Air Pollution Engineering Manual

Air Pollution Engineering Manual

Second Edition

AIR & WASTE MANAGEMENT
ASSOCIATION
◆
SINCE 1907

Edited by
Wayne T. Davis

A Wiley-Interscience Publication
JOHN WILEY & SONS, INC.
New York • Chichester • Weinheim • Brisbane • Singapore • Toronto

This book is printed on acid-free paper. ♾

Published simultaneously in Canada.

For ordering and customer service, call 1-800-CALL-WILEY.

Library of Congress Cataloging in Publication Data:

Air pollution engineering manual/Air & Waste Management Association; edited by Wayne T. Davis — 2nd ed.
p. cm.
Includes bibliographical references and index.
ISBN 0-471-33333-6 (alk. paper)
1. Air–Purification–Equipment and supplies. 2. Gases, Asphyxiating and poisonous–Environmental aspects. 3. Particles–Environmental aspects. 4. Factory and trade waste–Purification–Equipment and supplies. I. Davis, Wayne T. II. Air & Waste Management Association.

TD889.A39 2000
628.5′3–dc21 99-056637

Printed in the United States of America.

10 9 8 7 6 5 4 3 2

Contents

Foreword

The *Air Pollution Engineering Manual*, since its inception in the early 1960s, has been recognized by practitioners, engineers, academics, and students as a fundamental and practical source of information on the control of air pollution from stationary sources. The manual was originally published by the Los Angeles County Air Pollution Control District under the editorship of John A. Danielson with assistance from personnel of the agency.

In 1973 a revised second edition was published by the U.S. Environmental Protection Agency as AP-40. The editorial and technical content of the second edition was developed exclusively by the District under the direction of Robert G. Lunche, Eric Lemke, and John A. Danielson (as editor).

The manual, with its many flowcharts, photographs, and practical descriptions of various sources and control of air pollution, provided a valuable, if not indispensable, insight to practitioners throughout the 1970s and 1980s. However, many technological changes occurred in the late 1980s and early 1990s as a result of standards such as the PM-10 air quality standard, which placed more emphasis on the need to characterize emissions from source by particle size. Further, the Clean Air Act Amendments of 1990 mandated an aggressive review of 189 hazardous air pollutants and resulted in a major effort to develop Maximum Achievable Control Technology Standards (MACT standards). In light of the past significance of the manual and the increasing need for an updated manual, The Air & Waste Management Association and its Technical and Education Councils proposed and received partial funding from the U.S. Environmental Protection Agency in 1989 to update and publish the *Air Pollution Engineering Manual*. A&WMA oversaw the development of the manual, with coeditors Mr. Anthony J. Buonicore and Dr. Wayne T. Davis, who represented the Technical and Education Councils of the association, respectively. The manual was published in 1992 by Van Nostrand Reinhold.

This latest edition of the *Air Pollution Engineering Manual*, with Dr. Wayne T. Davis as editor, was initiated in 1998 by A&WMA, and the draft was provided to the publisher, John Wiley & Sons, Inc., in late August of 1999. Numerous authors from industry, federal, state, and local government agencies, trade associations, consulting firms, universities, and research organizations volunteered their efforts to produce the manual. We can only hope that this latest edition, inspired by its predecessors, will continue to be a practical and valuable source of information on air pollution control.

Acknowledgment

The editor would like to express his appreciation to the many authors who volunteered their time to make this manual possible, including those unnamed authors who contributed to the sections excerpted or adapted from the U.S. EPA Compilation of Air Pollutant Emission Factors (AP-42). It is clearly the volunteer efforts of the authors that have made the update of the manual a reality. The editor would also like to express his appreciation to the Air and Waste Management Association and to its staff and members of the Technical and Education Councils for their roles in initiating and continuing the effort to update this latest revision of the manual. Special thanks are expressed to Ms. Lisa Zavacky, Book Publishing and Sales Coordinator, and to Ms. Pam Groff, Book Production Coordinator (A&WMA) for their persistence and many long hours in coordinating the communication between the editor and the authors and for keeping the manual on schedule.

Wayne T. Davis, QEP
Fellow Member, A&WMA

Contributor List

1
Introduction

Air Pollution Control Engineering
Wayne T. Davis
Department of Civil and Environmental Engineering
University of Tennessee/Knoxville
Knoxville, TN

Anthony J. Buonicore
Environmental Data Resources
Southport, CT

Louis Theodore
Manhattan College
Manhattan College Parkway
Bronx, NY

Regulatory Aspects of Air Pollution in the United States
Leo Stander, Jr.
U.S. EPA, OAQPS
IT&PID, OPG (MD-12)
Research Triangle Park, NC

2
Control of Gaseous Pollutants

Absorption
Robert Jennings Heinsohn
Pennsylvania State University
University Park, PA

Adsorption
Kenneth E. Noll
Illinois Institute of Technology
Chicago, IL

Condensation
Anthony J. Buonicore
Environmental Data Resources
Southport, CT

Vapor Incineration
C. David Cooper
University of Central Florida
Orlando, FL

Biological Control Systems
Sarena J. Ergas
Department of Civil and Environmental Engineering
University of Massachusetts
Amherst, MA

Kerry A. Kinney
Department of Civil Engineering
University of Texas
Austin, TX

3
Control of Particulate Matter

Cyclones and Inertial Separators
Wayne T. Davis
Department of Civil and Environmental Engineering
University of Tennessee/Knoxville
Knoxville, TN

Timothy J. Herrick
Oakridge National Laboratory
Knoxville, TN

Wet Scrubbers

Kenneth E. Noll
Illinois Institute of Technology
Chicago, IL

Michael Pilat
Department of Civil Engineering
University of Washington
Seattle, WA

Electrostatic Precipitators

John D. McKenna
ETS, Inc.
Roanoke, VA

James H. Turner
Cary, NC

William M. Vatavuk
USEPA, OA QPS
MD-15
Research Triangle Park, NC

Phil A. Lawless, Toshiaki Yamamoto, and
David W. Coy
Research Triangle Institute
Research Triangle Park, NC

Gary P. Greiner
ETS, Inc.
Roanoke, VA

Fabric Filters

John D. McKenna
ETS, Inc.
Roanoke, VA

Arthur B. Nunn III
ETS, Inc.
Roanoke, VA

Dale A. Furlong
Clean Air Technologies
El Toro, CA

4
Furgitive Dust Emissions

John G. Watson, Judith C. Chow, and Thompson G. Pace
Desert Research Institute
Reno, NV

5
Odors

Willliam H. Prokop
Prokop Environmental Consulting
Deerfield, IL

6
Ancillary Equipment for Local Exhaust Ventilation Systems

Howard D. Goodfellow
Goodfellow Consultants, Inc.
Mississauga ON L5N 7G2
Canada

7
Combustion Sources

Coal

Arijit Pakrasi
IT Corporation
Monroeville, PA

Wayne T. Davis
Department of Civil and Environmental Engineering
University of Tennessee/Knoxville
Knoxville, TN

Fuel Oil

Arijit Pakrasi
IT Corporation
Monroeville, PA

Wayne T. Davis
Department of Civil and Environmental Engineering
University of Tennessee/Knoxville
Knoxville, TN

Natural Gas

Arijit Pakrasi
IT Corporation
Monroeville, PA

Wayne T. Davis
Department of Civil and Environmental Engineering
University of Tennessee/Knoxville
Knoxville, TN

Wood Waste

Arijit Pakrasi
IT Corporation
Monroeville, PA

Anthony J. Buonicore
Environmental Data Resources, Inc.
Southport, CT

Internal Combustion Sources

Michael J. Barboza, Michael J. Cannon
Nicholas J. Charno, and P. Steven Oliver
Paulus, Sokolowski and Sartor, Inc.
Warren, NJ

Stationary Gas Turbines

Michael J. Barboza, Michael J. Cannon
Nicholas J. Charno and P. Steven Oliver
Paulus, Sokolowski, and Sartor, Inc.
Warren, NJ

Reciprocating Internal Combustion Engines
Michael J. Barboza, Michael J. Cannon
Nicholas J. Charno, and P. Steven Oliver
Paulus, Sokolowski and Sartor, Inc.
Warren, NJ

8
Waste Incineration Sources

Municipal Waste Combustion
James R. Donnelly
Kvaerner Metals
San Ramon, CA

Hazardous Waste
Mackenzie L. Davis
Michigan State University
Department of Civil Engineering
East Lansing, MI

Clyde R. Dempsey and E. Timothy Oppelt
Risk Reduction Engineering Laboratory
Cincinnati, OH

Sewage Sludge
Calvin R. Brunner
Incinerator Consultants, Inc.
Reston, VA

Drum Reconditioning
Patricia Kay Rykowski
Stillwater Technologies, Inc.
Orlando, FL

9
Semiconductor Manufacturing

Peter F. Hess
Bay Area Air Quality Management District
San Francisco, CA

10
Evaporative Emissions and Surface Coatings

Organic Liquid Storage Tanks
Mark A. Morris
U.S. EPA, OAQPS
Research Triangle Park, NC

Evaporative Loss Sources, Organic Solvent Cleaning Operations
Mohamed A. Serageldin
US EPA, OAQPS MD-13
Research Triangle Park, NC

Paul Almodovar
USEPA
MD-13
Research Triangle Park, NC

Susan Miller
Harding-Lawson Associates
Marrisville, NC

Surface Coating
Ashok Kumar
University of Teledo
Teledo, OH

Taramati Shenoy
University of Toledo
Department of Civil Engineering
Toledo, OH

11
Graphic Arts Industry

Nonheatset Web Printing
C. Nelson Ho
Graphic Arts Technical Foundation
Sewickley, PA

H. Wilson Cunningham
American Newspaper Publishers Association
Reston, VA

Heatset Web Offset
C. Nelson Ho
Graphic Arts Technical Foundation
Sewickley, PA

Jerry Bender
Park Forest, IL

Robert M. Birkland
Lebanon Valley Offset
Annville, PA

William Schaeffer
Retired Graphics Arts Technical Foundation
Pittsburgh, PA

Graphic Arts Industry: Flexography
Doreen M. Monteleone
Flexographic Technical Association
Ronkonkoma, NY

John D. Serafano
Western Michigan State University
Kalamazoo, MI

Sheetfed Offset Printing
C. Nelson Ho
Graphic Arts Technical Foundation
Sewickley, PA

Gerald Bender
Park Forest, IL

David Johnson
Princeton Polychrome
Princeton, NJ

Frank Little
NIES/Aircraft
St. Louis, MO

William Schaeffer
Retired Graphics Arts Technical Foundation
Pittsburgh, PA

Gravure Printing
Gregory Tyszka
Gravure Association of America
Rochester, NY

Screen Printing
Marcia Y. Kinter
Screenprinting and Graphic Imaging Association
Fairfax, VA

12
Chemical Process Industry

Acrylonitrile
Ronald D. Bell
Petrocon Technologies, Inc.
Houston, TX

Alan V. Mechtenberg
Radian International LLC
Austin, TX

Carbon Black
Barry R. Taylor
Cabot Corporation
Billerica, MA

Charcoal/Activated Carbon
Mark P. Cal
New Mexico Tech.
MSEC Room 258
Socorro, NM

William D. Byers
CH2M Hill, Inc.
Corvallis, OR

Chlor-Alkali
Kenneth S. Walborn
PPG Industries, Inc.
New Martinsville, WV

Ethylene Oxide
George P. Partridge, Jr.
Penn State Harrisburg
Middletown, PA

R. Scott Huebner, Kevin S. McLeary
James F. White, and William M. Stayer
ChemDyn
Harrisburg, PA

Explosives
Institute of Makers of Explosives
Washington, DC

Phosphoric Acid Manufacturing
Gardon F. Palm (Deceased)

Phthalic Anhydride
Herbert P. Dengler
Exxon Chemical Co.
Baton Rouge, LA

Printing Inks
Templeton A. Elliott, Jr.
Sun Chemical
Winston Salem, NC

Soaps and Detergents
Richard C. Scherr
Chemical Waste Management
Oak Brook, IL

Sodium Carbonate
Michael J. Barboza
Paulus, Sokolowski and Sartor, Inc.
Warren, NJ

Stephen T. Coldwell
Paulus, Sokolowski and Sartor, Inc.
Warren, NJ

Neil M. Haymes
Paulus, Sokolowski and Sartor, Inc.
Warren, NJ

Sulfur
Bruce Scott
Bruce Scott Inc.
San Rafael, CA

Sulfuric Acid
Thomas Muller
E.I. Dupont de Nemours & Co. Inc.
Deepwater, NJ

Synthetic Rubber
USEPA public domain information only.

Wayne T. Davis
Department of Civil and Environmental Engineering
University of Tennessee/Knoxville
Knoxville, TN

Terephthalic Acid
Ronald D. Bell
Petrocon Technologies, Inc.
Houston, TX

Alan J. Mechtenberg
Radian International LLC
Austin, TX

Kevin McQuigg
IT McGill
Tulsa, OK

Natural Fiber Textile Industry
Anthony J. Buonicore
Environmental Data Resources
Southport, CT

Thermoplastic Resins
John A. Fey and Mitchell Jenkins, III
Union Carbide
South Charleston, WV

Thermosetting Resins
John A. Gannon
Consultant Ciba-Geiqy Corporation
Ardsley, NY

13
Food and Agricultural Industry

Bread Baking
Peter F. Hess
Bay Area Air Quality Management District
San Francisco, CA

Coffee Processing
USEPA public domain information only.

Grain Handling and Processing
Dennis Wallace
Midwest Research Institute
Cary, NC

Distilled Spirits
USEPA public domain information only.

Wines and Brandy
USEPA public domain information only.

Fish Processing
Anthony P. Bimbo
A.P. Bimbo & Associates
Kilmarnock, VA

Livestock Feedlots
John M. Sweeten
Texas A & M Agricultural Research Center
Amarillo, TX

Calvin B. Parnell, Bryan W. Shaw
and Ronald E. Lacy
Texas A & M University
College Station, TX

Brent W. Auvermann
Texas A & M University
Amarillo, TX

Rendering Plants
William H. Prokop
Deerfield, IL

14
Metallurgical Industry

Primary Aluminum Industry
Maurice W. Wei
Aluminum Company of America
Pittsburgh, PA

Metallurgical Coke
George Bradely
Indiana Harbor Coke Company
East Chicago, IN

Kelly R. Carmichael and Allen Ellis
Indiana Harbor Coke Company
East Chicago, IN

John Carson
Radian International
Oakridge Technical Center
Oak Ridge, TN

Thomas W. Easterly
Bethlehem Steel Corporation
Bethlehem, PA

Stefan P. Shoup and Dennis P. Kaegi
Inland Steel Company
East Chicago, IN

Copper Smelting
James R. Donnelly
Kvaerner Metals
San Ramon, CA

Jacques Moulins
Montreal PQ H3A 3JG
Canada

Ferroaolloy Industry Particulate Emissions
Steven Stasko
Air & Waste Management Association
Pittsburgh, PA

Steel Industry
Bruce A. Steiner
American Iron and Steel Institute
Washington, DC

Miscellaneous Fugitive Emission Sources
Robert E. Sistek
Pittsburgh, PA

Frank Pendelton
Midwest Research Institute
Kansas City, MO

Blast Furnace
Marek S. Klag
Dofasco, Inc.
Hamilton ONT L8N 3J5
Canada

Basic Oxygen Furnace Shops
S.S. Felton
AK Steel
Middletown, OH

Edward Cocchiarella and Murray S. Greenfield
Dofasco, Inc
Hamilton ONT L8N 3J5
Canada

Electric Arc Furnaces/Argon-Oxygen Decarburization Process
Manfred Bender
Bender Corporation
Beverly Hills, CA

Casters
Manfred Bender
Bender Corporation
Beverly Hills, CA

Ladle Metallurgy Vacuum Degassing
Edward Cocchiarella
Dofasco, Inc.
Hamilton ONT L8N 3J5
Canada

Fluxed Iron Ore Pellet Production
Gus R. Josephson
Inland Steel Mining Co.
Virginia, MN

Pickling
R.M. Hudson (Retired)
USS Group of USX Corporation
Pittsburgh, PA

Direct Reduction
Murray S. Greenfield
Dofasco, Inc.
Hamilton ONT L8N 3J5
Canada

Rolling
Michael T. Unger
Inland Steel
East Chicago, IN

Scarfing
S.J. Manganello
USS Technical Center
Monroeville, PA

Sinter Plants
S.S. Felton
AK Steel
Middleton, OH

L.M. Stuart
Bethlehem Steel Corporation
Bethlehem, PA

Primary Lead Smelting
USEPA public domain information only.

Zinc Smelting
James R. Donnelly
Kvaerner Metals
San Ramon, CA

Secondary Aluminum
Charles A. Licht
Charles Licht Engineering Associates
Olympia Fields, IL

Secondary Brass and Bronze Melting Processes
Charles A. Licht
Charles Licht Engineering Associates
Olympia Fields, IL

Iron Foundaries
American Foundrymen's Society
Air Quality Committee (10-E)*

Secondary Lead Smelting
Charles A. Licht
Charles Licht Engineering Associates
Olympia Fields, IL

Secondary Zinc
Charles A. Licht
Charles Licht Engineering Associates
Olympia Fields, IL

Steel Foundaries
Mackenzie L. Davis
Michigan State University
Department of Civil Engineering
East Lansing, MI

Souad A. Benromdhane
Michigan State University
Department of Civil Engineering
East Lansing, MI

15
Mineral Products Industry

Hot Mix Asphalt Mixing Facilities
George P. Partridge, Jr.
ChemDyn
Harrisburg, PA

Mark J. Miller
Penn State Harrisburg
Middletown, PA

R. Scott Huebner
ChemDyn
Harrisburg, PA

Portland Cement
Ann Dougherty
Portland Cement
Skokie, IL

Walter L. Greer
Ash Grove Cement Company
Overland Park, KS

Douglas M. Sweeney
BHA Group
Kansas City, MO

Ceramic and Brick Manufacturing
Dennis A. Brosnan
Center for Engineering Ceramic Manufacturing
Anderson, SC

Coal Processing
USEPA public domain information.
Larry Simmons and Lisa Lambert
Energy and Environmental Management
Murrysville, PA

Concrete Batching
John E. Yocom
West Simsbury, CT

Michael C. Shum and Michael D. Errickson
HRP Associates, Inc.
Plainville, CT

Fiberglass Operations
Aaron J. Teller
Air Water Technologies Corporation
Ft. Lauderdale, FL

Joseph Y. Hsieh
R-C Environmental Services & Technologies
Branchburg, NJ

Glass Manufacturing
Aaron J. Teller
Air Water Technologies Corporation
Ft. Lauderdale, FL

Joseph Y. Hsieh
R-C Environmental Services & Technologies
Branchburg, NJ

Lime Manufacturing
USEPA public domain information only.

Sand and Gravel Processing
Brian Palm
Air Control Techniques
Cary, NC

John Richards
Air Control Techniques
Cary, NC

Stone Crushing
Brian Palm
Air Control Techniques
Cary, NC

John Richards
Air Control Techniques
Cary, NC

Todd Brozell
Air Control Techniques
Cary, NC

16
Pharmaceutical Industry

Industry Characterization
Michael J. Barboza
Paulus, Sokolowski and Sartor, Inc.
Warren, NJ

Theodore V. Chleboski
Paulus, Sokolowsk and Sartor, Inc.
Warren, NJ

Richard V. Crume
Combustion Group/ESD/OAQPS (MD-13)
USEPA
Research Triangle Park, NC

Jeffrey W. Portzer
Research Triangle Institute
Cary, NC

Bulk Materials Processing—Bulk Pharmaceutical Production

Michael J. Barboza
Paulus, Sokolowski and Sartor, Inc.
Warren, NJ

Theodore V. Chleboski
Paulus, Sokolowski and Sartor, Inc.
Warren, NJ

Richard V. Crume
Combustion Group/ESD/OAQPS (MD-13)
USEPA
Research Triangle Park, NC

Jeffrey W. Portzer
Research Triangle Institute
Cary, NC

Pharmaceutical Product Formulation

Michael J. Barboza
Paulus, Sokolowski and Sartor, Inc.
Warren, NJ

Theodore V. Chleboski
Paulus, Sokolowski and Sartor, Inc.
Warren, NJ

17
The Petroleum Industry

J. Eldon Rucker
Robert P. Strieter
American Petroleum Institute
Washington, DC

USEPA public domain information.

18
Wood Processing Industry

Chemical Wood Pulping

Arun V. Someshwar
Senior Research Engineer
NCASI
Southern Regional Center
Gainesville, FL

John E. Pinkerton
Director, Air Quality
NCASI
Research Triangle Park, NC

Mechanical Pulping

Arun V. Someshwar
Senior Research Engineer
NCASI
Southern Regional Center
Gainesville, FL

John E. Pinkerton
Director, Air Quality
NCASI
Research Triangle Park, NC

Recovered Paper Pulping

Arun V. Someshwar
Senior Research Engineer
NCASI
Southern Regional Center
Gainesville, FL

John E. Pinkerton
Director, Air Quality
NCASI
Research Triangle Park, NC

Paper and Paperboard Manufacture

Arun V. Someshwar
Senior Research Engineer
NCASI
Southern Regional Center
Gainesville, FL

John E. Pinkerton
Director, Air Quality
NCASI
Research Triangle Park, NC

19
Soil/Solid Waste/Water Sources

Groundwater and Soil Treatment Process

Edward S. Alperin
IT Corporation
Knoxville, TN

Robert D. Fox
London, TN

Duane K. Root
London, TN

Drinking Water Treatment

Michael J. Barboza
Paulus, Sokolowski and Sartor, Inc.
Warren, NJ

Jack H. van Sciver
Paulus, Sokolowski and Sartor, Inc.
Warren, NJ

Municipal Solid Waste Landfill Gas Emissions

Michael J. Barboza
Paulus, Sokolowski and Sartor, Inc.
Warren, NJ

Wastewater Treatment Facilities

Michael J. Barboza
Paulus, Sokolowski and Sartor, Inc.
Warren, NJ

Gary Comerford
Paulus, Sokolowski and Sartor, Inc.
Warren, NJ

20
Accidental and Catastrophic Releases

George P. Partridge, R. Scott Huebner, Kevin S. McLeary
William M. Stayer, and James F. White
ChemDyn
Harrisburg, PA

Air Pollution Engineering Manual

1

INTRODUCTION

Air Pollution Control Engineering
Wayne T. Davis, Anthony J. Buonicore, and Louis Theodore
Regulatory Aspects of Air Pollution Control in the United States
Leo H. Stander, Jr.

AIR POLLUTION CONTROL ENGINEERING

WAYNE T. DAVIS, ANTHONY J. BUONICORE, AND LOUIS THEODORE

INTRODUCTION

In the past three decades that have transpired since the passage of the Clean Air Act Amendments of 1970 in the United States and the first publication of the Air Pollution Engineering Manual (early 1960s), the engineering profession has been heavily influenced by its responsibility to society. This responsibility has been directed towards the protection of public health and welfare and is guided by a host of environmental regulations. The extent to which the engineering and environmental professions respond to this challenge depends largely on the limits imposed by three principal considerations:

1. Legal limitations imposed for the protection of public health and welfare
2. Social limitations imposed by the community in which the pollution source is or will be located
3. Economic limitations imposed by marketplace constraints

Careful evaluation within the framework of all three limitations is now essential and often integral to corporate strategic planning processes.

The control strategy for environmental impact assessment often focuses on five alternatives whose purpose would be the reduction and/or elimination of pollutant emissions:

1. Elimination of the operation entirely or in part
2. Modification of the operation
3. Relocation of the operation
4. Application of appropriate control technology
5. Combinations thereof

In view of the relatively high costs often associated with pollution control systems, engineering professionals today are directing considerable effort towards pollution prevention by process modification to eliminate as much of the pollution problem as possible *at the source*. This includes evaluation of alternative manufacturing and production techniques, substitution of raw materials, and improved process control methods. Unfortunately, if there is no alternative, the application of pollution control equipment must be considered. Considering the relatively high costs, proper selection of this equipment is essential. The equipment must be designed to comply with regulatory emission limitations on a continual basis, with interruptions subject to severe penalty depending on the circumstances. The requirement for design performance on a continual basis places very heavy emphasis on operation and maintenance practices and equipment reliability. In fact, it is not unusual that favorable operation and maintenance requirements associated with a particular piece of equipment can strongly influence its selection, despite the fact that its capital cost may be higher. The rapidly escalating costs of energy, labor, and materials can make operation and maintenance considerations even more important than original cost.

FACTORS IN CONTROL EQUIPMENT SELECTION

There are a number of factors to be considered prior to selecting a particular air pollution control system and associated hardware.[1] In general, these can be grouped into three categories: environmental, engineering, and economic.

Environmental

1. Equipment location
2. Available space
3. Ambient conditions
4. Availability of adequate utilities (i.e., power, water, etc.) and ancillary system facilities (i.e., waste treatment and disposal, etc.)
5. Maximum allowable emission standards such as New Source Performance Standards (NSPS) and Maximum Achievable Control Technology (MACT) standards (air pollution control regulations)
6. Aesthetic considerations (i.e., visible steam or water vapor plume, impact on scenic vistas, odor, etc.)
7. Contribution of the air pollution control system to wastewater and solid waste
8. Contribution of the air pollution control system to plant noise levels

Engineering

1. Contaminant characteristics (i.e., physical and chemical properties, concentration, particulate shape and size distribution—in the case of particulates, chemical reactivity, corrosivity, abrasiveness, toxicity, etc.)
2. Gas stream characteristics (i.e., volume flow rate, temperature, pressure, humidity, composition, viscosity, density, reactivity, combustibility, corrosivity, toxicity, etc.)
3. Design and performance characteristics of the particular control system (i.e., size and weight, fractional efficiency curves—in the case of particulates, mass transfer and/or contaminant destruction capability—in the case of gases or vapors, pressure drop, reliability and dependability, turndown capability, power requirements,

utility requirements, temperature limitations, maintenance requirements, flexibility with regard to complying with more stringent air pollution regulations, etc.)

Economic

1. Capital cost (equipment, installation, engineering, etc.)
2. Operating cost (utilities, maintenance, etc.)
3. Expected equipment lifetime and salvage value

Prior to the purchase of control equipment, experience has shown that the following points should be emphasized:

1. Control equipment should not be purchased without reviewing *certified independent test data* on its performance in a similar application. The manufacturer should be asked to provide performance information and design specifications.
2. In the event that sufficient performance data are unavailable, the equipment supplier can often provide a small pilot plant test unit for evaluation under existing conditions.
3. Participation of the local control authorities in the decision-making process is strongly recommended.
4. A detailed and precise set of specifications is essential. A strong *performance guarantee* from the manufacturer should be obtained to ensure that the control equipment will meet all applicable local, state, and federal codes at specific process conditions.
5. Process and economic fundamentals should be closely reviewed. The possibility for emission tradeoffs (offsets) and/or applying the "bubble concept" should be assessed. The bubble concept permits a plant to find the most efficient way to control its emissions as a whole, rather than having the U.S. Environmental Protection Agency (EPA) regulate the emissions from individual sources. Reductions at a source where emissions can be lessened for the least cost can offset emissions of the same pollutant from another source in the plant.
6. A careful material balance study should be made before authorizing an emission test or purchasing control equipment.
7. Equipment should not be purchased until *firm* installation cost estimates have been added to the equipment cost. Escalating installation costs are the rule rather than the exception.
8. Operation and maintenance costs should be given high priority on the list of equipment selection factors.
9. Equipment should not be purchased until a *solid* commitment from the utility supplier(s) is obtained. Every effort should be made to ensure that the new system will utilize fuel, controllers, filters, motors, and so on that are compatible with those already available at the plant.
10. The specification should include written assurance of *prompt* technical assistance from the equipment supplier. This, together with a complete operating manual (with parts list and full schematics), is essential and is too often forgotten in the rush to get the equipment operating.
11. Schedules, particularly for projects being completed under a court order or consent judgment, can be critical. In such cases, delivery guarantees should be obtained from the manufacturers and penalties identified.
12. The air pollution equipment should be of fail-safe design with builtin indicators to show when performance is deteriorating.
13. A portion of the purchase price (10–15%) should be withheld until compliance is clearly demonstrated.

The usual design–procurement–construction–startup problems can be further compounded by any one or combination of the following:

1. Lack of familiarity of process engineers with air pollution engineering
2. New and changing air pollution regulations
3. New suppliers with unproven equipment
4. Lack of industry standards in some key areas
5. Inaccurate interpretations by control agency field personnel
6. Compliance schedules that are too tight
7. Vague specifications
8. Weak guarantees for the new control equipment
9. Unreliable delivery schedules
10. Variability; unreliable process operation

Proper selection of a particular system for a specific application can be extremely difficult and complicated. In view of the multitude of complex and often ambiguous pollution control regulations, it is in the best interest of the prospective user to work closely with regulatory officials as early in the process as possible. Finally, previous experience on a similar application cannot be overemphasized.

GENERALIZED DESIGN REVIEW PROCEDURE

Design reviews for air pollution control equipment are performed for a variety of reasons, including:

1. To anticipate compliance with applicable air pollution emission standards and regulations
2. To estimate performance of existing control equipment
3. To evaluate the feasibility of a proposed equipment design
4. To assess the effect on control equipment of process modification and the effect if any of the air pollution control equipment on the process operation

A typical generalized design review approach is presented in Figure 1. The design review investigation is an activity performed early in the evaluation process. Other activities that must be accomplished before final compliance is achieved are presented in Figure 2.

COMPARING CONTROL EQUIPMENT ALTERNATIVES

The final choice in equipment selection is usually dictated by that equipment capable of achieving compliance with regulatory codes at the lowest uniform annual cost (amortized capital investment plus operation and maintenance costs).

Table 1. Advantages and Disadvantages of Cyclone Collectors

Advantages

1. Low cost of construction
2. Relatively simple equipment with few maintenance problems
3. Relatively low operating pressure drop (for degree of particulate removal obtained) in the range of approximately 2 to 6 in. water column
4. Temperature and pressure limitations imposed only by the materials of construction used
5. Dry collection and disposal
6. Relatively small space requirements

Disadvantages

1. Relatively low overall particulate collection efficiencies, especially on particulates below 10 μm in size
2. Inability to handle tacky materials

In order to compare specific control equipment alternatives, knowledge of the particular application and site is essential. A preliminary screening, however, may be performed by reviewing the advantages and disadvantages of each type of air pollution control equipment. For example, if water or a waste treatment system is not available at the site, this may preclude use of a wet scrubber system and instead focus particulate removal on dry systems, such as cyclones, baghouses, and/or electrostatic precipitators. In this same circumstance gaseous pollutant removal may have to focus on activated carbon adsorption or a thermal oxidizer instead of a wet scrubber (absorption). If auxiliary fuel is unavailable on a continuous basis, it may not be possible to combust organic pollutant vapors in an incineration system. If the particulate-size distribution in the gas stream is relatively fine, cyclone collectors most probably would not be considered. If the pollutant vapors can be reused in the process, control efforts may be directed to adsorption systems. There are many more situations in which knowledge of the capabilities of the various control options, combined with common sense, will simplify the selection procedure. General advantages and disadvantages of some of the most popular types of air pollution control equipment for gases and particulates are presented in Tables 1 through 9.

SELECTING CONTROL EQUIPMENT FOR SPECIFIC INDUSTRIES

The basic types of emission control devices are mechanical collectors, wet scrubbers, baghouses, electrostatic precipitators, combustion systems and thermal oxidizers, condensers, absorbers, adsorbers, and biological degradation. All these have been used to some extent to control emissions from a variety of processes, with the selection procedure almost always dictated by experience.

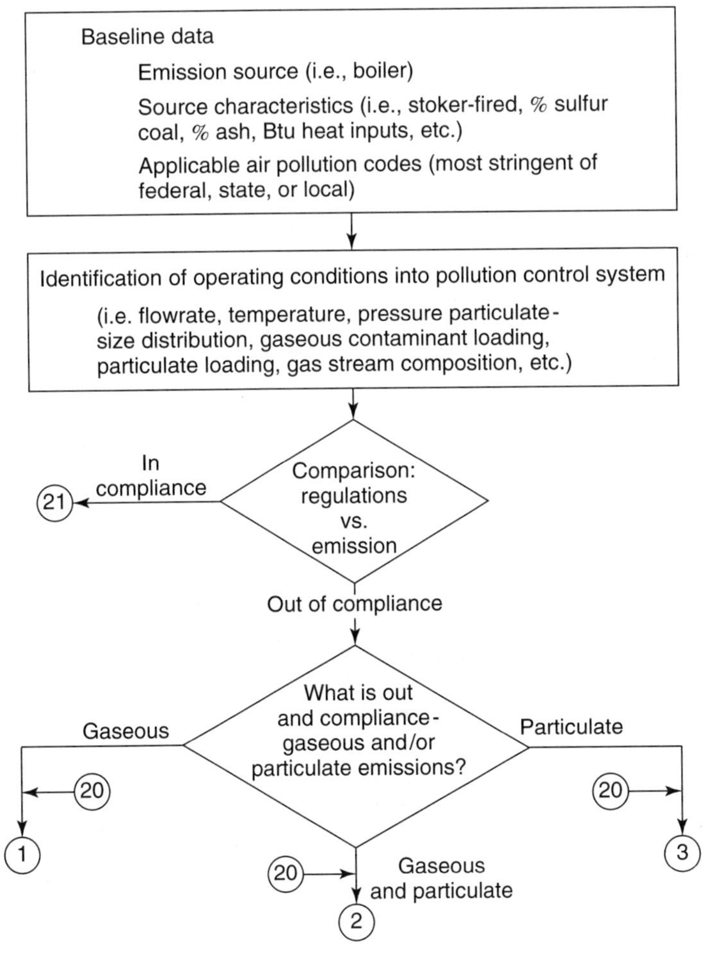

Figure 1. Typical Generalized Design Review Approach.

The chapters in this manual provide detailed information on (1) the basic principles of operation of each of the control devices, (2) a variety of industries affected by air pollution control regulations, (3) the general process descriptions of the industries, (4) the specific emission standards (NSPS, MACT, etc.) applicable to each industry, and (5) control systems used in each industry.

The regulations applicable to any industry may change as more advanced air pollution control technologies and new regulations are developed. This is especially true for the wide range of industries and pollutants that are addressed with the MACT standards, which are discussed and listed in the next section on regulations. It was the intent that each of the sections in the manual would address MACT standards that are applicable to each industry. An excellent source of the status and detailed regulations associated with the latest MACT standards can be found on the U.S. EPA Technology Transfer Network (TTNWeb) on the Unified Air Toxics Website

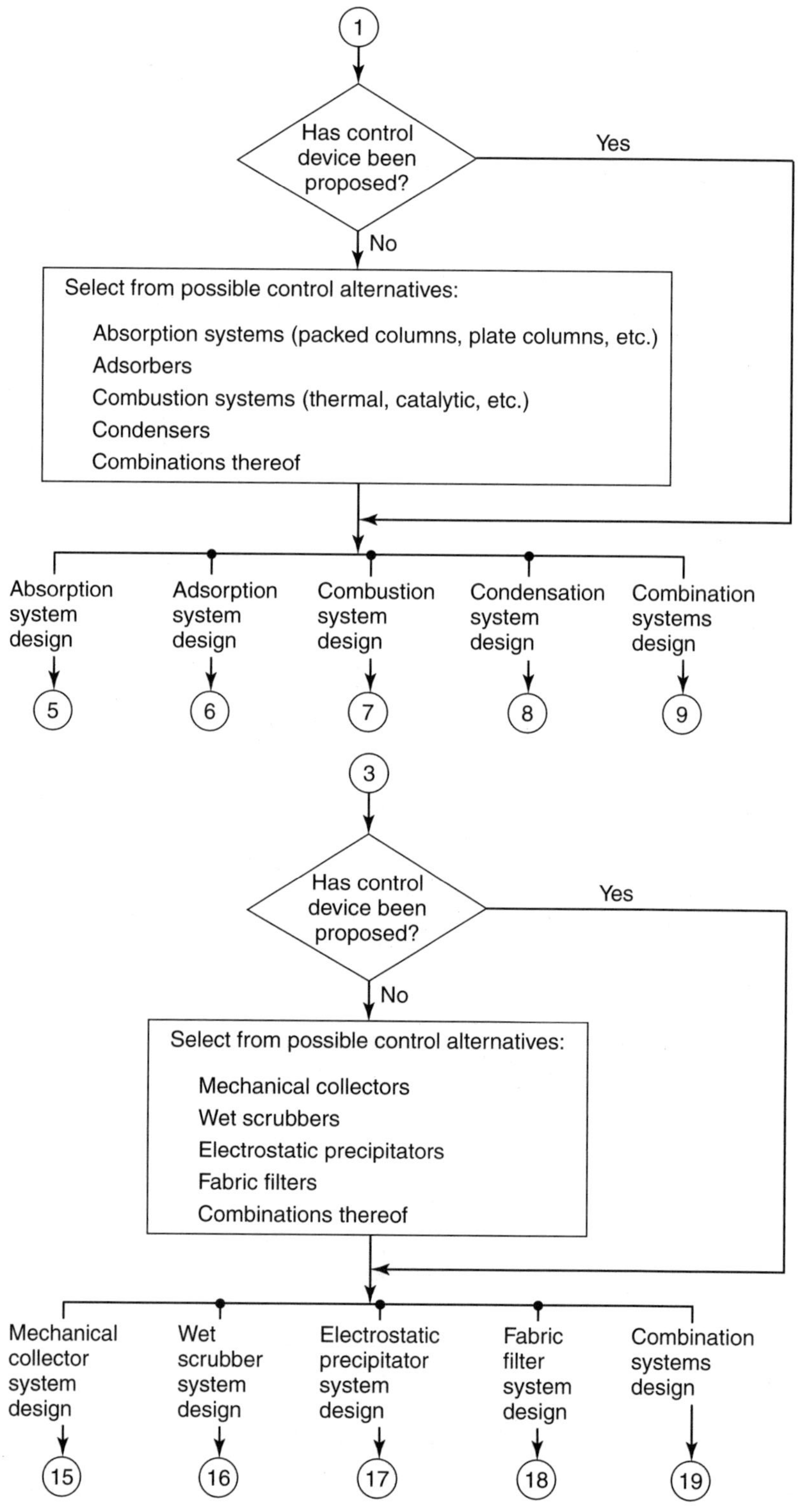

Figure 1. (*Continued*).

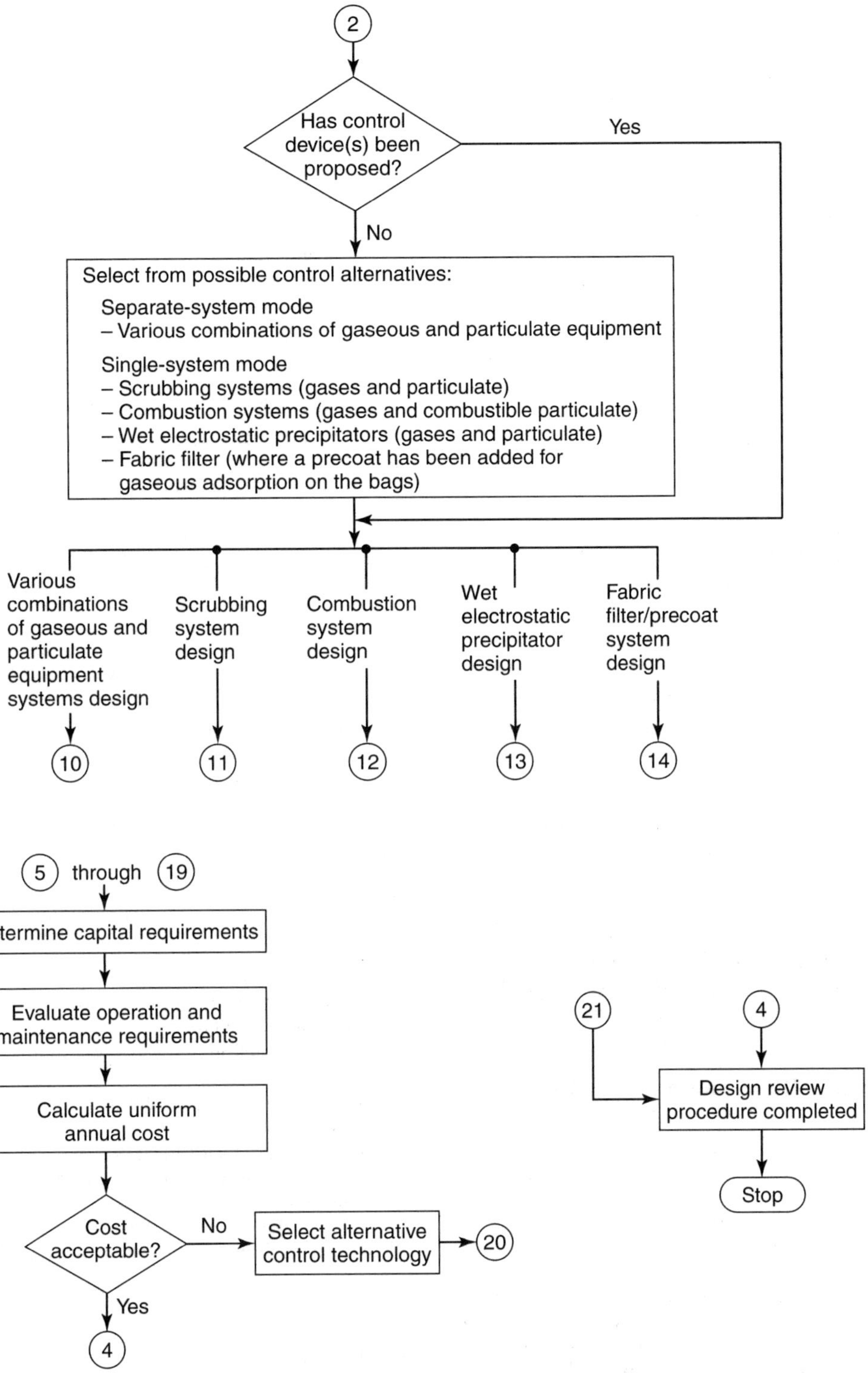

Figure 1. (*Continued*).

(UATW). While website addresses tend to change with time, the TTNWeb is currently found at http://www.epa.gov/ttn. Other Websites found at this same location that are of particular relevance to this manual are the following:

CATC/RBLC: Clean Air Technology Center/RACT/BACT/LAER Clearinghouse

CHIEF: Clearinghouse for Inventories and Emission Factors

FUTURE DIRECTION OF AIR POLLUTION CONTROL ENGINEERING

The future direction of regulations and air pollution control efforts is clearly towards significantly reducing the emissions to the environment of a broad range of compounds, including:

1. Volatile organic compounds (VOCs) and other ozone precursors (CO and NO_x)

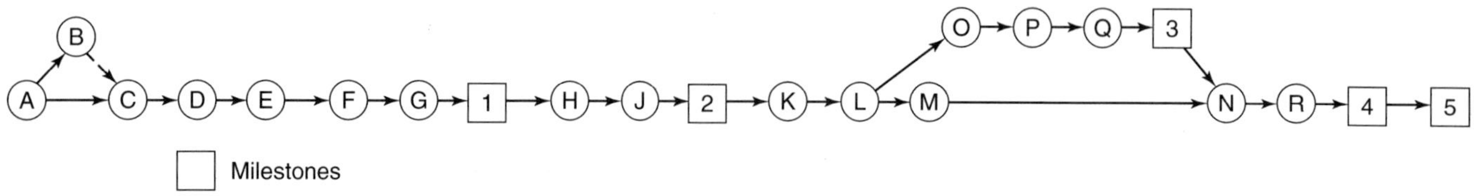

Milestones

1 Date of submittal of final control plan to appropriate agency.
2 Date of award of control device contract.
3 Date of initiation of onsite construction or installation of emission control equipment.
4 Date by which onsite construction or installation of emission control equipment is completed.
5 Date by which final compliance is achieved.

Activities

Designation		Designation	
A-C	Preliminary investigation	K-L	Review and approval of assembly drawings
A-B	Source tests, if necessary	L-M	Vendor prepares fabrication drawings
C-D	Evaluate control alternatives	M-N	Fabricate control device
D-E	Commit funds for total program	L-O	Prepare engineering drawings
E-F	Prepare preliminary control plan and Compliance schedule for agency	O-P	Procure construction bids
F-G	Agency review and approval	P-Q	Evaluate construction bids
G-1	Finalize plans and specifications	Q-3	Award construction contract
1-H	Procure control device bids	3-N	Onsite construction
H-J	Evaluate control device bids	N-R	Install control device
J-2	Award control device contract	R-4	Complete construction (system tie-in)
2-K	Vendor prepares assembly drawings	4-5	Startup, shakedown, source test

Figure 2. Compliance Activity and Schedule Chart.

2. Hazardous air pollutants, including carcinogenic organic emissions and heavy metals emissions
3. Acid rain precursors, including SO_x and NO_x

In addition, the development of the PM_{10} NAAQS and the more recently promulgated $PM_{2.5}$ NAAQS will continue to place emphasis on quantifying and reducing particulate emissions in the less than 10 μm and less than 2.5 μm

Table 2. Advantages and Disadvantages of Wet Scrubbers

Advantages

1. No secondary dust sources
2. Relatively small space requirements
3. Ability to collect gases as well as particulates (especially "sticky" ones)
4. Ability to handle high-temperature, high-humidity gas streams
5. Low capital cost (if wastewater treatment system is not required)
6. For some processes, the gas stream is already at high pressures (so pressure-drop considerations may not be significant)
7. Ability to achieve high collection efficiencies on fine particulates (however, at the expense of pressure drop)

Disadvantages

1. May create water disposal problem
2. Product is collected wet
3. Corrosion problems are more severe than with dry systems
4. Steam plume opacity and/or droplet entrainment may be objectionable
5. Pressure drop and horsepower requirements may be high
6. Solids buildup at the wet–dry interface may be a problem
7. Relatively high maintenance costs

Table 3. Advantages and Disadvantages of Electrostatic Precipitators

Advantages

1. Extremely high particulate (coarse and fine) collection efficiencies can be attained (at a relatively low expenditure of energy)
2. Dry collection and disposal
3. Low pressure drop (typically less than 0.5-in. water column)
4. Designed for continous operation with minimum maintenance requirements
5. Relatively low operating costs
6. Capable of operation under high pressure (to 150 psi) or vacuum conditions
7. Capable of operation at high temperatures (to 1300°F)
8. Relatively large gas flow rates can be effectively handled

Disadvantages

1. High capital cost
2. Very sensitive to fluctuations in gas stream conditions (in particular, flow rates, temperatures, particulate and gas composition, and particular loadings)
3. Certain particulates are difficult to collect due to extremely high or low resistivity characteristics
4. Relatively large space requirements required for installation
5. Explosion hazard when treating combustible gases and/or collecting combustible particulates
6. Special precautions required to safeguard personnel from the high voltage
7. Ozone is produced by the negatively charged electrode during gas ionization
8. Relatively sophisticated maintenance personnel required

Table 4. Advantages and Disadvantages of Fabric Filter Systems

Advantages

1. Extremely high collection efficiency on both coarse and fine (submicrometer) particulates
2. Relatively insensitive to gas stream fluctuation; efficiency and pressure drop are relatively unaffected by large changes in inlet dust loadings for continuously cleaned filters
3. Filter outlet air may be recirculated within the plant in many cases (for energy conservation)
4. Collected material is recovered dry for subsequent processing or disposal
5. No problems with liquid waste disposal, water pollution, or liquid freezing
6. Corrosion and rusting of components are usually not problems
7. There is no hazard of high voltage, simplifying maintenance and repair and permitting collection of flammable dusts
8. Use of selected fibrous or granular filter aids (precoating) permits the high-efficiency collection of submicrometer smokes and gaseous contaminants
9. Filter collectors are available in a large number of configurations, resulting in a range of dimensions and inlet and outlet flange locations to suit installation requirements
10. Relatively simple operation

Disadvantages

1. Temperature much in excess of 550°F require special refractory mineral or metallic fabrics that are still in the developmental stage and can be very expensive
2. Certain dusts may require fabric treatments to reduce dust seeping or, in other cases, assist in the removal of the collected dust
3. Concentrations of some dusts in the collector (~50 gm^3) may represent a fire or explosion hazard if a spark or flame is admitted by accident; fabrics can burn if readily oxidizable dust is being collected
4. Relatively high maintenance requirements (bag replacement, etc.)
5. Fabric life may be shortened at elevated temperatures and in the presence of acid or alkaline particulate or gas constituents
6. Hygroscopic materials, codensation of moisture, or tarry adhevise components may cause crusty caking or plugging of the fabric or require special additives
7. Replacement of fabric may require respiratory protection for maintenance personnel
8. Medium pressure-drop requirements, typically in the range of 4 to 10 in. water column

particle size range. Particle-size-specific emission factors have been developed for many sources, and size-specific emission standards have been developed in some states. These standards are addressing concerns related to the increased health effects of these smaller particles, sulfate and nitrate deposition of the particles, and visibility reduction and effects on scenic vistas, all of which are generally associated with submicrometer particles.

Although it is not possible to predict the future, it is possible to prepare for it and to influence it. It is highly recommended that maximum flexibility be designed into new air pollution control systems to allow for the increasingly stringent emissions standards for both particulate matter and gases that are on the horizon for both stationary point sources, area sources, and mobile sources. Further, it is everyone's

Table 5. Advantages and Disadvantages of Absorption Systems (Packed and Plate Columns)

Advantages

1. Relatively low pressure drop
2. Standardization in fibreglass-reinforced plastic (FRP) construction permits operation in highly corrosive atmospheres
3. Capable of achieving relatively high mass-transfer efficiencies
4. Increasing the height and/or type of packing or number of plates can improve mass transfer without purchasing a new piece of equipment
5. Relatively low capital cost
6. Relatively small space requirements
7. Ability to collect particulars as well as gases

Disadvantages

1. May create water (or liquid) disposal problem
2. Product collected wet
3. Particulars deposition may cause plugging of the bed or plates
4. When FRP construction used, it is sensitive to temperature
5. Relatively high maintenance costs

Table 6. Comparison of Plate and Packed Columns

Packed column

1. Lower pressure drop
2. Simpler and cheaper to construct
3. Preferable for liquids with high foaming tendencies

Plate column

1. Less susceptible to plugging
2. Less weight
3. Less of a problem with channeling
4. Temperature surge will result in less damage

Table 7. Advantages and Disadvantages of Adsorption Systems

Advantages

1. Product recovery may be possible
2. Excellent control and response to process changes
3. No chemical disposal problem when pollutant (product) is recovered and returned to process
4. Capability of systems to provide fully automatic, unattended operation
5. Capability to remove gaseous or vapour contaminants from process streams to extremely low levels

Disadvantages

1. Product recovery may require an exotic, expensive distillation (or extraction) scheme
2. Adsorbent progressively deteriorates in capacity as the number of cycles increases
3. Adsorbent regeneration requires a stream or vacuum source
4. Relatively high capital cost
5. Prefiltering of gas stream may be required to remove any particulate capable of plugging the adsorbent bed
6. Cooling of the gas stream may be required to get to the usual range of operation (less than 120°F)
7. Relatively high steam requirements to desorb high-molecular-weight hydrocarbons

Table 8. Advantages and Disadvantages of Combustion Systems

Advantages

1. Simplicity of operation
2. Capability to provide steam generation or heat recovery in other forms
3. Capability for high destruction effciency of organic contaminants

Disadvantages

1. Relatively high operating costs (particularly associated with fuel requirements)
2. Potential for flashback and subsequent explosion hazard
3. Catalyst poisoning (in the case of catalytic incineration)
4. Incomplete combustion can create potentially worse pollution problems

Table 9. Advantages and Disadvantages of Condensers

Advantages

1. Pure product recovery (in the case of indirect-contact condensers)
2. Water used as the coolant in an indirect contact condenser (i.e., shell-and-tube heat exchanger) does not contact the contaminated gas stream and can be reused after cooling

Disadvantages

1. Relatively low removal efficiency for gaseous contaminants (at concentrations typical of pollution control application)
2. Coolant requirements may be extremely expensive

responsibility to provide a thorough review of existing and proposed new processes and to make every attempt to identify economical pollution prevention approaches, including process modifications and/or material substitutions that reduce or, in some cases, eliminate both the emissions to the environment and the overdependency on retrofitted or new end-of-pipe control systems.

REFERENCES

1. L. Theodore and A. J. Buonicore, *Industrial Air Pollution Control Equipment for Particulates*, Chap. 6., CRC Press, West Palm Beach, FL, 1976.
2. M. Weiss and J. Palmisano, "Emissions trading gives flexibility in meeting clean-air laws," *Power*, 55–58 (March 1985).
3. A. J. Buonicore, "Air pollution control," *Chem. Eng.* **87**(13), 81–101 (June 30, 1980).

REGULATORY ASPECTS OF AIR POLLUTION CONTROL IN THE UNITED STATES

Leo H. Stander, Jr.

Engineering, as a profession, is heavily influenced by its responsibility to society. This responsibility is directed towards the advancement and protection of public health and welfare and is guided by a host of environmental regulations.

In this section various requirements that have an impact on sources of air pollution will be discussed.

REGULATORY ASPECTS

Laws provide the vision, scope, and authority for environmental protection and restoration. They are usually enacted by a legislative body and serve as a basis for administrative implementation.

Regulations establish general requirements that must be met by a regulated community. These requirements generally apply at a national, state, or local level and are usually adopted by an administrative entity that has received an authorization from a law. These regulations generally interpret the enabling legislation and explain how the law is to be implemented.

Policy is developed by an administrative entity to explain further how laws and regulations are to be implemented and to resolve issues and conflicts of interpretation.

History of Requirements

Although air pollution has been around for a long time, it has only been during the past 50 years that the governments of this country and many others have taken active roles in controlling it. In the United States, the first efforts at regulating air quality were principally at the individual community level in cities such as Cincinnati, Ohio, Pittsburgh, Pennsylvania, and Los Angeles, California. State requirements for controlling air pollution on a statewide basis came later.

At the federal level, the Clean Air Act defines the national policy for air pollution abatement and control in the United States. It establishes goals of protecting health and natural resources and delineates what is expected of federal, state, and local governments to achieve those goals. From its modest beginnings as an effort to determine the causes of illnesses and deaths that occurred in such places as Donora, Pennsylvania, or in New York City to the comprehensive document that exists today, the evolution of the Clean Air Act is an example of the country's efforts to protect people and the environment.

The federal effort began with the first national air pollution control legislation, called the Air Pollution Control Act of 1955. This legislation was enacted after epidemiological studies began showing that the effects of air pollution had broader implications than previously believed and that injuries and even death were the result of high concentrations of air pollutants in the atmosphere. The Air Pollution Control Act provided the Public Health Service of the Department of Health, Education and Welfare with limited authority to conduct air pollution research and to provide technical assistance to state and local governments. Amendments to this legislation, enacted in 1960 and 1962, authorized special studies to evaluate the health effects of emissions from motor vehicles.

The Clean Air Act of 1963 encouraged state, regional, and local programs to abate and control air pollution and enabled the U.S. Public Health Service to intervene in interstate problem areas. It also encouraged additional research for pollution resulting from motor vehicles and the development of air quality criteria to guide states in establishing ambient air quality standards and emission standards for stationary sources. The Clean Air Act was amended in 1965 to authorize emission standards for motor vehicles.

In 1967, the Clean Air Act was again amended and was renamed the Air Quality Act of 1967. The 1967 amendments provided the basic framework for many of our current efforts to control air pollution. Specifically, states were required to develop and implement ambient air quality standards on a fixed time schedule after the Secretary of Health, Education, and Welfare issued air quality criteria and control technique information.

Because of the slow process in developing ambient standards and the continuing deterioration of air quality across the country, the Clean Air Act of 1970 was enacted. These amendments required the newly formed Environmental Protection Agency (EPA) to adopt uniform national ambient air quality standards to protect health and welfare and required that these standards be attained and maintained across the country. Other aspects of the 1970 amendments included a significant federal enforcement authority; the development of state implementation plans to achieve the ambient air quality standards; more stringent motor vehicle emission standards; requirements for EPA to establish standards of performance for new and modified sources of air pollution; requirements for establishing national emission standards for hazardous air pollutants; and provisions for citizen suits.

In 1977, the Clean Air Act was again amended to enable EPA to make some corrections in the country's efforts to achieve and maintain ambient standards. Additional time was allowed for achieving the National Ambient Air Quality Standards in the dirtiest areas, and the concept of Prevention of Significant Deterioration was made part of the law. Additionally, EPA was given the flexibility to allow for some growth in dirty areas.

The 1990 Amendments to the Clean Air Act are the most recent changes in effect. Since 1977, the country had been unsuccessful in its efforts to achieve the ambient standards for many pollutants and EPA had been slow in identifying and requiring controls of hazardous air pollutants. The 1990 Amendments to the Clean Air Act provided a listing of hazardous air pollutants (Title III) and required EPA to establish emission standards for certain industrial and commercial sources of these pollutants. Under the 1990 Amendments to the Clean Air Act, EPA was authorized to require states to develop an operating permits program (Title V), to develop a program for reducing emissions of sulfur dioxide from power plants (Title IV), and to establish more stringent control requirements in those areas that continue to have dirty air (Title I). The EPA was also authorized to control and reduce those substances that are considered stratospheric ozone-depleting substances (Title VI) and to reduce emissions from motor vehicles (Title II).

The Clean Air Act (as amended in 1990) is composed of six subchapters, each of which establishes requirements for a portion of the nation's air pollution control effort. Subchapter I requires controls for stationary sources (including sources of toxic air pollutants as modified by Title III of the 1990 amendments to the Clean Air Act). Subchapter II establishes requirements for mobile sources. Subchapter III includes definitions and administrative provisions for development of regulations and for public involvement. In Subchapter IV, the acid rain provisions are prescribed. Subchapter V (referred to as Title V) covers the operating permits program, and Subchapter VI provides requirements for protection of stratospheric ozone and implements the Montreal Protocol, an international agreement to reduce emissions and ultimately eliminate the use of chlorofluorocarbons and other ozone-depleting chemicals.

PROGRAMS REQUIRING CONTROLS ON STATIONARY SOURCES

There are several programs that require the installation and operation of controls on stationary sources of air pollution. These will be briefly described in this section. Before we discuss these programs, a brief discussion of several regulatory terms is provided.

Definition of Terms

1. Air Pollution. Air pollution is defined as the addition to the atmosphere of any material that may have a deleterious effect to life on our planet. It results from natural processes or from anthropogenic, or manmade, sources. The legal definition for air pollutant is "any air pollution agent or combination of such agents, including any physical, chemical, biological, radioactive (including source material, special nuclear material, and by-product material) substance or matter which is emitted into or otherwise enters the ambient air. Such term includes any precursors to the formation of any air pollutant,. . ."[1]

2. Ambient Air[2]. A legal definition for ambient air is that portion of the atmosphere, external to buildings, to which the general public has access. Ambient air generally refers to the air out of doors and beyond the fences that surround a facility. Ambient air is considered a public property, and as such, the public is charged with maintaining its quality.

3. Indoor Air. Indoor air is considered to be the air inside a building or within the boundaries of a factory. Indoor air is considered as a private property and is subject to different controls and regulations.

4. Primary and Secondary Pollutants. These classifications indicate how the various pollutants are formed. Primary pollutants are those that have the same form (state and chemical composition) in the ambient atmosphere as when emitted from sources. Secondary pollutants are those that have changed in form after leaving the source due to oxidation or decay or reaction with other primary pollutants.

Primary pollutants exist in the atmosphere in a solid, liquid, or gaseous form with a variety of physical and chemical characteristics. Chemical characteristics range from inert particles to the highly reactive and acidic gases. Some examples of primary pollutants are particulate matter, lead, sulfur dioxide, carbon monoxide, volatile organic compounds, and oxides of nitrogen. Sources of these primary pollutants include combustion, chemical, and industrial processes and miscellaneous natural and human activities (volcanoes, farming, forest fires, etc.).

Secondary pollutants, like primary pollutants, are either solid or gaseous in form. They enter the atmosphere as one of the primary pollutants but react in an atmospheric chemical process to become something else. Examples of secondary pollutants include ozone and acid rain. Ozone is generated in the atmosphere by the chemical reaction between volatile organic compounds and oxides of nitrogen in the presence of sunlight. Acid rain, the common term for acid deposition, is produced when the primary pollutants of sulfur dioxides and oxides of nitrogen react with moisture in the atmosphere to form sulfuric and nitric acids and return to the earth as rain, snow, fog, dry particles, or gases.

5. Criteria Pollutants. *Criteria pollutant* is a legal term in the United States that classifies those substances, or common air pollutants, in the atmosphere for which health criteria have been established and for which National Ambient Air Quality Standards have been promulgated. Criteria pollutants are widely distributed across the country and are substances for which federal, state, and local emission limitations have been enacted and are the pollutants for which State Implementation Plans have been developed.

Criteria pollutants are emitted to the atmosphere by numerous and diverse stationary and mobile sources. The effects of varying concentrations of these pollutants in the atmosphere have been evaluated. The results of these studies have been described in documents referred to as "criteria documents." The studies show that varying quantities of these pollutants have identifiable impacts on public health and welfare.

The criteria pollutants are particulate matter, carbon monoxide, lead, ozone, nitrogen dioxide, and sulfur dioxide. Carbon monoxide, lead, nitrogen dioxide, and sulfur dioxide are primary pollutants. Ozone is a secondary pollutant, and particulate matter can be either a primary or secondary pollutant. All except carbon monoxide represent a class or family of pollutants with a similar composition.

Particulate matter consists of solid particles or liquid droplets such as bits of dust, smoke, soot, and mists ranging in size between 0.1 and 100 μm in diameter. The particles with a diameter of less than 2.5 μm are of the greatest concern due to their impact on the respiratory system. Particulate matter may consist of inert and nontoxic substances, acid mists, heavy metals, or toxic compounds. Even though the particles themselves may be inert and nontoxic, molecules of the more toxic gaseous pollutants may become attached and enter the body via the respiratory system. Primary particulate matter includes solid, liquid, or gaseous material at the pressure and temperature in the stack that would be expected to become a particulate at ambient temperature and pressure. Secondary particulate matter is formed from gaseous matter that has converted to particulate matter through atmospheric chemical reactions.

Carbon monoxide is a colorless, odorless, and tasteless gas and is a product of incomplete combustion. Because of its strong affinity for the oxygen bonding sites of the hemoglobin in blood, its effects are apparent and measurable. At lower concentrations, effects include headaches and drowsiness. Death can occur with exposures to high concentrations.

Lead is a term used for any lead-containing particle in the atmosphere. Lead particles range in size of from 0.1 to 100 μm in diameter; however, most lead particles are around 1 μm in diameter. The pollutant lead may exist in elemental form but consists mostly of complex lead salts and lead oxide. Once entering the body, lead concentrates in the bones and soft tissues. Accumulations of lead poisons the blood-forming, nervous, and renal systems in the body. It is particularly damaging to children. Lead is also considered a hazardous air pollutant.

Ozone is a secondary pollutant formed in chemical reactions of hydrocarbons and nitrogen oxides in the presence of sunlight. Ozone is a faintly bluish gas with a distinctive sharp odor. It is a powerful oxidizing agent that breaks down quickly. Ozone irritates the respiratory system, causing coughing, choking, and shortness of breath. Prolonged exposure results in a gradual deterioration of lung tissue.

Nitrogen dioxide represents a large complicated "family" of oxides of nitrogen, including nitric oxide, nitrogen oxide, nitrous oxide, and nitrogen dioxide. Nitrogen oxide is a colorless and odorless gas and is a common product of combustion. It, like some of other oxides of nitrogen, is easily oxidized to nitrogen dioxide, which is a red-brown gas with a pungent odor. Nitrogen dioxide is quite reactive and easily combines with water vapor or steam to produce nitrous acid droplets and mist. Nitrogen dioxide contributes to bronchitis, pneumonia, and generally lowered infection resistance. It is very toxic at high concentrations.

Sulfur dioxide also represents a complicated "family" of oxides of sulfur, including sulfur dioxide and sulfur trioxide. It is a product of fuel combustion and some manufacturing processes. Sulfur dioxide is a colorless, nonflammable, nonexplosive gas that has a detectable odor and taste at higher concentrations. Sulfur dioxide easily oxidizes to sulfur trioxide and combines with water to form sulfurous acid and sulfuric acid mists and droplets. In the body sulfur dioxide appears to increase resistance of the airways and increases the load on the operations of the heart and lungs.

6. Hazardous Air Pollutants. The term *hazardous air pollutants* refers to those air pollutants whose presence in the atmosphere is known to result in a variety of adverse health effects, including cancer, reproductive effects, birth defects, and respiratory illness or a variety of adverse environmental effects. Other terms often used interchangeably with hazardous air pollutants include *toxic air pollutants*, *air toxics*, *noncriteria pollutants*, and *regulated substances*. The Clean Air Act, in Section 112(b), and the Federal regulations in 40 CFR Part 61 identify substances that are considered hazardous air pollutants. The Environmental Protection Agency and state and local air pollution control agencies may add other substances to this list for their jurisdictions. These pollutants are all primary pollutants and may be present in the atmosphere as either particulate matter or gases. Some are pesticides, herbicides, or fungicides, and many are also volatile organic hydrocarbons.

7. Ozone-Depleting Substances (ODS)[3]. Evidence is strong that the ozone layer in the upper atmosphere that surrounds the earth and shields it from damaging ultraviolet radiation is being destroyed by the release to the atmosphere of *ozone-depleting substances*. These chemicals include chlorofluorocarbons (used in refrigerators, air conditioners, and foam blowers and as cleaning solvents), halons (used in fire extinguishers), carbon tetrachloride, and methyl chloroform. These chemicals are listed in Section 602 of the Clean Air Act. ODS are generally very stable in the troposphere and only degrade under intense ultraviolet light in the stratosphere. When they break down, they release chlorine or bromine atoms, which then deplete ozone, allowing greater ultraviolet (UV) radiation to reach the earth's surface. Increases in UV radiation exposure can lead to skin cancer, cataracts, immune deficiencies, and crop damage.

8. Greenhouse Gases. Combustion of fossil fuels and other human activities have increased the concentrations of carbon dioxide and other "greenhouse gases" in the atmosphere. Like the glass in a greenhouse, these gases allow radiation from the sun to penetrate the atmosphere to warm the earth's surface and then trap the infrared radiation reflected from the surface. In addition to carbon dioxide, greenhouse gases include methane, chlorofluorocarbons, and nitrous oxide.

9. Indoor Air Contaminants. For the most part, air contaminants that are considered indoor air contaminants are the same as those listed for the ambient air, except the sources differ somewhat and the gases exist in the indoor environment in higher concentrations. The concentrations of contaminants in the indoor environment are greater than for outside air, as the volume of air available for mixing and diluting the air contaminants in the outdoor environment is significantly greater than is available in the indoor environment. Other indoor contaminants include: vapors from commercial and industrial cleaning solutions; pesticides; formaldehyde and other hydrocarbon gases from carpets, furniture, insulation, paints, and office structures; radon; asbestos; and second-hand tobacco smoke. Biological contaminants such as viruses, bacteria, mold, mildew, pollen, and animal dander also are considered indoor air contaminants.

10. National Ambient Air Quality Standards[4]. National ambient air quality standards (NAAQS) are defined as those levels of air quality that are judged, by the Environmental Protection Agency, as necessary to protect the public health (primary standards), including the health of sensitive populations such as asthmatics, children, and the elderly, and to protect the public welfare (secondary standards), including protection against decreased visibility and damage to animals, crops, vegetation, and buildings. National ambient air quality standards for the United States can be found in the Federal Regulations at 40 CFR Part 50.

11. Emission Limitation. Emission limitations are requirements established to limit the quantity, rate, or concentration of emissions of air pollutants to the atmosphere.

12. Reasonably Available Control Technology (RACT). Reasonably available control technology refers to those devices, systems, process modifications, or other apparatus or techniques that are considered as reasonably available taking into account social, economic, and environmental impacts. When discussing RACT, this usually refers to controls on existing sources of air pollutants.

13. Best Available Control Technology (BACT). Any major source subject to the requirements for prevention of significant deterioration, in attainment areas, must conduct an analysis to ensure the application of Best Available Control Technology. BACT is an emissions limitation based on the maximum degree of reduction for each pollutant subject to regulation under the Clean Air Act which would be emitted in significant amounts as defined by EPA regulations from any proposed major stationary source or major modification. BACT is to be determined on a case-by-case basis and is to be achievable through application of production processes or available methods, systems and techniques, including fuel cleaning or treatment, or innovative fuel combustion techniques. In no event shall application of best available control technology result in emissions of any pollutant that would exceed the emissions allowed by any applicable standard under 40 CFR Parts 60 (Standards of Performance for New Stationary Sources) and 61 (National Emission Standards for Hazardous Air Pollutants).[5]

14. Lowest Achievable Emission Rate (LAER). LAER refers to the level of control required of a major source subject to the new source review requirements for nonattainment areas. This requirement applies only to those criteria pollutants for which the area is designated as nonattainment. The term means "for any source, the more stringent rate of emissions based on the following:

- The most stringent emissions limitation that is contained in the implementation plan of any state for such class or category of stationary source, unless the owner or operator of the proposed stationary source demonstrates that such limitations are not achievable; or
- The most stringent emissions limitation that is achieved in practice by such class or category of stationary sources. This limitation, when applied to a modification, means the lowest achievable emissions rate for the new or modified emissions units within a stationary source. In no event shall the application of the term permit a proposed new or modified stationary source to emit any pollutant in excess of the amount allowable under an applicable new source standard of performance."[6]

15. Maximum Achievable Control Technology (MACT). EPA's MACT standards are emissions limits for sources of one or more of the toxic air pollutants listed in Section 112(d) of the Clean Air Act. These limitations are based on the best demonstrated control technology or practices by similar sources or facilities. Such standards rely on a performance-based approach that is reasonable and effective in reducing toxic air emissions. For existing sources, MACT is defined as a stringency equivalent to the average of the best-performing 12% of the sources in the category. For new sources, MACT is defined as the best-controlled system.

Air Pollution Control Programs

The following paragraphs describe some of the principal air pollution control programs and activities in existence in the United States.

1. National Ambient Air Quality Standards (NAAQS)

a. National Ambient Air Quality Standards are levels of air quality concentrations determined by the U.S. Environmental Protection Agency (USEPA) as necessary to protect public health and public welfare.

   The U.S. national ambient air quality standards are summarized in Table 1.

b. State Implementation Plans (SIPs). A State Implementation Plan is a description of a state's strategies, efforts, and activities to achieve and maintain national ambient air quality standards. Similar plans are required for Indian country. These plans include the establishment

Table 1. National Ambient Air Quality Standards[7]

	Primary (Health Related)		Secondary (Welfare Related)	
Pollutant	Type of Average	Standard Level Concentration	Type of Average	Standard Level Concentration
Carbon Monoxide (CO)	8-hour[b]	9 ppm (10 mg/m^3)[a]	No secondary standard	
	1-hour[b]	35 ppm (40 mg/m^3)	No secondary standard	
Lead (Pb)	Maximum quarterly average	1.5 μg/m^3	Same as primary standard	
Nitrogen Dioxide (NO_2)	Annual arithmetic mean	0.053 ppm (100 μg/m^3)	Same as primary standard	
Ozone (O_3)[i]	1-hour[c,h]	0.12 ppm (235 μg/m^3)	Same as primary standard	
	8-hour[d]	0.08 ppm (157 μg/m^3)	Same as primary standard	
PM_{10}[i]	Annual arithmetic mean	50 μg/m^3	Same as primary standard	
	24-hour[e]	150 μg/m^3	Same as primary standard	
$PM_{2.5}$[i]	Annual arithmetic mean	15 μg/m^3	Same as primary standard	
	24-hour[e]	65 μg/m^3	Same as primary standard	
Sulfur Dioxide (SO_2)	Annual arithmetic mean	0.03 ppm (80 μg/m^3)	3-hour[b]	0.50 ppm (1300 μg/m^3)
	24-hour[b]	0.14 ppm (365 μg/m^3)		

[a]Parenthetical value is an approximate equivalent concentration.
[b]Not to be exceeded more than once per year.
[c]Not to be exceeded more than once per year on average.
[d]3-year average of annual fourth-highest concentration.
[e]The pre-existing form is exceedance based. The revised form is the ninety-ninth percentile.
[f]Spatially averaged over designated monitors.
[g]The form is the ninety-eighth percentile.
[h]The ozone 1-hour standard applies only to areas that were designated nonattainment when the ozone 8-hour standard was adopted in July 1997.
[i]Implementation pending action on court decision [American Trucking Associations, Inc. v. USEPA, Nos. 97-1440 and 97-1441 (D.C. Cir. May 14, 1999)].
Source: 40 CFR Part 50.

and enforcement of emission limitations for existing stationary sources (commonly referred to as RACT), description of procedures and programs to monitor ambient air concentrations and to monitor emissions from stationary sources, regulation of modification and construction of stationary sources [new source review for attainment areas (PSD program) and nonattainment areas], and enforceable strategies for reducing emissions and impact from mobile sources. The requirements for such plans differ for those areas that have air quality cleaner than the limits established for the NAAQS (attainment areas) than for those areas with air quality poorer or dirtier than NAAQS (nonattainment areas). The requirements for such plans are published at 40 CFR Part 51. Such plans are adopted by states and submitted to EPA for review and approval. Following EPA approval, the SIP becomes enforceable by citizens and the U.S. Government. If a SIP does not adequately comply with federal requirements, EPA is required to promulgate its own Federal Implementation Plan and implement it in those areas that lack approved SIPs. State plans as approved or Federal Implementation Plans as adopted are promulgated and published at 40 CFR Part 52.

2. Standards of Performance for New Stationary Sources (NSPS). The NSPS program was authorized in the Clean Air Act of 1970 in Section 111. These standards regulate emissions of criteria and other pollutants from new and modified sources in specific source categories. By focusing activities on new sources and modifications rather than existing sources of pollution, the NSPS prevents new air pollution problems and results in long-term improvements in air quality as the older existing plants are being replaced by new ones. The provisions of the NSPS are designed to ensure that new stationary sources are designed, built, equipped, operated, and maintained so as to keep emissions to a minimum. NSPS were promulgated for those categories of stationary sources that significantly cause or contribute to air pollution that could reasonably be anticipated to endanger public health or welfare. These standards are modified from time to time. The current standards can be found at 40 CFR Part 60.

A standard of performance is a standard for emissions of air pollutants that reflects the degree of emission limitation achievable through the application of the best system of emission reduction that (taking into account the cost of achieving such reduction and any non-air-quality health and environmental impact and energy requirements) the EPA determines has been adequately demonstrated. A list of source categories for which performance standards have been established is found in Table 2.

3. National Emission Standards for Hazardous Air Pollutants (NESHAP). The 1970 Amendments to the Clean Air Act included a provision to address air toxics. Subsequently, NESHAPs were promulgated for a small number of hazardous air pollutants. The Clean Air Act, as amended in 1990, directed EPA to set standards requiring companies to reduce routine emissions of toxic air pollutants. These amendments required EPA to establish and phase in specific performance-based, or MACT, standards for the industries that emit one or more of the listed toxic air pollutants in significant quantities.

Table 2. New Source Performance Standards Source Categories[8]

Municipal waste combustors

Municipal solid waste landfills

Sulfuric acid production units

Fossil-fuel-fired steam generators

Electric utility steam-generating units

Industrial–commercial–institutional steam-generating units

Small industrial–commercial steam-generating units

Incinerators

Portland cement plants

Nitric acid plants

Sulfuric acid plants

Hot asphalt facilities

Petroleum refineries

Storage vessels for petroleum liquids

Volatile organic liquid storage vessels

Secondary lead smelters

Secondary brass and bronze production plants

Primary emissions from basic oxygen process furnaces

Secondary emissions from basic oxygen process steel-making facilities

Sewage treatment plants

Primary copper smelters

Primary zinc smelters

Primary lead smelters

Primary aluminum reduction plants

Phosphate fertilizer industry: wet-process phosphoric acid plants

Phosphate fertilizer industry: Superphosphoric acid plants

Phosphate fertilizer industry: Diammonium phosphate plants

Phosphate fertilizer industry: Triple superphosphate plants

Phosphate fertilizer industry: Granular triple superphosphate storage facilities

Coal preparation plants

Ferroalloy production facilities

Steel plants: Electric arc furnaces

Steel plants: Electric arc furnaces and argon–oxygen decarburization vessels

Kraft pulp mills

Glass manufacturing plants

Grain elevators

Surface coating of metal furniture

Stationary gas turbines

Lime manufacturing plants

Lead-acid battery manufacturing plants

Metallic mineral processing plants

Automobile and light-duty truck surface coating operations

Phosphate rock plants

Ammonium sulfate manufacture

Graphic arts industry: Publication rotogravure printing

Pressure-sensitive tape and label surface coating operations

Industrial surface coating: Large appliances

Metal coil surface coating

Asphalt processing and asphalt roofing manufacture

Equipment leaks in synthetic organic

Chemical manufacturing industry

Beverage can surface coating industry

Bulk gasoline terminals

New residential wood heaters

Rubber tire manufacturing industry

Polymer manufacturing industry

Flexible vinyl and urethane coating and printing

Equipment leaks in petroleum refineries

Synthetic fibre production facilities

Synthetic organic chemical manufacturing industry air oxidation unit processes

Petroleum dry cleaners

Equipment leaks from onshore natural gas processing plants

Onshore natural gas processing; SO_2 emissions

Synthetic organic chemical manufacturing industry distillation operations

Nonmetallic mineral processing plants

Wool fiberglass insulation manufacturing plants

Petroleum refinery wastewater systems

Synthetic organic chemical manufacturing reactor processes

Magnetic tape coating facilities

Industrial surface coating: Surface coating of plastic parts for business machines

Calciners and dryers in mineral industries

Polymeric coating of supporting substrates facilities

Municipal solid waste landfills

Table 3. Hazardous Air Pollutants[9]

Chemical Abstract Service (CAS) Number	Chemical Name
75070	Acetaldehyde
60355	Acetamide
75058	Acetonitrile
98862	Acetophenone
53963	2-Acetylaminofluorene
107028	Acrolein
79061	Acrylamide
79107	Acrylic acid
107131	Acrylonitrile
107051	Allyl chloride
92671	4-Amino biphenyl
62533	Aniline
90040	*o*-Anisidine
1332214	Asbestos
71432	Benzene (including benzene from gasoline)
92875	Benzidine
98077	Benzotrichloride
100447	Benzyl chloride
92524	Biphenyl
117817	bis(2-*E*thylhexyl)phthalate (DEHP)
542881	bis(*C*hloromethyl)ether
75252	Bromoform
106990	1,3-Butadiene
156627	Calcium cyanamide
133062	Captan
63252	Carbaryl
75150	Carbon disulfide
56235	Carbon tetrachloride
463581	Carbonyl sulfide
120809	Catechol
133904	Chloramben
57749	Chlordane
7782505	Chlorine
79118	Chloroacetic acid
532274	2-Chloroacetophenone
108907	Chlorobenzene
510156	Chlorobenzilate
67663	Chloroform
107302	Chloromethyl methyl ether
126998	Chloroprene
1319773	Cresols/cresylic acid (isomers and mixture)
95487	*o*-Cresol
108394	*m*-Cresol
106445	*p*-Cresol
98828	Cumene
94757	2,4-D, salts and esters
3547044	DDE
334883	Diazomethane
132649	Dibenzofurans
96128	1,2-Dibromo-3-chloropropane
84742	Dibutylphthalate
106467	1,4-Dichlorobenzene (*p*)
91941	3,3-Dichlorobenzidene
111444	Dichloroethyl ether (bis(2-chloroethyl)ether)
542756	1,3-Dichloropropene
62737	Dichlorvos
111422	Diethanolamine
121697	N,N-Diethyl aniline (N,N-dimethyl aniline)
64675	Diethyl sulfate
119904	3,3-Dimethoxybenzidine
60117	Dimethyl aminoazobenzene
119937	3,3′-Dimethyl benzidine
79447	Dimethyl carbamoyl chloride
68122	Dimethyl formamide
57147	1,1-Dimethyl hydrazine
131113	Dimethyl phthalate
77781	Dimethyl sulfate
534521	4,6-Dinitro-*o*-cresol, and salts
51285	2,4-Dinitrophenol
121142	2,4-Dinitrotoluene
123911	1,4-Dioxane (1,4-diethylene oxide)
122667	1,2-Diphenylhydrazine
106898	Epichlorohydrin (l-chloro-2,3-epoxypropane)
106887	1,2-Epoxybutane
140885	Ethyl acrylate
100414	Ethyl benzene
51796	Ethyl carbamate (urethane)
75003	Ethyl chloride (chloroethane)
106934	Ethylene dibromide (dibromoethane)
107062	Ethylene dichloride (1,2-dichloroethane)
107211	Ethylene glycol
151564	Ethylene imine (aziridine)
75218	Ethylene oxide
96457	Ethylene thiourea
75343	Ethylidene dichloride (1,1-dichloroethane)
50000	Formaldehyde
76448	Heptachlor
118741	Hexachlorobenzene
87683	Hexachlorobutadiene
77474	Hexachlorocyclopentadiene
67721	Hexachloroethane
822060	Hexamethylene-1,6-diisocyanate
680319	Hexamethylphosphoramide
110543	Hexane
302012	Hydrazine
7647010	Hydrochloric acid
7664393	Hydrogen fluoride (hydrofluoric acid)
123319	Hydroquindne
78591	Isophorone
58899	Lindane (all isomers)
108316	Maleic anhydride
67561	Methanol
72435	Methoxychlor
74839	Methyl bromide (bromomethane)
74873	Methyl chloride (chloromethane)
71556	Methyl chloroform (1,1,1-trichloroethane)
78933	Methyl ethyl ketone (2-butanone)
60344	Methyl hydrazine
74884	Methyl iodide (iodomethane)
108101	Methyl isobutyl ketone (hexone)
624839	Methyl isocyanate
80626	Methyl methacrylate
1634044	Methyl *tert*-butyl ether

(Continued)

Table 3. *(Continued)*

Chemical Abstract Service (CAS) Number	Chemical Name
101144	4,4-Methylene bis(2-chloroaniline)
75092	Methylene chloride (dichloromethane)
101688	Methylene diphenyl diisocyanate (MDI)
101779	4,4′-Methylenedianiline
91203	Naphthalene
98953	Nitrobenzene
92933	4-Nitrobiphenyl
100027	4-Nitrophenol
79469	2-Nitropropane
684935	N-Nitroso-N-methylurea
62759	N-Nitrosodimethylamine
59892	N-Nitrosomorpholine
56382	Parathion
82688	Pentachloronitrobenzene (quintobenzene)
87865	Pentachlorophenol
108952	Phenol
106503	*p*-Phenylenediamine
75445	Phosgene
7803512	Phosphine
7723140	Phosphorus
85449	Phthalic anhydride
1336363	Polychlorinated biphenyls (aroclors)
1120714	1,3-Propane sultone
57578	β-Propiolactone
123386	Propionaldehyde
114261	Propoxur (baygon)
78875	Propylene dichloride (1,2-dichloropropane)
75569	Propylene oxide
75558	1,2-Propylenimine (2-methyl aziridine)
91225	Quinoline
106514	Quinone
100425	Styrene
96093	Styrene oxide
1746016	2,3,7,8-Tetrachlorodibenzo-*p*-dioxin
79345	1,1,2,2-Tetrachloroethane
127184	Tetrachloroethylene (perchloroethylene)
7550450	Titanium tetrachloride
108883	Toluene
95807	2,4-Toluene diamine
584849	2,4-Toluene diisocyanate
95534	*o*-Toluidine
8001352	Toxaphene (chlorinated camphene)
120821	1,2,4-Trichlorobenzene
79005	1,1,2-Trichloroethane
79016	Trichloroethylene
95954	2,4,5-Trichlorophenol
88062	2,4,6-Trichlorophenol
121448	Triethylamine
1582098	Trifluralin
540841	2,2,4-Trimethylpentane
108054	Vinyl acetate
593602	Vinyl bromide
75014	Vinyl chloride
75354	Vinylidene chloride (1,1-dichloroethylene)
1330207	Xylenes (isomers and mixture)
95476	*o*-Xylenes
108383	*m*-Xylenes
106423	*p*-Xylenes
0	Antimony compounds
0	Arsenic compounds (inorganic, including arsine)
0	Beryllium compounds
0	Cadmium compounds
0	Chromium compounds
0	Cobalt compounds
0	Coke oven emissions
0	Cyanide compounds[a]
0	Glycol ethers[b]
0	Lead compounds
0	Manganese compounds
0	Mercury compounds
0	Fine mineral fibers[c]
0	Nickel compounds
0	Polycyclic organic matter[d]
0	Radionuclides (including radon)[e]
0	Selenium compounds

Note: For all listings herein that contain the word *compounds* and for glycol ethers, the following applies: Unless otherwise specified, these listings are defined as including any unique chemical substance that contains the named chemical (i.e., antimony, arsenic, etc.) as part of that chemical's infrastructure.

[a]X′CN, where X = H′ or any other group where a formal dissociation may occur. For example, KCN or $Ca(CN)_2$.

[b]Includes mono- and diethers of ethylene glycol, diethylene glycol, and triethylene glycol $R\text{-}(OCH_2CH_2)_n\text{-}OR'$, where $n = 1, 2,$ or 3; R = alkyl or aryl groups; and R′ = R, H, or groups that, when removed, yield glycol ethers with the structure $R\text{-}(OCH_2CH)_n\text{-}OH$. Polymers are excluded from the glycol category.

[c]Includes mineral fiber emissions from facilities manufacturing or processing glass, rock, or slag fibers (or other mineral-derived fibers) of average diameter 1 μm or less.

[d]Includes organic compounds with more than one benzene ring, and which have a boiling point greater than or equal to 100°C.

[e]A type of atom that spontaneously undergoes radioactive decay.

Following promulgation of these standards, EPA is required continue to analyze the impact of the remaining or residual emissions on human health and to determine if additional control measures are necessary to ensure that there is no significant risk to human health or adverse environmental effect.

a. List of Pollutants. Pollutants for which national emission standards for hazardous air pollutants have been established include: arsenic, asbestos, beryllium, benzene, mercury, radionuclides, and vinyl chloride.

Table 4. Categories of Sources of Hazardous Air Pollutants[10]

Industry group	Source category[a]
Fuel combustion	
	Engine test facilities
	Industrial boilers
	Institutional/commercial boilers
	Process heaters
	Stationary internal combustion engines
	Stationary turbines
Nonferrous metals processing	
	Primary aluminum production
	Primary copper smelting
	Primary lead smelting
	Primary magnesium refining
	Secondary aluminum production
	Secondary lead smelting
Ferrous metals processing	
	Coke by-product plants
	Coke ovens: Charging, top side, and door leaks
	Coke ovens: Pushing, quenching, and Battery stacks
	Ferroalloys production
	Integrated iron and steel manufacturing
	Iron foundries
	Steel foundries
	Steel pickling—HCl process
Mineral products processing	
	Alumina processing
	Asphalt concrete manufacturing
	Asphalt processing
	Asphalt roofing manufacturing
	Asphalt/coal tar application—metal pipes
	Chromium refractories production
	Clay products manufacturing
	Lime manufacturing
	Mineral wool production
	Portland cement manufacturing
	Taconite iron ore processing
	Wool fiberglass manufacturing
Petroleum and natural gas production and refining	
	Oil and natural gas production
	Natural gas transmission and storage
	Petroleum refineries—catalytic cracking (fluid and other) units, catalytic reforming units, and sulfur plant units
	Petroleum refineries—other sources not distinctly listed
Liquids distribution	
	Gasoline distribution (Stage 1)
	Marine vessel loading operations
	Organic liquids distribution (non-gasoline)
Surface coating processes	
	Aerospace industries
	Auto and light duty truck (surface coating)
	Flat wood paneling (surface coating)
	Large appliance (surface coating)
	Magnetic tapes (surface coating)
	Manufacture of paints, coatings, and adhesives
	Metal can (surface coating)
	Metal coil (surface coating)
	Metal furniture (surface coating)
	Miscellaneous metal parts and products (surface coating)
	Paper and other webs (surface coating)
	Plastic parts and products (surface coating)
	Printing, coating, and dyeing of fabrics
	Printing/publishing (surface coating)
	Shipbuilding and ship repair (surface coating)
	Wood furniture (surface coating)
Waste treatment and disposal	
	Hazardous waste incineration
	Municipal landfills
	Off-site waste and recovery operations
	Publicly owned treatment works (POTW) emissions
	Sewage sludge incineration
	Site remediation
Agricultural chemicals production	
	4-Chloro-2-methylphenoxyacetic acid production
	2,4-D Salts and esters production
	4,6-Dinitro-*o*-cresol production
	Butadiene-furfural cotrimer (R-11) production[b]
	Captafol production[b]
	Captan production[b]
	Chlorone production[b]
	Chlorothalonil production[b]
	Dacthal(TM) production[b]
	Sodium pentachlorophenate production
	Tordon(TM) acid production[b]
Fibers production processes	
	Acrylic fibers/modacrylic fibers production
	Ryan production
	Spandex production
Food and agriculture processes	
	Baker's yeast manufacturing
	Cellulose food casing manufacturing
	Vegetable oil production
Pharmaceutical production processes	
	Pharmaceuticals Production[b]
Polymers and resins production	
	Acetal resins production
	Acrylonitrile–butadiene–styrene production
	Alkyd resins production
	Amino resins production
	Boat manufacturing
	Butyl rubber production
	Carboxymethyl cellulose production
	Cellophane production
	Cellulose ethers production
	Epichlorohydrin elastomers production
	Epoxy resins production
	Ethylene–propylene rubber production
	Flexible polyurethane foam production

(Continued)

Table 4. Categories of Sources of Hazardous Air Pollutants[10] *(Continued)*

- Hypalon$^{(TM)}$ Production[b]
- Maleic anhydride copolymers production
- Methyl cellulose production
- Methyl methacrylate–acrylonitrile–butadiene–styrene production[b]
- Methyl methacrylate–butadiene–styrene terpolymers production[b]
- Neoprene production
- Nitrile butadiene rubber production
- Nitrile resins production
- Non-nylon polyamides production
- Phenolic resins production
- Polybutadiene rubber production[b]
- Polycarbonates production[b]
- Polyester resins production
- Polyether polyols production
- Polyethylene terephthalate production
- Polymerized vinylidene chloride production
- Polymethyl methacrylate resins production
- Polystyrene production
- Polysulfide rubber production[b]
- Polyvinyl acetate emulsions production
- Polyvinyl alcohol production
- Polyvinyl butyral production
- Polyvinyl chloride and copolymers production
- Reinforced plastic composites production
- Styrene–acrylonitrile production
- Styrene–butadiene rubber and latex production[b]

Production of inorganic chemicals

- Ammonium sulfate production—caprolactam by-product plants
- Antimony oxides manufacturing
- Carbon black production
- Chlorine production[b]
- Cyanide chemicals manufacturing
- Fume silica production
- Hydrochloric acid production
- Hydrogen fluoride production
- Phosphate fertilizers production
- Phosphoric acid manufacturing
- Uranium hexafluoride production

Production of organic chemicals

- Ethylene processes
- Quaternary ammonium compounds production
- Synthetic organic chemical manufacturing

Miscellaneous processes

- Aerosol can-filling facilities
- Benzyltrimethylammonium chloride production
- Carbonyl sulfide production
- Chelating agents production
- Chlorinated paraffins production[b]
- Chromic acid anodizing
- Commercial dry cleaning (perchloroethylene)—transfer machines
- Commercial sterilization facilities
- Decorative chromium electroplating
- Dry cleaning (petroleum solvent)
- Ethylidene norbornene production[b]
- Explosives production
- Flexible polyurethane foam fabrication operations
- Friction products manufacturing
- Halogenated Solvent Cleaners
- Hard chromium electroplating
- Hydrazine Production
- Industrial cleaning (perchloroethylene)—dry-to-dry machines
- Industrial dry cleaning (perchloroethylene)—transfer machines
- Industrial process cooling towers
- Leather tanning and finishing operations
- OBPA/1,3-diisocyanate production[b]
- Paint stripper users
- Photographic chemicals production
- Phthalate plasticizers production
- Plywood/particle board manufacturing
- Pulp and paper production
- Rocket engine test firing
- Rubber chemicals manufacturing
- Semiconductor manufacturing
- Symmetrical tetrachloropyridine production[b]
- Tetrahydrobenzaldehyde production
- Tire production

Categories of area sources[c]

- Chromic acid anodizing
- Commercial dry cleaning (perchloroethylene)—dry-to-dry machines
- Commercial dry cleaning (perchloroethylene)—transfer machines
- Commercial sterilization facilities
- Decorative chromium electroplating
- Hard chromium electroplating
- Secondary lead smelting
- Halogenated solvent cleaners

[a]Only sources within any category located at a major source shall be subject to emission standards under Section 112 unless a finding is made of a threat of adverse effects to human health or the environment for the area sources in a category. All listed categories are exclusive of any specific operations or processes included under other categories that are listed separately.

[b]Equipment handling specific chemicals for these categories or subsets of these categories are subject to a negotiated standard for equipment leaks contained in the Hazardous Organic NESHAP (HON), which was promulgated on April 22, 1994. The HON includes a negotiated standard for equipment leaks from the SOCMI category and 20 non-SOCMI categories (or subsets of these categories). The specific processes affected within the categories are listed in Section XX.X0(c) of the March 6, 1991 Federal Register notice (56 FR 9315).

[c]A finding of threat of adverse effects to human health or the environment was made for each category of area sources listed.

In addition to those substances identified by EPA, the list of hazardous air pollutants in Table 3 were established within the Clean Air Act Amendments of 1990.

4. Acidic Deposition[11] Acidic deposition, or acid rain as it is commonly known, occurs when emissions of sulfur dioxide (SO_2) and oxides of nitrogen (NO_x) react in the atmosphere with water, oxygen, and oxidants to form various acidic compounds. These compounds fall to the earth in either dry form (such as gas and particles) or wet form (such as rain, snow, and fog).

Electric utility plants account for about two-thirds of annual SO_2 emissions and one-third of NO_x emissions in the United States. Mobile sources also contribute significantly to NO_x emissions. Overall, over 20 million tons of SO_2 and NO_x from all sources are emitted into the atmosphere each year.

Acid rain causes acidification of lakes and streams and contributes to damage of trees at higher elevations (for example, red spruce trees above 2000 feet in elevation). In addition, acid rain accelerates the decay of building materials and paints. Also, when in the atmosphere, SO_2 and NO_x gases and their particulate matter derivatives, sulfates and nitrates, contribute to visibility degradation and impact public health.

Title IV of the Clean Air Act requires the reduction of annual SO_2 emissions by 10 million tons below 1980 levels. To achieve these reductions, the law requires a tightening of the restrictions placed on fossil-fuel-fired power plants. The initial requirements began in 1995 (Phase I) and affected the larger coal-burning electric utility plants (263 units) located in 21 Eastern and Midwestern states. Phase II begins in 2000 and will encompass approximately 2000 sources throughout the country. Federal regulations for the acid rain program can be found at 40 CFR Parts 73 to 78.

Title IV of the Clean Air Act provides for the use of market-based techniques to aid in achieving the required reductions. These market-based techniques include an allowance trading system that harnesses the incentives of the free market to reduce pollution. Under this system, affected utility units were allocated allowances based on their historic fuel consumption and a specific emissions rate. Each allowance permits a unit to emit 1 ton of SO_2 during or after a specified year. For each ton of SO_2 discharged in a given year, one allowance is retired; that is, it can no longer be used.

These allowances may be bought, sold, or banked. Any person may acquire allowances and participate in the trading system. However, regardless of the number of allowances a source holds, it may not emit at levels that would violate other federal or state emission limitations. To firmly restrict emissions and ensure that environmental benefits will be achieved and maintained, the Clean Air Act sets a permanent ceiling (or cap) of 8.95 million allowances for total annual allowance allocations to utilities.

The NO_x component of the Acid Rain Program is more traditional than "cap and trade" in that it establishes an emissions rate limit for all affected utilities. Flexibility is introduced to this command and control measure, however, through compliance options such as emissions averaging, whereby a utility can meet the standard emission limitations by averaging the emission rates of two or more boilers. This allows utilities to overcontrol at units where it is technically easier to control emissions, thereby achieving emission reductions at a lower cost.

Phase I of the NO_x program began in 1996 and affects dry-bottom, wall-fired boilers and tangentially fired boilers. Starting in 2000, Phase II will also include cell burner, cyclones, wet-bottom, and vertically fired boilers and will achieve a 2 million ton reduction in NO_x per year compared to levels without the program.

5. Air Pollution Permits. In air pollution control there are basically two types of permitting programs. The first is a program for reviewing and permitting proposed projects that will result in air pollution emissions. This program is known as the new source review (NSR) or preconstruction permitting program. The second program is for permitting the ongoing operation of sources that emit air pollutants. The permits issued under this program are known as *operating permits*.

The *New Source Review program* requires an air pollution source owner or operator to apply for, and obtain, a permit to construct prior to breaking ground for an entirely new facility, a new addition to an existing facility, or the modification of operations at an existing facility. The permitting agency reviews the impact that the increases in emissions of air pollutants from the facility will have on the ambient air and determines what air pollution controls should be required on the new activity that will be constructed. The NSR permit, which contains the air pollution control requirements that the new or modified facility must meet, is then issued allowing construction.

The New Source Review Program is an integral part of the State Implementation Plan for achieving and maintaining ambient air quality standards. The NSR requirements for a particular source differ depending on the air quality of the area in which a source is located.

The EPA set national ambient air quality standards (NAAQS) to protect the health of the general public. The pollutants for which these ambient standards have been set are sulfur dioxide (SO_2), nitrogen dioxide (NO_2), particulate matter [nominally below 2.5 μm in diameter ($PM_{2.5}$)], lead, carbon monoxide (CO), and ground-level ozone, which is controlled by limiting emissions of volatile organic compounds (VOC) and nitrogen oxides (NO_x). Areas of the country that have air quality that is better than EPA's ambient standards are called *attainment areas*. Areas whose air quality does not meet EPA ambient standards are designated *nonattainment*. For attainment areas and for areas that cannot be classified as either attainment or nonattainment, Part C of the Clean Air Act sets up the prevention of significant deterioration (PSD) program to maintain the air quality and to prevent its deterioration. In these situations, major sources of air pollution must utilize best available control technology for new construction and modifications and ensure that their presence does not cause the air quality in the vicinity of the facility to be significantly deteriorated. PSD also provides special protection for areas such as National Parks and National Wilderness Areas. The requirements established to allow the construction of new or modified major sources in nonattainment areas are mandated under Part D of the Clean Air Act. Major sources may be constructed or modified in nonattainment areas if "lowest achievable emission rate" technology is used, if the increases in air pollution emission resulting from the operation of the new or modified facility are offset by reductions of comparable pollutants in the affected area, and if other specific requirements identified in the Clean Air Act are addressed.

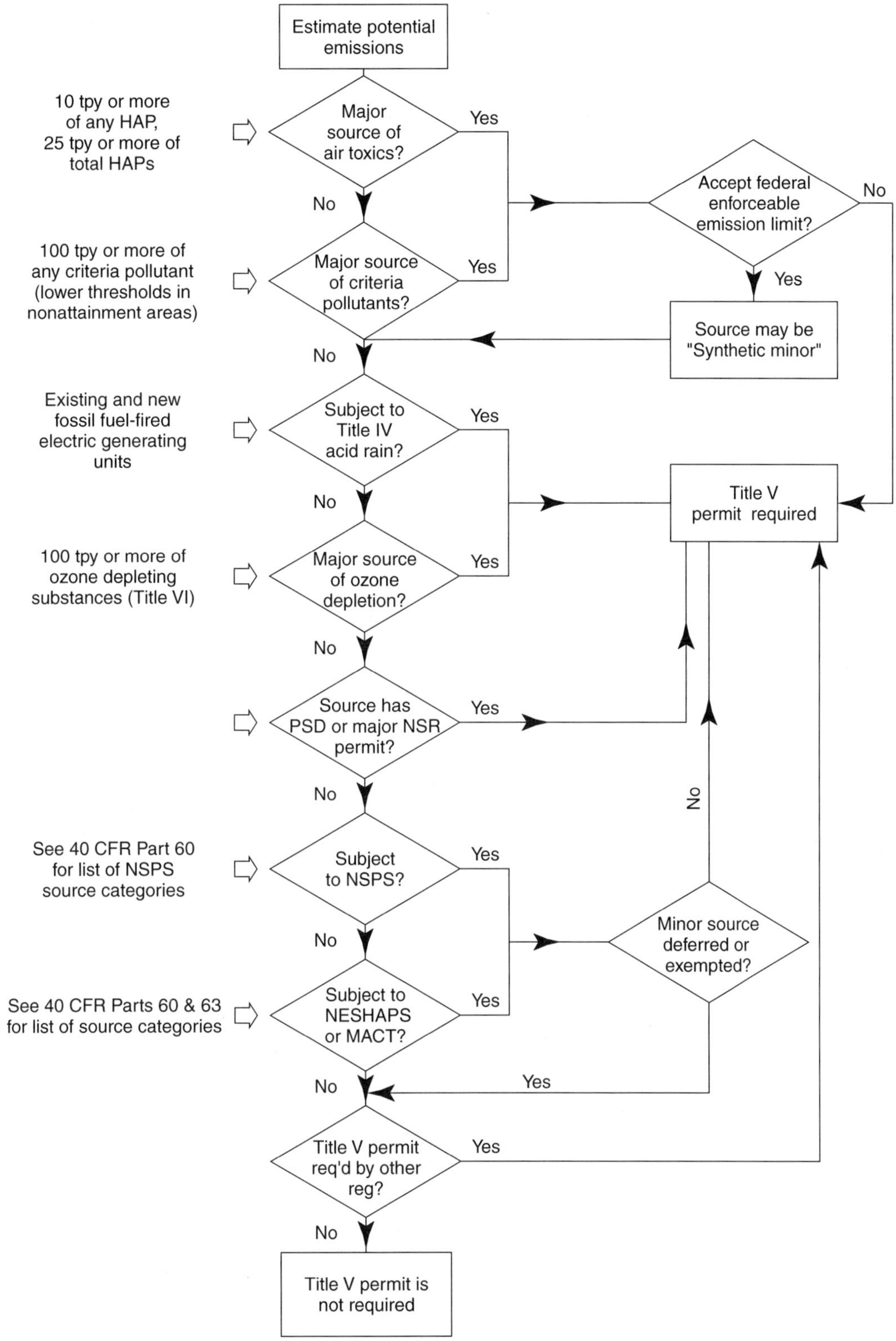

Figure 1. Sources Required to Obtain a Title V Permit.

The EPA has promulgated regulations to address the requirements of Parts C and D of the Clean Air Act. These regulations, which can be found in 40 CFR Parts 51 and 52, focus on permitting of "major" new sources or modifications of major sources. For minor changes to a source or for construction of "smaller" sources of air pollution, state and local permitting authorities administer minor NSR programs. Although the minor NSR program is completed prior to construction, it is generally less complicated and rigorous than the NSR program for major sources.

The *operating permits program*, required by Title V of the Clean Air Act, was designed to streamline the way federal, state, tribal, and local authorities regulate air pollution. This is done by consolidating all air pollution control requirements

Table 5. ODP and CAS Numbers of Class I Ozone-Depleting Substances[12]

Chemical Name Ozone-depleting substances are identified using an international numbering system, referred to as the ASHRAE Standard 34 scheme.		ODP	CAS Number
Group I (from Section 602 of the CAA)			
CFC-11	Trichlorofluoromethane	1.0	75-69-4
CFC-12	Dichlorodifluoromethane	1.0	75-71-8
CFC-113	1,1,1-Trichlorotrifluoroethane	0.8	354-58-5
	1,1,2-Trichlorotrifluoroethane	0.8	76-13-1
CFC-114	Dichlorotetrafluoroethane	1.0	76-14-2
CFC-115	Monochloropentafluoroethane	0.6	76-15-3
Group II (from Section 602 of the CAA)			
Halon 1211	Bromochlorodifluoromethane	3.0	353-59-3
Halon 1301	Bromotrifluoromethane	10.0	75-63-8
Halon 2402	Dibromotetrafluoroethane	6.0	124-73-2
Group III (from Section 602 of the CAA)			
CFC-13	Chlorotrifluoromethane	1.0	75-72-9
CFC-111	Pentachlorofluoroethane	1.0	354-56-3
CFC-112	Tetrachlorodifluoroethane	1.0	76-12-0
CFC-211	Heptachlorofluoropropane	1.0	422-78-6
CFC-212	Hexachlorodifluoropropane	1.0	3182-26-1
CFC-213	Pentachlorotrifluoropropane	1.0	2354-06-5
CFC-214	Tetrachlorotetrafluoropropane	1.0	29255-31-0
CFC-215	Trichloropentafluoropropane	1.0	1599-41-3
CFC-216	Dichlorohexafluoropropane	1.0	661-97-2
CFC-217	Chloroheptafluoropropane	1.0	422-86-6
Group IV (from Section 602 of the CAA)			
CCl_4	Carbon tetrachloride	1.1	56-23-5
Group V (from Section 602 of the CAA)			
Methyl chloroform	1,1,1-trichloroethane	0.1	71-55-6
Group VI (40 CFR Part 82.13 Appendix A)			
CH_3Br	Methyl bromide	0.7	74-83-9
Group VII (40 CFR Part 82.13 Appendix A)			
$CHFBr_2$		1.0	
CHF_2Br (HBFC-22B1)		0.74	
CH_2FBr		0.73	
C_2HFBr_4		0.3–0.8	
$C_2HF_2Br_3$		0.5–1.8	
$C_2HF_3Br_2$		0.4–1.6	
C_2HF_4Br		0.7–1.2	
$C_2H_2FBr_3$		0.1–1.1	
$C_2H_2F_2Br_2$		0.2–1.5	
$C_2H_2F_3Br$		0.7–1.6	
$C_2H_3FBr_2$		0.1–1.7	
$C_2H_3F_2Br$		0.2–1.1	
C_2H_4Br		0.07–0.1	
C_3HFBr_6		0.3–1.5	
$C_3HF_2Br_5$		0.2–1.9	
$C_3HF_3Br_4$		0.3–1.8	
$C_3HF_4Br_3$		0.5–2.2	
$C_3HF_5Br_2$		0.9–2.0	
Group VII (Continued)			
C_3HF_6Br		0.7–3.3	
$C_3H_2FBr_5$		0.1–1.9	
$C_3H_2F_3Br_4$		0.2–2.1	
$C_3H_2F_3Br_3$		0.2–5.6	
$C_3H_2F_4Br_2$		0.3–7.5	
$C_3H_2F_5Br$		0.9–1.4	
$C_3H_3FBr_4$		0.08–1.9	
$C_3H_3F_2Br_3$		0.1–3.1	
$C_3H_3F_3Br_2$		0.1–2.5	
$C_3H_3F_4Br$		0.3–4.4	
$C_3H_4FBr_3$		0.03–0.3	
$C_3H_4F_2Br_2$		0.1–1.0	
$C_3H_4F_3Br$		0.07–0.8	
$C_3H_5FBr_2$		0.04–0.4	
$C_3H_5F_2Br$		0.07–0.8	
C_3H_6FBr		0.02–0.7	

into a single, comprehensive "operating permit" that covers all aspects of a source's day-to-day air pollution related activities. This has reduced and eliminated uncertainty about the various emission control regulations and has helped resolve any questions on the requirements that apply to emission units. Operating permits programs are implemented by state and local permitting authorities except in Indian Country where the U.S. Environmental Protection Agency is implementing the program. Federal regulations for the operating permits program can be found at 40 CFR Part 70. Federal regulations for permits in Indian Country and in those locations outside the jurisdiction of EPA approved state or local permitting programs can be found at 40 CFR Part 71.

A significant aspect of the operating permits program is that the operating permit must prescribe methods for determining compliance with each air pollution control requirement, and

Table 6. ODPs and CAS Numbers of Class II Ozone-Depleting Substances[13]

Note: Only commonly used HCFCs are listed. However, all HCFCs are included in Class II, and all are covered by the accelerated phaseout. The full list of Class II substances is available in Section 602 of the Clean Air Act.

Chemical Name		ODP	CAS Number
HCFC-22	Chlorodifluoromethane	0.05	75-45-6
HCFC-123	2,2-Dichloro-1,1,1-trifluoroethane	0.02	306-83-2
HCFC-124	2-Chloro-1,1,1,2-tetrafluoroethane	0.02	2837-89-0
HCFC-141b	1,1-Dichloro-1-fluoroethane	0.1	1717-00-6
HCFC-142b	1-Chloro-1,1-difluoroethane	0.06	75-68-3

based on these methods, the source owner or operator must certify compliance with the permit every year. These compliance determination methods may include the use of emissions monitoring devices or may involve the maintenance of records detailing operating conditions that demonstrate compliance (e.g., hours of operation, sulfur content of fuel, combustion temperature of a unit incinerating pollutant emissions).

6. Stratospheric Ozone Protection. Ozone in the stratosphere serves as a protective shield for the earth by filtering out harmful ultraviolet rays. Exposure to these ultraviolet rays have been linked to the development of eye damage (cataracts) and skin cancer. Because of a concern about measured reduction in the concentrations of ozone in the stratosphere due to usage of chlorofluorocarbons (CFCs) and other ozone-depleting substances, nations of the world have begun to reduce the amounts of such substances. In the United States, the 1990 amendments to the Clean Air Act set a schedule for ending production and importation of these ozone-depleting substances (ODS).

Ozone-Depleting Substances. As defined in Section 602 of the Clean Air Act, a Class I substance is any chemical with an ozone-depletion potential of 0.2 or greater. Class I substances (see Table 5) cause the most damage to the ozone layer and are the first ODS to be phased out. Class II substances (see Table 6) include all the hydrochlorofluorocarbons (HCFCs). In the United States, the production of all listed CFCs and HCFCs is to end by January 2030.

Ozone-depletion potential (ODP) is a factor established by the Environmental Protection Agency to reflect the ozone-depletion potential of a substance, on a mass per kilogram basis, as compared to chlorofluorocarbon-11 (CFC-11). These factors are based upon the substance's atmospheric lifetime, the molecular weight of bromine and chlorine, and the substance's ability to be photolytically disassociated.

REFERENCES

1. Clean Air Act, 42 USC 7602(g).
2. 40 CFR 50.1(e).
3. Ozone Depletion Glossary, EPA website *http://www.epa.gov/ozone/defns.html#ods.*
4. Clean Air Act, 42 USC 7409(b).
5. 40 CFR 52.21(b)(12).
6. Clean Air Act, 42 USC 7501(3).
7. *National Air Quality and Emissions Trends Report, 1997*, EPA 454/R-98-016, U.S. Environmental Protection Agency, Research Triangle Park, NC, 1998.
8. 40 CFR Part 60.
9. Clean Air Act, 42 USC 7412(b).
10. *Federal Register*, February 12, 1998, **63**(29), 7155–66.
11. Acid Rain Program—Program Overview, EPA website *http://www.epa.gov/acidrain/overview.html.*
12. Ozone-depleting substances —EPA website *http://www.epa.gov/ozone/ods.html.*
13. *Ibid.*

2

CONTROL OF GASEOUS POLLUTANTS

Absorption
Robert Jennings Heinsohn
Adsorption
Kenneth E. Noll
Condensation
Anthony J. Buonicore
Vapor Incineration
C. David Cooper
Biological Control Systems
Sarina J. Ergas and Kerry A. Kinney

ABSORPTION

Robert Jennings Heinsohn

Wet scrubber is the generic name of an air pollution control device that uses the process of absorption to separate the pollutant from a process gas stream. Absorption is the physical (sometimes augmented chemically) process in which the pollutant leaves the gas phase and is dissolved in the (scrubbing) liquid phase. Wet scrubbers will be grouped as follows:

- Packed-bed counterflow scrubbers (random or structured packing) (Figure 1)
- Cross-flow (random or structured packing) scrubbers (Figure 2)
- Bubble-plate and tray scrubbers (Figure 3)

Only a counterflow, random packed-bed scrubber will be described in this section. An elementary approach will be presented that contains the essential elements found in more sophisticated techniques. More sophisticated approaches are given in the professional literature[1–3] and sophisticated proprietary programs created by industry. While the approach taken in this section is "plain vanilla," as it were, mastering it will enable users to understand more advanced techniques.[4–6]

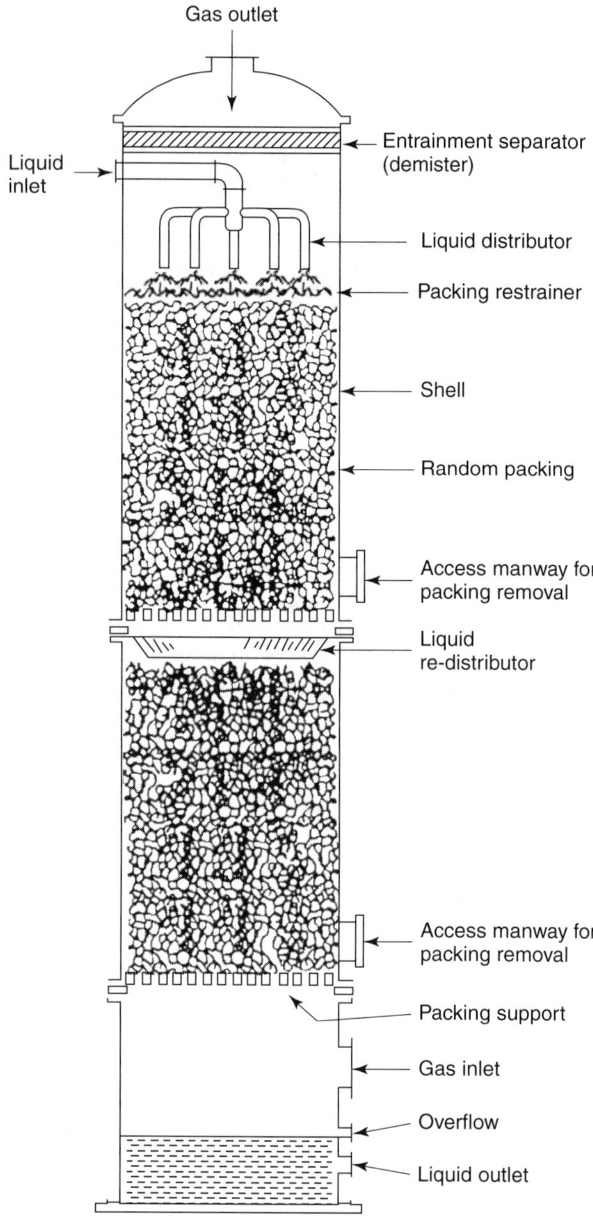

Figure 1. Countercurrent Random Packed-Bed Scrubber (Ref. 3).

GOALS OF DESIGN

It will be presumed that the volumetric flow rate of the gas to be cleaned and the inlet pollutant concentration are given conditions dictated by the process to be controlled. It will also be presumed that the outlet pollutant concentration is dictated by standards set by the state or local air pollution control agency or more stringent standards set by the company. Thus engineers who design or select an absorption system will be asked to recommend:

- Choice of scrubbing liquid
- Packed-bed height (Z), diameter (D)
- Type of packing
- Molar flow rate of a scrubbing liquid (L_s)
- Pressure drop across the bed

CONCEPTS OF ABSORPTION

Gas absorption is the removal of one or more pollutants from a contaminated gas stream by allowing the gas to come into intimate contact with a liquid that enables the pollutants to become dissolved by the liquid. The principal factor dictating performance is the solubility of the pollutants in the absorbing liquid. The rate of transfer in the liquid

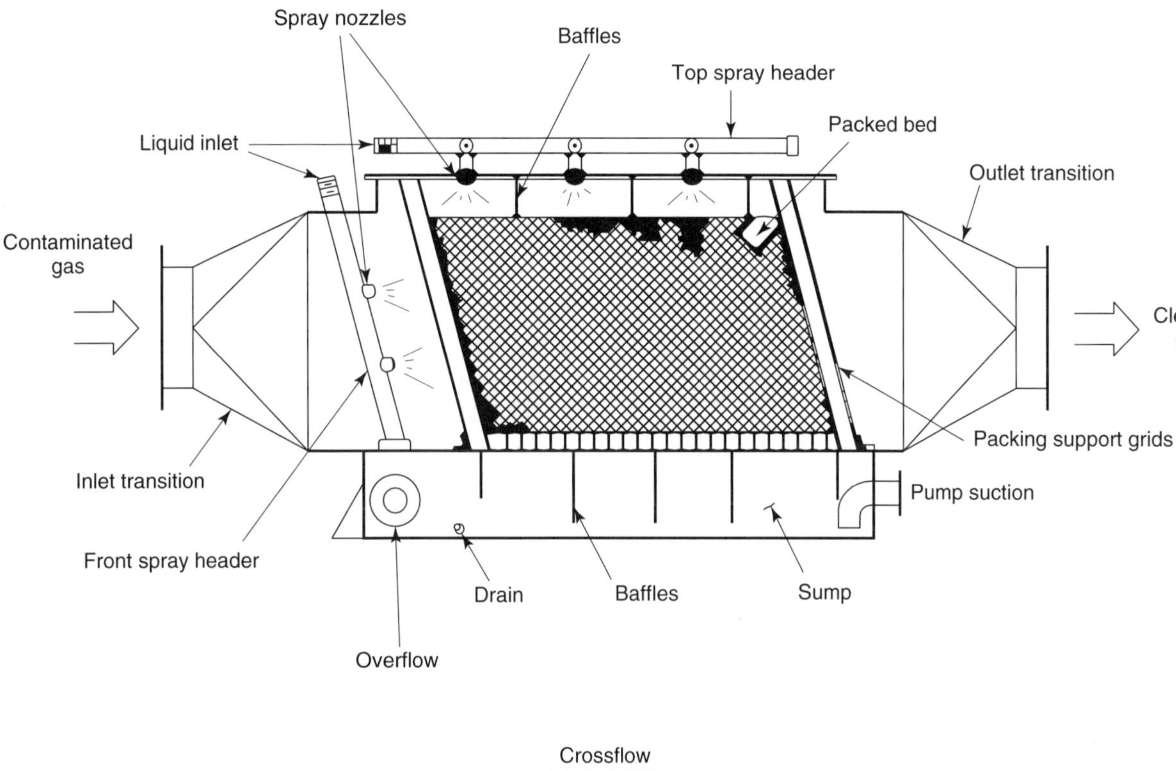

Figure 2. Cross-Flow Random Packed-Bed Scrubber (Ref. 6).

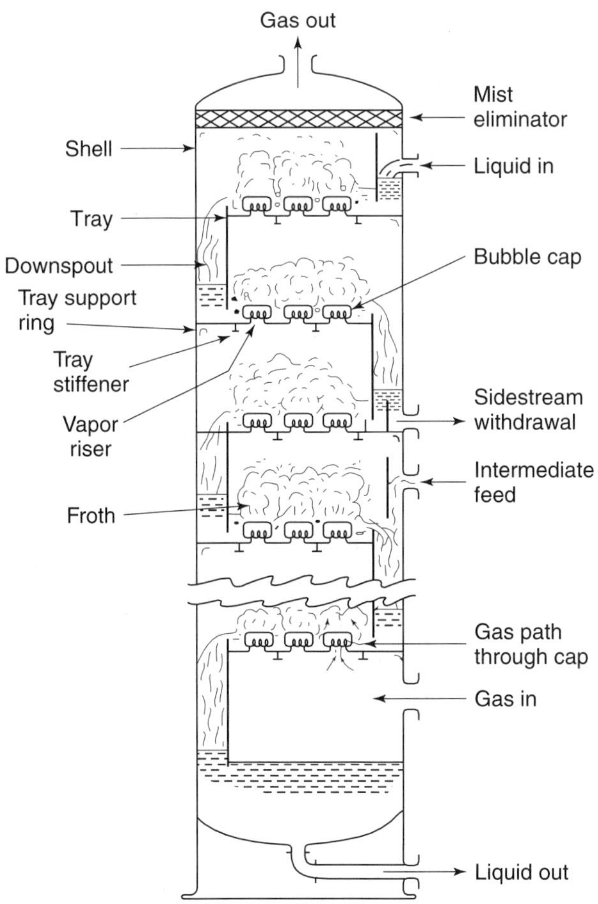

Figure 3. Bubble-Cap Plate Column Scrubber (Ref. 3).

is dictated by the diffusion processes occurring on each side of the gas–liquid interface. Pollutant molecules in the gas phase reach the gas–liquid interface by processes affecting diffusion through the gas-phase boundary layer (film). Once at the interface, the pollutant molecules diffuse into the liquid by processes affecting diffusion through a liquid-phase boundary layer (film). Often the mass transfers through one of these films faster than through the other film, in which case the slower diffusion process is said to "control" the overall mass transfer process.

Pollutants transferred to the scrubbing liquid need to be removed before the liquid is recycled. Thus scrubbers require equipment, storage vessels, additives, etc. to treat the scrubbing liquid. In addition to air permits, users will need to obtain water permits for storage ponds and discharge permits for the transfer of treated liquid to a municipal wastewater system.

Water is commonly used as the scrubbing fluid, but there are many processes in which other fluids are used. If water is chosen, the engineer should review the literature to determine the pollutant's solubility in water. If the pollutant's solubility in water is very small, another scrubbing fluid should be chosen if absorption is to be practical.

The physical principles governing absorption are given in terms of the absorption equilibrium line. On a graph y_j (or P_j) versus x_j (Figure 4) absorption will occur if the coordinates lie above the equilibrium line. So long as the coordinates lie above the equilibrium line, mass will be transferred from the gas to the liquid. If the coordinates (x_j, y_j) lie on the equilibrium line, mass transfer will cease. For low concentrations associated with air pollution, the absorption equilibrium can be expressed

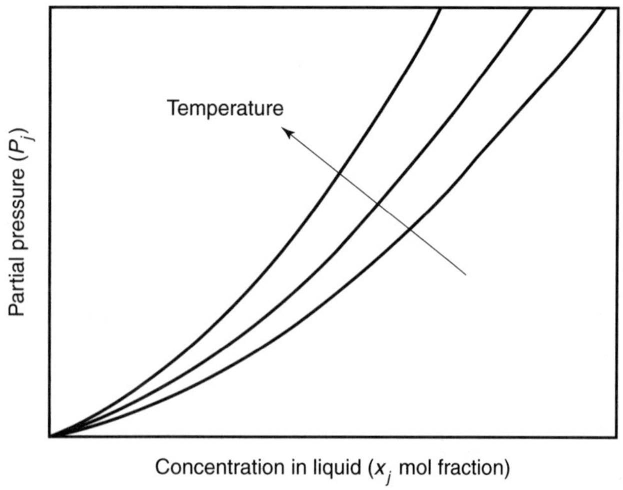

Figure 4. Absorption Equilibrium Isotherms for Contaminant in Air and Water.

by Henry's law in any one of the following forms.

$$P_j = Hc_j \tag{1a}$$

$$P_j = Py_j = Hx_jc_{\rm L} = H'x_j \tag{1b}$$

$$Hc_{\rm L} = H' \tag{1c}$$

$$y_j = (Hc_L/P)x_j = (H'/P)x_j \tag{1d}$$

Values of H' can be obtained from web pages for thermodynamic and physical properties on the Internet, for example, *http: / / www.uic.edu / ~mansoori / Thermodynamic.Data.and. Property_html*, or from Refs. 7–10. There is no unanimity on how to express Henry's law nor what units to ascribe to H. Readers can expect to see the following forms of the law used in the literature: pressure and concentration

$$P_j = Hc_j \tag{2a}$$

pressure and mol fraction:

$$P_j = (Hc_L)x_j = H'x_j \tag{2b}$$

where $H' = Hc_{\rm L}$; mol fraction and mol fraction:

$$y_j = P_j/P = (Hc_{\rm L}/P)x_j = (H'/P)x_j = H''x_j$$

where $H'' = (H'/P) = Hc_{\rm L}/{\rm P}$.

In air pollution, pollutant concentrations are usually quite small, and Henry's law is an acceptable representation of gas–liquid equilibrium. For components present in large proportion, Raoult's law is usually more accurate. In general, for nonideal solutions of any concentration, methods exist for determining absorption equilibria.[1]

NOMENCLATUARE AND ASSUMPTIONS

For the most part the material presented in this section will use SI units, but there are instances that commonly used empirical equations will be used that require engineering units. When terms pertaining to the scrubbing liquid and carrier gas accommodate SI units, they will be denoted with subscripts "s" and "c." When engineering units (lbm ft^2 h^{-1}) must be used, their terms will be denoted by L' and G'.

The total inlet pollutant concentration ($y_{j,\text{inlet}}$) and inlet molar flow rate ($G_{\rm m}$) of the process gas stream (containing the pollutant) are fixed by the industrial process the engineer is asked to control. The outlet pollutant concentration ($y_{j,\text{outlet}}$) can be no larger than a value specified by regulatory agencies or a more stringent value selected by a company. In the analysis that follows the following assumptions will be made:

- The only spatial variable is in the vertical direction; transverse gradients will be neglected.
- The liquid and gas velocities are uniform across the bed.
- No scrubbing fluid evaporates or is discharged from the scrubber exit.
- No carrier gas is absorbed in the scrubbing liquid.
- The exothermic nature of absorption will be neglected, and the flow will be assumed to be isothermal.
- The pollutant mol fraction entering the scrubber is small, e.g., $0 < y_j <<< 1$.

The bottom of the bed is designated by the subscript (1) and the top of the bed by the subscript (2).

PACKED BED SCRUBBER

Figures 1 and 5 show the elements of a random packed-bed scrubber. The scrubbing liquid passes downward through the packing, where it encounters the process gas stream moving upward. Spray nozzles are not needed, but it is important that the liquid and process gas stream are distributed uniformly across the packing. Packing material is inert solid material that produces a large surface area per volume of the bed that facilitates mass transfer. Shown in Table 1 are examples of

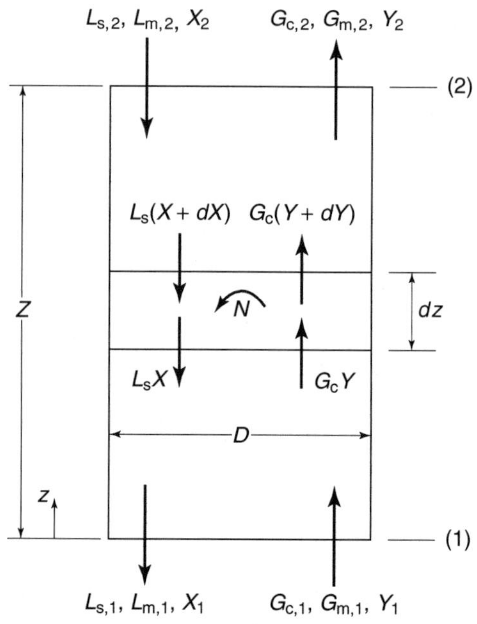

Figure 5. Schematic Diagram of a Packed-Bed Scrubber.

Table 1. Some Typical Packings and Applications (Ref. 6)

Packings	Application Features
Raschig rings 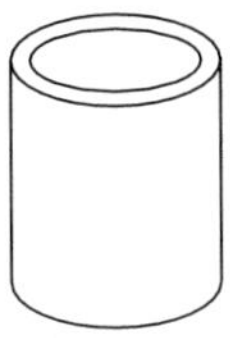	Most popular type, usually chapter per unit cost, but sometimes less efficient than others; available in widest variety of materials to fit service; very sound structurally; usually packed by dumping wet or dry, with larger 4- to 6-in. sizes sometimes hand stacked; wall thickness varies among manufactures, also some dimensions; available surface changes with wall thickness; produces considerable side thrust on tower; usually has more internal liquid channeling and directs more liquid to walls of column.
Berl saddles 	More efficient than Raschig rings in most applications, but more costly; packing nests together and creates "tight" spots in bed, which promotes channeling, but not as much as Raschig rings; do not produce much side thrust and have lower unit pressure drops with higher flooding point than Raschig rings; easier to break in bed than Raschig rings.
Intalox saddles 	One of the most efficient packings, but more costly; very little tendency or ability to nest and block areas of bed; gives fairly uniform bed; higher flooding limits and lower pressure drop than Raschig rings or Berl saddles; easier to break in bed than Raschig rings.
Pall rings 	Lower pressure drop (less than half) than Raschig rings; higher flooding limit; good liquid distribution, highly capacity; considerable side thrust on column wall; available in metal, plastic, and ceramic.
Spiral rings 	Usually installed as stacked, taking advantage of internal whirl of gas–liquid and offering extra contact surface over Raschig rings. Lessing rings, or cross–partition rings, available in single, double,and triple internal spiral designs, higher pressure drop, wide variety of performance data not available.
Teller rosette (Tellerette) 	Available in plastic: lower pressure drops, higher flooding limits than Raschig rings or Berl saddles; very low limit unit weight, low side thrust; relatively expensive.
Cross-partition rings 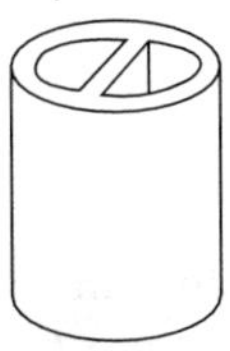	Usually used stacked, and as first layers support grids for smaller packing above; pressure drop relatively low, channeling reduced for comparative stacked packings: no side-wall thrust.
Lessing rings 	Not many performance data available, but in general slightly better than Raschig rings; pressure drop slightly higher; high side-wall thrust.

Table 1. (*Continued*)

Packings	Application Features
Ceramic balls 	Tend to fluidize in certain operating ranges, self-cleaning, uniform bed structure, higher pressure drop, and better contact efficiency than Raschig rings; high side thrust; not many commercial data.
Goodloe packing and wire mesh packing 	Available in metal only, used in large and small columns for distillation, absorption, scrubbing, liquid extraction; high efficiency, low pressure drop.

random packing. Random packing should provide a bed of uniform porosity and should not fracture due to its own mass. The packing is supported by structures that distribute the gas and liquid uniformly across the bed.

The base of the scrubber is a sump that collects the scrubbing liquid prior to its treatment and recycling. The inlet gas header distributes the flow uniformly so that the gas velocity entering the bed is distributed uniformly. At the scrubber outlet, a *demister* or *mist eliminator* collects small particles (*carryover*) of the scrubbing liquid entrained in the exit stream and prevents the liquid particles from leaving the scrubber.

MASS TRANSFER WITHIN THE PACKED BED

It is useful to describe the gas flow rates by *carrier gas molar flow rate* (G_c) and the *total molar gas flow rate* (G_m)

$$G_m = G_c + G_j \quad (3)$$

where G_j is the molar flow rate of pollutant species j in the gas phase. Thus G_c is the molar flow rate of the uncontaminated carrier gas entering or leaving the bed, while G_m is the total molar flow rate of all species in the gas phase entering or leaving the bed. In a similar fashion, for the liquid phase

$$L_m = L_s + L_j \quad (4)$$

where L_m is the *total molar liquid flow rate* entering or leaving the bed, L_j is the molar flow rate of pollutant species j in the liquid phase, and L_s is the *scrubbing liquid molar flow rate* of the uncontaminated scrubbing liquid. For the assumptions made above,

$$L_{s,1} = L_{s,2} = L_s \quad \text{and} \quad G_{c,1} = G_{c,2} = G_c \quad (5)$$

It is useful to describe the pollutant concentration by the *mol ratio* (X_j and Y_j). which is related to the mol fraction (x_j and y_j). Physically, X_j and Y_j represent the moles of molecular species j per mole of uncontaminated scrubbing liquid and uncontaminated carrier gas, respectively. Hence, for the gas phase:

$$Y_j = y_j/(1-y_j) = (n_j/n_t)/(1-n_j/n_t) \quad (6a)$$

$$y_j = Y_j/(1+Y_j) \quad (6b)$$

For the liquid phase:

$$X_j = x_j/(1-x_j) = (n_j/n_t)/(1-n_j/n_t) \quad (6c)$$

$$x_j = X_j/(1+X_j) \quad (6d)$$

Since the composition variables that will be used are the mol ratios X and Y, it is necessary to express Henry's law in these variables. Using Eqs. 1 and 6,

$$y_j = \left(\frac{Hc_L}{P}\right) x_j = H'' x_j$$

$$\frac{Y_j}{1+Y_j} = H'' \frac{X_j}{1+X_j} \quad (7)$$

$$Y_j = H'' X_j \left(\frac{1+Y_j}{1+X_j}\right) = mX_j$$

The absorption equilibrium line on the X–Y axes may be redrawn as

$$Y_j = mX_j \quad (8)$$

where,

$$m = H''(1+Y_j)/(1+X_j) \cong H'', \quad \text{since } 0 < X_j \ll 1 \text{ and } 0 < Y_j \ll 1 \quad (9)$$

In this analysis it will be assumed that the equilibrium curve is linear and that its slope (m) in the X–Y coordinates can be taken as constant.

The total number of mols of pollutant lost by the gas stream is equal to the total number of mols of pollutant gained by the scrubbing liquid:

$$L_s(X_1 - X_2) = G_c(Y_1 - Y_2) \quad (10)$$

From the above the, *liquid-to-gas ratio*, also called the *reflux ratio*, can be defined as:

$$L_s/G_c = (Y_1 - Y_2)/(X_1 - X_2) \quad (11)$$

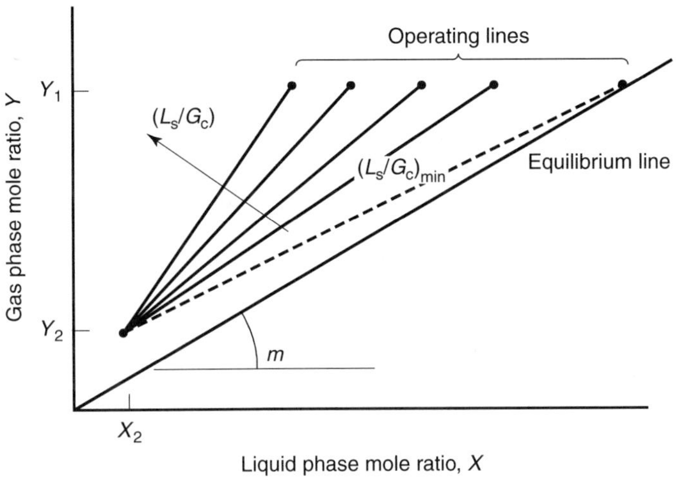

Figure 6. Operating Lines for Different (L_s/G_c) ratios, $L_s/G_c = (Y_1 - Y_2)/(X_1 - X_2)$.

In Eq. 11, the reflux ratio is computed on the basis of mols. The reflux ratio can also be expressed on the basis of mass or on the basis of gallons of water per 1000 ft^3 of gas cleaned. Readers must be careful to know when they have a choice of units and when empirical equations are used that compell the use of specific engineering units.

The molar flow rate of the carrier gas (G_c) and inlet pollutant concentration Y_1 are known quantities fixed by the process to be controlled, and Y_2 is dictated by the pollution regulations or company standards that need to be satisfied. The input concentration in the scrubbing liquid (X_2) is an independent parameter fixed by the performance of the water treatment process. Consequently there are two unknowns in Eq. 11 (i.e., X_1 and L_s). Designers select L_s and Eq. 11 determines the value of X_1.

Figure 6 is a Y–X diagram on which the equilibrium line satisfies Eq. 5 and values of X and Y satisfy Eq. 11. Each line expressing Eq. 11 depends on a particular value of (L_s/G_c) and is called an *operating line*. Readers must realize that while the operating line is a straight line, it does not imply that the mol ratios (X, Y) vary linearly with height from the bottom to the top of the packed bed. It is useful to compute the minimum value of the reflux ratio. The minimum value of mol ratio at point (1) is equal to its equilibrium value, $X_{1,\min} = Y_1/m$. Figure 6 indicates that $(L_s/G_c)_{\min}$ is set by thermodynamics, since the coordinates (X_1, Y_1) can never lie below the equilibrium line if pollutant is to transfer spontaneously from the gas phase to the liquid phase. Thus the minimum value of the reflux ratio, $(L_s/G_c)_{\min}$, corresponds to the value of X_1 that intersects the equilibrium line at the coordinates (Y_1, X_1).

$$(L_s/G_c)_{\min} = (Y_1 - Y_2)/[(Y_1/m) - X_2] \tag{12}$$

MASS TRANSFER COEFFICIENTS

Figure 5 depicts the contact between liquid and gas passing over an element of packing and the values in mol ratios X and Y (the species j is implicit and will hereafter be omitted) that exist inside each of the liquid and gas layers. Now consider a differential volume $(A\,dz)$ in the bed and the molar transfer of pollutant (N, kmol/s m^2) from the gas to the liquid at some elevation (z) inside the bed. The molar transfer rate (N) can be expressed by the following:

$$N = k_X(X_i - X) = K_X(X^* - X) \tag{13a}$$

$$N = k_Y(Y - Y_i) = K_Y(Y - Y^*) \tag{13b}$$

The variables at X_i and Y_i are the values of the mol ratio that exist at the interface where vapor–liquid equilibrium occurs. Thus X_i and Y_i correspond to some unknown point on the equilibrium line shown in Figure 7. Because the values of X_i and Y_i are unknown, it is useful to express the molar mass transfer rate in the alternative form on the right-hand side of Eq. 13, which involve new definitions:

K_X, the overall mass transfer coefficient in the liquid phase

K_Y, the overall mass transfer coefficient in the gas phase

X^*, the (star state) hypothetical equilibrium liquid-phase mol ratio corresponding to the actual gas-phase mol ratio (Y)

Y^*, the (star state) hypothetical equilibrium gas-phase mol ratio corresponding to the actual liquid-phase mol ratio (X)

The geometrical relationship between the star states and actual mol ratios is shown in Figure 7. The star states X^* and Y^* are the hypothetical mol ratios that would exist at equilibrium with the actual mol ratios Y and X. Based on the values of X and Y, the star states may be computed by a graphical construction or expressed in mathematical terms using Henry's law constant

$$Y^* = mX \quad \text{and} \quad X^* = Y/m \tag{14}$$

The analysis of a scrubber does not require the calculation of both K_X and K_Y, thus only the overall gas-phase hypothetical mass transfer coefficient (K_Y) will be computed. From Figure 7, the quantity $(Y - Y^*)$ can be expressed as

$$(Y - Y^*) = (Y - Y_i) + (Y_i - Y^*) \tag{15}$$

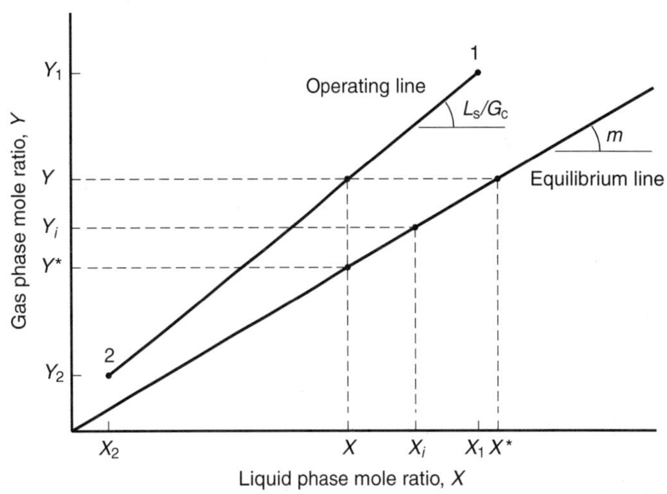

Figure 7. Relationship between the Star State and Actual State.

Using Eq. 13, the right- and left-hand terms of the above can be written as

$$\begin{aligned} N/K_Y &= N/k_Y + (mX_i - mX) \\ &= N/k_Y + m(X_i - X) \\ &= N/k_Y + mN/k_X \end{aligned} \tag{16}$$

or

$$1/K_Y = 1/k_Y + m/k_X \tag{17}$$

Since the slope of the equilibrium line (m) is known in terms of Henry's law constant, and since k_X and k_Y can be obtained from independent equations involving mass transfer properties, Eq. 17, expresses the overall mass transfer coefficient K_Y in terms of known parameters. Equations for the mass transfer coefficients k_X and k_Y are functions of the packing and the liquid and gas mass flow rates. Empirical equations for these mass transfer coefficients, k_X, k_Y, can be obtained from the manufacturers of packing material or from correlations in the research literature.[1–3,10]

An instructive way to understand Eq. 17 is to imagine that the mass transfer is proportional to a driving potential $(Y - Y^*)$ and inversely proportional to an *overall film resistance* (R_m) that is the sum of a gas-phase film resistance (R_X) and a liquid-phase film resistance (R_Y).

$$N = K_Y(Y - Y^*) = \frac{(Y - Y^*)}{\dfrac{1}{K_Y}} = \frac{(Y - Y^*)}{\left(\dfrac{1}{k_Y} + \dfrac{m}{k_X}\right)}$$

$$= \frac{(Y - Y^*)}{(R_Y + R_X)} = \frac{(Y - Y^*)}{R_m} \tag{18}$$

$$\begin{aligned} R_m &= 1/K_Y = R_Y + R_X \\ R_Y &= 1/k_Y \\ R_X &= m/k_X \end{aligned} \tag{19}$$

If we examine the consequences of choosing a scrubbing liquid that has a large or small value of the Henry's law constant (m), we find the following.

Small value of m: A low value of m produces a shallow slope of the equilibrium line in Figure 4, which implies that small values of Y (what is expected in air pollution) correspond to large values of X. Thus from Eq. 11, the value of (L_s/G_c) is small, and a large volumetric flow rate of scrubbing liquid will be required. The selection of such a scrubbing liquid is not what is desired; that is, large amounts of scrubbing liquid will be needed to heat a gas. Under these conditions, K_Y is essentially equal to k_Y providing, $\dfrac{1}{k_Y} \gg \dfrac{m}{k_X}$, which means that the gas-phase film (resistance) controls the mass transfer. Consequently the concentration gradient in the liquid phase is small compared to the concentration gradient in the gas phase.

Large value of m: Arguments used above can now be reversed. A large value of m implies a steep slope of the absorption line and a small value of (L_s/G_c). A small volumetric flow rate of liquid will be required. This is precisely what engineers want. Such a liquid is an attractive choice for a scrubber.

HEIGHT OF THE PACKED BED

The simplified analysis that follows is valid for an equilibrium curve that can be represented by a straight line, that is, m is constant and an operating line that satisfies Eq. 11:

$$Y = (L_s/G_c)X + Y_2 - (L_s/G_c)X_2 \tag{20}$$

For equilibrium lines that are not straight and for a more detailed analysis, readers should consult Refs.[1–3]. Consider a mass balance within the differential volume $(A\,dz)$ shown in Figure 6:

$$YG_c = G_c(Y + dY) + Na_pA\,dz \tag{21}$$

where N is the molar transfer rate, A is the cross-sectional area of the bed, and a_p, the packing parameter, which is equal to the total surface area of the packing per unit volume of the bed. Replace the molar transfer rate N by Eq. 19, solve for dz, and integrate assuming K_Y is constant:

$$\int_0^Z dz = Z = \left(\frac{G_c}{a_pAK_Y}\right)\int_{Y_2}^{Y_1} \frac{dY}{(Y - Y^*)} = (\text{HTU})(\text{NTU}) \tag{22}$$

The term before the integral sign is called the *overall height of a transfer unit* (HTU) and has units of length

$$\text{HTU} = G_c/a_pAK_Y \tag{23}$$

The integral is unitless and is called the *number of transfer units* (NTU). Estimating the number of transfer units is a task engineers need to perform. Manufacturers of packing may provide graphs of HTU as a function of gas and liquid flow rates, packing, and bed diameter for different chemical systems. Their clients can use these graphs in the design of absorption systems. More generally, engineers need to estimate the value of the HTU.

NUMBER OF TRANSFER UNITS

Figure 8 shows that the term $(Y - Y^*)$ in Eq. 22 is the vertical distance between the operating and equilibrium lines at any value of X. The integral is merely the reciprocal of this distance integrated over the difference $(Y_1 - Y_2)$. For a straight

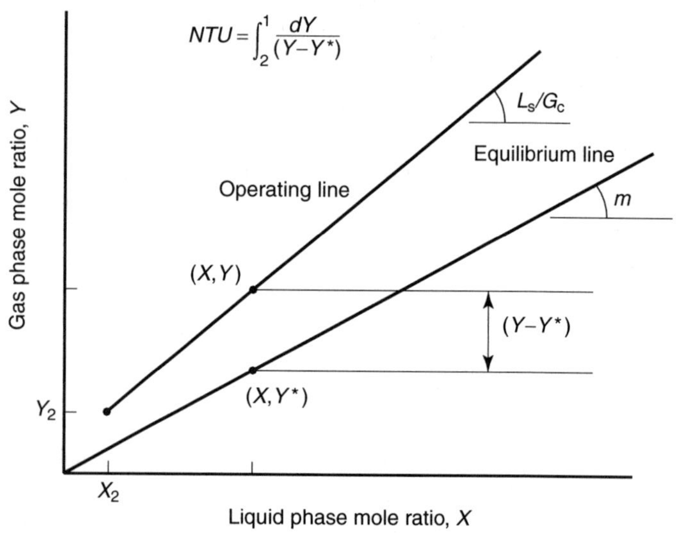

Figure 8. Diagram Used to Compute the Number of Transfer Units (NTU).

equilibrium line of slope m, the integral can be evaluated in closed form:

$$Y - Y^* = Y - mX \tag{24}$$

From Eq. 11,

$$X = \frac{Y - Y_2 + X_2\left(\frac{L_s}{G_c}\right)}{\left(\frac{L_s}{G_c}\right)} \tag{25}$$

Thus

$$Y - Y^* = Y - m\left(\frac{G_c}{L_s}\right)\left[Y - Y_2 + X_2\left(\frac{L_s}{G_c}\right)\right]$$

$$Y - Y^* = \omega Y + \theta \tag{26}$$

where

$$\omega = 1 - m\left(\frac{G_c}{L_s}\right)$$

$$\theta = m\left(\frac{G_c}{L_s}\right)\left[Y_2 - X_2\left(\frac{L_s}{G_c}\right)\right] = m\left[\left(\frac{G_c}{L_s}\right)Y_2 - X_2\right] \tag{27}$$

The integral in Eq. 22 can now be integrated:

$$\int_{Y_2}^{Y_1} \frac{dY}{(Y - Y^*)} = \int_{Y_2}^{Y_1} \frac{dY}{(\omega Y + \theta)} = \frac{1}{\omega}\ln\left(\frac{\omega Y_1 + \theta}{\omega Y_2 + \theta}\right) = \text{NTU} \tag{28}$$

An alternative method to evaluate the integral in Eq. 22 is available if the equilibrium and operating lines are parallel, or nearly so. In this case $(Y - Y^*)$ is constant, and its value can be taken directly from the graph

$$NTU = \int_{Y_2}^{Y_1} \frac{dY}{(Y - Y^*)} = \frac{1}{(Y - Y^*)} = \int_{Y_2}^{Y_1} dY = \frac{(Y_1 - Y_2)}{(Y - Y^*)} \tag{29}$$

The NTU can be evaluated from Figure 9 by graphical means. The transfer units are "stepped off" beginning at point (2) (smallest mol ratios, Y_2 and X_2) by drawing horizontal and vertical lines alternately between the operating and equilibrium lines, working up to point (1) (larger mol ratios, Y_1 and X_1). The number of horizontal lines represents the number of transfer units.

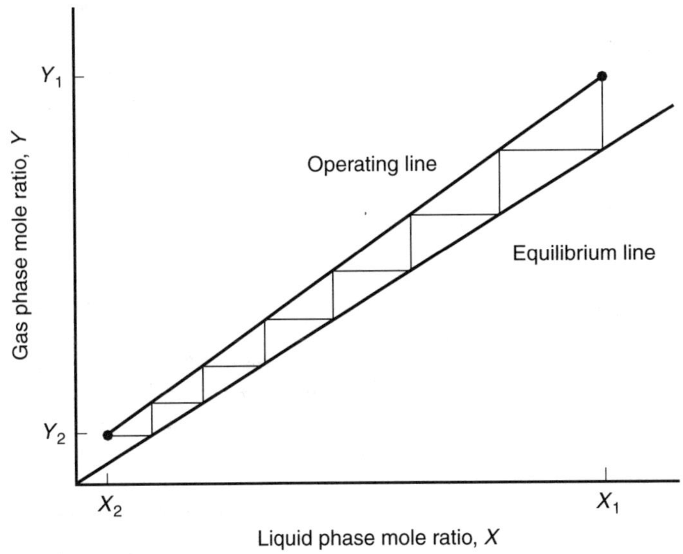

Figure 9. Graphical Method to Compute the Number of Transfer Units.

Equation 28 affords an opportunity to express the overall collection efficiency (η) as a function of the operating variables of the packed bed. Replace the value of Y_2 in Eq. 28 by

$$Y_2 = (1 - \eta)Y_1 \tag{30}$$

After simplifying, Eq. 28 can be written as

$$\eta = 1 - \frac{Y_2}{Y_1} = \left(1 - \frac{mX_2}{Y_1}\right)\left(\frac{e^{(1-\lambda)\text{NTU}} - 1}{e^{(1-\lambda)\text{NTU}} - (mG_c/L_s)}\right) \tag{31}$$

where

$$\lambda = (H'/P)(G_c/L_s) = m(G_c/L_s) \tag{32}$$

Using Eq. 31, one can estimate scrubber efficiency if all the other operating parameters of the packed bed are known. Admittedly Y_2 has to be known before a value of NTU can be determined accurately, but a crude trial and error process can be pursued in which an initial value of NTU can be estimated graphically assuming $Y_2 \approx 0$. Substituting this value of NTU into Eq. 31, a new value of Y_2 is obtained and the value of NTU is recomputed graphically. The process can be repeated until satisfactory convergence is obtained. By this procedure the efficiency can be estimated from a packed bed of known proportions (Z and A) and operated at a known (L_s/G_c).

OVERALL HEIGHT OF A TRANSFER UNIT

Multiply each term in Eq. 17 by (G_c/a_pA) and, in addition, multiply the last term by (L_s/L_s):

$$\left(\frac{1}{K_Y}\right)\left(\frac{G_c}{a_pA}\right) = \left(\frac{1}{k_Y}\right)\left(\frac{G_c}{a_pA}\right) + \left(\frac{m}{k_x}\right)\left(\frac{G_c}{a_pA}\right)\left(\frac{L_s}{L_s}\right)$$

$$\left(\frac{G_c}{a_pAK_Y}\right) = \left(\frac{G_c}{a_pAk_Y}\right) + \left(\frac{mG_c}{L_s}\right)\left(\frac{L_s}{a_pAk_X}\right) \tag{33}$$

The left-hand term corresponds to the overall height of a gas-phase transfer unit (HTU) in Eq. 23 and the remaining terms can be grouped as follows:

$$\text{HTU} = \frac{G_c}{a_pAk_Y} + \left(\frac{mG_c}{L_s}\right)\left(\frac{L_s}{a_pAk_X}\right)$$

$$= \text{HTU}_c + (mG_c/L_s)\text{HTU}_s \tag{34}$$

where L_s and G_c are expressed as molar flow rates:

$$\text{HTU}_c = \text{height of a gas-phase transfer unit} = G_c/a_pAk_Y \tag{35}$$

$$\text{HTU}_s = \text{height of a liquid-phase transfer unit} = L_s/a_pAk_X \tag{36}$$

Consequently HTU is the sum of the height of the gas-phase transfer unit (HTU_c) and the height of the liquid-phase transfer unit (HTU_s) multiplied by the slope of the equilibrium line divided by the reflux ratio expressed in mols.

The height of the liquid (HTU_s) and gas (HTU_c) transfer units are given in terms of the individual mass transfer coefficients, k_X and k_Y, in the gas and liquid phases and can be determined from empirical equations.[1,3,12] Since the coefficients depend on the geometry of the packing, this can be a difficult task. Alternatively, vendors who sell packing may provide such data. Typical of the data that an engineer can expect to be provided are the following in which in engineering units must be used.[11]

$$\mathrm{HTU_s} = L_s/a_p A k_X = \phi(L'/\mu_L)^{\eta}\mathrm{Sc}_L^{0.5} \tag{37}$$

$\mathrm{Sc_L}$ = liquid Schmidt number $(\mu/D\rho)_L$

L'(lbm/h ft^2) = superficial liquid mass flow rate $= L_s M_s/A$

μ_L = liquid viscosity (lbm/ft h)

$$\mathrm{HTU_c} = G_c/a_p A k_Y = \alpha(\mathrm{Sc_c})^{0.5}(G')^{\beta}/(L')^{\gamma} \tag{38}$$

$\mathrm{Sc_c}$ = carrier gas Schmidt number$(\mu/D\rho)_c$

G'(lbm/h ft^2) = superficial gas mass flow rate $= G_c M_c/A$

where G' and L' are called the superficial mass flow rates. Values of ϕ, α, β, η, γ are shown in Table 2 for several types of packing.

If random packing is used for which the empirical data shown in Table 2 are not available, engineers may wish to estimate the liquid and gas-phase mass transfer coefficients (k_L and k_G) by the commonly used Onda correlations. Dvorak et al.[10] found that the Onda correlations are good predictors for mass transfer coefficients for large ($D_c > 5$ cm) random packing, although they tend to underpredict mass transfer at high gas flow rates and when the gas-film resistance is large.

$$k_G(\mathrm{m/s}) = 5.23\left(\frac{Q_G\rho_G}{Aa_p\mu_G}\right)^{0.7}\left(\frac{\mu_G}{\rho_G D_G}\right)^{0.33}\frac{a_p D_G}{(aD_c)^2}$$

$$k_L(\mathrm{m/s}) = 0.0051\left(\frac{Q_L\rho_L}{Aa_p\mu_L}\right)^{0.67}\left(\frac{\mu_L}{\rho_L D_L}\right)^{-0.5}(a_p D_c)^{0.4}\left(\frac{\rho_L}{\mu_L g}\right)^{-0.33} \tag{39}$$

where the terms above and their units are defined as follows:

a_p, total packing surface area/overall bed volume (m^{-1})
D_G, D_L, diffusion coefficients of the pollutant in gas and liquid phases (m^2/sec)
Q_G, Q_L, volumetric flow rates of gas and liquid (m^3/sec)
μ_G, μ_L, viscosity of gas and liquid (kg/ms)
ρ_c, ρ_s densities of gas and liquid (kg/m^3)
A, packed bed cross sectional area (m^2)
$g = 9.8$ m/sec^2
D_c, characteristic diameter of the packing (m)

A full discussion of methods to estimate the mass transfer coefficients and other parameters in the mass transfer realm is beyond the scope of this book and readers are encouraged to consult Hines and Maddox.[12]

PRESSURE DROP IN PACKED BEDS

The pressure drop depends on the type of packing, height (Z) and diameter (D) of the packed bed, and the superficial liquid (L') and gas (G') mass flow rate per unit area, lbm ft^{-2} h^{-1}. Figure 10 typifies how the pressure drop per height ($\delta P/Z$) varies with the volumetric gas flow rate for a particular type of packing. It is important for readers to understand that Figure 10 contains three flow domains.

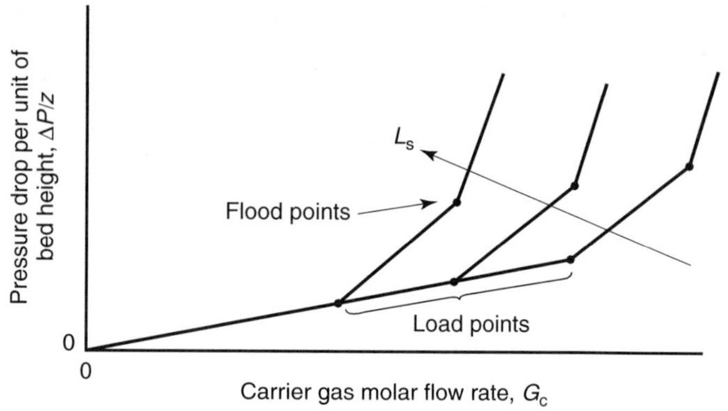

Figure 10. Pressure Drop in Packed Beds as a Function of the Gas and Liquid Molar Flow Rates.

1. At low values of L_s, $\delta P/Z$ increases with G_c as if the packing were dry. It is only mildly dependent on L_s since the liquid adheres to the packing surface and the void space in the packing is nearly constant.
2. At a particular value of G_c called the *load point*, portions of liquid become airborne (called *holdup*), the packing void space becomes smaller, and the slope of the curve is steeper. As the value of L_s increases, the value of G_c at the load point decreases.
3. At values of G_c above the load point, large amounts liquid (holdup) remain airborne and may even accumulate to the top of the packing. The packed bed is now said to be *flooded*. The value of G_c at which flooding occurs is called the *flood point*. Typically, flooding produces a pressure drop of 2–3 in. of water/ft of packing. The slope of the curve at values of G_c above the flood point is quite steep.

As a practical matter, packed beds are operated above the load point but at 40–70% of the flood point. To achieve operation at 40–70% of flooding the engineer will need to select the diameter of the packed bed carefully. Figure 11 illustrates the relationship between the pressure drop and the superficial liquid (L', lbm ft^{-2} s^{-1}) and gas (G', lbm ft^{-2} s^{-1}) flow rates. Figure 11 compels the use of specific engineering units. The viscosity (μ_L) must be in centipoise. The viscosity of water at 20°C is 1 centipoise (0.01 g cm^{-1} sec^{-1}).

EXAMPLE: DESIGN OF COUNTER-FLOW PACKED-BED SCRUBBER

The management of your plant believes that a packed-bed scrubber using fresh water as the scrubbing liquid is an effective way to remove phenol ($M = 94.1$) vapor from a 30,000 ACFM (14.164 m^3/sec) air stream at 30°C. Fresh water will enter the scrubber and the spent scrubbing liquid will be used elsewhere in the plant. The inlet phenol concentration in air (y_1) is 200 ppm. Management requests that the reflux ratio be 30% greater than the minimum value and that the scrubber operate at 60% the flood point. Owing to the modest solubility of phenol in water (9%), you believe that a wet scrubber using water is apt to be very tall and produce a large pressure drop. Nonetheless, you are instructed to prepare a preliminary

Table 2. Packing Constants (Ref. 6)

Packing	Liquid-Phase HTU_s Φ	η	L' (lbm/h ft^2)
Raschig rings			
3/8 in.	0.00182	0.46	400–15,000
1/2 in.	0.00357	0.35	400–15,000
1 in.	0.0100	0.22	400–15,000
1-1/2 in.	0.0111	0.22	400–15,000
2 in.	0.0125	0.22	400–15,000
Berl saddles			
1/2 in.	0.00666	0.28	400–15,000
1 in.	0.00588	0.28	400–15,000
11/2 in.	0.00625	0.28	400–15,000
Partion rings			
3 in.	0.0625	0.09	3000–14,000
Spiral rings (stacked, staggered)			
3 in. single spiral	0.00909	0.28	400–15,000
3 in. triple spiral	0.0116	0.28	3000–14,000

Packing	Gas-Phase HTU_c α	β	γ	G' (lbm/h ft^2)	L' (lbm/h ft^2)
Raschig rings					
3/8 in.	2.32	0.45	0.47	200–500	500–1500
1 in.	7.00	0.39	0.58	200–800	400–500
	6.41	0.32	0.51	200–600	500–4500
1 1/2 in.	17.30	0.38	0.66	200–700	500–1500
	2.58	0.38	0.40	200–700	1500–4500
2 in.	3.82	0.41	0.45	200–800	500–4500
Berl saddles					
1/2 in.	32.40	0.30	0.74	200–700	500–1500
	0.81	0.30	0.24	200–700	1500–4500
1 in.	1.97	0.36	0.40	200–800	400–4500
1-1/2 in.	5.05	0.32	0.45	200–1000	400–4500
Partion rings					
3 in.	6.50	0.58	1.06	150–900	3000–10,000
Spiral rings (stacked, staggered)					
3 in. single spiral	2.38	0.35	0.29	130–700	3000–10,000
3 in. triple spiral	15.60	0.38	0.60	200–1000	500–3000

Packing Factor F for Random Packing, Porosity (ε), and Surface Area to Bed Volume (a_p)

Packing	Nominal size (in.) 1/4	3/8	1/2	5/8	3/4	1	1-1/4	1-1/2	2
Raschig rings									
ceramic	1600	1000	580	380	255	155	125	95	65
ε	0.73	0.68	0.63	0.68	0.73	0.73	0.71	0.74	0.74
a_p (ft^{-1})	240	155	111	100	80	58	45	38	28
metal									
1/32 in. wall	700	390	300	170	155	115			
ε	0.69		0.84		0.88	0.92			
a_p (ft^{-1})	236		128		83.5	62.7			
1/16 in. wall			410	290	220	137	110	82	57
ε			0.73		0.78	0.85	0.87	0.90	0.92
a_p (ft^{-1})			118		71.8	56.7	49.3	41.2	31.4
Pall rings									
plastic				97		52		40	25
ε				0.88		0.90		0.905	0.91
a_p (ft^{-1})				110		63		39	31

Table 2. *(continued)*

Packing Factor F for Random Packing, Porosity (ε), and Surface Area to Bed Volume (a_p)									
	Nominal size (in.)								
Packing	1/4	3/8	1/2	5/8	3/4	1	1-1/4	1-1/2	2
metal				70		48		28	20
ε				0.902		0.938		0.953	0.964
a_p (ft^{-1})				131.2		66.3		48.1	36.6
Intalox ceramic saddles	725	330	200		145	98		52	40
ε	0.60		0.63		0.66	0.69		0.75	0.72
a_p (ft^{-1})	274		142		82	76		44	32
Berl saddles	900		240		170	110		65	45
ε	0.60		0.63		0.69	0.69		0.75	0.72
a_p (ft^{-1})	274		142		82	76		44	32

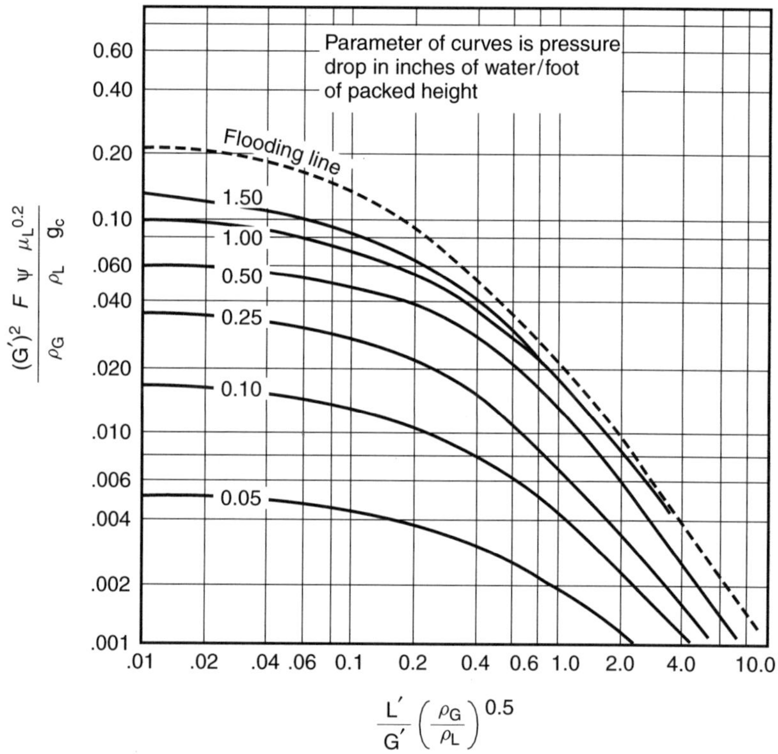

Figure 11. Generalized Pressure-Drop Correlation to Estimate Column Diameter. Packing Factor (F) Selected from Table 2. Users Must Use the Following Units: G', gas mass flow rate per unit area (lbm ft^{-2} s^{-1}); L', liquid mass flow rate per unit area (lbm ft^{-2} s^{-1}); Ψ, ratio of density of water to scrubbing liquid; μ_L, liquid viscosity (centipoise); μ_G, carrier gas viscosity (centipoise); g_c, 32.2 (lbm ft lbf^{-1} sec^{-2}); ρ_L, density of scrubbing liquid (lbm/ft^3); ρ_G, density of carrier gas (lbm/ft^3) (Ref. 6).

design of a random packed-bed scrubber using 1-in. Berl saddles to remove 95% of the phenol (exit phenol concentration in air is 10 ppm). Finally, you have been instructed to discuss the viability of removing the phenol by scrubbing with water.

ASSUMPTIONS

Since the phenol concentrations in air and water are small, assume that the carrier gas is only air and the scrubbing liquid is only water. Assume also that the pressure and temperature inside the scrubber are essentially 1 atm and 30°C.

INLET CONDITIONS

Air (carrier gas, subscript c or G)

$T_c = 30°C$, $\rho_c = 1.1614$ kg/m^3, $\rho_G = 0.07249$ lbm/ft^3

$Q_c = 30{,}000$ SCFM, $\dot{m}_c = 16.45$ kg/s (36.23 lbm/sec),

$G_c = 0.5678$ kmol/sec

$\mu_c = 1.865 \times 10^{-5}$ kg m^{-1} sec^{-1}

$\mathscr{D}_c$(phenol in air) $= 0.88 \times 10^{-5}$ m^2/sec

Schmidt number

$(Sc_c = \mu_c/\rho_c\mathscr{D}_c) = 1.865 \times 10^{-5}$ kg m^{-1} s^{-1}/$(0.88 \times 10^{-5}$ m^2/sec$)$ $\times$ (1.1614 kg/m^3) $= 1.824$

$y_1 = 200$ ppm $= 200 \times 10^{-6}$

Water (scrubbing liquid, subscript s or L)

$\rho_s = 1000$ kg/m^3, $\rho_L = 62.4$ lbm/ft^3

$\mu_s = 0.7978 \times 10^{-3}$ kg m^{-1} s^{-1}, $\mu_L = 1.928$ lbm h^{-1} ft^{-1}, 0.8908 centipoise

$\mathscr{D}_s$(phenol in water) $= 0.66 \times 10^{-9}$ m^2/sec

Schmidt number

$(Sc_s = \mu_s/\rho_s\mathscr{D}_s) = 0.7978 \times 10^{-3}$ kg m^{-1} s^{-1}/$(0.66 \times 10^{-9}$ m^2/sec)

$\times$ (1000 kg/m^3) $= 34.64$

SOLUTION

Compute the mol ratio (Y)

$$Y_1 = y_1/(1 - y_1) = 200 \times 10^{-6}/(1 - 200 \times 10^{-6}) = 200 \times 10^{-6}$$

Similarly, Y_2 will equal 10×10^{-6}. A company employee used one of the selections on the web page

http://www.uic.edu/~mansoori/Thermodynamic.Data.and.Property_html

and reported that the Henry's law constant was $H' = 4 \times 10^3$ N/m^2. Draw the equilibrium isotherm on a graph of X versus Y.

$$Y = (H'/P)X = mX = (4 \times 10^3 \text{ N/m}^2)/(1.01 \times 10^5 \text{ N/m}^2) = 0.04\,X$$

where m (0.04) is the slope of the equilibrium isotherm.

The minimum reflux ratio, $(L_s/G_c)_{min}$ (kmol water/kmol air), corresponds to a line (on the X–Y graph) whose slope is defined by the coordinates at the top of the bed, $X_2 = 0$, $Y_2 = 10 \times 10^{-6}$, the coordinates at the bottom of the bed defined by $Y_1 = 200 \times 10^{-6}$, $X_1 = Y_1/m = 200 \times 10^{-6}/0.04 = 5 \times 10^{-3}$. From Eq. 12, the minimum reflux ratio is,

$$(L_s/G_c)_{min} = (Y_1 - Y_2)/[(Y_1/m) - X_2] = 190 \times 10^{-6}/5 \times 10^{-3} = 0.038$$

The actual reflux ratio is 30% larger than the minimum value

$$(L_s/G_c)_{actual} = (1.3)(0.038) = 0.0494 \text{ kmol water/kmol air}$$

On the basis of mass, $(L_s/G_c)_{actual}$, is 0.0303 (which is equivalent to 0.28 gal/1000 ft^3 air). The actual flow rate of scrubbing liquid is

$$L_s(\text{actual}) = (0.0494)(0.5678) = 0.028 \text{ kmol/s (1.11 lbm/sec)}$$

The abcissa of Figure 12 can be computed from the following even though the cross sectional area (A) is not known.

$$\frac{L'}{G'}\left(\frac{\rho_c}{\rho_s}\right)^{1/2} = \frac{(L_s/A)}{(G_c/A)}\left(\frac{\rho_c}{\rho_s}\right)^{1/2} = \frac{1.11 \text{ lbm/s}}{36.23 \text{ lbm/s}}\left(\frac{0.07249 \text{ lbm/ft}^3}{62.4 \text{ lbm/ft}^3}\right)^{1/2} = 1.044 \times 10^{-3}$$

Figure 12 shows the equilibrium and operating lines for this Example problem. The diameter of the packed bed (D) can be

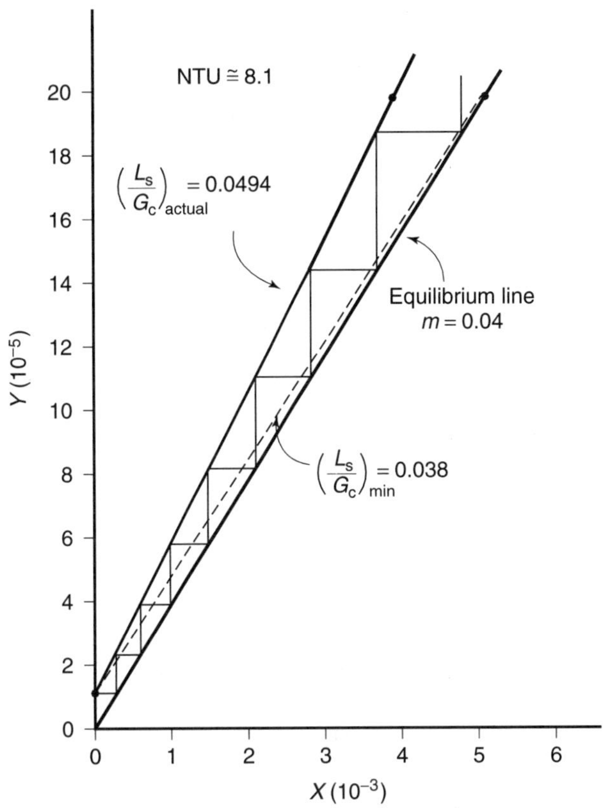

Figure 12. Graphical Method to Compute the NTU in a Packed-Bed Scrubber that Removes Airborne Phenol with Water.

determined from the requirement that the scrubber operate at 60% of the flood point. The flood point condition can be ascertained from Figure 12. The value of the abscissa parameter is 1.044×10^{-3}.

This value is situated to the left of the vertical axis in Figure 11, but since the curve for the pressure drop associated with flooding is nearly horizontal near the vertical axis, the curve will be extrapolated to 0.00104. The value of the ordinate parameter at flooding is approximately 0.25. The diameter of the packed bed at flooding can be found from the following, in which $F = 110$, a_p 76 ft^{-1} (249.3 m^{-1}) for 1-in. Berl saddles, and the remaining terms are expressed in engineering units. Thus

$$[(G')^2F(\mu_L)^{0.2}/g_c\rho_G\rho_L)]_{flood} \simeq 0.25$$

$$G'_{flood} = (G/A_{flood}) = 0.5818 \text{ lbm sec}^{-1}\text{ft}^{-2}$$

$$A_{flood} = 36.23/0.5818 = 62.27 \text{ ft}^2$$

Since the design cross-sectional area must be 60% the flood value,

$$0.6 = A_{flood}/A_{design}$$

$$A_{design} = 62.27/0.6 = 103.78 \text{ ft}^2 (9.6 \text{ m}^2)$$

$$D_{design} = 11.49 \text{ ft} (3.5 \text{ m})$$

It should be noted that $L' = 38.5$ lbm h^{-1} ft^{-2} and $G' = 1256.8$ lbm h^{-1} ft^{-2} are outside of the range suggested in Table 2, nonetheless the computations will proceed and the results interpreted in light of these limitations. The height of the packed bed (Z) will be computed from the expression $Z = \text{(NTU)(HTU)}$, where HTU is the height of transfer unit,

$$\text{HTU} = \text{HTU}_c + m(G_c/L_s)\text{HTU}_s$$

and G_c and L_s have the units kmol/sec. For 1-in. Berl saddles, Treybal[3] indicates. HTU_c and HTU_s will be evaluated from material from Wark and Warner[11]

air: $\alpha = 1.97, \beta = 0.36$, and $\gamma = 0.4$

water: $\phi = 0.0058, \eta = 0.28$

Thus HTU_c (ft) and HTU_s are,

$$\begin{aligned}\text{HTU}_c &= \alpha(\text{Sc}_c)^{1/2}(G')^{\beta}/(L')^{\gamma}\\ &= (1.97)(1.824)^{1/2}(1256.7)^{0.36}/(38.5)^{0.4} = 8.07\text{ ft}(2.46\text{ m})\\ \text{HTU}_s &= \phi(L'/\mu_s)^{\eta}(\text{Sc}_s)^{1/2}\\ &= 0.0058(38.5/1.928)^{0.28}(34.64)^{1/2} = 0.08\text{ ft}(0.02\text{ m})\end{aligned}$$

where G' and L' must be expressed in the units lbm ft^{-2} h^{-1}. The overall height of a transfer unit (HTU) is

$$\begin{aligned}\text{HTU} &= \text{HTU}_c + m(G_c/L_s)\text{HTU}_s\\ &= 2.46 + (0.04)[(0.5678/0.0280)](0.02) = 2.47\text{ m}(8.12\text{ ft})\end{aligned}$$

From Eqs. 37 and 38, the gas-phase and liquid-phase resistances, R_Y and R_X, can be computed from HTU_c and HTU_s

$$\begin{aligned}R_Y &= 1/k_Y = (a_pA/G_c)\text{HTU}_c\\ &= (249.3\text{ m}^{-1})(9.64\text{ m}^2)(2.46\text{ m})/0.5678\text{ kmol/sec}\\ &= 10{,}412.1\text{ sec/kmol}\\ R_X &= m/k_X = m(a_pA/L_s)\text{HTU}_s\\ &= (0.04)(249.3\text{ m}^{-1})(9.64\text{ m}^2)(0.02\text{ m})/0.028\text{ kmol/sec}\\ &= 68.7\text{ sec/kmol}\end{aligned}$$

The gas-phase resistance is much larger than the liquid phase resistance. Since m is small (0.04) owing to the small value of Henry's constant (H'), the diffusion of phenol from the gas to the liquid is much slower than the diffusion of phenol through the liquid.

Using the "stepping-off" procedure, Figure 12 indicates that NTU $\simeq$ 8.1. This value should be considered a crude estimate because the equilibrium and operating lines are not parallel. A more accurate computation for NTU can be obtained from,

$$\text{NTU} = (1/\omega)\ln[(\omega Y_1 + \theta)/(\omega Y_2 + \theta)]$$

where

$$\begin{aligned}\omega &= [(L_s/G_c) - m]/(L_s/G_c) = 1 - m(G_c/L_s)\\ &= 1 - (0.04)(0.5678/0.028) = 0.189\end{aligned}$$

$$\begin{aligned}\theta &= [m(L_s/G_c)][Y_2 - (L_s/G_c)X_2]\\ &= (0.04)(0.5678/0.028)(10 \times 10^{-6} - 0) = 8.11 \times 10^{-6}\end{aligned}$$

and L_s and G_c are expressed in kmol/s. Upon substitution, NTU = 8.06. Using this value, the height of the packed bed (Z) is

$$Z = \text{(NTU)(HTU)} = (7.96)(2.47) = 19.66\text{ m}(64.48\text{ ft})$$

To estimate the overall pressure drop, determine $\delta P/Z$ with Figure 10 at the actual coordinates under which the scrubber operates. The abscissa value does not change, $(L'/G')(\rho_G/\rho_L)^{1/2} = 1.04 \times 10^{-3}$, but the ordinate parameter has to be reevaluated at the actual operating conditions. The actual value can be computed by modifying the flooding value computed earlier:

$$\begin{aligned}&[(G')^2F\mu_L^{0.2}/g_c\rho_G\rho_L]_{\text{actual}} = (0.25)\\ &\times [(36.23\text{ lbm sec}/103.78\text{ ft}^2/(0.5818\text{ lbm ft}^{-2}\text{ sec}^{-1})]^2 = 0.09\end{aligned}$$

The pressure drop ($\delta P/Z$) corresponding to this value is approximately 1.0 in. water/ft of packing. Thus the pressure drop across the entire packed bed is:

$$\delta P = (1.0\text{ in water/ft bed})(64.48\text{ ft}) \simeq 65\text{ in. water}$$

DISCUSSION

It is easily seen that the height of a packed bed to remove 95% of the phenol with fresh water is large. Note also that this is only the height of the packing itself. The total height of the scrubber must include header space at the top and bottom of the bed. At the bottom of the scrubber, space must be provided to collect the scrubbing water and distribute the inlet air flow uniformly across the face of the bed. At the top of the scrubber space must be provided to accommodate a weir (or piping) system to distribute the water uniformly across the bed and a mist eliminator to prevent fine mist of scrubbing liquid from entering the atmosphere. Thus the scrubber that's been analyzed might be over 100 ft tall. Management should consider another way to remove phenol, i.e. perhaps change the scrubbing liquid, selection of the proposal of operating at 30% above the minimum reflux value and/or 60% the flood point. Thus your original suspicions is correct.

OPERATION AND MAINTENANCE

Scrubber Pressure Drop

After its installation, the overall pressure drop across the scrubber may be larger than intended, and may even increase with time. There are several reasons why this may occur. Packing may settle or fracture and make one side of the bed more porous than the other. Solid particles in the gas stream or suspended in the scrubbing liquid or crystallized salts may clog narrow passages in the packing. If the height of the packing decreases due to settling, additional packing should be added to return the packing to the intended level. The mist eliminator (demister) may also become clogged with solid matter from the gas stream or solids suspended in the scrubbing liquid.

Scrubber Liquid Flow Rate

Provision must be made to measure and control the scrubbing liquid volumetric flow rate. Maintaining the desired scrubbing liquid volumetric flow rate cannot be left to guess work or indirect measurement such as inferring it from the pump RPM or pressure drop across a pump. The volumetric flow rate of the scrubbing liquid must be measured directly. The scrubbing liquid should be distributed uniformly across to the top of the packed bed in ways that solids suspended in the scrubbing liquid or corrosion in pipes will not interfere with its distribution. A weir or pipes of large diameter are sufficient. Once added to the bed, the packing will distribute the scrubbing liquid. There is no need to use small-diameter nozzles to distribute the scrubbing liquid to the bed. Using packing support plates (distribution plates) located at different elevations in the packing not only supports the packing but also redistributes the liquid and gas uniformly across the bed. One must be careful that the openings in the support plates do not become clogged with solid matter. Not only will such blockage increase the overall pressure drop, but if the clogging is skewed to one side of the packing, it will also skew the flow of gas and liquid. Such clogging may even produce an uneven distribution "channeling" of the flow of gas and liquid. Channeling can not be tolerated because it will reduce the overall pollutant removal efficiency.

Treatment of the Scrubbing Liquid

Solid material in the scrubbing liquid will accumulate over time. The solids may be particles from the gas stream or sludge formed in the scrubbing liquid. Solid particles should be removed from the gas stream with filters before being introduced to the scrubber. If, however, oil vapors or oil droplets are contained in the gas stream, engineers can expect aggravating sludge to accumulate on packing support plates, sumps, strainers, pressure and pH sensors. Solids should not be allowed to accumulate in the sump. Sludge should be removed by filtration and discarded in ways commensurate with its toxicity. Depending on the pollutants to be removed, it may be necessary to maintain a certain pH in the scrubbing liquid. To control the pH of the scrubbing liquid, it may be necessary to add chemicals continuously in metered amounts.

Demister (Entrainment Separator)

Demisters prevent fine droplets of scrubbing liquid, "carryover" from being discharged to the atmosphere. Since the scrubbing liquid accumulates solids and other objectionable material, the carryover should not be discharged to the atmosphere. Indeed the pollutants contained in the carryover may be as hazardous (if not worse) than the pollutants removed from the gas stream. Depending on the design of the demister, suspended solids in the scrubbing liquid are apt to clog the demister. If the porosity of the demister is small, these solids may affect seriously the overall pressure drop in the system. If the scrubber is not operated continuously and perhaps shut down overnight or on weekends, particularly if the scrubber is installed outdoors in wintertime, condensed vapors may be very serious. If the fan is not sized properly, such clogging may reduce the gas volumetric flow rate.

Fans, Duct Work, Dampers

It is wise to install the fan on the downstream of the scrubber and put the scrubber under negative pressure so that leaks will cause ambient air to enter the scrubber rather than process gas stream escape to the indoor or outdoor atmosphere. Second, placing the fan downstream of the scrubber introduces a cleaned air stream to the fan and minimizes the accumulation of material on the fan blades. Mounting downstream of the scrubber will, however, introduce a moist gas stream to the fan and may be an issue over a period of time requiring replacement of the fan rotor, if not the entire fan. Duct work upstream of the scrubber is apt to collect solids suspended in the gas stream and condensates of the vapors in the gas stream. Duct work downstream of the scrubber will contain vapors of the scrubbing liquid and depending on the pH, may initiate corrosion. The scrubber should be designed with access doors appropriately located so that packing can be checked and removed and replaced if necessary. Duct work upstream and downstream of the scrubber should be equipped with appropriately located access doors to inspect and perhaps clean the ducts. Finally, provision should be made if vapors in the inlet and outlet gas stream are apt to condense inside the duct. Thus the pitch of the ducts must be chosen carefully and provision made for condensate drains.

REFERENCES

1. R. H. Perry, and C. H. Chilton, *Chemical Engineers' Handbook*, 6th edition, McGraw Hill, New York, 1984.
2. T. K. Sherwood, R. L. Pigford and C. R. Wilke, *Mass Transfer*, McGraw-Hill, New York, 1975.
3. R. E. Treybal, *Mass-Transfer Operations*, 3rd edition, McGraw-Hill, New York, 1980.
4. M. Crawford, *Air Pollution Control Theory*, McGraw-Hill, New York, 1976.
5. S. Calvert and H. M. Englund, *Handbook of Air Pollution Technology*, Wiley-Interscience, New York, 1984.
6. R. J. Heinsohn and R. L. Kabel, *Sources and Control of Air Pollution*, Prentice Hall, Upper Saddle River, NJ, 1998.
7. R. J. Lewis, Sr, "Sax's Dangerous Properties of Industrial Materials", 8th edition, Van Nostrand Reinhold, New York, 1994.
8. C. T. Norback, "Hazardous Chemicals on File", J C Norback, New York 1988.
9. R. P. Pohanish and S. A. Greene, "Hazardous Substances Resources Guide", Detroit Gale Research, Detroit, MI, 1993.
10. B. I. Dvorak, D. F. Lawle, J. R. Fair and N. E. Handler, "Evaluation of the Onda correlations for mass transfer with large random packings," *Environmental Science and Technology*, **30**, (3), 945–53, 1996.
11. K. Wark, and C. F. Warner, *Air Pollution*, Harper & Row, Publishers, New York, 1981.
12. A. L. Hines, and R. N. Maddox, *Mass Transfer Fundamentals and Applications*, Prentice-Hall, Englewood Cliffs, NJ, 1985.

NOMENCLATURE

Symbol	Description	Dimensions[a]
[j]	concentration of species j	M/L^3
a_p	total surface area of packing per volume of bed	L^{-1}

(Continued)

NOMENCLATURE *(Continued)*

Symbol	Description	Dimensions[a]
A	cross sectional area	L^2
a_p	packing parameter, total surface area of packing per unit volume of bed	
c, c_j	general concentration, concentrations of molecular species j	M/L^3
c_t	total molar concentration	
c_L	total molar concentration in the liquid phase, mass concentration in liquid phase	N/L^3, M/L^3
D	diameter of a packed bed	L
$\mathscr{D}$	diffusion coefficient	L^2/t
D_c	characteristic diameter of packing	
g	9.8 m/s^2	
G'	superficial total gas mass flow rate	M/L^2t
G_c	carrier gas molar flow rate	N/L^2t
G_j	molar flow rate of molecular species j in the gas phase	N/L^2t
G_m	total molar gas flow rate	N/L^2t
H, H', H''	Henry's law constant	variable
HTU	overall height of a transfer unit	L
HTU_c	height of gas-phase transfer unit	L
HTU_s	height of a liquid-phase transfer unit	L
k_X	mass transfer coefficient in the liquid phase	N/L^2t
K_X	overall mass transfer coefficient, liquid phase	N/L^2t
k_Y	mass transfer coefficient in gas phase	N/L^2t
K_Y	overall mass transfer coefficient, gas phase	N/L^2t
L'	superficial total liquid mass flow rate	M/L^2t
L_s	molar flow rate of a scrubbing liquid	N/t^-
L_j	liquid molar flow rate of molecular species j	N/t
L_m	total molar liquid flow rate	N/t
m	slope of the equilibrium line (X–Y coordinates)	—
M_s, M_c	molecular weight of the scrubbing liquid and carrier gas	M/N
n_t	total number of mols	
n_j	number of mols of molecular species j	
N	pollutant molar transfer rate	N/L^2t
NTU	number of transfer units	—
P_j	partial pressure of molecular species j	F/L^2
P	total pressure	F/L^2
Q	actual volumetric flow rate	L^3/t
Q_G, Q_L	gas and liquid volumetric flow rate	L^3/t
R_m	overall film resistance	
R_X	gas-phase film resistance	
R_Y	liquid-phase film resistance	
T	temperature	T
x_j	mol fraction of molecular species j in the liquid phase	—

(Continued)

Symbol	Description	Dimensions[a]
X_j	mol ratio of molecular species j in the liquid phase	—
X^*	hypothetical mol ratio in liquid phase	—
X_i	liquid mol ratio at gas-liquid interface	—
y_i	mol fraction of molecular species j in the gas phase	—
Y_i	gas mol ratio at gas-liquid interface	M/Nt
Y_j	mol ratio of molecular species j in the gas phase	—
Y^*	hypothetical mol ratio in gas phase	—
z	vertical distance	L
Z	height of a packed bed	L
	Subscript	
a	air	
0	inlet conditions	
1	base of a packed bed	
2	top of a packed bed	
c	carrier gas	
g	gas phase	
i	gas-liquid interface	
j	molecular species j	
L	liquid	
m	total molar	
s	scrubbing liquid	
t	total	
min	minimum	
flood	flood state	
design	design conditions	
actual	actual condition	
	Greek	
α	constant in HTU equation (units dictated by equation)	
β	constant in HTU equation (units dictated by equation)	
γ	constant in HTU equation, (units dictated by equation)	
η	capture efficiency	
λ	parameter	
ω	parameter	
Θ	parameter	
μ	viscosity	Ft/L^2
μ_G	viscosity of gas	
μ_L	viscosity of liquid	
ρ_g	gas (air and pollutant) density	M/L^3
ρ	density	M/L^3
Φ	constant in HTU equation (units dictated by equation)	
	Abbreviations	
ACFM	actual cubic feet per minute	L^3/t
		L
		—
PPM	parts per million	—
Sc_L, Sc_c	Schmidt number	—

[a] Q, energy; N, mols; t, time; T, temperature; F, force; M, mass; L, length.

ADSORPTION

Kenneth E. Noll

INTRODUCTION

During adsorption, gaseous components are removed from an effluent gas stream by adhering to the surface of a solid. The gas molecules being removed are referred to as the *adsorbate*, while the solid doing the adsorbing is called the *adsorbent*. Adsorbents are highly porous particles.

The application of adsorption technology for pollution control usually deals with the control of organic compounds. In general, any organic compound having a molecular weight greater than 45 is likely to be a good adsorbate. Adsorption is used to control emissions in many solvent-using operations such as drycleaning, degreasing, surface coating, rubber processing, and flexographic and gravure printing. Adsorption systems have also been used to control toxic or odorous vapors discharged from food processing plants, rendering plants, sewage treatment plants, and many chemical manufacturing processes (such as producing fuels, fertilizers, and pharmaceutical products).

Several materials are used effectively as adsorbing agents. The most common adsorbents used in industry are activated carbon, silica gel, activated alumina (alumina oxide), and zeolites (molecular sieves).[1] Adsorbents are characterized by their chemical nature, extent of their surface area, pore size distribution, and particle size. One of the most important characteristics in distinguishing between adsorbents is surface polarity. Polarity determines the type of vapors a particular adsorbent will have the greatest affinity. Of the above adsorbents, activated carbon is the primary nonpolar adsorbent. Polar adsorbents will preferentially adsorb any water vapor that may be present in the gas stream. Since moisture is present in most pollutant air streams, the use of polar adsorbents is severely limited for air pollution systems.

Activated carbon is produced from a variety of feedstocks such as wood, coal, and petroleum-based products. The term *activated* as applied to adsorbent materials refers to the increased internal and external surface area imparted by special treatment processes. The activation process, first, involves carbonization. Carbonization is the heating (in the absence of air) of the material to a temperature high enough to drive off all volatile material. To increase the surface area, the carbon is then "activated" by using steam, air, or carbon dioxide at higher temperatures. The temperatures involved and the type of feed stock all affect the adsorption qualities of the carbon.

Carbons used in gas-phase adsorption systems are manufactured in granular form, usually between 4×6 to 4×10 mesh in size. Bulk density of the packed bed can range from 5 to 30 lb/ft^3 depending on the internal porosity of the carbon. The surface area of the carbon can range from 3×10^6 to 8×10^6 ft^2/lb. This is equivalent to having the surface area of 2 to 5 football fields in one gram of carbon.

In discussing the fundamentals of adsorption, it is useful to distinguish between physical adsorption, involving only relatively weak intermolecular forces, and chemisorption, which involves essentially the formation of a chemical bond between the sorbate molecule and the surface of the adsorbent. Physical adsorption does not involve the sharing or transfer of electrons and thus always maintains the individuality of interacting species. The interactions are fully reversible, enabling desorption to occur. Physical adsorption is not site specific; the adsorbed molecules are free to cover the entire surface. Chemisorption involves chemical bonding and is irreversible. Chemisorption is characterized mainly by large interaction potentials that lead to high heats of adsorption that approach the value of chemical bonds. Chemical adsorption is not easily reversible.

All adsorption processes are exothermic, whether adsorption occurs from chemical or physical forces. This is because, in adsorption, the fast-moving gas molecules lose their kinetic energy of motion to the adsorbent in the form of heat. In chemisorption, the heat of adsorption is comparable to the heat evolved from a chemical reaction, usually over 10 kcal/mol. The heat given off by physical adsorption is much lower, approximately 100 cal/mol, which is comparable to the heat of condensation. For these reasons, chemisorption is not used extensively in air pollution control systems.

For physical adsorption processes, the capacity of an adsorbent decreases as the temperature of the system increases (Figure 2) As a general rule, adsorber temperatures are kept below 130°F to ensure adequate bed capacities. Temperatures above this limit can be avoided by cooling the exhaust stream that is to be treated. At low concentrations (below 100 ppm) the heat release is minimal and is dissipated by the air flow through the bed. At higher concentrations (approximately 5000 ppm) considerable heating of the bed can occur, which if not removed can cause the adsorber efficiency to decrease rapidly. In addition, granular carbon is a good insulator, which inhibits heat dissipation from the interior of the bed. In some cases, the temperatures rise can cause autoignition of carbon bed. Monitoring of bed temperatures and leaving the bed slightly wet after steam generation are techniques used to avoid bed fires.

ADSORPTION ISOTHERMS

Adsorption occur by a series of steps. In the first step, the contaminant diffuses from the major body of the air stream to the external surface of the adsorbent particle. In the second step, the contaminant molecule migrates from the relatively small area of the external surface (a few m^2/g) to the pores within each adsorbent particle. The bulk of adsorption occurs in these pores because the majority of available surface area is there (hundreds of m^2/g). In the third step, the contaminant molecule adheres to the surface in the pore. Figure 1 illustrates this overall diffusion and adsorption process.[2]

Due to these varied mechanisms, mass transfer rates in the pores are extremely difficult to predict. Unless extensive data are available concerning the specific adsorption application, determining the rate-controlling step is impossible. Therefore, empirical design procedures based on adsorption equilibrium conditions are the most common method used to predict adsorber size and performance.

The most common and useful adsorption equilibrium data are the adsorption isotherm. The isotherm is a plot of the adsorbent capacity versus the partial pressure of the adsorbate at a constant temperature. Adsorbent capacity is usually given in weight percent expressed as gram of adsorbate per 100 grams of adsorbent. Figure 2 is a typical example of an adsorption isotherm for carbon tetrachloride on activated carbon.

EXAMPLE 1

A dry cleaning process exhausts a 15,000 scfm air stream containing 680 ppm carbon tetrachloride. Given Figure 2 and

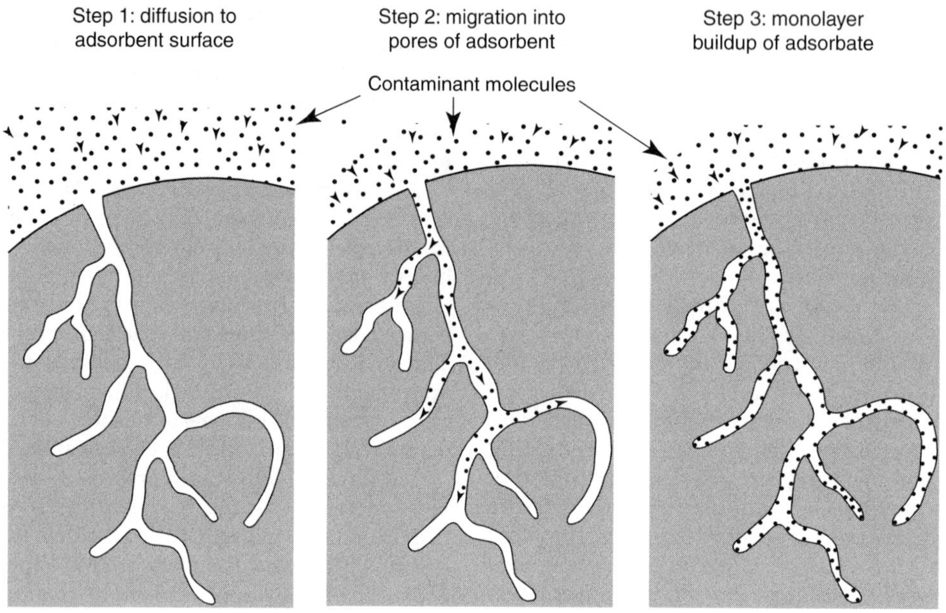

Figure 1. Mechanism of Adsorption.

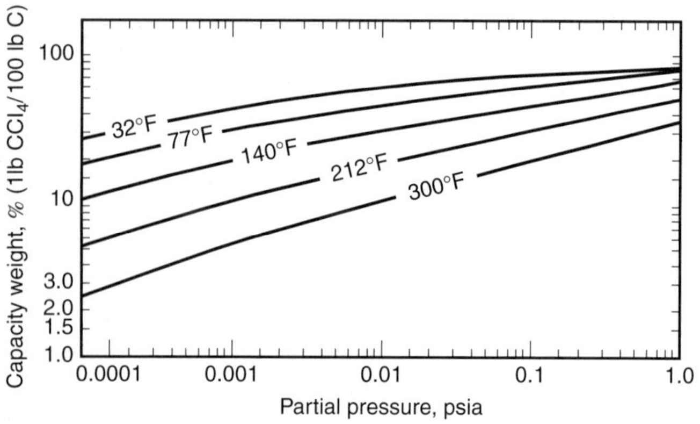

Figure 2. Adsorption Isotherm for Carbon Tetrachloride on Activated Carbon.

assuming the exhaust stream is at approximately 140°F and 14.7 psia, determine the saturation capacity of the carbon.

SOLUTION

In the gas phase, the mole fraction (Y) is equal to the percent by volume.

$$Y = \%\text{ volume} = 680\text{ ppm}$$
$$= 680/10^6$$
$$= 0.00068$$

Obtaining the partial pressure:

$$p = YP$$
$$= (0.00068)(14.7\text{ psia})$$
$$= 0.01\text{ psia}$$

From Figure 2, at a partial pressure of 0.01 psia and a temperature of 140°F, the carbon capacity is 30%. This means that at saturation, 30 lbs of vapor are removed per 100 lbs of carbon in the adsorber.

ADSORPTION EQUILIBRIUM RELATIONSHIPS

Attempts have been made to develop generalized equations that can predict adsorption equilibrium from physical data. This is very difficult because adsorption isotherms take many shapes depending on the forces involved. Isotherms may be concave upwards, concave downwards, or "S" shaped. To date, most of the theories agree with data only for specific adsorbate systems and are valid over limited concentration ranges.

One of the most useful mathematical models to describe adsorption equilibria is the Langmuir isotherm.[1,3] It is based on the following assumptions: (1) the adsorbed phase is a unimolecular layer, and (2) at equilibrium, the rate of adsorption is equal to the rate of desorption from the surface. f is the fraction of the total solid surface occupied by adsorbate molecules. The rate of adsorption, r_a, is proportional to the partial pressure of the adsorbate, p, and to the fraction of the solid surface area available for adsorption, $(1 - f)$. Therefore,

$$r_a = C_a\, p(1 - f) \tag{1}$$

where C_a is a constant. Conversely, the rate of desorption, r_a, is proportional to the fraction of the surface area occupied by the adsorbate:

$$r_d = C_d\, f \tag{2}$$

where C_d is a constant. At equilibrium, the rate of adsorption is equal to the rate of desorption. The fraction of the surface covered is, then, given by

$$f = \frac{C_a p}{C_d + C_a p} \tag{3}$$

Because the adsorbed phase is a unimolecular layer, the mass of adsorbate per unit mass of adsorbent, m, is also proportional to the surface covered:

$$m = C_m\, f \tag{4}$$

where C_m is a constant. Combining Eqs. 3 and 4,

$$m = \frac{k_1 p}{k_2 p + 1} \tag{5}$$

where $k_1 = C_a C_m / C_d$, and $k_2 = C_a / C_d$. Equation 5 is known as the Langmuir isotherm.

Equation 5 can be rearranged to give

$$\frac{p}{m} = \frac{1}{k_1} + \frac{k_2}{k_1} p \tag{6}$$

This equation is a straight line when plotted on rectangular coordinates as p/m versus p.

At very low adsorbate equilibrium partial pressure $k_2 p$ is near zero, and Eq. 5 becomes

$$m = k_1 p \tag{7}$$

Conversely, at high equilibrium partial pressure

$$m = \frac{k_1}{k_2} \tag{8}$$

Hence, over an intermediate range of partial pressures:

$$m = kp^n \tag{9}$$

where k is a constant and n, a constant with a value between 0 and 1. Equation 9 is known as Freundlich isotherm. The values of k and n are obtained from experimental data. Table 1 gives values of the Freundlich isotherm parameters for some adsorbates on activated carbon (4 × 10 mesh). Note that these isotherms may not be extrapolated outside the partial pressure ranges shown.

EXAMPLE 2: EQUILIBRIUM ADSORPTION ON ACTIVATED CARBON

Calculate the equilibrium adsorptivity of toluene (in air) on activated carbon (4 × 10 mesh). The temperature is 298 K and the total pressure is 101.3 kPa. The equilibrium concentration of toluene is 1000 ppm.

SOLUTION

Calculate the equilibrium partial pressure of toluene:

$$p = 1000(101.3)/10^6 = 0.1013 \text{ kPa} = 101.3 \text{ Pa}$$

According to Table 1, Freundlich isotherm applies for this partial pressure. Substituting the appropriate parameters from the table into Eq. 9,

$$m = 0.20842(101.3)^{0.110} = 0.346 \frac{\text{kg toluene}}{\text{kg carbon}}$$

ADSORPTION SYSTEM DESIGN; ADSORPTION BED CAPACITY

The most efficient arrangement for conducting adsorption operation is a packed bed of adsorbent through which the stream under treatment is passed. As the polluted stream travels through this bed, adsorption of the contaminants takes place and a purified effluent exists the column (Figure 3).

The portion of the bed close to the inlet is continuously contacted by the concentrated feed stream, whereas the subsequent portions are exposed to adsorbate not adsorbed

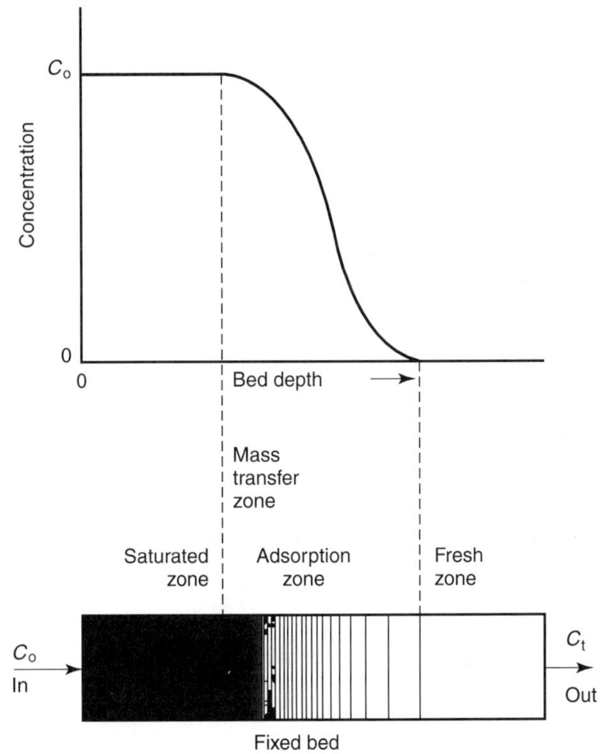

Figure 3. Concentration Profile along Column.

Table 1. Freundlich Isotherm Parameters for Some Adsorbates[a]

Adsorbate	Temperature (K)	$k \times 100$	n	Partial Pressure (Pa)
Acetone	311	1.234	0.389	0.69–345
Acrylonitrile	311	2.205	0.424	0.69–103
Benzene	298	12.602	0.176	0.69–345
Chlorobenzene	298	19.934	0.188	0.69–69
Cyclohexane	311	7.940	0.210	0.69–345
Dichloroethane	298	8.145	0.281	0.69–276
Phenol	313	22.116	0.153	0.69–207
Toluene	298	20.842	0.110	0.69–345
Trichloroethane	298	25.547	0.161	0.69–276
m-Xylene	298	26.080	0.113	0.69–6
m-Xylene	298	28.313	0.0703	6.9–345

Source: EAB Control Cost Manual, 3rd. ed. U.S. Environmental Protection Agency, Research Triangle Park, NC, 1987.
[a]The amount of adsorbed is expressed in kg adsorbate/kg adsorbent. The equilibrium partial pressure is expressed in Pa. Data are for the adsorption on Calgon activated carbon (4 × 10 mesh).

by an earlier portion. Thus the adsorbate becomes fully loaded at the inlet first, and then downstream.

The part of the bed that displays a gradient in solid concentration from zero to equilibrium is called the mass transfer zone (MTZ). As the name indicates, this is the active part of the bed where adsorption actually takes place. The concentration of pollutants changes continuously throughout this part of the bed, from a value close to zero at the beginning of the MTZ to the feed concentration at the end.

As the saturated part of the bed increases, the MTZ travels downstream and eventually exists the bed. This gives rise to a typical effluent concentration versus time profile, which is called the *breakthrough curve* (Figure 4). At breakthrough, the effluent concentration starts to rise and eventually reaches the effluent inlet concentration. Two important terms are used to describe the capacity of the adsorbent bed. The breakthrough capacity is defined as the capacity of the bed at which unreacted vapors begin to be exhausted. The saturation capacity is the maximum amount of vapors that can be adsorbed per unit weight of carbon (this is the capacity read from the adsorption isotherm).

When the effluent quality becomes unacceptable, or after a predetermined time interval, the operation of the column is stopped and the spent adsorbent is regenerated. Regeneration of an adsorbent with simultaneous recovery of the adsorbate can be achieved if the process is reversed by increasing the temperature. This is accomplished in air adsorption systems by the use of steam. Continuous operation of an adsorption system requires the involvement of two or more beds. One of them undergoes regeneration, while the rest contact the system.

EXAMPLE 3

Assume the same conditions as stated in Example 1. Estimate the amount of carbon that would be required if the adsorber were to operate on a 4-h cycle. The molecular weight of CCl_4 is 154 lb/lb mole.

SOLUTION

From Example 1 we know that the carbon used will remove 30 lbs of vapor for every 100 lbs of carbon at saturation conditions.

First compute the flow rate of CCl_4:

$$Q_{CCl_4} = 15{,}000 \text{ scfm } \times 0.00068$$
$$= 10.2 \text{ scfm } CCl_4$$

Converting to pounds per hour:

$$10.2\ \frac{ft^3}{min} \times \frac{lb\ mole}{359\ ft^3} \times \frac{154\ lb}{lb\ mole} \times \frac{60\ min}{h} = 262.5 \frac{lb\ CCl_4}{h}$$

For a 4-hour cycle there are: $4 \times 262.5 = 1050$ lb CCl_4.

The amount of carbon (at saturation) requires:

$$1050\ lb\ CCl_4 \times \frac{100\ lb\ carbon}{30\ lb\ CCl_4} = 3500\ lb\ carbon$$

ADSORBENT REGENERATION

Because it is simple and relatively inexpensive, steam stripping is the most common desorption technique. The steam heats the bed so that the adsorption capacity is reduced to a low level. The adsorbate leaves the surface of the carbon and is removed from the vessel. During the initial heating period no vapors are desorbed. This is because a fixed amount of steam is first required to raise the temperature of the cold bed to the

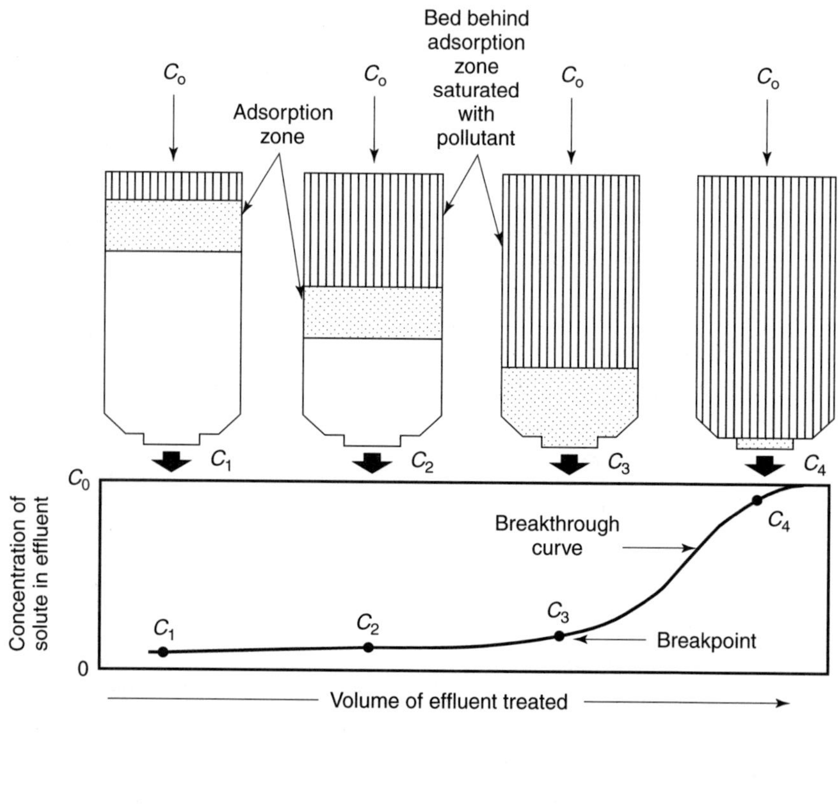

Figure 4. Adsorption Wave Front.

desorption temperature. After this initial period a substantial amount of adsorbate vapor is released. Figure 5 shows that the steam consumption per pound of solvent varies with time and the solvent adsorbed. The minimum point in the curves in Figure 5 represents the optimum steam requirements and steam time.[4] As the steam strip time is extended, more steam per pound of solvent recovered is required and a point is reached at which steam cost exceeds recovery benefits.

In Figure 5, the ratio of the pounds of the steam used per pound of perchloroethylene recovered is plotted for 15-min intervals. This reaches a minimum of about 4.7 pounds after an elapsed time of 90 min and then rises sharply. The pounds of solvent recovered reaches a maximum at this time and then decreases. In Figure 5 the desorbing of toluene follows the same pattern except that the steam consumption is higher. This is to be expected since its latent heat is greater. It is economical to operate the strip cycle to retrieve only part of the adsorbed solvent—that part recovered generally in less than 90 min, leaving a "heel" of solvent within the bed. A well-designed system will have steam consumption in the range of 1 to 4 pounds of steam per pound of recovered solvent supplied at pressures ranging from 3 to 15 psig.

The solvent vapors and steam that emerge from the bed during stripping are condensed in a water-cooled condenser. The cooling water will typically have an inlet temperature of 60°F. The temperature rise in the condenser will be approximately 40°F, making the outlet water temperature about 100°F. High cooling-water temperatures or low water-flow rates will slow the condensation process, thereby increasing the steam-strip cycle time and adding unnecessary cost to the process.

Separators are installed following the condenser to separate the contaminant from the water. Separators work on the principle of gravitational forces where the heavier material to be separated is removed from the bottom of the decanters and the lighter material is removed at the top. If the solubility of the condensed VOC in water is low enough, decanting is sufficient; otherwise an additional separation unit, such as distillation, is required.

For the capture of vapors in a continuous operation, a minimum of two adsorption units is required. With this arrangement, one unit is adsorbing while the other is being stripped of solvent and regenerated. Sufficient time must be available to cool this unit to nearly ambient temperature before it is returned to service. After the solvent is stripped, the carbon is not only hot but is saturated with water. Cooling and drying are usually done by blowing solvent-free air through the carbon. The ensuing evaporation of the moisture is helpful in removing the heat in the carbon.

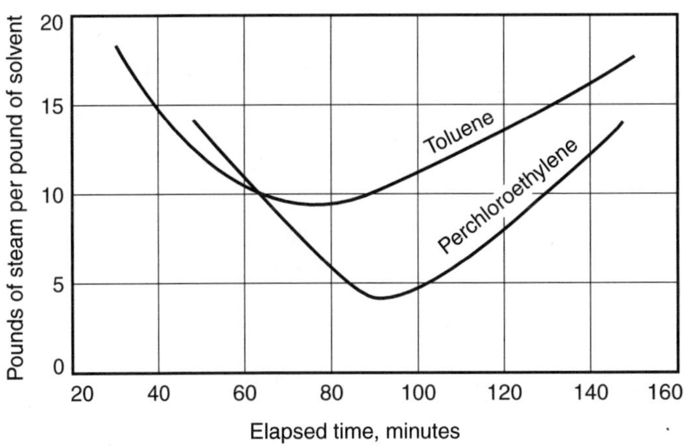

Figure 5. Steam Consumption per Pound of Solvent Recovered.

MAXIMUM ADSORBENT BED DEPTH

Often the adsorbent bed is sized to the maximum depth allowed by the pressure drop across the bed. Bed depth is controlled by the attrition (wear) losses of the granular activated carbon caused by excessive velocity (crushing velocity) and pressure during operation. The gas velocity is usually limited to 100 ft/min. The pressure drop across the bed is usually limited to 18–20 in. H_2O. Data on the pressure drop per foot of bed depth for typical carbons are presented in Figure 6.[5] The pressure drop per foot of bed depth is plotted versus the gas flow rate, with the carbon mesh size as a parameter. From the figure, an adsorber with a flow rate of 80 ft/min using 4×10 mesh carbon will have a pressure drop of approximately 6 in. H_2O per foot of bed depth. Therefore, if the pressure drop across the bed is limited to 18 in. H_2O then the local bed depth should not exceed 3 ft.

Carbon adsorption systems are designed for a maximum air flow velocity of 100 ft/min through the adsorber. A lower limit of at least 20 ft/min is maintained to avoid flow distribution problems, such as channeling. With the above-stated maximum and minimum flow rates, the allowable pressure drop usually dictates the maximum bed depth. By specifying a maximum velocity through the adsorber, the minimum cross-sectional area is also specified. For example, if 300 ft/min of contaminant gas is to be treated, and the maximum velocity through the adsorber is to be 30 ft/min, then the adsorber must have a cross-sectional area of at least 10 ft^2. The restriction on maximum bed depth caused by pressure-drop considerations leads to two types of fixed bed orientation, vertical and horizontal.

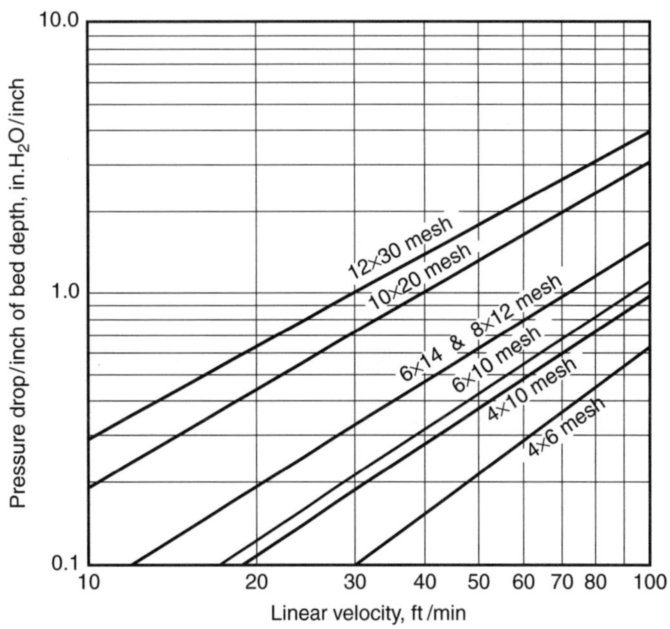

Figure 6. Pressure Drop versus Flow Rate through Granular Carbon Beds.

MINIMUM ADSORBENT BED DEPTH

Providing a minimum depth of adsorbent is very important in achieving efficient gas removal. If the adsorber bed depth is shorter than the required mass transfer zone, breakthrough will immediately occur, rendering the system ineffective. Fortunately, the MTZ is less than 3 in. for most air pollution application and represents only a small portion of the total bed depth.[3] Computing the length of the MTZ is very difficult since it depends upon six factors: the adsorbent particle size, gas velocity, adsorbate concentration, fluid properties of the gas stream, temperature, and pressure of the system. The MTZ can be estimated from experimental data (obtain from pilot adsorber column) using Eq. 10.[2]

$$\mathrm{MTZ} = \frac{1}{1 - X_S} D \left(1 - \frac{C_B}{C_S}\right) \tag{10}$$

where D is the bed depth; C_B, the breakthrough capacity %; C_S, the saturation capacity %; X_S, the degree of saturation in the MTZ (%, usually assumed to be 50%); and MTZ, the length of MTZ.

Equation 10 is used mainly as a check to ensure that the proposed bed depth is longer than the MTZ. Actual bed depths are usually many times longer than the length of the MTZ. The additional bed depth allows for cycle times required to regenerate the bed. As has already been discussed, minimum regeneration times are typically 2 h, with 60–90 min allowed for steaming and 30 min to 1 h for cooling and drying.

The length of the MTZ is directly proportional to the velocity; usually, the higher the velocity, the longer the unsaturated zone. An example is shown in Figure 7, where the effects of velocity on the length of MTZ is demonstrated for ethanol on activated carbon.[6] The data in Figure 7 are typical of most adsorption systems for VOC control (i.e., the maximum length of the MTZ for adsorption is typically less than 3 in.). This can be compared to water adsorption systems, where the MTZ is often a foot or more. This difference is due to the much higher diffusivities in the air adsorption systems.

SERVICE TIME

The service time is one of the prime factors to be considered in sizing adsorption equipment. Service time can be calculated as follows;

$$\tau_B = \frac{C_B}{V_0 C_0} \tag{11}$$

where τ_B is the service time (adsorption time) (min); V_0, the total gas volume flow rate (ft^3/min); C_0, inlet gas concentration (#/ft^3; and C_B breakthrough capacity.

Equation 10 can be rearranged to solve for the breakthrough capacity required in Eq. 1. The total capacity is the sum of the capacity in the saturated bed plus the amount adsorbed in the MTZ.

(active zone) + (saturated zone)

$$C_B = \frac{X_S C_S \mathrm{MTZ} + C_S(D - \mathrm{MTZ})}{D} \tag{12}$$

The total amount of adsorbent (C_S) is usually determined from the adsorption isotherm, as illustrated in Example 2.

The usual procedure in practice is to work with a term defined as the working charge (or working capacity) rather than C_B. It provides a numerical value for the actual adsorbing capacity of the bed under operating conditions. If experiment data are available, the working charge (WC) may be estimated from:

$$\mathrm{WC} = C_S \left[\left(\frac{D - \mathrm{MTZ}}{D}\right) + 0.5\left(\frac{\mathrm{MTZ}}{D}\right)\right] - \mathrm{HEEL} \tag{13}$$

where the MTZ is assumed to be 50% saturated and HEEL is the residual adsorbate present in the bed following regeneration. This heel accounts for a large portion of the difference between the saturation and the working capacity.

When deep adsorbent beds are used, as in solvent recovery, the MTZ is small compared to the overall bed depth and D, the adsorption capacity of the saturation zone, essentially determines the service time. With usual solvent-recovery practices, the solvent air velocity through the bed is maintained at near 100 fpm. In these cases, the MTZ is of the order of 3 in. and D, the adsorbent bed depth, can range from 16 to over 36 in. The capacity of the bed can be calculated with a considerable degree of accuracy if an adsorption isotherm has been determined from

$$W_C = C_S - \mathrm{HEEL}\ \text{(deep bed)} \tag{14}$$

In thin beds of 2 in. and less, flow velocities near 40 fpm are usually employed. In these applications. The MTZ and D are nearly the same and the W_C is essentially:

$$W_C = 0.5C_S - \mathrm{HEEL}\ \text{(thin bed)} \tag{15}$$

The adsorption phenomenon becomes somewhat more complex if the gases or vapor to be adsorbed consist of several compounds. The adsorption of organic compounds having higher molecular weights will tend to displace those having lower molecular weights. Lighter compounds will tend to be separated or partitioned from the heavier compounds and will pass through the bed at a faster rate. This will increase the mass transfer zone and may require additional carbon bed depth or shorter operating cycle.

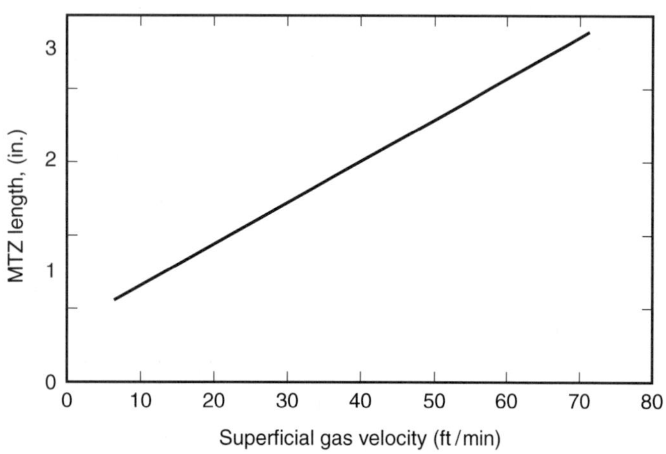

Figure 7. Effect of Gas Velocity on Length of MTZ.

EXAMPLE 4

You are asked to determine the required height of adsorbent for an adsorber that treats a degreaser ventilation stream contaminated with trichloroethylene (TCE) given the following operating and design data. The adsorption column cycle is set at 4 h in the adsorption mode, 2 h in heating and desorbing, 1 h in cooling, and 1 h in standby. The adsorber recovers 99.5% by weight of TCE. A horizontal unit with an inside width of 6 ft and length of 15 ft is to be used. Volumetric flow rate of contaminated air stream is 10,000 scfm; standard conditions are 60°F, 1 atm; operating temperature is 70°F; operating pressure is 20 psia; adsorbent is activated carbon; saturated capacity is 28 pounds TCE per 100 pounds carbon; inlet concentration of TCE is 2000 ppm (by volume); and molecular weight of TCE is 131.5.

SOLUTION

The total and TCE actual volumetric flow rates are first calculated.

$$\begin{aligned} Q_{\text{act}} &= 10{,}000[(70+460)/(60+460)](14.7/20) \\ &= 7491 \text{ acfm} \\ &= 4.5 \times 10 \text{ acfh} \\ Q_{\text{TCE}} &= (Y_{\text{TCE}})(Q_{\text{act}}) \\ &= (2000 \times 10^{-6})(4.5 \times 10^{5}) \\ &= 900 \text{ acfh} \end{aligned}$$

The mass flow rate of the TCE, W, is then

$$\begin{aligned} W &= Q\rho = Q(PM/RT) \\ &= (20)(131.05)(900)/(10.73)(70+460) \\ &= 416.2 \text{ lb/h} \\ \text{TCE adsorbed} &= (W)(0.995)(4) \\ &= (416.2)(0.995)(4) \\ &= 1656.6 \text{ pounds} \end{aligned}$$

Activated carbon required:

$$\begin{aligned} &= (\text{TCE to be adsorbed})(100 \text{ pounds carbon}/ \\ &\quad 28 \text{ pounds TCE adsorbed})/(\text{bulk density}) \\ &= (1656.6)(100/28)/(36) \\ &= 164 \text{ ft}^3 \\ \text{Height of adsorbent} &= (\text{activated-carbon volume})/ \\ &\quad (\text{cross-sectional area}) \\ &= (164)/(6)(15) \\ &= 1.83 \text{ feet} \end{aligned}$$

EXAMPLE 5

Design an adsorber system to collect 500 ppm benzene from an air–benzene mixture at 1.1 atm, 50°F, flowing at 500 acfm. Use 4 × 6 mesh activated carbon. Steam is available twice per day at the plant, and a steam–solvent ratio of 3.0 is to be employed. Thus each unit is to be operated for 24 h per total cycle (one on/one off). What are the pressure drops for this process? Also, calculate the steam flow per regeneration for this system. Pertinent data are as follows:

$$\begin{aligned} \rho_{\text{B}}, \text{ carbon} &= 30 \text{ lb/ft}^3 \\ \text{CAP} &= 0.45 \text{ pound benzene per pound carbon} \\ \text{HEEL} &= 0.03 \text{ pound benzene per pound carbon} \\ \text{velocity} &= 75 \text{ fpm} \end{aligned}$$

SOLUTION

Key calculations are as follows:

$$\begin{aligned} V &= (500 \text{ ft}^3/\text{min})(60)(12 \text{ h}) = 360{,}000 \text{ ft}^3 \text{ per } 12 \text{ h} \\ V_{\text{B}} &= 360{,}000(500/10) = 180 \text{ ft}^3 \\ \rho_{\text{B}} &= (78)(1.1)/(10.73)(510) = 0.2304 \text{ lb/ft}^3 \\ M_{\text{B}} &= (180)(0.2304) = 41.47 \text{ pounds benzene} \\ \text{WC} &= \text{CAP} - \text{HEEL} = 0.45 - 0.03 \\ &= 0.42 \text{ pound benzene per pound carbon} \\ M_{\text{AC}} &= 41.47/0.42 = 98.74 \text{ pounds activated carbon} \\ V_{\text{AC}} &= 98.74/30 = 3.291 \text{ ft}^3 \\ A &= 500/75 = 6.666 \text{ ft}^2 \\ \text{If vertical, } D &= 2.91 \text{ ft} \\ H &= 3.291/6.667 = 0.4937 \text{ ft} \\ &= 5.9245 \text{ in.} = 6 \text{ in.} \end{aligned}$$

From Figure 5:

$$\begin{aligned} \Delta P &= 0.4 \text{ in. water/ in. bed} \\ \Delta P_{\text{total}} &= (0.4)/6.0) = 2.4 \text{ in.} \\ M_{\text{steam}} &= (3.0)(41.5) \\ &= 124.5 \text{ pounds steam during regeneration} \end{aligned}$$

Extra carbon should be provided because the MTZ (2 in.) will only be 50% saturated.

REFERENCES

1. D. Ruthven, *Principles of Adsorption and Adsorption Processes*, Wiley, New York, 1984.
2. G. Joseph and D. Beachler, *Control of Gaseous Emissions*, Northrop Services, Inc., Research Triangle Park, NC, EPA 450/2-81-005, 1981.
3. K. Noll, V. Gounaris, and W. Hou, *Adsorption Technology for Air and Water Pollution Control*, Lewis Publishers, Chelsea, Mich., 1992.
4. J. Danielson (Ed.) *Air Pollution Engineering Manual*, AP40-USEPA, 1973.
5. A. Turk, "Adsorption," *Air Pollution, Vol. IV, Engineering Control of Air Pollution*, A. Stern (Ed.), Academic Press, New York, 1977.

6. T. Buonicore and W. Davis (Eds.), *Air Pollution Engineering Manual*, Air and Waste Management Assoc., Van Nostrand Reinhold, New York, 1992.

CONDENSATION

ANTHONY J. BUONICORE

Condensation is the process of converting a gas or vapor to liquid. Any gas can be reduced to a liquid by sufficiently lowering its temperature and/or increasing its pressure. The most common approach is to reduce the temperature of the gas stream, since increasing the pressure of a gas can be expensive.

Condensers are simple, relatively inexpensive devices that normally use water or air to cool and condense a vapor stream. Since these devices are usually not required or capable of reaching low temperatures (below 100°F), high removal efficiencies of most gaseous pollutants are not obtained unless the vapors will condense at high temperatures. Condensers are typically used as pretreatment devices. They can be used ahead of adsorbers, absorbers, and incinerators to reduce the total gas volume to be treated by more expensive control equipment. Employed in this manner, this device may help reduce the overall cost of the control system.

When a hot vapor stream contacts a cooler medium, heat is transferred from the hot gases to the cooler medium. As the temperature of the vapor stream is lowered, the average kinetic energy of the gas is reduced. Ultimately, the gas molecules are slowed down and crowded so closely together that the attractive (van der Waals) forces between the molecules cause them to condense to a liquid. The two conditions that aid condensation are low temperatures, so that the kinetic energy of the gas molecules is low, and high pressures, so that the molecules are brought closer together. The actual conditions at which a particular gas molecule will condense depends on its physical and chemical properties. Condensation occurs when the partial pressure of the pollutant in the gas stream equals its vapor pressure as a pure substance at the operating temperature.

As described, condensation of a gas can occur in three ways: (1) At a given temperature, the system pressure is increased (compressing the gas volume) until the partial pressure of the gas equals its vapor pressure; (2) at a fixed pressure, the gas is cooled until the partial pressure equals its vapor pressure; or (3) in a combined technique, the gas is compressed and cooled until its partial pressure equals its vapor pressure.

DESCRIPTION OF EQUIPMENT

As indicated in the previous section, condensation can be accomplished by increasing pressure or decreasing temperature (removing heat). In practice, air pollution control condensers operate by extracting heat. Condensers differ in the means of removing heat and the type of device used. The two different means of condensing are direct contact, where the cooling medium with vapors and the condensate are intimately mixed and combined, and indirect (or surface), where the cooling medium and vapor/condensate are separated by a surface area of some type.

Contact condensers are simpler, are less expensive to install, and require less auxiliary equipment and maintenance. The condensate/coolant from a contact condenser has a volume 10 to 20 times that of a surface condenser. This condensate often cannot be reused and may pose a waste disposal problem unless the dilution of the pollutant is sufficient to meet regulatory requirements. Some typical contact condensers are shown in Figure 1.

Surface condensers form the bulk of the condensers used for air pollution control. Among the applicable types of surface condensers are shell and tube, double pipe, spiral plate, flat plate, air cooled, and various extended-surface tubular units. This discussion focuses on shell-and-tube condensers, because they are so widely used in industry and have been standardized by the Tubular Exchanger Manufacturers Association (TEMA).

Condensing can be accomplished in either the shell or the tubes. The economics, maintenance, and operational ramifications of the allocation of fluids are extremely important, especially if extended-surface tubing is being considered. The designer should be given as much latitude as possible in specifying the condenser. Details of TEMA designations for shell-and-tube heat exchangers are shown in Figure 2.

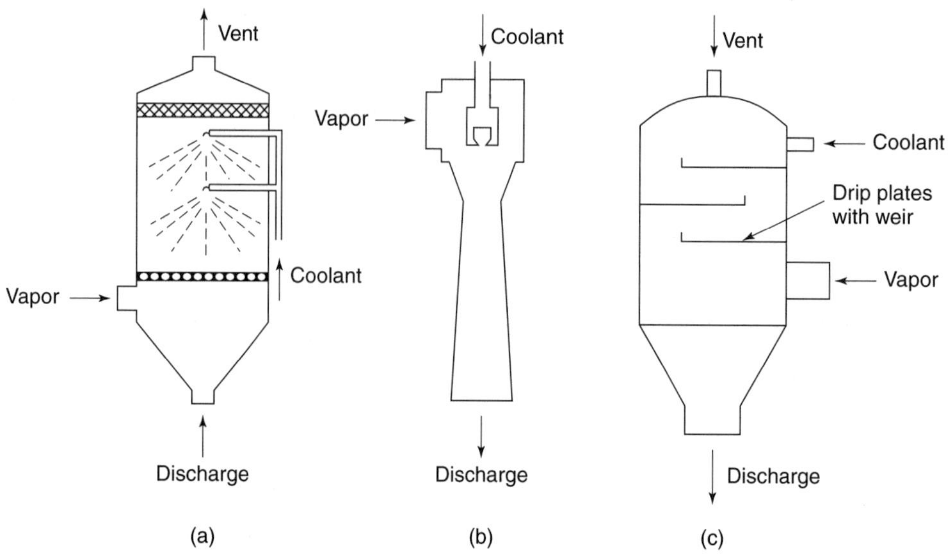

Figure 1. Contact Condensers: (a) Spray; (b) Jet; (c) Barometric.

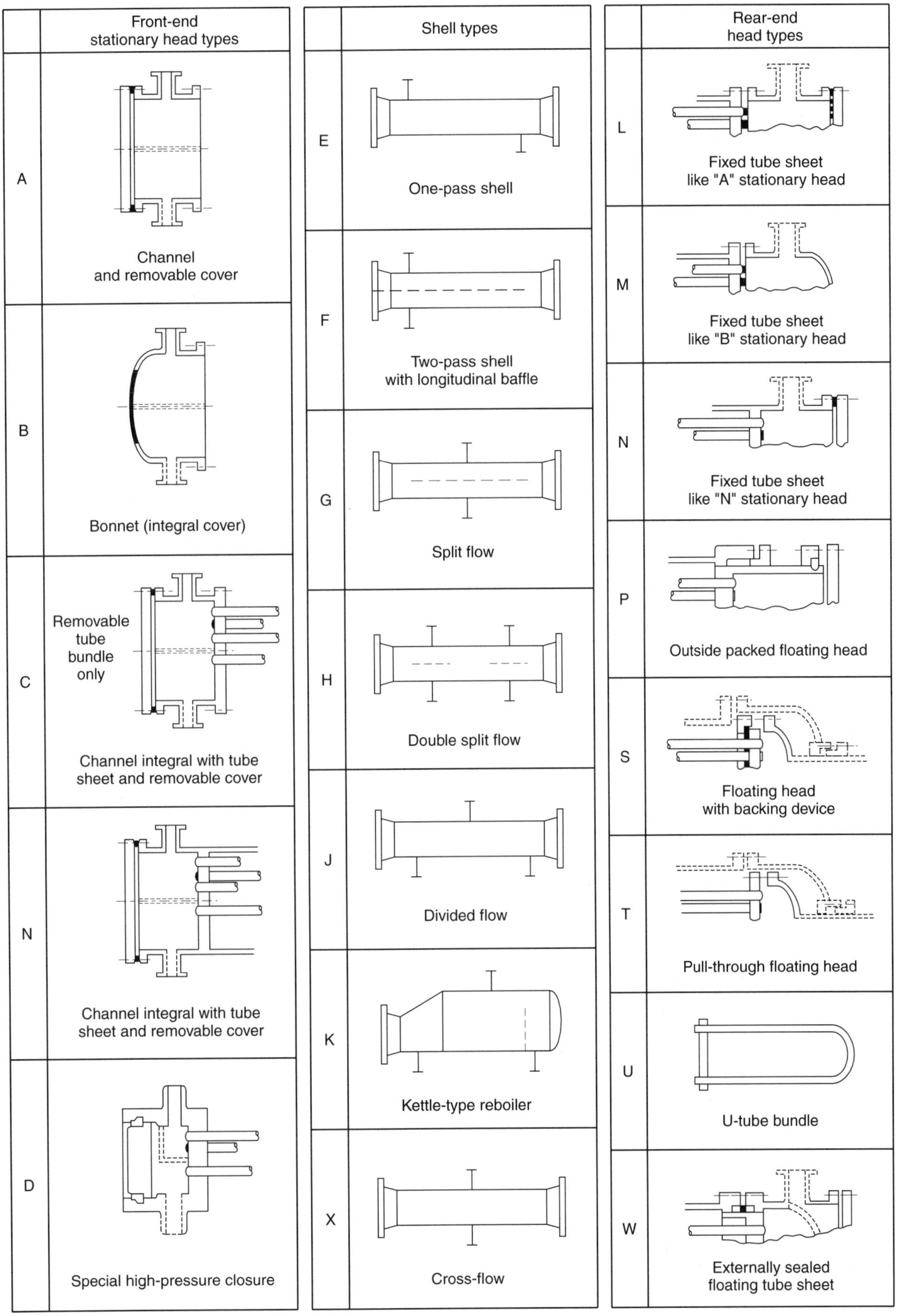

Figure 2. Shell-and-Tube Condensers (© Tubular Exchange Manufacturers Association).

Air-cooled condensers consist of a rectangular bank of high-finned tubes, a fan, a plenum for even distribution of air to the rectangular face of the tube bank, a header for vapor inlet and condensate outlet, and a steel supporting structure. Fins are usually aluminum, 0.5 to 0.625 in. high, applied to a bare tube by tension winding or soldering, or cold-extruded from the tube itself. Bimetallic tubes can be used to provide process corrosion resistance on the inside and aluminum extruded fins on the outside. Headers for the tube side frequently contain removable gasketed plugs corresponding to each tube end and are used for access to the individual tube-to-tube sheet joint for maintenance.

Nearly all tubular exchangers employ roller expanding of the tube ends into tube holes drilled into tube sheets as a means of providing a leakproof seal. According to TEMA requirements, tube holes must be of close tolerance and contain two concentric grooves. These grooves give the tube joint greater strength, but do not increase sealing ability and may actually decrease it. Research is currently being done to clarify what the best procedure is to follow for a given set of materials, pressures, temperatures, and loadings. Welding or soldering of tubes to tube sheets may be performed for additional leak tightness or in lieu of roller expanding.

DESIGN AND PERFORMANCE EQUATIONS

Design of contact condensers involves calculating the quantity of coolant required to condense and subcool the vapor and proper sizing of discharge piping and hotwell. Calculation of coolant flow rate is as follows:

$$M_c = [M_v H_v + M_l C_{pl}(T_i - T_d)]/[C_{pc}(T_d - t_i)] \quad (1)$$

where M_v, M_l, M_c are the flow rate of vapor, liquid condensate, and coolant, respectively (lb/in.); C_{p1}, C_{pc}, the heat capacity of condensate and coolant (Btu/lb °F); T_i, T_d, the inlet temperature of the vapor and the discharge temperature (°F); t_i, the inlet temperature of coolant (°F); and H_v, the latent enthalpy of vaporization of condensate (Btu/lb).

Use of a contact condenser requires consideration of coolant availability and liquid waste disposal or treatment facilities. Contact condensers are relatively efficient scrubbers as well as condensers and have the lowest equipment capital cost. Contaminants that are not ordinarily condensable or even particulates can be removed from the vapor stream. Reputable manufacturers of the various types of contact condensers should be consulted for sizing and layout recommendations.

The heat transfer for a surface condenser is governed by the following relationship:

$$Q = U_o A T_m \quad (2)$$

$$A = Q/(U_o T_m) \quad (3)$$

where Q is the total heat load (Btu/h); U_o, the overall heat-transfer coefficient (Btu/h °F ft^2); T_m, the mean temperature difference driving force (°F); and A, the surface area, (ft^2).

The first determination is the heat load. In the simplest case, it is the latent heat to condense plus the subcooling load. The next step is specifying the coolant flow rate and temperatures to balance the vapor heat load. Temperature crosses where the outlet coolant temperature is higher than the vent and/or condensate outlet temperature should be avoided.

After the terminal temperatures of the vapor and coolant have been determined, a mean temperature difference can be established. For isothermal or linear condensing and liquid subcooling, the logarithmic mean temperature difference (LMTD) is applicable. The LMTD is calculated as follows:

$$\text{LMTD} = (t_g - t_l)/[\ln(t_g/t_l)] \quad (4)$$

where t_g is the greater temperature difference between the hot and cold streams; t_l the lesser temperature difference between the hot and cold streams; and ln, the natural log.

Fluids are usually in counterflow, but in some cases, as with a fluid near its freezing point, parallel flow is used to afford a greater degree of safety from freezing the condenser. Frequently, there is a combination of parallel flow and counterflow as in a two-pass tube side, single-pass shell tubular condenser. The procedure is to apply a correction to the LMTD. The correction factors have been reduced to a series of graphs available in most heat-transfer texts.[1] Strictly, the correction factor applies only to the subcooling or desuperheating zones, but it is sometimes applied to condensers.

If condensation is nonlinear and has a range exceeding 20°F, the LMTD is not applicable, and a condensing curve or calculations at several points over the temperature range is required to determine a correct T_m. A mixture of miscible vapors or a vapor and noncondensables are examples of where the LMTD is not applicable.

The remaining unknown to calculate the required surface is the overall heat-transfer coefficient, U_o. The overall heat-transfer coefficient is the inverse of the sum of the resistances to heat transfer and is given by the following relationship:

$$U_o = 1/(1/h_o + 1/h_{io} + r_w + f_o + f_{io}) \quad (5)$$

where h_o is the outside or shell-side coefficient (Btu/h °F ft^2); h_{io}, the inside or tube-side coefficient corrected to the outside area; r_w, the resistance of the surface area or tube wall (h °F ft^2/Btu); f_o, the fouling resistance on the outside or shell side of the surface area; and f_{io}, the fouling resistance on the inside corrected to the outside of the surface area. Some typical industrial values of U_o are given in Table 1.

Table 1. Typical Overall Heat Transfer Coefficients in Tubular Heat Exchangers

Condensing Vapor (Shell Side)	Cooling Liquid (Tube Side)	U (Btu/°F ft^2 h)
Alcohol vapor	Water	100–200
High-boiling hydrocarbons (vacuum)	Water	20–50
Low-boiling hydrocarbons	Water	80–200
Organic solvents	Water	100–200
Organic solvents/high percent of noncondensables present	Water or brine	20–80
Naphtha	Water	50–75
Stabilizer reflux vapors	Water	80–120
Sulfur dioxide	Water	150–200
Tall oil derivatives, vegetable oil vapors	Water	20–50
Steam	Feed water	400–1000

If there is noncondensable gas present or a vapor mixture of miscible and immiscible components with a large condensing range, diffusion of the condensing vapor through another vapor or gas must be considered. The presence of a noncondensable gas in a vapor stream reduces the film coefficient; the amount of reduction is related to the size of the gas cooling load relative to the total load. The greater the ratio of sensible heat to total, the closer the heat-transfer coefficient approaches that of cooling the gas only. The calculation procedure is essentially that established by Colburn and Hougen.[2] It is a trial-and-error method performed for at least five points along the condensing curve. Typical condensing coefficients (Btu/°F ft^2 h) are given in Table 2.

The choice of a coolant will depend on the particular plant and the efficiency required of the condenser. The most common coolant is the primary plant coolant, usually a cooling tower or river water. It has been shown[3] that the vapor outlet temperature is critical to the efficiency of a condenser. In some instances, use of a chilled brine or a boiling refrigerant can achieve a collection efficiency that will be sufficient without additional control devices. It is not unusual to specify multiple-stage condensing, where condensers, usually two, are connected in series use cooling media with successively lower temperatures. For example, a condenser using cooling-tower water can be used prior to a unit using chilled water or brine—thereby achieving maximum efficiency while minimizing the use of chilled water.

Careful consideration should be given to the fouling resistances f_o and f_{io} applied to the condenser design. The purpose of adding a fouling resistance is to provide consistent performance without excessive cleaning. Theoretically, increased fouling factors will provide greater safety allowance. In practice, however, they can actually decrease the safety allowance. Consider, for example, a typical application where a maximum pressure drop and overall length are specified. The most efficient design will use the length and pressure drop to the limit specified. If the fouling resistance specified is so large that this is not possible, velocities of the respective streams will be lower than desirable, and fouling will be greater than in a design using a lower value of fouling resistance. Fouling resistances used indiscriminately can actually be "self-fulfilling." The most effective design feature to minimize fouling is high velocity and efficient use of pressure drop.[4]

Limited information is available in the literature to calculate the (vapor) collection efficiency of the condenser. The following calculational procedure is recommended. If F, V, and L represent the molar flow rates entering, vented, and (condensate) discharged, respectively, then:

$$F = V + L \tag{6}$$

If z_i, y_i, and x_i are the mole fraction of organic i in the feed, vent, and condensate streams, respectively, then:

$$Fz_i = Vy_i + Lx_i \tag{7}$$

Assuming the vent and condensate streams leave the unit in equilibrium:

$$K_i = y_i/x_i \tag{8}$$

The simultaneous solution of Eqs. 6–8 can provide information on the amount of organics (condensables) remaining in the gas stream:

$$\sum_i Fz_i/[V + (L/K_i)] = 1.0 \tag{9}$$

Table 2. Typical Film Coefficients (Btu/h °F ft)

Fluid	h
Steam	1500
Steam/10% gas	600
Steam/20% gas	400
Steam/40% gas	220
Pure light hydrocarbons	250
Mixed light hydrocarbons	175
Medium hydrocarbons	100
Medium hydrocarbons	125
Pure organic solvents	250

OPERATION AND MAINTENANCE[5]

The maximum allowable working pressures and temperatures are indicated on the nameplate of the condenser. These values must not be exceeded. Special precautions should be taken if any individual part of the condenser is designed for a maximum temperature lower than that of the condenser as a whole. The most common example is some copper-alloy tubing with a maximum allowable temperature lower than the actual inlet gas temperature. This is done to compensate for the low strength levels of some brasses or other copper alloys at elevated temperatures. An adequate flow of the cooling medium must be maintained at all times.

Condensers are designed for a particular fluid throughput. Generally, a reasonable overload can be tolerated without causing damage. If operated at excessive flow rates, erosion or destructive vibration could result. Erosion could occur at normally acceptable flow rates if other conditions, such as entrained liquids or particulates in a gas stream or abrasive solids in a liquid stream, are present. Evidence of erosion should be investigated to determine the cause.

In addition to flow overloads, vibration can be propagated by improper design, fluid maldistribution, or the corrosion/erosion of internal flow-directing devices such as baffles. Considerable study and research has been undertaken in recent years to develop a reliable analysis procedure to predict or correct damaging vibration. At this point, the developed correlations are considered "state of the art," yet most manufacturers have the capability of applying some type of vibration check when designing a condenser. Vibration can produce severe mechanical damage, and operation should not be continued when an audible vibration disturbance is evident.

Condensers should be warmed up slowly and uniformly; the higher the temperature ranges, the slower the warmup should be. This is generally accomplished by introducing the coolant and bringing the flow rate to design level and gradually adding the vapor. For fixed-tube-sheet units with different shell-and-tube material, consideration should be given to differential expansion of shell and tubes. As fluids are added, the respective areas should be vented to ensure complete distribution. Using another procedure could cause large differences in temperature between adjacent parts of the condenser and result in leaks or other damage. It is recommended that gasketed joints be inspected after continuous full-flow operation has been established. Handling, temperature fluctuations, and yielding of gaskets or bolting may necessitate retightening of the bolting.

Cooling down is generally accomplished by shutting off the vapor stream first and then the cooling stream. Again, fixed-tube-sheet condensers require consideration of differential expansion of the shell and tubes. Condensers containing flammable, corrosive, or high-freezing-point fluids should be thoroughly drained during prolonged outages.

The recommended maintenance of condensers requires regular inspection to ensure mechanical soundness of the unit and a level of performance consistent with the original design criteria. A brief general inspection should be performed on a regular basis while the unit is operating. Vibratory disturbance, leaking gasketed joints, excessive pressure drop, decreased efficiency indicated by higher gas outlet temperature or lower condensate rates, and intermixing of fluids are all signs that thorough inspection and maintenance procedures are required.

Complete inspection requires shutting down the condenser for access to internals and pressure testing and cleaning. Scheduling can only be determined from experience and general inspections. Tube internals and exteriors, where accessible, should be visually inspected for fouling, corrosion, or damage. The nature of any metal deterioration should be investigated properly to determine the anticipated life of equipment or the corrective action to take. Possible causes of deterioration include general corrosion, intergranular corrosion, stress cracking, galvanic corrosion, impingement, and erosion attack.

Fouling of condensers is the deposition of foreign material on the interior or exterior of tubes. Evidence of fouling in operation is increased pressure drop and general decreases in performance. Fouling can be so severe that tubes are completely plugged, resulting in thermal stresses and subsequent mechanical damage of equipment.

The nature of the deposited foul determines the method of cleaning. Soft deposits can be removed by steam, hot water, various chemical solvents, or brushing. Plant experience can determine which method to use. Chemical cleaning should be performed by contractors specialized in the field who will consider the deposit to be removed and the materials of construction. If the cleaning method involves elevated temperature, consideration should be given to thermal stresses induced in the tubes; steaming out individual tubes can loosen the tube-to-tube sheet joints.

Mechanical methods of cleaning are useful for soft and hard deposits. There are numerous tools for cleaning tube interiors: brushes, scrapers, and various rotating cutter-type devices. The condenser manufacturer or suppliers of tube tools can be consulted in the selection of the correct tool for the particular deposit. When cutting or scraping deposits, care should be exercised to avoid damaging tubes.

Cleaning of tube exteriors is generally performed using chemicals, steam, or other suitable fluids. Mechanical cleaning is performed, but it requires that the tubes be exposed, as in a typical air-cooled condenser, or are capable of being exposed, as in a removable bundle shell-and-tube condenser. The layout pattern of the tubes must provide sufficient intersecting empty lanes between the tubes, as in square pitch. Mechanical cleaning of tube bundles, if necessary, requires the utmost care to avoid damaging tubes or fins.

Proper maintenance requires testing the condenser to check the integrity of the tubes, tube-to-tube sheet joints, welds, and gasketed joints. The normal procedure consists of pressuring the shell with water or air at the nameplate-specified test pressure and viewing the shell welds and the face of the tube sheet for leaks in the tube sheet joints or tubes. Water should be at ambient temperature to avoid false indications due to condensation. Pneumatic testing requires extra care because of the destructive nature of a rupture or of explosion or fire hazards when residual flammable materials are present. Condensers of straight-tube, floating-head construction use a test gland. Tube bundles without shells are tested by pressurizing the tubes and viewing the length of the tubes and back face of the tube sheets.

Corrective action for leaking tube-to-tube sheet joints requires expanding the tube end with a suitable roller-type tube expander. Good practice calls for an approximate 8% reduction in wall thickness after metal-to-metal contact between the tube and tube hole. Tube expanding should not extend beyond 1/8 in. of the inner tube-sheet face to avoid cutting the tube. Care should be exercised to avoid overrolling the tube, which can cause work hardening of the material, an insecure seal, and/or stress-corrosion cracking of the tube.

Defective tubes can be either replaced or plugged. Replacing tubes requires special tools and equipment. The user should contact the manufacturer or a contractor qualified in repair. Plugging of tubes, although a temporary solution, is acceptable provided that the percentage of the total number of tubes per tube pass to be plugged is not excessive. The type of plug to be used is a tapered one-piece or two-piece metal plug suitable for the tube material and inside diameter. Care should be exercised in seating plugs to avoid damaging the tube sheets. If a significant number of tube or tube joint failures are clustered in a given area of the tube layout, their location should be noted and reported to the manufacturer. A concentration of failures is usually caused by something other than corrosion (e.g., impingement, erosion, or vibration).

MAINTAINING AND IMPROVING EQUIPMENT PERFORMANCE[5]

Within the constraints of the existing system, improving operation and performance refers to maintaining operation and original or consistent performance. There are several factors previously mentioned that are critical to the design and performance of a condenser: operating pressure, amount of noncondensable gases in the vapor stream, coolant temperature and flow rate, fouling resistance, and mechanical soundness. Any pressure drop in the vapor line upstream of the condenser should be minimized. Deaerators or similar devices should be operational where necessary to remove gases in solution with liquids. Proper and regular venting of equipment and leakproof gasketed joints in vacuum systems is necessary to prevent gas binding and alteration of condensing equilibrium. Coolant flow rate and temperatures should be checked regularly to ensure that they are in accordance with the original design criteria. The importance of this can be illustrated merely by comparing the winter and summer performance of a condenser using cooling-tower or river water. Decreased performance due to fouling will generally be exhibited by a gradual decrease in efficiency and should be corrected as soon as possible. Mechanical malfunctions can also be gradual, but will eventually be evidenced by a near total lack of performance.

Fouling and mechanical soundness can only be controlled by regular and complete maintenance. In some cases, fouling is much worse than predicted and requires frequent cleaning regardless of the precautions taken in the original design. These cases require special designs to alleviate the problems associated with fouling. A leading manufacturer of polyvinyl chloride found that carryover of polymer reduced the efficiency of its monomer condenser and caused frequent downtime. The solution was providing polished internals and high condensate loading in a vertical-downflow shell-and-tube condenser. A major pharmaceutical intermediate manufacturer had catalyst carryover to a vertical-downflow shell-and-tube condenser that accumulated on the tube internals. The solution was to recirculate condensate to the top of the unit and spray it over the tube-sheet face to create a film descending down the tubes to rinse the tubes clean.

Most condenser manufacturers will provide designs for alternative conditions as a guide to estimating the cost of improving efficiency via alternate coolant flow rates and temperatures, as well as alternative configurations (i.e., vertical, horizontal, shell side, or tube side).

EXAMPLE 1

Steam at atmospheric pressure is being condensed in a heat exchanger with water that enters and exits the unit at 80 and 115°F, respectively. Assuming no subcooling of the condensed steam, calculate the log mean temperature driving force for the exchanger.

Solution

Saturated steam at 1.0 atm is at 212°F. The log mean temperature difference is given by Eq. 4.

$$\text{LMTD} = (t_g - t_l)/\ln(t_g/t_l)$$

Substituting temperatures gives

$$t_g = 212 - 80 = 132$$

$$t_l = 212 - 115 = 97$$

$$\text{LMTD} = (132 - 97)/\ln(132/97) = 113.6^\circ\text{F}$$

EXAMPLE 2

The discharge gases from a meat-rendering plant contain essentially all steam with a small fraction of odor-carrying gases. This steam is typically condensed (to remove the water) prior to treatment with an adsorber or incinerator. Estimate the size of a condenser to treat 60,000 lb/h of this exhaust gas. Assume condensing occurs in a heat exchanger operating with the temperature driving force calculated in Example 1 and an overall heat-transfer coefficient of 135 Btu/h ft^2 °F.

Solution

Assuming that the discharge gas is all steam with an enthalpy of vaporization of approximately 1000 Btu/lb, the condenser heat load becomes

$$Q = (m)(\Delta H)$$
$$= (60{,}000)(1{,}000)$$
$$= 6.0 \times 10^7 \text{ Btu/h}$$

The condenser area is as follows:

$$A = Q/[(U)(\text{LMTD})]$$
$$= 6.0 \times 10^7/[(135)(113.6)]$$
$$= 3912 \text{ ft}^2$$

One should consult a vendor if information on the number of tubes and the tube diameter and length is required.

REFERENCES

1. D. Q. Kern, *Process Heat Transfer*, McGraw-Hill., New York, 1950.
2. A. P. Colburn and O. A. Hougen, *Ind. Eng. Chem.* **26**(11):1178 (1934).
3. W. F. Connery et al., "Energy and the environment," *Proclamation Third National Conference AIChE*, 1975, p. 276.
4. L. Theodore and A. J. Buonicore (Eds.), *Air Pollution Control Equipment: Selection, Design, Operation and Maintenance*, Prentice-Hall, Englewood Cliffs, NJ, 1982.
5. W. F. Connery, in *Air Pollution Control Equipment: Selection, Design, Operation and Maintenance*, L. Theodore and A. J. Buonicore, Eds., Prentic-Hall, Englewood Cliffs, NJ, 1982.

VAPOR INCINERATION*

C. David Cooper

INTRODUCTION

Incineration is a term that most people understand to mean the destruction of a waste by combustion. If, however, we restrict the use of the word *combustion* to an active self-sustained burning process producing flames, then incineration is better defined as a high-temperature oxidation process in which oxygen in the air attacks waste molecules, converting them into simpler substances such as carbon dioxide and water. Incineration is often the only viable method for controlling highly odorous, toxic, or otherwise dangerous chemicals (such as pesticides or war gases) that exhibit those odorous or toxic properties because of their organic molecular structure. Incineration is not appropriate for wastes that are problematic because of their elemental content (such as an industrial sludge containing heavy metals).

For many people in the general public, incineration is linked with the disposal of a solid or hazardous waste, and is perceived as a noisy, odorous, unsightly process posing a considerable risk to the local community, and which produces a significant

* Much of the material in this chapter has been taken/adapted from Chapter 11 of *Air Pollution Control—A Design Approach (2nd ed.)*, C. D. Cooper and F. C. Alley, Waveland Press (1994). Used by permission.

quantity of ash that must be discarded. However, when applied to the control of gases or vapors (here we may use the term *vapor incineration*), the process is much less intrusive, odors are destroyed, little or no ash is formed, and the process does not generate much public opposition. Because of the physical and public perception differences between solid or liquid waste incineration and vapor incineration, and because of the negative connotation of the word *incineration*, vapor incineration often is called *afterburning* or *thermal oxidation*, and the equipment is referred to as a *thermal oxidizer* instead of an incinerator.

Whatever the name, the fundamental principle of incineration remains the same: During the oxidation process the organic waste molecules are attacked by oxygen and destroyed, and simpler substances such as carbon dioxide and water are formed. In order to use incineration, the gases or vapors to be controlled must be substances that can be oxidized: typically, volatile organic compounds (VOCs) and/or carbon monoxide (CO). In general, vapor incineration is applied to a dilute stream of VOCs in air (such as may be found exiting a printing press or paint-baking oven), or to the incompletely combusted exhaust gases from another combustion process (such as the gases coming from the primary combustion chamber of a solid or liquid waste incinerator). Occasionally, some processes (especially in petroleum refineries) generate intermittent releases of nearly pure streams of VOCs. Such releases are often disposed of by burning in an elevated flare, another form of incineration.

In this section, we review the chemical principles of oxidation, some physical principles of applying this control technology, the engineering data and calculations supporting the process design of thermal oxidizers, and the regulations and objectives for control by vapor incineration, and give some examples of the calculations made to size thermal oxidizers.

OXIDATION CHEMISTRY

Consider the situation where a stream of air containing a low concentration of VOCs is heated to a high temperature. The concentration of VOCs is too low to support combustion (as defined previously), but the VOCs will become oxidized given the right conditions of temperature, time, and turbulence.

The oxidation of a hydrocarbon in air can be represented by the following stoichiometric equation for complete oxidation.

$$C_xH_y + bO_2 + 3.76bN_2 \rightarrow xCO_2 + \left(\frac{y}{2}\right) H_2O + 3.76bN_2 \quad (1)$$

where C_xH_y is the general formula for any hydrocarbon; $b = x + (y/4)$, the stoichiometric number of moles of oxygen required per mole of C_xH_y; and 3.76, the number of moles of nitrogen present in air for every mole of oxygen. In Eq. 1, we have included the nitrogen as a reminder that when oxidation occurs in air, much additional gas (the nitrogen) is always present. For simplicity, in the equations that follow, the nitrogen will not be shown.

Incomplete oxidation of a hydrocarbon results in CO, smaller-molecular-weight hydrocarbons, and some partially oxidized hydrocarbons being emitted as pollutants. In the case of highly concentrated streams of organic wastes being burned, then poor combustion may result in significant amounts of CO and VOCs being emitted. These emissions may even include compounds that were not in the wastes to begin with; compounds such as benzene, butadiene, formaldehyde, and others *can be formed* in the combustion process. However, it must be emphasized that a properly designed and operated thermal oxidizer can and does destroy 99.9% or more of these compounds.

If there are other elements in the waste besides carbon and hydrogen, they will be oxidized as well, forming end products that may be pollutants in and of themselves. The more common elements and the products they form are:

$$Cl \rightarrow HCl \quad \text{and} \quad F \rightarrow HF$$

$$S \rightarrow SO_2$$

$$N \rightarrow N_2 \quad \text{and} \quad NO$$

$$P \rightarrow P_2O_5$$

The kinetics of oxidation reactions are extremely important. The actual pathway or steps of the combustion reaction (the mechanism) can be very complex, and the oxidation does not occur in a single step as might be inferred from Eq. 1. Even the combustion of simple methane involves a branching chain reaction, the main reactions of which are depicted in Figure 1.[6] For heavier VOCs, the mechanisms are more complex owing to the larger number of possible intermediates.

Because of the necessity to simplify kinetic models for the design of thermal oxidizers, several authors (Refs. 4, 7 and 8) have developed "global models." A global model ignores many of the detailed steps that are required in mechanistic models, and ties the kinetics to the main stable reactants and products. Since carbon monoxide is a very stable intermediate, the simplest global model for the oxidation of a hydrocarbon is a two-step model as follows:

$$C_xH_y + \left(\frac{x}{2} + \frac{y}{4}\right) O_2 \rightarrow xCO + \left(\frac{y}{2}\right) H_2O \quad (2)$$

$$xCO + \left(\frac{x}{2}\right) O_2 \rightarrow xCO_2 \quad (3)$$

Again, the reactions do not actually occur in single steps, as might be inferred from Eqs. 2 and 3.

$CH_4 + O_2 \rightarrow CH_3 + HO_2$	chain initiation
$CH_3 + O_2 \rightarrow CH_2 + OH$ $OH + CH_4 \rightarrow H_2O + CH_3$ $OH + CH_2O \rightarrow H_2O + CHO$	chain propagation
$CH_2 + O_2 \rightarrow HO_2 + HCO$	chain branching
$HCO + O_2 \rightarrow CO + HO_2$ $HO_2 + CH_4 \rightarrow H_2O_2 + CH_3$ $HO_2 + CH_2O \rightarrow H_2O_2 + HCO$	chain propagation
$OH \rightarrow$ wall $CH_2O \rightarrow$ wall	chain termination

(Adapted from Ref. 6, as portrayed in Refs.1 and 3)

Figure 1. Major Reactions Involved in the Oxidation of Methane.

Many authors have shown the importance of atoms and radicals such as O, H, OH, CH_3, CH_2, and HO_2 in flame combustion. There is no reason to believe that these radicals, if they could be formed, would not also play an important role in the thermal oxidation of dilute VOCs in air. Research has shown that such oxidation reactions can be enhanced by injection of H_2O_2 or O_3 into the hot gases downstream of the flame zone. Hydrogen peroxide breaks apart into radicals even at temperatures as low as 500°C, which increases the extent and rate of the oxidation of the VOCs.[2,5]

PRINCIPLES OF OPERATION

The oxidation of VOCs occurs because of the interaction of the organic molecules with oxygen at a high temperature. This fact leads to the commonly cited "three Ts of incineration"—temperature, time, and turbulence. A sufficiently high temperature is crucial because the oxidation reactions require an activation energy in order to begin, and then must proceed very rapidly to promote the nearly complete destruction of the VOCs. Because the reactions proceed at a fast but finite rate, sufficient time must be provided at the desired temperature. The third key is turbulence in the gases, which promotes good mixing between the VOC molecules and the oxygen molecules. Consider, for example, a situation where 99.99% of the VOCs must be destroyed. If even 1% of the VOCs are never mixed with the high-temperature gases and thus exit the system unreacted, then the overall destruction efficiency is only 99%, and the emissions are 100 times more than the design requirement! Without temperature, time, and turbulence, the oxidizer cannot achieve the high degree of completion required.

In 1973, Danielson suggested that for good destruction, an afterburner (a thermal oxidizer following a primary combustion chamber) should be designed for temperatures of 1000–1500°F, for residence times of 0.3–0.5 sec, and for flow velocities (to promote turbulent mixing) of 20–40 ft/sec. More recently, these general guidelines have been modified to include higher temperatures (1200–2000°F) and longer residence times (0.5–2.0 sec) in order to promote more complete destruction of the VOCs.

It has become standard practice to include an afterburner as part of a hazardous waste incinerator (HWI) facility. U.S. EPA regulations for HWIs require a minimum of 99.99% destruction and removal efficiency (DRE) of the principal organic hazardous constituents (POHCs) being burned. In order to ensure such complete destruction, the final operating permits issued by the EPA almost always require a high temperature in the afterburner (the value of the required temperature depends on the specific compounds being burned, but 2000°F is not uncommon) and a minimum of 2.0 sec residence time in the afterburner. Hospital (biomedical) waste incinerators often utilize afterburners as well, and a regulatory requirement of 1600°F and 1.0 sec is typical for an afterburner on a hospital waste incinerator.

In a mathematical sense, the three Ts are related to three characteristic times—a chemical time, a residence time, and a mixing time—given by the following equations:

$$\tau_C = 1/k \tag{4}$$

$$\tau_r = V/Q = L/u \tag{5}$$

$$\tau_m = L^2/D_e \tag{6}$$

where τ_C, τ_r, τ_m are chemical, residence, and mixing times, respectively (sec); Q, the volumetric flow rate (at the temperature in the afterburner) (m^3/sec); L, the length of the reaction zone (m); u, the gas velocity in the afterburner (m/sec); and D_e the effective (turbulent) diffusion coefficient (m^2/sec).

Barnes, Putnam, and Barrett have noted[1] that if the ratio τ_m/τ_r is large and τ_r/τ_C is small, then mixing is the rate-controlling process in the afterburner. But if τ_m/τ_r is small and τ_r/τ_C is large, then the chemical kinetics are rate controlling. At the temperatures of most afterburners, as long as a reasonable flow velocity is maintained, the chemical kinetics, and not the mixing, will be the limiting factor. However, as the temperature is raised, the chemical time decreases rapidly, and at some point the overall rate is limited by the mixing within the device.

During normal operations, it is desirable to maintain the equipment running at steady state, a condition in which all flows, temperatures, concentrations, and other process parameters are held constant with time. Modern control instrumentation is used to help achieve steady-state operations, and stack gas analyzers help the operators monitor the process. Usually, the flow rates of the inlet VOC-in-air stream and the input fuel gas, the stack gas temperature, and the concentrations of oxygen, CO, and total hydrocarbons in the stack gas are measured continuously. Stack gas tests for specific compounds may also be conducted annually, semiannually, or more often to ensure that the emissions are being kept below permit limits.

Thermal oxidizers are important components of facilities designed for the incineration of hazardous wastes. Current regulations require a destruction and removal efficiency of 99.99% for POHCs. However, for a mixture of wastes, determining which POHCs will "control" the overall DRE of the mixture is not trivial. The minimum temperature required to achieve the target DRE in a thermal oxidizer is a function of the POHC that is the most difficult to destroy. A number of compounds have been ranked for their gas-phase thermal stability; the ranking is an indication of the difficulty of incineration. Selection of the critical POHC in a mixed waste is often based on this ranking. A partial listing is given in Table 1.

Because the compounds that are being oxidized can be explosive, safety considerations dictate that the stream of VOCs in air being fed into a thermal oxidizer be dilute. In fact, insurance companies require that the concentration of VOCs be below 25% of their lower explosive limits (LEL). Some lower explosive limits are presented in Table 2. But, if the stream is too dilute, this increases the size and cost of the equipment and the fuel needed to heat the stream of mostly air. For example, if the stream contains only 5% of the LEL of a certain compound, and if the dilution air can be reduced somewhere upstream in the process to increase the VOC concentration to 25% of the LEL, we would decrease the volume of air to be incinerated (and the fuel costs) by 80%!

DESIGN CONSIDERATIONS

Thermal Oxidizers

The design/specification of a thermal oxidizer requires first that a temperature of operation and a residence time be specified. Also a minimum flow velocity inside the chamber must be selected. Then the size of the device and the fuel flow rate

Table 1. Thermal Stability Ranking for Selected POHCs

POHC	T_{99} (°F)[a]	ΔH_c (kcal/g)[b]
Benzene	1150	10.03
Naphthalene	1070	9.62
Chlorobenzene	990	6.60
Acrylonitrile	985	5.57
Chloromethane	950	3.25
Toluene	895	10.14
Trichloroethene	865	1.74
Ethyl Cyanide	770	4.57
Acetyl Chloride	765	2.77
Isobutyl Alcohol	715	7.62
1,1,2,2-Tetrachloroethane	690	1.39
Methyl Ethyl Ketone	650	8.07
Tetrachloromethane	645	0.24
Trichloromethane	625	0.75

[a] Higher thermal stabilities indicated by higher temperatures for 99% destruction (T_{99}). All T_{99}s reported in this table were experimentally determined.
[b] ΔH_c, heat of combustion.

Source: Adapted from Ref. 3.

Table 2. Lower Explosive Limits (LELs) for Selected Organics

Organic	LEL (Percent by Volume in Air)
Acetone	2.15
Benzene	1.4
n-Butane	1.9
n-Butanol	1.7
Cyclohexane	1.3
Ethane	3.2
Ethanol	3.3
Ethyl acetate	2.2
Heptane	1.0
Hexane	1.3
Isobutane	1.8
Isopropanol	2.5
Methane	5.0
Methanol	6.0
Methyl ethyl ketone	1.8
Propane	2.4
Toluene	1.3
Xylene	1.0

Source: Adapted from Ref. 3.

required to achieve the specified time and temperature can be determined.

Material and Energy Balances

A labeled schematic diagram of a vapor incinerator is presented in Figure 2. In a typical design problem, the time, temperature, and target efficiency of destruction of the VOC are specified by regulations or can be selected from experience. Detailed kinetic studies for each compound are usually not warranted. The equations needed for calculating the fuel flow rate and for sizing the afterburner to achieve the desired temperature and time are developed below based on the diagram shown in Figure 2.

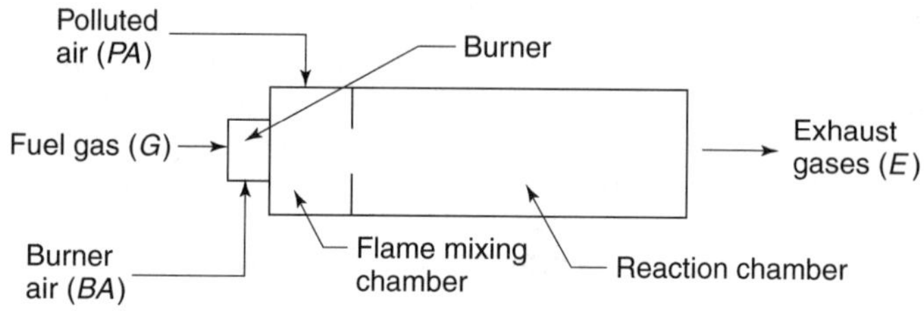

Figure 2. Schematic Diagram of a Vapor Incinerator (Source: Ref. 3. Used with permission.).

The steady-state material balance reduces to

$$0 = M_{PA} + M_{BA} + M_G - M_E \tag{7}$$

where M denotes the mass flow rates (in kg/min or lb_m/min), and the subscripts PA, BA, G, and E refer to the streams identified in Figure 2.

The steady-state enthalpy balance (other forms of energy are insignificant) is as follows:

$$0 = M_{PA}h_{PA} + M_{BA}h_{BA} + M_G h_G - M_E h_E + M_G \Delta H_{cG} \tag{8}$$

where h is the specific enthalpy (kJ/kg or Btu/lb_m); and ΔH_c, the net heat of combustion (lower heating value; kJ/kg or Btu/lb_m). In Eq. 8, any gain of heat from the combustion of the pollutants has been ignored, as well as any loss of heat through the walls. Often these two effects counterbalance each other. However, if the stream being incinerated is very dilute, then it may be desirable to include a heat loss term as a simple percentage of the heat released by burning the fuel. That is, the heat losses from the device can be taken into account by decreasing the rated heating value of the fuel gas by a small percentage.

Often, the enthalpies of all streams are assumed to be similar to those for pure air. Equation 8 then reduces to:

$$0 = M_{PA}h_{T_{PA}} + M_{BA}h_{T_{BA}} + M_G h_{T_G} - M_E h_{T_E} + M_G \Delta H_{cG}(1 - f_L) \tag{9}$$

where f_L is the heat loss from the device, as a fraction of heat input; and h_{T_i}, the enthalpy of air at temperature T_i (kJ/kg or Btu/lb_m).

If the burner is forced to use a portion of the polluted air stream to support the combustion of the fuel gas, then the second term in Eq. 7 and in Eq. 9 collapses into the first term. However, more often, the burner draws in fresh ambient air in a preset ratio (R_B) to the fuel gas. In this case, $M_{BA} = R_B M_G$. Substituting that relationship into Eq. 7, and assuming that $T_{BA} = T_G$, then Eq. 9 is solved as follows:

$$M_G = \frac{M_{PA}(h_{T_E} - h_{T_{PA}})}{(\Delta H_c)_G(1 - f_L) - (R_B + 1)(h_{T_E} - h_{T_{BA}})} \tag{10}$$

Note that in Eqs. 9 and 10, all enthalpies are coupled and appear as enthalpy differences. A numerical value of enthalpy is referenced to a given temperature (and the reference temperature varies with different tables of data). But when two enthalpies are subtracted, the effect of the reference temperature drops out. All items on the right sides of Eqs. 9 and 10 are known or can be found in tables of data (see Table 3).

Table 3. Enthalpy of Air at Various Temperatures

Temperature (°F)[a]	Enthalpy of Air (Btu/lb)
60	0.0
100	9.6
150	21.6
200	33.6
250	45.7
300	57.8
400	82.1
500	106.7
600	131.6
700	156.7
800	182.2
900	211.4
1000	234.1
1200	287.2
1400	341.5
1600	396.8
1800	452.9
2000	509.5
2500	654.3
3000	802.3

[a]In this table, 60°F is the reference temperature.

Source: Adapted from Ref. 3

Sizing the Device

The sizing calculations are relatively easy once the heat and material balances are done, and the quantity of fuel gas determined. Turbulent flow in the device is required to ensure good mixing of the polluted air and the hot combustion gases. Therefore, a linear gas velocity inside the afterburner is set at 10 m/sec or so. The velocity that is chosen mandates the length of the device. As mentioned earlier, a residence time of 2.0 sec or longer is required for a hazardous waste incinerator, and 1.0 sec or longer is often mandated for a biomedical waste incinerator.

The relations among time, length, and velocity are presented in the following. The length of the reaction chamber is given by:

$$L = u\tau_r \tag{11}$$

The volumetric flow rate of the exhaust gas is:

$$Q_E = \frac{M_E R T_E}{P(\mathrm{MW})_E} \tag{12}$$

The diameter of the reaction chamber is:

$$D = \sqrt{\frac{4Q_E}{\pi u}} \tag{13}$$

Catalytic Oxidation

A catalyst is a substance that promotes the desired reactions without itself being permanently changed. The catalysts used for oxidation promote the oxidation reactions and reduce the required temperature by hundreds of degrees. A very common example of a catalytic oxidizer is the catalytic converter in an automobile. Industrially, catalytic oxidizers reduce both the size of the equipment and the fuel usage, resulting in savings of as much as several hundred thousand dollars per year for large systems. In most cases, the gases are heated by a small auxiliary burner, and then are passed directly into the catalyst bed, which is contained in the same unit as the burner.

In air pollution control work, the catalyst is usually a noble metal such as palladium or platinum (other metals are used, including Cr, Mn, Cu, Co, and Ni) deposited on an alumina support in such a way so as to produce a small pressure drop. Pressure drop can be critical in the design of a catalytic oxidizer. A honeycomb arrangement typically results in a pressure drop of 0.05–0.5 in. H_2O/in. of bed depth, whereas a packed bed of 1/8-in.-diameter pellets results in pressure drops of 1.0–10 in. H_2O/inch of bed.[10] In addition to exhibiting good reactivity and low pressure drop, the catalyst must be able to resist attrition (crumbling, breakage, or other mechanical wear), withstand high-temperature excursions, and last a reasonable length of time in service.

The length of catalyst bed is usually from 6 to 24 in. for 99%+ conversion. The total amount of catalyst surface area provided is a very important parameter in the design. Surface area values are typically given per unit gas volumetric flow rate, or as the gas flow rate per unit volume of catalyst. This latter form has units of inverse time and is known as *space velocity*. In terms of space velocity, Snape[10] recommends 50,000–100,000 h^{-1} for honeycomb supports, and 30,000 h^{-1} for pelletized supports. Note that the inverse of space velocity is space–time, which is similar to residence time.

Comparing typical space–times in catalytic oxidizers (0.03 to 0.1 sec) to typical residence times in thermal oxidizers (0.3–1.0 sec), it is noted that the time required for oxidation is an order of magnitude longer in thermal systems than in catalytic systems.

There are four main disadvantages of catalytic oxidation. First, the catalysts often can be poisoned by other chemicals or elements (such as Cl, S, or Pb) in the waste gases. Second, the catalyst beds can become plugged by particulate matter. Third, the catalysts can be damaged by high-temperature excursions if the VOC concentration increases temporarily. Finally, catalytic oxidizers often have a higher capital cost than thermal oxidizers, and this may drive the investment decision. Despite the disadvantages, catalytic oxidizers are used in many industrial situations.

HEAT RECOVERY

In these days of energy efficiency, heat recovery is an essential consideration in the design and operation of thermal oxidizers. Since heat is provided by burning fuel gas (which is like burning money), it is desirable to recover heat from a vapor incinerator. At an energy cost of \$3.00/10^6 Btu, recovering just *half* of the heat content of 10,000 cfm of air at 1400°F results in a savings of about \$250 per day.

A heat exchanger is device used to recover some of the heat from an incinerator exhaust gas. The heat that is recovered can be used in several ways: to preheat the incoming waste gas, to heat and boil water, or to heat a different stream of material that would otherwise require the burning of fuel to heat. In these uses, the heat exchangers are often called *recuperators*. In recuperators, energy recoveries of 40–60% are common, and recoveries of 80% are often practical.[9] When the hot exhaust gases are used to preheat the incoming VOC-in-air stream, the thermodynamic properties of the two streams are so similar that the energy recovery can be approximated by a simple equation, as follows:

$$E = \frac{\Delta T \text{ recovered}}{\Delta T \text{ available}} \times 100\%$$
$$= \frac{(T_2 - T_1)}{(T_3 - T_1)} \times 100 \quad (14)$$

where E is the percent heat recovery; and $T_{1,2,3}$ are the temperatures of the VOC-in-air stream prior to preheat, of the stream after preheat but before incineration, and of the hot afterburner exhaust, respectively.

Recovering heat can result in economic savings. However, air-to-air heat exchange tends to be slow, requiring large, expensive heat exchangers. In addition, for higher percentages of energy recovery, more gas–metal contact is required; thus pressure drop is increased, increasing operating costs. Purchase of a heat exchanger must be economically justified on a case-by-case basis such that the cost of the heat exchanger is paid back by net energy savings (fuel gas savings less increased cost of running the fan). Recovery of a large portion of the heat is often easily justified, but recovery of the "last 10%" of the heat energy is usually very expensive.

In the past two decades an alternative to the traditional heat exchanger has evolved: *regenerative* heat recovery. This technology utilizes two chambers packed with ceramic heat transfer media, and joined by an insulated duct with a small burner installed in it. The duct serves as the thermal oxidizer, and as much as 80–95% of the heat can be recovered in the ceramic media. The burner is small because so much of the heat is recovered that very little additional fuel must be burned to achieve the desired operating temperature in the oxidizer duct. The external ducts and valves are arranged so that the cool incoming waste-gas stream can enter from either side, and the hot exhaust gases exit in either direction (see Figure 3).

The flow through the packed beds is alternated, flowing first in through the hot bed, through the oxidizer duct, and the out through the cool bed. As the cool polluted air flows through the hot bed, the air becomes heated and the bed cools off. As the hot exhaust gases flow out through Bed 2, which initially is cool, the gases give up heat to the ceramic media, heating the media and cooling the gases. After a period of time, when Bed 1 has become cool and Bed 2 hot, the valves are switched and the direction of flow is reversed. Vertical flow-through beds are preferred because they provide more uniform gas flow distribution (especially when the ceramic packing settles), and because the bottom face of the packing is always cool, avoiding the need to support hot packing. Pressure drops for these systems range from 15 in. H_2O (for 80% heat recovery) to 20 in. H_2O (for 95% recovery).

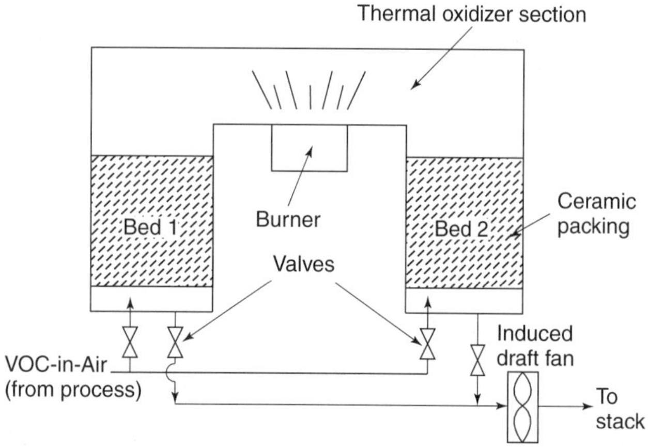

Figure 3. Thermal Oxidizer with Regenerative Heat Recovery (Source: Ref. 3. Used with permission).

Flares

A *flare* is a device designed to isolate and burn an intermittent or emergency release of hydrocarbons from within a petroleum or petrochemical processing plant. Flare tips are usually elevated a hundred or more feet above the ground in order to burn the gases safely. At night, from a distance, they look a little like giant candles. Because the stream being burned is essentially pure fuel, once ignited, combustion occurs continuously as long as the hydrocarbons continue to flow. Air is supplied by the open atmosphere. The flare is equipped with steam jets at the tip to promote turbulent mixing of atmospheric air with the hydrocarbon release. Steam injection rates are based on a mass ratio of steam to hydrocarbons of 0.1–0.5 to 1. If the flow of steam is too low, the mixing becomes inadequate and the flare may produce a large amount of black smoke. If operated properly, flares can provide destruction efficiencies of 99% or more.

EXAMPLE 1

Determine the mass flow rate of natural gas required to heat 10,000 acfm of polluted air (at $T = 200°F$ and 1 atm) from 200 to 1800°F. Assume 5% heat losses, and no heat benefits from oxidizing the VOCs in the polluted air. The afterburner draws in outside air (at 60°F) at a ratio of 14 lb of air per lb of fuel gas. The natural gas is almost pure methane and has a lower heating value of 21,500 Btu/lb. Also, solve for the volumetric flow rate of fuel gas, in standard cubic feet per minute (in the natural gas industry, STP is defined as 60°F and 1 atm).

SOLUTION

First solve for the mass flow rate of polluted air using the ideal gas law, and assuming that the molecular weight of the polluted air is the same as pure air:

$$M_{PA} = Q(P \times MW/RT)$$
$$= 10{,}000 \text{ ft}^3/\text{min} \times [(1 \text{ atm} \times 29 \text{ lb/lbmol})/$$
$$[(0.73 \text{ atm ft}^3/\text{lbmol °R} \times 660°\text{R}]$$
$$= 601.9 \text{ lb/min}$$

Using Eq. 10, and reading enthalpy values for air at 1800 and 200°F from Table 3, next solve for the mass of fuel gas:

$$M_G = 601.9(452.9 - 33.6)/[21500(1 - 0.05) - 15(452.9 - 0)]$$
$$= 18.5 \text{ lb/min of natural gas}$$

Finally, solve for the volumetric flow rate of natural gas using the ideal gas law, assuming that the molecular weight of natural gas is 16 (the same as methane):

$$Q_G = M(RT/P \times \text{MW})$$
$$= 18.5(0.73 \times 520)/(1 \times 16)$$
$$= 439 \text{ scfm}$$

EXAMPLE 2

Size the afterburner for the problem of Example 1. Assume that the gas velocity inside the device will be 20 feet per second, and the residence time must be 1.0 sec.

SOLUTION

First calculate the total exhaust gas flow rate in mass units using Eq. 7.

$$M = 601.9 + 14 \times 18.5 + 18.5$$
$$= 879.4 \text{ lb/min}$$

Next calculate the volumetric flow rate of exhaust using Eq. 12 and convert to cubic feet per second:

$$Q_E = [(879.4 \times 0.73 \times 2260)/(1 \times 29)]/60$$
$$= 833.8 \text{ ft}^3/\text{sec}$$

Next, calculate the diameter using Eq. 13:

$$D = [(4 \times 833.8)/(3.14 \times 20)]^{1/2}$$
$$= 7.28 \text{ ft (round to 7 ft 3 in.)}$$

Finally, the length of the device is determined using Eq. 11:

$$L = 20 \text{ ft/sec} \times 1.0 \text{ sec}$$
$$= 20 \text{ ft}$$

EXAMPLE 3

Calculate the value of the heat recovered from the exhaust gases in the preceding example problems if we can use a heat exchanger to cool the exhaust gases to 500°F and recover all of that enthalpy. Assume the process runs 24 hours per day, and that the heat is valued at $3.00/MMBtu. Give your answer in dollars per day.

SOLUTION

First calculate the enthalpy recovered by cooling the hot exhaust gas from 1800 to 500°F.

$$\Delta H = 879.4 \text{ lb/min} \times (452.9 - 106.7)\text{Btu/lb}$$
$$= 3.04 \times 10^5 \text{ Btu/min}$$

Next calculate the monetary value of this amount of heat:

$$\text{value} = 3.04 \times 10^5 \text{ Btu/min} \times 1440 \text{ min/d} \times \$3.00/10^6 \text{ Btu}$$
$$= \$1315/\text{day}$$

REFERENCES

1. R. H. Barnes, A. A. Putnam, and R. E. Barrett, *Chemical Aspects of Afterburner Systems*, a topical report to the U.S. Environmental Protection Agency (Contract No. 68-02-2629), Battelle, Columbus, OH, August 15, 1978.
2. C. A. Clausen, C. D. Cooper, M. Hewett, and A. Martinez, "Enhancement of organic vapor incineration using ozone," *J. of Hazardous Materials* **31** (1992).
3. C. David Cooper and F. C. Alley, *Air Pollution Control—A Design Approach* (2nd ed), Waveland Press, Prospect Heights, IL, 1994.
4. C. D. Cooper, F. C. Alley, and T. J. Overcamp, "Hydrocarbon vapor incineration kinetics," *Environmental Progress* **1**, 2 (May 1982).
5. C. D. Cooper, C. A. Clausen, D. Tomlin, M. Hewett, and A. Martinez, "Enhancement of organic vapor incineration using hydrogen peroxide," *J. of Hazardous Materials* **27** (1991).
6. I. Glassman, *Combustion*. Academic Press, New York, 1977.
7. K. H. Hemsath, and P. E. Susey, "Fume incineration in theory and practice," paper presented at the 71st National Meeting of the American Institute of Chemical Engineers, Dallas, TX, February 20–23, 1972.
8. K. C. Lee., N. Morgan, J. L. Hansen, and G. M. Whipple, "Revised model for the prediction of the time–temperature requirements for thermal destruction of dilute organic vapors and its usage for predicting compound destructability," a paper presented at the 75th Annual Meeting of the Air Pollution Control Association, New Orleans, LA, June 20–25, 1982.
9. J. H. Mueller, "Heat recovery," Chapter 19, in *Air Pollution Control and Design Handbook*, P. N. Cheremisinoff and R. A. Young, Eds., Dekker, New York, 1977.
10. T. H. Snape, "Catalytic incineration," Chapter 21, in *Air Pollution Control and Design Handbook*, P. N. Cheremisinoff and R. A. Young, Eds., Dekker, New York, 1977.

BIOLOGICAL CONTROL SYSTEMS

SARINA J. ERGAS AND KERRY A. KINNEY

Biological systems can be used to control air emissions contaminated with low to moderate concentrations of biodegradable volatile organic compounds (VOCs), hazardous air pollutants (HAPs), and odorous compounds. Biological control systems include three types of process designs:

- Biofilters
- Biotrickling filters
- Bioscrubbers

In biofilters and biotrickling filters, moist biofilms are attached to a porous media in a packed-bed bioreactor. The contaminated gas stream is passed through this biologically active packed bed, where the pollutants are degraded by the microorganisms. The principal difference between biofilters and biotrickling filters is that in biotrickling filters, the liquid phase moves (trickles) through the porous media, whereas the liquid phase is stationary in biofilters. In bioscrubber systems, absorption of the pollutants takes place in one unit, a packed column, spray tower, or bubble column scrubber, while biodegradation primarily takes place in a liquid-phase bioreactor, similar to an activated sludge reactor.

Biofilters are the most common type of biological control system currently in use for odor and VOC control applications in the United States. Numerous pilot-scale biotrickling filter systems have also been tested in the United States, and several full-scale systems are currently operating in Europe. Bioscrubbers, while also found in Europe, are not as common in the United States, and thus the discussion to follow will focus exclusively on biofilters and biotrickling filters.

A conceptual model of the three phases (gas, biofilm, solid media) found in a microscopic section of a packed-bed biological control system is shown in Figure 1. The contaminated gas stream is humidified and blown through the packing media. The packing media is composed of compost, peat, soil, ceramic, plastic packing, or other material and is surrounded by a moist biofilm. Soluble compounds in the gas stream partition into the biofilm, where they are transported by diffusion to the microorganisms. The contaminants then serve as a growth substrate for the attached microbial population.

Biofilters have been in use since the mid-1950s for *odor* control at wastewater treatment plants, composting plants, slaughterhouses, rendering plants, confined animal facilities, and industrial plants. Biological systems have been used in VOC control applications since the early 1980s. Industries utilizing biological control for VOCs include[1–3] adhesives production, breweries and bakeries, chemical manufacturing, electronics, film coating, flavors and fragrances, iron foundries, landfill gas extraction, painting and spraying operations, petroleum refining, plastics manufacturing, printing, pulp and paper manufacturing, soil and groundwater remediation, textiles, and waste oil recycling. Compounds that have been controlled include ammonia, hydrogen sulfide, methyl mercaptan, dimethyl sulfide, dimethyl disulfide, carbon disulfide, ethanol, methanol, butanol, phenol, formaldehyde, acetaldehyde, acetone, methyl ethyl ketone, methyl isobutyl ketone, ethyl acetate, dichloromethane, gasoline and jet fuel vapors, benzene, toluene, xylene, and styrene. Removal efficiencies greater than 95% are achievable with a properly designed and maintained system.

Biological systems have several advantages over conventional air pollution control alternatives including:

1. Low operating costs. Biological control systems operate at ambient temperatures and pressures so power consumption is minimal. Although power must be provided for a blower, pressure drops across biological control systems are generally less than 10 cm of water column per meter of packed bed depth. Pumping may be required for operation of humidification and supplemental water addition systems. Nutrients and/or buffers are also required in biotrickling filters, and the porous media will require periodic replacement in biofilters.
2. Absence of residuals and by-products. Compounds are usually mineralized to harmless products such as CO_2, H_2O, and inorganic ions such as Cl^-, SO_4^{2-}, and NO_3^-. Other gas-phase control alternatives produce residuals such as spent activated carbon and chemical sludges requiring further treatment or disposal.
3. Unlike thermal oxidation processes, biooxidation processes produce minimal CO_2 and NO_x emissions.

Disadvantages of biological control systems include relatively high area requirements and moderate to high capital costs. In addition, not all gas streams are amenable to biological control. Factors to be considered when choosing a biological control system include:

1. Pollutant biodegradability. The compound must be biodegradable under aerobic conditions. VOCs are generally used as carbon and energy sources for a

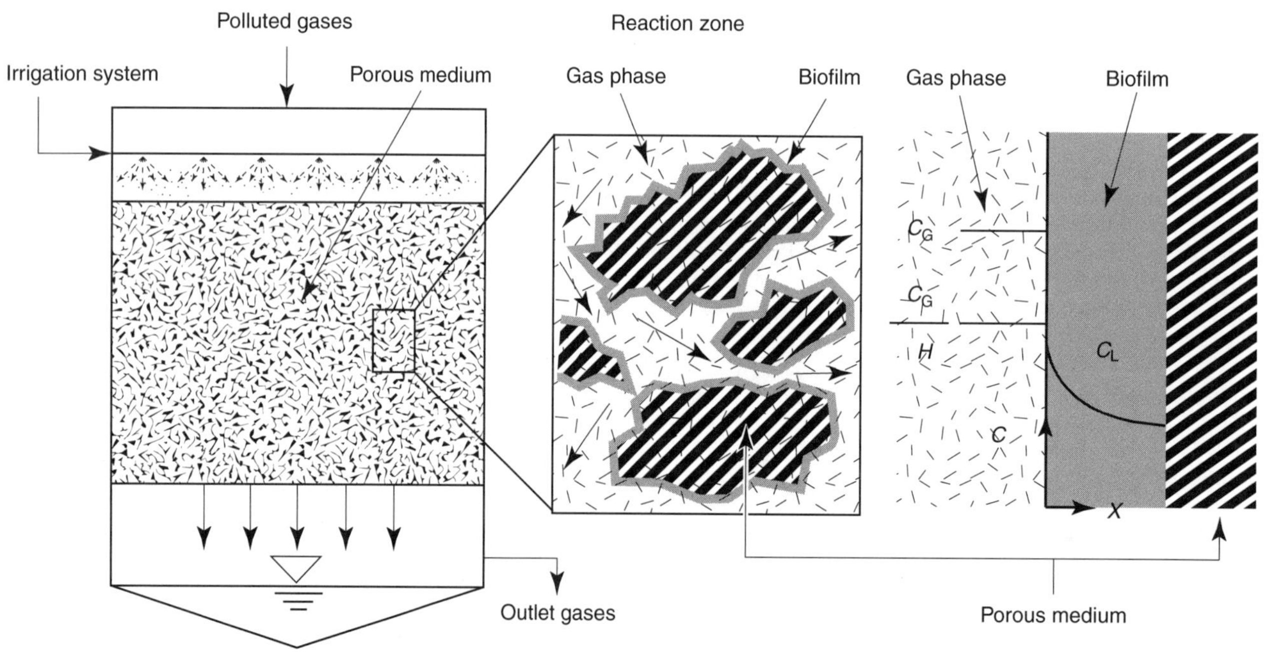

Figure 1. Conceptual Model of the Three Phases of a Packed-Bed Biological Control System. Contaminated Air (C_G) Is Passed through the Porous Medium. Soluble Compounds in the Gas Stream Partition into the Biofilm According to Henry's Law ($C_L = C_G/H$) where They are Transported by Diffusion and Biodegraded.

consortium of heterotrophic bacteria and fungi. H_2S and NH_3 are used as electron donors for chemolithotrophic metabolism. Molecules with certain structural groups such as amine, methoxy, sulfonate, and nitro groups; extensive halogenation; benzenes substituted in the meta position; ether linkages; and branched carbon chains may be recalcitrant or require excessively long residence times to biodegrade.

2. Compound solubility. The Henry's law constant H (dimensionless) provides a measure of the degree to which a compound partitions between the gas phase and the biofilm according to the following:

$$C_{\mathrm{L}} = \frac{C_{\mathrm{G}}}{H} \tag{1}$$

where C_{L} is the biofilm phase concentration (mg L^{-1}) and C_{G} is the gas-phase concentration (mg L^{-1}). Henry's law constants for VOCs controlled in biofilters vary by as much as five orders of magnitude. Pollutants with large Henry's constants (e.g., hexane, $H = 74$) are more difficult to treat and require longer residence times that those with small Henry's constants (e.g., ethanol, $H = 2.6 \times 10^{-4}$).

3. Pollutant loading rate and concentration. Biological control systems are generally limited to inlet VOC loading rates of less than 100 to 150 g $\mathrm{m}^{-3}\,\mathrm{h}^{-1}$. At high VOC loading rates, microbial growth results in clogging of media pore spaces, resulting in high pressure drops and uneven gas distribution. In some cases, high concentrations of VOCs are toxic to microorganisms and inhibit biodegradation.

4. Gas temperature. Most common microorganisms in biofilters are mesophiles with an optimum temperature range of 25–35°C. In very cold climates, inlet gases may need to be heated and the reactors insulated for stable year-round operation. Very hot waste gases may need to be cooled prior to biological control. Thermophilic operation of biofilters (50–65°C) has been successfully applied in a few cases.

5. Incomplete biodegradation. Biodegradation does not always result in the complete mineralization of an organic compound. An environmentally acceptable level of degradation is the biological transformation of the compound to the extent where toxicity or other undesirable characteristics of the compound are removed.

6. Acidic by-product formation. In many cases, biodegradation results in the formation of acidic metabolites. The oxidation of H_2S produces large amounts of H_2SO_4; however, sulfur oxidizing bacteria are acidophilic (acid loving) and can function well at pH values as low as 1. Dicloromethane (DCM) degraders produce two moles of HCl for each mole of DCM oxidized. Organic acid formation has also been observed in biofilter systems overloaded with ethanol.[4] Both DCM and ethanol degraders function best at a pH range between 6 and 8. Near neutral pH can be maintained through addition of buffers to the packing media of a biofilter or to the liquid phase of a biotrickling filter.

7. Acclimation. A period of days to several weeks is generally required for the microbial population to acclimate to the pollutant and for the population of VOC degraders to grow to the point where the system reaches maximum removal efficiency. For some pollutants, most notably H_2S, high removal efficiencies can be achieved immediately due to adsorption and chemical oxidation. Inoculation of the media with a previously acclimated microbial culture or with mixed liquor suspended solids (MLSS) from an activated sludge system can significantly shorten the acclimation period.

8. Transient conditions. A period of reacclimation may be required after shutting down a biological control system for several days or weeks, particularly if the air flow is not maintained though the media. In applications with frequent shutdowns and transients, however, the microbial population adapts to these conditions, and reacclimation periods of less than an hour have been observed. Shock loadings of toxic compounds should be avoided.

9. Mixtures of compounds. Microbial metabolism often operates in such a way that compounds that are more readily degradable are removed before less degradable compounds. Thus a readily degraded compound, such as ethyl acetate, may be inhibited by the presence of ethanol in the feed stream. In these cases, sufficient bed volume must be provided so that the more degradable compound is removed near the inlet to the bed and the less degradable compound can be removed deeper in the bed. The presence of compound mixtures makes prediction of system performance difficult without pilot tests.

10. Dust and grease. Particulate matter or grease in waste gases will clog the biofilter media, causing excessive pressure drops and the need for frequent media replacement. This has been a problem in the wood products industry and emissions from bakeries. Where high particulate matter concentrations are present in the inlet gases, biological air pollution control systems need to be preceded by a particulate matter control system, such as a wet collector or electrostatic precipitator.

DESCRIPTION OF EQUIPMENT

In biofilters, compost, soil, or peat are mixed with a bulking agent and used as the porous packing medium. Other amendments, such as nutrients and pH buffers, may be added to the medium. The medium serves as both a support for microbial attachment and a source of nutrients for the microbial population. Gases are prehumidified and the media are periodically moistened by spraying with water or buffer/nutrient solution. Compost biofilters are generally used to treat low mass loading rates of VOC and odorous emissions (less than 40 to 50 g $\mathrm{m}^{-3}\,\mathrm{h}^{-1}$) and for processes where pH control is not a concern. The high surface area available for microbial attachment and adsorption of pollutants, low cost and availability of the compost media, low pressure drop, and low chemical costs make biofilters the preferred option for these applications.

A schematic of an enclosed biofilter system used for VOC control is shown in Figure 2. Open biofilter systems are generally used for odor control systems, where control over environmental conditions is not as critical as in VOC control applications. An innovative, modular system that has been

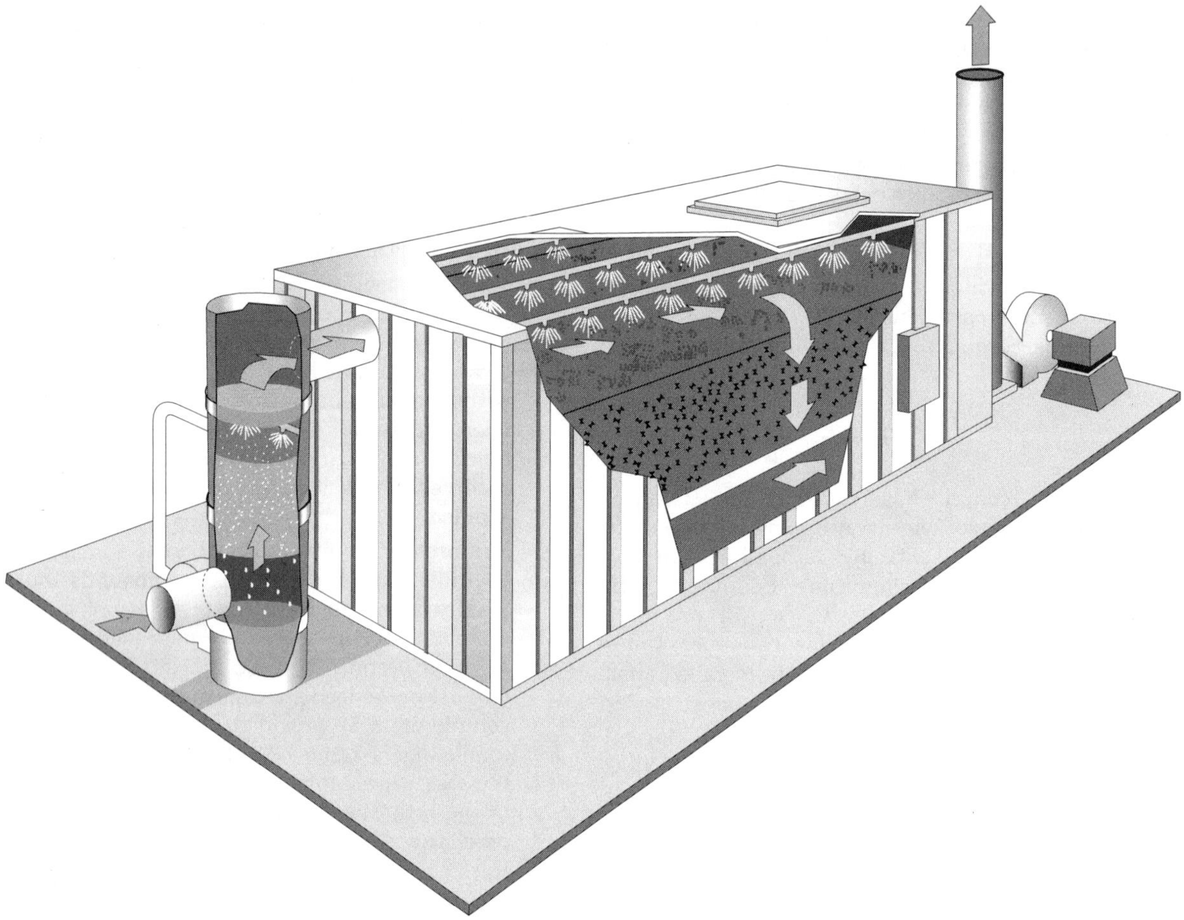

Figure 2. Schematic Diagram of a Fully Enclosed Compost Biofilter System (Courtesy of Monsanto Enviro-Chem, St. Louis, MO).

used to control high-concentration, low flow-rate emissions, such as off-gases from glycol dehydrators and aerosol can puncturing operations, is shown in Figure 3.

In biotrickling filters, an inert material such as porous ceramic pellets, granular activated carbon, plastic or fibrous packing, or polyurethane foams are used as the porous media. A recirculating nutrient/buffer solution is continuously or intermittently trickled down over the porous media. A portion of the recirculating solution is wasted continuously or semicontinuously and automated control of pH and/or nutrient concentrations is possible. A schematic of a biotrickling filter is shown in Figure 4. Biotrickling filters are generally applied

Figure 3. A Modular Biofilter System Used to Control Low-Volume High-Concentration Emissions from a Glycol Dehydrator (Courtesy of Bio-Reaction Industries, Inc., Tualatin, OR).

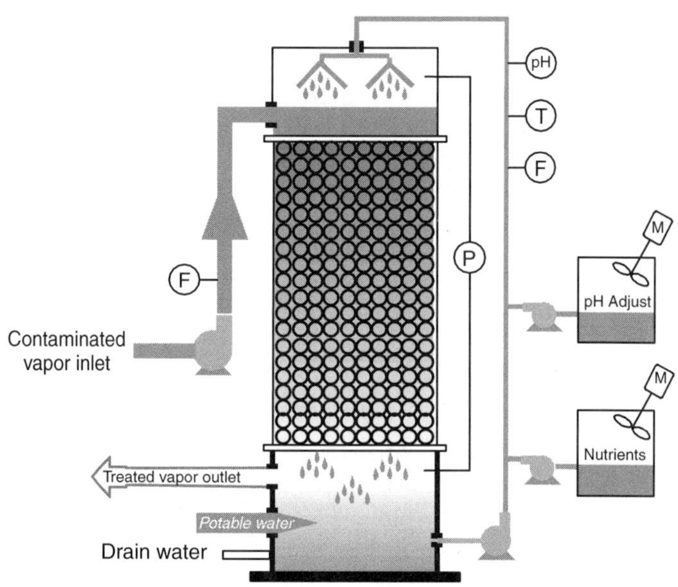

Figure 4. Schematic Diagram of a Biotrickling Filter (Courtesy of Envirogen, Inc., Lawrenceville, NJ).

at higher inlet loading rates (up to 100 to 150 g $m^{-3}h^{-1}$) and/or where continuous pH control is required.

Biofilter Media

Compost. Compost mixtures are preferred over other materials as a biofilter medium, particularly for the control of low-concentration emissions of odorous compounds and VOCs. Compost media can provide the necessary requirements for good overall biofilter performance including: high specific surface area, low pressure drop, high nutrient composition, high microbial diversity, good water holding capacity, and low cost.

Properties of different types of compost vary considerably and have a major impact on biofilter elimination capacity and long-term system performance. Many different types of raw materials and several different composting process configurations are used to produce compost. Major categories of organic substrates used in compost production include municipal biosolids, industrial sludges, yard wastes, animal manures, food and agricultural crop residues, and municipal solid waste.[5] During the composting process, organic substrates are degraded and stabilized, usually under thermophilic conditions. A maturation phase, characterized by lower temperatures, allows time for degradation of refractory organics and reestablishment of mesophilic microbial populations.

Compost stability has been shown to be a key parameter affecting long-term compost biofilter performance.[6] In unstable compost, degradation of bioavailable carbon may result in a decrease in media volume and compaction. In addition, heat released during biooxidation of the compost results in drying of the media and subsequent acidification and proliferation of fungi. Oxygen uptake rates below 200 mg O_2 kg total $solids^{-1}$ h^{-1} are indicative of stable compost.

Peat. Peat is a highly stable natural deposit of plant remains that has been protected from degradation by a high water table. Peat tends to be somewhat hydrophobic or resistant to wetting, making moisture control in these systems more difficult. Peat is naturally acidic and has a low nutrient content and a low indigenous microbial population. Successful biofilter systems have utilized peat media; however, supplemental nutrients, buffers, and microbial inoculum were required.

Soil. Some of the earliest biofilter systems were soil beds. Soil is inexpensive and has a high indigenous microbial population.[7] Soil is more structurally stable than compost or peat media, making deeper beds possible without compaction. In addition, soil tends to be hydrophilic, making moisture control less problematic in soil beds. Major disadvantages of soil media are their low degradative capacity and low air permeability. For these reasons, soil bed biofilters are usually much larger than compost-based biofilters and are generally restricted to systems where large area requirements are not a concern.

Bulking Agents. A coarse fraction of material is commonly added to biofilter media to improve the structural properties of the media, prevent compaction, promote drainage, and decrease pressure drop. The pressure drop per unit depth of biofilter media can be determined by:

$$\frac{\Delta P}{z} = v\frac{\mu}{k} \tag{2}$$

where ΔP is the pressure drop across the media (Pa), z is the height of the media (m), v is the superficial velocity (m^3 m^{-2} s^{-1}), μ is the dynamic viscosity of the fluid (N sec m^{-2}), and k is the permeability (m^2). Permeability is a function of the size distribution of the particles and pores that make up the media. A larger fraction of coarse material will increase the permeability, but decrease the surface area available for microbial attachment and adsorption of pollutants. In addition, bulking agents typically have lower nutrient content than compost materials. Bulking agents used in biofilter systems include wood chips and bark (2 to 6 cm diameter), polystyrene beads (BIOTON® packing), pearlite, vermiculite, and shredded tires. The fraction of bulking added to biofilter packing has been reported between 20 and 80% by volume,[8] with 50 to 60% common.

Biotrickling Filter Media

The selection of an appropriate packing media for biotrickling filters is essential to overall bioreactor performance. Random-packed plastic media such as Pall™ rings are one common type of packing for biotrickling filters. These packings are commonly manufactured for gas-scrubbing applications and are thus readily available and relatively inexpensive. Other packing-media types currently in use or under investigation include polyurethane foam cubes, fibrous packing materials, ceramic pellets, granular activated carbon, porous lava rocks, and structured plastic or ceramic media. Desirable properties for the biotrickling filter packing material include:

1. High surface area per unit volume of packing (e.g., >200 m^2/m^3). A high available surface area is necessary to support biofilm growth, minimize packed-bed volume, and achieve high pollutant-elimination capacities. It is important when selecting packing materials to consider how much of the media surface area is actually available to the microorganisms. For instance, a considerable fraction of the surface area in some activated carbon materials may be found in pore spaces too small to allow microbial penetration.
2. High void volume. A high void volume is desirable in biotrickling filters to maintain low pressure drops and minimize clogging as biomass accumulates in the system. This void volume requirement must be balanced against the need for high surface area, since the greater the void volume, the smaller the available surface area. In general, for random dumped packing media, packing materials between 1 and 2 in. in size should provide the appropriate balance between surface area and void volume requirements.
3. Chemical resistance. The selected packing material should be resistant to degradation by the gas stream contaminants, microbial by-products generated within the biofilm (including acids such as HCl), and the nutrient solution that is trickled over the packing material.
4. Lightweight. In general, it is advantageous to use lightweight packing material such as plastics or foams

for ease of handling and to avoid having to reinforce the biotrickling filter column. However, in some specific applications (e.g., odor control), heavy materials such as porous lava rock have been used successfully.

5. Inexpensive. The packing material can represent a significant fraction of the total system cost; so it is important to use inexpensive packing material. Although biotrickling filter packing material is generally more expensive than biofilter packing media, its useful life is considerably (e.g., up to 5 times) longer, thus greatly reducing replacement costs.

Amendments

Nutrients. Nutrient availability is a key factor affecting biological control system performance. All the chemical constituents of cell components must be available from the environment for cell growth to occur. Major nutrients required for microbial growth include nitrogen, phosphorus, sulfur, potassium, calcium, magnesium, and iron. A number of other elements (e.g., copper, selenium, zinc) are also required in trace amounts. In biofilters, the nutrient requirements are either met by the compost media or by adding nutrient solutions to the media. In biotrickling filters, a nutrient solution is continuously or semicontinuously trickled over the porous media.

The nitrogen requirements for biofilter media have not been clearly established; however, biosolids compost has been successfully used in biofilter systems. Total nitrogen content in a biosolids compost solid matrix typically has a range of 20–30 g/kg total solids. Several researchers[9,10] have reported an increase in VOC elimination capacity by adding slow-release nutrient granules or NO_3^- solutions to the media or gas-phase ammonia to the inlet gases. However, long-term improvement in biofilter performance due to nutrient addition has not been reported, and it is possible that systems run at these higher elimination capacities will quickly fail due to biomass accumulation. Addition of ammonium salts to biofilter media has not been effective, since ammonia is quickly stripped out in the gas stream.

pH Control. Agricultural liming materials such as slaked lime [$Ca(OH)_2$], unslaked lime (CaO), limestone ($CaCO_3$), crushed oyster shell ($CaCO_3$), dolomite [$CaMg(CO_3)_2$], marl (soil + $CaCO_3$), and slags ($CaSiO_3$) can be used to neutralize acidic by-products in biofilter media. Differences in buffering capacity and speed of reaction depend on the molecular composition and fineness of the particles. Finely crushed oyster shells are commonly mixed with compost media at approximately 10% of the dry weight of the media as a pH buffer. Larger amounts of buffer are required where acidic by-products are a concern. Feedback control of pH is often used to inject a controlled amount of caustic into biotrickling filter solutions to maintain near-neutral pH.

Humidification, Irrigation, and Leachate Collection

Moisture is provided to biofilters by humidifying the inlet gases and by spraying water over the surface of the bed, either manually or automatically. In uncovered systems, irregular irrigation of the bed also takes place by precipitation. Automated systems for control of moisture in biofilter media are usually based on timers or on total bed weight.

A humidification system is usually located upstream of the biofilter, either before or after the blower. If the blower is located downstream of the humidification system, the relative humidity of the gases will decrease due to the input of mechanical energy by the blower. The humidification system should be designed to maintain >95% relative humidity at the inlet to the bed. Simultaneous heating or cooling of the gas stream and/or removal of particulate matter is possible. Humidification systems used in biofilter applications have included[8] steam injection systems, spray chambers, venturi scrubbers, and packed-bed towers. Prehumidification of the inlet gases may not be required if the biofilter is used to treat humid gases, such as off-gases from activated sludge aeration basins.

Prehumidification is generally not sufficient to maintain optimal bed moisture content in biofilter systems due to: (1) incomplete humidification — gases entering the biofilter at less than 100% relative humidity will evaporate water from the media until they reach saturation; (2) biooxidation — an exothermic process that increases the temperature of the media and raises the water capacity of the gases; (3) metabolic water production — biodegradation of VOCs results in formation of water as a by-product of microbial metabolism; and (4) heat exchange with the environment. Irrigation systems should be designed to provide uniform distribution of a controlled amount of water throughout the bed. The frequency of supplemental moisture addition will be discussed later. Biofilter irrigation systems include[8] over-bed sprinkler systems, soaker hoses, and pressure-compensated soaker hoses. Clogging and freezing of sprinkler heads and soaker hose orifices should be avoided.

A leachate collection system is generally installed in the bottom of biofilter beds to collect any leachate that drains from the media. The bottom of the bed is usually excavated with a slight gradient toward a sump, where leachate is collected and pumped to a nearby sewer.

Gas Distribution and Flow

To maintain high elimination capacities, it is essential to uniformly distribute the contaminated gas stream as it flows through the biofilter or biotrickling filter packing material. Short-circuiting up the side of the bioreactor will result in contaminant breakthroughs and reduced elimination capacities. A smoke test is one useful way to test a full-scale system for short-circuiting problems.

For open biofilter systems, a network of slotted PVC pipe (or other chemically resistant pipe) buried in a layer of gravel at the base of the biofilter is often used as the gas distribution system. Specially designed concrete sinter blocks can also be arranged at the base of the bioreactor in such a way as to distribute the gas flow; these systems have the advantage that heavy equipment can be driven over them after they have been installed. For fully enclosed bioreactor systems, the inlet gas stream can be distributed in a properly designed plenum located at either the top of bottom of the bioreactor.

Biofilters can either be operated in an upflow or downflow mode. Downflow operation of biofilters has the advantage that moisture addition at the surface of the biofilter can be carried down throughout the packing material. However, any acidic or toxic products produced near the inlet will also be carried down and may damage the packed bed. Upflow mode has the advantage that any undesirable by-products produced near the inlet or particulates from the inlet gas stream can be more easily removed from the system.[8] However, it is difficult to add moisture to the system while it is operated in upflow mode. At a high superficial velocity, any water that is added to the top of an upflow system may be held up in the system, increasing the pressure drop across the packed bed and creating anaerobic zones. One potential solution to this problem is to design the bed so that it can operate in downflow or upflow mode. During moisture addition, the system could be operated in downflow mode to aid in the distribution of moisture throughout the bed while during regular operation, the system could be operated in an upflow mode. Periodic switches in the flow direction through a biofilter can also help distribute biomass growth more uniformly throughout the length of the bed, will be discussed later. Biotrickling filters are usually operated in with gas flow co-current with the liquid flow (downflow) to avoid stripping VOCs from the recirculating liquid into the outlet gases.

DESIGN AND PERFORMANCE EQUATIONS

A number of efforts have been directed at modeling biofilter processes.[11–14] These models generally assume equilibrium conditions at the gas/biofilm interface, transport of contaminants through the biofilm by diffusion, and simplified biodegradation kinetics. In general, these models have been used to understand the effect of various system parameters, such as media surface area, biofilm thickness, and biodegradation kinetic constants on contaminant removal. For the most part, these models have not been used in system design and operation.

Design and operation of biofilters and biotrickling filters are based on four parameters: empty-bed residence time, gas flux, contaminant mass loading rate, and elimination capacity. The empty-bed residence time, t_r (sec), represents the theoretical average time that a gas molecule would reside in the empty bed:

$$t_r = \frac{V}{Q} \tag{3}$$

where V is the bed volume (m^3) and Q is the gas flow rate ($m^3\ s^{-1}$). The true residence time is much shorter than the empty-bed residence time since the gases flow through the media pores and not the entire bed volume. The true residence time can be calculated by multiplying the empty-bed residence time by the porosity, n (volume of the voids/volume of the media). Empty-bed residence times in biological control systems vary from 20 sec to several minutes depending on the degradability and solubility of the compound. Empty-bed residence times provide an important way of comparing systems treating similar contaminant streams, since, for a given gas flow rate, t_r will be dependent only on the volume of the media and have a direct impact on the cost of the system.

The gas flux v ($m^3\ m^{-2}\ sec^{-1}$), also referred to a the superficial velocity or surface loading rate, represents the velocity of the gases as they approach the media:

$$v = \frac{Q}{A} \tag{4}$$

where A (m^2) is the cross-sectional area of the media. The gas flux has a significant impact on the pressure drop across the media, as shown in Eq. 2. The mass loading rate R ($g\ m^{-3}\ sec^{-1}$) represents the mass of contaminant entering the system per unit volume of bed material per unit time:

$$R = \frac{QC_i}{V} = \frac{C_i}{t_r} \tag{5}$$

where C_i is the inlet concentration ($g\ m^{-3}$). The elimination capacity, EC ($g\ m^{-3}\ sec^{-1}$), is calculated as the mass of contaminant degraded in the system per volume of biofilter media per unit time as:

$$\mathrm{EC} = \frac{Q(C_i - C_o)}{V} = \eta R \tag{6}$$

where C_o is the outlet concentration ($g\ m^{-3}$) and η is the removal efficiency. The relationship between mass loading rate and elimination capacity for a typical biofilter system is shown in Figure 5. Note that data obtained during the acclimation phase are typically not included in this graph. The 100% removal efficiency line (EC = R) is also shown. At low inlet-mass-loading rates, the pollutants are almost completely degraded in the media. In this region, the inlet-loading rate may be increased, either by increasing the pollutant concentration or increasing the gas flow rate; however, the contaminant removal efficiency remains constant. Ergas et al.[12] developed a first-order kinetic expression that is used to describe mass transfer and biodegradation at low inlet-loading rates:

$$\ln\left(\frac{C_o}{C_i}\right) = \frac{-k_1 n A_s L}{H}\frac{\tanh\phi}{\phi} t_r \tag{7}$$

where $k_1 4$ (sec^{-1}) is the first-order biodegradation rate coefficient, A_s ($m^2\ m^{-3}$) is the surface area per unit volume of the media, L (m) is the biofilm thickness, H is the dimensionless Henry's law constant, ϕ is the Thiele number ($\phi = \sqrt{k_1 L^2/D}$), and D is the diffusion coefficient ($m^2\ s^{-1}$). For constant media properties and biodegradation kinetics, Eq. 7 can be rewritten as:

$$\eta = 1 - \exp(-K_1 t_r) \tag{8}$$

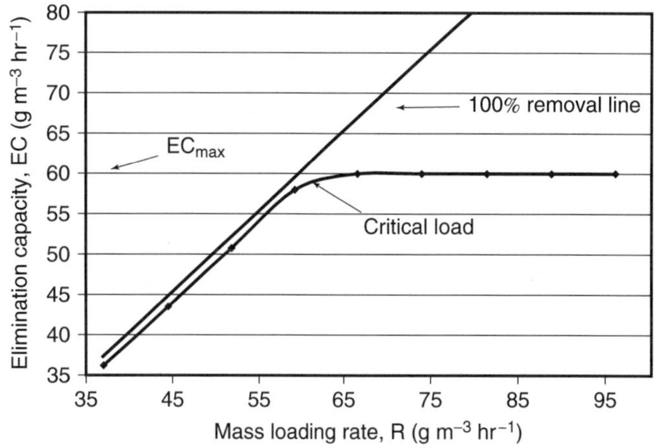

Figure 5. Typical Elimination Capacity (EC) versus Mass Loading Rate (R) Curve for a Packed-Bed Biological Control System. The Slope Equals 1 (EC = R) at 100% Removal Efficiency.

where K_1 (sec^{-1}) is a constant. Therefore, in this low-loading-rate region, removal efficiency is dependent only on gas residence time.

As the mass-loading rate on the system is increased, a critical EC will be reached (EC_{crit}), where the removal efficiency significantly decreases. At this point either the biomass cannot degrade the available substrate (biodegradation rate limitations) or there is insufficient residence time for the pollutant to diffuse into the biofilm (mass transfer rate limitations).

By increasing the mass loading rate further, a maximum elimination capacity is reached (EC_{max}). Ottengraf and van den Oever[11] developed a zero-order kinetic expression that is used to describe mass transfer and biodegradation at these high inlet loading rates:

$$\frac{C_o}{C_i} = 1 - \frac{k_o n A_s L}{C_i} t_r \quad (9)$$

where k_o (g m^{-3} sec^{-1}) is the zero-order biodegradation rate coefficient. For constant media properties and biodegradation kinetics, Eq. 9 can be rewritten as:

$$\eta = \frac{K_o}{C_i} t_r \quad (10)$$

where K_o (g m^{-3} sec^{-1}) is a constant. In the case of toxic contaminants or toxic metabolic by-products, increases in mass loading rate may result in decreases in elimination capacity.

Unless past research has been performed with the media, environmental conditions (temperature, pH) and contaminants of concern, design of full-scale biofilter systems is usually based on pilot- or lab-scale tests. Running the pilot system with either a slip-stream of the off-gas to be treated or a synthetic gas stream at varying mass loading rates generates a curve, similar to Figure 5. Plots of EC versus R are generated for each contaminant, and EC_{max} is determined for each compound to be controlled. The maximum elimination capacity determined for the most difficult to degrade compound is then used to determine the bed volume:

$$V = \frac{(SF)\eta C_i Q}{EC_{max}} \quad (11)$$

where SF is the safety factor (SF $>$ 1) and C_i and Q are the design inlet concentration and gas flow rate, respectively. Once the bed volume is determined, the cross-sectional area is obtained by dividing the volume by the bed height. Biofilter beds are typically 1–1.5 m in height. Deeper bed heights tend to undergo compaction quickly.

OPERATION AND MAINTENANCE

Acclimation Period

Unlike physical/chemical control systems, biofilters and biotrickling filters require an acclimation period following startup to reach full contaminant elimination capacity. This acclimation period generally ranges from days to weeks depending on the system and compounds being treated. Reduced loading rates or backup air pollution control systems may be necessary during this period. Processes that occur during the acclimation period include: (1) induction of the appropriate pollutant-degrading enzymes within the microbial population; and (2) growth and establishment of the pollutant-degrading biofilm throughout the bioreactor packing. One way to shorten the acclimation period is to inoculate the packing media with a microbial culture that is already acclimated to and actively degrading the contaminant(s) found in the gas stream. Microorganisms isolated from an existing biofilter or biotrickling filter treating a similar contaminant feed mixture are the most efficient inoculum source. For a general inoculum that contains a wide variety of microorganisms, mixed liquor suspended solids (MLSS) from an activated sludge system can also be used. Inoculation is almost always required for biotrickling filters, since these systems are packed with inert packing material, which is initially void of a microbial community. Inoculation is easily accomplished in these systems by recirculating the microbial solution throughout the packing material. Biofilters, on the other hand, do not necessarily require inoculation since the packing material contains an indigenous microbial population that can usually adapt to the contaminant feed stream. However, inoculation can shorten the acclimation period considerably and, for some recalcitrant pollutants (e.g., MTBE), inoculation with a specific microbial population may be absolutely necessary.

Response to Transients

Most biofilters in field and industrial applications are subjected to transient feed conditions, including process shutdowns and restarts as well as spikes in inlet contaminant concentration. During process shutdowns or periods of low contaminant feed concentrations, the biomass in the biofilter or biotrickling filter undergoes endogenous decay, and the pollutant-degrading activity decreases with time. The longer the period of nonuse, the greater the reacclimation period required following restart to fully restore bioreactor removal efficiency. In one study, a biofilter required 9–24 h to recover full activity after a 4 day shutdown.[15] In applications with frequent shutdowns and transients, however, the microbial population can usually adapt to these conditions, and shorter reacclimation periods are possible. During process shutdowns, it is often beneficial to keep the air flowing through the system to maintain aerobic conditions within the biofilter to minimize damage to the aerobic biofilm population. Unstable or transient feed conditions during startup are generally the most detrimental because they can inhibit the establishment of the biofilm within the system and result in excessively long acclimation periods.

Spikes in inlet concentration can also detrimentally affect biofilter performance, leading to contaminant breakthroughs or, in extreme cases, toxic inhibition or death of the microbial population within the bioreactor. For moderate and sustained spikes in inlet concentration, the microbial population eventually adapts to the higher concentration, but an intermediate acclimation period ranging from hours to days is usually required before full elimination capacity is achieved.[16–18]

Moisture Monitoring and Control

Inadequate moisture control has been cited as the most frequent cause of failure in biofilter systems. Biotrickling filter systems, however, are much less prone to drying problems, because a liquid phase is continuously or semicontinuously trickled over the packing material. Media moisture content is a critical parameter affecting biodegradation rates, microbial ecology, pH, mass transfer rates, and compost physical structure. Optimal moisture content, MC (%), in compost biofilter media ranges from 45 to 65% by mass:

$$MC = \frac{\text{mass of water}}{\text{mass of media} + \text{water}} 100\% \qquad (12)$$

When insufficient moisture is available, the liquid film surrounding the biofilm dries out, resulting in a loss of microbial activity. Too much moisture, however, can limit mass transport of oxygen, VOCs, and waste products in and out of the biofilm. Lack of oxygen can result in the development of anaerobic zones, which can lead to acidification and odor formation.

Hysteresis results in difficulty rewetting biofilter media that have undergone significant drying. This can result in the development of hydrophobic zones (areas of the media that are difficult if not impossible to rewet) in peat and compost media. This phenomenon is aggravated because dry conditions select for fungi, which tend to have hydrophobic surfaces, thus increasing the hydrophobicity of the media. Biofilter media that have undergone significant drying due to failure of the humidification or irrigation system may need to be completely replaced.

The most common method of measuring moisture content in biofilters is by performing spot checks on the media. A bucket auger is used to remove medium cores from the bed, which are analyzed for water content gravimetrically. Results of moisture measurements are used to adjust the frequency of supplemental moisture addition via the irrigation system. The size and number of samples that can be taken are limited since taking many core samples from the bed will cause channeling.

Several vendors have designed systems that measure bulk moisture content by weighing the entire bed using load cells. Losses in bed weight are assumed to be due to losses of moisture from the bed, although periodic recalibration may be done to account for media degradation and washout. Data obtained from these measurements are processed in a programmable logic controller that determines the frequency of moisture addition.[19] A disadvantage of this method is that moisture is not uniformly distributed though the bed; thus the average moisture content in the bed may be adequate, while some sections of the bed are extremely dry.

A number of probes have been developed for monitoring moisture content in agricultural soils, including electrical resistance sensors, thermocouple psychrometers, time-domain reflectometry probes, and tensiometers. Use of these systems for moisture monitoring and control in biofilters is an active area of current research.

Biomass Accumulation and Control

In the biofiltration process, microorganisms in the biofilm convert VOCs into biomass, CO_2, and other metabolic by-products. The fraction of the organic contaminant feed that is converted into biomass for a given mass of contaminant degraded is known as the *biomass yield*. Biomass yield varies significantly from one system to the next and is a strong function of the microbial community present, the contaminant(s) being degraded, as well as the nutrient and oxygen supply available to the microorganisms. The greater the mass loading rate and average biomass yield in a biofilter, the greater the potential to accumulate excess biomass within the packing material and develop clogging problems. Aerobic biodegradation of organic compounds results in greater biomass yields than do either anaerobic conditions or the biodegradation of inorganic compounds such as H_2S or NH_3.

In general, biofilters and biotrickling filters with an adequate nutrient supply that are continuously subjected to contaminant loadings in excess of 30 to 50 g VOC $m^{-3}\,h^{-1}$ may accumulate excess biomass within the packing material. This biomass accumulation is usually greatest near the inlet of the unit, where the contaminant feed concentrations are greatest. Biomass buildup and clogging may also occur in the plates used to support the packing material within the bioreactor. The biomass buildup reduces the effective porosity of the packed bed as well as the residence time and surface area per unit volume within the packed material.

Ultimately, the pressure drop across the bed rises, channeling develops, and contaminant removal efficiencies decline. Most biomass accumulation problems are found in biotrickling filters treating moderate to high concentrations of VOCs. Clogging problems are less common in biofilters because they generally treat lower contaminant concentrations and nutrient availability may limit biomass growth.

Several methods have been developed to control biomass accumulation in biotrickling filters. These methods include:

1. Limiting the supply of a key nutrient such as nitrogen or potassium to the microorganisms in the biotrickling filter
2. Physical removal of the biomass by backwashing and/or stirring of the packing material
3. Intermittent operation of the biofilter so as to more evenly distribute the contaminant degradation throughout the packed-bed volume

Regulating the nutrient supply to the microorganisms can control biomass accumulation in biotrickling filters; however, a 25 to 50% reduction in the pollutant removal efficiency may also result if the nutrient becomes limiting.[20,21] The composition of the nutrient feed to a biotrickling filter can be monitored and adjusted as necessary to deliver a limiting amount of nutrients to the biofilm. However, some recycling of the nutrients will occur within the biofilm itself, and this must be considered in the nutrient balance. Limiting the nitrogen supply limits biomass growth effectively but also has the most detrimental effect on pollutant degradation. Potassium, on the other hand, may be a more suitable choice since it can control biomass accumulation while only suppressing contaminant degradation to a moderate degree.

For biotrickling filters, excess biomass can also be physically removed from the packing material by periodically washing the packed bed with water or some other rinse solution such as 0.1 M sodium hydroxide.[21,22] Backwashing with water is a fairly straightforward method for biomass removal, although backwashing cycles as often as twice per week may be required for high contaminant loads. Backup air pollution control systems may also be required during each backwashing event. When a chemical rinse is used, fewer backwashing events may be required, although a longer period of time may be needed before the biofilm recovers full degradation capacity.

Another method for biomass control that may be used in conjunction with physical removal and nutrient limitation methods is to enhance the distribution of contaminant degradation throughout the packing material. One way to do this is periodically to switch the direction of the contaminant feed through the packing material. This feeding strategy more evenly distributes contaminant degradation and biomass throughout the packing material and prevents rapid clogging of the inlet section.

Media Replacement

Biofilter media are usually replaced every 2–5 years with fresh media. At a number of installations, media replacement is scheduled every 2 years regardless of system performance to maintain optimal elimination capacity. At other installations, media are replaced when a decrease in elimination capacity, an increase in pressure drop, decrease in media pH, or significant compaction is observed. Reasons for these changes in media performance include utilization, volatilization and/or leaching of nutrients, acidification, increases in media hydrophobiciy, biomass accumulation, degradation of the compost media, and gravitational forces. Factors that can accelerate the rate of media deterioration are inadequate moisture control, high-temperature operation, and high VOC loading rates.

ILLUSTRATIVE EXAMPLES

A compost biofilter is to be used to control ethanol (MW = 46) emissions from a chemical manufacturing plant. A pilot system was used with a slipstream from the process. The gas flow rate to the pilot system was 6 m³/min. The temperature was 30°C, and the pressure drop across the packed bed was 6 cm of water column. The dimensions of the rectangular bed were 3 m × 1.5 m × 1 m depth. The data of Table 1 were obtained from the pilot system.

EXAMPLE 1

Prepare a plot of elimination capacity versus gas loading rate (both should be in units of g/m³ h) for the pilot system.

The residence time and superficial velocity for the pilot system are determined using Eqs. 3 and 4, respectively:

$$t_r = \frac{(3\text{ m})(1.5\text{ m})(1\text{ m})}{6\text{ m}^3/\text{min}} = 0.75\text{ min}$$

$$v = \frac{(6\text{ m}^3)/\text{min}}{(3\text{ m})(1.5\text{ m})} = 1.3\text{ m/min}$$

The mass loading rate to the pilot system is determined by dividing the concentration, in g/m³, by the residence time (Eq. 5). The elimination capacity is determined by multiplying the mass loading rate by the removal efficiency (Eq. 6). For the 250-ppm inlet concentration, the calculations are:

$$R = \frac{(250\text{ ppm})(1\text{ atm})(46\text{ g/mol})}{8.21\times10^{-5}\dfrac{\text{m}^3\text{atm}}{\text{mol K}}(303\text{ K})(0.75\text{ min})}10^{-6}$$

$$= 0.62\text{ g/m}^3\text{min} = 37\text{ g/m}^3\text{ h}$$

$$\text{EC} = (0.98)37\text{ g/m}^3\text{ h} = 36.2\text{ g/m}^3\text{ h}$$

The graph generated from the data in Table 1 is shown as Figure 5. The maximum elimination capacity is read off the graph as 60 g/m³ h.

Table 1. Results of a Compost Biofilter Pilot Study for Ethanol Removal

Inlet Ethanol Concentration ppm	Removal efficiency (%)	Inlet Ethanol Concentration ppm	Removal efficiency (%)
250	98	500	81
300	98	550	74
350	98	600	68
400	98	650	62
450	90		

EXAMPLE 2

Give dimensions for a full-scale system to achieve 95% removal at an inlet ethanol concentration of 500 ppm. Assume a design gas flow rate of 30 m³/min and safety factor of 1.3.

The bed volume is determined using Eq. 11, with the concentration in the appropriate units:

$$V = \frac{(1.3)(0.95)(0.92)\text{ g/m}^3(30\text{ m}^3)/\text{min}(60\text{ min/h})}{60\text{ g/m}^3\text{ h}} = 34\text{ m}^3$$

A 1-m depth ×7-m × 5-m bed is suitable for this application.

EXAMPLE 3

Calculate the empty bed residence time, t_r, and superficial velocity, v, for the full-scale system:

$$t_r = \frac{(1\text{ m})(7\text{ m})(5\text{ m})}{30\text{ m}^3/\text{min}} = 1.2\text{ min}$$

$$v = \frac{(30\text{ m}^3/\text{min})}{35\text{ m}^2} = 0.86\text{ m/min}$$

EXAMPLE 4

Calculate the expected pressure drop for the full-scale system.

By rearranging Eq. 2, the term k/μ (permeability/viscosity) is determined for the pilot system:

$$\frac{k}{\mu} = \frac{vz}{\Delta P} = \frac{(1.33\text{ m/min})(1\text{ m})}{(6\text{ cm w.c.})} = 0.22\text{ m}^2/\text{mincm w.c.}$$

Equation 2 is then used to predict the pressure drop for the full-scale biofilter. Note that this assumes the same media mixture will be used in the full-scale system:

$$\Delta P = \frac{vz}{k/\mu} = \frac{(0.86\text{ m/min})(1\text{ m})}{(0.22\text{ m}^2/\text{min cm w.c.})} = 4\text{ cm w.c.}$$

Note that the pressure drop across the media may change over time due to changes in moisture content, biomass accumulation or compaction of the media.

REFERENCES

1. J. B. Eweis, S. J. Ergas, D. P. Y. Chang, and E. D. Schroeder, *Bioremediation Principles*, McGraw-Hill, NY, 1998.

2. R. J. Heinsohn and R. L. Kabel, *Sources and Control of Air Pollution*, Prentice Hall, NJ, 1999.
3. G. Leson and A. M. Winer, "Biofiltration: An innovative air pollution control technology for VOC emissions," *J. Air & Waste Magmt. Assoc.* **41**:1045–54 (1991).
4. G. Leson, R. Chavira, A. Winer and D. Hodge, "Experiences with a full-scale biofilter for control of ethanol emissions," *Proceedings 88th Annual Meeting of the Air & Waste Manag. Assoc.*, San Antonio, TX, 1995.
5. E. Epstein, *Compost Science and Technology*, Technomic, Lancaster, PA, 1997.
6. B. Cárdenas-González, S. J. Ergas, and M. S. Switzenbaum, "Characterization of compost biofilter media," *Proceedings 90th Annual Meeting of the Air & Waste Management Association*, Toronto, Ontario, 1997.
7. H. Bohn, "Biofilter media," *Proceedings of the 89th Annual Meeting of the Air & Waste Management Association*, Nashville, TN, 1996.
8. J. S. Devinny, M. A. Deshusses, and T. S. Webster, *Biofiltration for Air Pollution Control*, Lewis Publishers, Boca Raton, 1999.
9. R. Auria, M. Morales, M. Acuña, F. Perez, and R. Revah, "Biofiltration of toluene vapors: Start up and gaseous ammonia addition," *Proceedings of the USC-TRG Conference on Biofiltration*, Los Angeles, CA, 1996.
10. E. Morgenroth, E. D. Schroeder, D. P. Y. Chang, and K. M. Scow, "Nutrient limitation in a compost biofilter degrading hexane," *J. Air & Waste Manag. Assoc.* **46**:300–308 (1996).
11. S. P. P. Ottengraf and A. H. C. van den Oever, "Kinetics of organic compound removal from waste gases with a biological filter," *Biotechnology and Bioengineering* **25**:3089–102 (1983).
12. S. J. Ergas, E. D. Schroeder, and D. P. Y. Chang, "Control of air emissions of dichloromethane, trichloroethene, and toluene by biofiltration," *Proceedings of the 86th Annual Meeting of the Air and Waste Management Association*, Denver, CO, 1993.
13. Z. Shareefdeen and B. C. Baltzis "Biofiltration of toluene vapor under steady-state and transient conditions: Theory and experimental results," *Chemical Engineering Science* **49**(24A):4347–4360 (1994).
14. D. S. Hodge, and J. S. Devinny, "Modeling removal of air contaminants by biofiltration," *J. Environmental Engineering* **121**(1):21–32 (1995).
15. F. J. Martin and R. C. Loehr, "Effect of periods of non-use on biofilter performance," *J. Air & Waste Management Association* **46**:539–46 (1996).
16. C. B. Parks and R. C. Loehr, "Biofilter design for VOC remediation," *Proceedings of the 1994 Petroleum Hydrocarbons and Organic Chemicals in Groundwater Prevention, Detection and Remediation Conference* Houston, Texas, 1994.
17. H. M. Tang, S. J. Hwang and S. C. Hwang, "Dynamics of toluene degradation in biofilters," *Hazardous Waste & Hazardous Materials* **12**(3):207–19 (1995).
18. L. P. Seed and R. L. Corsi, "Biofiltration of BTEX-contaminated gas streams; laboratory studies," *Proceedings of the 87th Annual Meeting of the Air and Waste Management Association*, Cincinnati, Ohio, 1994.
19. C. van Lith, G. Leson, and R. Michelsen, "Evaluating Design Options for Biofilters," *J. Air & Waste Management Association* **47**:37–48 (1997).
20. P. Holubar, C. Andorfer, and R. Braun, "Prevention of clogging in trickling filters for purification of hydrocarbon-contaminated waste air," *Proceedings of the 1995 Conference on Biofiltration*, University of Southern California, Los Angeles, CA, 1995.
21. F. J. Weber, and S. Hartmans, "Prevention of clogging in a biological trickling bed reactor removing toluene from contaminated air," *Biotechnology and Bioengineering* **50**:91–97 (1996).
22. G. A. Sorial, F. L. Smith, M. T. Suidan, and P. Biswas, "Evaluation of a trickle bed biofilter media for toluene removal," *J. Air & Waste Management Association* **45**:801–10 (1995).

3

CONTROL OF PARTICULATE MATTER

Cyclones and Inertial Separators
Timothy J. Herrick and Wayne T. Davis
Wet Scrubbers
Michael Pilat and Kenneth E. Noll
Electrostatic Precipitators
James H. Turner, Phil A. Lawless, Toshiaki Yamamoto, David W. Coy, Gary P. Greiner, John D. McKenna, and William M. Vatavuk
Fabric Filters
John D. McKenna, Arthur B. Nunn III, and Dale A. Furlong

CYCLONES AND INERTIAL SEPARATORS

Timothy J. Herrick and Wayne T. Davis

Inertial separators are widely used for the collection of medium-sized and coarse particles. Their relatively simple construction and absence of moving parts mean that both the capital and the maintenance costs are lower than for baghouses and electrostatic precipitators. However, the efficiency is not as high, and inertial separators are usually used as precleaners upstream of these other devices, to reduce the dust loading and to remove larger, abrasive particles.

The general principle of inertial separation is that the particulate-laden gas is forced to change direction. As the gas changes direction, the inertia of the particles causes them to continue in the original direction and be separated from the gas stream.

Cyclones, where the gas is forced to spin in a vortex through a tube, are the most common type of inertial separators (see Figure 1). They can be used singly, or in a multiple-tube arrangement for large volumes. Centrifugal separators use the same circular motion, but have a rotating vane in the collector itself to both move the air and separate the dust (Figure 2). Baffle chambers and louver collectors also use the inertial principle (Figure 3). In a louver-type collector, the louvers cause a sudden change in the direction of the gas. As shown in Figure 4, the air enters through the air inlet, and as the air passes the louver, the direction of flow is reversed. By virtue of their inertia, particles move away from the high-velocity gases and are either collected in a hopper or trap or withdrawn in a concentrated dust stream. The concentrated dust stream is cleaned using a cyclone separator or a high-efficiency collector.

The baffle chamber also relies on an abrupt change of the direction of the gas to drop out the particles. The change of direction is caused by fixed baffle plates within the unit (Figure 5). Inertial collectors are occasionally used to control mist. In some applications, the interior of the collector may be irrigated to prevent reentrainment of dry dust and to remove soluble deposits.

Cyclone separators and centrifugal collectors are utilized in the following industries: chemical, coal mining and handling, combustion fly ash, metal melting, metal working, metal mining, rock products, plastics, and wood products. Common uses are the collection of grinding, crushing, conveying, buffing, machining, mixing, sanding, blending, and materials handling dust and for particle collections.[1]

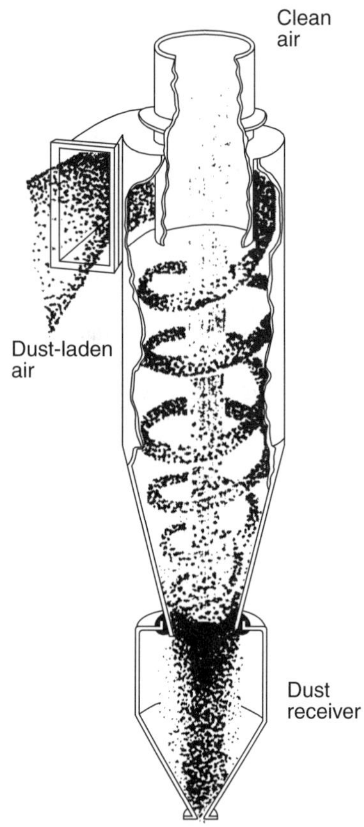

Figure 1. Operation of Cyclone (from Ref. 10).

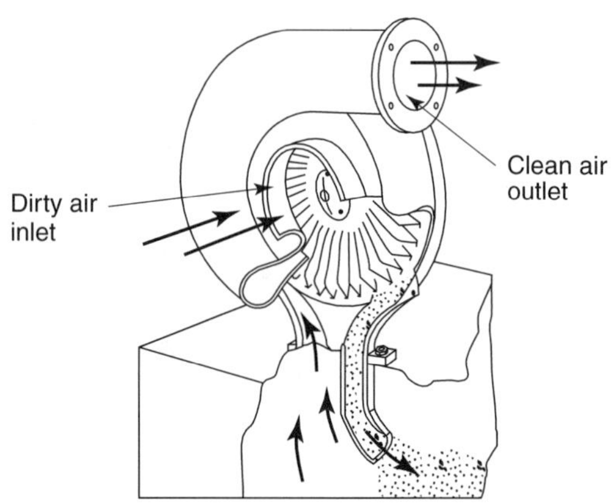

Figure 2. Centrifugal Collector (Reprinted by permission of the American Society of Heating, Refrigeration, and Air Conditioning Engineers, Atlanta, Georgia, from the *1996 ASHRAE Handbook-Systems and Equipment*).

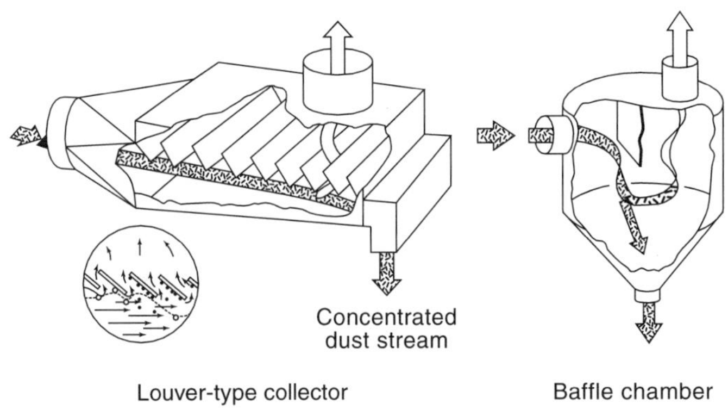

Figure 3. Typical Louver and Baffle Collectors (Reprinted by permission of the American Society of Heating, Refrigeration, and Air Conditioning Engineers, Atlanta, Georgia, from the *1996 ASHRAE Handbook-Systems and Equipment).*

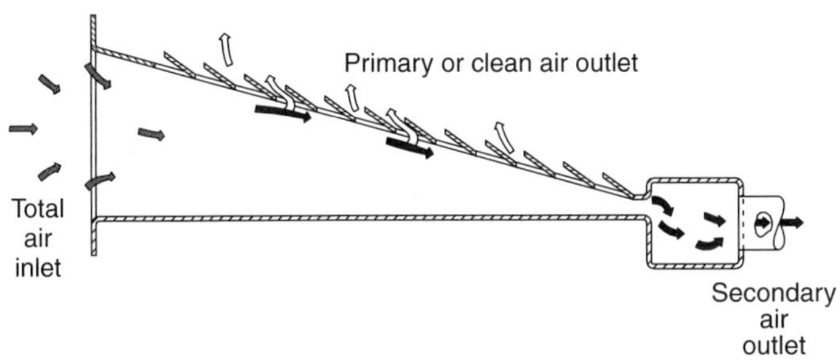

Figure 4. Cross-Sectional View of Dust Louver (from Ref. 13).

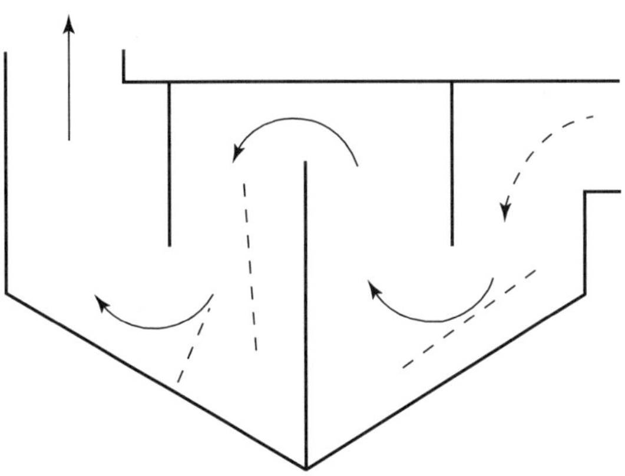

Figure 5. Multiple Baffle Chamber (from Ref. 13).

CYCLONES

Cyclones provide a low-cost, low-maintenance method of removing larger particulate matter from gas streams. On their own, they are not usually sufficient to meet stringent emissions standards, but they serve well as precleaners for other, more expensive control devices and for dry-product recovery. A cyclone reduces dust loading and removes larger abrasive particles, extending the life of the fabric filter, which may be used as the final collector. Multiple-tube cyclone arrangements are used on fossil-fuel boilers, where they handle the large volumes effectively as precleaners.

Particles enter the cyclone suspended in the gas stream, which is forced into a vortex by the shape of the cyclone. The inertia of the particles resists the change in direction of the gas, and they move radially outward across the gas streamlines (Figure 6). The gas inlet is usually at the top of the cyclone, and the gas is forced into a cyclonic path by the shape of the inlet itself, the outer wall of the device and the cylindrical tube extending down from the top to the cyclone known as the vortex finder. The gas spirals in the cylindrical section of the cyclone and the particles move outward under the influence of centrifugal force until they strike the body of the cyclone. There they are caught in the thin laminar layer of air next to the wall and are carried downward by gravity to be collected in the dust hopper. When the gas reaches the cone-shaped section, increased rotational gas velocity helps to keep the dust against the wall. The direction of the vortex reverses near the bottom of the cone, and the cleaned gas rises to exit through the central tube of the vortex finder (Figure 7).

Cyclones are usually designed with geometric similarity such that the ratios of the dimensions remain constant at different diameters, and these dimensions can be expressed

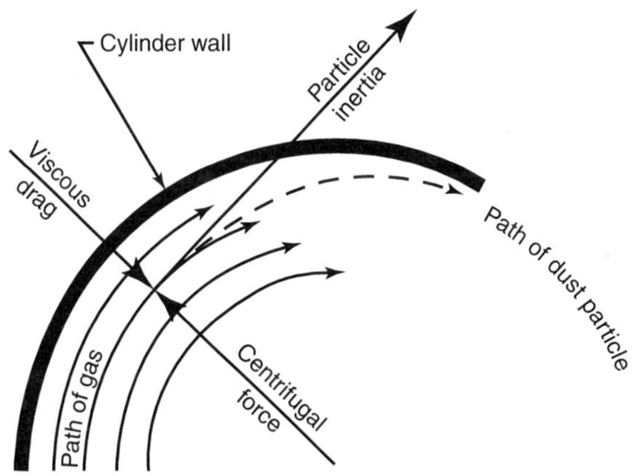

Figure 6. Movement of Particles Across Gas Streamlines (from Ref. 15).

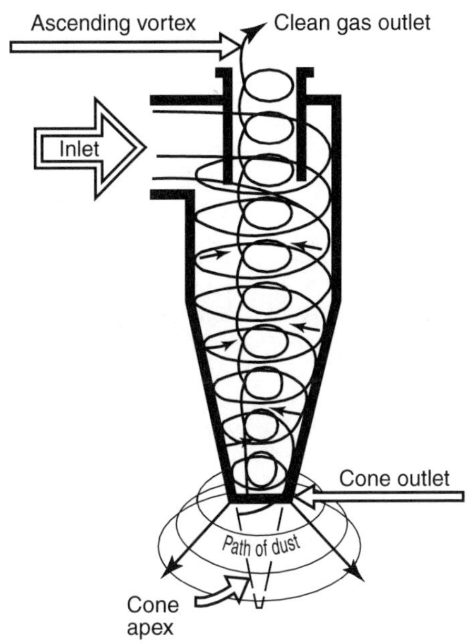

Figure 7. Movement of Gas Stream in a Cyclone (from Ref. 15).

in terms of the body diameter D (Figure 8). The values of the ratios determine whether the cyclone has conventional proportions or is a high-efficiency or high-throughput type (see Table 1).

The number of turns of the vortex, N_e, can be estimated from the cyclone dimensions, since it depends on the height of one turn of the vortex (i.e., the height of the inlet) and the length of the cyclone:

$$N_e \cong \frac{1}{H}\left(L_b + \frac{L_c}{2}\right) \quad (1)$$

Cyclones can be optimized for high collection efficiency by using small diameters, long cylinders, and high inlet velocities or for high throughput with converse dimensions. Small diameters, however, increase the pressure drop and cost of operation.

Different shapes of the inlet are used, all with the aim of merging the incoming gas with that already in the cyclone. Deflector vanes may also be used to improve the gas flow and to minimize turbulence (Figure 9). The inlet may be near the

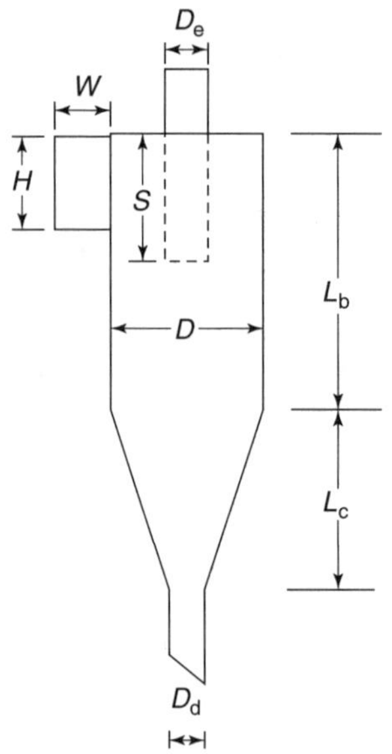

Figure 8. Standard Cyclone Dimensions (from Ref. 2).

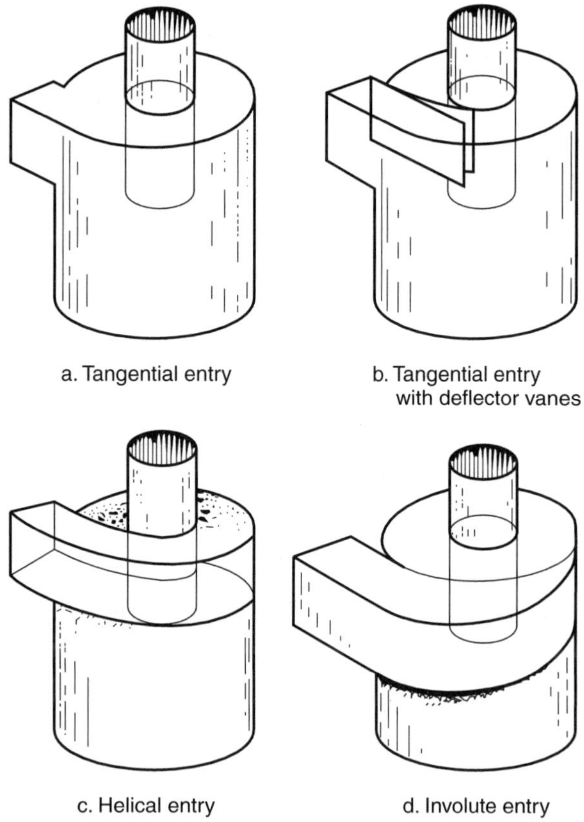

Figure 9. Types of Cyclone Inlets (from Ref. 11).

Table 1. Characteristics of Cyclones

	Cyclone Type[a]					
	High Efficiency		Conventional		High Throughput	
	(1)	(2)	(3)	(4)	(5)	(6)
Body diameter D/D	1.0	1.0	1.0	1.0	1.0	1.0
Height of inlet H/D	0.5	0.44	0.5	0.5	0.75	0.8
Width of inlet W/D	0.2	0.21	0.25	0.25	0.375	0.35
Diameter of gas exit, D_e/D	0.5	0.4	0.5	0.5	0.75	0.75
Length of vortex finder, S/D	0.5	0.5	0.625	0.6	0.875	0.85
Length of body L_b/D	1.5	1.4	2.0	1.75	1.5	1.7
Length of cone L_c/D	2.5	2.5	2.0	2.0	2.5	2.0
Diameter of dust outlet, D_d/D	0.375	0.4	0.25	0.4	0.375	0.4

[a]Columns 1 and 5 from Ref. 11; columns 2, 4, and 6 from Ref. 12; and column 3 from Ref. 5.

Source: Ref. 10.

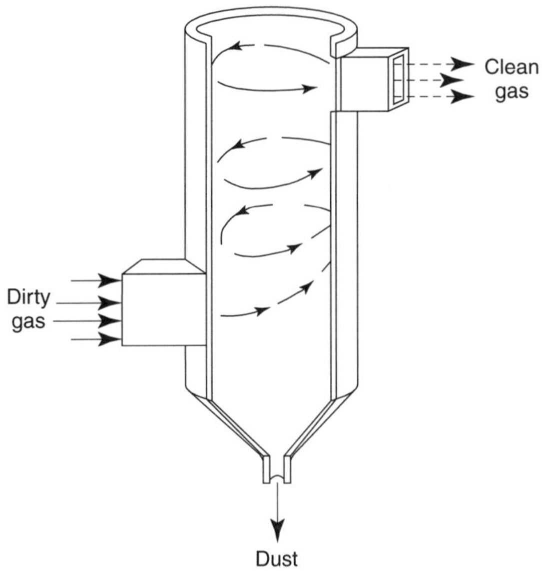

Figure 10. Bottom Inlet Cyclone Separator (from Ref. 11).

base of the cyclone, with the clean gas exiting from the top (Figure 10). This arrangement is often used for cyclones after wet scrubbers to collect the particulate-laden droplets from the wet collector.

Multiple-Tube Cyclones

When high efficiency (which requires small cyclone diameter) and large throughput are both desired, a number of cyclones can be operated in parallel. In a multiple-cyclone separator, the housing contains a large number of tubes that have a common gas inlet and outlet in the chamber. The gas enters the tubes through axial inlet vanes that impart a circular motion (Figure 11).

The arrangement of the tubes, as well as their diameter, affects the overall collection efficiency of a multitube cyclone. There is a small pressure drop in the incoming gas stream for each row of tubes it must pass; a decreasing length of outlet tube may be used to compensate for this, producing an organ-pipe arrangement. Alternatively, tubes can be arranged in groups, each with its own outlet plenum, and space in between the groups, which allows even and unimpeded gas flow to all tubes (Figure 12).

Collection efficiency can be improved by hopper evacuation or slip streaming, in which a small portion (about 15%) of the total gas flow is drawn off through the hopper. This suction considerably reduces dust reentrainment into the cyclone tubes. An increase of 40–50% in collection efficiency is possible.[2] The dust may be reinjected into the combustion zone in the case of fossil-fuel boilers or cleaned by a fabric filter and returned to the exit with the cleaned air from the cyclone tubes. Such systems have been used as a retrofit when higher efficiencies were necessary (Figure 13).

An alternative solution to the problem of dust reentrainment is to use a straight-through type of cyclone, rather than the reverse-flow type already discussed. Multiple-tube arrangements of these are common in older coal-fired boilers (Figure 14). The tubes are arranged horizontally, and the air is drawn axially into the tube. The fixed vanes impart a cyclonic motion to the gas, causing the particles to move to the

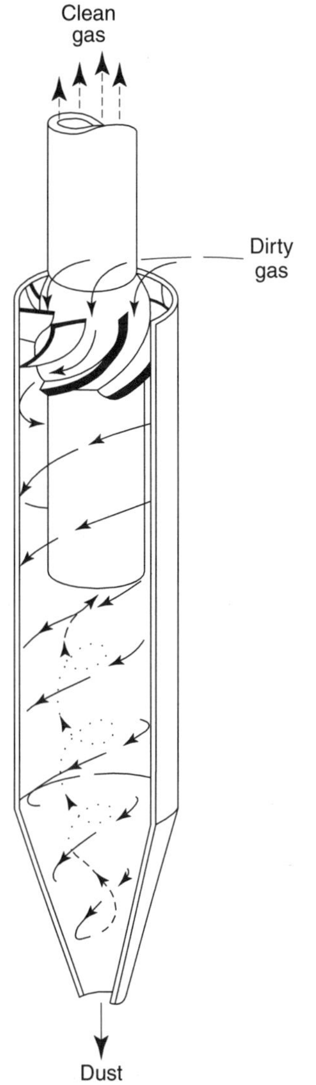

Figure 11. Vane Axial Entry Cyclone (from Ref. 11).

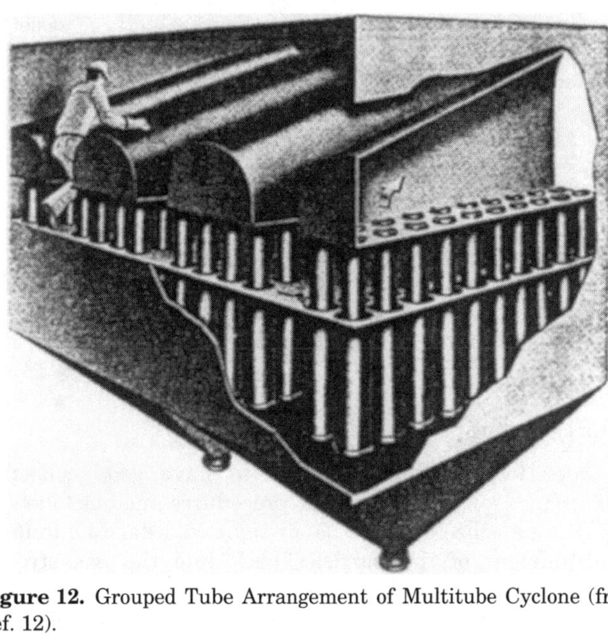

Figure 12. Grouped Tube Arrangement of Multitube Cyclone (from Ref. 12).

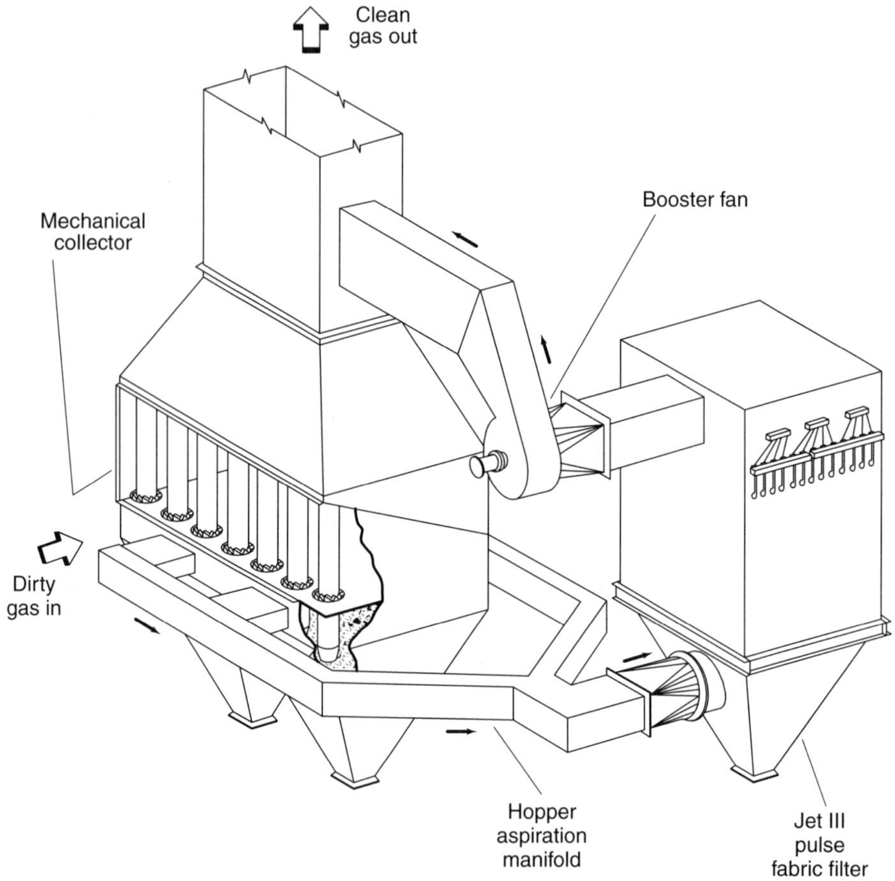

Figure 13. Hopper Aspiration System (from Ref. 1).

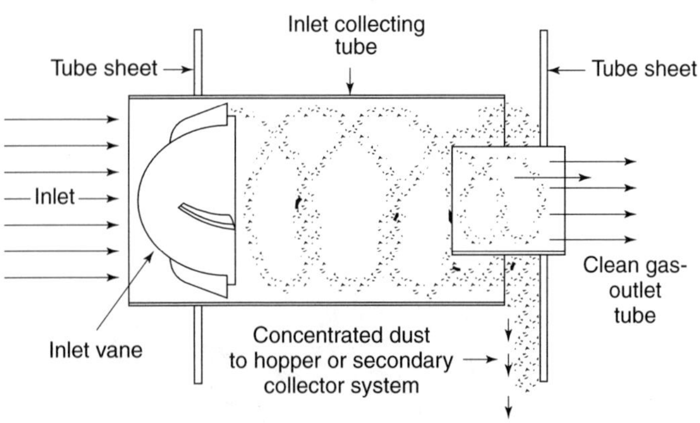

Figure 14. Fixed-Impeller Straight-Through Cyclone (from Ref. 13).

edges of the gas stream. In this case, the peripheral dust-laden gas stream is removed through an annular channel, and the cleaned gas passes out axially through the tube. The concentrated dust stream is then recirculated or further separated into a hopper.

Factors Affecting Cyclone Performance

In order for a cyclone initially to have good collection performance, proper installation procedures must be followed. The cyclone collector must be air-tight in order to eliminate reentrainment of the particles back into the gas stream. Therefore, when installing equipment such as access doors, inlet and outlet plenums, and dust disposal features, these areas must be completely sealed. Any leakage in a cyclone collector can cause a 25% or more loss in collection efficiency.[3]

Erosion and the fouling of cyclones are problems that can be encountered during operation and maintenance activities. Erosion in cyclones is caused by the striking or rubbing of dust particles on the inside wall of the cyclone. Erosion is self-accelerating once started. Erosion will increase with high dust loadings, high inlet velocities, high particle specific gravity values, and the strike angle. Erosion tends to be found in concentrated areas in the cone of the cyclone at light dust loadings, with the entire cone being affected at high dust loadings. Welded seams in the cyclone design are also areas that tend to be susceptible to erosion because of surface irregularities. The area of the shell opposite the inlet may experience excessive wear if the gas contains large dust particles. Erosion can be controlled by choosing the proper cyclone diameter size. Using thicker material in the cone area and abrasion-resistant removable wear plates (linings) at the strike zone are good design options that can control erosion.[3,4]

Fouling of a cyclone collector usually occurs by plugging the dust outlet or dust buildup on the cyclone wall. Plugging of the dust outlet occurs by large pieces of material becoming lodged in the outlet, forming an obstruction about which small particles can build or by the overfilling of the dust bin. These conditions can lead to reentrainment of the dust into the gas stream. Dust plugs remaining in the outlets can occur when a hopper that was once overfilled enough to plug has been emptied. For large-diameter cyclones, an axial cleanout opening with a bolted cover plate in the top of the outlet pipe or scroll can be provided so a rod can be inserted to clear a blockage. Material buildup on the cyclone wall is a function of the dust. Soft, fine dust has a tendency to build up on the wall. Particles below 3 μm in diameter possess inherently greater cohesive and adhesive forces. Condensation of moisture on the cyclone walls also contributes to the accumulation of material on the walls. Wall smoothness can help reduce the amount of material buildup on the walls. Electropolishing walls has been successful in minimizing buildup. If condensation is a problem, it must be eliminated by insulating the cyclone or other appropriate methods.[3,4]

Several other factors can play a major role in affecting the collection efficiency of a cyclone. Collection efficiency is a strong function of particle size and increases with increasing particle size. Also, the efficiency of dust collection is greater for particles of high density than for low density. Dust collection efficiency increases with dust loading. At very high loadings, all cyclones have relatively high efficiencies even though the gas discharge still contains progressively higher dust loadings. However, small-diameter cyclones are seldom used for heavy dust loads because erosion and plugging can occur. As will be shown later, the pressure drop theoretically varies with the square of the gas velocity. Collection efficiency does increase with an increasing inlet velocity, but at a lower rate than the increase in pressure drop. There is a limiting value of inlet velocity above which turbulence causes the reentrainment of particles into the gas stream and a decrease in the collection efficiency. Physical properties of the gas can also have some effect on the collection efficiency of a cyclone. Increasing the gas temperature decreases its density and increases its viscosity. The direct effect on efficiency by changes in gas density can be neglected, since the gas density is so much smaller than the density of the particles. If the viscosity of the gas that carries the dust particles to the cyclone increases, the collection efficiency will decrease with all other factors being constant.[3,4]

PREDICTION OF COLLECTION EFFICIENCY

Collection efficiency, as previously stated, is a strong function of particle size and increases with increasing particle size. The collection efficiency of a single particle size can be determined either by a semiempirical approach, developed by Lapple in 1951,[5] or by one of several theoretical equations that have since been developed. Determination of the overall collection efficiency requires knowledge of the particle size distribution of the dust particles. Collection efficiencies for each size range are then weighted accordingly and the values summed.

Theoretical Approach (Laminar Flow)

The basis of the simple theoretical derivation of efficiency is as follows: Particles enter with the gas stream, but tend to move outward under the influence of centrifugal force. This is resisted by the drag of the particles moving radially through the gas, and the resultant terminal or radial velocity of the particles is found by equating the centrifugal and drag forces. To be collected, particles must reach the outer wall before the gas leaves the outer vortex. The time and distance are both known: The time is the gas residence time, which depends on gas inlet velocity, radius of the cyclone, and number of turns in the vortex, and the maximum value of the distance to be traveled is the length from the inner edge of the inlet to the outer wall. Assuming laminar flow, an expression is derived that relates the collection efficiency to the cyclone parameters and operating conditions:

$$\eta = \frac{\pi N_e \rho_p d_p^2 V_g}{9\mu W} \tag{2}$$

where η is the efficiency (fraction); N_e, the effective number of turns; ρ_p, the particle density; d_p, the particle diameter; V_g, the gas velocity; and μ, the gas viscosity.

This model indicates the strong dependence of efficiency on particle diameter (squared), the dependence on the number of vortex turns (related to the dimensions of the cyclone) and inlet velocity, and the inverse dependence on cyclone inlet width, which is proportional to body diameter.

However, the model predicts a finite value of particle diameter, d_{pcrit} above which collection efficiency is 100% ("critical size"), whereas experimental evidence shows that efficiency approaches 100% asymptotically with increasing particle diameter.

$$d_{pcrit} = \left(\frac{9\mu W}{\pi N_e \rho_p V_g}\right)^{1/2} \tag{3}$$

Cut Diameter Approach

The semiempirical approach developed by Lapple[5] used this laminar-flow treatment, but introduced the concept of a cut size, d_{p50}, defined as the size of particle that is collected with 50% efficiency. The value is a characteristic of the control device and operating conditions, not of the size range of the dust particles. Using experimental particle collection efficiency versus particle-size data, correlated for cyclones of standard proportions, he plotted a generalized curve, as shown in Figure 15.

A particle whose collection efficiency is to be found is characterized by the ratio of its diameter d_p to the cut size d_{p50}, which gives the value of the abscissa of the graph, and the collection efficiency is read from the ordinate. The value of the cut size is calculated from Eq. 2 by setting efficiency equal to 0.5 and solving for d_p, which is now d_{p50} by definition:

$$d_{p50} = \left(\frac{9\mu W}{2\pi N_e \rho_p V_g}\right)^{1/2} \tag{4}$$

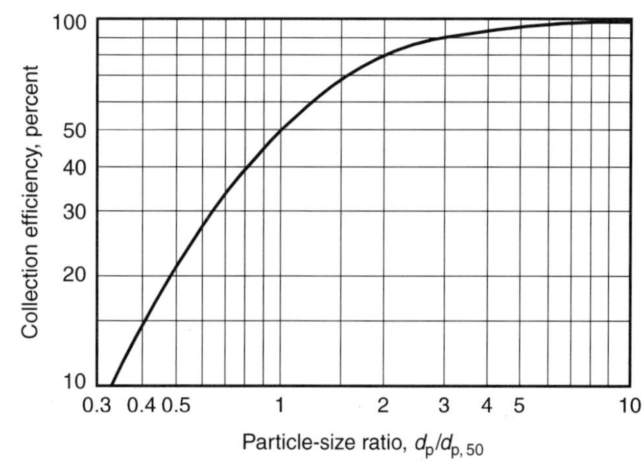

Figure 15. Cyclone Efficiency versus Particle-Size Ratio (from Ref. 2).

Lapple's graph has been fitted to an algebraic equation by Theodore and De Paola,[6] which makes it more convenient for computer applications:

$$\eta_j = 1/[1 + (d_{p50}/d_{pj})^2] \quad (5)$$

where d_{p50} is the particle cut diameter; d_{pj}, the particle diameter in range j; and η_j, the fractional efficiency in range j.

More Recent Approaches

The laminar-flow model has limitations, as gas flow in a cyclone is not simply laminar (nor is it fully turbulent, because the boundary layer has a significant depth).

Leith and Licht[7] and Leith and Mehta[8] developed an equation for collection efficiency from theoretical considerations that takes into account the back-mixing of uncollected particles and determines an appropriate average residence time for the gas in the cyclone. The equation is of the form

$$\eta = 1 - \exp(-Ad_p^B) \quad (6)$$

which is a common form for efficiency expressions, and indicates the asymptotic approach of efficiency to 100%. It uses a parameter C, which is a function of the cyclone's dimensional ratios, and Ψ, which is a modified inertial impaction parameter.

$$\eta = 1 - \exp(-2(C\Psi)^{1/2n+2}) \quad (7)$$

where:

$$n = 1 - \frac{(D)^{0.14}}{2.5}\left(\frac{T}{283}\right)^{0.3}$$

with cyclone diameter D in inches and T in degrees kelvin.

$$C = \frac{\pi D^2}{WH}\left\{2\left[1-\left(\frac{D_e}{D}\right)^2\right]\left(\frac{S}{D}-\frac{H}{2D}\right) + \frac{1}{3}\left(\frac{S+k-L_b}{D}\right)\left[1+\frac{d}{D}+\left(\frac{d}{D}\right)^2\right] + \frac{L_b}{D} - \left(\frac{D_e}{D}\right)^2\frac{k}{D} - \frac{S}{D}\right\} \quad (8)$$

where k is the farthest distance that the vortex extends below the gas exit duct,

$$k = 2.3D_e\left(\frac{D^2}{WH}\right)^{1/3}$$

and d is the diameter of the conical section at k

$$d = D - (D - D_d)(S + k - L_b)/L_c$$

In addition,

$$\Psi = \frac{\rho_p d_p^2 V_g}{18\mu_g D}(n+1) \quad (9)$$

The efficiency expression in Eq. 7 produces a curve of collection efficiency versus single-particle diameter, which, like the experimental data, is everywhere concave downwards, in contrast to the S-shaped curve of some of the many other theoretical models, as shown in Figure 16. The expression was shown to fit the experimental data to within 4%.

As an example of the application of the Leith and Licht expression, overall particle-collection efficiency has been calculated, using Eq. 7, for cyclones with diameters from 6 to 36 in., with an inlet velocity of 60 fps at 20°C for a Lapple conventional cyclone (Table 1, column 3). The particle size distribution was assumed to be log-normal with a standard deviation of 2.0. Figure 17 shows the theoretical relationship between the overall collection efficiency and the aerodynamic mass mean diameter of the dust assuming a particle density of 1.0 g/cm. The aerodynamic mass mean diameter is defined as

$$d_{A(MMD)} = d_{MMD}(\rho_p)^{1/2} \quad (10)$$

where $d_{A(MMD)}$ is the aerodynamic mass mean diameter; d_{MMD}, the physical mass mean diameter; and ρ_p, the particle density

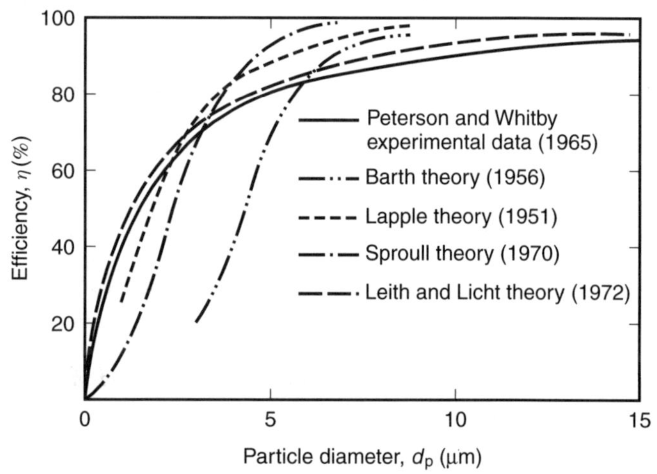

Source: Leith and Mehta, 1973.

Figure 16. Comparison of Experimental Data with Theoretical Predictions (from Ref. 6).

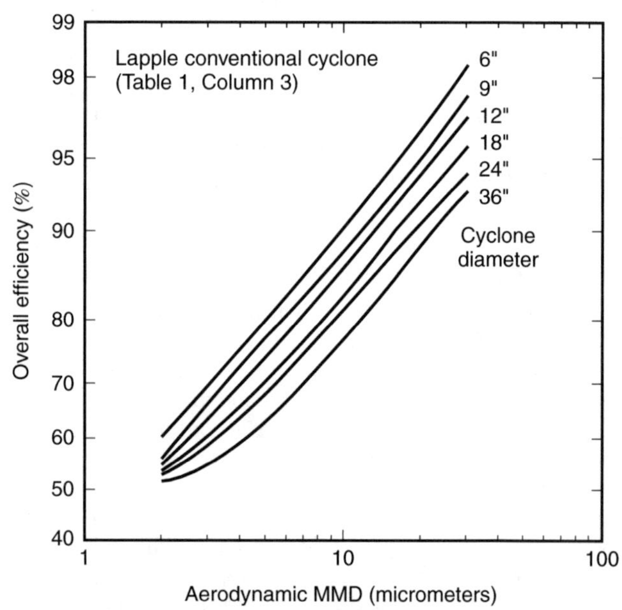

Figure 17. Overall Collection Efficiency versus Aerodynamic Mass Mean Diameter for a Range of Cyclone Sizes at 70°F.

(g/cm^3). The aerodynamic mass mean diameter is the value that is frequently reported based on sampling with a cascade impactor. In cases where the physical volume or mass mean diameter is given, simply calculate the aerodynamic mass mean diameter using the appropriate density value and particle physical mass mean diameter. Then read the overall collection efficiency for the required cyclone diameter from the curves in Figure 17. As shown in Figure 17, efficiency is seen to increase with increasing aerodynamic mass mean diameter and with decreasing cyclone diameter. For cyclones with relative dimensions different from those used, it would be necessary to recalculate the overall efficiency using Eqs. 4 and 5 or 7–9.

PREDICTION OF PRESSURE DROP

Pressure drop is an important parameter because it relates directly to operating costs. Higher efficiencies for a given cyclone can be obtained by higher inlet velocities, but this also increases the pressure drop, and a tradeoff must be made. Shepherd and Lapple[9] found empirically that pressure drop (expressed as the number of inlet velocity heads) depends inversely on exit diameter squared:

$$H_v = K\frac{HW}{D_e^2} \tag{11}$$

where H_v is the number of velocity heads and K is an empirical constant with a value of 16 for a tangential inlet cyclone and 7.5 for one with an inlet vane. The pressure drop is as follows:

$$\Delta P = \tfrac{1}{2}\rho_g V_g^2 H_v \tag{12}$$

where ΔP is the pressure drop (Pa; N/m^2); ρ, the gas density (kg/m^3); and V_g, the gas velocity (m/sec).

Although other equations have been derived from theoretical considerations, they have generally not been found to be more accurate than the equation developed by Lapple, which stands as the most useful one for calculating pressure drop.[8]

Pressure drop is a function of the square of inlet velocity; so too high a velocity will cause excessive pressure drop. On the other hand, too low a velocity would cause a low efficiency. A very high velocity would also actually decrease efficiency because of increased turbulence and saltation/reentrainment of particles; generally it is found that the best operating velocity is around 60 fps. Common ranges of pressure drops are as follows:

Low-efficiency cyclones	2–4 in. water
Medium-efficiency cyclones	4–6 in. water
High-efficiency cyclones	8–10 in. water

REFERENCES

1. *1996 ASHRAE Handbook: Heating, Ventilating, and Air Conditioning Systems and Equipment*, American Society of Heating, Refrigeration, and Air Conditioning Engineers, Atlanta, GA, 1996.
2. W. L. Olson, "Upgrading mechanical collector performance with hopper aspiration," Wheelabrator Air Pollution Control Technical Paper TP88-123, 1988.
3. A. J. Buonicor and L. Theodore, *Air Pollution Control Equipment: Selection, Design, Operation, and Maintenance*, Prentice Hall, New Jersey, 1982.
4. Air Pollution, Vol. IV—Engineering Control of Air Pollution, 3rd ed., A.C. Stern, Ed., Academic Press, New York, 1977.
5. C. E. Lapple, "Processes use many collector types," *Chem. Eng.* **58**:144–51 (1951).
6. L. Theodore and V. DePaola, "Predicting cyclone efficiency," *J. Air Pollution Cont. Assoc.* **30**(10), (1980).
7. D. Leith and W. Licht, "The collection efficiency of cyclone type particle collectors—A new theoretical approach," American Institute of Chemical Engineers, *Symposium Series* 126, **68**:196–206 (1972).
8. D. Leith and D. Mehta "Cyclone performance and design," *Atmospheric Environ.* **7**:527–549 (1973).
9. C. B. Shepherd and C. E. Lapple, "Flow pattern and pressure drop in cyclone dust collectors," *Ind. Eng. Chem.* **31**(8), (1939); **32**(9), (1940).
10. C. D. Cooper and F. C. Alley, *Air Pollution Control: A Design Approach*, Second Edition, Waveland Press, Illinois, 1994.
11. C. J. Stairmand, "The design and performance of cyclone separators," *Trans. Ind. Chem. Eng.* **29**, (1951).
12. P. Swift, "Dust control in industry," *Steam Heating Eng.* **38**, (1969).
13. M. Heumann, "Understanding cyclone dust collectors," *Plant Eng.* (May 26, 1983).
14. *Control of Particulate Emissions*, Course 413 Student Manual, EPA Air Pollution Training Institute.
15. Mechanical Collectors, Marsulex Environmetal Technologies, LLC, Lebanon, Pa.
16. Bulletin MC 581, Zurn Industries, Inc.

BIBLIOGRAPHY

L. W. Briggs, *Trans. AIChE* **42**(3):511–26 (1946).

W. Strauss, *Industrial Gas Cleaning*, International Series of Monographs in Chemical Engineering, Vol. 8, Pergamon Press, New York, 1966.

L. Theodore and A. J. Buonicore, *Air Pollution Control Equipment, Volume I—Particulates*, CRC Press, 1988.

K. Wark, C. Warner, and W. Davis, *Air Pollution—Its Origin and Control*, 3rd ed., Addison Wesley, CA, 1998.

WET SCRUBBERS

Michael Pilat and Kenneth E. Noll

INTRODUCTION

The control of particulate air pollutant emissions with wet scrubbers involves contacting or scrubbing the gases with a liquid (usually water or an aqueous solution). The aerosol particles are transferred from their suspension in a gaseous medium to the surface of the scrubbing liquid via mechanisms of inertial impaction, gravitational settling, Brownian diffusion, diffusiophoresis, electrostatics, and thermophoresis. For particles greater than about 0.5 μm diameter, inertial impaction is usually the primary collection mechanism, and for particles smaller than about 0.05 μm diameter, Brownian diffusion is the primary collection mechanism, depending on the relative velocity between the aerosol particle and the collecting liquid "object." The form or geometric shape of the scrubbing liquid "object" includes droplets, wetted walls, liquid sheets, and bubbles.

AREAS OF APPLICATIONS

Wet scrubbers are in use in many industries, some of which are presented elsewhere in this book. In general, wet scrubbers are used for:

1. Sticky, wet, corrosive, or liquid particles that are not easily removed from dry surfaces.
2. Explosive or combustible particles.
3. Collecting particles while simultaneously absorbing soluble gases such as SO_2.

FUNDAMENTALS OF PARTICLE COLLECTION IN WET SCRUBBERS

Particle Collection by Single Drops

The aerodynamic capture of aerosol particles by single liquid drops involves the collection mechanisms of inertial impaction, Brownian diffusion, electrostatics, diffusiophoresis, thermophoresis, and gravitational forces. A schematic of the trajectory of a single aerosol particle of radius r_p approaching a spherical droplet of radius R_d is shown in Figure 1. In the illustration, the particle is shown to just touch or graze the side of the droplet and is assumed to be collected by the drop. Note that the trajectory is of the center of the aerosol particle and although the particle center does not touch the droplet, the outer surface of the particle does touch. The single droplet particle collection efficiency η_{drop} is defined as the area swept free of aerosol particles divided by the projected cross-sectional area of the droplet or

$$\eta_{drop} = \frac{\text{area swept free of particles}}{\text{droplet cross-sectional area}} \tag{1}$$

Inertial Impaction. For inertial impaction an empirical equation for the single drop particle collection efficiency η_{drop} for potential flow of the gas around the drop (potential flow corresponds to high collector drop Reynolds numbers, Re_{drop} greater than 50) based on theoretical numerical calculations of the trajectories of particle centers (i.e., particles having no finite diameter) approaching spheres was reported for Stokes number Stk greater than 0.2 by Langmuir in 1948.

$$\eta_{drop,\ inertial} = \frac{(\text{Stk})^2}{(\text{Stk} + 0.5)^2} \tag{2}$$

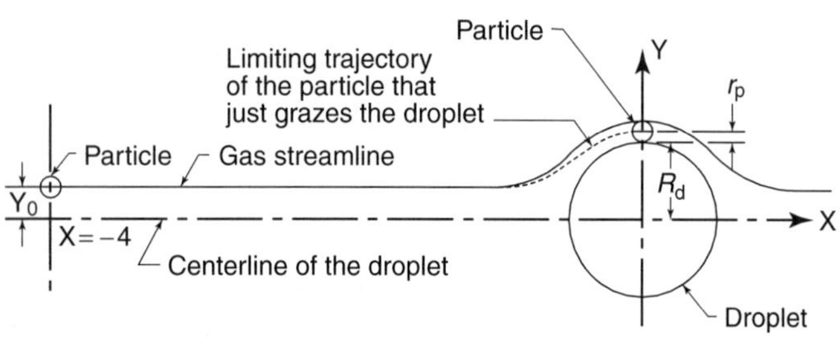

Figure 1. Trajectory of Particle Collected by a Single Spherical Droplet.

The Stokes number is a dimensionless group equal to the particle stop distance divided by the characteristic dimension of the fluid flow system (which in this case the system is the flow of a gas around a spherical drop). Historically, the characteristic dimension L of the drop has been either the drop radius or the drop diameter, and early researchers selected the drop radius. Because the literature calls both dimensionless groups using L as the drop radius or as the drop diameter the Stokes number, there is some confusion as to which definition of Stokes number is being used when comparing calculated and measured particle collection efficiencies. And unfortunately there are some equations using Stokes numbers in the published literature that may be in error—probably because of confusion regarding the definition. Here the Stokes number Stk will use the droplet radius as the characteristic dimension—in agreement with the early researchers, including Langmuir,[1] (1948) Richardson,[2] (1960) Herne,[3] (1960) Davies,[4] (1960) Walton and Woolcock,[5] (1960) Fuchs,[6] (1964) Sparks and Pilat,[7] (1970) and Calvert.[8] (1970)

$$\text{Stk} = \frac{\text{particle stop distance}}{\text{characteristic dimension}} = \frac{mu_o}{6\pi r_p \mu_g L} = \frac{2\rho_p u_o r_p^2}{9\mu_g R_d} = \frac{\rho_p u_o d_p^2}{9\mu_g D_d} \tag{3}$$

where m is the particle mass, ρ_p the particle density, r_p the particle radius, d_p the particle diameter, μ_g the gas dynamic viscosity, D_d the drop diameter, and u_o the gas velocity relative to the drop. Note that the early definitions of the Stokes number did not include the Cunningham slip correction factor (the Cunningham slip correction factor C does not have much effect on particles larger than about 5 μm at atmospheric pressure but becomes increasingly important as the particle diameter decreases). The experimental measured data of Walton and Woolcock[5] (1960) generally agrees with the Langmuir Eq. 2; the experimental particle collection efficiency data points are somewhat less than the Langmuir equation values in the Stokes number range of 0.3 to 3. A modification of the Langmuir equation to fit the measured data experimental curve reported by Walton and Woolcock[5] (1960) is given as:

$$\eta_{drop,\ inertial} = \frac{(\text{Stk})^2}{(\text{Stk} + 0.7)^2} \tag{4}$$

Walton and Woolcock used monodisperse methylene blue particles (1.25 to 24 μm diameter) and both stationary water drops and free-falling "gravity" monodisperse droplets in the 100 to 1000 μm diameter size range. The drop Reynolds numbers ranged from 1.5 to 100, which means that the low-drop Reynolds numbers do not meet the potential flow situation (i.e., Re of 1 to 10 is in the transition region between viscous and potential flow of the gas around the drop). The relatively good agreement between the measured and calculated single droplet particle collection efficiencies can probably be attributed to Walton and Woolcock's careful attention to having monodisperse aerosol particles, monodisperse water drops, and drops moving at known constant velocity relative to the air stream, which enabled clear identification of the correct Stokes number for each measured particle collection efficiency.

Brownian Diffusion. The Brownian diffusion mechanism of particle collection becomes increasing significant as the particle diameter decreases and the particle diffusion coefficient or diffusivity Δ increases. The flux or diffusion of the aerosol particles to the surface of the scrubber collection liquid takes time, and hence the gas velocity past the liquid needs to be slow enough for the particle to diffuse to the liquid surface. Accordingly, gas velocities will be lower and gas residence times greater in scrubbers that utilize the Brownian diffusion mechanism for fine particle collection. Johnstone and Roberts[9] (1949) reported an equation (based on a heat transfer analogy) for the single spherical droplet particle collection efficiency due to Brownian diffusion (which is an important mechanism for particles of less than 0.1 μm diameter) as

$$\eta_{\text{drop, Brownian diffusion}} = \frac{\pi}{(\text{Re})(\text{Sc})}\left[\frac{1}{\pi} + 0.557\sqrt{\text{Re}}(\text{Sc}^{3/8})\right] \quad (5)$$

where Re is the drop Reynolds number and Sc the particle Schmidt number:

$$\text{Re} = \frac{D_d u_o \rho_g}{\mu_g} = \frac{D_d u_o}{\nu_g}; \qquad \text{Sc} = \frac{\nu_g}{\Delta}$$

where D_d is the drop diameter, u_o the gas velocity relative to the drop, ρ_g the gas density, μ_g the gas dynamic viscosity, ν_g the gas kinematic viscosity, and Δ the particle diffusivity in the gaseous fluid. Balasubramanian and Meisen[10] (1975) reported that the Wilson and Geankoplis[11] theoretically derived equation for a single sphere Brownian diffusion collection efficiency of aerosol particles at low collector's Reynolds number (viscous flow of the gas around the spherical collector) agreed very well with Gebhart et al.[12] (1973) aerosol particle measurements using packed beds of uniform spherical glass beads, and this equation (converted to a form with the spherical drop Re and particle Sc) is in general agreement with the Johnstone and Roberts equation.

$$\eta_{\text{sphere, Brownian diffusion}} = \frac{2.180}{\text{Re}^{2/3}\text{Sc}^{2/3}} \quad (6)$$

Interception. Interception is commonly referred to as a particle collection mechanism for scrubbing with droplets, but interception originated from early mathematical calculations of the aerosol particle center trajectories around a single spherical droplet (i.e., assumed the particle had no finite diameter). The mathematical correction for the finite diameter of the aerosol particle was named "interception." All aerosol particles with finite diameters have this "interception" property because they are not merely dimensionless points (which were convenient for mathematical calculations — especially for numerical calculations using early computers in the 1940s and 1950s). Figure 1 illustrates this interception situation with the particle of radius r_p being collected as it just grazes the droplet when the trajectory of the particle center is a distance of r_p from the drop outer radius R_d. Ranz and Wong[13] (1952) reported an equation for the particle collection in potential flow (drop Re greater than 50) by interception on single spheres as

$$\eta_{\text{interception}} = (1 + d_p/D_d)^2 - \left(\frac{1}{1 + d_p/D_d}\right) \quad (7)$$

Note that *measured* particle collection efficiencies will include the so called "interception" phenomenon because all aerosol particles have finite diameters. Hence Eq. 7 for the particle collection efficiency of a drop due to interception should only be used to correct the single drop collection efficiencies obtained from *theoretical calculations* of the aerosol particle point center trajectories around a drop (and with today's computers it is possible to include the aerosol particle finite diameter in the trajectory calculations to obtain the drop or object particle collection efficiency, and accordingly the use of Eq. 7 to correct for this mathematically created "interception" situation should decrease).

Single Drop Collection with Combined Mechanisms. Assuming the theoretical single drop particle collection equations for inertial impaction (Langmuir) and Brownian diffusion are independent and apply for the collection of dimensionless particles (i.e., their point centers) and accordingly the correction for finite particle diameters via the interception (Ranz and Wong) equation is appropriate, the particle collection efficiency of a single collecting object such as a droplet can be calculated with

$$\eta_{\text{inertial \& Brownian diffusion}} = 1 - (1 - \eta_{\text{inertial}})(1 - \eta_{\text{Brownian diffusion}}) \times (1 - \eta_{\text{interception}}) \quad (8)$$

However, when obtaining the single drop particle collection efficiency from *measured* data equations such as the Walton and Woolcock equation for inertial impaction including interception, the particle collection efficiency should be calculated with

$$\eta_{\text{inertial \& Brownian diffusion}} = 1 - (1 - \eta_{\text{inertial}})(1 - \eta_{\text{Brownian diffusion}}) \quad (9)$$

Some calculated single droplet particle collection efficiencies reported by various authors are presented in Figure 2. The magnitude of these calculated (numerical calculations of particle trajectories to obtain the single drop particle collection efficiencies) single drop collection efficiencies are similar to those calculated using the combined mechanism Eq. 9 along with the Walton and Woolcock equation for the inertial mechanism and the Johnstone and Roberts equation for the Brownian diffusion mechanism. Figure 2 shows that the single drop collection efficiency is low for particles in the 0.1–1.0 μm diameter range for 100 μm diameter drops at their terminal gravity settling velocity, illustrating the need for a larger single drop collection efficiency in order for a scrubber to have the desired overall particle collection efficiency. The single drop particle collection efficiency can be increased by increasing the relative velocity between the drop and the aerosol particles (i.e., increased liquid velocities from spray nozzles, inject liquid into high velocity gases in venturi scrubber, use larger-diameter drops with higher gravity settling velocities) and by using other particle collection mechanisms such as diffusiophoresis, as reported by Pilat and Prem[14] (1977) or electrostatics.

Diffusiophoresis. Diffusiophoresis utilizes the water vapor migrating from the bulk gas stream toward a colder scrubber liquid upon which the water vapor condenses. If the gas stream contains a relatively high water vapor content (greater than 15% by gaseous volume water vapor) and the scrubbing liquid can be maintained at a somewhat cool temperature such that condensation occurs, diffusiophoresis can be a significant collection mechanism. Menoher[15] (1982) reported that using 80°F water in a 9-ft-high countercurrent packed-bed cooling tower to cool from 170 to 100°F the gases exhausting from an industrial incinerator removed 60% of the fine particles

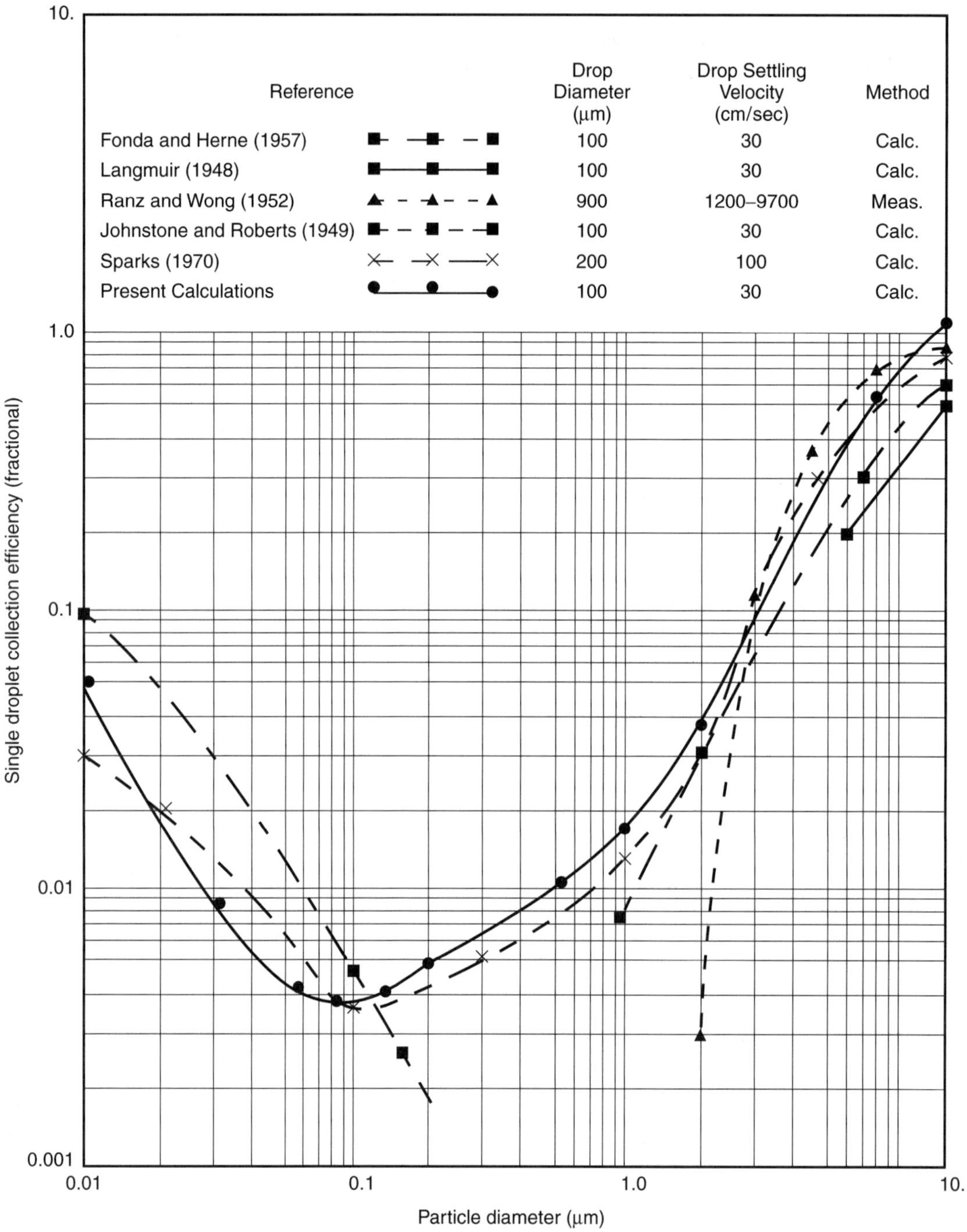

Figure 2. Calculated Single 100 μm Diameter Drop Collection Efficiencies.

downstream of a 30-in. H_2O pressure drop venturi (the venturi removed about 90% of the particles, cooled the gases to 170°F, saturated the gases with water vapor, and then the cooling tower removed 60% of the remaining fine particles).

Electrostatics. The use of electrostatics to improve the particle collection efficiency of wet scrubbers was proposed and demonstrated by Penney in the 1940s. Theoretical calculations of particle trajectories around single drops show significant increases in the collection efficiency upon charging the aerosol particles and drop to opposite polarities. Pilat et al.[16 (1974)] reported a measured increase from 68.8% at uncharged to 93.6% electrostatically charged overall particle collection (1.05 μm mass median diameter) with a 140-acfm pilot plant. For the particles charged to negative and the drops to positive polarities, the measured particle collection efficiency as a function of particle diameter is shown in Figure 3. The effect of charging is greatest for the smallest aerosol particles. The commercially available ionizing wet scrubber charges the aerosol particles at the inlet to the scrubber and utilizes a cross-flow packed tower scrubber to collect the particles (collecting charged aerosol particles with uncharged liquid droplets uses electrostatic mirror image forces to increase the particle collection). Wet electrostatic precipitators have been successfully used for aerosol particle collection for about 90 years and are a good configuration for the electrostatic

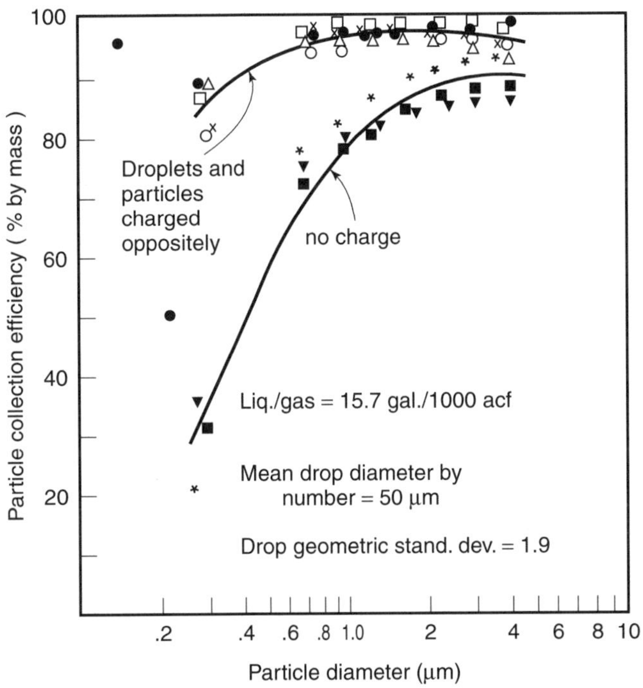

Figure 3. Particle Collection Efficiency of Electrostatic Spray Scrubber.

wet scrubber application because the electric field strength between the collecting liquid and the aerosol particle is much greater than in either the electrostatic spray scrubber (aerosol particle and liquid drop charged to opposite polarity) or the ionizing wet scrubber (aerosol particles charged and scrubbing liquid uncharged). Electrostatic sprays have been used to suppress emissions of fugitive dusts.

Particle Collection Efficiency of Spray Droplet Scrubbers

The overall particle collection efficiency η_o of a spray scrubber (Pease–Anthony cyclonic scrubber) was related (based on fundamental mathematics and chemical engineering) to the scrubber parameters with an exponential equation published in 1939 by Kleinschmidt.[17 (1939)] Almost all the wet scrubber particle collection efficiency equations developed after 1939 are in this exponential format (which can also be related to the number of mass transfer units N_t, commonly used in gaseous absorption calculations). The Kleinschmidt equation parameters include the spray liquid flow rate L, the gas flow rate G, the droplet diameter D_d, the distance the drop travels with respect to the gas R, and the single droplet particle collection efficiency η_{drop}.

$$\eta_o = 1 - e^{-(3RL/2D_dG)(\eta_{drop})} = 1 - e^{-f\eta_{drop}} = 1 - P_t = 1 - e^{-Nt} \quad (10)$$

where P_t is the overall particle penetration (fractional) through the wet scrubber. The fraction of gas swept by the droplets f (f is frequently greater than 1; so being named "fraction" may be misleading) is defined as the gas volume swept by the collecting drops divided by the total gas volume, as shown by Eq. 11. The fraction of gas swept by the droplets f is the ratio of the gas volumetric flow rate swept by the particle collecting objects (in this case liquid droplets) divided by the total gas volumetric flow rate through the wet scrubber. Perhaps f could be named the ratio of gas swept by the droplets to recognize that f is usually greater than unity.

$$f = \frac{\text{(gas volumetric flow rate swept by droplets)}}{\text{(total gas volumetric flow rate through scrubber)}} = \frac{3RL}{2D_dG} \quad (11)$$

The gas volumetric flow rate swept by the cross-sectional area of the monodisperse spherical droplets of diameter D_d and liquid flow rate L is given by

$$\text{gas volumetric flow rate swept by droplets} = \frac{3RL}{2D_d} \quad (12)$$

The distance the droplet travels with respect to the gas R is given by:

$$R = \text{(drop velocity with respect to gas)} \times \text{(drop residence time in scrubber)} \quad (13)$$

The droplets emit from the spray nozzles at velocities in the 10–100 ft/sec range and slow down because of the gaseous drag resistance. This initial droplet velocity can be estimated from the nozzle cross-sectional open liquid flow area and the liquid flow rate (recognize this is only an estimate as the droplets entrain adjacent gas so that the relative velocity between the gas and the drop is less than this estimated initial velocity). The drop then slows down until it either reaches the scrubber wall or slows down to its terminal gravity settling velocity with respect to the gas. A rigorous calculation of R requires dividing the drop trajectory into small distance segments, each segment at some drop velocity. A less rigorous R calculation may use an "average" drop velocity obtained from (initial drop velocity + terminal settling drop velocity)/2. Assuming the droplets travel at their gravitational terminal settling velocity immediately upon being emitted from the spray nozzle is an easy approach for obtaining R for vertical co-current or countercurrent spray towers. Then R is given by:

$$R = \text{(drop terminal settling velocity)} \times \text{(drop residence time in scrubber)} \quad (14)$$

For a co-current spray tower (gases and droplets flowing downward) R is given by:

$$R = \text{(drop terminal settling velocity)} \times \left(\frac{\text{height of scrubber zone}}{\text{drop term. settling vel + gas vel.}}\right) \quad (15)$$

For a co-current spray tower (gases and droplets flowing upward) R is given by:

$$R = \text{(drop terminal settling velocity)} \times \left(\frac{\text{height of scrubber zone}}{\text{gas vel.- drop term. settling vel.}}\right) \quad (16)$$

Note that mist eliminators are needed in the co-current spray towers to remove the scrubbing liquid droplets that are carried along with the gases.

For a countercurrent spray tower (droplets falling through rising gases) R is given by:

$$R = (\text{drop terminal settling velocity}) \times \left(\frac{\text{height of scrubber zone}}{\text{Drop term. settling vel.} - \text{gas vel.}}\right) \quad (17)$$

With the height of the scrubber zone H, the gas velocity v, and the drop terminal settling velocity V_d, the equation for f for a countercurrent tower assuming terminal settling is:

$$f = \frac{3RL}{2D_d G} = \left(\frac{3L}{2D_d G}\right)\left(\frac{V_d H}{V_d - V_g}\right) \quad (18)$$

Measurements have shown that in general larger spray scrubbers have higher particle collection efficiencies for the same liquid/gas flow-rate ratio, same droplet size distribution, and same particle size distribution, and this is probably because the droplets remain in the gas stream longer (before they hit the scrubber sides or are removed by a mist eliminator at the tower outlet), and hence the distance the droplet travels with respect to the gas is larger, providing for a larger magnitude of the fraction of gas swept f.

Venturi Scrubbers

The concept of venturi scrubbers was patented in the 1920s and put into practice in the 1940s. Johnstone and Roberts[9 (1949)] reported on their studies of aerosol particle collection, SO_2 absorption, and gas humidification in a rectangular venturi at air velocities up to 240 ft/sec. Both the SO_2 absorption and the aerosol particle collection efficiency measurements correlated with the scrubbing liquid specific surface area (ft^2 drop surface area/ft^3 gas volume). Johnstone et al.[18 (1954)] reported that inertial impaction was the predominant particle collection mechanism and that the measured particle collection efficiencies correlated with an exponential equation in a format similar to that reported by Kleinschmidt[17 (1939)]

$$\eta_o = 1 - e^{-k(l)/(g)(\Psi^{1/2})} = 1 - P_t = 1 - e^{-Nt} \quad (19)$$

where k is a constant, L/G is the liquid/gas flow ratio, and Ψ is the inertial impaction parameter and $\Psi = (\text{Stk}/2)$. For L/G in units of gallons/1000 ft^3 gas, Brink and Contant[19 (1958)] report k has a magnitude of 0.200 with units of 1000 ft^3/gallons, based on measured data from an industrial (6 m × 34 in. rectangular throat venturi) scale scrubber. Comparison with the Kleinschmidt equation shows that the $\Psi^{1/2}$ term is closely related to the single drop particle collection efficiency and the kL/G terms include the fraction of gas swept by the drops and some unit conversions. Identification of R (the distance the drop travels with respect to the gas) and of η_{drop} (the single drop particle collection efficiency) is complex, recognizing that the drop velocity relative to the gas is changing (drop rapidly accelerates in the venturi throat and then decelerates downstream in the venturi diverging section).

Calvert[8 (1970)] reported an equation for the overall particle penetration P_t of a venturi scrubber in the Kleinschmidt exponential form for standard air and water. Calvert et al.[20 (1972)] reported on the revision of his earlier equation for use in the interpretation of experimental data and introduced the variable f as an experimental coefficient. Earlier it was recommended to use an f of 0.25, and later Calvert[21 (1984)] reported that $f = 0.5$ for measurements on large-scale venturi scrubbers.

$$P_t = \exp\left(\frac{\rho_L u_G L D_d}{55\mu_g G}(F)\right) \quad (20)$$

$$F = \left(\frac{1}{\text{Stk}}\right)\left[1.4 \ln\left(\frac{\text{Stk} f + 0.7}{0.7}\right) + \frac{0.49}{0.7 + \text{Stk} f} - \text{Stk} f - 0.7\right] \quad (21)$$

where ρ_L is the liquid density, u_G the gas velocity in the venturi throat, L the liquid flow rate, D_d the drop diameter, μ_g the gas viscosity, and G the gas flow rate. Yung et al.[22 (1978)] extended the Calvert[8,20 (1970,1972)] solution of the differential equation for particle collection in the venturi throat and reported an exponential equation for the penetration P_t that does not have empirical parameters such as f. This equation takes into account the length of the venturi throat. Rudnick et al.[23 (1986)] reported on a comparison of measured and predicted venturi scrubber particle collection efficiencies using gas velocities in the 69–525 ft/sec range, an oil aerosol in the 0.32–2.8 μm diameter range, L/G in the 1.4–32 gal/1000 ft^3 range, venturi circular throat diameters in the 1.25–3.0 in. range, and mean drop diameters predicted by the Nukiyama–Tanasawa equation. They concluded that the Yung equation predicted the measured data better than the Calvert equation (with $f = 0.25$) or the Boll[24 (1973)] equation. All the particle collection efficiency equations tended to give better predictions as the venturi throat size increased.

The average droplet diameter (Sauter or droplet surface mean diameter) can be estimated from an empirical equation reported by Nukiyama and Tanasama[25 (1939)] for pneumatic atomization as

$$D_d = \frac{58{,}600}{u_g}\left(\sqrt{\frac{\sigma}{\rho_L}}\right) + 597\left(\frac{\mu_L}{\sqrt{\rho_L \sigma}}\right)^{0.45}\left(\frac{1000L}{G}\right)^{1.5} \quad (22)$$

where D_d is in μm, σ the liquid surface tension, μ_L the liquid viscosity, ρ_L the liquid density, and the parameters are in consistent cgs units. For typical conditions with air and water at standard conditions, simplified versions of the Nukiyama and Tanasama equation for the droplet diameter D_d with the parameters u_G and L/G in English units and in metric units are:

$$D_d(\mu\text{m}) = \frac{16{,}400}{(u_G, \text{ft/sec})} + 1.45\left(\frac{L}{G}, \frac{\text{gal}}{1000\ \text{ft}^3}\right)^{1.5} \quad (23)$$

$$D_d(\text{cm}) = \frac{50}{(u_G, \text{cm/sec})} + 91.8\left(\frac{L}{G}, \frac{\text{m}^3}{\text{m}^3}\right)^{1.5} \quad (24)$$

Venturi Scrubber Pressure Drop

A fairly large pressure drop (10–150 in. H_2O) across a venturi scrubber is needed to provide the high gas velocities. Licht[26 (1988)] reviewed the development of venturi pressure drop equations using a momentum balance over the venturi. A pressure drop equation reported by Hesketh[27 (1974)] is

$$\Delta P(\text{in. } H_2O) = \frac{v^2 \rho_g A^{0.133}(L/G)^{0.78}}{1270} \quad (25)$$

where v is the gas velocity in the venturi throat (ft/sec), A the throat cross-sectional area (ft^2), L/G the liquid-to-gas flow-rate ratio, and ρ_g the gas density (lb/ft^3). In metric units the equation is:

$$\Delta P(\text{cm H}_2\text{O}) = \frac{v^2 \rho_g A^{0.133} (L/G)^{0.78}}{3870} \tag{26}$$

where v is the gas velocity in the venturi throat (cm/sec), A the throat cross-sectional area (cm^2), L/G the liquid/gas flow-rate ratio (L/m^3), and ρ_g the gas density (g/cm^3).

Hesketh[28] (1992) correlated measured particle collection efficiencies at a specific particle diameter in the 0.1 to 10 μm range and pressure drop in the 6.8 to 50 in. H_2O range for rectangular venturi scrubbers, and his results are presented in Figure 4.

Contact Power Theory

The contact power theory of wet scrubbers reported by Lapple and Kamack[29] (1955) and extended by Semrau[30] (1960)

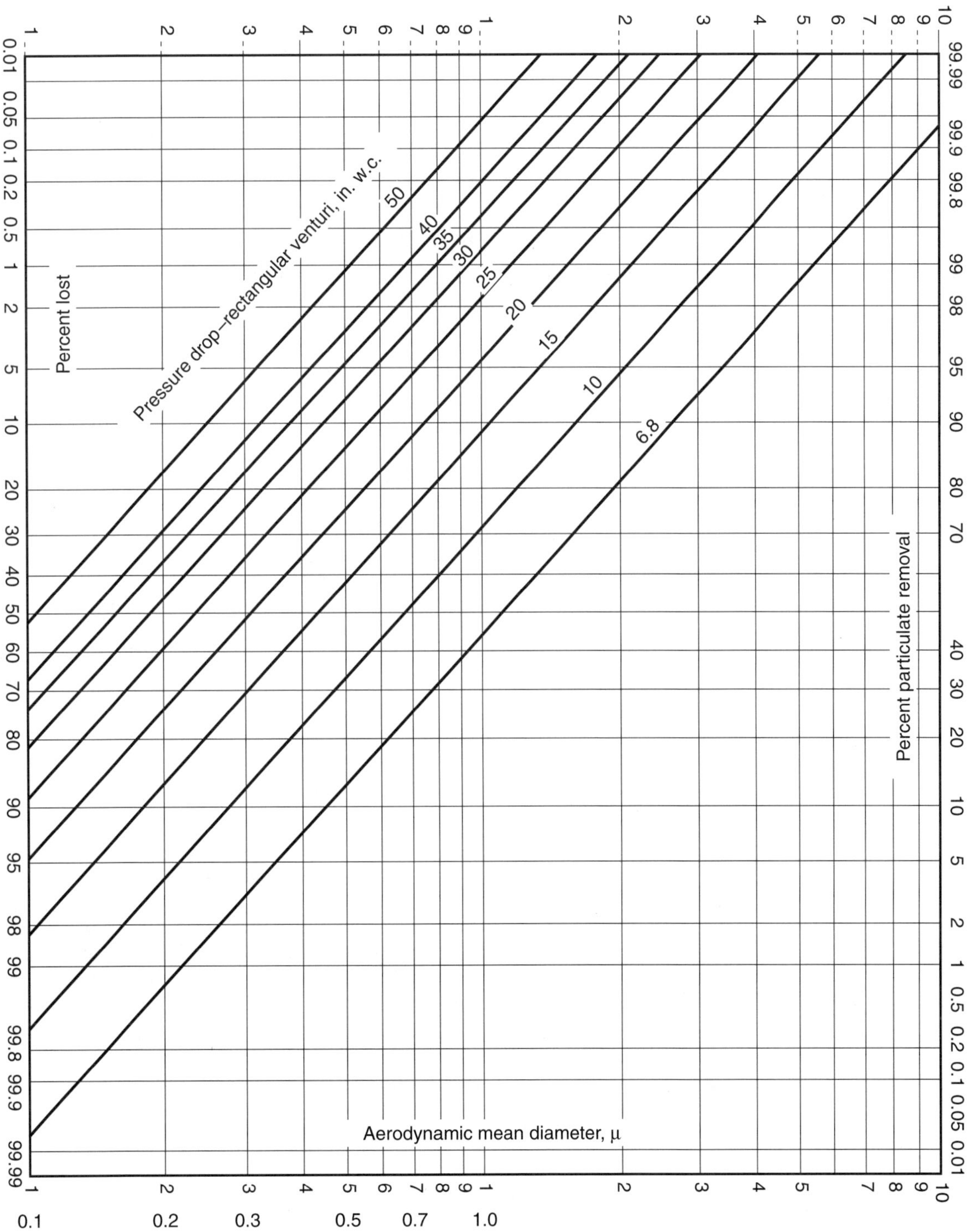

Figure 4. Collection Efficiencies of Particles in the 0.1–10-μm-Diameter Range at Venturi Pressure Drops in the 6.8–50 in. H_2O Range.

states that all scrubbers give the same degree of particle collection of a given dispersed dust at the same level of power consumption regardless of whether the pressure drop (or power consumption) is obtained in the gas stream or in the liquid stream. The overall wet scrubber particle collection efficiency is related to the number of mass transfer units (as shown before in Eq. 19)

$$\eta_o = 1 - P_t = 1 - e^{-Nt} \tag{27}$$

The number of mass transfer units is empirically related to the total contacting power P_T in units of horsepower/1000 acfm gas flow by

$$N_t = \alpha P_T^{\beta} \tag{28}$$

where α and β are empirical coefficients and the total contacting power P_T is the sum of the gas stream power input P_G and the liquid stream power input P_L.

$$P_T = P_G + P_L \tag{29}$$

The gas stream contacting power P_G and the liquid stream contacting power P_L are:

$$P_G(\text{hp/1000 acfm}) = 0.1575(\Delta P_G, \text{in. } H_2O) \tag{30}$$

$$P_L, \left(\frac{\text{hp}}{1000 \text{ acfm}}\right) = 0.583\left(\Delta P_L, \frac{lb}{\text{in.}^2}\right)\left(\frac{L}{G}, \frac{\text{gal}}{\text{ft}^3}\right) \tag{31}$$

where ΔP_G is the gas pressure drop (in. H_2O) and ΔP_L the liquid pressure drop (psig). In Table 1 are presented the measured magnitudes of α and β for wet scrubber control of aerosol particles emitted from a number of industries.

Table 1. Contact Power Scrubber Correlation Parameters

Aerosol	Scrubber	α	β
Lime kiln dust	venturi and cyclonic	1.47	1.05
Prewashed lime kiln dust	venturi, pipe line, and cyclonic	0.915	1.05
Talc dust	venturi	2.97	0.362
Talc dust	Orifice and pipe line	2.70	0.362
Phospheric acid mist	venturi	1.33	0.647
Foundry cupola dust	venturi	1.35	0.621
Open hearth steel furnace	venturi	1.26	0.569
Talc dust	cyclone	1.16	0.655
Ferrosilicon furnace	venturi and cyclonic spray	0.870	0.459
Odorous mist	venturi	0.363	1.41

TYPES OF WET SCRUBBERS FOR PARTICULATE CONTROL

There are many types of wet scrubbers, with the differences related to their gas–liquid contacting method and the geometrical shape, including venturi scrubbers, spray towers, cyclonic scrubbers, impingement or impaction scrubbers, wetted filters or grids, self-induced draft scrubbers, and packed towers. The wet scrubber design and operating parameters include gas pressure drop, liquid pressure drop, liquid/gas flow rate ratio (L/G), scrubber geometrical shape, location of water sprays or injection points, gas residence time, droplet size distribution, gas velocities, water temperature, gas temperature, water vapor content, and particle solubility in water. Three of the operating parameters and the scrubber cut diameters (aerodynamic diameter of particles collected with 50% efficiency by this scrubber) for some of these scrubbers are shown in Table 2. In general, the venturi scrubber has the lowest cut diameter (i.e., the highest particle collection efficiency at a given particle diameter).

Spray Towers (Preformed Sprays)

A spray tower uses liquid droplets formed by flowing the liquid through spray nozzles to produce the “preformed drops.” The aerosol particles are collected on these liquid drops. The size distribution and spray pattern of the drops are related to the nozzle configuration, the liquid being sprayed, the liquid pressure at the nozzle, and the liquid flow rate through the nozzle. Horizontal and vertical gas flow paths have been used with the spray introduced so that the liquid drops travel in a countercurrent, co-current, or cross-flow direction with respect to the gas direction. And in some spray scrubbers, all these droplet directions occur—depending on the spray pattern, nozzle orientation, and droplet size distribution. For the spray scrubber illustration presented in Figure 5 with the gas flowing vertically upward, large drops with sufficient gravity settling velocity will travel downward; smaller drops with settling velocities less than the upward gas velocity will travel upward to the mist eliminator. Spray nozzle types and sizes and spray nozzle locations are important to the successful operation of the spray tower.

Venturi Scrubbers

Venturi scrubbers utilize a constricted gas flow section or “throat,” which causes the gas to increase in velocity (accelerate) followed by a diverging section where the gases decrease in velocity (decelerate). The scrubbing liquid is injected upstream of the throat or directly into the throat. The liquid injection methods include pressurized spray nozzles and flow through straight tubes or pipes pointed toward

Table 2. Operating Parameters for Particulate Wet Scrubbers

Scrubber Type	Pressure Drop in. (H_2O)	Liquid/Gas Ratio (gal/1000 acf)	Liquid Pressure (psig)	Cut Diameter (μm)
Spray Tower	0.5–3	0.5–20	10–400	2–8
Cyclonic	2–10	2–10	10–400	2–3
Venturi	10–150	2–20	0.5–20	0.2

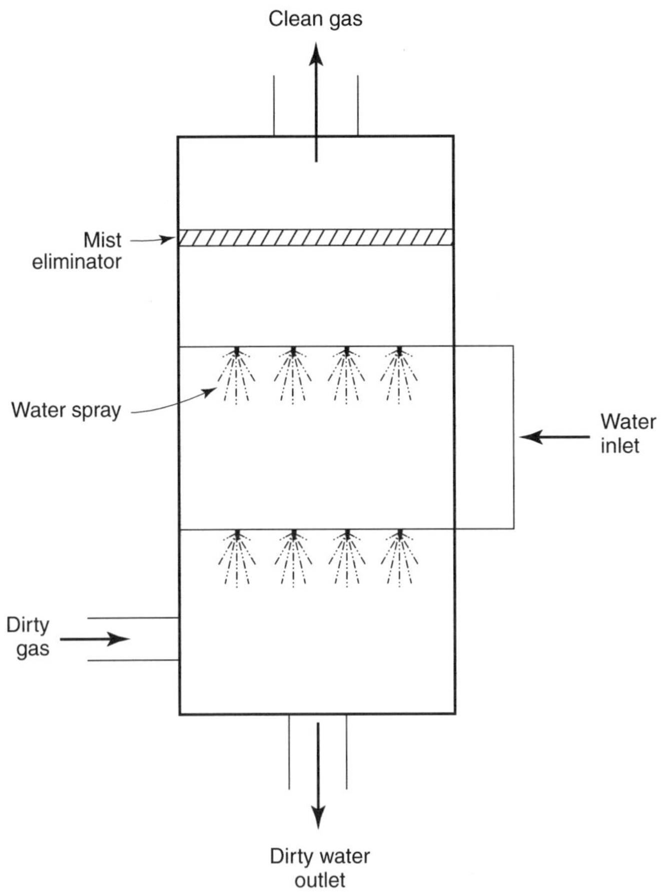

Figure 5. Spray Tower Wet Scrubber.

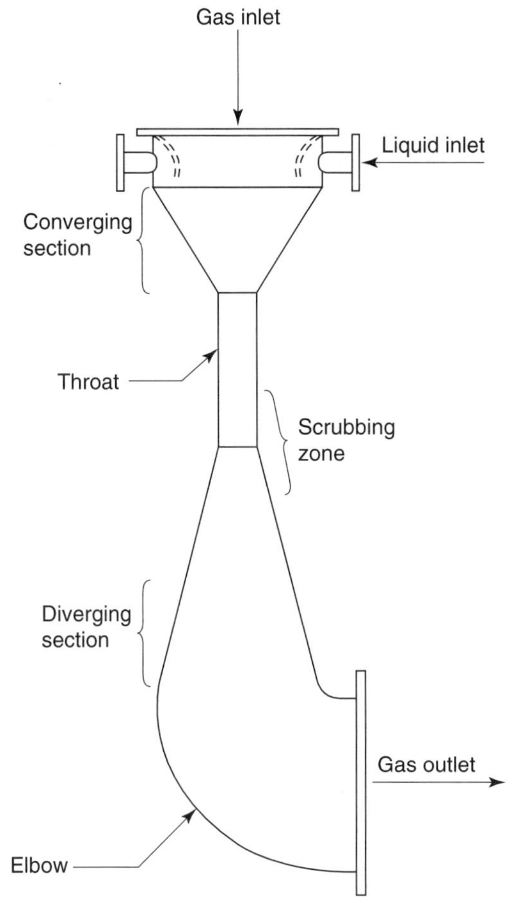

Figure 6. Venturi Scrubber.

the center (perpendicular to the direction of gas flow). It is important to have a uniform droplet distribution across the venturi cross section so as to have droplets properly located to sweep the incoming gases and aerosol particles evenly. The gases and aerosol particles moving at velocities in the 100–400 ft/sec range impact upon the slower-moving liquid droplets and the inertial impaction particle collection mechanism predominates. The throat configuration can be either circular or rectangular. Adjustable venturi throat openings enable the variation of the gas velocity and gas pressure drop (which accordingly affects the particle collection efficiency) and are able to adjust for variations in the total gas volumetric flow rate. Venturi scrubbers occupy the smallest volume of wet scrubbers and accordingly have the smallest gas residence time. However, some methodology to remove the entrained scrubbing liquid droplets is required, and the mist droplet removal device is probably much larger than the upstream venturi scrubber throat section. Variations of the venturi scrubber concept include educator or ejector venturis (gas flows due to high-velocity liquid flow) and multirod venturis (gas flows through closely spaced rods—the space between the rods being the flow constriction). An illustration of a venturi scrubber is shown in Figure 6.

Cyclonic Scrubbers

Cyclonic scrubbers are wet cyclones usually with the inlet gas flow through a tangential entry similar to the classic cyclone configuration. The scrubbing liquid can be injected at a number of locations, including through a center axial spray manifold (Pease–Anthony scrubber), from sprays located on outer walls of a cylindrical spray chamber, and from sprays evenly spaced throughout the tower chamber. The circular rotating gases with the entrained droplets and the resulting centrifugal force on the droplets cause them to migrate toward the outer scrubber walls. The droplet velocities relative to the gas stream are higher compared to gravity spray towers, and this increases the inertial impaction particle collection mechanism (which increases the single drop particle collection efficiency η_{drop}) but may reduce the distance the droplet travels with respect to the gas (which reduces the fraction of gas swept f). An illustration of a cyclonic wet scrubber is shown in Figure 7.

Wetted Filter Scrubber

Wetted filter scrubbers are wet filters and are useful for the collection of liquid particles or water soluble particles. The fibers or wires in the filter collect the particles. At higher gas velocities in the 1–20 ft/sec range, inertial impaction is the primary particle collection mechanism. At lower gas velocities (2–4 ft/min) and with fiber diameters in the 10–20 μm diameter range (such as with the Brink mist eliminator) the Brownian diffusion collection mechanism is effective for particles in the 0.01–0.5-μm-diameter size range. A wetted filter scrubber system is illustrated in Figure 8.

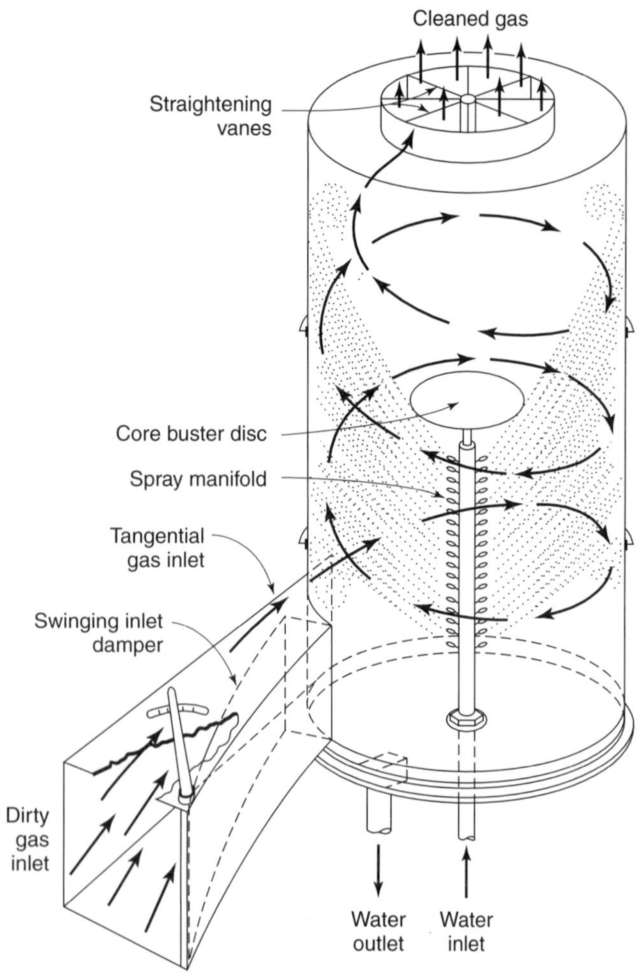

Figure 7. Cyclonic Wet Scrubber.

Plate and Tray Scrubber

Plate scrubbers are commonly named after the type of plate used. Sieve plate scrubbers use perforated plates, with the gas flowing vertically upward and the liquid flowing countercurrently downward. Impingement baffles located immediately downstream of the sieve plate orifice can be used as an impingement surface to collect particles (impingement tray scrubbers). The liquid flows downward through the plate perforations and liquid downcomers (similar to distillation towers). With some liquid depth on top of the perforated plate or tray plate, the gas flowing upward will form bubbles and foam. Particle collection occurs in these bubbles in the foam layer. The depth of the liquid and foam is dependent on the liquid flow rate, downcomer weir height, and other hydraulic parameters.

EXAMPLE CALCULATIONS

These example calculations are presented to illustrate the use of some of equations for calculating the particle collection efficiencies of two wet scrubbers, venturi scrubbers and gravity spray tower scrubbers. Such calculations will only provide estimates of the particle collection efficiencies. Pilot plant data and full-scale scrubber data are much more useful. However, the theoretical equations and empirical equations can be useful for the interpretation of measured data from pilot plants and full-scale scrubbers.

VENTURI SCRUBBER

Given:

$$\text{gas velocity} = 260 \text{ ft/sec} = 7,924.8 \text{ cm/sec};$$

$$\text{gas temp.} = 160^\circ\text{F} = 71^\circ\text{C}$$

$$\text{liquid/gas flow rate ratio} = L/G = 7.0 \text{ gal/1000 ft}^3$$

$$= 0.00093576 \text{ m}^3/\text{m}^3$$

$$\text{particle diameter} = 2.0\ \mu\text{m};\ \rho_\text{p} = 1.0 \text{ g/cm}^3;$$

$$\mu_\text{G} = 0.0002045 \text{ g/cm sec}\ f = 0.3;$$

$$\text{throat cross-sectional area} = 6 \text{ ft}^2;$$

$$\rho_\text{g} = 0.0639 \text{ lb/ft}^3$$

Find: (1) Mean drop diameter, (2) particle collection efficiency and penetration P_t for $d_\text{p} = 2.0\ \mu\text{m}$, and (3) pressure drop.

Mean drop diameter:

$$D_\text{d}(\mu\text{m}) = \frac{500{,}000}{(7924.8 \text{ cm/sec})} + 918{,}000\left(0.00093576\frac{\text{m}^3}{\text{m}^3}\right)^{1.5}$$

$$= 63.09 + 26.28 = 89.37$$

Particle collection efficiency and penetration for $d_\text{p} = 2.0\ \mu\text{m}$:

$$C = 1 + \frac{(0.00973)(\sqrt{344.44 \text{ K}})}{(2.0\ \mu\text{m})} = 1.0903$$

$$\text{Stk} = \frac{\rho_\text{p} u_\text{o} d_\text{p}^2}{9\mu_\text{g} D_\text{d}} = \frac{(1.0 \text{ g/cm}^3)(7924.8 \text{ cm/sec})(0.0002 \text{ cm})^2}{(9)(0.0002045)(0.008937 \text{ cm})}$$

$$= 21.0115$$

$$F = \left(\frac{1}{21.0115}\right)\left[1.4 \ln\left(\frac{6.30345 + 0.7}{0.7}\right) + \frac{0.49}{0.7 + 6.30345} - 6.30345 - 0.7\right] = -0.17653$$

Note that one needs to use many (perhaps 6 or more) significant figures in this equation for F, with the number of significant figures needed increasing as the particle diameter decreases.

$$P_\text{t} = \exp\left(\frac{\dfrac{(0.993 \text{ g/cm}^3)(7925 \text{ cm/sec})}{(0.0009358)(0.008937 \text{ cm})}}{55(0.0002045 \text{ g/cm sec})}(-0.1765)\right)$$

$$= 0.3560$$

$$\eta = 1 - P_\text{t} = 1 - 0.35597 = 0.6440 = 64.4\%$$

for 2 μm diameter particles

Similar calculations show the particle collection efficiency as 11.95% for 0.5 μm diameter and 38.59% for 1.0 μm diameter particles.

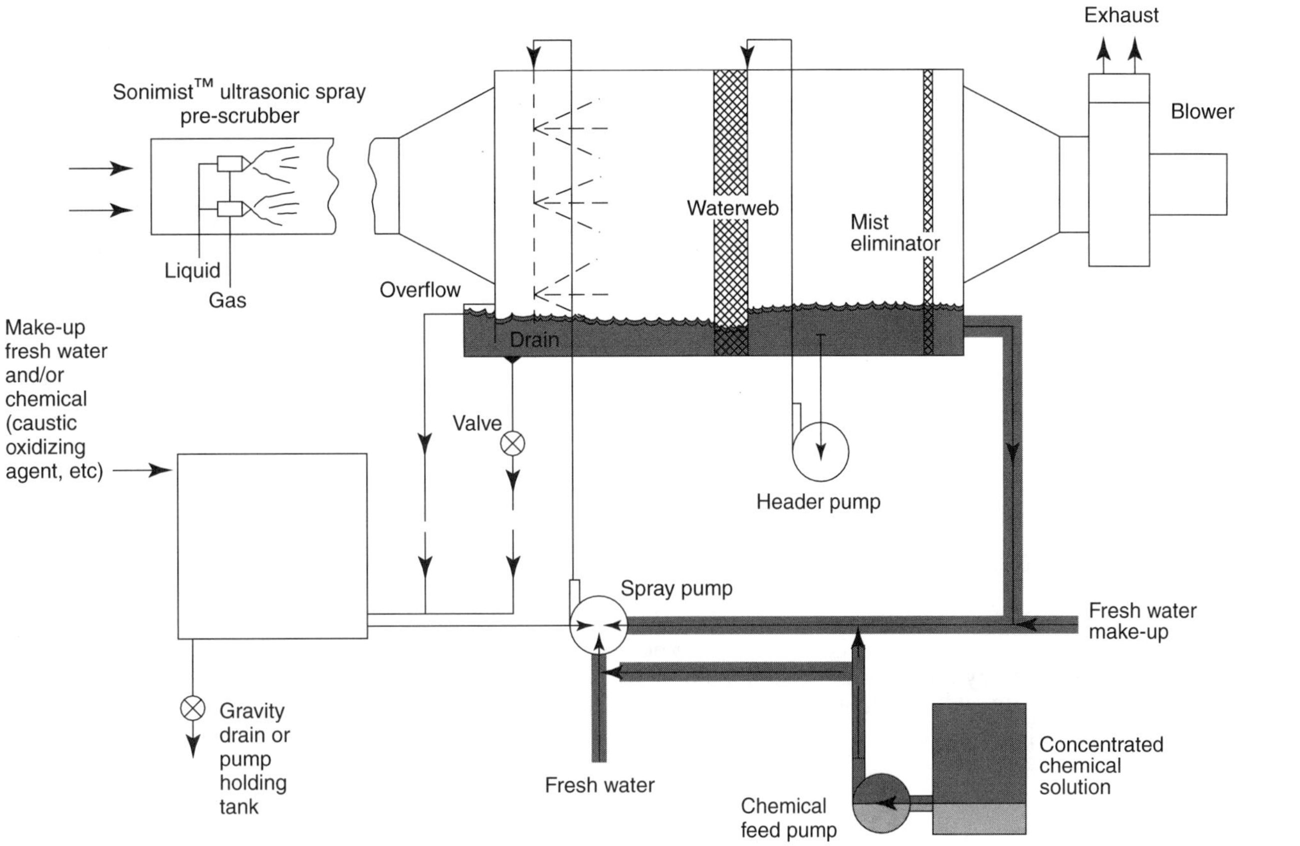

Figure 8. Wetted Filter Scrubber.

Pressure Drop

$$\Delta P\ (\text{in. H}_2\text{O}) = \frac{v^2 \rho_g A^{0.133} (L/G)^{0.78}}{1270}$$

$$= \frac{(260)^2 (0.0639)(6^{0.133})(5)^{0.78}}{1270} = 15.15$$

Using the equation reported by Hesketh,[27 (1974)] the calculated venturi pressure drop is 15.15 in. water. The graph in Figure 4 shows the particle collection efficiency is about 98.5% for 2 μm diameter particles at 15 in. water pressure drop compared to the predicted 64.4% calculated in this example. This illustrates the differences that may occur between calculated and measured particle collection efficiencies.

DROPLET MEAN SAUTER DIAMETER OF VENTURI SCRUBBER

Given:

$$\text{gas velocity} = 300\ \text{ft/sec} = 9144\ \text{cm/sec};$$
$$T = 20^\circ\text{C} = 68^\circ\text{F}; P = 1\ \text{atm}$$
$$L/G = 5\ \text{gal/1000 ft}^3 = 0.00066840\ \text{m}^3/\text{m}^3$$

Find the Mean Sauter diameter using English and metric unit equations.

$$D_d(\mu\text{m}) = \frac{16{,}400}{300\ \text{ft/sec}} + 14.5\left(5\frac{\text{gal}}{1000\ \text{ft}^3}\right)^{1.5}$$

$$= 54.67 + 16.21 = 70.88\ \mu\text{m}$$

$$D_d(\mu\text{m}) = \frac{500000}{(9144\ \text{cm/sec})} + 918000\left(0.0006684\frac{\text{m}^3}{\text{m}^3}\right)^{1.5}$$

$$= 54.68 + 15.86 = 70.54\ \mu\text{m}$$

The difference between the two equations is due to unit conversion factors.

COUNTERCURRENT SPRAY TOWER WITH 500 μM DIAMETER DROPLETS

Given:

gas temp. = 160°F = 71°C = 344.44 K; $\rho_g = 0.001023$ g/cm^3 = 0.0639 lb/ft^3

liquid/gas flow rate ratio = L/G = 7.0 gal/1000 ft^3 = 0.00093576 m^3/m^3

particle diameter = 2.0 μm; $\rho_p = 1.0$ g/cm^3;

$\mu_G = 0.0002045$ g/cm sec

drop diameter = 500 μm; drop terminal settling vel. = u = 202 cm/sec = 6.63 ft/sec Re = 63.9

gas flow of 1200 lb/ft^2 h(gas vel. = v = 5.22 ft/sec relative to tower wall)

droplets fall at terminal vel. from top of H = 20 ft high tower (countercurrent flow)

Inlet gases are saturated with water vapor so the condensation and evaporation effects can be neglected.

Find: (1) Single drop collection eff. for particles of 2.0 μm diameter; (2) fraction of gas swept by falling droplets for countercurrent flow; (3) overall particle collection efficiency for countercurrent flow.

Calculating the Stokes number:

$$\mathrm{Stk} = \frac{\rho_p u_o d_p^2}{9\mu_g D_d} = \frac{(1.0\ \mathrm{g/cm^3})(202\ \mathrm{cm/sec})(0.0002\ \mathrm{cm})^2}{(9)(0.0002045)(0.0500\ \mathrm{cm})}$$
$$= 0.08780$$

Calculating the single drop inertial impaction collection efficiency using the Walton and Woolcock experimental data fitted equation (drop Re is 63.9, but this equation is from experimental data):

$$\eta_{\mathrm{drop,\ inertial}} = \frac{(0.0878)^2}{(0.0878 + 0.7)^2} = 0.01242$$

The Brownian diffusion particle collection mechanism is ignored, as the 2-μm-diameter particles are too large and the velocity of the gases flowing by the settling 500-μm-diameter drop is too fast for the diffusion mechanism to contribute significantly to the particle collection.

$$R = (u)\left(\frac{\text{ht of scrubber zone}}{u - v}\right) = \left(6.88976\frac{\mathrm{ft}}{\mathrm{sec}}\right)$$
$$\times \left(\frac{20\ \mathrm{ft}}{6.88976 - 5.22\ \mathrm{ft/sec}}\right) = 82.524\ \mathrm{ft}$$

$$f = \frac{3RL}{2D_d G} = \frac{3(82.5238\ \mathrm{ft})(0.000935764\ \mathrm{ft^3 liq/ft^3 gas})}{2(0.05\ \mathrm{cm}(1\ \mathrm{ft}/30.48\ \mathrm{cm})}$$
$$= 70.6125$$

Substituting into the Kleinschmidt equation gives

$$\eta_o = 1 - e^{-(70.6125)(0.012421)} = 1 - e^{-0.87708} = 1 - 0.4160 = 58.4$$

The estimated overall collection efficiency for 2-μm-diameter particles is 58.4%.

COCURRENT SPRAY TOWER WITH 100-μM-DIAMETER DROPLETS

Given:

gas temp. = 160°F = 71°C = 344.44 K;

$\rho_g = 0.001023\ \mathrm{g/cm^3} = 0.0639\ \mathrm{lb/ft^3}$

liquid/gas flow rate ratio = $L/G = 7.0\ \mathrm{gal}/1000\ \mathrm{ft^3}$
$= 0.00093576\ \mathrm{m^3/m^3}$

particle diameter = 2.0 μm; $\rho_p = 1.0\ \mathrm{g/cm^3}$;

$\mu_G = 0.0002045$ g/cm sec

drop diameter = 100 μm; drop terminal settling vel. = u
= 24.6 cm/sec = 0.8071 ft/sec

gas flow of 1200 lb/ft^2 h(gas vel. = v
= 5.22 ft/sec relative to tower wall)

droplets injected at bottom of tower of z = 20 ft high tower

Inlet gases are saturated with water vapor; so the condensation and evaporation effects can be neglected.

Find: (1) Single drop collection eff. for particles of 2.0 μm diameter; (2) fraction of gas swept by falling droplets for countercurrent flow; (3) overall particle collection efficiency for countercurrent flow.

$$\mathrm{Stk} = \frac{\rho_p u_o d_p^2}{9\mu_g D_d} = \frac{(1.0\ \mathrm{g/cm^3})(24.6\ \mathrm{cm/sec})(0.0002\ \mathrm{cm})^2}{(9)(0.0002045)(0.0100\ \mathrm{cm})}$$
$$= 0.04463$$

Using the Walton and Woolcock data fitted equation

$$\eta_{\mathrm{drop,\ inertial}} = \frac{(0.04463)^2}{(0.04463 + 0.7)^2} = 0.003592$$

This single drop collection efficiency is somewhat below the theoretically predicted values for 100-μm-diameter droplets shown in Figure 2.

$$R = (u)\left(\frac{\text{ht of scrubber zone}}{v - u}\right) = \left(0.8071\frac{\mathrm{ft}}{\mathrm{sec}}\right)$$
$$\times \left(\frac{20\ \mathrm{ft}}{5.22 - 0.8\ \mathrm{ft/sec}}\right) = 3.1411\ \mathrm{ft}$$

$$f = \frac{3RL}{2D_d G} = \frac{3(3.1411\ \mathrm{ft})(0.000935764\ \mathrm{ft^3\ liq/ft^3\ gas})}{2(0.01\ \mathrm{cm})(1\ \mathrm{ft}/30.48\ \mathrm{cm})}$$
$$= 13.4386$$

The fraction of gas swept is also less than for the 500-μm-diameter drop and countercurrent flow.

$$\eta_o = 1 - e^{-(13.4386)(0.003592)} = 1 - e^{-0.0483} = 1 - 0.9529$$
$$= 0.0471 = 4.71\%$$

These two spray droplet tower examples illustrate that care needs to be taken with regard to the droplet size distribution, relative velocities between the drops and the gases, and locations of spray droplet injection into the spray towers. For these two examples, the 500-μm-diameter droplet falling down through rising gases (countercurrent flow) gives a higher 2-μm-diameter particle collection efficiency because of the larger single drop collection efficiency and the larger R distance the drop travels with respect to the gas. Considering the low efficiency for the 2-μm-diameter particles, this co-current 100-μm-diameter droplet scrubber is not appropriate for adequate control of 2-μm-diameter particles.

For additional information and greater detail on particle collection by wet scrubbers, see the books written by Licht,[26] (1988) Dullien,[31] (1989) and Noll.[32] (1999)

REFERENCES

1. I. Langmuir, "The production of rain by chain reaction in cumulus clouds at temperatures above freezing," *J. of Meteorology* **5**:175–92 (1947).
2. E. G. Richardson, Editor, *Aerodynamic Capture of Particles*, Proceedings of conference at British Coal Utilization Research Assoc., Leatherhead, Surrey, England, Pergamon Press, 1960.
3. H. Herne, "The classical computation of the aerodynamic capture of particles by spheres" *Aerodynamic Capture of Particles*, E. G. Richardson, Ed., Pergamon Press, pp. 26–43 (1960).
4. C. N. Davies, "The particle resistance/inertia parameter," *Aerodynamic Capture of Particles*, E. Richardson, Ed., Permagon Press, pp. 199–200 (1960).
5. W. Walton and A. Woolcock, "The suppression of airborne dust by water spray," *Int. J. Air Poll. Control* **3**:129 (1960).
6. N. A. Fuchs, *The Mechanics of Aerosols*, Dover, 1964.
7. L. Sparks, and M. Pilat, "Effect of diffusiophoresis on particle collection by wet scrubbers," *Atmospheric Environment* **4**:651–60 (1970).
8. S. Calvert, "Venturi and other atomizing scrubbers efficiency and pressure drop," *AIChE Journal* **16**:392–96 (1970).
9. H. Johnstone and M. Roberts, "Deposition of aerosol particles from moving gas streams," *Industrial and Engineering Chemistry* **41**:2417–23 (1949).
10. M. Balasubramanian and A. Meisen, "A note on the diffusional deposition of aerosol particles in packed beds," *J. Aerosol Science* **6**:461–63 (1975).
11. Wilson and Geankopolis
12. J. Gebhart, C. Roth and W. Stahlhofen, "Filtration properties of glass bead media for aerosol particles in the 0.1–2 μm size range," *J. Aerosol Science* **4**:355–71 (1973).
13. W. Ranz and J. Wong, "Impaction of dust and smoke particles," *Industrial and Engineering Chemistry* **44**:1371–81 (1952).
14. M. Pilat and A. Prem, "Effect of diffusiophoresis and thermophoresis on the overall particle collection efficiency of spray droplet scrubbers," *Air Pollution Control Assoc. J.* **27**:982–88 (1977).
15. C. Menoher, "An incinerator scrubber that works: A case study," *Third Symposium on the Transfer and Utilization of Particulate Control Technology: Vol. III Particulate Control Devices*," EPA Report EPA-600/9-82-006a, pp. 111–18, 1982.
16. M. Pilat, S. Jaasund and L. Sparks, "Collection of aerosol particles by electrostatic droplet spray scrubbers," *Environmental Science & Technology* **8**:360–62 (1974).
17. R. V. Kleinschmidt, "Factors in spray scrubber design," *Chemical and Metallurgical Engineering* **46**:487–90 (1939).
18. H. Johnstone, R. Feild and M. Tassler, "Gas absorption and aerosol collection in a venturi atomizer," *Industrial and Engineering Chemistry* **46**:1601–8 (1951).
19. J. Brink and C. Contant, "Experiments on an industrial venturi scrubber," *Industrial and Engineering Chemistry* **50**:1157–60 (1958).
20. S. Calvert, D. Lundgren, and D. Menenta, "Venturi scrubber performance," *J. Air Pollution Control Assoc.* **22**:529–532 (1972).
21. S. Calvert, "Particle control by scrubbing," *Handbook of Air Pollution Technology*, John Wiley & Sons, S. Calvert and H. Englund, Ed., pp. 215–248 (1984).
22. S. Yung, S. Calvert, and H. Barbarika, "Venturi scrubber performance model," *Environmental Science and Technology* **12**:456–59 (1978).
24. R. H. Boll, "Particle collection and pressure drop in venturi scrubbers," *Industrial Engineering Chemistry Fundamentals* **12**:40–49 (1973).
25. S. Nukiyama and Y. Tanasawa, *Trans. Soc. Mech. Engrs. (Japan)* **5**(18):68 (1939).
26. W. Licht, *Air Pollution Control Engineering: Basic Calculations for Particulate Collection*, 2nd Ed., Marcel Dekker, 1988.
27. H. E. Hesketh, "Fine particle collection efficiency related to pressure drop, scrubbant and particle properties, and contact mechanism," *Air Pollution Control Assoc. J.* **24**:939–42 (1974).
28. H. Hesketh, "Wet scrubbers," *Air Pollution Engineering Manual*, Buonicore and Davis, Ed., Van Nostrand Reinhold, pp. 80–81, 1992.
29. Lapple & Kamack 1955
30. K. T. Semrau, "Correlation of dust scrubber efficiency," *Air Pollution Control Assoc. J.* **10**:200–7 (1960).
31. F. A. Dullien, *Introduction to Industrial Gas Cleaning*, Academic Press, 1989.
32. K. E. Noll, *Fundamentals of Air Quality Systems: Design of Air Pollution Control Devices*, American Academy of Environmental Engineers, 1999.

Nomenclature:

A	venturi throat cross-sectional area
C	cunningham slip correction factor
d_d	particle diameter
D_d	droplet diameter
f	fraction of gas swept by collecting objects
f	experimental coefficient in Calvert equation
F	parameter used in Calvert equation
G	Liquid flow rate
H	height of spray scrubber zone
k	parameter in Johnstone and Roberts equation
K	degrees kelvin
L	liquid flow rate
L	characteristic dimension of collecting object
m	mass of aerosol particle
N_t	number of mass transfer units
P_t	penetration through wet scrubber
r_p	radius of aerosol particle
R_d	radius of droplet
R	distance droplet travels with respect to gas
Re	Reynolds number of droplet
Sc	Schmidt number of aerosol particle
Stk	Stokes number
u	gas velocity relative to collecting object
v	gas velocity relative to scrubber tower wall
V_d	drop terminal gravity settling velocity

Greek Letters

η_{drop}	single drop particle collection efficiency
η_o	overall particle collection efficiency of scrubber
ρ_g	gas density
ρ_p	particle density
ρ_L	liquid density
σ	liquid surface tension
μ_g	gas dynamic viscosity
ν_g	gas kinematic viscosity
Δ	particle diffusivity
Ψ	inertial impaction parameter = Stk/2

ELECTROSTATIC PRECIPITATORS

James H. Turner, Phil A. Lawless, Toshiaki Yamamoto, David W. Coy, Gary P. Greiner, John D. McKenna, and William M. Vatavuk

An electrostatic precipitator (ESP) is a particle control device that uses electrical forces to move the particles out of the flowing gas stream and onto collector plates. The particles are given an electric charge by forcing them to pass through a corona, a region in which gaseous ions flow. The electrical field that forces the charged particles to the walls comes from electrodes maintained at high voltage in the center of the flow lane.

Once the particles are collected on the plates, they must be removed from the plates without reentraining them into the gas stream. This is usually accomplished by knocking them loose from the plates, allowing the collected layer of particles to slide down into a hopper, from which they are evacuated. Some precipitators remove the particles by intermittent or continuous washing with water.

Figure 1 shows an ESP with its various components identified. Figure 2 shows two variations of charging electrode/collector electrode arrangements used in ESPs.

TYPES OF ESPs

Electrostatic precipitators are configured in several ways. Some of these configurations have been developed for special control action, and others have evolved for economic reasons. The types that will be described here are (1) the plate-wire precipitator, the most common variety; (2) the flat-plate precipitator; (3) the tubular precipitator; (4) the wet precipitator, which may have any of the previous mechanical configurations; and (5) the two-stage precipitator.

Plate-Wire Precipitator

Plate-wire ESPs are used in a wide variety of industrial applications, including coal-fired boilers, cement kilns, solid waste incinerators, paper mill recovery boilers, petroleum refining catalytic cracking units, sinter plants, basic oxygen furnaces, open hearth furnaces, electric arc furnaces, coke oven batteries, and glass furnaces.

In a plate-wire ESP, gas flows between parallel plates of sheet metal and high-voltage electrodes. These electrodes are long wires weighted and hanging between the plates or are supported there by mastlike structures (rigid frames). Within each flow path, gas flow must pass each wire in sequence as it flows through the unit.

The plate-wire ESP allows many flow lanes to operate in parallel, and each lane can be quite tall. As a result, this type of precipitator is well suited for handling large volumes of gas. The need for rapping the plates to dislodge the collected material has caused the plate to be divided into sections, often three or four in series with one another, which can be rapped independently. The power supplies are often sectionalized in the same way to obtain higher operating voltages, and further electrical sectionalization may be used for increased reliability.

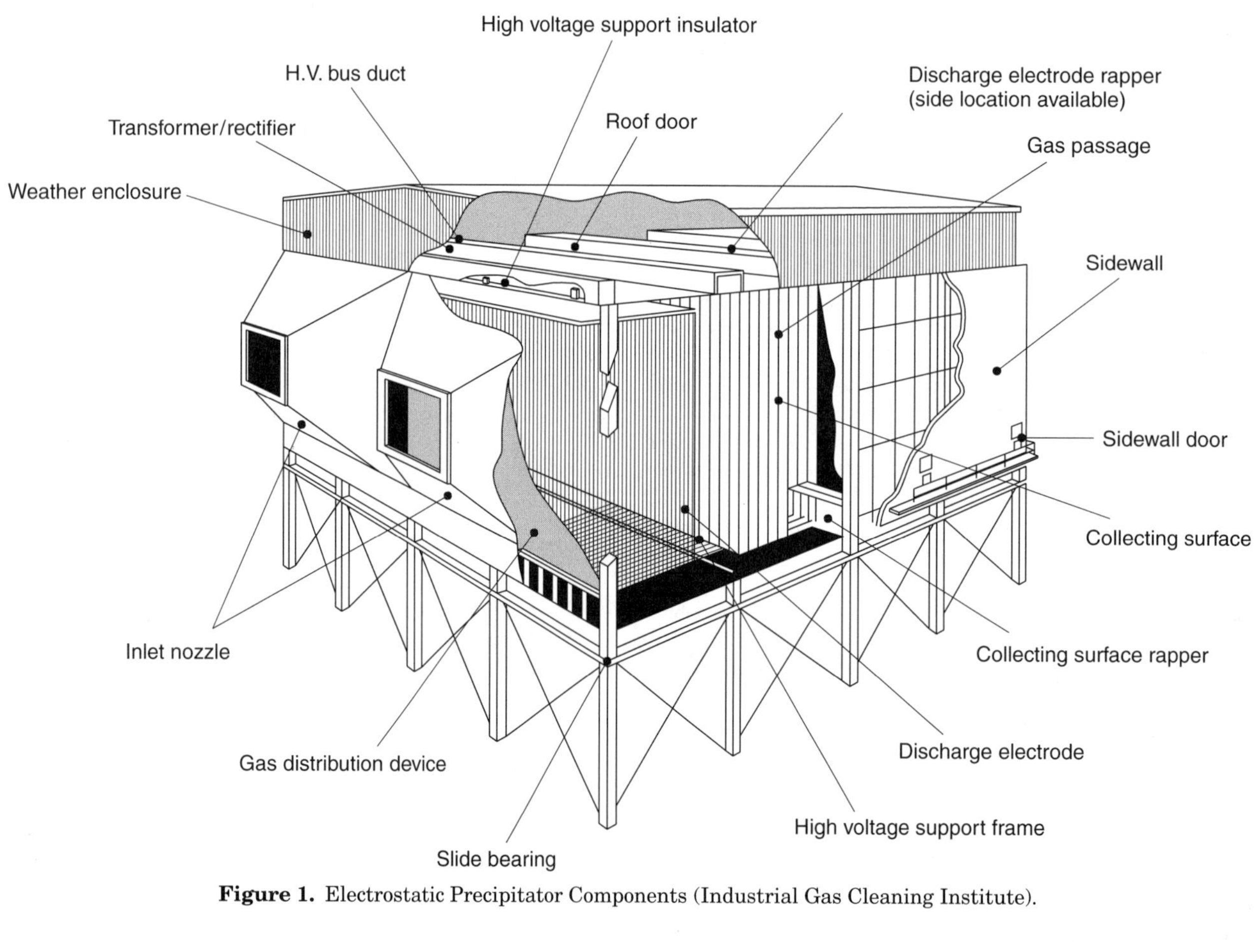

Figure 1. Electrostatic Precipitator Components (Industrial Gas Cleaning Institute).

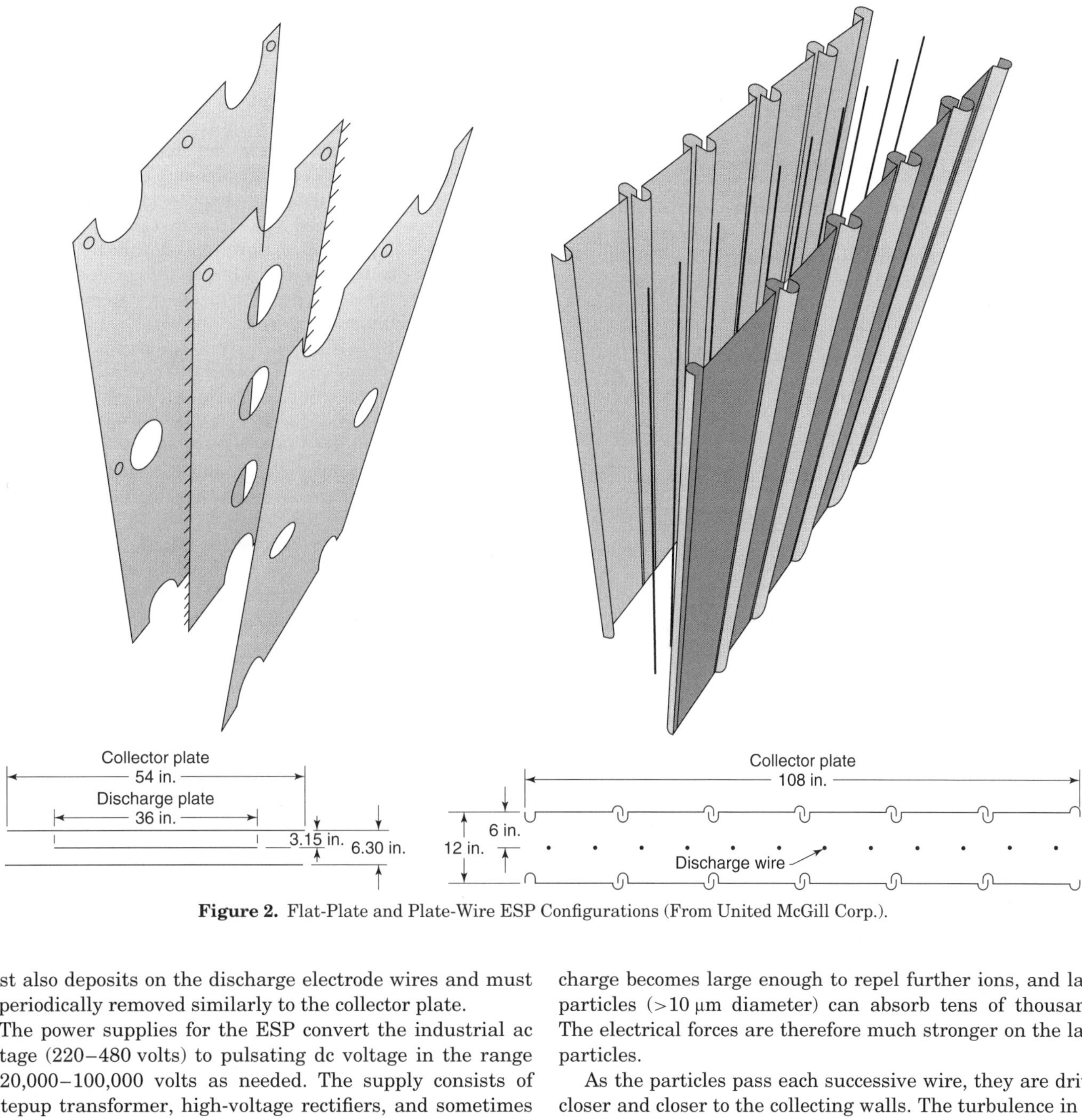

Figure 2. Flat-Plate and Plate-Wire ESP Configurations (From United McGill Corp.).

Dust also deposits on the discharge electrode wires and must be periodically removed similarly to the collector plate.

The power supplies for the ESP convert the industrial ac voltage (220–480 volts) to pulsating dc voltage in the range of 20,000–100,000 volts as needed. The supply consists of a stepup transformer, high-voltage rectifiers, and sometimes filter capacitors. The unit may supply either half-wave or full-wave rectified dc voltage. There are auxiliary components and controls to allow the voltage to be adjusted to the highest level possible without excessive sparking and to protect the supply and electrodes in the event a heavy arc or short circuit occurs.

The voltage applied to the electrodes causes the gas between the electrodes to break down electrically, an action known as a "corona." The electrodes usually are given a negative polarity because a negative corona supports a higher voltage than does a positive corona before sparking occurs. The ions generated in the corona follow electric field lines from the wires to the collecting plates. Therefore, each wire establishes a charging zone through which the particles must pass.

Particles passing through the charging zone intercept some of the ions, which become attached. Small aerosol particles (<1 μm diameter) can absorb tens of ions before their total charge becomes large enough to repel further ions, and large particles (>10 μm diameter) can absorb tens of thousands. The electrical forces are therefore much stronger on the large particles.

As the particles pass each successive wire, they are driven closer and closer to the collecting walls. The turbulence in the gas, however, tends to keep them uniformly mixed with the gas. The collection process is therefore a competition between the electrical and dispersive forces. Eventually, the particles approach close enough to the walls so that the turbulence drops to low levels and the particles are collected.

If the collected particles could be dislodged into the hopper without losses, the ESP would be extremely efficient. The rapping that dislodges the accumulated layer also projects some of the particles (typically 12% for coal fly ash) back into the gas stream. These reentrained particles are then processed again by later sections, but the particles reentrained in the last section of the ESP have no chance to be recaptured and so escape the unit.

Practical considerations of passing the high voltage into the space between the lanes and allowing for some clearance above the hoppers to support and align electrodes leave room for part of the gas to flow around the charging zones. This is called "sneakage" and amounts to 5–10% of the total flow. Antisneakage baffles usually are placed to force the sneakage flow to mix with the main gas stream for collection in later sections. But, again, the sneakage flow around the last section has no opportunity to be collected.

These losses play a significant role in the overall performance of an ESP. Another major factor is the resistivity of the collected material. Because the particles form a continuous layer on the ESP plates, all the ion current must pass through the layer to reach the ground plates. This current creates an electric field in the layer, and it can become large enough to cause local electrical breakdown. When this occurs, new ions of the wrong polarity are injected into the wire-plate gap, where they reduce the charge on the particles and may cause sparking. This breakdown condition is called "back corona."

Back corona is prevalent when the resistivity of the layer is high, usually above 2×10^{11} ohm cm. For lower resistivities, the operation of the ESP is not impaired by back corona, but resistivities much higher than 2×10^{11} ohm cm considerably reduce the collection ability of the unit because the severe back corona causes difficulties in charging the particles. At resistivities below 10^8 ohm cm, the particles are held on the plates so loosely that rapping and nonrapping reentrainment become much more severe. Care must be taken in measuring or estimating resistivity because it is strongly affected by such variables as temperature, moisture, gas composition, particle composition, and surface characteristics.

Flat-Plate Precipitators

A significant number of smaller precipitators (100,000–200,000 acfm) use flat plates instead of wires for the high-voltage electrodes. The flat plates (United McGill Corp. patents) increase the average electric field that can be used to collect the particles, and they provide an increased surface area for the collection of particles. Corona cannot be generated on flat plates by themselves; so corona-generating electrodes are placed ahead of and sometimes behind the flat-plate collecting zones. These electrodes may be sharp-pointed needles attached to the edges of the plates or independent corona wires. Unlike plate-wire or tubular ESPs, this design operates equally well with either negative or positive polarity. The manufacturer has chosen to use positive polarity to reduce ozone generation.

A flat-plate ESP operates with little or no corona current flowing through the collected dust, except directly under the corona needles or wires. This has two consequences. The first is that the unit is somewhat less susceptible to back corona than conventional units are because no back corona is generated in the collected dust, and particles charged with both polarities of ions have large collection surfaces available. The second consequence is that the lack of current in the collected layer causes an electrical force that tends to remove the layer from the collecting surface; this can lead to high rapping losses.

Flat-plate ESPs seem to have wide application for high-resistivity particles with small (1–2 μm) mass median diameters (MMDs). These applications especially emphasize the strengths of the design because the electrical dislodging forces are weaker for small particles than for large ones. Fly ash has been successfully collected with this type of ESP, but low-flow velocity appears to be critical for avoiding high rapping losses.

Tubular Precipitators

The original ESPs were tubular, like the smokestacks on which they were placed, with the high-voltage electrode running along the axis of the tube. Tubular precipitators have typical applications in sulfuric acid plants, coke oven by-product gas cleaning (tar removal), and iron and steel sinter plants. Such tubular units are still used for some applications, with many tubes operating in parallel to handle increased gas flows. The tubes may be formed as a circular, square, or hexagonal honeycomb with gas flowing upward or downward. The length of the tubes can be selected to fit conditions. A tubular ESP can be tightly sealed to prevent leaks of material, especially valuable or hazardous material.

A tubular ESP is essentially a one-stage unit and is unique in having all the gas pass through the electrode region. The high-voltage electrode operates at one voltage for the entire length of the tube, and the current varies along the length as the particles are removed from the system. No sneakage paths are around the collecting region, but corona nonuniformities may allow some particles to avoid charging for a considerable fraction of the tube length.

Tubular ESPs make up only a small portion of the ESP population and are most commonly applied where the particulate is either wet or sticky. These ESPs, usually cleaned with water, have reentrainment losses of a lower magnitude than do the dry particulate precipitators.

Wet Precipitators

Any of the precipitator configurations discussed above may be operated with wet walls instead of dry. The water flow may be applied intermittently or continuously to wash the collected particles into a sump for disposal. The advantage of the wet-wall precipitator is that it has no problems with rapping reentrainment or with back corona. The disadvantage is the increased complexity of the wash and the fact that the collected slurry must be handled more carefully than a dry product, adding to the expense of disposal.

Two-Stage Precipitators

The previously described precipitators are all parallel in nature; that is, the discharge and collecting electrodes are side by side. The two-stage precipitator invented by Penney is a series device with the discharge electrode, or ionizer, preceding the collector electrodes. For indoor applications, the unit is operated with positive polarity to limit ozone generation.

Advantages of this configuration include more time for particle charging, less propensity for back corona, and economical construction for small sizes. This type of precipitator is generally used for gas-flow volumes of 50,000 acfm and less and is applied to submicrometer sources emitting oil mists, smokes, fumes, or other sticky particulates because there is little electrical force to hold the collected particulates on the plates. Modules consisting of a mechanical prefilter, ionizer, collecting-plate cell, after-filter, and power pack may be placed in parallel or series–parallel arrangements. Preconditioning of gases is normally part of the system. Cleaning may be by water wash of modules removed from the system up to automatic, in-place detergent spraying of the collector followed by air-blow drying.

Two-stage precipitators are considered to be a separate and distinct type of device, as compared with large, high-gas-volume, single-stage ESPs. The smaller devices are usually sold as preengineered package systems.

AUXILIARY EQUIPMENT

Typical auxiliary equipment associated with an ESP system is shown schematically in Figure 3. Along with the ESP itself, a control system usually includes the following auxiliary equipment: a capture device (i.e., hood or direct exhaust connection); ductwork; dust removal equipment (screw conveyor, etc.); fans, motors, and starters; and a stack. In addition, spray coolers and mechanical collectors may be needed to precondition the gas before it reaches the ESP. Capture devices are usually hoods that exhaust pollutants into the ductwork or are direct exhaust couplings attached to a combustor or process equipment. These devices are usually refractory lined, water cooled, or simply fabricated from carbon steel, depending on the gas-stream temperatures.

Refractory or water-cooled capture devices are used where the wall temperatures exceed 800°F; carbon steel is used for lower temperatures. The ducting, like the capture device, should be water cooled, refractory, or stainless steel for hot processes and carbon steel for gas temperatures below approximately 1150°F (duct wall temperatures <800°F). The ducts should be sized for a gas velocity of approximately 4000 fpm for the average case to prevent particle deposition in the ducts. Large or dense particles might require higher velocities, but rarely would lower velocities be used. Spray chambers may be required for processes where the addition of moisture, or decreased temperature or gas volume, will improve precipitation or protect the ESP from warpage. For combustion processes with exhaust gas temperatures below approximately 700°F, cooling would not be required, and the exhaust gases can be delivered directly to the precipitator.

When much of the pollutant loading consists of relatively large particles, mechanical collectors, such as cyclones, may be used to reduce the load on the ESP, especially at high inlet concentrations. The fans provide the motive power for air movement and can be mounted before or after the ESP. A stack, normally used, vents the cleaned stream to the atmosphere. Screw conveyors or pneumatic systems are often used to remove captured dust from the bottom of the hoppers.

Wet ESPs require a source of wash water to be injected or sprayed near the top of the collector plates either continuously or at timed intervals. The water flows with the collected particles into a sump from which the fluid is pumped. A portion of the fluid may be recycled to reduce the total amount of water required. The remainder is pumped directly to a settling pond or passed through a dewatering stage, with subsequent disposal of the sludge.

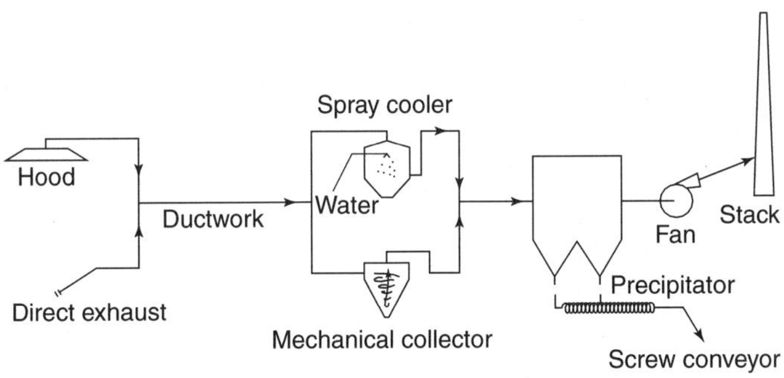

Figure 3. ESP Control System and Typical Auxiliary Equipment.

Gas conditioning equipment to improve ESP performance by changing dust resistivity is occasionally used as part of the original design, but more frequently it is used to upgrade existing ESPs. The equipment injects an agent into the gas stream ahead of the ESP. Usually, the agent mixes with the particles and alters their resistivity to promote higher migration velocity and thus higher collection efficiency. However, electrical properties of the gas may change, rather than dust resistivity. For instance, cooling the gas will allow more voltage to be applied before sparking occurs. Significant conditioning agents that are used include SO_3, H_2SO_4, sodium compounds, ammonia, and water, but the major conditioning agent by usage is SO_3. A typical dose rate for any of the gaseous agents is 10–30 ppm by volume.

The equipment required for conditioning depends on the agent being used. A typical SO_3 conditioner requires a supply of molten sulfur. It is stored in a heated vessel and supplied to a burner, where it is oxidized to SO_2. The SO_2 gas is passed over a catalyst for further oxidation to SO_3. The SO_3 gas is then injected into the flue gas stream through a multioutlet set of probes that breaches a duct. In place of a sulfur burner to provide SO_2, liquid SO_2 may be vaporized from a storage tank. Although more expensive in annualized cost, such systems have lower capital cost and are easier to operate.

Water or ammonia injection requires a set of spray nozzles in the duct, along with pumping and control equipment.

Sodium conditioning is often done by coating the coal on a conveyor with a powder compound or a water solution of the desired compound. A hopper or storage tank is often positioned over the conveyor for this purpose.

ELECTROSTATIC PRECIPITATION THEORY

The theory of ESP operation requires many scientific disciplines to describe it thoroughly. The ESP is basically an electric machine. The principal actions are the charging of particles and forcing them to the collector plates. The amount of charged particulate matter affects the electrical operating point of the ESP. The transport of the particles is affected by the level of turbulence in the gas. The losses mentioned earlier, sneakage and rapping reentrainment, are major influences on the total performance of the system. The particle properties also have a major effect on the operation of the unit.

The following subsections will explain the theory behind (1) electrical operating points in the ESP, (2) particle charging, (3) particle collection, and (4) sneakage and rapping reentrainment. General references for these topics are White[1] or Lawless and Sparks.[2]

Electrical Operating Point

The electrical operating point of an ESP section is the value of voltage and current at which the section operates. As will become apparent, the best collection occurs when the highest electric field is present, which roughly corresponds to the highest voltage on the electrodes.

In this work, the term *section* represents one set of plates and electrodes having a common power source. In the direction of flow, this unit is commonly called a *field*, and a section or "bus section" represents a subdivision of a field perpendicular to the direction of flow. In an ESP model, and in sizing applications, the two terms *section* and *field* are used equivalently because the subdivision into bus sections should have no effect on the model. This terminology has probably arisen because of the frequent use of the word *field* to refer to the electric field.

The lowest acceptable voltage is the voltage required for the formation of a corona, the electrical discharge that produces ions for charging particles. The (negative) corona is produced when an occasional free electron near the high-voltage electrode, produced by a cosmic ray, gains enough energy from the electric field to ionize the gas and produce more free electrons. The electric field for which this process is self-sustained has been determined experimentally. For round wires, the field at the surface of the wire is given by:

$$E_c = 3.126 \times 10^6 \left[d_r + 0.0301 \frac{d_r}{r_w} \right] \tag{1}$$

where E_c is the corona onset field *at the wire surface* (V/m); d_r, the relative gas density, referred to 1 atm pressure and 20°C; and r_w, the radius of the wire (m).

This is the field required to produce "glow" corona, the form usually seen in the laboratory on smooth, clean wires. The glow appears as a uniform, rapidly moving diffuse light around the electrode. After a period of operation, the movement concentrates into small spots on the wire surface, and the corona assumes a tuftlike appearance. The field required to produce "tuft" corona, the form found in full-scale ESPs, is 0.6 times the value of E_c.

The voltage that must be applied to the wire to obtain this value of field, V_c, is found by integrating the electric field from the wire to the plate. The field follows a simple $1/r$ dependence in cylindrical geometry. This leads to a logarithmic dependence of voltage on electrode dimensions. In the plate-wire geometry, the field dependence is somewhat more complex, but the voltage still shows the logarithmic dependence. V_c is given by:

$$V_c = E_c r_w \ln \frac{d}{r_w} \tag{2}$$

where V_c is the corona onset voltage (V); d, the outer cylinder radius in a tubular ESP (m); or $d = 4/\pi \times$ wire–plate separation for plate–wire ESP (m).

No current will flow until the voltage reaches this value, but the amount of current will increase steeply for voltages above this value. The maximum current density (A/m^2) on the plate or cylinder directly under the wire is given by:

$$j = \mu \varepsilon V^2 / L^3 \tag{3}$$

where j is the maximum current density (A/m^2); μ, the ion mobility (m^2/V s); E, the free-space permittivity, 8.845×10^{-12} F/m; V, the applied voltage (V); and L, the shortest distance from wire to collecting surface (m).

For tuft corona, the current density is zero until the corona onset voltage is reached, when it jumps almost to this value of j within a few hundred volts.

The region near the wire is strongly influenced by the presence of ions there, and the corona onset voltage magnitude shows strong spatial variations. Outside the corona region, it is quite uniform.

The electric field is strongest along the line from wire to plate and is approximated very well, except near the wire, by:

$$E_{max} = V/L \tag{4}$$

where E_{max} is the maximum field strength (V/m). When the electric field throughout the gap between the wire and the plate becomes strong enough, a spark will occur, and the voltage cannot be increased without severe sparking occurring. The field at which sparking occurs is not sharply defined, but a reasonable value is given by:

$$E_s = 6.3 \times 10^5 \left[\frac{273}{T} P \right]^{1.65} \tag{5}$$

where E_s is the sparking field strength (V/m); T, the absolute temperature (K); and P, the gas pressure (atm). This field would be reached at a voltage of, for example, 35,000 V for a wire–plate spacing of 11.4 cm (4.5 in.) at a temperature of 149°C (300°F). The ESP will generally operate near this voltage in the absence of back corona. E_{max} will be equal to or less than E_s.

Instead of sparking, back corona may occur if the electric field in the dust layer, resulting from the current flow in the layer, reaches a critical value of about 1×10^6 V/m. Depending on conditions, the back corona may enhance sparking or may generate so much current that the voltage cannot be raised any higher. The field in the layer is given by:

$$E_1 = j\rho \tag{6}$$

where E_1 is the electric field in dust layer (V/m); and p, the resistivity of the collected material (ohm m).

Particle Charging

Charging of particles takes place when ions bombard the surface of a particle. Once an ion is close to the particle, it is tightly bound because of the image charge within the particle. The "image charge" is a representation of the charge distortion that occurs when a real charge approaches a conducting surface. The distortion is equivalent to a charge of opposite magnitude to the real charge, located as far below the surface as the real charge is above it. The motion of the fictitious charge is similar to the motion of an image in a mirror, hence the name. As more ions accumulate on a particle, the total charge tends to prevent further ionic bombardment.

There are two principal charging mechanisms: diffusion charging and field charging. Diffusion charging results from the thermal kinetic energy of the ions overcoming the repulsion of the ions already on the particle. Field charging occurs when ions follow electric field lines until they terminate on a particle. In general, both mechanisms are operative for all sizes of particles. Field charging, however, adds a larger percentage of charge on particles greater than about 2 μm in diameter, and diffusion charging adds a greater percentage on particles smaller than about 0.5 μm.

Diffusion charging, as derived by White,[1] produces a logarithmically increasing level of charge on particles, given by:

$$q(t) = rkT/e \ln(1 + \tau) \tag{7}$$

where $q(t)$ is the particle charge (C) as function of time (sec); r, the particle radius (m); k, Boltzmann's constant (j/k); T, the absolute temperature (K); e, the electron charge, 1.67×10^{-19} C; and τ, dimensionless time given by:

$$\tau = \pi r v N e^2 \theta / kT \quad (8)$$

where v is the mean thermal speed of the ions (m/sec); N, the ion number concentration near the particle (No./m^3); and θ, the real time (exposure time in the charging zone; sec).

Diffusion charging never reaches a limit, but it becomes very slow after about three dimensionless time units. For fixed exposure times, the charge on a particle is proportional to its radius.

Field charging also exhibits a characteristic time dependence, given by:

$$q(t) = q_s \theta / (\theta + \tau') \quad (9)$$

where q_s is the saturation charge (charge at infinite time; C); θ, real time (sec); τ', another dimensionless time unit.

The saturation charge is given by:

$$q_s = 12\pi\varepsilon r^2 E \quad (10)$$

where ε is the free-space permittivity (F/m); and E, the external electric field applied to the particle (V/m).

The saturation charge is proportional to the square of the radius, which explains why field charging is the dominant mechanism for larger particles. The field charging time constant is given by:

$$\tau' = 4\varepsilon / Ne\mu \quad (11)$$

where μ is the ion mobility (all other terms as defined previously).

Strictly speaking, both diffusion- and field-charging mechanisms operate at the same time on all particles, and neither mechanism is sufficient to explain the charges measured on the particles. It has been found empirically that a very good approximation to the measured charge is given by the sum of the charges predicted by Eqs. 7 and 9 independently of one another:

$$q_{\text{tot}}(t) = q_{\text{d}}(t) + q_{\text{f}}(t) \quad (12)$$

where q_{tot} is the particle charge due to both mechanisms; q_{d}, the particle charge due to diffusion charging; and q_{f}, the particle charge due to field charging.

Particle Collection

The electric field in the collecting zone produces a force on a particle proportional to the magnitude of the field and to the charge:

$$F_e = qE \quad (13)$$

where F_e is the force due to electric field (N); q, the charge on particle (C); and E, the electric field (V/m).

Because the field-charging mechanism gives an ultimate charge proportional to the electric field, the force on large particles is proportional to the square of the field, which shows the advantage of maintaining as high a field as possible.

The motion of the particles under the influence of the electric field is opposed by the viscous drag of the gas. By equating the electric force and the drag force component due to the electric field (according to Stokes' law), we can obtain the particle velocity:

$$v(q, E, r) = q(E, r)EC(r)/6\pi\eta r \quad (14)$$

where v is the particle velocity (m/sec); q, the particle charge (C); C, the Cunningham correction to Stokes' law (dimensionless); and η, the gas viscosity (kg/m sec). The particle velocity is the rate at which the particle moves along the electric field lines, that is, toward the walls.

For a given electric field, this velocity is usually at a minimum for particles of about 0.5 μm diameter. Smaller particles move faster because the charge does not decrease very much, but the Cunningham factor increases rapidly as radius decreases. Larger particles have a charge increasing as r^2 and a viscous drag only increasing as r^1; the velocity then increases as r.

Equation 14 gives the particle velocity with respect to still air. In the ESP, the flow is usually very turbulent, with instantaneous gas velocities of the same magnitude as the particles' velocities, but in random directions. The motion of particles toward the collecting plates is therefore a statistical process, with an average component imparted by the electric field and a fluctuating component from the gas turbulence.

This statistical motion leads to an exponential collection equation, given by:

$$N(r) = N_0(r) \exp[-v(r)/v_0] \quad (15)$$

where N is the particle concentration of size r at the exit of the collecting zone (No./m^3); N_0, the particle concentration at the entrance of the zone (No./m^3); v, the size-dependent particle velocity (m/sec); and v_0, the characteristic velocity of the ESP (m/sec), given by:

$$v_0 = Q/A = 1/\text{SCA} \quad (16)$$

where Q is the volume flow rate of the gas (m^3/sec); A, the plate area for the ESP collecting zone (m^2); and SCA, the specific collection area (area per unit volume; sec/m).

When this collection equation is averaged over all the particle sizes and weighted according to the concentration of each size, the Deutsch equation results, with the penetration (fraction of particles escaping) given by:

$$p = \exp(-w_e\,\text{SCA}) \quad (17)$$

where p is the penetration (fraction); and w_e, the effective migration velocity for the particle ensemble (m/sec). The efficiency is given by:

$$\text{eff} = 100(1 - p) \quad (18)$$

This is the number most often used to describe the performance of an ESP.

Sneakage and Rapping Reentrainment

Sneakage and rapping reentrainment are best considered on the basis of the sections within an ESP. Sneakage occurs when a part of the gas flow bypasses the collection zone of a section. Generally, the portion of gas that bypasses the zone is thoroughly mixed with the gas that passes through the zone before all the gas enters the next section. This mixing cannot always be assumed, and when sneakage paths exist around several sections, the performance of the whole ESP is seriously affected. To describe the effects of sneakage and rapping reentrainment mathematically, we first consider sneakage by itself and then consider the effects of rapping as an average over many rapping cycles.

On the assumption that the gas is well mixed between sections, the penetration for each section can be expressed as:

$$p_s = S_N + (1 - S_N)p_c(Q') \quad (19)$$

where p_s is the section's fractional penetration; S_N, the fraction of gas bypassing the section (sneakage); and $p_c(Q')$, the fraction of particles penetrating the collection zone, which is functionally dependent on Q', the gas volume flow in the collection zone, reduced by the sneakage (m^3/sec).

The penetration of the entire ESP is the product of the section penetrations. The sneakage sets a lower limit on the penetration of particles through the section.

To calculate the effects of rapping, we first calculate the amount of material captured on the plates of the section. The fraction of material that was caught is given by:

$$m/m_0 = 1 - p_s = 1 - S_N - (1 - S_N)p_c(Q') \quad (20)$$

where m/m_0 is the mass fraction collected from the gas stream.

This material accumulates until the plates are rapped, whereupon most of the material falls into the hopper for disposal, but a fraction of it is reentrained and leaves the section. Experimental measurements have been conducted on fly ash ESPs to evaluate the fraction reentrained, which averages about 12%.

The average penetration for a section, including sneakage and rapping reentrainments, is:

$$p_s = S_N + (1 - S_N)p_c(Q') + \mathrm{RR}\{(1 - S_N)[1 - p_c(Q')]\} \quad (21)$$

where *RR* is the fraction reentrained. This can be written in a more compact form as:

$$p_s = \mathrm{LF} + (1 - \mathrm{LF})p_c(Q') \quad (22)$$

by substituting LF (loss factor) for $S_N + \mathrm{RR} - S_N \times \mathrm{RR}$. These formulas can allow for variable amounts of sneakage and rapping reentrainment for each section, but there is no experimental evidence to suggest that it is necessary.

Fly ash precipitators analyzed in this way have an average S_N of 0.07 and an RR of 0.12. These values are the best available at this time, but some wet ESPs, which presumably have no rapping losses, have shown S_N values of 0.05 or less. These values offer a means for estimating the performance of ESPs whose actual characteristics are not known, but about which general statements can be made. For instance, wet ESPs would be expected to have RR = 0, as would ESPs collecting wet or sticky particles. Particulate materials with a much smaller MMD than fly ash would be expected to have a lower RR factor because they are held more tightly to the plates and each other. Sneakage factors are harder to account for; unless special efforts have been made in the design to control sneakage, the 0.07 value should be used.

ESP DESIGN PROCEDURE

Specific Collecting Area

Specific collecting area (SCA) is a parameter used to compare ESPs and roughly estimate their collection efficiency. The SCA is the total collector plate area divided by gas volume flow rate and has the units of sec/m or sec/ft. It is often expressed as $m^2/(m^3/sec)$ or ft^2/kacfm (thousand acfm). It is also one of the most important factors in determining the capital and several of the annual costs (for example, maintenance and dust disposal costs) of the ESP because it determines the size of the unit. Because of the various ways in which SCA can be expressed, Table 1 gives equivalent SCAs in the different units for what would be considered a small, medium, and large SCA.

The design procedure is based on the loss-factor approach of Lawless and Sparks[2] and considers a number of process parameters. It can be calculated by hand, but it is most conveniently used with a spreadsheet program. For many uses, tables of effective migration velocities can be used to obtain the SCA required for a given efficiency. In the following subsection, tables have been calculated using the design procedure for a number of different particle sources and for differing levels of efficiency. If a situation is encountered that is not covered in these tables, then the full procedure that appears at the end of the subsection should be used.

Migration Velocity. If the migration velocity is known, then Eq. 17 can be rearranged to give the SCA:

$$\mathrm{SCA} = -\ln(p)/w_e \quad (23)$$

A graphical solution to Eq. 23 is given in Figure 4. The migration velocities have been calculated for three main precipitator types: plate-wire, flat-plate, and wet-wall ESPs of the plate-wire type. The following three tables, keyed to design efficiency as an easily quantified variable, summarize the velocities under various conditions. In Table 2, the migration velocities are given for conditions of no back corona and severe back corona; temperatures appropriate for each process have been assumed.

In Table 3, the migration velocities calculated for a wet-wall ESP of the plate-wire type assume no back corona and no rapping reentrainment.

In Table 4, the flat-plate ESP migration velocities are given only for no back-corona conditions because they appear to be less affected by high-resistivity dusts than the plate-wire types.

It is generally expected from experience that the migration velocity will decrease with increasing efficiency. In Tables 2–4, however, the migration velocities show some fluctuations. This is so because the number of sections must be increased as the efficiency increases, and the changing sectionalization

Table 1. Small, Medium, and Large SCAs as Expressed by Various Units

Units[a]	Small	Medium	Large
ft^2/kacfm	100	400	900
sec/m	19.7	78.8	177
sec/ft	6	24	54

[a] ft^2/kacfm = (sec/m) × 5.079.

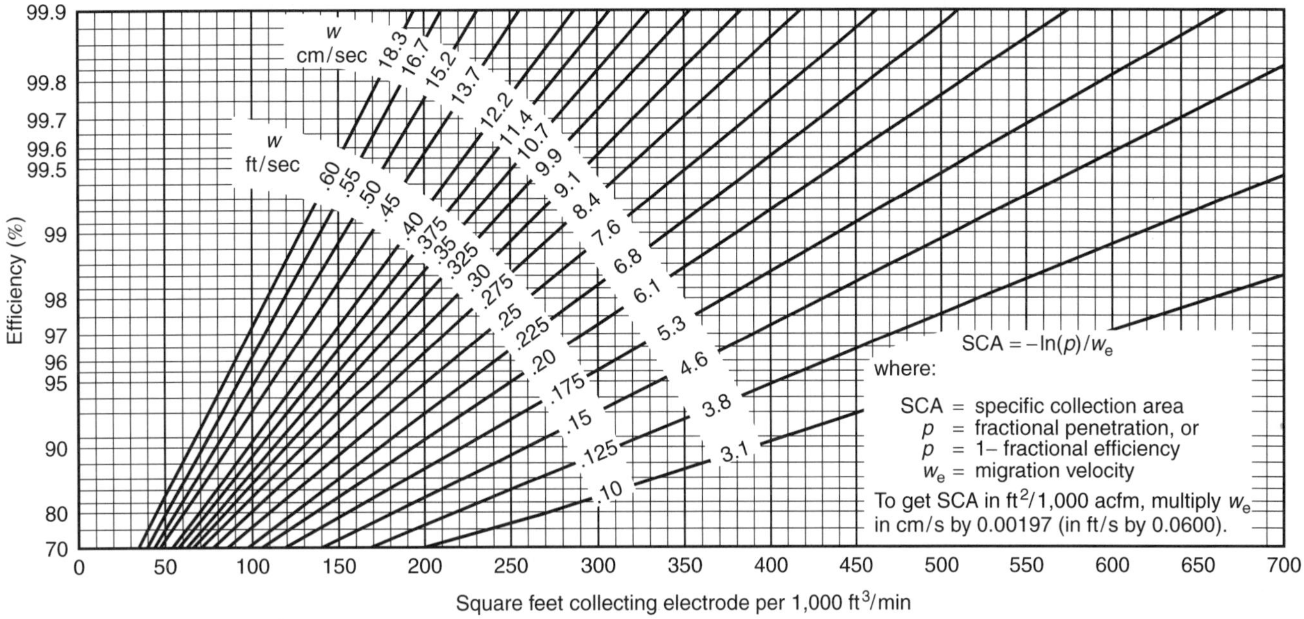

Figure 4. Chart for Finding Specific Collection Area (SCA).

Table 2. Plate-wire ESP Migration Velocities (cm/sec [fps])[a]

Particle Source	Design Efficiency (%) 95	99	99.5	99.9
Bituminous coal fly ash[b]				
(no BC)	12.6 (0.41)	10.1 (0.33)	9.3 (0.31)	8.2 (0.27)
(BC)	3.1 (0.10)	2.5 (0.082)	2.4 (0.079)	2.1 (0.069)
Subbituminous coal fly ash in tangential-fired boiler[b]				
(no BC)	17.0 (0.56)	11.8 (0.39)	10.3 (0.34)	8.8 (0.29)
(BC)	4.9 (0.16)	3.1 (0.10)	2.6 (0.085)	2.2 (0.072)
Other coal[b] (no BC)	9.7 (0.32)	7.9 (0.26)	7.9 (0.26)	7.2 (0.24)
(BC)	2.9 (0.095)	2.2 (0.072)	2.1 (0.069)	1.9 (0.062)
Cement kiln[c] (no BC)	1.5 (0.049)	1.5 (0.049)	1.8 (0.059)	1.8 (0.059)
(BC)	0.6 (0.020)	0.6 (0.020)	0.5 (0.016)	0.5 (0.016)
Glass plant[d] (no BC)	1.6 (0.052)	1.6 (0.052)	1.5 (0.049)	1.5 (0.049)
(BC)	0.5 (0.016)	0.5 (0.016)	0.5 (0.016)	0.5 (0.016)
Iron/steel sinter plant dust with mechanical precollector[b]				
(no BC)	6.8 (0.20)	6.2 (0.20)	6.6 (0.22)	6.3 (0.21)
(BC)	2.2 (0.072)	1.8 (0.059)	1.8 (0.059)	1.7 (0.056)
Kraft-paper-recovery boiler[b]				
(no BC)	2.6 (0.085)	2.5 (0.082)	3.1 (0.10)	2.9 (0.095)
Incinerator fly ash[e] (no BC)	15.3 (0.50)	11.4 (0.37)	10.6 (0.35)	9.4 (0.31)
(no BC)	6.2 (0.20)	4.2 (0.14)	3.7 (0.12)	2.9 (0.095)
Copper converter[g] (no BC)	5.5 (0.18)	4.4 (0.14)	4.1 (0.13)	3.6 (0.12)
Copper roaster[h] (no BC)	6.2 (0.20)	5.5 (0.18)	5.3 (0.17)	4.8 (0.16)
Coke plant combustion stack[c] (no BC)	1.2[j] (0.039)	—	—	—

[a] To convert cm/sec to fps, multiply cm/sec by 0.0328, but computational procedure uses SI units. To convert cm/sec to m/sec, multiply cm/sec by 0.01. Assumes same particle size as given in full computational procedure.

[b] At 300°F. Depending on individual furnace/boiler conditions, chemical nature of the fly ash, and availability of naturally occurring conditioning agents (e.g., moisture in the gas stream), migration velocities may vary considerably from these values. Likely values are in the range from back corona to no back corona. BC = back corona.

[c] At 600°F.

[d] At 500°F.

[e] At 250°F.

[f] 450 to 570°F.

[g] 500 to 700°F.

[h] 600 to 660°F.

[i] 360 to 450°F.

[j] Data available only for inlet concentrations in the range of 0.02 to 0.2 g/sec m^3 and for efficiencies less than 90%.

Table 3. Wet-Wall Plate-Wire ESP Migration Velocities (cm/sec [fps])[a]

Particle Source[b]	Design Efficiency (%) 95	99	99.5	99.9
Bituminous coal fly ash (no BC)	31.4 (1.03)	33.0 (1.08)	33.8 (1.11)	24.9 (0.82)
Subbituminous coal fly ash in tangential-fired boiler (no BC)	40.0 (1.31)	42.7 (1.40)	44.1 (1.45)	31.4 (1.03)
Other coal (no BC)	21.1 (0.69)	21.4 (0.70)	21.5 (0.71)	17.0 (0.56)
Cement kiln (no BC)	6.4 (0.21)	5.6 (0.18)	5.0 (0.16)	5.7 (0.19)
Glass plant (no BC)	4.6 (0.15)	4.5 (0.15)	4.3 (0.14)	3.8 (0.13)
Iron/steel sinter plant dust with mechanical precollector (no BC)	14.0 (0.46)	13.7 (0.45)	13.3 (0.44)	11.6 (0.38)

[a]To convert cm/sec to fps, multiply cm/sec by 0.0328, but computational procedure uses SI units. To convert cm/sec to m/sec, multiply cm/sec by 0.01. Assumes same particle size as given in full computational procedure. BC = back corona.
[b]All sources assumed at 200°F.

Table 4. Flat-Plate ESP Migration Velocities (cm/sec [fps])[a]

Particle Source[b]	Design Efficiency (%) 95	99	99.5	99.9
Bituminous coal fly ash[b] (no BC)	13.2 (0.43)	15.1 (0.50)	18.6 (0.61)	16.0 (0.53)
Subbituminous coal fly ash in tangential-fired boiler[b] (no BC)	28.6 (0.94)	18.2 (0.60)	21.2 (0.70)	17.7 (0.58)
Other coal[b] (no BC)	15.5 (0.51)	11.2 (0.37)	15.1 (0.50)	13.5 (0.44)
Cement kiln[c] (no BC)	2.4 (0.079)	2.3 (0.075)	3.2 (0.11)	3.1 (0.10)
Glass plant[d] (no BC)	1.8 (0.059)	1.9 (0.062)	2.6 (0.085)	2.6 (0.085)
Iron/steel sinter plant dust with mechanical precollector[b] (no BC)	13.4 (0.44)	12.1 (0.40)	13.1 (0.43)	12.4 (0.41)
Kraft-paper-recovery boiler[b] (no BC)	5.0 (0.16)	4.7 (0.15)	6.1 (0.20)	5.8 (0.19)
Incinerator fly ash[e] (no BC)	25.2 (0.83)	16.9 (0.55)	21.1 (0.69)	18.3 (0.60)

[a]To convert cm/sec to fps, multiply cm/sec by 0.0328, but computational procedure uses SI units. To convert cm/sec to m/sec, multiply cm/sec by 0.01. Assumes same particle size as given in full computational procedure. These values give the grounded collector plate SCA, from which the collector plate area is derived. In flat-plate ESPs, the discharge or high-voltage plate area is typically 40% of the ground-plate area. The flat-plate manufacturer usually counts all the plate area (collector plates plus discharge plates) in meeting an SCA specification, which means that the velocities tabulated above must be divided by 1.4 to be used on the manufacturer's basis. BC = back corona.
[b]At 300°F.
[c]At 600°F.
[d]At 500°F.
[e]At 250°F.

affects the overall migration velocity. This effect is particularly noticeable, for example, in Table 4 for glass plants. When the migration velocities in the tables are used to obtain SCAs for the different efficiencies in the tables, the SCAs will increase as the efficiency increases.

The full procedure for determining the SCA for large ESPs is given here. The smaller, two-stage precipitators are generally sold on the basis of a nominal ft³/min rating for packaged modules.

Procedure

1. Compute the design efficiency, $E(\%)$: $= 100 \times$ (1-outlet load/inlet load).
2. Compute design penetration, p: $= 1 - (E/100)$.
3. Compute or obtain the operating temperature: T_f, T_c, and/or T_k (°F, °C, K). T_k will be needed in the calculations.
4. Determine whether severe back corona is present. Severe back corona usually occurs for dust resistivities above 2×10^{11} or 3×10^{11} ohm cm. Its presence will greatly increase the size of the ESP required to achieve a certain efficiency.
5. Determine the MMD of the inlet particle distribution MMDi (μm). If this is not known, assume a value from the following table:

Source	MMDi (μm)
Bituminous coal	16
Subbituminous coal, tangential boiler	21
Subbituminous coal, other boiler types	10–15
Cement kiln	2–5
Glass plant	1
Wood burning boiler	5
Sinter plant,	50
with mechanical precollector	6
Kraft process recovery	2
Incinerators	15–30
Unknown	1
Copper reverberatory furnace	1
Copper converter	1
Coke plant combustion stack	1

6. Assume values for sneakage, S_N, and rapping reentrainment, RR, from the following table:

$S_N = 0.07$	Plate-wire ESPs
$S_N = 0.05$	Wet wall
$S_N = 0.10$	Flat plate
ESP type	RR
Coal fly ash, or not known	0.124
Wet wall	0.0
Flat plate with gas velocity >1.5 m/sec (not glass or cement)	0.15
Glass or cement	0.10

7. Assume values for the most penetrating size, MMDp (i.e., the MMD of the size distribution emerging from a very efficient collecting zone), and rapping puff size, MMDr (i.e., the MMD of the size distribution of rapped/reentrained material):

$$\text{MMDp} = 2\ \mu\text{m}$$

$$\text{MMDr} = 5\ \mu\text{m for ash with MMDi} > 5\ \mu\text{m, or}$$

$$\text{MMDr} = 3\ \mu\text{m for ash with MMDi} < 5\ \mu\text{m}$$

8. Use or compute the following factors for pure air:

$$\varepsilon_0 = 8.845 \times 10^{-12}, \text{ free space permittivity (F/m)}$$

$$\eta = 1.72 \times 10^{-0.5}(T_k/273)^{0.71}, \text{ gas viscosity (kg/m sec)}$$

$$E_{bd} = 630{,}000(273/T_k)^{1.65}, \text{ electric field at sparking (V/m)}$$

For plate-wire ESPs:

$$E_{avg} = E_{bd}/1.75, \text{ average field with no back corona}$$

or

$$E_{avg} = 0.7E_{bd}/1.75, \text{ average field with severe back corona}$$

For flat-plate ESPs:

$$E_{avg} = E_{bd} \times 5/6.3, \text{ average field, no back corona, positive polarity}$$

$$E_{avg} = 0.7E_{bd} \times 5/6.3, \text{ average field, severe back corona, positive polarity}$$

$$\text{LF} = \text{SN} + \text{RR} - S_N \times \text{RR}, \text{ loss factor (dimensionless)}$$

9. Assume the smallest number of sections for the ESP, n, such that $\text{LF}^n < p$. Suggested values of n are:

$E\%$	n
<96.5	2
<99	3
<99.8	4
<99.9	5
>99.9	6

These values are for an LF of 0.185, corresponding to a coal fly ash precipitator. The values are approximate, but the best results are for the smallest allowable n.

10. Compute the average section penetration, $p_s = p^{1/n}$.
11. Compute the section collection penetration, p_c: $= (p_s - \text{LF})/(1 - \text{LF})$. If the value of n is too small, then this value will be negative, and n must be increased.
12. Compute the particle size change factors, which are constants used for computing the change of particle size from section to section, D and MMDrp:

$$D = p_s = S_N + p_c(1 - S_N) + \text{RR}(1 - S_N)(1 - p_c)$$

$$\text{MMDrp} = \text{RR}(1 - S_N)(1 - p_c)\text{MMDr}/D$$

13. Compute a table of particle sizes for section 1 through n:

Section	MMDs
1	$\text{MMD1} = \text{MMDi}$
2	$\text{MMD2} = \{\text{MMD1} \times S_N + [(1 - p_c) \times \text{MMD}_p + p_c \times \text{MMD1}] \times p_c\}/D + \text{MMDrp}$
3	$\text{MMD3} = \{\text{MMD2} \times S_N + [(1 - p_c) \times \text{MMD}_p + p_c \times \text{MMD2}] \times p_c\}/D + \text{MMDrp}$
⋮	
n	$\text{MMDn} = \{\text{MMDn} - 1 \times S_N + [(1 - p_c)\text{MMD}_p + p_c \times \text{MMDn} - 1] \times p_c\}/D + \text{MMDrp}$

14. Calculate the SCA for sections 1 through n, using MMDn, η, ε, E_{avg}, and p_c:

$$SCA1 = -(\eta/\varepsilon) \times (1 - S_N) \times \ln(p_c)/(E_{avg}^2 \times \text{MMD1} \times 1 \times 10^{-6})$$

$$\vdots$$

$$SCAn = -(\eta/\varepsilon) \times (1 - S_N) \times \ln(p_c)/(E_{avg}^2 \times \text{MMDn} \times 1 \times 10^{-6})$$

where the factor 1×10^{-6} converts micrometers to meters. Note that the only quantity changing in the expressions is MMDx; therefore, the following relation can be used:

$$\text{SCA}n + 1 = \text{SCA}n \times \text{MMDn/MMDn} + 1$$

15. Calculate the total SCA and the English SCA:

$$\text{SCA} = \text{SCA1} + \text{SCA2} + \cdots + \text{SCA}n\text{(sec/m)}$$

$$\text{ESCA} = 5.079 \times \text{SCA}(\text{ft}^2/\text{kacfm})$$

This sizing procedure works best for p_c values less than the value of LF, which means the smallest value of n. Any ESP model is sensitive to the values of particle diameter and electric field. This one shows the same sensitivity, but the expressions for electric field are based on theoretical and experimental values. The SCA should not be strongly affected by the number

Table 5. Tubular ESP Migration Velocities (cm/sec [fps])[a]

Particle Source[b]	Design Efficiency (%)	
	90	95
Cement kiln (no BC)	2.2–5.4 (0.072–0.177)	2.1–5.1 (0.069–0.167)
(BC)	1.1–2.7 (0.036–0.089)	1.0–2.6 (0.033–0.085)
Glass plant (no BC)	1.4 (0.046)	1.3 (0.043)
(BC)	0.7 (0.023)	0.7 (0.023)
Kraft-paper-recovery boiler (no BC)	4.7 (0.154)	4.4 (0.144)
Incinerator 15 μm MMD, (no BC)	40.8 (1.34)	39 (1.28)
Wet, at 200°F MMD (μm)		
1	3.2 (0.105)	3.1 (0.102)
2	6.4 (0.210)	6.2 (0.203)
5	16.1 (0.528)	15.4 (0.505)
10	32.2 (1.06)	30.8 (1.01)
20	64.5 (2.11)	61.6 (2.02)

[a]BC = back corona. These rates were calculated on the basis of: $S_N = 0.015$, RR = 0, one section only. These are in agreement with operating tubular ESPs; extension of results to more than one section is not recommended.

of sections chosen; if more sections are used, the SCA of each section is reduced.

Specific Collecting Area for Tubular Precipitators

The procedure given above is suitable for large plate-wire or flat-plate ESPs, but must be used with restrictions for tubular ESPs. Values of $S_N = 0.015$ and RR = 0 are assumed, and only one section is used.

Table 5 gives migration velocities that can be used with Eq. 23 to calculate SCAs for several tubular ESP applications.

Flow Velocity

A precipitator collecting a dry particulate material runs a risk of nonrapping (continuous) reentrainment if the gas velocity becomes too large. This effect is independent of SCA and has been learned through experience. For fly ash applications, the maximum acceptable velocity is about 1.5 m/sec (5 fps) for plate-wire ESPs and about 1 m/sec (3 fps) for flat-plate ESPs. For low-resistivity applications, design velocities of 3 fps or less are common to avoid nonrapping reentrainment. The frontal area of the ESP must be chosen to keep gas velocity low and to accommodate electrical requirements (e.g., wire-to-plate spacing), while also ensuring that total plate area requirements are met. This area can be configured in a variety of ways. The plates can be short in height, long in the direction of flow, with several in parallel (making the width narrow). Or the plates can be tall in height, short in the direction of flow, with many in parallel (making the width large). After selecting a configuration, the gas velocity can be obtained by dividing the volume flow rate, Q, by the frontal area of the ESP:

$$v_{\text{gas}} = Q/WH$$

where v_{gas} = gas velocity (m/sec); W, the width of ESP entrance (m); and H, the height of ESP entrance (m).

When meeting the above restrictions, this value of velocity also ensures that turbulence is not strongly developed, thereby assisting in the capture of particles.

Pressure Drop Calculations

The pressure drop in an ESP is due to four main factors:

- Diffuser plate (usually present) (perforated plate at the inlet)
- Transitions at the ESP inlet and outlet
- Collection plate baffles (stiffeners) or corrugations
- Drag of the flat collection plate

The total pressure drop is the sum of the individual pressure drops, but any one of these sources may dominate all other contributions to the pressure drop. Usually, the pressure drop is not a design-driving factor, but it needs to be maintained at an acceptably low value. Table 6 gives typical pressure drops for the four factors. The ESP pressure drop, usually less than about 0.5 in. H_2O, is much lower than for the associated collection system and ductwork. With the conveying velocities used for dust collected in ESPs, generally 4000 fpm or greater, system pressure drops are usually in the range of 2–10 in. H_2O.

The four main factors contributing to pressure drop are described briefly below.

The diffuser plate is used to equalize the gas flow across the face of the ESP. It typically consists of a flat plate covered with round holes of 5–7 cm in diameter (2–2.5 in.) having an open area of 50–65% of the total.

Pressure drop is strongly dependent on the percent open area, but is almost independent of hole size.

The pressure drop due to gradual enlargement at the inlet is caused by the combined effects of flow separation and wall

Table 6. Components of ESP Pressure Drop

Component	Typical Pressure Drop (in. H_2O)	
	Low	High
Diffuser	0.010	0.09
Inlet transition	0.07	0.14
Outlet transition	0.007	0.015
Baffles	0.0006	0.123
Collection plates	0.0003	0.008
Total	0.09	0.38

friction and is dependent on the shape of the enlargement. At the ESP exit, the pressure drop caused by a short, well-streamlined gradual contraction is small.

Baffles are installed on collection plates to shield the collected dust from the gas flow and to provide a stiffening effect to keep the plates aligned parallel to one another. The pressure drop due to the baffles depends on the number of baffles, their protrusion into the gas stream with respect to electrode-to-plate distance, and the gas velocity in the ESP.

The pressure drop of the flat collection plates is due to friction of the gas dragging along the flat surfaces and is so small compared with other factors that it may usually be neglected in engineering problems.

Particle Characteristics

Several particle characteristics are important for particle collection. It is generally assumed that the particles are spherical or spherical enough to be described by some equivalent spherical diameter. Highly irregular or elongated particles may not behave in ways that can be easily described.

The first important characteristic is the mass of particles in the gas stream, that is, the particle loading. This quantity usually is determined by placing a filter in the gas stream, collecting a known volume of gas, and determining the weight gain of the filter. Because the ESP operates over a wide range of loadings as a constant-efficiency device, the inlet loading will determine the outlet loading directly. If the loading becomes too high, the operation of the ESP will be altered, usually for the worse.

The second characteristic is the size distribution of the particles, often expressed as the cumulative mass less than a given particle size. The size distribution determines how many particles of a given size there are, which is important because ESP efficiency varies with particle size. In practical terms, an ESP will collect all particles larger than 10 μm in diameter better than ones smaller than 10 μm. Only if most of the mass in the particles is concentrated above 10 μm would the actual size distribution above 10 μm be needed.

In lieu of cumulative mass distributions, the size distribution is often described by log-normal parameters. That is, the size distribution appears as a probabilistic normal curve if the logarithm of particle size used is the abscissa. The two parameters needed to describe a log-normal distribution are the mass median (or mean) diameter and the geometric standard deviation.

The MMD is the diameter for which one-half of the particulate mass consists of smaller particles and one-half is larger (see step 5 of the procedure). If the MMD of a distribution is larger than about 3 μm, the ESP will collect all particles larger than the MMD at least as well as a 3-μm particle, representing one-half the mass in the inlet size distribution.

The geometric standard deviation is the equivalent of the standard deviation of the normal distribution: It describes how broad the size distribution is. The geometric standard deviation is computed as the ratio of the diameter corresponding to 84% of the total cumulative mass to the MMD: It is always a number greater than 1. A distribution with particles all of the same size (monodisperse) has a geometric standard deviation of 1. Geometric standard deviations less than 2 represent rather narrow distributions. For combustion sources, the geometric standard deviations range from 3 to 5 and are commonly in the 3.5–4.5 range.

A geometric standard deviation of 4–5, coupled with an MMD of less than 5 μm, means that there is a substantial amount of submicrometer material. This situation may change the electrical conditions in an ESP by the phenomenon known as "space-charge quenching," which results in high operating voltages but low currents. It is a sign of inadequate charging and reduces the theoretical efficiency of the ESP. This condition must be evaluated carefully to be sure of adequate design margins.

Gas Characteristics

The gas characteristics most needed for ESP design are the gas volume flow and the gas temperature. The volume flow, multiplied by the design SCA, gives the total plate area required for the ESP. If the volume flow is known at one temperature, it may be estimated at other temperatures by applying the ideal gas law. Temperature and volume uncertainties will outweigh inaccuracies of applying the ideal gas law.

The temperature of the gas directly affects the gas viscosity, which increases with temperature. Gas viscosity is affected to a lesser degree by the gas composition, particularly the water vapor content. In lieu of viscosity values for a particular gas composition, the viscosity for air may be used. Viscosity enters the calculation of SCA directly, as seen in step 14 of the procedure.

The gas temperature and composition can have a strong effect on the resistivity of the collected particulate material. Specifically, moisture and acid-gas components may be chemisorbed on the particles in a sufficient amount to lower the intrinsic resistivity dramatically (orders of magnitude). For other types of materials, there is almost no effect. Although it is not possible to treat resistivity here, the designer should be aware of the potential sensitivity of the size of the ESP to resistivity and the factors influencing it.

The choice of power supplies' size (current capacity and voltage) to be used with a particular application may be influenced by the gas characteristics. Certain applications produce gas whose density may vary significantly from typical combustion sources (density variation may result from temperature, pressure, and composition). Gas density affects corona starting voltages and voltages at which sparking will occur.

Cleaning

Cleaning the collected materials from the plates often is accomplished intermittently or continuously by rapping the plates severely with automatic hammers or pistons, usually along their top edges, except in the case of wet ESPs that use water. Rapping dislodges the material, which then falls down the length of the plate until it lands in a dust hopper. The dust characteristics, rapping intensity, and rapping frequency determine how much of the material is reentrained and how much reaches the hopper permanently.

For wet ESPs, consideration must be given to handling wastewaters. For simple systems with innocuous dusts, water with particles collected by the ESP may be discharged from the ESP system to a solids-removing clarifier (either dedicated to the ESP or part of the plant wastewater treatment system) and then to final disposal. More complex systems may require skimming and sludge removal, clarification in dedicated equipment, pH adjustment, and/or treatment to remove dissolved solids. Spray water from the ESP preconditioner may be treated separately from the water used to flood the ESP collecting plates so that the cleaner of the two treated waters may be returned to the ESP. Recirculation of treated water to the ESP may approach 100%.

The hopper should be designed so that all the material in it slides to the very bottom, where it can be evacuated periodically, as the hopper becomes full. Dust is removed through a valve into a dust-handling system, such as a pneumatic conveyor. Hoppers often are supplied with auxiliary heat to prevent the formation of lumps or cakes and the subsequent blockage of the dust-handling system.

Construction Features

The use of the term *plate–wire geometry* may be somewhat misleading. It could refer to three different types of discharge electrodes: weighted wires hung from a support structure at the top of the ESP, wire frames in which wires are strung tautly in a rigid support frame, or rigid electrodes constructed from a single piece of fabricated metal. Using wire frames or rigid discharge electrodes in place of weighted wire discharge electrodes (particularly in coal-fired boiler applications) is stimulated by the user's desire for increased ESP reliability. The wire frames and rigid electrodes are less prone to failure by breakage and are readily cleaned by impulse-type cleaning equipment.

Other differences in construction result from the choice of gas passage (flow lane) width or discharge-electrode-to-collecting-electrode spacing. Typically, discharge-to-collecting-electrode spacing varies from 11 to 19 cm (4.3–7.5 in.). Having a large spacing between discharge and collecting electrodes allows higher electric fields to be used, which tends to improve dust collection. To generate larger electric fields, however, power supplies must produce higher operating voltages. Therefore, it is necessary to balance the cost savings achieved with larger electrode spacing against the higher cost of power supplies that produce higher operating voltages.

Most ESPs are constructed of mild steel. The ESP shells are constructed typically of 3/16- or 1/4-in. mild steel plate. Collecting electrodes are generally fabricated from lighter-gauge mild steel. A thickness of 18 gauge is common, but it will vary with the size and severity of the application.

Wire discharge electrodes come in various shapes, from round to square or barbed. A diameter of 2.5 mm (0.1 in.) is common for weighted wires, but other shapes used have much larger effective diameters, for example, 64-mm (0.25-in.) square electrodes.

Stainless steel may be used for corrosive applications, but it is uncommon except in wet ESPs. Stainless-steel discharge electrodes have been found to be prone to fatigue failure in dry ESPs with impact-type electrode cleaning systems.[3]

Precipitators used to collect sulfuric acid mist in sulfuric acid plants are constructed of steel, but the surfaces in contact with the acid mist are lead lined. Precipitators used on paper mill black-liquor-recovery boilers are steam jacketed. Of these two, recovery boilers have by far the larger number of ESP applications.

Example Problem

Assume an ESP is required for controlling fly ash emissions from a coal-fired boiler burning bituminous coal. The flue gas stream is 50 kacfm at 325°F and has an inlet ash loading of 4 g/ft^3. Analysis of the ash shows an MMD of 7 μm and a resistivity of less than 2×10^{11} ohm-cm. Assume that the ESP operates for 8640 hours per year (360 days) and that an efficiency of 99.9% is required.

Design SCA

The SCA can be calculated from Eq. 23. Assuming that a flat-plate ESP design is chosen, the fly ash migration velocity is 16.0 cm/sec (see Table 4). Then:

$$\text{SCA} - \ln(1 - 0.999)/16.0 = 0.432 \text{ sec/cm} = 43.2 \text{ sec/m}$$

Converting to English units (see "Procedure," step 15):

$$\text{ESCA} = 5.079 \times 43.2 = 219 \text{ ft}^2/\text{kacfm}$$

Total collector plate area is then:

$$219 \text{ ft}^2/\text{kacfm} \times 50 \text{ kacfm} = 10{,}950 \text{ ft}^2$$

To obtain a more rigorous answer, we can follow the steps of the "Procedure" previously given:

1. Design efficiency is required as 99.9.
2. Design penetration: $1 - (99.9/100) = 0.001$
3. Operating temperature in degrees kelvin:

$$(325°\text{F}-32°\text{F})5/9 + 273°\text{C} = 436 \text{ K}$$

4. Because dust resistivity is less than 2×10^{11} ohm-cm, no severe back corona is expected and back corona = 0.
5. The MMD of the fly ash is given as 7 μm.
6. Values for sneakage and rapping reentrainment (from the table presented in step 6, "Procedure," are as follows:

$$S_N = 0.10$$

$$\text{RR} = 0.124 \text{ (assuming gas velocity } <1.5 \text{ m/sec)}$$

7. The most penetrating particle size, from step 7 of the Procedure is:

$$\text{MMDp} = 2 \text{ μm}$$

The rapping puff size is:

$$\text{MMDr} = 5 \text{ μm}$$

8. From the Procedure:

$$\varepsilon 0 = 8.845 \times 10^{-12}$$

$$\eta = 1.72 \times 10^{-5}(436/273)^{0.71} = 2.40 \times 10^{-5}$$

$$E\text{bd} = 6.3 \times 10^5(273/436)^{1.65}$$

$$= 4.33 \times 10^5 \text{ V/m}$$

$$E\text{avg} = E\text{bd} \times 5/6.3 = 2.31 \times 10^5$$

$$LF = S_N + RR - S_N \times RR = 0.1 + 0.124 - 0.1 \times 0.124 = 0.212$$

9. Choose the number of sections for $LF^n < p, p = 0.001$. Try four sections:

$$LF^n = 0.212^4 = 0.002$$

This value is larger than p. Try five sections:

$$LF^n = 0.212^5 = 0.000428$$

This value is smaller than p and is acceptable.

10. Average section penetration is:

$$p_s = p^{1/n} = 0.001^{1/5} = 0.251$$

11. Section collection penetration is:

$$p_c = (p_s - LF)/(1 - LF) = (0.251 - 0.212)/(1 - 0.212) = 0.0495$$

12. Particle size change factors are:

$$D = p_s = S_N + p_c(1 - S_N) + RR(1 - S_N)(1 - p_c) = 0.10 + 0.0495(1 - 0.1) + 0.124(1 - 0.1) \times (1 - 0.0495) = 0.251$$

13. Particle sizes for each section are:

Section	MMD, μm
1	MMD1 = MMDi = 7
2	MMD2 = {MMD1 × S_N + [(1 − p_c) × MMDp + p_c × MMD1] × p_c}/D + MMDrp = {7 × 0.1 + [(1 − 0.0495) × 2 + 0.0495 × 7] × 0.0495}/0.251 + 2.11 = 5.34
3	MMD3 = {5.34 × 0.1 + [(1 − 0.0495) × 2 + 0.0495 × 5.34] × 0.0495}/0.251 + 2.11 = 4.67
4	MMD4 = {4.67 × 0.1 + [(1 − 0.0495) × 2 + 0.0495 × 4.67] × 0.0495}/0.251 + 2.11 = 4.39
5	MMD5 = {4.39 × 0.1 + [(1 − 0.0495) × 2 + 0.0495 × 4.39] × 0.0495}/0.251 + 2.11 = 4.28

14. SCAs for each section are:

Section	SCA, s/m
1	SCA1 = $-(\eta/\varepsilon 0) \times (1 - S_N) \times \ln(p_c)/(E\text{avg}^2 \times \text{MMD1} \times 1 \times 10^{-6})$ $= -(2.40 \times 10^{-5}/8.845 \times 10^{-12})(1 - 0.1) \ln 0.0495/[(3.44 \times 10^5)^2(7 \times 1 \times 10^{-6})]$ = 19.51
2	SCA2 = SCA1 × MMD1/MMD2 = 8.86(7/5.34) = 25.55
3	SCA3 = SCA2 × MMD2/MMD3 = 11.61(5.34/4.67) = 29.24
4	SCA4 = SCA3 × MMD3/MMD4 = 13.28(4.67/4.39) = 31.07
5	SCA5 = SCA4 × MMD4/MMD5 = 14.13(4.39/4.28) = 31.89

15. Total SCA = 19.51 + 25.55 + 29.24 + 31.07 + 31.89 = 137.26 s/m

English SCA = 5.079 × 137.26 = 697.15 ft^2/kacfm

Note that the more rigorous procedure calls for an SCA that is considerably higher than the value found by using Eq. 23. This discrepancy is caused by the considerably smaller particle size used in the example problem than is assumed for Table 4. In this case, the shorter method would lead to an unacceptably low cost estimate.

Total collector plate area is:

$$697.15\ \text{ft}^2/\text{kcafm} \times 50\ \text{kacfm} = 34{,}857\ \text{ft}^2$$

Acknowledgments

The authors gratefully acknowledge C. G. Noll, United McGill Co., for extensive review and the following companies for contributing data to this article: Research-Cottrell; Joy Industrial Equipment Co., Western Precipitation Division; and Environmental Elements.

REFERENCES

1. H. J. White, *Industrial Electrostatic Precipitation*, Addison-Wesley, Reading, MA, 1963.
2. P. A. Lawless and L. E. Sparks, "A review of mathematical models for ESPs and comparison of their successes," *Proceedings of Second International Conference on Electrostatic Precipitation*, S. Masuda, Ed., APCA, 1984, pp. 513–22.
3. R. L. Bump, "Evolution and design of electrostatic precipitator discharge electrodes," presented at the 75th APCA Annual Meeting, New Orleans, LA, June 1982.

FABRIC FILTERS

John D. McKenna, Arthur B. Nunn III, and Dale A. Furlong

Fabric filters remove dust from a gas stream by passing the stream through a porous fabric. Dust particles form a more or less porous cake on the surface of the fabric. It is normally this cake that actually does the filtration.

The manner in which the dust is removed from the fabric is a crucial factor in the performance of the fabric filter system. If the dust cake is not adequately removed, the pressure drop across the system will increase to an excessive amount. If too much of the cake is removed, excessive dust leakage will occur while fresh cake develops. The selection of design parameters is crucial to the optimum performance of a fabric filter system.

Fabric filter systems are frequently referred to as baghouses, since the fabric is usually configured in cylindrical "bags." The two most common baghouse designs are the reverse-air and the pulse-jet types. These names describe the cleaning system used with the design.

Reverse-air baghouses operate by directing the dirty flue gas into the inside of the bags; therefore, the collection of dust is on the inside surface of the bags. The bags are cleaned periodically by reversing the flow of air, causing the previously collected dust cake to fall from the bags into a hopper below. Since this cleaning procedure is accomplished at relatively low gas velocity, the fabric is not exposed to violent movement, and so the reverse-air cleaning technique normally results in maximum bag life. In a variation of reverse-air baghouse design and the forerunner of the reverse-air baghouse (i.e., the shaker baghouse), the bags are shaken during the reverse-air cleaning interval.

Pulse-jet baghouses are designed with internal frame structures, called *cages*, to allow collection of the dust on the outside of the bags. The dust cake is periodically removed by a pulsed jet of compressed air into the bag causing a sudden bag expansion; dust is removed primarily by inertial forces when the bag reaches its maximum expansion. This bag cleaning technique is quite effective; however, the vigorousness of the technique, and frequently the bag-to-cage fit, tend to limit bag life and also tend to increase dust migration through the fabric, thus decreasing dust collection efficiency.

The selection of the fiber material and fabric construction is important to baghouse performance. The fiber material from which the fabric is made must have adequate strength characteristics at the maximum gas temperature expected and adequate chemical compatibility with both the gas and the collected dust. Felted fabric construction generally gives better removal of fine dust particles as compared with woven fabrics. However, not all fiber materials can be felted into a fabric of adequate strength, and hence most filtration fabrics are constructed, at least in part, of filaments and/or fibers that are first twisted into yarns, and then woven or knitted into a fabric.

The published literature for fabric filtration is quite extensive. The reader is referred to *Fabric Filter—Baghouses I*, available from ETS, Inc., for one listing of references and for a more extensive discussion of subjects briefly included here.[1]

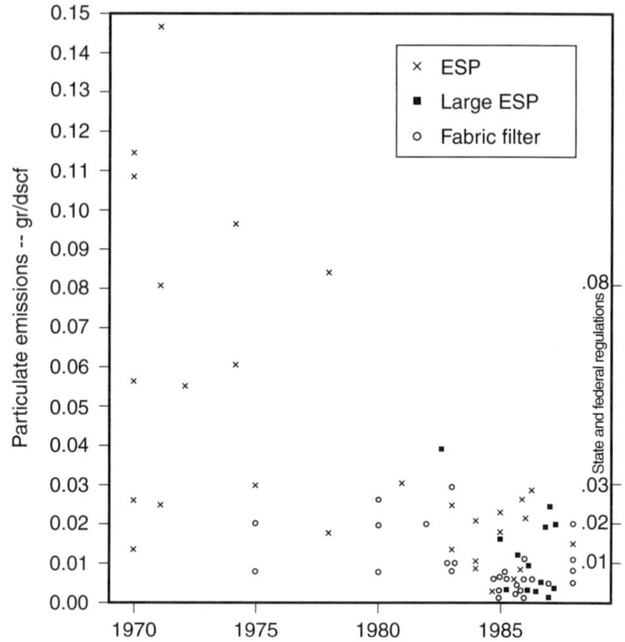

Figure 1. Particulate Emissions from MSW Incineration, 1970–1988.

CONTROL OF TOXIC AIR POLLUTANTS

Spray adsorption, or dry scrubbing, followed by a fabric filter is a rapidly maturing technology for simultaneously controlling particulate matter and acidic gases emitted from combustion processes. In addition, this combination is used to control emissions of other toxic air contaminants—heavy metal compounds and products of incomplete combustion such as chlorinated hydrocarbons (including dioxins), for example—that may be emitted from solid waste incinerators.

Dry scrubbing followed by a fabric filter is considered the best available technology for controlling emissions from solid waste incinerators for two basic reasons:

1. The dry scrubber both reacts with acidic gases to form solid particulate and causes other toxic vapors, including most heavy metals and chlorinated hydrocarbons, to condense. The condensed substances have been shown to collect preferentially on the very fine dust particles owing to the greater available surface area of the fine particulate.
2. The fabric filter is the best currently available technology for high-efficiency removal of fine particulate.

Baghouses are generally considered to be a superior choice, relative to electrostatic precipitators (ESPs), for fine-particulate control. A multifield precipitator with a very large plate area is required to provide comparable fine-particulate collection performance; therefore, an ESP of comparable performance is usually more expensive. Figure 1 illustrates the greatly improving trend in the control of particulate emissions by both ESPs and fabric filters since 1970. Figure 2 indicates that for typical two-field ESPs, the removal efficiency for fine, submicrometer particles is considerably less than that achieved in a fabric filter.

As experience with the use of baghouses has increased, their reliability also has increased as a result of the availability of different fibers/fabrics and improvements in the design of bag fabrics and in cleaning techniques. These measures have extended bag life to an average of five years, or more in some cases. Well-designed and operated baghouses have been shown to be capable of reducing overall particulate emissions to less than 0.010 g/dscf, and in a number of cases, to as low as 0.001–0.005 g/dscf. Based on the potential for greater removal efficiencies overall and in the submicrometer particle-size range, a number of states, including California, Connecticut, and Michigan, as well as Canada, prefer the use of scrubber/baghouses for solid-waste resource-recovery plants.

The concerns about air toxics control have been increased as a result of the more than 180 hazardous air pollutants targeted by the 1990 Clean Air Act Amendments. In the case of combustion sources, these tend to concentrate in the fine particle fraction. A review of the issue and existing performance data available for baghouses is provided by Dennis Helfritch.[2] A similar paper discussing the capabilities of both baghouses and ESPs relative to the control of fine particle toxics is provided by Kumar and Feldman[3] and concludes that both ESPs and fabric filters are capable of a high-level (>99%) trace element particulate control.

FILTRATION PROCESSES

Mechanisms

Several particle collection mechanisms are normally responsible for filter efficiency. Theoretical equations exist for the capture efficiency of each mechanism based on single particles approaching single fibers. For an operating fabric filter, however, the fabric is covered with a dust cake and the dust cake is of continually varying thickness. Once the dust cake has reformed after each cleaning, sieving is probably the dominant mechanism. As particles approach the porous mass of dust that constitutes the cake, they either will strike one or more surface particles or enter a pore. If the particle is larger than the pore it attempts to enter, it will be sieved out. If the particle is smaller than the pore it enters, it will continue traveling through the

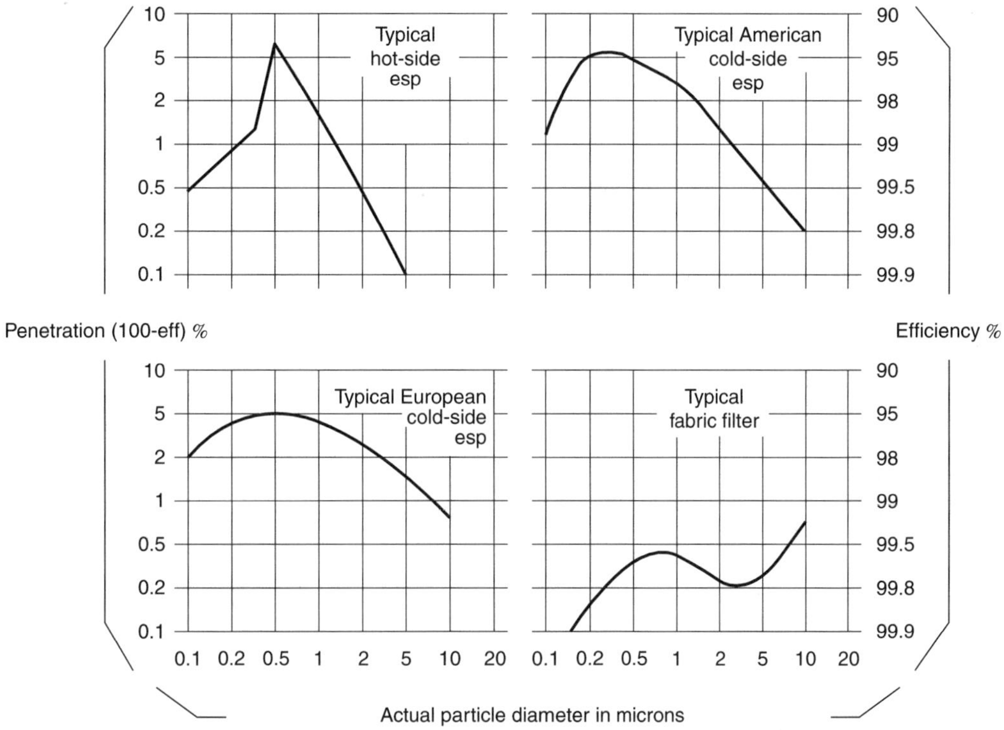

Figure 2. Typical Fractional Efficiencies for Existing Collectors (from *Electric Power Research Institute Economics of Fabric Filters vs. Precipitators*, Denver, CO June 1978).

pore until it touches the pore wall and adheres (rather than bounding or rolling along the wall); or until the pore narrows to dimensions smaller than the particle, causing the particle to be sieved out; or until the particle passes through the dust pore and a fabric pore and exits on the clean-air side of the air filter. Ordinarily, only one out of 1000, or an even 10,000, particles finds its way through the filter.

One might expect that larger particles would be sieved out with greater efficiency than smaller particles; that is, that the particles leaving the filter would have a smaller median diameter than the particles entering the filter. Experimentation has shown, however, that size distribution across the filter surface changes only slightly. The reason for the lack of change in size distribution stems from the manner in which most particles find their way through the filter. Most particles that transit the filter do so by a leakage process.[4]

The influence of electrically charged particles, a condition that is most common, on oppositely charged fibers of the filter has been shown to enhance attraction to an extent that particle-to-particle agglomeration is increased.

Consider a freshly cleaned bag: As filtration resumes, the cleaned areas of the fabric present pores of various sizes to the oncoming dust cloud. Individual particles strike the edges of pores or attach and begin to form chains or dendrites, a condition encouraged by the attraction produced as a result of opposing charge polarities. Before long, the smaller pores are bridged over by the chains and eventually become completely covered by porous cake. As time passes, more and more pores become covered with cake and the gas velocity through the remaining, uncovered pores becomes higher. Instead of a face velocity of a few feet per minute, we now have a pore velocity of up to several thousand feet per minute. Eventually, the velocity through the few remaining uncovered pores becomes so high and the pores so large that they cannot be bridged. For the remainder of the filtration cycle, these relatively free pores will be leakage points in the filter, and most of the particles passing through the filter will pass through these leakage points.

Billings and Wilder[5] report work by Tomaides suggesting that particles can bridge over a gap in a filter about 10 particle diameters wide. Presumably, the adhesive forces holding the chain together are exceeded by the aerodynamic forces trying to rupture the chain if its length exceeds 10 particle diameters.

Gas-to-Cloth Ratio

Gas-to-cloth ratio, G/C, is a measure of the amount of gas driven through each square foot of fabric in the baghouse, and is given in terms of the number of cubic feet of gas per minute passing through 1 square foot of cloth [$(ft^3/min)/ft^2$ or fpm]:

$$G/C = \frac{ft^3 \text{ per minute of gas}}{ft^2 \text{ of fabric}}$$

G/C can correctly be considered a superficial gas velocity. This is not the actual velocity through the openings in the fabric, but rather the apparent velocity of the gas approaching the cloth, although it is often referred to as "the gas velocity" and reported as feet per minute or meters per minute.

As the gas-to-cloth ratio increases, pressure drop (ΔP) also increases. Pressure drop is normally measured and reported as inches of water (in. H_2O, also written as inches water gauge, in. WC), the convention in the United States.

Pressure Drop

During the mid-1800s, Darcy formulated the following law for flow of fluid through a porous bed.

$$\Delta P = Lu_f V/K$$

where ΔP is the pressure difference across the bed; L, the bed thickness; u_f, the fluid viscosity; V, the superficial fluid velocity; and K, the bed permeability. This equation assumes that the fluid is essentially incompressible and steady, the fluid viscosity is Newtonian, and the velocity is low enough that only viscous effects occur. Over the past 100 years, investigators have been trying to find ways to predict K and to refine the Darcy equation.

The basic Darcy equation can be used to predict the pressure drop for an operating fabric filter with dust cake accumulating on the fabric.

$$\Delta P = S_E V + K_2 C_i V^2 t \tag{1}$$

where ΔP is the pressure drop (in. H_2O); S_E, the effective residual drag (in. H_2O/fpm); V, the velocity (fpm); K_2, the specific cake coefficient; C_i, the inlet dust concentrations (lb/cubic foot); and the filtration time (min).

Energy loss through a fabric filter is composed of two parts. The first, $S_E V$, represents drag or energy expended in pumping system gas through the cleaned equilibrium fabric of the fabric filter. The second part of the equation, $K_2 C_i V^2 t$, represents energy required to pump gas through the filter cake that builds up on the surface of the fabric. Gas velocity appears in both terms, but because it is squared for the cake portion of the equation, it is especially important for describing the energy consumed in pumping gas through the filter cake. Another important part of the equation is K_2, the specific cake coefficient. This term is characteristic of the dust, varies for different dusts, and is a measure of how rapidly pressure drop will build up in a system.

A fabric filter in stable, cyclic operation will normally reach a point of constant drag characteristics. That is, the resistance to gas flow of the freshly cleaned fabric is the same at the beginning of successive filtration cycles. In practice, the value may change as the fabric ages. Residual drag is a measured value. There is no useful predictive equation for residual drag.

Experimental Measurements of K_2

Many researchers have conducted laboratory and pilot-scale fabric filter tests to measure K_2. Billings and Wilder[5] reported an extensive field survey of K_2 as a function of the air–cloth ratio (filtration velocity) and particle size. In this early work, K_2 was determined from the reported values of operating air–cloth ratio (V), dust loading (C_i), filtration time (t), and residual and maximum pressure drops (ΔP_R, ΔP_m). While this earlier work was quantitative, the wide range of dusts, quality of reported data, and configuration of the individual systems (single compartment, multiple compartment, type of cleaning, etc.) led to considerable scatter. The relationship showed order-of-magnitude variations in K_2 at a given particle size.

In more recent tests, obtained under controlled conditions, the relationships among K_2, particle size, and velocity have been shown more clearly.

Data from Dennis et al.[6] and Davis and Kurzynske[7] are shown in Figures 3 and 4, respectively. The solid lines represent each researcher's best fit to the data, where available. The data reported by Dennis et al. were summarized from eight different sources for fly ash, mica, and talc at 2–6 fpm; the data by Davis and Kurzynske were on talc dusts at a velocity of 4 fpm. Both sets of data clearly indicate a strong dependence of K_2 on the particle size.

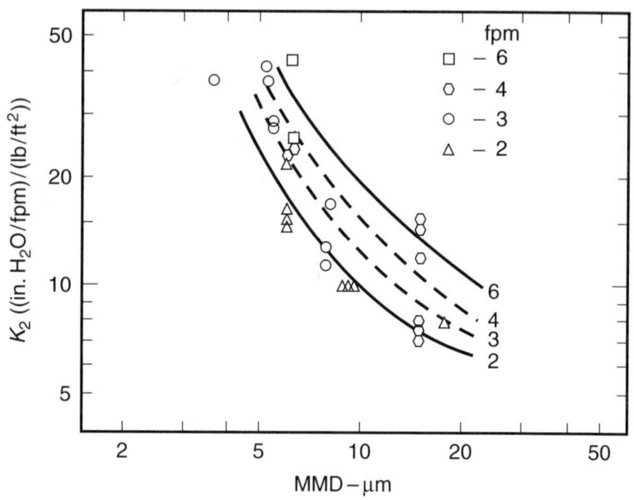

Figure 3. K_2 versus MMD and Face Velocity (Copyright ASTM. Reprinted with permission).

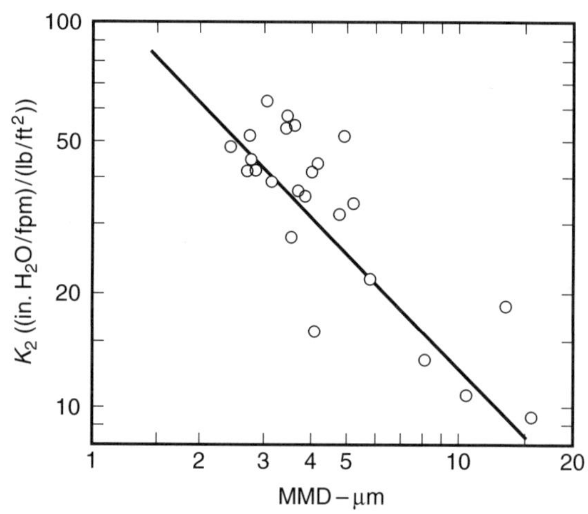

Figure 4. K_2 versus MMD for Talc at 4 fpm (Copyright ASTM. Reprinted with permission).

It is evident from these data that velocity also has an effect on K_2. While this observed effect may be partially attributed to the effect of velocity on dust cake packing and/or Reynolds number, most researchers have reported that K_2 is a function of velocity such that

$$K_2 = kV^x$$

Dennis et al.[6] reported that x had a value of 0.5 for fly ash and varied from 0.5 to 1.0. Davis and Frazier,[8] in a series of tests on fly ash using 11 different filter materials, reported an average value of 0.7 for fly ash. The data in Figures 3 and 4, data by Davis and Frazier,[8] and data by Frazier and Davis[9] were normalized to a velocity of 3 fpm and replotted in Figure 5 assuming an average value for x of 0.6. The normalized data show that there is a well-defined relationship between K_2 and the particle size.

A best-fit equation was determined for the data:

$$K_2 = 118.4 \text{ MMD}^{-1.10}$$

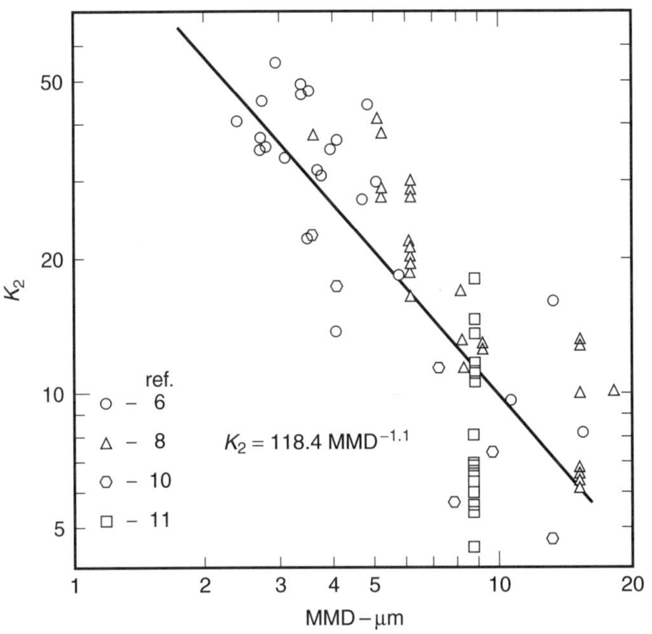

Figure 5. K_2 Normalized to 3 fpm versus MMD (Copyright ASTM. Reprinted with permission).

where K_2 is measured in the English system and MMD is in micrometers. The best-fit equation predicts the K_2 value within a factor of two. The agreement between various sets of data is excellent, considering that measurements obtained under carefully controlled laboratory conditions for a constant particle size distribution have shown a factor-of-two variation within a single laboratory.[8]

Pressure Drop in Multicompartment Baghouses

Equation 1 describes the instantaneous pressure drop after some definite period of filtration through an area of fabric. When the area is distributed across several compartments operated in parallel, the compartments are generally cleaned sequentially. The total pressure drop across the baghouse, in terms of drag, is analogous to a set of electrical resistances in parallel. Robinson et al.[10] used Hemeon's equation:

$$\frac{N}{S_T} = \frac{1}{S_1} + \frac{1}{S_2} + \cdots + \frac{1}{S_N} \tag{2}$$

where S_T is the total baghouse drag ($= \Delta P/V_T$); V_T, the total gas volume ($= V_1 + V_2 + \cdots V_N$); or

$$S_T = N\left(\sum_i \frac{1}{S_i}\right)^{-1}$$

and N is the number of compartments filtering. Subscripts refer to individual compartments.

If one knows the drag in each compartment at a given time, the instantaneous drag (and, therefore, pressure drop) could be easily calculated for the entire system. Equation 2 lends itself primarily to computer applications.

Efficiency

Efficiency is a measure of how well a filter separates dust from gas. For an operating baghouse, overall efficiency is calculated from:

$$\text{efficiency} = \frac{C_i - C_o}{C_i} = n$$

where C_o is the outlet concentration; and C_i, the inlet concentration. Penetration (1—efficiency) is also used as a measure of performance.

Efficiency and penetration may be measured or calculated for specific particle sizes or size ranges. Inertial impactors are used to measure efficiency by measuring particle concentration over several size ranges on the inlet and outlet streams of a baghouse.

There is no satisfactory set of published equations that allows a designer to calculate efficiency for a prospective baghouse.

FILTRATION FABRICS

Fiber Types

This section discusses fibers and fabric properties that will help the user select and specify fabrics for particular fabric filter applications. Fabrics made of natural fibers such as cotton or wool are still employed for many filter applications; the development of synthetic fibers, however, has greatly extended the possible range of applications for fabric filters. Continuing developments in fiber and fabric technology may be anticipated.

Synthetic fibers are widely used for filtration fabrics because of their low cost, better temperature- and chemical-resistance characteristics, and small fiber diameter. Synthetics used include acetates, acrylics, polyamides, polyesters, polyolefins, and polyvinyl chlorides. Specialty fibers for high-temperature use such as Teflon, Ryton, P84, and carbon fibers have been developed; however, the synthetic fiber most used for high-temperature applications is glass.

The properties of glass fiber, such as good acid resistance, good heat resistance, and high tensile strength, solve many of the problems inherent in baghouses.

Fiberglass has the following characteristics:

- It is noncombustible because it is completely inorganic.
- It has zero moisture absorption; therefore, it is not subject to hydrolysis. It has dimensional stability (low coefficient of linear expansion).
- It has very high tensile strength, but poor resistance to flex and abrasion. There are some chemical surface treatments (e.g., silicone, graphite, Teflon B) that improve the flex-abrasion characteristics of glass.
- It has good resistance to acids, but is attacked by hydrofluoric, concentrated sulfuric, and hot phosphoric acids.
- It has poor resistance to alkalis; hot solutions of weak alkalis attack glass.
- It has poor resistance to acid anhydrides and metallic oxides (e.g., fluorides and sulfur oxides). For this reason, glass baghouses should not be operated at or below the dew point.

Table 1 lists the major fiber alternatives for gas filtration and gives some of the important properties of these fibers.

Table 1. Fabric Selection Chart

Fabric	Maximum Temperature (°F)	Acid Resistance	Fluoride Resistance	Alkali Resistance	Flex Abrasion Resistance
Cotton	180	Poor	Poor	Good	Very Good
Polypropylene	200	Excellent	Poor	Excellent	Very good
Polyester	275	Good	Poor to fair	Good	Very good
Nomex	400	Poor to fair	Good	Excellent	Excellent
Teflon	450	Excellent	Poor to fair	Excellent	Fair
Fiberglass	500	Fair to good	Poor	Fair to good	Fair

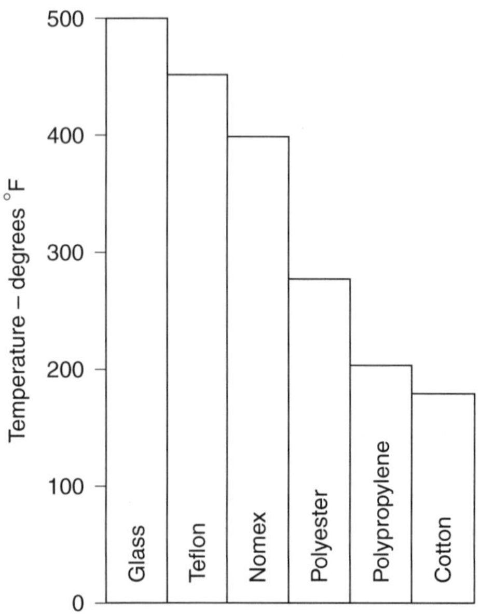

Figure 6. Recommended Maximum Operating Temperatures for Fabrics.

The first, and probably the key, inlet condition that must be identified prior to selecting a filtration medium is the temperature the fabric will experience. Figure 6 compares the recommended operating temperatures for the most-often-used filtration fabrics. Note that as temperature increases, the fabric choices become fewer and fewer. In 1991, the maximum temperature for which economical filter media were commercially available was about 500°F (260°C), although this may have changed with the advent of ceramic and metallic filters. Throughout the 1990s, the commercial development of high-temperature filter media, such as ceramics and metal filters, improved their economic competitiveness. Three papers that provide comprehensive information about high-temperature filtration media technology and costs are provided by Gregg,[11] Pethick,[12] and Jeffery.[13] When confronted with higher temperatures, the traditional approach has been to cool the gas down, at least to the vicinity of 500°F (260°C). Once a preliminary selection of filtering fabric has been made, the media supplier can usually provide additional information that should be considered in finalizing the fabric choice.

Important Fiber Characteristics

When selecting a fiber for gas filtration, attention must be paid to the following factors that interact and thus must be considered together.

1. Temperature. The fiber must have a maximum continuous service temperature higher than the normal temperature of the application. If temperature surges above the normal range occur, the ability of the fiber to withstand the expected conditions of surge temperature and duration must be considered.
2. Corrosiveness. The ability of the fiber to resist physical degradation from the expected application levels of acids, alkalis, solvents, or oxidizing agents must be considered.
3. Hydrolysis. Effects of the expected level of humidity must be taken into account.
4. Dimensional stability. If the fiber is expected to shrink or stretch in the application environment, the effects of such a change must be tolerable.
5. Cost. As with any engineering product, the least costly selection that will meet overall requirements is usually the best selection.

Media Selection

Baghouse operating costs are reduced if the baghouse has a high gas-to-cloth ratio, a low pressure drop, and a long life. In each case, the key to operation at minimum cost is the media selected for bag construction. This selection is crucial, but not easy, because many, often conflicting, requirements must be met.

The primary media selection criteria are the compatibility of the selected fiber with the gaseous environment and the physical configuration of the fiber and resulting fabric as it affects filtration performance. Usually, the selection criteria interact so much, or are not well enough understood, that the best selection is not apparent without long-term testing, except for bag-life determination, where experimental bag tests can provide specification data. Sometimes it is possible to equip complete baghouse compartments with various media[14] and observe the results for future selections. In 1985, ETS, Inc., introduced an individual bag flow monitor that is useful in making side-by-side comparisons of alternative media.[15]

The satisfactory performance of a fabric filter in a specific application requires the selection not only of a fiber material that is compatible with the gas–particle environment, but also that is of a fabric design appropriate to the dust collector geometry and collector cleaning requirements. *Fiber, yarn, and fabric parameters all influence the filtration process.*

Fabric strength, stability, and flexibility all are important parameters in determining the ability of the fabric to resist wear caused by abrasion. The term *abrasion* is defined as an eroding of fabric fibers or fiber surface material as a result of moving contact between the fiber and dust particles or between adjacent fibers. The flexibility of a filtration fabric is important for at least two reasons: cleanability and durability. Removal of the dust deposit may be improved by flexing of the fabric substrate, but such flexure may degrade the fabric. Thus fabric flexibility may be both necessary and harmful.

Permeability of the fabric must be considered when selecting a filter media. It must be understood, however, that the permeability of filtration fabrics is so reduced by a residual dust deposit that the permeability of clean fabric appears to have little relationship to permeability during use. The objective in fabric design is to maintain a highly permeable combination of residual dust and fabric while allowing a minimal amount of dust to pass through. To meet this objective, the pores through a fabric must be closely controlled. They must not exceed a certain bridging diameter, but if they are too small, they will either plug or pass too little gas. In an ideal filtration fabric, probably all the fabric pores should be of the same size, that size depending on properties of the dust and gas, fabric characteristics, and so on. Yarn texturizing and the postweave surface treatments contribute greatly to the permeability of clean fabric, and perhaps also to the permeability of residual dust deposit.

The last, but certainly not least, criterion for media selection is the ability of the fabric to *release* collected dust during the cleaning cycle. This ability depends largely on the mode and intensity of cleaning, but also on the adhesive character of the fabric. The way in which fabric construction relates to deposit release has not been completely determined, but it is known that a smooth fabric surface releases dust more readily than does a fuzzy surface. Dust may agglomerate on loose fibers and move away from the surface during cleaning, only to return once filtration resumes. This action, whose outcome is sometimes referred to as "dingleberries," can result in poor cleaning. Some fabric surface treatments are specifically intended to enhance cake release. Dust release also depends on the electrical resistances of selected fibers. Electrical resistance is known to depend on humidity, which itself has a marked effect on filtration fabric performance.[16] An extensive table of electrical properties for a variety of fabrics is given by Frederick.[17] "Electrical effects" is a factor that has rarely been quantified, but is known to influence pressure drop and efficiency. This is one of the factors that can cause major variations in baghouse performance.

In about 90% of the baghouses currently operating on coal-fired boilers, the bags are fabricated from glass fibers. Glass fibers are tiny filaments as small as 0.00015 in. (4 μm) in diameter that are extremely flexible and thus may be woven into fabric. Before 1930, discontinuous "glass wool" fibers were the only form produced commercially. The new technology was first used to develop continuous fibers for high-temperature insulation of fine electrical wires, designated "E"-glass because of its unique electrical insulation properties, and which was industrially practical because an effective lubricating finish protects the filaments. Today, E-glass is used in nearly all glass-fiber applications, ranging from printed circuit boards to boat hulls and filtration fabrics. Other formulations were also developed by altering percentages of the glass composition to obtain special performance characteristics. "C"-glass is particularly resistant to chemical attack; "S"-glass possesses outstanding strength characteristics. Thus far, production constraints and economics have limited glass filtration fabrics to E-glass.

Woven Fabric

Most filtration fabrics are either completely or partially made by weaving. Even felted or so-called "nonwoven" fabrics include a base (scrim) of woven fabric. Baghouses in which the gas flow is from the inside of the bags to the outside, such as reverse-air and shaker-cleaned baghouses, use woven fabrics almost exclusively. These baghouses generally operate at lower gas flow rates where the flow restriction of the fabric is not so significant. Woven fabrics usually have greater flow restriction but greater strength (for a given fabric weight) than do comparable nonwoven fabrics and thus are usually chosen for reverse-air and shaker applications.

Pulse-jet-cleaned baghouse designs offer increased cleaning energy and operate at higher gas flow rates. The tendency of woven fabrics to "bleed," resulting in low filtration efficiency when clean, usually restricts the use of woven fabrics for pulse-cleaned applications. Woven fabrics made with texturized yarn or with membrane films applied to the upstream surface offset the tendency to bleed when cleaned, and these types of fabrics are used in pulse-cleaned applications.

Woven fabric is manufactured by weaving together fibers that previously had been made into yarns. During weaving, longitudinal yarns (the warp) are interlaced at right angles with transverse yarns (the fill) by means of a loom.

Woven fabrics are formed by interlacing yarns at right angles on the loom, after which the raw (greige) fabric may be further treated. While there are many patterns of interlacing, the fabrics most commonly used in gaseous filtration are twill, satin, or plain weaves. These three weave patterns are depicted in Figure 7. Both the twill and the satin weaves have fabric sides where warp or fill yarns predominate, and those sides are referred to as warp face or filling face.

In its basic construction, satin weave is similar to twill, but generally uses a pattern of fill yarns going under one, then over 4 to 12 warp yarns. Satin differs in appearance from twill because the diagonal of satin weave is not visible; it is purposely interrupted in order to contribute to the flat, smooth, lustrous surface desired. There is no visible design on the face of the fabric because the yarns that are to be thrown to the surface are greater in number and finer in count than the yarns that form the reverse of the fabric.

Plain weave is sometimes referred to as tabby, homespun, or taffeta weave. It is the simplest type of construction

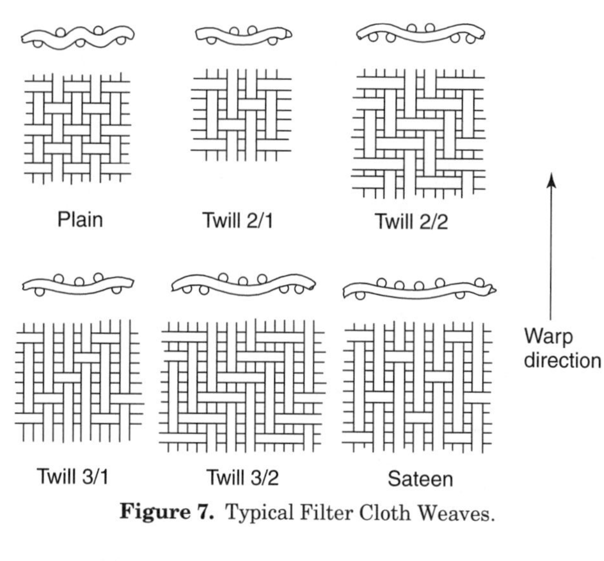

Figure 7. Typical Filter Cloth Weaves.

and consequently the least expensive. Each fill yarn goes alternately under and then over the warp yarns across the width of the fabric. On its return, the yarn alternates the pattern of interlacing. If the yarns are close together, plain weave has a high thread count, often a requirement for suitable efficiency.

Nonwoven Fabrics

Nonwoven fabrics generally are defined as sheet or web structures made by bonding and/or interlocking fibers, yarns, or filaments by mechanical, thermal, chemical, or solvent means.

The International Non-woven and Disposables Association (INDA), a trade association of the nonwoven fabrics industry, has given this definition:

> A non-woven is any fabric made directly from a web of fiber without the yarn preparation required for weaving or knitting. Natural or synthetic fibers can range in length from 0.5–15 cm from crimped staple products, up to continuous filaments for spun-bonded products. The fibers may be oriented in one direction or may be deposited randomly. This web is given structural integrity by (1) mechanical fiber intertangling, (2) thermal or chemically induced fusing of the fibers, or (3) application of any of several adhesives or resins.

Perhaps the most significant feature underlying all nonwoven fabric production, and the one that contributes most to its economic appeal, is the speed at which the fabric is produced. Many nonwoven production units in place today can produce fabric at speeds of 400 fpm (200 cm/sec) or more. Some advanced units operate at speeds over 1000 fpm (500 cm/sec). These rates may be compared with knitting machine speeds of about 5.0 fpm (2.5 cm/sec) and the even slower rates of weaving machines.

A needlepunched fabric is produced by introducing a fibrous web already formed by carding or air-laying into a machine equipped with groups of specially designed, barbed needles (see Figure 8). While the web is trapped between a bed plate and a stripper plate, the needles punch through it and reorient the fibers so that mechanical bonding is achieved among the individual fibers. The needlepunching process is generally used to produce fabrics that have high density and yet retain some bulk. Fabric weights usually range from 1.7 to 10 oz/yd^2 (58–340 g/m^2) and thicknesses from 15 to 160 mils (0.38–4.1 mm).

Needlepunching is often used to combine two or more layers of fiber into a feltlike fabric. Usually one layer is a woven fabric, called a *scrim*, for strength, while the other(s) may consist of fibers of almost any description or combination. Thus considerable control over separate properties of the finished material is possible. For example, the scrim may contribute to desired dimensional stability, while the top, or batting, layer contributes appropriate properties for filtration.

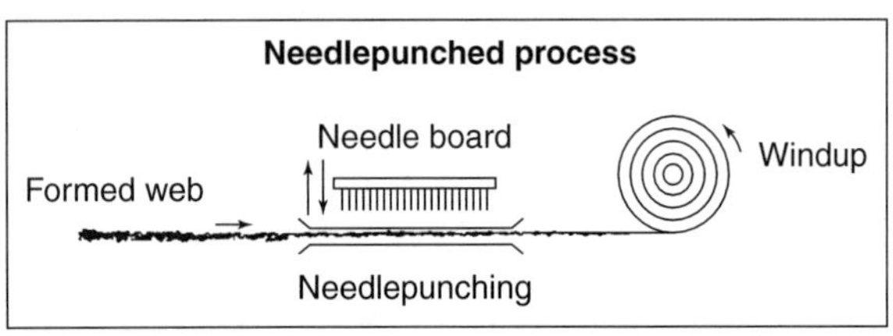

Figure 8. Needlepunched Process (from Ref. 12).

Needlepunching is accomplished by punching needles with forward barbs from the batting side into or through the scrim, and the batting fibers thus laced into the scrim remain behind when the needles are withdrawn. Variations on the needling process include changing the needle angle or number of repetitions or using two-sided needling. When a shrinkable scrim is used, the needled material may later be felted in various ways to produce a denser and more uniform material.

To improve the filtration characteristics, primarily the collection efficiency of the fabric surface, the surface may be napped. This is done by teasels (woody, thistlelike parts of weed plants), whose barbs pluck fibers from the surface of the felt. When enough nap has been raised in this manner, it may be singed or otherwise trimmed to the desired thickness.

Needlepunched fabrics, sometimes called needlefelts, are employed in many filtration applications. Modification of the basic structure produced by the needling process determines filtration performance. When punch density is increased during needling, one of the results is an increase in fabric density. Eventually, an optimum punch level is attained above which the density falls off. The maximum density possible from needlepunching alone would be insufficient to produce an efficient filtration medium. Additional processes are required to give the necessary densification and closing up of the structure. Heat treatments are normally employed at temperatures suited to the thermal characteristics of the particular fiber in use. Suitable processes include the squeezing of the fabric between rollers or a roller and a stationary bed. An increase in fabric density increases filtration efficiency, up to a point.

Successful fabrics are also required to provide good discharge of the dust cake. Good discharge is usually achieved by means of a surface modification in the form of singeing to remove individual fibers protruding from the surface. Good cake release surfaces can also be formed by heating the fabric surface to such an extent that fiber melting and fusion occur, although this will also increase the flow resistance of the fabric.

Relaxation of tensions built into the fabric structure by the needling operation can be brought about by passage through an oven. Such treatment is insurance against dimensional changes occurring when the fabric first encounters operating conditions.

Knitted fabrics, both plain and bulk, are being used (especially the latter) in high-ratio collectors.

Treatments and Finishes

"Finishing" includes those processes that improve the appearance or serviceability of the fabric after it leaves the weaving machine. Greige (unfinished) fabrics intended for use as filtration fabrics are treated and/or finished after weaving to improve their filtration characteristics and cleaning (release) characteristics. Fabric life and strength may also be affected. Treatments are defined as postweaving processes that affect the entire fabric, whereas finishes are postweaving processes that affect only the surface of the fabric.

One fabric treatment of significance for filtration fabrics is heat setting. In the process of heat setting, the fabric is exposed to temperatures exceeding those experienced in service. This treatment is done on a machine known as a *pin tenter*. The fabric is held under tension in both the warp and fill directions and passed through a heated oven.

Special finishes have been developed for the glass fabrics used in high-temperature filtration. For glass filter bags, organic materials, such as the starch binder that the yarn producer applies and the warp sizing (starch and mineral oil) applied to facilitate weaving, must be removed before a finish is applied to the fabric. This is necessary because these organic lubricants would not be stable at the process temperature, and also because they would interfere with the application of the desired finish. It is desirable that the applied finish have as close a contact as possible with the bare glass filaments, penetrating the yarn bundle and encapsulating individual filaments.

Organic sizings are removed thermally by a process called *coronizing*, which also sets the crimp in the glass yarns. Crimp is the bend in a yarn caused by interlacing during weaving. The yarn is heated to the softening point, and the crimp is set as the yarn cools. It is believed that setting the crimp in the yarn maximizes the flex life of the fabric.

The finish used for glass fabrics must be thermally stable at process temperatures (500–550°F or 260–290°C) and chemically resistant to the gas environments found in fiberglass filter bag applications. The basic purpose of the finish is to protect the glass fibers from abrading themselves, but it can also enhance dust-release characteristics.

In addition to providing lubricity to extend bag life, glass fabric finishes also help to promote dust release from the fabric and offer varying degrees of protection from chemical attack. The success of glass fabric as a viable filter medium depends to a large degree on the quality of the finish. Finish development has occurred in roughly three stages, with finishes making up primarily three groups:

I. Silicones. A glass-to-silicone coupler is the basic prefinish required before subsequent organic finishes are applied. These "couplers" ensure the complete individual fiber envelopment needed for effective protection.
II. Silicones and graphite with small amounts of fluorocarbons.
III. Fluorocarbon compounds.

All three groups are still in use, although Group I, the second-stage silicone finish, has largely faded from the scene. Group II finishes are divided into tricomponent and acid-resistant groups.[18]

Tricomponent
- Graphite (natural or synthetic) Teflon
- Silicones
- Agents to assist application

Acid-resistant
- Graphite (natural or synthetic)
- Polymers
- Binders
- Silicone
- Teflon
- Agents to assist application

Teflon is used in both Group II and III finishes. Group III finishes consist of Teflon, binders, and agents to assist application.

Most of the experience to date with Group III finishes has been at 10% loading; for example, 10% of the finished fabric weight is Teflon B. Amounts of finish vary greatly, but, in general, the Teflon finish is 7–10% weight added per weight of cloth, the acid-resistant finish is 4–6%, and the tricomponent finish is 1–2%.[19] Some experience indicates that Group II finishes are more resistant to chemical attack, while other experience favors the use of Group III finishes. There has been progress in using Group II finishes to protect Nomex fabrics from acid exposure in coal-fired boiler applications.

A special surface treatment gaining increased use in gaseous filtration is the application of a Gore-Tex™ membrane to the fabric surface.[19] The Gore-Tex membrane is expanded polytetrafluoroethylene (PTFE) deposited as a thin, fibrillated film. Figure 9 shows a photomicrograph of the fibrillated film of

Figure 9. Gore-Tex® Membrane Compared with Polyester Felt.

PTFE on the fabric surface. A cross section through the fabric is shown in Figure 10. The coarse woven fibers are seen on the right, and a thin PTFE film covers the left side. Gore-Tex membranes have been applied to many available backing materials, including woven polyester, Nomex, glass, and Teflon. Usually, the backing fabric is quite porous. Gore-Tex filter bags are constructed for pulse-jet, reverse-air, and shaker collectors. Gore-Tex membrane filter cartridges are also available.

The measured properties of fabric finished with a Gore-Tex membrane show the tensile and burst strengths characteristic of the woven fabric, but permeability is lower. Although lower permeability might suggest higher-than-desirable pressure drops and drags, that often is not the case, because the membrane improves cleanability and reduces residual dust buildup in the fabric.

In addition to Gore, a number of other suppliers such as Tetratec, Feasterville, PA,[20] and BHA Group, Inc., Kansas City, MO, also provide PTFE membranes.

CLEANING TECHNIQUES

The primary way to categorize a fabric filter is by the method used to clean it. In general, accumulated dust is separated from the fabric by some combination of the following effects.

1. Deflection of the fabric/dust cake, tending to fracture the cake and separate it from the fabric.
2. Acceleration of the fabric/dust cake, yielding separation forces.
3. Gas flow in the reverse direction, yielding aerodynamic forces that separate the dust from the fabric and subsequently move the dust toward the collecting hopper.

Four cleaning methods have evolved, each of which generates some combination of these fabric–dust-cake separation effects. The majority of baghouses currently in use employ one or more of these cleaning methods.

1. Shaker-cleaned baghouses. In shake cleaning the tops of the bags are shaken, preferably horizontally, resulting in deflections and acceleration forces throughout the bag. Zero or reverse flow is normally combined with the shaking.
2. Reverse-air-cleaned baghouses. In reverse-air cleaning, a combination of bag deflection (inward collapse) and reverse flow is used to remove dust from the fabric. This process, which results in very low stresses on the fabric, was developed specifically for easily damaged fabrics such as fiberglass.
3. Pulse-jet-cleaned baghouses. Pulse-cleaned baghouses use outside-in flow, where the fabric collapses against a wire cage during filtration. During cleaning, a pulse of high-pressure air is directed into the bag (the reverse-flow direction), inflating the bag and causing fabric/cake

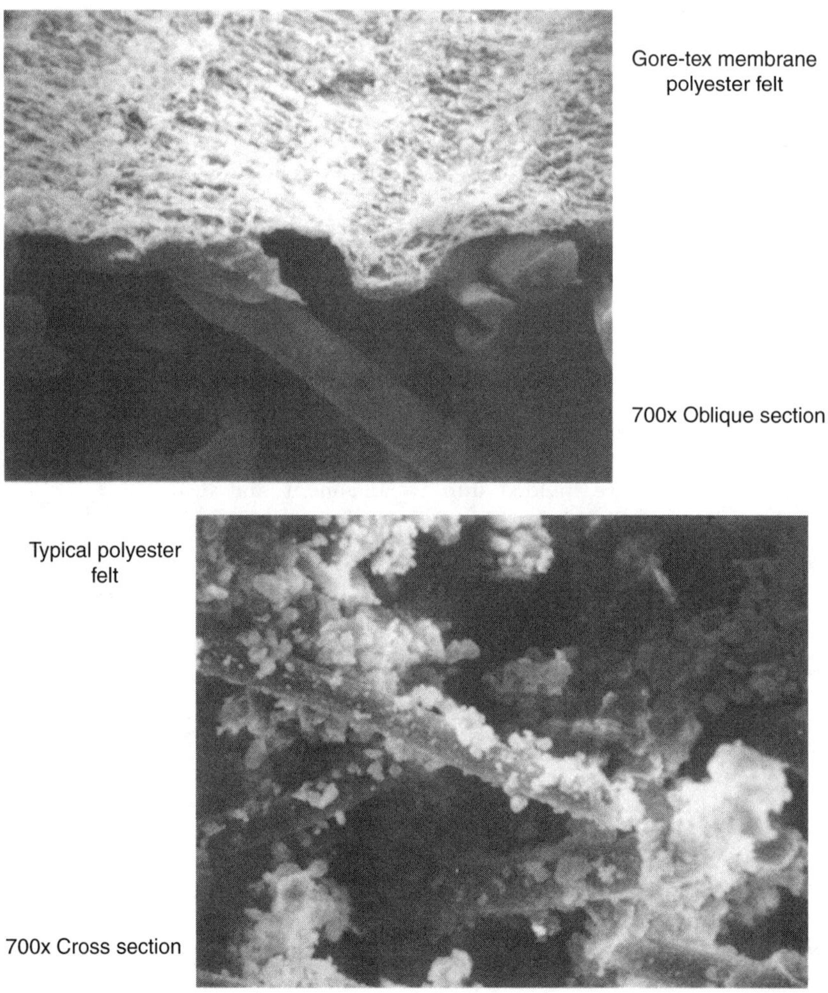

Figure 10. Cross Sections of a Gore-Tex® Membrane Filter Bag and a Typical Polyester Felt Filter Bag in Alumina Dust Collection.

deflection and high inertial forces that separate the dust from the bag. Although reverse airflow is involved, it is thought to have a minor effect on cleaning.
4. Sonic cleaning. Sonic cleaning, if used, usually augments another cleaning method. Sonic energy is normally introduced into the baghouse by air-powered horns. Although the process is not well understood, the sonic air shock waves apparently generate acceleration forces that tend to separate the dust from the fabric.

The significant parameters of shaker, reverse-air, and pulse-jet cleaning are given in Tables 2–4.

Table 2. Shake Cleaning—Parameters

Frequency	Usually several cycles/second; adjustable
Motion type	Simple harmonic or sinusoidal
Peak acceleration	1–10 g
Amplitude	Fraction of an inch to few inches
Mode	Off-stream
Duration	10–100 cycles, 30 sec to few minutes
Common bag diameters	5, 8, 12 inc.

Source: Reference 22.

Table 3. Reverse-Air Cleaning—Parameters

Frequency	Cleaned one compartment at a time, sequencing one compartment after another; can be continuous or initiated by a maximum-pressure-drop switch
Motion	Gentle collapse of bag (concave inward) upon deflation; slowly repressurize a compartment after completion of back-flush
Mode	Off-stream
Duration	1–2 min, including valve opening and closing and dust settling periods; reverse-air flow itself normally 10–30 sec
Common bag diameter	8, 12 in.; length 22, 30 f
Bag tension	50–75 pounds typical, optimum varies; adjusted after on-stream

Source: Reference 23.

Table 4. Pulse-Jet Cleaning—Parameters

Frequency	Usually, a row of bags at a time, sequenced one row after another; can sequence such that no adjacent rows clean one after another; initiation of cleaning can be triggered by maximum-pressure-drop switch or may be continuous
Motion	Shock wave passes down bag; bag distends from cage momentarily
Mode	On-stream: in difficult-to-clean applications such as coal-fired boilers, off-stream compartment cleaning being studied
Duration	Compressed-air (100 psi) pulse duration 0.1 sec; bag row effectively off-line
Common bag diameter	5–6 in.

Shaker

Although shaker-type baghouses are generally considered to be the oldest known form of fabric filter, they still have a significant place in present-day technology. It is known that in the smelter industry, when bag filters were developed in the mid-1800s and during the early 1900s, filter bags were cleaned by hand shaking. Shake cleaning has subsequently progressed through stages, from manually operated racks to today's devices that are automated for either motor or air operation. Many mechanisms have been developed to impart motion to the filter bags to clean them. The motion has been either vertical, horizontal, or some combination of the two, although shakers have been developed that twist or otherwise move the bags. In essence, all the mechanisms impart energy to the filter fabric in such a way that a change of direction allows inertial forces to remove the collected filter cake from the bags. The important parameters affecting the efficiency of cleaning are frequency, oscillation, and amplitude.

The one constant that must be provided in all shaker-type baghouses, regardless of the type of action imparted to the fabric, is that flow in the positive direction must be absent during cleaning. Forward differential pressure across the bags of less than 0.05 in. (12 Pa) has been observed to retard bag cleaning significantly.[21] Conversely, a slight reverse flow through the bags during shaking can be beneficial.

Shaking is usually accomplished by the use of a motor driving an eccentric (Figure 11), which, in turn, moves a rod connected to the bags. This bag cleaning technique is currently employed over the full range of baghouse capacities, from very small off-the-shelf units to extremely large structural design units.

The bag is generally open at the bottom and closed at the top, fixed in the tube sheet at the bottom, and attached to the shaking mechanism at the top (Figure 12). With this configuration, dust is collected on the inside of the bag. The bag normally contains no rings or cages. The flow of dirty gas to the bags is stopped during the cleaning process.

In the United States, shakers normally employ woven cloth at gas-to-cloth ratios below 4:1. Attempts thus far to use shaker cleaning in combination with domestic felts have led either to ineffective cleaning, and thus high pressure drop, or to the filter medium being shaken apart.

Reverse Air

Reverse-air cleaning is the gentlest cleaning method. Dust is removed from the bags by back-flushing with low-pressure (a few inches water gauge) reversed flow. In high-temperature applications, the just-cleaned hot gas is employed for back-flush, rather than ambient-temperature air. Woven filter media are generally employed with reverse-air cleaning.

Cleaning airflow is provided by a separate cleaning fan that is normally much smaller than the main system fan, since only a fraction of the total system is cleaned at any one time. In the case of a negative-pressure system, one often can clean without a reverse-flow fan. The flow rate of cleaning gas is normally about equal to that of the dirty gas (see Table 3).

Reverse-flow systems most often include a number of isolatable compartments, or modules. The gas-to-cloth ratios usually employed are less than 4:1. Dust is normally collected on the inside of the bag, the bag being open at the bottom and closed at the top. The bag contains rings to keep it from collapsing completely during flow reversal. Complete collapse

Figure 11. AAF's Shaker Support Logs (Photos Courtesy Snyder General Corp., Manufacturer and Marketer of American Air Filter Products).

Figure 12. AAF's Rugged Shaker Mechanism (Photos Courtesy Snyder General Corp., Manufacturer and Marketer of America American Air Filter Products).

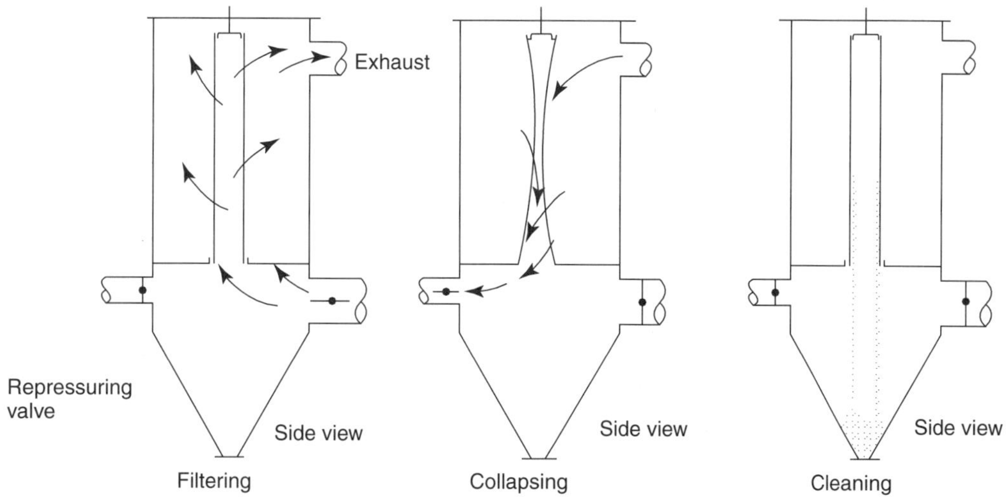

Figure 13. Reverse-Air Cleaning (from Ref. 23).

would, of course, prohibit cleaning because the dust particles could not fall down within the bag to the hopper. Cleaning is accomplished both with and without flexing (partial collapsing) of the bag. Reverse flow without flexing can be employed when the dust is very easily dislodged from the bag surface. Figure 13 shows a typical reverse-air cleaning cycle. In the on-stream gas filtering mode, the compartment and outlet dampers are open, and dirty gas enters the bag at the bottom. Dust collects on the inside of the bag, and the cleaned gas exits the outlet damper. During the bag cleaning cycle, the flow is reversed by closing the outlet plenum and opening a third damper that allows cleaned gas to enter the compartment on the clean side of the bags, thus back-flushing the bags and exiting the compartment through the inlet damper. This now-dirty gas progresses to the balance of the on-stream compartments. It should be noted that this process increases the system gas-to-cloth ratio by adding to the total gas volume the volume of gas employed in the back-flushing process.

One baghouse manufacturer studied the reverse-air cleaning process both analytically and experimentally[24] and concluded that the improved cleaning frequently observed when a thick dust cake is allowed to build up can be attributed to the pistonlike action of a falling plug of dust cake. The manufacturer theorizes that the cascading dust cake plug both scours that cake ahead of it and causes significant evacuation, and hence additional reverse flow behind it. The concept is shown in Figure 14.

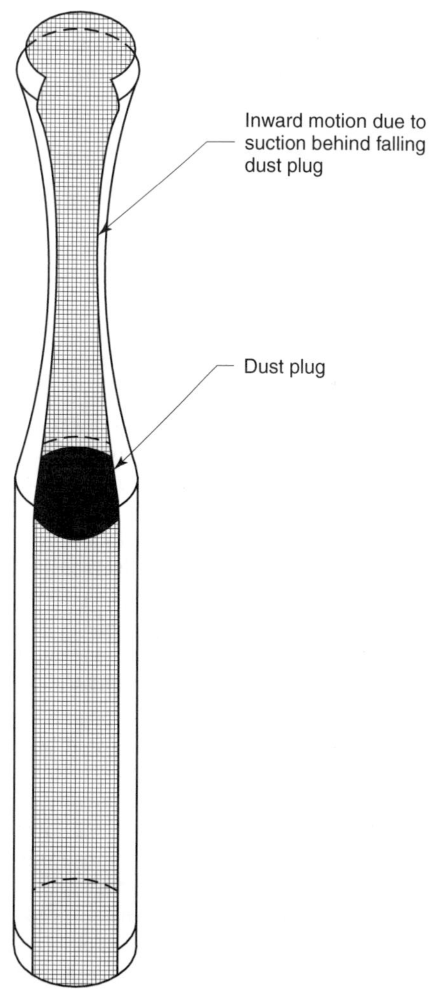

Figure 14. Idealization of the Falling Dust Cake upon Cleaning.

Reverse airflow can serve to flush out loosened particles from fabric interstices and carry the dislodged agglomerates toward the collecting hopper. Based on gravimetric measurements of the filter resistance characteristics, there is, however, little evidence to suggest any significant removal of dust particles by aerodynamic action alone. The above findings are in agreement with those of Larson,[25] who states that air velocities of the order of 200 fpm (102 cm/sec) are required to remove a single 20-μm particle from a fiber; and with Zimon,[26] who indicates that air velocities sweeping tangentially over a layer of dust must be in the range of 400 fpm (203 cm/sec) before any appreciable dust removal is accomplished. Because significant dust removal is attained in reverse-flow systems, one must conclude that separating forces other than aerodynamic drag are involved. According to the drag theory, the adhesive forces between adjacent particles are actually increased as dirty gas moves radially through the dust cake. During reverse airflow, any dust dislodgement is more likely to follow a spallation process, with the adhesive bond failure probably occurring close to the dirty face of the filter. Aerodynamic drag can be expected to flush out loosened particles.

Another adjunct to dust separation is the flexure produced in the fabric when the flow is reversed. In most systems, sufficient bending of the fabric surface occurs to cause a significant spallation at the dust–fabric interface. This effect is most pronounced for reverse-air-cleaned systems because, with the low gas-to-cloth ratios normally used, a large fraction of the collected dust appears as a superficial layer on the surface of the fabric.

If dust releases readily from the fabric, a reversal of flow alone may suffice for adequate cleaning. Because of its structural depth, and hence greater dust retentivity, a felted fabric is not usually cleaned by reverse flow. When reverse flow is used as the sole method of cleaning, the rate of flexure is probably the controlling factor with respect to fabric (or fiber) failure. Thus a gradual inflation or deflation process is unlikely to cause any serious fabric damage. Increased bag tension and reduced reverse flow rates also minimize the degree of flexure, as well as prevent a complete flattening of the bag. If the bag is flattened, there is no opportunity for loosened dust to fall to the hopper, nor is there a pathway for the reverse-flow air. The insertion of restraining rings or a supporting cage eliminates complete bag collapse, but introduces a potential problem of fabric chafing. Sewing the rings to the bag minimizes the problem of chafing or attrition.

Pulse Jet

Pulse-jet cleaning employs high-pressure (60–120-psi) compressed air, with or without a venturi, to back-flush the bags vigorously. This cleaning method creates a shock wave that travels down the bag, knocking dust away from the fabric.[27] Normally, this method is employed in conjunction with felted or bulk knitted filter media, and the gas-to-cloth ratio is generally higher than in shake and reverse-air cleaning methods. Typical gas-to-cloth ratios are given in Table 5. The duration of cleaning is shorter than for the other two methods; generally, the pulse lasts only a fraction of a second. The baghouse is often not subdivided when pulse-jet cleaning is employed.

The usual pulse-jet configuration has the bag closed at the bottom and open at the top, as shown in Figure 15. A metal cage is used inside the bag to keep it from collapsing. In the normal mode of operation, dirty gas enters the hopper and proceeds to the bags. Dust collects on the outside of the bags, and cleaned gas exits through the top of the bags and baghouse. Usually, a row of bags is cleaned simultaneously by introducing compressed air briefly at the top of each bag. The shock wave created drives the dust off the outside of the bag and down into the hopper. Continuous discharge of dust from the hopper is often employed. Cleaning parameters are given in Table 4. It is noteworthy that this system allows for removal of the bags from the clean side of the house, since they are usually connected only at the top.

The pulsed jets of compressed air commonly used to clean nonwoven fabrics are relatively inefficient at removing deposited dust. Measurements on a pilot-scale pulse-jet fabric filter using fly ash test dust indicate that typically less than 1% of the dust on a bag is removed to the hopper by a cleaning pulse.[24]

The following factors are important to the design and/or operation of a pulse-cleaned baghouse.

1. The location of the pulse-jet nozzle must be considered.
2. Bag material should be flexible, lightweight, and inelastic to obtain maximum acceleration for dust removal during the pulse. The fabric should have sufficient weight (i.e., number of fibers per unit area) to present many targets for dust collection. The pore structure should be as uniform as possible.
3. A large housing and hopper volume on the dirty side of the filter bag will minimize the pressure buildup in this region during the pulse and thus enlarge the magnitude of the pulse differential.
4. The pulse delivered to the bag should begin as abruptly as possible, with sufficient inflating flow to subject the entire bag length to a sudden pressure differential.
5. The back flow of air through the filter that accompanies the pulse assists cleaning in several ways. It flushes from the pore structure agglomerates loosened by the acceleration. It can itself loosen agglomerates if the shock alone was insufficient, although this appears to be a

Table 5. Typical[a] Gas-to-Cloth Ratios for Various Industries

	Shaker/Woven Reverse-Air/Woven	Pulse-Jet/Fest Reverse-Air/Felt
Alumina	2.5	8
Asbestos	3.0	10
Bauxite	2.5	8
Carbon black	1.5	5
Cement	2.0	8
Clay	2.5	9
Coal	2.5	8
Cocoa, chocolate	2.8	12
Cosmetics	1.5	10
Enamel frit	2.5	9
Feeds, grain	3.5	14
Feldspar	2.2	9
Fertilizer	3.0	8
Flour	3.0	12
Fly ash	2.5	5
Graphite	2.0	5
Gypsum	2.0	10
Iron ore	3.0	11
Iron oxide	2.5	7
Iron sulfate	2.0	6
Lead oxide	2.0	6
Leather dust	3.5	12
Lime	2.5	10
Limestone	2.7	8
Mica	2.7	9
Paint pigments	2.5	7
Paper	3.5	10
Plastics	2.5	7
Quartz	2.8	9
Rock dust	3.0	9
Sand	2.5	10
Sawdust (wood)	3.5	12
Silica	2.5	7
Slate	3.5	12
Soap, detergents	2.0	5
Spices	2.7	10
Starch	3.0	8
Sugar	2.0	7
Talc	2.5	10
Tobacco	3.5	13
Zinc oxide	2.0	5

[a] Generally safe design values; application requires consideration of particle size and grain loading.

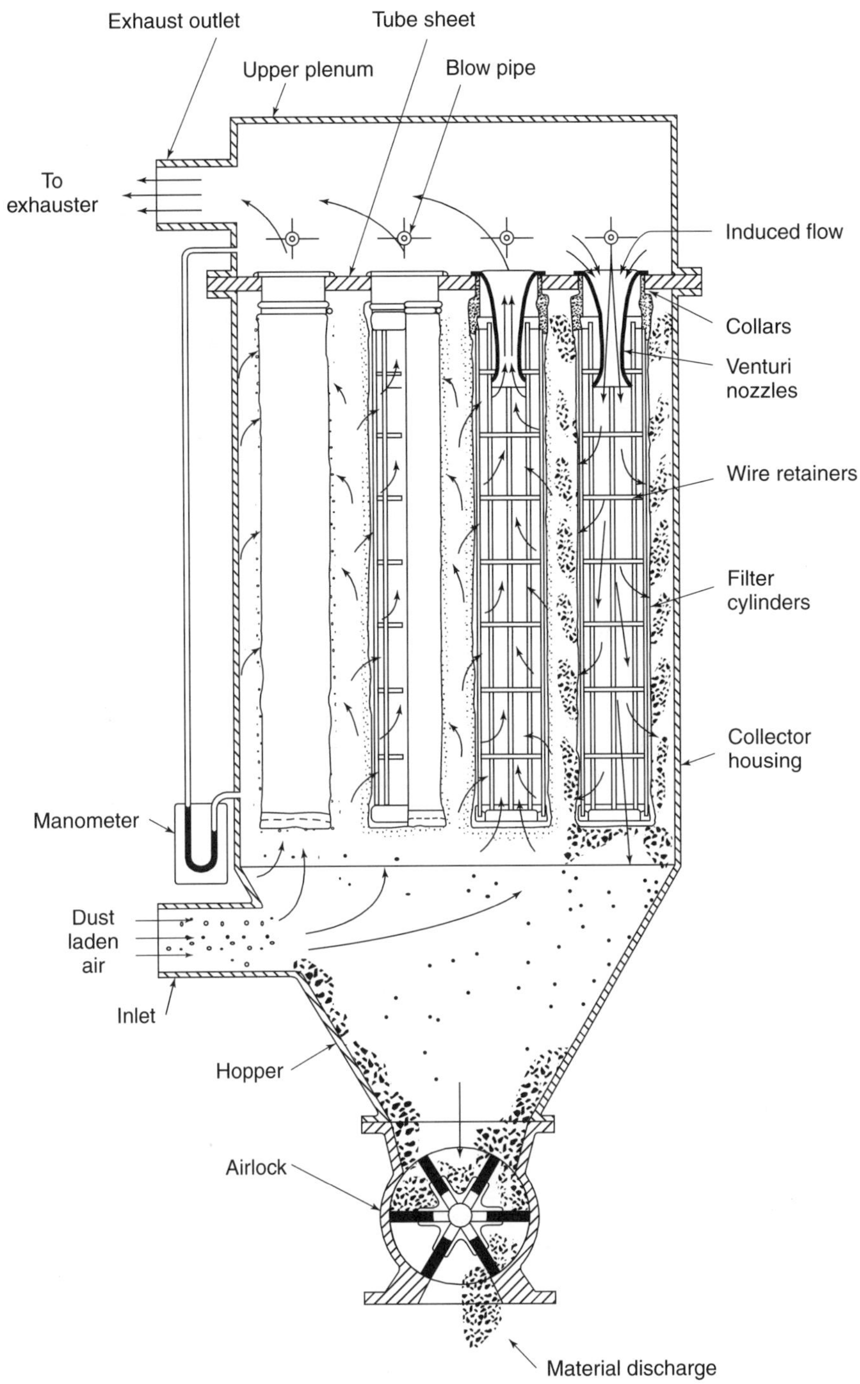

Figure 15. Pulse-Air-Jet Type of Bag Filter (from Ref. 29).

very inefficient use of compressed air. It also accelerates agglomerates that have already left the felt surface, helping to convey them to the hopper.

6. Pulse intensity should be as low as can be tolerated to save on compressed air (and reduce power needs) but sufficiently high to maintain equilibrium in the cleaning process.
7. Pulse duration should ordinarily be as short as possible.
8. Dry, clean (oil-free) air must be used.

Pulse-jet cleaning is sometimes described as the rapid passage of an air bubble through the bag, a concept depicted in Figure 16. During the pulse's descent, localized fabric distension is assumed to loosen the dust layer, while the transient reverse flow completes dust dislodgement from the upstream filtering surface. Although there is some truth to this concept, it overlooks the fact that typical pulse durations (of at least 0.06 sec) far exceed the time for the shock wave to travel to the bottom of a conventional felt bag (roughly 0.01 sec for a 10-foot-long bag). Donovan et al.[21] propose that, in reality, while the pressure front is advancing, the continued arrival of pulse air, at successively increasing rates until the solenoid valve is fully open, maintains the inflating pressure behind the front. Because pulse air entry continues well after the bottom of the bag is fully pressurized, the concept of an advancing bubble is questioned. It is suggested that parts (b), (c), and (d)

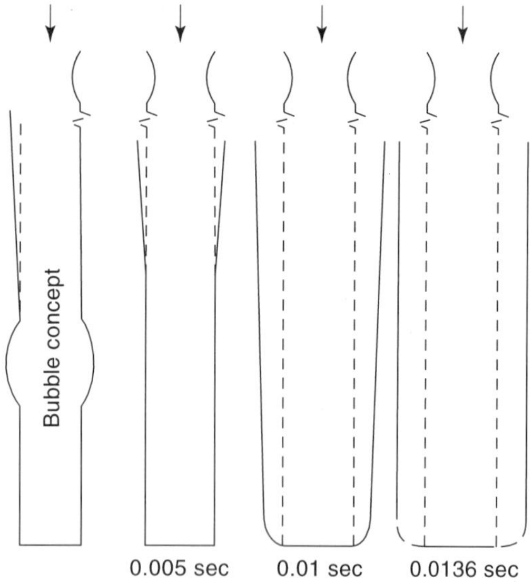

Figure 16. Bag Profiles versus Time After Pulse Initiation for a 4-Foot-Long Bag (from Ref. 28).

of Figure 16 are more descriptive of the movement of bag fabric during a pulse.

Sonic Horns

Sonic horns are increasingly being used to augment shaker-cleaned and reverse-air-cleaned baghouses. The horns are usually powered by compressed air, and acoustic vibration is introduced by a vibrating metal plate that periodically interrupts the airflow. A cast metal horn bell is normally used for cleaning. Typically, one to four horns are installed in the ceiling of a baghouse compartment containing several hundred bags.

Sonic frequencies of 150–550 Hz have been tested; Cushing, Pontius, and Carr[30] got the best results with horns that concentrated most of their energy at the lower frequencies. Relative sound pressure levels of 120–140 dB inside the baghouse are used.

Probably the most significant effect of sonic cleaning is on the weight of residual dust load on the bags. Menard and Richards[31] found that, before the sonic horns were operated, bag weights ranged from 34 to 55 pounds, with an average weight of approximately 46 pounds. After extended use of the sonic horns, bag weights ranged from 12.5 to 25 pounds, with an average weight of approximately 18 pounds. A new bag weighs approximately 9 pounds. The sonic air horns have thus reduced the amount of residual cake on the filtering elements by an average of 76%.

Cushing et al.[30] concluded:

> Sonic horns are an effective method for enhancing fabric filter performance. Their generally low cost and simple construction make them attractive additions to the available methods for cleaning fabric filter bags at coal-fired utility baghouses. Tests of six commercial sonic horns at the EPRI Fabric Filter Pilot Plant have demonstrated that, under appropriate conditions, reverse-air cleaning with sonic assist can be effective in reducing operating pressure losses. The test results have shown that overall sound pressure levels and the output frequency are important factors in determining whether a particular sonic horn application will enhance baghouse performance. One penalty encountered with sonic assist may be higher particulate penetration due to less residual dust cake on the bag surface. It is possible, however, that less frequent applications of horns will result in low ΔP without a significant increase in emissions.

Other Bag and Baghouse Technologies

One approach developed to provide excellent filtration performed at lower costs is the Compact Hybrid Particulate Collector (COHPAC). Test results on both pilot- and full-scale units have provided performance results at levels of <0.1 lb/MMBTU on coal-fired boilers. This technology involves the use of a pulse jet collector operating at high filtration rates in series with an energized electrostatic precipitator.[32,33]

A process combining the removal of SO_2, NO_x, and particulate in one unit, a high-temperature baghouse, has also been investigated. SO_2 is removed by injection of a sorbent upstream of the baghouse. NO_x is removed by selective catalytic reduction. NH_3 is also injected upstream of the baghouse, and the catalyst is contained within the bags.[34]

Baghouses that employ cartridge-based bags rather than the traditional tubular bags and cages have become popular where overhead space restrictions apply. One interesting case is where a shaker baghouse was retrofitted with cartridge pulse-jet technology, doubling the capacity of the unit, and the modified unit was reported to reduce maintenance cost as well.[35]

Other developments that have made progress in recent years include bag modifications such as significantly longer tubular bags, shape changes such as star and pleated bags, and broadened applications of materials such as Ryton,[36] P84,[37] and microdenier fibers.[38] The potential requirement for 2.5-μm removal is driving an R&D focus on fine-particle performance. The USEPA has initiated an Environmental Technology Verification program that now includes an effort devoted to verifying the performance of fine-particle filtration.[39]

BAGHOUSE MAINTENANCE

Why Keep Records?

Traditionally, baghouse operators do not keep records of routine baghouse operation. There are several reasons to reconsider this policy, including the following:

1. Records permit the operator to be aware of continuing normal operation.
2. Records permit the operator to be aware of abnormal operation, such sudden failures or slowly changing parameters (e.g., a slow rise in residual pressure that, if allowed to continue, would limit the ability of the process to reach full load).
3. Records provide a historical record that is useful when troubleshooting problems.

What Records to Keep

Pressure drop is often the only parameter monitored in a baghouse, usually by means of a gauge mounted on the baghouse wall. In multicompartment baghouses, the pressure drops of individual compartments may be monitored along with the total baghouse pressure drop. Unfortunately, over time, the pressure-sensing lines become clogged with dust or the gauge becomes unreadable owing to dust accumulation on the instrument face. An obvious first step in maintenance is to assure that the instrumentation is both functional and accurate.

An instrument reader who routinely records process data should probably also routinely record available baghouse data. A better scheme is to record the data as part of a data-logging system. Automated data-recording and parameter alarm systems are becoming more common. If such a system is a part of the process itself, perhaps baghouse data could also be recorded.

Other types of baghouse records, in addition to pressure-drop information, should be considered:

1. Flow rate. Pressure-drop information cannot be interpreted properly unless the flow rate is known. A record of flow rate may be useful in identifying a developing leak in the ducting or in the baghouse itself.
2. Opacity. If a continuous opacity monitor is incorporated into the system, its output should be recorded. If such opacity instrumentation is not available, visual opacity readings recorded manually should still prove useful. The cause of any change in opacity should be pursued and understood.
3. Temperature. The baghouse outlet temperature should certainly be monitored, even if other temperature records are not kept.
4. Dust removal. At least one parameter related to the quantity of dust removed from each baghouse compartment should be monitored and recorded. A change in dust quantity may be indicative of baghouse failure or of process changes.

In order to better monitor and record baghouse operating parameters and relate these to the process operation, an innovative monitoring, early warning, and diagnosing system has been developed and demonstrated. This system not only provides operational feedback but also guides the operator with information necessary to improve baghouse reliability and performance.[40,41]

The key to baghouse maintenance is frequent and routine inspection. It is essential that a regular program of routine maintenance be established and followed. Records (a log) should be kept of all inspections and maintenance. Inspection intervals will depend on the type of baghouse, the manufacturer's recommendation, and the process on which the unit is installed. The important thing is to be sure that the checks are performed regularly and as frequently as necessary, that no components are neglected, and that all pertinent information is logged for future reference.

When problems are located and isolated during routine or other inspection, it is important that corrections be made as quickly as possible to avoid possible equipment downtime or excess emissions from bypassing the control system. When there is a baghouse failure, the unit is usually shut down and/or bypassed and the malfunction corrected. Plant managers should expect considerable maintenance time to be expended on troubleshooting and correction of baghouse malfunctions. Maintenance personnel must learn to recognize the symptoms that indicate potential problems, and then to determine the cause of the difficulty and remedy it, either by in-plant action or by contact with the manufacturer or some other outside resource. High pressure drop across the system exemplifies one symptom for which there are many possible causes; for example, difficulties with the bag-cleaning mechanism, low compressed-air pressure, weak shaking action, loose bag tension, or excessive reentrainment of dust. Many other factors can cause excessive pressure drop, and several options are usually available for corrective action. Thus the ability to locate and correct malfunctioning baghouse components requires a thorough understanding of the system. The critical influence of moisture, in the baghouse, in the cleaning air, or with the particulate, should be noted and emphasized.

The operating experience of baghouses in conjunction with electric utility boilers has probably been more extensively documented than any other application. The Electric Power Research Institute (EPRI) currently supports an extensive research program intended to optimize the design, operation, and reliability of baghouse technology for the electric utility industry. Pilot- and full-scale evaluations are performed to measure, interpret, and predict baghouse performance. Results show that baghouses routinely achieve a clear stack and particulate matter collection efficiencies well in excess of 99.9%.

REFERENCES

1. J. D. McKenna and J. H. Turner, *Fabric Filter—Baghouses I*, ETS, Inc., Roanoke, VA, 1989.
2. D. J. Helfritch, "The fine particle control capability of fabric filters," Paper A848, Annual Meeting of the Air and Waste Management Association, Nashville, TN, June 1996.
3. K. S. Kumar and P. L. Feldman, "Assessing the technologies and costs of trace element control in coal burning plants," Paper A848, Annual Meeting of the Air and Waste Management Association, Nashville, TN, June 1996.
4. R. Dennis and H. A. Klemm, *Fabric Filter Model Format Change, Vol. I: Detailed Technical Report*, EPA-600/7-79 0432, U.S. Environmental Protection Agency, Research Triangle Park, NC, 1979, National Technical Information Service (NTIS) PB-293-551.
5. C. E. Billings and J. Wilder, *Handbook of Fabric Filter Technology, Vol. I: Fabric Filter Systems Study*, GCA Corp., Bedford, MA.; PB-200-648, NTIS, Springfield, VA, December 1970.
6. R. Dennis, R. W. Cass, D. W. Cooper, et al., *Filtration Model for Coal Flyash with Glass Fabrics*, EPA-600-7-77084, August 1977.
7. W. T. Davis and R. F. Kurzynske, "The effect of cyclonic precleaners on the pressure drop of fabric filters," *Filtration Separation*, **16**(5).
8. W. T. Davis and W. F. Frazier, "A laboratory comparison of the filtration performance of eleven different fabric filter materials filtering resuspended flyash," *Proceedings of the 75th Annual Meeting of the Air Pollution Control Association*, 1982.
9. W. F. Frazier and W. T. Davis, *Effects of Flyash Size Distribution on the Performance of a Fiberglass Filter*, Symposium on the Transfer and Utilization of Particulate Control Technology. Vol. II: Particulate Control Devices, EPA-600/9-82-OOSOc, July 1982, pp. 171–80.
10. J. W. Robinson, R. E. Harrington, and P. W. Spaite, "A new method for analysis of multicompartmented fabric filtration," *Atmospheric Environ.* **1**(4):499 (1967).
11. W. W. Gregg, "Survey of available technology for high temperature dry filtration to 2,000°F," *Proceedings, The User and Fabric Filtration Equipment VI Specialty Conference*, Toronto, Ontario, Canada, September 1992.
12. F. K. Pethick, "High Temperature Baghouse Operation Using a Unique Metallic Fiber Filter Design," *Proceedings, The User and Fabric Filtration Equipment VI Specialty Conference*, Toronto, Ontario, Canada, September 1992.

13. A. B. Jeffery, "High temperature filtration using cleanstac[TM1] filters," *Proceedings Baghouse Operation & Maintenance User and Fabric Filtration IX Specialty Conference*, Toronto, Ontario, Canada, September 1998.
14. A. A. Reisinger and W. T. Grubb, "Fabric evaluation program at Coyote Unit #1, operating results update," *Proceedings Second EPRI Conference on Fabric Filter Technology for Coal-Fired Power Plants, Electric Power Research Institute*, Palo Alto, CA., EPRI CS03257, November 1983.
15. G. P. Greiner, J. C. Mycock, and D. S. Beachler, "The IBFM, a unique tool for troubleshooting and monitoring baghouses," Paper 85-54.2, Annual Meeting of the Air Pollution Control Association, Detroit, MI, June 1985.
16. J. F. Durham and R. E. Harrington, "Influence of relative humidity on filtration resistance and efficiency of fabric dust filters," *Filtration Separation* **8**:389–93 (July/August 1971).
17. E. R. Frederick, *Electrostatic Effects in Fabric Filtration: Volume II Triboelectric Measurements and Bag Performance* (annotated data), EPA-600/7-78-142b, NTIS PB-287-207, July 1978.
18. International Non-Woven and Disposable Association, New York.
19. C. B. Hotchkiss and L. F. Cox, "Fabric and finish selection, manufacturing techniques and other factors affecting bag life in the coal-fired boiler applications," *Proceedings Second EPRI Conference on Fabric Filter Technology for Coal-Fired Power Plants*, Electric Power Research Institute, Palo Alto, CA, EPRI CS-3257, November 1983.
20. T. McKenna, "Expanded PTFE membrane for surface filtration the solution to many baghouse problems," *Proceedings, The User and Fabric Filtration Equipment VII Specialty Conference*, Toronto, Ontario, Canada, September 1994.
21. R. P. Donovan, B. E. Daniel, and J. H. Turner, *EPA Fabric Filtration Studies, 3. Performance of Filter—Bags Made from Expended PTFE Laminate*, EPA-600/2-76-168c, U.S. Environmental Protection Agency, Research Triangle Park, NC, December 1976, NTIS PB-263-132.
22. R. L. Adams. "Shaker type filter," *Proceedings, APCA Specialty Conference on the User and Fabric Filter Equipment*, Buffalo, NY, October 1973, Air Pollution Control Association, Pittsburgh, PA.
23. C. E. Billings, Ed., *The Fabric Filter Manual*, McIlvaine Co., Northbrook, IL.
24. M. M. Ketchuk, A. Walsh, O. F. Fortune, et al., "Fundamental strategies for cleaning reverse-air baghouses," *Proceedings Fourth Symposium on the Transfer and Utilization of Particulate Control Technology, Vol. I: Fabric Filtration*, EPA-600/9-84-025a, U.S. Environmental Protection Agency, Research Triangle Park, NC, November 1984.
25. R. I. Larson. "The adhesion and removal of particles attached to air filter surface," *AIHA J.* **19**, (1958).
26. A. D. Zimon, *Adhesion of Dust and Powder*, Plenum Press, New York, 1969, p. 112.
27. M. J. Ellenbecker and D. Leith, "Dust removal from nonwoven fabrics," Annual Meeting of the Air Pollution Control Association, Montreal, Que., June 22–27, 1980.
28. R. Dennis and L. S. Hovis, "Pulse-jet filtration theory—A state-of-the-art assessment," *Proceedings Fourth Symposium on the Transfer and Utilization of Particulate Control Technology, Vol. I: Fabric Filtration*, EPA-600/9-84-0252, U.S. Environmental Protection Agency, Research Triangle Park, NC, November 1984.
29. C. Orr, *Filtration, Principles, and Particles*, Part 1, Marcel Dekker, New York, 1977.
30. K. M. Cushing, D. H. Pontius, and R. C. Carr, "A study of sonic cleaning for enhanced baghouse performance," Annual Meeting of the Air Pollution Control Association, San Francisco, June 1984.
31. A. R. Menard and R. M. Richards, "The use of sonic air horns as an assist to reverse-air cleaning of a fabric filter dust collector," *Proceedings Fourth Symposium on the Transfer and Utilization of Particulate Control Technology, Vol. I*: Fabric Filtration, EPA-600/9-84-025a, U.S. Environmental Protection Agency, Research Triangle Park, NC, November 1984.
32. R. Miller, W. Harrison, K. Cushing, and R. Chang, "Recent compact hybrid particulate collector data (COHPAC) for fine particulate matter and air toxics removal from coal-fired power plants."
33. B. H. Easom, S. F. Burlatsky, R. F. Altman, and R. Chang, "Particulate control technologies for power generation," *Pollution Engineering*, pp. 40–42 (July 1999).
34. "SO_x–NO_x Rox Box[TM] (SNRB[TM]) flue gas cleanup demonstration," *Clean Coal Technology*, Topical Report Number 13, p. 9, May 1999.
35. L. Palazzo, J. Woolston, and Peter Ristevski, "Retrofitting shaker baghouses to cartridge pulse jet technology," *Proceedings the User and Fabric Filtration Equipment VII Specialty Conference*, Toronto, Ontario, Canada, September 1994.
36. B. Schaller, D. Miller, and Bob Brandt, "Comparison of baghouse bag material options on two identical circulating fluidized bed boilers at Iowa State University," *Proceedings, The User and Fabric Filtration Equipment V Specialty Conference*, Toronto, Ontario, Canada, September 1990.
37. A. Turkson, "Case histories involving P84/polyester versus other fabrics in dust collectors in the steel industry," *Proceedings, The User and Fabric Filtration Equipment V Specialty Conference*, Toronto, Ontario, Canada, September 1990.
38. J. W. Griffin, "Filtering with microdenier fibers or felts," ETS, Inc., 29th Fabric Filter Baghouse Seminar, Roanoke, VA, April 1996.
39. J. C. Mycock, J. H. Turner, D. W. VanOsdell, J. R. Farmer, and T. G. Brna, "Baghouse filtration products verification," *Proceedings, Baghouse Operation & Maintenance User and Fabric Filtration IX Specialty Conference*, Toronto, Ontario, Canada, September 1998.
40. S. Schliesser, E. Handel, and G. Greiner, "A candidate baghouse enhanced monitoring protocol," *Proceedings, The User and Fabric Filtration Equipment VII Specialty Conference*, Toronto, Ontario, Canada, September 1994.
41. S. Schliesser, T. Barham, and M. Furgurson, "Innovative baghouse performance monitoring system," Environment Virginia '96 Symposium, Lexington, VA, April 1996.

4

FUGITIVE DUST EMISSIONS

John G. Watson, Judith C. Chow, and Thompson G. Pace

INTRODUCTION

Fugitive dust[1] refers to small particles of geological origin that are suspended into the atmosphere from nonducted emitters. Open fields and parking lots, paved and unpaved roads, agricultural fields, construction sites, unenclosed storage piles, and material transfer systems are the major sources of fugitive dust. Large dust plumes are often noticed over these sources when wind speeds are high or when vehicles are moving. These visible plumes do not necessarily correspond to significant amounts of PM_{10} or $PM_{2.5}$ (particles with aerodynamic diameters less than 10 and 2.5 μm, respectively) measured by nearby air-quality monitors.[2] The PM_{10} and $PM_{2.5}$ size fractions are regulated by National Ambient Air Quality Standards.[3] Total suspended particulate (TSP, mass of particles with aerodynamic diameters less than ~30 μm) is also commonly used as a measure of fugitive dust emissions. Although PM_{10} and $PM_{2.5}$ dust particles scatter and absorb light,[4] they are too small to be seen with the naked eye.

PM_{10} and $PM_{2.5}$ fugitive dust emissions for 1997[5] include wind erosion; agriculture and forestry operations such as tilling and harvesting; roads and construction; and industrial processes such as cement production, quarrying, mining, and other mineral industries. Emissions from these fugitive dust source categories constitute 89% of the 33,574 thousand short tons per year (tpy) of PM_{10} and 66% of the 8,288 thousand tpy of $PM_{2.5}$ emitted into the air.[5] In contrast, on-road and off-road vehicle exhaust accounts for 2.4%, open and residential vegetative burning accounts for 1.8%, and other emitters (mostly industrial and residential coal, oil, and natural gas combustion) account for ~7% of total PM_{10} emissions. For $PM_{2.5}$ emissions, 7.5% derive from on-road and off-road vehicle exhaust, 6.9% come from open and residential vegetative burning, with the rest (~19%) deriving from the other emitters.

Figure 1 shows the categories and amounts of fugitive dust emissions.[5] Industrial emissions are a small fraction of the total. Unpaved roads, paved roads, construction, and wind erosion together constitute more than 80% of PM_{10} and 75% of $PM_{2.5}$ fugitive dust emissions. These emissions are from both urban and nonurban areas. Nearly all the agricultural emissions are from crop and livestock activities in nonurban areas. Most of the paved road dust and construction emissions are from urban areas. Unpaved road dust emissions are from both urban and nonurban sources, with a higher proportion from nonurban surroundings. Wind erosion is primarily nonurban, and the estimates are almost exclusively for agricultural operations.[6] These breakdowns give the impressions that unpaved road dust is the most important fugitive dust contributor on a nationwide basis and that most fugitive dust emissions derive from nonindustrial sources.

PM_{10} and $PM_{2.5}$ source apportionment studies use chemical profiles from different source types and from ambient particle samples to estimate source contributions.[7–11] These studies[12,13] show that, on average, fugitive dust contributes ~40% to ~60% of PM_{10} and ~5% to ~20% of $PM_{2.5}$ measured in the atmosphere. Fugitive dust contributions to ambient measurements are often overestimated by dispersion models that simulate contributions to receptor concentrations (e.g., Chow et al.,[14] Venkatram et al.[15]).

Resolving these discrepancies between emissions estimates and ambient source contributions is an important consideration in designing, applying, and evaluating control strategies intended to reduce fugitive dust emissions. The knowledge base needed to accomplish this resolution is incomplete and needs to be expanded. Fugitive dust emissions estimates contain a high amount of variability owing to the meteorological, physical, and chemical factors on which they are based. These factors can vary widely on a national or smaller regional basis. Systematic errors probably dominate over random errors for spatially and temporally averaged emissions rates, such as those of Figure 1. Random error may equal or exceed systematic biases for smaller areas and shorter time intervals.

This survey of fugitive dust emissions explains the processes that cause dust to be injected into and remain in the atmosphere. These processes depend on surface properties as well as the activities conducted on those surfaces. The survey summarizes fugitive dust emission factors and identifies indicators of the activities to which they are applied. Emissions reduction measures are described and quantitative studies of their effectiveness are reviewed. This survey complements, rather than replaces, detailed procedures for estimating emissions in "AP-42: Compilation of Air Pollutant Emission Factors"[16] that are regularly updated by U.S. EPA. Current procedures should be verified prior to use at *http://www.epa.gov/ttn/chief/*. Representative references are cited that can be consulted for more detailed information for each topic.

PROCESS DESCRIPTION

Fugitive dust emissions depend on particle sizes, surface loadings, surface conditions, wind speeds, atmospheric and surface moisture, and dust-suspending activities. Emission rates and control measures are also closely related to these properties.

Particle Sizes

Figure 2 shows the size distribution of suspended particles measured from common emission sources, including fugitive dust as determined by resuspension of dust samples and acquisition on filters through size-selective inlets similar to those used for ambient samples.[17–19] Construction dusts, road dusts, and soil dusts result from grinding down of larger particles and are predominantly in the coarse particle size range, larger than 2.5 μm. Combustion particles, on the other hand, dominate the $PM_{2.5}$ size fraction.

Similar size distributions are found in atmospheric particles, where the coarse mode is dominated by elements

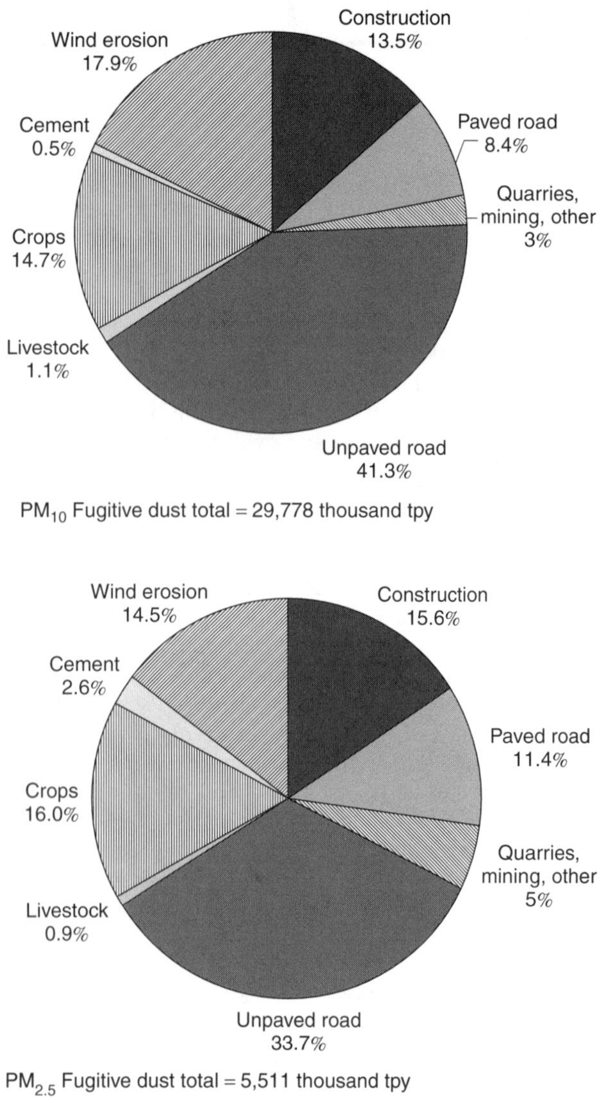

Figure 1. Fugitive Dust PM_{10} and $PM_{2.5}$ Emissions for Unpaved Roads, Paved Roads, Construction, Wind Erosion, Crops, Livestock, and Industrial Sources.[5]

indicative of geological origin. Pollen and spores also inhabit the coarse particle size range, as do ground-up trash, leaves, and tires. Particles larger than 30 μm deposit to the surface soon after suspension unless they are injected to high altitudes. This deposition effectively limits atmospheric concentrations for very large particles. The peak of the coarse mode in atmospheric size distributions has been measured between ~6 and 25 μm.[20] This peak shifts toward larger particle sizes when fugitive dust emissions are close to the measurement location, and toward smaller particles at nonurban locations far from recent dust emissions.

Laboratory resuspension is not practical for large-scale field surveys. The "silt" fraction of surface dust is most often used as a surrogate for supendable particles. Silt consists of particles with a geometric diameter <75 μm as determined by sieving dried soil samples acquired from surface loading tests. The 75-μm geometric diameter corresponds to an aerodynamic diameter of ~120 μm because the aerodynamic diameter varies inversely with the square root of the density,[21] which is ~2.65 g/cm^3 for minerals. Similarly, a 10-μm aerodynamic diameter dust particle has a ~6-μm geometric diameter, and a 2.5-μm aerodynamic diameter dust particle has a geometric diameter of ~1.5 μm. Little is known about the PM_{10} and $PM_{2.5}$ in surface dust deposits as these fractions are too small to be determined by simple sieving methods.

Particle-size distributions have been determined by sieving samples from different types of soils and recording these in soil surveys. These surveys have been used for agriculture and construction/engineering purposes since the early 1900s in many parts of the United States and are commonly available at county agricultural extension offices. The particle sizing procedure[22,23] most commonly followed for soil surveys creates a soil/water suspension in which soil aggregates are broken into their component parts prior to sieving. While the particle size distribution of the disaggregated sediment is useful for agricultural, construction, and other land uses, it is not entirely applicable for estimating air pollution emissions. This sieving method does not estimate the size of the dust aggregates that are entrained and suspended by surface winds or human activities. The silt fraction is determined by dividing the weight of material passing through the 75-μm sieve by the weight of material presented to a stack of sieves.

Gillette et al.[24] applied two methods to determine the particle and aggregate sizes in soil that might be entrained by winds. The first method ("gentle sieve") consists of drying a soil sample and sieving it gently with about twenty circular gyrations parallel to the plane of the sieve. The second method ("hard sieve") consists of up to one-half hour of vigorous shaking (usually using a shaking machine). Cowherd et al.[25] used a rotary sieving procedure described by Chepil[26] for estimating the modal aggregate size of sediment samples removed from unpaved roads. The gentle sieve method is assumed, without quantitative validation, to be a more suitable approach for determining the potential suspension properties of a soil because it attempts to sample the sediment with its in situ characteristics intact. Silt fractions and amounts determined by the hard sieve method probably provide a reasonable indicator of small particles from roads where vehicle tires abrade the surface. Threshold suspension velocities[24] for windblown dust apply to soil characteristics obtained by the gentle sieve method. Appendix C-2 of AP-42[16] provides detailed procedures for the hard sieve method applicable to other emissions.

The size distribution of dust particles affects the suspension process. A flat bed of particles with diameters less than 20 μm is difficult to suspend by wind. Bagnold[27] showed that fine Portland cement could not be entrained by wind velocities in excess of 1 m/sec at the surface. In this situation, there is no large cross section for wind to act on. In addition, adhesive forces such as van der Waals, electrostatic, and surface tension of adsorbed liquid films[21] increase the force required to entrain the particles.

Suspension of PM_{10} and $PM_{2.5}$ is reduced by larger nonerodible particles. Particles that exceed 840 μm in size shelter smaller particles in their lee.[28] Gillette and Stockton[29] sprinkled glass spheres with diameters ranging from 2,400 to 11,200 μm onto a bed of glass spheres with sizes from 107 to 575 μm and found major reductions in the horizontal flux of the smaller particles. However, Logie[30] found that erosion of a sand surface was enhanced when low concentrations of larger nonerodible obstructions were present on the surface. Logie[30] hypothesized that the increased erosion was due to wind acceleration around the isolated obstructions that scoured the

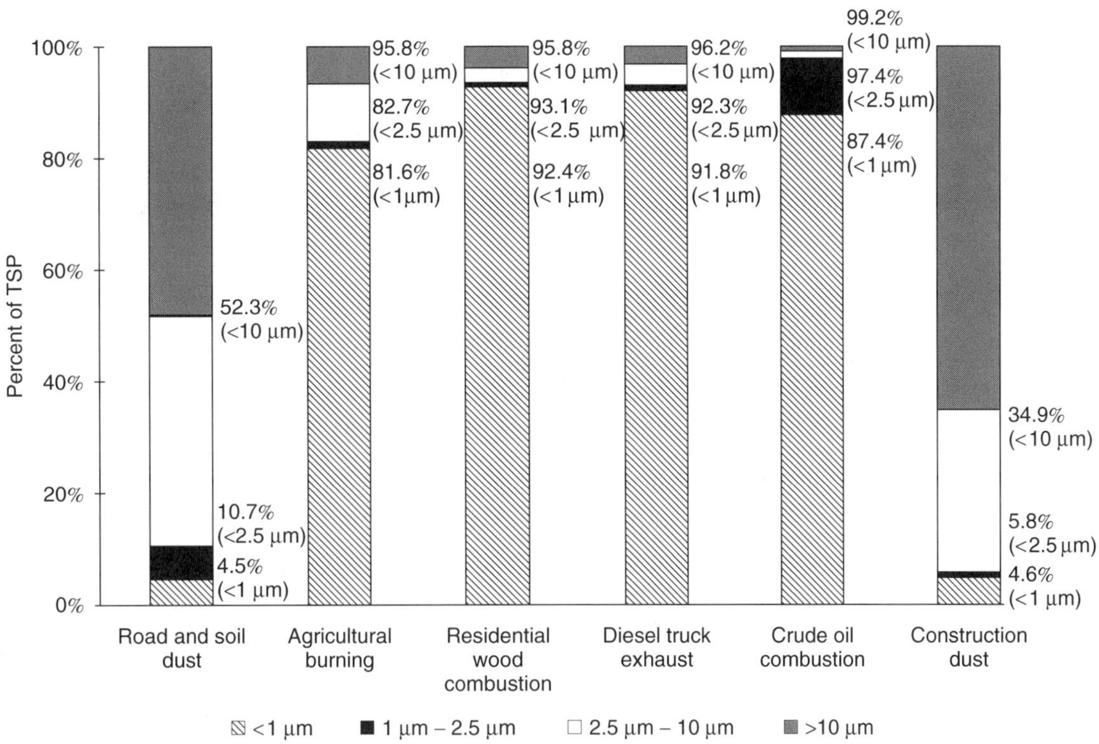

Figure 2. Size Distributions of Several Particulate Source Emissions[17–19].

loose sand. Bagnold[31] estimated that 800-μm particles can be separated from surfaces under high winds, although their large masses cause them to settle to the surface very rapidly.

Rosbury and Zimmer[32] and Flocchini et al.[33] observed higher dust emissions from unpaved road surfaces with lower silt contents and higher gravel content. When acted on by vehicle tires, larger gravel particles replenish the quantity of smaller suspendable particles by enhancing surface abrasion.

Carvacho et al.[34] have developed a method that directly measures the reservoir of PM_{10} available for suspension in a bulk soil sample. Pulses of air are run through a fluidized bed containing the sample, and the suspended dust is sampled through a size-selective inlet onto a series of parallel filters. After a few pulses, one of the filters stops sampling. After a few more pulses, another filter stops sampling and so on. Each filter is separately weighed, and the collected mass is plotted as a function of the aggregate agitation (as determined by the number of pulses). A maximum mass loading is reached after many pulses when the suspendable PM_{10} is depleted. This maximum varies between different types of dusts much more than it varies for repeated samples of the same dust. Shimp et al.[35] have related this PM_{10} suspension potential to the silt measurements in California soil surveys to improve their PM_{10} emissions estimates.

Silt fraction or quantity appear as explicit variables in many of the emissions factors cited below. The processes related to particle size indicate that actual emissions of PM_{10} and $PM_{2.5}$ are influenced by more detailed size distributions above and below the 75-μm geometric diameter that specifies silt content.

Surface Loading

Most soil surfaces are limited reservoirs; suspendable dust is depleted after a short time in the absence of direct abrasion. This depletion is represented as a negative exponential[36,37] or inverse[38–40] function of time.

As noted above, depletion of fine particles often results in the exposure of larger nonerodible sediments that shield the suspendable particles from the wind. The larger nonerodible elements also absorb momentum, thereby decreasing the wind's ability to erode the surface.[41,42] When surfaces are continually disturbed by very intense winds, by vehicular movement, or by other human activities, unlimited reservoirs are created that emit dust whenever winds exceed threshold suspension velocities. Suspendable dust loadings may vary substantially, even over periods of a few minutes, when there are no mechanisms to replenish the reservoir.

Surface loadings are determined by sweeping or vacuuming loose particles from several specified areas within a fugitive dust source. These samples are dried and weighed, then divided by the area sampled to determine the mass of dust per unit area. Appendix C-1 of AP-42[16] provides detailed procedures. The same samples are usually sieved to determine the silt fraction and the quantity of larger particle sizes. Surface loadings vary substantially with space and time. Zimmer et al.[43] measured silt loadings that varied by a factor of twenty at different times for the same paved road in Denver, CO. South Coast Air Quality Management District[44] reported paved road silt loadings from 0.112 to 1.83 g/m^2 for different areas with similar road and traffic conditions in the Los Angeles area.

Surface Conditions

Surface conditions refer to the landform shape and cohesion of a potential dust reservoir. The effects of landform shape on dust suspension are embodied in the concept of "surface roughness." Surface roughness is related to the heights of obstructions within and around exposed dust areas. Agriculturists often orient their furrows perpendicular to prevailing winds, or plant rows of trees upwind of their fields, to minimize soil losses from wind erosion by increasing surface roughness. Larger surface roughness decreases the force exerted by the wind on suspendable surface particles, thereby decreasing emissions. However, larger surface roughness increases vertical turbulence that can mix suspended particles higher into the atmosphere for longer transport distances.

The aerodynamic roughness length is the apparent distance above the surface at which the average wind speed approaches zero. In reality, wind speed does not become zero at this level, but it deviates from the logarithmic increase of wind speed with height that is commonly found in the atmosphere. Aerodynamic surface roughness is one-eighth to one-thirtieth of height of obstructions in and around an exposed area,[45] but it is site-specific and quantified with wind speed measurements taken at different elevations between ~1 and ~10 m above ground level. The ratio of wind speeds at lower levels is plotted against the logarithm of the height of the measurement and extrapolated to a wind velocity of zero. The intersection with the elevation axis at wind speed equals zero is the surface roughness. In practice, estimates from many hours of wind measurements are averaged to determine typical surface roughness. The slope of this relationship is termed the *friction velocity* and indicates the wind shear forces near an erodible surface.[46]

The presence of large particles or obstructions that increase surface roughness attenuates wind erosion by absorbing a significant fraction of the downward momentum flux from the air flow above.[47] While the principles are known, the effects of changes in surface roughness on fugitive dust emissions are not well quantified.

The same surface roughness may be associated with many degrees of surface cohesion. Natural weathered soil in arid environments tends to develop a crust over time that is highly resistant to suspension.[48] This surface is easily broken by footsteps, let alone by vehicular traffic and livestock grazing.[49,50] A large reservoir of suspendable material is often available beneath this crust. Desert pavements require many decades to reform, as evidenced by still-visible wagon tracks along emigrant trails blazed during the 1840s and 1850s.

Gillies et al.[51] estimated the strength and resilience of unpaved road surfaces by measuring the vertical force (kg/cm^2) needed for penetration with a Proctor Penetrometer®. This is similar in concept to the "rupture moduli" applied by Gillette et al.[50] Force measurements were taken across the road width every 0.25 m on different occasions and for different dust suppression treatments. Hard surfaces, similar to but stronger than natural desert crusts, experienced a brittle failure after application of a large force. The surface shattered, creating small aggregates and holes. Loose particles were exposed, as were edges of the surface that could be further ground down by tire wear. A flexible surface created by a polymer emulsion was penetrated at a relatively low force, but the size of the penetration was limited to the diameter of the Penetrometer, and no aggregates were created. While surface strength may be a good indicator of emissions potential for brittle surfaces, the nature of penetration is just as important. Vegetated and moist surfaces exhibit elastic characteristics similar to those of the polymer emulsion.

Wind Speeds

High wind speeds provide the energy needed to suspend loose particles from a surface; turbulence associated with these winds elevates particles to high altitudes where they can transport over long distances.[52–62] Wind erosion occurs in urban as well as non-urban areas. Figure 3 compares PM_{10} concentrations averaged for different wind speeds near a construction site in Las Vegas, NV, and at a location in the nearby non-urban desert. The urban site shows higher concentrations at wind speeds of 0 to 2 m/sec owing to accumulation of nearby emissions under stagnant conditions. PM_{10} levels begin to increase at wind speeds of 4 to 5 m/sec, attributable to windblown dust emissions from nearby land

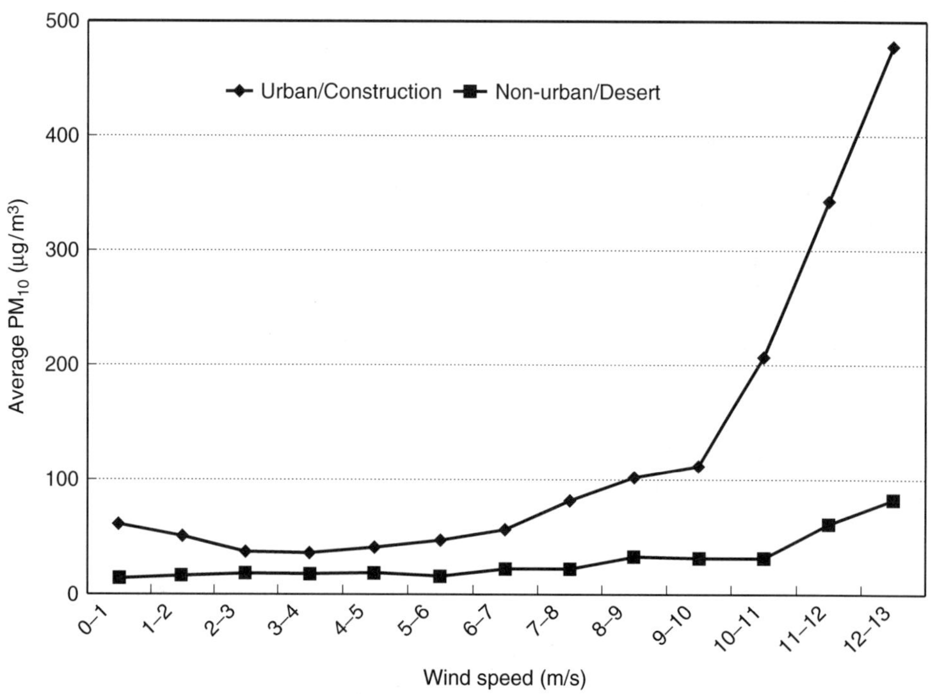

Figure 3. Average PM_{10} Classified by Wind Speed from Hourly Beta Attenuation Monitor Measurements at an Urban/Construction and a Non-Urban/Desert Site Near Las Vegas, NV During 1995.[63] Wind Speeds were Measured at 10 m agl.

parcels denuded for new construction.[14,63] Large increments in PM_{10} are not seen, however, until wind speeds exceed 7 m/sec with concentrations increasing rapidly for wind speeds in excess of 10 m/sec. Fortunately, these high winds are infrequent, occurring for only 83 h during 1995. Several of these hours were consecutive, however, and were associated with 24-h average PM_{10} exceeding 150 μg/m^3.

The nonurban desert site experienced a similar frequency of high winds, but it does not show significant increases in emissions until these speeds exceed 11 m/sec. A slight increase in PM_{10} is noticeable for wind speeds above 8 m/sec. This example refutes the argument that most urban dust derives from natural surfaces. Only when wind speeds are high and persistent do undisturbed areas provide measurable, though not major, fractions of PM_{10} or $PM_{2.5}$ in the atmosphere. Nonurban surfaces disturbed by off-road traffic, agricultural operations, and construction should not be classified as natural sources because they were created by human intervention.

Chepil and Woodruff[64] and Gillette and Hanson[65] show that the amount of dust suspended by wind depends on particle size distributions, wind speed at the surface, surface roughness, the relative fractions of erodible (<2 mm diameter) and nonerodible (>2 mm diameter) material, and the cohesion of the soil particles with one another. Values for each of these variables affect other variables. For example, a higher moisture content increases cohesion among particles and shifts the size distribution to larger particles. Larger agglomerations of small particles increase surface roughness, thereby decreasing wind speeds at the surface.

The effects of all these variables are embodied in a threshold friction velocity that is experimentally determined by placing a wind tunnel over an example of the affected soil and measuring the surface speed at which soil movement first becomes visible.[24,40,54,66–75] With the more common use of continuous particulate monitors,[76] threshold friction velocities might be inferred from hourly PM_{10}, $PM_{2.5}$, and wind-speed measurements at ambient sampling sites, as demonstrated in Figure 3. Averaging times of 1–5 min would provide more precise estimates than the hourly averages used in Figure 3.

Gillette[53] shows threshold friction velocities that vary from 0.19 to 1.82 m/sec for soils with different degrees of disturbance. Most ambient wind speed measurements such as those in Figure 3 are made at elevations between 5 and 10 m above ground level, and these must be translated to surface friction velocities to determine suspension. This is done using estimates of surface roughness and friction velocities from the actual or similar sites.[16] For this range of surface threshold values, emissions will be initiated at ambient wind speeds (measured at 7 m above the ground level, the height of most National Weather Service wind sensors) between 7 and 10 m/sec (26 to 37 km/h). Even though emissions begin at these velocities, the wind force contains insufficient energy to suspend very much of the erodible soil mass. The amount of dust suspended increases at approximately the cube of the wind speed above the threshold velocity. This is consistent with the measurements in Figure 3.

Particles suspended into the atmosphere are acted upon by gravity in a downward direction and by atmospheric resistance in an upward direction. Every particle attains an equilibrium between these forces at its terminal settling velocity. The settling velocity increases as the square of the particle diameter, and linearly with particle density.[77] For very small particles (<10 μm diameter), turbulent air movements in wind storms can counteract the gravitational settling velocity, and such particles can remain suspended for long times.[78–80] Transport distance depends on the initial elevation of a particle above ground level, the horizontal wind velocity component at the particle elevation, and the gravitational settling velocity.

Pye[81] shows vertical profiles for different-sized particles that might be elevated through a 100 m depth during a wind storm. The particles smaller than 10 μm are nearly uniformly distributed through this depth, while the larger particles exhibit much higher concentrations closer to the surface. Terminal settling velocities dominate most situations for >10 μm particles in the absence of violent winds.

Moisture Content

Water adhering to soil particles increases their mass and surface tension forces, thereby decreasing suspension and transport. Cohesion of wetted particles often persists after the water has evaporated due to the formation of aggregates and surface crusts. Soil moisture content is determined from representative samples swept or vacuumed from the potentially emitting surface and stored in an airtight container. A portion of the sample is weighed before and after 24-h heating at ~110°F in a laboratory drying oven, with the difference being equal to the evaporated water.[82,83] Low temperatures are used to minimize volatilization of materials other than water that might affect the weight change. Appendix C-2 of AP-42[16] provides detailed procedures.

Rosbury and Zimmer[32] found that moisture content affects the ejection of particles by vehicles, as well as the strength of the road bed and hence its ability to deform under vehicle weight. The addition of water to create surface moisture contents exceeding 2% resulted in >80% reductions for PM_{10} emissions compared to a control surface with an average moisture content of 0.56%.[33] Road surface-moisture content enhances the strength characteristics of surface crusts and the stability of aggregates.[84,85]

Kinsey and Cowherd[86] show how watering reduces emissions at a construction site. Significant dust control benefits are derived initially by doubling the area that is watered; however, emissions reductions are less pronounced as more water is applied to the site. Ultimately, control efficiency is limited because grading operations are continually exposing dry earth and burying the moistened topsoil.

Excessive moisture causes dust to adhere to vehicle surfaces so that it can be carried out of unpaved roads, parking lots, and staging areas. Carryout also occurs when trucks exit heavily watered construction sites.[87] This dust is deposited on paved (or unpaved) roadway surfaces as it dries, where it is available for suspension far from its point of origin. Fugitive dust emissions from paved roads are often higher after rainstorms in areas with many unpaved accesses, even though the rain may have flushed existing dust from the paved streets.

The same amount of moisture affects different dust surfaces in different ways. Moisture capacities of different geological materials are documented in soil surveys. Soil surveys include several indicators of the ability of soils to absorb moisture, with the most common being the "plastic limit." The plastic limit is the moisture content at which a soil changes from a semisolid to a plastic and is determined by adding water to a dry soil sample until it can be rolled into a coherent cylinder. Soil surveys also report liquid limits (the quantity of water required to create a slurry with the consistency of water), the infiltration rate (the movement of water through soil layers), and field moisture capacity.

The actual moisture content at a given time or place is not recorded and must be estimated. Thornthwaite[88] proposed the ratio of precipitation to evaporation as an indicator of the availability of moisture for soils. Thornthwaite's major concern was the agricultural potential of land in different areas. The precipitation–evaporation effectiveness index (P–E index) is ten times the sum of the monthly precipitation to evaporation ratios. Thornthwaite[88] classified North America as wet (P–E index > 128), humid (64 < P–E index < 128), subhumid (32 < P–E index < 64), semiarid (16 < P–E index < 32), or arid (P–E index < 16). Much of the Western United States is in the arid and semiarid categories. The P–E index has been used to estimate the moisture content of different soils, as an input to calculate emission factors for different surface types.

Precipitation events and the P–E index are crude methods of estimating soil moisture, which is likely to be affected by snow, fogs, and high humidity and to decrease over several days following heavy precipitation. An alternative model uses hourly rainfall, snowfall, relative humidity, and traffic volumes with monthly average Class A Pan evaporation rates to estimate soil moisture content.[89] Cowherd et al.[90] provide a map of average evaporation rates for 1946–1955.

The moisture content of soils varies throughout the year depending on the frequency and intensity of precipitation events, irrigation, and relative humidity and temperature of the surrounding air. Large amounts of rain falling during one month of a year will not be as effective in stabilizing dust as the same amount of rain interspersed at intervals throughout the year. Vehicle traffic enhances moisture evaporation by increasing air movement among the surface particles and exposing dry soil below the moist surface. Trees and other natural or manmade formations that evapo-transpire or cast shadows can also enhance or retain soil moisture content.

Vehicular Movement

The most common dust-suspending activity is vehicular movement on paved roads, unpaved roads, parking lots, and construction sites. This movement creates a reservoir of particles as well as providing the energy to inject them into the atmosphere.

Vehicle shape, speed, weight, number of wheels, as well as its previous history (e.g., dust acquisition for trackout) interact with different road surfaces to change the particle size, surface loading, wind effects, and surface moisture. Vehicular traffic in these areas adds to particle suspension because tire contact creates a shearing force with the road that lifts particles into the air.[91] Moving vehicles also create turbulent wakes that act much like natural winds to raise particles.[92] Natural crusts are often disturbed by vehicular movement, increasing the reservoir available for wind erosion.

Dust on paved roads must be continually replenished; minimizing the deposition of fresh dust onto these surfaces is a viable method for reducing their PM emissions. Dust loadings on a paved road surface build up by being tracked out from unpaved areas such as construction sites, unpaved roads, parking lots, and shoulders; by spills from trucks carrying dirt and other particulate materials; by transport of dirt collected on vehicle undercarriages; by wear of vehicle components such as tires, brakes, clutches, and exhaust system components; by wear of the pavement surface; by deposition of suspended particles from many emissions sources; and by water and wind erosion from adjacent areas.[1,93]

The relative contribution from each of these sources is unknown. Axetell and Zell[94] estimated typical deposition rates of 67.8 kg/km over a 24-h period for particles of all sizes from the following sources: (1) 42% from mud and dirt carryout; (2) 17% from litter; (3) 8% from biological debris; (4) 8% from ice control compounds (in areas with cold winters); (5) 8% from erosion of shoulders and adjacent areas; (6) 7% from motor vehicles; (7) 4% from atmospheric dustfall; (8) 4% from pavement wear; and (9) less than 1% from spills. Axetell and Zell[94] cite these fractions without describing the methodology used to estimate them.

Unpaved roads and other unpaved areas with vehicular activity are unlimited reservoirs when vehicles are moving. When regularly traveled, these surfaces are always being disturbed, and wind erosion seldom has an opportunity to decrease surface loadings or increase the surface roughness sufficiently to attenuate particle suspension. The grinding of particles by tires against the road surface shifts the size distribution toward smaller particles, especially those in the PM_{10} fraction. Pinnick et al.[95] found the distribution of particle sizes within a vehicle-created dust plume to be bimodal, with one mode peaking at ~50 μm and another mode peaking at ~2.5 μm. Patterson and Gillette[96] reported a similar distribution for naturally generated windblown dust plumes, with fewer large particles in the natural plume than in the vehicle-generated plume. The bimodal distribution was attributed to grinding processes caused by tires for the vehicle dust[95] and to a sandblasting process for wind-generated dust.[96] Nicholson et al.[91] show particle size and emissions rate increasing with vehicle velocity, consistent with higher energy transfer through surface contact and turbulent wakes at higher speeds.

Dyck and Stukel[97] hypothesized that vehicle weight and road type influence dust emissions. Mollinger et al.[98] found the shape of vehicles to have a large impact on the amount of dust suspension; a cylinder, an elliptical solid, and a rectangular solid were mounted on a pendulum that swung back and forth over dust-covered test areas. After twenty passes by the cylinder and elliptical solid, 65% and 45% of the dust remained in the test area, respectively. After twenty passes by the rectangular solid traveling at the same velocity, less than 20% of the dust remained. Moosmuller et al.[92] found that only high-profile vehicles, such as semi tractor trailers, produced sufficient turbulence to suspend dust along an unpaved shoulder next to a paved road. Normal passenger vehicles and pickup trucks produced negligible shoulder emmissions when traveling at speeds of 80 to 96 km/h.

Figure 4 shows the effect of several variables, especially surface loading, on emissions from a paved roadway. PM_{10} (light scattering equivalent) measured behind a vehicle tire on a clean-looking roadway was 10 to 100 times the PM_{10} in the surrounding air. PM_{10} measured behind the tire increased by another order of magnitude when the vehicle passed over a visible deposit from construction site trackout. Figure 4 demonstrates the variability of road emissions over short distances. It also suggests a method to evaluate roadway emissions potential that would couple high time-response PM_{10} measurements with geographic positioning system (GPS) tracking. These continuous measurements may be more closely related to actual emissions than sporadic silt content and surface loading samples.

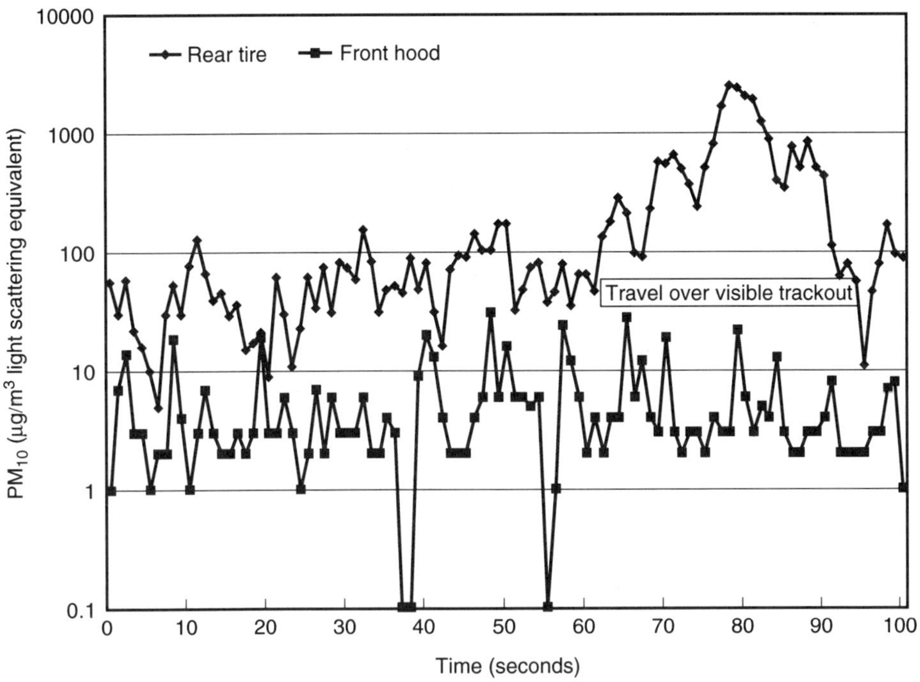

Figure 4. PM_{10} Equivalent Concentrations on a Paved Road on the Front Hood (~ 1 m Above Ground Level) and Behind the Right Front Tire (~ 0.25 m agl) of a Vehicle Moving at 10 mph. Concentrations Behind the Tire Increased Considerably with Travel Over a Surface with Visible Trackout from a Construction Site. One-Second PM_{10} Concentrations are Estimated Using a DUSTRAK® Photometer Calibrated with Arizona Road Dust.

Industrial Processes and Construction

In addition to paved and unpaved roads and disturbed areas, many industrial and construction sites that use mineral products have storage pile, material conveyance, and loading, digging, dozing, grading, scraping, and blasting activities. These activities create dust reservoirs, mechanically inject dust into the atmosphere, and present targets for wind erosion. Most of these operations are affected by the same variables described above, but the frequency, spatial extent, and magnitudes differ from those of urban and nonurban dust sources. Since most industrial operations are confined to private property, the combined PM_{10} and $PM_{2.5}$ emissions that travel beyond the property limits are of greatest concern to attainment of National Ambient Air Quality Standards. Visible clouds and deposition of large particles (>30 µm) on neighboring communities are often nuisances that need to be addressed.

Most modern industrial sites have designed the placement of their operations to minimize both emissions and transport of dust outside of property boundaries. Large blending domes, for example, are becoming more widely used in the cement industry to obtain a more homogeneous composition of the feed material to the kiln, thereby making better use of raw materials and minimizing upsets. Underground transport systems and covered conveyers are commonly used to minimize product loss. A fortunate by-product of these technological advances is to reduce fugitive dust from materials handling and plant upsets while increasing product volume and quality.

Other Dust-Emitting Activities

Botsford et al.[100] examined emissions from landfills, leafblowers, and equestrian facilities, and found them minor compared to those from paved roads, unpaved roads, construction, and wind erosion. These sources may be important within certain neighborhoods, however.

Botsford et al.[100] estimated landfill emissions to range from less than one ton per year for a small landfill about 650,000 m^2 and handling 1,150 m^3 of waste material to nearly 38 tons per year for a larger landfill about 5,120,000 m^2 in size and handling about 6,100 m^3 per day. For California and Nevada, it was concluded that landfills accounted for less than one percent of the local urban area fugitive dust emissions inventories.

Botsford et al.[100] made an aerodynamic calculation of potential dust suspension from leafblowers used in the Los Angeles area. With ~420,000 units used one day per week to clean an area of ~1,000 m^2, fugitive dust emissions could be almost 9 tons per day. This represented about 2% of the 1993 PM_{10} inventory for the South Coast Air Quality Management District.

Botsford et al.[100] used a box model to estimate emissions from horse-riding activities in the Los Angeles area. Based on surveys of equestrian centers in the Los Angeles area, it was concluded that about 13,000 horses were associated with equestrian centers, and that PM_{10} emissions from horse-riding activities were less than 0.2 tons per day, or less than 0.1% of the local fugitive dust inventory.

FUGITIVE DUST EMISSIONS CHARACTERIZATION

Fugitive dust emissions rates are determined by the following relationship:

$$E_{jkl} = R_{jkl}K_{jkl}A_{jkl}(1 - P_{jkl}) \tag{1}$$

where E_{jkl} is the emissions rate from source type j over time period k and area l; R_{jkl}, the rate of emissions (emissions factor) for a specific size fraction per unit of activity for source type j over time period k and area l; K_{jkl}, the particle size modifier applied to R_{jkl} when E_{jkl} is intended to represent a particle size fraction different from that represented by R_{jkl} (e.g., when $PM_{2.5}$ emissions are desired and emissions factors are only available for PM_{10} or TSP); this factor may be different for different source types j, time periods k, and area l; A_{jkl}, the activity that engenders dust emissions for source type j over time period k and area l; and P_{jkl}, the fractional reduction due to emissions controls applied to source j over time period k and area l.

Subscripts for source type, averaging time, and area are explicit in Eq. 1 to emphasize that these must be specified when estimating emissions. Some form of aggregation is always necessary owing to the static nature of most activity data. Averaging periods and spatial extent are usually much larger than those related to specific emissions events that might affect an ambient concentration. Chemical characterization of source and ambient samples provides an alternative for determining contributions to excessive 24-h PM_{10} or $PM_{2.5}$ concentrations, as discussed below.

U.S. EPA[16] provides regular updates on emissions factors, R_{jkl}, and methods for their application. The factors described below are consistent with 1999 practices, but emissions factors and their methods of application are continually being revised as new information becomes available.

Emissions factors are derived from experiments on specific sources that are believed to represent the general behavior of a source category. Most of the U.S. EPA emissions factors[16] are based on horizontal fluxes from an emitting area such as a road or vacant lot.[101] This is accomplished by locating sampling systems with the desired size-selective inlet (TSP, PM_{10}, or $PM_{2.5}$) at various elevations downwind of the dust-emitting area. Monitors located upwind are used to determine the flux into the emitting domain. This is most easily accomplished along roadways or open areas that have consistent traffic and confined boundaries.

Each of the downwind samplers is used to represent the amount of dust that is carried by the wind component perpendicular to a plane parallel to the source. Both the wind speed and concentration vary with height above ground level; so the horizontal flux is calculated through an area that extends above and below each sampler. These fluxes are added to obtain the aggregate emission rate from the source, after the flux of particles into the emitting area has been subtracted. Figure 5 shows a typical configuration for making these measurements.[102] Sampler elevations, distances from the source, and measurement devices vary among the different studies that have been conducted.

Figures 6 through 8 show the cumulative horizontal emissions fluxes at different elevations above ground level for paved roads, unpaved roads, and bare soil/construction sites.[99] These plots result from many of the same downwind profile tests that were used to derive commonly used PM_{10} emissions factors. The most noticeable feature from these plots is that 60% to 90% of the horizontal emissions flux is detected at elevations less than 1 to 2 m above ground level.

A 10-μm-aerodynamic-diameter particle has a settling velocity of ~0.3 cm/sec, and would therefore deposit to the surface within ~5 min after achieving an elevation of 1 m

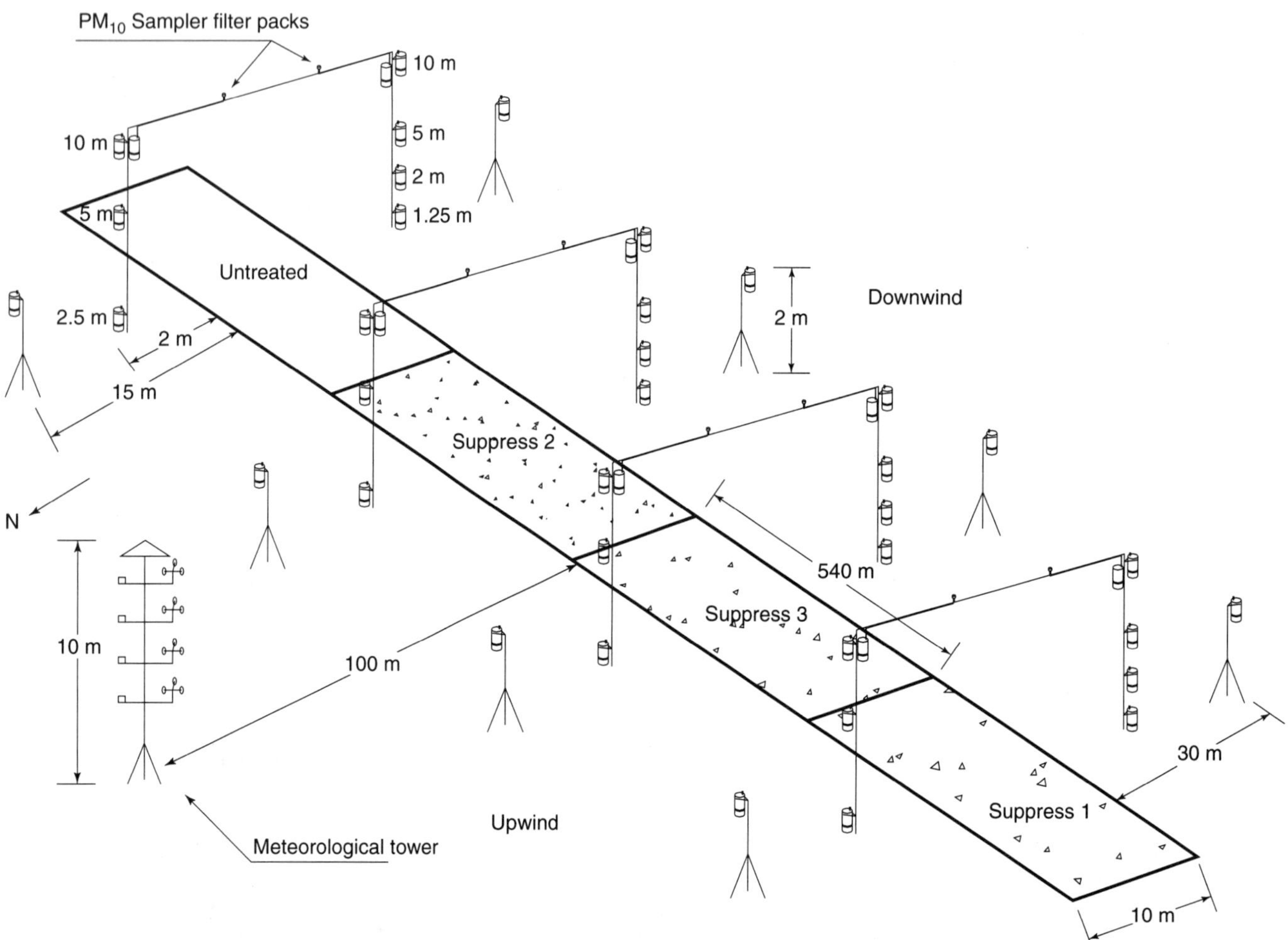

Figure 5. Example of Horizontal Flux Measurement System Around an Unpaved Road Treated with Different Suppressants[102].

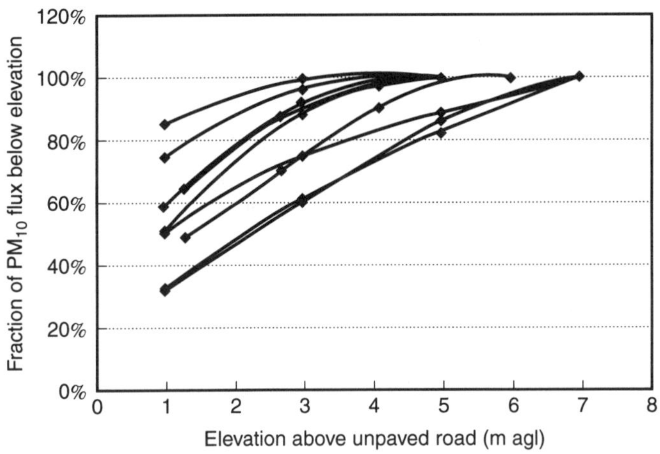

Figure 6. Cumulative Horizontal PM_{10} Flux at Different Downwind Elevations Above Different Unpaved Roads (Data from Cowherd et al.[99]).

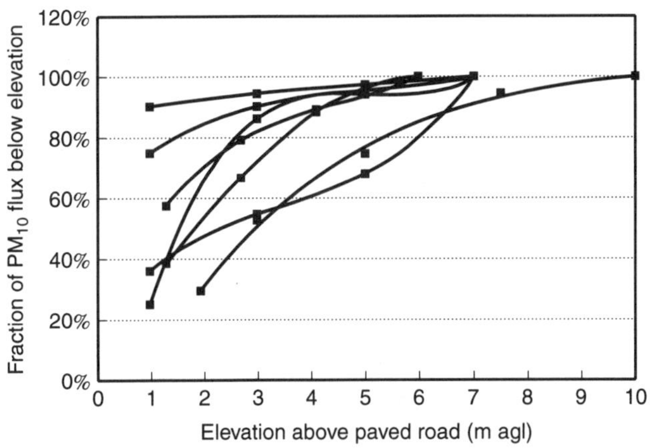

Figure 7. Cumlative Horizontal PM_{10} Flux at Different Elevations Above Different Paved Roads (Data from Cowherd et al.[99]).

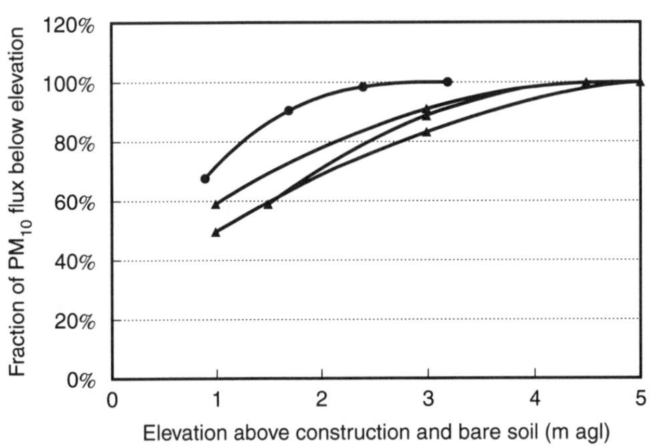

Figure 8. Cumulative Horizontal PM_{10} Flux at Different Downwind Elevations Above Different Construction and Bare Soil Sites. (Data from Cowherd et al.[99]).

above ground level. This corresponds to a travel distance of not more than 1 km in a 3-m/sec wind. Travel distances would be 0.25 km for a 20-μm particle and 4 km for a 5-μm particle under similar wind conditions. This assumes that these near-surface particles do not encounter low-level obstructions in their pathways. The rapid attenuation of PM_{10} concentrations due to deposition and vertical dispersion downwind of an unpaved road is illustrated in Figure 9.[102] While there is substantial scatter in the ratios of downwind to roadside concentrations, it is clear that PM_{10} is attenuated by ~90% within only 50 m from an unpaved roadside.

Dispersion models usually consider fugitive dust emissions at elevations higher than 2 m agl. Effective emissions heights of 5 to 10 m agl are often used so that a plume can develop downwind of a site. Most models are unable to resolve terrain variations of 1 to 2 m, and are therefore unable to account for the effects of horizontal impaction on gently rolling terrain and obstructions such as trees, shrubbery, and buildings.

There is little published information on the quantitative effect of nearby obstructions on low-level dust emissions, but it is common practice to place obstructions such as greenbelts[103,104] near visible dust emitters. Slinn[80] shows that dust deposition velocities through eucalyptus forest can be five to ten times higher than gravitational settling velocities with wind speeds of 5 to 10 m/sec.

Vertical flux is an alternative to horizontal flux as a method to estimate fugitive dust emissions.[105,106] Vertical flux is proportional to particle density, surface friction velocity, and the difference between particle concentrations at different elevations above ground level. Only a fraction of the horizontal flux moves upward, depending on meterological, aerosol, and surface variables. These variables can be obtained with a configuration similar to that of Figure 5.

Currently used emissions factors based on integrated horizontal flux adequately represent suspendable dust, the amount that leaves a dust-generating surface. These factors do not adequately represent transportable dust, the fraction of suspendable dust that is likely to travel more than a few hundred meters from the emitter. Future revisions to AP-42 may consider factors for horizontal fluxes above certain elevations that can be associated with appropriate zones of influence. For example, fluxes below 2 m would not be used to estimate impacts at distances larger than 1 to 10 km from the source, but would be available to estimate impacts on nearby receptors. Vertical flux estimates might be more appropriate for determining fugitive dust impacts over urban and larger spatial scales.

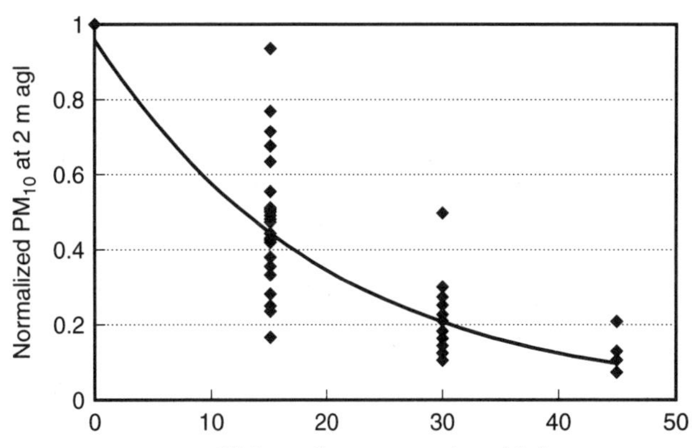

Figure 9. Attenuation of PM_{10} Concentrations with Distance from an Unpaved Road.[102].

Unpaved Road Dust Emissions

Results from field tests were used to develop unpaved road emissions factors. The PM_{10} emissions factor for unpaved roads is:

$$R_{ur.TSP} = 5.9(\text{silt}/12)(\text{speed}/30)(\text{weight}/3)^{0.7}(\text{wheels}/4)^{0.5} \quad (2)$$

where $R_{ur,PM_{10}}$ is the PM_{10} unpaved road dust emission factor for all vehicle classes combined (pounds/vehicle mile traveled); silt, the silt content of the surface material (% mass); speed, average speed (mph); weight, the average weight of all vehicle types combined (tons); and wheels, average number of wheels for all vehicle types.

Particle size modifiers are $K_{ur.PM_{10}} = 0.36$ and $K_{PM_{2.5}} = 0.0525$. For conversion to metric units of grams per vehicle kilometer traveled, multiply $R_{ur.PM_{10}}$ by 281.9.

Later emissions factors did not include variables for speed and number of wheels contacting the unpaved road surface, because these were not found to be statistically significant in the reanalysis of test data. U.S. EPA[16] provides ranges of average silt loadings for different source types from 5.1% to 24%. Individual sample silt loadings range from 0.1% to 68%. Equation 2 overestimates average emissions for vehicle speeds lower than 24 km/h.

A control effectiveness modifier due to rain is often appended to Eq. 2 to estimate annual averages: $P_{annual,county} =$ precidays/365, where precidays is the number of precipitation days per year with greater than 0.01 in. of rain as determined by the National Climatic Data Center. These data are usually sufficient to obtain spatial resolution for different counties within a state. This modifier assumes a 100% emissions reduction effectiveness for each day with more than 0.01 in. of precipitation.

Paved Road Dust

Results from regression analysis on data from 65 paved road dust tests that measured PM_{10} were used to derive the following paved road dust emissions factor:[3]

$$R_{pr,PM_{10}} = 0.016(\text{siltload}/2)^{0.065}(\text{weight}/3)^{1.5} \quad (3)$$

where $R_{pr,PM_{10}}$ is the PM_{10} paved road dust emission factor for all vehicle classes combined (pounds per vehicle mile traveled); siltload, the silt loading of the surface material (g/m^2); and weight, the average weight of all vehicle types combined (tons).

Particle size modifiers are $K_{pr,PM_{10}} = 1.0$ and $K_{pr,PM_{2.5}} = 0.247$. For conversion to metric units of grams per vehicle kilometer traveled, multiply $R_{ur,PM_{10}}$ by 287.5. Equation 3 is qualified for the range of variables that encompasses the tests from which it was derived. These are silt loadings between 0.02 and 400 g/m^2, mean vehicle weights of 2 to 42 tons, and mean vehicle speeds of 10 to 55 mph. Default silt loadings for normal roads are 0.1 g/m^2 for roads with more than 5,000 vehicles per day and 0.4 g/m^2 for roads with less than 5,000 vehicles per day. For dirty roads, such as those with visible carryout or road sand on them, the default values are 0.5 g/m^2 for more than 5,000 vehicles per day and 3 g/m^2 for fewer than 5,000 vehicles per day. These defaults may differ widely from actual values; great improvements in emissions estimates are gained with a few simple silt loading measurements.

Construction

The only available construction emission factor is a fixed value for total suspended particulate matter (TSP):

$$R_{const,TSP} = 1.2 \text{ tons/acre/month of activity} \quad (4)$$

Multiply $R_{const,TSP}$ by 2.24 to obtain emissions in mg/hectare/mo. No particle size modifiers are available. The acres of land under construction are usually estimated from the dollars spent on construction. The construction dust source category includes the building of residential structures, commercial structures, and roads. Emissions result from individual construction operations such as scraping, grading, loading, digging, compacting, light-duty vehicle travel, and other operations. Only the heavy earthmoving portion of a construction project approaches the emissions indicated by Eq. 4.[107] A better approach to estimating construction emissions is to break the project into its composite operations and apply individual emissions factors for unpaved roads, paved roads, and wind erosion.

Active Storage Piles

Active storage piles generate dust when material in dropped onto them from above in an unenclosed area. Material can be added in batches or continuously. The material is carried away from the pile by wind; so this is an important variable in emissions estimation. The emission factor for storage piles is:

$$R_{pile,PM_{10}} = 0.0011(\text{windspeed}/5)^{1.3}(\text{moisture}/2)^{-1.4} \quad (5)$$

where $R_{pile,PM_{10}} =$ PM_{10} storage pile emission factor (pounds/ton of material dropped); windspeed, the mean wind speed (mph); and moisture, the material moisture content (%).

Particle size modifiers are $K_{pile,PM_{10}} = 1.0$ and $K_{pile,PM_{2.5}} = 0.314$. Multiply $R_{pile,PM_{10}}$ by 0.5 to obtain units of kg/Mg of material dropped. Silt content was not found to be a significant variable in the regression analysis used to generate Eq. 5. The range of conditions for the emissions tests that generated Eq. 5 were silt contents of 0.44% to 19%, moisture contents of 0.25% to 4.8%, and wind speeds of 1.3 to 15 mph (21 to 24 km/h).

Wind Erosion

Since wind erosion depends on wind speeds at the surface exceeding the threshold velocity, hourly wind speed measurements at a known height above ground level are needed. Gusts or "fastest mile" winds are used in the estimation method. Gusts are the highest speed recorded for 60 sec or less during an hour. The fastest mile is reported by the National Weather Service as the speed corresponding to the fastest wind within an hour to traverse a full mile. The emission factor for wind erosion is:

$$R_{eros,PM_{10}} = 0.5\Sigma(58(u^* - u_t^*)^2 25(u^* - u_t^*) \quad (6)$$

Sum only for $(u^* - u_t^*) > 0$, and where $R_{eros,PM_{10}}$ is the PM_{10} emission factor (g/m^2 of available reservoir); u^*, the friction velocity at the surface (m/sec); and u_t^* the threshold suspension velocity at the surface (m/sec).

For a typical configuration with 0.05 surface roughness and measurements from a 10-m meteorological tower, the surface friction velocity is ~0.053 times the gust or fastest mile speed measured at 10 m. U.S. EPA[16] provides several examples and default values for surface roughnesses that can be used to adapt Eq. 6 to a variety of landforms. Particle size modifiers are $K_{eros,PM_{10}} = 1.0$ and $K_{pr,PM_{2.5}} = 0.30$.

CHEMICAL COMPOSITION

Fugitive dust contributions to PM_{10} and $PM_{2.5}$ concentrations in ambient air are easily distinguished from other source contributions by their chemical compositions. Hundreds of fugitive dust source profiles from many different areas have been measured.[108] Chemical source profiles are the fractional mass abundances of measured chemical species relative to primary PM_{10} or $PM_{2.5}$ mass in source emissions.

Source profile compilations (e.g., Watson and Chow,[4] Cowherd,[99] Watson,[109] Sheffield and Gordon,[110] Core and Houck,[111] Cooper et al.,[112] Houck et al.,[17, 113–115] Chow and Watson,[63, 116, 117] and Watson et al.[118–120]) include chemical abundances of elements, ions, and carbon for geological material (e.g., paved and unpaved road dust, soil dust, storage pile), motor vehicle exhaust (e.g., diesel-, leaded-gasoline-, and unleaded-gasoline-fueled vehicles), vegetative burning (e.g., wood stoves, fireplaces, forest fires, and prescribed burning), industrial boiler emissions, and other aerosol sources. More modern, research-oriented profiles include specific organic compounds or functional groups, elemental isotopes, and microscopic characteristics of single particles.[121,122]

Chow et al.[123] found that geological profiles typically contain large abundances of aluminum (Al), silicon (Si), potassium (K), calcium (Ca), and iron (Fe). The abundance of total potassium (K) is six to ten times the abundance of soluble potassium (K^+). The abundances of aluminum (Al), silicon (Si), potassium (K), and iron (Fe) are similar among fugitive dust profiles. Lead (Pb) is often enriched in paved road dust, even though leaded fuels are no longer used in the United States. Carbon constitutes 5% to 15% of road dust and some agricultural soils. Soluble ions such as sulfate (SO_4^{2-}), nitrate (NO_3^-), and ammonium (NH_4^+) are generally low, in the range of 0.1% to 0.2%. Sodium (Na) and chloride (Cl^-) are also low, except for situations when salt is used as a deicing agent.

Vehicle exhaust and burning are constituted mostly by carbonaceous material with elemental levels much less than 1% of mass emissions. Soluble potassium (K^+/K) ratios[124] are typically 0.80 to 0.90 in vegetative burning profiles, in contrast to the low soluble-to-total potassium ratios found in geological material.

Ducted industrial source emissions often contain a variety of different elements. Primary particle emitted by coal-fired power generators have the largest overlap in composition with fugitive dust sources. Crustal element abundances for silicon (Si), calcium (Ca), and iron (Fe) in coal-fired boiler profiles are 30% to 50% of the corresponding abundances in geological material. Aluminum (Al) abundances are similar to or higher than those found in geological material. Other elements such as phosphorus (P), potassium (K), titanium (Ti), chromium (Cr), manganese (Mn), strontium (Sr), zirconium (Zr), and barium (Ba) may be found in coal and certain soils. Selenium (Se) is not common in soils, but is usually found in coal-fired boiler emissions that do not pass through dry sulfur dioxide scrubbers. Watson et al.[119] found that limestone scrubbers processing hot exhaust gases removed selenium vapor before it could condense on fly ash particles.

The different sources of fugitive dust are not easily separated from each other because their chemical profiles are similar, but some attempts have been made to do so. Gatz and Prospero[125] applied Si/Al, Ca/Al ratios to separate transported North African dust from other sources. Gatz[126] and Gatz et al.[127] separated agricultural soil and unpaved roads based on their Ca/K ratios. Bruns et al.[128] identify different microbial species in different types of agricultural soils. Houck et al.[17] showed clear distinctions in Na, Cl, SO_4^{2-}, CO_3^{2-} abundances for alkaline lake beds relative to other sources. Davis et al.[129] identified clay and other specific minerals in a variety of soils. Finding additional markers that help to distinguish among fugitive dust emitters would improve source contribution estimates.

AIR POLLUTION CONTROL MEASURES

Mitigation measures include some combination of reducing suspendable dust reservoirs, preventing its deposit, stabilizing it, enclosing it, and reducing the activities that suspend it. These methods are applied with various degrees of effectiveness and diligence, often in response to local nuisance complaints rather than as part of a comprehensive emissions reduction strategy. Control effectiveness estimates, P_{jkl} in Eq. 1, vary considerably, and there is no single value appropriate for all situations. Table 1 summarizes several of these measures and the dust sources to which they can be applied. Ranges of effectiveness are discussed below.

Surface Watering

Surface watering is often applied on disturbed land such as construction sites or unpaved surfaces to reduce particle resuspension by vehicles. Flocchini et al.[33] found that the addition of sufficient water to increase the surface moisture content from 0.56% to 2% can achieve greater than 86% reduction in PM_{10} emissions. Kinsey and Cowherd[86] found immediate dust reductions at construction sites as a result of surface watering; however, the effectiveness of this measure did not increase as more water was applied to the site.

Chemical Suppression

The application of chemical suppressants on unpaved surfaces can reduce fugitive dust emissions. Watson et al.[102] enumerate commercially available dust suppressants. These products are classified into six categories according to their chemical composition and the suppressant mechanism they employ:

- Surfactants. Chemicals that reduce water surface tension and allow available moisture to more effectively wet the particles and aggregates in the surface layer.
- Salts. Hygroscopic compounds such as magnesium chloride or calcium chloride that adsorb water as ambient relative humidity exceeds 50%. Since salts are water soluble, precipitation tends to wash them away.
- Polymers. Long-chain molecular compounds that act as adhesives to bond soil particles together. Polymers may be able to stick to more particles than ordinary resins.
- Resin or petroleum emulsions. Non-water-soluble organic carbon compounds that are emulsified or suspended in water. When these emulsions are sprayed onto soil, they

Table 1. Fugitive Dust Emissions Control Methods

Control Method	Sources Controlled	Description
Street cleaning	Paved roads	Uses mechanical brushes, vacuum suction, regenerative-air suction, or blow-air/suction recirculation to remove street debris, litter, and dirt
Water flashing	Paved roads	Uses pressurized sprays of water or water with added surfactant to dislodge road dust and transport it into a drain system
Resurfacing	Paved roads	Repaving with nonerodible materials minimizes pavement cracks which trap and accumulate dust and minimizes pavement abrasion
Wet suppression	Unpaved roads	Water agglomerates small dust particles into larger entities that adhere to the surface, resist resuspension, or rapidly deposit after suspension
Windscreens	Disturbed open fields, storage piles, construction	Fabric, wood, or other light-weight material is placed upwind of an area to reduce surface winds, or downwind to intercept particles
Traffic controls	Paved roads, unpaved roads	Lower vehicle travel speeds, limited road usage, restriction of heavy-duty vehicle traffic, and providing parking opportunities and public transit reduces emissions levels
Carryout reduction	Construction, unpaved roads, paved roads	Wheel and truck washing on leaving a dusty site limits distribution of suspendable particles
Enclosures	Storage piles, open-haul trucks	Covering piles with tarps or building structures limits wind erosion
Chemical Stabilization	Unpaved roads, construction/demolition sites, storage piles, open fields	Surfactant or foaming agents bind fine particles into larger aggregates that reduce surface resuspension; some chemicals increase the surface dust moisture content to enhance particle agglomeration and fallout
Vegetative cover/ stabilization	Open fields	Vegetation reduces wind velocity at the surface and binds soil particles to the surfaces
Tilling implements, orientations, and frequencies	Agricultural fields	Wide-span planting equipment, no-till and minimum till planting procedures, land leveling, and sprinkler irrigation limit suspension of surface dust
Electrostatic fogger	Agricultural tilling, industrial processes	Electrostatically charged water droplets agglomerate suspended particles, thereby increasing particle size and deposition velocity
Skilled planting	Agricultural fields	Plug or punch planting, aerial seeding, and double cropping limit disturbance of surface materials

stick the soil particles together, and eventually harden to form a solid mass. Several emulsion products are based on tree resin, petroleum, or asphalt compounds.

- Bitumens. Materials such as asphalt or road oil that act as adhesives to bond soil particles together.
- Lignin sulfonate. A wood by-product from paper manufacture that forms a sticky but water-soluble layer on unpaved surfaces.

Most suppressants require repeated application at frequencies on the order of weeks or months. The effectiveness of chemical suppressants depends on road surface conditions, soil composition, application intensity, traffic volume, vehicle weight, and environmental factors such as precipitation and temperature. Prior to suppressant application, the road surface often needs to be graded or wetted. Most products can be dispensed as liquids by a truck equipped with a tank and spray bar. The spraying process injects the suppressant into the road material. Solid materials can be spread and mixed into the soil or road bed with a grader.

Several studies have examined the effectiveness of different chemical suppressants on various types of unpaved roads.[51,102] Rosbury and Zimmer[32] found considerable variability in dust emission rates for untreated and chemical-suppressant-treated surfaces. Much of this variation was attributed to the effects of ambient meteorological conditions (especially precipitation), the types of vehicles traveling on the road, and the initial road conditions. Control efficiency varied from 7% to 54% for well-mixed suppressant applications, from insignificant to 80% for topical suppressant applications.

Muleski and Cowherd[130] evaluated the effectiveness of chemical suppressants on unpaved roads and found emission rates similar to those found by Rosbury and Zimmer.[32] Average control efficiencies of ~50% or more were found within the first 30 days after suppressant application.

Grau[131] prepared suppressant-treated soil specimens under controlled laboratory conditions to determine their performance when subjected to simulated field conditions. The screening tests included repeated air impingement tests (1-min blasts from 80- and 160-km/h air jets at 20° from the horizontal) with simulated rainfall or jet fuel spills. Fortynine suppressants were tested, and suppressant effectiveness was determined based on visual observations. Only eleven suppressants were found acceptable.

Flocchini et al.[33] found that surface watering and road oiling reduced PM_{10} emissions by $87 \pm 6\%$ and $59 \pm 12\%$, respectively. Reducing vehicle travel speeds on unpaved roads from 40 to 16–24 km/h can effectively reduce PM_{10} emissions by $58 \pm 3\%$ and $42 \pm 35\%$, respectively.

Few studies evaluate the effectiveness of chemical suppressants over a long time period. Gillies et al.[51] and Watson et al.[102] describe a year-long study to assess changes in suppressant effectiveness in reducing PM_{10} emissions from unpaved public roads in the San Joaquin Valley, CA. Vehicle-induced vertical PM_{10} concentration profiles were measured. Three different types of chemical suppressants were applied on the unpaved roads: an acrylic co-polymer, a bitumen with copolymer additive, and a "biocatalyst." Gillies et al.[51] found that within a week of application the bitumen and acrylic copolymer showed over 95% emission reduction as compared to the untreated surface. The "biocatalyst" had an effectiveness of only 39%. After 3 months, the efficiency of the bitumen and biocatalyst had decreased by 20% and 30%, respectively, while the efficiency of the acrylic copolymer remained constant. After 11 months, the control effectiveness had been reduced to 53% for the bitumen, 85% for the acrylic copolymer and 0% for the "biocatalyst."[102]

Moosmüller et al.[92] applied several chemical suppressants to unpaved shoulders that were intended for stabilizing surfaces with minimal activity. They found that unpaved shoulders experience at least as much activity as unpaved roads, owing to vehicles leaving the roadway, mail carriers and school buses, and driveway entrances. None of the applied suppressants was effective a week after they were applied.

Street Sweeping

Mechanical broom and vacuum street sweeping have been applied in urban areas to remove street debris, litter, and dirt. Water droplets are often sprayed onto the road surface prior to sweeping to minimize dust resuspension induced by the sweeper. Sweeping schedules vary from weekly to monthly. Sweeping for aesthetic purposes is confined to the curb lanes in commercial and/or residential areas. Major highways and freeways are rarely swept.

Mechanical broom sweepers use large rotating brooms to lift the material from the street onto a conveyer belt. The conveyer discharges street debris into a collection hopper. Circular gutter brooms direct the debris into the path of the rotating broom. Mechanical broom sweeping has been questioned as a means of air pollution control on paved urban streets.[94,132,133] Chow et al.[93] found that the brushes resuspend as many small particles as they remove.

Commercially available vacuum sweepers use pure vacuum suction, regenerative-air suction, or blow-air/suction recirculation.[90,134–136] Vacuum sweepers use a gutter broom to loosen dirt and debris from the road surface and direct this material to a vacuum nozzle that sucks it into a hopper. The hopper usually consists of a chamber into which particles are collected by gravitational settling. The air passing through this chamber can be exhausted directly to the environment, through a bag-filter or precipitator, or to the collection nozzle for recirculation.

Pure vacuum sweepers create a strong vacuum within the pickup head that draws air from outside the head, through a duct, and into a hopper. The air movement across the road surface removes particles from the pavement and entrains them in the air flow. The vacuumed air is exhausted to ambient air after a brief residence time in the hopper. This residence time is usually insufficient to allow gravitational settling of PM_{10} or $PM_{2.5}$.

Regenerative-air vacuum sweepers direct the exhaust air back to one end of the pickup head at velocities between 50 and 200 m/sec. This blast air is directed perpendicular to the pavement where it is intended to dislodge particles. The blast air and its entrained particles move across the pickup head to a suction nozzle that transports the debris to the collection hopper. The pickup head must seal with the pavement using a flexible rubber curtain to prevent blast air from escaping the collection nozzle and to maintain a negative air pressure within the nozzle.

The blow-air/suction recirculation sweeper directs a portion of the exhaust air to a blast nozzle located immediately behind the pickup head. This blast air is directed at an angle to the pavement to blow particles from the road surface into the airstream caused by suction through the head. The suction flow rate exceeds the blast flow rate so that ambient air is always being drawn into the pickup head, thereby minimizing the escape of recirculated particles to the air. The nonrecirculated portion of the exhaust air is vented into a separate settling chamber before it escapes to ambient air.

Seton Johnson and Odell[137] described a 6-month street-sweeping study conducted in Portland, OR. Geological source contributions to chemically speciated TSP, PM_{15}, and $PM_{2.5}$ concentrations at a nearby sampling site were compared for sweeping and nonsweeping periods. No reductions in geological source contributions were detected with daily vacuum sweeping of the curb lane of industrial-area streets.

Hewitt[138] reported a 4-year study conducted in Bangor, ME, where ambient particle concentrations measured for 2 years without sweeping were compared with measurements from 2 years with sweeping. Nearly 800 TSP measurements were obtained and a 20% reduction in TSP as a result of vacuum street cleaning was reported. PEDCo Environmental[139] also measured TSP, PM_{15}, and $PM_{2.5}$ concentrations near paved roads in Denver, CO, at two sites that had been sanded during snowstorms. One site had the sand removed by sweeping, while the other site had no sand removal. Though a slight decrease in TSP and $PM_{2.5}$ was observed at the site where sand was removed, the difference was not statistically significant.

Cuscino et al.[140] made size-resolved vertical profile measurements near paved roads before and after sweeping. For PM_{15} mass concentrations, emissions reductions were estimated to be 16% 2 hours after sweeping, 51% 3 hours after sweeping, 0% 4 hours after sweeping, and 58% 24 h after sweeping. Cuscino et al.[140] note that meteorological variability may have played a large role in the differences in ambient mass concentrations among the measurement periods. In earlier studies, Cowherd[141] reported efficiencies of about 45% for PM_{15} and 35% for $PM_{2.5}$ when vacuum sweeping was used. Both of these vacuum-sweeping studies were performed on industrial paved roads, where initial street loadings and amount of material tracked on is typically much higher than on public streets.

Chow et al.[93] applied a receptor model to separate the contributions from fugitive road dust and vehicle exhaust. The ratio of these two contributions is less sensitive to meteorological and emission rate differences between sweeping and nonsweeping study periods than the absolute source contributions from either source. Chow et al.[93] found that daily street sweeping with a regenerative-air vacuum sweeper resulted in no detectable reductions in geological contributions to ambient PM_{10} measured in the sweeping area and that the street-sweeper design used in this study could not possibly reduce PM_{10} on pavements.

Fitz and Bumiller[142] investigated the effectiveness of street sweepers in reducing loose "blowsand" (material deposited on roadways after large-scale wind erosion events) on paved roads. They found that some street sweepers can successfully remove 97% to 99% of the loose surface material, whereas other sweepers resuspended as much PM_{10} as they removed.

While recently improved sweepers seem effectively to deplete the reservoir of material from which particles can be generated, none of these studies conclusively demonstrates the effectiveness of street sweeping on ambient concentrations of suspended PM_{10}. Suspended particle concentrations are affected by many variables, and differences caused by street sweeping may not be detectable at nearby monitoring sites. The variables that appear to influence emissions reduction efficiencies are: (1) loading of dirt on the street before and after sweeping; (2) particle size distribution of dirt on the street; (3) sweeper efficiency in removing dust from the street surface; (4) sweeper exhaust emission rates for small particles; (5) portion of roadway that is swept; (6) length of roadway that is swept; (7) sweeping frequency; and (8) meteorological variables such as precipitation, wind speed, wind direction, and relative humidity.

Water Flushing

Water flushing uses pressurized sprays from a water truck to dislodge road dust and transport it to the curb, where much of the particulate is washed into storm drains. A combination of water flushing followed immediately by broom sweeping has been used as a means of street cleaning.

Water flushing generally results in more consistent and higher efficiencies than sweeping, with 30% to 80% control effectiveness and 0 to 18 μg/m^3 reduction in nearby TSP measurements.[94,132,140,141] Flushing can be expected to reduce particle resuspension dramatically while the road surface is still wet, but its effect on average emissions during a typical cleaning cycle cannot be ascertained from the data reported.

The combination of water flushing and sweeping has the highest reported control effectiveness, from 47% to 90% for PM_{15} and 48% to 83% for $PM_{2.5}$.[140,141,143] The time frame of these measurements is either not specified or very short — less than 3 hours, when the road surface may still be wet. Large changes in control effectiveness over a short time period (e.g., 30% difference over a 12-min period) suggest that measurement uncertainty is high. Measurements were made on industrial paved roads; applicability to public roads has not been demonstrated.

Vehicle Washing

Washing vehicles as they depart from work areas can minimize dust trackout and carryout.[94,144] Carryout is the sediment attached to the vehicle that may eventually fall off and become resuspended by other vehicles. Axetell and Zell[94] found that TSP concentrations (~84 μg/m^3) increased by 40 to 60 μg/m^3 due to uncontrolled mud trackout. Immediate cleanup with shovel and broom reduced TSP by 10 to 20 μg/m^3, and daily cleanup reduced TSP by about 5 to 10 μg/m^3. These values correspond to control efficiencies of about 30% for immediate cleanup and 15% for daily cleanup.

Surface Roughness Alterations

Wind erosion emissions can be mitigated by: (1) altering the surface (such as increasing its roughness with tillage implements and practices), (2) vegetative stubble or mulch coverage, or (3) windbreaks. Applying tillage implements to ridge the soil or create large nonerodible clods is a standard practice in wind erosion control where the establishment of vegetative cover is difficult. This method results in large nonerodible roughness elements that absorb momentum from the erodible soil and trap the erodible soil in larger cracks and crevices.

No-till and minimum-till planting procedures minimize topsoil erosion, conserve water, and reduce dust emissions.[145] New seeds can be planted into stubble from the last crop or into a cropped field by cutting through the surface vegetation and inserting the seed directly into the ground. This type of low- or no-till farming reduces wind-generated fugitive emissions because vegetative cover (e.g., standing stubble or mulch) is left on the soil surface. The fraction of vegetative cover along with its shape and density of coverage determine the effectiveness of dust control.[146] Annual crops can be interplanted in narrow strips or rows to minimize ground-level soil movement. Interplanting is often supplemented with other practices such as protecting the strips with other vegetative coverage more efficiently to reduce dust emissions.

Windbreaks or other surface barriers absorb or deflect wind energy and minimize soil movement or particle resuspension from unprotected surfaces.[103,104,147] The length of a windbreak needs to be six to ten times its height to mitigate dust emissions effectively. In general, the shape, width, height, and porosity of the barrier, along with the surface wind speed and wind direction, affect the efficiency of the windbreak.

Minimization of Activity

Fugitive emissions caused by construction and agricultural activity can be reduced by minimizing activity levels. Vehicle-related resuspended dust can be reduced by regulating vehicle travel speeds, prohibiting road use by vehicles with more than four wheels, and reducing vehicle volume by providing perimeter parking and mass transit to industrial sites. Traffic controls are often supplemented with other control measures, such as surface watering and wheel washing.

SUMMARY AND CONCLUSIONS

Nonducted fugitive dust emissions arise from multiple sources that are often intermittent. National estimates show that unpaved roads, paved roads, construction projects, agricultural operations, and wind erosion are the largest contributors to fugitive dust. Industrial fugitive dust emissions are relatively small on a national basis, but they may be important in local circumstances.

The variables that affect emissions are complex and interrelated. Lacking a quantitative theoretical basis, emissions factors have been derived from empirical tests on representative sources that are intended to represent the entire population of emitters. The emission factors derived from these tests are applied to variables and activities that are specific to a source area. These variables include measures of surface particle loading, particle size, moisture content, and the type of activity.

Suspendable particles are not transportable particles. Sixty to ninety percent of PM_{10} mass is suspended within 1 to 2 m above ground level. These particles deposit to the surface or impact on nearby vertical structures within a few minutes after suspension. Although horizontal fluxes commonly used in empirically derived emission factors represent the mass of dust particles suspended from a surface, they do not represent the mass entrained into the atmosphere and transported over relevant distances of 0.1 to >10 km. The fraction of dust within the lowest 2 m above ground level probably deposits the surface or horizontally impacts on nearby obstructions and does not travel far from its source.

Fugitive dust control measures involve combinations of reducing suspendable dust reservoirs, preventing its deposit, stabilizing it, enclosing it, and reducing the activities that suspend it. Demonstrations studies on representative emitters provide a wide range of control effectiveness. The effectiveness of controls depends on the specific situation as well as the diligence with which it is applied. Tests are needed to demonstrate fugitive dust control effectiveness for specific circumstances.

DISCLAIMER

The material presented here represents the research and interpretation of the authors and does not represent the policy of the Desert Research Institute, the University and Community College System of Nevada, or the U.S. Environmental Protection Agency. The mention of commercial products does not constitute an endorsement or recommendation of those products.

Acknowledgments

Information presented in this chapter results from several fugitive dust studies undertaken at the Desert Research Institute since 1985. These projects were sponsored by the State of Nevada, Clark County Department of Comprehensive Planning, Washoe County District Health Department, California Air Resources Board, San Joaquin Valley Unified Air Pollution Control District, South Coast Air Quality Management District, Southern California Edison, and the U.S. Environmental Protection Agency. Useful information and contributions to these projects were obtained from Hampden Kuhns, Jack Gillies, Hans Moosmüller, Fred Rogers, Mark Green, and Nick Lancaster of the Desert Research Institute; David James of the University of Nevada at Las Vegas; Will Cates of the Clark County Department of Comprehensive Planning; Duane Ono from the Great Basin Unified Air Pollution Control District; William Barnard of E. H. Pechan and Associates; William Benjey, Ronald Myers, William Kuykendal, and Thomas Rosendahl of the U.S. Environmental Protection Agency; John Core of Core Environmental Consulting; Chatten Cowherd of the Midwest Research Institute; Dennis Fitz of the College of Engineering, Center for Environmental Research and Technology, University of California, Riverside; Donald Gatz of the Illinois State Water Survey; Dale Gillette of the National Oceanic and Atmospheric Administration; William Nickling of the University of Guelph, Canada; Dale Shimp of the California Air Resources Board; Melvin Zeldin of the South Coast Air Quality Management District; and George Slinn from Richland, WA.

REFERENCES

1. J. C. Chow and J. G. Watson, Fugitive emissions add to air pollution, *Environ. Protect.* **3**, 26–31 (1992).
2. D. L. Coe and L. R. Chinkin, "The use of a day-specific source activity database to augment CMB source apportionment modeling," in *Proceedings, $PM_{2.5}$: A Fine Particle Standard*, J. C. Chow and P. Koutrakis, Eds., Air & Waste Management Association, Pittsburgh, PA, pp. 463–74, 1998.
3. U.S. EPA, "National ambient air quality standards for particulate matter: Final rule." *Federal Register* **62**, 38651–760 (1997).
4. J. G. Watson and J. C. Chow, "Clear sky visibility as a challenge for society." *Annual Rev. Energy Environ.* **19**, 241–66 (1994).
5. U.S. EPA, "National air pollutant emission trends, 1997," Report No. EPA-454/R-98-016. U.S. Environmental Protection Agency, 1998.
6. W. R. Barnard and M. A. Stewart, "Modification of the 1985 NAPAP wind erosion methodology to assess annual PM_{10} emissions for EPA's Emissions Trends Report" in *Transactions, PM_{10} Standards and Non-Traditional Particulate Source Controls*, J. C. Chow and D. M. Ono, Eds., Air and Waste Management Association, Pittsburgh, PA, pp. 121–30, 1992.
7. J. G. Watson, "Overview of receptor model principles." *JAPCA* **34**, 619–23 (1984).
8. J. G. Watson, J. A. Cooper, and J. J. Huntzicker, "The effective variance weighting for least squares calculations applied to the mass balance receptor model." *Atmos. Environ.* **18**, 1347–55 (1984).
9. J. G. Watson, N. F. Robinson, J. C. Chow, R. C. Henry, B. M. Kim, T. G. Pace, E. L. Meyer, and Q. Nguyen, "The USEPA/DRI chemical mass balance receptor model, CMB 7.0." *Environ. Software* **5**, 38–49 (1990).
10. J. G. Watson, J. C. Chow, and T. G. Pace, "Chemical mass balance" in *Receptor Modeling for Air Quality Management*, P. K. Hopke, Ed., Elsevier, New York, NY, pp. 83–116, 1991.
11. J. G. Watson, N. F. Robinson, C. W. Lewis, C. T. Coulter, J. C. Chow, E. M. Fujita, D. H. Lowenthal, T. L. Conner, R. C. Henry, and R. D. Willis, Chemical Mass Balance Receptor Model Version 8 (CMB) User's Manual. Prepared for U.S. Environmental Protection Agency, Research Triangle Park, NC, by Desert Research Institute, Reno, NV, 1997.
12. J. C. Chow and J. G. Watson, "Summary of particulate data bases for receptor modeling in the United States," in *Transactions: Receptor Models in Air Resources Management*, J. G. Watson, Ed., Air & Waste Management Association, Pittsburgh, PA, pp. 108–33, 1989.
13. J. G. Watson and J. C. Chow, "Data bases for PM_{10} and $PM_{2.5}$ chemical compositions and source profiles," in PM_{10} *Standards and Nontraditional Particulate Source Controls*, J. C. Chow and D. M. Ono, Eds., Air & Waste Management Association, Pittsburgh, PA, pp. 61–91, 1992.
14. J. C. Chow, J. G. Watson, M. C. Green, D. H. Lowenthal, D. W. DuBois, S. D. Kohl, R. T. Egami, J. A. Gillies, C. F. Rogers, C. A. Frazier, and W. Cates, "Middle- and neighborhood-scale variations of PM_{10} source contributions in Las Vegas, Nevada." *JAWMA* **49**, 641–54 (1999).
15. A. Venkatram, D. Fitz, K. Bumiller, S. Du, M. Boeck, and C. Ganguly, "Using a dispersion model to estimate emission rates of particulate matter from paved roads." *Atmos. Environ.* **33**, 1093–102 (1999).
16. U.S. EPA, Compilation of Air Pollutant Emission Factors, AP-42, 5th ed. U.S. Environmental Protection Agency, Research Triangle Park, NC, 1999. The methods for estimating the emissions from unpaved roads is currently being reviewed. EPA's updates are on their Technology Transfer Network's CHIEF page at http://www.epa.gov/ttn/chief/. Also, the introduction to AP-42 provides substantial discussion on the appropriate use of the EPA's emission estimation procedures. This information can be viewed at http://www.epa.gov/ttn/chief/ap42.html#chapter by opening the file Introduction to AP-42, Volume 1, Fifth Edition.
17. J. E. Houck, J. C. Chow, and M. S. Ahuja, "The chemical and size characterization of particulate material originating from geological sources in California," in *Transactions, Receptor Models in Air Resources Management*, J. G. Watson, Ed., Air & Waste Management Association, Pittsburgh, PA, pp. 322–33, 1989.
18. M. S. Ahuja, J. J. Paskind, J. E. Houck, and J. C. Chow, "Design of a study for the chemical and size characterization of

particulate matter emission from selected sources in California," in *Transactions, Receptor Models in Air Resources Managements*, J. G. Watson, Ed., Air & Waste Management Association, Pittsburgh, PA, pp. 145–58, 1989.

19. J. E. Houck, J. M. Goulet, J. C. Chow, J. G. Watson, and L. C. Pritchett, "Chemical characterization of emission sources contributing to light extinction," in *Transactions, Visibility and Fine Particles*, C. V. Mathai, Ed., Air and Waste Management Association, Pittsburgh, PA, pp. 437–46, 1990.
20. D. A. Lundgren and R. M. Burton, "Effect of particle size distribution on the cut point between fine and coarse ambient mass fractions," *Inhalation Toxicology* **7**, 131–48 (1995).
21. W. C. Hinds, *Aerosol Technology: Properties, Behavior, and Measurement of Airborne Particles*, John Wiley & Sons, New York, NY, 1982.
22. American Society for Testing and Materials, "Standard test method for amount of material in soils finer than the No. 200 (75 μm) sieve," Method D1140–97. ASTM, West Conshohocken, PA, 1997.
23. American Society for Testing and Materials, "Standard test method for particle-size analysis of soils," Method D422–63 (1998). ASTM, West Conshohocken, PA, 1998.
24. D. A. Gillette, J. B. Adams, A. Endo, D. Smith, and R. Kihl, "Threshold velocities for input of soil particles into the air by desert soils," *J. Geophys. Res.* **85**, 5621–30 (1980).
25. C. Cowherd, P. J. Englehart, G. E. Muleski, J. S. Kinsey, and K. D. Rosbury, "Control of fugitive and hazardous dusts." Noyes Data Corporation, Park Ridge, NJ, 1990.
26. W. S. Chepil, "Improved rotary sieve for measuring state and stability of dry soil structure," *Soil Science Society of America Proceedings* **16**, 1113–17 (1952).
27. R. A. Bagnold, "The transport of sand by wind," *Geog. J.* **89**, 409–38 (1937).
28. W. S. Chepil, "Measurement of wind erosiveness by dry sieving procedure," *Sci. Agr.* **23**, 154–60 (1942).
29. D. A. Gillette and P. H. Stockton, "The effect of nonerodible particles on wind erosion of erodible surfaces," *J. Geophys. Res.* **94**, 12885–85 (1989).
30. M. Logie, "Influence of roughness elements and soil moisture of sand to wind erosion," *Catena Supp.* **1**, 161–73 (1982).
31. R. A. Bagnold, *The Physics of Blown Sand Desert Dunes*, Methuen, London, 1941.
32. K. D. Rosbury and R. A. Zimmer, "Cost-effectiveness of dust controls used on unpaved haul roads, Volume I—Results, analysis, and conclusions." Prepared for Bureau of Mines, U.S. Dept. of the Interior, by PEDCo Environmental, Inc., 1983.
33. R. G. Flocchini, T. A. Cahill, R. T. Matsumura, O. Carvacho, and Z. Lu, "Study of fugitive PM_{10} emissions from selected agricultural practices on selected agricultural soils." University of California, Davis, CA, 1994.
34. O. F. Carvacho, L. L. Ashbaugh, R. T. Matsumura, R. J. Southard, and R. G. Flocchini, "Measurement of PM_{10} potential from agricultural soils using a dust resupension test chamber." International Conference on Air Pollution from Agricultural Operations, Kansas City, MO, 1996.
35. D. Shimp, S. Campbell, and S. Francis, "Spatial distribution of PM_{10} emissions from agricultural tilling in the San Joaquin Valley," in *Proceedings, Geographic Information Systems in Environmental Resources Management*, Air & Waste Management Association, Pittsburgh, PA, 1996.
36. L. R. Anspaugh, J. H. Shinn, P. L. Phelps, and N. C. Kennedy, "Resuspension and redistribution of plutonium in soils," *Health Phys.* **29**, 571–82 (1975).
37. G. S. Linsley, "Resuspension of the trans-uranium elements—a review of existing data," Report No. NRPB-R75. HSMO, London, 1978.
38. J. A. Garland, *Precipitation, Scavenging, Dry Deposition and Resuspension*, H. R. Pruppacher, R. G. Semonin, and W. G. N. Slinn, Eds., pp. 1087–97, 1983.
39. M. W. Reeks, J. Reed, and D. Hall, "The long-term suspension of small particles by a turbulent flow—Part III: Resuspension for rough surfaces." Report No. TPRD/B/0640/N85, CEGB Berkeley Nuclear Labs, Gloucestershire, U.K., 1985.
40. K. W. Nicholson, "Wind tunnel experiment on the resuspension of particulate material," *Atmos. Environ.* **27A**, 181–88 (1993).
41. J. K. Marshall, "Drag measurements in roughness arrays of varying density and distribution," *Agricultural Meteorology* **8**, 269–92 (1971).
42. M. R. Raupauch, "Drag and drag partition on rough surface," *Boundary Layer Meteorology* **60**, 375–95 (1992).
43. R. A. Zimmer, W. Reeser, and P. Cummins, "Evaluation of PM_{10} emission factors for paved streets," in *Transactions: PM_{10} Standards and Nontraditional Particulate Source Controls*, J. C. Chow and D. M. Ono, Eds., Air & Waste Management Association, Pittsburgh, PA, pp. 311–23, 1992.
44. South Coast Air Quality Management District, "Inventory of PM_{10} emissions from fugitive dust sources in the South Coast Air Basin," South Coast Air Quality Management District, El Monte, CA, 1991.
45. R. Greeley and J. D. Iversen, *Wind as a Geological Process*, Cambridge University Press, Cambridge, 1985.
46. F. Pasquill and F. B. Smith, *Atmospheric Diffusion*, Ellis Horwood, Chichester, England, 1983.
47. J. D. Wilson, J. J. Finnigan, and M. R. Raupach, "A first-order closure for disturbed plant-canopy flows, and its application to winds in a canopy on a ridge," *Q. J. R. Meteorol. Soc.* **124**, 705–32 (1998).
48. J. Belnap and D. A. Gillette, "Disturbance of biological soil crusts: Impacts on potential wind erodibility of sandy desert soils in southeastern Utah," in *Land Degradation and Rehabilitation*, pp. 355–62, 1997.
49. D. A. Gillette, J. B. Adams, A. Endo, and D. Smith "Threshold friction velocities on typical Mojave Desert soils, undisturbed and disturbed by off-road vehicles," in *Proceedings, Int. Powder and Bulk Solids Handling and Processing*, Ind. and Sci. Conf. Manage., Chicago, IL, 1979.
50. D. A. Gillette, J. B. Adams, D. R. Muhs, and R. Kihl, "Threshold friction velocities and rupture moduli for crusted desert soils for the input of soil particles into the air," *J. Geophys. Res.* **87**, 9003–15 (1982).
51. J. A. Gillies, J. G. Watson, C. F. Rogers, D. Dubois, J. C. Chow, R. Langston, and J. Sweet, "Long term efficiencies of dust suppressants to reduce PM_{10} emissions from unpaved roads," *JAWMA* **49**, 3–16 (1999).
52. R. A. Duce, C. K. Unni, B. J. Ray, J. M. Prospero, and J. T. Merrill, "Long-range atmospheric transport of soil dust from Asia to the tropical North Pacific: Temporal variability," *Science* **209**, 1522–24 (1980).
53. D. A. Gillette, "Major contributors of natural primary continental aerosols: Source mechanisms," in *Aerosols, Anthropogenic and Natural Sources and Transport*, The New York Academy of Sciences, New York, NY, pp. 348–58, 1980.
54. D. A. Gillette, "Production of dust that may be carried great distances," in *Desert Dust: Origin, Characteristics, and Effect on Man*, T. Pewe, Ed., Geological Society of America, Boulder, CO, pp. 11–26, 1981.
55. D. A. Gillette and I. H. Blifford, "Composition of tropospheric aerosols as a function of altitude," *Journal of the Atmospheric Sciences* **28**, 1199–99.
56. D. A. Gillette, I. H. Blifford, and C. R. Fenster, "Measurements of aerosol-size distributions and vertical fluxes of aerosols on land subject to wind erosion," *J. Appl. Meteorol.* **11**, 977–87 (1972).

57. D. A. Gillette, R. N. Clayton, T. K. Mayeda, M. L. Jackson, and K. Sridhar, "Tropospheric aerosols from some major dust storms of the southwestern United States," *J. Appl. Meteorol.* **78**, 832–45 (1978).

58. D. A. Gillette, R. N. Clayton, T. K. Mayeda, M. L. Jackson, and K. Sridhar, "Tropospheric aerosols from some major dust storms of the southwestern United States," *J. Geophys. Res.* **87**, 9003–15 (1978).

59. J. M. Prospero, E. Bonatti, C. Schubert, and T. N. Carlson, "Dust in the Caribbean atmosphere traced to an African dust storm," *Earth Planet. Sci. Lett.* **9**, 287–93 (1970).

60. J. M. Prospero, R. A. Glaccum, and R. T. Nees, "Atmospheric transport of soil dust from Africa to South America," *Nature* **289**, 570–72 (1981).

61. J. M. Prospero, M. Uematsu, and D. L. Savoie, "Mineral aerosol transport to the Pacific Ocean," in *Chemical Oceanography*, J. P. Riley, R. Chester, and R. A. Duce, Eds., Academic, San Diego, CA, pp. 188–218, 1989.

62. J. M. Prospero and T. N. Carlson, "Vertical and aerial distribution of Saharan dust over the western equatorial North Atlantic Ocean," *J. Geophys. Res.* **77**, 5255–65 (1972).

63. J. C. Chow and J. G. Watson, "Fugitive dust and other source contributions to PM_{10} in Nevada's Las Vegas Valley," Report No. DRI 4039.1F, prepared for Clark County Department of Comprehensive Planning, Las Vegas, NV, by Desert Research Institute, Reno, NV, 1997.

64. W. S. Chepil and N. P. Woodruff, "The physics of wind erosion and its control," *Adv. Agron.* **15**, 211–302 (1963).

65. D. A. Gillette and K. Hanson, "Spatial and temporal variability of dust production caused by wind erosion in the United States," *J. Geophys. Res.* (1989).

66. S. Borrman and R. Jaenicke, "Wind tunnel experiments on the resuspension of submicrometer particles from a sand surface," *Atmos. Environ.* **21**, 1891–98 (1987).

67. D. A. Braaten, "Wind tunnel experiments of large particle reentrainment—Decomposition and development of large particle scaling parameters," *Aerosol Sci. Technol.* **21**, 157–69 (1994).

68. P. Giess, A. J. H. Goddard, G. Shaw, and D. T. Allen, "Resuspension of monodisperse particles from short grass swards: A wind tunnel study," *J. Aerosol Sci.* **25**, 843–58 (1994).

69. D. A. Gillette, "A wind tunnel simluation of the erosion of soil: Effect of soil texture, sandblasting, wind speed, and soil consolidation on dust production," *Atmos. Environ.* **12**, 1735–35 (1978).

70. D. A. Gillette, "Tests with a portable wind tunnel for determining wind erosion threshold velocities," *Atmos. Environ.* **12**, 2309–13 (1978).

71. D. A. Gillette, "Production of dust which may be carried great distances," in *Desert Dust: Origins, Characteristics, and Effect on Man*, Geological Society of America, 1982.

72. N. Rajendran and F. Fraftz, "A portable wind tunnel for sampling entrainable PM_{10} dust from soils and materials piles," J. C. Chow and D. M. Ono, Eds., Air & Waste Management Association, Pittsburgh, PA, 1992.

73. M. R. Raupauch, A. S. Thom, and I. Edwards, "A wind tunnel study of turbulent flow close to regularly arrayed rough surfaces," *Boundary Layer Meteorology* **18**, 373–97 (1980).

74. G. T. Visser, "A wind-tunnel study of the dust emissions from the continuous dumping of coal," *Atmos. Environ.* **26A**, 1453–60 (1992).

75. A. W. Zingg, "A portable wind tunnel and dust collector developed to evaluate the erodibility of field surfaces," *Agron. J.* **43**, 189–91 (1951).

76. J. G. Watson, J. C. Chow, H. Moosmüller, M. C. Green, N. H. Frank, and M. L. Pitchford, "Guidance for using continuous monitors in $PM_{2.5}$ monitoring networks, Report No. EPA-454/R-98-012, U.S. Environmental Protection Agency, Research Triangle Park, NC, by Desert Research Institute, Reno, NV, 1998.

77. S. K. Friedlander, *Smoke, Dust, and Haze*, Wiley and Sons, New York, NY, 1977.

78. G. A. Sehmel, "Particle resuspension: A review," *Environ. Int.* **4**, 107–27 (1980).

79. G. A. Sehmel, "Deposition and resuspension," in *Meteorology and Power Production*, D. Randerson, Ed., U.S. Department of Energy, 1984.

80. W. G. N. Slinn, "Predictions for particle deposition to vegetative canopies," *Atmos. Environ.* **16**, 1785–94 (1982).

81. K. Pye, *Aeolian Dust and Dust Deposits*, Academic Press, London, England, 1987.

82. T. Weems, "Survey of moisture measurement instruments," *Irrigation Journal* **41**, 12–16 (1991).

83. T. W. Ley, "An in-depth look at soil water monitoring and measurement tools," *Irrigation Journal* **44**, 8–20 (1994).

84. J. M. Bradford and R. B. Grosman, "In-situ measurements of near-surface soil strength by the fall-cone device," *Soil Science Society of America Journal* **46**, 685–88 (1982).

85. G. A. Lehrsch and P. M. Jolley, "Temporal changes in wet aggregate stability," *Transactions of the ASAE* **35**(2), 493 (1992).

86. J. S. Kinsey and C. Cowherd, "Fugitive emissions," in *Air Pollution Engineering Manual, First Edition*, A. J. Buonicore and W. T. Davis, Eds., Van Nostrand Reinhold, New York, pp. 133–46, 1992.

87. P. J. Englehart and J. S. Kinsey, "Study of construction related mud/dirt carryout," U.S. Environmental Protection Agency, Region V, Chicago, IL, 1983.

88. C. W. Thornthwaite, "The climates of North America according to a new classification," *Geographical Review* **21**, 633–55 (1931).

89. J. G. Watson and J. C. Chow, "Reconciling urban fugitive dust emissions inventory and ambient source contribution estimates: Summary of current knowledge and current research," prepared for U.S. Environmental Protection Agency, Research Triangle Park, NC, by Desert Research Institute, Reno, NV, 1999.

90. C. Cowherd, G. E. Muleski, and J. S. Kinsey, "Control of open fugitive dust sources," Report No. EPA-450/3-88-008, prepared for Office of Air Quality Planning and Standards, U.S. Environmental Protection Agency, by Midwest Research Institute, Kansas City, MO, 1988.

91. K. W. Nicholson, J. R. Branson, P. Giess, and R. J. Cannell, "The effects of vehicle activity on particle resuspension," *J. Aerosol Sci.* **20**, 1425–28 (1989).

92. H. Moosmüller, J. A. Gillies, C. F. Rogers, D. W. DuBois, J. C. Chow, J. G. Watson, and R. Langston, "Particulate emission rates for unpaved shoulders along a paved road," *JAWMA* **48**, 398–407 (1998).

93. J. C. Chow, J. G. Watson, R. T. Egami, C. A. Frazier, and Z. Lu, "Evaluation of regenerative-air vacuum street sweeping on geological contributions to PM_{10}," *JAWMA* **40**, 1134–42 (1990).

94. K. Axetell and J. Zell, "Control of reentrained dust from paved streets," Report No. EPA-907/9-77-007, U.S. Environmental Protection Agency, Region VII, Kansas City, MO, 1977.

95. R. G. Pinnick, G. Fernandez, B. D. Hinds, C. W. Bruce, R. W. Schaefer, and J. D. Pendleton, "Dust generated by vehicular traffic on unpaved roadways: Sizes and infrared extinction characteristics," *Aerosol Sci. Technol.* **4**, 99–121 (1985).

96. E. M. Patterson and D. A. Gillette, "Measurements of visibility vs. mass-concentration for airborne soil particles," *Atmos. Environ.* **11**, 193–96 (1977).

97. R. I. Dyck and J. J. Stukel, "Fugitive dust emissions from trucks on unpaved roads," *Environ. Sci. Technol.* **10**, 1046–48 (1976).

98. A. M. Mollinger, F. T. M. Nieuwstadt, and B. Scarlett, "Model experiments of the resuspension caused by road traffic," *Aerosol Sci. Technol.* **19**, 330–38 (1993).

99. C. Cowherd, "Profiling for open fugitive dust sources," prepared for U.S. Environmental Protection Agency, Emissions Factors Inventory Group, Research Triangle Park, NC, by Midwest Research Institute, Kansas City, MO, 1999.

100. C. Botsford, D. Lisoski, and W. Blackman, "Fugitive dust study characterization of uninventoried sources," prepared for South Coast Air Quality Management District, Diamond Bar, CA, Aero Vironment, Monrovia, CA, 1996.

101. C. Cowherd and P. J. Englehart, "Paved road particulate emissions: Source category report," Report No. EPA-600/7-84-077, U.S. Environmental Protection Agency, Research Triangle Park, NC, 1984.

102. J. G. Watson, J. C. Chow, J. A. Gillies, H. Moosmüller, C. F. Rogers, D. Dubois, and J. C. Derby, "Effectiveness demonstration of fugitive dust control methods for public unpaved roads and unpaved shoulders on paved roads," Report No. 685-5200.1F, prepared for San Joaquin Valley Unified Air Pollution Control District, Fresno, CA, by Desert Research Institute, Reno, NV, 1996.

103. V. K. Gupta and R. K. Kapoor, "Attenuation of air pollution by green belt—Optimization of density of tree plantation," in *Precipitation Scavenging and Atmosphere–Surface Exchange. Vol. 3—The Summers Volume: Applications and Appraisals*, S. E. Schwartz and W. G. N. Slinn, Eds., Hemisphere Publishing Corp., Washington, pp. 1265–76, 1992.

104. R. K. Kapoor and V. K. Gupta, "A pollution attenuation coefficient concept for optimization of green belt," *Atmos. Environ.* **18**, 1107–13 (1984).

105. D. A. Gillette, "Fine particulate emissions due to wind erosion," in *Transactions of the American Society of Agricultural Engineers*, American Society of Agricultural Engineers, St. Joseph, MI, pp. 890–97, 1977.

106. W. Nickling and C. Houser, "Estimation of PM_{10} dust fluxes from dust emission factors for different dust sources," prepared for U.S. Environmental Protection Agency, Research Triangle Park, NC, by MND Associates, Toronto, Canada, 1999.

107. Midwest Research Institute, "Improvement of specific emission factors" (BACM Project No. 1), Final report, South Coast AQMD Report No. 95040, prepared for BACM Work Group, by Midwest Research Institute, Kansas City, MO, 1996.

108. J. C. Chow and J. G. Watson, "Work plan to improve PM_{10} fugitive dust emissions inventory," Report No. DRI 0736.1D, prepared for South Coast Air Quality Management District, Diamond Bar, CA, by Desert Research Institute, Reno, NV, 1994.

109. J. G. Watson, "Chemical element balance receptor model methodology for assessing the sources of fine and total particulate matter in Portland, Oregon," Ph.D. Dissertation, Oregon Graduate Center, Beaverton, OR, 1979.

110. A. E. Sheffield and G. E. Gordon, "Variability of particle composition from ubiquitous sources: Results from a new source-composition library," in *Transactions, Receptor Methods for Source Apportionment: Real World Issues and Applications*, T. G. Pace, Ed., Air Pollution Control Association, Pittsburgh, PA, pp. 9–22, 1986.

111. J. E. Core and J. E. Houck, Pacific Northwest Source Profile Library Sampling and Analytical Protocols. Oregon Department of Environmental Quality, Portland, OR, 1987.

112. J. A. Cooper, D. C. Redline, J. R. Sherman, L. M. Valdovinos, W. L. Pollard, L. C. Scavone, and C. R. West, "PM_{10} source composition library for the South Coast Air Basin. Volume I: Source profile development documentation," Final report, South Coast Air Quality Management District, El Monte, CA, 1987.

113. J. E. Houck, J. M. Goulet, J. C. Chow, J. G. Watson, and L. C. Pritchett, "Chemical characterization of emission sources contributing to light extinction," in *Transactions, Visibility and Fine Particles*, C. V. Mathai, Ed., Air and Waste Management Association, Pittsburgh, PA, pp. 437–46, 1990.

114. J. E. Houck, J. C. Chow, J. G. Watson, C. A. Simons, L. C. Pritchett, J. M. Goulet, and C. A. Frazier, "Determination of particle size distribution and chemical composition of particulate matter from selected sources in California," Prepared for California Air Resources Board, Sacramento, CA, by OMNI Environmental Services, Inc., Beaverton, OR, 1989.

115. J. E. Houck, J. M. Goulet, J. C. Chow, and J. G. Watson, "Chemical source characterization of residential wood combustion emissions in Denver, Colorado; Bakersfield, California; and Mammoth Lakes, California," 82nd Annual Meeting of the Air & Waste Management Association, Anaheim, CA, 1989.

116. J. C. Chow and J. G. Watson, "Contemporary source profiles for geological material and motor vehicle emissions," Report No. DRI 2625.2F, prepared for U.S. EPA, Office of Air Quality Planning and Standards, Research Triangle Park, NC, by Desert Research Institute, Reno, NV, 1994.

117. J. C. Chow and J. G. Watson, "Imperial Valley/Mexicali Cross Border PM_{10} Transport Study," Report No. 4692.1D1, prepared for U.S. Environmental Protection Agency, Region IX, San Francisco, CA, by Desert Research Institute, Reno, NV, 1997.

118. J. G. Watson, J. C. Chow, Z. Lu, E. M. Fujita, D. H. Lowenthal, and D. R. Lawson, "Chemical mass balance source apportionment of PM_{10} during the Southern California Air Quality Study," *Aerosol Sci. Technol.* **21**, 1–36 (1994).

119. J. G. Watson, D. L. Blumenthal, J. C. Chow, C. F. Cahill, L. W. Richards, D. Dietrich, R. Morris, J. E. Houck, R. J. Dickson, and S. R. Andersen, "Mt. Zirkel Wilderness Area reasonable attribution study of visibility impairment—Volume II: Results of data analysis and modeling," prepared for Colorado Department of Public Health and Environment, Denver, CO, by Desert Research Institute, Reno, NV, 1996.

120. J. G. Watson, E. M. Fujita, J. C. Chow, B. Zielinska, L. W. Richards, W. Neff, and D. Dietrich, "Northern Front Range Air Quality Study," Final report, prepared for Colorado State University, Cooperative Institute for Research in the Atmosphere, Fort Collins, CO, by Desert Research Institute, Reno, NV, 1998.

121. J. J. Schauer, W. F. Rogge, M. A. Mazurek, L. M. Hildemann, G. R. Cass, and B. R. Simoneit, "Source apportionment of airborne particulate matter using organic compounds as tracer," *Atmos. Environ.* **30**, 3837–55 (1996).

122. B. Zielinska, J. McDonald, T. Hayes, J. C. Chow, E. M. Fujita, and J. G. Watson, "Northern Front Range Air Quality Study. Volume B: Source measurements," prepared for Colorado State University, Cooperative Institute for Research in the Atmosphere, Fort Collins, CO, by Desert Research Institute, Reno, NV, 1998.

123. J. C. Chow, J. G. Watson, J. E. Houck, L. C. Pritchett, C. F. Rogers, C. A. Frazier, R. T. Egami, and B. M. Ball, "A laboratory resuspension chamber to measure fugitive dust size distributions and chemical compositions," *Atmos. Environ.* **28**, 3463–81 (1994).

124. C. P. Calloway, S. M. Li, J. W. Buchanan, and R. K. Stevens, "A refinement of the potassium tracer method for residential wood smoke," *Atmos. Environ.* **23**, 67–69 (1989).

125. D. F. Gatz and J. M. Prospero, "A large silicon–aluminum aerosol plume in central Illinois: North African desert dust?, *Atmos. Environ.* **30**, 3789–800 (1998).

126. D. F. Gatz, "Source apportionment of rain water impurities in central Illinois," *Atmos. Environ.* **18**, 1895–904 (1984).

127. D. F. Gatz, W. R. Barnard, and G. J. Stensland, "Dust from unpaved roads as a source of cations in precipitation," 78th Annual Meeting of the Air Pollution Control Association, Detroit, MI, 1985.

128. M. A. Bruns, K. J. Graham, K. M. Scow, and T. VanCuren, "Biological markers to characterize potential sources of soil-derived particulate matter," Paper No. 98-TP43.02(A994) presented at the 1998 Annual Meeting of the Air & Waste Management Association, Pittsburgh, PA, 1998.

129. B. L. Davis, L. R. Johnson, R. K. Stevens, D. F. Gatz, and G. J. Stensland, "Observation of sulfate compounds on filter substrates by means of x-ray diffraction," in *Heterogeneous Atmospheric Chemistry*, Geophysical Monograph Series, **26**, American Geophysical Union, pp. 149–156, 1982.

130. G. E. Muleski and C. Cowherd, "Evaluation of the effectiveness of chemical dust suppressants on unpaved roads," Report No. EPA-600/2-87-102, Air and Energy Engineering Research Laboratory, U.S. Environmental Protection Agency, Research Triangle Park, NC, 1987.

131. R. H. Grau, "Evaluation of methods for controlling dust," Report No. GL-93-25. U.S. Army Corps of Engineers, 1993.

132. U.S. EPA, "Control techniques for particulate emissions from stationary sources," Volume 2, Report No. PB83-127480, U.S. EPA, Research Triangle Park, NC, 1982.

133. D. F. Gatz, S. T. Wiley, and L. C. Chu, "Characterization of urban and rural inhalable particulates," Illinois Department of Energy and Natural Resources, Division of Policy and Planning, Springfield, IL, 1983.

134. S. Calvert, H. Brattin, S. Bhutra, and D. M. Ono, "Improved street sweepers for controlling urban inhalable particulate matter," Report No. EPA-600/7-84-021, U.S. Environmental Protection Agency, Research Triangle Park, NC, 1984.

135. M. Duncan, R. Jain, S. C. Yung, and R. K. Patterson, "Performance evaluatoin of an improved street sweeper," Report No. EPA-600/7-85-008, U.S. Environmental Protection Agency, Research Triangle Park, NC, 1985.

136. C. Cowherd and J. S. Kinsey, "Identification, assessment and control of fugitive particulate emissions," Report No. EPA-600/8-86-023, U.S. Environmental Protection Agency, Research Triangle Park, NC, 1986.

137. Seton Johnson and Odell, Inc., "Portland road dust demonstration project—final report," Prepared for City of Portland Dept. of Public Works, Portland, OR, 1983.

138. T. R. Hewitt, "The effectiveness of street sweeping for reducing particulate matter background concentrations," Sirrine Environmental Consultants, Research Triangle Park, NC, 1981.

139. PEDCo Environmental Inc., "Study of street cleaning impact on particulate levels in Kansas City, Kansas," prepared under Contract No. 68-02-2535 for the U.S. Environmental Protection Agency, Kansas City, MO, by PEDCo Environmental Inc., 1981.

140. T. Cuscino, G. E. Muleski, and C. Cowherd, "Determination of the decay in control efficiency of chemical dust suppressants," in *Proceedings—Symposium on Iron and Steel Pollution Abatement Technology for 1982*, U.S. Environmental Protection Agency, Research Triangle Park, NC, 1983.

141. C. Cowherd, "Particulate emission reductions from road paving in California oil fields," 75th Annual Meeting of the Air Pollution Control Association, New Orleans, New Orleans, LA, 1982.

142. D. R. Fitz and K. Bumiller, "Determination of PM_{10} emission rates from street sweepers," 89th Annual Meeting of the Air & Waste Management Association, Nashville, TN, 1996.

143. T. Cuscino, G. E. Muleski, and C. Cowherd, "Iron and steel plant open source fugitive emission control evaluation," Report No. EPA-600/2-83-110, U.S. Environmental Protection Agency, Research Triangle Park, NC, 1983.

144. E. T. Brookman, "Linn County, Iowa, non-traditional fugitive dust study," Report No. EPA-907/9-83-002, U.S. Environmental Protection Agency, Research Triangle Park, NC, 1983.

145. R. W. Rice, "Fundamentals of no-till farming," American Association for Vocational Instructional Materials, Athens, GA, 1983.

146. J. F. Leys, "Towards a better model of the effect of prostate vegetation cover on wind erosion," *Vegetatio* **91**, 49–49 (1991).

147. M. A. Al-Afifi, I. Haffar, and S. Itani, "Use of date-fronds mat fence as a barrier of wind erosion control, 2. Effect of barrier density on microclimate and vegetation," *Agricul. Ecosyst. & Environ.* **33**, 47–47 (1990).

5

ODORS

William H. Prokop

SENSORY PROPERTIES OF ODOR

An odor is defined as a sensation resulting from the reception of a stimulus by the olfactory sensory system. The types of human responses to be evaluated depend on the particular sensory property to be measured, including odor intensity, detectability, character, and hedonic tone (pleasantness/unpleasantness). The combined effect of these properties is related to the annoyance that may be caused by an odor.

Intensity

Odor intensity is the strength of the perceived odor sensation and is related to the odorant concentration, which is an entirely different category of measurement. The intensity of an odor is perceived directly without any knowledge of the odorant concentration or of the degree of air dilution of the odorous sample needed to eliminate the odor.

The following equation defines the relationship between the odor intensity (I) and concentration (C), where k is a constant and n is the exponent.

$$I\ (\text{perceived}) = k(C)^n$$

or

$$\log I = \log K + n \log(C)$$

This is known as Stevens' law or the power law. For odors, n ranges from about 0.2 to 0.8, depending on the odorant. For an odorant with n equal to 0.2, a tenfold reduction in concentration decreases the perceived intensity by a factor of only 1.6, whereas for an odorant with n equal to 0.8, a tenfold reduction in concentration lowers the perceived intensity by a factor of 6.3. This is an important concept that is related to the basic problem of reducing the odor intensity of a substance by air dilution or other means.

Figure 1 shows data[1] for different chemicals as a log–log plot, where the concentration, in parts per billion, as the abscissa varies with the right-side ordinate, shown as relative odor intensity. The spacing of the numbers on the relative odor-intensity scale is based on data available for methyl sulfide and the four italicized odorants (IIT Research Institute data). The slope of the straight line is equal to n. It is evident that the effect of dilution on the odor intensity of methyl sulfide is much less than that for hydrogen sulfide.

Detectability

The detectability or threshold of an odor is a sensory property referring to the minimum concentration that produces an olfactory response or sensation. This threshold usually is determined by an odor panel consisting of a specified number of people; the numerical result is typically expressed as occurring when 50% of the panel correctly detect the odor.

At odor intensity levels at or just above "threshold," odors become difficult to perceive. As a result, the actual values depend on the type of sensory test, the panelist selection, the detectability criterion, and other factors. It can be defined only within the context of these parameters. The odor threshold is not a precisely determined value, as is, for example, vapor pressure. It is highly dependent on the sensitivity of the odor panelists, the method of presenting the odor stimulus to the panelists, including flow rate, and often the purity of the chemical odorant being tested.

An odor detection threshold relates to the minimum odorant concentration required to perceive the existence of the stimulus, whereas an odor recognition threshold relates to the minimum odorant concentration required to identify the stimulus. The detection threshold occurs at a lower concentration than the recognition threshold.

Data in the literature on odor threshold concentrations for any particular compound may differ significantly; in many cases, by tenfold or more. This is particularly true of the earlier values because of inadequate equipment or methods, too small a panel, or too large a stepwise change in odorant concentration.

Character

Odor character or quality is that property that identifies an odor and differentiates it from another odor of equal intensity. Odors are classified on the basis of odor descriptor terms. Odor character is evaluated by comparison with other odors, either directly or through the use of descriptor words (see Table 1).

The odor character is described by methods known as multidimensional scaling or profiling. In these methods, the odor is characterized by either the degree of its similarity to (or dissimilarity from) a set of reference odors or the degree of applying a scale of various descriptor terms to it. The result is an odor profile.

Hedonic Tone

Hedonic tone is a property of an odor relating to its pleasantness or unpleasantness. A distinction should be made between the acceptability and the hedonic tone of an odor. Acceptability is usually a judgment made by a specific person in the context of a specific situation and with specific expectations. For example, an otherwise pleasant odor may be unacceptable if it persists as part of an air pollution problem in a residential area and originates from a perfume factory instead of from a flower garden.

When an odor is evaluated in the laboratory for its hedonic tone in the neutral context of an olfactometric presentation, the panelist is exposed to a controlled stimulus in terms of intensity and duration. The degree of pleasantness or unpleasantness is determined by each panelist's experience and emotional associations. The responses among panelists may vary depending on odor character; an odor pleasant to many may be declared highly unpleasant to some.

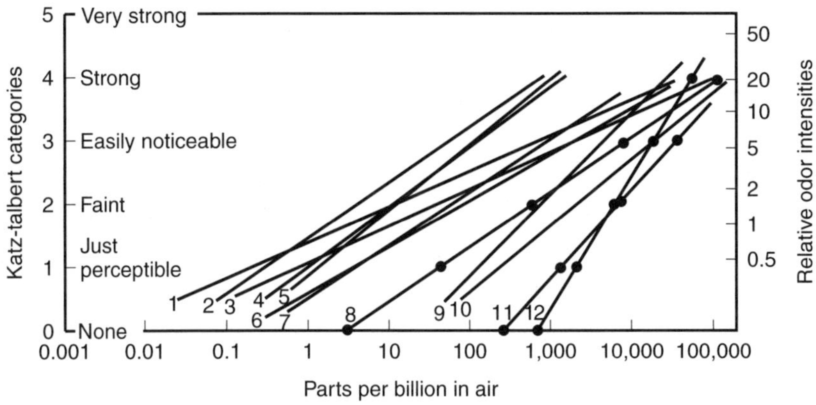

Odor intensity vs. concentration plots for several odorants. Individual points are shown for 8, 11, and 12 to illustrate the linearity. Legend: 1. Thiophenol; 2. Ethyl selenomercaptan; 3. Ethyl sulfide; 4. Phenyl isocyanide; 5. Ethyl selenide; 6. *Methyl sulfide*; 7. *Coumarin*; 8. Methylmercaptan; 9. Hydrogen sulfide; 10. *Pyridine*; 11. *Allyl alcohol*; 12. *Nitrobenzene*.

Figure 1. Odor Intensity versus Concentration.

Adaptation

Adaptation or olfactory fatigue is a phenomenon that occurs when people with a normal sense of smell experience a decrease in perceived intensity of an odor if the stimulus is received continually. Depending on the intensity of the stimulus, this self-adaptation and sensory recovery after removal of the stimulus both take place during a relatively short time. Adaptation to a specific odorant generally does not interfere with the ability of a person to detect other odors. Another phenomenon known as habituation or occupational anosmia occurs when a worker in an industrial situation experiences a long-term exposure and develops a higher threshold tolerance to the odor.

ODOROUS COMPOUNDS

Odorous substances that are emitted from industrial sources include both inorganic and organic gases and particulate. Hydrogen sulfide and ammonia are examples of inorganic gases. Many odorous compounds result from biological activity or are present in emissions from chemical processes. Most of the odorous substances derived from anaerobic decomposition of organic matter contain sulfur and nitrogen.

Table 1 lists some of the odorous compounds emitted from industrial sources and their odor detection and recognition thresholds. Most of the odorous substances are gaseous under normal atmospheric conditions or at least have a significant volatility. The volatility is shown in the table as parts per million (v/v) and is equal to the vapor pressure (mm Hg at 25°C) multiplied by 1316 (1 million ppm per 760 mm Hg). The molecular weights of these substances generally range from 30 to 150. Usually, the lower the molecular weight of a compound, the higher is its vapor pressure and potential for emission to the atmosphere. Substances of high molecular weight are normally less volatile and thus normally have less impact as a cause for odor complaints.

The reduced sulfur compounds, such as the mercaptans and organic sulfides, tend to be the most odorous, based on their relatively low odor threshold concentrations. This also applies to the nitrogen-bearing amines, but to a lesser extent.

ODOR SENSORY MEASUREMENT

An odor emission often consists of a complex mixture of many odorous compounds. As a result, odor sensory methods, instead of instrumental methods, are normally used to measure such odors. Analytical monitoring of individual chemical compounds present in such odors is usually not practical. These odor sensory methods depend on the olfactory response of individuals who serve on panels.

Odor sensory methods are available to monitor odors both from source emissions and in the ambient air. It is important to have both options. For example, an industrial source may be interested in evaluating the performance of odor control equipment by monitoring its odor emission and also may desire to know the downwind odor level in the ambient air beyond its property line to establish compliance with an ambient odor regulatory limit. These two diverse circumstances require different approaches for measuring odor. For example, the collection of odor samples is more easily accomplished for a source emission than for an odor in the ambient air. Also, due to atmospheric dilution, the odor in the ambient air is usually much lower in intensity than it is at the source. Thus the sensitivity of the odor sensory method must be significantly greater for measuring ambient odors than for source odor emissions.

The following discussion on odor dilution to threshold and suprathreshold (intensity scale) measurement methods is based on their treatment by Dravnieks[6,7] and Turk.[8]

Odor Dilution to Threshold

Odor dilution to threshold sensory results are expressed as a dimensionless ratio termed Z.[6,9] If volume v of an odorous sample is diluted to a total volume of V (where V equals the volume of odor diluted to threshold level), the ratio $Z = V/v$. As long as the same volume units are used to measure v and V, the value of Z is independent of the volumetric units.

Numerically, Z is equal to the term "odor units per cubic foot" (or per cubic meter), which has been in use for many years. The use of these units in this context is actually incorrect dimensionally. Other terms that are numerically equivalent to Z include ED_{50} (effective dosage at 50% panel) and D/T (dilution to threshold ratio).

Table 1. Odorous Compounds in Industrial Source Emissions

Compound Name	Formula	Molecular Weight	Volatility at 25°C, ppm, v/v	Detection[a] Threshold (ppm, v/v)	Recognition[a] Threshold (ppm, v/v)	Odor Descriptors
Acetaldehyde	CH_3CHO	44	Gas	0.067	0.21	Pungent, fruity
Allyl mercaptan	$CH_2{:}CHCH_2SH$	74		0.0001	0.0015	Disagreeable, garlic
Ammonia	NH_3	17	Gas	17	37	Pungent, irritating
Amyl mercaptan	$CH_3(CH_2)_4SH$	104		0.0003	—	Unpleasant, putrid
Benzyl mercaptan	$C_6H_5CH_2SH$	124		0.0002	0.0026	Unpleasant, strong
n-Butyl amine	$CH_3(CH_2)NH_2$	73	93,000	0.080	1.8	Sour ammonia
Chlorine	Cl_2	71	Gas	0.080	0.31	Pungent, suffocating
Dibutyl amine	$(C_4H_9)_2NH$	129	8,000	0.016	—	Fishy
Diisopropyl amine	$(C_3H_7)_2NH$	101		0.13	0.38	Fishy
Dimethyl amine	$(CH_3)_2NH$	45	Gas	0.34	—	Putrid, fishy
Dimethyl sulfide	$(CH_3)_2S$	62	830,000	0.001	0.001	Decayed cabbage
Diphenyl sulfide	$(C_6H_5)_2S$	186	100	0.0001	0.0021	Unpleasant
Ethyl amine	$C_2H_5NH_2$	45	Gas	0.27	1.7	Ammoniacal
Ethyl mercaptan	C_2H_5SH	62	710,000	0.0003	0.001	Decayed cabbage
Hydrogen sulfide	H_2S	34	Gas	0.0005	0.0047	Rotten eggs
Indole	$C_6H_4(CH)_2NH$	117	360	0.0001	—	Fecal, nauseating
Methyl amine	CH_3NH_2	31	Gas	4.7	—	Putrid, fishy
Methyl mercaptan	CH_3SH	48	Gas	0.0005	0.0010	Rotten cabbage
Ozone	O_3	48	Gas	0.5	—	Pungent, irritating
Phenyl mercaptan	C_6H_5SH	110	2,000	0.0003	0.0015	Putrid, garlic
Propyl mercaptan	C_3H_7SH	76	220,000	0.0005	0.020	Unpleasant
Pyridine	C_5H_5N	79	27,000	0.66	0.74	Pungent, irritating
Skatole	C_9H_9N	131	200	0.001	0.050	Fecal, nauseating
Sulfur dioxide	SO_2	64	Gas	2.7	4.4	Pungent, irritating
Thiocresol	$CH_3C_6H_4SH$	124		0.0001	—	Skunk, rancid
Trimethyl amine	$(CH_3)_3N$	59	Gas	0.0004	—	Pungent, fishy

[a]References 2–5.

The total emission of odorous substances from a source is then $Q = ZW$, where W is the volumetric flow rate in either cubic meters per minute or cubic feet per minute. As a result, Q has the same units as W and quantifies the odor emission to the atmosphere. If this emission is not to be detected downwind at ground level, the volumetric emission W must be diluted each minute to a volume either equal to or greater than Q expressed in either cubic meters or cubic feet.

Sampling of Source Odor Emissions

The sampling of source odor emissions for later evaluation by panels to determine odor thresholds requires the proper choice of sampling equipment, collection container, and sampling procedure. As a general rule, a sample should be evaluated by a panel as soon as possible after collection. In some cases, storage tests with various plastic bags of specific classes of odorous compounds at higher concentrations have indicated that a sample does not degrade within 24 hours after collection. In other cases, storage tests have indicated that a much shorter time is required for other classes of compounds, especially at lower concentrations.

Therefore, it is advisable to pretest the storage of a particular type of odorous sample in the proposed bags if previous data are not available.

A peristaltic pump is often used with essentially nonodorous plastic tubing to deliver the sample into a plastic bag. A preflushing technique is used whereby the bag is filled initially and then discharged to allow for the adsorption of the sample to equilibrate with the inside surfaces of the bag and tubing. The peristaltic method allows for a new piece of tubing to be used for each sample without having to clean the pump.

For samples with a lower odor intensity, such as ambient air, it usually is undesirable to have the sample pass through a pump because of adsorption losses. Instead, a bag located in an airtight, stainless-steel cylinder that is evacuated by means of a vacuum pump allows the sample to be drawn into the bag. Again, a preflush is used. Pressure can be applied to the cylinder to discharge the sample from the bag. Both sampling methods are shown in Figure 2.

Plastic bags are the most common containers used for collecting and storing samples of odorous gases. The bags should be free of pores, be mechanically strong, have little background odor, and minimize the loss of odorants by diffusion through the bag walls. For source emission samples, thick-walled collapsible polyethylene bags have been used successfully for various classes of odorous compounds. Also, they are relatively inexpensive and can be discarded after use. For weaker odors, the more expensive Tedlar or heat-sealed FEP Teflon bags are normally used because they contain less background odor. These bags may be reused by repeated flushing with odorless air.

Static versus Dynamic Methods

Static and dynamic methods of dilution to threshold measurement have significantly different characteristics. Olfactometry consists of receiving the odorous sample, diluting it in various proportions with odorless air, and presenting the diluted samples for evaluation by the panelists.

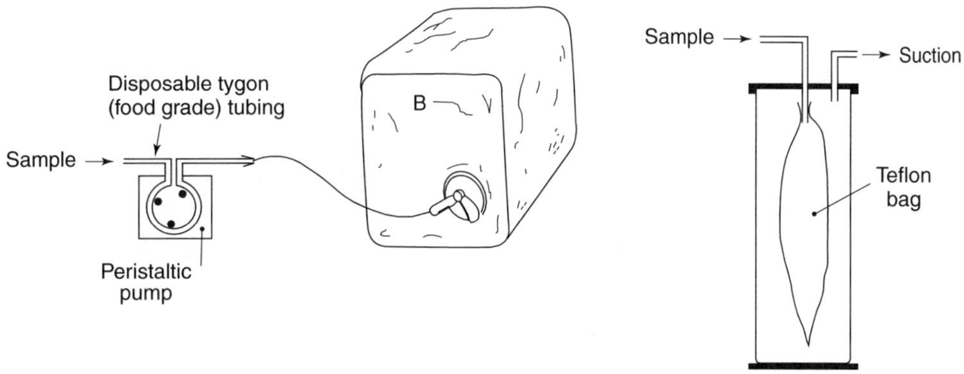

Figure 2. Methods for Sampling Source Odor Emission and Ambient Air.

Earlier methods developed for odor threshold measurement include the American Society for Testing and Materials (ASTM) syringe method,[10] which originally was issued in 1957. The 1978 revision of the ASTM D-1391 syringe method was withdrawn by the ASTM E-18 Committee on March 29, 1985. As a result, no official ASTM syringe method is in effect today. This method is cited in a number of state and local agency odor regulations and is known as a static olfactometric procedure. It consists of diluting the odorous sample in a container (100-mL syringe) and then expelling the entire diluted sample within a few seconds by directing the tip of the syringe toward the nostrils.

In contrast, a dynamic olfactometer provides a continuous and constant diluted odor stimulus by mixing controlled flows of sample and odorless air. With dynamic methods, larger samples are used and dilutions are presented at more reproducible flow rates and for a longer duration for panelists to evaluate.

At relatively low Z values, comparison tests with rendering odors showed that dynamic methods produced results that were numerically three- to fivefold greater than those obtained by the ASTM syringe method.[11] The lower ASTM values are considered to be due to adsorption losses on the glass syringes and the other factors previously discussed. Odor scientists recommend the use of dynamic olfactometric methods instead of static methods.[7] Factors in dynamic olfactometry that should be considered include the following:

1. Dynamic dilution
2. Delivery of diluted sample at olfactometer–nose interface
3. Scheduled presentation of various dilutions
4. Responses from panelists
5. Calculation of average panel threshold from raw experimental data
6. Number and selection of panelists

The sample and odorless airflows are mixed and then supplied to the olfactometer–nose interface. The olfactometer should be designed to minimize odor losses due to adsorption on interior surfaces. Also, the design should provide for strong odors to be readily flushed with odorless air or for the sample and diluted flowlines to be easily replaced. It is desirable that the flow rate to the olfactometer–nose interface be constant at all dilutions.

A large variety of methods exist for delivering the odor stimulus to the nose. These include the use of masks, sniffing ports, and nose cones. Each has advantages and disadvantages that relate to the potential for a residual odor to be sensed at the nose interface, and also to minimize any dilution by the surrounding ambient air of the stimulus before reaching the panelist's nose.

The most significant variable in the delivery of the odor stimulus to the nose is the flow rate from the olfactometer. For different olfactometers, it varies from 0.5 to 200 L/min. Higher flow rates produce greater numerical values of dilution to threshold. For example, a 100-fold variation in flow rate can produce a tenfold variation in the reported dilution ratio or suprathreshold intensity of the same odorant by the same panel. However, lower flow rates consume less sample per test, tending to improve the portability of an olfactometer and to have less impact on producing background odors in the test room for evaluation.

The Draft European CEN Standard[12] adopted a minimum flow rate requirement of 20 L/min with a minimum face velocity of 40 fpm (0.2 m/sec). O'Brien et al.[13] investigated the effect of sample flow rate in the determination of odor thresholds. Their experimental results indicated that a flow rate between 5 L/m and 10 L/m to each sniffing port with a face velocity ranging from a minimum of 0.012 m/s to a maximum of 0.103 m/s provides repeatable measurements most characteristic of odor perception in ambient air. The lower odor thresholds resulting from significantly higher flows are not necessarily "more correct" than those obtained at lesser flows.

The presentation of odorous sample dilutions to panelists and their responses depend on three sensory effects: judgment criterion, anticipation, and adaptation. The judgment criterion determines how the panelist is to respond when asked whether or not an odor is sensed. This is the case particularly when a single stimulus is presented and a yes or no answer is requested as to the sensation of odor. The anticipation effect is a tendency to expect an odor to occur when odorless or weak samples are consecutively presented. The adaptation effect is a temporary loss of sensitivity after smelling an odor. When a weak odor is detected initially, the same odor may not be detected again after smelling a stronger odor unless the panelist has had sufficient time to recover his or her olfactory sense.

Most of these problems are resolved by adopting a procedure in which a forced-choice multiple stimulus method is combined with a systematic ascending order of odorous sample concentrations (corresponding decrease in dilution). The diluted odor stimulus is presented along with two blanks of odorless air, and the panelist is asked to choose the one that contains the stimulus. The ASTM in Method E-679[14] recommends this approach when performing odor threshold measurements.

Data treatment consists of calculating from sets of individual responses an average threshold value representative of the panel as a whole. In most dynamic olfactometer procedures, the sample dilution is changed in discrete steps. For example, the dilution may be changed threefold per step. If a panelist does not detect odor at dilution $3D$, but detects it at dilution D (triple the odorant concentration) and at lesser dilutions, the dilution D does not necessarily represent the panelist's threshold. Instead, it might have been in the interval between $3D$ and D. Statistical considerations indicate that the overall error due to this factor is minimized if it is assumed that the threshold is the geometric mean of the two dilutions, $(3DD)^{1/2} = 1.7D$.

Individuals differ in their odor sensitivity, and panelists who have abnormally low or high sensitivities should be excluded. Panelists should be chosen with the objective in mind that they represent the average sensitivity of the population. Selection of the more sensitive half or third of the prospective panelists produces a data bias toward obtaining higher Z values.

The number of panelists selected for odor sensory evaluation is directly related to the percent confidence level desired to minimize the probability of error occurring by chance. For most odor sensory testing that requires decision making for odor control systems, a 99% confidence level is usually desired. This requires a panel of nine or more people to obtain a significant number of data points if each panelist evaluates the collected sample only once. If fewer panelists are used, they should evaluate the same sample at least twice.

The forced-choice, triangle olfactometer[15] was designed by the IIT Research Institute (IITRI) to implement the odor evaluation principles described under ASTM Method E679. The ASTM includes the IITRI method as an example of applying this standard. This particular version of the IITRI olfactometer is used to evaluate source emission odors.

Figure 3 shows this olfactometer with its various component parts coded. In the upper diagram, cup C contains three sniffing ports, one with the diluted odorous sample and the other two with odorless air. The panelist chooses the odorous port by depressing a pushbutton I at the selected port and a light bulb signals the choice at panel box S. The peristaltic pump P delivers the odorous sample to the olfactometer from sample bag B. F and R are flowmeters used to calibrate the sample and dilution airflows. Room air is used for dilution.

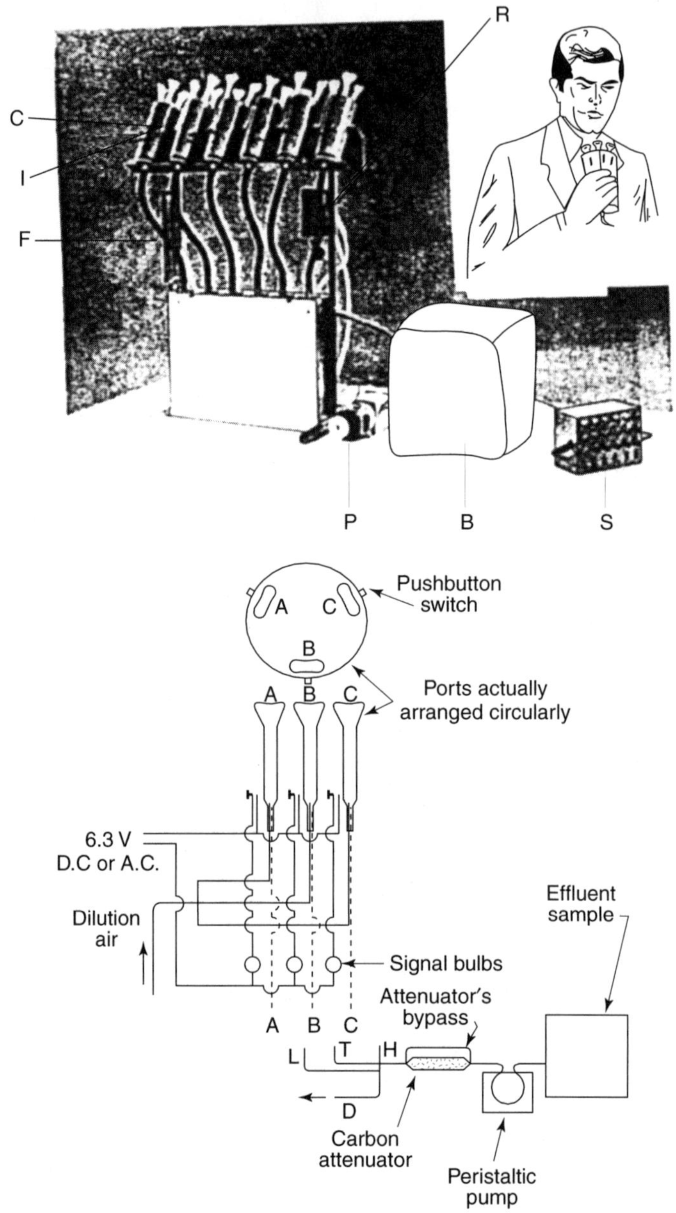

Figure 3. Dynamic Forced-Choice Triangle Olfactometer with Six Dilution levels.

There are six dilution levels, ranging typically from 4500× to 15×, or from 80,000× to 450× by use of the attenuator. Adjoining dilution levels differ in concentration by a factor of three. All required dilutions and blanks continuously discharge from the sniffing ports at an essentially uniform flow of 500 mL/min. This flow rate was adopted to prevent an odor buildup in a normally ventilated room.

Another commercially available olfactometer is known as the Hemeon Odor Meter.[16] It uses a forced-choice principle requiring the panelist to decide whether the diluted odor stimulus is (1) stronger than the previous presentation, or (2) has the same strength, or (3) has no odor (blank). The odorous sample is diluted through a series of valves and flowmeters with room air and is delivered at a rate of 150 L/min through each of three sniffing ports to three panelists simultaneously. The test begins with sufficient dilution air to start at an odor concentration below threshold. As the odor concentration is increased sequentially in steps using 50% less dilution air, the panelist rates the odor intensity according to a category scale of 0 to 4. Category 1 relates to detection and category 2 relates to recognition of the odor obtained from the parts-per-million values for the following experimentally derived equation.

$$I = 0.26(C)^{0.66}$$

where I is the perceived odor intensity of the sample that is matched against butanol; and C the concentration in parts per million of 1-butanol (in air emitted at 160 mL/min) that matches the sample's odor intensity.

The butanol odor intensity reference scale has the potential for use as a measurement of the odor intensity in the ambient air. Panelists can be used to match the perceived intensity of an ambient odor by comparison with the butanol reference scale. Panelists' judgments could be reinforced by referring to the butanol standards in between measurements of the ambient air for perceived odor intensity (Figure 4).

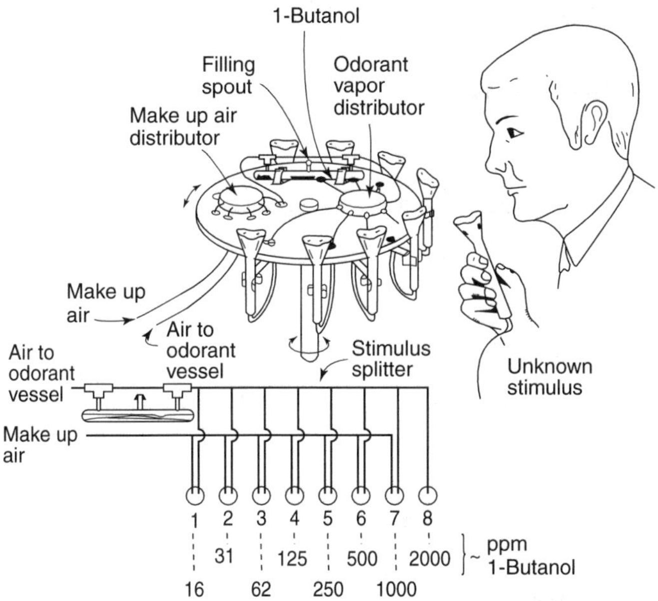

Figure 4. Dynamic Odor Intensity Reference Scale Based on 1-Butanol.

DYNAMIC OLFACTOMETRY IN EUROPE

In 1999, Harreveld et al.[17] conducted a review of the development of standards for odor concentration measurement by dynamic olfactometry in Europe. Issues covered are the development of instruments, standards, guidelines for methodology and calculation of odor concentration as well as validation of results through interlaboratory comparisons.

Individual national efforts led to establishing the Committee Europeen de Normalisation (CEN). A European standard was drafted and validated in 1996 with 19 laboratories from five countries participating. This CEN draft standard is expected to be introduced in 1999 and to become mandatory in the near future in the member countries of the European Union.

The unit of odor measurement was defined and related to an accepted reference value of a reference odorant, such as *n*-butanol. The unit of measurement is the European odor unit (ou_E) which is that amount of odorant(s) that, when evaporated into 1 m^3 neutral gas at standard conditions, elicits a physiological response from a panel (detection threshold) equivalent to that elicited by 1 European reference odor mass (EROM) evaporated into 1 m^3 neutral gas at standard conditions.[18] The European reference odor mass is the accepted reference value for the ou_E and is equal to 123 ug *n*-butanol, which when evaporated into 1 m^3 neutral gas, produces a concentration of 0.040/v ppm *n*-butanol. This value is obtained by a strict protocol for panel selection, which has been the main contributing factor in improving the reproducibility of olfactometry.

ODOR CONTROL METHODS

The methods available for controlling odor emissions from industrial sources include the following:

1. Condensation
2. Incineration
3. Wet scrubbing with chemical solutions
4. Activated-carbon adsorption
5. Biofiltration
6. Odor modification
7. Air dilution

Moisture-laden or organic vapor emissions should first be passed through a condenser that is capable of removing a high percentage of the water vapor or condensible volatile organic compounds (VOCs). This minimizes the odorous gas stream to be treated or allows recovery of a costly solvent, providing a more economical solution to solving an odor emission problem. Usually, an indirect condenser, such as a shell and tube or an air-cooled, finned-tube unit, is used to isolate the condensate from the cooling medium.

Thermal incineration of odor emissions is particularly effective at specific temperatures and residence times, normally resulting in the essentially complete removal of odor. The use of separate afterburners and incinerators, even with heat recovery, may be expensive, in terms of both initial investment and fuel operating cost. The use of a boiler firebox for steam generation often is an economical approach to incinerating odor emissions, particularly when the volumetric flow rate is not excessive. Catalytic incineration at lower temperatures is a possible method for controlling odors. However, its use is less prevalent than thermal incineration owing to the potential for fouling the catalyst.

Wet scrubbers with chemical solutions for the absorption of and reaction with odorous compounds are used extensively. Multistage scrubber systems are often used to treat process emissions, whereas single-stage scrubbers are used to treat plant ventilating air. Entrainment separators, venturi scrubbers, and spray-type scrubbers are normally used as the first stage in the multistage system to remove particulate and aerosols from gaseous streams so that the subsequent scrubber stages will be more effective in achieving odor removal. Usually, chemical oxidants, such as sodium hypochlorite and chlorine dioxide, are more effective scrubbing agents than the common bases and acids, such as caustic soda and sulfuric acid. However, the selection of the chemicals to use for scrubbing depends on the odorous compounds present in the source emission.

Activated-carbon adsorption columns are used for the treatment of odor emissions from sewage sources that release highly odorous compounds, such as hydrogen sulfide. Caustic-impregnated activated carbons are used that can be regenerated in situ. Biofilters, including soil-bed systems, have become prominent in the control of odors from various industrial and municipal sources consisting of chemical (VOCs), wastewater treatment, composting, and rendering facilities.

Odor modification involves the addition of an odorous substance to a given odor emission in order to reduce the odor intensity or to change the odor character to one that is less objectionable. Masking agents change the character of an odor, but also increase its resultant intensity. Counteractants are chemicals that may change the character and also reduce the intensity of an odor emission. Vapor-phase reactions may also occur with certain chemical compounds that reduce the odor intensity by producing less odorous constituents, similar to reactions with chemical oxidants.

Dilution of an odor emission with atmospheric air is a possible method of odor control that may be successfully applied under special conditions of meteorology and surrounding topography. However, odor emissions from point sources are normally treated first by an acceptable control method before being discharged to the atmosphere. An important part of the overall design of an odor control system is related to the stack design to ensure that the height and exit velocity provide the desired dispersion characteristics.

For more details on condensation, incineration, scrubbing and adsorption, refer to Chapters 2 and 3. Also, additional discussions of odor measurement and control methods are available in two Air and Waste Management Association conference proceedings.[19,20]

REFERENCES

1. A. Dravnieks, "Odor perception and odorous air pollution," *J. TAPPI* **55**, 737–42 (1972).
2. *Odor Thresholds for Chemicals with Established Occupational Health Standards*, American Industrial Hygiene Association, Akron, OH, 1989.
3. J. E. Moore and R. S. O'Neill, "Odor as an aid to chemical safety: Odor thresholds compared with threshold limit values and volatilities for 214 industrial chemicals in air and water dilution," *J. Appl. Toxicol.* **3**, 6 (1983).
4. *Odor Control for Wastewater Facilities*, Manual of Practice No. 22, Water Pollution Control Federation, Washington, DC, 1995.
5. R. J. Sullivan, *Preliminary Air Pollution Survey of Odorous Compounds*, U.S. Department of Health, Education and Welfare Public Health Service, Raleigh, NC, October 1969.
6. A. Dravnieks, "Measurement methods," *Odors from Stationary and Mobile Sources*, National Academy of Sciences, Washington, DC, January 1979, Chapter 4.
7. A. Dravnieks and F. Jarke, "Odor threshold measurement by dynamic olfactometry: Significant operational variables," *J. APCA* **30**(12), 1284 (1980).
8. A. Turk, E. D. Switala, and S. H. Thomas, "Suprathreshold odor measurements by dynamic olfactometry—principles and practice," *J. APCA* **30**(12), 1289 (1980).
9. A. Turk, "Expressions of gaseous concentration and dilution ratios," *Atmospher. Environ.* **7**, 967 (1973).
10. *Standard Test Method for Measurement of Odor in Atmospheres (Dilution Method)*, ASTM D-1391, American Society for Testing and Materials, Philadelphia, PA, 1978.
11. J. P. Wahl, R. A. Duffee and W. A. Marrone, *Evaluation of Odor Measurement Techniques: Vol. I—Animal Rendering Industry*, EPA 650/2-74-008-a, U.S. Environmental Protection Agency, January 1974.
12. *CEN/TC264/WG2 Draft European CEN Standard: Odour Concentration Measurement by Dynamic Olfactometer*, CEN: Delft, the Netherlands, Revised 1995.
13. M. O'Brien, R. A. Duffee and N. Ostijic, "Effect of Sample Flow Rate in the Determination of Odor Thresholds", *Proceedings of Odors: Indoor and Environmental Air*, Air & Waste Management Assoc., Pittsburgh, PA, 1996.
14. *Standard Practice for the Determination of Odor and Taste Threshold by the Forced-Choice Ascending Concentration Series Method of Limits*, ASTM E-679, American Society for Testing and Materials, Philadelphia, PA, 1979.
15. A. Dravnieks and W. H. Prokop, "Source emission odor measurement by a dynamic forced-choice triangle olfactometer," *J. APCA* **25**(1), 28–35 (1975).
16. W. Hemeon, "Technique and apparatus for quantitative measurement of odor emissions," *J. APCA* **18**(3), 166–170 (1968).
17. A. P. van Harreveld, P. Heeres, and H. Harssema, "A review of 20 years of standardization of odor concentration measurement by dynamic olfactometry in Europe," *JAWMA* **49**, 705–15 (1999).
18. *Standard Recommended Practices for Referencing Suprathreshold Odor Intensity*, ASTM E-544, American Society for Testing and Materials, Philadelphia, PA, 1975.
19. *Transactions of AWMA International Specialty Conference on Recent Developments and Current Practices in Odor Regulations, Controls and Technology*, Air and Waste Management Association, Pittsburgh, PA, 1991.
20. *Proceedings of Odors: Indoor and Environmental Air*, Air and Waste Management Association, Pittsburgh PA, 1996.

6

ANCILLARY EQUIPMENT FOR LOCAL EXHAUST VENTILATION SYSTEMS

Howard D. Goodfellow

FUNDAMENTALS

Introduction

Ventilation is one of the most important techniques available to control contaminant levels in the workplace. As will be discussed later, the control of a potentially hazardous contaminant can be achieved in two ways: by dilution of the concentration of the contaminant before it reaches the workers' breathing zone by mixing it with uncontaminated air (called industrial ventilation or dilution ventilation) or by the removal of the contaminant at or near its source or point of generation to prevent its release into the workplace environment. This latter approach is called local exhaust ventilation (LEV), and its elements, including design and the associated ancillary equipment, are the focus here.

Ventilation systems can be designed for commercial/residential or industrial applications. The design of ventilation systems for the residential/commercial fields is well presented in numerous publications of the American Society of Heating, Refrigerating, and Air Conditioning Engineers.[1] These references also contain valuable information on associated ancillary equipment (e.g., ducts, fans, heating/cooling units) that is of the same generic type as for the industrial ventilation systems. Industrial ventilation systems, however, require unique and specific design features for the ancillary equipment that must be taken into account in order for the design to be successful. The technology for the design and operation of industrial ventilation systems is not as well developed, but significant progress is being made.

A simple definition of ventilation is "the control of the environment with airflow." Ventilation systems must be designed as an integral part of the process. This approach, called the "systems approach," is essential to ensure a successful design of the industrial ventilation system. A textbook entitled *Advanced Design of Ventilation Systems for Contaminant Control*[1] outlines the steps required to implement the systems approach to solving ventilation problems for the processing and manufacturing industries.

The application areas for industrial ventilation systems (both general ventilation and LEV) are as follows:

1. Control of contaminants to acceptable levels
2. Control of heat and humidity for comfort
3. Prevention of fires and explosions

For industrial applications, all these factors are important considerations at the design stage. In general, the controlling design criterion will be to achieve acceptable contaminant levels in the workplace. The following are four steps to be considered in solving industrial ventilation problems.

1. Process modifications. This should always be the first approach to establish whether the process can be modified and the contaminant can be controlled to an acceptable level. For example, a processing plant handling silica using belt conveyors may solve a silica contaminant exposure problem by replacing the belt conveyor with a totally enclosed conveyor. The result of this process modification is that there is no need for a ventilation system.

2. Local exhaust ventilation. With the exception of process modifications, LEV systems will almost always be the most cost-effective technology for controlling contaminant levels in the workplace. Contaminants are controlled at the source with low volumes of gas.

3. Process building or general work area (i.e., general or dilution ventilation). The concept of general or dilution ventilation is based on supplying or exhausting large volumes of air to an occupied space or the complete processing/manufacturing complex. The ventilation air can be supplied and exhausted by natural ventilation or mechanical ventilation. A subsequent section will describe the features of general ventilation for processing and manufacturing facilities.

4. Personal protective equipment. For some specific applications, it may be necessary to use personal protective equipment to reduce the exposure for workers. In general, this is only applicable if process modification or ventilation is not feasible. Personal protective equipment can be considered for nonroutine maintenance work or for processes that do not have feasible engineering control solutions.

Conservation of Mass

A basic principle for airflow consideration is the continuity equation or the conservation of mass. This general physical law states that the mass, m, of a particle of the system remains constant with time, t, which means that the system equation can be written: the rate of change of mass (m) with time (t) must be zero.

$$\frac{dm}{dt} = 0$$

For industrial ventilation applications, gas flow in the ductwork can be treated as one dimensional if the proper averages are used for the variables across the section. A simplified equation can be written for any two points in a fluid system, as shown in Figure 1 based on one-dimensional compressible flow with only one entrance and one exit in the control surface. This equation is as follows:

$$\rho_1 A_1 V_1 = \rho_2 A_2 V_2$$

where ρ is the specific density of the fluid (lb/ft^3); A, the cross-sectional area of the duct (ft^2); and V, the average gas velocity (fpm).

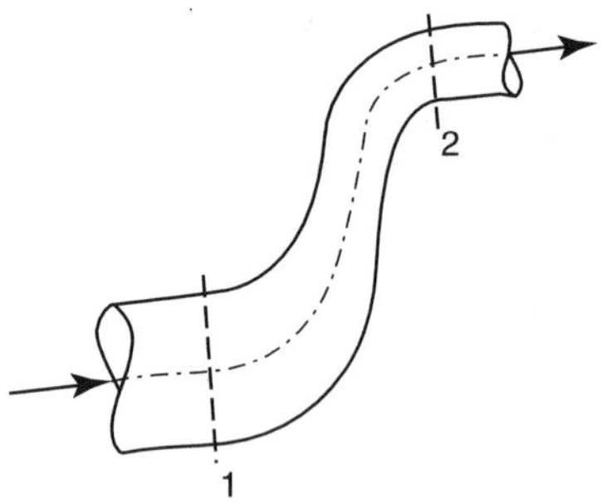

Figure 1. A One-Dimensional Fluid System.

For most industrial ventilation applications, the absolute pressure remains relatively constant and hence the specific density of the gas remains relatively incompressible.

For this case, $\rho_1 = \rho_2 =$ constant and the above equation becomes:

$$V_1 A_1 = V_2 A_2$$

or

$$Q_1 = Q_2$$

where $Q = AV$, the volumetric flow rate (ft^3/min).

The above equations must be modified if there is more than one inlet area, A_1, or outlet area, A_2.

Conservation of Energy

Bernoulli developed a general energy equation that relates kinetic energy, flow energy, and potential energy. This general equation can be simplified for the case of a steady flow of a frictionless, incompressible fluid along a streamline to give:

$$\frac{V^2}{2g_c} + \frac{P}{\delta} + Z = \text{const.}$$

where the constant is different for each streamline; V is the velocity (fps); g_c, the gravitational constant (= 32.12 ft/sec^2); P, the absolute pressure (lb/ft^2); δ, the specific density (lb/ft^3); and Z, the elevation above an arbitrary datum point (ft). Each of the above terms represents a manometric height of fluid (e.g., feet of water).

By taking the above equation and applying it to two points on a streamline and rearranging, the equation becomes:

$$\frac{V_2^2 - V_1^2}{2g_c} + \frac{P_2 - P_1}{\delta} + (Z_2 - Z_1) = 0$$

The equation shows that it is the differences in kinetic energy, flow energy, and potential energy that are significant. For most industrial ventilation applications, the change in potential energy is usually relatively small and is ignored. Also, the change in pressure, is only a small percent of the absolute pressure and the gas may be considered incompressible with an average specific weight. In real fluids, the common approach is to add a term to the Bernoulli equation to account for friction losses. For points 1 and 2 in a system, the equation for flow from point 1 to point 2 can be written:

$$\frac{V_1^2}{2g_c} + \frac{P_1}{\delta} + Z_1 = \frac{V_2^2}{2g_c} + \frac{P_2}{\delta} + Z_2 + \text{losses}_{(1-2)}$$

Pressures (Total, Static, Velocity)

Total pressure represents the head required to start and maintain flow for air moving through the duct. Total pressure is the algebraic sum of the static and velocity pressure (always positive) terms and is given by:

$$\text{TP} = \text{SP} + \text{VP}$$

or total pressure equals static pressure plus velocity pressure. Static pressure, sometimes called "frictional" or "resistance" pressure, can be positive or negative and is the potential pressure exerted in all directions by fluid at rest that tends to expand or compress the fluid. Static pressure exists even if there is no air motion. It produces the initial air velocity and overcomes the frictional resistance of the air against the duct surface, as well as the resistance resulting from any flow restrictions. For most ventilation systems, static pressure is negative in sign upstream from the fan and positive in sign downstream from the fan.

Velocity pressure is the kinetic pressure in the direction of flow necessary to cause a fluid at rest to flow at a given velocity. It exists only when air is in motion, acts in the direction of the air motion, and is always positive. Under certain conditions of fan operations, it is possible for either the velocity or static pressure to be equal to zero, but both can never equal zero at the same time. The relationship between the velocity of air and air pressure can be determined from the general Bernoulli equation and is as follows:

$$V = \sqrt{2g_c h}$$

where h is the head of air (ft).

For standard dry air at sea level (air density is 0.075 lb/ft), velocity can be determined using:

$$V = 4005\sqrt{\text{VP}}$$

where V here is in fpm, and is in VP is in. w.g.

Figure 2 shows the relationship among total pressures, static pressures, and velocity pressures for an open duct under both negative and positive pressure with a velocity of 4005 fpm.

If the velocity through the duct system increases, a part of the available static pressure is converted to additional velocity pressure to accelerate the fluid flowing through the duct. Conversely, if the velocity through the duct system is decreased at some point, a portion of the velocity pressure is converted to static or potential pressure. This conversion is always accompanied by a net loss of total pressure as a result of turbulence and friction losses. For any ducted system, it is important to be able to plot graphically pressures (TP, VP, SP) as a function of position for the ducted system. Figure 3 is a graphical plot of TP, VP, SP for a typical ducted system.

Friction and Dynamic Losses

Energy losses (friction plus dynamic) always occur for the flow of gases in industrial ventilation systems. An ideal fluid having zero viscosity and flowing in a straight horizontal pipe of constant diameter would have the same pressures all along the duct. Friction losses occur at the point of movement between the fluid particles and at the boundary layer adjacent

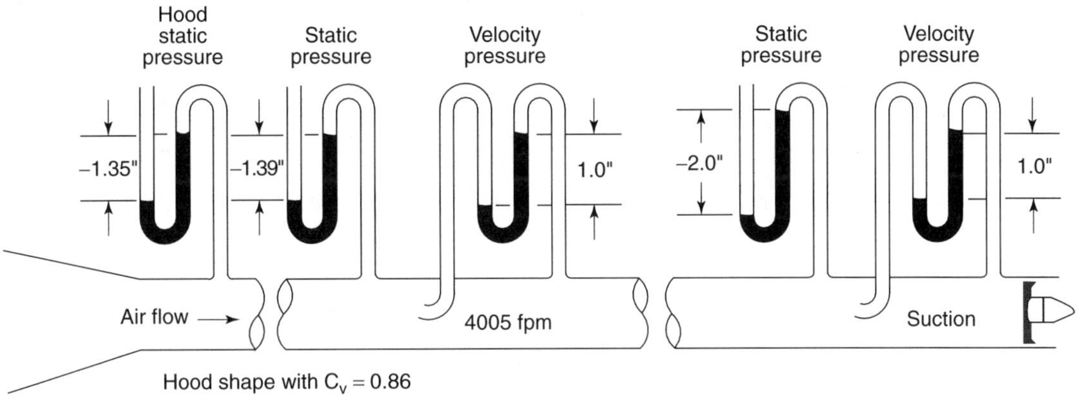

Figure 2. Typical Duct under Positive Pressure with a Velocity of 4005 fpm.

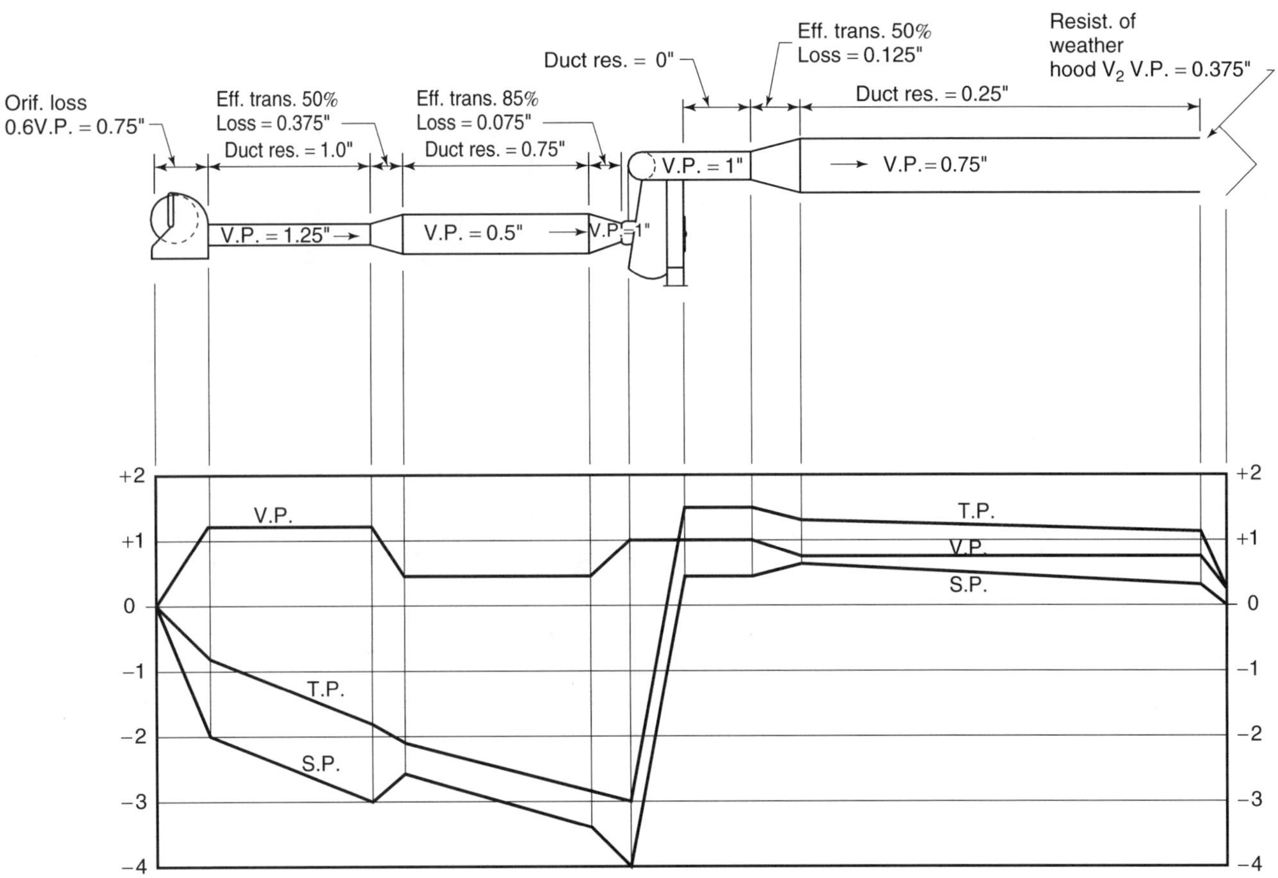

Figure 3. Plot of TP, VP, and SP for Typical Ducted System.

to the pipe wall. Dynamic losses occur as a result of turbulence caused by a change in direction or velocity within a duct, such as at a hood entry, elbows, or fittings. The design of a ventilation system requires the establishment of the total energy losses (friction plus dynamic) through the ductwork system for the design flow rate. As will be shown later, the fan static pressure and flow characteristics must be matched to the ductwork system static pressure and flow-rate characteristics. The design flow for the ductwork is established based on the process or equipment requirements and the required level of performance for the ventilation system. Once a system flow rate has been established, a standard design procedure can be developed to determine the friction and static pressures for the ductwork system.

Frictional loss in a duct varies directly as the length, inversely as the diameter, and directly as the square of the velocity of air flowing through the duct. The loss can be calculated from a Moody diagram as shown in Figure 4, which relates friction factor to Reynolds number ($Re = vd/\nu$, where Re is the Reynolds number; v, the velocity in fps; d, the pipe diameter in feet, and ν, the kinematic viscosity in ft^2/sec), duct material, and type of construction.

For calculation purposes, Loeffler[3] developed a simplified relationship that calculates friction factors over the range of

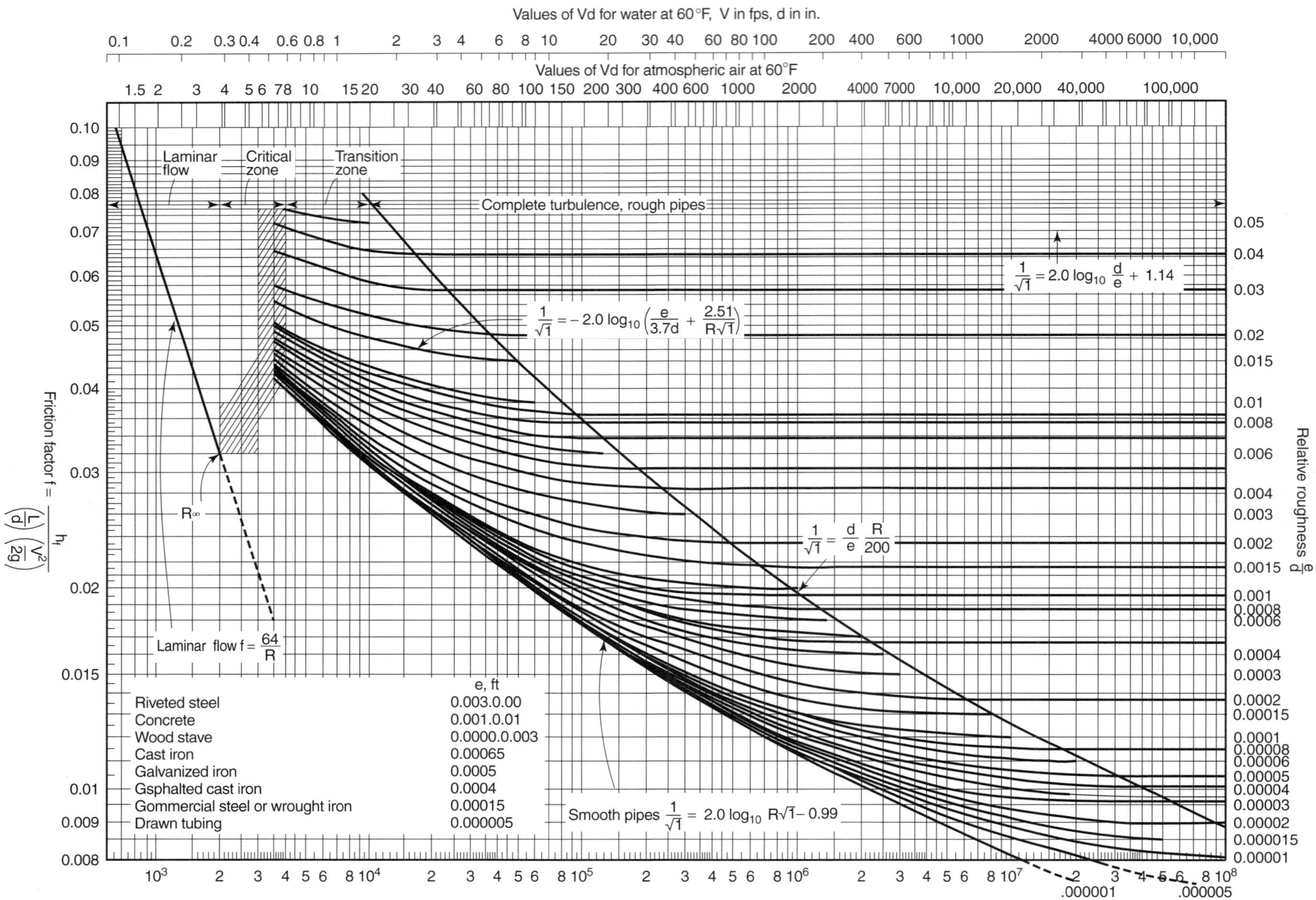

Figure 4. Moody Diagram Relating Friction Factor to Reynolds Number.

likely operating conditions for round, galvanized metal ducts:

$$H_{\rm f} = 0.0307\frac{V^{0.533}}{Q^{0.612}} = \frac{0.4937}{Q^{0.079}D^{1.066}}$$

Standard ventilation references can also be consulted and friction loss values read directly from charts. For gas conditions that are different from standard air, conversion tables exist that allow corrections for relative roughness, gas temperatures, and effects of elevation or gas density.

Dynamic losses occur because of turbulence as a result of a change in velocity or direction within a duct. The resulting pressure drop in a duct system due to dynamic losses increases with the number of elbows and angles and the number of velocity changes within the system. Dynamic losses occur as a gas enters a hood or passes through fittings, such as elbows, expansions and contractions, branch entries, and duct exits. Energy losses at hoods and duct fittings depend on different parameters and the design procedures and equations will be discussed separately.

Hoods are shaped inlets designed to capture contaminated air and conduct it to the exhaust duct system. The classification of hoods and details on hood design parameters and performance evaluation are discussed in detail under "Ancillary Equipment." For example, design procedures to calculate minimum exhaust design volumes are outlined. Based on a given hood volume flow rate, the designer must establish the design approach for a minimum entry loss (dynamic losses) for the selected hood.

For flow into a hood, the cross-sectional area of the flow begins to contract (vena contracta) before expanding again to fill the duct. The energy loss at the vena contracta of the hood is defined as the coefficient of entry ($C_{\rm e}$), which is the ratio of the actual flow to the theoretical flow, if there had been no vena contracta. For an ideal hood with no loss, $C_{\rm e}$ is 1.0.

The hood static pressure ($\mathrm{SP_h}$) measured a short distance downstream from the hood, is a direct measurement of the energy (VP) required to accelerate the fluid from rest to the duct velocity plus the hood entry turbulence losses ($h_{\rm e}$), which are a function of the hood shape. The entry loss plus the acceleration energy required to move the air at a given velocity (1 VP) make up the hood static pressure. The hood entry losses can be expressed as a number of velocity pressures using a hood pressure loss factor ($K_{\rm h}$).

The $\mathrm{SP_h}$, which can be measured directly at a short distance downstream from the hood entrance, is expressed algebraically as follows:

$$\mathrm{SP_h} = (1 + K_{\rm h})\mathrm{VP}$$

The coefficient of entry ($C_{\rm E}$) is related to the velocity pressure loss factor $K_{\rm h}$ by the following equation.

$$K_{\rm h} = \left(\frac{1}{C_{\rm E}^2} - 1\right)$$

For compound hoods, it is possible to add pressure losses for each hood. Novel or complex hoods may require experimental tests in a laboratory to establish hood entry losses. Figure 5 provides pressure loss factors for typical hoods.

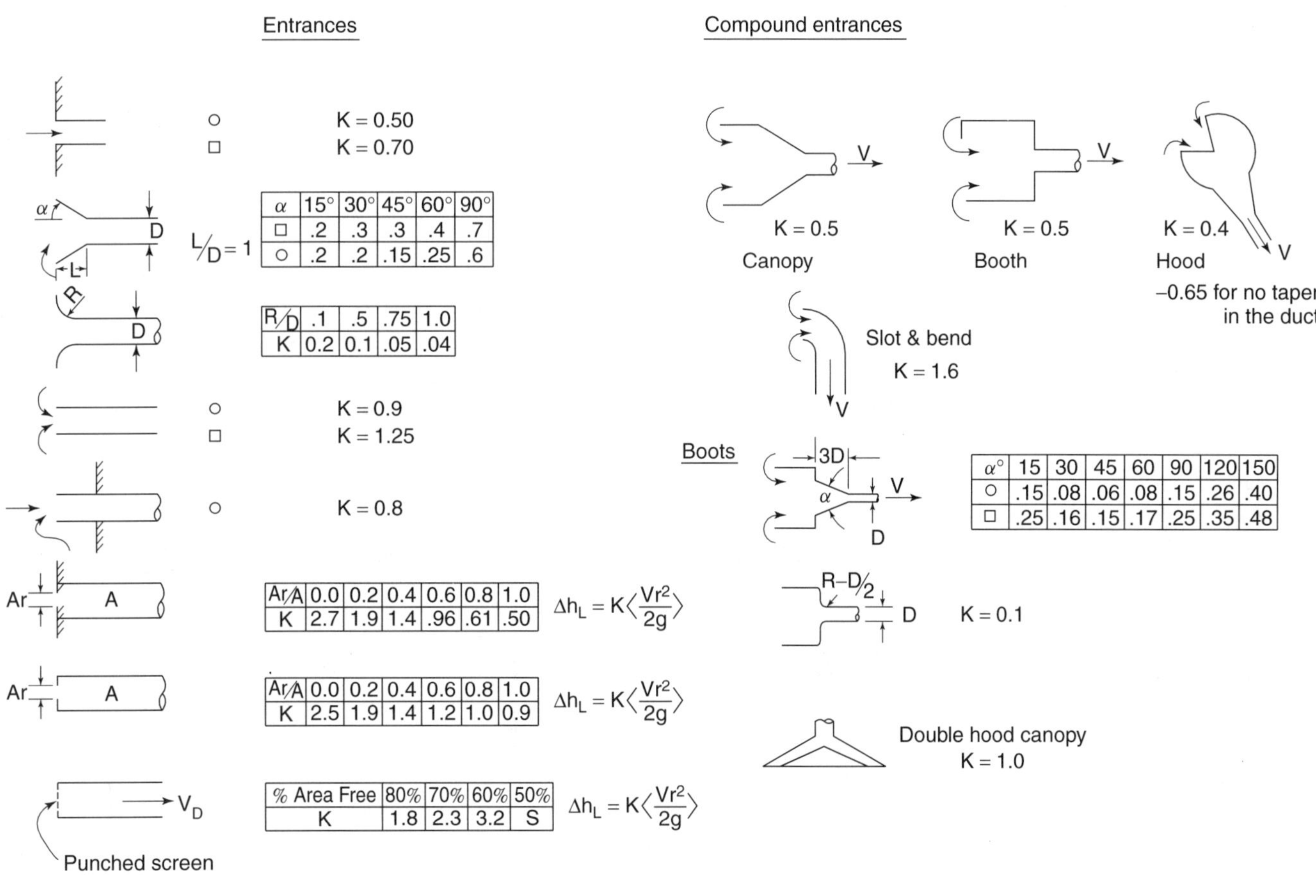

Figure 5. Pressure Loss Factors for Typical Hoods (from Ref. 2).

Fittings in a ducted system will result in additional static pressure losses because of the turbulence produced. The types of fittings include bends, expansions, contractions, dividing flows, combining flows, sucking flows, manifolds, and so on. For design purposes, their losses are expressed as a number of velocity pressures (VP), where K is the loss factor and VP is the velocity pressure in the duct. Figure 6 gives K factors for many common fittings for circular ducting. In other fittings and ducting of rectangular cross section, the necessary design calculations and correction factors can be obtained from the literature.[2,4]

The total static pressure loss for a system is calculated by summing up the friction and dynamic losses. An equation can be written as follows:

Explanation

The system resistance is determined by adding the total pressure drops across successive sections of the system of airways conveniently classified under the headings 1.6. This equals the total fan pressure required. Each item of loss. P, is given in terms of the velocity pressure. P_V, and the factor K given in the appropriate section.

$$P = K \times P_V \text{ in. w.g.}$$

P_V is calculated from the average velocity V at the section shown by the arrow in the diagrams, using the formula:

$$P_V = \left(\frac{V}{3970}\right)^3 \times \left(\frac{\text{Actual gas density}}{\text{Standard air density}}\right)$$

$$\text{Total fan pressure} = P = P_1 + P_2 + P_3 + P_4 + P_5 + P_6$$

Symbols

- L = Length of duct in feet
- D = Diameter of duct in feet
- P = Drop in total pressure in w.g.
- P_V = Velocity pressure, in w.g.
- V = Average velocity, ft./min.
- P = Gas density, lb, per cu. ft.
- P_2 = 0.0764 lb, per cu. ft.
- K = Factor defined in each section

1. Loss at entry to system $P_1 = K \times P_V$

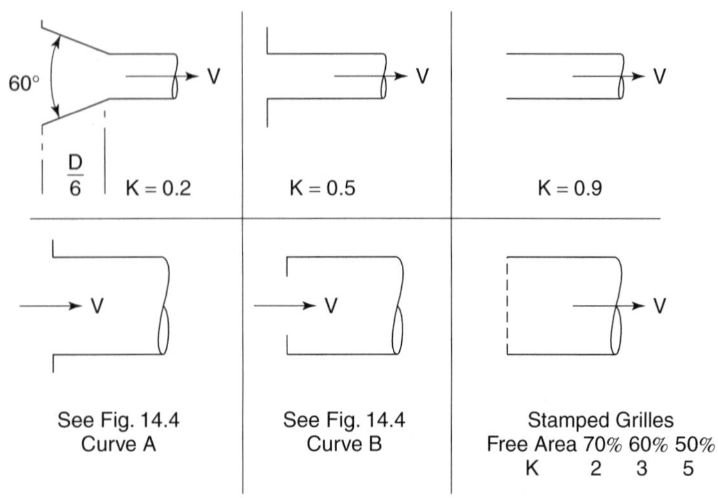

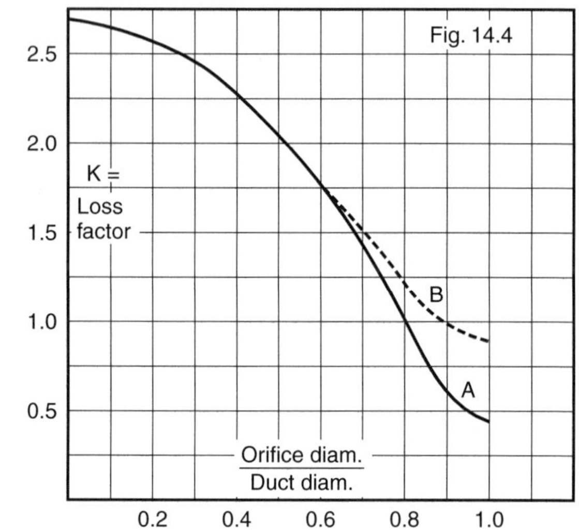

2. Losses along straight ducts $P_2 = K \times P_V$

Approximate loss formulas adequate for short ducts:

Circular Section

$$K = 0.02 \times \frac{L}{D}$$

Rectangular Section, A × B ft.

$$K = 0.01\ \frac{A + B}{AB} \times L$$

Other Sections

$$K = 0.005\ \frac{\text{Perimeter ft.}}{\text{Area sq. ft.}} \times L$$

Chart based on Friction Factor method for use with sheet metal ducts. Fig. 13-4 is for circular ducts. If duct is rectangular convert to circular equiv. using Table 13-1.

Based on e = 5

Friction Factor Method:

For maximum accuracy with long rough walled circular ducts:

$$K = f \times \frac{L}{D}$$

Determine f from Fig. 14.5 estimating surface roughness e (in units of 0.001 in.) using table below as guide:

Typical Surface	e
Drawn Tube	0.1
Rolled Sheet	1
Galvanised Sheet	5
Cement Facing	20
Cast Concrete	80
Heavily riveted	200
Rough Brickwork	500

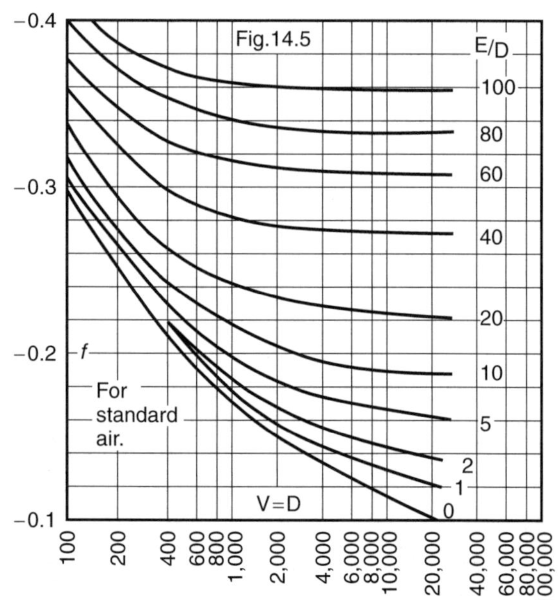

Figure 6. *K* Factors for Common Fittings for Circular Ducting.

3. Losses at heaters, filters, etc. $P_4 = K \times P_V$

Obtain drop in TOTAL PRESSURE across component at required volume rate from manufacturer's date. This equals the drop in STATIC PRESSURE if inlet and outlet areas are equal. Fig. 14.6 gives approximate values of K for banks of small gilled tubes. P_V is based on average FACE velocity For banks of plain tubes use formula:

$K = Cn$ where n = number of rows

Arrangements of Tubes: d = diameter of tube

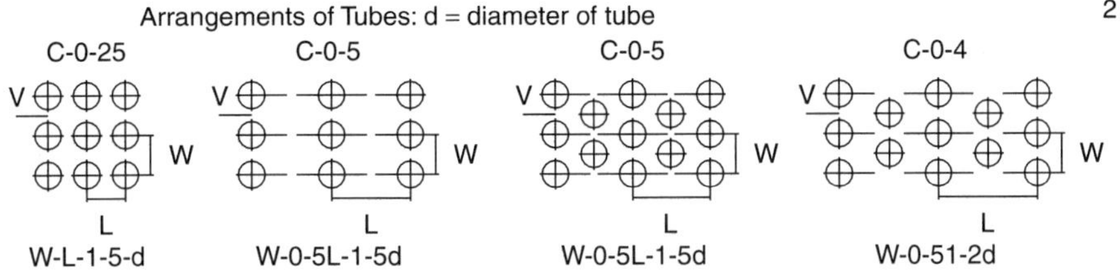

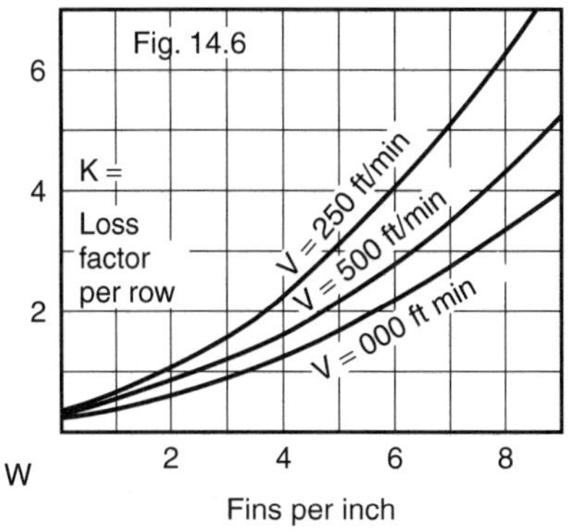

4. Losses at expansions and contractions $P_4 = K \times P_v$

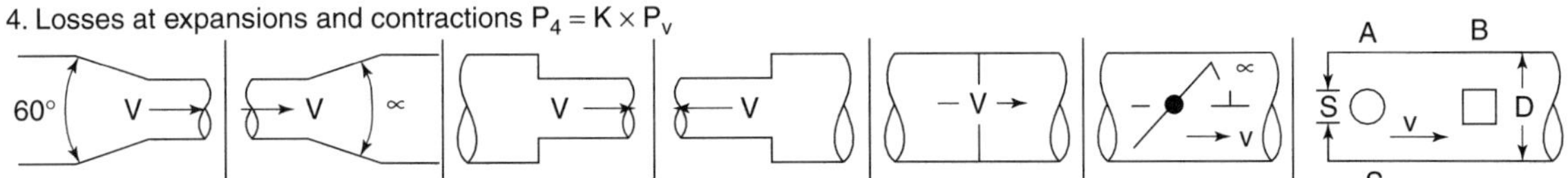

K = 0.0.6	See fig. 14.8	See fig. 14.8 Curve B	See fig. 14.8 Curve A	See fig. 14.7 Curve A	∝ / K	S/D, K_A, K_B
					9° 0.24	0.1 0.2 0.7
					10° 0.32	0.25 0.4 1.6
					20° 3.9	0.3 2.0 4.0

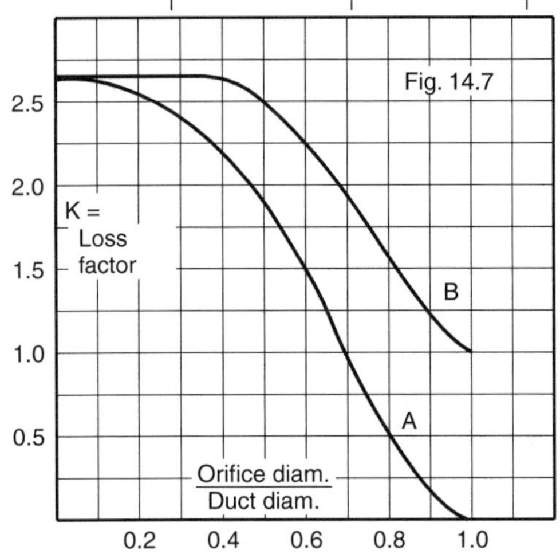

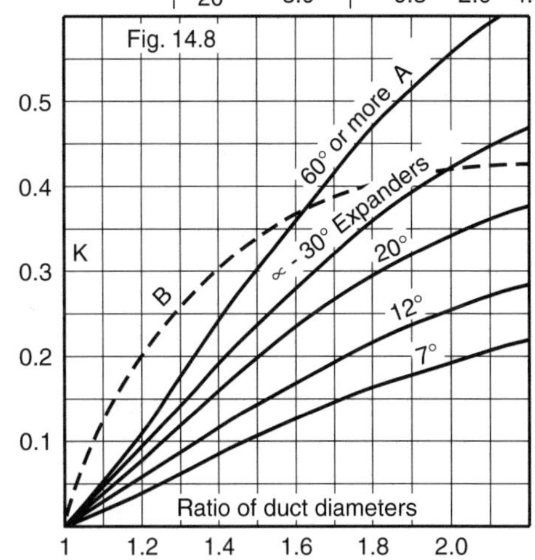

5. Losses at change of direction $P_3 = K \times P_v$

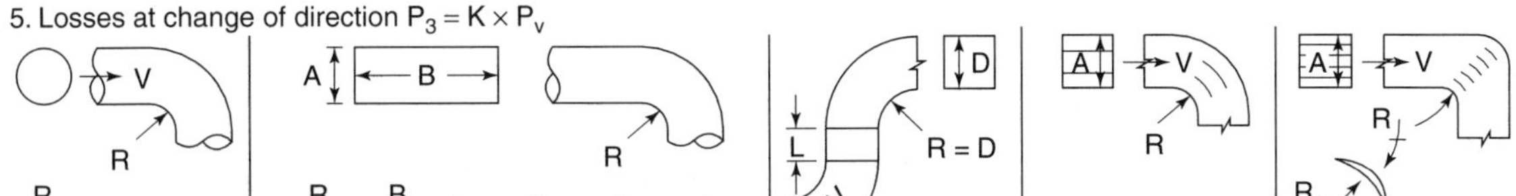

R/D	K
0	0.8
0.25	0.4
0.5	0.25
1.0	0.16

R/A	K: B/A = 1	2	3	4
0	1.0	0.9	0.8	0.73
0.25	0.4	0.4	0.39	0.32
0.5	0.2	0.2	0.19	0.16
1.0	0.13	0.13	0.13	0.10

L	X
O	0.62
D	0.68

Elbows vaned K = 0.19

Vanes No.	R/A	K
1	0.25	0.13
2	0.2	0.12
3	0.15	0.10

$Ri = \frac{Ro}{2} = \frac{A}{6}$

K = 0.1

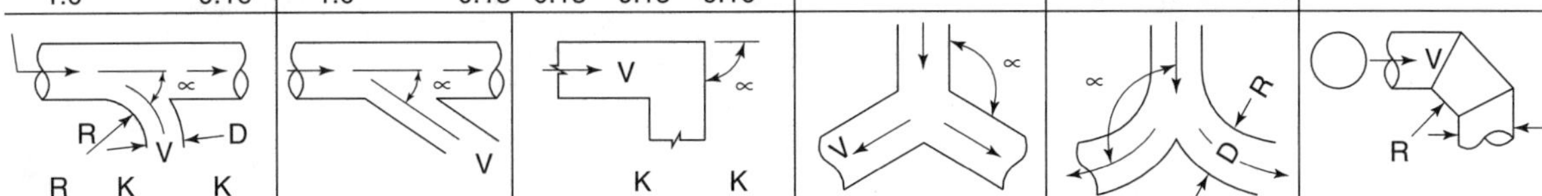

R/D	K ∝ = 90°	K ∝ = 45°
0	1.0	0.6
0.25	0.5	0.35
0.5	0.3	0.2
1.0	0.2	0.15

∝	K
90°	1.9
60°	0.5
45°	0.3
30°	0.2

∝	K ○	K □
90°	0.85	1.25
60°	0.65	0.95
45°	0.44	0.63
30°	0.22	0.3

∝	K ○	K □
90°	1.9	1.5
135°	0.6	0.9

R/D	K ∝ = 90°	K ∝ 135°
0.25	0.43	0.3
0.5	0.28	0.18

R/D	K 5 piece	K 3 piece
0.25	0.5	0.6
0.5	0.3	0.4
1.5	0.2	0.3

Figure 6. *(Continued).*

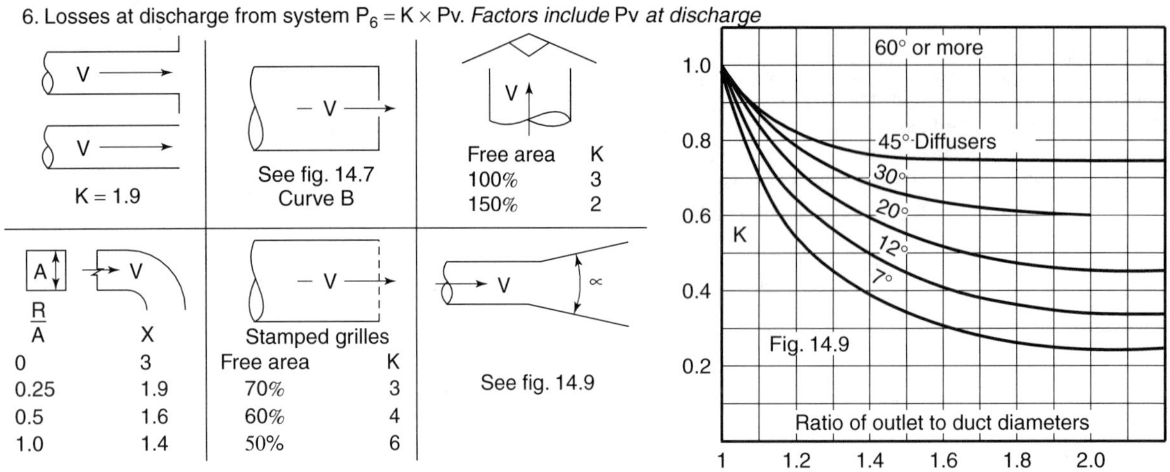

Figure 6. (*Continued*).

$$\text{total energy losses} = \text{friction losses} + \text{dynamic losses}$$
$$= (\text{straight duct} + \text{hood} + \text{fittings})\ \text{losses}$$
$$= (f(L/D) + (1 + K_h) + \Sigma K)\text{VP}$$

This general equation is used to calculate the total static pressure losses for an LEV system. This static pressure loss can be combined with the velocity pressure term to give the total fan pressure required. The details of the static pressure loss calculation procedures are presented in a subsequent section.

Measurement Techniques

There are numerous instruments and techniques that can be used to measure flows and pressures for a ventilation system. Detailed descriptions of the different instruments and their applications are available in standard references. An instrument that has wider applications for the measurement of pressures and temperatures is a *pitot tube*. A brief description of the pitot tube, and its operation and applications, follows.

The pitot tube with an inclined manometer is the standard air velocity meter. It is of rugged design, is relatively low in cost, and is available in many different sizes for different duct diameters. Figure 7 shows the proportional dimensions for a standard pitot tube. No calibration curves are required if the tube is carefully made and the velocity pressure readings obtained are considered accurate. Standard pitot tubes are made of Type 304 stainless steel and can be used at gas temperatures up to 1000°F. Above this temperature, special alloy steels or water-cooled pitot tubes are required.

The pitot tube consists of two concentric tubes. The opening of the inner tube is axial to the flow and measures total pressure; the opening of the outer tube is circumferential and measures static pressure. The velocity pressure is the

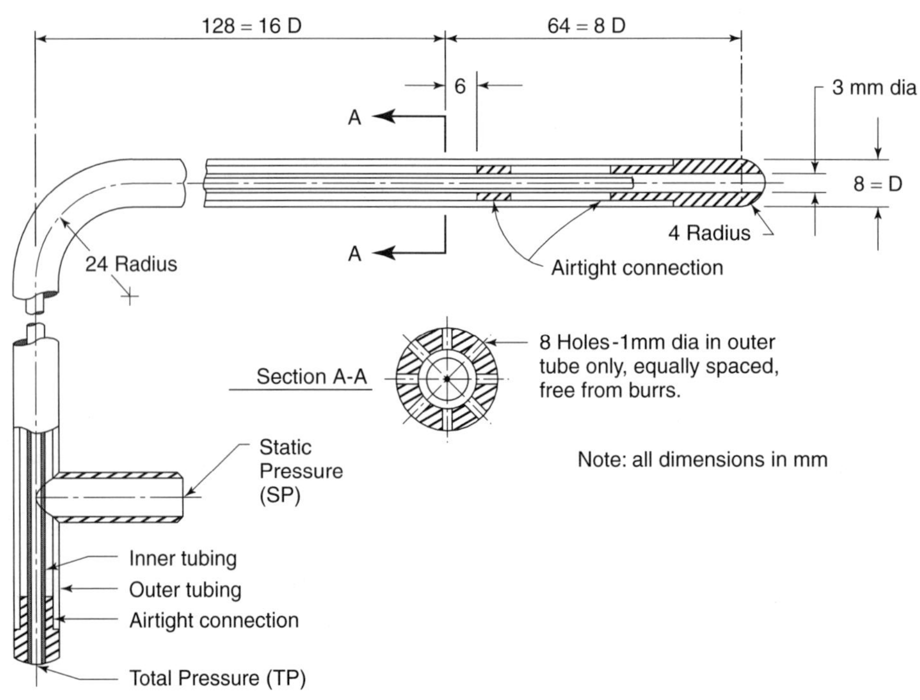

Figure 7. A Standard Pitot Tube.

difference between the total pressure and static pressure. Velocity pressure measurements are made with the pitot tube by connecting the static and total pressure tops as shown in Figure 8. By disconnecting one of the hoses, it is possible to perform static or total pressure measurements for a system.

Before taking any measurements with a manometer, the instrument must be calibrated to ensure that the manometer fluid and scales are correct. A simple field check is to use a U-tube or plastic tubing filled with water and the scale reads directly in the pressure units of inches of water gage (in. w.g.) In general, airflow is not uniform throughout the cross section of the duct. Traverses are used to obtain the average

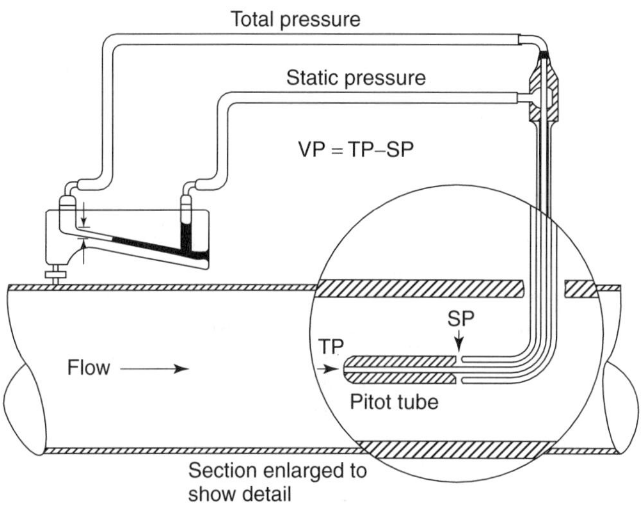

Figure 8. Pitot Tube Connected to an Inclined Monometer.

by measuring the velocity pressures at points in a number of equal areas in the cross section, as shown in Figure 9.

To minimize flow distribution measurement problems, it is recommended that measurements be taken at least ten diameters downstream and two diameters upstream from any obstructions or flow disturbances.

The design equation developed for a pitot tube from the general Bernoulli equation is as follows:

$$V = 1096\sqrt{\frac{\text{VP}}{(0.075)(d_f)}}$$

where V is the velocity (fpm); VP, the velocity pressure (in. w.g.); and d_f, the density factor:

$$d_f = \frac{530}{460 + t}\,\frac{B}{29.92}\,\text{(dry air)}$$

where t is the air temperature (°F); and B, the barometric pressure (in. Hg).

When the moisture content in the gas is above 0.02 lb/lb of dry air, the density factor should be obtained directly from psychometric charts. For dry air at standard conditions (70°F and 29.92 in. Hg), the density factor becomes unity and the equation becomes:

$$V = 4005\sqrt{\text{VP}}\ \text{(in fpm)}$$

To calculate the air volume flow rate from pitot tube readings, it is necessary to calculate velocities for each measured point and then to calculate average velocities. The measured air volume at the temperature in the duct is, then,

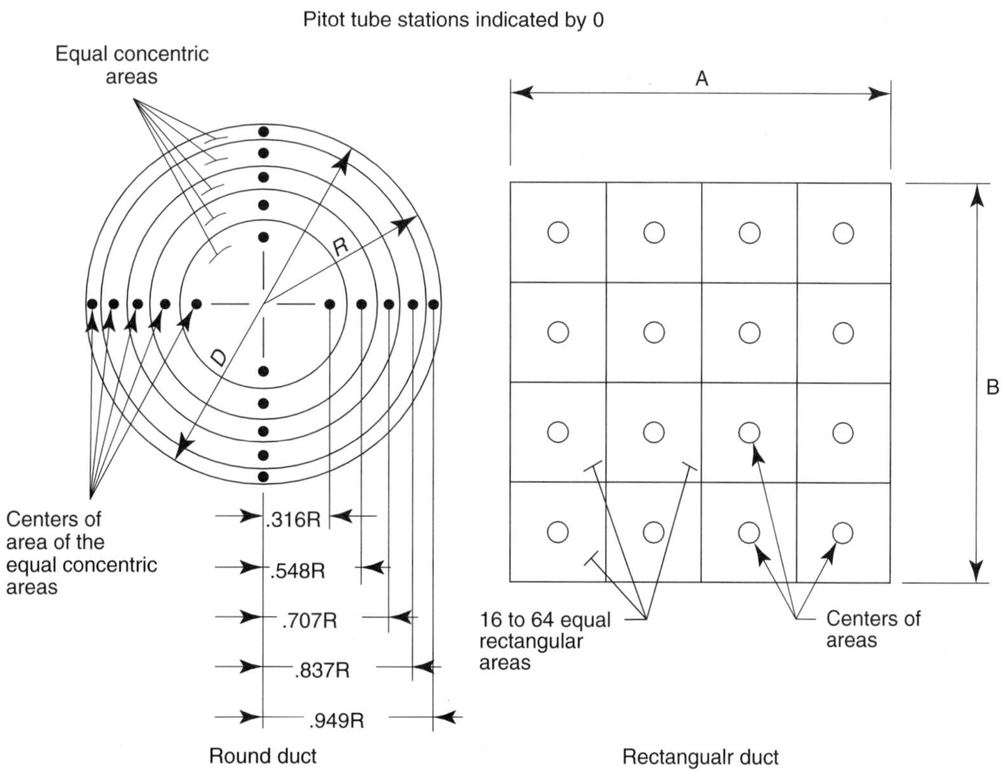

Figure 9. Traverse of a Round and Rectangular Duct Area.

the average velocity measured by the cross-sectional area of the duct.

LOCAL EXHAUST VENTILATION

Elements of LEV Systems

Figure 10 is a schematic of a typical LEV system used for the capture of a contaminant from an emission source. The major components of an LEV system are the hood, duct, gas cooling, fan, pollution control equipment, and stacks. Chapter 2 describes the pollution control equipment used for gases (absorption, adsorption, condensation, and incineration). Chapter 3 describes the pollution control equipment used for particulates (mechanical collectors, scrubbers, electrostatic precipitators, and fabric filters). Ancillary equipment described here will focus on the other components of an LEV system.

The term *hood* is applied to a shaped inlet that is designed to capture the contaminant from an emission source and conduct it into the exhaust duct system. Hoods can be of any shape or size. A subsequent section will describe a hood classification system, outline hood design methodologies, show schematics of different types of hoods within the groups, and present typical equations and criteria for hood design. Properly designed hoods should meet the design criteria of maximum efficiency of collection of the contaminant with minimal exhaust volumes and minimal interference with the process and the operators.

The contaminated air is exhausted through the ductwork, which provides a channel for the flow that moves the contaminant from the emission source to be discharged into the atmosphere at the stack. Ducted systems are typically classified as dust control, fume control, and mist/gaseous control. Although these ducted systems have many common features, a successful operation depends on a proper design by an experienced engineer who will incorporate many unique and detailed design features into the specific ducted system. A technical description and design parameters for ducted systems are provided in a later section.

Gas cooling technology applicable to LEV systems includes water-cooled ducting, radiation, forced-draft, evaporative, and air dilution. Each of these gas cooling technologies will be discussed later.

As one of the most important elements to ensure successful operation of an LEV system, fans provide the energy to move the air containing the contaminant through the system. A successful operation requires a fan to deliver the required volumetric flow rate in order to ensure capture of the contaminant at the source and ultimately to discharge the cleaned gas up the stack and into the surrounding atmosphere. A technical description of fans and the pertinent fan law equations for design are presented later in this discussion.

Stacks for LEV systems provide an elevated discharge point for the cleaned gas. The actual height of a stack is determined based on keeping contaminants within a specific contaminant concentration level at the point of impingement. A following section outlines a technical description of stacks and identifies applicable design equations to establish stack diameters and stack heights.

Design Parameters

Local exhaust ventilation is one of the standard "engineering methods of control" used to control contaminant levels in the workplace. Many manufacturing and processing facilities generate contaminants that would present worker exposure problem if the contaminants were not captured. Table 1 identifies the five major types of ducted air systems and shows typical hood velocities, duct velocities, and dust loadings. As also shown in Table 1, design parameters for LEV systems fall in between those for conventional heating, ventilation, and air conditioning (HVAC) systems and pneumatic conveying systems.

Factors to be considered in the design of LEV systems are as follows:

1. Capture of the contaminant at the source (if possible)
2. Provision of an adequate but not excessive capture velocity

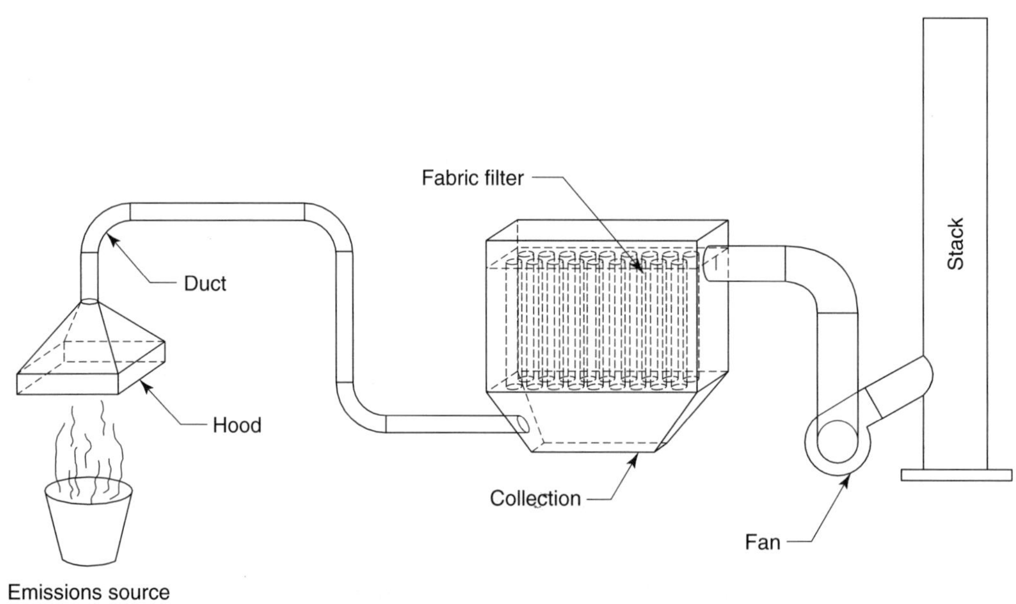

Figure 10. Schematic of a Typical LEV System.

Table 1. Ducted Air System Categories

System	Typical Operating Parameters		
	Hood Velocities (fpm)	Duct Velocities (fpm)	Dust Loading (grains/ft^3)
1. Heating and ventilation		2000	Insignificant
2. Local exhaust ventilation			
Vapour control	2000	2500	N/A
Fume control	2000–4000	2000–4000	0.05–0.5
3. Local exhaust ventilation			
Dust control	250	4000	5
4. Dedusting or classifying	1250	5000	100
5. Pneumatic conveying		7000	500

3. Incorporation of containment whenever possible
4. Use of gravitational force in hood design
5. Provision of adequate conveying velocity for the contaminant
6. Use of a design that minimizes pressure losses
7. Design of the fan so as to accommodate all operating modes
8. Rigorous application of the systems approach

The LEV systems have many advantages over general ventilation, some of which are as follows:

1. Level of control of contaminant can be very high (more significant as allowable contaminant levels decrease).
2. Volume of exhaust air is much lower (more economical design).
3. Contaminant level in exhaust gas is more concentrated (more economical design for pollution control equipment).
4. Performance is not grossly affected by cross-drafts.
5. Housekeeping is improved because large particles are collected.
6. Auxiliary equipment can be protected from heat, dust, corrosion, and so on.

The design parameters for LEV systems that are important are the size and shape of the hood; capture velocities and exhaust flow rates for the hood; materials of construction and diameter of ducts; type and size of gas cooling equipment; type of fan and parameters, such as fan speed, design gas flow, and system static pressure; type of pollution control equipment; and diameter and height of stack. Details on the technical specifications and design parameters will be found under "Ancillary Equipment."

Preparation of Technical Specifications

The LEV system is an integral part of the processing/manufacturing facility, and it is essential that the system designer develop the proper design parameters and technical specifications. Goodfellow[2] has described in detail the essential steps and provides a checklist for the preparation of better technical specifications. Typical standard technical specifications for all the major ventilation system components are presented. Components that are included are louvers, roof ventilators, ducted systems (including accessories), gas cooling, fans, gas cleaning, stacks, and dust-disposal systems. The technical specifications should be considered as guidelines or checklists that will assist in the preparation of equipment specifications for particular projects. For successful projects, technical specifications should be prepared only by well-qualified and experienced engineers.

Before preparing detailed technical specifications, a decision must be made as to whether the LEV system will be an "engineered system" or a "turnkey system." The current trend is to purchase a conventional LEV system on a turnkey basis, but any LEV system of a specialized nature will be purchased as an engineered system. The major advantages of an engineered system are that the client can select the best components based on competitive bidding, experienced process specialists can be employed, complex layouts can be integrated into the total project, and major capital expenditures can be deferred. It would appear, then, that emerging technology in the ventilation field will continue to favor the purchase of engineered systems for LEV.

The preparation of technical specifications must focus on communication, criteria, and conciseness. The specification to the bidders must communicate the user's requirement or criteria with clarity and brevity. A checklist of important steps to prepare better technical specifications would include the selection of a qualified engineer, a clear definition of the scope of work, a defined performance requirement, the supplying of all pertinent process and test data, and the attachment of a technical questionnaire. A well-prepared specification will help to ensure that the project will be successful and cost effective.

VENTILATION FOR PROCESSING/MANUFACTURING FACILITIES

Design Methodology

Significant progress has been made in developing a systematic design procedure for industrial ventilation systems to be accepted on a worldwide basis. The initial research and development effort was started in Finland in 1991 with the launch of the INVENT program and an investment of more than $20 million U.S. dollars. In 1995, a decision was made at a workshop in Zurich to start preparation of an international design guide book for industrial ventilation. An analysis of the current state-of-the-art identified the following issues:

1. No scientific basis for many design applications
2. No harmonization of design equations from different countries or researchers
3. Gaps in technical literature not defined and no roadmap identified for future technological developments
4. Many books are out of date
5. No longer acceptable to overdesign ventilation systems
6. No handbook in the industrial ventilation field
7. No accepted design methodology based on a rigorous scientific approach
8. Ventilation field fragmented on a global basis so need collaboration by a team of international experts
9. Excellent opportunity to collate worldwide research and development efforts into a single handbook
10. INVENT program has generated momentum and a critical mass to make project successful

An International research and development program is underway, with more than 25 countries participating, with the objective of preparing a design guide book for industrial ventilation. Two key elements of the design guide book which have been developed by the research and development team are design methodology and target levels.

Figure 11 is an outline of the systematic design methodology procedure for industrial ventilation. A brief description of the different design steps follows below. (More details can be found in the Scientific Report entitled "Design Methodology of Industrial Air Technology" listed in the bibliography.)

Step 1: Given data

- Identify and collect data that are site specific (i.e., climatic conditions, site elevation, etc.)

Step 2: Process description

- Understand the industrial process and identify subprocesses
- Identify possible emission sources, occupational areas, effects of environmental parameters on production, needs for enclosure and ventilation equipment
- Divide process in such parts that their inputs and outputs to the environment can be defined

Step 3: Building layout and structures

- Collect properties of building layout, structures and building envelope, openings
- Complete zoning of building based on division of the process and building layout
- Identify layout requirements and structures required for ventilation systems

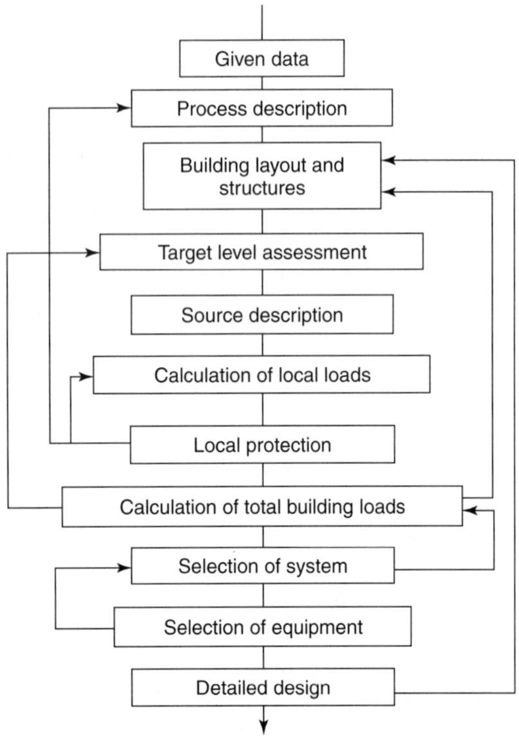

Figure 11. Design Methodology for Industrial Ventilation Systems.

Step 4: Target-level assessment

- Define target levels for indoor (zones) and outdoor (exhaust) conditions based on human risk assessment and technology
- Specify design conditions for which the target levels is to be met
- Define target levels for ventilation system based on reliability, energy consumption, investment and life-cycle costs, etc.

Step 5: Source description

- Identify emission sources by type (chemical and physical properties)
- Review requirements for head load calculations

Step 6: Calculation of local loads

- Calculate loads from individual sources
- Use emission data or calculation models

Step 7: Local protection

- Examine feasibility of local source control
- Calculate working conditions for different options
- Compare to target levels of local zone

Step 8: Calculation of total building loads

- Calculate total loads (heat, humidity, contaminants) from different subprocesses and building for specific zones of the building
- Account for time dependency of emissions
- Evaluate system performance and establish if target levels or building layout and structures need to be revised

Step 9: Selection of the system

- Select acceptable systems based on target levels
- Identify feasibility of different options selected
- Select the most effective ventilation system

Step 10: Selection of equipment

- Identify system specifications
- Select acceptable equipment based on performance characteristics
- Equipment recommendations
- Prepare a technical specification for selected equipment

Step 11: Detailed Design

- Calculate dimensioning requirements for selected systems
- Detailed layout ventilation systems
- Design control system
- Consider special issues such as thermal insulation, condensation risks, fire protection sound and vibration damping, maintenance, etc.
- Develop construction and commissioning plan

Target Levels

It is difficult to achieve a good indoor environment in an industrial facility because of the lack of a scientific basis for defining acceptable design levels for contaminants in an industrial setting. A detailed database for existing contaminant exposure is available from the Finnish Institute of Occupational Hygiene (FIOH).[16] Figure 12 shows a recommended approach for assessing target levels of industrial air quality.

Target concentrations of air contaminants can be developed based on both human risk assessment and technology. Health risk estimates at low concentrations (e.g., below one-tenth of the current occupational exposure limits) are, in general, inaccurate and rather unreliable. Consequently, technology-based approaches for the assessment of target concentrations have been emphasized. The control technology approach is based on information on "current" concentrations achieved by standard practices and "benchmark" concentrations achieved by the advanced ventilation and production technologies.

Existing contaminant exposure data banks, such as available from FIOH and other international health and safety bodies can be utilized for assessing concentration levels that are achievable. The "benchmark" air quality has been determined by measuring concentrations in factories equipped with the advanced technologies. This information on the "benchmark" air quality, the "current" concentration levels and the health and comfort effects form the foundation for setting target levels for industrial air quality. This target level becomes the design input for establishing the industrial ventilation system design parameters. A detailed description of the use of target levels of air quality as a valuable design tool is presented in a paper by Niemela et al.[16] Tahti[17] provides details of a case history of the application of target level and design methodology for paper machine room ventilation.

Industrial Ventilation

Most industrial ventilation problems for processing and/or manufacturing facilities are complex. The development of cost-effective solutions requires an experienced and qualified ventilation engineer. The design of a ventilation system must be incorporated into the plant design and layout at the earliest conceptual stage of the project. Decisions must be made on the level and requirements for general ventilation, dilution ventilation, and/or LEV. The ventilation engineer must work closely with the project design team at all stages of the project since the ventilation system can have a major impact on the type of process to be used, the plot plan, the building profile, and equipment layout within the plant. Cost-effective ventilation systems can only be developed if the ventilation system design is an integral part of the project planning and design activities from the preliminary conceptual stage.

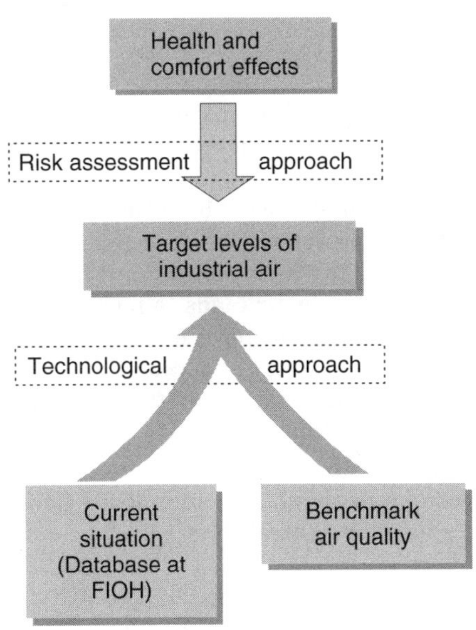

Figure 12. Target Levels for Industrial Air Quality.

In many applications, the ideal ventilation scheme may be in conflict with other environmental design criteria, such as noise control and fume control. For example, a building designed for hot processes may be based on the concept of a wide-open building to enhance natural ventilation; this is in conflict with noise control considerations that require the plant to be enclosed as much as possible to reduce neighborhood noise. Another example of this conflict is the use of canopies for remote fume capture. Canopies perform best if the building is totally enclosed to prevent cross-drafts from disturbing the rising plume. Totally enclosed buildings reduce the effectiveness of dilution ventilation.

The methodology for the design of a ventilation system for a new facility can be based on implementing a checklist of activities as summarized in Table 2.

Ventilation Design Equations

The design calculations for industrial ventilation systems for processing or manufacturing buildings are based on:

1. Heat-release calculations
2. Air set-in-motion calculation
3. Neutral plane calculations

Detailed heat-release calculations can be carried out for the different operations within the process building. Heat is released from surfaces by convective forces that cause thermal updrafts, producing the primary ventilation flows. Radiant heat that is not ultimately converted into convective heat does not affect the primary ventilation flow. Both the steady and the intermittent heat release are calculated. For design purposes, it is recommended that an average heat release be established to be somewhat higher than the steady heat release.

For convective heat losses, the heat release from equipment surfaces is calculated by:

$$q_c = h_c A(T_s - T_a)$$

where q_c is the convective heat loss (Btu/h); h_c, the natural convective heat transfer coefficient (Btu/h ft^2 °F); A, the surface area of hot source (ft^2); T_s, the surface temperature (°F); and T_a, the surrounding air temperature, (°F). The convective heat transfer coefficient h_c can be readily estimated from well-established formulas that can be found in standard heat-transfer reference books.

Air set-in-motion calculations account for major air and fume vertical flows that are produced by the total convective portion of heat release from hot processes. (The convective portion may include some radiant heat loss that is converted to convecting heat.) This hot air is heated by contact with the surfaces of the hot equipment and rises as a result of thermal

Table 2. Design Steps for Ventilation Systems for Greenfield Processing/Manufacturing Facilities

Design Step	Description of Activities
1. Obtain published ventilation design	Literature search to obtain published ventilation design data for similar processing operations
2. Obtain pertinent data on site condition	Obtain local meteorological data (minimum of three to five years on a monthly basis); data to include temperature, relative humidity, precipitation, and wind speed and direction
3. Develop a catalog of major emission	For each emission source, establish type of contaminant, gas composition of particulate, particle size of particulate, and dust loading
4. Develop ventilation design objectives	Review list of potential contaminants and prioritize Design must comply with applicable occupational health and safety regulations Client to review and approve design objectives
5. Develop a schedule of building openings	Prepare an isometric drawing Tabulate schedule of building openings
6. Perform analytical work	Heat release calculations Air set in motion Heat and flow balances Preliminary sizing of ventilation equipment
7. Define the design basis for the ventilation system	Use design objectives from above Prepare specific design bases for the ventilation system
8. Develop alternative ventilation system concepts	Preliminary layouts and flow sheets Performance evaluation using computer or fluid dynamic modelling
9. Perform a technical/economic evaluation of the alternatives	Detailed analysis of alternatives Develop a decision matrix (cost/benefits)
10. Prepare an engineering report with recommendations	Detailed description of system/design calculations Recommended ventilation system
11. Management approval of ventilation systems	Approval of recommended ventilation system
12. Detailed design	Prepare all necessary technical specification for major equipment Detailed design and drawings
13. Installation	Ventilation equipment to be installed in compliance with all detailed drawings, specifications, etc.
14. Startup and testing	Commissioning in compliance with startup and operating procedure manuals Testing of ventilation system Documentation of system performance

buoyancy. The size and velocity of this hot air column are a function of the heat release rate and the distance between the source and the roof level.

The heated airstream originating from the surfaces of hot bodies mixes turbulently with the surrounding air as it moves upward. An airstream at a hot body source may have a flow of 1000 cfm. After some vertical travel, this hot airstream may set in motion and become mixed with (i.e., entrained in) an additional flow of 100,000 cfm of surrounding air. To avoid recirculation, roof ventilators should, as a minimum, accommodate this total flow arriving at the roof level.

The plume flow rate can be calculated using

$$Q = 0.166(Z)^{5/3}(F)^{1/3}$$

where Q is the plume flow rate (cfm); Z, the distance from virtual origin (ft); and F, the buoyancy flux (ft^4/sec^3).

This equation shows that the plume flow rate depends very strongly on its distance from the source (five-thirds power), but only weakly on the buoyancy or thermal forces (one-third power). Therefore, the accuracy of the calculated plume flow is not highly dependent on the buoyancy flux, which is difficult to determine precisely.

Neutral plane calculations are based on the classical work of Ernswiler in 1926.[5] He defined the concept of a neutral zone or plane of neutral pressure as the elevation within a building at which neither the outside air tends to move into the building nor the inside air tends to move out. Figure 13 shows the plane of neutral pressure as defined by Ernswiler. Ernswiler developed two theorems that apply to industrial ventilation and form the basis for design.

Theorem 1. The sum of flow rates into the building below the neutral zone must equal the sum of the flow rates out of the building above the neutral zone.

$$\sum_{i=1}^{n} A_i V_i + \sum_{j=1}^{n} Q_j = 0$$

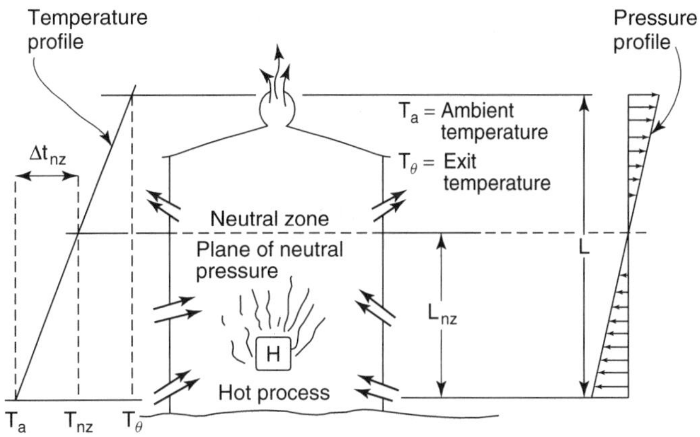

Figure 13. Plane of Neutral Pressure as Defined by Emswiler (from Ref. 2).

where A_i is the area of opening (ft^2); v_i, the velocity through the opening (fpm); and Q_j, the powered ventilation flow rate (cfm).

Theorem 2. The driving force at each opening is related to the vertical distance between the opening and the neutral zone and the temperature difference between the opening and the neutral zone.

$$Q_{\rm N} = A_{\rm N}\sqrt{L_{\rm nz}\,\Delta t_{\rm nz}/RT}$$

where $Q_{\rm N}$ is the flow rate (cfm); $A_{\rm N}$, the area of the opening (ft^2); $L_{\rm nz}$, the distance from the centerline of the opening to the neutral zone (ft); $\Delta t_{\rm nz}$, the temperature difference between the opening and the neutral zone (°F); R, the resistance of the opening; and T, the absolute temperature of the air at the opening (K).

The elevation of the neutral zone is determined by trial and error, using the preceding equations.

The temperature rise through a hot-process building is a function of the amount of heat released in the building and the ventilation flow rate. The heat balance equation can be written as

$$\Delta t = q/\rho C_p Q$$

where Δt is the temperature rise through the building (°F); q, the heat release rate (Btu/min); ρ, the air density (lb/ft^3); C_p, the heat capacity of air (Btu/lb °F); and Q, the ventilation flow rate (cfm).

For design of ventilation systems, the criteria of mass and heat balances must apply. The amount of air carried into the shop per unit of time is equal to the amount of air leaving the shop in the same period. For the heat balance equation, the quantity of heat removed from the shop must equal the sum of heat brought into the shop by the outdoor air plus the surplus heat given off in the shop in the same unit of time.

For natural ventilation, wind has a significant effect on building ventilation rates. Wind on a building manifests itself as an increase in pressure on the building's windward side and as suction on the leeward size. Pressure coefficient is the dimensionless ratio of the pressure at any arbitrary plane on a building to the ambient pressure. Experiments have shown that pressure coefficients remain constant on geometrically similar buildings. Thus pressure coefficients for buildings are found by wind-tunnel tests on geometrically similar models. The wind pressures are obtained by multiplying the pressure coefficient for the points concerned by the velocity pressure of the wind as

$$\rho = k_{\rm w}(\rho v^2/2g_{\rm c})$$

where p is the wind pressure (lb/ft^2); $k_{\rm w}$, the wind pressure coefficient (dimensionless); ρ, the air density (lb/ft^3); v, the velocity (fps); and $g_{\rm c}$, the acceleration due to gravity (32.2 ft/sec^2).

The amount of ventilation air required to meet allowable contaminant concentrations can be determined using the preceding equations. These equations also allow the areas of the vents in the walls and roof to be calculated. For systems with mechanical ventilation as well as natural ventilation, the quantity of air that is supplied or removed by mechanical ventilation is written into the air balance equation on the inlet or outlet side.

Dilution Ventilation

The terms *general ventilation* and *dilution ventilation* are often used interchangeably. In the current context, dilution ventilation refers to the dilution of contaminated air with uncontaminated air for the purpose of health hazard control.

Starting with a fundamental differential material balance, dilution ventilation requirements can be related to the generation and removal rates of a contaminant as

$$\underset{\text{rate of accumulation}}{V\,dt} = \underset{\text{rate of generation}}{G\,dt} - \underset{\text{rate of removal}}{Q'C\,dt}$$

where V is the volume of the room or enclosure (ft^3); C, the concentration of gas or vapor at time t (ppm); G, the rate of generation of contaminant (cfm); t, the time (sec); $Q' = Q/K$, the effective rate of ventilation corrected for incomplete mixing, (cfm); and K, the design distribution constant or mixing factor, allowing for incomplete mixing.

Ventilation of contaminated air in a space would be simple if the outside air could enter the room in a laminar fashion without turbulence and remove the contaminated air in a piston fashion. For this condition, one air change would be required to remove the contaminated air completely ($K = 1$). In practice, however, the introduction of the fresh air results in turbulent mixing with the contaminated air, and after one air change, the room still contains a dilution mixture of fresh and contaminated air. For most industrial ventilation applications, K varies from 3 to 10.

The material balance equation can be solved for different conditions. Table 3 presents the design equations for three typical conditions.

For any specific contaminant, the design criteria may be based on regulated limits, comfort limits, or odor threshold concentrations. For many industrial solvents, the most stringent design criterion for a dilution ventilation system is based on the odor threshold concentration. Computer models have been developed for calculating contaminant concentrations for different conditions.

Displacement ventilation, a concept developed and tested in laboratories in Norway, is based on ideal unidirectional flow

Table 3. Dilution Ventilation Equations

Case	Description	Equation	Unit Definition
A	Rate of contaminant concentration buildup ($C_1 = 0$ at $t_1 = 0$)	$t = -\frac{V}{Q'} \ln\left(\frac{(G - Q'C)}{G}\right)$	t = time, min V = room volume, ft^3 K = mixing factor Q' = effective ventilation rate, cfm G = rate of generation of contaminant, cfm C = contaminant concentration, ppm
B	Maintaining acceptable concentrations at steady state	$Q = \frac{KG}{C}$	Q = room ventilation rate, cfm
C	Rate of purging	$t_2 = -\frac{V}{Q'} \ln \frac{C_2}{C_1}$	C_2 = concentration at time t_2 C_1 = concentration at time t_1

(plug flow). No short-circuiting takes place, and the residence time for the room air is exactly the transit time (ideal case). A stratification takes place between the clean supply air and the contaminated air, as shown in Figure 14. Systems can be designed to utilize thermal or density stratification to create a tendency to unidirectional flow, that is, to create displacement flow.

The benefits of displacement flow are that it improves the air renewal and contaminant removal speed and that it assists in maintaining favorable concentration gradients of the contaminants generated in the room. Calculation procedures based on mathematical two-zone models have been established.

Fluid Dynamic Modeling

The technique of small-scale modeling for fluid flow problems (i.e., fluid-dynamic modeling) has been used extensively for a wide variety of industrial ventilation applications.[2]

The fluid-dynamic modeling technique is a cost-effective, flexible, and powerful tool for the design of ventilation systems for a greenfield plant and for the elimination of ventilation problems in existing plants. Some of the applications of scale modeling include:

- Finalizing building ventilation flow rates and schemes
- Examining internal flow patterns and contaminant concentrations at any location
- Examining external flow patterns, including quantitative measurements of downwash and transport of contaminants to other buildings
- Establishing the effectiveness of source hoods

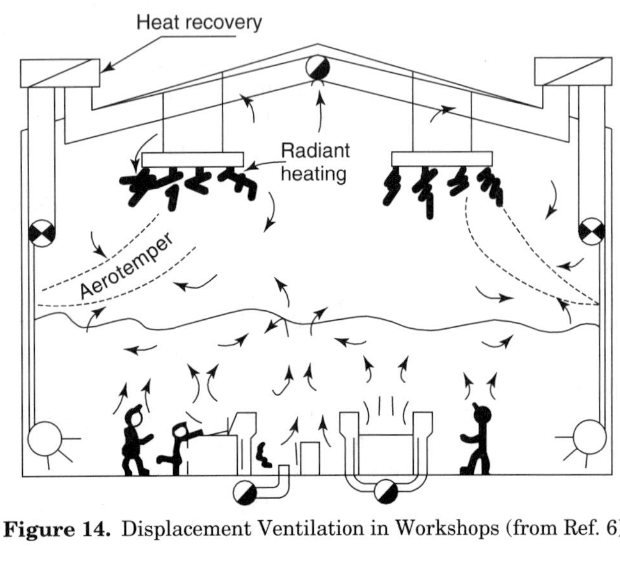

Figure 14. Displacement Ventilation in Workshops (from Ref. 6).

For new process buildings, the recommended sequence is to use the mathematical models and equations to develop the overall ventilation concepts and architectural constraints. At this stage, the project design team can continue with the design of the structural steel while the small-scale model is being constructed and tested. The results of the small-scale modeling are used to refine the ventilation concept and to finalize all requirements.

For an existing plant, the fluid dynamic model is a valuable tool in modeling the problems and developing alternative cost-effective solutions.

Figure 15 is a typical flowchart of activities required for a ventilation modeling study. The first step is to define the contaminant and the source characteristics. Parameters to be defined are the size of the process building and details of the source flux (i.e., heat and contaminant release rates, etc.). Information is required on the major sources of heat in order to calculate heat balances and volumes of air set in motion. Data are required on the external and internal flow conditions for the prototype. Information may be required on site conditions, such as wind speed, direction, and frequency.

The design of the model system requires an examination of the scaling parameters and the type of fluid medium to be used. The common media used are air and water, but the working or buoyancy-driven fluids for models can vary widely. Fluid systems that have been used include air and heated air, water and saltwater, water and carbon tetrachloride, and mercury and carbon tetrachloride. Air models are usually the simplest and least expensive models to build and test, but usually must be large models to ensure fully turbulent flow. Because water has a smaller kinematic viscosity than air, a small model is required to ensure a high Reynolds number and turbulent flow. How visualization is also easier with water-based models because velocities are lower than in air. A detailed analysis of the scaling parameters and possible fluid systems is required in order to select the best modeling technique for the specific ventilation problem. Data measured in the model flow can be related quantitatively to the full-scale prototype flow by establishing dynamic similarity (geometric and kinematic) between the model and the prototype.

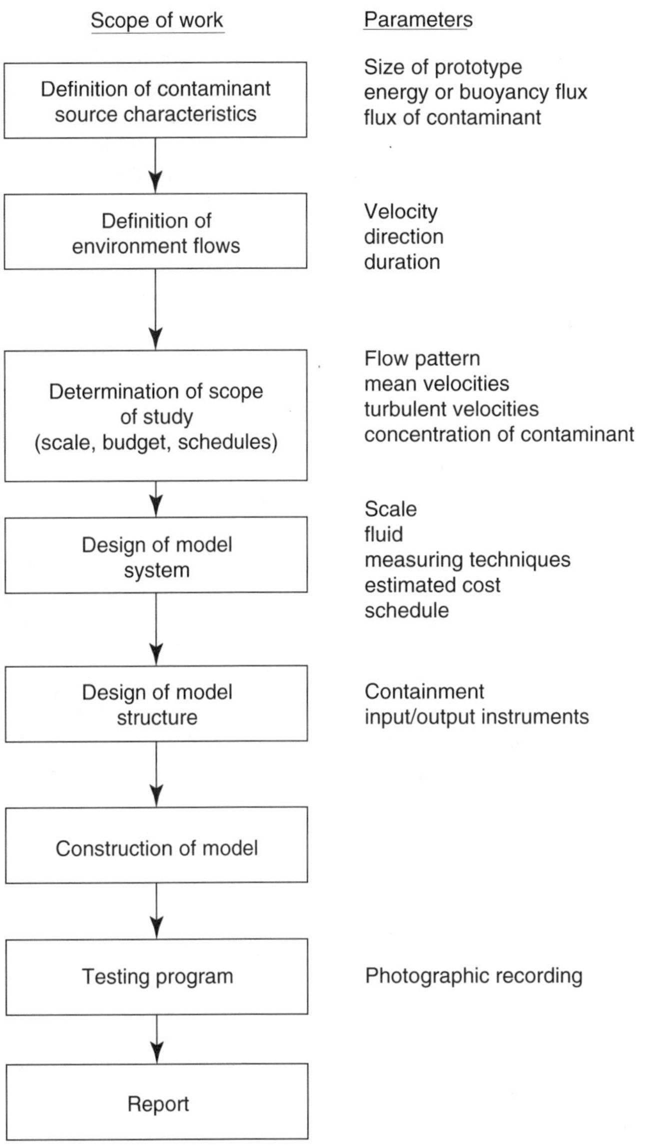

Figure 15. Typical Flowchart for Ventilation Modeling Study (Adopted from Ref. 2).

The flow required in the prototype is given by

$$Q_{\mathrm{p}} = Q_{\mathrm{m}} S^{5/3} (q_{\mathrm{p}}/q_{\mathrm{m}})^{1/3}$$

where Q is the volumetric flow rate (cfm); q, the heat flow rate (Btu/min); S, the model scale (e.g., for a 1 : 10 scale model, $S = 10$); the subscript p identifies the prototype parameter; and the subscript m identifies the model parameter.

The measuring technique for the contaminant of concern must also be considered since it may have a significant impact on the level of accuracy and the cost of the testing program. For any model testing program, extensive use should be made of photographic and video recording equipment. The photographs and films will be invaluable in analyzing the test results and for subsequent presentation of solutions to management.

Computer Modeling

Available computer programs, based on the application of the design equations in the earlier section, can be applied for solving industrial ventilation problems. Figure 16, a flow diagram of the computer ventilation model, shows a typical input, the initialization and iteration loop, and the output. Figure 17 shows that the overall agreement between predicted and measured flow rates is good.

Once correct modeling of observed ventilation flow rates has been established, the computer program can be used

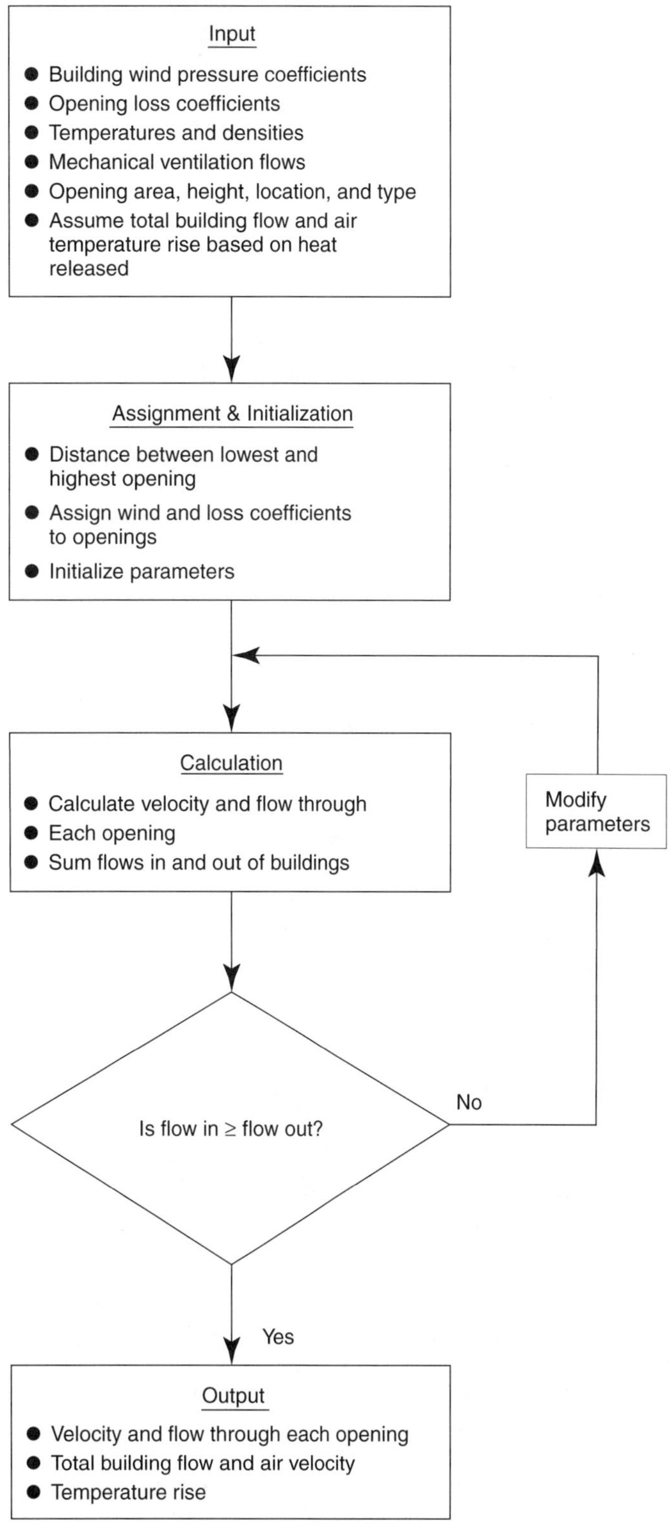

Figure 16. Computer Ventilation Model Flow Diagram (from Ref. 2).

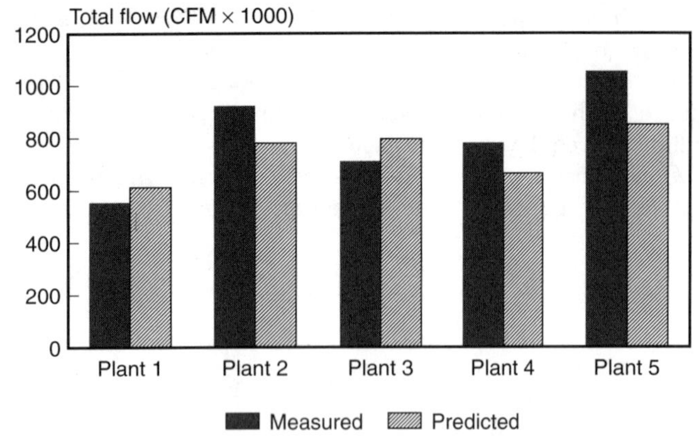

Figure 17. Total Ventilation Flow.

to study the effects of different atmospheric conditions (summer/winter, wind direction and speed, etc.) on the ventilation characteristics for the shop. It is also a very useful tool in the evaluation of proposed ventilation improvement schemes. Different ventilation schemes can be analyzed and cost-effective solutions developed and implemented to solve the ventilation problems.

Computer ventilation models can reliably predict the gross ventilation rates for complex process buildings. The use of high-speed computers provides the designer with the capability to examine the impact of architectural changes, wind conditions, or process changes on the performance of the proposed ventilation scheme. Problems such as contamination resulting from cross drafts or high temperatures in the work environment can be identified quickly and corrective measures taken.

Typical ventilation questions that may arise during the planning and design of ventilation systems for a new facility or to improve the environment in an existing facility include the following:

1. What are the internal flow patterns under different layout and operating conditions?
2. What are the effects of intermittent peak heat releases on ventilation flow characteristics?
3. Can the flow fields be represented satisfactorily as a two-dimensional flow?
4. What happens to contaminated plume that misses a hood?
5. What effect do cross drafts have on the workplace environment?
6. Where does the fresh air enter the building?
7. What are the predicted contaminant concentrations in the breathing zone?
8. What is the source of contamination in a specific area?

Computer programs are available that can operate on a microcomputer and solve three-dimensional complex ventilation problems.[7] Conferences such as Ventilation '88 and RoomVent '90 included many technical papers on computer programs to calculate contaminant levels as a function of position and time in a workplace environment.

The practical application of these computer programs represents a significant advancement in the science and technology of ventilation.

Field Testing Methodology

For most complex ventilation problems, it is recommended that both analytical calculations and a field measurement program be carried out in order to develop a proper, cost-effective solution. Experience has shown that the field sampling program must be organized and developed in detail prior to going into the field.

Ventilation flows, design parameters, and field testing protocol are unique for each specific processing or manufacturing operation, but many elements of the field testing program are common to all ventilation problems. Although there may be minor variations for any specific ventilation project, the eight activities listed in Table 4 represent the major activities for a field testing program and the sequence for performing this work. A brief description of the tasks and scope of work required for each activity is included.

As a starting point for the field testing program, all pertinent reports and drawings concerning the plant should be obtained for review and study. Specific information is required on the current and future operating practices, as well as on the nature of the in-plant contaminant concerns (chemical or heat/cold stress). A visit to the site should include a walk-through ventilation survey using a data sheet. The ventilation engineer should develop the questionnaire with experienced process and operating personnel for the specific industry. The prime purpose of the questionnaire is to identify the operating practices (present and future), which will then be used as the design basis for the ventilation system.

It is necessary to develop the details of a field testing program and all necessary field data log sheets prior to the actual field testing. The development of a proper field testing program is the most important step in ensuring a successful ventilation study. This program should be reviewed and approved by plant operating personnel before the actual field testing is carried out. Because the ventilation flow in large process buildings is usually complex, an inexperienced sampling team may collect insufficient data for subsequent analytical calculations or will expend considerable effort collecting field data that are irrelevant. A good approach is to perform preliminary calculations on all sources and then to identify clearly where information gaps exist and what essential data must be collected.

The measurement of air velocities through all the openings must recognize the need to have a "representative" velocity for the survey. It is not acceptable to "be on the safe side" by rounding readings to a higher value. Gusty wind conditions will cause velocity readings to fluctuate. Sufficient time must be allowed to elapse to enable the person monitoring the instrument to arrive at a representative velocity by "integrating" the observed values in his or her mind.

A temperature measurement is required for each velocity measurement. For air entering the building, this would generally be the ambient temperature, which should be recorded on an hourly basis. It will change during the day and could be higher in the wake of the building. The thermometer or temperature probe should not be exposed to sunlight or radiant heat from hot objects. The temperature of air leaving the melt shop will vary considerably. For heat balance calculations, good temperature readings are important.

Temperatures associated with in-plant flow patterns must also be recorded. Sufficient temperature data must be available

Table 4. Engineering Activities for a Ventilation Field Testing Program

Activity	Specific Tasks
1. Information gathering	Obtain drawings, reports, operating procedures Review existing data and studies Plant visit Define problem (summer, winter, heat/cold stress, chemical, supply exhaust)
2. Data on ventilation openings	Isometric drawings: plans, sections Schedule of openings (type, size, location)
3. Develop/complete plant questionnaire	Develop process flow sheet and general layout Identify significant heat sources in building Identify typical operating practice Identify gaps in data to be filled in by field testing program
4. Develop details of field testing program	Building ventilation flow rates In-plant flows Field data log sheets for ventilation measurements, weather conditions, plant operating records, etc.
5. Field testing program	Field measurements (volume, temperature, pressure, etc.)
6. Data analysis and calculations	Plant ventilation flow balance In-plant flow patterns Heat balance calculations (total plant/ventilation) Air set-motion volumes
7. Computer simulation of ventilation flows (natural ventilation)	Calibrate using field test data Run program for different conditions (summer, winter, etc.)
8. Field testing report	Summarize test conditions and test results Report with recommendations

to allow evaluation of the air density distribution within the shop.

Mean surface temperatures of hot surfaces must be recorded for subsequent heat release and air set-in-motion calculations. The location of the equipment on the floor plan and equipment surface temperature should be recorded on a separate sheet.

Weather data should be measured at the site and obtained from the nearest airport or meteorological station as well. Data to be recorded include ambient temperature, relative humidity, and wind speed and direction. During the test period, this information should be recorded on at least an hourly basis.

A record of plant activities during the testing program is required. Data to be included are the status of operation of all major process and environmental equipment, as well as production levels. Plant records and charts from process operation should be obtained.

An in-plant survey will be required to establish such parameters as levels of contaminants (dust, gases, etc.) and heat stress. These measurements would be concurrent with the ventilation survey. An industrial hygienist should work with the ventilation engineer to establish the scope and extent of the industrial hygiene sampling program.

Ventilation computer models should be used to determine ventilation flow rates for different operating conditions and to estimate contaminant concentrations in the workplace. Data from the computer model calculations can be compared with the measured ventilation and industrial hygiene sampling data.

For any ventilation field testing program, it is essential to prepare a proper engineering report, which includes all the field data, calculations, and test results. Using the results of this field testing program, an experienced ventilation engineer can develop cost-effective solutions for any plant ventilation problem.

ANCILLARY EQUIPMENT FOR LEV SYSTEMS

Hoods

Source Emission Characterization. As described earlier, hoods are the most important element of the LEV system. If the hood fails to capture the contaminant at the source, the performance and efficiency of the rest of the system becomes meaningless. The *Technical Manual: Hood System Capture of Process Fugitive Particulate Emissions*[8] describes design procedures for hoods for numerous industrial processes. The manual shows that each application must be studied for the specific process. A new approach stressed in the development of the hood design methodology is to develop a complete understanding of the source–hood interactions.

Although it is impractical to develop simple standard design procedures for hoods for different plant/process applications, significant progress is being made in the development of technology related to hood design equations. Numerous technical papers have been published at international conferences such as Ventilation '85 and Ventilation '88 that cover both the theoretical and practical design equations for hoods.

Factors known to influence hood performance are as follows:

1. Construction, size, and shape of hood (e.g., degree of enclosure of hood)
2. Suction velocity of exhaust hood
3. Position of hood relative to the source
4. Cross drafts

5. Physical/chemical characteristics of the emission sources
6. Velocity contours for gas and particulate emission sources

Hood design procedures will depend on whether the application is a greenfield or a retrofit plant. In a greenfield plant, the designer has many more options, and it is essential that the design be incorporated as an integral part of the development of the plant layout. It is recommended that the hood designer evaluate different alternatives using computer models or fluid-dynamic models. For retrofit plants, the layout may already dictate the hood configurations that are acceptable. The opportunity exists to perform field evaluations to collect data on existing hood performance or at least to characterize the emission source.

The emission sources can be characterized by parameters such as the following.

1. Buoyant or nonbuoyant source
2. Continuous or intermittent plume
3. Plume flow rate and geometry
4. Source heat flux and geometry
5. Physical/chemical characteristics of the particulates
6. Gas composition and temperature
7. Layout of the plant

Using this information, the emission source can be completely characterized as a function of time and position. This becomes the starting point for the establishment of hood design flow rates.

Hood Classification. The four hood classifications[9] are as follows:

Group A Enclosures
Group B Booths
Group C Captor hoods
Group D Receptor hoods

Figure 18 shows typical configurations for the different hood types. Table 5 summarizes characteristics of the four types of hoods, including operating principles, description of types, design considerations, design methodology, and typical industrial applications. Goodfellow[2] has presented further details on the hood classification scheme.

Design Parameters. Hood design methodologies are as follows:

1. Precedent
2. Rule of thumb
3. Analytical methods
4. Diagnosis of an existing site
5. Fluid-dynamic modeling

Precedent is based on a similar hood design on an existing application. The rule of thumb is to take data from a standard handbook that gives typical hood volume requirements for similar applications. Table 6 gives recommended control velocities and exhaust rates for enclosures of various materials-handling operations. *Industrial Ventilation*[4] contains more than 100 sketches of hoods for different processing operations. Analytical methods are based on using theoretical or empirical equations developed for different operations. In retrofit plants, it is possible to collect field data and to determine hood performance as a function of flow rate and different hood geometry. A newer approach is based on the use of scale models to simulate the source–hood interactions. This approach has the benefit of permitting hood performance evaluations to be established quantitatively.

Table 7 outlines typical equations and criteria for hood design for contaminant control. (See Table 7A for corresponding nomenclature.)

Hood Performance Evaluation. The evaluation of hood system performance is an important aspect of the technology of the design of LEV systems. Several techniques have been developed for an evaluation of hood system performance. The hood design manual[8] describes the techniques in detail. A brief description of the techniques follows.

Tracer gas techniques[10] are now well established as a procedure to measure hood performance. The technique is to inject a tracer at a known rate at the process source. The quantity captured by the hood source is a measure of the hood performance.

Goodfellow and Bender[11] have described three techniques, which are a more direct means of evaluating hood system performance for the actual process sources. For remote capture of buoyant plumes, plume flow rates can be measured by movie scaling, stopwatch clocking of the plume, and anemometer measurements at the roof trusses. Using this information on the estimated plume flow rates, it is possible to calculate the hood system performance for a given hood configuration and plant layout. Field measurements can be correlated with opacity measurements to establish performance relative to opacity criteria and overall hood performance evaluation.

Fluid modeling techniques, as described earlier for solving industrial problems, represent a sophisticated and rigorous approach to establishing hood system performance. This type of technique is based on building a scale model of the source–hood configuration in a water tank or air system. For a valid test, the flow in the scale model must be turbulent, and the Froude number similarity criteria must be met. Scale modeling permits convenient testing of hood performance for existing hood designs for a proposed modification to an LEV system.

For dust control systems, hood performance can be evaluated using tracer gas techniques, industrial hygiene-type sampling using area monitoring at sources, or a dust lamp. The dust lamp, developed in England, uses a high-intensity light beam utilizing the Tyndall effect, in which light is scattered forward from the particles in the atmosphere.[2] This technique is extremely sensitive, and photographs of the path of the dust clouds can be taken. In addition to its use in research and development applications for new or novel hoods, the dust lamp can be used in conjunction with other measuring equipment to establish hood performance data in the field for existing LEV systems.

Ducts Plus Accessories

Technical Description. Ducted systems are defined as the actual ducts plus accessories, such as expansion joints, explosion vents, flanges, and manholes, which are required to make a functional LEV system. Standard designs have been

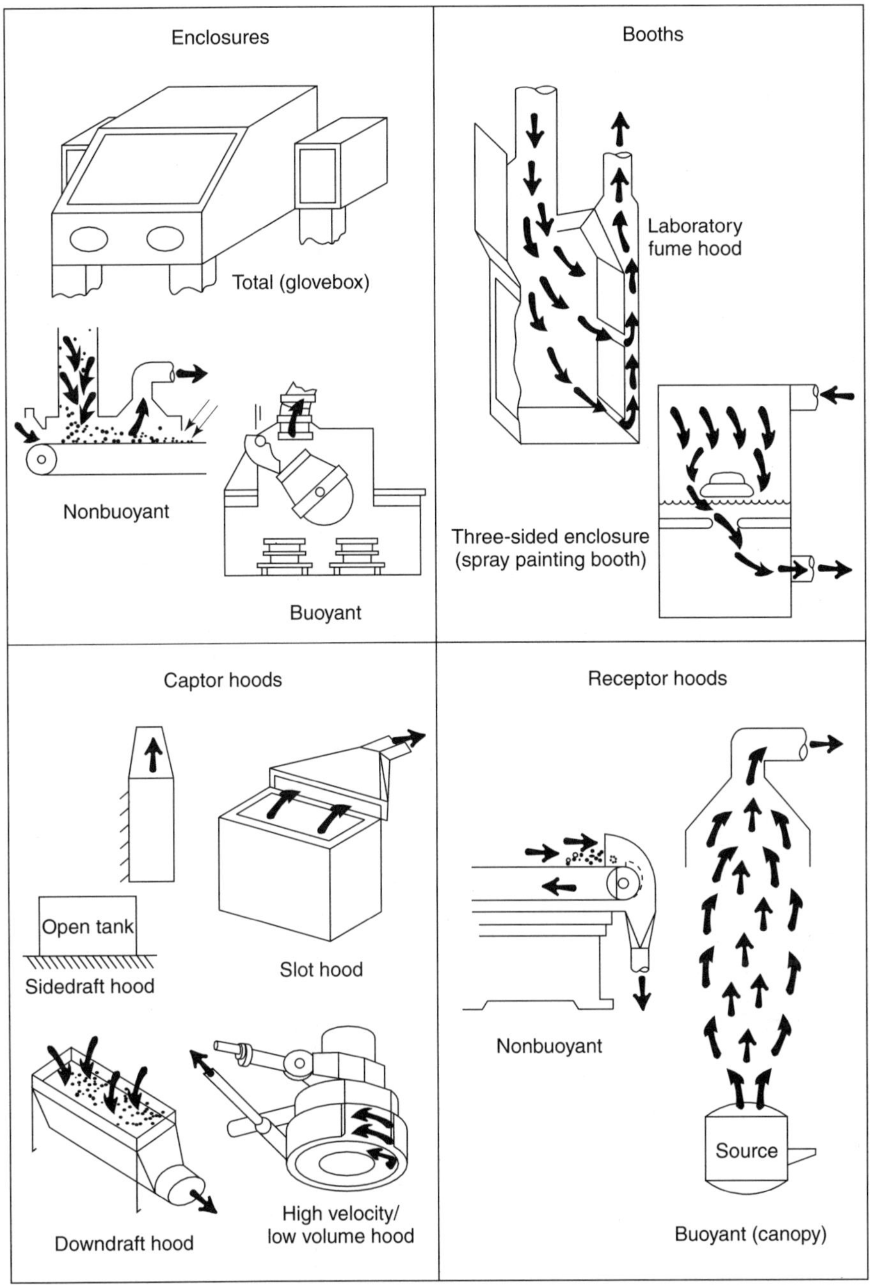

Figure 18. Four Different Hood Types (from Ref. 2).

developed that have been well proven in the field and are applicable for LEV systems.

Ducted systems should be constructed with materials suitable for the application and installed in a permanent and good workmanlike condition. The interior of all ductwork should be smooth and free of obstructions. Typical materials of construction range from sheet metal and mild steel platework to specialized materials, such as abrasion resistant steel, high-temperature alloys, plastics, and other corrosion-resistant materials.

For conventional dust control systems, materials of construction are black iron that is welded, flanged, and gasketed or galvanized sheet steel that is riveted and sealed. Corrosive conditions require the selection of plastics, special corrosion-resistant materials, or protective coatings. Galvanized sheet steel is not recommended for gas temperatures above 400°F. Material thickness of at least 16 gauge is recommended for arc welding.

Industrial Ventilation[4] has identified four classifications for ducts for LEV systems for noncorrosive applications.

Class 1 Light duty for nonabrasive applications such as replacement air and general ventilation

Class 2 Medium duty for applications with moderately abrasive particulates in light concentrations such as woodworking and grain dust

Class 3 Heavy duty for applications with highly abrasive particulates in low concentrations such as abrasive cleaning and sand handling

Class 4 Extra heavy duty for applications with highly abrasive particles in high concentrations such as canopying systems in heavy industrial plants

Table 5. Characteristics of the Four Types of Hoods

	Group A Enclosures (Total or Partial)	Group B Booths	Group C Captor Hoods	Group D Receptor Hoods
1. Operating principles	1. Complete enclosures of emission source. 2. Enclosures should be large enough to avoid direct impingement of particulate cloud on the walls.	1. Emission source inside of enclosure. 2. Enclosures should be large enough to avoid direct impingement of particulate cloud on the walls.	1. Operate at a distance from the source. Contaminant would not naturally enter the hood. 2. Fan supplies the power to establish the airflow patterns at the hood.	1. Positioned to catch the stream of contaminants or contaminated air thrown out in a given direction by an industrial process. 2. Hood exhaust volume is based on the volume of contaminated air that is pumped in by the process.
2. Description of types	*Total* 1. Not used except for radioactive materials. *Partial* 1. Most so-called total enclosures have small openings for entry and egress of materials. 2. Applicable for nonbuoyant sources.	1. Booths are enclosures from which one wall or a large part of one wall has been removed so that articles can be moved inside the enclosure.	*Side-draft hood* 1. Hood located at side as close as possible to emission. *Slot hood* 1. Slot in a narrow opening along the edge of the tank. *Downdraft hood* 1. Hood located under the source of contaminant. *High-velocity, low-volume hood* 1. Use extremely high velocities (>12,000 fpm) to capture contaminants at the source.	*Nonbuoyant sources* (local) 1. Local receiving hoods can be designed to capture particles thrown off. *Buoyant sources* 1. Heated air rising vertically from a hot source can be captured remotely using a canopy hood.
3. Design considerations	1. Total enclosures require materials inside to be handled by remote manipulators. 2. Total enclosures must be dusttight gastight. 3. For partial enclosures, total surface area of all openings must be small compared with total area of the walls.	1. Booth should be as deep as possible to ensure an even flow at the open face. 2. Operator should never stand between the work and the airstream. 3. Design of offtake may be important to ensure a high capture with an even gas flow at the face.	1. Since air velocities drop off rapidly with increasing distance from the hood, captor hoods must be located as close to the contaminant source as possible. 2. Captor hood performance depends on proper capture velocity and proper distribution of velocity over zone of contamination.	1. Receptor hoods are designed to prevent spillage because of overloading due to high flows or due to random air curtains which cause contaminated air to miss the hood. 2. Remote hoods only practical for buoyant sources.

4. Design methodology	1. Exhaust rate greater than rate at which contaminated air would escape. 2. For airtight design and minimal air disturbance, exhaust inlet airflow at each opening is 100 fpm. 3. Total flow = v (design) × A (opening). 4. Ventilation rate must be determined for each specific application.	1. Exhaust rate greater than rate at which contaminated air would escape. 2. For minimal air disturbance, exhaust inlet airflow at each opening is 100 fpm. For every small booths, an inlet velocity 400 fpm may be required.	1. Capture velocities range from 100 to 200 fpm. 2. Summay of design procedures: a. Examine behavior of contaminant. b. Determine capture velocity. c. Establish distance between hood and source. d. Determine flow required. e. Choose a suitable fan.	1. Design of hoods for nonbuoyant sources described in Table 7. 2. Design of hoods (i.e., canopy) for buoyant sources described in Table 7.
5. Typical industrial applications	1. Electric are furnace enclosure. 2. Pneumatic oxygen vessel enclosure. 3. Cooper–nickel converter enclosure. 4. Radioactive glove box. 5. Clamshell unloading enclosure. 6. Belt conveyor transfer point. 7. Screening operation. 8. Bin filling.	1. Spray painting booth. 2. Laboratory fume hood. 3. Portable grinding, polishing, or buffing operations.	*Side-draft hood* 1. Welding hoods. 2. Side shakeout hood. *Slot hood* 1. Narrow open tanks in electroplating. 2. Bench welding. *Down-draft hood* 1. Shakeout hood. 2. Bench soldering 3. Bench torch cutting *High-velocity / low-volume hood* 1. Power hand tools. 2. Manual arc welding.	*Nonbuoyant sources* 1. Grinding hood. 2. Belt sanders. *Buoyant sources* 1. Electric are furnace canopy hood.

Table 6. Tabulation of Recommended Control Velocities and Exhaust Rates for Materials-Handling Operations

Operation	Velocity (fpm)	Basis of Control Velocity	Exhaust Rate (cfm)	Basis of Exhaust Rate
Screens vibrating				
Flat deck	150–200	Openings in enclosure	25–30	Per square foot of screen area
Cylindrical (rotating)	400	Openings in enclosure	100	Per square foot of circular cross-sectional area of screen
Bag filling				
Paper	100	Openings in enclosure		
Cloth	200	Openings in enclosure		
Nontoxic	<400	Openings in enclosure	400–500	
Toxic	>400	Openings in enclosure	1000–1500	
Belt conveyors				
Transfer points	>150	Per square foot of openings	350	Per foot belt width (speed <200 fpm)
	>200	Per square foot of openings	500	Per foot belt width (speed >200 fpm)
Along length			350	Per foot belt width at 30-ft intervals
Bag emptying	40–100	Openings in insertion enclosures	400–1050	Multiplied by area of openings
Drum and barrel filling	100	Point of origin	100	Per square foot of barrel semitop enclosed
Bins and hoppers	150–200	Openings in enclosures	0.5	Per square foot of bin volume
Bucket elevators			100–200	Per square foot of casing cross section
Mixers	100–200	Openings in enclosures	2100–3100	Per foot of open area
Crushers	200	Through crusher gap in direction of material flow		

Round ducts are usually preferred for LEV systems because of their inherent strength and their ability to minimize the settling of airborne dust. The actual metal thickness for round industrial ducts will vary with the classification, static pressure, reinforcement, and span between supports. The Sheet Metal and Air Conditioning Contractors National Association (SMACNA)[12] gives the range of metal thicknesses shown in Table 8.

If rectangular ducts are necessary, the sections should be as nearly square as possible. All ducts should be constructed in accordance with the "Round or Rectangular Industrial Duct Construction Standards" as published by SMACNA.

Design Parameters. Design parameters to be determined for ducted systems include duct diameter (established from flow and velocity parameters), friction losses, dust filling criteria, materials of construction, duct thickness, structural support of duct, expansion joints, explosion protection, damper requirements, and such ductwork accessories as pitot tube connections and cleanout doors. For any application, the need for explosion prevention must be established at the preliminary design stage. Calculation procedures for establishing system pressure losses are outlined in previous sections. Duct thickness and type of reinforcing systems can be determined by using the three-step procedure outlined in the SMACNA duct manual as follows:

Step 1. Duct thickness

Step 2. Size and spacing of reinforcing member and weld specification

Step 3. Flange connection size and bolt requirements

Ducts and their ancillary equipment are key components that must be properly selected to ensure a satisfactory LEV system. Ducted systems should be designed and specified only by experienced engineers and must incorporate designs that will accommodate necessary maintenance features.

Gas Cooling

Technical Description of Types. There are many methods of cooling hot process off-gases. The five common cooling methods for LEV systems are as follows:

1. Dilution with ambient air
2. Natural convection and radiation from ductwork
3. Forced-draft cooling
4. Quenching with water (evaporative cooling)
5. Use of water-cooled ductwork

Table 9 compares the five different gas cooling techniques based on applicable gas temperature ranges and industrial experience.[2] A detailed technical and economic analysis of the alternative systems is required for any specific application before a selection can be made. A brief description of each system follows.

Design Parameters

Dilution with Ambient Air. Dilution of off-gas with ambient air is a simple method of reducing off-gas temperature.

Dilution cooling does not remove or absorb heat from the off-gas. Its effect can be examined by considering the off-gas volumes produced. Adding dilution air increases the volume of off-gas to be handled by the air pollution equipment. For example, if the air pollution control equipment is a baghouse, the cloth area must be increased to maintain reasonable air-to-cloth ratios. The adequacy of the existing fan must also be evaluated. The large volumes of air introduced when dilution air volumes are increased usually cannot be moved by the existing fan.

Table 7. Typical Equations and Criteria for Hood Design for Contaminant Control

	Total enclosure	Partial enclosure, nonbuoyant[a]	Partial enclosure, buoyant[b]	
Enclosures, Group A	$V = 40$ to 60 fpm $Q = VA$ or using a rule of thumb: $Q = 50$ cfm per square foot of open area	$Q_1 = 0.043\left(\frac{WH^2A_s^2}{P_d d}\right)$ $Q_2 = W/P_d$ $Q_3 = AV$; $V = 100$ (well-protected sources[c]) to 200 fpm (vigorous operations[d]) $Q = Q_1 + Q_2 + Q_3$	Volumetric flow rate after dilution $Q = \frac{q}{\rho c_p(T_{amb} - T_s)}$	
	Lab fume hood	Three-sided enclosure		
Booths, Group B	50 cfm per square foot for face velocities with good air supply distribution 150 cfm per square foot for poor air distribution	Airflow volume required per cross-sectional area 75 cfm per square foot 100 cfm per square foot 200 cfm per square foot	Size of booth[e] >150 ft^2 < 150 ft^2 < 4 ft^2	
	Side draft	Slot	Downdraft	High velocity/low volume
Captor hoods, Group C	$Q = V_u\left(\frac{T_a}{T_u}\right)^{0.5}\left(\frac{x}{y}\right)^{-0.2554} A_cE$ where: $V_u = 0.09V_{\max}(0.63 + 0.36y)$	$Q = V_u\left(\frac{T_a}{T_u}\right)^{0.5}\left(\frac{x}{y}\right)^{-0.2554} A_cE$ where: $V_u = 0.09V_{\max}(0.63 + 0.36y)$	$Q = (10X^{0.5} + A)V$ where: $V = 100$ to 200 fpm by rule of thumb or $Q = 150$ to 250 cfm per square foot of bench area	$Q = 25$ to 60 cfm per inch diameter Slot velocity $= 30{,}000$ to $39{,}000$ fpm Flexible hose at 1 to 2 inches diamter
	Nonbuoyant[1]	Buoyant[2]		
Receptor hoods, Group D	Belt width (in.) / Q exhaust <6 in. incl. / 440 cfm >6 to 9 in. incl. / 550 cfm >9 to 14 in. incl. / 800 cfm >14 in. / 1100 cfm	$Q = 0.166(Z)^{1.67}(F)^{0.33}$ where: $Z = 2D + Y$ $F = \frac{gq}{C_{pm}\rho T_m}$	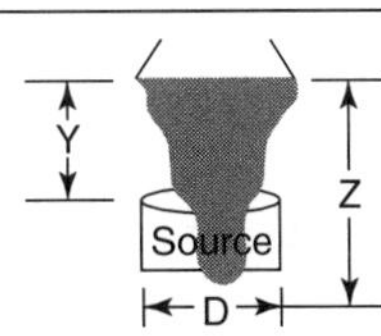 (Applies to group D, buoyant)	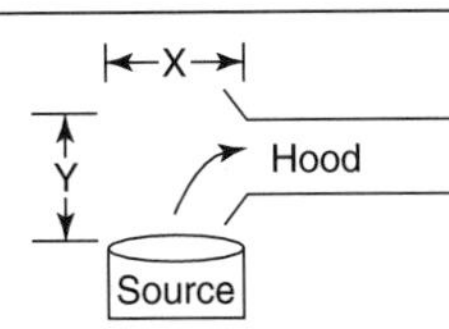 (Applies to group C; side-draft, slot, downdraft)

[a]Nonbuoyant, contaminant source at ambient temperature.
[b]Buoyant, contaminant source at elevated temperature.
[c]Protected from significant cross-drafts.
[d]Significant turbulence at the plume source caused by the source or by external factors.
[e]Open area at the front of the both enclosure.
Source: Reference 2.

The mixed temperature after dilution with ambient air can be calculated from:

$$T_m = \frac{Q_g c_{p_g}(T_g - T_{ref})] + [Q_d c_{p_d}(T_d - T_{ref})]}{Q_m c_{p_m}} + T_{ref}$$

where Q_m is the mixed gas volume (scfm); Q_d, the dilution air volume (scfm); Q_g, the off-gas volume before dilution (scfm); c_{p_g}, the mean molar heat capacity of off-gas before dilution (Btu/lb mol °F); c_{p_d}, the mean molar heat capacity of dilution off-gas (Btu/lb mol °F); c_{p_m}, the mean molar heat capacity of mixed off-gas (Btu/lb mol °F); T_g, the temperature of off-gas, (°F); T_d, the temperature of dilution air (°F); T_m, the temperature of mixed gases (°F); and T_{ref}, the reference temperature (°F).

Natural Convection and Radiation from Ductwork. Gas cooling by natural convection such as radiation from ductwork is a direct function of the outside area of the ductwork, the duct temperature, and ambient conditions, such as wind velocity and air temperatures. In radiation coolers, the off-gas passes through an assembly of U-shaped cooling columns, and heat is transferred by radiation and free convection. Radiation coolers of mild steel construction are suitable for cooling off-gas from about 1000 to 400°F. At lower off-gas temperatures, radiation coolers are not efficient. Pressure losses through radiation coolers are high. The ability of an existing system fan to accommodate the pressure requirements of a radiation cooler must be evaluated. Tube blockage and duct erosion are common problems on the bends of radiation coolers. Design must be amenable to regular inspection and ease of cleaning and maintenance.

Table 7A. Nomenclature for Table 7

A	control cross-sectional area, ft^2
A_c	control surface area calculated using Fletcher's formula and depending on hood configuration, ft^2
A_s	cross-sectional area of falling stream, ft^2
b_h	plume length scale at hood face, ft
C_p	specific heat or air at T_s, Btu/lb/°F
C_{pm}	specific heat or air at T_m, Btu/lb/°F
D	diameter of plume source, ft
d	particle mass median diameter, ft
E	1.6 when the vessel is not full and flow separation causes eddy formation
e	eccentricity, ft
F	buoyancy flux, ft^4/sec^3
g	gravitational constant, ft/sec^2
H	drop height, ft
L	length, ft
M	momentum, ft-lb $mass/min^2$
q	heat transfer rate, Btu/min
Q	total exhaust volume, cfm
Q_1	volumetric flow rate of induced air, cfm
Q_2	flow rate of displaced air, cfm
Q_3	flow rate for control velocity through openings, cfm
Q_p	peak plume flow rate, cfs
Q_s	hood exhaust flow rate, cfs
S	scale parameter (ratio of prototype to model)
T_a	ambient fluid temperature, R
T_{amb}	ambient dilution air temperature, °F
T_m	absolute temperature, R
T_s	specified air temperature after dilution, °F
T_u	plume fluid temperature, R
t	time, min
t_d	plume surge time, sec
U_{max}	plume centerline velocity taken at hood face, fpm
V	control velocity, fpm
V_{cross}	cross flow velocity, fpm
V_{max}	centerline velocity at one source diameter above the hood, fpm
V_u	updraft velocity of the plume above the source, fpm
W	material flow rate, lb/min
X	distance from fume source to hood, ft
Y	height to hood face, ft
y	vertical distance between source and hood in source diameters
Z	effective height from the virtual plume origin to hood face, ft
p	air density at T_s, lb/ft^3
P_d	density of dust or other particulate matter, lb/ft^3
α	entrainment constant (0.093 for a point plume source)

Table 8. Ducts and Gauges

Diameter of Straight Ducts	Range of Metal Thickness U.S. Standard Gauge for Steel Duct[a]			
	Class 1	Class 2	Class 3	Class 4
4 to 8 in.	22–20	22–18	16	14
Over 8 to 18 in.	22–12	22–12	16–11	14–11
Over 18 to 30 in.	18–7	16–7	16–6	14–6
Over 30 in.	14–2	14–2	12–2	12–2

[a]Thickness varies with classification, pressure, reinforcement, and span requirements. *Note*: 24- and 26-gauge metal cannot be welded.

Source: Reference 7.

The temperature reduction of off-gas through a radiation cooler can be estimated from:

$$T_o = T_{amb} + (T_i - T_{amb}) \exp\left(\frac{-UA \times 387}{Qc_p \times 60}\right)$$

where T_o is the off-gas temperature out of section of radiation cooler (°F); T_{amb}, the ambient air temperature (°F); T_i, the off-gas temperature into sections of the radiation cooler (°F); Q, the off-gas flow rate through the radiation cooler (scfm); A, the heat transfer area of the radiation cooler section (ft^2); c_p, the specific heat capacity of off-gas (Btu/lb mol °F); and U, the overall heat transfer coefficient at a location along the radiation cooler (Btu/h ft^2 °F).

The overall heat transfer coefficient typically varies between 1.5 and 3 Btu/h ft^2 °F, depending mainly on off-gas temperature, off-gas velocity, and ambient conditions.

Forced-Draft Coolers. In a typical forced-draft cooler design, the hot off-gas passes through banks of vertical tubes, and axial fans blow ambient air across the outside of the tubes to cool the off-gas. Forced-draft coolers of mild steel construction are suitable for cooling off-gas from about 1000 to 275°F, as required by polyester filter fabric. Special alloy steel tubes are sometimes used to operate at higher temperatures (e.g., in the ferroalloy industry). Pressure losses through forced draft coolers can be high, typically 4–6 in. w.g. Typical overall heat transfer coefficients range from 3 to 6 Btu/h ft^2 °F. In retrofit applications, the ability of the existing system fan to accommodate the pressure requirements introduced by the cooler must be evaluated. The main advantages of forced-draft coolers over radiation coolers are as follows:

1. Reduced surface area with direct cooling to 2750°F
2. Compact design
3. Potential for heat recovery
4. Lower maintenance cost because downstream fan and air pollution control equipment operate at a more uniform temperature.

Evaporative Cooling. In evaporative cooling, fine water droplets are injected into the off-gas. The droplets evaporate, absorbing the heat from the gas. Heat is not removed from the system; sensible heat is changed to latent heat, reducing the temperature. Evaporative cooling usually does not increase the volume of off-gas significantly. Unlike forced-draft coolers and radiation coolers, minimal pressure loss is added to the system by evaporative coolers, making their use a good option in a retrofit application. The disadvantage of evaporative cooling is that fairly sophisticated controls may be required to ensure that correct quantities of moisture are injected into the off-gas. Over spraying can lead to condensation of water vapor in the baghouse. Blinding of fabric filter bags will result, and condensation on metal surfaces will accelerate corrosion. Also, high off-gas moisture content reduces bag life. The degree of cooling that can be afforded by evaporative coolers is limited by the allowable off-gas moisture levels. However, when evaporative cooling systems are properly engineered, they provide the most cost-effective method of dealing with increased heat loads.

Water-Cooled Ductwork. Water-cooled ductwork provides an effective mechanism for heat removal from the off-gas. Temperature reduction of the off-gas through water-cooled

Table 9. Gas Cooling Techniques

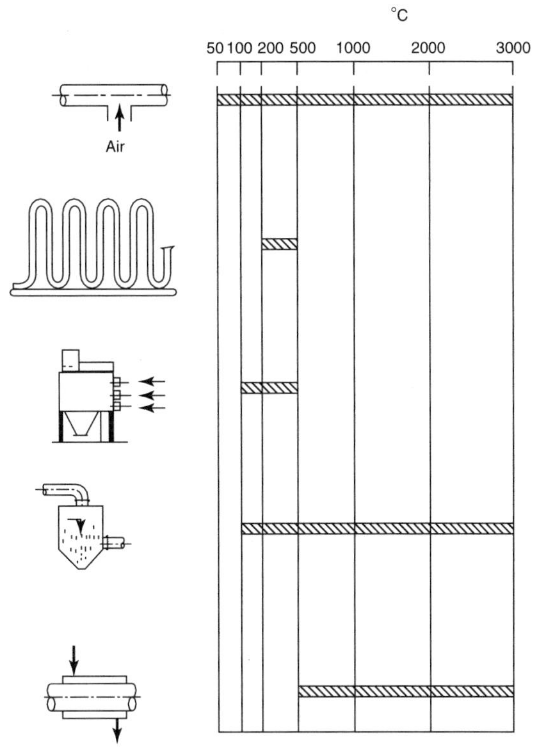

Cooling Technique	Schematic	Applicable Gas Temperature Range	Comments
1. Dilution with ambient air			Applicable over whole temperature range.
2. Natural convection and radiation from ductwork (radiation cooler or U-tube)			Results in smaller gas volumes. No direct operating costs.
3. Forced-draft cooling			Results in smallest gas volumes to be cleaned. Requires energy for cooling.
4. Quenching with water (evaporative cooling)			Potential for heat recovery. Works best for a steady heat flow. Dewpoint considerations important at low end.
5. Water-cooled ductwork			Prime purpose is to protect the ducting from high gas temperatures. High cost for heat removal.

Source: Reference 2.

ductwork can be estimated from:

$$T_o = T_w + (T_i - T_w)\exp - \left(\frac{UA \times 387}{Qc_p \times 60}\right)$$

where T_o is the off-gas temperature out of the section of water-cooled ductwork (°F); T_w, the average temperature of cooling water (°F); T_i, the off-gas temperature into section of water-cooled ductwork (°F); Q, the off-gas flow rate through ductwork (scfm); A, the heat transfer area of the water-cooled ductwork section (ft^2); c_p, the specific heat capacity of the off-gas (Btu/lb mol °F); and U, the overall heat transfer coefficient at location along the water-cooled ductwork (Btu/h ft^2 °F).

From the estimated temperature reduction, heat removal from the water-cooled ductwork can be calculated. The equation is valid for heat exchangers in which one of the fluids does not change temperature significantly relative to the temperature change of the other fluid. It provides good results in water-cooled ductwork heat transfer calculations because the temperature change of the water is very small compared with the temperature change of the off-gas.

The overall heat transfer coefficient varies significantly, depending mainly on the temperature of the off-gas. As the temperature of the off-gas decreases, the overall heat transfer coefficient also decreases. As a result, the heat transfer from the water-cooled ductwork decreases significantly at lower off-gas temperatures. Therefore, upstream sections of water-cooled ductwork are more effective in removing heat than are downstream sections of water-cooled ductwork. This should be considered when evaluating the length of ductwork to be installed, because the cost effectiveness of ductwork is reduced as the length of the duct increases. The addition of excess volumes of combustion air will quench the off-gas and will defeat the effectiveness of the ductwork. Water-cooled ductwork is usually sized for maximum velocities of 9000 to 10,000 fpm. High velocities are tolerable because pressure losses through ductwork are less severe at high temperatures.

Fans

Technical Description. Fans have been used in industry for about 100 years. Some of the earlier applications were in metallurgy, mines, and other manufacturing operations. Information covering the design, operation, and maintenance of fans for specific types of systems is widely available.

Fans are the primary moving devices used in industrial applications (such as LEV systems). They are used to convey the air contaminants from the source to a control device, which removes the contaminant from the airstream prior to discharge of the clean air to the atmosphere. The present section will describe the different types of air moving devices, design equations, fan selection procedures, and fan installations.

Fans can be classified into two main types: centrifugal and axial. A centrifugal fan consists of a wheel or rotor mounted on a shaft that rotates in a scroll-shaped housing. The air enters at the eye of the scroll, makes a right-angle turn, and is forced through the blades of the rotor by centrifugal force into the scroll-shaped housing. These fans have four basic impeller designs: airfoil, backward curved, radial, and forward curved. Figure 19 identifies the common terminology for centrifugal fan components. Figure 20 identifies the common terminology for inlet and outlet fans. Figure 21 identifies the drive arrangements for centrifugal fans.

The performance of a fan is characterized by the volume of gas flow, the pressure at which this flow is produced, the speed of rotation, the power required, and the efficiency.

Testing methods as specified by the Air Movement and Control Association (AMCA)[13] are used by the fan manufacturers to establish the relationships of these parameters. The test method consists of mounting a fan on the end of a straight length of duct and operating it with various-sized orifices (typically about 10) in the duct. For each test, the volume flow is measured, along with pressure, velocity, and power input. Test results are then plotted to obtain the fan characteristic curves. From the volume and pressure, the air horsepower is computed. The efficiency based on total pressure is called mechanical efficiency, while the efficiency based on the friction losses is called static efficiency. Figure 22 identifies the impeller design, housing design, performance curves, performance characteristics, and applications for centrifugal fans.

Axial fans include all those fans in which the air flows through the impeller in a path approximately parallel with the shaft upon which the impeller is mounted. The three

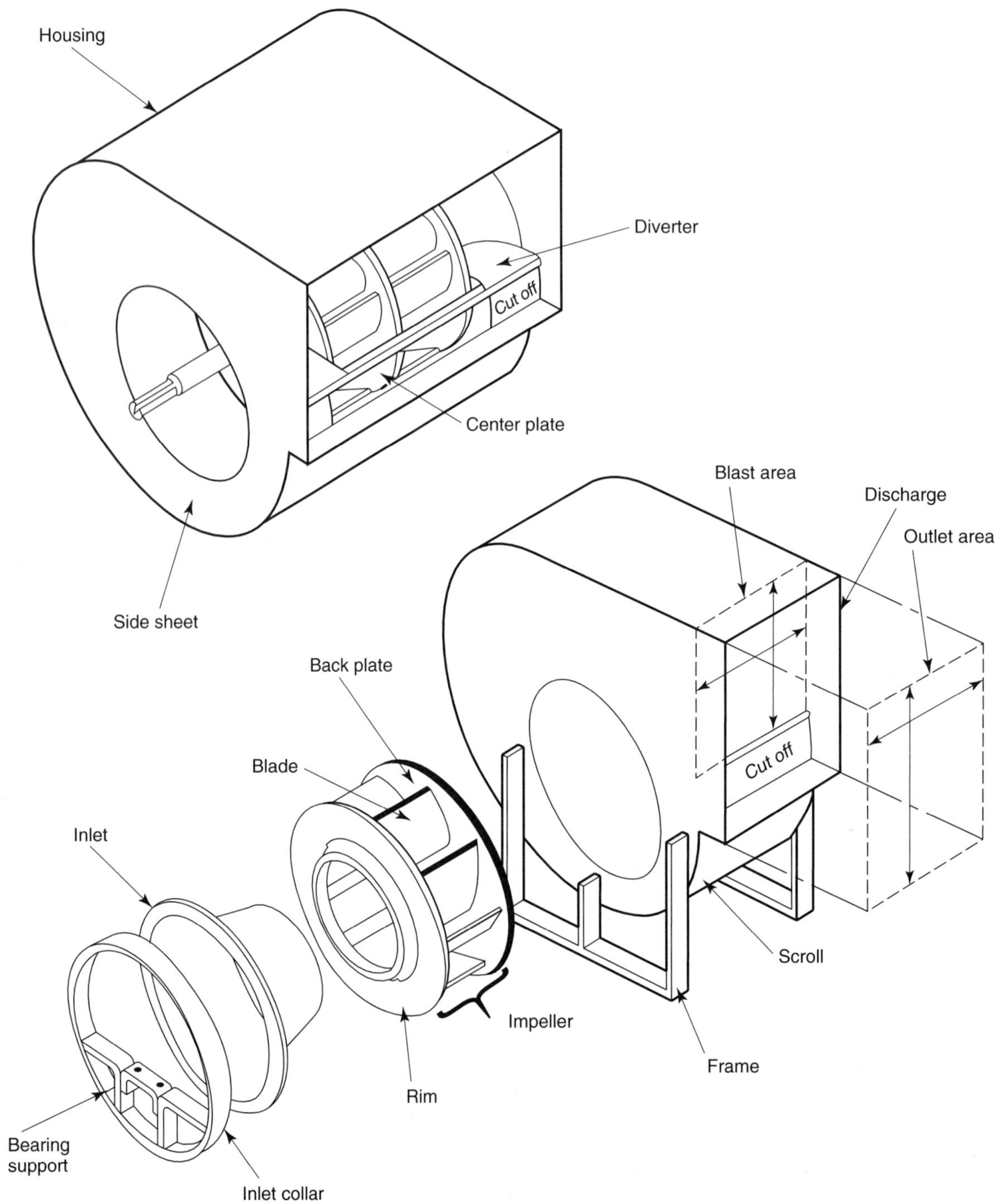

Figure 19. Common Terminology for Centrifugal Fans (from Ref. 12).

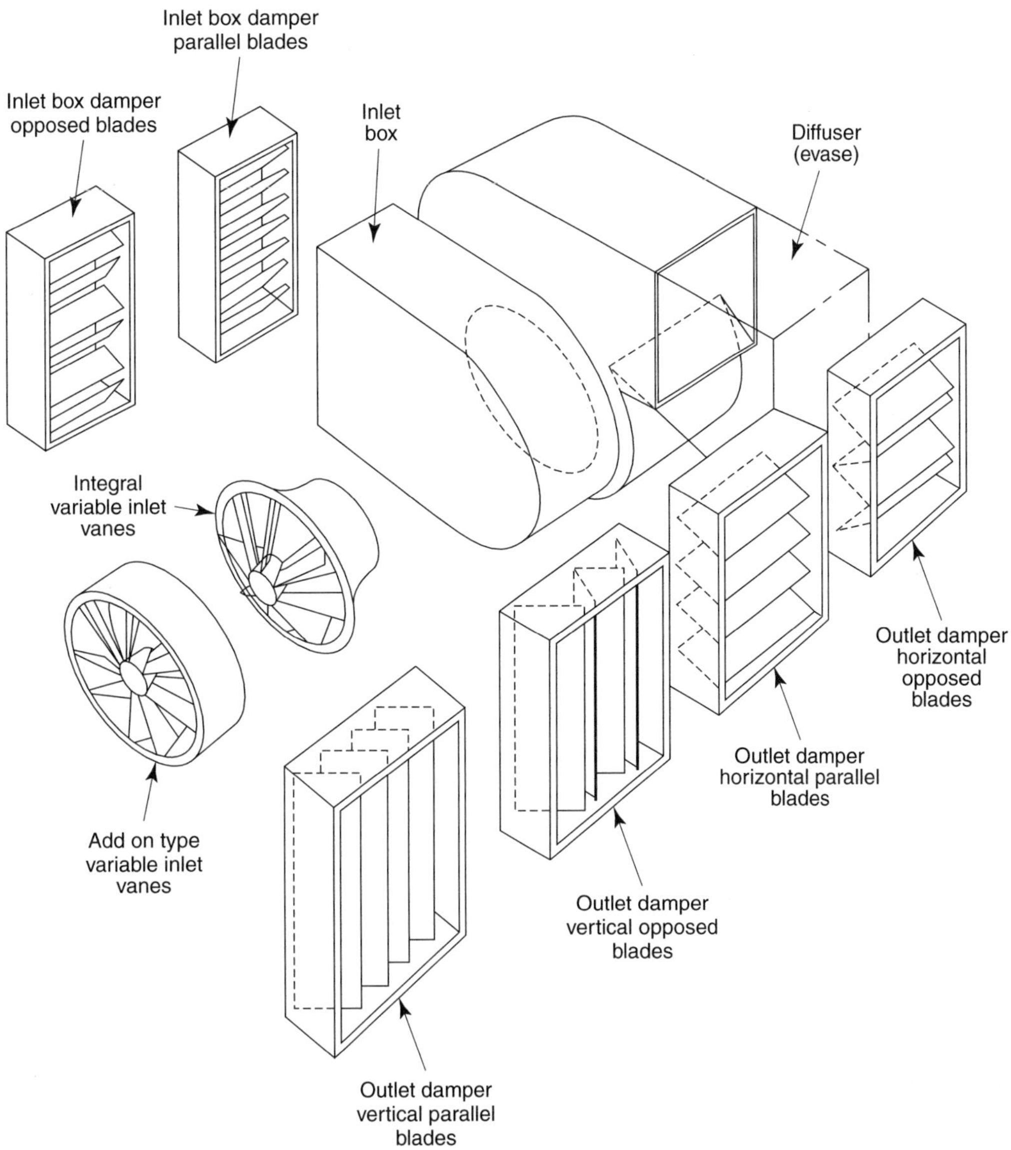

Figure 20. Common Terminology for Inlet and Outlet Fans (from Ref. 12).

basic types of axial fans are propeller, tube axial, and vane axial. Figure 23 presents the common terminology for axial fan components. Figure 24 identifies the impeller design, housing design, performance curves, performance characteristics, and typical applications for axial fans. It is generally recognized that axial fans are best suited for handling high volumes of a relatively clean gas at a lower static pressure. These fans tend to be of higher efficiency with an essentially flat and self-limiting horsepower curve.

Design Parameters. The starting point for the design of fans is the well-known "fan law" based on the classical theories of fluid mechanics. Simplifying assumptions are made that ignore considerations of Reynolds number, Mach number, kinematic viscosity, dynamic viscosity, surface roughness, impeller blade thickness, relative clearances, and so on. Although the accuracy of the fan law is sufficient for most cases, the AMCA is conducting investigations into the above effects with the aim of developing practical adjustments to the fan law that would give more accurate projections of fan performance. Preliminary results indicate that the correction factors being developed are different for each type of fan and may vary significantly among different manufacturers' designs. Based on these observations, it is recommended that the fan manufacturer be consulted before implementing any major changes in the speed or point of operation of an existing fan installation.

The equations based on the fan law enable the results of fan tests at one speed to be used to calculate performance at other speeds and other gas densities and for a complete series of different-sized, but geometrically similar (homologous) fans. Each set of equations for changes in speed, size, or density applies only to the same point of rating, and all equations in the set must be used to define the converted condition. "Point of rating" is defined by AMCA as a percentage of wide-open volume flow rate and a percentage of shutoff pressure, or a point of constant efficiency.

The fan law equations for any homologous series of fans at the same relative point of operation on any two performance curves are:

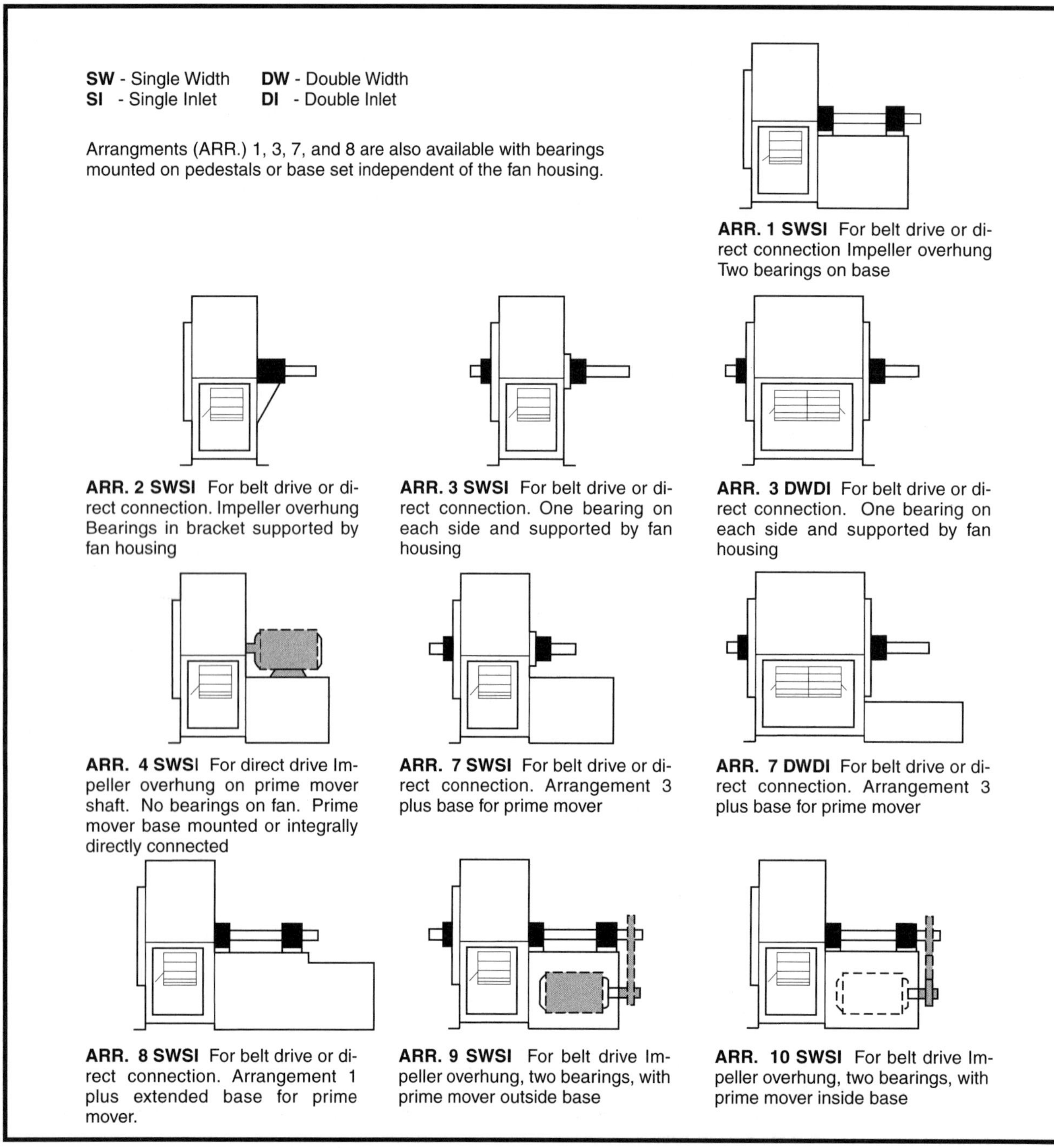

Figure 21. Drive Arrangements for Centrifugal Fans (from Ref. 4).

$$\frac{Q_2}{Q_1} = \frac{\text{RPM}_2}{\text{RPM}_1}\left(\frac{D_2}{D_1}\right)^3$$

$$\frac{\text{TP}_2}{\text{TP}_1} = \frac{\text{VP}_2}{\text{VP}_1} = \frac{\text{SP}_2}{\text{SP}_1} = \left(\frac{\text{RPM}_2}{\text{RPM}_1}\right)^2\left(\frac{D_2}{D_1}\right)\frac{\rho_2}{\rho_1}$$

$$\frac{\text{BHP}_2}{\text{BHP}_1} = \left(\frac{\text{RPM}_2}{\text{RPM}_1}\right)^3\left(\frac{D_2}{D_1}\right)^5\frac{\rho_2}{\rho_1}$$

where Q is the gas flow rate; RPM, the revolutions per minute; D, the wheel diameter; TP, the total pressure; VP, the velocity pressure; SP, the static pressure; ρ, the gas density; and BHP, the brake horsepower. Any convenient units may be employed as long as they are consistent, since the equations involve ratios of the variables.

The range of limitations of applicability of the fan law is as follows:

1. Gas is thermodynamically incompressible, and flow through fan is adiabatic.
2. T_2/T_1 is in absolute pressure units <1.036.
3. $D_2/D_1 < 3 : D_2 > D_1 : D_2/D_1) \times (\text{RPM}_2/\text{RPM}_1) \leq 3$.
4. $\text{RPM}_2/\text{RPM}_1 \leq 3$.
5. $(D_2/D_1)(\text{RPM}_2/\text{RPM}_1) \leq 3$.
6. $\text{Re}_1 \geq \text{Re}_2$.

In practice, the principles expressed by the fan law equations are normally applied to determine the effect of changing only one variable. Table 10 shows equations that can be used to determine the effect of varying speed, changing density, and increasing wheel-diameter size.

The pressure loss in the ducted system is obtained by calculating pressure losses for each of the components of the system. At a fixed volume flow through a given ducted system, a corresponding pressure loss will exist. If the flow changes, the new static pressure (SP) is closely approximated using the following parabolic relationship.

$$\mathrm{SP} = aQ^2$$

where a is a constant; and Q, the volume flow rate (scfm). The reason for this relationship is that nearly all losses in the system are functions of the velocity squared. Table 11 presents velocity–flow relationships for different types of losses in ducted systems.

Figure 25 illustrates typical plots of the resistance to flow versus volume flow rate for three different and arbitrary fixed-duct systems (A, B, and C). For a fixed system, an increase or decrease in system resistance results from an increase or decrease in the volume flow rate along the given system curve only.

Duct system A has a system design point at 100% volume and 100% duct system resistance. An increase in flow to 120% of the design volume results in an increase in duct system resistance of 150% from design. A decrease in volume flow to 50% results in a decrease to 30% of the design resistance. Similar relationships also hold for duct system B and duct system C, as shown in Figure 25.

The system resistance depends on the density of the gas flowing through the system. The standard gas density is 0.075 lb/ft^3. The dependence of fan performance on gas density is obtained from the fan law equations. The fan pressure and power vary directly as the ratio of the gas density at the fan inlet to standard density. The density ratio must always be

	Type	Impeller design	Housing design
Centrifugal fans	Airfoil	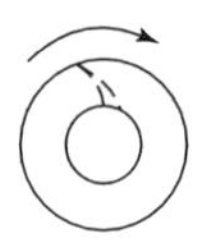 Highest efficiency of all centrifugal fan designs 9 to 16 blades of airfoil contour curved away from the direction of rotation. Air leaves the impeller at a velocity less than its tip speed and relatively deep blades provide for efficient expansion within the blade passages. For given duty, this will be the highest speed of the centrifugal fan designs.	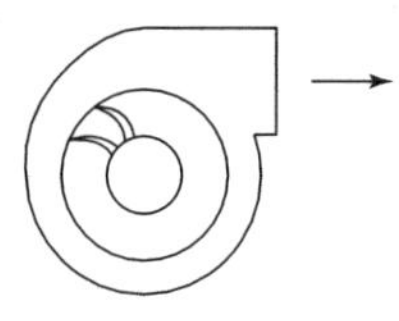 Scroll-type, usually designed to permit efficient conversion of velocity pressure to static pressure, thus permitting a high static efficiency, essential that clearance and alignment between wheel and inlet bell be very close in order to reach the maximum efficiency capability. Concentric housings can also be used as in power roof ventilators, since there is efficient pressure conversion in the wheel.
	Backward-inclined backward-curved	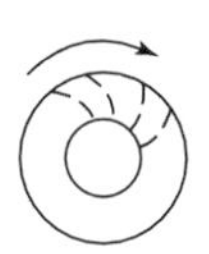 Efficiency is only slightly less than that of airfoil fans. Backward-inclined or backward-curved blades are single thickness. 9 to 16 blades curved or inclined away from the direction of rotation. Efficient for the same reasons given for the airfoil fan above.	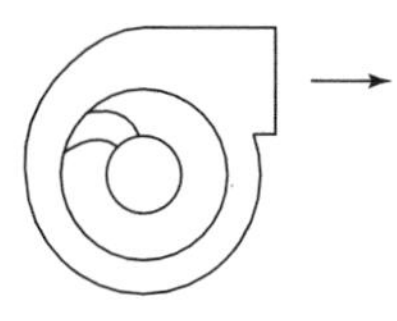 Utilizes the same housing configuration as the airfoil design.
	Radial	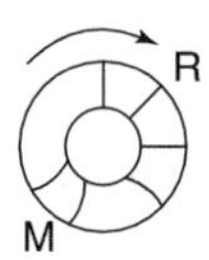 Simplest of all centrifugal fans and least efficient. Has high mechanical strength and the wheel is easily repaired. For a given point of rating, this fan requires medium speed. This classification includes radial blades (R) and modified radial blades (M), usually 6 to 10 in number.	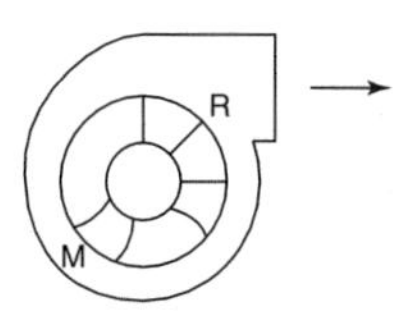 Scroll-type, usually the narrowest design of all centrifugal fan designs described here because of required high velocity discharge. Dimensional requirements of this housing are more critical than for airfoil and backward-inclined blades.
	Forward-curved	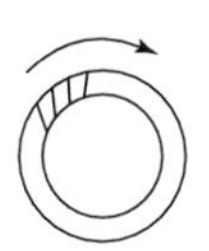 Efficiency is less than airfoil and backward-curved bladed fans. Usually fabricated of lightweight and low cost construction. Has 24 to 64 shallow blades with both the heel and tip curved forward. Air leaves wheel at velocity greater than wheel. Tip speed and primary energy transferred to the air is by use of high velocity in the wheel. For given duty, wheel is the smallest of all centrifugal types and operates at lowest speed.	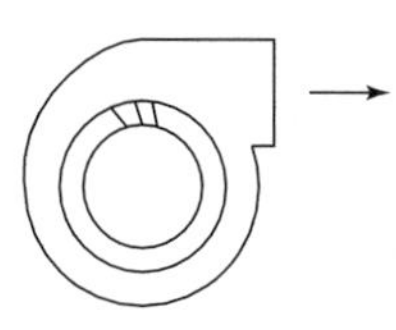 Scroll is similar to other centrifugal-fan designs. The fit between the wheel and inlet is not as critical as on airfoil and backward-inclinded bladed fans Uses large cut-off sheet in housing.

Figure 22. Characteristics of Centrifugal Fans (from Ref. 4).

Performance curves	Performance characteristics*	Applications
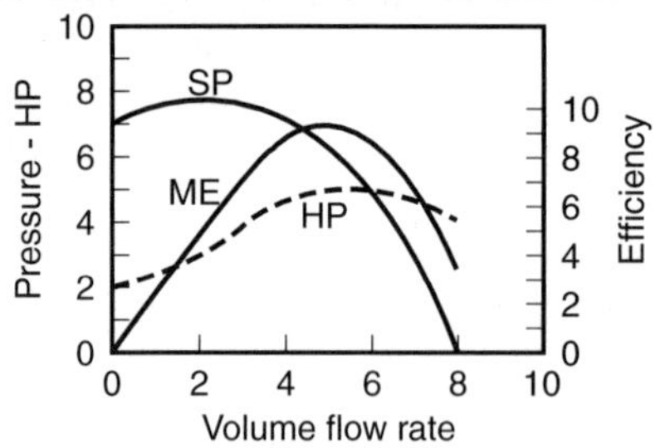	Highest efficiencies occur 50 to 65% of wide open volume. This is also the area of good pressure characteristics, the horsepower curve reaches a maximum near the peak efficiency area and becomes lower toward free delivery, a self-limiting power characteristic as shown.	General heating, ventilating and air-conditioning systems. Used in large sizes for clean air industrial applications where power savings are significant.
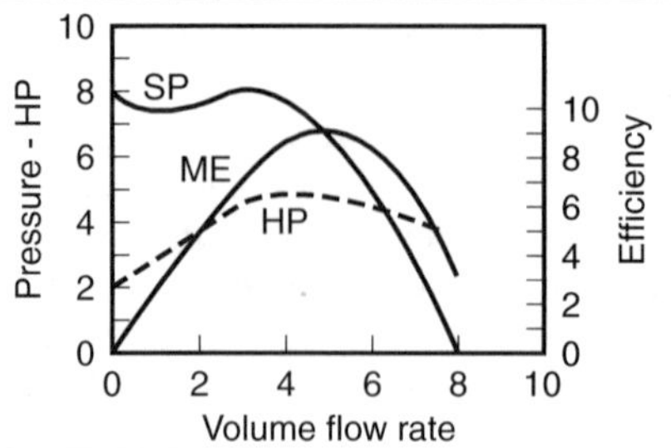	Operating characteristics of this fan are similar to the airfoil fan mentioned above. Peak efficiency for this fan is slightly lower than the airfoil fan. Normally unstable left of peak pressure.	Same heating, ventilating and air-conditioning applications as the airfoil fan. Also used in some industrial applications where the airfoil blade is not acceptable because of corrosive and/or erosion environment.
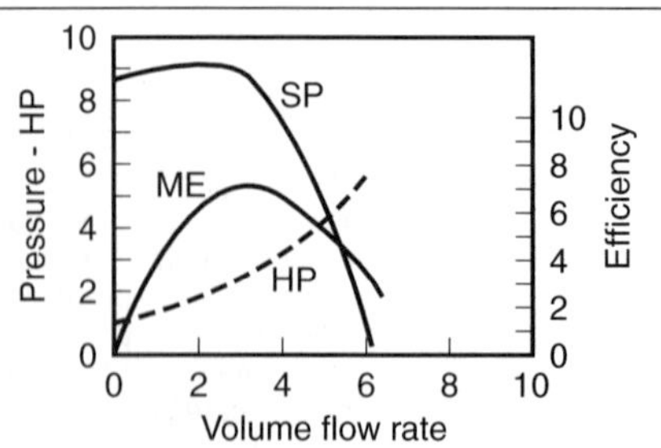	Higher pressure characteristics than the above mentioned fans. Power rises continually to free delivery.	Used primarily for material-handing applications in industrial plans. Wheel can be of rugged construction and is simple to repair in the field. Wheel is sometimes coated with special material. This design also used for high-pressure industrial requirements. Not commonly found in HVAC applications.
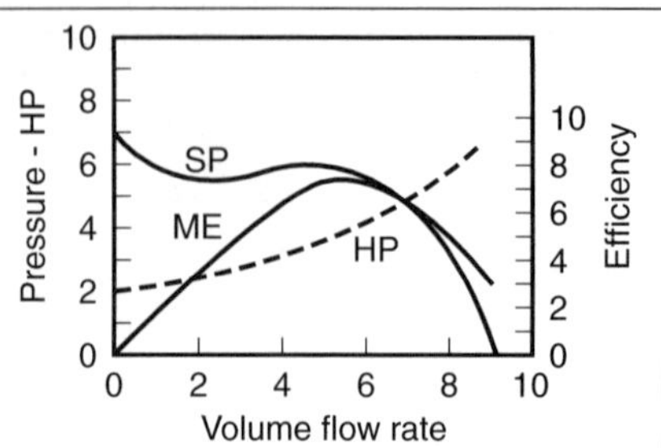	Pressure curve is less steep than that of backward-curved bladed fans. There is a dip in the pressure curve left of the peak pressure point and highest efficiency occurs to the right of peak pressure. 40 to 50% of wide open volume. Fan should be rated to the right of peak pressure. Power curve nses continually toward free delivery and this must be taken into account when motor is selected.	Used primarily in low-pressure heating, ventilating and air-conditioning applications such as domestic furnaces, central station units and packaged air-conditioning equipment from room air-conditioning units to roof top units.

Figure 22. *(Continued).*

considered when selecting fans from manufacturers' catalogs or curves.

The calculation sheets for obtaining system pressure losses are based on static pressures. Most industrial fan rating tables are based on fan static pressures, which cannot be read directly from the calculation sheet. Fan static pressure is defined by AMCA 201 as the total pressure for the fan minus the fan velocity pressure.

$$\begin{aligned} SP_{fan} &= TP_{fan} - VP_{fan} \\ &= SP_{outlet} - SP_{inlet} - VP_{inlet} \end{aligned}$$

Fan power is calculated from the equation

$$P_i = \frac{Q \times TP_{fan}}{\eta_T}$$

where P_i is the fan input power; Q, the volume flow rate; TP_{fan}, the fan total pressure; and η_T, the fan total efficiency.

A fan performance curve is a graphical representation of the performance of a fan. Typical fan curves cover the range from no delivery (an airtight system with no air flowing) to free delivery with no obstructions to flow. Information stated on a fan curve should include gas density, impeller size, and speed. Fans selected at nonstandard density require knowledge of the actual volumetric flow rate at the fan inlet, the actual pressure requirement, and the density of the gas at the fan inlet.

Fan Selection. A successful fan installation requires the proper selection of a fan to match the required flows and static pressures for the LEV system. The following is a checklist of factors to be specified or considered:

1. Actual inlet airflow/unit time
2. Fan static pressure
3. Density of gas at the inlet
4. Dust characteristics and loading
5. Barometric pressure

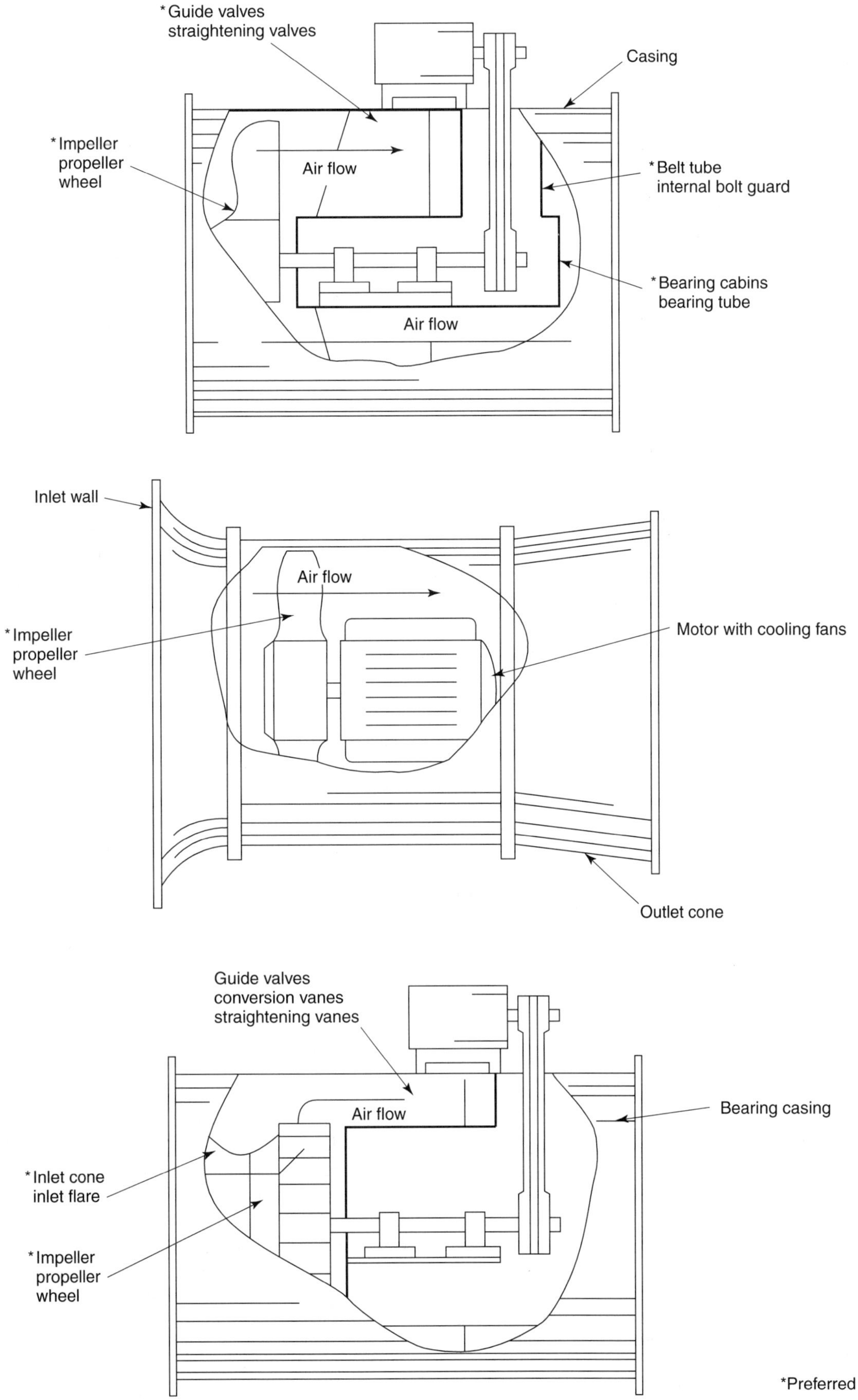

Figure 23. Common Terminology for Axial Fans (from Ref. 12).

6. Desired fan speed
7. Explosive or inflammable material
8. Whether direct drive or belt drive
9. Noise requirements
10. Fan efficiency
11. Layouts and space limitations
12. Capital and operating cost considerations

Performance is based on fan inlet conditions and the negative pressure developed by the fan must be taken into consideration in flow, pressure, and density calculations. Fan specifications are often based on volume flow units, but sometimes mass flow units are used. The expected range of operating conditions (minimum/maximum/average) must be specified, along with any expected sudden temperature changes, so that the fan design can accommodate those

Type			Impeller design		Housing design	
Axial fans	Propeller		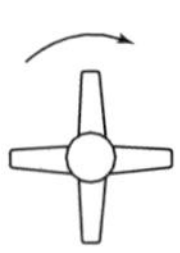	Efficiency is low. Impellers are usually of inexpensive construction and limited to low pressure applications. Impeller is of 2 or more blades, usually of single thickness attached to relatively small hub. Energy transfer is primarily in form of velocity pressure.	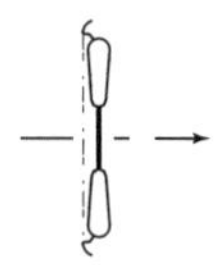	Simple circular ring, orifice plate, or venturi design. Design can substantially influence performance and optimum design is reasonably close to the blade tips and forms a smooth inlet flow contour to the wheel.
Axial fans	Tubeaxial			Somewhat more efficient than propeller fan design and is capable of developing a more useful static pressure range. Number of blades usually from 4 to 8 and hub is usually less than 50% of fan tip diameter. Blades can be of airfoil or single thickness cross section.	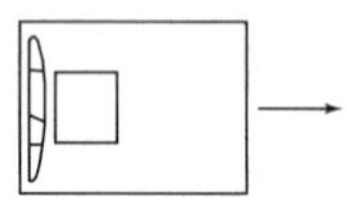	Cylindrical tube formed so that the running clearance between the wheel tip and tube is close. This results in singnificant improvement over propeller fans.
Axial fans	Vaneaxial		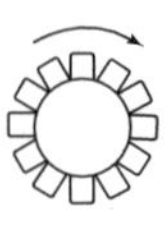	Good design of blades permits medium- to high-pressure capability at good efficiency. The most efficient fans of this type have airfoil blades. Blades are fixed or adjustable pitch types and hub is usually greater than 50% of fan tip diameter.	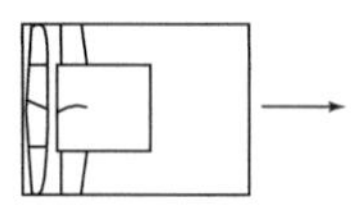	Cylindrical tube closely fitted to the outer diameter fo blade tips and fitted with a set of guide vanes. Upstream or downstream from the impeller, guide vanes convert the rotary energy imparted to the air and increase pressure and efficiency of fan.
Special designs	Tubular	Centrifugal	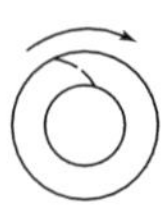	This fan usually has a wheel similar to the airfoil backward-inclined or backward-curved blade as described above. (However, this fan wheel type is of lower efficiency when used in fan of type.) Mixed flow impellers are sometimes used.	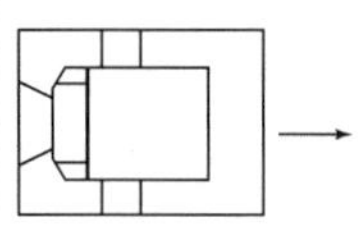	Cylindrical shell similar to a vaneaxial fan housing, except the outer diameter of the wheel does not run close to the housing. Air is discharged radially from the wheel and must change direction by 90 degrees to flow through the guide vane section.
Special designs	Power roof ventilators	Centrifugal	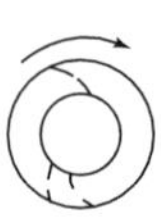	Many models use airfoil or backward-inclined impeller designs. These have been modified from those mentioned above to produce a low-pressure, high-volume flow rate characteristic. In addition, many special centrifugal impeller designs are used, including mixed-flow design.	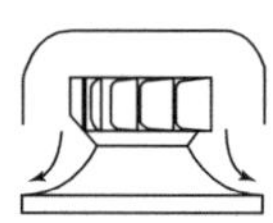	Does not utilize a housing in a normal sense since the air is simply discharged from the impeller in a 360 degree pattern and usually does not include a configuration to recover the velocity pressure component.
Special designs	Power roof ventilators	Axial		A great variety of propeller designs are employed with the objective of high-volume flow rate at low pressure.	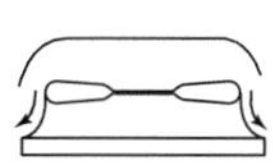	Essentially a propeller fan mounted in a supporting structure with a cover for weather protection and safety considerations. The air is discharged through the annular space around the bottom of the weather hood.

Figure 24. Characteristics of Axial Fans (from Ref. 4).

Performance curves	Performance characteristics*	Applications
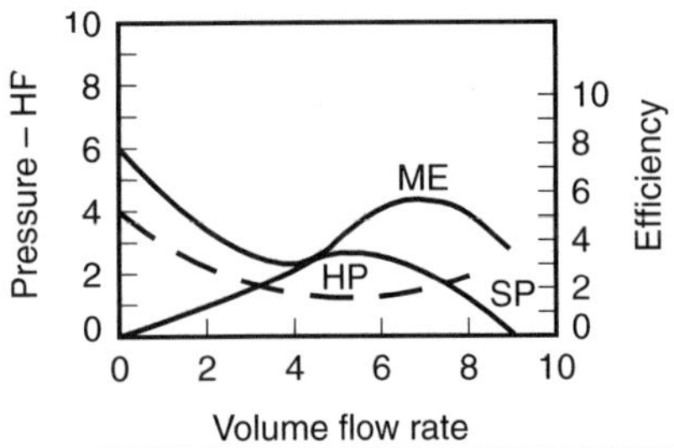	High flow rate but very low-pressure capabilities and maximum efficiency is reached near free delivery. The discharge pattern of the air is circular in shape and the air stream swirls because of the action of the blades and the lack of straightening facilities.	For low-pressure, high-volume air moving applications such as air circulation within a space or ventilation through a wall without attached duct work. Used for replacement air applications.
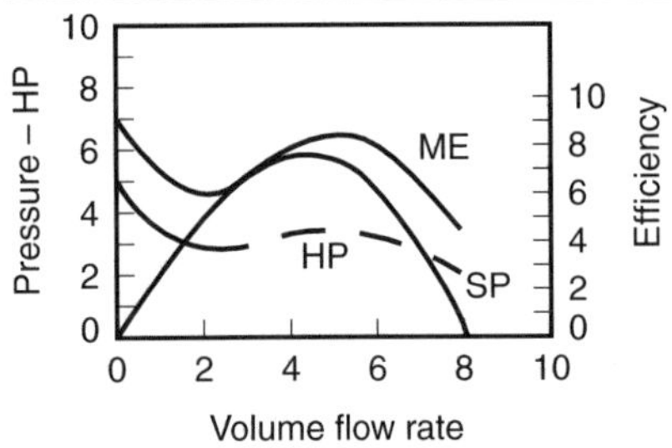	High flow-rate characteristics with medium-pressure capabilities. Performance curve includes a dip to the left of peak pressure which should be avoided. The discharge air pattern is circular and is rotating or whirling because of the propeller rotation and lack of guide vanes.	Low, and medium-pressure ducted heating, ventilating, and air-conditioning applications where air distribution on the downstream side is not critical. Also used in some industrial applications such as drying ovens, paint spray booths, and fume exhaust systems.
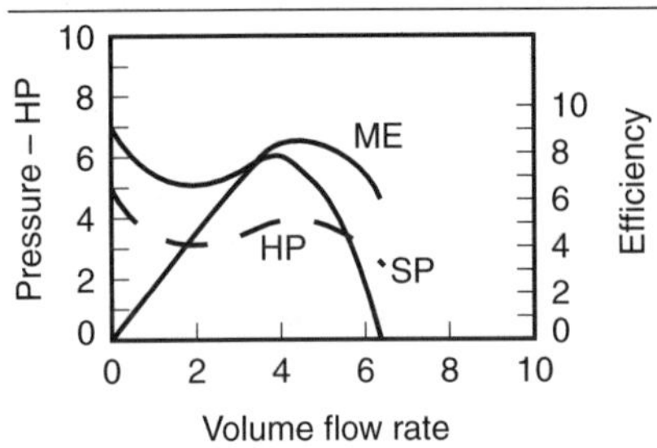	High-pressure characteristics with medium volume flow rate capabilities. Performance curve includes a dip caused by aerodynamic stall to the left of peak pressure which should be avoided. Guide vanes correct the circular motion imparted to the air by the wheel and improve pressure characteristics and efficiency of the fan.	General heating, ventilating and air-conditioning systems in low-, medium-, and high-pressure applications is of advantage where straight-through flow and compact installation are required, air distribution on downstream side is good. Also used in industrial application similar to the tubeaxial fan. Relatively more compact than comparable centrifugal-type fans for same duty.
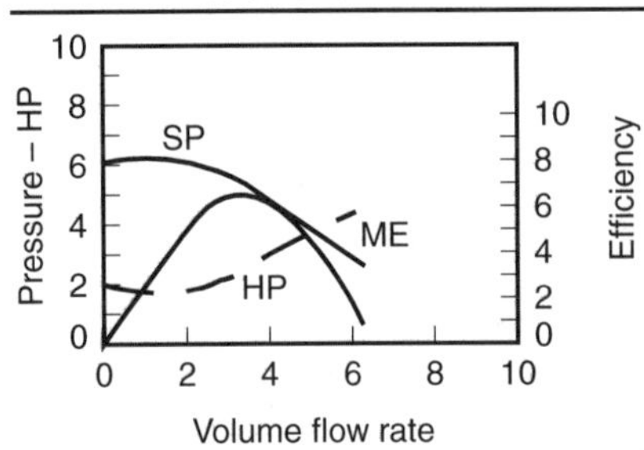	Performance is similar to backward-curved fan, except lower capacity and pressure because of the 90 degree change in direction of the air flow in the housing. The efficiency will be lower than the backward-curved fan. Some designs may have a dip in the curve similar to the axial-flow fan.	Used primarily for low-pressure return air systems in heating, ventilating, and air-conditioning applications. Has straight-through flow configuration.
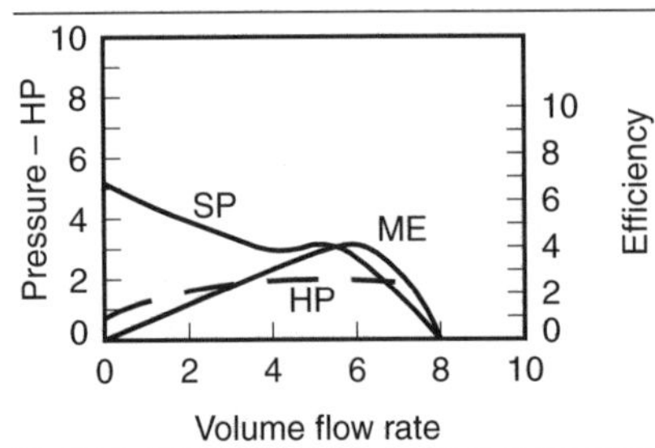	Usually intended to operate without attached ductwork and therefore to operate against a very low-pressure head. It is usually intended to have a rather high-volume flow rate characteristic. Only static pressure and static efficiency are shown this type of product	For low-pressure exhaust systems such as general factory, kitchen, warehouse, and commercial installations where the low-pressure rise limitation can be tolerated. Unit is low in first cost and low in operating cost and provides positive exhaust ventilation in the space which is a decided advantage over gravity-type exhaust units. The centrifugal unit is somewhat quieter than the axial unit described below.
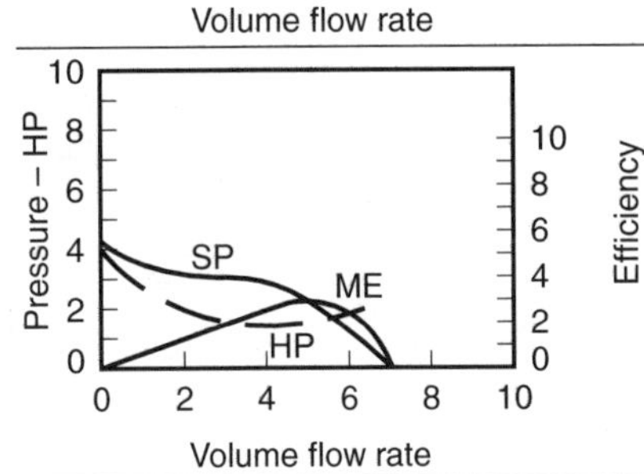	Usually intended to operate without attached ductwork and therefore to operate against very low-pressure head. It is usually intended to have a high-volume flow rate characteristic. Only static pressure and static efficiency are shown for this type of product.	For low-pressure exhaust systems such as general factory, kitchen, warehouse, and commercial installations where the low-pressure rise limitation can be tolerated. Unit is low in first cost and low in operating cost and provides positive exhaust ventilation in the space which is a decided advantage over gravity-type exhaust units.

Figure 24. *(Continued).*

Table 10. Fan Laws Equations

Fan Parameter	Effect of Varying Speed	Effect of Changed Density	Effect of Increased Size
	$(D_2/D_1 = \rho_2/\rho_1 = 1)$	$(\mathrm{RPM_2/RPM_1} = 1, \mathrm{D_2/D_1} = 1)$	$(\mathrm{RPM_2/RPM_1} = 1, \rho_2/\rho_1 = 1)$
Q	$Q_2 = Q_1 \times \left(\frac{\mathrm{RPM_2}}{\mathrm{RPM_1}}\right)$	$Q_2 = Q_1$	$Q_2 = Q_1\left(\frac{D_2}{D_1}\right)^3$
TP	$\mathrm{TP_2} = \mathrm{TP_1} \times \left(\frac{\mathrm{RPM_2}}{\mathrm{RPM_1}}\right)^2$	$\mathrm{TP_2} = \mathrm{TP_1} \times \left(\frac{\rho_2}{\rho_1}\right)$	$\mathrm{TP_2} = \mathrm{TP_1} \times \left(\frac{D_2}{D_1}\right)^2$
VP	$\mathrm{VP_2} = \mathrm{VP_1} \times \left(\frac{\mathrm{RPM_2}}{\mathrm{RPM_1}}\right)^2$	$\mathrm{VP_2} = \mathrm{VP_1} \times \left(\frac{\rho_2}{\rho_1}\right)$	$\mathrm{VP_2} = \mathrm{VP_1} \times \left(\frac{D_2}{D_1}\right)^2$
SP	$\mathrm{SP_2} = \mathrm{SP_1} \times \left(\frac{\mathrm{RPM_2}}{\mathrm{RPM_1}}\right)^2$	$\mathrm{SP_2} = \mathrm{SP_1} \times \left(\frac{\rho_2}{\rho_1}\right)$	$\mathrm{SP_2} = \mathrm{SP_1} \times \left(\frac{D_2}{D_1}\right)^2$
BHP	$\mathrm{BHP_2} = \mathrm{BHP_1} \times \left(\frac{\mathrm{RPM_2}}{\mathrm{RPM_1}}\right)^3$	$\mathrm{BHP_2} = \mathrm{BHP_1} \times \left(\frac{\rho_2}{\rho_1}\right)$	$\mathrm{BHP_2} = \mathrm{BHP_1} \times \left(\frac{D_2}{D_1}\right)^5$

conditions. The type, size, and expected concentration of the entrained dust should be included so that these factors can be considered in the wheel design and selection.

Any information known on acceptable materials of construction for the fan components or the specific environmental conditions of the gas stream should be specified by the user.

The system designer preparing specifications should use as a reference the fan publication manuals[13] prepared by the AMCA entitled:

Table 11. Velocity and Flow Relationships for Different Types of Losses

Type of Loss	Equation[a]	Relationships: Velocity	Relationships: Flow[a]
Hood	$h_e = F \times \mathrm{VP}$	$h_e = V^2$	$h_e = Q^2$
Fitting	$h_f = K \times \mathrm{VP}$	$\mathrm{Loss} = V^2$	$\mathrm{Loss} = Q^2$
Friction	$h_f = K \times Q^{1.9}$	$\mathrm{Loss} = V^{1.9}$	$\mathrm{Loss} = Q^{1.9}$
Baghouse			$\mathrm{Loss} = aQ^2 + bQ^{1.6}$

[a] F, K, a, and b are constants.

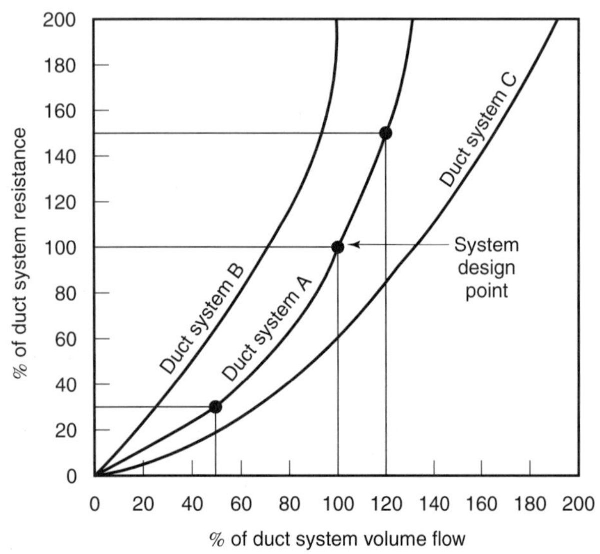

Figure 25. Flow System Resistance for Three Different Ducted Systems.

Publication 201	Fans and Systems (fan testing in laboratories and calculation of performance ratings)
Publication 202	Troubleshooting
Publication 203	Field Performance Measurement
Publication 210	Laboratory Methods of Testing Fans for Rating Purposes

The topics covered by these publications give the designer a realistic estimate of fan-system performance for any specific application.

AMCA prescribes minimum static pressure and air velocity performances for Class I, II, and III fans. There are separate standards for backward-inclined, single- and double-width fans; forward-curved, single- and double-width fans; and backward-inclined fans. As an example, the class standard for backward-inclined fans is as follows:

Class I	5 in. w.g. at 2300 fpm to 2.5 in. w.g. at 3200 fpm.
Class II	8.5 in. w.g. at 3000 fpm to 4.25 in. w.g. at 4150 fpm
Class III	12.5 in. w.g. at 3850 fpm to 6.5 in. w.g. at 5250 fpm.
Class IV	Above Class III minima

The quality of construction is specified based on minimum gauge thickness and not on class.

Fans may be operated in series or in parallel. Figure 26 shows combined characteristic curves for two fans in series and for two fans in parallel. The combined volume–pressure curve for two fans operating in series is obtained by adding the fan pressures at the same air volume flow. For fans in parallel, the combined volume–pressure curve is obtained by adding the air volume flows at the same fan pressures.

The effect of the duct system curve not being at the design point is shown in Figure 27. The design system pressure (point 1) is overestimated from the actual system pressure (point 2). A common mistake made by designers is the belief that an overestimation of system pressure losses is a conservative approach. As shown in Figure 27, an overestimation of the design system pressure loss results in the actual fan power's requirements (point 4) being higher than the design fan power (point 5). The consequence of this situation is that the fan motor may be undersized for the specific application.

The effect of the fan and its inlet and outlet duct connections will often have a significant effect on the flow conditions at the fan inlet and outlet. The result of these disturbances or

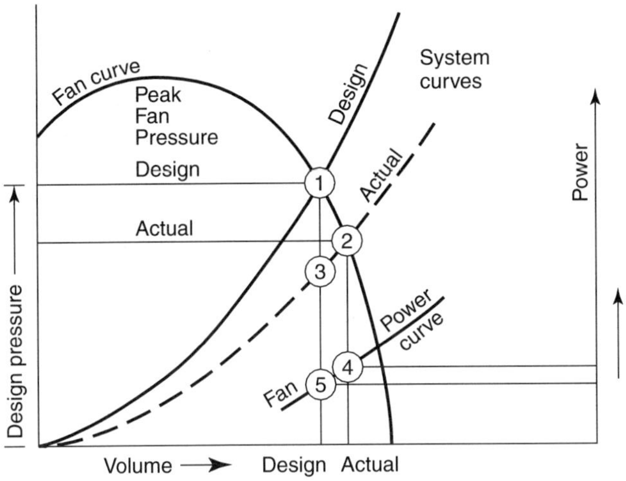

Figure 26. Duct System Curve Not at Design Point.

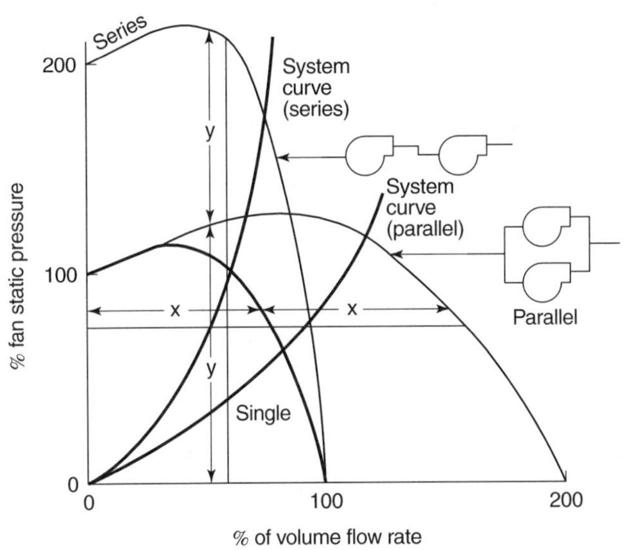

Figure 27. Combined Characteristic Curves for Fans in Series or in Parallel.

system effects is that the fan does not perform as expected. Publication 203 of the AMCA has developed an approach to quantifying the system effect losses using a concept called the "system effect factor." A family of system effect curves can be established that gives system effect losses as a function of velocity for different configurations. This system effect factor, which must be determined separately for the inlet and the outlet, is added to the estimated total system effect pressure losses to give the system design point of operation. This total estimated system effect pressure loss is used to determine the fan power requirements from the manufacturer's catalog data. Information related to specific types and models of fans may be available directly from the manufacturer.

The following are the four most common causes of system effects:

1. Eccentric flow into fan
2. Spinning flow into the fan
3. Improper outlet ductwork
4. Obstructions at the inlet or outlet

A designer must take system effects into account in order to ensure successful fan and ventilation system performance.

Fan Installation. The proper installation of a fan is critical to ensure a successful startup and ongoing operation. Some of the important features are as follows:

1. Complete a detailed inspection of fan installation before startup.
2. Complete the AMCA 202 troubleshooting lists.
3. Make sure that the foundation is sound.
4. Make sure that the fan is firmly secured to the foundation.
5. Check that the impeller is installed correctly.
6. Make sure that inlet sleeves to the wheel are properly located and secure.
7. Check the installation of the damper and its operation.
8. Check the level of the fan shaft and plumb of the casing.
9. Ensure correct bearing installation, including alignment.
10. Ensure proper bearing lubrications.
11. Check the alignment of the fan to the drive for both parallelism and concentricity.
12. Bump the fan for correct rotation.
13. After startup, check for overall smoothness of operation, along with overall operation of the bearing.

An important aspect of fan performance is fan balancing. Technology development in this field is quite recent and is well described in the literature. The techniques for static and dynamic balancing of fan wheels and shaft assemblies, prior to shipment, are described by Myers.[14] The use of vibration measurements and analyses allows the detection and diagnosis of problems in fans before they become serious.

Satisfactory performance of a ventilation system is dependent on matching the fan with the system. It is essential that proper homework and calculations be done when fan specifications are being prepared and duct layouts are being developed. Field modifications to fans to correct system performance problems are always expensive and very often unsatisfactory. It is common to have the fan operating away from its point of maximum efficiency and at increased noise levels. The careful selection and specification of the fan and its associated duct system are essential to ensure that the ventilation system operates at its required performance level.

Stacks

Technical Description. In the not-so-distant past, stacks were used as a method of pollution control, the idea being that the best solution to pollution was "dilution." In other words, the taller the stack, the greater the dispersion of contaminants and, therefore, the more effective the "control." Now, as the direct result of changes in environmental legislation and public attitudes, stacks are the final component of sometimes highly complex air pollution control or ventilation systems.

The design of stacks involves rigorous procedures that employ air dispersion modeling techniques to predict downwind contaminant loadings. Stack heights and diameters have to adhere to "good engineering practice" and also to regulatory constraints.

Stacks are used in a multitude of applications and, consequently, their design is invariably specific to the site and process involved. Considering the large capital outlay required, it is important that a proper engineering feasibility study be carried out in order to evaluate all the alternatives so that the most cost-effective solution can be implemented.

There are four important parameters to be considered in specifying a stack.

- Materials of construction
- Liners
- Structural support
- Accessories

Stacks can be made of many different materials, including steel, brick, fiberglass-reinforced plastic (FRP), and reinforced concrete. In order properly to assess which material would best suit the application, information on the physical and chemical properties of the gas stream is very important. Such properties as the corrosiveness and acidity or basicity of the gas must be determined. The anticipated temperature differential between the gas and the outside air is also an important parameter for materials selection.

Stacks can be loosely categorized into two types, lined and unlined. Liners are usually recommended for most designs because they are able to withstand the chemical composition and temperature of the gas. Also, a liner allows the airspace between it and the outer column to act as insulation. Liners can be replaced at much lower cost than can complete stacks, should they be exposed to particularly corrosive gases. Liners exposed to such conditions should be made of stainless steel, brick, or FRP. Where corrosion is not a problem, carbon-steel liners are adequate. For unlined stacks, coatings can be applied to the inside of the stack for protection; however, these tend to be expensive and, in general, do not last as long as liners.

Stack supports can be free standing, intermittently supported, or guyed, although the last is applicable only to short stacks. Regardless of the type of support structure, the designer must design the stacks with wind loads and seismic zones in mind.

Stack accessories include a cleanout or access door, a sampling platform, ladders, lightning protection systems, and aircraft warning lights. The cleanout door is needed to allow the removal of any accumulated materials at the bottom of the stack, and in the case of lined stacks, it should also provide access to the space between the liner and the outer shell. Regulatory authorities require sampling platforms for all stacks for auditing and compliance purposes, and these should be located at least 10 stack diameters downstream of the inlet duct. Access ladders are an important part of sampling and maintenance activities. Lightning protection systems are important to protect a sizable investment. The requirement for aircraft warning lights is based on site-specific aviation requirements.

Since stacks form an integral part of air pollution control systems, complete drawings detailing the above should be developed and included in the required permit application.

Design Parameters. The minimum information required for the proper and successful design of a stack is as follows:

1. Detailed description of the process generating the off-gas, including performance characteristics of the pollution control devices located upstream of the stack
2. Gas flow rates and inlet temperatures for current, future, and design conditions, including maximum, minimum, and average values
3. Draft required at the stack opening
4. Site-specific data, including elevation above sea level and ambient temperature fluctuations
5. Topographical description and information on potential for seismic activities
6. Plan and elevation layout of the plant property and buildings
7. Meteorological data, of which wind data are especially important

Using this process and plant information, the important stack parameters that need to be calculated and/or specified are as described in the following.

1. *Stack height.* Stack heights are designed according to good engineering practice (GEP). Accepted GEP is a stack height equal to 2.5 times the height of the nearest building. The U.S. Environmental Protection Agency (EPA) has defined a building within a distance of $5L$ (where L is the lesser of the building height or diagonal width of the building) to be a building that affects the stack plume, and consequently the stack height must be designed according to GEP. The EPA has also defined GEP to be calculated as: GEP = bldg. height + $1.5L$, or a maximum of 65 m (213 ft). No credit for stacks of a height greater than 65 m will be given in modeling calculations.
2. *Effective stack height.* The effective height of a stack is defined as the stack height plus the plume rise. Plume rise is a function of plume velocity and temperature differential between the off-gas and the ambient air. A 1°F differential is expected to produce a corresponding increase in effective stack height of about 25 feet. EPA-approved models such as the ISC model contain algorithms for determining the effective stack height.
3. *Duct entry size.* The diameter or width of the entry duct should not be greater than 50% of the liner diameter or 33% of the outer shell diameter when there is no liner.
4. *Gas exit velocity.* Gas exit velocities are an important parameter in stack design because of their relationship to downwind contaminant concentrations. High stack velocities increase plume height and, therefore, the effective stack height. Low stack velocities would allow the gas to be drawn into the eddy zones created by the wind flowing over the stack and may, in fact, cause a downwash. This, in turn, would cause the off-gas to enter the building cavity and introduce contaminants into the building, as shown in Figure 28. Consequently, the stack exit velocity should be approximately one to five times that of the wind. Exit velocities of between 3000 and 4000 fpm have been found to provide an adequate margin of safety in design. However, in order to be sure, average, maximum, and minimum wind velocities should be obtained for the site.

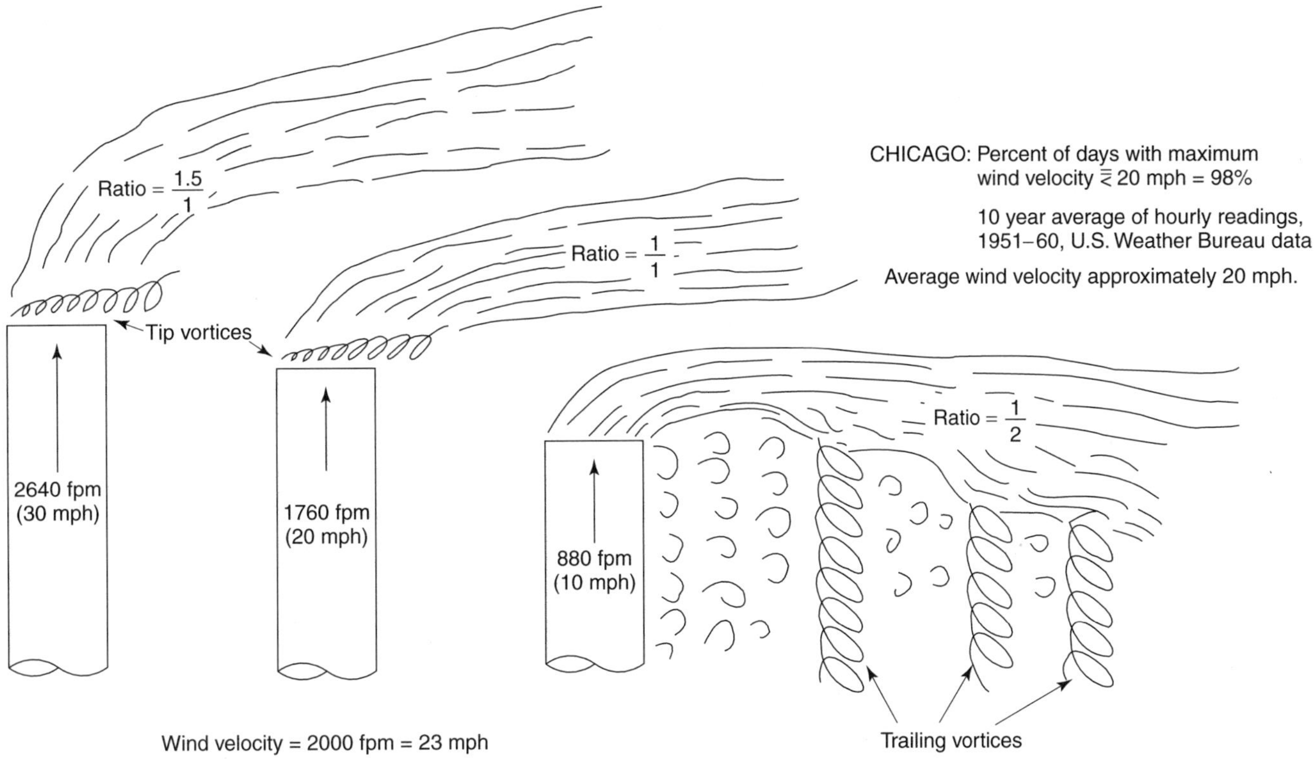

AIR FLOW over stacks is similar to that over buildings. Eddies in the form of tip and trailing vortices will bring the effluent down the stack unless the discharge velocity is high enough. Such downwash may bring the discharge into the building cavity and cause re-entry. Good discharge is obtained with a stack to wind velocity ratio of 1.5:1. Marginal flow occurs with a ratio of 1:1, and very poor flow occurs with a ratio of 1:2. For a 20 mph design wind velocity for Chicago, the stack velocity should be a minimum of 30 mph, or approximately 3000 fpm.

Figure 28. Relationship between Stack Exit Velocity and Wind Velocity (from Ref. 11).

5. *Stack effect.* The stack effect or stack draft is the draft created by the difference in temperature between the off-gas and the ambient air. Stack draft can be calculated by using the following equation:

$$P_{\mathrm{d}} = 0.0343 L P_{\mathrm{T}} \left(\frac{1}{T_{\mathrm{amb}}} - \frac{1}{T_{\mathrm{mg}}} \right)$$

where P_{d} is the theoretical stack draft (in. w.g.); L, the stack height above breeching (ft); P_{T}, the barometric pressure (in. w.g.); T_{amb}, the ambient temperature (°R); and T_{mg}, the average gas temperature (°R).

For stacks with cool gases, the stack effect need not be considered. However, for stacks handling hot gases, the stack effect may be as great as 1–2 in. w.g.

Other important design concerns for stacks include:

- Thickness of stack outer shell
- Resonance problems
- Reinforcement of openings in the outer shell
- Design of base plate and anchor bolts
- Design of stack stiffness

The above are described in detail in the *Guide for Steel Stack Design and Construction*[15] and are beyond the scope of this manual.

DESIGN AND OPERATION OF LOCAL EXHAUST VENTILATION SYSTEMS

System Parameters

Design Procedures for Sizing LEV Systems. In the previous sections, details of the design, sizing, and selection of the individual components were discussed (i.e., hoods, ducts, gas cooling equipment, fans, air cleaning devices, and stacks). Using this information, the steps required to design an LEV system will be outlined. It is important that the designer accept and use the "systems approach" in order to develop cost-effective solutions. The systems approach requires the integration of the process, the work methods, and the LEV system. The size, cost, and effectiveness of the LEV system applied to any process operation will depend on how well the designer understands the process and how well the LEV system has been integrated into the process.

The preliminary data required for the design of an LEV system are as follows:

1. Process details of the operation (e.g., type of contaminant, level of toxicity, continuous or intermittent)
2. Site conditions (e.g., minimum/maximum/average temperatures, relative humidity, wind speed and direction)
3. Performance level for contaminant control
4. Layout of operation

5. Preliminary sketch of each hood
6. Line sketch, including elevations of the ductwork layout, air cleaner location, fan location, stack location (isometric sketch recommended)

The design of an LEV system is an iterative process. The starting point is defining the parameters or practical ranges from the velocities or flow rates for each hood and through the duct network. The design for the LEV system continuously changes through the course of the project as equipment and plant layouts change. The final design for LEV systems must be based on approved layout drawings and certified drawings of the specific mechanical equipment.

A task to be addressed at the start of the design of an LEV system is the establishment of the nature and extent of the contaminant control problem. As described previously, the design objectives for the LEV system must be clearly defined in terms of expected hood performances or contaminant levels in the workplace environment. The recent development of high-speed computers provides the designer with the capabilities to predict contaminant levels using three-dimensional models for different performance levels of LEV systems in an industrial setting (e.g., RoomVent '90).

Before starting the design, two major decisions are required: the type of system (centralized versus distributed) and the balancing method (balanced system or plenum method). Figure 29 shows a flowchart for the ductwork design. A centralized system is based on the use of a single air-cleaning device using a complex network of properly designed ductwork to convey all the contaminated air from the different sources to a central air-cleaning unit. For some applications, it is advantageous to have a multiplicity of units serving different parts of the plant.

The balanced system is the conventional approach and is based on the ducts' being sized for balance and adequate transport velocity. In the plenum approach, the plenum can be either vertical or horizontal, with ducts simply connected to the large plenum, which has a low velocity and a low resistance. For dust or fume control systems, the plenum approach requires that a dust/fume removal system be included in the design.

The procedure for designing an LEV system is as follows:

1. Select and design each exhaust hood.

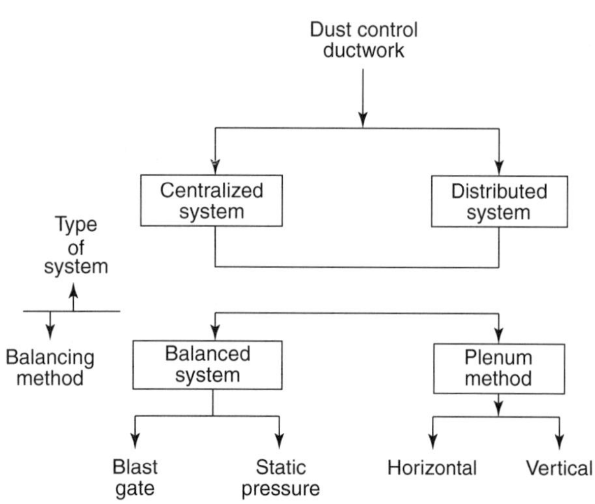

Figure 29. Flowchart for Ductwork Design for LEV Systems.

2. Determine the volume of air that will be required at each hood enclosure in the system in order to ensure adequate contaminant capture.
3. Determine the minimum duct velocity based on the required transport velocity (for systems capturing particulate matter).
4. Determine the branch duct size by dividing the design flow rate by the minimum duct velocity.
5. Determine the design length for each duct segment and the number and type of any special fittings and elbows that are required.
6. Calculate the pressure losses for the entire exhaust system. (Two methods that can be used for these calculations are the velocity–pressure method and the equivalent foot method.
7. Select the fan on the basis of the final volumetric flow rate and the calculated system assistance.
8. Determine makeup air requirements for building air balance.

The procedure for selecting hood type and design exhaust flow rates was outlined in a previous section. The final design exhaust flow rate will be based on the specific hood application and the required level of contaminant control required for the operation.

The calculation of the duct diameter is based on the minimum transport velocity for the given material, as outlined in Table 12. The air velocity in the ductwork must be high enough to ensure that the contaminant does not drop out in the ductwork. A velocity much higher than the minimum transport velocity will result in excessive pressure losses, and for abrasive ducts, it will result in excessive erosion and high maintenance costs. Duct areas, and hence duct diameters, can be determined by dividing the exhaust flow rate by the minimum design velocity.

The general principles for ductwork layout are as follows:

1. Duct layout should be an integral part of the process equipment and plant layout (i.e., simple duct routing).
2. It should minimize horizontal runs.
3. Headers should be located as close to equipment as possible.
4. Lengths of ducts and number of bends should be minimized.
5. Small branches should join the header as close to the fan as possible.
6. Ductwork should provide for ease of cleanout and incorporate necessary maintenance features.

Static Pressures and System Balancing. The required exhaust hood flow rates for an LEV system can be achieved only if the selected fan can provide the required airflow against the static pressure losses for the system. Hence the calculation of the static pressure or resistance losses through the system is a critical element in the design of LEV systems.

The two main procedures for the calculation are the velocity–pressure method and the equivalent-length method. The velocity–pressure method for calculating pressure losses is based on the fact that all frictional losses in ducts and fittings are proportional to the velocity pressure in the duct. The velocity–pressure method is generally more rapid, more

Table 12. Transport Velocities for Different Materials

Material, Operation, or Industry	Minimum Transport Velocity (m/sec)	(fpm)
Abrasive blasting	17.8–20.3	3500–4000
Aluminum dust, coarse	20.3	4000
Asbestos carding	15.2	3000
Bakelite molding powder dust	12.7	2500
Barrel filling or dumping	17.8–20.3	3500–4000
Belt conveyors	17.8	3500
Bins and hoppers	17.8	3500
Brass turnings	20.3	4000
Bucket elevators	17.8	3500
Buffing and polishing		
Dry	15.2–17.8	3000–3500
Sticky	17.8–22.9	4000
Cast iron boring dust	22.9	4000
Ceramics, general		
Glaze spraying	12.7	2500
Brushing	17.8	3500
Fettling	17.8	3500
Dry pan mixing	17.8	3500
Dry press	17.8	3500
Sagger filling	17.8	3500
Clay dust	17.8	3500
Coal (powdered) dust	20.3	4000
Cocoa dust	15.2	3000
Cork (ground) dust	12.7	2500
Cotton dust	15.2	3000
Crushers	15.2	3000 or higher
Flour dust	12.7	2500
Foundry, general	17.8	3500
Sand mixer	17.8–20.3	3500–4000
Shake-out	17.8–20.3	3500–4000
Swing grinding booth exhaust	15.2	3000
Tumbling mills	20.3–25.4	4000–5000
Grain dust	12.7–15.2	2500–3000
Grinding, general	17.8–22.9	3500–4500
Portable hand grinding	17.8	3500
Jute		
Dust	12.7–15.2	2500–3000
Lint	15.2	3000
Dust shaker waste	16.3	3200
Pickerstock	15.2	3000
Lead dust	20.3	4000
with small chips	25.4	5000
Leather dust	17.8	3500
Limestone dust	17.8	3500
Lint	10.2	2000
Magnesium dust, coarse	20.3	4000
Metal turnings	20.3–25.4	4000–5000
Packaging, weighing, etc.	15.2	3000
Downdraft grille	17.8	3500
Pharmaceutical coating pans	15.2	3000
Plastics dust (buffing)	19.3	3800
Plating	10.1	2000
Rubber dust		
Fine	12.7	2500
Coarse	22.9	4000
Screens		
Cylindrical	17.8	3500
Flat deck	17.8	3500
Silica dust	17.8–22.9	3500–4500
Soap dust	15.2	3000
Soapstone dust	17.8	3500
Soldering and tinning	12.7	2500
Spray painting	10.1	2000
Starch dust	15.2	3000
Stone cutting and finishing	17.8	3500
Tobacco dust	17.8	3500
Woodworking		
Wood flour, light dry sawdust and shaving	12.7	2500
Heavy shavings, damp sawdust	17.8	3500
Heavy wood chips, waste, green shavings	20.3	4000
Hog waste	15.2	3000
Wool	15.2	3000
Zinc oxide fume	10.1	2000

Material	Minimum Transport Velocity (m/sec)	(fpm)
Very fine, light dusts	10.2	2000
Fine, dry dusts and powders	15.2	3000
Average industrial dusts	17.8	3500
Coarse dusts	20.3–22.9	4000–4500
Heavy or moist dust loading	22.9 and up	4500 and up

Source: Reference 2.

flexible, and more amenable to the use of computers for calculations than other methods.

For simple systems operating at ambient conditions, the American Conference of Governmental Industrial Hygienists (ACGIH)[4] developed a simplified calculation procedure with design charts, based on the equivalent length method. This technique involves the addition of losses by elbows, entries, fittings, and so forth, expressed as the equivalent length of straight duct and added to the centerline length of straight duct. For conditions other than ambient and for different duct roughness factors, the ACGIH supplies correction tables for density and roughness. Standard calculation sheets are used for the design of LEV systems. The *Industrial Ventilation* manual contains many worked examples for conventional applications, and the reader should refer to the manual for specific details of design.

For an LEV system that has multiple branches, it is necessary to provide a means of distributing airflow among the branches, since air will always take the path of least resistance. At each junction, the exhaust volumetric flow rate will distribute itself according to the pressure losses of the available flow paths. If the system is not designed correctly, the design flows at each hood may not be achieved. In order to provide the proper airflow distribution, the designer must ensure that all flow paths (ducts) entering a junction will have equal calculated static pressure requirements. Two methods are available to do this: the balance-by-design method and the blast gate method. The advantages and disadvantages of these two methods are shown in Table 13.

In the balance-by-design method, duct sizes are chosen so that the static pressure balance at each junction will achieve

Table 13. Relative Advantages and Disadvantages of the Two Balancing Methods

Balance-by-Design Method	Blast Gate Method
1. Volumetric flow rates cannot be changed easily by workers or at the whim of the operator.	1. Volumetric flow rates may be changed relatively easily. Such changes are desirable where pickup of unnecessary quantities of material may affect the process.
2. There is little flexibility for future equipment changes or additions. The duct is "tailor-made" for the job.	2. Depending on the fan and motor selected, there is somewhat greater flexibility for future changes pr additions.
3. The choice of exhaust volumetric flow rates for a new operation may be incorrect. In such cases, some duct revision may be necessary.	3. Correcting improperly estimated exhaust volumetric flow rates is relatively easy within certain ranges.
4. No unusual erosion or accumulation problems will occur.	4. Partially closed blast gates may cause erosion, thereby changing resistance or causing particulate accumulation.
5. Duct will not plug if velocities are chosen wisely.	5. Duct may plug if the blast gate insertion depth has been adjusted.
6. Total volumetric flow rate may be greater than design due to higher air requirements.	6. Balance may be achieved with design volumetric flow rate; however, the net energy required is usually greater than for the balance-by-design method.
7. The system layout must be in complete detail with all obstructions cleared and length of runs accurately determined. Installation must follow layout exactly.	7. Moderate variations in duct layout are possible.

Source: Reference 4.

the desired air volume in each branch duct and hood. With the blast gate method, blast gates are installed that must be adjusted after installation in order to achieve the desired flow at each hood.

The design procedures for the two balancing approaches are outlined in the following.

Method A: Balanced System Design (Static Pressure Balance Method)

1. Begin in branch of greatest resistance.
2. Based on design data, calculate total loss from exhaust hood to junction of the next branch.
3. At each junction point, the static pressure necessary to achieve desired flow in one air stream must match the static pressure *(SP)* in the joining airstream.
4. Where differences in SP are greater than 20%, the branch with the lower resistance should be redesigned to increase its pressure drop (e.g., decrease duct size).
5. If SPs are greater than 5%, but less than 20%, balance can be obtained by increasing airflow through the run with lower resistance. Increase in flow can be quickly calculated by:

$$\text{corrected flow} = \text{design flow} \times \sqrt{\frac{\text{SP run with larger SP loss}}{\text{SP run with smaller SP loss}}}$$

6. If SP is within 5%, it is usual to ignore the small error and treat it as if the paths were in complete balance.

Method B: Blast Gate Adjustment (Velocity Method)

1. Begin in branch of greatest resistance.
2. Calculate pressure losses as for Method A.
3. At each junction, the desired volume of that branch is added to the main flow. No attempt is made to balance the static pressure in the joining airstreams. Joining branches are merely sized to give the desired minimum duct velocity at the desired flow.
4. Blast gates are adjusted after installation in order to achieve the desired airflow at each hood.

The final step is to select a suitable fan to give the design flow rates and static pressures (i.e., point of operation) at an acceptable efficiency. A preceding section described the different types of fans and their operating characteristics. Both the fans and the system have variable operating characteristics, but the actual "point of operation" will be the one single point at the intersection of the fan curve and the system curve. The designer's responsibility is to match the fan performance and system requirement. For complex LEV systems, the fan selection is the critical component to ensure a satisfactory performance. The experienced designer must work closely with the process operating engineers and the fan manufacturer's technical staff.

It is important to balance a system after the installation in order to ensure that it is performing according to the design specifications. Modifications to the system may be necessary to guarantee adequate contaminant capture. The objectives of checking an exhaust system are as follows:

1. To determine the exhaust volume and capture velocity at each pickup point and evaluate the adequacy of contaminant capture
2. To determine the total exhaust volume and evaluate the size and performance of the air cleaning device
3. To determine the system's static pressure and evaluate the fan capacity, speed, and horsepower requirements
4. To determine the temperature at all points in the system in order to evaluate the materials of construction of the ductwork and the collector

Testing of LEV Systems

System Evaluations. Testing of LEV systems should be carried out both for new systems, to verify original design

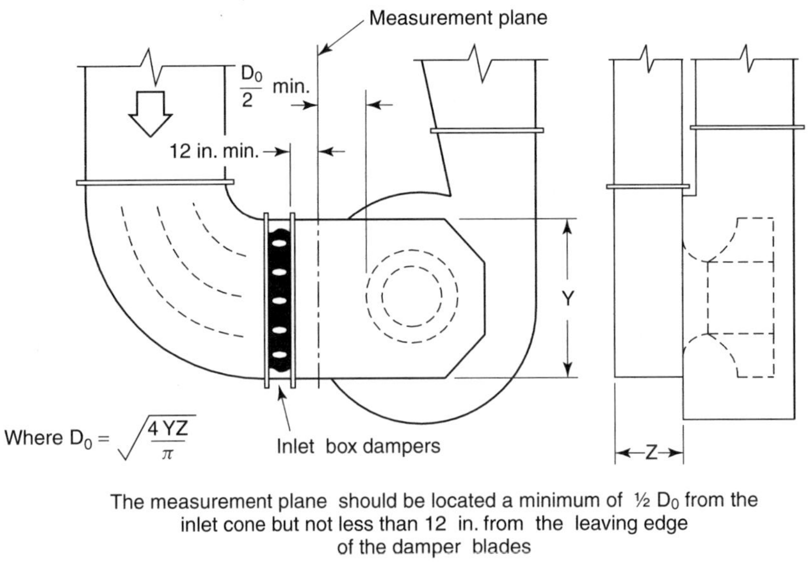

Figure 30. Measurement Plane at Fan Inlet Box (from Ref. 12).

data, and for existing systems that have had or need modifications to perform satisfactorily. For more specific details on measurement equipment and testing procedures, the reader is referred to the standard ventilation references,[2,4] which identify the initial tests and periodic field tests, including test data sheets, to be completed. For each LEV system, static pressure and velocity measurements should be made in each branch and the main duct. The LEV ductwork design should include appropriate testing points and suitable access to these points. Static pressures and velocity measurements should be made upstream and downstream of the fan and the air-cleaning equipment. Details of the procedures to follow for the performance field testing of fans are given in the next section.

Once the system has been commissioned and properly balanced, it may not be necessary to carry out the rigorous testing procedure described. Instead, the routine approach should be to measure static pressures at hoods and to compare them with the original or design values. A marked reduction in hood static pressure could be caused by such factors as the following:

1. Decrease in fan performance
2. Accumulation of dust in ductwork
3. Holes in ductwork or hood
4. Addition of further exhaust points to the system
5. Increased pressure drop across gas cleaning unit
6. Modifications to the system or fan dampers

The reduction in hood static pressure would indicate the need for detailed testing and review of the LEV system performance.

Field Performance Testing of Fans. Fans, being the critical component of LEV systems, require rigorous testing for the commissioning of new systems and identification of problems or lack of performance for existing installations.

Publication 203-90 of the AMCA[13] on field performance measurements of fan systems describes fan performance as a function of fan flow rate, fan total or static pressure, and fan power input at a stated fan speed and fan air density. The major difficulty in making reliable field performance tests is selecting suitable locations for measurements. This section provides the general calculation procedures for making such tests and provides guidance in selecting suitable measurement locations. The procedure is based on those outlined in AMCA publication 203-90.

Fan Flow Rate. Publication 203-90 describes the conditions for an acceptable measurement of fan flow rate as follows:

1. The velocity distribution should be uniform throughout the traverse plane. The uniformity of distribution is considered acceptable when more than 75% of the velocity–pressure measurements are greater than 1/10 of the maximum measurement.
2. The flow stream should be at right angles to the plane.
3. The cross-sectional shape of the airway in which the plane is located should not be irregular.
4. The cross-sectional shape of the airway should be uniform throughout the length of the airway in the vicinity of the plane.
5. The plane should be located in such a way as to minimize the effects of leaks in the portion of the system that is located between the plane and the fan.

Figure 30 is a guide for locating the measurement plane within an inlet box.

To determine fan flow rate, the velocity pressure is measured at the traverse location. Figures 31 and 32 show

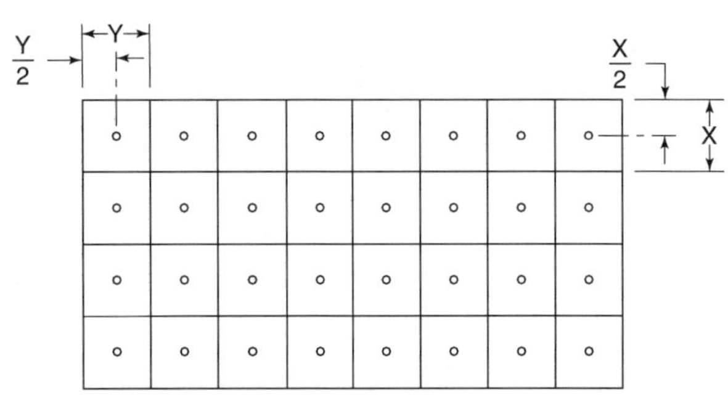

Figure 31. Traverse Points for Rectangular Ducts (from Ref. 31).

In order to obtain a representative average velocity in a duct, it is necessary to locate each traverse point accurately. It is recommended that the number of traverse points increase with increasing duct size. The distributions of traverse points for circular ducts, as indicated below, are based on log-linear Pitot traverse method.

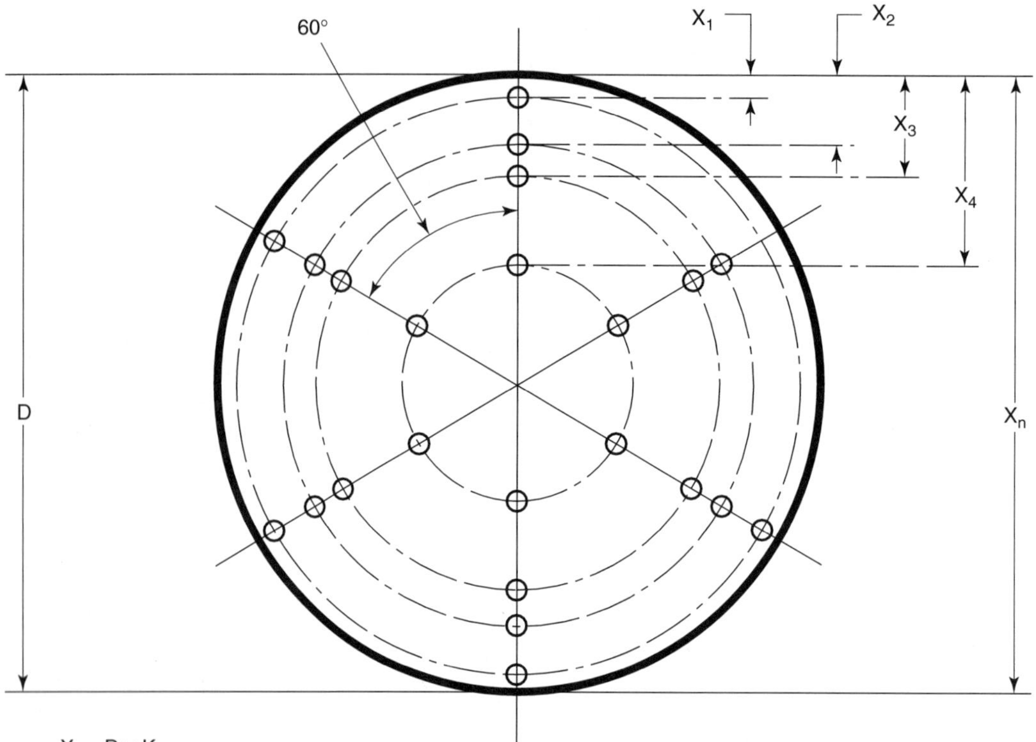

$X_a = D \times K_a$

where D is the inside diameter of the duct and K_a is the factor corresponding to the duct size and the traverse point location as indicated in the table below.

Inside diameter of duct	Number of traverse points in each of 3 diameters	K_1	K_2	K_3	K_4	K_5	K_6	K_7	K_8	K_9	K_{10}	K_{11}	K_{12}	K_{13}	K_{14}	K_{15}	K_{16}
Less than 8 ft	8	.021	.177	.184	.345	.655	.816	.883	.979	–	–	–	–	–	–	–	–
8 ft thru 12 ft	12	.014	.075	.114	.183	.241	.374	.626	.759	.817	.886	.925	.986	–	–	–	–
Greater than 12 ft	16	.010	.055	.082	.128	.166	.225	.276	.391	.609	.724	.775	.834	.872	.918	.945	.990

Figure 32. Traverse Points for Circular Ducts (from Ref. 12).

the recommended data collection points on the traverse plane for rectangular and circular ducts.

The fan flow rate is calculated as follows:

$$Q = VA$$

where Q is the flow rate (acfm); A, the area of the traverse plane (ft^2); and V, the average velocity at the traverse plane (fpm):

$$V = 1096(P_v/\rho)^{0.5}$$

where P_v is the root-mean-square velocity pressure at the traverse plane (in. w.g.) = [Σ(velocity pressure readings)$^{0.5}$/number of readings]^2; and ρ, the gas density at the traverse plane (lb/ft^3).

Fan Static Pressure. Static pressure is measured at the fan inlet and the fan outlet. The requirements for selecting a traverse plane for measurements of fan static pressure are the same as those used for the selection of a traverse plane for flow rate.

The static pressure at a traverse plane (P_s) is calculated from:

$$P_s = \frac{\sum \mathrm{Ps_r}}{\text{number of readings}}$$

where $\mathrm{Ps_r}$ is the static pressure reading (in. w.g.).

The fan static pressure (P_s) is given by:

$$P_s = P_{s2} - P_{s1} - P_{v1}$$

where P_{s2} is the fan static pressure at the fan outlet (in. w.g.); P_{s1}, the fan static pressure at the fan inlet (in. w.g.); and P_{v1}, the velocity pressure at fan inlet (in. w.g.). P_{s1} and P_{s2} usually cannot be measured directly. In most cases, the measurements will be made a short distance upstream of the fan inlet and downstream of the fan outlet.

The downstream static pressure of the fan should be measured a short distance from the outlet because of the turbulence in the vicinity of the outlet. Figure 33 is a guide to selecting the location for static pressure measurements downstream of the fan.

The static pressure measurement upstream of the fan inlet should be measured a minimum of one-half of the equivalent diameter from the inlet. Figure 30 is a guide for locations within the inlet box.

The static pressure at the fan inlet (P_{s1}) can be calculated from:

$$P_{s1} = P_{s4} + P_{v4} - P_{v1} - \Delta P_{4,1}$$

where P_{s4} is the static pressure at the measurement plane; P_{v4}, the velocity pressure at the measurement plane; and $\Delta P_{4,1}$, the pressure losses between the inlet and the measurement plane from friction, fittings, etc.

The static pressure at the fan outlet (P_s) can be calculated from:

$$P_{s2} = P_{s5} + P_{v2} + \Delta P_{2,5}$$

where P_{s5} is the static pressure at the measurement plane; P_{v5}, the velocity pressure at the measurement plane; P_{v2}, the velocity pressure at the fan inlet; and $\Delta P_{2,5}$, the pressure losses between the outlet and the measurement plane from friction, fittings, etc.

The pressure losses $\Delta P_{4,1}$ and $\Delta P_{2,5}$ must be calculated. When the pressure losses between the measurement planes and the planes of interest are negligible, the equations reduce to:

$$P_{s1} = P_{s4} + P_{v4} - P_{v1}$$

$$P_{s2} = P_{s5} + P_{v5} - P_{v2}$$

In addition, when the cross-sectional areas between the measurement planes and the planes are equal, the equations reduce to:

$$P_{s1} = P_{s4}$$

$$P_{s2} = P_{s5}$$

For more accurate fan static pressure calculations, conditions affecting fan performance should be calculated when the fan is installed. For example, fan inlet and outlet restrictions will reduce fan performance. To account for these effects, *system effect factors* must be used in the analysis. Refer to AMCA publications 203-90 and 201-90 for this information.

Fan Power Input. Several measurement methods are available to determine fan power input, including the following:

1. Phase current method
2. Typical motor performance data
3. Calibrated motors methods

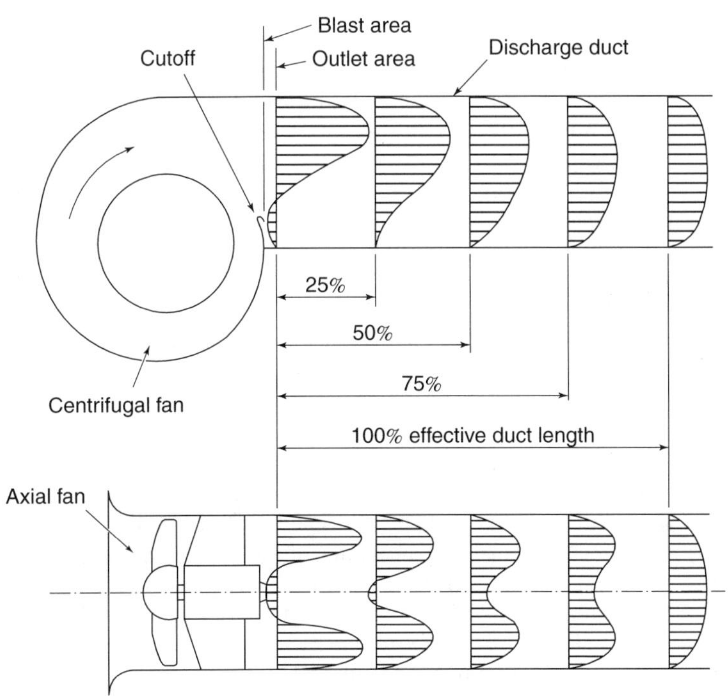

To calculate 100% effective duct length, assume a minimum of 2-1/2 duct diameters for 2500 fpm or less. add 1 duct diameter for each additional 1000 fpm.

Example: 5000 fpm = 5 equivalent duct diameters

If duct is rectangular with side dimensions a and b. the equivalent duct diameter is equal to $(4ab/\pi)^{0.5}$

Figure 33. Velocity Profiles in Outlet Duct (from Ref. 12).

In this section, only the phase current method is discussed. This method is applicable to three-phase motors. For a detailed discussion of other methods, refer to AMCA Publication 203-90.

The method requires measurement of the phase currents and voltages supplied to the motor. The closer the phase current is to the full-load current, the greater will be the accuracy. Two equations are used. Equation A:

$$H_{\mathrm{mo}} = \mathrm{NPH}\left(\frac{\text{measured amps}}{\mathrm{FLA}}\right)\left(\frac{\text{measured volts}}{\mathrm{NPV}}\right)$$

where H_{mo} is the motor power output (hp); NPH, the nameplate horsepower; FLA, the full-load amperes; NPV, the nameplate volts; measured volts, the average of measured phase volts; and measures amperes, the average of measured phase amperes.

Equation B:

$$H_{\mathrm{mo}} = \mathrm{NPH}\left(\frac{\text{measured amps} - \mathrm{NLA}}{\mathrm{FLA} - \mathrm{NLA}}\right)\left(\frac{\text{measured volts}}{\mathrm{NPV}}\right)$$

where NLA is the average of measured phase values of no-load amperes, which must be determined with the motor operating and disconnected from the fan shaft.

Equation A is used to estimate the H_{mo} with fan motors greater than 5 hp operating at greater than 90% of FLA. The average of equations A and B is used when the fan is operating at lower than 90% FLA.

Fan Speed. A minimum of three measurements should be averaged to determine fan speed.

Operation and Maintenance

Introduction. The proper operation and maintenance of LEV air pollution control systems are important. The promulgation of more stringent air pollution legislation has necessitated a much higher performance level from these systems. Regulations in many countries require good operation and maintenance practices (including record keeping) and a comprehensive preventative maintenance program. It is necessary to approach maintenance not only of the actual air pollution control equipment, but also of the entire LEV system, in the same scientific manner that has been adopted for other plant functions.

It is still common to find LEV systems that are badly neglected and receive only minimal maintenance attention. The reasons for this lack of maintenance are as follows:

1. Process equipment has a much higher priority than pollution control equipment.
2. Pollution control equipment is dusty and difficult to maintain.
3. The equipment is often bought on a low-bid basis; attention to maintenance features is not of paramount concern.
4. The owner does not keep an inventory of parts for repairs.
5. No specific individual is responsible for equipment operation.
6. No proper training is conducted for operation and maintenance of equipment.
7. Equipment is complex and not easy to troubleshoot.
8. Equipment is inadequately designed for ease of maintenance.

Successful operation of an installation requires that a proper approach be adopted at the early design stage of the project. Details of the important steps for system design are summarized as:

1. Prepare technical and economic analyses of alternative types of environmental control systems.
2. Develop technical and economic specifications for major equipment for competitive bidders.
3. Prepare technical and economic bid analyses for each bid.
4. Develop detailed design and layouts to accommodate the operation and maintenance requirements.
5. Develop detailed startup and operating manuals and procedures, including operator training programs.
6. Implement a preventative maintenance program.

Items 1 through 4 of the preceding list are covered in detail in previous sections. For establishing correct operation and maintenance procedures, we will focus on items 5 and 6.

Startup and Commissioning. The startup and commissioning of any system are critical. System operation must be established and performance adjusted to obtain design performance. Since LEV systems often work in conjunction with general or dilution ventilation systems, by design they must function in a specified operating range. The steps involved in a startup and commissioning program can be outlined as follows:

1. Obtain process flow instrumentation drawings of the overall system. Review of this information with the goal of reviewing key operating and ventilation requirements is necessary.
2. Obtain and review detailed operating and maintenance manuals provided on each specific piece of equipment by vendors. For large and significant components, schedule manufacturer's training sessions prior to startup.
3. Prepare an overall commissioning and startup schedule, starting from the completion of installation and proceeding through to system turnover. Ensure that adequate tradespeople are on hand during commissioning to allow immediate corrections to any deficiencies that may become apparent.
4. Establish sufficient measurement points in the system to obtain field flow and static pressure measurements. Ensure that access points for measurements have been installed.
5. Perform mechanical checkout on all moving components to ensure freedom of motion throughout range of travel and verify correct installation of drives and linkages.
6. Perform electrical checkout of individual components of system manually (i.e., isolated from interlocks). This may require overrides on some automated systems. Ensure that each piece of equipment is responding to electrical and instrumentation controls.
7. Calibrate all instruments and instrumentation.
8. Verify that all interlocks and process control automation are functioning correctly.
9. Obtain field flow and static pressure measurements throughout system. Compare with original design values and make adjustments as necessary. Compare final field settings and system flows for operating records.

Preventative Maintenance Program. Operating and maintenance problems may develop with equipment and components in the system. The main equipment components and a brief discussion of the potential problems follows. Further details on air pollution control equipment can be found elsewhere in this manual.

Baghouses. Experience with baghouses has identified high-maintenance problems in the following decreasing order of frequency.

1. Filter bags
2. Flow control (blast gases, dampers)
3. Hoppers (screw conveyors, air locks, dust disposal)
4. Cleaning system maintenance

The pressure drop across the bags should be checked frequently during normal operation. This is the most important parameter indicating overall system performance.

Scrubbers. For scrubbers, high-maintenance items are related to the physical and chemical characteristics of the fluids being handled in the ducts, housing, and piping. The highly reactive scrubbing liquids can produce corrosion, erosion, and plugging. The scrubbing liquids must usually be filtered and treated chemically before being reused. A common problem for scrubbers is dust buildup at the wet–dry interface. The solution to this problem may require a redesign of the scrubber inlet ducting or a change in the humidification of the incoming gas. Pipe clogging, buildup, and erosion can occur in the scrubber liquor-handling system. Improperly specified pipes, elbows, reducers, and other fittings can lead to plugging and abrasion. It is also common to find valves misused to throttle flow, which results in high erosion and plugging. Mist eliminators are common sources of plugging and can result in increased system pressure drop and lower system flows.

Electrostatic Precipitators. Experience has shown the following to be typical of users of electrostatic precipitators.

1. The majority of users are satisfied with the precipitator as a functioning piece of equipment.
2. The principal sources of malfunction are the discharge electrodes and design improvements are required.

Fans. The successful operation and maintenance of fans were discussed in earlier sections.

For fume control, the high-temperature gases containing dust-laden materials often present a severe erosion and corrosion problem. Fan balancing is also a common problem and requires careful attention.

Field Inspection Checklist. Troubleshooting to solve problems for existing LEV systems can be organized into the following five steps.

Step 1. Identifying the problem

Step 2. Developing alternative solutions and recommendations

Step 3. Implementation

Step 4. Testing and sampling

Step 5. Doing preventative maintenance

By maintaining thorough, accurate records of system and equipment performance, the system operation can be tracked and changes in performance recorded. Table 14 provides a checklist of field inspection items and a suggested frequency for an LEV system using a baghouse for the air pollution control equipment. A similar list could be developed for other types

Table 14. Field Inspection Checklist

Inspection Item	Frequency
1. Dust inspection for dust settlement (and removal if required)	Quarterly, yearly
2. Bag compartment—painting, patching, regasketing, recaulking	Yearly
3. Supporting structure and framing—painting, bolt tightening, weld inspection	Yearly
4. Heating equipment for baghouse—operating temperature sensors and controls, insulation	Daily
5. Fan drives, electrical motors, gear trains—checks for wear, lubrication	Daily, weekly
6. Repressure fan (for reverse air cleaning) check for temperature and vibration	Daily
7. Compartment isolation and bypass dampers—mechanical drive, jamming, leakage	Daily
8. Bag (fabric) inspection, clean side—wear, damage, dust seepage, surface deposits, bag clamps	Monthly, quarterly
9. Lubrication of damper valves, shaker mechanism	Weekly, monthly
10. Access doors—jamming, insulation sealing, hinges, safety locks	Quarterly
11. Hopper—dust level, bridging, transfer valve material balance on dust recovery	Daily, yearly
12. Bag suspensions—frames, drive attachment hoods, tension adjustment system	Quarterly, yearly
13. Baghouse—check for floor deposits, gross leakage	Weekly, quarterly, yearly
14. Control panel—indicator lights, switch position, dial displays	Daily
15. Baffle plate and guide vanes—erosion or dust deposition	Yearly
16. Shaking mechanism—linkage, slack, amplitude or frequency adjustment	Yearly
17. Compressed air supply—pressure, oil and water separators	Daily, weekly
18. Fire detectors, sprinklers, explosion panels	Semiannually
19. Instrumentation—direct display, continuous monitors	
a. Baghouse temperature	Daily
b. Overall baghouse pressure loss	Daily
c. Individual compartment pressure loss	Daily
d. Fan static pressure	Daily
20. Stack emissions—visual estimates, automatic in-stack opacity meters	Daily
21. Fabric cleaning controls—timing circuits, pressure sensor	Daily
22. Manometer—check and record pressure drop readings. Watch for trends	Daily
23. Collector—observe all indicators on control panel and listen to system for properly operating subsystems	Daily
24. Check all bolts and welds. Inspect entire collector thoroughly, clean, and touch up paint where necessary	Yearly

of pollution control equipment (e.g., scrubbers, electrostatic precipitators).

Acknowledgment

The Department of Chemical Engineering and Applied Chemistry of the University of Toronto provided financial support for the preparation of this chapter.

REFERENCES

1. *ASHRAE Handbook—Heating, Ventilating, and Air Conditioning Systems and Application*, American Society of Heating, Refrigerating and Air Conditioning Engineers, Atlanta, GA, 1995.
2. H. D. Goodfellow, *Advanced Design of Ventilation Systems for Contaminant Control*, Elsevier, New York, 1985.
3. J. J. Loeffler, "Simplified Equations for HVAC Duct Friction Factors," *ASHRAE J.*, 76–79 (January 1980).
4. *Industrial Ventilation*, 20th ed., American Conference of Governmental Industrial Hygienists, Cincinnati, OH, 1988.
5. J. E. Ernswiler, "The Neutral Zone in Ventilation," *J. Am. Soc. Heating Ventilating Eng.* **32**(1): 1–16 (January 1926).
6. H. D. Goodfellow, Ed., *Ventilation '85, Proceedings of the 1st International Symposium on Ventilation for Contaminant Control*, Elsevier, New York, 1986.
7. G. C. Cawkwell and H. D. Goodfellow, "Multiple Cell Ventilation Model with Time-Dependent Emission Sources," *Proceedings of Engineering Aero and Thermo Dynamics of Ventilated Room 2nd International Conference*, A 1–9, Norsk VVS, Oslo, Norway, June 13–15, 1990.
8. E. R. Kashdon et al., *Technical Manual: Hood System Capture of Process Fugitive Particulate Emissions*, EPA 1600/786/016, Environmental Protection Agency, Research Triangle Park, NC, April 1986.
9. H. D. Goodfellow, "Hood Design for Ventilation Systems," *Heating, Piping/Air Conditioning*, 60–67 (February 1987).
10. V. Hampl, "Evaluation of Local Exhaust Hood Efficiency by a Tracer Gas Technique," *Am. Indus. Hygiene Assoc. J.* **45**(7), 485–90 (July 1984)
11. H. D. Goodfellow and M. Bender, "Design considerations for Fume Hoods and Process Plants," *Am. Indus. Hygiene Assoc. J.*, **41**, 473–84 (1980).
12. *HVAC Duct Construction Standards, Metal and Flexible*, Sheet Metal and Air Conditioning Contractors National Association, Vienna, VA. 1985.
13. AMCA Publication 201: *AMCA Fan Application Manual—Part 1—Fans and Systems*, 1973; AMCA Publication 202: *AMCA Fan Application Manual—Part 2—Troubleshooting*, 1972; AMCA Publication 203: *AMCA Fan Application Manual—Part 3—Field Performance Measurements*, 1976; AMCA Publication 210: *AMCA Standard 210-74: Laboratory Methods of Testing Fans for Rating Purposes*, 1974. Air Movement and Control Association, Arlington Heights, IL.
14. R. C. Myers, "Industrial Fans—guidelines for a Successful Installation," *Iron Steel Eng.*, 38–44 (October 1976).
15. *Guide for Steel Stack Design and Construction*, Sheet Metal and Air Conditioning Contractors National Association, Vienna, VA, 1983.
16. R. Niemela et al., "Target Levels of Air Quality—A Valuable Design Tool," pp. 133–40, Ventilation '97. Proceedings of the 54th International Symposium on Ventilation for Contaminant Control, Ottawa, Canada, September 14–17, 1997.
17. E. Tahti et al., "Target Level and Design Methodology for Paper Machine Room," pp. 511–12, Ventilation '97. Proceedings of the 54th International Symposium on Ventilation for Contaminant Control, Ottawa, Canada, September 14–17, 1997.

BIBLIOGRAPHY

AMCA Publication 203-90, *Field Performance Measurements of Fan Systems*, Air Movement and Control Association, Arlington Heights, IL.

M. O. Amdur, J. R. Anticaglia, E. C. Barnes, et al., *The Industrial Environment—Its Evaluation and Control*, U.S. Department of Health and Human Services, Washington, DC, 1973.

W. Burgess, M. Ellenbecker, and R. Treitman, *Ventilation for Control of the Work Environment*, John Wiley & Sons, New York, NY, 1989.

H. Goodfellow and E. Tahti, *Ventilation '97—5th International Symposium on Ventilation for Contaminant Control*, Canadian Environment Industry Association, Ottawa, ON, 1997.

K. Hagstrom, K. Pontinen, J. Railio, and E. Tahti. *Design Methodology of Industrial Air Technology INVENT TECHNOLOGY Program Report 55*, April 1996, Fimet, Helsinki, Finland.

T. Hayashi, R. H. Howell, M. Shibata, et al., *Industrial Ventilation and Air Conditioning*, CRC Press, Boca Raton, FL, 1987.

R. Hughes, H. Goodfellow, and G. Rajhans, *Ventilation '91—3rd International Symposium on Ventilation for Contaminant Control*, American Conference of Governmental Industrial Hygienists, Inc., Cincinnati, OH, 1993.

A. Jansson and L. Olander, *Ventilation '94—4th International Symposium on Ventilation for Contaminant Control*, National Institute of Occupational Health, Stockholm, Sweden, 1994.

R. Jorgensen, *Fan Engineering*, 8th ed., Buffalo Forge Co., Buffalo, NY, 1983.

Research and Education Association, *Modern Pollution Control Technology*, Vol. 1, New York, 1980.

J. H. Vincent, Ed., *Ventilation '88—2nd International Symposium on Ventilation for Contaminant Control*, The Annals of Occupational Hygiene, Pergamon Press, New York, 1989.

R. A. Wadden and P. A. Scheff, Engineering Design for the *Control of Workplace Hazards*, McGraw-Hill, New York, 1987.

7

COMBUSTION SOURCES

COAL

Arijit Pakrasi and Wayne T. Davis

INTRODUCTION

The most common example of solid fuel is coal. Coal-fired boilers accounted for 68% of total megawatts generated in the United States in 1994, and this trend is expected to continue in the future. Different ranks of coal are found in nature, depending on the conditions under which the coal was formed. Classification by rank is based on the volatile matter, moisture content, and fixed carbon content of the coal, as shown in Table 1. The natural transition of coal with time is as follows, with each step in the process enriching the carbon content:

$$\text{peat} \rightarrow \text{lignite} \rightarrow \text{subbituminous} \rightarrow \text{bituminous} \rightarrow \text{anthracite}$$

The composition of coal varies with the rank and also within the rank. Coal compositions are reported in a standard format based on American Society for Testing and Materials (ASTM) standard analyses referred to as "proximate" and "ultimate" analyses. Proximate analyses report coal composition in the broad subgroups of moisture content (MC), volatile matter (VM) content, fixed carbon (FC) content, and ash content (noncombustibles). The volatile matter is that portion of the coal that is driven off (exclusive of H_2O vapor) when it is exposed to a temperature of 950°C for 7 min (ASTM D-3175). The fixed carbon is the combustible residue that is left after the coal has been subjected to the moisture and volatile matter tests. The noncombustible residue is referred to as the ash.

The ultimate analysis reports the carbon, hydrogen, sulfur, oxygen, nitrogen, and water content of the coal and is used to compute the air requirements and approximate composition of flue gas products. An overall average formula for the coal can be determined from the ultimate analysis of the coal. Table 2 shows typical proximate and ultimate analyses based on the rank of the coal.

Coal contains many noncombustibles collectively termed "ash." The major constituents of ash are various minerals and mineral oxides. During combustion, minerals and oxides partition into either the bottom ash or the fly ash, depending on such factors as combustion temperature, heat-to-fluid flow pattern in combustion, vapor pressure characteristics of the minerals, fusion properties, and the sintering properties of the ash. The fly ash in the flue gas varies in composition and loading (i.e., mass of fly ash per unit volume of flue gas) for each type of coal, combustor, and process operation.

It is possible to conduct a detailed metals analysis of the coal ash using ASTM D-3174. The results are reported as the mass percent for the metal oxides as follows:

$$SiO_2 + Al_2O_3 + Fe_2O_3 + CaO + MgO + Na_2O + K_2O + TiO_2 + P_2O_5 + SO_3 = 100\%$$

While the coal ash properties, as measured by conducting a laboratory ash analysis of the coal sample, may be different from those of the fly ash emitted from the combustion chamber, the coal ash analysis is frequently used to determine slagging characteristics in the boiler and to predict the electrical resistivity of the fly ash. This parameter is a significant one used in the design of electrostatic precipitators (ESPs). A model for the prediction of fly ash resistivity based on coal ash properties was developed by Bickelhaupt. Table 3 shows the broad range of variation in chemical composition of coal ashes.

Although coal ash properties are frequently used to predict the behavior of the ash, the fly-ash properties are also important with regard to the design of particulate control systems. The fly-ash particle-size distribution and surface area directly affect the pressure drop across fabric filter dust collectors and the efficiency of all the control devices. Vann Bush et al. showed that the resistance to flow through fabric filters (baghouses) increased as the surface area increased and the size of the particles decreased. Table 4 gives a summary of key fly-ash characteristics for 24 coal-fired plants. Specific descriptions of each parameter and its effect on performance were reported. Figure 1 shows typical scanning electron microscope (SEM) photographs of the fly-ash dust cake collected from four different coal-fired power plant ashes, ranging from an ash composed of nearly spherical fly-ash particles with a low surface area of 0.85 m^2/g (Figure 1a) to 2.23, 3.23, and 14.44 m^2/g for Figures 1b–d, respectively. The mass mean diameter of the 24 fly ashes ranged from 3.6 to 9.4 μm.

PROCESS DESCRIPTIONS*

There are three major coal combustion techniques — suspension firing, grate firing, and fluidized bed. Suspension firing is the

* Taken in part from AP-42.

Table 1. Classification of Coals by Rank

Class	Fixed Carbon Limits (%)[a]		Volatile Matter Limits (%)[a]		Calorific Value Limits (Btu/lb)[b]	
	Equal or Greater Than	Less Than	Equal or Greater Than	Less Than	Equal or Greater Than	Less Than
			I. Anthracitic			
Meta-anthracite	98	—	—	2	—	—
Anthracite	92	98	2	8	—	—
Semianthracite[c]	86	92	8	14	—	—
			II. Bituminous			
Low-volatile bituminous coal	78	86	14	22	—	—
Medium-volatile bituminous coal	69	78	22	31	—	—
High-volatile A bituminous coal	—	69	31	—	14,000[c]	—
High-volatile B bituminous coal	—	—	—	—	13,000[c]	14,000
High-volatile C bituminous coal	—	—	—	—	11,500	13,000
					10,500	11,500
			III. Subbituminous			
Subbituminous A coal	—	—	—	—	10,500	11,500
Subbituminous B coal	—	—	—	—	9,500	10,500
Subbituminous C coal	—	—	—	—	8,300	9,500
			IV. Lignitic			
Lignite A	—	—	—	—	6,300	8,300
Lignite B	—	—	—	—	—	6,300

[a] Dry mineral-free basis.
[b] Moist mineral-free basis.
[c] Coals having 69% or more fixed carbon shall be classified by fixed carbon, regardless of calorific value.

Source: Reference 1.

Table 2. Analyses of Typical U.S. Coals, as Mined

Coal Type	% Proximate Analysis				% Ultimate Analysis						HHV	Air[a]
	H_2O	VM	FC	Ash	H_2O	C	H_2	S	O_2	N_2	(Btu/lb)	(lb/10^6 Btu)
Anthracite	2.5	5.7	83.8	8.0	2.5	83.9	2.9	0.7	0.7	1.3	13.72	787
Bituminous	3.3	20.5	70.0	6.2	3.3	80.7	4.5	1.8	2.4	1.1	14.31	765
Sub-bituminous	23.2	33.3	39.7	3.8	23.2	54.6	3.8	0.4	13.2	1.0	9.42	757
Lignite	34.8	28.2	30.8	6.2	34.8	42.4	2.8	0.7	12.4	0.7	7.21	750

[a] Theoretical air required for combustion under stoichiometric conditions (no excess air).

Source: Adapted from Reference 2.

Table 3. Coal Ash Properties

Component	Percentage
SiO_2	10–70
Al_2O_3	8–38
Fe_2O_3	2–50
CaO	0.5–30
MgO	0.3–8
Na_2O	0.1–3
K_2O	0.1–3
TiO_2	0.4–4
SO_3	0.1–30

Source: Reference 2.

primary combustion mechanism in pulverized-coal-fired and cyclone-fired units. Grate firing is the primary mechanism in underfeed and overfeed stoker-fired units. Both mechanisms are employed in spreader stokers. In a fluidized bed combustor (FBC), the coal is introduced to a bed of either sorbent or inert material (usually sand), which is fluidized by an upward flow of air.

Pulverized-coal furnaces are used primarily in utility and large industrial boilers. In these systems, the coal is pulverized in a mill to the consistency of talcum powder (i.e., at least 70% of the particles will pass through a 200-mesh sieve). The pulverized coal is generally entrained in primary air before being fed through the burners to the combustion chamber, where it is fired in suspension. Pulverized-coal furnaces are classified as either dry or wet bottom, depending on the ash removal technique. Dry-bottom furnaces fire coals with high ash fusion temperatures, and dry ash removal techniques are used. In wet-bottom (slag tap) furnaces, coal with a low ash fusion temperature is used, and molten ash is drained from the bottom of the furnace. Pulverized coal furnaces are further classified by the firing position of the burners, that is, single (front or rear) wall, horizontally opposed, vertical, tangential (comer fired), turbo, or arch fired. Figure 2 shows

Table 4. Fly-Ash Characteristics for 24 Coal-Fired Plants

Fabric Filter	Chemical Characteristics of Dust-Cake Ashes (% wt)												Soluble SO_4^a	Equilibrium pH
	Li_2O	Na_2O	K_2O	MgO	CaO	Fe_2O_3	Al_2O_3	SiO_2	TiO_2	P_2O_5	SO_3	LOI		
Monticello	0.02	0.49	0.99	2.20	9.70	3.4	20.4	60.5	1.90	0.08	0.44	0.29	0.48	9.81
Scholz HSFP	0.04	0.58	2.83	0.94	2.72	21.1	22.4	46.3	1.42	0.55	0.65	11.10	3.65	4.77
Brunner Island	0.06	0.32	2.70	0.93	2.33	11.7	28.5	49.6	1.53	0.71	0.36	6.10	3.10	4.26
IPP	0.03	2.10	1.46	2.68	10.72	6.1	19.6	53.0	1.10	0.61	1.56	0.64	1.95	10.68
Nixon	0.03	2.60	1.10	1.55	6.00	4.8	28.4	50.2	0.96	1.65	1.02	2.40	2.00	7.15
Shawnee Unit 5	0.06	0.40	2.00	0.84	1.80	4.5	31.5	53.8	2.70	0.19	0.34	6.30	1.83	4.38
Cameo	0.03	0.75	1.20	1.95	6.35	5.0	28.4	50.5	1.70	1.55	1.35	1.73	1.35	8.39
Harrington	0.01	1.55	0.40	6.05	29.00	6.3	18.8	30.5	1.25	1.45	2.60	0.53	3.15	11.05
Escalante	0.02	0.50	1.20	1.60	5.30	6.4	23.5	58.3	0.92	0.07	0.70	0.89	0.90	10.00
Arapahoe Unit 4	0.02	1.60	1.30	1.80	6.30	5.2	25.7	53.4	0.83	1.30	0.76	2.60	2.00	7.80
Arapahoe FFPP	0.02	2.80	1.30	1.60	4.90	4.1	26.4	55.5	1.10	1.00	0.78	1.80	1.50	7.27
Arapahoe FFPP (downstream of multiclone)	0.02	1.90	1.40	2.50	7.60	6.6	25.8	48.0	1.00	2.00	1.30	3.10	2.10	7.40
Arapahoe Ecolaire	0.02	1.40	1.30	2.00	6.20	5.7	26.3	52.0	0.92	1.60	0.71	3.00	1.80	7.40
Crane Unit 1	0.04	1.20	2.10	1.10	4.30	18.6	24.3	44.3	1.20	0.86	1.50	15.50	6.30	4.40
Comanche Ecolaire	0.01	1.20	0.18	6.60	35.90	5.3	18.8	21.8	1.80	2.00	3.30	0.55	4.20	11.40
Cherokee	—	0.57	1.10	1.80	5.60	4.6	28.2	52.1	1.50	1.70	0.95	1.60	1.07	—
Eraring	0.02	0.53	1.70	0.73	1.30	4.4	26.8	61.5	0.92	0.17	0.17	2.00	0.57	5.77
Tallawarra	0.02	0.11	1.20	0.52	0.95	4.9	20.1	70.5	0.84	0.19	0.13	3.00	0.65	5.45
TVA AFBC	0.03	1.33	1.05	1.85	34.67	10.0	11.4	22.4	0.53	0.27	15.78	11.10	16.58	11.47
Plant A CFBC	—	—	—	—	—	—	—	—	—	—	—	—	—	—
Plant B CFBC	0.01	0.28	0.97	1.20	29.20	23.5	7.3	19.2	0.42	0.84	16.30	6.50	18.60	11.32
Nucla CFBC	0.03	0.18	0.93	0.81	16.40	3.8	23.9	45.1	1.00	0.04	5.80	8.80	6.60	11.40
EPRI High Sulfur Test Center	0.61	0.40	0.65	0.70	31.00	4.1	7.4	15.2	0.35	0.24	39.10	5.80	43.60	10.30
Plant C Spray Dryer	0.02	1.70	1.20	1.30	12.20	4.4	18.0	48.4	0.91	0.05	10.60	2.50	12.10	10.40

a typical pulverized-coal-fired boiler. Approximately 90% of utility units were either wall-fired or tangential-fired units in 1994.

Cyclone furnaces burn coal of low ash fusion temperature crushed to a 4-mesh size. The coal is fed tangentially, with primary air, to a horizontal cylindrical combustion chamber. In this chamber, small coal particles are burned in suspension, while the larger particles are forced against the outer wall. Because of the high temperatures developed in the relatively small furnace volume, and because of the low fusion temperature of the coal ash, much of the ash forms a liquid slag, which is drained from the bottom of the furnace through a slag tap opening. Cyclone furnaces are used mostly in utility and large industrial applications (see Figure 3). Approximately 8.5% of all utility units were cyclone-fired units in 1994.

Stoker-fired systems account for the vast majority of coal-fired water-tube boilers for industrial applications. In spreader stokers, a flipping mechanism throws the coal into the furnace and onto a moving fuel bed. Combustion occurs partly in suspension and partly on the grate. Because of significant carbon in the particulate, fly ash reinjected from mechanical collectors is commonly employed to improve boiler efficiency. Ash residue in the fuel bed is deposited in a receiving pit at the end of the grate. In overfeed stokers, coal is fed onto a traveling or vibrating grate, and it burns on the fuel bed as it progresses through the furnace. Ash particles fall into an ash pit at the rear of the stoker. The term *overfeed* applies because the coal is fed onto the moving grate under an adjustable gate. Conversely, in *underfeed* stokers, coal is fed into the firing zone from underneath by mechanical rams or screw conveyers. The coal moves in a channel, known as a *retort*, from which it is forced upward, spilling over the top of each side to form and to feed the fuel bed. Combustion is completed by the time the bed reaches the side dump grates, from which the ash is discharged to shallow pits. Underfeed stokers include single-retort units and multiple-retort units, the latter having several retorts side by side. Figure 4 shows a schematic of a stoker-fired boiler.

FBCs have gained popularity in the past decade for smaller boilers used in cogenerators, independent power producers, and utilities. FBCs operating under atmospheric pressure are more common than FBCs operating under pressure. Both bubbling bed design and circulating bed designs are used, the difference being in the fluidization velocity. In bubbling bed design, the fluidization velocities are low, which minimizes solids carryover. In the circulating bed design, high fluidization velocities are used to promote circulation of solids, and uncontrolled particulate emissions are consequently higher.

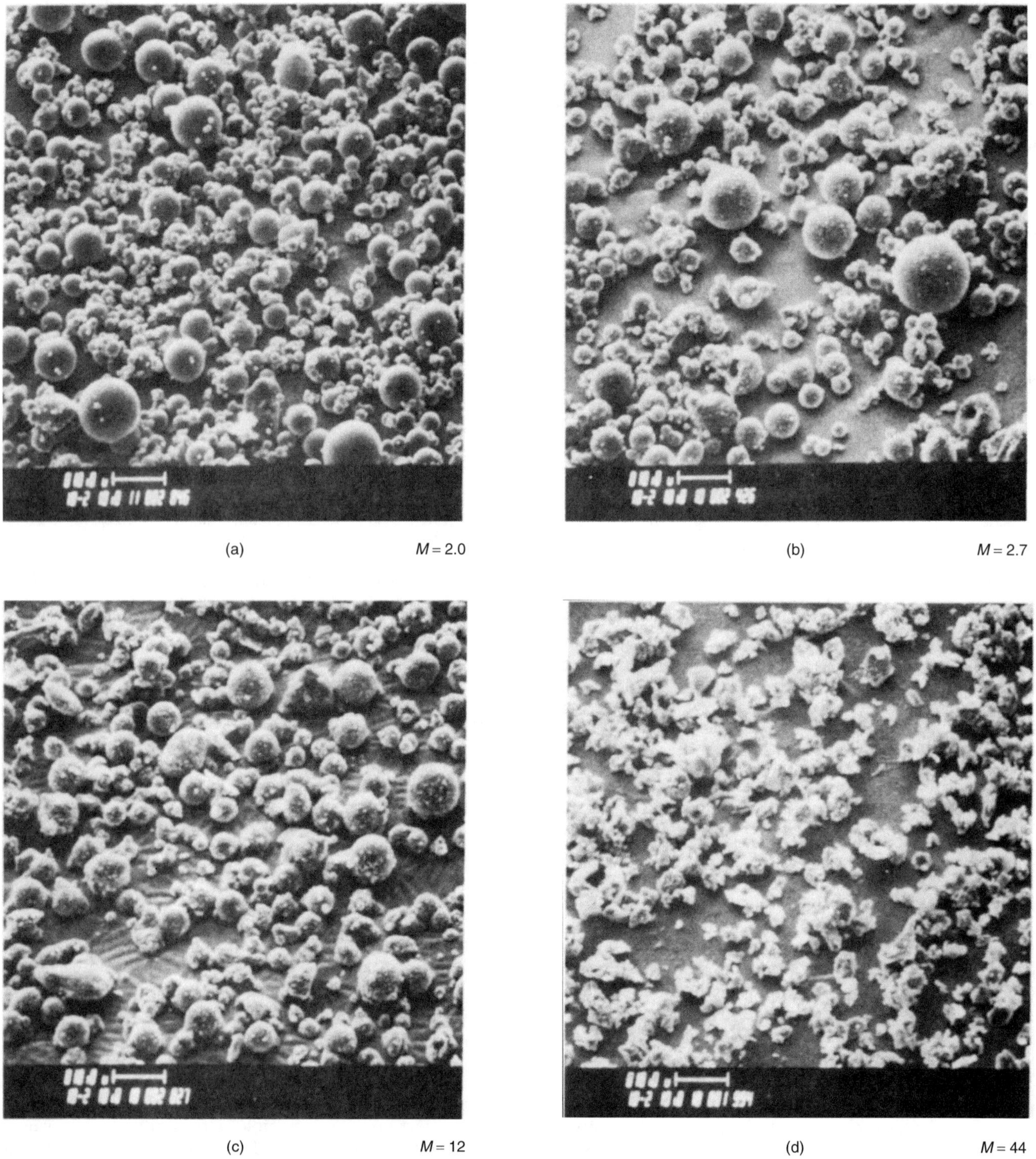

(a) $M = 2.0$ (b) $M = 2.7$

(c) $M = 12$ (d) $M = 44$

Figure 1. SEM Photographs Comparing the Appearance and Morphology Factor, M, of Dust-Cake Ashes from (a) Monticello, (b) Escalante, (c) Scholz HSFP, and (d) TVA Atmospheric Fluidized-Bed Combustor (from Ref. 4, P. 233).

AIR EMISSION CHARACTERIZATION

The major pollutants of concern from external coal combustion are particulate, sulfur oxides, and nitrogen oxides. Some unburned combustibles, including numerous organic compounds and carbon monoxide, are generally emitted even under proper boiler operating conditions.

Particulate Emissions

Particulate composition and emission levels are a complex function of firing configuration, boiler operation, and coal properties. In pulverized-coal systems, combustion is almost complete, and thus the particulate is largely composed of the inorganic ash residue. In wet-bottom, pulverized-coal units and

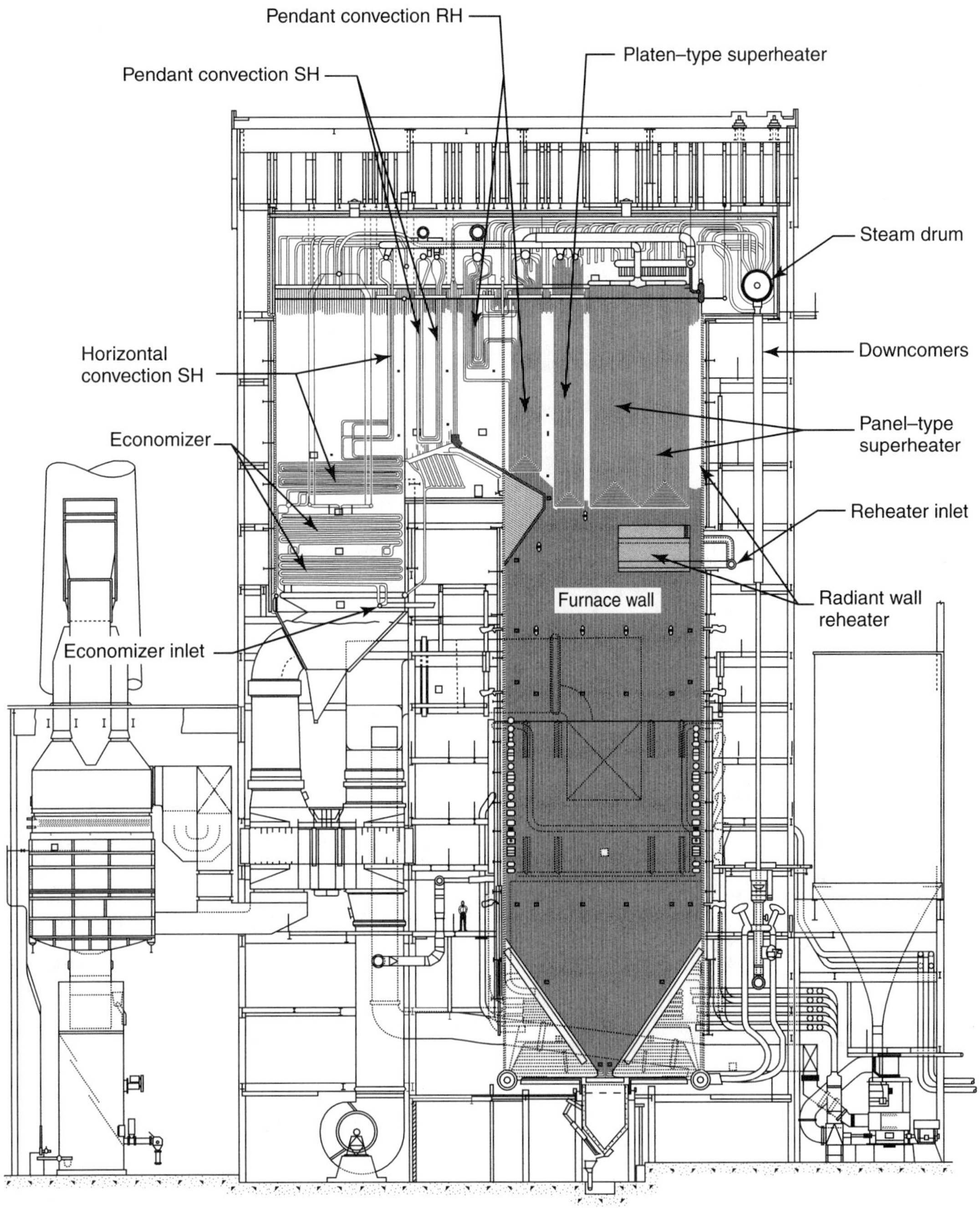

Figure 2. Side Elevation of Kansas Power and Light Gas Services, Lawrence Energy Center Unit No. 5 as Originally Built.

cyclones, the quantity of ash leaving the boiler is less than in dry-bottom units, since some of the ash liquefies, collects on the furnace walls, and drains from the furnace bottom as molten slag. To increase the fraction of ash drawn off as wet slag, and thus to reduce the fly-ash disposal problem, fly ash may be reinjected from collection equipment into slag tap systems. Dry-bottom unit ash may also be reinjected into wet-bottom boilers for the same purpose.

Because a mixture of fine and coarse coal particles is fired in spreader stokers, significant unburned carbon can be present in the particulate. To improve boiler efficiency, fly ash from collection devices (typically multiple cyclones) is sometimes reinjected into spreader stoker furnaces. This practice can dramatically increase the particulate loading at the boiler outlet and, to a lesser extent, at the mechanical collector outlet. Fly ash can also be reinjected from the boiler, air heater, and economizer dust hoppers. Fly-ash reinjection from these hoppers does not increase particulate loadings nearly so much as that from multiple cyclones.

Uncontrolled overfeed and underfeed stokers emit considerably less particulate than do pulverized-coal units and spreader stokers, since combustion takes place in a relatively quiescent fuel bed. Fly-ash reinjection is not practiced in these kinds of stokers.

Variables other than firing configuration and fly-ash reinjection also can affect emissions from stokers. Particulate loadings will often increase as load increases (especially as full load

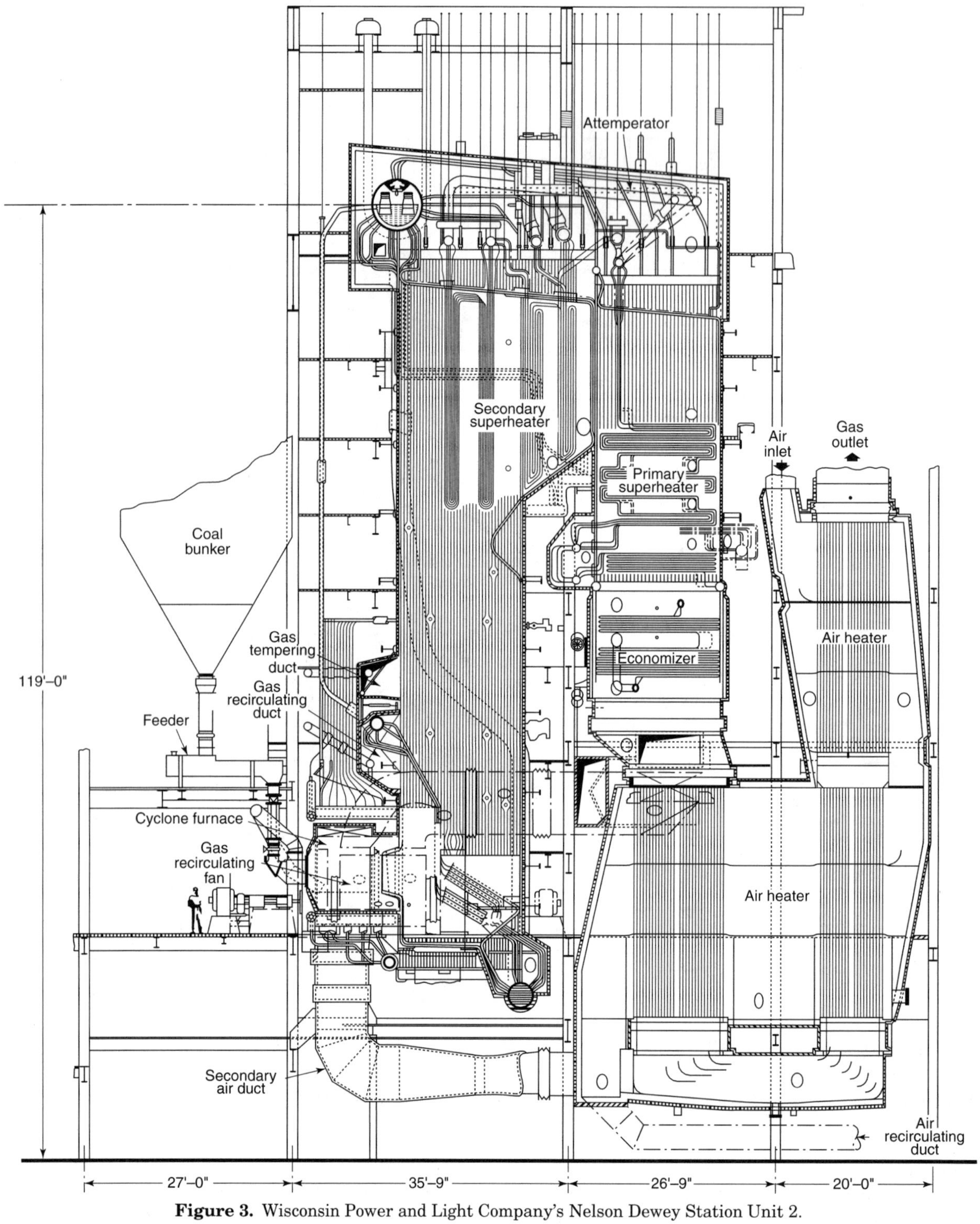

Figure 3. Wisconsin Power and Light Company's Nelson Dewey Station Unit 2.

is approached) and with sudden load changes. Similarly, particulate can increase as the ash and fines contents increase. ("Fines," in this context, are coal particles smaller than about 1.6 mm, or 1/16 in., in diameter.) Conversely, particulate can be reduced significantly when overfire air pressures are increased.

Sulfur Oxides Emissions

Gaseous emissions of sulfur oxides from external coal combustion are largely sulfur dioxide (SO_2) and, to a much smaller extent, sulfur trioxide (SO_3) and gaseous sulfates. These compounds form as the organic and pyritic sulfur in the coal is oxidized during the combustion process. On average, 95% of the sulfur present in bituminous coal will be emitted as gaseous sulfur oxides, whereas somewhat less will be emitted when subbituminous coal is fired. The more alkaline nature of the ash in some subbituminous coal causes some of the sulfur to react to form various sulfate salts that are retained in the boiler or in the fly ash. Generally, boiler size, firing

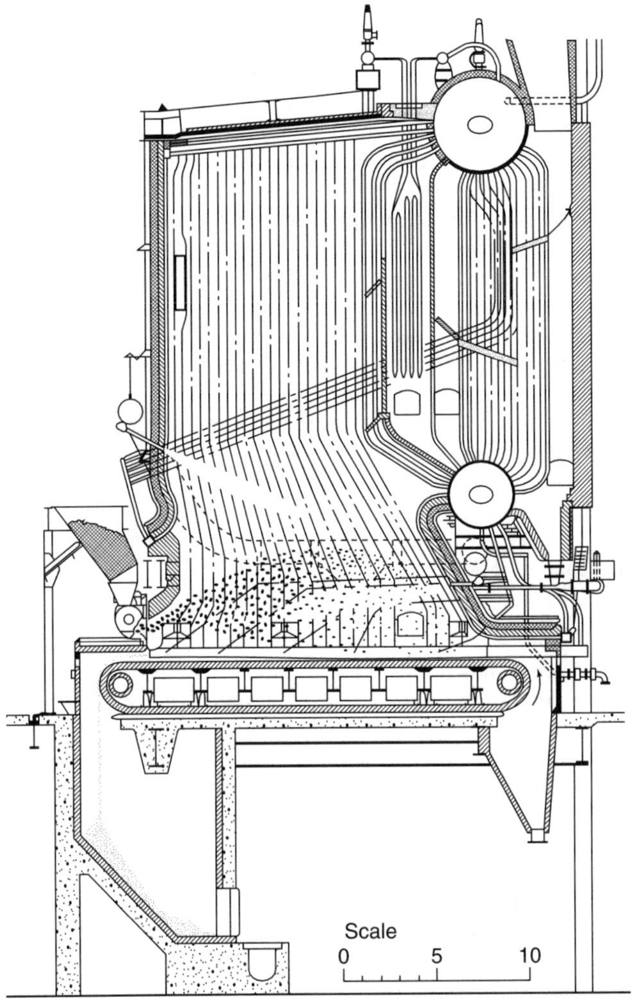

Figure 4. A Steam-Generating Unit Equipped with Continuous-Discharge Type of Spreader Stoker. Capacity, 150,000 Pounds of Steam per Hour at 450 psi and 750°F When Burning Midwest Bituminous Coal. (from *Combustion Engineering*, 5–25, 1948).

configuration, and boiler operations have little effect on the percent conversion of fuel sulfur to sulfur oxides.

Nitrogen Oxides Emissions

Nitrogen oxides (NO_x) are produced in the combustion process by at least two different mechanisms: (1) from the molecular nitrogen in the combustion air (thermal NO_x, and (2) from the nitrogen in the fuel being burned. (fuel NO_x). The relative contributions of these two formation pathways depend on the combustion conditions, the type of boiler, and the type of fuel being burned. Figure 5 shows typical uncontrolled NO concentrations in the flue gas for various types of coal-fired boilers.

The thermal NO_x formation rate is markedly dependent on gas temperature and becomes rapid at gas temperatures of 3000–3600°F. The currently accepted model for thermal NO_x formation was developed by Zeldovich (1946). The most important reactions in the Zeldovich model are as follows:

$$N_2 + O \rightarrow NO + N$$

$$N + O_2 \rightarrow NO + O$$

$$N + OH \rightarrow NO + H$$

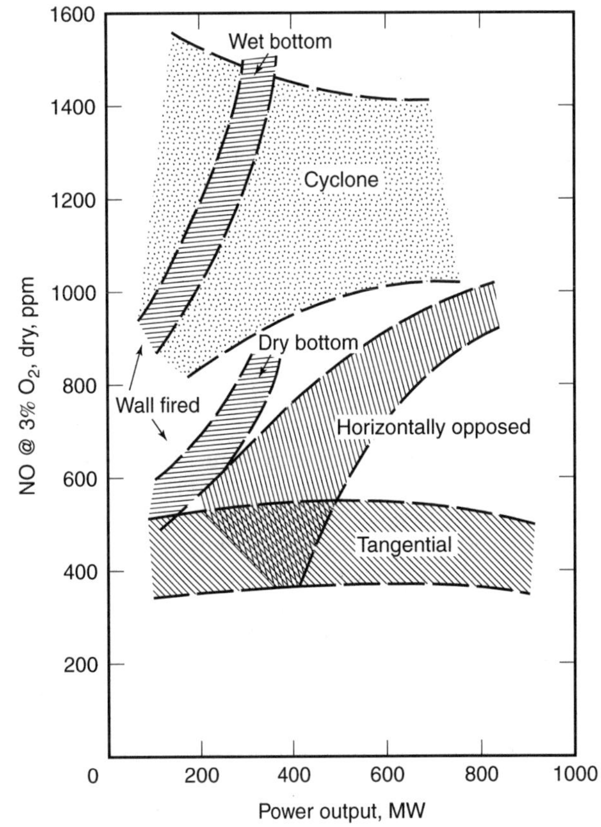

Figure 5. Uncontrolled NO Concentrations for Coal-Fired Boilers (from Ref. 11, Chapter 15).

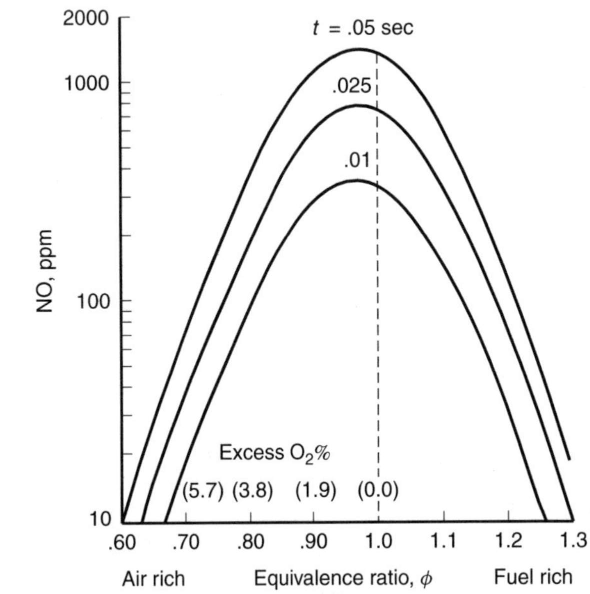

Figure 6. NO versus Fuel Equivalence Ratio and Time (stoichiometric A/F = 16.3, Air Preheat = 650°F (from Ref. 11, Chapter 15, p. 460).

It has been observed that the most important operating variables in the formation of thermal NO_x are (1) peak gas temperature, (2) gas residence time in peak temperature, and (3) fuel–air stoichiometric ratio. In general, a lower operating temperature, gas residence time, and off-stoichiometric fuel–air ratio reduce NO_x formation. Figure 6 shows the simulated effects of these operating variables on the formation of NO_x in a combustion system.

The nitrogen oxides can also be formed in the combustion zone through an alternative mechanism in which molecular nitrogen in the combustion reacts with free hydrocarbon radicals available from the fuel in the combustion zone, as shown in the following reaction.

$$N_2 + CH \rightarrow HCN + N$$

The HCN then combines with free OH radicals to form CN, which is then oxidized to NO via a number of intermediate steps. The nitrogen oxide formed through the pathway is also called "prompt NO." The contribution of prompt NO to the overall NO_x formation is rather small.

If the fuel contains organically bound nitrogen as in the case of most coals (0.5–2.0%) and residual fuel oils (0.1–0.5%), reactions with the oxygen in the combustion air produce NO_x. The amount of nitrogen in the fuel is relatively small compared with that in the combustion air. However, this nitrogen is highly reactive; depending on combustion conditions, the contribution of fuel nitrogen can be significant.

A schematic representation of the relative contributions of the three mechanisms of NO_x formation to the overall NO_x concentration is shown in Figure 7.

Bituminous and subbituminous coals usually contain from 0.5 to 2 weight percent nitrogen, mainly present in aromatic ring structures. Fuel nitrogen can account up to 80% of the total NO_x from coal combustion.

VOC and CO Emissions

Volatile organic compounds (VOCs) and carbon monoxide (CO) are unburned gaseous combustibles that are also emitted from coal-fired boilers, but generally in quite small amounts. However, during startups, temporary upsets, or other conditions preventing complete combustion, unburned combustible emissions may increase dramatically.

The VOC and CO emissions per unit of fuel fired are normally lower from pulverized-coal or cyclone furnaces than from smaller stokers and hand-fired units, where operating conditions are not so well controlled. Measures used for NO_x control can increase CO emissions; so to reduce the risk of explosion, such measures are applied only to the point at which CO in the flue gas reaches a maximum of about 200 ppm. Other than maintaining proper combustion conditions, control measures are not applied to control VOCs and CO.

Organic emissions are either due to constituents present in the coal or are formed as products of incomplete combustion (PIC). The emissions include alkanes, aldehydes, alcohols, benzene, toluene, xylene, and ethyl benzene. Other organic emissions of interest are polychlorinated dibenzo-*p*-dioxins and polychlorinated dibenzofurans (PCDD/PCDF), which are formed through reactions in downstream air pollution control devices. The formation of PCDD/PCDF depends primarily on the flue gas temperature, with maximum potential in the temperature window of 450–650°F.

Organic emissions in condensed phase from coal combustion can be classified either as polycyclic organic matter (POM) or a subset of compounds called polynuclear aromatic hydrocarbon (PAH).

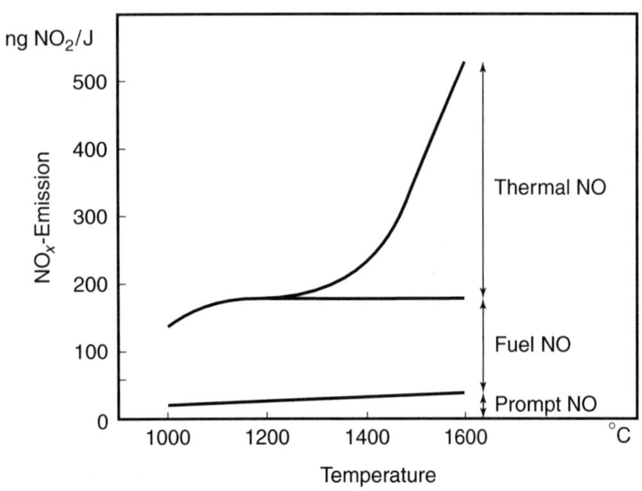

Figure 7. Schematic Representation of Distribution of Formation of NO between Different Formation Mechanisms when Burning Coal, as a Function of Temperature (from Ref. 12).

Trace Metals

Metals are constituents of coal and are emitted as part of fly ash during combustion. Trace metal emissions include antimony, arsenic, beryllium, cadmium, chromium, cobalt, lead, magnesium, manganese, mercury, nickel, and selenium. Some of these metals and metallic compounds are considered hazardous air pollutants (HAP) under Title III of the Clean Air Act Amendment of 1990. Partitioning of the metals in the bottom ash, ash collected in the air pollution control devices, and in the fly ash depends on the volatility of the metals, temperature profile across the system, and available oxygen concentration. The metals can be present in both particulate and gaseous phases.

Currently, mercury emissions from coal-fired utility and industrial boilers are of special concern. Mercury tends to accumulate in certain media and pose hazard to human through multimedia pathways. Mercury emissions from combustion of coal depend on the concentration in coal. Due to its high volatility, almost all mercury in the coal is entrained in the flue gases. Mercury is emitted both as elemental mercury [Hg(0)] and in the divalent oxidized form [Hg(II)]. Also, mercury is emitted both as vapor and bound in particulate. EPA estimates that 20% of the total mercury is emitted as elemental mercury in the vapor phase, 60% of the total mercury is emitted as divalent mercury in the vapor phase, and the balance 20% of the total mercury is emitted as divalent mercury in the particle-bound phase.

Hydrogen Chloride and Hydrogen Fluoride

Combustion of coal also results in emission of chlorine and fluorine, primarily in the form of hydrogen chloride (HCl) and hydrogen fluoride (HF). Free chlorine and fluorine are also present in the flue gases in smaller amounts. Some of these acid gases are adsorbed on the surface of the fly ash and are responsible for leaching of metals in the fly ash.

Greenhouse Gases

The most significant greenhouse gas from coal combustion is carbon dioxide, and nearly 99% of the fuel carbon is converted to carbon dioxide. The conversion is independent of firing configuration but depends on combustion conditions. Poor combustion conditions tend to generate CO instead of CO_2.

Nitrous oxide (N_2O) is also generated during the combustion process through a complex series of reactions, which depends on many factors, such as temperature in the combustion zone and availability of oxygen. Generally, lower than 1600°F and greater than 1% oxygen concentration results in higher N_2O formation. The formation also depends on firing configuration, and FBCs tend to generate significantly higher N_2O than any other type of firing configuration.

Methane emissions are generally due to incomplete combustion and are usually higher during upset conditions or during startups and shutdowns.

Tables 5 through 12 present the emission factors for air pollutants from coal combustion for different types of boilers and firing configurations. The factors are mostly taken from AP-42. Emission factor ratings indicated on the tables reflect the quality of the data with following categorizations (per AP-42): A = excellent; B = above average; C = average; D = below average; and E = poor.

AIR EMISSION CONTROL MEASURES

Particulate Control

The primary kinds of particulate control used for coal combustion are (1) multiple cyclones, (2) ESPs, (3) fabric filters, and (4) venturi scrubbers. The application of particular equipment will depend on the type of furnace, coal properties, and operating conditions.

Prior to the Clean Air Act of 1970, multiple cyclones were generally utilized to control fly-ash emissions. These have generally been replaced in recent years by more efficient control systems or retained in the system as a precleaner upstream of the more efficient ESPs or fabric filters, with the intent of reducing the loading of the abrasive fly-ash particles.

Electrostatic precipitators have been used widely for the control of emissions from steam generation for nearly 60 years, principally on coal-fired boilers. For many years, all fly-ash ESPs were installed downstream of the air heaters at temperatures of 270–350°F (referred to as cold-side ESPs). As a result of the increased combustion of low-sulfur coals with higher resistivities, ESPs have also been installed upstream of the air heaters in a temperature range of 600–750°F to take advantage of the lower fly-ash resistivity at higher temperatures. These units are referred to as hot-side ESPs. Figure 8 illustrates a plan view of a typical cold-side ESP. Applications on coal-fired boilers include ESPs that use the weighted wire electrodes (commonly referred to as the "American" design) and the rigid frame-supported electrodes (referred to as the "European" design).

In the early 1970s, the ESP was the generally preferred choice for a higher-efficiency particulate control device over fabric filters or wet venturi scrubbers. In 1994, it was reported that ESPs accounted for about 91% of all utility particulate controls in the United States. In general, the ESP was a more economical option than the baghouse, particularly on high-sulfur-coal applications where the ESP performance was unaffected by the resistivity of the dust.

The new Clean Air Act of 1990 requires electric utilities to reduce their SO_2 emissions by approximately 10 million tons per year by the year 2000. If utilities choose to do this by using wet scrubbing techniques (downstream of the ESP), the changes may not affect existing ESPs. However, a number of utilities will likely retrofit existing plants with dry SO_2 technologies or switch to low-sulfur coals. Either of these choices will increase dust loadings and change the resistivity of the fly ash, making it more difficult to collect the materials with ESPs.

Offen and Altman[14] concluded in 1991 that:

> The selection of particulate control equipment by U.S. utilities and independent power producers in the 1990s will depend largely on regulatory trends and the technical and commercial success of ongoing particulate control research and development efforts. Well-designed, modem ESPs theoretically can meet emission targets somewhat lower than the current New Source Performance Standards (NSPS). Therefore, if new particulate emission limits are not significantly lower than the current NSPS, utilities will be able to continue to choose either ESPs or baghouses based on considerations such as unit size, fuel type, and boiler type. Requirements for greater particulate reductions or stringent emission limits based on respirable particulate (PM_{10}) may limit the number of sites where an ESP would be acceptable.

Options for upgrading existing ESPs to meet the greater demands are on the horizon and are likely to influence the future application of ESPs.

Electric utilities have made significant progress in recent years in designing and operating fabric filter collectors (baghouses) for the collection of coal fly ash. As a result, baghouses have become an accepted, and sometimes preferred, alternative to ESP. Interest in baghouses will continue to increase because of the increasing stringency of air emission standards, concerns over fine particulates, and the use of dry SO_2 control technologies (duct injection and spray drying). A comparison of capital and levelized costs for ESPs versus baghouses is shown in Figures 9 and 10. In general, the capital costs favored ESPs on high-sulfur coals with 1971 emission standards and favored baghouses in the lower-sulfur applications, with 1979 emission standards.

Cushing et al.[16] reported that as of December 1989, there were 99 baghouses in operation on utility boilers with 21,359 MW of generation capacity, ranging in size from 6 to 860 MW. The largest operating baghouses were two 739-MW units at the Four Corners utility, with 19,872 filter bags in each unit. The oldest unit was reported to be the Sunbury installation, commissioned in 1973. Figure 11 illustrates the rapid growth in the application of baghouses on coal-fired utility boilers since 1972. In 1994, 7% of all coal-fired units were equipped with baghouses. Table 13 summarizes the primary operating characteristics of the baghouses applied to coal-fired boilers. The majority of the baghouses are cleaned by reverse gas cleaning and, in some cases, use sonic horns to assist in the cleaning. Shaking and shake/deflate are also used. Figure 12 shows a schematic of a typical reverse-gas-cleaned baghouse. These baghouses have air–cloth ratios typically ranging from 1.5 to 2.0 acfm/ft^2 with operating pressure drops of 3.5–9.0 in. H_2O with an average of 6.4 in. H_2O.

Design efficiencies for the baghouses ranged from 98% to over 99.9% emission levels from the baghouses with reverse air cleaning ranged from 0.005 to 0.03 lb/MBtu; with reverse air with sonic horn assistance, from 0.008 to 0.125 lb/MBtu; and with shake and deflate cleaning, from 0.007 to 0.07 lb/MBtu. All of the baghouses reported opacities of less than 5.0%.

Table 5. Emission Factors for SO_x, NO_x, and CO from Bituminous and Subbituminous Coal Combustion[a]

	SO_x[b]			NO_x[c]			CO[d,e]		
Firing Configuration	Emission Factor (lb/ton)	Emission Factor (kg/mg)	Emission Factor Rating	Emission Factor (lb/ton)	Emission Factor (kg/mg)	Emission Factor Rating	Emission Factor (lb/ton)	Emission Factor (kg/mg)	Emission Factor Rating
PC, dry bottom, wall-fired,[f] bituminous pre-NSPS[g]	38*S*	19*S*	A	22	11	A	0.5	0.25	A
PC, dry bottom, wall-fired,[f] bituminous pre-NSPS[g] with low-NO_x burner	38*S*	19*S*	A	11	5.5	A	0.5	0.25	A
PC, dry bottom, wall-fired,[f] bituminous NSPS[g]	38*S*	19*S*	A	12	6.0	A	0.5	0.25	A
PC, dry bottom, wall-fired,[f] subbituminous Pre-NSPS[g]	35*S*	17.5*S*	A	12	6.0	C	0.5	0.25	A
PC, dry bottom, wall-fired,[f] subbituminous NSPS[g]	35*S*	17.5*S*	A	7.4	3.7	A	0.5	0.25	A
PC, dry bottom, cell burner fired, bituminous	38*S*	19*S*	A	31	15.5	A	0.5	0.25	A
PC, dry bottom, cell burner fired, subbituminous	35*S*	17.5*S*	A	14	7.0	E	0.5	0.25	A
PC, dry bottom, tangentially fired, bituminous, pre-NSPS[g]	38*S*	19*S*	A	15	7.5	A	0.5	0.25	A
PC, dry bottom, tangentially fired, bituminous, pre-NSPS[g] with low-NO^x burner	38*S*	19*S*	A	9.7	4.85	A	0.5	0.25	A
PC, dry bottom, tangentially fired bituminous, NSPS[g]	38*S*	19*S*		10	5.0	A	0.5	0.25	A
PC, dry bottom, tangentially fired, subbituminous, pre-NSPS[t]	35*S*	17.5*S*	A	8.4	4.2	A	0.5	0.25	A
PC, dry bottom, tangentially fired subbituminous, pre-NSPS[g]	35*S*	17.5*s*	A	7.2	3.6	A	0.5	0.25	A
PC, wet bottom, wall-fired[f], bituminous, pre-NSPS[g]	38*S*	19*S*	A	31	15.5	D	0.5	0.25	A
PC, wet bottom, tangentially fired, bituminous, NSPS[g]	38*S*	19*S*	A	14	7.0	E	0.5	0.25	A
PC, wet bottom wall-fired, subbituminous	35*S*	17.5*S*	A	24	12.0	E	0.5	0.25	A
Cyclone furnace, bituminous	38*S*	19*S*	A	33	16.5	A	0.5	0.25	A
Cyclone furnace, subbituminous	35*S*	17.5*S*	A	17	8.5	C	0.5	0.25	A
Spreader stoker, bituminous	38*S*	19*S*	B	11	6.5	B	5	2.5	A
Spreader stoker, subbituminous	35*S*	17.5*S*	B	8.8	4.4	B	5	2.5	A
Overfeed stoker[h]	38*S* (35*S*)	19*S* (17.5*S*)	B	7.5	3.75	A	6	3.0	B
Underfeed stoker	31*S*	15.5*S*	B	9.5	4.75	A	11	6.5	B
Hand-fed units	31*S*	15.5*S*	D	9.1	4.55	E	275	137.5	E
FBC, circulating bed	—[i]		E	5.0	2.5	D	18	9.0	E
FBC, bubbling bed	—[i]		E	15.2	7.6	D	18	9.0	D

[a] *Source*: AP-42. Factors represents uncontrolled emissions unless otherwise specified and should be applied to coal feed, as fired.

[b] Expressed as SO_2, including SO_2, SO_3, and gaseous sulfates. Factors in parentheses should be used to estimate gaseous SO_x emissions for subbituminous coal. In all cases, S is weight % sulfur content of coal as fired. Emission factor would be calculated by multiplying the weight percent sulfur in the coal by the numerical value preceding S. For example, if fuel is 1.2% sulfur, then $S = 1.2$. On average for bituminous coal, 95% of fuel sulfur is emitted as SO_2, and only about 0.7% of fuel sulfur is emitted as SO_3 and gaseous sulfate. An equally small percent of fuel sulfur is emitted as particulate sulfate. Small quantities of sulfur are also retained in bottom ash. With subbituminous coal, about 10% more fuel sulfur is retained in the bottom ash and particulate because of the more alkaline nature of the coal ash. Conversion to gaseous sulfate appears about the same as for bituminous coal.

[c] Expressed as NO_2. Generally, 95 vol % or more of NO_x present in combustion exhaust will be in the form of NO, the rest NO_2. To express factors as NO, multiply factors by 0.66. All factors represent emissions at baseline operation (i.e. 60 to 110% load and no NO_x control measures).

[d] Nominal values achievable under normal operating conditions. Values 1 or 2 orders of magnitude higher can occur when combustion is not complete.

[e] Emission factors for CO_2 emissions from coal combustion should be calculated using lb CO_2/ton coal $= 72.6C$, where C is the w % carbon content of the coal. For example, if carbon content is 85%, then C equals 85.

[f] Wall-fired includes front and rear wall-fired units, as well as opposed wall-fired units.

[g] Pre-NSPS boilers are not subject to any NSPS. NSPS boilers are subject to Subpart D or Subpart Da. Subpart D boilers are boilers constructed after August 17, 1971, and with a heat input rate greater than 250 million Btu per hour (MMBtu/h). Subpart Da boilers are boilers constructed after September 18, 1978, and with a heat input rate greater than 250 MMBtu/h.

[h] Includes traveling grate, vibrating grate, and chain grate stokers.

[i] SO_2emission factors for fluidized bed combustion are a function of fuel sulfur content and calcium-to-sulfur ratio. For both bubbling bed and circulating bed design, use: lb SO_2/ton coal $= 39.6(S)(\mathrm{Ca/S})^{-1.9}$. In this equation, S is the weight percent sulfur in the fuel and Ca/S is the molar calcium-to-sulfur ratio in the bed. This equation may be used when the Ca/S is between 1.5 and 7. When no calcium-based sorbents are used and the bed material is inert with respect to sulfur capture the emission factor for underfeed stokers should be used to estimate the SO_2 emissions. In this case, the emission factor ratings are E for both bubbling and circulating units.

Table 6. Uncontrolled Emission Factors for PM and Pm$_{10}$ from Bituminous and Subbituminous Coal Combustion[a]

Firing Configuration	Filterable PM[b] Emission Factor (lb/ton)	Filterable PM[b] Emission Factor (kg/mg)	Filterable PM[b] Emission Factor Rating	Filterable PM$_{10}$ Emission Factor (lb/ton)	Filterable PM$_{10}$ Emission Factor (kg/mg)	Filterable PM$_{10}$ Emission Factor Rating
PC-fired, dry bottom, wall-fired	10*A*	5*A*	A	2.3**A**	1.15*A*	E
PC-fired, dry bottom, tangentially fired	10*A*	5*A*	B	2.3*A*[c]	1.15*A*	E
PC-fired, wet bottom	7*A*[d]	3.5*A*	D	2.6*A*	1.3*A*	E
Cyclone furnace	2*A*[d]	1*A*	E	0.26*A*	0.13*A*	E
Spreader stoker	66[e]	33	B	13.2	6.6	E
Spreader stoker, with multiple cyclones, and reinjection	17	8.5	B	12.4	6.2	E
Spreader stoker, with multiple cyclones, no reinjection	12	6.0	A	7.8	3.9	E
Overfeed stoker[f]	16[g]	8	C	6.0	3.0	E
Overfeed stoker, with multiple cyclones[f]	9[h]	4.5	C	5.0	2.5	E
Underfeed stoker	15[j]	7.5	D	6.2	3.1	E
Underfeed stoker, with multiple cyclones	11[h]	6.5	D	6.2[j]	3.1	E
Hand-fed units	15	7.5	E	6.2[k]	3.1	E
FBC, bubbling bed	—		E	—		E
FBC, circulating bed	—		E	—		E

[a]*Source*: AP-42. Factors represent uncontrolled emission unless otherwise specified and should be applied to coal feed, as fired.
[b]Based on EPA Method 5 (front half catch). Where particulate is expressed in terms of coal ash content, the *A* factor is determined by multiplying w % ash content of coal (as fired) by the numerical value preceding the *A*. For example, if coal with 8% ash is fired in a PC-fired, dry bottom unit, the PM emission factor would be 10 × 8, or 80 lb/ton.
[c]No data found; emission factor for PC-fired dry bottom boilers used.
[d]Uncontrolled particulate emissions, when no fly-ash reinjection is employed. When control device is installed, and collected fly ash is reinjected to boiler, particulate from boiler reaching control equipment can increase up to a factor of 2.
[e]Accounts for fly-ash settling in an economizer, air heater, or breaching upstream of control device or stack. (Particulate directly at boiler outlet typically will be twice this level.) Factor should be applied even when fly ash in reinjected to boiler form air heater or economizer dust hoppers.
[f]Includes traveling grate, vibrating grate, and chain grate stokers.
[g]Accounts for fly-ash settling in breaching or stack base. Particulate loadings directly at boiler outlet typicallty can be 50% higher.
[i]Accounts for fly-ash settling in breaching downstream of boiler outlet.
[k]No data found; emission factor for underfeed stoker used.
[m]No date found; use emission factor for spreader stoker with multiple cyclones and reinjection.

Table 7. Condensable Particulate Matter Emission Factors for Bituminous and Subbituminous Coal Combustion[a]

Firing Configuration[b]	Controls[c]	CPM-TOT[d,e] Emission Factor (lb/MMBtu)	CPM-TOT[d,e] Emission Factor Rating	CPM-IOR[d,e] Emission Factor (lb/MMBtu)	CPM-IOR[d,e] Emission Factor Rating	CPM-ORG[d,e] Emission Factor (lb/MMBtu)	CPM-ORG[d,e] Emission Factor Rating
All pulverized coal-fired boilers	All PM controls (without FGD controls)	$0.1S-0.03$[f]	B	80% of CPM-TOT emission factor[e]	E	20% of CPM-TOT emission factor[e]	E
All pulverized coal-fired boilers	All PM controls combined with an FGD controls	0.02	E	ND		ND	
Spreader stoker, traveling grate overfeed stoker, underfeed stoker	All PM controls, or uncontrolled	0.04	C	80% of CPM-TOT emission factor[g]	E	20% of CPM-TOT emission factor[g]	E

[a]*Source*: AP-42. All condensable PM is assumed to be less than 1.0 μm in diameter.
[b]No data are available for cyclone boilers or for atmospheric fluidized bed combustion (AFBC) boilers. For cyclone boilers, use the factors provided for pulverized coal-fired boilers and applicable control devices. For AFBC boilers, use the factors provided for pulverized coal-fired boilers with PM and FGD controls.
[c]FGD = flue gas desulfurization.
[d]CPM-TOT, total condensable particulate matter; CPM-IOR, inorganic condensable particulate matter; CPM-ORG, organic condensable particulate matter; ND, no data.
[e]Factors should be multiplied by fuel rate on a heat input basis (MMBtu), as fired. To convert to lb/ton of bituminous coal, multiply by 26 MMBtu/ton. To convert to lb/ton of subbituminous coal, multiply by 20 MMBtu/ton.
[f]S = coal sulfur percent by weight, as fired. For example, if the sulfur percent is 1.04, then S = 1.04. If the coal sulfur percent is 0.4 or less, use a default emission factor of 0.01 lb/MMBtu rather than the emission equation.

Table 8. Emission Factors for Various Organic Compounds from Controlled Coal Combustion[a]

Pollutant[b]	Emission Factor (lb/ton)	Emission Factor (kg/mg)	Emission Factor Rating
Acetaldehyde	5.7E-04	2.9E-04	C
Acetophenone	1.5E-05	7.5E-06	D
Acrolein	2.9E-04	1.5E-04	D
Benzene	1.3E-03	6.5E-03	A
Benzyl chloride	7.0E-04	3.5E-04	D
Bis(2-ethylhexyl)phthalate (DEHP)	7.3E-05	3.7E-05	D
Bromoform	3.9E-05	2.0E-05	E
Carbon disulfide	1.3E-04	6.5E-03	D
2-Chloroacetophenone	7.0E-06	3.5E-06	E
Chlorobenzene	2.2E-05	1.1E-05	D
Chloroform	5.9E-05	3.0E-05	D
Cumene	5.3E-06	2.7E-06	E
Cyanide	2.5E-03	1.3E-03	D
2,4-Dinitrotoluene	2.8E-07	1.4E-07	D
Dimethyl sulfate	4.8E-05	2.4E-05	E
Ethyl benzene	9.4E-05	4.7E-05	D
Ethyl chloride	4.2E-05	2.1E-05	D
Ethylene dichloride	4.0E-05	2.0E-05	E
Ethylene dibromide	1.2E-06	6.0E-07	E
Formaldehyde	2.4E-04	1.2E-04	A
Hexane	6.7E-05	3.4E-05	D
Isophorone	5.8E-04	2.9E-04	D
Methylbromide	1.6E-04	8.0E-05	D
Methyl chloride	5.3E-04	2.7E-04	D
Methyl ethyl ketone	3.9E-04	1.95E-04	D
Methyl hydrazine	1.7E-04	8.5E-05	E
Methyl methacrylate	2.0E-05	1.0E-05	E
Methyl *tert*-butyl ether	3.5E-05	1.75E-05	E
Methylene chloride	2.9E-04	1.5E-04	D
Phenol	1.6E-05	8.0E-06	D
Propionaldehyde	3.8E-04	1.9E-04	D
Tetrachloroethylene	4.3E-05	2.2E-05	D
Toluene	2.4E-04	1.2E-04	A
1,1,1-Trichloroethane	2.0E-05	1.0E-05	E
Styrene	2.5E-05	1.3E-05	D
Xylenes	3.7E-05	1.9E-05	C
Vinyl acetate	7.6E-06	3.8E-06	E

[a] *Source*: AP-42. Factors were developed from emissions data from ten sites firing bituminous coal, eight sights firing subbituminous coal, and from one site firing lignite. The emission factors are applicable to boilers using both wet limestone scrubbers or spray dryers and an electrostatic precipitator (ESP) or fabric filter (FF). In addition, the factors apply to boilers utilizing only an ESP or FF.

[b] Pollutants sampled for but not detected in any sampling run include: carbon tetrachloride, 2 sites; 1,3-dichloropropyene, 2 sites; N-nitrosodimethylamine, 2 sites; ethylidene dichloride, 2 sites; hexachlorobutadiene, 1 site; hexachloroethane, 1 site; propylene dichloride, 2 sites; 1,1,2,2-tetrachloroethane, 2 sites; 1,1,2-trichloroethane, 2 sites; vinyl chloride, 2 sites; and hexachlorobenzene, 2 sites.

[c] Emission factor should be applied to coal feed, as fired.

Table 9. Emission Factors for Hydrogen Chloride (HCl) and Hydrogen Fluoride (HF) from Coal Combustion[a] Emission Factor Rating: B

Firing Configuration	HCl (lb/ton)/(kg/mg)	HF (lb/ton)/(kg/mg)
PC-fired, dry bottom, all types	1.2 / 0.6	0.15 / 0.75

[a] *Source*: AP-42. The emission factors were developed from bituminous coal, subbituminous coal, and lignite emissions data.

The average bag life was reported to vary from 2.9 years on high-sulfur applications to 4.6 years for low-sulfur applications. Seven of 22 of the systems reported a bag life of more than 6 years. The majority of the installations used woven fiberglass bags with fabric coatings of Teflon®, silicon graphite, or a proprietary acid-resistant finish.

Figures 13 to 18, taken from AP-42, show the extent of particulate emission control achieved by multiple cyclones, ESPs, baghouses, and wet scrubbers for the various types of furnaces. Uncontrolled emission factors are also shown

Table 10. Emission Factor Equations for Trace Elements from Coal Combustion[a] Emission Factor Equation Rating: *A*[b]

Pollutant	Emission Equation ($lb/10^{12}Btu$)[c]
Antimony	$0.92\ [(C/A)PM]^{0.63}$
Arsenic	$3.1\ [(C/A)PM]^{0.85}$
Beryllium	$1.2\ [(C/A)PM]^{1.1}$
Cadmium	$3.3\ [(C/A)PM]^{0.5}$
Chromium	$3.7\ [(C/A)PM]^{0.58}$
Cobalt	$1.7\ [(C/A)PM]^{0.69}$
Lead	$3.4\ [(C/A)PM]^{0.80}$
Manganese	$3.8\ [(C/A)PM]^{0.60}$
Nickel	$4.4\ [(C/A)PM]^{0.48}$

[a]*Source*: AP-42. The equations were developed from emissions data from bituminous coal combustion, subbituminous coal combustion, and from lignite combustion. The equations may be used to generate factors for both controlled and uncontrolled boilers. The emission factor equations are applicable to all typical firing configurations for electric generation (utility), industrial, and commercial/industrial boilers firing bituminous coal, subbituminous coal, and lignite.
[b]AP-42 criteria for rating emission factors were used to rate the equations.
[c]The factors produced by the equations should be applied to heat input. To convert from $lb/10^{12}$ Btu to kg/joules, multiply by 4.31×10^{-16}. C = concentration of metal in the coal, parts per million by weight (ppmwt); A = weight fraction of ash in the coal. For example, 10% ash is 0.1 ash fraction; PM = site-specific emission factor for total particulate matter, $lb/10^6$ Btu.

for comparison. As a result of the concern over emissions in various size ranges, the emissions are presented as a function of particle size and are referred to as cumulative size-specific emission factors.

Sulfur Oxides Control

Several techniques are used to reduce sulfur oxides from coal combustion. One way is to switch to lower-sulfur coals, since sulfur oxides emissions are proportional to the sulfur content of the coal. This alternative may not be possible where lower-sulfur coal is not readily available or where a different grade of coal cannot be fired satisfactorily. In some cases, various cleaning processes may be employed to reduce the fuel's sulfur content. Physical coal cleaning removes mineral sulfur such as pyrite, but is not effective in removing organic sulfur. Chemical cleaning and solvent refining processes are being developed to remove organic sulfur.

The primary development of sulfur dioxide control technology has occurred as a result of the need to control emissions from the coal-fired electric utility power industry and is referred to as flue gas desulfurization (FGD). In 1994, the electric utility power industry in the United States had a generating capacity of approximately 850 GW. Approximately 55% was produced by coal-fired units (see Table 14).

Table 15 presents a summary of FGD systems by process type for 1989 (actual capacity) and 2010 (projected capacity). The three dominant processes are all throwaway-product systems, including the two wet scrubbing systems (limestone

Table 11. Emission Factors for CH_4, TNMOC, and N_2O from Bituminous and Subbituminous Coal Combustion[a]

	CH_4[b]			TNMOC[b,c]			N_2O		
Firing Configuration	Emission Factor (lb/ton)	Emission Factor (kg/mg)	Emission Factor Rating	Emission Factor (lb/ton)	Emission Factor (kg/mg)	Emission Factor Rating	Emission Factor (lb/ton)	Emission Factor (kg/mg)	Emission Factor Rating
PC-fired, dry bottom, wall fired	0.04	0.02	B	0.06	0.03	B	0.03	0.015	B
PC-fired, dry bottom, tangentially fired	0.04	0.02	B	0.06	0.03	B	0.08	0.04	B
PC-fired, wet bottom	0.05	0.025	B	0.04	0.02	B	0.08	0.04	E
Cyclone furnace	0.01	0.005	B	0.11	0.055	B	0.09[c]	0.045	E
Spreader stoker	0.06	0.03	B	0.05	0.025	B	0.04	0.02	E
Spreader stoker, with multiple cyclones, and reinjection	0.06	0.03	B	0.05	0.025	B	0.04	0.02	E
Spreader stoker, with multiple cyclones, no reinjection	0.06	0.03	B	0.05	0.025	B	0.04	0.02	E
Overfeed stoker	0.06	0.03	B	0.05	0.025	B	0.04	0.02	E
Overfeed stoker, with multiple cyclones	0.06	0.03	B	0.05	0.025	B	0.04	0.02	E
Underfeed stoker	0.8	0.4	B	1.3	0.65	B	0.04	0.02	E
Underfeed stoker, with multiple cyclone	0.8	0.4	B	1.3	0.65	B	0.4	0.02	E
Hand-fed units	5	2.5	E	10	5	E	0.04	0.02	E
FBC, bubbling bed	0.06[h]	0.03	E	0.05[h]	0.025	E	3.5	1.75	B
FBC, circulating bed	0.06	0.03	E	0.05	0.025	E	3.5	1.75	B

[a]*Source*: AP-42. Factors represent uncontrolled emissions unless otherwise specified and should be applied to coal feed, as fired.
[b]Nominal values achievable under normal operating conditions; values 1 or 2 orders of magnitude higher can occur when combustion is not complete.
[c]TNMOC are expressed as C_2 to C_{16} alkane equivalents. Because of limited data, the effects of firing configuration on TNMOC emission factors could not be distinguished. As a result, all data were averaged collectively to develop a single average emission factor for pulverized coal units, cyclones, spreaders, and overfeed stokers.

Table 12. Default CO_2 Emission Factors for U.S. Coals[a] Emission Factor Rating: C

Coal Type	Average %C[b]	Conversion Factor[c]	Emission Factor[d] lb/ton coal	Emission Factor[d] kg/mg coal
Subbituminous	66.3	72.6	4810	2405
High-volatile bituminous	75.9	72.6	5510	2755
Medium-volatile bituminous	83.2	72.6	6040	3020
Low-volatile bituminous	86.1	72.6	6250	3125

[a] *Source*: AP-42. This table should be used only when an ultimate analysis is not available. If the ultimate analysis is available, CO_2 emissions should be calculated by multiplying the % carbon (%C) by 72.6. This resultant factor would receive a quality rating of B.

[b] Based on average carbon contents for each coal type (dry basis) based on extensive sampling of U.S. coals.

[c] Based on the following equation:

$$\frac{44 \text{ ton CO}_2}{12 \text{ ton C}} \times 0.99 \times 2000 \frac{\text{lb CO}_2}{\text{ton CO}_2} \times \frac{1}{100\%} = 72.6 \frac{\text{lb CO}_2}{\text{ton \%C}}$$

where 44 = molecular weight of CO_2; 12 = molecular weight of carbon; and 0.99 = fraction of fuel oxidized during combustion (Ref. 16).

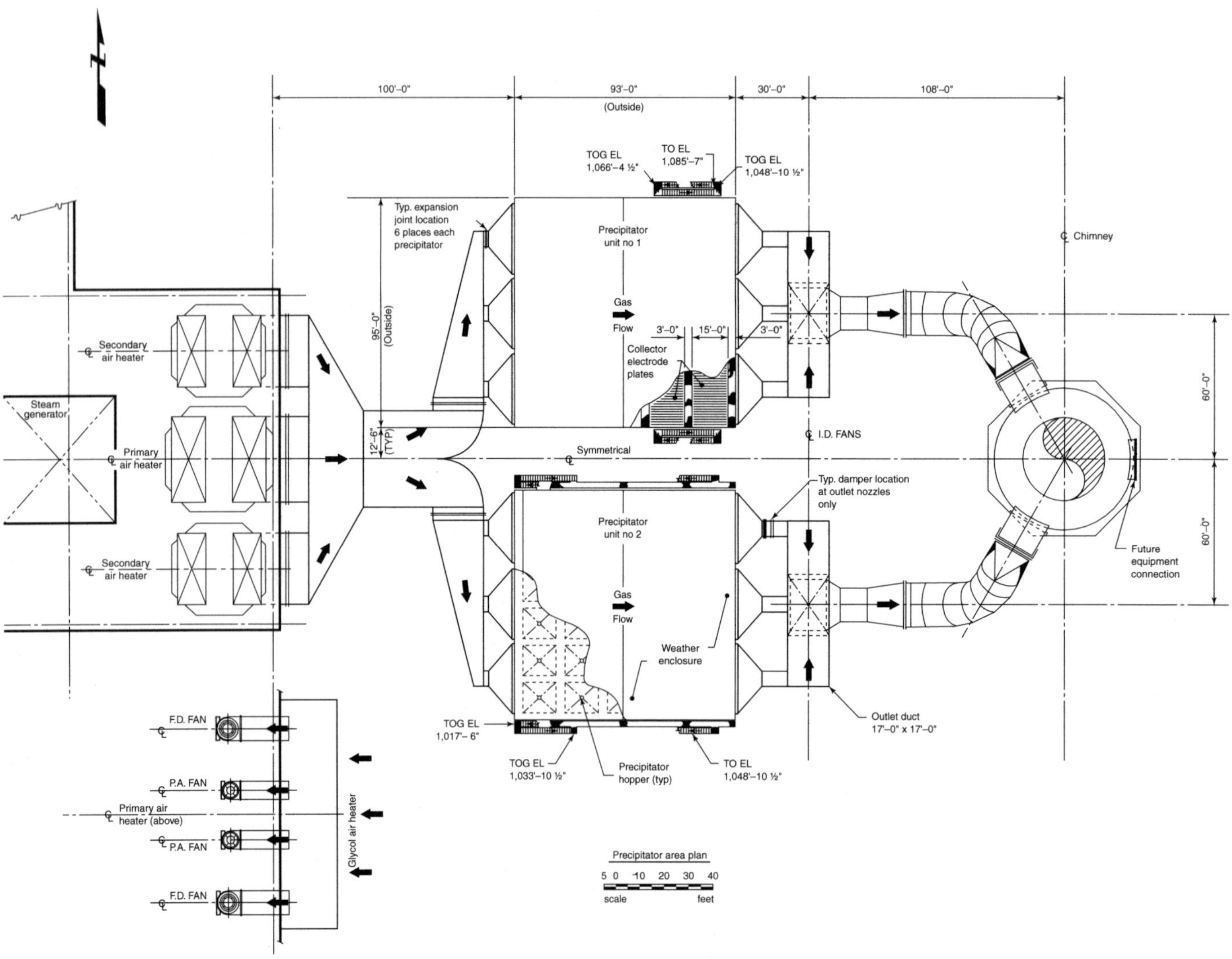

Figure 8. European Coal-Side ESP, Plan View (from Ref. 13).

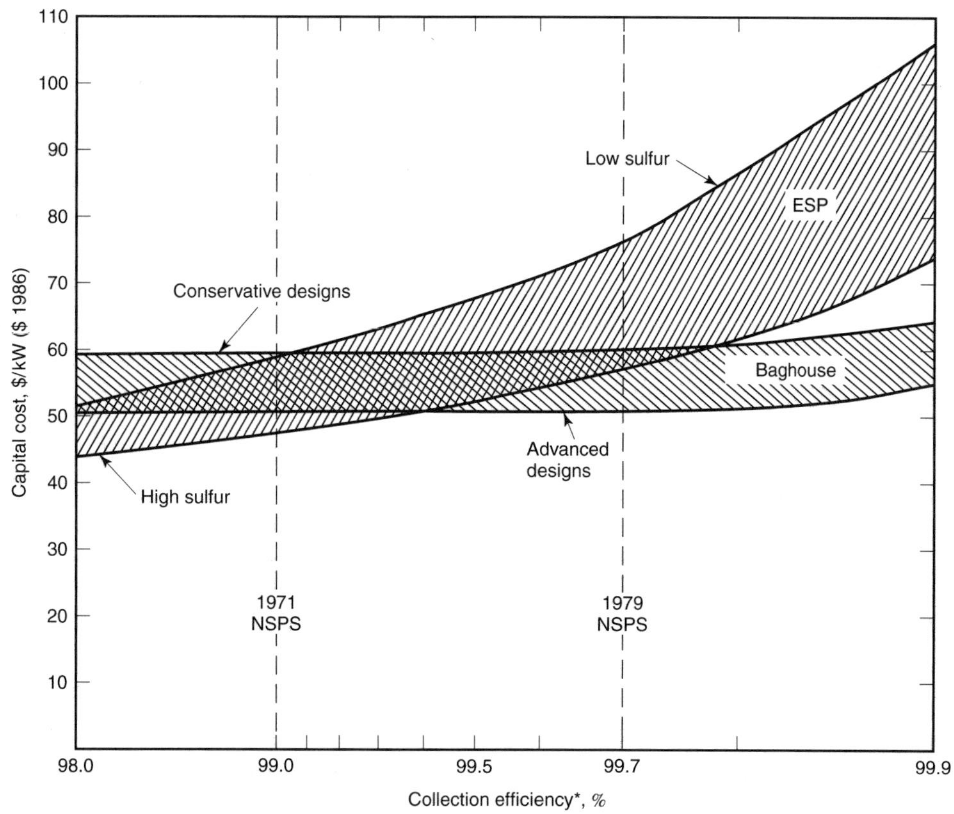

* Based on a coal rated at 10,000 Btu/lb and containing 10% ash. the 1971 NSPS is 0.1 lb/10^6 Btu, and the 1979 NSPS is 0.03 lb/10^6 Btu.

Figure 9. Capital Costs for Baghouses and ESPs, Expressed as a Function of Nominal Collection Efficiency (from Ref. 15, p. I-5).

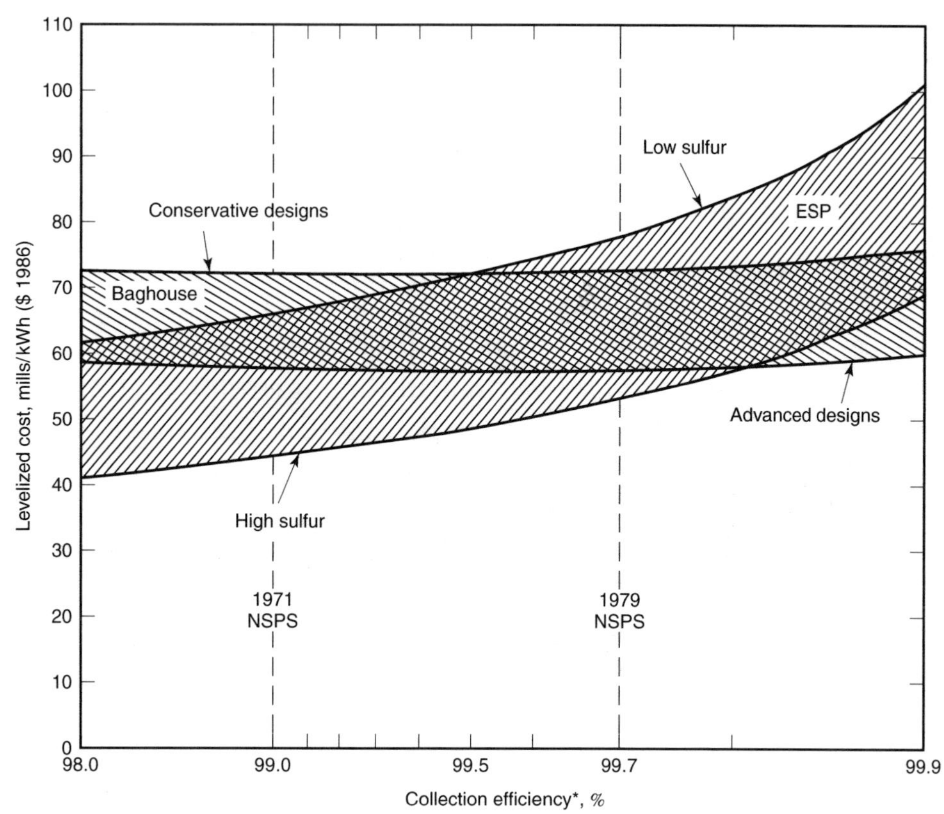

* Based on a coal rated at 10,000 Btu/lb and containing 10% ash. the 1971 NSPS is 0.1 lb/10^6 Btu, and the 1979 NSPS is 0.03 lb/10^6 Btu.

Figure 10. Levelized Cost for Baghouses and ESPs, Expressed as a Function of Nominal Collection Efficiency (from Ref. 15, p. I-6).

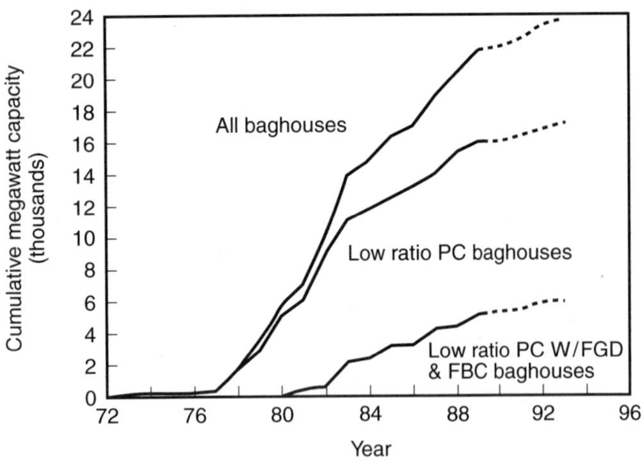

Figure 11. Chronological History of Cumulative Megawatt Capacity of Utility Baghouses. Data Are Presented for All Baghouses and Separately for Low-Ratio Baghouses Downstream from Pulverized-Coal Boilers and Low-Ratio Baghouses Downsteam from Dry FGD Systems of Pulverized-Coal Boilers and Fluidized-Bed Boilers. Data for Baghouses in Design or Construction Are Shown beyond 1989 (from Ref. 16).

and lime) with 48.7% and 16.3% of the capacity and the lime spray drying system with 8.8% of the capacity.

The current NSPS for coal-fired utility boilers requires from 70% up to 90% for SO_2 removal, depending on the sulfur content of the coal. In 1990, the average design efficiency for new and retrofit systems was 82% and 76%, respectively.

The principal types of absorbers used in the wet scrubbing systems include venturi scrubber absorbers, static packing scrubbers, moving-bed absorbers, tray tower absorbers, and spray tower absorbers. In the case of the venturi scrubber absorber, the venturi (for particulate removal) is also an integral part of the absorber. The particulate collection (using either an ESP or baghouse) precedes the wet scrubber to minimize operational problems. Table 16 provides a summary of typical design parameters for lime scrubber/absorbers. A flowchart for a typical lime FGD system (wet scrubber) is shown in Figure 19. The limestone scrubber is similar except that it does not involve the lime slaking step, but instead includes limestone handling and crushing (see Figure 20).

In applications of spray dryer technology for FGD, the particulate collector is downstream and is considered an integral part of the FGD system. The major equipment items found in a typical spray dryer scrubbing system are the spray dryer absorber, the particulate-collection system, reagent and slurry preparation and handling equipment, solids transfer and recycle equipment, and process control and instrumentation.

Figure 21 shows a typical utility application that recycles both spray dryer and baghouse (or ESP) products. Applications may or may not use partial recycle of the reacted products. The spray dryer absorber provides the initial contact between the atomized reactive alkali and the acid-gas contaminants. There are two types of atomizers: rotary disks or wheels and dual-fluid pneumatic nozzles. In each case, the slurry is atomized as droplets into the dryer, reacts with the SO_2, and dries to a fine powder, which is then carried over into the ESP or fabric filter and removed. The particulate collector also serves as an additional contact point between the dried reactants and SO_2, providing additional removal.

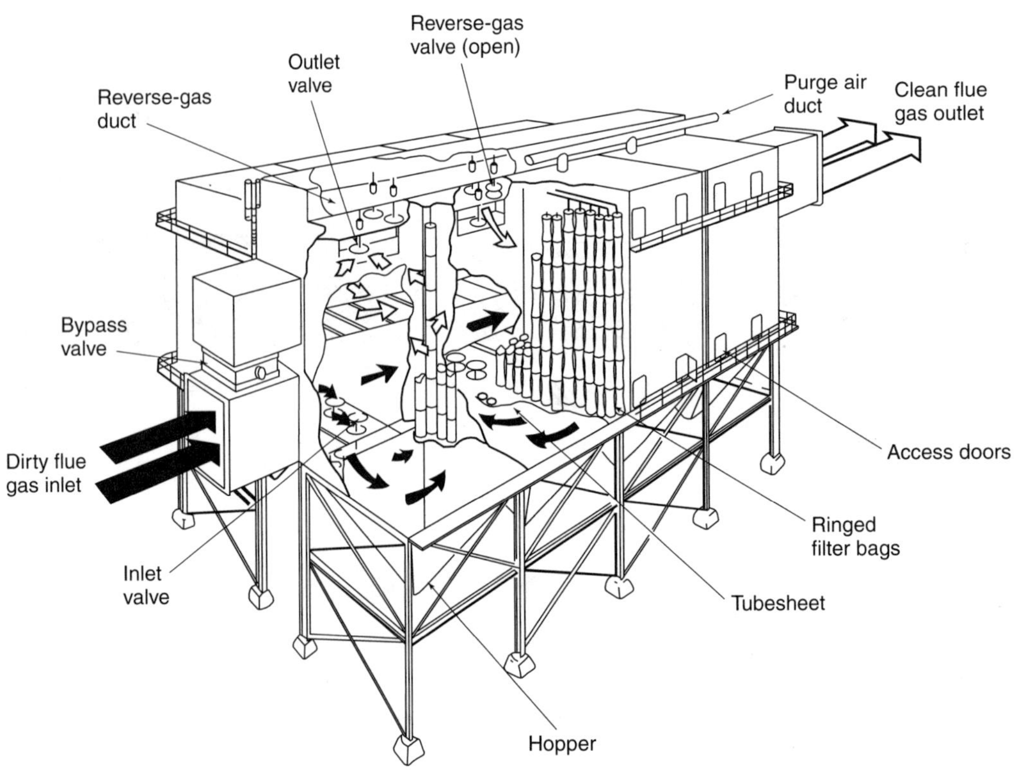

Figure 12. Cutaway View of a Typical Reverse-Gas-Cleaned Baghouses; the Compartment in the Foreground is Filtering Flue Gas While the Adjacent Compartment is Being Cleaned (from Ref. 15, p. 2–5).

Table 13. Baghouse Performance Data

Plant Generating Capacity (MW)	Coal Type[a]	Coal Sulfur Content (%)	Bag Cleaning Method[b]	Gas Temp. (°F)	Flange/Flange Pressure Drop (in. H_2O)	Tube-Sheet Pressure Drop (in. H_2O)	Gross Air/Cloth Ratio[c]	Dust-Cake Areal Density (lb/ft^2)	Emission Rate (lb/MBtu)	Stack Opacity (%)
					Pulverized-coal boilers					
150	WS	0.24	RG	325	7.5–8.0	—	1.95	—	—	—
85	WS	0.37	RG	—	6.0	—	1.77	—	—	—
223	WS	0.37	RG	270	7.0	6.0	1.58	—	0.012	—
223	WS	0.37	RG	282	6.0	—	1.81	—	—	—
405	WS	0.41	RG	267–305	5.0–5.5	4.6–4.8	1.65	0.78	—	0.5–2.0
447	WS	0.43	RG	273–306	3.5–5.0	2.5–3.5	1.72	0.35	0.0045	—
840	WS	0.43	RG	260–280	5.0–6.0	3.8–4.5	1.89	0.24	—	2.0–3.0
245	WS	0.47	RG	320	4.0–5.0	—	1.46	—	0.01	—
24	WS	0.49	RG	360	6.0	—	1.65	—	—	—
110	WS	0.52	RG	290	6.0–7.1	—	1.80	—	0.015	—
150	WS	0.52	RG	283–296	5.0–5.2	4.2–4.7	1.49	0.86	0.013	3.0–4.0
295	WS	0.52	RG	309	4.0–5.0	3.5–4.5	1.97	0.35	—	3.0
30	WS	0.61	RG	290	6.0–7.0	—	1.90(*D*)	—	—	—
565	WS	0.3	RG/S	275	8.0	5.6–5.8	1.7	0.35	0.03	3.0–5.0
565	WS	0.3	RG/S	275	8.0	—	1.7	0.28	0.023	3.0–5.0
254	WS	0.33	RG/S	230	5.5–6.8	4.5–5.5	1.98(*D*)	0.29	—	1.0
570	WS	0.45	RG/S	325	5.5–6.5	4.0–5.0	1.91	1.19	0.008	1.0–2.0
44	WS	0.52	RG/S	290	6.5–8.2	—	2.09(*D*)	—	0.016	—
100	WS	0.52	RG/S	290	4.2–6.0	—	1.98(*D*)	—	—	—
166	WS	0.6	RG/S	315	5.5–6.5	—	2.0(*D*)	—	—	—
44	WS	0.61	RG/S	290	6.0–8.0	—	1.93(*D*)	—	—	—
739	WS	0.69	RG/S	240–280	4.8–5.5	4.0–4.4	1.50	0.64	0.023	—
185	EB	0.85	RG/S	301	5.0–6.5	2.5–3.5	1.76	0.32	0.209	3.0
185	EB	0.86	RG/S	305	5.7	2.7	1.87	—	0.018	3.0–5.0
185	EB	0.87	RG/S	300	5.0–6.5	2.5–3.5	1.91	—	0.036	3.0–5.0
79	AP	1.79	RG/S	325	6.0	—	1.71	—	—	—
350	EB	1.83	RG/S	—	5.0–9.0	4.0–8.0	1.83(*D*)	—	—	—
191	EB	2.2	RG/S	304	7.0	—	1.5	—	0.039	—
191	EB	2.4	RG/S	303	7.0	—	1.5	—	0.125	—
87.5	AP	2.6	RG/S	400	3.5	—	1.89(*D*)	—	0.085	—
87.5	AP	2.7	RG/S	400	3.5	—	1.89(*D*)	—	0.085	—
384	WS	0.35	S/D	305	9.0	8.0	3.2	0.23	0.3	2.0–4.0
384	WS	0.36	S/D	320	7.5	—	2.8	—	0.051	—
593	TL	0.43	S/D	350	9.0–13.0	7.0–11.5	2.6	—	—	—
593	TL	0.49	S/D	350	9.0–13.0	7.0–11.5	2.6	—	—	—
(79)	AP	1.79	S/D	350	6.0	—	1.9	—	0.01–0.07	—
					Pulverized-coal boilers with dry FGD systems					
279	WS	0.31	RG	185	4.0	—	1.58	—	—	—
319	WS	0.36	RG	165	6.0	—	1.60(*D*)	0.09	—	—
44	WS	0.52	RG	180	6.0	—	1.54(*D*)	—	0.03	—
860	WS	0.6	RG	165	9.8	—	2.00(*D*)	—	0.024	—
415	NDL	1.08	S/D	200	4.0–8.0	—	2.24(*D*)	—	0.018	—
					Fluidized-bed combustion boilers.					
160	EB	0.33	RG	290	7.2	6.8	1.53	—	<0.03	—
110	WS	0.39	S/D	294	5.0–6.5	3.7–5.2	2.4–2.9	0.23	0.0072	—

[a]WS = western subbituminous, WB = western bituminous, AP = anthracite/petroleum coke, TL = Texas lignite, EB = eastern bituminous, NDL = North Dakota lignite.

[b]RG = reverse gas, RG/S = reverse gas with sonic assistance, S/D = shake/deflate.

[c]*D* = design air-to-cloth ratio.

Source: Reference 16.

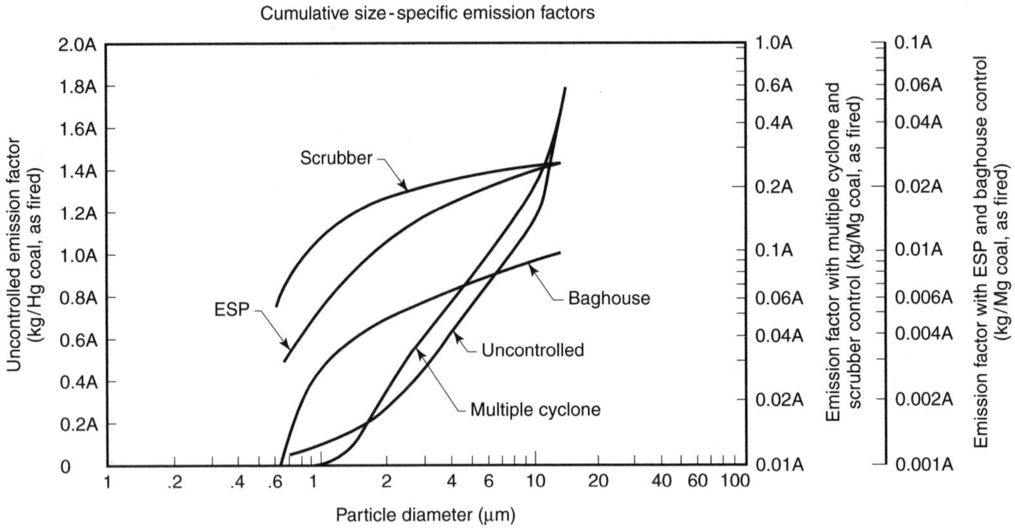

Figure 13. Emission Factors for Dry Bottom Boilers Burning Pulverized Bituminous Coal (from Ref. 6).

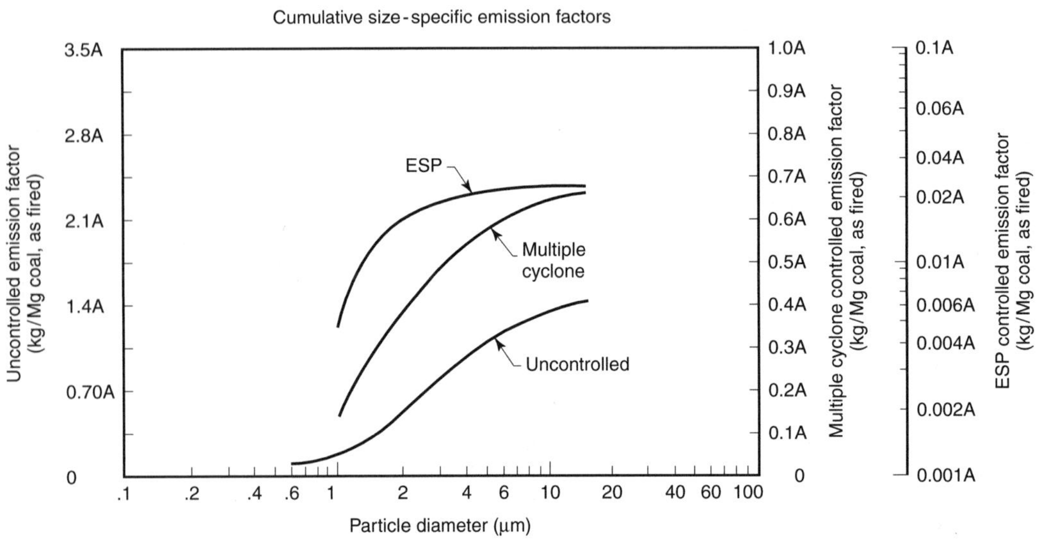

Figure 14. Emission Factors for Wet Bottom Boilers Burning Pulverized Bituminous Coal (from Ref. 6).

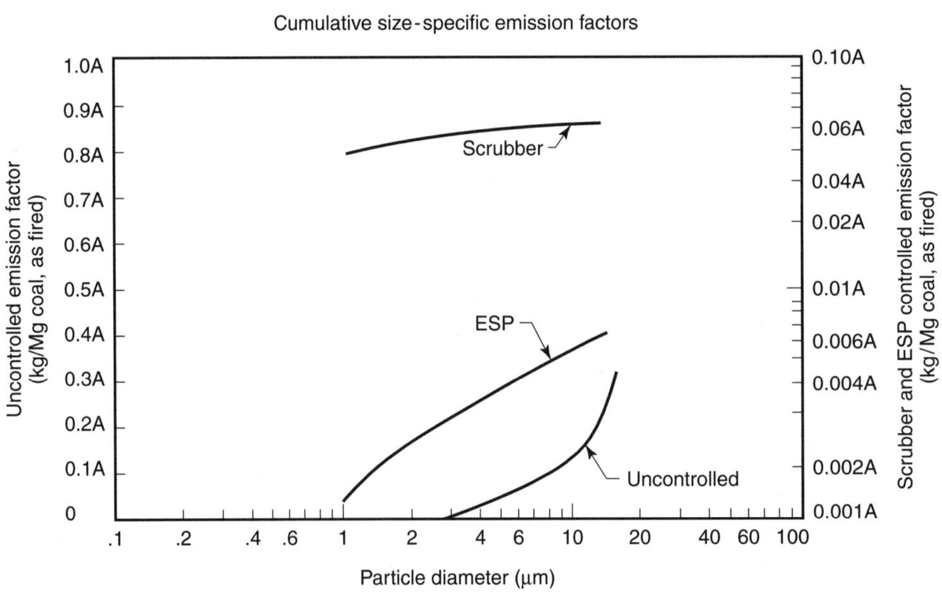

Figure 15. Emission Factors for Cyclone Furnaces Burning Bituminous Coal (from Ref. 6).

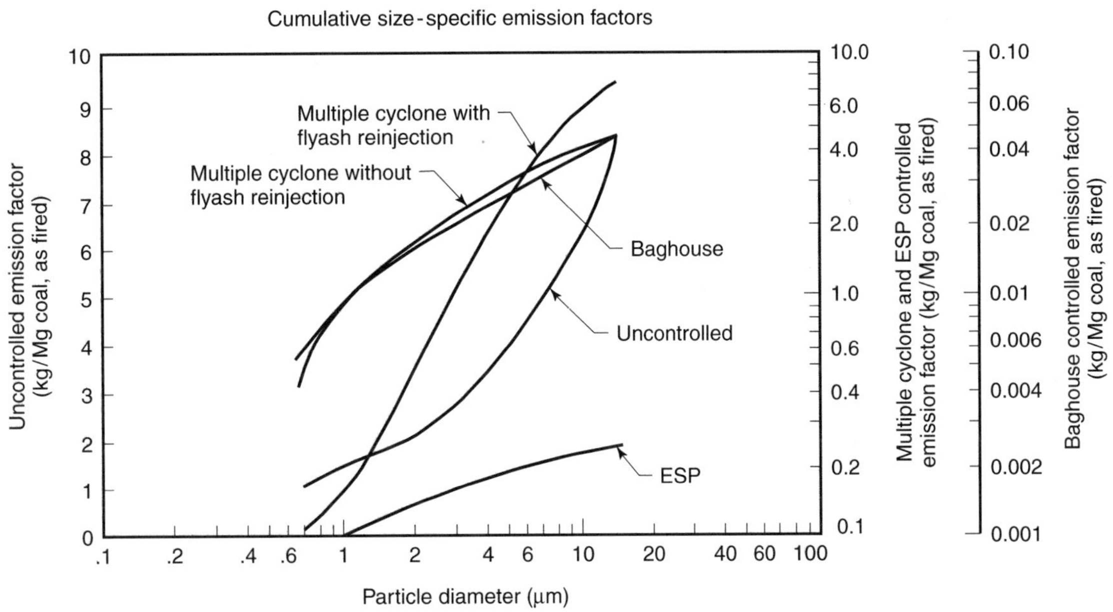

Figure 16. Emission Factors Spreader Stokers Burning Bituminous Coal (from Ref. 6).

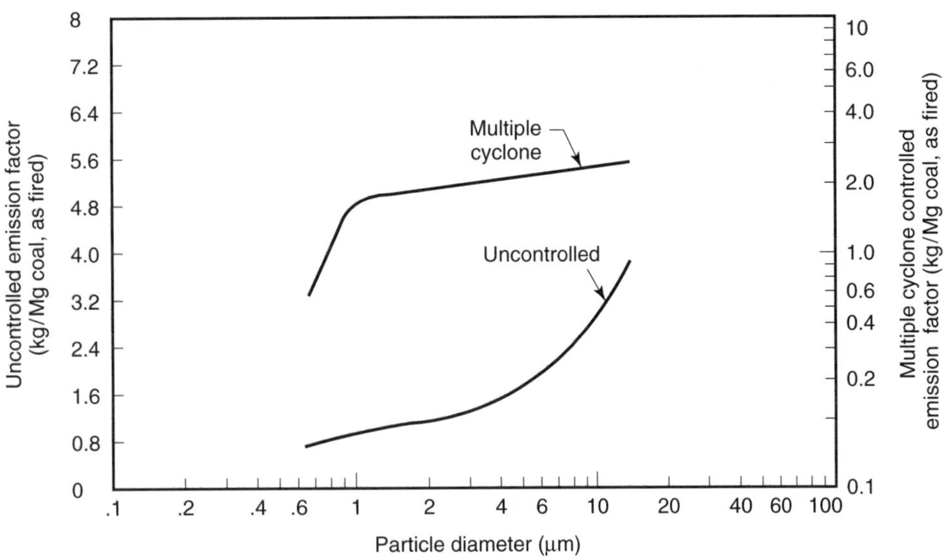

Figure 17. Emission Factors for Overfeed Stokers Burning Bituminous Coal (from Ref. 6).

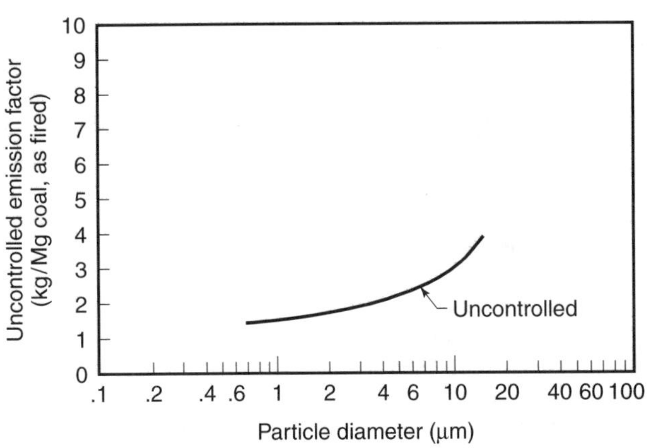

Figure 18. Emission Factors for Underfeed Stokers Burning Bituminous Coal (from Ref. 6).

Table 14. Fossil-Fuel Power Generation Sources

	Coal	Oil	Gas	Other	Total Approximate Capacity (GW)
1994	55%	19%	7%	19%	850

The chosen atomization method affects the design of the spray dryer absorber vessel, including the physical dimensions. For a rotary-atomizer type of spray dryer, which projects the droplets radially outward and perpendicular to the gas flow, the length-to-diameter ratio of the dryer (L/D) is typically 0.8 to 1. Figure 22A illustrates two typical configurations of rotary atomizer spray dryers. The droplets (which range in size from 25 to 150 μm) decelerate rapidly owing to the drag forces of the

Table 15. Summary of FGD Systems by Process (Percentage of Total MW)

Process	By-product	December 1989	December 2010
	Throwaway product		
Wet scrubbing			
Dual alkali		3.4	2.3
Lime		16.3	13.5
Lime/alkaline fly ash		7.0	4.9
Limestone		48.2	43.9
Limestone/alkaline fly ash		2.4	1.6
Sodium carbonate		4.0	3.3
Spray drying			
Lime		8.8	7.9
Sodium carbonate		0	0.5
Reagent type not selected		0.7	2.1
Dry injection			
Lime		0.2	0.1
Sodium carbonate		0	1.6
Reagent type not selected		0	0.2
Process not selected		0	2.2
	Saleable product		
Wet scrubbing			
Lime	Metals/fly ash/other	<0.1	<0.1
Limestone	Gypsum	4.1	4.6
Magnesium oxide	Sulfuric acid	1.4	1.0
Wellman Lord	Sulfuric acid	3.1	2.1
Spray drying			
Lime	Dry scrubber waste	0	0.3
Process undecided		0	7.8
Total		100.0	100.0

Source: Reference 17.

downward-moving flue gas and eventually attain the speed and direction of the flue gas. To avoid wall deposition, the designed radial distance between the atomizer and the dryer wall must be sufficient to allow for adequate drying of the largest droplets. This is accomplished by proper choice of the L/D, droplet size, and residence time.

For a two-fluid pneumatic nozzle type of spray dryer, which atomizes the droplets in the direction of the gas flow (downward), the L/D is typically 2 : 1. In this case, sidewall deposition is a minor problem. Two examples of these spray dryers are shown in Figure 22B. Experience has shown that with the above L/D ratios, spray dryers on flue-gas systems can be operated with design residence times of 10–12 sec, based on the actual outlet flow rate conditions [i.e., residence time (min) = dryer volume/acfm]. Although the designed residence time for most spray dryers is 10–12 sec, few systems operate at 100% of the design flow rate. The actual operating residence time of most systems is 12–15 sec.

Table 17 summarizes selected spray dryer applications and the operating characteristics of each. Typically, industrial boiler spray dryers have diameters of 25–30 ft, whereas utility spray dryers have diameters of 40–50 ft. In most cases, multiple spray dryers are required to handle the large volume of flue gas found in utility boilers.

The spray dryer FGD system is incomplete without an appropriate means of particulate-matter collection. Not only is a well-designed particulate-matter control system needed to meet particulate-matter and opacity emissions requirements, but it can help to meet acid-gas-removal requirements. Acid gases are removed when the flue gas comes into contact with lime-containing particles and encounters the collected particulate matter in the fabric filter or ESP.

The four types of particulate-matter collectors most commonly employed in the removal of the spray-dried products and fly ash downstream of the spray dryer are fabric filters with reverse-air cleaning, fabric filters with reverse-air cleaning and assistance from shaking or from sonic horns, fabric filters with pulse-jet cleaning, and ESPs. The selection of the type of dust collector depends on many factors, such as particulate-matter emission limits, overall acid-gas removal requirements, client and/or vendor preference, and the system flow rate. Each offers advantages and disadvantages that must be evaluated for each site.

The fabric filter has been used in the majority of new installations, whereas ESPs have been used in a few large utility plants and in one retrofit plant. The high percentage of fabric filters found in new installations of dry scrubbers is due to the inherently greater ability of the fabric filter than that of the ESP to remove acid gases. The fabric filter has been observed to remove 15–30% of the SO_2, depending on the operating conditions of the spray dryer and the filter.

Although the characteristics of the fly ash, the presence of spray dryer products, and the flue-gas temperature exiting the spray dryer are considerably different from conditions in a system without a spray dryer, many systems have bypass provisions or may need to be operated without cooling in the spray

Table 16. Summary of Scrubber/Absorber Designs at Various Operating Units

Plant Name	No. of Modules per Unit	% Capacity per Module	Type of Module	L/G(gpm/1000 acfm) Presaturator/ Scrubber	L/G(gpm/1000 acfm) Absorber	Gas Flow per Module (acfm)	åP (in. H_2O)	Gas Velocity (fps)	Materials of Construction Presaturator/ Scrubber	Materials of Construction Absorber	Materials of Construction Internals
Pleasants 1 and 2	4	25	Venturi/tray tower	14–17	35	600,000	6	10.4	CS, inorganic lined	Cs, rubber lined	316L SS
Four Corners 1,2, and 3	2	50	Venturi	25	—	407,000 (1 & 2)	12	NR	CS, 316L SS with polyester lining	—	316L SS
R.D. Green 1 and 2	2	50	Spray tower	NR[a]	45	500,000	3	9.2	CS, inorganic lined	CS, inorganic lined	316L SS
Conesville 5 and 6	2	60	Moving bed	0.55	57	500,000	6	10	Carpenter 20	CS, rubber lined	CS, rubber lined
Coal Creek 1 and 2	4	33	Spray tower	—	60	685,000	4.5	10.6	—	316L SS	316L SS
Elrama Station	5 (total)	25	Venturi	—	40	550,000	11	NR	—	CS, polyester lined	316L SS
Phillips Station	5 (total)	20	Venturi	—	40	547,000	16	40	—	CS, polyester lined	316L SS
Hawthorn 3 and 4	2	60	Marbel bed	26	—	250,000	12	10	316L SS and CS polyester lined	—	316L SS
Green River Station	1 (total)	100	Venturi/moving bed	34	34	288,000	7/4	14	CS, inorganic lined	CS, inorganic lined	316L SS
Cane Run 4	2	50	Venturi/moving bed	NR	60	368,000	4	11.5	CS, inorganic lined	CS, inorganic lined	316L SS
Mill Creek 3	4	30	Venturi/moving bed	5	60	400,000	11	10	CS, 316L SS	CS, inorganic lined	316L SS
Paddy's Run 6	2	50	Marbel bed	—	16.5	175,000	11.5	9	—	CS, polyster lined	316L SS
Clay Boswell 4	4	33	Venturi/spray tower	20	50	640,000	20	12	CS, rubber lined	CS, rubber lined	CS, rubber lined and 316L SS
Milton R. Young 2	2	56	Spray tower	—	80	859,000	8	9	—	CS, polyester lined	—
Colstrip 1 and 2	3	40	Venturi/spray tower	15	18	426,000	17/1	200/8.7	CS, polyester and acid brick lined	CS, polyester lined	316L SS
Bruce Mansfield 1 and 2	6	17	Venturi/venturi	23	20	558,000	18/6	200/100	CS, polyester lined	CS, polyester lined	316L SS
Bruce Mansfield 3	5	25	Horizontal spray chamber	0.7	88	992,000	2.8	22	Inconel 625	CS, polyester lined	316L SS
Hunter 1 and 2	4	25	Spray tower	—	70	330,000	5–6	9.6	—	CS, polyester lined	FRP
Huntington 1	4	25	Spray tower	—	70	330,000	5–6	9.6	—	CS, polyester lined	FRP

[a]NR = Not Reported

Note: gpm/1000 acfm = 0.1337 m^3/km^3, acfm = 0.00047 m^3/s; in. H_2O = 0.2488 kPa. ft/sec = 0.3048 m/sec.

Source: Reference 19.

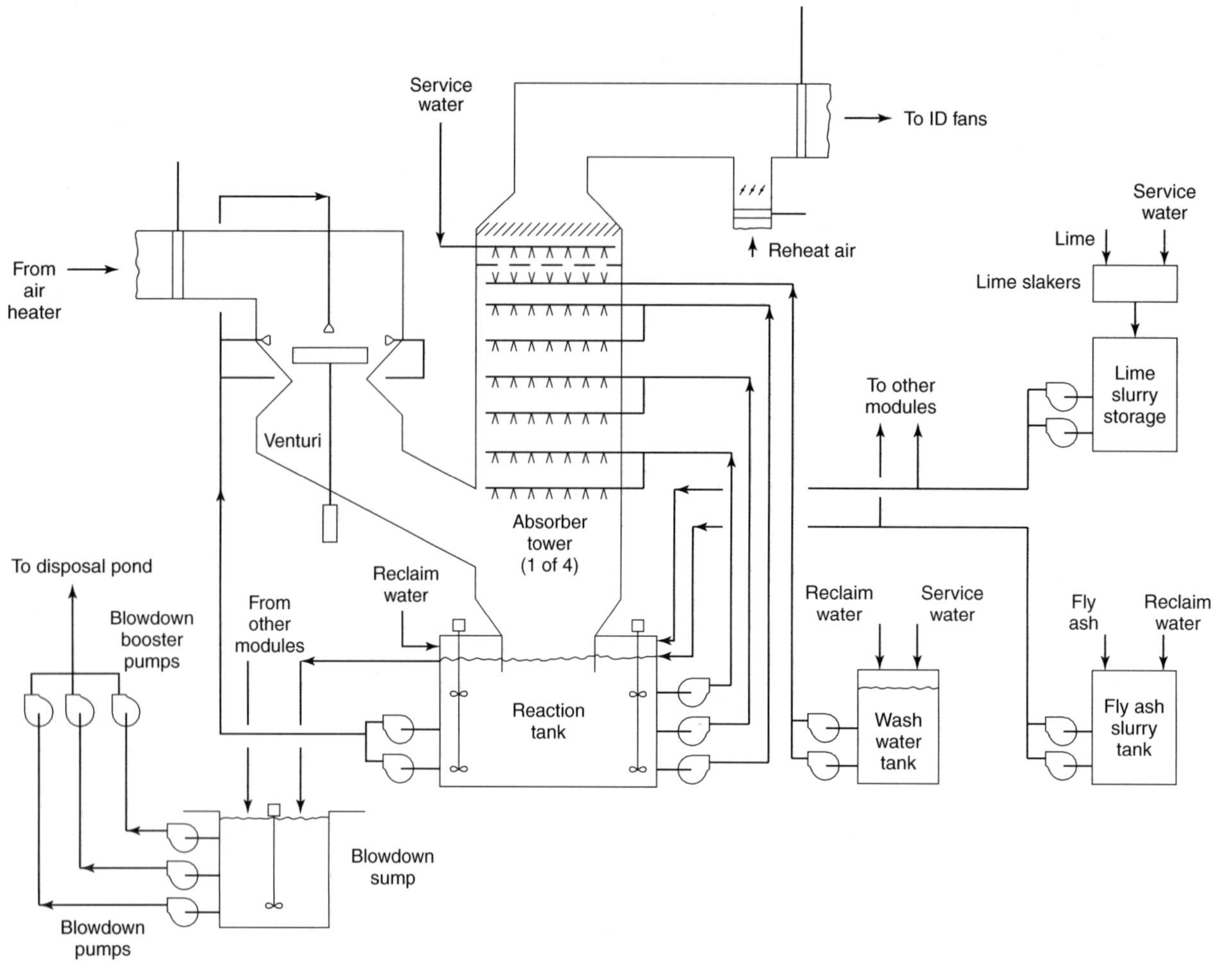

Figure 19. Clay Boswell Station Unit 4 Process Flow Diagram (from Ref. 19, p. 4-1–17).

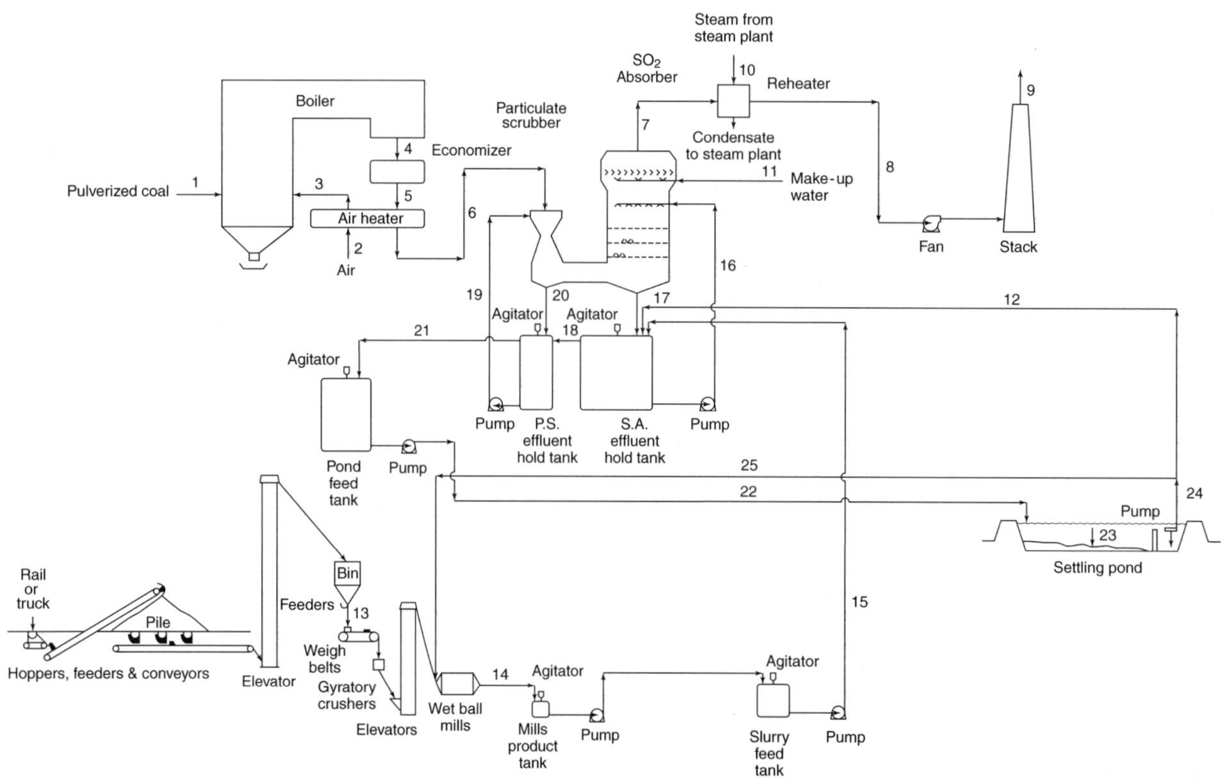

Figure 20. Limestone Slurry Process. Flow Diagram (from Ref. 20, p. 322).

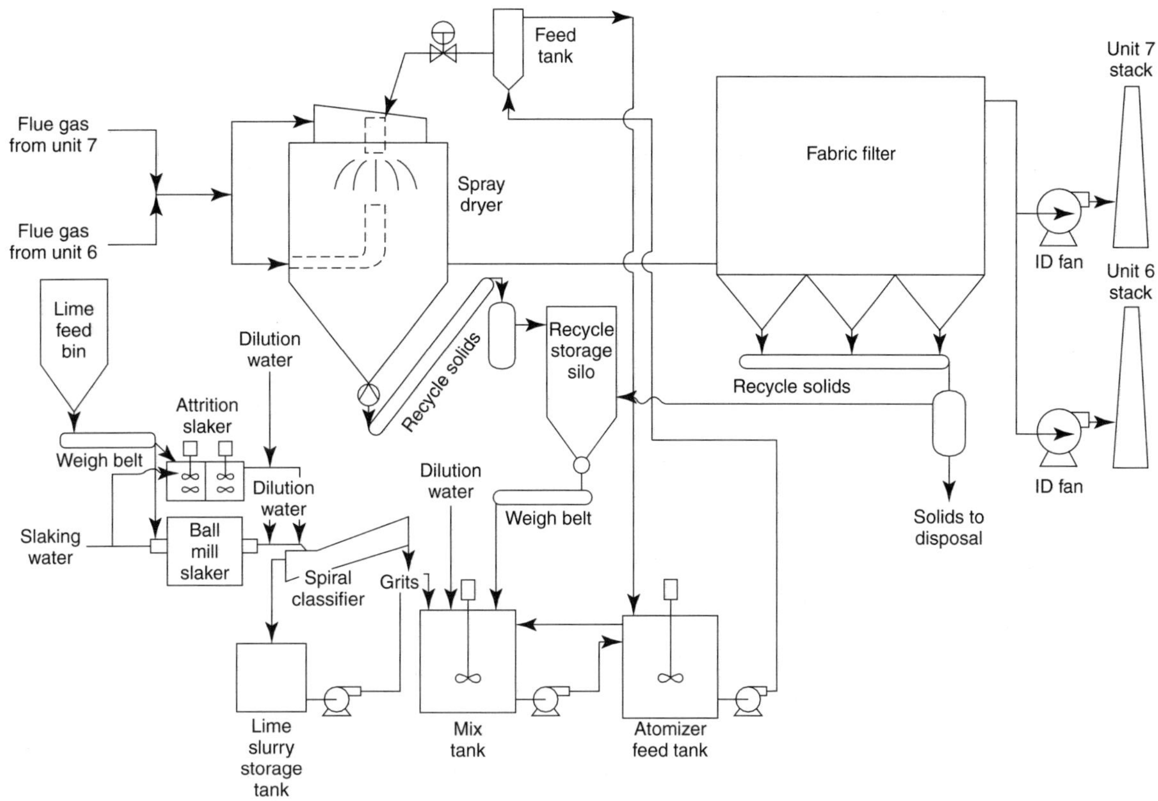

Figure 21. Diagram of a Utility Spray Dryer Scrubbing System, with Recycle (from Refs. 21 and 22).

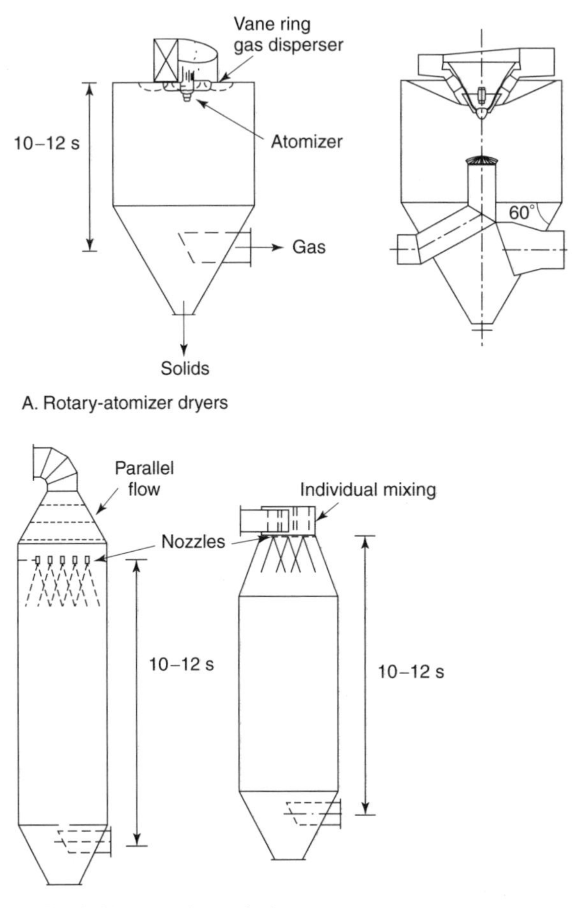

Figure 22. Spray Dryer Arrangements (from Refs. 23 and 24).

dryer because of the spray dryer downtime. Therefore, systems are generally designed with dust collector specifications that meet both operating conditions. Otherwise, the entire system (SO_2 control and particulate-matter control) would need to be bypassed if a spray dryer component were to fail. This may or may not be an acceptable practice, depending on the local air pollution regulations and the time involved. In addition, it is not uncommon to start up the system (bypassing the spray dryer) until the system has been operating for some time.

The primary system operating conditions affected by the application of the spray dryer (which, in turn, affect the performance of the particulate-matter collector) are dust loading, particle size distribution, temperature, and gas-flow rate. For boilers burning coal without SO_2 control, the fly-ash load may vary from as little as 0.5 g/ft^3 to in excess of 4 g/ft^3, depending on (1) the presence of cyclonic precleaners (commonly found on industrial-size boilers), (2) the type of boiler, and (3) the ash content of the coal. The particle size distribution varies similarly, with mass-mean diameters ranging from 6 to 15 μm. Flue-gas temperatures are typically 275–350°F at the particulate collector.

In contrast to the foregoing conditions, the application of a spray dryer results in (1) flue-gas temperatures of 150–170°F, (2) approximately 20% less gas flow through the particulate-matter collector due to lower gas temperatures, (3) a particle size distribution with a mean diameter of 20–30 μm, and (4) a significantly increased dust loading. The increased particle size distribution and decreased flow rate are beneficial, but the decreased temperature and increased dust loading tend to increase pressure drops across fabric filters and to decrease efficiency across ESPs. The increased dust loading caused by the reacted products ranges from as little as 1.5 g/acf at

Table 17. Summary of Spray Dryer Applications at Various Operating Units

Industry Type	Size (MW)	Unit	Atomizer Type[a]	Atomizers per Dryer/ Number of Dryers	RT[b] sec	T[c] °F	δT[d] °F	Spray Dryer Diameter (ft)
Utilities	440 (each)	Antelope Valley 1,2 Basin Electric	R	1/5	12	310	20	46
	110	Riverside 6,7 Northern States Power	R	1/1	10	350	Var.[e]	46
	60	Stanton 10 United Power	R	3/1	10	323	20	38
	450	Craig 3 Colorado-Ute	N	12/4	NR[f]	NR	25	NR
	280	Rawhide 1 Platte River Power	R	1/3	11	276	NR	46
	320	Holcomb 1 Sunflower Coop.	R	1/3	10.6	249	50	51
	44	Shiras 3 City of Marquette	R	1/1	10	265	25	36
	270	North Valmy Sierra Pacific Power Idaho Power	R	3/3	NR	260–300	30	NR
	570	Laramie River 3 Basin Electric	N	12/4	8	286	23	55
	370	Springerville 1,2 Tucson Electric	R	1/3	12	256	20	46
	575	GRDA	R	3/4	12	310	20	52
Industrial (size in acfm)	75,000	Argonne National Lab. Argonne, IL	R	1/1	12	330–340	≈20	25
	90,500	Container Corp. Philadelphia, PA	R	1/1	NR	350	NR	NR
	46,500 (3 units)	Fairchild Air Force Base Spokane, WA	R	1/1	NR	375	≈25	20
	167,000	General Motors Buick Div., Flint, MI	R	1/1	NR	300	NR	32
	48,600	Griffis Air Force Base Rome, NY	R	1/1	NR	400	NR	22
	44,400	Malstrom Air Force Base Great Falls, MI	R	1/1	NR	325	≈ 35	20
	40,000	Strathmore Paper Woronoco, MA	N	4/1	NR	NR	NR	NR
	62,000	University of Minnesota	N	1/1				
	96,000		N	1/1	12	375	20	24
	97,000	Rockwell Int 1. Columbus, OH	R	1/1	12	450	30	30
	205,000	M.M.Carbon Long Beach, CA	R	1/1	10	405	90–120	36
	81,710	Ohio State Univ. Columbus, OH	N	NR/1	NR	400	NR	24

[a]R = rotary; N = nozzle.
[b]Residence time.
[c]Flue-gas temperature at entrance.
[d]Approach to saturation at exit.
[e]Varies.
[f]Not reported.

Source: Reference 21.

500 ppm of SO_2 (with no recycling) to as high as 14 g/acf at 3000 ppm (with a 50% recycle) of reacted products. This increased loading, added to the fly-ash loading, results in overall dust loadings of 2–20 g/acf, depending on the overall conditions. Therefore, increased dust loading must be taken into account in the design of the particulate-matter collector and the associated dust-handling support systems.

As a result of the higher fraction of $Ca(OH)_2$, calcium sulfites, calcium sulfates, increased moisture (5–15%) in the product, and closer approach to saturation, it is important to minimize the heat loss in the particulate collector to prevent cold-end corrosion. Operating experience and related problems have been summarized by a number of vendors. Most of the operational problems have been a result of the

Table 18. Summary of Baghouse Applications at Various Operating Units

Industry Type	Size	Unit	A/C (acfm/ft^2)	ΔP^a (in. H_2O)	Cloth,[b] Material	Bag Life (years)	Type of Cleaning	Cleaning Cycle (h)	No. of Compartments
Utilities (size in MW)	440	Antelope Valley 1,2 Basin Electric	2.2	4	FG/TC (10)	2	Reverse gas	2	14 (2 units)
	110	Riverside 6.7 Northern States Power	1.70	4	FG	4	Reverse gas	1	12
	60	Stanton 10 United Power	1.61	6	FG/SGT (14)	2	Reverse gas	1	NR[c]
	450	Craig 3 Colorado-Ute	2.3	6	NR[c]	2	NR	NR	
	280	Rawhide 1 Platte River Power	1.78	3.5	FG/TC (10)	2	Reverse gas	1	12 (2 units)
	320	Holcomb 1 Sunflower Coop.	1.55	3.5	FG/TC (10)	2	Reverse gas	2 (or 4 in.)	14 (2 units)
	44	Shiras 3 City of Marquette	2.0	6.8	FG/AR	2	Reverse gas	8	NR
	270	North Valmy Sierra Pacific Power Idaho Power	1.8	5.8	FG/TC	NR	Reverse gas	0.5	10
Industrial (size in acfm)	75,000	Argonne National Lab. Argonne, IL	3.0	3.5	FG/TC (16)	2	Pulse	1.5	4
	90,500	Container Corp. Philadelphia, PA	4.5	5.5	FG/TC	NR	Pulse	NR	8
	46,500	Fairchild Air Force Base Spokane, WA	2.0	NR	NR	NR	Reverse gas	NR	6
	167,000	General Motors Buick Div., Flint, MI	2.0	NR	NR	NR	Reverse gas	NR	8
	48,600	Griffis Air Force Base Rome, NY	4.0	9.0	FG/TC	NR	Pulse	NR	4
	44,400	Malstrom Air Force Base Great Falls, MI	2.0	NR	NR	NR	Reverse gas	NR	5
	40,000	Strathmore Paper Woronoco, MA	NR	NR	Acrylic	NR	Pulse	NR	NR
	62,000	University of Minnesota	2.5	2.9	FG/TC	>6	Reverse gas	5	8
	96,000		1.7	3.0	FG/TC	>6	Reverse gas	15	8
	97,000	Rockwell Int'l. Columbus, OH	2.1	3.5	FG/TC	2	Reverse gas	30	8
	205,000	M.M. Carbon Long Beach, CA	1.9	4	FG/TC	NR	Reverse gas	24	12
	81,710	Ohio State Univ. Columbus, OH	5.3	8	FG/TC	NR	Pulse	NR	6

[a]Baghouse pressure drop (ΔP).
[b]FG/TC (10) = fiberglass with Teflon® coat (10 oz/yd^2); FG/AR = fiberglass with acid-resistant finish; FG/SGT = fiberglass with silicon–graphite–Teflon® finish.
[c]NR = not reported.
Source: Reference 21.

lower operating temperatures near the water dew point and associated moisture-acid condensation. Five methods that have been employed or recommended to minimize these problems involve the use of:

1. Insulation (usually a minimum of 4 in.)
2. Control of air in-leakage
3. Hopper heating
4. Improved operating procedures
5. Protective coatings on inner surfaces
6. A crusher (delumper) on the spray dryer hopper

To date, most spray dryer applications have relied on a fabric filter collector for final particulate-matter control. Table 18 summarizes the baghouse applications at several installations where fabric filter design and operating in formation was available.

Generally, reverse-air collectors have been used for large utility boilers, while both reverse-air and pulse-cleaned collectors have been used on small industrial boilers. As shown in Table 18, the air-to-cloth ratios have ranged from 1.6 to 2.5 acfm/ft^2 for the reverse-air collectors and from 3 to 5 acfm/ft^2 for the pulse types of collectors.

The filter material used in reverse-air collectors is typically a woven fiberglass with a Teflon B® coating with a filter weight of 9.5–10 oz/yd^2. Although other materials may be better suited for the low-temperature operation downstream of the FGD spray dryer, the bags must also be able, to operate at temperatures in excess of 350°F and up to 500°F (i.e., during startup without the spray dryer). Fiberglass bags with the Teflon® coating have also been used in pulse collectors, although the filter material generally weighs 16–22 oz/yd^2. The heavier weight is required be cause of the mechanical wear associated with the higher-energy pulsed cleaning, the greater abrasion resulting from the increased air-to-cloth ratios, and the greater potential for abrasion associated with the internal bag cages.

The major advantage of the pulse type of collectors is that they operate at relatively higher filter velocities than the reverse-gas collector (typically 4 fpm, compared with 2 fpm for the reverse-air collectors), thus resulting in a substantially smaller collector. Ironically, the higher velocity tends to reduce bag life, increase outlet emissions, and increase pressure drop. Although these problems can be minimized by the choice of filter material and design changes, there is a cost for these factors that must be considered in evaluating design tradeoffs.

Electrostatic precipitators have been used in only a few large utility applications and, in one case, where units were already in existence. The primary disadvantage of the ESP is the decreased sorbent utilization that occurs when the fly ash and reaction products, collected on the plates of the precipitator, are removed from the gas stream. Several units have been installed, including Grand River Dam Authority (575 MW), Pacific Power and Light—Wyodak 1 (300 MW), and Basin Electric's Laramie River Plant (570 MW). The specific collector area (SCA) of the existing ESPs, which have spray dryers upstream, are 685 and 700 $ft^2/10^3$ acfm for the Grand River Dam Authority and Wyodak, respectively, based on the actual cubic feet per minute at the outlet of the spray dryers.

Nitrogen Oxides Control

The methods for controlling nitrogen oxides can be divided into (1) combustion control methods, in which operating conditions for combustion are modified to reduce the formation of NO_x; and (2) postcombustion control methods, in which NO_x is removed from the gas stream after formation.

The underlying concepts of the combustion control methods are as follows:

1. To reduce peak temperatures in the combustion zone by operating the primary flame zone under fuel-rich conditions, cooling the flame at a high rate, and decreasing the adiabatic flame temperature by dilution
2. To reduce the gas residence time in the high-temperature zone
3. To operate on an off-stoichiometric ratio by using a rich fuel–air ratio in the primary flame zone and lower overall excess air conditions

The actual method of combustion control used depends on the type of boiler, especially the method of firing the fuel. For example, in the case of pulverized-coal-fired boilers (which accounted for 70% of all utility NO_x emissions in the United States in 1989), low-NO_x burners (LNB) and overfire air (OFA) have been successfully applied to tangentially and wall-fired units, whereas reburning is the only currently available option for cyclone boilers.

The uncontrolled NO_x emissions from tangentially fired and wall-fired units at full load are estimated to range from 0.4 to 1.1 and from 0.5 to 1.4 lb/MBtu, respectively. In the United States, Europe, and Japan, most utilities incorporate LNB systems with OFA for NO_x control through combustion modifications. The LNB combustion scheme is shown in Figure 23. After the initial combustion zone, a pyrolitic zone is formed in which the liberation of volatile nitrogenous species takes place. In the next stage, a fuel-rich combustion zone is formed to limit the formation of NO_x. This region is followed by a burnout zone in which completion of combustion is allowed to take place.

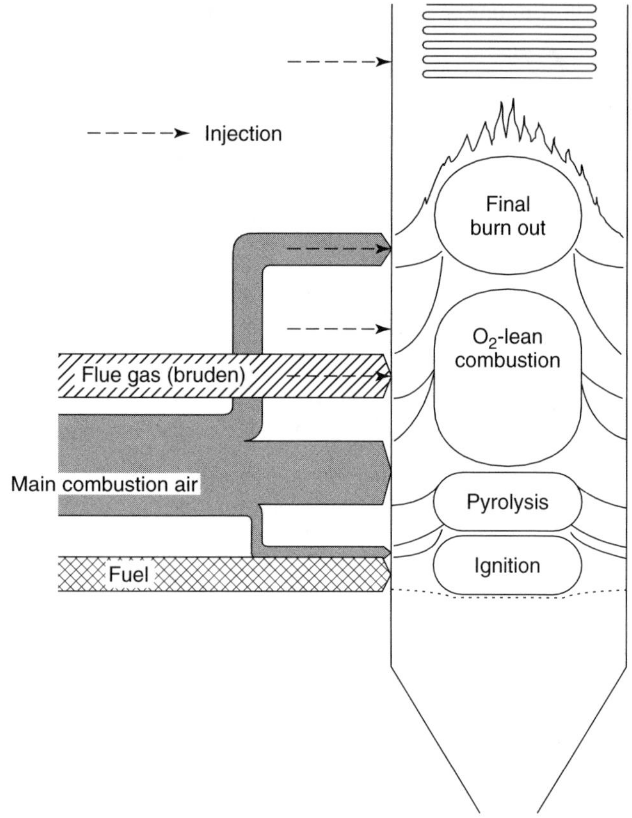

Figure 23. Low NO_x-Combustion (scheme) (from Ref. 27).

The actual design of the LNBs differs substantially between the two boilers. Figure 24 shows the schematic of an LNB for a wall fired unit; Figure 25 compares the configurations of a conventional burner and an LNB for a tangentially fired unit.

One of the major findings during recent retrofit LNB demonstrations was the importance of OFA in NO_x reduction. In fact, OFA could be used as a standalone modification in some cases to meet the emission standards. Results have indicated that up to 30% NO_x reduction can be achieved by conventional OFA systems that stage 10–20% of the combustion air. Advanced OFA systems, staging 20–30% of the combustion

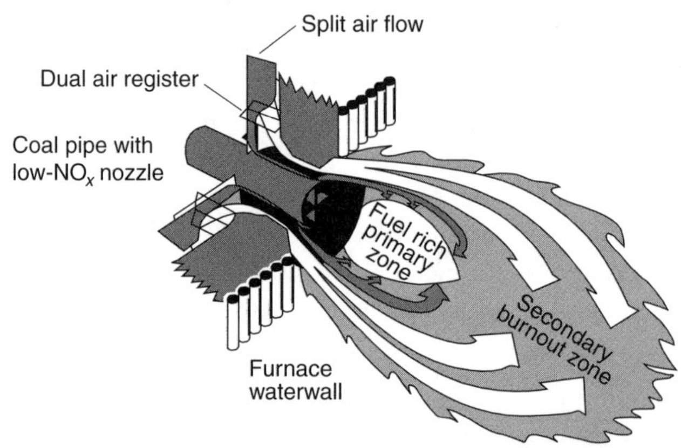

Figure 24. Schematic of a Low-NO_x Burner for Wall-Fired Boilers (from Ref. 26, p. 18).

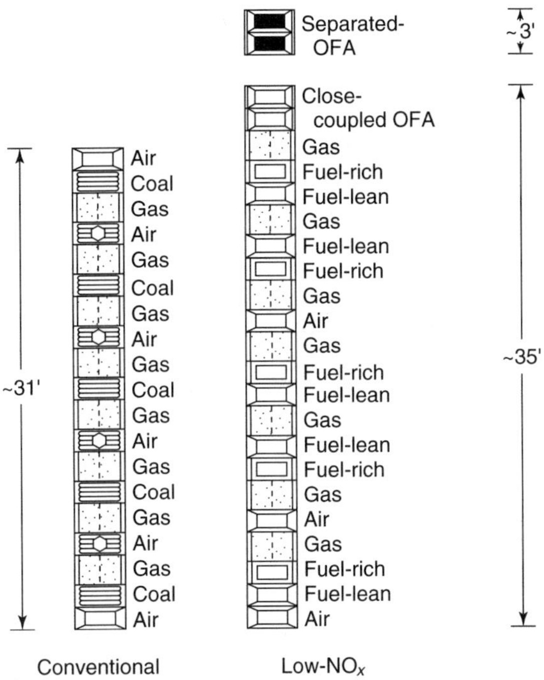

Figure 25. Comparison of Conventional and PM Burner Configurations (from Ref. 28, p. 2–77).

air, have a potential of higher reductions in NO_x. Figure 26 shows the dependence of NO_x emissions on the percentage of combustion air staged in the OFA nozzles, and Figure 27 shows a furnace view of an OFA nozzle in a 400-MW wall-fired boiler.

Reductions in NO_x of up to 50% have been demonstrated in LNB systems incorporating OFA. Capital costs for conventional and advanced OFA alone were estimated to be \$5 per kilowatt and \$10 per kilowatt in 1991. For LNB with OFA, the capital costs were estimated to range between \$15 and \$25 per kilowatt in the same period. Operating costs in all cases are expected to be low, but have not been estimated.

As of 1989, there were 105 operating cyclone-fired utility boilers with a capacity of 26,000 MW (approximately 14% of the pre-NSPS coal-fired generating capacity). However, these units contributed to approximately 21% of the total NO_x emissions, largely because of their high operating temperatures, which are conducive to NO_x formation. Uncontrolled NO_x emissions from cyclone-fired boilers, shown in Figure 5, normally range between 660 and 1400 ppm.

The typical configuration of the cyclone furnaces prevents the successful application of standard LNB technology. The cyclone furnace consists of a cyclone burner connected to a horizontally water-cooled cylindrical chamber, as shown in Figure 28. Fuel (crushed coal) and air are introduced into the chamber by the cyclone burner. The large coal particles are forced onto the molten slag layer formed on the chamber wall by the cyclonic effect of the injection. Combustion of these large particles takes place in this slag layer. The finer particles are held in suspension by the gas stream and are burned in the primary combustion zone within the cylindrical chamber. The flue gases and the remaining ash leave the cyclone and enter the main furnace, which is a separate part of the boiler. Since the combustion occurs outside the main furnace, modification of the furnace, such as with LNBs, is not suitable for these applications. Other conventional NO_x

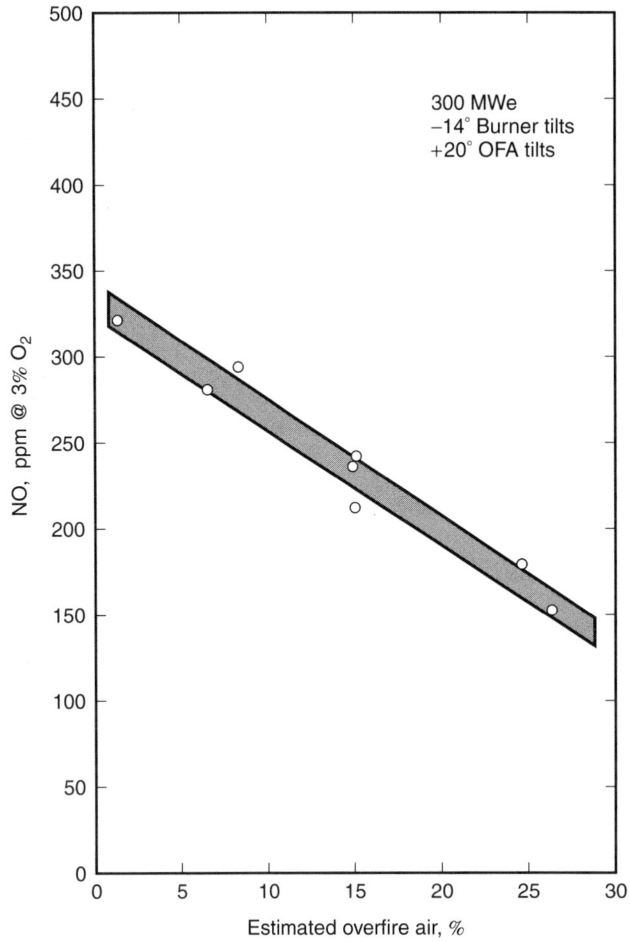

Figure 26. NO_x Dependence on Overfire Air Flow (from Ref. 28, p. 2–82).

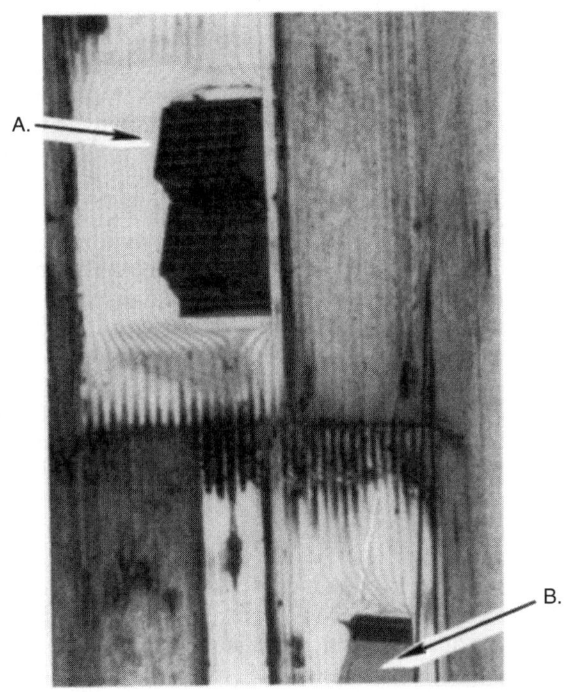

Figure 27. Furnace View of Separated OFA Nozzles (Offset) (from Ref. 28); A. Separated Overfire Air Nozzles; B. Top of Close Coupled Overfire Air Nozzles.

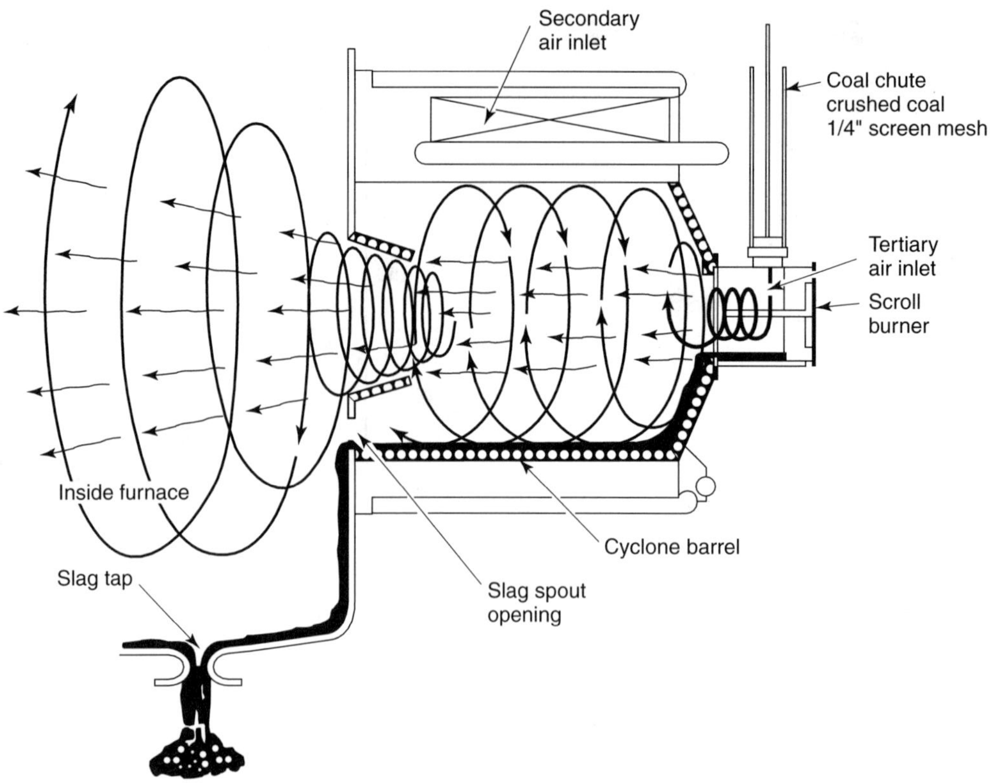

Figure 28. Cyclone Furnace (from Ref. 29, p. 3–16).

control methods, such as staged combustion, have been shown to create cyclone corrosion problems. Reburning remains the only available promising combustion control method for NO_x control in these types of boilers. At present, this technology is still in the pilot/field-demonstration stage.

The concept of reburning technology is shown in Figure 29. This technology involves injection of a second fuel in the furnace above the main cyclonic combustion zone. This section is maintained at reducing conditions by a controlled supply of combustion air. The reducing atmosphere converts NO_x to molecular nitrogen, thus destroying a portion of the NO_x formed in the main combustion zone. This secondary combustion zone is maintained at a lower temperature by recirculation of flue gas to prevent additional NO_x formation in this zone. Complete burnout of the fuel takes place in a subsequent burnout zone, in which additional combustion air is used through OFA ports. The fuel–air ratio and the heat release rates in the various zones need to be optimally maintained for the greatest control of NO_x.

From an economic standpoint, coal is favored as the secondary fuel for coal-fired utilities. However, the use of oil and natural gas is being actively explored, especially in light of the better burnability of these fuels. An additional advantage of these fuels is the proportional reduction of SO_x, particulate, and NO_x emissions achieved over coal.

Reductions in NO_x of 40–60% have been measured in pilot-scale and early short-term demonstration tests. Data obtained by Babcock & Wilcox in studies sponsored by EPRI and GRI using a pilot-scale cyclone boiler showed NO_x reductions of 50–75% from a baseline of about 1000 ppm. More data were expected to be available from ongoing studies. From the preliminary studies in the pilot-scale unit mentioned above, it was found that the type of reburn of fuel used, the percent of flue gas recirculated in the reburning zone, and the fuel–air ratio maintained in the cyclone burner zone affected the level of NO_x control achievable with this control method. Depending on the size of the boiler and the type of secondary fuel used in the reburn chamber, the capital costs for installing reburning technology are estimated to be \$35 to \$45 per kilowatt-hour depending largely on the type of reburn fuel used.

Currently, the most developed and widely applied postcombustion NO_x control technology is selective catalytic reduction (SCR). Selective noncatalytic reduction (SNCR) using ammonia- or urea-based compounds is in the developmental stage and may be available in the future as another postcombustion NO_x control method. In the SCR process, ammonia, used as the reducing agent, is injected into the NO_x-laden flue gas, usually upstream of the air heater. This location provides a temperature window of 550–750°F, which is optimum for the reduction reactions. The NO_x is reduced to molecular nitrogen in a separate reactor vessel containing a catalyst, which is usually a mixture of titanium dioxide, vanadium pentoxide, and tungsten trioxide. The schemes of the reducing reactions are as follows:

$$4NO + 4NH_3 + O_2 \rightarrow 4N_2 + 6H_2O$$

$$2NO_2 + 4NH_3 + O_2 \rightarrow 3N_2 + 6H_2O$$

A schematic of the SCR process is shown in Figure 30. The location of the catalytic converter in the overall equipment layout determines the space velocities and the catalyst geometry used. In “high-dust” SCR systems, in which the converter is located between the boiler and a cold-side ESP, the mesh size of the catalyst has to be kept higher to avoid dust buildup and catalyst fouling. This results in lower catalyst

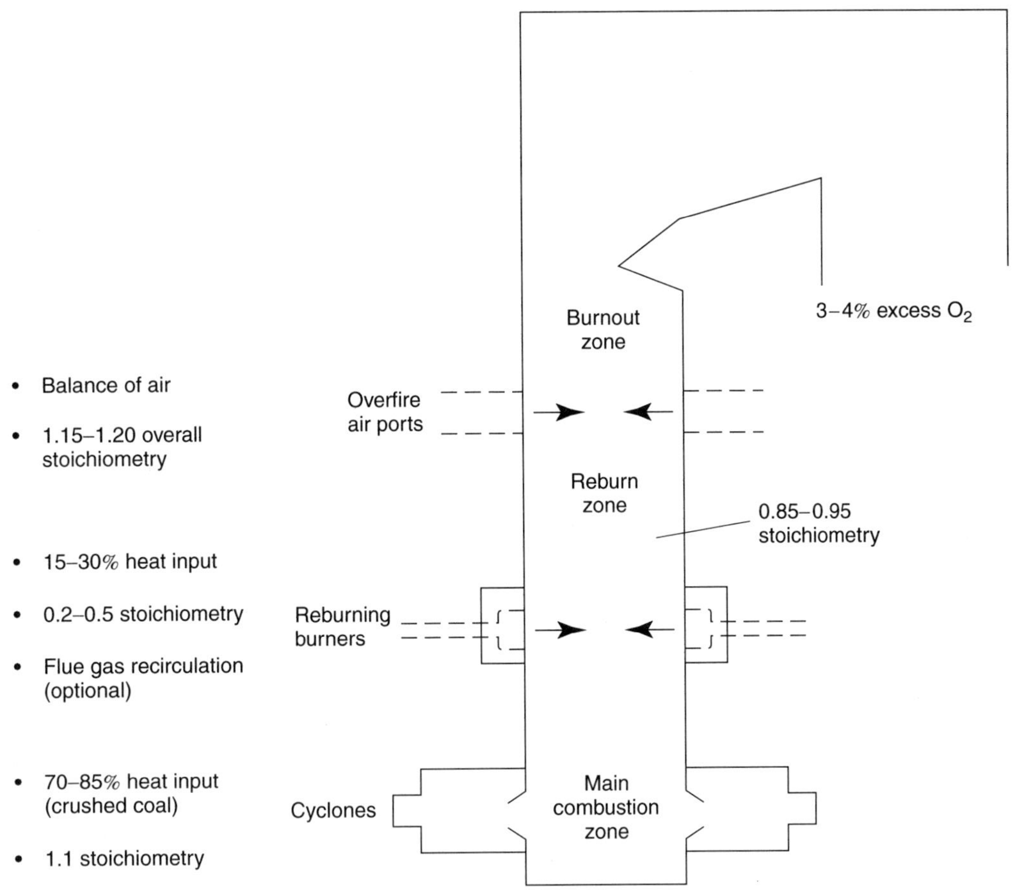

Figure 29. Reburning Technology (from Ref. 29, p. 3–17).

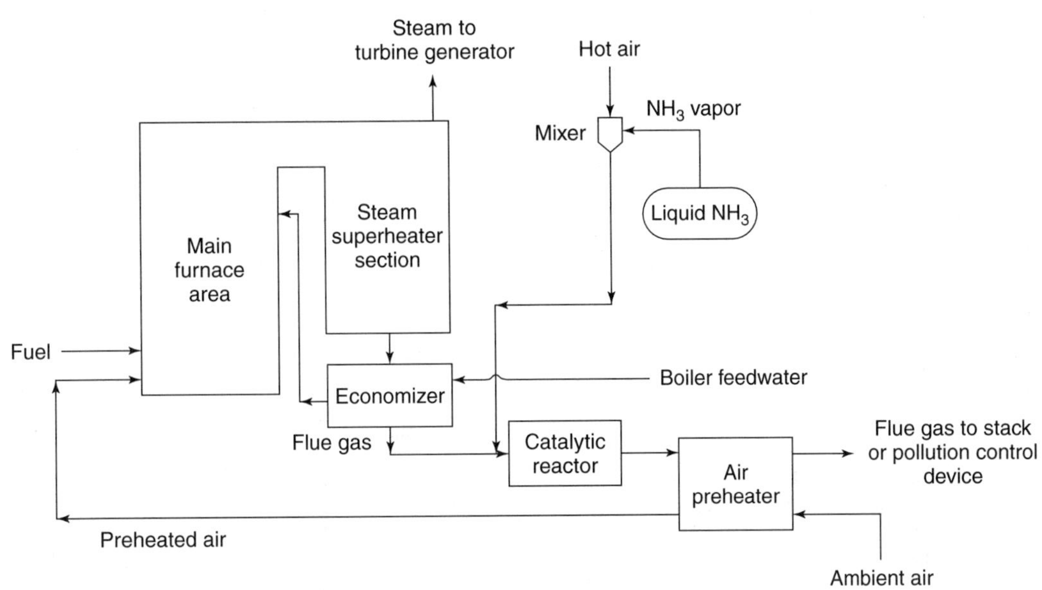

Figure 30. Schematic Flow Diagram for the Selective Catalytic Reduction Method of NO_x Control (from Ref. 11).

specific surface area. In the "low-dust" systems, in which the converter is located after a hot-side ESP, finer mesh catalysts with higher specific areas can be used. Figure 31 shows the two configurations.

Pressure drop and catalyst space velocities are important considerations in the design of SCR systems. Pressure drops in the catalytic converters of between 5 and 7 mbar[31] have been reported, and depend on the type of catalyst geometry used, that is, plate type (lower pressure drop) or honeycomb type. For the desired NO_x-removal efficiencies, space velocities of 2000–2500 per hour are reported to be necessary. Since catalyst costs dominate the SCR system costs, maximum

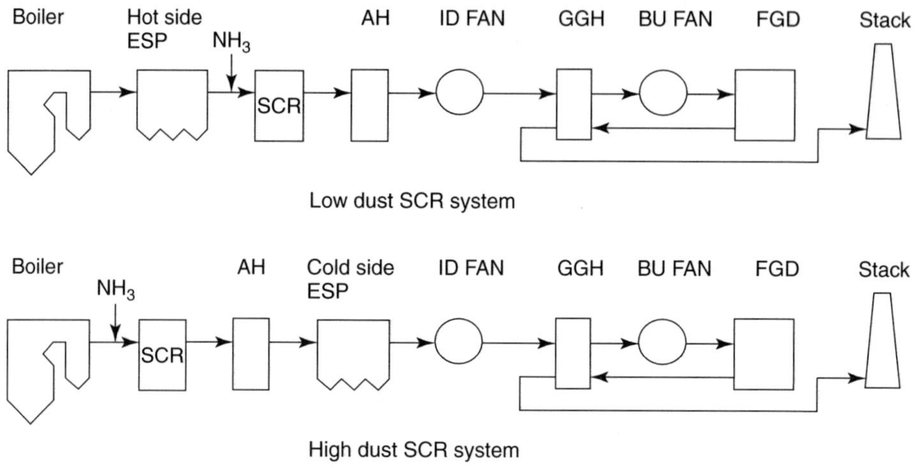

Figure 31. Type of SCR System (from Ref. 30).

allowable space velocities are desired from an economic standpoint.

Catalyst deactivation and residual NH_3 in the flue gas are the two key operating considerations in an SCR system. Catalyst activity decreases with operating time due to fouling of the catalyst. Thus a low-dust SCR system in which a hot-side ESP precedes the SCR reactor is expected to have a longer life, though the capital costs and operating costs may also be higher.

Unreacted NH_3 passing through the reactor forms ammonium sulfate and causes clogging in downstream equipment such as air heaters. With increasing catalyst lifetime and consequent deactivation, the residual NH_3 in the flue gas or the "NH_3 slip" increases. The particulate ammonium sulfate is also known to deposit on the induced-draft fan and to create vibrational problems. From operating experiences in Japanese and European installations, the maximum allowable NH_3 slips seems to be 5 ppmv.

Reductions of 60–90% in NO_x have been reported in full-scale SCR systems internationally and in pilot-scale systems in the United States. Retrofit capital costs are expected to range from \$75 to \$150 per kilowatt, with levelized operating costs ranging between 5 and 9 mills per kilowatt-hour.

Several issues remain to be resolved before SCR can be extensively used in the United States. One of the main issues is whether the data obtained from low-sulfur alkaline-coal-fired boilers in Japan and Europe can be directly extrapolated to the use in the United States of higher-sulfur coals. The other potential concern is that the NH_3 absorbed by the flue gas in this process could lower the resale value of the fly ash by increasing the leachability of metals.

In the SNCR process, urea- or ammonia-based chemicals (either alone or with additives) are injected as the reducing agent to convert NO to molecular nitrogen. However, the reaction is allowed to take place at a higher temperature of 1600–2000°F to obtain a high activation energy and thereby eliminate the use of catalysts. This means that the injection point is usually within the furnace. Based on current pilot-scale and fullscale experiences, 30–70% NO_x reductions are expected for SCNR systems. Capital costs are estimated to range between \$10 and \$20 per kilowatt. Operating costs have not been documented as of 1991.

Mercury Emissions Control. Normally, specific control devices are not installed for controlling mercury emissions from coal-fired units. However, several particulate control devices and FGD scrubbing systems provide some degree of control of mercury emissions. Average reductions of mercury from ESP, baghouses, and FGD systems are 26%, 39%, and 45%, respectively.[33] Mechanical collectors do not provide significant control of mercury emissions.

REFERENCES

1. *Classification of Coals by Rank*, ASTM Standard D-388.
2. J. G. Singer, Ed., *Combustion—Fossil Power Systems*, Combustion Engineering, Inc., 1981.
3. R. E. Bickelhaupt, *A Technique for Predicting Fly Ash Resistivity*, EPA-600/7-79-204, NTIS PB80-102379, U.S. Environmental Protection Agency.
4. P. Vann Bush, T. R. Snyder, and R. L. Chang, "Determination of Baghouse Performance from Coal and Ash Properties: Part I," *J APCA* **39**(2), 228 (1989).
5. P. Vann Bush, T. R. Snyder, and R. L. Chang, "Determination of Baghouse Performance from Coal and Ash Properties: Part II," *J. APCA* **39**(3), 361 (1989).
6. *Compilation of Air Pollutant Emission Factors, Volume I: Stationary Point and Area Sources*, AP-42, U.S. Environmental Protection Agency 1985.
7. *Steam*, 38th ed., Babcock & Wilcox, New York, 1975.
8. H. Farzan et al., "Pilot Evaluation of Reburning for Cyclone Boiler NO_x Control," presented at the 1989 Symposium on Stationary Combustion Nitrogen Oxide Control, Vol. 1, EPRI Publication No. GS-6423, 1989, p. 2–1.
9. R. D. Lewis et al., "Retrofit and Boiler Performance Evaluation of its Low-NO_x PM Firing System at Kansas Power and Light," presented at the 1989 Symposium on Stationary Combustion Nitrogen Oxide Control, Vol. 1, EPRI Publication No. GS-6423, 1989, p. 2–87.
10. *Control Techniques for Nitrogen Oxide Emission from Stationary Sources*, 2nd ed., EPA-450/3-83-002, U.S. Environmental Protection Agency, 1983.
11. C. D. Cooper and F. C. Alley, *Air Pollution Control: A Design Approach*, PWS Engineering, 1988, p. 452.
12. M. Hupa, R. Backman, and S. Bostrom, "Nitrogen Oxide Emission of Boilers in Finland," *J APCA* **39**(11), 1496 (1989).
13. *Electrostatic Precipitator Reference Manual*, EPRI Publication No. CS-2809, 1983.

14. G. R. Offen and R. F. Altman, "Issues and Trends in Electrostatic Precipitation Technology for U.S. Utilities," *J. AWMA* **41**(2), 222 (1991).

15. C. Bustard et al., *Fabric Filters for the Electric Utility Industry, Vol. I-General Concepts*, EPRI Publication CS 5161, 1988.

16. K. M. Cushing, R. L. Merritt, and R. L. Chang, "Operating History and Current Status of Fabric Filters in the Utility Industry," *J. AWMA* **40**(7), 1051 (1990).

17. S. B. Hance and J. L. Kelly, "Status of Flue Gas Desulfurization Systems," Paper No. 91-157.3, 84th Annual Meeting of the Air and Waste Management Association, 1991.

18. *Flue Gas Desulfurization Information System (FGDIS), Computerized Database*, U.S. Department of Energy, Energy Information Administration, Forrestal Building, Washington, D.C.

19. *Lime FGD Systems Data Book — Second Edition*, EPRI Publication No. CS-2781, Black & Veatch Consulting Engineers, 1983.

20. K. E. Noll and W. T. Davis, *Power Generation: Air Pollution Monitoring and Control*, Ann Arbor Science, Ann Arbor, MI, 1976.

21. *Spray-Dryer Flue-Gas-Cleaning Systems Handbook*, Publication No. ANL/ESD-7, Energy Systems Division, Argonne National Laboratory, U.S. Department of Energy, 1988.

22. G. M. Blythe et al., *Field Evaluation of a Utility Dry Scrubbing System — Project Summary*, EPA/600/s7-85/020, U.S. Environmental Protection Agency, 1985.

23. J. R. Donnelly et al., "Spray Dryer FGD Experience: Joy–Niro Installation," *Proceedings of Ninth Symposium on Flue Gas Desulfurization*, U.S. Environmental Protection Agency/EPRI, Cincinnati, OH, 1985.

24. P. J. Kroll and P. Williamson, "Application of Dry Flue Gas Scrubbing to Hazardous Waste Incineration," Paper No. 8610.4, 79th Annual Meeting of the Air Pollution Control Association, 1986.

25. D. Eskinazi et al., "Stationary Combustion NO_x Control — A Summary of the 1989 Symposium," *J. APCA* **39**, 1131 (1989).

26. D. Eskinazi, "Retrofit NO_x Control Strategies for Coal Fired Utility Boilers," Paper No. 91-94.5, 84th Annual Meeting of the Air and Waste Management Association, 1991.

27. K. G. Hein, "The Application of Combustion Modifications for NO_x Reduction to Low Rank Coal Fired Boilers," presented at the 1989 Symposium on Stationary Combustion Nitrogen Oxide Control, Vol. 1, EPRI Publication No. GS-6423, 1989, p. 2–1.

28. R. E. Thompson et al., "NO_x Emission Results for a Low NO_x PM Burner Retrofit," Presented at the 1989 Symposium on Stationary Combustion Nitrogen Oxide Control, Vol. 1, EPRI Publication No. GS-6423, 1989, p. 2–67.

29. H. Farzan et al., "Pilot Evaluation of Reburning for Cyclone Boiler NO_x Control," presented at the 1989 Symposium on Stationary Combustion Nitrogen Oxide Control, Vol. 1, EPRI Publication No. GS-6423, 1989, p. 3–1.

30. B. Schonbucher, "Reduction of Nitrogen Oxides from Coal Fired Power Plants by Using the SCR Process. Experience in the Federal Republic of Germany with Pilot and Commercial Scale $DeNO_x$ Plants," presented at the 1989 Symposium on Stationary Combustion Nitrogen Oxide Control, Vol. 1, EPRI Publication No. GS-6423, 1989, p. 6A–1.

31. T. Mori and N. Shimuzu, "Operating Experience of SCR Systems at EPDC's Coal Fired Power Station," presented at the 1989 Symposium on Stationary Combustion Nitrogen Oxide Control, Vol. 1, EPRI Publication No. GS-6423, 1989, p. 6A–85.

32. P. Necker, "Experience Gained by Neckerwerke from Operation of SCR $DeNO_x$ Units," presented at the 1989 Symposium on Stationary Combustion Nitrogen Oxide Control, Vol. 1, EPRI Publication No. GS-6423, 1989, p. 6A–19.

33. *Emission Factor Documentation for AP-42 Section 1.1, Bituminous and Subbituminous Coal Combustion*, USEPA, Research Triangle Park, NC, April 1993.

34. *Study of Hazardous Air Pollutant Emissions from Electric Utility Steam Generating Units — Final Report to Congress*, EPA, February 1998.

FUEL OIL

Arijit Pakrasi

INTRODUCTION

The most common liquid fuel used in utilities and industrial processes is fuel oil derived from crude petroleum. Fuel oils are broadly classified into (1) distillates–kerosene used mostly for domestic and small commercial applications and (2) residual fuel oils used in utility and industrial boilers. Fuel oils are classified into six grades, depending on the physical and chemical properties of the oils. Table 1 shows typical analyses and properties for each grade of fuel oil.

Pumpability and flowability are important considerations in using oils as fuel. These properties are characterized by the pour point and temperature–viscosity relationship of the oil. Figure 1 shows a typical relationship between the temperature and viscosity of No. 6 fuel oil. Other important properties are the heating value and the residual carbon content. The residual carbon content is an indicator of the particulate emission expected from the fuel oil.

PROCESS DESCRIPTION

The major boiler configurations for fuel-oil-fired combustors are watertube and firetube design. Boilers are classified according to design and orientation of heat transfer surfaces, burner configuration, and size. These factors can all strongly influence emissions as well as the potential for controlling emissions.

Watertube boilers are used in a variety of applications ranging from supplying large amounts of process steam to providing space heat for industrial facilities. In a watertube boiler, combustion heat is transferred to water flowing through tubes that line the furnace walls and boiler passes. The tube surfaces in the furnace (which houses the burner flame) absorb heat primarily by radiation from the flames. The tube surfaces in the boiler passes (adjacent to the primary furnace) absorb heat primarily by convective heat transfer.

Firetube boilers are used primarily for heating systems, industrial process steam generators, and portable power boilers. In firetube boilers, the hot combustion gases flow through the tubes while the water being heated circulates outside the tubes. At high pressures and when subjected to large variations in steam demand, firetube units are more susceptible to structural failure than watertube boilers. This is because the high-pressure steam in firetube units is contained by the boiler walls rather than by multiple small-diameter watertubes, which are inherently stronger. As a consequence, firetube boilers are typically small and are used primarily where boiler loads are relatively constant. Nearly all firetube boilers are sold as packaged units because of their relatively small size.

Table 1. Typical Analyses and Properties of Fuel Oils

Grade	No. 1 Fuel Oil	No. 2 Fuel Oil	No. 4 Fuel Oil	No. 5 Fuel Oil	No. 6 Fuel Oil
Type	Distillate (Kerosene)	Distillate	Very Light Residual	Light Residual	Residual
Color	Light	Amber	Black	Black	Black
API gravity, 60°F	40	32	21	17	12
Specific gravity, 60/60°F	0.8251	0.8654	0.9279	0.9529	0.9861
Pounds/U.S. gallon, 60°F	6.870	7.206	7.727	7.935	8.212
Viscosity, centistokes, 100°F	1.6	2.68	15.0	50.0	360.0
Viscosity, Saybolt univ., 100°F	31	35	77	232	—
Viscosity, Saybolt Furol, 122°F	—	—	—	—	170
Pour point (°F)	Below zero	Below zero	10	30	65
Temperature for pumping (°F)	Atmospheric	Atmospheric	15 min.	35 min.	100
Temperature for atomizing (°F)	Atmospheric	Atmospheric	25 min.	130	200
Carbon residue (%)	Trace	Trace	2.5	5.0	12.0
Sulfur (%)	0.1	0.4–0.7	0.4–1.5	2.0 max.	2.8 max.
Oxygen and nitrogen (%)	0.2	0.2	0.48	0.70	0.92
Hydrogen (%)	13.2	12.7	11.9	11.7	10.5
Carbon (%)	86.5	86.4	86.10	85.55	85.70
Sediment and water (%)	Trace	Trace	0.5 max	1.0 max.	2.0 max.
Ash (%)	Trace	Trace	0.02	0.05	0.08
Btu/gallon	137,000	141,000	146,000	148,000	150,000

Sources: Combustion—Fossil Power Systems, J.G Singer, Ed., published by Combustion Engineer, Inc.

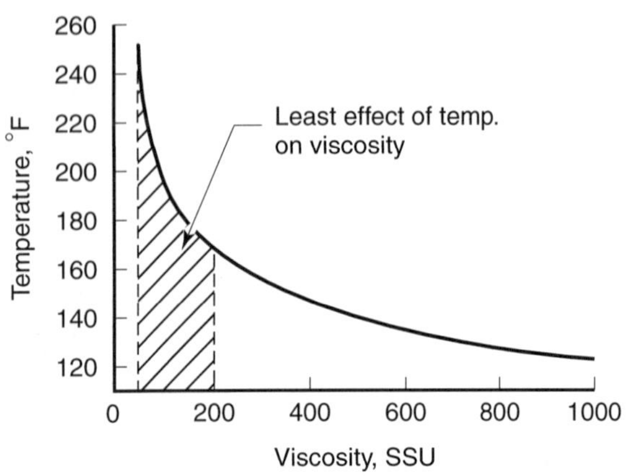

Figure 1. Viscosity versus Temperature, No. 6 Fuel Oil (from *Combustion—Fossil Power Systems*, J. G. Singer, Ed.; published by Combustion Engineering, Inc.).

AIR EMISSION CHARACTERIZATIONS*

Emissions from fuel-oil combustion depend on the grade and composition of the oil, the type and size of the boiler, the firing practices used, and the level of equipment maintenance. Table 2 presents emission factors for pollutants from fuel-oil combustion. Cumulative size distribution data and size-specific emission factors for particulate emissions from fuel-oil combustion and uncontrolled and controlled size-specific emission factors are presented in Figures 2–5. Distillate and residual oil categories are given separately, because these produce significantly different particulate, sulfur dioxide (SO_x), and nitrogen oxide (NO_x) emissions.

* Taken in part from AP-42.

Particulate Matter

Particulate emissions depend on the grade of oil fired. The lighter distillate oils result in particulate emission significantly lower than with heavier residual oils. Among residual oils, Nos. 4 and 5 usually produce less particulate than does the heavier oil, No. 6.

In boilers firing No. 6 oil, particulate emissions can be described, on the average, as a function of the sulfur content of the oil. As shown in Table 2, particulate emissions can be reduced considerably when low-sulfur No. 6 oil is fired. This is so because low-sulfur No. 6, either refined from naturally low-sulfur crude oil or desulfurized by one of several current processes, exhibits substantially lower viscosity and reduced asphaltene, ash, and sulfur, which results in better atomization and cleaner combustion.

Boiler load can also affect particulate emissions in units firing No. 6 oil. At low-load conditions, particulate emissions may be lowered 30–40% from utility boilers and by as much as 60% from small industrial and commercial units. No significant particulate reductions have been noted at low loads from boilers firing any of the lighter grades. At too low a load condition, proper combustion conditions cannot be maintained and particulate emissions may increase significantly. It should be noted, in this regard, that any condition that prevents proper boiler operation can result in excessive particulate formation.

Trace elements are also emitted from the combustion of fuel oil. The quantity of trace elements emitted depends on the combustion temperature, the fuel feed mechanism, and the composition of the fuel. The temperature determines the degree of volatilization of specific compounds contained in the fuel. The fuel feed mechanism affects the separation of trace element (metals) emissions into bottom ash and fly ash.

Sulfur Oxides

Total SO_x emissions are almost entirely dependent on the sulfur content of the fuel and are not affected by boiler size,

Table 2. Uncontrolled Emission Factors for Fuel Oil Combustion (Emission Factor Rating: A)

Boiler Type[a]	Particulate[b] Matter (kg/10^3 L)	Particulate[b] Matter (lb/10^3 gal)	Sulfur Dioxide[c] (kg/10^3 L)	Sulfur Dioxide[c] (lb/10^3 gal)	Sulfur Trioxide (kg/10^3 L)	Sulfur Trioxide (lb/10^3 gal)	Carbon Monoxide[d] (kg/10^3 L)	Carbon Monoxide[d] (lb/10^3 gal)	Nitrogen Oxide[e] (kg/10^3 L)	Nitrogen Oxide[e] (lb/10^3 gal)	Volatile Organics[f] Nonmethane (kg/10^3 L)	Volatile Organics[f] Nonmethane (lb/10^3 gal)	Volatile Organics[f] Methane (kg/10^3 L)	Volatile Organics[f] Methane (lb/10^3 gal)
Utility boilers														
Residual oil	—[g]	—[g]	19S	157S	0.34S	2.9S	0.6	5	8.0 (12.6)(5)[h]	67 (105)(42)[h]	0.09	0.76	0.03	0.28
Industrial boilers														
Residual oil	—[g]	—[g]	19S	157S	0.24S	2S	0.6	5	6.6[i]	55[i]	0.034	0.28	0.12	1.0
Distillate oils	0.24	2	17S	142S	0.24S	2S	0.6	5	2.4	20	0.024	0.2	0.006	0.052
Commercial boilers														
Residual oil	—[g]	—[g]	19S	157S	0.24S	2S	0.6	5	6.6	55	0.14	1.13	0.057	0.475
Distillate oil	0.24	2	17S	142S	0.24S	2S	0.6	5	2.4	20	0.04	0.34	0.026	0.216
Residential furnaces														
Distillate oil	0.3	2.5	17S	142S	0.24S	2S	0.6	5	2.2	18	0.085	0.713	0.214	1.78

[a]Boilers can be approximately classified according to their gross (higher) heat rate:
Utility (power plant) boilers: $>106 \times 10^9$ j/h ($>100 \times 10^6$ Btu/h)
Industrial boilers: 10.6×10^9 to 106×10^9 j/h (10×10^6 to 100×10^6 Btu/h)
Commercial boilers: 0.5×10^9 to 10.6×10^9 j/h (0.5×10^6 to 10×10^6 Btu/h)
Residual furnaces: $<0.5 \times 10^9$ j/h ($<0.5 \times 10^6$ Btu/h)

[b]References 3–7 24, and 25. Particular matter is defined in this section as that material collected by EPA Method 5 (front half catch).

[c]References 1–5. S indicates that the wt % of sulfur in the oil should be multiplied by the value given.

[d]References 3–5 and 8–10. Carbon monoxide emissions may increase by factors of 10 to 100 if the unit is improperly operated or not well maintained.

[e]Expressed as NO_2. Test results indicate that at least 95% by weight of NO_x is NO for all boiler types except residential furnaces, where about 75% is NO.

[f]Volatile organic compound emissions are generally negligible unless boiler is improperly operated or not well maintained, in which case emissions may increase by several orders of magnitude.

[g]Particulate emission factors for residual oil combustion are, on average, a function of fuel oil grade and sulfur content;
Grade 6 oil: 1.25(S) + 0.38 kg/10^3 L [10(S) + 3 lb/10^3 gal] where S is the wt % of sulfur in the oil. This relationship is based on 81 individual tests and has a correlation coefficient of 0.65.
Grade 5 oil: 1.25 kg/10^3 L (10 lb/10^3 gal)
Grade 4 oil: 0.88 kg/10^3 L (7 lb/10^3 gal)

[h]Use 5 kg/10^3 L (42 lb/10^3 gal) for tangentially fired boilers, 12.6 kg/10^3 L (105 lb/10^3 gal) for vertical fired boilers, and 8.0 kg/10^3 L (67 lb/10^3 gal) for all others, at full load and normal (>15%) excess air. Several combustion modifications can be employed for NO_x emissions 40–70% but may increase emissions of ammonia. Combinations of these modifications have been employed for further reductions in certain boilers.

[i]Nitrogen oxide emissions from residual oil combustion in industrial and commercial boilers are strongly related to fuel nitrogen content, estimated more accurately by the empirical relationship:
kg $NO_2/10^3$ L $= 2.75 + 50(N)^2$ [lb $NO_2/10^3$ gal $= 22 + 400(N)^2$], where N is the wt % of nitrogen in the oil. For residual oils having high (>0.5 wt %) nitrogen content, use 15 kg $NO_2/10^3$ L (120 lb $NO_2/10^3$ gal) as an emission factor.

Source: Reference 31.

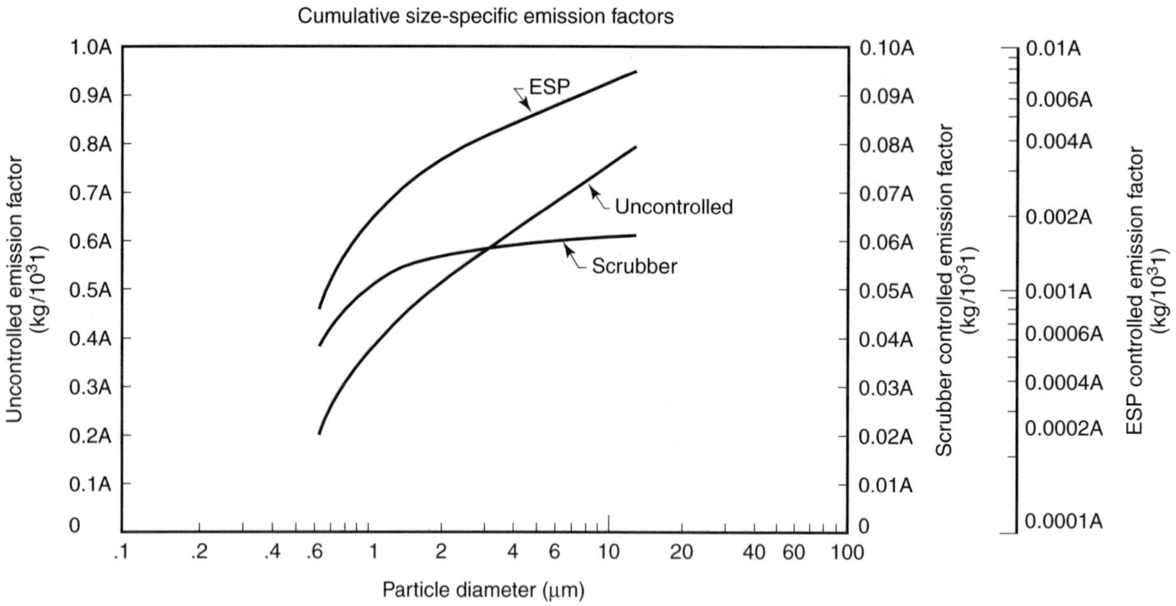

Figure 2. Emission Factors for Utility Boilers Firing Residual Oil (from Ref. 31).

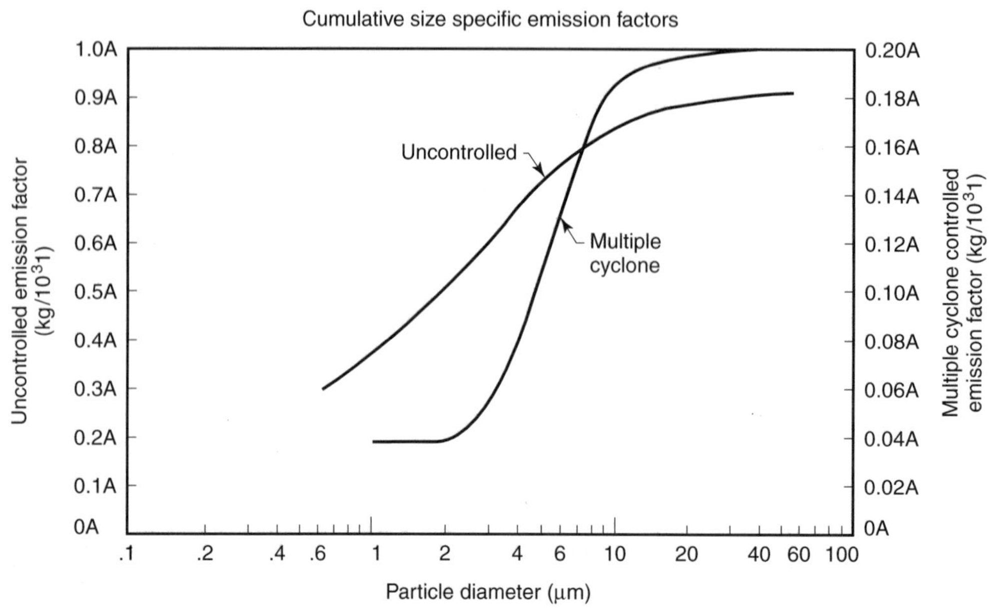

Figure 3. Emission Factors for Industrial Boilers Firing Residual Oil (from Ref. 31).

burner design, or grade of fuel being fired. On the average, more than 95% of the fuel sulfur is emitted as SO_2, about 1–5% as SO_3, and about 1–3% as sulfate particulate. The SO_3 readily reacts with water vapor (in both air and flue gases) to form a sulfuric acid mist.

Nitrogen Oxides

Two mechanisms form NO_x: oxidation of fuel-bound nitrogen and thermal fixation of the nitrogen in combustion air. Fuel NO_x is primarily a function of the nitrogen content of the fuel and the available oxygen. On average, about 45% of the fuel nitrogen is converted to NO, but this may vary from 20% to 70%. Thermal NO_x is largely a function of peak flame temperature and available oxygen, factors that depend on boiler size, firing configuration, and operating practices. Thermal NO_x also depends on fuel nitrogen concentration.

Fuel nitrogen conversion is the more important NO_x-forming mechanism in residual oil boilers. Except in certain large units having unusually high peak flame temperatures or in units firing a low-nitrogen residual oil, fuel NO_x will generally account for over 50% of the total NO_x generated. Thermal fixation, on the other hand, is the dominant NO_x forming mechanism in units firing distillate oils, primarily because of the negligible nitrogen content in these lighter oils. Because distillate oil-fired boilers usually have low heat-release rates, however, the quantity of thermal NO_x formed in them is less than that of larger units.

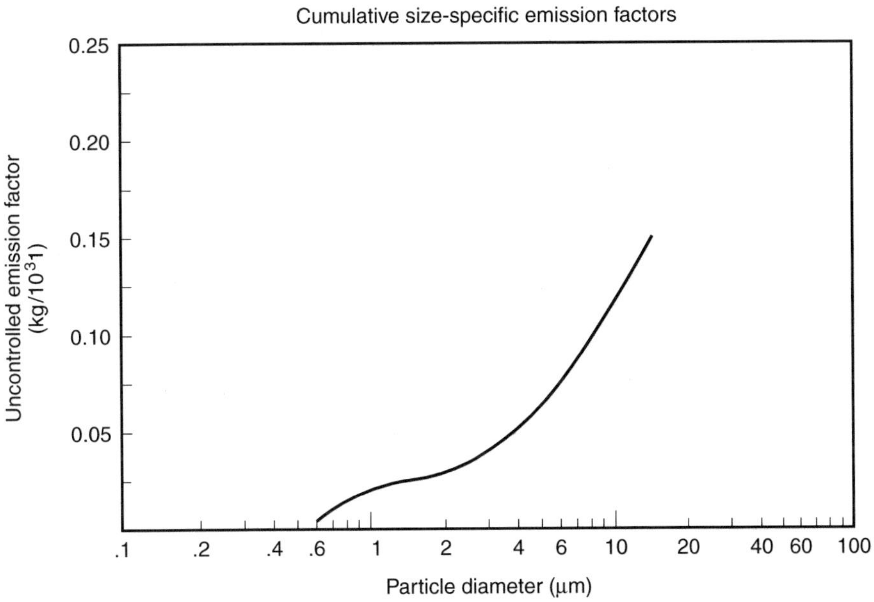

Figure 4. Emission Factors for Uncontrolled Industrial Boilers Firing Distillate Oil (from Ref. 31).

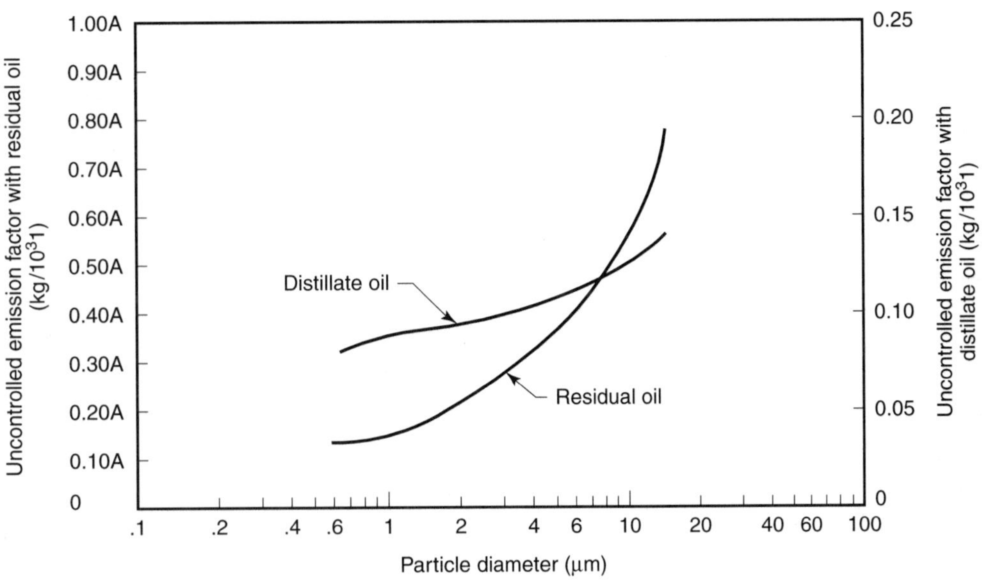

Figure 5. Emission Factors for Uncontrolled Commercial Boilers Burning Residual and Distillate Oil (from Ref. 31).

A number of variables influence how much NO_x is formed by these two mechanisms. One important variable is firing configuration. Nitrogen oxide emissions from tangentially fired boilers are, on the average, less than those from horizontally opposed units. Also important are the firing practices employed during boiler operation.

Carbon Monoxide and VOCs

As a rule, only minor amounts of volatile organic compounds (VOCs) and carbon monoxide are emitted from the combustion of fuel oil. The rate at which VOCs are emitted depends on combustion efficiency. Organic compounds present in the flue gas of boilers include aliphatic and aromatic hydrocarbons, esters, ethers, alcohols, carbonyl, carbolic acids, and polycyclic organic matter. The last includes all organic matter having two or more benzene rings.

The presence of VOCs and CO in the exhaust gases are mainly due to incomplete combustion. Small boilers tend to emit more CO and VOCs than larger boilers due to lower combustion efficiency. Some of the VOCs are classified as hazardous air pollutants (HAP) under Title III of the 1990 Clean Air Act Amendments.

Formaldehyde is formed during combustion of fuel oil and is subject to oxidation/decomposition at the high temperatures in the combustion zone. Larger units with more efficient combustion emit less formaldehyde than smaller, less efficient combustion units.

Toxic Metals

Several trace metals, considered as HAP under Title III of the 1990 Clean Air Act Amendments are emitted from combustion of fuel oil. The trace metal emissions depend on concentration in the fuel oil, combustion temperature, fuel feed mechanism, and the composition of the fuel. Combustion temperature determines the degree of volatilization of the metals, and the fuel feed mechanism affects the percentage entrained in flue gas as fly ash. Particle enrichment, which depends on the type of metal, is also an important factor in partitioning of the metals and consequently on the emissions.

Greenhouse Gases

Carbon dioxide (CO_2), methane (CH_4), and nitrous oxide (N_2O) emissions are all produced during fuel-oil combustion. Nearly all the fuel carbon (99%) in fuel oil is converted to CO_2 during the combustion process. This conversion is relatively independent of firing configuration. Although the formation of CO acts to reduce CO_2 emissions, the amount of CO produced is insignificant compared to the amount of CO_2 produced. The majority of the fuel carbon not converted to CO_2 is due to incomplete combustion in the fuel stream.

Formation of N_2O during the combustion process is governed by a complex series of reactions, and its formation is dependent upon many factors. Formation of N_2O is minimized when combustion temperatures are kept high (above 1475°F) and excess air is kept to a minimum (less than 1%). Additional sampling and research is needed fully to characterize N_2O emissions and to understand the N_2O formation mechanism. Emissions can vary widely from unit to unit, or even from the same unit at different operating conditions. Average emission factors based on reported test data have been developed for conventional oil combustion systems.

Methane emissions vary with the type of fuel and firing configuration, but are highest during periods of incomplete combustion or low-temperature combustion, such as the startup or shut-down cycles. Typically, conditions that favor formation of N_2O also favor emissions of CH_4.

Emission factors for air pollutants for oil-fired boilers are presented in Tables 2–6. Emission factor ratings reflect the quality of the data (as per AP-42) as follows: A = Excellent, B = Above Average, C = Average, D = Below Average, E = Poor.

Table 3. Emission Factors for Speciated Organic Compounds from Fuel Oil Combustion[a]

Organic Compound	Average Emission Factor (lb/10^3 gal)	Average Emission Factor (kg/10^3 L)	Emission Factor Rating
Benzene	2.14E-04	2.57E-05	C
Ethylbenzene	6.36E-05	7.63E-06	E
Formaldehyde	3.30E-02	3.96E-03	C
Naphthalene	1.13E-03	1.36E-04	C
1,1,1-Trichloroethane	2.36E-04	2.83E-05	E
Toluene	6.20E-03	7.44E-04	D
o-Xylene	1.09E-04	1.31E-05	E
Acenaphthene	2.11E-05	2.53E-06	C
Acenaphthylene	2.53E-07	3.04E-08	D
Anthracene	1.22E-06	1.46E-07	C
Benz(a)anthracene	4.01E-06	4.81E-07	C
Benzo(b,k)fluoranthene	1.48E-06	1.78E-07	C
Benzo(g,h,I)perylene	2.26E-06	2.71E-07	C
Chrysene	2.38E-06	2.86E-07	C
Dibenzo(a,h)anthracene	1.67E-06	2.00E-07	D
Fluoranthene	4.84E-06	5.81E-07	C
Fluorene	4.47E-06	5.36E-07	C
Indo(1,2,3-cd)pyrene	2.14E-06	2.57E-07	C
Phenanthrene	1.05E-05	1.26E-06	C
Pyrene	4.25E-06	5.10E-07	C
OCDD	3.10E-09	3.72E-10	E

[a]Data are for residual oil-fired boilers.

Source: AP-42.

AIR POLLUTION CONTROL MEASURES

Since air emissions from oil-fired boilers are substantially lower than those from coal-fired boilers, little or no control

Table 4. Emission Factors for Metals from Uncontrolled No. 6 Fuel-Oil Combustion[a]

Metal	Average Emission Factor (lb/10^3 gal)	Average Emission Factor (kg/10^3 L)	Emission Factor Rating
Antimony	5.25E-03	6.30E-04	E
Arsenic	1.32E-03	1.58E-04	C
Barium	2.57E-03	3.08E-04	D
Beryllium	2.78E-05	3.34E-06	C
Cadmium	3.98E-04	4.78E-05	C
Chloride	3.47E-01	4.16E-02	D
Chromium	8.45E-04	1.01E-04	C
Chromium VI	2.48E-04	2.98E-05	C
Cobalt	6.02E-03	7.22E-04	D
Copper	1.76E-03	2.11E-04	C
Fluoride	3.73E-02	4.48E-03	D
Lead	1.51E-03	1.81E-04	C
Manganese	3.00E-03	3.60E-04	C
Mercury	1.13E-04	1.36E-05	C
Molybdenum	7.87E-04	9.44E-05	D
Nickel	8.45E-02	1.01E-02	C
Phosphorus	9.46E-03	1.14E-03	D
Selenium	6.83E-04	8.20E-05	C
Vanadium	3.18E-02	3.82E-03	D
Zinc	2.91E-02	3.49E-03	D

[a]Data are for residual oil-fired boilers.
[b]18 of 19 sources were uncontrolled and 1 source was controlled with low efficiency ESP.
[c]For oil/water mixture, reduce factors in proportion to water content of the fuel (due to dilution). To adjust the listed values for water content, multiply the listed value by 1-decimal fraction of water (ex: for fuel with 9 percent water by volume,multiply by 1 − 0.09 = 0.91).

Source: AP-42.

Table 5. Emission Factors for Trace Metals from Distillate Fuel Oil Combustion Source

Metal	Emission Factor (lb/10^2 Btu)	Emission Factor Rating
As	4	E
Be	3	E
Cd	3	E
Cr	3	E
Cu	6	E
Pb	9	E
Hg	3	E
Mn	6	E
Ni	3	E
Se	15	E
Zn	4	E

Source: AP-42.

Table 6. Carbon Dioxide Emission Factors for Liquid Fuel Combustion[a] Emission Factor Rating: B

Fuel Type	%C[b]	Density (lb/gal)	Emission Factor ($lb/10^3$ gal)	Emission Factor ($kg/10^3$ L)
No. 1 (kerosene)	86.25	6.88	21,500	2,580
No. 2	87.25	7.05	22,300	2,676
Low Sulfur No. 6	87.26	7.88	25,000	3,000
High Sulfur No. 6	85.14	7.88	24,400	2,928

[a]Based on 99% conversion of fuel carbon content to CO_2.
[b]Based on an average of fuel carbon contents.

of the emissions normally is required, except in large oil-fired boilers. Large oil-fired boilers are often equipped with mechanical collectors to control particulates generated during soot blowing, upset conditions, or when very dirty fuel oil is fired. For these situations, high-efficiency mechanical collectors can achieve up to 85% control efficiency. The particulates emitted during normal operation are too fine to be collected efficiently by mechanical collectors, and in such cases, electrostatic precipitators (ESPs) are used. Up to 90% collection efficiencies have been reported for these ESPs in oil-fired-boiler applications. Fabric filter can be used for oil-fired boilers. The particulate removal efficiency depends on a variety of particle and operational characteristics including particle distribution, particle cohesion characteristics, and particle electrical resistivity. The structure of the fabric filter, filter composition, and bag properties also affect collection efficiency. Collection efficiencies of fabric filters may be greater than 99%.

Particulate emission control for small boilers is usually accomplished by improving burner servicing, oil atomization, and combustion aerodynamics.

Wet scrubbing and spray dryers have been used to reduce sulfur oxide emissions in boilers using high-sulfur oil. Wet systems are more commonly used with alkaline slurries as the absorbent medium. Typical sulfur oxide control efficiencies are in the range of 90% and depend not only on the design of the scrubber but also on the operational variables.

Nitrogen oxide emissions from oil-fired boilers are usually low because of the low amount of nitrogen present in the fuel. However, since many of these are located in large metropolitan areas, where ozone nonattainment and other health issues related to acid rain are of great concern, these units are also subjected to significant reductions of NO_x The NO_x-control methods for oil-fired boilers are similar to those for coal-fired boilers. Combustion control methods, including limited excess air firing, flue gas recirculation, staged combustion, or some combination thereof, may result in NO_x reductions of 5–60%. In Japan, however, selective catalytic reduction technology is more common for oil-fired-boiler NO_x control.

Load reduction can likewise decrease NO_x production. Nitrogen oxide emissions may be reduced from 0.5% to 1% for each percentage reduction in load from full-load operation. It should be noted that most of these variables, with the exception of excess air, influence the NO_x emissions only of large oil-fired boilers. Limited excess air firing is possible in many small boilers, but the resulting NO_x reductions are not nearly as significant.

Postcombustion nitrogen oxide control includes selective noncatalytic reduction (SNCR) and selective catalytic reduction (SCR). SNCR involves injecting ammonia or urea into specific temperature zones to react with nitrogen oxides to produce nitrogen and water. The control efficiency depends on several factors, such as the temperature at which the reagents are injected, mixing between the reagents and the flue gas, residence time within the required temperature window, and the quantity of the reagent injected. Based on limited data available from commercial applications, NO_x reductions of 25–40% have been achieved by SNCR.

SCR involves injection of ammonia into the flue gas in the presence of a catalyst to reduce NO_x to nitrogen and water. The performance of SCR is affected by flue gas temperature, fuel sulfur content, ammonia to NO_x ratio, inlet NO_x concentration, space velocity, and catalyst condition. Control efficiency 75–85% has been achieved by SCR systems on oil-fired boiler operating in the United States.

One U.S. utility noted, in a study, that the particulate emissions tended to increase with NO_x controls. Further studies have been planned, and emphasis is being placed on improving atomizer design and on staging of air to reduce NO_x without increasing the particulate emissions.

Table 7. Postcombustion SO_2 Controls Fuel Oil Combustion Source

Technology Control	Process	Typical Control Efficiencies	Remarks
Wet scrubber	Lime/limestone	80–95+%	Applicable to high-sulfur fuels; wet sludge product
	Sodium carbonate	80–98%	5–430 MMBtu/h typical application range; high reagent costs
	Magnesium oxide/hydroxide	80–95+%	Can be regenerated
	Dual alkali	90–96%	Uses lime to regenerate sodium-based scrubbing liquor
Spray drying	Calcium hydroxide slurry, vaporizes in spray vessel	70–90%	Applicable to low- and medium-sulfur fuels; produces dry product
Furnace injection	Dry calcium carbonate/hydrate injection in upper furnace cavity	25–50%	Commercialized in Europe; several U.S. demonstration projects underway
Duct injection	Dry sorbent injection into duct, sometimes combined with water spray	25–50+%	Several R&D and demonstration projects underway; not yet commercially available in the U.S.

Source: AP-42.

Table 8. NO_x Controls Option for Oil-Fired Boilers

Control Technique	Description of Technique	NO_x Reduction Potential (%) Residual Oil	Distillate Oil	Range of Application	Commercial Availability/ R&D Status	Comments
Low excess air (LEA)	Reduction of combustion air	0 to 28	0 to 24	Generally excess O_2 can be reduced to 2.5%, representing a 3% drop from baseline	Available for boilers with sufficient operational flexibility	Added benefits included efficiency; limited by increase in CO_2 HC, and smoke emissions
Staged combustion (SC)	Fuel-rich firing burners with secondary combustion air ports	20 to 50	17 to 44	70–90% burner stoichiometries can be used with proper installation of secondary air ports	Technique is applicable on packaged and field-erected units; however, not commercially available for all design types	Best implemented on new units; retrofit is probably not feasible for most units, especially packaged ones
Burners out of service (BOOS)	One or more burners on air only; remainder of burners firing fuel-rich	10 to 30	ND	Most effective on boilers with 4 or more burners in a square pattern	Available.	Requires careful selection of BOOS pattern and air flow; may result in boiler derating unless fuel delivery system is modified
Fuel gas recirculation (FGR)	Recirculation of portion of flue gas to burners	15 to 30	58 to 73	Up to 25–30% of, flue gas recycled; can be implemented on most design types	Available; best suited for new units	Requires extensive modifications to the burner and windbox; possible flame instability at high FGR rates
Flue gas recirculation plus staged combustion	Combined techniques of FGR and staged combustion	25 to 53	73 to 77	Maximum FGR rates set at 25% for distillate oil and 25% for residual oil	Available for boilers with sufficient operational flexibility	May not be feasible on all existing boiler types; best implemented on new units
Load reduction (LR)	Reduction of air and fuel flow to all burners in service	33% decrease to 25% increase in NO_x	31% decrease to 17% increase in NO_x	Applicable to all boiler types and sizes; load can be reduced to 25% of maximum	Available in retrofit applications	Technique not effective when it necessitates an increase in excess O_2 levels; LR possibly implemented in new designs as reduced combustion intensity (i.e., enlarged furnace plan area)

Low NO_x burners (LNB)	New burner designs with controlled air/fuel mixing and increased heat dissipation	20 to 50	20 to 50	New burners described generally applicable to all boilers	Commercially available	Specific emissions data from industrial boilers equipped with LNB are lacking
Reduced air preheat (RAP)	Bypass of combustion air preheater	5 to 16	ND	Combustion air temperature can be reduced to ambient conditions	Available	Application of this technique on new boilers requires installation of alternate heat recovery system (e.g., an economizer)
Selective noncatalytic reduction (SNCR)	Injection of NH_3 or urea as a reducing agent in the flue gas	40 to 70	40 to 70	Applicable for large packaged and field-erected watertube boilers; may not be feasible for fire-tube boilers	Commercially offered but not widely demonstrated on large boilers	Elaborate reagent injection, monitoring, and control system required; possible load restrictions on boilers and air preheater fouling when burning high-sulfur oil; must have sufficient residence time at correct temperature
Conventional selective catalytic reduction (SCR)	Injections of NH_3 in the presence of a catalyst (usually upstream of air heater)	Up to 90% (estimated)	Up to 90% (estimated)	Typically large boiler designs	Commercially offered but not widely demonstrated	Applicable to most boiler designs as a retrofit technology or for new boilers

Source: AP-42. ND = no data.

Retrofit capital costs for installing LNB and OFA systems in oil-fired boilers are estimated to be \$20 to \$40 per kilowatt, based on 1989 estimates.

REFERENCES

1. W. S. Smith, *Atmospheric Emissions from Fuel Oil Combustion: An Inventory Guide*, 999-AP-2, U.S. Environmental Protection Agency, Washington, D.C., 1962.
2. *Air Pollution Engineering Manual*, J. A. Danielson, Ed.; 2nd ed., AP-40, U.S. Environmental Protection Agency, Research Triangle Park, NC, 1973 (out of print).
3. A. Levy et al., "A Field Investigation of Emissions from Fuel Oil Combustion for Space Heating," API Bulletin 3099, Battelle Columbus Laboratories, Columbia, OH, 1971.
4. R. E. Barrett et al., *Field Investigation of Emissions from Combustion for Space Heating*, EPA-R2-73-084a, U.S. Environmental Protection Agency, Research Triangle Park, NC, 1973.
5. G. A. Cato et al., *Field Testing: Application of Combustion Modifications to Control Pollutant Emissions from Industrial Boilers–Phase I*, EPA-650/2-74-078a, U.S. Environmental Protection Agency, Washington, D.C., 1974.
6. G. A. Cato et al., *Field Testing: Application of Combustion Modifications to Control Pollutant Emissions from Industrial Boilers–Phase II*, EPA-600/2-76-086a, U.S. Environmental Protection Agency, Washington, D.C., 1976.
7. *Particulate Emission Control Systems for Oil Fired Boilers*, EPA-450/3-74-063, U.S. Environmental Protection Agency, Research Triangle Park, NC, 1974.
8. W. Bartok et al., *Systematic Field Study of NO, Emission Control Methods for Utility Boilers*, APTD-1163, U.S. Environmental Protection Agency, Research Triangle Park, NC, 1971.
9. A. R. Crawford et al., *Field Testing: Application of Combustion Modifications to Control NO_x Emissions from Utility Boilers*, EPA-650/2-74-066, U.S. Environmental *Protection Agency*, Washington, D.C., 1974.
10. J. F. Deffner et al., "Evaluation of Gulf Econojet Equipment with Respect to Air Conservation," Report No. 731RC044, Gulf Research and Development Co., Pittsburgh, PA, December 18, 1972.
11. C. E. Blakeslee and H. E. Burbach, "Controlling NO_x Emission from Steam Generators," *J. APCA*, **23** 37–42 (1973).
12. C. W. Siegmund, "Will Desulfurized Fuel Oils Help?" *ASHRAE J.* **11** 29 (1969).
13. F. A. Govan et al., "Relationships of Particulate Emissions versus Partial to Full Load Operations for Utility-Sized Boilers," *Proceedings of Third Annual Industrial Air Pollution Control Conference*, Knoxville, TN, 1973.
14. R. E. Hall et al., *A Study of Air Pollutant Emissions from Residential Heating Systems*, EPA-650/2-74-003, U.S. Environmental Protection Agency, Washington, D.C., 1974.
15. "Flue Gas Desulfurization: Installations and Operations," PB 257721, National Technical Information Service, Springfield, VA, 1974.
16. *Proceedings: Flue Gas Desulfurization Symposium — 1973*, EPA-650/2-73-038, U.S. Environmental Protection Agency, Washington, D.C., 1973.
17. R. J. Milligan et al., *Review of NO_x Emission Factors for Stationary Fossil Fuel Combustion Sources*, EPA-450/4-79021, U.S. Environmental Protection Agency, Research Triangle Park, NC, 1979.
18. N. F. Suprenant et al., *Emissions Assessment of Conventional Stationary Combustion Systems, Vol. 1: Gas and Oil Fired Residential Heating Sources*, EPA-600/7-79-029b, U.S. Environmental Protection Agency, Washington, D.C., 1979.
19. C. C. Shih et al., *Emissions Assessment of Conventional Stationary Combustion Systems, Vol. 11: External Combustion Sources for Electricity Generation*, EPA Contract No. 68-02-2197, TRW, Inc., Redondo Beach, CA, 1980.
20. N. F. Suprenant et al., *Emissions Assessment of Conventional Stationary Combustion Systems, Vol. IV: Commercial Institutional Combustion Sources*, EPA Contract No. 68-022197, GCA Corp., Bedford, MA, 1980.
21. N. F. Suprenant et al., *Emissions Assessment of Conventional Stationary Combustion Systems, Vol. V: Industrial Combustion Sources*, EPA Contract No. 68-02-2197, GCA Corp., Bedford, MA, 1980.
22. *Fossil Fuel Fired Industrial Boilers — Background Information for Proposed Standards* (Draft EIS), Office of Air Quality Planning and Standards, U.S. Environmental Protection Agency, Research Triangle Park, NC, 1980.
23. K. J. Lim et al., *Technology Assessment Report for Industrial Boiler Applications: NO, Combustion Modification*, EPA600/7-79-178f, U.S. Environmental Protection Agency, Washington, D.C., 1979.
24. *Emission Test Report*, Docket No. OAQPS-78-1, Category 11-1-257 through 265, Office of Air Quality Planning and Standards, U.S. Environmental Protection Agency, Research Triangle Park, NC, 1972 through 1974.
25. *Primary Sulfate Emission from Coal and Oil Combustion*, EPA Contract No. 68-02-3138, TRW, Inc., Redondo Beach, CA, 1980.
26. C. Leavitt et al., *Environmental Assessment of an Oil Fired Controlled Utility Boiler*, EPA-600/7-80-087, U.S. Environmental Protection Agency, Washington, D.C., 1980.
27. W. A. Carter and R. J. Tidona, *Thirty-Day Field Tests of Industrial Boilers: Site 2 — Residual-Oil-Fired Boiler*, EPA600/7-80-085b, U.S. Environmental Protection Agency, Washington, D.C., 1980.
28. G. R. Offen et al., *Control of Particulate Matter from Oil Burners and Boilers*, EPA-450/3-76-005, U.S. Environmental Protection Agency, Research Triangle Park, NC, 1976.
29. *Inhalable Particulate Source Category Report for External Combustion Sources*, EPA Contract No. 68-02-3156a, Acurex Corp., Mountain View, CA, 1985.
30. D. Eskinazi, J. E. Cichanowicz, W. P. Linak, et al., "Stationary Combustion NO_x Control — a summary of the 1989 Symposium," *J. APCA* **39** 1131 (1989).
31. *Compilation of Air Pollutant Emission Factors, Volume I, Stationary Point and Area Sources*, AP-42, U.S. EPA, 1985.

NATURAL GAS*

Arijit Pakrasi

INTRODUCTION

Natural gas is used for power generation, and also for heat/steam generation in the industry. The primary component of natural gas is methane, though varying amounts of ethane and trace compounds such as nitrogen, helium, and carbon dioxide are also present. The average gross heating value of natural gas is approximately 1020 Btu/scf, usually varying between 950 and 1050 Btu/scf. Typical properties of natural gas are presented in Table 1.

* Taken in part from AP-42

Table 1. Characteristics of Typical Natural Gas

% by Volume								Density	HHV		A at Zero Excess
CO_2	N_2	H_2S	CH_4	C_2H_6	C_3H_8	C_4H_{10}	C_5H_{12}	(lb/ft^3)	(Btu/ft^3)[b]	(Btu/lb)	Air (lb/10^6Btu)
5.50	—	7.00	77.73	5.56	2.40	1.18	0.63[a]	0.05621	1061	18,880	738
3.51	32.00	0.50	52.54	3.77	2.22	2.02	3.44[a]	0.06610	874	13,220	729
26.2	0.7	—	59.2	13.9	—	—	—	0.06747	849	12,580	732
0.17	87.69	—	10.50	1.64	—	—	—	0.07120	136	1,907	732
0.20	0.60	—	99.20	—	—	—	—	0.04491	1006	22,410	732
—	0.60	—	—	79.40	20.00	—	—	0.08812	1935	21,960	735
—	0.50	—	—	21.80	77.70	—	—	0.11079	2389	21,560	738

[a]All hydrocarbon heavier than C_5H_{12} are assumed to be C_5H_{12}.
[b]If gas is saturated with moisture at 60°F and 30.0 in. Hg, reduce by 1.74%.

Source: Combustion—Fossil Power Systems, by J. G. Sunger, Ed., Combustion Engineering, Inc.

PROCESS DESCRIPTION

In commercial, industrial, and utility applications, both water-tube and fire-tube boilers are used with natural gas fuel. In water-tube boilers, the water passes through the inside of the tubes, which are heated by direct contact with combustion gases and by radiant heat transfer. Packaged units are available up to a heat transfer rate of 100 MMBtu/h and field-erected units can be as large as 250 MMBtu/h. Field-erected units have multiple or single burners located on a single or opposing walls (wall-fired units). Burners may also be located on the four corners of the combustion chamber (tangential-fired). The customized location of burners in field-erected units allows greater operational flexibility and nitrogen oxides (NO_x) control options.

Fire-tube boilers are designed to pass hot combustion gases through the tubes and water outside the tubes. The applications are limited to process steam generation and space heating. Since these boilers are almost always packaged units, the operational flexibility and NO_x control options are limited.

AIR EMISSIONS CHARACTERIZATION

Natural gas is considered a clean fuel compared to coal and fuel oil. Most of the air emissions are NO_x, carbon monoxide (CO), and greenhouse gases such as carbon dioxide (CO_2). Other air emissions include small amounts of particulate matter (PM) and sulfur oxides (SO_x).

The principal mechanism of formation of NO_x during natural gas combustion is thermal NO_x. Thermal NO_x is produced in the high-temperature region of the boilers through thermal dissociation and subsequent reaction of nitrogen (N_2) and oxygen (O_2) molecules in combustion air. The thermal NO_x formation increases with greater oxygen concentration, higher peak temperature in the combustion chamber, and longer exposure time at peak temperature.

Emissions of CO and volatile organic compounds (VOCs) are primarily due to incomplete combustion and are generally higher for improperly operating boilers. Application of low-NO_x burners and flue gas recirculation (FGR) decreases combustion efficiency in some cases resulting in higher CO and VOC emissions. In addition to constituents of natural gas, most of the VOC emissions are formaldehyde or benzene, which are present in natural gas in trace amounts.

Emission of SO_x depends on the sulfur content of the natural gas, which is rarely above 2000 grains per million cubic feet. Small amounts of SO_x may result from oxidation of mercaptans, which are intentionally added to the gas for detection of leaks. In general, the SO_x emission from natural gas combustion is insignificant.

Particulate matter from natural gas combustion is generally higher-molecular-weight hydrocarbons, which are not fully combusted due to poor fuel–air mixing and improper combustion. The PM has both filterable and condensable fractions and are usually less than 1 μ in diameter.

CO_2 is the most insignificant greenhouse gas produced during combustion of natural gas by oxidation of the fuel carbon. Other greenhouse gases from natural gas combustion are nitrous oxide (N_2O) and methane (CH_4). N_2 emissions increase with lower temperature and higher excess oxygen in the combustion zone. Methane emissions are generally due to incomplete combustion.

Both organic and inorganic hazardous air pollutants (HAP) are produced during combustion of natural gas. Organic HAPs are formed primarily due to incomplete combustion. Though reliable data are unavailable at the present time, hexane and benzene are considered to be the two most significant organic HAPs from combustion of natural gas. Other organic HAP emissions are usually below detection limits. Also, the organic HAP emissions are believed to be independent of the design of the boiler and calorific value of the fuel. Inorganic HAP emissions are metals (or metal compounds) such as arsenic, beryllium, cadmium, chromium, cobalt, manganese, mercury, nickel, and selenium.

Typical emissions from natural gas combustion in boilers are shown in Tables 2–5.

AIR POLLUTION CONTROL MEASURES

Under improper operating conditions of the boilers, such as poor air–fuel mixing, insufficient air, and so on, significant amounts of particulate emissions may result. However, gas-fired boilers are usually not equipped with any particulate collectors.

Sulfur dioxide emissions from gas-fired boilers depend on the sulfur content of the gas. Normally, natural gas is free from sulfur. In the processing stage, however, small amounts of mercaptans are intentionally added to the gas to permit detection; these will produce small quantities of SO_x in

Table 2. Uncontrolled and Controlled Emission Factors for Nitrogen Oxides (NO_x) and Carbon Monoxide (CO) from Natural Gas Combustion[a]

	NO_x^b	
Combustor Type (MMBtu/h Heat Input)	Emission Factor ($lb/10^6 scf$)	Emission Factor ($kg/10^6 m^3$)
Large wall-fired boilers		
(>100)		
Uncontrolled (pre-NSPS)	190–280	3040–4480
Controlled — low-NO_x burners	140	2240
Controlled — flue gas recirculation	100	1600
Small boilers		
(<100)		
Uncontrolled	100	1600
Controlled — low-NO_x burners	50	800
Controlled — low-NO_x burners/flue gas recirculation	32	512
Tangential-fired boilers	170	2720
(all sizes)	76	1216
Uncontrolled		
Controlled — flue gas recirculation	94	1504
Residential furnaces		
(<0.3)		
Uncontrolled		

[a] Emission factors are based on an average natural gas higher heating value of 1,020 Btu/scf. To convert from $lb/10^6$ scf to lb/MMBtu, divide by 1,020. The emission factors in this table may be converted to other natural gas heating values by multiplying the given emission factor by the ratio of the specified heating value to this average heating value.

[b] Expressed as NO_2. For large and small wall-fired boilers with SNCR control, apply a 24% reduction to the appropriate NO_x emission factor. For tangential-fired boilers with SNCR control, apply a 13% reduction to the appropriate NO_x emission factor.

Source: AP-42.

Table 3. Uncontrolled Emission Factors for Criteria Pollutants, Organic Compounds, and Greenhouse Gases from Natural Gas Combustion[a]

Pollutant	Emission Factor ($lb/10^6$ scf)	Emission Factor ($kg/10^6$ m^3)
Criterial Pollutants and Organic Compounds		
Lead	0.0005	0.008
PM (total)[b]	7.6	121.6
PM (condensable)[c]	5.7	91.2
PM (filterable)[c]	1.9	30.4
SO_2^d	0.6	9.6
VOC	5.5	88
TOC	11	176
Greenhouse Gases		
CO_2^c	120,000	1,920,000
N_2O (uncontrolled)	2.2	35.2
Methane	2.3	36.8

[a] Data are for all natural gas combustion sources. To convert from $lb/10^6$ scf to lb/MMBtu, divide by 1,020. The emission factors in this table may be converted to other natural gas heating values by multiplying the given emission factor by the ratio of the specified heating value to this average heating value. TOC, total organic compounds; VOC, volatile organic compounds.

[b] All PM (total, condensible, and filterable) is assumed to be less than 1.0 μ in diameter. Therefore, the PM emission factors presented here may be used to estimate PM_{10}, $PM_{2.5}$ or PM_1 emissions. Total PM is the sum of the filterable PM and condensible PM. Condensible PM is the particulate matter collected using EPA Method 202 (or equivalent). Filterable PM is the particulate matter collected on, or prior to, the filter of an EPA Method 5 (or equivalent) sampling train.

[c] Based on approximately 100% conversion of fuel carbon to CO_2. $CO_2(lb/10^6$ scf$) = (3.67)$ (CON) (C)(D), where CON is the fractional conversion of fuel carbon to CO_2, C, is the carbon content of fuel by weight (0.76), and D, the density of fuel, 4.2×10^4 $lb/10^6$ scf.

[d] Based on 100% conversion of fuel sulfur to SO_2. Assumes sulfur content is natural gas of 2,000 grains/10^6 scf. The SO_2 emission factor in this table can be converted to other natural gas sulfur contents by multiplying the SO_2 emission factor by the ratio of the site-specific sulfur content (grains/10^6 scf)

Source: AP-42.

Table 4. Uncontrolled Emission Factors for Organic HAPs and Other Compounds from Natural Gas Combustion[a]

Pollutant	Emission Factor (lb/10^6 scf)	Emission Factor (kg/10^6/m^3)
2-Methylnaphthalene[b,c]	2.4E−05	3.84E−04
3-Methylchloranthrene[b,c]	<1.8E−06	<2.88E−05
7,12-Dimethylbenz(a)anthracene[b,c]	<1.6E−05	<2.56E−04
Acenaphthene[b,c]	<1.8E−06	<2.88E−05
Acenaphthylene[b,c]	<1.8E−06	<2.88E−05
Anthracene[b,c]	<2.4E−06	<3.84E−05
Benz(a)anthracene[b,c]	<1.8E−06	<2.88E−05
Benzene[b]	2.1E−03	1.936E−02
Benzo(a)pyrene[b,c]	<1.2E−06	<1.92E−05
Benzo(b)fluoranthene[b,c]	<1.8E−06	<2.88E−05
Benzo(g,h,i)perylene[b,c]	<1.2E−06	<1.92E−05
Benzo(k)fluoranthene[b,c]	<1.8E−06	<2.88E−05
Butane	2.1E+00	3.336E+01
Chrysene[b,c]	<1.8E−06	<2.88E−05
Dibenzo(a,h)anthracene[b,c]	<1.2E−06	<1.92E−05
Dichlorobenzene[b]	1.2E−03	1.92E−02
Ethane	3.1E+00	4.96E+01
Fluoranthene[b,c]	3.0E−06	4.8E−05
Fluorene[b,c]	2.8E−06	4.48E−05
Formaldehyde[b]	7.5E−02	1.2E+00
Hexane[b]	1.8E+00	2.88E+01
Indeno(1,2,3-cd)pyrene[b,c]	<1.8E−06	<2.88E−05
Naphthalene[b]	6.1E−04	9.76E−03
Pentane	2.6E+00	4.16E+01
Phenanathrene[b,c]	1.7E−05	2.72E−04
Propane	1.6E+00	2.56E+01
Pyrene[b,c]	5.0E−06	8.0E−05
Toluene[b]	3.4E−03	5.44E−02

[a]Data are for all natural gas combustion sources. To convert from lb/10^6 scf to lb/MMBtu, divide by 1,020. Emission factors preceded with a less-than symbol are based on method detection limits.
[b]Hazardous air pollutant (HAP) as defined by Section 112(b) of the Clean Air Act.
[c]HAP because it is polycyclic organic matter (POM). POM is a HAP as defined by Section 112(b) of the Clean Air Act.
Source: AP-42.

the combustion process. Usually, the emissions of SO_x are sufficiently small that a control system is not warranted.

Nitrogen oxides are the major pollutants in natural-gas-fired boilers. The amount of NO_x emission is dependent on the combustion temperature and the rate of cooling of the combustion products. The NO_x control strategies for gas-fired boilers are all based on combustion modifications and are similar to the strategies described for coal-fired boilers in the section on coal combustion. These include staged combustion, off-stoichiometric firing, flue gas recirculation, and low-NO_x burner systems with overfire air. These control strategies can reduce the NO_x emissions by 5–50%. Currently, FGR and low-NO_x burners are the two most common technologies used for NO_x control. FGR reduces formation of thermal NO_x by reducing peak temperatures and limiting availability of oxygen. Low-NO_x burner reduced formation of thermal NO_x by delayed combustion (staging) resulting in a cooler flame. Both staged air burners and staged fuel burners have been used in natural gas fired boilers. Properly operated low-NO_x burners can achieve NO_x emission reduction of 40–85%. In conjunction with FGR, the burners can achieve NO_x emission reduction of 60–90%.

Table 5. Uncontrolled Emission Factors for Metals from Natural Gas Combustion[a]

Pollutant	Emission Factor (lb/10^6 scf)	Emission Factor (μg/10^6 m^3)
Arsenic[b]	2.0E-04	3.2E-03
Barium	4.4E-03	7.04E-02
Beryllium[b]		<1.92E-04
Cadmium[b]	<1.2E-05	1.76E-02
Chromium[b]	1.1E-03	2.24E-02
Cobalt[b]	1.4E-03	1.344E-03
Copper	8.4E-05	1.36E-02
Manganese[b]	8.5E-04	6.08E-03
Mercury[b]	3.8E-04	4.16E-03
Molybdenum	2.6E-04	1.76E-02
Nickel[b]	1.1E-03	3.36E-02
Selenium[b]	2.1E-03	<3.84E-04
Vanadium	<2.4E-05	3.68E-02
Zinc	2.3E-03	4.64E-01
	2.9E-02	

[a]Data are for all natural gas combustion sources. Emission factors preceded by a less-than symbol are based on method detection limits. To convert from lb/10^6 scf to lb/MMBtu, divide by 1,020.
[b]Hazardous air pollutant as defined by Section 112(b) of the Clean Air Act.
Source: AP-42.

Several recent NO_x compliance studies at gas-fired boilers have concluded that emission control strategies for these boilers are very site specific. For example, for tall boilers, flue gas recirculation and staging of the air can be more economical than low-NO_x burners. In the case of short furnaces or where operational flexibility is required, low-NO_x burners seem to be more appropriate.

Though not commonly used, both selective noncatalytic reduction (SNCR) and selective catalytic reduction (SCR) can be used for postcombustion NO_x emission control for natural-gas-fired boilers. In SNCR, ammonia or urea is injected at a specific temperature zone, and NO_x control efficiencies between 25–40% have been documented. The SCR involves injection of ammonia in the presence of a catalyst. Currently, no data are available on performance of SCR systems, although estimated NO_x control efficiency range between 80–90%.

REFERENCES

1. W. Bartok et al., *Systematic Field Study of NO_x Emission Control Methods for Utility Boilers*, APTD-1163, U.S. Environmental Protection Agency, Research Triangle Park, NC, 1971.
2. F. A. Bagwell et al., "Oxides of Nitrogen Emission Reduction Program for Oil and Gas Fired Utility Boilers," *Proceedings of the American Power Conference*, 1970, pp. 683–693.
3. H. E. Dietzmann, "A Study of Power Plant Boiler Emissions," Final Report No. AR-837, Southwest Research Institute, San Antonio, TX, 1972.
4. R. E. Barrett et al., *Field Investigation of Emissions from Combustion Equipment for Space Heating*, EPA-R2-73-084, U.S. Environmental Protection Agency, Research Triangle Park, NC, 1971.
5. C. E. Blakesless and H. E. Burbock, "Controlling NO_x Emissions from Steam Generators," *JAPCA* **23**, 37–42 (1979).
6. J. W. Bradstreet and R. J. Fortman, "Status of Control Techniques for Achieving Compliance with Air Pollution Regulations by the Electric Utility Industry," presented at the Third Annual

Industrial Air Pollution Control Conference, Knoxville, TN, March 1973.

7. *Study of Emissions of NO_x from Natural Gas Fired Steam Electric Power Plants in Texas, Phase II, Vol. H*, Radian Corp., Austin, TX, 1972.
8. N. F. Suprenant et al., *Emissions Assessment of Conventional Stationary Combustion Systems, Vol. I: Gas and Oil Fired Residential Heating Sources*, EPA-600/7-79-029b, U.S. Environmental Protection Agency, Washington, DC, 1979.
9. C. C. Shih et al., *Emission Assessment of Conventional Stationary Combustion Systems, Vol. 111: External Combustion Sources for Electricity Generation*, EPA Contract No. 68-02-2197, TRW, Inc., Redondo Beach, CA, 1980.
10. N. F. Suprenant et al., *Emission Assessment of Conventional Stationary Combustion Sources, Vol. IV: Commercial Institutional Combustion Sources*, EPA Contract No. 68-02-2197, GCA Corp., Bedford, MA, 1980.
11. N. F. Suprenant et al., *Emissions Assessment of Conventional Stationary Combustion Systems, Vol. V: Industrial Combustion Sources*, EPA Contact No. 68-02-2197, GCA Corp., Bedford, MA, 1980.
12. R. J. Milligan et al., *Review of NO_x Emission Factors for Stationary Fossil Fuel Combustion Sources*, EPA-450/4-79 021, U.S. Environmental Protection Agency, Research Triangle Park, NC, 1979.
13. W. H. Thrasher and D. W. Dewerth, *Evaluation of the Pollutant Emissions from Gas Fired Forced Air Furnaces*, Research Report No. 1503, American Gas Association, Cleveland, OH, 1975.
14. G. A. Cato et al., *Field Testing: Application of Combustion Modification to Control Pollutant Emissions from Industrial Boilers, Phase I*, EPA-650/2-74-078a, U.S. Environmental Protection Agency, Washington, DC, 1974.
15. G. A. Cato et al., *Field Testing: Application of Combustion Modification to Control Pollutant Emissions from Industrial Boilers, Phase II*, EPA-600/2-76-086a, U.S. Environmental Protection Agency, Washington, DC, 1976.
16. W. A. Carter and H. J. Buening, *Thirty-Day Field Tests of Industrial Boilers—Site 5*, EPA Contract No. 68-02-2645, KVB Engineering, Inc., Irvine, CA, 1981.
17. W. A. Carter and H. J. Buening, *Thirty-Day Field Tests of Industrial Boilers—Site 6*, EPA Contract No. 68-02-2645, KVB Engineering, Inc., Irvine, CA, 1981.
18. *ECS Update*, Summer 1991, No. 23, published by EPRI, Palo Alto, CA, 1991.

BIBLIOGRAPHY

C. E. Blakesless and H. E. Burbock, "Controlling NO_x Emissions from Steam Generators," *J. APCA* **23**, 37–42 (1979).

R. L. Chass and R. E. George, "Contaminant emissions from the combustion of fuels," *J. APCA* **10**, 34–43 (1980).

Confidential Information, American Gas Association Laboratories, Cleveland, OH, 1970. *ECS Update*, Summer 1991, No. 23, published by EPRI, Palo Alto, CA, 1991.

G. C. England et al., *Hazardous Air Pollutant Emissions from Gas-Fired Combustion Sources: Emissions and Effects of Design and Fuel Type*, Presentation at the Air and Waste Management Association Annual Meeting and Exhibition, June 1998, San Diego, California.

H. H. Hovey et al., *The Development of Air Contaminant Emission Tables for Non-Process Emission*, New York State Department of Health, Albany, NY, 1965.

D. M. Hugh et al., *Exhaust Gases from Combustion and Industrial Processes*, EPA Contract EHSD 71-36, Engineering Science, Inc., Washington, D.C., 1971.

L. K. Jain et al., *State of the Art for Controlling NO_x Emissions: Part 1, Utility Boilers*, EPA Contract No. 68-02-0241, Catalytic, Inc., Charlotte, NC, 1972.

K. J. Lim et al., *Technology Assessment Report for Industrial Boiler Applications: NO_x Combustion Modification*, EPA Contract No. 68-02-3101, Acurex Corp., Mountain View, CA, 1979.

J. H. Perry (Ed.), *Chemical Engineers' Handbook*, 4th ed., McGraw-Hill Book Co., New York, 1963.

H. J. Taback et al., *Fine Particle Emissions from Stationary and Miscellaneous Sources in the South Coast Air Basin*, California Air Resources Board Contract No. A6-191-30, KVB, Inc., Tustin, CA, 1979.

W. H. Thrasher and D. W. Dewerth, *Evaluation of the Pollutant Emissions from Gas Fired Water Heaters*, Research Report No. 1507, American Gas Association, Cleveland, OH, April 1977.

Unpublished data on domestic gas fired units, U.S. Environmental Protection Agency, Cincinnati, OH, 1970.

WOOD WASTE

ARIJIT PAKRASI AND ANTHONY J. BUONICORE

INTRODUCTION

Wood has always been an important boiler fuel in the forest products industry. Each year considerable quantities of timber are harvested and millions of tons of wood products manufactured. In the process, very large quantities of wood "wastes" or residues are generated. Logging operations alone produce billions of cubic feet of residues. In fact, in conventional logging, nearly 40% of the harvested tree is left behind on the forest floor, as well as diseased and rotted trees. Primary processing and manufacture of the logs create additional supplies of residues, including trimmings, edging, sawdust, sander dust, bark, and low-grade chips.

Such residues can represent as much as 50% of the original log that reaches the mill. These wastes also represent a considerable opportunity for energy recovery. Heating values typically range between 7500 and 9000 Btu/lb on a dry basis, with as-fired heating values between 4000 and 5000 Btu/lb.

Local governments also add to the available supply of wood residues. Large volumes of wood "wastes" are developed as communities construct roads, sewers, transmission lines, drainage basins, wastewater treatment plants, and reservoirs. Cleanup operations after storms add to the wood waste supply, as do the normal maintenance of parks and tree-lined streets and local construction and demolition activities. Traditionally, municipalities have discarded these wastes in dumps or landfills. However, as disposal sites become less available, alternatives must be investigated. Wood waste-to-energy projects have been developed and continue to be proposed as one such alternative.

The type of wood waste depends on the industry. Generally, bark is the major wood waste in the pulp industry. A mixture of wood waste and bark or wood waste alone is commonly burned in the lumber, furniture, and plywood industries.

PROCESS DESCRIPTION

Various boiler firing configurations are used in burning wood waste. One common type in smaller operations is the Dutch oven, or extension type of furnace with a flat grate. This unit is widely used because it can burn fuels with very high moisture.

Fuel is fed into the oven through apertures atop a firebox and is fired in a cone-shaped pile on a flat grate. The burning is done in two stages: drying and gasification and the combustion of gaseous products. The first stage takes place in a cell separated from the boiler section by a bridge wall. The combustion stage takes place in the main boiler section. Unfortunately, the Dutch oven is not adequately responsive to changes in steam load, and it provides poor combustion control.

In another type, the fuel cell oven, fuel is dropped onto suspended fixed grates and is fired in a pile. Unlike the Dutch oven, the fuel cell also uses combustion air preheating and repositioning of the secondary and tertiary air injection ports to improve boiler efficiency.

In many large operations, more conventional boilers have been modified to burn wood waste. These units may include spreader stokers with traveling grates, vibrating grate stokers, and so on, as well as tangentially fired or cyclone-fired boilers. The most widely used of these configurations is the spreader stoker. Fuel is dropped in front of an air jet, which casts the fuel out over a moving grate, spreading it in an even, thin blanket. The burning is done in three stages in a single chamber: (1) drying, (2) distillation and burning of volatile matter, and (3) burning of carbon. This type of operation has a fast response to load changes, has improved combustion control, and can be operated with multiple fuels. This is done to maintain constant steam when the wood waste supply fluctuates and/or to provide more steam than is possible from the waste supply alone.

An alternative of spreader-stoker fired boiler is the suspension-fired boiler. In this type, small-sized wood waste (generally less than 2 mm) are combusted by supporting in air rather than on fixed grates. This type of boiler can respond very well to rapid changes in steam demand.

Fluid-bed combusters using a sand-bed medium have also been developed to burn wood residues. These units are able to combust fuels with very high moisture content (up to approximately 65% by weight). The waste fuel is typically hogged to minus 3 in. Some of the sand in the bed with ash is continuously removed and passed over a vibrating screen. Material suitable for reinjection is returned to the bed. The material removed from the bed is usually fused ash or gravel and rocks initially present in the hogged fuel.

Sander dust is often burned in various boiler types at plywood, particleboard, and furniture plants. Sander dust contains fine wood particles with low moisture content (less than 20% by weight). It is fired in a flaming horizontal torch, usually with natural gas as an ignition aid or supplementary fuel.

AIR EMISSIONS CHARACTERIZATION

In wood-fired boiler plants, the air pollutant of primary concern is particulate matter. Removal of sulfur oxides is required only when large amounts of high-sulfur coal or oil, together with wood waste, are burned. Nitrogen oxide emissions are considerably less for wood-fired steam generators than for coal-fired boilers of comparable size, despite the fact that larger quantities of excess air are normally used in burning wood. Generally, nitrogen oxide emissions can be controlled by boiler design; however, in some locations, regulations have required the installation of nitrogen oxide emission control systems.

If demolition debris is part of the wood waste fuel, additional air pollution control equipment may be required, depending on what and how much is burned. In addition to waste wood, demolition debris can contain plastics, paint, creosote-treated wood, glues, synthetics, wire, cable, insulation, and so forth. Such materials can contribute to acid gas emissions (such as HCl), heavy metal emissions (including lead, cadmium, chromium, copper, and zinc), inorganic emissions (including arsenic, cyanides, and asbestos), and organic emissions (including formaldehyde, dioxins, and furans).

Actual emissions will depend on many variables, including:

- Composition of the wood waste fuel and its variability
- Type and amount of fossil fuel burned in combination with the wood waste (if any)
- Firing method and type of furnace
- Extent of carbon reinjection
- Air pollution control system

The composition of wood waste depends largely on the industry from which it originates. Pulping operations, for example, produce great quantities of bark that may contain more than 70% by weight moisture, along with sand and other noncombustibles. Because of this, bark boilers in pulp mills may emit considerable amounts of particulate matter to the atmosphere unless they are well controlled. On the other hand, some operations, such as furniture manufacturing, produce a clean dry wood waste, 5–50% by weight moisture, with relatively low particulate emissions when properly burned. Still other operations, such as sawmills, burn a varying mixture of bark and wood waste that results in particulate emissions somewhere between these two extremes.

Furnace design and operating conditions are particularly important when firing wood waste. For example, because of the high moisture content that can be present in this waste, a larger than usual area of refractory surface is often necessary to dry the fuel before combustion. In addition, sufficient secondary air must be supplied over the fuel bed to burn the volatiles that account for most of the combustible material in the waste. When proper drying conditions do not exist or when secondary combustion is incomplete, the combustion temperature is lowered, and increased particulate, carbon monoxide, and hydrocarbon emissions may result. The lowering of combustion temperature generally means decreased nitrogen oxide emissions. Also, emissions can fluctuate over short periods if significant variations occur in fuel moisture content.

Fly-ash reinjection, which is common to many large boilers to improve fuel efficiency, has a considerable effect on particulate emissions. Because a fraction of the collected fly ash is reinjected into the boiler, the dust loading from the furnace, and consequently from the collection device, increases significantly per unit of wood waste burned. It is reported that full reinjection can cause a tenfold increase in the dust loadings of some systems, although an increase of 1.2 to 2 times is more typical for boilers using 50–100% reinjection. A major factor affecting this dust loading increase is the extent to which the sand and other noncombustibles can be separated from the fly ash before reinjection to the furnace.

Although reinjection increases boiler efficiency from 1% to 4% and reduces emissions of uncombusted carbon, it increases boiler maintenance requirements, decreases average fly-ash particle size, and makes particulate collection more difficult. Properly designed reinjection systems should separate sand and char from the exhaust gases, reinject the larger carbon particles to the furnace, and divert the fine sand particles to the ash disposal system.

Uncontrolled particulate emissions of wood waste-fired boilers generally range between 0.5 and 5 g/dscf or 1.5–6.0 pounds per million Btu heat input, or 7–50 pounds per ton of wood waste burned. Approximately 80–95% of the ash in wood waste becomes fly ash off the boiler. The particle size distribution of this fly ash may range from 10% to 30% by weight less than 10 μm. In some cases, up to 90% of the particles by mass may be less than 10 μm. The particulate emission problem becomes more complex if the wood waste fuel contains salt residue resulting from seawater-floated logs. The salt is discharged as a submicrometer-sized fume much more difficult to capture in any air pollution control equipment.

Uncontrolled nitrogen oxide emissions are typically less than 200 ppmv, sulfur dioxide emissions less than 10 ppmv, carbon monoxide emissions less than 200 ppmv, and total hydrocarbons less than 10 ppmv. Reasonable emission factors for modem boiler designs are limited to 0.165 lbs/MM Btu for nitrogen oxides and 0.06–0.10 lb/MM Btu for hydrocarbons. Sulfur dioxide emissions rarely exceed 0.02 lb/MM Btu with wood waste alone.

Most of the greenhouse gases from wood waste combustion is carbon dioxide (CO_2) though traces of methane (CH_4) and nitrous oxide (N_2O) may be also present depending on combustion conditions. Common toxic air emissions are dioxins, furans, polycyclic organic matters (POMs), metals, and other organic compounds.

Emission factors for of air pollutants from wood waste combustion are presented in Tables 1 to 4.

Table 1. Emission Factors for Particulate Matter (PM), Particulate Matter Less Than 10 μm (Pm-10), and Lead (Pb) from Wood Waste Combustion[a]

Source Category	PM Emission Factor (lb/ton)	PM-10 Emission Factor (lb/ton)	Pb Emission Factor (lb/ton)
Bark-fired boilers			
Uncontrolled	47	17	2.9E-03
Mechanical collector with flyash reinjection	14	11	ND
without flyash reinjection	9.0	3.2	ND
Wet scrubber	2.9	2.5	ND
Wood/bark-fired boilers			
Uncontrolled	7.2	6.5	ND
Mechanical collector with flyash reinjection	6.0	5.5	3.2E-04[c]
without flyash reinjection	5.4	1.7	3.2E-04[c]
Wet scrubber	0.48	0.47	3.5E-04
Electrostatic precipitator	0.04	ND	1.6E-05
Wood-fired boilers			
Uncontrolled	8.8	ND	ND
Mechanical collector without flyash reinjection	4.2	2.6[b]	3.1E-04
Electrostatic precipitator	0.17	ND	1.1E-03

[a]Units are lb of pollutant/ton of wood waste burned. To convert from lb/ton to kg/Mg, multiply by 0.5. To convert from lb/ton to lb/MMBtu, multiply by 0.11. Emission factors are based on wet, as-fired wood waste with average properties of 50 wt % moisture and 4500 Btu/lb higher heating value. PM-10 = particulate matter less than 10 μm. ND = no data. NA = not applicable.

[b]Based on one test in which 61% of emitted PM was less than 10 μm in size.

[c]Due to lead's relative volatility, it is assumed that flyash reinjection does not have a significant effect on lead emissions following mechanical collectors.

Source: AP-42.

Table 2. Emission Factors for Nitrogen Oxides (NO_x), Sulfur Oxides (SO_x), and Carbon Monoxide (CO) from Wood Waste Combustion[a]

Source Category	NO_x^b Emission Factor (lb/ton)	SO_x^c Emission Factor (lb/ton)	CO^d Emission Factor (lb/ton)
Fuel cell/Dutch oven boiler	0.38 (0.0033–1.5)	0.075 (0.01–0.2)	6.6 (0.65–21)
Stoker boilers	1.5 (0.66–3.6)	0.075 (0.01–0.2)	13.6 (1.9–80)
FBC boilers	2.0	0.075 (0.01–0.2)	1.4 (0.47–24)

[a]Units are lb of pollutant/ton of wood waste burned. To convert from lb/ton to kg/Mg, multiply by 0.5. To convert from lb/ton to lb/MMBtu, multiply by 0.11. Emission factors are based on wet, as-fired wood waste with average properties of 50 wt % moisture and 4500 Btu/lb higher heating value. FBC = fluidized bed combustion.

[b]NO_x formation is primarily a function of wood nitrogen content. Values in parentheseis represent the range of emission factors.

[c]Lower limit of the range (in parentheses) should be used for wood and higher values for bark.

[d]Values in parenthesis represent the range of emission factors.

Source: AP-42.

Table 3. Emission Factors For Total Organic Compounds (TOC), Methane (CH_4), Nitrous Oxide (N_2O), and Carbon Dioxide (CO_2) From Woodwaste Combustion[a]

Pollutant	Emission Factor[b] (lb/ton)
TOC[c]	0.18–0.22
CH_4	0.01–0.1
N_2O	0.04–0.2
CO_2	1800–2000

[a]Units are lb of pollutant/ton of wood waste burned. To convert form lb/ton to kg/Mg, multiply by 0.5. To convert from lb/ton to lb/MMBtu, multiply by 0.11. Emission factors are based on wet, as-fired wood waste with average properties of 50 wt % moisture and 4500 Btu/lb higher heating value. FBC = fluidized bed combustion.

[b]Emission factor range based on type of boiler.

[c]Emissions measured as total hydrocarbons.

Source: AP-42.

AIR POLLUTION CONTROL MEASURES

Since the air pollutant of primary concern is particulate matter, air pollution control equipment designed to remove particulates will be the principal focus.

Particulate control equipment used on wood-fired boilers generally falls into one of four categories:

1. Mechanical collectors
2. Wet scrubbers
3. Electrostatic precipitators (ESPs) — dry and wet
4. Filters

Multitube cyclonic mechanical collectors have traditionally been the sole source of particulate control for many existing wood-fired boiler plants. These consist of a series of small-diameter (typically 9-in.-inner-diameter) cyclonic collector tubes operating in parallel at a total gas pressure drop typically in the 3 in. w.c. range. Unfortunately, mechanical collectors alone are not capable of meeting the current New Source Performance Standard (NSPS) for wood-fired boilers. Consequently, these systems are most often used in series with scrubbers, ESPs, or fabric filters. Large pieces of char and much of the sand can be removed from the gas stream before it flows to these secondary dust collectors. Reinjection of the char by mechanical collectors can increase boiler efficiency by as much as 4%.

The most common type of scrubber used for particulate removal on wood-fired boilers is the venturi scrubber, operating

Table 4. Emission Factors of Toxic Air Pollutants (Uncontrolled) for Industrial Wood-Fired Boilers[a]

Pollutant	Emission Factor	Remarks
Dioxins		Salt-laden wood
Heptachlorodibenzo-*p*-dioxins	39 ng/dscm @ 3% O_2	
Hexachlorodibenzo-*p*-dioxins	49 ng/dscm @ 3% O_2	
Octachlorodibenzo-*p*-dioxins	11 ng/dscm @ 3% O_2	
Pentachlorodibenzo-*p*-dioxins	48 ng/dscm @ 3% O_2	
2,3,7,8-Tetrachlorodibenzo-*p*-dioxins	0.28 ng/dscm @ 3% O_2	
Tetrachlorodibenzo-*p*-dioxins	47 ng/dscm @ 3% O_2	Not including 2,3,7,8-isomer
Furans		Salt-laden wood
Heptachlorodibenzo-*p*-furans	6.5 ng/dscm @ 3% O_2	
Hexachlorodibenzo-*p*-furans	15 ng/dscm @ 3% O_2	
Octachlorodibenzo-*p*-furans	0.92 ng/dscm @ 3% O_2	
Pentachlorodibenzo-*p*-furans	23 ng/dscm @ 3% O_2	
2,3,7,8-Tetrachlorodibenzo-furans	1.8 ng/dscm @ 3% O_2	
Tetrachlorodibenzo-furans	37 ng/dscm @ 3% O_2	Not including 2,3,7,8-isomer
Polycyclic organic matter	0.00044–0.0062 lb/ton dry wood	Gaseous and particulate POM
Acenaphthene	0.0000022–0.00011 lb/ton	Particulate control with mechanical collector
Benzene	0.00995	As above
Acetaldehyde	0.00192	As above
Formaldehyde	0.0082	As above
Naphthalene	0.00339	As above
Anthracene	0.0000044 lb/ton	As above
Barium	0.007 lb/ton	As above
Benzo(k)fluoranthene	0.0000044 lb/ton	As above
Chromium	0.0003 lb/ton	As above
Chrysene	1× 10E−9 lb/ton	As above
Cobalt	0.0000176–0.0002 lb/ton	As above
Copper	0.0004–0.001 lb/ton	As above
Fluoranthene	0.0000066–0.000028 lb/ton	As above
Lead	0.00022–0.00040 lb/ton	As above
Naphthalene	2 × 10E−7 lb/ton	As above
Nickel	0.0052–0.0064 lb/ton	As above
Phenanthrene	0.000044–0.00032 lb/ton	As above
Phenol	0.000022–0.0002 lb/ton	As above
Pyrene	0.0000044–0.000013 lb/ton	As above
Silver	0.00032–0.034 lb/ton	As above

[a]*Source*: AP-42.

at gas pressure drops ranging between 6 and 25 in. w.c. The venturi scrubber normally is provided with an adjustable throat to maintain constant pressure drop (and, therefore, particulate collection efficiency) at variable boiler load. Liquid-to-gas ratios in the venturi system typically range between 8 and 10 gallons of water per 1000 acfm saturated. Solids buildup in the recirculation loop rarely is allowed to exceed 5% by weight, and wastewater treatment facilities may be necessary to treat the scrubber blowdown. Particulate collection efficiency of 90% have been reported.

High-carbon ash is not easily collected by an ESP. Hence, the highly conductive (very low-resistivity) ash off wood-fired boilers often necessitates a very conservative design. Specific collections areas (SCA) in the 300–500 ft^2/1000 acfm area and power requirements at the 150–400 watts per 1000 acfm level would be reasonable, depending on the efficiency required and the nature of the fuel fired.

Superficial gas velocities typically range between 3 and 5 fps. While the use of a wet scrubber eliminates concern about fires, the potential for a fire must be considered in the design if a dry ESP is to be used. For example, glowing char can cause hopper fires if air is allowed to leak in via the hopper discharge port. For this reason, the induced-draft fan is installed between the mechanical collector and precipitator. In this way, the precipitator is slightly pressurized. Figure 1 shows an ESP installed on a wood-fired boiler. Collection efficiencies of 93 to 99.8% have been reported for ESPs operating on wood-fired boilers.

To eliminate the fire concern, ESPs can operate in the wet mode (Figure 2). In such systems, collection plates and internal parts are wetted continuously with water. Because of lower gas volumes, a wet precipitator would be smaller in size than the equivalent dry version. A prequench is typically used to saturate the gas stream and ensure that "condensable" pollutants have condensed into particulate and can be collected in the unit. Water is recirculated around the precipitator, but not completely, since a small blowdown stream is necessary. Wet precipitators on wood-fired boilers are capable of achieving outlet grain loadings similar to dry precipitators, that is, 0.014.02 g/dscf corrected to 12% CO_2 at power levels typically in the 30–150 watts per 1000 acfm range.

Fabric filters have only limited experience on wood-fired boilers, despite the advantage of very high removal efficiency for fine particulates. The majority of these systems are operating on the West Coast, with more than half on boilers burning bark from logs transported in saltwater. The filters were installed to collect the chloride fumes that resulted from the bark combustion. Operating experience indicates that collected chloride can act as a fire inhibitor. The principal concern with the use of a fabric filter is the potential for fire damage resulting from burning cinders, temperature excursions, and/or operating upsets. Both glass and Nomex bags have been used, with bag life generally not exceeding 18 months to 2 years. Reverse air-cleaned fabric filters generally operate at an air-to-cloth ratio of 1–2 (acfm/ft^2). Pulse jet-cleaned fabric filters generally operate with an air-to-cloth ratio in the range of 4 to 5. Particulate collection efficiency up to 98% has been reported.

Gravel-bed filters have been developed to eliminate the potential for fire hazard. Instead of a fabric medium, gravel-bed filters have a slowly moving bed of granular "rock" as the filtration medium through which the flue gas must travel. To enhance efficiency, these systems have been electrostatically augmented (10–20 watts/1000 acfm). An electrical conductor configured in the form of a cage is positioned within the medium bed. A high voltage (50 kV) is applied to this conductor, and the electrical field generated between the conductor and the inlet and outlet louvers enhances particulate collection. Gas pressure drops are typically in the 5–6 in. w.c. range.

Performance experience with particulate removal air pollution control equipment is presented in Table 5. For a major new wood-fired source that must comply with a 0.1 lb/MM Btu particulate emission limitation, the alternatives include a

Figure 1. Electrostatic Precipitator on a 300,000-lb/h Wood-Fired Boiler (Courtesy PPC Industries).

Figure 2. Wet Electrostatic Precipitator on a Wood-Waste-Fired Boiler (Courtesy United McGill Corp.).

high-energy venturi scrubber, an ESP, an electrostatically augmented moving granular-bed filter, or a fabric filter. The principal disadvantages of the venturi scrubber are the relatively high energy utilization and the need for disposal or utilization of the scrubber blowdown stream. Electrostatic precipitators are relatively expensive, but have a satisfactory operating history on wood-waste-fired boilers. The potential for fires in fabric filters may make it necessary to provide a bypass (for temperature excursions and operating upsets), which may complicate the permitting process.

If nitrogen oxides removal is required and combustion modifications (such as staged combustion and lowering the excess air) for existing units or design techniques for new units are not sufficient, ammonia injection (for example, the Exxon DeNOX process) can be used. Such systems have already been used on wood-waste-fired boilers in California. The process requires the injection of ammonia into the hot flue gases off the boiler at a point where the temperature is in the range of 1700°F to 1800°F. Nitrogen oxide removal efficiencies in the 50–70% range are possible with this technology.

Table 5. Performance Expectation of Air Pollution Control Equipment on New Wood — Waste-Fired Boilers

		Outlet Particulate Loading	
Device	Efficiency	(lb/MMBtu)	(g/dscf at 12% CO_2)
1. Mechanical-collector multitube, cyclonic	85–92	0.30–0.60	0.15–0.25
2. Two mechanical collectors in series	90–94	0.20–0.30	0.08–0.15
3. Venturi scrubber			
$\Delta P = 6–8$ in. w.c.	95–96	0.10–0.15	0.04–0.06
$\Delta P = 10–15$ in. w.c.	97–98	0.06–0.08	0.025–0.03
$\Delta P = 20–25$ in. w.c.	99	0.03–0.05	0.015–0.02
4. Electrostatic precipitator	99–99.5	0.025–0.05	0.01–0.02
5. Granular-bed (moving) filter	90–95	0.20–0.30	0.10–0.15
6. Electrostatically augmented granular-bed (moving) filter	98–99.2	0.03–0.1	0.01–0.04
7. Fabric filter	99.5–99.9	0.01–0.03	0.004–0.015

Should acid-gas removal equipment be required because of the use of high-sulfur auxiliary fuel or the addition of demolition debris containing plastics or other chloride-containing matter, the following alternatives are possible:

- Use of limestone in a fluid-bed boiler
- Addition of a spray dryer acid-gas reactor using a lime slurry upstream of the fabric filter or ESP
- Injection of lime/limestone in the ductwork upstream of the ESP or fabric filter
- Addition of a packed-tower absorber using a sodium hydroxide (or some other alkaline-based) scrubbing solution downstream of the venturi or other particulate removal device

REFERENCES

1. *Atmospheric Emissions from the Pulp and Paper Manufacturing Industry*, EPA-450/1-73-002, U.S. Environmental Protection Agency, Research Triangle Park, NC, September 1973.
2. S. F. Galeano and K. M. Leopold, "A Survey of Emissions of Nitrogen Oxides in the Pulp Mill," *J. Tech. Assoc. Pulp Paper Indus.* **56**(3),74–76 (March 1973).
3. *Control of Particulate Emissions from Wood Fired Boilers*, EPA-340/1-77-026, U.S. Environmental Protection Agency, Research Triangle Park, NC, 1978.
4. *Wood Residue Fired Steam Generator Particulate Matter Control Technology Assessment*, EPA-450/2-78-044, U.S. Environmental Protection Agency, Research Triangle Park, NC, October 1978.
5. H. S. Oglesby and R. O. Blosser, "Information on the Sulfur Content of Bark and its Contribution to SO_2 Emissions when Burned as a Fuel," *J. Air Pollution Control Assoc.* **30**(7), 769–72 (1980).
6. *A Study of Nitrogen Oxides Emissions from Wood Residue Boilers*, Technical Bulletin No. 102, National Council of the Paper Industry for Air and Stream Improvement, New York, November 1979.
7. A. Nunn, *NO_x Emission Factors for Wood Fired Boilers*, EPA-600/7-79-219, U.S. Environmental Protection Agency, September 1979.
8. *Nonfossil Fueled Boilers—Emission Test Report*, Weyerhaeuser Company, Longview, Washington, EPA-80-WFB10, Office of Air Quality Planning and Standards, U.S. Environmental Protection Agency, Research Triangle Park, NC, March 1981.
9. *A Study of Wood Residue Fired Power Boiler Total Gaseous Non-methane Organic Emissions in the Pacific Northwest*, Technical Bulletin No. 109, National Council of the Paper Industry for Air and Stream Improvement (NCASI), New York, September 1980.
10. *Volatile Organic Carbon Emissions from Wood Residue Fired Power Boilers in the Southeast*, NCASI, Technical Bulletin No. 455, New York, April 1985.
11. *A Study of Factors That Affect Nitrogen Oxides and Carbon Monoxide Emissions from Wood-Residue Fired Boilers*, NCASI, Technical Bulletin No. 448, New York, December 1984.
12. *Carbon Monoxide Emissions from Selected Combustion Sources Based on Short-Term Monitoring Records*, NCASI, Technical Bulletin No. 416, New York, January 1984.
13. *Nonfossil Fuel Fired Industrial Boilers—Background Information*, EPA-450/3-82-007, March 1982.
14. W. O. Dameworth, "Application of an ESP for the Control of Particulate Emissions from a Wood-Residue-Fired Boiler," *J. TAPPI*, 225 (July 1989).
15. A. Pope, et al., *Toxic Air Pollutant Emission Factors—A Compilation for Selected Air Toxic Compounds and Sources*, 2nd ed., EPA-450/2-90-01 1, U.S. Environmental Protection Agency, October 1990.
16. *National Dioxin Study: Report to Congress*, EPA-530/SW87-025, Environmental Protection Agency Office of Solid Waste and Emergency Response, Washington, D.C., August 1987.
17. G. W. Brooks, et al., *Locating and Estimating Air Emissions from Sources of Polycyclic Organic Matter (POM)*, EPA450/4-84-007p, U.S. Environmental Protection Agency, Research Triangle Park, NC, May 1988.
18. *Environmental Assessment of a Wood Waste-Fired Industrial Watertube Boiler*, ORD Project Summary, EPA-600/S7-87/ 012, AEERL, U.S. Environmental Protection Agency, Research Triangle Park, NC, May 1987.

BIBLIOGRAPHY

A. J. Buonicore and D. Scerbo, "Air Pollution Control Alternatives for Wood Waste Combustion Facilities," Air and Waste Management Association Conference on Air Quality Issues Related to Power Production, Enfield, CT, October 24–25, 1989.

L. Theodore and A. J. Buonicore, *Industrial Air Pollution Control Equipment for Particulates*, CRC Press, Boca Raton, FL, 1976.

INTERNAL COMBUSTION SOURCES

Michael J. Barboza, Michael J. Cannon,
Nicholas J. Charno, and P. Steven Oliver

Internal combustion sources include stationary gas turbines, sometimes referred to as combustion turbines, and reciprocating internal combustion engines (RICE). Internal combustion sources are often used in applications similar to those associated with external combustion sources (e.g., boilers). For example, combustion turbines and/or boiler/steam turbine combinations are commonly used for cogeneration (concurrent production of steam and electricity). The ability or inability to burn certain fuels will often dictate whether an internal or external combustion source will be utilized for a given industrial application (e.g., cogeneration). The paper industry typically utilizes boiler/steam turbine combinations (external combustion devices) because of their ability to burn wood wastes and other solid by-product fuels generated during paper production while combustion turbines, burning natural gas and by-product gases generated by process operations, are commonly used for cogeneration at chemical and petroleum refining facilities. Compared to boiler/steam turbine installations, combustion turbines are typically smaller in size, mass, and initial capital cost. However, boiler/steam turbine installation advantages, relative to combustion turbines, include higher cycle efficiencies and the ability to burn solid fuels.[1]

Stationary gas turbines and reciprocating engines are also used to provide shaft horsepower for a variety of mechanical applications such as generating electric power, pumping gas or other fluids, or compressing air for pneumatic machinery. Selection of gas turbines versus reciprocating engines is generally determined by site specific criteria such as installed capital costs and operating/maintenance costs which typically include: emissions control costs, fuel costs, and installation/structural costs. Operating experience, compatibility with existing equipment and availability are also important considerations in selection of this type of equipment.[2]

STATIONARY GAS TURBINES

Stationary gas turbines find application in the gas and oil industry, emergency power generation facilities, independent electric power producers, electric utilities, and other industrial applications. Oil and gas industries typically use simple cycle gas turbines with a size range of under 300 hp to over 20,000 hp to provide shaft horsepower for oil and gas extraction and transmission. Emergency power generation sites also utilize simple cycle gas turbines, usually under 5,000 hp; to provide backup or emergency power to critical equipment. Independent electrical power producers (IPP) include large industrial complexes and refining facilities that generate their own power and sell excess power to other neighboring sites. Most IPP installations use combined cycle gas turbines ranging from 1,000 hp to over 100,000 hp, however, simple and regenerative gas turbines can also be found at IPP installations. Electric utilities use simple cycle gas turbines ranging from 20,000 hp to over 200,000 hp mostly as peaking units for meeting power demand peaks imposed by large commercial and industrial users on a daily or seasonal basis. Other industrial applications for gas turbines include pulp and paper, chemical, and food processing, where gas turbines are used to produce steam and electricity.[3] Advantages of combustion turbines include: low vibration, high reliability, no requirement for cooling water, suitability for remote operation, lower capital costs than reciprocating engines and lower capital costs than boiler/steam turbine electric power generating plants.[2]

RECIPROCATING INTERNAL COMBUSTION ENGINES

Reciprocating engines combust gaseous and liquid fuels to provide mechanical shaft power for a variety of applications. Large stationary engines are used for electric power generation (both base load and standby service), oil and gas pumping/transport, oil and gas exploration and production, irrigation and nuclear power plant emergency cooling water pump operation. Smaller engines are used in stationary applications including generators, pumps, compressors and nonroad applications. Nonroad engines[4] are generally distinguished from highway engines (mobile sources) in that nonroad engines are used in:

- Motive equipment that propels itself in addition to performing an auxiliary function (e.g., bulldozers grading construction sites)
- Equipment that is intended to be propelled as it performs its function (e.g., lawnmowers)
- Equipment that is stationary but portable (e.g., generators or compressors)
- Motive equipment that propels itself, but is primarily used for offroad functions (e.g., offhighway trucks)

Nonroad engines[4] are also different from other mobile source categories because nonroad engines include:

- A wider range of engine sizes and power ratings
- Equipment applications in which the engines are used are extremely diverse
- Engines that can be used in widely varying equipment applications (e.g., the same engine used in a backhoe can also be used in a drill rig or in an air compressor)

Engines for nonroad equipment range from capacities of less than 37 to over 560 kW. Figure 1 shows the distribution of annual sales of nonroad engine equipment, from 1991 through

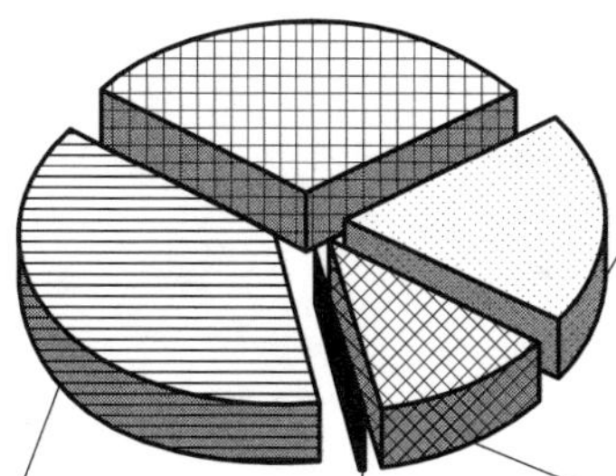

Figure 1. Nonroad Engine Sales Distribution (1991–1995).

Table 1. Selected Nonroad Equipment Applications[4]

Category	Applications		
Agriculture	Tractor	Sprayer	Skidder
	Baler	Swather	Combine
Construction	Backhoe	Off-highway truck	Rubber-tired loader
	Bore/drill rig	Paver	Scraper
	Cement mixer	Paving equipment	Signal board
	Crawler tractor	Plate compactor	Skid-steer loader
	Excavator	Roller	Trencher
	Grader	Rubber-tired dozer	Feller/buncher
General industrial	Concrete saw	Oil field equipment	Scrubber/sweeper
	Crushing equipment	Refrigeration/AC	Rail maintenance
Lawn and garden	Garden tractor	Rear engine mower	Chippers/grinder
Utility	Air compressor	Pump	Irrigation set
	Hydro power unit	Generator set	Welder
	Pressure washer	Aircraft support	
Material handling	Aerial lift	Forklift	Rough-terrain forklift
	Crane	Terminal tractor	
Marine <27 kW	Propulsion	Auxiliary	

1995, by engine size, and indicates that 67% of these sales are less than 75 kW.[4]

Nonroad equipment can be grouped into the following categories: agriculture and logging, construction, general industrial, lawn and garden, utility, material handling, and small marine. Nonroad engines are also used in locomotives, large marine applications (rated over 37 kW), aircraft, and underground mining equipment. Table 1 contains examples of the types of nonroad equipment in each category.[4]

REFERENCES

1. M. M. El-Wakil, *Powerplant Technology*, McGraw-Hill, Inc., New York, 1984.
2. *Alternative Control Techniques Document—NO_x Emissions from Stationary Gas Turbines*, EPA 53/R-93-007, January 1993.
3. *Emission Factor Documentation for AP-42 Section 3.1 Stationary Gas Turbines*, Draft Document, U.S. Environmental Protection Agency, Office of Air Quality Planning and Standards, EPA Contract No. 1.68-D7-0070, Alpha-Gamma Technologies Inc., Research Triangle Park, NC, May 1998.
4. *Final Regulatory Impact Analysis: Control of Emissions from Nonroad Diesel Engines*, EPA, August 1998.

STATIONARY GAS TURBINES

Michael J. Barboza, Michael J. Cannon, Nicholas J. Charno, and P. Steven Oliver

Gas turbines are internal combustion engines that operate with rotary rather than reciprocating motion. Gas turbines are used in a broad scope of applications including electric power generation, cogeneration, natural gas transmission, and for various process applications. Gas turbines are available with power outputs ranging in size from 300 to over 268,000 hp, with an average size of 40,200 hp. The primary fuels used in gas turbines are natural gas and distillate (No. 2) fuel oil. Other fuels such as landfill gas and digester gas can also be burned in gas turbines.[1–4]

PROCESS DESCRIPTION[1–3]

Gas turbines are essentially composed of three major components: compressor, combustor, and power turbine. Figure 1 presents a cutaway view of a dry, low-NO_x, combined cycle combustion turbine with a silo combustor. In the compressor section, ambient air is drawn in and compressed up to 30 times ambient pressure and directed to the combustor section where fuel is introduced, ignited, and burned. Combustors can either be annular, can-annular, or silo. An annular combustor is a doughnut-shaped, single, continuous chamber that encircles the turbine in a plane perpendicular to the air flow. Can-annular combustors are similar to the annular; however, they incorporate several can-shaped combustion chambers rather than a single continuous chamber. A silo combustor has one or more combustion chambers mounted external to the gas turbine body.[1]

The combustion process in a gas turbine can be classified as diffusion flame combustion, or lean premix staged combustion. In the diffusion flame combustion, the fuel/air mixing and combustion take place simultaneously in the primary combustion zone. This generates regions of near-stoichiometric

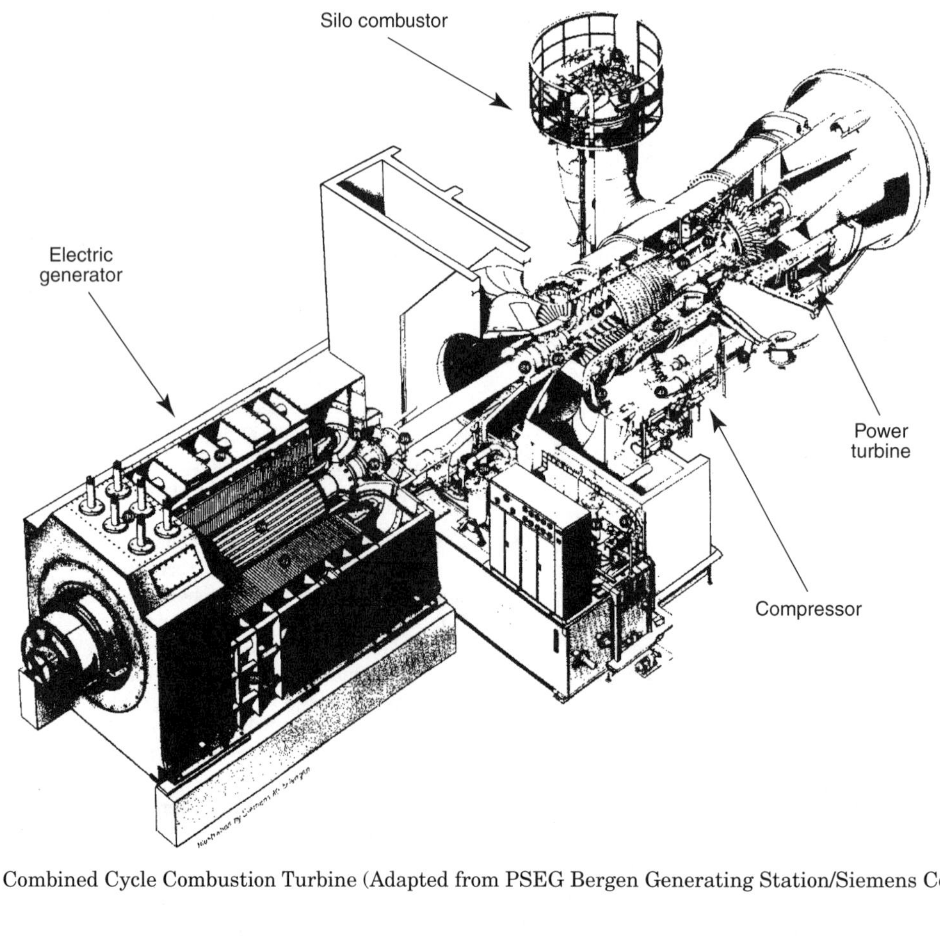

Figure 1. Combined Cycle Combustion Turbine (Adapted from PSEG Bergen Generating Station/Siemens Corporation).

fuel/air mixtures where the temperatures are very high. For lean premix combustors, fuel and air are thoroughly mixed in an initial stage resulting in a uniform, lean, unburned fuel/air mixture which is delivered to a secondary stage where the combustion reaction takes place. Manufacturers use different types of fuel/air staging, including fuel staging, air staging, or both; however, the same staged, lean premix principle is applied. Gas turbines using staged combustion are also referred to as dry, low-NO_x combustors.[1]

Hot gases from the combustion section are diluted with additional air from the compressor section and directed to the power turbine section at temperatures up to 2350°F. Energy from the hot exhaust gases, which expand in the power turbine section, are recovered in the form of shaft horsepower. More than 50% of the shaft horsepower is needed to drive the internal compressor and the balance of recovered shaft horsepower is available to drive an external load. Gas turbines may have one, two, or three shafts to transmit power between the inlet air compression turbine, the power turbine, and the exhaust turbine.[1,2]

Operating Cycles

Gas turbines are classified by the following types of operating cycles:

- Simple cycle
- Regenerative cycle
- Cogeneration cycle
- Combined cycle

The heat content of the exhaust gases exiting the turbine can either be discarded without heat recovery (simple cycle); recovered with a heat exchanger to preheat combustion air entering the combustor (regenerative cycle); recovered in a heat recovery steam generator to raise process steam, with or without supplementary firing (cogeneration); or recovered, with or without supplementary firing, to raise steam for a steam turbine Rankine cycle (combined cycle).[1]

Simple Cycle. The simple cycle is the most basic operating cycle of gas turbines with a thermal efficiency of 32%. Figure 2 presents a schematic of a simple cycle gas turbine. The cycle thermal efficiency is defined as the ratio of the percent of useful shaft energy to fuel energy input. Simple cycle gas turbines are typically used for shaft horsepower without recovery of exhaust heat.[1]

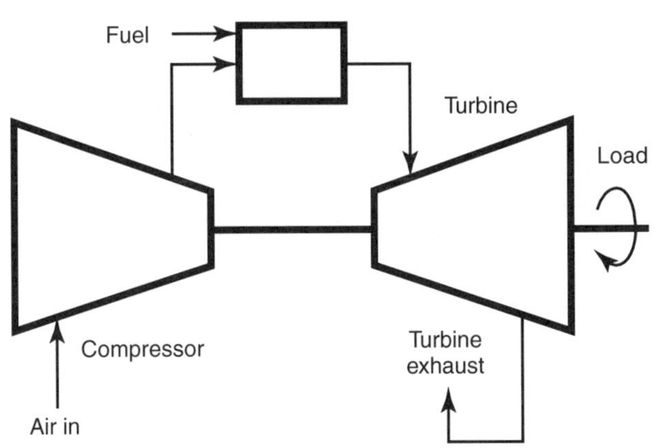

Figure 2. Simple Cycle.

Regenerative Cycle. A regenerative cycle is a simple cycle gas turbine with an added heat exchanger. Figure 3 presents a schematic of a regenerative gas turbine. The heat exchanger preheats the combustion air, which reduces the amount of fuel required to reach combustor temperatures. The thermal efficiency of a regenerative cycle is around 35%; however, the added efficiency may not be sufficient to justify the capital cost of the heat exchanger.[1]

Cogeneration Cycle. A cogeneration cycle consists of a simple cycle gas turbine with a heat recovery steam generator (HRSG). Figure 4 presents a schematic of the cogeneration cycle utilizing a gas turbine. The cycle thermal efficiency can be as high as 75%. In a cogeneration cycle, the steam generated by the HRSG can be delivered at a variety of pressures and temperatures to other thermal processes at the site. For situations where additional steam is required, a supplementary burner, or duct burner, can be placed in the exhaust duct upstream of the HRSG to increase steam production.[1]

Combined Cycle. A combined cycle gas turbine is a gas turbine with an HRSG applied at electric utility sites. The gas turbine drives an electric generator, and the steam from the HRSG drives a steam turbine that also drives an electric generator. Figure 5 presents a schematic of a combined cycle. A supplementary burner can also be used to increase steam

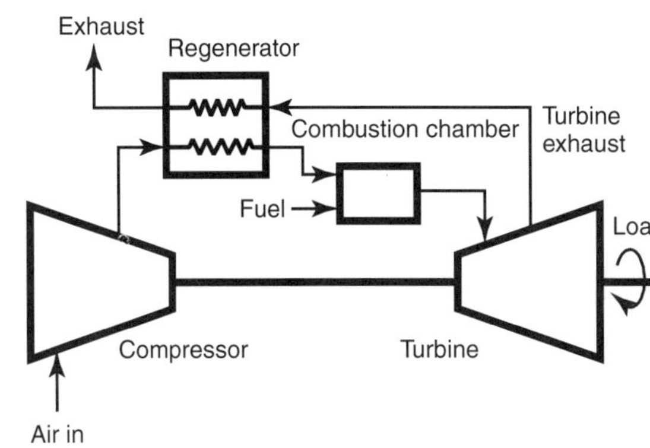

Figure 3. Regenerative Cycle.

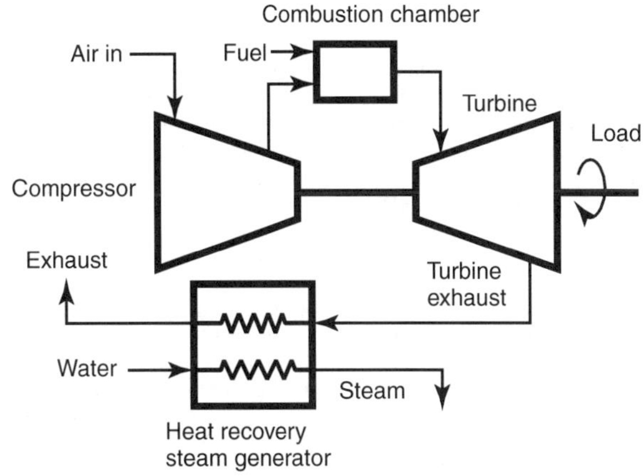

Figure 4. Cogeneration Cycle.

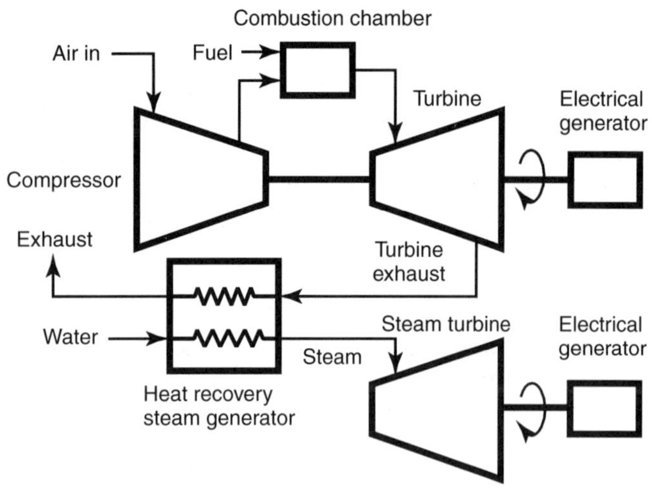

Figure 5. Combined Cycle.

production/electricity generation. The thermal efficiency of a combined cycle gas turbine is approximately 50%.[1]

AIR EMISSIONS CHARACTERIZATION

The primary pollutants from gas turbine engines are nitrogen oxides (NO_x), carbon monoxide (CO), and to a lesser extent, volatile organic compounds (VOC), and particulate matter (PM). NO_x formation is strongly dependent on the high temperatures developed in the combustor. CO, VOC, HAPs (hazardous air pollutants), and PM are primarily the result of incomplete combustion. Ash and metallic additives in the fuel may also contribute to PM in the exhaust. Trace to low amounts of HAPs and SO_x are emitted from gas turbines. Emissions of oxides of sulfur (SO_x), mainly sulfur dioxide (SO_2), are directly related to the sulfur content of the fuel. Significant SO_x emissions can be generated if heavy oils are fired in the turbine. Table 1 presents a compilation of air contaminant emission

Table 1. Emission Factors for Stationary Combustion Turbines

	Emission Factors[a]							
	Natural Gas-Fired Turbines		Distillate Oil-Fired Turbines		Landfill Gas-Fired Turbines		Digester Gas-Fired Turbines	
Pollutant	(lb/10^6 scf)[b]	Emission Factor Rating	(lb/10^3 gal)[c]	Emission Factor Rating	(lb/10^6 scf)[d]	Emission Factor Rating	(lb/10^6 scf)[e]	Emission Factor Rating
CO_2[f]	110,000	B	22,000	C	20,000	C	16,000	C
Lead	<0.016[g]	E	0.0015	C	ND	ND	<0.0020[g]	D
PM[h]	7.4[i]	B	1.6[i]	C	9.3	B	7.2	D
PM (condensible)[i]	4.0	B	1.0	C	ND	ND	ND	ND
PM (filterable)[i]	3.4	B	0.60	C	ND	ND	ND	ND
SO_2	2.3	B	92	D	18	C	3.9	E
THC[j]	8.5	B	ND	ND	ND	ND	ND	ND
VOC[j]	2.8	E	0.51	E	ND	ND	ND	ND
NMHC[j]	0.68	E	ND	ND	5.4	B	3.5	E
NO_x								
Uncontrolled	410	A	93	C	36	B	98	E
Water/steam injection	160	A	35	B	—	—	—	—
Lean-premix	81	B	—	—	—	—	—	—
CO								
Uncontrolled	88[k]	C	2	D	240	B	10	E
Water/steam injection	7.6	B	11	D	—	—	—	—
Lean-premix	5.1	B	—	—	—	—	—	—

[a]Emission factors based on fuel input. Only tests with 90–100% loads were considered for CO and NO_x. NO_x emissions decrease at lower loads, while CO emissions increase at lower loads. Test data indicate that as much as 30% NO_x reduction and 200% CO increase are possible when the unit load is reduced to 75%. ND = No Data.

[b]Emission factors based on an average natural gas heating value of 1020 Btu/scf at 60°F. To convert from (lb/10^6 scf) to (lb/MMBtu), divide by 1020. These emission factors can be converted to other natural gas heating values by multiplying the given emission factor by the ratio of the specified heating value to this heating value.

[c]Emission factors based on an average distillate oil heating value of 139 MMBtu/10^3 gallons. To convert from (lb/10^3 gal) to (lb/MMBtu), divide by 139.

[d]Emission factors based on an average landfill gas heating value of 400 Btu/scf at 60°F. To convert from (lb/10^6 scf) to (lb/MMBtu), divide by 400.

[e]Emission factors based on an average digester gas heating value of 600 Btu/scf at 60°F. To convert from (lb/10^6 scf) to (lb/MMBtu), divide by 600.

[f]Based on 99.5% conversion of the fuel carbon to CO_2 for natural gas and 99% conversion of the fuel carbon to CO_2 for distillate oil. CO_2 (lb/10^6 scf) = (3.67)(%CON)(C)(D) where %CON = weight percent conversion of fuel carbon to CO_2, C = carbon content of fuel by weight, and D = density of fuel. For natural gas, the carbon content is assumed at 75%, and the gas density is assumed at 4.1E + 04 lb/10^6 scf. For distillate oil, the carbon content is assumed at 87%, and the fuel density is assumed at 6.9E + 03lb/10^3 gal. For landfill gas and digester gas, CO_2 is presented in test data as volume percent of the exhaust stream (4 to 4.5%).

[g]Emission factors based on one half the detection limit and not on measured values. Expected emissions are lower than the presented emission factor.

[h]PM emission factor determined using EPA Method 5.

[i]All PM (total, condensible, and filterable) is assumed to be less than 2.5 μm in diameter. Therefore, the PM emission factors presented here may be used to estimate PM-2.5 or PM-10 emissions. Total PM is the sum of the filterable PM and condensible PM. Condensible PM is the particulate matter collected using EPA Method 202 (or equivalent). Filterable PM is the particulate matter collected on, or prior to, the filter of an EPA Method 5 (or equivalent) sampling train.

[j]Only tests with 90–100% loads were considered for VOC, THC, and NMHC. In cases where emissions data are limited, NMHC emission are presented.

[k]CO emissions varied widely for the units tested. The range of emission factors calculated for CO was 1 to 500 lb/10^6 scf with O_2 levels of 14.8% to 17.5%.

Source: Reference 1.

factors for stationary combustion turbines firing natural gas, distillate oil, landfill gas, and digester gas.[1]

Gas turbines firing distillate oil may emit trace metals carried over from the metals content of the fuel. If the fuel analysis is known, the metals content of the fuel ash should be used for flue gas emission factors assuming all metals pass through the turbine.[1]

Digester and landfill gases are by-products of wastewater treatment and landfilling municipal refuse. These gases are formed through anaerobic decomposition of organic materials and are composed principally of methane (50–65%) and carbon dioxide (35–50%). Trace quantities of other compounds including hydrogen sulfide, ammonia, VOC, and PM are also present. Digester and landfill gases are similar in their composition, and their emissions after combustion are very similar to natural gas. However, emissions from digester and landfill gas combustion may also include hydrogen chloride and SO_x emissions (generated from combusting chlorinated and sulfur-bearing compounds), which are typically not generated in significant quantities when combusting natural gas.

If the HRSG is not supplementary fuel fired, the simple cycle input-specific emission factors [pounds per million standard cubic feet ($lb/10^6 scf$) or pounds per thousand gallons ($lb/10^3$ gal)] will also apply to cogeneration/combined cycle systems. The output-specific emission factors [pounds per horsepower-hour (lb/hp h)] are only appropriate for simple cycle turbines and would not be an accurate representation of combined cycle turbine emissions due to differences in cycle efficiencies. If the HRSG is supplementary fired, the emissions attributable to the supplementary firing must also be considered to estimate total stack emissions.[1]

Nitrogen Oxides[1]

Nitrogen oxides formation occurs by three fundamentally different mechanisms. The principal mechanism with turbines firing gas or distillate fuel is thermal NO_x, which arises from the thermal dissociation and subsequent reaction of nitrogen (N_2) and oxygen (O_2) molecules in the combustion air. Most thermal NO_x is formed in high-temperature stoichiometric flame pockets downstream of the fuel injectors, where combustion air has mixed sufficiently with the fuel to produce the peak temperature fuel/air interface. The second mechanism, called prompt NO_x, is formed from early reactions of nitrogen molecules in the combustion air and hydrocarbon radicals from the fuel. Prompt NO_x forms within the flame and is usually negligible when compared to the amount of thermal NO_x formed. The third mechanism, fuel NO_x, stems from the evolution and reaction of fuel-bound nitrogen compounds with oxygen. Natural gas has negligible chemically bound fuel nitrogen (although some molecular nitrogen is present). Essentially all NO_x formed is thermal NO_x. Distillate oils have low levels of fuel-bound nitrogen. Fuel NO_x from distillate oil-fired turbines may become significant in turbines equipped with a high degree of thermal NO_x controls. Otherwise, thermal NO_x is the predominant NO_x formation mechanism in distillate oil-fired turbines.

The maximum thermal NO_x formation occurs at a slightly fuel-lean mixture because of excess oxygen available for reaction. The control of stoichiometry is critical in achieving reductions in thermal NO_x. Thermal NO_x formation also decreases rapidly as the temperature drops below the adiabatic flame temperature, for a given stoichiometry. Maximum reduction of thermal NO_x can be achieved by control of both the combustion temperature and the stoichiometry. Gas turbines operate with high overall levels of excess air, because turbines use combustion air dilution as the means to maintain the turbine inlet temperature below design limits.

In older gas turbine models, where combustion is in the form of a diffusion flame, most of the dilution takes place downstream of the primary flame, which does not minimize peak flame temperature and suppress thermal NO_x formation. Diffusion flames are characterized by regions of near-stoichiometric fuel/air mixtures where temperatures are very high and significant thermal NO_x is formed. Water vapor in the turbine inlet air may lower the peak flame temperature, thereby reducing thermal NO_x emissions. Thermal NO_x can also be reduced in diffusion-type turbines through water or steam injection. The injected water/steam acts as a heat sink, lowering the combustion zone temperature, and therefore thermal NO_x. Newer model gas turbines use lean, premixed combustion where the fuel is typically premixed with more than 50% theoretical air, which results in lower flame temperatures, thus suppressing thermal NO_x formation.

Ambient conditions can affect emissions and power output from gas turbines more than from external combustion systems. The operation at high excess air levels and at high pressures increases the influence of inlet humidity, temperature, and pressure. Variations of emissions of 30% or greater have been exhibited with changes in ambient humidity and temperature. Humidity acts to absorb heat in the primary flame zone through the sensible heat and, if condensation occurs during compression, the latent heat of vaporization, thus suppressing thermal NO_x formation. For a given fuel firing rate, lower ambient temperatures will cause lower peak flame temperatures, resulting in reducing NO_x significantly. Lower barometric pressure will also lower the temperature exiting the compressor which will lower NO_x.

Carbon Monoxide and Volatile Organic Compounds[1]

CO and VOC emissions both result from incomplete combustion. CO results when there is insufficient residence time at high temperature to complete the final step in fuel carbon oxidation. The oxidation of CO to carbon dioxide (CO_2) at gas turbine temperatures is a slow reaction compared to most hydrocarbon oxidation reactions. In gas turbines, failure to achieve CO burnout may result from quenching by dilution air. With liquid fuels, this can be aggravated by carryover of larger droplets from the atomizer at the fuel injector. In gas turbines, CO emissions are usually higher when the unit is operated at low loads.

The pollutants commonly classified as VOCs can encompass a wide spectrum of volatile organic compounds. VOC emissions are emissions of total organic compounds (TOC) less methane and ethane emissions. These organic compounds are discharged into the atmosphere when some of the fuel remains unburned or is only partially burned during the combustion process. With natural gas, some organics are carried over as unreacted, trace constituents of the gas, while others may be pyrolysis products of the heavier hydrocarbon constituents. With liquid fuels, large droplet carryover to the quench zone accounts for much of the unreacted and partially pyrolized volatile organic emissions.

Available data indicate that emission levels of HAPs are lower for gas turbines than for other combustion sources. This is due to the high combustion temperatures reached during normal operation. For No. 2 distillate oil-fired gas turbines, the exhaust stream may include certain metallic HAPs that are carried over from the fuel constituents. No specific control technology is currently implemented for HAP emissions for gas turbines.

Particulate Matter[1,5]

PM emissions from turbines primarily result from carryover of noncombustible trace constituents in the fuel. PM emissions are minimal with natural gas firing and marginally significant with distillate oil firing because of the low ash content. PM emissions can be classified as "filterable" or "condensible" PM. Filterable PM is that portion of the total PM that exists in the stack in either the solid or liquid state and can be measured on an EPA Method 5 filter. Condensible PM is that portion of the total PM that exists as a gas in the stack but condenses in the cooler ambient air to form particulate matter. Condensible PM exists as a gas in the stack; so it passes through the Method 5 filter and is typically measured by analyzing the impingers, or "back half" of the sampling train. The collection, recovery, and analysis of the impingers is described in EPA Method 202 of Appendix M, Part 51 of the Code of Federal Regulations. Condensible PM is composed of organic and inorganic compounds and is typically less than 1 μm in aerodynamic diameter.

Greenhouse Gases[1,7–13]

Carbon dioxide, methane (CH_4), and nitrous oxide (N_2O) emissions are all produced during natural gas and distillate oil combustion in gas turbines. Nearly all of the fuel carbon is converted to CO_2 during the combustion process. This conversion is relatively independent of firing configuration. Although the formation of CO acts to reduce CO_2 emissions, the amount of CO produced is insignificant compared to the amount of CO_2 produced. The majority of the fuel carbon not converted to CO_2 is due to incomplete combustion.

Formation of N_2O during the combustion process is governed by a complex series of reactions, and its formation is dependent upon many factors. Formation of N_2O is minimized when combustion temperatures are kept high (above 1475°F) and excess air is kept to a minimum (less than 1%).

Methane emissions vary with the fuel, combustion temperature, and firing configuration, but are highest during periods of incomplete combustion or low-temperature combustion, such as during the startup or shutdown cycle. Typically, conditions that favor formation of N_2O also favor emissions of CH_4.

AIR POLLUTION CONTROL MEASURES[1,14]

There are three generic types of emission controls in use for gas turbines: wet controls using steam or water injection to reduce combustion temperatures for NO_x control; dry controls using advanced combustor design to suppress NO_x formation and/or promote CO burnout; and postcombustion catalytic control to selectively reduce NO_x and/or oxidize CO formed in the turbine. Other recently developed technologies (e.g., flameless combustion) promise significantly lower levels of NO_x and CO emissions from diffusion combustion-type gas turbines. These technologies are currently being demonstrated in several installations.

Water/Steam Injection

Water or steam injection is a technology that has been demonstrated to effectively suppress NO_x emissions from gas turbines. The effect of steam and water injection is to increase the thermal mass by dilution and thereby reduce peak temperatures in the flame zone. With water injection, there is an additional benefit of absorbing the latent heat of vaporization from the flame zone. Water or steam is typically injected at a water-to-fuel weight ratio of less than one. Depending on the initial NO_x levels, such rates of injection may reduce NO_x by 60% or higher. Water or steam injection is usually accompanied by an efficiency penalty (typically 2 to 3%) but an increase in power output (typically 5–6%). The increased power output results from the increased fuel flow required to maintain turbine inlet temperature at manufacturer's specifications. Both CO and VOC emissions are increased by water injection dependent on the amount of water injection.

Dry Controls

Since thermal NO_x is a function of both temperature (exponentially) and time (linearly), the basis of dry controls are to either lower the combustor temperature using lean mixtures of air and/or fuel staging, or decrease the residence time of the combustor. A combination of methods may be used to reduce NO_x emissions such as lean combustion, reduced combustor residence time, and staged combustion (two-stage lean/lean combustion or two-stage rich/lean combustion).

Lean combustion involves increasing the air-to-fuel ratio of the mixture so that the peak and average temperatures within the combustor will be less than that of the stoichiometric mixture, thus suppressing thermal NO_x formation. Introducing excess air not only creates a leaner mixture, but it also can reduce residence time in the combustor.

Two-stage lean/lean combustors are essentially fuel-staged, premixed combustors in which each stage burns lean. The two-stage lean/lean combustor allows the turbine to operate with an extremely lean mixture while ensuring a stable flame. A small stoichiometric pilot flame ignites the premixed gas and provides flame stability. The NO_x emissions associated with the high-temperature pilot flame are insignificant. Low NO_x emission levels are achieved by this combustor design through cooler flame temperatures associated with lean combustion and avoidance of localized "hot spots" by premixing the fuel and air.

Two-stage rich/lean combustors are essentially air-staged, premixed combustors in which the primary zone is operated fuel rich, and the secondary zone is operated fuel lean. The rich mixture produces lower temperatures (compared to stoichiometric) and higher concentrations of CO and hydrogen (H_2), because of incomplete combustion. The rich mixture also decreases the amount of oxygen available for NO_x generation. Before entering the secondary zone, the exhaust of the primary zone is quenched (to extinguish the flame) by large amounts of air, and a lean mixture is created. The lean mixture is preignited, and the combustion completed in the secondary zone. NO_x formation in the second stage are minimized through combustion in a fuel-lean, lower-temperature environment. Staged combustion is identified through a variety of names, including dry, low NO_x (DLN), dry low emissions (DLE), or SoLoNO$_x$.

Selective Catalytic Reduction Systems

Selective catalytic reduction systems selectively reduce NO_x emissions by injecting ammonia (NH_3) into the exhaust gas stream upstream of a catalyst. Nitrogen oxides, NH_3, and O_2 react on the surface of the catalyst to form nitrogen and water. The exhaust gas must contain a minimum amount of oxygen and be within a particular temperature range (typically 450–850°F) in order for the SCR system to operate properly. The range is dictated by the catalyst, typically made from noble metals, base metal oxides such as vanadium and titanium, or zeolite-based material. Exhaust gas temperatures greater than the upper limit (850°F) cause NO_x and NH_3 to pass through the catalyst unreacted. Ammonia emissions, called NH_3 slip, may be a consideration when specifying an SCR system.

Ammonia, either in the form of liquid anhydrous ammonia or aqueous ammonia hydroxide, is stored on site and injected into the exhaust stream upstream of the catalyst. Although an SCR system can operate alone, it is typically used in conjunction with water/steam injection systems to reduce NO_x emissions to their lowest levels (less than 10 ppm at 15% oxygen for SCR and wet injection systems).

The catalyst and catalyst housing used in SCR systems tend to be very large and dense (in terms of surface-area-to-volume ratio) because of the high exhaust flow rates and long residence times required for NO_x, O_2, and NH_3, to react on the catalyst. Most catalysts are configured in a parallel-plate, "honeycomb" design to maximize the surface-area-to-volume ratio of the catalyst. Some SCR installations incorporate CO catalytic oxidation modules along with the NO_x reduction catalyst for simultaneous CO/NO_x control.

Both digester gas and landfill gas contain a family of silicon-based gases collectively called siloxanes. Siloxanes are found in many cosmetics and cleaning solutions that are disposed of in either landfills or sewers. Combustion of siloxanes forms compounds that can foul fuel systems, combustion chambers, and postcombustion catalysts. The fouling renders catalysts inoperable within a very short time period. Because of this problem, catalytic technology has not been demonstrated to work effectively on combustion devices that burn diesel and either digester gas or landfill gas.

Other Catalytic Reduction Technologies[1,15,16]

Other catalytic reduction technologies have been developed and are currently being commercially demonstrated for gas turbines. Such technologies include, but are not limited to, the $SCONO_x$ and the XONON systems, both of which are designed to reduce NO_x and CO emissions. The $SCONO_x$ system is applicable to natural gas turbines. It is based on a unique integration of catalytic oxidation and absorption technology. CO and NO are catalytically oxidized to CO_2 and NO_2. The NO_2 molecules are subsequently absorbed on the treated surface of the $SCONO_x$ catalyst. The system manufacturer guarantees CO emissions of 1 ppm and NO_x emissions of 2 ppm. Emissions data for gas turbines equipped with the $SCONO_x$ catalyst are not currently available. The $SCONO_x$ system does not require the use of ammonia, eliminating the potential of ammonia slip conditions evident in existing SCR systems.

The XONON system is applicable to diffusion and lean premix combustors and is currently being demonstrated with the assistance of leading gas turbine manufacturers. The system utilizes a flameless combustion system where fuel and air reacts on a catalyst surface, preventing the formation of NO_x while achieving low CO and unburned hydrocarbon emission levels. Emissions data from gas turbines equipped with a XONON catalyst are not currently available. The catalyst manufacturer claims that gas turbines equipped with the XONON catalyst emit NO_x levels below 3 ppm and CO and unburned hydrocarbons levels below 10 ppm. The system is totally contained within the combustor of the gas turbine and is not a process for cleanup of the turbine exhaust. The overall combustion process consists of the partial combustion of the fuel in the catalyst module followed by completion of the combustion downstream of the catalyst. The partial combustion within the catalyst produces no NO_x, and the combustion downstream of the catalyst occurs in a flameless homogeneous reaction that produces almost no NO_x.

REFERENCES

1. *Emission Factor Documentation for AP-42 Section 3.1 Stationary Gas Turbines*, Draft Document, U.S. Environmental Protection Agency, Office of Air Quality Planning and Standards, Research Triangle Park, NC, May 1998.
2. *Alternative Control Techniques Document—NO_x Emissions from Stationary Gas Turbines*, EPA 53/R-93-007, January 1993.
3. C. C. Shih et al., *Emissions Assessment of Conventional Stationary Combustion Systems, Vol. II: Internal Combustion Sources*, EPA-600/7-79-029c, U.S. Environmental Protection Agency, Cincinnati, OH, February 1979.
4. *Final Report—Gas Turbine Emission Measurement Program*, GASLTR787, General Applied Science Laboratories, Westbury, NY, August 1974.
5. *In-Stack Condensible Particulate Matter Measurement and Permitting Issues for Maryland Power Plants*, Maryland Department of Natural Resource, Prepared by Versar, Inc., January 1998.
6. *Standards Support and Environmental Impact Statement, Volume 1: Proposed Standards of Performance for Stationary Gas Turbines*, EPA-450/2-77-017a, U.S. Environmental Protection Agency, Research Triangle Park, NC, September 1977.
7. L. P. Nelson et al., *Global Combustion Sources of Nitrous Oxide Emissions*, Research Project 2333-4 Interim Report, Sacramento: Radian Corporation, 1991.
8. R. L. Peer et al., *Characterization of Nitrous Oxide Emission Sources*, U.S. Environmental Protection Agency, Office of Research and Development, Research Triangle Park, NC, 1995.
9. S. D. Piccot et al., *Emissions and Cost Estimates for Globally Significant Anthropogenic Combustion Sources of NO_x, N_2O, CH_4, CO, and CO_2*, U.S. Environmental Protection Agency, Office of Research and Development, Research Triangle Park, NC, 1990.
10. G. Marland and R. M. Rotty, *Carbon Dioxide Emissions from Fossil Fuels: A Procedure for Estimation and Results for 1951–1981*, DOE/NBB-0036 TR-003, Carbon Dioxide Research Division, Office of Energy Research, U.S. Department of Energy, Oak Ridge, TN, 1983.
11. G. Marland and R. M. Rotty, "Carbon Dioxide Emissions from Fossil Fuels: A Procedure for Estimation and Results for 1950–1982," *Tellus* **36B**, 232–61, 1984.
12. *Inventory of U.S. Greenhouse Gas Emissions and Sinks: 1990–1991*, EPA-230-R-96-006, U.S. Environmental Protection Agency, Washington, D.C., November 1995.
13. *IPCC Guidelines for National Greenhouse Gas Inventories Workbook*, Intergovernmental Panel on Climate Change/Organization for Economic Cooperation and Development, Paris, France, 1995.
14. L. M. Campbell and G. S. Shareef, *Sourcebook: NO_x Control Technology Data*, Radian Corp., EPA-600/2-91-029, Air and Energy

Engineering Research Laboratory, U.S. Environmental Protection Agency, Research Triangle Park, July 1991.

15. *Catalysts for Power Generation, the $SCONO_x$ System.* Goal Line Environmental Technologies, *www.glet.com/gl_prod_sconox.htm*, March 1998.
16. Information from Chuck Solt of Catalytica Combustion System, Inc., to EPA, XONON Flameless Combustion, January 1996.

BIBLIOGRAPHY

D. C. Brandt, *GE Turbine Design Philosophy*, General Electric Company, Schenectady, New York, presented at the 33rd GE Turbine State-of-the-Art Technology Seminar for Industrial, Cogeneration, and Independent Power Turbine Users, 1989.

C. Castaldini, *Evaluation of Water Injection Impacts for Gas Turbine NO_x Control at Compressor Stations*, prepared by Acurex Corp. for the Gas Research Institute, GRI-90/0138, July 1990.

Y. A. Cengel and M. A. Boles, *Thermodynamics—An Engineering Approach, WCB*, McGraw-Hill, New York, 1998.

M. M. El-Wakil, *Powerplant Technology*, McGraw-Hill, Inc., New York, 1984.

Engines, Turbines, and Compressors Directory, Catalog #XF0488, American Gas Association, Arlington, VA, 1985.

J. M. Kovick, *Cogeneration Application Considerations*, General Electric Company, Schenectady, New York, presented at the 33rd GE Turbine State-of-the-Art Technology Seminar for Industrial, Cogeneration, and Independent Power Turbine Users, 1989.

H. Maghon and A. Kreutzer (Siemens Product Group KWU, Muelheim, Germany), and H. Termuehlen (Utility Power Corporation, Bradenton, Florida), "The V84 Gas Turbine Designed for Base-Load and Peaking Duty," presented at the American Power Conference, Chicago, April 18–20, 1988.

"Repowering Old Plants Gains Favor," *Power Engineering* **94**, 25–27, May 1990.

R. W. Smock, "Need Seen for New Utility Capacity in the '90's," *Power Engineering* **93**, 29–31, April 1990.

R. W. Smock, "Coalgas-Fired Combined Cycle Projects Multiply," *Power Engineering* **103**, 32–34, February, 1991.

R. Williams and E. Larson (Princeton University), *Expanding Roles for Gas Turbines in Power Generation*, prepared for Vattenfall Electricity with the support of the Office of Energy of the U.S. Agency for International Development, December 1985.

RECIPROCATING INTERNAL COMBUSTION ENGINES

Michael J. Barboza, Michael J. Cannon, Nicholas J. Charno, and P. Steven Oliver

Reciprocating internal combustion engines (RICE) are utilized in various industrial applications to provide mechanical shaft power. The following engine design characteristics can be used to distinguish the major types of RICE[1]:

- Ignition system (compression ignition or spark ignition)
- Air scavenging cycle (4-stroke or 2-stroke)
- Air-to-fuel ratio (rich-burn or lean-burn)

Ignition systems and air scavenging cycles are not interchangeable for existing RICE, and operation in rich or lean-burn mode is principally fixed by engine design. Engine design characteristics affect the combustion process (fuel and air mixing, ignition, flame propagation, and quenching), as well as the type and quantity of air contaminants generated by the engine. Engine design characteristics also can affect the viability of air pollution control options by affecting the constituents in the RICE exhaust stream and the exhaust temperature.[1]

Figure 1 illustrates the relative quantity of engines according to engine design characteristics and shows that the majority of reciprocating engines are spark ignition, natural gas-fired engines. Available data indicate that about two-thirds of these engines are 4-stroke and approximately two-thirds are lean-burn. The following describes three major categories of RICE.[1]

Heavy-duty Pipeline Compressor Engines—Natural Gas-fired (800 to 5000 bhp)[2]

Natural gas-fired internal combustion engines are used in the natural gas industry at storage facilities and pipeline compressor stations to provide mechanical shaft power. At storage facilities, reciprocating engines are used to help inject the natural gas into high-pressure underground cavities (e.g., empty oil fields) to create natural gas storage fields. At pipeline compressor stations, the engine is used to help move natural gas from station to station. Pipeline compressor stations also use stationary gas turbines to provide mechanical shaft power to drive compressors (see section on stationary gas turbines for additional details). Pipeline population statistics show a nearly equal installed capacity of turbines and reciprocating engines. For reciprocating engines at pipeline compressor stations, 2-stroke designs contribute approximately two-thirds of installed capacity.

Although they can operate at a fairly constant load, pipeline engines must be able to operate under varying pipeline requirements. The size of these engines ranges from 800 brake horsepower (bhp) to 5000 bhp.

Industrial Engines—Gasoline (up to 250 hp) And Diesel (up to 600 hp)[3]

These engines cover a wide variety of industrial applications of both gasoline and diesel internal combustion (IC) engines such as aerial lifts, fork lifts, mobile refrigeration units, generators, pumps, industrial sweepers/scrubbers, material handling equipment (such as conveyors), and portable well-drilling equipment. Gasoline and diesel-fired engines can also be used to drive nonroad equipment. Figure 2 presents a diagram of a skid-mounted electric generator set utilizing a RICE to drive the electric generator.

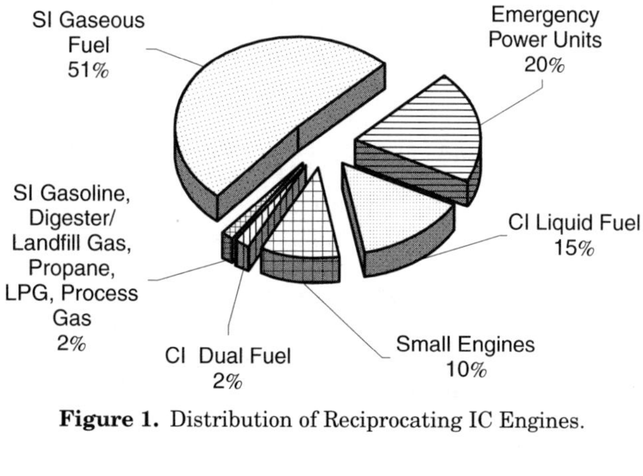

Figure 1. Distribution of Reciprocating IC Engines.

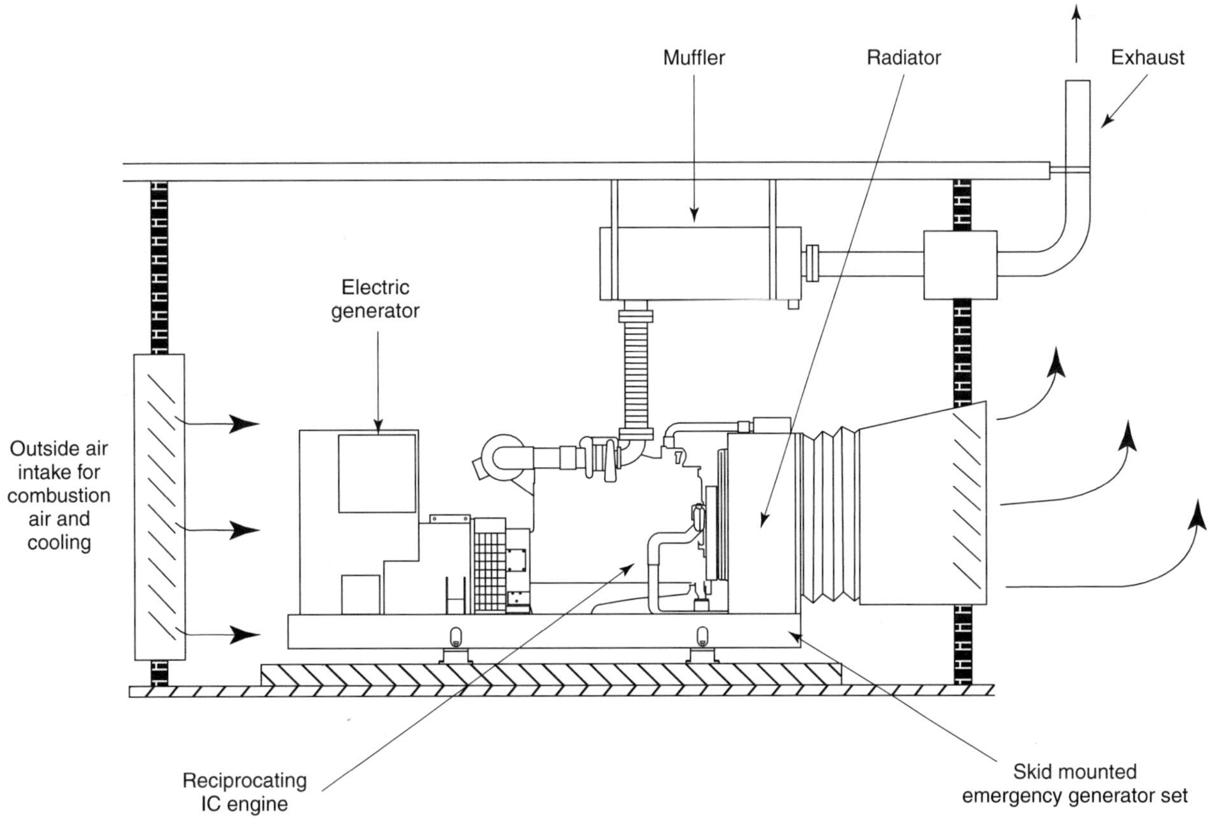

Figure 2. Typical Skid-Mounted Emergency Generator Set Layout (Adapted from Onan Corporation Application Manual, 1993).

Table 1. Emission Factors for Uncontrolled Natural Gas-Fired Heavy-Duty Pipeline Compressor Engines (800 to 5000 bhp)[a]

	2-Stroke Lean-Burn		4-Stroke Lean-Burn		4-Stroke Rich-Burn	
Pollutant	Emission Factor (lb/hp h) (power output)	Emission Factor ($lb/10^6$ scf) (fuel input)	Emission Factor (lb/hp h) (power output)	Emission Factor ($lb/10^6$ scf) (fuel input)	Emission Factor (lb/hp h) (power output)	Emission Factor ($lb/10^6$ scf) (fuel input)
NO_x (uncontrolled)	2.7E−02	3.1E+03	3.3E−02	3.7E+03	4.6E−02	4.4E+03
NO_x (with control)[b]	ND[c]	ND[c]	7.6E−03	8.5E+02	5.1E−03	4.8E+02
CO (uncontrolled)	2.7E−03	3.1E+02	2.7E−03	3.1E+02	1.6E−02	1.8E+03
CO (with control)[b]	ND[c]	ND[c]	ND[c]	ND[c]	5.0E−03	5.6E+02
CO_2[c]	8.9E−01	1.1E+05	9.7E−01	1.1E+05	1.1E+00	1.1E+05
SO_2[d]	5.0E−06	5.7E−01	5.0E−06	5.7E−01	5.5E−06	5.7E−01
TOC	1.8E−02	2.1E+03	1.0E−02	1.1E+03	3.4E−03	3.5E+02
VOC	2.1E−03	2.3E+02	6.9E−04	7.8E+01	4.8E−04	2.9E+01
Methane	1.5E−02	1.8E+03	9.0E−03	9.8E+02	2.8E−03	3.1E+02
PM-10 (total)	ND[c]	ND[c]	8.0E−05	1.0E+01	9.8E−05	1.3E+01
PM-10 (filterable)	ND[c]	ND[c]	6.2E−07	7.8E−02	5.5E−06	7.5E−01
Inorganic condensible PM[e]	ND[c]	ND[c]	4.4E−05	5.6E+00	3.7E−05	5.0E+00
Organic condensible PM[e]	ND[c]	ND[c]	3.5E−05	4.5E+00	5.5E−05	7.5E+00

[a]Factors are based on large population of engines. Factors for individual engines from specific manufacturers may vary. To convert from lb/hp h to kg/kW h, multiply by 0.608. To convert from $lb/10^6$ scf to ng/J, multiply by 0.422.

[b]Control for 4-stroke lean burn is SCR and NSCR for 4-stroke rich burn.

[c]Based on 99.5% conversion of the fuel carbon to CO_2. CO_2 (lb/MMBtu) = (3.67)(%CON)(C)(D), where %CON is the percent conversion of fuel carbon to CO_2; C, the carbon content of fuel by weight (0.75); and D, the density of fuel, 4.1×10^{-4} $lb/10^6$ scf.

[d]Based on 100% conversion of the fuel sulfur to SO_2. Assumes sulfur content of natural gas is 2000 $g/10^6$ scf.

[e]PM considered less than or equal to 1 μm in aerodynamic diameter.

Source: Reference 2.

The three primary fuels for reciprocating IC engines are gasoline, diesel fuel oil (No. 2), and natural gas. Gasoline is used primarily for mobile and portable engines. Diesel fuel oil is the most versatile fuel and is used in IC engines of all sizes. The rated power of these engines covers a rather substantial range, up to 250 hp for gasoline engines and up to 600 hp for diesel engines. Understandably, substantial differences in engine duty cycles exist.

Table 2. Emission Factors for Uncontrolled Gasoline (<250 hp) and Diesel (<600 hp) Industrial Engines[a]

	Gasoline Fuel		Diesel Fuel	
Pollutant	Emission Factor (lb/hp h) (power output)	Emission Factor (lb/MMBtu) (fuel input)	Emission Factor (lb/hp h) (power output)	Emission Factor (lb/10MMBtu) (fuel input)
NO_x	0.011	1.63	0.031	4.41
CO	0.439	62.7	6.68E-03	0.95
SO_x	5.91E−04	0.084	2.05E−03	0.29
PM-10[b]	7.21E−04	0.10	2.20E−03	0.31
CO_2[c]	1.08	154	1.15	164
Aldehydes	4.85E−04	0.07	4.63E−04	0.07
TOC				
Exhaust	0.015	2.10	2.47E−03	0.35
Evaporative	6.61E−04	0.09	0.00	0.00
Crankcase	4.85E−03	0.69	4.41E−05	0.01
Refueling	1.08E−03	0.15	0.00	0.00

[a]When necessary, an average brake-specific fuel consumption (BSFC) of 7000 Btu/hp h was used to convert from lb/MMBtu to lb/hp h. To convert from lb/hp h to kg/kW h, multiply by 0.608. To convert from lb/MMBtu to ng/J, multiply by 430.
[b]PM-10, the particulate matter less than or equal to 10 μm in aerodynamic diameter. All particulate is assumed to be less than or equal to 1 μm in size.
[c]Assumes 99% conversion of carbon in fuel to CO_2 with 87% carbon (by weight) in diesel, 86% carbon (by weight) in gasoline, average BSFC of 7000 Btu/hp h, diesel heating value of 19,300 Btu/lb, and gasoline heating value of 20,300 Btu/lb.
Source: Reference 3.

Table 3. Emission Factors for Large Stationary Diesel (>600 hp) and Stationary Dual-Fuel Engines[a]

	Diesel Fuel		Dual Fuel[b]	
Pollutant	Emission Factor (lb/hp h) (power output)	Emission Factor (lb/MMBtu) (fuel input)	Emission Factor (lb/hp h) (power output)	Emission Factor (lb/MMBtu) (fuel input)
NO_x				
Uncontrolled	0.024	3.2	0.018	2.7
Controlled	0.013[c]	1.9[c]	ND	ND
CO	5.5E−03	0.85	7.5E−03	1.16
SO_x[d]	8.09E−03 (S1)	1.01 (S1)	4.06E−04 (S1)+9.57E−03 (S2)	0.05 (S1)+0.895 (S2)
CO_2[e]	1.16	165	0.772	110
PM	0.0007[c]	0.1[c]	ND	ND
TOC (as CH_4)	7.05E−04	0.09	5.29E−03	0.8
Methane (CH_4)	—[f]	—[f]	3.97E−03	0.6
TNMOC	—[f]	—[f]	1.32E−03	0.2[g]

[a]Based on uncontrolled levels for each fuel. When necessary, the average heating value of diesel was assumed to be 19,300 Btu/lb with a density of 7.1 lb/gal. The power output and fuel input values were averaged independently from each other, because of the use of actual brake-specific fuel consumption (BSFC) values for each data point and of the use of data possibly sufficient to calculate only one of the two emission factors (e.g., enough information to calculate lb/MMBtu, but not lb/hp h). Factors are based on averages across all manufacturers and duty cycles. The actual emissions from a particular engine or manufacturer could vary considerably from these levels. To convert from lb/hp h to kg/kW h, multiply by 0.608. To convert from lb/MMBtu to ng/J, multiply by 430.
[b]Dual-fuel assumes 95% natural gas and 5% diesel fuel.
[c]Controlled NO_x is by ignition timing retard.
[d]Assumes that all sulfur in the fuel is converted to SO_2. S1 is the % sulfur in fuel oil; S2 is the % sulfur in natural gas. For example, if fuel oil sulfur content is 1.5%, then S1 = 1.5.
[e]Assumes 100% conversion of carbon in fuel to CO_2 with 87% carbon in diesel, 70% carbon in natural gas, dual-fuel mixture of 5% diesel with 95% natural gas, average BSFC of 7000 Btu/hp h, diesel heating value of 19,300 Btu/lb, and natural gas heating value of 1050 Btu/scf.
[f]Based on data from one engine, TOC is by weight 9% methane and 91% nonmethane.
[g]Assumes that nonmethane organic compounds are 25% of TOC emissions from dual-fuel engines. Molecular weight of nonmethane gas stream is assumed to be that of methane.
Source: Reference 4.

Large Stationary Engines—Diesel (>600 hp) and Stationary Dual fuel[4]

The primary domestic use of large stationary diesel engines greater than 600 hp is in oil and gas exploration and production. These engines, in groups of 3 to 5, supply mechanical power to operate drilling (rotary table), mud pumping, and hoisting equipment, and may also operate pumps or auxiliary power generators. Another frequent application of large stationary diesels is electric power generation for both base and standby service. Smaller uses include irrigation, hoisting, and nuclear power plant emergency cooling water pump operation.

Dual-fuel engines were developed to obtain compression ignition performance and the economy of natural gas, using a minimum of 5–6% diesel fuel to ignite the natural gas. Large dual-fuel engines have been used almost exclusively for prime electric power generation. Dual-fuel engines that burn diesel and either digester gas or landfill gas are common within the wastewater treatment and landfill industries.

PROCESS DESCRIPTION

All reciprocating IC engines operate by the same basic process. A combustible mixture is first compressed in a small volume between the head of a piston and its surrounding cylinder. The mixture is then ignited, and the resulting high-pressure products of combustion push the piston through the cylinder. This movement is converted from linear to rotary motion by a crankshaft. The piston returns, pushing out exhaust gases, and the cycle is repeated.[3–7]

Ignition Methods

There are two ignition methods used in stationary reciprocating IC engines, *compression ignition* (CI) and *spark ignition* (SI). In CI engines, combustion air is first compression heated in the cylinder, and diesel fuel oil is then injected into the hot air. Ignition is spontaneous because the air temperature is above the autoignition temperature of the fuel. SI engines initiate combustion by the spark of an electrical discharge. Usually the fuel is mixed with the air in a carburetor (for gasoline) or at the intake valve (for natural gas), but occasionally the fuel is injected into the compressed air in the cylinder.[3–7]

All diesel-fueled engines are compression ignited and all gasoline- and gas-fueled engines are spark ignited. Gaseous fuels (natural gas, landfill/digester gas) can be used in a CI engine if a small amount of diesel fuel is injected into the compressed gas/air mixture to burn any mixture ratio of gas and diesel oil; hence the name dual-fuel. The concentration of diesel oil in dual-fuel engines ranges from 6 to 100%.[4–7]

CI engines usually operate at a higher compression ratio (ratio of cylinder volume when the piston is at the bottom of its stroke to the volume when it is at the top) than SI engines because fuel is not present during compression; hence there is no danger of premature autoignition. Since engine thermal efficiency rises with increasing pressure ratio (and pressure ratio varies directly with compression ratio), CI engines are more efficient than SI engines. This increased efficiency is gained at the expense of poorer response to load changes and a heavier structure to withstand the higher pressures.[1,3–5]

Air Scavenging Cycles[1]

Reciprocating engines are separated into two design classes: 2-stroke and 4-stroke. Each of these has design differences that affect both baseline emissions as well as the potential for emissions control. Two-stroke engines complete the power cycle in a single crankshaft revolution as compared to the two crankshaft revolutions required for 4-stroke engines.

Two-stroke engines have the advantage of higher horsepower-to-weight ratio compared to 4-stroke engines when operated at the same speed. If ports are used instead of valves, the mechanical design of the 2-stroke engine can also be simplified relative to the 4-stroke valve design. However, combustion can be better controlled in a 4-stroke engine and excess air is not needed to purge the cylinder. Therefore, 4-stroke engines tend to be slightly more efficient and typically generate less air emissions, particularly unburned hydrocarbons, than 2-stroke engines.

Two-Stroke Cycle. In a 2-stroke engine, the air/fuel charge is injected with the piston near the bottom of the power stroke. The intake ports are then covered or closed, and the piston moves to the top of the cylinder, thereby compressing the charge. Following ignition and combustion, the power stroke starts with the downward movement of the piston. Exhaust ports or valves are then uncovered to exhaust the combustion products, and a new air/fuel charge is injected. Two-stroke engines may be turbocharged using an exhaust-powered turbine to pressurize the charge for injection into the cylinder and to increase cylinder scavenging. Nonturbocharged engines may be either blower scavenged or piston scavenged to improve removal of combustion products. Figure 3 illustrates the various steps in the 2-stroke cycle.

4-Stroke Cycle. Four-stroke engines use a separate engine revolution for the intake/compression cycle and the power/exhaust cycle. These engines may be either naturally aspirated, using the suction from the piston to entrain the air charge, or turbocharged, using an exhaust-driven turbine to pressurize the charge. Turbocharged units produce a higher power output for a given engine displacement, whereas naturally aspirated units have lower initial cost and maintenance. Rich-burn engines operate near the stoichiometric air/fuel ratio with exhaust excess oxygen levels less than 4%. Lean-burn engines may operate up to the lean flame extinction limit, with exhaust oxygen levels of 12% or greater. Figure 4 illustrates the various steps in the 4-stroke cycle.

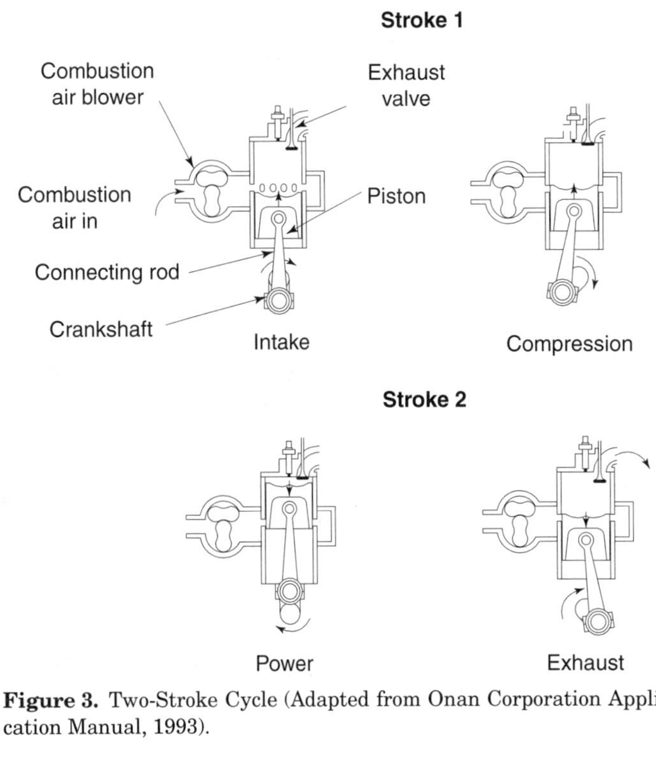

Figure 3. Two-Stroke Cycle (Adapted from Onan Corporation Application Manual, 1993).

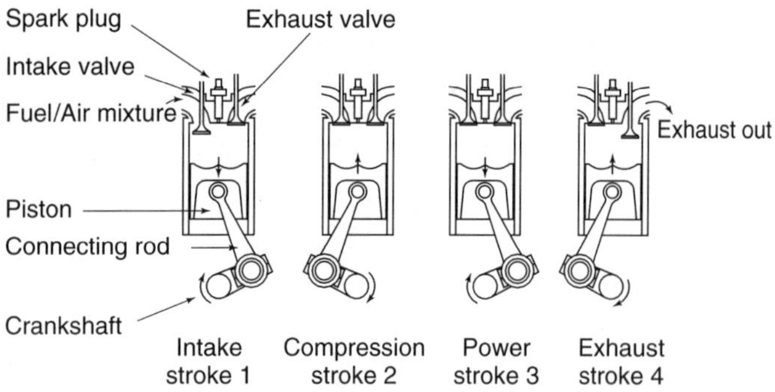

Figure 4. Four-Stroke Cycle (Adapted from Onan Corporation Application Manual, 1993).

Air-to-Fuel Ratio ("Rich" or "Lean")[1]

Reciprocating engines operate with various air-to-fuel ratios. In general, air-to-fuel ratios may be classified as either rich or lean of stoichiometry, the theoretical air-to-fuel ratio required for complete combustion. All stationary CI engines are lean-burn engines, usually utilizing turbochargers and intercoolers to achieve the desired fresh air density. SI engines may be either rich-burn engines or lean-burn engines.

A common method used to differentiate between "rich-burn" and "lean-burn" engines is the percentage oxygen in the exhaust stream. Several regulatory agencies have adopted a value of 4% oxygen in the exhaust as the defining limit for "rich-burn" engines. An engine with more than 4% exhaust oxygen is classified as "lean-burn", while most "lean-burn" engines manufactured today are designed to have at least 7% oxygen in the exhaust.

AIR EMISSIONS CHARACTERIZATION

Natural Gas-Fired Engines (800 to 5000 hp)

The primary pollutants from natural gas-fueled reciprocating engines are nitrogen oxide (NO_x), carbon monoxide (CO), and total organic compounds (TOC)/hydrocarbons (HC). Nitrogen oxides formation is strongly dependent on the high temperatures developed in the cylinder. The other pollutants, CO and HC species, are primarily the result of incomplete combustion. Trace amounts of metals and noncombustible inorganic material may be carried over from the lubricating oil, from engine wear, or from trace constituents in the gas. Sulfur oxides are very low since most of the sulfur compounds are removed from natural gas in the gas treatment plant prior to entry into the pipeline.[2]

Nitrogen Oxides. NO_x formation occurs by two fundamentally different mechanisms. The principal mechanism with gas-fired engines is thermal NO_x, which arises from the thermal dissociation and subsequent reaction of nitrogen (N_2) and oxygen (O_2) molecules in the combustion air. Most thermal NO_x is formed in high-temperature regions in the cylinder, where combustion air has mixed sufficiently with the fuel to produce the peak temperature fuel/air interface. A component of thermal NO_x, called prompt NO_x, is formed from early reactions of nitrogen intermediaries and hydrocarbon radicals from the fuel. The prompt NO_x forms within the flame and is usually negligible compared to the amount of thermal NO_x formed. The second mechanism, fuel NO_x, stems from the evolution and reaction of fuel-bound nitrogen compounds with oxygen. Natural gas has negligible chemically fuel-bound nitrogen (although some molecular nitrogen), and essentially all NO_x formed is thermal NO_x. The formation of prompt NO_x can form a significant part of total NO_x only under highly controlled situations where thermal NO_x is suppressed. It is more prevalent with rich-burn engines. The rates of these reactions are highly dependent upon the stoichiometric ratio, combustion temperature, and residence time at the combustion temperature.[2]

The maximum thermal NO_x production occurs at a slightly lean fuel/air mixture ratio because of the excess availability of oxygen for reaction. The control of stoichiometry is critical in achieving reductions in thermal NO_x. Premixing with lean-burn reciprocating engines is effective in suppressing NO_x relative to rich-burn engines. The thermal NO_x generation decreases rapidly as the temperature drops below the adiabatic temperature. Thus, maximum reduction of thermal NO_x generation can be achieved by control of both the combustion temperature and the stoichiometry.[2]

Carbon Monoxide and Total Organic Compounds. Carbon monoxide and total organic compound emissions both result from the products of incomplete combustion. Carbon monoxide results when there is insufficient residence time at high temperature to complete the final step in hydrocarbon oxidation. The oxidation of CO to carbon dioxide is a slow reaction compared to most hydrocarbon oxidation reactions. In reciprocating engines, CO emissions may indicate early quenching of combustion gases on cylinder walls or valve surfaces. CO emissions are usually higher when the unit is run at low loads.[2]

Total organic compounds can encompass a wide spectrum of volatile and semivolatile organic compounds. TOC emissions include methane (CH_4) and nonmethane organic compounds (NMOC) fractions. TOCs are discharged into the atmosphere when some of the gas remains unburned or is only partially burned during the combustion process. With natural gas, some organics are carryover, unreacted, trace constituents of the gas, while others may be pyrolysis products of the heavier hydrocarbon constituents. Partially burned hydrocarbons can occur because of poor air/fuel homogeneity due to incomplete mixing prior to, or during, combustion; incorrect air/fuel ratios in the cylinder during combustion due to maladjustment of the engine fuel system; or low cylinder temperature due to excessive cooling through the walls or early cooling of the gases by expansion of the combustion volume caused by piston motion before combustion is completed.[2]

Particulate Matter. Particulate matter (PM) emissions from reciprocating engines are typically nondetectable with conventional protocols unless the engines are operated in a sooting condition. Otherwise, PM could arise from carryover of noncombustible trace constituents in the gas or from lube oil.[8]

Gasoline (up to 250 hp) and Diesel-Fired Engines (up to 600 hp)

Most of the pollutants from IC engines are emitted through the exhaust. However, some TOCs/HCs escape from the crankcase as a result of blowby (gases that are vented from the oil pan after they have escaped from the cylinder past the piston rings) and from the fuel tank and carburetor because of evaporation. Nearly all the TOCs from diesel CI engines enter the atmosphere from the exhaust.[3,4]

Crankcase blowby is minor because TOCs are not present during compression of the charge. Evaporative losses are insignificant in diesel engines due to the low volatility of diesel fuels. In general, evaporative losses are also negligible in engines using gaseous fuels because these engines receive their fuel continuously from a pipe rather than via a fuel storage tank and fuel pump.[4]

The primary pollutants from internal combustion engines are NO_x, HC and other organic compounds, CO, and PM, which include both visible (smoke) and nonvisible emissions. NO_x formation is directly related to high pressures and temperatures during the combustion process and to the nitrogen content, if any, of the fuel. The other pollutants, HC, CO, and smoke, are primarily the result of incomplete combustion. Ash and metallic additives in the fuel also contribute to the PM content of the exhaust. Sulfur oxides also appear in the exhaust from IC engines. The sulfur compounds, mainly sulfur dioxide (SO_2), are directly related to the sulfur content of the fuel.[3,4,6]

Nitrogen Oxides. NO_x formation occurs by two fundamentally different mechanisms. The predominant mechanism with internal combustion engines is thermal NO_x which arises from the thermal dissociation and subsequent reaction of nitrogen (N_2) and oxygen (O_2) molecules in the combustion air. Most thermal NO_x is formed in the high-temperature region of the flame from dissociated molecular nitrogen in the combustion air. Some NO_x, called prompt NO_x, is formed in the early part of the flame from reaction of nitrogen intermediary species, and HC radicals in the flame. The second mechanism, fuel NO_x, stems from the evolution and reaction of fuel-bound nitrogen compounds with oxygen. Gasoline, and most distillate oils, have no chemically bound fuel nitrogen, and essentially all NO_x formed is thermal NO_x.[3,4]

Total Organic Compounds. Total organic compounds are composed of a wide variety of organic compounds and are discharged into the atmosphere when some of the fuel remains unburned or is only partially burned during the combustion process. TOC emissions include methane and nonmethane organic compounds fractions. Most unburned hydrocarbon emissions result from fuel droplets that were transported or injected into the quench layer during combustion. This is the region immediately adjacent to the combustion chamber surfaces, where heat transfer outward through the cylinder walls causes the mixture temperatures to be too low to support combustion.[3,4]

Partially burned hydrocarbons can occur because of poor air and fuel homogeneity due to incomplete mixing, before or during combustion; incorrect air/fuel ratios in the cylinder during combustion due to maladjustment of the engine fuel system; excessively large fuel droplets (diesel engines); and low cylinder temperature due to excessive cooling (quenching) through the walls or early cooling of the gases by expansion of the combustion volume caused by piston motion before combustion is completed.[3,4,6]

Carbon Monoxide. Carbon monoxide is a colorless, odorless, relatively inert gas formed as an intermediate combustion product that appears in the exhaust when the reaction of CO to CO_2 cannot proceed to completion. This situation occurs if there is a lack of available oxygen near the hydrocarbon (fuel) molecule during combustion, if the gas temperature is too low, or if the residence time in the cylinder is too short. The oxidation rate of CO is limited by reaction kinetics and, as a consequence, can be accelerated only to a certain extent by improvements in air and fuel mixing during the combustion process.[3,4,6,7]

Particulate Matter. White, blue, and black smoke may be emitted from IC engines. Liquid particulates appear as white smoke in the exhaust during an engine cold start, idling, or low load operation. These are formed in the quench layer adjacent to the cylinder walls, where the temperature is not high enough to ignite the fuel. Blue smoke is emitted when lubricating oil leaks, often past worn piston rings, into the combustion chamber and is partially burned. Proper maintenance is the most effective method of preventing blue smoke emissions from all types of IC engines. The primary constituent of black smoke is agglomerated carbon particles (soot).[3,4,6]

Sulfur Oxides. SO_x emissions are a function of only the sulfur content in the fuel rather than any combustion variables. In fact, during the combustion process, essentially all the sulfur in the fuel is oxidized to SO_2. The oxidation of SO_2 gives sulfur trioxide (SO_3), which reacts with water to give sulfuric acid (H_2SO_4). Sulfuric acid reacts with basic substances to give sulfates, which are fine particulates that contribute to PM emissions (smoke) and visibility reduction. Sulfur oxide emissions also contribute to corrosion of the engine parts.[3,4,6,7]

AIR POLLUTION CONTROL MEASURES

Three generic control techniques have been developed for reciprocating engines: parametric controls (timing and operating at a leaner air/fuel ratio for reciprocating engines); combustion modification such as advanced engine design for new sources or major modification to existing sources (clean-burn reciprocating head designs); and postcombustion catalytic NO_x reduction [selective catalytic reduction (SCR) for lean-burn reciprocating engines and nonselective catalytic reduction (NSCR) for rich-burn engines].[2]

Control measures to date are primarily directed at limiting NO_x and CO emissions since they are the primary pollutants from these engines. From an NO_x control viewpoint, the most important distinction between different engine models and types of reciprocating engines is whether they are rich-burn or lean-burn. Rich-burn engines have an air-to-fuel ratio operating range that is near stoichiometric or fuel-rich of stoichiometric and as a result the exhaust gas has little or no excess oxygen. A lean-burn engine has an air-to-fuel operating range that is fuel-lean of stoichiometric; therefore, the exhaust from these engines is characterized by medium to high levels of O_2. The most common NO_x control technique for diesel and dual-fuel engines focuses on modifying the combustion process. However, selective catalytic reduction and nonselective catalytic reduction, which are postcombustion techniques, are becoming available. Controls for CO have been partly adapted from mobile sources.[3,4,9]

Combustion modifications include injection timing retard (ITR), preignition chamber combustion (PCC), air-to-fuel ratio adjustments, and derating. Injection of fuel into the cylinder of a CI engine initiates the combustion process. Retarding the timing of the diesel fuel injection causes the combustion process to occur later in the power stroke when the piston is in the downward motion and combustion chamber volume is increasing. By increasing the volume, the combustion temperature and pressure are lowered, thereby lowering NO_x formation. ITR reduces NO_x from all diesel engines; however, the effectiveness is specific to each engine model. The amount of NO_x reduction with ITR diminishes with increasing levels of retard.[3,4,9]

Improved swirl patterns promote thorough air and fuel mixing and may include a precombustion chamber (PCC). A PCC is an antechamber that ignites a fuel-rich mixture that propagates to the main combustion chamber. The high exit velocity from the PCC results in improved mixing and complete combustion of the lean air/fuel mixture, which lowers combustion temperature, thereby reducing NO_x emissions.[3,4,9]

The air-to-fuel ratio for each cylinder can be adjusted by controlling the amount of fuel that enters each cylinder. At air-to-fuel ratios less than stoichiometric (fuel-rich), combustion occurs under conditions of insufficient oxygen, which causes NO_x to decrease because of lower oxygen and lower temperatures. Derating involves restricting the engine operation to lower than normal levels of power production for the given application. Derating reduces cylinder pressures and temperatures, thereby lowering NO_x formation rates.[3,4,9]

SCR is an add-on NO_x control placed in the exhaust stream following the engine and involves injecting ammonia (NH_3) into the flue gas. The NH_3 reacts with NO_x in the presence of a catalyst to form water and nitrogen. The effectiveness of SCR depends on engine duty cycle (load fluctuations) and fuel quality. Load fluctuations can cause variations in exhaust temperature and NO_x concentration, which can create problems with the effectiveness of the SCR system. Contaminants in the fuel may poison or mask the catalyst surface, causing a reduction or termination in catalyst activity. Both digester gas and landfill gas contain a family of silicon-based gases collectively called *siloxanes*. Siloxanes are found in many cosmetics and cleaning solutions that are disposed of in either landfills or sewers. Combustion of siloxanes forms compounds that can foul fuel systems, combustion chambers, and postcombustion catalysts. The fouling renders catalysts inoperable within a very short time period. Because of this problem, catalytic technology has not been demonstrated to work effectively on internal combustion engines and dual-fuel engines that burn diesel and either digester gas or landfill gas.

NSCR is often referred to as a three-way conversion catalyst system because the catalyst reactor simultaneously reduces NO_x, CO, and HC and involves placing a catalyst in the exhaust stream of the engine. The reaction requires that the O_2 levels be kept low and that the engine be operated at fuel-rich air-to-fuel ratios.[3,4,9]

Natural Gas-Fired Engines

Control Techniques for Rich-Burn Reciprocating Engines[2]

Nonselective Catalytic Reduction. This technique uses the residual hydrocarbons and CO in the rich-burn engine exhaust as a reducing agent for NO_x. In NSCR, hydrocarbons will be oxidized by O_2 and NO_x, hence the designation "nonselective." This is in contrast to ammonia injection for SCR, where ammonia selectively reacts with NO_x. The excess hydrocarbons and NO_x pass over a catalyst, usually a noble metal (platinum, rhodium, or palladium), which reduces the reactants to nitrogen (N_2), carbon dioxide, and water.

The NSCR technique is effectively limited to engines with normal exhaust oxygen levels of 4% or less. This includes 4-stroke naturally aspirated engines and some 4-stroke turbocharged engines. Engines operating with NSCR require tight air/fuel control to maintain high reduction effectiveness without high hydrocarbon emissions. To achieve optimum NO_x reduction performance, the engine may need to be run in a richer fuel condition than normal.

Prestratified Charge. Prestratified charge combustion is a retrofit system that is limited to 4-stroke carbureted natural gas engines. In this system, controlled amounts of air are introduced into the intake manifold in a specified sequence and quantity. This stratification provides both a fuel-rich ignition and rapid flame cooling, resulting in reduced formation of NO_x.

Control Techniques for Lean-Burn Reciprocating Engines[2]

Lean Combustion. Lean combustion techniques use increased bulk air/fuel ratios to lower peak flame temperature and reduce NO_x formation. Typically, air/fuel ratios are increased from normal levels of 20 to 35 up to controlled levels of 45 to 50. The upper limit is constrained by the onset of misfiring at the lean limit. This condition also increases CO and HC emissions.

To maintain acceptable engine performance at lean conditions, high-energy ignition systems have been developed that promote flame stability at very lean conditions. With high-energy ignition, a rich mixture is ignited in a small ignition cell located in the cylinder head. The ignition cell flame passes to the cylinder, where it provides a uniform ignition source. The technique can be retrofit to existing turbocharged 2- and 4-stroke engines. With new engine designs, NO_x reductions of 80–90% have been achieved compared to conventional spark ignition engine designs. In most cases, the NO_x reductions have been accompanied by increases in power output and increased fuel economy.

Selective Catalytic Reduction. SCR is applicable to lean-burn engines. NH_3 is injected upstream of a noble metal, metal oxide or zeolite catalyst to give an $NH_3:NO_x$ ratio of about 1:1. The mixture of NH_3 and NO_x is selectively reduced over the catalyst within a temperature range of 600–900°F depending on the catalyst. The major system components are the catalyst and associated housing, the ammonia storage and delivery system, and the control system. The performance has been less acceptable than NSCR with rich-burn engines. The primary difficulty with lean-burn engines has been maintaining air/fuel control, very limited automatic controls, and engine performance and the inherent variety of engine loading while achieving the necessary exhaust temperature window for efficient SCR operation.

REFERENCES

1. *Subcategories and MACT Floor Determinations for Existing Stationary Reciprocating Internal Combustion Engines (RICE)*, Reciprocating Internal Combustion Engine Work Group of the Industrial Combustion Coordinated Rulemaking, August 1998.
2. *Compilation of Air Pollutant Emission Factors*, AP-42, Section 3.2, Natural Gas-Fired Reciprocating Engines, U.S. Environmental Protection Agency, Office of Air Quality Planning and Standards, RTP NC, Draft 2/97.
3. *Report on Revisions to 5th Edition AP-42 Section 3.3 Gasoline and Diesel Industrial Engines*, U.S. Environmental Protection Agency, Office of Air Quality Planning and Standards, Research Triangle Park, NC, September 1996.

4. Compilation of Air Pollutant Emission Factors, AP-42, Section 3.4, *Large Stationary Diesel and All Stationary Duel Fuel Engines*.
5. H. I. Lips et al., *Environmental Assessment of Combustion Modification Controls for Stationary Internal Combustion Engines*, EPA-600/7-81-127, U.S. Environmental Protection Agency, Cincinnati, OH, July 1981.
6. *Standards Support and Environmental Impact Statement, Volume 1: Stationary Internal Combustion Engines*, EPA-450/2-78-125a, U.S. Environmental Protection Agency, Research Triangle Park, NC, July 1979.
7. M. Hoggan et al., *Air Quality Trends in California's South Coast and Southeast Desert Air Basins, 1976–1990, Air Quality Management Plan, Appendix II-B*, South Coast Air Quality Management District, July 1991.
8. *Limiting Net Greenhouse Gas Emissions in the United States, Volume II: Energy Responses*, Report for the Office of Environmental Analysis, Office of Policy, Planning and Analysis, Department of Energy (DOE), DOE/PE-0101 Volume II, September 1991.
9. R. B. Snyder, *Alternative Control Techniques Document. NO_x Emissions From Stationary Reciprocating Internal Combustion Engines*, EPA-453/R-93-032, U.S. Environmental Protection Agency, Research Triangle Park, July 1993.

BIBLIOGRAPHY

Air Emission from Stationary Diesel Engines for the Alaska Rural Electric Cooperative Association, Environmetrics, 1992.

Air Pollution Source Testing for California AB2588 on an Oil Platform Operated by Chevron USA, Inc., Platform Hope, California, Chevron USA, Inc., Ventura, CA, August 29, 1990.

Air Pollution Source Testing for California AB2588 of Engines at the Chevron USA, Inc., Carpinteria Facility, Chevron USA, Inc., Ventura, CA, August 30, 1990.

C. Castaldini and L. R. Waterland, *Environmental Assessment of a Reciprocating Engine Retrofitted with Nonselective Catalytic Reduction*, EPA-600/7-84-073B, U.S. Environmental Protection Agency, Cincinnati, OH, June 1984.

C. Castaldini and L. R. Waterland, *Environmental Assessment of a Reciprocating Engine Retrofitted with Selective Catalytic Reduction*, EPA Contract No. 68-02-3188, U.S. Environmental Protection Agency, Research Triangle Park, NC, December 1984.

C. Castaldini, *Environmental Assessment of NO_x Control on a Spark-ignited Large Bore Reciprocating Internal Combustion Engine*, EPA-600/7-86-002A, U.S. Environmental Protection Agency, Cincinnati, OH, January 6, 1986.

C. Castaldini, *Environmental Assessment of NO_x Control on a Compression Ignition Large Bore Reciprocating Internal Combustion Engine, Volume I: Technical Results*, EPA-600/7-86/001a, U.S. Environmental Protection Agency, Cincinnati, OH, April 1984.

C. Castaldini, *NO_x Reduction Technologies for Natural Gas Industry Prime Movers*, GRI-90/0215, Gas Research Institute, Chicago, IL, August 1990.

Compliance Measured Particulate Emissions from an Emergency Diesel Generator, Silorsky Aircraft, United Technologies, Stratford, CT, TRC Environmental Consultants, 1987.

Compliance Test Report for Particulate Emissions from a Caterpillar Diesel Generator, St. Mary's Hospital, Waterburg, CT, TRC Environmental Consultants, 1987.

Compliance Test Report for Particulate Emissions from a Cummins Diesel Generator, Colonial Gold Limited Partnership, Hartford, CT, TRC Environmental Consultants, 1988.

Compliance Test Report for Particulate Emissions from a Cummins Diesel Generator, CIGNA Insurance Company, Bloomfield, CT, TRC Environmental Consultants, 1988.

Definition of "Rich Burn Engine" for the Reciprocating Internal Combustion Engine (RICE) MACT Standard, Reciprocating Internal Combustion Engine Work Group of the Industrial Combustion Coordinated Rulemaking, September 1998.

Engines, Turbines, and Compressors Directory, American Gas Association, Catalog #XF0488.

Final Regulatory Impact Analysis: Control of Emissions from Nonroad Diesel Engines, EPA, August 1998.

R. E. Fanick et al., *Emissions Data for Stationary Reciprocating Engines and Gas Turbines in Use by the Gas Pipeline Transmission Industry—Phases I & II*, Project PR-15-613, Pipeline Research Committee, American Gas Association, Arlington, VA, April 1988.

Final Report for an Emission Compliance Test Program on Two Standby Generators Located at American Car Company, Greenwich, CT, York Services Corp., 1987.

Final Report for an Emission Compliance Test Program on a Standby Diesel Generator at South Central Connecticut Regional Water Authority, West Haven, CT, York Services Corp., 1988.

C. T. Hare and K. J. Springer, *Exhaust Emissions from Uncontrolled Vehicles and Related Equipment Using Internal Combustion Engines, Part 5: Farm, Construction, and Industrial Engines*, APTD-1494, U.S. Environmental Protection Agency, Research Triangle Park, NC, October 1973.

Inventory of U.S. Greenhouse Gas Emissions And Sinks: 1990–1991, EPA-230-R-96-006, U.S. Environmental Protection Agency, Washington, D.C., November 1995.

IPCC Guidelines for National Greenhouse Gas Inventories Workbook, Intergovernmental Panel on Climate Change/Organization for Economic Cooperation and Development, Paris, France, 1995.

G. Marland and R. M. Rotty, *Carbon Dioxide Emissions from Fossil Fuels: A Procedure for Estimation and Results for 1951–1981*, DOE/NBB-0036 TR-003, Carbon Dioxide Research Division, Office of Energy Research, U.S. Department of Energy, Oak Ridge, TN, 1983.

G. Marland and R. M. Rotty, "Carbon Dioxide Emissions from Fossil Fuels: A Procedure for Estimation and Results for 1950–1982," *Tellus* **36B**, 232–261, 1984.

N. L. Martin and R. H. Thring, *Computer Database of Emissions Data for Stationary Reciprocating Natural Gas Engines and Gas Turbines in Use by the Gas Pipeline Transmission Industry Users Manual (Electronic Database Included)*, GRI-89/0041, Gas Research Institute, Chicago, IL, February 1989.

W. E. Osborn and M. D. McDannel, *Emissions of Air Toxic Species: Test Conducted under AB2588 for the Western States Petroleum Association*, CR 72600-2061, Western States Petroleum Association, Glendale, CA, May 1990.

Pooled Source Emission Test Report: Gas Fired IC Engines in Santa Barbara County, ARCO, Bakersfield, CA, July 1990.

Pooled Source Emission Test Report: Oil and Gas Production Combustion Sources, Fresno and Ventura Counties, California, ENSR 7230-007-700, Western States Petroleum Association, Bakersfield, CA, December 1990.

A. Rosland, *Greenhouse Gas Emissions in Norway: Inventories and Estimation Methods*, Oslo: Ministry of Environment, 1993.

Sector-Specific Issues and Reporting Methodologies Supporting the General Guidelines for the Voluntary Reporting of Greenhouse Gases under Section 1605(b) of the Energy Policy Act of 1992 (1994) DOE/PO-0028, Volume 2 of 3, U.S. Department of Energy.

Standards Support and Environmental Impact Statement, Volume I: Stationary Internal Combustion Engines, EPA-450/2-78-125a, U.S.

Environmental Protection Agency, Office of Air Quality Planning and Standards, Research Triangle Park, NC, July 1979.

Technical Feasibility of Reducing NO_x and Particulate Emissions from Heavy-Duty Engines, CARB Contract A132-085, California Air Resources Board, Sacramento, CA, March 1992.

C. Urban, *Compilation of Emissions Data for Stationary Reciprocating Gas Engines and Gas Turbines in Use by American Gas Association Member Companies*, prepared by Southwest Research Institute Pipeline Research Committee of the American Gas Association, Project PR-15-86, May 1980.

8

WASTE INCINERATION SOURCES

Municipal Waste Combustion
James R. Donnelly
Hazardous Waste
Mackenzie L. Davis, Clyde R. Dempsey, and E. Timothy Oppelt
Sewage Sludge
Calvin R. Brunner
Drum Reconditioning
Patricia Kay Rykowski

MUNICIPAL WASTE COMBUSTION

JAMES R. DONNELLY

Environmentally sound disposal of municipal and industrial refuse is a major issue faced by communities in developed countries over the past two decades. The volume of municipal waste generated continues to grow annually, and traditional disposal methods (landfilling or ocean dumping) have become less acceptable because of cost and environmental concerns. Incineration of refuse in modem high-efficiency combustors is being employed for a growing fraction of waste streams to achieve significant reductions in refuse volume, while, in many instances, also achieving energy recovery in the form of steam or electricity. This increase in waste combustion has been coupled with an increase in the complexity and efficiency of air-pollution controls to limit the associated potential emissions. This discussion focuses on commonly employed municipal waste combustion devices and the air-pollution controls applied to them.

REFUSE INCINERATION

Refuse or municipal solid waste includes nonhazardous waste generated in households, institutions (excluding hospital wastes), and commercial and light industrial facilities, as well as agricultural wastes and sewage sludge. In 1988, the United States generated approximately 180 million tons of refuse.[1] Table 1 lists the major components of refuse. In addition to the major components listed in the table, several toxic trace metals are found in refuse. These include arsenic, cadmium, chromium, lead, and mercury.

Table 1. U.S. Municipal Solid Waste Stream (*Annual Generation Rate—180,000 Tons/Year*)

Composition	%
Paper and paperboard	41.0
Glass	8.2
Metals	8.7
Plastics	6.5
Rubber, leather, textiles, and wood	8.1
Food wastes	7.9
Yard wastes	17.9
Miscellaneous inorganic wastes	1.6

Source: Reference 2.

Refuse per capita generation rate in the United States is the highest in the world and has shown an annual growth over the past several decades. This same growth phenomenon is also observed in other industrial nations. Because of this continual refuse growth, environmental concerns with landfilling or ocean dumping, and the lack of availability and cost of landfills, alternate methods have been developed to handle this refuse. These include composting, recycling, and incineration (or municipal waste combustion). Of these alternatives, municipal waste combustors (MWCs) have shown the ability to achieve the greatest reduction in refuse volume (70–90%) and can be used in conjunction with the other two alternatives to achieve even greater reductions in disposal volume. MWCs currently treat about 15% of the refuse stream in the United States and even higher levels in many European countries.

The increased use of MWCs has led to a concomitant increase in concern over air pollutant emissions from these sources, which, in turn, has led to the promulgation of air emissions standards and the application of modern air-pollution controls to limit these emissions. During the 1990s several countries either established or revised downward emission standards for MWCs. The U.S. Environmental Protection Agency (EPA) promulgated New Source Performance Standards for Municipal Waste Combustors on December 19, 1995.[2] Germany revised their emission standards in 1990,[3] and the Netherlands revised their standards in 1993.[4] The European Community (EC) prepared a revised draft directive in 1994.[5] Table 2 presents a summary of key elements of these standards.

Although each standard is unique in its specific emission limits and averaging times, there are also many similarities. All require significant reduction in the uncontrolled emissions of acid gases (HCl, SO_2, HF, and NO_x), particulates, and heavy metals. Each standard includes limits on carbon monoxide (CO) emissions as a measure of good combustion practices. Although not shown in Table 2, the standards also include provisions to ensure proper combustion and requirements for operator training and/or certification.

The United States has established emission limits for total polychlorinated dibenzyl-dioxins (PCDD) plus polychlorinated dibenzyl-furans (PCDF). European emission limits are based on international toxic equivalents (ITEQ). Both are set at levels that are deemed to be protective of human health.

In the United States a number of states and local air-pollution control districts have developed their own sets of emission standards for new MWC sources. Many of these standards (or permit conditions) are more stringent than the U.S. EPA standards.

MUNICIPAL WASTE COMBUSTION SOURCES

Several different technologies are employed in municipal waste combustion. These include mass burn combustors (modular, traveling grate, and rotary combustors); refuse-derived, fuel-fired combustors; and, to a lesser extent, fluid-bed combustors.

Table 2. Municipal Waste Combustor Emissions Standards and Guidelines

Pollutant	U.S. EPA, NSPS 12/95	German[a] 17th BIMSch V 11/90	Netherlands[a] BLA 1993	EU Draft[a] Directive 1994
		Gases		
HCl	25 ppm or 95% reduction[b]	60 mg/N m^3	10 mg/N m^3	5 mg/N m^3
SO_2 ($SO_2 + SO_3$)	30 ppm or 80%	200	40	25
HF	—	4	1	1
CO	50, 100, or 150 ppm[c]	100	50	50
NO_x (as NO_2)	150 ppm	400	70	—
Particulates	24 mg/dscm[d]	20	5[e]	5
		Metals		
Cadmium	0.020 mg/dscm[d]		0.05[e]	
Mercury	0.080 mg/dscm[d]	0.05[f]	0.05[e]	
Lead	0.20 mg/dscm[d]			
Class I		0.05[f]		0.05[f]
Class III		0.5[f]	1.0[e]	0.5[f]
Dioxins/furans	13 ng/dscm[d]	0.1 ng/N m^3 ITEQ[e]	0.1 ng/N m^3 ITEQ[e]	0.1 ng/N m^3 ITEQ[e]
Measurement Conditions	Corrected to 7% O_2 Daily average	Corrected only when $O_2 > 11\%$ 30 min avg. unless noted	Corrected to 11% O_2 11 h avg. unless noted	Corrected only when $O_2 > 11\%$ 30 min avg. unless noted

[a]Values are mg/N m^3 (0°C, 1 atm) unless otherwise noted.
[b]Whichever is less stringent.
[c]Specific emission limit based on type of combustor.
[d]Dry standard cubic meter (20°C, 1 atm).
[e]8 h avg.
[f]For greater that 30 min.
Source: References 3 and 4.

Mass Burn Combustors

Mass burn combustors are the predominant type of incinerators currently being employed for refuse. These are characterized by their accepting refuse that has undergone very limited preprocessing other than removal of oversized items and, in many instances, recyclable materials. Mass burn combustors may be of the modular (or starved-air) type, the traveling grate type, or the rotary combustor type.

Modular Combustors

Modular combustors are typically small, two-chamber, starved-air systems with capacities between about 5 and 100 tons/d. Figure 1 shows a cutaway view of a modular incinerator.

Typically, refuse is moved from a tipping floor to a ram feeder, where it is pushed into the primary combustion chamber. Here the refuse is combusted under starved-air conditions, which leads to the formation of combustible gases and ash. These combustible gases pass into the secondary chamber, where they are fired with auxiliary fuel to complete combustion and raise the temperature to above 1000°C. Flue gases, leaving the secondary chamber, often pass through a heat-recovery boiler and economizer, where they are cooled to approximately 250°C before entering the flue gas cleaning system. Ash formed in the primary chamber is pushed through the chamber and typically exits the end of the chamber via an ash quench system. Two or more modular combustors are typically employed for incinerating refuse at smaller facilities. Modular mass burn incinerators typically involve lower capital costs than do traveling grate or rotary combustors.

Traveling Grate Combustors

Traveling grate mass burn combustors are typically much larger than modular combustors and have capacities ranging from about 150 to 750 tons/d. Figure 2 shows a traveling grate combustor with water walls, a burnout zone, a heat recovery boiler, and an economizer.

Typically, refuse is dumped into a tipping bay, where it is mixed and oversized objects are removed by an overhead crane. An overhead crane dumps the mixed refuse into a feed chute, where it is fed into the grate system by a hydraulic ram. Traveling grates move the refuse through the various zones of the combustion chamber in a tumbling motion. Air is added at multiple points through the grate to assist in refuse dry out, combustion, and burnout. Additional combustion air may be added above the grate to assist in flue gas mixing and complete combustion.

Residual ash falls from the grate into a wet quench system, where it is removed and sent to a landfill. Flue gases pass upward to a burnout zone, where the temperature is maintained at about 1000°C to ensure complete destruction of organic compounds. The flue gas then passes through the boiler and economizer, where its temperature is reduced to 180–250°C prior to its entering the flue gas cleaning system.

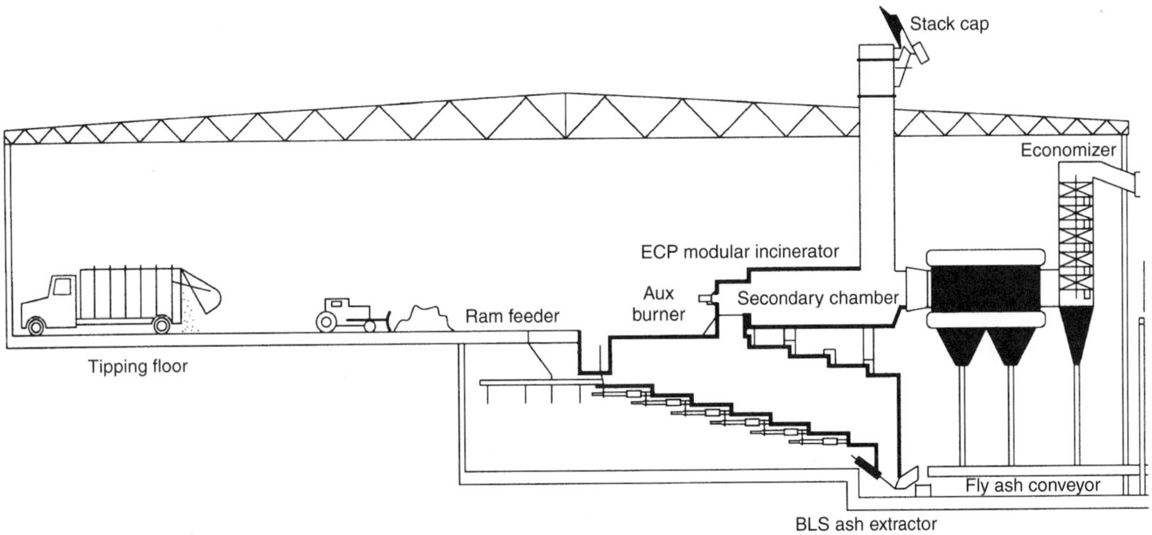

Figure 1. Modular Mass Burn Incinerator (Joy Technologies, Inc.).

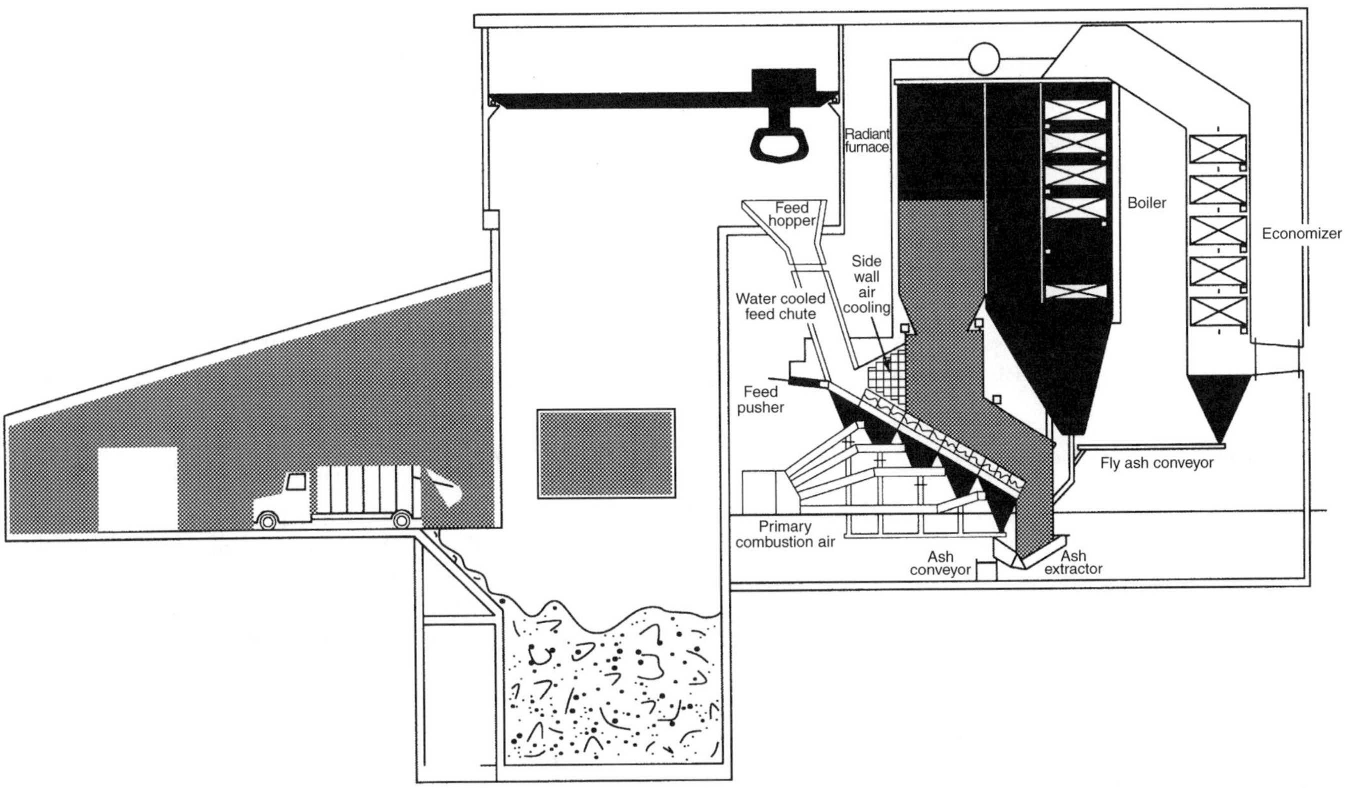

Figure 2. Travelling Grate Mass Burn Incinerator (Joy Technologies, Inc.).

Rotary Combustors

Some mass burn municipal waste combustors employ water-cooled rotary combustors instead of traveling grates. These systems typically range in size from approximately 200 to 450 tons/d. Figure 3 shows a water-cooled rotary combustor and boiler.

Refuse is fed to the feed chute via an overhead crane and pushed into the combustor by a hydraulic ram. The rotating combustor barrel is mounted at a slight angle and rotates at 10 to 20 revolutions per hour. Preheated combustion air is fed to the combustor through typically six air zones (four longitudinal and two transverse) to provide good combustion control. The rotation and angle of the combustor barrel cause the refuse to tumble and slowly move through the combustor onto a water-cooled afterburning grate, where remaining combustibles are consumed.

Residue ash is then discharged via a quench system and sent to a landfill. The flue gases at 1000°C leave the combustor barrel, pass up through a radiant zone where the temperature is reduced to between 450 and 650°C, and then pass through a boiler and economizer before discharging to the air-pollution control system. Because of the air-staging arrangement and good combustion control for this type of combustor, NO_x,

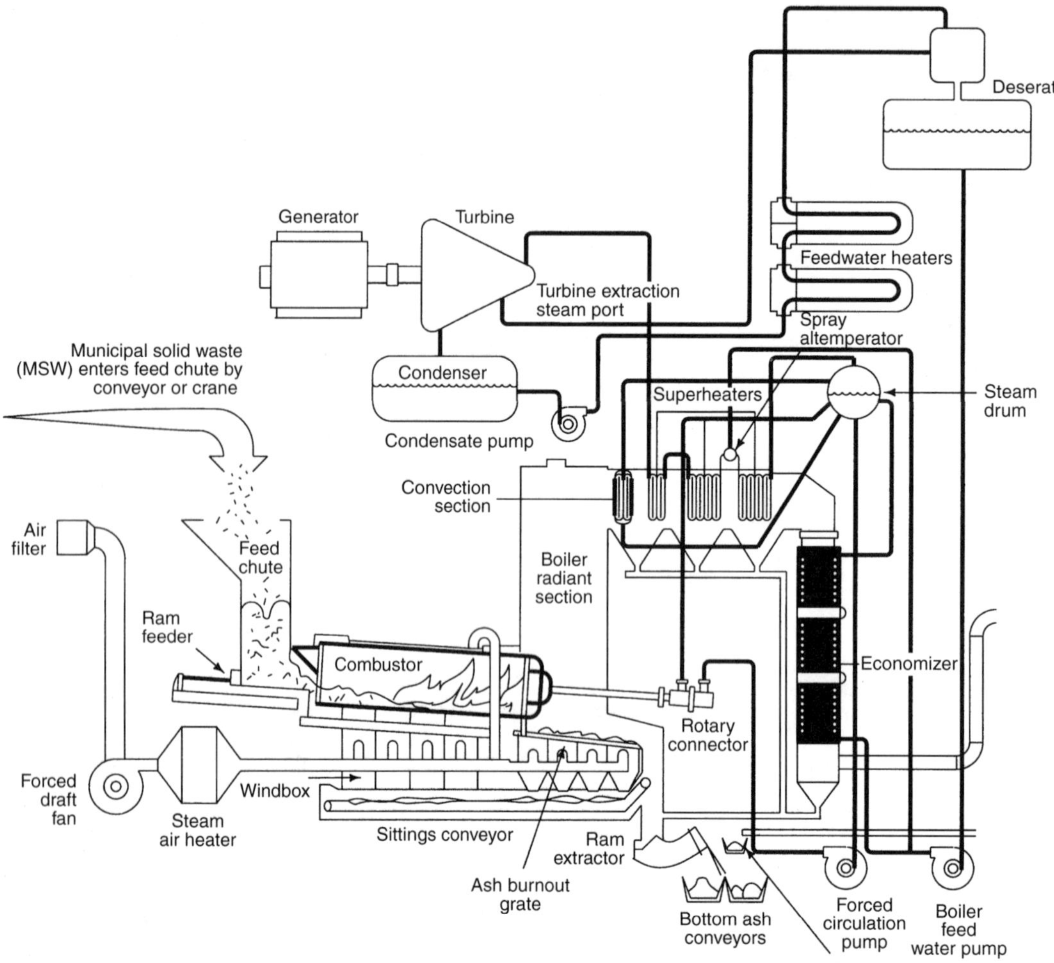

Figure 3. Water-Cooled Rotary Combustor and Boiler (Westinghouse Electric Corp.).

generation is typically somewhat lower than for a traveling grate combustor.

Refuse-Derived Fuel Combustors

Refuse-derived fuel (RDF)-fired combustors are designed to burn a fuel that has been preprocessed to produce a fuel with a more uniform size, composition, and heat rate. A variety of combustor designs can be employed with RDF, depending on the degree of preprocessing. The simplest form and most commonly employed method of RDF processing is shredding of refuse followed by magnetic separation to remove ferrous metals, and, in some cases, by air classification to remove ash. This type of fuel is often burned in a spreader stoker combustor or is suspension fired over a stoker.

RDF may be further processed to produce:

- A densified fuel by pelletizing
- A recovery-prepared RDF in which a larger portion of metals and glass is removed
- A fluff RDF for co-firing, with coal, in suspension-fired combustors

RDF combustors range in capacity from approximately 400 to 1,000 tons/d. Because of the nature of the fuel and firing, particulate matter carryover to the air pollution control system is generally much higher than for a mass burn combustor.

EMISSIONS CHARACTERIZATION

Municipal waste combustion has the potential of emitting a wide range of pollutants to the environment. These potential emissions arise from compounds present in the refuse stream, are formed as a part of the normal combustion process, or are formed as a result of incomplete combustion. Table 3 lists principal potential municipal-waste-combustion emissions and the prime source for each.

Table 3. Principal Municipal-Waste-Combustion Emissions and Sources

Pollutant	Principal Source
Particulate matter	Ash in waste stream
Acid gases	
HCl	Chlorinated plastic in waste stream
SO_2	Sulfur compounds in waste stream
SO_3	Oxidation of SO_2 in flue gas
HF	Fluorocarbons in waste stream
NO_x	Air and fuel nitrogen conversion
CO	Incomplete combustion
Heavy metals (arsenic, cadmium, lead, mercury)	Metal compounds in waste stream
Organic compounds (dioxins furans)	Products of incomplete combustion or contained in waste stream

Particulate matter consists primarily of noncombustible inorganic material entrained in the flue gas, and typically ranges in size from less than 1 μm to about 50 μm. Uncontrolled particulate matter emission rates vary substantially with the different types of MWCs. Modular combustors produce the lowest levels of uncontrolled emissions, with RDF-fired units having the highest.

Table 4 presents emission factors for a number of pollutants for modular, mass burn, and RDF-fired combustors. Acid gases, hydrogen chloride (HCl), sulfur dioxide (SO_2), and hydrogen fluoride (HF), are formed from the combustion of chloride-, sulfur-, and fluoride-containing compounds found in the waste stream. A small fraction of the SO_2 (approximately 1–5%) in the flue gas is oxidized to sulfur trioxide (SO_3). These gases, in the presence of water or water vapor, react to form hydrochloric, sulfurous, hydrofluoric, or sulfuric acid.

Nitrogen oxides (NO_x) are found predominantly in the form of NO and are formed primarily through the conversion of fuel-bound nitrogen, although some nitrogen in the combustion air may also be converted. Carbon monoxide (CO) is formed from the incomplete combustion of organic compounds in the waste stream and is used as an indicator of combustion conditions.

Heavy-metal compounds of concern emitted from MWCs include oxides and chlorides of arsenic, cadmium, lead, and mercury. These compounds are formed from the combustion of heavy-metal-containing components of the waste stream, such as batteries, plastics, paper products, and metal alloys. A number of these compounds have boiling points or sublime at temperatures below the 1000°C typical of incineration systems and are thus vaporized into the flue gas. As the flue gas temperature cools, they tend to condense out and are concentrated on fine particulate matter in the flue gas. For mercury and lead compounds, a significant fraction may remain in the vapor state at typical combustor exit gas temperatures.

Organic emissions are formed as a result of incomplete combustion of compounds found in the waste stream. The prime organic compounds of concern are PCDD and PCDF. These emissions can arise from incomplete thermal destruction of PCDD- and PCDF-containing materials in the waste stream, from incomplete thermal destruction of other organic compounds that produce PCDD/PCDF precursors, and through chemical reactions that occur at relatively low temperatures downstream of the combustor.

As shown in Table 4, uncontrolled particulate emissions and the fine particulate fractions vary widely for the three major types of combustors. These variations are a function of the turbulence in the combustion zone, the flue gas velocity through the combustor, and the fineness of the fired fuel. RDF-fired boilers have substantially higher uncontrolled total and fine particulate fractions, reflecting these conditions. Variations of other uncontrolled pollutant emissions rates are much less for the various types of combustors and are more a function of waste-stream composition and combustor operating conditions.

Table 4. Emission Factors for Municipal Waste Combustion *(All Values are Pounds per Ton, Uncontrolled)*

	Type of Incinerator		
	Modular Starved Air	Mass Burn	Refuse-Derived Fuel
Particulate matter			
Total	1.4	14	44
PM_{10}	0.12	0.18	0.13
Lead	0.12	0.18	0.13
SO_2	1.7	1.7	1.7
NO_x	4.4	3.6	5.0
CO	3.4	2.2	3.6

Source: Reference 6.

Table 5. Typical Refuse Incinerator Uncontrolled and Controlled Emissions

Pollutant	Uncontrolled Emissions	Controlled Emissions	Percent Reduction
Particulate matter, gr/dscf	0.5–4.0	0.002–0.015	99.5+
Acid gases, ppmdv			
HCl	400–1000	10–50	90–99+
SO_2	150–600	5–50	65–90+
HF	10–20	1–2	90–95+
NO_x	150–300	60–180	30–65[a]
Heavy metals, mg/nm^3			
Arsenic	<0.1–1	<0.01–0.1	90–99+
Cadmium	1–5	<0.01–0.5	90–99+
Lead	20–100	<0.1–1	90–99+
Mercury	<0.1–1	<0.1–0.7	10–90+
Total PCDD/PCDF, ng/nm^3	20–500	<1–10	80–99

Note: Reference conditions: dry gas at 12% CO_2.
[a] Reduction associated with nonselective catalytic reduction.

Table 5 shows typical uncontrolled and controlled emissions for a number of pollutants of concern from municipal waste combustion. Percent reduction ranges typical of levels being achieved utilizing the best available control technologies are also shown for each pollutant.

Modern MWCs achieve very low emissions as a result of the proper application and operation of available air-pollution control systems. Average MWC emission levels for all pollutants has been decreased substantially over the past decade as more modern installations with current emissions controls have been brought into service.

AIR-POLLUTION CONTROL SYSTEMS

MWC air-pollution control systems can be classified by the pollutant they control or their operating principles. Table 6 presents a list of the common pollutants of concern found in MWC flue gas and the methods used to control their emissions.

Often more than one control device will be used in series to control a number of pollutants. The most common examples of this are the use of an electrostatic precipitator followed by a wet scrubber or a spray dryer absorption system including an electrostatic precipitator or fabric filter. As emission control requirements have gotten more stringent the complexity and number of components of these systems have increased substantially. In some installations as many as five or more control devices may be installed in series to control all the pollutants of concern and simultaneously minimize generation of waste.

Table 6. Incinerator Emissions and Controls

Carbon monoxide (CO)	Good combustion practices
Nitrogen oxides (NO_x)	Staged combustion selective noncatalytic reduction (SNCR)
Particulate matter	Electrostatic precipitator (ESP) Pulse jet fabric filter (PJF) Reverse air fabric filter (RAF) Wet scrubber (WS)
Acid gases (HCl, SO_2, SO_3, HF)	Dry sorbent injection (DSI) Spray dryer absorption (SDA) Wet scrubbing
Heavy metals	ESP, PJF, RAF, SDA, WS
PCDD/PCDF	Good combustion practices ESP, PJF, RAF, SDA, WS

Carbon Monoxide Controls

Carbon monoxide emissions are controlled through the use of "good combustion practices." These practices include operational and combustor design elements to control the amount and distribution of excess air in the flue gas to ensure that there is enough oxygen present for complete combustion. Design of modern efficient combustors is such that there is adequate turbulence in the flue gas to ensure good mixing, a high-temperature zone (greater than 1000°C) to complete burnout, and a long enough residence time at the high temperature (1–2 sec) for complete burnout.

Feed to the combustor is controlled to minimize fuel spikes that lead to fuel-rich firing. Combustors are equipped with adequate instrumentation and combustion air controls to adjust for rapid changes in fuel conditions.

Good combustion practices also limit PCDD/PCDF emissions exiting the combustor. This is accomplished by maintaining firing conditions that destroy PCDD/PCDFs found in the fuel and also destroy PCDD/PCDF precursors that may be formed from the combustion of other chlorinated organic compounds.

Nitrogen Oxide Controls

Nitrogen oxide emissions are controlled by limiting their formation in the incinerator using staged combustion or applying selective noncatalytic reduction to reduce the NO_x content in the flue gas. In some instances where very low NO_x emission levels must be achieved, a tail-end selective catalytic reduction (SCR) may be employed.

Staged combustion is accomplished by splitting up the introduction of combustion air into the combustor so that areas of fuel-rich and fuel-lean firing are established. This lowers the peak flame temperatures and limits the amount of oxygen available to react with nitrogen in the air at the peak temperature. Introduction of additional secondary air downstream in the combustor ensures complete combustion and minimizes CO formation. Generally, staged combustion is effective in reducing NO_x formation due to air–nitrogen conversion, but is not very effective for conversion of fuel-bound nitrogen to NO_x.

NO_x present in the flue gas can be reduced through the use of either a selective catalytic or noncatalytic reduction process. Selective catalytic reduction utilizes ammonia injection upstream of a catalytic reactor, at about 315–350°C, to reduce NO_x to nitrogen. Selective catalytic reduction has been applied to a wide range of combustion sources with 80–85% NO_x reduction demonstrated. However, because of the nature of the compounds found in refuse incinerator flue gas, the successful application of SCR requires installation downstream of the acid-gas and particulate control systems with subsequent reheat to the reactor operating temperature.

Selective noncatalytic reduction (SNCR) reduces flue gas NO_x, through a reaction with ammonia in a temperature range of 900–1050°C. Ammonia may be supplied as anhydrous ammonia, aqueous ammonia, or urea. At flue gas temperatures above 1100°C, oxidation of ammonia to NO_x increases, and SNCR can actually result in an increase in overall NO_x. At temperatures below about 900°C, NO_x reduction falls off rapidly and ammonia breakthrough increases, leading to the potential for a visible ammonium chloride plume and increased fouling by ammonia bisulfite deposits.

Ammonia injection, also known as thermal de-NO_x, has been applied to many different combustion sources, including mass burn combustors. Reductions in NO_x levels of up to 65% have been demonstrated at an ammonia-to-NO_x ratio of about two, with ammonia breakthrough as low as 5 ppm(v). This corresponds to an NO_x emission level of approximately 60 ppm(v). Thermal de-NO_x operates most efficiently under steady-state operating conditions. Changes in fuel feed rate, excess air rate, or incinerator load can significantly change flue gas conditions at the ammonia-injection point, leading to a major change in control efficiency.

Urea injection has been demonstrated full scale on refuse combustors in the United States and Europe. Urea injection offers the advantage of not requiring a hazardous material for operation. At the injection temperatures employed (900–10,500°C), the urea quickly breaks down to form the active reagent. In some cases, reaction enhancers are added to the urea to expand the effective temperature window to as low as 650°C. Tests with urea injection have achieved greater than 65% NO_x reduction with very low (approximately 5 ppm) ammonia slip.[7]

Particulate Matter Controls

Particulate emissions are primarily controlled by electrostatic precipitators (ESPs) or fabric filters, although wet scrubbers are sometimes used on small combustors or in series with ESPs for additional control. ESPs are installed either alone, to control particulate emissions, or after a spray dryer, as a part of an acid-gas cleaning system. Fabric filters are typically installed downstream of a quench tower or spray dryer, where the conditions of increased flue gas moisture and lowered temperature aid in protecting filter bags from hot embers.

Electrostatic Precipitators

Electrostatic precipitators collect particulate matter by introducing a strong electrical field in the flue gas, which imparts a charge to the particulates present. These charged particles are then collected on large plates, which have an opposite charge applied to them. The collected particulate is periodically removed by rapping the collection plates. The agglomerated particles fall to a hopper, where they are removed. Key design parameters for ESPs include the following:

- Particulate composition, density, and resistivity
- Flue gas temperature and moisture content

- Inlet particulate loading and collection efficiency
- Specific collection area (SCA, square meters of collecting surface per m^3/sec of flue gas)
- Number of fields, flue gas velocity, and collector plate spacing
- Rapping frequency and intensity
- Transformer rectifier power levels

Table 7 presents sizing parameters typical for ESPs applied for MWC particulate emissions control. The ranges in parameters shown reflect straight ESP particulate control applications and ESP applications as part of an acid-gas cleaning system. Although the inlet particulate loading to the ESP is much higher as part of an acid-gas cleaning system, the number of fields and specific collecting areas required to achieve a similar outlet emission do not change significantly. This is due to lower ash resistivity values and increased flue gas moisture contents, which improve the ESP's performance. MWCs that have had spray dryers retrofitted in front of existing ESPs have been able, in most cases, to maintain the same level of particulate emissions (e.g., 5–15 mg/N m^3 at 12% CO_2).

Weighted-wire, rigid-frame, and rigid-electrode types of precipitators are employed for MWC applications; however, rigid-frame and rigid-electrode types predominate. This is related to the corrosive gas conditions and sticky nature of the fly ash being collected. Electrode failures associated with rigid-frame and rigid-electrode systems are less frequent than for weighted wire electrodes. This is especially true where higher rapping forces are needed to dislodge the sticky fly ash. For rigid-frame systems, high-alloy (e.g., Incoloy 825) spring-wound electrodes are also used to minimize electrode corrosion problems.

Insulator compartment ventilation system must be designed to minimize the effects of the corrosive nature of the flue gas and fly ash stickiness. A pressurized ventilation system employing heated air is recommended to maintain clean insulators and reduce potential electrical tracking problems.

Fabric Filters

Both the reverse-air and pulse-jet types of fabric filters are used for MWC particulate emission control. Each type offers advantages that should be evaluated on a site-specific basis. Both types are capable of achieving particulate emissions of the order of 5–15 mg/N m^3 at 12% CO_2. Table 8 presents design parameters typical of incinerator fabric filter applications.

Table 7. Electrostatic Precipitator-Design Parameters

Particulate loading, gr/acf		0.5–9
Required efficiency, %		98–99.9
Number of fields		3–4
SCA, ft^2/1000 acfm		400–550
Average secondary voltage, kV		35–55
Average secondary current, mA/1000 ft^3		30–50
Gas velocity, ft/s		3.0–3.5
	Particulate	Acid-Gas Control
Flue gas temperature, °F	350–450	230–300
Flue gas moisture, % vol.	8–16	12–20
Ash resistivity, ohm-cm	10^9–10^{12}	10^8–10^9

Source: Reference 8.

Table 8. Fabric Filter—Design Parameters

	Reverse Air	Pulse Jet
Operating temperature, °F	230–450	
Type of fabric	Woven fiberglass	
Fabric coating	10% Teflon B or acid Resistant	
Fabric weight, oz/yd^2	9.5	16 or 22
Bag diameter, inches	8	6
Net air-to-cloth ratio	1.5–2.0:1	3.5–4.0:1
Minimum compartments	6	4
Overall pressure drop, in. w.g.	4–6	8–10
Estimated Bag Life, Years	3–4	1.5–2

Temperature ranges shown represent operation after both a dry quench chamber (175–230°C) and a spray dryer (110–150°C). For these temperature ranges, woven fiberglass is typically used as the bag material, although Nomex fabric is also used. A 10% Teflon B coating is the most commonly specified, with acid-resistant coating also used.

Bag sizes differ substantially for the two types of filters. Reverse-air filters generally employ bags 200 mm in diameter by up to 7 m long. Pulse-jet bags are usually 150 mm diameter by 3.5–4.5 m long. However, some vendors offer a low-pressure pulse filter with bags up to 7 m long. The major differences in operating parameters are in the air-to-cloth ratio and system pressure drop. Pulse-jet filters generally operate at double the air-to-cloth ratio than reverse-air filters and at nearly double the pressure drop. This results in more frequent bag cleaning and a substantially shorter bag life.

Main advantages of a pulse-jet fabric filter are a lower capital cost and a smaller footprint. However, because of the shorter bag life and higher pressure drop, the pulse-jet filter generally has a higher total evaluated cost for plants exceeding 15 years of life. A reverse-air filter typically has lower particulate emissions when compared with a pulse-jet filter.

The majority of fabric filter applications are as part of an acid-gas cleaning system and incorporate specific design features for operating after a spray dryer. Flue gas, after a spray dryer has been cooled (110–150°C), has a high moisture content (12–20%), is closer to the dew point (25–70°C), and may have a higher particulate loading. These conditions can lead to severe corrosion and baghouse plugging.[8]

Corrosion control is accomplished by insulation design, control of air in-leakage into the filter, hopper heating, and, in some instances, coating of the fabric filter internals with an acid-resistant material. Insulation specifications usually require a minimum of 100 mm with double lapping on side panels and with the insulation extending into the hopper crotch areas. Air in-leakage is controlled by good quality control during erection and by minimizing the number of openings into the filter. Hopper heating is used to maintain the hopper skin temperature at the flue gas temperature to prevent cold spots and to aid in maintaining product flowability.

As part of an acid-gas cleaning system, the fabric filter also acts as a reactor to aid in acid-gas absorption, especially for sulfur dioxide. Sulfur dioxide in the flue gas is absorbed by alkaline material in the filter cake on the bags. Therefore, when a bag is freshly cleaned, SO_2 absorption decreases. In order to minimize this impact on overall absorption, the number of bags being cleaned simultaneously should be minimized. This can be accomplished by increasing the number of compartments. A minimum of six compartments is generally specified for acid-gas cleaning systems.

Wet Scrubbers

Wet scrubbers are typically employed as part of a two-stage flue gas cleaning system downstream of an ESP. They function as a particulate-removal polishing stage and as an acid-gas absorber. A venturi scrubber followed by a packed or tray tower is commonly used; however, other types of wet scrubbers, such as charged droplet scrubbers, are also used. Figure 4 shows a typical wet scrubber design used for both particulate and acid-gas control.

Typically, water is recycled in the venturi stage to achieve particulate removal. Hydrogen chloride present in the gas would also be removed in this stage. Additional particulate and acid-gas removal can take place in the second scrubbing stage. Absorption of SO_2 is enhanced in this stage by maintaining a recirculating solution pH in the range of about 6.5 to 9 through addition of caustic (sodium hydroxide). A blowdown stream is maintained for each stage to control the recirculating solution solids content. Typical design parameters for refuse-incinerator wet scrubber applications are presented in Table 9.

The venturi section is subjected to severe corrosive conditions due to the low circulating solution pH, the high hydrochloric acid concentration, and the presence of small amounts of sulfuric, nitric, and hydrofluoric acids. The scrubber inlet temperature may be as high as 250°C, which preclude the use of corrosion-resistant resins; therefore, high-alloy steels are typically specified as the materials of construction. In cases where inlet flue gas contains high levels of particulate matter, the venturi section may be lined with a corrosion-resistant material such as bricks. The venturi section typically is equipped with a set of emergency quench nozzles to ensure that flue gas temperatures leaving the venturi are maintained at an acceptable level for the absorber-stage materials of construction.

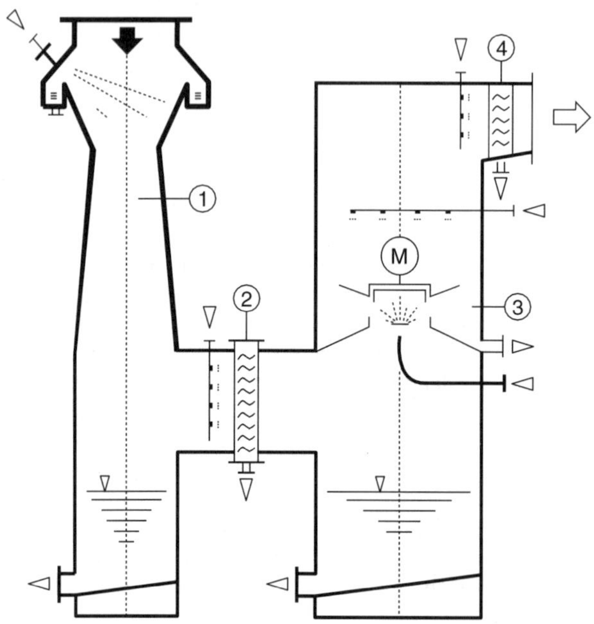

1 saturator venturi/1st scrubbing stage
2 lamellar droplet separator
3 radial-flow scrubber/2nd scrubbing stage
4 lamellar droplet separator

Figure 4. Saturator Venturi with Radial-Flow Scrubber (Courtesy Lurgi Corp. Reprinted with permission).

Table 9. Wet Scrubber-Design Parameters

	Venturi Stage	Absorber Stage
Gas velocity, ft/s	90–150	6–10
Pressure drop, in. w.c.	40–70	4–8
L/G. Gal/Kacfm	10–20	20–40
Scrubbing medium	Water	Caustic
Solution pH	<1–2	6.5–9
Materials of construction	High-alloy steel (e.g., Inconel, Hastelloy)	FRP Lined carbon steel

The absorber stage may be a packed tower, a tray tower, or a radial flow tower, as shown in Figure 4. Materials of construction for the absorber are typically fiberglass-reinforced plastics (FRP), although carbon-steel vessels lined with rubber or a corrosion-resistant resin material are also used.

ACID-GAS CONTROLS

Control of refuse incinerator acid-gas (HCl, SO_2, SO_3, and HF) emissions is achieved by dry sorbent injection, spray dryer absorption, or wet scrubbing. Each of these types of technologies has been successfully applied to meet existing emissions regulations; however, as emissions limitations become more stringent, the trend is toward spray dryer absorption, wet scrubbing, and/or hybrid systems, which incorporate both technologies.

Dry Sorbent Injection

Dry sorbent injection (DSI) involves the addition of an alkaline material—usually hydrated lime, $Ca(OH)_2$, or soda ash, $Na_2(CO_3)$ to the gas stream to react with acid gases present, thus producing a salt that is collected in a particulate-collection device. This very simple process can capture up to 90% of the HCl present in the flue gas and about 50% of the SO_2. However, stoichiometric ratios (equivalents of alkali added per equivalents of acid in the flue gas) are high, typically of the order of 2–4. Therefore, because of the high reagent costs associated with this system, simple DSI applications are normally limited to small facilities with moderate emissions control requirements.

The overall acid-gas control efficiency of DSI can be improved and reagent consumption decreased by:

- Increasing flue gas humidity
- Recycling reaction products into the flue gas stream

Increasing the flue gas relative humidity can be accomplished by cooling the flue gas using heat exchangers or by quenching the flue gas using water sprays. Both approaches are commercially applied; however, the use of quench chambers predominates. Figure 5 presents a simplified process flow diagram for a humidification DSI process.

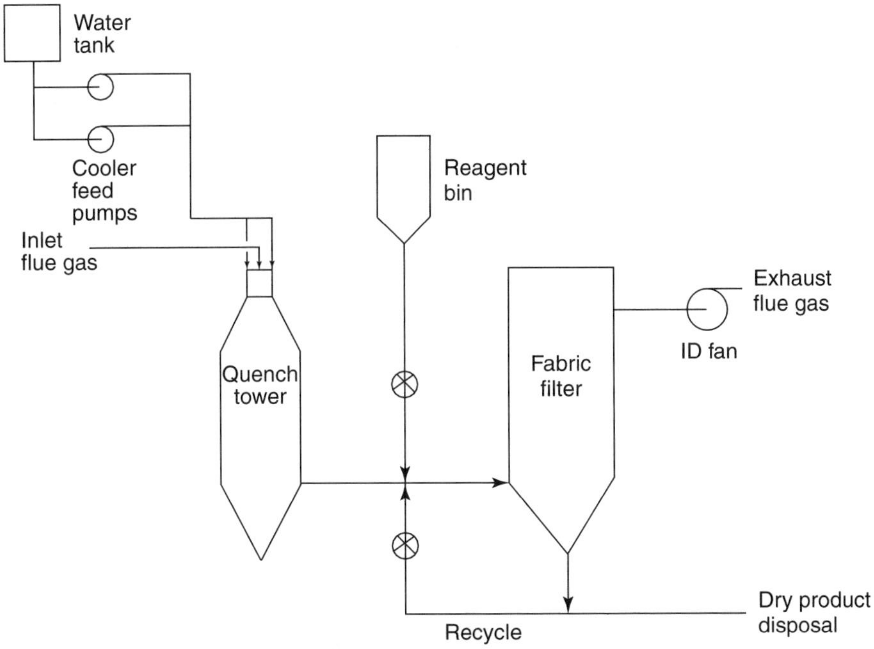

Figure 5. Humidification—Dry Sorbent Injection Process.

Flue gas from the MWC enters a 3–5-sec retention time cooling tower (or dry quench chamber), where water is sprayed into the gas to lower the temperature. The flue gas temperature leaving the cooling tower is controlled at a temperature high enough to ensure that all water droplets evaporate (150–175°C). Dry reagent is mixed with the flue gas via pneumatic transport systems or eductor venturis. The reagent reacts with acid gases prior to removal in a dust collector (typically a fabric filter). A portion of the collected reaction products in some cases is reinjected to increase acid-gas removal and decrease reagent consumption. Humidification and reagent injection steps can also be carried out together in specially designed reactors. This type of process can achieve greater than 95% HCl removal and 90% SO_2 removal at stoichiometric ratios between 1 and 2.

Spray Dryer Absorption

Spray dryer absorption (SDA) has been widely applied for MWC emissions control and has been specified as best available control technology (BACT) in a number of air permits. The SDA process combines a spray dryer with a dust collector. Reagent addition, flue gas humidification, and some acid-gas absorption take place in the spray dryer. Additional acid-gas absorption and collection of the dry fly-ash/reaction product mixture take place in the dust collector. The SDA process is capable of achieving very high removal efficiencies for all acid gases (99+% HCl, 95% SO_2, 99+% SO_3, 95% HF), as well as for the removal of trace metals and organic compounds at stoichiometric ratios between 1 and about 1.8. Figure 6 presents a simplified flow diagram for the SDA process.

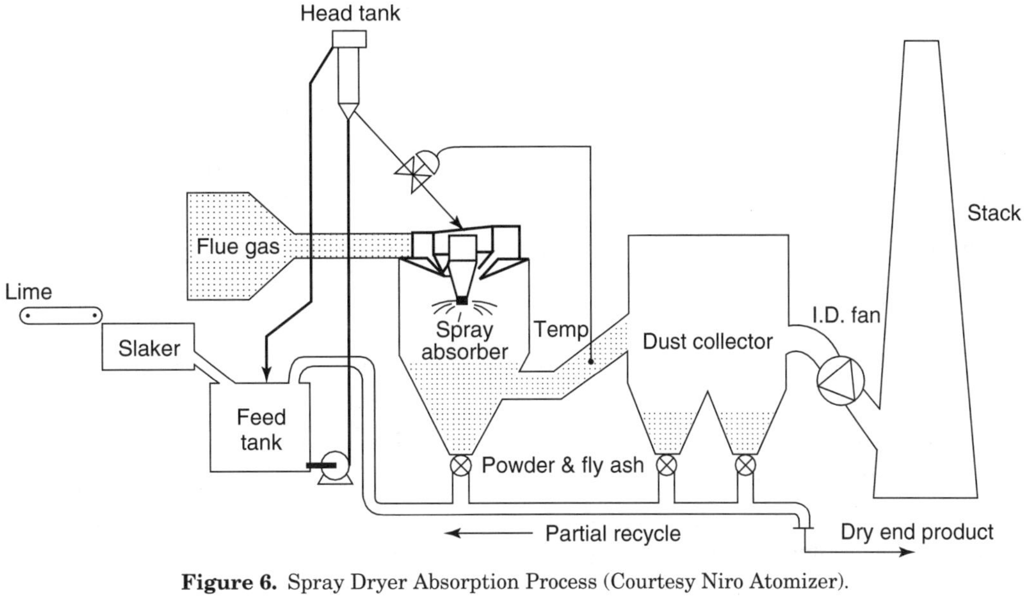

Figure 6. Spray Dryer Absorption Process (Courtesy Niro Atomizer).

Flue gas enters the spray dryer, where it is contacted by a cloud of finely atomized droplets of reagent (typically, a hydrated lime slurry). The flue gas temperature is decreased and the flue gas humidity is increased as the reagent slurry simultaneously reacts with acid gases present and evaporates to dryness. In some systems, a portion of the dried product is removed from the bottom of the spray dryer, while in others, it is carried over to the duct collector. Collected reaction products are sometimes recycled to the feed system to reduce reagent consumption.

Several different spray-dryer design concepts have been employed for MWC applications. These include single rotary, multiple rotary, and multiple dual fluid nozzle atomization; downflow, upflow, and upflow with a cyclone precollector spray dryers; and single and multiple gas inlets. Flue gas retention times range from 10 to 18 sec and flue gas temperatures leaving the spray dryer range from 110°C up to 150°C.

Generally, the lower the spray dryer outlet temperature, the more efficient will be the acid-gas absorption. The minimum reliable operating outlet temperature is a function of the spray dryer and dust collector design and the composition of the dry fly ash–reaction product mixture. The spray dryer outlet temperature must be maintained high enough to ensure complete reagent evaporation and the production of a free-flowing product. Low-outlet-temperature operation requires efficient reagent atomization, good gas dispersion and mixing, adequate residence time for drying, and design of the dust collector to minimize heat loss and air in-leakage.

Dust collectors downstream of the spray dryer may be ESPs, reverse-air baghouses, or pulse-jet type baghouses. The selection of a specific type of dust collector is dependent on such site-specific factors as particulate emission limits, overall acid-gas removal requirements, and project economics. Each of these dust-collection devices offers process advantages and disadvantages that are evaluated on a site-specific basis. Generally, where high acid-gas control is required (95+% HCl, 85+% SO_2), a baghouse is utilized as it is a better reactor than an ESP.

Whether a fabric filter or an ESP is selected as the dust collector, minimization of heat loss from the dust collector to avoid corrosion and increased product stickiness is a prime design consideration. Four methods employed to achieve this are as follows:

- Insulation, to control heat loss
- Control of air in-leakage, to minimize cold spots
- Hopper heating, to maintain product temperature
- Operating procedures to maintain product flowability and minimize cold areas

The end product of the SDA process is a fine hygroscopic material with a significant soluble fraction. It tends to be stickier than MSW fly ash and more difficult to convey and store. Major end-product constituents include:

Fly ash	Calcium sulfite
Calcium hydroxide	Calcium sulfate
Calcium chloride	Calcium fluoride
Calcium carbonate	Moisture

8Calcium chloride formed at typical spray dryer outlet temperatures is a mixture of mono and dihydrates ($CaCl_2{\cdot}H_2O$ and $CaCl_2{\cdot}2H_2O$) and at lower temperatures will absorb moisture until it reaches the hexahydrate form ($CaCl_2{\cdot}6H_2O$) and melts. Therefore, it is necessary to keep the product from being exposed to cold and/or moist air. This is accomplished by proper design of the product conveying and storage systems.

Wet Scrubbing

Wet scrubbing systems are capable of achieving high acid-gas removal efficiencies and have been applied to a large number of installations in Europe. Typical wet scrubbing applications include two-stage scrubbers located downstream of an ESP. The first stage is used for HCl removal and the second for SO_2 removal. Water is used to capture the HCl and either caustic or hydrated lime is used for SO_2 capture. Figure 4 shows a typical two-stage wet scrubber, and Figure 7 shows a process flow diagram for an application of wet scrubbing with fly-ash treatment.

In this process, the HCl stream from the first scrubbing stage is pumped to a fly-ash leaching tank, where it is used to leach out heavy metals from the fly ash collected in the dust collector. After leaching, residual fly-ash solids are either disposed of or used in construction applications. The heavy-metals-bearing HCl stream is then treated alone or with the sodium sulfite–sulfate solution from the second scrubber stage in a neutralization/precipitation stage to concentrate the heavy metals and produce salt-containing wastewater for disposal. When lime is used in the SO_2 absorption section of the scrubber, the calcium sulfite slurry can be oxidized to calcium sulfate (gypsum) for utilization.

Wet scrubbers offer a number of process advantages:

- They are relatively inexpensive to install and require relatively small plot space.
- They are capable of achieving very high removal efficiencies for acid gases (99+% HCl, 95+% SO_2).
- They are capable of high removal efficiencies for many volatile trace compounds.
- They require the lowest reagent stoichiometrics (1.0–1.2) of any of the alternatives considered.

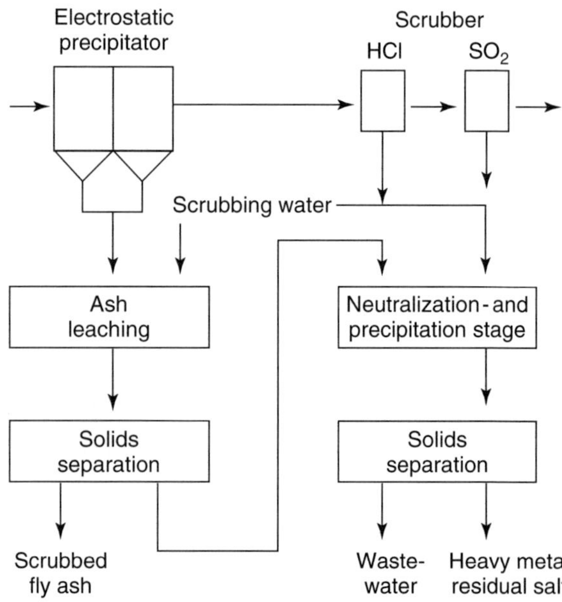

Figure 7. Wet Scrubbing with Ash Treatment.

Wet scrubbers also have some disadvantages:

- They produce a wet effluent that requires additional treatment with complex effluent treatment systems.
- Economics and space requirements are not as attractive as for the other alternatives.
- Wet scrubbers are more prone to corrosion problems and may require expensive materials of construction.
- Historically, wet scrubbers have experienced more operating problems and higher maintenance requirements than the alternatives.

HEAVY-METALS CONTROL

The primary heavy metals of concern from MWCs (arsenic, cadmium, lead, and mercury) are collected in the particulate control device or in the acid-gas control system. Most of these metals exist as solid particulates at incinerator-exit flue gas temperatures and are collected as particulate matter. However, some arsenic, lead, and mercury compounds exist in the vapor state at MWC gas exit temperatures, and these compounds must be collected by condensation through cooling of the flue gas. This can be accomplished with either an SDA or wet scrubbing process.

In the SDA process, the flue gas cooling takes place rapidly in a cloud of finely atomized droplets. These droplets serve as sites on which metals can condense or into which they can be absorbed. The condensed metal is then removed with the reaction products in the downstream dust collector. Collection efficiencies for arsenic and lead at typical SDA system operating temperatures are greater than 90%.

A significant fraction of mercury remains in the vapor phase, even at SDA system outlet temperatures of 250°F. Additions of small amounts of powdered activated carbon or sodium sulfide upstream of the spray dryer have been used to enhance mercury control and greater than 90% capture has been achieved.[9]

Wet scrubbers following a dust collector operate at saturated flue gas temperatures (65–80°C) and can achieve greater than 90% removal of mercury. They can also remove a major fraction of the other metals that may escape the particulate control device. Wet scrubbers operating in a subadiabatic temperature mode can achieve even higher removal efficiencies.

Activated carbon packed-bed adsorbers, located downstream of other gas cleaning devices, are seeing greater application in Europe, where mercury emission limits are quite stringent. They are capable of achieving very low mercury emissions.

PCDD/PCDF CONTROL

The PCDD/PCDF emissions are controlled by good combustion practices that inhibit their formations and by particulate and acid-gas controls. Combustion temperatures above 1000°C for more than 2 sec are specified as a method of destroying PCDD/PCDF found in the waste stream and their precursors formed from the combustion of other organic and chlorine-containing compounds. However, some PCDD/PCDF compounds may still form downstream of the MWC, on the surface of fly ash, at temperatures from 250°C to 375°C.

Control of PCDD/PCDF compounds found in the flue gas leaving the incinerator is achieved by ESPs operating below 230°C or by acid-gas control systems. Acid-gas control systems achieve a higher PCDD/PCDF capture efficiency because of their reduced outlet temperatures and the large droplet surface area available for adsorption to take place. Capture efficiencies of PCDD/PCDF of up to 99% can be achieved, and total emissions can be reduced to less than about 10 ng/N m^3.

REFERENCES

1. *EPA Characterization of Municipal Solid Waste in the United States 1960–2000* (update 1988), Final Report, U.S. Environmental Protection Agency, Office of Solid Waste and Emergency Responses, Franklin Associates Ltd., March 30, 1988.
2. *Decision-Makers Guide to Solid Waste Management*, U.S. Environmental Protection Agency Office of Solid Waste and Emergency Responses, EPA/530-500-89-072, November 1989.
3. *Subpart Eb—Standards of Performance for Municipal Waste Combustors for Which Construction is Commenced after September 20, 1994*, 40 CFR Part 60, 60 FR 65419, Dec. 19, 1995.
4. A. Licata, H. U. Hartenstein, and L. Terracciano, "Comparison of U.S. EPA and European Emission Standards for Combustion and Incineration," •••.
5. Incineration brochures, Joy Energy Systems, Inc., Charlotte, NC.
6. D. G. Jones et al., "Two stage De NO_x, Process Test Data from Switzerland's Largest Incineration Plant," International Conference on Municipal Waste Combustion, Hollywood, FL, April 1989.
7. W. Panknins, "Research into Activated Carbon Technology on Harmful Organic Substances, Heavy Metals and NO_x Control," Air and Waste Management Association 83rd Annual Meeting, Pittsburgh, June 1990.
8. B. Brown, J. R. Donnelly, T. D. Tarnok, et al., "Dust Collector Design Considerations for MSW Acid Gas Cleaning Systems," EPA/EPRI 7th Particulate Symposium, Nashville, March 1988.
9. J. R. Donnelly, "Design Considerations for MSW Incinerator APC Systems Retrofit," Air and Waste Management Association 83rd Annual Meeting, Pittsburgh, June 1990.

BIBLIOGRAPHY

T. G. Brna and J. D. Kilgroe, "The Impact of Particulate Emissions Control on the Control of Other MWC Air Emissions," *J. AWMA* **40**(9):1324–29 (1988).

M. J. Clarke, "Technologies for Minimizing the Emission of NO, from MSW Incineration," International Conference on Municipal Waste Combustion, Hollywood, FL, April 1989.

D. J. Collins and P. W. Guletsky, "Operating Experiences with Spray Dry Scrubbing Systems: Case Histories of Three Industrial Plants," IGCI Forum 90, Baltimore, March 1990.

J. R. Donnelly and K. S. Felsvang, "Joy/Niro SDA-FGC Systems North American and European Operating Results," International Conference on Municipal Waste Combustion, Hollywood, FL, April 1989.

R. Emery, R. Gregporini, and A. Berst, "Design and Startup of the Imperial Resource Recovery Spray Reactor/Fabric Filter System," IGCI Forum 90, Baltimore, March 1990.

S. Greene, *Municipal Waste Combustion Study: Report to Congress*, U.S. Environmental Protection Agency, NTIS PB 87206074, 1987.

K. S. Felsvang, T. S. Holms, and B. Brown, "Control of Mercury and Dioxin Emissions from European MSW Incinerator by Spray Dryer Absorption Systems," AIChE Summer National Meeting, San Diego, August 1990.

J. H. Hahn, H. P. Von Dem Fange, and D. S. Sofaer, "Recent Air Emissions Data from Ogden Martin Systems, Inc., Resource

Recovery Facilities Which Became Operational during 1988," Air and Waste Management Association 83rd Annual Meeting, Pittsburgh, June 1989.

K. Masters, *Spray Drying Handbook*, 3rd ed., George Goodwin Ltd., London, 1979.

Municipal Waste Combustion Study: Sampling and Analysis of Municipal Waste Combustors, U.S. Environmental Protection Agency, NTIS PB 87-206124, 1987.

Municipal Waste Combustion Study: Characterization of the Municipal Waste Combustion Industry, U.S. Environmental Protection Agency, NTIS PB 87-206140, 1987.

L. P. Nelson, P. Shindley, and J. D. Kilgroe, "Development of Good Combustion Practices to Minimize Air Emissions from Municipal Waste Combustors," International Conference on Municipal Waste Combustion, Hollywood, FL, April 1989.

P. Schindler, *Municipal Waste Combustion Study: Emissions Data Base for Municipal Waste Combustors*, U.S. Environmental Protection Agency, NTIS PB 87-206082, 1987.

W. K. Seeker, W. S. Lanier, and M. P. Heap, *Municipal Waste Combustion Study: Combustion Control of MSW Combustors to Minimize Emission of Trace Organics*, U.S. Environmental Protection Agency, NTIS PB 87-206090, 1987.

E. Wheless, "Air Emission Testing at the Commerce Refuse-to-Energy Facility," Environment Canada MSW Incinerator Workshop, Toronto, October 1987.

HAZARDOUS WASTE

Mackenzie L. Davis, Clyde R. Dempsey, and E. Timothy Oppelt

The following discussion has been excerpted from Ref. 1.

PROCESS DESCRIPTION

Different incineration technologies have been developed for handling the various types and physical forms of hazardous waste. The four most common incinerator designs are liquid injection (sometimes combined with fume incineration), rotary kiln, fixed hearth, and fluidized bed incinerators. While a detailed examination of design principles is beyond the scope of this discussion, a generalized review of the most prominent features of incineration systems and important design factors will be helpful in understanding incinerator operation and emissions performance.

The four major subsystems of a hazardous waste incineration system are: (1) waste preparation and feeding, (2) combustion chamber(s), (3) air pollution control, and (4) residue/ash handling. The selection of the appropriate combination of these components is primarily a function of the physical and chemical properties of the waste stream or streams to be incinerated.

Waste Preparation and Feeding

The physical form of the waste determines the appropriate feed method. Liquids are blended, then pumped into the combustion chambers through nozzles or via specially designed atomizing burners. Wastes containing suspended particles may need to be screened to avoid clogging of small nozzle or atomizer openings. While sustained combustion is possible with waste heat content as low as 4,000 Btu/lb, liquid wastes are typically blended to a net heat content of 8,000 Btu/lb or greater. To incinerate lower heating value wastes, supplementary fuel will normally be required. Blending may be achieved by either mixing the wastes before they are fed to the combustion chamber or by using separate nozzles for different types of waste, wherein the mixing occurs in the combustion chamber. Blending is also used to control the chlorine content of the waste fed to the incinerator. Wastes with a chlorine content of 70% and higher can be incinerated in specially designed incinerators. Blending provides better combustion control and limits the potential for periodic formation of high concentrations of free-chlorine gas in the combustion gases.

Sludges are typically fed using progressive cavity pumps and water-cooled lances. Bulk solid wastes may require shredding for control of particle size. They may be fed to the combustion chamber via rams, gravity fed, air-lock feeders, vibratory or screw feeders, or belt feeders. Containerized waste is typically gravity or ram fed.

Combustion Chambers

The physical form of the wasted and its ash content determine the type of combustion chamber selected. Table 1 provides

Table 1. Applicability of Major Incinerator Types to Wastes of Various Physical Form

	Liquid Injection	Rotary Kiln	Fixed Hearth	Fluidized Bed
Solids				
Granular, homogeneous		X	X	X
Irregular, bulky (pallets, etc.)		X	X	
Low melting point (tars, etc.)	X	X	X	X
Organic compounds with fusible ash constituents		X	X	X
Unprepared, large, bulky material		X	X	
Gases				
Organic vapor laden	X	X	X	X
Liquids				
High-organic-strength aqueous wastes	X	X	X	X
Organic liquids	X	X	X	X
Solids/liquids				
Waste contains halogenated aromatic (2200°F minimum)	X	X	X	
Aqueous organic sludge		X		X

general selection considerations for the four major combustion chamber (incinerator) designs as a function of wastes of different forms.[2] Most incineration systems derive their names from the type of combustion chamber employed.

Liquid injection incinerators or combustion chambers are applicable almost exclusively for pumpable liquid waste. These units (Figure 1) are usually simple, refractory-lined cylinders (either horizontally or vertically aligned) equipped with one or more waste burners. Liquid wastes are injected through the burner(s), atomized to fine droplets, and burned in suspension. Burners, as well as separate waste injection nozzles, may be oriented for axial, radial, or tangential firing. Improved utilization of combustion space and higher heat release rates, however, can be achieved with the utilization of swirl or vortex burners or designs involving tangential entry.

Good atomization is critical to achieving high destruction efficiency in liquid combustors. Nozzles have been developed to produce mists with mean particle diameters as low as 1 micrometer (μm), compared to typical oil burners, which yield droplets in the 10 to 50 μm range. Atomization may be attained by low-pressure air or steam [1–10 pounds per square inch gauge (psig)], high-pressure air or steam (25–100 psig), or mechanical (hydraulic) means using specially designed orifices (25–450 psig).

Vertically downward oriented liquid injection incinerators are preferred when wastes are high in inorganic salts and fusible ash content, while horizontal units may be used with low ash waste. In the past, the typical capacity of liquid injection incinerators was roughly 30×10^6 Btu/h heat release. However, units as high as 210×10^6 Btu/h are now in operation.

Rotary kiln incinerators (Figure 2) are more versatile in the sense that they are applicable to the destruction of solid wastes, slurries, and containerized waste as well as liquids. The rotary kiln is a horizontal cylindrical refractory-lined shell that is mounted on a slight slope. Rotation of the shell provides for transportation of waste through the kiln as well as enhanced mixing of the burning solid waste. The waste may move either concurrent or countercurrent to the gas flow. The residence time of waste solids in the kiln is generally 0.5–1.5 h. This is controlled by the kiln rotation speed (typically 0.5 to 1.0 revolutions per minute), the waste feed rate, and in some instances, the inclusion of internal dams to retard the rate of waste movement through the kiln. The feed rate is also generally adjusted to limit the amount of waste being processed in the kiln to at most 20% of the kiln volume.

The primary function of the kiln is to convert solid wastes to gases, which occurs through a series of volatilization, destructive distillation, and partial combustion reactions. An afterburner is necessary, however, to complete the gas-phase combustion reactions. The afterburner is connected directly to the discharge end of the kiln, where the gases exiting the kiln are directed to the afterburner chamber. Some more recent systems have installed a "hot cyclone" between the kiln and afterburner to remove solid particles that might otherwise create slagging problems in the afterburner. The afterburner itself may be horizontally or vertically aligned, and essentially functions as a liquid injection incinerator. In fact, many facilities also fire liquid hazardous waste through separate waste burners in the afterburner. Both the afterburner and kiln are usually equipped with an auxiliary fuel firing system to bring the units up to temperature and to maintain the desired operating temperatures. On the other hand, some operators make it a practice of firing their aqueous waste streams into the afterburner as a temperature control measure. Rotary kilns have been designed with a heat release capacity as high as 150×10^6 Btu/h in the United States. On average, however, units are typically around 60×10^6 Btu/h.

Fixed hearth incinerators, also called controlled air, starved air, or pyrolytic incinerators, are the third technology in use for hazardous waste incineration today. These units employ a two-stage combustion process, much like rotary kilns (Figure 3). Waste is ram fed or pumped into the first stage or primary chamber, and burned at roughly 50–80% of stoichiometric air requirements. This starved-air condition causes most of the volatile fraction of the waste to be vaporized by the endothermic heat provided by the oxidation of the fixed carbon fraction. The resultant smoke and pyrolytic products consisting primarily of methane, ethane, and other hydrocarbons; carbon monoxide; and products of combustion pass to the second

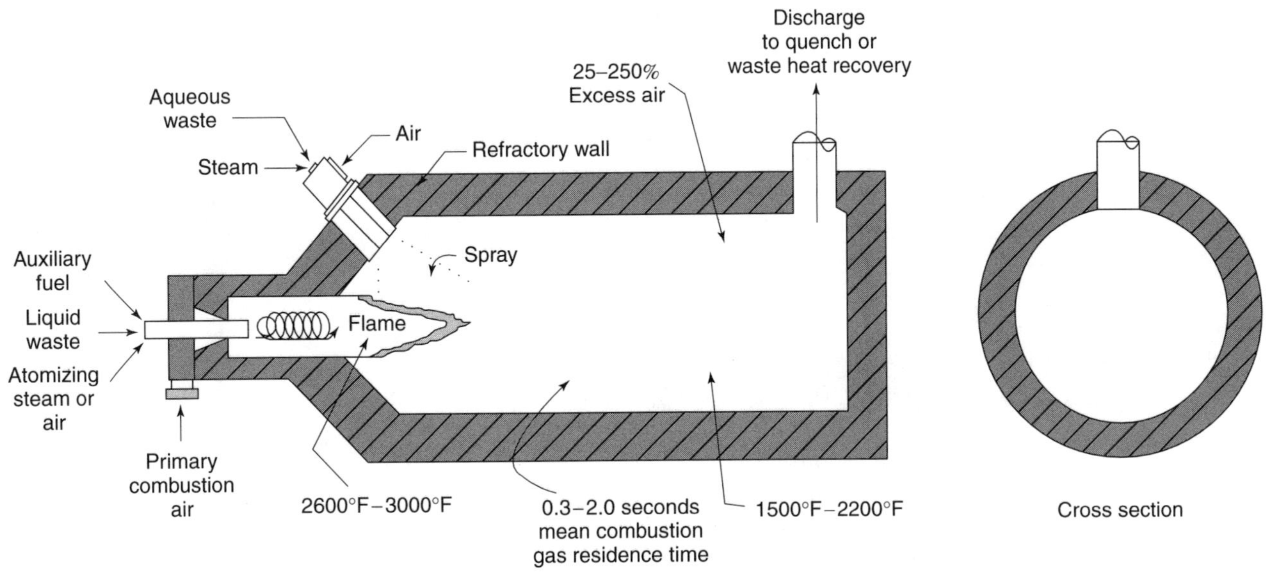

Figure 1. Typical Liquid Injection Conbustion Chamber.

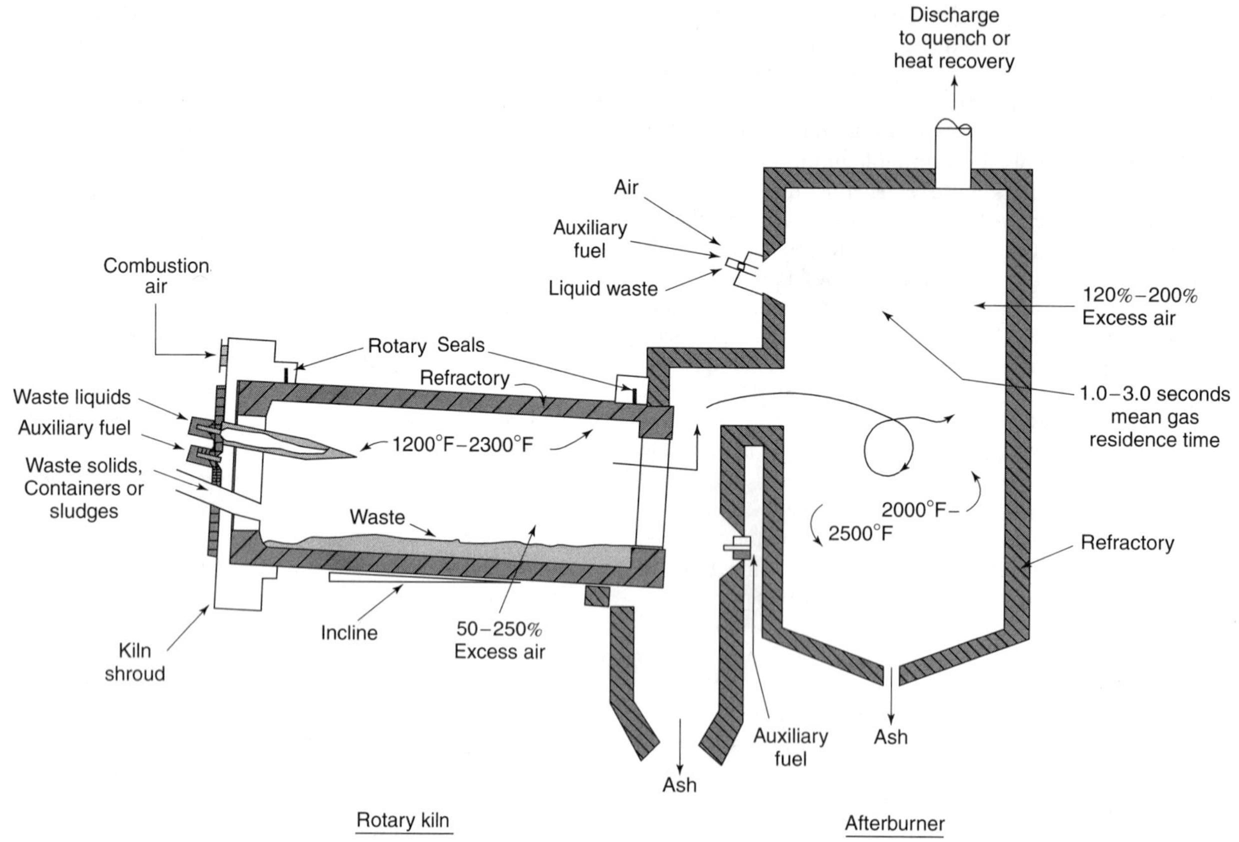

Figure 2. Typical Rotary Kiln/Afterburner Combustion Chamber.

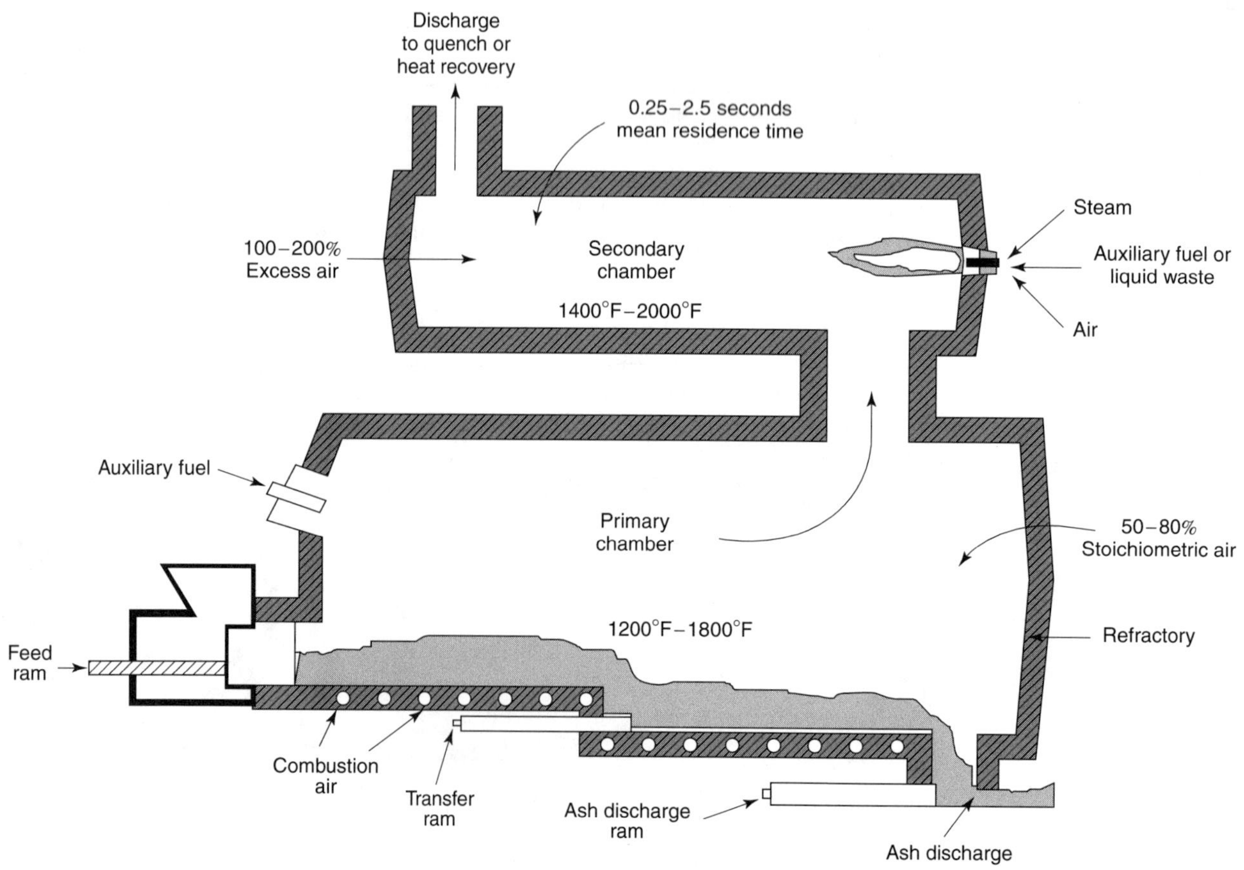

Figure 3. Typical Fixed Hearth Combustion Chamber.

stage, or secondary chamber. Here, additional air is injected to complete the combustion. Supplementary fuels may be added to ensure complete combustion. The primary chamber combustion reactions and turbulent velocities are maintained at low levels by the starved-air conditions to minimize particulate entertainment and carryover. With the addition of secondary air, total excess air for fixed hearth incinerators is in the 100–200% range.

Fixed hearth units tend to be of smaller capacity than liquid injection or rotary kiln incinerators because of physical limitations in ram feeding and transporting large amounts of waste material through the combustion chamber. These lower the capital costs and potentially reduced particulate control requirements, thus making them more attractive than rotary kilns for smaller on-site installations.

Fluidized bed incinerators may be either circulating or bubbling bed designs. Both types consist of a single refractory-lined combustion vessel partially filled with particles of sand, alumina, calcium carbonate, or similar materials. Combustion air is supplied through a distributor plate at the base of the combustor (Figure 4) at a rate sufficient to fluidize (bubbling bed) or entrain part of the bed material (circulating bed). In the circulating bed design, air velocities are higher and the solids are blown overhead, separated in a cyclone, and then returned to the combustion chamber. Operating temperatures are normally maintained in the 1400–1600°F range, and excess air required ranges from 25–150%.

Fluidized bed incinerators are primarily used for liquids, sludges, or shredded solid materials, including soil. To allow for good distribution of waste materials within the bed and removal of solid residues from the bed, all solids generally require prescreening or crushing to a size less than 0.2 in. in diameter. Fluidized bed incinerators offer: high gas-to-solids ratios, high heat transfer efficiencies, high turbulence in both, gas and solid phases, uniform temperatures throughout the bed, and the potential for in situ acid-gas neutralization by lime, limestone, or carbonate addition. Fluidized beds also have the potential for solids agglomeration in the bed, especially if salts are present in waste feeds.

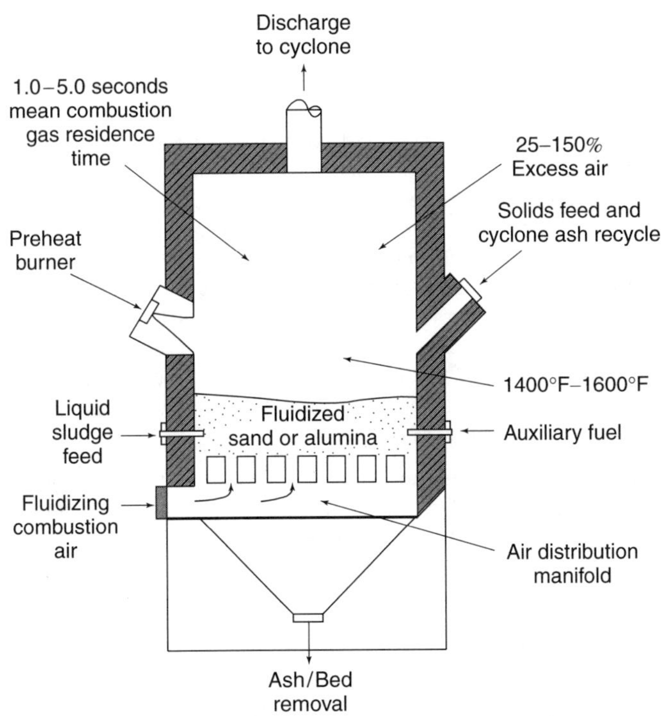

Figure 4. Typical Fluidized Bed Combustion Chamber.

AIR EMISSIONS CHARACTERIZATION

While incineration devices are designed to optimize the chances for complete oxidation of the waste, they never completely attain the ideal. Rather, small quantities of a multitude of other products may be formed, depending on the chemical composition of the waste and the combustion conditions encountered. These products, along with potentially unreacted components of the waste, comprise the emissions from the incinerator.

Hydrogen chloride and small amounts of chlorine, for example, are formed from the incineration of chlorinated hydrocarbons. Hydrogen fluoride (HF) is formed from the incineration of organic fluorides, and both hydrogen bromide (HBr) and bromine (Br_2) are formed from the incineration of organic bromides. Sulfur oxides, mostly as SO_2, but also including 1–5% sulfur trioxide (SO_3), are formed from the sulfur present in the waste material and auxiliary fuel. Highly corrosive phosphorus pentoxide (P_2O_5) is formed from the incineration of organophosphorus compounds. In addition, oxides of nitrogen may be formed by fixation of nitrogen from nitrogen compounds present in the waste material or in the combustion air. Suspended particulate emissions are also produced and include particles of mineral oxides and salts from the mineral constituents in the waste material. A wide range of organic compounds may also be formed in trace amounts from the incomplete thermal destruction of organic compounds in the waste and auxiliary fuel.

Numerous test programs and trial burns have been conducted for hazardous waste incinerators in compliance with the RCRA permitting requirements. The following sections discuss the available information in four areas:

1. Destruction and removal efficiency (DRE), particulate emissions, and HC control
2. Metal emissions
3. Combustion by-product emissions
4. Dioxin and furan emissions

DRE, Particulate Emissions, and HC Control

Table 2 summarizes waste destruction efficiency, HC, and particulate emissions results for incinerators tested by EPA in the early 1980s. Information from 14 trial burns is also included in Table 2.

These data reveal that well-operated incinerators are capable of achieving 99.99% (the RCRA performance standard) to greater than 99.999% DRE. In many cases the target principal organic hazardous constituent (POHC) was below the detection limit and the calculated DRE was a maximum value assuming the stack gas concentration of the POHC equaled the detection limit. All the incinerators tested by EPA achieved this level of performance for candidate POHC compounds in concentrations greater than 1200 parts per million (ppm) in the waste feed. Candidate POHC compounds between 200 and 1200 ppm frequently were not destroyed to a 99.99% DRE, and no compounds below 200 ppm in the waste feed met the RCRA DRE limit. In fact, regression analysis of the pooled data

Table 2. Incinerator Performance and Stack Emissions Data (Data Reported as Averages for Each Facility)

Facility Type	O_2 (%)	CO (ppm)	HC (ppm)	DRE (%)	Particulate (mg/m^3)	HCl control (%)
Commercial rotary kiln/liquid incinerator	10.5	6.2	1.0	99.999	152	99.4
Commercial fixed hearth, two-stage incinerator	11.4	6.9	1.0	99.994	400	98.3
On-site two-stage liquid incinerator	8.1	9.4	6.0	99.994	143	99.7
Commercial fixed hearth, two-stage incinerator	11.0	327.7	18.7	99.997	60	—[b]
On-site liquid injection incinerator	13.2	11.9	1.0	99.999	186	—[b]
Commerical two-stage incinerator	10.2	1.1	1.3	99.998	902	—[b]
On-site rotary kiln incinerator	9.7	554.0	61.7	99.999	23	99.9
Commercial two-stage fixed hearth incinerator	13.4	26.8	1.8	99.996	168	98.3
On-site rotary kiln	—[c]	794.5	NA[a]	99.998	184	99.7
On-site liquid injection incinerator	9.7	66.3	7.8	99.994	95	—[b]
On-site rotary kiln incinerator	10.7	5.8	NA	99.996	404	99.9
On-site rotary kiln incinerator	14.1	323.0	NA	99.996	NA	99.8
On-site liquid injection incinerator	12.4	31.9	1.9	99.999	163	98.6
On-site liquid injection incinerator	9.3	1.0	NA	99.996	40	—[b]
On-site fluidized bed incinerator	3.6	67.4	NA	99.996	259	—[b]
On-site fixed hearth incinerator	12.9	ND[d]	NA	99.999	93	—[b]
On-site liquid injection incinerator	4.5	358.0	NA	99.995	99	—[b]
On-site liquid injection incinerator	3.6	28.4	NA	99.998	12	99.3
Commercial rotary kiln incinerator	9.4	8.0	0.5	99.999	172	99.9
On-site liquid injection incinerator	3.1	779.3	NA	99.999	88	99.6
On-site liquid fumace incinerator	6.4	56.3	NA	99.999	4	99.9
On-site fixed hearth incinerator	13.5	5.0	NA	99.999	150	98.4

[a] NA = Not available.
[b] HCl emissions <4 lb/h.
[c] Reported only as a range (3.1–16.7%).
[d] ND = not detected.

Sources: U.S. EPA, *Performance Evaluation of Full-Scale Hazardous Waste Incineration*, 5 volumes, NTIS PB 85-129500, 1984; U.S. EPA, *Permit Writers Guide to Test Burn Data—Hazardous Waste Incineration*, EPA-625/6-86/012, 1986.

suggested that statistically significant correlations (correlation coefficients were 0.76 and 0.84) existed between compound penetration (1-DRE) and compound feed concentration, showing that DRE increased with waste feed concentration.

All incinerators tested met or approached the RCRA HC removal standard of 99%.

Particulates from hazardous waste incinerators are composed mostly of oxides of silicon, iron, calcium, and aluminum. Typically from 1 to 10% of the particulate is RCRA-regulated metals. Achieving the RCRA particulate emission standard of 180 mg/dscm was a problem for a number of the incinerators tested by EPA. Four of the eight units tested failed to meet the RCRA standard. Two of those facilities were marginally above the emission limit and could likely meet the standard with minor operating adjustments. The remaining two facilities appeared to need significant design and/or operational changes. In some cases, failure of the particulate emission standard may be attributed to dissolved neutralization salts in mist carryover from alkaline acid-gas scrubbers. It is clear, however, that the particulate emission standard of 180 mg/dscm is achievable if proper air pollution control is provided, since at least 150 hazardous waste incinerators have now demonstrated this during their trial burns.

Metal Emissions

Metals are of possible concern in waste incineration because of their presence in many hazardous wastes and because of possible adverse health effects from human exposure to emissions. Those metals posing carcinogenic risk include arsenic, cadmium, chromium, and beryllium. Among the noncarcinogenic toxic metals are antimony, barium, lead, mercury, silver, and thallium. It is anticipated that the noncarcinogenic toxic metals, nickel and selenium, will also be regulated in the near future.

Incineration may change the form of metal fractions in waste streams, but it will not destroy the elemental metals. As a result, metals are expected to emerge from the combustion zone essentially in the same total quantity as the input. The principal environmental concern is to identify where and in what physical or chemical form the metals exit the combustion system, i.e., bottom ash, APCD residues, or stack emissions. Unless emission testing is conducted, it is conservatively assumed that 100% of the metals in the feed are emitted to the air. Whether or not this feed rate is economically or practically feasible must be determined on an individual basis. Treatability tests of the waste may be warranted.

Combustion By-Product Emissions

The current RCRA incineration standards regulate destruction and removal only for the major hazardous compounds in the waste. One of the concerns expressed by some scientists and environmentalists regarding hazardous waste thermal destruction is the possible impact on human health and the environment of potentially hazardous products of incomplete combustion (PIC) emissions. While many of the incinerator field tests conducted to date have attempted to quantify by-product emissions, these data have been criticized as being incomplete and insufficient for the purposes of a full risk assessment. Testing has focused largely on identifying Appendix VIII organic compounds only. Appendix VIII is a list of approximately 400 organic and inorganic hazardous chemicals first published in Part 261 of the May 19, 1980, Federal Register. This list is updated semiannually in 40 CFR 26. POHC selection is covered by 40 CFR 264.342(b), which suggests that Appendix VIII constituents, which are in the highest concentration in the waste feed and are the most difficult to incinerate, are the most appropriate to be selected as POHCS. Comparison of total hydrocarbon emissions with the total quantity of specific organic compounds identified in the emissions has usually revealed that only a relatively small percentage of the total hydrocarbon emissions may have been identified.

Dioxin and Furan Emissions

EPA full-scale hazardous waste tests have examined dioxin/furan emissions at six incinerators. Data are also available from test burns at one incinerator burning pentachlorophenol (PCP) waste and three PCB incinerators. Dioxin/furan emission data are also available from a pilot-scale treatability test of PCB-contaminated soil from a Superfund site. Dioxin/furan emissions from all these sources are summarized in Table 3. In addition, these emissions are expressed in I-TEQs/89 in both concentration and mass per year.

No 2,3,7,8-TCDD was detected in any of the ten full-scale hazardous waste incinerators burning hazardous or PCB waste.

AIR POLLUTION CONTROL MEASURES

Following the incineration of hazardous wastes, the combustion gases need to be further treated in an air pollution control system. The presence of chlorine or other halogens in the waste will generally signal a need for a scrubbing or absorption step for combustion gases to remove HC and other haloacids. Ash in the waste is not destroyed in the combustion process. Depending on its composition, ash will exit as bottom ash, at the discharge end of a kiln or hearth, for example, and/or as particulate matter suspended in the combustion gas stream (fly ash). Particulate emissions from most hazardous waste combustion systems generally have particle diameters down to less than 1 μm and require high-efficiency collection devices to meet the RCRA or state emission standards.

One of the most commonly employed air pollution control systems for hazardous waste facilities is a quench (gas cooling and conditioning) followed by high-energy venturi scrubber (particulate removal), a packed tower absorber (acid-gas removal), and a demister (visible vapor plume reduction). Facilities handling low-ash, low-halogen-content liquid waste streams have been able to operate without any control. The major advantage of venturi scrubbers is their reliability and relative simplicity of operation. On the other hand, maintaining the significant pressure drop across the venturi throat (160–120 in. of water column) required for efficient hazardous waste combustion particulate matter control represents a significant percentage of the total cost of operation of incineration facilities employing venturi scrubbing. Also, venturi scrubbers may not be very effective in controlling the emission of fine particulates such as metal aerosols.

Acid-gas removal is generally accomplished in packed-bed or plate-tower scrubbers. Packed-bed or plate-tower scrubbers are commonly used at liquid injection incinerator facilities, where absorption of soluble gaseous pollutants (HC and sulfur oxides) is often most important and particulate control is less critical. At rotary kiln or fixed hearth facilities, or liquid injection facilities where high-ash-content wastes are incinerated, however, venturi scrubbers are often used in series with packed-bed or plate-tower scrubbers.

Many designs have begun to incorporate waste heat boilers as a substitute for gas quenching and as a means of energy recovery. Wet electrostatic precipitators (ESP), ionizing wet scrubbers (IWS), collision scrubbers, spray dryer absorbers (SDA), and fabric filters (FF) are also being incorporated into newer systems. This is largely due to their high removal efficiencies for small particles and lower pressure drop.

Table 3. Dioxin/Furan Emissions from Thermal Destruction Facilities (ng/dscm @ 7% O_2)

Facility Type[a]	Sample (Waste)[b]	2378-TCDD	PCDD	PCDF	I-TEQs/89[c]	
					(ng/dscm)	(g/yr[d])
HWI (commercial, Rotary kiln, liquid injection)	FG[e] (HW)	ND[f]	ND	ND	ND	ND
HWI (confidential)	FG/FA (HW)	ND	22	70	17.7	1.95
HWI (on-site liquid injection)	FG(HW)	ND	ND	7.3	0.93	0.02
HWI (on-site liquid injection)	FG (HW)	ND	ND	ND	ND	ND
HWI (commercial, two chamber, liquid injection, and hearth)	FG/FA (HW)	ND	ND	1.7	0.57	0.02
HWI (on-site kiln and liquid, injection in parallel)	FG (HW)	ND	ND	ND	ND	ND
HWI (liquid injection, incinerator ship)	FG/FA (PCB)	ND	ND	ND	0.3	0.16
HWI (fixed hearth)	FG/FA (PCP)	ND	ND	ND	ND	ND
HWI (liquid injection)	FG/FA (PCB)	ND	0.64	9.9	1.63	0.81
HWI (rotary kiln/liquid injection)	FG/FA (PCB)	ND	0.18	2.1	0.39	0.12
HWI (pilot-scale rotary kiln)	FG/FA (PCB)	0.003	0.108	3.18	0.073	0.001

[a] HWI, hazardous waste incinerator.
[b] Information in parentheses describes waste feed; HW, hazardous waste; PCB, polychlorinated biphenlys; PCP, pentachlorophenol waste.
[c] Calculated by the international toxicity equivalency factor/89 (I-TEF/89) method. If isomer specific data were not available, homologue data were considered to be composed of the most toxic isomers.
[d] Assumes 8160 operating hours per year.
[e] FG, flue gases analyzed; FA, flue gas particulate analysed.
[f] ND, not detected.

In a test of a full-scale PCB incinerator equipped with a spray dryer/baghouse followed by an ionizing wet scrubber, particulate emissions averaged 20 mg/dscm for one test condition and 109 mg/dscm for a second test condition.[3] In another test of a full-scale hazardous waste incinerator equipped with a venturi/packed-bed wet scrubber system, the particulate emissions averaged 48 mg/dscm.[4] The particulate emissions from a pilot-scale hazardous waste incinerator[5] ranged from 17 to 64 mg/dscm when it was equipped with an ESP.[5] When this system was equipped with a Hydro-Sonic[R] wet scrubber, the particulate emissions averaged 10 mg/dscm.

Residue and Ash Handling

The inorganic components of hazardous wastes are not destroyed by incineration. These materials exit the incineration system either as bottom ash from the combustion chamber, as contaminants in scrubber waters and other air pollution control residues, and in small amounts in air emissions from the stack. Residues generated from the incineration of hazardous waste must be managed carefully.

Ash is commonly either air-cooled or quenched with water after discharge from the combustion chamber. From this point, ash is frequently accumulated on-site in storage lagoons or in containers prior to disposal in a permitted hazardous waste land disposal facility. Dewatering or chemical fixation/stabilization may also be applied to meet the land disposal restriction (LDR) regulations prior to disposal.

Air pollution control residues are generated from the combustion gas quenching, particulate removal, and acid-gas absorption steps in an incineration system. These residues are typically aqueous streams containing entrained particulate matter, absorbed acid gases (usually as HC), salts, and trace amounts of organic contaminants. These streams are often collected in sumps or recirculation tanks, where the acids are neutralized with caustic and returned to the process. Eventually, a portion or all these waters must be discharged for treatment and disposal. Many facilities discharge neutralized waters to settling lagoons or to a chemical precipitation step to allow for suspended contaminants to be concentrated and ultimately sent to land disposal. Depending upon the nature of the dissolved contaminants and their concentration after treatment, waters may either be returned to the process or discharged to sewers. One alternative to the management of aqueous residue streams is to use dry scrubber systems, which do not generate any wastewater.

REFERENCES

1. C. R. Dempsey and E. T. Oppelt, "Incineration of Hazardous Waste: A Critical Review Update," *J. AWMA* **43**(1):25–73 (1993)
2. U.S. EPA, *Engineering Handbook on Hazardous Waste Incineration*, SW-889, NTIS PB81-248163, Sept. 1981.
3. U.S. EPA, *A Performance Test on a Spray Dryer, Fabric Filter and Wet Scrubber System; APTUS Incinerator at Coffeyville, KS*, Radian Corp., NTIS PB 90120544, Oct. 1989.
4. U.S. EPA, *Measurement of Particulates, Metals, and Organics at a Hazardous Waste Incinerator; Midwest Research Institute*, NTIS PB 89230668, Nov. 1988.
5. U.S. EPA, *Pilot Scale ESP and Scrubber Parametric Tests for Particulates, Metals, and HC Emissions: John Zink Company Research Facility, Tulsa, OK*, Radian Corp., NTIS PB90129362, June 1989.

SEWAGE SLUDGE

Calvin R. Brunner

Incineration has been utilized for the treatment and disposal of sewage sludge for over 50 years. The Herreshoff multiple-hearth furnace was first used for this application in 1932. The initial use of a fluid-bed furnace for sewage sludge incineration occurred in 1952. Among other types of incineration systems used for sewage sludge are the cyclone furnace, electric (or radiant heat) furnace, and codisposal systems.

Other methods of treatment and disposal include land application, land burial, composting, and soil amendment (fertilizers) application. These technologies were found to be less costly than incineration in the past, but as land application is becoming more strictly regulated, the relative cost of incineration is decreasing.

With the potential for heat recovery and low fuel costs, incineration of sewage sludge is growing in popularity. Its capital cost is high, but with appropriate selection of upstream equipment to obtain dryer, or autogenous burning sludge cake, the annual fuel cost can be significantly reduced. There are many unique advantages to incineration processes. These include:

- Incineration achieves a substantial reduction in the weight, volume, and toxicity of the initial waste stream.
- Less material remains for ultimate disposal.
- Residue is often a sterile ash that has potential for use in road surfacing, building block, metals reclamation, and so on.
- Equipment is available to meet the most stringent air emissions requirements.
- The incineration process is compact and requires nominal land area as compared with land-intensive disposal techniques such as land-filling.
- Incineration is immediate. Months or years are not required as with other treatment or disposal methods, such as burial or land application.
- Incineration takes place at the point of generation of sludge, eliminating the need to transport sludge to a distant disposal site. (Often the air emissions resulting from truck hauling of sludge are far greater than those of an incinerator discharge for the same sludge quantity.)
- The potential for recovery of heat exists.

Incineration systems have a useful life of 10–15 years before replacement or major overhaul is necessary. Public acceptance of a continued incinerator presence is easier to obtain than acceptance of necessary searches for new land disposal sites, particularly when a distant area is asked to accept exported sludge.

- Incinerators are stationary, fixed-point dischargers. As such, they can be readily observed and monitored as compared with discharges from nonpoint sources such as trucks or landfills. This feature can provide the public with greater confidence in the assessment of incinerator operation.

The disadvantages of incineration as a sludge treatment method include the following:

- An external source of fuel is always required, if not for maintaining combustion, then to bring the furnace to operating temperature.
- A relatively high investment in capital equipment is necessary.
- Skilled operators are required.
- Metals concentration in the residue (ash) may classify the ash as a "special" or hazardous waste. This could represent a relatively high cost for its ultimate disposal.

THE INCINERATION PROCESS

Sewage sludge is normally fired as a sludge cake produced from dewatering equipment such as a centrifuge, belt filter press, vacuum filter, or plate-and-frame press. It normally contains from 55% to 85% moisture and from 40% to 80% combustibles (as a percent of dry weight). It cannot start to burn until its moisture content has been reduced to less than 30%. The ultimate purpose of incineration is to reduce its moisture content until it will burn, to allow the sludge to burn out to a sterile residue or ash, and to condition the exhaust gas stream to an innocuous discharge to the atmosphere.

Sludge incineration systems in use today include the following:

- Multiple-hearth incinerator
- Fluid-bed incinerator
- Electric furnace
- Cyclonic furnace
- Codisposal systems

All these systems rely on the excess-air process, in which air in excess of the ideal or stoichiometric requirement is provided to the burning sludge to assure that all the sludge is effectively and efficiently burned. A number of years ago, substoichiometric (starved-air or so-called pyrolysis) processes were attempted in an effort to reduce equipment size and supplemental fuel requirements. However, this technology was not successful anywhere in the United States, and these substoichiometric processes were abandoned in favor of conventional burning.

Air requirements for sludge incinerators range from 20% in excess of the stoichiometric requirement for the electric furnace to over 100% of the stoichiometric requirement for multiple-hearth furnaces.

Multiple-Hearth Incinerators

The multiple-hearth furnace is the most common incinerator for the disposal of sewage sludge in this country. It was developed at the turn of the century for ore roasting and has been adapted to carbon regeneration, recalcining, and many other industrial applications.

Multiple-hearth furnaces range from 6 to 25 ft (183 to 762 cm) in diameter and from 12 to 65 ft (366 to 1981 cm) in height. The number of hearths depends on the waste feed and processing requirements, but it generally varies between 5 and 12. Waste retention time is controlled by the rabble tooth pattern and the rotational speed of the center shaft.

Sludge cake is introduced at the top of the furnace (see Figure 1). The furnace interior consists of a series of circular refractory hearths, one above the other. The hearths are arch structures, self-supporting off the refractory-lined cylindrical wall of the furnace. They are numbered No. 1 as the top hearth, No. 2 as the next to top, and so on. Where the top hearth is used as an afterburner section, with sludge deposited on the hearth beneath the top hearth, the top hearth may be referred to as a "zero" hearth. In this case, the next to top hearth would be referred to as hearth No. 1, and so on.

A vertical shaft is positioned in the center of the furnace. Rabble arms are attached to the center shaft above each hearth. The center shaft, along with the rabble arms, rotates relatively slowly, approximately one revolution per minute. Teeth on the rabble arms are positioned to move sludge across the hearth and through the furnace.

Every other hearth has a large annular opening between the hearth and the center shaft. These are termed "in" hearths. The teeth on the rabble arms of these hearths will "rabble" or wipe sludge to the center of the hearth, where the sludge will fall off the edge of the inner refractory ring, landing on the hearth below, an "out" hearth.

There are a series of openings on the outside or periphery of the out hearth. The inside of an out hearth is fairly close to the center shaft. A "lute ring cover" is a collar located above the hearth, attached to and moving with the center shaft, preventing sludge from dropping adjacent to the center shaft. Teeth on the rabble arms on the out hearths move sludge to the drop holes on the outside of the hearth, where the sludge drops to the hearth below, an in hearth. This process repeats until sludge, or ash, reaches the bottom hearth, the floor of the furnace, where it discharges from the furnace. Teeth on each hearth agitate the sludge, exposing new surfaces of the sludge to the gas flow within the furnace. As sludge falls from one hearth to another, it again has new surfaces exposed to the hot gas.

The upper hearths of the furnace provide a drying zone where the sludge cake gives up moisture (evaporation) while cooling the hot flue gases. Flue gas exits the top hearth of the furnace at from 800–1200°F (427–649°C). The center hearths make up the burning zone, where gas temperatures can reach 1600°F (871°C). Burnout of sludge to ash occurs in the lower hearths of the furnace.

The center shaft and rabble arms are hollow. Air is passed through the center shaft, which is constructed to distribute this air to each of the rabble arms and discharge it through the top of the shaft. After passing through the rabble arms and center shaft, this cooling airflow reaches temperatures of 300–450°F (149–232°C). This heated air often is recycled into the furnace as preheated sludge combustion air.

The temperature above at least two hearths should be maintained at approximately 1600°F (871°C) at all times when burning sludge cake. The off-gas temperature can range from 800°F to 1400°F (427–760°C). In general, the exit temperature of the multiple-hearth furnace is from 400–600°F (204–316°C) lower than the maximum furnace temperature. This is an important consideration. The maximum allowable temperature anywhere within the furnace is 1800°F (982°C). Above this temperature the sludge ash fusion temperature will be approached, and ash clinkering will occur. The formation of clinker and slag can clog drop holes, interfere with rabble-arm motion, and create other problems within the furnace that can result in shutdown and damage.

Occasionally, odor problems can exist (unburned hydrocarbon emissions), which may require the installation of afterburning equipment. This is of particular concern where there is a significant industrial component to the waste stream,

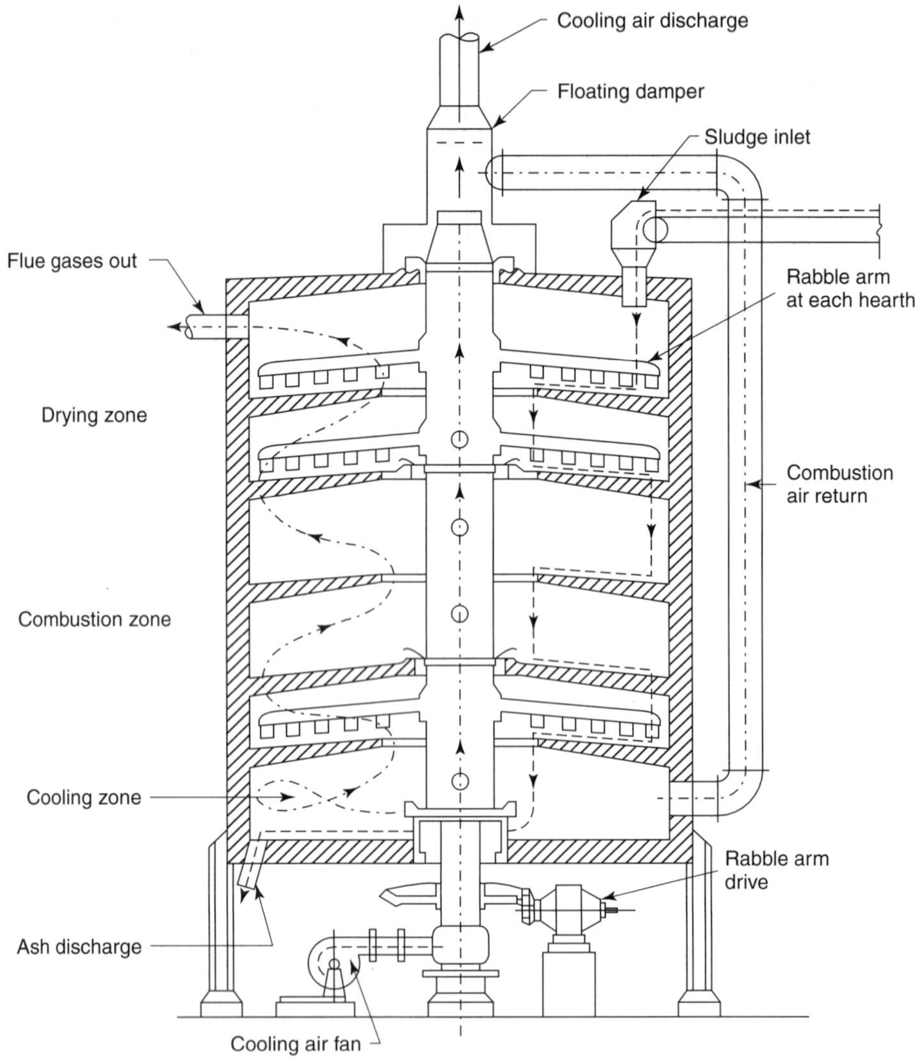

Figure 1. Multiple-Hearth Furnace (from Ref. 4, p. 230).

in excess of 20% of the total influent flow to the treatment plant.

Multiple-hearth incinerators are designed to burn materials with a low gross (as received) heating value, such as sewage sludge and other high-moisture-content wastes. Their design includes drying as well as burning sections. Materials with less moisture content, or higher heating values, such as coal or solid waste, will start to burn too high in the furnace. There would be insufficient residence time above these top hearths, and the temperatures in this volume of the furnace would be too low for effective burnout. Generally, solid fuel or solid waste is not compatible with a multiple-hearth incinerator system and should not be introduced into it.

Figure 2 is a diagram of a generalized multiple-hearth incinerator system. It includes a fume incinerator (afterburner) separate from the multiple-hearth furnace and a waste-heat recovery boiler. The boiler is placed in the flue gas stream whenever the fume incinerator is in operation. The fume incinerator is designed to bring the temperature of the gases exiting the incinerator from 800°F (427°C) to the range of 1400–1500°F (760–816°C), which represents a significant amount of heat. The waste-heat boiler can reclaim over 50% of the heat input from the supplemental fuel.

Ash exits this system dry, from beneath the furnace. It can be collected and disposed of dry, or it can be dropped into a wet hopper, mixed with water, and pumped to a lagoon or settling tank for dewatering and ultimate disposal. The multiple-hearth furnace is flexible enough to allow wet or dry ash handling.

Feeding sludge to a multiple-hearth furnace is usually by gravity; it is dropped onto the top hearth. Sludge also can be deposited on the top hearth or onto a lower hearth by a screw conveyor. Attempts to feed sludge to a multiple-hearth furnace pneumatically have met with failure. Pneumatic systems necessarily inject air into the furnace in strong discrete bursts, which makes the maintenance of constant furnace draft difficult, if not impossible.

Generally, grease (scum) should not be added to sludge feed. If grease is to be incinerated in a multiple-hearth furnace, it should be added at a lower hearth (a burning hearth) through a separate nozzle(s). If it is mixed with or placed on the sludge fed to the furnace, it will tend to volatilize upon entering the furnace at too high a hearth. The volatilized grease would not experience a high enough temperature or sufficient residence time for effective destruction.

The multiple-hearth furnace should always be run at a negative pressure (draft) to prevent external leakage of hot,

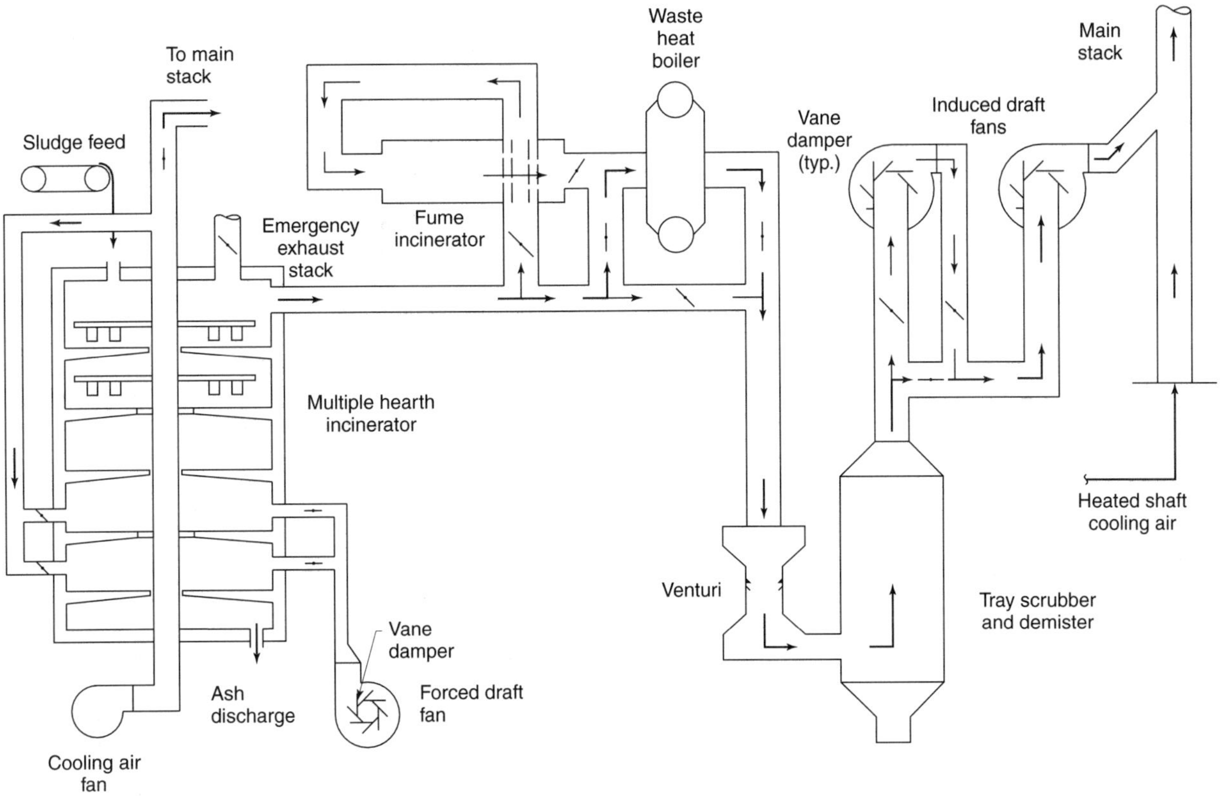

Figure 2. Multiple-Hearth Sludge Incinerator System (from Ref. 5).

toxic flue gas. As shown in Figure 2, an induced draft (ID) induces, or pulls, gases through the furnace system. An inlet damper on the ID fan is normally controlled by furnace draft. The position of this damper automatically modulates to maintain constant draft (constant negative pressure) within the furnace. If a separate afterburner is needed, and if an emergency discharge stack is provided, the stack should always be located downstream of the afterburner.

Incinerator instrumentation should include a temperature strip chart that at least records temperatures on each hearth. A chart recording scrubber pressure drop and fraction oxygen in the flue gas should also be provided. Carbon monoxide indication is a useful tool in the control of sludge combustion air. Opacity meters should not be used downstream of a wet scrubber, as entrained water from the scrubber will appear opaque and the meter will give a false reading.

Fluid-Bed Incineration

The fluid-bed furnace was developed for catalyst recovery by the Exxon Corp. during World War II. The first fluid bed furnace used for incineration of sewage sludge was installed in 1962.

As illustrated in Figure 3, the fluid-bed furnace is cylindrical refractory-lined shell with a grid structure, or bed plate (either a refractory arch or an alloy steel plate), above a wind box to support a sand bed. Air is introduced at the fluidizing air inlet at pressures in the range of 3.5–5 psig (72 mb). The air passes through openings (tuyeres) in the bed plate supporting the sand and generates a high degree of turbulence within the sand bed. The sand undulates and has the appearance of a fluid in motion.

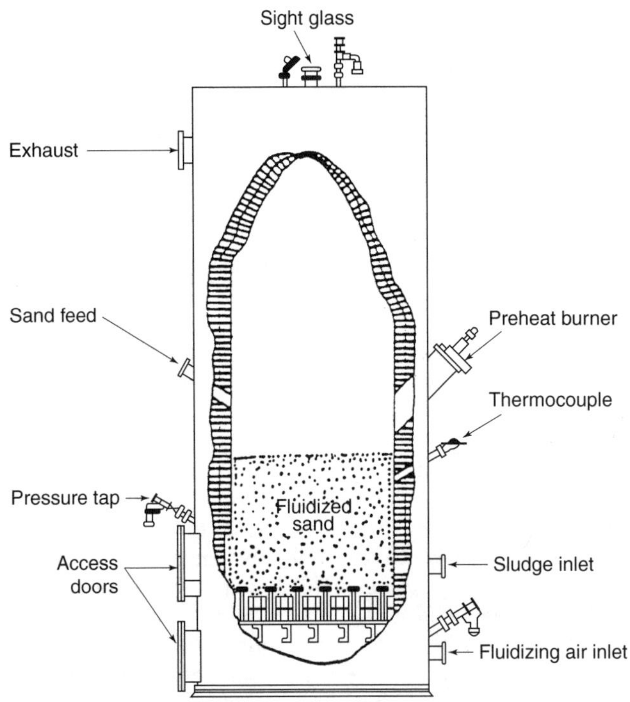

Figure 3. Fluid-Bed Furnace (from Ref. 4, p. 232).

Air can be introduced to the wind box cold or, as is usually the case, preheated by the exiting flue gas (see Figure 4). The sand bed is normally maintained at approximately 1300–1400°F (704–760°C). It expands from 30% to 60% in volume when fluidized.

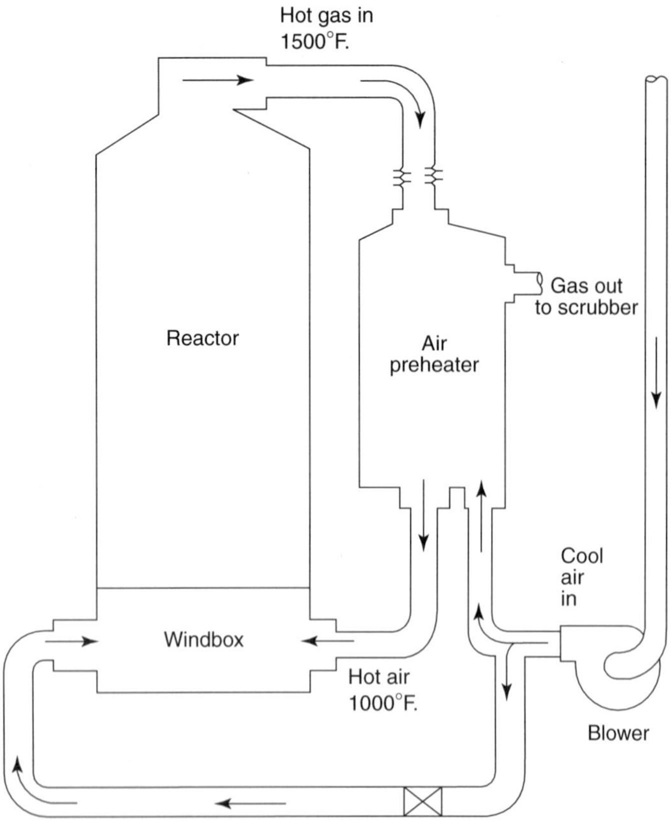

Figure 4. Fluid-Bed Furnace Incineration System (from Ref. 4, p. 233).

Sludge cake is normally introduced within or just above the fluid bed. Fluidization provides maximum contact of air with the sludge cake particles for optimum burning. The drying process is practically instantaneous. Moisture flashes into steam upon entering the hot bed.

Most of the fluid-bed furnaces used for sludge incineration have one major item of air-moving equipment, the forced-draft fan (or fluidizing air blower). The fan is sized to move (push) air/flue gas through the heat exchanger, the fluid bed, the hot gas side of the heat exchanger, and then through the gas scrubbing system. This requires that the reactor be pressurized and that it be airtight to prevent leakage of flue gas from the incinerator.

The large amount of sand within the furnace is a heat sink that provides a significant thermal inertia within the system. This allows the furnace to be shut down with minimal heat loss. It is a relatively tight system; for example, the sand will retain heat to allow quick startup after an overnight shutdown. A weekend shutdown will normally require only two or three hours of heating on a Monday morning. A multiple-hearth furnace, by comparison, cannot effectively maintain temperature when not burning sludge without the firing of supplemental fuel.

All the sludge ash and some sand become airborne and exit the furnace within the flue gas stream. The gas cleaning system has to be sized for this relatively high particulate loading.

As a result of the intimate mixing of air and sludge in the fluid sand bed, excess air requirements are low, generally 40%. The large volume (termed freeboard) within the furnace above the sand bed is maintained at 1500–1600°F (816–871°C). Residence time of the flue gases in the freeboard can be as high 6 sec or more, which, at these temperatures, is normally sufficient to obtain complete burnout and to eliminate odor.

Fuel is used for startup, for reheat, and, depending on the net heat content of the sludge, for maintenance of temperature during incineration. It can be injected directly into the bed or sprayed on top of the bed when the bed is fluidized and is at a temperature high enough to ensure ignition.

The fluid-bed furnace is sensitive to waste consistency. Wastes containing soluble alkalis or phosphates, lead, or low-melting-point eutectics may cause seizure or fusion of the bed. A major issue associated with a fluid-bed furnace is its ability to handle a wide variety of waste streams. Certain wastes, particularly those containing salts, will tend to slag (seize) the bed, preventing fluidization. Test burns on materials that have not previously been fired in a fluid-bed furnace should be made to determine the bed reaction to that material. If bed seizure is indicated, bed additives may be available to help eliminate this problem by changing the physical properties (eutectic) of the waste feed. If additives are not available, a fluid-bed furnace may not be appropriate.

A related issue is agglomeration. Waste materials may build up on individual sand particles within the bed. With changes in operation (such as continuous bed withdrawal), this can be controlled. If not addressed, agglomeration can result in defluidization. This is of particular concern with sewage sludge generated in shore (coastal) communities where the sludge will have a relatively high salt content. The residual sodium chloride and other sea salts will bring the melting temperature of sludge below 1400°F (760°C), which can cause both agglomeration and seizure of the bed. Note that the multiple-hearth furnace is not as sensitive as the fluid-bed furnace to salts and other low-melting-point materials. In a multiple-hearth furnace, the wastes and their residuals are continually being wiped out of the furnace. By contrast, in a fluid-bed furnace, the wastes are resident within the bed, and their high residence time encourages agglomeration and seizure.

An internal water-spray system is often employed in fluid-bed systems to help protect the air preheater from high-temperature excursions. Water normally is automatically injected into the freeboard when the heat exchanger inlet temperature is above a preset figure, generally no higher than 1600°F (871°C).

A fluid-bed furnace requires a minimum amount of air to maintain bed fluidization, regardless of the waste feed. When the bed is operating at a feed rate below the design capacity, approximately the same air quantity is required at the design load to maintain fluidization. This results in a system that may have good fuel consumption at its design point, but a relatively high fuel consumption at lesser feeds.

Ash from a fluid-bed furnace normally exits with the flue gas. Some materials, such as sewage grit, when burned in a fluid-bed furnace, will not be airborne, but will build up within the bed. This requires periodic bed tapping. Provisions should be included in the fluid-bed system to accommodate the possibility of excessive bed buildup, which would include automatic and continuous bed-tapping and sand injection systems.

The air emissions control system must be designed to remove 100% of the sludge ash and elutriated sand from the gas stream. This requires high-energy venturi scrubbing systems with higher pressure requirements than many other incineration systems. Typical fluid-bed systems need scrubbers rated from 30 to 60 in. w.c. (16–31 mb).

Feeding of a fluid-bed furnace requires particular attention. Except for furnaces that use an ID fan where negative pressure is maintained in the freeboard and where waste can be dropped into the furnace by gravity, most fluid-bed furnaces are under positive pressure. Waste must be force-fed into the furnace, and gravity feed cannot be utilized. Positive displacement pumps or screws are normally used for feeding fluid-bed furnaces. Sludge feeding is a potential problem area because of the tendency of the sludge within the feed lines to dry during periods when the furnace is maintained hot without sludge feed (hot standby conditions). Hard piping should be used from the pumps to the furnace, and a hot-water purge system should be included to help prevent caking and to provide clearing of the piping when not in use.

Electric Furnace

The electric or radiant heat (or infrared) furnace is basically a conveyor belt system passing through a long, rectangular refractory-lined chamber, as shown in Figure 5. An ID fan maintains a negative pressure within the system.

Combustion air is introduced at the discharge end of the belt, shown in Figure 5 as the viewport. Air will pick up heat from the hot burned sludge as sludge and air travel countercurrent to each other. Supplemental heat is provided by electric infrared heating elements within the furnace above the belt. Cooling air is injected into the incinerator chamber to prevent local hot spots in the immediate vicinity of the heaters and is used as secondary combustion air within the furnace.

The conveyor belt is continuous woven wire mesh made of high-temperature steel. The refractory is not brick, but ceramic felt fiber. It does not have a high capacity for holding heat and can be started up from a cold condition relatively quickly (within one to two hours).

Sludge cake is fed by gravity onto the belt, and it is immediately leveled to a depth of approximately 1 in. (2.5 cm). There is no other sludge contact or sludge-handling mechanism. The belt speed and travel are sized to provide burnout of the sludge without agitation. This feature results in a relatively low level of particulate emissions.

Supplemental fuel, in this case, electricity, is required for startup. Electric power also is used to provide the heat required to maintain combustion temperatures. Unfortunately, the power needed for startup results in a large connected load.

Electric energy costs approximately four times that of fossil fuel on a heat-value basis. Therefore, unless the sludge will burn autogenously, an electric furnace may not be cost effective—and even with an autogenously burning sludge, the cost of installed kilowatts (demand charge) may be prohibitive.

Cyclonic Furnace

In the cyclonic furnace, the hearth moves and the rabble teeth are stationary (see Figure 6). Sludge is rabbled toward the center of the hearth, where, as ash, it is discharged. This type of incinerator is built in relatively small sizes and has been used in shipboard and shore (Navy base) applications.

Air is introduced at tangential burner ports on the shell of the furnace. The furnace is a refractory-lined cylindrical shell with a domed top. The air, heated with the introduction of supplemental fuel, creates a violent swirling pattern that provides good mixing of air with the sludge feed. The air, later flue gas, swirls up vertically in cyclonic flow through the discharge flue in the center of the domed roof. Sludge is fed into the furnace with a screw feeder and deposited on the periphery of the rotating hearth. A progressive cavity pump also can be used to feed sludge.

A variation of the cyclonic furnace is shown in Figure 7. This is a horizontal skid-mounted cyclonic reactor system. Ash

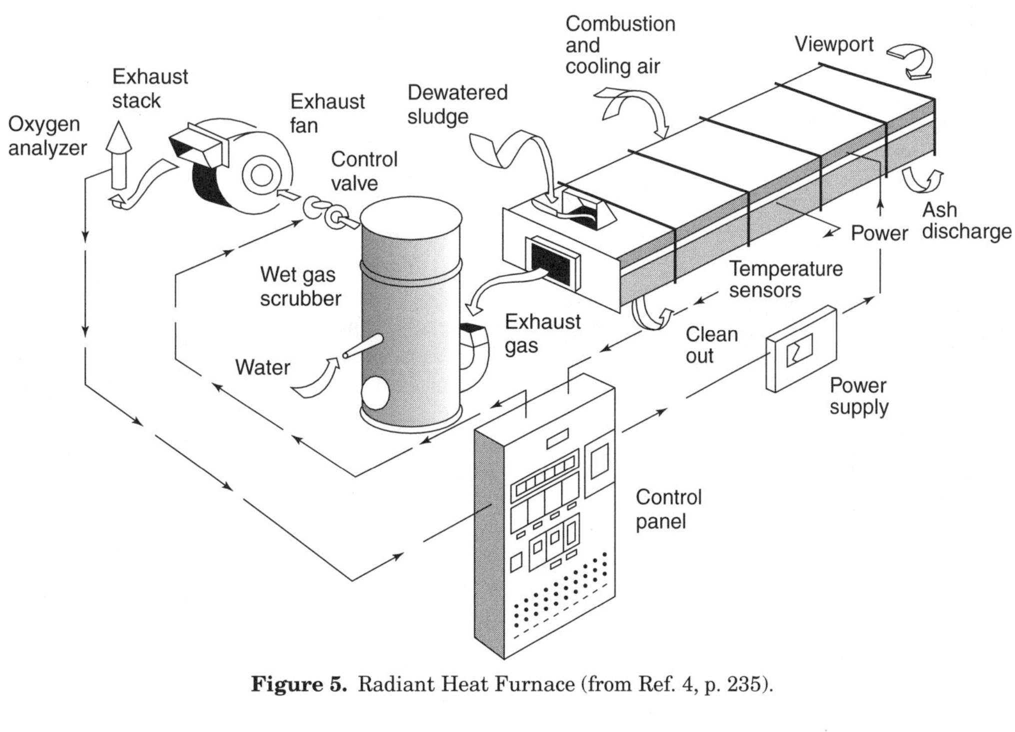

Figure 5. Radiant Heat Furnace (from Ref. 4, p. 235).

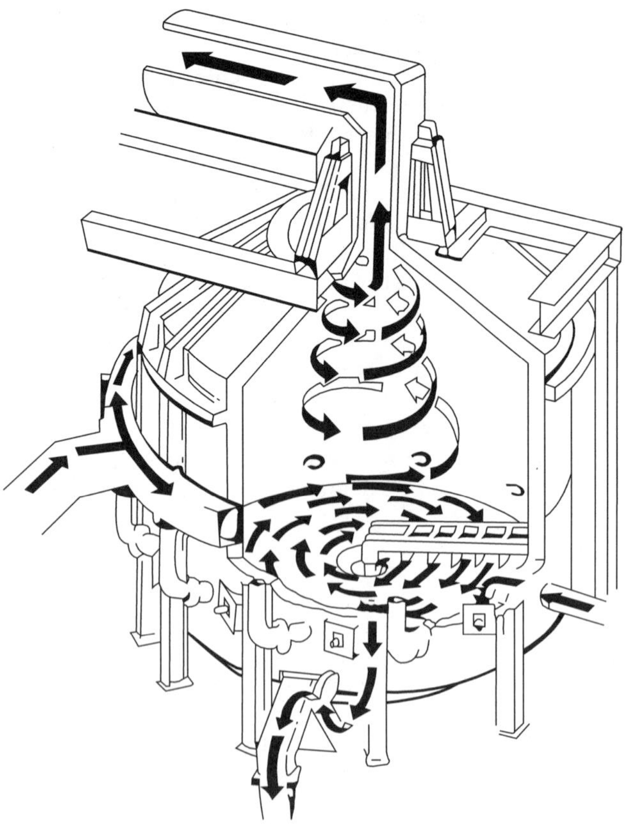

Figure 6. Cyclonic Furnace (from Ref. 4 p. 237).

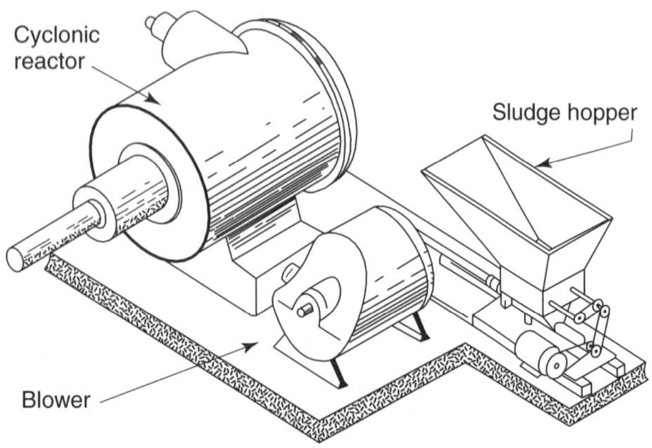

Figure 7. Skid-Mounted Cyclonic Reactor System (from Ref. 4, p. 237).

is discharged from the furnace within the flue gas. Sludge is pumped into the furnace tangentially from the furnace wall. Air is introduced at tangential burner ports, creating a cyclonic effect.

There is no hearth, only the furnace shell and refractory. The sludge detention time is no greater than 10 sec in this furnace. The products of combustion exit the furnace in vortex flow at 1500°F (816°C), and complete combustion is assured. Temperatures within the furnace are from 1500°F to 1600°F (816–871°C). These furnaces are relatively small and can be placed in operation, at operating temperature, within an hour's time.

Cyclonic furnaces are suited for sludge generated from relatively small influent flows, 2 million gallons per day and less. They are relatively inexpensive, mechanically simple units. These systems can be purchased skid mounted for installation as a complete independent package, requiring only utility and feed connections and a stack. Many packaged units like this are sold for on-site sludge and other liquid waste disposal and have been used on shipboard.

Codisposal

A small number of installations have combined sludge and refuse disposal in a single unit. This technique has not met with much success where refuse is charged into a sludge incinerator. Refuse cannot be effectively fired in a multiple-hearth furnace. As noted previously, adding a solid fuel will result in burning too high in the furnace. Feed preparation also is required to size the refuse and to remove low-melting-point materials.

With codisposal in a fluid-bed incinerator, the materials within the refuse stream with low melting temperatures (such as aluminum cans and some glass products) must be removed from the charge prior to injection into the furnace. This is necessary to avoid slagging and subsequent seizure of the bed. The process of preparing the refuse, and of sizing it to less than a 1-inch (2.5-cm) particle size for effective feeding to the bed, has been maintenance intensive, unreliable, and costly.

The charging of sludge into a refuse incinerator has been found to be successful in a limited number of installations. The refuse moisture content is approximately 25%, whereas that of the sludge cake is approximately 75%. To assure that the sludge will burn out as the refuse burns within the incinerator, its moisture content must be reduced to that of the refuse. Drying equipment is used that generally utilizes heat from the gas stream exiting the refuse incinerator to dry the sludge to from 80% to 90% solids content. This high-solids sludge is mixed with sludge cake entering the facility (at 75% moisture content), generally in a pugmill, and the resulting sludge consistency is approximately 75% solids (or 25% moisture). The sludge cake is then introduced into the incinerator, either on the ceiling of the incinerator firing chamber or above the refuse as it enters the incinerator. The moisture-laden stream leaving the drying sludge is directed back to the refuse incinerator, where elutriated organics are destroyed.

The nature of emissions from sludge burning is different from that from refuse burning. In the former, the emissions are predominantly inorganic, whereas unburned organics may be the major emission of concern with the latter. While water scrubbers may be the control system of choice for the sludge incinerator, fabric filters or electrostatic precipitators (ESPs) may be required to capture the small organic constituents of the refuse exhaust stream. These two duties, that is, sludge burning and refuse burning, require emissions control devices that are incompatible. For this reason, the sludge feed rate in codisposal systems has been held to no more than 10% (by weight) of the refuse feed throughput. At this ratio, there has been no noticeable effect of sludge burning on emissions from a refuse incinerator.

The need for drying equipment and associated sludge receiving, storage, and handling facilities adds further operating requirements and costs to codisposal facilities. As a result, very few of these facilities are in operation today.

SLUDGE INCINERATION OVERVIEW

Past experience has indicated that the multiple-hearth furnace is a reliable and predictable sludge incinerator. However, it is relatively complex and requires a high degree of skill and dedication to operate properly. The fluid-bed furnace requires fewer operational decisions to operate satisfactorily, and is a far less complex machine than the multiple-hearth furnace. The fluid-bed furnace, however, does not have universal application. Not all sludges can be incinerated in this furnace. The presence of salts or other materials with low melting temperatures will agglomerate the sand bed and can cause bed seizure. Fuel usage in the fluid-bed furnace is generally lower than in a multiple-hearth furnace at the design point, but at lower feed rates, the fluid-bed furnace becomes less efficient.

Preventive maintenance requires more attention in a multiple-hearth than in a fluid-bed furnace; however, when a failure occurs in a fluid-bed furnace system, it can be more serious (replacing a recuperator heat exchanger section, for instance).

Other types of equipment discussed here have limited application. The electric power rate structure may eliminate the electric furnace from consideration, and the cyclonic furnace is applicable only to smaller systems and may not be able to meet air emission requirements.

The selection of an appropriate sludge incinerator or thermal treatment system must be evaluated on a case-specific basis.

AIR EMISSIONS CHARACTERIZATION

The emissions from a sewage sludge incinerator, besides the carbon dioxide and water vapor produced from the burning process, will be both organic and inorganic and will generally include the constituents discussed in the following.

Inorganic Gases

Carbon Monoxide. Carbon monoxide (CO) is generated in relatively small quantities when any organic material is burned. In incinerators, it is generated through the burning of the sludge feed and firing of supplemental fuel.

A well-operated incinerator will have no more than 100 ppm CO in its stack gas. The CO generation rate will increase with a decrease in air supply and with a decrease in combustion temperature. The lower the CO concentration, the more effective and complete is the combustion process.

Nitrogen Oxides. Nitrogen oxide (NO) is a minor product of combustion. It is considered harmless; however, it will slowly oxidize to nitrogen dioxide (NO_2) and this gas is a problem. NO is the predominant nitrogen oxide from the combustion process; NO_2 will be generated at approximately 5% of the quantity of NO generated. Nitrogen oxides contribute to gross atmospheric effects such as haze and smog.

The generation of nitrogen oxides will increase with increasing temperature and, to a lesser extent, with an increase in the air supply. The sludge incinerator is generally limited to a temperature both within and outside the furnace of 1800°F (982°C). At this temperature, the generation of nitrogen oxides is relatively low and is not a major concern. The burner usually operates at a temperature much higher than the furnace temperature, in excess of 2200°F (1204°C). Often, nitrogen oxides are generated primarily by the supplemental fuel burner.

Another source of nitrogen oxides in a sludge incinerator is in the burning of the sludge itself. Organic nitrogen is present in the sludge and can be as high as 15% of the weight of sludge volatiles. As the sludge volatiles burn, this organic nitrogen will oxidize and generate nitrogen oxides.

Sulfur Oxides. Sulfur is a constituent of sewage sludge, representing up to 2% of the weight of the sludge volatiles. In the burning process, sulfur will oxidize to sulfur dioxide (SO_2) and sulfur trioxide (SO_3). Less than 5% of the sulfur oxides will be SO_3, which is a less desirable compound. It is readily soluble in water and will form sulfuric acid, which is highly corrosive. All sulfur oxides have gross atmospheric effects, including the generation of smog and tree damage.

Hydrogen Chloride. When a chlorine-containing organic compound (organic chlorine) is burned, the chlorine will combine with hydrogen from the water produced by the burning material to generate hydrogen chloride, HCl. The HCl is soluble in water and will readily form hydrochloric acid. As with SO_2, HCl is highly corrosive and does contribute to gross atmospheric effects.

Particulate Matter

Ash. A fraction of the sludge ash will elutriate into the gas stream. In a fluid-bed incinerator, the entire ash load, plus some bed sand, will enter the gas stream, whereas only from 5% to 10% of the sludge ash will enter into the gas stream in an electric furnace. In a multiple-hearth furnace, from 10% to 20% of the ash produced by the burning sludge will enter the gas stream.

The amount of ash elutriated into the exhaust gas is a function of the turbulence the sludge experiences in the burning process. In a fluid-bed furnace, the sludge experiences maximum turbulence. The bed is designed with turbulence in mind, which is the mechanism utilized for effective sludge burning. In the electric furnace, the sludge lies on the conveyor belt, and no motion is applied to the sludge at all.

Heavy Metals. Heavy metals will generally concentrate in a sludge. They enter the plant with the influent, and the amount and nature of these metals are related to the industrial profile of the service area of a plant. In general, the more industrial users that feed the plant, the higher will be the heavy metal loading.

Metals may enter the sludge as part of organic compounds, as metals, or as metal oxides. Within the incinerator, they will oxidize and either stay with the ash or exit in the exhaust as a particulate or as a vapor. As the furnace temperature increases, the release of heavy metals to the incinerator hot gas stream increases. One limitation in furnace operation is in the temperature above the burning hearths. At a temperature in excess of 1650°F (899°C), the release of heavy metals will significantly increase. For this reason, temperatures within the furnace should not be allowed to exceed 1650°F (899°C).

The heavy metals of greatest concern in the sludge incineration process are arsenic, cadmium, chromium, lead, mercury, and nickel. Most of these metals and their oxides (and other compounds) have been found to be toxic when ingested in relatively large quantities.

Table 1. Heavy Metals Removal by Wet Scrubbing

Metal	% Removed
Arsenic	96%
Beryllium	99%
Cadmium	65%
Chromium	96%
Lead	67%
Mercury	0%
Nickel	95%

Source: Reference 1.

Metal oxides will either be adsorbed onto other particulate matter or be free within the gas stream. When adsorbed onto particulate, they will be removed from the gas stream as the particulate is removed. As small free oxides, or as metallic oxides in a gaseous form that has condensed through a quench or other low-temperature process, they are generally less than 1/2 μm mean particle size. This is too small for many conventional scrubbing systems to remove, and they will likely pass through a wet scrubbing system and discharge through a stack.

The effectiveness of removal of these metals through a water scrubber is noted in Table 1.

Organic Compounds

Volatile organic compounds (VOCs) are undesirable in a furnace off-gas. They can be visible, generating a smoke, and can result in the presence of odor. Some of these organics will be condensible, that is, they will condense out of the gas stream when the stream is reduced in temperature to below 400°F (204°C).

The presence of these organics results from two factors. First, the lighter organic fraction will tend to vaporize as the sludge first enters the incinerator. If the temperature at the sludge entrance is too high, the release of VOCs will increase. If there is insufficient residence time at a temperature in excess of 1400°F (760°C), the temperature required for destruction of the majority of VOCs, these compounds will be present.

The second factor contributing to the generation of these compounds is the lack of sufficient oxygen in the incinerator. The less oxygen there is, the more likely it is that incomplete combustion will result, and incomplete combustion is, almost by definition, a designation of the presence of organics (VOCs) in the furnace exhaust. In a multiple-hearth furnace, the oxygen level in the exhaust should not be allowed to fall below 6%. In a fluid-bed furnace, the exhaust oxygen should always be above 4% (measured on a dry volume basis).

Although not always the case, operation of the multiple-hearth furnace may result in release of a greater quantity of VOCs than will a similarly sized fluid-bed furnace. This can occur as a result of higher temperatures than desired on the top hearth of the furnace. At times, the top-hearth temperatures may be difficult to control. When skimmings are mixed with the sludge, or if industrial wastewater going to the plant generates additional volatiles in the sludge feed, a higher release of VOCs will take place.

An afterburner may be required in a multiple-hearth installation to control VOC destruction. Many multiple-hearth furnaces have an expanded top hearth where residence time is designed for a minimum of 1 sec. This residence time is limited by the geometry of the furnace. Another limitation is the temperature to which the top hearth of the furnace can be brought. At higher than 1400°F (760°C), sludge wiped across, or even dropped through, the top hearth will begin to heat up, and by the time the sludge reaches the burning zone of the furnace, it may be hot enough to slag. In this case, an external afterburner is desirable. With the off-gas from the incinerator at or less than 1200°F (649°C), a separate chamber at the outlet of the top hearth of the incinerator would be designed for a minimum 1400°F (760°C) temperature maintained for 2 sec, minimum, to assure the complete destruction of VOCs without affecting sludge burning within the furnace.

In a fluid-bed furnace, VOC release will increase if the temperature within the bed is too low or the bed has insufficient height, or with too high a velocity (excessive fluidizing airflow) through the bed, VOCs are released into the freeboard. Rather than provide additional afterburning equipment, the operation of the fluid-bed furnace can usually be adjusted to minimize the release of VOCs into the exhaust.

Polycyclic organic matter (POM), which includes polychlorinated biphenyls (PCBs), dioxins, and furans, are suspected carcinogens. As techniques for detecting these compounds are becoming more and more sophisticated, they can be found in concentrations of one part per trillion or less. It appears that at these low levels, POM is generated in practically every combustion process, from cigarette smoking to automobile exhaust and from charcoal broiling to incineration.

The generation of POM from a sludge incinerator has not been found to result in a significant health risk. The control of these compounds is through efficient combustion. When the generation of carbon monoxide is low and the generation of VOCs is low, the POM generation rate is also likely to be low.

Opacity

Opacity is a measure of the light interference of an emission. As shown in Figure 8, light interference (a measure of opacity) will vary with particle size. For a given particulate emission rate, the smaller the particle size, the less visible it will be. At the same rate (lb or kg per hour) of emissions, for larger particles, fewer particles will be present and their visibility

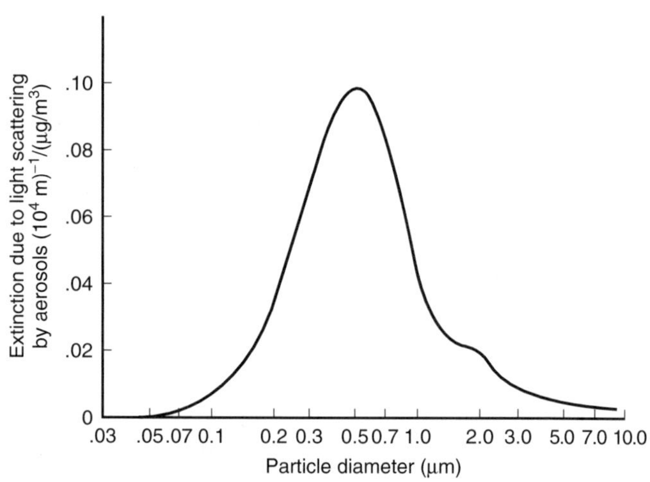

Figure 8. Light Scattering as a Function of Particle Diameter (from Ref. 4, p. 318.

less. In the range of 0.3 to 0.7 μm mean particle size, opacity will be greatest. Noting these properties, it is not possible to relate opacity to an emissions quantity without knowing the particle size distribution of that emission. A low emission might be visible if the 0.5-μm particles predominate; however, if the particle size were less than 0.2 μm, the visibility could be nil.

AIR EMISSIONS CONTROL

Air emissions control is critical to an incinerator system. Particulates, including metals emissions, hydrocarbon emissions, and both organic and inorganic gaseous discharges, are of concern, as are opacity and odor.

Control devices can limit particulate, metals, and acid gas emissions from the stack discharge, but the control of other gaseous, condensible, and noncondensible hydrocarbon emissions is generally a function of the combustion process and may not be specifically affected by gas scrubbing or conditioning.

Metals and other inorganics (ash), as well as some noncondensible organics, will form into or become attached to particulate matter within the exhaust gas stream. When removing particulate from the gas stream, therefore, most of the metals and many of the organic constituents, as well as most of the elutriated ash, will be removed.

The Hot Gas Stream

The gas stream from an incinerator has a very high moisture content. The sludge moisture, which is approximately 80% of the sludge weight, will eventually be discharged in the incinerator off-gas. The gas exiting the incinerator also will contain organic gases, some of which will be condensible. When the gas is reduced in temperature, these condensibles will drop out of the gas and will produce a highly corrosive liquid.

The high moisture content of the gas stream discourages the use of a fabric filter (baghouse) for emissions control. The effectiveness of a baghouse requires that the accumulated particulate matter on the bags be cleaned from them effectively and regularly. With the high moisture content of the exhaust gas, the particulate will tend to cake on the bags and blind them. It would be extremely difficult, if not impossible, to keep the bags clean and functioning (filtering the gas stream) in this moist and dirty gas environment.

With regard to other dry systems for air emissions control, these systems lack the means to capture and remove condensibles from the gas stream. They must be used after the gas stream is wetted. Such wetting will remove the condensibles from the gas, but also will produce a gas stream saturated with moisture.

For these reasons, the conventional baghouse or the dry ESP is not applicable to sludge incineration systems. In all sewage sludge incinerations systems in this country, wet scrubbing is employed, although additional air emissions control systems are used downstream of the scrubber in some installations.

Particle-Size Analysis

The efficiency of any particulate control device will vary as a function of the particulate size. Particulate matter is normally designated by mean diameter, measured in micrometer (μm, 1 millionth of a meter).

The particulate size distribution of an emission cannot be calculated, but can only be inferred from prior data based on measurement. As would be expected, data on particle size distribution are rare. It is a function of waste properties, furnace design, and furnace operation, and will vary from test to test. It is virtually impossible to define particle size distribution with any degree of confidence prior to construction and operation of the incinerator in question. As an estimate of the type of distribution that can be expected, note Figures 9–11, for multiple-hearth, fluid-bed, and infrared incinerators. These are only estimates, and should not be used as a basis for systems or equipment design.

Wet Scrubbing

Wet gas scrubbing with plant effluent is used almost universally for sludge incinerator exhaust gas treatment. Wet scrubber operation is described in the chapter on particulate removal. Table 2 lists typical emissions from sewage sludge incinerators using wet scrubbers.

The effectiveness of a scrubbing system is usually directly related to the pressure drop across the scrubber. The higher the pressure drop, the greater is the turbulence/mixing, and,

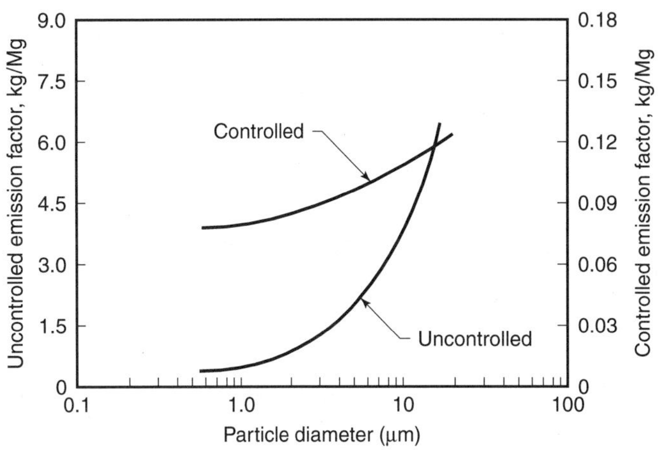

Figure 9. Cumulative Particle Size Distribution and Size-Specific Emissions Factors for Multiple-Hearth Incinerators (from Ref. 2).

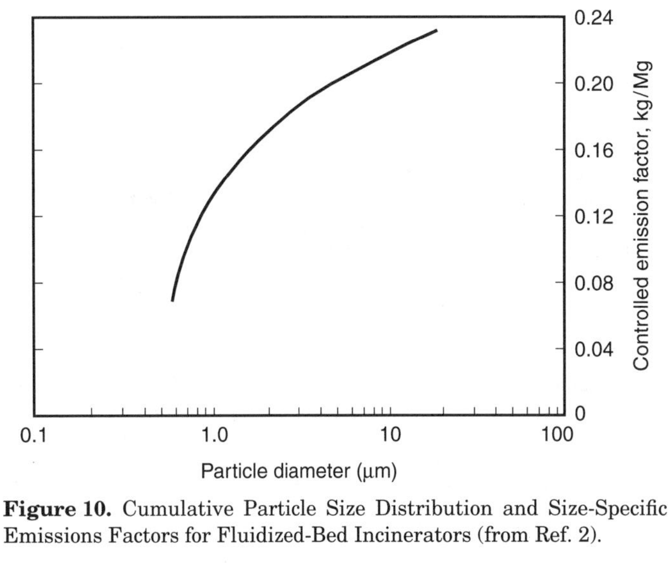

Figure 10. Cumulative Particle Size Distribution and Size-Specific Emissions Factors for Fluidized-Bed Incinerators (from Ref. 2).

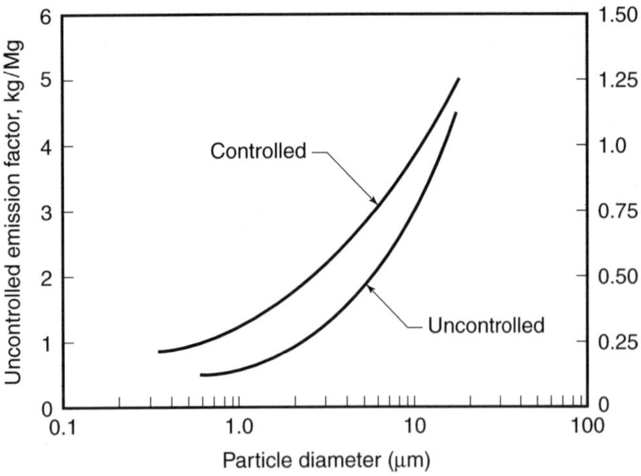

Figure 11. Cumulative Particle Size Distribution and Size-Specific Emissions Factors for Electric (Infrared) Incinerators (from Ref. 2).

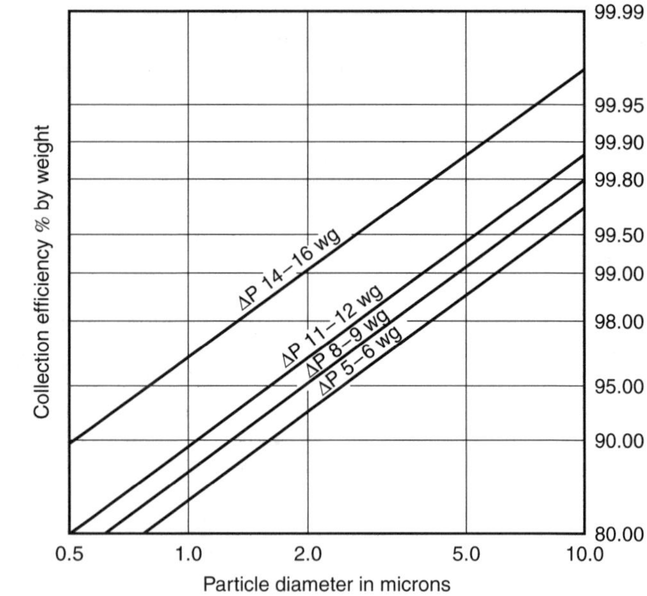

Figure 12. Collection Efficiency Versus Particle Size (from Ref. 4, p. 331).

therefore, the more effective is the scrubbing action. This feature is illustrated by the graph in Figure 12. For a 2-μm-diameter particle, for instance, a pressure differential of less than 3 in. w.c. (1.6 mb) will provide a removal efficiency of 99%, whereas a 20-in. w.c. (10.5-mb) differential is necessary for 99% removal of a 0.8-μm particle from the gas stream.

Scrubbing systems are often described by system pressure drop. A low-energy system is normally defined as one producing less than 12 in. w.c. (6.3 mb) for particulate removal, whereas high-energy systems will have significantly higher pressure differentials, from 20 to over 60 in. w.c. (10 to over 31 mb). Likewise, medium-energy systems operate with a 7–20-in. w.c. (3.7–10.5-mb) differential.

Gases exiting a sludge incinerator will contain hydrogen chloride and sulfur oxides. These constituents result in acid formation in the gas scrubbing liquid, as do condensible organic residuals. The use of nonmetallic chambers or nonmetallic coated carbon steel is a growing method of corrosion control. These materials, such as flake polyester, fiberglass, neoprene, and epoxy, have good anticorrosion properties; however, they are temperature sensitive. In most cases, they cannot withstand temperatures above the range of 160–200°F (71–93°C). Use of such materials must include provisions for emergency quenching to provide cooling if the normal cooling water supply is lost.

Wet gas cleaning equipment floods a gas with scrubbing liquid. Water droplets usually are carried off by the gas stream exiting the scrubber. A mist eliminator is a passive component that removes most of the entrained water droplets from the gas stream. This is desirable to remove collected particulate that is sorbed onto the water particles and to reduce the size of the plume exiting the stack (excess water in off-gas will appear as a white plume at the stack discharge), and will reduce the moisture collected in the ID fan and other downstream equipment. Decreasing this liquid accumulation will also decrease attendant corrosion.

One of the more common mist eliminators in use on sludge incinerators is the chevron type (Figure 13). The moisture-laden gas stream travels through a series of baffles. Gas exits the baffles and the heavier moisture particles, because of their inertia, hit the chevron-shaped baffles and fall to the bottom of the chamber, leaving the gas stream. Other mist eliminators include impingement plate designs and packed towers.

Table 2. Emission Factors for Sewage Sludge Incinerators [values in pounds per dry ton (kg/Mg dry)]

	Multiple Hearth		Fluid Bed		Electric	
Pollutant	Uncontrolled	After Scrubber[a]	Uncontrolled	After Scrubber[a]	Uncontrolled	After Scrubber[a]
Particulate	80 (40)	0.80 (0.40)	NA[b]	6.0 (3.0)	20 (10)	4.0 (2.0)
Lead	0.10 (0.05)	0.06 (0.03)	NA	0.006 (0.003)	NA	NA
Sulfur dioxide	20 (10)	4.0 (2.0)	20 (10)	4.0 (2.0)	20 (10)	4.0 (2.0)
Nitrogen oxides	11 (5.5)	5.0 (2.5)	NA	2.0 (1.0)	8.6 (4.3)	6.0 (3.0)
Carbon monoxide	60 (30)	4.0 (2.0)	NA	4.2 (2.1)	NA	NA
VOC—methane	4.6 (2.3)	4.6 (2.3)	1.6 (0.80)	1.6 (0.80)	NA	NA
—nonmethane	1.7 (0.85)	1.7 (0.85)	NA	NA	NA	NA

[a]Scrubbers are either venturi or venturi plus impingement tray.
[b]NA, not available.

Source: Reference 2.

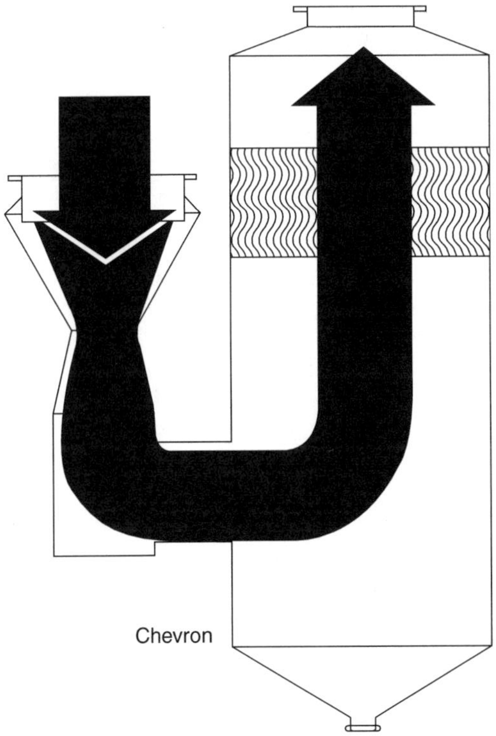

Figure 13. Chevron-Type Mist Eliminator (from Ref. 4, p. 328).

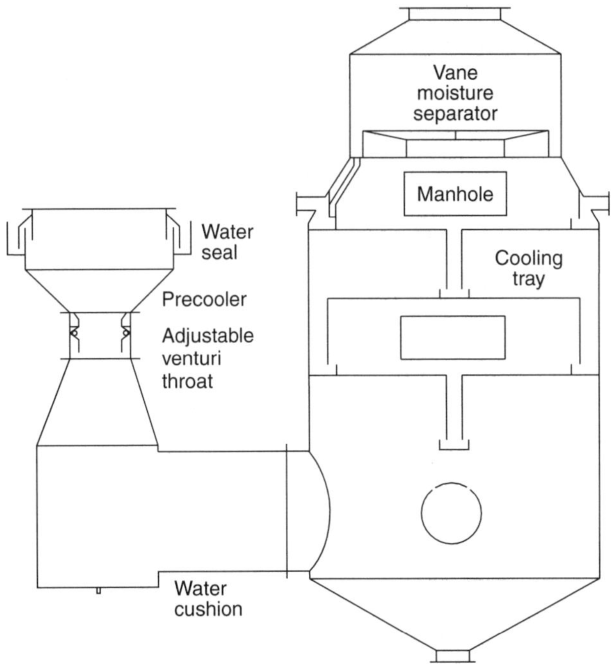

Figure 14. Variable Throat Venturi Scrubber (from Ref. 4, p. 334).

Venturi Scrubber

Venturi scrubbers are widely used, where water is readily available, as high-efficiency, high-energy gas-cleaning devices. The heart of the system is a venturi throat where gases pass through a contracted area, such as that shown in Figure 14, reaching velocities of 200–600 fps (6096–18,288 cm/sec), and then pass through an expansion section. From the expansion section, the gas enters a large chamber for separation of particles or for further scrubbing.

Impingement Tray Scrubbers

Impingement tray scrubbers are essentially perforated plates with target baffles. Tray scrubbers have no large gas-directing baffles, but are simply perforated plates within a tower, usually immediately downstream of a venturi. A water level is maintained above the trays (there are usually two or more trays). The geometrical relationship of the tray thickness, hole diameter, and spacing, as well as the impinger details, results in a high-efficiency device for the removal of small particulate of less than 2 μm mean diameter.

Wet Electrostatic Precipitator

Wet ESPs have been found to be effective devices for the removal of airborne particulate matter. Most wet ESP installations utilize them after a conventional scrubber to reduce the scrubber particulate and heavy metals emissions discharge. One of these systems is shown schematically in Figure 15 and operates as follows:

1. The hot gas stream is initially wetted either by upstream equipment or by the water sprays within the wet ESP.
2. The gas stream passes through a series of discharge electrodes. These electrodes are negatively charged, in the range of 1,000–6,000 volts dc. This voltage creates a corona around the electrode. A negative charge is induced in the particulate matter passing through the corona.
3. A grounded surface, or collector electrode, surrounds the discharge electrode. Charged particulate will collect on the grounded surface.

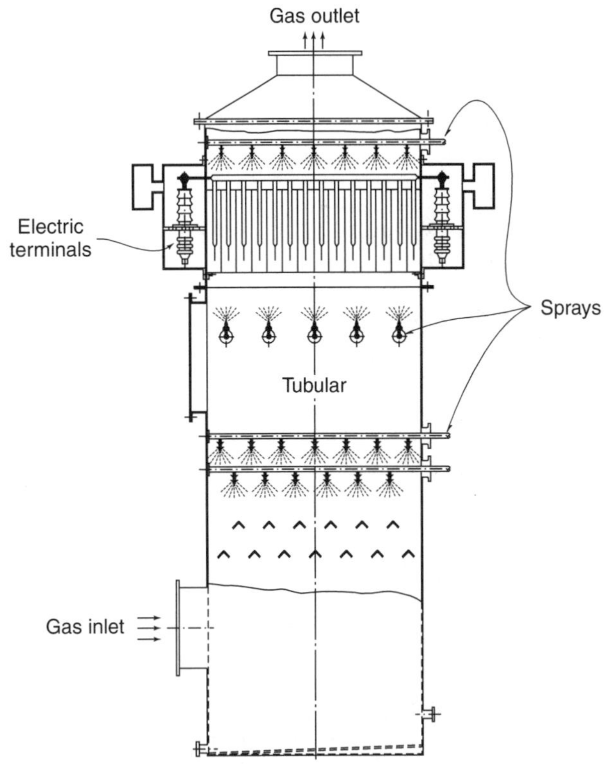

Figure 15. Wet Electrostatic Precipitator (Courtesy Beltran Associates Inc., Brooklyn, N.Y.).

Table 3. Particulate Discharge after Wet Electrostatic Precipitator Maryville, Tenn., Wastewaster Treatment Plant

	Particulate Discharge [1b/dry ton (kg/Mg)]
Test 1	0.49 (0.25)
Test 2	0.23 (0.12)
Test 3	0.21 (0.11)
Average	0.31 (0.16)

Source: Reference 3.

4. The particulate matter will be removed from the collector surface by intermittent water sprays directed within the electrode grounding surface assembly.

The wet ESP is very efficient in the collection of small particulate, down to the submicrometer range. Table 3 lists the results of particulate emissions tests (three replications) measured downstream of a wet ESP. The precipitator is located after a conventional venturi—impingement-tray scrubber on a multiple-hearth sludge incinerator.

REFERENCES

1. *The Federal Register*, February 6, 1989, p. 5890.
2. Draft report. Emission Factor Documentation for AP-42, Section 2.5-Sewage Sludge Incineration, Monitoring and Data Analysis Division, Office of Air Quality Planning and Standards, U.S. Environmental Protection Agency, Research Triangle Park, NC, September 1987.
3. *Maryville Wastewater Treatment Plant Sewage Sludge Incinerator Test Report*, EMI Consultants, Knoxville, TN, September 18, 1986.
4. C. R. Brunner, *Incineration Systems Handbook*, Incinerator Consultants Inc., Reston, VA, 1996.
5. C. R. Brunner, "Sewage Sludge Incineration at the Cleveland Southerly Wastewater Treatment Center," Proceedings of the 16th Intersociety Energy Conversion Conference, Atlanta, 1980.

BIBLIOGRAPHY

C. R. Brunner, *Design of Sewage Sludge Incineration Systems*, Noyes, Park Ridge, NJ, 1980.

C. R. Brunner, *Hazardous Air Emissions from Incineration*, 2nd ed., Chapman & Hall, New York, 1986.

C. R. Brunner, *Medical Waste Disposal*, Incinerator Consultants Incorporated, Reston, VA, 1996.

C. R. Brunner, "Sludge Incineration Applications and Processes." *TAPPI Journal*, November 1992 (Technical Association of the Pulp and Paper Industry, Atlanta, GA).

C. R. Brunner, "Biological Sludge Incineration." *Waste Management Journal*, October 1991.

C. R. Brunner, "Sludge Incineration," Paper, April 1991, London, England.

C. R. Brunner, "Incineration of Sludge Waste," Proceedings, Incineration Conference, San Diego, CA, May, 1990.

C. R. Brunner, "Sludge Incineration at Atlantic City," Proceedings, Incineration Conference, Air & Waste Management Association, Atlantic City, NJ, October 1989.

C. R. Brunner, "Industrial Sludge Incineration," *Environmental Progress*, August 1989.

C. R. Brunner, "Waste Disposal by Incineration," Article in *Encyclopedia of Chemical Processing and Design*, Marcel Dekker, 1988.

C. R. Brunner, "Sludge Incinerator Practice," in *Sludge Incineration*, WPCF, Alexandria, VA, November 1988.

C. R. Brunner, "Hospital Waste Disposal by Incineration." *Journal of the Air Pollution Control Association*, October 1988.

C. R. Brunner, "Incineration of an Industrial Sludge Waste," Monograph, AIChE Summer Conference, Denver, August 1988.

R. Coulter, "Smoke, Dust, Fumes Closely Controlled in Electric Furnaces," *Iron Age* **173**, 107 (1954).

T. D. Ellis and C. R. Brunner, *Industrial Hazardous Waste Incineration Course Notes*, AIChE Short Course, New York, 1990.

M. Kraus, "Baghouses," *Chem. Eng.* (1979).

National Air Pollution Control Administration, *Control Techniques for Particulate Pollutants*, AP-51, 1981.

Water Pollution Control Federation, *Incineration Manual of Practices*, OM-11, 1988.

DRUM RECONDITIONING

PATRICIA KAY RYKOWSKI

In the early 1900s the demand for wooden barrels had already established the need for business that collected used barrels, cleaned and refurbished them for reuse. As the steel drum replaced the wooden barrel as the most common industrial shipping container, the business of refurbishing wooden barrels evolved into the collection of processes known today as drum reconditioning.

The steel drum is now used for every type of liquid and semiliquid material that is shipped in bulk in commerce, including petroleum products, industrial chemicals, paints, varnishes, inks, adhesives, resins, food and agricultural products, cleaning products, pesticides, and industrial wastes. The current industry demand is for approximately 80 million steel drums. To meet this demand, approximately 42 million reconditioned steel drums are provided, compared to only 36 million new drums. The role of drum reconditioning is thus vitally important to every industry that ships or uses materials packaged in steel drums.[1]

In the life cycle of a steel drum, illustrated in Figure 1, a single drum may be reconditioned many times, resulting in multiple reuse by industry before being processed for steel scrap recycling. The drum reconditioning industry is the critical component of this life cycle, facilitating an overall net savings in energy, raw materials, air emissions, water emissions, and solid waste volume.[2]

PROCESS DESCRIPTION

Drum Types

There are two types of steel drums: tight-head (also called closed-head) and open-head. Tight-head drums have threaded bung openings (2- and 3/4 in.) in the top (head). They are generally used to transport flowable materials, including oils, solvents, and other liquids. Open-head drums have fully removable heads (lids) and are used to contain more viscous materials, such as petrochemicals and adhesives.[3]

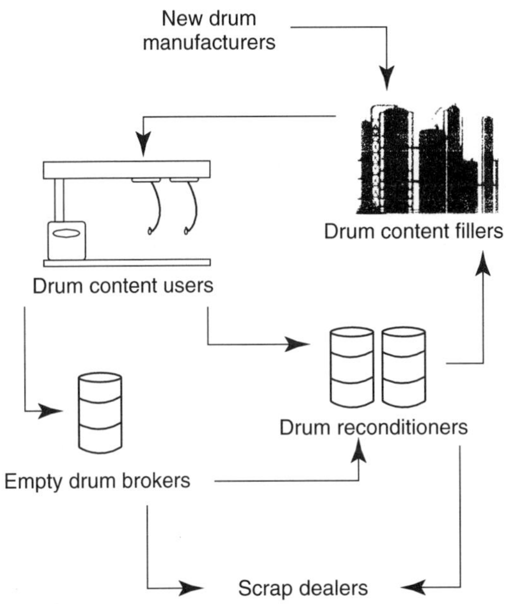

Figure 1

Process Overview

The reconditioning process is different for each drum type. Tight-head drums are reconditioned using a series of high-pressure alkaline and acid flush-and-rinse stations. The drums are reshaped using pneumatic methods to remove dents and restore chimes, and then pressure-tested as required by Department of Transportation (DOT) regulations. The drum exterior is refinished by removing the old coating through shot blasting or other methods, and repainting to meet customer specifications. If a closed-head drum contains hardened residue that cannot be removed through washing, the head is removed and the drum is converted to an open-head drum for reconditioning.[3]

Open-head drums are reconditioned using a drum reclamation furnace to pyrolize the exterior coating, and any interior lining and residual content. After processing in the furnace, the drum's interior and exterior are stripped to bare metal through shot blasting. Just like closed-head drums, the open-head drums are reshaped using pneumatic methods to remove dents and restore chimes, and then pressure-tested as required by Department of Transportation (DOT) regulations. If necessary to meet customer demand, an open-head drum may be "remanufactured" into a tight-head drum by reseaming with a newly manufactured head. The drum exterior is refinished by removing the old coating through shot blasting or other methods, and repainting to meet customer specifications.[3]

Both tight-head and open-head drums may include a removable plastic liner to isolate the contents from the metal to avoid product contamination or corrosion. Depending on customer requirements, open-head drums may be lined with a chemically resistant coating.

Drum Reclamation Furnace

The drum reclamation furnace can generally be described as a continuous tunnel furnace, as illustrated in Figures 2 and 3. Figure 4 provides a photograph of a typical furnace installation. As stated above, the purpose of the furnace is to pyrolize the drums. This pyrolosis is to change the previous coatings, linings, and any residual contents of the drums into a dry and friable condition, which allows them to be easily removed in the subsequent abrasive blasting operation.

The drums are conveyed through the furnace in their inverted position by means of a drag-chain conveyor. The lids, rings, and plugs are included by placing them on top of the inverted drums; they may also be pyrolyzed in a separate furnace. The furnace feed rates may range from two or three drums per minute to 10 to 12 drums per minute.

Drum-reclamation furnaces are constructed in a wide variety of sizes and configurations. A typical design consists of a refractory-lined tunnel furnace with a number of natural-gas-fired burners mounted in the sides of the tunnel. Furnaces may also use propane or fuel oil for burner firing.

The furnace can be viewed as being divided into three zones. After entering the tunnel, the drums pass through the preheat zone, where they are heated by radiation from the furnace flame area. They then pass through the ignition zone, where

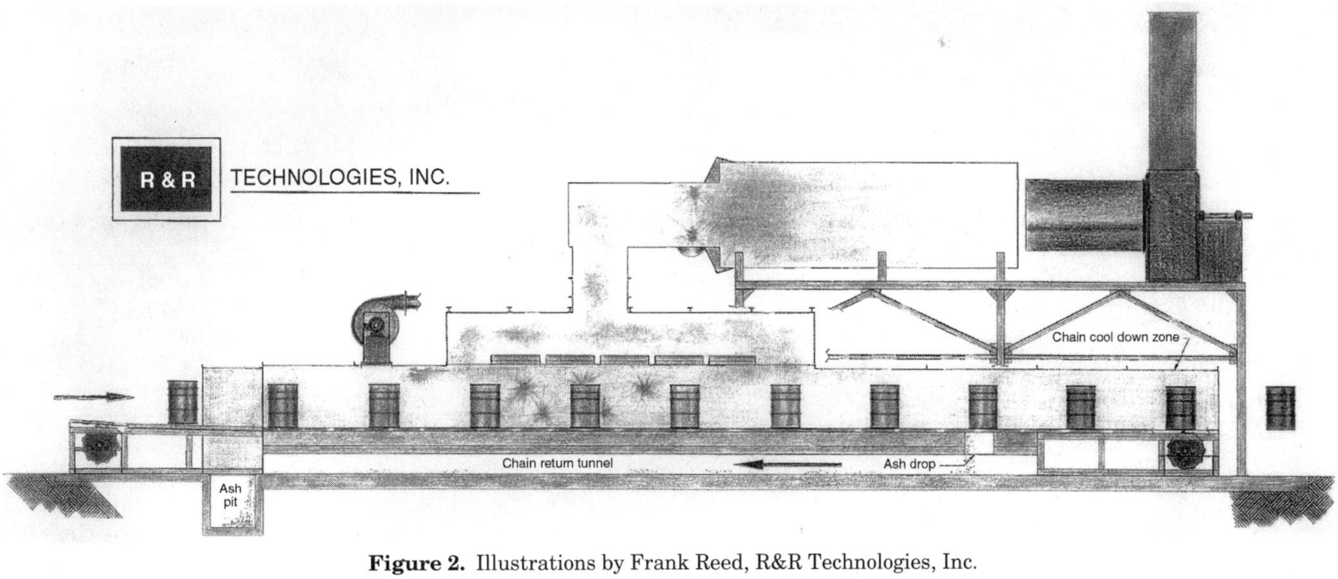

Figure 2. Illustrations by Frank Reed, R&R Technologies, Inc.

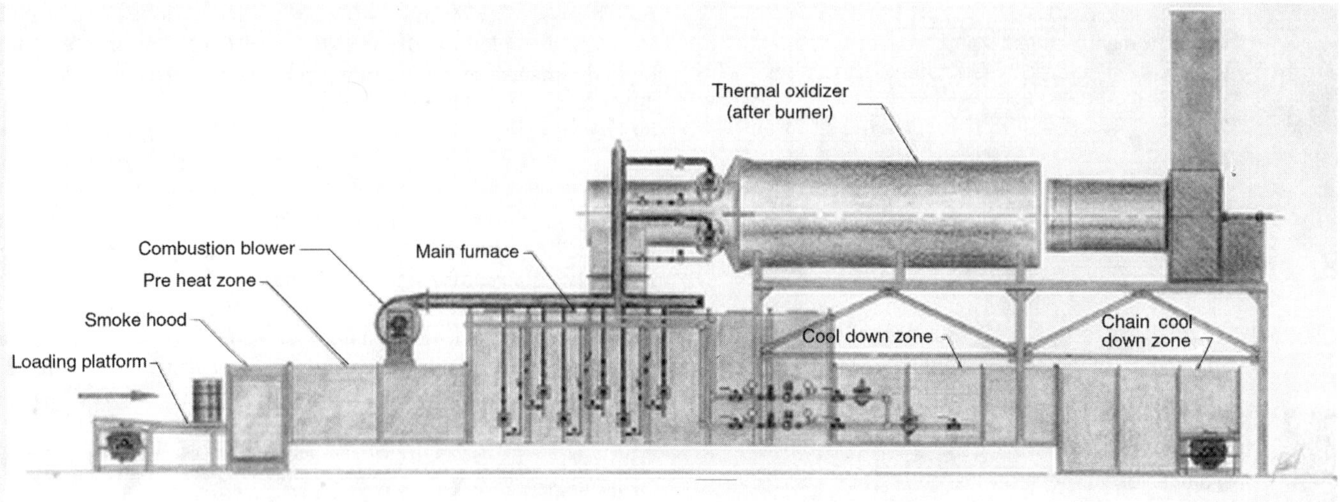

Figure 3. Illustrations by Frank Reed, R&R Technologies, Inc.

Figure 4. Photo Courtesy of Drum Service Co. of Florida and Palex Container Systems.

combustibles in contact with furnace flames ignite and burn. Last, they pass through the cooling zone, where a small amount of burning continues and the drums are cooled somewhat by the induced air. The exhaust gases from the furnace pass into a thermal oxidizer, where carbon monoxide and organic vapors are more completely combusted.

Table 1 provides a summary of the range of process conditions typical of drum furnaces operating today.

Table 1. Summary of Typical Drum Furnace Operating Parameters[2,4]

Process throughput	100 to 600 drums/h
Furnace operating temperature	850–1500°F
Thermal oxidizer temperature	800–1800°F
Thermal oxidizer gas residence time	0.4–1.7 sec
Stack gas flow rate	50–150 scfh per drum/h

Source Classification

There are no known state regulations that have been established specifically for drum reclamation furnaces. In general, drum furnaces have been regulated based on general combustion rules included in the relevant state or local program. Drum reconditioning furnaces are properly classified as industrial process furnaces. Regulatory permit writers at the state and local level may sometimes incorrectly classify a drum reconditioning furnace as an incinerator and subsequently may apply regulations or other requirements specifically developed for incinerators. This is an improper regulatory classification. It is important to note that the purpose of a drum furnace is to pyrolize the used drum in preparation for shot blasting. The drum reconditioning facility is not a waste disposal operation.

Thermal Oxidizer

Most drum reclamation furnaces in operation today are equipped with a thermal oxidizer for secondary combustion. The thermal oxidizer, when designed as an integral part of the drum furnace structure, is typically a rectangular chamber located above the furnace tunnel, the roof of the tunnel being the base of the thermal oxidizer chamber. Alternatively, the thermal oxidizer can be a separate, free-standing chamber connected to the furnace exhaust outlet by a refractory-lined breeching.

The thermal oxidizer provides control of organic compounds, carbon monoxide, and particulate matter. To achieve optimum performance, adequate temperature, gas residence-time, and turbulence must be provided. Thermal oxidizers in operation today throughout the industry vary in design and operation. Gas residence times range from 0.4 to 1.7 sec, with operating temperatures from 800°F to 1800°F.[2] Adequate turbulence, or mixing of exhaust gases (organic vapors) with combustion air (oxygen), is achieved largely through design, by means of baffles and abrupt right-angle turns of the flow path or by inducing cyclonic flow patterns.

Waste-Heat-Recovery Boiler

Many drum furnaces are equipped with waste-heat-recovery boilers. The boiler installations may be designed for either side-stream or integrated (full-flow) operation. In the side-stream design, some of the flue gases from the thermal oxidizer are drawn through the boiler by means of an induced-draft (ID) fan and the balance of the flue gases exhaust up the stack by natural draft. In the integrated design, the entire flue gas stream is drawn through the boiler by the ID fan. The choice of side-stream versus integral design depends on many factors. In the side-stream design, the rate of withdrawal of hot flue gases is dictated by the facility's steam demand, whereas in the integrated design, all the recovered heat is converted to steam and any excess steam above demand is vented to the atmosphere.

The Combustion Process

The combustion process in a drum furnace system is quite complex. There are at least four sources of air for combustion: the air induced into the entrance and exit openings of the furnace tunnel and the forced air provided by the combustion air blowers for the fuel burners in the furnace and the thermal oxidizer. Fuel sources consist of the furnace and thermal oxidizer burners. Depending on the type and nature of the drums being processed, some fuel value may be available from any drum content residuals volatized during furnace operation.

Example Furnace Design

The analysis below provides an example of typical drum furnace design, including three major components: (1) the primary ignition (furnace) chamber; (2) the thermal oxidizer; and (3) the induced draft fan. This analysis is based on calculations first presented in the *Air Pollution Engineering Manual*, 1st Edition (AP-40), and has been updated to reflect current methodology used by industry experts.[5,6]

PROBLEM

Design a continuous-tunnel-type drum furnace for processing 300 standard 55-gallon, steel, open-top drums per hour and an afterburner operating at a temperature 1800°F with a 1-sec retention time.

GIVEN

Combustible material attached to each drum varies from near zero (0) to four (4) pounds. All combustible material on the drums is considered to have a composition equivalent to that of U.S. Grade #6 fuel oil.

SOLUTION

1. Primary ignition chamber

A. Induced air through openings at 200 fpm and 60°F:

opening width = 28 in.

opening height = 50 in.

$$\text{opening cross-sectional area} = \frac{(28)(50)(2)}{(144)} = 19.44\ \text{ft}^2$$

$$\begin{aligned}\text{Induced air} &= (19.44)(200)\\ &= 3889\ \text{scfm}\\ &= \text{use } 4500\ \text{scfm}\end{aligned}$$

B. Size of ignition burners to raise effluent to 1000°F.

Design ignition burners for most severe operation, that of negligible combustible material per drum. Burners must raise temperature of drums to 900°F. Density of air at 60°F and 14.7 psi is 0.076 lb/ft^3. Average specific heat of products of combustion is 0.26 Btu/lb °F.

i. Heat required to raise induced air from 60°F to 1000°F:

$$Q_1 = W_a C_{pa}(T_2 - T_a)$$

where Q_1 is the heat required (Btu/h); W_a, the weight of air (lb/h); C_{pa}, the average specific heat over the temperature range; T_2, the final temperature (°F); and T_a, the ambient air temperature (°F).

$$\begin{aligned}Q_1 &= (4500)(60)(0.076)(0.26)(100^\circ\text{F} - 60^\circ\text{F})\\ &= 5{,}015{,}088\ \text{Btu/h}\end{aligned}$$

ii. Heat required to raise temperature of drums from 60°F to 900°F:

The specific heat of steel for this temperature range is 0.12 Btu/lb°F

$$Q_2 = W_d C_{pd}(T_2 - T_1)$$

where Q_2 is the heat required (Btu/h); W_d, the weight of drums (lb/h); C_{pd}, the specific heat of steel (Btu/lb °F); T_1, the initial temperature; and T_2, the final temperature.

$$Q_2 = (250)(55)(0.12)(900°\text{F} - 60°\text{F})$$
$$= 1,663,200 \text{ Btu/h}$$

iii. Total heat required in ignition zone:
Assume heat losses through radiation, storage, etc. are 10% of the total gross heat input.

$$= \frac{5,015,088 + 1,663,200}{0.90}$$
$$= 6,678,288 \text{ Btu/h}$$

iv. Natural gas capacity of primary burners:
The caloric value of one (1) SCFM of natural gas with 20% excess air is 746.2 Btu @ 1,000°F.

$$\text{total capacity} = \frac{6,683,700 \text{ Btu/h}}{736.2 \text{ Btu/scf}}$$
$$= 9071 \text{ scfh}$$

v. Individual burner capacity:
Install 12 burners, 6 per side

$$\text{burner capacity} = \frac{9071 \text{ scfh}}{12}$$
$$= 756 \text{ scfh}$$

C. Excess primary combustion air:
Assume all air for burning the material on drums is induced through the tunnel openings.

i. Maximum design burning rate:

$$(300)(4) = 1200 \text{ lb/h}$$
$$= 20 \text{ lb/min}$$

ii. Total combustion air available through tunnel openings

$$\text{air} = \frac{4500 \text{ scfm}}{20 \text{ lb/min}}$$
$$= 225 \text{ scf/lb}$$

1 lb U.S. Grade 6 fuel oil requires 177 scf air 30% saturated @ 60°F.

$$\%\text{ excess air available} = \frac{225 \text{ scf} - 177 \text{ scf}(100)}{177 \text{ scf}}$$
$$= 27.1\%$$

D. Volume of ignition zone:
Assume a heat release factor of 22,000 Btu/h, which is similar to heat release factor for oil-fired furnace fireboxes operating at less than 1800°F. Assume the drums contain negligible combustible materials.

$$\text{volume} = \frac{(9071 \text{ scfh})(1100 \text{ Btu/scf})}{22,000 \text{ Btu/h ft}^3}$$
$$= 453.5 \text{ ft}^3$$

E. Assume width = 30 in., height = 87 in. including conveyor

$$\text{minimum length} = \frac{453.5 \text{ ft}^3}{(3.3 \text{ ft})(7.3 \text{ ft})}$$
$$= 19 \text{ ft}$$
$$\text{actual length} = 28 \text{ ft}$$

F. Cooling zone length:
Design ignition zone and cooling zone for a total residence time of 4 min. Assume a drum spacing of 27 in.

$$\text{conveyor speed} = \frac{(300)(2.2)}{60}$$
$$= 11 \text{ fpm}$$

Length of ignition and cooling zone:

$$\text{total length} = (11 \text{ ft/min})(4 \text{ min})$$
$$= 44 \text{ ft}$$
$$\text{cooling zone length} = 44 - 28 = 16 \text{ ft}$$

G. Preheat zone length:

$$\text{design length} = 16 \text{ ft}$$

2. Secondary combustion chamber (thermal oxidizer)

A. Design the gas burners for the most severe operation. Thermal oxidizer will raise the temperature of products of combustion from ignition zone from 1200°F to 1800°F.

i. Weight of the products of combustion from ignition zone (furnace):

$$\text{induced air} = (4500)(0.076)(60)$$
$$= 20,520 \text{ lb/h}$$
$$\text{natural gas} = (9071)(0.999 \text{ lb/scf})$$
$$= 9,062 \text{ lb/h}$$
$$\text{Total} = 29,582 \text{ lb/hr}$$

ii. Heat required to raise temperature of products of combustion to 1800°F.
Average specific heat of products of combustion is 0.26 Btu/lb °F over the given temperature range.

$$Q_3 = W + C_{pc}(T_2 - T_1)$$

where W is the weight of products of combustion; C_{pc}, the average specific heat (Btu/lb °F); T_1, the initial temperature; and T_2, the final temperature.

$$Q_3 = (29{,}582)(0.26)(1800 - 1200°\text{F})$$
$$= 4{,}614{,}792 \text{ Btu/h}$$

iii. Total heat required in thermal oxidizer:

Assume heat losses by radiation, convection, and so on are 10% of gross heat supply at 1800°F.

$$\text{total heat} = \frac{4{,}164{,}792 \text{ Btu/h}}{0.90}$$
$$= 5{,}127{,}546 \text{ Btu/h}$$

iv. Total capacity of secondary burners:

The calorific value of 1 scf of natural gas is 615.4 Btu at 1800°F with 20% excess air.

$$\text{natural gas} = \frac{5{,}127{,}546 \text{ Btu/h}}{615.4 \text{ Btu/scf}}$$
$$= 8332 \text{ scfm of natural gas}$$

v. Individual secondary burner capacity:

Install four burners @ 90° apart

Capacity of each burner $= 2{,}083 \text{ ft}^3/\text{h}$

B. Cross-sectional area:

Design the afterburner for a cross-section velocity of 20 fps maximum at 1800°F. From Appendix resources, there are 13.53 scf products of combustion from 1 scf of natural gas with 20% excess air.

i. Volume of products of combustion when drums are burned with negligible combustibles:

induced air:

$$(4500 \text{ scfm})(60) = 270{,}000 \text{ scfh}$$

primary burners:

$$(9071 \text{ scfh})(13.53) = 122{,}731 \text{ scfh}$$

secondary burners:

$$(8332 \text{ scfh})(13.53) = 112{,}732 \text{ scfh}$$
$$\text{total} = 505{,}463 \text{ scfh}$$
$$= 8424 \text{ scfm}$$
$$= 140 \text{ scfs}$$

ii. Internal cross-sectional area and dimensions:

$$\text{area} = \frac{140 \text{ scfs}(2{,}260°\text{F})}{20 \text{ fps}(520°\text{R})}$$
$$= 30 \text{ ft}^2$$
$$\text{diameter} = 30/0.785$$
$$= 72'' \text{ in.}$$
$$\text{length} = (20 \text{ fps})(1 \text{ sec}) = 20 \text{ ft}$$

3. Fan capacity (without waste heat boiler)

A. Total air volume with cooling air:

Assume maximum temperature at exit of the afterburner is 1800°F.

$$Q_1 = (1800)(0.076)(0.26)(60)(8424)$$
$$= 17{,}977{,}489 \text{ Btu/h}$$

Assume air temperature entering the fan to be 650°F.

$$\text{scfm} = \frac{17{,}977{,}489 \text{ Btu/h}}{(650)(0.076)(0.26)(60)}$$
$$= 23{,}328 \text{ scfm}$$
$$\text{cfm} = \frac{(22{,}461)(1110)}{520}$$
$$= 47{,}945 \text{ cfm @ } 650°\text{F}$$

B. Static pressure of system is 6 in. water column.

AIR EMISSIONS CHARACTERIZATION

The types and concentrations of exhaust gas contaminants from drum reclamation furnaces vary depending on the age and design of the furnace, the operation of the furnace (combustion conditions), the types of drums processed, and drum management procedures.

Furnace Design

It is estimated that the first drum reclamation furnaces were constructed in the 1940s. However, these systems were primitive in comparison to current furnace technology. Advances in burner systems, combustion air supply, and process instrumentation/controls have resulted in a more energy-efficient, high-performance drum furnace design. The furnaces in operation today vary in age, and include both older and modern technologies. While sufficient data are not available to quantify and compare emissions from old and new furnaces, it is expected that newer systems with controlled combustion air supply and automatic draft control will result in less entrainment of particulate matter in combustion gases.

Types of Drums Processed

The types of drums processed are largely dependent on the customer base served by the reconditioning facility. As an example, in certain regions the customer base may be predominantly agricultural, resulting in processing a high percentage of drums that have contained food products. The type of drums processed can affect the release of air emissions for pollutants such as acid gases and metals. Emission rates for these compounds are generally directly proportional to the feed concentration into the furnace and thus vary widely depending on the quantity of drums processed with acid or metal constituent residues. Chlorinated solvents and sulfur-containing compounds are usually shipped in tight-head drums, but may occur in open-head drums typically processed in drum furnaces.

Drum Management Procedures

The U.S. Environmental Protection Agency (EPA), under the authority of the Resource Conservation and Recovery Act (RCRA), regulates the use of steel drums for the storage and disposal of hazardous wastes. Under RCRA, a reconditioner cannot legally accept drums that do not meet the RCRA definition of an empty container unless it is a permitted treatment, storage, or disposal facility (TSDF). This regulation, promulgated in 1980, is commonly referred to as the "one-inch rule", however, this is misleading, in that an empty container can have a residue level of 1 in. only in very limited cases involving nonflowable products.

The drum reconditioning industry, through the Reusable Industrial Packaging Association (RIPA, formerly the Association of Container Reconditioners) advocates an in-plant Responsible Container Management (RCM) policy that surpasses the minimum requirements of RCRA. This policy states that all containers (including steel drums) accepted for reconditioning must be "drip-dry," indicating that all materials that can be removed (using normal methods, like pouring) have been removed. Implementation of this policy effects a pollution prevention strategy of emissions control for a variety of pollutants, including acid gases, metals, particulates, organics, and carbon monoxide.[7] Thus emissions of these pollutants may vary depending on the efficacy of RCM implementation at the reconditioning facility.

Emissions Data Resources

Limited data are available to characterize air emissions generated during drum furnace operation. The most significant published resource available is EPA's AP-42 Fifth Edition, Section 4.8 Tank and Drum Cleaning.[8] Although most recently published in 1995, this section was authored in 1980 and is based on data from single report from 1978. These data reflect operation prior to implementation of the RCRA empty container requirements and the industry RCM policy, and thus is not considered to represent current drum furnace operating conditions. One additional publication documents tests conducted in 1990[9] and provides only limited emissions data for some pollutants. Table 2 provides a summary of the range of emission rates typical of current drum furnace operation with thermal oxidizer control. It should be noted that this summary does not include specific data for PM10. Particle-size distribution analyses of the particulate emissions from a typical drum reconditioning furnace indicate that 80–90% of the total particulate emissions are in the PM10 range.

AIR-POLLUTION CONTROL MEASURES

There are three primary control strategies used throughout the industry to limit drum furnace emissions. Pollution prevention (P2), good combustion practices (GCP), and addon emissions control equipment.

Table 2. Summary of Typical Drum Furnace Emissions[4,9]

Total particulate	0.01–0.17 g/dscf @ 12% CO_2 0.0026–0.17 lb/drum
Carbon monoxide (CO)	0–50 ppm @ 7% O_2
Nitrogen oxides (NO_x)	70–150 ppm @ 7% O_2

Pollution Prevention (P2)

Drum management practices, especially RCM, as described above constitute pollution prevention (P2) as defined by USEPA. Consistent and rigorous implementation of the RCM drum management policy is an effective strategy for minimizing emissions of acid gases and metals.

Good Combustion Practices (GCP)

GCP incorporates a variety of issues, including combustion system design, operator training, operation and maintenance procedures. Table 3 provides a summary of GCP techniques and practices applicable to drum reclamation furnace operation. This table is adapted from guidance developed by EPA for industrial combustion sources.[10]

Addon Control Equipment

As discussed above, the thermal oxidizer is commonly used for control of particulates, organics, and carbon monoxide. Performance testing of existing drum reclamation furnaces suggest that a thermal oxidizer can reliably achieve 95% thermal destruction efficiency for control of organic compounds.

Use of additional control equipment for particulates, such as a baghouse or wet scrubber, is not typical throughout the drum reconditioning industry. There are currently no federal (and no known state) emission limiting standards requiring

Table 3. Summary of Good Combustion Practices (GCP)[10]

Good Combustion Technique	Examples of Practice
Operator qualifications	Training on applicable equipment and procedures
Operator practices	Documented operating procedures, updated as required for equipment or practice change Startup, shutdown, and malfunction procedures Operating logs/recordkeeping
Maintenance knowledge	Training on applicable equipment and procedures
Maintenance practices	Documented maintenance procedures, updated as required for equipment or practice change Routinely scheduled evaluation, inspection, overhaul as appropriate for equipment Maintenance logs/recordkeeping
Furnace/thermal oxidizer stoichiometric (fuel/air) ratio	Burner and control adjustment based on visual checks Burner and control adjustment based on continuous temperature monitoring Safety interlocks
Thermal oxidizer residence time, temperature, turbulence	Residence time by design Turbulence/mixing by design Minimum combustion chamber temperature continuous monitoring

such control specifically for drum furnaces. However, in cases where a state or local general particulate emission limit is applied, the need for particulate control will depend on the specific performance of the drum furnace and thermal oxidizer and the nature of drums processed.

Use of addon control equipment for acid gases and metals is also not typical throughout the drum reconditioning industry. The need for such control, which may include dry scrubber, packed-bed, or packed tower scrubber technologies, is best evaluated on a case-by-case basis, as the types of drums processed and drum management procedures greatly affect the potential for emissions of these compounds. There are currently no federal (and no known state) emission limiting standards requiring such control specifically for drum furnaces.

Internet Resources

Information on drum reconditioning can be obtained from the Reusable Industrial Packaging Association (formerly the Association of Container Reconditioners) at *http://www.reusablepackaging.org*.

Additional resources are available from R&R Technologies, Inc., at *http://rrconsultant.com/info.html* (or see the technical resources link at the main page *http://rrconsultant.com*).

REFERENCES

1. *Air Pollution Engineering Manual*, Air & Waste Management Association, 1992.
2. Franklin Associations, *Life Cycle Inventory of Single-Trip and Multi-Trip Steel Drum Systems in the U.S., Europe, and Japan*, January 1999.
3. D. Barber, *Steel Drums*, Association of Container Reconditioners.
4. Miscellaneous Stack Test Reports, R&R Technologies, Inc.
5. *Air Pollution Engineering Manual*, 2nd ed., John A. Danielson, Ed.; AP-40, U.S. Environmental Protection Agency, Research Triangle Park, NC, 1973.
6. *Drum Furnace Design & Burner Sizing Calculations*, R&R Technologies, Inc.
7. *Empty Container Collection and Management*, Reusable Industrial Packaging Association, 1998.
8. T. R. Blackwood et al., *Source Assessment: Rail Tank Car, Tank Truck, and Drum Cleaning, State of the Art*, EPA-600/278-004g, U.S. Environmental Protection Agency, Cincinnati, OH, April 1978. (Cited in Compilation of Air Pollutant Emission Factors, 5th ed., AP-42, U.S. Environmental Protection Agency, Research Triangle Park, NC, 1995).
9. Technical Report: *Assessment of Combustion Sources that Emit Polychlorinated Dioxins and Furans, Polycyclic Aromatic Hydrocarbons, and Other Toxic Compounds*, California Air Resources Board, 1992 (test date 1990).
10. *Good Combustion Practices Guidance*, USEPA, March 31, 1998.

BIBLIOGRAPHY

1. P. Terry, *Fifty-Five Gallons: The History of Steel Drum Reconditioning*, NABADA—The Association of Container Reconditioners, 1991.
2. *Air Pollution Engineering Manual*, Air & Waste Management Association, 1992.

9

SEMICONDUCTOR MANUFACTURING

Peter F. Hess

Semiconductor manufacturing refers to the series of manufacturing processes, which produce packaged integrated circuits. A separate manufacturer assembles most packaged integrated circuits onto printed circuit boards. However, some packaged integrated circuits, such as microprocessor upgrades and memory modules, are used without preassembly onto a circuit board. Printed circuit board assembly will not be reviewed in this chapter.

Semiconductor manufacturing is unique for two reasons: the requirements for an ultraclean work environment that is necessary to produce the product and because the processes are structured into varying-sized operating units within a campuslike setting. The manufacturing process requires that the wafer and semiconductor fabrication areas be free of dirt and chemical contamination, called an "ultraclean room environment." The "fab area" structure is also unique to semiconductor fabrication facilities. For example, all of a semiconductor manufacturing facilities equipment may be grouped into a single fab area, or multiple fab areas may be established to correspond to product lines or clean room environments. Frequently the campus is located within or adjacent to a residential area. Semiconductor manufacturing uses a wide array of chemicals that generate criteria, toxic or hazardous air pollutants (HAP). The proximity of the semiconductor manufacturing facility near the populace and the chemicals used in the manufacturing process call for a careful review of the air emissions and a strong good neighbor policy.

PROCESS DESCRIPTIONS

The semiconductor manufacturing processes may be divided into three major categories:

1. Blank wafer production, where blank wafers are produced, usually at dedicated facilities that perform no semiconductor fabrication or packaging
2. Semiconductor fabrication, where integrated circuits integrated circuits are produced on the wafers
3. Assembly and packaging, where the wafers are cut into individual integrated circuits that are then mounted into a package for assembly on a printed circuit board

Semiconductor wafers may be made from a variety of substrates. The most common is silicon; others include gallium arsenide (GaAs) and indium phosphide (InP). In general, all substrates will have the same processing steps and similar emissions of typically regulated pollutants such as precursor organic compounds, hazardous air pollutants, ozone-depleting compounds, and fine particulate matter (PM_{10} and $PM_{2.5}$). However, different substrates require the use of different processing materials for etching, doping, and layering operations, resulting in different hazardous air pollutant emissions. A HAP is any compound listed in the Federal Clean Air Act, Title V, Section 112(b). Although the following discussion focuses on silicon substrate processing, the engineering guidelines and emission calculation methods are applicable to any substrate.[1]

Blank Wafer Production

Molten silicon is introduced into a mold with a seed crystal of silicon. As the molten silicon cools, it crystallizes around the seed to "grow" a single crystalline ingot. Crystal growth is not in itself a source of air contaminants. The crystals or the wafers are cleaned with solvents; these cleaning solvents are typically regulated and called miscellaneous cleaning operations. Miscellaneous cleaning operations may result in precursor organic or inorganic emissions. Individually, cleaning operations may have very low emissions, but many cleaning operations at one location may stimulate emission control requirements.

The cooled silicon ingot is shaped and sliced into round wafers that are mechanically polished ("lapped"). These steps generate particulate emissions. When the lapping operations are vented to the atmosphere, they are subject to emission limits and are typically controlled by simple abatement devices. After polishing, wafers are etched in a chemical bath to remove surface imperfections. Depending on the type and concentration of etchants used, mists and aerosols may be produced. The mists and aerosols generated by this process are classified as a particulate matter and are abated by a low-pressure drop scrubber.

Finally, each wafer is polished to a smooth finish. Adhesives are used to fix wafers in position during polishing. The amount and composition of organic compounds used in the adhesives may be subject to local or state regulations and should be reviewed for compliance with hazardous air pollutant regulations.

Miscellaneous cleaning operations during wafer manufacturing may result in organic or inorganic emissions. Abatement may be at each solvent cleaning station or may be accomplished by control of the manufacturing building.

Semiconductor Fabrication

The process used to form integrated circuits on the wafer is illustrated in Figure 1 and includes the following steps:

1. Oxidation of the wafer is the formation of an inert layer of silicon dioxide on the wafer by exposing the wafer to a hot oxygen environment.
2. Photoresist application, exposure, and development is the uniform application of solvent-based, light-sensitive resin solutions to the wafer and then processing to leave a pattern of cured photoresist on the wafer that corresponds to the circuit image, while removing the nonimage coating. Photoresist applicators may also apply non-light-sensitive coatings (spin-on-glass, antireflective coatings) and solvents (edge bead remover, hexamethyldisilazane).

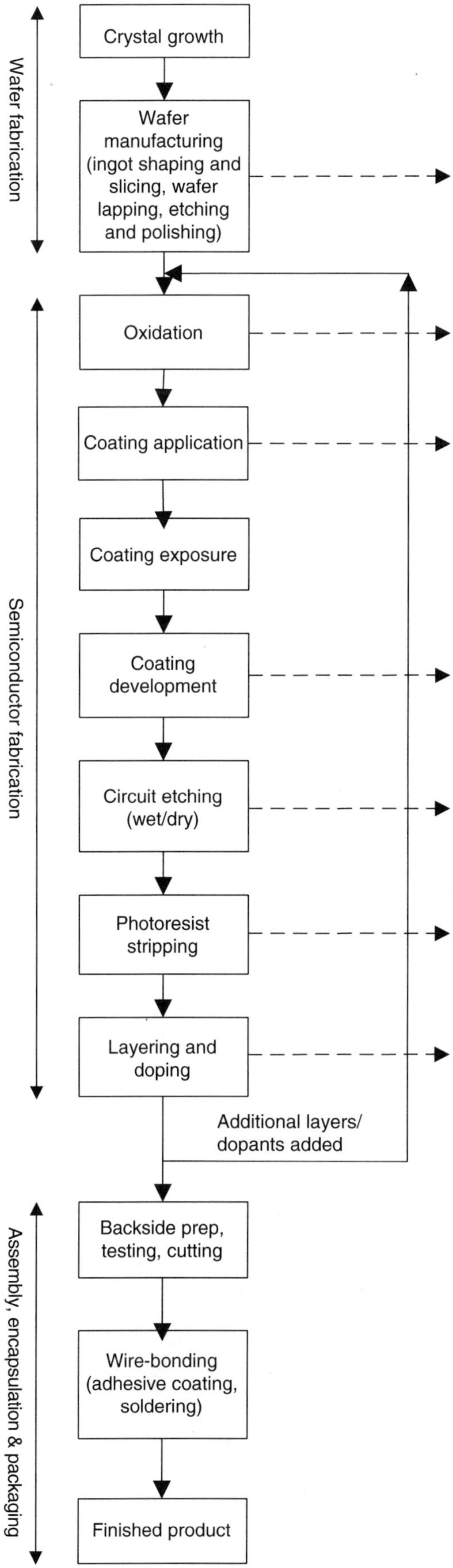

Figure 1

3. Etching, is the use of reactive gases or liquids to remove the silicon dioxide layer from the wafer surface, where it is not protected by cured photoresist, thereby exposing the underlying silicon for further processing; liquid etching is classified as a wet chemical station.
4. Photoresist stripping is the removal of cured photoresist from the wafer after it has under gone selective surface processing. The photoresist stripping solutions may be organic or inorganic depending on the composition of the underlying wafer surface, and therefore may be classified as either solvent stations or a wet chemical station, as will be described.
5. Doping (diffusion, ion implant) is the exposing of the wafer to impurities that penetrate into the exposed silicon patterns to modify selectively the electrical conductivity of the silicon, thereby producing electronic components and circuits.
6. Layering (epitaxial growth, metallization films, chemical vapor deposition) is the covering of a doped wafer with either a uniform layer of silicon to form a base for additional circuit layers or metal to form a conductive connection between the circuit layers and the external integrated circuit package.
7. Chemical mechanical polishing is the polishing of wafer surfaces to maintain wafer flatness during processing.
8. Solvent stations is the cleaning of wafers or tools by immersion in a solvent liquid or vapor or by being sprayed with a solvent liquid.
9. Wet chemical stations are where wafers or tools are cleaned or etched by immersion in an inorganic solution or by being sprayed with such a solution.
10. Wipe cleaning is the cleaning of tools and work surfaces in place or at a dedicated station.

Assembly and Packaging

After circuit production is complete, the noncircuit wafer side is mechanically milled down to improve heat dissipation and reduce the possibility of breakage ("backside prep"). After each integrated circuit on the wafer is tested, the individual integrated circuits on the wafer are cut apart. During "wire bonding" the integrated circuits is mounted in a ceramic or metal package, or in metal lead frames that will be encapsulated in plastic. Wires are soldered between connection points on the integrated circuits and leads on the mounting package or lead frame.

AIR EMISSIONS CHARACTERIZATION

Semiconductor manufacturing operations may generate many of the regulated classes of air pollutants. The emissions are generated from the numerous manufacturing steps necessary for the formation of an integrated circuit on a wafer. The type and quantity of emissions is not uniform in each manufacturing step. The pollutants generated are: nitrogen oxides (NO_x), sulfur dioxide (SO_2), precursor organic compound (POC), nonprecursor organic compound (NPOC), particulate matter smaller than 10 μm diameter (PM_{10}), and carbon monoxide (CO), as well as a variety of hazardous air pollutants and other toxic compounds, as summarized in Table 1.

Oxidation is the exposing of silicon wafers to oxygen at high temperature to form a layer of silicon dioxide on the

Table 1. Air Contaminants Generated by Specific Semiconductor Sources

Pollutant	Source
NO_x	Combustion of fuel at abatement devices
POC, NPOC[a]	Evaporation of wipe cleaning solvents
	Evaporation of solvent station solvents
	Evaporation of wafer coating solvents
	Evaporation of developer/stripper solvents
	Drying/curing of wafer coatings
	Combustion of fuel at abatement devices
PM_{10}	Aerosols from fuming/agitated wet chemical station chemicals
	Combustion of fuel at abatement devices
CO/SO_2	Combustion of fuel at abatement devices
HAP	POC, NPOC, and PM_{10} emissions may also be a HAP
	Reactive gas emissions from gaseous operations may be HAP
	Metal emissions from sputtering operations may be HAP

[a]*Note*: NPOC is a generic classification for a series of organic compounds that do not have ozone-generation potential, but may have ozone-depleting or global-warming capacity

wafer surface. This layer is selectively removed in subsequent processing steps, allowing the underlying silicon to be modified in a pattern that corresponds to electronic components and circuits. The oxygen source used in oxidation may either be steam (wet oxidation) or oxygen (dry oxidation).

Oxidation is not itself a source of air emissions and is not subject to emission limitations. However, sometimes a chlorine source (chlorine gas, hydrochloric acid, trichloroethylene, or trichloroethane) is used during oxidation to modify the oxide characteristics. Oxidation furnaces may also be cleaned with chlorine vapor. The chlorine emissions used in this operation should be evaluated by using published toxic risk screening methodologies. The California Air Pollution Control Officers Association[2] has published an excellent reference for the evaluating air toxic emissions. Standard industry practice is to abate oxidation chambers with wet or dry scrubbers.

Photoresist coating application is where photoresist, a solution of light-sensitive resin and solvent, is applied on a "spin track." Modern spin tracks are integrated tools where a variety of different materials such as photoresist, edge bead remover, spin-on-glass, may be applied. On the spin track, a fixed amount of photoresist is metered onto the wafer, which is then spun at a very high speed on a rotating element to uniformly coat the wafer surface. Photoresist may have a high precursor organic compound content, in which case the source may be subject to emission control requirements. Some photoresist solution contains less than 1% precursor organic compound by weight, which is a cost-effective substitution for abatement equipment. Negative photoresist has a higher precursor organic compound content than positive photoresist.

Part of the photoresist solution applied on the spin track is flung from the wafer into the spin track tool and recovered as waste. The coated wafer is then heated to evaporate photoresist solvent, either on the spin track tool or in a separate oven. Therefore, all the solvent in the photoresist, which remains on the wafer, is emitted, and some portion of the solvent in the excess photoresist may also be emitted depending on the design of the spin track and waste collection system. Spin-on-glass and edge bead remover is applied in the same way as photoresist. The non-light-sensitive coating called spin-on-glass acts as a protective layer that provides a good adhesive surface for subsequent layers. Edge bead remover is a solvent that is spun onto a coated wafer to remove the photoresist bead that forms on the outside edge of a coated wafer. Hexamethyldisilazane may be applied to a wafer to improve adhesion of subsequent coating layers.

Spin tracks may utilize automatic cleaning where solvent, typically isopropyl alcohol (IPA), acetone, or propylene glycol methyl ether acetate (PGMEA), is dispensed onto the spin track to prevent photoresist buildup. Manual wipe cleaning of spin tracks is also regularly performed, typically with an IPA solution.

Coating application equipment is typical candidates for control by abatement equipment due to their precursor organic emissions. State and local ozone regulations are triggered by the quantity of precursor organic compounds used in this process. Manufacturing plants can contain many banks of coating equipment. The organic compound emissions from coating equipment can be cost effectively captured and controlled to 90% efficiency. The type of control equipment used to abate these operations should be selected very carefully. A change of the type of organic compounds in the photoresist may impact the efficiency of the abatement equipment. The residual emissions should be evaluated using published toxic risk screening methodologies. Spin tracks where negative photoresist is applied should also be evaluated for compliance with air pollution regulations. The Bay Area Air Quality Management District (BAAQMD) Regulation 8, Rule 30 provides an excellent guideline for regulating semiconductor coating application equipment.[3,4]

Photoresist exposure occurs after a wafer that has been coated with photoresist and the photoresist has been cured is exposed to a light source through a template called a *mask*. The mask is a glass plate on which an image of the circuit has been produced. Exposing a wafer through the mask causes a portion of the photoresist coating, corresponding to the circuit image, to react. "Positive photoresist" will soften relative to the unexposed photoresist. "Negative photoresist" will harden relative to unexposed photoresist. Depending on which type of photoresist is used, either the image or nonimage portion of the photoresist is removed in a photoresist developer. Because photoresist solvent is evaporated during application and curing, exposure of coated wafers does not result in additional emissions. Some equipment associated with exposure, such as the masks, may require solvent cleaning, typically in a solvent station. Solvent stations are subject to state and local emission limitations.

Photoresist development is the process of removing soft photoresist on the wafer (either the image or nonimage portion depending on the type of photoresist used) to expose a portion of the silicon dioxide layer corresponding to the circuit image. Development may be performed on spin tracks, in sinks, or in enclosed spray units. Development of positive photoresist is typically performed with mild caustic solutions. Negative photoresist developers may use a precursor organic or an inorganic developer solution. The emissions from photoresist development should be evaluated using published toxic risk screening methodologies.

Developer emissions from organic developers are similar to photoresist emissions when development is performed on a spin track. When sinks or enclosed units are used, emissions are similar to those from solvent stations. Photoresist developers are candidates to be abated. The organic gasses from the developer can be cost effectively captured and controlled by a 90% efficiency abatement device.

Wet etching is the process of removing exposed silicon dioxide in the pattern created by photoresist exposure and development with a liquid, typically with an inorganic acid. Wet etchers are classified as wet chemical stations and may be exempt from air pollution regulations. However, the inorganic acid emissions may cause a negative impact at ground level or the building superstructure. Standard industry practice is to abate wet chemical stations by the use of wet scrubbers.

Dry etching is where reactive gas plasma is used in place of a wet etching solution. Dry etching provides a higher resolution than wet etching and therefore is more likely to be used as circuit elements become smaller. Dry etching generally produces less "undercutting" of the wafer substrate under photoresist, often provides more control over etching rate, and may be necessary where the etched layer is resistant to liquid etchants. Gaseous etching is generally performed with halogenated compounds such as, depending on the wafer substrate, bromine-, chlorine-, fluorine-, or iodine-based gases in a carrier gas.[4] Some of these etchants may be hazardous air pollutants. Also, plasma operations will break the halogenated compounds into related compounds. For example, use of chlorine (Cl_2) plasma with a hydrogen carrier will produce chlorine and hydrochloric acid (HCl) emissions. Standard industry practice is to abate dry etchers by the use of wet scrubbers. All dry etching process emissions should be evaluated using published toxic risk assessment methodologies.

Photoresist stripping occurs after etching where the remaining photoresist on the wafer is removed in a photoresist stripper. Photoresist strippers are classified as solvent stations if they use organic solvents, or wet chemical stations if they use acids or hydrogen peroxide solutions. Strippers have a variety of physical designs, ranging from manual or automated dip sinks to more sophisticated units where wafers are immersed in or sprayed with stripper solution. Standard industry practice is to abate strippers that use inorganic solutions by the use of wet scrubbers.

Doping is performed through diffusion by either exposing the wafer to a high-temperature environment of dopant vapor in a carrier gas (gaseous diffusion), or by covering the surface to be doped with a temporary layer of dopant oxide (nongaseous diffusion). The most common dopants include arsenic, boron, and phosphorus. Others include aluminum, antimony, beryllium, gallium, germanium, gold, magnesium, tellurium, and tin.

Dopant gases such as arsine, silane, phosphine, and diborane are highly toxic. All these gases, except arsine, are also "pyrophoric"—igniting on contact with atmospheric oxygen. Silane has the property of forming silica particles when ignited. Standard industry practice is to vent pyrophoric gases to chambers where they are ignited under controlled conditions, and then vented to wet scrubbers, and to abate nonpyrophoric gases at wet scrubbers. Due to the highly toxic nature of the contaminants, the storage and transfer of the dopant gases should be evaluated under a leak-and-release scenario using published toxic risk screening methodologies.

Ion implantation provides a more controlled doping mechanism than diffusion. In ion implantation, dopant material is ionized in a vacuum chamber, then accelerated and imbedded into the wafer with an implanter. Standard industry practice is to abate ion implanters at combustion chambers (when pyrophoric gases are used) by the use of wet scrubbers. The residual emissions should be evaluated using published toxic risk screening methodologies.

Epitaxial growth is the process of re-establishing a fresh silicon layer on a wafer using exposed silicon on the wafer surface as a seed for additional silicon crystal growth. Epitaxial growth occurs in an epitaxial reactor (or "siliconizing reactor"), where the wafer is exposed to silane and dopant gases in a high-temperature environment. Silicon layer growth is also performed with molecular beam epitaxy (MBE), in which silicon and dopants are evaporated and deposited on the wafer in a vacuum environment. Standard industry practice is to abate epitaxial growth processes at combustion chambers with wet scrubbers. The residual emissions should be evaluated using published toxic risk screening methodologies.

Sputtering is the process of impinging ionized gas atoms, typically argon, on a metal target in order to generate microscopic metal fragments that are deposited on the wafer as a thin film. Sputtered metals include titanium, platinum, gold, molybdenum, tungsten, nickel, and cobalt. Standard industry practice is to abate sputtering chambers at wet scrubbers. The residual emissions should be evaluated using published toxic risk screening methodologies.

Chemical vapor deposition (CVD) is a process in which a metal or silicon compound is vaporized and deposited onto the wafer as a thin film. Some chemical vapor deposition operations are operated within a vacuum environment. Standard industry practice is to abate exhaust gases from chemical vapor deposition chambers by wet scrubbers. The residual emissions should be evaluated using published toxic risk screening methodologies.

Chemical mechanical polishing (CMP) is a polishing step performed after layering or other coating processes in order to re-establish the wafer flatness required for subsequent processing steps. Chemical mechanical polishing is usually performed with inorganic abrasive slurries. Standard industry practice is to vent chemical mechanical polishing tool exhaust to wet scrubbers. The residual emissions should be evaluated using published toxic risk screening methodologies.

Solvent station refers to any process used to clean wafers or remove photoresist from wafers with an organic solvent solution (excluding developer solutions applied at spin tracks, which are classified as developers), as well as processes used to clean tools (excluding manual wipe cleaning). The most common organic solvents used in solvent stations are isopropyl alcohol, acetone, and *n*-methyl pyrrolidone (NMP). State and local regulations regulate solvent stations. The type of cleaning device and the capacity for holding organic solvents usually determines the degree of control. For the purpose of determining solvent station capacity, the following guidelines should be used:

For solvent sinks, the capacity is the maximum volume of liquid that the sink will hold during use. This may be less than the actual volume of the sink. Devices where liquid solvent is sprayed or where a chamber is cyclically flooded with liquid solvent are considered to have a solvent capacity equal to the liquid solvent holding capacity of the device, excluding reservoirs that hold fresh solvent or waste solvent that will not be reused.

Devices that heat solvents above their boiling points (such as solvent vapor dryers) should be considered to have a "solvent capacity" exceeding 100 gallons because of their potential to produce much higher emission levels than liquid-based stations. Therefore, the use of a vapor-based station renders all solvent stations in the same fab area subject to abatement requirements. The grouping of these operations within a fab area makes the abatement of the emissions practical and efficient. The use of combustion or adsorption techniques is a typical a means of abatement.

Solvent stations within a fab area should be evaluated as an aggregate group. Generally, if the aggregate capacity of the solvent stations within a fab area exceeds 100 gallons of precursor solvents, the emissions are abated. The use of combustion or adsorption is a typical means of abatement.

Wet chemical station refers to any process in which inorganic solutions are used to clean or etch wafers, or to clean tools. Inorganic solutions include caustic or acidic solutions as well as unheated organic solutions with less than 10 wt % unheated and heated solutions with less than 2.5 wt % ozone precursors. Some wet chemical stations, such as wet etchers, may use air-agitation of solutions to increase or control the etching rate.

Wet chemical stations are a more complex emission problem than solvent stations. Many materials used in wet chemical stations are classified as hazardous air pollutants. Some of these materials produce odorous or irritating fumes in the form of $PM_{10/2.5}$ emissions. Table 2 lists some chemicals commonly used at wet chemical stations.

Tool/wipe cleaning may occur at solvent and wet chemical stations. These stations may be exempt from permit requirements in accordance with the criteria for solvent stations and wet chemical stations. Wipe cleaning refers to manual wipe cleaning of work surfaces or tools which is performed either in place or at a dedicated cleaning station. Depending on the quantity used at a fab area, wipe cleaning may be subject to general solvent surface cleaning regulations. Abatement of manual wipe cleaning may be required if total fab area emissions exceed the state or local emission limit applicability level.

Assembly and packaging occurs after circuit production is complete and the noncircuit wafer side is mechanically milled down to improve heat dissipation and reduce the possibility of breakage ("backside prep"). After each integrated circuit on the wafer is tested, the individual integrated circuits on the wafer are cut apart. During "wire bonding" the integrated circuit is mounted in a ceramic or metal package, or in a metal lead frame that will be encapsulated in plastic. Wires are soldered between connection points on the integrated circuits and leads on the mounting package or lead frame.

Wafer processing during backside prep, testing, cutting, and wire bonding are not considered emission sources. Assembly and packaging do produce some precursor organic solvent emissions due to wipe cleaning. However, abatement of these other operations may be required if estimated emissions exceed the state or local miscellaneous solvent regulation thresholds. These emissions are considered separately from blank wafer production and semiconductor fabrication processes.

Table 2. Properties of Wet Chemical Station Chemicals

Material	Properties
Aqua regia (nitrohydrochloric acid), hydochloroic acid, nitric acid ("fuming"), sulfuric acid ("fuming," "oleum")	Fuming acids that generate irritating PM_{10} aerosols; aqua regia, hydochloric acid, hydrofluoric acid, and nitric acid are HAPs
Ammonia, ammonia hydroxide, and phosphoric acid	HAP, irritants
Hydrogen peroxide, acetic acid	Irritants

Miscellaneous operations encompass chemical handling and laboratory operations that may be found in any of the three categories of semiconductor manufacturing processes. Chemical handling includes the preparation of reagents and distribution of solvents from bulk containers. Laboratory operations, including failure analysis and quality assurance testing, will produce minor emissions, but should be reviewed for possible abatement.[5]

Calculation Methods

Emissions from semiconductor operations may be calculated by mass balance, by applying an emission factor derived from a process-specific source test, or by applying a generic emission factor.

Mass balances provide the most accurate emissions estimates and should be used whenever possible.[6] Mass balance emission calculations require more records than other methods and may not be preferred by applicants for this reason. Also, a mass balance may not be possible for processes that commingle waste streams. However, source operators should always be given the option to perform mass balance emission calculations.

Process-specific source tests are costly because even identical process equipment may have widely variable emissions at different processing rates or if different process materials are used. Therefore, accurate characterization of an entire fab with process-specific source tests may require that a large number of tests be performed and may require retests when process variables change. Also, the high ventilation airflow used in clean room environments typically result in very dilute pollutant concentrations, which may make a source test impractical. Where process-specific emission factors are used, permit conditions for the fab area should clearly limit the applicability of such factors to appropriate processes.

Generic emission factors are average factors derived from a large number of related processes.[7] Generic emission factors require the minimum amount of records, although the fact that emissions may be substantially overestimated may increase source operator costs by triggering abatement requirements. The source of generic emission factors should be considered whenever they are applied, and such factors should only be used where they will provide conservative emission estimates. It should be noted that EPA has not developed emission factors for semiconductor operations in AP-42 ("Compilation of Air Pollution Emission Factors"). No other authoritative emission factors are known to exist. When generic emission factors are used, permit conditions should identify those factors. Table 3 is a summary of recommended emission calculation methods.

It should be noted that Santa Clara County, California, where many semiconductor fabrication facilities are located, has adopted a "toxic gas ordinance"[8] that is intended to minimize the impact of accidental releases of toxic gases. The toxic gas ordinance does not apply to normal production process emissions, which the ordinance intends to be regulated directly by the air pollution agency. Therefore, it should be assumed that a municipality or other local authority would require the use of gas scrubbers or other control equipment at process gas emission points.

Table 3. Recommended Emission Calculation Methods

Operation (Pollutant)	Emission Calculation Method (in Order of Highest Preference)
Any material applied at a wafer coating applicator, including coatings, developers, dispensed cleanup solvents, processing solvents (POC, NPOC, HAP)	1. Mass balance 2. Generic emission factor of 90% emission of total solvent content of material[a]
Wipe cleaning solvent (POC, NPOC, HAP)	Mass balance
Solvent station chemicals used at solvent sinks or other solvent stations (POC, NPOC, HAP)	1. Mass balance 2. Generic emission factors of 30% of total solvent content of all chemicals[a]
Fuel combustion at abatement devices (NO_x, NPOC, CO)	Source test (emission factors from EPA AP-42; Tables 1.4-1 and 1.4-3 may be used as a preliminary estimate)
Fuel combustion at abatement devices (SO_2, PM_{10})	Emission factors from EPA AP-42, Tables 1.4-1 and 1.4-2 (emissions are too low to justify a source test)
Liquid chemicals used at wet chemical stations (PM_{10}) and (HAP)	Assume negligible where equipment is abated by a wet scrubber, otherwise estimate with simplified "spill" correlation in Ref. 8 as discussed in Ref. 9
Gases from dry etching, layering (CVD, epitaxial growth), doping (ion implant) (HAP)	Mass balance[b]
Sputtered metals (HAP)	Mass balance[b]

[a] Reference 4.

[b] Emissions of gases and metals, which are HAPs, should be calculated in the most conservative manner possible. As a first estimate, emissions should be assumed to be equal to total gross usage amounts and compared to toxic emission risk screening trigger levels. For those materials where the trigger levels are exceeded, either a risk screening should be performed to verify an insignificant risk, or else a more refined second estimate should be made that takes into account abatement efficiency, etc. This process should be repeated until an insignificant toxic risk is established or the emission estimate has been refined as much as possible.

AIR POLLUTION CONTROL MEASURES

General Limitations

The regulatory requirements for semiconductor manufacturing operations are dependent upon the location of the facility. A model for the abatement of emissions of semiconductor manufacturing operations can be found in San Francisco Bay Area Regulation 8, Rule 30. Depending on the size of the semiconductor manufacturing facility, the EPA Control Technique Guideline for Solvent Surface Cleaning Operations may apply to solvent cleaning operations. The EPA NESHAP for Halogenated Solvent Cleaning Operations may apply if perchloroethylene or trichloroethane is used as a cleaning solvent. After the year 2000 the EPA is scheduled to develop a MACT for Semiconductor Manufacturing Operations that will be similar to the San Francisco Bay Area Regulation 8, Rule 30.[10]

The state and local rules pertaining to visible emission mass particulate loading and grain loading will be applicable to the fine particulate matter and aerosol emissions from semiconductor manufacturing operations. Regulations pertaining to criteria pollutants emitted from boilers may be applicable to the manufacturing plants.

Many areas of the country have risk-based toxic material regulations. These regulations provide insight to the allowable health impacts due to the emission of toxic compounds to the atmosphere.

Photoresist Operations Using Solvent-Based Developer

If the solvent-based developer is used to process a photoresist within a specific fab area, then all exhaust gases containing precursor or nonprecursor organic compounds from both the solvent-based development operations in that fab area should be vented to an air pollution control device that captures and abates at least 90% (by weight) of the organic compounds evaporated at the application and development operations. This measure applies to developer applied by any method other than by immersion in a solvent sink. The abatement techniques generally applied to photoresist operations are adsorption or combustion.

Major semiconductor manufactures have numerous photoresist applicators and developers in their fab areas. The exhaust from the photoresist machines can be combined and vented to a single abatement device. Photoresist machines are typically enclosed and readily applicable for exhaust to an abatement device.

The inlet organic gas concentration dictates the control efficiency of the abatement device. For inlet organic gas concentrations greater than 2000 ppmv, the outlet concentration should be below 10 ppmv or abatement efficiency greater than 98.5%. When the inlet concentration is between 2000 and 200 ppmv, a control efficiency of 97% is achievable. When the inlet concentration drops below 200 ppmv, 90% control efficiency is recommended. Source testing using EPA method 25 or 25A can be used to determine the compliance of photoresist operations.[3]

Solvent Spray Station, Solvent Vapor Station, Solvent Sink, and Fab Area Wipe Cleaning

A variety of abatement strategies are available for the control of solvent emissions from semiconductor cleaning operations (see Chapter 9). Source controls such as low-volatility solvents

Table 4. Typical Scrubber Abatement Efficiency[a,b]

Hydrogen chloride	99%
Hydrogen sulfide	92%
Chlorine	86%
Hydrogen bromide	92%
Diborane	91%
Hydrogen fluoride	99%
Ammonia	99%

[a]Efficiency is based on 500 fpm face velocity, 5 ft of packing, and 5 gpm scrubber liquor recirculation rate per ft^2 of packing surface area.
[b]Acid gas injection is necessary to attain a high efficiency for an ammonia gas stream; caustic injection is necessary to attain a high efficiency for an H_2S, Cl gas stream.[12]

and chilled freeboards at degreasers are the most cost effective in reducing emissions.[11] Based upon the number and type of solvent cleaning sources in a fab area, a control device is sometimes employed by a facility.

Aerosol and Fine Particulate Scrubbers

The abatement of aerosol and fine particulate matter emissions generated during the semiconductor manufacturing process can be accomplished by a low-pressure-drop combination scrubber and mist eliminator. A scrubber with a 2–4-in. water column pressure drop is sufficient to meet most emission limitations. Scrubbers that accommodate a flow rate from 525 to 61,000 cfm, with a water recirculation rate from 5 to 585 gpm and liquor makeup rate of 0.3 to 29 gpm, are available to achieve the abatement efficiencies contained in Table 4.

The specifications for control are highly influenced by the exhaust flow rate from the fab area plus the contaminants emitted from the process device. If there are numerous particulate and gaseous emitting sources located within a large fab area, it is usually practical to vent all the sources to a single scrubber rather than having individual scrubbers for each source. Abatement equipment assists in providing a clean room atmosphere that is critical to the semiconductor manufacturing process.

REFERENCES

1. M. K. Lee and J. Elliot, Bay Area Air Quality Management District, *Permit Handbook Section 7, Chapter 4—Semiconductor Manufacturing*, May 1998.
2. California Air Pollution Control Officers Association, *CAPCOA Air Toxic Hotspots Program, Revised 1992 Risk Assessment Guidelines*, October 1993.
3. Bay Area Air Quality Management District, *Regulation 8, Rule 30, Semiconductor Manufacturing Operations*, October 1998.
4. Bay Area Air Quality Management District, *BACT/TBACT Workbook*, June 1995.
5. S. Hill, Bay Area Air Quality Management District, *Emissions from Semiconductor Manufacturing in the Bay Area*, April 1987.
6. P. Halloway and G. McQuire, *Handbook of Compound Semiconductors*, Noyes Publications, 1995.
7. T. Higgs, *Emissions Estimates for a Semiconductor Manufacturing Facility*, presented at Air & Waste Management Association Specialty Conference, New Orleans, September 1996.
8. Santa Clara County Fire Chief's Association, Toxic Gas Model Ordinance, June 29, 1997.
9. N. Sax, and R. Lewis, *Hawley's Condensed Chemical Dictionary*; eleventh edition; Van Nostrand Reinhold, 1987.
10. U.S. Environmental Protection Agency, *Preferred and Alternative Methods for Estimating Air Emissions from Semiconductor Manufacturing*, Review Draft, April 1997.
11. U.S. Environmental Protection Agency, *Estimating Releases and Waste Treatment Efficiencies for the Toxic Chemical Release and Inventory Form,* EPA 560/4-88-002, 1987, p. 6–4.
12. M. Doherty, Harrington Environmental Engineering Technical Data Sheet, 1999.

10

EVAPORATIVE EMISSIONS AND SURFACE COATINGS

Organic Liquid Storage Tanks
Mark A. Morris
Evaporative Loss Sources, Organic Solvent Cleaning Operations
Mohamed A. Serageldin, Paul Almodovar, and Susan Miller
Surface Coating
Ashok Kumar and Taramati Shenoy

ORGANIC LIQUID STORAGE TANKS

Mark A. Morris

PROCESS DESCRIPTION

Storage vessels containing organic liquids can be found in many industries, including: (1) petroleum producing and refining, (2) petrochemical and chemical manufacturing, (3) bulk storage and transfer operations, and (4) other industries consuming or producing organic liquids. Organic liquids in the petroleum industry, usually called petroleum liquids, generally are mixtures of hydrocarbons having dissimilar true vapor pressures (for example, gasoline and crude oil). Organic liquids in the chemical industry, usually called volatile organic liquids, are composed of pure chemicals or mixtures of chemicals with similar true vapor pressures (for example, benzene or a mixture of isopropyl and butyl alcohols).

The most common storage tank types are fixed roof (vertical and horizontal), external floating roof, and internal floating roof. Vertical fixed roof tanks consist of a cylindrical steel shell with a permanently affixed roof, which may vary in design from cone- or dome-shaped to flat. Horizontal fixed roof tanks are not discussed in this chapter. Fixed roof tanks are either freely vented or equipped with a pressure/vacuum vent. The latter allows the tanks to operate at a slight internal pressure or vacuum to prevent the release of vapors during very small changes in temperature, pressure, or liquid level. Of current tank designs, the fixed roof tank is the least expensive to construct and is generally considered the minimum acceptable equipment for storing organic liquids.

An external floating roof tank (EFRT) consists of an open-topped cylindrical steel shell equipped with a roof that floats on the surface of the stored liquid. The floating roof consists of a deck, fittings, and rim seal system. Floating decks that are currently in use are constructed of welded steel plates and are of two general types: pontoon and double-deck. A pontoon-type EFRT is shown in Figure 1. In an EFRT, the roof rises and falls with the liquid level in the tank. Numerous fittings pass through or are attached to floating roof decks to accommodate structural support components or allow for operational functions. Floating decks are equipped with a rim seal system, which is attached to the deck perimeter and contacts the tank wall. The purpose of the floating roof and rim seal system is to reduce evaporative loss of the stored liquid. Some annular space remains between the seal system and the tank wall. The seal system slides against the tank wall as the roof is raised and lowered.

An internal floating roof tank (IFRT) has both a permanent fixed roof and a floating roof. There are two basic types of IFRTs: tanks in which the fixed roof is supported by vertical columns within the tank, and tanks with a self-supporting fixed roof and no internal support columns. A column-supported IFRT is shown in Figure 2. The typical IFRT is freely vented by circulation vents at the top of the fixed roof. The vents minimize the possibility of organic vapor accumulation in the tank vapor space in concentrations approaching the flammable range. IFRTs without vents, or closed IFRTs, are not discussed in this chapter.

AIR EMISSIONS CHARACTERIZATION

Emissions from organic liquid storage tanks occur because of evaporation of the liquid during its storage and as a result of changes in the liquid level. The emission types and magnitudes vary with tank design and stored liquid. Emissions from fixed roof tanks consist of evaporative losses during storage (known as breathing losses or standing storage losses) that result from barometric pressure and/or ambient temperature changes, and evaporative losses during filling and emptying operations (known as working losses). Emissions from EFRTs and IFRTs consist of evaporative losses that occur during standing storage and withdrawal of liquid from the tank. Standing storage losses are a result of evaporative losses through rim seals, deck fittings, and/or deck seams. Withdrawal losses result from the evaporation of the liquid that clings to the tank wall when the floating roof is lowered.

The equations in this section are from the U.S. Environmental Protection Agency AP-42 publication, which contains storage tank descriptions and equations to estimate storage tank emissions.[1] The equations are based on emission estimation procedures developed by the American Petroleum Institute (API) Environmental Technical Advisory Group and published in API's Manual of Petroleum Measurement Standards.[2] AP-42 and the API publications on which it is based include more detailed descriptions of storage tanks and equation variables.

Fixed Roof Tanks

Total losses from fixed roof tanks are equal to the sum of the standing storage loss and working loss:

$$L_T = L_S + L_W \tag{1}$$

where L_T are total losses (lb/yr); L_S, standing storage losses (lb/yr); and L_W, working losses (lb/yr).

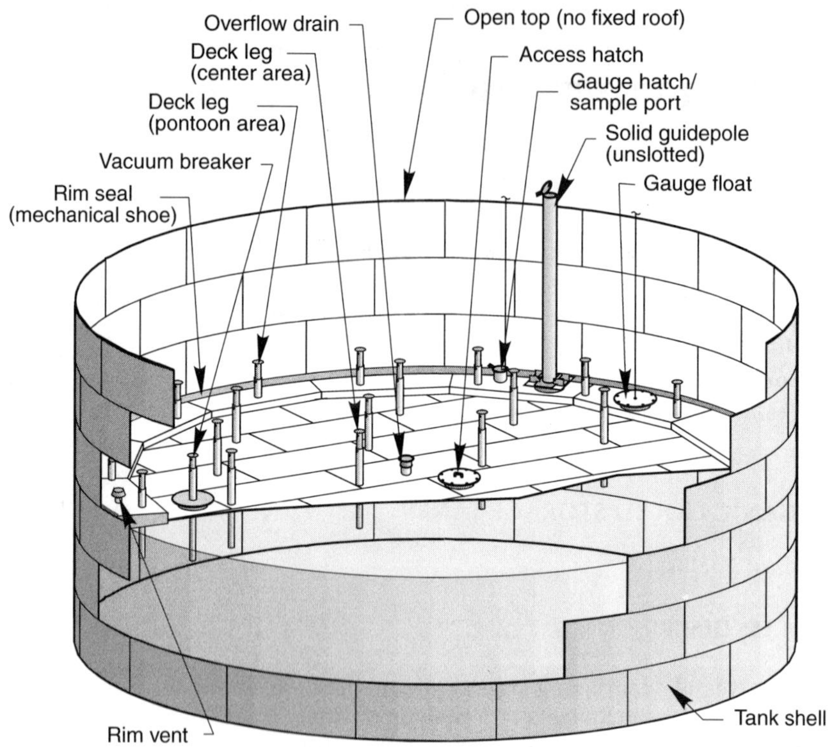

Figure 1. External Floating Roof Tank (Pontoon Type).

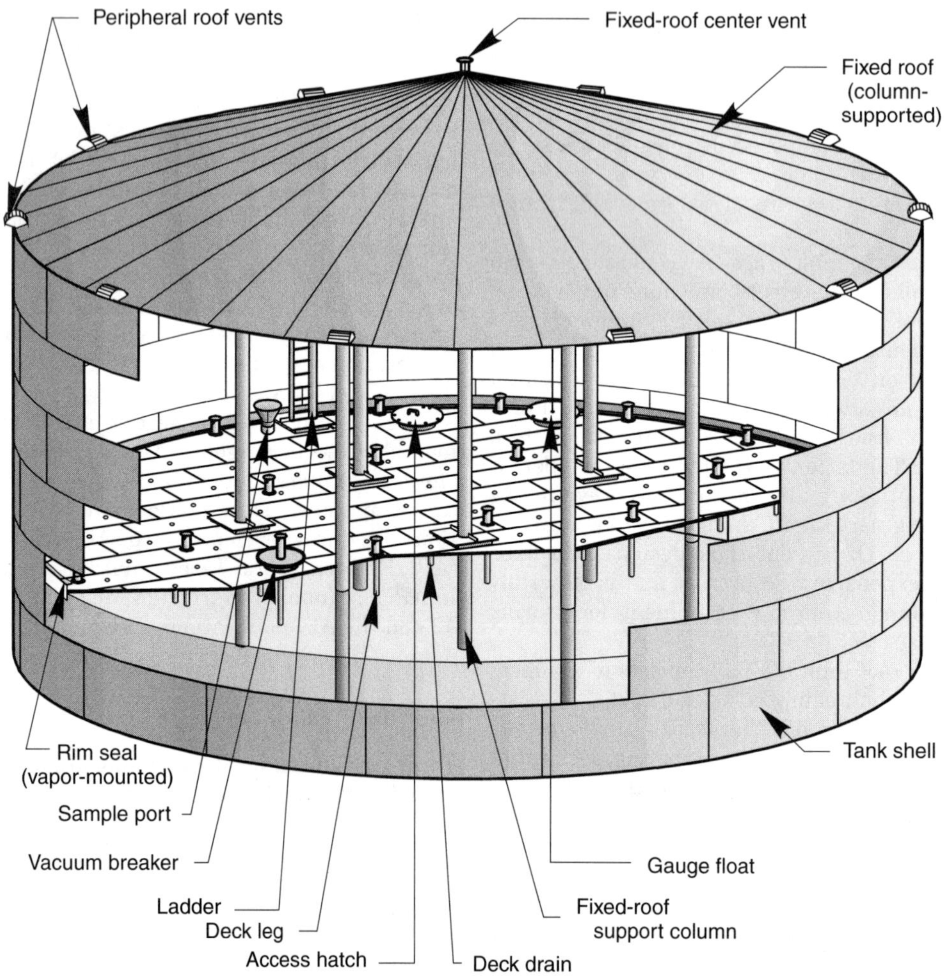

Figure 2. Internal Floating Roof Tank.

Standing Storage Loss

Fixed roof tank standing storage losses can be estimated using the following equation:

$$L_S = 365 V_V W_V K_E K_S \quad (2)$$

where V_V is the vapor space volume (ft^3); W_V, the vapor density (lb/ft^3); K_E, the vapor-space expansion factor (dimensionless); and K_S, the vented vapor saturation factor (dimensionless).

Tank Vapor Space Volume, V_V. The tank vapor-space volume is calculated using the following equation:

$$V_V = \frac{\pi}{4} D^2 (H_S - H_L - H_{RO}) \quad (3)$$

where D is the tank diameter (ft); H_S, the tank shell height (ft); H_L, the liquid height (ft); and H_{RO}, the roof outage (ft); see Note 1 for a cone roof or Note 2 for a dome roof.

Notes:

1. For a cone roof, H_{RO} is calculated as follows:

$$H_{RO} = \tfrac{1}{3} S_R R_S$$

where S_R is the tank cone roof slope (ft/ft); if unknown, a standard value of 0.0625 ft/ft is used; and R_S the tank shell radius (ft).

2. For a dome roof, H_{RO} is calculated as follows:

$$H_{RO} = H_R \left[\frac{1}{2} + \frac{1}{6} \left(\frac{H_R}{R_S} \right)^2 \right]$$

where H_R is the tank roof height (ft); and R_S the tank shell radius (ft).

Vapor Density, W_V. The density of the vapor is calculated using the following equation:

$$W_V = \frac{M_V P_{VA}}{R T_{LA}} \quad (4)$$

where M_V is the vapor molecular weight (lb/lb mole); P_{VA}, the vapor pressure at daily average liquid surface temperature (psia); R, the ideal gas constant (10.731 psia ft^3/lb mole °R); and T_{LA}, the daily average liquid surface temperature (°R); see Note 1.

Notes:

1. If T_{LA} is unknown, it can be estimated using the following equation:

$$T_{LA} = 0.44 T_{AA} + 0.56 T_B + 0.0079 \alpha I$$

where T_{AA} is the daily average ambient temperature (°R); see Note 2; T_B, the liquid bulk temperature (°R); see Note 3; α, the tank paint solar absorptance (dimensionless); see Table 1; and I, the daily total solar insolation factor (Btu/ft^2 d); see Table 2.

Table 1. Paint Solar Absorptance for Fixed Roof Tanks

		Paint Factors (α) Paint Condition	
Paint Color	Paint Shade or Type	Good	Poor
Aluminum	Specular	0.39	0.49
Aluminum	Diffuse	0.60	0.68
Gray	Light	0.54	0.63
Gray	Medium	0.68	0.74
Red	Primer	0.89	0.91
White	NA	0.17	0.34

2. T_{AA} is calculated using the following equation:

$$T_{AA} = (T_{AX} + T_{AN})/2$$

where T_{AX} is the daily maximum ambient temperature (°R); and T_{AN}, the daily minimum ambient temperature (°R).

3. T_B is estimated using the following equation:

$$T_B = T_{AA} + 6\alpha - 1$$

where T_{AA} and α are as previously defined.

Vapor-Space Expansion Factor, K_E. The vapor-space expansion factor is calculated using the following equation:

$$K_E = \frac{\Delta T_V}{T_{LA}} + \frac{\Delta P_V - \Delta P_B}{P_A - P_{VA}} \quad (5)$$

where ΔT_V is the daily vapor temperature range (°R); see Note 1; ΔP_V, the daily vapor pressure range (psi); see Note 2; ΔP_B, the breather vent pressure setting range (psi); see Note 3; P_A, the atmospheric pressure (psia); and P_{VA}, T_{LA} are as previously defined.

Notes:

1. ΔT_V is calculated using the following equation:

$$\Delta T_V = 0.72 (T_{AX} - T_{AN}) + 0.028 \alpha I$$

where T_{AX}, T_{AN}, α, and I are as previously defined.

2. ΔP_V is calculated using the following equation:

$$\Delta P_V = P_{VX} - P_{VN}$$

where P_{VX} is the vapor pressure at the daily maximum liquid surface temperature (psia); and P_{VN}, the vapor pressure at the daily minimum liquid surface temperature (psia).

3. ΔP_B is calculated using the following equation:

$$\Delta P_B = P_{BP} - P_{BV}$$

where P_{BP} is the breather vent pressure setting (psig); and P_{BV}, the breather vent vacuum setting (psig).

Table 2. Solar Insolation Factor (Btu/ft^2 d) for Selected U.S. Locations

Location	Monthly Averages												Annual
	Jan.	Feb.	Mar.	Apr.	May	June	July	Aug.	Sept.	Oct.	Nov.	Dec.	Average
Birmingham, AL	707	967	1296	1674	1857	1919	1810	1724	1455	1211	858	661	1345
Montgomery, AL	752	1013	1341	1729	1897	1972	1841	1746	1468	1262	915	719	1388
Homer, AK	122	334	759	1248	1583	1751	1598	1189	791	437	175	64	838
Phoenix, AZ	1021	1374	1814	2355	2677	2739	2487	2293	2015	1577	1151	932	1869
Tucson, AZ	1099	1432	1864	2363	2671	2730	2341	2183	1979	1602	1208	996	1872
Fort Smith, AR	744	999	1312	1616	1912	2089	2065	1877	1502	1201	851	682	1404
Little Rock, AR	731	1003	1313	1611	1929	2107	2032	1861	1518	1228	847	674	1404
Bakersfield, CA	766	1102	1595	2095	2509	2749	2684	2421	1992	1458	942	677	1749
Long Beach, CA	928	1215	1610	1938	2065	2140	2300	2100	1701	1326	1004	847	1598
Los Angeles, CA	926	1214	1619	1951	2060	2119	2308	2080	1681	1317	1004	849	1594
Sacramento, CA	597	939	1458	2004	2435	2684	2688	2368	1907	1315	782	538	1643
San Francisco, CA	708	1009	1455	1920	2226	2377	2392	2117	1742	1226	821	642	1553
Santa Maria, CA	854	1141	1582	1921	2141	2349	2341	2106	1730	1353	974	804	1608
Denver, CO	840	1127	1530	1879	2135	2351	2273	2044	1727	1301	884	732	1568
Grand Junction, CO	791	1119	1554	1986	2380	2599	2465	2182	1834	1345	918	731	1659
Wilmington, DE	571	827	1149	1480	1710	1883	1823	1615	1318	984	645	489	1208
Atlanta, GA	718	969	1304	1686	1854	1914	1812	1709	1422	1200	883	674	1345
Savannah, GA	795	1044	1399	1761	1852	1844	1784	1621	1364	1217	941	754	1365
Honolulu, HI	1180	1396	1622	1796	1949	2004	2002	1967	1810	1540	1266	1133	1639
Chicago, IL	507	760	1107	1459	1789	2007	1944	1719	1354	969	566	402	1215
Springfield, IL	585	861	1143	1515	1866	2097	2058	1806	1454	1068	677	490	1302
Indianapolis, IN	496	747	1037	1398	1638	1868	1806	1644	1324	977	579	417	1165
Wichita, KS	784	1058	1406	1783	2036	2264	2239	2032	1616	1250	871	690	1502
Louisville, KY	546	789	1102	1467	1720	1904	1838	1680	1361	1042	653	488	1216
Baton Rouge, LA	785	1054	1379	1681	1871	1926	1746	1677	1464	1301	920	737	1379
Lake Charles, LA	728	1010	1313	1570	1849	1970	1788	1657	1485	1381	917	706	1365
New Orleans, LA	835	1112	1415	1780	1968	2004	1814	1717	1514	1335	973	779	1437
Detroit, MI	417	680	1000	1399	1716	1866	1835	1576	1253	876	478	344	1120
Grand Rapids, MI	370	648	1014	1412	1755	1957	1914	1676	1262	858	446	311	1135
Minneapolis, MN	464	764	1104	1442	1737	1928	1970	1687	1255	860	480	353	1170

Jackson, MS	754	1026	1369	1708	1941	2024	1909	1781	1509	1271	902	709	1409
Billings, MT	486	763	1190	1526	1913	2174	2384	2022	1470	987	561	421	1325
Las Vegas, NV	978	1340	1824	2319	2646	2778	2588	2355	2037	1540	1086	881	1864
Newark, NJ	552	793	1109	1449	1687	1795	1760	1565	1273	951	596	454	1165
Roswell, NM	1047	1373	1807	2218	2459	2610	2441	2242	1913	1527	1131	952	1810
Buffalo, NY	349	546	889	1315	1597	1804	1776	1513	1152	784	403	283	1034
New York, NY	548	795	1118	1457	1690	1802	1784	1583	1280	951	593	457	1171
Cleveland, OH	388	601	922	1350	1681	1843	1828	1583	1240	867	466	318	1091
Columbus, OH	459	677	980	1353	1647	1813	1755	1641	1282	945	538	387	1123
Toledo, OH	435	680	997	1384	1717	1878	1849	1616	1276	911	498	355	1133
Oklahoma City, OK	801	1055	1400	1725	1918	2144	2128	1950	1554	1233	901	725	1461
Tulsa, OK	732	978	1306	1603	1822	2021	2031	1865	1473	1164	827	659	1373
Astoria, OR	315	545	866	1253	1608	1626	1746	1499	1183	713	387	261	1000
Portland, OR	310	554	895	1308	1663	1773	2037	1674	1217	724	388	260	1067
Philadelphia, PA	555	795	1108	1434	1660	1811	1758	1575	1281	959	619	470	1169
Pittsburgh, PA	424	625	943	1317	1602	1762	1689	1510	1209	895	505	347	1069
Providence, RI	506	739	1032	1374	1655	1776	1695	1499	1209	907	538	419	1112
Columbia, SC	762	1021	1355	1747	1895	1947	1842	1703	1439	1211	921	722	1380
Sioux Falls, SD	533	802	1152	1543	1894	2100	2150	1845	1410	1005	608	441	1290
Memphis, TN	683	945	1278	1639	1885	2045	1972	1824	1471	1205	817	629	1366
Amarillo, TX	960	1244	1631	2019	2212	2393	2281	2103	1761	1404	1033	872	1659
Corpus Christi, TX	898	1147	1430	1642	1866	2094	2186	1991	1687	1416	1043	845	1521
Dallas, TX	822	1071	1422	1627	1889	2135	2122	1950	1587	1276	936	780	1468
Houston, TX	772	1034	1297	1522	1775	1898	1828	1686	1471	1276	924	730	1351
Midland-Odessa, TX	1081	1383	1839	2192	2430	2562	2389	2210	1844	1522	1176	1000	1802
Salt Lake City, UT	639	989	1454	1894	2362	2561	2590	2254	1843	1293	788	570	1603
Richmond, VA	632	877	1210	1566	1762	1872	1774	1601	1348	1033	733	567	1248
Seattle, WA	262	495	849	1294	1714	1802	2248	1616	1148	656	337	211	1053
Charleston, WV	498	707	1010	1356	1639	1776	1683	1514	1272	972	613	440	1123
Huntington, WV	526	757	1067	1448	1710	1844	1769	1580	1306	1004	638	467	1176
Cheyenne, WY	766	1068	1433	1771	1995	2258	2230	1966	1667	1242	823	671	1491

If specific information on the breather vent pressure setting and vacuum setting is not available, assume 0.03 psig for P_{BP} and −0.03 psig for P_{BV} as typical values. If the fixed roof tank is of bolted or riveted construction in which the roof or shell plates are not vapor tight, assume that $\Delta P_B = 0$, even if a breather vent is used.

Vented Vapor Saturation Factor, K_S. The vented vapor saturation factor is calculated using the following equation:

$$K_S = \frac{1}{1 + 0.053 P_{VA}(H_S - H_L + H_{RO})} \quad (6)$$

where P_{VA}, H_S, H_L, and H_{RO} are as previously defined.

Working Loss

The working loss can be estimated using the following equation:

$$L_W = 0.0010 M_V P_{VA} Q K_N K_P \quad (7)$$

where Q is the annual net throughput [tank capacity (bbl) times annual turnover rate] (bbl/yr); K_N, the turnover factor (dimensionless); see Note 1; K_P, the working loss product factor, (dimensionless); $K_p = 0.75$ for crude oils; $K_p = 1$ for all other organic liquids; and M_V and P_{VA} are as previously defined.

Notes:

1. For turnovers >36, $K_N = (180 + N)/6N$; for turnovers ≤36, $K_N = 1$

$$N = \frac{7.149Q}{D^2 H_{LX}}$$

where N is the number of turnovers per year (dimensionless); H_{LX}, the maximum liquid height (ft); and Q and D are as previously defined.

Floating Roof Tanks

Total floating roof tank losses are the sum of rim seal, withdrawal, deck fitting, and deck seam losses:

$$L_T = L_R + L_{WD} + L_F + L_D \quad (8)$$

where L_T is the total loss (lb/yr); L_R, the rim seal loss (lb/yr); L_{WD}, the withdrawal loss (lb/yr); L_F, the deck fitting loss (lb/yr); and L_D, the deck seam loss (internal floating roof tanks only) (lb/yr).

Rim Seal Loss

Rim seal losses from floating roof tanks can be estimated using the following equation:

$$L_R = (K_{Ra} + K_{Rb} V^n) D P^* M_V K_C \quad (9)$$

where K_{Ra} is the zero-wind-speed rim seal loss factor (lb mole/ft yr); see Table 3; K_{Rb}, the wind-speed-dependent rim

Table 3. Rim Seal Loss Factors, K_{Ra}, K_{Rb}, and n, for Floating Roof Tanks

	Average-Fitting Seals		
Tank Construction and Rim Seal System	K_{Ra} (lb mole/ft yr)	K_{Rb} [lb mole/(mph)^n ft yr]	n (dimensionless)
	Welded Tanks		
Mechanical-shoe seal			
Primary only[a]	5.8	0.3	2.1
Shoe-mounted secondary	1.6	0.3	1.6
Rim-mounted secondary	0.6	0.4	1.0
Liquid-mounted seal			
Primary only	1.6	0.3	1.5
Weather shield	0.7	0.3	1.2
Rim-mounted secondary	0.3	0.6	0.3
Vapor-mounted seal			
Primary only	6.7[b]	0.2	3.0
Weather shield	3.3	0.1	3.0
Rim-mounted secondary	2.2	0.003	4.3
	Riveted Tanks		
Mechanical-shoe seal			
Primary only	10.8	0.4	2.0
Shoe-mounted secondary	9.2	0.2	1.9
Rim-mounted secondary	1.1	0.3	1.5

Note: The rim-seal loss factors K_{Ra}, K_{Rb}, and n may only be used for wind speeds below 15 miles per hour.

[a] If no specific information is available, a welded tank with an average-fitting mechanical-shoe primary seal can be used to represent the most common or typical construction and rim-seal system in use for external and domed external floating roof tanks.

[b] If no specific information is available, this value can be assumed to represent the most common or typical rim-seal system currently in use for internal floating roof tanks.

seal loss factor [lb mole/(mph)^n ft yr]; see Table 3; V, the average ambient wind speed at tank site (mph); n, the seal-related wind speed exponent (dimensionless); see Table 3; P^*, the vapor-pressure function (dimensionless); see Note 1

$$P^* = \frac{P_{VA}/P_A}{\{1 + [1 - (P_{VA}/P_A)]^{0.5}\}^2}$$

K_C, the product factor (dimensionless); $K_C = 0.4$ for crude oils; $K_C = 1$ for all other organic liquids; and D, M_V, P_{VA}, and P_A are as previously defined.

Notes:

1. The API recommends using the stock liquid temperature instead of the liquid surface temperature to calculate P_{VA} for use in the vapor-pressure function. If the stock liquid temperature is unknown, API recommends the following equations to estimate the stock temperature:

Tank Color	Average Annual Stock Temperature, T_s(°F)
White	$T_{AA} + 0^a$
Aluminum	$T_{AA} + 2.5$
Gray	$T_{AA} + 3.5$
Black	$T_{AA} + 5.0$

[a] T_{AA} is the average annual ambient temperature in degrees Fahrenheit.

Withdrawal Loss

The withdrawal losses from floating roof storage tanks can be estimated using the following equation:

$$L_{WD} = \frac{(0.943)QCW_L}{D}\left(1 + \frac{N_C F_C}{D}\right) \quad (10)$$

where C is the shell clingage factor (bbl/1000 ft^2); see Table 4; W_L, the average organic liquid density (lb/gal); N_C, the number of fixed roof support columns (dimensionless); see Note 1; F_C, the effective column diameter (ft) [column perimeter (ft)/π]; see Note 2; and Q and D are as previously defined.

Notes:

1. For a self-supporting fixed roof or an external floating roof tank, $N_C = 0$. For a column-supported fixed roof, use tank-specific information or see Table 5.
2. Use tank-specific effective column diameter or:

$F_C = 1.1$ for 9×7-in. builtup columns

$F_C = 0.7$ for 8-in.-diameter pipe columns

$F_C = 1.0$ if column construction details are not known

Table 4. Average Clingage Factors, C[a] (bbl/10^3 ft^2)

Product Stored	Shell Condition		
	Light Rust	Dense Rust	Gunite Lining
Gasoline	0.0015	0.0075	0.15
Single-component stocks	0.0015	0.0075	0.15
Crude oil	0.0060	0.030	0.60

[a] If no specific information is available, the values in this table can be assumed to represent the most common or typical condition of tanks currently in use.

Table 5. Typical Number of Columns as a Function of Tank Diameter for Internal Floating Roof Tanks With Column-Supported Fixed Roofs[a]

Tank Diameter Range D, (ft)	Typical Number of Columns, N_C
$0 < D \leq 85$	1
$85 < D \leq 100$	6
$100 < D \leq 120$	7
$120 < D \leq 135$	8
$135 < D \leq 150$	9
$150 < D \leq 170$	16
$170 < D \leq 190$	19
$190 < D \leq 220$	22
$220 < D \leq 235$	31
$235 < D \leq 270$	37
$270 < D \leq 275$	43
$275 < D \leq 290$	49
$290 < D \leq 330$	61
$330 < D \leq 360$	71
$360 < D \leq 400$	81

[a] This table was derived from a survey of users and manufacturers. The actual number of columns in a particular tank may vary greatly with age, fixed roof style, loading specifications, and manufacturing prerogatives. Data in this table should not be used when actual tank data are available.

Deck Fitting Loss

Deck fitting losses from floating roof tanks can be estimated by the following equation:

$$L_F = F_F P^* M_V K_C \quad (11)$$

where F_F is the total deck fitting loss factor (lb mole/yr); see Note; and P^*, M_V, and K_C are as previously defined.

Note:

The total deck fitting loss factor is given by:

$$F_F = [(N_{F_1} K_{F_1}) + (N_{F_2} K_{F_2}) + \cdots + (N_{F_{n_f}} K_{F_{n_f}})]$$

where N_{F_i} is the number of deck fittings of a particular type ($i = 0, 1, 2, \ldots, n_f$) (dimensionless); see Tables 6–9; K_{F_i}, the deck fitting loss factor for a particular type of fitting (lb mole/yr) = $K_{Fa_i} + K_{Fb_i}(K_V V)^{m_i}$, where K_{Fa_i} is the zero-wind-speed loss factor for a particular type of fitting (lb mole/yr); see Table 6; K_{Fb_i}, the wind-speed-dependent loss factor for a particular type of fitting [lb mole/(mph)^m yr]; see Table 6; K_V, the fitting wind-speed correction factor (dimensionless); $K_V = 0.7$ for external floating roof tanks; $K_V = 0$ for internal floating roof tanks; V, the average ambient wind speed (mph); m_i, the wind-dependent exponent for a particular type of deck fitting (dimensionless); see Table 6; and $i = 1, 2, \ldots, n_f$ (dimensionless); and n_f, the total number of different types of fittings (dimensionless).

Deck Seam Loss

Neither welded deck internal floating roof tanks nor external floating roof tanks have deck seam losses. Internal floating

Table 6. Deck-Fitting Loss Factors, K_{Fa}, K_{Fb}, and m, and Typical Number of Deck Fittings, N_F

Fitting Type and Construction Details	Loss Factors K_{Fa} (lb mole/yr)	K_{Fb} (lb mole/(mph)^m yr)	m (dimensionless)	Typical Number of Fittings, N_F
Access hatch (24-in. diameter well)				1
Bolted cover, gasketed[a]	1.6	0	0	
Unbolted cover, ungasketed	36[b]	5.9	1.2	
Unbolted cover, gasketed	31	5.2	1.3	
Fixed roof support column well[c]				N_C (Table 5)
Round pipe, ungasketed sliding cover	31			
Round pipe, gasketed sliding cover	25			
Round pipe, flexible fabric sleeve seal	10			
Builtup column, ungasketed sliding cover	47[b]			
Builtup column, gasketed sliding cover	33			
Unslotted guide-pole and well (8-in.-diameter unslotted pole, 21-in.-diameter well)				1
Ungasketed sliding cover[a]	31	150	1.4	
Ungasketed sliding cover w/pole sleeve	25	2.2	2.1	
Gasketed sliding cover	25	13	2.2	
Gasketed sliding cover w/pole wiper	14	3.7	0.78	
Gasketed sliding cover w/pole sleeve	8.6	12	0.81	
Slotted guide-pole/sample well (8-in.-diameter slotted pole, 21-in.-diameter well)				*d*
Ungasketed or gasketed sliding cover	43	270	1.4	
Ungasketed or gasketed sliding cover, with float[e]	31	36	2.0	
Gasketed sliding cover, with pole wiper	41	48	1.4	
Gasketed sliding cover, with pole sleeve	11	46	1.4	
Gasketed sliding cover, with pole sleeve and pole wiper	8.3	4.4	1.6	
Gasketed sliding cover, with float and pole wiper[e]	21	7.9	1.8	
Gasketed sliding cover, with float, pole sleeve, and pole wiper[f]	11	9.9	0.89	
Gauge-float well (automatic gauge)				1
Unbolted cover, ungasketed[a]	14[b]	5.4	1.1	
Unbolted cover, gasketed	4.3	17	0.38	
Bolted cover, gasketed	2.8	0	0	
Gauge-hatch/sample port				1
Weighted mechanical actuation, gasketed[a]	0.47	0.02	0.97	
Weighted mechanical actuation, ungasketed	2.3	0	0	
Slit fabric seal, 10% open area	12[b]			
Vacuum breaker				N_{vb} (Table 7)[g]
Weighted mechanical actuation, ungasketed	7.8	0.01	4.0	
Weighted mechanical actuation, gasketed[a]	6.2[b]	1.2	0.94	
Deck drain (3-in. diameter)				N_d (Table 7)
Open	1.5	0.21	1.7	
90% closed[a]	1.8	0.14	1.1	
Stub drain (1-in. diameter)[h]	1.2[b]			N_d (Table 9)
Deck leg (3-in. diameter)				N_l (Tables 8, and 9)
Adjustable, internal floating deck[b]	7.9			
Adjustable, pontoon area ungasketed[a]	2.0	0.37	0.91	
Adjustable, pontoon area gasketed	1.3	0.08	0.65	
Adjustable, pontoon area sock	1.2	0.14	0.65	
Adjustable, center area ungasketed[a]	0.82	0.53	0.14	
Adjustable, center area gasketed[i]	0.53	0.11	0.13	
Adjustable, center area sock[i]	0.49	0.16	0.14	

Table 6. *(Continued)*

Fitting Type and Construction Details	Loss Factors K_{Fa} (lb mole/yr)	K_{Fb} (lb mole/(mph)^m yr)	m (dimensionless)	Typical Number of Fittings, N_F
Adjustable, double-deck roofs	0.82	0.53	0.14	
Fixed	0	0	0	
Rim vent[j]				1
Weighted mechanical actuation, ungasketed	0.68	1.8	1.0	
Weighted mechanical actuation, gasketed[a]	0.71	0.10	1.0	
Ladder well[c]				1[c]
Sliding cover, ungasketed	76[b]			
Sliding cover, gasketed	56			

Note: The deck-fitting loss factors, K_{Fa}, K_{Fb}, and m, may only be used for wind speeds below 15 miles per hour.
[a]If no specific information is available, this value can be assumed to represent the most common or typical deck fitting currently in use for external and domed external floating roof tanks.
[b]If no specific information is available, this value can be assumed to represent the most common or typical deck fitting currently in use for internal floating roof tanks.
[c]Column wells and ladder wells are not typically used with self-supported fixed roofs.
[d]A slotted guide-pole/sample well is an optional fitting and is not typically used.
[e]Tests were conducted with floats positioned with the float wiper at and 1 in. above the sliding cover. The user is cautioned against applying these factors to floats that are positioned with the wiper or top of the float below the sliding cover ("short floats"). The emission factor for such a float is expected to be between the factors for a guidepole without a float and with a float, depending upon the position of the float top and/or wiper within the guidepole.
[f]Tests were conducted with floats positioned with the float wiper at varying heights with respect to the sliding cover. This fitting configuration also includes a pole sleeve that restricts the airflow from the well vapor space into the slotted guidepole. Consequently, the float position within the guidepole (at, above, or below the sliding cover) is not expected to significantly affect emission levels for this fitting configuration, since the function of the pole sleeve is to restrict the flow of vapor from the vapor space below the deck into the guidepole.
[g]$N_{vb} = 1$ for internal floating roof tanks.
[h]Stub drains are not used on welded contact internal floating decks.
[i]These loss factors were derived using the results from pontoon-area deck legs with gaskets and socks.
[j]Rim vents are used only with mechanical-shoe primary seals.

Table 7. External Floating Roof Tanks: Typical Number of Vacuum Breakers, N_{vb}, and Deck Drains, N_d^a

Tank Diameter D (feet)[b]	Number of Vacuum Breakers, N_{vb}: Pontoon Roof	Double-Deck Roof	Number of Deck drains, N_d
50	1	1	1
100	1	1	1
150	2	2	2
200	3	2	3
250	4	3	5
300	5	3	7
350	6	4	ND
400	7	4	ND

[a]This table was derived from a survey of users and manufacturers. The actual number of vacuum breakers may vary greatly depending on throughput and manufacturing prerogatives. The actual number of deck drains may also vary greatly depending on the design rainfall and manufacturing prerogatives. For tanks more than 350 feet in diameter, actual tank data or the manufacturer's recommendations may be needed for the number of deck drains. This table should not be used when actual tank data are available. ND, no data.
[b]If the actual diameter is between the diameters listed, the closest diameter listed should be used. If the actual diameter is midway between the diameters listed, the next larger diameter should be used.

roof tanks with bolted decks may have deck seam losses. Deck seam loss can be estimated using the following equation:

$$L_D = K_D S_D D^2 P^* M_V K_C \qquad (12)$$

where K_D is the deck seam loss per unit seam length factor (lb mole/ft yr); $K_D = 0.0$ for welded decks; $K_D = 0.14$ for bolted decks; S_D, the deck seam length factor (ft/ft^2), $= L_{seam}/A_{deck}$; if unknown, use Table 10, where L_{seam} is the total length of deck seams (ft); A_{deck}, the area of deck (ft^2), $= \pi D^2/4$; and D, P^*, M_V, and K_C are as previously defined.

AIR POLLUTION CONTROL MEASURES

The emission estimation procedures in the previous section can be used to evaluate how tank type and equipment affects storage tank emissions. Floating roof technology is usually the most economical control technology for storage tanks. The

Table 8. External Floating Roof Tanks: Typical Number of Roof Legs, N_l^a

Tank Diameter, D (ft)[b]	Pontoon Roof		Number of Legs on Double-Deck Roof
	Number of Pontoon Legs	Number of Center Legs	
30	4	2	6
40	4	4	7
50	6	6	8
60	9	7	10
70	13	9	13
80	15	10	16
90	16	12	20
100	17	16	25
110	18	20	29
120	19	24	34
130	20	28	40
140	21	33	46
150	23	38	52
160	26	42	58
170	27	49	66
180	28	56	74
190	29	62	82
200	30	69	90
210	31	77	98
220	32	83	107
230	33	92	115
240	34	101	127
250	35	109	138
260	36	118	149
270	36	128	162
280	37	138	173
290	38	148	186
300	38	156	200
310	39	168	213
320	39	179	226
330	40	190	240
340	41	202	255
350	42	213	270
360	44	226	285
370	45	238	300
380	46	252	315
390	47	266	330
400	48	281	345

[a]This table was derived from a survey of users and manufacturers. The actual number of roof legs may vary greatly depending on age, style of floating roof, loading specifications, and manufacturing prerogatives. This table should not be used when actual tank data are available.
[b]If the actual diameter is between the diameters listed, the closest diameter listed should be used. If the actual diameter is midway between the diameters listed, the next larger diameter should be used.

Table 9. Internal Floating Roof Tanks: Typical Number of Deck Legs, N_l, and Stub Drains, N_d

Deck fitting type	Typical Number of Fittings, N_F
Deck leg or hanger well[a]	$\left(5 + \frac{D}{10} + \frac{D^2}{600}\right)$
Stub drain (1-in. diameter)[a,b]	$\left(\frac{D^2}{125}\right)$

[a]D, is the tank diameter (ft).
[b]Not used on welded contact internal floating decks.

major equipment options that affect emissions from floating roofs include the rim seal system and the type of deck fittings. The most effective rim seal system is a liquid-mounted primary seal with a rim-mounted secondary seal. Deck fitting losses can be controlled by gasketing or sealing the openings in the deck, and by substituting lower-emitting fitting types.

Pressure vessels were not discussed in the previous sections, but such vessels are a viable control option if the pressure setting is high enough; however, it is a costly method of control.

The use of addon vapor recovery or destruction techniques, such as refrigerated condensation and incineration, is an alternative method for emission control. For example, vapor

Table 10. Deck Seam Length Factors (S_D) for Typical Deck Constructions for Internal Floating Roof Tanks[a]

Deck Construction	Typical Deck Seam Length Factor, S_D (ft/ft^2)
Continuous sheet construction[b]	
5 ft wide	0.20[c]
6 ft wide	0.17
7 ft wide	0.14
Panel construction[d]	
5 × 7.5 ft rectangular	0.33
5 × 12 ft rectangular	0.28

[a]Deck seam loss applies to bolted decks only.
[b]$S_D = 1/W$, where W is the sheet width (ft).
[c]If no specific information is available, this value can be assumed to represent the most common bolted decks currently in use.
[d]$S_D = (L + W)/LW$, where W, is the panel width (ft) and L, the panel length (ft).

recovery may be used for chlorinated solvents because they cause corrosion of some floating roof materials.

REFERENCES

1. *Compilation of Air Pollutant Emission Factors*, AP-42, 5th ed., U.S. Environmental Protection Agency, Office of Air Quality Planning and Standards, Research Triangle Park, NC, 1995.
2. *Manual of Petroleum Measurement Standards*, 1st ed., American Petroleum Institute, Washington, D.C., April 1997.

EVAPORATIVE LOSS SOURCES, ORGANIC SOLVENT CLEANING OPERATIONS*

Mohamed A. Serageldin†, Paul Almodovar, and Susan Miller

Organic solvents, solvent blends, or their vapors are used by many industries to remove water-insoluble soils such as grease, oils, waxes, carbon deposits, fluxes, and tars from metal, plastic, fiberglass, printed circuit boards, and other surfaces. Typically, organic solvent cleaning is performed prior to such operations as painting, plating, inspection, repair, assembly, heat treatment, and machining. A portion of all the used solvent evaporates during cleaning. Because organic cleaning solvents are used in large amounts, the U.S. Environmental Protection Agency (EPA) and many states have enacted regulations and developed guideline documents to help reduce emissions of pollutants such as volatile organic compounds (VOCs) and hazardous air pollutants (HAPs) from their use in industrial operations.

This chapter provides an overview of the different cleaning operations that constitute the universe of cleaning. It will also emphasize the importance of tracking and accounting solvent usage at the operational level as part of a management system for setting targets and meeting objectives. The approach described below was shared by the U.S. EPA with the public in 1992 before publication of a final document in 1994.[1] Only control techniques for parts cleaners, which are also referred to as cleaning machines (degreasers) will be detailed. A good overall discussion of spray gun cleaning and spray booth cleaning including control aspects are given elsewhere.[1,2] The last section of this chapter will provide general recommendations for reducing emissions, waste, and risk from the use of organic solvents in industry. The material accounting and management approach discussed below has general applicability in a manufacturing process.[4]

* The views expressed in this chapter are those of the authors and do not necessarily reflect those of their affiliation.
† To whom communications should be directed.

CLEANING TERMINOLOGY[1,3]

A basic requirement for an inventory system that identifies solvent usage and emissions accurately is to know where the solvent is being used and what needs to be cleaned and to have a formalized methodology for estimating emissions. The term *unit operation* is used for an industrial operation, classified or grouped according to its function in an operating environment. It will be used to refer to a spraybooth, a parts (machine) cleaner, and a gun cleaner or the gun itself when a gun cleaner is not used. The term *cleaning activity* is used for actions such as wiping, flushing, or brushing that are performed during cleaning of a unit operation. For characterizing the emissions from cleaning the term *unit operation system* (UOS) is used to describe the ensemble of equipment and cleaning tools around which a material balance is performed. Thus the unit operation system defines the boundary of the "source" at the operational level for determining emissions, waste, cost of cleaning, level of risk, or for determining' cost-effectiveness of an alternative action.

The *work practice* measure is reserved for specific human activities within industry that lead to reduction in pollution, level of risk, or costs. The activities include increased operational training, management directives, segregation of the waste solvent, and practices that lead to a reduction in cleaning frequency. It does not include the use of specialized equipment, such as solvent dispensers.

EXAMPLE OF UNIT OPERATION SYSTEMS

Three subcategories of parts cleaning unit operation systems are illustrated to explain the convention for reporting emission and waste information in the figures. Figure 1 includes a storage container or tank for virgin or clean solvent, a batch parts cleaner, a waste solvent container, and addon control unit operation. F is the makeup feed to the virgin solvent tank to compensate for the volume of solvent transferred to the parts cleaner. V_1 is the evaporative emissions from the virgin solvent, which is not necessarily dedicated tank to this operation.

The broken lines indicate that the units are not connected by pipes, meaning that the volumes of clean and waste solvent are transferred manually. Fugitive emissions are likely to occur during these steps; however, they would be counted as part of the other emission points: V_2, from the parts cleaner; and V_3 from the waste solvent:

$$\text{emissions} = V_1 + V_2 + V_3 = F - L \tag{1}$$

In Figure 1, the UOS boundary for the material balance is drawn around the parts cleaner and does not include the impacts from the clean solvent or waste solvent containers. We are adding clean, L', solvent or removing used solvent, L, from

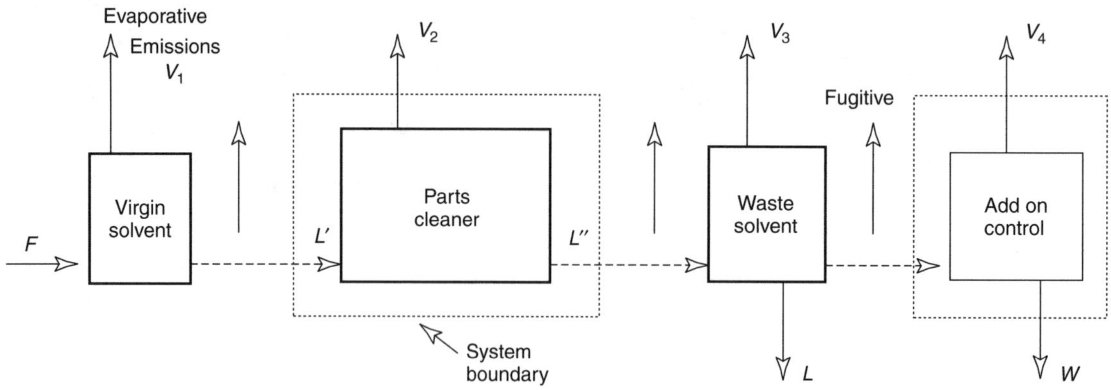

Figure 1. Parts Cleaner Unit Operation System (UOS) (Ref. 5) (Primary and Secondary Unit Operations Depicted).

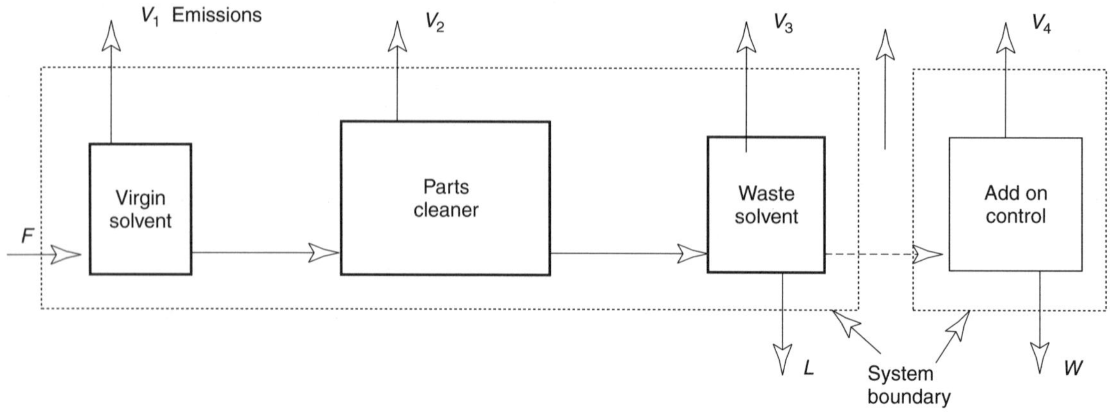

Figure 2. Parts Cleaner Unit Operation System (with Dedicated Feed and Waste Tanks).

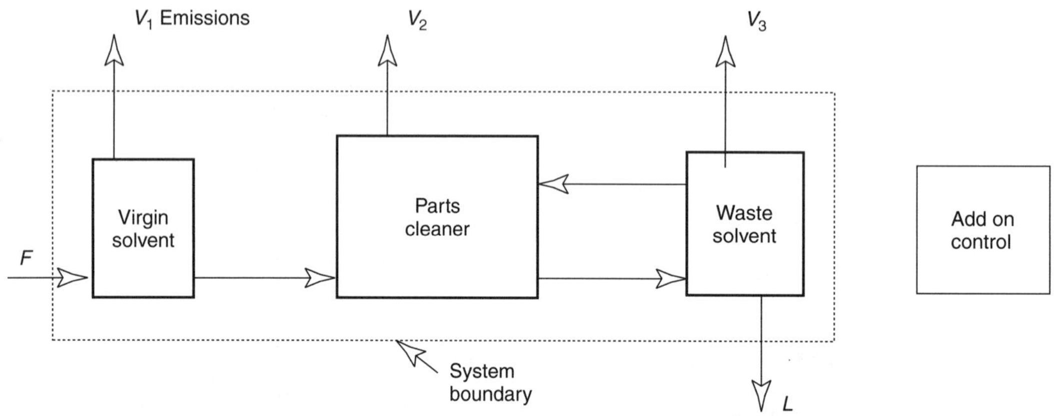

Figure 3. Parts Cleaner UOS (with Closed-Loop Recycling).

the parts cleaner itself. Losses during transfer or size of the inventory will not impact the emissions as in Eq. 1. Hence,

$$V_1 = L' - L'' \tag{2}$$

This is the representation around which will revolve much of the discussion in this chapter.

The stream V_4 represents the emissions from an addon control unit operation. In this presentation the emission stream V_2 from the parts cleaner is not directed to the addon control unit but is released to the atmosphere. Only the waste solvent, L, collected from the parts cleaner is distilled producing a waste stream W. Here, one is assuming that the used solvent is being processed at the facility. However, some facilities may have this step done by the solvent supplier or another entity that will periodically replace or recycle the used solvent.[6]

Figure 2 depicts a unit operation system for a parts cleaner with a dedicated supply of solvent. The waste solvent from the cleaner is removed via a pipe to the waste container. Such an arrangement will result in lower emissions $(V_1 + V_2 + V_3)$ compared with the one shown in Figure 1. However, the spent solvent stream L is still significant and will need to be treated in an addon unit operation. This step will result in evaporative emissions, V_4 and a waste stream, W.

Figure 3 depicts a closed system. Both the evaporative emissions $(V_1 + V_2 + V_3)$ from the UOS and waste solvent stream L are negligible compared with a conventional batch or vapor (machine) cleaners. The integrated recycling unit should bring the quality level of waste solvent to that of the virgin solvent. This representation can be used for low-emission vapor cleaners, which are completely enclosed and air tight.[5] Different technologies that fit this description are briefly discussed elsewhere.[6,7]

THE UNIVERSE OF CLEANING

Cleaning operations are one of the many operations in a manufacturing process. For example, in the coating industry a surface will need to be cleaned of contaminants before a coating is applied to ensure good adhesion of the coating. This is sometimes referred to as surface preparation and often involves the use of organic cleaning solvents resulting in emission of volatile organic compound and hazardous air pollutant materials. Cleaning solvents are also used to clean tanks used for mixing coatings, lines used to pump material from the mixing area to the application area or spray-booth, and application equipment and other tools that need to be reused during the manufacturing process. Other industries such as the chemical and pharmaceutical industries use organic solvents to clean tanks and other containers before a new batch of

Table 1. Cleaning Unit Operation Systems (UOS defines the mass-balance boundary around a unit operation)

Focus of Cleaning	Unit Operation	Unit Operation System (UOS) (boundary for material balance)	Comments
Exterior surface	Spray-booth	1. Spray booth cleaning: solvent is applied mostly by spraying or wiping of surfaces inside the booth. The surfaces include the walls, floor, and grating.	Emissions are generated from the solvent container, the applicator, the surface cleaned, and the used applicator and any contaminated solvent.
	Manufacturing of large components	2. Large manufactured components cleaning: solvent is applied mostly by wiping. The surfaces cleaned include furniture and large appliances components, auto bodies, and metal sheets prior to stamping.	Emissions are generated from the solvent container, the applicator, the surface cleaned, and the used applicator and any contaminated solvent.
	Manufacturing of small components	3. Small manufactured components cleaning: solvent is applied mostly by wiping. The surfaces include switches, machined parts, and circuit breaker cases.	Emissions are generated from the solvent container, the applicator, the surface cleaned, and the used applicator and any contaminated solvent.
	Equipment units and cleaning tools	4. Equipment cleaning: solvent is applied by wiping, brushing, and dipping. The surfaces cleaned include pumps, machine conveyors, packaging machines, bearings, punch press and coating applicators such as rollers, coaters, and ink templates.	Emissions are generated from solvent container, the applicator, the surface cleaned, and the used applicator and any contaminated solvent.
	Floor (other than spray-booth)	5. Floor cleaning: solvent is applied using a cloth or a mop to floor surfaces in a plant other than those of spray booths.	Emissions are generated from solvent container, the applicator, the surface cleaned, and the used applicator and any contaminated solvent.
Interior surface	Line (pipe lines) of the operation	6. Line cleaning: solvent is applied by flushing through a system of pipes and associated tanks to a closed waste container. The system may include a closed-loop recycling system.	Emissions are generated from mixing tanks and from any leaks and any wiping or brushing performed.
	Tank	7. Tank cleaning: solvent is applied by spraying and manual brushing in open tank cleaning. Spent solvent is sometimes used to clean short connecting lines.	Emissions are generated from the interior surface of the tank when the tank is manually cleaned. These emissions are reduced significantly when the tanks are of the type that can be cleaned automatically with the lid closed.
Exterior/interior surface	Parts (machine) cleaner	8. Parts (machine) cleaning: solvent is applied by spraying or by dipping.	Emissions are generated from the parts being cleaned and from the spent solvent container.
	Spray gun (if manual); spray gun cleaner (if automated)	9. Spray gun cleaning: solvent is applied by purging, wiping, and brushing when cleaning manually. It is applied by spraying in automated cleaners.	During manual cleaning, emissions are generated from the solvent container, during transfer of solvent to the gun being cleaned, the gun, transfer to the used solvent container, and from the container. Losses from automated cleaning are greatly reduced.

Source: Adapted from Ref. 1.

material is prepared to prevent cross-contamination of the different batches. Common organic solvent cleaners include alcohols, aliphatic solvents, aromatic solvents, chlorinated solvents, ester solvents, ketones, and fluorinated solvents.

Nine Categories of Cleaning Operations[1]

Cleaning operations can be identified by nine categories: (1) spray-booth cleaning, (2) large manufactured parts (and assemblies) cleaning, (3) small manufactured parts cleaning, (4) equipment cleaning, (5) floor cleaning, (6) line cleaning, (7) tank cleaning, (8) parts (machine) cleaning, and (9) spray gun cleaning. A description of these cleaning operations is presented in Table 1. Sub-categories of each of the nine identified here may be found and will be a reflection of the level of the technology in use within a facility, as was illustrated in Figures 1–3. The first five unit operation systems represent situations where the solvent is brought to the external surface being cleaned. For the sixth unit operations system, solvent in a covered mixing tank is pumped to purge transfer lines from coatings. Most of the surface area that needs to be cleaned is that of the lines rather than the associated mixing tanks; hence the name line cleaning. On the other hand, for tank cleaning (No. 7), the interior surface of the tanks makes up most of the area of such systems, rather than the associated lines that supply or interconnect the tanks. This is an important distinction. In the case of tank cleaning, most of the solvent is applied by spraying and flushing, Wiping of the external surface of a tank may be done occasionally. However, for parts cleaners (No. 8) the parts are brought to the solvent and cleaned by dipping or soaking in the liquid solvent or by exposing the parts to hot solvent vapor. The cleaning efficiency is augmented by mechanical action or by applying ultrasonic waves to the liquid solvent. The last category, spray gun cleaning UOS, will vary from one industry to another depending whether the cleaning is done manually or automatically in a spray gun cleaner. This grouping was adopted by EPA to help obtain emission and waste data that are specific and well characterized. In the next sections two types of parts cleaners (which will also be referred to as cleaning machine cleaners) will be discussed in some detail. Hand wiping using rags or brushes is done as necessary to remove dirt of small parts or areas. The impact of theses cleaning activities should be counted under the applicable UOS identified in Table 1.

DESCRIPTION OF PARTS (MACHINE) CLEANERS

This section describes in detail

- Two most commonly used organic solvent parts cleaners
- How parts are cleaned
- Types of cleaning applications for which each parts cleaner is used

Batch cleaners (cold and vapor), and in-line cleaner (cold and vapor) machines are widely used because they result in dry surfaces and have a low cleaning cost on a parts basis. Cold cleaning machines use organic solvents, either at room temperature or heated below the solvent's boiling point, to clean or dry the parts. Vapor cleaning machines boil the liquid solvent to generate solvent vapors for cleaning or drying the parts. The two basic types of solvent cleaning machines are batch cleaners and in-line or continuous cleaners (cold and vapor). The vast majority of halogenated solvent use is in vapor cleaning, both batch and in-line. The four commonly used halogenated solvents are methylene chloride (MC), perchloroethylene (PCE), trichloroethylene (TCE), and 1,1,1,-trichloroethane (TCA). The primary solvents used in batch cold cleaners are mineral spirits, Stoddard solvents, and alcohols. Very little halogenated solvent use has been identified in batch cold cleaning.

Batch Cleaners[8]

In a batch cleaning machine individual parts or sets of parts move through the entire cleaning cycle before new parts are introduced into the solvent cleaning machine. An open-top vapor cleaning (OTVC) machine is the most common type of batch cleaning machine. Solvent cleaning machines, such as Ferris wheel and cross-rod cleaners that clean multiple batch loads simultaneously and are manually loaded, are also considered batch cleaning machines. In a cross-rod cleaner the parts baskets are suspended on rods and moved by power-driven chains. Batch vapor cleaners are used primarily in metalworking operations and other manufacturing facilities. They are seldom used for ordinary maintenance cleaning because batch cold cleaners using petroleum distillate solvents can usually perform this type of cleaning at a lower cost. Exceptions include maintenance cleaning of electronic components, small equipment parts, and aircraft parts, where a high degree of cleanliness is needed.

There are two basic types of batch cold cleaners: immersion cleaners and remote reservoir cleaners. In an immersion cold cleaner parts are immersed in the solvent when being cleaned. In a remote reservoir cold cleaner liquid solvent is pumped to a sinklike work area that drains solvent back into an enclosed container while parts are being cleaned, which prevents used solvent from draining in the work area. Properly designed batch cold cleaners are fitted with tightly fitting covers and a water layer or a tightly fitting cover and a freeboard ratio of 1.0, in order to minimize evaporation losses. Other options are indicated in Table 2. The use in cold cleaners of a water layer over the solvent substantially limits evaporation losses because very little organic solvent will come into contact with the air and is recommended for cleaning in which rust or water damage to the part are not of concern.

Figure 4 shows a basic batch open-top vapor cleaner as a tank designed to generate and contain solvent vapor. At least one section of the tank is equipped with a heating system that uses steam, electricity, hot water, or heat pumps to boil liquid solvent. As the solvent boils, dense solvent vapors rise and displace the air inside the tank . The solvent vapors rise to the level of the primary condenser and on the surface that needs to be cleaned.

In-Line (or Continuous) Cleaners[8]

In-line cleaners (also called continuous cleaners) use automated parts handling systems, typically a conveyor, automatically to provide a continuous supply of parts to be cleaned. In-line cleaners can be either cold or vapor cleaning machines. The majority of in-line machines using halogenated solvents are vapor cleaners. A continuous or multiple-batch loading system greatly reduces manual parts handling associated with batch cleaning. The same cleaning techniques are used during

Table 2. Control Options for Batch Vapor Solvent Cleaners

Vapor Cleaner Interface Area	Opt. Ref. 9	Working Mode Cover	Reduce Draft	Freeboard 1.0	Freeboard Refrigeration Device	Superheated Vapor
Less than	1	X		X		X
1.2 m^2	6		X		X	
(13 ft^2)	7			X	X	
Greater than						
1.2 m^2	1			X	X	X
(13 ft^2)	6		X	X	X	

Source: Adapted from Ref. 9.

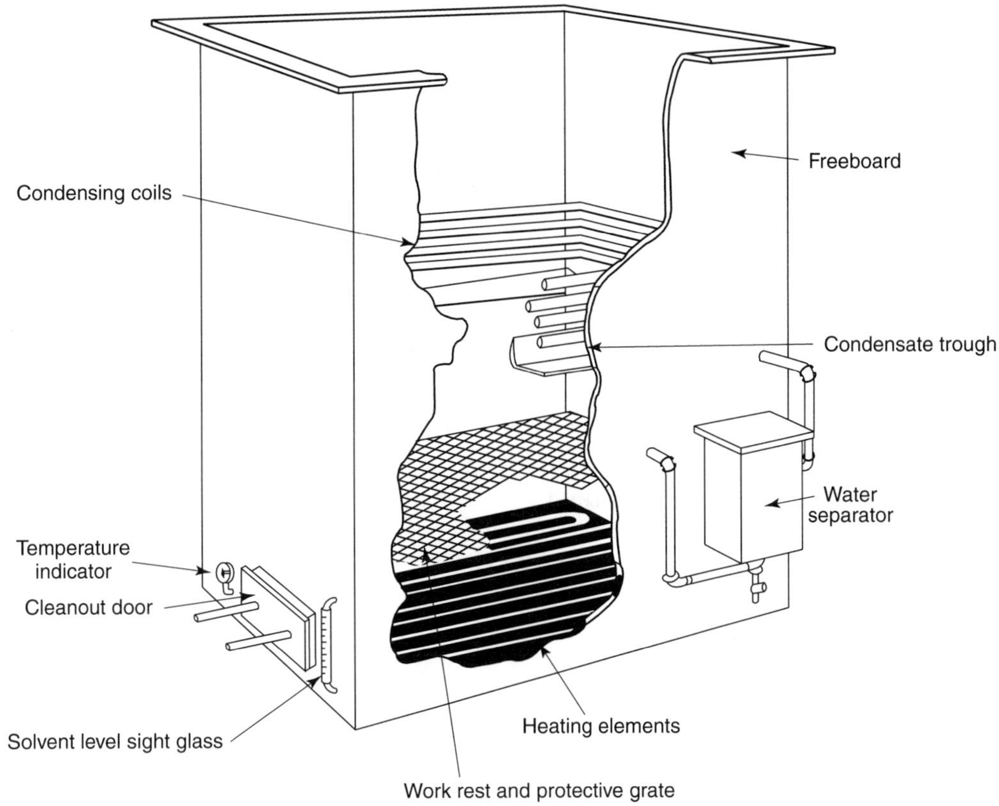

Figure 4. Schematic of an Open-Top Vapor Cleaner (Ref. 8).

in-line cleaning, but usually on a larger scale than with batch units. In-line cleaners are nearly always enclosed, except for parts/conveyor inlet and exit openings, to help control solvent losses from the system. In-line cleaners are most often found in plants where there is a constant stream of parts to be cleaned. There are five main types of in-line cleaners that use halogenated solvents: monorail, belt, strip, printed-circuit-board processing equipment (photoresist strippers, flux cleaners, and developers), and continuous web cleaners.

A monorail vapor cleaner (Figure 5) is usually chosen when the parts to be cleaned are being transported between manufacturing operations on a monorail conveyor. The monorail cleaner is well suited to automatic cleaning with solvent spray and vapor. It can be of the straight-through design illustrated or can incorporate a U-turn within the machine so that parts exit through an opening parallel to the entrance. The U-turn monorail cleaner benefits from lower vapor loss because the design eliminates the possibility of drafts flowing through the machine. Both the belt cleaner and the strip cleaner are designed to allow simple and rapid loading and unloading of parts. A belt cleaner conveys parts through a long, narrow boiling chamber in which the parts are cleaned, either by the condensing vapor or by immersion in the solvent sump. The strip cleaner is similar to the belt cleaner except that the strip itself is the material being cleaned. Several equivalent emission control options for in-line solvent cleaners are presented in Table 3.

LOSSES FROM PARTS CLEANERS[8–10]

There are several different operating modes that exist with a typical solvent cleaning machine, each with its own particular emission characteristics:

- Working: This is the time where a cleaning machine is actively being used to clean a part.

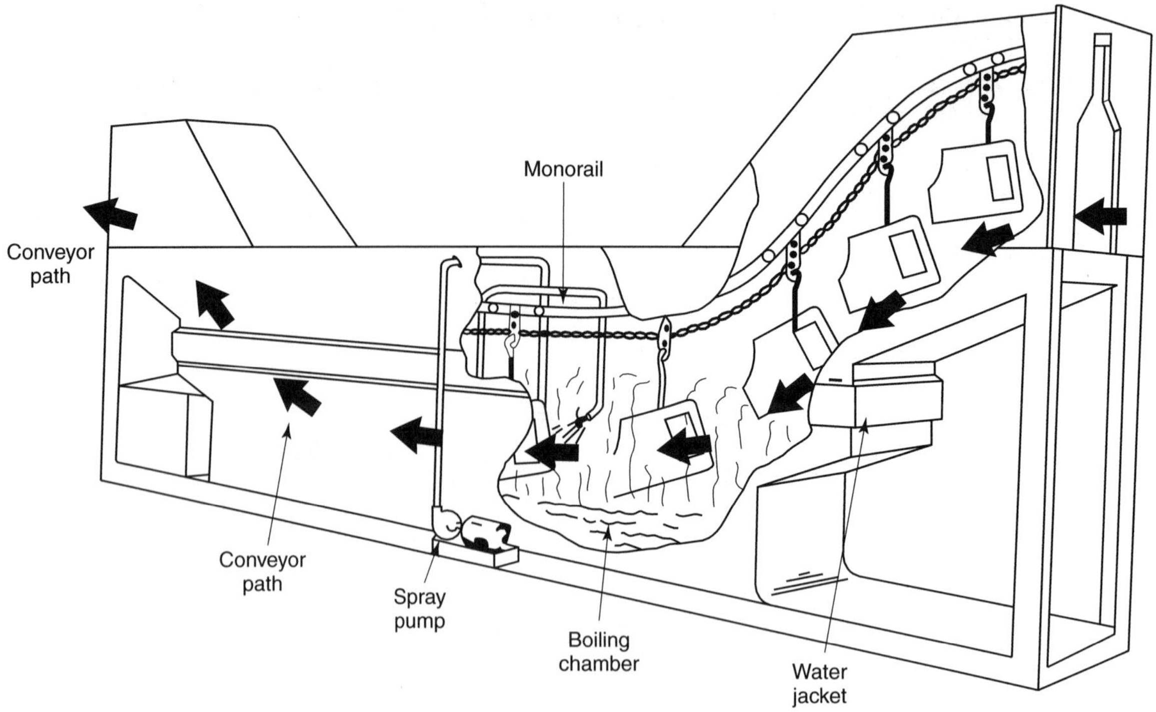

Figure 5. Schematic of a Monorail In-Line (Continuous) Cleaner (Ref. 8).

Table 3. Equivalent Control Options for In-Line Solvent Cleaners

Vapor Cleaner	Opt. Ref. 9	Freeboard 1.0	Freeboard Refrigeration Device	Superheated Vapor	Addon Control
Existing	1	X		X	
	2	X	X		
	3		X		X
New	1		X	X	
	2		X		X
	3			X	X

Source: Adapted from Ref. 9.

- Idling: This includes all time that the cleaning machine is turned on (i.e., sump heat and cooling coils are on) and the machine is ready for cleaning, but is not currently cleaning any part. It should be noted that in cold cleaning machines, downtime and idling modes are virtually synonymous.
- Downtime: This is the time that the cleaning machine is turned off (i.e., sump heat is off on vapor cleaning machines), and the cleaning machine is not ready for cleaning.

The emissions during each of these modes occur from multiple mechanisms. An understanding of these mechanisms is critical to developing the optimal control strategy for any solvent cleaning machine. Optimal control will result in lower costs of operation and solvent usage, thus reducing emissions and waste.

Design-Specific Losses[8–13]

This is the first of three categories of losses from parts cleaners and constitutes of three elements.

Working-Mode Losses. When a cleaning machine is actively being used to clean parts, evaporative losses occur at the solvent–air interface, due to diffusion and convection effects and as liquid and vapor are carried out on parts. Solvent in the vapor zone diffuses into the ambient air above the vapor zone (i.e., in the freeboard area) because of the concentration difference. The effect is enhanced by the temperature difference, which is responsible for convective losses. The interface losses during the working mode can be significantly higher than the interface losses during the idling mode due to the greater potential for disturbances. Losses can be reduced by purchasing machines (or retrofitting existing machines when possible) with superheated vapor. Raising the temperature of the vapor zone above the normal boiling-point temperature will reduce dragout and diffusion emissions considerably as the parts will dry rapidly in the vapor zone. Other operational options for reducing emissions are shown in Table 4.

In batch machines, a primary disturbance is caused by the introduction and removal of parts and parts baskets through the vapor zone. The magnitude of this loss depends on the

Table 4. Operational and Monitoring Options for Reducing Emissions

Solvent Losses	Batch Loaded Cleaner (Operated below Solvent Boiling Point)	Batch Open-Top Vapor Parts Cleaner	Open-Top Vapor Conveyorized Parts Cleaner
Splashing due to spraying: increases losses by diffusion of solvent from air–solvent vapor interface	Use cover to reduce startup or idling losses. Maintain agitation pressure below kg/m^2 (2 lb/in.^{2}). Direct a spray or liquid stream used to clean parts downwards to avoid turbulence at air–solvent interface. Position solvent transfer pipes below solvent surface to avoid splashing losses.	Use cover to reduce startup or idling losses. Maintain agitation pressure below kg/m^2 (2 lb/in.^{2}). Direct a spray or liquid stream used to clean parts downward to avoid turbulence at air–solvent interface. Position solvent transfer pipes below solvent surface to avoid splashing losses.	Use cover to reduce startup or idling losses. Direct a spray or liquid stream used to clean parts downward to avoid turbulence at air–solvent interface. Position solvent transfer pipes below solvent surface to avoid splashing losses
Drafts/Ventilation above opening of a parts (machine) cleaner: promote diffusional and convective losses	During startup, turn on the primary condenser, then the sump heater. During shut down, turn off the heater first. Do not add solvent should above fill line. Use tightly fitting covers and keep cover closed and turn off fans when not using the cleaner (i.e., at downtime). Minimize drafts when cover is open.	During startup, turn on the primary condenser, then the sump heater. During shutdown, turn off the heater first. Do not add solvent should above fill line Use tightly fitting covers and keep cover closed and turn off fans when not using the cleaner (i.e., at downtime). Minimize drafts when cover is open. Maximum temperature cutoff of primary condenser coils should be monitored. Refrigerated freeboard chiller (outlet temperature should not exceed 4.4°C).	During startup, turn on the primary condenser, then the sump heater. During shut down, turn off the heater first. Do not add solvent above fill line. Use tightly fitting covers and keep cover closed and turn off fans when not using the cleaner (i.e., at downtime). Minimize drafts when side covers are open. Exhaust should be sent to an addon control unit or use freeboard refrigeration. Maximum temperature cutoff of primary condenser coils should be monitored. Chilled air blanket should be less than 40% of solvent b.p. or 50% of b.p. for solvents that form azeotrope.
Piston effect: disturbs vapor layer at the air–solvent vapor interface, promotes diffusional losses and carryout of solvent on parts being cleaned.	Use a programmable hoist: speed of raising/lowering parts vertically should not exceed 3.4 m/min (11 ft/min).	Work load area must not exceed 1/2 area of air–solvent vapor interface. Use a programmable hoist: speed of raising/lowering parts vertically should not exceed 3.4 m/min (11 ft/min).	Use a programmable hoist: speed of raising/ lowering parts vertically should not exceed 3.4 m/min (11 ft/min).
Dripping parts: results in carryout (dragout) of solvent on parts cleaned.	Parts must not be removed when dripping; preferably remove when dry. Proper racking reduces amount of solvent trapped in parts or basket. Orient basket so that solvent drains freely. Do not clean porous materials.	Parts must not be removed from vapor zone when dripping; preferably remove when dry. Proper racking reduces amount of solvent trapped in parts or basket. Orient basket so that solvent drains freely. Do not clean porous materials	Parts must not be removed from vapor zone when dripping, preferably remove when dry. Proper racking reduces amount of solvent trapped in parts or basket. Temperature within a superheated zone should be >5°C above b.p. of solvent. Orient basket so that solvent drains freely.

Source: Adapted from Refs. 8 and 10–13.

speed of the basket[8,10,12] as well as the characteristics of the parts being cleaned.[5] The vapor layer can be disturbed as a portion of the layer is displaced by the piston-type effect as the parts/parts basket are lowered into the machine. Thus the stop-and-go technique is used[12] to reduce the amount of solvent vapor loss when the part is lowered or raised through the vapor zone of batch vapor cleaner. Dragout emissions can be reduced by using a programmable transport system capable of moving at a low and steady speed parts/baskets from the initial loading of parts to the point of removal of parts. The low and steady speeds reduces disturbance to the vapor zone and keeps the parts/basket for the required length of time in the freeboard zone to drain solvent carried out.

Idling Mode Losses. These losses occur mainly at the solvent–air interface (interface losses), which include diffusion and convection losses. The largest source of idling emissions in most vapor solvent cleaning machines is from diffusion. Cleaners with 100% freeboard (and open) will have less emissions than one with a 75% freeboard. One can reduce idling losses considerably for higher-boiling solvents by installing secondary condenser coils (0°C) just above the primary condenser coils (45°C). Covers will also help reduce emissions when a parts cleaner is exposed to fans or drafts in an industrial environment (Table 4). They should be free of cracks or other defects to keep in the solvent.

Downtime Losses. They occur when the heat to the sump is turned off and the machine is not operating because of evaporation of solvent from the liquid solvent surface and subsequent diffusion into the ambient air. The use of a tight-fitting cover during downtime will help contain the vapors within the cleaning machine, which will reduce idling losses.

Work-Practice-Type Losses[8,10]

In addition to working, idling, and downtown losses the following activities and mechanisms can contribute to overall solvent losses from an organic solvent parts cleaner. The magnitude of these emissions relative to total losses is dependent on the machine design, machine integrity, and operating and maintenance procedures.

Leaks. Such losses can occur continuously (depending on where the leak is located), whether the machine is off or on. Leaks can result from manufacturing defects or from machine wear. They can occur from piping connections, cracks in the machine or tank, and gasketed portholes or viewing windows. Leaks are often difficult to detect because the solvent will evaporate quickly when it reaches the atmosphere and may not leave telltale drips or wet areas. Because solvent typically has a low surface tension, it can escape through cracks that may not be easily visible. These characteristics magnify the chance of leaks becoming a serious source of solvent loss. Many manufacturers leak-test their machines before that are sold, but cracks can occur during shipping, and therefore the operator should routinely check for leaks as part of a work practice schedule.

Filling/Draining Losses. These losses occur from a parts cleaner are minimized by adopting good work practices. Hence, an operator should avoid filling or emptying a cleaner using open buckets or barrels. This practice could result in both the high emissions and operator exposure. Submerged filling through the use of leakproof couplings will provide the best results. If solvent spills occur during filling or draining, the operator may be subject to Comprehensive Environmental Response, Compensation, and Liability Act (CERCLA) regulations, which require notification of all spills above reportable quantities.[14]

Startup/Shutdown Losses. They occur during the transition from downtime to steady-state idling or working or from operating back to downtime. Startup losses are due to pumpout of solvent-laden air within the machine after the sump heat has been activated and as the solvent layer is being established. Shutdown losses are due to the evaporation of hot liquid from the sump after the heat has been turned off and the vapor layer has collapsed. If a cleaning machine is not controlled (i.e., through the use of a cover during this time), shutdown losses can be significant since the solvent is near its boiling point at the beginning of the shutdown period.

Solvent Decomposition. This will occur if the solvent is not properly monitored and is allowed to become acidic due to the presence of water. Not only will the solvent need to be replaced, but also dangerous fumes (e.g., chlorine gas, hydrochloric acid) can be emitted form solvent decomposition. The waste solvent would be subject to hazardous waste guidelines under the Resource Conservation and Recovery Act (RCRA).[15] Stabilizers are sometimes added to prevent the solvent from turning acidic.

Other Losses

Losses may occur from other than the parts cleaner, the primary unit operation, as shown in Figure 1. They may include streams from one or more associated or secondary unit operations.

Solvent Storage Losses. They occur by diffusion from tanks and other containers. The emissions will depend on the type of container and the number of containers or size of the solvent inventory.

Distillation Losses/Sludge Disposal. They occur when spent solvent is regenerated through on-site distillation (see Figure 1). Solvent lost during this operation stems from the evaporation during the transfer to and from the distillation unit or, if a piping system is used, from leaks in the equipment. Solvent may also evaporate form distillation sludge that is removed for disposal.

Water Separator Losses[8,10]. These occur when one is recovering solvent from the solvent–water mixture that condenses at the primary condenser of vapor machines. Water separators are typically minor sources of solvent losses from solvent cleaning machines. The solvent loss occurs when the water is decanted from the separator containing a slight amount of solvent depending on the solubility of the specific solvent in water. Operators must ensure that only the water is decanted from these units and that no pure solvent is removed. This will ensure that the water separator remains a minimal source of emissions. Water impacts could also occur if carbon adsorbers are used, due to the condensation of steam used to desorb these carbon beds.

SOLVENT ACCOUNTING/MANAGEMENT

When a company has a good knowledge of the fate and costs of material it is using, it is likely to take action to minimize pollution and cost of doing business. Many companies do not yet properly account for the cost of cleaning and may not realize the potential for savings in cost and therefore take no action towards that goal.[1,3]

Facility-wide emissions from the use of organic cleaning solvents can be determined from annual usage of solvent for the purpose of cleaning. This may be difficult to do if a facility uses some of the purchased solvent for thinning of coatings, unless the facility has already set up an accounting system that distinguishes between the two kinds of uses. Hence, a facility will need to know where it is using the solvent. When the facility knows, at the operational level, how much solvent it is using and where the solvent is being used, it will be in a position to take the next step, which is to identify the opportunities for minimizing emissions and waste as well as address risk and other impacts under consideration.

The solvent emissions should preferably be determined by mass balance on a well-defined system boundary. Hence,

$$\text{input} - \text{output} = \text{losses} \tag{3}$$

What one includes within the boundary of the system for the mass balance will define the limits and magnitude of the environmental impacts under consideration. Compare, for instance, the sources defined in Figures 1–3. The accuracy of the mass balance will depend on whether or not one is using operational-specific data. If actual usage data are used, it will most likely be more accurate than if purchase data or records were the source of the information. One should base the analysis when possible on the amount used at a given unit operation site—the unit operation system—rather than on the amount purchased for a cost center.[1,3] A solvent reduction plan should address actions to be taken for other cleaning unit operation system category or subcategory. Three subcategories of parts cleaners are defined in Figures 1–3. This information is important for matching operational usage against the overall facility-wide usage of solvents.

The difference between total solvent issued for cleaning and the sum of solvent usage accounted by unit operation systems provides a measure of the uncertainty in the data, which is often attributable to the procedures used to collect the information and possible transfer losses that were not accounted for.[3,16]

A Standardized Accounting System

First, let us examine the purposes of having a standardized accounting system based on unit operation systems. If corporate management wants to reduce the usage of cleaning solvents, it can require all facilities to reduce their usage by a fixed percentage across the board, without consideration of differences in the individual cleaning operations. Alternatively, it can give facilities the opportunity to track the usage of solvents for each of the categories described in Table 2 to determine the locations of the unit operations that are a source of large emissions and waste from cleaning and those pose the highest level of risk. It is important at this point that a facility compares total usage of cleaning solvent to that accounted for by unit operation system, to determine the difference as a percentage of usage as was explained before. If a facility is unable to identify where the missing solvent was used, it could readjust proportionally the amount used at the UOS level so that the total cleaning solvent issued to the operators during a given period of time is accounted for. That step should be easy to make and will help provide more accurate impact and cost values. Therefore, instituting a standardized accounting system such as the one described here is a first step in a solvent management plan that promotes data comparability. Once this is done it is much easier to compare usage, emissions, or risk from like unit operation systems within an industry and across industries.[3,16]

ENVIRONMENTAL PERFORMANCE INDICATORS

Many facilities today are developing "indicators" or "metrics" for reducing emissions, waste, or risk levels from their important operations. Those indicators are developed to measure the level of progress towards attaining the objectives and goals of that organization.[16–19] The objectives could be local or global. An organization will select indicators, set targets and other performance criteria on the basis of policy objectives. These objectives may be influenced by economic justifications or driven by the need to comply with regulatory standards.

A possible list of local indicators applicable to subcategories of parts cleaners depicted in Figures 1–3 are: (1) amount of solvent used; (2) amount of evaporative emissions; (3) amount of solvent waste generated; (4) level of toxic compounds/environmental risk; (5) resultant water pollution and soil pollution; and (6) energy consumption.

If one focuses on parts cleaners, one would find that some cleaning machines are very large and some are not. The level of emissions from a cleaner will depend on the type of machine, type of solvent used, and work practices adopted for minimizing emissions. As part of the overall solvent management system, one should also evaluate associated emissions at the virgin or clean solvent storage site and waste disposal site. In order to generate an environmental performance indicator for a part cleaner, it is important that one understands the relationship between operating modes and the emissions from the system which we seek to improve. For a parameter such as emissions to be meaningful it has to be "normalized." Such values are shown in Table 5. The normalized emissions rates are presented by operating mode for both batch and in-line (continuous) cleaning machines with solvent–air interface area. The most significant losses are vapor–air interface (interface losses) and workload-related losses (workload losses).

The relative amounts of emissions per operating mode will depend on the design of the parts (machine) cleaner and on the operating schedule followed. It is also dependent on mass and shapes of the parts cleaned.[5] The total emissions from a small and medium uncontrolled batch vapor cleaner operating schedule of 2 h working, 6 h idling, and 16 h downtime for 260 d and 24 h downtime for 105 d is 3755 kg/m² of interface surface in a year.[20] This value is the sum of two values in Eqs. (4) and (5).

$$(T_w F_w + T_i F_i + T_d F_d) D_{op} = 3146 \text{ kg/m}^2 \tag{4}$$

$$T_d F_d D_d = 609 \text{ kg/m}^2 \tag{5}$$

Table 5. Typical Emission Rates for Batch and Continuous Cleaning Machines[20]

	Emission Rates for Typical Cleaning Machines, kg/(m² h) [lb/(ft² h)]		
Cleaning Machine	Working Emissions	Idling Emissions	Downtime Emissions
Batch vapor	1.95 [0.40][a]	0.73 [0.15][a]	0.24 [0.03][b]
Continuous vapor[d]	1.12 [0.23][b]	0.34 [0.07][b]	0.24 [0.03][b]

[a] EPA 1991 ACT for halogenated solvent cleaners.
[b] Estimates in 1993.

where the subscripts w, I, d represent working, idling, and downtime, respectively; T is the operating time (h); D_d, the number of downtime days; and D_{op}, the number of operating days.

If we assume that the maximum control for the operating schedule of 2 h working (520 h/yr) and 6 h idling (1560 h/yr) is 60%, the value in Eq. (4) above becomes 1755 kg/m^2 and the total annual emissions are (1755 + 609) equal to 2364 kg/m^2 of interface surface.

In practice the emissions for a batch or line cleaner should be determined by test data using a material balance. For a parts cleaner shown in Figure 1, an operator would need to know the mass of

- Solvent added to the cleaner
- Clean solvent removed
- Solvent removed from the cleaner as liquid waste
- Solvent removed from the cleaner in solid waste
- Solvent–air interface area.

For cleaning machines with no interface such as completely enclosed vapor cleaners (Figure 3), one can estimate emissions, E_i (mass/time), based on the cleaning capacity of the cleaner using an expression such as:

$$E_i = a(\text{volume})^b \tag{6}$$

Values of $a = 330$ and $b = 0.6$ are used by U.S. EPA to define emission limits from cleaners that have a cleaning volume or capacity greater than 2.95 cubic meters.[9,10] They allow a 3-monthly rolling average to determine the average monthly emissions. This mean that the emissions averaged over 3 months (kg/month) must not exceed those calculated from the equation when the appropriate constants are used. The procedure for developing this form of the equation is discussed elsewhere.[21] It is important, when we compare results for cleaners that the boundary of the source and the time period used for averaging emissions, waste, risk, or cost must be kept the same. For example, the material balance boundaries, the unit operation systems in Figures 1 and 3 are not the same, because the technologies are different. Thus, to provide an accurate comparison, one must account for the impacts of the secondary operations on the total emissions, risk, waste, and costs associated with the primary operation.[3,16]

OPTIONS FOR SOLVENT REDUCTION

Many organic solvents used in cleaning operations fall in the category of hazardous air pollutants. A number are photochemically reactive and contribute to ground-level ozone formation. Several have ozone-depleting potential. Therefore, an organization that uses such solvents needs to manage the use of solvents to minimize the impact on the environment and operating cost. The latter cost involves training for worker protection and meeting regulatory requirements. Another element is the cost of the solvent used.

There are opportunities for reducing the amount of organic solvent used[6,7,11,22–24]

Options With a Focus on Parts Cleaners

Use a parts cleaner that is tailored for the work load. In-line cleaners are usually specifically designed. If a facility does not need to operate continuously during a shift, it should plan to purchase a batch cleaner rather than an in-line (continuous) cleaner.

Maximize cleaning efficiency. Removing using a brush caked-on solids from parts before placing in solvent parts cleaner will increase the life of the solvent in a cold cleaner. Avoid operating conditions that will expose solvent to water in vapor cleaners.

Use waste solvent recycled when possible. This will reduce the amount of virgin or clean solvent used and solvent waste that will have to be disposed. Segregation of solvent will affect its reuse and recyclability. The economics of solvent recycling is dependent on several factors, including the type of solvent used, the operation, and the level of quality assurance and quality control maintained.[23]

Options for All Cleaning Operations

Reduce the need to clean with organic solvent. Making the operation process solventless may be an option in some cases. This requires planning before purchasing a technology, e.g., using powder coatings rather than solvent based coatings, reduces the need to clean application equipment and application area. Do not use organic solvent to clean a shop floor, and adopt good work practices to reduce losses during storage and transfer of solvent. Alternative cleaning materials or fluids are available.

Substitute with a water-based cleaning material.[6,7,22,24] Aqueous cleaning materials are increasingly being used to remove oils, grease, and dirt from parts when exposure to water is not a problem. Water-soluble cutting fluids (used to dissipate heat during machining) have been developed and can be used instead of solvent-based fluids. The cleaning fluid is then removed in another operation and recycled back to the cutting operation.

Aqueous cleaning machines eliminate the solvent use and associated hazards; however, they generally require more space than solvent-based machines and use more energy. They include automated aqueous cleaners (usually for cleaning small parts), aqueous power washers (cleaning large and small parts), and ultrasonic cleaners (part size is determined by the energy required for cleaning).

Use pollution prevention technologies.[7,22,23] These include the use of alternative materials in an operation such as the use of low solids fluxes, which eliminates the need for cleaning; the use of alternative cleaning materials, such as alkali and citric acid-based materials, which are suited for many cleaning needs when moisture is not a problem; and closed-loop parts cleaners, which emit very little solvent and generate low amounts of solvent waste. Other technologies that have the potential to reduce the organic cleaning solvents are being developed. Reference 24 gives information on a number of useful online sources of information on solvent substitution.

REFERENCES

1. *Alternative Control Techniques Document—Industrial Cleaning Solvents*, U.S. Environmental Protection Agency, Office of Air Quality Planning and Standards, Research Triangle Park, NC, February 1994; EPA-453/R-94-015; NTIS: PB94-156791.
2. *Automobile Assembly Plant Spray Booth Cleaning Emission Reduction Technology Review*, Environmental Protection Agency, Office of Air Quality Planning and Standards, Research Triangle Park, March 1994; EPA-453/R-94-029; NTIS: PB94-206257.

3. M. A. Serageldin, "Standardized Accounting for a Formal Environmental Management and Auditing System," in *Waste Minimization through Process Design*, A. P. Rossiter, Ed. McGraw-Hill, New York, April 1995, pp. 289–303.
4. M. A. Serageldin, "Multi-Media Accounting and Management: Application of the Unit Operation System Concept to the Chemical Industry," American Institute of Chemical Engineers, 1994 Spring National Meeting, paper 36 D, Atlanta, April 17–21, 1994.
5. M. A. Serageldin, *Determining the Performance of a Parts Cleaner*, National Technical Information Services, Springfield, VA. PB97-109466, September 1995.
6. A. R. Gavaskar, "Process Equipment for Cleaning and Degreasing," in *Industrial Pollution Prevention Handbook*, M. F. Freeman, Ed., Mc Graw-Hill, Inc., New York, 1995, pp. 467–81.
7. *Guide to Cleaner Technologies: Cleaning and Degreasing Process Changes*, U.S. Environmental Protection Agency, Office of Research and Development, Washington, D.C., EPA/625/R-93/017, February 1994.
8. B. T. Mark, and V. C. Richard, "Organic Solvent Cleaning (Degreasing)," in *Air Pollution Engineering Manual*, Air & Waste Management Association, Pittsburgh, PA 1992, pp. 345–60.
9. *National Emission Standards for Halogenated Solvent Cleaning—Code of Federal Regulations, Protection of the Environment*, Title 40, Part 63, Subpart T; U.S. Government Printing Office, Washington, D.C.
10. *Guidance Document for the Halogenated Solvent Cleaner NESHAP*, U.S. Environmental Protection Agency, Office of Air Quality Planning and Standards, Research Triangle Park, EPA-453/R-94-081, April 1995.
11. J. Hoffman, "Saving Solvent," *Product Finishing*, November 1989, pp. 101–5.
12. "Vapor Degreasing," in *Tool and Manufacturing Engineering Handbook, Vol. III, Materials Finishing and Coating*, Chapter 18, Fourth Edition, C. Wick, and R. F. Veilleux, Eds., Society of Manufacturing Engineers, Dearborn, 1985, pp. 8–15 to 18–25.
13. *Reducing Solvent Emissions from Vapor Degreasers*, Fact Sheet, (7/91-70); Minnesota Technical Assistance Program; Minnesota Office of Waste Management, Minneapolis, Minnesota, 1991, 6 pp.
14. *The Comprehensive Environmental Response, Compensation, and Liability Act (CERCLA or Superfund)—U.S.C.A.*, Title 42, Public Health and Welfare, Section 9601 et seq. (1980), U.S. Government Printing Office, Washington, D.C.
15. *The Resource Conservation and Recovery Act (RCRA)—U.S.C.A.*, Title 42, Public Health and Welfare, Section 321 et seq. (1976); U.S. Government Printing Office, Washington, D.C.
16. M. A. Serageldin, *Unit Operation System for Measuring Environmental Performance*, National Technical Information Services, Springfield, VA, PB98-124399INZ, November 1996.
17. ISO 14031: *Environmental Performance Evaluation*, 1996.
18. M. A. Serageldin, *Pollution Prevention: An Informal Discussion (with supporting examples)*, National Technical Information Services, Springfield, VA, PB98-124381 INZ, November 1996.
19. *Developing and Using Production-Adjusted Measurements of Pollution Prevention*, U.S. Environmental Protection Agency, Office of Research and Development, Washington, D.C., EPA/R-97/048, September 1997.
20. *National Emission Standards for Halogenated Solvent Cleaning—Docket: A-92-39*; Document No. IV-D-40, January 28, 1994; U. S. Environmental Protection Agency; Air and Radiation Docket and Information Center (Mail Code 6102), Washington, D.C.
21. *National Emission Standards for Halogenated Solvent Cleaning—Docket: A-92-39*; Document No. IV-B-02, July 22, 1994; U.S. Environmental Protection Agency; Air and Radiation Docket and Information Center (Mail Code 6102), Washington, D.C.
22. "Waste Minimization Options for Automotive Repair Shops," in *Guide to Pollution Prevention, The Automotive Repair Industry, Section 3 & 4*; Document No. 1214 [Online]: URL. *http://www.ccar-greenlink.org/documents/cat1200/doc1214.html#table1* (accessed February 23, 1999).
23. "On-Site Reuse and Recycle of Solvents," in *Solvent Waste Reduction*; Pollution Technology Review No. 193; U.S. Environmental Protection Agency, ICF Consulting Associates Inc.; Noyes Data Corporation, New Jersey, 1990, pp. 78–89.
24. *Understanding Solvent Cleaning and Air Quality* [Online]: URL. *http://www.scapca.org/solvent.html*, 6 pp. (accessed March 22, 1999).

SURFACE COATINGS

Ashok Kumar and Taramati Shenoy

INTRODUCTION

Surface-coating operations involve the application of wet or dry coating material to a surface. The coating materials are applied for decorative and/or protection against damage or corrosion on items manufactured by transportation, furniture, appliances, cans, and computer industries. Different surface coating methods are used in various industries (see Table 1). The surfaces can consist of a wide range of substrates such as metal, wood, paper, and plastic. The products and shapes coated vary from automobiles to zippers.

However, each surface-coating operation can result in the emission of volatile organic compounds (VOCs) and other hazardous air pollutants (HAPs). The HAPs are too many and may target many organs in the body. Their effects occur after long, latent periods. The percentage of VOC emissions emitted during the surface coating operations for selected industries are tabulated in Table 2. VOCs are compounds of carbon. When these are emitted into the atmosphere, they may take part in the photochemical reaction with oxides of nitrogen to produce ozone, the major ingredient in photochemical smog. Ozone is a strong irritant that can cause a number of human health problems, besides damaging crops and property. Hence the calculation and control of VOC and other emissions, which occur during the surface-coating operations, is important in air pollution.

This chapter has been broadly divided into three sections. The first deals primarily with the surface-coating operation, the various kinds of coatings, their application techniques, and equipment used for the various processes. The second part involves the mathematical characterization of the air emissions. The third part deals with the control of air pollution, which occurs as a result of the surface-coating operations. This includes the design of an enclosed and ventilated paint spray booth, in which painting operations are usually carried out in an automobile industry to ensure a good finish on a painted surface, protect against fire and explosion hazards, arising from chemicals contained in the paints, and reduce possible health problems.[7] The last section includes subsidiary subjects such as the Alternative Emission Control Plans in California and the Maximum Achievable Control Technology (MACT) standards being developed by the U.S. Environmental Protection Agency (EPA).

Table 1. Surface-Coating Application Methods for Various Industries

Surface Coating Method	Coil Coating	Metal Furniture	Auto and Light-Duty Truck	Large Appliances	Can	Auto Refinish	Traffic Marking	Wood Building Products Industry	Fabric Industry
Dip		X		X				X	X
Flow		X		X				X	
Roller	X				X			X	X
Electrodeposition	X		X	X					
Spray									
Air-atomized spray			X	X	X	X	X	X	X
Airless spray				X		X		X	X
Electrostatic		X	X	X		X		X	X
High-volume, low-pressure spray			X			X		X	X
Electrostatic bell and disk spray				X				X	X

Source: Refs. 1, 2, 3.

Table 2. Percentage of VOC Emissions Emitted during Surface Coating Operations for Selected Industries

	Percentage of Total VOC Emissions	
Industry	Spray Booth or Application Area and Flashoff	Bake Oven
Metal furniture	70	30
Automobile and light-duty truck	85–90	10–15
Large appliance	80	20
Coil coating[a]	8	90

[a]Remaining VOC emissions (2%) come from the quench section after the baking/curing oven.

Source: Refs. 4, 5, 6.

PROCESS DESCRIPTION

The surface-coating process consists of the following steps:

1. Surface preparation:
   a. Cleaning the object's surface with an industrial detergent and water to remove dirt, oils, and grease.
   b. Wiping the object surface clean with solvents in order to remove dirt, oils, heavy machine greases, or milling agents.
2. Application of coatings.
3. Curing of coatings.

A brief description is given in the following sections.

Surface Preparation

The preparation of a surface for coating requires cleaning of the surface. Generally sources use six treatment methods[8]:

1. Cleaning the surface with an industrial detergent in water to remove dirt and oils.
2. Running through a degreaser or wiping clean with solvents in order to remove dirt, heavy machine greases, or milling agents.
3. Washing with an acid solution to cleanse the part and provide surface etching to provide better coating adhesion properties.
4. Treating with alkaline/caustic solutions.
5. Phosphating, that is, treating in a surface conversion process.
6. Blasting with abrasives such as sand, metal slag, and plastic beads.

These treatments improve the coating adhesion and corrosion resistance. The cleaned articles are dried prior to coating.

Surface Coating

The surface coating consists of solvents and solids. The solvents lower the viscosity of the paint by acting as a carrier for the solids as well as being solvent for solid resins.[9] Solvents include volatile organic compounds, water, and non-VOC solvents. Coating components vary among basic coating systems, specific formulations, and color variations.

Coatings are either applied directly or are mixed to a standard mix ratio or viscosity. For solvent-borne coatings, additional solvents are often used to reduce the coating to the viscosity desired.[10] Table 3 enlists the various coatings used in coil coating.

Emission of VOCs occur during mixing and application of the coatings, cleanup of mixing equipment, and storage of

Table 3. Coatings Used in Coil Coating

Coatings	Volatile Content (wt%[a])
Acrylics	40–45
Adhesives	70–80
Alkyds	50–70
Epoxies	45–70
Fluorocarbons	55–60
Organosols	15–45
Phenolics	50–75
Plastisols	5–30
Polysters	45–50
Silicones	35–50
Vinyls	60–75
Zincromet™	35–40

[a]The volatile content by volume is generally 5–10% greater than the volatile content by weight.

Source: Ref. 11.

coatings and solvents. The various types of coatings discussed in the following section are[12]:

1. High-solids coatings
2. Water-borne coatings
3. Powder coatings

High-Solids Coatings

High-solids coatings contain a greater proportion of film-producing solids than conventional coatings. The solvent content is reduced by the increased solid content. It is because of the increase of solids that the coating will cover more surface area than a conventional coating and hence less amount of coating is required. The control efficiencies for high-solids coatings have been tabulated in Table 4.

The high-solids coatings are further classified into single- or multicomponent systems. Single-component coatings take longer to polymerize a process that is hastened by oven baking. The multicomponent coatings have a mix of coatings and a catalyst so that the oven baking process is not required.

Water-Borne Coatings

Water-borne coatings use water as a major solvent or dispersant. These coatings contain VOCs as a co-solvent. The VOC concentration ranges from 5 to 40% by weight of coating. Water-borne coatings exist in three basic forms: water-soluble dispersions, water-soluble polymers, and emulsions. The water-soluble dispersions are usually transparent coatings, water-soluble polymers are translucent, and emulsions are opaque. Water-borne paint has a higher initial gloss and superior weathering properties. The chalking and fade resistance are much better than the solvent-borne paint.[13]

The advantages of water-borne coatings include:

- Reduced fire hazard
- Lower toxicity
- Reduced VOC emissions from coating
- Reduced VOC cleanup

Table 4. Control Efficiencies for High-Solids Coatings

High-Solids Coatings	Emission Reduction by Weight or Volume (%)[a]
60% by volume solids	
Flat surface	62
Complex surface	61
65% by volume coatings	
Flat surface	69
Complex surface	69
70% by volume coatings	
Flat surface	75
Complex surface	75
80% by volume coatings	
Flat surface	85
Complex surface	85

[a]Some round off error (approximately 2%) may exist for the calculations.

Source: Ref. 1.

The disadvantages of water-borne coatings are:

- Better surface preparation
- Greater temperature and humidity control during curing
- Stainless steel application equipment
- Special care while using electrostatic equipment
- Electrical grounding of application to reduce shock hazards

The typical organic solvent contents for conventional solvent-borne coatings used in various industries are given in Table 5. The HAP emissions from facilities with coating operations are given in Table 6.

Powder Coatings

Powder coatings are composed of fine, dry particles of paint solids; they usually contain less than 4% VOCs. These particles are applied to the surface using electrostatic attraction and are then baked to fuse the particles into a smooth continuous film. The resins used in the powder coatings are different from the resins used in the liquid coatings. Liquid coatings require low-viscosity and low-molecular-weight resins, that are soluble

Table 5. Typical Organic Solvent Contents for Conventional Solvent-Borne Coatings Used in Various Industries

Industry	Solvent-Borne Coating	Typical % Solvent by Volume
Metal furniture	Not specified	65
Automobile and light-duty truck	Enamel	67–76
	Laquer	82–88
Automobile refinishing	Enamel	72–76
	Laquer	87–91
Large appliance	Not specified	70
Traffic marking	Alkyd	50

Source: Ref. 4, 5, 6, 14, 15.

Table 6. HAP Emissions from Facilities with Coating Operations

Coating System	HAP Emitted	Emissions (TPY)
Solvent-based	MEK	1–25
	Methylene chloride	50–163
	Methyl isobutyl ketone	0.2–47
	Toulene	0.1–85
	Xylene	0.3–315
	Vinyl acetate	5
	Vinyl chloride	5
Water-based	Ethyl acrylate	5
	Ethylene glycol	3
	Formaldehyde	1
	Glycol ethers	75
	Methanol	15
	Phenol	8
	Styrene	4
	Vinyl acetate	1.6
	Xylene	1.31

Source: Ref. 2.

Table 7. Emission Reduction via Powder Coatings

Coating	Emission Reduction %
Thermosetting powders[a]	
Epoxy	97–99
Acrylics	99
Polyster (urethane)	96–98
Thermoplastic powders[b]	
Polyster (others)	99
Acrylics	99
Polyvinyl chloride and cellulose acetate	
Butyrate	90–95

[a]VOC emissions from bake oven only.
[b]VOC emissions from application and cool-down areas.
Source: Ref. 1.

in organic solvents or water. Powder coatings require resins that are solid at room temperature, to facilitate application, and melt during the baking process.[16] The advantages of the powder coatings are:

- Better physical properties than conventional coatings for chemical and abrasion resistance properties
- Lower air flow required
- Touchup of parts is easy prior to baking
- Excess powder can be easily discarded

The emission reduction via powder coatings has been enlisted in Table 7.

The disadvantages of powder coatings are:

- High capital costs of the system including powder reclamation booth
- High baking temperature required to form a film
- Difficulty in mixing of powder coatings to avoid removal of all initial powder

Alternative Coatings

To assist coating users in identifying and learning about lower-emitting coatings, Research Triangle Institute (RTI) is working in cooperation with the U.S. Environmental Protection Agency's Air Pollution Prevention and Control Division (APPCD) to develop a coatings guide for use on a personal computer. This guide, called the Coatings Alternatives Guide (CAGE),[17] contains an expert system and information source. CAGE provides technically innovative, cost-effective and low-polluting coatings for specific applications to the users of coatings and technical assistance providers. Initially, the focus of CAGE was to provide information on low-emitting coatings for metal parts and products painting to those coating users. Future versions of CAGE will include nonmetallic substrates such as wood and plastics.

APPLICATION OF COATINGS

After the preparation of the surface, one or more coatings are applied. The first coating consists of a primer, which provides corrosion resistance, fills the surface imperfections, and provides a bonding surface for the top coat.[18] The top coat determines the color of the article.

There are many ways of application of coatings. They are discussed in the following.

Dip Coating. Dip coating involves the immersion of individual pieces into a tank for coating. After withdrawal, the parts are held over the tank allowing the excess coating to drip back into the tank. Very often the parts are dipped into the tank on hangars or trays. They can be dipped in batches or by conveyor line. This simple equipment allows for the coating of many shapes with high transfer efficiency provided the waste collects back into the tank. It is important to control the circulation, filtering, temperature, and viscosity control of the coatings. Specialized dip coating operations, such as electro/autodeposition systems may also be used.

Flow Coating. In flow coating metal parts or products are conveyed over a closed sink and a pumped stream of coating is allowed to flow gently over the surface of the part. The excess coating is filtered as it drains into the sink and is then pumped back to a holding tank for reuse.

Roller Coating. Roller coating is the application of a coating from a paint trough to a flat surface by a mechanical series of rollers. The machine typically has three or more power-driven rollers. One roller runs partially immersed in the coating and transfers the coating to a second, parallel roller. The strip or sheet to be coated is run between the second and third roller and is coated by the transfer of coating from the second roller. The distance between the rollers establishes the quantity of coating applied to the sheet or strip.

Electrodeposition. Electrodeposition (EDP) involves the application of a dc voltage between the coating bath (or carbon or stainless steel electrodes in the bath) and the part to be coated. The part acting as the cathode or the anode is dipped into the bath. Coating particles, being oppositely charged, are attracted from the bath to the part, yielding an extremely even coating. The coatings used in the EDP tanks are water-borne solutions. Transfer efficiencies for EDP are commonly above 95%.

Spray Coating. Spraying operations generate two types of air pollutants. These include VOCs and particulate matter (PM). PM is affected by the transfer efficiency of the operation and control efficiency of the filters that may be present. VOC emissions in this type of operations are usually controlled.[19] In an automobile plant air-atomized spray coatings are used in airless spray, electrostatic spray and high-volume, low-pressure spray. To protect the coated surface from dirt and to provide a well-ventilated area that protects the workers from solvent vapors, the coatings are sprayed onto the metal surface. Different kinds of equipment are available for the application of coating, and some of them are explained below. The transfer efficiency as a function of spraying method and sprayed surface is shown in Table 8.

Air-Atomized Spray. In an air-atomized spray the paint stream is dispersed by forced air. Compressed air is supplied to the spray gun at pressures up to 60 psi and to the container (pressure pot) at fluid pressures of 10 to 30 psi. As the coating stream enters the application mechanism (spray gun), the high-pressure air mixes with the coating, forcing the stream to break up into small droplets in a turbulent mist toward the part to be coated. Air spray is often employed in the painting

Table 8. Transfer Efficiency as a Function of Spraying Method and Sprayed Surface

Method of Spraying	Flat Surface	Table Leg Surface	Bird Cage Surface
Air atomized	50	15	10
Airless	75–80	10	10
Electrostatic			
Disk	95	90–95	90–95
Airless	80	70	70
Air atomized	75	65	65

Source: Ref. 20, Table 232, which shows overspray percentages.

operations in spite of its low transfer efficiency. A very fine finish or appearance can be achieved by adjusting the air, the fluid, and the fan pattern.

Airless Spray. Instead of air atomization, airless spray uses hydraulic atomization. The coating is supplied to the gun under high fluid pressure between 1500 and 3000 psi and then forced through a small orifice. As the coating stream exits the tip into the air, the sudden decrease in pressure atomizes the coating. The advantage of the airless spray is:

- It can apply coatings to large surfaces efficiently.

The disadvantages of the airless spray are:

- The high fluid pressure causes paint to bounce back.
- The fluid tip wear increases.
- The life of the pump is decreased.

Air-Assisted Airless Spray. Air-assisted airless spray combines air and airless features. An airless type fluid tipped spray atomizes the paint and shapes the fan pattern at fluid pressures around 800 to 1000 psi. Lower-pressure air from 10 to 30 psi combines at the spray cap to adjust the fan shape to eliminate heavy edges. Air-assisted airless spray provides for greater pattern adjustment than airless spray and often gives a much better finish. Reducing the air pressure reduces the surface turbulence resulting in reduced paint bounce back vis-á-vis atomization.

High-Volume, Low Pressure (HVLP) Spray. This type of spray equipment uses a high volume of low-pressure air to atomize and carry the coating. The air pressure is approximately 4 to 6 psi, and the air passes through an atomizing hole in laminar flow. This process increases the temperature of the atomizing air, resulting in decreased viscosity of the coating. The advantages of using this type of equipment are:

- Increased transfer efficiency
- Less fluid wear on parts
- Reduced fogging
- Decreased viscosity because of heated air created by the turbine units

The disadvantages of using this type of equipment are:

- High costs
- Relative unfamiliarity of operators with the new technology

Flame Spray Coating. In flame spraying (combustion spraying) finely divided metallic or nonmetallic materials are deposited onto a surface in a molten or semimolten state. Usually a powder, wire, or rod has to be deposited. It is fed at a controlled rate to the flame spray gun, which heats the material to the molten or semimolten state with combustible gases or electricity. Compressed gas then propels the particles to the surface at a speed of 10–20 m/sec. The coating is formed when the particles hitting the surface flatten and conform to the surface. Generally, no additional heating or curing is required. This process is compatible with most surfaces.

The advantages of this process are:

- No primer is required when coating thermoplastics.
- Sharp edges and welds are well covered.
- The coating can be used as soon as it cools.
- There are no runs, drips, sags, or incomplete cure.
- Coatings have high chemical, impact, and abrasion resistance.

Electrostatic Spray. Electrostatic spray is the application of a coating to a surface by applying a charge to the coating and providing a grounded surface for deposition. In electrostatic spray operations, the object to be coated is grounded while the paint is given an induced negative charge at the spray applicator. The negatively charged particles are electrostatically attracted to the grounded surface, and the pigment particles are deposited in the same manner as in a normal air spray. Consequently, there is reduced paint overspray because the coating is forcibly attracted by the substrate; therefore, fewer pigments miss the target.

The advantages of electrostatic spray include:

- Reduced VOC emissions
- Better quality control
- Reduced paint wastage

The disadvantages of the electrostatic spray are:

- Potential for electric shocks
- Potential for increased fire hazards

The transfer efficiency of the gun used to apply a coating is an important factor as far as pollutant emission is concerned. The transfer efficiency of a spray gun is defined as the ratio of the amount of solids adhering on to the surface to the amount of solids applied, expressed as a percentage. The use of application techniques such as electrostatic attraction, high-volume, low-pressure spray, and flow or dip coating significantly reduce the volume of coatings used, thus reducing the total emissions. The transfer efficiency of the spray gun is dependent upon numerous parameters such as type of spray equipment used, the particle size and geometry of the paint solids, air temperature and humidity, spray booth air velocity, electrochemical properties of the paint, viscosity of the paint, the time between mixing and application of the paint, the experience of the painter, speed and precision of application, and the maintenance of the equipment.

Electrostatic Bell and Disk. Other electrostatic methods use bells or disks. In these methods, the coating is atomized to a small extent by the centrifugal forces associated with rapid spinning of the bell or disk and to a greater extent by the high voltage applied to repel the particles from the disk or bell and from each other. In addition, the bell or disk housing may reciprocate up and down or back and forth allowing a complete coating of the object. The coating particles passing across the negatively charged surface of the bell or disk get a negative charge. These particles are then attracted to the positively grounded parts as controlled by the applied voltage and the centrifugal force of the system.

CURING OF COATINGS

Large ovens are usually used individually to dry and bake the primer and the top-coat applications. The automobile industry uses multibake ovens, usually after each coating is applied. Depending upon the kind of coating used, the oven temperature can range from 200 to 500°F. With no exhaust systems, in the bake ovens, the solvent concentrations can increase to an explosive level, as the solvent concentrations increase above the lower explosive limit (LEL). The value of LEL for a solvent can be obtained from the chemical's material safety and data sheet (MSDS).

AIR EMISSIONS CHARACTERIZATION

The surface coating process is comprised of several steps as discussed above. The following section consists of a general theory for the calculations of air contaminant emissions from surface coating operations.

General Coating Calculations

The liquid coating consists of solids and solvents in which the solvents can be a mixture of one or more volatile organic compound, water, and non-VOC solvent. Applying the mass balance equation to mass of coating, one obtains

$$M_{\mathrm{m}} = M_{\mathrm{sol}} + M_{\mathrm{voc}} + M_{\mathrm{w}} + M_{\mathrm{nVOC}} \tag{1}$$

where M_{m} is the mass of the coating; M_{sol}, the mass of solids; M_{voc}, the mass of VOCs; M_{w}, the mass of water; and M_{nVOC}, the mass of nonvolatile compound solvents.

The coating volume, including water and nonvolatile organic compound solvents, is the sum of all the compound volume in liters or gallons.

Density of Coating

The density of coating material (ρ_{m}) is the measure of mass per unit volume (i.e., grams per liter). The coating density depends on both the mass and volume of each solvent and solid present in the coating

$$\rho_{\mathrm{m}} = M_{\mathrm{m}}/V_{\mathrm{m}} \tag{2}$$

where ρ_{m} is the density of the coating material; M_{m}, the mass of the coating material; and V_{m}, the volume of the coating material. In many processes the volume of coating applied may be computed directly from the type of device used.

Actual Emissions of VOCs

The actual emissions are the emissions from the facility in a specified period of time.

$$E = P\,(\%\mathrm{VOC}) \tag{3}$$

where E is the VOC emissions from the coating; P, the amount of coating used per unit time; and %VOC, the percent VOCs present in the coating (expressed as a fraction).

The actual emissions (E_{a}) including transfer efficiency and the efficiency of the addon control equipment is

$$E_{\mathrm{a}} = P\,(\%\mathrm{VOC})\,(1\text{-tr eff})\,(1\text{-OE}) \tag{4}$$

where E_{a} is the actual VOC emissions from coating; tr eff, the transfer efficiency of the spray gun; and OE, the overall efficiency of the control equipment.

The hazardous air pollutant emissions from the VOCs emitted can also be calculated. The HAP contents of the paint can be obtained from the MSDS, in which they are denoted as weight percentage of the VOC.

Actual Emissions of Particulate

The emission rate of the particulate matter is given by:

$$\mathrm{PM} = P\,(\%\ \text{solids})\,(1\text{-tr eff})\,(1\text{-OE}) \tag{5}$$

where PM is the particulate matter emissions; and % solids is the percent solids present in the coating (expressed as a fraction).

An alternative method for computing emission rates is to use the available values of emission factors from different sources. For example, some emission factors are given in the FIRE database maintained by the U.S.E.P.A. (Web site: *http://www.epa.gov/ttn/chief/*). A comparison of five different methods for computing emission rates from painting operations is given by Dasgupta and Trapani.[19]

AIR POLLUTION CONTROL MEASURES

The emissions of VOCs and particulate matter are of much concern for the health and safety of the personnel directly involved with the surface-coating applications and also other external receptors. As explained in the introductory passage, emission of VOCs can result in the formation of ozone, which can cause much damage. Hence air pollution control measures and equipment become an integral part of any surface-coating operation, as discussed in the following paragraphs. Air pollution control measures that can be applied to surface-coating operations can generally be categorized into three areas: reduced-VOC coatings, higher-transfer-efficiency equipment, and addon air pollution control devices (thermal and catalytic incinerators, carbon adsorbers, and condensers). The control efficiencies for the addon air pollution control equipment are shown in Table 9. The industry-specific data on emission reduction efficiency associated with applicable air pollution control measures as applied have been presented for the following industries: metal furniture coating, coil coating, automobile and light-duty truck manufacturing, large-appliance manufacturing, automobile refinishing, and the application of traffic markings in Tables 10–13. At times, facilities in these industries may be using a different baseline coating (other than the conventional solvent-borne coating) as a result of regulatory requirements or for technical reasons. Therefore, the potential emission reduction efficiencies for these facilities will be different from those shown here.[1]

Table 9. Control Efficiencies for Addon Air Pollution Control Equipment

Addon Control Device	Process Being Controlled	Control Efficiency for Control Device % by Weight
Carbon adsorber	Spray booth	90
Carbon adsorber	Entire coating line	80
Thermal incinerator	Bake oven	96
Catalytic incinerator	Bake oven	90

Source: Ref. 1

Air pollution control equipment often consists of control devices such as an incinerator or an emission collection system (VOCs or particulate or both). EPA-required pollution abatement systems collect most of the fugitive VOC emissions from the drying oven and spray paint booth exhaust. This treatment system is connected to the spray painting booth, where the pollutants originate.

The efficiency of an air pollution control unit depends upon the efficiency of the abatement device and the collection system efficiency. The ratio of the outlet to the inlet VOC concentration determines the efficiency of the abatement device. Collection efficiency is the ratio of the emissions that enter an abatement device to those that arise from an uncontrolled operation. The important operations involved in the control operations are incineration and adsorption.

Incineration

Incineration destroys VOCs through thermal oxidation producing water vapor and carbon dioxide. Combustion of VOC-contaminated air requires an auxiliary fuel, such as natural gas. Incineration efficiency depends upon the time, temperature, and turbulence in the combustion. The time during which

Table 10. Theoretical Emission Reduction Potential Associated with Various New Coating Materials for Use as Automotive Body Coatings

Coating Type and Percent Solids by Volume	Application Method	Transfer Efficiency %	Solvent/Dry Solids Applied l/l	Potential Percent Emission Reduction When Replacing: Laquer 16 Vol. % Solids	Potential Percent Emission Reduction When Replacing: Enamel 28 Vol. % Solids
Solvent-based enamel, 28 vol %	Air spray	40	6.4	51	—
Solvent-based laquer, 16 vol. %	Air spray	40	13.1	—	—
Powder Coating, 97 vol %	Electrostatic spray	98	0.03	99+	99+
Water Based	Electrodeposition	96	0.12	99	98
Water-based,[a] 25 vol %	Air spray	40	1.3	90	79
Water-based,[b] 25 vol %	Electrostatic spray	87	0.39	97	94

[a]Assumed 82–18 water-organic-solvent ratio by volume.
[b]Assumed 88–12 water-organic-solvent ratio by volume.

Source: Ref. 4.

Table 11. Common Methods of Reducing VOCs in the Large-Appliance Surface-Coating Industry

Control Method	Means of Reduction	% Reduction over Uncontrolled Conventional Coatings
High-solids coatings (0.340 kg VOC/l)	Lower organic-solvent content	70[a]
Powder coatings	No use of organic solvent	99[a]
EDP (0.040 kg VOC/l)	Lower organic solvent content	94[a]
Water-based coatings (0.140 kg VOC/l)	Lower organic-solvent content	47[a]
Carbon adsorption[b,c,d]	Adsorption of hydrocarbon emissions on a carbon bed	Top-coat spray booth, 33
		Top-coat oven, 15
Incineration[b,c]	Catalytic or thermal oxidation of hydrocarbon emissions	Top-coat spray booth, 33
		Top-coat oven, 15

[a]Calculated by determining the difference in the organic-solvent content of the coating containing 0.61 kg VOC per liter (30% solids).
[b]81% overall control efficiency.
[c]Based on 40% emissions occurring in the spray booth, 40% in the flash off, and 20% in the curing oven.
[d]In addition to control of VOCs, solvents maybe recovered through carbon adsorption.

Source: Ref. 5.

Table 12. Matrix of VOC Emissions Reduction Alternatives for a Medium-Sized Automobile Refinishing Shop and Estimated Emission Reductions

Coating Alternatives	VOC Emission Reduction Alternatives: Solvent Recovery		Transfer Efficiency %		Emissions (tons/yr)	Emission Reduction (tons/yr)	%
	Yes	No	35	65			
Use of laquers and enamels (current practice)							
Baseline		X	X		3.63	NA	NA
Alternative 1	X		X		3.08	0.54	15
Alternative 2		X		X	2.46	1.17	32
Alternative 3	X			X	1.91	1.72	47
Replace laquer and enamel primers with waterborne primers							
Alternative 4		X	X		2.73	0.90	25
Alternative 5	X		X		2.34	1.28	35
Alternative 6		X		X	1.83	1.80	50
Alternative 7	X			X	1.44	2.19	60
Replace laquer and enamel clears with higher-solids clears							
Alternative 8		X	X		2.82	0.81	22
Alternative 9	X		X		2.28	1.35	37
Alternative 10		X		X	2.08	1.61	44
Alternative 11	X			X	1.48	2.15	59
Replace laquers with enamels							
Alternative 12		X	X		2.27	1.36	37
Alternative 13	X		X		1.73	1.90	52
Alternative 14		X		X	1.55	2.08	57
Alternative 15	X			X	1.00	2.63	72
Replace lacquers and enamels with urethanes							
Alternative 16		X	X		1.89	1.73	48
Alternative 17	X		X		1.35	2.28	63
Alternative 18		X		X	1.52	2.11	58
Alternative 19	X			X	0.98	2.65	73

Source: Ref. 14.

the VOCs are in the high-temperature zone must be sufficient to complete their combustion. The incinerator temperature also must be high enough to completely oxidize the VOCs. The turbulence must ensure adequate mixing of combustion air, auxiliary fuel, and the VOCs.

The two different types of incinerators are:

- Direct flame incinerators
- Catalytic incinerators

Direct flame incinerators depend on flame contact and high temperatures in the range of 1200 to 1500°F for complete VOC oxidation. Direct flame incinerators consist of refractory-lined chambers, burners, controllers, safety equipment, and heat recovery equipment.

Catalytic incinerators promote VOC oxidation at lower temperature through the use of a catalyst. An increase in temperature occurs across the catalyst bed as the VOCs are oxidized. The catalyst is often a platinum series metal deposited in a porous form onto a structure, such as honeycomb, or on pellets. The advantage of catalytic incinerator over a direct flame incinerator is that fuel savings occur due to a lower incinerator operating temperature.

The disadvantages of incinerators are:

- High capital costs
- Ongoing auxiliary fuel costs to facilitate VOC destruction should be less
- Reduced destruction efficiencies due to catalyst poisoning
- Production of additional gaseous pollutants such as oxides of nitrogen and carbon monoxide

Adsorption

Adsorption is the other unit operation used for the removal of VOCs. It is a surface-controlled phenomenon. The material mainly used as an adsorbent is activated carbon. Carbon adsorbs VOCs from the gaseous stream on to its surface. Carbon's very large surface area is the result of many pores present throughout it. The adsorption capacity of a unit is a function of the following:

- Amount of carbon
- Carbon characteristics

Table 13. VOC Emission Reductions for Alternative Traffic Marking Materials from Baseline

Traffic Marking Material	Typical Annual VOC Emissions (lb/mile yr)	Typical Annual VOC Reduction from Baseline (lb/mile yr)	% Reduction from Baseline
Solvent-borne paints	69	NA[a]	NA[a]
Waterborne paints	13	56	81
Thermoplastics	0	69	100
Preformed tapes			
Without adhesive primer	0	69	100
With adhesive primer	58	11	16
Field related	0	69	100
Permanent markers	0	69	100

[a]NA, not applicable.

Source: Ref. 15.

- Gas stream temperature and humidity
- VOC concentration
- Inherent chemical characteristics of the VOC

Carbon adsorption units contain either fixed or fluidized beds of carbon. The feed containing VOCs is pretreated to remove solids, liquids, and the longer-chain high-boiling organic compounds that would coat the carbon. A concentration gradient develops and a band of carbon becomes saturated with VOC as the air progresses through the bed. The entire bed becomes saturated over a period of time. The bed is regenerated in place by stripping the VOCs from the bed with steam or high-temperature nitrogen. The stripped VOCs are either recovered or disposed of.

The advantages of the carbon adsorption process are:

- High efficiency

The disadvantages of the carbon adsorption process are:

- High capital cost
- High operation costs
- Carbon regeneration

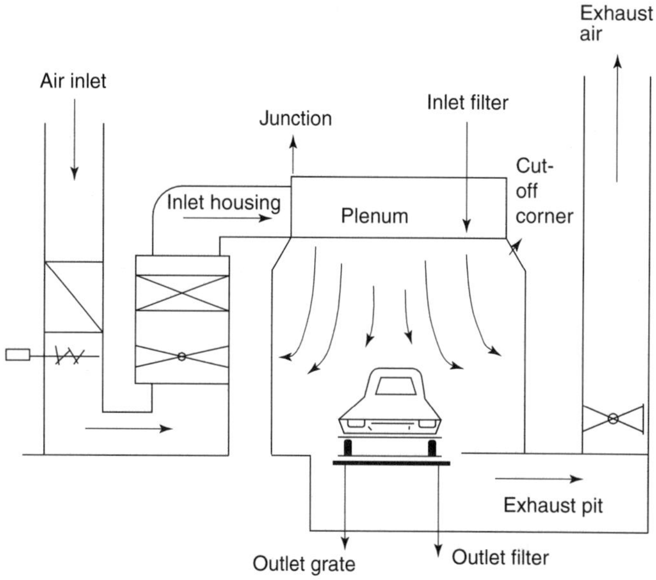

Figure 1. Schematic Painting Booth Configuration Performance of Painting Booths Equipped with Down-Draft Ventilation (*Source*: Ref. 7).

Condensation and absorption control devices could potentially be used at metal parts and products coating operations; however, neither technology is used to a large extent. Condensers reduce the temperature of the VOC stream to a temperature at which the VOC condenses out into a liquid. Absorbers usually pass a VOC-containing air countercurrent to a liquid stream, so that the liquid absorbs the VOC. The VOC must then be removed from the liquid.

Besides these, provision for proper enclosures in which to conduct the surface coating operations can also be regarded as an effective emission control measure.

An example of a paint spray booth design for emission control is explained in the following section.

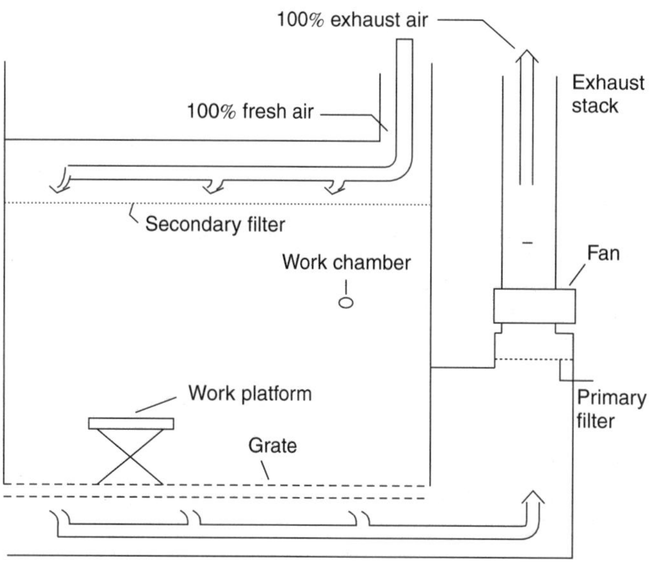

Figure 2. Conventional Spray Booth (*Source*: Ref. 21).

PAINT SPRAY BOOTH DESIGN

The general schematics of a paint spray booth and its details are presented in Figures 1[7] and 2,[21] respectively. The paint spray booth is an enclosed space, which is well ventilated for the purpose of containing the spray application of a surface coating onto a workpiece. The ventilating air is frequently heated or cooled and then drawn through the entire cross-section of the booth as uniformly and free of turbulence as possible at a velocity at least equal to the applicable OSHA minimum (60, 100 or 150 ft). The ventilating air entrains the overspray (wasted paint that forms a fine dust after failing to adhere to the workpiece) and solvent vapors and carries them through the exhaust stack to the atmosphere. The elimination of overspray greatly improves the quality of the finish; if overspray settles on the workpiece, the appearance of the finish will be unacceptably rough. The harmful overspray and toxic solvent vapors are greatly reduced if not eliminated from the worker's breathing zone by the ventilating air. The concentration of solvent vapor is kept low by the dilution effect of the ventilating air, thereby reducing the risk of fire and explosion.

The spray booths are either side-draft or down-draft. In a side-draft booth, the fresh ventilating air enters through one side of the booth, traverses the spray booth work area, and exits as exhaust air through the opposite side of the booth. In a down-draft booth, the fresh ventilating air enters through the ceiling, traverses the spray booth work area and exits through grates that form the floor. In both booths, a diffuser grill or filter section is often used to distribute the ventilating air uniformly across the cross-section of the booth to establish the laminar flow. The down-draft booth has the advantage of gravity. A disadvantage of the down-draft booth is increased expense in fabrication and installation. The downdraft booth is used for large workpieces such as automobiles, trucks, and aircraft, especially when a finish is required. The side-draft booth is typically used for smaller pieces such as furniture, and for components of larger products.

The size of the paint spray booth is determined by the largest workpiece plus space for the worker to maneuver while applying the spray paint. The volume of the ventilating air required is proportional to the size of the booth. The workpieces may enter and leave the booth on a continuous basis using a conveyor or they may enter periodically on a batch basis. The painted workpieces may be left in the booth to air dry or moved to a curing oven to force dry. The spray paint booths alone cannot capture all the VOCs; however, they help substantially in containing them. A spray paint booth is equipped with two supplemental units:

1. Application equipment: a device for application of coatings using different spray techniques.
2. Air pollution control equipment: a device/devices used for destruction/removal of air pollutants emanating from spray paint booths as discussed in the preceding sections.

Dry baffle spray booths use baffle plates to force the air to change directions and to provide a surface for the particulate to adhere to. Particulate removal efficiencies are comparatively low and depend on the particle size and how tacky the overspray is. Laquers, which dry rapidly, will not be collected well on dry baffles. Dry baffles require frequent cleaning. Dry filter spray booths use filter pads, usually of a woven mesh, through which air flow passes. The particulate is trapped on the filters. When the filters become loaded with paint, the pressure drop across the filters increases and the air flow through the filters decreases. Unless the filters are changed, the visibility in the work area will decrease.

Manometers enable the operator to determine the pressure drop provided they are filled with fluid, zeroed, and properly installed and operated. Filters, which contain activated carbon, are also being used. Additionally, powder spray booths often contain a special filter and reclaiming system to prevent the powder from settling on nearby surfaces where they can flow and set up on a hot day.

Water-wash spray booths may include wet curtains and spray nozzles. Water (or occasionally oil) flows down a baffle plate, which enhances the capture of particles and the flow washes the particulate down to a circulating tank. Additionally spray nozzles are often used to knock down the particulate. Greater particulate removal efficiencies can be achieved without the need to purchase filters. However surface-active agents are added to water to aid in the removal of paint from the tank. The proper maintenance of the booth and of the chemical additives is very important to ensure adequate removal efficiencies. Additionally, sludge is generated from water-wash bottles. Health services and water control agencies have additional requirements for the discharge of wastewater and for the disposal of the sludge.

ALTERNATIVE EMISSION CONTROL PLANS IN CALIFORNIA

Many local districts in California allow facilities to demonstrate compliance with surface-coating regulations for certain operations through an alternative emission control plan (AECP).[8] Under AECPs, sources must demonstrate that the emissions from the source are equal to or less than the emissions that would have resulted if they would have complied with the specific emission limit. If the emissions are offset through the use of coatings with less VOC, the sources are usually allowed to use noncomplying coatings. Only emissions from operations subject to a single prohibitory rule are allowed to be included in an AECP. In addition, emission reductions used must be surplus, permanent, quantifiable, and enforceable.

Several districts allow varied alternative emission control plans. Industries may use any means necessary to control emissions from their operation to an equivalent level, as would be expected to be achieved if all the processes complied with the requirements of the rule. The following general requirements are stated in the AECP policy and must be met with[8].

1. Facilities operating under an AECP must demonstrate that they comply with daily limits of the regulation through alternative means on a coating solids (production) basis.
2. Emission averaging periods in nonattainment areas may not exceed 24 h.
3. On a daily basis, the operations included in the emission trade must provide a 20% emission reduction from a daily weighted baseline calculated from the previous two operating years.
4. Emission reduction credits must be quantifiable by approved test methods, which eliminates existing credits for improved transfer efficiency in all categories, except automobile manufacturing topcoat applications.
5. Emission reduction credits must be enforceable, which eliminates existing credits for the substitution of exempt solvents.
6. Detailed daily usage of records must be maintained for all operations and must demonstrate compliance on a daily basis.

7. All AECPs must be submitted to the EPA as a source specific revision to the State Implementation Plan (SIP).
8. A State Assurance document must accompany the SIP revision certifying that the AECP has been administered properly, is technically sound and serves the intent of the SIP.

MAXIMUM ACHIEVABLE CONTROL TECHNOLOGY STANDARDS

The Emission Standards Division (ESD) of the U.S. EPA is currently developing national emission standards for hazardous air pollutants (NESHAP) for twelve industrial surface coating and composite categories.[22] The twelve categories are:

1. Auto and light-duty truck
2. Boat manufacturing
3. Fabric printing, coating, and dyeing
4. Large appliance
5. Metal can
6. Metal coil
7. Metal furniture
8. Miscellaneous metal parts
9. Paper, film, Foil, and others
10. Plastic parts and products
11. Reinforced plastic composites manufacturing
12. Wood building products

Under the Clean Air Act of 1990 (CAA), the U.S. EPA is required to regulate sources of hazardous air pollutants, which are known or suspected to cause serious health problems such as cancer. The U.S. EPA published a list of industry groups (source categories) that emit one or a combination of these air toxics, which contains several industrial surface coating and composites categories. Facilities that emit or have the potential to emit at least 10 tons/yr of a listed HAP or at least 25 tons/yr of a combination of HAP, termed "major sources," have to abide by the restrictions posed. The CAA directs the U.S. EPA to apply stringent air pollution reduction measures known as maximum achievable control technology in developing its standards. During development of these regulations, the U.S. EPA will evaluate whether nonmajor (also called "area") sources of the same type should also be regulated in accordance with the law.

In addition to the MACT standards, to reduce the volatile organic compound emissions as required under section 183(e) of the CAA, for eight of the twelve surface coating and composite categories, the U.S. EPA is also developing national rules or control technique guidelines (CTG). A CTG may affect facilities in ozone nonattainment areas with smaller amounts of emissions than the 10–25 tons/yr for major HAP sources. This may affect more small businesses. Also, even small businesses can have emissions that qualify them as a "major source"; so small entities can be affected by both the VOC and HAP standards.

The coating and composites MACT standards are all scheduled for final rules by November of 2000. Each MACT standard specifies the dates by which the owner or operator must comply with the requirements. In general, the maximum amount of time allowed by the statute for the CAA for compliance with MACT standards is 3 years after the promulgation of the final rule for existing sources, and upon startup for new sources. The section 183(e) national rules or CTGs for these categories are planned for issuance in roughly the same time frame as the MACT standards.

REFERENCES

1. M. B. Turner, *Surface Coating*, Air Pollution Engineering Manual, Air and Waste Management Association and John Wiley and Sons, 1992.
2. U.S. EPA Web site: *http://www.epa.gov/ttn/uatw/coat/fabric/pic-fabr/pdf*, 1999.
3. U.S. EPA Web site: *http://www.epa.gov/ttn/uatw/coat/flatw/wbppic.pdf*, 1999.
4. *Automobile and Light-Duty Truck Surface Coating Operations—Background Information for Proposed Standards*, Draft EIS, U.S. Environmental Protection Agency, Research Triangle Park, NC, EPA-450/3-79-030, 1979.
5. *Industrial Surface Coating: Appliances Background Information for Proposed Standards*, Draft EIS, U.S. Environmental Protection Agency, Research Triangle Park, NC, EPA-452/3-80-037, 1980.
6. *Surface Coating of Metal Furniture—Background Information for Proposed Standards*, Draft EIS, U.S. Environmental Protection Agency, Research Triangle Park, NC, EPA-450/3-88-007, 1988.
7. N. Goyer, "Performance of Painting Booths Equipped with Down-Draft Ventilation," *Journal of American Industrial Association* **56**, 1995.
8. *Metal Parts and Products Surface Coating Operations, Compliance Assistance Program*, California Air Resources Board Compliance Division, 1989.
9. W. J. Athanson, *New World Automotive Chemical Coatings—Solvents Glossary, Autobody Repair and Paint Guide*, D. Van Nostrand Co., New York, NY, 1995.
10. *Annual Book of ASTM Standards, Paint, Related Coatings and Aromatics*, American Society for Testing and Materials, Vol. 06.01, D-16-83a, 1984.
11. *Metal Coil Surface Coating Industry—Background Information for Proposed Standards*, Draft EIS, U.S. Environmental Protection Agency, Research Triangle Park, NC, EPA-450/3-80-035a, 1980.
12. S. Paul, *Surface Coatings, Science and Technology*, John Wiley and Sons, Inc., 1985.
13. T. W. Mc Daniel, and L. L. Pickens, "Reducing Emissions from Truck Chassis Painting in an Existing Paint Spray Booth," Paper #95-RP138.01, Air and Waste Management Association, Proceedings of the 88th Annual Meeting and Exhibition, San Antonio, Texas, June 18–23, 1995.
14. *Reduction of Volatile Organic Compound Emission from Automobile Refinishing*, U.S. Environmental Protection Agency, Research Triangle Park, NC, EPA-450/3-88-009, 1988.
15. *Reduction of Volatile Organic Compound Emission from the Application Traffic Markings*, U.S. Environmental Protection Agency, Research Triangle Park, NC, EPA-450/3-88-007, 1989.
16. G. L. Schneberger, *Understanding Paint and Painting Processes*, Third Edition, Hitchcock Publishing Co., New York, NY, 1985.
17. D. R. Cornstubble, J. N. Baskir, and M. Kosusko, "A Personal Guide for Selecting Alternative Coatings for Metal Parts and Products Painting," Paper #96-WA4D.01, Air and Waste Management Association, Proceedings of the 89th Annual Meeting and Exhibition, Nashville, Tennessee, June 23–28, 1996.
18. B. J. Suther and U. Potasku, *Controlling Pollution from the Manufacturing and Coating of Metal Products—Metal Coating Air Pollution Control*, U.S. Environmental Protection Agency, Vol. 1, EPA-625/3-77-009 1977.

19. S. Dasgupta, and M. F. Trapani, "Emission Estimation from Spray Painting Operation," Iowa Waste Reduction Center, University of Northern Iowa, Paper #95-MP1.04, Proceedings of the 88th Annual Meeting and Exhibition of Air and Waste Management Association, San Antonio, Texas, 1995.
20. J. A. Danielson, Ed., *Air Pollution Engineering Manual*, AP-40, 2nd ed., Air Pollution Control District of Los Angeles, Environmental Protection Agency, Office of Air and Water Programs, and Office of Air Quality Planning and Standards, Research Triangle Park, NC, 1973.
21. C. M. Smith, and W. E. Brown, "Conventional Spray Booth, Elimination of VOC Emissions from Surface Coating Operations," *Air and Waste*, **43**, 1018, 1993.
22. U.S. EPA Web site: *http://www.epa.gov/ttn/uatw/coat/smbizpg.html*, 1999.

11

GRAPHIC ARTS INDUSTRY

NONHEATSET WEB PRINTING

H. Wilson Cunningham and C. Nelson Ho

In web printing, a rotary press is used to print an image on a continuous web of paper that may exceed 8 miles (12 km) in length. After printing at speeds between 600 and 2100 fpm (180–670 m/min), the web is mechanically cut and manipulated into its final form.

The nonheatset or cold-set web printing process can be distinguished from other web printing processes by its unique ink and paper requirements. Nonheatset web inks are semifluid materials whose function depends on the rapid absorption of the liquid component of the ink vehicle by the surface of the paper. The capillary action of the paper surface draws the liquid into the paper surface and "sets" the ink. This process does not require the application of heat to cure the resin or evaporate the liquid component.

Newsprint, which accounts for most of the paper used for nonheatset web printing, is a lightweight paper stock that weighs 44–53 g/m^2. The paper must have enough absorptive capacity to allow rapid penetration of the ink vehicle. This requirement excludes the use of coatings and permits a minimum of filler materials or machine finishing that smoothes the surface and decreases the rate of absorption.

Offset lithography and letterpress printing represent the nonheatset web printing processes. The flexographic process, another such process, is covered in another section of this chapter.

PROCESS DESCRIPTION

Letterpress printing uses a relief plate similar to a rubber stamp to transfer ink from its surface directly to the surface of the paper (Figure 1). Newspaper printing accounts for 90% of letterpress production, and book and specialty printing represent the remainder.[1] Letterpress news inks are relatively simple inks consisting of an oil vehicle, resinous binder, and solid pigment (Table 1). Letterpress printing, the only mode of newspaper printing up to 1960, represented less than 31% of the daily circulation in 1987.[2] Conversion to offset lithography—and, since 1987, to aqueous flexography—has caused the decrease in newspaper letterpress printing.

Nonheatset web printing represents about 35% of the diverse offset lithography classification of printing. The 1400 daily newspapers that use offset printing[2] account for most of the nonheatset web offset market.[1] Offset lithography differs from letterpress printing in that a planographic printing surface is used that has been chemically modified into a hydrophobic (oleophilic) ink-receptive image area and water-loving nonimage area. To keep the hydrophobic ink film from migrating into the nonimage region of the plate, a second fluid, water, is applied to the plate in conjunction with the ink. A thin film of ink is transferred from the plate to an intermediate blanket cylinder before being transferred to the moving paper web (Figure 1).

The required presence of an immiscible liquid and the amount of ink transferred from a planographic plate have a profound effect on the composition of nonheatset web offset (NHWO) ink (Table 1). An offset print receives about 30% less ink than a corresponding print made from a relief plate.

This thinner ink film necessitates an increased pigment concentration to provide equal print density. Suspending and properly dispersing more solid pigment within the liquid vehicle require an increase in the amount of resinous binders. Solubilization of the resinous binder affects the types and amounts of hydrophobic liquids in NHWO inks. The choice of the resin depends on its ability to wet the pigment surface for dispersion and its ability to release the liquid component only after the ink has been applied to the surface of the paper. In NHWO inks, this release process, which "sets" the resin and pigment into the paper, cannot be dependent on the evaporation of the liquid by the addition of heat. Finally, the resinous binders cannot dramatically increase the adhesion of the ink film that will pick paper fibers from the uncoated paper surface. The chemical characteristics of the resin should not drastically increase the amount of water emulsified by the ink.

The dampening system is the portion of an offset press that applies water to the surface of the planographic plate. The contact dampening system applies a film of water directly to the plate by some type of roller or misting system. The ink train dampening system applies the water solution to the ink roller train, where an emulsion forms. This ink–water emulsion is then transferred to the plate surface, and the shear of the printing nip causes the emulsion to break into its hydrophobic ink component and nonimage-area-protecting water component. Both types of dampening systems are open to the atmosphere. Refrigeration of the dampening system in NHWO is extremely rare because the use of low-molecular-weight isopropyl alcohol is not common.

The water component, referred to as fountain solution or etch, used in the lithography printing process is a dilute chemical mixture. The chemicals are added to lower the

Table 1. Composition of a Typical Nonheatset Web Printing Ink

	Component[a]	Typical[b] Range (%) Letterpress	Offset
Pigment	Carbon black	8–15	15–20
	Organic color pigment	6–12	8–15
	Rheological solid[c]	0–1	0–10
Resinous binder[d]	Alkyd resin	0–7	0–20
	Thermoplastic resin	0–7	0–20
Hydrophobic vehicle[e]	High-viscosity oil, 310–550°C bp	0–85	0–80
	Low-viscosity oil, 230–300°C bp	0–10	0–50
	Vegetable oil	0–85	0–80

[a]Typical materials used in formulation. Materials in this category may be used singly or in combinations.
[b]Maximum and minimum amounts of component that may be found in a typical formulation.
[c]Rheological solids are inert materials, such as a clays, used to adjust viscosity and tack of ink.
[d]Resinous binders are solid thermoplastic materials solubilized in the hydrophobic liquid to create the ink vehicle.
[e]Hydrophobic liquid is a hydrotreated or severely solvent-refined mineral or vegetable oil. Low- and high-viscosity petroleum oils can be distinguished by boiling-point ranges and viscosity range. High-viscosity oils have viscosities between 750 and 2400 Saybolt universal seconds (SUS). Low-viscosity oil has a 50–100 SUS range.
Source: Reference 3.

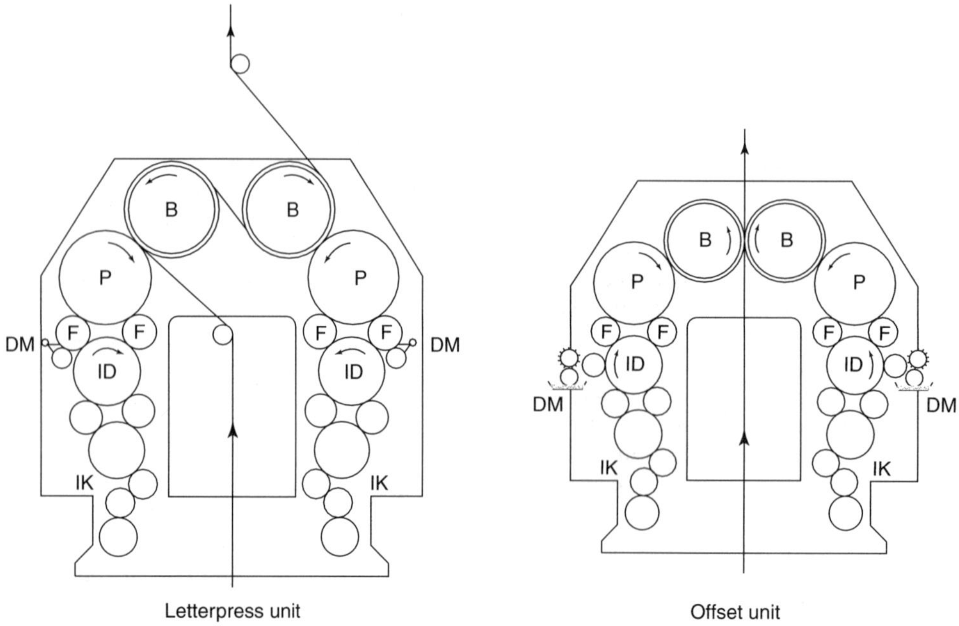

Figure 1. Typical Components of Offset and Letterpress Nonheatset Web Printing Units. Identification of parts: IK, ink train roller system; ID, ink drum roller; F, form roller; P, plate cylinder; B, blanket cylinder for offset and impression cylinder for letterpress; DM, ink train type of dampening system.

surface tension of the water and provide a controlled acidity level (Table 2). Fountain solutions used in NHWO printing can be distinguished by the lack of low-molecular-weight isopropyl alcohol. The low tack and viscosity of the NHWO inks do not require the aggressive wetting power provided by the addition of low-molecular-weight isopropyl alcohol. Instead, alcohol substitutes, high-boiling-point glycols and glycol ethers, have the advantage of providing lubrication and improved release characteristics at the blanket-plate printing nip. In recent years, several manufacturers of fountain solution have developed different types of fountain etch used in NHWO printing, especially in business form printing, without using any isopropyl alcohol or glycol ether–based alcohol substitutes. In other words, those types of fountain solution are literally "alcohol free" as well as "0" VOC.

Table 2. Chemical Additives Listed on Material Safety Data Sheets for Fountain Solution[a] Used in Nonheatset Web Offset Printing

Phosphate salts
Silicates
Colloid materials such as gum arabic or dextrins
Organic surfactants
Alcohol substitutes

[a]Fountain solution is a dilute water solution with specific conductance between 1000 and 2500 mho that may contain some or all of the listed components. Typical fountain solutions are either acidic (pH 3.0–5.5), neutral (pH 6.5–8.5), or alkaline (pH 9.5–11). Phosphate salts are used to maintain the pH of the fountain solution. Alcohol substitutes are high-molecular-weight glycols, polyols, and possibly glycol ethers in concentrations below 0.5%. Use of low-molecular-weight alcohols such as isopropanol is not common.

AIR EMISSIONS CHARACTERIZATION

Ink, fountain solution, platemaking, and press-cleaning operations represent the four potential sources of air emission from the nonheatset web printing process. Unfortunately, actual emission factors are not known at the time of this writing, and most data must be extrapolated.

Ink

Potential emissions from the ink used in nonheatset web printing could occur during the printing process. The design of a nonheatset web printing press requires the ink to travel through a series of ink-distribution rollers, known as the ink train, that control the application of ink to the plate. The potential evaporation from the operating press would depend on the temperatures of this roller system and the volatile components of the ink.

In 1989, the American Newspaper Publishers Association (ANPA) used an infrared pyrometer to measure ink-train temperatures for large and small offset and letterpress printing operations. The measurements were taken in February and repeated in July. The maximum temperature observed during the production cycles was 42°C. As it occurred at press stoppage, only a small amount of ink would be exposed to this temperature. The temperature of an operating press was about 3°C lower. No difference was observed between the summer and winter operating temperatures.[4]

Exposure to temperatures that are relatively low in comparison with the boiling point of the liquids used to manufacture nonheatset web inks would suggest that emissions during press operations are minimal. The ANPA, in cooperation with the National Association of Printing Ink Manufacturers (NAPIM), measured the emissions during press operations for three different formulations of NHWO inks (Table 3). Air samples from the ink train and the paper web as it exited the press were taken using standard National Institute for Occupational Safety and Health (NIOSH) procedures for measuring hydrocarbon exposure. Emissions, determined by gas chromatography, were between 0.6 and 5.7 mg/m^3/h. The observed differences in the emissions were a result of ink formulation. Extrapolation of these data to a nine-unit press operating for 24 h predicts an emission of less than 1.3 g of hydrocarbon into the atmosphere.

These emission data support the theory that all materials transferred to the paper penetrate the surface and are retained within the product. Measurement of the loss of any retained materials from the printed sheet is hampered by the low concentration of ink (about 1% to 2% by weight) on a printed page and the very hygroscopic nature of uncoated paper. Even under controlled humidity and temperature, the gain or loss of water exceeds the accuracy of any gravimetric procedure to detect emissions of organic components from the printed product.

Laboratory oven tests such as Method 24 of the American Society for Testing and Materials (ASTM) were developed to accelerate the evaporation of volatile components from products designed to dry by evaporation. The tests were not designed to be applied to products such as NHWO inks where the drying mechanism is not evaporation. However, certain regulatory authorities have chosen to apply Method 24 to nonevaporative systems in an attempt to predict the potential loss of material over its extended lifetime.[5] This type of testing does not take into consideration the reduction of emissions that can occur in a nonheatset web product as a result of the pacification of the liquid component within the "set" resin and pigment complex or the absorption of the material within the cellulose fiber. Until this reduction in emissions can be determined quantitatively, the severe conditions of Method 24 will continue to overestimate the emissions from nonheatset web inks.

Method 24 demonstrates that different nonheatset inks have different potential emissions (Table 4). Most of the differences seen in this table reflect the ratio of low-viscosity to high-viscosity mineral oil required to formulate the performance of a given ink or the substitution of vegetable oils for the petroleum oil. Nonheatset web ink formulated with vegetable oil (e.g., soya oil) has significantly lower emission potential than inks formulated with petroleum oils. The choice of different types of petroleum oils not traditionally used in ink formulations may have a similar effect of lowering the potential emissions from nonheatset printing inks.

In the past decade, many business form printers, part of the NHWO family, have elected to use UV-curing nonheatset web inks because the final products have better and shinier look. This type of inks is mainly composed of UV-sensitive alkyl acrylate esters, trimethylpropane triacrylate, methyldiethanolamine, and other UV-sensitive chemicals. Unlike the soya-based inks that still have a small amount of VOCs, the UV-cured inks are basically "VOC free" or have very minuscule amount VOCs.

Air Pollution Control Measures. The low volatility of the components used in nonheatset web printing inks limits the

Table 3. Volatile Organic Component Emission Rates Measured with Three Different NHWO Inks

Sampling Location	Ink Type: Low-Rub[a] Black	No-Rub[b] Black	Petroleum[c] Black
Method 24 VOC	10%	19%	5%
Right ink form roller (mg/m^3/h)	3.1	5.0	2.6
Left ink form roller (mg/m^3/h)	3.7	5.2	3.2
Pipe roller (mg/m^3/h)	0.6	5.7	1.1

[a]Low-rub formulation contained 60% petroleum oil that is a mixture of low-viscosity and high-viscosity oils.
[b]No-rub formulation composed of 45% low-viscosity oil and 10% high-viscosity oil.
[c]Regular black contained about 75% high-viscosity oil.
Source: Reference 4.

Table 4. Content of Four Classes of Printing Inks Determined by ASTM Method D-2369-81 (Method 24)

Ink Type	Mean (%)	Range (%)
Color offset[a]	9	2–20
Black offset	9	2–13
Color letterpress	12	10–14
Black letterpress	5	5–10

[a]The lower range of these samples represents samples formulated with soya oil. No soya oil formulations of black offset, color letterpress, or black letterpress were available for testing.
Source: Reference 4.

control technology options. Applications of drier systems would be expensive and would have little effect on the removal of the boiling-point petroleum component from the printed page. The heat from the drier system also would reduce the viscosity of the ink vehicle and would allow the materials to penetrate the paper instead of being isolated near the surface. This penetration would destroy the visual integrity of the product. The operating-press emission data demonstrate that press enclosures and air purification systems would have little benefit in reducing the release of volatile compounds into the environment.

If nonheatset web inks were considered a significant source of volatile materials, the best control technology would be in ink formulation. The data indicate that the inks[4] are not a major source of volatile materials during printing.

Fountain Solutions

Nonheatset letterpress printing does not require a dampening system; therefore, emission from the fountain solution does not apply to this printing process. Fountain solutions used in NHWO printing are generally purchased in concentrated form and diluted with water to create a "working solution." The working solutions are at least 99.5% water by volume. High-boiling-point glycols and glycol ethers represent the remainder of the working-solution liquid composition. These compounds, in combination with organic surfactants, have effectively replaced isopropyl alcohol as the primary wetting agent in the fountain solution.

The mechanical design of the dampening systems on most presses requires that the working solution be open to the atmosphere. Any evaporation from these systems may result in a measurable worker exposure to any organic compound in the fountain solution. Since 1985, ANPA has conducted about 45 chemical air-sampling studies in various nonheatset web printing operations. Concentrations of glycols and glycol ethers were below 5 mg/m^3 at all sites tested. These data suggest that the potential emission from the dampening system of an NHWO press operating for 24 h would be less than 1.2 g/d.

Air Pollution Control Measures. The limited amount of information that is available from manufacturers' Material Safety Data Sheets suggests that the NHWO dampening system is not a major source of emission. If the printer adds isopropyl alcohol to the dampening system, refrigeration of the system or replacement with less volatile alcohol substitutes may be required.

Plate Making

Nonheatset web letterpress plate-making chemistry is very dependent on the manufacturer's polymer system. The chemistry of these systems is highly proprietary, and only limited information is available.

The NHWO plates use an aqueous diphenylamine azide coating as the photoactive polymer. After polymerization of the image area by ultraviolet light, the image is developed by an acidified asphalt–water emulsion. Traditionally, this emulsion contained an alcohol–ketone component to stabilize the emulsion. Concerns about wastewater discharges and worker exposure have all but eliminated this volatile component from the plate developer. Normally, there are very little VOC emissions involved in plate making and film making, if any.

Air Pollution Control Measures. All systems are vented, and if the emissions are considered a problem, air purification systems could be installed. Obviously, any exhaust containing significant levels of air contaminants could be processed through air pollution devices such as activated-charcoal filters.

Press-Cleaning Operations

Press cleaning, the removal of the built up layer of ink and paper dust from rollers and blankets, is a necessary maintenance operation for the nonheatset web press. The cleaning of a nonheatset web press, however, differs from other printing processes in that the inks do not readily dry or oxidize into a solid film. These "soft" films do not require aggressive degreasers for removal. Traditionally, the press-cleaning solvent of choice has been petroleum distillates such as deodorized mineral spirits and kerosene. These materials sometimes are mixed with water and detergents to form a cleaning solution mainly to remove paper lint and paper dust.

The best available data from the ANPA chemical air sampling in nonheatset web operations would suggest that press cleaning is the source of 90–95% of all emissions from the nonheatset process.[4] The total amount of emissions is hard to estimate because of the different cleaning procedures used in different printing locations.

Air Pollution Control Measures. Viable control technologies to reduce the emission from this periodic cleaning action are to reduce the amount of solvent in the cleaning solution required for cleaning or to use less volatile solvents. The water extension is a common means of reducing the volume of solution used. The unique nature of nonheatset web inks may also permit the use of an aqueous-based cleaner. Various manufacturers are currently testing these products, but their functionality has not been proved as yet.

REFERENCES

1. *The Rauch Guide to the U.S. Ink Industry*, Rauch Associates, Bridgewater, NJ, 1988.
2. *Facts about Newspapers '90*, American Newspapers Publishers Association, Reston, VA, 1990.
3. P. Volpe, National Association of Printing Ink Manufacturers, Harrison, NY, personal communication, 1990.
4. C. S. McLendon and H. W. Cunningham, *Volatile Organic Content of Non-Heatset Newspaper Printing Inks*, American Newspaper Publishers Association, Reston, VA, 1989.
5. J. Berry, Chief of Chemical Applications Section, U.S. Environmental Protection Agency, Research Triangle Park, NC, personal communication, 1990.

BIBLIOGRAPHY

N. R. Eldred and T. Scarlett, *What the Printer Should Know About Ink*, 2nd ed., Graphic Arts Technical Foundation, Pittsburgh, 1990.

R. H. Leach, *The Printing Ink Manual*, 4th ed., Northwood Books, London, 1988.

HEATSET WEB OFFSET

Robert M. Birkland, Gerald Bender, William Schaeffer, and C. Nelson Ho

Heatset web offset lithography is the process that produces over 75% of magazines, catalogs, books, and tabloids. This process is similar to other offset lithographic processes in the type of plates used and the requirement for fountain solution to maintain a printing image. Differences include the web speed, the inks that require energy to dry, energy-application devices (dryers) to dry the ink, web cooling devices (chill rolls), and ink oil emission control devices, such as catalytic, thermal, and regenerative afterburners.

PROCESS DESCRIPTION

The heatset web offset printing process is a complex process that varies depending on the type of product to be produced. This process is used to produce a vast array of materials, from complex technical manuals with highly illustrated images to simple one-color text. As a result, the process can vary from the use of one-color printing stations to complex multicolor presses utilizing up to eight or more printing units. In its basic form, the web offset process consists of an unwind stand, in-feed, printing units, dryer, chill stand, folder, sheeter, and delivery table. A typical web offset printing press is shown in Figure 1.

The reel is the mechanism that allows the continuous unwinding of rolls of paper for subsequent printing in the printing units. This mechanism has the capability of splicing the expiring roll of paper to a new roll of paper without stopping, thus maximizing the operational efficiency of the press. The in-feed tension device keeps a constant tension throughout the press and isolates the press from the tension upsets that occur when a splice is made. Without this device, web offset printing would be extremely difficult to accomplish because the physical appearance of the printed material will vary depending on the tension applied to the paper.

After the paper web unwinds on the roll stand or reel, it may pass through a heated web conditioner before entering the first printing unit. Ink and fountain solution is applied to the web, typically on both sides, as shown in Figure 2. Depending on the number of colors being printed, the web will pass either into the dryer or to a number of additional printing units before entering the dryer. The dryer will apply energy to cause the ink oil to evaporate. Because the web has been heated by the dryer, it is passed onto the chill rolls, where it is cooled to about 20 degrees above ambient temperature before folding and cutting.

Dryers for heatset web offset-type inks are typically recirculating hot air systems, although some direct flame impingement and infrared dryers are still in use. Usually, air, flame, or radiation is applied to both sides of the web within the dryer to raise the web temperature to about $275 \pm 50°F$. The volatile portions of the ink are exhausted from the dryer, either to an emission control device or to the atmosphere. Evaporation from the hot web exiting the dryer is collected by either a tunnel or a hood extended from the dryer over the chill rolls.

In addition to the inks that dry by evaporation, others cure by chemical reaction among components. Reactive-cure ink systems include three major categories:

1. Ultraviolet-radiation cure (UV)
2. Electron-beam-radiation cure (EB)
3. Thermal catalytic cure

Reactive cure inks contain only a nonvolatile reactive binder or film former and pigment. Dryers for chemical reaction inks are usually very-high-energy UV lamps, electron-beam generators, or thermal ovens. The absorption of this energy within the ink causes the film former to cure and harden. Very

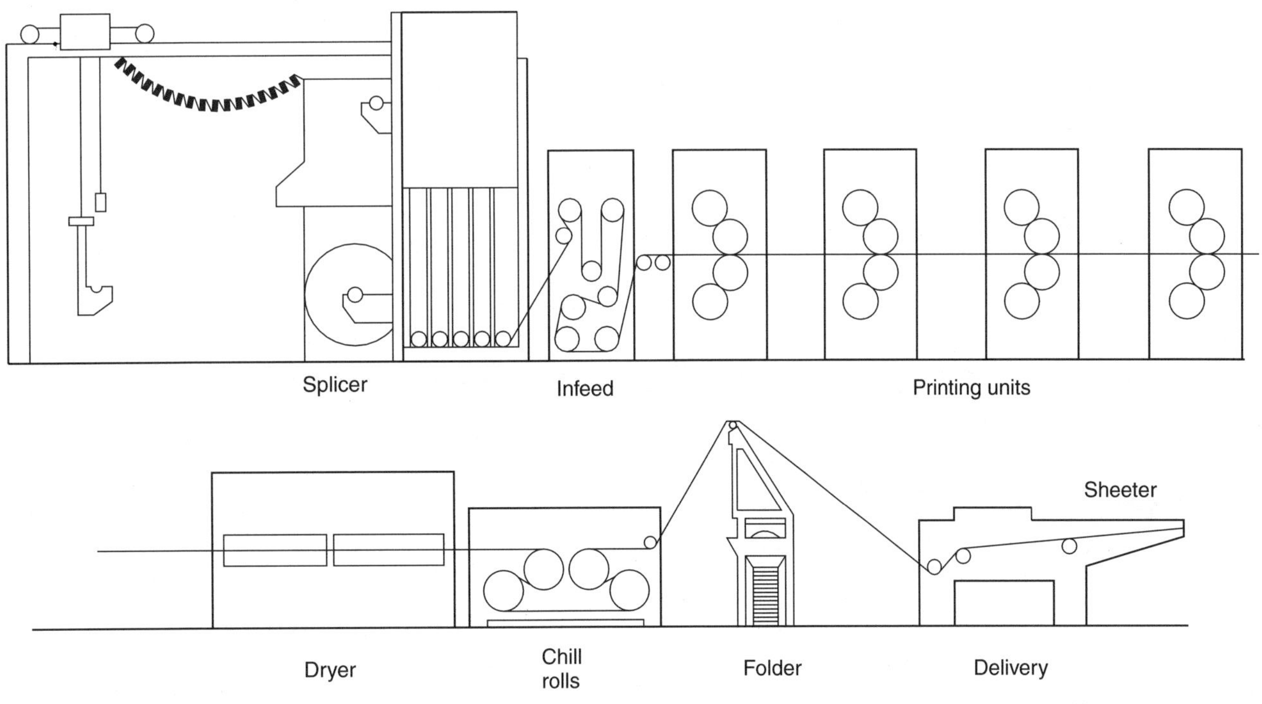

Figure 1. A Four-Unit Blanket-to-Blanket Web Offset Press with Optional Delivery to a Folder or Sheeter.

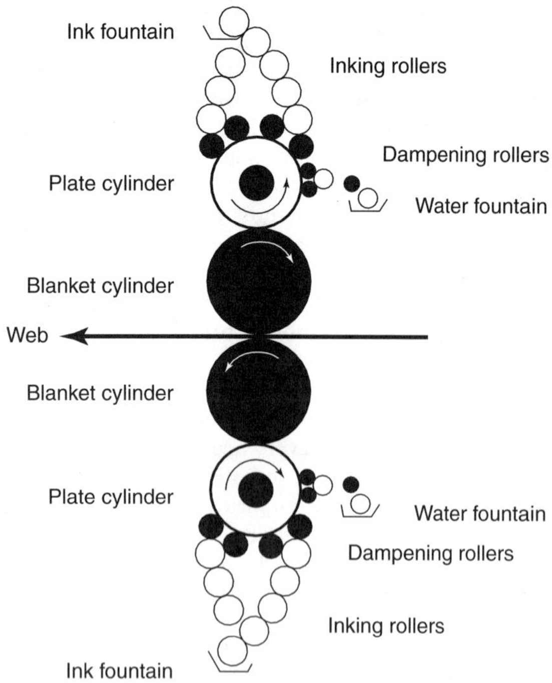

Figure 2. Diagram of a Web Offset Printing Unit.

few, if any, volatiles are evaporated. The emissions, if any, are associated with the dryer energy sources, not the ink. A low level of ozone may be present during warmup of UV lamps and decreases as the lamps reach their operating temperature.

AIR EMISSIONS CHARACTERIZATION

Air emissions from heatset web offset lithography are primarily from the three sources—the ink, the fountain solution, and the cleaning solvents. Ink oils used in heatset inks have vapor pressures usually less than 0.1 mm of mercury and, when in the vapor state, have a smog-generating reactivity similar to that of ethane. The ink oil is a condensable organic material that, if emitted in high concentrations, will condense into an opaque plum. Higher concentrations may require controls to reduce opacity to acceptable limits as specified by individual state air pollution control regulations. In fact, most of the existing air pollution control regulations enacted by individual states have specified the VOC emissions threshold as well as emission control requirements for heatset web offset printing operations.

The fountain solution and the cleaning solvents contain constituents that are defined as volatile organic compounds (VOCs). Fountain solution is essentially water with nonvolatile additives, but may contain isopropyl alcohol or replacements, such as 2-butoxyethanol or ethylene glycol. Isopropyl alcohol concentration in the fountain solution ranges from 5% to 25%, while the alcohol substitutes are present at 2–5% total concentration. Emission factors vary greatly, depending on the additives involved. As much as 75% of the isopropyl alcohol may evaporate into the room, while up to 25% of 2-butoxy ethanol may evaporate. Ethylene glycol is even less volatile. The amount remaining in the dried web is also variable, ranging from essentially zero to as much as 5%.

Manual cleaning solvent emissions are limited to the amount of solvent transferred to blankets, plates, rollers, or paper webs, if the cleaning cloths are placed in sealed containers after use. Cleaning solvents for heatset inks usually consist of mixtures of C7 to C10 hydrocarbons, with some aromatic components. These mixtures have flash points from under 80°F to over 140°F. Cleaning solvents for reactive cure inks usually are alcohol systems rather than aliphatic hydrocarbon types.

The dryer exhaust emissions from heatset inks are primarily a mixture of hydrocarbons of about an average C15 chain length. Typical concentrations would range from 2% to 12% of the lower flammable limit. Some heat distortion (oxidation) can occur as a result of contact with either very hot air or flame. Depending on the rate of ink use and type of paper being printed, approximately 40% to 90% of the input ink oil are evaporated in the dryer. The balance of the input ink is retained in the printed paper web. A low ink-input rate usually results in a low percentage of ink oil being evaporated. Other materials present would be either alcohol or alcohol substitute from the fountain solution. Fewer heatset printers use isopropyl alcohol because of cost and environmental concerns, but isopropyl alcohol does provide great operational flexibility. In most of the existing air pollution control regulations, isopropyl alcohol is not forbidden to be used in fountain solutions except NHWO printing operations in several states, such as Illinois. However, the concentration of isopropyl alcohol in the fountain solution is usually limited, usually less than 5% by volume. Alcohol substitutes usually include 2-butoxyethanol and/or ethylene glycol. Both are listed as air toxics by SARA 111, Section 313.

Presses with automatic blanket washers also will have cleaning solvent vapors in the dryer exhaust during the wash cycle. Relatively small amounts of solvent are used because the wash cycle is very short and concentrations are closely controlled to maintain a 4:1 safety factor below the lower flammable limit.

The U.S. EPA has proposed two guidelines, Control Technique Guidelines for Lithographic Printing (CTG) and Alternative Control Technique Guidelines (ACT) to quantify the rate of emission and retention of ink oil, fountain solution additives, and washup solvents (from automatic blanket washers only) based on studies completed by the printing industry. The emission and retention rates are listed in the following:

Sheetfed and Nonheatset Web Printing:

- 5% of ink oil emitted as fugitive emission, and 95% of ink oil is retained on the paper.
- 100% of fountain solution additives emitted as fugitive emissions.

Heatset Web Printing:

- 80% ink oil emitted through the dryer, and 20% retained on the paper.
- 70% of isopropyl alcohol or alcohol substitutes emitted through the dryer and 30% emitted as fugitive emissions.
- 40% of the washup solvent used in the automatic blanket washer emitted through the dryer and 60% emitted as fugitive emission.

AIR POLLUTION CONTROL MEASURES

Air pollution control equipment has been installed only to control the emissions from evaporative-style ink dryer exhausts.

Three methods of hydrocarbon emission control—carbon adsorption, incineration, and condensation/removal—theoretically are applicable. Adsorption has been tried, but desorption of ink oil from carbon requires extreme conditions of elevated temperature and low pressure, and one-time use of carbon is prohibitively expensive. The primary methods of ink oil emission control are fume incinerators (afterburners) and condenser-droplet-removal systems. Afterburners are the usual method of emission control, with 95–99% of the controls being of this type.

Afterburners are divided into catalytic and thermal types. Most afterburners often incorporate heat exchanger to reduce fuel consumption if operating hours are sufficient and airflow volume is large enough to provide an economic advantage.

A catalytic afterburner consists of a preheating section, a temperature indicator–controller, a chamber containing the catalyst, safety equipment, and, usually, heat-recovery equipment. Figures 3 and 4 show two arrangements. A catalyst is a substance that changes the rate of a chemical reaction and does not appear to change chemically in doing so. In the case of afterburners, the catalyst functions to promote the oxidation reactions at a somewhat lower temperature than occurs in a thermal afterburner. The catalyst often is platinum combined with other metals and deposited in porous form on an inert substrate. Metallic oxide catalysts are usually homogeneous granules. Catalyst afterburners typically control air temperature into the catalyst at 650–800°F. Typical hydrocarbon destruction efficiencies range from 90% to 98%, depending on the ratio of the volume of catalyst used to the air-flow volume. Catalysts may be supported on granular particles or on rigid structures (monoliths). Catalysts are susceptible to loss of performance as a result of coating (masking) and reaction with contaminants (poisoning). Dryer exhaust gases containing silicones, tars, resins, and dusts will cause masking, and those containing high levels of phosphorus compounds, heavy metals, halogens, or sulfur will cause poisoning. Excessive temperatures can cause crystal growth of the catalyst support and loss of activity. The upper temperature limit appears to be 1400°F for platinum catalysts and lower for metallic oxide catalysts. Granular catalysts are susceptible to abrasion, which may extend the useful life of the catalyst by removing masking and surface poisons. However, it may ultimately cause failure by converting the catalyst into fines, which may eventually pass through the retaining screens. The preheating section may have either electric or gas fuel heaters. Liquid fuels are avoided as inefficient burning will contribute to masking of the catalyst. Annual efficiency test of the catalyst is usually required for catalytic thermal afterburners.

A thermal afterburner has functional components similar to those of a catalytic afterburner, except that the catalyst bed is replaced with a larger combustion chamber, heat recovery is greater and nearly mandatory economically, and construction materials are changed to handle higher temperatures. An effective thermal afterburner design must provide for an adequate combination of (1) a sufficiently high temperature, (2) a high enough residence time (usually above 0.5 sec), and (3) adequate mixing or turbulence in the combustion chamber. A small deficiency in any one component may be compensated for by an increase in another. Direct flame contact is not required, although the extremely high temperature in the flame even for a short period is beneficial. Heat recovery is standard owing to the high temperatures involved and has resulted in two types of systems—recuperative and regenerative heat recovery. Recuperative units continually transfer heat from a hot stream to a colder stream in a countercurrent flow arrangement. Figure 5 illustrates a typical shell-and-tube recuperative unit. Regenerative units cyclically store and transfer heat between

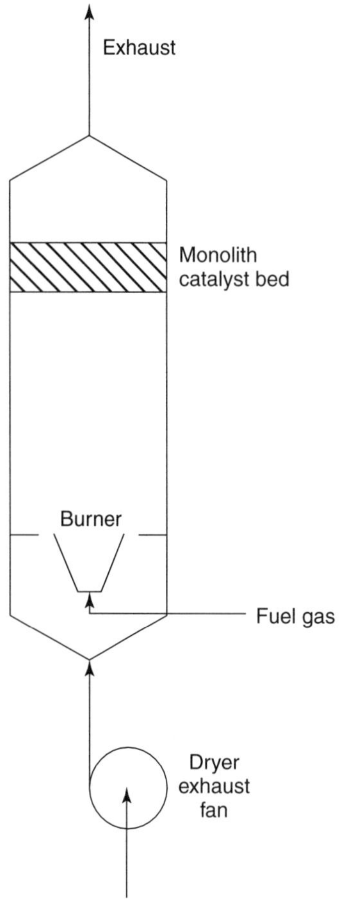

Figure 3. Catalytic Afterburner without Heat Recovery.

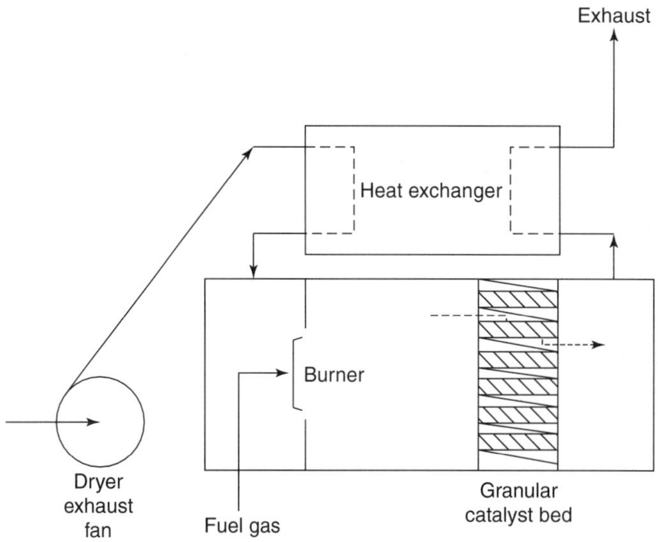

Figure 4. Catalytic Afterburner with Heat Recovery.

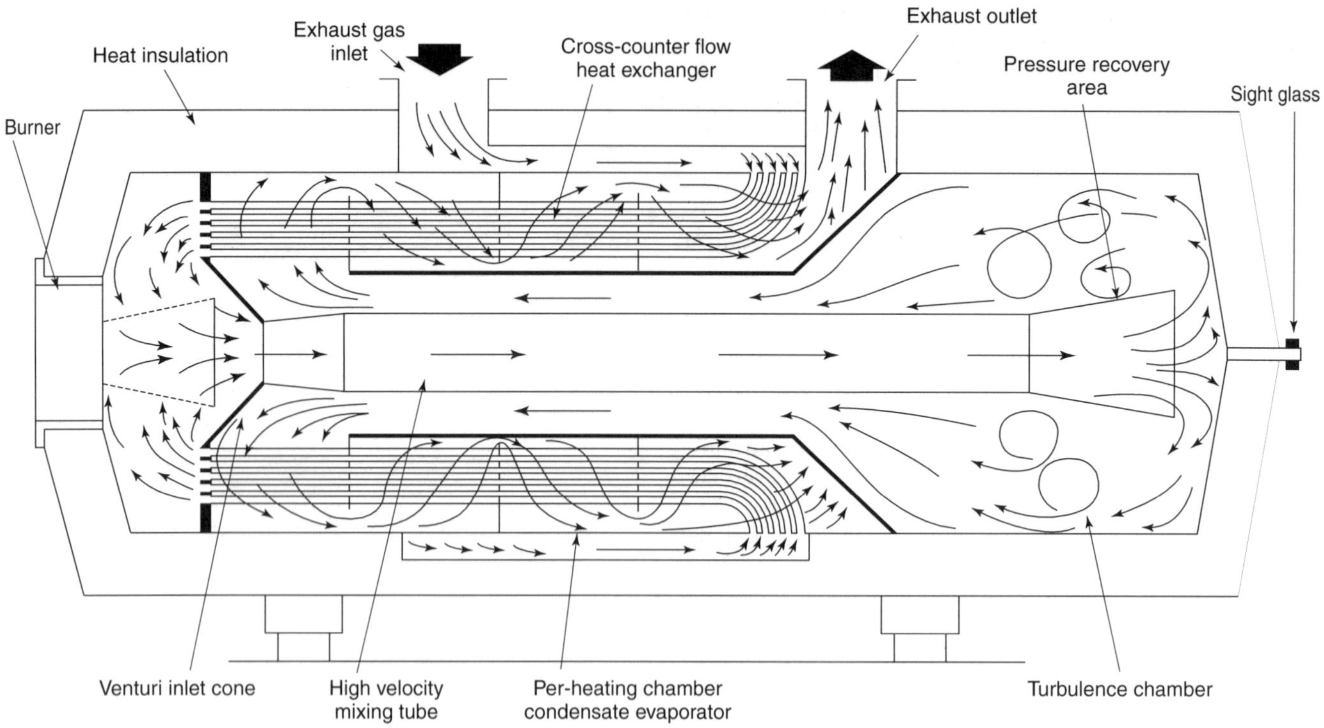

Figure 5. Recuperative-Type Thermal Afterburner.

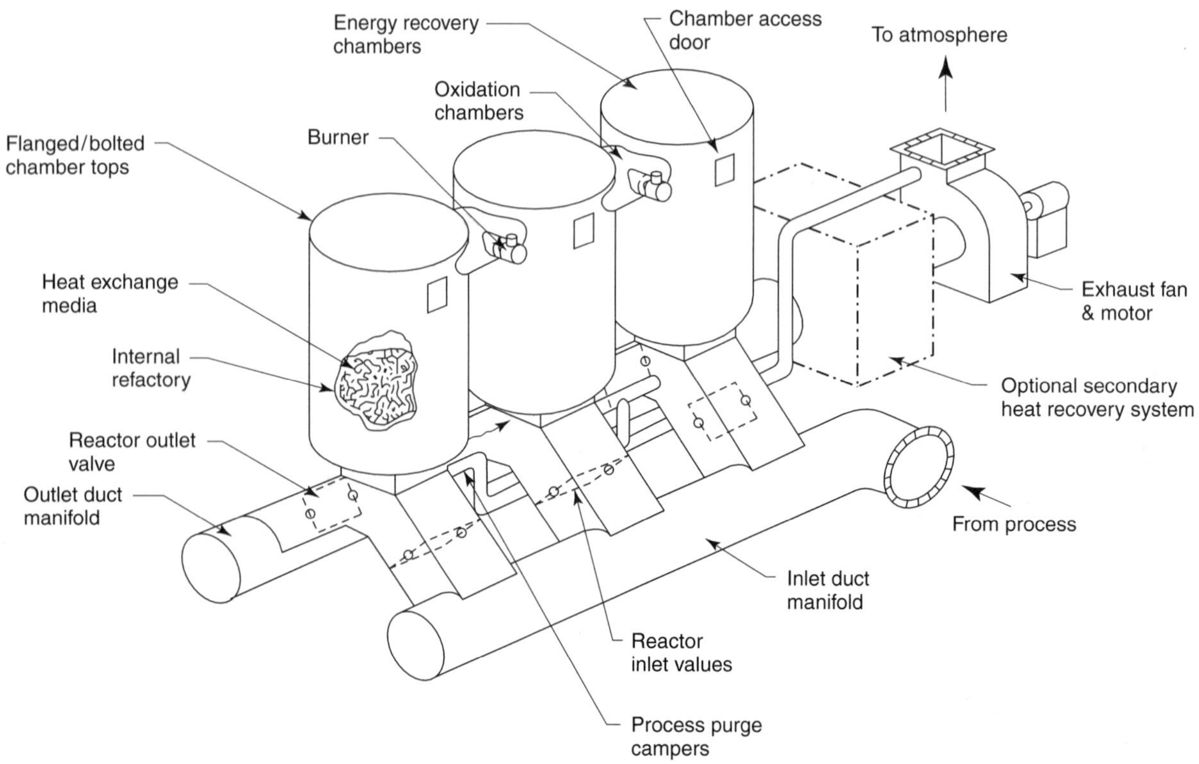

Figure 6. Regenerative-Type Thermal Afterburner.

streams, and temperatures are always changing. Figure 6 is an example of a regenerative thermal afterburner in which the temperature storage unit consists of ceramic pieces. Operating temperatures of recuperative units range between 1250°F and 1450°F. Regenerative units operate in the range of 1400°F to 1800°F. Hydrocarbon destruction efficiencies range from 97% to 99.8% for the recuperative units and from 95% to 99% for the regenerative units. Thermal recovery effectiveness typically ranges from 45% to 76% for recuperative units and from 80% to 95% for regenerative units. The use of heat exchangers reduces fuel consumption, which results in fewer NO_x compounds.

Cooling and condensing type of control consists of heat exchangers followed by either filters or electrostatic precipitators to remove oil droplets. The exchangers are often followed by an oil–water separator. Optional equipment often includes additional mist eliminators to reduce droplet loading or after treatment by a carbon filter further to reduce hydrocarbon vapor. The heat exchangers cool the dryer exhaust to a range of 100–120°F. Hydrocarbon oils with concentrations above a dew point of these temperatures will condense. Condensate will be divided between liquid that drains from the heat exchangers and an aerosol fog that must be removed from the air stream. The droplet-removal system usually involves fiberglass filters or electrostatic precipitators, as the droplets are very small in size (0.1 Å or less). Because cooling is usually accomplished using ambient air, lower temperatures are achieved in winter months. Hydrocarbon removal efficiencies may vary from 5% to 75%, depending on the condensing temperature reached and the input hydrocarbon concentration. Cooling and condensing control device emissions have a relatively constant output concentration and can be a mixture of uncontrolled oil droplets and some uncondensed oil vapor. The condensed liquid is a water and oil mixture, with most of it being water. It usually goes through an oil–water separator where the separated oil is recovered. Heat exchangers, droplet-removal systems, and oil–water separators must be cleaned regularly because of the buildup of residues.

STACK TESTING

Stack testing is often required to determine the emissions entering and leaving a control device. The U.S. Environmental Protection Agency (EPA) Method 25 is not suitable for testing hydrocarbon emissions that consist of a mixture of aerosol droplets and vapor concentrations. It is very difficult to obtain consistent results from operations that are as intermittent as heatset web offset and are further complicated by the installation of multiple press dryers into a single emission-control device. If aerosols are known to be present, the appropriate test is EPA Method 5. For complex emission sources, a continuous readout hydrocarbon monitor is more appropriate. Hydrogen flame ionization detectors provide sufficient sensitivity, and heated sampling lines promote sampling accuracy. Frequently, dryer exhaust concentrations of hydrocarbons can be less than 500 ppm carbon. If 90% control is mandated, it is difficult to demonstrate compliance using EPA Method 25, which has not been found to produce accurate or reproducible results below 50 ppm carbon. The alternative methods described above should be considered in these situations.

BIBLIOGRAPHY

Air Pollution Control in the Heatset Web Offset Printing Industry, Report 8403, Tech Systems, W. R. Grace Co., DePere, 1989.

D. Crous and R. J. Schneider, *Web Offset Press Operating*, 3rd ed., Graphic Arts Technical Foundation, Pittsburgh, 1989.

The Lithographers Manual, 8th ed., R. Blair, Ed., Graphic Arts Technical Foundation, Pittsburgh, 1988.

United States Environmental Protection Agency, *Control Technique Guideline for Offset Lithographic Printing*, Office of Air Quality, Research Triangle Park, NC, September 1993.

United States Environmental Protection Agency, *Alternative Control Techniques for Offset Lithographic Printing*, Office of Air Quality, Research Triangle Park, NC, November 1993.

GRAPHIC ARTS INDUSTRY: FLEXOGRAPHY

Doreen M. Monteleone and John D. Serafano

Flexography is a simple and economical print method capable of print quality rivaling alternative, historically superior, printing techniques such as offset lithography and gravure. Flexography uses a printing plate with a raised image, called a *relief plate*. Flexographic plates are soft and flexible, made from molded or engraved rubber or imaged from a light-sensitive synthetic mater called *photopolymer*. Plates are mounted on a rotary cylinder on a press equipped with anywhere form one to twelve color stations.

Flexographic printing ink is liquid and fast drying. Unique to flexography, the ink applied by way of a finely engraved rotary cylinder called an *anilox roll*. The anilox delivers ink to the printing plate from microscopic cells with uniform size and spacing. Excess ink is removed from the anilox roll with a doctor (or metering) blade placed at a reverse angle to the revolving anilox cylinder.

The printed image in flexo is transferred directly from the inked plate to the substrate. Flexography can print almost any size job, from small labels to large refrigerator boxes. Flexographic presses can print on virtually any type of substrate, from corrugated cardboard to flexible plastic film, to textiles and lightweight paper. Flexo presses are often equipped with in-line finishing processes, such as coating, laminating, slitting, and diecutting.

MARKET PROFILE

Flexography is the world's fastest growing print technology (Table 1). Today, flexography routinely prints jobs that the general public comes in contact with on a daily basis in the supermarket, at the warehouse store, at the shopping mall, and even at the local newsstand. Much of the reputation and endorsement of flexographic printing is a result of the need for shorter production runs, the ability to address promotional and marketing strategies on a more timely basis, particularly those aimed at the varying demographic shifts, and the economics of fast-changing consumer demands. Flexography's versatility, simplicity, economics, and major strides in quality are driving the market to new highs.

Flexographic printing is seen in everyday life. Flexo is used to print newspapers and comics. The mail received daily is in envelopes printed by flexo. Houses are full of packaging for store-bought items printed by flexo: computer software, toys, and appliances, for example. Grocery store shelves are stacked with food packaging printed by flexo: cereal boxes, shampoo bottle labels, soda bottle wrap and cartons, frozen food bags, bread bags, and milk cartons. The bags groceries are carried home in, whether they are paper or plastic, are printed flexographically. At fast-food restaurants, food is wrapped in and boxed in flexo-printed packaging, and drinking cups are printed by flexo.

Flexography is the youngest and the most dynamic printing process. With rapid advances in flexo technology and society's

Table 1. 1998 Market Trends for Flexography

Flexible film packaging	70% market share of a $18.8 billion segment
Paper bags and multiwall sacks	Dominates a $2.5–3.0 billion segment
Labels and tags	Growth trend in a $9 billion segment
Corrugated and preprint	100% market share of corrugated and 50% market share of preprint; total segment revenue is $26 billion
Folding carton	Strong growth
Envelope	60% market share of $3.8 billion segment
Newspaper	50 companies to date and growing

Source: Reference 1.

increasing reliance on packaging, flexo will continue to grow at an unprecedented rate.

PROCESS DESCRIPTION

Publication and packaging products printed with flexography vary considerably from application to application. Flexo is capable of printing a variety of substrate widths and production speeds. Relatively low-viscosity liquid inks are used almost exclusively in flexographic printing and can have either an organic solvent or water vehicle. The liquid ink is transported and metered precisely through the printing process, resulting in a thin uniform ink film deposition of 7–25 μm in thickness on the substrate. Once applied, the ink film gives up between 40 and 65% of its liquid vehicle volume by direct contact of heat and evaporation, and/or absorption. Drying is an integral part of the flexo process. Each ink color must be dried before subsequent colors can be applied. Printing color over color is called "trap," and flexo is a dry-trap process. A typical flexographic press contains the following functional sections:

1. Unwind—substrate roll handling
2. Printers—individual color units or "decks"
3. Dryers—individual for each color and main tunnel oven
4. Rewind—substrate roll handling
5. Converting

Important flexo characteristics are as follows:

- Press speeds of a few feet per minute (fpm) to more than 2200 fpm
- Plates from 0.019 to 0.250 in. thick
- Substrates from kraft paper to board to cellophane to plastic film to newsprint to filled and coated paper
- Web widths from 2 in. to more than 100 in.
- Inks that are hydrocarbon based, alcohol based, or water-based; pigment based or dye based; that vary from high viscosity to low, from expensive to relatively inexpensive, and from simple to formulate to complex
- Anilox cylinders with fine engravings, coarse engravings, or no engravings; large cell volumes or small cell volumes; random cell placement or very precise cell placement; and electroplated surfaces or ceramic plasma-coated surfaces
- Ink applicator techniques that vary from a fountain roller (two-roll system) to a single-doctor-bladed system to a hybrid system (two rolls with doctor blade) to a double-doctor-bladed system (chambered doctor blade)
- Print repeat lengths that vary from a few inches to more than 100 in.
- Ink drying needs that vary from none to extensive (utilizing many technologies).

Flexographic Press Designs

Flexographic press designs fall into one of the following categories, depending on the application:

- Common or central impression (CI)
- In-line
- Stack
- Newspaper unit
- Dedicated four-, five-, or six-color unit.

The common or central impression (CI) design uses one impression cylinder common to all printing stations (called couples in publication printing). The number of stations using the same impression cylinder varies from two through eight. The advantages of a common impression design include compactness, excellent registration of multi-colored images and excellent control of the substrate. One drawback to the CI press is its drying capacity between the colors, which is limited by color deck to color deck distance. The dryers used to dry the ink between colors is called the "BCD", or between color dryers. A lengthy tunnel dryer is used at the end of the last applied color for final drying of all colors; this dryer is referred to as the "Main Dryer". Figure 1 shows a typical CI design for a packaging application.

With the in-line design, the printing stations are separate, discrete units mounted in-line to one another (Figure 2). With this design, any number of stations can be supplied and the printing width and repeat length can be selected from a wide range of possibilities. The distance between colors is much greater than the CI press, allowing for longer drying lengths and usually better drying capacity. Color registration on thin flexible materials can be difficult to control on an in-line press.

In a stack press design (Figure 3), individual color stations are stacked one over another (usually on one or both sides of a main press frame). There are two primary advantages of a stack-type press. First, it is usually possible to reverse the web to allow both sides to be printed during one pass through the press. Second, the printing stations are usually quite accessible, which facilitates easy changeover from one job to another and easy washup.

A newspaper printing unit (Figure 4) consists of two dedicated printing couples (stations) located back-to-back in a common pair of frames. This permits the printing of black and white images on both sides of the web using a dedicated lead. Multiple units are arranged into a pressline to print the many pages required of a large newspaper (Figure 5). Color decks (often called half-decks), each consisting of one dedicated

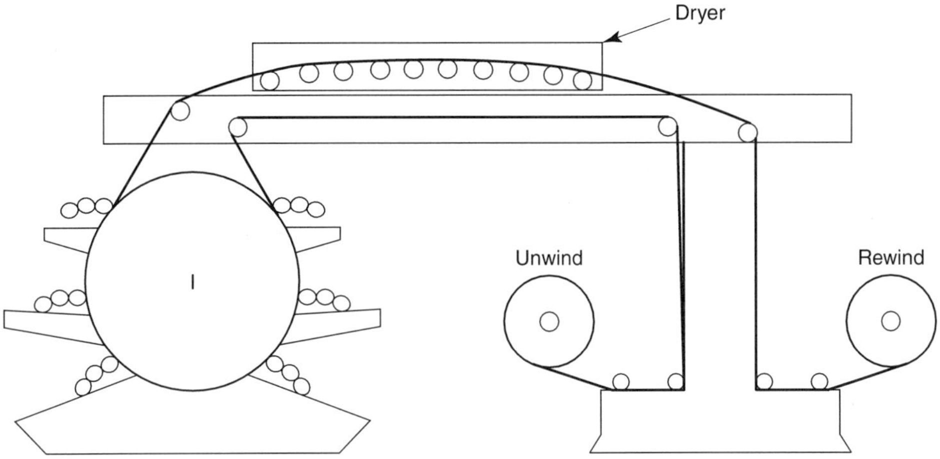

Figure 1. Packaging Flexographic Central Impression Press.

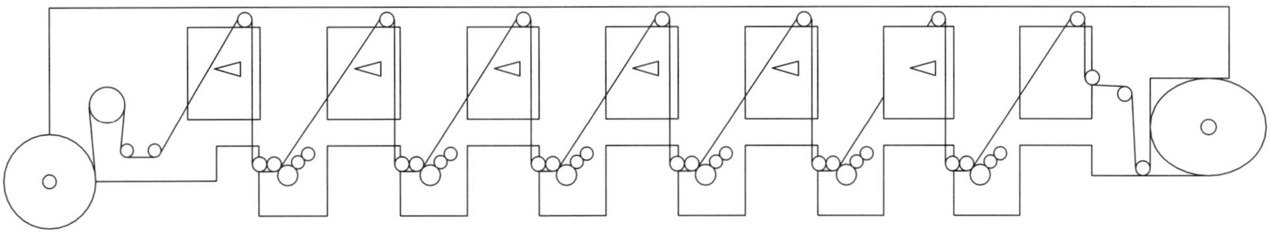

Figure 2. Packaging Flexographic In-Line Press.

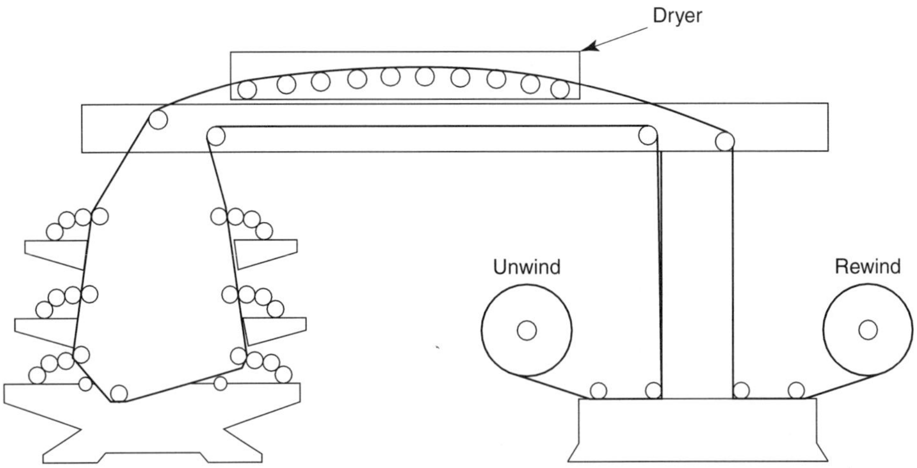

Figure 3. Packaging Flexographic Stack Press.

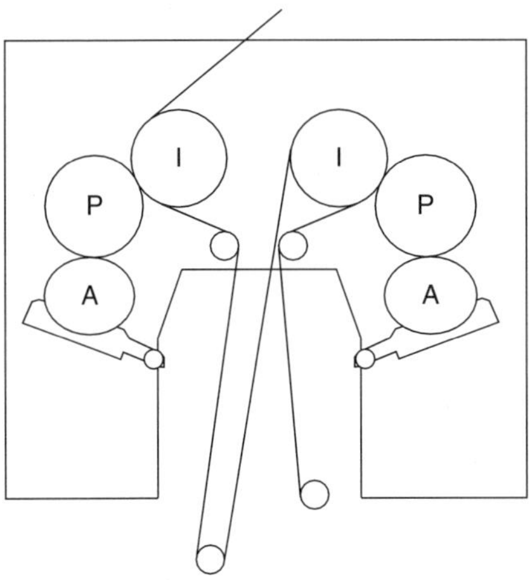

Figure 4. Newspaper Flexographic Printing Unit.

printing couple, are positioned above those unit, positions where the publisher wants to offer (single) spot color capability.

Commercial publication flexographic presses utilize dedicated four-color units (Figure 6). Two units are combined into one press for printing four colors on both sides of the web. The advantages of this design include wide web capability, high operating speed, and compact design. In most publication designs, dryers, usually infrared, are included to ensure adequate drying of the water-based ink after each side of the web is printed.

Inking Systems

The anliox roll is the primary ink control device in a flexo inking system. The engraved cells in the anilox surface, either chrome or ceramic, hold the designed ink volume for the product to be printed. Applied ink volume and ink film thickness

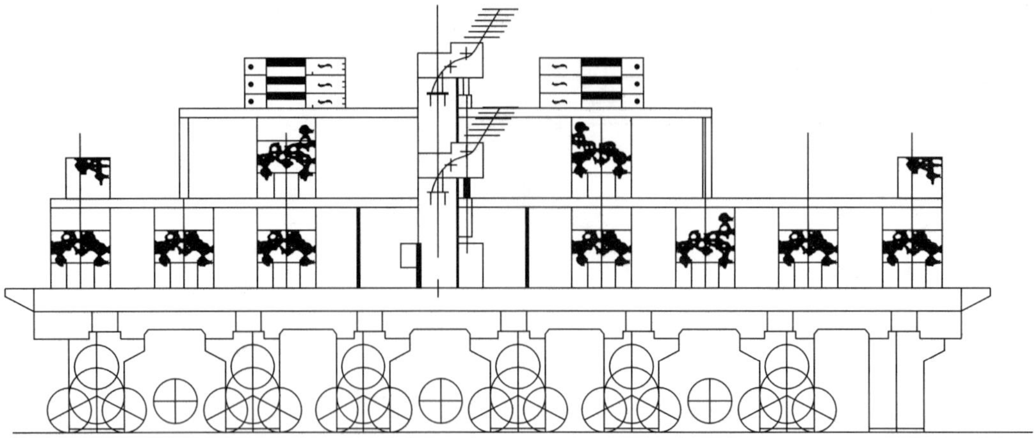

Figure 5. Newspaper Flexographic Printing Press.

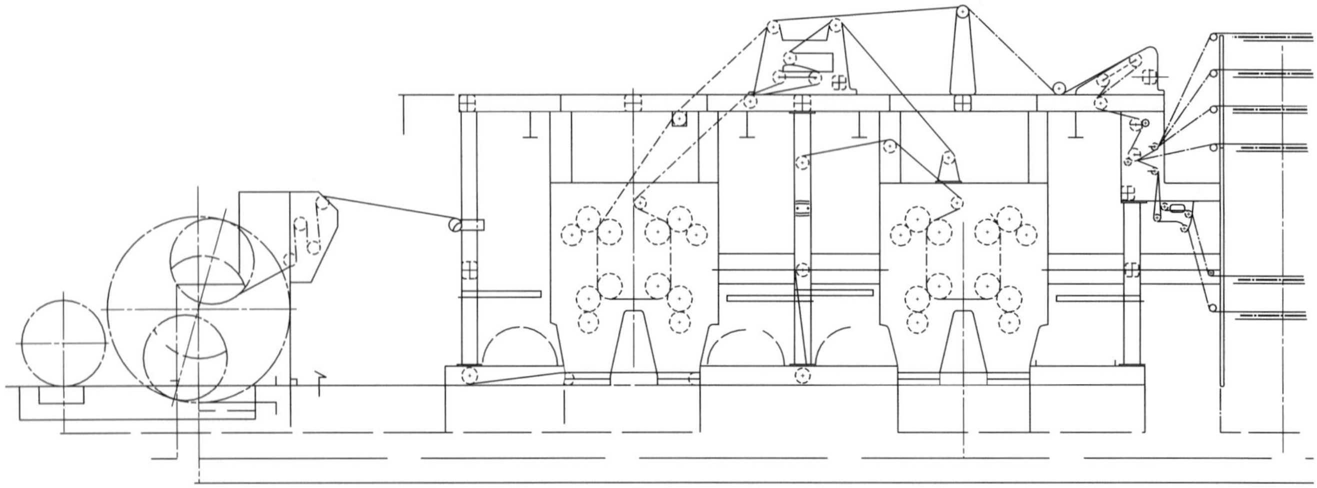

Figure 6. Commercial Publication Flexographic Press.

are controlled by the anilox roll. Volumes for these rolls are measured in BCM (billion cubic micrometers)/square inch, and range from less than 1 BCM to more than 15 BCM, depending on color and functional ink film properties.

There are several systems employed to meter the ink from the anilox roll. In a "two-roll system," a fountain roller rotating in an ink bath transfers the ink to the anilox cylinder, which, in turn, transfers it to the printing plates and web (Figure 7). The ink film thickness efficiency is controlled by the dynamics of the nip between the rubber-covered fountain roller and the anilox. The metering action of the fountain roll is controlled by the hardness of the roll, its diameter, rotational speed, and pressure. This type of metering system deposits the heaviest volume of ink to the substrate.

The second and most common design used today utilizes an enclosed ink chamber and two doctor blades (Figure 8). Ink is pumped into the chamber and doctored (wiped) by one of the blades. Ink containment and easy color change of the printing station (couple) are two of this system's advantages. The double-doctor-bladed ink chamber is often used where both high-speed and high-quality printing are requirements. Chambered doctor blade systems are required where high graphics quality and fine screen rulings are necessary.

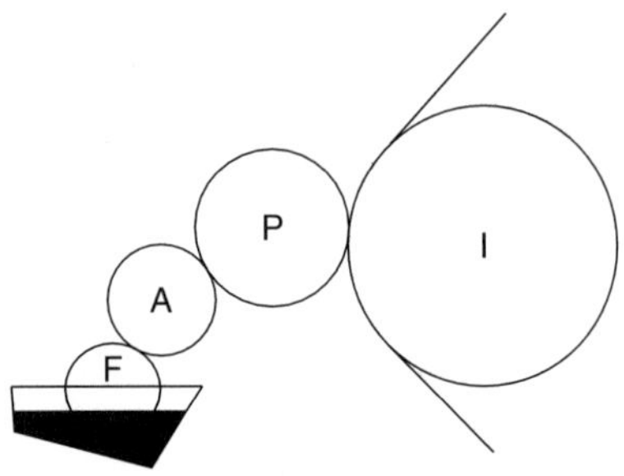

Figure 7. Fountain Roller Style of Printing Couple.

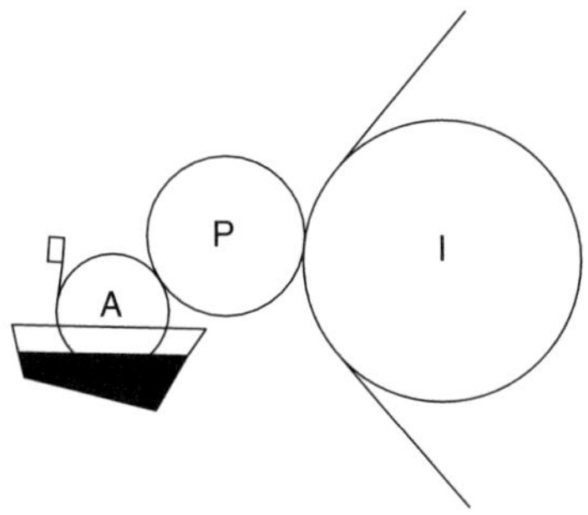

Figure 8. Single-Doctor-Bladed Style of Printing Couple.

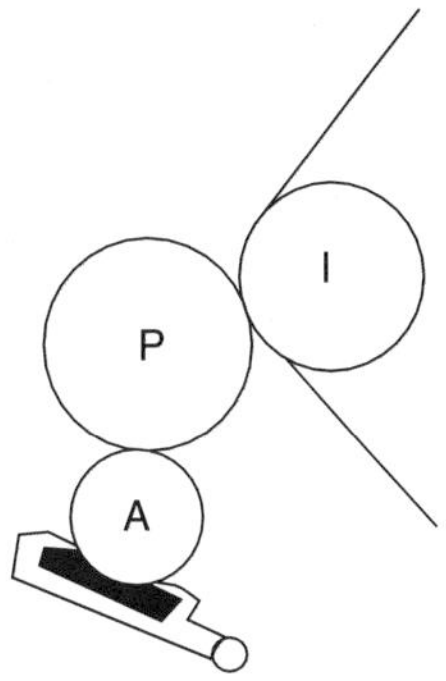

Figure 9. Double-Doctor-Bladed Style of Printing Couple.

A third system uses a single reverse-angle doctor blade to scrape or doctor the excess ink off of the anilox surface (Figure 9). With this system, the anilox rotates in the open ink fountain directly in contact with the ink. A doctor blade is located on the entrance side of the anilox-plate cylinder nip to scrape off the excess ink.

Other Press Components

In addition to printing units, all presses have a number of other components. These components vary depending on the application. They include:

- Unwind and rewind stand, or reel—source for unprinted and printed substrate rolls can be used with a reel/tension/paster (RTP) for continuous nonstop production.
- Folder—to convert printed web into partially or completely finished product.
- A press drive and a control system.
- Web tension control devices are utilized on both the unwind and rewind of the press.
- Other converting equipment including: slitters, dryers, chill rollers, manual and/or automatic register controls, and die cutters.

Dryers are an important part of most packaging flexo presses and some publication presses. They are often necessary to cause the ink to dry fast enough to make the process economically viable. Dryer types are frequently infrared or hot air; other types are rarely applied.

AIR EMISSIONS CHARACTERIZATION

Air emissions from the flexographic printing process come from inks and solvents used for cleanup and dilution. There are three ink systems common to flexography, water-based, organic solvent-base, and energy curable. Although Reference Method 24 does not accurately determine VOC content in low-VOC or water-based inks, the U.S. Environmental Protection Agency utilizes Method 24 to characterize VOC content in printing inks.

Flexo inks dry by four mechanisms: evaporation, absorption, chemical, or radiation. Each ink is specially formulated to meet specific substrate and drying requirements. Most packaging inks used in wide web flexo printing dry by evaporation, while flexo news inks dry primarily by absorption. Higher production speeds are generally possible with organic solvent inks because of their low boiling point and quick solvent release from the ink film layer. Water inks require more energy (air flow and volume) to facilitate adequate drying to meet production requirements. Energy-curable inks such as UV (ultraviolet) and EB (electron beam) dry by direct radiation of the ink film via an energy source such as a lamp or emitter. UV and EB inks have unique benefits, including very fast drying and quality image formation. These inks, however, are currently limited to narrow web flexo printing, such as labels, and relatively few wide web specialty products.

Because of these widely varying job requirements, ink formulators have developed a wide variety of inks to meet them. Consequently, widely diverse resin systems are utilized in the vehicles of flexographic inks. Each resin system requires specific solvent characteristics to maintain solubility and to meet the drying requirements of the application.

Solvents commonly used in flexographic printing include ethanol, isopropanol, N-propanol, hexane, toluene, isopropyl acetate, N-propyl acetate, glycols, glycol ethers, and water. These solvents are often co-blended in various percentages to enhance or control drying rates and facilitate end use properties and resin compatibility.

Water-based inks were first applied in the 1950s. Their use began to expand in the late 1970s with the development of better inks and in response to environmental pressures. High-solids ink technology was attempted beginning in the late 1970s in response to these pressures, but these efforts were largely unsuccessful, although they continue today.

Organic solvent-based inks are consumed in greater quantities than are water-based inks, though water inks have made significant market advancements and represent nearly half of all flexo inks used today. Water-based inks work best where substrates are hydrophilic and porous. The more hydrophobic the substrate and the higher the press speeds, the less suitable are water-based inks. Film and foil are two examples of substrates to which water-based inks are less adaptable. Water-based ink formulations have been successful in printing some films, but cannot be universally applied.

Water-based inks and organic solvent-base inks have different drying requirements. Water inks need high airflow volume exchanges within the dyer and high airflow impingement out of the dryer nozzle on to the web to ensure adequate drying and color to color trap of the ink film. Typical air-flow requirements for drying solvent inks are 3000–7000 CFM (cubic feet per minute). Water inks require higher air flow, typically from 6,000 to 12,000 FCM.

Newsprint is an excellent substrate for water-based inks. Thus the products where newsprint is the substrate of choice are the ones with the highest application of water flexo. These products include newspapers, Sunday comics, newspaper inserts, and telephone directories. The inks used for newspaper printing have low viscosity and absorb quickly into the absorbent newsprint, minimizing the need for high-volume air dryers.

Water-based inks may also contain VOCs (although in smaller amounts than solvent inks). These are required to optimize the solubility of the resins in water and improve flowout of the ink film on the substrate. The VOCs used in water-based inks are one or more of the following: amines, alcohols, glycols, and glycol ethers (the latter two may or may not be VOCs depending on the specific compounds and test methods used).

In publication flexo, amines and glycols are the only compounds that meet or could meet the definition of VOC. Alcohols are specifically not used. Amines and glycols are utilized in quantities that never exceed 5% of the ink formulation (prior to dilution with water).

In packaging flexo, VOC content may be higher than 5%, particularly if alcohols are used as co-solvents. In such cases, the VOC content is limited so that the ink is exempt from emission controls either by virtue of the specific VOC content or because it meets the definition of a high-solids ink. A high-solids water-based ink is one that contains 60% solids and 40% solvent after the water portion has been extracted under U.S. Environmental Protection Agency testing guidelines.

In packaging flexo, organic solvents are used to clean systems that use solvent-based inks. Where water-based inks are successfully used, nonorganic cleaning techniques utilizing high-alkaline solutions are applicable. In some cases, mechanical cleaning, such as with baking soda or dry ice blasting, can be used, resulting in no emissions.

Emissions from flexographic printing arise from organic solvent that may be present in the ink and from solvent that may be used for press cleaning. Emissions are limited in all cases to volatile organic compounds (VOCs). The types of VOCs emitted from a specific operation depend on the ink type and its formulation, the specific job requirement, and may be influenced by drying capabilities of the press equipment.

Flexo inks also contain other functional components such as plastisizers, waxes, silicones, and fatty acids, which are used to add physical functional properties to the ink. Some of these materials, such as silicone, may be detrimental to pollution control systems.

AIR POLLUTION CONTROL MEASURES

In operations utilizing water-based inks exclusively, typically no emission controls are necessary. Locations where energy-curable inks have been successfully adapted are also exempted. However, solvent-based inks are still widely used because of their compatibility with nonporous substrates.

In flexo pressrooms utilizing solvent-based inks, the applicability of emission controls is related to a number of factors. These include ink consumption rate, statutory emission control requirements, and the degree of success in applying high-solids ink technology (if attempted). The type of emission controls employed is based on other factors, including the type(s) of solvents used in the ink formulations, the need to change inks from press run to press run as job requirements change, the number of inks used in a given pressroom, the statutory requirements, and the relative cost effectiveness of competing alternatives.

Were there no technological stumbling blocks, solvent recovery strategies using such technologies as adsorption, absorption, and condensation theoretically would be the preferred emission control strategy, as is the case in publication gravure. Since these difficulties do exist, solvent-recovery systems and the subsequent reuse of flexographic solvents have been applied to only a few flexographic printing operations. As many of the flexo solvents are alcohols and acetates, and because many packaging operations do, in fact, utilize multiple inks and solvents in a single pressroom, attempts to recover those solvents for reuse have been mostly or wholly unsuccessful.

The reason for this lack of success is that these solvents are miscible with water and form an azeotropic mixture that is nearly impossible to separate, even with expensive thin-film evaporation techniques. Furthermore, printability and adhesion can only be optimized with inks whose physical characteristics — and, therefore, chemical makeup — are held within a narrow range. Most attempts to reuse mixed solvents collected from a recovery system as diluents for the flexographic ink have met with failure, and the others have had only limited success. Thus the used solvents are valuable only for their fuel value.

A second problem with solvent recovery is that many flexo inks contain some high-molecular-weight components. These compounds are sometimes difficult to desorb from the carbon bed. Bed efficiency would drop precipitously from such contamination. For the above reasons, the addon emission control strategy most frequently used in solvent-based flexographic printing operations is oxidation. Biofiltration, a relatively new technology is being used in some facilities. The two types of oxidizers are catalytic and thermal (see Figures 10 and 11). The catalytic category consists of fluidized-bed catalytic, fixed-bed catalytic, and precious metal catalytic using either metal-on-metal or metal-on-ceramic monolith. The thermal oxidizer category can be divided into direct thermal, thermal regenerative, and thermal recuperative.

The thermal oxidation process relies on high temperature (1400–1500°F), turbulence, and adequate retention time (>0.5 sec) to facilitate the combustion of organic vapors into carbon dioxide and water. Each of the three types of thermal oxidizers utilizes a different approach to the destruction of the organic vapors. All require a separate fuel, often natural gas, to maintain continuous combustion as the solvent concentration in the waste stream varies depending on press utilization, press speeds, and ink coverage on the substrate.

The catalytic oxidation process relies on a catalyst operating in the activation range of 500°F to 750°F to convert the organic vapors to carbon dioxide and water. Uniform flow presentation, uniformity of inlet temperature across the bed, and retention time (referred to as space gas hourly velocity, SGHV) directly affect not only the destruction efficiency, but the life of the catalyst itself. The specific mechanical and catalyst technology utilized in each type of catalytic oxidizer is different.

No system has an inherent advantage that makes it universally superior to another. The decision by a flexographer to purchase one system rather than another (or one type versus another) would depend more on such issues as system design, purchase price, ease of installation, available space, installation costs, operating costs, and source and quantity of emissions to be controlled.

In very general terms, catalytic systems are well suited for processes that have a minimum potential for poisoning or masking the catalyst. Well-designed catalytic units can achieve 90–98% conversion efficiency. Conversion efficiency of a catalytic system depends on properly applying the catalyst to the manufacturing system to be controlled and protecting the catalyst from damage. This is especially critical when striving for high-end conversion efficiencies.

Thermal systems are better suited for operations where poisoning and masking would be a problem with a catalytic system or where large process exhaust flows are involved. Thermal units normally achieve 95–98% conversion efficiency.

Thus, unlike the selection of a particular system within each category, choosing between categories (thermal or catalytic)

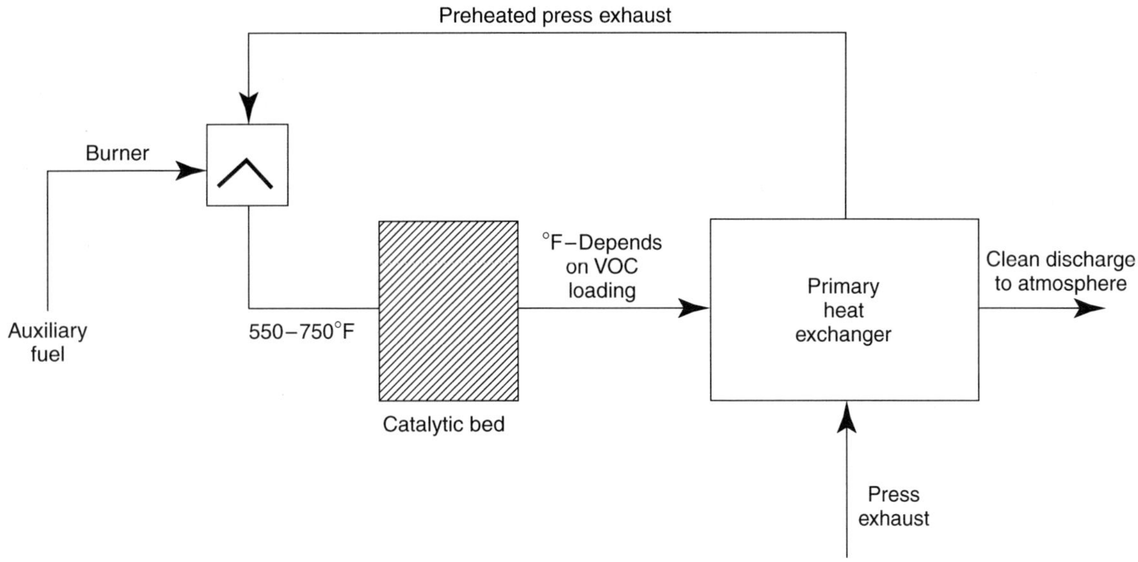

Figure 10. Catalytic Incinerator with Heat Exchanger.

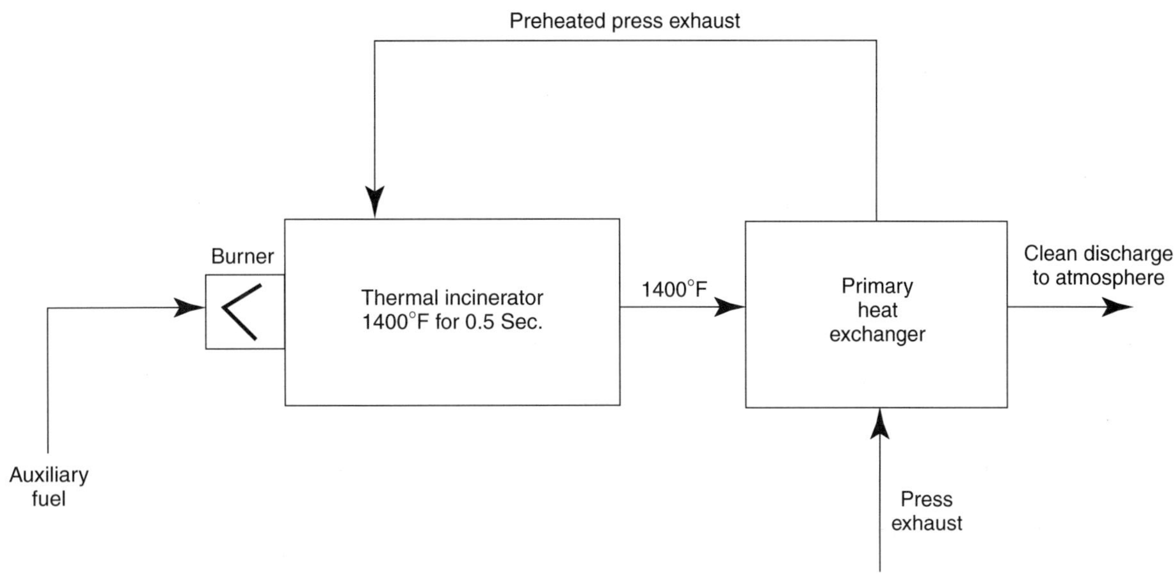

Figure 11. Thermal Incinerator with Heat Exchanger.

involves analyzing all factors relevant to a specific printing operation. This initial decision is often dictated by the sum of those factors. Also, reduction in the press exhaust flow rate (by recirculation and/or cascading between stations) can have a major, beneficial effect in reducing both capital and operating costs.

Biofiltration is an air pollution control technology in which VOCs are oxidized into carbon dioxide and water using microorganisms. Process air is passed through a pretreatment humidifier to saturate the gas stream before it enters the biofilter. The humidified air then flows through the biofilters, where the VOCs are absorbed into an aqueous layer surrounding the filter material. The microorganisms contained in the filter material use the VOCs as their primary carbon source and convert the VOCs to CO_2 and water.

Biofiltration is not applicable to all gas streams, but in situations where it does apply, the economics indicate that it should be seriously considered as an alternative to other control technologies.

Inherent in an efficient control strategy is the control of fugitive emissions. Proper control here involves different enclosures for ink sumps, fountains, and tanks; efficient, properly enclosed dryers; negative pressroom air pressure; good operating practices, such as keeping doors and tank lids closed; efficient press cleaning using minimal amounts of solvent; and a preventative maintenance program for the control device and ancillary equipment. Periodic hose and damper inspection, keeping lids closed on ink containers, selecting presses that use enclosed ink chambers, and using enclosed cleaning tanks either connected to the control device or that utilize low-evaporation-rate cleaning solvents also minimize fugitive emissions. Improving capture efficiency should be a high priority in planning an installation, as control efficiency is improved and supplemental fuel consumption is often minimized. This is even more important where heat recovery is practiced and used to supplement process and room heating needs. System components such as powered pickups, floor sweeps, and efficient dryers improve capture efficiency.

REFERENCES

1. *1999 Business Plan*, Flexographic Technical Association, 1998.

BIBLIOGRAPHY

Foundation of Flexographic Technical Association, *Flexography: Principles and Practices*, Fifth Edition, 1999.

USEFUL URLS

Flexographic Technical Association: *www.fta-ffta.org*

Printers' National Environmental Assistance Center: *www.pneac.org*

U.S. Environmental Protection Agency: *www.epa.gov*

SHEETFED OFFSET PRINTING

JERRY BENDER, C. NELSON HO, DAVID JOHNSON, FRANK LITTLE, AND WILLIAM SCHAEFFER

Sheetfed offset lithography is the process used by most small- to medium-sized printers because of its versatility. Among the establishments that employ the process are quick printers, in-plant printing operations, commercial printers, paperboard converters, and metal decorators, and the products printed range in size from individual envelopes, newsletters, advertising brochures, books, magazines, and folding cartons to billboards. While paper is the dominant substrate, significant volumes of paperboard, tin plate, and plastic are printed with the process. Inasmuch as it uses inks formulated with drying oils, resins, and high-boiling-point, petroleum-based oils that form dry films under ambient conditions through oxidation and polymerization, the air emissions generated are normally very small.

PROCESS DESCRIPTION

The sheetfed lithographic printing press consists of a feeder, a printing unit, and a delivery unit. The feeder supplies sheeted paper, paperboard sheets, or other substrates to the printing unit. A typical printing unit usually consists of the following.

1. An inking system containing multiple metal and polymeric rollers to produce and deliver, at high speed, a very thin, uniform ink film to the printing plate.
2. A dampening system that delivers fountain solution to the printing plate by one of a number of different mechanical systems and from which the nonprinting areas of the printing plate are preferentially wet and maintained during printing runs.
3. A plate cylinder on which the printing plate with its printing images is mounted and from which only the inked images are transferred from the plate to the blanket cylinder.
4. A blanket cylinder on which an elastomeric rubber covered, multiple-tiered lithographic blanket is mounted, and which receives the inked images and passes them onto the substrate.
5. An impression cylinder that controls the force imposed on the substrate in contact with the lithographic blanket so that the inked images are transferred properly from the blanket to the substrate.

Figure 1 illustrates one configuration of a typical lithographic printing unit. As in all other offset lithographic processes, printing is done by offsetting an inked image from the printing plate onto the blanket cylinder and then transferring the image from the blanket cylinder onto the substrate (paper sheet). Any number of printing units can be installed sequentially, with a six-unit printing press being the most common printing configuration used today. Depending on the product to be printed and the type of ink to be used, the delivery unit may contain various accessories, such as coating devices, radiation sources for ink drying and curing [infrared, ultraviolet (UV), and electron beam], antisetoff powder applicators, and a substrate-handling system for delivering and stacking the printed paper sheets. Figure 2 depicts a modem six-color sheetfed offset lithographic printing press.

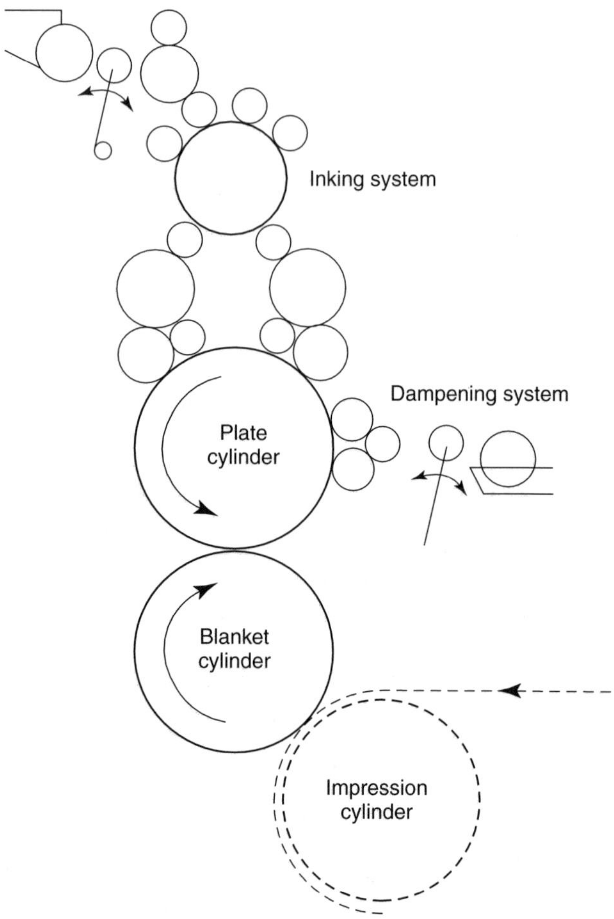

Figure 1. Configuration of a Typical Lithographic Printing Unit (From Ref. 1, p. 14:9).

Figure 2. Miller TP104PLUS Six-Color Press (Roland and Miller Sheetfed Press Systems, Pittsburgh).

AIR EMISSIONS CHARACTERIZATION

A variety of chemicals, including cleaning solvents, fountain solution additives, printing inks, and coating materials, can be used in the sheetfed offset printing process, a number of which may be sources of air emissions. Due to the nature of the printing process, most of the air emissions are fugitive, evaporative losses and are very small in quantity. The emissions can be categorized as described in the following.

Prepress Operations

Most sheetfed lithographers have photographic operations, almost exclusively for processing images on photographic film to develop the desired images for lithographic reproduction. Trace amounts of air emissions may be generated from the use of isopropyl alcohol and other solvents in film developing, color proofing, and some printing-plate processing.

Dampening System

The key to successful sheetfed lithographic printing is the proper maintenance of image boundaries by the application of aqueous fountain solution to the printing plate. This technique is known as dampening, and it is done either by applying the fountain solution directly onto the printing plate through a series of rollers or by applying it to the surface of the inkform rollers.

The fountain solution is usually composed of 80–95% water, 5–20% isopropyl alcohol or alcohol substitutes (usually ranges from 2% to 5%), and many other additives in smaller quantities, such as gum and phosphoric acid to maintain the pH of the fountain solution. Air emissions are usually generated during the dampening process from the evaporation of the isopropyl alcohol or its substitutes from the fountains, the dampening rollers, and the printing plates.

A small amount of the fountain solution is transferred to the substrate during printing, where subsequently a minimal amount of alcohol carried over by the fountain solution may evaporate and cause emission. Some minor emission also may occur from the fountain solution transferred to the substrate and emulsified in the ink on the inking roller trains. The amounts of these emissions are very small and of a fugitive type. Some presses use dampening systems that function with fountain solutions that do not require the addition of isopropyl alcohol or alcohol substitutes. These presses, frequently serving the one- or two-color end of the market, do not generate any emissions from their dampening systems.

Inking System

Most sheetfed offset inks used today for general commercial printing are formulated to set quickly, usually within minutes after they are applied. This enables the printer to print the other side of the sheet within a few hours instead of waiting overnight. The main components of a typical sheetfed offset printing ink normally include drying oil (about 30%), such as linseed oil, tung oil, or soy oil; resins (about 30%); pigments (about 15%); petroleum distillates (about 15%); waxes; and other additives.

The ink-setting mechanism involves the penetration of the distillate–resin phase of the ink into the substrate and deposition of the film-forming drying oil–resin phase with the pigments onto the substrate surface for subsequent oxidation and polymerization.

Minor emissions may result from the evaporation of petroleum distillates from the ink fountain, from the distribution of ink on the ink roller in the inking system, and from the ink in and on the substrates after the ink has been applied. The amount of emission is very small because of the low volatility of the petroleum distillate and the low operating temperature encountered during printing. Inks also may be formulated without petroleum distillates and those organic emissions will not occur.

Ink Roller, Plate, and Blanket Washing

Blankets are cleaned either manually or automatically during press stops, between runs, and between jobs with solvent formulations (principally petroleum based) or solvent and water emulsions and mixtures. Plates and ink rollers are normally manually cleaned with cleaning rags or paper wipes soaked with solvent between runs or ink color changes. The emissions are of a fugitive type and are often generated while these solvents or solvent mixtures are being applied, and the amounts are relatively small. In some applications, blankets and rollers must be cleaned frequently using lower-boiling-point volatile solvents so that less solvent residue remains on the rollers and blankets. In these situations, considerable quantities of the solvent may become fugitive air emissions. The used cleaning rags and wipes are usually placed in metal containers with lids that curtail the amount of solvent vapor emitted into the pressroom.

In-Line Coating Systems

Printers frequently utilize in-line coating finishes following printing, either as the last unit on the printing press or as auxiliary equipment. The coating provides a glossy or dull finish for surface protection of the final product. There are three main types of coating compositions: varnish, aqueous coating, and UV-cured coating. Press varnish is essentially sheetfed offset ink without pigments and may generate very small amounts of ink oil emission. Water-based coatings may also emit a minimal amount of emissions owing to the use of a small amount of alcohol in their formulation. A UV-cured coating normally does not generate any emissions, except for a small amount of ozone from the use of the UV lamps.

Antisetoff Agents (Spray Powders)

Atomizers are used to deposit antisetoff agents of controlled particle sizes (usually 15 to 60 μm) on the surface of the printed sheets to prevent contact of printed sheets in the pile. Antisetoff agents normally consist of chemically modified cornstarch. They are applied onto each printed sheet to create a gap between sheets so that the applied inks can go through oxidization and polymerization quicker or to "set" properly. Excess antisetoff agents usually are collected in the delivery unit by a vacuum device and transported to a cyclone and/or bag filters. Printers often use antisetoff agents only on printing jobs with high ink coverage.

AIR POLLUTION CONTROL MEASURES

Since most air emissions from sheetfed offset printing are fugitive, addon devices to control the emissions are not practical. The only feasible way to minimize emissions is to reduce evaporative losses during the printing operations. The methods routinely used by sheetfed printers include the following:

1. Refrigeration of the fountain solution. Refrigeration devices to control fountain solution temperature in the 55–60°F range are normally used by sheetfed offset printers. By lowering the temperature of the fountain solution, the evaporation rate of isopropyl alcohol in the dampening systems is considerably reduced. But reducing the temperature of the fountain solutions to below the 55–60°F range is not recommended because random water condensation may form on the printing plate and the condensation may affect print quality.
2. Isopropyl alcohol substitutes in the dampening solution. Dampening agents other than isopropyl alcohol have been employed to reduce emissions. Glycol ethers have been found to be the most acceptable substitutes. These substitutes are considerably less volatile and can be used in much lower concentration (usually about 1–3%). This combination of factors can reduce emissions effectively. The use of alcohol substitutes does require much better control over press operating parameters, as well as a well-balanced fountain solution. Many printers still need to use fountain solutions with isopropyl alcohol for special printing jobs. Nevertheless, substitutes have been gaining acceptance within the printing industry.
3. Use of organic solvents with less volatile organic compound (VOC) content. A wide variety of degreasing, cleaning, and washing organic solvents are used in sheetfed offset printing operation, with most of the organic solvents used to clean press rollers and blankets. The amount of VOC emissions that these cleaning solvents contribute is second to the amount contributed by the isopropyl alcohol in the dampening solution. A large number of printers have successfully tried several different types of cleaning solvents available on the market with less VOC content and have achieved the same cleaning effectiveness. This substitution has helped many printers in several states to maintain compliance with stringent state VOC emission regulations.

REFERENCES

1. *The Lithographers Manual*, 8th ed., Ray Blair, Ed., Graphic Arts Technical Foundation, Pittsburgh, 1988.

BIBLIOGRAPHY

A. Glassman, *Printing Fundamentals*, TAPPI Press, Atlanta, 1985.

V. Strauss, *The Printing Industry*, Printing Industry of America, Washington, D.C., 1967.

GRAVURE PRINTING

Gregory Tyszka

The gravure printing industry consists of three branches, namely, publication, packaging, and product gravure. Publication gravure plants print magazines, catalogs, and free-standing advertising material. Presses in the United States can accommodate paper webs up to 139 in. wide and can run at speeds up to 3000 fpm.

Packaging gravure is used to print cartons, paper wraps and labels, plastic film, and aluminum foil. Some presses operate in-line with laminating or extrusion equipment, producing sophisticated structures of paper, film, and foil with barrier properties for water vapor and oxygen to preserve the freshness of the contents of the package. Press widths range from 12 to 64 in., and press speeds typically range from 300 to 1000 fpm and higher.

Product gravure is used to print gift-wrap paper, wall coverings, floor coverings, wood grains for laminates, some postage stamps, and other specialty products. Press widths and press speeds in product gravure printing vary considerably, with floor-covering presses up to 192 in. in width.

PROCESS DESCRIPTION

Almost all gravure is web fed; that is, the printing substrates, such as paper, paperboard, plastic film, aluminum foil, or laminates of these materials, are supplied in roll form rather than in the form of sheets. During printing, the substrate rolls are unwound in a reel stand. The web may be preconditioned to increase its dimensional stability before entering the printing section of the press.

For each application of primer coatings, printing inks, or top lacquers, a separate printing unit (sometimes called printing station) is used, into which a gravure cylinder has been installed prior to the press run (Figure 1).

In rotogravure printing, the image areas (the areas from which inks or coatings are transferred) consist of small, recessed cells engraved into the surface of the gravure cylinder. The material that is engraved is typically an electroplated layer of copper. After engraving, the copper is protected from wear by a very thin electroplated layer of chromium.

As the cylinder rotates around a horizontal axis, its lower portion is flooded with liquid ink. The rotation carries the ink-covered surface past a doctoring station, where a thin doctor blade removes the ink from the nonprinting surface of the cylinder, leaving ink only in the recessed cells. The rotation then carries the inked and doctored surface of the cylinder to the printing nip, where ink transfer onto the substrate is accomplished by pressing the substrate web against the cylinder with a rubber-covered impression roll.

Each printing station is followed by a high-volume air dryer, before the next ink or coating is applied. The ink dries by evaporation of its liquid phase at temperatures that typically range from 80 to 150°F for publication gravure and up to 280°F for packaging gravure (Figure 1).

Gravure inks have to be highly fluid with a viscosity in the 7–30-centipoise range to allow for proper flooding of the cylinder by ink, for doctoring, and for transfer of the ink from the cells to the substrate that is printed. The low ink viscosity and fast drying are traditionally achieved by the use of low-boiling-point organic solvents in the ink. The ink in the press fountains can contain as much as 75% solvent by weight.

To maintain ink viscosity during the press run, further solvent is usually added to compensate for solvent that evaporates from the inking system.

In publication gravure, the printing stations are followed by a folder that delivers either cut and folded signatures for

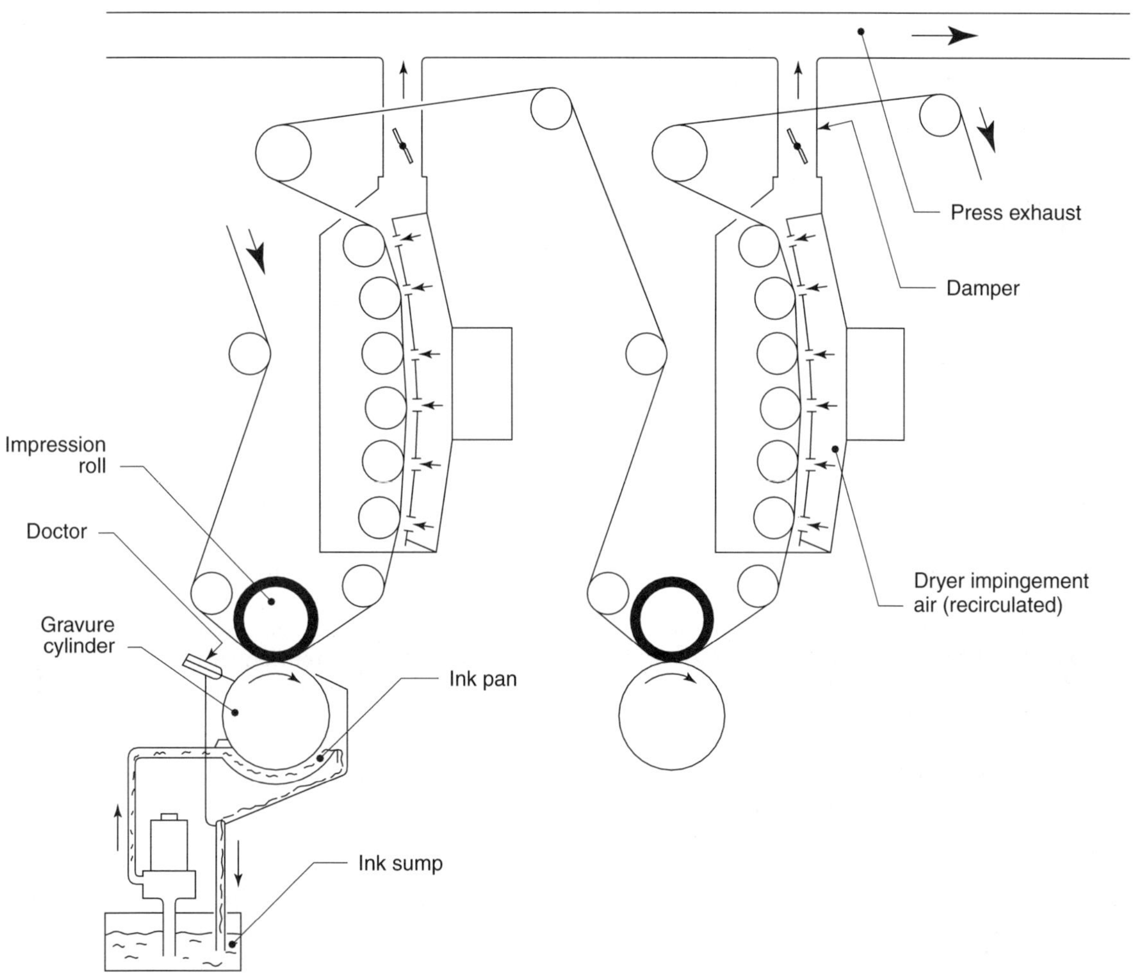

Figure 1. Section of a Gravure Press.

bindery use or products that may be trimmed in-line with the press. In packaging and product gravure, the printed web is rewound, sheeted, or die-cut into carton blanks.

AIR EMISSIONS CHARACTERIZATION

The solvents used in gravure printing inks and coatings are defined by the U.S. Environmental Protection Agency (EPA) as volatile organic compounds (VOCs). Most of the solvents are evaporated in the press dryers, although some solvents (up to 7%) may be retained in the web. Some solvent evaporates from exposed portions of the cylinder, the inking system, and the web, either before or after it has passed through the dryers. Measurements have shown that most of these fugitive vapors are drawn into the dryers, provided there is sufficient dryer exhaust flow. The dryer inlet slots are within about 4 ft of the printing area proper and the fountains are enclosed. Some solvents used for press cleanup may also evaporate and escape as fugitive emissions.

In addition, VOC emissions may result from breathing and working losses from bulk ink and solvent tanks. These emissions are small and normally uncontrolled, although, in some cases, they may be vented to the pollution control device.

Ancillary sources of emissions connected with gravure printing include the boilers used to produce steam for steam-heated dryers or, in plants with solvent-recovery systems, to strip the solvent from the carbon beds.

In plants that operate chrome-plating tanks, the chromic acid mist generated is normally controlled by a demister that, in some cases, is preceded by a wet scrubber. Where paper trimmings and other paper waste are pneumatically transported for collection, the paper is mechanically separated from the airstream using a cyclone. In cases where high loadings of paper and clay dust are present in the airstream, the dust may be removed from the cyclone exhaust by using a baghouse fabric filter. No VOC emissions result from these operations.

AIR POLLUTION CONTROL MEASURES

The following options to control solvent vapor emissions for compliance with regulations arising from the Clean Air Act are recognized:

- Solvent recovery
- Solvent vapor incineration
- Use of water-based inks and coatings
- Use of high-solids inks and coatings

Solvent recovery and incineration require pollution control systems, while the other control options employ low-VOC-content materials to achieve compliance. The high-solids compliance option could not be implemented in gravure printing.

Capture, Control Device, and Overall Control Efficiencies

For solvent recovery and incineration, overall emission control efficiency is the product of the capture efficiency and the control device efficiency. In plants with solvent recovery, overall efficiency is typically determined as the ratio of solvent recovered to solvent used.

The amount of solvent used is the sum of the solvent contained in the ink and varnishes that are received from the ink maker plus the solvent added by the printer to reduce the ink viscosity to the level required for printing. The solvent content of the ink can be obtained from the ink maker. Alternatives to determine the solvent level of ink are Methods 24 and 24A in the appendix to 40 CFR (Code of Federal Regulations) 60, for inks that contain water and inks that contain no water, respectively. As the methods specify temperatures of 110°C and 120°C to evaporate the volatile phase of the inks, they can indicate higher values of VOC content than are likely to evaporate in practical situations.

The control device efficiency of a solvent-recovery or incineration system is given by one minus the ratio of the VOC content of the exhaust airstream of the device over the content at the device inlet. Control device efficiencies are typically very high, and hence the overall efficiency of a control system is normally limited by the capture efficiency.

Control devices operate most efficiently at reasonably high vapor concentration levels. All publication and the newer packaging and product gravure presses use two dryer air systems, namely, a recirculating impingement air system and a dryer exhaust system (Figure 1). This cuts down on the amount of air that has to be exhausted and, therefore, increases the exhaust concentration. However, solvent vapor concentration levels are limited by safety considerations, since gravure solvents are flammable. Peak concentrations in dryers and ducts are typically limited to 25% of LFL (lower flammable limit) or 40–50% of LFL, when continuous monitors are employed to provide automatic warning and equipment shutdown. Dryer exhaust dampers must be set to keep concentrations below these limits at all times.

Solvent use is roughly proportional to press speed. When a press is running slowly, the vapor concentrations in the exhaust tend to be low. In packaging and product gravure, there are wide differences in the ink coverage among different jobs, a factor that leads to lower average solvent concentration levels in the captured air. This, in turn, can lead to reduced control device efficiency.

Capture efficiency is the ratio of solvent in the airstream going to the control device compared to the solvent used. Methods to measure solvent concentration and airflow volume are also specified in the appendixes to 40 CFR 60. One hundred percent capture efficiency is assumed for plants designed with fully enclosed pressrooms at specified levels of negative air pressure and with all exhaust flow through a control device. Fire- and explosion-hazard protection dictates the provisions for emergency venting to atmosphere in the event of equipment malfunction or breakdown.

Low capture efficiencies may result from fugitive losses as well as solvent retention in the substrate. When capture efficiency is measured on presses with gas-heated dryers drawing pressroom air, allowances should be made for solvents that are burned. Depending on the age of the plant, plant layout, and the plant air-handling system design, fugitive losses can be minimized by selective enclosures of parts of the presses, high dryer exhaust rates, the proper location of makeup air inlets, and similar measures. Efficient capture of fumes is important not only in maximizing overall efficiency, but also in ensuring that pressroom solvent fume levels are kept within Occupational Safety and Health Administration (OSHA) exposure levels.

Emission Control Options

Solvent Recovery. Solvent recovery is used in all publication gravure printing plants and in some packaging gravure printing plants, where the solvent can be reused without too much further processing.

In the operation of a solvent-recovery system, the exhaust air from the press dryers and, where practical, from other exhausts near the production presses is collected and passed through large beds of activated carbon (Figure 2). The carbon adsorbs all (>95%) but trace amounts of the solvent vapor. When a bed is saturated, the exhaust stream is diverted to a nonsaturated bed, and the saturated bed is regenerated by steaming. The resulting steam–solvent vapor mixture is condensed. The solvent and the condensed water separate by gravity in a decanter vessel. In newer systems, a brief cooling cycle follows steaming. This avoids steam plumes and solvent loss that would occur when solvent-laden air (SLA) is passed over a hot, wet carbon bed and then exhausted.

Solvent-recovery systems are equipped with control instrumentation not shown in Figure 2. For example, a sensitive vapor concentration monitor on the exhaust of the adsorbers will initiate the steaming cycle when a bed is saturated and solvent "breakthrough" occurs. This system is usually backed up by a system that schedules steaming on a time basis.

For efficient operation, it is essential that the air that enters the solvent recovery be properly conditioned. The temperature of the SLA that enters the adsorbers should be below 110°F. The air must be filtered to avoid particulates that can clog the carbon beds. Compounds with high boiling points will stay on the carbon permanently and reduce its working capacity. The use of chlorinated solvents anywhere in the building can lead to corrosion and should be avoided.

Publication gravure solvents are composed of either toluene or mixtures of toluene and similar-boiling-point aliphatic hydrocarbons. These solvents are miscible with water only in trace amounts and can be separated from water in the steam–solvent condensate by gravity in the decanter. The recovered solvents are used within the plant for ink dilution and viscosity control, and some solvent is returned to the ink supplier.

Solvents used for packaging and product gravure are usually more water miscible. These solvents are usually mixtures that contain acetates, ketones, and alcohols. Such recovered solvents may require water removal or redistillation before reuse. In any case, solvent traces may have to be removed from the decant water before it can be reused or discarded.

The ratio of pounds of steam to pounds of recovered solvent, which ranges from 1:1 to 5:1, is a major criterion in operating costs. However, given the relative costs of solvent and steam, plants are usually operated to recover as much solvent as possible.

Solvent Vapor Incineration. Solvent vapor incineration oxidizes the solvent vapors into harmless carbon dioxide and water vapor. By and large, incinerators are employed where a variety of solvents and water-miscible solvents are used that would require redistillation after solvent recovery. This applies mainly to packaging and product gravure plants where complex end-use requirements mandate the use of special ink resins and solvents such as acetates, alcohols, and ketones, as well as aliphatic and aromatic hydrocarbons. In all solvent incineration systems, high overall VOC destruction requirements will increase energy consumption and may shorten control equipment life.

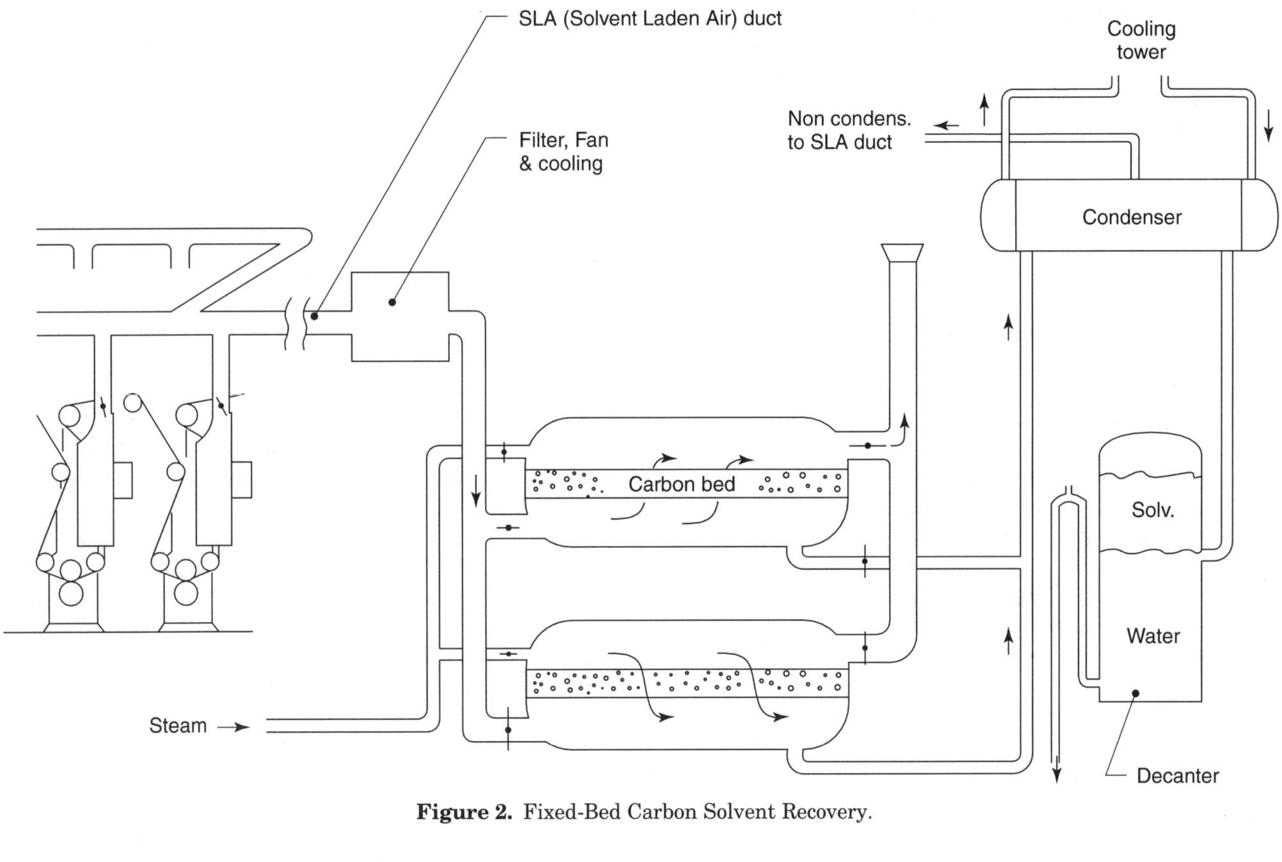

Figure 2. Fixed-Bed Carbon Solvent Recovery.

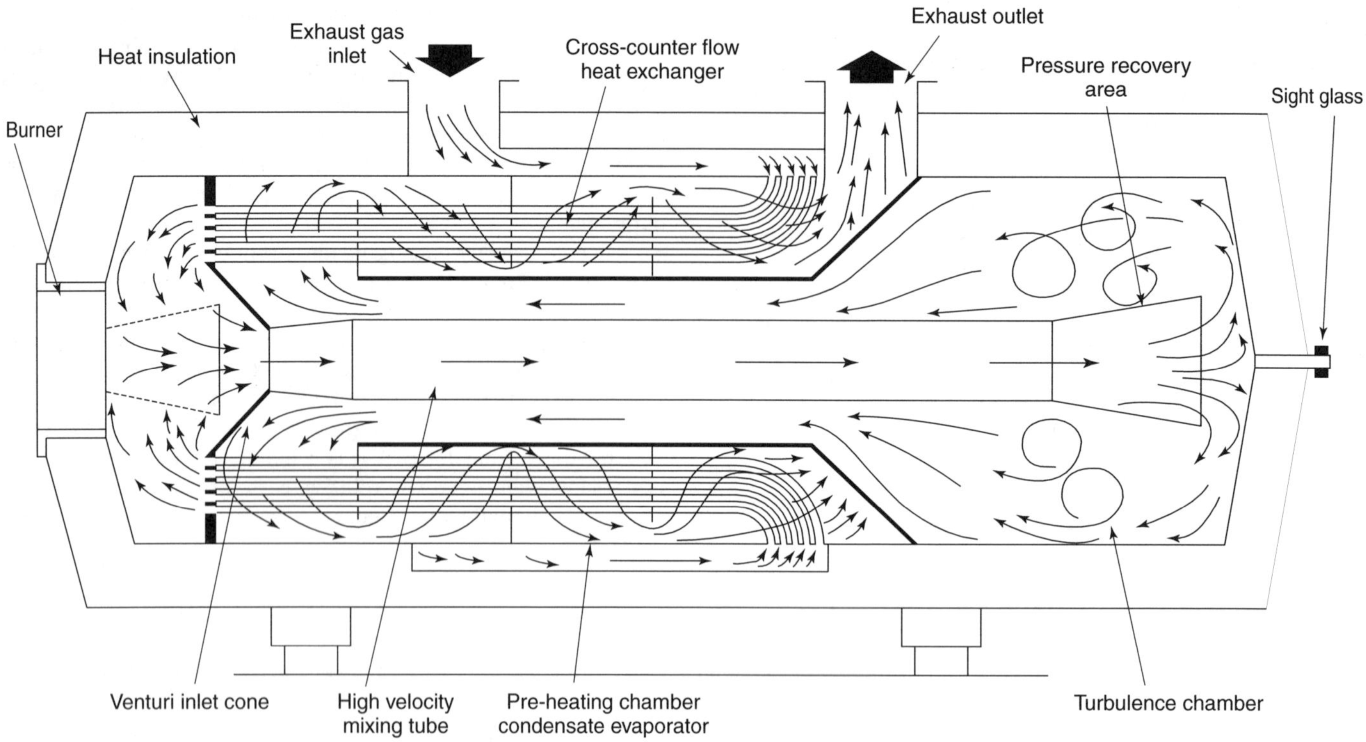

Figure 3. Recuperation-Type Thermal Afterburner.

Thermal oxidation with heat recovery is appropriate for large, steady exhaust streams. Operating at 1325–1600°F, destruction of over 99% of the captured VOCs can be achieved. To reduce energy requirements, heat recovery is usually employed. Recuperative systems incorporate primary and sometimes secondary heat exchangers. Primary heat exchangers are used to preheat the SLA before it enters the combustion chamber. Secondary heat exchangers provide return heat to the building or for other purposes in the plant (Figure 3).

Regenerative systems use a number of chambers filled with a heat-absorbing medium (usually ceramic) to preheat the airstream before it enters the combustion chamber. When needed, a burner further elevates the temperature in the combustion chamber where oxidation of the VOCs occurs. The airstream is redirected every few minutes so that air going to the combustion chamber is passed through chambers that have previously been heated by the high-temperature exhaust air from the combustion chamber (Figure 4). Regenerative systems are more frequently used to control exhaust streams above 15,000 scfm.

Catalytic incinerators operate in the 500–750°F temperature range, with device efficiencies that range from 90% to 98% (Figure 5). They are usually employed for exhaust streams below 15,000 scfm and where press operations are frequently interrupted by job changes. Due to their lower operating temperatures, catalytic units are better able to withstand the thermal stress placed on the equipment by cycling between standby and active status. Care has to be taken to avoid materials that can poison or mask the catalyst.

Concentrator devices have been introduced to raise afterburner inlet VOC concentrations by a factor of about ten by adsorbing VOCs from low-concentration airstreams by carbon adsorption and subsequent desorption by relatively small volumes of hot air for subsequent incineration. This increases energy and destruction efficiency by reducing the volume of air to be treated and increasing the energy value of the SLA stream.

Water-Based Inks. Water-based ink and coating systems have been under development in the gravure industry for many years. Certain packaging and product gravure operations are successfully printing with water-based systems routinely. Although most current water-based systems contain some VOCs, they appear to be the alternative to control solvent vapor emissions without the need for capital and energy-intensive control devices. Water-based systems also reduce safety hazards and the complications arising from the need to transport, store, and handle large amounts of flammable solvents. The inherently higher surface tension and the slower drying rate of water are major obstacles. Extensive work on ink formulation, cylinder engraving, press operations, and dryer design has achieved success in some areas.

Control by high-efficiency solvent-recovery systems is employed throughout publication gravure. Water-based ink systems are seldom used in this segment of the industry because of rough printing, paper distortion, and press-speed limitations. However, some limited success has been achieved with uncoated paper on slow presses.

In packaging gravure, some paperboard and paper products are successfully printed at speeds up to 1000 fpm. Water-based usage is expected to increase in this area, although problems still exist at higher press speeds. Water-based systems are used to print a number of aluminum foil products. On plastic film, there has been little success.

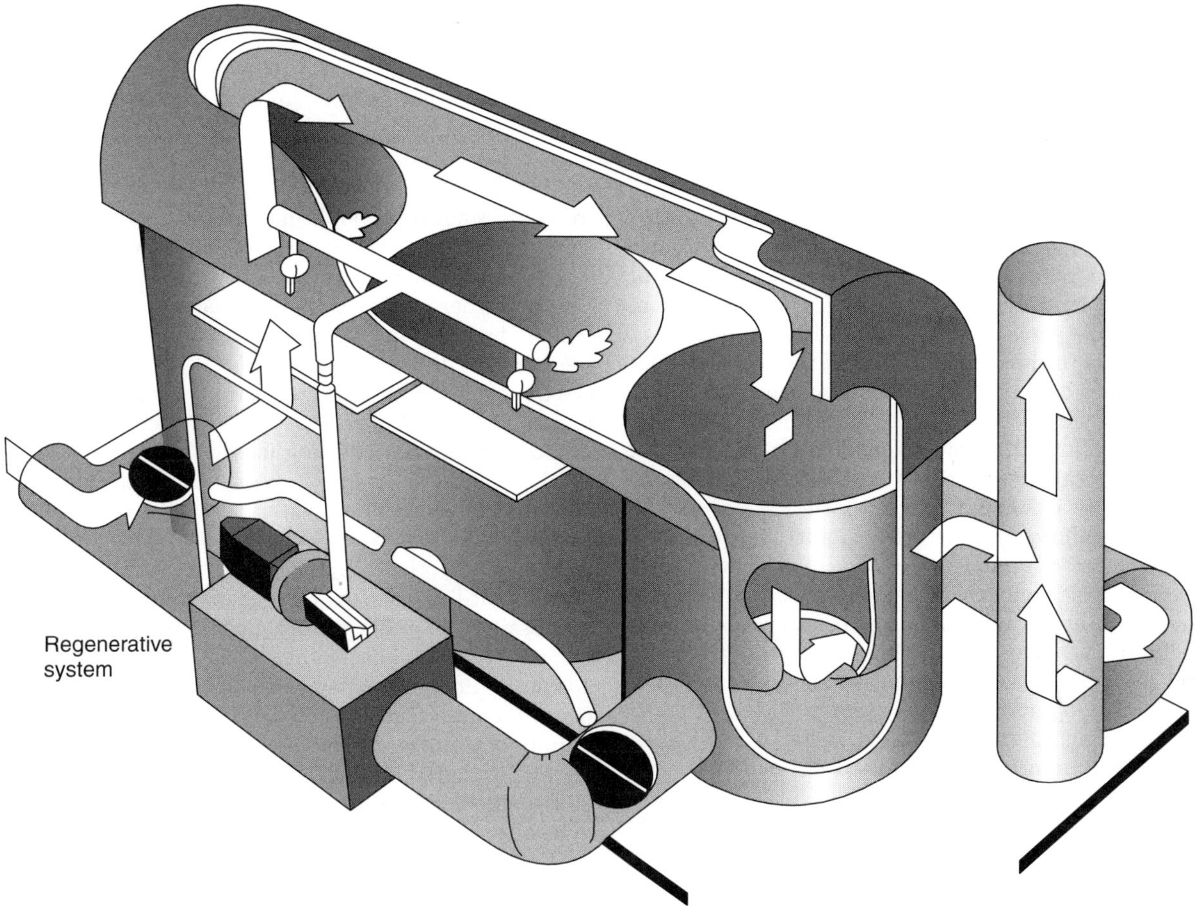

Figure 4. Regenerative Afterburner.

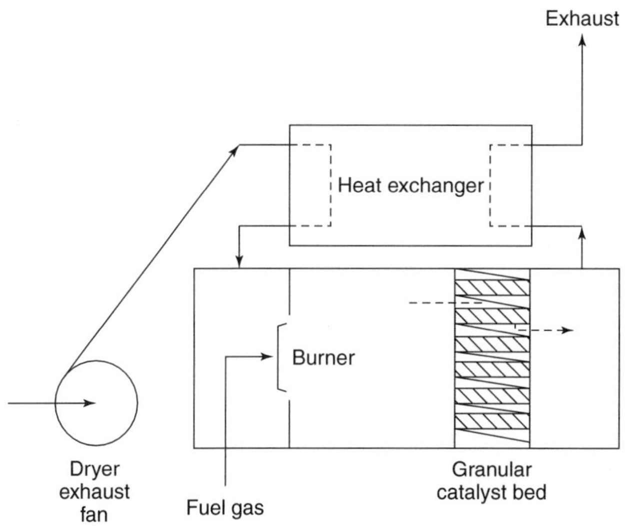

Figure 5. Catalytic Afterburner with Heat Recovery.

Product gravure involves a great variety of printing substrates and end-use requirements, so that no generalizations can be made. Water-based systems have been very successful on medium-weight paper, but on lightweight paper, problems with paper distortion and curl are encountered. Water-based systems are used successfully on some nonabsorbent substrates at low press speeds.

NESHAP FOR PRINTING AND PUBLISHING INDUSTRY

(update by Dave Salman (U.S. EPA) and Wayne Davis, editor)

National Emission Standards for Hazardous Air Pollutants associated with the printing and publishing industry were proposed by the U.S. EPA in March 1995. The final rule for the industry was published on May 30, 1996. The final rule is applicable to all existing and new rotogravure and wide-web flexographic facilities that are major sources of HAP or are located at plant sites that are major sources of HAP. Section 112(a) of the CAA defines a major source as a source, or group of sources, located within a contiguous area and under common control that emits or has the potential to emit, considering controls, 9.1 Mg/yr (10 tpy) or more of any individual HAP or 22.7 Mg/yr (25 tpy) or more of any combination of HAP. The HAP emissions include toluene, xylene, ethylbenzene, methanol, methyl ethyl ketone, methyl isobutyl ketone, ethylene glycol, and certain glycol ethers.

The "compliance date" for existing sources for this regulation was May 30, 1999. New major sources are required to comply with the NESHAP requirements upon start-up or the promulgation date, whichever is later. The final standards will reduce organic HAP emissions from rotogravure and wide-web flexographic printing operations by 6,700 Mg/yr (7,400 tpy) from a baseline level of 21,700 Mg/yr (23,900 tpy).

Publication rotogravure facilities subject to this rule must limit emissions of organic HAP to no more than eight percent of the total HAP volatile matter used each month. The emission limitation may be achieved by capture and control of at least 92 percent of organic HAP used, by substitution of non-HAP materials for organic HAP, or by a combination of capture and control technologies and substitution of materials. Product and packaging rotogravure and wide-web flexographic printing facilities subject to this rule must limit organic HAP emissions to no more than five percent of the organic HAP applied each month, or to no more than four percent of the mass of inks, coatings, varnishes, adhesives, primers, solvents, reducers, thinners, and other materials applied each month, or to no more than 20 percent of the solids applied each month, or to an equivalent allowable mass based on the as-applied solids contents of the materials applied each month.

The regulation allows sources the flexibility to select from various options for compliance. Sources may reduce HAP usage and emissions through conversion to water-borne, lower HAP solvent-borne or ultraviolet/electron beam cure materials. Alternatively, sources may install or upgrade existing capture and control devices to meet the proposed standard. Finally, sources have the option to comply by a combination of lower-HAP materials and capture and control. Each facility may select its compliance strategy based on facility specific considerations.

The reader should refer to the final rule for more details as well as to any subsequent information that may be published.

REFERENCES

1. *National Emission Standards for Hazardous Air Pollutants; Final Standards for Hazardous Air Pollutant Emissions from the Printing and Publishing Industry* (40 CFR Subpart KK), U.S. EPA, 61 FR 27131, May 30, 1996 (available at *http://www.epa.gov/ttn/uatw/print/printpg.html*).

BIBLIOGRAPHY

A. Glassman, *Printing Fundamentals*, TAPPI Press, Atlanta, 1985.

V. Strauss, *The Printing Industry*, Printing Industry of America, Washington, DC, 1967.

Technical Guide for the Gravure Industry, 3rd ed., Gravure Technical Association, New York, 1975.

SCREEN PRINTING

Marcia Y. Kinter

The screen printing industry, although often considered the smallest arm of the graphic arts industry, is the most versatile. The flexibility of the screen printing process allows the controlled deposit of inks onto a substrate in a manner that provides the user with the needed legibility, color opacity, and durability. This method is known for its ability to impart relatively heavy deposits of ink onto practically any type of surface, in a controlled pattern, with few limitations on the size and shape of the object being printed.

Today, there are over 30,000 screen printing operations located throughout the United States. It has been estimated that in 1986, the screen printing industry posted gross sales of $48 billion. This printing process is an integral part of every industry that produces products that need imprinted instructions, identification, or color application. Over the past two decades, the rate of growth of the screen printing industry has exceeded that of any other type of printing process.

The average screen printing company has 21 employees, in both production and management. Nearly half of the screen printing facilities produce imprinted textile garments. This category includes clothing, caps, children's wear, and towels. Other major categories of screen-printed products include banners and billboards for outdoor advertising, point-of-purchase displays, posters (both single and multisheet), electronics such as circuit boards, containers of all types and sizes, and pressure-sensitive decals. End products produced by screen printing include in-store displays, decals, labels, emblems, banners, flags, pennants, binders, transit advertising, nameplates, wallpaper, and product-identification markings.

Unlike the other major printing processes, screen printing is able to utilize a wide variety of substrates. All substrates can be classified as either absorbent or nonabsorbent. Substrates for screen printing include all types of plastics (acrylic, epoxies, vinyl, topcoated and nontopcoated polyester, and polycarbonate); fabric (both natural and synthetic); metals (aluminum, brass, copper, lacquer-coated metals, and steel); papers (uncoated, coated, corrugated coated fiberboard, poster, and cardboard); and other substrates, such as leather, masonite, wood, and electronic circuit boards.

This listing of substrates and end products graphically illustrates the differences between screen printing and all other major printing processes. It needs to be noted at this point that the majority of screen printers do not restrict their operations to printing on one substrate or to the production of one end product.

PROCESS DESCRIPTION

In the most simplistic terms, the screen printing process involves the flow of ink through a porous screen mesh to which a stencil has been added to define the image. The ink flows through the imaged screen by virtue of hydraulic pressure that is initiated by the action of a flexible rubber or synthetic blade known as a squeegee. The squeegee blade sweeps across the surface of the pretensioned, preimaged screen, pressing the ink through those areas of the screen not blocked by the stencil and onto the substrate in the pattern defined by the stencil image. The substrate is then either manually placed onto drying racks or mechanically/manually placed onto a conveyor transport system for conveyance into a drying unit.

Mesh

Many different material supplies are necessary for the completion of the screen printing process; however, mesh is a major player when discussing the emission of volatile organic compounds (VOCs). Mesh is available in the following materials: stainless steel, natural silk, monofilament nylon, monofilament polyester, multifilament polyester, metalized monofilament polyester, and calandered (one side flattened) monofilament polyester. All these fabrics are available in differing numbers of threads per inch and thread diameters, which affect the fabric thickness, the size of the mesh open areas, and subsequently the volume of ink deposited.

The selection of the screen fabric is critical to the screen printing process. The diameter of the mesh thread determines the overall mesh thickness. This, in combination with the number of threads woven per inch, determines the amount of ink deposited onto the substrate. Ink can be deposited onto a substrate in a thickness of 0.04 to 10 mils.

Prepress Applications

The most significant advance in screen printing is the replacement of the traditional analog prepress systems with digital prepress systems. Analog technologies consist of manual labor and machine-assisted operations not fully executed by computers. Examples of analog technologies include the use of reprographic cameras, contact proofs, and cutting of film.

Digital technologies are those that are completely executed by a computer or computer-operated equipment. Using digital applications in the prepress area reduces or eliminates the use of traditional photographic processes. Computer-to-screen digital systems utilized digital output devices with thermal, laser, or inkjet imaging to image the screen directly without films.

Printing Equipment

Most commercial and industrial screen printing is produced on automated printing presses. These presses typically consist of an automatic squeegee, an indexing system to place the screen and printing base in register each time a substrate is loaded, and some type of substrate transport method. Within the industry, there are manually operated, semiautomatic, three-quarters automatic, and fully automatic presses. All types are considered power-assisted presses. The automation refers to the substrate transport method and the mechanical printing device mechanism.

Within the industry, flatbed, cylinder, and web presses are used, with the flatbed press probably the most common. The print bed of such presses is horizontal and parallel to the screen. The cylinder press prints the substrate on a cylinder as the screen, substrate, and cylinder move together.

Also found within the screen printing industry are web presses. On a web press, the substrate unwinds off a roll, moves through various print and dry stations, and finishes by passing through a die cutter or slitter or is rewound onto another spool. Web presses can be of either the flatbed or cylinder type.

The press is typically configured in-line with a conveyorized drying oven. This oven takes the freshly printed substrates and subjects them to the energy level necessary for quick and effective solidification of the printed image. Absorbent substrates are typically printed with all colors consecutively applied prior to the ink-drying step. However, nonabsorbent substrates, such as plastics, require ink drying between each individual color application.

Energy sources used to dry the wet printed ink include infrared (IR), convection, and ultraviolet (UV) light. Each of these methods must be suited to the chemical composition of the ink used. All drying/curing units include adjustable temperature/wattage controls, variable belt speeds, and exhaust systems vented to the outside to remove emissions from the workplace. These vents are viewed as air emission sources. Also, in some cases, the printed product is air dried on racks prior to final packaging.

The most common type of drying system utilized is the hot-air convection oven. The key to a convection oven is the recirculation of hot air. Within this type of drying oven, typically 80% of the air is recirculated to maintain operating efficiencies, while 20% is exhausted. Temperatures are absolutely controllable.

Infrared systems use inductive, as compared with convection, energy to create heat within the material and effect the drying process. The IR energy meets the substrate and excites the molecules, which creates heat within the material. Heat is produced through electrically conductive alloys with 3–7-μm wavelengths being the most useful for the curing process. The IR spectrum can be shifted to suit the absorption characteristics of the ink and/or substrate.

The IR elements are contained in an insulated and heated section that also contains the controls for the elements, airflow, and exhaust devices. The heated section separates the heating elements from the environment and provides for a controllable airflow and exhaust for the system.

The accuracy of the conveyor belt speed is a major contributor to the accuracy of the entire curing process. The amount of energy absorbed by the material depends on the amount of time spent under the IR element. Air is circulated within the heating system; however, very little is exhausted to the outside. The air is swept over the substrate and recirculated within the system.

Ultraviolet curing units provide relatively energetic photons of 200–400 nm. Photons are emitted from mercury vapor lamps within the curing unit. The intensity of commonly used lamps is 200 W/in.^{2}. Either the lamp is housed in an extruded aluminum reflector with highly polished walls for maximum UV reflectance or its energy is focused onto the substrate by parabolic reflectors.

The UV energy provided by the lamps is absorbed by the photoinitiators present within the liquid formulation of the ink. Free radicals are produced that attack the double bonds of the resin molecules, increasing the molecular weight as the particles add to themselves and form a cross-linked chain. This chain produces a solid polymer film. This process takes place in a fraction of a second. The UV curing units are 10 to 12 times more energy efficient than the traditional thermal curing units.

Ink Systems

The basic chemistry of ink systems revolves around three major elements: pigments that are the colorants for impact, identification, and contrast; binders that hold the colorants together in a continuous film and to the substrate onto which the film is printed; and solvents that dissolve the resins so that they are put in a fluid state and, together with the pigments, make a pliable and screenable material.

Resins are the important element of any ink system. They are the solids of the ink and bond the pigment to the surface. These resins have to be dissolved by the solvents. Ink manufacturers develop products from a host of resin systems to meet end-product demands for substrate compatibility, opacity, flexibility, abrasion resistance, and durability. For example, to print on a vinyl substrate, an appropriately formulated ink with a vinyl resin system would be required to assure durability. The volatile content of the ink is dependent on the demands of the resin system.

A variety of ink systems are available for use by the screen printing industry. The selection of a given ink system for a particular job is determined in large part by the end use of the product and the substrate being printed.

Within the screen printing ink family, four different major types of ink systems currently are utilized: UV inks (UV curables), water-based inks, plastisols, and the solvent-based ink systems.

Ultraviolet inks consist of pigments, monomers, oligomers, additives, and modifiers. The photoinitiator reacting to the UV light source produces free radicals that initiate a chain reaction, resulting in polymerization of the monomers and oligomers. Some UV inks are suitable for outdoor usage because they exhibit resistance to abrasion, solvents, chemicals, and other environmental menaces. They do have limitations in the area of flexibility and are not recommended for use on most corrugated surfaces.

Water-based ink systems use water as part or all of their solvent component. These ink systems all contain organic pigments, resins, and additives, such as flow promoters, retarders, and other performance enhancers. Water-based systems require the use of water-soluble resins and contain up to 65–70% solids. These inks can be used on a wide variety of substrates. A distinction must be made between water-reducible and water-based ink systems. Not all water-reducible systems are water based. Water-reducible systems can contain very little water, as low as 10%, but as long as the system can be reduced or thinned with water, it is considered water reducible. Water-based systems have replaced the traditional solvent with water. This is a major difference between true water-based systems and a water-reducible system that needs to be recognized.

Plastisol inks are generally used in the textile market. This ink system consists of two basic ingredients: polyvinyl chloride (PVC) resins and the plasticizer. This ink system is considered to be 100% solid with little or no VOC emission. Because of their chemical makeup, plastisols fuse rather than dry or cure; they must be exposed to heat in order to fuse. The PVC resin is present as discrete particles dispersed in the plasticizer. Only when the appropriate fusing heat (320°F) is applied to the ink does this condition change.

Conventional solvent-based ink systems are still widely used within the screen printing industry. These ink systems consist of pigments, resins, solvents, and additives. They are dried by the evaporation of the solvent from the system, leaving behind a thin film of resins and pigments. These ink systems are used on virtually all substrates. The amount of volatiles emitted by these different ink systems depends on the resin system utilized and the amount of volatile additive required.

Inks are rarely used as supplied by the ink manufacturers. Because of variations in temperature and humidity, the screen printer will generally use ink additives. These additives perform various functions. Thinners hold the resin in the binding agent and reduce the viscosity of the inks. Retarding agents are used to slow the drying time of the ink, so that the ink will not block the mesh openings of the screen. These additives are similar to thinners in that they hold the solid elements of the ink system soluble.

Generally, solvents are used as ink additives. The three general categories of solvents used in the screen printing industry are aliphatic hydrocarbons, aromatic hydrocarbons, and oxygenated solvents that are miscible in water.

Ultraviolet and plastisol ink additives are not solvent based and are considered nonvolatile materials. As previously mentioned, water-based ink systems are reduced with a combination of water and solvent, depending on the substrate. Solvent-based ink systems are reduced with VOC materials.

Screen/Stencil Reclamation

Depending on the ink system utilized during the printing process, screen reclamation activities can be undertaken with organic solvents or water-based systems. During the screen reclamation process, the ink residue is removed after all excess ink is removed from the screen. Water-soluble ink degradents or the appropriate solvents are used for this purpose. The fabric/stencil must then be degreased. Any chemical or solvent residue left on the stencil will impede the effectiveness of the stencil remover. After the stencil remover has been applied, the emulsion begins to dissolve and is completely removed using a high-pressure washer.

After repeated uses, hardened ink and stencil material removed appear as a ghost image from previous stencil applications. Ghost/haze removers are used to remove this unwanted condition. After using the haze removers, the screen fabric must be degreased again, to prepare it for reuse.

AIR EMISSIONS CHARACTERIZATION

The major emission problem in the screen printing industry remains the emissions of VOCs. Very little information is available concerning the release of air toxic or particulate air emissions.

It has been estimated that 95% of the VOCs emitted in a screen printing facility are fugitive emissions. These emissions result from the ink and solvents used throughout the printing process — two elements that account for most air emission sources in a screen printing facility. Petroleum-based solvents are used in both the ink formulation and screen reclamation operations. The ink systems themselves may contain up to 60% solvent. The previous discussion on the ink systems in the screen printing industry makes it very clear that they differ dramatically from the inks used in other printing processes.

The current method utilized for the calculation of VOCs in inks is Reference Method 24, which has been accepted by the U.S. Environmental Protection Agency (EPA) for the analysis of most printing inks. However, because of the wide variety of ink systems used by the screen printing industry, this test method does not apply for several very important reasons.

The test method calculation requires the removal of the water from the product in order to determine the actual VOC content. The water-based ink systems developed for screen printing use have replaced VOCs with water. Although water-reducible systems are used, the industry also uses true water-based ink systems. In utilizing Reference Method 24 to determine the VOC content of these inks, the industry is penalized for technological improvements designed to reduce the amount of VOCs emitted.

Although Reference Method 24 has been amended to included test method changes for non-thin-film reactive cure coatings, the reactive cure, or UV ink systems, used in screen printing are thin film, and thus the new methodology is not applicable. Since RM 24 calls for the baking of the product to determine the VOC content, erroneous VOC readings are obtained through the test method. This occurs most notably with the UV-curable inks systems that are cured, not baked. The baking requirement seriously distorts the actual VOC content of these ink systems and again penalizes the industry for the utilization of low-VOC emitting systems.

For these reasons, it is advocated that a test method that evaluates the VOC content in grams per liter of applied solids be used by the screen printing industry until Reference Method 24 has been amended for thin-film reactive-cure coatings. At this time if Reference Method 24 is used for the UV-curable ink systems, then it is recommended that a cure cycle be inserted before the bake cycle to simulate actual use.

AIR POLLUTION CONTROL MEASURES

Air pollution control equipment with a capture efficiency of 90% and an overall efficiency of 95% currently is not available for the screen printing industry. Both thermal and catalytic afterburners have been installed in several screen printing facilities; however, the average capture-control efficiency level for this equipment ranges between 70% and 80%. The air pollution control equipment either can be retrofitted to existing equipment or new drying units that include the control equipment can be purchased. The cost of an afterburner unit ranges from $12,000 to $2,000,000, depending on the configuration and the size of the facility and whether or not new drying units are purchased. Annual operating costs range as high as $500,000 per year.

The major air pollution control method currently applied by the screen printing industry involves the use of low-VOC-content or low-vapor-pressure products. Generally, the VOC content of inks is expressed in either grams per liter, as applied, or pounds per gallon, as applied. Because of the relatively small amounts of solvents used during the screen reclamation process, these VOC products are generally regulated by a vapor pressure figure, that is, 5 mm Hg at 20°C.

Several of the ink systems discussed previously do offer the screen printer the option of utilizing low-VOC-content inks. The UV-curable inks have very low emissions of VOCs. True water-based ink systems also are viewed as viable low-VOC-content inks, as are plastisols. However, these low-VOC-content inks are not suitable for use on all substrates or for all end products. Considering the end use of the product, such as outdoor graphics, or the type of substrate, it may still be necessary to use an ink system with a relatively high solvent content.

Due to the versatility of the screen industry, there are still certain end uses, regardless of the substrate, that require an ink with a higher solvent content. These situations include ceramic inks, inks used to manufacture ceramic decals, conductive inks used for circuitry printing, and inks for products that are designed to withstand any of the following: more than two years outdoor exposure, exposure to chemicals, solvents, acids, detergents, oil products, or cosmetics, temperatures in excess of 170°F, vacuum forming, embossing, or molding. Ink manufacturers are still in the process of verifying low-solvent-content inks for durability.

BIBLIOGRAPHY

A. Kosloff, *Screen Printing Techniques*, Signs of the Times Publishing Co., Cincinnati, 1981.

B. Magee, *Screen Printing Primer*, Graphic Arts Technical Foundation, Pittsburgh, 1985.

A. Peyskens, *The Technical Fundamentals of Screen Making*, SAATI, Como, Italy, 1989.

Technical Guidebook of the Screen Printing Industry, Vols. I–VI, Screen Printing Association International, Fairfax, VA, 1989.

MANUALS

Manual on Determination of Volatile Organic Compounds in Paints, Inks and Related Coatings, J. J. Brezenski, Ed., American Society for Testing Materials, Philadelphia, 1989.

1998 Industry Profile Study, Summary Report, Screen Printing Association International, Fairfax, VA, 1987.

Screen Printing Fact Book, J. Letuner, Ed., Screen Printing Network, Denver, 1990–1991, pp. 33–72.

ARTICLES

H. Bubley and F. Matthews, "A Comparison of Conveyorized Drying Systems," *Tech. Guidebook Screen Printing Ind.* **4**, Q1–Q8 (1978).

J. Clark, "Plastisol Inks: Construction and Application," *Screen Printing Mag.* 58–60 (August 1988).

S. Ducilli, "Making a Splash (Again): Waterbased Textile Inks Stage a Comeback," *Screen Printing Mag.* 126–31, 133–45 (August 1989).

S. Ducilli, "UV Braves the Elements, Newer Formulations Find Acceptance Outdoors," *Screen Printing Mag.* 63–66, 153 (August 1988).

T. Frescka, "Infrared Drying and Equipment," *Screen Printing Mag.* 60–69 (June 1987).

K. Gilleo, "Rheology and Surface Chemistry for Screen Printing," *Screen Printing Mag.* 128–33 (February 1989).

P. R. Herman, "Inks as Applied to Close Tolerance," *Tech. Guidebook Screen Printing Ind.* **2**, H1–H11 (1978).

A. Kosloff, "Ultraviolet Curing or Drying," *Tech. Guidebook Screen Printing Ind.* R1–R4 (1978).

C. Moret, "Pigmented UV Ink Review—An Incurable Craving of Curable Color," *Screen Printing Mag.* 133–35, 154 (August 1988).

C. Moret, "Waiting in the Wings," *Screen Printing Mag.*, 124, 125, 132, 142, 143 (August 1989).

W. J. Ramler, "Ultraviolet Curing: For Now and for the Future," *Screen Printing Mag.* 58 (September 1989).

PERSONAL COMMUNICATIONS

W. Pfirrmann, KIWO, Inc., Seabrook TX, Personal Communication, December 1998.

12

CHEMICAL PROCESS INDUSTRY

ACRYLONITRILE

Ronald D. Bell and Alan V. Mechtenberg

Acrylonitrile, $CH_2{=}CH{-}C{\equiv}N$ was initially prepared in 1893 by the dehydration of acrylamide or ethylene cyanohydrin using phosphorus pentoxide; however, it did not gain importance until shortly before World War II, when it was discovered that it could be used as a copolymer to increase the resistance of synthetic rubbers to oil and solvents. While a number of synthesis methods have been used to produce acrylonitrile (ACN) in the laboratory, relatively few have proven to be of commercial interest. Some of the early production methods included the addition of hydrogen cyanide to acetylene using a cuprous chloride catalyst, a catalytic reaction of propylene with nitric oxide, and a reaction of ethylene oxide with hydrogen cyanide followed by catalytic dehydration of ethylene cyanohydrin.[1] Acrylonitrile has several synonyms and trade names, including propenenitrile, vinyl cyanide, cyanoethylene, Acrylon®, Carbacry®, Fumigrain®, and Ventox®.

The commercial production of ACN has increased rapidly since the 1960s, when technical advances involving the introduction of ammoxidation processes revolutionized the economics of its production. Reduced production costs stimulated widespread research into new applications, and the resulting market development spurred tremendous growth in production capacity worldwide. Production costs have declined steadily since the introduction of ammoxidation processes because of the continuous development and use of improved catalysts and the increasing size of production units. Table 1 lists the ACN production sites in the United States, with corresponding annual production capacities based on 1990 data. A comparison of the total 1990 U.S. ACN capacity of over 3 billion pounds with 1960 U.S. ACN production of less than 200 million pounds[2] provides an indication of the growth that has occurred during this period.

Currently, the major end use of ACN is in the production of acrylic fibers for clothing and textiles, which make up more than half of current ACN monomer demand. The ACN-containing plastics, such as acrylonitrile–butadiene–styrene (ABS) and styrene acrylonitrile (SAN), constitute another major consumer of ACN. Other uses include the application of ACN monomer in the production of nitrile rubbers and nitrile barrier resins, adiponitrile (an intermediate in the manufacture of nylon), and acrylamide.[2]

Air emissions from the prevailing processes for ACN monomer production primarily consist of propane, propylene, acetonitrile, acrylonitrile, and hydrogen cyanide. One previous study of a model ACN plant states that approximately 86% of all air emissions consist of propane and propylene.[4] The study also states that ACN production has been identified by

Table 1. U.S. Production Capacity of Acrylonitrile[a]

Manufacturer	Location	Annual Capacity (Millions of Pounds)
American Cyanamid Co.	Avondale, LA	350
BP Chemicals America, Inc.	Greenlake, TX	720
	Lima, OH	380
E.I. du Pont de Nemours and Co., Inc.	Beaumont, TX	375
Monsanto Chemical Co.	Alvin, TX	500
Sterling Chemicals, Inc.	Texas City, TX	700
Total capacity		3025

[a] 1990 data from Ref. 3.

preliminary studies as causing the highest emissions of volatile organic compounds (VOCs) of all processes in the synthetic organic chemicals industry. Acrylonitrile is defined in 29 CFR 1910.1200 (d)(4) as a carcinogen.

PROCESS DESCRIPTION

Nearly all the ACN produced in the world today is produced using the Sohio process for ammoxidation of propylene and ammonia. The overall reaction takes place in the vapor phase in the presence of a catalyst according to:

$$CH_2{=}CH{-}CH_3 + NH_3 + \tfrac{3}{2}O_2 \rightarrow CH_2{=}CH{-}C{\equiv}N + 3H_2O$$

Reference 1 provides a detailed discussion and illustration of the reaction mechanisms and catalyst interaction involved.

Figure 1 shows a typical simplified process flow diagram for an uncontrolled Sohio process. Streams are designated as main product flow and other flow streams and are described in Table 2. The table separates vent/discharge streams that would result in air emissions from other product and flow streams.

The primary by-products of the process are hydrogen cyanide, acetonitrile, and carbon oxides. The recovery of these by-products depends on such factors as market conditions, plant location, and energy costs. Hydrogen cyanide and acetonitrile, although they carry a market value, are usually incinerated, indicating that the production of these by-products has little effect on the economics of producing ACN.

In the process represented in Figure 1, by-product hydrogen cyanide and acetonitrile are routed to an incinerator. As previously discussed, variations within the Sohio process may provide for purification, storage, and loading facilities for these recoverable by-products. Reference 1 provides further details regarding basic by-product recovery processes. Other variations of the Sohio process include the recovery of ammonium sulfate from the reactor effluent to allow for biological treatment of a wastewater stream and variations in catalysts and reactor conditions.[2]

In the standard Sohio process, air, ammonia, and propylene are introduced into a fluid-bed catalytic reactor operating at 5–30 psig and 750–950°F. Ammonia and air are fed to the reactor in slight excess of stoichiometric proportions because excess ammonia drives the reaction closer to completion and air continually regenerates the catalyst. A significant feature of the process is the high conversion of reactants on a once-through basis with only a few seconds residence time. The heat generated from the exothermic reaction is recovered via a waste-heat-recovery boiler.

The reactor effluent is routed to a water quench tower, where sulfuric acid is introduced to neutralize any unconverted ammonia. The product stream then flows through a countercurrent water absorber–stripper to reject inert gases and recover reaction products. The operation yields a mixture of ACN, acetonitrile, and hydrogen cyanide, which is distilled

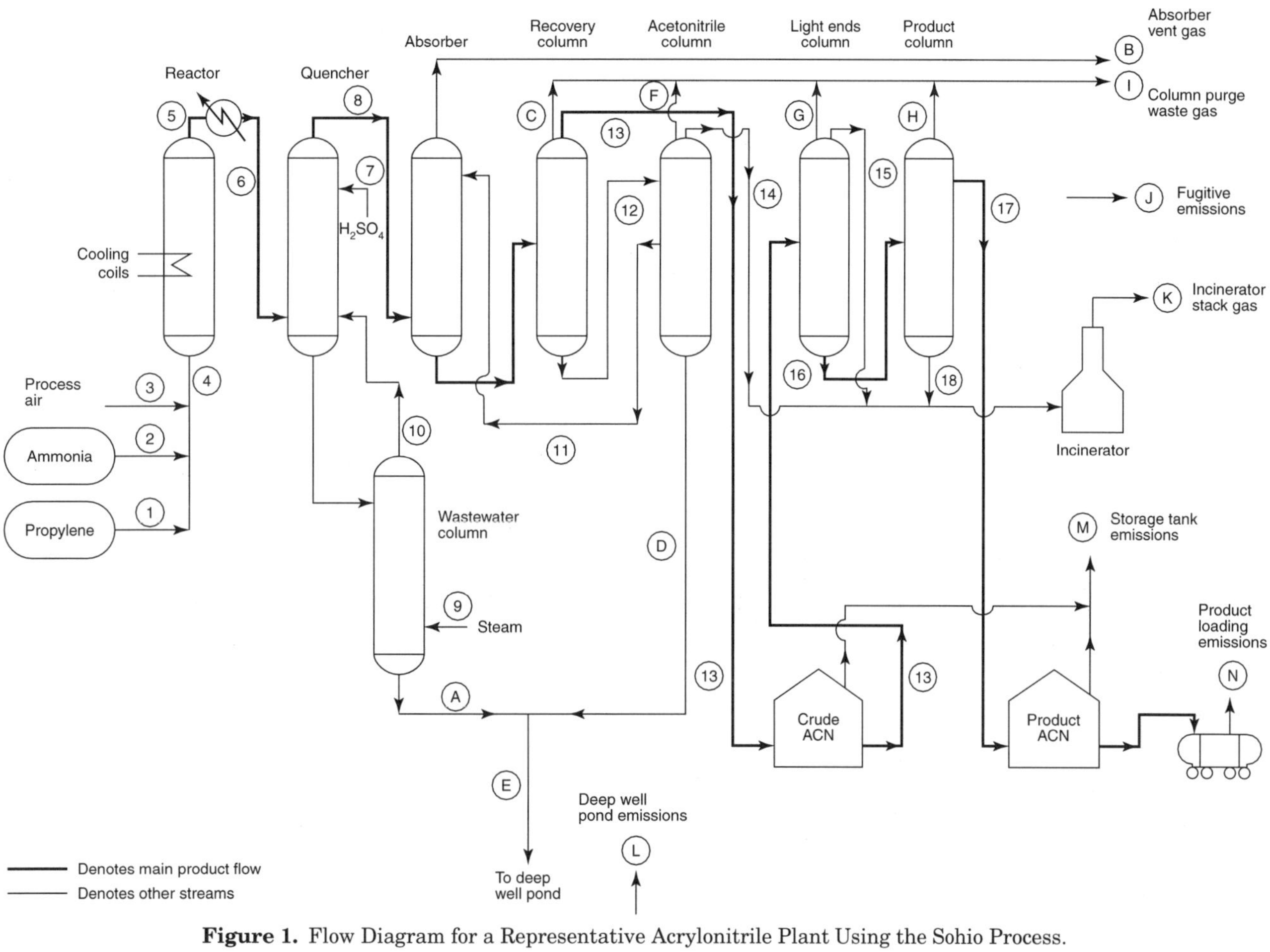

Figure 1. Flow Diagram for a Representative Acrylonitrile Plant Using the Sohio Process.

Table 2. Stream, Vent, and Discharge Descriptions for Figure 1

Stream/Vent/Discharge	Discharge
	Stream
1	Propylene feed
2	Ammonia feed
3	Process air
4	Reactor feed
5	Reactor product
6	Cooled reactor product
7	Sulfuric acid
8	Quenched reactor product
9	Stripping steam
10	Wastewater column volatiles
11	Water recycle
12	Crude acetonitrile
13	Crude acrylonitrile
14	Acetonitrile
15	Hydrogen cyanide
16	Light-ends column bottoms
17	Product acrylonitrile
18	Heavy ends
	Vent/Discharge
A	Wastewater column bottoms
B	Absorber vent gas
C	Recovery column purge vent
D	Acetonitrile column bottoms
E	Acrylonitrile plant wastewater
F	Acetonitrile column purge vent
G	Light-ends column purge vent
H	Product column purge vent
I	Column purge waste gas
J	Fugitive emissions
K	Incinerator stack gas
L	Deep well/pond emissions
M	Storage tank emissions
N	Product transport loading facility vent

to remove acetonitrile and water and then is sent to a fractionator to remove hydrogen cyanide. The final two steps involve the drying of the ACN stream and the final distillation to remove heavy ends. The fiber-grade ACN obtained from the process is 99+% pure.[5]

Several fluid-bed catalysts have been used since the inception of the Sohio ammoxidation process. Catalyst 49, which represents the fourth major level of improvement, is currently recommended in the process. Catalyst 49 replaced bismuth–molybdenum Catalyst 41, which was preceded by uranium Catalyst 21 and the older bismuth–molybdenum Catalyst A.

AIR EMISSIONS CHARACTERIZATION

Air emissions of ACN and other VOCs resulting from ACN production via the Sohio process may come from several sources, including the absorber vent, column purge vents, storage tanks, transport and loading facilities, and volatilization from deep wells and ponds. Fugitive emissions also result from valves, compressor seals, pumps, and flanges throughout the process. The ACN itself represents a relatively small fraction of the total VOCs emitted from these sources, with propane and propylene making up the majority of VOC emissions.[6]

Description of Sources

Figure 1 and Table 2 identify 14 waste streams and volatilization points that contribute to emissions from an uncontrolled ACN facility. Seven of these streams/points (vent/discharge B, 1, J, K, L, M, and N in Figure 1) are considered the primary air pollutant emission points for the typical ACN plant. Table 3 lists these emission points and provides a checklist of uncontrolled pollutants that may be found in each of these streams. The checklist was compiled using a material balance for a representative ACN plant[7] that did not include estimates of nitrogen oxides (NO_x), carbon monoxide (CO), and carbon dioxide (CO_2) emissions from the by-product incinerator.

Table 3. Characterization of Pollutants from Air Emission Point for Acrylonitrile Production Facilities[a]

	Stream Identifier						
	B	I	J	K	L	M	N
	Description						
	Absorber Vent Gas	Column Purge Waste Gas	Fugitive Emissions	Incinerator Stack Gas[b]	Deep Well/Pond Emissions	Storage Tank Emissions	Product Transport Loading Facility Vent
Carbon dioxide	X	X					
Propylene	X	X	X				
Propane	X		X				
Ammonia			X				
Carbon monoxide	X						
Acrylonitrile	X	X	X			X	
Hydrogen cyanide	X	X		X		X	X
Hydrocarbons (general)					X		

[a]Data taken from material balance for a representative ACN plant (Ref. 7).
[b]The material balance cited did not include combustion products such as CO_2, CO, and NO_x.

Emissions of ACN during startup are substantially higher than during normal operation. During startup, the reactor is heated to operating temperature before the reactants (propylene and ammonia) are introduced. Effluent from the reactor during startup begins as oxygen-rich, then passes through the explosive range before reaching the fuel-rich zone that is maintained during normal plant operation. To prevent explosions in the lines to the absorber, the reactor effluent is vented to the atmosphere until the fuel-rich effluent mixture can be achieved. The ACN emissions resulting from this startup procedure have been estimated to exceed 10,000 lb/h.[7]

The absorber vent gas (stream B) contains nitrogen and unconverted oxygen from the air fed to the reactor, propane and unconverted propylene from the propylene feed, product ACN, by-product hydrogen cyanide and acetonitrile, other organics not recovered from the absorber, and some water vapor.

The ACN content of the combined column purge vent gases (stream 1) is relatively high, about 50 wt % of the total VOCs emitted from the recovery, acetonitrile, light ends, and product columns.[6] The rest of the vent gases consist of noncondensibles that are dissolved in the feed to the columns, the VOCs that are not condensed, and, for the columns operating under vacuum, the air that leaks into the column and is removed by the vacuum jet systems.

For the ACN process illustrated in Figure 1, by-product hydrogen cyanide and acetonitrile are incinerated, along with product column bottoms. The primary pollutant problem related to the incinerator stack is the formation of NO_x from the fuel nitrogen of the acetonitrile stream (stream 14) and hydrogen cyanide (stream 15). Carbon dioxide and lesser amounts of CO are emitted from the incinerator stack gas.[7]

Other emission sources involve the volatilization of hydrocarbons through process leaks (fugitive emissions) and from the deep well ponds, breathing and working losses from product storage tanks, and losses during product loading operations. The fugitive and deep well/pond emissions primarily consist of propane and propylene, while the storage tank and product loading emissions consist primarily of ACN.

Emission Factors

Emission factors have been developed for ACN and total VOC emissions from a model ACN plant utilizing the Sohio process.[4] The model plant has an ACN capacity of 180 billion grams per year, based on 8760 hours of operation annually. The model plant, unlike that represented in Figure 1, is arranged either to incinerate or to recover and sell by-product hydrogen cyanide and acetonitrile; therefore, emission factors are provided for storage and handling losses of these by-products through tank and loading emission points. Emission factors are based on the assumption that the plant incinerates 8.64 billion grams of the 21.6 billion grams of hydrogen cyanide product and all of the 5 billion grams of the acetonitrile produced. The emission factors resulting from this study are listed in Table 4.

Emissions from the absorber vent (stream B in Figure 1) are listed for normal operation and startup conditions involving the venting of the reactor effluent to the atmosphere until normal fuel-rich conditions are reached. Startup emissions are listed as yearly averaged values based on four startups per year.

Table 4. Uncontrolled Emissions of Acrylonitrile and Total Volatile Organic Compounds from Model Plant Using Sohio Process[a,b]

Source	Stream Designation (Figure 1)	Emission Ratio (g/kg)[c]		Emission Rate (kg/h)	
		Acrylonitrile	Total VOC	Acrylonitrile	Total VOC
Absorber vent, normal[a]	B	0.1000	100.0000	2.0500	2050.0000
Absorber vent, startup[d]	B	0.1870	0.2490	3.8400	5.1100
Column vents	I	5.0000	10.0000	103.0000	205.0000
Storage vents					
Crude acrylonitrile	M	0.0480	0.0480	0.9860	0.9860
Acetonitrile			0.0210		0.4300
Hydrogen cyanide	N/A[e]		0.0410		0.8420
Acetonitrile run tanks	N/A[e]	0.1280	0.1280	2.6200	2.6200
Acrylonitrile storage	M	0.5310	0.5321	10.9000	10.9000
Handling					
Hydrogen cyanide[f]	N/A[e]		0.0260		0.5400
Acetonitrile[f]	N/A[e]				
Acrylonitrile[f]	N	0.1670	0.1670	3.4400	3.4400
Acrylonitrile[g]	N/A[e]	0.1500	0.1500	2.1600	2.1600
Fugitive[h]	J	0.4630	0.9470	9.5200	19.4700
Secondary					
Incinerator	K		0.3600		7.4000
Deep well/pond	L		13.0000		267.0000
Total		6.7700	126.0000	139.0000	2567.0000

[a] Reference 4.
[b] Model plant has an annual ACN capacity of 180 billion grams. Factors are based on 8760 h of operation annually.
[c] Grams of emission per kilogram of acrylonitrile produced.
[d] Average rate for entire year, based on eight startups per year lasting 1 h each.
[e] Streams are not shown on Figure 1.
[f] By tank car.
[g] By barge.
[h] Reference 8.

Storage tanks are assumed to be fixed-roof tanks with atmospheric vents for all materials except ammonia and propylene reactants, which are stored in pressure vessels. Details on the number and dimensions of the tanks and other parameters used to determine the listed storage loss factors are available in Ref. 4.

The fugitive emission factors listed in Table 4 represent the component counts used in the model study; however, the various uncontrolled emission factors were reestablished based on the more recent average uncontrolled fugitive emission factors for the synthetic organic chemical manufacturing industry.[8] Component counts used for the model plant include 50 pumps in light-liquid service, 200 pipeline valves in gas/vapor service, 1000 valves in light-liquid service, 80 relief valves in gas/vapor service, and two compressor seals.

A second study was conducted with emphasis on ACN emissions only.[6] The study utilized the uncontrolled emission factors for ACN from the source referenced in Table 4 and applied hypothetical control devices and assumed efficiencies to develop controlled emission factors for ACN only.

AIR POLLUTION CONTROL MEASURES

Table 3 characterizes the pollutants from air emission points for an ACN facility. Seven point sources are included:

1. Absorber vent gas
2. Column purge waste gas
3. Fugitive emissions
4. Incinerator stack gas
5. Deep well/pond emissions
6. Storage tank emissions
7. Product transport loading facility vent

Absorber Vent Gas

The absorber vent gas stream contains nitrogen, oxygen, unreacted propylene, hydrocarbon impurities from the propylene feed stream, CO, CO_2, water vapor, and small quantities of ACN, acetonitrile, and hydrogen cyanide. Two control methods are used to treat this stream: thermal incineration and catalytic oxidation.

The thermal incineration units have demonstrated VOC destruction efficiencies of 99.9% or greater, while most catalytic units can only achieve destruction efficiencies in the 95–97% range. Destruction efficiencies in the 99% and greater range can be achieved with catalytic oxidizers, but these are not achieved on a long-term basis because of deactivation of the catalyst by a number of causes. The advantage of catalytic oxidation is low fuel usage, but emissions of NO_x formed in the reactors and not destroyed across the catalyst can pose problems.

The thermal units require additional fuel to maintain operating temperature and usually have some form of preheat and energy recovery in the form of steam generated by a heat recovery boiler. In addition to higher VOC destruction efficiency, the thermal incinerators can treat other streams, such as the by-product hydrocyanic acid (HCN) stream and the crude aceto stream. These streams, however, generate high concentrations of NO_x when treated by a standard incinerator. To minimize the formation of NO_x, staging of the combustion air stream and the absorber off-gas stream is practiced. In some units, ammonia is injected to reduce the level of NO_x in the incinerator stack exhaust.

If the incinerator or the catalytic oxidizer has an emergency shutdown, the absorber vent stream is routed to a flare on a temporary basis.

Column Purge Waste Gas

Waste gas releases from the recovery column, light-ends column, product column, and the acetonitrile column are frequently tied together and vented to a flare. The estimated VOC destruction efficiency of the flare is 98–99% for all streams with a heat content of 300 Btu/scf or greater. The use of a flare is ideally suited for streams that are intermittent and have heating values in excess of 300 Btu/scf.

Fugitive Emissions

Fugitive emissions from piping, valves, pumps, and compressors are controlled by periodic monitoring by leak checking with a VOC detector and a directed maintenance program.

Incinerator Stack Gas

Staged combustion and ammonia injection are used to control the emissions of NO_x from the incinerator that treats the absorber off-gas vent, the crude aceto waste liquid stream, and the by-product liquid HCN stream. Staged combustion suppresses the formation of NO_x by operating under fuel-rich conditions (less than stoichiometric air) in the flame zone where most of the NO_x is formed and oxygen-rich conditions downstream at lower temperatures where NO_x is not appreciably formed. The staging is achieved by splitting the air stream and absorber off-gas stream; one portion of each stream is introduced near the flame zone, while the remainder is introduced downstream. Staging the absorber off-gas not only enhances the "fuel-rich/fuel-lean" condition, but also lowers the temperature since this stream is overall endothermic.

Ammonia injection reduces NO_x by selectively reacting ammonia with NO_x. The reaction occurs at temperatures in the range of 1600–1800°F, and, as such, the ammonia must be injected in the postflame zone of the combustion chamber. Residence times of 0.5–1.0 sec are required for NO_x destruction efficiencies in the range of 80%, which is compatible with the residence time required for VOC destruction. Selective catalytic reduction using ammonia injection can also be used to control NO_x emissions, but no such units are in service at this time. For this control, the catalyst bed and the ammonia injector are placed downstream of the heat recovery boiler, because the operating temperature must be in the 600–700°F range.

Deep Well/Pond Emissions

Emissions of acrolein and other odorous components in vents from wastewater treatment steps are controlled with water scrubbers. In some cases, pond emissions are controlled by adding a layer of a low-vapor-pressure oil on the surface of the pond to limit volatilization.

Storage Tank Emissions

Product storage tank emissions are controlled with double-seal floating roofs or, in some cases, water scrubbers. Field experience indicates that a removal efficiency of 99% can be achieved with water scrubbing.

Product Transport Loading

Product transport loading vents are gathered and sent to a flare or incinerator for VOC control. Destruction efficiencies of 98–99% are achieved using the flare and greater than 99% using incineration.

REFERENCES

1. W. Löwenbach and J. Schlesinger, *Acrylonitrile Manufacture: Pollutant Prediction and Abatement*, MTR-7752, MITRE Technical Report, McLean, VA, 1978.
2. *Kirk-Othmer Encyclopedia of Chemical Technology*, 3rd ed., Vol. 1, John Wiley & Sons, New York, 1978.
3. *1990 Directory of Chemical Producers United States of America*, SRI International, Menlo Park, CA, 1990.
4. *Organic Chemical Manufacturing, Vol. 10: Selected Processes*, EPA-450/380-028e, U.S. Environmental Protection Agency, Research Triangle Park, NC, 1980.
5. *Hydrocarbon Processing, Petrochemical Handbook '85*, Gulf Publishing Co., Houston, 1985.
6. Radian Corp., *Locating and Estimating Air Emissions from Sources of Acrylonitrile*, U.S. Environmental Protection Agency, Research Triangle Park, NC, 1983.
7. M. T. Anguin and S. Anderson, *Acrylonitrile Plant Air Pollution Control*, EPA-600/2-79-048, U.S. Environmental Protection Agency, Washington, D.C., 1979.
8. *Protocols for Generating Unit-Specific Emission Estimates for Equipment Leaks of VOC and VHAP*, EPA-450/3-88-010, U.S. Environmental Protection Agency, Research Triangle Park, NC, 1988.

BIBLIOGRAPHY

T. W. Hughes and D. A. Hom, *Source Assessment: Acrylonitrile Manufacture (Air Emissions)*, EPA-600/2-77-107, U.S. Environmental Protection Agency, Research Triangle Park, NC, 1977.

Industrial Process Profiles for Environmental Use: Chapter 6: The Industrial Organic Chemicals Industry, EPA-600/2-77-023f, U.S. Environmental Protection Agency, Cincinnati, OH, 1977.

W. A. Schwartz et al., *Engineering and Cost Study of Air Pollution Control for the Petrochemical Industry, Vol. 2: Acrylonitrile Manufacture*, EPA-450/3-73-006-b, U.S. Environmental Protection Agency, Research Triangle Park, NC, 1975.

CARBON BLACK

BARRY R. TAYLOR

Carbon black is a very finely divided form of industrial carbon with median stokes diameters ranging from approximately 30 to 600 nm. The principal uses of carbon black are as a reinforcing agent in rubber compounds (e.g., for tires, belts, and hoses); as a black pigment in printing inks, surface coatings, paper, and plastics; and as a conductive additive for plastics used in the wire and cable industry. There are six major producers of carbon black in the United States operating a total of 21 plants: Cabot Corp. (four plants), Columbian Chemicals Division of Phelps Dodge (five plants), Degussa (three plants), J. M. Huber (three plants), Sid Richardson Carbon & Gasoline Co. (three plants), and Continental Carbon Division of Witco Corp. (three plants). Typical plant production capacities range from 50 to 300 million pounds per year of carbon black, with total U.S. capacity estimated at approximately 3.3 billion pounds annually.

PROCESS DESCRIPTION

Types of Processes

Carbon black is a product of endothermic hydrocarbon pyrolysis. It can be produced by partial combustion processes involving flames or by purely thermal decomposition processes in the absence of flames. Thermal cracking processes, namely, the thermal and acetylene processes, are still used to produce small quantities of special carbon black grades having performance properties that have not been duplicated with the oil furnace process. However, among the major producers, the oil furnace process, which is a partial combustion process, currently accounts for more than 98% of the total production of carbon black in the United States and is the subject of further discussion.

The Oil Furnace Process

Figure 1 is a flow diagram of a typical oil furnace process. The reactor inputs are preheated air, auxiliary fuel (natural gas or oil), and preheated feedstock. The feedstock is preferably a highly aromatic petrochemical or carbochemical heavy oil preheated to 150 to 250°C to achieve a viscosity appropriate for atomization. The feedstock is injected, frequently as an atomized spray, into a highly turbulent high-temperature zone, which is achieved by burning the auxiliary fuel with air. The oxygen in the system is in stoichiometric excess with respect to the auxiliary fuel, but is insufficient for complete combustion of the feedstock. Consequently, a major portion of the feedstock is pyrolyzed to form carbon black.

The carbon-containing combustion products are quenched in the reactor with water sprays and pass through heat exchangers that preheat the combustion air and, in some instances, the feedstock oil. The product stream is then cooled further to a temperature in the range of 230–260°C before flowing to the main unit filter, where the carbon black is separated from the gas-phase combustion products (tail gas).

The carbon black, in a finely divided fluffy form, is conveyed pneumatically to a process filter, from which the black drops into a grinder and subsequently into a surge tank. From the surge tank, the carbon black may be packed in fluffy form or may be fed to a pelletizer, where it is compacted into pellets for ease of handling. The majority of rubber blacks are pelletized by a wet process in which carbon black, water, and small quantities of a binder are mixed in pelletizers to form wet pellets. The wet pellets are dried in continuous-flow, rotating drum dryers heated indirectly by burning tail gas or natural gas.

Some grades of carbon black are pelletized by a dry process (not shown in Figure 1) in which the fluffy carbon black is fed to a rotating drum along with a recycle stream of pelletized black. The recycled pellets act as seed material for the formation of new pellets and, in concert with the rotation of the drum, act to densify the new pellets.

Carbon black pellets are transported from the pelletization process by elevators, screw conveyors, and belt conveyors to

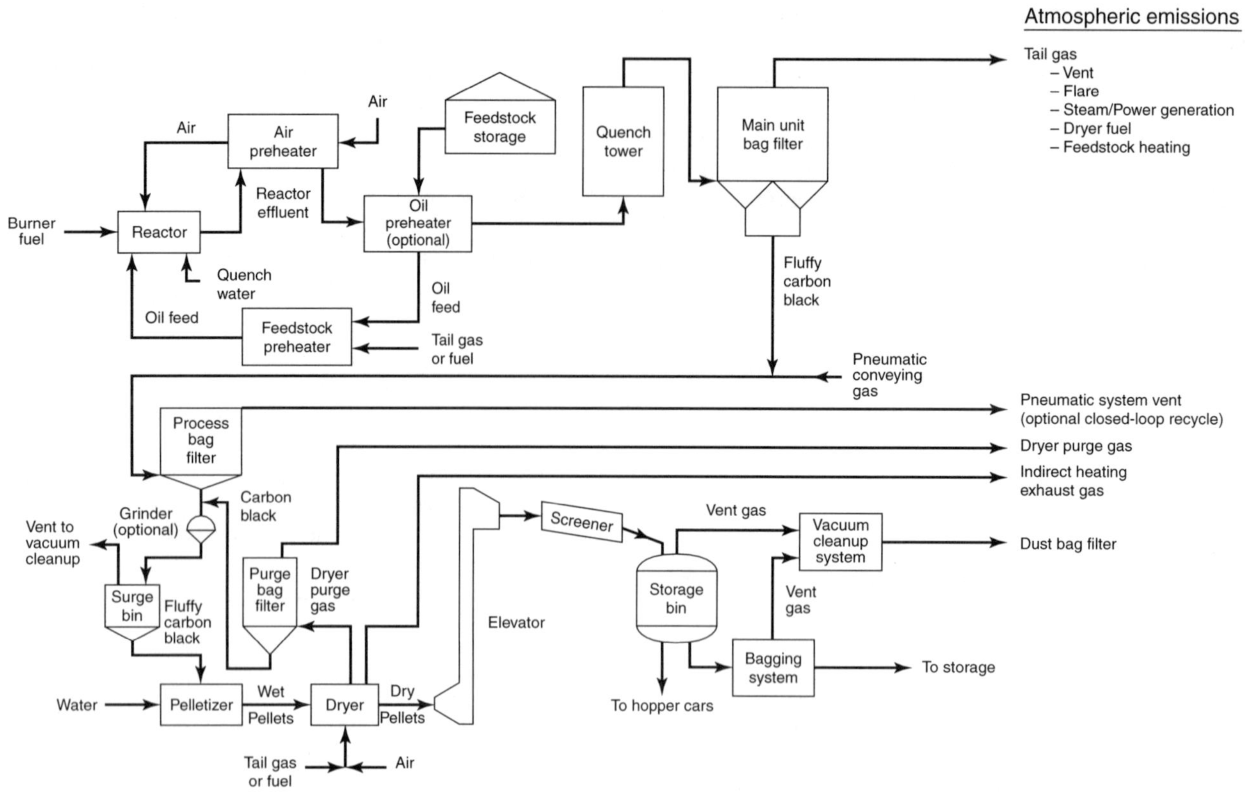

Figure 1. Schematic Flow Diagram for Typical Oil Furnace Process.

bulk storage tanks for subsequent loading into railcars and trucks or packing in bags.

AIR EMISSIONS CHARACTERIZATION

Emissions from the manufacture of carbon black include particulate matter, carbon monoxide, hydrocarbons, and sulfur compounds.

Particulate Matter

Carbon black is the main particulate emitted in a carbon black plant. It is a relatively innocuous material with an Occupational Safety and Health Administration permissible exposure limit of 3.5 mg/m^3.[1] As a black, finely divided material that is readily transported by the wind, it can be a nuisance if allowed to disperse. Consequently, sources of emission must be identified and controlled.

Since fluffy carbon black cannot be completely removed from a gas stream, carbon black is emitted when main process gas streams in the manufacturing process are discharged to the atmosphere. The principal process gas stream, reactor tail gas, as well as gases used for dryer purge and open-loop pneumatic conveying, fall into this category. Emission factors (in pounds of particulate per ton of carbon black produced) used by the U.S. Environmental Protection Agency (EPA) to characterize particulate emissions for carbon black production are 6.53, 0.24, and 0.58 for tail gas, dryer purge gas, and conveying gas, respectively. The carbon black content in tail gas can be reduced when the tail gas is burned; consequently, an emission factor of 2.70 has been established for tail gas that has been flared and of 2.07 for tail gas that has been incinerated.[2]

Intermittent sources are another category of carbon black emissions. During some portions of the reactor warmup cycle, and in some instances, during operation in a mode to maintain the reactor temperature without making carbon black (referred to as heat load), the oxygen-bearing combustion products that flow through the reactor are vented to the atmosphere, and some entrained carbon black will be emitted if not controlled. Intermittent emissions of entrained carbon black also occur during the venting of carbon black storage silos because of displacement of air during filling operations and the expansion of air in the head space upon heating. Carbon black can also be emitted during the bulk loading of railcars and trucks owing to the entrainment of carbon black in the air that is displaced from the vessel being loaded.

Fugitive sources constitute a final category of carbon black emissions. Quantitative information on fugitive emissions is limited. In a survey conducted in 1974, average fugitive emissions of carbon black at one plant were estimated to be 0.1 pound per ton of product.[3] The emission factor currently used for fugitive emissions by the EPA is 0.2 pound per ton of product.[2]

Fugitive emissions of carbon black can occur through imperfect seals on grinders, rotary screens, bucket elevators, pellet conveying systems, and bag packers. Other potential sources of fugitive emissions include ports for equipment access and product sampling, leaks on process equipment, and accidental spills of carbon black that may take place during equipment cleanout operations or from broken packages.

Gas-Phase Emissions

Reactor tail gas is the major source of gas-phase pollutants generated during the production of carbon black. Because

Table 1. Tail Gas Emission Information

Component	Data for ASTM N330[a] Concentration (vol %)	Data for ASTM N330[a] Emissions (lb/ton Carbon Black)	Range of Emissions[a] (lb/ton Carbon Black)	EPA Emission Factor[b] (lb/ton Carbon Black)
Nitrogen	40.	17,700	9,150–26,800	—
Argon	0.48	302	133–456	—
Oxygen	0.95	480	59–1,280	—
Carbon dioxide	2.0	1,400	838–3,440	—
Water	42.	12,000	5,330–25,900	—
Carbon monoxide	6.3	2,800	1,400–4,380	2,800
Hydrogen	7.5	240	104–366	—
Methane	0.20	50	22–120	50
Acetylene	0.22	90	12–258	90
Hydrogen sulfide	0.11	60	Trace–60	60
Carbon disulfide	0.05	60	No analysis	60
Carbonyl sulfide	0.02	20	No analysis	20
Sulfur dioxide	—[c]	0	Trace–24	0

[a]Reference 3.
[b]Reference 2.
[c]Not detectable at 1 ppm.

carbon black is produced by a fuel-rich combustion process, reactor tail gas contains significant quantities of hydrogen and carbon monoxide, along with small quantities of light hydrocarbons, primarily methane and acetylene. Table 1 shows these species, as well as the other major components and pollutants found in carbon black tail gas. The table also shows tail gas composition (vol %) for a typical grade of carbon black [American Society for Testing and Materials (ASTM) grade N330] and corresponding emission factors (weight of emissions per unit weight of carbon black) for each component. Another column in the table contains data on the range of emission factors measured for tail gas components in surveys of operating plants conducted for the EPA. These data help to show that concentrations of various components in tail gas, as well as the emissions per unit weight of product, are strongly dependent on the grade of carbon black being produced, the particular processing equipment and conditions that are being used to produce it, and the composition of the feedstock used. The final column in the table shows average emission factors currently used by the EPA to characterize emissions of gas-phase pollutants from carbon black tail gas.

Sulfur enters the carbon black manufacturing process via sulfur-bearing feedstocks. Sulfur exiting the process is distributed between the carbon black and the tail gas, with 50–80% of the sulfur ending up in the tail gas, depending on the grade of carbon black being produced, the equipment and the operating conditions used for production, and the level of sulfur in the feedstock. As with other tail gas species, the partitioning of sulfur among species in the tail gas depends on the processing equipment and operating conditions employed.

Since the overall environment in the reactor is a reducing one, the sulfur in the tail gas is expected to be in a reduced state. Sulfur is present as hydrogen sulfide (H_2S), carbon disulfide (CS_2), carbonyl sulfide (COS), and, in smaller quantities, sulfur dioxide (SO_2). Measurements of H_2S concentrations have accounted for approximately 40% to more than 99% of the sulfur in the tail gas; generally, more than half of the sulfur in the tail gas is H_2S. Measured fractions of sulfur as CS_2 in the tail gas have ranged from less than 1% to approximately 60%. Carbonyl sulfide has always been detected in the tail gas, but has not been observed to exceed 10% of the total tail gas sulfur. The emission factor for sulfur oxides in vented tail gas currently used by the EPA is zero pounds per ton of carbon black.[2] In published emission surveys of a wide variety of operating conditions, measured SO_2 concentrations have been relatively low, but have varied from undetectable levels (<1 ppmv) to 210 ppmv.[3]

AIR-POLLUTION CONTROL MEASURES

Particulate

High-performance bag filters are used to remove carbon black from the main process gas streams. The collection efficiency of these filters during normal operation is enhanced significantly by the buildup of a filter cake of carbon black on the filter fabric. New bags typically require up to 24 h of operation to achieve maximum efficiency. Performance testing indicates that effluent carbon black concentrations of 6 mg/nm^3 (wet basis) are at the upper end of capture efficiency capabilities and that 30 mg/nm^3 (wet basis) is an effluent concentration that is consistently achievable for a broad range of carbon black grades with a bag filter system maintained in good operating condition. Using an effluent concentration of 30 mg/nm^3 and the data provided in the Monsanto Research Corp. report[4] on tail gas compositions for various operations, particulate emissions attributable to vented tail gas are calculated to be less than 2.0 pounds of particulate per ton of carbon black. This is significantly less than the emission factor of 6.5 pounds per ton currently used by the EPA.

Vent scrubber systems, which consist of a venturi scrubber, a cyclone, and a spray tower with demisters, can be used effectively to remove the entrained carbon black from reactor effluent gases during intermittent reactor warmup and heat-load operations. Bin vent filters, which consist of a filter bag connected to the vent, are used to remove entrained carbon black from the vent gases of carbon black storage silos. To control the emission of carbon black during the loading of railcars and trucks, air displaced from the vessels is collected in a plant vent system and discharged to the atmosphere through a bag filter.

Fugitive emissions of carbon black through imperfect seals on process equipment are prevented by connecting the pieces of equipment to a vacuum system and keeping them under reduced pressure. Air collected by the vacuum system is discharged to the atmosphere through a bag filter. Emissions from product sampling ports, process equipment leaks, and accidental spills are minimized through proper maintenance of operating equipment and aggressive housekeeping practices. Housekeeping equipment includes sweeping machines to clean smooth-surfaced portions of the plant and plantwide vacuum systems to enable the cleanup of isolated spills of carbon black before they are dispersed by the wind.

Gas-Phase Emissions

Because of the high concentrations of hydrogen and carbon monoxide in carbon black tail gas, it can be used as a low-Btu fuel. In carbon black plants that have wet pelletization units, some of the tail gas is burned to supply thermal energy to the dryers. In situations where it is not economical to burn the tail gas to generate process steam or electric power, the gas can be flared to convert reduced tail gas species to others more compatible with the environment.[2]

The amount of sulfur released to the atmosphere in North America through the production of carbon black (estimated to be 210,000 tons per year as SO_2 using emission factors of 60 pounds H_2S, 60 pounds CS_2, and 20 pounds COS per ton of carbon black and an annual carbon black production rate of 3.3 billion pounds) is small relative to the estimated total quantity of anthropogenic sulfur emissions (22.8 million tons per year as SO_2.[5] In carbon black plants where the emission of sulfur species to the atmosphere must be reduced, control is accomplished through the use of low-sulfur feedstocks. Tail gas that is not burned in pellet dryers or in power-generating systems can be flared to oxidize reduced sulfur species to less objectionable sulfur oxides.[2] Converting reduced sulfur species to SO_2 during combustion has little impact on sulfur oxides in the environment because species such as H_2S are oxidized to SO_2 in the atmosphere over time.

REFERENCES

1. Occupational Safety and Health Administration, *OSHA Safety and Health Standards: General Industry Standards* (revised), publication no. 29CFR 19 10.1000, U.S. Department of Labor, Washington, D.C., 1983.
2. Office of Air Quality Planning and Standards, *Compilation of Air Pollutant Emission Factors, Volume 1: Stationary Point and Area Sources*, 4th ed., publication AP-42, U.S. Environmental Protection Agency, Research Triangle Park, NC, 1985.
3. Office of Air Quality Planning and Standards, *Engineering and Cost Study of Air Pollution Control for the Petrochemical Industry, Vol. 1*, report EPA-450/3-73-006-a, U.S. Environmental Protection Agency, Research Triangle Park, NC, 1974.
4. Monsanto Research Corp., *Source Assessment: Carbon Black Manufacture*, NTIS no. P13-273-068, U.S. Department of Commerce, Springfield, VA, 1977.
5. Office of Air Quality Planning and Standards, *National Air Quality and Emissions Trends Report, 1988*, report EPA-450/4-90-002, U.S. Environmental Protection Agency, Research Triangle Park, NC, 1990.

BIBLIOGRAPHY

Dannenberg, E. M., "Carbon Black," *Encyclopedia of Chemical Technology*, 3rd ed., Wiley-Interscience, New York, 1978.

CHARCOAL/ACTIVATED CARBON

William D. Byers and Mark P. Cal

Charcoal is the solid carbon residue following the pyrolysis (carbonization and destructive distillation) of carbonaceous raw materials. The principal raw materials for charcoal are hardwoods (beech, birch, maple, hickory, and oak), softwoods (long leaf and slash pine), nutshells, fruit pits, coal, vegetable wastes, and paper mill residues. Charcoal is used primarily as a fuel for outdoor cooking. In some instances, charcoal manufacture may be considered as a solid waste disposal technique and may also be used in forest management for disposal of refuse.

Activated carbon is produced by the pyrolysis and subsequent activation of carbonaceous material, and it is a widely used material in chemical and environmental processes. Gas-phase applications for activated carbon include: gas separation, odor control, purification of indoor air, recovery of valuable solvents, and removal of hazardous air pollutants from a variety of industrial sources (chemical manufacturing, food processing, rendering plants, sewage treatment plants, and pharmaceutical plants). Activated carbon is also used extensively for liquid-phase applications including: removal of organic and inorganic pollutants from drinking water, decolorizing food products, removal of taste and odor from drinking water, and for combined physical/biological processes. The demand for activated carbon has increased sharply in response to more rigorous environmental regulations. The U.S. consumption of activated carbon in 1997 was reported at 365 million pounds, and it is predicted to rise to 440 million pounds in 2002.[1] Costs for activated carbon range from about 25 cents per pound to several dollars per pound. Many of the major U.S. activated carbon manufacturing companies (Calgon Carbon, Norit, Westvaco, etc.) produce carbons for specialized applications, such as mercury removal, H_2S removal, odor control, or gas or liquid filtration. Activated carbons with catalytic properties can also be produced.

Almost any carbonaceous material can be converted to activated carbon, if properly treated. Raw materials for activated carbon include: wood, coals, petroleum, coke, peat, coconut shell, fruit, and nuts, and more recently, polymers have been used to make materials such as activated carbon cloth. Anthracite, lignite, and bituminous coals are major sources of activated carbon. The manufacturing of activated carbon generally includes the following steps: material preparation, pelletization, carbonization, and activation. Activated carbon is generally grouped into two broad size ranges. Granular activated carbon (GAC) has a particles sizes predominantly greater than 80 mesh, while powdered activated carbon (PAC) has particles smaller than 80 mesh. PAC is typically used for liquid-phase applications due to higher diffusion rates. PAC is normally not used in adsorption columns because PAC packs much more densely than GAC, increasing pressure drop and lowering flow rates. GAC is used in both gas- and liquid-phase adsorption column applications. Recently, activated carbon has been produced from polymeric materials, such as phenolic resins. These materials are usually called activated carbon cloth (ACC) or fibers (ACF) and are present in the form of a woven or nonwoven cloth or matting. These materials usually have a higher degree of microporosity than GAC or PAC, leading to high mass transfer rates. Also, due to the homogeneous polymeric starting material, it is generally easier to carefully control the pyrolysis and activation steps.

PROCESS DESCRIPTION

Process Overview

The manufacture of charcoal and activated carbon begins with a carbonization operation that converts organic material to primary carbon or "char." In the manufacture of charcoal, the char is mixed with a binder and pressed to form briquettes as the final product. In activated-carbon production, the char is "activated" physically, by contact with hot gases, or chemically, by the addition of certain inorganic chemicals ($ZnCl_2$, H_2SO_4). In the 15 years prior to 1978, there were over 190 patents applicable to the manufacture and reactivation of activated carbon.[2]

Carbonization

During the carbonization operation, organic material is reduced by the slow application of heat in the (relative) absence of oxygen to form a char. Carbonization is accomplished in a four-step operation:

1. Heat is applied to the organic material and, as the temperature approaches 100°C, water and highly volatile organics are driven off.
2. As the moisture is driven off, the temperature rises to approximately 170°C, causing degradation with the evolution of CO, CO_2, and organic (acetic and other) acids.
3. At temperatures of 270–280°C, the process becomes exothermic, with the formation of substantial amounts of hydrocarbon distillates, including methanol, tars, and other by-products.
4. At a temperature of about 350°C, exothermic pyrolysis ends, and heat is applied to raise the temperature to the 400–600°C range. At this temperature, some of the less volatile tarry materials are removed from the char.

The overall yield from the carbonization process varies widely depending on the raw material and the actual process conditions. Yields of 25–30% are reported for the production of charcoal from wood.

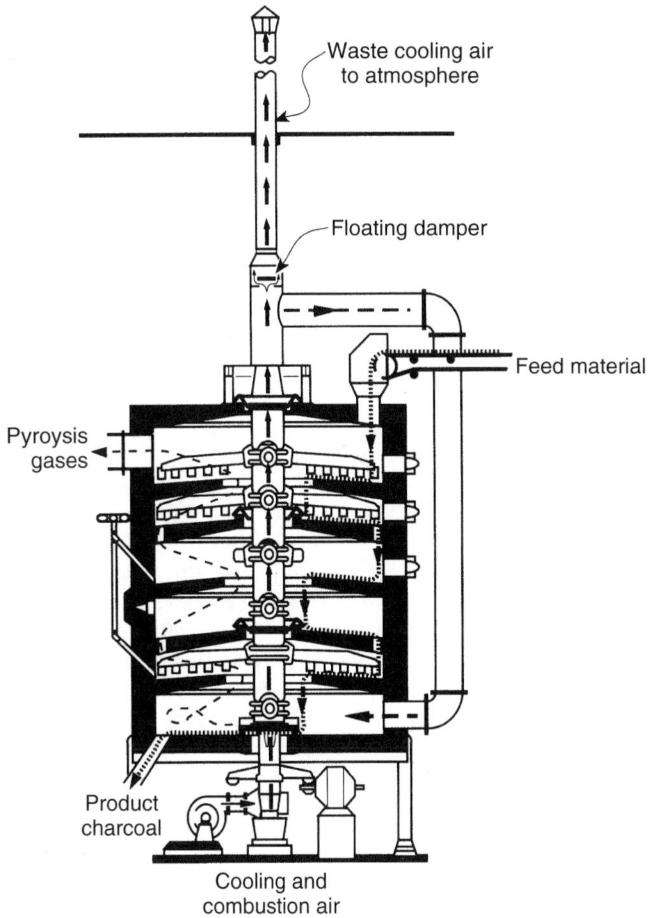

Figure 1. Typical Cross Section of Multiple-Hearth Furnace.

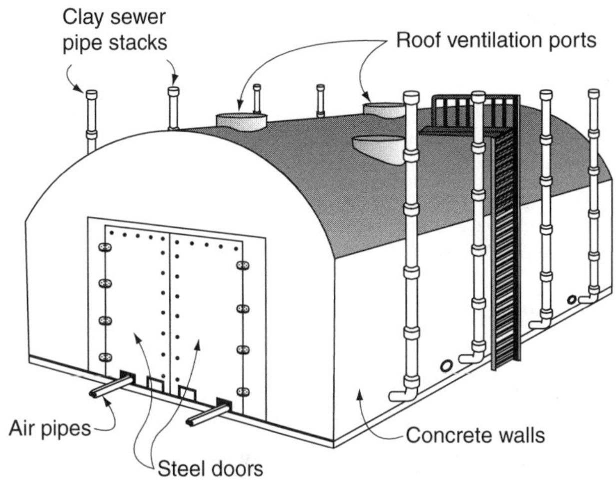

Figure 2. The Missouri-Type Charcoal Kiln.

The carbonization of wood for charcoal production is most commonly accomplished as a continuous process in large, cylindrical, multiple-hearth furnaces, where the charge is stirred and moved from one hearth to the next lower one by rotating rabble arms. Figure 1 shows a typical cross section of a multiple-hearth furnace.[3] Batch kilns are also used for the carbonization of wood for charcoal production. The most common style of batch kiln is the "Missouri-type" kiln shown in Figure 2.[3]

Charcoal Manufacture

Briquetting. The fabrication of briquettes from raw charcoal involves hammer-milling or crushing the char, mixing the char with a starch binder, forming the briquettes in a press, and drying the briquettes at approximately 135°C to achieve a product moisture content of about 5%.

Activated-Carbon Manufacture

Activation. Activation is essentially a two-phase process requiring the burnoff of tars formed in carbonization plus the enlargement of pores in the carbonized material. Two types of activation processes are used — chemical and physical. Chemical activation involves the addition of a chemical agent such as phosphoric acid or zinc chloride. Physical activation involves the use of a hot gas such as steam or CO_2 at temperatures of 800–1000°C.

Physical activation is the more commercially important process, as well as the process of greater interest from an air emission standpoint. It is accomplished in continuous internally or externally fired rotary retorts or in multiple-hearth furnaces. The activation process uses sufficient oxygen to burn off residual volatile carbon remaining on the char after carbonization without burning excessive amounts of char. The activation process is exothermic, using the heating value of the volatile carbon plus heat supplied from supplemental fuel.

The carbonization and activation processes can be carefully controlled to produce an activated carbon with a desired pore structure and mechanical strength. Through postactivation chemical processes, the surface chemistry of the carbon can also be modified to enhance or inhibit the adsorption of particular species or to produce catalytic properties; for example, chlorine addition has been shown to decrease water vapor adsorption while sulfur addition has been shown to increase mercury adsorption. Activated carbons are typically highly microporous ($d < 2$ nm) materials with the majority of their pores being less than 5 nm in diameter, and they have a large surface area of between 500 and 2000 m^2/g.

Reactivation. Carbon that has been used for its intended purpose can be reactivated using essentially the same process as for the original activation. Organics adsorbed on the carbon during its use are burned off in the reactivation process. In addition, a portion of the carbon particles is consumed during reactivation. Reactivation may be performed at the site where the carbon was used (on-site reactivation) or conducted at a commercial manufacturing or reactivation facility. For smaller-scale on-site reactivation operations, fluidized-bed and infrared furnaces are gaining acceptance, in addition to the conventional multiple-hearth furnace.

Furnace types used for activated-carbon reactivation, in the order of their total on-line capacity, are shown in Table 1.[4] Also listed are the approximate number of operating reactivation furnaces of each type.

Table 1. Types of Equipment Used for Activated-Carbon Reactivation

Furnace Type	No. of Units
Multiple hearth	>100
Fluidized bed	<20
Indirect fired rotary kiln	>50
Direct fired rotary kiln	<30
Vertical tube type	<30
Infrared horizontal	<5
Infrared vertical	4

AIR EMISSIONS CHARACTERIZATION

Five types of products result from the manufacture of charcoal and activated carbon: the solid charcoal or activated carbon; noncondensible gases (carbon monoxide, carbon dioxide, methane, and ethane); pyroacids (primarily acetic acids and methanol); tars and heavy oils; and water. The products and product distribution are varied, depending on the raw materials and the process conditions used.

Particulate Matter

Particulate matter emissions result from handling the charcoal or activated carbon in transfer, size reduction/screening, briquetting, and related activities. In addition, tars emitted from the process, if not combusted, may solidify to form particulate emissions and pyroacids may form aerosol emissions.

Gases

Noncondensible gases are products of the carbonization process (charcoal and activated-carbon manufacture) and the activation process (activated-carbon manufacture). The quantity and relative concentrations of noncondensible gases depend heavily on the character of the raw material and the operating conditions of the process used.

Uncontrolled emissions from carbonization in batch kilns have been reported as shown in Table 2.[3]

Average noncondensible gas composition ranges from batch kilns are presented in Table 3.[5]

Acetic acid is seen to be 4–7% of the condensible products, methanol is 3–6%, and insoluble tars are 8–13%.[5]

Air Toxics

The uncontrolled emissions from carbonization and activation contain several constituents classified as air toxics (e.g., methanol). Reactivation has an even greater potential for causing air toxics emission, since the carbon being reactivated has often been employed in adsorbing compounds that would be classified as air toxics.

Of special interest is the potential for formation of polychlorinated dibenzodioxins (PCDDs) and polychlorinated dibenzofurans (PCDFs) in the high-temperature, low-oxygen environment of the activation (or reactivation) furnace. One test program found no evidence of PCDDs or PCDFs emitted from the reactivation of virgin carbon, but detected the presence of both families of compounds when reactivating

Table 2. Uncontrolled Emission Factors for Charcoal Manufacturing[a]

Pollutant	Average Emission Factor (kg/Mg)	(lb/ton)
Total PM[b]	160	310
CO	140	290
NO_x	12	24
CO_2	560	1,100
VOC[c]	140	270
Methane	54	110
Ethane	26	52
Methanol	76	150
POM	0.0047	0.0095

[a]Factor units are kg/Mg and lb/ton of product. All factors are for uncontrolled emissions, Manufacturing process used a charcoal kiln. The data have been given an EPA rating of E (poor).
[b]Includes condensibles and consists primarily of tars and oils; approximately 74% consists of organic material.
[c]Consists primarily of methanol acetic acid, formaldehyde, pyroacids, and unsaturated hydrocarbons.

Table 3. Composition Range for Noncondensible Products of Charcoal Manufacture

Product	Percent of Noncondensibles
Carbon dioxide	50–60
Carbon monoxide	22–33
Methane	3–18
Hydrogen	1–4
Higher hydrocarbons[a]	1–6

[a]"Higher hydrocarbons" are assumed to be nonmethane, noncondensible hydrocarbons.

spent carbon from water treatment operation. This would tend to indicate that the by-products were formed from the adsorbed organics on the spent carbon rather than from impurities in the virgin carbon.[6]

AIR POLLUTION CONTROL MEASURES

Control of particulates from material sizing, handling, and briquetting operations is typically achieved by the use of centrifugal or fabric filter collectors. Collection efficiencies of 65% for centrifugal collection and 99% for fabric filtration have been reported.[5]

Emissions from carbonization, activation, and reactivation retorts are generally controlled by afterburners followed by water scrubbers.[7] Typical afterburner operating conditions include combustion temperatures of 1800°F (980°C) or greater and residence times in excess of 2 sec. There are no data available on destruction-removal efficiency (DRE) for such a control system in this application. The configuration and conditions are similar to those used for controlling hazardous waste incinerator emissions where DREs of 99.99% are expected.

In one application, where data have been published on PCDD and PCDF emissions from reactivation systems using afterburners, a fluidized-bed reactivation system was fitted

Table 4. Total Dioxin and Furan Equivalent Emissions as 2,3,7,8-TCDD[a]

Pollutant	2,3,7,8-TCDD Toxic Equivalence Factor	Average Concentration, mg/dscm	Average 2,3,7,8-TCDD Toxic Equivalent Concentration, ng/dscm
2,3,7,8-TCDD	1.00	ND (0.023)[b]	ND (0.023)[b]
Other TCDD	0.01	0.012	0.00012
Penta-CDD	0.5	ND (0.024)[b]	ND (0.012)[b]
Hexa-CDD	0.04	0.448	0.0179
Hepta-CDD	0.001	0.032	3.2×10^{-5}
Octa-CDD	0.000	0.250	0.000
2,3,7,8-TCDF	0.1	0.068	0.0068
Other TCDF	0.001	0.189	1.89×10^{-4}
Penta-CDF	0.1	0.075	0.0075
Hexa-CDF	0.01	0.03	3×10^{-4}
Hepta-CDF	0.001	0.043	4×10^{-5}
Octa-CDF	0.000	0.033	0.000
Total			0.068

[a]Annual emissions for a 196-dry standard m^3/h combustion gas glow rate, 7000 hours per year facility operation — 100 μg per year.
[b]ND — not detected; values in parentheses denote the average detection limit, used in lieu of actual data in the risk assessment.

with an afterburner designed to provide a 2-sec residence time at a temperature of 2400°F (1316°C). In another application, a horizontal infrared reactivation furnace was fitted with an afterburner sized for a 0.3-min (20-sec) residence time at 1850°F (1010°C). Table 4 shows the measured PCDD and PCDF emissions from the latter installation.[6] The operating rate of the system was 215 lb/h of spent activated carbon.

REFERENCES

1. Chemical Market Reporter, "Chemical Profile: Activated Carbon," Vol. 253, Iss. 2, January 12, 1998.
2. A. Yehaskel, *Activated Carbon Manufacture and Regeneration*, Noyes Data Corp., Park Ridge, NJ, 1978.
3. AP-42, U.S. Environmental Protection Agency, Research Triangle Park, NC, September, 1995.
4. W. G. Schuliger and L. G. Knapil, "Reactivation Systems," Sunday Seminar, American Water Works Association Annual Conference, June 1990.
5. C. M. Moscowitz, *Source Assessment: Charcoal Manufacturing State of the Art*, EPA-600/2-78-004z, U.S. Environmental Protection Agency, Cincinnati, OH, December 1978.
6. W. E. Koffskey and B. W. Lykins, Jr., "GAC Adsorption and Infrared Reactivation: Case Study," *J. AWWA*, 48–56 (January 1990).
7. L. Sorrento, AtoChem North America, Philadelphia, personal communication, June 5, 1991.

CHLOR-ALKALI

Kenneth S. Walborn

Carl Wilhelm Scheele, a young Swedish apothecary, is generally recognized as the first person to isolate pure chlorine gas in the year 1774. The major coproduct, caustic soda, has an older but more obscure history. Caustic soda was produced in ancient times by reacting lime and sodium carbonate for use as a cleanser.

The first commercial use for chlorine was in the bleaching of textiles in France in the year 1789 and the bleaching of textiles, paper, and flour remained its major applications until the early 1900s, when it was used to sterilize the Jersey City water supply. The ability of chlorine to destroy lower forms of organic life resulted in its rapid acceptance as a treatment for sewage sludge and drinking water, thus virtually eliminating typhoid fever epidemics. The production history of chlorine is characterized by rapid growth to match its uses in the manufacture of inorganic chemicals, plastics, organic chemicals, solvents, pulp, and paper, and for water treatment.

The growth of caustic soda paralleled that of chlorine and the production volume of caustic soda by the electrolytic chlor-alkali processes passed that of the lime soda process in 1940. The uses of caustic soda have grown from soaps and cleaners to include the manufacture of organic chemicals, detergents, and inorganic chemicals, and its utilization in the pulp and paper, petroleum, aluminum, and water treatment industries.

The growth of the chlor-alkali industry is shown in Figure 1. The units of production are in electrochemical units (ECUs), with 1 ECU equal to 1 ton of chlorine plus 1.1 tons of caustic soda (the ratio of formation).

PROCESS DESCRIPTION

Chlorine and caustic soda are produced concurrently by the electrolysis of salt dissolved in water in either a diaphragm cell, a mercury cell, or a membrane cell. In a diaphragm cell (Figure 2), chlorine gas is produced at the positive electrode (anode) and caustic soda and hydrogen are produced at the negative electrode (cathode). The overall reaction can be expressed as:

$$2NaCl + 2H_2O \xrightarrow{\text{electricity}} Cl_2 + 2NaOH + H_2$$

The chlorine produced at the anode is physically separated from the coproducts by an asbestos diaphragm that allows passage of sodium ions to the cathode, where caustic soda and hydrogen gas are produced.

In the mercury cell, two liquid layers flow through a long cell (Figure 3). The bottom layer is liquid mercury, which serves as the cathode for the cell. The upper layer is a saturated brine solution. Anodes protrude into the brine layer, and when an electric current is passed through the cell, chlorine is produced at the anode and is withdrawn from the cell. The sodium ions are drawn to the cathode, where a sodium–mercury amalgam is formed. This amalgam is removed from the cell and reacted in a secondary cell with purified water to form caustic soda and hydrogen.

The membrane cell is an improvement on the diaphragm cell that was made possible with the development of stable

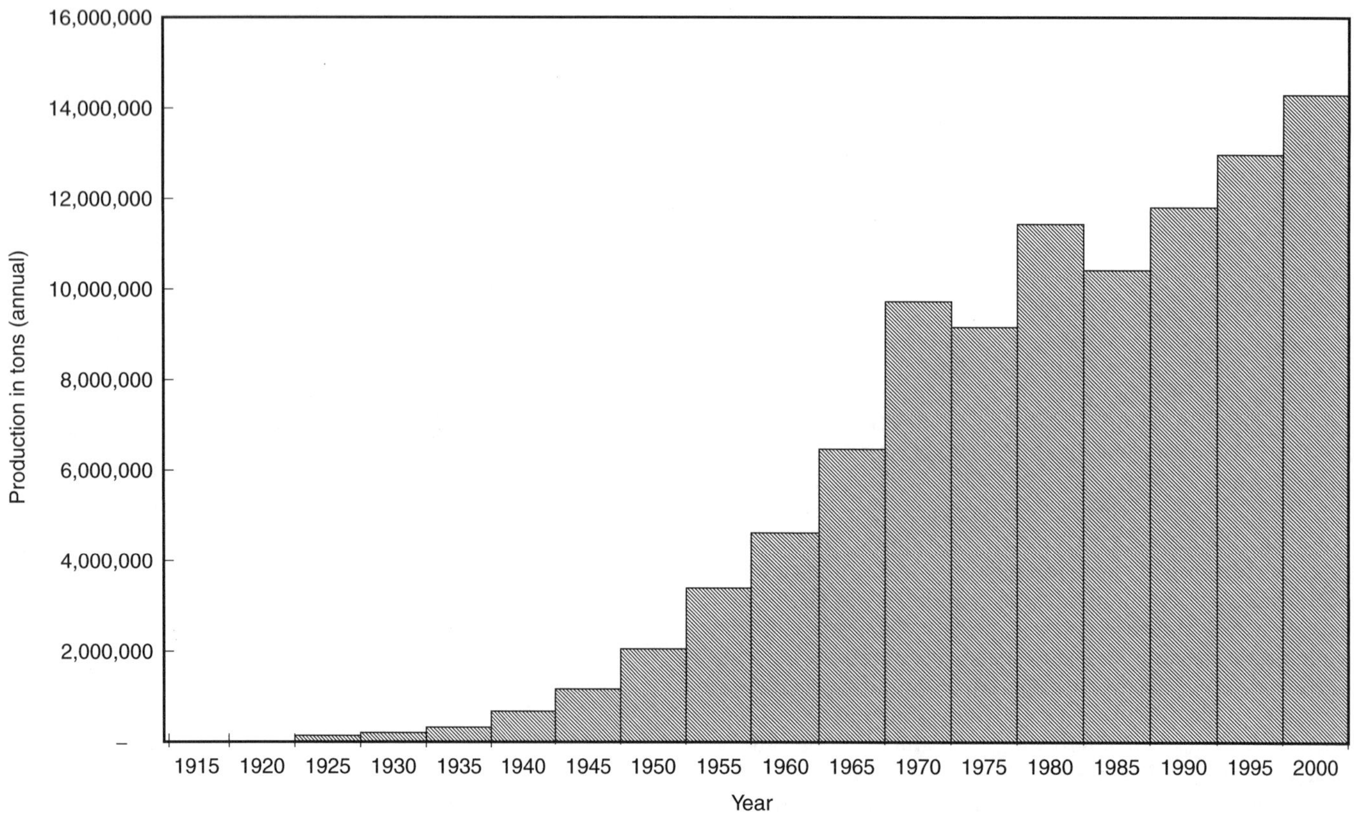

Figure 1. United States Chlor-Alkali Production in ECUs for the Years 1915 through 2000 Projected (*Source*: Chlorine Institute).

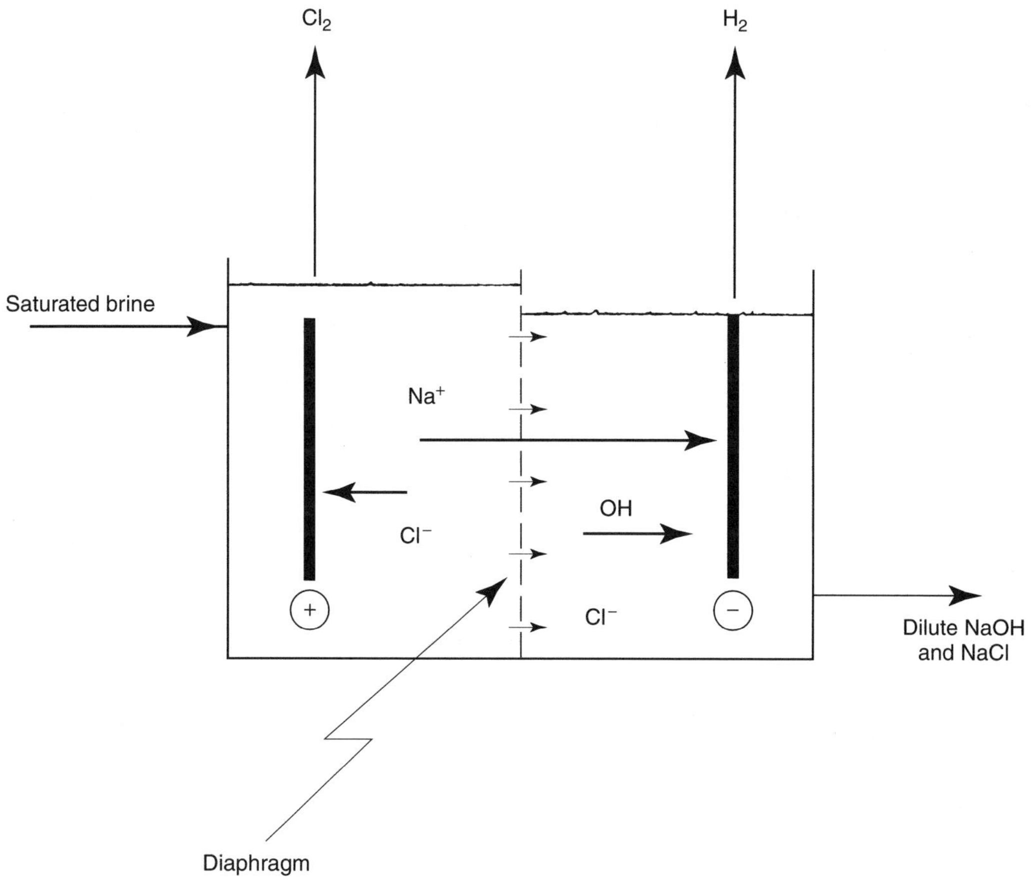

Figure 2. Simplified Diaphragm Cell.

membranes in the late 1960s that allowed the elimination of asbestos diaphragms (Figure 4). The passage of chloride through the membrane is reduced, allowing the production of high-quality caustic soda directly from the cell. The brine from the anode side of the cell is depleted of its chloride and sodium values and is recycled for reconcentration. Pure water is added to the cathode side of the cell to produce pure sodium hydroxide.

The choice of cell type or diaphragm material depends on space limitations, salt balance, brine purity, steam and power costs and availability, and caustic-soda strength and purity requirements. The cells are installed in groups operating in electrical series, with production of products in parallel.

The caustic soda from diaphragm cells is low in caustic concentration and contains unreacted salt, and it must be concentrated in evaporators to increase its strength and to recover salt. The caustic from mercury cells is produced at commercial strength and high purity and requires no further treatment before sale. Membrane cells produce high-purity caustic of somewhat less strength than its mercury cells.

The hydrogen gas produced in the cells is cooled and scrubbed prior to sale or use. Hydrogen from mercury cells must receive further chemical treatment or carbon absorption to remove all traces of mercury before it is used.

The chlorine gas produced in the cells is saturated with water and must be cooled and dried prior to compression. Cooling is normally done in noncontact heat exchangers, and the drying is carried out with sulfuric acid in a series of direct-contact scrubbers. The moisture level must be reduced to eliminate corrosion of conventional (steel) equipment that is employed as the construction material in the system from the compressor on. After the chlorine gas is dried, it is compressed and cooled. The high-pressure gas (35–150 psig) is liquefied in several stages of refrigeration. The liquefied chlorine is stored under pressure in tanks for shipment in tank cars, trucks, and barges. The chlorine gas stream entering the refrigeration units contains small quantities of "noncondensibles," primarily as a result of air in-leakage at the cells. The noncondensible gases leaving the refrigeration units still contain chlorine gas, which must be removed by recovery units or scrubbers or used in processes capable of utilizing low-chlorine-concentration streams. A schematic of a simplified chlorine system is shown in Figure 5.

AIR EMISSIONS CHARACTERIZATION

The air emissions to be controlled from diaphragm-, mercury-, and membrane-cell chlorine plants include chlorine gas, mercury vapors, asbestos, and hydrogen.

Chlorine

Chlorine gas is listed as an extremely hazardous substance under SARA Section 302, and any person handling in excess

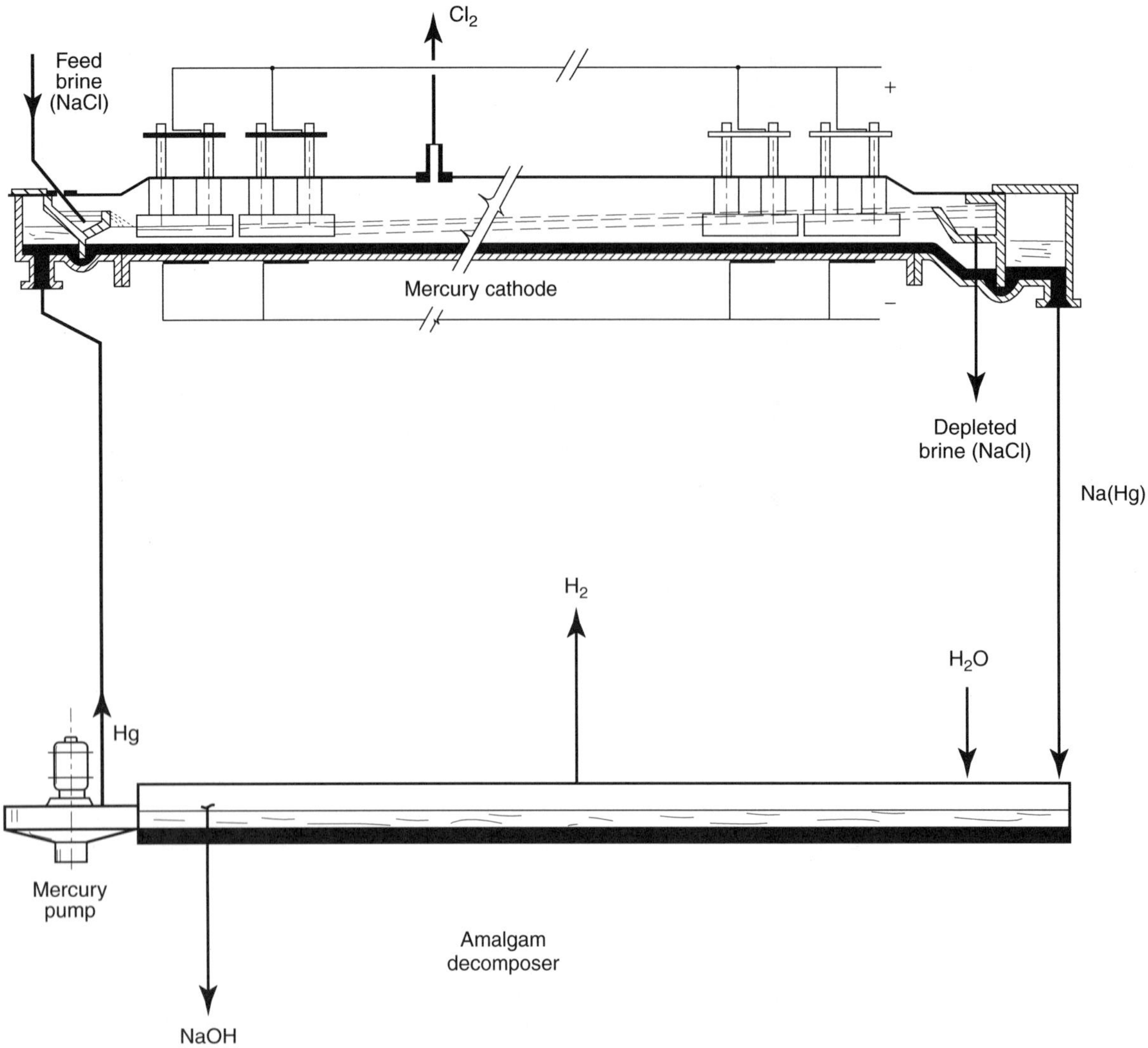

Figure 3. Typical Mercury Cell Construction.

of 100 pounds of the gas must take certain emergency precautions. It is also subject to the release-reporting requirements of SARA Section 304, and noncontinuous releases exceeding 10 pounds in 24 h must be reported to the National Response Center (NRC), the Local Emergency Planning Commission (LEPC), and the State Emergency Response Commission (SERC). Continuous releases that exceed the reportable quantity must also be reported to the same agencies, by both telephone and letter, with a 1-year followup letter required. Under the requirements of SARA 313, detailed emission data must be reported annually to all groups if a manufacturer produces in excess of 25,000 pounds per year or a consumer uses in excess of 10,000 pounds per year.

The potential sources of the various emissions are:

1. From the cell flanges, connections, and relief devices
2. From the low-pressure headers during normal operations, startups and shutdowns, and emergency operations
3. From chlorine dissolved in the sulfuric acid and then released to the atmosphere
4. From chlorine compressor seals
5. From high-pressure piping and liquefaction equipment
6. Tail gas from recovery units and scrubbers
7. From tanks and equipment being loaded

Chlorine emission factors[1,2] for uncontrolled plant processes and, to some extent, specific control practices are shown in a table of emission factors (Table 1).

Mercury

Mercury liquid or vapor is included on the SARA hazardous chemical list and spills or releases in excess of 1 pound in 24 h must be reported to the NRC, LEPC, and SERC. Annual reporting under the terms of Section 313 is also required. Chlor-alkali mercury emissions are also covered by promulgated National Emission Standard for Hazardous Air Pollutants (NESHAP) limits, which impose a limit of 2300 grams a day for total emissions.

The potential sources of the various emissions are:

1. From the cells during operation and maintenance
2. In the vent gas from the cells
3. In the hydrogen from the cells
4. From piping leaks

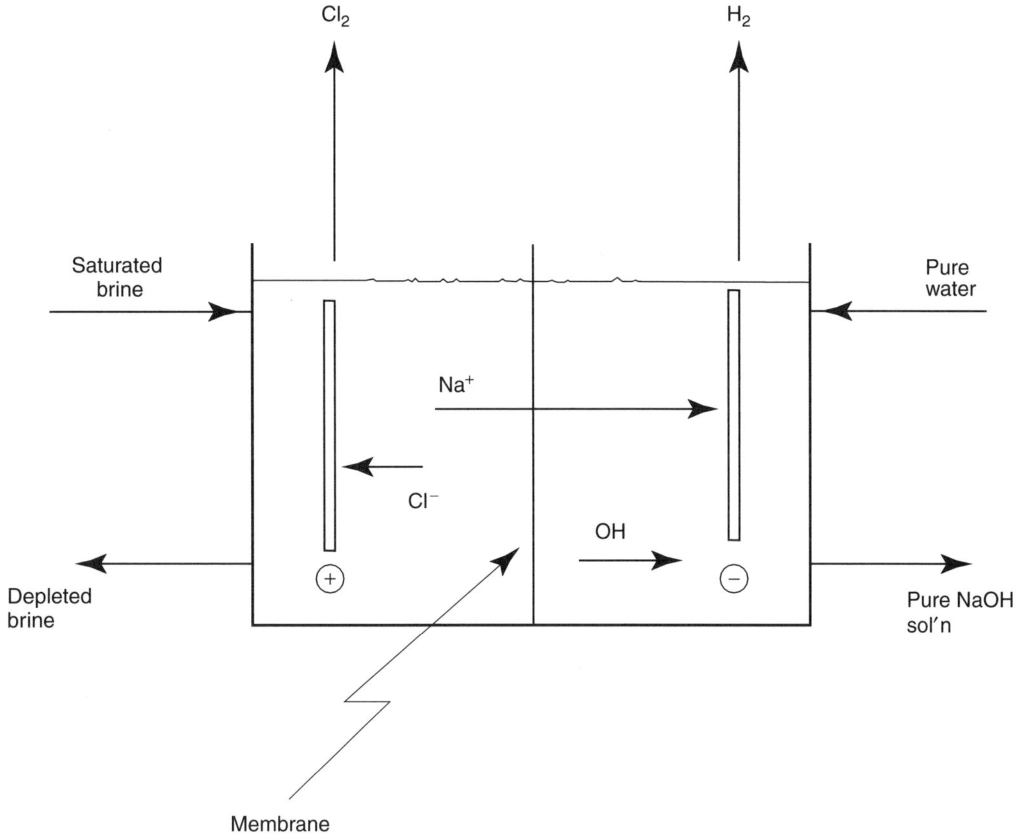

Figure 4. Simplified Membrane Cell.

Asbestos

Friable asbestos (dry) is included on the SARA hazardous chemicals list and spills or releases of dry asbestos in excess of 1 pound in 24 h must be reported to the NRC, LEPC, and SERC. Annual reporting under the terms of Section 313 is required as well. Handling of friable asbestos is also covered under NESHAP regulations, which set standards and emission limits.

The potential sources of the various emissions are:

1. Receiving and handling
2. Diaphragm manufacture
3. Disposal methods

Hydrogen

This very light gas does not appear on the regulatory lists, and spill and release reporting is not required. Hydrogen is an extremely flammable gas, and extreme care and good equipment design are required in its handling to avoid fires and explosions.

Among the potential sources of the various emissions are:

1. From the cells during operation
2. From vent leak-through

AIR POLLUTION CONTROL MEASURES

The control and elimination of fugitive and point-source emissions depend on plant design, operating philosophy, and the maintenance program.

Chlorine

Emissions from diaphragm cells and low-pressure headers are controlled or eliminated by the use of corrosion-resistant materials, the use of elastomer gaskets retained by bolts or heavy compression springs, and operation at atmospheric pressure or under a very slight vacuum. All headers and major equipment are vented through safety relief systems to caustic scrubbers during startups, shutdowns, or emergency conditions to eliminate releases to the atmosphere. These scrubbers must have sufficient capacity to neutralize full production capacity for the time necessary to start up with inerts or to shut down and evacuate the system.

A control system capable of maintaining a consistent pressure throughout the system under varying rates and stream compositions is extremely important to successful operation. Operation of the low-pressure system above atmospheric pressure can result in the release of fugitive chlorine, and operation at deep vacuums can result in the intrusion of atmospheric air, which causes high noncondensible flows in the high-pressure system. Cells, piping headers, and major equipment are evacuated to a vacuum system and purged prior to opening them for repairs.

The design and operation of the sulfuric acid drying system are extremely important to long-term successful operation. The rate of corrosion of steel used in the high-pressure system is accelerated to unacceptable levels as the moisture level increases. The sulfuric acid used to dry the chlorine will contain dissolved chlorine, which must be recovered and recycled by vacuum stripping or neutralized with sulfides to prevent

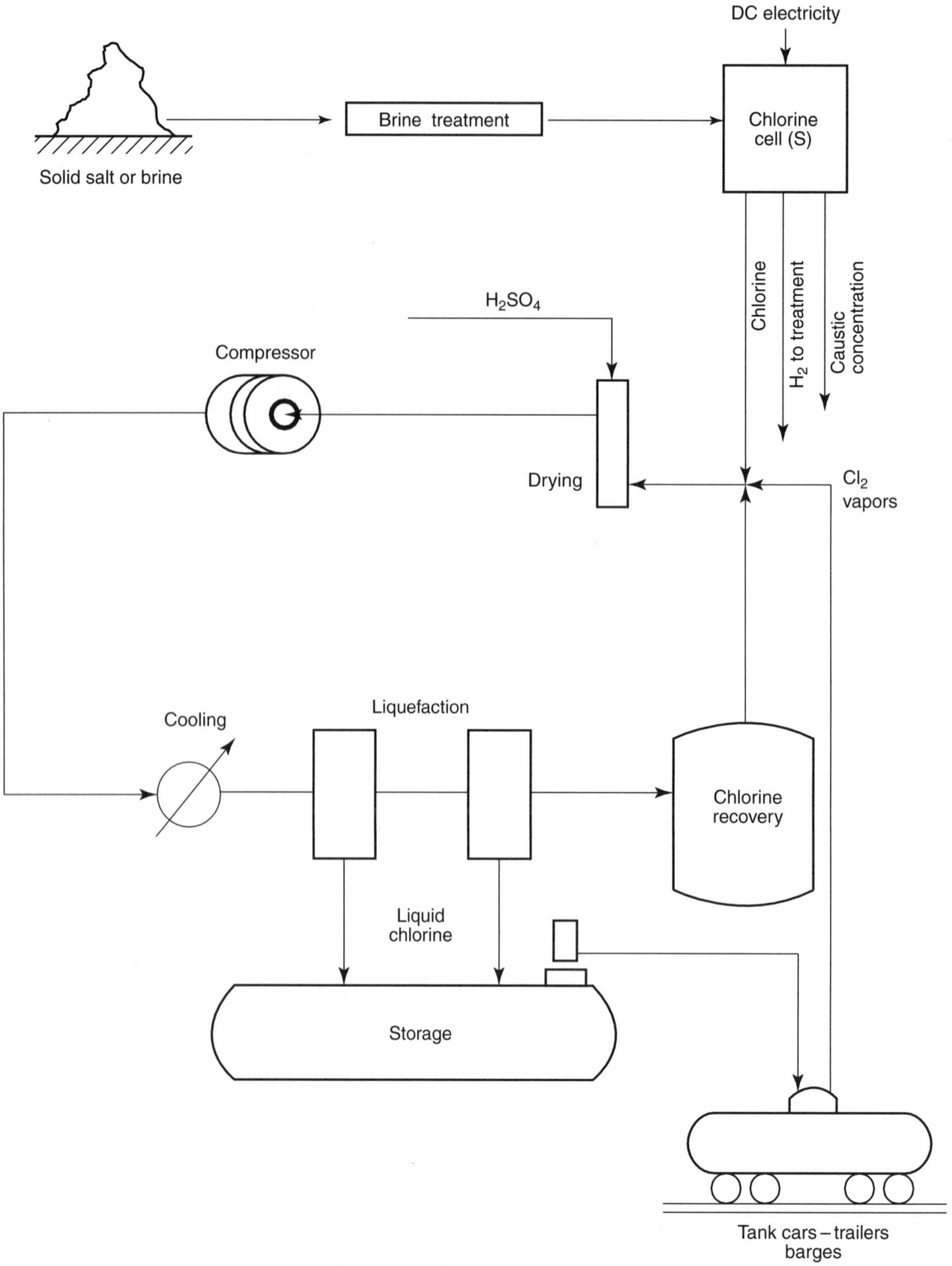

Figure 5. Simplified Chlorine Manufacturing Flow Sheet.

release to the atmosphere after discharge and neutralization of the spent acid.

Most manufacturers now use multistage centrifugal compressors, although reciprocal and sulfuric acid wetted-lobe compressors have been used. Double seals with a dry air purge between the seals are used to prevent seal leakage. It is important to design all high-pressure piping and equipment with the minimum number of flanges, and no leakage is acceptable. Because of the rapid corrosion of steel when it is exposed to wet chlorine, any small leak will rapidly become a large leak. High-pressure equipment is also vented through safety relief valves to a caustic scrubber with sufficient capacity to handle full production rates for the period required to shut down and evacuate the equipment.

The high-pressure chlorine is condensed in liquefaction equipment at temperatures ranging from $+75^\circ$F to -50°F, depending on the operating pressure of the system and the chlorine removal desired. The tail gas from the system, which consists of the noncondensibles and residual chlorine, can be sent to chlorination processes that can use low-concentration streams or to recovery systems. Recovery systems are capable of complete recovery and recycling of the chlorine in the tail gas. The design of recovery systems must recognize the presence of low levels of hydrogen that can react violently with the chlorine

Table 1. Emission Factors for Chlor-Alkali Plants[a] — Emission Factor

Type of Source	Chlorine Gas	
	(1b/100 tons)	(kg/100 MT)
Liquefaction blow gas		
Diaphragm cell	2,000–10,000	1,000–5,000
Mercury cell[b]	4,000–16,000	2,000–8000
Water absorber[c]	25–1,000	12.5–500
Caustic or lime scrubber	1	0.5
Loading of chlorine		
Tank car vents	450	225
Storage tank vents	1,200	600
Air blowing of mercury-cell brine	500	250

[a] References 1 and 2.
[b] Mercury cells lose about 1.5 pounds per 100 tons (0.75 kg/100 MT) of chlorine liquefied.
[c] Control devices.

in the system. The hydrogen level must be monitored closely and steps taken when required to ensure that it is kept below the explosive range. These systems consist of multiple stages of scrubbing, absorption/reaction, and refrigeration of the stream, with the chlorine being recovered by desorption/reaction from the scrubbing medium. The capability to divert the tail gas to a caustic scrubber should be included in the process to handle upsets or outages.

The liquid chlorine is transferred to scale tanks or spheres for storage prior to shipment. All liquid lines are equipped with expansion chambers between the valves to allow for the expansion of trapped liquid without pipeline rupture. The storage tank vapors are equalized to the drying tower system for recovery. Vapors from transportation equipment (i.e., tank cars, tank trucks, and barges) are recovered by a vacuum system. Pumps are equipped with double seals with a dry air pad to eliminate chlorine leakage at the seals. Valves and cocks are of a special leak-free design with bellows or multiseal construction.

Mercury

The control of mercury emissions is dependent on equipment and facility design, a good preventive maintenance program, and observation of appropriate operations procedures.

The primary cell is operated at a slight vacuum imposed by the chlorine compressor to eliminate emissions. The end boxes, which are the mercury inlet and outlet chambers, should have vapor-tight covers. In addition, water is added to the end boxes to wash, cool, and control vapor emissions from the mercury. A slight vacuum is pulled on the end boxes by a fume system further to reduce the possibility of mercury release. These vent streams are cooled, scrubbed, and/or passed through treated carbon to ensure a mercury-free discharge. Mercury pump tanks, pumps, and seal pots are treated in the same way: vacuum operation, cover water, and tight covers.

Cell maintenance procedures and program quality are extremely important to the prevention of the escape of mercury vapors. Leaks in pipes, flanges, and other equipment must be repaired immediately, and a preventive maintenance program should be in place to prevent their occurrence. Any time that the cell cover or top must be removed for maintenance, the cell should be cooled and vented and, when possible, the bottom kept covered with water or brine to reduce emissions. The time that the cell is open and the frequency of such openings should be held to a minimum.

A potentially major source of mercury loss from the system is with the hydrogen generated in the decomposer or denuder cell. The hydrogen from the cells should be cooled at each individual cell and the pipes arranged to return condensed mercury directly to the cell, thus reducing handling of the mercury in later recovery steps. The hydrogen product is further cooled in refrigerated coolers and/or scrubbed with chemical scrubbers in series to recover the maximum quantity of mercury. Liquid mercury recovered is recycled directly to the cells, and the mercury recovered in scrubbers can be returned with the brine.

The floors in the cell room area should be smooth with no cracks and all seams sealed, so that no mercury or mercury solutions can be held for later release. Trenches should be sloped to prevent mercury puddles and a layer of water kept on sumps. Where possible, wastewaters are recycled, and when this is not possible, they are treated to remove mercury before they are discharged.

A good inspection program involving daily inspections and floor washdown can further limit emissions. Special vacuum cleaners equipped with carbon filters are available for the cleanup of mercury droplets discovered by the inspection program.

Operator training and compliance with operating standards are an important part of the maintenance of minimum emissions. They involve housekeeping, spill control, personal hygiene, and control of operating parameters that affect emission levels.

A waiver from cell room emissions measurement is available if the requirements of Regulation 40CFR61.53(c) (4) — which generally spell out design, maintenance, inspection, and operation requirements for mercury-cell rooms — are met. The Environmental Protection Agency assumes that a facility meeting these requirements will emit less than 1300 g/d of mercury from the cell room itself. Well-designed systems can reduce emissions to less than 100 g/d from end-box systems and to less than 10 g/d from the hydrogen system. Actual testing is required to meet the 1000 gram per day emission limit for point sources. As previously mentioned, the total emissions NESHAP limit is 2300 g/d.

Asbestos

The control of asbestos (friable) emissions starts with the receipt of the asbestos, which arrives in airtight plastic bags, shrink wrapped with plastic, on pallets in boxcars or trailer trucks. The pallet loads are inspected before removal is started. If any tears or spills are noted, they are repaired with tape and the spilled material cleaned up using a vacuum cleaner. Workers wear special clothing and masks during all phases of unloading. The asbestos bags should be stored in a dedicated area with appropriate warning signs. If water is used for the cleanup of spills, care must be taken to limit pressure to ensure that fibers are not broadcast into the air.

Periodically, a pallet of asbestos bags is transferred to the cell repair area. Access to this area is limited, and appropriate warning signs are posted. When it is necessary to add asbestos to the liquid bath used to produce the asbestos diaphragms, a bag is placed inside a glove box through a hinged Plexiglas door. The box is equipped with rubber gloves hermetically sealed through openings into the box. These gloves are used for all operations inside the glove box. A vacuum is pulled on the box system through the vacuum mix tank, which ensures that no fibers escape the box system. The asbestos bags can be opened inside the glove box and the asbestos transferred to the mix tank through a transfer pipe. From this point on, the asbestos is handled in the wet (nonfriable) state, eliminating exposure. The empty asbestos bags are transferred to a clean disposal bag attached to an opening in the glove box. Before adding the next bag, the box can be cleaned with a vacuum wand located inside the box. All spills of asbestos slurry are hosed down as required to prevent any later generation of fibers.

Cell diaphragms eventually must be replaced because of the gradual plugging caused by trace impurities in the brine. At that time, the wet asbestos is washed from the cathode into a sump. It is then recovered in a filter operation and, while still in the wet form, is placed in plastic-lined boxes for secure landfill. Cleanliness is important, and any spilled material should be washed to the sump for recovery.

Hydrogen

Hydrogen is a very light flammable gas that does not appear on regulatory lists. Hydrogen from mercury cells can contain mercury, for which the relevant control measures are as discussed under "Mercury." The major emphasis for hydrogen is on control systems to provide safety in the handling of this extremely flammable gas.

Cell and suction header pressures are maintained near atmospheric pressure to reduce losses from flanges. Because of the difficulty in preventing hydrogen from leaking through isolation and vent valves, the use of water seals is recommended where pressure permits. Double-block and bleed arrangements are recommended for higher-pressure service. High-pressure valves can be checked for leak-through by venting through seal tanks that can be filled with water to determine valve tightness.

REFERENCES

1. *Atmospheric Emissions from Chlor-Alkali Manufacture*, Publication AP-80, U.S. Environmental Protection Agency, Air Pollution Control Office, Research Triangle Park, NC, January 1971.
2. A. L. Duprey, *Compilation of Air Pollutant Emission Factors*, PHS Publication 999-AP-42, U.S. Department of Health, Education, and Welfare, PHS, National Center for Air Pollution Control, Durham, NC, 1968, p. 49.

BIBLIOGRAPHY

A. E. Dungan, *Chlorine Institute*, Washington D.C., personal communication, December 1990.

P. M. King, PPG Industries, Inc., Pittsburgh, PA, personal communication, August 1990.

R. J. Samelson, PPG Industries, Inc., Pittsburgh, PA, personal communication, August 1990.

ETHYLENE OXIDE

R. Scott Huebner, Kevin S. McLeary, George P. Partridge, Jr., William M. Stayer, and James F. White

INTRODUCTION

Ethylene oxide (C_2H_4O; CAS #75-21-8) is a member of the class of petrochemicals called *epoxides*. It is known as oxirane by the International Union of Pure and Applied Chemistry; however, it is also called dihydrooxirene; dimethylene oxide; 1,2-epoxyethane; oxacyclopropane; oxane; oxidoethane; and a,b-oxidoethane.[1,8]

Ethylene oxide is a molecule with a highly strained ring, and is very reactive. Because the ring can be opened easily, it is a very versatile chemical intermediate that is mainly used as an intermediary in the manufacture of solvents, antifreeze, adhesives, detergents, polyurethane foam, textiles, and more. Minor uses of ethylene oxide (EO) include commercial sterilization and fumigation. It is used on items that are sensitive to heat or moisture, such as some food products, cosmetics, library materials, and hospital sterilization of certain equipment and devices. Although its use as a sterilant accounts for less than 1% of its production, such facilities are widespread throughout the United States.[1] This section first discusses air emissions and controls for EO industrial manufacturing processes and then addresses those for EO sterilant and fumigant operations.

Health Risks

EO is a suspected carcinogen and poses an occupational risk during its manufacture and use.[2] Pathways include inhalation, ingestion, and dermal contact. Its inhalation can cause nausea, vomiting, and, in extremes, neurological disorders. Contact with EO in solution may irritate the eyes, skin, and lungs. Long-term exposure to EO fumes may lead to eye cataracts. Occupational health studies have indicated elevated cases of leukemia, stomach, and pancreatic cancers, and Hodgkin's disease, although the related study results are at present deemed inconclusive due to limited data. Animal studies have shown lung, gland, and uterine tumors caused by inhalation exposure.

Characteristics[3,4]

EO is a colorless gas at room temperature and has a sweet odor. It is highly reactive and flammable. It presents a severe explosion hazard when exposed to heat or flame. Its vapor is heavier than air and may travel to a source of ignition, and then flash back. It dissolves readily in water and solvents. Physical characteristics are given in Table 1.

Table 1. Physical Characteristics of Ethylene Oxide

Molecular weight	44.06
Boiling point	51.3°F, 10.7°C
Vapor pressure	1,095 mm Hg at 20°C
Melting point	−170.5°F, −112.5°C
Enthalpy of combustion	1306.1 kJ/mol at 298.15 K
Thermodynamic properties (gas)	(at 298.15 K, 100 kPa)
Enthalpy of formation	−166.2 kJ/mol
Gibbs energy of formation	−132.8 kJ/mol
Standard entropy	263.7 J/mol K
Heat capacity	55.3 J/mol K
Flash point	Between −18°C and 0°C
Autodecomposition temperature	Approx. 500°C at 1 atm

Careful attention must be given to process and storage vessel design, and process operations must be closely monitored. EO autodecomposition, possibly caused by a hot spot due to polymerization in a reactor in a production plant in 1991, resulted in an explosion and injuries.[5] EO attacks organic materials such as plastics. Stainless and carbon steel should be used in system components. Empty storage tanks should be filled with inert gas until used and should be cleansed of iron oxide. The latter precaution is necessary to avoid runaway polymerization reaction that may cause failure in the tank.[7]

INDUSTRIAL MANUFACTURE OF ETHYLENE OXIDE

Almost all EO production is associated with ethylene glycol manufacture.[6] In 1997 the United States, Western Europe, and Japan accounted for 57% of worldwide EO process capacity, which was 27 billion pounds. This ratio was in flux as more Asian and Pacific countries constructed EO facilities to support their own ethylene glycol production.

PROCESS DESCRIPTON[7]

EO was first synthesized in the mid-nineteenth century and was initially commercially produced in Germany during World War I using the chlorohydrin process. This process used hypochlorous acid to react with ethylene, and was brought to the United States in the mid 1920s. It was replaced by the more economical direct oxidation method introduced in the United States in the late 1930s, which did not require the large amounts of chlorine and the accompanying plant structure.

Commercial direct oxidation processes react ethylene with air or oxygen at a pressure of 10–20 atm and a temperature of 220–280°C over a silver-based catalyst. The reactor gas stream contains the product as well as by-products, which must be separated and disposed of. The reaction produces large amounts of heat that must be dissipated, but may be captured for energy recovery and plant operation.

The ethylene oxide chemical reaction is given by:

$$C_2H_4 + \tfrac{1}{2}O_2 \xrightarrow{Ag} C_2H_4O$$

By-products in the form of carbon dioxide and water are formed by:

$$C_2H_4 + 3O_2 \longrightarrow 2CO_2 + 2H_2O$$

and

$$C_2H_4O + \tfrac{5}{2}O_2 \longrightarrow 2CO_2 + 2H_2O$$

The reaction may produce trace amounts of formaldehyde.[8] It is also possible for the ethylene oxide to isomerize to acetylaldehyde over the silver catalyst.[9] Catalyst activity can be enhanced by the addition of alkali metals or alkaline earth metals. Halide, a catalyst inhibitor, may be added to suppress the secondary reaction without interfering with the primary reaction.[8]

The direct oxidation process uses one of two methods. The first of these is the air-based method, which was commercially developed in the United States by Union Carbide in 1937. It uses compressed air or air enriched with oxygen. The second, or pure oxygen, method is the predominant process used today. It uses an oxygen gas feed instead of air. The two variations of the direct oxidation process are outlined below in general terms. Specific plant processes will differ in some degree due to proprietary process designs.[7]

Air-Based Direct Oxidation Process

The air-based process has three general parts: the primary reaction stage, the recovery stage, and the purification stage (see Figure 1).

Stage 1 is the primary reactor stage. Here ethylene and compressed air feed separately into one or more shell-and-tube reactors (see Figure 2) connected in parallel to the process train. The reactor vessels contain several thousand steel tubes, measuring approximately 6–12 m in length, which are surrounded by water or oil acting as coolant. Inside the tubes silver catalyst material is mounted on carriers. Ethylene reacts with oxygen as previously stated to produce EO, CO_2, and H_2O. The considerable heat released by the reaction is removed from the vessel by the coolant circulation system and can be moved to a steam generator for energy recovery. Estimated EO conversion per pass is 20–30%. Vapor-phase oxygen inhibitors are added to reduce CO_2 formation. The effluent gas is cooled before moving on to Stage 2.

Stage 2 is the recovery stage. EO is removed from the product gas in absorber towers using cold water streams. The EO dissolves together with the CO_2 and N_2, as well as traces of ethylene, ethane, and aldehydes. The exiting water stream then moves to a desorber. A large portion of the desorbed gas recirculates to the primary reactors, and a smaller portion moves through heat exchangers to raise its temperature, and then on to purge reactors. Additional air may be added to the purge reactor to increase the oxygen content. The product gases from the purge reactor go to the purge absorber, where EO and O_2 are again removed. Multiple purge reactors and absorbers may function in series to increase EO yield. The scrubbed gas from the final absorber may be vented to the product storage vessels. These gases may be processed for valuable energy recovery to generate steam and power for plant operations.

Stage 3 is the purification stage. This stage reclaims further EO from primary and purge absorber–desorber units. The effluent streams pass through a heat exchanger to reduce their temperatures and then enter desorbers, where the gas is steam stripped. The lean water is then returned to the primary and purge absorbers while the vent gases may be moved through an additional stripper and then transferred to product storage vessels.

Oxygen-Based Direct Oxidation Process

The oxygen-based process differs chiefly in that it does not require the purge reactor system used by the air process (see Figure 3). This is because the primary reactor yields a higher product per pass due to the higher partial concentration of oxygen.

The reaction vessel design and process is the same as that for the air-based process. Inert gas presence in the system is reduced. While carbon dioxide in both processes must be removed, argon is an additional concern in the oxygen process. It must be removed carefully, since its departure renders the remaining gas stream considerably more flammable. The oxygen content must be reduced at the same time.

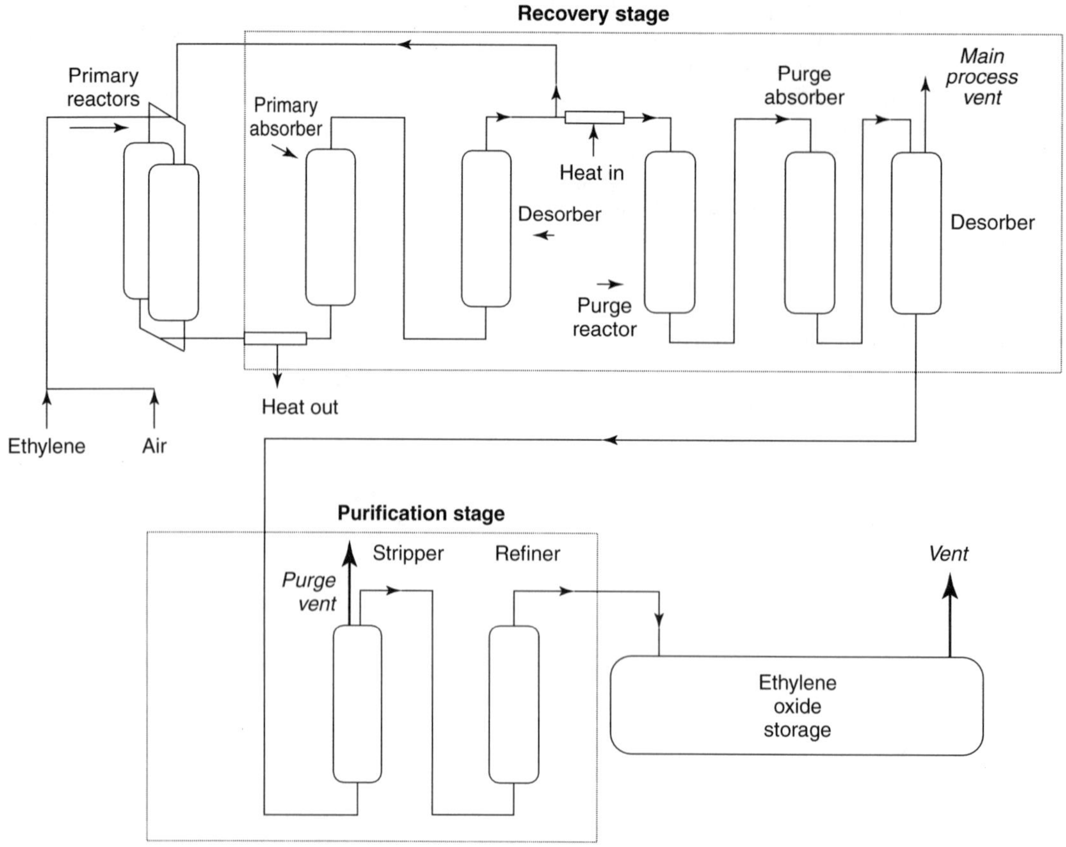

Figure 1. Schematic of an Air-Based Direct Oxidation Process.

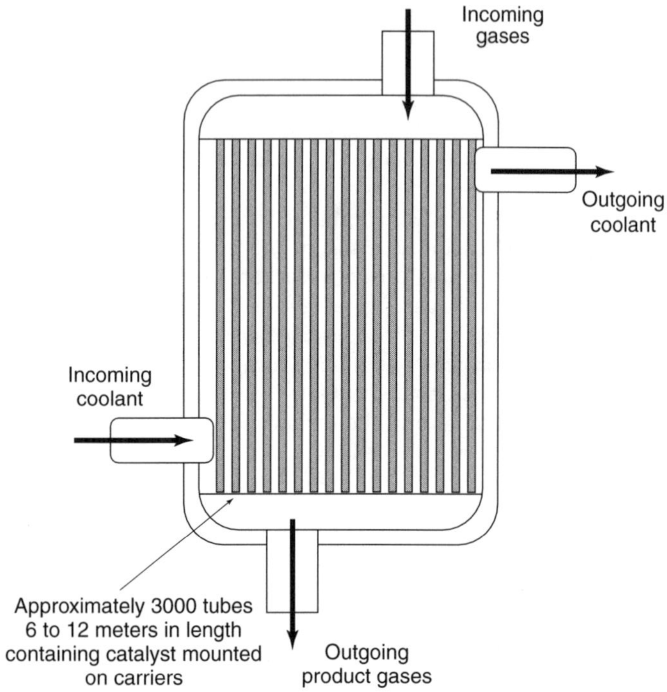

Figure 2. Schematic of Shell-and-Tube Reactor.

EMISSIONS

The primary air emissions from EO manufacture are ethylene, EO, CO_2, and ethane. Small amounts of SO_x and NO_x may also emanate from the process machinery, environmental devices, and supporting equipment. The largest source of emissions will be from the process cycle purge. In the air process this takes place at the last purge reactor. The vent produces a high-volume stream with small amounts of hydrocarbon as constituents, resulting in an emission with low energy-conversion value. In the oxygen process the comparatively low-volume stream is high in hydrocarbon content (approximately 85 mol %) and highly eligible for energy conversion as pump or heater unit fuel.

Further emissions come from the EO purification and recovery stages. CO_2 with small amounts of ethylene and EO may be vented from the oxide distillation system. The carbon dioxide regenerator vent in the oxygen process releases CO_2 again with small amounts of ethylene and EO. In addition to the process stream vents, small amounts of emissions can emanate from incomplete combustion in gas-fired turbines, which use process gases for fuel, as well as from valves, flanges, and other fittings. Standby flare apparatus can also be sources when these devices are placed in operation.

The joints between sections of pipe or fittings can be sources for fugitive emissions of ethylene oxide. Pump and valve seals can be other sources of fugitive emissions. These emissions can vary depending on equipment design and construction and on maintenance practices. EPA has published a methodology for predicting fugitive emissions, *Protocol for Equipment Leak Emission Estimates*, EPA-453/R-95-017. Emissions from valves make up approximately 60% of fugitive emissions within the chemical and petrochemical industries. Leaks from relief valves make up 15% of the fugitive emissions, while tanks make up approximately 10%. Pumps represent 10% of the

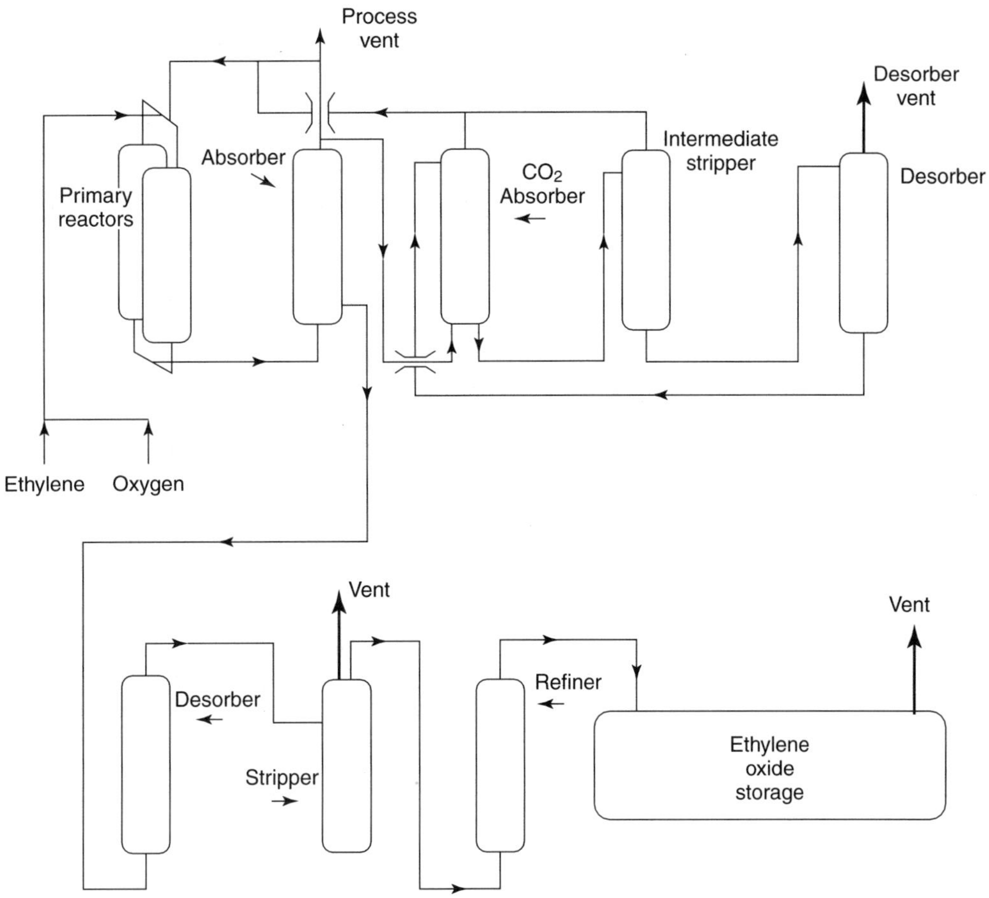

Figure 3. Schematic of an Oxygen-Based Direct Oxidation Process.

fugitive emissions, and flanges make up 5% of the fugitive emissions.[10]

EMISSION CONTROLS[11]

EO emissions are reduced or eliminated from both the air-based and the oxygen-based processes through the use of scrubbers, catalytic and thermal oxidizers, and, for standby use, flares. Processes seek to recover EO for the production stream through stripping. Vent gas streams, where EO cannot easily be recovered, are sent through thermal or catalytic oxidizers to destroy the EO and other emissions. EO, at some stages, can be used to provide fuel to plant heating and power generation operations.

In the air-based operation the main vent occurs after the last purge absorber. The components of the vent stream include EO, nitrogen, oxygen, ethane, and ethylene. The stream is processed through a catalytic or thermal oxidizer and may be used for energy recovery. The next vent in the air-based system occurs after the scrubbing and stripping operation. The scrubbed vent stream is generally completely destroyed in a fire box. The third and final process vent takes place as the product EO is pumped into the storage vessels under nitrogen, thereby displacing the inert gas. The vent stream can be scrubbed and combusted to remove any escaping EO that has mixed with the nitrogen.

In the oxygen-based system the main venting takes place in the argon purge vent, which can be incinerated to destroy organic compounds or can be sent to boiler fire boxes for energy recovery. The next vent occurs at the CO_2 desorber and consists almost entirely of CO_2 and water. The CO_2 may be recovered for sale or other use, while the bottoms stream is sent to a flasher for destruction. The third vent, the purge vent, can be sent for reabsorber wash for further EO recovery, and the end gas stream can be thermally destroyed. Finally, as with the air-based system, thermal destruction can be used to destroy any EO moving through the inert gas displacement operation in the product storage tanks and out through the final vent.

Robert F. Dye, in the 1992 edition of this book, recommended several equipment adaptations to reduce fugitive EO emissions emanating from flanges, valves, and pumps. He cited the 1988 study that found a range of between 8.7 and 40.4 kg/d of these types of emissions from U.S. facilities. He also emphasized the importance of safety considerations in EO processes. His safety recommendations remain pertinent and are given in Table 2.

Centrifugal Pump Shaft Seals

Fugitive emissions from pump seals are minimized by the use of magnetic drive (sealless) pumps, or by the use of double mechanical seals.[9,12] For double mechanical seals, the seal fluid used should be relatively nonreactive in contact with ethylene oxide. A 50% aqueous solution of ethylene glycol is often used for this purpose.[9] If the seal fluid is maintained at a higher pressure than the pumped fluid, the seal will be nearly 100% effective in preventing fugitive emissions. Sealless pumps are

Table 2. Safety Practices for EO Process Operations[11]

1. Install EO and flammable gas leak detectors at critical pump seals, flanges, etc. with regular monitoring and sample analysis.
2. Make routine gasket replacements during planned maintenance turnarounds.
3. Use all welded construction, where feasible, to minimize the number of flange joints.
4. Provide upstream rupture disks for safety/relief valves.
5. Use closed-loop sample systems.
6. Use a pressurized buffer gas in labyrinth shaft seals of centrifugal EO compressors.
7. Equip EO handling pumps with double seals of the tandem type.
8. Provide alarms and automatic pump shutdown switches for high temperature and seal failure.
9. Have plant personnel make routine inspections for leaks, with immediate leak repair.

100% effective in preventing emissions, since there are no seals to leak.[12] Packed pumps and single mechanical seals should not be used for ethylene oxide service.[9]

Centrifugal Compressor Shaft Seals

Like pump shaft seals, double mechanical seals with a buffer gas may be used to minimize fugitive emissions from compressor seals. The buffer gas may be nitrogen for air-based systems or oxygen-based plants that use nitrogen as a ballast gas. Methane may be used for the buffer gas in oxygen-based plants that use methane as a ballast gas.[11]

Valves

Emissions from valve stem seals may be eliminated entirely by the use of sealless, bellows-seated valves. These valves contain a bellows assembly, which is welded to the bonnet assembly at the top of the valve and to the stem at the bottom.[11]

Recent Development

John Crane Mechanical Seals (*www.johncrane.com*) has developed a gas-lubricated, noncontacting mechanical double seal for use in pumps. These seals eliminate the potential of damage from face wear, since the faces do not contact. This seal may also eliminate downtime, which may be required to clean conventional double mechanical seals.[10] These seals work at pressures up to 600 psi.[13]

Flowserve (*www.flowserve.com*) manufactures a non-contacting, gas-lubricated double-bellows seal for applications involving volatile hazardous substances.[10]

A&A Environmental Seals, Inc., has developed a secondary seal to capture emissions from the primary seal. The system works according to the Coanda effect, which is the tendency of a free fluid jet to follow any nearby curved or inclined surface upon leaving the jet nozzle.[14]

Seal manufacturers are also providing "split" seals, which lets the seal users replace the seals without disassembling their equipment.[13]

For the latest information on pump seals and manufacturers, refer to the Pump World web site at *www.pumpworld.com*, the Fluid Sealing Association web site at *www.fluidsealing.com*, or the European Sealing Association web site at *www.europeansealing.com*.

COMMERCIAL USE OF ETHYLENE OXIDE AS A STERILANT AND FUMIGANT[15]

EO is used as a commercial and special application sterilant of items that are sensitive to heat and moisture. It is highly effective and can penetrate paper and cardboard packaging and certain plastic wrapping materials. Approximately 200 large-scale EO sterilization or fumigation facilities are in operation at dispersed locations throughout the United States as of 1997. This does not include small-scale use sites such as hospitals, veterinary facilities, labs, and museum curation sites. EO is used on materials such as surgical instruments and gloves, pharmaceuticals, cosmetics, spices, and rare or old books, and other artifacts.

EO can be used in pure form or in a mixture with other gases. Two commonly used mixtures are EO at 12% with dichlorodifluoro methane, and EO at 10% with carbon dioxide. EO is stored in liquid for bulk fumigation operations and is vaporized during its application.

Sterilization and fumigation operations fall into two categories: single item sterilization, or bulk. Single item processes generally involve the item being placed in an envelope or small container. EO gas is then injected into the container and allowed to persist for a specified period of time.

Bulk sterilization and fumigation takes place in chambers ranging in size from approximately 100 cubic feet to 1,000 cubic feet (see Figure 4). The chambers operate at less than ambient pressure by using vacuum pumps in order to create optimum dispersion conditions, and, later, to evacuate the EO after application.

PROCESS DESCRIPTION

Chamber sterilization processes follow six general steps. First, the subject materials are loaded into the chamber and the pressure is reduced below ambient. Both temperature and humidity are adjusted to optimize EO effectiveness. Evacuation of air also reduces the threat of EO ignition.

Second, liquid EO (in mixture or pure form) is vaporized and injected into the chamber, where it resides for approximately 4–6 h. Third, the EO is evacuated by vacuum pump in a span of about 10 min. Fourth, the chamber is flushed with nitrogen or carbon dioxide four or more times to remove remaining EO. This step is called the "sterilization chamber vent." Fifth, the chamber gases are exhausted through a vent, a step called the "chamber exhaust vent," as the entrance door is cracked in order to remove any remaining gas before workers enter. The final step is the removal of the bulk items to an aeration space, where they will remain for hours or days to permit additional diffusion of EO. The off-gassing of EO from the product takes place here, in what is called the "aeration vent." All three venting operations produce EO streams from the process.

EMISSIONS

The principle sources of EO emissions from bulk commercial sterilization operations are the gas/liquid separator on the system vacuum pumps, the vacuum pump drain, the chamber's exhaust vent, and the aeration room vent.

The EPA issued a national emission standard for hazardous air pollutant (NESHAP) for EO on 6 December 1994 in light of health concerns regarding EO and the chronic

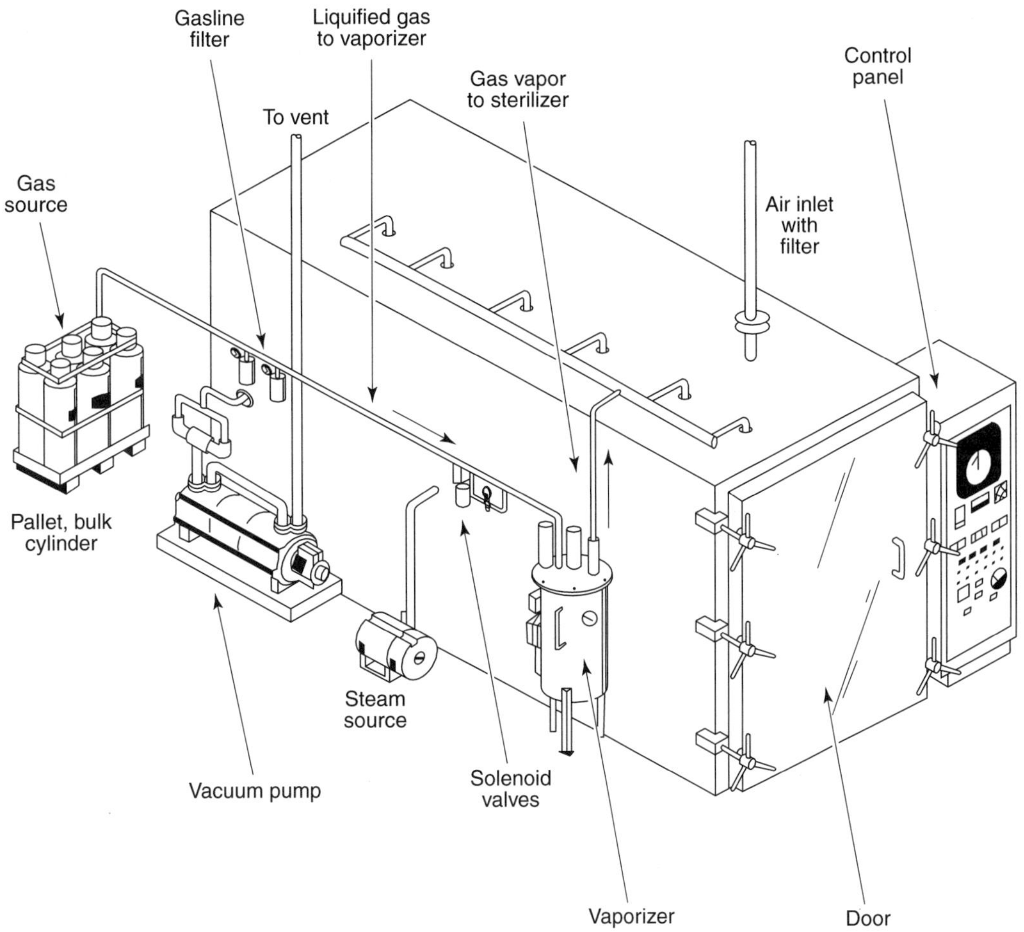

Figure 4. Schematic of a Commercial Ethylene Oxide Sterilization Chamber.

exposure of workers at commercial fumigation facilities. The rule exempted beehive fumigators, medical and veterinary facilities, laboratories, research activities, and certain other operations.

EMISSION CONTROLS

The EPA promulgated rules for commercial ethylene oxide fumigant and sterilization facilities in the Federal Code.[16] The rules regulated facilities that used one or more tons of EO during a consecutive 12-month period and set limits on the amount of EO emissions from the three types of vent procedures.

EPA estimated that 150 of the 200 large-scale commercial operations in 1997 would be affected nation wide. The EO emission standards are given in Table 3.

The rule required operators to conduct initial performance tests for their new or existing facilities to determine the efficiency of the EO emission control devices. The tests would also establish parameters for determining compliance. Operators had to measure the maximum ethylene glycol concentration and the maximum liquor tank level for the acid-water scrubber control system. They had to determine baseline temperatures for the catalytic and thermal oxidizers. After these parameters were established, operators would submit periodic reports on performance as well as reports of violations when the parameters were exceeded. Operators were also required to obtain a Title V permit.

EPA Recommended Control Technologies

EPA recommended three types of control techniques to reduce emissions: acid-water scrubbing, catalytic oxidation, and thermal oxidation. EPA also allowed for industry use of other unspecified technology. Initial testing and reporting parameters for the other-than-EPA recommended technology were to be approved by the Administrator. The three EPA recommended control technologies are described in the following.

Acid-Water Scrubber

This system uses countercurrent flow through a packed tower, in which EO gas passes through an acidic water solution (generally aqueous sulfuric acid). Cleaned nitrogen or carbon dioxide is vented from the top of the tower (see Figure 5). The EO joins the solution and then moves with it to a reactor tank where hydrolysis is completed. The EO completes reaction to ethylene glycol [$CH(OH)_2$]. The solution is returned to the tower for more EO capture until a predetermined weight of glycol is achieved, signaling decreased removal efficiency. The spent solution is then removed for sale as an industrial ingredient or for disposal.

Table 3. EPA Standards for Ethylene Oxide Commercial Sterilization and Fumigation Operations

Source Size	Sterilization Chamber Vent	Aeration Chamber Vent	Chamber Exhaust Vent
Less than 1 ton/yr	None (minimum record keeping)	None (minimum record keeping)	None (minimum record keeping)
1 to less than 10 tons/yr	99% EO emission reduction	No controls	max conc. 5,300 PPM prior to activation of exhaust
10 or more tons	99% EO emission reduction	1 ppm conc. at outlet or 99% EO emission reduction, whichever is less stringent	99% EO emission reduction

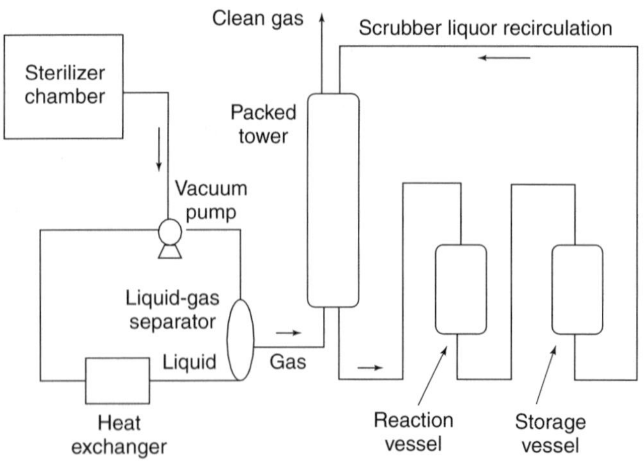

Figure 5. Schematic of Acid Water Scrubber System.

Catalytic Oxidation Unit

In this process EO is passed through a catalytic reactor for oxidation and final venting to the atmosphere (see Figure 6). Prior to the reaction the EO may be diluted to prevent excess catalyst bed temperatures, which can cause damage to the catalyst. The EO gas is then filtered to remove particulate matter, and its temperature is increased to optimize its reactivity with the catalyst. Both the acid-water scrubber and the catalytic oxidation processes are said to achieve 99% removal efficiency.

Thermal Oxidation

This method destroys EO by flaring it to the atmosphere (see Figure 7). The drawback of this method is that supplemental

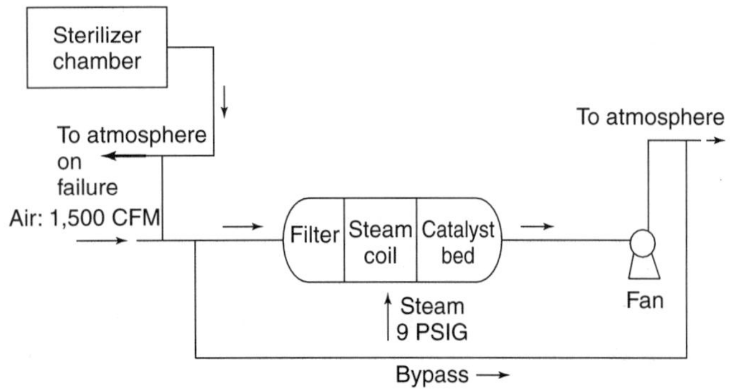

Figure 6. Schematic of Catalytic Oxidizer Unit.

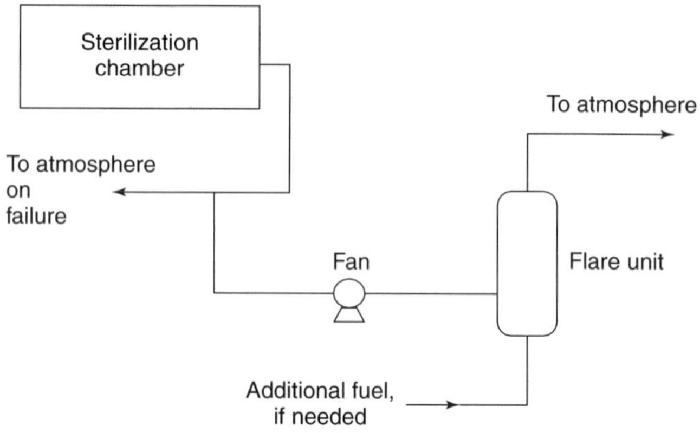

Figure 7. Thermal Oxidation Unit.

fuel would have to be provided to the oxidizer in many cases, since EO flow might not be sufficient to sustain the burn. This method achieves 98% removal.

Suspension of Control Technology Requirements

In 1997 the EPA announced that it would delay implementation of the emission control requirements deadline for 1 year.[17] The reason was four explosions or fires reported at commercial EO fumigation facilities in Massachusetts, Virginia, Wisconsin, and Indiana. The Indiana event resulted in one fatality and over 60 injuries. Firms were given permission to vent EO without controls. In December 1998 the EPA reinstated the requirement for emission controls for the sterilization chamber vent operation, but extended the deadline for implementing the controls on the other two vents, the chamber exhaust vent and the aeration room vent, until December 1999. After investigation of the previous explosions and consultation with the Ethylene Oxide Sterilization Association, the EPA concluded that some of the accidents appeared to be the result of an "abnormal activation of the chamber back vent [chamber exhaust vent]," at high concentrations of ethylene oxide causing the catalytic oxidizer units to be overfed with highly volatile EO.[18] EPA stated that operators should conduct comprehensive process hazard analyses (PHA) for each sterilizer process before putting the controls into operation.

REFERENCES

1. "Public Health Statement: Ethylene Oxide," Agency for Toxic Substances and Disease Registry, U.S. Department of Health

and Human Services, December 1990, *http://atsdr1.atsdr.cdc.gov:8080/toxprofiles/phs9016.html.*

2. "Ethylene Oxide," U.S. Environmental Protection Agency Technology Transfer Network, Unified Air Toxics Website, 26 May 1998, *www.epa.gov/ttn/uatw/hlthef/ethylene.html.*
3. U.S. EPA Chemical Profile: Ethylene Oxide, U.S. Environmental Protection Agency, *www.epa.gov/swercepp/ehs/profile/75218p.txt.*
4. Lide, David R. (Ed.), *CRC Handbook of Chemistry and Physics*, 74 ed., Boca Raton, CRC Press.
5. G. A. Viera, L. L. Simpson, B. C. Ream, "Lessons Learned from the Ethylene Oxide Explosion at Seadrift Texas," *Chemical Engineering Progress*, August, 1993, pp. 66–75.
6. "Ethylene Oxide/Glycol Producers 'Shifting,'" *Hydrocarbon Processing*, p. 27 (August 1997).
7. "Ethylene Oxide," *Encyclopedia of Chemical Technology*, 4th ed., Vol. 9, John Wiley, New York, 1991, pp. 915–958.
8. *Locating and Estimating Air Emissions from Sources of Ethylene Oxide*, EPA-450/4-84-007L, U.S. Environmental Protection Agency, Office of Air Quality Planning and Standards, Research Triangle Park, NC, September, 1986.
9. *Ethylene Oxide User's Guide*, Hoechst Celanese Polyester Intermediates, Occidental Chemical Corporation, Shell Chemical Company, and Sun Company, Inc., 1995.
10. *Chemical Engineering Magazine* **105**(7), McGraw-Hill, New York, NY (July 1998).
11. R. F. Dye, "Ethylene Oxide," *Air Pollution Engineering Manual*, Air and Waste Management Association, Pittsburgh, PA, 1992, pp. 422–433.
12. *Protocol for Equipment Leak Emission Estimates*, EPA-453/R-95-017, U.S. Environmental Protection Agency, Office of Air Quality Planning and Standards, Research Triangle Park, NC, November, 1995.
13. *Chemical Engineering Magazine* **105**(6), 121, McGraw-Hill, New York, NY (June 1998).
14. *Chemical Engineering Magazine* **105**(4), McGraw-Hill, New York, NY (April 1998).
15. *Ethylene Oxide Commercial Sterilization and Fumigation Operations NESHAP Implementation Document*, Midwest Research Institute, for the Office of Air Quality Planning and Standards, U.S. Environmental Protection Agency, September 1997.
16. "Rules for Commercial Ethylene Oxide Sterilization and Fumigation Facilities," *Code of Federal Regulations*, Subpart O, 40 CFR 63 (§63.360–63.367).
17. "Citing Explosions, EPA Delays Ethylene Oxide Controls," *Environmental Science and Technology* **31**(10), 451 (October 1997).
18. "Interim Final Rule: National Emission Standards for Hazardous Air Pollutants for Ethylene Oxide Commercial Sterilization and Fumigation Operations," *Federal Register* **63**(233), 4 December 1998, p. 66990.

EXPLOSIVES

Institute of Makers of Explosives

Explosives were introduced into commerce when Marco Polo brought the art of making black powder back to Italy from China. Explosives are designed and manufactured to detonate and today's safe manufacturing processes and practices are the result of experience gained from over 500 years of manufacturing. These practices include siting explosive manufacturing facilities in accordance with the American Table of Distances, intraplant or interline separation distances, or other quantity/distance tables. Buildings are often of frangible construction protected, by barricades, and posted with personnel and product limitations. History has dictated that the manufacturing of explosives is one of minimization. They are manufactured in the smallest possible quantity to reduce the hazards to the minimum with practicality.

The commercial explosives industry is involved in manufacturing a wide variety of explosive products, which the industry generally defines as "explosive materials." Commercial explosives are typically used for blasting dirt or rock, including excavating for road building or construction, removing overburden at a strip mine, fragmenting ore in an underground mine, and quarrying. Some situations require the use of specialty products, such as oil well charges (which are used for perforating oil well casings and cladding metal parts together), seismic activities, and demolition. Because there are numerous applications for commerical explosives, there are many types of explosives, each with its own manufacturing process.

Commercial explosives can be divided into the following groups: (1) dynamite, (2) emulsion, (3) ammonium nitrate fuel oil mixtures, (4) cast boosters, (5) detonators, (6) detonating cord, (7) initiating explosives, (8) black powder, (9) water gels and slurries, (10) specialty explosives. Each of these processes uses different raw materials, methods, and equipment to manufacture the explosives. Most production plants might produce one or more of these products, but there are no plants that produce all of them at one location.

Only the manufacturing processes in the commerical explosives industry will be covered here. The discussion will not include explosives that are manufactured under the control of the Department of Defense, and also will not cover workplace air contaminants that are subject to Occupational Safety and Health Administration regulation.

DYNAMITE

Process Description

Dynamites are combinations of nitrate esters, nitrate salts such as ammonium nitrate, and wood pulp or similar absorbent materials. Nitrocellulose is also used as a plasticizer or gelatinizer, particularly in semigelatin and gelatin dynamites. The dynamite is then packaged in paper, plastic, or cardboard shells and tubes of varying diameters, lengths, and weights.

The first step in the manufacture of dynamites is the nitration of organic alcohols to produce nitrate esters that sensitize the dynamite, such as nitroglycerin (NG), ethylene glycol dinitrate (EGDN), diethylene glycol dinitrate (DEGDN), and metriol trinitrate (MTN). The nitration process involves reacting the organic substrate (usually glycerin and ethylene glycol) with a mixture of nitric and sulfuric (oleum) acids. Nitration can be conducted in batches or in a continuous process. The by-products of the nitration are nitrate esters, spent acids, and water. There are no significant air emissions from the nitration process.

Manufacture of NG

$$H_2SO_4/HNO_3 + C_3H_8C_3 \rightarrow C_3H_5N_3O_9 + HNO_3 + H_2SO_4 + H_2O$$

Manufacture of NG/EGDN

$$H_2SO_4/HNO_3 + C_3H_8O_3 + C_2H_6O_2 \rightarrow C_3H_5N_3O_9 + C_2H_4N_2O_6 + HNO_3 + H_2SO_4 + H_2O$$

Dry carbonaceous materials such as corn flour, rice hulls, and sawdust are blended together and are called "dopes" in the industry. These dopes are used to absorb the NG/EDGN. All dry ingredients are received in containers and stored in separate hoppers in the "dope house." The various formulae are mixed on a batch basis by metering portions of each ingredient from its respective storage hopper into transport containers. Particulates may be trapped within the material-handling systems or dope-house air emissions may be controlled by the use of exhaust fans with dry cyclone dust collectors, baghouses, or both.

The dopes and nitrated esters are then blended together in the "mix house" to make dynamite, which has either a dry, damp, or paste consistency. The dynamite is then transferred to pack houses, where it is loaded into the various packages. Dust generated may be contained within the process equipment and the building. If it is not, mix- and pack-house air emissions are controlled with conventional particulate-control technology. The loaded sticks or tubes are placed in fiberboard cases either at the pack house or at a casing house. No air emissions result from the packaging process.

Air Emissions Characterization

The spent acid generated during the nitration process contains weak nitric acid, weak sulfuric acid, water, and some trace amounts of NG/EGDN. To remove the trace amounts of NG/EGDN, the spent acids are passed through an absorption tower to separate the two acids from the water and to destroy the trace amounts of NG/EGDN. In the commercial industry, this is called a "denitrating tower." The weak acids are reclaimed and are cycled back to the acid plants, where they are concentrated into their proper strengths. Nitrogen oxide fumes generated during this process are recovered with high efficiency in the absorption tower as weak nitric acid. The other by-product of the denitration process is steam, which is condensed into water.

Denitration of NG/EGDN

$$H_2SO_4 + HNO_3 + C_3H_4N_2O_6 + C_3H_5N_3O_9 \rightarrow HNO_3 + H_2SO_4 + H_2O + NO_x + CO + CO_2$$

Air Pollution Control Measures

Dynamite plants have extensive engineering controls installed to correct the dust and fumes in the workplace environment. Ventilation of buildings for industrial hygiene purposes produces minimal air pollution. Air pollution control is sometimes provided to control particulate dust emissions into the environment. In these cases, the pollution control technology utilized is conventional wet scrubbers, cyclones, or filters.

EMULSIONS

Process Description

Commercial explosive emulsions are typically water-in-oil emulsions of a concentrated aqueous solution of one or more oxidizer salts (typically ammonium nitrate, calcium nitrate, and sodium nitrate) and about 6% of a water-insoluble organic fuel mixture (e.g., mineral oil, fuel oil, or waxes), which are emulsified to form a product with a soft to firm consistency. These emulsions are less sensitive than dynamites, and their processed densities cannot be detonated by a blasting cap. In order to make them cap sensitive, it is common to add sensitizers. These compounds may be explosives, aluminium, solid fuel, and/or such density-control materials as hollow glass, ceramic, or plastic beads.

The fuels and oils are usually heated to about 70–110°C in their separate tanks. A solution of nitrate salts is prepared in another heated tank and kept at 70–105°C. The formulated amounts of the fuels and oils are then either batched or continuously forced through a high-shear mixer to form a thick, high-viscosity emulsion. This is either packaged in paper or plastic cartridges or prepared for use in bulk applications.

Air Emissions Characterization

Prior to blending, the oils, waxes, solutions, fuels, emulsifiers, and oxidizer salts are usually stored in metal or plastic tanks, which have submerged fill lines for the liquids. Fugitive emissions from these bulk tanks are minimal and are not controlled because of the low volatility of these liquids. Once the process has started, it is common to have all of the ingredients in a closed-loop system, The pumping, heating, blending, and packaging of the explosive emulsion are usually done in the same room of the plant. This allows for the collection of dust, scrap, and waste materials within the plant. The processed emulsions are packaged in plastic film or paper cartridges or produced for shipment in bulk quantities. Particulate emissions from the handling of aluminium and density-control materials are minimal. Handling takes place in buildings, and, in some cases, ventilation may be required for industrial hygiene purposes.

Air Pollution Control Measures

Since most of the process systems are closed loops and air emissions from tanks and buildings are minimal, air pollution control systems are not warranted. Submersible fill pipes in fuel oil tanks provide emission control. Lids and control valves are used on ammonium nitrate bins.

AMMONIUM NITRATE-FUEL OIL MIXTURES

Process Description

Ammonium nitrate-fuel oil mixtures are classified as blasting agents by the Department of Transportation. These blasting agents are usually mixtures of industrial-grade ammonium nitrate and No. 2 fuel oil. Aluminium, bagasse, nitroparaffins, or plastic spheres are sometimes added to afford more energy or to control the density of the packaged afford more energy or to control the density of the packaged product. Often a red dye is added to the fuel for quality-control purposes. Most blasting agent plants are modest installations and are the most common explosive plants in the United States.

Prilled ammonium nitrate is usually received in bulk by truck or railcar and unloaded into storage bins. The fuel oil is stored in tanks and would have the typical fugitive emissions of No. 2 fuel oil in storage tanks. The explosive is produced by metering the raw materials into a mixer/blender in the proper proportions, where, after mixing, it is either packaged or put into bulk bins for delivery directly to the consuming site.

Air Emissions Characterization

The ammonium nitrate prill and fuel oil are usually stored in large tanks and bulk bins outside the plants. The fuel oil would have the same emissions characteristics as would any common above-ground storage tank with a submerged fill pipe and which is vented to the atmosphere. Ammonium nitrate prill bins use containment lids and valves to limit the generation of dust. The mixing equipment usually is enclosed, and this limits air emissions. Fugitive emissions from production are usually contained within the building and do not escape to the atmosphere.

Air Pollution Control Measures

Emission control measures include submersible fill pipes in the fuel oil tanks, lids and flow control valves on the ammonium nitrate prill bins, and enclosed mixing systems in the production buildings. These measures limit the small amount of fugitive air emissions from storage and handling. Fugitive emissions from production are minimal and typically stay within the buildings.

CAST BOOSTERS

Process Description

Cast boosters are formulated by adding pentaerythritol tetranitrate (PETN), trimethylentrinitramin (RDX), or other, similar high-brisance explosives with trinitrotoluene (TNT) and casting the mix into small charges. The TNT is melted in a steam-jacketed kettle that has an agitator to blend the materials. Various proportions of TNT, PETN, and/or RDX are used, depending on market conditions and the physical performance required. The molten explosives are cast into a variety of forms and shapes, but they are typically cylindrical with one or more channels through the axis of the cylinder to accept either a blasting cap, a detonating cord, or both. The weights of these units range from about 50 to 2500 g, and they are usually packaged in fiberboard boxes.

Air Emission Characterization

They drying of the PETN and RDX may generate a minimum amount of dust that is typically contained within the dryer system. No chemical by-products are produced in the blending of these materials. The explosives are very sensitive to heat, friction, impact, and shock, and thus safety considerations are of paramount importance in the collection and retention of this dust.

Air Pollution Control Measures

The fumes generated by the mix kettles and the cooling operations are typically exhausted through a water scrubber that is contained in or is a part of the product building. The cleaning of the exterior of the cast booster and packaging results in the generation of small amounts of dust and chips, which are contained within the building.

DETONATORS

Process Description

Detonators (also known as blasting caps) are copper or aluminium tubes containing a very small quantity of initiating explosive(s). They are used to provide an energetic impulse capable of initiating a detonation wave in larger charges of high explosives. The two most prevalent types of detonators are electric and nonelectric. Both types are fairly similar in their methods of construction and differ substantially only in the type of energy input and the method of ignition involved.

An explosive base charge of granular dry PETN is press-loading remotely into the metal cap shell. A flame-sensitive priming charge of lead azide, diazodinitrophenol, or hexanitromannitol (or mixtures thereof) is remotely loaded atop the base charge. One or more delay elements are usually next introduced on top of the priming charge. These provide a specific time delay to the detonator and contain various proprietary delay powders (fuel/oxidizer mixtures) encased in a steel or lead sheath.

Electric detonators typically utilize either of two configurations above the delay column: a bridge wire surrounded by a proprietary flash powder or an electric match of proprietary composition. Both configurations are connected to two plastic-jacketed wires (either copper or iron) that exit the top of the shell and are used to deliver the energy input to ignite the bridgewire or match head.

Nonelectric detonators incorporate a hollow plastic tube above the delay column. This plastic tube contains a low dose of proprietary energetic material and provides the nonelectric alternative energy input to the delay column. Another form of nonelectric detonator is the fuse-cap type, which contains only the base charge and the priming charge and uses the flame from a safety fuse to provide the energy input.

Air Emissions Characterization

The materials utilized in the manufacturing process are combinations of dry powders. These raw materials remain process captive during the drying, blending, and loading operations and do not give rise to fugitive air emissions. Mixing of fuel-oxidizer raw materials for detonator manufacture may involve the use of a liquid mix vehicle (i.e., isopropanol), which is discharged to air as the mixtures are dried.

Air Pollution Control Measures

There are minimal fugitive emissions, and, therefore, no pollution controls are employed.

DETONATING CORD

Process Description

Detonating cord is a flexible linear product containing a center core of high explosive and is used to initiate other explosives or other lines of detonating cord. The explosive core loading is usually PETN, but other explosives such as RDX and cyclotetramethylene tetranitramine (HMX) can also be used. The detonating cord is made by either a dry- or wet-loading process. The dry-loading process is generally more popular because of the lower production costs for the ordinary types of cord.

Dry-process manufacturing requires that the PETN be dried to a very low moisture content. Additives such as flow enhancers, water-repellent compounds, or desensitizers may also be added if required. The PETN is placed in a hopper leading to an orifice on a spinning machine. A continuous strip of plastic is folded around the core of PETN as the plastic is drawn through a loading die. The tube thus formed, which contains relatively loose PETN, is overspun with several layers of yarns and has a layer of plastic extruded over it. Additional yarns may be overspun on the cord, and wax may be applied to hold these yarns in place. The finished cord is usually coiled onto spools and packaged in fiberboard boxes.

The wet-process detonating-cord manufacturing system uses a fine-granulation explosive such as PETN. The explosive is mixed to a fluid paste in water with a suspending agent. This paste is then pressure loaded into a core surrounded by a braid of textile fiber and subsequently dried. The process is finished by countering with various yarns, overextruding with a plastic jacket, and final overspinning and waxing, if desired. The slow, limited speed of the braiding process makes this method of manufacturing more expensive than the dry method.

Air Pollution Characterization

Both methods of manufacturing employ explosive powder as feedstock. Only a very small amount of dust is generated during drying, and it is contained within the dryer system. Any dust generated during the braiding or spinning process is process captive within the building.

Air Pollution Control Measures

There are no fugitive or point source emissions from this manufacturing process, and, therefore, no pollution controls are employed.

INITIATING EXPLOSIVES

Process Description

Initiating explosives, as the name implies, are used to provide a localized point-source impulse to trigger the detonation of other explosives. An example of this function of an initiating explosive is found in a detonator (blasting cap). The principal initiating explosives used commercially are lead azide, diazodinitrophenol (DDNP), and hexanitromanite (HNM). Pentaerythritol tetranitrate, the explosive of choice for use in detonating cord, is also an initiating explosive.

Lead azide, $Pb(N_3)_2$, is synthesized by combining aqueous solutions of sodium azide and lead nitrate (or lead acetate). Lead azide precipitates as a which solid from the combined solutions. The liquid is filtered away from the crystalline product, and the lead azide is washed with water. The process uses no solvents, nor are there any gaseous by-products of the reaction. There are no fugitive airborne emissions from this process. The wash water is collected, treated, and released.

Pentaerythritol tetranitrate (PETN) is manufactured by either a batch or a continuous nitration process. Pentaerythritol, $C(CH_2OH)_4$, a white, granular solid, is added to concentrated (98%) nitric acid. The nitration conversion to the tetranitrate is rapid. This produces a crude PETN that is separated from the spent nitric acid by filtration while being washed with water. The crude PETN can be used in its present state or it can be recrystallized further after being dissolved in acetone in the next downstream process.

Air Emissions Characterization

The principal airborne process-related fugitive emissions are oxides of nitrogen (NO_x) and acetone vapor. Sources of NO_x fumes include nitration reactors, drown tanks (specific to batch nitrators), spent acid filters, spent acid storage tanks, and strong nitric (98%) storage tanks. Wet scrubbing systems are employed to capture NO_x vapors from these various sources. Many scrubber designs can be used to accomplish this goal. The difficulty in the removal of oxides of nitrogen is a combination of low solubility of nitric oxide and the conversion of nitrogen dioxide (NO_2) into nitric oxide (NO). When NO_2 is absorbed, it converts to nitrous acid (HNO_2) and NO. As a result, without any other consideration, one-third of the NO_2 is lost as NO because of the low solubility of NO. One chemical that will interfere with this quick conversion is sodium sulfide, and it is utilized in some scrubbers. Here, the NO is converted to NO_2 in the first of two towers. The NO_2 is reduced in tower 2 by the presence of sodium sulfide to nitrogen gas (N_2). The reaction of sodium sulfide is fast enough to interfere with the normal conversion of part of the NO_2 to NO.

As noted above, two different sets of reactions take place in the scrubber columns. The following equations show which reactions take place in each tower.

Tower 1

$$2NO + NaClO_2 \rightarrow 2NO_2 + NaCl$$

Tower 1

$$3NO_2 + H_2O \rightarrow 2HNO_3 + NO$$

Towers 1 and 2

$$HNO_3 + NaOH \rightarrow NaNO_3 + H_2O$$

Towers 1 and 2

$$2NO_2 + 2Na_2S + H_2O \rightarrow N_2 + Na_2S_2O + 2NaOH$$

Tower 2

$$2NO_2 + Na_2S_2O + 2NaOH \rightarrow N_2 + 2Na_2SO_4 + H_2O$$

The scrubber chemicals are used in two packed towers in series. The exact liquid rates and the residence time help to ensure that the NO_x removal is accomplished. The acetone vapors generated at the dissolvers, crystallizers, and final filtration locations are commonly recondensed by shell-and-tube condensers using chilled (40°F) water as the condensing medium.

Nitromannite (HNM) is produced by the mixed-acid (sulfuric and nitric) nitration of mannitol ($C_6H_{14}O_6$). The powdered solid mannitol is fed into agitated mixed acid in a nitrator. After the nitration phase is completed, the liquid mixture, composed essentially of suspended HNM and spent acids, is discharged into drowning tanks containing water. The resultant suspension is then separated in a centrifuge. The retained solid is washed to remove residual surface acids. Solid HNM is then dissolved in acetone, at which the point the residual acid is neutralized. The solution separates into layers, and the water and acid are drawn off; the acetone layer is diluted with water in a continuous precipitator to form a slurry mixture and then filtered and washed. Both the acetone–water mixture and the filtrate wash water are collected for acetone recovery by distillation.

Diazodinitrophenol (DDNP) is produced by the diazotization of picramic acid. Sodium picramate and a DDNP seed mixture are reacted with sodium nitrate and nitric acid over a 4-h period in a cooled agitated reactor. After brief settling, the liquor and the very fine waste DDNP are pumped off, leaving the product in the reactor. Wash water is added, the DDNP is resuspended with agitation, and the fine-grain-size DDNP is washed and pumped to a vacuum filter. The coarse fraction of DDNP is again suspended in water and filtered. The liquor, waste DDNP fines, DDNP wash water, and building wastewater are discharged into a digestion tank. This waste is treated with hydrogen peroxide, ferrous sulfate, and sodium carbonate to oxidize all phenol compounds to water and carbon dioxide. The by-product fugitive emissions are water and carbon dioxide, and, therefore, no controls are required for entrapment or recovery of gases or vapors.

Air Pollution Control Measures

The air pollution controls for the initiating explosives are outlined in the process description or the "Air Emission Characterization" section.

BLACK POWDER

Process Description

Black powder is a deflagrating or low explosive of an intimate mixture of potassium (or sodium) nitrate, charcoal, and sulfur. The raw materials are stored in either bins or containers prior to usage. The potassium or sodium nitrate is crushed and batch weighted to formulation. The sulfur and charcoal are batch weighted by formula and pulverized to a very fine powder. These ingredients are remotely incorporated, plasticized with a small amount of water, and mechanically kneaded to produce a relatively loose, low-density, and moist mass known as "wheel cake." The wheel cake is pressed to the desired density and then chipped into the maximum dimension of choice for the grade to be produced. The chips are granulated through a system of crushing rollers and screens. The rate of feed, the spacing of the rolls, and then screen mesh produce the desired sizes of "green grains." The rough edges of the green grains are rounded through the friction process within the glazing cycle. They are the dried and graphited to add polish and luster. Some black powder is not glazed by choice.

Air Emission Characterization

The production buildings are designed to contain most of the dust generated during the different stages of the process.

Air Pollution Control Measures

As there are minimal fugitive emissions from this manufacturing process, no air pollution controls are employed.

WATER GELS AND SLURRIES

Process Description

Water gels and slurries comprise a wide variety of explosive materials used for blasting. As manufactured, they have varying degrees of sensitivity proportion of ammonium nitrate, some of which is in a water solution. These products may be sensitized with metals such as aluminum, fuels, or molecular explosives. A metal cross-linker, typically an antimony or chromium compound, is responsible for controlling the viscosity and gelling consistency of the materials.

The process is usually started by making a heated solution of ammonium nitrate and water. This solution can then be further processed by either a continuous or batch method. Both systems use the same general methods. The solution is added to the process stream in measured amounts with fuel and sensitizer metals, if required. The addition of gelling agents, which are activated by the heat of the product, causes the material to form a gelled mass that can be from a loose to a very firm form. The consistency is chosen to match the method of packaging or bulk delivery desired.

Air Emission Characterization

The manufacturing process of water gels and slurries is a closed loop, and very few air emissions are generated. Liquid raw materials and the products are nonvolatile. Minimal fugitive air emissions result from handling of aluminum and dry oxidizer ingredients.

Air Pollution Control Measures

The fuel oils are stored in large above-ground storage tanks that usually have submerged fill pipes. The solid ammonium nitrate is usually stored in large hopper bins that have lids covering the openings.

SPECIALTY EXPLOSIVES

Process Description

Specialty explosives are those that are used in a specific application and usually are in small quantities. There are two general categories of these products: sheet explosives and oil-field specialty explosives. There may be other classifications of commercial explosives, but they would be so small that they are not categorized by the industry.

Sheet explosives' commercial applications lie principally in the area of shock-hardening of metals. These explosives are pliable, flat material most commonly manufactured in thin sheets, although they can also be made in various other shapes, such as cylindrical. The process is typically composed of three steps: mixing of components, homogenization, and extrusion into finished form.

The mixing operation combines the explosive (usually PETN or RDX) with a matrix material and plasticizer (nitrocellulose and acetyl tributyl citrate for PETN-based explosives) in an aqueous slurry in a mixer. The process water is then removed from the mixture, usually by a stationary vacuum filter.

The mixture generally is further homogenized in an intensive mixer, at which point residual process water from the mixing step is removed. The final step is processing the uniformly mixed explosive and matrix system through an extruder to form the final desired shapes.

Oil-field specialty explosives are a wide variety of products manufactured for commercial oil-well-completion applications. The preforating charges make up the principal product line, and they consist of a charge case, a main charge, a high-explosive primer, and a cone-shaped charge liner. The charge is usually a hollow, machined steel cup. It is the receptacle into which the explosive components are loaded. The most common explosives used are RDX, PETN and HMX.

A primer charge (a pure form of the main charge explosive) is first added to the 5% of the net explosive weight. The main charge is a pressed granular powder (usually RDX, PETN or HMX with a waxlike coating around the individual particles).

The conically shaped charge liner is a hollow cone previously formed by pressing a powdered copper and lead mixture that has been pressed into the explosive-loaded charge case.

Air Pollution Characterization

The materials utilized in the manufacturing process are combinations of dry powders. These raw materials remain process captive during the blending, loading and pressing operations and do not give rise to fugitive air emissions.

Air Pollution Control Measures

There are no fugitive emissions, and, therefore, no pollution controls are required.

BIBLIOGRAPHY

American National Standards Institute, *Safety Requirements for the Transportation, Storage, Handling and Use of Commercial Explosives and Blasting Agents in the Construction Industry*, Standard A10.7, American National Standards Institute, New York, 1970.

Atlas Powder Co., *Handbook of Electric Blasting*, 1985 ed., Atlas Powder Co., Dallas, 1985, 74 pp.

A. Bailey and S. G. Murray, *Explosives, Propellants, and Pyrotechnics*, Brassey's, United Kingdom, 1989.

D. G. Borg, R. F. Chiapetta, R. C. Morhard, et al., *Explosives and Rock Blasting*, Atlas Powder Co., Dallas, 1987, 662 pp.

G. H. Damon, C. M. Mason, N. E. Hanna, et al., *Safety Recommendations for Ammonium Nitrate-Based Blasting Agents*, Information Circular 8746, U.S. Bureau of Mines, Washington, DC.

Design Considerations for Toxic Chemicals and Explosives Facilities, R. A. Scott and L. J. Doemeny, Eds.; ACS Symposium Series #345, American Chemical Society, Washington, DC, 1987.

R. A. Dick, L. R. Fletcher, and D. V. D'Andrea, *Explosives and Blasting Procedures Manual*, Information Circular 8925, U.S. Bureau of Mines, Washington, DC, 1983.

R. A. Dick, *Factors in Selecting and Applying Commercial Explosives and Blasting Agents*, Information Circular 8405, U.S. Bureau of Mines, Washington, DC.

Encyclopedia of Explosives and Related Items, B. Federoff and S. M. Kaye, Eds.; PATR 2700, U.S. Army ARRADCOM, Dover, NJ, 1960–1983 (10 vols.).

Engineering Design Handbook: Explosives Series—Properties of Explosives of Military Interest, ACMP 706–177, U.S. Army Material Command, Washington DC, 1971.

Explosive Materials Regulations, Title 27, Code of Federal Regulations, Parts 55 and 181, U.S. Bureau of Alcohol, Tobacco and Firearms, Washington, DC.

Explosives Technologies International, *Blaster's Handbook*, 16th ed., Explosives Technologies International, Wilmington, DE, 1980.

H. D. Fair and R. F. Walker, *Energetic materials*, Plenum Press, New York, 1977.

S. Fordam, *High Explosives and Propellants*, Macmillan Co., New York, 1980.

C. E. Gregory, *Explosives for North American Engineers*, Trans Tech Publications, Rockport, MA, 1979.

Hazardous Materials Regulations, Title 49, Code of Federal Regulations, Part 174.16, U.S. Department of Transportation, Washington, DC.

C. H. Johanssen and P. A. Persson, *Detonics of High Explosives*, Acamedic Press, London 1970, 326 pp.

R. Meyer, *Explosives*, 3rd ed., Verlag Chemie, Weinheim/New York,m 1987.

T. Urbanski, *Chemistry and Technology of Explosives*, Macmillan Co., New York, 1984 (four vols.).

PHOSPHORIC ACID MANUFACTURING

Gordon F. Palm
(with update by Editor)

There are two basic types of processes for the production of phosphoric acid (H_3PO_4): wet processes and furnace (or thermal) processes. In 1987, 96% of the 9.4 million megagrams (Mg) of phosphoric acid produced utilized the wet process for fertilizer production, and this trend is likely to continue in the future. The other 4%, produced by the furnace processes, include the blast furnace process and the electric furnace process. The blast furnace is no longer used; however, the electric furnace process is used extensively to make elemental phosphorus, most of which is converted to phosphoric acid for nonfertilizer uses. It is unlikely that the electric furnace process will become competitive for producing phosphoric acid for fertilizer use.

Wet processes may be classified according to the acid used to decompose the rock.[1] Sulfuric, nitric, and hydrochloric acid have been used in commercial processes. However, processes involving nitric acid are no longer used in the United States, and those using hydrochloric acid are not competitive for fertilizer production. Therefore, the discussion will be limited to the sulfuric acid processes.

PROCESS DESCRIPTION

Wet-Process Phosphoric Acid

Wet-process phosphoric acid is produced, as shown in Figure 1, by digesting ground phosphate rock, a fluorapatite, with an amount of sulfuric acid approximately stoichiometrically equal to the calcium oxide in the rock.[2] About 97% of the sulfate is precipitated as calcium sulfate. Some 2% to 3% appears as unreacted sulfate (free acid) in the phosphoric acid. For efficient operation the free acid content must be closely controlled. The concentration of phosphorus pentoxide (P_2O_5) in the product and the solids concentration in the reaction slurry are both controlled by the quantity and concentration of recycle acid returned from the filter. Gypsum from the filter is slurried with water and pumped to a gypsum stack for storage. The water is recycled through a surge-cooling pond to the phosphoric acid process.

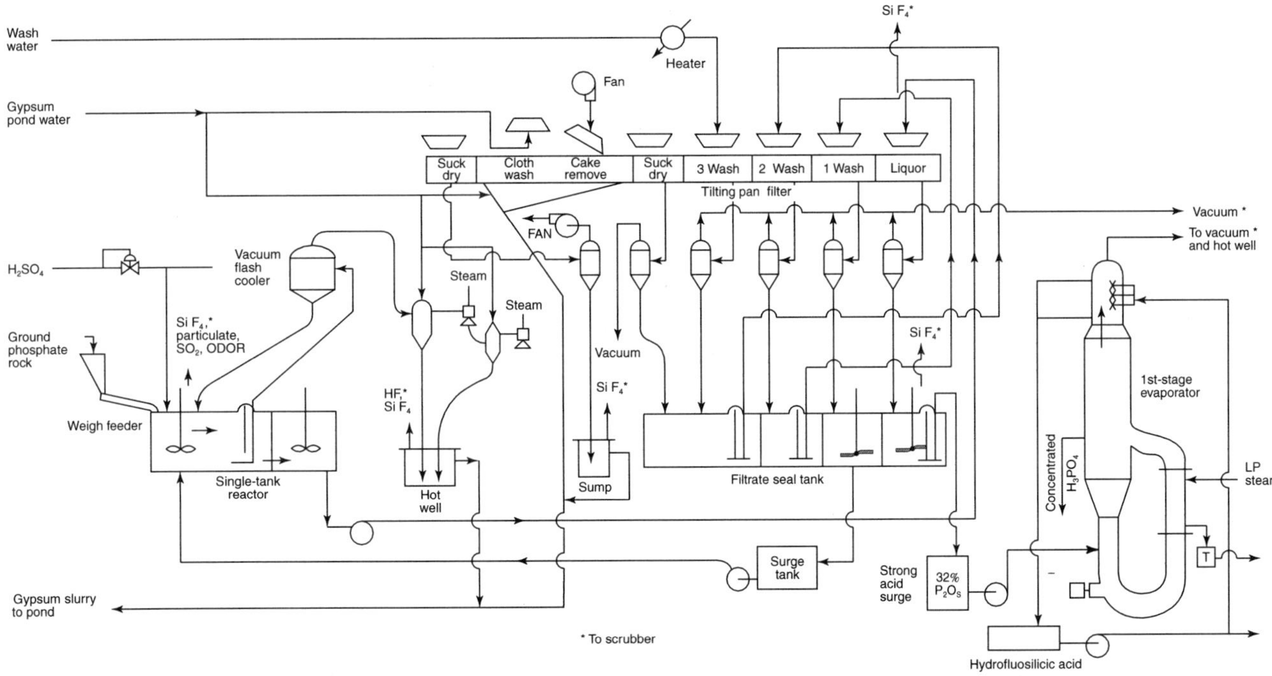

Figure 1. Diagram of Wet Process Phosphoric Acid Plant (AP-42, *Compilation of Air Pollution Factors*, U.S. Environmental Protection Agency, revised February 1980, p. 5, 41-1).

Table 1. Commercial Wet Process in Use

Crystal Form(s)	Number of Separation Steps[a]	Usual Concentration of Acid (% P_2O_5)	Usual Temperature (°C) Reactor	Usual Temperature (°C) Recrystallizer
Dihydrate	1	26–32	70–85	
Hemihydrate	1	40–50	85–100	
Hemihydrate–dihydrate	1	26–30	90–100	50–60
Hemihydrate–dihydrate	2	40–50	90–100	50–65
Dihydrate–hemihydrate	2	35–38	65–70	90–100

[a] Filtration of centrifuging steps.

Commercial wet processes may be classified according to the hydrate form in which the calcium sulfate crystallizes.[3]

Anhydrite: $CaSO_4$

Hemihydrate: $CaSO_4 \cdot \frac{1}{2}H_2O$

Dihydrate: $CaSO_4 \cdot 2H_2O$

At present there is no commercial use of the anhydrite process, mainly because the required reaction temperature is high enough to cause severe corrosion difficulties. Processes in commercial use are listed in Table 1.

Straight dihydrate processes are the most popular today. Recently, some have been converted to a one-step hemihydrate process that has the advantage of producing phosphoric acid of a relatively high concentration and with lower impurity levels. Hemihydrate–dihydrate processes without intermediate filtration and hemihydrate–dihydrate processes with two separation steps are not used in the United States.

Furnace (or Thermal) Process

The raw materials used in the thermal process for production of phosphoric acid are elemental (yellow) phosphorus, air, and water, and the process includes three major steps: (1) combustion, (2) hydration, and (3) demisting.

In the combustion process, the elemental phophorus is burned in ambient air in a chamber at temperatures of 1650–2760°C (3000–5000°F) to form phophorus pentoxide as follows:

$$P_4 + 5O_2 \rightarrow 2P_2O_5 \quad (1)$$

The phosphorus pentoxide is then hydrated to produce strong phosphoric acid:

$$2P_2O_5 + 6H_2O \rightarrow 4H_3PO_4 \quad (2)$$

The phosphoric acid mist is then removed from the combustion gas stream before the gases are released to the atmosphere using high-efficiency/pressure-drop demisters. The concentration of phophoric acid normally ranges from 75 to 85% and is required for high grade chemical production and other nonfertilizer uses.

CHEMISTRY

In the wet process, digestion is considered to take place in several stages.[2] First, tricalcium phosphate in the rock is attacked by phosphoric acid to form monocalcium phosphate. The monocalcium phosphate then reacts with sulfuric acid to yield additional phosphoric acid and gypsum. The overall reaction is as follows:

$$Ca_{10}(PO_4)_6F_2CaCO_3 + 11H_2SO_4 \rightarrow 6H_3PO_4 + 11CaSO_4 \cdot nH_2O + 2HF + CO_2 + H_2O \quad (3)$$

where fluoroapatite and sulfuric acid yield phosphoric acid, gypsum and hydrogen fluoride, carbon dioxide, and water. Simplified versions of these reactions are as follows:

$$Ca_2(PO_4)_2 + 4H_3PO_4 \rightarrow 3CaH_4(PO_4)_2 \quad (4)$$

$$3CaH_4(PO_4)_2 + 3H_2SO_4 + 6H_2O \rightarrow 3CaSO_4 \cdot 2H_2O + 6H_3PO_4 \quad (5)$$

Depending on the temperature and phosphoric acid strength maintained during digestion, dihydrate, hemihydrate, or even the anhydrite of calcium sulfate may result.[3]

A number of minor reactions occur between fluorides, silica, and water in the scrubbing of gaseous fluorides from the digester. These reactions are discussed under "Air Pollution Control Measures."

AIR EMISSION CHARACTERISTICS

Wet-Process Fluorides

In the digester, about 3–7% of the fluoride in the phosphate rock is evolved.[2,4,5] At the higher P_2O_5 concentrations and temperature of the hemihydrate and hemihydrate–dihydrate processes, more silicofluoride fumes are emitted from the attack tank than in the dihydrate process.[5] The digesters are vented to wet scrubbers for fluoride removal.

The relative quantities of fluorides in the filter acid and gypsum depend on the type of rock and the operating conditions.[6] The main factors appear to be the following:

1. The quantity of sodium or potassium salts present, which will precipitate insoluble fluorine compounds
2. The reaction temperature at which digestion occurs (increasing temperature results in increasing gaseous fluorides)
3. The concentration of the product phosphoric acid

Filter acid contains about 67% of the fluoride in the phosphate rock,[7] much of which is volatilized during evaporation. The gaseous fluorides are removed by scrubbing to prevent pollution. In some cases, saleable products are produced[8] (fluosilicic acid, fluosilicates, cryolite, or aluminum fluoride). The amount of fluoride evolved during concentration from 30% to 54% P_2O_5 may be 70–80%[8] of that originally present in the acid. This is equivalent to 50% of the fluoride in the phosphate rock. In the production of superphosphoric acid (69–72% P_2O_5) by concentrating wet process acid, most of the fluoride is volatilized[8] so the acid contains only 0.2–0.3% as F. The addition of reactive silica during evaporation enhances silicon tetrafluoride volatization and reduces fluoride content to about 0.1% as F.

The evaporators used to concentrate phosphoric acid operate under a vacuum. The steam from the evaporated acid and the evolved fluorides goes to a total condenser, where they are directly contacted with cool process water. The steam is condensed and the fluorides are absorbed in a large quantity of process water. Gaseous fluoride emissions from the condenser waters are negligible.

Wet process phosphoric acid[9] is defluorinated for use in animal feed products. The stripping of the fluoride from the acid is more easily attained when the fluoride is present in the silicon tetrafluoride form,[10] which is more volatile than hydrogen fluoride. Silicon dioxide is added as needed to the phosphoric acid to allow the chemical reactions necessary for producing silicon tetrafluoride. The various methods of defluorinating are as follows:

1. Air stripping by blowing air through hot 54% P_2O_5 acid (the exit air must be scrubbed)
2. Steam stripping 54% P_2O_5 phosphoric acid by circulating through a stripping column with steam blown countercurrent to the acid (exit steam must be condensed to remove fluorides)
3. Evaporation method by diluting 54% P_2O_5 acid and then reconcentrating the acid to 54%; uses conventional vacuum evaporation equipment
4. Concentrating to superacid under vacuum will give a P:F (weight ratio) of 80 at 70% P_2O_5 acid. Reactive silica, such as diatomaceous earth, when added to the acid, allows it to be stripped of fluorides in the evaporator to a P:F ratio of 100 or higher.

Acid used in animal feed products must have a minimum P:F of 100. If there is fluoride in the limestone used for feed-grade dicalcium phosphate, the acid P:F must be about 125. Fifty-four percent P_2O_5 acid has a P:F of about 15; 30% P_2O_5 acid is hard to defluorinate.

Wet-Process Particulates

In the older plants, dry ground phosphate rock is generally used. It is transported to the phosphoric acid plant digester by conveyor belt, screw conveyor, and/or airveying systems. The handling of the dry ground rock results in the generation of fugitive dust, which is collected by ventilation ductwork and removed by bag collectors, which discharge the dust to the rock feed bin. In addition to the gaseous fluoride in the digester vent, there is usually a small amount of rock dust generated by the mechanical handling of the rock.[11] This dust is removed in the wet scrubber designed for the removal of fluorides.

In newer plants, there are no particulate emissions, since wet rock grinding is used.[12] The rock is ground as a slurry in a ball mill to about 65% solids. The advantages of wet rock grinding are a 30–40% reduction in horsepower in the grinding area, elimination of the rock dryer, elimination of atmospheric pollution by dust, and a savings in the cost of fuel for drying. Many older plants have been converted to wet rock grinding to realize these significant environmental and cost benefits.

Gypsum Stacks and Process Water Ponds

The gypsum stacks and process water ponds are an integral part of a typical phosphoric acid plant.[13] The gypsum stacks serve primarily as storage for stacking of the by-product gypsum. Adjoining these stacks are process water ponds that provide surge volume for wet and dry weather conditions to avoid water treatment and water discharge and to conserve on fresh water makeup. The water in the ponds, commonly called "pond water," is recirculated for process cooling, recovery of P_2O_5 and sulfate values, control of effluents, scrubbing of process emissions, process makeup water, and hydraulic transport of by-product gypsum. The tops of the gypsum stacks are frequently used for supplemental cooling of the pond water. During the manufacturing process, the pond water comes in direct contact with raw materials, intermediate products, by-products, waste products, or finished products. The cooling ponds, including the cooling water on top of the gypsum stacks, also serve as surge equalization ponds for the rainfall runoff from the gypsum stacks.

Pond water is used mainly in the phosphoric acid processes to wash the filter cake, in scrubbers, in barometric condensers, and for slurrying waste gypsum. After settling of gypsum solids in the gypsum stack, the pond water is allowed to cool by natural evaporation and is recirculated. Pond water is acidic, has a pH of about 1.5–2.0, and contains weak sulfuric, phosphoric, and fluosilicic anion components. Phosphate analysis, reported as P_2O_5, varies from about 1% to 2% P_2O_5. Fluoride analysis, reported as percent of fluorine, varies from about 1% to 1.8% F.

Hydrogen fluoride (HF) and silicon fluoride (SiF_4) are prevented from escaping from the facility into the atmosphere by the use of wet scrubbers. These recycle pond water as the scrubbing medium.[13] In the scrubbers, HF and SiF_4 are removed from the air, forming H_2SiF_6.

Low-level fluoride emissions occur from the pond water on top of the gypsum stack and pond water cooling ponds. Measurements of these levels are extremely difficult to correlate because of changing air velocity and direction and the concentration variation of the fluorides along the elevation over the pond water surface. Estimations based on fluoride vapor pressure are also difficult to make because of the wide variations in pond water composition and the lack of available data.[14]

The U.S. Environmental Protection Agency, in discussing the verification of predicted fluoride emissions, made the following statement[13]: "Based on our findings concerning the emissions of fluoride from gypsum ponds, it was concluded that no investigator had as yet established the fluoride emission rate from gypsum ponds."

Cooling towers are mentioned for cooling condenser pond water[15] rather than a pond water cooling pond. However, if a cooling tower is used, surge equalization provisions must also be considered to avoid excessive treatment and the discharge of treated water. Cooling towers are used for cooling gypsum pond water in four U.S. locations instead of pond water cooling ponds. Fluoride emissions from the cooling towers are a function of the fluoride vapor pressure of the pond water. The fluoride emissions are low but must meet limits of the state regulatory authorities.

Thermal Process Emissions

The major source of emissions from the thermal process are the phoshoric acid mist emissions contained in the gas stream from the hydrator with particle sizes ranging from 1.4 to 2.6 micrometers. It is not uncommon that as much as 50% of the total P_2O_5 is present as acid mist and must be captured, both from an air emissions standpoint as well as from an economic product recovery standpoint. Control equipment commonly includes venturi scrubbers, cyclonic separators with wire mesh mist eliminators, fiber mist eliminators, high-energy wire mesh mist eliminators, and wet electrostatic precipitators.[19]

EMISSION FACTORS

Emission factors for wet phosphoric acid production and for thermal process production are shown in Table 2. These values were obtained from appropriate tables found in the U.S. EPA AP-42 dated July 1993.[19] In addition, the reader is referred to the last section, which discusses the new NESHAP for phosphoric acid production, which will result in changes in the emission factors in the future as existing and new plants are required to meet the NESHAP.

AIR POLLUTION CONTROL MEASURES

Wet-Process Fluoride Emissions Control

Gaseous fluoride compounds are removed from the vent air of the digester, filter, and various tanks. Wet scrubbers have been used exclusively for this service in the United States.

Wet scrubbing combines the ability to remove particulates (rock dust) from gas streams by impaction of the particulates on the surface of liquid droplets and the ability to absorb gaseous constituents into the liquid phase.[16] Both of these functions are limited by the characteristics of the scrubbing liquor, the properties of the materials to be removed, and sometimes by the two in combination. In the phosphoric acid plant wet scrubbers, the pond water and the fluoride-containing gases from the wet-process phosphoric acid plant reactor can produce a gelatinous silica precipitate. This plugs the packing in the scrubbers and limits the types of scrubbing equipment that can be used.

The basic chemistry of the compounds of fluoride, silica, and water must be considered to characterize the application. In the reactor, the fluoride contained in the fluorapatite or fluorspar goes into solution according to the following reactions[16]:

$$\underset{\text{calcium fluoride}}{CaF_2} + \underset{\text{sulfuric acid}}{H_2SO_4} \rightleftharpoons \underset{\text{calcium sulfate}}{CaSO_4} + \underset{\text{hydrogen fluoride}}{2HF} \tag{6}$$

or

$$\underset{\text{hydrogen fluoride}}{2HF} + \underset{\text{silicon tetrafluoride}}{SiF_4} \rightleftharpoons \underset{\text{fluosilicic acid}}{H_2SiF_6} \tag{7}$$

A vapor equilibrium is set up between the reactants in Eq. 7, which may be regarded as the reverse of the following equation:

$$\underset{\text{fluosilicic acid}}{H_2SiF_6} \rightleftharpoons \underset{\text{hydrogen fluoride}}{2HF} + \underset{\text{silicon tetrafluoride}}{SiF_4} \tag{8}$$

High temperature drives the reaction to the right, increasing the vapor pressure of both the HF and SiF_4, and increases the relative significance of SiF_4 as the fluorine-containing species.[16] These vapor pressures set the lower limit of fluoride concentration in the gas phase leaving the scrubber.

In addition to the reactions given, hydrolysis of SiF_4 occurs when the concentration of this component is higher than the

Table 2. Controlled Emission Factors for Wet Phosphoric Acid Production[a] Emission Factor Rating: B (except as noted) (Emission Factors—Metric and English Units

Source	Fluoride kg/Mg P_2O_5 Produced	Fluoride lb/ton P_2O_5 Produced
Reactor[b] (SCC 3-01-016-01)	1.9×10^{-3}	3.8×10^{-3}
Evaporator[c] (SCC 3-01-016-99)	0.022×10^{-3}	0.044×10^{-3}
Belt filter (SCC 3-01-016-99)	0.32×10^{-3}	0.64×10^{-3}
Belt Filter vacuum pump (SCC 3-01-016-99)	0.073×10^{-3}	0.15×10^{-3}
Gypsum settling and cooling ponds[c,d] (SCC 3-01-016-02)	Site-specific	Site-specific

Uncontrolled Emission Factors for Wet Phosphoric Acid Production[a]
Emission Factor Rating: C (except as noted)

Source	Nominal Percent Control Efficiency	Fluoride kg/Mg P_2O_5 Produced	Fluoride lb/ton P_2O_5 Produced
Reactor[e] (SCC 3-01-016-01)	99	0.19	0.38
Evaporator[c] (SCC 3-01-016-99)	99	0.00217	0.0044
Belt filter (SCC 3-01-016-99)	99	0.032	0.064
Belt filter vacuum pump (SCC 3-01-016-99)	99	0.0073	0.015
Gypsum settling and cooling ponds[c,d] (SCC 3-01-016-02)	ND	Site-specific	Site-specific

Controlled Emission Factors for Thermal Phosphoric Acid Production[a]
Emission Factor Rating: E

Source	Nominal Percent Control Efficiency	Fluoride kg/Mg P_2O_5 Produced	Fluoride lb/ton P_2O_5 Produced
Packed tower (SCC 3-01-017-03)	95.5	1.07	2.14
Venturi scrubber (SCC 3-01-017-04)	97.5	1.27	2.53
Glass fiber mist eliminator (SCC 3-01-017-05)	96–99.9	0.35	0.69
Wire mesh mist eliminator (SCC 3-01-017-06)	95	2.73	5.46
High-pressure drop mist (SCC 3-01-017-07)	99.9	0.06	0.11
Electrostatic precipitator (SCC 3-01-017-08)	98–99	0.83	1.66

[a] SCC, source classification code.
[b] Emission Factor Rating: A
[c] Site-specific. Acres of cooling pond required: ranges from 0.04 hectare per daily Mg (0.10 acre per daily ton) P_2O_5 produced in the summer in the southeastern U.S. to 0 in the colder locations in the winter months when the cooling ponds are frozen.
[d] No investigator had as yet established experimentally the fluoride emission from gypsum ponds.
[e] Emission factor rating: B

equilibrium values, according to the following reaction.

$$\underset{\text{silicon tetrafluoride}}{3SiF_4} + \underset{\text{water}}{4H_2O} \rightleftharpoons \underset{\text{silicic acid}}{Si(OH)_4} + \underset{\text{fluosilicic acid}}{2H_2SiF_6} \quad (9)$$

Reaction 9 occurs as the temperature of a gas stream is reduced in the presence of water and leads to the formation of gelatinous deposits of polymeric silica, which plug scrubber packings. This problem limits the use of conventional packed countercurrent absorbers in this service, as well as other contacting devices that have small gas passages that might plug up.

Transfer Unit Concept. Removal of fluorine compounds from effluent gases is achieved by absorption in water. With the decrease of allowable emissions in the past few years, scrubber efficiencies in the range of 99+% are required.[17] Scrubber emissions are now expressed as pounds per hour of fluorides emitted. If once-through or neutralized water can be used, at least 4.6 transfer units are required to achieve this efficiency. If recycled pond water is used, the number of transfer units required for the same efficiency has been as many as eight units in several installations. This depends on the partial pressure of fluorides in the incoming water. The number of transfer units required is established for these lean systems as shown in the following.

It is first assumed that enough absorbing liquid is used to prevent the concentration of fluoride in the liquid from varying significantly as a result of the absorption. Then,[17]

$$N_{og} = \ln[(y_1 - a)/(y_2 - a)] \quad (10)$$

where N_{og} is the number of transfer units required; y_1, the inlet concentration of fluorine-containing compound in gas phase; y_2, the outlet concentration of fluorine-containing compound in gas phase; and a, the concentration of fluorine-containing compounds in gas phase in equilibrium with scrubbing liquid.

All concentrations are expressed in the same units, which may be volume percent, mole percent, or partial pressure. For very dilute systems, expressing concentrations in weight percent or weight per unit volume loses little accuracy.

In those cases where the vapor pressure of fluorides in the scrubbing liquid, inlet and outlet, is negligible, then

$$N_{og} = \ln(y_1/y_2) \tag{11}$$

Scrubbers. Scrubbing systems used in phosphoric acid plants are venturis, wet cyclonic, and semi-cross-flow[18] scrubbers. Figure 2 is an artist's sketch of a cyclonic scrubber with outside wall-mounted spray boxes. This design feature allows the spray nozzles, which are mounted in a number of boxes, to be cleaned or replaced while the scrubber is in operation. Since the nozzles tend to plug with silica, this feature improves the operating time of the scrubber. Figure 3 shows a two-stage cyclonic scrubber recently installed on a phosphoric acid reactor.

In the current design of semi-cross-flow scrubbers, pond water is sprayed countercurrent to the gas stream in several rows of spray headers in the front of the scrubber. Then spray headers are used to spray concurrent with the gas stream onto two or more Kimre tower packing sections (multiple woven pads) in series followed by a dry Kimre tower packing for drift elimination. Several older spray cross-flow packed scrubbers have been modified because of plugging of the initial type of tower packing used. In some plants, the packing was removed and the scrubber converted by installing Kimre tower packing. In other plants, the Kimre tower packing was installed prior to the initial tower packing and a dry Kimre tower packing installed before the fan suction for drift elimination. Figure 4 presents a sketch of a phosphoric acid scrubber showing these changes.

Figure 3. Two-Stage Cyclonic Scrubber Installed on a Wet Process Phosphoric Acid Reactor (Hydronics Engineering Corp., Midland Park, NJ).

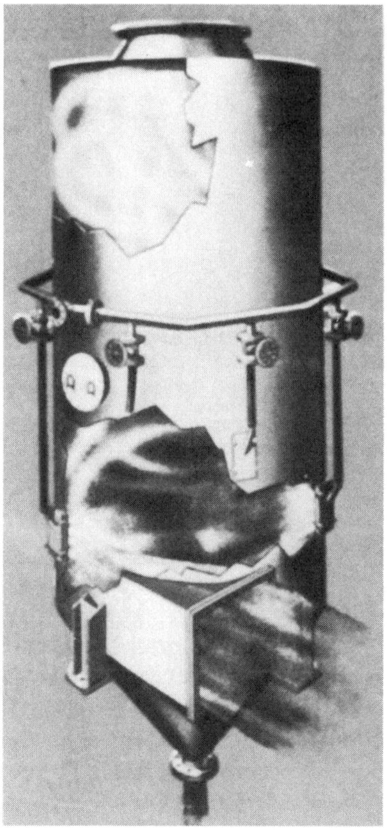

Figure 2. Artist's Sketch of Cyclonic Scrubber with Wall-Mounted Spray Boxes Showing Flows of Internal Gases and Scrubbing Liquid (used with permission Hydronics Engineering Corp., Midland Park, NJ).

NESHAP REGULATIONS (PROMULGATED IN 1999)

A NESHAP for new and existing major sources in phosphoric acid manufacturing and phosphate fertilizers production plants was proposed in the *Federal Register* on December 27, 1996 (61 FR 68430). EPA's final decision on the rule was signed in April 1999. This action promulgated national emission standards for hazardous air pollutants (NESHAP) for new and existing major sources in phosphoric acid manufacturing and phosphate fertilizers production plants (SIC 2874). Hazardous air pollutants (HAPs) emitted by the facilities covered by this rule include hydrogen fluoride (HF); arsenic, beryllium, cadmium, chromium, manganese, mercury, and nickel (HAP metals); and methyl isobutyl ketone (MIBK) in the case of one plant where purified phosphoric acid was being manufactured. Existing sources must comply with the NESHAP within 3 years after publication of the final rule. Sources constructed after December 27, 1996, were required to meet the standard as of the publication of the final rule. The full text of the ruling can be downloaded from

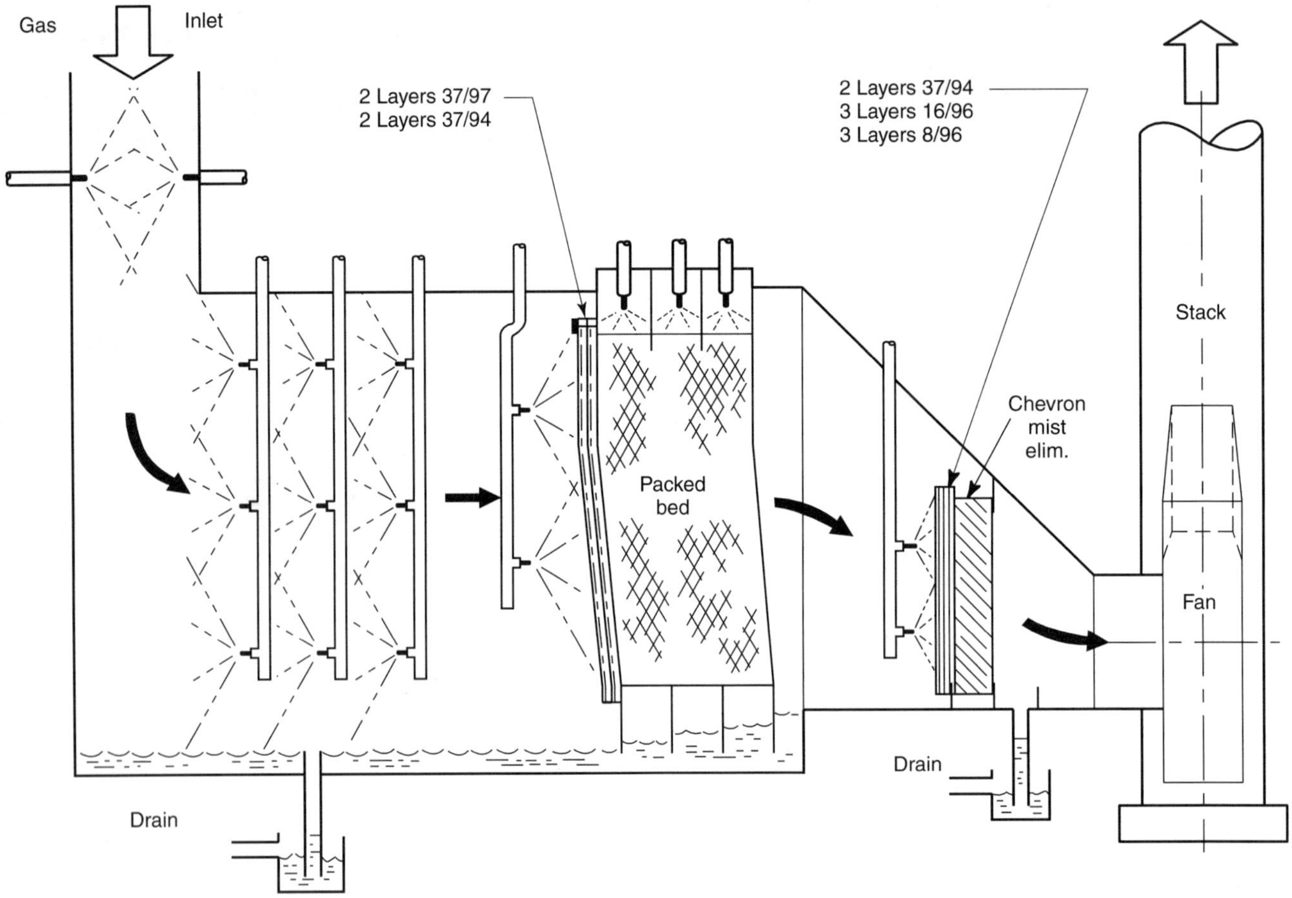

Figure 4. Phosphoric Acid Fluoride Scrubber after Changes (Kimre, Inc., Perrine, FL).

Table 3. Emissions Limitations for Existing Phosphoric Acid Manufacturing Plants and Phosphate Fertilizers Plants

Class of Source	Pollutant	Emission Limit
Wet-process phosphoric acid process line	Total fluorides	0.020 lb. total fluoride (F^-) per ton P_2O_5 feed
Superphosphoric acid process line	Total fluorides	0.010 lb. F^- per ton P_2O_5 feed
Diammonium and/or monoammonium phosphate process line	Total fluorides	0.060 lb. F^- per ton P_2O_5 feed
Granular triple superphosphate process line	Total fluorides	0.150 lb. F^- per Ton P_2O_5 Feed
Granular triple superphosphate storage buildings	Total fluorides	5.0×10^{-4} lb. F^- per hour per ton of P_2O_5 stored
Phosphate rock dryers	Particulate matter	0.2150 lb. PM per ton of rock feed
Phosphate rock calciners	Particulate matter	0.060 grains PM per dry standard cubic foot
Purified phosphoric acid process line	MIBK	Implement Part 63, Subpart H, Leak Detection and Repair Program

http://www.epa.gov/ttn/oarpg/t3/fr_notices/pfertfr.wpd, which is on the EPA TTN bulletin board or by searching on "phosphoric acid" on the TTN in the case that the URL has changed.

The NESHAP emissions limits for existing and new sources are given in Tables 3 and 4, respectively. An annual performance test is required to demonstrate compliance with each applicable numerical limit for total fluorides or particulate

Table 4. Emissions Limitations for New Phosphoric Acid Manufacturing Plants and Phosphoric Fertilizers Plants

Class of Source	Pollutant	Emission Limit
Wet-process phosphoric acid process line	Total Fluorides	0.01350 lb. total fluoride (F^-) per ton P_2O_5 feed
Superphosphoric acid process line	Total Fluorides	0.00870 lb. F^- per ton P_2O_5 feed
Diammonium and/or monoammonium phosphate process line	Total Fluorides	0.0580 lb. F^- per ton P_2O_5 feed
Granular triple superphosphate process line	Total Fluorides	0.1230 lb. F^- per ton P_2O_5 feed
Granular triple superphosphate storage buildings	Total Fluorides	5.0×10^{-4} lb. F^- per hour per ton of P_2O_5 stored
Phosphate rock dryers	Particulate Matter	0.060 lb. PM per ton of rock feed
Phosphate rock calciners	Particulate matter	0.040 grains PM per dry standard cubic foot
Purified phosphoric acid process line	MIBK	Implement Part 63, Subpart H, Leak Detection and Repair Program

matter. The monitoring provisions require an owner or operator using a wet scrubbing device to continuously monitor the pressure drop and liquid flow rate of scrubbing devices used to control total fluorides or particulate matter. The feed rate of raw materials to the processes must also be continuously monitored. During the performance test, the owner/operator must record the scrubber pressure drop and liquid flow rate to establish baseline levels. Following the performance test, it is an operating requirement that the owner/operator must maintain scrubber pressure drop and liquid flow rate within ±20% of the values recorded during the performance test or within a range established upon the basis of prior successful tests. Any exceedance of the operating range averaged over 24 h is a violation of the operating requirement.

For PPA plants that emit MIBK, the standards require implementation of a leak detection and repair program, continuous monitoring of chiller stack temperature and daily monitoring of MIBK concentrations at two points in the process.

As required by the NESHAP General Provisions (40 CFR part 63, subpart A), the owner or operator must develop and implement a startup, shutdown, and malfunction plan. Most notification, recordkeeping, and reporting requirements in the general provisions apply to phosphoric acid manufacturing and phosphate fertilizers production facilities. These include but are not limited to: (1) initial notification(s) of applicability, notification of performance test, and notification of compliance status; (2) a report of performance test results; (3) a startup, shutdown, and malfunction plan with semiannual reports of reportable events (if they occur); and (4) semiannual reports of excess emissions. If excess emissions are reported, the owner or operator must report quarterly until a request to return the reporting frequency to semiannual is approved.

The NESHAP General Provisions also require that records be maintained for at least 5 years from the date of each record. The owner or operator must retain the records on site for at least 2 years but may retain the records offsite the remaining 3 years.

REFERENCES

1. International Fertilizer Development Center, *Fertilizer Manual*,Muscle Shoals, AL, 1981, p. 163.
2. A. V. Slack, *Phosphoric Acid*, Vol. 1, Marcel Dekker, New York, 1969, p. 183.
3. International Fertilizer Development Center, *Fertilizer Manual*, Muscle Shoals, AL, 1981, p. 164.
4. A. V. Slack, *Phosphoric Acid*, Vol. 1, Marcel Dekker, New York, 1969, p. 197.
5. J. Gobbitt, "Hemihydrate Phosphoric Acid Plant Retrofits at Geismar and Chinhae," AIChE Central Florida Section, Annual Clearwater Meeting, 1990, p. 5.
6. A. V. Slack, *Phosphoric Acid*, Vol. 2, Marcel Dekker, New York, 1969, p. 745.
7. A. V. Slack, *Phosphoric Acid*, Vol. 2, Marcel Dekker, New York, 1969, p. 765.
8. International Fertilizer Development Center, *Fertilizer Manual*, Muscle Shoals, AL, 1981, pp. 170–171.
9. W. E. Rushton, "Defluorination of Wet Process Acid," *Chem. Bibliography Eng. Prog.*, 52 (November 1978).
10. A. C. Perkins, "Acid Defluorination for Animal Feeds," AIChE Central Florida Section, Annual Clearwater Meeting, 1988, p. 2.
11. P. A. Boys, *Air Pollution Control Technology and Costs in Seven Selected Areas*, EPA-450/3-73-01 0, U.S. Environmental Protection Agency, Research Triangle Park, NC, 1973, p. 38.
12. S. V. Houghtaling, "Wet Grinding of Phosphate Rock Hold Down Dollars, Dust and Fuel," *Eng. Mining J.*, 94–96 (January 1975).
13. A. A. Linero and R. A. Baker, *Evaluation of Emissions and Control Techniques for Reducing Fluoride Emissions from Gypsum Ponds in the Phosphoric Acid Industry*, EPA-600/278-124, U.S. Environmental Protection Agency, Research Triangle Park, NC, 1978, pp. 1–219.
14. P. Beker, *Phosphates and Phosphoric Acid*, Marcel Dekker, New York, 1983, p. 486.
15. P. Tiberghien and F. L. Prado, "The Rhone-Poulenc Phosphoric Acid Process and the Environment," AIChE Central Florida Section Annual Clearwater Meeting, 1986, pp. 2–5.

16. P. A. Boys, *Air Pollution Control Technology and Costs in Seven Selected Areas*, EPA-450/3-73-0 10, U.S. Environmental Protection Agency, Research Triangle Park, NC, 1973, pp. 38–47.
17. A. V. Slack, *Phosphoric Acid*, Vol. 2, Marcel Dekker, New York, 1969, pp. 753–754.
18. G. C. Pederson, "Design Procedure for Fluorine Scrubbers in the Fertilizer Industry," AIChE Central Florida Section Meeting, September 8, 1987, p. 15.
19. *Phosphoric Acid*, U.S.EPA AP 42, 5th Edition, Chapter 8, Section 9, July 1993, *http://www.epa.gov/ttn/chief/ap42c8.html*.

PHTHALIC ANHYDRIDE

HERBERT P. DENGLER

The demand for phthalic anhydride (PAN) in the United States hovered around a billion pounds per year in the decade of the 1980s. A record peak of 1050 million pounds per year was reached in 1988 compared with 950 million pounds per year in 1989. Demand growth rates in the early 1990s were expected to accelerate up to 2.5% per year. Over the past two decades, the number of producers has dwindled to six, all with capacities in the range of 150–250 million pounds a year. Total capacity was estimated at 1120 million pounds per year as of mid-1989.[1]

Current production is obtained by catalytic air oxidation in fixed-bed units. Five of the six U.S. plants use *o*-xylene as feedstock. The sixth uses naphthalene derived from coal tar for at least a portion of its feedstock. As of mid-1989, the price of *o*-xylene was about 21 cents per pound, and molten PAN was selling in the range of 40–44 cents per pound.[2] The price of flaked and bagged PAN in the United States is typically 4 cents per pound higher than for molten PAN.

The major end use for PAN is in the manufacture of plasticizers. For the last quarter of a century, domestic end uses in the United States have been 50% to plasticizers, 25% to polyester resins, 20% to alkyd resins, and the remaining 5% to a myriad of uses such as insecticides and dyestuffs. Exports have ranged to as high as 5% of the production in some years.

PROCESS DESCRIPTION

Commercially, PAN is made by the catalytic air oxidation of either *o*-xylene or naphthalene (Figure 1). *o*-Xylene is obtained from petroleum sources and typically is about 98% pure, with the major impurity being *p*-xylene. The naphthalene is obtained from coal tar sources, and the purity typically exceeds 95%.[3] The catalyst is a vanadium pentoxide–titanium dioxide mixture containing small quantities of promoters.[4] This active material is surface coated on a support that is inert at the operating temperature levels. The reaction is strongly exothermic. To facilitate heat removal in commercial plants, the reaction is carried out in nominal 1-in.-diameter vertical tubes in a reactor with circulating molten heat-transfer salt in the shell.[5]

The major reactions are as follows:

$$o\text{-xylene} + 3O_2 \rightarrow \text{PAN} + 3H_2O + \sim 265 \text{ kcal/g mol}$$

$$\text{naphthalene} + 4.5O_2 \rightarrow \text{PAN} + 2H_2O + 2CO_2 + \sim 427 \text{ kcal/g mol}$$

Temperatures of molten salt in the reactor are in the 630–730°F range. Process temperatures are limited to less than 900°F to avoid damage to the catalyst, especially because surface temperatures can be considerably higher than bulk temperatures.

The air oxidation process is carried out at pressures only slightly above atmospheric pressure. Process air is compressed to approximately 10 psig and preheated. Preheated *o*-xylene or naphthalene is mixed with the air in carefully controlled concentrations. Typically, the *o*-xylene concentration is in the range of 60–70 g of *o*-xylene feed per cubic meter of wet air and is lower for naphthalene. The mixture is less than 1.5 vol % *o*-xylene, but it is flammable. In the event of an ignition, the equipment is protected by rupture disks vented to the atmosphere. Because ignitions are very infrequent and the concentrations of hydrocarbons are low, atmospheric emissions from this source are insignificant.

The residence time in the reactor is less than 1 sec. Essentially complete conversion of the feed occurs. The molar selectivity for the conversion of *o*-xylene to PAN has improved steadily over the years and is now about 80% of theoretical, while that of naphthalene is in the 90–95% range. The corresponding weight yield for *o*-xylene is 1.1 pound of PAN per pound of 100% *o*-xylene. Naphthalene weight yields are about 10% lower. Theoretical weight yields are 1.4 for *o*-xylene and 1.16 for naphthalene (pounds of PAN per pound of 100% pure feedstock).

The effluent from the reactor is primarily nitrogen and oxygen. Reactor effluent oxygen concentrations are in the 12–15 vol % range because less than half of the oxygen in the air is consumed in the reaction. In descending order of volumetric concentration, the effluent contains nitrogen, oxygen, water, carbon dioxide, PAN, carbon monoxide, maleic anhydride, and other oxygenated by-products. The other oxygenated by-products are about 1 wt % of the hydrocarbon feed when using *o*-xylene and slightly higher when naphthalene is the feed, depending on feed purity.

The by-products with lower boiling points than that of PAN are primarily maleic anhydride and benzoic acid. With naphthalene feedstocks, naphthaquinone production can reach 0.1–0.5 wt % on feed. The naphthaquinone must be removed from the final product by distillation and/or chemical treatment to much lower levels than those of benzoic acid and maleic anhydride because it has a detrimental effect on product color. The by-products with higher boiling points include phthalide, trimellitic anhydride, and a variety of carboxylic acids, together with a small quantity of unidentified tars and carbon particles.

After exiting the reactor, the process stream is cooled and 99% of the PAN is removed as a solid by desublimation in switch condensers. These condensers alternately are switched onstream to a cooling/desublimation service and isolated from the process stream for heating up and melting out. During melting, the crude PAN is recovered as a liquid.

The crude PAN is then subjected to a heat treatment of 12–24 h at 450–550°F, which improves product quality. The pressure level ranges from about 10 mm Hg vacuum to slightly above atmospheric pressure. In the latter case, a vacuum jet would not be required. The pretreatment step is frequently called decomposition because one of the reactions that occurs is the decomposition of phthalic acid to PAN and water vapor. The phthalic acid concentration in the crude can be as high as 1 wt %, depending on the operating conditions in the switch condensers. Chemicals such as sodium

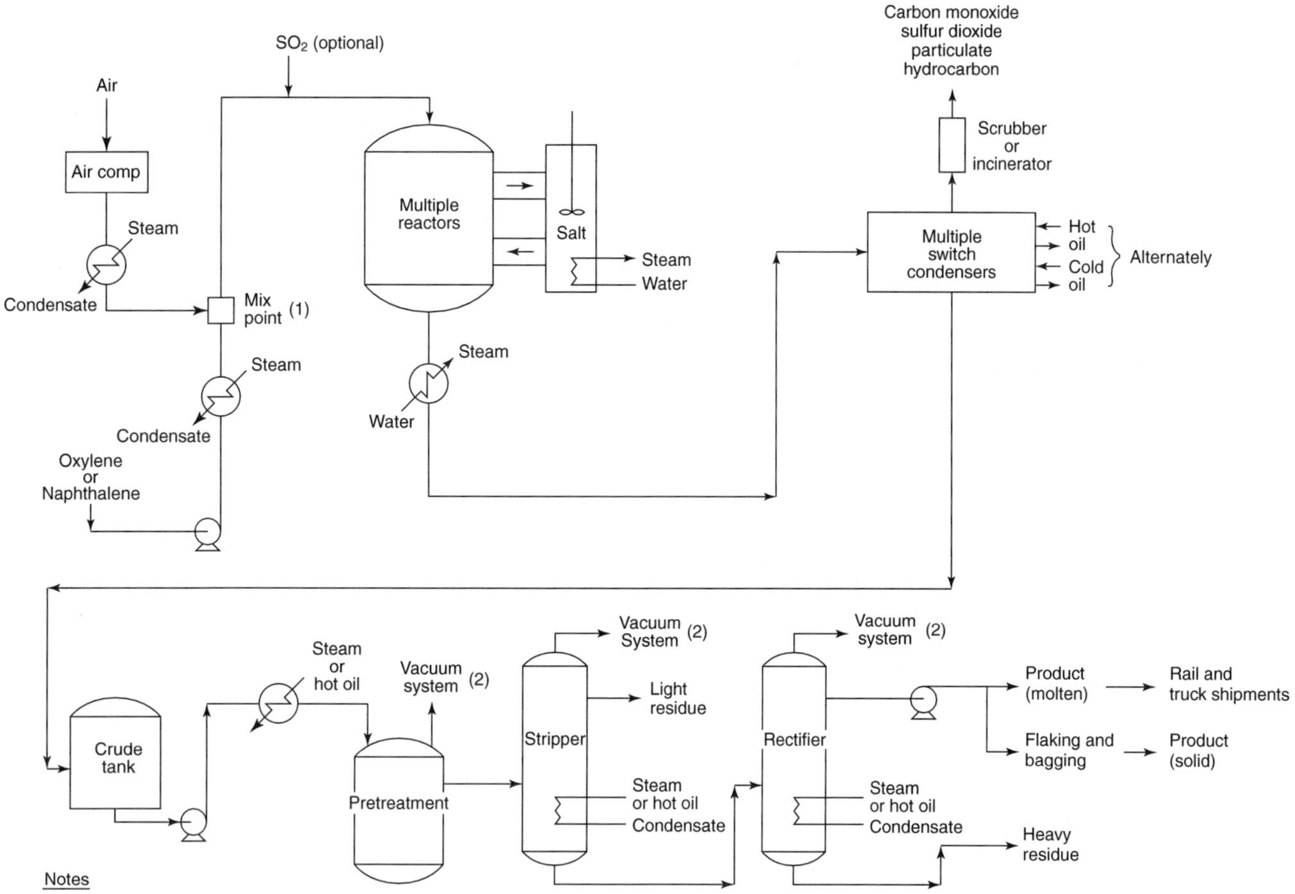

Notes

(1) Oxylene is typically sprayed into the air. Naphthalene is usually vaporized by passing a slipstream of heated air through the liquid; the unvapored residue is removed from the process.

(2) Vacuum obtained by one of the following:

1. Steam jets barometric condensers (vent composition similar to scrubber effluent)
2. Steam or oil powered jets with combustion of effluent as furnace fuel
3. Air powered jets with recycle to switch condensers

Figure 1. Phthalic Anhydride Process.

carbonate are sometimes added during the decomposition step to improve product quality further, and, in the case of naphthalene feeds, especially to reduce naphthaquinone levels in the product.

After the heat treatment, the light ends are removed from the crude product by vacuum distillation. The pure product is then taken overhead in a second vacuum distillation tower, leaving behind the higher-boiling-point materials. The light and heavy cuts are about equal in volume, depending on the purity of the feedstock. The PAN content of these cuts is about 50%, but can be lower if additional distillation facilities are provided.

The light and heavy residues air usually mixed together and typically are used as fuel to generate steam or to heat oil in a high-temperature heat-transfer oil circuit. The gases leaving the switch condensers contain light ends, unrecovered PAN, carbon oxides, and water. This stream is either incinerated or water scrubbed to remove the organic materials.

The molten PAN product is a water-white liquid with fluid properties similar to those of water. Most of the PAN used in the United States is transferred in its molten state. This requires keeping it above its freezing point of 268°F. Shipments are made in insulated railcars or tank trucks, most of which are fitted with steam coils to melt cargo that has partly solidified during transit. The material that is flaked and bagged is used for long-term storage or exported; this is the only method currently available for overseas shipments.

AIR EMISSIONS CHARACTERIZATION

Sources

Air emission sources from the process include the following:

1. Incinerator or scrubber effluent
   a. Oxygenated hydrocarbons.
   b. Particulates (*Note:* most are organic solids that eventually vaporize on exposure to air at ambient conditions).
      i. Primarily due to PAN, which desublimates when cooled below its freezing point of 268°F.
      ii. Minor quantities of ash, primarily from corrosion of carbon steel heat-transfer equipment.

iii. In the case of scrubbers, liquid particles from entrainment of the scrubber solution.

iv. Residue particulates if residue is incinerated.

c. Sulfur dioxide (SO_2) is optionally injected into the reactor as a catalyst promoter: it passes essentially unchanged to the incinerator or scrubber. Also, SO_2/SO_3 is produced by combustion of sulfur-containing fuels used in process heaters and thermal incinerators.

d. Carbon monoxide (CO) is produced in the process reaction, as well as in thermal incinerators and process heaters.

e. Nitrogen oxides (NO_x are not produced in the process, but are produced in the thermal incineration and in process furnaces from the combustion of the fuel used to provide the heat).

2. Vacuum system vents

a. Steam jets with barometric condensers—similar to scrubber effluent, but with very little CO and maleic anhydride present.

b. Vents discharged to a furnace—part of furnace emissions.

c. Recycle to switch condensers part of switch condenser effluent.

d. If the pretreatment is operated above atmospheric pressure, a vacuum system is not required to vent it to, for example, the switch condenser effluent circuit.

3. Storage tanks

a. Crude and product storage and transfers—particulates (PAN).

b. Feed tankage—*o*-xylene or naphthalene hydrocarbon emissions.

c. Heat transfer oils—weathering of light ends if degradation occurs.

4. Fugitive emissions valves, flanges, and so on. Very small quantities are involved because most of the process operates close to atmospheric pressure or under a vacuum. Values are based on heavy liquid because of the low vapor pressure of PAN. Hydrocarbon emissions also occur from the *o*-xylene–naphthalene feed system up to the reactor catalyst bed.

5. Emissions from flaking operations—particulates (PAN).

Emission Factors

Table 1 is a revision of Table 6.5.1 from AP-42[6] updated on the following basis:

1. The portion of Table 6.5.1 entitled "oxidation of naphthalene," which was largely based on the older fluid-bed processes, was deleted because PAN is currently produced only in fixed-bed units. Because the fixed-bed units are similar, whether *o*-xylene or naphthalene is used as the feedstock, only one set of values for both feedstocks is shown.

2. Since the completion of the 1975 engineering study, which set the base values for AP-42, technological changes in the industry have permitted operations at higher-weight concentrations of feedstock in the process air.[5] Considerable energy savings have resulted because it has been possible to increase production at constant air capacity. Typical concentration increases were from 40 g of *o*-xylene per cubic meter of air to about 66. While the main process stream concentrations of SO_2 and CO are increased by the capacity ratio of 66:40, this is offset by the increased plant production ratio of 66:40. Thus the SO_2 and CO values reported in Table 6.5.1 are unchanged because they were based on a unit weight of product. Although it is unlikely that all the plants are running at the 66 g/m^3 air level, this value was chosen to reflect an average of current and near-future trends in the industry.

3. For particulates, the recovery systems have been improved with the increased production so that losses per ton have remained constant. Incineration and scrubbing technology has been improved so that 98% instead of 95% destruction of the particulates is achieved, as shown in the table.

4. Only incineration technology is shown because it is the best of current technology. Combinations of scrubbers and incinerators were deleted because it is currently more likely that an incinerator would be used to replace a scrubber than be added to a scrubber.

5. NO_x emissions have been added. The process produces nil NO_x, but NO_x is produced in the thermal incineration and in any process furnaces. NO_x can also be produced in catalytic incineration prior to startup. Values shown are based on fuel gas firing in a thermal incinerator.

AIR POLLUTION CONTROL MEASURES

The incinerator or scrubber is an integral part of the process because of the large volumes of air used. Although about 99% of the phthalic is recovered in the switch condensers, the remaining phthalic and by-products have to be removed. Water scrubbing recovers the materials for possible sales or for combustion as a liquid solution. Scrubbing has practical limitations in the recovery of the oxygenated hydrocarbons and does not convert CO to carbon dioxide. Therefore, the preferred choice, currently, is either catalytic or thermal incineration.

In thermal incinerators, the effluent is heated to about 1500°F by burning additional fuel. Residence times are about 1 sec at this temperature, after which the gases are cooled, usually by generating steam.[8] Catalytic incinerators operate in the 500–800°F temperature range and, depending on the amount of organics in the waste gas, may not require auxiliary fuel, except for startup and end-of-run operations.[9] Both types achieve a destruction efficiency of the hydrocarbons and CO of 98%. Increased destruction efficiency can be obtained in a thermal incinerator by increasing residence time and/or temperature. Similar results can be obtained in a catalytic system by increasing either inlet temperature or the amount of catalyst (i.e., residence time in the catalyst bed). Higher temperature in a thermal incinerator increases NO_x formation and requires more steam generation to remain efficient. Catalytic incinerators are limited on catalyst bed outlet temperatures because of potential catalyst damage at elevated temperatures.

Table 1. Emission Factors for Phthalic Anhydride[a,b] (Emission Factor Rating: B)

	Particulate[c] (Revised)		NO_x (Added)	
	(g/kg)	(lb/ton)	(g/kg)	(lb/ton)
Main process Stream[d]				
Uncontrolled	69	138.0	0.0	0.0
With incinerator[e]	1.4	2.8	0.5	1[f]
Pretreatment (vacuum system)				
Uncontrolled	6.4	13.0	0.0	0.0
With incinerator[e]	0.1	0.3	0.0	0.0
Distillation (vacuum system)				
Uncontrolled	45	89.0	0.0	0.0
With incinerator[e]	0.9	1.70	0.0	0.0

The following factors are unchanged from AP-42

	SO_x		Nonmethane VOC[g]		CO	
	(g/kg)	(lb/ton)	(g/kg)	(lb/ton)	(g/kg)	(lb/ton)
Main stream						
Uncontrolled	4.7[h]	9.4[h]	0.0	0.0	151.0	301.0
With incinerator[e]	4.7	9.4	0.0	0.0	3.0	6.0
Pretreatment						
Uncontrolled	0.0	0.0	0.0	0.0	0.0	0.0
With incinerator[e]	0.0	0.0	0.0	0.0	0.0	0.0
Distillation						
Uncontrolled	0.0	0.0	1.2[c,e]	2.4[c,e]	0.0	0.0
With incinerator[e]	0.0	0.0	<0.1	0.1	0.0	0.0

[a]Revised Table 6.5.1 in reference 6, which was based on data in reference 7. Factors are in grams of pollutant per kilogram of PAN produced (or in pounds per ton).
[b]Control devices listed are those currently being used by PAN plants.
[c]Consists of PAN, maleic anhydride, benzoic acid, and for naphthalene feedstocks, naphthaquinone.
[d]Main process stream includes reactor and multiple switch condensers as vented through condenser unit.
[e]Normally a vapor, but can be present as a particulate at low temperature.
[f]Typical value for a thermal incinerator operating on a fuel gas that does not contain nitrogen compounds. Varies considerably with design, operating temperature, and fuel composition.
[g]Emissions contain no methane.
[h]Value shown is for a relatively fresh catalyst that requires an SO_2 promoter. The requirement can change with age and can be 9.5–13 kg/mg (19–25 lb/ton) for aged catalysts. Catalysts that do not require SO_2 are available, but with debits of shorter life and lower yield of product.

Vacuum Systems

Because PAN solidifies at 268°F and reacts with water to form solid phthalic acid, vacuum pumps with water seals normally are not used to provide a vacuum. Instead, vacuum jets are used. The effluent from the jets can be treated in one of three ways: scrubbed with water and the resultant solution treated in a biox system, burned as fuel, or recycled back to the process. In the last case, the noncondensibles in the stream are eventually treated in the switch condenser effluent incinerator or scrubber. Alternatively, condensers can be installed ahead of the jets to reduce the organic levels to the point where further treatment of the jet effluent is not required.

Tank Vents and Transportation

Vent control consists of two types: desublimators or a vacuum system using jets (as described above and including any vents at elevated pressure that can bypass the jets and go directly to the downstream treatment steps). The simplest type of desublimator is a large sheet-metal box that provides sufficient surface area to cool the vented gases. On cooling, the PAN desublimes from the vapor and collects in the box. Suitable baffles are provided to prevent the vapor from exiting the box until cooled and for periodic removal and recycling of the solid PAN crystals. The vent gas composition corresponds to the vapor pressure of PAN at the exit temperature. Because PAN has a low vapor pressure (0.0035 mm Hg at 100°F), the concentration is low. The total amount of PAN vented will depend on tank filling rates and the rate of the gas blanketing on the tank.

Desublimators mounted on the top of the tank and equipped with heating and cooling facilities also are used. These can be operated to lower outlet temperatures depending on the cooling-medium temperature and do not require manual transfer of the recovered solid phthalic.

REFERENCES

1. *Chemical Marketing Reporter*, July 24, 1989, p. 58.
2. *Chemical Marketing Reporter*, July 24, 1989, p. 15.
3. H. P. Dengler, *Phthalic Anhydride. Vol. 36, Encyclopedia of Chemical Processing and Design*, J. J. McKetta and W. A. Cunningham (Eds.), Marcel Dekker, New York, 1990, p. 33.

4. R. Y. Saleh and I. E. Wachs, U.S. Patent 4,582,912, April 15,1986 (to Exxon Research & Engineering Co.).
5. W. Gierer and O. Weideman, "Phthalic Anhydride Made with Less Energy," *Chem. Eng.* **86**(3), 26 (1979).
6. *Compilation of Air Pollutant Emission Factors, Chapter 6 Section 5, Phthalic Anhydride*, AP-42, U.S. Environmental Protection Agency, 5th Edition, Research Triangle Park, NC, 1983, p. 6.5-6.
7. *Engineering and Cost Study of Air Pollution Control for the Petrochemical Industry, Vol. 7, Phthalic Anhydride Manufacture from Ortho-xylene*, EPA-450/3-73-006g, U.S. Environmental Protection Agency, Research Triangle Park, NC, July 1975.
8. T. F. McGowan and R. D. Ross, "Incineration: Flexible and Efficient," *Chem. Proc.* **53**(11), 79 (1990).
9. R. G. McInnes, S. Jelinek, and V. Putsche, "Cutting Toxic Organics," *Chem. Eng.* **97**(9), 108 (1990).

PRINTING INKS

TEMPLETON A. ELLIOTT, JR.

Printing inks can be classified by the printing process in which they are used: (1) letter press, which uses a paste type of ink; (2) lithographic or offset, which also uses a paste type of ink; (3) flexographic, which uses a fluid ink; and (4) gravure, which also uses a fluid ink. Other printing processes, which consume less ink on an annual basis than the four major processes, include screen-printing, intaglio, ink jet printing, and xerography. Printing inks are used for the printing of publications, packaging, and certain specialty products. While letterpress and offset inks are used for almost every type of publication and packaging, gravure inks are used principally for magazines, advertising circulars, packaging, and specialty products. Flexographic inks are used almost exclusively for package printing, although there is a trend towards increased use of the process to print newspapers. Products such as gift wrap, wall coverings, and floor coverings are usually printed using the gravure process, with lesser amounts being printed using the flexographic process. Approximately 2.1 billion pounds of ink were used in 1995, with an estimated 2.5 billion pounds to be consumed annually in the year 2000.[1] Figure 1 provides a breakdown by type of ink manufactured in the United States.

All inks consist of pigments, which are the colorants, dispersed in a vehicle, which is the component that binds the pigments to the receiving surface. Receiving surfaces include paper, paperboard, foil, and plastic films. Generally, while the pigments used in all four types of printing ink are similar, the vehicles are quite different. Letterpress and lithographic processes use paste inks that incorporate a thick, viscous vehicle, while gravure and flexographic inks are fluid and utilize a much lower-viscosity vehicle. Letterpress inks and lithographic inks dry by oxidation of the vehicle or vehicle absorption into the printing surface, while fluid inks dry by solvent evaporation. Oil and paste inks are less important emissions sources because the processes, that utilize them, lithography and letterpress printing, do not normally use volatile organic compounds (VOCs).[2] One exception is a variation of lithographic ink known as web offset inks. These inks can contain over 50% by weight VOC but because the solvents used have very low vapor pressures, VOC emissions from the production of these inks are low. Another exception, which is a significant source, occurs where vehicle (varnish) for paste-type inks is cooked on-site. This is the primary source of VOC emissions from lithographic and letterpress ink manufacturing.[3] While many companies purchase the vehicle after cooking, emissions estimates are included under "Air Emissions Characterization," for use when applicable.

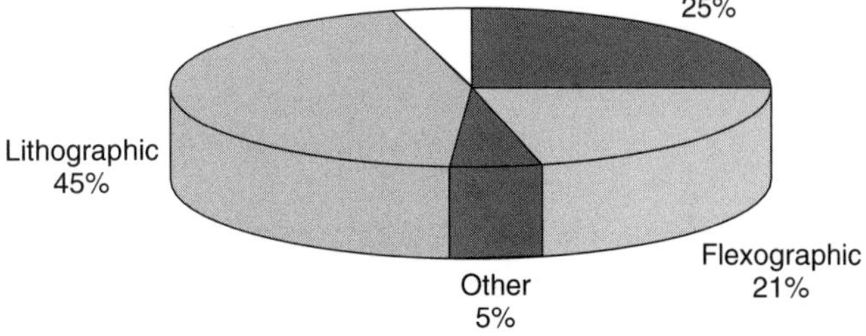

Figure 1. Ink Production in the United States. Percent of Total Production by Weight.

Flexographic and rotogravure inks have many properties in common: particulates (pigments) dispersed in a vehicle, low viscosities, and drying after application by the evaporation of volatile solvents and/or water. These inks are manufactured by a roughly similar process sequence. This sequence may be best described as a simple blending of prepared components. Preparation of components consists dissolving (cutting) of resins into solvent mixtures or water in the case of water-based inks, and wet grinding (dispersing) pigments in vehicles.

The principal difference between flexographic and rotogravure inks relates directly to their respective printing processes. Flexographic inks are used in a printing process where the design is transferred from a rubber or synthetic relief plate onto the receiving surface. The solvents used must not degrade the surface of the relief plate. Principal solvents used in the flexographic process are alcohols or mixtures containing alcohols and small amounts of other solvents such as acetates or aliphatics. The gravure printing process differs from the flexographic process in that the design to be printed is etched or engraved into the surface of a metal printing cylinder. Gravure ink formulations include solvents that would degrade flexible printing plates. Various hydrocarbon-based solvents such as toluene and other solvents such as acetates and ketones can be used in gravure formulations.

Since the mid-1970s, much of the flexographic ink for printing of corrugated containers and folding cartons has been formulated with water-soluble materials. The Hull Report estimated that a total of 342 million pounds of water-based flexographic ink was produced in 1995.[4] The emissions from the manufacture of these inks contain few, if any, VOCs. In the 1980s, water-based gravure inks for packaging increased in usage, reaching an estimated 68 million pounds annually in 1995.[5] Future increases in the production and use of water-based fluid ink is limited by the higher surface tension and slower evaporation rate of water as compared with VOC solvents. Ink manufacturers continue to attempt to overcome these problems, which would result in the increased manufacture of water-based inks and a corresponding decrease in the production of solvent-based ink.

Both flexographic and rotogravure inks for packaging printing are typically formulated for specific customer applications and to match a myriad of different colors. Variations in color, adhesion, drying, printing equipment, substrate, operator preference, and regulatory requirements all can result in complex formulation changes. As a result, the flexographic and rotogravure ink manufacturing processes for packaging printing can be best described as batch operations that must be highly adaptable to customer demands. Gravure inks used for publication printing can be produced in more or less continuous operations because only four standard colors normally are manufactured.

PROCESS DESCRIPTION

Raw-Materials Receiving

A flow diagram for the ink manufacturing process is shown in Figure 2. Most raw materials are purchased from outside sources rather than being manufactured by the ink company. However, some paste-ink manufacturers do manufacture their own vehicles from oils and other materials. Raw materials are received in tank trucks, portable tanks, drums, and bags. Materials include resins, dry pigments, solvents (such as ketones, alcohols, aromatic and aliphatic hydrocarbons, and acetates), and other processing chemicals. Except for accidental spills and leaks, the only emissions from material receiving occur during tank truck unloading, when vapors are displaced to the atmosphere as the storage tanks are filled (working losses).

Storage

Storage of solvents in tanks may result in breathing losses. Since breathing losses result from diurnal temperature variation of the stored material, there are negligible emissions of this type from underground tanks; only aboveground tanks located outside of buildings are affected. Other materials stored in portable tanks, drums, and bags emit air pollutants only while undergoing transfers into process containers.

Vehicle Preparation

Vehicles for paste inks are prepared by mixing oils such as petroleum oils and reactive oils with various resins to impart suitable tack and flow characteristics. Some paste inks incorporate reactive oils such as linseed, soybean, or china wood oil that cause the ink to dry by oxidation. By chemical modification of these oils, alkyd vehicles are produced that are much superior in drying and certain other properties such as hardness. Many of these vehicles must be heated or cooked during preparation. Most paste-ink manufacturers purchase these vehicles from unaffiliated suppliers, but some companies

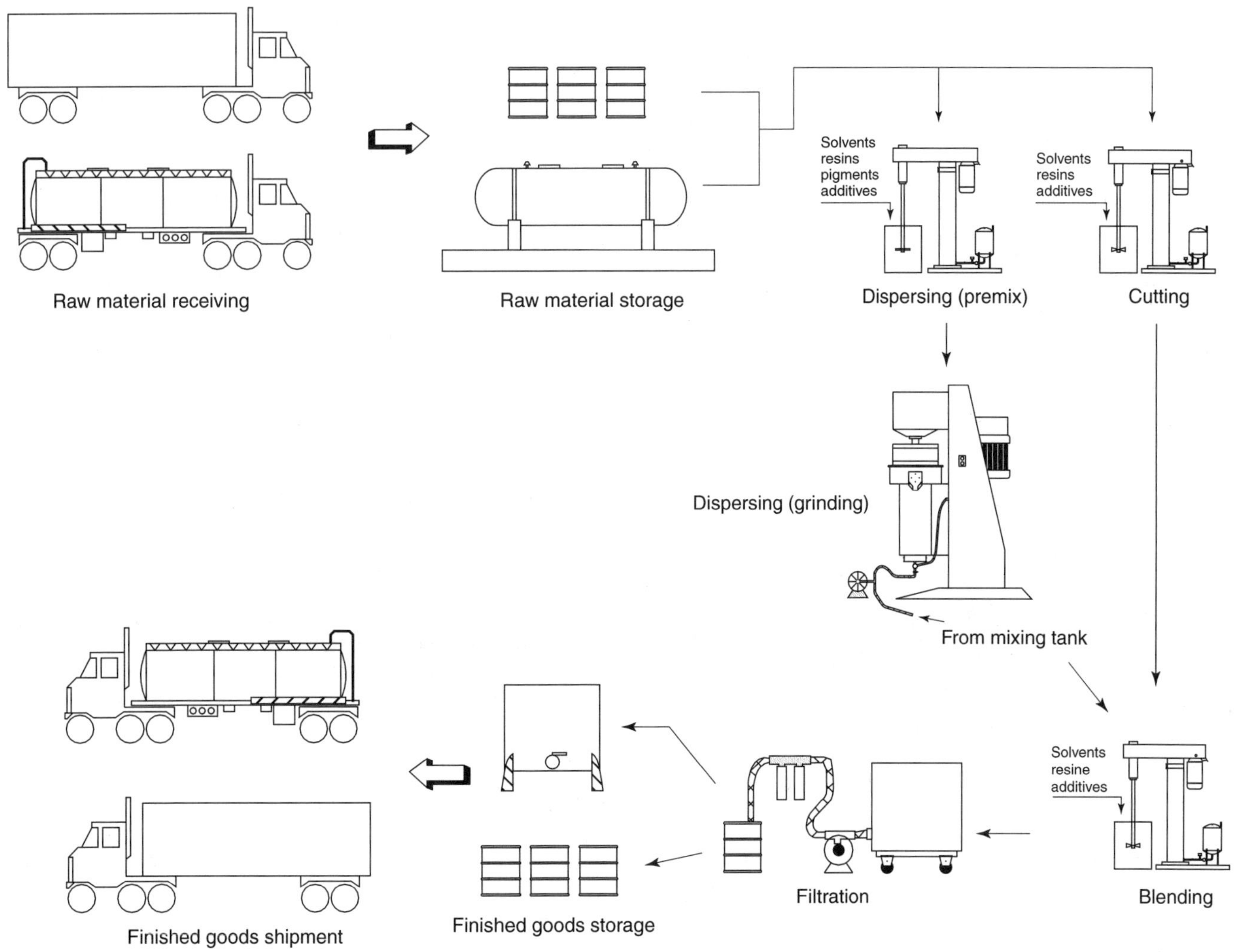

Figure 2. Ink Manufacturing Process Flow Diagram.

have vehicle cooking capability on the same site as that of ink manufacturing. Primary emissions are the oils and solvents that evaporate during the cooking process. When closed vessels are used for vehicle cooking, few vapor escape.

Vehicles for fluid inks are prepared by the cutting operation described below.

Cutting

Cutting, or dissolving, is the primary production step for vehicles used in fluid inks. These solutions may be used as intermediates, vehicles, and extenders or as finished products such as varnishes. Processing these unpigmented solutions involves adding solvents to a receiving vessel and then adding dry resins and other chemicals. Mixing is accomplished by a mechanically driven shear blade equipped with slotted teeth, a propeller-type blade or a turbine-type blade. This agitation speeds the dissolving of the resin in the solvent. During mixing, some frictional heating usually occurs.

Emissions are a function of the frictional heating, mix time, the extent to which the vessel is closed, and losses associated with liquid transfers (displaced vapor). The degree of containment of air emissions can vary by available equipment and operator technique.

Dispersing

Dispersing is the primary production step in making an intermediate, usually referred to as a base or dispersion. A base is a colored concentrate used to produce a finished ink. The first step in dispersing is the preparation of a "premix." A premix involves the addition of vehicles (either manually or automatically), dry pigments (manually), and other chemicals into a mixing container. The batch is then mixed mechanically by the use of a special slotted blade in the case of fluid inks, or by the use of a slower-speed, higher-torque mixing blade in the case of paste inks. Frictional heating may occur, depending on the duration of the mixing, the peripheral speed of the mixing blade, and the viscosity of the premix. Following the premix stage, wet grinding is accomplished by any of several technologies (see Figure 2). These technologies utilize spherical media or the abrasiveness of the mixture itself to reduce the agglomerated pigment to the micron or sub-micron size.

Common technologies operate as follows:

1. *Dispersion via mixing.* For certain applications and for easily dispersed pigments, the premix step, which occurs in a large mixing vessel, may be continued until dispersion is accomplished. Emissions result from the displacement of vapors when materials are added to the vessel and from the surface of the moving liquid as it is being mixed. Normally metal covers or plastic sheeting are used to reduce emissions.
2. *Rotor stator devices.* These are devices in which a high-speed turbine or plate rotates inside a stator and disperses pigment by forcing the pigment and vehicle through the close clearances between the rotating rotor and stationary stator. These devices may be fixed inside a large vessel or mounted on a lift and lowered inside portable vessels. Usually there is no premix step and the pigment is added directly to the vehicle in the vessel while the rotor is turning. Thus premix and dispersion are achieved simultaneously. These devices generate considerable frictional heat. However, since they are usually used with covers, emissions to air result only from the displacement of vapors when loading and from losses during discharge from the vessel. Some of these vessels are jacketed and use cooling water to prevent over heating of the materials.
3. *Ball mills.* Ball mills consist of a water-jacketed steel cylinder partially filled with ceramic or steel balls. The raw materials may be added directly into the ball mill or they may be premixed and then added. Raw materials and balls are tumbled within this rotating cylinder, mixing and grinding the ingredients for periods up to 24 h. After the milling has been completed, the base mixture is strained into containers, through a screening hatch, which retains the balls within the mill. Air emissions result when vapors within the cylinder are displaced to atmosphere while the unit is being filled and evaporation during transfer when product is discharged.
4. *Small media mills.* Small media mills are the most prevalent types of machinery used to disperse pigments for use in the fluid ink industry. Several variations are in common use. The oldest of the small media mills is the sand or shot mill in which the premix is pumped through a vertical chamber containing sand, steel shot, or other small spherical media which ranges from 0.5 to 2.0 mm in diameter. Mechanically driven disks within the chamber churn the media, dispersing the pigment particles in the vehicle by breaking apart agglomerates. A screen at the top of the chamber retains the media in the mill and allows the product to be discharged in a trough. Most emissions from these mills occur at the screen face and discharge trough. An improvement in this type of mill is the enclosed mill, in which internal, narrow gaps or a rotating screen separates the media from the product and retains the media inside the mill. These mills may be vertical or horizontal and always include a water jacket to dissipate frictional heating. The discharge from these enclosed mills is through a pipe or hose so air emissions from evaporation are minimal.
5. *Double-arm mixers.* Dispersion in a double-arm mixer requires ingredients to be added directly to the unit in the same fashion as loading into a ball mill. The ingredients contain little liquid and result in an extremely viscous product. Within this mixer, two counter-rotating arms force the pastelike mixture against itself, causing dispersion and grinding. No separate media are required in this process. This vessel is completely closed, and few emissions occur except for transfer losses.
6. *Two- and three-roller mills.* Two- and three-roller mills are composed of two or three horizontal, closely spaced steel rollers that rotate in opposite directions at different speeds. The premix is fed to the rear roller from a tub by pouring or by hand. As the ink transfers between the rollers, the pigment is dispersed by the shearing action that occurs at the nips between the rollers. The ink or dispersion is removed from the last roller with a scraper blade. Because the mill rolls are exposed, the potential for evaporative losses is great depending on the evaporation rate of the solvents (if any) employed. Three roller mills are used primarily for making paste inks. Two roller mills utilize the same principle but evaporate most of the solvent used in the premix and discharge a solid product

that is broken into chips or pulverized into powder. The solvent emissions may be released to the atmosphere or captured by collection devices and treated.

Once the base or dispersion has been reviewed for quality, it may be used directly, shipped to customers, or placed into intermediate storage. This intermediate storage can be in drums or in bulk tanks. Air emissions can result during transfer operations.

Further Steps

Blending. Blending is a batch process in which intermediates and raw materials are mixed in specific quantities to produce specific finished ink. The blending process sequence and ingredients often have subtle variations that impart specific qualities to the final ink product. These are usually considered trade secrets. Emissions include transfer losses during material transfer and evaporative losses during mechanical mixing.

Filtering. Because of customer and quality-related concerns, most inks are filtered prior to or during packaging. This is accomplished using cloth or bag filters, vibrating screens, or porous resin cartridge filter systems. Filtration is an extra handling step that may lead to some transfer and evaporative losses.

Packaging. After filtration the finished product is pumped or poured into the final shipping container (cans, pails, drums, portable tanks, or tank trucks). Emissions occur during transfer operations.

Shipment. Finished ink products are shipped in cans, pails, drums, portable tanks, or tank trucks, and under normal conditions, no emissions result during shipment.

Cleaning. Blending vessels, mills, filter screens, and other equipment are routinely cleaned after the production of an ink formulation is complete. However, in a few large plants, only one color or a limited number of colors are produced, and each piece of equipment is dedicated to a single color, which eliminates the need to clean between batches. Cleaning is done with either solvents or a hot water and caustic mixture and may be done manually with rags and brushes, by closed washing machines, or by immersion in soak tanks. When using solvents for cleaning, a final rinse that is compatible with the next ink formulation to be produced in the equipment is used. Washing machines use either solvent or hot caustic water solutions to remove ink from vessels and equipment. All are closed, and solvent emissions occur only when opening or closing the washing machine. No solvent emissions result when hot caustic solutions are used except from solvents contained in residual ink clinging to the vessel before cleaning. Soak tanks or pits utilize hot caustic solutions and usually have open tops. Emissions from these devices are mostly water vapor and small amounts of VOC from inks clinging to the equipment when it is initially being submerged in the pit.

AIR EMISSIONS CHARACTERIZATION

Emissions from printing ink manufacturing consist of volatile organic compounds from solvents employed and dusts from the addition of dry pigment and other fine solids to mixing vessels.

Because of the small size of most ink manufacturing plants, few detailed emissions surveys exist. Theoretically, solvent emissions can be calculated for sequential process steps, starting with product receiving and storage and carrying through to product packaging. The U.S. EPA and the Emission Inventory Improvement Program commissioned the compilation of numerous methods used for estimation emissions from ink manufacturers.[9] One actual measurement of emissions from a cutting operation showed emissions to be 0.1% of solvent (toluene) employed when using covers and 0.8% when uncontrolled.[6] Storage tank breathing losses can be estimated using algorithms published by the U.S. Environmental Protection Agency (EPA). Emissions from processing steps can be estimated using engineering principles. Often, many simplifying assumptions must be made, and final estimates may be off by a considerable margin.

Stack testing can be used to supplement engineering estimates. This is difficult to implement because of the batch nature of ink manufacturing, the wide array of solvents being used in the ink products, and the often complex and diverse nature of the emissions capture and collection systems employed. In short, accurately estimating emissions from an ink manufacturing facility is a challenging task requiring considerable planning.

For typical printing-ink manufacturing facilities, solvent losses to air as a percentage of total usage are between 0.3% and 1.6% for newer facilities and can exceed 2% for older facilities or for batch operations requiring frequent manual cleaning because of equipment changeover. These ranges reflect differences in facility age, equipment, operating, and maintenance methods, and the degree to which water-based production capacity has replaced solvent-based production capacity. Flexographic ink producers with a high percentage of water-based products may tend to be at the low end of the range, or even well below it at individual facilities.

Dry pigment varies in particle size. This characteristic directly affects potential emissions, along with the following parameters: method of addition to the mixing vessel (manual versus automated), product mix (e.g., clear varnishes use no pigments), ventilation, and operating practices. This includes the use of covers on mixers, the size of pigment bags, how empty drums or bags are handled, and even how bags are split open for emptying, lengthwise versus at one end.

To represent a variety of conditions, an emission factor of 2 pounds per ton of pigment mixed has been developed.[7] It is felt that this represents the upper end of actual emissions and that good practice, such as the use of mixer covers while emptying bags, splitting bags endwise, and emptying them thoroughly before disposal, would reduce emissions further.

Cooking an ink vehicle is a special case. This step is more prevalent in lithographic and letterpress ink manufacturing than in rotogravure and flexographic ink manufacturing facilities. Vehicle cooking is not applicable to facilities that purchase a varnish or other clear solution that has already been heated, or that make varnish or other clear solutions without heating them at all. Vehicle cooking involves heating a vessel containing resins, solvents, and other additives at temperatures up to 340°C (650°F) for an average of 8 to 12 h.[8] Where it is used, this process is the primary potential source of VOC emissions at a facility.[2] Emissions result from decomposition or evaporation of varnish ingredients at peak temperatures or from purging the reactor vessel with air or inert gas. The quantity, composition, and rate of emissions from the varnish cooking step are a function of cooking temperature and time, the ingredients, the way additives are introduced, the degree of stirring, and the extent of air/inert gas purging. Uncontrolled VOC releases from this operation is estimated at from 40 to 160 pounds per ton for various varnish types.[7]

AIR POLLUTION CONTROL MEASURES

Air emissions from the manufacturing of printing ink consist of dusts and solvent vapors (VOCs). Controlling them at the source can be cost effective and goes a long way towards minimizing air pollution.

Source Control

For pigments, good manufacturing practice would include splitting bags at their ends, minimizing the vertical drop of the pigment to the extent practicable, and emptying bags and drums thoroughly to reduce fugitive emissions when disposing of empty bags in a compactor. The use of intermediate bulk containers, which are large (2000 lbs.) sacks for shipment and addition of dry materials, also reduces dust emissions.

For vehicle cooking, where applicable, the use of retractable hoods, enclosed reactors, and/or add-on controls will substantially reduce emissions. Besides condenser systems, potentially applicable add-on controls are reported to include thermal and catalytic incineration and carbon absorption. However, this is based on a study involving vehicles working in an allegedly similar operation in the paint industry.[8]

For solvents, equipping vessels with well-fitting covers and paying attention to housekeeping help to minimize solvent losses. "Proper" manufacturing techniques are a function of each facility, reflecting its age, layout, type of equipment, and personnel experience and training. However, in general, best manufacturing practice, when using solvents, would include using well-fitting covers on mixing vessels, keeping vessel transfers to a practical minimum, closure of mixer cover openings immediately after addition of ingredients, cleaning of mixers and tanks by using a closed washing machine or a tank or pit using caustic water solutions; and good housekeeping in general. Improvements in initial plant design or in retrofits can also reduce VOCs. For example, equipping aboveground storage tank vent lines with conservation vents to reduce breathing losses and equipping bulk tanks with vapor return lines can reduce VOC emissions.

End-of-Line Controls

Dusts, usually particulates associated with the transfer of resins and pigments, are collectible by fabric or cartridge filters (bag houses). Depending on pigment particle size, this control equipment may achieve up to the 98–99% collection efficiencies observed for other processes. In theory, collected material may be reused within the process. However, because of cross-contamination and customer specifications, most material is ultimately disposed of as waste.

A unique source of solvent vapors is resin cooking, which, while performed by some ink manufacturers, should be considered a chemical process because chemical reactions take place. Emissions from this high-temperature process are reduced by the use of closed vessels and by the use of reflux condensers to return vapors to the process as condensate.

Solvent emissions from other steps in the ink manufacturing process are usually a mixture of solvents depending on product mix and, if not controlled at the source become fugitive emissions and present a much more difficult control problem. Three factors make it difficult to control the solvent vapors in the exhaust from ink plants:

1. The concentration of fugitive VOCs is quite low (5–50 ppm is typical).
2. The VOC concentration is highly variable because of the nature of batch operations.
3. The VOC in the exhaust is a highly variable mixture of chemicals, ranging from polar to nonpolar species and high- to low-boiling-point compounds.

Because of these factors, a number of control technologies can be eliminated as unfeasible to control fugitive emissions. Vapor refrigeration, while effective when used on individual vessels, is generally not reasonable for control of fugitive emissions owing to the dilute concentration of VOCs and the resulting low operating temperature required in the condenser to chill the exhaust stream containing fugitive VOCs below its dew point. Because of the wide range in physical properties of the chemicals in the exhaust, carbon absorption may have limited effectiveness and scrubber systems may not be feasible.

Oxidizing systems (thermal or catalytic) are technologically feasible control processes. Ink manufacturing tends to be a relatively small-scale operation, and any control technology found to be technologically feasible must be evaluated for economic viability. It is quite common for all control options to prove prohibitively expensive in terms of cost per ton of VOC removed. In these instances, control of emissions at the source must be emphasized.

REFERENCES

1. *Printing Ink U.S. 1995–2000, A Multi Client Study*, Hull & Co. Greenwich, CT, April 1996, p. 33.
2. S. V. Capone and M. Petruccia, *Guidance to State and Local Agencies in Preparing Regulations to Control Volatile Organic Compounds from Ten Stationary Source Categories*, EPA-450/2-79-004, U.S. Environmental Protection Agency, Research Triangle Park, NC, 1989, p. 155.
3. R. Blaszczak, "A Preliminary Review of 19 Source Categories of VOC Emissions," unpublished document, U.S. Environmental Protection Agency, Research Triangle Park, NC, May 1988, p. 8–2.
4. *Printing Ink U.S. 1995-2000, A Multi Client Study*, Hull & Co. Greenwich, CT, April 1996, p. 67.
5. Ibid. p. 91.
6. *Control of VOC Emissions from Ink and Paint Manufacturing Processes*, EPA-450/3-92-013, U.S. Environmental Protection Agency Control Technology Center, Research Triangle Park, NC, April 1992, Appendix C.
7. *Compilation of Air Pollutant Emission Factors*, Vol. I, Chapter 6, 5th ed. AP-42. U.S. Environmental Protection Agency. September 1991, *http://www.epa/ttn/chief/ap42c6.html*.
8. *Air Pollution Control Engineering and Cost Study of the Paint and Varnish Industry*, EPA-450/3-74-03 1, U.S. Environmental Protection Agency, June 1974.
9. *Preferred and Alternative Methods for Estimation Air Emissions from Paint and Ink Manufacturing Facilities*, Vol. II, Chapter 8, Eastern Research Group, Inc., Morrisville, NC, March 1998, *http://www.epa/ttn/chief/eiip/techrep.htrn#pointsrc*.

Acknowledgment is made to Stephen W. Paine and Stephen K. Harvey, authors of the Printing Inks chapter in the first edition of *Air Pollution Engineering Manual*, much of whose information is used in this second edition.

SOAPS AND DETERGENTS

RICHARD C. SCHERR
(with update by Editor)

The term *synthetic detergent products* applies broadly to cleaning and laundering compounds containing surfactants along with other ingredients formulated for use in aqueous solutions. These products are marketed as heavy- or light-duty granules or liquids, cleansers, and laundry or toilet bars. The heavy-duty granules and liquids represent the major portion of all products manufactured, with the light-duty liquid and light-duty granules produced to a much lesser degree. Heavy-duty powders and liquids for home and commercial laundry, laundry detergent, make up 60% to 65% of the U.S. soap and detergent market and were estimated at 2.6 million metric tons in 1990.[1] The manufacture of all detergent products incorporates equipment and processes similar to those used for manufacturing soap products. The manufacture of the granular products via spray drying is of paramount interest, with more severe air pollution problems than those encountered with soap granule spray drying. The manufacture of liquid detergent and bar products is of lesser importance, with little or no difference from similar soap products in either process equipment or air pollution potential, and will not be discussed.

Laws requiring zero or low phosphate content in detergents and the emergence of heavy-duty liquid detergents have had a significant impact on the production of granule laundry detergents. Until the early 1970s, almost all laundry detergents sold in the United States were heavy-duty powders. Public concerns regarding the role of phosphates in the eutrophication of lakes resulted in the banning of phosphates in detergents in some parts of the United States and Canada in the early 1970s. Nitrilotriacetic acid (NTA), a leading substitute builder for phosphates, was eliminated for use in the United States because of health concerns. Manufacturers, forced to formulate granule laundry products without an adequate builder, used sodium carbonate and zeolites. This significantly decreased cleaning performance and also resulted in a product that was difficult to dissolve. The introduction of heavy-duty laundry liquids utilizing sodium citrate and sodium silicate offered the consumer superior cleaning performance and improved solubility at only a slightly increased cost. While the performance of zero-phosphate granules has improved in recent years, heavy-duty liquids now account for 40% of the laundry detergents sold in the United States, up from 15% in 1978.[2] This growth in liquid share and consequent reduction in the demand for granule production, combined with improvements in manufacturing techniques, resulted in the shutdown of over half of the spray-drying facilities for laundry granule production since 1970. Consumption of laundry detergents has grown approximately 4% per year since 1980, reaching 7 billion pounds in 1989. Future growth is forecasted at between 2% and 3%.[1]

RAW MATERIALS

Components of detergents fall into three major categories: surfactants, to remove dirt and other unwanted materials from clothes; builders, to treat the water to improve the efficiency of the surfactants by softening the water; and additives, to improve cleaning or physical properties. Additives may include bleaches, bleach activators, antiredeposition agents, optical brighteners, antistatic agents, fabric softeners, and fillers. Each particular formulation depends on the ultimate design for consumer use. Table 1 illustrates typical detergent formulations for the United States, Europe, and Japan.[2]

Surfactants

The surfactants used in formulating synthetic detergent products are either anionic, cationic, nonionic, or amphoteric. Anionic surfactants derive their detergency from the large anion part of the molecule. Anionic surfactants are alkyates of benzene and propylene trimer or tetramer. Typical detergent surfactants are linear alkylbenzene sulfonates (LAS), alcohol sulfates, alcohol ether sulfates, and alkylglyceryl ether sulfonates. Cationic surfactants derive their detergency from the cation. Most cationic surfactants are derivatives of amines. Nonionic surfactants are hydrophobic molecules made water soluble by the addition of ethylene oxide groups. Typical hydrophobes include fatty acids, alcohols and acids, amines, and amides. Typical detergent nonionics include alkyl polyethylene glycol ethers and nonylphenyl polyethylene glycol ethers. Amphoterics are surfactants that contain both a cationic and an anionic end group.

Detergent surfactants include alkyl sulfonics, alkyl sulfates, and alcohol sulfates, discussed above, and almost the entire range of anionic and nonionic surfactants, including soap. Anionics and nonionics are the predominant surfactants used in laundry detergents. Cationics and amphoterics are generally used in specialty products because of their poor cleaning efficiency and high cost, respectively.[2] The most widely used surfactant for laundry detergents is LAS, followed closely by the alcohol sulfonates. Some manufacturers combine these surfactants, with the balance between the two being determined by their relative cost.[1]

Anionic surfactant materials such as LAS and alcohol sulfonates are sometimes produced at detergent plants by sulfonation and sulfation or, as with LAS, received as liquid sulfonic acids. In either case, the "acid mix" is neutralized with 30% sodium hydroxide in an exothermic reaction to form the final surfactant. Plants manufacturing their own sulfonic and sulfate surfactants also use other surfactants in some or all of their products. The detergents are received and handled mostly as liquid solutions of varying strength, but some surfactants are received as flake or powder. The exact method and processing vary for each surfactant, but anionic surfactants are principally mixed with the slurry before drying, while some nonionic surfactants can be sprayed on the finished granule.

Amides of various types can be used as supplementary surfactants in many formulations. They improve the detergency of the sulfonic and sulfate surfactants and act as foam boosters or stabilizers. Amides used include the higher fatty amides (e.g., cocomonethanolamide), ethanolamides, dialkyl and alkylol (hydroxyalkyl) amides, morpholides, and nitriles, as well as the lower acyl derivatives of higher fatty amines. These

Table 1. Household Detergent Formulations in the United States, Western Europe, and Japan

Component	Examples	Composition (%)					
		United States		Western Europe		Japan	
		With builders	Without builders	With builders	Without builders	With builders	Without builders
		Heavy-Duty Powdered Detergents					
Anionic surfactants	Alkylbenzene sulfonates	0–15%	0–20%	5–10%	5–10%	5–15%	5–15%
	Fatty alcohol sulfates	—	—	1–3	—	0–10	0–10
	Fatty alcohol ether sulfates	0–12	0–10	—	—	—	—
	Alpha-olefin sulfonates	—	—	—	—	0–15	0–15
Nonionic surfactants	Alkyl and nonylphenyl poly(ethylene glycol) ethers	0–17	0–17	3–11	3–6	0–2	0–2
Suds-controlling agents	Soaps, silicon oils, paraffins	0–1.0	0–0.6	0.1–3.5	0.1–3.5	1–3	1–3
Foam boosters	Fatty acid monoethanol amides	—	—	0–2	—	—	—
Chelators (builders)	Sodium tripolyphosphate	23–55	—	20–40	—	10–20	—
Ion exchangers	Zeolite 4A, poly(acrylic acids)	—	0–45	2–20	20–30	0–2	10–20
Alkalis	Sodium carbonate	3–22	10–35	0–15	5–10	5–20	5–20
Cobuilders	Sodium citrate, sodium nitrilotriacetate	—	—	0–4	0–4	—	—
Bleaching agents	Sodium perborate	0–5	0–5	10–25	20–25	0–5	0–5
Bleach activators	Tetraacetylethylenediamine	—	—	0–5	0–2	—	—
Bleach stabilizers	Ethylenediaminetetraacetate	—	—	0.2–0.5	0.2–05	—	—
Fabric softeners	Quaternary ammonium compounds	0–5	0–5	—	—	—	0–5
Antiredeposition agents	Cellulose ethers	0–0.5	0–0.5	0.5–1.5	0.5–1.5	0–2	0–2
Enzymes	Proteases, amylases	0–2.5	0–2.5	0.3–0.8	0.3–0.8	0–0.5	0–0.5
Optical brighteners	Stilbene derivatives	0.05–0.25	0.05–0.25	0.1–0.3	0.1–0.3	0.1–0.8	0.1–0.8
Anticorrosion agents	Sodium silicate	1–10	0–25	2–6	2–6	5–15	5–15
Fragrances		a	a	a	a	a	a
Dyes and blueing agents		a	a	a	a	a	a
Formulation aids		0–1.0	0–1.0	—	—	—	—
Fillers and water	Sodium sulfate	Balance	Balance	Balance	Balance	Balance	Balance
		Heavy-Duty Liquid Detergents					
Anionic surfactants	Alkylbenzene sulfonates	5–17	0–10	5–7	10–15	5–15	—
	Fatty alcohol ether sulfates	0–15	0–12	—	—	5–10	15–25
	Soaps	0–14	—	—	10–15	10–20	—
Nonionic surfactants	Alkyl poly(ethylene glycol) ethers	5–11	15–35	2–5	10–15	4–10	10–35
Subs-controlling	Soaps	—	—	1–2	3–5	—	—
Foam boosters	Fatty acid alkanolamides	—	—	0–2	—	—	—
Enzymes	Proteases	0–1.6	0–2.3	0.3–0.5	0.6–0.8	0.1–0.5	0.2–0.8
Builders	Potassium diphosphate, sodium tripoly-phosphate	—	—	20–25	—	—	—
	Sodium citrate, sodium silicate	6–12	—	—	0–3	3–7	—
Formulation aids	Xylene sulfonates, ethanol, propylene glycol	7–14	5–12	3–6	6–12	10–15	5–15
Optical brighteners	Stilbene derivatives	0.1–0.25	0.1–0.25	0.15–0.25	0.15–0.25	0.1–0.3	0.1–0.3
Stabilizers	Triethanolamine	—	—	—	1–3	1–3	1–5
Fabric softeners	Quaternary ammonium salts	0–2	0	—	—	—	—
Fragrances		*a*	*a*	*a*	*a*	*a*	*a*
Dyes		*a*	*a*	*a*	*a*	*a*	*a*
Water		Balance	Balance	Balance	Balance	Balance	Balance

[a]Component may be present in very small concentrations.

materials are handled as liquids and received in tank cars or barrels. In granule manufacture, they can be incorporated in the slurry before drying, but are usually blended with the detergent granules after drying to avoid emission problems from the spray dryer.

Builders

Sodium tripolyphosphate (STP) or tetrasodium pyrophosphate (TSPP) is incorporated in granular formulations as a "builder" or sequestering agent where phosphates are not prohibited. These agents serve to eliminate interference with the detergent action by the calcium and magnesium ions (hardness) in the water used in the wash solution. Both STP and TSPP may be used in powder, prill, or granule form and are received in car lots. These ingredients are most often blended into the slurry before spray drying.

Nitrilotriacetic acid and its sodium salts are used in Canada in zero- and low-phosphate detergents. The acid is a crystalline powder and the salts (disodium or trisodium) are powders.

They are received in car lots. The NTA is added to the slurry mix before drying.

Zero- and low-phosphate detergents in the United States use sodium carbonate (soda ash), sodium silicate, and calcium carbonate, or a combination of these compounds, with sodium aluminosilicates (zeolites) as builders. Sodium carbonate and calcium carbonate are powders received in bulk. Sodium silicates and zeolites are generally received as liquids. Like other builders, these materials are added to the slurry before drying.

Other Ingredients

Fillers, usually sodium sulfate or sodium carbonate, are incorporated in granule products. They are either powders or crystalline powders and are added in bulk form to the slurry before drying.

Trisodium phosphate (TSP) is used in detergent granule formulations such as dishwasher compounds and wall cleaners that are designed to clean hard surfaces. Considered functionally as an alkali rather than a sequestering agent, TSP is usually handled as a crystalline powder and is received in car lots, drums, or bags.

Carboxy methylcellulose (CMC—sodium cellulose glycolate) usually is added to heavy-duty granules and serves to prevent redeposition onto the fabric of the dirt removed by the detergent. This chemical is received in bags or drums as a powder or granule. It is added to the slurry mixture before drying.

In most synthetic detergent formulations, sodium silicate is added to inhibit the surfactant's tendency to corrode metal. It also is used to overcome production and packaging problems encountered with detergent granules. It is functionally used as a primary detergent alkali in compounds designed for hard-surface washing (e.g., machine dishwashing compounds). It can serve to retain uniform viscosity in the mixing and pumping of the slurry before it is dried, and it reduces the "tackiness" of the dried granules, facilitating their handling and reducing caking of the product after packaging. The sodium silicates are received in tank cars as water solutions.

Optical brighteners are added to many formulations. These are usually fluorescent dyes that absorb ultraviolet rays and reflect them as visible light. The dyes are received as powders in bags or as liquids in drums, and they are usually blended in the slurry before drying. Also, perfumes are added to almost all detergent products to overcome unpleasant odors and impart a pleasing scent to laundered fabrics. The perfumes are added by spraying onto the dried granules or by mixing with the liquid detergents. They are handled as liquids in small containers or drums.

Bleaches of various kinds are frequently incorporated in heavy-duty detergents. Sodium perborate, along with magnesium silicate as a stabilizer, is commonly employed. They are received as powders or crystals in boxes or bags and are added to the granules after drying. Because of lower washing temperatures, perborate activators are being introduced in some U.S. heavy-duty laundry products. These compounds are more commonly used in European laundry products. Tetraacetylethylenediamine (TAED) is the predominant bleach activator used.

In Europe and Japan, 90% of the detergents contain enzymes. Enzymes are limited to 45–50% of the heavy-duty laundry detergents in the United States because of the high use of chlorine bleaches that destroy enzymes; however, lower washing temperatures and increasing use of perborate bleaches by consumers are leading to increased enzyme usage.[1] Enzymes assist in the removal of protein-based stains from fabrics. The enzymes, which are received as prills (coated powders) in bags or drums, are heat sensitive and are destroyed if heated to 212°F. Most manufacturers blend the enzymes into the detergent granules after drying.

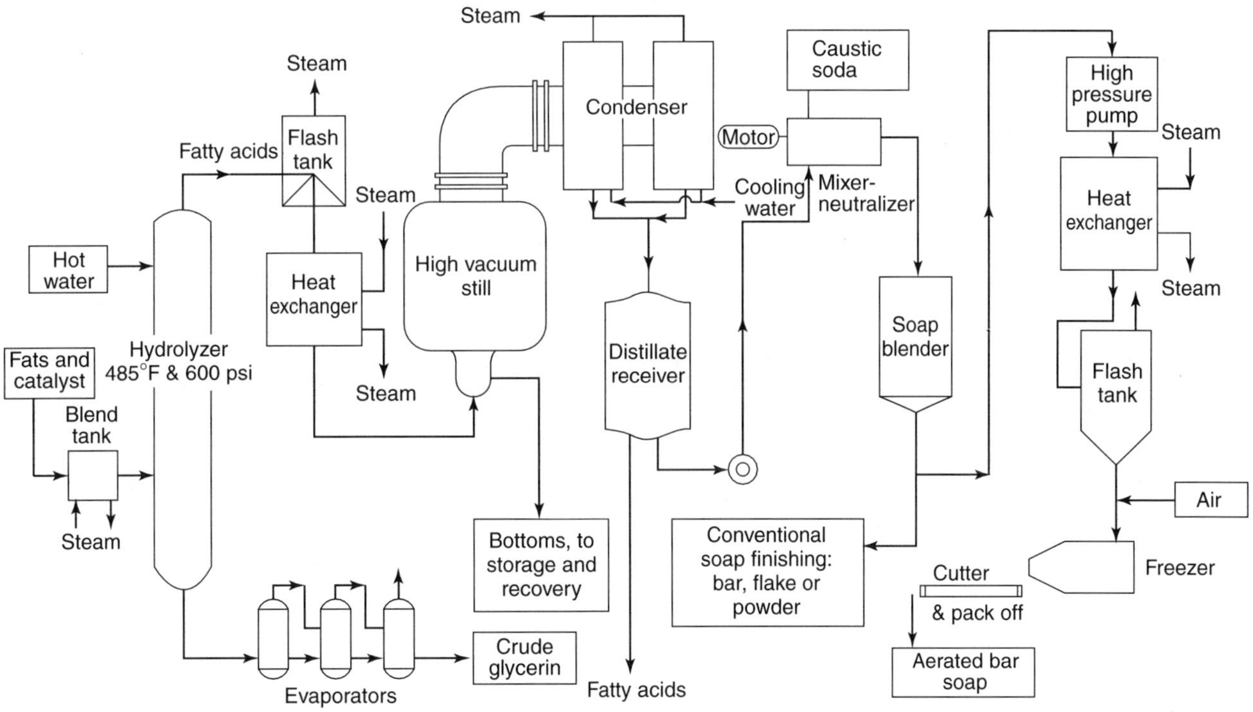

Figure 1. Continuous Process for Fatty Acids and Soaps (Ref. 4).

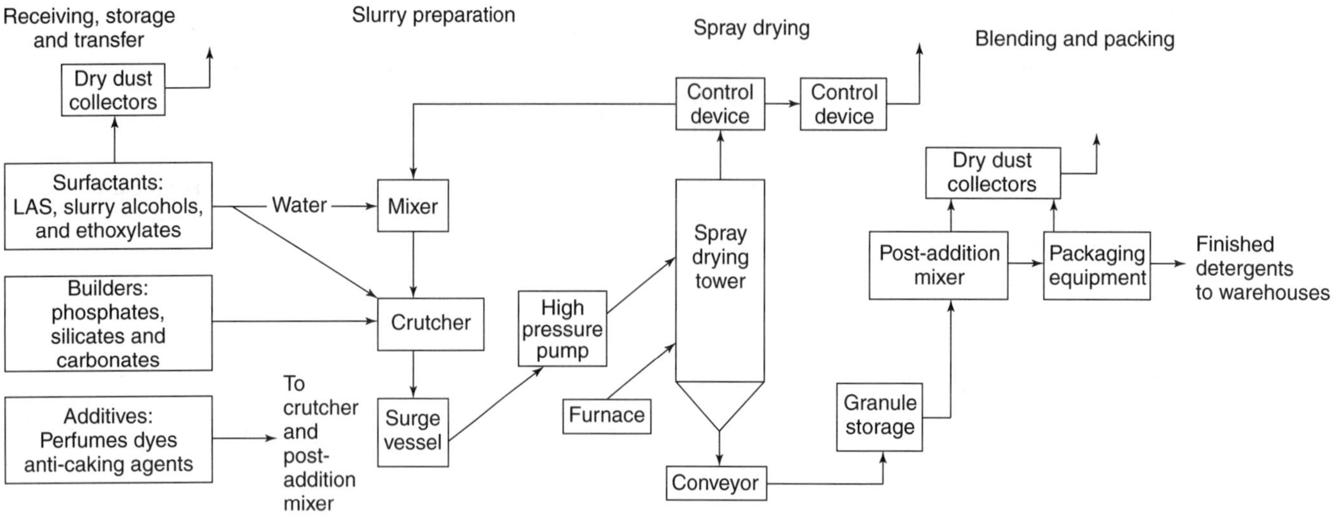

Figure 2. Manufacture of Spray-Dried Detergents (Ref. 4).

Many other compounds may be incorporated in various products. Preservatives, antioxidants, foam suppressors, and other types of additives are used. The scouring cleansers are composed principally of finely pulverized silica, active detergent, small amounts of phosphates, and frequently a bleach.

PROCESSES

The first process discussed is that of soap manufacturing. Up until the early 1940s, soap was manufactured by an alkaline hydrolysis reaction called saponification.[4] Soap was made in huge kettles where fats, oils, and caustic soda were heated to a brisk boil. After cooling for several days, salt was added, causing the "neat" soap to separate from the undrerlying spent lye and water. The soap was pumped to a closed mixing tank (crutcher) where builders, perfumes, and other ingredients were added. The soap was then rolled into flakes, cast or milled into bars, or spray-dried into a powder. Since the 1940s soap has generally been made by the direct hydrolysis of fats by water at high temperatures. This permits fractionation of the fatty acids, which are neutralized to soap in a continuous process shown in Figure 1. The direct hydrolysis process provides better control of the soap concentration and and chemical content. After the soap is recovered, it is pumped to the crutcher and treated the same as the product from the kettle process.

The second process is that of manufacturing of detergent granule formulations incorporating spray-drying processes and is discussed in more detail in the following sections. All other products are produced in processes, such as drum drying, similar to soap production. The manufacture of detergent granules incorporates three separate steps: slurry preparation, spray drying, and granule handling (including cooling, additive blending, and packaging). Figure 2 illustrates the various operations.[4]

SLURRY PREPARATION

Process Description

The formulation of slurry for detergent granules requires the intimate mixing of various liquids, powdered, and granulated materials. The soap crutcher is almost universally used for this mixing operation. Premixing of various minor ingredients is performed in a variety of equipment prior to charging to the crutcher or final mixer. The slurry, mixed in batch operations, is then held in surge vessels for continuous pumping to the spray dryer.

Air Emissions Characterization

The receiving, storage, and hatching of the various dry ingredients create dust emissions. Pneumatic conveying of fine materials causes dust emissions when conveying air is separated from the bulk solids. Many detergent products require raw materials with a high percentage of fines. Typical specifications for some raw materials include the following percentage of fine materials passing a 200-mesh screen: sodium sulfate, 12%; TSPP, 74%; STP, 53%.

The storage and handling of the liquid ingredients, including the sulfonic acids, sulfonic salts, and sulfates, do not cause emission problems other than mild odors. In the batching and mixing of the fine dry ingredients to form slurry, dust emissions are generated at scale hoppers, mixers, and the crutcher. Liquid-ingredient addition to the slurry creates no visible emissions, but may cause odors.

Air Pollution Control Measures

Control of dusts generated from pneumatic or mechanical conveying or from discharge of fine materials into bins or vessels is described in Chapter 3. No unique problems occur in hooding or exhaust systems for controlling dust emissions from conveying and slurry preparation.

Baghouses are employed not only to reduce and eliminate the dust emissions, but also for the salvage of raw materials. None of the dusts causes any serious corrosion problems. Filter fabrics should be selected that have good resistance to alkalis. Filter ratios for baghouses with intermittent shaking cleaning mechanisms should be under 3 cfm/ft^2.

SPRAY DRYING

Process Description

All spray-drying equipment designed for detergent granule production incorporates the following components:

spray-drying tower, air heating and supply system, slurry atomizing equipment, slurry pumping equipment, product cooling equipment, and conveying equipment. The towers are cylindrical with cone bottoms and range in size from 12 to 24 ft in diameter and 40 to 125 ft in height. Single towers may vary in diameter, being larger at the top and smaller at the bottom. Air is supplied to the towers from direct-heated furnaces fired with either natural gas or fuel oil. The products of combustion are tempered with outside air to lower temperatures and then blown to the dryer under forced draft. The towers are usually maintained under slightly negative pressure, between 0.05 and 1.5 in. of water column, with exhaust blowers adjusted to provide this balance. Most towers designed for detergent production are of the countercurrent type, with the slurry introduced at the top and the heated air introduced at the bottom. A few towers of the concurrent type are used for detergent spray drying, with both hot air and slurry introduced at the top. Some towers are equipped for either mode of operation, as illustrated in Figure 2.

In most towers today, the slurry is atomized by spraying through a number of nozzles rather than by centrifugal action. The slurry is sprayed at pressures of 600–1000 psi in single-fluid nozzles and at pressures of 50–100 psi in two-fluid nozzles. Steam or air is used as the atomizing fluid in the two-fluid nozzles.

Tower operations vary widely among manufacturers and among products. Heated air supplied to the tower varies from 350 to 750°F. Temperatures of air supplied to countercurrent towers are generally lower and most often range from 500 to 650°F. Concurrent tower temperatures are somewhat higher. Solids content of slurries for detergent spray drying varies from 50% to 65% by weight, with some operations as high as 70%. Moisture content of the dried product varies from 10% to 17%. Towers are designed for specific airflow rates, and these rates are maintained throughout all phases of operation. Slurry temperatures may vary, but in most formulations they do not exceed 160°F and are frequently as low as 80°F. Exit gas temperatures range from 150 to 250°F, with wet-bulb temperatures of 120 to 150°F. Air velocities in concurrent towers are usually higher than velocities in countercurrent towers. The concurrent towers produce granules that are mostly hollow beads of light specific gravity (0.05 to 0.20). Countercurrent towers produce multicellular, irregularly shaped granules that have higher specific gravities ranging from 0.25 to 0.45.

In countercurrent towers, with lower air velocities and droplets descending against a rising column of air, most of the dried granules fall into the cone at the bottom of the tower. They are discharged through a star valve, or regulated opening, while still hot. Cooling of the granules is discussed below with other granule processing procedures. Unlike in other product spray-drying operations (e.g., powdered milk), the desired detergent granule product is comparatively large. The specifications for some well-known granular products require 50% by weight or more to be retained on a 28-mesh screen. A certain amount of the product is dried to a comparatively small size. This amount depends on tower feed rates, the liquid droplet size in slurry atomization, the paste viscosity, the particular product, and other variables. Usually, the exhaust air entrains 7–10% of that portion of the granular product that is too fine to settle at the base of the tower.

Concurrent towers operate with higher air velocities than do countercurrent towers. The air is vented just above the bottom of these towers through a baffle that causes violent changes of direction to the exhaust air dynamically to separate the dried granules, which then fall to the cone bottom for discharge. Concurrent towers producing very-low-gravity granules vent air still conveying the product to auxiliary equipment for separation. The loss of detergent fines entrained in the exhaust air stream will be somewhat higher from concurrent towers than from countercurrent towers.

Air Emissions Characterization

The exhaust air from detergent spray-drying towers contains two types of air contaminants. One is the fine detergent particles entrained in the exhaust air discussed above; the second consists of organic materials vaporized in the higher-temperature zones of the tower.

The detergent particles entrained in the exhaust air are relatively large in size. Over 50% by weight of these particles are over 40 μm. These particles constitute over 95% of the total weight of contaminants in the exhaust air.[3] They consist principally of detergent compounds, although some of the particles are uncombined phosphates, sulfates, and other mineral compounds.

The second type of air contaminant, organic compounds, originates primarily from the surfactants included in the slurry. Various organic components in the slurry vaporize in the tower. The amount vaporized depends on many variables, such as tower temperatures and volatility of the organics in the slurry mixture. Volatility of the organics is a function of the chain length of surfactant compounds and the reaction completeness in the sulfonation and sulfation of anionic surfactants. The vaporized organic materials condense upon cooling in the tower exhaust air stream into micrometer- and submicrometer-sized droplets or particles.

The variety of possible detergent compounds is almost infinite, and manufacturers are continually introducing new formulations or reformulating older ones. It is not always possible to predict how certain organic compounds in a slurry mixture will affect stack emission. Work in the early 1970s indicated that if amides are present in the slurry in amounts greater than 0.5% by weight, emission problems will occur. Source tests of exhaust air from an air-pollution control scrubber with amides present in the slurry being dried in the spray tower revealed 0.08 grain of organic particulate per standard cubic foot of exhaust gas. The presence of this relatively low concentration of submicrometer-sized aerosols causes water vapor plumes to persist for long distances. Following the break or end of the water vapor plume, a highly visible contaminant plume persists for even greater distances. The amide emission rate increases as tower operating temperatures rise. Many tower operating variables affect air contaminant emissions. The most significant variables are the formulation's final granule temperature and moisture, the dryer inlet air temperature, and temperature profiles in the dryer.

Slurry formulations containing alcohol ethoxylate surfactants or alcohol sulfates with high levels of unsulfated alcohols cause similar aerosol emissions. Source tests of the aerosol leaving the scrubber indicate a particle size range of 0.2 to 1 μm, where ethoxylated alcohol surfactants were added to the crutcher. It is believed that very small amounts of unreacted alcohols coat water droplets like amides, inhibiting the evaporation of the droplets and resulting in gray water vapor plumes that persist for long distances.

Air Pollution Control Measures

The collection of air contaminants not only provides for the economic return of detergent fines to the process, but it also provides for control of submicrometer particles and aerosols to ensure compliance with air pollution prohibitory rules.

Manufacturers producing detergent granules have developed several separate approaches for capturing the detergent fines in the spray-dryer effluent for return to process. Dry cyclones and cyclonic impingement scrubbers are the primary approaches used. Cyclonic impingement scrubbers return a slurry to the crutcher, while dry cyclones return detergent fines to the crutcher. The dry cyclone separators can remove 90% or more by weight of the detergent product fines in the tower exhaust air. The detergent dust remaining in the effluent vented from the cyclones consists of particles 95% by weight less than 10 μm.[4] Particulate concentrations vary from 0.1 to 1.0 g/scf. The cyclones are designed for relatively high efficiencies and operate at pressure drops of from 8 to 10 in. of water column.[3]

Secondary collection equipment is used to collect fine dust and organic aerosols that pass the primary collectors. Mist eliminators are used after cyclonic impingement scrubbers. Cyclones are generally followed by scrubbers, scrubber/precipitators, or fabric filters. Several types of scrubbers can be used following the cyclone collectors. Venturi scrubbers operating at 8 to 10 in. of water pressure drop (using water at 8 to 10 psig distributed through nozzles in the throat, throat velocities of 8500 fpm, and water supplied to the throat at a ratio of 4.5 to 5.0 gal per 1000 ft^3 of effluent) have been used following the cyclones. Venturi scrubbers have been replaced largely by packed bed scrubbers operating at 0.5 to 2.0 in. of water pressure drop and water rates of 1 to 3 gal per 1000 acfm. Packed-bed scrubbers are usually followed by wetpipe-type electrostatic precipitators constructed immediately above the packed bed in the same vessel. Fabric filters are sometimes used after cyclones, but their application is limited to spray dryers having low drying loads and drying products with low surfactant concentrations. On efficient spray dryers with exhaust conditions near the dew point drying products with more than 10–15% surfactant, condensing water vapor and condensing organic aerosols will bind the filter fabrics.

Particulate emissions, as measured by the Environmental Protection Agency's Method 5, vary slightly with formula and operating conditions. Typical control efficiencies and emission factors for various control options are shown in Table 2.

Table 2. (Metric and English Units) Particulate Emission Factors for Detergent Spray Drying[a,4] Emission Factor Rating: E[b]

Control Device	Efficiency (%)	Particulate kg/Mg of Product	Particulate lb/ton of Product
Uncontrolled	NA		
(SCC 3-01-099-01)		45	90
Cyclone	85	7	14
Cyclone with			
Spray chamber	92	3.5	7
Packed scrubber	95	2.5	5
Venturi scrubber	97	1.5	3
Wet scrubber	99	0.544	1.09
Wet scrubber/ESP	99.9	0.023	0.046
Packed bed/ESP	99[c]	0.47	0.94
Fabric filter	99	0.54	1.1

[a]Some type of primary collector, such as a cyclone, is considered integral to a spray-drying system. NA, not applicable; ESP, electrostatic precipitator; SCC, Source Classification Code.
[b]Emission factors are estimations and are not supported by current test data.
[c]Emission factor has been calculated from a single source test. An efficiency of 99% has been estimated.

As previously mentioned, small amounts of organic, especially paraffin alcohols and amides, can result in a highly visible plume that can persist after the condensed water vapor plume has dissipated. The precipitator/scrubber appears to provide the best control for these compounds, although no conclusive data exist. Differences in formulation among manufacturers and differences in drying systems make comparison impossible. Fabric filters operating at higher temperatures will not collect organic aerosols, some of which will be caught on the Method 5 filter.

Opacity and the organics emitted are influenced by granule moisture and temperature at the end of drying, temperature profiles in the dryer, and formulation of the slurry. An alternative method for controlling visible emissions from the dryer caused by volatile organics in the slurry is to reformulate the slurry to eliminate these offending organic compounds. Sometimes the purity of raw materials such as alcohol ethoxylates can be improved by stripping short-chain molecules from the raw material. When amide compounds were identified as causing the emission problems, some manufacturers developed other formulations or methods for adding the amides to the spray-dried granules after the dryer to achieve a comparable product.

When reformulation is not possible, the tower production rate may be reduced, permitting operation at lower air inlet temperatures and lower exhaust gas temperatures. When tower temperatures are reduced, lower temperatures may result in lower granule temperature and higher granule moisture, thus reducing organic emissions.

GRANULE HANDLING

Process Description

Many manufacturers discharge hot granules from the spray tower into mixers where dry or liquid ingredients are added. The granules are usually mechanically conveyed away from the tower or mixer discharge and then are air-conveyed to storage and packaging. Air conveying cools the granules and elevates them for gravity flow through further processing equipment to storage and packaging. Air conveying of low-density granules usually is designed for 50–75 scfm air per pound of granules conveyed. At the end of the conveyor, gravity or centrifugal separators remove granule product from the conveying air. Some manufacturers mechanically lift the granules from the spray tower to aeration bins, where they are cooled or aged by injecting air at the bottom of the bin. This air percolates upward through the scrubber.

The cooled granules are screened to deagglomerate the large granules and to remove undersize or oversize particles. Further mixing or blending may be performed to add heat-sensitive compounds such as enzymes to the detergent products. Many manufacturers do not store the finished granules, but convey them directly to packaging equipment. Some detergent products are held in storage, in either large fixed bins or small-wheeled buggy bins, and then are charged to packaging equipment. Packaging is done with either scale or volumetric filling machines.

Air Emissions Characterization

Conveying, mixing, packaging, and other equipment used for granules can cause dust emissions. The granule particles, which are hollow beads, are crushed during mixing and conveying and generate fine dusts. With continuous exposure dusts emitted from screens, mixers, bins, mechanical conveying equipment, and air-conveying equipment are quite irritating to eyes and nostrils. Some additive materials, such as enzymes, bleaches, bleach activators, and brighteners, also can cause health problems. Equipment involving enzymes requires very efficient ventilation in addition to proper dust collection. Dust emissions in most cases represent a significant product loss, and their collection and return to process (usually as an ingredient of the slurry to be spray dried) is necessary for cost-efficient plant operation, as well as for air pollution control.

Air Pollution Control Measures

Dust generated by granule processing, conveying, and storage equipment does not create unique air pollution control problems. Usually, baghouses provide the best control. Collection efficiencies for baghouses are high; efficiencies can exceed 99%. No extreme conditions of temperature or humidity have to be met, but filter fabrics selected must show good resistance to alkaline materials. Baghouses utilizing intermittent shaking mechanisms should not have filtering velocities exceeding 3 fpm. Baghouses with continuous cleaning mechanisms may have filtering velocities as high as 6 fpm.

CURRENT DEVELOPMENTS

The recent introduction of superconcentrated detergents and a new push for more environmentally friendly products and packaging may cause a change equal to or greater than the shift to liquids sparked by phosphate bans. Manufacturers are developing more biodegradable surfactants from natural oils. The introduction of polymers for builder systems offers improved performance for zero- and low-phosphate detergents. Superconcentrated powders offer the environmental benefit of reducing packaging material and shipping energy by eliminating fillers and compressing the detergent to achieve two-to-one and three-to-one volume reductions. These factors may push the market back toward powder detergents.[1] While this may increase the demand for spray-drying capacity, newer formulations might also allow other manufacturing processes with less pollution potential. As of 1992, only two detergent plants had been built in the United States since 1970, and both were completed after 1988. One of those plants was the typical spray-drying plant described here; the other utilized a low-temperature fluid-bed dryer to dry the detergent with a fabric filter to control emissions. While the second plant is still limited as to the type and amount of surfactant used in the formulation, changes in types of surfactant and surfactant sulfation technology may shift significant granule production away from spray-drying towers.

REFERENCES

1. "Soaps and Detergents, New Opportunities in a Mature Business," *Chem. Wk.*, 20–64 (January 31, 1990).
2. B. F. Greek and P. L. Layman, "Higher Costs Spur New Detergent Formulations," *Chem. Eng. News* **67**(4), 29 (1989).
3. A. H. Phelps, "Air Pollution Aspects of Soap and Detergent Manufacture," *J. Air Poll. Cont. Assoc.* **17**(8), 505 (1967).
4. *Compilation of Air Pollution Emission Factors, Chapter 6, Section 8, Soaps and Detergents*, 5th ed., U.S. Environmental Protection Agency, Research Triangle Park, NC, 1993. Also available at *http://www.epa.gov/ttn/chief/ap42c6.html.*

SODIUM CARBONATE

Michael J. Barboza, Stephen T. Coldwell, and Neil M. Haymes

Sodium carbonate (Na_2CO_3), or soda ash, is a white, crystalline, alkaline, hygroscopic powder used primarily in the production of glass (49% in 1997) and in the chemical (26%) and cleaning agent (12%) industries, as well as in the pulp and paper and water treatment industries. Sodium carbonate is used to synthesize many different sodium compounds, including sodium bicarbonate (baking soda), sodium silicate and sodium tripolyphosphate (used in detergents) sodium hydroxide (caustic soda), sodium chromate and sodium dichromate (used in chrome plating), sodium aluminate (used in refining aluminum), and sodium cyanide (for electroplating). Demand depends to great extent upon the price of, and environmental issues surrounding, caustic soda, which is interchangeable with soda ash in many uses and is widely coproduced with chlorine.[1]

Soda ash is produced in density grades from 545 to 1250 kg/m^3 (34 to 78 lb/ft^3), with the lighter grade being preferred in chemical, detergent, and water treatment applications and the dense grade preferred in the glass industry. Anhydrous sodium carbonate slowly absorbs moisture and carbon dioxide from the atmosphere to form sodium sesquicarbonate ($Na_2CO_3 \cdot NaHCO_3 \cdot 2H_2O$). Sodium carbonate monohydrate ($Na_2CO_3 \cdot H_2O$), which contains 85.48% Na_2CO_3 and 14.52% water of crystallization, separates as small crystals from saturated solutions above 35.4°C (95.7°F) or may be formed by wetting soda ash with the calculated quantity of water at or above this temperature. Sodium carbonate decahydrate ($Na_2CO_3 \cdot 10H_2O$), commonly called "sal soda" or "washing soda," forms large transparent crystals containing 37.06% Na_2CO_3 and 62.94% water of crystallization; it may be crystallized from saturated solutions in the temperature range of −2.1 to 32°C (28.2 to 89.6°F) or may be formed by wetting soda ash with the calculated quantity of water in this temperature range.[2]

Although soda ash is not highly toxic, it is alkaline. Contact with the eyes may be harmful, prolonged contact with the skin may be irritating, and ingestion may be corrosive to the lining of the stomach. Soda ash is not flammable, and soda ash dust is not explosive.[2]

Table 1. U.S. Sodium Carbonate Plants

Company	Plant Location	Plant Nameplate Capacity million Mg/yr	million tons/yr	Process Type
IMC Chemicals, Inc.	Trona, CA	1.31	1.45	Direct carbonation
FMC Wyoming Corp.	Green River, WY	3.22	3.55	Sesquicarbonate, Monohydrate
General Chemicals (Soda Ash) Partners	Green River, WY	2.18	2.40	Monohydrate
Solvacy Minerals, Inc.	Green River, WY	2.09	2.30	Monohydrate
OCI Chemical Corp.	Green River, WY	2.09	2.30	Monohydrate
Tg Soda Ash, Inc.[a]	Granger, WY	1.18	1.30	Monohydrate
	Total	12.07	13.30	

[a]Plant and mine formerly owned by Texasgulf, Inc., subsidiary of Elf Aquitaine of France. Purchased by FMC in 1999.

Source: Refs. 1, 4, and 5.

The leading world producing regions, in descending order, are North America, Asia, Eastern Europe and Western Europe. Nine countries have the capacity to produce more than one million megagrams (1.1 million tons) annually. They are, in descending order, the United States, China, Russia, India, Germany, France, Japan, Poland, and the United Kingdom.[1]

Sodium carbonate is produced by *synthetic* processes such as the Solvay process and the Japanese ammonium chloride (NH_4Cl) coproduction process, or from *natural* sources by the sesquicarbonate, monohydrate, and direct carbonation processes.[3]

Most of the world manufactures sodium carbonate synthetically via the Solvay process, first commercialized in 1865. Because of increasing fuel costs, the use of the Solvay process in the United States declined since the mid-1960s, and it is no longer used in this country, although it is still used in Canada. In the United States, sodium carbonate is produced from natural sources. In 1947, soda ash was first produced from trona ore (sodium sesquicarbonate) in the Green River Basin of Wyoming by what is now FMC Corporation. Approximately 90% of domestic production originates from trona ore mined in this region of southwest Wyoming. The remainder is produced from Searles Lake brines in California.[1,3]

As of July 1999 there were six producers in the United States, with a total nameplate capacity of approximately 12 million megagrams per year (13.3 million tons per year), as listed in Table 1. The following sections will describe the natural sodium carbonate processes in detail and only briefly discuss the Solvay process, which is no longer used in the United States.

INDUSTRIAL PROCESSES[3,6,7]

The processes currently used to produce natural soda ash differ only in the recovery and primary treatment of the raw material used. The raw material for Wyoming soda ash is mined trona ore, while California soda ash is derived from sodium carbonate-rich brine extracted from Searles Lake.

Wyoming trona ore is mined by dry underground methods known as room-and-pillar, longwall, and shortwall. In the past it has also been mined by *solution mining*, and is now being mined by a variation of solution mining called "flooding extraction" mining. For the dry underground methods, the conventional blasting agent is prilled ammonium nitrate (spherical particles of ammonium nitrate) and fuel oil (ANFO). Beneficiation is accomplished with either the monohydrate or the sesquicarbonate process.

Monohydrate Process

The monohydrate process, shown in Figure 1, is the most commonly used process in the United States. Some producers use a single crushing/screening step, while others use two stages of crushing. The crushed and sized ore is heated to approximately 200°C (392°F) in the calciner. Carbon dioxide and water vapor are removed, leaving crude sodium carbonate. The crude sodium carbonate, which also contains insoluble impurities, is fed into leach tanks, or dissolvers, where the sodium carbonate dissolves. The liquor is passed to a clarifier, where suspended solids are allowed to settle.

Multiple effect evaporators are used to crystallize sodium carbonate monohydrate from the clear liquor. The mechanism of crystallization involves an increase in the concentration of dissolved sodium carbonate monohydrate until the liquor becomes supersaturated and crystallization is initiated. The liquor is returned to the process, and the sodium carbonate monohydrate crystals are transferred to product dryers. In the product dryers, both free and chemically bound moisture are evaporated from the sodium carbonate monohydrate at approximately 120–180°C (248–350°F). At some facilities, air clarifiers or rotary tubes with external cooling water are used

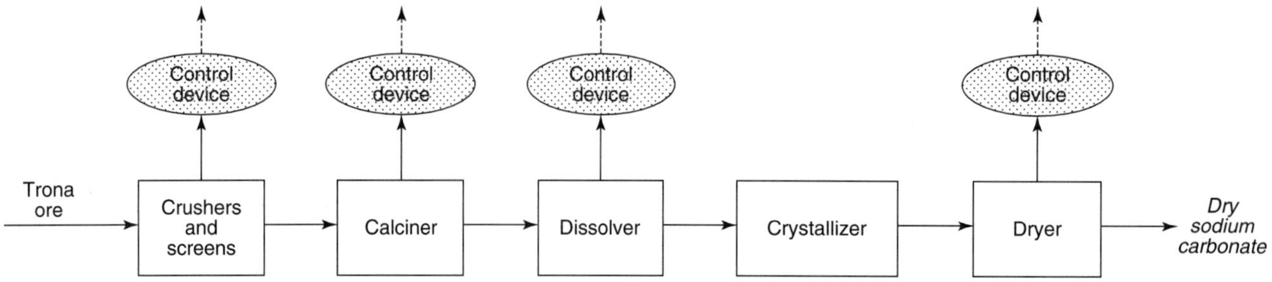

Figure 1. Sodium Carbonate Production by Monohydrate Process.

to cool the product. The product is conveyed to intermediate storage silos and then to loading facilities.

Sesquicarbonate Process

The first plant to begin processing trona used the sesquicarbonate process. The processing steps in this process are very similar to those in the monohydrate process, but the order in which they occur is different. Here the raw trona ore is first dissolved in water and then treated as brine. The liquid is filtered to remove insoluble impurities before the sodium sesquicarbonate is precipitated by using vacuum crystallizers. The result is centrifuged to remove remaining water, and can be sold as a finished product or further calcined to yield soda ash of light to intermediate density. Because the calcination step follows purification, direct firing of the calciners with coal or high-sulfur oil would contaminate the product with coal ash and sulfur. Steam tube calciners or gas-fired calciners are thus used in the sesquicarbonate process.

Direct Carbonation Process

Direct carbonation is used in refining brines containing sodium sesquicarbonate, sodium carbonate, and other salts, extracted from below Searles Lake in California. In the past, the direct carbonation process has also been used in Wyoming in conjunction with solution mining, which involves injecting dilute caustic soda (sodium hydroxide, NaOH) into the trona to dissolve it.

Sodium carbonate is produced from the brine or solution in the direct carbonation process. A flow diagram of this process is presented in Figure 2. Initially, carbon dioxide is used to partially carbonate the brine in precarbonation towers. Further carbonation of the brine occurs in primary and secondary carbonation towers. This carbonation converts the sodium carbonate to sodium bicarbonate ($NaHCO_3$). Vacuum crystallizers are used to recover sodium bicarbonate from the brine. The sodium bicarbonate filter cake enters steam-heated predryers, where some of the moisture is evaporated. The temperature in these predryers is kept below approximately 50°C (122°F) so that no carbon dioxide is evolved. The partially

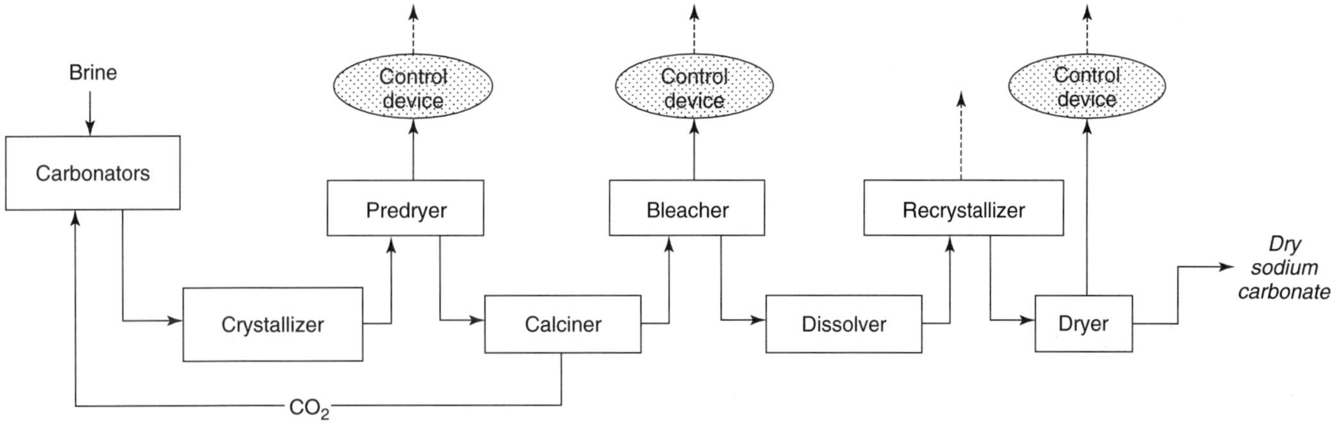

Figure 2. Sodium Carbonate Production by Direct Carbonation Process.

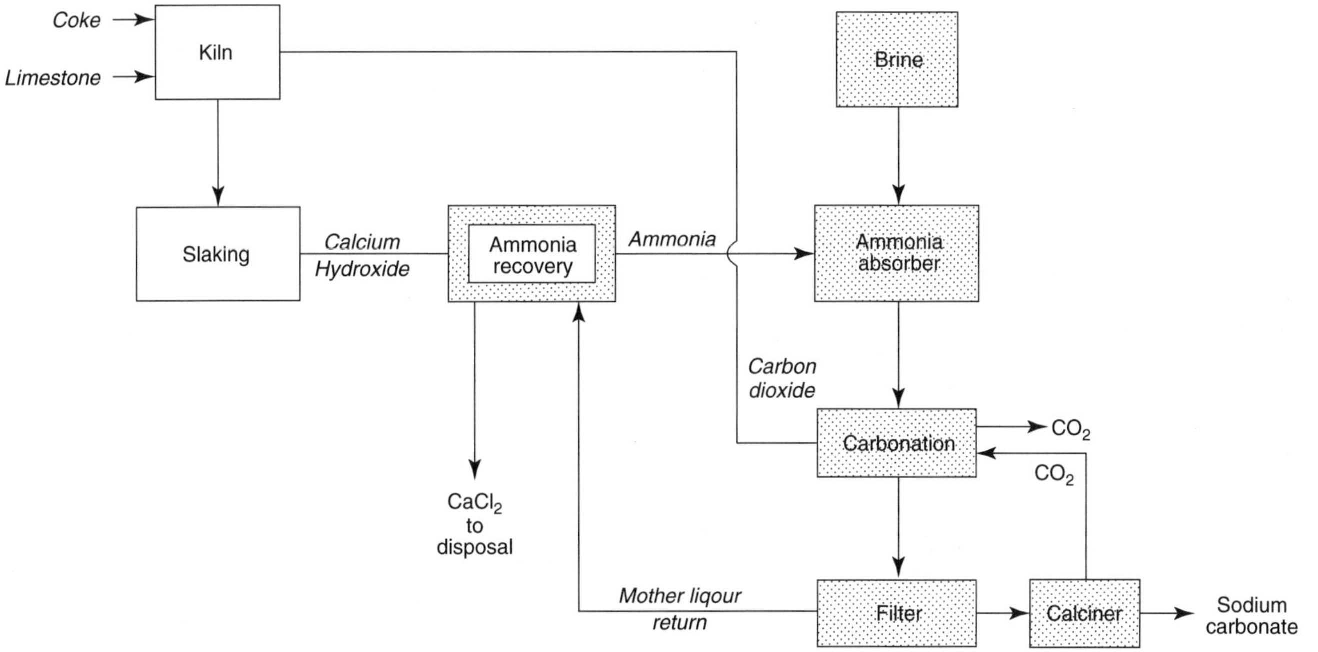

Figure 3. Sodium Carbonate Production by Solvay Process.

dried sodium bicarbonate is then further heated in a steam-heated calciner. Carbon dioxide and remaining water vapor are removed, forming impure sodium carbonate, which is bleached with sodium nitrate to burn off discoloring materials. The light sodium carbonate from the bleacher is recrystallized to sodium carbonate monohydrate. The monohydrate crystals are transferred to product dryers where free and bound moisture are evaporated.

In 1996, FMC developed a new above-ground processing method that is used in combination with "flooding extraction" mining. The technique extracts trona left in a worked-out portion of the mine, 1500 ft underground. FMC gets double duty from an aqueous slurrying agent used for carrying soda-ash process tailings down into the cavity for disposal. After depositing the tailings, the solution contacts the remaining trona, then the saturated brine is pumped to the surface. The low dissolution temperature of 20–22°C (68–72°F) saves energy costs associated with hot-solution mining. After filtration to remove insolubles, the liquor is heated to 100–110°C (212–230°F) to evaporate water and drive off carbon dioxide, then cooled to 5–15°C (41–59°F), to crystallize sodium carbonate decahydrate. These crystals can be reacted to make other sodium-containing chemicals or processed to make soda ash. Impurities are concentrated in the final mother liquor, minimizing disposal requirements.[4,8]

Solvay Process

Light sodium carbonate is made by carbonating a sodium chloride brine in the Solvay process. The carbon dioxide is produced by firing coke and limestone. Ammonia is used as a catalyst for the carbonation reaction as shown in Figure 3.

AIR EMISSIONS CHARACTERIZATION[3,6,7]

There are a number of emission sources within the natural sodium carbonate industry. Those judged most significant include calciners, bleachers, dryers, and predryers. These are process emission sources that have the potential to emit large quantities of particulate matter.

Most of the recent, high quality emission test data available represent emissions after control equipment. Table 2 presents a summary of controlled particulate emission factors for natural process sodium carbonate plants. Limited emission data are available for predryers and bleachers in the direct carbonation process, based upon a small number of test runs conducted on both controlled and uncontrolled process equipment in the late 1970s. Uncontrolled particulate emission rates and emission factors for predryers and bleachers are presented in Table 3.

Calciners

Calciners are the largest source of particulate emissions from plants using the monohydrate process. These particulates consist of sodium carbonate and inerts. The exit gas from coal-fired calciners also contains fly ash. Particulate emissions from calciners are affected by the gas velocity and the particle size distribution of the ore feed. As the gas velocity increases, the rate of increase in the particulate emissions steadily increases. Thus coal-fired calciners may have higher particulate emissions than gas-fired calciners because of higher gas flow rates. Additionally, small particles are more easily entrained in a moving stream of gas than are larger particles.

Sulfur oxides are produced from fuel combustion. The quantities produced depend upon the sulfur content of the fuel. Organics are also emitted from calciners, some of which are present in the feed in the form of oil shale. At the calcination temperatures used, organics may vaporize or be partially combusted. In addition, some organics may result from partial or incomplete combustion of the fuel.

Nitrogen oxides are emitted from direct-fired process heating units such as calciners and soda ash dryers, but very little emission test data are available for these

Table 2. Controlled Emission Factors for Particulate Matter from Sodium Carbonate Production Using Natural Process

Process	Filterable Emissions[a]			Total Emissions[b]			Nominal Control Efficiency (%)[d,e]
	kg/Mg of product	lb/ton of product	Emission Factor Rating[c]	kg/Mg of product	lb/ton of product	Emission Factor Rating[c]	
Ore mining[e]	0.0016	0.0033	C	ND	ND	NA	99.9
Ore crushing screening[e]	0.0010	0.0021	D	0.0018	0.0035	C	99.9
Ore transfer[e]	0.00008	0.0002	E	0.0001	0.0002	E	99.9
Monohydrate process: rotary ore calciner	0.091	0.18	A	0.12	0.23	B	99.9
Sesquicarbonate process: rotary calciner	0.36	0.72	B	0.36	0.73	C	99
Sesquicarbonate process: fluid-bed calciner	0.021	0.043	C	ND	ND	NA	99
Rotary soda ash dryers	0.25	0.50	C	0.25	0.52	D	99
Fluid-bed soda ash dryers/coolers	0.015	0.030	C	0.019	0.39	D	99
Soda ash screening	0.0097	0.019	E	0.013	0.026	E	99.9
Soda ash storage/loading and unloading[e]	0.0021	0.0041	E	0.0026	0.0051	E	99.9

[a]filterable particular matter is that material collected in the probe and filter of a method 5 or method 17 sampler.
[b]Total particulate matter includes filterable particulate and inorganic condensable particulate. ND, no data; NA, not applicable.
[c]EPA rating of emission factor reliability and accuracy (A through E, with A the best).
[d]For processes operating at significantly greater than ambient temperatures, emission factors for uncontrolled total particulate matter can be calculated by applying the nominal control efficiency to the controlled filterable particulate emission factors.
[e]For these ambient temperature processes, all particulate matter emissions can be assumed to be filterable at ambient conditions; however, particulate sampling according to EPA Refernce Method 5 involves the heating of the front half of the sampling train. For consistency, particulate matter measured as condensable according to Method 5 is reported as such. For these processes, emission factors for total particulate matter can be calculated by applying the nominal control efficiency to the controlled total particulate emission factors.

Source: Ref. 3.

Table 3. Uncontrolled Particulate Emissions from Bleachers and Predryers in the Direct Carbonation Process

Operation	Feed Rate[a] (Mg/h) [ton/h]	Particulate Emission Factor[a] (kg/Mg Feed) [lb/ton Feed]	Particulate Emission Rate (1000 kg/h) [1000 lb/h]	Particulate Concentration (g/dNm^3) [g/dscf]	Exit Gas Flow Rate (1000 dNm^3/min) [1000 dscf/min]
Bleacher	82	45.5–228	3.7–19.0	105–380	0.059–0.82
	[90]	[90.1–455]	[8.1–41.0]	[46–166]	[21.0–29.0]
Predryer	59	0.377–3.21	0.0222–0.189	0.261–1.49	1.4–2.4
	[65]	[0.754–6.42]	[0.049–0.417]	[0.14–0.653]	[46.0–78.0]

[a]Reported as dry impure sodium carbonate feed to the bleachers. The impurity content ranges from 0 to 15% for bleachers and 0 to 10% for predryers.

Source: Ref. 6.

processes. Reported emission factors are 0.0008 kg/Mg of feed (0.0016 lb/ton) for one calciner with a cyclone and scrubber, and 0.056 kg/Mg (0.111 lb/ton) from another calciner with a cyclone and ESP.

Dryers, Predryers, and Bleachers

Three types of dryers are used for product drying in the monohydrate and direct carbonation processes: rotary steam tube, rotary gas fired, and fluid-bed steam tube. Sodium carbonate fines are emitted from each of these dryers.

Particulate emissions from dryers are affected by the gas velocity of the feed. As the gas velocity increases, the rate of increase in the total emission rate of particulates steadily increases. Therefore, because of higher gas flow rates, and higher gas velocities, fluid-bed steam tube dryers and rotary gas-fired dryers have higher emission rates than rotary steam tube dryers.

In the direct carbonation process, rotary steam-heated predryers are used to lower the water content of wet sodium bicarbonate crystals before they are calcined. Particulates of sodium bicarbonate are the primary type of emissions from predryers.

Emissions from bleachers consist mainly of particulates of sodium carbonate. Small amounts of compounds formed from the reactions of sodium nitrate may also be present in the particulates.

Synthetic Process Emissions

Both the Solvay process and the ammonium chloride coproduction process result in ammonia emissions, as well as particulate emissions. Because these processes are not used in the United States, and emission data are not readily available, their emissions are not discussed here.

AIR POLLUTION CONTROL MEASURES[3,6]

Particulate emission control techniques applicable to sources in sodium carbonate plants include:

- Centrifugal separation
- Wet scrubbing
- Electrostatic precipitation
- Fabric filtration

Centrifugal separators, or cyclones, rely on centrifugal forces to effect particulate separation from the gas stream. Cyclones are frequently used upstream of scrubbers or electrostatic precipitators (ESPs).

Scrubbers rely mainly on inertial impaction of particles with water droplets to effect particulate separation from the gas stream. Particles are contacted with wetted surfaces or atomized liquid droplets. As the gas stream diverges to pass such obstructions, the inertia of particles in the gas stream carries the particles into the water droplets or wetted surfaces. The particulate laden liquid is then separated from the gas stream, and either recycled to the production process or discharged as waste.

Electrostatic precipitators generate an electrical field by applying a high voltage to a discharge electrode system consisting of rows of vertical wires. The strength of the field depends in part on the gas composition. The subsequent migration of the charged particles to the collected plates depends on the particle size, resistivity, gas velocity and distribution, rapping, and field strength. The collecting electrodes are rigid plates that are baffled. Electromagnetic or pneumatic hammers are used to rap the electrodes, dislodging the collected particles, which then fall into hoppers. Baffling on the collecting electrodes provides shielded air pockets that reduce re-entrainment of particles after rapping.

Fabric filtration is a process where dust particles in a gas stream are filtered out and collected.

Ore and Product Handling

Particulate emissions from ore and product handling operations, such as conveyor transfer points, crushing, and product sizing, are typically controlled by either venturi scrubbers or fabric filters (baghouses). These control devices are an integral part of the manufacturing process, capturing raw materials and product for economic reasons.

Calciners and Bleachers

Calciners and bleachers are typically controlled by cyclones in series with ESPs. Venturi scrubbers, also used to control emissions from calciners, typically achieve lower removal efficiencies than ESPs. Higher removal efficiencies may be achieved with sufficient pressure drop. It appears that a pressure drop of 154 cm (60 in.) of water may be required to achieve a removal efficiency comparable to that achieved with a four-stage ESP, based on the removal efficiency achieved from EPA test data from a gas-fired calciner with a scrubber pressure drop of about 85 cm (33.5 in.) of water.[6]

Fabric filters (baghouses) have not been reported to be in use to control emissions from calciners or bleachers. The sticky, hygroscopic nature of sodium carbonate could lead to bag blinding and/or caking.

Dryers and Predryers

Venturi scrubbers are the devices used to control emissions from rotary steam tube dryers. Cyclones in series with venturi scrubbers are used to control emissions from fluid-bed steam tube dryers and rotary steam-heated predryers. Both venturi scrubbers and ESPs have been used to control emissions from gas-fired dryers.

The exhaust gas from both rotary and fluid-bed steam tube dryers and predryers is well suited to control by wet scrubbing. The removed sodium carbonate particles are quite soluble and hygroscopic. These characteristics enhance the removal of sodium carbonate particles in wet scrubbers. However, these characteristics coupled with the high water content of the dryer exit gas can result in operating problems for ESPs or baghouses. Moisture in the exit gas can condense in ESPs or baghouses. Wet, sticky dust adheres to the electrodes and hoppers of ESPs or blinds and cakes the bags in baghouses. ESPs have been used to control emissions from rotary gas-fired dryers, but the exit gas from these dryers is at a higher temperature and lower relative humidity than gas from steam tube dryers.

RECENT AND FUTURE INDUSTRY TRENDS[3,6]

Sodium carbonate facilities are potential major sources of particulate emissions. These emissions can be effectively controlled through the proper use and maintenance of conventional add-on particulate control techniques. Tests conducted at sodium carbonate plants along with industry data contributed to the selection of ESPs as the best system of emission reduction for calciners and bleachers and venturi scrubbers as the best system for dryers and predryers.

Small amounts of sulfur oxides and organics are also emitted from direct-fired calciners; however, source tests have indicated that these emissions are very low relative to uncontrolled particulate emissions.

Environmental issues have contributed to the phaseout of the production of sodium carbonate by the Solvay process in the United States; for example, substantial quantities of aqueous waste, containing high concentrations of calcium chloride, are produced in the Solvay process. Another reason for the decline in the use of the Solvay process has been high fuel costs, since the Solvay process is more fuel intensive than any of the natural processes.

Most of the new sodium carbonate process lines constructed in Wyoming in the past decade use the monohydrate process. This process is likely to continue to be used in the future. Because of the proximity of the Wyoming trona mines to western coal mines, future plants are expected to make greater use of coal than older existing plants.

REFERENCES

1. D. S. Kostick, "Soda Ash," *Minerals Yearbook 1997*, Vol. I: Metals and Minerals, U.S. Department of the Interior, U.S. Geological Survey, 1998.
2. *Soda Ash: Technical/Handling*, General Chemical Corporation, November 1988.
3. *Compilation of Air Pollution Emission Factors, Volume I: Stationary Point and Area Sources*, AP-42, 5th ed., U.S. Environmental Protection Agency, Office of Air Quality Planning and Standards, Research Triangle Park, NC, January 1995.
4. D. S. Kostick, U.S. Geological Survey, personal communication, July 1999.
5. "FMC Completes Acquisition of Tg Soda Ash Inc.," FMC Corporation news release, July 1, 1999.
6. *Sodium Carbonate Industry—Background Information for Proposed Standards—Draft EIS*, EPA-450/3-80-029a, U.S. Environmental Protection Agency, Office of Air Quality Planning and Standards, Emission Standards Division, Research Triangle Park, NC, August 1980.
7. *Background Report: AP-42 Section 5.16, Sodium Carbonate Production*, prepared for U.S. Environmental Protection Agency, Office of Air Quality Planning and Standards by Pacific Environmental Services, Inc., Research Triangle Park, NC, January 1996.
8. G. Parkinson, "A New Soda Ash-From-Trona Route Cuts 30–40% from Standard Operating Costs," *Chemical Engineering*, April 1996.

BIBLIOGRAPHY

D. S. Kostick, "Soda Ash," *Mineral Commodity Summaries*, U.S. Department of the Interior, U.S. Geological Survey, annual.

Sodium Carbonate Emission Test Report F.M.C. Green River, Wyoming, EMB Report 79-SOD-2, U.S. Environmental Protection Agency, Office of Air Quality Planning and Standards, Research Triangle Park, NC, May 1979.

Sodium Carbonate Emission Test Report Kerr–McGee Chemical Corporation Trona, California, EMB Report 79-SOD-3, U.S. Environmental Protection Agency, Office of Air Quality Planning and Standards, Research Triangle Park, NC, March 1980.

Compilation of Air Pollution Emission Factors, Volume I: Stationary Point and Area Sources, AP-42, 4th ed., Supplement A, U.S. Environmental Protection Agency, Office of Air Quality Planning and Standards, Research Triangle Park, NC, October 1986.

40 CFR Part 60, Subpart OOO—Standards of Performance for Nonmetallic Mineral Processing Plants, *Code of Federal Regulations*, U.S. Government Printing Office, July 1, 1998.

J. R. Dyni and R. Jones, Eds., Proceedings of the First International Soda Ash Conference (June 11, 1997), Wyoming State Geological Survey, 1998.

BIBLIOGRAPHY—INTERNET

U.S. Geological Survey Minerals Information: Soda Ash Statistics and Information, *http://minerals.usgs.gov/minerals/pubs/commodity/soda_ash/index.html*.

Isonex, Inc., Chemical Engineers & Consultants, *http://www.isonex.com/*.

FMC Corporation, *http://www.fmc.com/*.

OCI Chemical Corp., *http://www.ocichemical.com/*.

Solvay Minerals, Inc., *http://www.solvay.com/na/smi/index.htm*.

SULFUR

Bruce Scott

Very-high-purity elemental sulfur is recovered as a by-product of the processing of crude oil and natural gas. This source of recovered sulfur, over 6.7 million tons in 1993, represents about three-quarters of the total elemental sulfur market in the United States, the remainder being derived from Frasch mining or imports.[1] The Frasch mining process has little direct impact on air emissions and will not be discussed further in this chapter. Elemental sulfur has a wide variety of industrial uses, the chief being the manufacture of sulfuric acid. From the standpoint of air pollution, the sulfur-recovery plants operated by petroleum refineries and natural gas plants dramatically reduce the emissions of sulfur compounds into the atmosphere from those facilities by converting noxious hydrogen sulfide into benign elemental sulfur.

All crude oils and most natural gases contain sulfur compounds in widely varying amounts. Crude oils may contain from as little as 0.1% by weight sulfur to as much as 5.0%. Natural gas is sometimes sulfur-free or may contain 15–20% or more of hydrogen sulfide (H_2S) or other sulfur compounds. These sulfur compounds are not tolerable in most finished petroleum products, and can prove to be corrosive in refinery processes used to make salable products.

While the sulfur in natural gas is usually hydrogen sulfide (H_2S), which can be directly removed by the sulfur-recovery processes described here, the removal of sulfur from raw crude oil is much more complex. In most modem refineries, this is usually accomplished by treating distillate cuts from crude oil with hydrogen in the presence of a catalyst at elevated temperature and pressure. This treatment converts the various sulfur compounds into H_2S. Some residual cuts are also desulfurized by this same means to produce low-sulfur feedstocks for further processing. Heavy, low-value cuts are usually cracked into lighter, higher-value fractions. The process of cracking normally converts organic sulfur compounds into H_2S. Being a gas at normal refinery conditions, H_2S tends to concentrate in the off-gas streams used for plant fuel. If this gas were to be burned as is, it would frequently result in unacceptable levels of atmospheric sulfur dioxide (SO_2) pollution.

PROCESS DESCRIPTION

The sulfur-recovery facilities used in most refineries and gas plants to prevent this pollution consist of three distinct steps: gas scrubbing to remove and concentrate the H_2S, sulfur recovery to convert the concentrated H_2S to elemental sulfur, tail gas treating to clean the effluent from the sulfur-recovery plant to levels appropriate for release to the atmosphere, and finally incineration to ensure that all the sulfur species released are oxidized. The refinery or gas plant operator has considerable flexibility in the choice of specific processes to be used to accomplish each of the three steps, as discussed in the following.

GAS SCRUBBING

The processes described here are used primarily for removing H_2S from gas streams, but are also used to scrub H_2S from liquid streams such as propane and butane. Application to liquids differs from gas scrubbing only in the engineering details and will not be described further.

Removal of H_2S from gas streams is accomplished by scrubbing with a water solution of an organic amine in a packed or trayed tower (scrubber or absorber). The amine solution is alkaline and the weakly acidic H_2S in the gas stream readily dissolves in it. The acid and base react to form a salt, as illustrated in the following reaction.

$$2R_2NH + H_2S \longleftrightarrow (R_2NH_2)_2S + \text{heat} \qquad (1)$$

It should be noted that carbon dioxide (CO_2) is also a weak acid in water and will react with amine by the same mechanism as reaction (1). Hydrocarbon gases, being insoluble in water and neutral rather than acidic, are not affected by the amine solution and so pass out of the scrubbing tower. As indicated by reaction (1), the chemical reaction of H_2S or CO_2 with an amine is exothermic. Reaction (1) is shifted to the right to promote scrubbing by low temperatures and to the left by high temperatures to promote regeneration of the scrubbing solvent.

In commercial plants, the rich amine solution, loaded with H_2S and CO_2 (acid gases), is sent to a stripping tower (regenerator), where it is boiled and the acid gases are steam stripped out. The amine solution, now containing very little acid gas, is cooled and returned to the scrubbing tower for reuse. The acid gases from the regenerator are cooled and sent to the sulfur-recovery plant for conversion to elemental sulfur. Heat exchangers, pumps, filters, and a flash drum for hydrocarbon removal complete the equipment list for a typical plant. Frequently, one regenerator serves several absorbers, with the amine being pumped around the refinery. In other applications the regenerators may be associated with individual absorbers and the concentrated acid gas is piped to a central sulfur recovery unit. Figure 1 illustrates a typical gas scrubbing plant.

A number of amines are in commercial use in gas scrubbing plants. The most common is diethanolamine (DEA), a secondary amine as illustrated in reaction (1). Monoethanolamine (MEA), a primary amine, has been used extensively and is still widely used where its higher reactivity is needed, particularly for CO_2 removal. Methyldiethanolamine (MDEA), a tertiary amine, is coming into greater favor with refiners because of its ability to absorb H_2S selectively and reject most of the CO_2 in the gas stream. MDEA may also offer improved energy consumption and increased treating capacity. There are other amines in use, and the above amines can be blended or otherwise promoted to enhance their performance in specialized applications.

Typical operating conditions in an amine plant are approximately as follows, although individual plants may vary greatly. Absorbers in refineries are operated at about 90–120°F and at pressures ranging from 80 to 150 psig in most towers to over 2000 psig in certain hydrotreating applications. Natural gas plant absorbers are operated at the same temperatures, but normally at higher pressures, typically 400–1000 psig, or more. Regenerators operate at low pressure, usually 10–20 psig, and at bottoms temperatures in the range of 245–265°F, set by the boiling point of the amine solution. Acid gases are typically cooled to below 120°F to minimize the water sent to the Claus unit.

In some cases, usually where there is a very low level of H_2S in the sour gas, other treatment processes may be used in lieu of amine treatment. These generally use a solid or liquid scrubbing agent that reacts chemically with H_2S, removing it from the gas stream. The spent scrubbing agent frequently is sent directly to disposal, although some can be regenerated and reused. There are a number of these scrubbing agents on

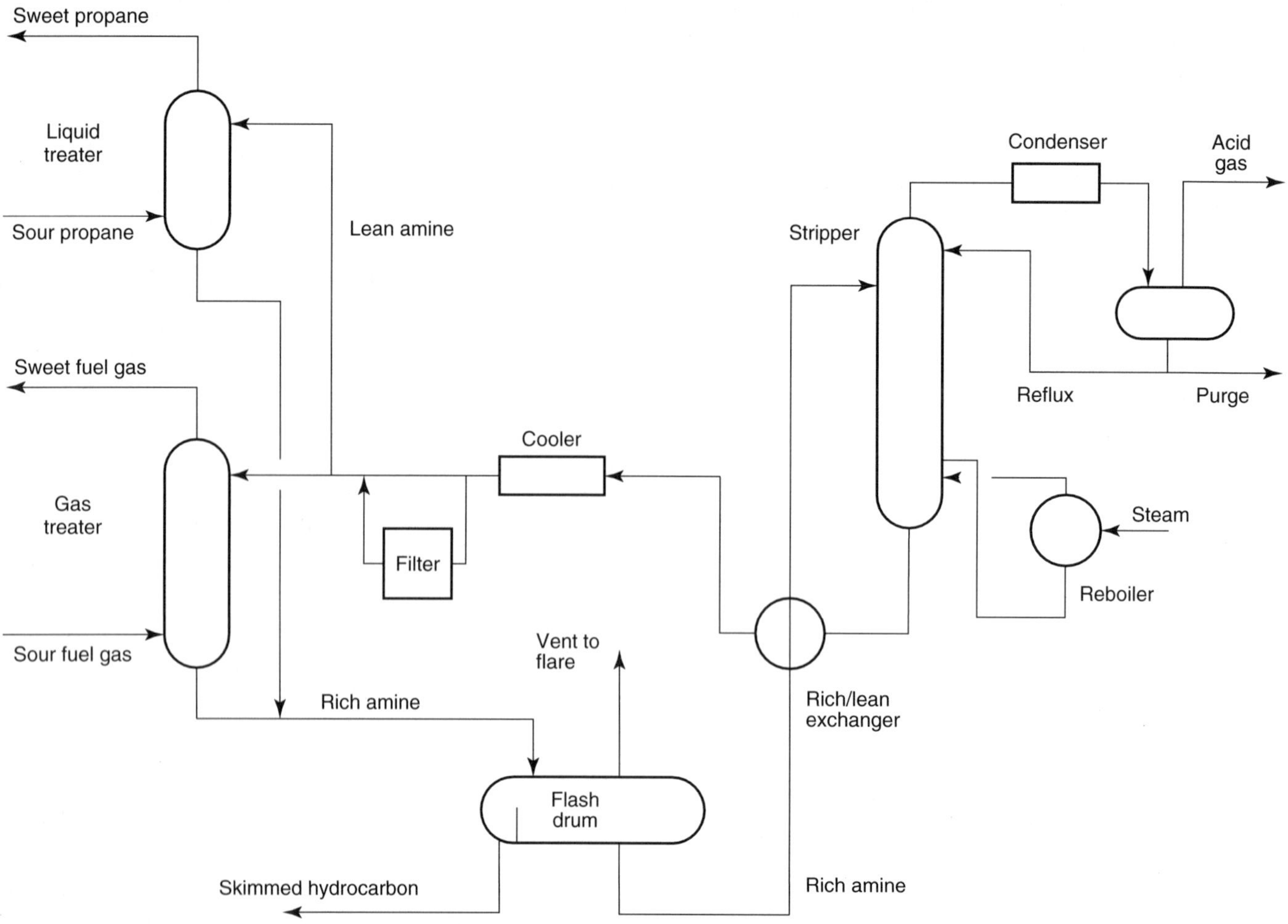

Figure 1. Typical Amine Scrubbing Plant.

the market and a detailed discussion will not be attempted here.

Sour gases are also sometimes scrubbed directly with liquid redox solutions. The H_2S is removed and converted directly to elemental sulfur, which can be recovered and the scrubbing solution regenerated for further use. These processes have also been widely used as tail gas cleanup processes; however, this application is declining.

SULFUR RECOVERY

Acid gases from the amine plant are processed in a sulfur-recovery plant, utilizing the modified Claus process in the vast majority of cases. In the Claus process, one-third of the H_2S in the acid gas stream is burned with air to form SO_2:

$$H_2S + \tfrac{3}{2}O_2 \rightarrow SO_2 + H_2O + \text{heat} \qquad (2)$$

The SO_2 reacts with the remaining H_2S to form sulfur, as follows:

$$2H_2S + SO_2 \longleftrightarrow 3\,S + 2H_2O + \text{heat} \qquad (3)$$

Reactions (2) and (3) take place in the high-temperature combustion part of the plant and accounts for something over half of the total plant conversion. The remaining sulfur is formed by reaction (3) in lower-temperature catalytic parts of the plant. Because eq. (3) represents an equilibrium chemical reaction, it is not possible for a Claus plant to convert all the incoming sulfur compounds to elemental sulfur. Each catalytic stage can recover half to two-thirds of the sulfur entering it. Some H_2S, SO_2, sulfur vapor and mist, and traces of other sulfur compounds formed in the combustion section escape with the inert gases from the tail end of the plant (tail gas). Thus it is frequently necessary to follow the Claus plant with a tail-gas cleanup plant.

Since the 2:1 ratio of H_2S:SO_2 indicated by Eq. (3) is essential to efficient operation of the plant, an extensive control system is usually provided to ensure the correct injection of air into the combustion chamber.[2] Many of the engineering differences among Claus plants are centered on the combustion chamber and the need for proper combustion conditions for a wide range of compositions and flammabilities of the acid gas.[3] Hot gas from the combustion chamber is quenched by passing through a waste-heat boiler, in which high- or medium-pressure steam is generated, before entering a sulfur condenser. Since reaction (3) is an equilibrium reaction, conversion to sulfur is improved by removing that product as the reactants move from stage to stage.

This step is followed by a catalytic stage, consisting of a gas reheater, a catalyst chamber, and a condenser to remove the product sulfur as a liquid. This catalytic stage can be repeated

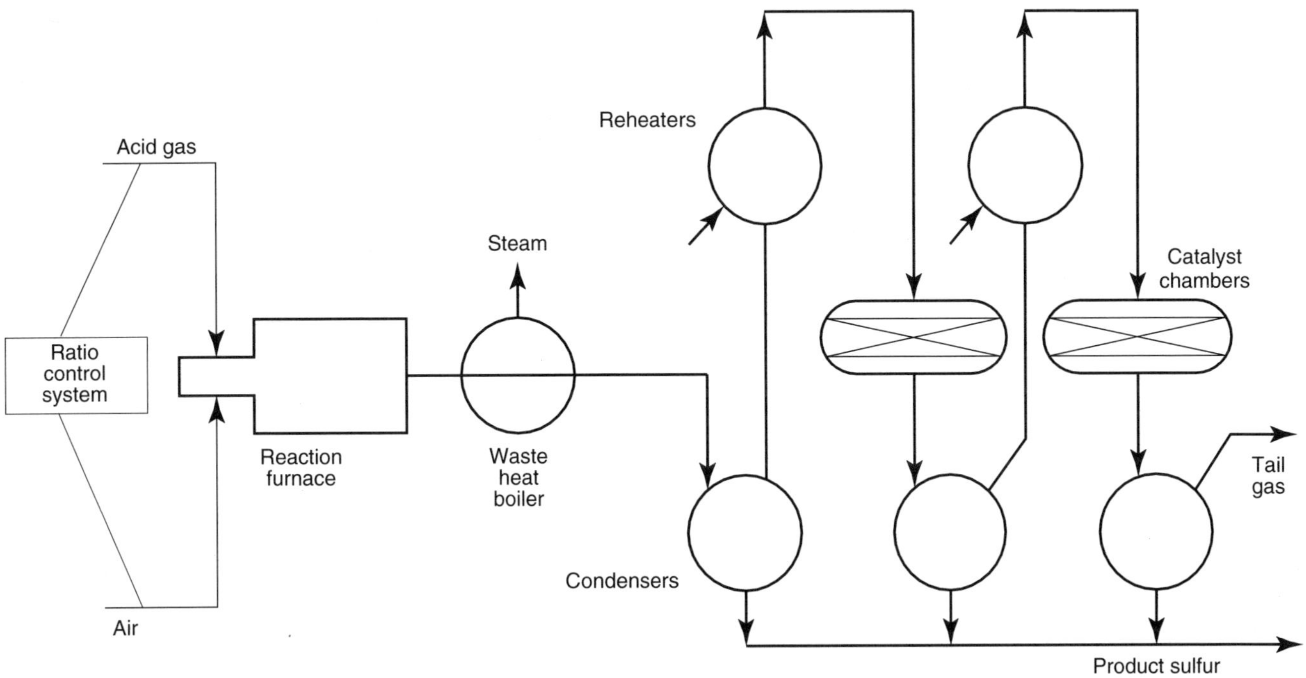

Figure 2. Typical Two-Bed Claus Plant.

one to four times, depending on the level of conversion desired. Most plants are now built with two stages (some jurisdictions require three), since the need for a tail-gas cleanup plant makes the third and fourth stages uneconomical. From the condenser of the final catalytic stage, the process stream passes to some form of tail-gas treatment process.

The liquid sulfur, at 99.95% or higher purity, runs from the condensers through a seal leg into a covered pit, from which it is pumped to trucks or rail cars for shipment to end users. Some plants, remote from end users, form molten sulfur into solid pellets or other shapes for shipment. Figure 2 illustrates a simple, two-stage Claus plant.

Claus plants operate at low pressures, rarely higher than 10 psig. Temperatures range from 1800 to 2800°F in the combustion chamber to 400–600°F in the catalytic reactors. Tail gas from the last condenser is at about 270°F, just above the freezing point of the sulfur, to minimize the carryover of sulfur vapor. The catalyst used is alumina, sometimes promoted to aid in the hydrolysis of COS and to resist sulfation.

Claus plants in refineries are frequently used to process gas from the treatment of refinery sour waters. This gas may contain significant quantities of ammonia, as well as H_2S. The ammonia can be efficiently converted to nitrogen and water in the Claus plant, if proper attention is given to maintaining operating conditions in the right range. Higher combustion temperatures (above 2200–2300°F) and adequate residence time in the combustion chamber are required to ensure complete destruction of ammonia. Note that feeding ammonia to the Claus unit will necessarily reduce the conversion efficiency of the unit slightly.

TAIL-GAS TREATMENT

Gas from the Claus plant still contains 0.5–1.5% of sulfur compounds, which must be removed before the gases can be safely vented to the atmosphere. This is done in a tail-gas treatment, or cleanup plant. The number and variety of tail-gas cleanup options are beyond our scope here, and only the two most common types will be described in any detail. These are the amine scrubbing and the subdewpoint processes. In jurisdictions requiring high levels of cleanup, the amine scrubbing process has displaced the liquid redox processes because of its ease of operation and high efficiency. Where an intermediate level of cleanup is acceptable, the subdewpoint process has become the dominant method for achieving these targets. There are a number of other, less frequently used, processes for acheiving intermediate levels of sulfur recovery.

AMINE SCRUBBING PROCESS

Tail gas from the Claus plant is heated to about 550°F, hydrogen is added (or generated in the heater, if necessary), and the gas is passed through a catalyst bed, where virtually all the sulfur compounds are converted to H_2S; for example:

$$SO_2 + 3H_2 \rightarrow H_2S + 2\,H_2O + \text{heat} \quad (4)$$

Other sulfur compounds are similarly hydrogenated. The hot gas is frequently cooled in a waste-heat boiler before passing to a quench column where excess water and other contaminants are removed. Figure 3 shows the above reaction section. The cool, dry gas stream flows to the bottom of an amine absorber, where the H_2S is removed in the same manner as indicated in the gas scrubbing section above. Acid gas is stripped from the amine, cooled to remove water, and recycled to the front of the Claus plant to be converted to sulfur. In its general layout, this amine plane is quite similar to that described above, although many engineering details are different. Gas from the amine absorber is routed to an incinerator (see below) before being vented to the atmosphere.

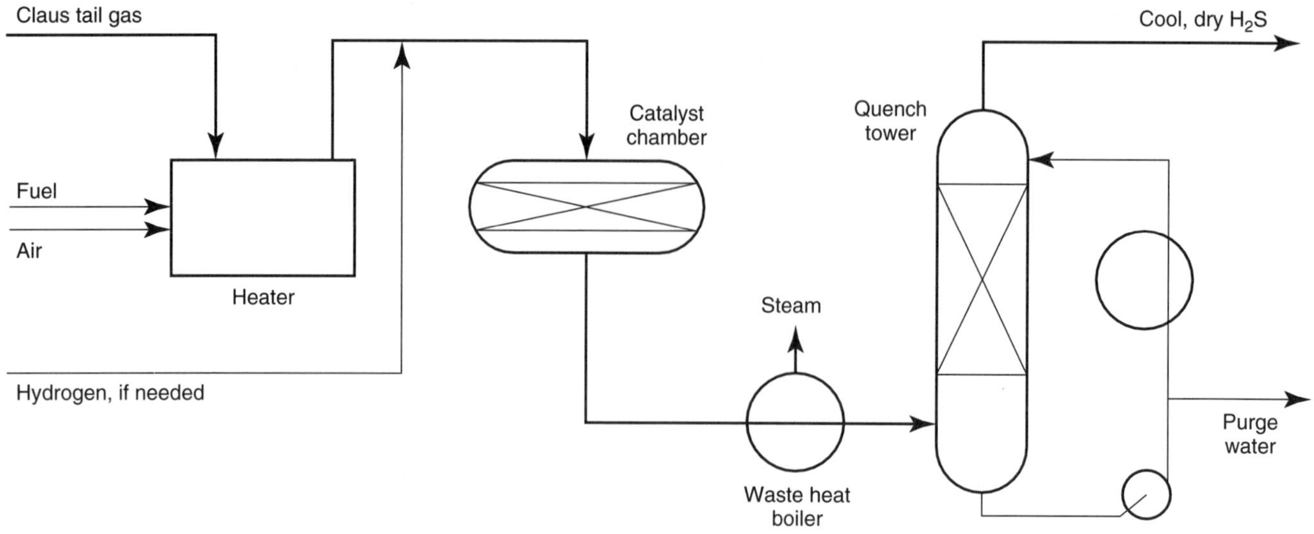

Figure 3. Tail Gas Treater, Hydrogenation Section.

Because of the low scrubbing pressure and the stringent emissions target, the need for careful control of the amine temperature and lean acid gas loading is much more important than in the main refinery treating system. Selective amines (MDEA or DIPA) are used exclusively because of the need to vent CO_2 from the system.

SUBDEWPOINT PROCESSES

In area where pollution regulations permit, outside the United States, subdewpoint processes are frequently used. The Claus reaction, reaction (3) above, has a higher conversion at lower temperatures. In the Claus process the catalyst temperature is kept above the sulfur dewpoint to prevent condensation of liquid sulfur in the catalyst bed. In the subdewpoint tail gas process two or more additional catalyst beds are added to the Claus unit. These are operated at temperatures just above the sulfur freezing point and sulfur is allowed to condense in the catalyst bed. The beds are cycled out of the Claus unit and regenerated by passing hot gas through the catalyst to vaporize the sulfur off the catalyst. The regenerated beds are then cycled back into the process.

OTHER PROCESSES

A number of other tail-gas treating processes that are used commercially are not discussed here because they are not widely used or would not normally meet the criterion of best available control technology. These include liquid redox processes that convert H_2S directly to elemental sulfur using a vanadium or iron chemistry and oxidation processes in which sulfur compounds are oxidized to SO_2, and recycled to the Claus plant. Other processes catalytically oxidize H_2S directly to sulfur and water at high efficiency.

While not strictly pollution control devices, most sulfur-recovery facilities end by routing the tail gas through a thermal (or sometimes catalytic) incinerator. These incinerators convert the small amount of H_2S remaining in the cleaned tail gas to SO_2. Nearly complete oxidation of sulfur compounds in the thermal incinerator require a temperature of 1100–1200°F with 1.0–1.5 sec residence time and adequate mixing. The oxygen level in the incinerator stack is typically 2–3%. Most stacks are fitted with continuous monitors to measure the level of SO_2 in the stack, frequently corrected for excess air.

AIR EMISSIONS CHARACTERIZATION

Air emissions are affected at both ends of the sulfur recovery train. The amine plant sets the sulfur level in the refinery fuel gas system, which in turn controls stack emissions from furnaces. The tail-gas treatment plant controls sulfur emissions from the sulfur-recovery complex itself.

Typical refinery amine plants are designed to limit the H_2S content of fuel gases to below 160 ppm, the current Environmental Protection Agency (EPA) limit, and generally are able to clean to well below that level, although total reduced sulfur content may be significantly higher than the H_2S level. Uncontrolled emissions from refineries and gas plants could be as high as several thousand parts per million, depending on the raw materials and processing equipment used.

Natural gas plants are able to treat gas to below 4 ppm H_2S, the normal pipeline specification. This is better than accomplished at refineries because of the higher treating pressure, which favors the scrubbing reaction, and the normally lower content of other trace sulfur compounds. The other pollution issue in amine treating plants involves occasional losses of amines into the plant effluent water system.

Older, or very small (less than 20 LTPD sulfur), Claus plants without tail-gas cleanup plants influence air pollution directly by their sulfur-recovery efficiency. This is a function of the number of catalytic stages in the plant and the concentration of H_2S in the acid gas fed to the plant. The higher the acid gas concentration and the more stages, the higher will be the recovery of elemental sulfur. Contaminants (e.g., carbonyl sulfide, hydrocarbons, water or CO_2) in the acid gas stream also reduce the expected recovery efficiency. While it is difficult to generalize, one would expect a two-catalytic-bed Claus plant to recover 94–96% of the inlet sulfur as liquid. The range covers acid gas H_2S concentrations from 35% to 90%. The equivalent recoveries from a three-bed plant are 96–97.5% and from a four-bed plant, 97–98.5%.[4] The liquid sulfur produced from a

Claus plant is saturated with H_2S, and, unless degassed, some low parts-per-million levels of H_2S and elemental sulfur will be released from pit and storage tank vents.

Tail-gas treatment plants are rated by the concentration of sulfur compounds in the treated vent gas, corrected for moisture and air dilution. Amine scrubbing plants normally release 50–250 ppm of sulfur compounds, although older plants may be designed for less efficient cleanup. Current EPA limits are 250 ppmv. Subdewpoint tail-gas treatment plants increase Claus recovery efficiency to 98.8–99.3% of the feed sulfur.

AIR POLLUTION CONTROL MEASURES

Measures taken by plant operators to maintain, or improve, the pollution control efficiency of their sulfur recovery facilities include the following.

Gas Scrubbing

The principal cause of excess emissions from an amine plant is foaming of the amine solution in the absorber or the regenerator. Foaming can be caused by excessive particulate contamination, the presence of liquid hydrocarbons or oils in a gas scrubber, and contamination by surface-active chemicals. Thus the operator's first concern must be for the cleanliness of the amine solution. Strongly acidic compounds in the gas stream, or oxygen contamination that forms acidic compounds, react with the amine to form salts that will not decompose in the stripper. These are called heat-stable salts and will gradually cause the performance of the amine to deteriorate. At levels above 4–5% in the amine solution, heat-stable salts promote corrosion, which can lead to foaming and eventual equipment failure. Keeping the amine solution cool and well-stripped is also important to the scrubbing efficiency of the process. Amines (and H_2S) are known to cause cracking in the heat-affected zone of welds in carbon-steel vessels. Care must be exercised in the fabrication and repair of steel equipment for this service, and stress relief of vessel and piping welds is required.

Sulfur Recovery

Careful control of the air injection rate into the Claus plant is the chief factor affecting the recovery efficiency. In addition to the front-end control system, most plants have a tail gas ratio analyzer to measure the concentrations of H_2S and SO_2 in the gas leaving the last condenser. Poor combustion control can cause the catalyst to "sulfate" and lose activity. Periodic rejuvenations can control this phenomenon temporarily. Because of the toxic nature of H_2S, Claus plants must be equipped with a safety interlock system to put the plant automatically into a safe condition in case certain critical variables are beyond preset limits.

More and more sulfur plant operators are adding oxygen to the Claus unit to increase the capacity of the plant. By attention to the engineering details it is possible to more than double the capacity of a Claus unit by using gaseous oxygen for the combustion reaction (2) above. This capacity is frequently used to provide redundancy for the sulfur recovery system in case of shutdown.

Some plants remove (degas) the H_2S dissolved in the liquid sulfur produced. This may reduce the emissions from the complex slightly, but its main value is in transportation and storage safety. It is possible to reach explosive levels of H_2S in the vapor space of sulfur pits and rail cars with undegassed sulfur.

Tail Gas Treating

The pollution control concerns in the amine scrubbing process discussed include the maintaining of proper air-to-fuel ratios in the heater, adequate hydrogen to accomplish the necessary hydrogenation, and control of the pH of the quench tower water. Concerns in the amine process parallel those in the amine gas scrubbing process, with the extra care to ensure adequate stripping and cooling of the amine to the scrubber. Amine stripping and temperature set the H_2S leak rate through the absorber.

Emissions from the subdewpoint plants are controlled primarily by maintaining the proper air-to-feed ratio and temperature profile in the catalyst beds and by adequate regeneration and cooling of the regenerated bed before putting it into service.

Incinerators must operate at a temperature of 1200°F or higher if all the H_2S is to be combusted. Proper air-to-fuel ratios are needed to eliminate pluming from the incinerator stack. Stacks should be equipped with analyzers to monitor the SO_2 level.

REFERENCES

1. "Industry Statistics," *Sulphur* **230** (1993).
2. "Control: The Key to Successful Claus Sulphur Recovery," *Sulphur* **202**, 24 (1989).
3. J. B. Hyne, "Optimum Furnace Configuration for Sulphur Recovery Units," *Sulphur* **198**, 24 (1988).
4. H. G. Paskall, *Capability of the Modified Claus Process*, Western Research, Calgary, 1979, p. 172.

BIBLIOGRAPHY

A. L. Kohl and F. C. Riesenfeld, *Gas Purification*, 4th ed., Gulf Publishing Co., Houston, 1988.

U. H. F. Sander, H. Fischer, et al., *Sulphur, Sulphur Dioxide and Sulphuric Acid*, British Sulphur Corp. Ltd., London, 1984.

"Sweetening of natural gas: I," *Sulphur* **192**, 30 (1987).

"Sweetening of natural gas: II," *Sulphur* **193**, 30 (1987).

"Gas Processes '98," *Hydrocarbon Processing* **77**(4), April 1998.

SULFURIC ACID

Thomas Muller

INTRODUCTION

Sulfuric acid is a basic raw material in an extremely wide range of industrial processes and manufacturing operations. Because of its widespread use and relatively low production cost versus shipping cost, sulfuric acid plants are scattered throughout the nation near every industrial complex.

The production of sulfuric acid involves the generation of sulfur dioxide (SO_2), its oxidation to sulfur trioxide (SO_3), and the reaction of SO_3 with water to form sulfuric acid. In practice, the reaction of SO_3 with water is accomplished by first absorbing the SO_3 in strong sulfuric acid and then adding water.

The combustion of elemental sulfur is the predominant source of SO_2 for on-purpose acid production. The combustion of hydrogen sulfide from waste gases, the thermal decomposition of spent sulfuric acid or other sulfur-containing materials, and the roasting of pyrites are also used as sources of SO_2. Since the mid 1970s, primarily for environmental reasons, many non-ferrous metal producers have built sulfuric acid plants to capture the large amounts of SO_2 generated in the smelting process. Although not currently a major source, SO_2 gases from power plants have also been used to produce sulfuric acid.[1,2]

PROCESS DESCRIPTION

The two main processes for sulfuric acid manufacture are the chamber process and the contact process. The chamber process uses the reduction of nitrogen dioxide (NO_2) to nitric oxide (NO) as the oxidizing mechanism to convert SO_2 to SO_3. The contact process uses a vanadium-based catalyst to oxidize SO_2 to SO_3. During the first part of this century, the contact process gradually replaced the chamber process. Continued improvements have made it highly efficient from the standpoint of conversion efficiency (SO_2 to SO_3) and energy recovery. Today, essentially all sulfuric acid is manufactured by the contact process. Therefore, further discussion will be restricted to the contact process.

Contact Process

Contact plants may be divided into two broad classifications, "hot gas" plants and "cold gas" plants. In hot gas plants the SO_2-containing gas, generally from burning sulfur, need only be cooled before entering the catalytic converter. Cold gas plants have dirty gas streams, from spent acid regeneration, pyrite roasting, smelting operations, etc., as their SO_2 source. These gas streams must be cooled and cleaned before entering the contact plant. So called "wet gas" plants generally use hydrogen sulfide (H_2S) as the sulfur source and are a special type of hot gas plant. The process gas in wet gas plants contains appreciable water, but is sent to the converter without drying.

SO_2 Oxidation

The heart of the contact process is the oxidation of SO_2 to SO_3. It is difficult to explain the contact process without first explaining the characteristics of this reaction. The reaction is written as:

$$SO_{2(g)} + \tfrac{1}{2}O_{2(g)} \longleftrightarrow SO_{3(g)} + \text{heat} \qquad (1)$$

$$\Delta H \cong -41,400 \text{ Btu/lb mole}$$

Two important characteristics of this reaction are that it is highly exothermic and reversible. Being reversible, higher SO_3 concentrations and higher temperatures (i.e., more heat) favor the reverse reaction. As the forward reaction proceeds, and the temperature and amount of SO_3 increase, the reverse reaction rate increases, slowing the *net* oxidation of SO_2. Removing either heat and/or SO_3 will decrease the reverse reaction and keep the forward reaction proceeding at a reasonable rate. This is an essential factor in converter design and in the *dual-absorption* process currently used in most modern sulfuric acid plants.

The reversible nature of this reaction can also be described using the concept of chemical *equilibrium*. When the forward and reverse reaction rates are equal, the overall reaction is said to be in equilibrium. The equilibrium state is a function of gas composition, temperature, and pressure.* Once the reaction reaches equilibrium, no additional conversion of SO_2 to SO_3 is possible unless the process gas conditions are changed. Cooling the gas moves the gas away from the existing equilibrium curve, creating a new driving force for the forward reaction. Removing SO_3 restores the forward reaction by defining a new equilibrium curve entirely.

An additional consideration when designing any chemical process is reaction rate. Many reactions proceed very rapidly without much help. Unfortunately, the oxidation of SO_2 to SO_3 is not one of them. Without the use of a catalyst, this reaction would proceed very slowly, if at all.

Even with a catalyst, at 400°C the reaction rate of SO_2 oxidation is too slow to be practical. Speeding up the reaction requires higher temperatures. Unfortunately, as we saw above, increased temperatures lead to unfavorable equilibrium conditions. Optimum converter design and operation requires a balance between reaction rate (which likes high temperature) and equilibrium (which likes low temperature).

Three primary factors influencing overall conversion are:

- Equilibrium considerations (gas strength and temperature)
- Minimum temperatures required for catalyst to operate
- The need to keep catalyst masses of practical size

Together, these factors limit conversion in a *single-absorption* plant to about 98%. *Dual-absorption* plants can achieve overall conversions greater than 99.7%. (See "Single-Absorption Process" and "Dual-Absorption Process" below.)

Single-Absorption Process

A flow diagram of a typical single-absorption sulfur-burning contact sulfuric acid plant is shown in Figure 1.† Combustion air is typically provided by a single-stage centrifugal blower. The blower is usually driven by a steam turbine, but electric drives are not uncommon, especially in enhanced energy recovery and co-generation plants. The pressure at the blower discharge may be as high as 6 psig, or 180 in. H_2O.

Combustion air passes through a drying tower to remove moisture before entering the sulfur burner. In the drying tower, moisture is removed from the air by countercurrent scrubbing with sulfuric acid. Drying tower acid strength is usually maintained at 93% or 98–99%, depending on product requirements. Since the acid in the drying tower picks up moisture from the air, some drying tower acid is

* A few high-pressure processes have been proposed, and at least one new high-pressure plant was actually built. However, most plants run at essentially atmospheric pressure.

† Although most new plants in the U.S. are *dual-absorption* plants, it is helpful to consider the *single-absorption* design first since dual-absorption plants are merely an extension of the same principles used in a single-absorption plant.

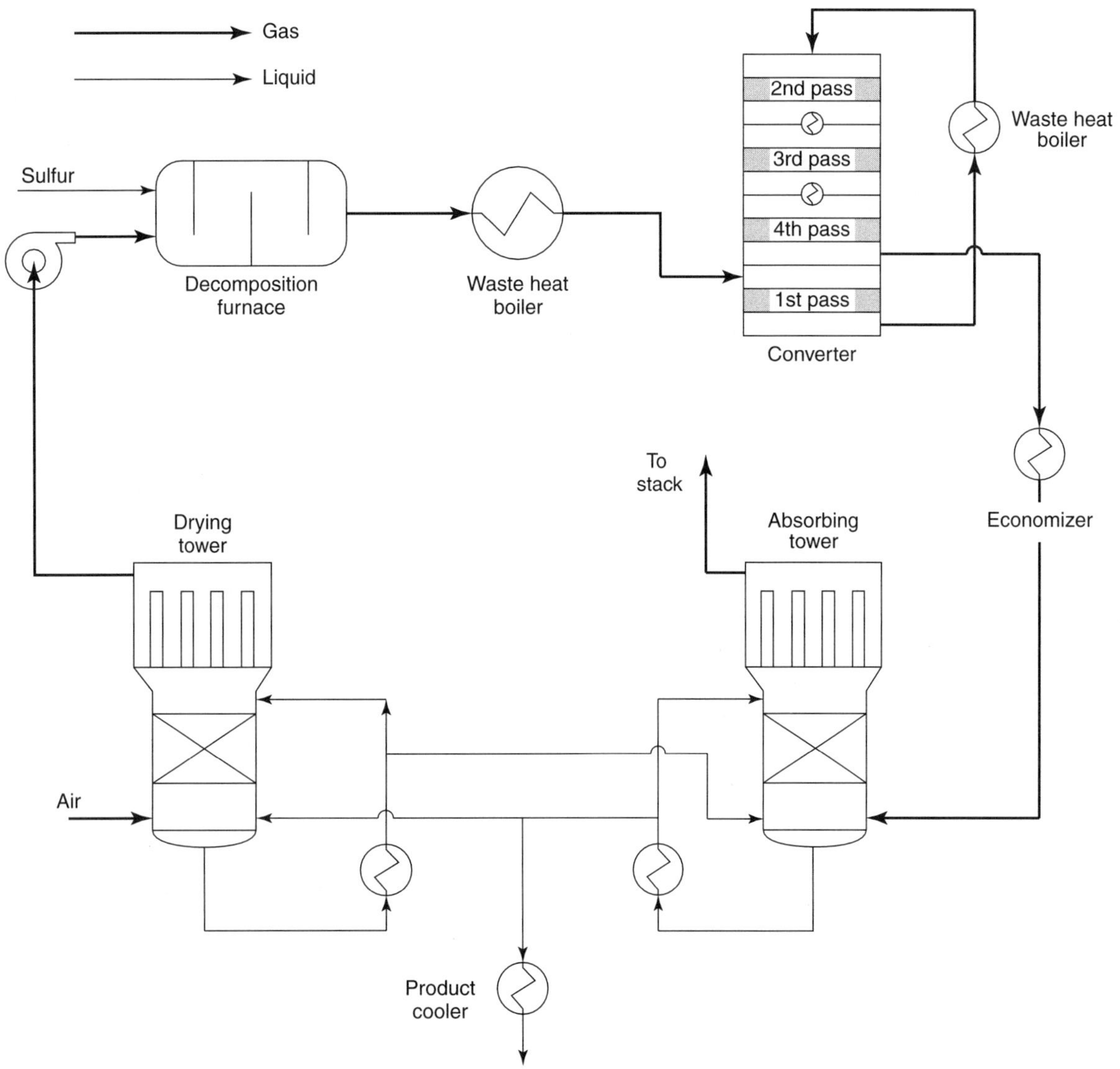

Figure 1. Typical Single-Absorption Sulfur Burning Acid Plant.

continuously removed and replaced with stronger acid from the absorbing tower. Drying towers can operate successfully over a wide range of acid temperatures, depending on selected acid concentration; however, 40–70°C is most common.

Drying towers are usually brick-lined steel or special alloy vessels filled with ceramic packing. To prevent mechanically generated spray and mist from carrying over into the rest of the process, drying towers usually have internal spray catchers or mist eliminators built into the top of the tower. The two most common types of mist and spray eliminators are the mesh pad and the packed fiber "candle" or tubular mist eliminator. These will be discussed later.

Molten sulfur is pumped to the furnace, where it is burned with the dry combustion air to form SO_2, according to the following reaction:

$$S_{(rhombic)} + O_{2(g)} \rightarrow SO_{2(g)} + \text{heat} \qquad (2)$$

$$\Delta H \cong -127{,}700 \text{ Btu/lb mole}$$

The SO_2 concentration leaving the furnace is typically 7–11% SO_2 by volume. Some SO_3 also may be produced in the sulfur burner.[3,4] If not accounted for, the SO_3 may be high enough to cause calculated SO_2 conversion to be low and emissions (lb_m SO_2 emitted per ton of acid produced) to be erroneously high. The combustion reaction is highly exothermic, and in the absence of reducing substances, is irreversible. The SO_2 and excess air leave the sulfur burner at 700–1000°C and must be cooled, typically, in a waste heat boiler to about 425°C before entering the converter.

Sulfur dioxide is converted to SO_3 in a multipass catalytic converter having three to five, but most often four, catalyst beds. Process gas enters the first pass of the converter to begin the oxidation of SO_2 to SO_3. In order to achieve acceptable conversion, the process gas is cooled between catalyst passes. Removing heat from the process gas allows the oxidation reaction to continue according to Eq. 1. (See "SO_2 Oxidation" above.)

Gas leaving the converter is cooled to 185–250°C in an economizer. This increases overall plant energy recovery and improves SO_3 absorption by lowering the process gas temperature entering the absorption tower.

From the economizer, the process gas enters the bottom of the absorbing tower, where SO_3 is absorbed in a recirculating stream of acid. To maintain acid strength as SO_3 is absorbed, water or weak acid is added to the absorbing tower. Acid from the drying tower also may be used for this purpose, although additional water probably will still be required.

Acid concentration entering the absorbing tower should be 98–99% H_2SO_4. This is near the azeotrope, where the sum of the partial pressures of SO_3, H_2O, and H_2SO_4 is lowest. Below 98% the partial pressure of water is high enough that it reacts with SO_3 in the gas to form H_2SO_4 mist. The partial pressure of SO_3 increases at strengths much above 99%, allowing some SO_3 to leave with the tail gas.

The temperature of acid entering the tower is typically 60–80°C but may be significantly higher or lower. In some newer towers designed for high energy recovery, temperatures may be as high as 200°C. Construction of the absorbing tower is similar to the drying tower. Process gas, with the SO_3 removed, passes through mist removal equipment and is discharged to the atmosphere through the stack.

Dual-Absorption Process

Concern over SO_2 emissions led to the development of the dual absorption process in the 1960s. The single-absorption process cannot economically achieve conversions much higher than about 98%. According to Eq. 1, however, additional SO_2 oxidation can occur if the SO_3 is removed from the process gas. (See "SO_2 Oxidation" above.) This is the principle used by the dual-absorption process.

Conceptually, the dual absorption process takes the tail gas from a single-absorption plant, reheats it, sends it to another catalyst pass, then absorbs the SO_3 in an absorption tower. Since the gas entering the dual-absorption part of the plant contains no SO_3, the second conversion step can achieve overall conversions of greater than 99.7%.

Figure 2 is a flow diagram of a dual-absorption sulfur-burning plant. The gas flow is essentially the same as for a single-absorption plant until the gas leaves the third catalyst pass. Following the third catalyst pass, the process gas is cooled and sent to an *interpass absorbing tower*, where the SO_3 is absorbed in sulfuric acid. SO_3-free gas leaves the interpass absorber, is reheated, and enters a fourth catalyst pass, where nearly all the SO_2 is converted to SO_3. Gas leaving the fourth catalyst pass is cooled before entering a final absorbing tower, where the remaining SO_3 is absorbed.

The use of two successive catalysis/absorption steps gives this process its name of "dual absorption." The terms "double absorption" or "double catalysis" are also used. Virtually all new plants built in the United States since the implementation of New Source Performance Standards (NSPS) have been dual-absorption plants.

Spent Acid Regeneration and Metallurgical Plants

Figure 3 is a flow diagram of a spent acid regeneration plant. In this plant the sulfur source may be waste sulfuric acid, hydrogen sulfide gas (H_2S), other sulfur-containing compounds, or a mixture of these. At temperatures of 1000–1050°C these materials decompose to produce SO_2. Carbon dioxide (CO_2) and water are the other two main products. With the exception of H_2S, these materials usually provide little heat of combustion; so natural gas or fuel oil is added to the spent furnace to provide the necessary heat. Elemental sulfur also may be burned in the spent furnace. Decomposition efficiency is aided by near stoichiometric ratio of oxygen to combustibles.

The gases leaving the spent furnace contain excess water and, depending on the spent materials burned, ash. Various equipment combinations, including waste heat boilers, humidification towers, reverse jet scrubbers, packed gas cooling towers, impingement tray columns, Karbate® coolers, and electrostatic precipitators, are used to cool and clean the gas before it is sent to the drying tower.

The balance of the spent plant is the same as the sulfur-burning plant, except that the gas leaving the drying tower must be heated to the reaction temperature (about 425°C) before it enters the converter. This is done using the heat of reaction in the converter as the heat source, with gas to gas heat exchangers replacing waste heat boilers and superheaters.

Metallurgical acid plants, that is, those receiving wet, dirty SO_2 from burning or smelting sulfide ores, have essentially the same flow diagram as the spent acid regeneration plant shown in Figure 2. The only major difference is that the smelter or ore roaster replaces the combustion chamber. In operation, they are similar, with metallurgical plants having continuously varying feed rates and concentrations, causing difficulty in achieving high conversions (of SO_2 to SO_3), that is, low SO_2 emissions.

Process Variations

There are many variations on the plants described above. Where oleum is desired, a separate oleum tower is used ahead of the primary absorbing tower. In cold-gas and dual-absorption plants, gas-to-gas heat exchangers are used instead of boilers and superheaters to cool the converter gas. In some older converters dried air is injected into the process gas following the second and third passes to provide cooling. Especially when older single-absorption plants are converted to dual absorption, other equipment combinations can be found.

AIR EMISSIONS CHARACTERIZATION

The only significant source of air emissions from a contact sulfuric plant is the tail gas leaving the final absorbing tower. This gas contains small amounts of SO_2 and even smaller amounts of SO_3, sulfuric vapor, and sulfuric acid mist.

SO_2 Emissions

SO_2 emissions and limits are reported either as pounds SO_2 per ton of 100% H_2SO_4 produced, or percent SO_2 converted to SO_3, the two being related by a simple formula. Some old "grandfathered" plants have only "in-stack" SO_2 limits. These typically limit emissions to 2000 ppm, or sometimes 3500 ppm for spent acid regeneration plants.

A minor source of air emissions from spent acid regeneration and metallurgical acid plants is the stack from a fuel-fired heater. These are typically fired with natural gas or light oils, and have characteristic emissions for those burners. These heaters are used to add heat indirectly to the process, and may be used only at startup, or continuously in some cases.

Single-absorption sulfur-burning plants are generally capable of meeting emission standards of 27 lb_m SO_2 per ton of 100% H_2SO_4, equivalent to 97.97% conversion. Dual-absorption plants can meet NSPS requirements of 4 lb_m SO_2 per ton of 100% H_2SO_4 produced, or 99.7% conversion. Dual absorption is well demonstrated and is the technology most frequently used today to reduce SO_2 emissions, although

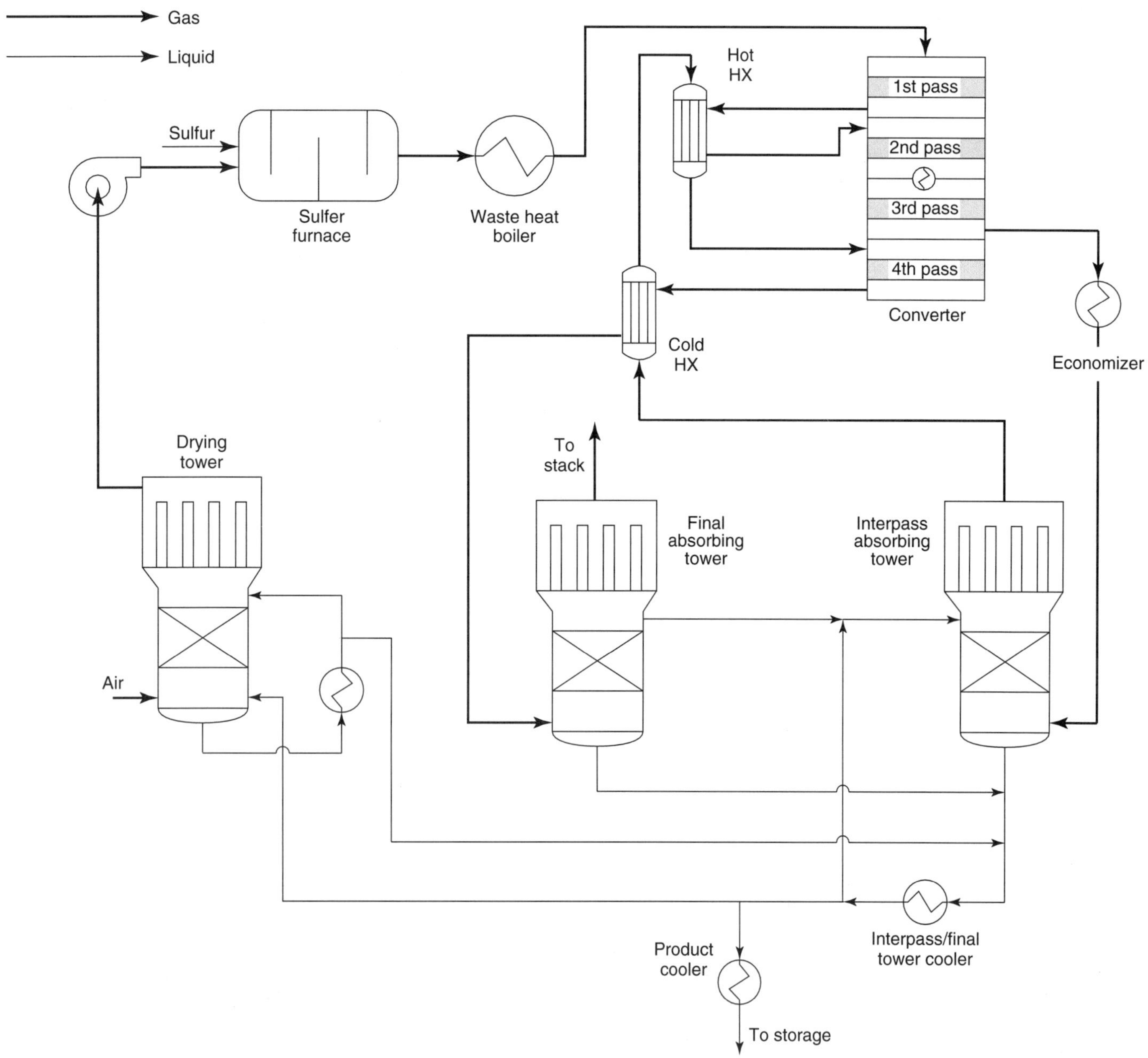

Figure 2. Typical Dual-Absorption Sulfur Burning Sulfuric Acid Plant.

various scrubbing technologies are also available. These will be briefly discussed below.

Conversion and SO_2 emissions are generally calculated from SO_2 entering the converter and SO_2 leaving the stack using the following formulas. These formulas are plotted in Figure 4.

$$C = \frac{A - B}{A - 1.5AB}$$

$$SO_2 \text{ emissions} = \frac{1306}{C - 1306}$$

where C is the conversion (fraction); A, the inlet SO_2 (fraction); B, the stack SO_2 (fraction); and SO_2 emissions equal lb_m SO_2 per ton 100% H_2SO_4.

The conversion equation has been used for years and is based on simple material balance. Fairlie[5] and Miles[6] both give its derivation. The emissions equation is simple algebra based on the fact that it takes 1306 lb_m of SO_2 to make a ton of sulfuric acid at 100% conversion efficiency.

The above formulas are valid for hot- or cold-gas plants but do not apply to plants that use air injection for cooling in the lower converter passes unless a correction is made to include the amount of air added. These formulas assume that the conversion in the gas entering the converter is zero; that is, no SO_3 is present. If significant SO_3 is present, as can happen in the combustion of sulfur, the measured inlet SO_2 will be low and the calculated SO_2 emissions will be high.

For hot-gas plants using only sulfur and air, a more precise way of measuring conversion and SO_2 emissions is based on SO_2 and oxygen (O_2) in the stack gas. Conversion is calculated from the following formula (the basis for its derivation is given by Miles[7]):

$$C = \frac{0.209 - Y - B}{0.209 - Y + 0.186B}$$

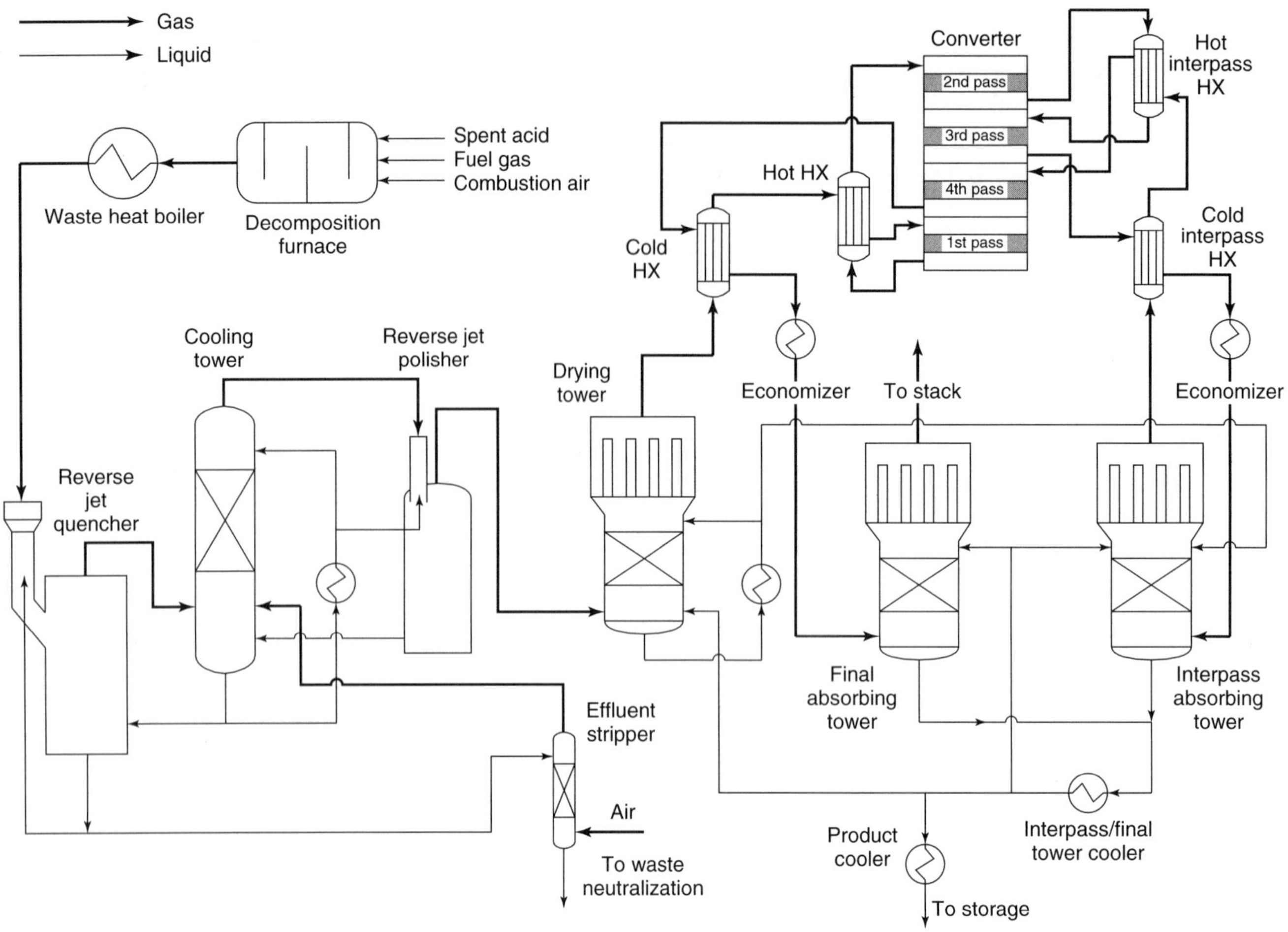

Figure 3. Typical Dual-Absorption Spent Acid Plant.

where Y is the stack O_2 (fraction). This formula is also applicable and especially useful to air injection plants if their only SO_2 source is sulfur and air. Trace amounts of H_2S and hydrocarbons in sulfur cause slight errors in this calculation since the hydrogen and carbon consume oxygen during combustion. These errors are usually negligible.

SO_3 and H_2SO_4 Mist

The gas leaving the stack contains small amounts of SO_3, H_2SO_4 vapors, and H_2SO_4 mist. SO_3 is normally combined with mist when determining acid plant emission factors because, upon leaving the stack, SO_3 rapidly reacts with atmospheric moisture to form H_2SO_4 mist. Hence, the resultant pollution problem is the same whether the initial cause is SO_3 or H_2SO_4 mist. However, the sources of SO_3 and mist emissions and their control are different. An increase in either SO_3 or mist emissions is usually first seen as increased stack opacity. Except in extreme cases, however, it is nearly impossible to tell the source of the problem by observing the stack plume. Hence, understanding the difference between SO_3 and mist is necessary to properly diagnose and control these emissions. Mist and SO_3 emission factors can be determined only by testing.

When reporting the emission factor for SO_3 and mist, SO_3 is calculated as 100% H_2SO_4 and added to the mist value. The NSPS requirement for combined SO_3 and mist is 0.15 lb_m/ton of acid. Some "grandfathered" plants have limits of 0.5 lb_m/ton. (The very small amount of H_2SO_4 vapor in the gas stream will be measured as SO_3 by existing test methods like EPA Method 8.)

SO_3 in the stack gas comes from a poorly operating final absorbing tower. Significant variables for proper tower operation are acid strength, acid temperature, circulation rate, and acid distribution over the top of the tower. The least important to good SO_3 absorption is temperature, although temperature can significantly affect internal mist generation. (See later discussion.) The SO_3 concentration leaving a properly operating tower will normally be <0.1 mg/scf.

Mist is formed within the process by mechanical means (splashing/entrainment) and physico-chemical means (reaction/condensation). Mechanically generated mist is very coarse and is easily collected. Reaction/condensation generated mist forms from water in the gas stream. Any water vapor in the gas stream will react with SO_3 to form H_2SO_4. If the gas temperature falls below the dew point, this H_2SO_4 will condense into mist particles. The dew point is a function solely of the H_2SO_4 concentration. Therefore, two ways to prevent mist condensation are to keep the gas temperature elevated and to keep the H_2SO_4 concentration low, that is, keep water out of the system.

Water vapor may come from poor drying tower performance, boiler tube or other steam leaks, or hydrocarbon impurities in the sulfur. Even with efficient gas drying and no steam leaking into the system, mist formation is impossible to prevent

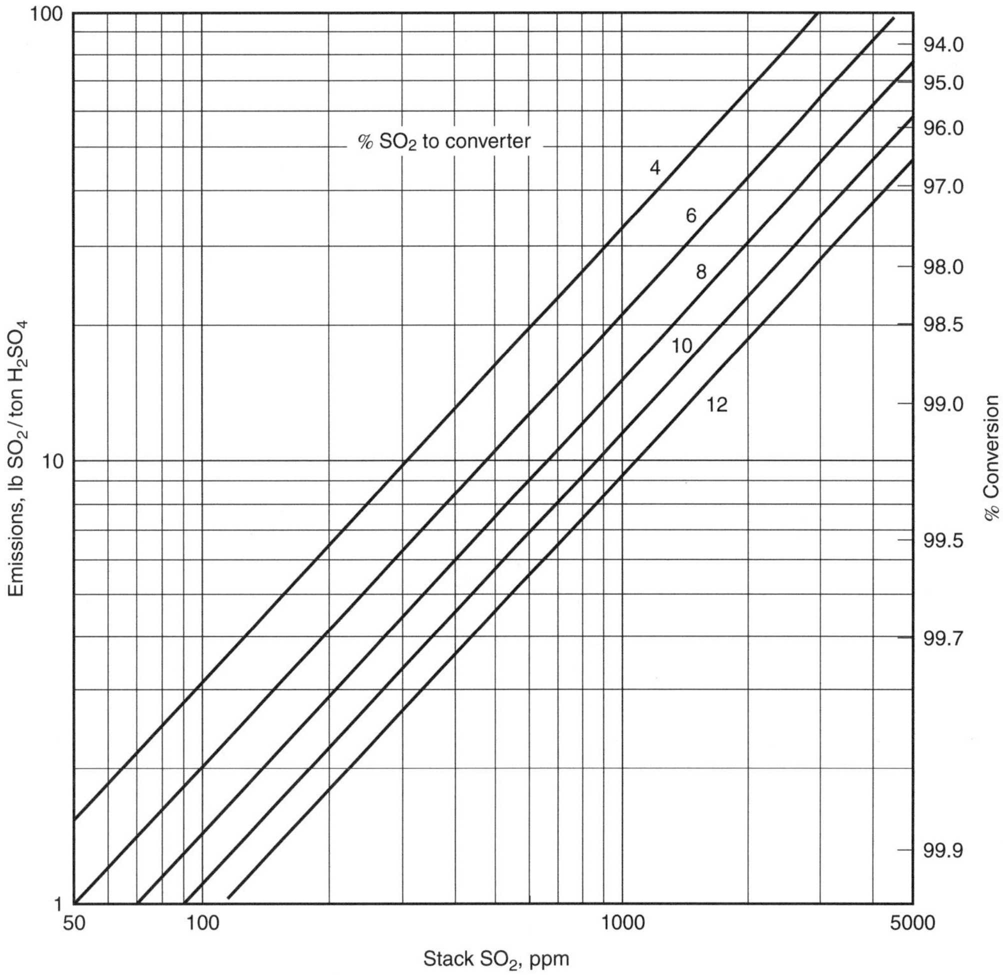

Figure 4. Sulfur Dioxide Conversion and Mass Emissions Versus Inlet and Exit Sulfur Dioxide Concentrations.

completely. Sufficient water vapor is always present in the process gas so that if the gas stream is cooled too quickly, before the SO_3 can be absorbed, a mist will be formed. This is particularly a problem in plants with oleum towers ahead of the final absorber, with bypass oleum towers producing more mist than full flow oleum towers. Acid mist formed by condensation is very difficult to collect. Because of its small particle size, sufficient mist will pass through the absorbing tower and, if unabated, will create a highly visible stack plume.

Typical uncontrolled mist levels leaving the final absorber are 15–50 mg/scf for dry-gas plants. Wet-gas plants may have uncontrolled mist levels of several thousand mg/scf.

The size of mist particles range from about 0.1 to greater than 10 μm. Mist particles greater than 5–10 μm are probably from mechanical entrainment. Plume visibility is a strong function of particle size. Acid mist composed of particles less than 10 μm can be visible even at concentrations less than 1 mg/scf (see Figure 5). As the particle size decreases, the plume becomes more opaque because of the greater light-scattering effect of smaller particles. Maximum light scattering occurs when the particle size approximates the wavelength of light, about 0.5 μm. Thus particles about 0.5 μm diameter are the most visible. In practice, these are also the most difficult particles to collect.

Sulfuric acid plants must meet maximum opacity limits on stack emissions. However, as seen from Figure 5, opacity is not an accurate indication of mass emissions. In one study of 24 variables affecting plume visibility, the angle of the sun caused a greater change in apparent opacity than any other variable.[8] Other significant factors listed included stack (plume) diameter, color of the sky, distance of observer from the stack, wind direction, particle density, particle index of refraction, and plume color. The last three variables are normally fixed for H_2SO_4 mists. However, these may change significantly if the use of tail-gas scrubbers changes the chemical nature of the emissions (such as with ammonia scrubbers).

AIR POLLUTION CONTROL MEASURES

Sulfur Dioxide

Sulfur dioxide emissions are determined primarily by overall plant design (e.g., number of catalyst passes, amount of catalyst, dual or single absorption, etc.) and not by the inclusion of pollution control processes or devices. New plants are usually designed to meet NSPS emissions limits using the dual-absorption process. In certain situations, plants can achieve better than NSPS limits with special designs. For example, in a metallurgical acid plant lower emissions can sometimes be achieved catalytically if the process gas from a smelter has a sufficiently high oxygen-to-SO_2 ratio. Proper

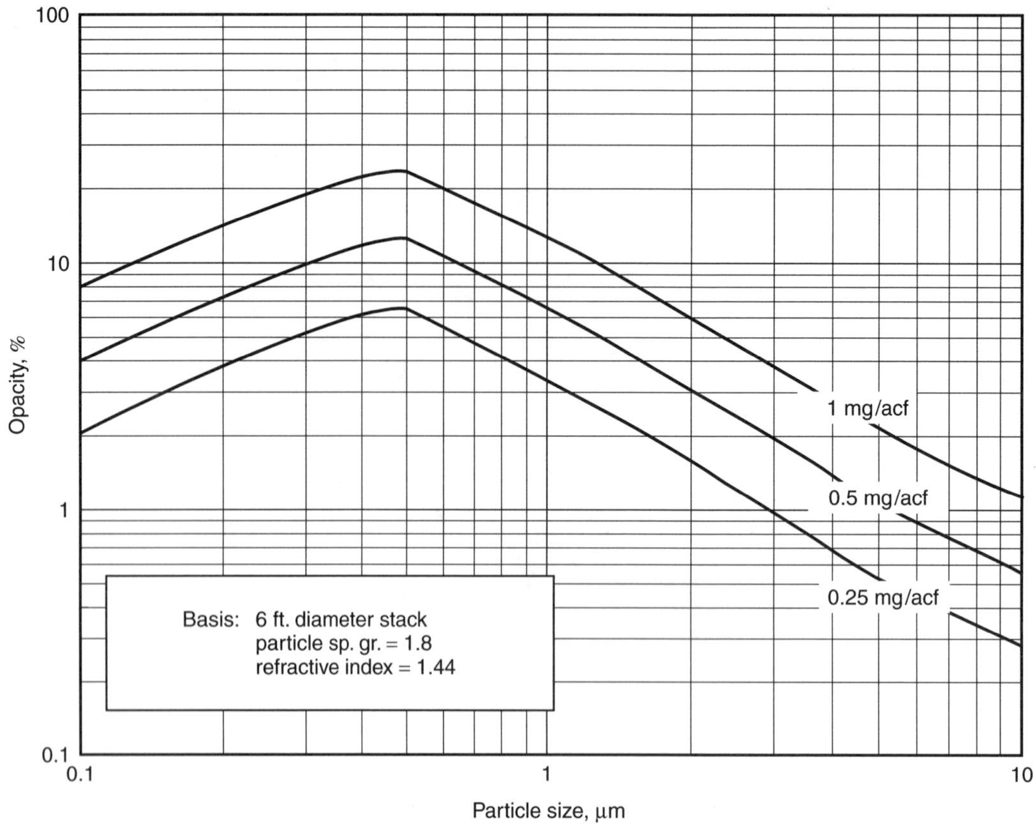

Figure 5. Effect of Particle Size on Stack Opacity (Ref. H. C. Van de Hulst, *Light Scattering by Small Particles*, John Wiley, New York, 1957; Courtesy of Monsanto Enviro-Chem. Used with permission).

catalyst volumes and interpass cooling must be incorporated into the initial design, however. Existing plants that are forced to reduce emissions are generally faced with converting to dual absorption or installing a tail-gas scrubber.

Dual Absorption. Dual absorption has been generally accepted as Best Available Control Technology (BACT) for meeting NSPS emission limits. Conceptually, dual absorption is the addition of another converter and absorbing tower on the tail end of a single-absorption plant (with appropriate heating and cooling of the gas stream) so there is no new technology involved. Only sulfuric acid is produced in the dual-absorption equipment. There are no by-products or waste scrubbing materials. This typically has the least economic impact of abatement methods available to the acid producer, especially for new plants. Converting an existing plant to dual absorption is rarely justified economically. A notable exception is when unused SO_2 allowances can be sold or used elsewhere. See "Process Description" section for detailed discussion of dual-absorption technology.

Tail-Gas Technologies. Various scrubbing technologies are available for removing SO_2 from gas streams. These are described in the literature. Scrubbers may be the "throwaway" type, where the scrubbing chemicals are used on a once-through basis and disposed of as by-products or discarded, or the "regenerative" type, where the SO_2 is removed from the gas and the active scrubbing solution is regenerated. Even with a regenerative process, some waste materials are usually produced that must be purged from the system. The economics of processes that produce significant by-products is usually determined by the market for the by-product or the cost of its disposal.

Tail-gas treatment is rarely used to achieve NSPS limits for new plants. A tail-gas process at the end of a dual-absorption plant may be the preferred technology where local regulations require substantially lower emissions than NSPS,

Tail-gas processes that produce a by-product that can be recycled to the acid plant (e.g., weak sulfuric acid) are of special interest because they eliminate the need for off-site by-product disposal. Two such processes will be briefly discussed: hydrogen peroxide scrubbing and SO_2 oxidation with activated carbon.

Hydrogen Peroxide Scrubbing. DuPont developed a hydrogen peroxide (H_2O_2) scrubbing process to reduce SO_2 emissions in the early 1970s. Dilute sulfuric acid and H_2O_2 are circulated over a packed bed countercurrent to the stream of SO_2 containing gas. Sulfur dioxide is absorbed in the solution where a rapid, high-yield reaction takes place to produce sulfuric acid. The acid produced in the scrubber becomes part of the plant's total production by blending with high-strength acid in the drying or absorbing towers. Thus there is no by-product or purge stream to dispose of with this process. The basic reaction is:

$$SO_2 + H_2O_2 \longrightarrow H_2SO_4$$

Lurgi and Süd-Chemie developed the similar Peracidox® process in Europe for scrubbing the tail gas from double-absorption plants.[9] This process uses either hydrogen peroxide or electrolytically produced peroxymonosulfuric acid (H_2SO_5) to oxidize SO_2 to H_2SO_4 according to the following reactions:

$$SO_2 + H_2O_2 \longrightarrow H_2SO_4$$

$$SO_2 + H_2SO_5 + H_2O \longrightarrow 2H_2SO_4$$

Both the DuPont process and the Peracidox® process have been successfully demonstrated. The high cost of hydrogen peroxide, or electric power to generate peroxymonosulfuric acid, makes this technology an unlikely candidate for meeting NSPS from a single-absorption plant.

SO_2 Oxidation with Activated Carbon. Lurgi developed its Sulfacid® process in the 1970s.[10] Monsanto Enviro-Chem introduced its MAC100® process (originally designated Centaur®) in 1996 as a potential low-cost alternative to dual absorption.[11] In both processes, the dry gas leaving the final absorbing tower is humidified then passed through a reactor filled with activated carbon. The activated carbon oxidizes the SO_2 to sulfuric acid, which is retained in the pores of the carbon. Clean, but wet, tail gas is discharged to the stack. Periodically, the carbon bed is regenerated by flushing with water. This produces a weak sulfuric acid stream that can be recycled back to the contact plant as dilution water. Both processes produce 20–30% sulfuric acid.

The primary difference between the Sulfacid® and MAC100® processes is the improved catalytic activity of the patented Centaur® activated carbon, manufactured by Calgon. Monsanto Enviro-Chem claims that the enhanced activity allows the use of significantly smaller vessels and smaller carbon beds, while providing less SO_2 breakthrough and longer carbon bed life.[12]

The Sulfacid® process has been successfully demonstrated at several plants. The first MAC100® process started up on the 1000 metric TPD single-absorption Philphos plant in the Philippines in 1998.[13]

The large amount of water that must be recycled back to the acid plant from both the Sulfacid® and MAC100® processes may be a problem to total spent regeneration plants and plants that produce large amounts of oleum or liquid SO_3.

SO_3 and Acid Mist Removal

Acid mist removal from sulfuric acid plant tail gases is accomplished almost exclusively with packed fiber mist eliminators or demister pads. Although a small portion of the SO_3 that leaves the final absorber will be absorbed in fiber mist eliminators and demister pads, SO_3 emission control depends primarily on proper plant operation.

A successful packed fiber tubular mist eliminator using treated glass fibers was developed by Monsanto Chemical Company in 1959. Figure 6 shows a diagram of this style mist eliminator. The original name, Brink® mist eliminator, is still used today. These units are available in two styles, the hanging (shown) and standup varieties. These devices capture particles using a combination of three different mechanisms: interception, impaction, and Brownian motion. Each mechanism operates most efficiently for a particular particle size. Together, they provide overall collection efficiencies that can exceed 99% depending on the inlet mist loading.

Packed fiber mist eliminators must be properly sized for the gas flow to provide optimum collection efficiency. Collection by the impaction mechanism suffers if gas velocity is too low. Re-entrainment becomes a problem if gas velocity is too high. Fiber mist eliminators typically operate with 6–8 in. H_2O pressure drop. While high-pressure drop can be an indication of fouling, there is no direct correlation between pressure drop and collection efficiency or mass emissions. The most frequent causes of poor mist eliminator performance are:

- Poor sealing at the mist eliminator flanges, which allows gas bypassing
- Loss of liquid in the seal leg (see Figure 6), which allows gas bypassing
- Fouling, which causes increased local velocities through the mist eliminators

Packed fiber mist eliminators are capable of meeting NSPS requirements of 0.15 lb_m/ton of acid.

Demister pads are mesh pads designed to capture larger mist particles by the interception and impaction mechanisms. Sometimes a coalescing pad is used ahead of the demister pad to provide higher collection efficiency.

Coalescer pads use high velocity to coalesce smaller particles into larger particles via the impaction mechanism. Because of the high gas velocity, the larger coalesced particles are almost always re-entrained into the gas stream leaving the coalescer pad. These re-entrained particles are then collected in the larger, low-velocity demister pad.

Demister pads, with or without a coalescer section, are not able to collect submicrometer particles as efficiently as packed fiber demisters. Coalescer pads also operate at much higher pressure drop than packed fiber demisters, up to 20 in. H_2O. Nevertheless, they require significantly less investment, and many plants have used them to achieve nearly invisible stack emissions. Successful use of demister pads requires careful

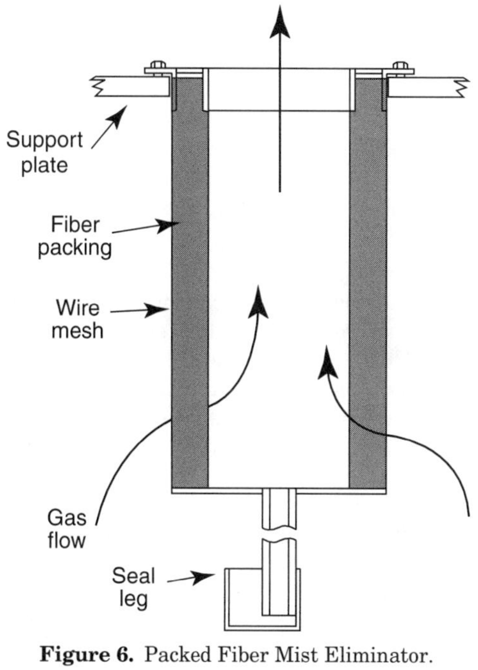

Figure 6. Packed Fiber Mist Eliminator.

control of plant operating parameters to minimize internal mist formation.

REFERENCES

1. O. Rud-Bendixen, "The Topsøe WAS-2 Process," British Sulphur Corporation, *Sulphur 1990* International Conference, Cancun, Mexico, April 1990.
2. J. W. Ward, "Topsøe's Wet Gas Sulfuric Acid (WSA) Process: An Alternative Technology for Recoverinf Refinery Sulfur," 1995 NPRA Annual Meeting, San Francisco, March 1995.
3. F. D. Miles, *The Manufacture of Sulphuric Acid*, D. Van Nostrand Company, New York, 1925, p. 29.
4. T. J. P. Pearce, "Sulphuric Acid: Physico-Chemical Aspects of Manufacture," in *Inorganic Sulphur Chemistry*, G. Nickless, Ed., Elsevier Publishing Co., New York, 1968, p. 538.
5. A. M. Fairlie, *Sulfuric Acid Manufacture*, Reinhold Publishing Corporation, New York, 1936, pp. 496–498.
6. F. D. Miles, *op. cit.*, pp. 187–189.
7. Ibid., p. 193.
8. A. Weir, Jr., D. G. Jones, L. T. Papay, et al., "Measurement of Particle Size and Other Factors Influencing Plume Opacity," International Conference on Environmental Sensing and Assesment, session 33, paper #2, September 14–19, 1975, Las Vegas, Nevada.
9. U. H. F. Sander, H. Fischer, U. Rothe, et al., *Sulphur, Sulphur, Dioxide and Sulphuric Acid*, Verlag Chemie International Inc., English edition, 1984, pp. 338–340.
10. Ibid., p. 337.
11. D. R. McAlister et al., "Centaur® SO_2 Removal Process: A New Approach to an Old Problem," Sulphur 96 (British Sulphur International Conference), Vancouver, B.C., October 20–23, 1996.
12. Ibid.
13. Private communication from Monsanto Enviro-Chem, July 20, 1998.

INTERNET WEB SITES

Sections of 40 C.F.R. 60 pertaining to sulfuric acid plant new source performance standards can be found and downloaded through *www.epa.gov* and other sites.

SYNTHETIC RUBBER

Excerpted from AP-42, U.S. Environmental Protection Agency

Two types of polymerization reaction are used to produce styrene butadiene copolymers, the emulsion type and the solution type. This discussion addresses volatile organic compound (VOC) emissions from the manufacture of copolymers of styrene and butadiene made by emulsion polymerization processes. The emulsion products can be sold in either a granular solid form, known as crumb, or in a liquid form, known as latex.

Copolymers of styrene and butadiene can be made with properties ranging from those of a rubbery material to those of a very resilient plastic. Copolymers containing less than 45 wt % styrene are known as styrene–butadiene rubber (SBR). As the styrene content is increased over 45 wt %, the product becomes increasingly more plastic.

PROCESS DESCRIPTION

Emulsion Crumb Process

As shown in Figure 1, fresh styrene and butadiene are piped separately to the manufacturing plant from the storage area. Polymerization of styrene and butadiene proceeds continuously through a train of reactors, with a residence time in each reactor of approximately one hour. The reaction product formed in the emulsion phase of the reaction mixture is a milky white emulsion called latex. The overall polymerization reaction is not carried out beyond a 60% conversion of monomers to polymer, because the reaction rate falls off considerably beyond this point and product quality begins to deteriorate.

Because recovery of the unreacted monomers and their subsequent purification are essential to economical operation, unreacted butadiene and styrene from the emulsion crumb polymerization process normally are recovered. The latex emulsion is introduced to flash tanks where, using vacuum flashing, the unreacted butadiene is removed. The butadiene is then compressed, condensed, and pumped back to the tank-farm storage area for subsequent reuse. The condenser tail gases and noncondensibles pass through a butadiene adsorber/desorber unit, where more butadiene is recovered. Some noncondensibles and VOC vapors pass to the atmosphere or, at some plants, to a flare system. The latex stream from the butadiene-recovery area is then sent to the styrene-recovery process, which usually takes place in perforated-plate, steam-stripping columns. From the styrene stripper, the latex is stored in blend tanks.

From this point in the manufacturing process, latex is processed continuously. The latex is pumped from the blend tanks to coagulation vessels, where dilute sulfuric acid (H_2SO_4 of pH 4 to 4.5) and sodium chloride solution are added. The acid and brine mixture causes the emulsion to break, releasing the styrene–butadiene copolymer as crumb product. The coagulation vessels are open to the atmosphere.

Leaving the coagulation process, the crumb and brine-acid slurry is separated by screens into solid and liquid. The crumb product is processed in rotary presses that squeeze out most of the entrained water. The liquid (brine/acid) from the screening area and the rotary presses is cycled to the coagulation area for reuse.

The partially dried crumb is then processed in a continuous belt dryer that blows hot air at approximately 93°C (200°F) across the crumb to complete the drying of the product. Some plants have installed single-pass dryers, where space permits, but most plants still use the triple-pass dryers that were installed as original equipment in the 1940s. The dried product is baled and weighed before shipment.

Emulsion Latex Process

Emulsion polymerization can also be used to produce latex products. These latex products have a wider range of properties and uses than do the crumb products, but the plants are usually much smaller. Latex production, shown in Figure 2, follows the same basic processing steps as emulsion crumb polymerization, with the exception of final product processing.

As in emulsion crumb polymerization, the monomers are piped to the processing plant from the storage area. The polymerization reaction is taken to near completion (98–99% conversion), and the recovery of unreacted monomers,

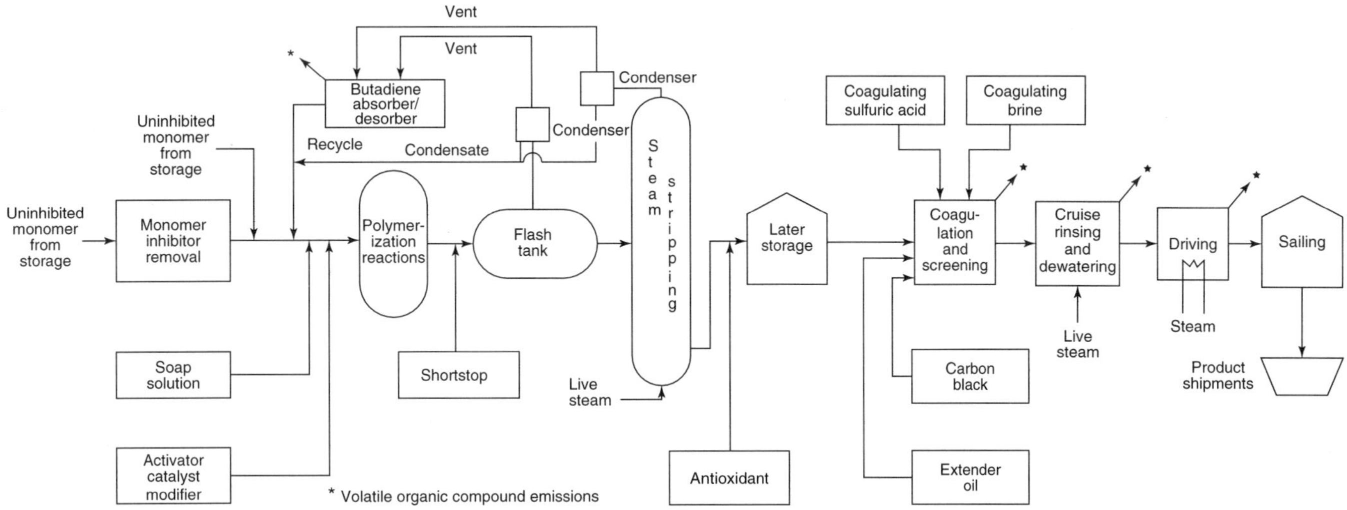

Figure 1. Typical Process for Crumb Production by Emulsion Polymerization.

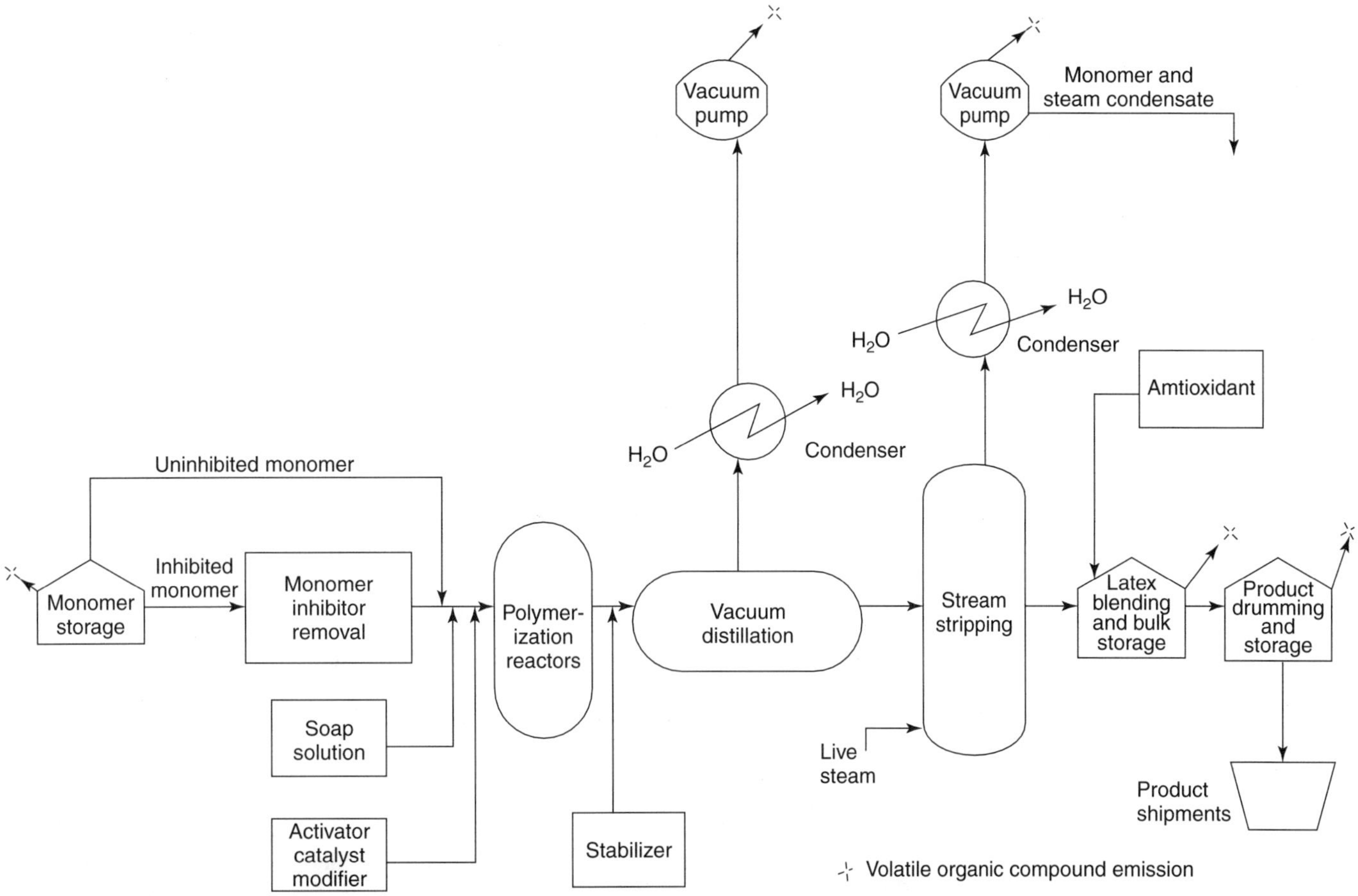

Figure 2. Typical Process for Latex Production by Emulsion Polymerization.

therefore, is uneconomical. Process economy is directed toward maximum conversion of the monomers in one process trip.

Since most emulsion latex polymerization is done in a batch process, the number of reactors used for latex production is usually smaller than for crumb production. The latex is sent to a blow-down tank where, under vacuum, any unreacted butadiene and some unreacted styrene are removed from the latex. If the unreacted styrene content of the latex has not been reduced sufficiently in the blow-down step to meet product specifications, the latex is introduced to a series of steam-stripping steps to reduce the content further. Any steam and styrene vapors from these stripping steps are taken overhead and sent to a water-cooled condenser. Any noncondensibles leaving the condenser are vented to the atmosphere.

After discharge from the blow-down tank or the styrene stripper, the latex is stored in process tanks. Stripped latex is

Table 1. Emission Factors for Emulsion Styrene–Butadiene Copolymer Production[a]

Process	Volatile Organic Emissions[b] (g/kg)	(lb/ton)
Emulsion crumb		
Monomer recovery, uncontrolled[c]	2.6	5.2
Absorber vent	0.26	0.52
Blend/coagulation tank, uncontrolled[d]	0.42	0.84
Dryers[e]	2.51	5.02
Emulsion latex monomer removal		
Condenser vent[f]	8.45	16.9
Blend tanks		
Uncontrolled[f]	0.1	0.2

[a]Nonmethane VOC, mainly styrene and butadiene. For emulsion crumb and emulsion latex processes only.
[b]Expressed as units per unit of copolymer produced.
[c]Average of three industry-supplied stack tests.
[d]Average of one industry stack test and two industry-supplied emission estimates.
[e]No controls available. Average of three industry-supplied stack tests and one industry estimate.
[f]EPA estimates from industry-supplied data, confirmed by industry.

passed through a series of screen filters to remove unwanted solids and is stored in blending tanks, where antioxidants are added and mixed. Finally, latex is pumped from the blending tanks to be packaged into drums or to be bulk loaded into railcars or tank trucks.

AIR EMISSIONS CHARACTERIZATION

Emission factors for emulsion styrene-butadiene copolymer production processes are presented in Table 1.

AIR POLLUTION CONTROL MEASURES

In the emulsion crumb process, uncontrolled noncondensed tail gases (VOCs) pass through a butadiene absorber control device, which is 90% efficient, to the atmosphere or, in some plants, to a flare stack.

No controls are employed at present for the blend tank and/or coagulation tank areas on either crumb or latex facilities. Emissions from dryers in the crumb process and the monomer-removal part of the latex process are not currently subject to control devices.

Individual plant emissions may vary from the average values listed in Table 1 with facility age, size, and plant modification factors.

BIBLIOGRAPHY

Control Techniques Guideline (Draft), U.S. Environmental Protection Agency Contract No. 68-02-3168, GCA, Inc., Chapel Hill, NC, April 1981.

Emulsion Styrene–Butadiene Copolymers: Background Document, U.S. Environmental Protection Agency Contract No. 68-023063, TRW Inc., Research Triangle Park, NC, May 1981.

C. Fabian, U.S. Environmental Protection Agency, Research Triangle Park, NC, confidential written communication to Styrene–Butadiene Rubber File (76/15B), July 16, 1981.

TEREPHTHALIC ACID

RONALD D. BELL, ALAN J. MECHTENBERG, AND KEVIN MCQUIGG

Terephthalic acid (TPA) produced in a purified form (P-TPA) is used almost exclusively in the manufacture of polyethylene terephthalate (polyester) fibers. A small percentage of P-TPA is used for the manufacture of polyester films, polybutylene terephthalate resins, and barrier resins for carbonated-beverage bottles. Table 1 provides a listing of the U.S. producers of P-TPA, the locations of their facilities, and the annual P-TPA capacities of these facilities. Production capacity in the United States has grown to 4.8 billion pounds in 1990 from a reported 2.9 billion pounds in 1978.[2]

Beginning in the United Kingdom in 1949 and the United States in 1953, polyester production initially involved the use of TPA in a technical grade as an intermediate to dimethyl terephthalate. The acid was produced by a dilute nitric acid oxidation of *p*-xylene, and a number of subsequent variations were made to the process that reduced the consumption of nitric acid by introducing air into the reaction. Neither of these nitric acid processes is currently in commercial application.

Prior to 1963, monomer produced from the direct esterification of TPA did not meet required standards for purity. This was primarily due to the difficulty in purifying TPA itself, since it is extremely insoluble in most organic solvents, as well as in water. Dimethyl terephthalate, on the other hand, was susceptible to most common methods of purification available at the time and was the source for transesterification to all monomer.

New methods of producing and purifying TPA from the liquid-phase oxidation of *p*-xylene were commercialized in 1963. Thus P-TPA is currently used interchangeably with dimethyl terephthalate in polyester production in which 1.17 g of dimethyl terephthalate as a reactant is equivalent to 1 g of P-TPA. While production of P-TPA requires two separate processes to synthesize and purify crude TPA, the use of P-TPA in polyester production via the direct esterification process offers distinct cost advantages to the producer. Specifically, the use of dimethyl terephthalate requires recovery of methanol, a by-product of the transesterification. Also, P-TPA requires less ethylene glycol for polyesterification, thereby improving final polymer quality.

Air emissions from the synthesis of P-TPA vary with respect to solvent-recovery portions of the process because of variations from facility to facility. Emissions of concern include *p*-xylene, acetic acid, carbon dioxide (CO_2), carbon monoxide (CO), methyl acetate, and TPA in particulate and

Table 1. U.S. Production Capacity of Purified Terephthalic Acid[a]

Manufacturer	Location	Annual Capacity (millions of pounds)
Amoco Corp.	Charleston, NC	1600
	Decatur, AL	1800
Cape Industries	Wilmington, NC	380
E. I. du Pont de Nemours & Co., Inc.	Wilmington, NC	598
Eastman Kodak Co.	Columbia, SC	400
Total capacity		4778

[a]1990 data from Ref. 1.

vapor form. One information source describes the emissions from the industry as a whole as "moderate."[3] This study also indicates that, compared with other commercial P-TPA processes, the currently dominant liquid-phase air oxidation process is one of the lowest polluters.

PROCESS DESCRIPTION

The predominant process for technical-grade TPA manufacture is the Amoco process for liquid-phase air oxidation of *p*-xylene. Of the domestic P-TPA producers listed in Table 1, the Eastman Kodak Co. and Cape Industries utilize different methods that account for 780 million pounds of the 4.8 billion pounds of P-TPA production capacity. The E. I. du Pont de Nemours & Co. and Eastman Kodak Co. facilities also produce crude TPA, which is subsequently esterified with methanol to produce dimethyl terephthalate. The purification processes for the production of P-TPA also vary within the industry; however, the Amoco process for catalytic hydrogenation of impurities is dominant in the United States.

A simplified process flow diagram of the Amoco process for production of crude TPA is depicted in Figure 1. The process represented is essentially that practiced by Amoco at the Charleston and Decatur facilities. Streams are designated as main product flow and other flow streams. Streams with letter designations are those that potentially result in air pollutant emissions.

The overall reaction in the Amoco process takes place in the liquid phase in the presence of a catalyst, according to the following reaction. The parallel reaction that results in the formation of carbon oxides is considered a minor reaction.

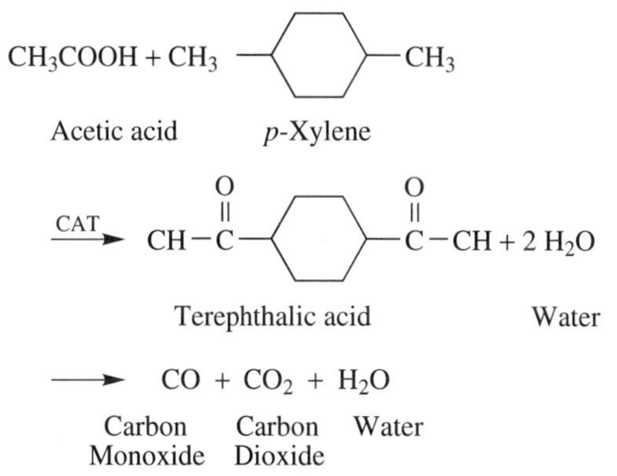

The primary by-products of the reaction system are products of partial oxidation of *p*-xylene, such as *p*-toluic: acid and *p*-formyl benzoic acid, which appear to some extent as impurities in the TPA product. Methyl acetate is also formed in significant amounts in the reaction.

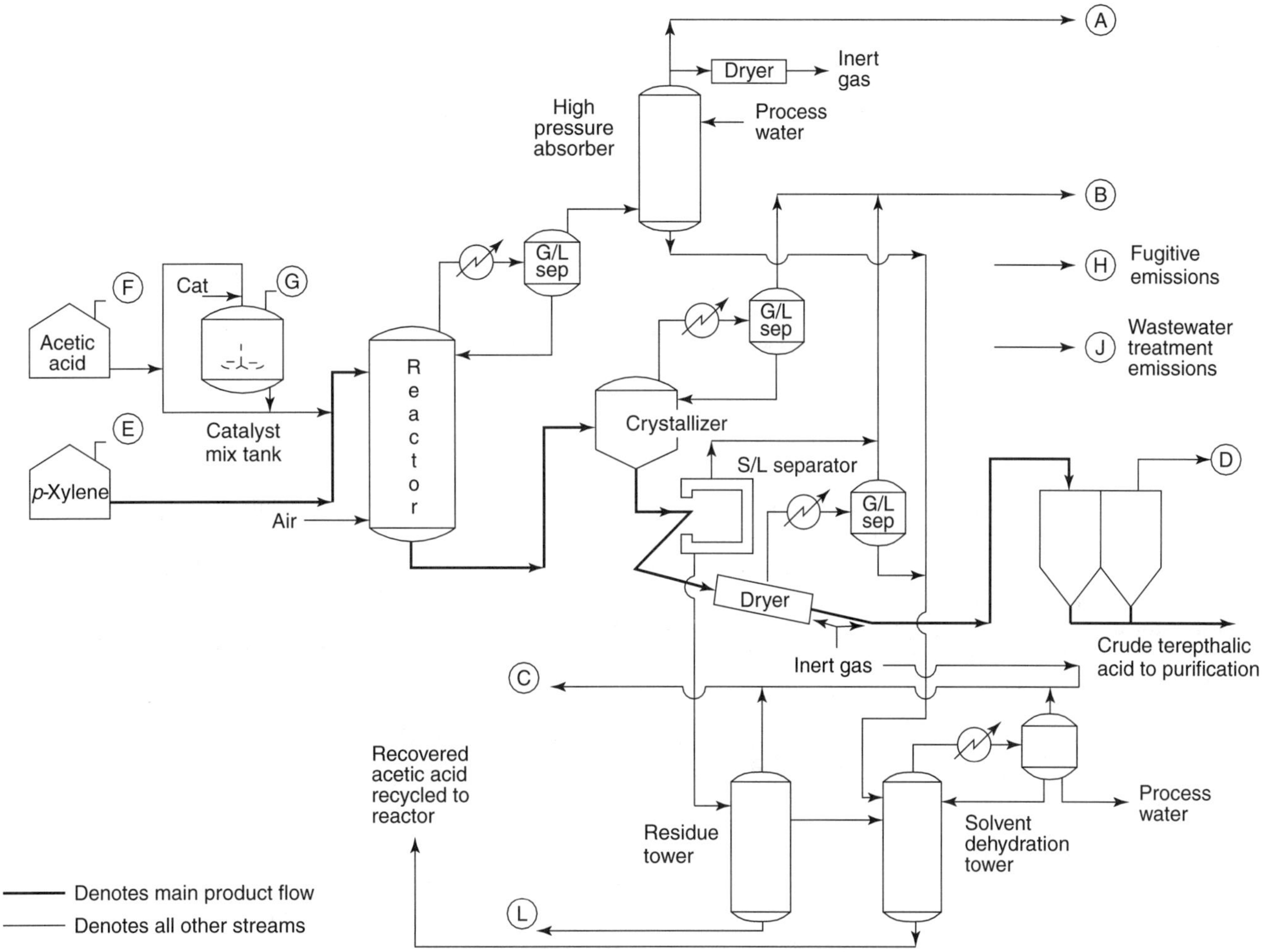

Figure 1. Flow Diagram for Crude TPA Production Using the Amoco Process.

Fresh and recycled acetic acid, *p*-xylene, and a heavy-metal catalyst (normally manganese or cobalt acetate and sodium bromide) are continuously fed to the reactor, which is maintained at about 175–230°C and 220–435 psi. The reaction temperature is controlled by regulating the pressure at which the reaction mixture is permitted to boil and form the vapor stream leaving the reactor. Air is added in greater than stoichiometric amounts to minimize the formation of by-products. The residence time in the reactor ranges from 30 min to 3 h, depending on process conditions. Reportedly, the conversion of *p*-xylene exceeds 95%, and the yield of TPA is at least 90 mol %.[4]

Inert gases, excess oxygen, CO, CO_2, and volatile organic compounds (VOCs) are routed through a gas–liquid separator to a high-pressure absorber, where the gas stream is scrubbed with process water to reduce the VOC content. Part of the stream is dried and used as a source of inert gas, and the remainder is vented to the atmosphere (vent A in Figure 1). The scrubbing liquor from the absorber is routed to the solvent dehydration tower for acetic acid recovery.

The effluent from the reactor is a slurry, since TPA is somewhat soluble in acetic acid. The slurry flows to a series of crystallizers, where the pressure is relieved and the liquid is cooled by the vaporization and return of condensed VOCs and water. Noncondensible inert gases containing some VOCs are vented from the condenser (vent B).

The impurities from the partial oxidation of *p*-xylene are somewhat soluble in acetic acid and remain in the mother liquor as the effluent from the crystallizers is routed to solid–liquid separators (normally centrifuges). The TPA recovered as wet cake is sent to a rotary dryer, where inert gases and heat are used to drive off the remaining water and VOCs. The hot VOC-laden vapors are routed to a condenser, where the noncondensibles containing some VOCs are again vented to the atmosphere (vent B), and the condensed liquids are routed to the solvent dehydration tower for acetic acid recovery.

Dry product TPA is transported to storage silos using inert gas; there the inert gas is vented to the atmosphere after passing through a particulate-recovery device. The collected solids are combined with the stored product for processing in the purification section.

The mother liquor from the solid–liquid separators flows to a residue still, where acetic acid, water, and methyl acetate are recovered overhead. The bottoms from the residue still contain tars, products of partial oxidation of *p*-xylene, catalyst residue, and some acetic acid. The overhead from the residue still, along with the condensed vapors from the dryer and the scrubbing liquor from the high-pressure absorber, is routed to the solvent dehydration tower that produces a relatively pure acetic acid, which is recycled to the reactor. A reflux accumulator at the column is used to recover condensed water and methyl acetate, which is used as the scrubbing liquor in the high-pressure absorber. Noncondensibles from the reflux accumulator and the residue column are vented to the atmosphere (vent Q, along with some VOCs and inert gas, which is used for blanketing and instrument purging.

The purpose of the purification process is to produce a product (P-TPA) that is suitable for polyester production. In the Amoco purification process presented in Figure 2, this is accomplished by converting certain impurities to a more water-soluble form via catalytic hydrogenation in an aqueous medium. The converted impurities remain in solution upon subsequent recrystallization of TPA.

The primary impurity formed by the partial oxidation of *p*-xylene in the reactor, *p*-formyl benzoic acid, is converted to the more water-soluble *p*-toluic acid, according to the following reaction.

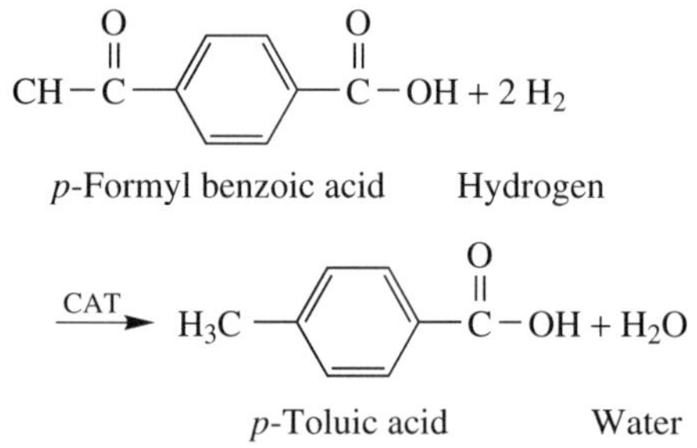

Crude TPA is sent to the feed slurry tanks and combined with hot water, resulting in the release of gases that had been trapped in the TPA crystals (vent A in Figure 2). The slurry is then routed to a dissolver, which is maintained under slight positive pressure and approximately 250°C, to achieve dissolution of TPA.

The dissolved TPA slurry and hydrogen in the amount of about 0.004 gram per gram of TPA, which represents a significant stoichiometric excess, are fed to the reactor, which is maintained at the same temperature as the dissolver. The reactor contains a noble-metal catalyst on a carbon support.

The reactor effluent flows to the crystallizers, where the temperature is reduced by allowing the water to boil at a controlled rate. Minor amounts of TPA in the vapor phase are released with the water vapor (vent B). Since TPA will sublime rather than melt, the released TPA vapors are cooled in the atmosphere and fall as particles.

The slurry from the crystallizers is sent to a solid–liquid separator (normally a centrifuge), and the mother liquor is sent to wastewater treatment. The wet cake in the centrifuge is then washed with hot water and the water is recycled to the feed slurry tanks. The wet cake is dried in a rotary drier and transported to product storage silos with inert gas; this gas is vented to the atmosphere after passing through a particulate-control device. The final concentration of *p*-formyl benzoic acid is less than 20 ppm, and the yield from the purification process is reportedly 98%.[4]

Several variations of the Amoco processes for crude and purified TPA production are used in the United States. The Cape Industries facility listed in Table 1 obtains P-TPA directly by hydrolyzing dimethyl terephthalate. The Eastman Kodak facility utilizes acetaldehyde to replace acetic acid solvent losses and a somewhat different catalyst in the crude TPA process. Eastman Kodak also employs further steps in the solvent-recovery sections of the process in which an azeotroping agent is applied and methyl acetate is recovered further to reduce VOC content in the wastewater from the solvent-recovery operations. The additional solvent-recovery steps are typical of those carried out in dimethyl terephthalate technology.[5] Variations within the purification operations do not result in a significant difference in air emissions from those resulting from the Amoco process.

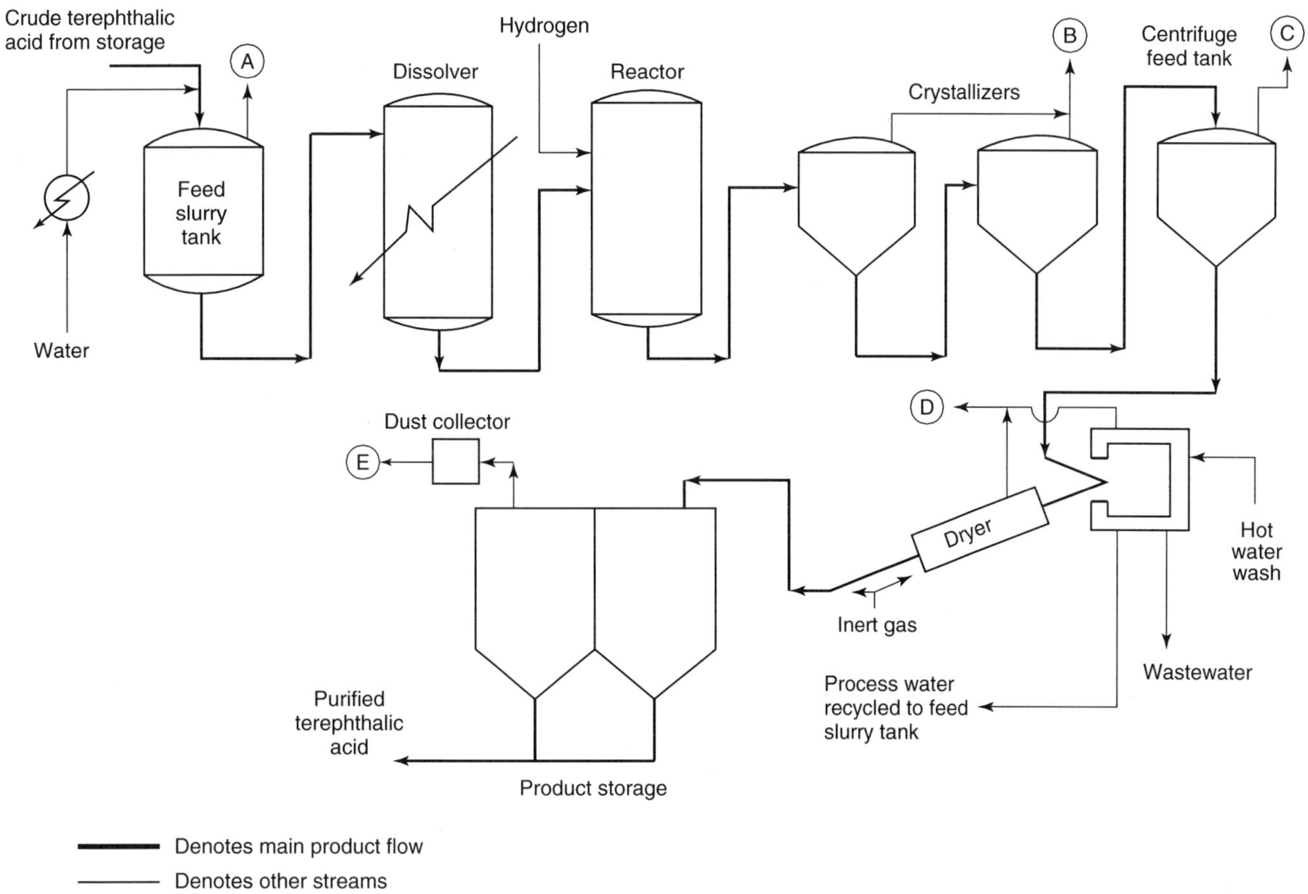

Figure 2. Flow Diagram for P-PTA Production Using the Amoco Purification Process.

AIR EMISSIONS CHARACTERIZATION

Air emissions resulting from crude TPA production may arise from several sources, including the reactor vent, the crystallization/separation/drying vent, the solvent-recovery vent, the product-transfer vent, material storage and handling vessel vents, and the wastewater treatment operations. Fugitive emissions from valves, compressor seals, pumps, and flanges throughout the process also result. In addition, TPA is also emitted in particulate form, primarily from the crude TPA storage areas.

Air emissions from the purification process consist primarily of TPA vapor and particulate matter from various processing vents. No VOCs are introduced or generated in the purification process; therefore, no VOCs are emitted to the atmosphere.

Description of Sources

Figure 1 identifies the waste streams and volatilization points that potentially contribute to air emissions from an uncontrolled crude TPA production process. Letter designations for each of these streams are referenced with the following discussion of emission characterization. The reactor vent gas (vent A) consists of the substances within the vapor-phase reactor effluent that are recovered via the high-pressure absorber. These substances include nitrogen from air oxidation, unreacted oxygen and *p*-xylene, CO, and CO_2. Acetic acid and methyl acetate are also present in certain amounts, depending on the removal efficiency of the absorber.

The crystallization/separation/drying vent (vent B) consists primarily of the noncondensible off-gases from the crystallization and drying portions of the process. Carbon dioxide and CO are not present in an appreciable amount; however, some VOCs carried by the noncondensibles are emitted through this vent.

The residue and solvent-recovery column vent (vent C) contains the noncondensible vapors from these operations, including some CO_2 and CO. Inert gases that are introduced for blanketing and instrument purging are released at this point and VOCs are carried with the vent gas stream. The VOC content within the discharge stream is reportedly very low.[2]

The product storage vent (vent D) releases the inert gases used to transport solid TPA product to the storage silos and contains the TPA particulate matter carried with these gases. The VOCs that are carried with recovered inert gases from the high-pressure absorber may also be present in this stream, where the recovered inert gases are used for product transfer.

The emissions from storage and handling vessels (vents E, F, and G) consist of *p*-xylene and acetic acid from the vessels in which they are present. Fugitive emissions throughout the process consist primarily of VOCs that are introduced and generated in the various process areas.

Wastewater treatment emissions (point J) consist of volatilized pollutants from the treatment area. The sources of wastewater from facilities that use the Amoco process as described include the blowdown from the high-pressure absorber (scrubbing liquor) and the effluent from equipment washdown and process drains. Where the process does not employ a methyl acetate recovery process, significant amounts of methyl acetate are present in this scrubbing liquor.

Producers of P-TPA may also discharge the bottoms from the residue still to a wastewater treatment facility, where emissions of VOCs through volatilization would occur. The residue still bottoms contain catalyst residue, tars, some acetic acid, and the products of incomplete oxidation of *p*-xylene.

The crude TPA purification process as shown in Figure 2 does not include VOCs in significant amounts, since VOCs are not introduced or generated. Very small amounts of VOCs are released in the dissolution process from the crude TPA crystals. The venting of excess hydrogen and water from the hydrogenation reactor carries some TPA in vapor form, which is cooled in the atmosphere and converted to TPA particulate matter. Additionally, some particulate P-TPA product is released from the product storage facility.

Emission Factors

Emission factors have been developed in terms of total VOC and CO emissions from an uncontrolled model plant that produces crude TPA.[2] The model plant has a crude TPA capacity of 507 million pounds per year based on 8760 h of annual operation.

The model plant is a combination of the Amoco process shown in Figure 1 and the solvent-recovery portion representative of dimethyl terephthalate technology as practiced at the Eastman Kodak facility. The model plant's solvent-recovery operation involves the use of *n*-propyl acetate as an azeotroping agent in the solvent dehydration tower and additional separation steps for methyl acetate and plant wastewater. The *n*-propyl acetate and methyl acetate emissions resulting from storage of these materials are, therefore, included in the factors. The recovered methyl acetate, along with the bottoms from the residue tower, is incinerated in the model plant; therefore, incinerator stack gas emission factors are listed. The emission factors resulting from the study are given in Table 2.

Storage tank emission factors were developed assuming two raw-material tanks and one in-process *p*-xylene tank, two storage tanks and one catalyst mix tank containing acetic acid, and four crude TPA tanks. The tanks are typical-size, fixed-roof tanks. Details regarding additional parameters used to determine the storage loss factors are available in Ref. 2.

The fugitive emission factors listed in Table 2 represent the component counts used in the model study; however, the various uncontrolled factors were reestablished based on the more recent average uncontrolled fugitive emission factors for the synthetic organic chemical manufacturing industry.[6] Component counts used for the model plant include 50 pumps, 900 process valves, and 40 pressure-relief devices in light-liquid VOC service.

The emission factors in Table 2 are considered uncontrolled factors, with the exception of the incinerator stack emissions, since the incinerator itself is considered a control device. The factor for vent A in Figure 1 is an uncontrolled factor since the high-pressure absorber is considered a part of the production process.

The study[2] also resulted in the development of solid TPA emission factors for the Amoco purification process represented in Figure 2, which were based on a screening study of particulate emissions from the purification process.[7] The factors are considered controlled emission factors since particulate emissions from each of the vents (vents A–D in Figure 2) were measured following the application of a water scrubber for particulate control. The emissions from the product storage vent (vent E in Figure 2) were measured following the passage of the gas stream through a dust collector. Controlled particulate TPA emission factors resulting from the screening study are listed in Table 3.

Table 2. Uncontrolled Emissions of VOC and CO from Model Plant for Crude TPA Production[a,b,c]

Source	Stream Designation (Figure 1)	Emission Ratio (g/kg)[d]		Emission Rate (kg/h)	
		VOC	CO	VOC	CO
Reactor vent	A	14.6	17	383.3	446
Crystallization, separation, and drying vent	B	1.9		49.9	
Distillation and recovery vent[e]	C	1.14		29.9	
Product transfer vent[f]	D	1.78	2	46.7	53
Raw material storage[g]	E,F	0.112		2.94	
Other storage[h]	G	0.006		0.017	
Fugitive emissions	H	0.50		13.02	
Incinerator stack[i]	N/A	0.00605		0.1583	
Wastewater treatment[j]		<0.004		<0.1	
Total		20.05	19	526.04	499

N/A = Not applicable.
[a]Reference 2.
[b]Model plant has an annual crude TPA capacity of 507 million pounds. Factors are based on 8760 h of annual operation.
[c]Factors are uncontrolled factors with the exception of the incinerator stack as noted in this table.
[d]Grams of emission per kilograms of crude TPA produced.
[e]Emission factors include methyl acetate and wastewater towers in the model plant, which are not shown in Figure 1.
[f]VOC and CO emissions originate from inert gases that are used to transport product and are recovered from the high-pressure absorber.
[g]Emission factors include emissions from *n*-propyl acetate storage in the model plant, which is not shown in Figure 1.
[h]Emission factors include emissions from methyl acetate storage in the model plant, which is not shown in Figure 1.
[i]Incinerator stack emission factors are based on the model plant, in which residue still bottoms and recovered methyl acetate are incinerated.
[j]Wastewater treatment emissions are based on the model plant, in which methyl acetate is removed from plant wastewater.

Table 3. Controlled TPA Particulate Emissions from Model Plant for TPA Purification Using the Amoco Purification Process[a,b,c]

Source	Stream Designation (Figure 2)	Emission Ratio (g/kg)[d]	Emission Rate (kg/h)
Feed slurry tank vent	A	0.088	5.08
Crystallizer vent	B	0.098	5.69
Atmospheric centrifuge feed tank	C	0.023	1.32
Dryer vent	D	0.0012	0.07
Silo dust collector vent	E	0.0017	0.10

[a] Reference 2.
[b] Model plant has an annual P-TPA capacity of 507 million pounds. Factors are based on 8760 h of annual operation.
[c] Control of source streams A–D includes a water scrubber.
[d] Grams of emission per kilogram of P-TPA produced.

AIR-POLLUTION CONTROL MEASURES

The types and quantities of pollutants from air emission sources within the TPA and P-TPA processes are shown in Table 2. Nine point sources are included as follows:

- Reactor vent
- Crystallization, separation, and drying vent
- Distillation and recovery vent
- Product transfer vent
- Storage and handling vents
- Fugitive
- Wastewater treatment
- Residue tower bottoms
- TPA purification

Reactor Vent and Product Transfer Vent

The reactor vent gas normally contains nitrogen, unreacted oxygen, unreacted *p*-xylene, acetic acid, CO, CO_2, methyl acetate, and water. The quantity of VOCs in the reactor vent varies depending on the absorber operating pressure and temperature. The product transfer vent gas has the same composition as the reactor vent gas, except that it includes water and TPA particulate matter. Particulate matter in the product transfer vent gas is typically removed by bag filters downstream of the product silos. Three technologies for control of VOCs are applicable to these vent streams: aqueous absorbers, carbon adsorption, and thermal oxidation.

In general, VOCs in the reactor vent and product transfer vent gases are controlled by the use of aqueous absorbers. The liquid streams from the absorbers are separated and returned to the process. The vent gas from the absorbers, which still contains from 3% to 10% of the original VOC and an unchanged amount of CO, is vented to the atmosphere.

Carbon adsorption can be expected to provide approximately 97% control of the VOCs in the vent streams. However, no change in the level of CO results. For continuous operation, a typical carbon-adsorption system utilizes two carbon beds: one in operation and one undergoing regeneration. Regeneration is generally achieved by steam stripping the VOCs from the carbon bed. The VOC-laden stripping steam is condensed and decanted. The organic layer is returned to the reactor section, and the aqueous layer is sent to distillation for recovery of water-soluble VOCs.

Aqueous absorbers and carbon adsorption may prove to be inadequate as regulations regarding VOC and CO emissions become more and more stringent. Thermal oxidation can reduce VOC and CO levels in the vent gases by 99% to 99.9%. Depending on the VOC and CO content of the stream, considerable supplemental fuel may be required to achieve the operating temperature required (about 875°C for 99% control to 1000°C for 99.9% control) to obtain this level of control efficiency. Supplemental fuel cost can be offset by adding a waste-heat boiler to provide steam for the TPA production process.

Crystallization, Separation, and Drying Vent

Noncondensible gases carrying VOCs are released during crystallization of the TPA and separation of crystallized solids from the solvent. Different methods incorporated into the process affect the amount of noncondensible gases and accompanying VOCs. Control of the VOCs can be obtained by carbon adsorption, thermal oxidation, or, as is typically practiced, by the use of aqueous absorbers. The use of aqueous absorbers to remove acetic acid will reduce VOC emissions from this vent stream by approximately 98%.

Distillation and Recovery Vent

This small discharge consists of the gases that are dissolved in the feed stream sent to distillation; the inert gases used for blanketing, instrument purging, and pressure control; and the VOCs carried by the inert gases. This vent gas is typically incorporated with gas from the crystallizer, separation, and drying vent, and sent to aqueous absorbers. Removal of acetic acid by the aqueous absorbers reduces VOC emissions from this vent stream by approximately 96%. Alternatively, the vent gas can be sent to a carbon-adsorption system or to a thermal oxidizer to achieve VOC control.

Storage and Handling Vents

Some emissions result from the storage of *p*-xylene and acetic acid. Most emissions occur only during tank filling, especially for the *p*-xylene storage tanks that are maintained at a constant temperature and are equipped with conservation vents. Emissions from acetic acid storage tank vents are typically controlled using aqueous absorbers. The resultant aqueous stream is returned to the process. Alternatively, these vents could be routed to a carbon-adsorption system or thermal oxidizer.

Fugitive Emissions

Fugitive emissions from piping, valves, pumps, and compressors are controlled by periodic monitoring by leak checking with a VOC detector and a directed maintenance program.

Wastewater Treatment

Currently, the small amount of VOC emissions from the various wastewater treatment steps is not controlled. If control of these emissions becomes necessary, aqueous absorbers would be the most easily applied technology. Other treatment alternatives might include carbon adsorption or air stripping of the wastewater. Although these technologies seem applicable, lack of experience would require some development to determine the effect of inorganic species in the wastewater on carbon adsorption and the efficiency of air stripping on the organic compounds in the wastewater stream.

Residue Tower Bottoms

The residue tower bottoms stream contains products of partial oxidation, tars, catalyst residue, and some acetic acid. If not treated, volatilization of light ends will result in VOC emissions. This stream is most effectively treated by sending it to a liquid-waste incinerator. The incinerator is capable of an efficiency of 99.99% in the destruction and removal of the organic compounds. Bromine salts and other inorganic solids contained in the still bottoms must be controlled by removal prior to incineration (usually very difficult) or by the treatment of the incinerator flue gas with some type of particulate-control device, such as a bag filter, electrostatic precipitator, or venturi scrubber.

Terephthalic Acid Purification

There are no VOC emissions associated with the purification of TPA, since the purification process does not involve the use of any VOC. Emissions of TPA, not considered a VOC, is a vapor result from the venting of excess hydrogen and water vapor at elevated temperatures. The TPA vapors sublime in the atmosphere, thereby creating particulate emissions. Water scrubbers are currently used to provide effective control of the TPA in the emissions from the purification process.

REFERENCES

1. *1990 Directory of Chemical Producers, United States of America*, SRI International, Menlo Park, CA, 1990.
2. *Organic Chemical Manufacturing, Vol. 7: Selected Processes*, EPA-450/3-80-028b, U.S. Environmental Protection Agency, Research Triangle Park, NC, 1980.
3. J. W. Pervier et al., *Survey Reports on Atmospheric Emissions from the Petrochemical Industry, Vol. II*, EPA-450/3-73-0056, U.S. Environmental Protection Agency, Research Triangle Park, NC, 1974.
4. *Kirk–Othmer Encyclopedia of Chemical Technology*, 3rd ed. Vol. 17, John Wiley & Sons, New York, 1978.
5. *Supplement B to Compilation of Air Pollutant Emission Factors, Vol. I: Stationary Point and Area Sources*, U.S. Environmental Protection Agency, Research Triangle Park, NC, 1988.
6. *Protocols for Generating Unit-Specific Emission Estimates for Equipment Leaks of VOC and VHAP*, EPA-450/3-88-010, U.S. Environmental Protection Agency, Research Triangle Park, NC, 1988.
7. L. M. Elkin, "Terephthalic Acid and Dimethyl Terephthalate," pp. 49–55 in *Report No. 9, A Private Report by the Process Economics Program*, Stanford Research Institute, Menlo Park, CA, 1966.

BIBLIOGRAPHY

D. F. Durocher et al., *Screening Study to Determine Need for Standards of Performance for New Sources of Dimethyl Terephthalate and Terephthalic Acid Manufacturing*, EPA Contract No. 68-02-1316, Task Order no. 18, 1976, p. 4.

*UNIPOL® is a trademark of Union Carbide Chemicals and Plastics Technology Corp.

NATURAL FIBER TEXTILE INDUSTRY

Anthony J. Buonicore

Textiles are any product derived from the processing of natural fibers such as wool and cotton, and/or from the processing of fibers synthesized and extruded from petrochemicals and modified wood pulp, such as polyester, nylon, rayon, and acetate. The processing that the raw fibers receive functions to remove natural impurities (in the case of cotton and wool) and to impart particular qualities of appearance, touch, and durability. The textile industry consists of companies that process natural and/or synthetic fibers into yarns, fabrics, and certain consumer products. The production of synthetic fibers, however, is not included in this discussion, as the nature of such plants is more closely identified with that of organic chemical plants.

The activities of the natural fiber textile industry (excluding fiber production) typically include:

- The production of yarns by spinning
- The production of fabrics by weaving, knitting, and other processes
- The dyeing and finishing of fibers, yarns, and fabrics
- The production of clothing, household linens, curtains, carpets, and other consumer products

The textile "complex" is shown schematically in Figure 1.[1] The industry uses many different raw materials and chemicals to produce a great variety of products. Many of the chemicals are not retained in the final textile product and are either recovered for reuse, discarded in the plant effluent, or driven off into the atmosphere.

PROCESS DESCRIPTION

The classical method of categorizing the industry involves grouping the manufacturing plants according to the fiber being processed, that is, cotton, wool, or synthetics. The modern approach to textile industry categorization, however, involves grouping the manufacturing plants according to their particular operation. Two broad categories can be identified: low-water-use processing and wet processing. Low-water-use processing includes the manufacture of yarns and unfinished fabrics ("greige" goods). Wet processing includes several categories according to the processes and raw materials employed in the production of the finished product. For example, such categories might include:

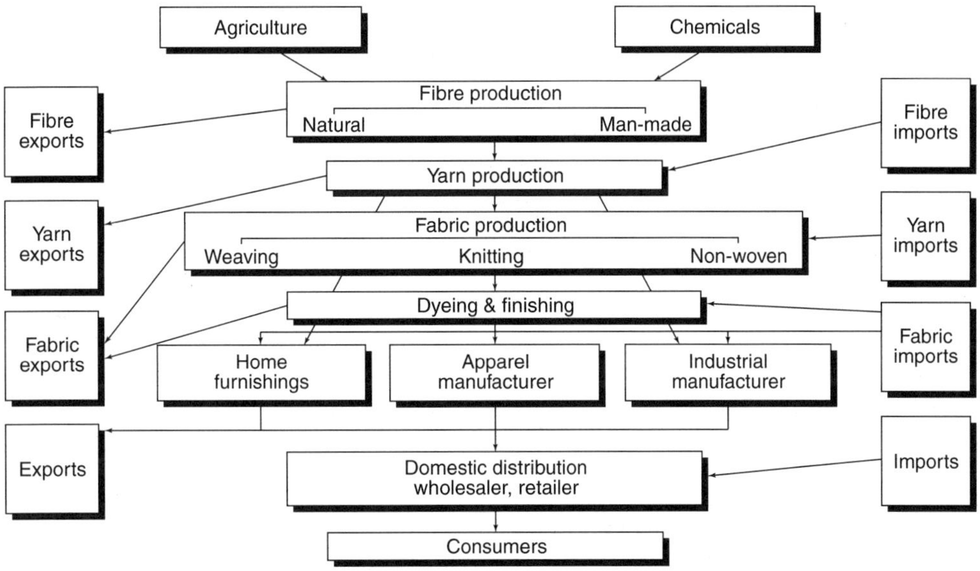

Figure 1. Schematic Diagram of the Textile Complex (Courtesy Canadian Textile Institute).

- Wool scouring
- Wool finishing
- Woven fabric dyeing and finishing
- Knit fabric dyeing and finishing
- Carpet finishing
- Stock and yarn finishing
- Nonwoven manufacturing
- Felted fabric finishing

The low-water-use process unit operations consist of carding, spinning, slashing (sizing), weaving, knitting, and tufting. The wet process unit operations include singeing, desizing, scouring, wool carbonizing, wool fulling, cotton mercerizing, bleaching, dyeing/printing, and finishing. These processes are briefly described below.[1]

Low-Water-Use Processing Unit Operations

1. Carding is the preliminary process in spun yarn manufacture. The fibers are separated, distributed, equalized, and formed into a thin web and condensed into a continuous, untwisted strand of fibers called a sliver. This process also removes impurities and a certain amount of short, broken, or immature fibers. The operation is performed on a card.
2. Spinning is the process of making yarn from fibers by a combined drawing out and twisting operation or from filament tow by the combination of cutting/breaking with drafting and twisting in a single series of operations.
3. Slashing (sizing) is the process of sizing warp yarns on a slasher to protect the yarns against injury during weaving. It also gives temporary added strength and abrasion resistance to the yarns, enabling them to be processed successfully for weaving. The entire warp is coated by immersion in the sizing material that may contain starch, polyvinyl alcohol (PVA), wax, gelatin oils, and/or other materials, and is dried before winding on the beam of the loom prior to weaving.
4. Weaving is the process of interlacing two yarns of similar materials so that they cross each other at right angles to produce woven fabric. The three basic weaving patterns are plain, twill, and satin.
5. Knitting is the process of constructing fabric by an interlocking series of loops of one or more yarns. The three classes of knitted fabrics are circular knit, flat knit, and warp knit.
6. Tufting is the process of making carpets and involves a wide multiple-machine needle process that sews pile yarns to a broad fabric backing.

Wet Processing Unit Operations

1. Singeing removes protruding surface fibers to reduce pilling. The fabric passes over an impinging flame, which burns the fibers from the fabric surface.
2. Desizing removes the sizing compounds applied to yarns to impart tensile strength. The starch sizing compounds are solubilized with alkali, acid, or enzyme, and the fabric is washed thoroughly. Alkaline desizing utilizes a weak alkaline solution to facilitate size removal, while acid desizing employs a dilute acid solution to hydrolyze the size and render it water soluble. Enzyme desizing utilizes vegetable or animal enzymes to decompose size (usually only starch size) to a water-soluble form. After solubilizing the size, the fabric is rinsed clean.
3. Scouring removes natural and acquired impurities from fibers and fabric. Synthetic fibers require less scouring than does cotton or light wool. Scouring agents include detergents, soaps, and various assisting agents, such as alkalis, wetting agents, foamers/defoamers, and lubricants. After scouring, the goods are thoroughly rinsed (or washed) to remove excess agents.
4. Wool carbonizing removes burrs and other vegetable matter from loose wool or woven fabric goods. The process consists of acid impregnation, baking, and mechanical agitation. A dilute solution of sulfuric acid is used to degrade cellulosic impurities to hydrocellulose without damaging the wool. The excess acid is squeezed from the wool, and the wool is baked to oxidize the contaminants to gases and a solid carbon residue. The material then

passes through pressure rollers to crush the solid residue and into a mechanical agitator to shake loose the crushed material. The acid content in the material remains high after agitation, requiring neutralization and rinsing before further processing.

5. Wool fulling gives woven woolen cloth a thick, compact feel, finish, or appearance. Cloth is mechanically worked in fulling machines to make the fibers felt together, causing shrinkage, area weight increase, and concealment of the woven threads of the cloth. The two common methods of fulling are acid and alkali. In acid fulling, an aqueous solution of sulfuric acid, hydrogen peroxide, and small amounts of metallic catalyst are used. In alkali fulling, soap or detergent, sodium carbonate, and a sequestering agent are added to provide the required lubrication. Acid fulling is usually followed by alkali fulling.
6. Mercerizing is usually used in cotton processing to increase the tensile strength, luster, dye affinity, and abrasion resistance of the goods. In this operation, the cotton, usually in fabric form, is impregnated with a cold sodium hydroxide solution. The solution causes the cotton fibers (cellulose) to swell. After the desired contact period, the alkali is thoroughly washed out, sometimes with the use of a dilute acid bath to ensure neutralization. It has been normal practice to recover the caustic solution for concentration and reuse in scouring or mercerization.
7. Bleaching is a common process used to whiten cotton, wool, and some synthetic fibers by removing the natural coloring. It is usually performed after scouring and prior to dyeing or printing. Bleaching chemicals include sodium hypochlorite, hydrogen peroxide, and sodium perborate, as well as optical brighteners. Batch bleaching is done in kiers (continuous processes use J-boxes), where fabric is stacked for a given period to allow the chemical to work before goods are withdrawn from the bottom of the box. Bleaching is followed by thorough rinsing.
8. Dyeing can be performed in the stock, yarn, or fabric state, and single- or multiple-fiber types can be dyed. Multiple-fiber types may require multiple or sequential steps.

   Stock dyeing is performed before the fiber is converted to the yarn state and can be a batch or continuous process. The batch process consists of placing the stock in a vat or pressure kettle, applying a sufficient amount of dye solution, providing optimum environmental conditions, allowing time for dye fixation, and rinsing. The stock is then processed into yarn.

   Yarn dyeing is performed on yarns used for woven goods, knit goods, and carpets. Usual methods include skein, package, and space dyeing. Skein dyeing is performed by placing turns of yarn on a frame and putting the frame in a dye bath under the required dyeing conditions. Package dyeing is the most common yarn dyeing process and is effected by placing yarn wound onto perforated tubes on a frame, placing the frame into a pressure vessel, circulating dye liquor in and out of the tubes under the required conditions, and then rinsing. Space dyeing is a specialty yarn dyeing process. Dye liquor is applied to yarns at a repeat or random interval by a roller-type dye pad. The dyed yarn then enters a subsequent step(s) for development and fixation of the color and is rinsed.

   Fabric dyeing is the most common method in use today because it can be continuous or semicontinuous, as well as a batch process. Methods employed include beck (winch), jet, jig, beam, and continuous range. Beck dyeing is accomplished with the fabric in rope form in either atmospheric or pressure vessels. The fabric is connected end-to-end and rotated over a large drum through the dye liquor. The length of the fabric is such that the fabric lies in a heap at the bottom of the vessel for a short time. As with all dyeing methods, the proper conditions of time and chemicals must be provided.

   Jet dyeing is also accomplished with the fabric in the rope form. Jet machines are similar to pressure becks, except that the cloth passes through a venturi tube, which helps transport the fabric. A pump circulates the dye liquor through the tubes and suction at the venturi causes the fabric to rotate. Jet machines have improved on certain deficiencies of beck dyeing by providing faster dyeing times, lower liquor-to-cloth ratios, reduced tangling, more uniform temperatures, and reduced elongation of fabric due to tension.

   Jig dyeing is performed with the fabric in an open-width configuration. Both atmospheric and pressure equipment are available. The fabric is wound onto rollers that pass through a shallow trough containing the dye liquor. The rollers rotate in clockwise and counterclockwise directions alternately and guide rolls ensure complete immersion of the cloth. Only a few meters of cloth are immersed at a time, and it is possible to work with a low liquor-to-cloth ratio.

   Continuous dyeing also is performed with the fabric in an open width on what are termed ranges. The ranges consist of a number of dip troughs, from which dyes and chemicals are applied; drying and fixation sections; wash boxes to remove excess dye liquor; steam chambers; and a final drying step.

   Beam dyeing is a batch process involving the use of a kettle where open-width fabric is wound onto a perforated drum with dye solution circulated through the material in a manner similar to package yarn dyeing. This is unlike the other fabric dyeing processes described, in which the fabric is circulated through the dye solution.

   Thermosol dyeing is a continuous process normally used for dyeing polyester and polyester-cotton blends. Dye is applied to the fabric in the pigment form and dried, causing a film of pigment to adhere to the fabric. The fabric is then heated to 350–400°F (near the melting point of polyester), causing the synthetic fibers to swell, and as the dispersed dye sublimes, it penetrates the polyester. The cotton portion is dyed by passing the fabric through an additional chemical pad and steamer. The fabric is then washed and dried.
9. Printing is similar to dyeing, except that print color is applied to specific areas of the cloth to achieve a planned design instead of the whole cloth being colored. Dyes and dyeing assists are similar to those used in fabric dyeing; however, the color application techniques are quite different. Textiles are usually wet-printed by roller, rotary screen, or flatbed screen methods.

Roller printing is accomplished by applying print paste to an etched or engraved roller and transferring the design to the fabric by contact with the rollers. After printing, the pattern is fixed by steaming, aging, or some other treatment.

Screen printing differs from roller printing in that the print paste is transferred to the fabric through openings in specially designed screens. The process can be manual, semiautomatic, or completely automatic. Automatic screen printing can be either flatbed or rotary, while manual and semiautomatic are flatbed processes only.

Screens are made by manually or photographically transferring the design onto the screen material. The area outside the design is specially treated so that the print paste is retained. The screens are securely stretched over a frame for correct positioning. A separate screen is required for each color in the design. Modern techniques for rotary screen production include the use of laser engravers, which etch the cotton in a coated screen.

In manual screen printing, the fabric is placed on a long table, the screen frames correctly positioned for the pattern, and the selected print paste forced through the screen mesh onto the fabric by squeegee. The fabric is then dried and given other finishing treatments. In the semiautomatic process, the fabric travels and the screens remain in place. Handling of the screens and application of the print paste are performed manually.

Automatic flatbed screen printing takes place on a machine that electronically performs and controls each step. The process is continuous in that the fabric moves along a table, the screens are positioned, and print paste is applied and squeezed through the screen onto the fabric. The fabric moves forward one frame between each screen, and when all color has been applied, it moves into a steaming and/or drying box.

In rotary screen printing, the color is applied to the fabric through lightweight metal-foil screens that resemble the cylinder rollers of the roller printing process. The fabric moves continuously under the cylinder screens, and print paste is forced from the inside of the screens through and onto the fabric. A separate screen is required for each color in the design.

10. Finishing operations are included as wet processes because the majority of treatments are applied in the liquid state. Wet finishing involves the application of a wide range of chemicals to add properties to a fabric. Finishes can be applied, for example, to make a fabric wrinkle resistant, crease retentive, water repellant, flame resistant, mothproof, mildew resistant, bacteriostatic, and/or stain resistant. Application is usually by means of a pad that contains immersion rollers in a pan with finish and a set of nip rolls that squeeze the finish excess and assure an even finish application. The fabric is then dried to cure the finish onto the fabric.

    Synthetic resins are most often used to achieve wrinkle resistance and crease retention (permanent press). The resins are adhesive and are permanently cross-linked with the fibers by curing with heat and a catalyst. Silicones and other synthetic materials are used to provide water repellency. The silicones successfully repel oily fluids as well. Flame-resistant finishes are applied to cellulosic fabrics (cotton) to prevent them from supporting combustion.

    Mothproofing finishes are typically applied to wool and other animal hair fibers to make them unfit as food for moth larva. The growth of mildew, mold, rot, and fungus is inhibited by the application of compounds that usually contain chlorinated phenols, metallic salts or zinc, copper, or mercury. Selected hygienic additives can also be used to inhibit the growth of bacteria.

    Soil-release finishes usually contain organosilicone compounds, fluoro compounds, or oxazoline derivatives. These compounds allow stains to be removed by ordinary washing.

Cotton Processing

Cotton processing consists of two basic steps—weaving (greige manufacturing) and finishing (wet processing) (refer to Figure 2).

Greige manufacturing entails mostly dry operations, as follows:

1. Trash and foreign matter are manually or mechanically removed from the raw fibers (opening, cleaning, carding, and combing).
2. The fibers are joined, straightened, drawn into yarn, and wound on spools (drawing, roving, spinning, and winding).
3. The yarn is run through a starch solution (sizing or slashing) and then dried so that it has the strength and stiffness required to withstand the abrasion and friction generated in the weaving operation. Yarn for knitting is waxed to provide lubrication for the knitting machine and reduce yarn-to-metal friction.
4. The strengthened or waxed yarn is woven into cloth (greige goods).

The finishing plant receives the greige goods and must process them to satisfy various client demands. Finishing is essentially wholly of the wet type, wherein:

1. Natural impurities (wax, pectins, alcohols, etc.) and acquired impurities (unremoved size agent, dirt, oil, grease, etc.) must be removed to make the cloth suitable for chemical processing.
2. Wet chemical processing is performed to impart to the cloth the desired properties and appearance.

Wool Processing

As in the case of cotton, wool must undergo both weaving (greige manufacturing) and finishing operations; the former is necessary in order to develop a continuous cloth from the bits of raw fiber, and the latter in order to impart the desired characteristics to the cloth.

Greige manufacturing entails numerous mechanical and manual dry operations (see Figure 3). Only two operations are of the wet type (scouring and slashing).

Greige operations include sorting and blending of raw fibers, scouring (to remove natural grease and foreign matter), drying, carding (opening and paralleling of fibers), rewashing, oiling (for reducing static electricity), gilling (paralleling of fibers), combing (to remove trash), roving (or drawing of fibers

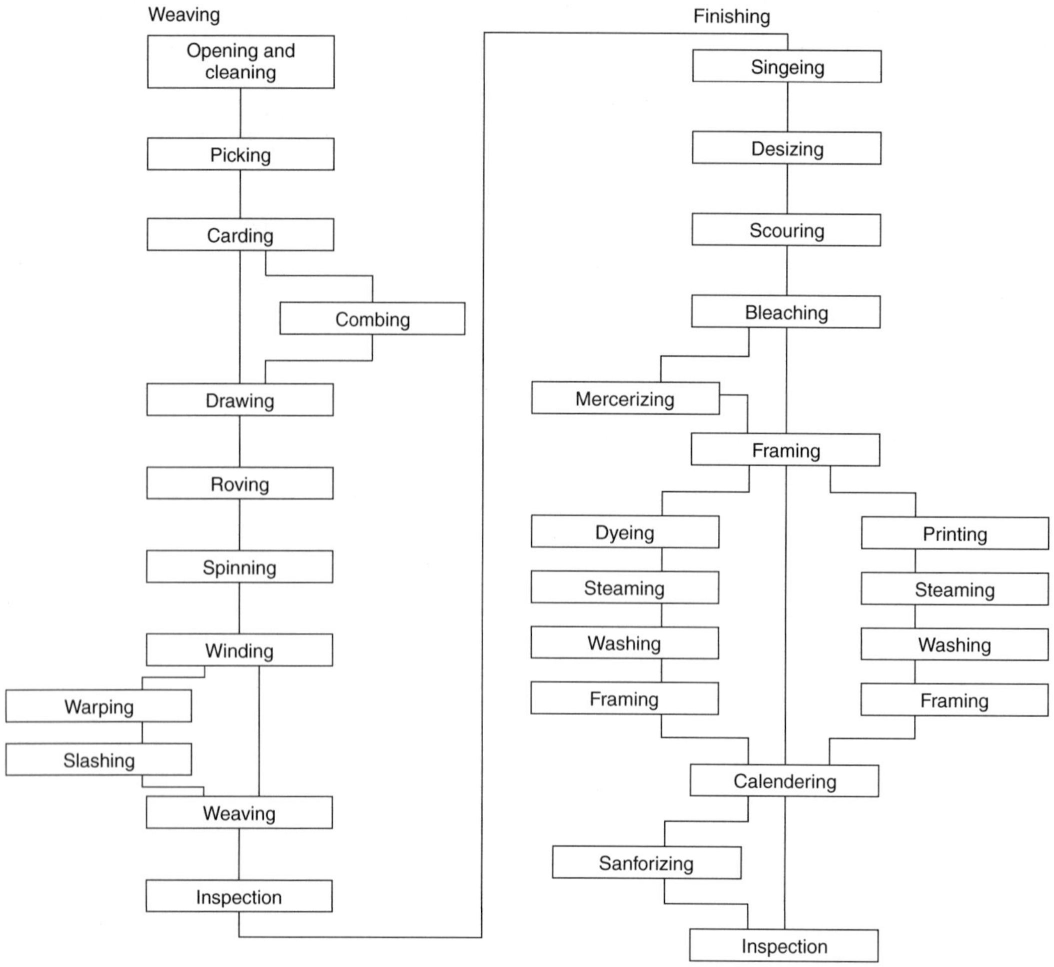

Figure 2. Cotton Processing Flow Diagram.

into rope form), spinning (or further drawing of fibers into yarns), slashing (starch sizing of wool yarn to protect against abrasion and friction generated in the weaving operation), weaving (yarns are woven into cloth), and mending. Scouring is performed by detergents or solvents.

Dyes and Chemicals

Dyes used by the textile industry are classified according to the application and include:

1. Acid dyes are sodium salts, usually of sulfonic acids or carboxylic acids, and are the main class of dyes used in wool dyeing and for dyeing nylon.
2. Direct dyes resemble acid dyes in that they are sodium salts of sulfonic acids and are often azo compounds. They are used primarily for the dyeing of cellulosic fibers (e.g., cotton).
3. Basic dyes are usually hydrochlorides or salts of organic bases. The pigment is found in the cation, and the dyes are often referred to as cationic dyes. They are mainly used on acrylic fibers.
4. Disperse dyes are suspensions of finely divided organic compounds with very slight aqueous solubility. These dyes were developed to dye hydrophobic fibers such as cellulose acetate and the synthetic fibers.
5. Mordant dyes include natural and synthetic dyes. They have no natural affinity for textile fibers, but can be applied if the fiber has been mordanted with a metallic oxide. The most commonly used mordant is chrome, and these dyes can be referred to as chrome dyes. The dyes are most commonly applied in a boiling acid dye bath after application of the chrome mordant to the wool. Alternatively, the order of application can be reversed.
6. Premetallized dyes are a variation on mordant dyes with the metallic oxide, again usually chrome, already included in the formulation. This eliminates the need for addition of the dichromate as a second step in the process.
7. Reactive dyes are generally referred to as fiber-reactive dyes. They react chemically with functional groups on the fiber, and because of this feature, possess good wash-fastness properties. They are primarily applied to natural and some synthetic fiber types (e.g., cotton, rayon, linen, wool, and silk). Fiber-reactive dyes are applied either by exhaust or continuous methods. A common method, utilizing a continuous system, is to impregnate the fiber with the dye and an alkali followed by steaming to promote the reaction between fiber and dye. Most fiber-reactive dyes are used to dye cellulose.
8. Sulfur dyes are complex organic compounds that contain sulfur linkages within the molecules. They are usually

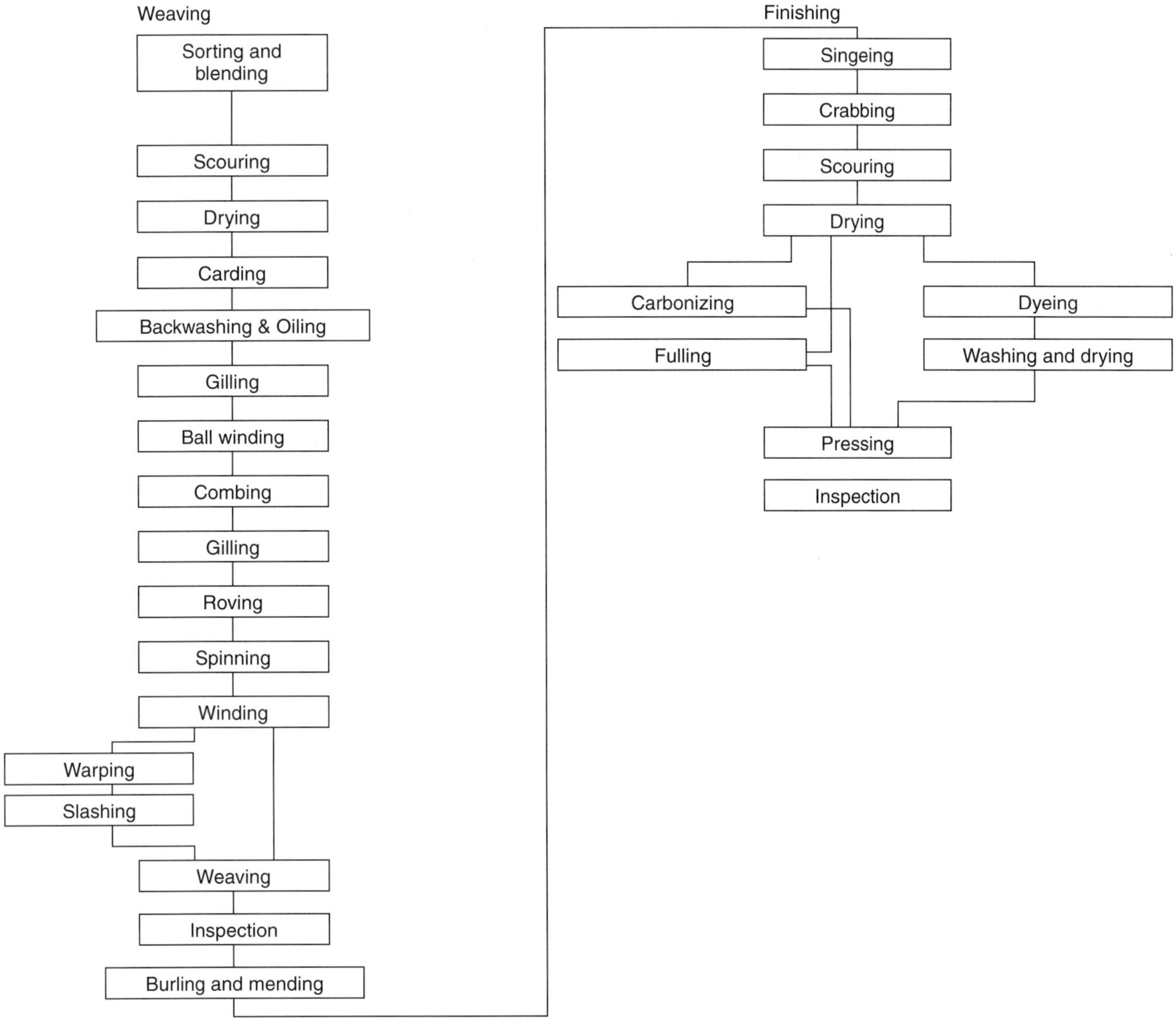

Figure 3. Wool Processing Flow Diagram.

insoluble in water, but will form soluble components in the presence of a reducing agent such as sodium sulfide. The soluble components have an affinity for cellulose (cotton), and once on the fiber, are oxidized back to the original insoluble form. These dyes have good resistance to washing and moderate resistance to sunlight, and are not very bright.

9. Vat dyes are made from indigo, anthraquinone, and carbazol, and are used mainly on cotton, linen, and rayon. They are insoluble in water, but when treated with reducing agents, they are converted into soluble compounds in the presence of alkali. The soluble compounds have an affinity for cellulose and reoxidize to the insoluble pigment within the fiber when exposed to air.

Auxiliaries typically recommended for satisfactory dyeing are listed in Table 1.[1] Other chemicals used in different textile processes are listed in Table 2.[1]

AIR EMISSIONS CHARACTERIZATION

Possible emissions from textile processing include:

1. Oil mists and organic emissions produced when textile materials containing knitting and lubricating oils, plasticizers, and other materials that can volatilize or be thermally degraded into volatile substances, are subjected to heat. Processes that can be the sources of oil mists include tentering, calendaring, heat setting, drying, and curing.
2. Acid mists produced during the carbonizing of wool.
3. Solvent vapors released during and after solvent processing operations such as dry cleaning.
4. Dust and lint produced by the processing of natural fibers and synthetic staple prior to and during spinning, as well as by napping and carpet shearing.

Table 1. Auxiliaries Required for Satisfactory Dyeing

Chemical Auxiliary	Type of Dye Acid	Direct	Basic	Disperse	Mordant	Premetallized	Reactive	Sulfur	Vat
Acetic acid	×		×	×	×	×	×	(or alternative acids)	×
Ammonium acetate	×								
Ammonium phosphate	×								
Ammonium sulfate	×				×	×			
Aromatic amines		×							
Buffer							×		
Defoamer				×					
Dispersing agents				×					
Formic acid	×		×		×	×			
Gelatin									×
Hydrochloric acid		×							
Hydrogen peroxide								×	×
Leveling or retarder agents	×		×	×					×
Oxalic acid			×	×					
Penetrating agents					×	×			
Potassium dichromate					×	×			
Sequestering agents		×		×					
Sodium acetate			×						
Sodium carbonate		×					×	×	
Sodium chloride		×					×	×	
Sodium dichromate						×		×	
Sodium hydrosulfite									×
Sodium hydroxide							×		×
Sodium nitrate		×							
Sodium sulfate (Glauber's salt)	×		×		×	×			
Sodium sulfide								×	
Soluble oil									×
Sulfuric acid	×				×	×			
Urea							×		
Wetting agent		×							

AIR POLLUTION CONTROL MEASURES

Emissions from finishing operations, including drying, curing, and heat setting, have been and continue to be one of the more significant air pollution problems in the textile industry. The blue haze and odor that are characteristic of these emissions have posed both technical and economic problems for large and small facilities alike. The amount of air pollution depends on the finishing processes used. Various kinds and amounts of oils and finishing resins can vaporize from the cloth in tenter frames (commonly used for drying, curing, and heat setting) operating at temperatures typically in the 300–400°F range. Some exhausts also can carry a significant amount of lint.

There are process modifications that can reduce some of this air pollution. For example, most of the oils can be removed by effectively prescouring the cloth. Alternatively, the selection of "less polluting" finishing chemicals may help, but cloth finishing quality and economics often limit the ability of such changes to correct the problem. Over the longer term, new technology, such as solvent-free, radiation-curable coatings, may prove to be a factor in overcoming the air pollution problem.[2] Such technology not only would eliminate the problem, but would result in significant energy savings, higher line speeds, and greatly reduced floor space. Unfortunately, lack of availability of radiation-curable coatings and high cost have been the major limitations to commercializing this technology.

Historically, the solution to the characteristic blue haze and odor problem has been the use of air pollution control equipment.[3–6] Wet (two-stage) electrostatic precipitators, fabric-type (glass fiber-wool) mist eliminators, and fume incinerators have traditionally been the equipment of choice. Wet precipitators and mist eliminators have been plagued by continuous and costly operation and maintenance problems. In general, their performance has been strongly dependent on the effectiveness of gas stream cooling prior to treatment in the control device. Experience has shown that temperatures generally should not exceed 120°F. Fume incinerators operating at 1200–1400°F with approximately a 0.3-sec gas residence time have been effective in eliminating blue haze and odor problems; however, capital and operating costs may have a significant impact on the facility.

Air pollution problems resulting from greige processing of cotton consist predominantly of dust generation in the high-speed mechanical operations, such as spinning, drawing, carding, and twisting. Dry dust collectors of the vacuum type are well suited to maintain a dust-free working environment.

Finely divided lint is present in the air in the vicinity of various operations associated with spinning the yarn and weaving the cloth. Most modern mills have air tunnels built under the floor to effect a downward flow of air around the machines that produce the largest volumes of waste lint (carding, roving, spinning, twisting, weaving, etc.). In the

Table 2. Other Chemicals Used in Different Textile Processes

Textile Process	Chemicals	
	Major	Minor
Printing	Dyestuffs	Ammonium
	Pigments	Sodium alginate
	Gums/thickeners	Soda ash
	Solvents	Hypochlorite
	Alcohols	Soaps and detergents
Bleaching	Hydrogen peroxide	Sodium phosphates
	Sodium hypochlorite	Sodium carbonate
	Sodium chlorite	Sulfur dioxide
	Sodium silicate	Sodium bisulfite
	Sodium hydroxide	Epsom salt
	Organic stabilizers	Calcium carbonate
	Surfactants, detergents	Sodium nitrite
	Chelating agents	
Slashing	Cornstarch	
	Polyvinyl alcohol	
	Carboxymethyl cellulose	
	Soaps	
Spinning	Wool oil	
Carpet finishing	Latex	
	Calcium carbonate	
	Aluminum hydrate	
	Thickeners	
	Styrene-butadiene rubber	
Desizing	Enzymes	Sodium persulfate
	Sodium hydroxide	Sodium bromite
	Hydrogen peroxide	Potassium perphosphate
	Surfactants, detergents	Glauber salt
	Chelating agents	Sodium chloride
	Sodium carbonate	Potassium chloride
	Epsom salt	Calcium chloride
Scouring	Sodium hydroxide	Sodium silicate
	Surfactants, detergents	Solvents
	Chelating agents	Salt
	Sodium phosphates	
	Sodium carbonate	
Fiber preparation	Oleic acid	
Special finishing	Softening agents (oils, waxes, soaps)	
	Stiffeners (PVA, ethyl cellulose)	
	Permanent press (resins)	
	Shrinkproofing (cross-linking resins)	
	Oil repellents (perfluorofatty acids and acrylates)	
	Water repellents (silicone resins)	
	Antistatic finishes	
	Fire retardants	
	Bactericides (trichlorophenol)	
	Mothproofer (dieldrin)	

opening and picking rooms in a cotton plant, stock is usually handled by overhead ductwork. Lint and dust are collected by the vacuum system as close as possible to their point of origin. Automatic traveling vacuum cleaning systems are frequently used in conjunction with the underfloor system. The suspended lint and dust are conveyed by air movement through ducts to a centrally located filter system for removal. Automatic dry-type traveling disposable air filters are extensively used for this purpose. Since any filtering medium tends to hold the lint fibers tenaciously, a low-cost, disposable paper medium is often used.

The fumes from wool carbonizing pose another problem that must be dealt with. This process generates very fine carbon particles that appear as smoke, as well as some fumes and odors. These fumes probably include residual sulfur oxides (if sulfuric acid has been used for carbonizing), in addition to organic decomposition products, and are generally very corrosive. Corrosion-resistant stainless steels and/or plastics have been used in the collection systems, which usually exhaust the fumes to the atmosphere. Since a significant quantity of particulate matter is in the submicrometer range, a visible emission is usually apparent unless there is a very efficient air pollution control system. The severity of the opacity problem will depend to a large extent on location, topography, and local meteorological conditions. Incinerators, wet scrubbers, and fabric filters have been used to resolve the problem. Wet scrubbing systems have the added advantage of reducing residual gases such as sulfur oxides, as well as particulate emissions.

REFERENCES

1. E. C. Chen, *Environmental Assessment of the Canadian Textile Industry*, Environment Canada Report EPS 5/TX/1, June 1989.
2. *Symposium Proceedings: Textile Industry Technology* (December 1978, Williamsburg, VA), F. A. Ayers, Ed., prepared for U.S. Environmental Protection Agency, Report No. EPA600/2-79-104, May 1979.
3. P. Miller and J. L. Wilki, "Removing Visible Oil Fumes from Tenter Frame Exhausts," *Mod. Textiles*, February 1974.
4. M. R. Beltran, "How to Recover Heat from your Oven Exhausts," *Am. Dyestuff Reporter*, April 1974.
5. W. Hebrank and L. Nelson, "Heat Recovery System Saves 45% Fuel at Cannon Mills," *Textile Indust.*, February 1980.
6. Textile finishing application literature from the following companies: wet electrostatic precipitators—Beltran (Brooklyn, NY), Trion (Sanford, NC, and United Air Specialists (Cincinnati, OH); fume filters—CVM (Wilmington, DE) and Monsanto's Brink's mist eliminator (St. Louis, MO); incinerators—Hirt Combustion (Montebello, CA), AER (Ramsey, NJ), and Smith Environmental (Duarte, CA).

THERMOPLASTIC RESINS

JOHN A. FEY AND J. MITCHELL JENKINS III

Thermoplastic resins are materials that soften and melt upon heating, with no chemical change and with no permanent change in physical properties. These resins are readily extruded or molded into film, fibers, bottles, and many types of molded articles. In the following sections, two of the major thermoplastic resins are discussed: polyethylene and polypropylene.

POLYETHYLENE

The earliest polyethylene product was low-density polyethylene (LDPE), which was first produced in the United States during World War II for use as radar cable insulation. It is made from ethylene by a polymerization process that operates at pressures as high as 50,000 psi.

During the 1950s, catalysts were discovered that allowed production of polyethylene at much lower pressures, a few hundred pounds per square inch. The polyethylene was found to have somewhat different physical properties than the LDPE produced by high-pressure processes. Because of its higher density, the material was called high-density polyethylene (HDPE).

Further catalyst development has allowed the copolymerization of ethylene with propylene, 1-butene, and other higher α-olefins. Incorporation of these higher α-olefins resulted in a decrease in the density of the polyethylene, along with other physical property changes. It was found, however, that this "new" material had properties similar to, but not identical with, those of the LDPE produced by the high-pressure processes. This new low-density polyethylene was called linear low-density polyethylene (LLDPE).

A large percentage of the HDPE and LLDPE produced worldwide is made by gas-phase fluidized-bed processes. The most widely used process of this type is the UNIPOL® polyethylene process, which was developed by Union Carbide Chemicals and Plastics Co. Inc. and has been licensed for use by polyethylene producers worldwide.

Process Description

Following is a brief description of facilities in each part of a typical gas-phase fluidized-bed polyethylene plant. References such as R-1 indicate process streams as shown on the block flow diagram Figure 1. Nomenclature for Figure 1 is identified in Table 1.

Ethylene Supply and Purification. The ethylene feed to the plant (R-1) is compressed or pumped to the required feed pressure, after which it is passed through a series of purifiers where trace quantities of impurities are removed. The purifiers are vented to a flare before and during regeneration. The fresh ethylene is combined with recycled ethylene and is injected into the reactor.

Raw-Materials Supply and Purification. Other reactor feed materials are identified as R-2 through R-8. R-2, R-3, and R-7 are passed through degassing columns and temporarily stored. R-2 (or alternatively R-7) and R-3 are passed through purifiers to remove impurities and are injected into the reactor. The purifiers are vented to the flare before and during regeneration. R-4 is passed through a series of purifiers, where trace quantities of impurities are removed, and then injected into the reactor. R-8 also is passed through a series of purifiers, where trace quantities of impurities are removed. A portion of R-8 is compressed and then routed to the reactor. The remaining R-8 is used throughout the plant. R-5 and R-6 are routed directly to the reactor.

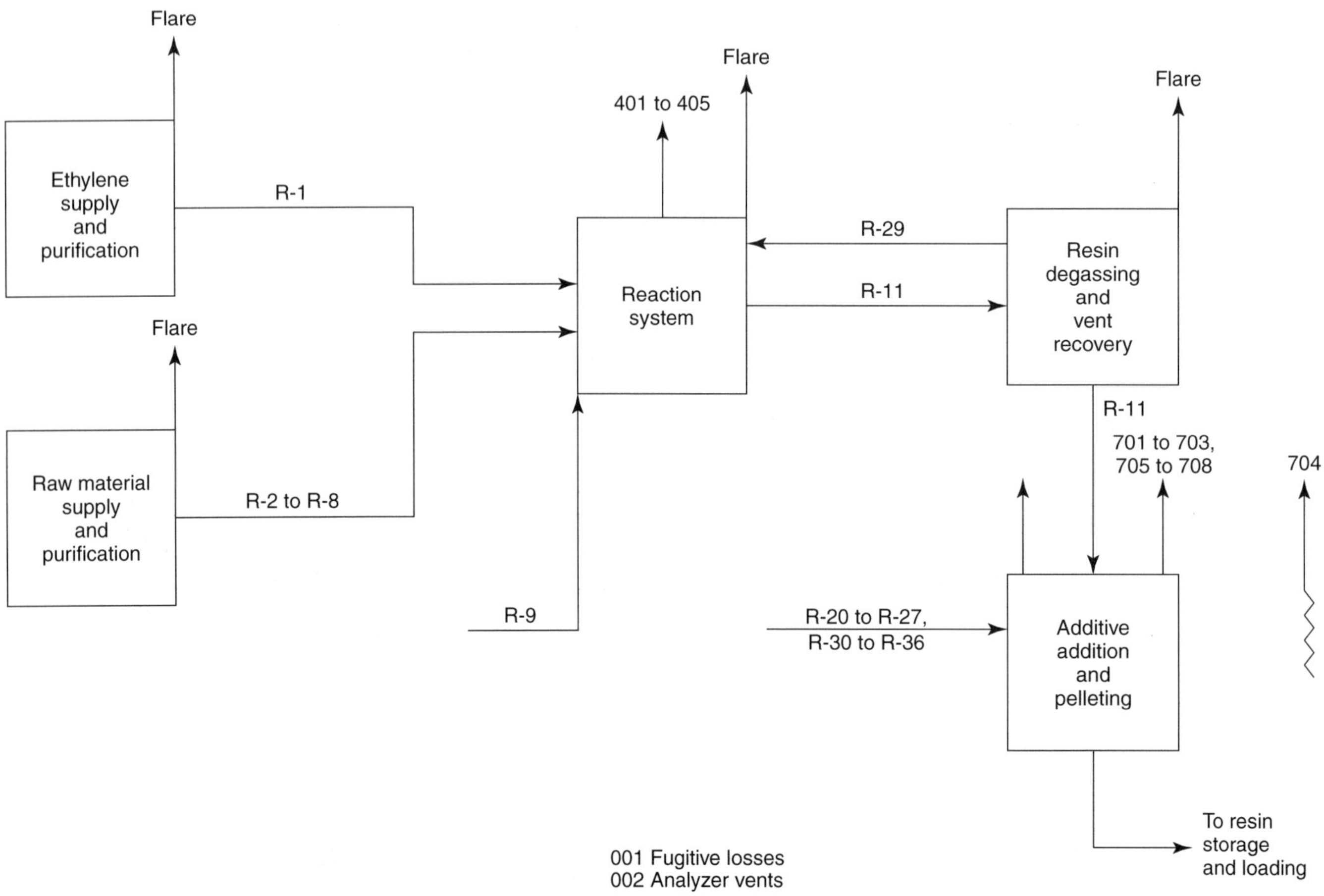

Figure 1. Polyethylene Plant Using Gas-Phase Fluidized-Bed Process.

Table 1. Nomenclature for Figure 1

001	Hydrocarbons (fugitive losses)
002	Hydrocarbons (from analyzers)
401	Cycle gas compressor seal and lube oil reservoir vent
402, 403	Catalyst feeder vents
404	Reactor seed bed vent
405	Catalyst vent filter
701	Dust collector exhaust
702	Pelleter dryer exhaust
703	Polyethylene fines while conveying to storage
704	Hydrocarbon evolution from storage
705	Resin surge bin vent
706	Additive dump hopper vent
707	Resin cooler vent
708	Seed bed resin rotary feeder vent

Reaction System. The reaction system consists of a fluidized-bed reactor, a cycle gas compressor and cooler, and product discharge tanks. Ethylene (R-1), R-2, R-3, and/or R-7, R-4, R-5, R-6, and catalyst (R-9) are fed continuously to the reactor. The unreacted ethylene is recycled. Polyethylene resin (R-11) is removed from the reactor and separated from the small amount of gas accompanying it in the discharge tank. The separated gas either is pumped back to the reactor as part of the recycle stream or is flared.

Resin Degassing and Vent Recovery

Resin from the reactor (R-11) is conveyed to a purger, where the dissolved hydrocarbons are stripped from the resin and returned to the process (R-29). Any excess gas not used in the process is vented to the flare. Resin from the purger flows to the additive system.

Additive System. Resin (R-11) and solid additives (R-21 to R-27, R-30 to R-36) are metered by gravimetric feeders and flow by gravity to the pelleting system.

Pelleting. The additives and resin are thoroughly mixed, melted, and pelleted in the pelleting system. The pellets are dried, cooled, and conveyed to product storage and packaging or shipping.

POLYPROPYLENE

Polypropylene was first produced commercially in the late 1950s. In the early plants, polypropylene was polymerized as a slurry in a hydrocarbon diluent. Later, processes were developed in which the polypropylene was polymerized as a slurry in liquid propylene. Mechanically stirred gas-phase processes also were developed. In the early 1980s, Union Carbide Chemicals and Plastics Co. Inc. and the Shell Chemical Co. [a division of Shell Oil Co. (USA)] jointly developed the UNIPOL® polypropylene process, a gas-phase fluidized-bed process. This process has been licensed for use by polypropylene producers worldwide.

Polypropylene products can generally be divided into three classes: homopolymers, random copolymers, and impact copolymers. Polypropylene homopolymers are crystalline polymers containing only propylene. Random copolymers are copolymers of propylene with small amounts of ethylene (or other α-olefins). Both homopolymers and random copolymers are made in a single reactor, and the same plant can produce both types of polymers as market demand may require.

Impact copolymers are a mixture of polypropylene and ethylene–propylene copolymer produced by two reactors operating in series. Catalyst is fed to the first reactor, where the polypropylene homopolymer is produced. The homopolymer, containing still-active catalyst, is transferred to a second reactor, where the ethylene–propylene copolymer is produced. Most UNIPOL® polypropylene plants are built with two reactors so that the full range of homopolymer, random copolymer, and impact copolymer products can be produced.

Process Description

Following is a brief description of facilities in each part of a typical gas-phase fluidized-bed polypropylene plant. References such as R-1 indicate process streams as shown on the block flow diagram, Figure 2.

Propylene Supply and Purification. Fresh propylene feed to the unit is passed through a degassing column where trace quantities of impurities are removed. The column is vented to a flare. This propylene (R-1) is pumped through a purifier, where additional impurities are removed. The purifier is vented to a flare before and during regeneration. R-1 is then routed to one or both reactors.

Raw-Materials Supply and Purification. The R-2, R-3, and R-4 are passed through separate series of purifiers, where trace quantities of impurities are removed. Each is supplied as required to various parts of the unit. All except one of the purifiers are vented to a flare before and during regeneration. R-5 is routed directly to the reactors.

Reaction System No. 1. Reaction system No. 1 consists of a fluidized-bed reactor, a cycle gas compressor and cooler, and product discharge tanks. Streams R-1 to R-6, catalyst (R-7), and R-11 from the vent recovery system are fed continuously to the reactor. Polypropylene (R-9) is removed from the reactor by the discharge tanks and is sent either to reaction system No. 2 or resin degassing. A purge (R-10) is sent to the vent recovery system.

Reaction System No. 2. Reaction system No. 2 also consists of a fluidized-bed reactor, a cycle gas compressor and cooler, and product discharge tanks. Streams R-1 to R-5, additives (R-8 and R-22), R-9 from reaction system No. 1, and R-14 from the vent recovery system are fed continuously to the reactor. Polypropylene R-13 is removed from the reactor by the discharge tanks and is sent to resin degassing. A purge (R-12) is sent to the vent recovery system.

Resin Degassing. Resin from either reaction system (R-9 or R-13) is conveyed to a receiver, where interstitial and some dissolved hydrocarbons are stripped from the resin and are sent to the vent recovery system (R-15). The resin flows to a purger, where additional hydrocarbons are removed. The purged resin (R-16) is sent to the pelleting system. The gas from the purger is flared.

Vent Recovery. The vent recovery system recovers as much hydrocarbon as possible from the streams sent to it (R-10, R-12, and R-15). The recovered materials are returned directly

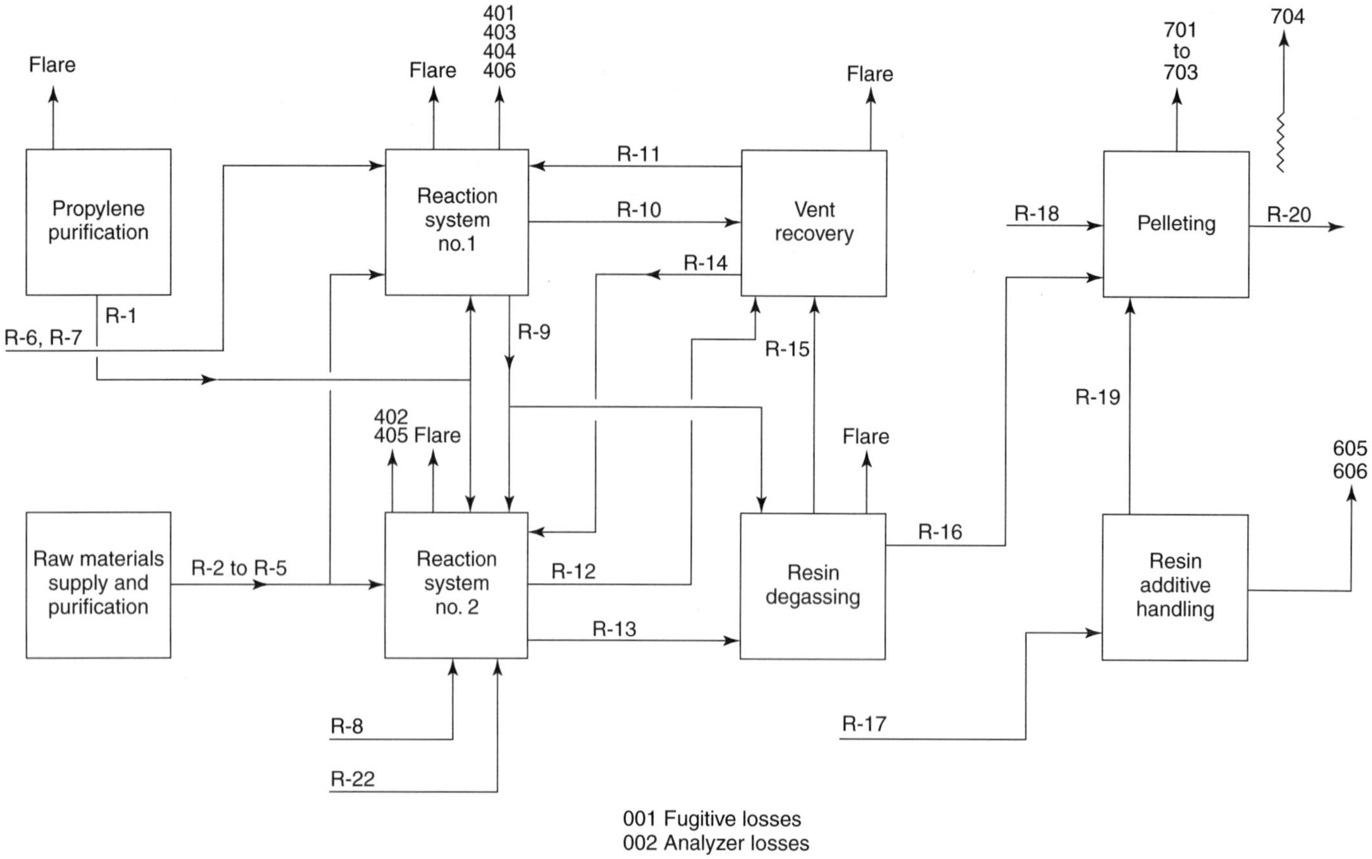

Figure 2. Polypropylene Plant Using Gas-Phase Fluidized-Bed Process.

to both reaction systems (R-11 and R-14). Any gas not returned to the process is vented to the flare.

Resin Additive Handling. Solid additives (R-17) are metered and sent to the pelleting system (R-19).

Pelleting. The resin (R-16), solid additives (R-19), and liquid additive (R-18) are mixed, melted, and pelleted in the pelleting system. The pellets (R-20) are dried, cooled, and sent to product storage and packaging or shipping.

Air Emissions Characterization

Table 2 characterizes the emissions from a UNIPOL® polyethylene plant, and Table 3 characterizes the emissions from a UNIPOL® polypropylene plant.

Table 2. Emissions Characterization for Polyethylene[a]

UNIPOL™ Polyethylene emissions	
Hydrocarbon	
Process	0.20 kg/ton
Fugitive	0.03 kg/ton
Particulate (resin)	0.02 kg/ton
Aqueous	None
Nonaqueous liquids	<0.1 ton/yr
Solid wastes	
Catalysts and treatment beds	<0.15 kg/ton
Polymer spills	Normally recovered and sold

[a]Data provided by Union Carbide Chemicals and Plastics Co. Inc.

Table 3. Emissions Characterization for Polyethylene[a]

UNIPOL™ Polyethylene emissions	
Hydrocarbon	
Process	0.16 kg/ton
Fugitive	0.03 kg/ton
Particulate (resin)	0.02 kg/ton
Aqueous	None
Solid wastes	
Catalysts and treatment beds	<0.1 kg/ton
Polymer spills	Normally recovered and sold

[a]Data provided by Union Carbide Chemicals and Plastics Co. Inc.

AIR POLLUTION CONTROL MEASURES

The gas-phase fluidized-bed processes for producing polyethylene and polypropylene are low-pressure, solventless processes with a low inventory of in-process hydrocarbons. This provides for very low hydrocarbon emissions to the atmosphere. The low operating pressure and virtual absence of polluting effluents enable this type of process to meet current government regulations for air, water, and solid wastes. Low operating pressure, a minimum of process steps, and absence of solvent combine to ensure compliance with hydrocarbon emission standards. Hydrocarbon discharges from occasional routine depressurization of the reactor and from purifier regeneration for polyethylene and sieve regenerations for polypropylene are routed to a high-efficiency flare. All but the last traces, less than 50 ppm, of hydrocarbons removed from the resin granules are recovered or flared. Polymer particulate emissions to the air are very low, with dust controlled by filters on gas exhausted from conveying stations. Surface drainage is collected and screened to remove polymer spills. Solid waste consists of a minimal amount of polymer scrap and spills that can be recovered and sold or incinerated. There are also small amounts of inactive catalyst and treatment beds that can be sold or disposed of. There are no aqueous process effluents whatsoever from the gas-phase process.

The simplicity of the gas-phase process allows for the minimum use of valves, pumps, and compressors, which are the primary source of fugitive emissions. Fugitive emissions, therefore, are minimal.

For polyethylene, less than 0.15 kg per ton and for polypropylene less than 0.1 kg per ton of waste treatment beds and, in both cases, 1.1 kg per ton of spilled and scrap polymer are generated each year. The spilled polymer is washed down to a collection pit, filtered from the water, and sold. Spilled polymer, therefore, does not normally contribute to the solid waste load.

As a result of the exceptionally low emissions, the low-pressure gas-phase fluidized-bed process has been considered best available control technology by regulatory agencies. Government permits have been obtained for UNIPOL® polyethylene and polypropylene plants in Texas, Louisiana, and Illinois, and in over a dozen countries around the world.

BIBLIOGRAPHY

Anonymous, "New Route to Low-Density Polyethylene," *Chem. Eng.*, 86:80 December 3, 1979.

I. D. Burdett, "Polypropylene: Lowest Cost Process," *Hydrocarbon Proc.*, 75 November 1986.

H. H. K. Ficker, G. L. Goeke, and G. W. Powers, "Gas-Phase Process for PP Surpasses other Methods," *Plast. Eng.*, 29 February 1987.

F. F. D. Hussein and T. L. Nemzek, *UNIPOL® PP-Innovation Through Combined Technologies*, AIChE 1989 Spring National Meeting, Houston, TX, April 2–6, 1989.

J. M. Jenkins III, "Polyethylene: Continuing Technological Advancement," in *Hydrocarbon Technology International*, P. Harrison, Ed.; Sterling Publications, London, 1987, pp. 93–95.

F. J. Karol, "Catalysis and the Polyethylene Revolution," in *History of Polyolefins*, R. B. Seymour and T. Cheng, Eds.; D. Reidel Publishing Co., Dordrecht, Holland, 1986, pp. 193–211.

F. J. Karol and F. I. Jacobson, "Catalysis and the UNIPOL® Process," *Stud. Surf. Sci. Catal.*, 25:323 1986.

S. Royse, "UNIPOL® Process Brings UC Continued Success," *Europ. Chem. News* **28**, May 28, 1990.

THERMOSETTING RESINS

John A. Gannon

Thermosetting synthetic resins constitute a variety of manufactured polymers characterized by conversion to an insoluble, infusible state brought about by a second-stage chemical reaction, usually employing heat. The process is often referred to as curing or hardening, and the result is a cross-linked, three-dimensional structure. Prior to curing, the resinous material can be processed, along with the curing or hardening agent, in a variety of applications, including molding, casting, powder coating, and embedding, during which the composition is cured to develop the physical and chemical properties required for the specific application.

In contrast to thermoplastics, the thermosetting synthetic resins depend on the curing reaction to provide the useful properties they possess. Prominent thermosetting resins are varieties of polyesters, including both unsaturated and saturated types with linear and branched varieties sometimes containing fatty acids (alkyds). A family of polyethers with oxirane groups represents the commonly known epoxy resins that can exist in many forms, but principally are based on epichlorohydrin and bisphenol-A.

Polyurethanes composed of diisocyanate reactants with various diols or polyols make up another prominent category of synthetic resins. An important family of thermosetting resins is also derived from the condensations of phenols, ureas, and melamine derivatives with formaldehyde to form useful commercial products upon conversion to a three-dimensional state or as curing agents for other synthetic resins.

This discussion will feature synthetic resin processes used in manufacturing epoxy resins.

Epoxy resins derived from epichlorohydrin and bisphenol-A exist in two major forms: liquid epoxy resins (base liquid resin, BLR) and solid epoxy resins derived from the liquid resin by further reaction with bisphenol-A ("advancement process"). Both the liquid and solid resins are thermoset or cured by a variety of reagents that react with the terminal epoxy groups of the resin and/or the pendant hydroxyl groups characteristic of advanced products.

Important multifunctional epoxy resins are prepared by glycidylation (epoxidation with epichlorohydrin) of polyfunctional amines or amino phenols with subsequent use of sodium hydroxide as a dehydrohalogenating agent. Such resins are tri- or tetrafunctional in epoxy content and upon curing with aromatic amines, for example, exhibit enhanced chemical resistance and thermal properties as a consequence of the higher cross-linking realizable with these resins.

Epoxy cresol novolak resins are derived from epoxidation of polyfunctional cresol-novolaks, which are, in turn, derived from *o*-cresol and formaldehyde in various ratios using an acid catalyst for the condensation reaction. A range of epoxy cresol-novolak resins is manufactured that includes resins of different molecular weights, functionalities, and softening points.

Curing of the multifunctional epoxy cresol-novolak resins again leads to products with enhanced electrical, chemical, and thermal properties. Similar resins are based on glycidylation of phenol novolaks derived from the condensation of phenol and formaldehyde in the presence of acidic catalysts.

PROCESS DESCRIPTION

The manufacture of solid epoxy resins and related solid epoxy resin solutions is accomplished by the so-called advancement process, wherein a liquid epoxy resin is further reacted in a polymerization process in the presence of a catalyst to form a higher-molecular-weight solid resin possessing terminal epoxy groups and a secondary hydroxyl group in each repeat unit of the polymer chain. A series of solid resins and solid resin solutions is manufactured to various epoxy values (epoxy equivalent weights) and resin viscosities (melt or solution viscosities at various temperatures) depending on the charge ratios of liquid epoxy resin to bisphenol-A, with the former employed in excess to provide epoxy terminal groups. In some instances, a monofunctional phenol (chain terminator) is also used in minor amounts relative to the bisphenol-A charge in order to exert better control of the product viscosity.

Isolation of the solid advanced resin from the reaction vessel is performed by transferring the bag-filtered molten mass to a flaker feed tank and then to a chilled-watercooled belt flaker. The flakes are sized and dropped into a jacketed blending hopper, from which they are fed to a packaging system. The dust generated from the sizing, flaking, and packaging is collected in a dust collector and blended into subsequent batches (see Figure 1).

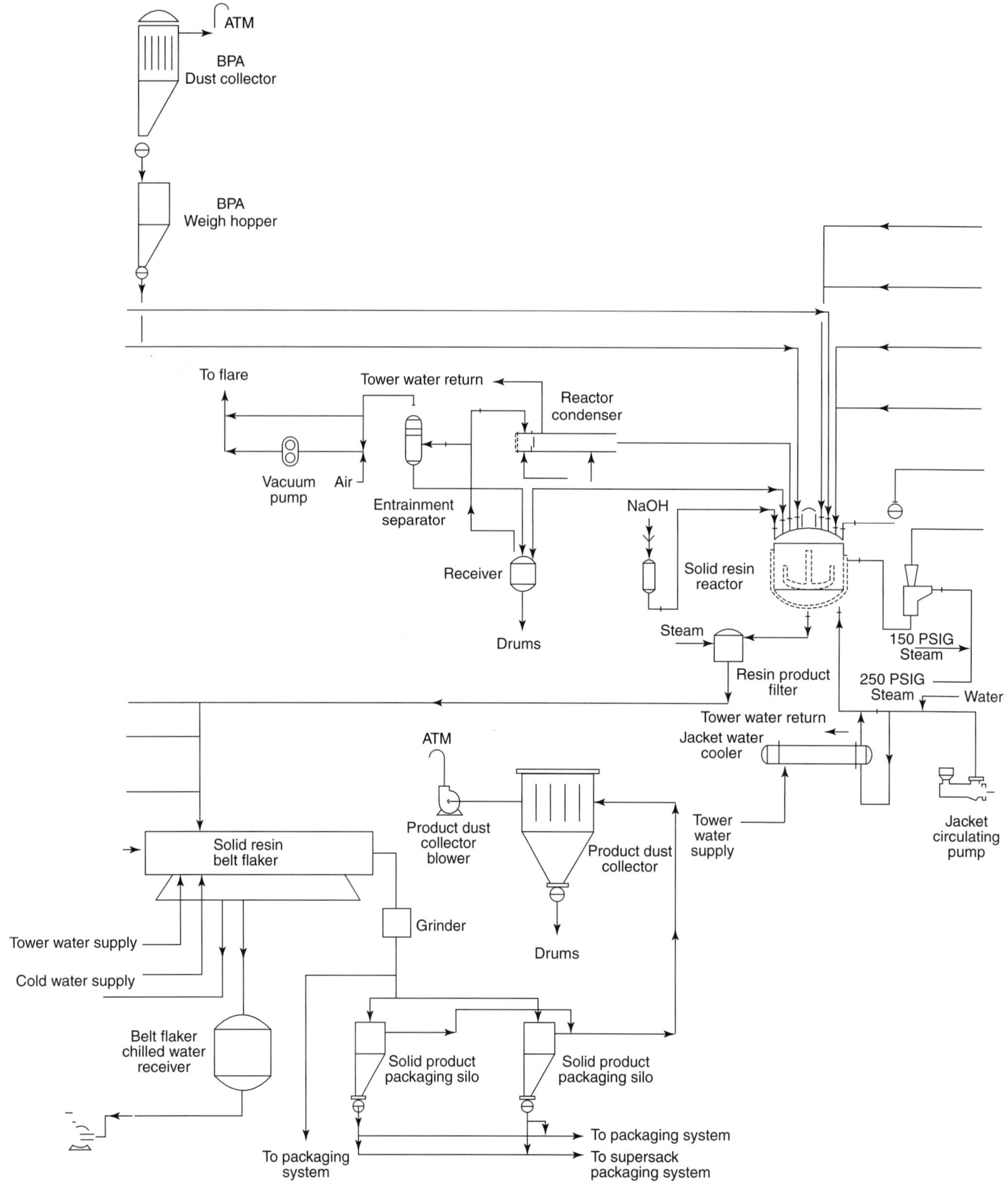

Figure 1. Solid Epoxy Resin Plant Flow Sheet.

Table 1. Emission Source Characteristics

Description	Pollutant	Emission Rate (g/sec)	Height (m)	Exit Gas Temperature (K)	Exit Gas Velocity (m/sec)	Stack Diameter (m)
Packing dust collector	Product dust	0.0210	9.1	289	23.2	0.508
BPA[a] dust collector	BPA[a]	0.161	38.1	294	14.0	0.254
BPA[a] dust collector	BPA[a]	0.161	35.1	294	14.0	0.254
Product blend or term.[b] bag	Term.[b]	0.0328	38.1	294	10.7	0.289
Charge station	Product dust	0.0328	38.1	294	10.7	0.289
Celite[c] and term.[b] bag	Term.[b]	0.0328	30.5	294	10.7	0.289
Charge station	Celite[c]	0.0328	30.5	294	10.7	0.289
Term.[b] and celite[c] bag	Term.[b]	0.0328	38.1	294	10.7	0.289
Charge station						
Vent gas combustor	Dowanol PM[d]	0.00756	48.8	1,144	27.95	0.254
	Dowanol PMA[e]	0.00171	48.8	1,144	27.95	0.254
	MAK[f]	0.00237	48.8	1,144	27.95	0.254
	MIBK[g]	0.0151	48.8	1,144	27.95	0.254
	Cyclosol-53[h]	0.00529	48.8	1,144	27.95	0.254
	n-Butanol	0.00277	48.8	1,144	27.95	0.254
	Isopropanol	0.107	48.8	1,144		
	Toluene	0.360	48.8	1,144	27.95	0.254
	Methanol	0.0464	48.8	1,144	27.95	0.254
	Acetone	0.0628	48.8	1,144	27.95	0.254
	MEK[i]	0.0788	48.8	1,144	27.95	0.254
	Xylene	0.00738	48.8	1,144	27.95	0.254

[a]Bisphenol-A.
[b]Chain terminator.
[c]Solid filter aid.
[d]Monomethyl ether of propylene glycol (Dow Chemical Co.).
[e]Monomethyl ether of propylene glycol acetate (Dow Chemical Co.).
[f]Methyl amyl ketone.
[g]Methyl isobutyl ketone.
[h]Aromatic hydrocarbon.
[i]Methyl ethyl ketone.

Table 2. Emission Inventory

				Pollutant Emissions with Controls		
Emission Point Description	Pollutant	Compound	Control Efficiency	Peak lb/h	Average lb/h	Annual tons/year
Packing dust collector	TSP	Product dust	99.%	0.167	0.167	0.74
BPA[a] dust collector	TSP	BPA[a]	99.%	1.28	0.2256	0.988
BPA[a] dust collector	TSP	BPA[a]	99.%	1.28	0.024	0.104
BPA[a] dust collector	TSP	BPA[a]	99.%	1.28	0.1095	0.475
Product blend or term.[b] bag	TSP	Term.[b]	99.%	0.26	0.001	0.0046
Charge station	TSP	Product dust	99.%	0.26	0.0034	0.0148
Term.[b] and celite[c] bag	TSP	Term.[b]	99.%	0.26	0.0014	0.0125
Charge station						
Term.[b] and Celite[c] bag	TSP	Term.[b]	99.%	0.26	0.0014	0.0125
Celite[c] and term.[b] bag	TSP	Term.[b]	99.%	0.26	0.0015	0.0066
Charge station	TSP	Celite[c]	99.%	0.26	0.00028	0.0012
Vent gas combustor	VOC	Isopropanol	98.%	0.8466	0.064	0.2803
	VOC	Toluene	98.%	2.8555	0.296	1.2964
	VOC	Methanol	98.%	0.3683	0.052	0.2277
	VOC	Acetone	98.%	0.4983	0.0238	0.1042
	VOC	MEK[d]	98.%	0.625	0.0108	0.0473
	VOC	Xylene	98.%	0.0586	0.049	0.0214

[a]Bisphenol-A.
[b]Chain terminator.
[c]Solid filter aid.
[d]Methyl ethyl ketone.

Table 3. Fugitive VOC Emission Inventory

Source	Number	Emission Factor (lb/h/source)	VOC Emissions without Controls (tons/year)	Control Efficiency (%)	VOC Emissions with Controls (tons/year)
Pumps					
Light liquid (mech. seal)	53	0.109	25.3	61[a]	9.9
Heavy liquid (mech. seal)	25	0.0472	5.2	61[a]	2.0
Light liquid (sealless)	40	0.109	19.1	100[b]	0
Valves					
Light liquid	2235	0.016	162.9	59[a]	66.8
Heavy liquid	625	0.00051	1.4	—	1.4
Compressors	0	0.503	0	—	0
Relief valves	6	0.229	6.0	50[a]	3.0
Flanges	4720	0.0018	37.2	—	37.2
Open-ended lines	148	0.0037	2.4	100[c]	0
Sampling connections	37	0.0331	5.4	100[d]	0
Totals			264.9		120.3

[a]Leak detection and repair program credit.
[b]Using sealless pumps as an equipment alternative results in no leaks.
[c]Using caps on open-ended lines as an equipment alternative results in no leaks.
[d]Using closed purge sampling systems as an equipment alternative results in no leaks.

The solid resin solutions are prepared by quenching (in reactors equipped with a suitable cooled reflux condenser) the molten advanced resin with solvent(s) from the solvent weigh tank via the tank farm or from drums. The resin solution, after analysis, is charged with a solid filter aid through a charging hopper and then filtered, after which it is transferred through a bag filter to a storage tank, tank truck, railcar, or drums.

Possible air emission sources in solid resin manufacture include the bisphenol-A and chain terminator storage and charging bins, as well as vents from vessels used in processing/purification of the resinous products and corresponding resin solutions.

AIR EMISSIONS CHARACTERIZATION

Table 1 details the air emission source characteristics for the various pollutants that are part of the solid epoxy resin advancement systems and the epoxy resin solution manufacture. As is evident from the table, solid bisphenol-A and chain terminator reactants are handled with multiple dust collectors or in charging stations with built-in dust collectors and the solvents used in the processing are vented to vent gas combustors from reactors and storage/processing vessels.[1]

Table 2 details the emissions inventory for solid resins and resin solutions, whereby control of total suspended particulates (TSP) and volatile organic compound (VOC) pollutant is effected.[2,3]

Table 3 summarizes the data obtained on fugitive VOC emissions, including emission factors at an epoxy resin plant.[4–7]

AIR POLLUTION CONTROL MEASURES

Two major emission control techniques are utilized in the epoxy resin manufacturing plant described here — recovery via baghouses and combustion via vent gas combustors. Display of the efficiency of these techniques is detailed in Table 2, and it can be seen that both control measures are effective in a plant devoted to solid epoxy resin and resin solution manufacture.

REFERENCES

1. Central Engineering Dept. CIBA-GEIGY Corp.
2. *Rules and Regulations*, Alabama Air Pollution Control Commission, revised September 21, 1989.
3. *Compilation of Air Pollutant Emission Factors*, EPA AP-42, 4th ed., U.S. Environmental Protection Agency, September 1985.
4. "Method 21-Determination of Volatile Organic Compounds Leaks," *Code of Federal Regulations*, Title 40, Part 60, Appendix A, U.S. Environmental Protection Agency, July 1, 1986.
5. *Fugitive Emission Sources of Organic Compounds—Additional Information on Emissions, Emission Reductions, and Costs*, EPA-450/2-82-010, U.S. Environmental Protection Agency, April 1982.
6. "Standards of Performance for New Stationary Sources: Volatile Organic Liquid Storage Vessels (Including Petroleum Liquid Storage Vessels)," *Federal Register* **52**(67), April 8, 1987.
7. "Subpart VV—Standards of Performance for Equipment Leaks of VOC in the Synthetic Organic Chemicals Manufacturing Industry," *Code of Federal Regulations*, Title 40, Part 60, July 1, 1986.

BIBLIOGRAPHY

"Performance Tests," *Code of Federal Regulations*, Title 40, Part 60 Section 8, July 1, 1986.

Guidelines on Air Quality Models (revised), EPA-450/2-78-027R, U.S. Environmental Protection Agency, July 1986.

P. C. Siebert, K. R. Meardon, and J. C. Serne, "Emission Controls in Polymer Production," *Chem. Eng. Prog.* **80**(1), 68 (1984).

Industrial Source Complex (ISC) Dispersion Model User's Guide, 2nd ed., Vol. 1, EPA-450/4-86-005a, U.S. Environmental Protection Agency, June 1986.

Addendum to Regional Workshops on Air Quality Modeling: A Summary Report, EPA-450/4-82-015, U.S. Environmental Protection Agency, revised October 1983.

Users Manual for Single Source (CRSTER) Model, EPA-450/277-013, U.S. Environmental Protection Agency.

13

FOOD AND AGRICULTURAL INDUSTRY

BREAD BAKING

PETER F. HESS

The control of ethanol emissions is spurred by the increasing need for precursor organic emissions to meet the 1-h and the new 8-h federal ozone ambient air quality standard in large metropolitan areas.[1] Bread baking operations are also facing the need to meet the requirements of New Source Review, Prevention of Significant Deterioration, and the federal operating permit program of Title V of the 1990 Clean Air Act Amendments.[2] Although bread baking is not viewed as a classical source of air pollution, there may be a significant amount of ethanol generated from bread-baking ovens. Bread-baking ovens can generate significant amounts of ethanol. The baking industry is large and decentralized with 2,636 wholesale bakeries in the United States in 1990. Because bread is a perishable and distribution delays undesirable, bakeries tend to correlate with population and be located in larger cities in all states.[3] Ethanol is generated from the fermentation process that takes place in yeast-leavened bakery products such as bread and bread-type rolls. Chemically leavened bakery products, such as cakes, cookies, cake doughnuts, and biscuits generate little if any ethanol emissions.

The amount of ethanol generated during the bread-baking process varies depending on the type of dough process employed, with those processes that use a higher percentage of yeast and have a longer fermentation time tending to generate the largest amount of ethanol. Several different types of dough processes are employed in large commercial bakeries, including sponge dough, straight dough, liquid and flour brews, and continuous-mix or "no time" dough. Sponge dough processes typically will have the highest percentage of yeast and longest fermentation time and will generate the greatest amount of ethanol.[4] But while the sponge dough process is used in the formulation of white breads, the popularity of and the demand for other bread varieties have resulted in the broad use of virtually all dough processes. However, the sponge dough process is still the prevalent process to produce bread.

Bakery ovens are predominantly directly fired by natural gas. In such ovens, the ethanol vapors driven from the bread during baking and the combustion product gases exit the oven through the same exhaust stacks. Combustion gases never come in contact with the bread in indirectly fired ovens, and are exhausted through separate stacks.

The exhaust flow rate from a bakery oven must be balanced with the requirements to produce a quality product. With bakery ovens there is a need to properly control the oven temperature to produce a product that meets with the demands of the public and government regulations. Improper heat transfer within bakery oven zones may result in the formation of unacceptable bread crust, texture, flavor, color, or shelf life. The application of abatement equipment to existing uncontrolled ovens presents a challenge to abatement equipment design engineers due to the sensitivity of the baking process.

PROCESS DESCRIPTION

Bread baking at large commercial bakeries is a highly automated process. When operating at full capacity, it is not unusual for a large commercial bread bakery to produce up to 300,000 pounds of several dozen different varieties of bread and other bakery products per day. All physical mixing and blending of ingredients, as well as the working and dividing of the dough is performed mechanically. The dough is continuously conveyed through each step of the process, from the initial dividing through the final slicing and bagging without manual handling.

There are four basic types of dough mixing processes: sponge dough, straight dough, brew, and continuous-mix ("no-time"). These processes vary in the manner in which the various dough ingredients are mixed, which determines the amount of fermentation time available. Large commercial bakeries most often use the sponge dough or brew methods to produce bread. The process starts with the mixing of flour, water, sugar, and yeast to form dough, initiating a long series of complex biochemical changes that ends in the oven where the bread is baked.

In producing bread from sponge dough, only a portion of the ingredients (typically yeast, yeast food, sweetener, part of the water, and two-thirds of the flour) are initially combined to form the sponge. The sponge is allowed to ferment for several hours. After the fermentation reactions are well established, the remaining ingredients, such as salt and the remaining flour and water, are mixed in to form the dough. The dough is then allowed to stand and "relax" before being molded, divided into individual loaves, and put into pans on a tray conveyor.

The panned dough is first conveyed to the proofing box, a warm (100°F), and humid chamber where the dough is allowed to rise to the volume desired for baking. The dough is automatically conveyed out of the proofing box and into the oven for baking.

The baking process actually occurs in the oven itself, which causes expansion of the loaf to final volume, crust formation, yeast and enzymatic activity inactivation, coagulation of dough proteins, partial gelatinization of starch, and reduction of loaf moisture, all of which are necessary to produce high-quality, saleable bread products. To accomplish all these product and process effects in the proper sequence, commercial bread ovens have from three to eight temperature gradient zones, which are maintained in critical temperature balance. In the initial zones of the oven there is oven rise, which determines the final loaf volume and internal texture. Thermal death of the yeast occurs as the internal bread temperature reaches 140–145°F, thereby stopping the fermentation process. In the middle zones the protein is denatured at between 140°F and 180°F. At the end of the process, browning and crust color develop, while ethanol and moisture are evaporated to cool the loaf and prevent the internal temperature from reaching the boiling point of water.[4] After baking the bread is depanned, allowed to cool, sliced, and bagged.

There are three fundamental types of large ovens: tunnel, tray, and spiral (Figures 1–3). In each design, the operator adjusts the oven to compensate for differences between batches and bread varieties based on visual inspection and experience. The operators adjust the ribbon burners, controlling the air-to-fuel ratio, and manually adjust the exhaust stack dampers to control the temperature in each zone. Tunnel ovens are long horizontal ovens in which dough enters at one end and is conveyed to the opposite end, where it exits as bread. Generally the oven has two or more exhaust stacks. Tray ovens are also horizontal, but the dough entering the oven exits on the same side as that on which it entered, after being conveyed the length of the oven, lowered to a second level, and conveyed to the exit near where it entered. Usually these ovens have two or three exhaust stacks. In spiral ovens, dough enters at the top corner of the oven and is conveyed in a downward spiral to the bottom comer of the oven, where it exits through an opening diagonally below where it entered the oven. Spiral, or Lanham ovens, require only a single exhaust stack. Each oven is also equipped with a purge stack for exhausting residual gases prior to burner ignition each day. The damper for this stack is normally closed during baking.[5]

AIR EMISSIONS CHARACTERIZATION

Stack tests performed by the Bay Area Air Quality Management District (BAAQMD) at numerous commercial bakery

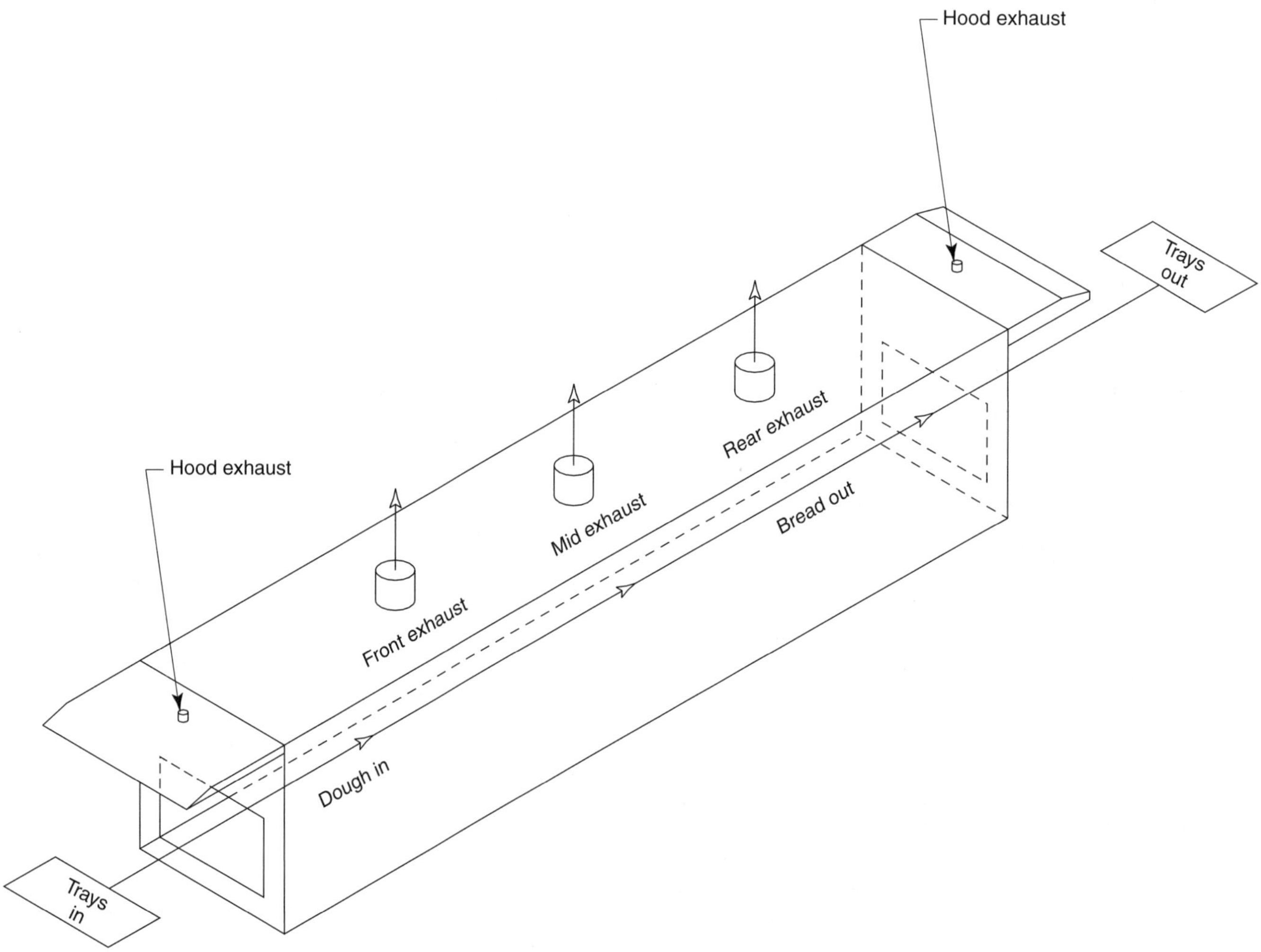

Figure 1. Tunnel Oven.

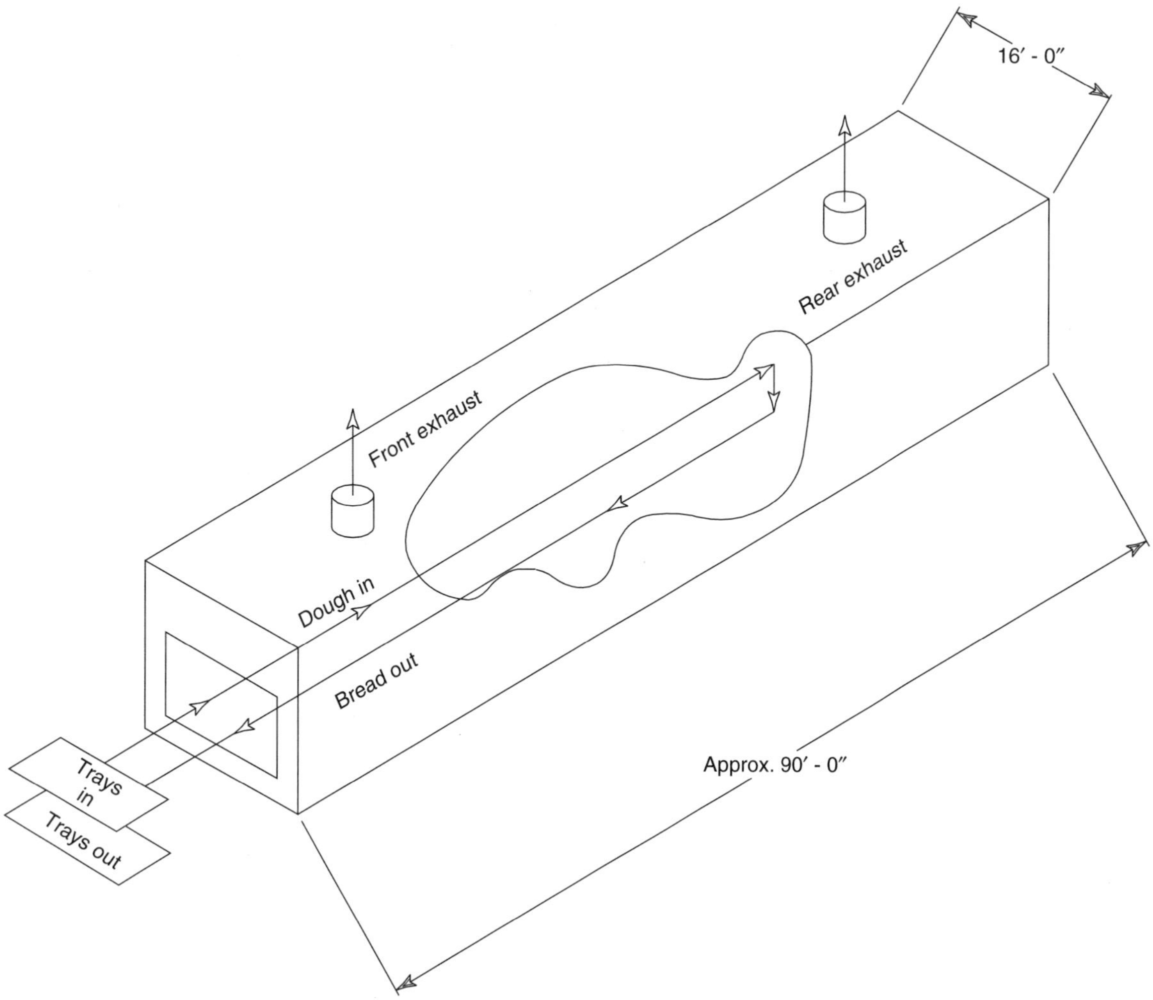

Figure 2. Tray Oven.

ovens demonstrated that ethanol from fermentation constitutes more than 99% of the volatile organic compound (VOC) emissions from bread ovens.[6] The ethanol is generated as a result of fermentation, which causes the sugars and starches to be converted to ethanol, carbon dioxide, and water. Fermentation begins immediately following the initial mixing of ingredients and continues until the yeast is killed in the oven. During the initial fermentation period, a skin forms on the top of the dough. The skin keeps the ethanol and carbon dioxide inside the dough, thereby allowing the dough to rise and minimizing fugitive emissions. As a result, most of the ethanol generated by fermentation is released in the oven during baking.[6,7] In multiple-stack tunnel and tray ovens, the BAAQMD stack tests have shown that 60–70% of the ethanol typically exits from one "principal" stack in the oven, although the exact percentages appear to vary by oven and bread type.[8]

The first attempt at estimating an emission factor was performed by Midwest Research Institute (MRI) under an EPA contract in 1977. MRI mixed, fermented, and baked four loaves of bread under a tent to capture all emissions from each stage of the bread-making process. An emission factor for sponge dough formulations of 5.6 lbs. ethanol/1000 lbs. bread was developed. However, the amount dough formulas were not typical of commercial operations.[9] The BAAQMD tested 16 ovens, some under varying operating conditions, resulting in a range of an emission factor from 0.3 to 7.0 lb. ethanol per 1000 pounds bread baked.[8]

The American Institute of Baking (AIB) analyzed the BAAQMD stack test results to develop ethanol emission factors for bread baking.[10] Based on that evaluation, as well as a separate chemical engineering evaluation, the AIB developed a formula for calculating ethanol emission factors based on the percent of the initial yeast in the dough and the total fermentation time. The AIB emission factor formula is set forth as follows:

$$\text{VOC E.F.} = 0.95Y_i + 0.195t_i - 0.51S - 0.86t_s + 1.90$$

where VOC E.F. is the pounds VOC per ton of baked bread; Y_i, the initial baker's percent of yeast to the nearest tenth of a percent; t_i, the total yeast action time in hours to the nearest tenth of an hour; S, the final (spike) baker's percent of yeast to the nearest tenth of a percent; and t_s, the spiking time in hours to the nearest tenth of an hour.

Although it appears that by changing a bread formula and increasing the amount of final yeast (S), it would be possible to obtain low or even a negative value for VOC emission estimates, a product of high quality would not be produced.

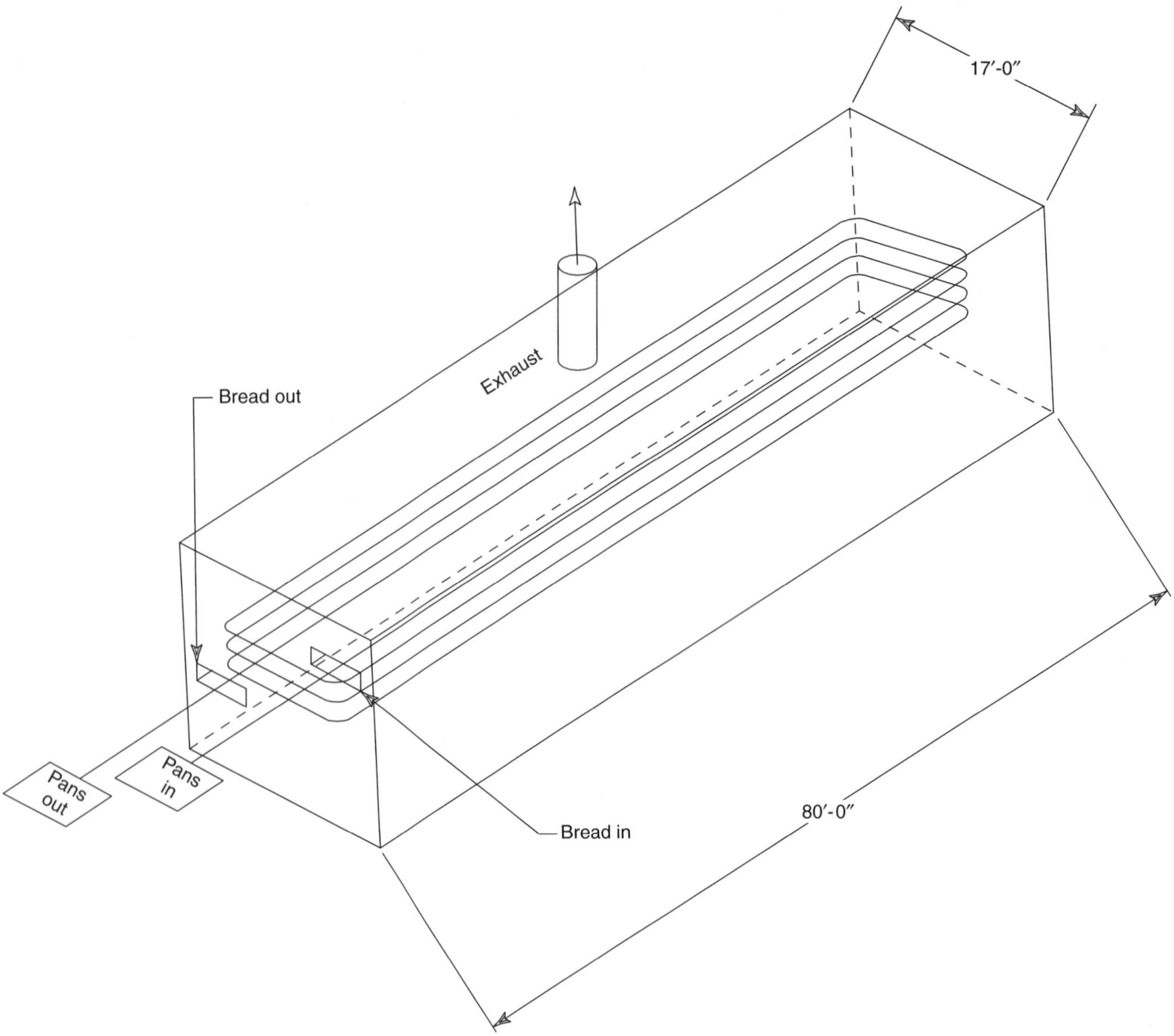

Figure 3. Spiral Oven.

Where no final yeast is added, the formula condenses to:

$$\text{VOC E.F.} = 0.95Y_{\mathrm{i}} + 0.195t_{\mathrm{i}} + 1.90$$

This predictive equation can be used for quantifying VOC emissions from bakery ovens. A baker knows the yeast concentrations and yeast action times for each variety baked. Those values can be inserted into this equation and pounds of VOC per ton of bread baked can be calculated. This number is multiplied by the tons of bread baked during a given time period, and the product is pounds of VOC emitted from the oven for that particular product for the given time period (typically per year). The following equation demonstrates this calculation:

$$\text{VOC emissions tons/yr} = \text{VOC E.F.} \times \text{BP} \times k$$

where BP is the bread production (tons/yr); and k, the conversion constant (ton/2000 lb). Using the ABI formula a separate emission factor can be developed for each variety of bread. Although this procedure is cumbersome, it is easier than conducting a separate source test for each of the numerous varieties of bread that may be baked in an oven. The most accurate means to determine emissions is with a source test.[3]

AIR POLLUTION CONTROL MEASURES

The sensitivity of the bread-baking process to air flow turbulence in the oven presents a challenge to the application of air pollution controls to bakery oven exhaust stacks, particularly retrofit applications. Turbulence in the exhaust airflow can cause unstable airflow, resulting in extinguished burner flames and nonuniform lateral heat distribution throughout the oven zone. This may cause in uneven, improperly baked bread with poor texture, crumb characteristics, and flavor. The balancing of airflow from a multiple stack operation demands engineering rigor and precision. Government labeling regulations require that the loaves must be within a small percent of the label weight. The physical size of the packaging limits loaf volume.

As bread is produced for human consumption, bakeries must maintain strict sanitary conditions at all times. The plant is cleaned daily, and 2 days per week it is shut down for a thorough floor-to-ceiling cleaning. The design of abatement equipment must avoid the buildup of fats and oils in exhaust ducts.[5,8]

Available Controls

There are several types of abatement equipment available for the control of ethanol emissions from bakery ovens. Although thermal incineration and catalytic incineration are the most widely used there are other techniques available for consideration as a means of control. These alternative techniques are scrubbing, carbon adsorption, refrigerator condensation, and the use of biofilters. However, these alternative techniques have one or more negative drawbacks.

The cost to control an existing bakery oven is generally higher than a new operation because bakery roofs are not commonly designed to support the weight of abatement equipment and many bakeries lack the area for a ground-level unit. Approximately 20% of the capital costs for roof-mounted abatement equipment is for the purchase and installation of support platforms. If floor space is adequate, a substantial capital cost savings can be realized. Using 1990 as a baseline year, the cost of control by use of a single or multiple catalytic incinerator ranges from \$0.72 to \$2.69 per pound of ethanol reduced.[11]

With either direct or catalytic incineration a combustion process is used to oxidize the organic compounds in the bakery oven exhaust to primarily carbon dioxide and water vapor. Burning natural gas or some other gaseous fuel provides auxiliary heat required to raise the oven exhaust to combustion temperatures. Temperatures up to 1450°F are required for conversion in direct incineration and up to 1000°F for catalytic incineration. In both cases, fuel consumption can be reduced by primary heat exchange of inlet and outlet gasses.

Incinerators are most effective in controlling exhaust streams with relatively high concentrations of organic compounds. Ethanol concentrations in the oven exhaust gas can substantially decrease during gaps in production. The incinerator may not be able to achieve 90% reduction during these periods.

Catalytic incinerators usually require very clean exhaust streams to avoid fouling the catalyst bed. After numerous years of operation, the concern that oils in bakery oven exhaust may foul the catalyst and reduce the destruction efficiency has been put to rest. Catalytic incinerators are the preferred means of control in the San Francisco Bay Area.

Carbon adsorption as a method of control offers the advantages of simplicity of operation and comparatively low energy and maintenance costs. Ethanol has a high affinity for carbon and is difficult to strip from the carbon beds. Incomplete stripping lowers bed capacity and reduces abatement efficiency. Fats and oils in the exhaust stream may clog the carbon pores, reducing bed capacity and life. The resulting ethanol/water mixture would require treatment and disposal. The incorporation of carbon adsorption as a means of control warrants careful review prior to use.

Because ethanol is readily soluble in water, scrubbing has been considered a means of control where wastewater disposal is not a problem. A packed tower scrubber with a 90% abatement efficiency would require roughly 80 gal/min of nonrecirculating water to abate a 3000 sdcfm exhaust from a typical large bakery oven. Unless ethanol recovery is practical, scrubbing may not be the most cost-effective means of abatement.

Condensation, where the exhaust gases are cooled below the dew point of ethanol also presents abatement challenges. Condensing the ethanol would require freon (as opposed to water) chilled coils to cool the very wet gas stream from 250°F to 50°F. Water in the exhaust would freeze the coils, insulating them and reducing the abatement efficiency. Fats and oils would condense more readily than ethanol, exacerbating any sanitation problems. The condensed ethanol will need to be dispose of in a practical manner. However, a recent improvement in condenser design and the potential sale of ethanol as a feed supplement has sparked the interest in condenser controls.

The use of biofilters, where the exhaust stream is passed through a bed of soil or peat moss that allows microorganisms naturally present in the medium to break down the organic, has several advantages not offered by other abatement options. The capital cost is sufficiently low to allow installation of separate beds for multistack oven operations. This would avoid any flow balance problems in the oven and would minimize the expense for additional ducting. Annual operating expense would be minimal. However, there are some difficulties in applying this technology to bakery ovens. Because each stack would require 600 ft^2 of peat bed or 13,000 ft^2 of soil bed, there may be a space problem. The incoming exhaust stream must be cooled to 170°F to avoid killing the microorganisms. This may cause fat condensation in the ducting, a potential sanitation problem. A biotrickling filter pilot study in Columbus Ohio during 1997–1998 showed an average control efficiency of 91%.[13]

Control Restraints

While the installation of controls on a new and existing bakery oven presents difficulties not found with other oven applications where the process is not as sensitive as the baking process, the application of controls has been successfully demonstrated in the San Francisco Bay Area and several other jurisdictions. The retrofitting of existing bakery ovens, particularly multiple-stack ovens, identifies challenges concerning the risk of interference with the baking process, increased risk of bacterial contamination, construction problems based on space, and sanitation constraints and bread production demands.[4,12]

The most significant challenge concerning the retrofitting of bakery ovens is the ability to maintain the airflow balance if a single control unit is used on an oven with multiple stacks. To assure a consistent quality of bake, the exhaust flows in each oven zone must be steady, and the operator must be able to adjust each zone independently of the other zones. Adjusting the exhaust damper in one stack may affect the airflow through another stack. These exhaust dampers may require small incremental adjustments, relying on the skill of the oven operator to assure an even bake. Turbulent airflow within the oven results in unstable ribbon burner flames, uneven lateral heat distribution, incorrect zone temperatures, and ultimately, improperly baked bread and unusable product.

The airflow balance problem can be solved in two ways: the damper controls for each stack to solve the airflow balance problem or a multiple-abatement system. Every bakery in the BAAQMD operating multistack ovens subject to control requirements has chosen the first option. The South Coast AQMD has a similar rule to require the control of existing ovens. Several bakeries that installed new ovens have been required to abate the ethanol emissions due to NSR regulations.

Another potential problem concerns the possible buildup of grease or biological growth in long runs of horizontal ductwork. Contamination can be minimized by installing insulated ducting, with the ductwork sloped toward the control device. This has reduced the potential for condensation and directs any condensate to flow toward the control device instead of the oven. Traps are another option to collect the condensate and to allow for easier cleaning.

Bakeries normally operate on a schedule that will permit the shutdown of equipment two days per week for sanitary purposes. But since bakeries must supply fresh bread for the marketplace, the ovens cannot be taken out of service for any period longer than that allowed by a bakery's normal schedule without interrupting customer supplies. Maintenance of tight sanitation standards is also of concern during construction because such standards must be observed even when the operating oven is being retrofitted with controls. Special construction techniques must be used to avoid bread contamination during retrofit.

Retrofitting abatement devices on existing bakery ovens poses significant engineering and logistical problems. However, all these problems are readily addressed by good engineering and sound construction. Currently, numerous ovens are being abated by afterburners; five of the control units have been in place for more than 8 years. Each oven has remained in operation without any significant problem. The experience in the Bay Area has demonstrated that bakery oven controls are a technically feasible and cost-effective means of reducing ozone precursor organic emissions.

REFERENCES

1. Bay Area Air Quality Management District ("BAAQMD") Rule 8-42, *Control of Organic Compounds from Bread Baking Ovens*, San Francisco, CA.
2. 42 U.S.C.A. Sections 7401–7671Q.
3. *Alternative Control Technology Document, Bakery Oven Emissions*, U.S. EPA, EPA 453/R-92-017, December 1992.
4. E. J. Pyler, *Baking Science and Technology*, 3rd ed., Sosland Publishing Co., Merriam, KS, 1988, p. 589.
4. J. W. Stitley, *Baking Technology, Oven Emissions and Control Devices*, American Institute of Baking, Manhattan, KS, 1986.
5. BAAQMD Staff Report supporting adoption of proposed Rule 8-42, July 1988.
6. BAAQMD bakery oven source test results, 1985 and 1986.
7. H. Weiss, Chief Environmental Services, Pennsylvania Department of Environmental Resources, 1989 field test results.
8. BAAQMD Supplemental staff report—supporting adoption of Rule 8-42, September 1988, San Francisco, CA.
9. R. Keller and D. Henderson, *Nonmethane Organic Emissions from Bread Producing Operations*, Research Triangle Park, EPA 450/4-79-001, 1978.
10. J. W. Stitley et al., *Bakery Oven Ethanol Emissions Experimental and Plant Survey Results, American Institute of Baking*, Technical Bulletin IX, 12, Manhattan, KS, 1987.
11. J. Cutino and S. Owen, *Technical Assessment Report for Regulation 8, Rule 42, Organic Compounds from Large Commercial Bakeries*, November 1988, San Francisco, CA.
12. *Draft Control of Ethanol Emissions from Ovens: Bakery, Demonstration Program*, American Bakers Association, Washington, D.C., 1988.
13. *Biotrickling Filter Pilot Study for Ethanol Emissions Control*, U.S. Department of Agriculture Cooperative State Research, Education, and Extension Service and the U.S. EPA partnering with the FOOD Manufacturing Coalition, July 1998.

COFFEE PROCESSING

Taken from AP-42 U.S. EPA

GENERAL

Most coffee is grown in Central and South America. After harvesting and drying at or near the coffee plantation, most "green" coffee beans are exported and further processed before sale to the consumer. Coffee processing in the United States consists essentially of cleaning, roasting, grinding, and packaging.[1]

Roasting is the key operation and produces most of the air contaminants associated with the industry. Roasting reduces the sugar and moisture contents of green coffee and also renders the bulk density of the beans about 50% lighter. An apparently desired result is the production of water-soluble degradation products that impart most of the flavor to the brewed coffee. Roasting also causes the beans to expand and split into halves, releasing small quantities of chaff.[1]

The coffee roasting industry involves the processing of green coffee beans into roasted coffee products, including whole and ground beans and soluble coffee products. The Standard Industrial Classification (SIC) code for coffee roasting is 2085.

PROCESS DESCRIPTION[1–6]

The coffee roasting process consists essentially of cleaning, roasting, cooling, grinding, and packaging operations. Figure 1 shows a process flow diagram for a typical coffee roasting operation. Bags of green coffee beans are hand- or machine-opened, dumped into a hopper, and screened to remove debris. The green beans are then weighed and transferred by belt or pneumatic conveyor to storage hoppers. From the storage hoppers, the green beans are conveyed to the roaster. Roasters typically operate at temperatures between 370 and 540°C (698 and 1004°F), and the beans are roasted for a period of time ranging from a few minutes to about 30 min. Roasters are typically horizontal rotating drums that tumble the green coffee beans in a current of hot combustion gases; the roasters operate in either batch or continuous modes and can be indirect- or direct-fired. Indirect-fired roasters are roasters in which the burner flame does not contact the coffee beans, although the combustion bases from the burner do contact the beans. Direct-fired roasters contact the beans with the burner flame and the combustion gases. At the end of the roasting cycle, water sprays are used to "quench" the beans. Following roasting, the beans are cooled and run through a "destoner." Destoners are air classifiers that remove stones, metal fragments, and other waste not removed during initial screening from the beans. The destoners pneumatically convey the beans to a hopper, where the beans are stabilized and dried (small amounts of water from quenching exist on the surface

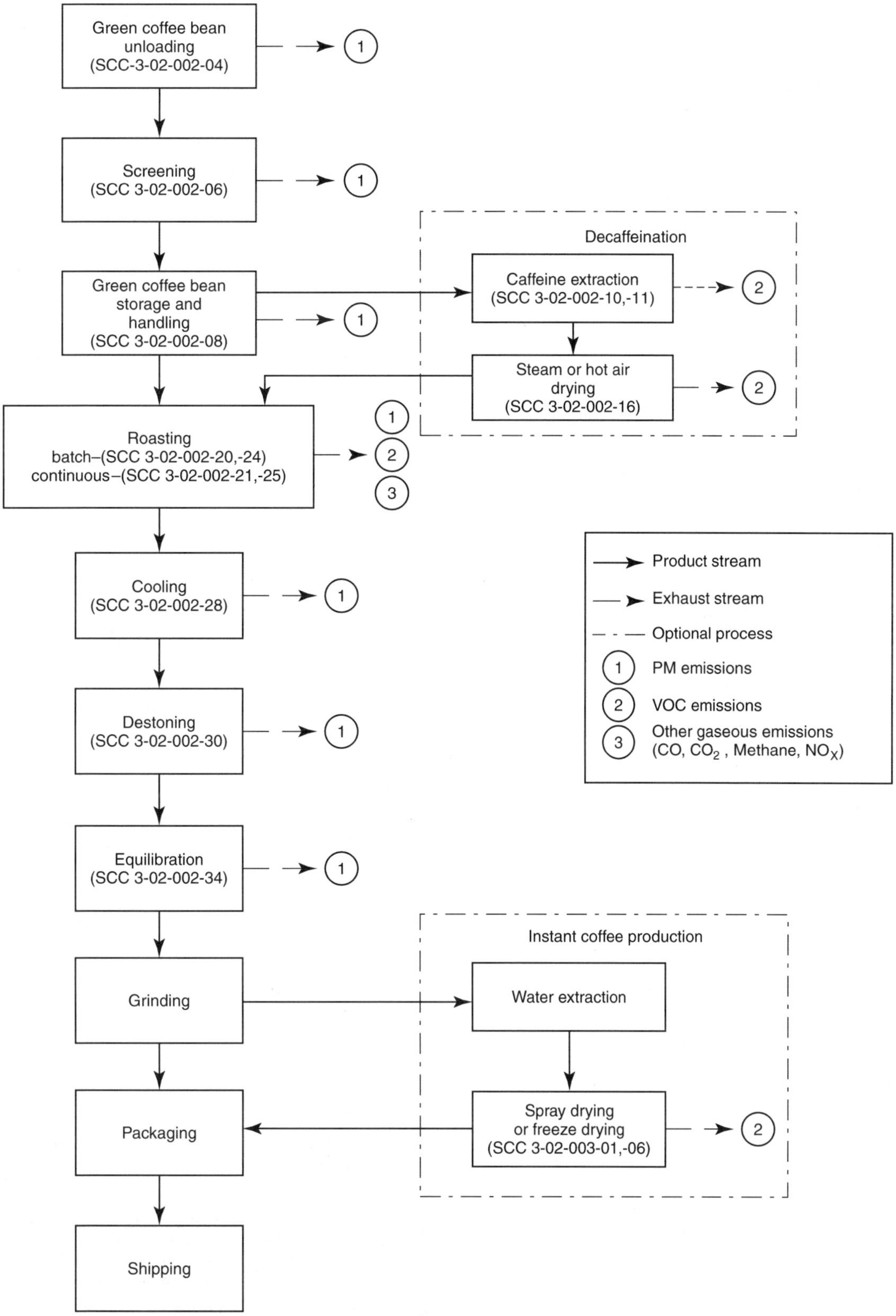

Figure 1. Typical Coffee Roasting Operation (Source Classification Codes in Parentheses).

of the beans). This stabilization process is called *equilibration*. Following equilibration, the roasted beans are ground, usually by multistage grinders. Some roasted beans are packaged and shipped as whole beans. Finally, the ground coffee is vacuum sealed and shipped.

Additional operations associated with processing green coffee beans include decaffeination and instant (soluble) coffee production. Decaffeination is the process of extracting caffeine from green coffee beans prior to roasting. The most common decaffeination process used in the United States is supercritical carbon dioxide (CO_2) extraction. In this process, moistened green coffee beans are contacted with large quantities of supercritical CO_2 [CO_2 maintained at a pressure of about 4000 pounds per square inch and temperatures between 90 and 100°C (194°C and 212°F)], which removes about 97% of the caffeine from the beans. The caffeine is then recovered from the CO_2, typically using an activated carbon adsorption system. Another commonly used method is solvent extraction, typically using oil (extracted from roasted coffee) or ethyl acetate as a solvent. In this process, solvent is added to moistened green coffee bans to extract most of the caffeine from the beans. After the beans are moved for the solvent, they are steam-striped to remove any residual solvent. The caffeine is then recovered for the solvent, and the solvent is reused. Water extraction is also used for decaffeination, but little information on this process is available. Decaffeinated coffee beans have a residual caffeine content of about 0.1% on a dry basis. Not all facilities have decaffeination operations, and decaffeinated green coffee beans are purchased by many facilities that produce decaffeinated coffee.

In the manufacture of instant coffee, extraction follows the roasting and grinding operations. The soluble solids and volatile compounds that provide aroma and flavor are extracted from the coffee beans using water. Water heated to about 175°C (347°F) under pressurized conditions (to maintain the water as liquid) is used to extract all the necessary solubles from the coffee beans. Manufacturers use both batch and continuous extractors. Following extraction, evaporation or freeze-concentration is used to increase the solubles concentration of the extract. The concentrated extracts are then dried in either spray dryers or freeze dryers. Information on the spray-drying and freeze-drying processes is not available.

Table 1. Emission Factors for Coffee Roasting Operations[a] Emission Factor Rating: D

Source	Filterable PM, lb/ton	Condensible PM lb/ton
Batch roaster with thermal oxidizer[b] (SCC 3-02-002-20)	0.12	ND
Continuous cooler with cyclone[c] (SCC 3-02-002-28)	0.028	ND
Continuous roaster[d] (SCC 3-02-002-21)	0.66	ND
Continuous roaster with thermal oxidizer (SCC 3-02-002-21)	0.092[e]	ND
Green coffee bean screening, handling, and storage system with fabric filter[f] (SCC 3-02-002-08)	0.059	ND
Destoner (SCC 3-02-002-30)	ND	ND
Equilibration (SCC 3-02-002-34)	ND	ND

[a]Emission factors are based on green coffee bean feed. Factors represent uncontrolled emissions unless noted. SCC, Source Classification Code; ND, no data; D-rated and E-rated emission factors are based on limited test data; these factors may not be representative of the industry.
[b]References 12 and 14.
[c]Reference 15.
[d]References 8 and 9.
[e]Reference 7–9, 11, and 15. Includes data from thermal catalytic oxidizers.
[f]Reference 16. Emission factor rating: E.

EMISSIONS AND CONTROLS

Particulate matter (PM), volatile organic compounds (VOC), organic acids, and combustion products are the principal emissions from coffee processing. Several operations are sources of PM emissions, including the cleaning and destoning equipment, roaster, cooler, and instant coffee drying equipment. The roaster is the main source of gaseous pollutants, including alcohols, aldehydes, organic acids, and nitrogen and sulfur compounds. Because roasters are typically natural gas-fired, carbon monoxide (CO) and carbon dioxide (CO_2) emissions are expected as a result of fuel combustion. Decaffeination and instant coffee extraction and drying operations may also be sources of small amounts of VOC. Emissions from the grinding and packaging operations typically are to vented to the atmosphere.

Particulate matter emissions from the receiving, storage, cleaning, roasting, cooling, and stoning operations are typically ducted to cyclones before being emitted to the atmosphere. Gaseous emissions from roasting operations are typically ducted to a thermal oxidizer or thermal catalytic oxidizer

Table 2. Emission Factors for Coffee Roasting Operations[a] Emission Factor Rating: D

Source	VOC[b], (lb/ton)	Methane (lb/ton)	CO (lb/ton)	CO_2 (lb/ton)
Batch roaster[c] (SCC 3-02-002-20)	0.86	ND	ND	180
Batch roaster with thermal oxidizer (SCC 3-02-002-20)	0.047[d]	ND	0.55[d]	530[e]
Continuous roaster (SCC 3-02-002-21)	1.4[f]	0.26[g]	1.5[h]	120[j]
Continuous roaster with thermal oxidizer (SCC 3-02-002-21)	0.16[k]	0.15[m]	0.098[k]	200[n]
Decaffeination: solvent or supercritical CO_2 extraction (SCC 3-02-002-10,11)	ND	ND	ND	ND
Steam or hot air dryer (SCC 3-02-002-16)	ND	ND	ND	ND
Spray drying (SCC 3-02-003-01)	ND	ND	ND	ND
Freeze drying (SCC 3-02-003-06)	ND	ND	ND	ND

[a]Emission factors are based on green coffee bean feed. Factors represent uncontrolled emissions unless noted. SCC, Source Classification Code; ND, no data; D-rated and E-rated emission factors are based on limited test data; these factors may not be representative of the industry.
[b]Volatile organic compounds as methane. Measured using GC/FID.
[c]Reference 14.
[d]Reference 12–14.
[e]References 12 and 14.
[f]References 8, 9, 11, and 15.
[g]References 8, 9, 11, and 15. Emission factor rating: E.
[h]References 8, 9, and 15.
[j]References 8, 9, 11, and 15. Emission factor rating: C.
[k]References 8, 9, 11, and 15. Includes data from thermal catalytic oxidizers.
[m]References 8, 9, 11, and 15. Includes data from thermal catalytic oxidizers. Emission factor rating: E.
[n]References 9, 11, and 15. Includes data from thermal catalytic oxidizers.

following PM removal by a cyclone. Some facilities use the burners that heat the roaster as thermal oxidizers. However, separate thermal oxidizers are more efficient because the desired operating temperature is typically between 650°C and 816°C (1200°F and 1500°F), which is 93°C to 260°C (200°F to 500°F) more than the maximum temperature of most roasters. Some facilities use thermal catalytic oxidizers, which require lower operating temperatures to achieve control efficiencies that are equivalent to standard thermal oxidizers. Catalysts are also used to improve the control efficiency of systems in which the roaster exhaust is ducted to the burners that heat the roaster. Emissions from spray dryers are typically controlled by a cyclone followed by a wet scrubber.

Table 1 presents emission factors for filterable PM and condensible PM emissions from coffee roasting operations. Table 2 presents emission factors for volatile organic compounds, methane, CO, and CO_2 emissions from roasting operations. Emissions from batch and continuous roasters are shown separately, but with the exception of CO emissions, the emissions for these two types of roasters appear to be similar.

REFERENCES

1. R. G. Ostendorf Ed., "Coffee Processing," *Air Pollution Engineering Manual*, Van Nostrand Reinhold, New York, NY, 1992.
2. M. N. Clifford and K. C. Willson, *COFFEE—Botany, Biochemistry and Prevention of Beans and Beverage*, The AVI Publishing Company, Inc., Westport, CT, 1985.
3. J. M. L. Penninger, *Supercritical Fluid Technology—Potential in the Fine Chemicals and Pharmaceutical Industry*, presented at the Workshop on Prevention of Waste and Emissions in the Fine Chemicals/Pharmaceutical Industry, Cork, Ireland, October 1993.
4. Telephone communication between B. Shrager, Midwest Research Institute, Cary, NC, and M. Wood, Tetley's Corporation, Palisades Park, NJ, December 20, 1994.
5. R. J. Clarke and R. MacRae, Eds., *Coffee, Volume 2: Technology*, Elsevier Science Publishing Company, Inc., New York, NY, 1987.
6. G. Wasserman et al., "Coffee," *Kirk-Othmer Encyclopedia of Chemical Technology*, 4th Ed., Volume 6, John Wiley & Sons, Inc., 1992.
7. *Source Test Report, Particulate Emissions, Premium Coffee, Wall, New Jersey*, Princeton Testing Lab, Princeton, NJ, January 1987.
8. *Compliance Stack Sampling Report for Hills Brothers Coffee, Inc., Edgewater, New Jersey*, Ambient Engineering, Inc., Parlin, NJ, September 23, 1988.
9. *Stack Sampling Report for Hills Brothers Coffee, Inc., Edgewater, New Jersey, On Thermal Oxidizer #22 Inlet/Outlet*, Ambient Engineering, Inc., Parlin, NJ, October 5, 1988.
10. *Compliance Stack Sampling Report for General Foods Corporation, Maxwell House Division, Hoboken, New Jersey, On Thermal Oxidizer Inlet and Outlet*, Recon Systems, Inc., Three Bridges, NJ, March 13, 1989.
11. *Nestle Food Corporation Compliance Emission Testing Report*, AirNova, Inc., Pennsauken, NJ, October 1990.
12. *Source Test Report for Particulate, Volatile Organic Compounds, and Carbon Monoxide Emissions from the Coffee Roaster 7D Thermal Oxidizer at General Foods—Maxwell House Division, Hoboken, New Jersey*, Air Consulting and Engineering, Inc., Gainesville, FL, December 20, 1990.
13. *Source Test Report for Volatile Organic Compounds and Carbon Monoxide Emissions from the Coffee Roaster 7D Thermal Oxidizer at General Foods—Maxwell House Division, Hoboken, New Jersey*, Air Consulting and Engineering, Inc., Gainesville, FL, December 20, 1990.
14. *Melitta USA, Inc.*, Blaw Knox Roaster Emission Compliance Test Program, AirNova, Inc., Pennsauken, NJ, February 1992.
15. *Nestle Beverage Co. Source Test Report, Coffee Roaster and Cooler*, Best Environmental, Inc., San Leandro, CA, October 1, 1992.
16. *Summary of Source Test Results*, Bay Area Air Quality Management District, San Francisco, CA, January 1991.

GRAIN HANDLING AND PROCESSING

DENNIS WALLACE (with update by Editor)

Grain handling and processing facilities move grain (wheat, corn, soybeans, rice, oats, etc.) from the farm, through storage and transfer locations (grain elevators), to the mills and plants that generate a variety of products (flour, starch, oil, animal feed, etc.).[1] A key characteristic of the industry is its diversity, which reflects the geographic dispersion of the raw-materials sources (farms), the variety of raw materials (different types and qualities of grain), and the wide range of final products. Consequently, the industry comprises a large number of facilities (as many as 9,000 grain elevators[2] about 500 grain mills,[3] and 15,000 feed manufacturing plants[4]) spread throughout the United States in areas with vastly different population densities. Individual facilities vary in size and use different unit processes. As a consequence of their diversity of location and operating characteristics, each facility has different emission levels, emissions control needs, and technical and economic feasibility of emissions control. This discussion includes a brief overview of the emission characteristics and control alternatives, but because the industry is so diverse, the information should be used very cautiously in assessing emission problems or control options at an individual facility or group of facilities.

In the broadest sense, the grain handling and processing industry can be defined as including all operations, from the point of production (the farm) to the points where final products are prepared for consumer use. This review will focus on a narrower segment of the industry in which facilities have similar air pollution problems. Specific segments that will be considered are grain elevators, grain milling (including flour milling and wet corn and rice milling), and animal feed manufacturing, segments that have many common unit processes. The remainder of this presentation provides a general description of each of these industry segments. The operating characteristics of the different unit processes and the mechanisms by which they generate emissions are described later.

Grain elevators transfer, condition, and store grains and soybeans as they are passed from the farmer to the grain processor or exporter. The industry divides elevators into two classes—country elevators and terminal elevators—as a function of their size, source of grain, and destination of shipments. Country elevators are those facilities that receive the bulk of their grain directly from the farm. Terminal elevators, those that ship grain directly to a processor or export grain, are often further divided into port terminals and inland terminals (or subterminals). The subterminals receive the bulk of their grain from country elevators, but also receive a small amount directly from farmers; they ship grain to port terminals or directly to processors. Although these different elevator classes vary greatly in size, all generally perform the same receiving, conditioning, internal handling and transfer, and storage and shipping operations.

A schematic of a traditional country elevator is shown in Figure 1.[5] Newer country elevators differ from the one shown in that they tend to have free-standing metal legs and metal storage bins. Most modern country and terminal elevators have eliminated the enclosed headhouse and gallery and employ a more open structural design. The typical subterminal or terminal elevator is much more complex than the facility shown in Figure 1. Figure 2 shows a process flow diagram for the various operations that occur at a grain elevator illustrating the location of potential emissions of particulate matter and VOCs.[18]

Grain is received at the elevator by truck, rail, or barge and is unloaded by one of a variety of receiving systems, which include gravitational processes, power shovels, and augers. As grain is unloaded, it is transferred by gravity to the heart of the elevator operations—an enclosed bucket elevator called the "leg." Almost all grain transfer within the elevator is accomplished by a combination of gravity flow, belt conveyor, and the elevator leg. Grain flows into the bottom, or "boot," of the leg and is picked up by the bucket elevator and carried up to the top of the leg—"the headhouse." Here, the grain is discharged to the distributor, which can discharge it directly to a loadout spout, to a holding bin for loadout, to the gallery belt for transfer to a storage bin, or to an ancillary operation such as weighing, cleaning, scalping, or drying. Grain moves by conveyor through an enclosed area called the gallery to the appropriate bin, where it is discharged from the belt into the bin by a tripper or is transferred directly from the distributor through a metal spout to a detached bin. When stored grain is to be shipped or conditioned further, it is discharged by gravity from the bottom of the bin onto the tunnel belt, a belt conveyor that transfers the grain to the leg for subsequent distribution.

Grain processing facilities have receiving, handling, and transfer operations comparable to those found in grain elevators. Each processing facility also has unit processes that are specific to its final products. Detailed process descriptions for different types of grain mills are presented in Ref.[6]; Figure 3 illustrates one type of mill, flour milling from wheat.[6] Although specific operations vary from plant to plant, grain mills generally use four major types of operation: a series of cleaning steps to remove light impurities (hulls, straw, dust, and metallic objects); tampering or conditioning to adjust the moisture content of the grain; milling operations such as grinding, sifting, rolling, aspirating, and drying; and packaging or bagging as illustrated in Figure 3.

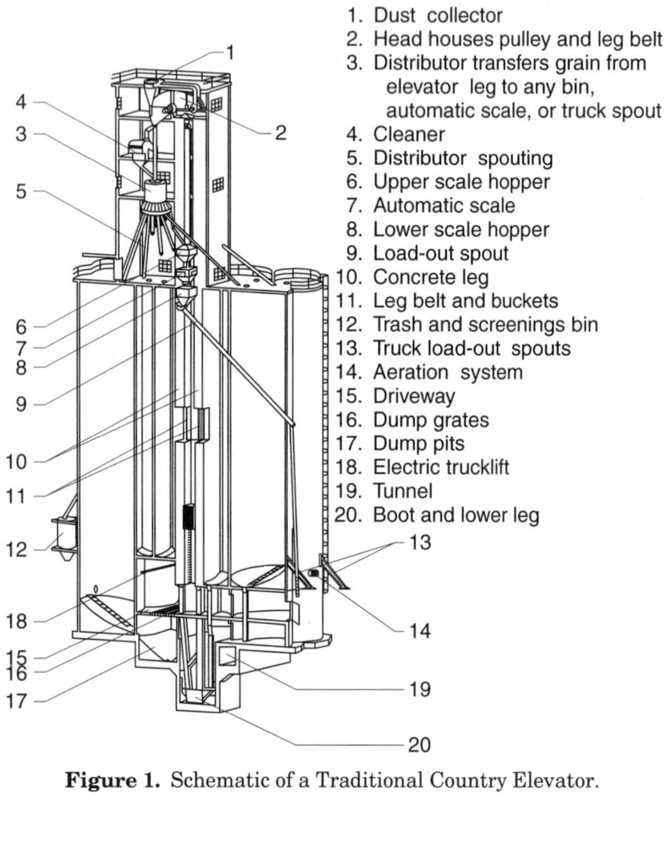

Figure 1. Schematic of a Traditional Country Elevator.

A detailed description of the animal feed manufacturing industry and the production processes used in the industry is presented in Ref. 7. A simplified diagram of a typical smaller plant is shown in Figure 4.[8] Again, these facilities use receiving and storage operations that are comparable to those used in grain elevators. As grain is received, it goes through scalpers to remove coarse materials such as sticks, husks, and straw. Then the grain is stored or transferred directly to the grinding area. The ground materials are mixed, generally in batch units. The material from the mixer is a meal, or mash, and may be bagged and marketed as is.[9] However, the material is often formed into pellets in the pellet mill. Further size reduction, if needed, is provided by the crumbler or granulator. Finally, screens are used to remove undersized and oversized materials, and the final product is transferred to the bagging area, to storage, or to bulk load-out.

PROCESS DESCRIPTION

Although each grain handling and processing facility is unique, many different facilities use the same unit operations. Further, different unit operations are similar in that they are either materials handling operations or operations such as grinding, sizing, and mixing that mechanically change grain characteristics. Consequently, common mechanisms generate emissions from these operations. Each process is a potential source of process fugitive emissions because of the dusty nature of grain and the dirt that is mixed with the grain during harvest and transport. Even processes involving "clean" grain are subject to emissions because additional grain dust is generated by abrasion and breakage as grain is handled and processed. Particulate emissions are generated by a combination of mechanical energy imparted to particles in handling and processing operations and particle entrainment by drag forces in the locally turbulent airflows that accompany these operations. These entrained particles are then emitted from the facility via natural airflows or process or building ventilation systems. The following briefly describes the more common grain handling and processing operations and identifies how emissions are generated from these processes.

Grain Receiving

Grain handling and processing facilities receive grain by truck, railroad hopper car, railroad boxcar, barge, and ship. Unloading procedures depend on the type of vehicle delivering the grain, and multiple options are available for unloading from each type of vehicle. However, most receiving procedures generate emissions by the same mechanisms. The two principal factors that affect emission levels during bulk unloading are local wind currents in the receiving area and the quantity of dust generated when the falling stream of grain strikes the

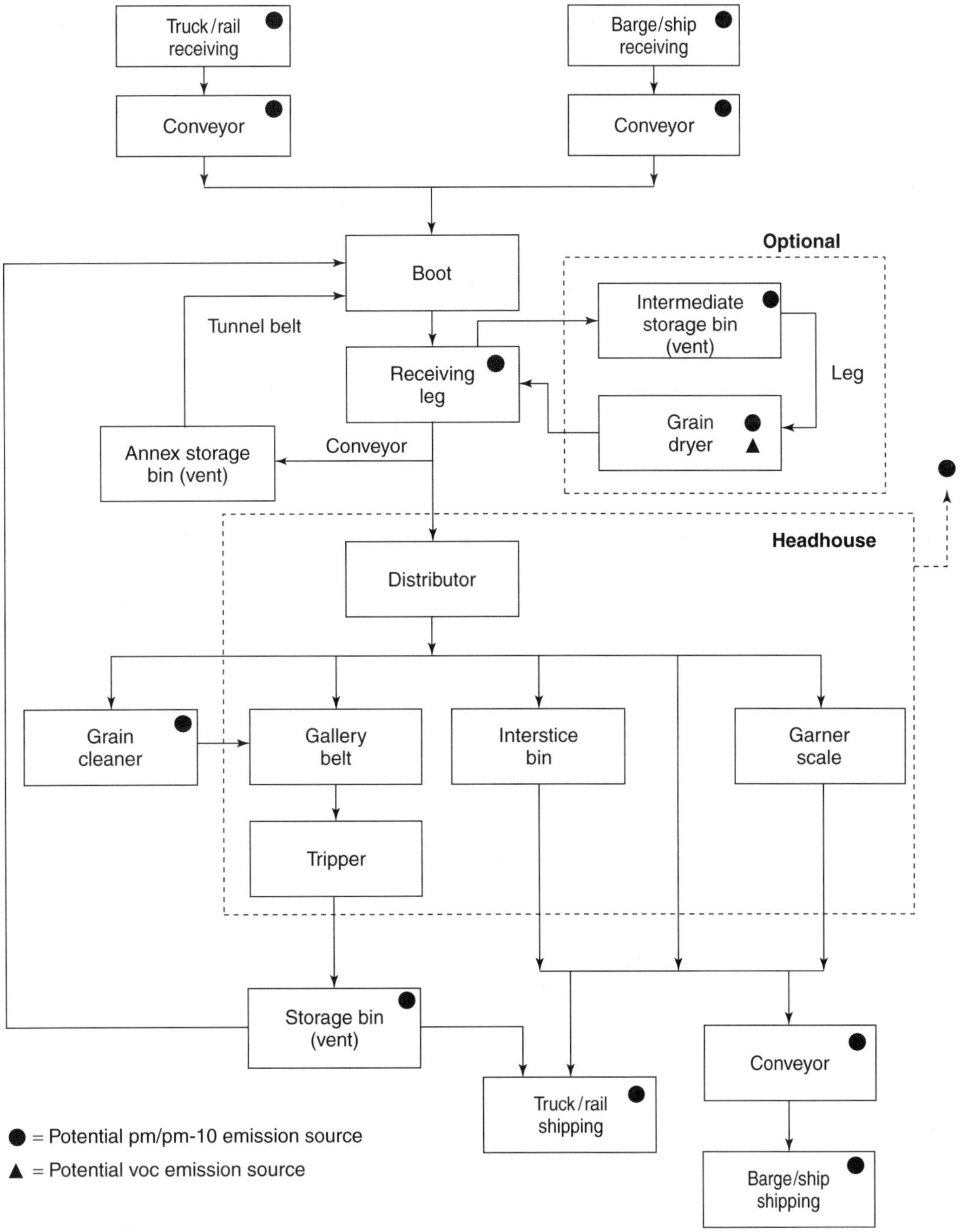

Figure 2. Major Process Operations at a Grain Elevator (Ref. 18).

receiving pit. When falling grain strikes an immovable object (e.g., the base of a pit), the energy expended causes extreme air turbulence. A violent generation of dust occurs from a combination of splashing of dust particles upon impact and entrainment of dust in the turbulent airstream.[10]

Trucks are generally unloaded at some type of dumping platform or dump pit. The area may be unenclosed, partially enclosed, or fully enclosed. Typically, the front of the truck is elevated and grain flows from the back of the truck through a steel grate into an underground pit. Emission levels are affected by drop height and wind currents in the pit area. Hopper cars are unloaded into a pit similar to the one used for truck unloading, but grain is unloaded through the bottom of the car. The dust emissions from this operation can be reduced by allowing the grain to flow at a velocity high enough that a cone forms around the hopper discharge at the receiving pit (i.e., choke unloading) and by limiting wind velocities in the pit area.

Two methods are commonly used for boxcar unloading. In the first, the inside grain door is broken, allowing a surge of grain to flow from the car into the receiving pit. After this initial surge stops, the remaining grain is unloaded into the pit with a power shovel or bobcat. Alternatively, some large facilities use a mechanical car dump, which clamps the car to a movable section of track and rotates and tilts it to dump grain from the car door into the receiving pit. Again, the primary factors that affect emissions are the drop pattern and wind currents in the receiving area. Barges and ships typically are

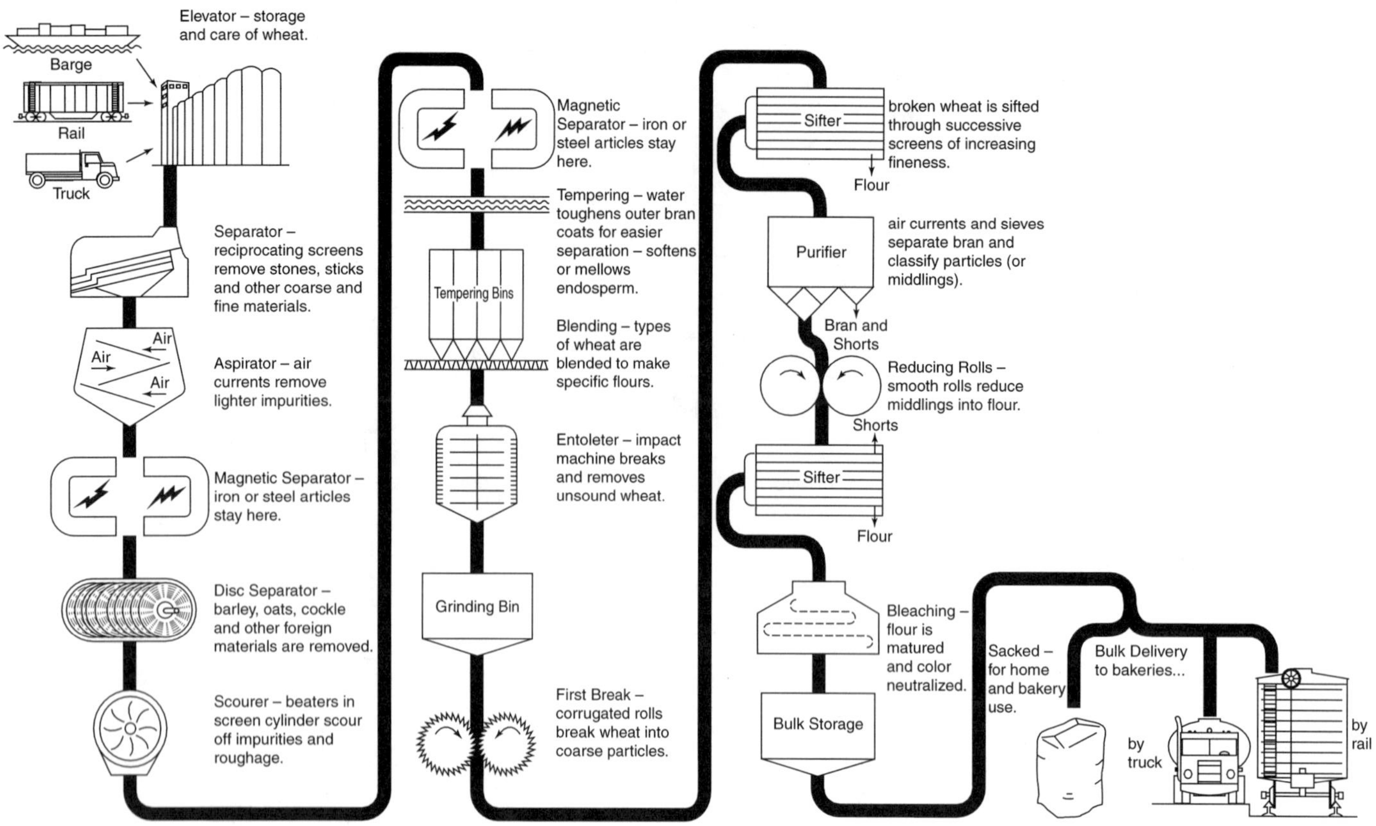

Figure 3. Grain Milling Process.

unloaded by either a bucket elevator that is lowered into the ship's hold or a pneumatic discharge system.[10]

Grain Transfer

The three primary operations used to transfer grain in both grain elevators and grain processing plants are by the bucket elevator (or leg), by belt conveyors, and by gravity flow through spouts. The moving pulley system and attached buckets in the legs pick up grain in the boot and transport it to the head of the leg. Dust is generated as grain is dumped into the boot, as buckets move through the boot to pick up grain, and as grain is discharged from the elevators at the head. Typically, the legs are aspirated at the top, and possibly at the boot, to relieve air pressure and remove dust generated by the moving elevator, but some systems are operated without venting.

Belt conveyors are used for the horizontal transfer of grain within grain elevators and grain processing plants. Typically, grain is transferred to the belt from a spout by gravity feed. Grain can be removed from the belt at a dump station at the tail pulley or by a tripper that removes the grain at an intermediate point in the belt. The primary source of emissions is falling grain as it is transferred onto the belt and discharged from the belt. Emissions are also generated when dust particles that have adhered to the belt drop off the return side of the belt. These particles may be entrained immediately in local air currents or may fall to an exposed floor or equipment surface and subsequently be reentrained. Frequently, transfer of the grain to storage and loadout is accomplished via gravity flow through spouts. The primary source of emissions is the "splash" of dusty material at the spout discharge. Such emissions are most pronounced from barges, ships, and loadout operations.

Grain Cleaning

Both terminal grain elevators and grain processing plants employ various cleaning devices to remove impurities from the grain. These devices rely on differences in size, shape, specific gravity, and material between the grain and the impurities to promote separation. All the different cleaning devices involve grain movement, potential for abrasion, and substantial airflows. Consequently, cleaning operation can emit significant levels of particulate matter (PM).

The primary cleaning operations are scalping, air aspiration, metals removal, and scouring. A preliminary step widely used at both elevators and processing plants is called scalping. In the scalper, grain is run through a coarse mesh shaker or reel screen to remove straw, sticks, stones, strings, or other coarse material. Low-velocity vibrating screens are also used to remove fine dust from the grain. In air aspiration systems, air is dispersed through falling grain in a cross-current or countercurrent flow to remove field dust, fibers, chaff, and light trash. Feed mills and feed manufacturing plants use magnetic drum separators to remove bolts and screws from the grain or from milled grain dust. Many mills also employ scouring as the final step in the cleaning room. A scourer is a machine in which beaters attached to a central shaft throw grain violently against a surrounding drum. This violent action buffs each kernel and breaks off loose material. Air currents are used to separate the detached dust and bran from the grain.[11,12] The exhaust gases from aspiration systems and scourers contain significant quantities of dust that will be emitted if these streams are not controlled. If grain velocities are kept low and the systems are tightly enclosed, the scalper and the screening operations will have essentially no emissions.[13] However,

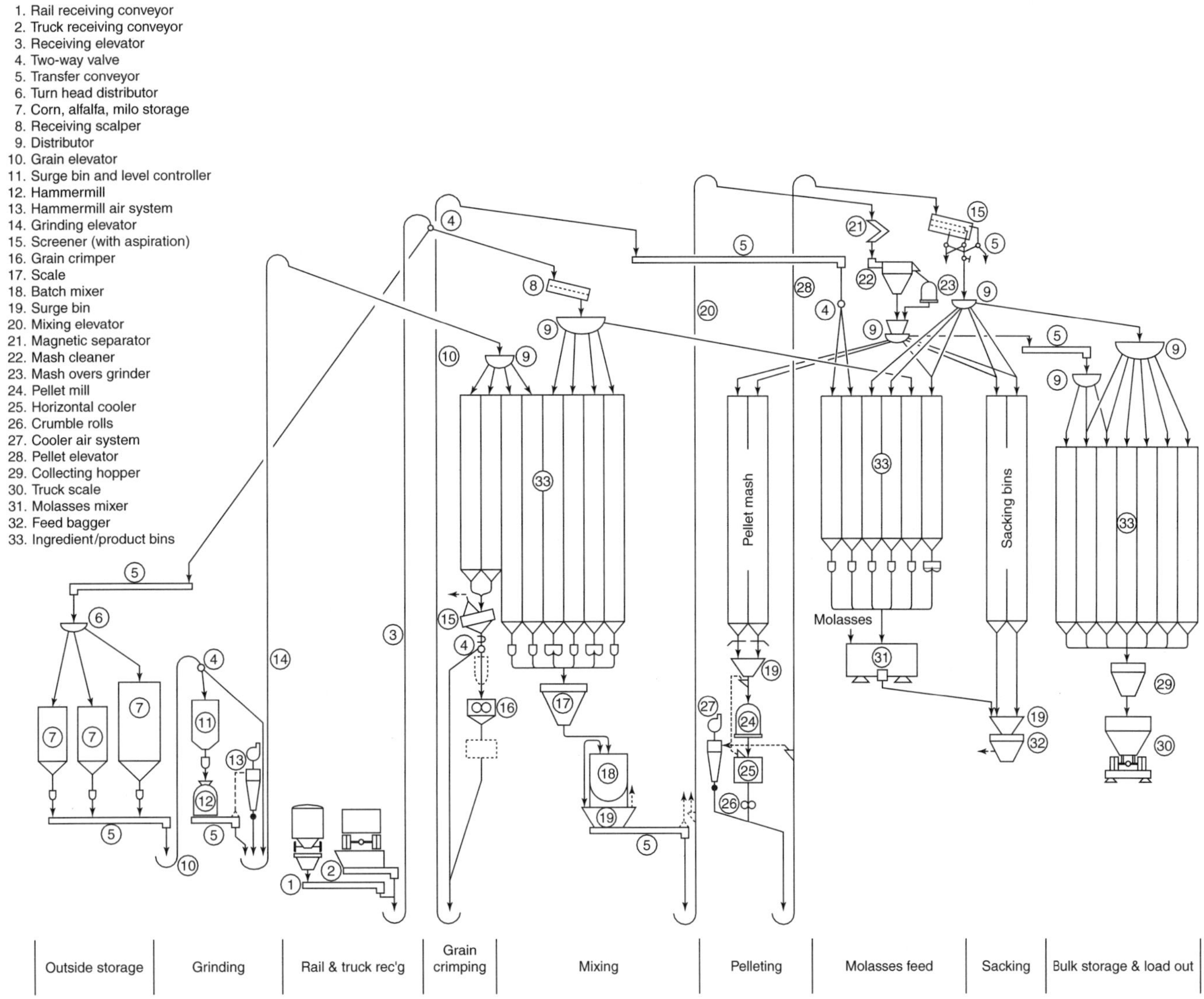

Figure 4. Flow Diagram for a Small Feed Manufacturing Process.

these operations can generate substantial fugitive emissions if grain flows are excessive or if the equipment is open and grain streams are exposed to local air currents.

Grain Milling and Processing Operations

In addition to the handling and cleaning steps defined above, grain mills and feed manufacturing plants use a series of milling and processing steps to convert the raw grain to a usable final product. All facilities use some type of milling operation to reduce the size of the grain material. Options for milling include hammer mills, attrition mills, burr mills, and roller mills. Other operations conducted at many facilities are proportioning and mixing, pelletizing and subsequent sizing, and grain drying.

Hammer mills are commonly used at grain mills and feed manufacturing plants to grind ingredients. These units accomplish both grinding and air conveyance. Grain is fed into the center of a high-speed rotor, where the pivoted hammers move it to the perimeter screen. The grain is forced through the screen, usually with the aid of supplemental air. The ground product is pneumatically conveyed to a cyclone separator or filter receiver that separates the air from the ground ingredient.[11] Attrition and burr mills are used in feed plants for preliminary size reduction. Attrition mills and burr mills comprise two disks, or plates. In the burr mill, one plate is stationary, the other rotating. The plates have rough surfaces, and different types of plate surfaces can be used for different materials or to produce different particle sizes. The material is fed between the plates and reduced by cutting, crushing, and shearing. The "double-runner" attrition mill is a heavy-duty precision plate mill, and each plate rotates at high speeds.[14]

Proportioning and mixing operations are the heart of any formula-feed manufacturing plant. Each ingredient has a single identity until it is mixed with one or more ingredients to form a formula feed. Key operations are moving both sacked and bulk ingredients from storage to the mixing center; proportioning these ingredients; mixing, conveying, scalping, and blending them; and conveying the mixed feed to processing bins, such as pelleting, sacking, or bulk loadout bins.[15] Dust is generated when the ingredients are dumped into the hopper scale and when dry materials are dumped from the scale into the mixer. The mixing operation itself produces minimal emissions if the system is tightly closed.

Pelletizing, extruding, and blocking are methods of producing moist agglomerated particles; they generate little or no dust emissions. However, following pelletizing and extruding, cooling or drying processes use high airflow rates that can generate dust. However, minimal dust is typically emitted because high-efficiency collection systems are used to prevent product loss. The pelletized or extruded product passes through a shaker to size particles before packaging, bulk storage, or bulk shipping. Some dust may be generated during this operation, depending on product moisture; therefore, a dust-tight enclosure is required to prevent product loss.[11]

Many grain elevators, grain mills, and feed manufacturing plants use grain dryers to reduce the moisture content of the grain. Almost all grain dryers can be classified as either batch or continuous-flow dryers. In batch dryers, the grain is essentially stationary during the drying and cooling process. The dryer does not operate during loading and unloading.[16]

In continuous-flow dryers, grain flows continuously downward through the dryer. Air temperature and grain flow rates are usually not adjustable. Continuous-flow dryers can be further classified by how air moves in relation to the grain column. If air flows across the column, the dryer is a cross-flow type. Cross-flow dryers are the easiest and simplest to design and are most common. If air moves in the same direction as the grain, the dryer is called a concurrent-flow model. This design allows the use of higher air temperature and produces better-quality grain, but it is more susceptible to fire and more complicated to build. If air moves opposite to the direction of grain flow, the dryer is a counterflow type. This version accomplishes the most drying in the least amount of space, but it is less efficient, more difficult to build, and produces the lowest quality of grain of the three types of dryers.[16]

All types of grain dryers present a substantial air pollution problem. The emission stream from grain dryers is characterized by large air volumes, a large cross-sectional area of the exhaust stream, low PM specific gravity, and high moisture content. The PM from dryers generally comprises particles of relative large diameter, but the individual particles are light and have proven difficult to collect. "Bees' wings," a light, flaky material that breaks off from the corn kernel during drying, is particularly difficult to control.

AIR EMISSIONS CHARACTERIZATION

Each of the handling and processing operations in this industry has the potential to release PM emissions. A recent study for the U.S. Environmental Protection Agency provided a comprehensive review of all available PM data.[17,18] Tables 1

Table 1. Particulate Emission Factors for Grain Elevators (lb/ton)[a]

Emission Source	Type of Control	Filterable[b] PM	Emission Factor Rating	PM-10[c]	Emission Factor Rating
Grain receiving (SCC 3-02-005-05)					
Trucks (unspecified) (SCC 3-02-005-50)	None	ND		ND	
Straight truck (SCC 3-02-005-51)	None	0.18[d]	E	0.059	E
Hopper truck (SCC 3-02-005-52)	None	0.035[d]	E	0.0078	E
Railcar (SCC 3-02-005-53)	None	0.032	E	0.0078	E
Barge (SCC 3-02-005-54)	None	—[e]		—[e]	
Ships (SCC 3-02-005-55)	None	—[e]		—[e]	
Grain cleaning (SCC 3-02-005-03)					
Internal vibrating (SCC 3-02-005-37)	Cyclone	0.075[f]	E	—[h]	
Stationary enclosed (SCC 3-02-005-38)	None	ND		ND	
Grain drying (SCC 3-02-005-04)					
Column dryer (SCC 3-02-005-27)	None	0.22[g]	E	—[h]	
Rack dryer (SCC 3-02-005-28)	None	3.0[g]	E	—[h]	
	Self-cleaning screens (<50 mesh)	0.47[g]	E	—[h]	
Headhouse and internal handling (SCC 3-02-005-30) (legs, belts, distributor, scale, etc.)	None	0.061	E	0.034	E
Storage bin (vent) (SCC 3-02-005-40)	None	ND		ND	
Grain shipping (SCC 3-02-005-06)					
Trucks (unspecified) (SCC 3-02-005-60)	None	0.86[d]	E	0.029	E
Railcar (SCC 3-02-005-63)	None	0.027	E	0.0022	E
Barge (SCC 3-02-005-64)		—[e]		—[e]	
Ships (SCC 3-02-005-05)		—[e]		—[e]	

[a]Factors are in units of lb/ton of grain handled or processed. lb/ton divided by 2 gives kg/Mg. SCC, Source Classification Code; ND, no data available.

[b]Weight of total filterable PM, regardless of size, per unit weight of grain throughput.

[c]Weight of PM $\leq$ 10 μm in aerodynamic diameter per unit weight of grain throughput.

[d]Mean of two values.

[e]No data are available that represent current loading and unloading practices for ships and barges.

[g]Mean of two D-rated data points.

[h]PM-10 test data are not available. PM-10 emission factors can be estimated by taking 25% of the filterable PM emission factor.

Table 2. Particulate Emission Factors for Grain Processing Facilities (lb/ton)[a]

Type of Facility/ Emission Source	Type of Control	Filterable[b] PM	Emission Factor Rating	PM-10[d]	Emission Factor Rating
Animal feed mills					
Grain receiving (SCC 3-02-008-02)	None	0.017[e]	E	0.0025[e]	E
Grain cleaning (SCC 3-02-008-07)	Cyclone	—[f]		—[f]	
Storage	None	ND		ND	
Grain milling (SCC 3-02-008-15)					
Hammermill (SCC 3-02-008-17)	Cyclone	0.067[h]	E	—[g]	
Flaker (SCC 3-02-008-18)	Baghouse	0.012[i]	E	—[f]	
	Cyclone	0.15	E	—[g]	
Grain cracker (SCC 3-02-008-19)	Cyclone	0.024	E	—[g]	
Mixer	None	ND		ND	
Conditioning	None	ND		ND	
Pelletizing					
Pellet cooler[j] (SCC 3-02-008-16)	Cyclone	0.36[k]	E	—[g]	
		0.059[l,c] (condensible)	E		
	High efficiency cyclone[n]	0.15[m]	E	—[g]	
Feed shipping (SCC 3-02-008-03)	None	0.0033[e]	E	0.0008[e]	E

[a]Factors are in units of lb/ton of grain handled or processed. lb/ton divided by 2 gives kg/Mg. SCC, Source Classification Code; ND, no data available.
[b]Weight of total filterable PM, regardless of size, per unit weight of grain throughput.
[c]Condensible PM is material collected in the impinger portion of a PM sampling train.
[d]Weight of PM ≤ 10 μm in aerodynamic diameter per unit weight of grain throughput.
[e]Feed shipping emission factor based on data for loading of bulk feed (not pellets) barges.
[f]See emission factors for grain elevators, Table 2.
[g]PM-10 test data are not available. PM-10 emission factors can be estimated by taking 50% of the filterable PM emission factor.
[h]Mean of two values.
[i]Mean of two B-rated values.
[j]Includes column and pan dryers.
[k]Mean of 11 A-, B-, and C-rated values.
[l]Mean of three B- and C-rated values.
[m]Mean of two B-rated values.
[n]Equivalent to triple cycle or modern high efficiency cyclone.
[o]Mean of two values. Value converted from bushels to tons using a conversion factor of 50 bu/ton.
[p]PM-10 test data fare not available. PM-10 emission factors can be estimated by taking 100% of the filterable PM emission factor.

and 2 present emission factors that were generated by that survey for grain elevators and grain processing facilities, respectively. The estimates in the tables represent average emissions in pounds of emissions per ton of grain handled or processed. The emissions in kg/Mg can be found by dividing the values in lb/ton by a factor of 2.0. Note that for a particular grain elevator, the amount of grain handled in relation to the amount of grain received or shipped varies from operation to operation. Data on the emissions from the various milling operations such as wheat, corn, rice, durum, rye, and oats are very limited. The receiving, drying, and handling of the grains at milling operations can be estimated from Tables 1 and 2 for similar operations. Limited data on mills can be found in Reference 18.

While these estimates are industry-wide averages, they do not necessarily provide reliable emission estimates for a particular facility. Earlier studies indicate that emissions exhibit large variability, both between facilities and within a given facility, over time.[20] Factors that contribute to this variability include the type of grain handled, quality or grade of the grain, moisture content of the grain (usually 10–30%), amount of foreign material in the grain (usually 5% or less), moisture content of the grain at harvest (hardness), amount of dirt harvested with the grain, degree of enclosure at loading and unloading areas, and type of cleaning and conveying.[20] Of particular note are the studies from a subterminal elevator that show a wide variability of emissions as a function of grain type with emissions from soybean handling consistently two to four times higher than emissions from corn, wheat, and milo handling.[19]

While a reasonable quantity of information is available on total PM emissions from grain handling and processing, data on the particle size distribution of these emissions and on the fraction of emissions that might be toxic air pollutants are quite limited. The 1973 version of this manual indicated that 70–98% of the particulate matter captured in the control devices associated with shipping and receiving operations was 10 μm in diameter or larger.[11] The more recent surveys shown in Table 1 indicate that about 70–80% of the particulate matter emitted from grain receiving operations such as trucks and railcars are 10 μm in diameter or larger. As seen in Tables 1 and 2, however, data are relatively limited on the particle size distribution of emissions from grain handling and processing. Although quite limited, the data suggest that PM emissions from grain handling and processing operations tend to have relatively small fractions of fine PM.

No quantitative or qualitative information is available on toxic air pollutant emissions from grain handling and processing. Generally, the potential appears to be relatively small. The only two exceptions appear to be contamination of entrained grain dust with pesticides or herbicides applied in the fields and the use of fumigants for rodent and fungus control in grain storage facilities. No data are available on the air toxic emissions potential of these processes. However, limited information is available on adverse health effects associated with grain dust emissions.

Data indicate that dust from grain handling operations have adverse effects on the skin, eyes, and respiratory systems of humans. Allergic skin reactions and eye irritation have occurred in workers exposed to high concentrations of grain dust, but the most severe effects are on the respiratory system. The two main hypotheses on mechanisms by which grain dusts affect the respiratory system are (1) that the fine particles in the grain dust may act as a mechanical irritant and (2) that the organic, or perhaps the inorganic, part of the dust may produce an antigenic reaction, with asthmatic reactions in the case of the former and fibrotic changes in the latter.[2] Detailed descriptions of the effects of grain dust on workers' respiratory systems are presented in Refs. 21–24.

Data also indicate that these same respiratory effects can be found in some persons exposed to airborne dusts emitted from grain elevators. In particular, a study done at the University of Minnesota over a 5 year period supported a hypothesis of asthma attacks induced by respiratory irritation from atmospheric particles generated by grain handling operations. Evidence also suggests that asthmatic attacks in the vicinity of a New Orleans elevator are partially attributable to airborne effluent from the large grain elevator. Generally, no evidence exists for adverse effects on healthy people from grain emissions at concentrations less than 100 $\mu g/m^3$. However, people with preexisting respiratory disorders may be affected by rapid increases above the seasonal mean concentration of particulate grain dust.[20]

AIR POLLUTION CONTROL MEASURES

The two general types of measures that are available to reduce emissions from grain handling and processing operations are process modifications that are designed to prevent or inhibit emissions and capture/collection systems. Detailed descriptions of both process controls and capture systems for specific operations are available in the literature.[25–29] These detailed descriptions are not repeated here, but Table 3 identifies the types of controls available for each source and directs the reader to specific documents for more detailed information. The following describes the general approaches to process controls and capture systems. The characteristics of the collection systems most frequently applied to grain handling and processing plants [mechanical collectors (or cyclones) and fabric filters] are then described, and common operation and maintenance problems found in the industry are discussed.

Because PM emissions from grain handling operations are fugitive emissions generated as a consequence of mechanical energy imparted to the dust by the operations themselves and of local air currents in the vicinity of the operations, an obvious control strategy is to modify the process or facility to limit the

Table 3. Process Control and Exhaust Systems for Grain Handling and Processing[a]

Grain Handling and Processing Operation	Control Mechanism	Source of Control System Information: Dust Control[b]	Feed Manufacturing[c]	Environmental Controls[d]	Emission Control[e]	Industrial Ventilation[f]
Receiving operations	Grain flow control	35–38				
	Exhaust	41–52	113–114	12–20	297–300	
Belt conveyors	Enclosure	17–19				
	Flow control	90				
	Hooding/exhaust	98–108	115–116	30–31		5-39
Elevator legs	Exhaust	80–86	115	29–30		5-38
Distributors	Exhaust					5-128
Cleaners	Enclosure/exhaust		117	26–27		5-40
Scales	Enclosure/exhaust	109–113		21–24		5-128
Grain driers	Screens	203–262			287–297	
Hammer mills	Exhaust		117–118	41–46		
Roller mills	Exhaust			48		
Mixers	Exhaust		118	56–61		5-128
Truck/rail loadout	Dust suppression	127–142				
	Exhaust	143–160			300–302	
Barge/ship loadout	Dust suppression	99				
	Exhaust	166–173		301–314		

[a]Numbers in table are page references in these documents. Dashes in columns 3 to 6 represent page ranges, while those in column 7 represent chapter and page number.
[b]Reference 29.
[c]Reference 27.
[d]Reference 26.
[e]Reference 25.
[f]Reference 30.

effects of those factors that generate emissions. The primary preventive measures that facilities have used are construction and sealing practices that limit the effect of air currents, and minimizing grain free-fall distances and grain velocities during handling and transfer. Some recommended practices of this type are to enclose the receiving area to the degree practicable, preferably with doors at both ends of a receiving shed; to specify dust-tight cleaning and processing equipment; to use lip-type shaft seals at bearings on conveyor and other equipment housings; to use flanged inlets and outlets on all spouting, transitions, and miscellaneous hoppers; and to fully enclose and seal all areas in contact with products handled.[30]

A substantial reduction in emissions from receiving, shipping, and handling and transfer areas can be achieved by reducing grain free-fall distances and grain velocities. Figure 5 illustrates a choke unloading procedure used to reduce free-fall distance during hopper car unloading. The same principle can be used to control emissions from grain transfer onto conveyor

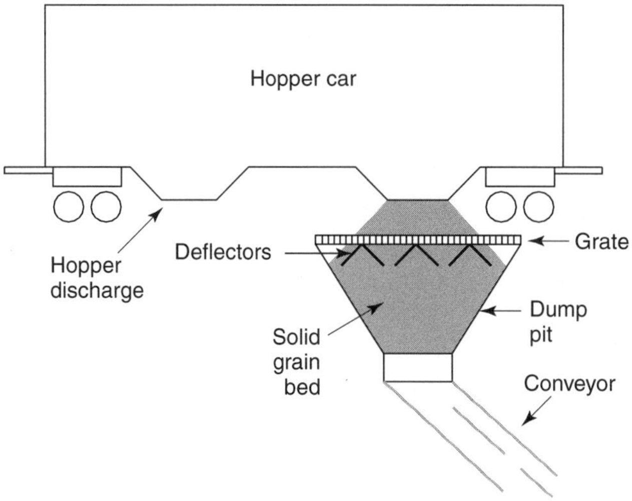

Figure 5. Chock Unloading for a Grain Receiving Process.

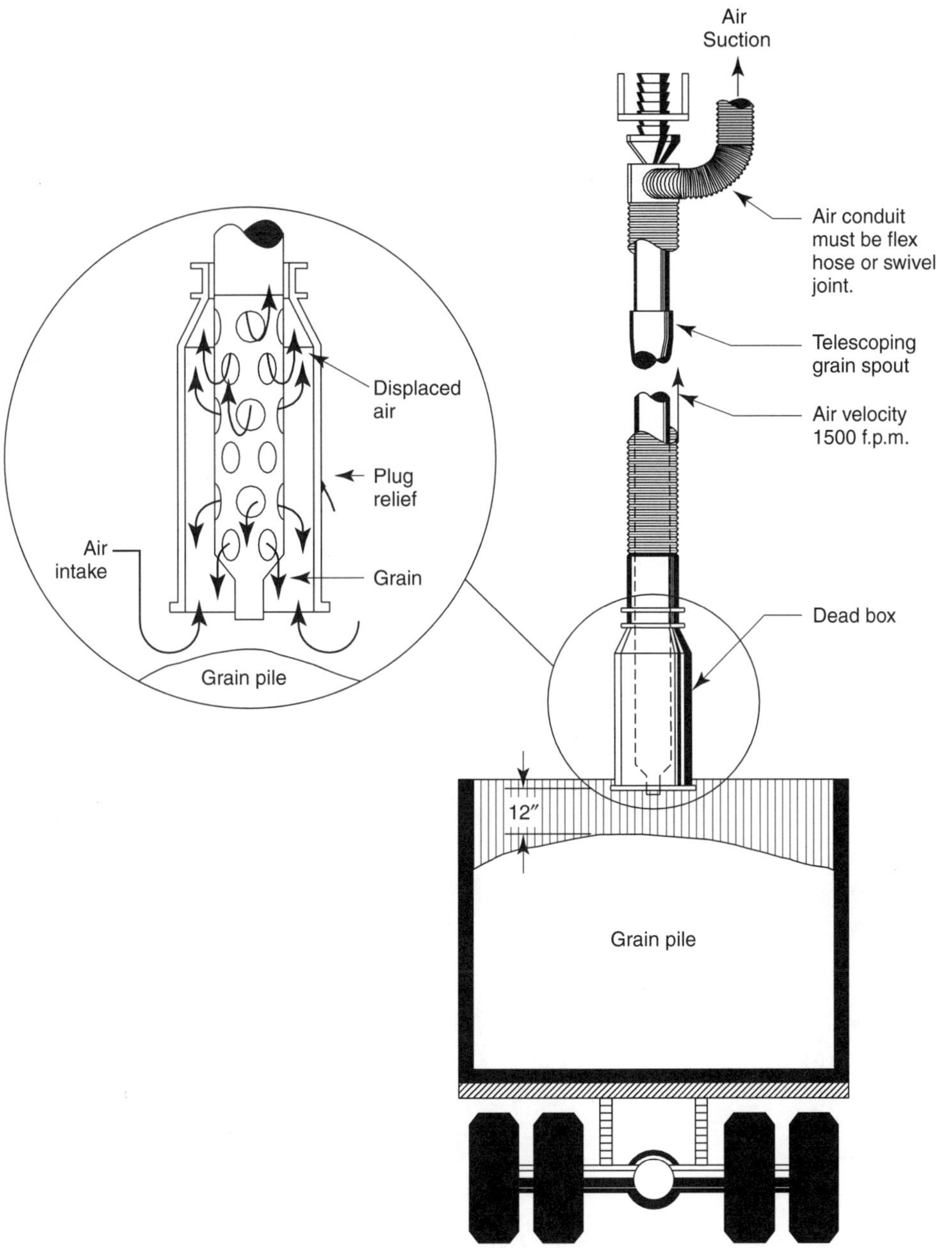

Figure 6. Dead Box for Reducing Loadout Emissions.

belts and from loadout operations. An example of a mechanism that reduces grain velocities is a dead box spout, which is used in loadout operations. The dead box spout slows down the flow of grain and stops the grain in an enclosed area (Figure 6). The dead box is mounted on a telescoping spout to keep it close to the grain pile during operation. In principle, the grain free-falls down the spout to an enclosed impact dead box, with grain velocity going to zero. It then falls about 12 in. to the grain pile. Typically, the entrained air and dust liberated at the dead box are aspirated back up the spout to a dust collection system.[31]

While the preventive measures described can minimize emissions, most facilities also require ventilation, or capture, systems to reduce emissions to acceptable levels. In fact, air aspiration (ventilation) is a part of the dead box system. Almost all grain handling and processing facilities, except relatively small grain elevators, use ventilation systems on the receiving pits, cleaning operations, and elevator legs. Generally, milling and pelletizing operations at processing plants are ventilated, and some facilities use hooding systems on all handling and transfer operations. An example of a ventilation system at a truck receiving station is illustrated in Figure 7. Details on capture systems can be found in the references noted in Table 3.

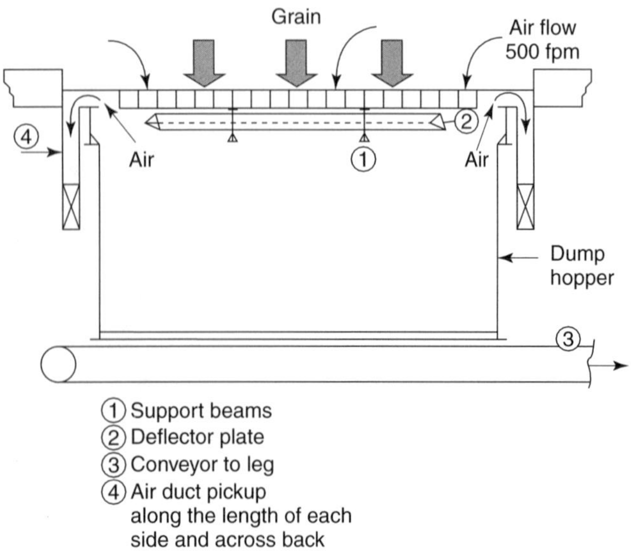

Figure 7. Receiving Pit Ventilation System.

The PM control devices typically used in the grain handling and processing industry are cyclones (or mechanical collectors) and fabric filters. Cyclones are used only on country elevators and small processing plants located in sparsely populated areas. Terminal elevators and processing plants in densely populated areas, as well as some country elevators and small processing plants, use fabric filters for PM control. Both of these systems can achieve acceptable levels of PM control for many grain handling and processing sources, as noted in Table 4.

Although cyclone collectors can achieve acceptable performance in some scenarios, and fabric filters are highly efficient, both devices are subject to failure if they are not properly operated and maintained. Also, malfunction of the ventilation system can lead to increased emissions at the source. Some typical malfunction problems at grain handling and processing facilities are described in the following.

Limited emission test data indicate that cyclone malfunctions can cause elevated emissions at country elevators. The efficiency of a cyclone depends on its design parameters and the velocity of the gas stream through it. Reduced gas velocities also result in decreased efficiency. Factors that cause an increased pressure drop through the control system, such as plugging in the cyclone, improper damper settings in the ductwork, or plugging in the ductwork, lead to decreased flow through the cyclone and a drop in cyclone efficiency. Periodic checks of cyclone pressure drop, exhaust fan current, the condition of cyclone walls and fan blades, and the exhaust rate at pickup points provide indications of any problems that might reduce efficiency. One other problem with cyclone operation that often occurs is leakage of air into the cyclone from the dust discharge. Therefore, dust discharge mechanisms should be checked regularly to control leakage.

Table 4. Cyclone and Fabric Filter Performance

	Cyclone Performance			Fabric Filter Performance		
		Outlet Levels			Outlet Levels	
Process Source	Efficiency (%)	gr/dscf	lb/ton	Efficiency (%)	gr/dscf	lb/ton
Grain receiving						
Truck			0.06[a]		0.002–0.01[a]	0.005[a]
Rail			0.1[a]		<0.002[a]	0.0002[a]
Not specified	85–95[b]	0.02–0.09[b]		99+[b]	0.002–0.006[b]	
Handling operations (legs, conveyors)			0.1[a]		<0–0.002[a]	0.0002[a]
Cleaning						
Elevators			0.3–0.6[a]		0.002–0.004[a]	0.003–0.014[a]
Grain mills	88–99[b]	0.03–0.1[b]		99+[b]	0.005–0.02[b]	
Grain milling	96–99[b]	0.04–0.2[b]		99+[b]	0.005–0.01[b]	
Pellet cooling	88–99[b]	0.02–0.1[b]				
Grain loadout						
Ship			0.03[a]		<0.002[a]	0.001[a]
Rail			0.06[a]		0.003–0.01[a]	0.001[a]

[a]Reference 33.
[b]Reference 34.

A major contributor to excess emissions from well-controlled grain elevators is the malfunction of fabric filters. Typical causes are torn bags that allow excess emissions from the control device outlet and plugging of the bags, which creates excess emissions at the source. Industry personnel, environmental control officials, and equipment vendors indicate that there is some evidence that operating problems may result in significant periods during which the control equipment is shut down or operating inefficiently.[32] Substantial bag blinding problems in areas of high humidity have been reported. Blinding problems also have been reported in cooler climates when the exhaust stream collected from handling operations for relatively warm, moist grain, which has been stored inside the elevator, is ducted to a control device located in a cooler outdoor atmosphere. Using baghouse heating systems and cleaning the bags thoroughly prior to shutdown can reduce these problems. Dust bridging in the baghouse hopper can be a major problem. Potential causes are poor design of hoppers and improper shutdown procedures. Problems can be reduced if systems are designed so that elevator personnel have easy access to the hopper for inspection and cleaning. Finally, malfunction of the bag cleaning mechanism can result in dust buildup on the bags and eventually in total plugging of the systems. At some facilities, the pulse-jet control panel is located on top of an outdoor baghouse and not in a weathertight enclosure. Accumulation of dust and moisture inside the control panel can cause short circuits in the solenoid valves for the pulse jet system, ultimately leading to system failure. To avoid such problems, general baghouse operation and maintenance practices should be routinely applied at grain handling and processing facilities.

Generally, emissions from multiple operations at grain handling and processing facilities are controlled by a single air pollution control system. If the ventilation system that connects these sources is not properly designed and operated, excess fugitive emissions are generated at the source. References 35 and 36 provide excellent information on designing, operating, and maintaining balanced ventilation systems.

REFERENCES

1. *Emission Factor Documentation for AP-42, Section 9.9.1 Grain Elevators and Grain Processing Plants—Final Report*, U.S. EPA contract 68-D2-0159, Midwest Research Institute. Available on the CHIEF bulletin board, May 1998.
2. D. Wallace and V. Ramananthan, *Review of Compliance Monitoring Programs with Respect to Grain Elevators*, EPA Contract No. 68-01-4139, WA 14, U.S. Environmental Protection Agency, Washington, D.C., 1980, p. 44.
3. U.S. Department of Commerce, *1987 Census of Manufacturers—Grain Mill Products*, U.S. Department of Commerce, Washington, D.C., 1990, pp. 200–7.
4. R. R. McElhiney, *Feed Manufacturing Technology III*, American Feed Manufacturing Association, Arlington, VA, 1985, p. 9.
5. L. J. Shannon, R. W. Girstel, P. G. Gorman, et al., *Emissions Control in the Grain and Feed Industry, Vol. 1—Engineering and Cost Study*, EPA-450-/3-73-003a, U.S. Environmental Protection Agency, Research Triangle Park, NC, 1973, p. 112.
6. *Ibid.*, pp. 203–69.
7. Reference 4, pp. 1–216.
8. *Ibid.*, pp. 36–37.
9. Reference 5, p. 159.
10. *Ibid.*, pp. 120–29.
11. J. A. Danielson, *Air Pollution Engineering Manual, AP–40* , U.S. Environmental Protection Agency, Research Triangle Park, 1973, NC, pp. 352–61.
12. Reference 5, p. 205.
13. Reference 4, p. 436.
14. *Ibid.*, p. 143.
15. *Ibid.*, p. 151.
16. *Ibid.*, p. 16.
17. Reference 1, 24 pp.
18. U.S. Environmental Protection Agency, *Compilation of Air Pollutant Emission Factors*, Grain Elevators and Processes, U.S. Environmental Protection Agency, Research Triangle Park, NC, 1998, pp. 9.9.1-1–27.
19. P. G. Gorman, *Potential Dust Emissions from a Grain Elevator in Kansas City, Missouri*, EPA Contract No. 68-02-0228, U.S. Environmental Protection Agency, Research Triangle Park, NC, 1974.
20. Reference 2, pp. 56–77.
21. N. Williams, A. Skoulas, and N. Merriman, "Exposure to Grain Dust: I. Survey of Effects," *J. Occupational Med.* **6**(8),319 (1964).
22. A. Skoulas, N. Williams, and J. Merriman, "Exposure to Grain Dust: II. A Clinical Study of the Effects," *J. Occupational Med.* **6**(9),359 (1964).
23. R. Gordon, *Dust Control for Grain Elevators*, National Grain and Feed Association, Washington, D.C., 1981, pp. 358–75.
24. T. Hurst and J. Dosman, "Characterization of Health Effects of Grain Dust Exposures," *Am. J. Ind. Medi.* **17**(l),27 (1990).
25. Reference 5, 544 pp.
26. W. Briggs, M. Shiver, and T. Stivers, *Environmental Controls for Feed Manufacturing and Grain Handling*, American Feed Manufacturer's Association, Chicago, 1971, 184 pp.
27. Reference 4, 608 pp.
28. Committee on Industrial Ventilation, *Industrial Ventilation, a Manual of Recommended Practice*, 19th ed, American Conference of Governmental Industrial Hygienists, Lansing, MI, 1986, 392 pp.
29. Reference 23, 466 pp.
30. Reference 26, pp. 8–9
31. Reference 23, p. 149.
32. Reference 2, pp. 15, 16.
33. Reference 5, pp. 293–94.
34. *Standards Support and Environmental Impact Statement, Vol. I: Standards of Performance for Grain Elevator Industry*, U.S. Environmental Protection Agency, Research Triangle Park, NC, 1977, pp. 4-1–30, 5-2.
35. Reference 23, pp. 274–306.
36. Reference 28, pp. 6-1–55.

DISTILLED SPIRITS

(Adapted from Section 9.12.3 AP-42)

GENERAL[1,2]

The distilled spirits industry includes the production of whisky, gin, vodka, rum, and brandy. The production of brandy is discussed in another section of this manual and also in AP-42 Section 9.12.2, "Wines and Brandy." Distilled spirits production also may include the production of secondary products such as distillers dried grains used for livestock feed and other feed/food components.

Distilled spirits, including grain spirits and neutral spirits, are produced throughout the United States.[1] The Bureau of Alcohol, Tobacco, and Firearms (BATF) has established "standards of identity" for distilled spirits products.[2]

PROCESS DESCRIPTION[3,4]

Distilled spirits can be produced by a variety of processes. Typically, in whisky production, grains are mashed and fermented to produce an alcohol/water solution, that is distilled to concentrate the alcohol. For whiskies, the distilled product

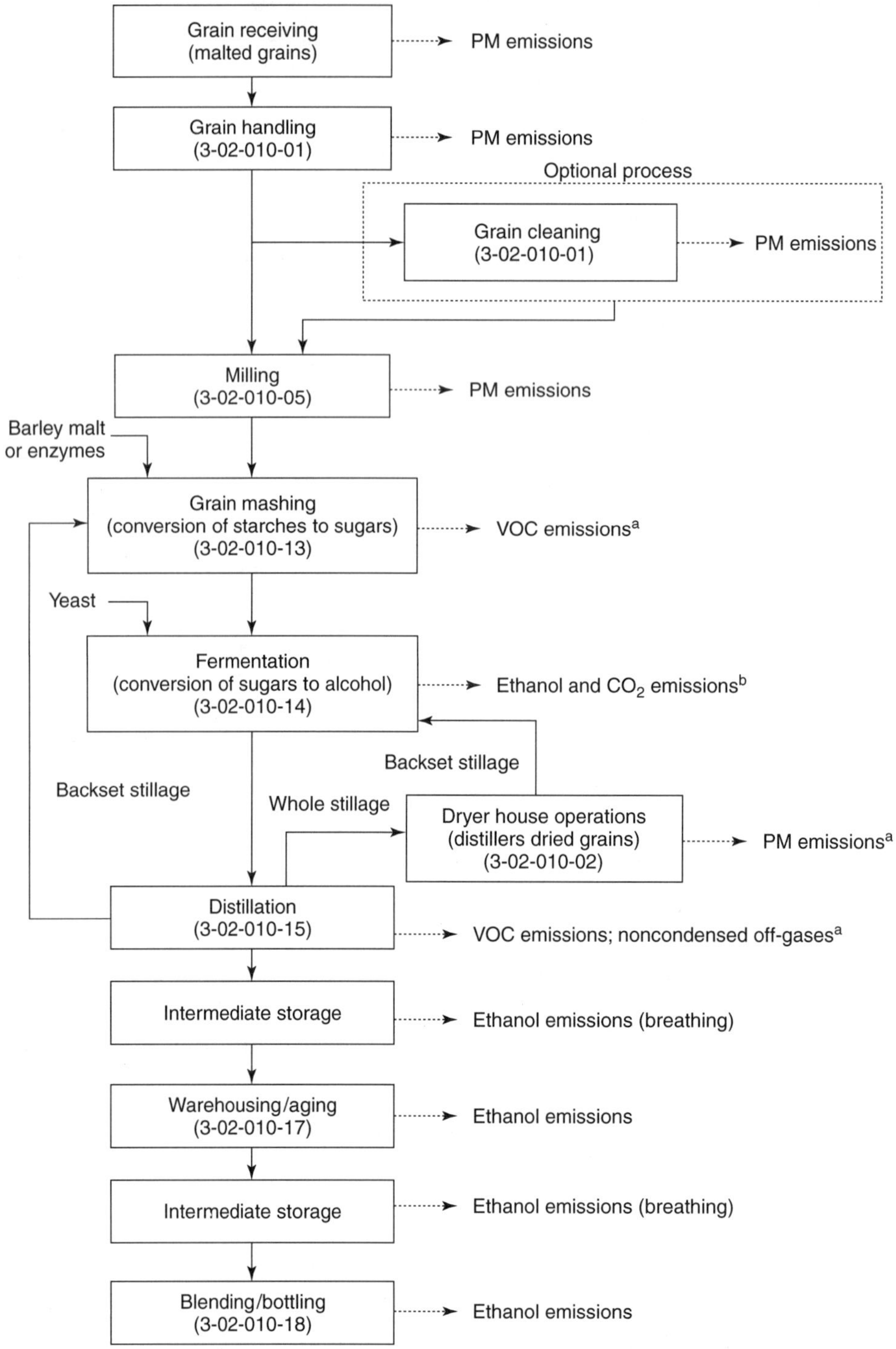

[a] Processes require heat. Emissions generated (e.g., CO, CO_2, NO_X, SO_2, PM, and VOCs) will depend on the source of fuel.
[b] Other compounds can be generated in trace quantities during fermentation including ethyl acetate, fusel oil, furfural, acetaldehyde, sulfur dioxide, and hydrogen sulfide. Acetaldehyde is a hazardous air pollutant (HAP).

Figure 1. Whisky Production Process (Source Classification Codes in Parentheses).

is aged to provide flavor, color, and aroma. This discussion will be limited to the production of Bourbon whiskey. Figure 1 is a simple diagram of a typical whisky production process. Emission data are available only for the fermentation and aging steps of whiskey production.

GRAIN HANDLING AND PREPARATION

Distilleries utilize premium cereal grains, such as hybrid corn, rye, barley, and wheat, to produce the various types of whisky and other distilled spirits. Grain is received at a distillery form a grain-handling facility and is prepared for fermentation by milling or by malting (soaking the grains to induce germination). All U.S. distillers purchase malted grain instead of performing the malting process onsite.

GRAIN MASHING

Mashing consists of cooking the grain to solubilize the starch from the kernels and to convert the soluble starch to grain sugars with barley malt and/or enzymes. Small quantities of malted barley are sometimes added prior to grain cooking. The mash then passes through a noncontact cooler to cool the converted mash prior to entering the fermenter.

FERMENTATION

The converted mash enters the fermenter and is inoculated with yeast. The fermentation process, which usually lasts 3–5 days for whiskey, uses yeast to convert the grain sugars into ethanol and carbon dioxide. Congeners are flavor compounds that are produced during fermentation as well as during the barrel aging process. The final fermented grain alcohol mixture, called "beer," is transferred to a "beer well" for holding. From the beer well, the beer passes through a preheater, where it is warmed by the alcohol vapors leaving the still, and then to the distillation unit. The beer still vapors condensed in the preheater generally are returned to the beer still as reflux.

DISTILLATION

The distillation process separates and concentrates the alcohol from the fermented grain mash. Whiskey stills are usually made of copper, especially in the rectifying section, although stainless steel may be used in some stills. Following distillation, the distilled alcohol spirits are pumped to stainless steel tanks and diluted with demineralized water to the desired alcohol concentration prior to filling into oak barrels and aging. Tennessee whiskey utilizes a different process from bourbon in that the distillate is passed through sugar maple charcoal in mellowing vats prior to dilution with demineralized water.

GRAIN AND LIQUID STILLAGE ("DRYER HOUSE OPERATIONS")

In most distilleries, after the removal of alcohol. Still bottoms (called whole stillage), are pumped from the distillation column to a dryer house. Whole stillage may be sold, land applied (with permitting), sold as liquid feed, or processed and dried to produce distillers dried grains (DDG) and other secondary products. Solids in the whole stillage are separated using centrifuges or screens; the liquid portion (thin stillage) may be used as a backset or concentrated by vacuum evaporation. The concentrated liquid may be recombined with the solids or dried. Drying is typically accomplished using either steam-heated or flash dryers.

WAREHOUSE/AGING

Aging practices differ from distiller to distiller, and even for the same distiller. Variations in the aging process are integral to producing the characteristic taste of a particular brand of distilled spirit. The aging process, which typically ranges from 4 to 8 years or more, consists of storing the new whisky distillate in oak barrels to encourage chemical reactions and extractions between the whisky and the wood. The constituents of the barrel produce the whisky's characteristic color and distinctive flavor and aroma. White oak is used because it is one of the few woods that holds liquids while allowing breathing (gas exchange) through the wood. Federal law requires all bourbon whisky to be aged in charred new white oak barrels.

The oak barrels and the barrel environment are key to producing distilled spirits of desired quality. The new whiskey distillate undergoes many types of physical and chemical changes during the aging process that removes the harshness of the new distillate. As whiskey ages, it extracts and reacts with constituents in the wood of the barrel, producing certain trace substances, called congeners, which give whisky its distinctive color, taste, and aroma.

Barrel environment is extremely critical in whisky aging and varies considerably by distillery, warehouse, and even location in the warehouse. Ambient atmospheric conditions, such as seasonal and diurnal variations in temperature and humidity, have a great affect on the aging process, causing changes in the equilibrium rate of extraction, rate of transfer by diffusion, and rate of reaction. As a result, distillers may expose the barrels to atmospheric conditions during certain months, promoting maturation through the selective opening of windows and doors and by other means.

Distillers often utilize various warehouse designs, including single- or multistory buildings constructed of metal, wood, brick, or masonry. Warehouses generally rely upon natural ambient temperature and humidity changes to drive the again process. In a few warehouses, temperature is adjusted during the winter. However, whisky warehouses do not have the capability to control humidity, which varies with natural climate conditions.

BLENDING/BOTTLING

Once the whisky has completed its desired aging period, it is transferred from the barrels into tanks and reduced in proof to the desired final alcohol concentration by adding demineralized water. Following a filtration process that renders it free of any solids, the whisky is pumped to a tank in the bottling house, bottled, and readied for shipment to the distributors.

EMISSIONS AND CONTROLS[3–6]

Emissions

The principal emissions from whisky production are volatile organic compounds (VOCs), principally ethanol, and occur primarily during the aging/warehousing stage. In addition to ethanol, other volatile compounds, including acetaldehyde (a HAP), ethyl acetate, glycerol, fusel oil, and furfural, may be produced in trace amounts during aging. A comparatively small source of ethanol emissions may results from the fermentation stage. Smaller quantities of ethyl acetate, isobutyl alcohol, and isoamyl alcohol are generated as well; carbon dioxide is also produced during fermentation. Particulate matter (PM) emissions are generated by the grain receiving, handling, drying, and cleaning processes and are discussed in more detail in AP-42 Section 9.9.1 (Grain Elevators and Processes). Other emissions, including SO_2, CO_2, CO, NO_x, and PM may be generated by fuel combustion from power production facilities located at most distilled spirits plant.

Ethanol and water vapor emissions result from the breathing phenomenon of the oak barrels during the aging process. This phenomenon of wood acting as a semipermeable membrane is complex and not well understood. The emissions from evaporation from the barrel during aging are not constant. During the first 6–18 months, the evaporation rate from a new barrel is low because the wood must become saturated (known as "soakage") before evaporation occurs. After saturation, the evaporation rate is greatest, but then decreases as evaporation lowers the liquid level in the barrel. The lower liquid level decreases the surface area of the liquid in contact with the wood and thus reduces the surface are subject to evaporation. The rate of extraction of wood constituents, transfer, and reaction depend upon ambient conditions, such as temperature and humidity, and the concentrations of the various whisky constituents. Higher temperatures increase the rate of extraction, transfer by diffusion, and reaction. Diurnal and seasonal temperature changes cause convection currents in the liquid. The rate of diffusion will depend upon the differences in concentrations of constituents in the wood, liquid, and air blanketing the barrel. The rates of reaction will increase or decrease with the concentration of constituents. The equilibrium concentrations of the various whisky components depend upon the humidity and air flow around the barrel.

Minor emissions are generated when the whisky is drained from the barrels for blending and bottling. Residual whisky remains in the used barrels both as a surface film ("heel") and within the wood ("soakage"). For economic reasons, many distillers attempt to recover as much residual whisky as possible by methods such as rinsing the barrel with water and vacuuming. Generally, barrels are refilled and reentered into the aging process for other distilled spirits at the particular distiller or sealed with a closure (bang) and shipped offsite for reuse with other distilled spirits. Emissions may also be generated during blending and bottle filing, but no data are available.

CONTROLS

With the exception of devices for controlling PM emissions, there are very few emissions controls at distilleries. Grain handling and processing emissions are controlled through the use of cyclones, baghouses, and other PM control devices (see AP-42, Section 9.9.1). There are currently no current control technologies for VOC emissions from fermenters because the significant amount of grain solids that would be carried out of the fermenters by air entrainment could quickly render systems, such as carbon adsorption, inoperable. Addon air pollution control devices for whisky aging warehouses are not used because of potential adverse impact on product quality. Distillers ensure that barrel construction is of high quality to minimize leakage, thus reducing ethanol emissions. Ethanol recovery would require the use of a collection system to capture gaseous emissions in the warehouse and to process the gases through a recovery system prior to venting them to the atmosphere.

EMISSION FACTORS

Table 1 provides uncontrolled emission factors for emissions of VOCs from fermentation vats and for emissions of ethanol from aging due to evaporation. Because ethanol is the principal VOC emission from aging, the ethanol emissions factors are reasonable estimates of VOC emissions for these processes. Emission factors for grain receiving, handling, and cleaning may be found in AP-42, Section 9.9.1, Grains Elevators and Processes and in the appropriate section of this manual. Emission factors are unavailable for grain mashing, distillation, blending/bottling, and spent grain drying. An emission factor for carbon dioxide from fermentation vats is also unavailable although carbon dioxide and ethanol are theoretically generated in equal molecular quantities during the fermentation process.

Recognizing that aging practices may differ from distiller to distiller, and even for different products of the same distiller, a method may be used to estimate total ethanol emissions from barrels during aging. An ethanol emission factor for aging (total loss emission factor) can be calculated based on annual emissions per barrel in proof gallons (PG). The term "proof gallon" refers to a U.S. gallon of proof spirits, or the alcoholic equivalent thereof, containing 50% of ethyl alcohol (ethanol) by volume. This calculation method is derived from the gauging of product and measures the difference in the amount of product when the barrel was filled and when the barrel was

Table 1. Emission Factors for Distilled Spirits[a] Emission Factor Rating: E

Source[b]	Ethanol	Ethyl acetate	Isoamyl Alcohol	Isobutyl Alcohol
Grain mashing (SCC 3 -02-010-13)	NA	NA	NA	NA
Fermentation vats (SCC 3-02-010-14)	14/2[c]	0.046[c]	0.013[c]	0.004[c]
Distillation (SCC 3-02-010-15)	ND	ND	ND	ND
Aging (SCC 3-02-010-17)				
Evaporation loss[d]	6.9[e]	ND	ND	ND
Blending/bottling (SCC 3-02-010-18)	ND	ND	ND	ND
Dryer house operations (SCC 3-02-010-02)	ND	ND	ND	ND

[a] Factors represent uncontrolled emissions. SCC, Source Classification Code; ND, no data available. To convert from lb to kg, divide by 2.2. NA, not applicable.
[b] Emission factors for grain receiving, handling, and cleaning processes are available in AP-42 Section 9.9.1, Grain Elevators and Processes.
[c] Reference 5. In units of pounds per 1,000 bushels of grain input.
[d] Evaporation losses during whisky aging do not include losses due to soakage.
[e] References 6 and 7. In units of lb/bbl/yr; barrels have a capacity of approximately 53 gal.

emptied. Fugitive evaporative emissions, however, are not the sole difference between these two amounts. During the aging period, product soaks into the barrel, test samples are drawn, and other losses (e.g., spillage, leakage) may occur. Estimates of ethanol loss due to evaporation during aging based only on the gauging of product will produce an overestimate unless soakage and sampling losses (very small losses) are subtracted. The emission factor for evaporation loss in Table 1 represents an overestimate because only data for soakage losses could be calculated; data for other losses were not available.

REFERENCES

1. Bureau of Alcohol, Tobacco, and Firearms (BATF), "Monthly Statistical Release—Distilled Spirits," Department of the Treasury, Washington, D.C., January 1995 through December 1995.
2. "Standards of Identity for Distilled Spirits," 27 CFR Part 1, Subpart C, Office of the Federal Register, National Archives and Records Administration, Washington, D.C., April 1, 1996.
3. J. E. Bujake, "Beverage Spirits, Distilled," *Kirk–Othmer Encyclopedia of Chemical Technology*, 4th Ed., Vol. 4, John Wiley & Sons, Inc., 1992.
4. *Cost and Engineering Study Control of Volatile Organic Emissions from Whiskey Warehousing*, EPA-450/2-78-013, Emission Standards Division, Chemical and Petroleum Branch, Office of Air Quality Planning and Standards, U.S. Environmental Protection Agency, Research Triangle Park, NC, April 1978.
5. R. V. Carter, and B. Linsky, "Gaseous Emissions from Whiskey Fermentation Units," *Atmospheric Environment* **8** 57–62, January 1974; also a preliminary paper of the same title by these authors (undated).
6. Written communication from R. J. Garcia, Seagrams America, Louisville, KY, to T. Lapp, Midwest Research Institute, Cary, NC, March 3, 1997. RTGs versus age for 1993 standards.
7. Written communication from L. J. Omlie, Distilled Spirits Council of the United States, Washington, D.C., to T. Lapp, Midwest Research Institute, Cary, NC, February 6, 1997. Ethanol emissions data from Jim Beam Brands Co.

WINES AND BRANDY

(Adapted from Section 9.12.2 AP-42)

Wine is an alcoholic beverage produced by the fermentation of sugars in fruit juices, primarily grape juice. In general, wines are classified into two types based on alcohol content: table wines (7–14%, by volume) and dessert wines (14–24%, by volume). Table wines are further subdivided into still and sparkling categories, depending upon the carbon dioxide (CO_2) content retained in the bottled wine. Still table wines are divided into three groups: red, rose (blush), and white, based on the color of the wine.

PROCESS DESCRIPTION[1–4]

The production of still table wines is discussed in the following paragraphs, followed by more concise discussions of the production of sweet table wines, sparkling wines, dessert wines, and brandy.

STILL TABLE WINES

The basic steps in vinification (wine production) include harvesting, crushing, pressing, fermentation, clarification, aging, finishing, and bottling. A simplified process diagram outlining the basic steps in the production of still table wines is shown in Figure 1.

Harvesting of grapes is usually conducted during the cooler periods of the day to prevent or retard heat buildup and flavor deterioration in the grape. Most wineries transport the whole grapes but some crush the grapes in the vineyard and transport the crushed fruit to the winery. Stemming and crushing are commonly conducted as soon as possible after harvest. These two steps are currently done separately using a crusher–stemmer, which contains an outer perforated cylinder to allow the grapes to pass through but prevents the passage of stems, leaves, and stalks. Crushing the grapes after stemming

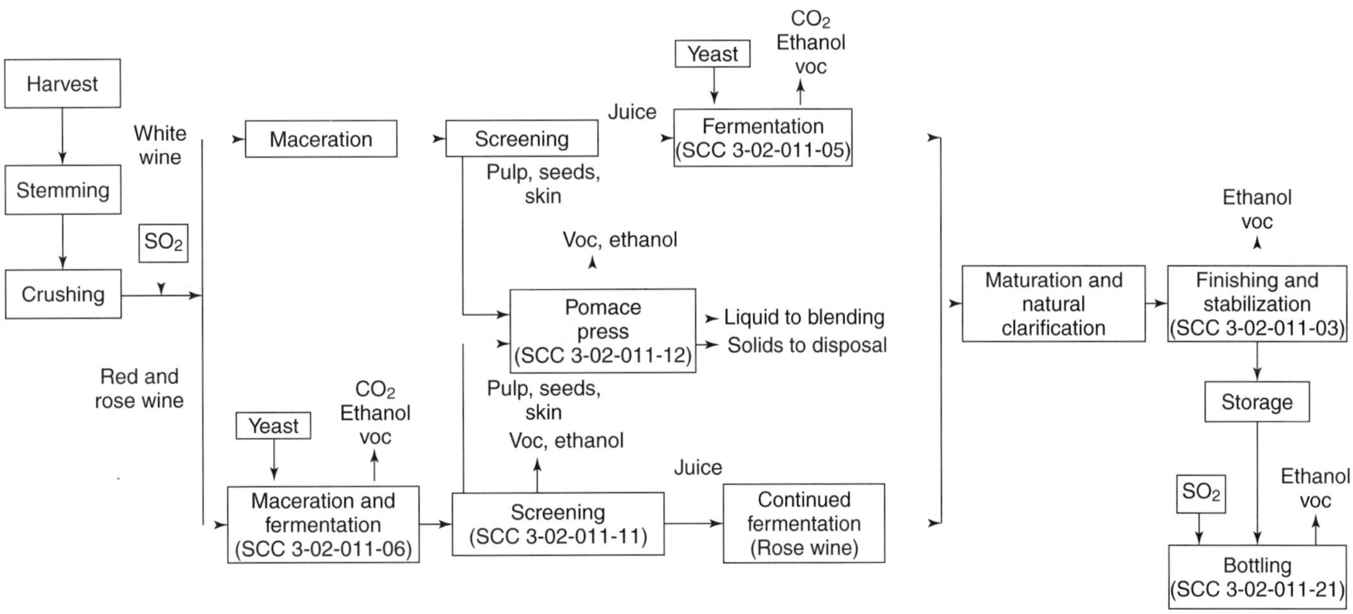

Figure 1. Basic Steps in Still Table Wine Production (Source Classification Codes in Parentheses).

is accomplished by any one of many procedures. The three processes generally favored are: (1) pressing grapes against a perforated wall; (2) passing grapes through a set of rollers; or (3) using centrifugal force. Generally, 25–100 milligrams (mg) of liquefied sulfur dioxide (SO_2) are added per liter of the crushed grape mass to control oxidation, wild yeast contamination, and spoilage bacteria.

Maceration is the breakdown of grape solids following crushing of the grapes. The major share of the breakdown results from the mechanical crushing but a small share results from enzymatic breakdown. In red and rose wine production, the slurry of juice, skins, seeds, and pulp is termed the "must." Thus the term *must* can refer to either the mixture of juice, seeds, skins, and pulp for red or rose wines or only the juice for white wines. Maceration is always involved in the initial phase of red wine fermentation. The juice from the grapes may be extracted from the "must" in a press. Additionally, gravity flow juicers may be used initially to separate the majority of the juice from the crushed grapes and the press used to extract the juice remaining in the mass of pulp, skins, and seeds (pomace). There are many designs of dejuicers, but, generally, they consist of a tank fitted with a perforated basket at the exit end. After gravity dejuicing has occurred, the pomace is place in a press and the remaining juice extracted. There are three major types of presses. The horizontal press is used for either crushed or uncrushed grapes. A pneumatic press can be used for either crushed or uncrushed grapes as well as for fermented "must." In the continuous screw press, the "must" is pumped into the press and forced in the pressing chamber where perforated walls allow the juice to escape. After pressing, white "must" is typically clarified and/or filtered prior to fermentation to retain the fruity character. The white juice is commonly allowed to settle for up to 12 h but may be centrifuged to speed the clarification.

Fermentation is the process hereby the sugars (glucose and fructose) present in the "must" undergo reaction by yeast activity to form ethyl alcohol (ethanol) and CO_2 according to the equation:

$$C_6H_{12}O_6 \longrightarrow 2C_2H_5OH + 2CO_2$$

In the United States the sugar content of the juice is commonly measured with a hydrometer in units of degree Brix (°B), which is grams (g) of sugar per 100 grams of liquid. Fermentation may be initiated by the addition of yeast inoculation to the "must." The fermentation process takes place in tanks, barrels, and vats of a wide variety of shapes, sizes, and technical designs. Tanks are different from vats in the tanks are enclosed, whereas vats have open tops. In most of the larger wineries, tanks have almost completely replaced vats. Since the 1950s, the move has been away from the use of wooden tanks, primarily to stainless steel tanks. Lined concrete tanks are also used, and fiberglass tanks are becoming more popular because of their light weight and lower cost.

The fermentation process is an exothermic reaction and requires temperature control of the fermenting "must." Red wines are typically fermented at 25 to 28°C (70° to 82°F) and white wines at 8° to 15°C (46° to 59°F). Almost all the fermentation is conducted by the batch process and continuous fermentors are rarely used in the United States. Size of the fermentors is based primarily on the volume of "must" to be fermented. During fermentation of red wines, the CO_2 released by the yeast metabolism becomes entrapped in the pomace (layer of skins and seeds) and causes it to rise to the top of the tank where it forms a cap. The pomace cap is periodically covered with the "must" to increase color, removal, aerate the fermenting "must," limit growth of spoilage organisms in the cap, and help equalize the temperature in the fermenting "must." For white wines, the main technical requirement is efficient temperature control. Temperature is one of the most influential factors affecting the fermentation process. During fermentation of both white and red "must," the CO_2, water vapor, and ethanol are released through a vent in the top of the tank. Malolactic fermentation sometimes follows the primary fermentation and results in a reduction in acidity and increased pH. There are very diverse opinions about this step because the fermentation, to varying degrees, can improve or reduce wine quality.

After fermentation, all wines undergo a period of adjustment (maturation) and clarification prior to bottling. The process of maturation involves the precipitation of particulate and colloidal material from the wine as well as a complex range of physical, chemical, and biological changes that tend to maintain and/or improve the sensory characteristics of the wine. The major adjustments are acidity modification, sweetening, dealcoholization, color adjustment, and blending. Following the fermentation process, a preliminary clarification step is commonly accomplished be decanting thee wine from one vessel to another, called racking, in order to separate the sediment (lees) from the wine. Currently racking practices range from manually decanting wine from barrel to barrel to highly sophisticated, automated, tank-to-tank transfers. In all cases, separation occurs with minimal agitation to avoid resuspending the particulate matter. The residue from racking may be filtered to recover wine otherwise lost with the lees or may be used "as is" for brandy production.

Stabilization and further clarification steps follow maturation and initial clarification to produce a permanently clear wine with no flavor faults. The steps entail various stabilization produces, additional clarification (fining), and a final filtration prior to bottling. The most common stabilization technique used for may red wines and some white wines is aging the wine for a period of months or years. Vessels used to store and age wine, termed cooperage, are produced in a wide range of sizes, depending on their intended use. White oak has traditionally been used for the barrels to age wine, but currently its usage is reserved primarily for the production of premium white and red wines and some fortified wines. Water and ethanol are lost through the barrel surfaces and partial vacuum develops in the space created by this loss. Each barrels periodically opened and topped off with wine to fill the void created by the ethanol and water loss. Cooperage constructed from materials other than wood has many advantages and is less expensive to maintain. Stainless steel is often preferred, but fiberglass and concrete are also used. In addition to aging, other stabilization procedures are used to prevent formation of potassium bitartrate or calcium tartrate crystals, haziness (casse) resulting from protein coalescence, casse resulting from oxidation of tannins present in the wine, and haziness due to metal ions such as iron and copper. Enzyme mixtures are used to remove polysaccharides which can cause filtration problems and haze formation. Most wines contain viable but dormant microorganism. Racking is used as an initial step in microbial stabilization but long-term stability frequently requires use of sulfur dioxide as the antimicrobial agent. Other methods include pasteurization and filter sterilization. Sulfur dioxide may be added at various stages in wine production to prevent microbial growth and oxidation. Finishing (fining) agents are commonly added to accelerate the precipitation of suspended material in wine. Prior to bottling, a final clarification step is used to remove any remaining suspended material and microbes in the wine. This step involves only physical methods of clarification, generally a filtration procedure.

Glass bottles are the container of choice for premium-quality wines and for sparkling wines. Because of disadvantages such as weight and breakage, glass bottles are sometimes being replaced by new containers, such as bag-in-box, for many standard-quality, high-volume wines. To protect the wine against microbial spoilage, and to limit oxidation, the SO_2 content in the wine is adjusted to a final level of 50 mg/L before filling. Precaution is taken to minimize contact with air during filling and thereby to reduce oxidation. This is done by either flushing the bottle with inert gas before filling or flushing the headspace with inert gas after filling.

SWEET TABLE WINES

The most famous of the sweet wines are those made from noble-rotted, *Botrytis*-infected grapes. These wines are produced to a limited extent in the United States. The *Botrytis* mold acts to loosen the grape's skin so moisture loss occurs rapidly and the sugar concentration increases in the grape. The grapes are then selectively picked, followed by pressing and fermentation. Fermentation is a slow process, however, because of the high sugar content and the use of SO_2 to retard the growth of undesirable molds and microorganisms. Nonbotrytized sweet wines are also produced by drying the grapes. Drying involves allowing the grapes to dehydrate on mats or trays in the shade for weeks or months and then crushing the grapes and fermenting the concentrated juice. Heating, boiling, or freezing is also used to concentrate juice for semisweet wines.

Sparkling Wines

Most sparkling wines obtain CO_2 supersaturation using a second alcoholic fermentation, typically induced by adding yeast and sugar to dry white wine. There are three principal methods of sparkling wine production: the *methode champagnoise*, the transfer method, and the bulk method. In the *methode champagnoise*, both red and white grapes may be used, but most sparkling wines are white. The grapes are harvested earlier than those used for still table wines and pressed whole without prior stemming or crushing to extract the juice with a minimum of pigment and tannin extraction. This is important for producing white sparkling wines from red-skinned grapes. Primary fermentation is carried out at approximately 15°C (59°F), and bentonite and/or casein may be added to aid the process and improve clarity. The blending of wines produced from different sites, varieties, and vintages distinguishes the traditional method. Before preparing the blend (*cuvee*), the individual base wines are clarified and stabilized. Aging typically takes place in stainless steel tanks but occasionally takes place in oak cooperage. The secondary fermentation requires inoculation of the *cuvee* wine with a special yeast strain. A concentrated sucrose solution is added to the *cuvee* just prior to the yeast inoculation. The wine is then bottled, capped, and stacked horizontally at a stable temperature, preferably between 10 and 15°C (50–59°F), for the second fermentation. After fermentation, the bottles are transferred to a new site for maturation and stored at about 10°C (50°F).

Riddling is the technique used to remove the yeast sediment (lees). The process involves loosening and suspending the cells by manual or mechanical shaking and turning, and positioning the bottle to move the lees toward the neck. Disgorging takes place about 1 or 2 years after bottling. The bottles are cooled and the necks immersed in an ice/$CaCl_2$ or ice/glycol solution to freeze the sediment. The disgorging machine rapidly removes the cap on the bottle, allowing for ejection of the frozen yeast plug. The mouth of the bottle is quickly covered and the fluid level is adjusted. Small quantities of SO_2 or ascorbic acid may be added to prevent subsequent in-bottle fermentation and limit oxidation. Once the volume adjustment and other additions are complete, the bottles are sealed with special corks, the wire hoods added, and the bottles agitated to disperse the additions. The bottles are then decorated with their capsule and labels and stored for about 3 months to allow the corks to set in the necks. The transfer method is identical to the *methode champagnoise* up to the riddling stage. During aging, the bottles are stored neck down. When the aging process is complete, the bottles are chilled below 0°C (32°F) before discharge into a transfer machine and passage to pressurized receiving tanks. The wine is usually sweetened, sulfited, clarified by filtration, and sterile filtered just before bottling.

Other methods of production of sparkling wine include the "rural" method and carbonation. The rural method involves prematurely terminating the primary fermentation prior to a second in-bottle fermentation. The injection of CO_2 (carbonation) under pressure at low temperatures is the least expensive and the least prestigious method of producing sparkling wines.

DESSERT WINES

Dessert wines are classified together because of their elevated alcohol content. The most common dessert wines are sherries and ports.

Baking is the most popular technique for producing sherries in the United States. Grapes are crushed and stemmed and SO_2 added as soon as possible to control bacteria and oxidation. The maximum amount of juice is separated from the skins and the juice is transferred to fermentors. The juice is inoculated with starter and fermented at temperatures of 25–30°C (77–86°F). The new wine is then pumped from the fermentor or settling tank to the fortification tank. High-proof spirits are added to the sherry material, or shermat, to raise the alcohol content to 27–189% by volume and then the wine is thoroughly mixed, clarified, and filtered before baking. Slow baking occurs when the wine is stored in barrels exposed to the sun. More rapid baking is achieved through the use of artificially heated storage rooms or heating coils in barrels or tanks. After taking, the sherry is cooled, clarified, and filtered. Maturation is then required and is usually carried out in oak barrels. Aging can last form 6 months to 3 years or more.

Port wines are produced by the premature termination of fermentation by addition of brandy. When the fermenting must is separated from the pomace by gravity, it is fortified with wine spirits containing about 77% alcohol, by volume. Most white ports are fortified when half the original sugar content has been fermented, except for semidry and dry white ports, which are fortified later. The type and duration of aging depend on the desired style of wine. Blending is used to achieve the desired properties of the wine. The final blend is left to mature in oak cooperage for several months prior to fining, filtration, stabilization, and bottling.

BRANDY PRODUCTION

Brandy is an alcoholic distillate or mixture of distillates obtained from the fermented juice, mash, or wine from grapes or other fruit (e.g., apples, apricots, peaches, blackberries, or boysenberries). Brandy is produced at less than 190 proof and bottled at a minimum of 80 proof. [In the United States, "proof" denotes the ethyl alcohol content of a liquid at 15.6°C (60°F), stated as twice the percent ethyl alcohol by volume.] Two types of spirits are produced from wine or wine residue: beverage brandy and "wine spirits."

In brandy production, the grapes are pressed immediately after crushing. There are major differences in the fermentation process between wine and brandy production. Pure yeast cultures are not used in the fermentation process for brandy. Brandy can be made solely from the fermentation of fruit or can be distilled either from the lees leftover from the racking process in still wine production or from the pomace cap that is leftover from still red wine fermentations.

In the United States, distillation is commenced immediately after the fermentation step, generally using continuous column distillation, usually with an aldehyde section, instead of pot stills. For a detailed discussion of the distillation and aging of distilled spirits, which include brandy and brandy spirit, refer to AP-42, Section 9.12.3, "Distilled and Blended Liquors." After distillation, the brandy is added in oak casks for 3 to 15 years or more. During aging, some of the ethanol and water seep through the oak and evaporate; so brandy is added periodically to compensate for this loss. Caramel coloring is added to give the brandy a characteristic dark brown color. After aging, the brandy may be blended and/or flavored, and then chilled, filtered, and bottled.

EMISSIONS AND CONTROLS[5–11]

Ethanol and carbon dioxide are the primary compounds emitted during the fermentation step in the production of wines and brandy. Acetaldehyde, methyl alcohol (methanol), *n*-propyl alcohol, *n*-butyl alcohol, *sec*-butyl alcohol, isobutyl alcohol, isoamyl alcohol, and hydrogen sulfide also are emitted but in much smaller quantities compared to ethanol emissions. In addition, a large number of other compounds are formed during the fermentation and aging process. Selected examples of other types of compounds formed and potentially emitted during the fermentation process included a variety of acetates, monoterpenes higher alcohols, higher acids, aldehydes and ketones, and organosulfides. During the fermentation step, large quantities of CO_2 are also formed and emitted.

Fugitive ethanol emissions also occur during the screening of the red wine, pressing of the pomace cap, aging in oak cooperage, and the bottling process. In addition, as a preservative, small amounts of liquified SO_2 are often added to the grapes after harvest, to the "must" prior to fermentation, or to the wine after the fermentation is completed; SO_2 emissions can occur during these steps. There is little potential for VOC emissions before the fermentation step in wine production.

Except for harvesting the grapes and possibly unloading the grapes at the winery, there is essentially no potential for particulate (PM) emissions from this industry.

Emission controls are not currently used during the production of wines or brandy. Five potential control systems have been considered, and three have been the subject of pilot-scale emission tests studies at wineries or universities in California. The five systems are (1) carbon

Table 1. Emission Factors for Wine Fermentation[a] Emission Factor Rating: E

Wine Type	Type of Control	Ethyl alcohol ($lb/10^3$ gal)	Methyl alcohol ($lb/10^3$ gal)	*n*-Propyl alcohol ($lb/10^3$ gal)	*n* Butyl alcohol ($lb/10^3$ gal)	sec-Butyl alcohol ($lb/10^3$ gal)	Isobutyl alcohol ($lb/10^3$ gal)	Isoamyl alcohol ($lb/10^3$ gal)	Acet-aldehyde ($lb/10^3$ gal)	Hydrogen sulfide ($lb/10^3$ gal)
Red (SCC 3-02-011-06)	None[b]	4.6[c]	0.0025	0.0034	5.5E-5	4.5E-5	0.0036	0.014	0.0027	0.0017
	Carbon adsorption[d]	0.17[c]	ND	ND	ND	ND	ND	ND	ND	ND
	Catalytic incineration[e]	1.1	ND	ND	ND	ND	ND	ND	ND	ND
	Wet scrubber[e]	0.056	ND	ND	ND	ND	ND	ND	ND	ND
White (SCC 3-02-011-05)	None[b]	1.8[c]	6.4E-4	0.0023	ND	ND	6.9E-4	0.0051	7.2E-5	0.0014
	Carbon adsorption[d]	0.092[c]	ND	ND	ND	ND	ND	ND	ND	ND
	Catalytic incineration[e]	0.15	ND	ND	ND	ND	ND	ND	ND	ND
	Wet scrubber[e]	0.083	ND	ND	ND	ND	ND	ND	ND	ND

[a] Emission factor units are lb/1000 gal of fermented juice produced. SCC, Source Classification Code; ND, no data.
[b] References 8–11.
[c] Emission Factor Rating: C
[d] References 8–10.
[e] Reference 8.

adsorption, (2) water scrubbers, (3) catalytic incineration, (4) condensation, and (5) temperature control. All the systems have disadvantages in either low control efficiency, cost effectiveness, or overall applicability to the wide variety of wineries.

Emission factors for VOC and hydrogen sulfide emissions from the fermentation step in wine production are shown in Table 1. The emission factors for controlled ethanol emissions and the uncontrolled emissions of hydrogen sulfide and other VOCs from the fermentation step should be used with caution because the factors are based on a small number of tests and fermentation conditions vary considerably from one winery to another.

The only emission factors for wine production processes other than fermentation were obtained for a 1982 test. These factors represent uncontrolled fugitive ethanol emissions during handling processes. The factor for fugitive emissions from the pomace screening for red wine (SCC 3-02-011-11) is 0.5 lb/1000 gal of juice. An ethanol emission factor for the pomace press is applicable only to red wine because the juice for white wine goes through the pomace press before the fermentation step. The emission factor for red wine (SCC 3-02-001-12) is 0.02 lb/ton of pomace. Although fugitive emissions occur during the bottling of both red and white wines, an emission factor is available only for the bottling of white wine. The factor for white wine bottling (SCC 3-02-011-21) is 0.1 lb/1000 gal of wine. All these factors are rated E. These emission factors should be used with extreme caution because they are based on a limited number of tests conducted at one winery. There is no emission factor for fugitive emissions from the finishing and stabilization step (aging).

There are no available data that can be used to estimate emission factors for the production of sweet table wines, dessert wines, sparkling wines, or brandy.

REFERENCES

1. R. S. Jackson, *Wine Science: Principles and Application*, Academic Press, San Diego, CA, 1994.
2. M. A. Amerine, "Wine," in *Kirk — Othmer Encyclopedia of Chemical Technology*, Third Edition, Volume 24, John Wiley and Sons, New York, 1984.
3. J. A. Heredia, "Technical Assessment Document on Ethanol Emissions and Control from California Wineries," Master of Science Dissertation, California Polytechnic State University, San Luis Obispo, CA, June 1993.
4. M. A. Amerine et al., *Technology of Wine Making*, Fourth Edition, AVI Publishing Company, Westport, CT, 1980.
5. G. C. Miller, et al., "Loss of Aroma Compounds in Carbon dioxide Effluent during White Wine Fermentation," *Food Technol. Aust.* **39**(5), 246–49, 1987.
6. Written communication from Dean C. Smith, California Air Resources Board, Sacramento, CA, to Mark Boese, San Joaquin Valley Unified Air Pollution Control District, Fresno, CA, November 1, 1994.
7. EAL Corporation, "Characterization of Ethanol Emissions from Wineries," Final Report, California Air Resources Board, Sacramento, CA, July 1982.
8. *Ethanol emissions and Control for Wine Fermentation and Tanks*, Report # ARB/ML-88-027, California Air Resources Board, April 1988.
9. D. F. Todd et al., "Ethanol Emissions Control from Wine Fermentation Tanks Using Charcoal Adsorption: A Pilot Study," California Air Resources Board, published by California Agricultural Technology Institute, March 1990.
10. *Ethanol Emissions Control from Wine Fermentation Tanks Utilizing Carbon Adsorption Technology*, Akton Associates, Martinez, CA, June 1991.
11. Written communication from Arthur Caputi, Jr., E&J Gallo Winery, Modesto, CA, to Maria Lima, San Joaquin Valley Unified Air Pollution Control District, Fresno, CA, December 14, 1992.

FISH PROCESSING

Anthony P. Bimbo

World landings of fish and shellfish have been growing at a rate of about 9% per year since 1953 and passed the 100 million ton mark in 1992. Upon closer examination of the landings, however, we find that ocean landings have actually leveled off and most of the growth in the world catch since 1984 is coming from aquaculture. World landings are characterized by the intended end use of the fish caught, which is broken down into 6 broad categories. The 10-year average disposition of the world catch by use is shown in Figure 1. In general terms, except for the fresh category, each of the other categories requires a certain level of processing, which in turn generates by-products that must be further processed.

The processing of fish is divided into two main categories. Freshly caught fish such as salmon and tuna are either frozen or canned for human consumption. The by-product material from these operations is processed by "fishmeal plants," which are somewhat similar to the rendering plants for meat and poultry processing. The products from these plants are used for pet food, land animal, and aquaculture feeds.

At fish freezing plants, the fresh fish is filleted and the pieces trimmed to size and packaged for freezing either in the raw state or after cooking. Fish canneries use one of the two basic processes: wet-fish and precooked fish.[1]

The wet-fish method is used to preserve salmon, mackerel, sardines, and similar species that are collected locally and are brought to the cannery quickly. The distinctive feature of the wet-fish process is the complete removal of heads, tails, and viscera before cooking. The raw, trimmed fish are cooked directly in unsealed cans. These cans are drained of any liquid, and then sealed and pressure cooked before shipment. In some cases, the drained liquid is replaced by oil, water, or a sauce.

The precooked process is used for larger species of fish such as tuna. Whole eviscerated fish are cooked in live-steam-heated chambers operated at about 5 psig pressure. After cooking, the flesh is cooled so that it becomes firm before further handling. A little more than one-third of the raw tuna weight is canned for humans and pets. The remaining skin, bone, viscera, and other waste parts are processed in fishmeal plants.

Since the odor emissions from plants that process fresh fish for human consumption and pet food are considerably less than those associated with fishmeal plants, the remaining discussion is devoted to the fishmeal process and associated odor problems.

PROCESS DESCRIPTION

Fish used for the production of fishmeal and oil comes from one of several categories: (1) fish caught primarily for reduction

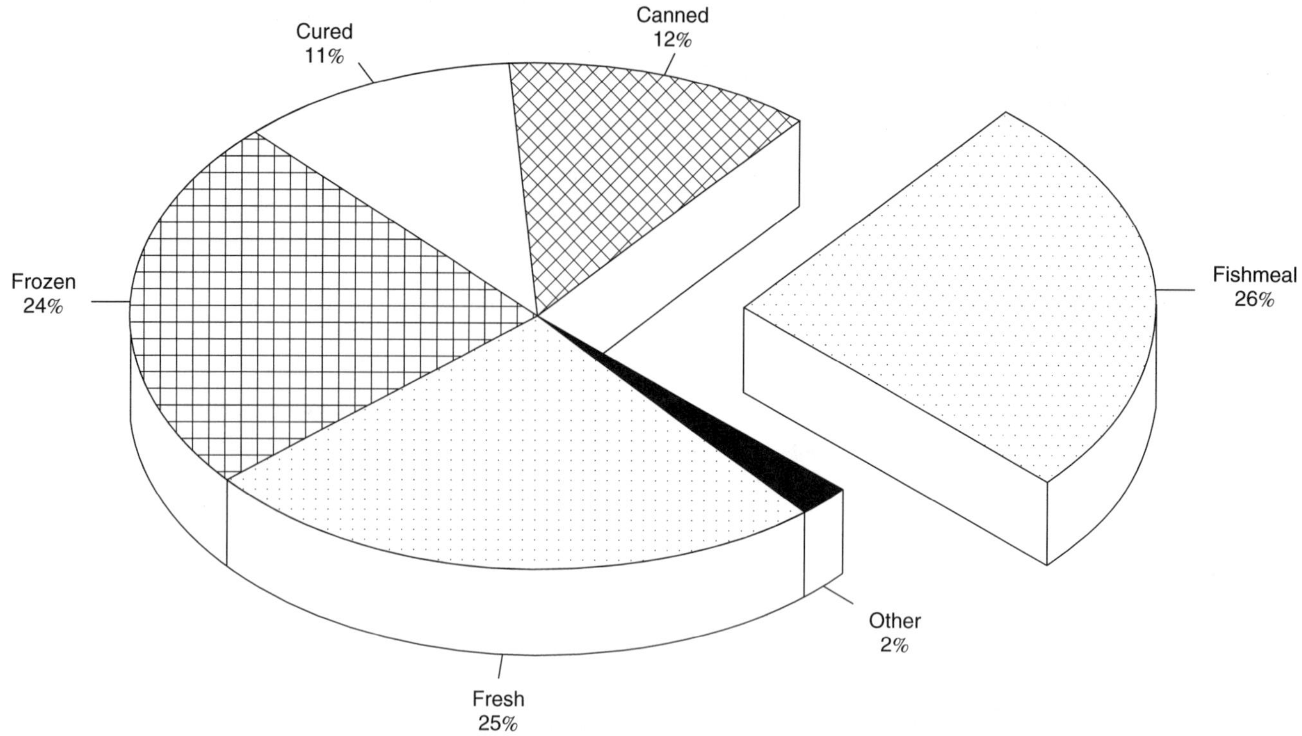

Figure 1. Ten-Year Average Disposition of Fish Landings by End Use.

to fishmeal and oil, such as the U.S. menhaden; (2) by-catch from other fisheries, such as the shrimp fishery; and (3) fish trimmings and cuttings from edible fisheries, such as cutting waste from filleting operations (catfish), roe operations, waste from surimi production, and tuna cannery waste. These categories are similar in that the fish are either considered industrial because of excessive bones or high oil content, or the trimmings cannot be used for edible purposes and present a disposal problem. In these circumstances, the raw material is converted into valuable feed and industrial products and often makes its way into the food supply indirectly by conversion to edible protein in livestock and farmed fish.[4] In the fish meal process, three main product streams result from the raw material: fish solids known as fishmeal, marine oil, and stickwater concentrate (solubles). It requires slightly over 3 pounds of raw fish to produce 1 pound of product, including the meal, oil, and stickwater concentrate (solubles).

Table 1 illustrates 6-year production statistics for fish meal, marine oil, and fish solubles in the United States.[2] The predominant fish processed in the United States is menhaden, which accounts for over 80% of the total. Shellfish account for less than 5%.

Figure 2 provides a flow diagram of the wet reduction (rendering) process for a fishmeal plant. As shown, the total operation includes the unloading of whole fish from fishing vessels into storage bins or silos for a short time before the various processing stages of cooking and pressing take place.

Unloading of whole fish from boats is divided into dry and wet methods. Dry unloading methods include the use of a bucket elevator and conveyor, an air suction method that moves fish in a current of air, and a vacuum method employing a vacuum pump and a rotary valve to maintain a proper seal. Direct pumping without water is also possible to convey fish in a sealed pipeline. The wet unloading method uses various types of pumps (centrifugal and positive displacement) as well as pressure-vacuum pumps with water as the transporting medium to convey the fish from the boat to the plant. The major difference between the pumps is the ratio of water to fish that is needed to move the fish to the plant. The water is then separated at the plant and recycled back to the vessel for reuse and eventual processing or it is discharged after screening.

Table 1. U.S. Production[a] of Industrial Fishery Products

Year	Fish Meal	Marine Oil	Stickwater Concentrate (Solubles)
1992	645	185	93
1993	751	293	127
1994	808	292	147
1995	667	242	90
1996	643	248	82
1997	725	283	144
1998	593	223	NA

[a] In million pounds.

Source: Fisheries of the United States 1998. NOAA National Marine Fisheries Service, U.S. Department of Commerce, Current Fishery Statistics No. 9800, 1999.

The fishmeal process typically is a continuous operating system. The following description is divided into two process phases, solids flow and liquid flow.[3,4]

Solids Flow

Cooking. Raw fish enter the cookers, which are horizontal cylindrical vessels equipped with steam-heated jackets and steam-heated screws with hollow flights. The cooking step coagulates the fish protein so that liquids and solids can be separated mechanically. Fat cells are also ruptured, releasing

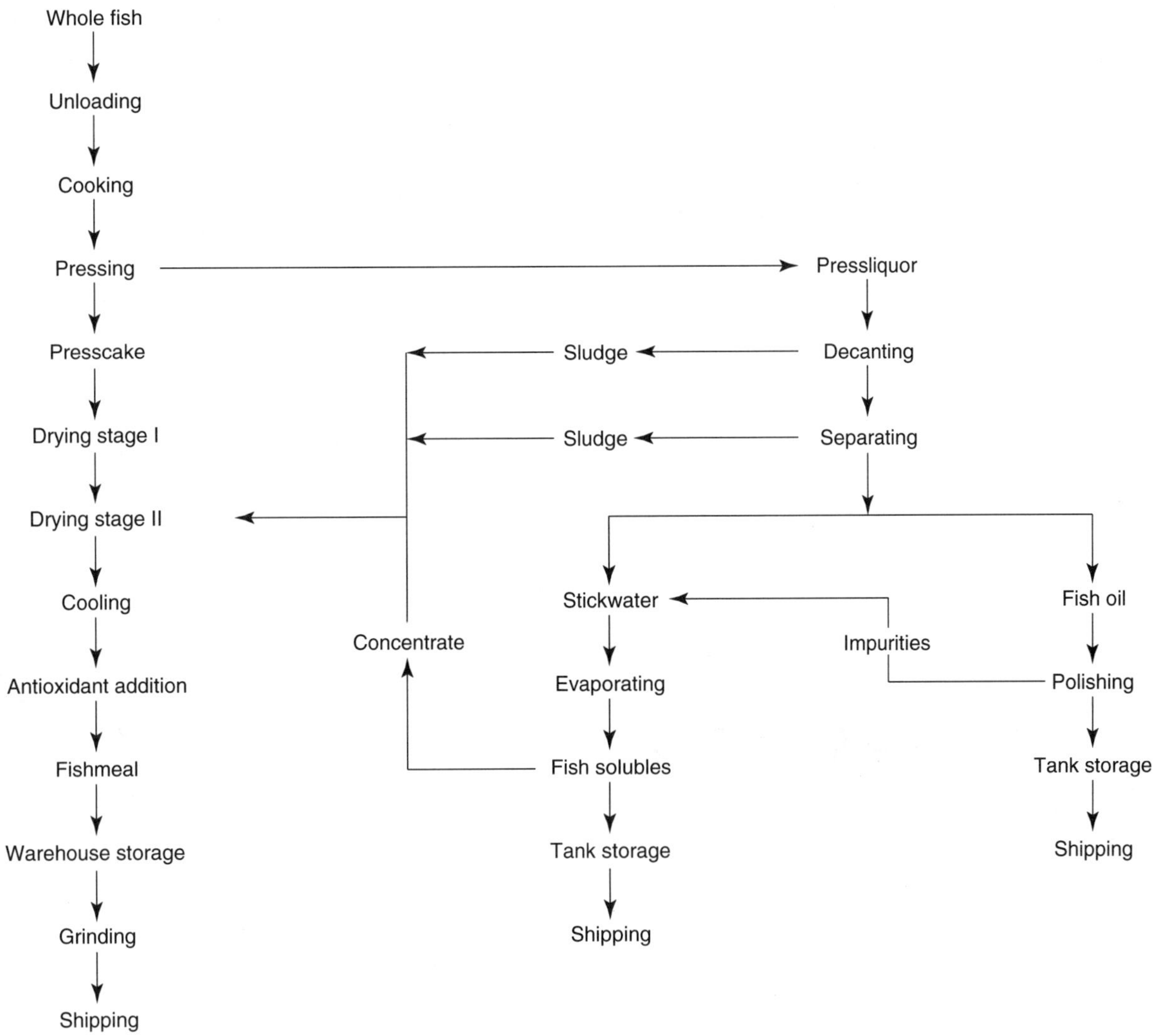

Figure 2. Typical Fishmeal Process.

the oil into the liquid phase. A typical cooking process heats the fish for 10–15 min at 90°C.

Pressing. The cooked fish are conveyed to the press by an inclined screw conveyor with perforations for dewatering the cooked material. The purpose of pressing is to separate the liquid phase from the remaining solids in the cooked fish. Both continuous single- and twin-screw presses are used in the fishmeal industry. The single-screw press is designed with a taper and exerts an increasing pressure on the cooked fish by reducing its volume as it passes through the press. A problem may be experienced with a single-screw press on soft fish, which can slip through the press without being processed. This difficulty may be minimized by the use of a twin-screw press.

Drying. The pressed fish solids are then dried to produce fish meal. Drying reduces the moisture content in order to prevent microbial decomposition. The presscake fed to the dryer normally averages about 50% moisture, and the dried fishmeal has a moisture content varying from 5% to 10%. Two main types of dryers are used: direct-fired rotary units and indirect rotary dryers.

The conventional direct-fired or flame-contact rotary dryer mixes the hot gases of combustion from the dryer firebox with a larger volume of fresh air before contact with the wet meal inside the rotating section. The exhaust air from this dryer usually passes through a cyclone to recover the entrained fish meal product. In this drying process, known as convection drying, the drying air and moisture laden solids flow concurrently, coming into direct contact with each other. The drying air inlet temperature is maintained below 600°C. The residence time of the solids in the dryer is about 15 min. This type of dryer is shown in Figure 3.

The indirect rotary dryer relies on heated surfaces that provide conduction drying for moisture removal. These heated surfaces consist of a series of parallel hollow disks or coils mounted vertically on a rotating hollow shaft through which passes either hot air or steam under 75–100 psig pressure. A small volume of fresh air passes countercurrent to the solids flow for moisture removal. The steam temperature generally ranges from 160 to 170°C and the temperature of the exiting product is usually 85–95°C. The retention time in this dryer is about 30 min. A typical rotary disk steam dryer is shown in Figure 4.

Another type of indirectly heated dryer combines certain features of both the direct-fired and the indirect rotary dryers.

Figure 3. A Direct-Fired or Flame-Contact Dryer.

Figure 4. A Rotary Disk Steam Dryer.

In this dryer, the combustion gases from the furnace pass through a heat exchanger to heat another stream of air that subsequently enters the rotary dryer, coming into direct contact with the solid particles, and removes moisture by convection. This type of indirect dryer can be referred to as an indirect rotary convection dryer to differentiate it from the indirect rotary conduction dryer previously described. This type of dryer system is shown in Figure 5.

Low-temperature, gentle drying systems have become more prevalent in the fishmeal industry during the past 8 years.

Figure 5. An Indirect Rotary Convection Dryer.

These improved drying methods include the increased use of indirect hot-air, vacuum, and fluid-bed dryers.[5] For example, the vacuum dryer, which is steam heated, is operated under a reduced pressure, resulting in correspondingly lower product temperatures of 50–60°C. Sand and Burt[5] compare the direct-fired dryer with various types of indirect dryers (steam, hot air, and vacuum) regarding certain criteria, including product quality, capital investment and operating costs, odor control, energy consumption, process control, maintenance, and space requirements. Wiedswang[6] compares the theoretical drying requirements for various types of dryers: direct-fired rotary, indirect rotary conduction, indirect rotary convection, and indirect vacuum.

Two stages of drying (I and II) are shown in Figure 2. A current trend exists to utilize both the direct-fired and or the indirect hot-air rotary dryer and steam dryers in two stages. The direct-fired or indirect hot air unit is used in the first stage to reduce the moisture in the press cake to 20–25% moisture. The steam unit in the second stage completes the moisture removal. The stickwater concentrate and the sludge removed by the decanter centrifuges (in the liquid-flow part of the process) are added to the feed inlet in the second-stage dryer, since this additional moisture load is better assimilated at this point than in the feed inlet to the first-stage dryer.

Cooling, Storage, and Shipping. The dried meal is normally cooled in a rotary unit through which ambient air is passed countercurrent to the meal flow. The meal is then ground to pass 100% through a U.S. No. 7 Standard Screen. Fishmeal must be stored in weatherproof, well-ventilated areas. Fishmeal can be stored in bulk in sheds of single- or multiunit construction with concrete floors and walls. Besides bulk storage, fishmeal may be stored and shipped in multiwall paper, burlap, or woven plastic bags. A new type of bulk bag is capable of holding about 1 ton of product.

Liquid Flow

During the pressing operation, two intermediate products result: presscake and pressliquor. As described under "Solids Flow," the presscake is dried to produce fishmeal. The pressliquor squeezed from the cooked fish contains coarse particles of fish and bone that must be removed before the liquor is centrifuged. These solids are removed by passing the liquor over a vibrating screen with 5–6-mm perforations. The recovered solids are sent back with the presscake to be dried.

Separation of the oil–water mixture is normally carried out in three separate stages, or types of centrifuges: (1) a decanter to remove suspended solids; (2) a separator to recover oil from the pressliquor, which becomes the stickwater; and (3) a polisher to remove water and other impurities from the oil. The stickwater is evaporated in a multiple-effect evaporator to produce a stickwater concentrate containing 50% dry substance. This concentrate is recycled back to the dryer for further moisture removal or is sold as a high-protein concentrate for animal feed.

Decanting (Solids Removal). The decanter centrifuge consists of a horizontal cylindrical bowl that rotates and contains a rotating screw conveyor or scroll. The press liquor is fed into the bowl, and the solids are forced to the outer wall. They are scraped off and conveyed out of the centrifuge by the scroll, which rotates at a slower speed than the bowl. Before discharge, the solids pass through a dewatering zone that tends to thicken or concentrate the solids. This sludge is returned to the dryers.

Separation. The liquid discharge from the decanter passes through a heat exchanger that raises its temperature to about 95°C before it enters the separator centrifuge. The separator is a three-phase centrifuge in which the feed liquor is separated into oil, stickwater, and a solids phase. The centrifuge consists of a series of cone-shaped disks rotating about a vertical axis and enclosed in a circular bowl. The feed liquor enters at the top center of the bowl, and a sensor in the bowl is able to control the separation cycle by automatically discharging the accumulated solids. Since the main purpose of the separator is to maximize the recovery of oil, its operation is adjusted accordingly.

Oil Polishing. The design of the polisher centrifuge is similar to that of the separator. However, its purpose is to polish or purify (ensure maximum removal of impurities from the oil). As a result, the operation of the polisher is different from that of the separator. In order to obtain the desired purity, hot water equal to 10% of the oil volume is added to the feed liquor. The water and solids removed by polishing are added to the stickwater.

Evaporation of Stickwater. The water phase from the separator centrifuge is known as stickwater and usually contains 6–12% dry substance consisting of soluble proteins, fat, minerals, and vitamins. The stickwater is concentrated by means of a multiple-effect evaporator that may consist of two, three, or four stages.

For a three-stage evaporator, the dilute stickwater enters the first stage, with a more concentrated solution passing on to the second stage and a final stickwater concentrate of 50% dry substance discharging from the third stage. Steam under pressure is injected into the jacket at the first stage. The vapor from evaporation in the first stage passes on to the second-stage jacket, and the vapor from the second stage passes on to the third-stage jacket. The vapor from the third stage is condensed and a vacuum is applied to the third-stage evaporator, which reduces the boiling points of evaporation in stages 2 and 3.

Most recent multiple-effect evaporators operate according to the failing film principle. A typical falling film evaporator is shown in Figure 6. Because the inside of the vertical evaporator tubes becomes gradually fouled due to deposits of proteins and calcium phosphate, it is necessary to circulate a caustic soda solution through the evaporator periodically in order to maintain production.

Since the raw material processed in fishmeal plants contains 70–80% moisture, considerable fuel energy is required to evaporate this water. The recent trend in European and many U.S. fishmeal plants has been to provide systems that require less energy.[7,8] This is accomplished by utilizing the exhaust vapors at atmospheric pressure from the indirect steam-heated or hot-air-heated dryers for moisture removal in the multiple-effect evaporator, which is operated under vacuum.

Fishmeal is used in the feeds of poultry, pigs, ruminants, fish, crustaceans, pets, and fur-bearing animals because it increases productivity and improves feed efficiency. Fish oil is a major source of energy and provides the essential omega-3 fatty acids that are required by many species. The omega-3 fatty acids also positively affect the immune system of animals, improve the fatty acid composition of eggs, and may have an effect on bone development. Some of these diseases cost the poultry industry several hundred million dollars in lost productivity per year. Currently, more fishmeal is produced under improved drying methods and process control. This type of product is used to feed salmon, trout, shrimp, and baby pigs. Fish oil is also used as an industrial drying oil, and in margarine, shortening, and bakery fats.

Figure 6. A Falling-Film Evaporator System.

AIR EMISSIONS CHARACTERIZATION

The largest odor emission source in fishmeal plants normally is at the dryer, especially direct-fired dryers that require significant volumes of air for moisture removal. Lesser emissions occur at the cookers, evaporators, and other parts of the process. Table 2 provides an estimate of the odor emissions from various stages of the fishmeal process.[9]

An extremely important consideration in the handling and storage of raw fish aboard a vessel concerns its temperature. The deterioration of raw fish begins rather quickly after the catch because of the bacteria and enzymes in the digestive tract. It has been determined for raw fish that the formation of volatile amines doubles with an increase in temperature of 6°C, and the odor unit profile increases tenfold.[7] It is essential

Table 2. Odor Emissions from the Fishmeal Process

Stage of Process	Total Emissions (%)
Dryers	60–80
Cookers	10–20
Raw material conveying	10–20
Pneumatic meal conveying	2–5

Source: Ref. 9.

to keep the raw fish at a low enough temperature to prevent this deterioration. As a result, fishing vessels are equipped with ice to be mixed with the fish or with refrigerated seawater or fresh water circulated through their holds to maintain the temperature of the fish as low as possible.

The storage of the raw material before processing is a definite source of odor and can vary from 1,000 to 100,000 odor units per cubic foot ASTM syringe method in the head space above the material, depending on its degree of deterioration.[10] These odor emissions from open areas, unless they are enclosed, are virtually impossible to control.

Where fugitive odor emissions from fish plants may affect the surrounding community more severely, a plant ventilating air system with subsequent odor control equipment may be necessary. Other potential sources of fugitive emissions include the leakage from certain process equipment, such as presses and open screens, and at the transfer point from one stage of the process to the next.

The odor emissions from the cookers may vary from 5,000 to 100,000 odor units per cubic foot (ASTM syringe method), depending on the state of decomposition of the raw material. The cooker is normally vented to odor control equipment at a volumetric emission rate varying from 100 to 1,000 ft^3/min.[1] Although the multiple-effect evaporator system may remove as much as 50% of the water evaporated from the total process, its operation normally is under vacuum, and the odor emission is slight.

Table 3 illustrates the odor emissions from several rotary direct-fired dryers without odor control measures being applied.[11] It should be noted for dryer B that the odor units per standard cubic foot increased from 1500 to 4000 when the dryer discharge temperature increased from 240°F to 300°F. This illustrates the importance of controlling process conditions adequately in order to minimize odor generation. The exhaust from direct-fired dryers normally averages about 200°F, and its moisture content ranges between 15% and 25% by volume.

Exhaust from indirect steam dryers is 30–45% lower in volume than from comparable direct-fired units. Steam dryers typically have a 25% moisture content in their exhaust, compared with about 15% moisture from a direct-fired unit processing the same material. Odor concentrations from steam dryers are generally in the same range as those from direct-fired units when fresh fish waste material is processed under proper operating conditions. Steam dryers are less likely to overheat the meal and result in excessive odor concentrations.

Table 3. Typical Odor Emissions from Rotary Fish Meal Dryers

Dryers	Feed Rate (tons/h)	Type of Fish	Temperature at Dryer Discharge (°F)	Exhaust Gas Volume (scf/min)[a]	Odor Units[b] per scf
A	10	Tuna	220	18,500	1500
A	15	Mackerel	220	18,500	1500
B	70	Tuna	220	9,000	700
B	10	Tuna	240	10,000	1500
B	14	Tuna	300	8,000	4000
C	9	Tuna	200	17,000	2500

[a]Standard cubic feet per minute.
[b]Odor units per standard cubic foot (70°F and 14.7 psia), ASTM syringe method.

In many European countries, direct-fired dryers that use relatively large volumes of air and higher temperatures have been replaced by indirect steam-heated dryers because a lesser volume of fresh air is circulated through this type of dryer.[10] As a result, the operation of drying equipment that reduces odor volumetric emissions from the process allows the odor control systems to be more effective.

The composition of the odorous compounds present in the fishmeal process is variable due to the various species of fish to be considered, as well as the degree of spoilage that has occurred and the actual process operating conditions being used. The concentration of gases in the head space above the storage of raw materials has been found to be quite high.[12] For example, inorganic gases such as hydrogen sulfide and ammonia were reported at levels up to 2000 and 1000 ppm, respectively. The threshold limiting values (TLVs) for H_2S and NH_3 are specified at 10 and 50 ppm, whereas their odor detection thresholds are reported at 0.00021 and 21.4 ppm, respectively.[13] Also present were organic amines such as trimethylamine, with an odor detection threshold of 0.00021 ppm.[13] Because of these low thresholds, even small volumes of these gases escaping to the surrounding atmosphere are capable of causing complaints.

With regard to dryer exhaust emissions, the use of gas chromatography and mass spectrometry[14] indicated the presence of sulfur compounds, such as hydrogen sulfide, carbon disulfide, carbonyl sulfide, and methyl and *n*-propyl mercaptans. In addition to ammonia, the only amine present was trimethylamine. Analyses of the cooker exhaust indicated that a number of sulfur compounds were present, but only one amine, trimethylamine.

Dust emissions from the fishmeal process are usually limited to the direct-fired dryers and to the grinding and pneumatic conveying of the dried fishmeal. There are relatively few fines in ground fishmeal. A sample of the dried product was collected from a pneumatic conveyor transporting ground fishmeal. Only 0.6% by weight of the sample was less than 5 μm in diameter and 1.4% was less than 10 μm in diameter.[1] The cyclone collecting the pneumatically conveyed fishmeal was found to be better than 99.9% efficient, with an exhaust dust concentration of less than 0.01 gr/scf.

AIR POLLUTION CONTROL MEASURES

Since the process emissions from the cookers and dryers contain considerable moisture at temperatures of about 200°F (94°C), the necessary means should be provided to remove most of this moisture and to cool the process exhaust air before further odor control treatment. Also, there may be dust particles in the dryer cyclone exhaust that should be removed before effective odor control measures can be applied. This is normally accomplished by either direct contact or indirect water-cooled condensers. The indirect type includes the shell-and-tube condenser. The direct contact type includes cocurrent-flow venturi scrubbers and countercurrent-flow spray-type scrubbers. They are often used where large volumes of seawater are available.

In addition to condensing water vapor and cooling the gas, a specific degree of odor reduction is obtained with direct-contact condensers since certain odorous compounds, such as

NH_3, H_2S and $N(CH_3)_3$, are quite water soluble. Table 4 shows the ability of direct-contact condensers to reduce the inlet odor level.[10]

Two basic methods of odor control have been applied to the process emissions from cookers and dryers after condensing the water vapor: boiler incineration by direct-flame oxidation and wet scrubbing by chemical oxidation or the use of other scrubbing agents.

Thermal oxidation of the odorous process air is readily achieved in a boiler firebox operated at a temperature ranging from 650°C to 800°C (1200°F to 1475°F). However, the amount of odorous air to be treated must be compatible with the combustion air required for the boiler.

Where indirect steam-heated dryers are used, it may be feasible to use the dryer exhaust as combustion air after condensing the water vapor. However, if a direct-fired dryer is used, it normally is more feasible to use wet scrubbing. The fishmeal industry in Europe uses the process air as a source of secondary combustion air in the boiler.[10] Table 5 illustrates the odor reduction obtained by boiler incineration.

As fuel costs have increased significantly, particularly in Europe, wet scrubbing systems have become more prevalent. Also, if fugitive odor emissions from the raw material storage area and the processing area are significant and cannot be captured by the exhaust duct system, a plant ventilating air system with adequate wet scrubbing equipment is required. Two wet scrubbing systems located in Norway and Denmark are described by Hansen.[15]

Depending upon the type of dryer (steam-heated disk units versus direct-fired units), the wet scrubbing system treats 500–1500 m^3 of dry air at standard conditions per ton of raw material processed. If plant ventilating air is required, the capacity of the scrubber system may increase to 3000–4000 m^3 of dry air at standard conditions per metric ton of raw material processed.

Table 4. Odor-Removal Ability of Direct Condensers

System	Odor Units/scf[a] Inlet	Odor Units/scf[a] Outlet	Percent Removal	Air Temperature (°C) Inlet	Air Temperature (°C) Outlet
1	138,000	19,000	86		
	18,000	3,600	80	92	35
2	52,000	18,000	65	79	28

[a]Odor units per standard cubic foot, ASTM syringe method.

Source: Ref. 10

Table 5. Odor-Removal Ability of Boiler Incineration

System	Odor Units/scf[a] Inlet	Odor Units/scf[a] Outlet	Percent Removal
1	144,000	5,600	96
	18,000	840	95
2	4,750	850	82
	3,600	600	83
3	18,000	1,200	93
	9,000	1,000	89

[a]Odor units per standard cubic foot, ASTM syringe method.

Source: Ref. 10

Seawater scrubbers are used to condense water vapor and cool the incoming gas stream. These may be venturi, packed-tower, or spray-type scrubbers and are normally used as the first stage of a two-stage wet scrubbing system. The exhaust air from this first-stage scrubber passes on to the second-stage chemical oxidation scrubber, which normally contains packing. In the past, the packing consisted of 1- or 2-in. Rasching rings. Currently, more efficient types of packing are available. The use of sodium hypochlorite solution as a wet scrubbing agent in the second stage is preferred over potassium permanganate and hydrogen peroxide. Table 6 illustrates the odor reduction results obtained with two stages of scrubbing.[15]

Another fishmeal plant in Norway is described by Onarheim and Utvik.[16] It uses two separate odor control systems—one exclusively for the dryer exhaust and the other for treating the remaining process odor emissions and the plant ventilating air. This plant processes 700 tons of raw fish (mackerel, herring, fish waste, etc.) every 24 h.

The exhaust air from three rotodisc driers passes to a packed tower scrubber operating at a linear velocity of 2 m/sec (6.5 fps) with 6 m^3/min (1600 U.S. gpm) of seawater. The scrubber outlet passes through an indirect heat exchanger, which uses the steam vapors from the dryers as a heating medium to preheat the scrubber exhaust. Most of this air is recirculated back through the dryers and the remainder is incinerated in the boiler firebox.

The second odor control system, which treats the remaining process emissions and plant ventilating air, consists of two stages of scrubbers. A packed-tower scrubber through which seawater is circulated precedes a second packed tower through which sodium hypochlorite solution is recirculated. The first-stage scrubber consumption of seawater is 2.5 m^3/min (660 U.S. gpm). In the second stage, the consumption of NaOCl is estimated at 13–19 pounds per 1000 metric tons of raw material. Amines and ammonia could not be detected analytically in the second-stage exhaust.

In summary, fishmeal plants in Europe and the United States have developed an odor control capability to treat their emissions and avoid becoming local nuisances. Process emissions are controlled by a combination of wet scrubbing with seawater and boiler incineration. Plant ventilating air is treated by wet scrubbing.

Biofilter technology developed during the past 15 years in Europe and elsewhere has been applied to the treatment of odor emissions from fishmeal plants.[17] Biofilters consist of large beds of porous media that are capable of adsorbing odorous gaseous compounds and reducing these by aerobic microbial action to nonodorous components. This technology has been tested by fishmeal plants in the United States, but so far the technology has not been applied.

Table 6. Odor Removal with Two Stages of Wet Scrubbing

Category of Scrubber	D/T[a]
Seawater scrubber inlet	200,000
Seawater scrubber outlet	20,000
Chemical scrubber outlet	1,000

[a]Odor dilution to threshold by IITRI dynamic olfactometer.

Source: Ref. 15

The Torry Research Station[18] in Scotland has developed a set of recommended practices to be used to reduce odor emissions during the production of fishmeal. These practices include the delivery and handling of raw material, processing, handling and storage of fishmeal, plant and equipment design, plant management, and maintenance.

One final thought on the subject should be mentioned. Quality of the raw material is the key element in both water and air pollution issues in fishmeal plants. As the fish age, the compounds developed during decomposition are volatile and some are water soluble. If not treated properly, these compounds are either discharged into the air or if scrubbed become part of the water discharge from the plants. In other words, an air pollution problem can be quickly changed into a water pollution problem.

Acknowledgments

The assistance of Mr. William H. Prokop, P.E, of Prokop Enviro Consulting, Deerfield, IL, is greatly appreciated.

REFERENCES

1. "Fish Canneries and Fish Reduction Plants," *Air Pollution Engineering Manual*, 2nd ed., AP-40, U.S. Environmental Protection Agency, Research Triangle Park, NC, May 1973, pp. 760–70.
2. Fisheries of the United States, 1998. NOAA National Marine Fisheries Services, U.S. Department of Commerce, Current Fishery Statistics No. 9800, 1999.
3. A. P. Bimbo, "Production of Fish Oil," *Fish Oils in Nutrition*, M. E. Stansby, Ed. Van Nostrand Reinhold, New York, 1990, p. 155.
4. A. P. Bimbo, "Fish Meal and Oil," *The Seafood Industry*, Roy Martin and George Flick, Eds. Van Nostrand Reinhold, New York, 1990, pp. 325–50.
5. G. Sand and J. Burt, "Which Kind of Dryer Is Best," *Processing Bulletin No. 2*, International Association of Fish Meal Manufacturers, Herts, England, July 1987, pp. 1–11.
6. H. Wiedswang, "Low Temperature Dryers," paper presented at the 29th Annual Conference of the International Association of Fish Meal Manufacturers, Lima, Peru, November 1988.
7. P. Fosbol, "Production of Fish Meal and Fish Oil," INFOFISH International, May 1988, pp. 26–32.
8. A. 0. Utvik, "Modern Fish Oil and Meal Processing," 75th Annual Meeting of American Oil Chemical Society, Dallas, TX, 1984.
9. W. Schmidtsdorff, Symposium, International Exhibition, Jonkoping, Denmark, 1974.
10. J. Wignall, "Odor Pollution in the Fish Meal and Oil Industry," *News Summary No. 44, Special Issues on Processing*, International Association of Fish Meal Manufacturers, Herts, England, September 1978, pp. 44–65.
11. J. L. Mills, J. A. Danielson, and L. K. Smith, "Control of Odors from Inedible Rendering and Fish Meal Reduction in Los Angeles County," 60th Annual Meeting of APCA, Cleveland, OH, June 1967.
12. J. B. Dalgaard, F. Denker, and B. Fallentin, *Br. J. Indus. Med.* **29**, 307 (1972).
13. G. Leonardos, D. Kendall, and H. Barnard, "Odor Thresholds for 53 Commercial Chemicals," *J. Air Poll. Control Assoc.* **19**(2), 91 (1969).
14. Y. Hoshika, S. Kadowaki, 1. Kojima, et al., *Oil Chemistry* (Japan), **24**(4), 233 (1975).
15. K. W. Hansen, "Odor Control in the Fish Meal Industry," *News Summary No. 44, Special Issues on Processing*, International Association of Fish Meal Manufacturers, Herts, England, September 1978, pp. 66–74.
16. R. Onarheim, and A. 0. Utvik, "A System of Complete Effluent Control," *News Summary No. 44, Special Issues on Processing*, International Association of Fish Meal Manufacturers, Herts, England, September 1978, pp. 75–80.
17. G. Leson and A. M. Winer, "Biofiltration: An Innovative Air Pollution Control Technology for VOC Emissions," *J. Air Waste Manag. Assoc.* **41**(8), 1048 (1991).
18. "Reducing Odour in Fish Meal Production," *Torry Advisory Note No. 72*, Torry Research Station, Aberdeen, Scotland, May 1979, pp. 1–7.

LIVESTOCK FEEDLOTS*

John M. Sweeten, Calvin B. Parnell, Brent W. Auvermann, Bryan W. Shaw, and Ronald E. Lacey

The purposes of this section are to identify and assess (1) the relationship of concentrated livestock production facilities to air pollutant concentrations and emissions, and (2) alternative technologies and management practices to reduce potential for air contamination from livestock and poultry operations.

MANURE PRODUCTION AND DISTRIBUTION IN THE UNITED STATES

Total U.S. manure production (dry basis) for all livestock and poultry species were calculated by Van Dyne and Gilbertson.[1] They estimated that total annual manure production was 112 million dry tons and the manure contained 8.2 billion pounds of nitrogen, 2.1 billion pounds of phosphorus, and 4.8 billion pounds of potassium. Volatilization, leaching, and runoff losses reduced the dry weight by 10%, total nitrogen by 36%, phosphorus by 5%, and potassium by 4%. Recoverable (collectible) manure and nutrients were estimated by subtracting the portion of materials voided on pasture areas. Collectible (economically recoverable) dry manure was estimated at 52 million tons, and the fractions from various animal species were dairy cattle, 39%; feeder cattle, 31%; hogs, 11%; laying hens, 6%; broilers, 5%; sheep, 3%; and turkeys, 2%.

The estimates[1] were based on an engineering practice standard adopted by the American Society of Agricultural Engineers[2] based on constituent production per unit weight of live animal. These standard values were updated in 1988 to reflect more recent research data.[3] In most cases, average values of dry manure and nutrients (pounds per day per 1000 pounds live weight) were revised upward, and standard deviations were calculated to reflect the degree of variability.

A more recent estimate was made of as-voided dry manure production in the United States from concentrated animal feeding operations (CAFOs), using the new ASAE standard values resulted in an estimate of 58 million dry ton/year.[4] This annual rate appeared to be divided almost equally among beef

* Revised from Chapter 13, "Food and Agricultural Industry," *Air Pollution Engineering Manual*, A. J. Buonicore and W. T. Davis, Eds., Air and Waste Management Association, Pittsburgh, PA.

cattle, dairy cattle, swine, and poultry, although large regional differences among species exist.[4]

PROCESS DESCRIPTION

Intensive Animal Production Systems

Intensive livestock production systems are regarded as "animal feeding operations," which are defined (for purposes of water pollution control) in U.S. Environmental Protection Agency (EPA) regulations for the feedlots point source category as areas where animals are "stabled or confined and fed or maintained for a total of 45 days or more in any 12 month period, and crops, vegetation, forage growth or post-harvest residues are not sustained in the normal growing season over any portion of the lot or facility."[5] The EPA definition is not specific as to animal species, type of confinement facility, or animal density, but essentially integrates these factors (along with climate and soils) into a single, visually determined criterion — absence of vegetation — which develops where manure production and/or animal traffic are sufficient to prevent germination or growth of forages during the normal growing season.

Major types of livestock and poultry production facilities, design information, and associated manure management systems have been described in other reports.[6–9] Roofed or total confinement facilities are common for poultry and swine and, to a lesser extent, dairy and beef production.[10] It is estimated that less than 2% of the cattle on feed in the United States are in housed lots.

Cattle Feedlots

Open feedlots (nonroofed) are the type of intensive confinement facilities most commonly used for beef cattle. They are also widely used for dairy, swine, and sheep production in the southwestern United States.

The United States had 10.0 ± 0.8 million beef cattle (monthly average) in feedlots being finished for slaughter during 1997.[11] Typically, these cattle range in live weight from 600 to 1200 pounds per head, averaging 900 pounds. Each animal fed in a normal 130- to 150-day fattening period produces approximately one dry ton of collectible manure solids, or about 2 dry tons of collected manure per year per head of feedlot capacity. The animal spacing per head varies according to rainfall and temperature, slope, and other factors from typically 100 to 125 ft^2 per head in arid areas (less than 15 in. annual rainfall) to 150–200 ft^2 per head in the southern central Great Plains (15 to 25 in. per year), which includes the largest cattle feeding states, and to 250 to 400 ft^2 per head in the eastern and northern Great Plains (25 to 35 in. per year).

Most of the manure deposited on the feedlot surface is compacted by cattle into a relatively moist manure pack of 35–50% moisture content (wet basis). At higher moisture contents, odors can develop, especially in warm weather.

The surface manure may become pulverized by cattle hooves during prolonged drying conditions to only 10–25% moisture wet basis. When surfaces are excessively dry, there is a potential for dust problems, and these problems have been most severe in the arid and semiarid cattle-feeding areas of Arizona, California, Texas, Oklahoma, New Mexico, and Kansas. Dust from cattle feedlot surfaces, alleys, and roads can annoy neighbors, irritate feedlot employees, create a traffic hazard on adjacent highways,[12] and possibly impair cattle performance.[13] Total particulate emissions are affected by feedlot area, cattle density in pens, wind speed, precipitation, and evaporation.[14] They may also be affected by feed rations, feeding frequency, and management programs.

Feedlot cattle can create high dust concentrations, especially for about 2–6 h around dusk when cattle activity increases and manure moisture content is low.[15] Dust is usually minimal in early morning. Under calm conditions, feedlot dust can drift over nearby highways and buildings.

Odors from Livestock Feeding Operations

Odors from livestock facilities are sometimes an annoyance and may affect the well-being of downwind neighbors. Public or private nuisance lawsuits can threaten the survival of an operation, and livestock producers need to minimize the emission of odorous compounds to the extent practical.

Odorous gases arise from feed materials (food-processing wastes and fermented feeds), fresh manure, and stored or decomposing manure. Research has been directed toward the control of volatile compounds released by the storage and treatment of manure. Odor is generally less objectionable from fresh manure than from anaerobically decomposing manure. Fresh manure evolves large quantities of ammonia, but it is generally not accompanied by other decomposition products that contribute the most objectionable characteristics. The types and concentrations of odorous compounds that evolve from manure treatment facilities are a function of the material as excreted, the configuration of the storage or treatment unit, presence or absence of oxygen, and the biological reactions occurring in the material.

Roofed confinement facilities typically have high odor-production potential, owing to the high animal density involved, the large inventory of manure frequently in storage, and limited rate of air exchange.[10] Manure-covered surfaces (e.g., building floors and animals), manure storage tanks beneath slotted floors, and anaerobic lagoons often used for manure storage and treatment are important odor sources.

Open feedlot surfaces that become wet, particularly in warm weather, support anaerobic decomposition with an associated large surface area for the evolution of odorous gases. Feedlot odor problems are most frequent in warm, humid areas and in feedlots constructed in areas of inadequate drainage or poor drying conditions.

Feedlot odor research conducted using a dynamic triangle forced-choice olfactometer for odor measurement determined that odor concentrations from wet feedlot manure pads were approximately 50–100 times greater than those of dry pads; odor concentrations and offensiveness peaked 2 days after the manure became wet; and wet pads produced more offensive odors than dry pads.[16–19] They concluded that a small area of wet feedlot pad can be responsible for generating a significant amount of odor, and poorly drained feedlots with manure buildup can create far more odor than well-drained feedlots that are regularly cleaned and dry rapidly.

AIR EMISSIONS CHARACTERIZATION

Particulate Emission Standards

In 1971, the EPA[20] promulgated primary and secondary national ambient air-quality standards for total suspended particulate matter (TSP). The primary standards were set at 260 μg/m^3 for 24-h average, not to be exceeded more than once per year, and an annual geometric mean of 75 μg/m^3. Secondary standards were set at 150 μg/m^3 for a 24-h sampling period, not to be exceeded more than once per year. The purpose of primary standards was protection of public health. As a consequence, primary standards were to be achieved regardless of cost within a specified period of time. The intent of secondary standards was to protect public welfare from known or anticipated adverse effects.

The EPA[20] replaced the TSP standards for all sources with a PM_{10} standard in 1987 based on particulate matter (PM) having mass median diameter (MMD) of 10 μm aerodynamic equivalent diameter (AED). In essence, the revision was based on the premise that finer dust particles needed to receive greater focus than coarser dust particles in protecting human health. The PM_{10} primary and secondary 24-h standards were changed to 150 μg/m^3 for a 24-h average with no more than one exceedance per year and 50 μg/m^3 annual arithmetic mean. EPA also replaced the secondary TSP standard with 24-h and annual PM_{10} standards that were identical to the aforementioned primary standards.

Two instruments (manufactured by Wedding and Associates and by Sierra Andersen) were accepted for PM_{10} measurement by the EPA. Other instruments or methods have been developed as well.[21]

A procedure developed by Raina and Parnell[22] involved use of a Coulter Counter to determine particle size distribution (PSD) of particulate collected with a high-volume sampler and, based on these measurements, mathematically derive the PM_{10} concentration. Their results suggest that, relative to direct measurement using a PM_{10} sampler, a more accurate concentration of PM_{10} can be obtained using a high-volume TSP sampler to measure the concentration of total suspended particulate matter and then multiplying this concentration by the fraction of PM_{10} determined from the mass versus particle diameter AED particle size distribution obtained using a Coulter Counter.

With increasing concerns for human health effects believed caused by fine particulate matter (respirable dust), the National Ambient Air Quality Standard (NAAQS) was again revised in July 1997 by EPA, adding new primary and secondary standards for $PM_{2.5}$ (particulate matter less than 2.5 μm, AED).[23] The new 24-h primary and secondary $PM_{2.5}$ standard is 65 μg/m^3, calculated as the 3-year average of the ninety-eighth percentile reading at each monitor. The annual standard is 15 μg/m^3 as the 3-year average of annual arithmetic means. In addition to the new $PM_{2.5}$ standard, the 1987 NAAQS for PM_{10} was left in place, except that the PM_{10} exceedance criterion for 24-h samples was changed to ninety-ninth percentile (i.e., fourth highest concentration) rather than one exceedance per year.

Care must be taken when comparing the NAAQS to net concentrations downwind from feedyards. An ambient standard is used to estimate the average 24-h exposure of the public in an area such as a city or town. Net concentrations downwind from feedyards, which are usually located in a rural area, will disperse rapidly downwind. Property line net concentrations do not correspond to the 24-h concentrations measured to determine compliance with the NAAQS in the city or town that may be located thousands of meters downwind.

Dust Emissions in Cattle Feedlots

Elam et al.[15] collected feedlot dust samples inside of 65 pens at 10 California feedlots using a Staplex high-volume air sampler operated in 1- to 3-h increments during 24-h continuous sampling periods. Peak particulate concentrations, collected between 1900 and 2200 Pacific Daylight Time (PDT), ranged from 1946 to 35,536 μg/m^3 and averaged 14,200 μg/m^3 (±11, 814 μg/m^3 standard deviation). Lowest concentrations occurred in early morning, when concentrations were only 130 to 250 μg/m^3 in some feedlots.

Algeo et al.[24] measured 24-h TSP concentrations both upwind and downwind of 25 California feedlots. Net particulate concentrations (downwind minus upwind) for 24-h continuous sampling time ranged from 54 to 1268 μg/m^3. The average value for all 25 feedlots was 654 ± 376 μg/m^3. Downwind concentrations exceeded upwind concentrations by 300%. Both upwind and downwind particulate levels usually exceeded the EPA ambient air-quality standards for TSP. Because feedlot particulates are commonly regarded as coarse dust, these measured property line concentrations are not indicative of ambient concentrations downwind.

Peters and Blackwood[14] cited several major limitations of the results of Algeo et al.[24] including: (1) All sampling was performed in the dry season; and (2) details such as feedlot size, cattle number, distances from samplers to feedpens, and climate conditions were not reported. Consequently, emission rates developed for cattle feedlots from their data were regarded as *worst-case projections*.[14]

Based on the Peters and Blackwood[14] treatment of the California data, the EPA[25] published so-called emission factors (AP-42) for cattle feedlots and termed them "crude estimates" at best. These estimates of particulates were indexed according to cattle feedlot size and cattle throughput as follows[25]:

1. Capacity basis (283 pounds particulates per day per 1000 head)
2. Throughput basis (27 tons particulates per 1000 head fed)

These mass particulate emission rates would place feedyards with more than 2,000 head above a projected TSP emission level of 100 tons per year, not including the feedmill, subject to the Title V program for Major Sources. Other emissions factors were similarly written for ammonia, amines, and total sulfur compounds.[25]

The EPA particulate emission factors[25] ignored the major climatic differences that exist between southern California and the larger cattle regions of the Great Plains and the Midwest, in which both total rainfall and seasonality of rainfall are different. California represents less than 4% of the U.S. cattle on feed as compared with the Southern Great Plains states of Texas, Nebraska, Kansas, Colorado, Oklahoma, and New Mexico, which combined have 81% of the 10.6 million head on feed as of January 1, 1999.[26]

To obtain a broader database, more representative of Southern Great Plains conditions, dust emissions were measured at three Texas cattle feedlots ranging from 17,000 to 45,000 head in 15 field experiments in 1987 to determine concentrations of feedlot dust measured both as TSP and PM_{10}.[27] Net 24-h feedlot dust (TSP) concentrations (downwind minus upwind) ranged from 16 to 1700 μg/m^3 and averaged 412 ± 271 μg/m^3 (which is 37% less than the earlier California data). Feedlot dust concentrations were generally highest in early evening and lowest in early morning, and upwind concentrations averaged 22% of downwind concentrations. For the experiments where TSP and PM_{10} samplers were co-located, TSP concentrations averaged 700 ± 484 μg/m^3, and PM_{10} averaged 285 ± 214 μg/m^3 (range of 11 to 866 μg/m^3).[28]

Using two types of PM_{10} samplers (Wedding and Andersen-321A), the PM_{10} dust concentrations were 19% and 40%, respectively, of mean TSP concentrations in direct comparisons.[27] There was good correlation between PM_{10} and TSP concentrations, with $r^2 = 0.634$ and 0.858 for the Wedding and the Andersen 321-A samplers, respectively.[27]

The mass median diameters obtained from the Coulter Counter PSDs[22] were 9.5 ± 1.4 μm (AED) for the PM captured with high-volume samplers (TSP) and 6.9 ± 0.8 μm for the PM captured with the Andersen PM_{10} sampler. The fraction of $PM_{2.5}$ from the Coulter Counter PSDs was less than 5% of the TSP.[28]

When the Wedding sampler was used for PM_{10} measurements, feedlots were below the new EPA standard for PM_{10}, and peak concentrations did not coincide with expected early-evening peaks caused by cattle activity. Hence, it would appear that a relatively small proportion of the TSP feedlot manure dust may have been captured in the Wedding PM_{10} instruments.

Analysis with a Coulter Counter resulted in aerodynamic particle size distribution curves for TSP and PM_{10} samplers (Figure 1).[28] Mass median diameters (MMD) of dust particles captured on high-volume samplers averaged 14.2 μm downwind as compared with values of 12.3 μm upwind of feedlots. The PM_{10} sampler oversampled particles larger than 10 μm. Most of the TSP sampled downwind of three feedlots was found to be relatively coarse dust as compared to current PM_{10} and $PM_{2.5}$ standards. Sweeten et al.[28] determined that the Coulter Counter method can be used to determine PSD for feedlot dust and supplement PM_{10} samplers.

Lesikar et al.[29] and Parnell[30] calculated a range of particulate emission factors for feedyards using field data from Sweeten et al.[27] The cattle feedyard emission factors were found to vary by season.[29] The emission factors for winter, spring, summer, and fall were 2.9, 11.6, 1.5, and 2.9 kg of TSP per 1,000 head per day. It was recommended that the EPA AP-42 emission factor[24] of 127 be changed to 4.5 kg of TSP per 1,000 head per day.[28,29] The latter figure represents only 3.5% of the EPA emission factor[24] for feedyards.

Odor from Livestock Facilities

Animal manure odor is composed of gaseous compounds that are intermediate and final products of biodegradation and includes these groups (Table 1): ammonia and amines, sulfides, volatile fatty acids, alcohols, aldehydes, mercaptans, esters, and carbonyls.[10,31–33] These odorous gases are not considered toxic at concentrations found downwind of livestock feeding facilities. Eaton[34] listed 170 compounds from swine manure odor associated with human olfactory response. Although concentrations of these compounds at downwind locations are low, some may exceed olfactory threshold values and create nuisance conditions. Olfactory threshold values detected by human panelists may span a range as great as 5 or 6 orders of magnitude for a single compound and range from as low as 7.5×10^{-8} ppm for skatole to as high as 12,000 ppm for formaldehyde.[34] Ammonia has reported odor threshold values ranging from 0.0317 to 37.8 ppm. Ammonia and propionic acid concentrations were correlated with odor intensity.[35]

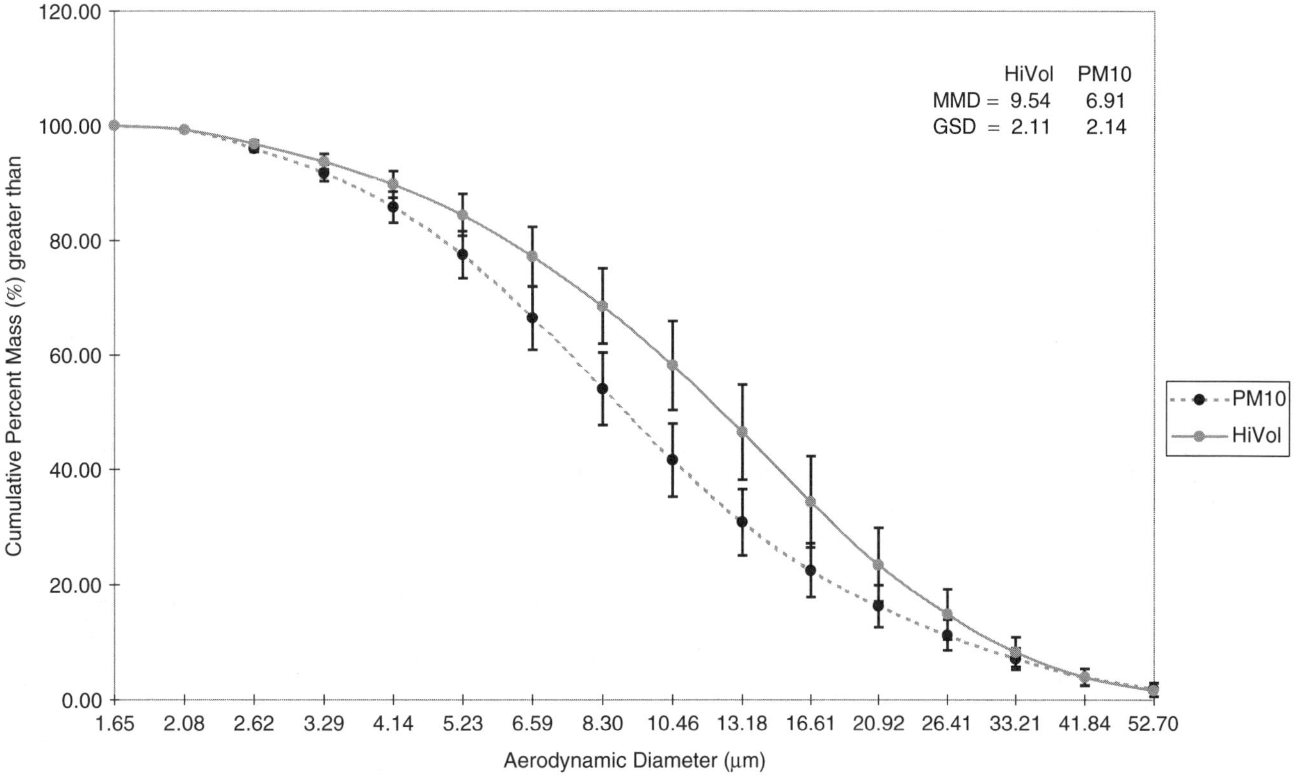

Figure 1. Mass Fraction of Feedlot Dust of Given Size Captured with High Volume and PM_{10} Samplers.

Table 1. Compounds Resulting from the Anaerobic Decomposition of Livestock and Poultry Manure

Alcohols	Amines
	Methylamine
Acids	Ethylamine
Butyric	Trimethylamine
Acetic	Diethylamine
Propionic	
Isobutyric	Esters
Isovaleric	
	Fixed gases
Carbonyls	Carbon dioxide (odorless)
	Methane (odorless)
Sulfur compounds	Ammonia
Hydrogen sulfide	
Dimethyl sulfide	Nitrogen heterocycles
Diethyl sulfide	Indole
Methylmercaptan	
Disulfides	Skatole

Correlations between odor intensity and gas concentrations were developed in two- and three-component mixtures of hydrogen sulfide, ammonia, and methyl amine.[36]

Odor characteristics that contribute to, and interact in, odor nuisance conditions are as follows[37]: (1) the intensity, concentration, or strength of the odor; (2) the odor frequency, or number of times detected during a time period; (3) the duration of the period in which the odor remains detectable; (4) the perceived offensiveness and character or quality of the odor. Odor frequency and duration are partly dictated by climatic conditions, including wind-direction frequency, atmospheric stability, and moisture conditions.

There is nearly universal acceptance of sensory approaches, involving the use of trained human panelists, for measurement of odor intensity or concentration. However, instruments and techniques to facilitate sensory odor measurement may vary. Odor measurement technology[38] applicable to livestock operations is used for determining:

1. Concentrations of specific compounds (ammonia, hydrogen sulfide, volatile organic acids, etc.)
2. Dilutions-to-threshold with a dynamic forced-choice olfactometer or scentometer
3. Equivalent concentration of a reference gas, such as butanol vapor, that matches the ambient odor intensity using a butanol olfactometer

Several states and municipalities have property-line odor standards based on these and other measurement methods.[39]

Odors caused by anaerobic decomposition of swine manure were determined by Meyer and Converse,[40] who found that H_2S and NH_3 production were, respectively, 218% and 118% higher at 73°F as compared with 60°F during a 23-d storage period. In European research,[41] the odor emission rate from swine houses with anaerobically stored manure increased 20-fold for each 10°F rise in manure temperature and, including ventilation rate influences, was over four times greater in summer than in winter. Emissions were 73% greater for fully slotted floors than for partially slotted floors.

Odor-intensity observations were made with a scentometer during eight experiments at three cattle feedlots at upwind and downwind locations alongside dust samplers.[27] Upwind odor intensities were typically in the range of 0 to 2 dilutions to threshold (DT), while downwind concentrations were 7, 31, or 170 DT alongside or a short distance from the feedlots. Arithmetic average downwind odor intensities ranged from 13 to 49 DT.

AIR POLLUTION CONTROL MEASURES

Dust Emissions from Cattle Feedlots

Guidelines on feedlot dust control[12,42,43] usually involve water sprinkling at strategic times and in proper amounts. Carroll et al.[44] compared two feedlots, one unsprinkled and the other sprinkled at the rate of 2 h on, 2.5 h off, 1.5 h on each day, and reported that sprinkling reduced dust emissions by at least half.

Elam et al.[15] reported that feedlot manure moisture content of 20–30% was needed for dust control. Particulate concentrations (24-h averages) increased from 3,150 to 23,300 μg/m^3 when daily water sprinkling was terminated for 7 d.

Sweeten et al.[27] found that feedlot dust concentrations decreased with increasing moisture content in the top 1 in. of feedlot surface. Regression equations indicated that the manure moisture needs to be 26–31% (wet basis) in the loose surface manure and 35–41% at 0- to 1-in. depths in order to control feedlot dust to TSP limits of 150 and 260 μg/m^3. Odor intensity (DT) increased as net dust concentrations decreased, perhaps because of moisture influences.

Dust control methods include watering unpaved roads, watering feedlot surfaces with mobile tankers or solid-set sprinklers, and controlling cattle stocking rate in relation to precipitation and evaporation.[12,45] The amount of manure moisture generated by cattle varies directly with liveweight and inversely with average stocking rate (ft^2/head). Large cattle on lower than normal animal spacing can easily double the effective "precipitation" on the feedlot surface, which has implications for both dust and odor control. A water-balance approach has been suggested.[45,46] Research by Texas A&M University researchers in the Texas High Plains is comparing stocking density in terms of average cattle liveweight per unit area as the basis for emissions comparisons.

Lott[46] fitted a three-stage drying model to the data for evaporation from pen surfaces at three feedlots. The first stage of manure surface drying represented a fast-rate linear function that is directly related to potential evaporation. In the second stage of drying, evaporation rate slows as manure restricts the release of water, and evaporation occurs at a very slow rate in the third stage. This drying rate function in analogous to drying curves for typical soils.[12]

Odor Control Methods

Specific measures have been devised to reduce odors from livestock manure,[31,32] including the development of a standard engineering practice.[33] These methods fall into three broad categories[39]:

1. Treatment of manure — aeration, anaerobic digestion, biochemical treatment.

2. Capture and treatment of odorous gases — containment, covered storage pits or lagoons, wet scrubbing, packed-bed adsorption, soil incorporation, soil filter fields, or absorption beds
3. Odor dispersion — site selection, separation distance, elevated source, topography, frequency of wind direction, and atmospheric stability

Research involving several of these methods will be discussed in the following section. A fourth approach of altering the animal's ration is beyond the scope of this subchapter.

Manure Treatment

Manure treatment methods for odor control include maintaining aerobic conditions during storage, aerobic treatment (aerated lagoons or composting), anaerobic digestion, or biochemical treatment.

For open lot surfaces, odor control can best be achieved by maintaining conditions for rapid drying, which includes frequent, uniform removal of surface manure and excellent drainage. Manure should be regularly harvested leaving a smooth, uniformly sloped pen surface with interfacial layer intact to maintain surface sealing. Wet manure on an open feedlot or dairy lot surface can be responsible for significant odor intensity and offensiveness.

Watts et al.[47] determined a 60-fold difference in measured odor concentration (in terms of odor units, measured with a dynamic forced-choice triangle olfactometer) between dry and wet feedlot surfaces. Odor generation peaked at 2–3d after rainfall and at a surface moisture content of 60–67% wet basis. Therefore, feedlots with wet anaerobic manure accumulation will create far greater odor than a well-drained and well-maintained feedlot. Cattle ration had less effect on odor concentration than moisture content. Odors were highest at mid-day.

A feedlot should be designed and managed to shed water. Pen slope of 3–5% away from feedbunks or feeding alleys is needed, with discrete drainage provided for each feed pen into a drainage channel that accelerates runoff away from the feedlot surfaces with minimal solids deposition. Potholes should be backfilled as soon as they develop, and overflows or leaks from cattle watering facilities onto the feedlot surface should be avoided. Proper stocking density in pens can ensure that moisture excretion by cattle plus rainfall does not exceed average seasonal evaporation. Where pen-to-pen drainage is unavoidable, pens should be maintained to prevent manure from accumulating under fence lines and impeding good drainage.

Controlled-rate anaerobic digestion of liquid swine manure at 32°C for methane gas production reduced the odor emission rate from land-applied digested slurry by 90% as compared with pit-stored slurry.[41] Anaerobic digestion also reduced the time for odor dissipation (in terms of 50% panelist detection threshold) from 72 to 24 h.

Anaerobic lagoons must have adequate capacity (i.e., low specific loading rate) to obtain relatively low odor production. Design criteria have been developed based principally on the volatile solids loading rate, which is inversely proportional to the treatment volume per pound of live weight[48–50].

Mechanical aeration of liquid manure in oxidation ditches or lagoons has been determined to be an effective odor-control method.[51] Aerating only the top third or half of swine lagoon contents proved successful and reduced power requirements as compared with complete mixing.[51] Converse et al.[52] used limited aeration of liquid swine manure without a measurable dissolved oxygen residual and reduced odor as compared with nonaerated storage. Phillips et al.[53] rapidly reduced hydrogen sulfide and methanol emissions from swine manure by aeration, but less volatile and less offensive compounds such as phenols persisted. Zhang[54] reduced odor concentration from surface-aerated laboratory digesters, but ammonia concentrations increased. Aeration just prior to land spreading could reduce odors from field application.

Frequent manure collection by flushing, cable scraping, or pit drainage/recharge is useful for absorbing odorous gases and eliminating anaerobic storage conditions in confinement buildings.[40,55,56] For example, Oosthoek and Kroodsma[57] determined a threefold reduction in ammonia emission rate by flushing the concrete floor in a free stall dairy barn, with minimal ammonia reduction from scraping the concrete floor.

Biochemicals for odor control include masking agents, counteractants, digestive deodorants, chemical deodorants, adsorbents, and feed additives.[58] Digestive deodorants are the most prevalent, and necessitate frequent additions to allow selected bacteria to become predominant. Potassium permanganate (100–500 ppm), hydrogen peroxide (100–125 ppm), and chlorine are oxidizing chemicals capable of controlling hydrogen sulfide emissions.

Warburton et al.[59] significantly reduced odors from anaerobic swine manure slurry with four treatments: aeration, chlorination, and two biochemical formulations. Numerous other chemicals, bacterial, or enzyme products that were tested were not effective for either odor control or solids reduction. Lindvall et al.[60] reduced odors of liquid swine manure with ammonia persulfate, and Miner and Stroh[61] determined that zeolites (clinoptilolite and erionite) were somewhat effective for odor reduction from a dirt-surfaced cattle feedlot.

Odor Capture and Treatment

Capture and treatment of odorous gases includes the use of covered storage pits or lagoons; soil incorporation of applied liquid or solid manure; and dry scrubbers for building exhaust gases, including soil absorption beds. Installing a cover on an outside manure storage pit, tank, or lagoon is an effective means of odor control because it effectively reduces the ventilation rate and hence the rate of odor emission. Safley and Westerman[62] demonstrated the use of a floating flexible membrane cover to capture and collect biogas (including odorants) produced from a primary treatment lagoon for a 150 cow free-stall dairy to fuel an internal combustion engine and electric generator. However, rigid covers are expensive, and flexible membrane covers over large surfaces are subject to photodegradation and wind damage.

Wet scrubbers that involve spraying exhaust air with water or oxidizing chemicals are widely used for industrial and food processing plant odors, and some researchers have adapted them to livestock confinement buildings. Van Geelen and Van Der Hoek[63] obtained an 88% reduction in odor concentration with wet scrubbing of exhaust from a swine house, although captured dust formed a sludge that caused difficulty in recirculating the scrubbing water. Schirz[64] cited problems with clogging of spray nozzles when scrubbing with recycled water and biological treatment was required. Licht and Miner[65] built a horizontal cross-flow packed-bed wet scrubber for swine confinement building exhaust air. They achieved 50% and 90% removal of particulates larger than 1 and 5 μm, respectively; ammonia reduction of 8–38%; and 82% reduction of odor intensity. A packed-bed dry scrubber filled with a zeolite (clinoptilolite) reduced ammonia emissions from a poultry house by 45% initially, but efficiency dropped to only 15% in 18 days.[66]

Soils and organic materials such as peat or wood chips readily absorb odorous gases and provide for aerobic decomposition of captured odorants. The soil is an excellent odor-scrubbing medium because of chemical adsorption, oxidation, and aerobic biodegradation of organic gases.[67] Lindvall et al.[60] determined that soil injection reduced odor emissions (measured as DT) from liquid swine manure by 90–99% as compared with surface spreading. Odor from soil-injected manure was about equal to that from a nonmanured soil surface. Disk harrowing or plowing of surface spread manure reduced odor threshold levels by 67–95%.

Soil filters with perforated pipe in a shallow soil bed have proved effective for scrubbing odors from a process or building exhaust air. Kowalewsky[68] removed 52–78% of the ammonia and 46% of the organic constituents from ventilation air from a swine-confinement building using a soil filter system. Prokop and Bohn[69] reported 99.9% odor reduction (ED_{50}) from a soil filter used to treat high-intensity odors in exhaust from rendering plant cookers. Soil filters required a moderately fine-textured soil, sufficient moisture, and pH between 7 and 8.5. A typical land area requirement is 2,500–4,600 ft^2 per 1000 cfm, depending on the airflow rate.[69] Sweeten et al.[70] measured 95–99% reduction in ammonia emissions and 30–82% reduction in odor intensity (matching butanol concentrations) using a 0.25-acre sand filter field to scrub air from a poultry manure composting operation.

Odor Dispersion

Sound site selection with adequate separation distance and, if necessary, elevated sources or mechanical turbulence will help achieve odor dispersion and avoid nuisance conditions. Odorants may be transformed between the source and the receptor, and this includes interactions with other odorous gases or particulates.[14] Ammonia and hydrogen sulfide are highly reactive, have relatively high odor thresholds and low molecular weights and disperse rapidly (i.e.,low persistence factor).[71]

Sound site selection often is the simplest and cheapest odor control strategy that protects investments in new concentrated animal feeding operations and surrounding real estate and helps avoid legal expense to defend nuisance lawsuits. To achieve good dispersion, operators should choose a remote site relative to neighbors; gently sloped topography without confining valley walls; low probability of wind direction toward nearby neighbors; and low probability of stable atmospheric conditions that retard dispersion in those directions.

Odorous gases become diluted with distance downwind of the source, depending on atmospheric turbulence and odorant reactions. Atmospheric dispersion models are sometimes used to predict the transport of odorous gases[72] and/or their impact on communities in terms of odor intensity. The use of dispersion models is limited to short distances and to nonreactive odorous gases.[10] Where odor modeling is used for regulatory applications, versions of the Gaussian diffusion model are normally used. The prediction models require that atmospheric stability, wind speed, and odor emission rates be known.

Emission rate is the product of odor concentration times airflow rate and is often used as a surrogate for mass emission rate.[73,74] For instance, Smith and Watts[73] used dynamic forced-choice triangle olfactometer measurement to calculate odor emission rates ranging from 5 OUm/sec for a dry feedlot pad to over 500 OUm/sec for a wet feedlot pad, and these data were used to model dispersion. It should be noted that the use of OU as a "dispersion" variable is subject to criticism because of the nonlinear relationship between OU and actual concentrations.

Odors, ammonia, and propionic acid traveled farther downwind from land-spreading liquid manure than from aboveground manure storage tanks or confinement buildings.[35] An odor panel observed a 90% reduction in odor intensity, as determined by a matching butanol olfactometer,[75] over a distance 0.5 mile downwind of a cattle feedlot in Texas.[76]

Klarenbeek[41] reported an ordinance in the Netherlands that stipulates a "required minimum distance" of swine finishing houses from residences. For example, required distances for a 1500-hog finishing operation were 1000 ft from residential areas or 550 ft from an individual nonagricultural residence.

Sweeten[77] linked separation distances for odor control at two typical types and sizes of swine facilities in the central United States with land area requirements for agronomic use of nitrogen and phosphorus in the manure when land spread or irrigated for nutrient recovery through the soil/plant system. He found that separation distances needed to reduce odor to 2 DT measured with a Barnebey–Sutcliffe Scentometer were similar to distances established by furnishing adequate land area to achieve a phosphorus balance for corn production in typical cases.

The Queensland Department of Primary Industries[78] developed guidelines for feedlot separation distance from residences or populated areas. Criteria included feedlot size, design, and management factors, together with terrain and degree of tree cover.

The nonlinear/nonadditive nature of odor emissions from contributing sources makes it difficult to accurately predict odor emission rates from complex sources, such as feedlots and dairies. Modeling of trace gases such as ammonia has appeal for simplicity,[79] but may not accurately characterize odor as perceived by humans.

In summary, modeling will be used in the future to predict odor impacts on surrounding land users in advance, before projects (agricultural or nonagricultural) are actually built.[38] However, much more research is needed before accurate odor models are developed, calibrated, and utilized with confidence.

REFERENCES

1. D. L. Van Dyne, and C. B. Gilbertson, *Estimating U.S. Livestock and Poultry Manure and Nutrient Production*, ESCS-12, Economics, Statistics and Cooperative Service, U.S. Department of Agriculture, Washington, D.C., 1978, 150 pp.
2. *Manure Production and Characteristics*, ASAE Data D384, American Society of Agricultural Engineers, St. Joseph, MI, 1976, 1 p.
3. *Manure Production and Characteristics*, ASAE Data D384.1, American Society of Agricultural Engineers, St. Joseph, MI, 1988, 4 pp.
4. J. M. Sweeten, "Livestock and Poultry Waste Management: A National Overview," National Livestock, Poultry and Aquaculture Waste Management, Proceedings of National Workshop, American Society of Agricultural Engineers, St. Joseph, MI, 1992, pp. 3–15.

5. U.S. Environmental Protection Agency, "State Program Elements Necessary for Participation in the National Pollutant Discharge Elimination System — Concentrated Animal Feeding Operations," 40 CFR 124.82, Federal Register, March 18, 1976, p. 11460. (See also 40 CFR 122.23, including Appendix B thereof.)
6. *Development Document for Proposed Effluent Limitations Guidelines and New Source Performance Standards for the Feedlots Point Source Category*, EPA-440/1-73/004, Washington, D.C., 1973, pp. 59–64.
7. R. K. White and D. L. Forster, *A Manual on Evaluation and Economic Analysis of Livestock Waste Management Systems*, EPA 600/2-78-102, U.S. Environmental Protection Agency, Robert S. Kerr Environmental Research Laboratory, Ada, OK, 1978, 302 pp.
8. *Beef Housing and Equipment Handbook*, MWPS-6, Midwest Plan Service, Iowa State University, Ames, IA, 1987.
9. J. Foster and V. Mayrose, *Pork Industry Handbook*, Cooperative Extension Service, Purdue University, West Lafayette, IN, 1987.
10. National Research Council, *Odors from Stationary and Mobile Sources*, National Academy of Sciences, Washington, D.C., 1979.
11. Texas Cattle Feeders Association, *Cattle Feeders Annual*, Amarillo, TX, 1998, p. 119.
12. J. M. Sweeten, *Feedlot Dust Control*, L-1340, Texas Agricultural Extension Service, Texas A&M University System, College Station, TX, 1982.
13. D. W. MacVean, D. K. Franzen, T. J. Keefe, and B. W. Bennett, Airborne Particulate Concentration and Meteorologic Conditions Associated with Pneumonia Incidence in Feedlot Cattle, *Am. J. Vet. Res.* **47**(12), 2676–82, 1986.
14. J. A. Peters and T. R. Blackwood, *Source Assessment: Beef Cattle Feedlots*, Monsanto Research Corp., EPA-600/2-77107, U.S. Environmental Protection Agency, Industrial Environmental Research Laboratory, Research Triangle Park, NC, 1977.
15. C. J. Elam, J. W. Algeo, T. Westing, et al. "Measurement and Control of Feedlot Particulate Matter," Bulletin C, *How to Control Feedlot Pollution*, California Cattle Feeders Association, Bakersville, CA, January 1971.
16. P. J. Watts, M. Jones, S. C. Lott, R. W. Tucker, and R. J. Smith, "Odor Measurement at a Queensland Feedlot," ASAE Paper 92-4516, Presented at International Winter Meeting of the American Society of Agricultural Engineers, Nashville, TN, December 15–17, 1992, 18 pp.
17. R. Tucker, "Odour Measurements from Simulated Feedlot Pads," in: *Odour Update '92, Proceedings of a Workshop on Agricultural Odours*, MRC Report No. DAQ 64/24, Department of Primary Industries, Toowoomba, Queensland, Australia, April 1992, pp. 103–21.
18. P. J. Watts and R. W. Tucker, *The Creation and Reduction of Odour at Feedlots*, Workshop on Agricultural Odours, Australian Water and Wastewater Association and Clean Air Society of Australia and New Zealand, 1993a, pp. 3.1–3.14.
19. P. J. Watts and R. W. Tucker, "The Effect of Ration on Waste Management and Odour Control in Feedlots," *Recent Advances in Animal Nutrition in Australia 1993*, University of New England, Armidale, New South Wales, Australia, 1993b, pp. 117–129.
20. U.S. Environmental Protection Agency, 40CFR50, Revisions to the National Ambient Air Quality Standards for Particulate Matter and Appendix J — Reference Method for the Determination of Particulate Matter as PM-10 in the Atmosphere, *Federal Register*, **52**(126), 24634–69, 1987.
21. D. J. Heber, and C. B. Parnell, "Comparison of PM-10 and High-Volume Air Samplers Using a Coulter Counter Particle Size Analyzer," Paper no. SWR 88-109, presented at 1988 Southwest Region Meeting of ASAE, Lubbock, TX, 1988.
22. M. Raina and C. B. Parnell, Jr., "Determination of PM_{10} Concentrations from the TSP Data Using the Coulter Counter Multisizer," ASAE Paper No. 944537, Paper presented at the Winter Meeting of the American Society of Agricultural Engineers, Atlanta, Georgia. 1994.
23. J. W. Anderson, "New Air Quality: Standards Are Tighter but Compliance Is Distant," *Resources*, Issue 129, Fall, 1997, p. 608.
24. J. W. Alego, C. J. Elam, A. Martinez, and T. Westing, "Feedlot Air, Water and Soil Analysis," Bulletin D, *How to Control Feedlot Pollution*, California Cattle Feeders Association, Bakersville, CA, June 1972, 75 pp.
25. U.S. Environmental Protection Agency, Supplement A to *Compilation of Air Pollution Emission Factors*, Section 6.15, Beef Cattle Feedlots (Stationary Point and Area Sources, Vol 1), AP-42, Office of Air Quality Planning and Standards, Research Triangle Park, NC, 1986.
26. Texas Cattle Feeders Association, USDA Cattle on Feed Report, TCFA Newsletter, Amarillo, Texas, January 22, 1999, p. 5.
27. J. M. Sweeten, C. B. Parnell, R. S. Etheredge, and D. Osborne, "Dust Emissions in Cattle Feedlots," Veterinary Clinics in North America, *Food Animal Practice* **4**(3), 557–78, November 1988.
28. J. M. Sweeten, C. B. Parnell, B. W. Shaw, and B. W. Auvermann, "Particle Size Distribution of Cattle Feedlot Dust Emission," *Transactions ASAE* **41**(5), 1477–81, 1998.
29. B. J. Lesikar, S. A. Parnell, and J. M. Sweeten, "Dispersion Modeling for Prediction of Emission Factors from Cattle Feedyards," Final Report, National Cattlemen's Association, Englewood, Colorado, November 17, 1994, 121 pp.
30. S. E. Parnell, "Dispersion Modeling for Prediction of Emission Factors from Cattle Feedyards," unpublished Master of Science thesis, Department of Agricultural Engineering, Texas A&M University, College Station, Texas, 1994.
31. J. R. Miner, "Management of Odors Associated with Livestock Production," in *Managing Livestock Wastes*, Proceedings of Third International Symposium on Livestock Wastes, American Society of Agricultural Engineers, St. Joseph, MI, 1975, pp. 378–80.
32. C. L. Barth, L. F. Elliot, and S. W. Melvin, "Using Odor Control Technology to Support Animal Agriculture," *Trans. ASAE*, **27**, 859–684, 1984.
33. "Control of Manure Odors," ASAE EP-379.2, ASAE Standards-1998, American Society of Agricultural Engineers, St. Joseph, MI, 1998, pp. 638–40.
34. D. L. Eaton, *Swine Waste Odor Compounds*, Pioneer Hi-Bred International, Livestock Environmental Systems, Des Moines, Iowa, 1996, 14 pp.
35. H. H. Kowalewsky, R. Scheu, and H. Vetter, "Measurement of Odor Emissions and Emissions," in *Effluents from Livestock*, J. K. R. Gasser, Ed. Applied Science Publishers, London, 1979, pp. 609–25.
36. D. T. Hill and C. L. Barth, "Quantitative Prediction of Odor Intensity," *Trans. ASAE*, **19** 939–44, 1976.
37. M. Jones, "Odour Measurement Using Dynamic Olfactometry," in *Odour Update 92: Proceedings of a Workshop on Agricultural Odors*, MRC Report No. DAQ 64/24, Department of Primary Industries, Toowoomba, Queensland, Australia, April 1992, pp. 17–32.
38. J. M. Sweeten, "Odor Measurement Technology and Applications: A State of the Art Review," in: *Proceedings of the International Symposium on Agricultural and Food Processing Wastes*, Chicago, Illinois, ASAE — The Society for Engineering in Agricultural, Food, and Biological Systems, 1995, pp. 214–29.
39. J. M. Sweeten, "Odor Measurement and Control for the Swine Industry," *J. Envir. Health* **50**(5), 286, 1988.
40. D. J. Meyer and J. C. Converse, "Gas Production vs. Storage Time on Swine Nursery Manure," Paper no. 81-4512, American Society of Agricultural Engineers, St. Joseph, MI, 1981.
41. J. V. Klarenbeek, "Odour Emissions of Dutch Agriculture," *Agricultural Waste Utilization and Management*, Proceedings

of the Fifth International Symposium on Agricultural Wastes, American Society of Agricultural Engineers, St. Joseph, MI, 1985, pp. 439–45.

42. P. D. Andre, "Sprinklers Solved This Feedlot Dust Problem," *Beef* 70–72, 74, 79–81, February 1985.
43. F. M. Simpson, "The CCFA Control of Feedlot Pollution Plan," Bulletin A, *How to Control Feedlot Pollution*, California Cattle Feeders Association, Bakersville, May 28, 1970.
44. J. J. Carroll, J. R. Dunbar, R. L. Givens, et al. "Sprinkling for Dust Suppression in a Cattle Feedlot," *Calif. Agr.* 12–13, March 1984.
45. J. M. Sweeten and S. C. Lott, "Dust Management," in: P. J. Watts and R. Tucker, Eds., *Designing Better Feedlots*, Queensland Department of Primary Industries, Toowoomba, Queensland, Australia, 1994, pp. 6.23–6.30.
46. S. C. Lott, "Australian Feedlot Hydrology, Part I (Data)," in: *Proceedings, Feedlot Waste Management Conference*, Queensland Department of Primary Industries, Gold Coast, Queensland, Australia, 1995, 29 pp.
47. P. J. Watts, M. Jones, S. C. Lott, R. W. Tucker, and R. J. Smith, "Feedlot Odor Emissions Following Heavy Rainfall," *Trans. ASAE* **37**(2), 629–36, 1994.
48. C. L. Barth, "A Rational Design Standard for Anaerobic Livestock Waste Lagoons," *Agricultural Waste: Utilization and Management*, Proceedings of the Fifth International Symposium on Agricultural Wastes, American Society of Agricultural Engineers, St. Joseph, MI, pp. 638–47.
49. F. J. Humenik and M. R. Overcash, *Design Criteria for Swine Waste Treatment Systems*, EPA-600/2-76-233, U.S. Environmental Protection Agency, Ada, OK, 1976, 291 pp.
50. "Design of Anaerobic Lagoons for Animal Waste Management," Engineering Practice EP403.1, Standards 1990, American Society of Agricultural Engineers, St. Joseph, MI, 1990, 1.
51. F. J. Humenik, R. E. Sneed, M. R. Overcash, et al. "Total Waste Management for a Large Swine Production Facility," *Managing Livestock Wastes*, Proceedings of the Third International Symposium on Livestock Wastes, American Society of Agricultural Engineers, St. Joseph, MI, 1975, pp. 168–71.
52. J. C. Converse, D. L. Day, J. T. Pfeffer, et al. "Aeration with ORP Control to Suppress Odors Emitted from Liquid Swine Manure System," *Livestock Waste Management and Pollution Abatement*, Proceedings of International Symposium on Livestock Wastes, American Society of Agricultural Engineers, St. Joseph, MI, 1971, pp. 267–71.
53. D. Phillips, M. Fattori, and N. R. Bulley, "Swine Manure Odors: Sensory and Physico-Chemical Analysis," Paper No. 79-4074, American Society of Agricultural Engineers, St. Joseph, MI, 19 pp.
54. R. Zhang, P. N. Dugba, and D. S. Bundy, "Laboratory Study of Surface Aeration of Anaerobic Lagoons for Odor Control of Swine Manure," Trans. *ASAE* **40**(1),185–90, 1997.
55. W. Korsmeyer, M. D. Hall, and T. H. Chen, "Odor Control for a Farrow-to-Finish Swine Farm—A Case Study," *Livestock Waste: A Renewable Resource*, Proceedings of the Fourth International Symposium on Agricultural Wastes, American Society of Agricultural Engineers, St. Joseph, MI, 1981, pp. 193–97, 200.
56. S. J. Raabe, J. M. Sweeten, B. R. Stewart, et al. "Evaluation of Manure Flush Systems at Caged Layer Operations," *Trans. ASAE* **27** 852–58, 1984.
57. J. Oosthoek and W. Kroodsma, "Ammonia Emission Reduction from Cowsheds," in: *Agricultural and Food Processing Waste*, Proceedings of the 6th International Symposium on Agricultural and Food Processing Wastes, American Society of Agricultural Engineers, 1990, pp. 442–48.
58. W. F. Ritter, "Chemical Odor Control of Livestock Wastes," Paper no. 80-4059, American Society of Agricultural Engineers, St. Joseph, MI, 1980, 16 pp.
59. D. J. Warburton, J. N. Scarbrough, D. L. Day, et al. "Evaluation of Commercial Products for Odor Control and Solids Reduction of Liquid Swine Manure," *Livestock Waste: A Renewable Resource*, Proceedings of the Fourth International Symposium on Livestock Wastes, American Society of Agricultural Engineers, St. Joseph, MI, 1981, pp. 309–13.
60. T. Lindvall, O. Noren, and L. Thyselius, "Odor Reduction for Liquid Manure Systems," *Trans. ASAE* **17** 508–12, 1974.
61. J. R. Miner and R. C. Stroh, "Controlling Feedlot Surface Odor Emission Rates by Application of Commercial Products," *Trans. ASAE* **19**, 533–38, 1976.
62. L. M. Safley and P. W. Westerman, "Psychrophilic Anaerobic Digestion of Dairy Cattle Manure," in: *Agricultural and Food Processing Wastes*, Proceedings of the 6th International Symposium on Agricultural and Food Processing Wastes, American Society of Agricultural Engineers, St. Joseph, MI, 1990, pp. 250–58.
63. M. A. Van Geelan and K. W. Van Der Hoek, "Odor Control with Biological Air Washers," *Agr. Envir.* **3** 217–22, 1972.
64. S. Schirz, "Odour Removal from the Exhaust of Animal Shelters," *Agr. Envir.* **3**, 223–28, 1977.
65. L. A. Licht and J. R. Miner, "A Scrubber to Reduce Livestock Confinement Building Odors," Paper No. PN-78-203, American Society of Agricultural Engineers, St. Joseph, MI, 1978, 12 pp.
66. J. K. Koelliker, J. R. Miner, M. L. Hellickson, et al. "A Zeolite Packed Air Scrubber to Improve Poultry House Environment," *Trans. ASAE* **23**, 157–61, 1980.
67. H. Bohn, "Soil Absorption of Air Pollutants," *J. Environ. Quality* **1**, 372–77, 1972.
68. H. H. Kowalewsky, "Odor Abatement through Earth Filters," *Landtechnik* **36**(l), 8–10, 1981.
69. W. H. Prokop and H. L. Bohn, "Soil Bed System for Control of Rendering Plant Odors," Paper 85-79.6, 78th Annual Meeting, Air Pollution Control Association, Pittsburgh, 1985, 17 pp.
70. J. M. Sweeten, R. E. Childers, J. S. Cochran, and R. Bowler, "Odor Control from Poultry Manure Composting Plant Using a Soil Filter," *Applied Engineering in Agriculture* **7**(4), 439–49, 1991.
71. W. Summer, *Odor Pollution of Air—Causes and Control*, The Chemical Rubber Company Press, Cleveland, Ohio, 1971, 310 pp.
72. K. A. Janni, "Modeling Dispersion of Odorous Gases from Agricultural Sources," *Trans. ASAE* **25**, 1721–23, 1982.
73. R. J. Smith and P. J. Watts, "Determination of Odor Emission Rates from Cattle Feedlots: Part 1, A Review," *Journal of Agricultural Engineering Research* **57**, 145–55, 1994.
74. A. M. S. McFarland, "Dispersion Modeling of Odor Transport from Open Lot Dairies," in: *Proceedings, Odors: Indoor and Environmental Air*, Air Waste Management Association Conference, Bloomington, MN, 1995, 13 pp.
75. J. E. Sorel, R. O. Gauntt, J. M. Sweeten, et al. "Design of a 1-Butanol Scale Dynamic Olfactometer for Ambient Odor Measurements," *Trans. ASAE* **26**, 1201–06.
76. J. M. Sweeten, D. L. Reddell, A. R. McFarland, et al. "Field Measurement of Ambient Odors with a Butanol Olfactometer," *Trans. ASAE* **26**, 1206–16, 1983.
77. J. M. Sweeten, "Separation Distances for Swine Odor Control in Relation to Manure Nutrient Balances," *Applied Engineering in Agriculture* **14**(5), 543–49, 1998.
78. Queensland Department of Primary Industries, *Queensland Government Guidelines for Establishment and Operation of Cattle Feedlots*, Brisbane, Queensland, Australia, September 1, 1989, 16 pp.
79. J. P. Kelly, "Ammonia and Odour Emission from Agricultural Sources: A Review," Faculty of Engineering and Surveying, University of Southern Queensland, Toowoomba, Queensland, Australia, 1995, 57 pp.

RENDERING PLANTS

William H. Prokop

Rendering plants process or "recycle" animal and poultry by-product materials in order to produce tallow, grease, and protein meals. The rendering industry is divided into two basic groups: integrated and independent. The integrated rendering plants are operated in conjunction with animal slaughterhouses or poultry processing plants. This type of plant is an integral part of the animal slaughterhouse or meat packing operation and processes all the residual material from these operations. This also applies to those rendering plants that process the offal and feathers from poultry processing plants.

Independent rendering plants collect their raw materials from a variety of sources that are off-site or separate from the plant itself. These renderers send out special-body route trucks to collect discarded fat and bone trimmings, meat scraps, restaurant grease, blood, feathers, offal, and entire animal carcasses from a variety of sources: butcher shops, supermarkets, restaurants, fast-food chains, poultry processors, slaughterhouses, farms, ranches, feedlots, and animal shelters. Independent renderers collect raw materials primarily in metropolitan areas. However, a number of these renderers collect raw material from a variety of regional areas and require the use of a transfer station, which receives the collected material for further transport to a rendering plant.

Animal rendering systems are divided into two classes: edible rendering of animal fatty tissue into edible fats and proteins for human consumption, and inedible rendering of animal by-product materials into fats and proteins for animal feed and nonedible applications. Edible rendering plants are normally operated in conjunction with meat packing plants under the inspection and processing standards established by the U.S. Department of Agriculture, Food Safety and Inspection Service (USDA/FSIS). Edible tallow or lard is produced from beef or swine fatty tissue.

Inedible rendering plants are operated by independent renderers to produce inedible tallow and grease. These are used in the feed rations for livestock and poultry, for soap production, and for fatty-acid manufacture. A number of different grades of inedible tallow and grease are produced for these different categories of use. For example, "yellow grease" or feed-grade animal fat is produced from waste cooking fats received from restaurants and deep fat fryers. This product is used as an ingredient in animal and poultry feeds. Protein meal includes animal and poultry meals that are used as feed supplements for livestock and poultry. Rendering plants also produce blood meal and feather meal. These products usually have their protein content specified.

The meat packing industry during the past 20 years has converted its meat production facilities almost totally from an animal carcass to a "boxed beef" and "boxed pork" operation.[1] This has resulted in more fat and bone being trimmed at the slaughterhouse and less raw material being available to the independent renderer. Also, there has been a recent downward trend in tallow and grease prices.

During the past 15 years or more, the number of independent rendering plants has been reduced significantly. Currently, an estimated 135 independent rendering plants are operating in the United States.[2] This consolidation also applies to integrated rendering, with an estimated 80 meat packing plants and 40 poultry processing plants also including rendering operations.

PROCESS DESCRIPTION

Current edible and inedible rendering systems are described in detail by Prokop.[3] These include the batch cooker, Duke continuous, and Stord waste heat dewatering systems for inedible rendering. The Sharples Trim-R process is described for edible rendering.

Edible Rendering

The current edible rendering process is continuous and consists of two stages of centrifugal separation. A typical feedstock of beef fat trimmings from USDA inspected meat processing plants consists of 14–16% fat, 60–64% moisture, and 22–24% protein solids.

The Sharples Trim-R Edible Fat Process is shown in Figure 1. Fat trimmings are ground through a Weiler grinder and belt conveyed to a melt tank equipped with an agitator and steam-heated jacket. The melted fatty tissue at 110°F is pumped to a Reitz disintegrator to rupture the fat cells. A Sharples Super-D-Canter Centrifuge separates the proteinaceous solids from the melted fat and moisture containing a small percentage of solids or fines.

A second-stage centrifuge is required to "polish" the edible fat, which is first heated to 200°F by a shell-and-tube heat exchanger with steam. The Westfalia De-Sludging Separator makes a two-phase separation where the polished edible fat discharges from the top and the water fraction containing the protein fines is discharged as sludge from the bottom. The edible fat is pumped to storage, whereas the sludge from the centrifuge is either transported to an inedible rendering plant or passed through a primary treatment system for wastewater.

Since no cooking vapors are emitted from current edible rendering processes and heat contact with the edible fat is minimal, odor emissions from these rendering plants are perceived as having little impact. This also is due to the freshness of the raw material being processed and the plant sanitation and housekeeping practices that have been established by the USDA/FSIS. As a result, the remainder of this section is devoted to inedible rendering processes and the control of their odor emissions.

Wet Rendering Versus Dry Rendering

Wet rendering is defined as a process of separating fat from raw material by boiling with water. This is normally accomplished by boiling the raw material in a tank of water. The products of wet rendering are fat, stick water containing glue, and wet tankage (protein solids). The wet rendering process involves the addition of water to the raw material and the use of live steam to cook the raw material and accomplish separation of the fat.

The wet rendering process no longer is used in the United States because of the high costs of energy and of an adverse effect on fat quality. The water added to the raw material must be evaporated, which increases fuel costs to generate additional steam for moisture removal. Also, contact of the fat with excess water under a boiling temperature tends to increase the free fatty acid content of the fat.

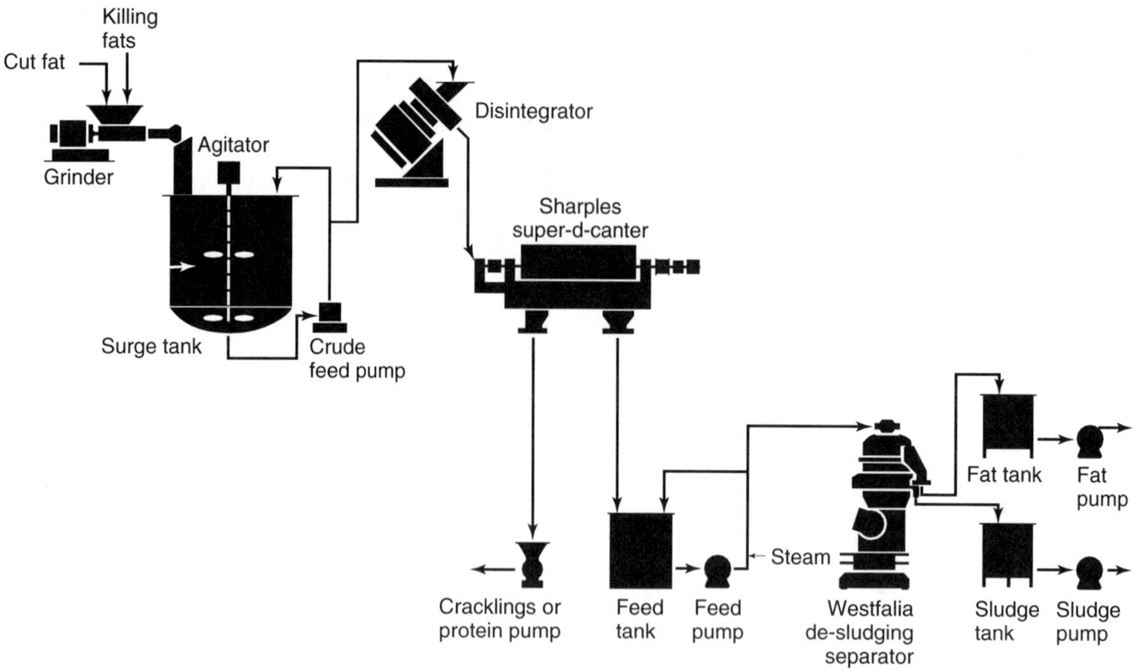

Figure 1. Edible Rendering System (From Ref. 3).

The wet rendering process has been exclusively replaced by dry rendering, which is defined as a process for releasing fat by dehydrating raw material in a batch or continuous cooker. After moisture removal in the cooker, the melted fat is separated from the protein solids. No excess water or live steam is added to the raw material in this process.

Raw Materials for Rendering

The integrated rendering plant for an animal slaughterhouse or poultry processor normally receives only one type of raw material. This simplifies the control of the processing conditions, which usually require only minor adjustments. Also, the raw material is relatively fresh, undergoing little or no noticeable deterioration. Conversely, the independent renderer often handles a variety of raw materials that require either the operation of multiple rendering systems in parallel or significant changes in the operating conditions for a single system to process variable raw material.

Table 1 provides specific raw-material yield data and the split between the fat and protein solids content of various categories of raw materials for inedible rendering.[1]

In comparing Table 1 with a similar table (Table 221) in the previous edition of this manual, we see that the fat content of steers and hogs has increased significantly as a result of improved breeding and the nutritional feeding of fat to livestock. However, the fat content of butcher shop fat and bone has decreased due to the production of "boxed beef/pork," since more of the available fat is trimmed at the packing house.

Basic Rendering Process — Batch Cooker System

Figure 2 from Ref. 3 illustrates the basic rendering process where batch cookers are used. These are multiple units arranged in a row or series of rows, depending on the size and arrangement of the rendering plant. Each cooker consists of a horizontal, steam-jacketed cylindrical vessel equipped with an agitator. This vessel is known as a batch cooker because it follows a repetitive cycle: the cooker is charged with the proper amount of raw material, the cook is made under controlled conditions, and, finally, the cooked material is discharged.

Table 1. Composition of Raw Materials for Inedible Rendering

Source	Tallow/Grease (wt %)	Protein Solids (wt %)	Moisture (wt %)
Packing house offal and bone			
Steers	30–35	15–20	45–55
Cows	10–20	20–30	50–70
Calves	10–15	15–20	65–75
Sheep	25–30	20–25	45–55
Hogs	25–30	10–15	55–65
Dead stock (whole animals)			
Cattle	12	25	63
Calves	10	22	68
Sheep	22	25	53
Hogs	30	28	42
Butcher shop fat and bone	31	32	37
Blood	None	16–18	82–84
Restaurant grease	65	10	25
Poultry offal	10	25	65
Poultry feathers	None	33	67

Source: Ref. 1.

The raw material from the receiving bin is screw conveyed to a crusher or similar device for size reduction. For batch cookers, the raw material is reduced in size to 1 or 2 in. to provide efficient cooking, which normally requires $1\frac{1}{2}$ to $2\frac{1}{2}$ hours. The raw material is quite variable, depending on the source, and adjustments in the cooking time and temperature may be required to process it. The final temperature of the cooked material ranges from 250°F to 275°F, depending on the type of raw material.

After the cooking process is completed, the cooked material is discharged to the percolator drain pan, which contains a perforated screen that allows the free-run fat to drain and

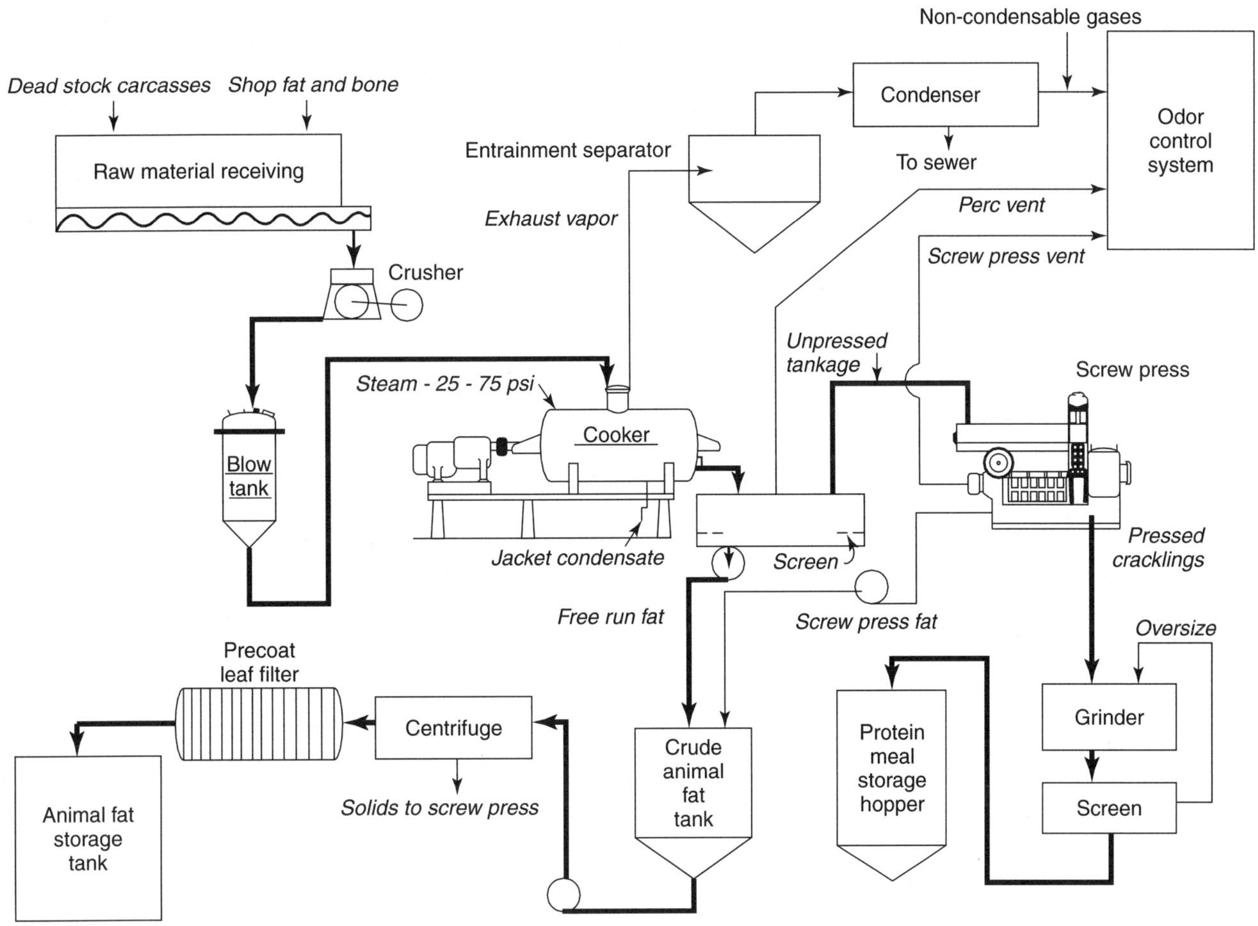

Figure 2. Batch Cooker Rendering Process.

be separated from the protein solids, which are known as "tankage." After 1 or 2 h of drainage, the protein solids still contain about 25% fat and are conveyed to the screw press, which completes the separation of fat from solids. The final protein solids have a residual fat content of 10%.

The solid protein material discharged from the screw press is known as "cracklings." It is normally screened and ground with a hammer mill to produce protein meal that essentially passes a 12-mesh screen. The fat discharged from the screw press usually contains fine solid particles that are removed by either centrifuging or filtration.

Continuous Rendering Systems

Since the 1960s, a variety of continuous rendering systems have been installed to replace the batch cooker systems. Continuous rendering is synonymous with continuous cooking. The raw material is fed continuously to the cooker, and the cooked material is likewise discharged at a constant rate.

A continuous rendering system normally consists of a single continuous cooker, whereas the batch cooker system consists of multiple cooker units. A continuous system usually has a higher capacity than the batch cooker system it replaces. This increased capacity provides for more efficient processing of the raw material by processing more material in less time.

Continuous rendering also has a number of other inherent advantages over the batch system. Since a continuous process requires less cooking time or exposure to heat, improved product quality normally results. Further, the continuous system occupies considerably less space than a batch cooker system with equivalent capacity, thus saving building construction costs. Finally, a single-cooker unit is inherently more efficient than multiple-cooker units in terms of steam consumption and achieves a significant savings in fuel usage by the boilers. Likewise, less electric power is consumed for agitation in the single continuous cooker unit.

The Duke continuous rendering system is manufactured by the Dupps Co., Germantown, Ohio. This rendering system is shown in Figure 3.[3] The Duke system is designed to provide a method of cooker operation similar to that of the batch cooker. The Equacooker is a horizontal steam-jacketed cylindrical vessel equipped with a rotating shaft to which are attached paddles that lift and move the material horizontally through the cooker. Steam also is injected into the hollow shaft to provide increased heat transfer.

The feed rate to the Equacooker is controlled by adjusting the speed of the variable-speed drive for the feed screw, which establishes the production rate for the system. The discharge rate for the Equacooker is controlled by the speed at which the control wheel rotates. The control wheel contains buckets, similar to those used in a bucket elevator, that pick up the cooked material from the Equacooker and discharge it to the drainer.

The drainer performs the same function as the percolator drain pan in the batch cooker process. It is an enclosed screw conveyor that contains a section of perforated trough for the free melted fat to drain through to the crude tallow tank. The protein solids containing residual fat are then conveyed to the

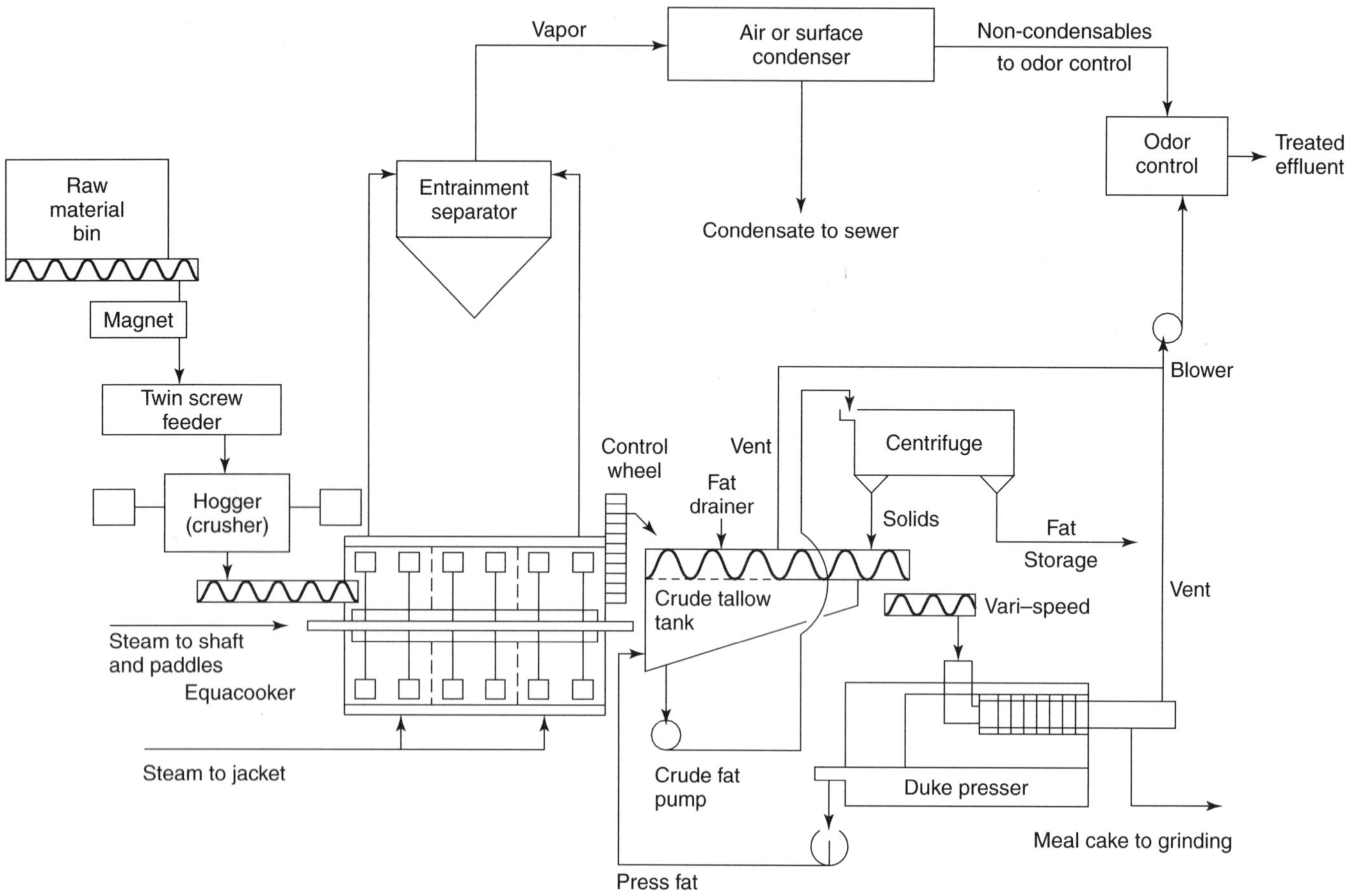

Figure 3. Duke Continuous Rendering System.

pressers for the additional separation of fat. The pressers and other components of the Duke continuous system are similar to those used for the batch cooker system. A central control panel is provided for the operator to consolidate the instruments required for process control of these continuous systems.

New Continuous Systems with Reduced Energy Usage

During the early 1980s, considerable effort was spent on developing new rendering systems that utilize the cooking vapors from either a batch cooker system or a continuous rendering system to obtain further moisture removal. Most of this new rendering technology has evolved in Europe, where energy costs have been significantly higher than in the United States.

The Stord waste-heat dewatering (WHD) system, for example, is manufactured by Stord, Inc., in Bergen, Norway. It is illustrated by Figure 4 from Ref. 3. The Stord WHD system consists of a preheater, twin-screw press, and evaporating system. It usually is installed in conjunction with an existing rendering system.

In this system, raw material is screw conveyed as usual from the raw material bin over an electromagnet and fed to a prebreaker for coarse grinding. This ground material passes through the preheater, which is a horizontal, steam-jacketed, cylindrical vessel with a rotating shaft and agitator to move the material through the vessel continuously and to improve heat transfer. The temperature of the raw material in the preheater is controlled and ranges from 65°C to 85°C (150°F to 190°F), depending on the type of raw material being processed. This heating step is necessary to melt the fat and condition the animal fibrous tissue properly for the subsequent pressing operation.

The twin-screw press consists of intermeshing, counter-rotating, twin screws moving inside a press cage assembly. It includes a perforated screen through which the liquid is pressed and a series of vertical support plates to secure the perforated screen in place. The perforated screen with heavy backing plate is constructed to follow the contour of the rotating flights of the twin screws. The feed material fills the free space between the screws and the press cage. At the feed end, the twin screws are constructed with a lower-diameter shaft and deeper flights to provide a larger volume of space. As the press screws rotate, this space decreases and the material is subjected to a steadily increasing pressure to squeeze out the liquid through the perforated screen.

The twin-screw press separates the heated and ground raw material into two phases, a press cake of solids containing fat and moisture and a liquid containing mostly the melted fat and water. The solids are screw conveyed to the existing cooker or dryer, which is steam heated to remove the moisture. Final separation of the fat from the solids is completed with a screw press.

Liquid from the twin-screw press is pumped from the feed tank to the evaporator, which consists of a tubular heat exchanger mounted vertically and integrally with the vapor chamber. The vapors from the existing cooker or dryer provide the heating medium for evaporation. The liquid pumped to the evaporator enters at the top of the heat exchanger and flows by gravity downward through the tubes discharging into the

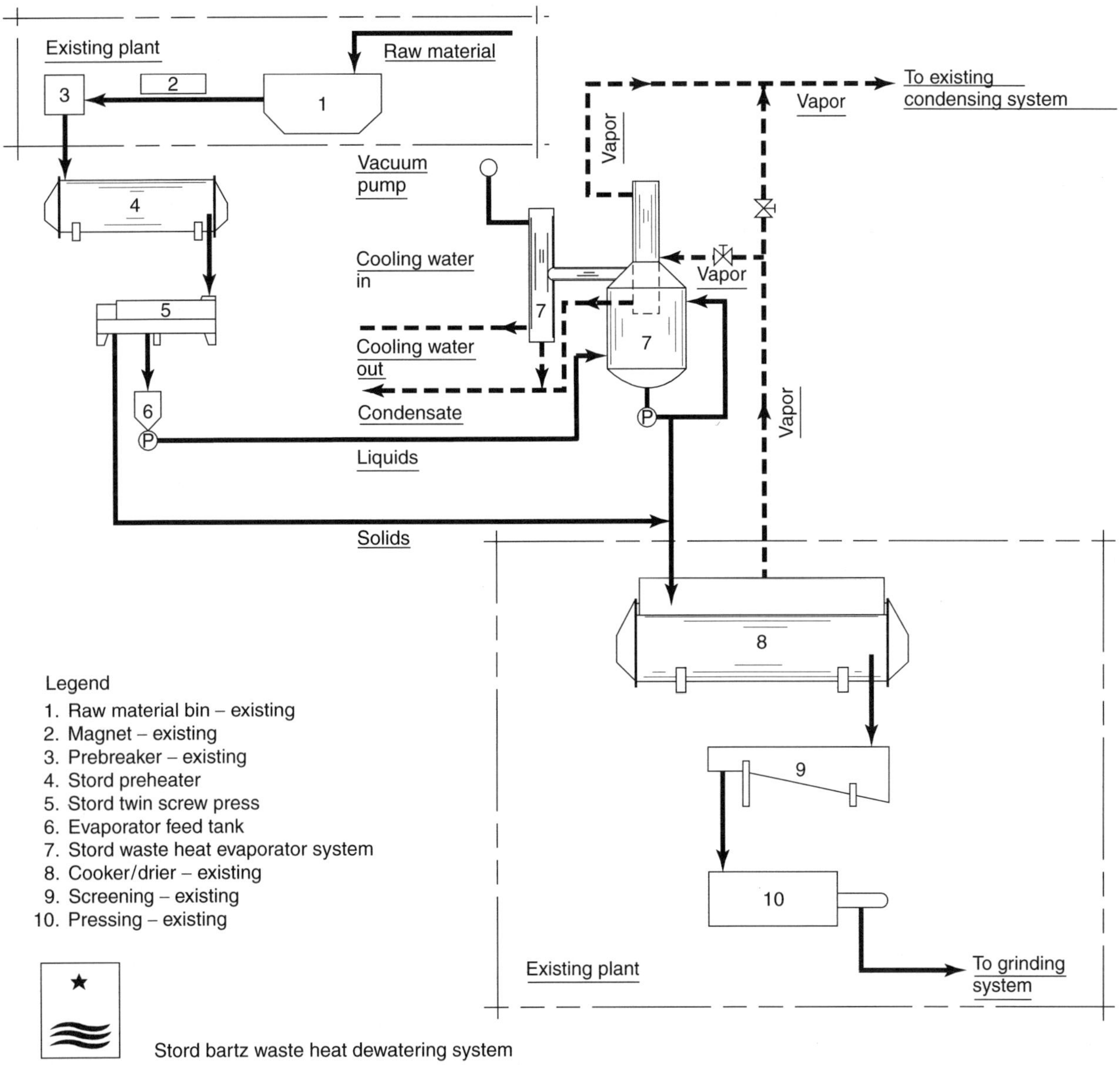

Figure 4. Stord Waste-Heat Dewatering System.

vapor chamber, which is maintained under a vacuum of 24 to 26 in. Hg provided by a vacuum pump. The temperature of the liquid ranges from 70°C to 90°C (160°F to 200°F) at which point the moisture is evaporated. The water vapor from the vapor chamber is condensed with a shell-and-tube condenser through which cooling water is circulated.

The basic concept of the Stord WHD system is to use the waste heat in the vapors from the existing cooker to evaporate the moisture from the liquid removed by pressing the raw material, thus converting the existing plant into a two-stage evaporation system. It is essential to balance the operation of this system so that sufficient vapors are available from the existing cooker to evaporate the moisture from the pressed liquid. Operation of the preheater and twin-screw press requires adjustment for different types of raw material with varying moisture content in order to achieve proper balance of this system.

In addition to reducing fuel costs by 30–40%, the operation of the Stord WHD system, in conjunction with the existing cooker system, is capable of increasing production throughput by 75%.

These new systems[4] incorporate the use of microcomputer control concepts that are capable of performing essentially all start-stop sequences, monitoring specific process elements, and recording process data to provide trend or deviation outputs. If a particular malfunction occurs, the control system automatically shuts down key operating elements that otherwise could cause serious damage or loss of production.

Blood Processing and Drying

Whole blood received from animal slaughterhouses contains 16–18% total protein solids. Of this amount, approximately 70% of the total protein is recovered as blood meal after steam coagulation and drying. The soluble protein fraction remains in the serum water.

In the past, batch cookers were used to coagulate and dry the blood. However, the dried blood becomes quite dusty and is easily entrained in the exit gases from the cooker. The batch cookers have been replaced by continuous drying processes that are more energy efficient and provide a product with improved quality. Blood meal is highly valued as an animal feed ingredient because of its relatively high lysine (amino acid) content. The continuous drying processes maintain the lysine content of blood nearly at the maximum available, whereas the batch process reduces this lysine availability by 30–50%.[5]

In the continuous blood process, whole blood passes through an inclined tubular vessel into which steam is injected to coagulate the blood solids. This slurry is pumped to a horizontal, solid bowl centrifuge, which separates the blood solids from the blood serum water. The blood solids at 50–55% moisture are fed into a continuous drying system: either a gas-fired, direct-contact ring dryer or steam tube, rotary dryer.

The Dupps Ring Dryer is manufactured by the Dupps Co. in Germantown, Ohio. The coagulated blood solids enter the dryer through a hammer-type mill and are air conveyed through the ring duct at a temperature of 200°F to the manifold. At this point, the moist product is separated from the dried product and is recirculated back to the ring duct for further drying. Heated air at 600°F from the direct fired furnace passes through the hammer mill and provides the heat necessary to dry the moist product. This heated air consists of 60% recycled exhaust air and 40% makeup air. The dried product is separated from the exhaust air by twin cyclone collectors with rotary valves. The exhaust air normally passes through a venturi scrubber and packed tower scrubber in series before discharging to atmosphere. This dryer is furnished in four different models with varying evaporative capacities, ranging from 1000 to 4000 pounds of water evaporated per hour.

The Anderson 72 Tube Rotary Steam Dryer is manufactured by Anderson International Corp. in Cleveland, Ohio. The moist feed material cascades over the rotating, steam-heated tubes. Fresh air passes through the dryer across the steam-heated tubes and exits from the dryer with the moisture released from the solids. The 72-tube unit is the only model available and has an evaporative capacity estimated to be 650 pounds of water evaporated per hour.

Poultry Feathers and Hog Hair Processing and Drying

The feathers and hog hair consist mostly of keratin, which is a long-chain, highly cross-linked, relatively indigestible protein. The rendering process converts the keratin by chemical hydrolysis, combining with water at elevated temperatures (280–300°F), into shorter-chain, more digestible amino acids. This hydrolyzation is accomplished by processing the feathers or hair in a batch cooker with an internal cooker pressure of 40–50 psig maintained for 30–45 min. The moisture content after hydrolyzation is approximately 50%. The feather hydrolysis process is discussed,[6] including the effect of the processing conditions on the quality of the feather meal product. The feathers or hair may be dried in the batch cooker. However, the drying operation normally is conducted with either the Dupps Ring Dryer or Anderson Rotary Steam Tube Dryer, both of which were described under "Blood Processing and Drying."

Grease Processing

The recent growth of the restaurant business, and of fast food chains in particular, has made the recycling of restaurant grease an important part of the rendering industry. In the past, 55-gal drums of grease were picked up at restaurants and unloaded manually by plant employees for processing at the grease plant. Currently, much of this grease is bulk loaded into specially designed and constructed vehicles, which transport and discharge it directly to the grease processing system without any manual labor.

The melted grease is screened to remove coarse solids, then heated to 200°F in vertical processing tanks, and stored for 36–48 h to separate the grease from the water and fine solids by gravity. Four phases of separation normally occur: (1) solids, (2) water, (3) emulsion layer, and (4) grease product. The solids settle to the bottom of the processing tank and are separated from the water layer above. The emulsion is processed through two stages of centrifuges: a horizontal, solid bowl type to remove solids and a vertical disk type to remove water and fines. The grease product is skimmed off the top. No air is introduced into the processing tanks for "blowing" or drying of the grease.

AIR EMISSIONS CHARACTERIZATION

Odor is the primary air pollutant emitted from the rendering process. There is a potential dust problem in a few instances, such as the exhaust from the Dupps Ring Dryer when processing blood or feathers. However, the provision of proper control equipment, such as cyclones and venturi scrubbers, is normally adequate to abate this type of emission.

Points of Emission

For the batch cooker rendering process, the primary sources of high-intensity odors include the noncondensibles from the cooker exhaust and emissions from the screw press, since the material in both cases is heated to temperatures of about 220°F. The processing of blood or poultry feathers normally results in high odor levels. Other sources of high-intensity odors include dryers, centrifuges, tallow processing tanks, and the perc pans that are open to the plant atmosphere and receive the discharge from the batch cookers. The hot, cooked material from the batch cooker not only releases odor, but also fat particles, which tend to become airborne and are deposited on equipment and building surfaces within the plant.

The raw material is another source of odor, but it normally is not significant when processed without delay.

However, the age of the raw material is important because older material that has deteriorated will result in substantially higher odors being generated during the cooking and pressing operations. Also, the type of raw material being cooked is a significant factor. For example, dead stock will tend to result in the emission of higher-intensity odors during the rendering process.

Table 2 illustrates the effect of raw material on the odor emission from an inedible batch cooker when discharged to the perc pan.[7]

An important trend during the 1970s and 1980s involved the replacement of batch cooker systems with continuous rendering systems that are essentially enclosed and are capable of confining the odors and fat aerosol particles within the equipment. By providing proper equipment seals and locating suction pickup vents at strategic points, a major percentage of the odor generated from the continuous rendering process can be confined and treated by a low-volume scrubber system or by boiler incineration.

Odor control measures have been applied successfully to the Duke continuous rendering system because of its ability to confine the process odors within the system. Recent innovations in rendering equipment technology have resulted

Table 2. Odor Concentration of Emissions from Inedible Batch Cookers during Discharge to Perc Pan

Type of Raw Material Cooked	Emission during Cooker Discharge (odor units/cf)[a]	Emission 5 min after Discharge (odor units/cf)[a]
Poultry feathers	200	20
Poultry and turkey offal	2,000	500
Slaughterhouse viscera and bones	150	150
Fresh meat and bone trimmings from beef slaughterhouse	100	70
Meat & bone trimmings with high percent rancid restaurant grease	25,000 40,000	3,000 200
Mixture of dead cats and dogs, fish scrap, poultry offal, etc.	1,000	1,500

[a]Odor units per cubic foot by ASTM syringe method. *Source*: Ref. 7.

in the replacement of screw conveyors delivering raw material to and cooked material from the Equacooker with specially designed pumping systems. This has provided a better seal at the Equacooker, preventing the leakage of high-intensity odors into the plant operating area.

The primary sources of high-intensity odors from the Duke continuous process are similar to those for batch cooker systems: the cooker exhaust noncondensibles and the press vents. In the past, only these two odor emissions from a Duke system were normally vented to an afterburner to incinerate them. However, experience has proven that additional pickup vents are required at the drainer and at the centrifuge to prevent leakage of high-intensity odor to the plant atmosphere. Depending on the leakage occurring at the feed end of the Equacooker, an odor pickup may be located at the inlet chute.

The degree of tightness of an enclosure at each pickup point is an important factor to consider in designing an exhaust duct system. A definite but regulated excess of air is needed at each pickup point to minimize leakage of odor from the rendering process. Often in the past, the exhaust system for an afterburner was undersized because it was desired to incinerate the least volume of air and so minimize fuel costs. A damper is provided at each pickup point to adjust and maintain each flow at the desired rate and also to balance the exhaust duct system.

Composition of Rendering Odors

Odor emissions from rendering plants are relatively complex mixtures of organic compounds. Samples of rendering plant odors have been analyzed by a combination of gas chromatograph and mass spectrometric methods. A total of 30 or more odorous compounds were identified.[8]

Further research work identified the odorous compounds present in rendering plant emissions. Improved analytical techniques provided a more complete list of components. The major compounds included organic sulfides, disulfides, C-4 to C-7 aldehydes, trimethylamine, and various C-4 amines, quinoline, dimethyl pyrazine and other pyrazines, C-3 to C-6 organic acids. Compounds of lesser significance included C-4 to C-7 alcohols, ketones, aliphatic hydrocarbons, and aromatic compounds. Odor panel tests were also conducted to relate odor intensity with the various important peaks identified by gas chromatography.[9]

Table 3 lists odor-detection and recognition threshold values for certain odorous compounds known to be present in emissions from rendering plants. Some of the compounds have extremely low odor threshold values and can be detected at concentrations as low as 1 ppb or less.

Odor Emission Data

It is essential that accurate odor emission data be available in order to design an odor control system that successfully abates the emission. If such data are not available, it may be necessary to conduct appropriate odor sensory measurements and obtain accurate gas-flow data. In this section, odor dilution to threshold values are provided for various high-intensity odor emissions from the rendering processes, together with their corresponding volumetric emission rates expressed in cubic feet per minute.

Odor emissions from batch rendering cookers are discussed in detail.[7] In a batch cooker, the rate of moisture removal rises initially, reaching a peak usually within 1 h, then decreases rapidly until the end of the cooking cycle. The cooking time for a batch cycle normally ranges from $1\frac{1}{2}$ to $2\frac{1}{2}$ h, depending on the initial moisture content and type of raw material being processed.

Table 3. Odorous Compounds in Rendering Plant Emissions (Refs. 7, 10, 11, 12)

Compound Name	Formula	Molecular Weight	Detection[a] Threshold (ppm, v/v)	Recognition Threshold (ppm, v/v)
Acetaldehyde	CH_3CHO	44	0.067	0.21
Ammonia	NH_3	17	17	37
Butyric acid	C_3H_7COOH	88	0.0005	0.001
Dimethyl amine	$(CH_3)_2NH$	45	0.34	—
Dimethyl sulfide	$(CH_3)_2S$	62	0.001	0.001
Dimethyl disulfide	CH_3SSCH_3	94	0.008	0.008
Ethyl amine	$C_2H_5NH_2$	45	0.27	1.7
Ethyl mercaptan	C_2H_5SH	62	0.0003	0.001
Hydrogen sulfide	H_2S	34	0.0005	0.0047
Indole	$C_6H_4(CH)_2NH$	117	0.0001	—
Methyl amine	CH_3NH_2	31	4.7	—
Methyl mercaptan	CH_3SH	48	0.0005	0.0010
Skatole	C_9H_9N	131	0.001	0.050
Trimethyl amine	$(CH_3)_3N$	59	0.0004	—

The average steam rate during the batch cooking cycle will normally vary from 450 to 900 ft^3/min. Shell-and-tube condensers through which cooling water is circulated or air-cooled, finned-tube condensers are used to condense the steam vapors and cool the condensate to normally below 120°F. The batch cooker noncondensibles range in odor intensity from 5000 to 1 million odor units/standard cubic feet by the ASTM syringe method, depending on the age and type of raw material. The volumetric emission rate of the noncondensibles may vary from 25 to 75 ft^3/min, depending on the tightness of the batch cooker top cover/discharge door openings and shaft seals.

Table 4 shows odor sensory data obtained[13] during a batch cooker cycle, with polyethylene bag samples being collected at half-hour intervals during the cook cycle. These samples were evaluated using the IITRI dynamic olfactometer. Readings were also obtained with an instrument measuring the concentration of total organic compounds (as methane) in parts per million.

Table 4. Batch Cooker Noncondensible Odor Emissions

Cook Cycle (h)	Cooker Temperature (°F)	Total Organics (ppm)	Odor Dilution[a] to Threshold
$\frac{1}{4}$	150		
$\frac{1}{2}$	220	180	40,000
1	245	1,000	45,000
$1\frac{1}{2}$	245	700	97,000
2	245	400	75,000
$2\frac{1}{2}$	245	260	93,000
3	245	200	127,000

[a]IITRI dynamic olfactometer. *Source*: Ref. 13.

Another series of six samples were taken simultaneously on a different day approximately 3 h after the batch cooker cycle began. The cooker temperature was recorded at 265°F and the noncondensibles temperature at 90°F, and the total organics varied from 230 to 330 ppm during the sampling. The odor dilution to threshold values (IITRI dynamic olfactometer) varied from 184,000 to 276,000. The batch rendering plant was known to process raw material such as dead stock, which emits unusually high-intensity odors during the cooking process.

The steam vapor emission rate from continuous rendering processes is relatively constant and can be calculated for a specific moisture content and tonnage rate of raw material being processed. The capacity of cookers and dryers usually is expressed in terms of evaporative capacity as pounds of water evaporated. For example, the Series 1800 Duke Equacooker with regular shaft has a rated evaporative capacity of 12,000 pounds of water evaporated per hour based on a steam pressure of 100 psig.

During the past 5 years, a number of Duke Series 1800 continuous systems have been evaluated to quantify the high-intensity odors emitted from the process. These emissions include the noncondensibles from the Equacooker, the drainer, centrifuge, and other sources of odor. Samples were taken of the noncondensibles alone and also of the total odor emissions from the Duke system or other rendering plant operations. All samples were taken during the late spring or summer months and were evaluated using the IITRI dynamic olfactometer. Also, these evaluations were conducted with varying types of raw material, including that from beef slaughterhouse, shop fat and bone, and restaurant grease and fish. Table 5 from Ref. 14 summarizes the data, which include the volumetric emission rates.

Table 5. Odor Emissions from Continuous Rendering Processes

Plant	Category of Renderer and Material	Rendering Process and Type of Emission	Odor Dilution[a] to Threshold	Emission Rate (ft^3/min)
A	Integrated beef slaughterhouse	Duke—noncond.	20,000–50,000	450
		Duke—total[b]	39,100–43,200	1,500
		Blood ring dryer[c]	300–1,000	2,000
B	Independent beef slaughterhouse and restaurant grease	Duke—noncond.	24,400–62,700	—
		Duke—total[d]	56,000–138,000	—
B	Shop fat and bone	Duke—noncond.	11,000–16,000	665
		Duke—total[d]	7,600–13,200	1,200
C	Independent beef slaughterhouse and restaurant grease	Duke—noncond.	36,100–39,800	800
		Duke—total[e]	21,600—73,700	1,700
D	Independent herring (fish)	Duke—noncond.	39,500	600
		Duke—total[f]	23,800	2,600
D	Meat scraps and beef slaughterhouse	Duke—noncond.	59,400–93,800	600
		Duke—total[f]	28,100–59,200	2,600

[a]IIRTI dynamic olfactometer.
[b]Includes noncondensibles, two presses, drainer, centrifuge, and blood coagulator and centrifuge. Tests conducted in mid-April in Texas.
[c]Venturi scrubber discharge after dryer exhaust.
[d]Both tests include noncondensibles, presses, drainer, and centrifuge. First test conducted during late July and second test conducted during late September in New England.
[e]Includes two presses, two tallow tanks, two centrifuges, drainer, and grease vapor noncondensibles. Test conducted in mid-August in New York.
[f]Includes two presses, centrifuge, drainer, meal product conveyor, and storage bins. Tests conducted in late may on West Coast.
Source: Ref. 14.

Table 6. Suggested Odor Pickup Flow Rates for Duke Process Emissions

Category of Odor Emission Pickup	Duct ϕ (in.)	Flow Rate (ft^3/min)
Equacooker inlet chute (if not sealed)	6	500
Equacooker noncondensibles	6	500
Drainer section	6	500
Equacooker unloading elevator	4	250
Duke presser (10- or 12-in.)	6	500
Centrifuge (24 × 60 in. or 24 × 38 in.)	4	250
Tallow processing tank	4	250

There has been a tendency in the past to have inadequate odor pickup flow rates for certain emissions from Duke continuous systems. Table 6 provides a guide for establishing the volumetric rate of pickup for various odor emissions from a Series 1800 Duke continuous system. These should not be considered as rigid requirements, but instead as guidelines.

No odor sensory data are available for the new continuous systems with reduced energy usage. However, it has been observed in these plants that the odor emissions from the rendering process are definitely lower in intensity. This no doubt can be attributed to the lower processing temperatures that are used, particularly in the evaporator where the liquid from the twin-screw press is under vacuum for moisture removal.

AIR POLLUTION CONTROL MEASURES

Rendering plant operation and maintenance considerations are of basic importance in developing odor control measures to abate plant emissions. Raw material received at the plant should be processed with a minimum of delay. Cooking and pressing operations should be conducted to prevent overheating and burning of the processed material. As an example, a Duke Equacooker normally operated at a temperature of 270°F experienced a process upset, causing an increase to 320°F. This resulted in a corresponding increase from 14,000 to 102,000 odor units per standard cubic foot (ASTM syringe method).[15]

Startup and shutdown operating procedures should ensure that all odor control equipment is operating properly while any raw material is being processed through the rendering system. Process equipment leaks to the floor or to the plant atmosphere should be corrected in a timely manner. Daily plant cleanup should normally follow shutdown of the rendering plant. Sanitation practices are crucial. A substantial amount of odor may be generated from within a plant building whose walls and ceiling have become permeated with fat aerosol odor emissions.

The basic purpose of providing odor control in a rendering plant is to reduce the odor emissions from the plant to a level that will result in the surrounding ambient air *not* containing odors that are a source of valid nuisance complaints. In designing a new control system or revising an old one, each individual plant situation must be evaluated separately based on a variety of factors, including proximity of neighbors to the plant, categories of neighbors present, surrounding topography, prevailing winds, plant building features, ability of rendering process to confine odors, residual plant odors, type of raw material, and seasonal climatic conditions.

The fundamental question often to be resolved is whether to treat the high-intensity odors only or also to treat a large volume of air that would be used to ventilate the operating area within the plant. A decision to treat only the high-intensity odors is usually predicated on the ability of the rendering process to confine these odors within the equipment. As discussed before, continuous rendering systems usually have this capability. Boiler incineration or multistage, low-volume scrubbing of the high-intensity odors is particularly compatible with this type of system.

CONDENSING OF COOKER VAPORS

The batch cooker and continuous rendering systems (Figures 2 and 3 on pp. 449 and 500) both evaporate the moisture in the raw material. This water vapor must be condensed to accomplish efficient removal of odors from the cooker emission by use of either boiler incineration or wet scrubbing. Condensers used in rendering plants consist of two basic types: a shell and tube heat exchanger or an air-cooled, fin-tubed heat exchanger.

The shell and tube condenser is shown in Figure 2 on page 45. The cooking vapors enter the inlet header and pass thru the tubes to the outlet header where the noncondensibles are separated from the condensate which discharges to the wastewater treatment system. The cooling water normally enters the shell near the outlet header and flows countercurrent to the vapor flow inside the tubes, exiting near the vapor inlet header. Baffles are provided at spaced intervals inside the shell to direct the water over the outside of the tubes in order to utilize the entire tube surface area. This type of condenser provides heat recovery in the form of hot water which is used for plant cleanup. If heat recovery is not desired, a cooling tower is normally provided to cool the hot water circulated thru the condenser. However, the hardness of the water is a factor to consider and may require considerable chemical treatment to minimize scaling of the condenser tubes.

An air-cooled condenser consists of a rectangular bank of finned tubes, a large volume, low static pressure fan(s), a plenum chamber for even distribution of ambient air across the rectangular face of the tube bank, a header for vapor inlet and a discharge manifold for removal of the noncondensibles from the condensate. The vapors and noncondensibles are pulled from the cooker thru the finned tubes by means of an exhaust fan at the discharge manifold. Usually, a forced draft fan(s) is located below the finned tubes and it forces the ambient air upward, passing across the tubes. This type of condenser is more prevalent in the rendering industry. If heat recovery is not required, the usage of water for cooling and the need for chemical treatment of the water are eliminated. However, the performance of an air-cooled condenser is limited by the temperature of the ambient air. In extremely warm climates, a shell and tube condenser may be the only logical choice.

Both types of condensers are normally provided with stainless steel tubes for corrosion resistance. In air-cooled condensers, the cooling fins attached to the tubes are usually constructed of aluminum. Stainless steel construction is used wherever the cooker vapors come into contact with a metal surface. Periodic inspection and cleaning of the condenser tubes, particularly the inside, is important to maintain condenser efficiency.

For odor removal, it is desired to have a noncondensibles temperature not exceeding 120°F. If this temperature is too high, the excess water vapor in this saturated stream may cause the burner system used for boiler incineration to malfunction; or, it may affect adversely the odor removal efficiency of the wet scrubber receiving the noncondensibles.

Boiler Incineration of Process Odors

The installation and operation of afterburners or incinerators solely for pollution control is relatively uncommon due to the capital investment and fuel costs required. Currently, boiler incineration of the high-intensity process odors is a regular practice throughout the United States, since all rendering plants require the generation of steam for the cooking and drying processes.

Two basic choices are available for the odorous air to be introduced into the boiler: primary combustion air (that mixed with fuel before ignition) or secondary combustion air (that mixed with the burner flame to complete combustion). The following factors should be considered for boiler incineration of high-intensity odors.

1. The volume of odorous air to be handled should be minimized. Ensure that odor pickup points in the rendering process are not pulling excessive quantities of air.
2. Maximum fuel economy is achieved by using the odorous stream as primary combustion air, whereas its use as secondary combustion air probably requires additional fuel. When used as primary combustion air, particular care must be taken to see that the air stream is essentially free of moisture and particulate, which can interfere with the operation of the burner and controls.
3. Cooker noncondensibles can be successfully used as primary combustion air for incineration, provided that proper precautions are taken to "clean" it up. This is accomplished with a combination scrubber and entrainment separator of proper design. A water spray is provided to cool the odorous air and condense out the moisture. Likewise, the solid and fat aerosol particles should be removed.
4. Any high-intensity odors used as secondary combustion air should also be "cleaned up" in a manner similar to that described. When used as secondary combustion air, it is essential that the odorous stream come into intimate contact with the burner flame to accomplish efficient odor removal by incineration. Merely ducting the odorous air to the firebox without achieving contact with the flame may result in unsatisfactory odor removal.
5. The boiler size and burner capacity should be compatible with the amount of odorous air to be incinerated. If multiple boilers are used, the odorous air can be split among the various boilers or a smaller single unit can be used to incinerate the odors. The boiler should be equipped with suitable burner controls to ensure that the minimum firing rate is sufficient to incinerate the volume of odorous air passing through the firebox, regardless of the steam demand. A temperature of 1200°F or more is usually obtained in the firebox at the minimum firing rate. The residence time in the boiler firebox at maximum fuel rate is normally more than 1 sec.
6. If an existing boiler is to be used, a thorough analysis should be made to establish that the combustion control and safety requirements are satisfied for incinerating odors. The boiler manufacturer should be consulted regarding any details to modify the unit. Likewise, the insurance company should be contacted to receive their approval of any such revisions.

Figure 5 illustrates the boiler incineration of process odors used as primary combustion air. A two-stage spray scrubber with tangential inlet and entrainment separator is shown.

Previously, Table 5 summarized odor emission data for high-intensity odors from the Duke continuous process for Plants B, C, and D. Table 7 illustrates the odor removal efficiency achieved by boiler incineration for these same plants. The odor dilution to threshold values shown in this table were obtained with the IITRI dynamic olfactometer. These results clearly show that boiler incineration is a very efficient method of odor control for treating the high-intensity odors from the rendering process. Two different incineration conditions are shown for Plant B to illustrate the higher stack exhaust odor levels resulting from the use of high-sulfur No. 6 fuel (2% sulfur) oil compared with those for natural gas. A strong, pungent, sulfur dioxide smell was observed in the exhaust stack emission. When natural gas or low-sulfur fuel oil is used as boiler fuel, the odor character of the stack exhaust is that of combustion gases only. In both cases, a rendering odor was not detected.

Wet Scrubbing of Process Odors

Multistage scrubber systems for treating the high-intensity odors provide an alternative to boiler incineration when the latter approach may not be feasible because of the boiler plant's being remotely located or for other reasons. These multistage systems have been successfully applied to rendering plant emissions since the early 1970s.

An evaluation[9] was made of various scrubbing agents to determine their comparative effectiveness to absorb and neutralize selected odorous compounds in an experimental packed column. These odorous compounds were synthetically generated in the laboratory. Their selection included two sulfides, two amines, an organic acid, a ketone, an aldehyde, and an alcohol. Included among the various chemicals studied as scrubbing agents were water, soda ash, caustic soda, hydrochloric acid, sodium bisulfite, hydrogen peroxide, potassium permanganate, and sodium hypochlorite. The chemical oxidizing agents were found to be the most effective. From an overall viewpoint of treating a variety of odorants present in rendering plant emissions, sodium hypochlorite was considered to be the most effective agent.

Sodium hypochlorite or the addition of chlorine gas to a caustic soda solution has been used since the early 1970s. More recently, chlorine dioxide generated on-site by the addition of chlorine gas to sodium chlorite solution has been used in a number of these scrubber systems. Chlorine dioxide is known to have a higher oxidizing potential than sodium hypochlorite.[16]

It is important to establish the design and operational criteria for a wet scrubbing system that uses a chemical oxidant solution, not only to obtain effective odor control, but also to achieve practical operation that is economical. This subject is discussed by Lundgren and others.[17]

The chemical oxidant solutions used in wet scrubbers normally are recirculated to conserve water and minimize chemical and wastewater treatment costs. The concept of balancing the oxidant chemical addition rate with the chemical use rate is important to achieve optimum usage and minimum cost. It is advisable to treat gas streams containing solid and fat aerosol particles with a preconditioning device, such as a low- to medium-pressure-drop venturi scrubber, to remove this particulate before passing to the chemical scrubber. Since the mass of the particulate matter suspended in the gas stream

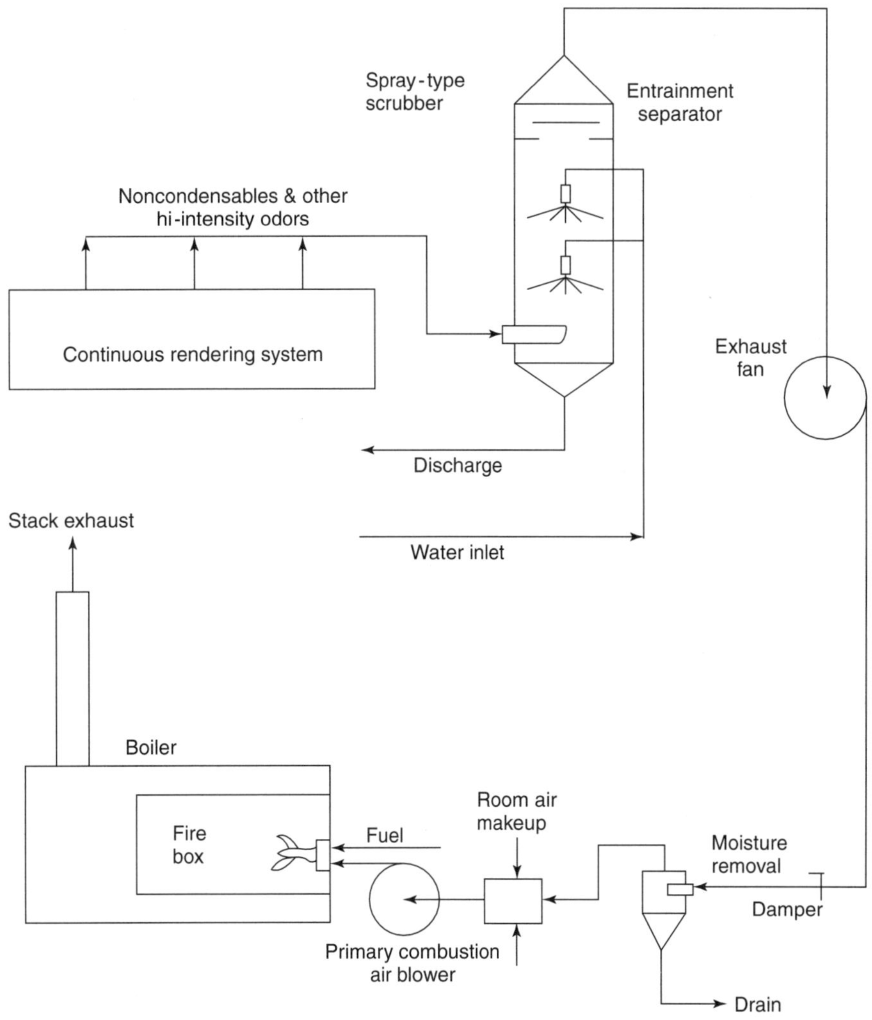

Figure 5. Boiler Incineration of Rendering Process Odors.

Table 7. Boiler Incineration of Rendering Process Odors

Plant	Fuel Used	Fire Box Temperature (°F)	Odor Dilution to Threshold Values[a] Boiler Inlet	Stack Exhaust	Odor Removal (%)
B	No. 6 oil	—	56,000–138,000	234–650	99.5
B	Natural gas	—	7,600–13,200	88–128	99.0
C	No. 6 oil	1,400	21,600–73,000	76–157	99.6
D	Natural gas	1,250	28,100–59,200	202–356	99.3

[a] IITRI dynamic olfactometer. *Source*: Ref. 14.

may exceed significantly that of the gaseous odor components, the rate of chemical oxidant consumption could rise quickly because the presence of the particles in the recirculating solution will tend to use up most of the oxidizing agent.

A two-stage scrubber system consisting of a venturi and a packed tower is described by Prokop.[8] This scrubber system treats high-intensity odors, including those from the screw press vents, blood dryer exhaust, raw feather receiving, feather noncondensibles, feather cooker discharge, and feather dryer exhaust. The scrubber system capacity is 32,000 ft^3/min. The scrubbing solutions consist of water circulating through the venturi at 110 gpm and sodium hypochlorite solution circulating through the packed tower at 350 gpm. Chemical consumption consists of 2 gph of 12% NaOCl solution.

A three-stage scrubber system consisting of a venturi and two packed towers in series is described by Prokop.[15] This scrubber system treats high-intensity odors, including those from the Equacooker noncondensibles, drainer, presser vents, centrifuge, and the steaming of grease barrels. The scrubber system capacity is 7500 ft^3/min. The scrubbing solutions consist of a weak solution of trisodium phosphate for circulation through the venturi at 22 gpm, a solution of phosphoric acid at 2–3 pH and recirculated through the first packed tower at 80 gpm, and a sodium hypochlorite solution at 10 pH and recirculated through the second packed tower at 80 gpm. The NaOCl solution is generated by an electrolytic cell from sodium chloride. The NaOH solution is added to control the pH of the NaOCl solution.

Figure 6 illustrates the operation of this multistage scrubber system. The high-intensity odors enter the throat of the low-energy venturi scrubber, which removes the particulate matter and cools and saturates the air with water vapor. A demister removes entrained droplets from the air before it passes upward through the first-stage counterflow packed-bed scrubber. Acid solution is recirculated through this stage to remove the amine-type odors, and NaOCl solution is recirculated throughout the second stage to remove the sulfide-type odors.

Table 8 summarizes the odor sensory data obtained for these two multiple-stage scrubber systems. The odor removal efficiency of these two systems is approximately 99%.

Wet Scrubbing of Plant Ventilating Air

This approach provides a more complete solution to an overall plant odor problem. Fugitive odors within the plant can be captured and treated in a uniform manner with this type of scrubber. It is particularly suited for rendering plants located near sensitive population areas, such as residential or commercial. For this application, it is essential to have adequate distribution and flow of air throughout the plant in order to pick up and capture in the ventilating air those fat aerosols and other particles that are emitted from the rendering process.

In addition to accomplishing effective odor control, these scrubbers provide proper ventilation of the plant operating areas, thereby maintaining satisfactory working conditions for the employees and improved compliance with the Occupational Safety and Health Administration standards. It is important to recognize that sufficient ventilating air must pass through the operating area during the summer months. Otherwise, doors and windows will be opened excessively, allowing the plant odors to escape from the building instead of being treated by the scrubber system. Ideally, a slight negative pressure should be maintained within the rendering plant.

Certain factors should be considered for the installation of a plant ventilating air scrubber.[19] These include the following.

1. Different types of operations should be separated. For example, protein meal grinding and storage should be isolated from rendering processes and raw material receiving in order to prevent bacterial contamination of the protein meal.
2. Ventilation air should flow from low-odor sources to high-odor sources. Thus air from the finished product storage area should flow toward the process operating area, where the scrubber is normally located. This also applies to the raw material receiving area.
3. Building openings need to be controlled in some positive manner. The overhead truck doors need to be kept closed whenever possible. A building fitted with gravity intake louvers that automatically close when a door is opened may be able to maintain a constant air volume exhausted from the building.

An important consideration in designing a plant ventilating air scrubber system concerns the number of room changes per hour for various categories of rendering plant operations. Table 9 provides a listing of these ventilation rates. A range of values is provided to allow sufficient flexibility in selecting an appropriate ventilation rate that takes into consideration the presence of an operator(s), emission of vapors or odors into the room atmosphere, ambient temperature (summer conditions), and raw material being processed.

A typical plant ventilation air scrubber is shown in Figure 7 as a single packed tower through which NaOCl solution is circulated. These scrubbers have capacities normally ranging from 40,000 to 80,000 ft^3/min.

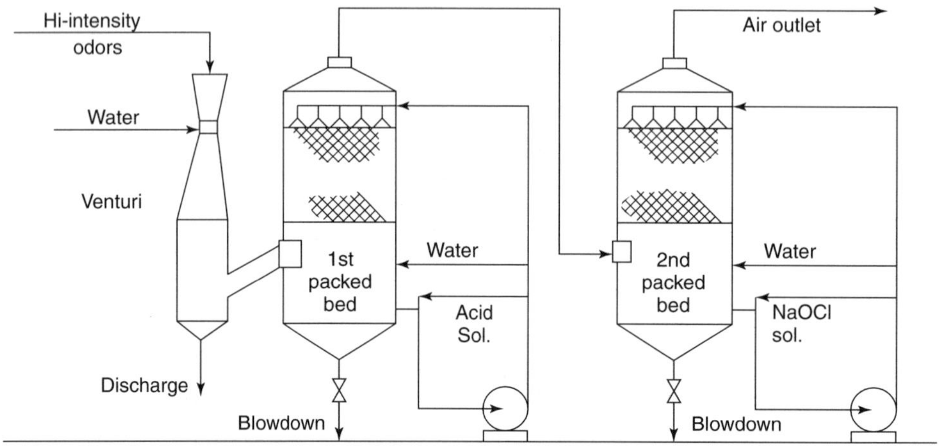

Figure 6. Multistage Scrubbing of Rendering Process Odors.

Table 8. Multistage Scrubbing of Rendering Process Odors

Scrubber Category	Exhaust Flow (ft^3/min)	Scrubber Solutions			Odors Units/scf[a]	
		First	Second	Third	Inlet	Outlet
Venturi and one packed tower	32,000	Water	NaOCl		5000–20,000	50–100
Venturi and two packed towers	7,500	Na_3PO_4	H_3PO_4	NaOCl	14,000	185

[a]ASTM syringe method. *Source*: Refs. 15 and 18.

Table 9. Ventilation Guidelines for Rendering Plant Processing Areas

Processing Area Category	Number of Room Changes per Hour
Batch rendering system operating area	20–40
Continuous rendering system (suitably enclosed)	10–20
Grease melting and processing	10–20
Raw-material storage and handling	5–15
Fat processing (filtering and bleaching)	5–10
Protein meal milling and conveying	5–10
Fat storage	3–5
Protein meal storage	3–5

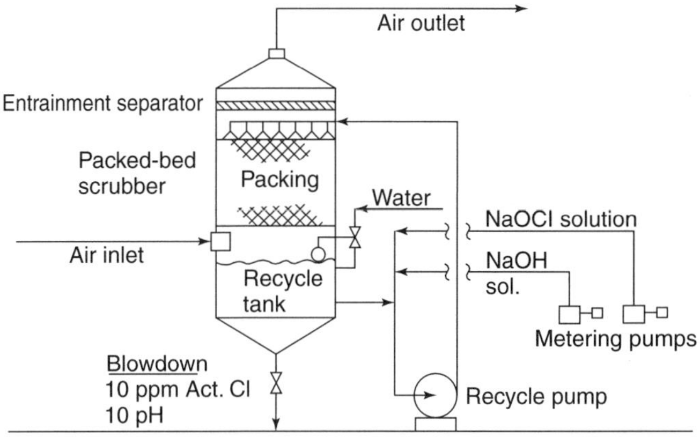

Figure 7. Packed-Bed Scrubbing of Plant Ventilating Air.

The addition of NaOCl and caustic soda is controlled by means of oxidation–reduction potential (ORP) and pH measurements. This type of scrubber is capable of achieving 95% or more of odor removal based on inlet values of 2000 to 5000 and exhaust values of 100 to 200 (IITRI dynamic olfactometer).

Other types of wet scrubbers that have been successfully applied to treating the plant ventilating air in rendering plants include the following.

1. Cross flow, packed-bed scrubber where the air flows horizontally and is contacted by the scrubbing solution as it flows downward by gravity through the packing. The capacity for this type of scrubber ranges above 100,000 cfm, and the scrubbing solution is recirculated at a rate equivalent to 8 gal/1000 ft^3 of odorous air. This type of scrubber is described by Frega and Prokop.[20]
2. Cross-flow, coarse-spray scrubber where the air flows horizontally past a series of baffles countercurrent to the spray pattern. The capacity for this type of scrubber ranges from 38,000 to 150,000 cfm, and the scrubber solution is recirculated through each stage of scrubbing at 7.5 gal/1000 ft^3 of odorous air. Either one or two stages are furnished where different scrubbing solutions are normally used in both stages. A two-stage scrubber of this type is described by Prokop.[18]
3. Mist spray, cocurrent flow scrubbers where very fine droplets are formed by atomizing nozzles normally supplied with compressed air. The mist spray at the top of the scrubber is directed downward in the same direction as the flow of odorous air. The capacity for this type of scrubber is available up to 75,000 cfm, with the once-through scrubbing solution delivered at 0.5 gal/10,000 ft^3 of odorous air.

Current practice for designing wet scrubbing systems to treat both the high-intensity process odor and plant ventilating air is to combine both types of scrubbers into a single system. A three-stage scrubber system is described[20] that consists of the following.

1. A venturi scrubber in series with a vertical countercurrent packed tower to treat the high-intensity odors, which include the Rotadisc noncondensibles, the drainer screw, and two Duke pressors. The scrubbing solutions consist of fresh water circulated through the venturi at 70 gpm and sulfuric acid solution at 2 pH recirculated through the packed tower at 75 gpm. This two-stage scrubber system treats 6300 cfm of odorous air.
2. A horizontal, cross-flow, packed-bed scrubber that receives the exhaust from the two-stage scrubber system described above and the plant ventilating air. Sodium hypochlorite solution at 9 pH is recirculated through this packed-bed scrubber at 575 gpm and its capacity is 70,000 cfm.

Table 10 summarizes the odor sensory data obtained with this three-stage scrubber system.

The overall reduction of the high-intensity odors by three stages of scrubbing (venturi, small packed tower, and large packed bed) exceeded 99%. The plant ventilating air scrubber had an odor removal efficiency ranging from 90% to 95%, depending on the inlet odor level.

Other Methods of Odor Control

Biofilter technology developed during the past 20 years in Europe and elsewhere has been applied to the treatment of rendering odors.[21,22] Biofilters consist of large beds of porous media that are capable of adsorbing odorous gaseous compounds and reducing these by aerobic microbial action to nonodorous components.

Two basic types of biofilter media influence flow rate. Compost, peat moss, heather, and other fibrous media, because of their greater porosity, allow high airflow rates ranging from 35 to 180 m^3/h/m^2 (2 to 10 cfm/ft^2) of bed surface without excessive pressure drop resulting. This category of biofilter is analogous to the plant ventilating air wet scrubbers described earlier. Rendering plants currently are operating such biofilters in Canada, Denmark, France, Holland, Germany, New Zealand, and the United States.

Table 10. Multiple-Stage Scrubbing of Process Odor and Plant Ventilating Air

	Odor Dilution to Threshold[a]	
	Average	Range
High-intensity odor scrubber		
Venturi scrubber inlet	112,000	42,900–175,000
Plant ventilating air scrubber		
Inlet	3,810	2,180–6,180
Outlet	332	165–628

[a] IITRI dynamic olfactometer. *Source*: Ref. 20.

Table 11. Soil-Bed Treatment of Cooker Noncondensibles

Date	4/11/84	4/12/84	6/26/85	6/27/85
Noncondensibles[a]	52,500	29,000	93,000	209,000
Soil-bed exhaust[a]	43	23	46	120
Odor removal (%)	99.9	99.9	99.9	99.9

[a]Odor dilution to threshold values (IITRI dynamic olfactometer). *Source*: Ref. 23.

Soil-bed systems are another type of biofilter. Less porous media such as soil result in lower airflow rates, ranging from 2 to 10 $m^3/h/m^2$ (0.1 to 0. 5 cfm/ft^2) of bed surface. Such a system is described[23] for a soil bed with a surface area of 4500 ft^2 that treats 650 cfm of Duke Equacooker noncondensibles. A bed depth of 60 cm (24 in.), was provided and the pressure drop across the bed was measured at 5 cm (2 in.) of water column. Two series of odor sensory tests were conducted approximately 14 months apart. Composite samples of soil-bed exhaust were evaluated, along with samples of the inlet noncondensibles. Table 11 summarizes the results, which are expressed as odor dilution to threshold (IITR1 dynamic olfactometer). These odor sensory performance results clearly indicate the potential for soil-bed treatment of high-intensity process odors of relatively low volume. Compared with wet scrubbing systems, soil beds require considerably less initial investment and have lower operating costs.

Other odor control methods include activated-carbon adsorption and catalytic oxidation. No actual odor sensory performance tests related to rendering plant emissions have been reported for these methods since the publication of the second edition of this manual.

A very important part of the overall design of an odor control system is related to the emission from the control system into the surrounding atmosphere. Specific precautions should be taken to ensure that the stack design takes into account the height and exit velocity needed to provide the desired dispersion characteristics.

In performing atmospheric dispersion calculations to predict downwind ground-level concentrations, it should be recognized that estimates based on averaging times of 10–60 min typical of traditional models may be in error. Objectionable odors are detected in much shorter time intervals and peak concentrations for these short times can be much higher than the long-term concentrations for traditional models. Jann and Cha[24] discuss the importance of the odor dispersion averaging times being expressed in seconds in order to be able to predict the maximum odor concentration at ground level.

REFERENCES

1. *U.S. Production, Consumption, Trade and Stocks of Tallow/Grease, Lard and Animal Protein Meals; Estimates and Projections*, P-318 (April 1980), prepared by Development Planning & Research Associates, Manhattan, KS, for the USDA Foreign Agricultural Service, Washington, D.C.
2. W. H. Prokop, "The Rendering Industry—a Commitment to Public Service," *The Original Recyclers*, National Renderers Association, VA, 1996, pp. 17–21.
3. W. H. Prokop, "Rendering Systems for Processing Animal By-Product Materials," *J. Am Oil Chem. Soc.* **62**(4), 805–11, April 1985.
4. W. H. Prokop, "Low Temperature Rendering by ATLAS System," *Meat Proc.*, p. 24, April 1986.
5. W. R. Boehme, "New Drying Techniques Preserve Lysine in Blood Meals," *Director's Digest No. 116*, Fats and Proteins Research Foundation, Fort Myers Beach, FL, February 1974. Feather Meal, National Renderers Association, Washington, D.C., revised March 1985, 11 pp.
7. R. T. Walsh, "Reduction of Inedible Rendering Matter," *Air Pollution Engineering Manual*, 2nd ed., J. A. Danielson, Ed., U.S. Environmental Protection Agency, Research Triangle Park, NC, Publication No. AP-40, May 1973, p. 823.
8. T. A. Burgwald, "Identification of Chemical Constituents in Rendering Industry Odor Emissions," Project No. C8172, IIT Research Institute, Chicago, IL, January, 1971, p. 39 (supported by Fats and Proteins Research Foundation).
9. R. H. Snow and H. G. Reilich, "Investigation of Odor Control in the Rendering Industry," Project No. C8210, IIT Research Institute, Chicago, August 1972, p. 117 (performed under Contract No. 68-02-0260 with U.S. Environmental Protection Agency and with the support of Fats and Proteins Research Foundation).
10. *Odor Thresholds for Chemicals with Established Occupational Health Standards*, American Industrial Hygiene Association, Akron, OH, 1989.
11. J. E. Moore and R. S. O'Neill, "Odor as an Aid to Chemical Safety: Odor Thresholds Compared with Threshold Limit Values and Volatilities for 214 Industrial Chemicals in Air and Water Dilution," *J. Appl. Toxicol.* **3** 6, 1983.
12. G. Leonardos, D. Kendall, and N. Bernard, "Odor Thresholds for 53 Commercial Chemicals," *J. AWMA* **19**(2), 91, 1969.
13. F. Jarke, "Investigation of High-Intensity Odors," Project C8563, IIT Research Institute, Chicago, June 1981 (supported by Fats and Proteins Research Foundation).
14. W. H. Prokop, "Control Methods for Treating Odor Emissions from Inedible Rendering Plants," Paper 91-146.8 at Annual Meeting of AWMA, Vancouver, B.C., Canada, 1991.
15. W. H. Prokop, "Wet Scrubbing of High Intensity Odors from Rendering Plants," *Proceedings AWMA Specialty Conference on Odor Control Technology II*, Pittsburgh, PA, March 1977, pp. 153–66.
16. G. C. White, *Handbook of Chlorination*, Van Nostrand Reinhold Co., New York, 1972, p. 600.
17. D. A. Lundgren, L. W. Rees, and L. D. Lehmann, "Odor Control by Chemical Oxidation: Cost, Efficiency and Basis for Selection," Paper 72-116 at Annual Meeting of AWMA, Miami Beach, FL, 1972.
18. W. H. Prokop, "Wet Scrubbing of Inedible Rendering Plant Odors," *Proceedings AWMA Specialty Conference on Odor Control Technology I*, Pittsburgh, PA, 1974, pp. 132–50.
19. J. M. Sweeten, G. I. Wallin, and J. A. Chittenden, "Control of Rendering Plant Odors: State of the Art," Paper 80-6004 at Annual Meeting of American Society of Agricultural Engineers, 1980.
20. V. Frega and W. H. Prokop, "Wet Scrubbing of Multiple Source Odors in a Rendering Plant," Paper 81-39.2 at Annual Meeting of APCA, Philadelphia, PA, 1981.
21. M. B. Rands, D. E. Cooper, C. P. Woo, et al., "Compost Filters for 1-12S Removal from Anaerobic Digestion and Rendering Exhausts," *J. WPCF* **53**, 185, 1981.
22. W. Koch, H. G. Liebe, and B. Striefler, "Betfiebserfahrungen mit Biofiltem zur Renduzierung geruchsin-tensiver Emissionen," *Staub-Reinhaultung der Luft* **43**, 488, 1982.
23. W. H. Prokop and H. L. Bohn, "Soil Bed System for Control of Rendering Plant Odors," *J. AWMA* **35**(12), 1332–38, 1985.
24. P. R. Jann and S. S. Cha, "Atmospheric Dispersion and Odor Modeling," *Odor Control for Wastewater Facilities*, Manual of Practice No. 22, Water Environment Federation, Washington, D.C., 1995.

14

METALLURGICAL INDUSTRY

PRIMARY ALUMINUM INDUSTRY

Maurice W. Wei
(with update by Editor)

The process used to make aluminum was discovered in 1886 almost simultaneously by Charles Martin Hall of Oberlin, Ohio, and Paul-Louis-Toussaint Héroult of France. Working independently, Hall and Hóroult discovered that alumina (Al_2O_3) dissolved in cryolite (Na_3AlF_6) will undergo electrolysis and produce aluminum at the carbon cathode. The Hall process was commercialized in 1888, when the first smelter to use this technology was built in Pittsburgh, Pennsylvania.

Aluminum is the second most common metallic element in the earth's crust, and economically recoverable quantities of aluminum oxide exist principally as boehmite and gibbsite in bauxite deposits found in many areas of the world. Major deposits are found in Australia, West Africa, Brazil, Jamaica, Surinam, and Venezuela. Aluminum oxide is recovered from bauxite, usually by the Bayer process. The world production of primary aluminum in 1990 was 17,832,000 tonnes (metric tons), of which 4,048,000 tonnes were produced in the United States.

Because the production of aluminum requires a significant amount of electricity, primary production facilities are generally located in countries with relatively low-cost energy and/or a good supply of bauxite. Thus most aluminum plants were originally built in North America, Latin America, and certain areas of Europe. With the discovery of bauxite in Australia and the availability of coal and gas there, the Australian aluminum industry has grown rapidly in the past few years. The countries formerly comprising Russia and the Peoples Republic of China are also major producers of primary aluminum.

Pure aluminum is a relatively soft metal and has low tensile strength. But by alloying aluminum with other metals, such as copper, magnesium, manganese, and silicon, several series of aluminum alloys have been developed with properties tailored to their specific applications. Some properties exhibited by these alloys are a high strength-to-weight ratio, corrosion resistance, and the ability to be extruded, formed, machined, and to accept certain finishes.

In the United States, the major uses for aluminum are in containers and packaging (28.3%), such as beverage and food cans, followed by transportation (17.3%), building and construction (15.9%), electrical (7.8%), consumer durables, machinery, and equipment (12.4%), and others (3.4%). Over 14% of the U.S. primary production was exported.

Aluminum scrap retains high residual value, resulting in a thriving secondary aluminum industry. In 1990, 2,393,000 tonnes of aluminum were recovered from scrap. Reclamation from aluminum cans accounted for about 877,300 tonnes. Over 63% of the estimated 83 billion cans produced in this country were recycled, and this percentage has been going up every year. The target for 1995 was 75%. As energy and other costs go up, and the public's concern over the recyclability of consumer and commercial products increases, the incentive for recycling increases. Recycling aluminum scrap saves 95% of the energy required to make new aluminum from bauxite.

PROCESS DESCRIPTION

Alumina Production

Aluminum is produced by the electrolysis of alumina, and the latter is produced from bauxite by the Bayer process. In the Bayer process, aluminum hydroxides or hydrates are selectively separated from the other components by extracting them with sodium hydroxide to form sodium aluminate. Alumina is then precipitated from the sodium aluminate solution, washed, and calcined (Figure 1).

Bauxite Preparation. To prepare bauxite for processing, mined ore is crushed and ground in ball mills to a finely divided state. Most bauxite, as mined, has a relatively low moisture

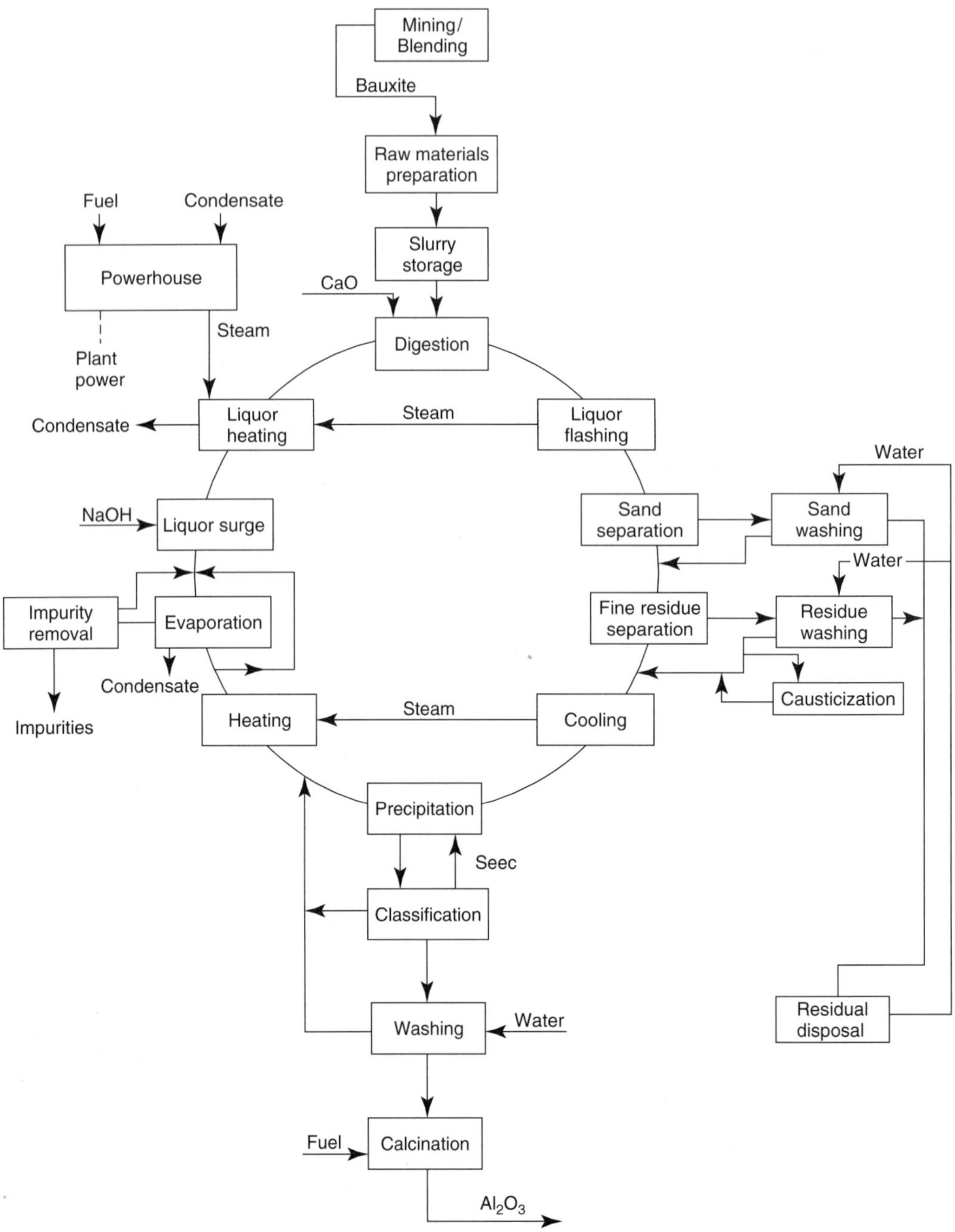

Figure 1. Bayer Process Flowsheet.

content. However, in certain parts of the world, the moisture content is very high, and the bauxite must be dried in rotary kilns before it is sent to storage. The drying operation emits significant amounts of bauxite dust, requiring dust control equipment. The types of control equipment used are usually mechanical collectors such as a cyclone followed by either a venturi scrubber or an electrostatic precipitator (ESP).

Digestion of Bauxite. To produce high-purity alumina suitable for aluminum reduction cells, finely divided bauxite containing 30–70% alumina is slurried with sodium hydroxide solution and reacted at a high temperature and pressure in reactors called digesters. Sodium aluminate is formed according to the general equation

$$Al_2O_3 \cdot x(H_2O) + 2NaOH \rightarrow 2NaAlO_2 + (1+x)H_2O$$

while leaving behind substantially all of the silicon, iron, titanium, and calcium oxides as insoluble components in the solid waste residue.

Clarification and Precipitation. The hot slurry is usually cooled by flash evaporation, producing steam that is used to preheat incoming slurry. Next, the slurry is processed in a series of clarification steps designed not only to separate the solid residue from the liquor, but also to rid the solution of the impurities that would lower the quality of the aluminum when the alumina later is electrolyzed. These impurities are mainly oxides of silicon and iron.

If the bauxite contains coarse material (mainly sand), it is first removed from the slurry in hydrocyclones, and starch or some settling agent is added to the overflow before it enters a clarifier. Starch promotes agglomeration and settling of the fine particles, and the relatively clear overflow from the clarifier is filtered in either filter presses or sand filters to remove any remaining suspended solids. The clarifier underflow is then washed in thickeners to remove as much caustic as possible before it is pumped to a bauxite residue pond. The filtrate containing sodium aluminate, now at 120°F to 160°F, is processed in precipitators where aluminum hydroxide crystallizes and precipitates as shown in the following.

$$2NaAlO_2 + 4H_2O \rightarrow 2Al(OH)_3 + 2NaOH$$

After precipitation, the slurry containing trihydrate crystals is sent to classification, where the product is separated according to crystal size. The coarse product is washed to remove excess sodium hydroxide, dewatered on vacuum filters, and sent to calcination. The finer crystals are recycled to the precipitation process as seed, thereby providing nucleation sites for precipitation to occur. Spent liquor from the precipitators is recycled to the digesters.

Calcination. The coarse alumina is calcined in rotary kilns or fluid-bed calciners at about 1800°F. During calcination, water (mechanically and chemically bound) is driven off. Calciners produce hot flue gases containing alumina and water vapor, and the usual types of control equipment consist of cyclonic separators followed by ESPs. The control equipment not only minimizes air pollution, but also recovers valuable product as well. Calcined alumina is stored for shipment to the smelter as feedstock for the aluminum electrolytic cells.

Aluminum Production

Electrolytic Reduction. Aluminum production is carried out in a semibatch manner in large electrolytic cells called pots with a dc input of up to 280,000 amperes at about 5 volts. The pot, a rectangular steel shell ranging in size from 30–50 ft long, 9–12 ft wide, and 3–4 ft high, is lined with a refractory insulating shell on which carbon blocks are placed to form the cathode. Figure 2 shows a cross-section of the commonly used center-worked prebaked cell. Steel collector bars are inserted in the bottom blocks to carry current from the pot. The insulation is designed to allow enough heat loss at the vertical walls that a protective layer of frozen electrolyte will form, but not on the bottom, where electrical continuity must be maintained

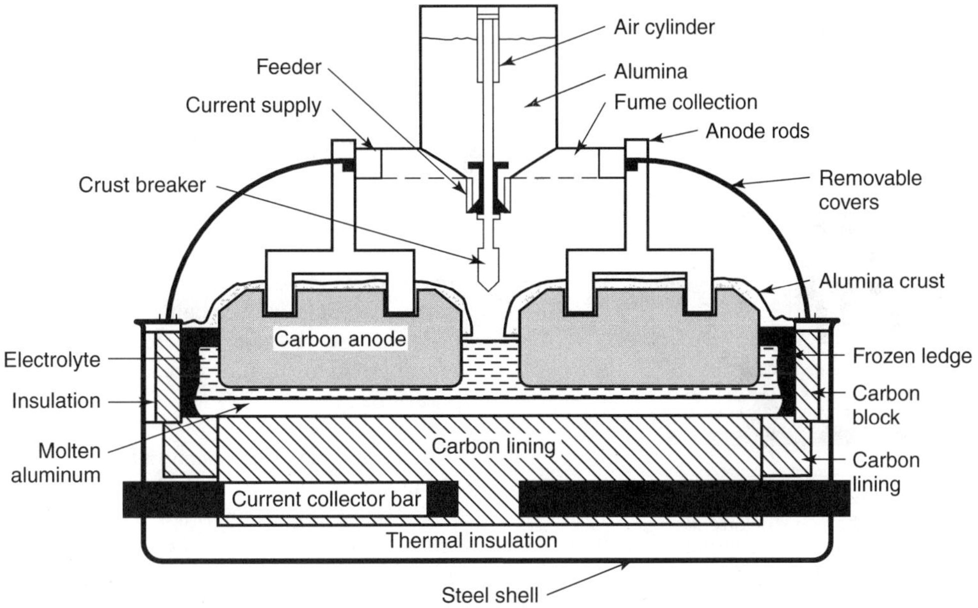

Figure 2. Aluminum Electrolyzing Cell—Prebaked Anode (*Kirk–Othmer Encyl. of Chem. Tech.*, Vol. 2, 4th ed.).

between molten aluminum and the cathode block. Molten cryolite is placed in the cavity formed by the cathode blocks, and anodes, also of baked carbon, are immersed in the cryolite, completing the electrical path. There are usually two rows of 20–30 closely spaced anodes hung on busbars along the full length of the pot that supplies the current.

Alumina is fed to the pots either from a crane-mounted hopper or from a hopper running the full length of the pot attached to the pot superstructure. A hole is first punched in the frozen cryolite crust on the pot by a crust breaker on the overhead crane or by breakers on the superstructure. Alumina is then metered to the opening in the crust. The anodes can be lowered or raised simultaneously or individually. As the anodes are consumed by combining with oxygen from the alumina, they are lowered into the bath and are replaced when most of the carbon is gone. The carbon butt from the spent anode is allowed to cool and the bath adhering is later removed, usually by shot blasting. The anode butt is then crushed and recycled to make more anodes. Operation is semicontinuous because the pots are fed periodically and tapped about once a day as the aluminum reaches a certain level.

In the Soderberg cells, continuous self-baking anodes are used instead of prebaked anodes (Figure 3). A paste consisting of coal tar pitch and petroleum coke is fed to the top of a steel casing and is baked into an anode by heat from the cell and also by heat produced by the current passing through the anode. Electricity is introduced by the spikes inserted into the anode either vertically or at an angle from the sides. As the lower portion of the anode is consumed, the mass of baked material moves downward through the casing. More paste is added to the top, and the spikes must be pulled and reset at a higher level.

Electrolysis of the molten bath involves several complex electrode reactions. However, in the overall reaction, molten aluminum is produced at the cathode, and oxygen, released at the anode, reacts with carbon to form carbon dioxide (CO_2).

$$Al_2O_3 + \tfrac{3}{2}C \rightarrow 2Al + \tfrac{3}{2}CO_2$$

About 0.4–0.5 lb of carbon is consumed for each pound of aluminum produced.

The major gaseous emissions besides CO_2 are hydrogen fluoride (HF) gas, sulfur dioxide (SO_2), and carbon monoxide (CO). Particulate emissions are particulate fluoride, carbon dust, and alumina. The HF gas is produced by the hydrolysis of fluoride compounds due to adsorbed water, to water of hydration associated with the alumina, and from hydrogen in the anodes. Particulate fluoride results from the vaporization of fluoride compounds from the molten bath and subsequent condensation in the cooler zone above the pot. Sulfur dioxide comes from the sulfur present in the coke and coal tar pitch binder in the anodes. The transportation and handling of the materials used in the process, such as alumina, coke, and pitch, are also sources of emissions.

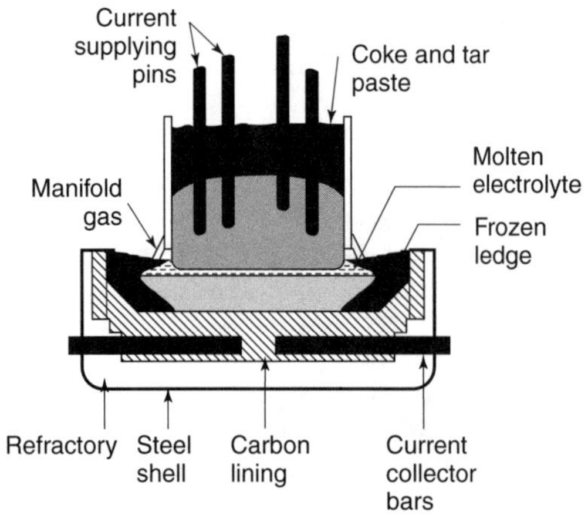

Figure 3. Aluminum Electrolytic Cell—Soderberg (*Kirk–Othmer Encycl. of Chem. Tech.*, Vol. 2, 4th ed.).

In the Soderberg cell, unlike the prebake cell, there is continuous evolution of hydrocarbon vapors from the anode paste that not only creates a less desirable working environment, but also makes collection and treatment of the cell gases more difficult. On the cells with vertical studs, gas burners are sometimes installed at the gas collection skirt to destroy some of the pitch fumes prior to the pollution control system. Although Soderberg cells avoid costly anode baking facilities, the majority of new smelters are designed with prebake cells because of their higher current efficiency and more manageable emission problems.

Anode Production

Green Mill. Anodes for prebake pots are prepared in two stages. The first step is the production of green anodes, where pitch and petroleum coke are mixed and heated to 140–180°F in either batch or continuous mixers. In the batch mixer, liquid or pencil pitch and coke are added batchwise, and each batch is heated and blended separately until the mixture is uniformly mixed and at the proper temperature. The batch of anode mix is then discharged and transported in a covered conveyor belt to the anode-forming machine. Coke dust and hydrocarbon vapors from pitch evolve during the mixing, discharging, and conveying of the mix to the anode former.

To produce good high-density anodes, an important parameter is the proportion of coarse and fine coke particles used in the mix. To produce different size fractions, the coke is sized in classifiers and transported to different tanks according to particle size. Coke dust is emitted in this operation.

Included with the coke used in this operation are other carbonaceous materials obtained from anode butts. These butts are first mechanically cleaned to remove most of the adhering bath material and then crushed to the required size fractions. Good butt cleaning to remove the fluoride compounds will be discussed in the next section, since it affects both fluoride emission and furnace performance.

With the continuous mixer, pitch and coke are fed continuously to the inlet of the mixer, and as the material travels through the mixer, it is heated, blended, and then transported by conveyor belt to the anode-forming area. At the anode former, the mix is metered into molds and compressed to form green anodes. Hydrocarbons are also emitted from this operation.

Anode Baking. The second step in anode production is the baking operation. Green anodes are placed in a baking furnace consisting of refractory-lined pits with flues on both sides of each pit. The anodes are then covered with coke and fired for 30 h or less at temperatures up to 2400°F. Hot flue gas from the firing section is used to preheat a section of packed green anodes located ahead of the fired section. Simultaneously, combustion air for the burners is drawn through the flues of the previously fired section that preheats the air and also cools the baked anodes.

Thus at any given moment, one section is being cooled while another is being fired and a third section is undergoing preheat. After firing is completed, the fired section is cooled, the preheat section is fired, and a new section is preheated. A movable waste-gas manifold is placed after the preheat section to collect the flue gases and direct them to the waste gas main, which connects to the fume control equipment. In a baking furnace operation, several fires are usually in progress at a given moment. Because firing of the anodes proceeds around the building in a circular manner, the furnace is usually referred to as a ring furnace.

Negative pressure created in the flues by the waste gas fans pulls air through the packing over the anodes into the flues, where it blends with products of combustion. Flue gas from the baking furnace contains HF, SO_2, CO_2, CO, and hydrocarbons from pitch. Particulate emissions are carbon dust and particulate fluoride. Gaseous and particulate fluorides are produced from residual bath material in the anode butts.

Peak hydrocarbon emission occurs when the firing frame containing the burners and the waste-gas manifold are moved. At this point, firing of the preheat section commences and green anodes in the preheat section are exposed to heat for the first time. The temperature of the flue gas will normally range from 300°F to 450°F, with the lower temperature usually occurring when a fire is moved.

As mentioned above, bath material adhering to the anode butts is removed before the butts are crushed and recycled. Thorough removal of the bath is important, since it not only will reduce the fluoride emissions, but will provide several other benefits. The HF acid aggressively attacks silica in the furnace refractory; therefore, with lower HF in the flue gases, the furnace flues will require less maintenance. Furnaces in good condition will permit better temperature control and higher finishing temperatures and thus higher-quality anodes. Higher flue gas temperatures also promote better incineration of hydrocarbons in the flue gas, which not only provides heat to the process, but also results in lower emissions.

In contrast, furnaces with poor flue conditions and high air in-leakage invariably produce lower-quality anodes and higher hydrocarbon emissions. This is a good example of the principle that a well-controlled process will usually produce a better product and less pollution.

There is another type of furnace such as the Riedhammer furnace, which by design will produce a tar by-product. The furnace is closed on top to prevent air in-leakage and to create a nonoxidizing environment. Tar in the off-gases is condensed, collected, and sold or reused as binder in green anode production. The environmental control system, therefore, must be capable of collecting tar as well as fluorides. Such a system is described later in the section on baking furnace emission control systems. The design of this furnace is more complex than the open ring furnace, but a more uniform temperature distribution is claimed for the closed furnace.

AIR EMISSIONS CHARACTERIZATION

Potroom

Potroom Emissions. An aluminum pot will typically emit 20–35 kg per tonne of gaseous and particulate fluoride and roughly an equal amount of particulate matter. The emissions factors for the various operations in primary aluminum production are summarized in Tables 1–4 reproduced from Tables 7.1-2 to 7.1-5 of *AP-42 Supplement A to Compilation of Air Pollutant Emission Factors*.[1] U.S. federal New Source Performance Standards (NSPS) require emissions of no more than 1 kg fluoride per tonne of aluminum from the Soderberg plant and 0.95 kg fluoride per tonne from prebake plants with 0.05 kg fluoride per tonne from the anode bake plant. The standard also requires less than 10% opacity. Each cell is equipped with a hooding and manifold system that captures and conveys the emission to the fume control equipment. The temperature of the pot exhaust gases entering the control equipment ranges from 150°F to 300°F, depending on the ambient temperature.

Pot Hooding. Pot hooding is an integral part of the smelter fume control system. The purpose of the hooding is to enclose the pot as tightly as possible to collect the cell off-gases and, at the same time, to allow the pot attendant to change anodes and perform other functions. This enclosure may consist of individual panels in close contact forming a continuous covering or large continuous covers on each side of the pot. Each panel is removable to allow access to the cell. Cell covers must perform many functions well. They must seal tightly as they lie side by side on the cell, and they should be rugged, but also fairly light, so that they can be easily removed by the operator.

From this hooding, the cell off-gases rise and enter a rectangular enclosure above the cell superstructure from which they enter an exhaust duct. The duct from each cell is connected to a manifold running the length of the potroom that connects to the fume control system. In a well-designed system, the off-gases are exhausted uniformly from the whole cell, and draft from each pot is adjusted so that there is equal volume from each cell.

In theory, hooding efficiency is the amount of off-gas produced by the electrolytic process that is collected by the hooding system as a percentage of the total off-gas generated by the cell.

$$\text{hooding efficiency (\%)} = \frac{\text{off-gas collected by hoods} \times 100}{\text{off-gas produced by cells}}$$

In actual practice, hooding efficiency is calculated as follows:

$$\text{hooding efficiency (\%)} = \frac{A \times 100}{A + B}$$

where A is the measured total fluoride to the fume treatment system; and B, the measured total fluoride escaping from the roof monitor. A good pot hooding system should be at least 95% efficient on a center-worked prebaked line.

Much effort has gone into the design of pot hood systems over the years, and new designs are being tested constantly in an effort to improve their performance and durability. It should be mentioned that a primary source of the fluoride emission from a smelter is from around the pot enclosure, either from leaks or when the covers are removed. Significant amounts of fluoride are also emitted to the environment when an anode is removed from the pot and placed in the aisle. The hot carbon continues to air burn for some time while fluorides evolve from the adhering bath material.

Table 1. Emission Factors for Primary Aluminum Production Processes[a,b] (Emission Factor Rating: A)

	Total Particulate[c]		Gaseous Fluoride		Particulate Fluoride	
Operation	kg/Mg	lb/ton	kg/Mg	lb/ton	kg/Mg	lb/ton
Bauxite grinding						
Uncontrolled	3.0	6.0	Neg	NA		
Spray tower	0.9	1.8	Neg	NA		
Floating-bed scrubber	0.85	1.7	Neg	NA		
Quench tower and spray screen	0.5	1.0	Neg	NA		
Aluminum hydroxide calcining						
Uncontrolled[d]	100.0	200.0	Neg	NA		
Spray tower	30.0	60.0	Neg	NA		
Floating-bed scrubber	28.0	56.0	Neg	NA		
Quench tower	17.0	34.0	Neg	NA		
ESP	2.0	4.0	Neg	NA		
Anode baking furnace						
Uncontrolled	1.5	3.0	0.45	0.9	0.05	0.1
Fugitive	NA	NA	NA	NA	NA	NA
Spray tower	0.375	0.75	0.02	0.04	0.015	0.03
ESP	0.375	0.75	0.02	0.04	0.015	0.03
Dry alumina scrubber	0.03	0.06	0.0045	0.009	0.001	0.002
Prebake cell						
Uncontrolled	47.0	94.0	12.0	24.0	10.0	20.0
Fugitive	2.5	5.0	0.6	1.2	0.5	1.0
Emissions to collector	44.5	89.0	11.4	22.8	9.5	19.0
Multiple cyclones	9.8	19.6	11.4	22.8	2.1	4.2
Dry alumina scrubber	0.9	1.8	0.1	0.2	0.2	0.4
Dry ESP plus spray tower	2.25	4.5	0.7	1.4	1.7	3.4
Spray tower	8.9	17.8	0.7	1.4	1.9	3.8
Floating-bed scrubber	8.9	17.8	0.25	0.5	1.9	3.8
Coated bag filter dry scrubber	0.9	1.8	1.7	3.4	0.2	0.4
Cross-flow packed bed	13.15	26.3	3.25	6.7	2.8	5.6
Dry plus secondary scrubber	0.35	0.7	0.2	0.4	0.15	0.3
Vertical Soderberg stud cell						
Uncontrolled	39.0	78.0	16.5	33.0	5.5	11.0
Fugitive	6.0	12.0	2.45	4.9	0.85	1.7
Emissions to collector	33.0	66.0	14.05	28.1	4.65	9.3
Spray tower	8.25	16.5	0.15	0.3	1.15	2.3
Venturi scrubber	1.3	2.6	0.15	0.3	0.2	0.4
Multiple cyclones	16.5	33.0	14.05	28.1	2.35	4.7
Dry alumina scrubber	0.65	1.3	0.15	0.3	0.1	0.2
Scrubber plus ESP plus spray screen and scrubber	3.85	7.7	0.75	1.5	0.65	1.3
Horizontal Soderberg stud cell						
Uncontrolled	49.0	98.0	11.0	22.0	6.0	12.0
Fugitive	5.0	10.0	1.1	2.2	0.6	1.2
Emissions to collector	44.0	88.0	9.9	19.8	5.4	10.8
Spray tower	11.0	22.0	3.75	7.5	1.35	2.7
Floating bed scrubber	9.7	19.4	0.2	0.4	1.2	2.4
Scrubber plus wet ESP	0.9	1.8	0.1	0.2	0.1	0.2
Wet ESP	0.9	1.8	0.5	1.0	0.1	0.2
Dry alumina scrubber	0.9	1.8	0.2	0.4	0.1	0.2

[a] For bauxite grinding, expressed as kg/Mg (lb/ton) of bauxite processed. For aluminum hydroxide calcining, expressed as kg/Mg (lb/ton) of alumina produced. All other factors are per Mg (ton) of molten aluminum product. NA = not available. Neg = negligible.

[b] Sulfur oxides may be estimated, with an emission factor rating of C, by the following calculations.

Anode baking furnace, uncontrolled SO_2 emissions (excluding furnace fuel combustion emissions):

$20(C)(S)(1 - 0.01K)$ kg/Mg [$40(C)(S)(1 - 0.01K)$ lb/ton]

Prebake (reduction) cell, uncontrolled SO_2 emissions:

$0.2(C)(S)(K)$ kg/Mg [$0.4(C)(S)(K)$ lb/ton]

where C is the anode consumption* during electrolysis, lb anode consumed/lb Al produced;

S, the % sulfur in anode before baking; and

K, the % of total SO_2 emitted by prebake (reduction) cells.

*Anode consumption weight is weight of anode paste (coke + pitch) before backing.

[c] Includes particulate fluorides.

[d] After multicyclone.

Source: Ref. 1.

Table 2. Uncontrolled Emission Factors and Particle Size Distribution for Roof Monitor Fugitive Emissions from Prebake Aluminum Cells (Emission Factor Rating: C)

Particle size[a] (μm)	Cumulative Mass (%) ≤ Stated Size	Cumulative Emission Factor kg/Mg Al	Cumulative Emission Factor lb/ton Al
15	65	1.62	3.23
10	58	1.45	2.90
5	43	1.08	2.15
2.5	28	0.70	1.40
1.25	18	0.46	0.92
0.625	13	0.33	0.67
Total	100	2.5	5.0

[a]Expressed as equivalent aerodynamic particle diameter.

Source: Ref. 1.

Table 3. Uncontrolled Emission Factors and Particle Size Distribution for Roof Monitor Fugitive Emissions from HSS Aluminum Cells (Emission Factor Rating: D)

Particle size[a] (μm)	Cumulative Mass (%) ≤ Stated Size	Cumulative Emission Factor kg/Mg Al	Cumulative Emission Factor lb/ton Al
15	39	1.95	3.9
10	31	1.55	3.1
5	23	1.15	2.3
2.5	17	0.85	1.7
1.25	13	0.65	1.3
0.625	8	0.40	0.8
Total	100	5.0	10.0

[a]Expressed as equivalent aerodynamic particle diameter.

Source: Ref. 1.

Table 4. Uncontrolled Emission Factors and Particle Size Distribution for Primary Emissions from HSS Reduction Cells (Emission Factor Rating: D)

Particle size[a] (μm)	Cumulative Mass (%) ≤ Stated Size	Cumulative Emission Factor kg/Mg Al	Cumulative Emission Factor lb/ton Al
15	63	30.9	61.7
10	58	28.4	56.8
5	50	24.5	49.0
2.5	40	19.6	39.2
1.25	32	15.7	31.4
0.625	26	12.7	25.5
Total	100	49.0	98.0

[a]Expressed as equivalent aerodynamic particle diameter.

Source: Ref. 1.

Finally, it should be pointed out that fume control systems are expensive, costing \$35–\$50 per actual cubic foot per minute of gas treated (1992). Minimizing the exhaust volume from each pot can produce considerable savings in the cost of the control equipment. A well-designed and well-maintained pot hooding will require less exhaust volume per pot and, therefore, a smaller, less expensive fume treatment system.

Data on potroom and anode baking operations, environmental control systems, regulations, and capital cost are summarized in the appendix.

Control Systems

Dry Systems. Most smelters constructed within the past 20 years use alumina to scrub the gaseous fluoride emission and the alumina containing fluoride is then fed to the cells. The two types of dry scrubbing systems in use today differ primarily in the manner in which the potroom gases are brought into contact with alumina. One system capitalizes on the ability of a fluid-bed reactor to promote intimate mixing between solids and fluids. In this process, potroom gases fluidize a bed of alumina. In the other system, the alumina is injected into a zone of high gas velocity and turbulence. Alumina is entrained and turbulent mixing promotes a reaction between HF and alumina.

These dry scrubbing systems are capable of removing over 99% of the HF gas and particulate matter. The alumina containing HF is then used to make aluminum. With the high recovery of fluorides and their subsequent reuse, consumption of fluoride has been reduced significantly in smelters with dry scrubbing systems. Because these dry scrubber systems are so efficient in capturing particulate and gaseous fluorides, the emissions from their stacks contribute very little to the overall plant fluoride emission.

Wet Systems. A few plants still use the older control system consisting of electrostatic precipitators to collect particulate emissions, followed by spray towers to scrub the gaseous fluoride. A small amount of particulate matter is also removed in the scrubber. The scrubber effluent is then neutralized, usually by the addition of lime to form calcium fluoride, and the water is recirculated to the scrubber. Wet systems have many disadvantages, such as corrosion by hydrofluoric acid, scaling, and the expense involved in recovering the fluoride values from calcium fluoride. The only benefit of a wet scrubber system is that it removes some of the sulfur oxide in the cell off-gases. Since new smelters no longer install wet systems, only dry scrubber systems will be covered here.

Dry Scrubber Chemistry. In potroom dry scrubbers, gaseous HF reacts rapidly with alumina through chemisorption to form a monolayer of HF gas on the Al_2O_3. The HF is then converted to AlF_3 when exposed to heat in the cell. To achieve high HF-removal efficiencies, it is necessary to use alumina with a high surface area. If the surface area is too low, HF will be adsorbed in excess of a monolayer and could later be desorbed in the pot. In this case, the HF concentration in the exhaust gas stream will build up as more HF is added to the gas stream than is removed by the alumina, and high fluoride emission will result from the scrubber system.

A surface area of 45 m^2/g or greater as measured by the BET method is generally considered to be satisfactory. For example, an alumina with a surface area of 50 m^2/g will adsorb 1.79% of its weight as HF. In some earlier dry scrubbing systems, to obtain high surface area, kiln-activated alumina (KAH) was ground and then used in injection systems. For the most part, these earlier systems have been converted to the fluid-bed process.

Dry Scrubbers

Fluid-Bed Scrubber. The heart of this system is a fluid-bed reactor roughly 12 ft wide by 50 ft long and 8 ft high with a fabric filter located on top of the reactor (Figure 4). Metal-grade alumina is fed into one end, usually with a rotary vane feeder, and exits at the other end of the reactor through a seal leg. Potroom gases, introduced at the bottom of the reactor through dribble plates, fluidize the alumina while, simultaneously, HF gas reacts with the alumina. The alumina bed in the nonfluidized state is usually 4–8 in. deep. The reaction between HF and alumina is very rapid, and HF removal efficiency is 99% or more. The pressure drop through the system ranges from 12 to 20 in. w.c.

The fabric filter on top of the reactor captures and returns the entrained alumina particles to the reactor below, as well as particulate fluoride and carbon particles produced by the cells. The fume control system (Figure 5) consists of a bank of reactors (only one reactor shown) fed by alumina supplied from storage tanks. Alumina leaving the reactors is stored in

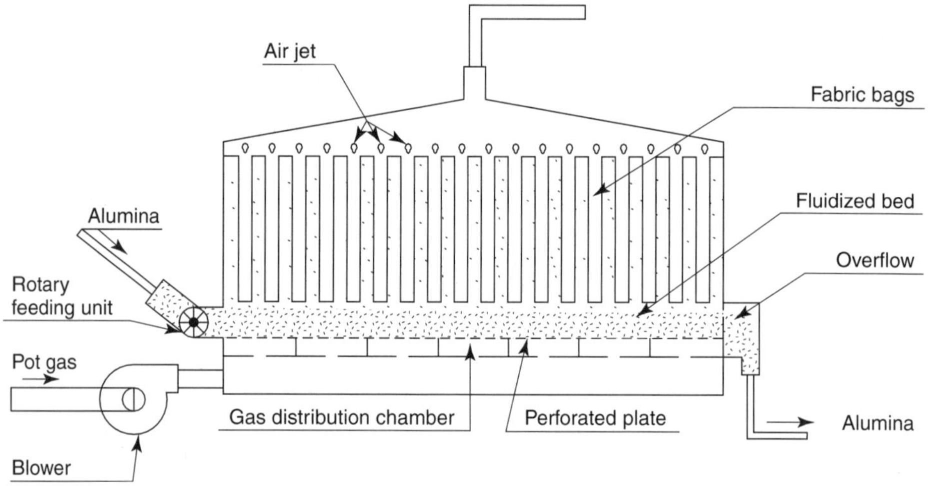

Figure 4. Fluid-Bed Reactor for Potroom Fumes (Aluminum Company of America, Pittsburgh, PA).

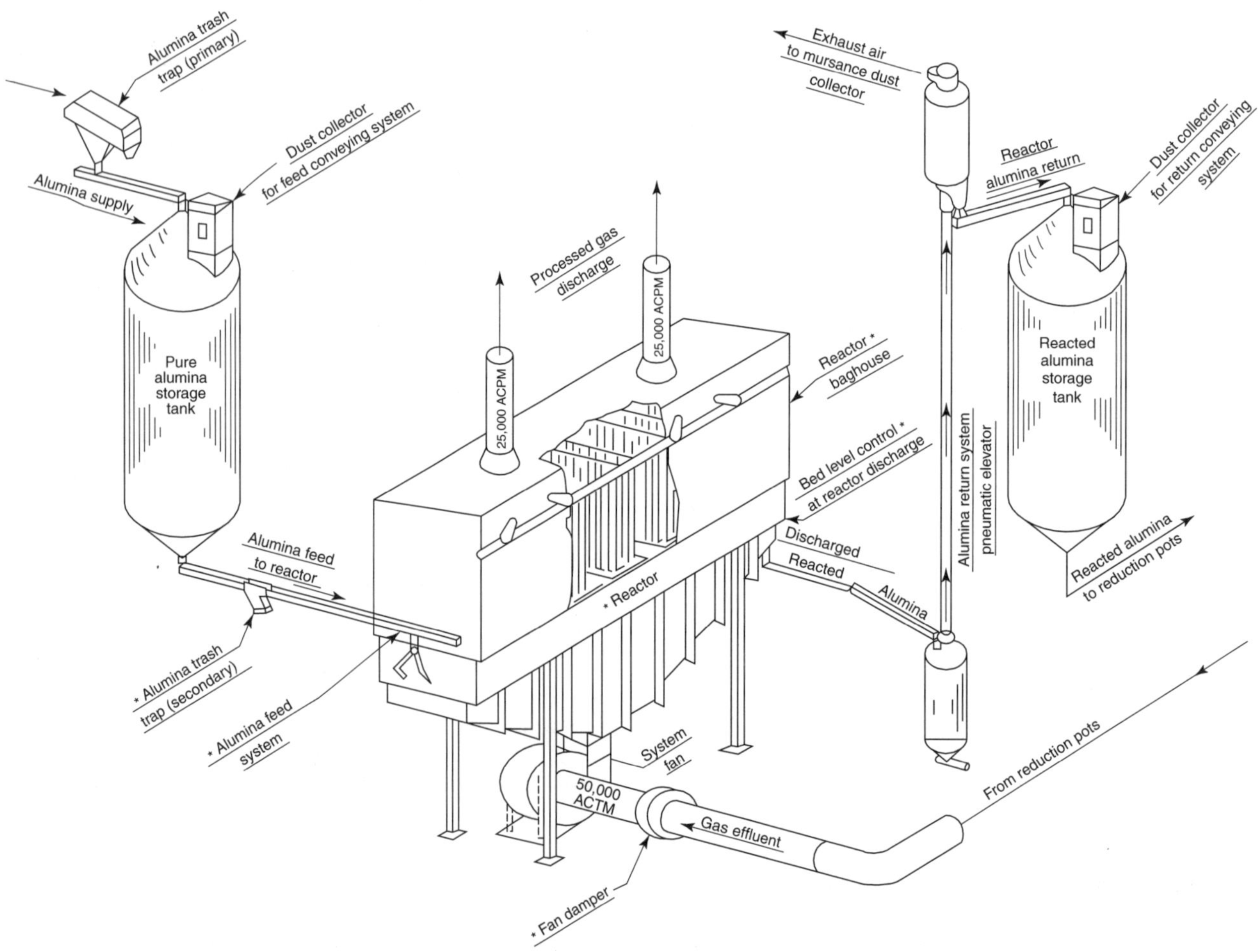

Figure 5. Fluid-Bed Scrubber System for Potroom Fumes (Aluminum Company of America, Pittsburgh, PA).

adjacent tanks that supply the cells. In a typical smelter, most of the alumina is used in the dry scrubber system before it goes to the potroom, but in some plants, a portion of the fresh ore may be segregated and used in cells where higher-purity aluminum is desired because reacted ore contains higher levels of trace impurities such as iron and silica. The anode baking operation could also use up to 10% of the fresh alumina in the dry scrubbing system.

The choice of fabric material for the filter is dictated by the gas temperature and resistance to HF and sulfur oxide gases. Polyester and acrylic fabrics are both suitable, except where the cell exhaust gas temperature exceeds 275°F. Nomex® is usually selected for higher temperatures. The filter bags in the baghouse must be cleaned periodically as the pressure drop builds up across the reactor. To clean the bags, the reactor–baghouse unit is isolated by shutting off the potroom gases, causing the bags to deflate. They are then cleaned by a shaker mechanism. After bag cleaning, the reactor is started up, and another reactor unit is taken off-line for cleaning. One or more spare reactor units thus are needed to pick up the load during bag cleaning, as well as when one is down for bag changing. The cleaning and reactor startup cycle is usually performed automatically.

Most fluid-bed installations on prebake cells use a shaker-type fabric filter, but in a few installations, reverse-pulse-type filters are used. The shaker collector, using woven filter bags, is usually designed with an air-to-cloth ratio of 2:1, while the reverse-pulse collectors with felt bags have air-to-cloth ratios ranging from 4.0:1 to 5.5:1. A set of woven filter bags in the shaker collector will provide up to 5 years' service before replacement is required, and felt bags in the reverse-pulse collector will provide a bag life of about 2 years.

The working environment for bag changing in a shaker collector is usually better than in the reverse-pulse unit, since the dirty side is on the inside surface of the bags, while with pulse collector bags, filtering occurs on the bag's outside surface.

The exhaust volume from a pot ranges from 2000 to 5000 acfm, depending on its size. With up to 200 pots per line, dry scrubber systems are designed to handle from 500,000 to over 1,000,000 cfm of exhaust gases. Since a reactor can handle up to 50,000 cfm, a large installation will consist of up to 25 reactor–baghouse units.

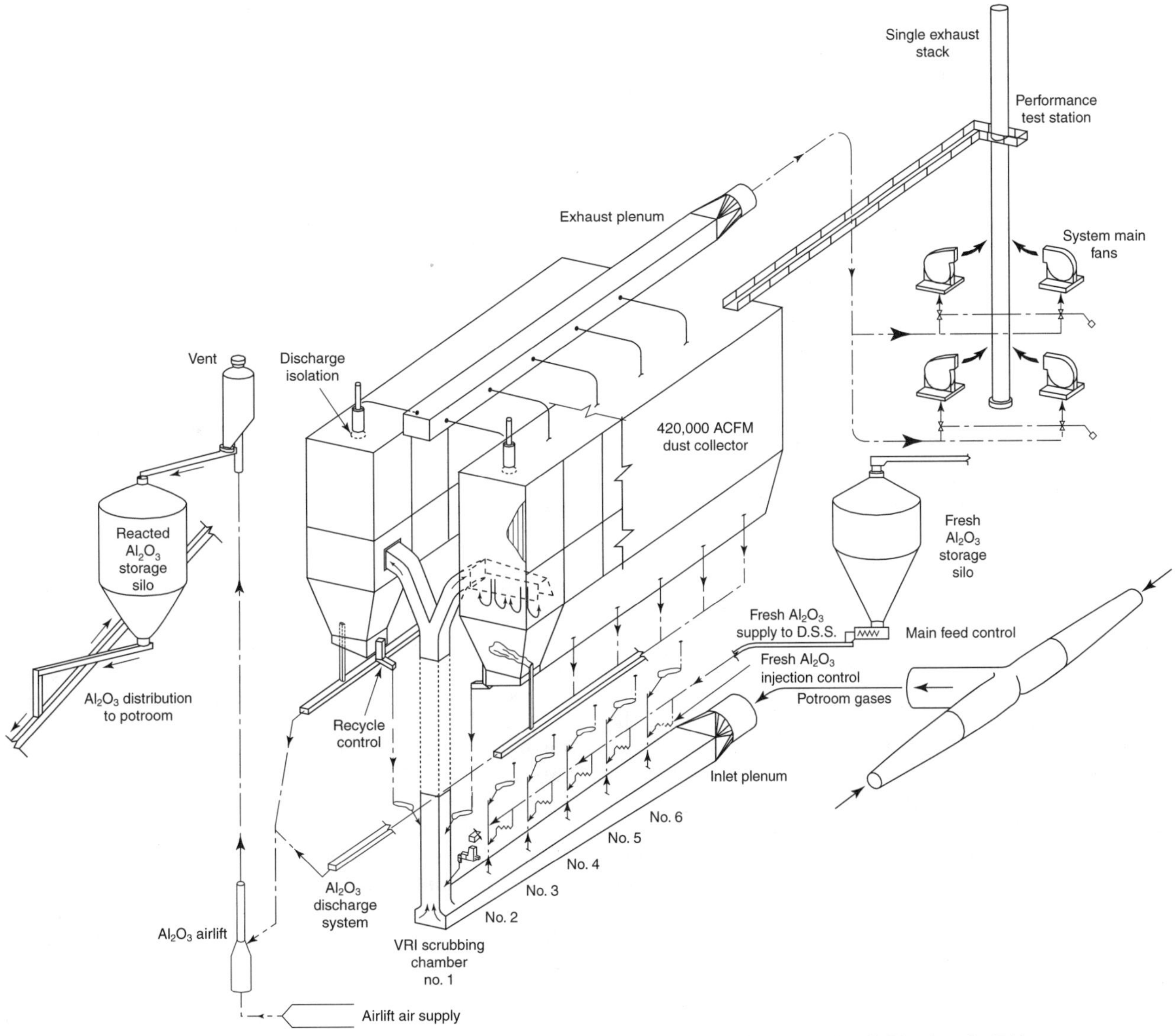

Figure 6. Dry Scrubber Injection System—Potroom Fumes (Busch International, Pittsburgh, PA).

Injection System. In the injection system (Figure 6), alumina is injected into the gas stream, where turbulent contact occurs between the alumina and the potroom gas, resulting in the adsorption of HF. Alumina particles are entrained in the gas stream and recovered in a fabric filter. There are variations in the method of injecting alumina into the gas stream. In some units, alumina is injected in a venturi throat, while others introduce the alumina through a specially designed pipe into a plenum. The main objectives of these designs are to create good mixing and to distribute the alumina evenly into the gas stream. To increase utilization of the alumina, a portion of it is usually removed from the fabric filter hopper and reinjected into the gas stream. The balance of the alumina is conveyed to the potroom storage feed tank.

The system is designed with two large multicompartment bag filters side by side. A reverse-pulse collector is the usual choice, although shaker collecters can be used. Compartmentalization of the baghouse allows maintenance of the collector and bag changes while the system is operating. With the reverse-pulse collector, which uses felt bags, it is possible to use a higher air-to-cloth ratio of 3.0:1 to 3.5:1, resulting in fewer bags and a smaller baghouse; thus it is a more economical design from a capital cost standpoint. However, the pulse collector uses a large amount of compressed air for bag cleaning. Shaker-type collectors, on the other hand, with woven bags require a lower air-to-cloth ratio of about 2.0:1, and so are larger and more costly. However, bag life is greater for the shaker because the cleaning method is gentler than with pulsing.

Baking Furnace

Baking Furnace Emissions. The amount of emissions produced by baking furnaces depends on many factors, such as the type of furnace, the condition of the flues, firing practice, and type of fuel used, that is, heavy oil or gas. Also, as mentioned earlier, the fluoride levels in the emissions are a function of the amount of fluorides retained with the recycled anode butts. Uncontrolled fluoride emissions will range from 0.25 to 1.5 kg per tonne alumina with similar amounts of coke dust and hydrocarbon vapors from the pitch binder. New Source Performance Standards (NSPS) for fluoride emissions from baking furnaces are 0.05 kg per tonne alumina and 10% maximum opacity.

Control Systems

Wet Systems. As with potrooms, wet scrubbers were originally used to reduce baking furnace emissions. These scrubbers ranged from simple cyclonic to high-energy venturi scrubbers. While cyclonic scrubbers were fairly effective in removing gaseous fluorides, they removed only some of the pitch fumes and the escaping matter consisted of very fine condensed tar particles, which could cause unacceptably high stack opacity. The high-energy venturi scrubber is a more efficient collector, but both scrubber systems generate large amounts of low-pH, high-fluoride water containing suspended tar, polynuclear aromatic compounds, and particulate matter. Under the best circumstances, this effluent is difficult to treat and expensive to dispose of.

Wet ESPs have also been used, and if properly maintained, will provide high collection efficiency on tar. The tar vapors must first be condensed in a quench chamber since the ESP will collect only particulate matter. The maintenance cost of wet ESPs is quite high because the collector plates must be cleaned regularly to maintain high collection efficiency and the water treatment problems are the same as with scrubbers. Wet ESPs are used where the emission from the closed-type furnace is basically a tar product and must be collected before the fluorides can be removed in the dry scrubbers. However, for the closed furnace emission, a dry system will be described in a later section for collecting tar vapor and fluorides.

Dry Scrubbers. Dry scrubbers, which are widely used today on anode baking furnaces, are similar to those used on potrooms, but quite different in their mission. Dry scrubbers for anode furnaces were originally developed to collect, primarily, the hydrocarbons that caused high stack opacity. As federal regulations evolved, however, the NSPS for aluminum smelters also limited fluoride emissions from the baking furnace; consequently, fluoride removal is now as important as hydrocarbon control.

The major difference between the baking furnace and potroom control systems is that the former is designed to cool the furnace gas to condense the tar vapors so that they can be collected on the alumina. Cooling also increases fabric filter bag life. The first dry scrubber installation used a fluid-bed reactor, but later the injection system used on smelters was adapted to the baking furnace by adding a gas cooler to the system.

The fluoride collection efficiency of these systems is similar to that obtained for potroom emissions—99% or greater. The collection efficiency for hydrocarbon emissions is about 90%, but this value can vary depending on the manner in which the baking furnace is operated, the amount of tar in the flue gas, and the operating temperature of the dry scrubber.

In some baking furnaces, the hydrocarbon level in the flue gas is so low that the scrubber system is used mainly to control fluoride emissions. In other plants where local regulations do not require hydrocarbon control and hydrocarbon levels are relatively low, the gas is cooled by convective cooling in long runs of ductwork, thus avoiding the evaporative coolers.

The cost of dry scrubber systems for anode furnace emissions ranges from \$60 to \$75 per actual cubic foot per minute (1992). The higher cost as compared with potroom scrubbers reflects the added cost for cooling the gas and the smaller gas volumes to be treated. The large potroom's treatment systems can take advantage of the economy of scale.

Fluid-Bed System. The fluid-bed reactor is quite different from that used on potroom emissions. The baking furnace flue gas cannot be used to fluidize the alumina because the tars would plug the small holes in the distribution or dribble plates. Therefore, the alumina is fluidized, as shown in Figure 7, by a small amount of ambient air, and the flue gas enters the reactor through a row of pipes above the dribble plate. Another difference, mentioned earlier, is the need to cool the flue gases to condense and capture the hydrocarbons. This is accomplished by spraying water into the fluidized bed. Fluidization provides efficient mixing among the water droplets, alumina, and pollutants, thus promoting good mass and heat transfer. The water droplets evaporate as they contact the hot alumina, causing the bed to cool, and condensation of the tar fume takes place on the cooled alumina particles. Fresh alumina is introduced at one end of the reactor bed, which is 12–15 in. deep, and is removed at the other end. The entrained alumina and other solids are separated in a bag filter located on the reactor, and the scrubbed flue gas exits from stacks on the baghouse.

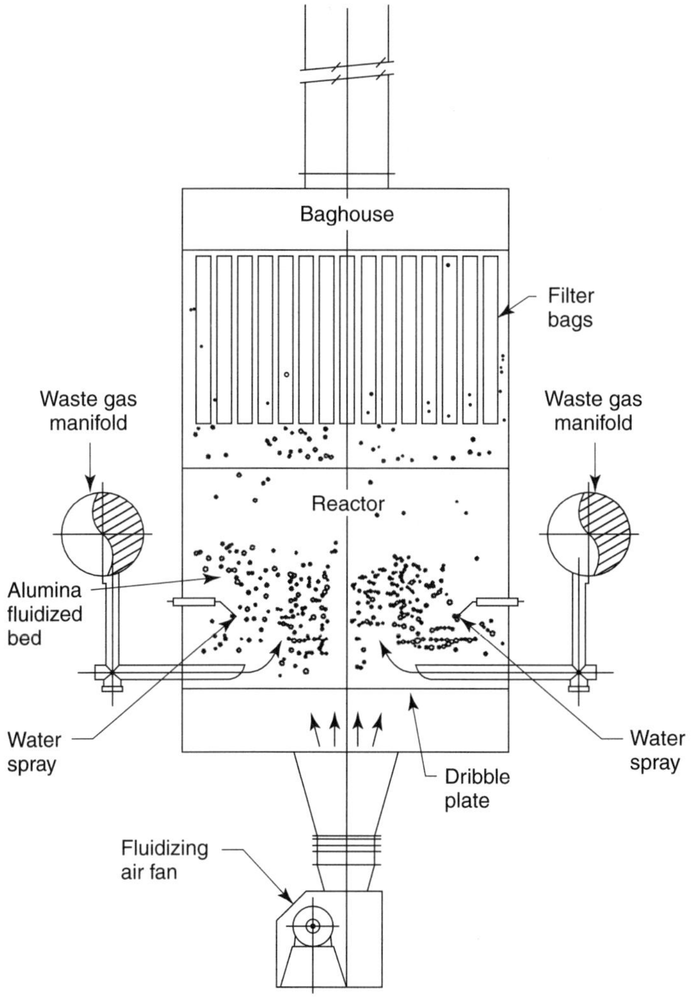

Figure 7. Dry Scrubber Fluid-Bed Reactor—Baking Furnace (Aluminum Co. of America, Pittsburgh, PA).

The reactors are designed to process 35,000–50,000 acfm of flue gas, with an alumina feed rate of 600–1000 lb/h per reactor. The bed is cooled to 150°F to 180°F, requiring from 10 to 15 gal–min of water, depending on the flue gas temperature. The material handling system is quite similar to that used in the potroom fluid-bed scrubber system except on a smaller scale, and reacted alumina is also sent to the potrooms.

In some installations, the reacted alumina containing coke and tar is heated to destroy the combustibles before the alumina is used in the pots. In this case, the alumina is fed to a regenerator, where it is fluidized and heated to 1200°F. The bed is about 55 in. deep and is fluidized by ambient air, which provides oxygen for combustion. Radiant tube gas-fired burners immersed in the bed supply the heat. Hot alumina emerging from the regenerator is cooled to about 200°F in a water-cooled heat exchanger before being transferred to storage. A typical system with the regenerator is shown in Figure 8.

Injection System. The injection system, as pointed out earlier, is similar to that used to control potroom emissions, with the addition of an evaporative cooler up front to cool the flue gas (Figure 9). The cooler will range in size from 12 to 15 ft in diameter and will be about 100 ft tall, depending on the gas volume to be treated and the desired residence time. Flue gas enters the cooler at the top, where a bank of nozzles spray water mist into the gas stream. The entrained water droplets travel co-currently with flue gas down the tower. Cooled gas exits the cooler bottom and enters the injection system.

The water spray is produced from two-fluid nozzles using compressed air to create the fine mist particles ranging in size from 50 to 100 μm. The tower is designed to provide several seconds of residence time in order to ensure complete evaporation of the water droplets. The operating temperature of the cooler is usually 200°F to 210°F, which is a practical limit. Attempts to lower the temperature can result in wetting of the inside wall, as well as carryover of water into the bag filter. Perhaps lower gas temperatures could be achieved with spray nozzles that produce smaller water droplets. Coke particles and condensed hydrocarbons settle out in the cooler and are removed continuously from the bottom.

After the cooler, alumina is injected into the gas stream in a contact chamber or in the venturi section in the ductwork, and adsorption of HF gas and tar deposition occurs on the alumina. The gases then enter the compartmentalized baghouse, where entrained alumina, together with coke particles and condensed hydrocarbons, are removed. Cleaned flue gas exits to a blower and out of the stack. As in the potroom system, a portion of the alumina from the bag filter hopper is recycled to the contact chamber to increase alumina utilization and the balance sent to the potroom.

The size of a system will vary with the volume of the flue gas to be treated and will range from about 50,000 to 150,000 cfm.

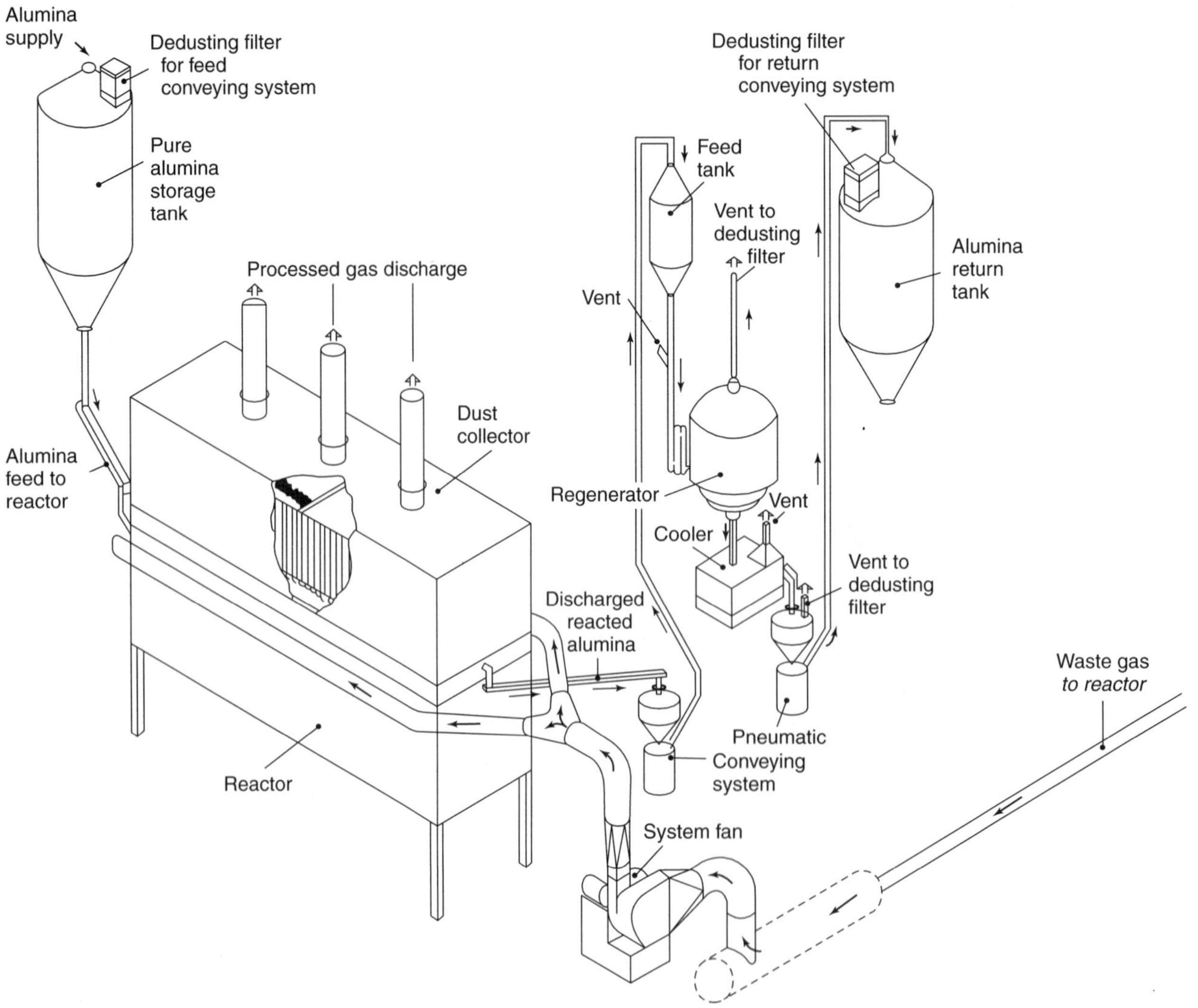

Figure 8. Fluid-Bed Reactor System—Baking Furnace (Aluminum Co. of America, Pittsburgh, PA).

As mentioned, the system is similar in design to the potroom dry scrubber. It has a multicompartment baghouse reverse-pulse bag cleaning or a shaker system that permits each compartment to be isolated for maintenance and bag change.

Tar Collection and Dry Scrubbing. As mentioned previously, there are closed baking furnaces that produce a flue gas low in oxygen and high in tar content. The control scheme for some of the earlier installations consisted of cooling the flue gas to 150°F to 160°F in an evaporative cooler to condense the tar into aerosol particles, which were then removed in an ESP. To achieve this relatively low temperature, it was necessary to operate the cooler with a wet bottom, that is, with excess water flowing from the cooler bottom. This effluent contained dissolved HF and suspended tar and required treatment before it could be disposed of.

To avoid the evaporative cooler and water treatment, a system was recently developed in which the flue gases are cooled in shell-and-tube water-cooled heat exchangers (Figure 10), with flue gas on the shell side and cooling water on the tube side. Through careful control of the water temperature in each cooling zone, the flue gas is cooled to 160°F without plugging the heat exchanger. Tar that condenses on the tubes drips to a heated sump below. This particular temperature was chosen because it is low enough to cause most of the tar vapor to condense, but warm enough for the tar to flow. The cooled gases then go to a plate-type ESP, where the remaining fine aerosol tar particles are captured on the plates and tar drips off to a heated hopper below. Finally, the gases are treated in a dry alumina injection system to remove fluoride.

In these systems, the detarring equipment must be insulated and heat traced to prevent tar from freezing in cold weather, and a fire protection system is also required.

Green Anode

Green Mill Control. As mentioned earlier, emissions produced in the various steps in green anode production are coke dust from the coke classifiers and hydrocarbon vapors from the mixing, conveying, and anode-forming equipment. Coke dust from the classifiers and conveying equipment is usually controlled by fabric filters.

Control of hydrocarbon emissions is more difficult. Since the gases are at relatively low temperature—150°F or less—tar vapors readily condense in the ductwork and, together with the coke dust, cause plugging. Heat tracing the ductwork will help to minimize the problem, but will not completely prevent

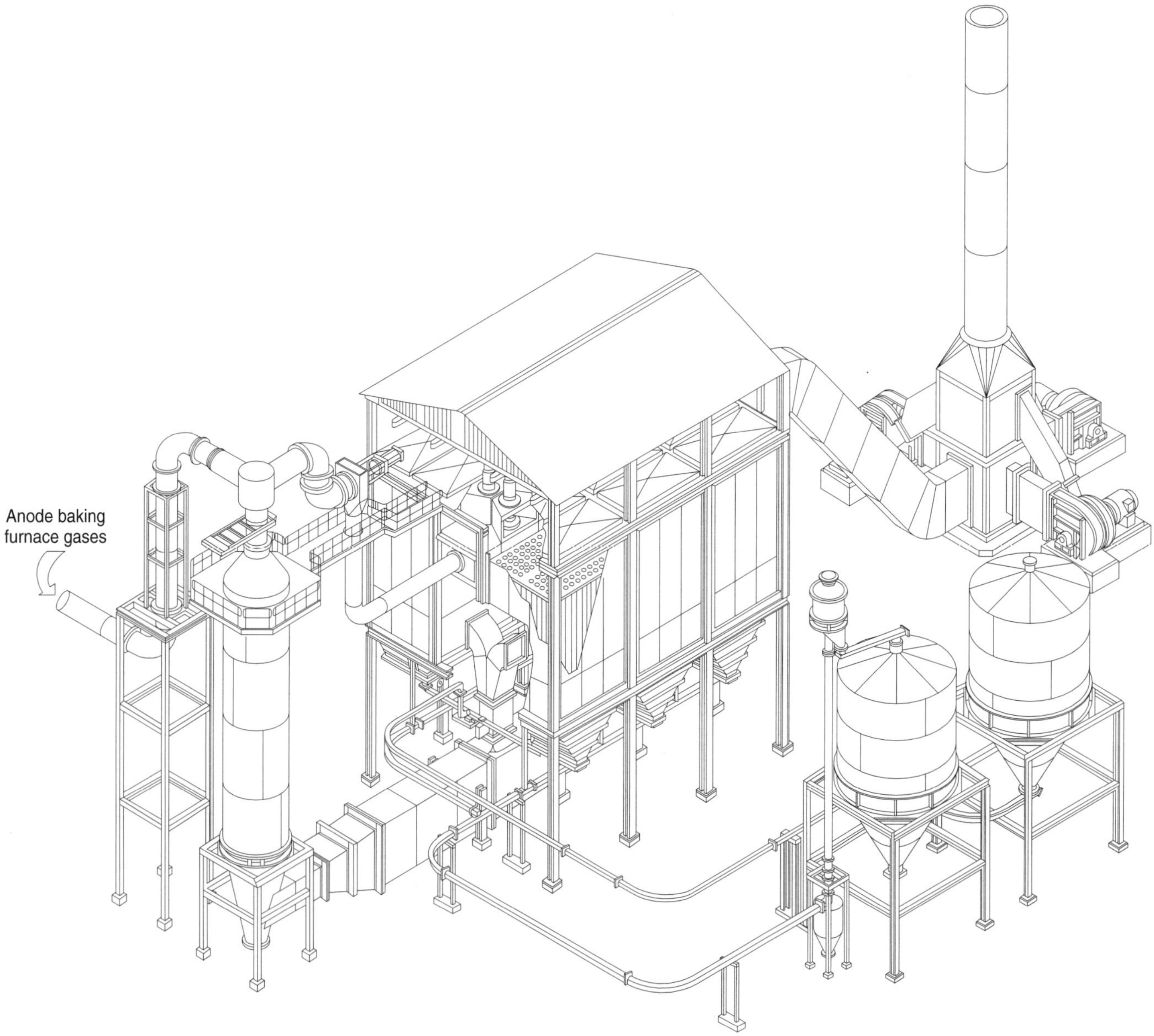

Figure 9. Dry Scrubber Injection System — Baking Furnace (Busch International, Pittsburgh, PA).

it. Gas velocity in the ductwork should be at least 3000 fpm to prevent coke particles from dropping out, and cleanout ports in the ductwork will facilitate cleaning.

An effective control technology for green mill emission is a precoat bag filter using coke dust as the precoat material, which not only absorbs/adsorbs the hydrocarbon vapors, but also prevents the tar from blinding the filter bags. Coke dust can be obtained from coke pneumatic conveying systems, coke classifiers, and other exhaust streams. The bags are precoated by injecting the coke dust into the ductwork prior to the bag house. Since the gas is relatively cool, the vapors are almost totally condensed at this stage and exist as finely condensed particles of tar together with a small amount of vapor. Some vapor adsorption probably occurs as well.

The amount of coke required will depend on the concentration of hydrocarbon vapors in the gas stream, but in a shaker collector, a minimum thickness of about 1/8 in. of precoat material is necessary for tar absorption and to protect the filter bags from blinding. Another operating parameter is to provide enough coke dust so as not to exceed 4–5% tar content on the coke dust. As a general rule, tar-removal efficiency will increase with the amount of coke used, since more coke will produce more available surface, as well as a thicker precoat layer for increased filtration efficiency.

Another possible method for controlling pitch fumes is to scrub the gases with a heavy oil. The oil absorbs the pitch fumes, and since the scrubbing oil has a low vapor pressure, it does not emit a significant amount of vapor when it is exposed to the gases being scrubbed. The scrubbing oil will eventually become saturated with hydrocarbons and it will have to be replaced.

OTHER ENVIRONMENTAL CONSIDERATIONS

The foregoing discussion has dealt largely with control systems used to abate air emissions from the primary

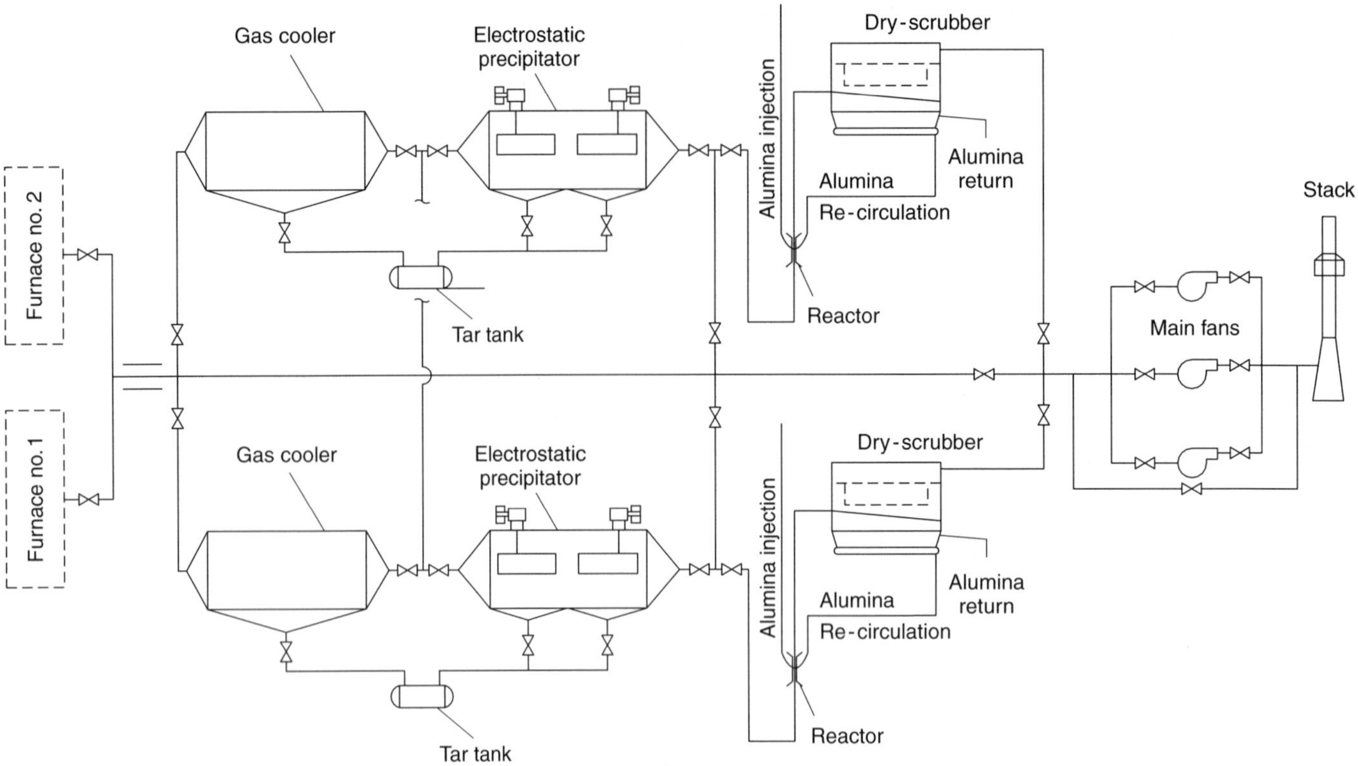

Figure 10. Tar Precipitation and Dry Scrubber System (Mikropul Env. Systems, Morris Plains, NJ).

aluminum industry. Historically, as environmental problems were identified, suitable control technology was applied or developed to correct the problems. This is the so-called end-of-the-pipe treatment approach. However, with mounting environmental costs and an increased awareness of the global effects of pollution, there is an emerging consensus that a broader approach must be taken to address these problems not only in the aluminum industry, but in other industries as well. Therefore, it is appropriate to mention some of the other programs and steps-long and short range-that will find wider acceptance as industry takes a more holistic view toward addressing these environmental issues.

Process Control

Good process control in baking furnace operation to reduce emissions and minimization of exhaust volume from pots through a well-designed hood system were mentioned earlier. The use of modem computer control systems has led to improved process control in the potrooms. For example, computerized process control can reduce emissions by minimizing the "anode effect." An anode effect is caused by a deficiency of alumina feed to a pot, resulting in high cell resistance, high temperature, and higher fluoride emissions. Computer controls, improved alumina feeders, and process monitoring can help to reduce anode effects.

Other process control items that can affect environmental costs are anode burn-off and premature potlining failure. Anode burn-off can be caused by poor-quality anodes or a bad connection at the carbon-anode rod interface. And the disposal of spent potlining removed from failed linings is currently receiving a great deal of attention in the industry.

Choice of Control Technology

The proper choice of control technology is also important. The example most often cited to illustrate a poor choice of technology is one where a wet system instead of a dry one was chosen because of lower capital cost, overlooking the high cost of waste treatment. Thus, an air pollution problem was merely exchanged for a worse one.

In many respects, the aluminum industry is fortunate to have developed dry environmental control systems for potroom and baking furnace emissions, thus replacing the wet systems. However, it should be mentioned that solving the major air emission problems resulted in solid and liquid waste problems with which to contend. The large number of used filter bags, some of which are impregnated with hydrocarbons from the anode baking operation, presents one problem, and the tar collection system used on the closed anode furnace generates tar contaminated with acids.

Equipment Operation and Maintenance

Good operation and maintenance of pollution control and process equipment cannot be overstressed. Poorly maintained pot hooding, fabric filters, and pneumatic conveyors, for instance, can cause major emissions. Poor flue maintenance in the baking furnace, and its relationship to good product quality, has been pointed out already. In the long run, well-maintained environmental control equipment not only emits less, but also provides longer service life.

Environmental control systems should also be integrated into the production process and not operated as an appendage. The fume control system for the baking furnace is a good example. It is an important part of the baking operation, providing draft for combustion, and must work in unison with the process. However, the two systems are sometimes operated and maintained as separate systems by different groups. As a result, the baking operation is often listed as the cause of poor performance of the control system, and vice versa. Because of this separation, the relationship between good process control and good performance of the fume control system often goes unrecognized.

Process Modification, TQM, Early Involvement, and Waste Minimization

Process modification, if it can be carried out without major expenditures, can be an inexpensive method for reducing and perhaps even eliminating an emission. To take advantage of process modification as a control scheme, continuous evaluation of the process, and input from operators, supervisors, and others familiar with the operation are necessary.

The approach of total quality management (TQM) could also be applied to environmental control to improve the operation of the systems in terms of reliability, cost of operation and maintenance, and reduction in emissions. This concept, which also embodies a goal of continuous improvement, is becoming more accepted in this country as an industrial management tool. It is a system that uses quality control tools, such as data gathering and statistical controls, and applies them in every step of an operation to produce the best-quality product in the most efficient manner. The terms operation and product are used in the broadest sense and can range from a highly sophisticated production process to the most mundane task.

Another long-range program is one of early involvement during process development. By identifying environmental problems in the early stages of development, they can be resolved before the process is in place and expensive end-of-the-pipe treatment has to be installed.

Finally, a program of waste minimization has been shown by many companies to be quite beneficial in terms of resource conservation, pollution reduction, and good payback. Like process modification and TQM, waste minimization is a never-ending program requiring plantwide support, environmental awareness, and a clear definition of the objectives of waste minimization.

These approaches are proactive in nature, and in order to succeed, will require some training, teamwork, and the commitment of an entire organization. However, they represent a more logical and integrated management of environmental issues, and in the long run, a combination of these programs with the best end-of-the-pipe solutions will prove to be less stressful to the environment and far more cost effective.

NESHAP FOR PRIMARY ALUMINUM REDUCTION PLANTS

On July 16, 1992 (57 FR 31576), the EPA published a list of major and area sources for which National Emission Standards for Hazardous Air Pollutants (NESHAP) were to be promulgated; primary aluminum production was one of the 174 categories of sources listed. There are 23 primary aluminum plants currently located in a total of 14 states. Many of these plants are concentrated in the Northwest in close proximity to hydroelectric power sources. The 23 plants have 91 potlines that produce aluminum, each plant has a paste production operation, and 17 of these plants have anode bake furnaces. The major HAPs emitted by these facilities are HF and POM. The NESHAP or MACT standards were first proposed in September 1996[2] with the final rule promulgated and effective as of October 7, 1997.[3]

The final rule established national emission standards for each new or existing potline, paste production operation, and anode bake furnace associated with primary aluminum reduction plants. The major hazardous air pollutants (HAPs) emitted by the facilities covered by the rule include hydrogen fluoride (HF) and polycyclic organic matter (POM). Polycyclic aromatic hydrocarbons (PAHs) are included in the chemical group POM. The standard applies to emissions of HF, measured using total fluorides (TF) as a surrogate, and POM (as measured by methylene chloride extractables) from each affected source associated with primary aluminum reduction. Affected sources are each potline of reduction cells, each anode bake furnace, and each paste production plant.

The standard is based on a combination of control techniques that either prevent the escape of emissions or capture the pollutants and return them to the process. These pollution prevention measures include work practices, equipment modifications, operating practices, housekeeping measures, and in-process recycling. The overall effect of the standard will be to raise the control performance of nearly half of the industry to the level of control achieved by the best performing plants. Currently, over 11,000 tpy of fluoride and POM are emitted nationwide; these emissions will be reduced by more than 50%, and higher reductions will be achieved at particular sites. Emissions of total particulate matter also would be reduced by 16,000 tpy.[2,3]

The EPA's analysis of existing aluminum production processes and operations resulted in the designation of seven subcategories for potlines, as shown in Table 5. For the subcategories of potlines, the distinctions are based primarily on differences in the process operation, process equipment, emissions, and the applicability of control devices. Additional information on the subcategorization is included in the Basis and Purpose Document for Primary Aluminum Reduction Plants available on the TTNWeb Unified Air Toxics website.

The control option for primary emissions from the reduction process for six of the seven subcategories of existing potlines and for all new potlines is the installation of a dry alumina scrubber (with a baghouse to collect the alumina and other particulate matter) at those plants that do not have them. The control option for prebake plants producing high-purity aluminum is a wet scrubber that removes impurities that would otherwise contaminate the aluminum. The MACT technology used to establish the floor of performance for potline secondary emission control involves the use of wet roof scrubbers for side-worked prebake potlines and one type of Soderberg potline. Work practice programs, inspection procedures, and maintenance programs for repairing or replacing damaged hoods and seals provide the most efficient control for secondary emissions from other types of existing and new potlines. Based on EPA's MACT floor analyses, the dry alumina scrubber also is the MACT floor technology for new and existing anode bake furnaces, and a capture system vented to a dry coke scrubber is the MACT floor technology for new and existing paste production operations. For the one bake furnace plant not located with a primary aluminum reduction plant, the MACT floor control technology (dry alumina scrubbers) does not apply because the plant does not have access to alumina as do other bake furnaces, and there are no potlines onsite to use the reacted alumina.

The emission levels shown in Table 5 for existing sources and Table 6 for new sources represent the MACT floor and MACT for each emission source. The limits for potlines are in the same format as the NSPS (40 CFR part 60, subpart S)—kilogram of pollutant per megagram of aluminum (kg/Mg) or pound of pollutant per ton of aluminum (lb/ton). A similar format, lb/ton of anode, is used for emission limits for anode bake plants.

An equipment standard requiring installation of a capture system and the routing of emissions through a closed system to a dry coke scrubber or equivalent alternative control device is required for paste production. If an alternative to the dry coke scrubber is used, the control device must achieve a POM removal efficiency of at least 95% for continuous paste mixing operations and at least 90% for batch operations. The capture system must be designed and operated to meet generally accepted engineering standards for minimum exhaust rates.

Table 6. Summary of Emission Limits for New Sources

Source	Emission Limit
Potlines	TF Emission Limit 0.6 kg/Mg (1.2 lb/ton) of aluminum produced POM Emission Limit 0.32 kg/Mg (0.63 lb/ton) of aluminum produced
Paste production	POM Emission Limit Install, operate, and maintain equipment for the capture of emissions and vent emissions to a dry coke scrubber
Anode bake furnace	TF Emission Limit 0.01 kg/Mg (0.02 lb/ton) of anode POM Emission Limit 0.025 kg/Mg (0.05 lb/ton) of anode

Additional information on the NESHAP can be found in the proposed and final rules.[2,3]

Table 5. Summary of Emission Limits for Existing Sources

Source	Emission Limit
Potlines	TF Emission Limits 0.95 kg/Mg (1.9 lb/ton) of aluminum produced for CWPB1[a] potlines 1.5 kg/Mg (3.0 lb/ton) of aluminum produced for CWPB2 potlines 1.25 kg/Mg (2.5 lb/ton) of aluminum produced for CWPB3 potlines 0.80 kg/Mg (1.6 lb/ton) of aluminum produced for SWPB potlines 1.1 kg/Mg (2.2 lb/ton) of aluminum produced for VSS1 potlines 1.35 kg/Mg (2.7 lb/ton) of aluminum produced for VSS2 potlines 1.35 kg/Mg (2.7 lb/ton) of aluminum produced for HSS potlines
	POM Emission Limits 2.35 kg/Mg (4.7 lb/ton) of aluminum produced for HSS potlines 1.2 kg/Mg (2.4 lb/ton) of aluminum produced for VSS1 potlines 1.85 kg/Mg (3.6 lb/ton) of aluminum produced for VSS2 potlines
Paste Production	POM Emission Limit Install, operate, and maintain equipment for capture of emissions and vent emissions to a dry coke scrubber.
Anode bake furnace (located with a primary aluminum plant)	TF Emission Limit 0.10 kg/Mg (0.20 lb/ton) of anode POM Emission Limit 0.09 kg/Mg (0.18 lb/ton) of anode

[a]Abbreviations defined:

CWPB1	Center-worked prebake potline with the most modern reduction cells; includes all center-worked prebake potlines not specifically identified as CWPB2 or CWPB3
CWPB2	Center-worked prebake potlines located at Alcoa in Rockdale, Texas; Kaiser Aluminum in Mead, Washington; Ormet Corporation in Hannibal, Ohio; Ravenswood Aluminum in Ravenswood, West Virginia; Reynolds Metals in Troutdale, Oregon; and Vanalco Aluminum in Vancouver, Washington
CWPB3	Center-worked prebake potline that produces very high-purity aluminum, has wet scrubbers as the primary control system, and is located at the primary aluminum plant operated by NSA in Hawesville, Kentucky
HSS	Horizontal stud Soderberg potline
SWPB	Side-worked prebake potline
VSS1	Vertical stud Soderberg potline at Northwest Aluminum in The Dalles, Oregon, or at Columbia Aluminum in Goldendale, Washington
VSS2	Vertical stud Soderberg potlines at Columbia Falls Aluminum in Columbia Falls, Montana.

Appendix. Summary of Smelter Environmental Data

Potroom	
• NSPS regulations	1 kg F/tonne Al (Soderberg pot) 0.95 F/tonne Al (prebake pot) 10% opacity
• Exhaust volume per pot	2000–5000 cfm
• Hooding efficiency	95% (approximately)
• HF removal efficiency — dry scrubbers	99%
• Minimum required surface area of alumina for dry scrubber	45 m^2/g (BET method)
• Carbon consumption	0.4–0.5 lb C/lb Al
• Capital cost — dry scrubber system	\$35–\$50/cfm
Carbon Baking Furnace	
• NSPS regulations	0.05 kg F/tonne Al 10% opacity
• Emission factors	
Fluoride	1–5 lb/tonne Al
Hydrocarbons + particulate	1–5 lb/tonne Al
• Dry scrubber efficiency	
Fluoride	99%
Hydrocarbon	90% or less[a]
Particulate	99%
• Capital cost — dry scrubber system	\$60–\$75/cfm

[a]Hydrocarbon removal efficiency will depend on many factors, such as gas temperature, alumina feed rate, and gas composition.

REFERENCES

1. *AP-42 Supplement A to Compilation of Air Pollutant Emission Factors*, October 1986. Vol. 1: Stationary Point and Area Sources. U.S. Environmental Protection Agency, Office of Air and Radiation, Research Triangle Park, NC.
2. *NESHAP for Source Categories: NESHAP for Primary Aluminum Reduction Plants: Proposed Rule*, September 26, 1996, U. S. EPA TTNWeb/UATW.
3. *NESHAP for Source Categories: NESHAP for Primary Aluminum Reduction Plants: Final Rule*, October 7, 1997. U.S. EPA TTNWeb/UATW.

BIBLIOGRAPHY

Air Pollution Control in the Primary Aluminum Industry, EPA-450/33-73-004A and EPA-450/33-73-004B, Singmaster and Breyer, New York, 1973.

Aluminum Statistical Review for 1990, Aluminum Association, Washington, D.C.

P. R. Atkins and K. P. Karsten, "In Plant Practices for Job Related Health Hazards Control: Aluminum," Chap. 4, Vol. 1, *Production Processes*, John Wiley & Sons, New York, 1989.

W. W. M. Boon and R. de Fluiter, "Leaning of Flue Gas from Anode Bake Furnaces," *Light Metals*, Metallurgical Society of AIME, 1980, pp. 721–733.

C. N. Cochran, W. C. Sleppy, and W. B. Frank. "Fumes in Aluminum Smelting: Chemistry of Evolution and Recovery," *J. Metals*, September 1970.

C. N. Cochran, "Recovery of Hydrogen Fluoride Fumes on Alumina in Aluminum Smelting," *Environ. Sci. Technol.*, **8**(1), 1974.

P. A. Comella and R. W. Rittmeyer, "Waste Minimization/Pollution Control," *Pollution Eng.* April 1990.

C. C. Cook, G. R. Swary, and J. W. Colpitts, "Operating Experience with the Alcoa 398 Process for Fluoride Recovery," *J. APCA* **21**(8), August 1971.

C. C. Cook and G. R. Swany, "Evolution of Fluoride Recovery Processes Alcoa Smelters," *Light Metals*, Metallurgical Society of AIME, Vol. 11, 1971, pp. 465–477.

Development Document for Effluent Guidelines and New Source Performance Standards—Primary Aluminum Smelting, EPA-440/1-74-019-d. U.S. Environmental Protection Agency, Washington, D.C.

J. Edwards, *The Immortal Woodshed*, Dodd, Mead & Co., New York, 1955.

J. C. Files, "Treatment of Bake Oven Exhaust Gases Using Unique System of Alumina Dry Scrubbing," *Light Metals*, Metallurgical Society of AIME, 1981, pp. 931–939.

L. K. Hudson, *Alumina Production*, Alcoa Research Laboratories, Alcoa Center, PA, 1982.

Primary Aluminum: Guidelines for Control of Fluoride Emissions from Existing Primary Aluminum Plants, EPA-450/2-78-049b, Office of Air Quality Planning Standards, Research Triangle Park, NC, 1979.

Review of New Source Performance Standards for Primary Aluminum Reduction Plants, EPA-450/3-86-0 10, U.S. Environmental Protection Agency, Research Triangle Park, NC.

P. D. Stobart. *Centenary of the Hall & Heroult Processes 1886–1986*, International Primary Aluminum Institute, London, 1986.

T. J. Robare and M. W. Wei, "Control of Pitch Volatiles in Green Electrode Plant," *Light Metals*, Metallurgical Society of AIME, Vol. II, 1984, pp. 1521–29.

D. Rush, J. C. Russel, and R. E. Iverson, "Air Pollution Abatement on Primary Aluminum Potlines: Effectiveness and Cost." *J. APCA* **23**(2), 1973.

Ullman's Encyclopedia of Industrial Chemistry, 5th ed., Vol. A., Verlagsgesellschaft mbH, Weinheim, Germany, 1985.

M. W. Wei, "The Alcoa 446 Process for the Control of Anode Baking Furnace Fumes." *Light Metals*, Metallurgical Society of AIME, Vol. II, 1975, pp. 261–268.

METALLURGICAL COKE

Kelly R. Carmichael, Allen Ellis, George Bradley, John Carson, Thomas W. Easterly, Stefan P. Shoup, and Dennis P. Kaegi

Metallurgical coke is manufactured by heating high-grade bituminous coal (low sulfur and low ash) to around 1050°C (1925°F) in an enclosed oven chamber without oxygen. The resulting solid material consists of elemental carbon and any minerals (ash) that were present in the coal blend that did not volatilize during the process.

PURPOSE OF METALLURGICAL COKE

There are two major categories of metallurgical coke: furnace coke and foundry coke. Furnace coke is used in steel mill blast furnaces, and foundry coke is used in foundry cupolas. In general, foundry coke is made from coals with a lower volatility, is heated longer, and is larger than furnace coke. Both types of coke can be made with the same equipment by changing the coals and the operating practices. In the United States in the early 1990s, over 90% of coke produced is furnace coke for use in steel mill blast furnaces.

Furnace coke is essential to the blast furnace method of iron production, where it serves three major functions. First, the coke provides the porous support for the other materials in the blast furnace burden (principally iron ore pellets and limestone), allowing the hot blast to blow through the furnace. Second, coke provides fuel to react with the blast air, which, in turn, provides the heat and reducing atmosphere required to refine the ore into iron. Third, the coke acts as a chemical reductant, removing oxides from the ore.

Because of the costs and environmental concerns associated with coke manufacturing, the steel industry has been working on processes to reduce the need for coke in blast furnaces. To date, there has been success in substituting oil, natural gas, and coal for some of the fuel value provided by the coke, but large amounts of coke are still needed both to provide a porous support to the blast furnace burden and to act as a chemical reductant. As less coke is used in the blast furnace per ton of hot metal produced, it is expected that the quality of coke will have to be improved, particularly its strength or stability.

TYPES OF BATTERIES

Two types of coke oven batteries are currently employed to produce metallurgical coke. The most common is the slot oven by-product type, which was used to produce over 99% of the metallurgical coke in 1990. In this type of battery, the volatiles driven off in the process are refined in the coal chemical plant to produce clean coke oven gas, tar, sulfur, ammonium sulfate, and light oil. Some 33% to 40% of the coke oven gas produced is used to heat the battery, while the rest can be utilized as a fuel at other facilities. Figures 1 and 2 show the general layout and sources of emissions of a typical by-product coke oven battery.

The other type is the nonrecovery battery, which is currently used to produce coke at two locations in the United States and is seriously being considered by a number of steel companies as an alternative to the slot oven by-product battery when present coke plants need to be reconstructed. A nonrecovery coke battery is one in which the coal volatiles driven off in

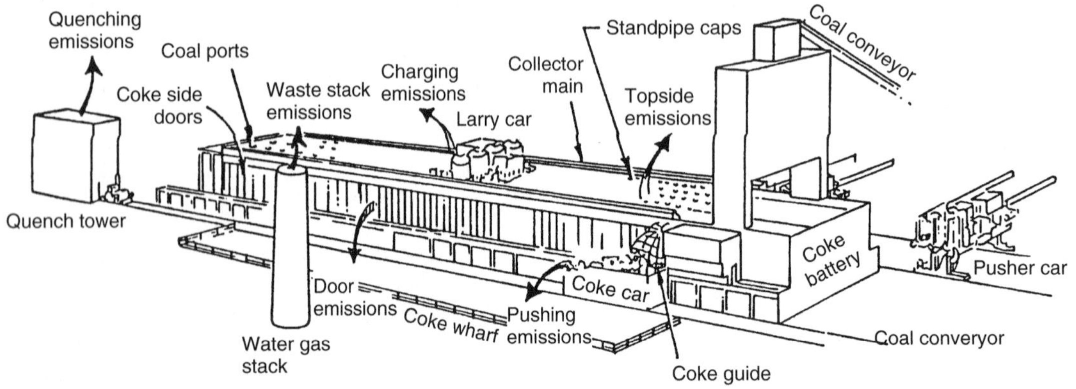

Figure 1. General Layout of By-Product Coke Oven Battery.

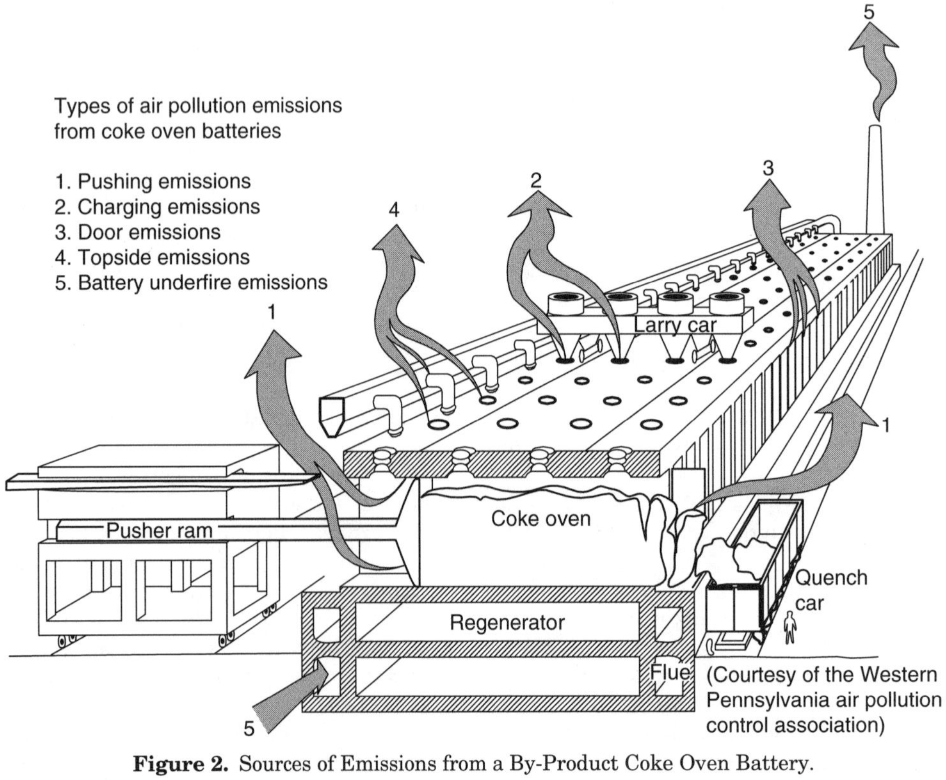

Figure 2. Sources of Emissions from a By-Product Coke Oven Battery.

the process are immediately combusted within and around the oven to provide the heat required for carbonization. This eliminates the need for associated by-products plants. Emission control for exhaust gas generally would consist of removal of particulate matter and sulfur dioxide prior to discharge through the waste gas stack. Figures 3 and 4 show a typical individual oven and layout of a nonrecovery battery facility.

An older type of coke oven that is no longer in use is the beehive oven. It was an environmentally uncontrolled and manually operated forerunner of the current design of nonrecovery ovens.

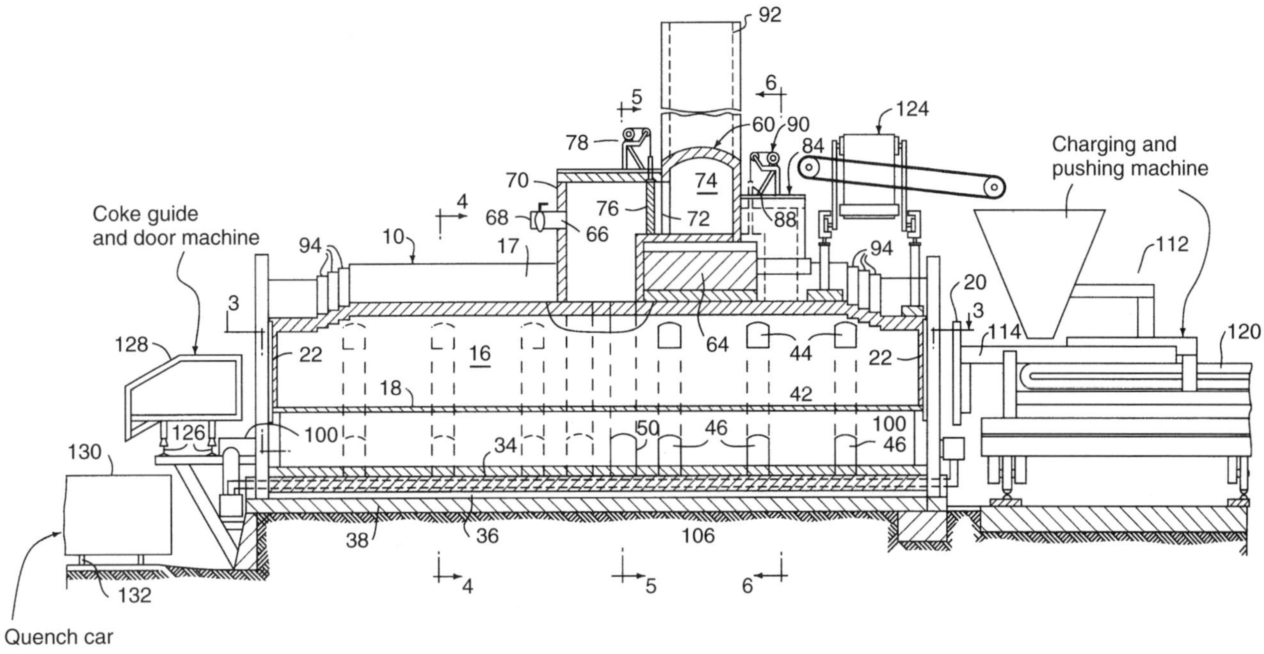

Figure 3. Side View of an Oven on a Nonrecovery Coke Battery.

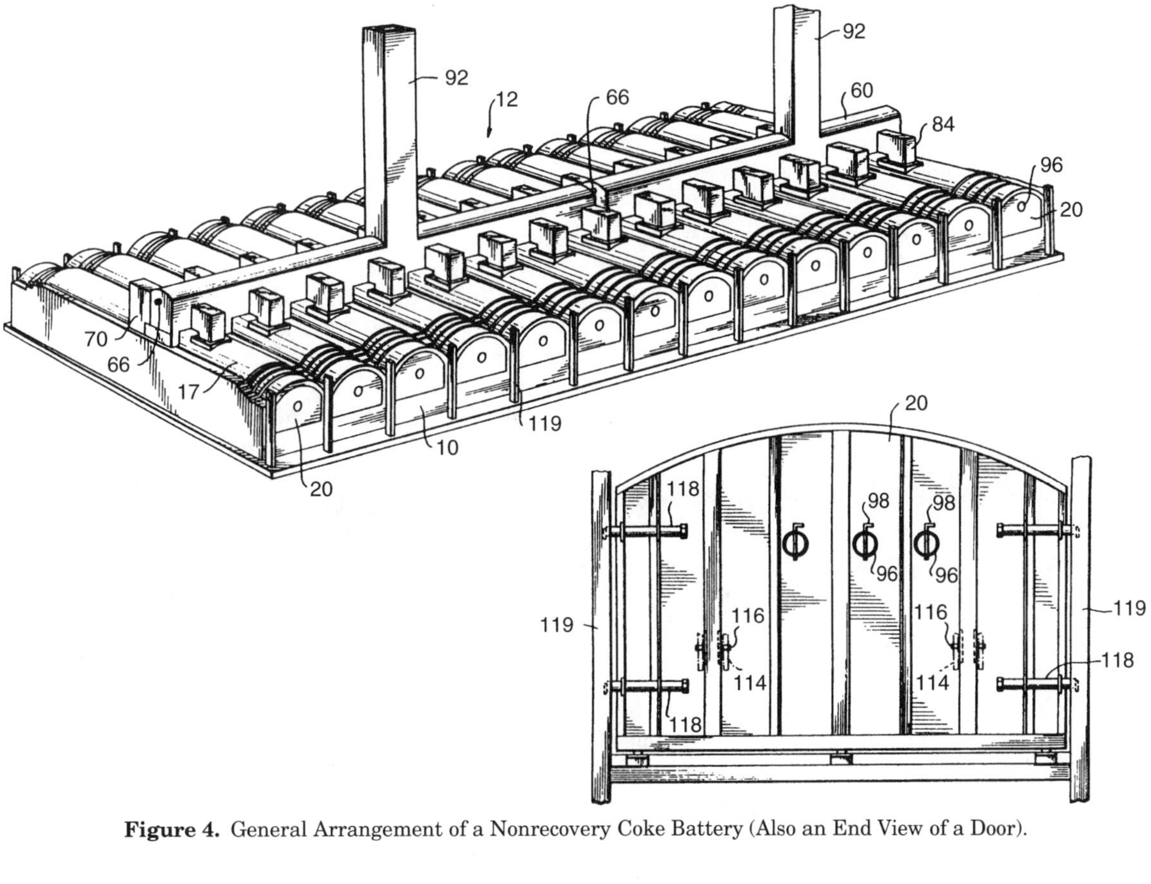

Figure 4. General Arrangement of a Nonrecovery Coke Battery (Also an End View of a Door).

PROCESS DESCRIPTION

By-Product Coke Production

Oven Size, Shape, and Capacity. In by-product coke production, the coke is produced in a series of narrow, 400–600-mm- (16–24-in.) wide, 12–18-m- (40–60-ft) long ovens that may be 3–6 m (10–20 ft) high, or even higher. The smallest ovens produce about 7.5 tons of coke per cycle, while the larger ones produce about 39 tons of coke per cycle. The new jumbo ovens (7.5 m high, 550 mm wide, 18 m long) operated by Mannesman in Germany produce 50.6 tons of coke per oven cycle. Depending on the size of the ovens and the desired production rate, there may be from 35 to over 100 ovens in a battery. The ovens are slightly wider on the coke side (the side of the battery where the coke comes out of the oven) than on the pusher side (the side of the battery on which the pusher machine is located) so that the coke will more easily move through the oven as it is being pushed.

Oven Heating. In a battery, the wall separating adjacent ovens, as well as each end wall, is made up of a series of heating flues. At any one time, half of the flues in a given wall will be combusting gas, while the other half will be conveying the heated products of combustion (waste heat) from the combustion flues to the combustion or waste-heat stack through the "checker brick" heat exchanger. The flame temperature is above the melting point of the battery brickwork. In order to avoid melting the battery brickwork, and to provide more uniform heating to the coke mass, every 20–30 min the battery "reverses." The former waste-heat flues become combustion flues, and the former combustion flues become waste-heat flues.

The maximum design heating rates of a slot-oven type (long, narrow, and high ovens) of coke battery, which charges wet coal, range from 1 to 1.25 in. of oven width per hour. Therefore, a 16-in.-wide oven may be pushed (pushing is the process of forcing the coke out of the oven using a ram that moves from the pusher side to the coke side of the battery) in about 13 h after being charged, while a 24-in.-wide oven may require 24 h per coking cycle, depending on the heating rate. Preheated-coal-charged batteries, where the coal is preheated to about 400°F before charging into the ovens, can heat the coal at rates of up to 1.5 in. of oven width per hour. Coke produced in an 18-in.-wide preheated-coal-charged battery can be pushed in 12 h.

A by-product battery may be fired with a number of fuels, including coke oven gas, natural gas, and blast furnace gas. Coke oven gas is the most common fuel because it is produced by the coking process in quantities that exceed the amount required to heat the battery.

Maintaining the proper heat distribution in a coke oven is a complex task. In a typical battery, there will be over 2000 individual flues with from one to four burners per flue. Each burner has a nozzle that must be of the correct size and must be kept clean. A number of factors affect battery heating, including the types of coal used, the moisture content of the coal, the bulk density of the coal, the level and volume of the coal charged, variations in the coke production schedule, the fact that the coke side of the oven is wider than the pusher side, and the order in which the ovens are charged and pushed.

Coke Oven Gas Control. During the coking process, about one quarter of the mass of the coal charged is volatilized and removed as gas. Collecting, cleaning, and distributing this gas properly is a major component of the emission control effort in the coke production process. Each oven is connected via offtake piping assemblies to one or two collector mains, which run the length of the battery. The gas reaching the collector main(s) is evacuated through the by-product recovery plant, where it is processed into clean gas and a number of by-products, including tar, light oil, naphthalene, ammonia, sulfur, and many other components.

The Coke Production Operation

Coal Handling. Most coke-making operations use a blend of from two to seven distinct coals selected to produce coke with the desired qualities of size, strength, reactivity, and so on. To reduce variations in the quality of the coke and the required coking times, it is desirable for the coal to be of consistent blend, moisture content, and bulk density at all times. Over a million tons of coal are processed at a typical coke plant in a year. Therefore, the coal grinding, blending, moisture control, and bulk density control steps of coal handling are critical to the successful operation of the coke plant. While they will not be discussed here, the coal mining, drying and cleaning, and quality assurance steps done at the mine are also critical to the successful operation of the coke plant.

Charging. With the oven doors in place and properly sealed, a larry car that contains the proper blend of pulverized coal is placed (spotted) over the oven to be charged. With the oven under a slight negative pressure via steam or liquor aspirated into the off-take piping, the charging hole ports or lids on top of the oven are removed and the coal is discharged into the oven until it is full. Near the end of the charging sequence, the leveler (chuck) door within and near the top of the pusher side door is opened and a bar about as long as the oven is inserted to level the charge and aid in filling the oven. When all the coal is in the oven and leveled, the leveler door is closed, the charging port lids are replaced and sealed, the steam or liquor aspiration is turned off, and the off-take piping assembly is adjusted so that the gas evolved from the coke reaches the collector main.

Note: A number of batteries have been built that utilize a pipeline, rather than a larry car, for charging the oven. Pipe line-charged batteries have proved to be very difficult to operate. Therefore, pipeline-charged batteries will not be discussed here.

Coking. The coal in the oven is heated by the walls, and the coke mass increases in temperature from the outside in until the volatile matter has been removed from the entire coke mass. During this time, all gas evolved is removed from the oven via the off-take piping to the collector main.

Pushing. The oven to be pushed is isolated (dampered off) from the collector main so that gas in the collector main is not discharged through the oven when the doors are removed. Both doors (coke side and pusher side) are then removed from the oven. The pusher ram, which has a face that is taller than the coke mass and is almost as wide as the narrow end of the oven, is inserted from the pusher side and pushed through the oven toward the coke side, pushing the coke out of the oven. The coke is typically collected in a quench car, which is mounted on rails and is used to transport the hot coke to the quench tower.

Quenching. The coke pushed from the oven is very hot (about 1050°C) and will burn when it contacts oxygen from the air. To keep the coke from burning, it is normally quenched with water. The quench car, which receives the hot coke pushed out of the oven, is driven to the quench tower, where the coke is sprayed with water to reduce its temperature to below the ignition point of the coke and below the temperature at which the wharf conveyor belts will burn. Ideally, the temperature of the quenched coke on the wharf will vary from 120°F to 400°F.

Nonrecovery Coke Production

While numerous nonrecovery coking processes are being discussed, only the Jewell Thompson heat recovery process is successfully operating in the United States. There are two facilities that are producing coke: one in Vansant, Virginia, and one in East Chicago, Indiana. The new 268-oven facility in East Chicago, Indiana, consists of four batteries designed to produce 1.2 million tons of screened coke per year. The waste gases generate steam that drives a turbine capable of generating 87 MW of electricity.

The Jewell Thompson Process. The Jewell Thompson oven, depicted in Figure 5, is constructed of high-silica refractory material. The oven cavity, walls, crowns, and sole flues are built with less than two dozen different brick shapes. The oven is supported with steel work I-beams (buck-stays) on the front and back of each wall. The buck-stays are bound front to back with tie rods to offer lateral support for the oven brick work. The rectangular carbonization chamber is 45 ft long, 12 ft wide and 8 ft high at the crown of the arch. Typical charging levels range between 34 and 38 in. corresponding to 40 and 45, tons respectively. Coal is introduced into the carbonization chamber by removing the front door of the oven and feeding the coal into the oven cavity via conveyor mounted to the pusher/charger machine. The refractory maintains heat from the previous coking cycle. This residual temperature in the oven initiates the gassing of the newly charged coal. Heat generated during the coking cycle is absorbed into the refractory and supports the next charge, making the process self-sustaining. The ovens are operated under negative pressure, and air is introduced into the oven through air ports in the doors of the oven. The ports are kept open enough to maintain an excess of air to ensure that all the gases are combusted. The gases generated during the coking cycle exit the carbonization chamber through gas ports in the walls located above the coal line. The gas moves through the gas ports (three ports on opposite walls) into the down comers and into the sole flue area. The hot gases move under the oven floor in the sole flues, where baffles cause the gases to make four passes under the coal bed. This sinuous passage of the hot gases under the coal bed heats it from below. The coking process can then occur from the top of the bed downward and from the bottom of the bed upward.

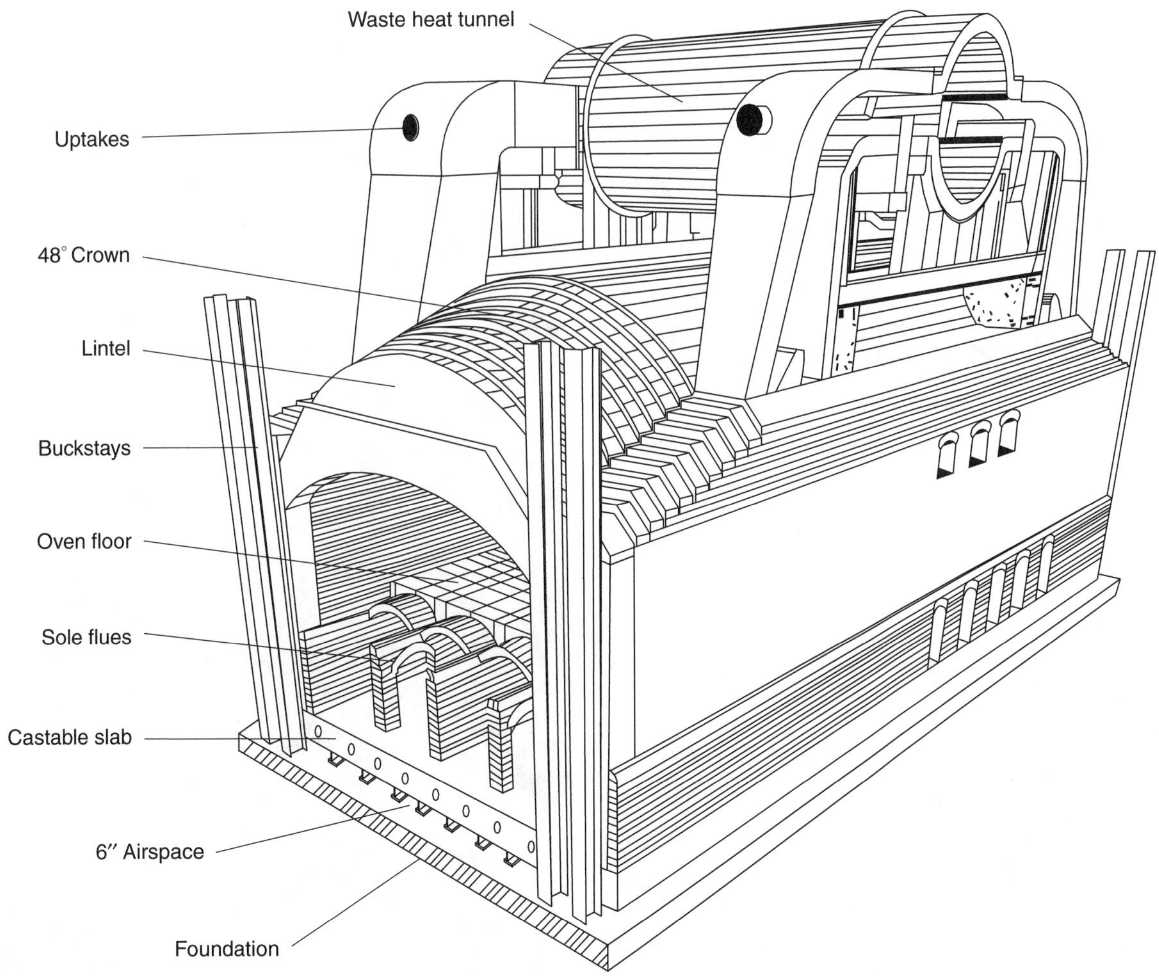

Figure 5. Jewell Thompson Oven.

The effect of heating from the top and the bottom of the coal bed is seen in Figure 6. There is a line of cleavage in the center of the coke profile, which shows that the coking process proceeded efficiently from both directions. The down comers serve identical sole flue systems on the coke side and the pusher side of the oven, allowing a more even distribution of the hot gas into the floor area of the oven (refer to Figure 5). The gases move from the sole flue into the uptakes toward the common tunnel. Before entering the common tunnel the gases are checked by the uptake damper. This damper mechanism regulates the pressure in the oven and allows operating personnel to regulate the coking rate of the coal in the oven. The gases on each battery of ovens are collected in a common gas tunnel and move into the waste heat boilers shown in Figure 7. The gases enter the waste heat boiler at 1750°F and exit the boiler at less than 400°F. The ovens are capable of a 45-ton coal charge and are coked for 48 h. Depending on the volatile matter in the coal blend charged into the oven, the charge weights, yield, and cycle rates will vary.

Coal Handling and Blending. Coal handling and blending are similar to conventional battery designs with many upgrades. The unloaded coals move from the dumper on a system of conveyors to the stacker, as shown in Figure 8. The stacker traverses the length of the coal field to coal pile locations and delivers the coal into its respective pile. The coals are fed from the piles by pile conveyors onto a blending belt. Vibrating feeders below the grizzlies at the base of each coal pile communicate weight to the mixing computer. Belt scales weigh the coal from the individual pile and will vary the speed of the vibrating feeders to deliver the correct weight ratio of coal onto the blending belt. This system is very reliable, operates with minimum disruptions, and delivers the blend consistently with less than a 2% error.

The coal is delivered by the blending conveyor to the grinders (Gundlach design) and mixed during grinding. The coals are controlled to 60% minus 1/8 in. The blended coal is then conveyed to the batteries for charging.

The East Chicago Facility consists of 268 ovens pushing and charging on a 48-h cycle. The ovens are grouped in four batteries, 67 ovens on each battery (Figure 9). The operating crews work two turns per day, each crew pushing 67 ovens per turn for a total of 134 ovens per day. The average charge is 42 tons per oven, producing 3450 tons per day, with an allowance for availability, or 1.22 million tons of screened coke per year.

The ovens are aligned two batteries in a row and opposite each other. The blended coal moves by conveyor to silos in the middle of the four batteries. The silos are arranged to

Figure 6. Pusher Side Ready to Be Pushed.

Figure 7. Waste Heat Boiler.

Figure 8. Stacker and Coal Field Piles.

Figure 9. Profile of Coke Battery.

feed two batteries each. At the base of each silo is a weigh feeder conveyor that delivers the specified amount of coal to the pusher/charger machine, shown in Figure 10. The coal is delivered to the charging machine via belt conveyor that runs the length of the battery. The pusher/charger machine fits on rails that run the length of the front side of the ovens on two batteries.

The machines are equipped with hydraulic drives that are powered by electric motors. The machines are equipped with a pusher ram, charging conveyor, and two door lifters that work immediately in front of the ram and conveyor. The support members carrying the pusher ram and charging conveyor are water cooled to prevent heat deformation during extended use. To charge the oven the charging conveyor is spotted in front of the oven to be charged and the door manual latches are released; the door is then lifted above the conveyor by hydraulic cylinders, and the conveyor begins to enter the oven. At the time the charging conveyor begins to move into the oven, coal is already moving down the battery conveyor to the coal hopper on the pusher machine located above the conveyor. The operator uses an onboard computer to specify the oven number and weight of the charge. The door is immediately placed on the oven, and the coking process begins.

The coking cycle is continuously monitored by the central control computer. The computer receives temperature data

Figure 10. Pusher/Charger Machine.

from thermocouples located in the crown and sole flues of each oven. Heating adjustments are made throughout the coking cycle to ensure the coke process is completed and over burning (yield loss) is minimized.

The combusted gases move into a common tunnel on each battery and are drawn into the waste heat boilers (WHB). There are four WHBs on each battery that convert the waste gases into steam. The turbine converts this energy into electricity. The spent gases exit the WHB and move into the cold gas duct and finally into the lime spray scrubbers to reduce the sulfur content of the gas. These scrubbers are of a cyclone wet spray design. The gas is then drawn into the baghouse and out the waste heat stack. Figure 11 depicts typical control of the spent gas.

The battery pushing schedule for the facility is designed to distribute the gas loads evenly across the entire plant. Every other oven on each battery is pushed every other day; that is, the even number ovens are pushed one day and the odds the next. The pushing is done by the pusher/charger machine. To push the oven the pushing ram is spotted in front of the oven to be pushed, the door manual latches are released, the door is then lifted above the ram by hydraulic cylinders, and the ram begins to enter the oven. At the same time a utility car operator removes the coke side door spots up the coke guide and calls out the oven is ready to be pushed. The utility cars are electric-powered hydraulic drive units, one unit for each battery.

The hot cars that receive the hot coke are powered by diesel locomotives. The hot coke is received in the hot car and carried to the quench station. The station utilizes conventional quench sprays designed to fit the hot car to maximize the even distribution of the quench. After quenching the coke is dumped into warfs for delivery to coke handling systems.

AIR EMISSIONS CHARACTERIZATION

By-Product Coke Batteries

At the time of this publication, the EPA was in the process of revising and updating emission factors for by-product and nonrecovery coke batteries. The majority of emission factors reported below reflect the original AP-42 listing in order to avoid reporting factors that are in the process of review and potential revision. Please refer to EPA's latest version of AP-42 for the most up-to-date emission factor listing.

Coke Oven Doors. Door leaks occur when there are gaps between the coke door and jamb or between the door jamb and the surrounding brickwork. Coke ovens are kept at a slight positive pressure (relative to atmospheric) during the coking cycle. Therefore, if there is an imperfection in the seal between the door and jamb or between the jamb and the brickwork, the material volatilized during the coking process is likely to leak from the door or jamb.

Emissions. The emissions from door leaks are referred to by the U.S. Environmental Protection Agency (EPA) as "coke oven emissions" and are listed as a hazardous air pollutant under Section 112 of the Clean Air Act. These emissions are essentially uncleaned coke oven gas. There have been few reported measurements of coke oven door emission rates or composition. What has been reported is that estimates of coke oven door emission rates vary by a factor of 10–100 or more.

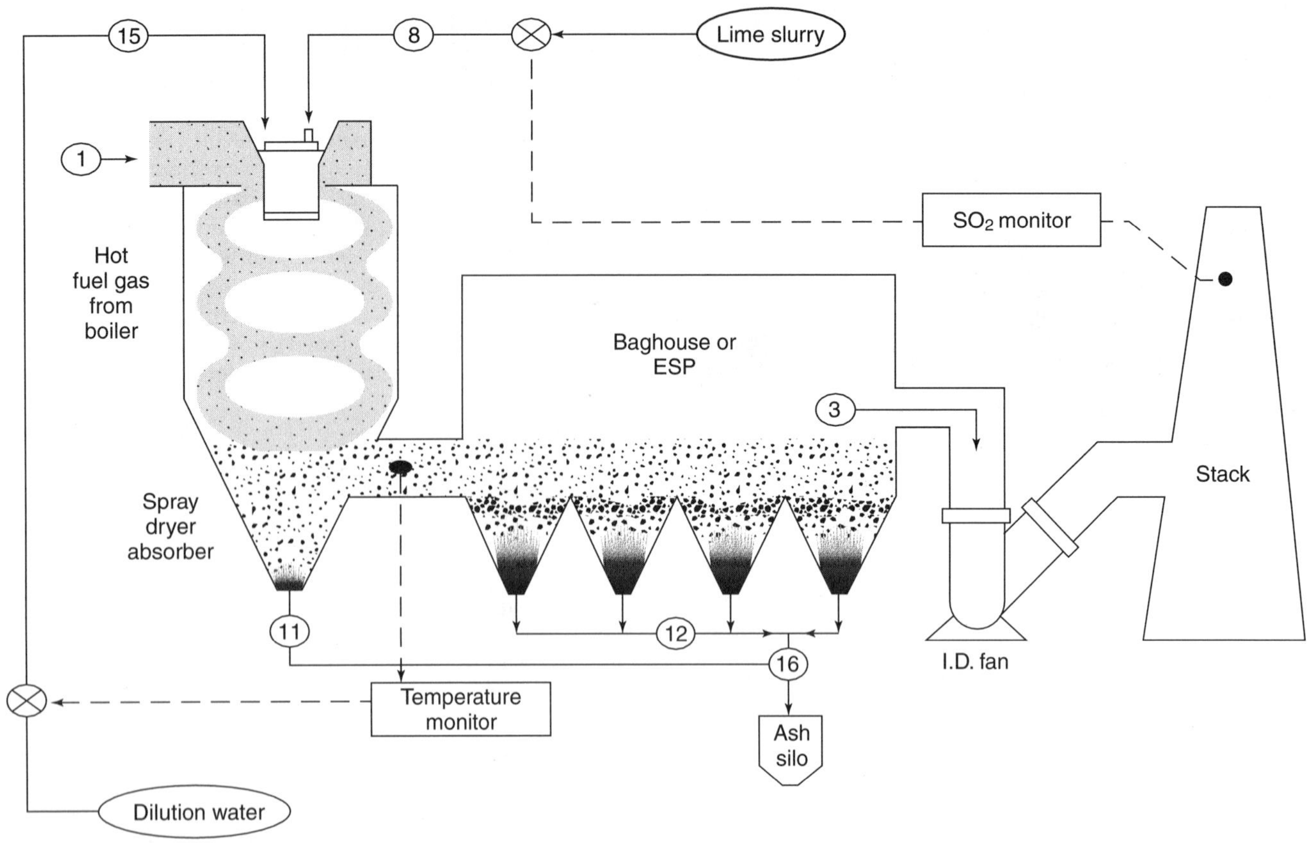

Figure 11. Lime Semidry FGD System.

Particulates. Since coke oven door emissions are actually uncleaned coke oven gas, the particulate emissions are the fractions of raw coke oven gas that condense out at the filter temperature (248°F or 120°C during a standard EPA Method 5 test). This condensed gaseous material is virtually all "benzene soluble organics" (BSO).

There is little information on the rate of emission generation. In the draft AP-42, the EPA reports 0.5 pound of particulate emissions per ton of coal charged for "uncontrolled" door leaks. However, in the EPA's *Coke Oven Emissions from Wet-Coal Charged By-Product Coke Oven Batteries—Background Information for Proposed Standards*,[1] door leak emissions for a model battery with 10% leaking doors are reported as 0.01 to 0.12 kg BSO per hour per leak. The EPA's data are based on two sets of measurements of emissions from leaking doors under a coke side shed. One of these measurements was taken on a battery with about 70% leaking coke side doors, and the other on a battery with about 30% leaking coke side doors. During the tests reported by the EPA, the amount of BSO per leak per hour from the battery with 70% door leaks was significantly lower than that from the battery with 30% leaking doors.

Recent industry measurements have generally confirmed the range of the EPA's reported door leak emissions and have provided further insight into the reason for the range. The numbers at the low end of the range are associated with the light wisp type of leaks now commonly found on most batteries, while values near the high end of the EPA's range are associated with the very large leaks that are now relatively rare. Most large leaks now observed have mass emissions closer to the midpoint of the EPA's reported range of emissions. There currently is no reported information on the size distribution of door leak particulates. However, since they are condensed gaseous material, the size of the particles for these emissions are expected to be very small (i.e., submicron).

Gases. Because door leaks are essentially raw coke oven gas, the gaseous portion of the emission is the lighter portion of the coke oven gas and can be approximated by the typical composition shown in Table 1.

Table 1. Typical Emissions from Coke Oven Door Leaks

Material	Typical Wt% Composition	Epected Emissions (g/h/leak)
Acetylene	0.11	0.01
Ammonia	1.90	0.16
Benzene	4.05	0.34
Butane	0.20	0.02
Butylene	0.85	0.07
Carbon dioxide	5.59	0.47
Carbon monoxide	18.28	1.53
Ethane	3.42	0.29
Ethylene	5.97	0.50
Hydrogen	9.58	0.80
Hydrogen cyanide	0.45	0.04
Hydrogen sulfide	2.90	0.24
Methane	40.92	3.43
Nitrogen	3.04	0.25
Oxygen	0.70	0.06
Propane	0.42	0.04
Propylene	1.02	0.09
Toluene	0.38	0.03
Xylene	0.22	0.02

Based on recent industry measurements of the condensible portion of an EPA Method 5 particulate emission test procedure, the emission rate for gaseous benzene from a leak appears to be essentially independent of the particulate (BSO) emission rate from a leak; it is about 0.34 g/h per leak. The other gases are expected to be present in the same relative proportions in which they are found in coke oven gas.

Toxics. As stated, the EPA has listed "coke oven emissions" as a hazardous air pollutant under the Clean Air Act. In addition, the gaseous portions of the door emissions contain the following chemicals that are identified as toxics under SARA: ammonia, benzene, ethylene, propylene, toluene, and xylene.

Coke Oven Lids and Off-Takes. Coke oven lids are normally closed and sealed except when the oven is being charged or pushed. During these periods, the oven is either at atmospheric pressure or at a negative pressure so there should be no emissions. Similarly, coke oven off-take piping assemblies are normally sealed closed, except when the oven is being decarbonized or pushed so that they should not leak. However, both lids and off-takes are under pressure during the coking cycle and can leak if sealing methods fail.

Emissions. The EPA has updated AP-42 to list emissions from lid leaks and off-take leaks. Uncontrolled BSO emissions from lid leaks and offtake leaks are estimated at 0.05 lb/ton. Post-NESHAP controls are estimated at 0.00060 and 0.0020 lb/ton, respectively.

Coke Oven Charging. Charging is the process of introducing coal into the oven through the charging ports on top of the oven.

Emissions. Emissions from charging are similar in character to those from door, lid, and off-take leaks. All these emissions are listed by EPA as the hazardous air pollutant "coke oven emissions." However, in addition to raw coke oven gas, charging emissions may contain coal dust. The EPA's AP-42 lists the emissions from wet coal charging shown in Table 2.

Unfortunately, AP-42 does not relate emissions to the typical charging standards of seconds of visible emissions for a series of charges. Based on measurements of door emissions, it would be more accurate to relate charging emissions to seconds of light, moderate, and significant emissions during the charge. The emissions of benzene-soluble organic particulate emissions would be about 0.1 g/sec for light visible emissions, 0.8 g/sec for moderate visible emissions, and 1.7 g/sec for heavy visible emissions. The difference between particulate emissions determined by this method and those predicted by other emission factors are likely to be coal-dust emissions of a fairly large particle size. The ratio of other gaseous emissions to benzene-soluble organic emissions would be the same as for door emissions.

In AP-42, the EPA has also estimated particle size distributions for stage charging operations. These estimates are given in Table 3.

Coke Oven Pushing. The process of pushing hot coke out of the oven is often one of the most spectacular and hard-to-control processes in the coke-making process. The hot coke breaks up as it falls out of the coke guide on its way to the quench car. As a result of this fall, the coke, which has newly exposed surfaces to the air and may contain volatile matter not removed during the coking cycle, often ignites violently. This ignition and the mechanical impact of falling coke hitting the quench car often release coke particles into the atmosphere. On the other hand, perfectly coked material can be pushed with virtually no flames and little smoke or other emissions. The variation in uncontrolled pushing emissions from battery to battery and from oven to oven on the same battery makes reliable control difficult.

Emissions. Coke pushing emissions are primarily carbon particulate. The EPA has published the particulate emission factors by particle size for uncontrolled coke pushing as shown in Table 4.

There is very little information on gaseous emissions from pushing. The EPA does report gaseous emissions from uncontrolled coke pushing Table 5. However, these emission factors are based on a 1968 United Nations report and are not well documented. For example, a recent industry measurement indicates that uncontrolled coke pushing volatile organic compound emissions may be 0.005 kg/mg or 0.01 lb/ton, which is less than one-tenth of that reported in AP-42.

Table 3. EPA Emission Factor and Particulate Size Distribution for Wet-Coal Larry Car Stage Charging

Particle Size, μm	Cumulative Mass % Less Than or Equal to Stated Size	Cumulative Mass Emission Factors kg/Mg	lb/ton
0.5	13.5	0.001	0.002
1.0	25.2	0.002	0.004
2.0	33.6	0.003	0.005
2.5	39.1	0.003	0.006
5.0	45.8	0.004	0.007
10.0	48.9	0.004	0.008
15.0	49.0	0.004	0.008
30.0	100.0	0.008	0.016

Table 2. EPA Emission Factors for Wet-Coal Larry Car Charging

Pollutant	Uncontrolled		Stage Charging		Scrubber Control	
	kg/Mg	lb/ton	kg/Mg	lb/ton	kg/Mg	lb/ton
Particulates	0.24	0.48	0.008	0.016	0.007	0.014
Sulfur dioxide	0.01	0.02				
Carbon monoxide	0.3	0.6				
Volatile organics	1.25	2.5				
Nitrogen oxide	0.015	0.03				
Ammonia	0.01	0.02				

Table 4. EPA Emission Factors and Particulate Size Distribution for Uncontrolled Coke Pushing Operations

Particle Sizes, μm	Cumulative Mass % Less Than or Equal to Stated Size	Cumulative Mass Emission Factors kg/Mg	Cumulative Mass Emission Factors lb/ton
0.5	3.1	0.02	0.04
1.0	7.7	0.04	0.09
2.0	14.8	0.09	0.17
2.5	16.7	0.10	0.19
5.0	26.6	0.15	0.30
10.0	43.3	0.25	0.50
15.0	50.0	0.29	0.58
30.0	100	0.58	1.15

Table 5. EPA Gaseous Emission Factors for Uncontrolled Coke Pushing Operations

Material	Emission Factor kg/Mg	Emission Factor lb/ton
Carbon monoxide	0.035	0.07
Volatile organics	0.1	0.2
Ammonia	0.05	0.1

Coke pushing is generally not considered a source of air toxics. The emissions from coke pushing are not included in the listed hazardous air pollutant "coke oven emissions."

Coke Oven Underfire Stack. The underfire (or combustion) stack's main purpose is to provide natural draft to the battery combustion system and to convey and emit the combustion products from heating the battery.

Emissions. The emissions from the underfire stack come from two sources. Every underfire stack conveys the emissions from fuel combustion of the gas used to heat the battery. In addition, there are often some unburned coal fines that make their way to the combustion stack via oven-to-flue leakage. These coal fines can often be traced to a recently charged oven, and thus stack opacity provides useful feedback to the operator concerning oven wall damage needing repair.

Coke battery stack emissions are no more toxic than emissions from other fuel-burning sources and are not included in the listed hazardous air pollutant "coke oven emissions." The effect of fuel type on particulate emissions is indicated by the data from the EPA's AP-42 uncontrolled emission factors shown in Table 6.

Table 6. EPA Emission Factors for Coke Battery Combustion Stacks

Fuel Type	Particulate Emission Factor kg/Mg	Particulate Emission Factor lb/ton	Sulfur Dioxide Emission Factor kg/Mg	Sulfur Dioxide Emission Factor lb/ton
Coke oven gas (undesulfurized)	0.234	0.47	2.0	4.0
Blast furnace gas	0.085	0.17		

In AP-42, EPA also gives the particle size estimates for uncontrolled emissions when using undesulfurized coke oven gas Table 7. While this emission information is from a reputable source, it should be remembered that coke battery underfire particulate emissions are highly dependent on the condition of the battery walls and on whether the underfire gas has been desulfurized. It is not uncommon for batteries fired on similar fuels to have particulate emissions that vary by a factor of 4 or 5. Similarly, the sulfur dioxide emission factor for underfiring with undesulfurized coke oven gas is based upon a number of assumptions. It is much more reliable to use the actual gas sulfur content and the quantity of fuel gas consumed to determine battery stack emissions.

Coke Quenching. Quenching is the process of spraying water on hot coke to reduce its temperature so that it will not burn.

Emissions. Emissions from quenching are typically larger-sized particulates created by the breakup of the hot coke when it is hit with water. This particulate is carried up the quench tower by the velocity of the steam plume. Quenching is normally conducted with relatively clean water (less than 1500 mg/L total dissolved solids); however, at some facilities, water containing over 5000 mg/L of total dissolved solids may be used (this is often called dirty water quenching). The typical control for conventional quenching is to install baffles in the quench tower. The EPA has published the size-specific emission factors shown in Table 8 for the various combinations of coke quenching practices.

Very few data are available about other emissions from quench towers. Some researchers have reported various volatile organic emissions from quenching, while some industry tests have been unable to duplicate these findings when measuring emissions.

Battery Venting. When the exhausters have insufficient suction to remove the gas being generated by the battery, emergency relief vents on the battery will open and the battery will vent raw coke oven gas.

Emissions. These emissions are essentially raw coke oven gas and are hazardous air pollutant "coke oven emissions." The composition of the emission is expected to be similar to that reported for coke oven door leaks. The quantity of emissions depends on the amount of gas not evacuated by the exhausters.

Coke By-Product Recovery Plants. As their name indicates, the original purpose of coke oven by-product recovery plants

Table 7. EPA Particulate Emission Factors and Size Distribution for Combustion Stack of Coke Batteries Fired with Undesulfurized Coke Oven Gas

Particle Sizes, μm	Cumulative Mass % Less Than Stated Size	Cumulative Mass Emission Factors kg/Mg	Cumulative Mass Emission Factors lb/ton
1.0	77.4	0.18	0.36
2.0	85.7	0.20	0.40
2.5	93.5	0.22	0.44
5.0	95.8	0.22	0.45
10.0	95.9	0.22	0.45
15.0	96	0.22	0.45
30.0	100	0.23	0.47

Table 8. EPA Emission Factors and Size Distribution for Coke Quenching Operations

Particle Size, µm	Cumulative Mass % Less Than or Equal to Stated Size				Cumulative Mass Emission Factors (lb/ton of Coal Charged to Oven) Divide by 2 to get kg/Mg			
	Clean Water		Dirty Water		Clean Water		Dirty Water	
	No Baffles	Baffles	No Baffles	Baffles	No Baffles	Baffles	No Baffles	Baffles
1.0	4.0	1.2	13.8	8.5	0.05	0.006	0.72	0.11
2.5	11.1	6.0	19.3	20.4	0.13	0.03	1.01	0.27
5.0	19.1	7.0	21.4	24.8	0.22	0.04	1.12	0.32
10.0	30.1	9.8	22.8	32.3	0.34	0.05	1.19	0.42
15.0	37.4	15.1	26.4	49.8	0.42	0.08	1.38	0.65
30.0	100	100	100	100	1.13	0.54	5.24	1.30

was to recover by-products such as tar, benzene, sulfur, and ammonium sulfate for sale. Because of current economics in North America, most by-products plants now function primarily to clean the coke oven gas for downstream users. The "by-products" are now often produced as a necessary part of cleaning the gas. This has resulted in the more recently constructed coal chemical plants being designed to destroy, rather than produce, many of the by-products historically produced. Older plants, which have the capability to refine the tar and light oil into many constituents, are now often operated to produce tar and light oil for sale as feedstocks to chemical manufacturers.

Because it is essentially a chemical plant, the coke oven by-products emission sources are similar to those found at other chemical manufacturing facilities. These sources are product storage tanks, decanters, reactors, and heat exchangers. Figure 12 shows the simplified flows at a typical coal chemical plant. Notice that flushing liquor, circulating liquor, and weak (or excess) ammonia liquor are similar. Similarly, light oil and BTX are essentially the same material.

Emissions. Product storage tanks for tar, light oil (or BTX), and excess ammonia liquor have emissions related to the vapor pressures of the materials and the operation of the tanks. Decanters and sumps also can have emissions related to the material they contain, flow rates, temperature, and decanter design. Cooling towers and process tanks for materials (such as final cooler cooling water and sulfuric acid) that come into

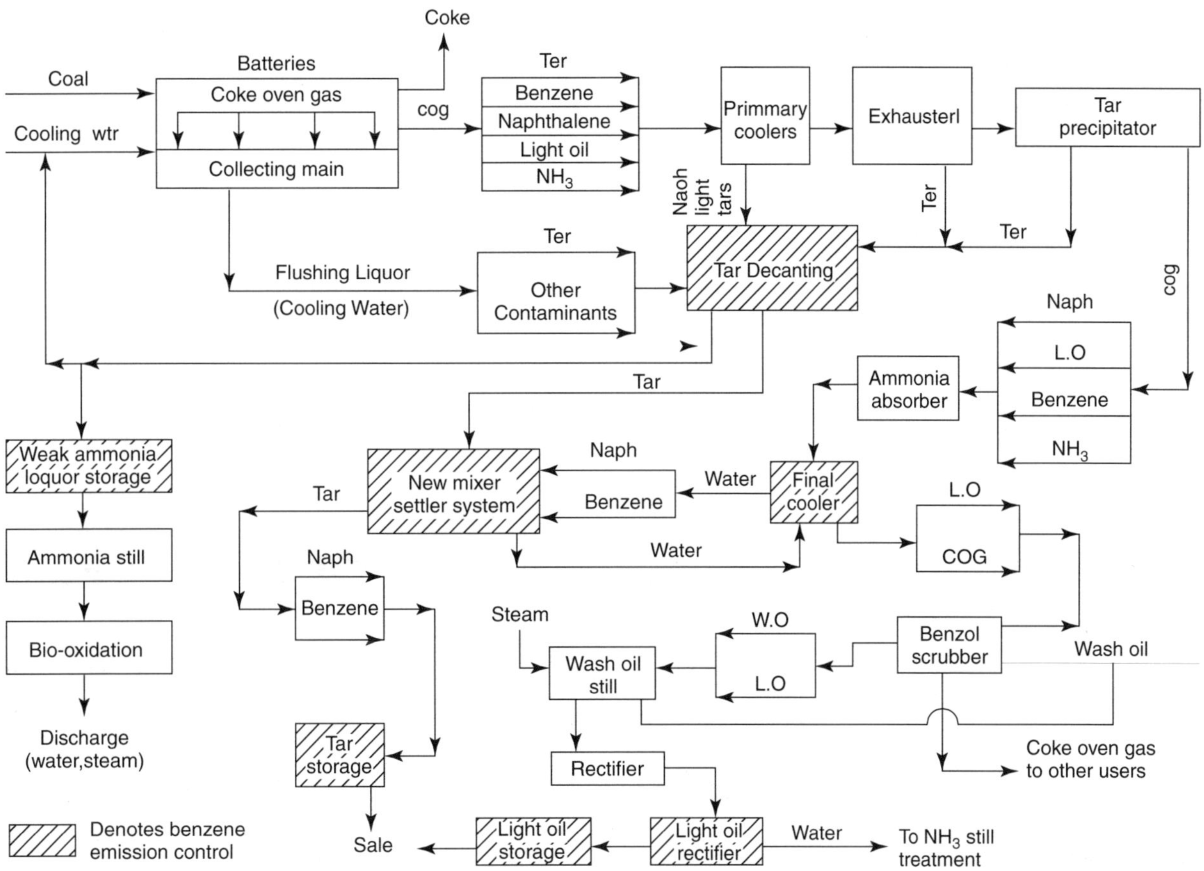

Figure 12. Simplified Flow Diagram of a Coke By-Product-Recovery Plant.

contact with the coke oven gas can emit materials that have been adsorbed from the gas.

There are normally no particulate emissions from the coal chemical plant. In addition, the only sulfur emissions are likely to occur from the "tail gas" of any sulfur-recovery plant that is used to convert hydrogen sulfide into elemental sulfur. The only nitrogen dioxide or carbon monoxide emissions will be from combustion sources such as those associated with the sulfur-recovery operation.

The most common emissions from coke oven by-product recovery plants are the various species of volatile organic compounds (VOCs), including benzene and other air toxics. Chemicals that may be found in typical coke oven gases and are listed as toxics under SARA Title III include ammonia, anthracene, benzene, biphenyl, dibenzofuran, ethylbenzene, ethylene, naphthalene, phenol, propylene, quinoline, styrene, toluene, 1,2,4-trimethylbenzene, and xylene.

Most of the effort in estimating emission rates from by-product recovery plants has focused on benzene emissions. Some work has also been done on estimating total VOC emissions. The data in Table 9 from the EPA's *Benzene Emissions from Coke By-Product Recovery Plant Background Information for Proposed Standards*[2] indicate the estimated range of uncontrolled benzene emissions from both furnace and foundry coke plants.

While the emission estimates in Table 9 are the best commonly available, they should be used with caution. Not all by-product plants have all these components. In addition, some of the emission estimates do not make sense. For example, the emission estimate for the storage of tar, which has a relatively small amount of benzene, is twice as high as the emission estimate for the storage of benzene (where all emissions are benzene). It is best to use these factors as a screening tool to determine the likely sources of benzene emissions at a facility and then to use facility-specific information to determine actual facility benzene emissions.

The other source of emissions at a by-product recovery plant is equipment leaks, primarily from equipment in the light-oil-recovery plant carrying liquids that contain significant concentrations of benzene. The EPA estimates of the emissions from such sources are given in Table 10.

Table 9. EPA Benzene Emission Factors for Uncontrolled Furnace and Foundry Coke By-product Plants (Grams of Benzene per Megagram of Coke)

Source	Furnace Plant, g/Mg	Foundry Plant, g/Mg
Direct- water final cooler cooling tower	270	197
Tar-bottom final cooler cooling tower	70	51
Naphthalene separation and processing	107	79
Light oil condenser vent	89	48
Tar intercepting sump	90	45
Tar dewatering	21	9.9
Tar decanter	77	36
Tar storage	12	5.6
Light-oil sump	15	8.1
Light oil, BTX, or benzene storage	5.8	3.1
Flushing liquor circulation tank	9	6.6
Excess ammonia liquor circulation tank	9	6.6
Wash-oil decanter	3.8	2.1
Wash-oil circulation tank	3.8	2.1

Nonrecovery Battery Sources

Coke Oven Charging, Pushing, and Quench

Emissions. Because of the oven design and operating practices (negative oven pressure), emissions from doors and uptakes do not exist.

Coke Oven Charging

Emissions. Limited testing at a point above the door during 3.5-min charges indicates the emission factors shown in Table 11.

Coke Oven Pushing and Quenching

Emissions. Pushing and quenching operations are similar to conventional by-product oven operations. Table 12 and 13 lists

Table 10. EPA Benzene Emission Factors for Equipment Leaks (Kilograms of Benzene per Leak per Day)

Source	VOC Emissions	Furnace Plant		Foundry Plant	
		LO[a]	Benzene[b]	LO[a]	Benzene[b]
Valves	0.26	0.18	0.22	0.16	0.20
Pumps	2.7	1.9	2.3	1.7	2.1
Exhausters	1.2	0.28	0.28	0.25	0.25
Pressure-relief devices	3.9	2.7	3.4	2.5	3.1
Sampling connections	0.36	0.25	0.31	0.23	0.28
Open-ended lines	0.055	0.038	0.047	0.035	0.043

[a]"LO" indicates that the plant recovers light oil. The amount of benzene in the light oil is assumed to be 70% at furnace plants and 63.5% at foundry plants.

[b]"Benzene" indicates that the plant recovers refined benzene. The amount of benzene averaged over the light oil and benzene portions of the plant is assumed to be 86% at furnace plants and 78% at foundry plants.

Table 11. Charging Emissions for Nonrecovery Batteries

Pollutant	Uncontrolled Emissions		Controlled Emissions[b]	
	kg/Mg	lb/ton	kg/Mg	lb/ton
Filterable PM[c]	0.013	0.027	0.0041	0.0081
TSO[d]	0.0013	0.0026	0.0011	0.0022
Benzene	1.8×10^{-5}	3.6×10^{-5}	1.8×10^{-5}	3.6×10^{-5}
Toluene	8.4×10^{-6}	1.7×10^{-5}	8.4×10^{-6}	1.7×10^{-5}
Xylene	3.4×10^{-6}	6.7×10^{-6}	3.4×10^{-6}	6.7×10^{-6}
Carbon disulfide	1.1×10^{-6}	2.1×10^{-6}	1.1×10^{-6}	2.1×10^{-6}
Chloromethane	1.0×10^{-6}	2.0×10^{-6}	1.0×10^{-6}	2.0×10^{-6}
Ethyl benzene	3.6×10^{-7}	7.3×10^{-7}	3.6×10^{-7}	7.3×10^{-7}
Naphthalene	1.2×10^{-5}	2.3×10^{-5}	1.2×10^{-5}	2.3×10^{-5}
Total PAHs[e]	1.4×10^{-5}	2.7×10^{-5}	1.2×10^{-5}	2.3×10^{-5}
Manganese	7.5×10^{-7}	1.5×10^{-6}	2.3×10^{-7}	4.6×10^{-7}
Arsenic	4.0×10^{-7}	7.9×10^{-7}	1.2×10^{-7}	2.4×10^{-7}
Nickel	2.5×10^{-7}	5.0×10^{-7}	7.5×10^{-8}	1.5×10^{-7}
Lead	1.7×10^{-7}	3.4×10^{-7}	5.0×10^{-8}	1.0×10^{-7}
Chromium	1.7×10^{-7}	3.4×10^{-7}	5.0×10^{-8}	1.0×10^{-7}
Cobalt	1.2×10^{-7}	2.4×10^{-7}	3.6×10^{-8}	7.1×10^{-8}
Beryllium	1.5×10^{-8}	2.9×10^{-8}	4.4×10^{-9}	8.7×10^{-9}
Mercury	1.3×10^{-9}	2.6×10^{-9}	4.0×10^{-10}	7.9×10^{-10}

Table 12. Pushing Emission Factors for Nonrecovery Batteries

Pollutant	Emission Factor	Ref.	Comment
TSP	0.09	*a*	EF includes baghouse control
PM_{10}	0.04	*a*	PM_{10} = 43.3% of TSP per Table 12.2-12 of Draft AP-42, Section 12.2
SO_2	0.05	*b*	
NO_x	0.016	*c*	
CO	0.077	*c*	
VOC	0.2	*c*	
Lead	0.077	*c*	
NH_3	0.1	*c*	

[a]Draft AP-42, Table 12.2-7.
[b]Table 2 of *T.W. Easterly, Measurement Based Coke Oven Battery Emission Factors,* Presented at the 88th Annual Meeting of the AWMA, San Antonio, TX, June 1995.
[c]Draft AP-42, Table 12.2-9.

Table 13. Quenching Emission Factors for Nonrecovery Batteries

Pollutant	Emission Factor (lb/tons wet coal)	Ref.	Comment
TSP	0.54	*a*	Dirty water (1200 mg/L), normal tower height and proper maintenance
PM_{10}	0.12	*b*	PM_{10} = 22.8% TSP per Draft AP-42, Table 12.2-10

[a]Draft AP-42, Table 12.2–7.

Table 14. Combustion Stack Emissions For Nonrecovery Batteries

Pollutant	Emission Factor kg/Mg	Emission Factor lb/ton
Filterable PM[b]	1.1	2.2
Condensible PM[c]	0.075	0.15
SO_2[d]	5.3	11
NO_x[d]	0.39	0.78
CO[d]	0.035	0.07
Benzene[e]	0.00026	0.00051
Toluene[f]	0.00026	0.00051
Naphthalene[g]	0.00017	0.00034
Phenol[h]	3.6×10^{-5}	7.1×10^{-5}
Antimony[g]	6.3×10^{-5}	0.00013
Arsenic[g]	0.00063	0.0013
Barium[g]	6.3×10^{-5}	0.00013
Beryllium[g]	1.0×10^{-5}	2.0×10^{-5}
Cadmium[g]	9.0×10^{-5}	0.00018
Chromium[g]	0.00032	0.00063
Copper[g]	0.0014	0.0028
Lead[d]	0.0017	0.0034
Manganese[g]	0.00015	0.00030
Mercury[g]	0.0017	0.00033
Nickel[g]	0.00029	0.00058
Phosphorus[g]	0.0070	0.014
Selenium[g]	0.00016	0.00032
Silver[g]	2.3×10^{-5}	4.5×10^{-5}
Thallium[g]	9.0×10^{-5}	0.00018
Zinc[g]	0.0026	0.0051

emissions factors for pushing and quenching on nonrecovery coke batteries. Limited testing inside a hot-car shed has provided some qualitative results. It was not possible to quantify exhaust gas volume during pushing.

Combustion Stack

Emissions. Stack emissions are, to a large extent, a function of the coal mix used in the coking process. Some testing has been done for specific coal mixes. Emission factors indicated were as given in Table 14.

None of the polycyclic aromatic hydrocarbons (PAHs) typically associated with coke oven gas emissions was detected (e.g., benzo-a-pyrene, anthracene, phenanthrene, chrysene, pyrene). Among the volatile and semivolatile organic compounds, only four were consistently present. They were benzene, bromomethane, methyl ethyl ketone (2-butanone), and carbon disulfide.

AIR POLLUTION CONTROL MEASURES

By-Product Coke Batteries

Coke Oven Doors

Equipment. *Door and Sealing Designs.* In the past, when batteries were shorter, it was typical to utilize rigid doors and jambs and to hand lute (seal) the opening between the door and jamb with a claylike material. More recently, door leakage control has relied on maintaining very close tolerances between the door sealing edge and the door jamb. The tars contained in the raw coke oven gas escaping from small leaks remaining between the door sealing edge and the door jamb will condense and seal the opening-and thus these doors are called self-sealing.

The task of minimizing gaps between the door knife (or sealing) edge and the door jamb is complicated by the tendency of the door jambs and door frames to warp over time. This tendency for warpage is greater on taller ovens that have been constructed since the early 1970s. A number of techniques, in addition to the continual replacement of doors and jambs that are not perfectly straight, have been employed to reduce leaks from coke oven doors. These include adding adjustment to the door knife (or sealing) edge to allow it to reconform to a jamb that is no longer straight and plumb and using sealants such as sodium silicate to plug gaps remaining after the door is seated.

In addition to simply increasing the adjustability of the door knife edge with various spring arrangements, a number of manufacturers are now offering "flexible" doors that are designed to bend. This allows them to seal with jambs that have many inches of warpage. These doors have not yet been fully tested in long-term operation. The initial results after replacing all doors and jambs on a battery with flexible doors are typically positive; that is, there is a significant decrease in leakage. However, based on experience to date and depending on battery and door operation and maintenance practices, the improvement has not always proved to be long lasting. When maintained similarly to conventional doors with adequate adjustability, the new flexible-door designs appear to have about the same leak performance.

The use of sodium silicate has been shown dramatically to reduce door leaks for short periods. However, over the longer term, use of this material tends to mask the underlying problem of the door and jamb's not sealing correctly. Typically, after less than a year's use, the gaps between the door sealing edges and the jambs become so large that even the sodium silicate no longer can seal the opening. After that time, the battery typically will have significant door leakage until the core problems are addressed. The use of door sealants such as sodium silicate also creates major door and jamb cleaning problems, which make proper door maintenance difficult.

Auxiliary Capture Devices. An alternative method of controlling emissions from door leaks is to capture the emissions and duct them to an appropriate control device. Two methods have been used to capture door leak emissions after they have been generated: door hoods and coke side sheds.

A few batteries have been constructed with evacuation hoods positioned over the doors to capture the door leak emissions as they rise above the battery. Reports on at least one of these batteries indicate that the hood did not have a significant impact on the control of door emissions. This was so because typical door leaks are so small in relation to the distance from the leak to the hood that they are often affected by winds and are blown away from the battery before they reach the hood.

There are also a number of batteries with coke side sheds. A properly designed coke side shed will capture virtually all leaks from coke side doors (which are typically the most difficult doors to keep from leaking). If the control device is capable of cleaning the door leak emissions, there will be no impact on the ambient air regardless of the amount of leakage. A number of the control devices used on the exhaust gases from coke side sheds have the potential to remove the benzene-soluble particulate emissions from the shed exhaust.

Work Practices. Work practices play a very significant part in door emission control. With the exception of the coke side shed, all the current methods of door emission control are highly dependent on the work practices discussed in the following.

Jamb and Brickwork Maintenance. The surface between the door jamb and the battery brickwork must be reasonably level and properly packed to prevent leakage there. The jamb also must be properly supported by the underlying brickwork or it will soon warp, making proper sealing difficult. In addition, if the oven walls are too close together or have excessive carbon buildup, the door plug will not fit fully into the oven. This will prevent the door sealing edge from fully mating with the door jamb, thus virtually ensuring that the door will leak.

There are significant differences in battery designs. On some batteries, changing door jambs is relatively routine, whereas on others, doing so requires a major effort, with the possibility that significant brickwork damage will occur that will necessitate major wall repairs coincident with the door jamb change. Thus while maintaining true and flat door jambs is an important part of some door leakage programs, other programs rely on being able to mate the doors to the jambs properly as they currently exist. In these programs, changing door jambs is a last resort.

Door Maintenance. In order for the door to seal properly, the sealing knife edge must not have any nicks or other imperfections and it must mate with the door jamb. This means that the door sealing edge and the springs that hold it in place must be properly maintained. There are many other parts of a door assembly that affect its leak performance, such as the door latches and door plug. To maintain these items in proper condition, regular rebuilding of each coke oven door is essential. In addition, doors that leak before their next scheduled rebuild, because they either are damaged or have some component failure, often need additional maintenance. The schedule for regular rebuilds varies depending on door designs, the age of the doors, and battery operational practices. To date, there is no consensus on an ideal routine door-rebuild frequency. (In the United Kingdom, the standard frequency is reported to be one door repaired or rebuilt per day for a two-battery operation or about once a year for each door.)

Door Adjusting. The sealing edges of most modern coke oven doors have a significant range of adjustment that can be used to ensure that the door mates with the jamb properly. In addition, each door has stop blocks that prevent the door from being pressed against the jamb so hard that the sealing edge is permanently deformed. Proper adjustment of the door to its jamb is essential to control door leaks. However, there is significant disagreement on the advisability of adjusting the door sealing edge after the door has been in service on a jamb. Some operators believe that door leaks can be controlled by routinely adjusting the sealing edge to stop observed leaks. Others believe that once the door is adjusted to the jamb, routine adjustments of the sealing edge are likely to cause additional leaks, and so it is better to determine the cause of the leak and make repairs as needed or even to rebuild the door completely if it starts leaking. There are no reported controlled long-term studies to determine which method of operation results in less door leakage.

Door Cleaning. In normal battery operation, the door seal gas channel, the door plug, both the outside and inside areas of the jamb, and the leveler door should be cleaned of tar and carbon buildup prior to every charge. This cleaning may be done manually (typically on short batteries) or by machine. However, this cleaning, particularly manual cleaning, is generally erratic in frequency and in quality, resulting in hard carbon buildup on the seal, plug, and jamb areas. Therefore, in most operations, each coke oven door is cleaned by sand blasting or high-pressure water blasting on a routine basis between door rebuilds. As with door rebuilds, there is no consensus on the ideal type of cleaning or the ideal timing of the cleaning. What seems to work at one facility often increases leaks when tried at another, similar facility. However, because clean doors and jambs are essential to good sealing performance, all new batteries are equipped with mechanical cleaners, and some older batteries are being retrofitted with mechanical cleaners on the door and pusher machines.

Sodium Silicate Sealing. As indicated above, sodium silicate can be used to seal a door leak when the opening causing the leak is too large to self-seal. However, in addition to the practical problems of properly applying the sealant to portions of the door that are more than 20 ft above the nearest supporting surface, use of this material often leads to more significant leakage later. This happens because sodium silicate will mask the underlying cause of the door leakage until the gap causing the leak becomes so large that even this sealant cannot stop the emission successfully. It becomes evident whenever there is a loss of gas pressure control on the battery and the higher pressure in the oven breaks all the sodium silicate seals.

At that time, there is often a major leak instead of the minor wisp leak that was originally controlled with the sealant. Thus while the limited application of sodium silicate, along with proper door maintenance, may be used to improve door leak performance, reliance on sodium silicate to the extent that the need for proper door maintenance is hidden will simply result in more significant leaks in the future.

Battery Back-Pressure Optimization. The collector main is maintained at a positive pressure designed to ensure that when the coal has completed coking and is no longer generating gas, the pressure at the bottom of an oven will not be negative. This is to ensure that air does not leak into the battery. Air leaking into the battery can cause the coke oven gas to burn, resulting in permanent damage to battery brick- and ironwork that will result in greatly increased emissions. In addition, too much air can cause the gas mixture in the collector main to become explosive, and possibly to explode, resulting in personal injury or major damage to the by-product plant, as well as very large emissions.

The rate of gas evolution from each single oven is not constant during the coking cycle. In addition, the gas flow through the collector main is affected by the oven charging process. Therefore, a damper system called an askania valve is used to regulate the battery back pressure. If this valve allows significant variations in back pressure, the tar that forms the final seal in a self-sealing door will be dislodged and the door will resume leaking. In addition, if the back pressure is set too high, the pressure on the door seals will be harder to contain within the oven. Similarly, if the back pressure is set too low, the fires associated with air infiltration will cause significant damage to the battery brickwork and door jamb, resulting in additional leakage.

Door Removal and Installation, Spotting, and Latching. The sealing edges of coke oven doors are easily damaged by contacting adjacent battery surfaces during their removal or installation. Frequently, the seals are damaged by hitting the latch hooks while the doors are being inserted into the oven. The removal and replacement of a door is a complex process involving unlatching, lifting, and twisting the door, and then reversing the process. If the machine performing this task is not properly adjusted or is not properly located in relation to the oven, the door sealing edge may contact the fixed portion of the door latches or buckstays and be damaged.

In addition to routine maintenance to ensure that the machines are properly adjusted, there must be a reliable method of "spotting" or properly locating the machine in front of each oven. Spotting methods range from the simple (such as aligning a pointer on the machine with a pointer on the battery) to sophisticated laser-alignment methods. All methods can be made to work. While the laser systems are more precise, all methods rely on the target on the battery being in the proper location. These targets are often located on, or are attached to, the battery buckstays. Over time, the buckstays move and warp. Therefore, no matter what spotting system is used, there must be routine realignment to ensure that the targets are properly located.

Door latching is important to ensure that the door sealing edge is held securely against the doorjamb, but is not under so much pressure that it is permanently deformed. As discussed above, the door stop blocks are often used as an indication that the door is properly secured. In addition, door latching pressure is routinely measured as an indication of proper door placement and latching.

Coke Oven Lids and Off-takes

Equipment Design. *Standpipe Caps.* The standpipe cap on the off-take piping is normally opened prior to each push. This means that the cap must be sealed to the standpipe in such a way that it can be opened, but that it must seal reliably during each coking cycle. The three most common methods of sealing standpipe caps are water seals, luting, and a sealing knife edge.

In a water seal cap, a lip on the cap is submerged in a water-filled trough. Major concerns with this design are keeping the standpipes horizontal so that the lip on the cap is submerged in the proper amount of water to maintain the seal and keeping the water inputs and drains on the troughs clean so that the troughs are filled with water, but do not overflow.

Many caps are designed to seal on a slant and must be sealed with a luting compound each time they are closed. The major concern with this type of system is ensuring that the luting compound is properly mixed and applied.

Caps with knife edge seals are similar in operation to the sealing edges of flexible coke oven doors. These caps require adequate levels of cleaning and maintenance to seal reliably.

Standpipe Slip Joints. Standpipes extend from the oven brickwork to the gooseneck, which is, in turn, connected to the collector main. In many batteries, there is a slip joint where the standpipe emerges from the battery top paving (in other batteries, the slip joint may be between the standpipe and the gooseneck or between the gooseneck and the collector main). These slip joints must be kept gastight to prevent leaks, but must slip to compensate for normal battery movement. They are often sealed with a packing material. However, especially when the joint is located where the standpipe exits from the battery top paving, currently available packing materials often burn out. Another sealing method is to use tar in the slip joint.

Charging Port Sealing. Charging port lids are normally luted whenever they are replaced after a charge. On some batteries, charging port lids are designed to seal without the need for luting.

Work Practices. Control of off-take and charging port lid leaks is mainly a function of paying attention to the details of cleaning, luting, and maintenance. There are occasions when castings need to be replaced and the refractory linings of the offtakes need to be repaired or replaced.

The operating practices of dampering the oven off from one of the collector mains partway through the coking cycle on a dual-collector main battery and of decarbonizing properly can reduce the need for casting and refractory repairs.

Coke Oven Charging

Equipment. *Stage Charging.* Virtually all larry car charging emission control systems now rely on stage charging. Stage charging is a process that uses steam aspiration at the gooseneck assembly of the off-take piping to create a vacuum in the oven being charged. The material that would otherwise be emitted is evacuated to the collector main and cleaned in the by-product plant.

During the stage charging process, prior to leveling, coal will tend to pile and extend to the roof of the oven, effectively blocking the gas channel. Therefore, it is important that there be evacuation from both ends of the oven being charged. In a dual-collector main battery, steam is aspirated into each off-take assembly so that emissions can be evacuated from each end of the oven. In a single-collector main battery, steam is aspirated into the off-take assembly of a nearby oven referred to as a "jumper oven." A jumper pipe, which may either be integral with the larry car or a separate assembly, is used to connect the charging port farthest from the collector main of the oven being charged to the jumper oven.

In addition to proper evacuation of the oven, stage charging relies on proper seals between the larry car drop sleeves and the charging ports, as well as a "smoke seal" on the leveler bar, to reduce the infiltration of clean air that would overwhelm the evacuation capabilities of the steam aspiration system. Stage charging also relies on proper placement of the coal in the oven to ensure that there are no empty spots in the oven and that the gas channel is not blocked in more than one place at any one time. In stage charging, generally, the coal is dropped out of the end larry car hoppers first and then out of the center hoppers after the end hoppers are empty. On many batteries, the charging hole lids are removed only when the coal is ready to be dropped and are replaced whenever the hopper is empty. When this method is used, no more than two charging hole lids are off the oven at any one time in order to maximize the emission control effect of the steam aspiration. Since batteries are of different sizes and have different numbers of charging ports and different methods of removing and replacing the charging hole lids, the coal charging portion of the stage charging process is custom designed for each battery.

Scrubber-Equipped Larry Cars. Early attempts to control charging emissions produced many types of controls, including the scrubber-equipped larry car. This car functioned by evacuating the emissions through a scrubber on the larry car. Because of weight limitations, the fans and scrubbers on these cars were often not sufficiently powerful to property evacuate and clean the emissions from the charging process. Therefore, virtually all batteries now use stage charging as the means of controlling charging emissions. However, there are still a few scrubber-equipped larry cars that are used as backups when the cars equipped for stage charging are out of service for maintenance or repair.

Evacuation to a Land-Based Control Device. In an attempt to overcome the problems with a mobile scrubber, some designs evacuated the charging emissions through a series of ducts to a fixed, land-based fan and control device. Because stage charging has proved to be simpler and more workable, only one land-based charging emission control system is currently in use in North America.

Pipeline Charging. Pipeline charging provided another means of controlling emissions from the charging process. This system had both production-related problems (e.g., inability to fill ovens, shortened battery life) and emission-related problems (e.g., sometimes lids were blown off the oven being charged owing to off-take plugging and often created large door leaks during charging). Therefore, all pipeline-charged batteries, except for Inland Battery No. 11, have been shut down or converted to larry-car-type charging. Inland reports that it is considering conversion of its No. 11 battery to a wet-charge battery using stage charging.

Work Practices. *Stage Charging.* Stage charging, which is as much a work practice as it is an equipment-related control, is described in detail above. For stage charging to control emissions effectively, the drop sleeves on the larry car must mate effectively with the charging ports on the battery. In addition, the steam nozzles must be clean, properly sized, properly oriented, and supplied with adequate steam pressure and volume; the smoke boot must be effective in reducing the infiltration of air around the leveler bar; and the gas channels from the oven to the collector main must be open.

Carbon and Tar Control. Proper evacuation of the ovens depends on the presence of a clear gas channel for the material being evacuated from the oven. This channel can be clogged by excessive oven roof carbon, the buildup of carbon in the standpipe, the buildup of carbon or tar in the gooseneck, and the buildup of tar in the collector main.

Carbon is built up as coke oven gas passes over hot coke oven surfaces and is cracked (i.e., broken into its simple components) into carbon and hydrogen. The carbon attaches to the hot surface and can grow to many inches in depth, blocking the gas channel. The growth of roof carbon is controlled by operating practices, by using carbon cutters and decarbonizing air on the pusher machine ram, and by letting the oven sit with its charging ports and standpipes open either just before or just after the push. Standpipes and goosenecks are typically cleaned before each charging operation to ensure that they have adequate gas-handling capacity. In addition, periodic water blasting or other cleaning methods are used to keep this equipment clean.

Coke Oven Pushing

Equipment. *Devices Connected to Land-Based Pollution Controls.* The following types of capture devices are ducted to fixed control systems located on land. The control devices in use include baghouses, venturi scrubbers, and wet electrostatic precipitators. As long as the design of the system protects the control device from possible failures (e.g., baghouses from high temperature or flame), all types of control devices can function acceptably for pushing emission control.

Sheds. Coke side sheds are typically large steel structures that are at least twice the height of the battery and extend the entire length of the battery from its coke side face across the quench car tracks. These sheds are constantly evacuated. Depending on the size of the battery and the design of the shed, the evacuation rate can be from 154,000 to 600,000 acfm. The fume rises to the top of the shed, where it is stored in a plenum until it is completely evacuated to the control device. Sheds can be very effective in capturing all pushing and travel emissions, as well as door leak emissions.

Traveling Hoods. A traveling hood is connected to a duct extending the length of the battery in such a way that the hood can be evacuated while it is moving. A typical design is the use of a hood with a flexible belt as the top member. The belt is lifted by, and moves on, rollers through the hood as the hood travels. Because of hood travel, these systems can control emissions during both coke pushing and quench car travel to the quench tower.

Typical designs for hood systems use from 6000 to 9000 acfm per ton of coke pushed. The fan is dampered, except during the pushing and travel cycle, resulting in less energy use than in a shed system. A major challenge in hood system design is capturing the emissions at the coke guide door jamb interface. Fume generated at this point has a high thermal drive in the vertical direction. Weight and space restrictions on the hood, when combined with the horizontal distance from the point of fume generation to the source of suction at the pollution control duct, make it extremely difficult to capture these emissions. A related concern with hood systems is controlling heating so that the emissions generated during the push do not overwhelm the system and spill from the hood, thereby escaping capture.

Indexing Hoods. Indexing hoods are similar to traveling hoods except that they are connected to the pollution control duct by inserting one to three connecting ducts from the hood into openings in the pollution control duct. Because they can only provide evacuation when mated with the pollution control duct at fixed locations, they cannot travel with the coke to the quench tower. An advantage of the indexing hood is that, because there is no airflow through the hood except when it is mated with the pollution duct, it is practical to maintain a "spare" hood that can be used to control emissions when the other hood is being repaired.

Coke Transfer Machines. A coke transfer machine is an integrated coke guide, door machine, and pushing emission control hood and is connected to a pollution control duct that extends the length of the battery. Because of the integration of all pushing-emissions-related equipment into a single unit, close tolerances are maintained and these devices provide excellent emission control at relatively low air flows (3000 to 7000 acfm per ton of coke pushed). Because it integrates three functions into one device, a coke transfer machine is relatively complex and heavy, and, therefore, it is costly. Concerns over the ability of existing battery foundations to accept the weight of these machines, along with the relatively high cost of providing a second machine as a backup for use when the primary machine is being maintained, have resulted in their limited use in North America.

Mobile Systems. *Mobile Scrubber Cars.* There are a number of proprietary-design pushing emission control cars that combine a mobile scrubber control device with the quench car in a single unit. These devices are generally used on batteries that are no larger than 4 m high. Mobile scrubber cars are typically used to control emissions from existing, rather than new, batteries. An advantage of these units is that with sufficient spare units, all pushes can be controlled without the need to construct a fixed system around an operating battery. Mobile scrubber cars can reduce pushing emissions to the same levels as those attained by land-based systems.

These devices put a scrubber, its energy source, and the water supply on wheels. They are heavy, use a lot of energy, and require a lot of maintenance. Owing to the close fit required between the machine and the battery to collect the pushing emissions successfully and the weight of the machines, substantial quench car track maintenance is typically required when these cars are used. Because of the high maintenance requirements associated with these cars, a number of them are being replaced with land-based systems.

Mobile Fume Suppression Systems. An attempt to overcome the maintenance and energy consumption problems associated with mobile scrubber cars has resulted in the development of the simpler mobile fume suppression car. These cars typically include pumps, water spray nozzles, and a water tank. They are much lighter and more reliable than mobile scrubber cars.

These systems work by spraying water on the coke as it is pushed. This can stop combustion, knock down particulates, and agglomerate particles into larger sizes that will fall out near the battery. These cars are typically used on small batteries with longer than normal coking times. Tests indicate that they can reliably reduce particulate emissions by about 50% over uncontrolled pushes.

Kress Indirect Dry Cooling (KIDC). This proprietary system works to prevent the generation of the emissions so that capture and control are not needed. The system consists of a number of steel boxes that are roughly the size of the ovens being pushed, a cooling rack, and mobile equipment. In operation, the hot coke is pushed into a box that mates directly to the door jamb. The oxygen in the box is quickly consumed, and since the coke moves as an oven-shaped mass that does not break up, emissions are not generated. Once the coke is in the box, the box is sealed and taken to the cooling rack, where it is cooled from the outside. When the coke in the box has cooled, it can be handled as dry, cool coke. There are essentially no pushing or quenching emissions from this system.

The KIDC system has operated as a prototype at National Steel's Granite City facility and Bethlehem Steel's Sparrows Point facility. The system has significant pollution control benefits; however, it has not yet been proven in production operation. The first full-production KIDC system will be tried at the Sparrows Point facility in 1991. If it is successful, the system will be extended to an adjacent battery at this facility.

Work Practices. *Heating Control.* The amount of volatile material left in the coke when it is pushed significantly affects the amount of uncontrolled pushing emissions. Control of the heating process to ensure that volatile matter and emission generation are minimized is most important when controls such as hoods, mobile scrubber cars, and mobile fume suppression are used. The control of coke volatile matter is also important to the blast furnace that uses the coke and where the presence of excess volatile material has an adverse impact on the coke's properties.

In order to heat the coal properly, the coal mixtures, coal bulk density, coal moisture, and oven charge levels must be as uniform as possible. Variations in these parameters will change the amount of heat needed to ensure proper coking. Because the coke battery is a massive structure that holds a lot of heat, it is virtually impossible to adjust coke heating to respond to short-term variations in the coal inputs to the process.

Ram Speed. Within narrow limits, reducing the ram speed may reduce the rate of emission generation.

Coke Oven Underfire Stack

Equipment (Add-On Controls). As stated above, the major function of a coke battery stack is to provide sufficient natural draft for optimum battery combustion operation. Since all the add-on controls interfere with the natural draft of the battery, they are employed on a small number of batteries where all other means of emission control have failed.

For the control of opacity caused by coal leaking from the oven to the flue, electrostatic precipitators have been used most often. In 1990, Inland Steel operated electrostatic precipitators on four of its Indiana Harbor Works coke batteries. Electrostatic precipitators can operate at the temperatures commonly encountered at the base of a stack and thus do not require the gas to be cooled. Precipitators are not effective against sulfate particulate because the sulfate condenses at a temperature lower than those found at the base of the stack. The main operating problems associated with precipitators include fires in the dust hoppers and the shutdown of the induced-draft fans during power failures. At least one battery is believed to have had its life substantially shortened because failures of both the induced-draft fan and the bypass dampers resulted in explosions in the heating system.

Other controls, such as baghouses, can effectively clean the gas with low-pressure drop, but would normally require the gas to be cooled before it reaches the baghouse. Scrubbers should also be effective in controlling emissions from coke underfire stacks, but would have the disadvantages of high energy consumption and the need for subsequent wastewater treatment. Both of these control devices would also have the same induced-draft problems that have been associated with precipitators.

Work Practices. Because a well-operated battery in good condition does not usually have excessive particulate emissions, the focus on battery stack emission problems is normally on operating and maintenance practices.

Carbon Control. While excessive carbon growth can cause charging emission problems, insufficient wall carbon leads to increased oven-to-flue leakage. Wall carbon is an effective means of sealing minor cracks between the oven and flue. Operating the battery to keep the proper amount of wall carbon is an important part of stack emission control.

Oven Dusting. Silica dusting—which involves blowing a dust, under pressure, into a hot empty oven before it is charged—can seal cracks that are too large to be controlled by normal carbon growth. Typical oven dusting programs involve dusting a certain number of ovens each week so that each oven is dusted on a routine recurring basis.

Battery Maintenance Practices. The most important battery work practices for the control of stack emissions are those relating to battery brickwork maintenance. If the brickwork's condition is adequate to prevent excessive oven-to-flue leakage, control of any remaining stack emission problems is a function of ensuring proper combustion.

Oven Spraying. A number of companies offer proprietary refractory spray products that are used to make effective, but temporary, repairs to small oven-wall imperfections. When used on a routine basis, these products can effectively control oven-to-flue leakage that does not require more extensive repairs.

Panel Patching. Panel patching involves the use of castable zero-expansion brick in place of conventional brickwork to replace damaged sections of battery brickwork. The advantages of this system include quicker repair time, fewer joints with the potential to leak, easier tie-in to the remaining brickwork, and no problems with expansion during heat-up of the repaired area. As this practice is still fairly new, the life of these repairs is not yet known, but it is over 4 years.

End Flue Repairs. Normal battery brickwork damage is usually confined to a few flues at each end of each oven. The general practice is to tear out the damaged flues and replace them with new bricks that are identical to the original bricks. When done properly, these repairs are effective in returning the battery to its original condition.

Through-Wall Repairs. If the damage extends further than the end flues, it is often possible to replace the entire oven wall with brickwork similar to that originally used to construct the battery. This type of repair is also effective in returning the battery walls to their original condition.

Combustion Control. *Combustion Air Control.* The combustion air to the battery is controlled by a series of finger-box adjustments for each oven flue wall. Although it is rare, it is possible for these adjustments to be sufficiently far off that they cause incomplete combustion and the emission of unburned carbon particulate.

Cleaning of Underfire Fuels. The combustion of properly cleaned fuels (e.g., properly treated coke oven gas) affects emissions in a number of ways. Combustion of undesulfurized gas often results in a sulfate plume. The sulfate also reports as particulate during the standard EPA Method 5 stack test.

Adequate removal of materials such as naphthalene from the coke oven gas is necessary to maintain the cleanliness of the nozzles that control the fuel flow to the heating flues. When the nozzles are not clean, it is difficult to maintain proper combustion conditions.

Coke Quenching

Equipment. *Baffles in Quench Tower.* According to the EPA's emission factors, the installation of baffles in the quench tower can reduce quenching emissions by at least 50%. This is the only control device for quenching emissions that is currently in widespread use.

Kress Indirect Dry Cooling. This proprietary pushing and dry quenching system (described under "Coke Oven Pushing") will completely eliminate quenching emissions. It has not yet been used in actual production, but is scheduled for a full production trial in late 1991.

Dry Quenching. Quenching of coke with inert gas under controlled conditions instead of water can also eliminate quenching emissions and recover heat energy from the hot coke. While this process has been used in Europe and Japan, it has not been tried in North America.

Work Practices. *Quench Water Quality.* Using relatively clean water for quenching can reduce quenching emissions. Most North American coke facilities use clean water for quenching.

Battery Venting

Equipment (Add-On Controls). *Flares.* When a battery is venting, the only effective control measure is to combust the emissions with a flare. There will still be sulfur dioxide emissions from the combustion of the undesulfurized gas.

Work Practices. *Exhauster Redundancy and Maintenance.* Battery venting incidents, which indicate a breakdown in the exhauster system, are best prevented. While prevention is never perfect, maintaining multiple exhausters, on different power systems (e.g., some on steam, some on electricity) and using power systems that are as independent of each other as possible can minimize battery venting incidents.

Change of Battery Operation during Venting Incident. The gas being released during venting incidents is generated from the heating of coal. When a venting incident is occurring, it is prudent to stop all coal charging. In addition, if the venting incident is expected to last for some time, reducing the heat input to the battery may be prudent (because the battery has a large thermal mass, this action will have no effect on emissions during short venting incidents).

Coke By-Product Recovery Plants

Process Changes to Eliminate Sources. In by-product recovery plants, major reductions in plant benzene (and VOC) emissions can be obtained by replacing the direct-water final cooler cooling tower and naphthalene separation and processing units with either a wash-oil final cooler or a tar-bottom mixer settler with indirect cooling. In many cases, this one process change will reduce facility emissions by over 90%.

The replacement of sumps with pipes and tanks is often a relatively simple means of reducing benzene (and VOC) emissions. The goal is to reduce the amount of material exposed to the air.

Gas Blanketing. Once process changes have been made, the major remaining sources of emissions are likely to be from storage and process tank vents. In late 1989, the EPA promulgated regulations requiring that all openings in such tanks be sealed and that they be vented to the coke oven gas main (or a control device). To allow the draw-down of product storage tanks connected to such systems, some provision must be made to replace the volume of the product with another material. The materials commonly chosen for this are nitrogen, coke oven gas, or natural gas.

There has been very little operating experience with these systems. Among concerns that have been expressed are tank rupture resulting from the plugging of gas lines by condensibles, transmission of fires and explosions from one tank to another through the gas manifold, and adverse effects on the quality of the coke oven gas (that accepts the material vented from the tanks). The EPA regulations require that these systems were to be in operation by September 1991. Since that time, all operators should have been gaining experience to determine which, if any, of the potential problems with the systems occur in operation.

Leak Detection and Repair. Once the two emission-reduction steps listed above have been taken, the remaining source of emissions is leaks from devices in liquid service that contain benzene or other VOCs and from the gas blanketing system and various tanks. The control of these leaks can be accomplished by routine monitoring and repair. The EPA has promulgated regulations requiring that the exterior of all valves, pumps, exhausters, and gas blanketing systems be sampled on a routine basis for VOC concentration levels above background. The agency has currently established 500 ppmv above background as the level of "no detectable emissions" and 10,000 ppmv as the level of a "leak." Once a leak is detected, there are time limits set for the first attempt at repair and in which to accomplish the final repair.

Nonrecovery Coke Batteries

Coke Oven Charging

Equipment. Charging of nonrecovery ovens is accomplished via horizontal conveyors through the oven door. Because there is suction on the oven, fugitive charging emissions are minimal. The pusher/charging machine has an on-board baghouse to capture and minimize emissions.

Work Practices. Minimizing emissions is not difficult since the ovens operate under negative pressure. Charging takes about 3.5 min on average using a retractable coal conveyor. Work practices consist of:

1. Maintaining and recording common tunnel duct temperature within a specified range to ensure VOC and NO_x limits are not exceeded.
2. Ensuring and recording that uptake dampers are fully open during pushing and charging.
3. Monitoring and recording emissions control equipment pressure drops and fan amps daily.
4. Monitoring and recording common tunnel pressures to ensure negative pressure is maintained at all times.

Coke Oven Pushing and Quenching

Equipment. Pushing emissions are controlled by a coke side shed or a shed and baghouse. Quenching emissions are controlled by baffles in the quench tower and maintaining a total dissolved solids level of less than 1100 mg/L.

Combustion Stack

Equipment. Air toxic compounds have been shown either not to be present or to be occasionally present in trace amounts. Therefore, off-gas controls for only SO_2 and particulate matter are indicated in the form of scrubbers and baghouses.

REFERENCES

1. *Coke Oven Emissions from Wet-Coal Charged By-Product Coke Oven Batteries—Background Information for Proposed Standards*, EPA-450/3-85-028a, U.S. Environmental Protection Agency, Research Triangle Park, NC, April 1987.
2. *Benzene Emission from Coke By-Product Recovery Plants—Background Information for Revised Proposed Standards*, EPA-450/3-83-016b, U.S. Environmental Protection Agency, Research Triangle Park, NC, June 1988.
3. *Compilation of Air Pollution Emission Factors*, AP-42 Supplement A, U.S. Environmental Protection Agency, Research Triangle Park, NC, October 1986, pp. 7.2-1–7.2-22.

COPPER SMELTING

JACQUES MOULINS, EDITOR
(with updated by James R. Donnelly)

Copper is one of the first metals used by man and today rates third in usage behind iron and aluminum. Copper is mined throughout the world, with significant deposits worked on all continents except Antarctica. Copper's value lies in its high electrical conductivity, ease of working, corrosion resistance, durability, and use in alloys.

Copper is found, in nature, principally as a sulfide and/or oxide ore. The primary sulfide ores are the mineral chalcopyrite ($CuFeS_2$) and chalcocite (Cu_2S). Copper is mined primarily using open-pit and/or underground mining techniques. Coppers ores generally contain about 1–4% copper. They also typically contain significant amounts of other metals, including antimony, arsenic, bismuth, gold, lead, molybdenum, nickel, selenium, silver, and zinc. Beneficiation at the mine (crushing, grinding, gravity separation, and flotation) is used to upgrade the ores to copper-rich concentrates (15–35% copper). Sulfur contents of copper concentrates generally are in the range of 25–35%, with iron representing the other main element (20–25%). Dewatered copper concentrates generally contain 8–10% moisture. Pyrometallurgical smelting is the primary route used to produce copper metal from sulfide ores.

PROCESS DESCRIPTION[1–3]

A traditional pyrometallurgical copper smelting process is illustrated in Figure 1. Key unit operations include roasting or drying, smelting, converting, and fire refining to produce anode copper (99.5%). Anode copper is further refined

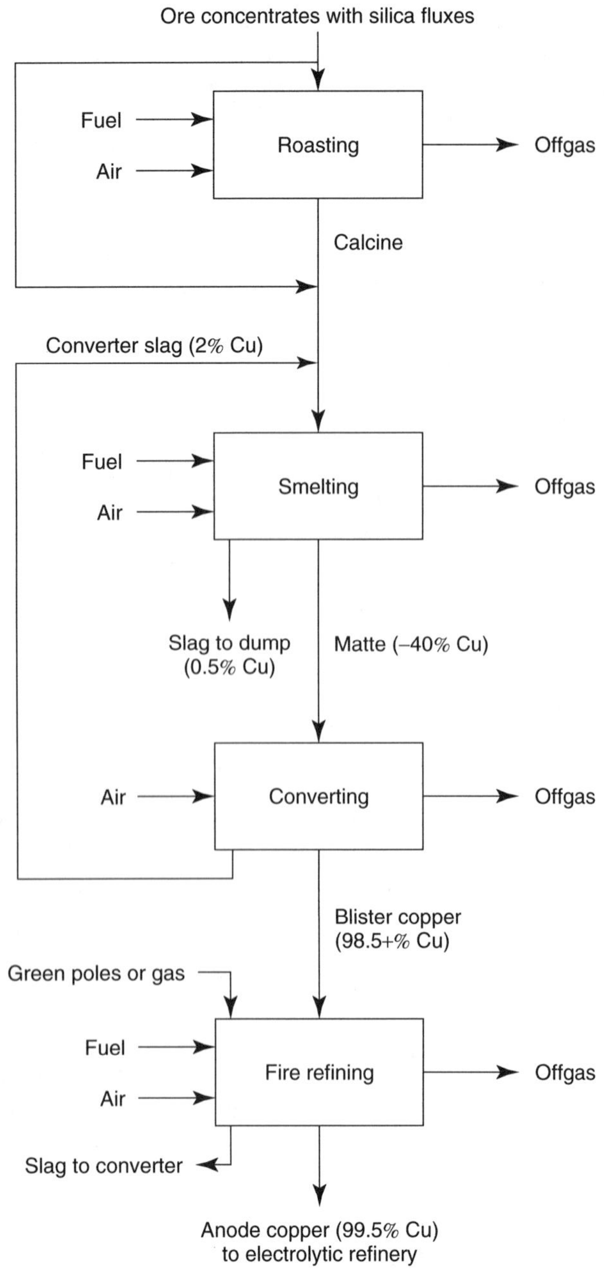

Figure 1. Typical Primary Copper Smelter Process.

hydrometallurgically to produce cathode copper (99.9%) and recover valuable metal values (gold, silver, etc.).

Roasting is only used in smelters, which employ reverbatory or electric smelting furnaces. Roasting is used to drive off a portion of the sulfur and some impurities. Copper concentrate is mixed with a siliceous flux (often a low-grade ore) and then heated in air to about 650°C in a multiple-hearth or fluid bed roaster. Approximately 25% of the sulfur is eliminated as sulfur dioxide (SO_2), some iron is converted to oxide, and portions of such impurities as antimony, arsenic, and lead are driven off. Roaster off-gas typically contains 5–15% SO_2 and is usually fed to an acid plant to recover sulfur values as sulfuric acid (H_2SO_4). The roasted product (calcine) serves as a dried and heated charge for the smelting furnace.

Multiple-hearth roasters accept moist concentrate, whereas fluid-bed roasters are fed finely ground, dried material (60% minus 200 mesh). With both of these types, the roasting is autogenous. Because there is less air dilution, higher SO_2 concentrations are present in fluidized-bed roaster gases than in multiple-hearth roaster gases.

Smelting converts the calcine or copper concentrate feed to copper matte (Cu_2S–FeS) containing from 35 to 65% copper. In the smelting process, hot calcine or raw unroasted concentrate is melted in the presence of air, at about 1000–1100°C, with siliceous flux in a smelting furnace to produce copper matte and a slag. Most of the iron and some of the impurities in the charge oxidize with the fluxes to form a slag that floats on top of the molten matte bath. This slag is periodically removed and discarded. Copper matte remains in the furnace until tapped for transfer to the converters. Heat for melting comes from the partial oxidation of the sulfide charge and from the burning of external fuel.

Copper smelting metallurgy is based on the greater affinity of sulfur to copper than for iron and most other common metals. The predominant reactions in the smelting step are summarized below:

$$FeS_2 + O_2 \rightarrow FeS + SO_2 \quad (1)$$

$$CuO_2 + 2FeS + O_2 \rightarrow CuS + 2FeO + O_2 \quad (2)$$

$$2Cu + FeS \rightarrow Cu_2S + Fe \quad (3)$$

$$2FeS + 3O_2 \rightarrow FeO + SO_2 \quad (4)$$

$$FeO + SiO_2 \rightarrow FeO \cdot SiO_2 \text{ (slag)} \quad (5)$$

Formation of FeS (1) ensures that copper present in forms other than sulfide will be converted to the sulfide form (2)+(3). When all of the copper is converted to sulfides, iron will oxidize (4) and react with the silica flux to form slag (5).

The most common types of smelting furnaces are reverbatory, electric, Noranda, Outokumpu (flash), Inco (flash), Mitsubishi, and Ausmelt.

Reverbatory furnace operation is a continuous process, with frequent charging of input materials and periodic tapping of matte and skimming of slag. Reverbatory furnaces have high heat loses due to the large volume of gases sweeping through the furnace. They require an outside source of fuel to keep the process going. They produce an off-gas stream with relatively low concentrations of SO_2 (1–3%) that is unsuitable for feed to most acid plants. For these reasons reverbatory furnaces have lost favor with the industry and are being phased out of service.

In electric arc smelting furnaces, heat is generated by the flow of an electric current in carbon electrodes lowered through the furnace roof and submerged in the slag layer of the molten bath. The feed generally consists of dried concentrates or calicoes. Charging of wet concentrates is avoided. The chemical and physical changes occurring in the molten bath are similar to those occurring in the molten bath of a reverbatory furnace. Also, the matte and slag tapping practices are similar at both furnaces. Electric furnaces do not produce fuel combustion gases; so flow rates are lower and SO_2 concentrations higher in the effluent gas than in that of reverberatory furnaces. Electric smelting furnaces depend on a low-cost reliable supply of electrical power to be competitive in the industry. As such they have been losing favor to other more economical processes.

Flash furnace smelting combines the operations of roasting and smelting to produce a high-grade copper matte from concentrates and flux. In flash smelting, dried concentrates and finely ground fluxes are injected, together with oxygen, preheated air, or a mixture of both, into a furnace of special design, where the temperature is maintained at approximately 1000°C (1830°F). Flash furnaces, in contrast to reverberatory and electric furnaces, use heat generated from the partial oxidation of their sulfide charge to provide much or all of the energy (heat) required for smelting. They also produce off-gas streams containing high concentrations of SO_2 (15–60% SO_2), which are fed to an acid plant.

Slag produced by flash furnace operations contains significantly higher amounts of copper than does that from reverberatory or electric furnace operations. As a result, flash furnace slags are treated in a slag-cleaning furnace, or by hydrometallurgical methods, to recover the copper. Slag cleaning furnaces usually are small electric furnaces. Flash furnace and converter slags are charged to a slag-cleaning furnace and are allowed to settle and separate, under reducing conditions, with the addition of coke or iron sulfide. Copper oxide in the slag is converted to copper sulfide, and subsequently removed from the furnace and charged to a converter with regular matte.

The Noranda process was originally designed to allow continuous production of blister copper in a single vessel by effectively combining roasting, smelting, and converting into one operation. However, for the operation of these reactors, the copper matte mode is preferred. As in flash smelting, the Noranda process takes advantage of the heat energy available from the copper ore. The remaining thermal energy is supplied by natural gas burners or by coal mixed with the ore concentrates. Mitsubishi reactors have recently entered the market for smelting and converting. They are similar in operation to Noranda reactors.

Copper matte is processed (converted) to blister copper in converters. The converting process is composed of two distinct phases, slag making and copper making. The approximate chemical equations for the phases are:
Slag Making

$$FeS + O_2 \rightarrow FeO + SO_2 \quad (6)$$

$$6FeO + O_2 \rightarrow 2Fe_3O_4 \quad (7)$$

$$2FeO + SiO \rightarrow 2FeO \cdot SiO_2 \text{ (slag)} \quad (8)$$

$$Cu_2O + FeS \rightarrow FeO + Cu_2S \quad (9)$$

Copper making

$$2Cu_2S + 3O_2 \rightarrow Cu_2O + SO_2 \quad (10)$$

$$Cu_2S + 2Cu_2O \rightarrow 6Cu + SO_2 \quad (11)$$

$$Cu_2S + O_2 \rightarrow 2Cu + SO_2 \quad (12)$$

In the slag-making phase iron and other metal sulfides are converted into ferrous oxide (6)+(7), which reacts with silica to form slag (8). An undesirable side reaction is the oxidation of ferrous oxide to hematite (9). Copper oxides formed in the slag-making step are readily reduced to sulfides as long as ferrous sulfide is present.

In the copper-making step, copper sulfides are oxidized to "blister" copper. Blister copper contains about 98–99% copper, 0.02–0.5% sulfur, and 0.5–0.8% oxygen.

The most common types of converters in use are Pierce–Smith converters. These are refractory-lined, cylindrical, steel shells mounted on trunnions at either end and rotated about the major axis for charging and pouring. An opening in the center of the converter functions as a mouth through which molten matte, siliceous flux, and scrap copper are charged and gaseous products are vented. During the slab-making step (slag blow) air or oxygen-rich air is blown through the molten matte. During this step additional copper matte is periodically added to the converter and the blowing process repeated until an adequate amount of relatively pure Cu_2S, called "white metal," accumulates in the bottom of the converter.

A renewed air blast (copper blow) oxidizes the copper sulfide sulfur to SO_2, leaving blister copper in the converter. The blister copper is subsequently removed and transferred to an anode furnace for fire refining or cast into ingot for transfer to a fire refining facility for further processing. SO_2 produced throughout the converting operation is vented to pollution control devices.

Another type of converter used in copper smelters is the Hoboken converter. The primary advantage of these converters lies in emission control. The Hoboken converter is essentially like a conventional Pierce–Smith converter, except that this vessel is fitted with a side flue at one end shaped as an inverted U. This flue arrangement permits siphoning of gases from the interior of the converter directly to the off-gas collection system. This leaves the converter mouth under a slight vacuum thus reducing secondary emissions.

Blister copper usually contains from 98.5% to 99.5% pure copper. Impurities may include gold, silver, antimony, arsenic, bismuth, iron, lead, nickel, selenium, sulfur, tellurium, and zinc. Blister copper is further refined using fire refining and electrolytic techniques. In fire refining, blister copper is placed in a fire-refining furnace (anode furnace), a flux is usually added, and air is blown through the molten mixture to remove sulfur and oxidize remaining impurities, which are removed as a slag. Following the oxidation step, the remaining metal bath is subjected to a reducing atmosphere to reconvert cuprous oxide to copper. The temperature in the furnace is around 1100°C.

Fire-refined copper is cast into anodes, which are transferred to an electrolytic refinery for further purification and recovery of valuable metals. Electrolytic refining separates copper from impurities by electrolysis in a solution containing copper sulfate and sulfuric acid. Metallic impurities precipitate from the solution and form a sludge, which is removed and treated to recover precious metals. Copper is dissolved from the anode and deposited at the cathode. Cathode copper is remelted and made into bars, ingots, or slabs for marketing purposes. Cathode copper is 99.95–99.97% pure.

AIR EMISSIONS CHARACTERIZATION

Particulate matter and SO_2 are the principal air contaminants emitted by primary copper smelters. These emissions are generated directly from the processes involved, as in the liberation of SO_2 from copper concentrate during roasting or in the volatilization of trace elements as oxide fumes. Fugitive emissions are generated by leaks from major equipment during material-handling operations.

Roasters, smelting furnaces, and converters are the principal sources of particulate matter, SO_2, and metals emissions. Copper and iron oxides are the primary constituents of the particulate matter, but other oxides, such as arsenic, antimony, cadmium, lead, mercury, and zinc, may also be present, with metallic sulfates and sulfuric acid mist. In multiple-hearth roasters and reverberatory furnaces fuel combustion products also contribute to the particulate and SO_2 emissions.

Fugitive Emissions

The process sources of particulate matter and SO_2 emission are also the potential fugitive sources of these emissions: roasting, smelting, converting, fire refining, and slag cleaning. Table 1 presents fugitive emission factors for these sources, while Tables 2 through 4 and Figures 2–7, taken from AP-24 "Compilation of Air Pollution Emission Factors" U.S. EPA, present cumulative size-specific particulate emission factors for fugitive emissions from reverberatory furnace matte, slag tapping, converter slag, and copper blow operations. Actual quantities of emissions from these sources depend on the type and condition of the equipment and on the smelter operating techniques. Although emissions from many of these sources are released inside a building, ultimately they are discharged to the atmosphere.

Fugitive emissions are generated during discharge and transfer of hot calcine from multiple-hearth roasters, with negligible amounts possible from the charging of these roasters. Fluid-bed roasting, a closed-loop operation, has negligible fugitive emissions.

Matte tapping and slag skimming operations are sources of fugitive emissions from smelting furnaces. Fugitive emissions can also result from the charging of a smelting furnace or from leaks, depending on the furnace type and condition. A typical single matte tapping operation lasts from 5 to 10 min, and a single slag skimming operation lasts from 10 to 20 min. Tapping frequencies vary with furnace capacity and type. In an 8-h shift, matte is tapped 5–20 times and slag is skimmed 10–25 times.

Each of the various stages of converter operation—charging, blowing, slag skimming, blister pouring, and holding—is a potential source of fugitive emissions. During blowing, the converter mouth is in stack (i.e., a close-fitting primary hood is over the mouth to capture off-gases). Fugitive emissions escape from the hoods. During charging, skimming, and pouring operations, the converter mouth is out of stack (i.e., the converter mouth is rolled out of its vertical position and the primary hood is isolated). Fugitive emissions are

Table 1. Fugitive Emission Factors for Primary Copper Smelters[a] (Emission Factor Rating: B)

Source of Emission	Particulate		Sulfur Dioxide	
	kg/Mg	lb/ton	kg/Mg	lb/ton
Roaster calcine discharge	1.3	2.6	0.5	1
Smelting furnace[b]	0.2	0.4	2	4
Converter	2.2	4.4	65	130
Converter slag return	NA	NA	0.05	0.1
Anode furnace	0.25	0.5	0.05	0.1
Slag cleaning furnace[c]	4	8	3	6

[a] References 16, 22, 25–32. Expressed as mass units per unit weight of concentrated ore processed by the smelter. Approximately four unit weights of concentrate are required to produce one unit weight of copper metal. Factors for flash furnace smelters and Noranda furnace smelters may be lower than reported values. NA, not available.
[b] Includes fugitive emissions from matte tapping and slag skimming operations. About 50% of fugitive particulate emissions and about 90% of total SO_2 emissions are from matte tapping operations, with the remainder from slag skimming.
[c] Used to treat slags from smelting furnaces and converters at the flash furnace smelter.

Table 2. Uncontrolled Particle Size and Size-Specific Emission Factors for Fugitive Emissions from Reverberatory Furnace Matte Tapping Operations[a] (Emission Factor Rating: D)

Particle Size[b] (μm)	Cumulative Mass (%) ≤Stated Size	Cumulative Emission Factors	
		kg/Mg	lb/ton
15	76	0.076	0.152
10	74	0.074	0.148
5	72	0.072	0.144
2.5	69	0.069	0.138
1.25	67	0.067	0.134
0.625	65	0.065	0.130
Total	100	0.100	0.200

[a] Reference 25. Expressed as units per unit weight of concentrated ore processed by the smelter.
[b] Expressed as aerodynamic equivalent diameter.

Table 3. Particle Size Distribution and Size-Specific Emission Factors for Fugitive Emissions from Reverberatory Furnace Slag Tapping Operations[a] (Emission Factor Rating: D)

Particle Size[b] (μm)	Cumulative Mass (%) ≤Stated Size	Cumulative Emission Factors	
		kg/Mg	lb/ton
15	33	0.033	0.066
10	28	0.028	0.056
5	25	0.025	0.050
2.5	22	0.022	0.044
1.25	20	0.020	0.040
0.625	17	0.017	0.034
Total	100	0.100	0.200

[a] Reference 25. Expressed as units per unit weight of concentrated ore processed by the smelter.
[b] Expressed as aerodynamic equivalent diameter.

Table 4. Particle Size Distribution and Size-Specific Emission Factors for Fugitive Emissions from Converter Slag and Copper Blow Operations[a] (Emission Factor Rating: D)

Particle Size[b] (μm)	Cumulative Mass (%) ≤Stated Size	Cumulative Emission Factors	
		kg/Mg	lb/ton
15	98	2.2	4.3
10	96	2.1	4.2
5	87	1.9	3.8
2.5	60	1.3	2.6
1.25	47	1.0	2.1
0.625	38	0.8	1.7
Total	100	2.2	4.4

[a] Reference 25. Expressed as units per unit weight of concentrated ore processed by the smelter.
[b] Expressed as aerodynamic equivalent diameter.

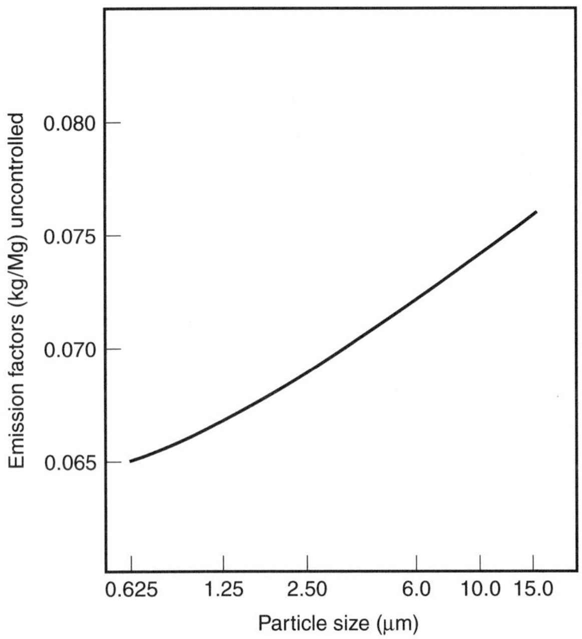

Figure 2. Size-Specific Fugitive Emission Factors for Reverberatory Furnace Matte Tapping Operations.

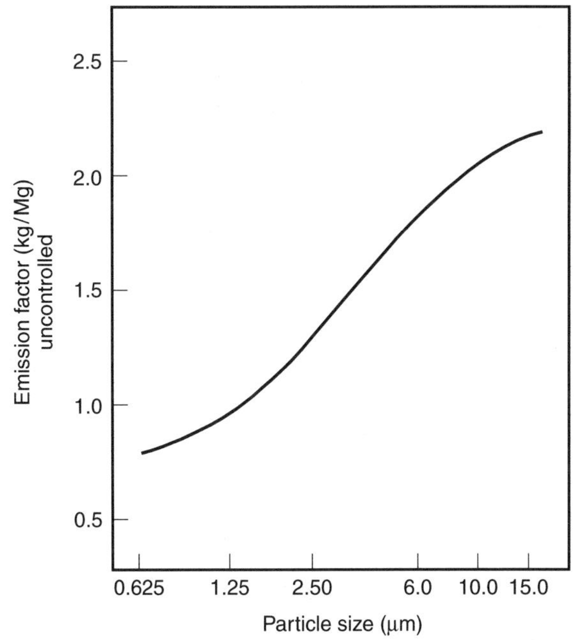

Figure 4. Size-Specific Fugitive Emission Factors for Converter Slag and Copper Blow Operations.

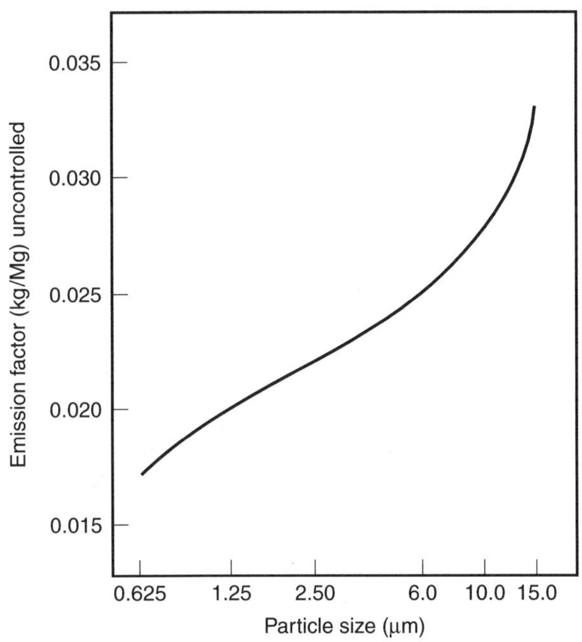

Figure 3. Size-Specific Fugitive Emission Factors for Reverberatory Furnace Slag Tapping Operations.

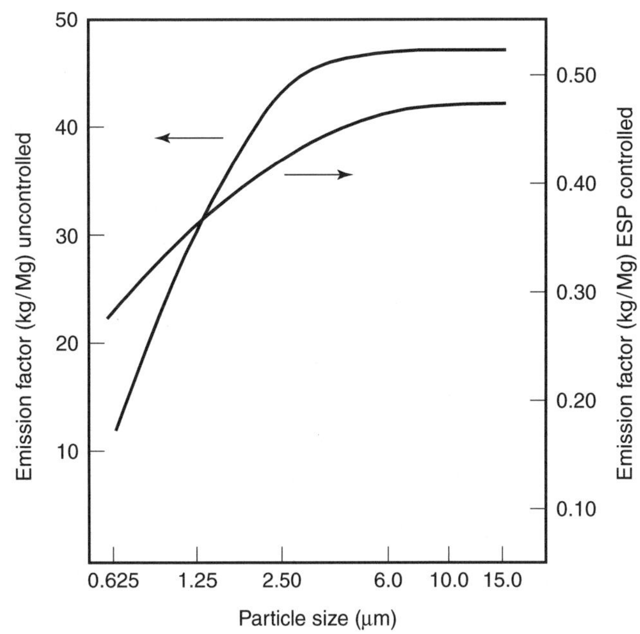

Figure 5. Size-Specific Emissions Factors for Multiple-Hearth Roaster and Reverberatory Smelter.

discharged during rollout. In a number of smelters secondary hoods are used to capture these emissions and route them to a pollution control device.

At times during normal smelting operations, slag or blister copper cannot be transferred immediately from or to the converters. This condition, a holding stage, may occur for several reasons, including insufficient matte in the smelting furnace or the unavailability of a crane. Under these conditions, the converter is rolled out of its vertical position and remains in a holding position, and fugitive emissions may result.

Metals Emissions

At primary copper smelters, both process emissions and fugitive particulate from various pieces of equipment contain oxides of many inorganic elements, including antimony, arsenic, bismuth, cadmium, gold, lead, mercury, molybdenum, nickel, selenium, silver, and zinc. The metals content of particulate emissions depends on both the metals content of the smelter feed and the process off-gas temperature. Many of the volatile metals emissions are effectively removed in particulate control systems operating at low temperatures, about 40–60°C.

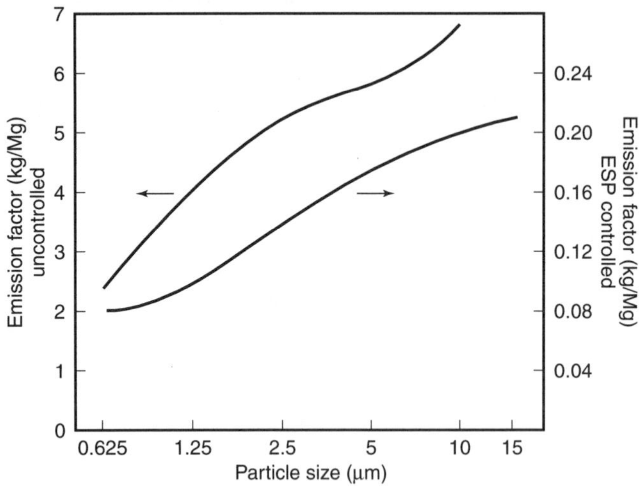

Figure 6. Size-Specific Emission Factors for Reverberatory Smelting.

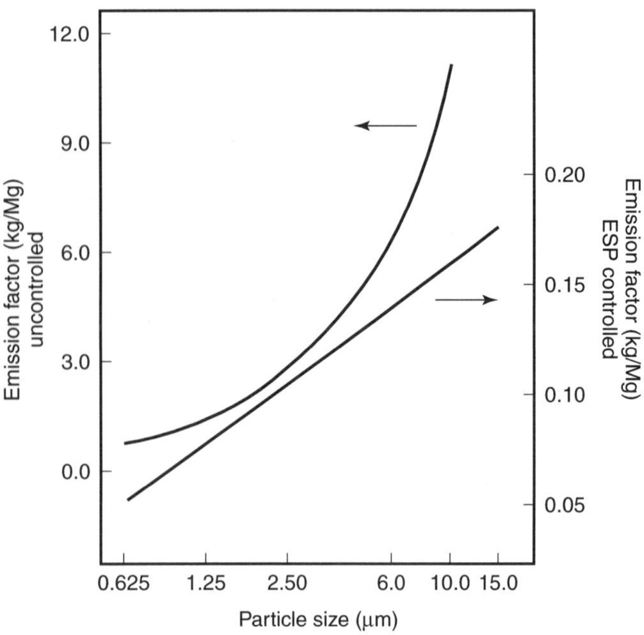

Figure 7. Size-Specific Emission Factors for Copper Converting.

Table 5 presents process and fugitive lead emission factors for various operations of primary copper smelters. Fugitive emissions are captured through the application of localized or general ventilation techniques. Once captured, emissions are vented directly to a collection device or, in some instances, combined with process off-gases before collection. Close-fitting exhaust hood capture systems are used for multiple-hearth roasters and hood ventilation systems for smelter matte tapping and slag skimming operations. For converters, secondary hood systems or building evacuation systems are used.

Actual emissions from a particular smelter unit depend on the configuration of the equipment in that smelting plant and its operating parameters. Table 6 gives emission factors for various smelter configurations, and Tables 5, 7, 8, and 9, taken from AP-42, "Compilation of Air Pollution Emission Factors," U.S. EPA, give size-specific emission factors for those copper production processes where information is available.

Table 5. Lead Emission Factors for Primary Copper Smelters[a] (Emission Factor Rating: C)

	Emission Factor[b]	
Operation	kg/Mg	lb/ton
Roasting	0.075	0.15
Smelting	0.036	0.072
Converting	0.13	0.27
Refining	NA	NA

[a] Reference 33. Expressed as units per unit weight of concentrated ore processed by smelter. Approximately four unit weights of concentrate are required to produce one unit weight of copper metal. Based on test data for several smelters with 0.1–0.4% lead in feed throughput. NA, not available.
[b] For process and fugitive emissions totals.

Table 6. Typical Sulfur Dioxide Concentrations in Off-Gases from Primary Copper Smelting Sources

Unit	SO_2 Concentration (Vol %)
Multiple-hearth roaster	1.5–3
Fluidized-bed roaster	10–12
Reverberatory furnace	0.5–1.5
Electric arc furnace	4–8
Flash smelting furnace	10–70
Continuous smelting furnace	5–15
Pierce-Smith converter	4–7
Hoboken converter	8
Single-contact H_2SO_4 plant	0.2–0.26
Double-contact H_2SO_4 plant	0.05

AIR POLLUTION CONTROL MEASURES

Control measures used in copper smelting operations are similar to the types of controls used in lead and zinc smelting. They include control measures to limit SO_2, acid mist, particulate, and metals emissions. The major emission sources in copper smelting are the roasters, smelters, and converters. Other point sources include dryers, anode furnaces, and revert furnaces.

Control of SO_2 emissions from smelter sources is most commonly performed in a single- or double-contact sulfuric acid plant. Use of a sulfuric acid plant to treat copper smelter effluent gas streams requires that gas be free from particulate matter and contain a minimum SO_2 concentration. Practical limitations have usually restricted sulfuric acid plant application to gas streams that contain at least 3% SO_2. Table 6 shows typical average SO_2 concentrations for the various smelter unit off-gases. Primary gases from flash, Noranda, Mitsubishi, and Ausmelt smelters and from converters are treated for SO_2 control in sulfuric acid plants. Gas effluents of some multiple-hearth roaster operations and of all fluid-bed roaster operations also are treated in sulfuric acid plants. Weak-SO_2-content gas effluents from reverberatory furnace operations are usually released to the atmosphere with no reduction of SO_2.

Table 7. Emission Factors for Primary Copper Smelters[a,b] (Emission Factor Rating: B

Configuration[c]	By Unit	Particulate kg/Mg	Particulate lb/ton	Sulfur Dioxide[d] kg/Mg	Sulfur Dioxide[d] lb/ton	References
Reverberatory furnace (RF)	RF	25	50	160	320	4–10
followed by converters (C)	C	18	36	370	740	9, 11–15
Multiple-hearth roaster (MHR)	MHR	22	45	140	280	4, 5, 16, 17
followed by reverberatory	RF	25	50	90	180	4–9, 18, 19
furnace (RF) and converters (C)	C	18	36	300	600	8, 11–13
Fluid-bed roaster (FBR) followed	FBR	NA	NA	180	360	20
by reverberatory furnace (RF)	RF	25	50	90	160	e
and converters (C)	C	18	36	270	540	e
Concentrate dryer (DC) followed	CD	5	10	0.5	1	21,22
by electric furnace (EF) and	EF	50	100	120	240	15
converters (C)	C	18	36	410	820	8, 11–13, 15
Fluid-bed roaster (FBR) followed	FBR	NA	NA	180	360	20
by electric furnace (EF) and	EF	50	100	45	90	15, 23
converters (C)	C	18	36	300	600	e
Concentrate dryer (CD) followed	CD	5	10	0.5	1	21, 22
by flash furnace (FF),	FF	70	140	410	820	24
cleaning furnace (SS), and	SS[f]	5	10	0.5	1	22
converters (C)	C[e]	NA[g]	NA[g]	120	240	22
Concentrate dryer (DC) followed	CD	5	10	0.5	1	21, 22
by Noranda reactors (NR) and	NR	NA	NA	NA	NA	NA
converters (C)	C	NA	NA	NA	NA	NA

[a]Expressed as units per unit weight of concentrated ore processed by the smelter. Approximately four unit weights of concentrate are required to produce one unit weight of blister copper. NA = not available.

[b]For particulate matter removal, gaseous effluents from roasters, smelting furnaces, and converters usually are treated in hot ESPs at 200–340°C (400–650°F) or in cold ESPs with gases cooled to about 120°C (250°F) before the ESP. Particulate emissions from copper smelters contain volatile metallic oxides that remain in vapor form at higher temperatures (120°C or 250°F). Therefore, overall particulate removal in hot ESPs may range from 20% to 80% and in cold ESPs may be 99%. Converter gas effluents and, at some smelters, roaster gas effluents are treated in single-contact acid plants (SCAP) or double-contact acid plants (DCAP) for SO_2 removal. Typical SCAPs are about 96% efficient and DCAPs are up to 99.8% efficient in SO_2 removal. They also remove over 99% of particulate matter. Noranda and flash furnace off-gases are also processed through acid plants and are subject to the same collection efficiencies as cited for converters and some roasters.

[c]In addition to sources indicated, each smelter configuration contains fire-refining anode furnaces after the converters. Anode furnaces emit negligible SO_2. No particulate emission data are available for anode furnaces.

[d]Factors for all configurations except a reverberatory furnace followed by converters have been developed by normalizing test data for several smelters to represent 30% sulfur content in concentrated ore.

[e]Based on the test data for the configuration of multiple-hearth roaster followed by reverberatory furnace and converters.

[f]Used to recover copper from furnace slag and converter slag.

[g]Since converters at flash furnace and Noranda furnace smelters treat high-copper-content matte, converter particulate emissions from flash furnace smelters are expected to be lower than those from conventional smelters with multiple-hearth roasters, reverberatory furnace, and converters.

Table 8. Particle Size Distribution and Size-Specific Emission Factors for Multiple-Hearth Roaster and Reverberatory Smelter Operations[a] (Emission Factor Rating: D)

	Cumulative Mass. % ≤Stated Size		Cumulative Emission Factors			
			Uncontrolled		ESP Controlled[c]	
Particle Size.[b] μm	Uncontrolled	ESP Controlled	kg/Mg	lb/ton	kg/Mg	lb/ton
15	100	100	47	95	0.47	0.95
10	100	99	47	94	0.47	0.94
5	100	98	47	93	0.46	0.93
2.5	97	84	46	80	0.40	0.80
1.25	66	76	31	72	0.36	0.72
0.625	25	62	12	59	0.29	0.59
Total	100	100	47	95	0.47	0.95

[a]Reference 25. Expressed as units per unit weight of concentrated ore processed by the smelter.

[b]Expressed as aerodynamic equivalent diameter.

[c]Nominal particulate removal efficiency is 99%.

Table 9. Particle Size Distribution and Size-Specific Emission Factors for Reverberatory Smelter Operations[a] (Emission Factor Rating: E)

	Cumulative Mass. % ≤ Stated Size		Cumulative Emission Factors			
			Uncontrolled		ESP Controlled[c]	
Particle Size,[b] μm	Uncontrolled	ESP Controlled	kg/Mg	lb/ton	kg/Mg	lb/ton
15	NR	83	NR	NR	0.21	0.42
10	27	78	6.8	13.6	0.20	0.40
5	23	69	5.8	11.6	0.18	0.36
2.5	21	56	5.3	10.6	0.14	0.28
1.25	16	40	4.0	8.0	0.10	0.20
0.625	9	32	2.3	4.6	0.08	0.16
Total	100	100	25	50	0.25	0.50

[a]Reference 25. Expressed as units per unit weight of concentrated are processed by the smelter. NR = not reported because of excessive extrapolation.

[b]Expressed as aerodynamic equivalent diameter.

[c]Nominal particulate removal efficiency is 99%.

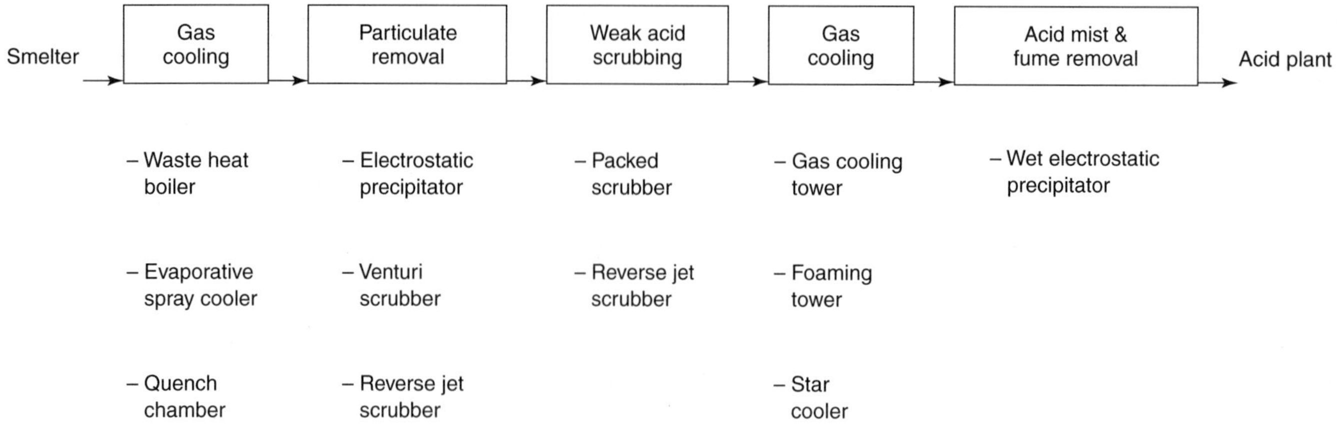

Figure 8. Metallurgical Acid Plant Gas Cleaning System; Unit Operations and Equipment Utilized.

Typically, single-contact acid plants achieve 92.5–98% conversion of SO_2 to acid, with approximately 2000 ppm SO_2 remaining in the acid plant effluent gas. Double-contact acid plants collect from 98% to more than 99% of the SO_2 and are currently being designed to limit SO_2 emissions to less than 100 ppm.

The efficient and reliable operation of an acid plant depends on a clean, relatively dry feed gas to the acid plant-drying tower. Figure 8 presents a schematic of a metallurgical acid plant gas cleaning system. This gas cleaning system functions to remove particulate matter, SO_3, and metal fumes to low levels and predries the gas entering the acid plant.

Gas cooling is needed to reduce the roaster off-gas temperatures to levels acceptable for use of a hot electrostatic precipitator (350–400°C) and/or to condense out volatile metals. For Outokumpu flash furnaces and some Pierce–Smith converters where an electrostatic precipitator (ESP) is used for particulate removal, a waste heat boiler is often used in order to recover heat energy in the form of steam. For other applications an evaporative gas cooler is used. Where particulate removal will be accomplished using a wet scrubber, a quench chamber is used to saturate and cool the inlet gas to its adiabatic saturation temperature.

Particulate removal is most often accomplished using a dry ESP. Metallurgical sulfuric acid plant gas cleaning ESPs are typically of the rigid frame or rigid electrode type with three to five electrical fields and specific collecting areas on the order of 100 to 200 m^2/m^3/sec. Outlet particulate loadings are generally on the order of 25–150 mg/Nm^3.

Wet scrubbers are also used for particulate removal. These are typically high-energy venturi scrubbers (1200–2500-mm pressure drop) or other particulate scrubbers, such as reverse jet scrubbers. Wet scrubbers are more often used when higher particulate loading is allowable to the acid plant and/or when wet handling/treatment of the collected dust is specified.

Weak acid scrubbers function to remove additional particulate and sulfuric acid mist, and condense volatile metals. Packed-bed and reverse jet scrubbers are typically used for this purpose. These scrubbers typically operate at a pH of less than 1 and therefore do not remove SO_2.

Gas coolers lower the gas temperature further, below the adiabatic saturation temperature, through the use of either gas/liquid heat exchangers or recirculating chilled scrubber fluid. Gas temperatures on the order of 35–40°C are common. These coolers function to condense out additional volatile metals and SO_3 and also reduce the moisture content of the gas. A number of different gas cooler designs are used.

Wet ESPs remove acid mist and condensed metal fumes from the gas stream. These ESPs may be of the wet tube or plate type. In some applications two wet ESPs may be used in series to achieve very high acid mist and particulate removal. The gas entering the acid plant typically is relatively dry and contains low levels of SO_3, metals, and particulate (on the order of 10–50 mg/Nm^3).

In some instances where the concentrate contains high levels of mercury and or very high acid quality is required, a mercury removal step follows the wet ESPs. Both calomel and sulfide-based technologies are used.

Rotary or fluid bed dryers off-gases contain low levels of SO_2, particulates, and some metals. Dryer emissions are typically controlled using electrostatic precipitators, fabric filters, and/or wet scrubbers.

Dryer ESPs are typically of multifield rigid-frame and rigid-electrode types operating at 200–350°C. They are capable of achieving high collection efficiencies for particulates (99+%) but are ineffective for metal fumes. These types of ESPs are selected based on the corrosive gas conditions and sticky nature of the fly ash being collected. Electrode failures associated with rigid-frame and rigid-electrode systems are less frequent than for weighted wire electrodes. This is especially true where higher rapping forces are needed to dislodge the sticky fly ash. For rigid-frame systems, high-alloy (e.g., Incoloy 825) spring-wound electrodes are also used to minimize electrode corrosion problems.

Due to more stringent particulate and visible emission limitations, fabric filters are being used more for dryer emissions control. Pulse-jet-type filters are the norm due to their low capital cost, however; at least one installation uses a reverse-air-type filter in this type of application. Fabric filters can achieve very high particulate collections (99.5+%) and are more effective than ESPs for metal fume control. The major potential problems with fabric filters is high-pressure drop due to bag blinding and fires where concentrates are pyrophoric. Gas coolers are often required in front of these fabric filters to reduce the inlet gas temperature to below about 150°C.

In the standard Pierce–Smith converter, flue gases are captured during the blowing phase by a primary hood over the converter mouth. To prevent the hood's binding to the converter with splashing molten metal, there is a gap between the hood and the vessel. During charging and pouring operations, significant fugitives may be emitted when the hood is removed to allow crane access. Secondary hoods are becoming standard to control fugitive emissions when the converter is out of stack. Converter off-gases are treated in ESPs to remove particulate matter and in sulfuric acid plants to remove SO_2.

The remaining smelter processes handle material that contains very little sulfur, and hence SO_2 emissions from these processes are relatively insignificant. Particulate emissions from fire refining operations, however, may be of concern. Sulfuric acid mist emissions from electrolytic refining can be a problem and require control. Crushing and grinding systems, used in ore, flux, and slag processing, also contribute to fugitive dust problems.

REFERENCES

1. *Background Information for New Source Performance Standards: Primary Copper, Zinc and Lead Smelters, Volume I, Proposed Standards*, EPA-450/2-74-002a, U.S. Environmental Protection Agency, Research Triangle Park, NC, October 1974.
2. *Arsenic Emissions from Primary Copper Smelters Background Information for Proposed Standards, Preliminary Draft*, EPA Contract No. 68-02-3060, Pacific Environmental Services, Durham, NC, February 1981.
3. *Background Information Document for Revision of New Source Performance Standards for Primary Copper Smelters*, EPA Contract No. 68-02-3056, Research Triangle Institute, Research Triangle Park, NC, March 31, 1982.
4. *Air Pollution Emission Test: Asarco Copper Smelter*, El Paso, TX, EMB-77-CUS-6, Office of Air Quality Planning and Standards, U.S. Environmental Protection Agency, Research Triangle Park, NC, June 1977.
5. W. F. Cummins, Inc., El Paso, TX, written communications to A. E. Vervaert, U.S. Environmental Protection Agency, Research Triangle Park, NC, June 1977.
6. AP-42 Background Files, Office of Air Quality Planning and Standards, U.S. Environmental Protection Agency, Research Triangle Park, NC, March 1978.
7. *Source Emissions Survey of Kennecott Copper Corporation, Copper Smelter Converter Stack Inlet and Outlet and Reverberatory Electrostatic Precipitator Inlet and Outlet, Hurley, NM*, EA-735-09, Ecology Audits, Inc., Dallas, TX, April 1973.
8. *Trace Element Study at a Primary Copper Smelter*, EPA-600/ 2-78-065a and 065b, U.S. Environmental Protection Agency, Research Triangle Park, NC, March 1978.
9. *Systems Study for Control of Emissions, Primary Nonferrous Smelting Industry, Vol. II: Appendices A and B*, PB 184885, National Technical Information Service, Springfield, VA, June 1969.
10. *Design and Operating Parameters for Emission Control Studies: White Pine Copper Smelter*, EPA-600/2-76-036a, U.S. Environmental Protection Agency, Washington, D.C., February 1976.
11. R. M. Statnick, *Measurements of Sulfur Dioxide, Particulate and Trace Elements in Copper Smelter Converter and Roaster/Reverberatory Gas Streams*, PB 238095, National Technical Information Service, Springfield, VA, October 1974.
12. AP-42 Background Files, Office of Air Quality Planning and Standards, U.S. Environmental Protection Agency, Research Triangle Park, NC.
13. *Design and Operating Parameters or Emission Control Studies*, Kennecott–McGill Copper Smelter, EPA-600/2-76036c, U.S. Environmental Protection Agency, Washington, D.C., February 1976.
14. *Emission Test Report (Acid Plant) of Phelps Dodge Copper Smelter, Ajo, AZ*, EMB-78-CUS-11, Office of Air Quality Planning and Standards, Research Triangle Park, NC, March 1979.
15. S. Dayton, "Inspiration's Design for Clean Air," *Eng. Mining J.* **175**, 6, June 1974.
16. *Emission Testing of Asarco Copper Smelter, Tacoma, WA*, EMB-78-CUS-12, Office of Air Quality Planning and Standards, U.S. Environmental Protection Agency, Research Triangle Park, NC, April 1979.
17. Written Communication from A. L. Labbe, Asarco, Inc., Tacoma, WA, to S. T. Cuffe, U.S. Environmental Protection Agency, Research Triangle Park, NC, November 20, 1978.
18. *Design and Operating Parameters for Emission Control Studies: Asarco–Hayden Copper Smelter*, EPA-600/2-76036j, U.S. Environmental Protection Agency, Washington, D.C., February 1976.
19. *Design and Operating Parameters for Emission Control Studies: Kennecott, Hayden Copper Smelter*, EPA-600/2-76036b, U.S. Environmental Protection Agency, Washington, D.C., February 1976.
20. R. Larkin, *Arsenic Emissions at Kennecott Copper Corporation, Hayden, AZ*, EPA-76-NFS-1, U.S. Environmental Protection Agency, Research Triangle Park, NC, May 1977.
21. *Emission Compliance Status, Inspiration Consolidated Copper Company, Inspiration, AZ*, U.S. Environmental Protection Agency, San Francisco, CA, 1980.
22. M. P. Scanlon, Phelps Dodge Corp., Hidalgo, AZ, written communication to D. R. Goodwin, U.S. Environmental Protection Agency, Research Triangle Park, NC, October 18, 1978.
23. G. M. McArthur, Anaconda Co., written communication to D. R. Goodwin, U.S. Environmental Protection Agency, Research Triangle Park, NC, June 2, 1977.
24. V. Katari, Pacific Environmental Services, Durham, NC, telephone communication to R. Winslow, Hidalgo Smelter, Phelps Dodge Corp., Hidalgo, AZ, April 1, 1982.
25. *Inhalable Particulate Source Category Report for the Nonferrous Industry*, Contract 68-02-3159, Acurex Corp., Mountain View, CA, August 1986.
26. *Emission Test Report, Phelps Dodge Copper Smelter, Douglas, AZ*, EMB-78-CUS-8, Office of Air Quality Planning and Standards, U.S. Environmental Protection Agency, Research Triangle Park, NC, February 1979.
27. *Emission Testing of Kennecott Copper Smelter, Magna, UT*, EMB-78-CUS-13, Office of Air Quality Planning and Standards, U.S. Environmental Protection Agency, Research Triangle Park, NC, April 1979.
28. *Emission Test Report, Phelps Dodge Copper Smelter, Ajo, AZ*, EMB-78-CUS-9, Office of Air Quality Planning and Standards, U.S. Environmental Protection Agency, Research Triangle Park, NC, February 1979.
29. R. D. Putnam, Asarco, Inc., written communication to M. O. Varner, Asarco, Inc., Salt Lake City, UT, May 12, 1980.
30. *Emission Test Report, Phelps Dodge Copper Smelter, Playas, NM*, EMB-78-CUS-10, Office of Air Quality Planning and Standards, U.S. Environmental Protection Agency, Research Triangle Park, NC, March 1979.

31. *Asarco Copper Smelter, El Paso, TX*, EMB-78-CUS-71 Office of Air Quality Planning and Standards, U.S. Environmental Protection Agency, Research Triangle Park, NC, April 25, 1978.
32. A. D. Church et al., "Measurement of Fugitive Particulate and Sulfur Dioxide Emissions at Inco's Copper Cliff Smelter," Paper A-79-51, Metallurgical Society, American Institute of Mining, Metallurgical and Petroleum Engineers, New York.
33. *Copper Smelters, Emission Test Report—Lead Emissions*, EMB-79-CUS-14, Office of Air Quality Planning and Standards, U.S. Environmental Protection Agency, Research Triangle Park, NC, September 1979.

FERROALLOY INDUSTRY PARTICULATE EMISSIONS

Steve Stasko

Adapted from AP-42, Compilation of Air Pollutant Emission Factors, U. S. Environmental Protection Agency with NESHAP update by Editor

A ferroalloy is an alloy of iron and one or more other elements, such as silicon, manganese, or chromium. Ferroalloys are used as additives to impart unique properties to steel and cast iron. The iron and steel industry consumes approximately 95% of the ferroalloy produced in the United States. The remaining 5% is used in the production of nonferrous alloys, including cast aluminum, nickel–cobalt base alloys, and titanium alloys, and in making other ferroalloys.

Three major groups—ferrosilicon, ferromanganese, and ferrochrome—constitute approximately 85% of domestic production. Subgroups of these alloys include silicomanganese, silicon metal, and ferrochromium. The various grades manufactured are distinguished primarily by their carbonol silicon content. The remaining 15% of ferroalloy production is specialty alloys, typically produced in small amounts and containing such elements as vanadium, columbium, molybdenum, nickel, boron, aluminum, and tungsten.

Ferroalloy facilities in the United States vary greatly in size. Many facilities have only one furnace and require less than 25 MW of electricity. Others consist of 16 furnaces, produce six different types of ferroalloys, and require over 75 MW.

PROCESS DESCRIPTION

A typical ferroalloy plant is illustrated in Figure 1. A variety of furnace types produce ferroalloys, including submerged electric arc furnaces, induction furnaces, vacuum furnaces, exothermic reaction furnaces, and electrolytic cells. Furnace descriptions and their ferroalloy products are given in Table 1. Ninety-five percent of all ferroalloys, including all bulk ferroalloys, are produced in submerged electric arc furnaces, the furnace type discussed here.

The basic design of submerged electric arc furnaces is generally the same throughout the ferroalloy industry in the United States. This furnace is composed of a cylindrical steel shell with a concave hearth. The interior of the shell is lined with two or more layers of carbon blocks. Raw materials are charged through feed chutes from above the furnace. The molten metal and slag are removed through one or more tapholes that extend through the furnace shell at the hearth level. Three carbon electrodes, arranged in a delta formation, extend downward through the charge material to a depth of 3–5 ft.

Submerged electric arc furnaces are of two basic types, open and covered, with about 80% of these furnaces in the United States of the open type. Open furnaces have fume collection hoods at least 3 ft above the top of the furnace. Movable panels or screens sometimes are used to reduce the open area between the furnace and hood to improve emissions-capture efficiency. Covered furnaces have a watercooled steel cover to seal the top,

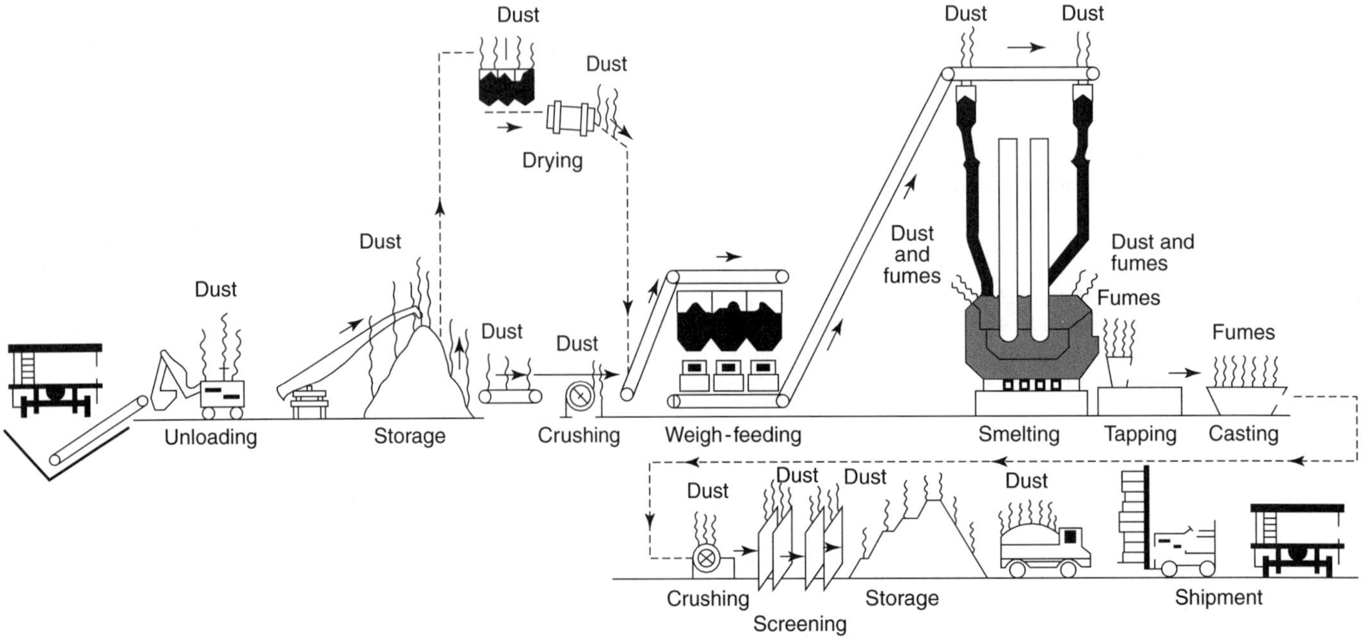

Figure 1. Typical Ferroalloy Production Process, Showing Emission Points.

Table 1. Ferroalloy Processes and Respectie Product Groups

Process	Product
Submerged arc furnace[a]	Silvery iron (15–22% Si)
	Ferrosilicon (50% Si)
	Ferrosilicon (65–75% Si)
	Silicon metal
	Silicon/manganese/zirconium (SMZ)
	High carbon (HC) ferromanganese
	Siliconmanganese
	HC ferrochrome
	Ferrochrome/silicon
	FeSi (90% Si)
Exothermic[b]	
Silicon reduction	Low-carbon ferrochrome, low-carbon ferromanganese, medium-carbon ferromanganese
Aluminum reduction	Chromium metal, ferrotitanium, ferrocolumbium, ferrovanadium
Mixed aluminothermal/ silicothermal	Ferromolybdenum, ferrotungsten
Electrolytic[c]	Chromium metal, manganese metal
Vacuum furnace[d]	Low-carbon ferrochrome
Induction furnace[e]	Ferrotitanium

[a]Process by which metal is smelted in a refractory-lined, cup-shaped steel shell by three submerged graphite electrodes.
[b]Process by which molten charge material is reduced, in exothermic reaction, by addition silicon, aluminum, or a combination of the two.
[c]Process by which simple ions of a metal, usually chromium or manganese in an electrolyte, are plated on cathodes by direct low-voltage current.
[d]Process by which carbon is removed from solid-state, high-carbon ferrochrome in vacuum furnaces maintained at temperature near the melting point of alloy.
[e]Process that converts electric energy without electrodes, into heat to melt metal charge in a cup or drum-shaped vessel.

with holes through it for the electrodes. The degree of emission containment provided by the covers is quite variable. Air infiltration sometimes is reduced by placing charge material around the electrode holes. This type is called a mix seal or semienclosed furnace. Another type is a sealed or totally closed furnace with mechanical seals around the electrodes and a sealing compound packed around the cover edges.

The submerged arc process is a reduction smelting operation. The reactants consist of metallic ores and quartz (ferrous oxides, silicon oxides, manganese oxides, chrome oxides, etc.). Carbon, usually as coke, low-volatility coal, or wood chips, is charged into the furnace as a reducing agent. Limestone also may be added as a flux material. After crushing, sizing, and, in some cases, drying, the raw materials are conveyed to a mix house for weighing and blending, and then by conveyors, buckets, skip hoists, or cars to hoppers above the furnace. The mix is fed by gravity through a feed chute either continuously or intermittently. At high temperatures in the reaction zone, the carbon sources react chemically with oxygen in the metal oxides to form carbon monoxide and to reduce the ores to base metal. A typical reaction illustrating 50% ferrosilicon production is as follows:

$$Fe_2O_3 + 2SiO_2 + 7C \rightarrow 2FeSi + 7CO$$

Smelting in an electric arc furnace is accomplished by converting electric energy to heat. An alternating current applied to the electrodes causes a current to flow through the charge between the electrode tips. This provides a reaction zone of temperatures up to 2000°C (3632°F). The tip of each electrode changes polarity continuously as the alternating current flows between the tips. To maintain a uniform electric load, electrode depth is continuously and automatically varied by mechanical or hydraulic means, as required. Furnace power requirements vary from 7 to over 50 MW, depending on the furnace size and the product being made. The average is 17.2 MW.

The molten alloy and slag that accumulate on the furnace hearth are removed at 1–5-h intervals through the taphole. Tapping typically lasts 10–15 min. Tapholes are opened with a pellet shot from a gun, by drilling, or by oxygen lancing. The molten metal and slag flow from the taphole into a carbon-lined trough, then into a carbon-lined runner that directs the metal and slag into a reaction ladle, ingot molds, or chills. Chills are low, flat iron or steel pans that provide rapid cooling of the molten metal. Tapping is terminated and the furnace resealed by inserting a carbon paste plug into the taphole.

When chemistry adjustments are necessary to produce a specified product, a reaction ladle is used. Ladle treatment reactions are batch processes and may include chlorination, oxidation, gas mixing, and slag metal reactions.

During tapping and/or in the reaction ladle, slag is skimmed from the surface of the molten metal. It can be disposed of as a raw material in a furnace or reaction ladle to produce a chemically related ferroalloy product.

After cooling and solidifying, the large ferroalloy castings are broken with drop weights or hammers. The broken ferroalloy pieces are then crushed, screened (sized), and stored in bins until shipment.

AIR EMISSIONS CHARACTERIZATION

Particulates are generated from several activities at a ferroalloy facility, including raw-material handling, smelting, and product handling. The furnaces are the largest potential sources of particulate emissions. The emission factors in Tables 2 and 3 reflect emissions from ferroalloy smelting furnaces.

Electric-arc furnaces emit particulates in the form of fume, accounting for an estimated 94% of the particulate emissions in the ferroalloy industry. Large amounts of carbon monoxide and organic materials also are emitted by submerged electric-arc furnaces. Carbon monoxide is formed as a by-product of the chemical reaction between oxygen in the metal oxides of the charge and carbon contained in the reducing agent (coke, coal, etc.). Reduction gases containing organic compounds and carbon monoxide continuously rise from the high-temperature reaction zone, entraining fine particles and fume precursors. The mass weight of carbon monoxide produced sometimes exceeds that of the metallic product. The chemical constituents of the heat-induced fume consist of oxides of the products being produced, carbon from the reducing agent, and enrichment by SiO_2, CaO, and MnO, if present in the charge.[14]

AIR POLLUTION CONTROL MEASURES

In an open electric arc furnace, all carbon monoxide burns with induced air at the furnace top. The remaining fume, captured

Table 2. Emission Factors for Particulate from Submerged ARC Ferroalloy Furnaces[a]

		Particulate Emission Factors Uncontrolled[c]						Particulate Emission Factors Controlled[c]				
Product[a]	Furnace Type	kg/Mg (lb/ton) Alloy	kg(lb)/MWh	Size Data	Notes	Emission Factor Rating	Control Device[d]	kg/Mg (lb/ton) alloy	kg (lb)/MWh	Size Data	Notes	Emission Factor Rating
FeSi (50%)	Open	35 (70)	7.4 (16.3)	Yes	*e, f, g*	B	Baghouse	0.9 (1.8)	0.2 (0.44)	Yes	*e, f*	B
	Covered	46 (92)	9.3 (20.5)		*h*	E	Scrubber					
							High energy	0.24 (0.48)	0.05 (0.11)		*h, j*	E
							Low energy	4.5 (9.0)	0.77 (1.7)		*h, j*	E
FeSi (75%)	Open	158 (316)	16 (35)		*k*	E	Scrubber					
	Covered	103 (206)	13 (29)		*h, j*	E	Low energy	4.0 (8.0)	0.5 (1.1)		*h, j*	E
FeSi (90%)	Open	282 (564)	24 (53)	Yes	*m*	E						
Si metal (98%)	Open	436 (872)	33 (73)	Yes	*n, p*	B						
FeMn (80%)	Open	14 (28)	4.8 (11)	Yes	*q, r*	B	Baghouse	16 (32)	1.2 (2.6)	Yes	*n, p*	B
							Baghouse	0.24 (0.48)	0.078 (0.172)	Yes	*q, r*	B
							Scrubber					
FeMn (1% Si)	Covered	6 (12)	2.4 (5.3)		*h, t*	E	High energy	0.8 (1.6)	0.34 (0.75)		*h, s*	E
	Sealed	37 (74)	17 (37)		*u, v*	E	High energy	0.25 (0.5)	0.10 (0.22)		*h, s, w*	C
FeCr (high carbon)	Open	78 (157)	15 (33)	Yes	*x, y*	C	Electrostatic precipitator	1.2 (2.3)	0.23 (0.5)	Yes	*x, y*	C
							Scrubber	2.1 (4.2)	0.44 (1.0)	Yes	*aa, bb*	C
SiMn	Open	96 (192)	20 (44)	Yes	*z, aa*	C	High energy	0.15 (0.30)	0.016 (0.035)		*v, w*	E
	Sealed	—(–)	—(–)									

[a]Factors are for main furnace dust collection system before and after control device. Where other emissions, such as leaks or tapping, are included or quantified separately, such is noted. Particulate sources not included: raw-material handling, storage, preparation; product crushing, screening, handling, packaging.
[b]Percentages are of the main alloying element in product.
[c]In most source testing, fugitive emissions not measured or collected. Where tapping emissions are controlled by primary system, their contribution to total emissions could not be determined. Fugitive emissions may vary greatly among sources, with furnace and collection system design and operating practices.
[d]Low-energy scrubbers are those with $\Delta P < 20$ in. H_2O; high energy, with $\Delta P > 20$ in. H_2O.
[e]Includes fumes captured by tapping hood (efficiency estimated near 100%).
[f]References 4, 5, 15.
[g]Factor is average of three sources, fugitive emissions not included. Fugitive emissions at one source measured an additional 10.5 kg/Mg alloy or 2.7 kg/MWh.
[h]References 4,5.
[j]Does not include emissions from tapping or mix seal leaks.
[k]References 18, 19.
[m]Reference 16.
[n]Estimated 60% of tapping emissions captured by control system (escaped fugitive emissions not included in factor).
[p]References 5, 7.
[q]Estimated 50% of tapping emissions captured by control system (escaped fugitive emissions not included in factor).
[r]References 4–6.
[s]Includes fume only from primary control system.
[t]Includes tapping fumes and mix seal leak fugitive emissions. Fugitive emissions measured at 33% of total uncontrolled emissions.
[u]Assumes tapping fumes not included in emission factor.
[v]Reference 8. Dash, no data.
[w]Does not include tapping or fugitive emissions.
[x]Tapping emissions included. Factor developed from two test series performed on the same furnace seven years apart. Measured emissions in latter test were 36% less than in former.
[y]References 2, 9–11.
[z]Factor is average of two test series. Tests at one source included fugitive emissions (3.4% of total uncontrolled emissions). Second test insufficient to determine whether fugitive emissions were included in total.
[aa]References 2, 12, 13.
[bb]Factors developed from two scrubber controlled sources, one operated at $\Delta P = 47$–57 in. H_2O, the other at unspecified ΔP. Uncontrolled tapping operations emissions are 2.1 kg/Mg alloy.

Table 3. Size-Specific Emission Factors for Submerged Arc Ferroalloy Furnaces

Product	Control Device	Particle size[a] (μm)	Cumulative Mass (%) ≤Stated Size	Cumulative Mass Emission Factor kg/ton Alloy	(lb/ton) Alloy	Emission Factor Rating
50% FeSi						
Open furnace	None[b,c]	0.63	45	16	(32)	B
		1.00	50	18	(35)	
		1.25	53	19	(37)	
		2.50	57	20	(40)	
		6.00	61	21	(43)	
		10.00	63	22	(44)	
		15.00	66	23	(46)	
		20.00	69	24	(48)	
		d	100	35	(70)	
	Baghouse	0.63	31	0.28	(0.56)	B
		1.00	39	0.35	(0.70)	
		1.25	44	0.40	(0.80)	
		2.50	54	0.49	(1.0)	
		6.00	63	0.57	(1.1)	
		10.00	72	0.65	(1.3)	
		15.00	80	0.72	(1.4)	
		20.00	85	0.77	(1.5)	
			100	0.90	(1.8)	
80% FeMn						
Open furnace	None[e,f]	0.63	30	4	(8)	B
		1.00	46	7	(13)	
		1.25	52	8	(15)	
		2.50	62	9	(17)	
		6.00	72	10	(20)	
		10.00	86	12	(24)	
		15.00	96	13	(26)	
		20.00	97	14	(27)	
		d	100	14	(28)	
80% FeMn						
Open furnace	Baghouse[e]	0.63	20	0.048	(0.10)	B
		1.00	30	0.070	(0.14)	
		1.25	35	0.085	(0.17)	
		2.50	49	0.120	(0.24)	
		6.00	67	0.160	(0.32)	
		10.00	83	0.200	(0.40)	
		15.00	92	0.220	(0.44)	
		20.00	97	0.235	(0.47)	
		d	100	0.240	(0.48)	
Si Metal[h]						
Open furnace	None[g]	0.63	57	249	(497)	B
		1.00	67	292	(584)	
		1.25	70	305	(610)	
		2.50	75	327	(654)	
		6.00	80	349	(698)	
		10.00	86	375	(750)	
		15.00	91	397	(794)	
		20.00	95	414	(828)	
		d	100	436	(872)	
	Baghouse	1.00	49	7.8	(15.7)	B
		1.25	53	8.5	(17.0)	
		2.50	64	10.2	(20.5)	
		6.00	76	12.2	(24.3)	
		10.00	87	13.9	(28.0)	
		15.00	96	15.4	(31.0)	
		20.00	99	15.8	(31.7)	
			100	16.0	(32.0)	

(*Continued*)

Table 3. *(Continued)*

Product	Control Device	Particle size,[a] μm	Cumulative Mass% ≤Stated Size	Cumulative Mass Emission Factor kg/ton Alloy	(lb/ton) Alloy	Emission Factor Rating
FeCr (HC)						
Open furnace	None[b,j]	0.5	19	15	(30)	C
		1.0	36	28	(57)	
		2.0	60	47	(94)	
		2.5	63[k]	49	(99)	
		4.0	76	59	(119)	
		6.0	88[k]	67	(138)	
		10.0	91	71	(143)	
		d	100	78	(157)	
	Electrostatic					C
	precipitator	0.5	33	0.40	(0.76)	
		1.0	47	0.56	(1.08)	
		2.5	67	0.80	(1.54)	
		5.0	80	0.96	(1.84)	
		6.0	86	1.03	(1.98)	
		10.0	90	1.08	(2.07)	
		d	100	1.20	(2.30)	
SiMn						
Open furnace	None[b,m]	0.5	28	27	(54)	C
		1.0	44	42	(84)	
		2.0	60	58	(115)	
		2.5	65	62	(125)	
		4.0	76	73	(146)	
		6.0	85	82	(163)	
		10.0	96[k]	92[k]	(177)[k]	
		d	100	96	(192)	
	Scrubber[m,n]	0.5	56	1.18	(2.36)	
		1.0	80	1.68	(3.44)	
		2.5	96	2.02	(4.13)	
		5.0	99	2.08	(4.26)	
		6.0	99.5	2.09	(4.28)	
		10.0	99.9[k]	2.10[k]	(4.30)[k]	
			100	2.10	(4.30)	

[a] Aerodynamic diameter, based on Task Group on Lung Dynamics definition. Particle density = 1 g/cm^3.
[b] Includes tapping emissions.
[c] References 4, 5, 15.
[d] Total particulate, based on Method 5 total catch.
[e] Includes tapping fume (capture efficiency 50%).
[f] References 4–6.
[g] Includes tapping fume (estimated capture efficiency 60%).
[h] References 5, 7
[j] References 1, 9–11.
[k] Interpolated data.
[m] References 2, 12, 13.
[n] Primary emission control system only, without tapping emissions.

by hooding about 3 ft above the furnace, is directed to a gas cleaning device. Baghouses are used to control emissions from 85% of the open furnaces in the United States. Scrubbers are used on 13% of the furnaces and electrostatic precipitators on 2%. Control efficiencies for well-designed and -operated control systems [i.e., baghouses with air-to-cloth ratios of 1:1 to 2:1 ft^3/ft^2 and scrubbers with a pressure drop from 14 to 24 kPa (55 to 96 in. H_2O)] have been reported to be in excess of 99%.[4] Air-to-cloth ratio is the ratio of the volumetric airflow through the filter medium to the medium area.

Two emission capture systems, not usually connected to the same gas cleaning device, are necessary for covered furnaces. A primary capture system withdraws gases from beneath the furnace cover. A secondary system captures fume released around the electrode seals and during tapping. Scrubbers are used almost exclusively to control exhaust gases from sealed furnaces. The gas from sealed and mix sealed furnaces is usually flared at the exhaust of the scrubber. The carbon monoxide–rich gas has an estimated heating value of 300 Btu/ft^3 and is sometimes used as a fuel in kilns and sintering machines. The efficiency of flares for the control of carbon monoxide and the reduction of organic emission has been estimated to be greater than 98% for steam-assisted flares with a velocity of less than 60 fps and a gas heating value of

300 Btu/scf.[17] For unassisted flares, the reduction of organic and carbon monoxide emissions is 98% with a velocity of less than 60 fps and a gas heating value greater than 200 Btu/scf.

Tapping operations also generate fumes. Tapping is intermittent and is usually conducted during 10–20% of the furnace operating time. Some fumes originate from the carbon lip liner, but most are a result of induced heat transfer from the molten metal or slag as it contacts the runners, ladles, casting beds, and ambient air. Some plants capture these emissions to varying degrees with a main canopy hood. Other plants employ separate tapping hoods ducted to either the furnace emission control device or a separate control device. Emission factors for tapping emissions are unavailable because of a lack of data.

A reaction ladle may be used to adjust the metallurgy after furnace tapping, by chlorination, oxidation, gas mixing, and slag metal reactions. Ladle reactions are an intermittent process, and emissions have not been quantified. Reaction ladle emissions often are captured by the tapping emissions control system.

Available data are insufficient to provide emission factors for raw-material handling, pretreatment, and product handling. Dust particulates are emitted from raw-material handling, storage, and preparation activities (see Figure 1), from such specific activities as unloading raw materials from delivery vehicles (ship, railcar, or truck), storing raw materials in piles, loading raw materials from storage piles into trucks or gondola cars, and crushing and screening raw materials. Raw materials may be dried before charging in rotary or other types of dryers, and these dryers can generate significant particulate emissions. Dust may also be generated by heavy vehicles used for loading, unloading, and transferring material. Crushing, screening, and storing ferroalloy products emit particulates in the form of dust. The properties of particulate emitted as dust are similar to the natural properties of the ores or alloys from which they originated, ranging in size from 3 to 100 μm.

Approximately half of all ferroalloy facilities have some type of control for dust emissions. Dust generated from raw-material storage may be controlled in several ways, including sheltering storage piles from the wind with block walls, snow fences, or plastic covers. Occasionally, piles are sprayed with water to prevent airborne dust. Emissions generated by heavy vehicle traffic may be reduced by using a wetting agent or paving the plant yard.[3] Moisture in the raw materials, which may be as high as 20%, helps to limit dust emissions from unloading and loading. Dust generated by crushing, sizing, drying, or other pretreatment activities is sometimes controlled by dust collection equipment such as scrubbers, cyclones, or baghouses. Ferroalloy product crushing and sizing usually require a baghouse. The raw-material emission collection equipment may be connected to the furnace emission control system.

MACT STANDARD FOR FERROALLOY PRODUCTION

The MACT rule for hazardous emissions from the ferroalloy industry was originally proposed under the title of "national emission standards for hazardous air pollutants for ferroalloys production." The final ruling, as published in the *Federal Register* on May 20, 1999, finalized the National Emission Standards for Hazardous Air Pollutants (NESHAP) for ferroalloys production and included only ferromanganese and silicomanganese production. Ferrochromium and ferronickel production facilities were excluded from the final rule.

The NESHAP applies to new and existing ferroalloy production facilities that manufacture ferromanganese and silicomanganese and are major sources of HAP emissions or are co-located at major sources of HAP emissions. The following HAP emission sources at a ferroalloy production facility are affected by the rule: submerged arc furnaces; metal oxygen refining (MOR) process; crushing and screening operations; and fugitive dust sources. The rule contains emission standards that limit particulate matter emissions from existing and new or reconstructed emission sources. The limits for the submerged arc furnaces depend on the product produced and furnace design. The rule also sets limits for the air pollution control devices associated with the MOR process and crushing and screening operations. The emission standards are summarized in Table 4. As of May 1999, only one facility was affected by the NESHAP (the Elkem Metals Company plant in Marietta, Ohio), and no new facilities were anticipated at the time of passage of the final rule. The reader is referred to the final rule for details on the MACT standard, which can be found at the UATW bulletin board on EPA's TTN bulletin board (*http://www.epa.gov/ttn/uatw/ferroa/fr20my99.txt*).

REFERENCES

1. F. J. Schottman, "Ferroalloys," 1980 Mineral Facts and Problems, Bureau of Mines, U.S. Department of the Interior, Washington, D.C., 1980.
2. J. O. Dealy and A. M. Killin, *Engineering and Cost Study of the Ferroalloy Industry*, EPA-450/2-74-008, U.S. Environmental Protection Agency, Research Triangle Park, NC, May 1974.

Table 4. Emission Standards for Particulate Matter (NESHAP) (as of May 20, 1999)

New or Reconstructed or Existing Source	Affected Source	Particulate Emission Standard
New or reconstructed	Submerged arc furnace (primary and tapping)	1. 0.23 kg/h/MW or 2. 35 mg/dscm
Existing	Open submerged arc furnace (prim. and tapping)	16.3 kg/h (silicomanganese) 6.4 kg/h (ferromanganese)
Existing	Semisealed submerged arc furnace (primary, tapping, and vent stacks)	11.2 kg/h
New, reconstructed, or existing	MOR process	69 mg/dscm
New or reconstructed	Individual equipment associated with crushing and screening operation	50 mg/dscm
Existing	Individual equipment associated with crushing and screening operation	69 mg/dscm

3. *Background Information on Standards of Performance: Electric Submerged Arc Furnaces for Production of Ferroalloys, Volume I: Proposed Standards*, EPA-450/2-74-018a, U.S. Environmental Protection Agency, Research Triangle Park, NC, October 1974.
4. C. W. Westbrook and D. P. Dougherty, *Level I Environmental Assessment of Electric Submerged Arc Furnaces Producing Ferroalloys*, EPA-600/2-81-038, U.S. Environmental Protection Agency, Washington, D.C., March 1981.
5. C. W. Westbrook, *Multimedia Environmental Assessment of Electric Submerged Arc Furnaces Producing Ferroalloys*, EPA-600/2-83-092, U.S. Environmental Protection Agency, Washington, D.C., September 1983.
6. T. Epstein et al., *Ferroalloy Furnace Emission Factor Development, Roane Limited, Rockwood, Tennessee*, EPA-600/ X-85-325, U.S. Environmental Protection Agency, Washington, D.C., June 1981.
7. S. Beaton et al., *Ferroalloy Furnace Emission Factor Development, Interlake Inc., Alabama Metallurgical Corp., Selma, Alabama*, EPA-600/X-85-324, U.S. Environmental Protection Agency, Washington, D.C., May 1981.
8. J. L. Rudolph et al., *Ferroalloy Process Emissions Measurement*, EPA-600/2-79-045, U.S. Environmental Protection, Agency, Washington, D.C., February 1979.
9. Joseph F. Eyrich, Macalloy Corp., Charleston, SC, written communication to GCA Corp., Bedford, MA, February 10, 1982, citing Airco Alloys and Carbide test R-07-7774-000-1, Gilbert Commonwealth, Reading, PA, 1978.
10. *Source Test, Airco Alloys and Carbide*, Charleston, SC, EMB-71-PC- I 6(FEA), U.S. Environmental Protection Agency, Research Triangle Park, NC, 1971.
11. Joseph F. Eyrich, Macalloy Corp., Charleston, SC, telephone communication with Evelyn J. Limberakais, GCA Corp., Bedford, MA, February 23, 1982.
12. Source test, Chromium Mining and Smelting Corporation, Memphis, TN, EMB-72-PC-05 (FEA), U.S. Environmental Protection Agency, Research Triangle Park, NC, June 1972.
13. Source Test, Union Carbide Corporation, Ferroalloys Division, Marietta, Ohio, EMB-71-PC-12 (FEA), U.S. Environmental Protection Agency, Research Triangle Park, NC, 1971.
14. R. A. Person, "Control of Emissions from Ferroalloy Furnace Processing," *J. Metals* **23**(4), 17–29, April 1971.
15. S. Gronberg, *Ferroalloy Furnace Emission Factor Development, Foote Minerals, Graham, W. Virginia*, EPA-600/X85-327, U.S. Environmental Protection Agency, Washington, D.C., July 1981.
16. *Air Pollutant Emission Factors*, Final Report, APTD-0923, U.S. Environmental Protection Agency, Research Triangle Park, NC, April 1970.
17. Leslie B. Evans, Office of Air Quality Planning and Standards, U.S. Environmental Protection Agency, Research Triangle Park, NC, telephone communication with Richard Vacherot, GCA Corp., Bedford, MA, October 18, 1984.
18. R. Ferrari, "Experiences in Developing an Effective Pollution Control System for a Submerged Arc Ferroalloy Furnace Operation," *J. Metals*, 95–104, April 1968.
19. Fredriksen and Nestaas, *Pollution Problems by Electric Furnace Ferroalloy Production*, UN Economic Commission for Europe, September 1968.
20. A. E. Vandergrift et al., *Particulate Pollutant System Study Mass Emissions*, P13-203-128, PB-203-522 and P-203-521, National Technical Information Service, Springfield VA, May 1971.
21. *Control Techniques for Lead Air Emissions*, EPA-450/2-77012, U.S. Environmental Protection Agency, Research Triangle Park, NC, December 1977.
22. W. E. Davis, *Emissions Study of Industrial Sources of Lead Air Pollutants*, 1970, EPA-APTD-1543, W. E. Davis and Associates, Leawood, KS, April 1973.
23. *Source Test, Foote Mineral Company, Vancoram Operations, Steubenville, Ohio*, EMB-71-PC-08 (FEA), U.S. Environmental Protection Agency, Research Triangle Park, NC, August 1971.

BIBLIOGRAPHY

1. F. J. Schottman, "Ferroalloys," *Minerals Yearbook, Vol. I: Metals and Minerals*, Bureau of Mines, U.S. Department of the Interior, Washington, D.C., 1980.
2. S. Beaton and H. Klemm, *Inhalable Particulate Field Sampling Program for the Ferroalloy Industry*, TR-80-115-G, GCA Corporation, Bedford, MA, November 1980.
3. G. W. Westbrook and D. P. Dougherty, *Environmental Impact of Ferroalloy Production Interim Report: Assessment of Current Data*, U.S. Environmental Protection Agency, Research Triangle Institute, Research Triangle Park, NC, November 1978.
4. K. Wark, C. F. Warner, and W. T. Davis, *Air Pollution: Its Origin and Control*, Addison Wesley Longman, Third Edition, 1998.
5. M. Szabo and R. Gerstle, *Operations and Maintenance of Particulate Control Devices on Selected Steel and Ferroalloy Processes*, EPA-600/2-78-037, U.S. Environmental Protection Agency, Washington, D.C., March 1978.
6. S. Gronberg et al., *Ferroalloy Industry Particulate Emissions: Source Category Report*, EPA-600/7-86-039, U.S. Environmental Protection Agency, Cincinnati, OH, November 1986.
7. R. W. Gerstle et al., *Review of Standards of Performance for New Stationary Air Sources—Ferroalloy Production Facility*, EPA-450/3-80-041, U.S. Environmental Protection Agency, Research Triangle Park, NC, December 1980.

STEEL INDUSTRY

Bruce A. Steiner

In the years that have passed since the first publication of the *Air Pollution Engineering Manual* (AP-40) in 1967, the iron and steel industry has changed dramatically in many respects. More change has occurred since the Manual's 1992 edition.

The number of steel-making facilities and the domestic capacity have been greatly reduced over the past 30 years. As an indication, steel industry employment has dropped from a high of over 500,000 persons to about 160,000 today. One reason is that increased imports have cut into domestic production. In 1967, imports represented only 13% of steel shipments, but imports have increased in recent years and in 1998 reached a record high of 41.5 million tons, or about 30% of apparent U.S. steel consumption. Productivity and yields have also increased substantially. For example, in 1967, the U.S. industry produced 127 million tons of raw steel and shipped 88 million tons. In 1998, the industry produced 109 million tons of raw steel but shipped 102 million tons. The increased ratio of shipped tons to raw steel production reflects substantial increases in productivity, accomplished in large part by the increased installation of continuous casting facilities, which now account for over 95% of the raw steel output. Moreover, the development of stronger and lighter steels and coated grades allow steel to be more efficiently utilized than in past years, and the tailoring of steel properties and shapes to the needs of consuming industries results in less scrap and greater yield of finished product.

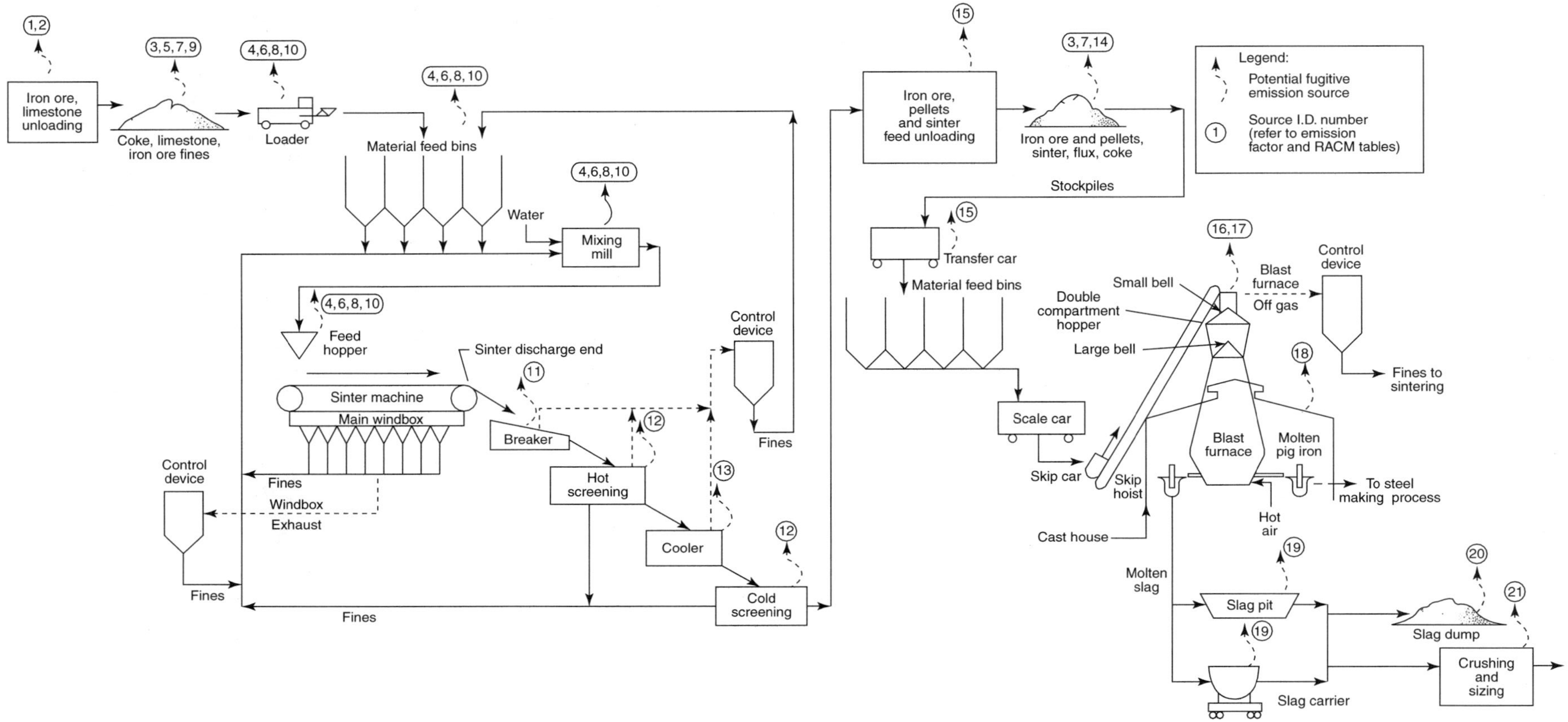

Figure 1. Process Flow Diagram of a Typical Iron Production Facility (From Ohio EPA, *Reasonably Available Control Measures for Fugitive Dust Sources*, September, 1980).

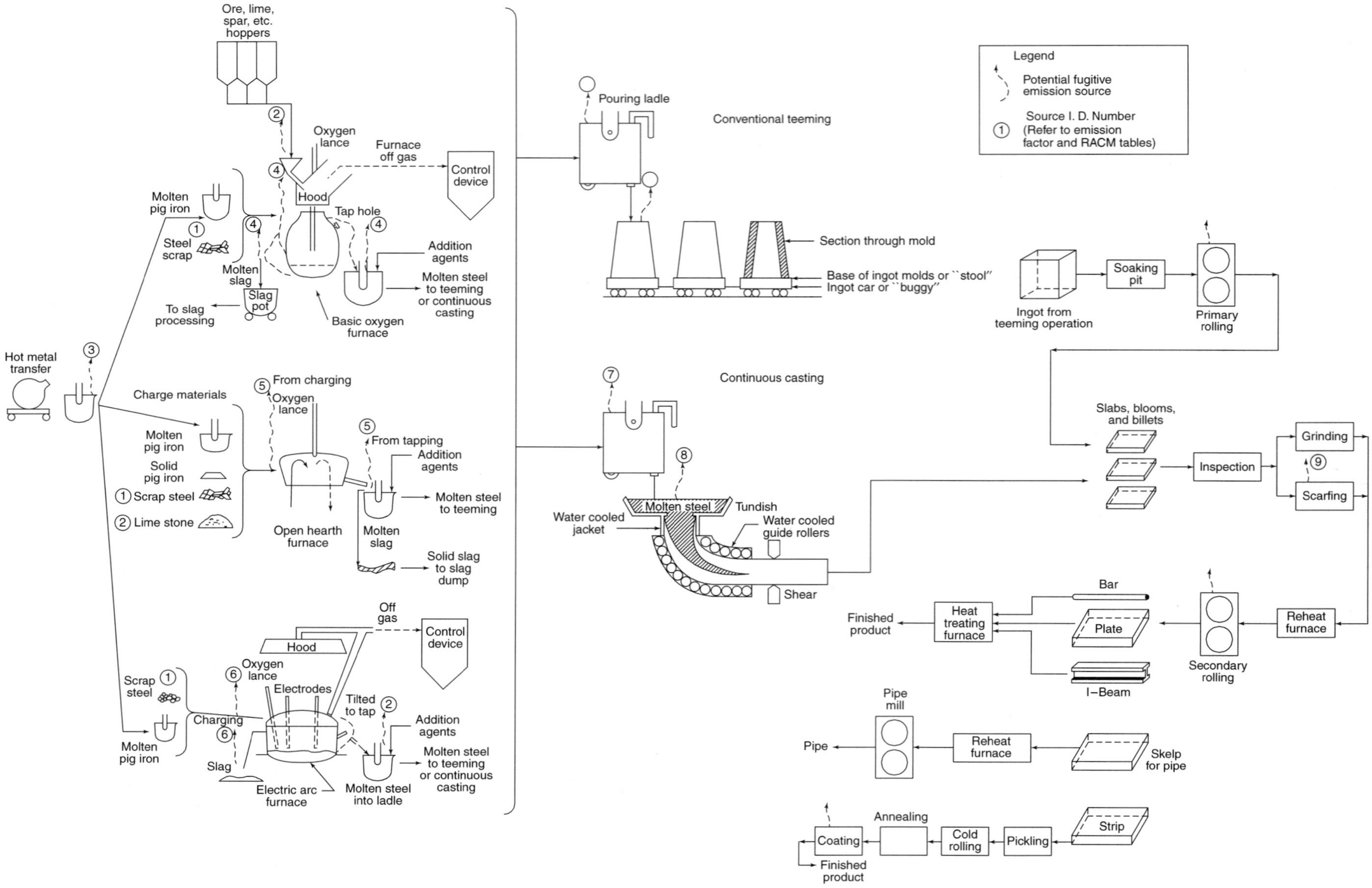

Figure 2. Process Flow Diagram of a Typical Steel Production Facility (From Ohio EPA, *Reasonably Available Control Measures for Fugitive Dust Sources*, September, 1980).

The changing face of the U.S. iron and steel industry has also brought about many process changes. In 1967, the dominant method of making steel was the open hearth furnace. Since then, however, all open-hearth operations have been discontinued, and the basic oxygen process as a percentage of the total production has climbed from 33% to 55% and electric furnace production has increased from 12% to 45%. Over that period, annual domestic coke consumption has fallen to less than half, due in part to the lower raw steel production, but also to improved blast furnace operations that require lower coke rates. Improved yields, improved processing technologies, increased utilization of recycled steel, and energy conservation programs have allowed the industry to reduce its energy consumption per ton of shipped steel by about 45% since 1975. New processing techniques also continue to emerge. Among the technologies now commonly employed, but rarely used or unknown in 1967, are hot metal desulfurization, argon–oxygen decarburization, electrogalvanizing, vacuum degassing, ladle metallurgy, and slag skimming.

With these numerous changes come new challenges for air pollution control technology. Emphasis cannot be placed solely on the selection and efficiency of the control device for criteria pollutants. Other factors need to be considered, including the full range of air pollutants that may be emitted, the capture efficiency of the entire system, and the energy efficiency, cost effectiveness, and ease of maintenance and reliability of the control system.

The production of iron and steel at an integrated iron and steel plant is accomplished using several interrelated processes. A flow diagram of a typical iron production facility and a typical steel production facility are shown in Figures 1 and 2, respectively. The major operations are: (1) coke production, (2) sinter production, (3) iron production, (4) iron preparation, (5) steel production, (5) semifinished and finished product preparation, (7) heat and electricity supply, and (8) handling and transport of raw, intermediate, and waste materials. As shown in Figure 1, iron production involves the sintering of raw materials into an agglomerated product (sinter), after which the sinter is charged to blast furnaces to make iron. The molten iron is then further processed into steel, as shown in Figure 2 in basic oxygen furnaces or electric arc furnaces, depending on the type of steel to be produced.

The following sections describe the individual operations associated with the iron and steel production and the potential air emissions and associated control technologies associated with each operation. Emission factors associated with each of the operations are included in each section.

MISCELLANEOUS FUGITIVE EMISSION SOURCES

ROBERT E. SISTEK AND FRANK PENDLETON
(with update by Editor)

Fugitive emissions in the iron and steel industry can be defined as those emissions that enter the atmosphere from some means other than a stack, chimney, or similar device. They are commonly classified as either process or nonprocess emissions. Such emissions occur when pollutants escape capture by a control device or are generated without employing control measures. Many of the significant sources of fugitive emissions in the iron and steel industry are identified, along with appropriate emission factors, in Sections 12.5 and 13.2.4 of the U.S. Environmental Protection Agency (EPA) publication AP-42,[1] as well as in the following sections.

SOURCES OF FUGITIVE EMISSIONS

Emissions of particulate matter, the most common fugitive emissions in the iron and steel industry, are generated by mechanical operations, process reactions, and fuel combustion. Gaseous pollutants are generated during certain processes, as well as during fuel combustion. Process emissions, which result from the chemical or physical alteration of a raw material, are normally generated inside a building or enclosure. Certain process emissions, such as oxygen-fuel cutting or lancing, can also occur as open-source emissions. Potential sources of fugitive emissions from iron and steel mill operations are indicated in Figures 1 and 2 of the previous section and identified in Table 1. Sources for which emission data are included in AP-42 are listed in Table 2. Several additional sources that are less significant or occur less frequently are addressed in the following.

Oxygen-Fuel Cutting and Lancing

Emissions from cutting and lancing are assumed to be similar to those from uncontrolled scarfing.[2] An emission factor of 0.0015 pound of particulate matter per pound of metal removed has been derived from the AP-42 emission factor for uncontrolled scarfing.[3] A torch operator typically removes 200 pounds of metal in 1 h and generates about 0.3 pounds of particulate matter emissions.[4] Characteristics of the emissions are expected to be similar to those from scarfing. Emissions can be controlled by enclosing the operation and exhausting the emissions to a control device. Some degree of control can be achieved by the enclosure alone and can be as much as 70% for certain process emissions (see Table 12.5-1 of Ref. 1 or the emission factors for each process in the following sections). Visible emissions from the operation can be reduced by injecting CO_2 into the oxygen-fuel mixture. Whether CO_2 injection actually reduces the mass emissions or simply prolongs the emission rate by extending cutting times is uncertain.

Kish Collection and Disposal

As the solubility of carbon in molten iron decreases during cooling, carbon precipitates out of the solution as a fine graphite particle commonly referred to as kish. Although kish is generated during several operations, most is produced during the desulfurization process, generally utilizing an efficient emission control system. Following desulfurization, the kish is skimmed from the surface of the ladle, along with other slag components, and transported to a processing site. Emissions generated during skimming can be controlled by an appropriately placed hood, which exhausts to a control device, such as a baghouse. Control of fugitive dust emissions from the dumping and subsequent processing of kish, as well as other slag components, can best be achieved by moving the entire process indoors. Emissions from uncontrolled disposal of kish can be estimated using the material-handling equation found later in this section. Because of the hot, dry nature of the material, a moisture content of 0.25% should be used, which reflects the lowest value included in the data set used in developing the equation. A typical two-vessel basic oxygen furnace (BOF) shop can be expected to produce about 480 tons of kish per day.[5]

Ladle and Tundish Repairs[5]

Emissions from ladle and tundish repairs are made up of dust and fumes from the removal and installation of refractory products, as well as combustion products from curing and preheating the repaired unit. Typically, a two-vessel BOF shop repairs an average of nine tundishes and 35 ladles each day. Removing the metallic residue (skulls) requires about 15 min of lancing or oxygen-fuel cutting per unit. Damaged refractory is either patched or replaced by spray methods. Tundish spraying often uses local hoods to control overspray. Curing and preheating typically require about 4100 ft^3 of natural gas per tundish and 1000 ft^3 per ladle. Repairs to tundishes and ladles are normally conducted indoors. Dumping the old refractory usually produces little, if any, dust emission. However, certain types of tundish construction can produce considerable dust during removal. In such cases, a hood with an appropriate control device can be utilized. Emissions from lancing and natural gas combustion can be estimated using the factor for lancing contained here and the AP-42 factors for fuel combustion.

Ground Casting[5]

Ground casting (pouring molten metal on the ground) results from misalignment of the torpedo or subcars at the blast furnace cast house or burnout of a ladle. It is normally considered a nonroutine operation or a malfunction. The practice is, however, employed at times to allow for continued iron production during BOF maintenance outages. Estimates of ground cast metal can vary from about 600 tons per year when used only for emergency situations to as high as 190,000 tons per year when used in conjunction with BOF outages. An estimate of uncontrolled fugitive dust emissions can be made using the AP-42 factor for hot metal transfer. However, because of the much greater emissions associated with ground casting, it is recommended that the factor be doubled. The most suitable method of controlling emissions from ground casting is to conduct the operation indoors.

Conditioning Slag Pots[5]

Slag from the BOF process is transported in thimble-shaped pots to a slag processing site. To aid in removal of the slag, a release agent (commonly hot solid slag) is placed in the pot prior to its use. Typically, about 2–3 cubic yards of this material are charged in each slag pot after it is emptied. Some facilities, however, use as much as 6 cubic yards per slag pot. A two-vessel BOF shop will use an estimated 24 slag pots per day. Emissions from this conditioning process can be estimated using the material-handling equation found later in this section. Because of the hot, dry nature of the material, a moisture content of 0.25% should be used, which reflects the lowest value included in the data set used in developing the equation. Fugitive dust emissions from this practice can be minimized by controlling the amount of dry slag being charged in the pots and by moving the entire process indoors.

REFERENCES

1. *Compilation of Air Pollutant Emission Factors*, AP-42, U.S. EPA TTN/CHIEF (*http://www.epa.gov/ttn/chief/ap42c12.html*) and (*http://www.epa.gov/ttn/chief/ap42c13.html*).
2. *Fugitive Emissions from Integrated Iron and Steel Plants*, EPA-600/2-78-050, Midwest Research Institute, March 1978, p. 2–1.
3. F. Pendleton and R. Sistek, "PM10 Characterization of Iron and Steel Sources Not Previously Included in Fugitive Emission Inventories,"presented at the 83rd Annual Meeting and Exhibition of AWMA, Pittsburgh, June 1990.
4. W. Hass, Heckett Corp., East Chicago, IN, personal communication, January 1990.
5. *Characterization of Sources Not Previously Included in LTV's PM10 Inventory for the Indiana Harbor Works*, Project 9415L, Midwest Research Institute, June 20, 1990.

Emissions from Aggregate Handling and Storage Piles (adapted from AP-42 Section 13.2.4)

The quantity of dust emissions from aggregate storage operations varies with the volume of aggregate passing through the storage cycle. Emissions also depend on three parameters associated with the pile: age of pile, moisture content, and proportion of fines.

When freshly processed aggregate is loaded onto a storage pile, the potential for dust emissions is at a maximum. Fines are easily disaggregated and released to the atmosphere. As the pile weathers, potential for dust emissions is greatly reduced. Moisture causes aggregation and cementation of fines to the surfaces of larger particles. Any significant rainfall soaks the interior of the pile, and then the drying process is very slow.

The silt content (particles equal to or less than 75 μm in diameter) is determined by measuring the portion of dry aggregate material that passes through a 200-mesh screen using ASTM-C-136 method.[1] Table 1 (taken from AP-42) summarizes measured silt and moisture values for various industries including iron and steel production.

Either adding aggregate material to a storage pile or removing it usually involves dropping the material onto a receiving surface either in a batch mode or a continuous mode. Truck dumping on the pile or loading out from the pile to a truck with a front-end loader are examples of batch drop equations. Adding material to a pile by a conveyor stacker is an example of a continuous drop operation.

The quantity of particulate emissions generated by either type of drop operation, per ton of material transferred, may be estimated, with a rating of A, using the following empirical expression[2]:

$$E = k(0.0016)\frac{(U/2.2)^{1.3}}{(M/2)^{1.4}} \text{ (kg/Mg)}$$
$$E = k(0.0032)\frac{(U/5)^{1.3}}{(M/2)^{1.4}} \text{ (lb/ton)} \qquad (1)$$

where E is the emission factor; k, the particle size multiplier (dimensionless); U, the mean wind speed, m/sec (mph); and M, the material moisture content (%). The particle size multiplier, k, varies with aerodynamic particle diameter, as shown in the Table 2.

The equation retains the assigned quality rating if applied within the ranges of source conditions that were tested in developing the equation, as given in Table 3. Note that silt content is included in Table 3, even though silt content does not appear as a correction parameter in the equation. While it is reasonable to expect that silt content and emission factors are interrelated, no significant correlation between the two was found during the derivation of the equation, probably because

Table 1. Typical Silt and Moisture Contents of Materials at Various Industries[a]

			Silt Content (%)			Moisture Content (%)		
Industry	No. of Facilities	Material	No. of Samples	Range	Mean	No. of Samples	Range	Mean
Iron and steel production	9	Pellet ore	13	1.3–13	4.3	11	0.64–4.0	2.2
		Lump ore	9	2.8–19	9.5	6	1.6–8.0	5.4
		Coal	12	2.0–7.7	4.6	11	2.8–11	4.8
		Slag	3	3.0–7.3	5.3	3	0.25–2.0	0.92
		Flue dust	3	2.7–23	13	1	—	7
		Coke breeze	2	4.4–5.4	4.9	2	6.4–9.2	7.8
		Blended ore	1	—	15	1	—	6.6
		Sinter	1	—	0.7	0	—	—
		Limestone	3	0.4–2.3	1.0	2	ND	0.2
Stone quarrying and processing	2	Crushed limestone	2	1.3–1.9	1.6	2	0.3–1.1	0.7
		Various limestone products	8	0.8–14	3.9	8	0.46–5.0	2.1
Taconite mining and processing	1	Pellets	9	2.2–5.4	3.4	7	0.05–2.0	0.9
		Tailings	2	ND	11	1	—	0.4
Western surface coal mining	4	Coal	15	3.4–16	6.2	7	2.8–20	6.9
		Overburden	15	3.8–15	7.5	0	—	—
		Exposed ground	3	5.1–21	15	3	0.8–6.4	3.4
Coal-fired power plant	1	Coal (as received)	60	0.6–4.8	2.2	59	2.7–7.4	4.5
Municipal solid waste landfills	4	Sand	1	—	2.6	1	—	7.4
		Slag	2	3.0–4.7	3.8	2	2.3–4.9	3.6
		Cover	5	5.0–16	9.0	5	8.9–16	12
		Clay/dirt mix	1	—	9.2	1	—	14
		Clay	2	4.5–7.4	6.0	2	8.9–11	10
		Fly ash	4	78–81	80	4	26–29	27
		Misc. fill materials	1	—	12	1	—	11

[a]Refs. 1–10. ND, no data.

Table 2. Aerodynamic Particle Size Multiplier *k* for Eq. 1 (Taken from AP-42 13.2)

<30 μm	<15 μm	<10 μm	<5 μm	<2.5 μm
0.74	0.48	0.35	0.20	0.11

Table 3. Range of Source Conditions for Eq. 1

Silt Content	Moisture Content	Wind Speed	
(%)	(%)	(m/sec)	(mph)
0.44–19	0.25–4.8	0.6–6.7	1.3–15

most tests with high silt contents were conducted under lower winds, and vice versa. It is recommended that estimates from the equation be reduced one quality rating level, if the silt content used in a particular application falls outside the range given in Table 3.

REFERENCES

1. C. Cowherd, Jr., et al., *Development of Emission Factors for Fugitive Dust Sources*, EPA-450/3-74-037, U.S. Environmental Protection Agency, Research Triangle Park, NC, June 1974.
2. R. Bohn et al., *Fugitive Emissions from Integrated Iron and Steel Plants*, EPA-600/2-78-050, U.S. Environmental Protection Agency, Cincinnati, OH, March 1978.
3. C. Cowherd, Jr., et al., *Iron and Steel Plant Open Dust Source Fugitive Emission Evaluation*, EPA-600/2-79-103, U.S. Environmental Protection Agency, Cincinnati, OH, May 1979.
4. R. Bohn, Evaluation of Open Dust Sources in the Vicinity of Buffalo, New York, EPA Contract No. 68-02-2545, Midwest Research Institute, Kansas City, MO, March 1979.
5. C. Cowherd, Jr., and T. Cuscino, Jr., Fugitive Emissions Evaluation, MRI-4343-L, Midwest Research Institute, Kansas City, MO, February 1977.
6. T. Cuscino et al., *Taconite Mining Fugitive Emissions Study*, Minnesota Pollution Control Agency, Roseville, MN, June 1979.
7. K. Axetell and C. Cowherd, Jr., *Improved Emission Factors for Fugitive Dust from Western Surface Coal Mining Sources*, 2 Volumes, EPA Contract No. 68-03-2924, PEI, Inc., Kansas City, MO, July 1981.
8. E. T. Brookman et al., *Determination of Fugitive Coal Dust Emissions from Rotary Railcar Dumping*, 1956-L81-00, TRC, Hartford, CT, May 1984.
9. G. A. Jutze et al., *Investigation of Fugitive Dust Sources Emissions and Control*, EPA-450/3-74-036a, U.S. Environmental Protection Agency, Research Park, NC, June 1974.

BLAST FURNACE

Marek S. Klag

The blast furnace is at the present time the most efficient method of producing hot metal (molten iron). Depending on the size, modern furnaces can produce up to 10,000 tons of hot metal daily. It is possible to achieve uninterrupted furnace operation for more than 10 years. Blast furnace process control systems have been developed to optimize operational efficiency, increase productivity, and protect the environment.

PROCESS DESCRIPTION

A blast furnace as shown in Figure 1 consists of a refractory-lined steel shaft in which the charge is fed into the top through a gas seal. Air heated from 871°C to 1100°C is blown through tuyeres into the lower part of the furnace. The combustion of coke provides the carbon monoxide (CO) to reduce the iron oxides to iron and provides additional heat to melt the iron and impurities. Auxiliary fuels such as coal, oil, natural gas, or tar may also be injected through the tuyeres. As the burden moves downward through the furnace, it is heated by the countercurrent upward flow of gases that exit at the top of the furnace. Molten iron and slag accumulate in the hearth of the furnace and a taphole is drilled into the hearth to drain the slag and iron accumulation into a trough. The trough is equipped with a skimmer and dam at the outlet end, allowing the iron to be drawn off separately from the slag. A typical charge required to produce a ton of hot metal consists of 1.55 tons of iron ore, 0.5 ton of coke, and 0.05–0.15 ton of limestone and dolomite. Higher productivity has been achieved through the use of high-purity iron units, sized raw materials, higher blast temperatures, blast humidity control, higher top pressures, and fuel and oxygen injection.

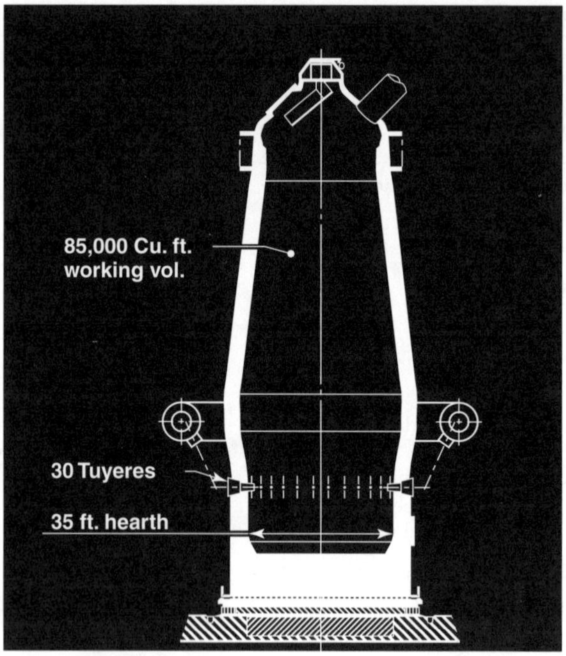

Figure 1. Modern Blast Furnace.

Pollution Control Equipment

Pollution control in the iron-making industry focuses on primary and secondary emissions. The blast furnace top gas and emissions problems associated with the iron production activities have been essentially solved. The difficulty in capturing dust often requires the extraction of a large volume of dilution air; particularly for secondary processes. Emissions contained in blast furnace top gas are generally handled through wet-collection systems. These include dust catchers, venturi scrubbers, and precipitators (Figure 2). The collection efficiency of wet systems can be related to the energy losses

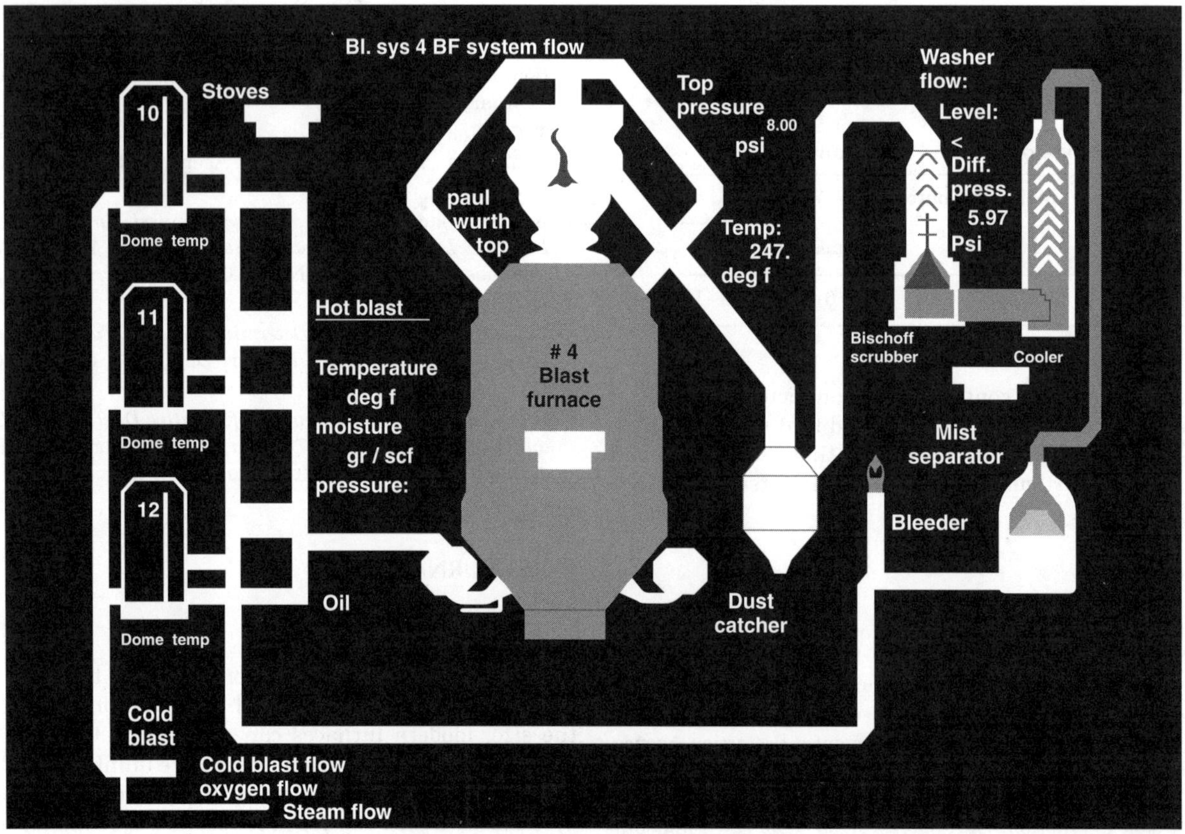

Figure 2. Blast Furnace System Flow.

resulting from the pressure drop of gas through the units. Disadvantages include high energy requirements and large volumes of scrubbing liquid to be treated. Collection efficiency is in the range of 99% for particulate larger than 1 μm.[1] Secondary emission collection (e.g., cast-house operation) is accomplished using a baghouse. The dirty gases are forced through filter bags that trap the particulate. Dust particulates are separated from the bags by gravity. The material from which the bags are woven depends on particle size, the gas temperature, acidity, alkalinity, and the desired life of the filter element. The collection efficiency for fabric filtration is greater than 99%. Reference 2 shows particulate emission factors and particulate size for iron and steel mills.

AIR EMISSIONS CHARACTERIZATION

Nature of Emissions

Top Gas

The top gas contains CO and particulate levels. The blast furnace gas has a relatively low heating value in the range of 3.7 MJ/M^3.[3,4] The top gas can be used as a fuel in steel plants. Submicrometer particles are formed when vapor from low-boiling-point materials such as sodium (Na), potassium (K), cerium (Ce), and zinc (Zn) compounds vaporize in the bottom of the furnace, where temperatures rise above 1650°C, and condense in the upper section of the furnace. The gas leaving the furnace contains trace quantities of hydrogen sulfide (H_2S) and sulfur dioxide (SO_2) due to the efficiency of the iron oxide and flux in removing sulfurous gases (see Table 1).

Tapping (Casting) Liquid Iron

Every 45–90 min, the taphole in the hearth of the furnace is drilled open. The slag metal mixture exits at a temperature of 1300–1500°C. The slag floats to the top of the trough and a dam system separates the hot metal and slag into two separate streams. The hot metal is drained through a runner system into a torpedo car mounted on railroad wheels. Hot metal cars typically hold 80–300 tons of liquid iron for delivery to oxygen steel-making facilities. The slag is drained into a pit or slag pots by slag runners. Casting emissions generated in the cast-house are shown in Table 2.

Iron oxides emissions are generated during casting either by the direct vaporization of compounds or the partial pressure of CO bursting bubbles at the metal atmosphere interface. The ejection of hot metal into the air quickly oxidizes to produce iron oxide fume (see Figure 3). If vaporization were the major mechanism, then enrichment of the manganese (Mn) level in the collected dust should occur. Since the Mn in the dust collected by the blast furnace cast-house baghouse is only 1.6 times the Mn analysis in the hot metal, very little evaporation takes place.[5] Droplet ejection must be the major mechanism for fume generation.

Further discussion on the bursting-bubble phenomenon can be found in Refs. 4 and 5. The particulate size analysis collected from a total building ventilation system is shown in Table 2. Sulfur dioxide and CO, collected at a local hood mounted over a tilting iron runner, amounted to 0.018 and 0.025 kg per ton, respectively. Particulate collected at the same location amounted to 0.14 kg per ton. Total particulate collected by a local collection system was reported at 0.65 kg per ton with 0.14 kg per ton less than 15 μm[6]

Table 1. Particulate, Energy Content, Gas Flow, Temperature, and Analysis of Blast Furnace Top Gas

		Chemical Analysis of Top Gas (wt %)	
Temperature	175–250°C	Fe	47
Gas quantity	1000–2000 nm^3/ton of iron	C	12
Gas analysis	CO: 23–40%	SiO_2	12
	CO_2: 15–22%	Ca	2.9
	H_2: 1.5–6.0%	Mg	1.2
	N_2: remainder	Mn	0.38
Energy content	3.7 MJ/m^3		
Dust content	Up to 30 g/nm^3		

Table 2. Tapping Emissions and Particle Size Distribution

Tapping Emission (wt%)		Particulate Size Distribution (μm[a])		Emission (kg/ton of Hot Metal)	
Fe	51	0.25	0.026	Taphole	0.06–0.11
C	33	0.50	0.083	Runner	0.04
SiO_2	4.5	0.75	0.110	Ladle spot	0.06–0.11
CaO	4.0	1.0	0.177	Slag	0.01
MgO	1.5				
MnO	0.5				

[a] Blast furnace cast-house total building ventilation.

Slag Processing

A common practice is to drain the molten slag, after separation from the hot metal, into a pit or pot adjacent to the furnace. The slag emissions contribute to the dust generated by oxidizing sulfur in the slag to SO_2 during a casting. The light-colored and low-mass slag emissions are estimated to contribute approximately 10% of the total emissions. Studies have correlated slag composition and slag emissions.[7] There is a direct relationship between lime content, lime–silica ratio,

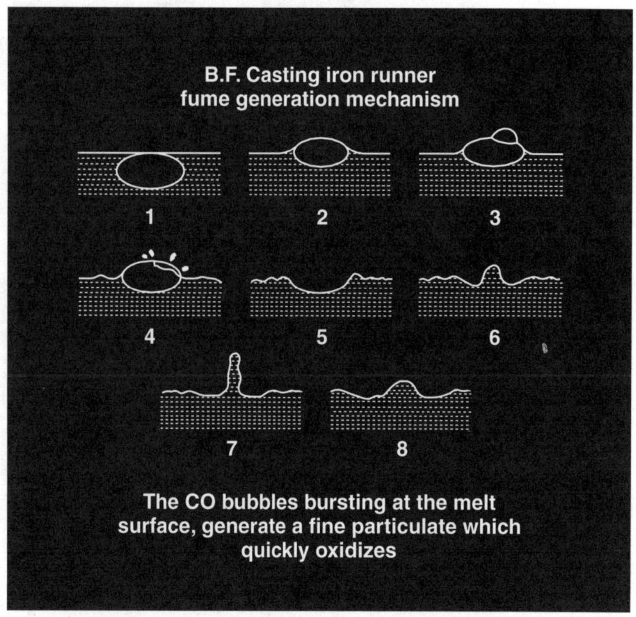

Figure 3. Proposed Mechanism of Fume Generation.

and slag basicity. An inverse relationship has been found with silica and potassium oxide content. The authors propose a mechanism that relates the behavior of potassium in the furnace to the slag emissions.

Additional emissions occur during slag processing. Water may be used to increase the cooling rate in the pit or produce granulated slag. The sulfur in the slag cast from the furnace is present as calcium sulfide (CaS).[8] The use of water to cool hot slag produces H_2S emissions. Hydrogen sulfide concentrations measured over a section of the slag pit being quenched peaked at 250 ppm H_2S by volume.[9] The emission factors shown in Table 3 are based on actual measurements.[10]

Hot-Metal Desulfurization

Blast furnaces can be operated to produce iron containing from 0.02 to 0.03% sulfur, which can be utilized directly in steel-making operations. Certain plants have increased blast furnace productivity and lowered operating costs by producing iron containing from 0.04% to 0.1% sulfur, which is externally desulfurized prior to use in steel-making by injecting calcium carbide (CaC_2), lime (CaO), magnesium (Mg), or soda-based reagents. Desulfurization station particulate emission characteristics are shown in Table 4. The particulate emission factor for a system injecting 5.2 kg (CaC_2) per ton of metal was 0.8 kg per ton.

Table 3. Hydrogen Sulfide Concentration at Source

Slag Processing	Average of 30-min Average (ppm)
Pit quenching	22
Expending (palletizing)	10

Table 4. Chemical Analysis of Calcium Carbide Desulfurization; Particulate Size Distribution

Desulfurization using CaC_2 wt%		Particulate Size Distribution (μm[a])
Fe	11	0.015
C	33	0.025
SiO_2	4	0.03
Ca	46	—
Mg	<1.0	—
Mn	<1.0	—

[a]Fraction smaller than.

FACTORS AFFECTING EMISSION

Top Gas

Most of the raw materials charged into a blast furnace are screened. Quality and size play the most decisive roles in influencing the gas distribution. Blast furnace operation is influenced by the reactions taking place:

- At the burden surface at the top of the furnace
- In the granular zone in the middle of the furnace
- In the cohesive zone in the lower shaft belly and bosh
- In the zone in front of the tuyeres
- In the hearth (see Figure 4)

These influences govern the counterflow of gas to solids or liquids. As the burden descends in the furnace, it is heated by the ascending reducing gases and is prereduced. The iron ores begin to soften and melt and the ore layers become impermeable to gas. The gas flows through the "coke channels" between the ore layers. Buildups or secretions may occur on

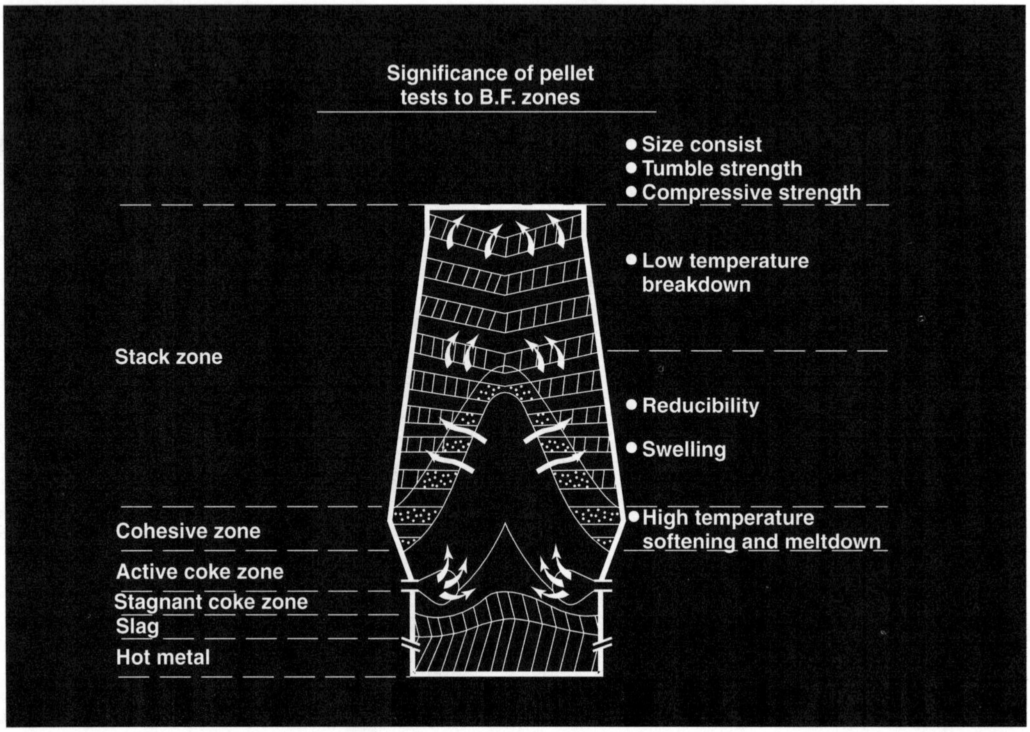

Figure 4. Significance of Pellets.

the walls of the furnace. This condition causes extremely slow, erratic, or stopped stock-line (raw materials) movement. When the burden movement exhibits a sudden large drop, there is a resulting increase in a furnace top pressure. The excess gas is vented through the safety release bleeder valve. Estimated emissions from slips can be found in Ref. 11 and 12. The shape of the cohesive zone is largely responsible for the behavior of the furnace and productivity (see Figure 5).

Tapping (Casting) Liquid Iron

The tapping temperature of hot metal is in the range of 1400°C. The hot metal is saturated with carbon. During furnace casting, turbulent conditions break up the slag layer, increasing oxygen transport, and reducing the temperature by exposing more metal surface to the air. With a drop in temperature, carbon is ejected from solution in the form of graphite flakes (kish). After slag separation, the iron becomes exposed to oxygen from the air, resulting in oxidation of molten iron (dust). One of the major problems with blast furnace cast-house fume emissions is the long length of the runners over which the fumes are generated. Trozzo et al.,[13] estimated iron runner emissions at 0.04 kg per ton and slag runner emissions at 0.01 kg per ton of hot metal. A number of factors increase emissions generated during casting:

- The turbulence at the trough, which is a function of the angle of the drill and the size of the hole (Figure 6)
- Hot-metal casting temperature and sulfur content
- Runner width-to-depth ratio

The blast furnace refractory material for runners, trough, and mud-gun clay may use organic coal-tar binders, closing the taphole or resulting in emissions during casting. After completing the cast, the liquid metal reservoir has been drained away. This will allow high-pressure gases inside the furnace to channel out through the taphole, generating large quantities of fume.

Slag Processing

There are two common blast furnace slag-handling processes:

- Discharge slag into a pit/pot
- Granulate/pelletize slag

The blast furnace slag discharge temperature is in the range of 1500°C. A typical cooling rate at 1.4 m below the surface is 2.8°C/h.[9] The outer layer of slag cools down between casts, but inner layers can remain molten. Therefore, the slag must be water-cooled to allow the pit to be excavated safely. This practice of water quenching generates high levels of H_2S (Table 3).

The method available for granulation of liquid slag utilizes the centrifugal force imparted by a rotating drum or plate and water jets. Hydrogen sulfide gas and fibers are generated.

Hot Metal Desulfurization

Sulfur is one of the most notorious impurities in steel. The hot metal may be desulfurized at the blast furnace, at a separate desulfurization facility, or at the steel-making operation. A common method of desulfurization is to inject a desulfurization reagent through a lance into the hot metal using a carrier gas (Figure 7). The bubbles from the carrier gas and the reacting agent cause rapid stirring, breaking the surface and creating emissions. The turbulence might also cause molten metals to be ejected from the ladle, creating excess emissions. Insufficient metal level in the ladle, which limits the required lance submergence, may also result in increased emissions.

The constituents of this reagent are typically CaC_2, which desulfurizes the hot metal, and calcium carbonate ($CaCO_3$), which provides the dioxide (CO_2) gas required to mix the hot metal and the desulfurizing powder. Another common injection material is salt-coated Mg granules containing approximately 85% Mg. Approximately 80% of the injected reagent is captured by the slag and the remaining salt escapes as a fume.[14] During desulfurization, Mg salt goes throughout the steps of boiling

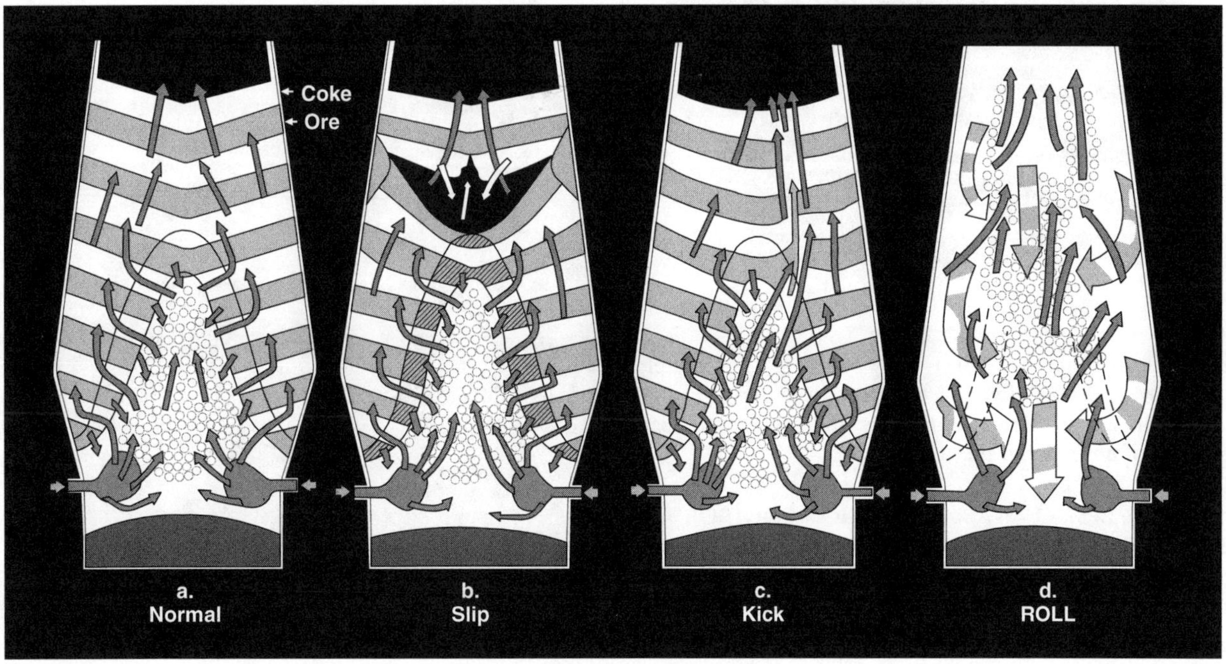

Figure 5. Blast Furnace Zones.

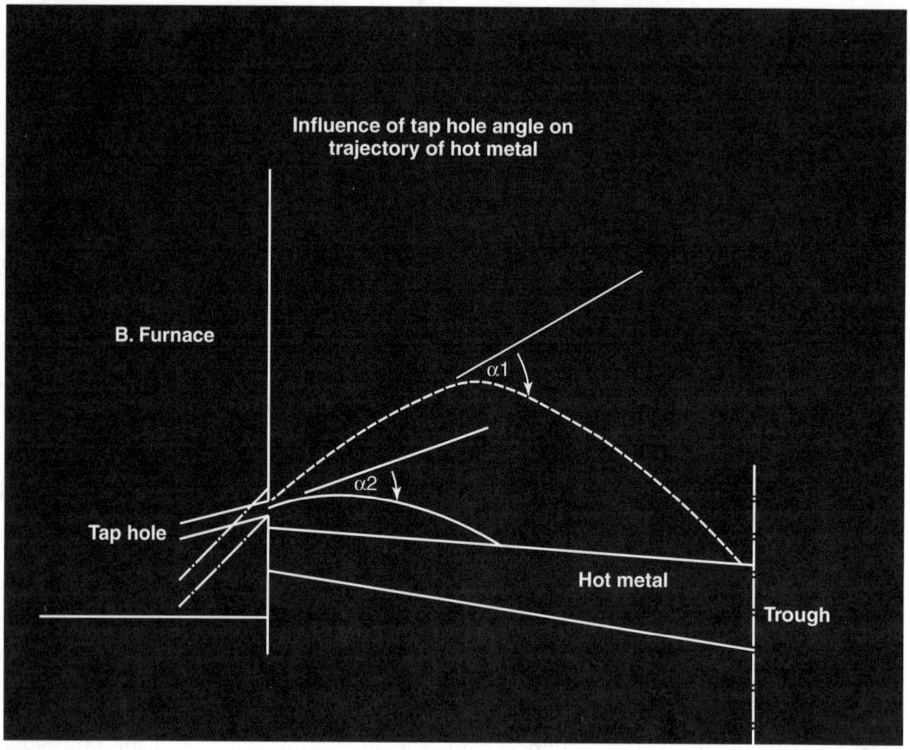

Figure 6. Influence of Taphole Angle on Trajectory of Hot Metal.

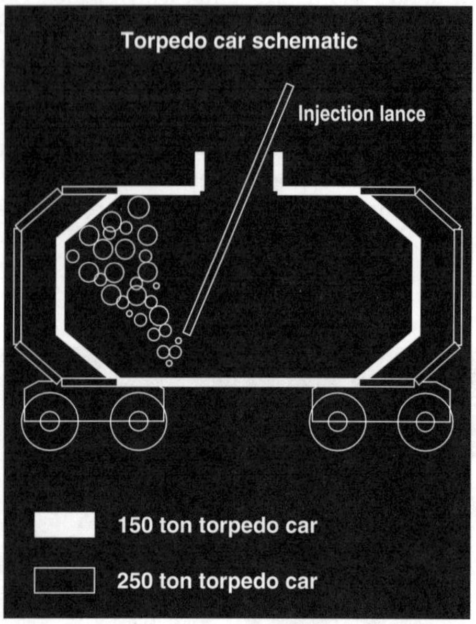

Figure 7. Torpedo Car Schematic.

dissolution. The Mg vapor is extremely volatile. The particle size of salt-coated material would be that of a vaporized fume due to the evaporation of the alkali salts and the evaporation and oxidation of nonreacted metallic Mg.

BLAST FURNACE BEST AVAILABLE POLLUTION CONTROL TECHNOLOGY

Top Gas

The productivity of a furnace is governed by the aerodynamics of gas flow from the tuyere to the top of the furnace. Due to the energy content, blast furnace top gas is used as fuel in blast furnaces, stoves, reheat furnaces, boilers, and so on. The practice of achieving low-maintenance burners and high-efficiency gas cleaning is widespread.

The gas permeability in the granular zone and the lower part of the furnace are very important. A modern charging installation (Paul Worth top) makes it possible to distribute coke and burden materials in separate layers or in any mixtures on the surface, compared with two bell-top systems that have batching charges (Figure 8). The permeability in the burden zone depends on the quality, size of raw materials,

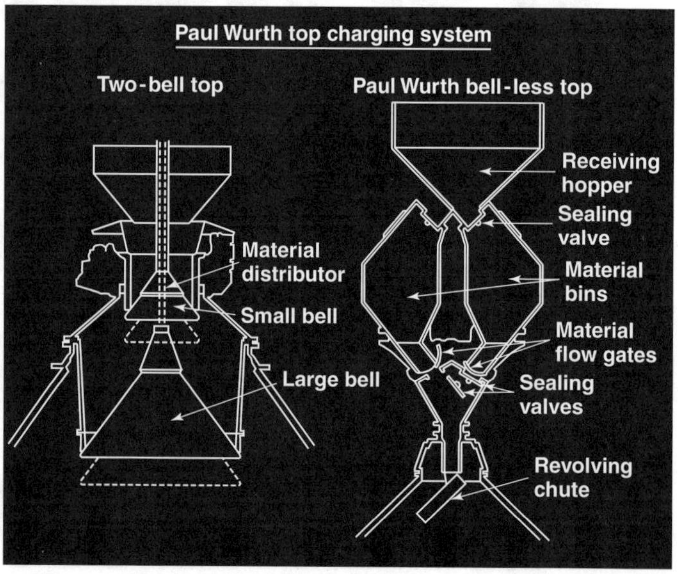

Figure 8. Paul Wurth Top Charging System.

their fines content, the reduction degradation of the iron-bearing materials, and the burden distribution. For large blast furnaces, the installation of an energy-recovery turbine can contribute to the works' electricity needs.[15] The installation of a gas expansion turbine, utilizing the excess pressure, therefore provides a useful method of recovering both pressure energy and some sensible heat energy from the gas (Figure 9).

Tapping (Casting) Hot Metal Liquid Iron

The modern blast furnace may be equipped with the following to control casting emissions:

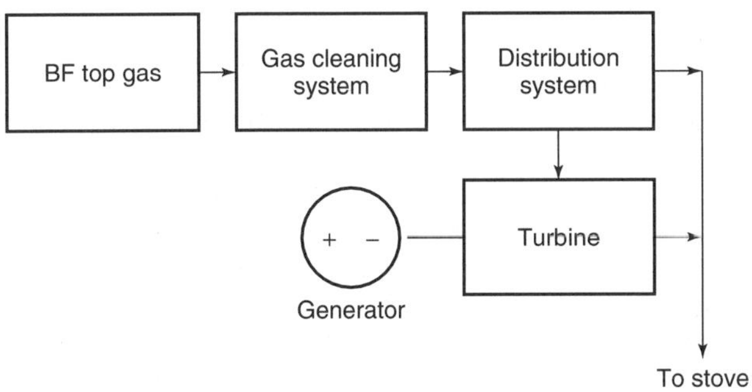

Figure 9. Blast Furnace Energy Recovery System.

- Local collections at the taphole, skimmer, and tilting runners
- Each hood with its own control system to adjust air flow independently
- Local hoods that require the evacuation rates indicated in Table 5

Tilting runners are used to reduce the runner length, and, therefore, to reduce surface iron oxide fume emissions. Dofasco's No. 4 blast furnace was originally fitted with runner length of over 60 ft with four ladle spots per cast-house. For production and environmental reasons, the runner length was reduced to 25 ft, with two ladle spots and tilting runners (Figure 10). The iron runner emissions were reduced from an estimated 0.04 to 0.017 kg per ton. The iron temperature losses were reduced from 27.2°C to 4.4°C.[16,17]

Table 5. Local Hoods Evacuation Rate[a]

Location	Rate (nm^3/min)
Taphole	3450
Dam skimmer	1100
Tilting iron runner	3450

[a] Dofasco No. 4 blast furnace cast-house local hoods rate.

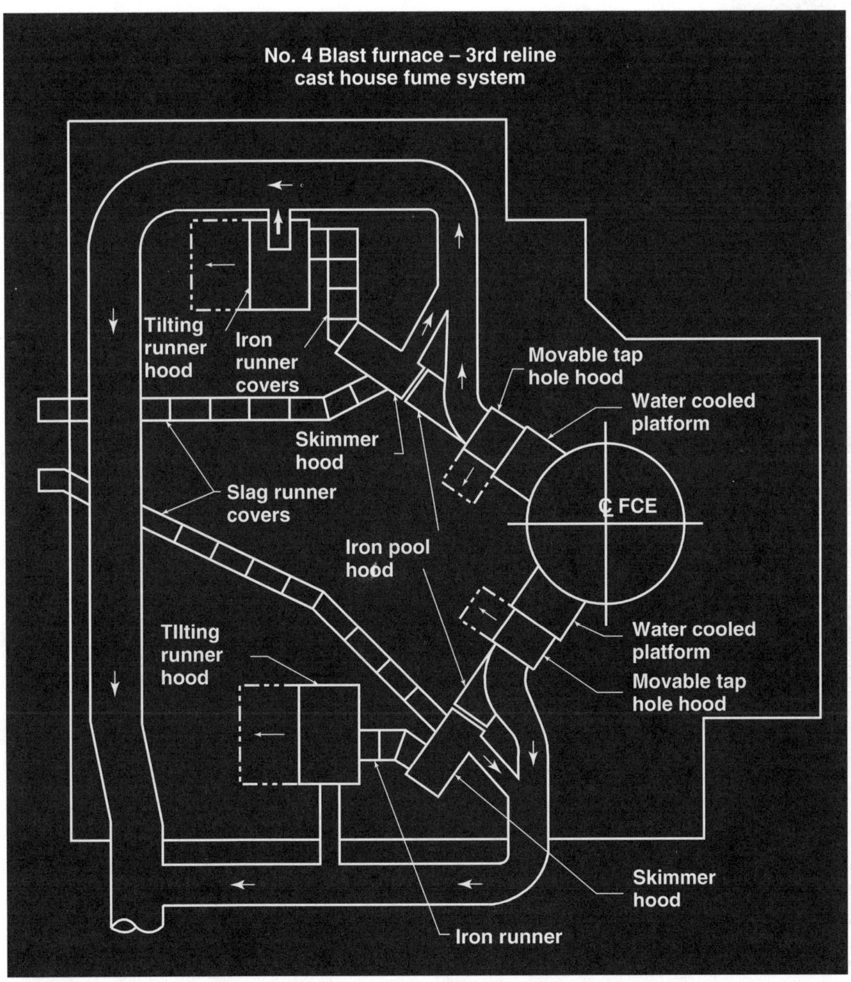

Figure 10. Casthouse Fume System.

Covering the runners is an effective way to reduce emissions (Figure 11). The evacuated runners generate emissions that are captured by the control device. A favored practice is tightly covered runners without evacuation. The use of covered runners also reduces the heat loss from iron in runners. The hot-metal temperature delivered to the steel making operation increases to 12.7°C if covers are used (see Table 6).[18] The emission temperature during casting does not usually exceed 90°C. Satisfactory gas cleaning installations for secondary emissions are baghouses.

The casting emissions can also be controlled using suppressant gases like methane (CH_4) and nitrogen (N_2). Certain North American companies such as USX and LTV have attained considerable success with gas suppression at the ladle spot, trough, and runner.[19] The principle of suppression is to eliminate oxygen from contact with liquid iron during casting. Figure 12 shows the arrangement for fume suppression during a trial period at Dofasco's blast furnaces.

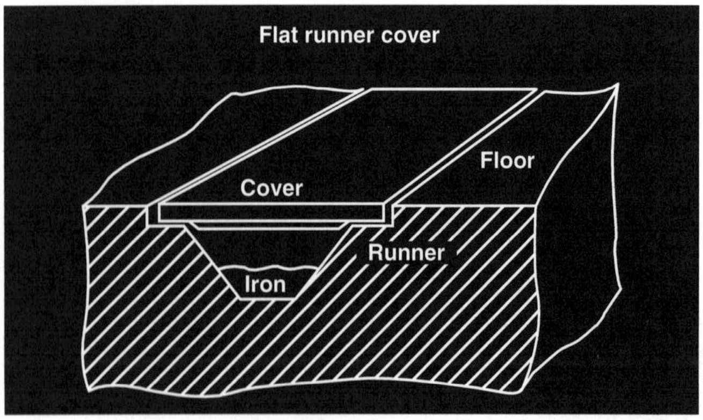

Figure 11. Flat Cover Runners.

Table 6. Hot Metal Temperature Loss with and without Runner Covers

	Sample Number	Average Temperature	Standard Deviation
Without covers	47	27.1°C	6.04
With covers	43	14.37°C	11.51

Slag Processing

Using water as a coolant for slag quenching results in H_2S emissions. A reduction in H_2S generation can be achieved by increasing the air cooling time as long as possible before water is used. Depending on the pit size and water cooling system, the process of slag quenching can allow effective control of H_2S emissions. Based on the adopted practice at Dofasco, up to 85% reduction of H_2S emissions can be anticipated. The pits with high surface area and shallow depth are favored.

A reduction in H_2S can also be attained by adding caustic or potassium permanganate ($KMnO_4$) to the quench water. A total reduction of 88% was attained by a combination of a 3-d cooling period before the quench as well as a 100-ppm addition of $KMnO_4$ to the quench water.[3] For the slag granulating process, a water-free system may be considered by individual plants (see Refs. 20 and 21).

Hot Metal Desulfurization

A collection volume of 1000 nm³/min for a hood located within 0.3 m of the ladle car will satisfactorily collect emissions from

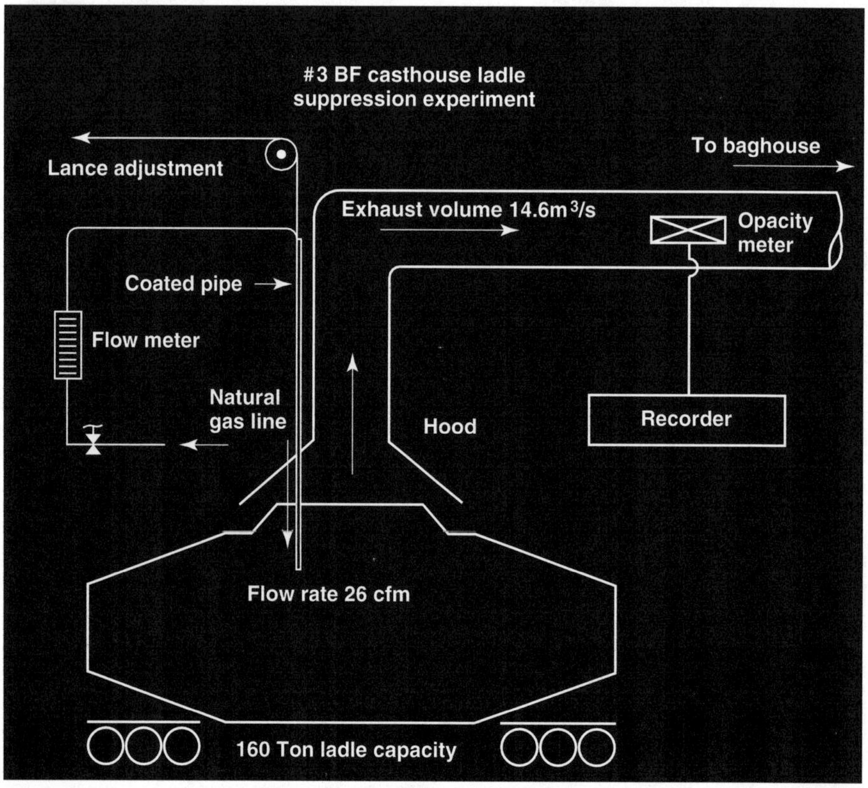

Figure 12. Supperssion Experiment.

the injection of 80 kg/min of a CaC_2–$CaCO_3$ mixture.[22] The average temperature of gases would be approximately 80°C and could peak much higher if a lance break occurred. Scrubbers, baghouses, and WESPs are satisfactory gas-cleaning devices. Provisions for the safe collection of unreacted CaC_2 or Mg reagents are required. Calcium carbide and salt-coated magnesium are the most prominent of the external treatment process agents. The treatment can reduce the sulfur level to less than 0.005%. The selection of the desulfurization techniques is governed by the characteristics and objective of each plant.[14]

REFERENCES

1. J. H. Seinfeld. *Air Pollution, Physical and Commercial Fundamentals*, McGraw-Hill, New York, 1975.
2. W. Davis, "Particulate Emission Factor for Iron and Steel Mills," Tables 7.5-1 and 7.5-2 of AP-42.
3. S. Calvert and H. M. Englund, *Handbook of Air Pollution Technology*, John Wiley & Sons, New York, 1984.
4. R. D. Pehlke et al., *BOF Steelmaking*, Theory Process Technology Division, Iron and Steel Society, AIME.
5. F. Goetz, "Mechanism of BOF Fume Formation," thesis for McMaster University, Hamilton, Ont., April 1984.
6. T. J. Maslany, "Blast Furnace Control Technology Update," Symposium of Iron and Steel Pollution Abatement Technology, Chicago, 1981.
7. A. W. Simon, and J. F. Kelly, *Characterization of Slag Emissions at a Blast Furnace Casthouse*, U.S. Steel Corp., Research Laboratory, Monroeville, PA.
8. R. A. Stoehr and J. P. Pezze, "Effect of Oxidizing and Reducing Conditions on the Reaction of Water with Sulfur Bearing Blast Furnace Slag," *J. APCA* **25**(11), 1975.
9. F. H. Remmus et al., "Control of H_2S Emissions during Slag Quenching," *J. APCA* **23**, October 1973.
10. M. Klag and S. Iaboni, "Determination and control of H_2S Emissions from Slag Processing Operations," Dofasco internal report, November 1987.
11. *Pollution Effects of Abnormal Operations in Iron and Steelmaking, Vol. III, Blast Furnace Iron Making Manual of Practice*, U.S. Environmental Protection Agency.
12. *Blast Furnace Slips and Accompanying Emissions as an Air Pollution Source*, U.S. Environmental Protection Agency.
13. D. L. Trozzo and C. F. Hoffmann, "Method for Determining Particulate Mass Emissions from Iron and Slag Runners," EPA Conference, June 1980.
14. *Blast Furnace Iron Making-Intensive Course*, Vols. II, III. McMaster University, Hamilton, Ont., May 1989.
15. D. Young, "A Project to Recover Energy from the Blast Furnace," *Steel Technol. Int.*, 1989.
16. "Minimizing Energy Costs in the Pollution Control System," *J. APCA*, 1985.
17. P. Spawn and R. Craig, "Status of casthouse control technology in the USA, Canada and West Germany in 1980," GCA Technology Division, Second Symposium on Iron and Steel Pollution Abatement Technology, Bedford, MA, November 1980.
18. A. M. McClure, *Hot Metal Temperature Loss with / without Runner Covers*, Dofasco Research, January 25, 1983.
19. "Molten Metal Fume Suppressions," U.S. Patent, No. 445,883, July 10, 1984.
20. M. Yoshinaga, K. Fujii, T. Shigematsu, et al., "Dry Granulation and Solidification of Molten Blast Furnace Slag," *Tetsu-ToHagane* **67** 917, 1981; (English version) Iron Steel Inst., *Jpn* **22**, 823–29, 1982.
21. S. J. Pickering, N. Hay, F. Roylance, et al., *New Process for Dry Granulation and Heat Recovery from Molten Blast Furnace Slag*, Institute of Metals, 1985.
22. C. Borgianni and A. Praitom, "Injection of Powdered Reagents into the Melt," *Steel Technol. Int'l*, 1988.

BASIC OXYGEN FURNACE SHOPS

S. S. Felton, E. Cocchiarella, and M. S. Greenfield

Basic oxygen steel-making was developed in Linz-Donawitz, Austria, in the 1950s and is a variation of the older Bessemer process. The basic oxygen process now accounts for most steel-making capacity worldwide and accounted for 58.4% of U.S. steel production in 1988.[1]

PROCESS DESCRIPTION

A basic oxygen process furnace (BOPF) is a large open-mouthed vessel lined with a basic refractory material.[2] The furnace (Figure 1) is mounted on trunnions to permit full rotation. A typical vessel can be 12–14 ft across and 20–30 ft high. Furnace capacities range from 15 to 400 tons. A BOPF receives a charge composed of approximately 30% scrap and 70% molten iron and converts it to molten steel. To do this, it utilizes a jet of high-purity oxygen at rates up to 30,000 cfm, which oxidizes the carbon and the silicon in the molten iron, removes these products, and provides heat for melting the scrap. After the oxygen blow is started, lime, usually in the form of pebble lime, is added to the top of the bath to provide a slag of the desired basicity. An entire heat cycle may take only 30 to 50 min.

Three types of furnaces are currently in use. The most common type is the top-blown furnace, in which oxygen is blown into the vessel through a water-cooled lance suspended above the mouth of the furnace. The second type of furnace is the bottom-blown Q-BOPF. In this furnace, oxygen is introduced into the vessel through tuyeres (openings) in the furnace bottom. The third type is the combined-blown K-OBM furnace where 30% of the oxygen is blown through the bottom.[3] Powdered lime, natural gas, nitrogen, and argon may also be injected through the bottom tuyeres.

Other process variations involve different techniques for hooding and combustion of the gases evolved from the basic oxygen process. Hoods can be open combustion, in which excess air is introduced in quantities from 10% to 300%, or they can be "suppressed" combustion, into which only 5% to 70% of the air theoretically required for combustion is allowed to infiltrate.[4] With open hoods, there is a gap between the hood and the furnace top into which air can be induced. With closed hoods, a movable skirt seals as tightly as feasible to the furnace top to discourage air inflow. Closed hoods have the advantages of reduced gas volume and reduced heat in the hood, reduced fume generation, and possible savings through increased metallic yield or offsetting credits for recovered carbon monoxide. Several partial (suppressed) combustion systems have also been developed independently and differ only slightly in the method of controlling infiltration and combustion levels at the interface of the vessel mouth and hood. These include the OG, IRSID-CAFL, Baumco, and Krupp systems.[4]

There are several ancillary operations associated with the basic oxygen process of making steel. The first is the handling

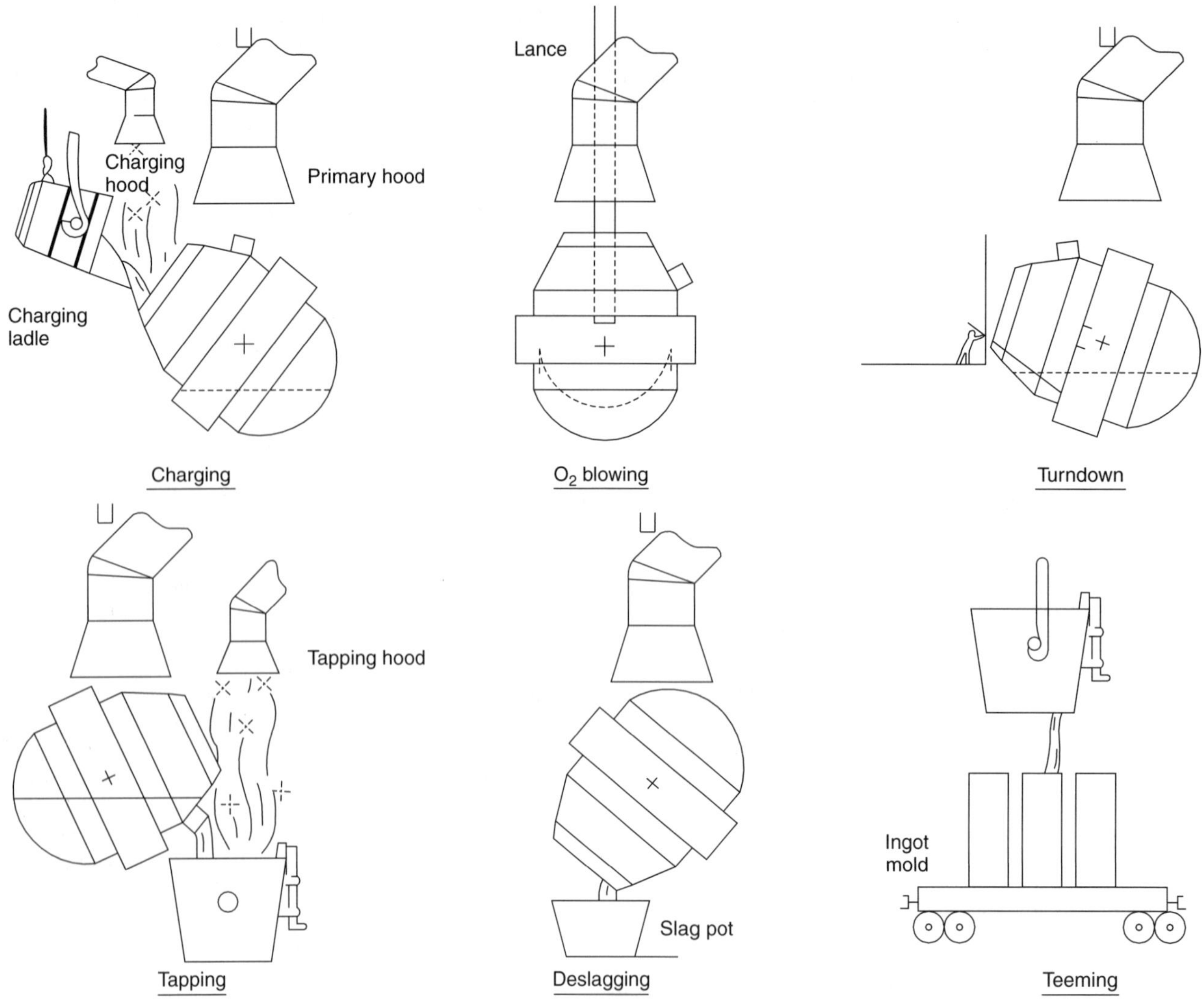

Figure 1. Steps for Making Steel by the Basic Oxygen Process (from Ref. 2).

of scrap. The next is the transfer of molten iron from the torpedo car to the charging ladle and from the charging ladle to the furnace itself. The handling of molten iron may include mechanically skimming slag from the top of the bath of iron. After the blowing period, there are emissions from tapping the furnace into teeming ladles. A fourth operation is the teeming of the finished steel into ingot molds or into continuous casting machines. Finally, there is the handling and treating of molten slag. After the slag solidifies, it is conveyed in trucks, railroad cars, or special slag pot carriers to recycling facilities. Screening and magnetic separation take place, providing recyclable or salable materials.

AIR EMISSIONS CHARACTERIZATION

The operations in the BOPF shop are directly responsible for two principal types of air pollution. The first is the direct result of the steel-making process itself (oxygen blow) and consists of dense emissions of fumes from the mouth of the basic oxygen vessel. These are commonly referred to as "primary" emissions. The fumes are mostly metallic oxides that result from the reaction between the jet of oxygen and the molten bath. Also included in these fumes are particles of slag. Carbon monoxide produced by the reaction of both carbon and oxygen is emitted as well.

The gases that leave the mouth of the furnace vessel, in addition to being dusty, are extremely hot. In a closed-hood system, temperatures are in the neighborhood of 1650°C (3000°F). In an open-hood system, carbon monoxide combustion takes place at the entrance of the hood, raising the temperature perhaps another 540°C (1000°F). Before the gases can be cleaned of their particulate matter, they must be cooled.

The second type of air pollution source comprises a variety of operations that are generally classified as "fugitive" or "secondary" emissions. A description of these sources follows.

- Reladling or "hot metal transfer" of molten iron from the torpedo railroad car to the charging ladle is accompanied by emissions of kish, a flat particle usually composed of a mixture of fine iron oxide particulates together with larger graphite particles.
- Desulfurizing molten iron may be accomplished with various reagents, such as calcium carbide, soda ash, lime, and magnesium. Injection of the reagents into the molten iron is done pneumatically, with either dry air or nitrogen.

Desulfurizing may take place at various locations within the iron- and steel-making facility; however, if the location is the BOPF shop, then it is often carried out at the reladling station to take advantage of the fume collection system at that location.

- Skimming slag from the ladle of molten iron keeps this source of high sulfur out of the steel-making process. Skimming results in the emissions of kish particles described above.
- Charging scrap and molten iron into the BOPF vessel can result in a large flame and a dense cloud of emissions. Emissions from the charging of scrap are particularly severe if the scrap is dirty or contaminated or has been preheated.
- Tapping the molten steel from the BOPF vessel into the ladle results in iron oxide fumes. The quantity of fumes is substantially increased by adding alloying materials, such as silicon and manganese, into the ladle. The pouring rate and tilt must be carefully controlled to ensure that local hooding captures most of the emissions.
- Turndown of the vessel for the purpose of taking samples or for pouring out the slag results in emissions. These emissions are particularly dense in the case of the Q-BOPF and K-OBM, because when the vessel is turned down, a flow of nitrogen must be maintained in the tuyeres in the bottom of the vessel in order to keep out the molten metal and slag.
- Slag handling may consist of transporting the ladle of molten slag from the shop to a remote dump area or dumping the molten slag on the ground at the end of the shop and allowing it to cool there. The dumping of slag and its subsequent removal by bulldozer is a very dusty operation that is generally uncontrolled.
- Teeming of steel from the ladle to the ingot mold or continuous caster results in emissions that are normally uncontrolled. With teeming into a continuous caster, emissions are usually much lower than with teeming into ingots because of the sophisticated shrouding techniques employed and the reduced air contact.
- Flux material handling is accomplished with a sophisticated system of receiving hoppers for accepting deliveries from trucks, railroad cars, a belt conveyor, large overhead bins, weigh hoppers, or special feeders. Particulate emissions result when the bulk materials fall into the vessel, as well as at all transfer points in the conveyor and bin system.
- Skull burning and ladle maintenance relate to the fact that the molten steel that remains in the ladle after teeming may cool and solidify between successive uses. After accumulating for some time, these skulls may interfere with proper ladle operation. To prevent this, they are burned out with oxygen lances. This ladle-lancing procedure results in the emission of iron oxide fumes. Ladles must also be relined at regular intervals to protect their steel shells. The ladles are turned upside down to dump loose material onto the shop floor. This generates fugitive dust.

As previously mentioned, particulate emissions generated from BOPFs are usually classified as either primary or secondary in nature and are produced mainly by the condensation of vaporized metal oxides and the coagulation of these particles to form agglomerates.

Generally, more than 50% of the primary particulate emissions generated during the oxygen blow phase is less than 1 μm in physical diameter. Particulates from bottom-blown BOPFs (Q-BOPFs) are smaller and generally estimated to have a mean diameter of about 0.1 μm.

A significant change in particle-size distribution appears to occur when BOPF primary emissions are collected in closed-hooded gas collection systems as compared with open-hooded gas collection systems. The closed-hooded exhaust gas particles are larger than those typically obtained from an open-hooded furnace.

The particulate emission rate in the basic oxygen process depends on several factors, including oxygen blow rate, carbon content of iron, percentage of scrap charged, quality of scrap charged, rate of additions, operational practices, and the condition of the refractory lining of the vessel. During the production cycle, the gas evolution rate and gas temperature vary considerably. Because of the resultant variations in the concentration of particulate matter and gas temperature and volume in the inlet gas stream, emissions are greater at the beginning of the blowing period than during the remainder of the oxygen blow and the rest of the cycle. About 30 pounds of particulates are produced per ton of raw steel during the entire oxygen blow segment of the BOPF cycle.

The sources of secondary emissions within the BOPF shop are hot-metal transfer, desulfurization, skimming, charging, turndown, tapping, deslagging, teeming, ladle maintenance, flux handling, and slag handling. AP-42 emission factors, particle-size distribution, and trace metals analysis have been developed for most steps within the BOPF cycle. This information is summarized in Tables 1–3.[5]

AIR POLLUTION CONTROL MEASURES

Just as there are primary and secondary sources of particulate emissions from BOPFs, control systems are also designated as either primary or secondary.

BOPF Primary Emission Collection Systems

Two types of primary emission collection systems are in common use. For open-hood systems, the gases that evolve during the oxidation process and the combustion of carbon monoxide are captured by the hood and enter a hood cooling section, where heat is extracted. The gas then passes through a conditioning chamber, where it is cooled to the required temperature for the gas-cleaning equipment. The gas-cleaning system usually consists of a precipitator or wet scrubber, fans, dust-handling equipment, and a stack for carrying away the cleaned gases. For closed-hood systems, the gases are collected in an uncombusted state, their volume is reduced as compared with those in the open hood, and the yield of the process is increased. Because the gases remain combustible in the closed-hood system, gas cleaning is generally performed with a wet scrubber, since precipitators present a potential source of explosions. The cleaned gas is then vented to a stack where any combustible gases are flared.

Although both the closed- and open-hood systems are capable of achieving 99+% particulate removal efficiencies, closed systems are becoming more popular for several reasons. First, they clean only the process gases; therefore, a lesser

Table 1. Particulate Emission Factors for Basic Oxygen Furnace Shops

Source	Units	Emission Factors		Emission Factor Rating	Size Data
Basic oxygen furnaces					
BOF melting and refining	kg/Mg (1b/ton) steel				
Uncontrolled		14.25	(28.05)	B	
Controlled by open hood vented to:					
ESP		0.065	(0.13)	A	
scrubber		0.045	(0.09)	B	
Controlled by closed hood vented to:					
scrubber		0.0034	(0.0068)	A	Yes
BOF charging	kg/Mg (1b/ton) hot metal				
At source		0.3	(0.6)	D	
At building monitor		0.071	(0.142)	B	
Controlled by baghouse		0.0003	(0.0006)	D	Yes
BOF tapping	kg/Mg (1b/ton) steel				
At source		0.46	(0.92)	D	
At building monitor		0.145	(0.29)	B	
Controlled by baghouse		0.0013	(0.0026)	D	Yes
BOF metal transfer	kg/Mg (1b/ton) hot metal				
At source		0.095	(0.19)	A	
At building monitor		0.028	(0.056)	B	
BOF monitor (all sources)	kg/Mg (1b/ton) steel	0.25	(0.5)	B	
Q-BOP melting and refining	kg/Mg (1b/ton) steel				
Controlled by scrubber		0.028	(0.056)	B	Yes
Hot metal desulfurization emissions	kg/Mg (1b/ton hot metal)				
Uncontrolled		0.55	(1.09)	D	Yes
Controlled by baghouse		0.0045	(0.009)	D	Yes

Source: Ref. 5.

quantity of gas requires cleaning and equipment can be smaller. Process off-gas temperatures are somewhat lower and, so, are not as damaging to the control system components. Also, less water is required for quenching and gas scrubbing.

There is a much higher percentage of larger particles in the closed-system exhaust gas. This is advantageous in that it is easier and less expensive to remove larger particles than fine ones. Thus the closed-hood system also achieves a lower mass emission rate than the open-hood system. Electricity consumption in a closed system is typically about 60% of that required in an open system. With the presence of flammable carbon monoxide in the closed hood off-gas, there is a potential energy recovery for use within or outside the plant.

The amount of gases exhausted from the furnace varies according to the type of fume collection system. An open-hood system with an electrostatic precipitator (ESP) utilizes the greatest volume of gas, approximately 2000 scfm per ton of steel. A supplementary benefit of the high- exhaust volume is that it facilitates the capture of emissions from the mouth of the vessel when it is tilted away from the hood at various steps in the BOPF cycle.

An open-hood system with a wet scrubber generally requires a much lower flow of gases than the ESP, with the amount usually in the vicinity of 900 scfm per ton of steel. The reduced volume results from the need to conserve energy in a scrubber system with a pressure drop in the range of 50 to 70 in. of water.

Most closed-hood systems operate in conjunction with a wet scrubber in order to reduce the hazard of igniting the carbon monoxide–laden exhaust gas. The volumetric flow through this closed wet system is considerably lower than in the above two systems at approximately 360 scfm per ton of steel.

Early performance data on the effectiveness of these two primary control systems rank them in terms of particulate control from lowest emission rate to highest as follows: the closed hood with a scrubber, the open hood with a scrubber, and the open hood with a precipitator.

BOPF Secondary Emission Collection Systems

Steps in the BOPF cycle that require the vessel to be tilted out from under the hood are scrap charging, hot-metal charging, sampling, tapping, and deslagging.

A large part of these secondary furnace emissions is captured by one or more of the following special techniques.

- Use of furnace enclosures
- Use of local hoods
- Building evacuation (full or partial)

A summary of use and success to date is presented in the following.

Furnace Enclosures

Furnace enclosures may partially (on at least two sides) or fully (on four sides plus the top) enclose a furnace vessel. Most of the BOPFs brought on-line in the United States since 1973 have been enclosed. Partial enclosures can shield the BOPF from most drafts, thus permitting hoods within or adjacent to the enclosure to be more effective at lower airflow rates.

Table 2. Size-Specific Emission Factors for Basic Oxygen Furnaces

Source	Emission Factor Rating, A–E	Particle Size (μm)	Cumulative Mass % Less Than Stated Size	Cumulative Mass Emission Factor kg/Mg Alloy	(1b/ton Alloy)
Basic oxygen furnaces					
BOF melting and refining controlled by closed hood and vented to:					
Scrubber	C	0.5	34	0.0012	(0.0023)
		1.0	55	0.0019	(0.0037)
		2.5	65	0.0022	(0.0044)
		5.0	66	0.0022	(0.0045)
		10	67	0.0023	(0.0046)
		15	72	0.0024	(0.0049)
		*	100	0.0034	(0.0068)
BOF charging at source	E	0.5	8	0.02	(0.05)
		1.0	12	0.04	(0.07)
		2.5	22	0.07	(0.13)
		5.0	35	0.10	(0.21)
		10	46	0.14	(0.28)
		15	56	0.17	(0.34)
		*	100	0.3	(0.6)
BOF charging controlled by baghouse	D	0.5	3	7.5×10^{-6}	(1.5×10^{-5})
		1.0	10	3.0×10^{-5}	(6.0×10^{-5})
		2.5	22	66×10^{-5}	(0.0001)
		5.0	31	9.3×10^{-5}	(0.0002)
		10	45	0.0001	(0.0003)
		15	60	0.0002	(0.0004)
		*	100	0.0003	(0.0006)
BOF tapping at source	E	0.5	N/A	N/A	N/A
		1.0	11	0.05	(0.10)
		2.5	37	0.17	(0.34)
		5.0	43	0.20	(0.40)
		10	45	0.21	(0.41)
		15	50	0.23	(0.46)
		*	100	0.46	(0.92)
BOF tapping Controlled by baghouse	D	0.5	4	5.210^{-5}	(0.0001)
		1.0	7	0.0001	(0.0002)
		2.5	16	0.0002	(0.0004)
		5.0	22	0.0003	(0.0006)
		10	30	0.0004	(0.0008)
		15	40	0.0005	(0.0010)
		*	100	0.0013	(0.0026)
Q-BOP melting and refining Controlled by scrubber	D	0.5	45	0.013	(0.025)
		1.0	52	0.015	(0.029)
		2.5	56	0.016	(0.031)
		5.0	58	0.016	(0.032)
		10	68	0.019	(0.038)
		15	85	0.024	(0.048)
		*	100	0.028	(0.056)
Hot metal desulfurization Controlled baghouse	D	0.5	8	0.0004	(0.0007)
		1.0	18	0.0008	(0.0016)
		2.5	42	0.0019	(0.0038)
		5.0	62	0.0028	(0.0056)
		10	74	0.0033	(0.0067)
		15	78	0.0035	(0.0070)
		*	100	0.0045	(0.009)
Uncontrolled	E	0.5	N/A		
		1.0	2	0.01	(0.02)
		2.5	11	0.06	(0.12)
		5.0	19	0.10	(0.22)
		10	19	0.10	(0.22)
		15	21	0.12	(0.23)
		*	100	0.55	(1.09)

Source: Ref. 5.

Table 3. Selected Metals Analysis in BOF and Q-BOP Fume

	Concentration in Gas (mg/nm^3)	
Element	BOF	Q-BOP
Aluminum	MC[a]	0.43
Antimony	0.006	<0.001
Arsenic	≤0.05	<0.02
Barium	>0.11	0.02
Bismuth	>0.11	0.02
Cadmium	0.077	0.002
Calcium	MC[a]	64
Chromium	0.84	0.26
Copper	0.18	0.1
Iron	>2.7	85.3
Lead	>0.02	0.41
Magnesium	>0.55	2.3
Manganese	>0.11	3.8
Mercury	0.0008	>0.0031
Nickel	0.31	0.18
Phosphorus	MC[a]	0.53
Selenium	0.087	<0.033
Silicon	>1.1	4.2
Strontium	0.016	<0.056
Sulfur	>0.06	7.9
Zinc	MC[a]	0.14

[a]MC, major component
Source: Ref. 5.

A partial enclosure is less expensive, easier to retrofit, and less likely to impede furnace operations. However, the trend in recent years has been toward "total" enclosure. Enclosures are capable of controlling other furnace emissions, such as puffing, turndown, hot-metal charging, and skimming.

Local Hoods

The use of localized hoods has become a popular method for capturing particulates from scattered sources throughout the BOPF plant; however, they present design problems as a result of complicated cross-drafts within the building, or they may get in the way of crane operations. Local hoods exhausting to a secondary system make up only a part of the secondary emission control system.

Building Evacuation

Another approach to capturing secondary emissions is to evacuate all air from within the building; however, this can be quite costly and has several disadvantages.

BOPF Secondary Gas Cleaning Systems

Whereas high-performance wet scrubbers and ESPs have traditionally been used for primary furnace emissions, baghouses are typically used for secondary furnace emissions. Sometimes, through the use of enclosures, secondary emissions are collected through the primary gas-cleaning system.

Additional BOPF Gas Cleaning

Baghouses also have been traditionally used to control emissions from hot-metal transfer, desulfurization, slag skimming, and flux material-handling operations. Occasionally, mechanical collectors have been used for these applications, but with limited success.

REFERENCES

1. "World Steel Statistics—1989," *Association of Iron and Steel Engineers Year Book—1989*, 416 (1989).
2. J. H. Lucas, *An Evaluation of Basic Oxygen Process Furnace Air Pollution Control Technology*, Mellon Institute, April 1984.
3. Edward Cocchiarella, DOFASCO, Inc., written communication to S. S. Felton, Armco Steel Co., October 10, 1990.
4. B. A. Steiner, "Air Pollution Control in the Iron and Steel Industry," *Intl. Metals Rev.* **21**, 171, 1976.
5. *Compilation of Air Pollutant Emission Factors*, AP-42, U.S. Environmental Protection Agency, October 1986.

BIBLIOGRAPHY

1. A. G. Nicola, "Fugitive Emission Control in the Steel Industry," *Iron Steel Eng.* **53**(7), 25, 1976.
2. J. Pearce, "Q-BOP Steelmaking Developments," *Iron Steel Eng.* **52**(2), 29, 1975.
3. C. M. Parker, "BOP Air Cleaning Experiences," pp. 66–100, presented at Annual Meeting of the Air Pollution Control Association, San Francisco, CA, June 1966.
4. B. A. Steiner, "Air Pollution Control in the Iron and Steel Industry," *Int. Metals Rev.* **21**, 171, 1976.
5. B. A. Steiner, "Ferrous Metallurgical Operations," Chapter 21 in *Air Pollution, Vol. IV*, A. C. Stem, Ed., Academic Press, New York, 1977.
6. *Background Information for Proposed New Source Performance Standards: Iron and Steel Plants*, Vol. 1, Main Text, Publication APTD-1352a, U.S. Environmental Protection Agency, Research Triangle Park, NC, June 1973.
7. *The Making, Shaping and Treating of Steel*. 9th ed., H. E. McGannon, Ed., U.S. Steel Corp., 1971.
8. *Background Information for Proposed New Source Performance Standards: Iron and Steel Plants*, Vol. 2, Appendix: Summaries of Test Data, Publication APTD-1352b, U.S. Environmental Protection Agency, Research Triangle Park, NC, June 1973.
9. *Iron and Steel Industry Particulate Emissions: Source Category Report*, EPA-600/7-86-036, U.S. Environmental Protection Agency, October 1986.
10. *Background Information for an Opacity Standard of Performance for Basic Oxygen Process Furnaces in Iron and Steel Plants*, U.S. Environmental Protection Agency, Research Triangle Park, NC, February 1976.
11. "Standards of Performance for New Stationary Sources—Basic Oxygen Process Furnaces: Opacity Standard," *Federal Register* **43**, 72, 15600 (April 13, 1978).
12. D. W. Coy et al., *Pollution Effects of Abnormal Operations in Iron and Steel Making: Vol. VI. Basic Oxygen Process, Manual of Practice*, EPA-600/2-78-118f, U.S. Environmental Protection Agency, Research Triangle Park, NC, June 1978.
13. *Revised Standards for Basic Oxygen Process Furnaces Background Information for Proposed Standards*. Preliminary Draft, U.S. Environmental Protection Agency, Research Triangle Park, NC, November 1980.
14. L. J. Goldman et al., "Performance of BOF Emission Control Systems," presented at USEPA Symposium on Iron and Steel Pollution Abatement Technology, Chicago, October 6–8, 1981.

15. *Revised Standards for Basic Oxygen Process Furnaces Background Information for Promulgated Standards*, Final EIS, EPA-450/3-82-005b, Office of Air Quality Planning and Standards, U.S. Environmental Protection Agency, December 1985.
16. M. Drabin and R. Helfand, *A Review of Standards of Performance for New Stationary Sources—Iron and Steel Plants/Basic Oxygen Furnaces*, EPA-450/3-78-116 (NTIS PB289877), Metrek Division, Mitre Corp., November 1978.
17. C. W. Westbrook, *Hot Metal Desulfurization, BOF Charging, and Oxygen Blowing: Level I Environmental Assessment*, EPA-600/2-81-036 (NTIS PB 179251), December 1981.
18. T. A. Cuscino, Jr., *Particulate Emission Factors Applicable to the Iron and Steel Industry*, EPA-450/4-79-028 (NTIS PB81-145914), September 1971.
19. J. Steiner and L. F. Kertcher, "Fugitive Particulate Emission Factors for BOP Operations," Proceedings: First Symposium on Iron and Steel Pollution Abatement Technology, Chicago, EPA-600/9-80-012 (NTIS PB80-176258), February 1980.
20. C. W. Westbrook, *Level 1 Assessment of Uncontrolled QBOP Emissions*, EPA-600/2-79-190 (NTIS PB80-100399), September 1979.
21. S. Gronberg and S. Piper, *Characterization of Inhalable Particulate Matter Emissions from a Q-BOP Furnace, Vols. I and 11*, EPA-600/X-85-329a-329b, GCA/Technology Division, August 1982.
22. C. W. Westbrook, "BOF and Q-BOP Hotel Metal Charging Emission Comparison," Proceedings Symposium on Iron and Steel Pollution Abatement Technology for 1980, EPA-600/981-017 (NTIS PB81-244808), March 1981.

ELECTRIC ARC FURNACES/ARGON–OXYGEN DECARBURIZATION PROCESS

MANFRED BENDER

By the year 2005, electric arc furnaces (EAFs) are expected to generate approximately 50% of the raw steel produced in the United States. The market share of EAFs is increasing because of the continued rapid development of EAF steelmaking technology.

Electric arc furnaces are the prime means of recycling steel scrap into liquid steel. They are, however, also increasingly being used to produce liquid steel from iron sources such as direct reduced iron (DRI) pellets or briquettes and blast furnace liquid molten steel or pig iron.

EAF steel is batch produced. Each batch is called a "heat." The furnaces generally range in capacity from 50 to over 200 tons. Smaller furnaces (5-ton capacity and up) are used to produce batches of special steels. The economical furnace size is limited by the power transformer size and the power the electric supply grid can provide. The largest transformers are rated at approximately 180 MVA. The most powerful furnaces approach a power-to-tapped-steel-weight ratio of 1 MVA per ton.

Because of electric power supply limits, electric power costs, and furnace efficiency considerations, oxygen, natural gas, and coal are increasingly used to augment the melting process. An EAF is only 55% energy efficient, with 45% of the total furnace energy input going into losses such as slag (10%), water-cooled panels (12%), the off-gas system (21%), and miscellaneous other losses (2%).

Most EAFs in use today supply ac power to the arc using three roof electrodes. Some furnaces now supply dc power through two electrodes (one through the furnace roof and one in the furnace hearth center) and through three electrodes. These furnaces may lead to new continuously charged, higher-efficiency designs.

The trend is to use EAFs as melting machines. The furnace is tapped once the desired percentage of carbon in the melt and the tapping temperature are established. The steel is tapped at a temperature lower than that required for refining and casting. Alloying is done in the ladle during tapping, in ladle metallurgy stations, or in ladle furnaces. Ladle furnaces are small EAFs that allow the temperature of the molten steel in the ladle to be raised to the temperature required for casting.

An introduction to EAF steelmaking can be found in Ref. 1. A critical look at EAF steelmaking is presented in Ref. 2.

PROCESS DESCRIPTION

Equipment

Multifurnace meltshops usually consist of a furnace or charging aisle and a tapping or casting aisle. The process flows for each furnace are generally kept totally independent of each other. In single-furnace shops, the furnace and caster are generally in the same aisle. The newest meltshop technology places the EAF into a closed building ("elephant house"). These are meltshops with an enclosed crane runway area of approximately 100 sqft/(t/h) of tapped steel production. Such a meltshop is just large enough to accommodate only the charging crane and the EAF with its operating floor and scrap bucket hoisting area (Figures 1 and 2). All heat released within the building is removed by the shop roof exhaust because there are no roof ventilators. The meltshop layout has a significant influence on the design of the furnace secondary emission control system.

The furnace consists of three parts: the hearth, the side walls, and the roof. The hearth is refractory lined and is sized to contain the molten steel capacity of the furnace. The side walls contain the scrap volume. The side walls have traditionally been refractory lined as well, but now are almost exclusively water-cooled. The same holds for the roof, except that the roof center, which accommodates the three electrode ports, is generally made from refractory.

The furnace rests on heavy concrete foundations on which the furnace can rock forward and back. Furnaces have the ability to rock 15 degrees backward for slagging or 20–40 degrees forward, depending on whether the furnace is a bottom tapping or spout furnace.

Most furnaces have such accessories as burners, limestone feeders, carbon injectors, and oxygen lances. These accessories are attached through ports in the side wall of the furnace. The side opposite to the tap hole has a slag door through which slag is allowed to flow out. The tapping spout directs the molten steel into the ladle. More and more furnaces are turned into eccentric bottom tapping (EBT) or offset bottom tapping (OBT) furnaces that discharge the molten steel vertically into a ladle placed underneath the furnace by a transfer car or a ladle turret.

The Process

The steelmaking process (the heat) begins with the first addition of scrap. The furnace roof swings open and the overhead crane operator opens the clam shell bottom of the scrap bucket and lets the scrap drop into the furnace. The roof

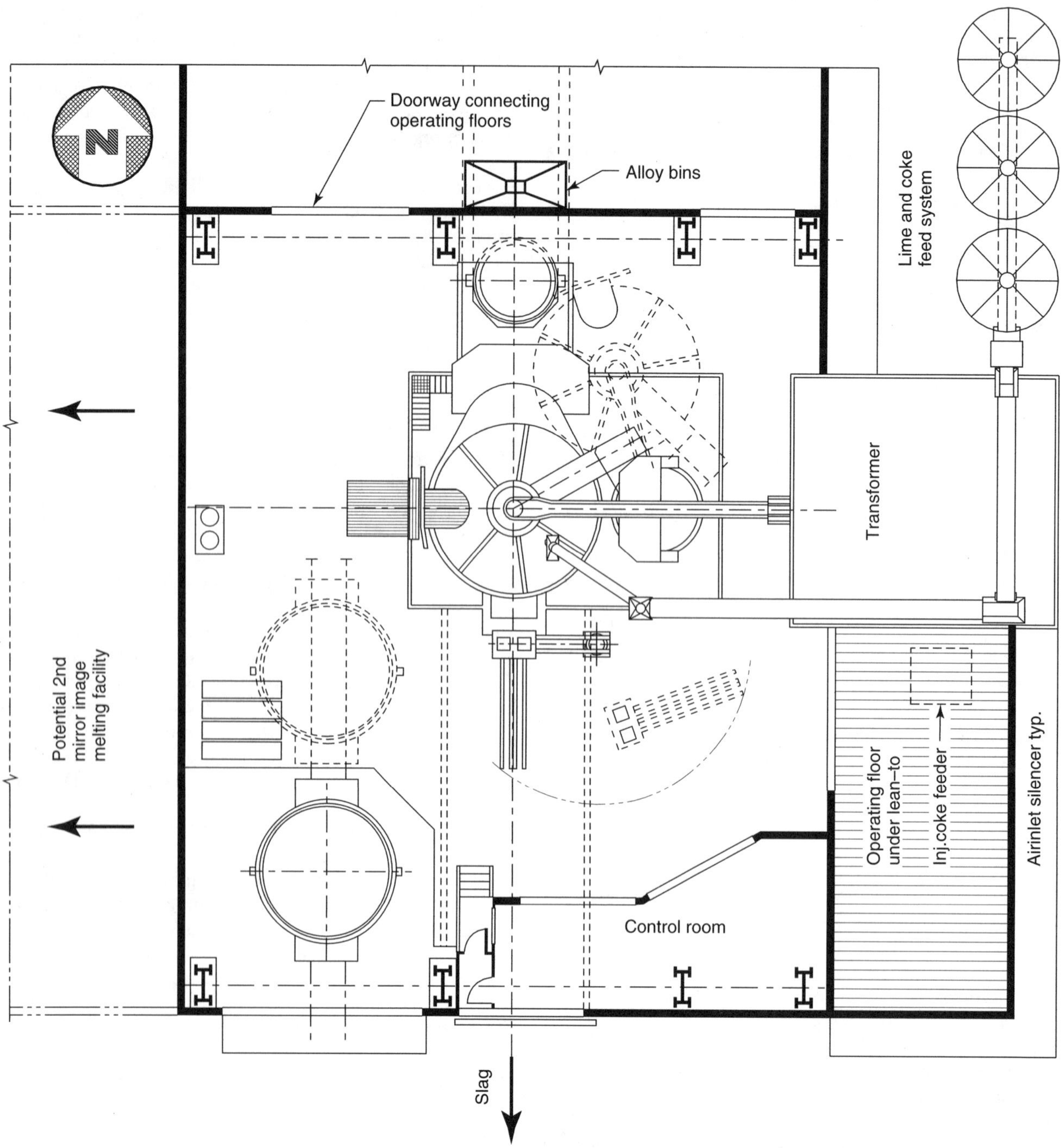

Figure 1. "Elephant House" Meltshop.

closes and power is applied. Most furnaces now use oxy-fuel burners. At least three oxy-fuel burners are turned on during the first 10 min of the heat in order to accelerate melting and add heat to the cold spots of the furnaces. Oxygen lances are used to make the scrap collapse into the melt and to burn combustibles in the furnace, thereby accelerating the scrap heating and melting process. Carbon and/or lime is often added with the scrap charge. The coke or coal provides the proper reducing environment for the process. The limestone removes impurities from the steel through the formation of slag.

Subsequent additions of scrap are made in a similar manner until the furnace capacity is achieved. Two or three scrap bucket loads are needed to make a full charge. As the molten steel temperature rises, impurities are removed by the production of slag. After the burners are turned off, carbon and oxygen are added to the slag in order to make the slag foam. Foamy slag improves the furnace power input, lowers hearth refractory wear by shielding arc radiation, and improves heat transfer to the cooler scrap zones near the furnace walls.

The floating slag rises as it foams. It is poured from the bath through the slag door into a pit or a slag pot below the furnace. Metallurgical tests indicate the progress of the process according to established goals for temperature, residual metals, carbon, sulfur, phosphorus, and so on.

When these goals are achieved, a preheated ladle is brought to the furnace and the furnace is tapped. Two general methods

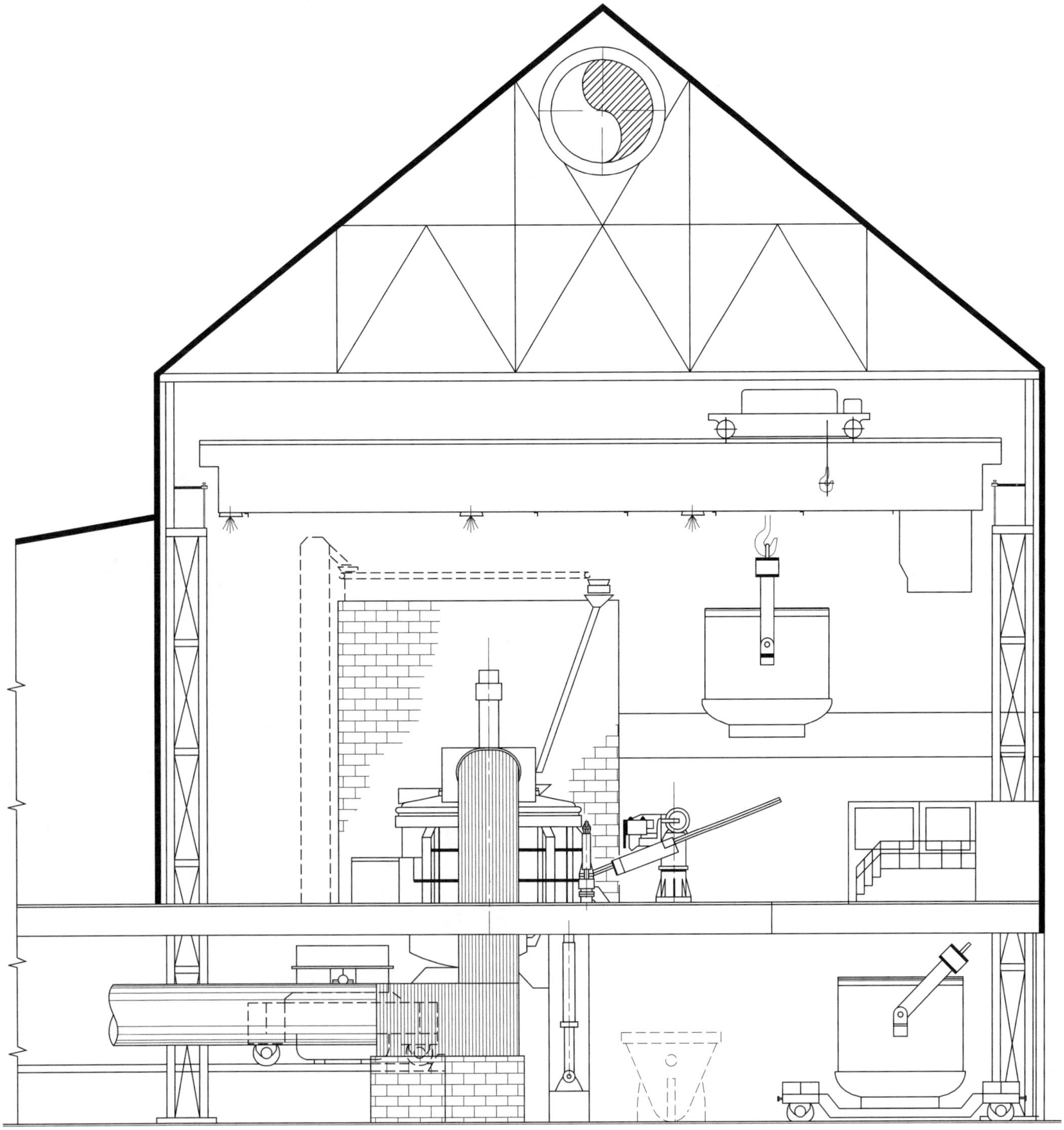

Figure 2. Section Through Shop in Figure 1.

of tapping furnaces are employed. A traditional spout furnace is tapped through a liquid metal runner (spout) attached to the furnace wall near the liquid steel elevation. The furnace is tilted to about 40 degrees to tap the hot metal into the ladle. The ladle is supported by either a tapping crane, a ladle stand, a turret, or a transfer car. The spout furnace ladle may become encrusted with slag and solidified steel. Almost all new EAFs now have EBT designs. This type of furnace taps through a hole in the bottom of a forehearth attached to the furnace. The advantage of this method is rapid, nearly slag-free tapping, molten steel temperature retention, and less tap hole and ladle maintenance.

Alloy additions, if needed, can be provided either directly to the furnace or into the ladle. Many meltshops homogenize the melt for temperature and chemistry by injecting inert gas into the ladle. Gas is bubbled through the ladle by either a lance or a porous bottom plug.

There are approximately 65 EAF steel mills in the United States operating 172 EAFs. Only $\frac{1}{3}$ of the steel mills have EAFs with tap-to-tap times of 60 min or less. A tap-to-tap time of 40 min with a power-on time of 30 min seams to be the fastest practical way of making a batch of steel in a traditional EAF. Many steelmakers are now starting to look for more cost-efficient novel EAF designs because melting fast does not necessarily mean melting cost-effectively.

Figure 3 shows that the loss of power (heat loss rate) to the off-gas system increases hyperbolically with a reduction in the power-on time. The higher energy input rates of faster melting EAFs with a constant contaminant level in the scrap cause this exponential increase in the heat loss to the off-gas system. As the power-on time approaches zero, the off-gas heat release must approach infinity. This means that fast-melting EAFs require an enormous amount of water-cooled off-gas ducting, a fact that fast-melting steelmakers know only too well. Some new EAF designs are therefore using off-gas heat to preheat scrap.

Batch preheating scrap using EAF off-gas presents a completely new set of process problems related to full combustion of the gas. Such systems are generally equipped with afterburners. Continuous scrap feed and preheating systems appear to be simpler and more promising in the long run. They also do not have the off-gas combustion problems that make batch scrap preheating systems so complicated and maintenance intensive. A more detailed description of these new scrap preheating EAF steelmaking processes is beyond the scope of this document.

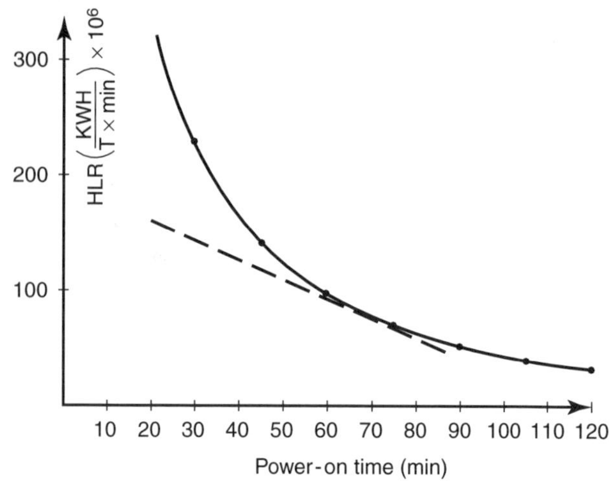

Figure 3. Average Heat Loss Rate (HLR) to EAF Off-Gas versus Power-on Time.

Figure 4. The Mushroom Cloud Generated by Electric Arc Furnace Charging Spreads around the Scrap Bucket and the Charging Crane on its way to the Canopy Fume Hood.

AIR EMISSIONS CHARACTERIZATION

The EAF emissions are generally characterized in three forms: charging, melting, and tapping. Each of these will be individually discussed.

Charging

Before the scrap bucket arrives to charge the furnace, the furnace roof swings open. Large volumes of hot fume-laden air rise into the meltshop roof trusses. Both the heat and the fume contained in the rising plume have to be exhausted from the meltshop. The evacuation system should be able to instantaneously extract these emissions as they are dispersed by the arriving scrap bucket and crane. Most plants keep oxy-fuel burners operational during this period, thereby adding more heat and fume to the rising plume.

The crane operator opens the clam-shell bottom of the scrap bucket and lets the scrap drop into the furnace. The drop generates a mushroom cloud and flames, which often engulf the crane (see Figure 4). Peak flow rates in these plumes on medium-sized furnaces have been estimated to exceed 1.5 million acfm.

The second or third bucket of scrap is charged in the same manner, except that the mushroom cloud is usually larger because of the presence of molten steel in the furnace. The volume flow rate and emission level in the mushroom cloud are also increased if the steelmaker places additives such as coal and lime into the scrap bucket.

Melting

With each charge, as the electrodes are lowered onto the scrap through each of the three ports, the arc is struck and the burners are turned to full power. An immense amount of heat is generated in the furnace. This heat melts the scrap, but it also produces much waste heat, gases, and fume. These are extracted from the furnace through a hole in the furnace roof (fourth hole). Some smaller furnaces extract the emissions from enlargements of the electrode ports using partly water-cooled or stainless-steel side-draft hoods or a hybrid of the side-draft hood and the fourth-hole concept.

Emissions generated by the EAF while the power is on and/or the roof is closed are called furnace primary emissions. The off-gas volume flow rate from furnaces is relatively low. A modern ultrahigh-power, 100-ton furnace with oxy-fuel burners using the foamy slag practice generates only 50,000 scfm of fourth hole off-gas, but can produce a peak waste heat flow rate equivalent to the electric power input.

The off-gas calculations not only require experience in process engineering, but also an appreciation of the large variety and ever-changing furnace operating practices.[3]

The amount of heat generated depends on many factors:

- The electric power input (up to 1 MVA per ton of molten steel).
- The percentage of combustibles in the scrap (normally 1.5% measured as oil and likely to be higher if the scrap contains more than 5% turnings).
- The amount of air infiltrating the furnace as a function of the furnace air infiltration areas and the pressure maintained by the exhaust system under the furnace roof. All nitrogen entering the furnace as part of infiltration air robs the furnace of valuable melting heat.
- The amount of scrap cutting oxygen injected (15–25 scfm per ton of molten steel).
- The amount of slag foaming and post-combustion oxygen injected (usually 20–45 scfm per ton of molten steel), together with the amount of foaming carbon added (generally 0.9 pounds/min of coal or coke per ton of molten steel).
- The amount of decarburization oxygen injected during refining (generally 25–35 scfm per ton of molten steel).
- The total oxy-fuel burner power input (generally about 15% of the furnace power).
- The electrode consumption rate (2–6 pounds per ton of tapped steel).
- The percentage of DRI fines exhausted by the fume system.
- The percentage of oxygen in oxygen carriers such as iron ore pellets, mill scale, recycled baghouse dust, and extra-rusty scrap (such as high-manganese turnings) charged.

The heat in the off-gas from an EAF greatly depends on the furnace operating practice. The off-gas is generally assumed to be air, but in off-gas calculations and emission control equipment selection consideration is given to the individual gaseous components and particulate in the gas stream.

Carbon monoxide (CO) can be generated in the production of steel in EAFs. Carbon-containing compounds in the additives, scrap contamination, and particularly the foamy slag practice are the source of these emissions.

As the furnace contents are heated to approximately 2950°F, any metals that volatilize below this temperature will be carried away by the furnace off-gases. The volatized metals recondense on particulate matter in the off-gases as the temperature is reduced. When coke, coal, or limestone is injected into the furnace, fine particulates of these commodities may be drawn into the off-gas system.

In EAF steelmaking, a fair amount of heavy solid particulate is injected into the off-gas. The quantities of heavy particulate are difficult to estimate because they can vary greatly from furnace to furnace. A furnace using the foamy slag practice can expect to collect 35 pounds of dust per ton of molten steel,[4] but one could expect to collect more with unfavorable oxygen injection practices or too small a fourth hole.

Assuming that the heavy particulate has been properly separated from the particulate carried to the baghouse, the gas contains particulate of the approximate size distribution and composition given in Table 1.

Table 1. Approximate Size Distribution and Composition of Baghouse Dust from an EAF Producing Carbon Steel

Particle Size		Composition[a]	
% less than	μm	Component	%
15	1	Fe	35.1
55	2	Fe(II)	1.7
80	5	Zn	15.4
90	10	Cd	0.028
99.5	50	Pb	1.5
		Mn	3.0
		Cr	0.38
		Ni	0.07
		SiO_2	5.3
		Al_2O_3	0.37
		CaO	4.8
		Na_2O	0.83
		C	0.44
		S	0.36
		F	0.9
		Cl	1.0
		Plus oxides	

[a]Ref. 4.

Tapping

Furnace spout and furnace bottom tapping produce similar emissions. The emissions are mostly iron oxide and slag particulate. However, almost all EAF steelmaking processes add alloying elements to the ladle while tapping. This procedure can significantly increase tapping fume evolution. The emissions, therefore, also contain particulate consisting of oxides of these additives.

Ladle Metallurgy

After tapping, the molten steel ladle is moved to the ladle furnace or other ladle metallurgy stations where the liquid steel temperature and composition are adjusted to meet the standard of the specific grade to be produced.

These processes all require emission control. The type of emission capture process used is affected by metallurgical considerations and the main emission control concept employed for the meltshop. The concept, therefore, can vary greatly from case to case. The ladle then moves to the casting area. For a description of casting processes see proceeding section.

Post-Production Emission Sources

After most of the scrap is melted, enough slag has accumulated in the furnace to cause it to flow over the slag door sill into the slag pit or into slag pots. Without the use of slag pots, the slag is usually removed by a bulldozer from the pit in a solid red-hot state in order to minimize dust generation and heating of the meltshop air by the cooling slag.

Before charging the first scrap load for the next heat, the furnace is inspected and the hearth is maintained. Hearth maintenance consists of coating the refractory at the slag line. In some meltshops, furnace melting additives are added. All this work is generally performed with the furnace roof open. Emissions from these sources consist of particulate matter and fumes.

AIR POLLUTION CONTROL MEASURES

The peak heat flow rate (Btu/min) through the fourth hole during the melting cycle usually determines the most critical operation for the design of the direct evacuation control (DEC) system.

A major factor affecting the combustion of furnace off-gas components is the amount of combustion air supplied to the gas leaving the furnace. The off-gas should receive at least 100% excess combustion air based on the period giving the largest heat release from the furnace. This combustion air is introduced for furnace DEC systems at the break flange between the furnace off-gas elbow and the entrance to the water-cooled duct and for side-draft hood systems with the gas cooling air around the electrodes above the furnace roof.

Furnace Direct Evacuation Control (DEC) Systems

The prime goal in the design of DEC systems is to make them inherently safe. The process should not allow the gas to reignite or explode even if combustible gas travels through the DEC system. A DEC system cannot be designed to ensure with absolute certainty that such a condition does not occur. Spark arrestors or other spray chambers should be equipped with explosion vents of a proven design. These vents should not be self-closing in order to prevent a subsequent even more severe explosion or detonation.

The DEC concept uses a hole in the furnace roof (a fourth hole) and a fourth hole duct elbow to direct the off-gas into the fixed water-cooled duct. The elbow preferably rests on the furnace roof ring and is attached temporarily to the furnace roof lifting beams for roof maintenance.

The inlet to the fixed duct is usually enlarged in order to ensure discharge of the elbow gas into the fixed duct as the furnace tilts forward and back within reasonable limits. The elbow is equipped with a flange that prevents excessive air from leaking into the enlarged duct.

Good furnace operation is achieved when the furnace discharges a nearly constant amount of fume (not flames!) from around the electrodes into the meltshop. The lack of electrode emissions is a sign of excessive infiltration of air into the furnace. This causes increased electrode consumption and excessive heat loss to the fume system.

The fourth hole gas flow rate is roughly proportional to the off-gas heat flow rate because during most phases of the power-on period, the off-gas temperature varies from 2400°F to 3000°F. This is a result of the draft control system limiting the volume flow rate of cool air drawn into the furnace. As another consequence, the combustion air infiltration rate at the break flange varies nearly in proportion to the off-gas volume flow rate.

The concept provides good combustion of the furnace off-gas with an average CO concentration during the power-on period of less than 800 ppm. Rapid changes in the furnace off-gas volume flow rate caused by scrap cave-ins, coal injection, oxygen injection, the oxy-fuel burner operation, and varying furnace power levels make it difficult for the draft control system to maintain the desired constant furnace pressure at all times. Carbon monoxide peaks of over 7000 ppm (0.7%) are, therefore, not uncommon. 7000 ppm corresponds to approximately 100 ppm in the baghouse exhaust when combined with canopy hood exhaust.

The following total process gas emission factors apply to most EAFs. EAFs producing specialty steels or EAFs using special operating practices can have emission factors as much as 50% lower or 150% higher. The data are given on a per ton of tapped steel basis:

PM_{10} particulate	0.20 lb/t
Carbon monoxide (CO)	2.00 lb/t
Volatile organic compounds (VOC)	0.10 lb/t excluding CH_4
Sulfur oxides (SO_x)	0.20 lb/t
Nitrogen oxides (NO_x)	0.50 lb/t

The heavy particulate in furnace off-gas is mostly slag and limestone dust. It is beneficial to shape the furnace water-cooled elbow and the subsequent fixed duct to prevent obstruction of the gas flow by slag settling in these components. At the same time, the water-cooled duct next to the furnace is most appropriately designed to contain a dropout box, where this material can be collected. In a properly designed water-cooled duct and dropout, the collected material is free from the scrap and frozen slag lumps that preclude the use of conveying equipment for the continuous removal of the dropout box catch.

Furnace draft control is accomplished by either a damper in the DEC duct system (which can sometimes be the fan inlet damper) or a damper in the canopy hood duct.

On older EAFs, a dilution air damper at the end of the water-cooled duct protects the uncooled duct from temperatures in excess of 1200°F (1400°F for peaks lasting less than 30 sec). The drawback of this design is that the moment the damper opens, the furnace starts to smoke and flame excessively into the meltshop. For modern EAFs operating with a foamy slag practice, uncontrolled slag boils can flood the furnace DEC duct with vast amounts of heat and carbon monoxide. With small amounts of dilution air added to the gas stream, be it from another EAF, a ladle furnace, or a dilution air damper, the gas can reignite or even explode. Both can cause severe damage to the emission control system.

If gas temperatures above 1200°F at the end of the water-cooled duct do not occur for more than about 10 min at a time, air-atomized water spray nozzles may be used to inject water into the duct for the purpose of peak shaving occasional temperature spikes. This concept maintains furnace draft and is also inherently safe.

As already outlined above, fast-melting EAFs require enormous amounts of water-cooled ducting. Also, the life of water-cooled ducts depends on the scrap quality and the number of hot–cold cycles the duct experiences in a given time. Poor scrap quality accelerates duct corrosion due to hydrochloric acid condensation on the cold duct wall. Steelmakers experiencing this problem go to water-cooled ducting made from stainless steel. In this case, stainless steel ducting is more cost-efficient than mild steel ducting.

The best-available control technology for fast-melting EAFs is to terminate the water-cooled duct in an evaporative cooler, that is, a spray chamber. Such a concept is shown in Figure 5. It is the technology of choice for fast-melting EAFs because it reduces the enormous amount of maintenance intensive water-cooled ducting by 30% to 50%. It also allows the steelmaker more freedom with the EAF operating practice without concern for overheating air-cooled ducting. Another important benefit is that the spray chambers act as spark arrestors and that they can be operated to serve as chlorinated hydrocarbon emission control devices in jurisdictions where the control

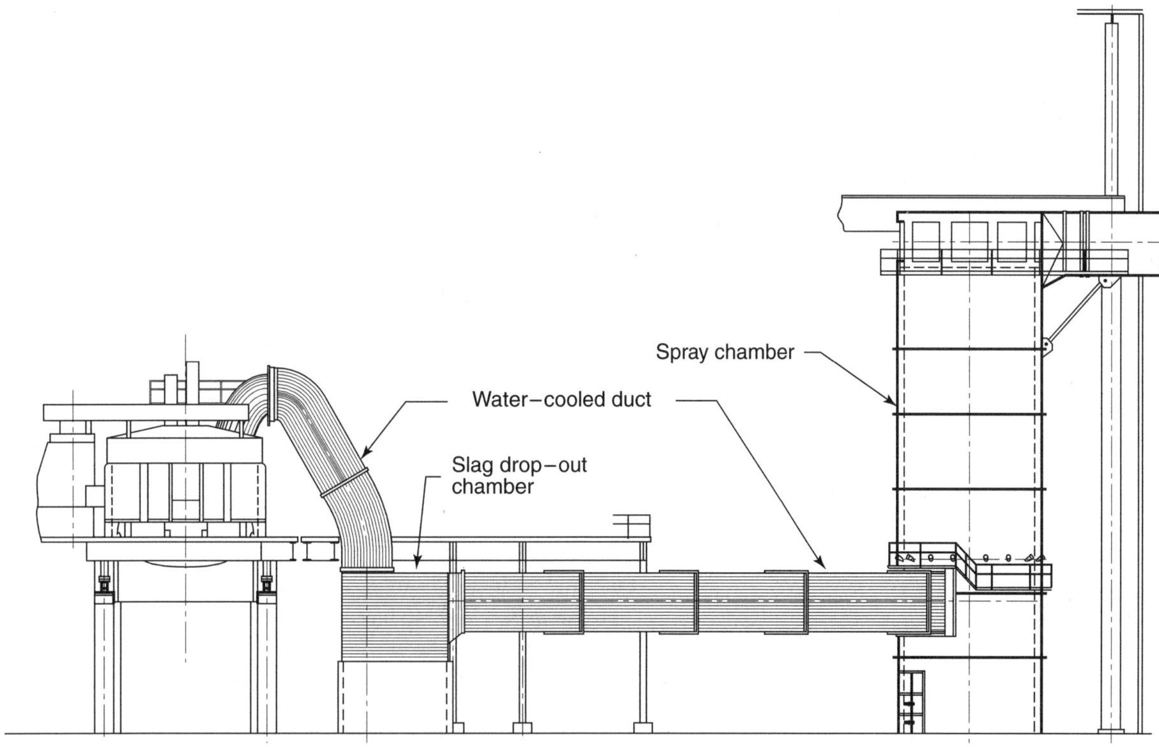

Figure 5. Electric Arc Furnace Off-Gas Direct Evacuation Control (DEC) System using Spray Chamber Technology to Reduce Water-Cooled Duct Requirements by 30 to 50%.

of these compounds is mandated. In the design of spray chambers for EAF off-gas systems it is important to size them so the baghouse does not experience excessive moisture levels keeping all other process moisture sources in mind as well.

Side-Draft Hood Systems

A few EAFs, and mostly EAFs of less than 50 tons in capacity, still use side-draft hood systems. This system consists of a hood, portions of which may be water-cooled, surrounding the electrodes and capturing emissions from the enlarged electrode openings. The concept does not maintain a negative pressure in the furnace.

Since temperatures downstream of the hoods are usually less than 1200°F, water-cooled ducting is not needed, but the duct system and gas flow rates become very large. Carbon monoxide combustion occurs readily because of the high ratio of ambient air to furnace gases. In some cases, a fourth hole exhaust is integrated into the side-draft hood.

Secondary-Emission Control Canopy Hood Systems

Secondary-emission control in EAFs includes the control of emissions generated by operations not associated with melting. The most significant of these emissions are those from furnace charging and tapping.

Charging and tapping emissions are generally captured by canopy hoods in the meltshop roof trusses. Tapping emissions are sometimes captured by low-level tapping hoods when the ladle is not crane held. Canopy hood shapes and hood exhaust flow rates are determined by experience and from plume rise formulas modified for the presence of obstructions such as cranes.

Because potential emission control investments are usually large, secondary-emission capture concepts, including canopy hoods and scavenging duct off-take systems, are often developed in a fluid dynamic model of the meltshop that takes into consideration the particular distribution of heat, fume, and dust sources prevailing in the plant.[5–7]

Charging canopy hoods are not designed to capture the furnace charging mushroom cloud instantaneously. Rather, the hood off-take flow rate is determined by the fume flow rate from the open furnace before and after charging. The hood is built deep and wide enough either to store the charging plume surge temporarily, as is the case in elephant-house-type meltshops where the entire meltshop roof acts as a canopy hood, or to spill the mushroom cloud into a space surrounding the primary hood. The spilled fume is drawn from the space by so-called scavenging ducts or is allowed to drift back into the hood.

When the DEC duct branch has a much higher pressure drop than the secondary emission control (canopy hood) branch, separate hot-gas fans are needed to move the gas to the canopy duct. In this case, it is advisable to provide a coarse particulate separator, such as a low-pressure drop cyclone, ahead of the hot gas fans in order to protect the fans against excessive erosion from abrasive particulate. Since the hot gas fans cannot usually withstand temperatures in excess of 700°F, it becomes advisable to carefully design the systems for proper gas cooling. For modern EAF off-gas systems, these hot-gas booster fans are a major maintenance concern. They also limit the DEC capacity to the performance given by the fans. It is best to avoid hot-gas booster fans.

Depending on the space constraints and the duct configuration selected, it is often possible to select a simpler system that artificially increases the pressure drop in the canopy branch in order to match it to the pressure drop of the DEC branch. In this case, the canopy branch exhaust fans are selected to also handle the gas from the DEC system, and it is best to design the DEC system to have a low-pressure drop and the canopy duct to have a high gas velocity.

In totally closed roof meltshops, the only means of heat removal is the air pollution control system. It is therefore important to keep heat from ladle preheaters and casters, etc. away from the melting area roof because the resulting high-temperature inversion can cause crane maintenance and in-plant plant visibility problems due to low-level fume stratification. It can also cause emissions to the environment from low-level plant openings. It is therefore very important that in closed-roof meltshops, the input of clean heat into the building is kept as low as possible. A plant air temperature rise between the ambient air and the air under the meltshop roof of 30–35°F is a good target to use.

In addition to the above, generally applicable plant ventilation criteria should be met in order to prevent emissions from reaching a low level or even escaping through low-level plant openings. When considering these aspects of EAF secondary emission control, special attention must be paid to the nonferrous metallic oxides and respirable particulate because of the applicable Occupational Safety and Health Administration regulated workplace air threshold limit values.

In naturally ventilated (open-roof) meltshops, a design consideration must be to prevent visible EAF emissions through roof ventilators. The desire to keep meltshop fume at a high elevation and the plant cool is motivating steel makers to remove ladle preheat and caster billet or slab heat from the meltshop through forced- or natural-draft exhaust systems separated from the meltshop by sheeting. Spout EAFs use tapping canopy hoods to capture emissions from tapping. These hoods are sometimes combined with the charging canopy hood.

EBT furnaces are sometimes equipped with local tapping hoods that are evacuated by the DEC system. A more and more common method for capturing tapping emissions for EAFs with ladles on transfer cars is to direct tapping fume into the charging canopy hood using tapping sheds. These are walls which start at the tapping ladle level and reach all the way up to the nearest edge of the charging canopy hood. For EAFs tapping across a column line, these sheds incorporate the crane runways with the charging crane traveling on one side of the shed wall and the ladle crane traveling on the other side.

For these furnaces, tapping canopy hoods are less cost-effective because they have to capture emissions dispersed by the furnace tilting platform. The same concept may be applied to short spout furnaces and side tapping furnaces that do not utilize crane-held tapping ladles. Sometimes, however, the tapping process generates so little fume as a result of low tapping temperatures that the canopy hood concept can be employed just as well.

Charging/tapping canopy hood off-take volume flow rates for well-designed hoods usually range from 250,000 to 500,000 acfm at 150°F, depending on the furnace size and the factors discussed above. Hoods should be sized to match the rising plume at the hood face taking into account the dispersion of emissions by objects such as scrap buckets, crane runways the cranes themselves, etc. Hoods should preferably have a minimum face velocity of approximately 120 ft/min. Smaller tapping canopy hoods without obstructions under their hood face need face velocities up to 350 ft/min for 100% capture of emissions. Larger canopy hoods sized for fume storage as used for furnace charging may have face velocities as low as 50 ft/min. Each scavenging off-take may have a volume flow rate ranging from 25,000 to 100,000 acfm at 150°F. Local tapping hoods have off-take volume flow rates from 50,000 to 100,000 acfm at 400°F. Most main fume fans handling unfiltered emissions have a standby fan and operate at speeds of 900 rpm.

The best available secondary emission control technology for fast-melting EAFs today is the meltshop elephant house. The combined DEC and canopy hood exhaust system capacity for such a concept preferably amounts to more than 300,000 acf/t [= acfm × 60/(hourly production)] of tapped steel and preferably gives the meltshop an updraft velocity of more than 50 ft/min. These criteria are consistent with the 100 sqft/(t/h) enclosed crane runway area guide referenced above.

Furnace Enclosures

Furnace enclosures are enclosures around the furnace just large enough to accommodate the furnace and its ancillary equipment. The enclosures have movable doors to permit scrap bucket access, tapping ladle access if necessary, access to the slag door, and access for maintenance.

Enclosures are generally used for EAF melting noise control in densely populated neighborhoods. They are also more popular for alloy- and stainless-steel-producing EAFs because the longer tap-to-tap times result in the enclosure being less of an interference with access to the furnace.

Many carbon steel–producing furnaces are now too fast for this type of emission capture concept to be practical. Their off-gas heat release is so high that the enclosure does not eliminate the need for a furnace DEC system. It has also become evident that, in most cases, a furnace enclosure does not make canopy hoods or meltshop roof fume off-takes unnecessary. This eliminates the major advantage of the enclosure concept.

A common problem with enclosures is that they have too many open areas around the bottom perimeter. As a result, a negative pressure cannot be maintained in the enclosure and emissions escape around doors near the top. Another problem that is sometimes encountered, because of the confined volume of the enclosure, is the occurrence of high baghouse temperatures during furnace charging. High-temperature bags or larger system capacities are usual solutions to these problems.

Air Pollution Control Equipment

Most EAFs have fabric filters (baghouses) for fume cleaning and collection. Both positive- and negative-pressure fabric filters are used. Cleaning methods of all types — reverse air, shaker, and pulse jet — are employed. The reverse air filter is the most commonly used type of filter in North America. Negative-pressure pulse-jet filters are more common in the rest of the world. They often initially cost less but require more maintenance in order to keep emissions low. Pulse-jet filters are preferred for smaller system capacities.

The advantages of negative-pressure filters include easier filter maintenance because the exterior of the filter usually remains cleaner. Also, use of higher-efficiency and lower-cost fume fans without the need for standby fans becomes possible. The advantage of positive-pressure filters is lower filter cost.

However, the price of negative-pressure filters is slowly dropping as a result of filter designs that use less steel for the 25 in. w.c. negative pressure. Negative-pressure filters are now finding wider use because of their significant operating benefits.

For reverse-air and shaker filters, the air-to-cloth ratio (filtration velocity) with one compartment down for cleaning (the net ratio) is usually selected to be 3 ft/min. Well-maintained reverse-air filters with bags selected for the specific steelmaking operation have operated at nct filtration velocities of up to 3.5 ft/min. Inlet-to-outlet fabric filter pressure drops of 6–8 in. w.c. are normal.

Pulse-jet baghouses operate with net filtration velocities ranging from 4.5 to 7 ft/min, depending on the dust load and the cleaning method employed. For reasonable maintenance, pulse-jet filters serving EAFs with DEC systems should have a net filtration velocity of approximately 4.5 ft/min. Off-line cleaning is preferred because it provides lower overall filter pressure fluctuations. The selection of pulse-jet fabric filters for EAF fume cleaning requires considerable expertise, but with care, often-encountered problems such as high filter pressure drops and high dust penetration can be avoided.

Emission control systems with separate ducts, baghouses, and fans for DEC and secondary emission control often experience high DEC system baghouse temperatures. These temperatures are either reduced with air-to-air heat exchangers called forced-draft coolers or are accepted by the fabric filter through the installation of high-temperature bags. Forced-draft coolers generally require considerable maintenance. High-temperature bags can handle up to 500°F as compared with the normally employed polyester fabrics, which have a continuous operation limit of 275°F. It is generally preferred to combine the DEC and canopy hood exhaust systems and invest the extra cost for forced-draft coolers and high-temperature bags in more baghouse and fan capacity. The higher capacity also benefits meltshop secondary emission control.

Most EAF fume systems carry furnace sparks to the baghouse where they tend to cause bag spark damage. Well-designed baghouse hopper inlets have the ability to drop these sparks out in the hopper before they have a chance to travel onto the bag surface. For furnaces using foamy slag practices, glowing carbon particles often pose a major spark problem. In many cases, spark arrestors in the DEC ducts are used to remove sparks from the gas stream. The units are settling chambers or preferably low-pressure drop cyclones. Sometimes the spark arrestor includes the canopy duct. In this case, the arrestor can also capture sparks from EAF charging, even though these sparks are rarely a problem.

Well-designed canopy hoods have a tendency to send the hot gas surge from the EAF charging fire ball to the baghouse without insufficient gas cooling. This can cause severe bag damage. It is advisable to place the baghouse sufficiently far away from the meltshop in order to allow the gas to be cooled by the thermal inertia of the duct system.

Collected dust should be removed from the baghouse hoppers as soon as possible in order to prevent dust re-entrainment and bridging of dust in baghouse hoppers.

As an alternative to the baghouse (dry) emission control system, scrubbers are still in use in rare cases today. These systems use water sprays to trap to the particulate matter. The water is collected and recirculated to the scrubber. Sludge is collected in a clarifier. This sludge must be dewatered prior to further handling. The high operating costs and low efficiencies of these systems make them unattractive for EAF emission control.

ARGON–OXYGEN DECARBURIZATION VESSELS

Argon–oxygen decarburization (AOD) vessels are used as metallurgical processing units for the production of specialty steels. They are charged with molten steel from EAFs. The molten steel is usually high in carbon and chromium as a result of charging low-cost, high-carbon ferrochromium into the EAF. The AOD is a converter-style vessel that is equipped with bottom gas injection ports (tuyeres). The process involves the injection of oxygen together with an inert gas (usually argon, but sometimes also nitrogen and air) for the purpose of decarburizing the heat without oxidizing the alloying components.[8,9]

Primary AOD emissions are generated when the process gases are blown through the vessel bottom into the molten steel. These emissions are captured in a local movable hood and ducted to the EAF fume system. Sometimes, AOD vessels are equipped with movable accelerator hoods that serve to prevent the growth and dispersion of the AOD off-gas plume on its way to a roof-mounted canopy hood. The concept requires large air pollution control systems because the gas is cooled using large volumes of dilution air.

An AOD primary off-gas hood can require excessive maintenance if it is not designed to withstand impingement by the vessel-mouth flame. In the initial section of AOD off-gas systems, the vessel off-gas, which consists of CO and the inert injection gas, is fully burned and cooled. Therefore, the hood and subsequent ducting are sometimes water cooled.

Canopy hoods are used to capture vessel charging and tapping emissions. These canopy hoods have design considerations similar to those of EAF canopy hoods. Again, it is recommended that the best canopy hood concept be developed in a fluid dynamic model of the facility if the potential emission control investment is large, if plume dispersion above the vessel is affected by a crane and unusually warm air under the plant roof, or if the plant roof structure is not deep enough to accommodate a normal canopy hood design.

For a more detailed investigation of AOD emission control concerns, see Ref. 10.

REFERENCES

1. R. J. Fruehan, "A Nontechnical Introduction to Electric Furnace Steelmaking," in the series "Keeping Current," *Iron Steelmaker*, 65–66, March 1989; 43–44, April 1989; 58–59, May 1989; 40–41, June 1989; 36–37, July 1989; 52–53 August 1989.
2. K.-H. Klein and G. Paul, "Reflections on the Possibilities and Limitations of Cost Savings in Electric Arc Furnace Steel Production," *Iron Steelmaker*, 25–34, January 1989.
3. M. Bender and L. F. Rostik, "Emission Control Aspects of Modem EAF Steelmaking," *Iron Steel Eng.* **64**(9), 22–26, 1987.
4. D. R. MacRae, "Electric Arc Furnace Dust; Disposal, Recycle and Recovery," Report 85-01, Center for Metals Production, April 1985.
5. M. Bender, "Fume Hoods, Open Canopy Type. Their Ability to Capture Pollutants in Various Environments," *Am. Ind. Hyg. Assoc. J.* **40**, 118–27, 1979.
6. D. L. Harmon, *Technical Manual: Hood System Capture of Process Fugitive Emissions*, Office of Research and Development, U.S. Environmental Protection Agency, January 1985.

7. M. Bender, G. Paul, and R. V. McCabe, "Meltshop Ventilation Optimization Using Fluid Dynamic Models," *Electric Furnace Conference Proceedings*, Vol. 45, 1987, pp. 161–66.
8. G.H.H. Sterkrade, "Production of Special Steels in AOD Converters with Plant Description," *Steel Times Int.*, 13–20, March 1980.
9. A. G. Williams and I. Ludlam, "The AOD Process and Equipment Design," in a series of case histories entitled: "Quo Vadis AOD," *Steel Times Int.*, 50–72, December 1978.
10. R. Iverse, *Background Information Document for the Review of the New Source Performance Standard for ElecSystem Capture of Process Fugitive Emissiotric Arc Furnaces and Argon Oxygen Decarburization Vessels in the Steel Industry*, U.S. Environmental Protection Agency, February 1982.

CASTERS

Manfred Bender

Casters are steel plant equipment used to process liquid steel into continuous lengths of solid steel of various cross sections. The most common sections are less than 8 in. square (billets); larger squares are blooms and rectangles averaging 48 in. in width (slabs). Special sections are rounds and so-called near net shapes, such as beam blanks and slabs or sheets as thin as 1 in.

PROCESS DESCRIPTION

Molten steel is tapped from the furnace into a refractory-lined ladle. The molten steel stream emerges from the slide gate in the ladle bottom and fills a tublike distribution vessel (tundish).

The number of drain holes in the tundish corresponds to the number of strands to be cast simultaneously. The molten steel streams emerging from the tundish bottom fill oscillating water-cooled copper molds in which the molten steel solidifies as it passes through the mold and emerges from its bottom. The molds provide immediate cooling of the outer skin of the steel. As the steel emerges from the mold, the center is still molten. After complete solidification, the continuous strip is cut into convenient lengths using torch machines or shears.

AIR EMISSIONS CHARACTERIZATION

Ladle emissions are negligible. The molten steel is sometimes given a cover of inert material such as rice hulls to provide thermal insulation. In addition, many modern casting facilities employ refractory-lined lids.

The molten steel stream from the ladle bottom generates almost no emissions, particularly when it is surrounded by a ceramic submersion tube. Some emissions are seen to emerge from the molds. The emissions are caused by mold powders and mold lubricating oils. These emissions are mostly a white-blue haze generated by vaporized oil condensate or mold powder combustion products. The emissions are greater if the molten steel stream is nitrogen shrouded because the nitrogen prevents the burnoff of combustibles.

The caster cutoff torches may also generate minor emissions because the red-hot steel section that emerges from the casting

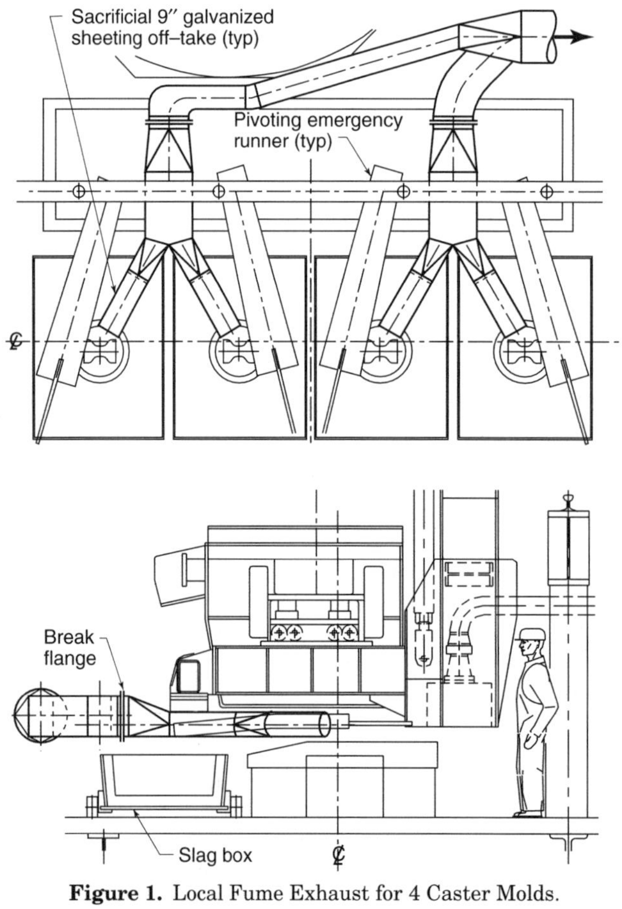

Figure 1. Local Fume Exhaust for 4 Caster Molds.

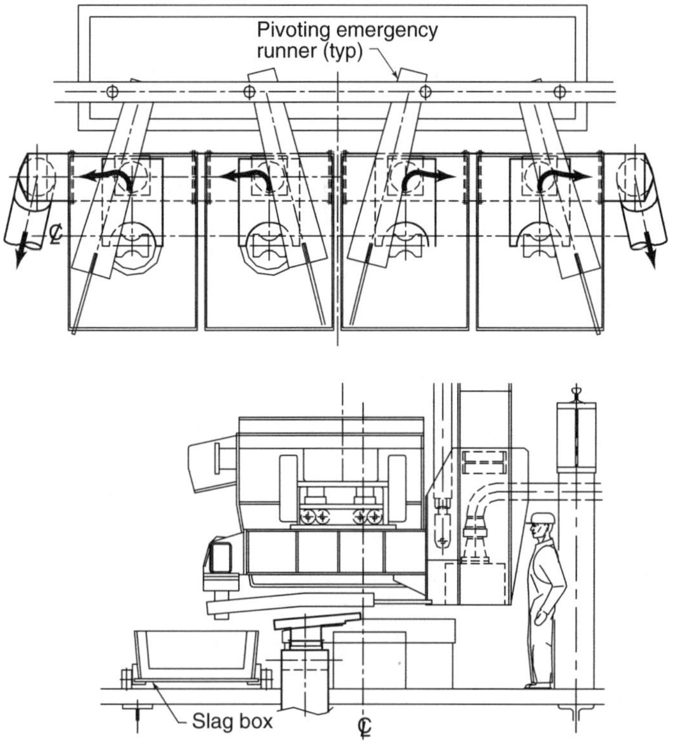

Figure 2. Integral Fume Exhaust for 4 Caster Molds.

machine is cut by a traveling torch with an oxidizing flame. Other emissions are generated during the maintenance of tundishes and casting ladles. These are mostly dust and iron oxide fumes from dumping and oxygen lancing sculls.

AIR POLLUTION CONTROL MEASURES

Casting emissions are generally too insignificant to be controlled at the source. In special circumstances, canopy hoods in the roof trusses above the casting machines capture these emissions. The exhaust is usually in the 100,000 acfm range. The emissions are then ducted to the electric furnace emission control system.

Caster mold emission can also be captured at the source. The exhaust per mold is in the 1500 acfm range regardless of whether local exhaust (Figure 1) or integrated exhaust (Figure 2) is used. The systems use small booster fans, which direct the emissions to the plant emission control system. These systems are not only more cost-effective than canopy hood installations, they also improve the air at the workplace.

Cutoff machine torches generate a strong vertical downward blast into a water flume below the caster runout. Most of the heavy particulate is washed away in the flume water. A slight amount of fine iron oxide particulate is generated. It is rapidly dispersed and is hardly visible, even in the immediate vicinity of the machine.

Ladle and tundish lancing emissions are generally allowed to rise into the plant's fugitive emission control system. In cases where this is not possible and where visible emissions from the process building do occur, custom-designed hooding systems and fabric filters are employed to control the emissions.

LADLE METALLURGY VACUUM DEGASSING

EDWARD COCCHIARELLA

Ladle metallurgy facilities (LMFs) are used primarily to adjust the composition and temperature of steel heat from a primary melting furnace, such as an electric-arc furnace (EAF) or basic-oxygen furnace (BOF). They include a series of processes to treat molten steel in the ladle before continuous casting or ingot teeming. The LMF is also a buffer between the primary melting furnace and a continuous-casting machine.[1]

Ladle metallurgy, also known as secondary metallurgy, is an increasingly important operation in both electric-arc and basic-oxygen steel production. The series of processes includes raking or skimming slag; injecting or adding fluxes, nonferrous metals, and gases; degassing by application of vacuum; and electric-arc reheating. These were developed to make cleaner steels in response to more stringent standards and difficult applications.[2,3] The combination of processes used at a particular plant depends on the products made, the quality desired, and the requirements of the melting furnace and continuous caster. At Dofasco Steel in Hamilton, Ont., Canada, for instance, the No. 2 Melt Shop has two deslagging stations, one ladle furnace, one vacuum degasser tank, two strong stir-powder-injection stations, and one wire injection station. The typical sequence of ladle metallurgy processes used between the K-OBM (a variation of the BOF steel-making process) and the continuous caster is shown in Figure 1.

PROCESS DESCRIPTION

In steel-making shops with LMFs, the EAF or BOF becomes primarily a melting tool. The heat is then refined at the LMF.

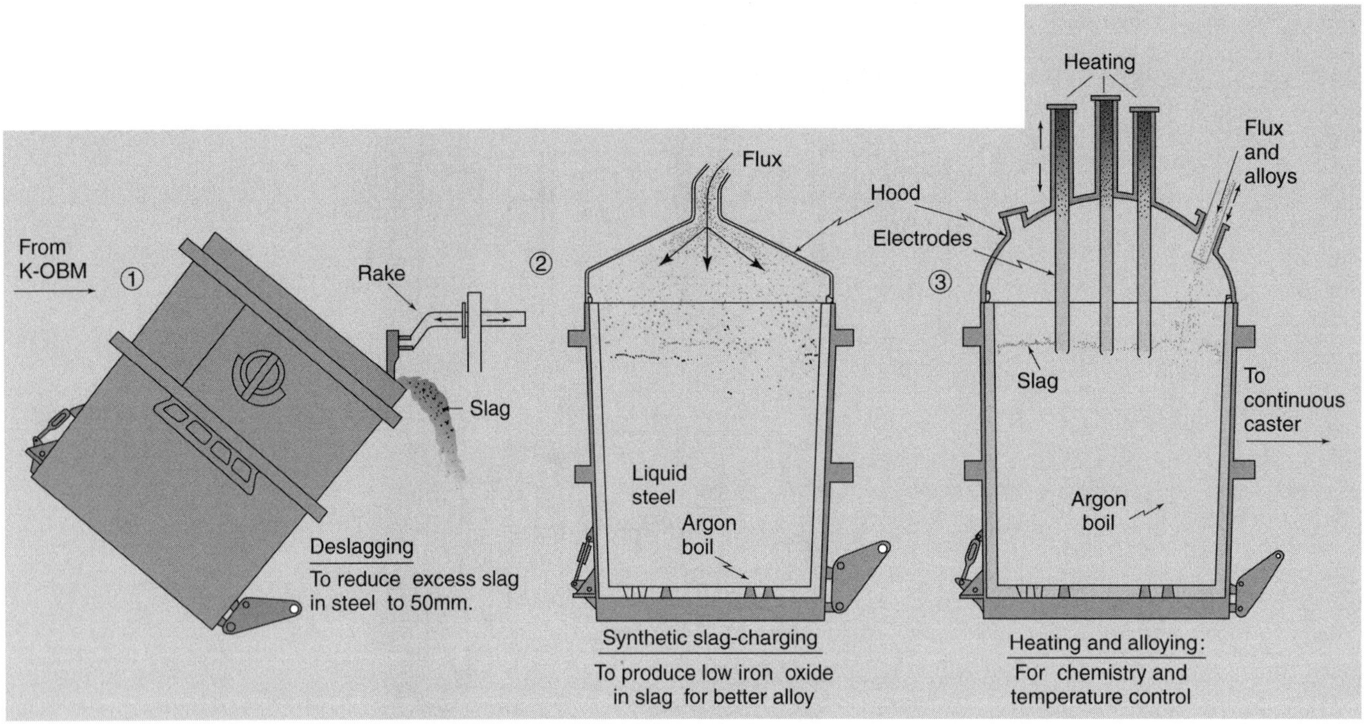

Figure 1. Typical Sequence of Ladle Metallurgy Processes.

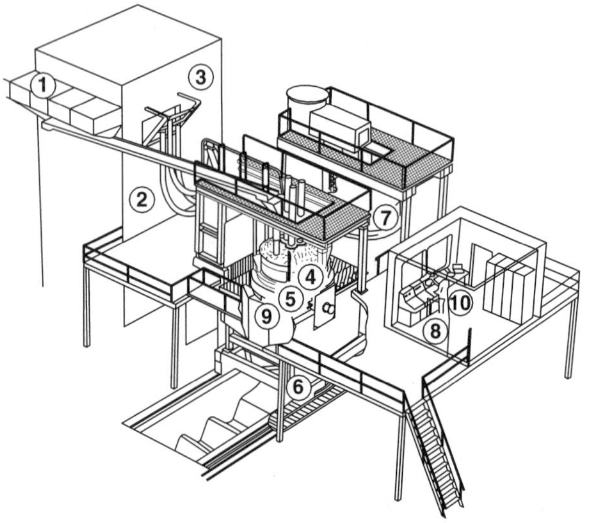

1 Alloy addition equipment
2 Hydraulic and electric rooms
3 Transformer house
4 Heating station
5 Ladle
6 Ladle transfer car
7 Degassing station
8 Main control room
9 Induction stirrer
10 ASEA master refiner

Figure 2. Layout of an ASEA Refining System.

This results in shorter process times, which may increase the shop capacity by some 20%. However, the additional operations produce new sources of emissions. Particulate emissions are generally collected by baghouses, except for vacuum degassing systems, which have steam ejectors and condensers.

The LMFs may have several stations connected by a track, a turntable, or a crane ladle transfer system. The stations are usually centered around a ladle furnace. Other stations may include slag raking, powder or wire injection, vacuum degassing, inert gas or magnetic stirring, alloy addition, and synthetic-slag additions. The different stations may incorporate fixed or movable hoods to capture emissions. Figures 2 and 3 show two examples of LMFs.[4,5]

Vacuum degassing is achieved by three main processes. The simplest is a tank degasser, where the entire ladle is placed in a tank, as in Figure 4. The tank is sealed and vacuum is applied. The steel is stirred by argon bubbling. The second is the Ruhrstahl Heraeus (RH) process, in which a vacuum vessel with two snorkels is lowered into the ladle, as shown in Figure 5. Argon bubbling is used to circulate the molten steel through the vacuum vessel. The RH degassers are prone to the buildup of solidified steel and slag, known as skulling, on the walls of the vessel caused by steel splashing and low refractory temperatures. The skulls are melted down by gas-oxygen burners.[6] The final process, ladle furnace/vacuum degassing (LF/VD), involves the application of a tight-fitting vacuum hood to a ladle furnace. Several other vacuum technologies are used for special steel applications. These include vacuum-arc degassing (VAD), vacuum-oxygen decarburization (VOD), vacuum-oxygen heating (VOH), and vacuum-induction degassing (VID).[1]

The vacuum for degassing processes is usually generated by multiple-stage steam ejectors. Condensers trap particulate and carbon monoxide in the cooling water. The carbon monoxide is released in the condenser sump and cooling tower. The sump vent and cooling tower are minor sources of carbon monoxide emissions.

Secondary sources of particulate emissions arise during the pneumatic or conveyor transport of alloys and fluxes. Emissions from transfer points, hoppers, and transport gases are collected by baghouses.

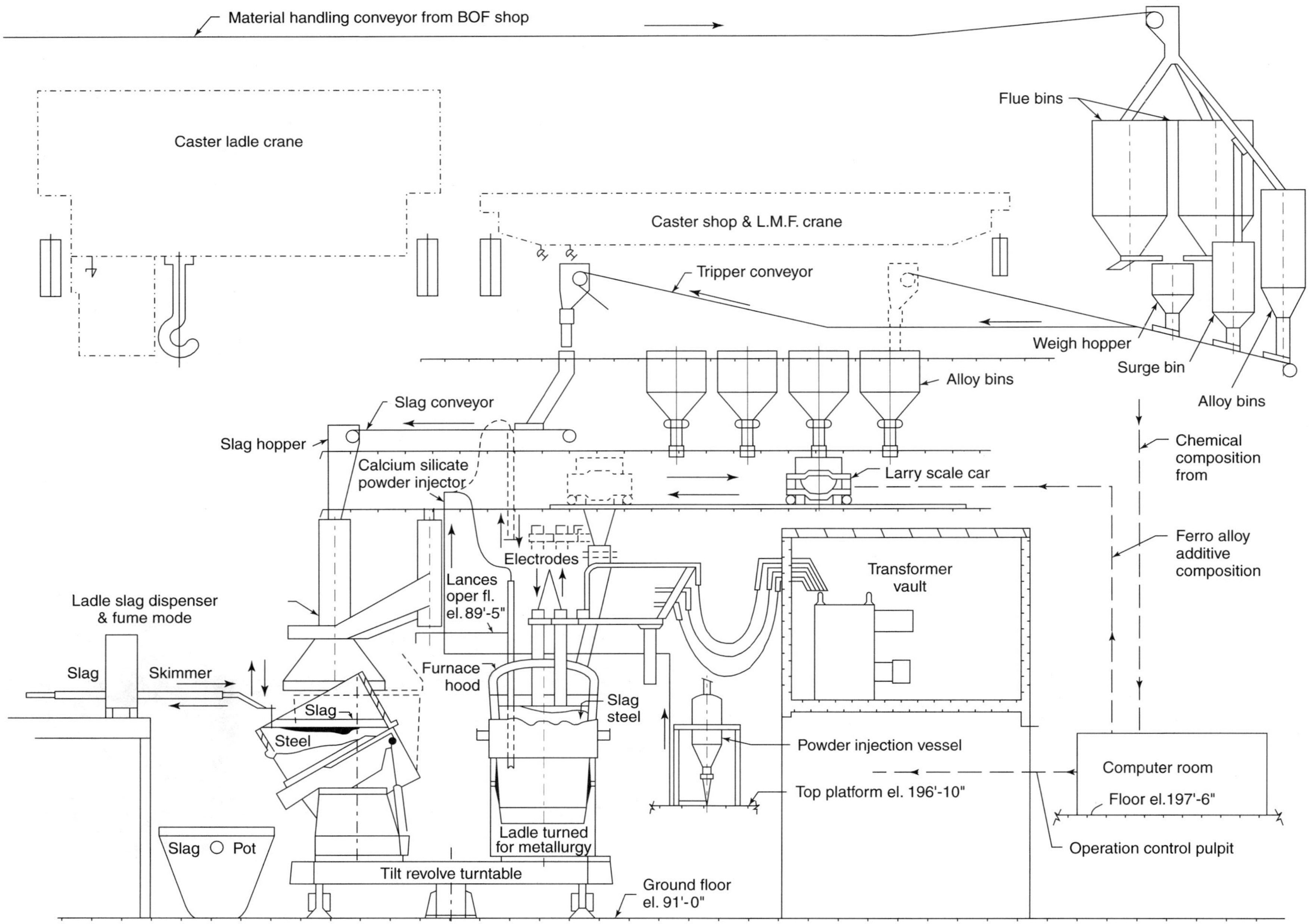

Figure 3. Schematic of a Ladle Metallurgy Facility (from Ref. 5).

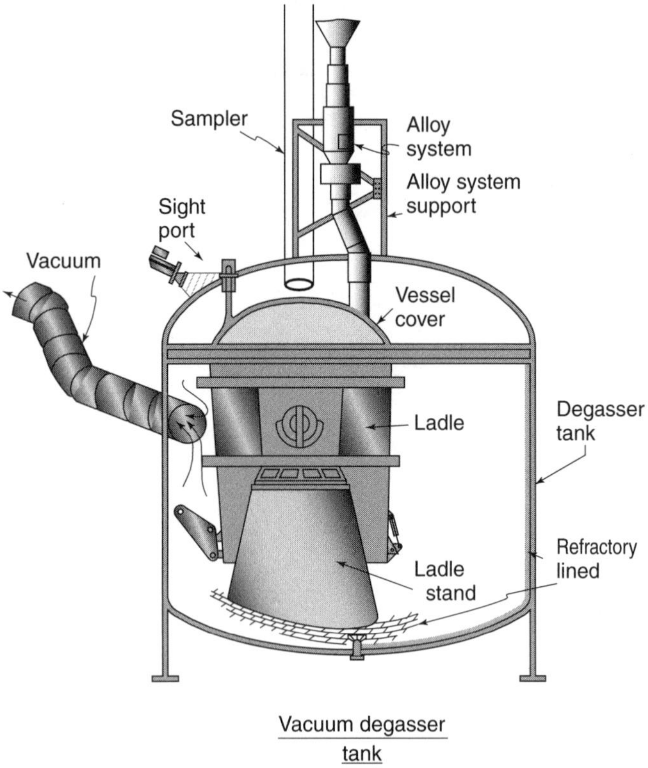

Figure 4. Schematic of a Vacuum Tank Degassing Process.

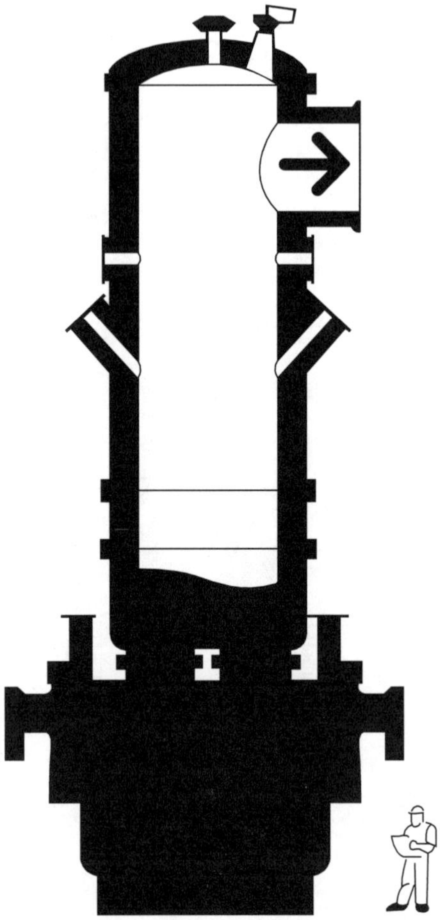

Figure 5. Schematic of the RH-OB Process (from Ref. 6).

AIR EMISSIONS CHARACTERIZATION

Ladle emissions can be either particulate or gaseous. Emission factors for ladle furnaces or other ladle metallurgy processes are not included in the U.S. Environmental Protection Agency's Emission Factor Manual (AP-42). Particulate emissions factors from two steel plants are listed in Table 1, with AP-42 emission factors for EAF operations shown for comparison.[7]

Particulate emissions include dusts from fluxes, slag, and various additives, and fumes from the oxidation of metals. Table 2 shows a typical LMF baghouse dust analysis for a shop with a ladle furnace, slag raking, and synthetic-slag facilities.[8] The analysis indicates that flux–slag compounds are the main components of the dust, since fine particulate is drawn into the hood as the granular fluxes break down. Handling, inert gas stirring, and electric-arc heating contribute to the formation of fine flux–slag particulate. The metal oxide fume is mainly formed when the molten steel vaporizes, oxidizes, and finally condenses. Operations that expose molten steel to air or oxygen promote the formation of metal oxide fume, as during slag raking, inert gas stirring, electromagnetic stirring, oxygen lancing, electric-arc heating, and pouring.

Another method of fume formation is by bubble bursting. This takes place when an entrained gas bubble erupts at the surface of the steel. Metal and slag droplets are ejected and

Table 1. Emission Factors for Ladle Metallurgy Facilities

Source	Emission Factor (lb/ton Steel)	Reference No.
Uncontrolled LMF		
Ca-Si heats		
Plant A	0.25	12, 13
Plant B	0.23	12, 13
Non–Ca-Si heats		
Plant A	0.012	12, 13
Plant B	0.016	
Controlled by baghouse	0.0077	12, 13
Fugitives at 99% capture efficiency	0.0016	12, 13
Uncontrolled EAF melting/ refining	38.0	7
Controlled EAF with direct shell evacuation and baghouse	0.043	7

Table 2. Typical Analysis of LMF Baghouse Dust

Parameter	Composition (%)
C	4.1
SiO_2	2.8
Al_2O_3	13.9
CaO	39.3
MgO	10.2
Fe_2O_3	9.3
K_2O	0.60
MnO	10.0
TiO_2	0.47
P_2O_3	0.10
Loss on ignition (1000°C)	4.4

Source: Ref. 8.

partially oxidize.[9] Operations that generate this type of fume include inert gas stirring, oxygen lancing, vacuum degassing, and the injection of certain materials that react vigorously, such as calcium silicate.

Gases emitted include carbon monoxide and carbon dioxide, which are formed during final decarburization of the steel by oxygen lancing or vacuum degassing at the LMF. Since only residual carbon is removed at the LMF, the generation of carbon monoxide is low.

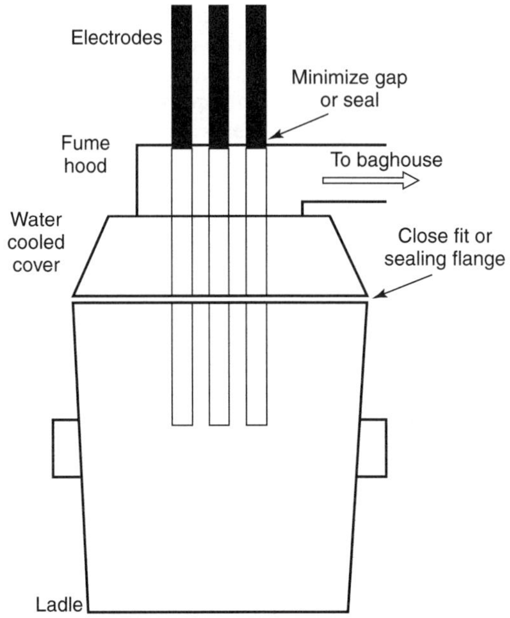

Figure 7. Ladle Furnace Cover.

AIR POLLUTION CONTROL MEASURES

The most effective measure is control at source. Contact of the molten steel with air should be minimized. Maintaining a slag layer, shrouding, and using ladle covers are examples of this type of control to reduce the formation of particulate emissions.

The degree of inert gas stirring may increase the fumes generated. High gas rates lead to vigorous agitation of the steel heat, exposing molten steel to the air. Figure 6 shows that certain practices of electromagnetic stirring can expose the steel surface.[10] Certain wire or powder additions, such as calcium silicate, greatly increase fume emissions owing to their high vapor pressures and high reactivity. When injected into the ladle, vigorous reactions result in the release of a white fume.

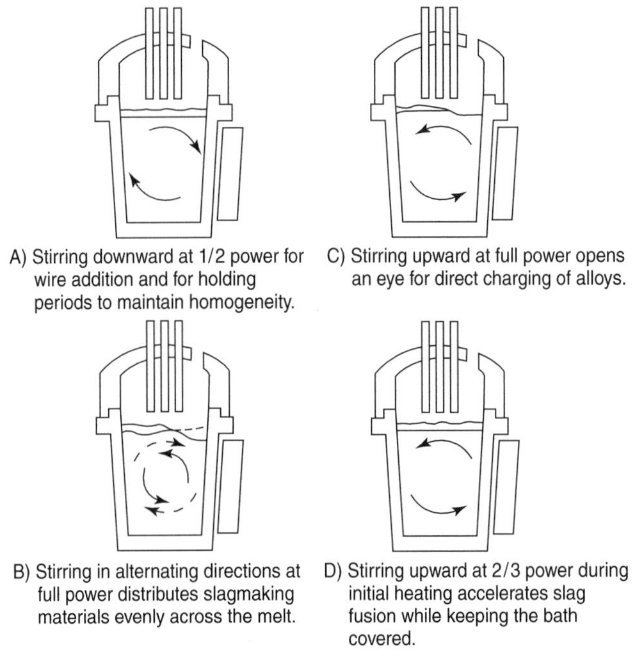

Figure 6. Electromagnetic Stirring Patterns (from Ref. 10, p. 338, Figure 5).

Direct-evacuation systems used in vacuum degassing have good particulate collection. Either the entire ladle is enclosed, as in tank degassing, or the exposed molten-steel surface is sealed from air contact by a snorkel or vacuum hood.

Close-capture hood systems are employed at various stations. Ladle furnace hoods are mainly water-cooled membrane designs similar to those used in EAFs. Refractory-lined hoods tend to experience more operational problems. Steel hoods can be damaged by heat cycling, causing poor fit and a loss of capture efficiency. Good features include a close fit to the ladle mouth and sealing flanges to maximize fume capture and minimize air infiltration (Figure 7).[11] Some close-capture hoods feature swing-away designs to allow for crane transfer of ladies. This allows lower exhaust volumes than would be required for side-draft or large canopy hoods above the crane rails.

The different stations, except for vacuum degassing, may be exhausted to a common baghouse. Also, the LMF may share a common baghouse with an EAF or other secondary emission sources in the shop. The duct system may have dampers to control flow and minimize baghouse size.

Basic information on a number of LMF fume collection systems is given in Table 3. The table shows that the baghouses

Table 3. Features of LMF Fume Collection Systems

Plant	Positive/ Negative System	Volume Flow (acfm)	Air-to-Cloth Ratio	Average Temperature (°F)	Cleaning Method	Bag Fabric	Bag Size (in.)
A	Negative	165,000	2.8:1	150	Shaker	Polyester	5 × 171
B	Negative	N/A	N/A	N/A	Pulse jet	Polyester	5 × 168
C	Negative	60,000	4.2:1	>375	Pulse jet	Nomex	5.75 × 144
D	Negative	50,000	5.5:1	145	Pulse jet	Polyester	5.75 × 147
E	Positive	65,000	6.0:1	330	Pulse jet	Nomex	4.63 × 145
F	Negative	85,000	4.5:1	130	Pulse jet	Polyester	5.75 × 146
G	Negative	100,000	4.9:1	est. 150	Pulse jet	Dacron felt	6 × 144

are predominantly negative-pressure (before the fan) units with pulse-jet cleaning and air-to-cloth ratios of over 4:1. Operating experience has shown that good features include a negative-pressure design, off-line pulse-jet cleaning, and an air-to-cloth ratio of 4:1 or less. Reported bag life ranges from 7 to 18 months.[14]

REFERENCES

1. G. Bruckmann, A. Choudhury, W. Dietrich, et al., "Vacuum Metallurgy and its Potential—A General Overview of Secondary Metallurgy Systems," *Metall. Plant Technol.* **12**(2), 8, 1989.
2. W. Hogan, "Ladle Metallurgy," *Iron Steel Eng.* **11**, 38, 1989.
3. H. Muller, "Secondary Metallurgy Plant Technology and Development," *Proceedings of the International Conference on Secondary Metallurgy*, ICS, Stahleisen, Aachen, September 21–23, 1987, pp. 85–94.
4. K. Larsudd and S. Olund, "ASEA Master Refiner-Ladle Furnace Computer Control System," *Fachberichte Huttenpraxis Metallweiterverarbeitung* **23**(8), 584, 1985.
5. G. McQuillis, J. Pruett, and J. Bell, "Ladle Metallurgy Experience at the Granite City Division of National Steel," *Steelmaking Conference Proceedings*, 1988, pp. 317–22.
6. D. Kittenbrink, W. Krajcik, and M. Szatkowski, "The Startup of the First RH-OB Vacuum Degasser in North America," *Steelmaking Conference Proceedings*, 1988, pp. 299–310.
7. *Compilation of Air Pollution Emission Factors, Vol. 1, Stationary Point and Area Sources*, AP-42, 4th ed., U.S. Environmental Protection Agency, September 1988.
8. "Analysis Report—Baghouse Dust from Ladle Metallurgy System," Internal Report, Dofasco, Inc., March 1988.
9. F. Goetz, "Mechanism of BOF Fume Formation," Master's thesis, McMaster University, Hamilton, Ont., April 1984.
10. M. Alavanja and M. Pacillas, "Inland's Ladle Metallurgy Facility—Equipment and Processing," AISE Yearbook, 1987, pp. 335–339.
11. H. Legrand and M. Amblard, "Technical Evolution of Ladle Furnaces," *Proceedings of the International Conference on Secondary Metallurgy*, ICS, Stahleisen, Aachen, September 21–23, 1987, pp. 449–60.
12. R. Sistek, "Comparative LMF Data for Two LTV Plants," unpublished data, 1987–1988.
13. F. Pendleton and R. Sistek, "PM10 Characterization of Iron and Steel Sources not Previously Included in Fugitive Emission Inventories," presented at the 83rd Annual Meeting and Exhibition of AWMA, Pittsburgh, June 1990.
14. B. Byrd and J. Jenkins, BHA Group, Inc., Kansas City, MO, Personal Communication, March 26, 1991.

FLUXED IRON ORE PELLET PRODUCTION

Gus R. Josephson

The following is an overall description of the fluxed iron ore pellet production process from the initial ground excavation to the final stage of shipping.

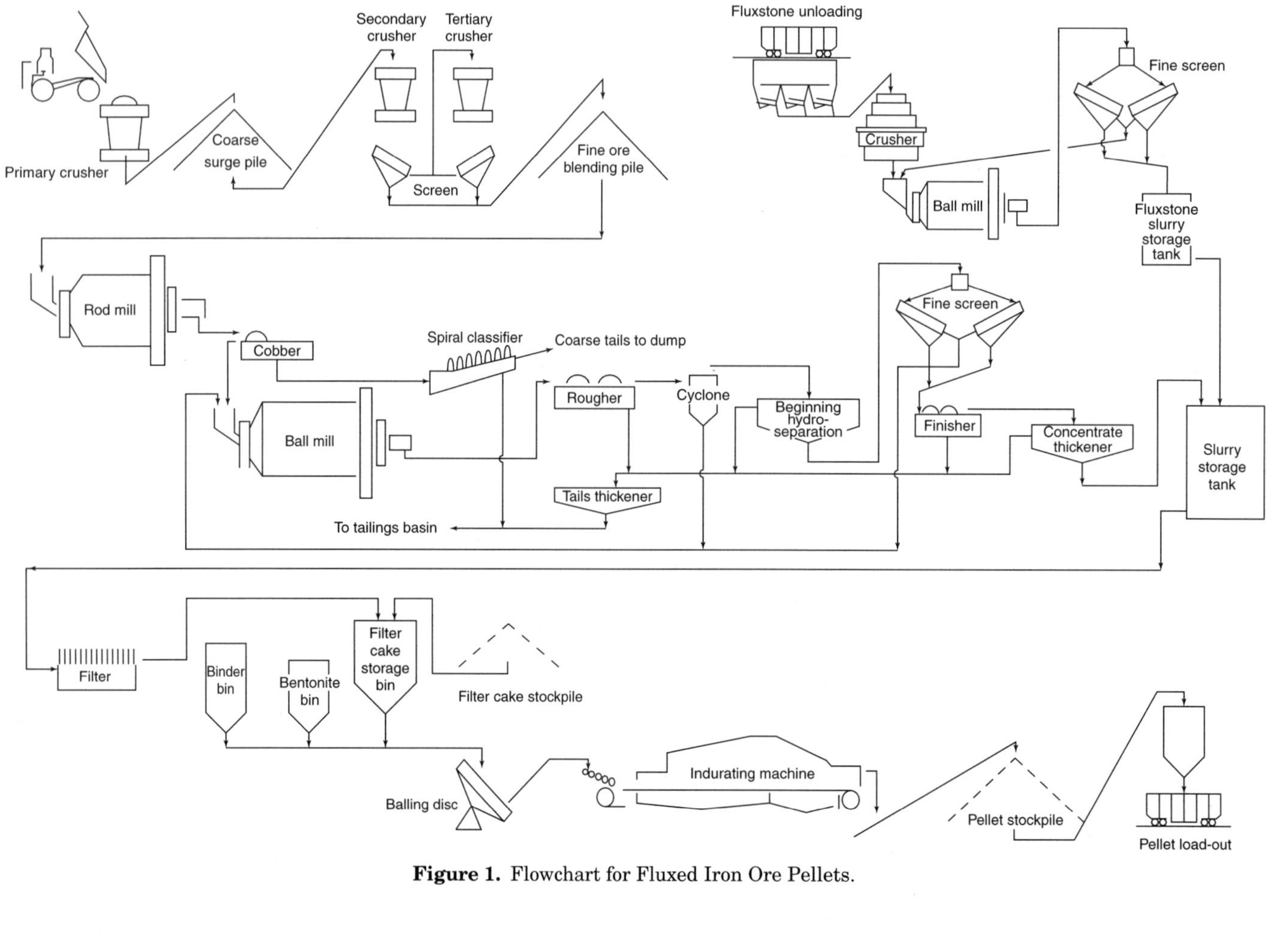

Figure 1. Flowchart for Fluxed Iron Ore Pellets.

PROCESS DESCRIPTION

The production of fluxed iron ore pellets in a traveling grate taconite plant is accomplished by a series of interrelated operations:

1. Mining the ore
2. Ore crushing
3. Ore concentrating
4. Fluxstone processing
5. Pelletizing
6. Shipping

The interrelation of these operations is depicted in Figure 1.

Mining

The mining process involves stripping overburden material to expose the ore body by using power shovels to load into the mine haulage trucks. The crude ore is then drilled using rotary drills and blasted so that it can be loaded into mine haulage trucks by power shovels for transport to the primary crusher.

Crushing

The crushing process involves reducing the size of the crude ore to less than $\frac{9}{16}$ in. through a series of gyratory and cone crushers.

Ore Concentrating

Water is added to the minus-9/16-in. crude ore, which is further ground in a rod mill to minus-eight mesh and discharged to rotating magnetic drums (magnetic separators), where the good magnetic ore is separated from the waste tailings. The good magnetic ore is next discharged to a ball mill, where it is ground to 90% minus-325 mesh. The ore concentrate and water mixture forms a slurry that is again magnetically separated over two more sets of drums to remove silica and other impurities. The ore concentrate slurry is then pumped to a storage tank.

Fluxstone Crushing and Grinding

Fluxstone (dolomite and limestone) of less than $\frac{11}{2}$ in. in size is crushed, milled, and screened to 50% minus 500. Water is added to the fluxstone to form a slurry, and the slurry is added to the iron ore concentrate slurry in the concentrate storage tank.

Pelletizing

The water is removed from the fluxed concentrate slurry with vacuum filters to about 9% moisture content. The fluxed concentrate is then mixed with a binder (organic or bentonite) and formed into $\frac{3}{8}-\frac{1}{2}$-in. pellets in a balling disk or drum much the same way snowballs form when rolling down hills. They are then fired with natural gas in a traveling grate

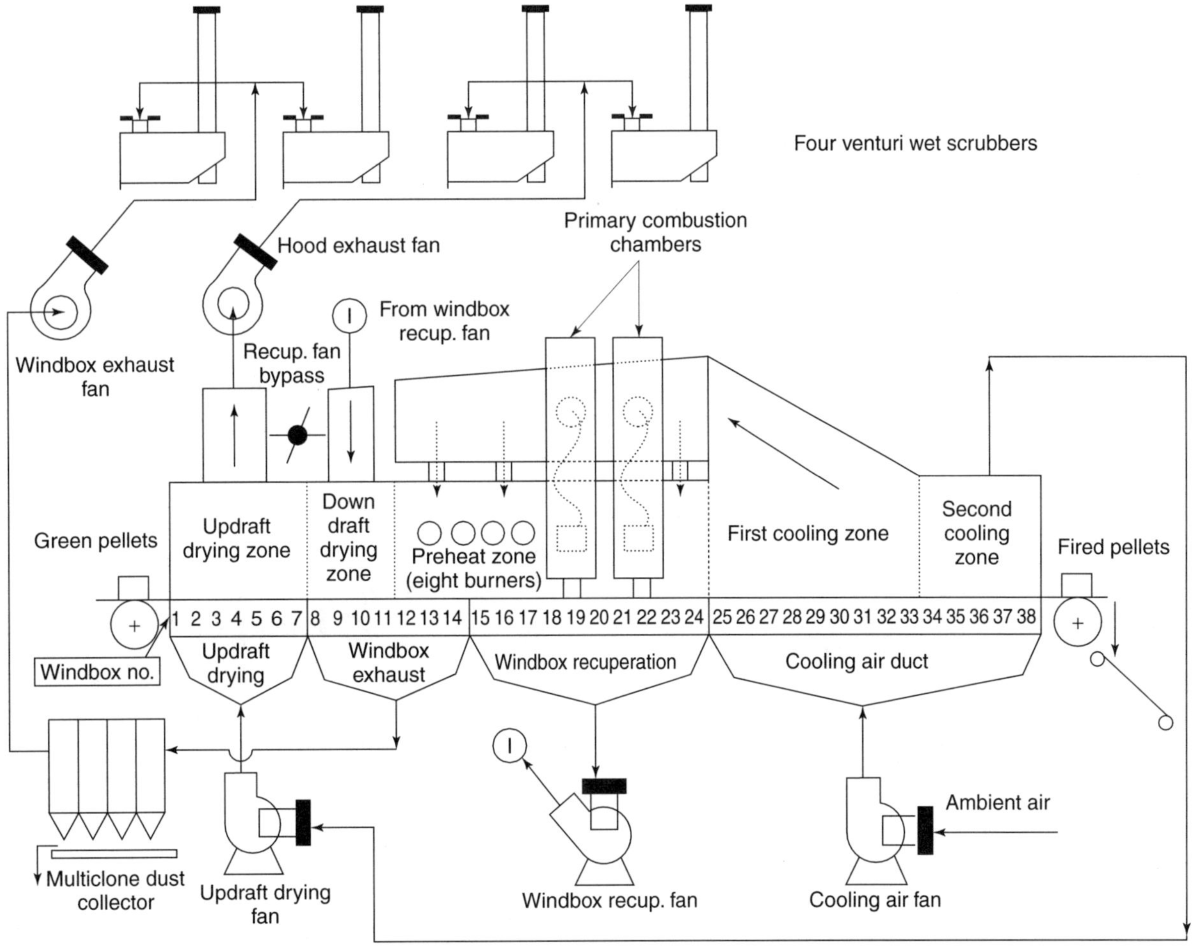

Figure 2. Fluxed Pellet Configuration for Traveling Grate Indurating Furnace.

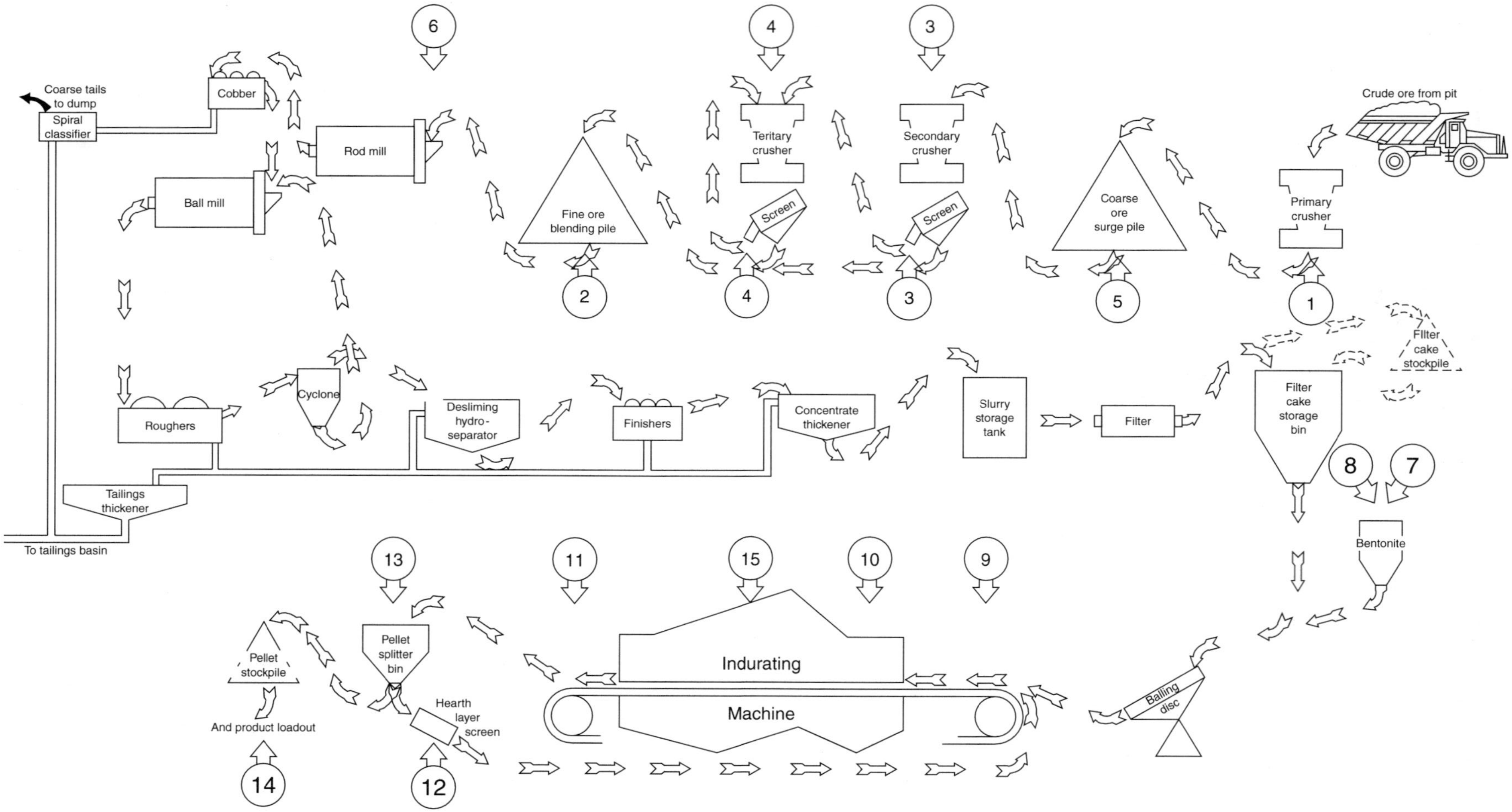

Figure 3. Taconite Simplified Flowchart, with Emission Control Points Indicated.

furnace. The green or unhardened pellets enter the furnace on top of a layer of hardened pellets and move on the traveling grate through the furnace, where they are indurated or heat hardened.

The indurating machine is divided into main drying zones consisting of wind-box sections as viewed in Figure 2. The furnace has two primary and preheat burners that provide the thermal energy to indurate the pellets.

The furnace's air distribution system is designed to preheat, indurate, and cool the taconite pellets. The induration and calcination processes occur in the combustion zone of the furnace. The above-bed air temperature reaches 2200°F in the preheat zone and 2400°F in the primary combustion zone. The combustion gases are next routed through the down-draft drying zone that is used to preheat the pellets. After passing through the down-draft drying zone, the combustion gases enter the wind-box exhaust duct, then a primary dust collector and process gas scrubber. Once the indurating process is complete, the taconite pellets are cooled in the furnace's cooling zones, as seen in Figure 2.

After cooling the pellets, the cooling air is split into the first and second cooling zone. The cooling air from the first cooling zone is distributed to the preheat and primary combustion zones. The air from the second cooling zone is recirculated to the up-draft drying zone, where the green pellets are preheated. The air is then routed to the hood exhaust fan, which exhausts the flow to the process gas scrubbers.

Shipping

The fired pellets are shipped by rail and boat to the steel mill.

AIR EMISSIONS CHARACTERIZATION AND CONTROL MEASURES

Mining

Emissions from the mining operation are generated by drilling, blasting, and hauling. Fugitive particulates are the main emission. Dust is controlled with wet drilling. Blasting emissions are minimized by utilizing controlled blasting techniques that monitor wind direction and temperature inversions. Hauling emissions are controlled by using dust suppression (water, lignosulfanate, chloride) on roads (see Figure 3).

Crushing

Emissions are generated from the crushing itself and the ore transfer (conveying). Emissions from the crushers, feeders, and transfer points on conveyors are captured by local hoods and ducted to a baghouse.

Concentrating

Emissions from grinding and concentrating occur as fugitive dust, where the ore is transferred to the rod mills. Emissions are controlled at these transfer points by local hoods venting to a wet scrubber.

Fluxstone Processing

Emissions from fluxstone processing occur primarily in crushing and conveying as fugitive dust. Emissions are controlled by hooded pickup points vented to a wet scrubber.

Pelletizing

Emissions from the pelletizing operation are nitrogen oxide, sulfur dioxide, carbon monoxide, and carbon dioxide particulates. Particulates are controlled by a dust collector and gaseous emissions are controlled by process gas scrubbers.

Shipping

Emissions from loading and shipping are from fugitive dust. Emissions are controlled by applying dust suppressants and water sprays.

BIBLIOGRAPHY

Fluxed Pellet Conversion Operating Manual, Dravo Corp., Pittsburgh, PA, July 1987.

PICKLING

R. M. Hudson
(with update by Wayne T. Davis, Editor)

During the hot rolling of steel in air, an oxide scale forms that must be removed before subsequent forming or processing operations. The process of removing scale by contacting steel with acid solutions is termed pickling.

PROCESS DESCRIPTION

Bars, rods, and tubing are usually immersed in batches in tanks containing solutions of sulfuric acid. After a sufficient time has elapsed for scale to be removed, the work is lifted out of the pickling tank, rinsed with fresh water, and dried. For alloy steels, scale removal may require the use of other acids, such as nitric or hydrofluoric acid. Scale is removed from hot band by continuously passing the strip through several successive tanks that contain solutions of either sulfuric or hydrochloric acid. After pickling, the strip is rinsed with water, dried, and usually oiled before recoiling.

Because of the shorter immersion times available during continuous strip pickling, higher temperatures and acid concentrations are required than in batch pickling to assure complete scale removal. To maintain satisfactory pickling rates, fresh acid additions are made to offset the lowered acid concentrations that result when scale dissolves to form iron salts and water. Inhibitors are generally added to the acid solutions to minimize acid attack on the base metal with formation of iron salts and hydrogen gas. Control of the pickling operation requires a knowledge of solution concentrations and temperatures.

AIR EMISSIONS CHARACTERIZATION

A spray comprising droplets of acid solution is formed during hydrogen evolution and the spray and water vapor are present in the air above the surface of pickling solutions. When hydrochloric acid is used, hydrogen chloride vapor is also present in concentrations that increase with solution temperature and with increases in the concentration of hydrochloric and/or ferrous chloride in the solution.

AIR POLLUTION CONTROL MEASURES

An air exhaust system and scrubber are used to recover acid constituents present; the scrubber water can be used for makeup additions to the pickling tanks. Waste acid solutions can be subjected to a regeneration process or treated to prepare iron salts for industrial applications.

Section 112(c) of the Clean Air Act Amendments of 1990 required the EPA to list each category of major and area sources, as appropriate, emitting one or more of the HAPs listed in section 112(b) of the Act. On July 16, 1992 (57 FR 31576), the EPA published a list of major and area sources for which NESHAPs were to be promulgated, followed by a schedule for promulgation of those standards (58 FR 63941, December 3, 1993). "Steel Pickling–HCl Process" was included on the list of major sources for which EPA was required to establish national emission standards. The term *major source* means a source emitting 10 tpy or more of any one HAP or 25 tpy or more of any combination of HAPs. The EPA proposed national emission standards for this source category on September 18, 1997, and the final rule became effective on June 22, 1999.

The final rule required that new and existing pickling lines that use hydrochloric acid as the primary pickling solution, hydrochloric acid regeneration plants, and acid storage tanks meet emission standards reflecting application of the maximum achievable control technology (MACT). Implementation of the rule was expected to reduce HAP emissions by more than 2,200 megagrams per year (Mg/yr) (2,500 tons per year, tpy) from levels that existed in 1999.[1,2]

Conditions that distinguish pickling from other operations such as cleaning or surface activation are now defined such that each new or existing pickling line (batch or continuous process) using an acid solution in any tank in which hydrochloric acid is at a concentration of 6% by weight or greater and has a temperature of 100°F or greater is subject to the rule. For the purposes of the rule, steel pickling is limited to hydrochloric acid pickling of carbon steels, which contain approximately 2% or less carbon, 1.65% or less manganese, 0.6% or less silicon, and 0.6% or less copper. An acid regeneration plant includes the collection of equipment and processes configured to reconstitute fresh hydrochloric acid pickling solution from spent pickle liquor using a thermal treatment process. A new or existing plant that regenerates only pickling solution other than HCl is not subject to the rule. The rule is not applicable to facilities that pickle only specialty steels. Specialty steel means a category of steel that includes silicon electrical, alloy, tool, and stainless steels. Specialty steels are pickled by a process that may include the use of hydrochloric acid but also includes the use of other acids, which may be mixed with hydrochloric acid in the same pickling bath or used in separate baths as part of a multiacid/multibath pickling sequence. The EPA has indicated that it will determine at a later date if the specialty steel pickling process should or should not be subject to the requirements of a rule that limits HCl emissions. The emission standards for the processes are shown in Table 1.

The compliance date for the operations of Table 1 is June 22, 2001.

Table 1. Emission Standards for Affected Sources

Affected Source	Emission Standard
Pickling line	
Existing	HCl concentration in air pollution control device or process exhaust gas no more than 18 parts per million by volume (ppmv) or air pollution control device minimum HCl collection efficiency of 97%
New	HCl concentration in air pollution control device or process exhaust gas no more than 6 ppmv for continuous lines and 18 ppmv for batch lines or air pollution control device minimum HCl collection efficiency of 99% for continuous lines and 97% for batch lines
Hydrochloric acid regeneration plant	
Existing	HCl concentration in air pollution control device or process exhaust gas no more than 25 ppmv and chlorine concentration in air pollution control device or process exhaust gas no more than either 6 ppmv or a source-specific maximum concentration limit
New	HCl concentration in air pollution control device or process exhaust gas no more than 12 ppmv and chlorine concentration in air pollution control device or process exhaust gas no more than 6 ppmv
Hydrochloric acid storage vessel	
Existing and new	Cover and seal all openings and route emissions to air pollution control device or alternative control system; and use enclosed line or local fume capture system vented to air pollution control device or alternative control system at each point where acid is exposed to atmosphere

REFERENCES

1. *Federal Register*, Vol. 62, No. 181, pp. 49051–75, 40 CFR Part 63, *National Emission Standards for Hazardous Air Pollutants for Source Categories; National Emission Standards for Hazardous Air Pollutants for Steel Pickling Facilities—HCl Process*; Proposed Rule, September 18, 1997.
2. 40 CFR Part 63, *National Emission Standards for Hazardous Air Pollutants for Steel Pickling—HCl Process Facilities and Hydrochloric Acid Regeneration Plants*, U.S. EPA, Final rule, June 22, 1999.

BIBLIOGRAPHY

ASM Metals Handbook, Vol. 5: Surface Cleaning, Finishing, and Coating, 9th ed., 1982, pp. 68–82.

DIRECT REDUCTION

Murray S. Greenfield

The blast furnace is the dominant method of smelting iron from oxides. Coke or charcoal is required to form a porous bed in the hearth of the furnace for the passage of the reducing gases. North American coke oven batteries are not being replaced when they are phased out. Steel producers using scrap-based electric-arc furnaces are moving upscale rolling products that require lower levels of residual elements available from direct reduced iron (DRI). The amounts of DRI available and the number of hot-metal facilities are predicted to grow substantially.

According to Table 1, in 1990, direct reduction plants produced 17,880,000 metric tons per year, or 2–3% of the world's steel production, and an additional 10 million tons[1] per year capacity is under construction. Direct reduction covers the nonblast furnace techniques for producing metallic iron from oxides. Some processes produce hot metal (molten iron), while others produce a metallized lump that is suitable as a scrap substitute for an electric-arc furnace.

The direct reduction techniques can be categorized into four main types:

1. The vertical shaft type using reformed gas as the reductant. Commercial variations in this category include Midrex, supplied by the Midrex Corp., and HYL, supplied by Hysla. In 1990, Midrex and HYL accounted for about 90% of direct reduction production.
2. The rotary kiln or rotary hearth type using coal as the reductant (for example, SLRN from Lurgi, CODIR from Mannesman Demag, Fastmet from the Midrex Corp., Inmetco from Inco).
3. The melter gasifier using coal as the reductant. Partially reduced iron oxides are smelted with noncoking coals in a melter gasifier. (Corex from Deutsche Voest-Alpine is of this type.)
4. Hot metal bath (HMB) using coal as the reductant. The bath may be contained by a hearth or converter. (This type includes HIsmelt from the HIsmelt Corp., NKK, DIOS, Nippon Steel, AISI, etc.)

Table 1. 1990 Production and Market Share by Process of Steel-Making–DRI[a]

Process	Fuel	Production (tonnes per year)	Total World Production (%)
Midrex	Gas	10.84	60.6
Hyl I	Gas	3.69	20.6
Hyl III	Gas	1.55	8.6
SL/RN	Coal	0.84	4.7
Fior	Gas	0.39	2.2
Codir	Coal	0.19	1.1
Accar	Coal	0.09	0.5
DRC	Coal	0.14	0.8
Tisco	Coal	0.05	0.3
Purofer	Gas	0.05	0.3
DAV	Coal	0.03	0.2
K-M	Coal	0.02	0.1
Total		17.88	100.0

[a]Preliminary figures as of January 31, 1991.

PROCESS DESCRIPTION

Vertical Shaft Using Reformed Gas

The vertical shaft using reformed gas produces over 86% of the world's DRI. As shown in Figure 1, heated reformed reducing gas at temperatures of up to 930°C is injected into the reduction zone of the vertical shaft through a refractory annular duct countercurrent to the descending pellet ore charge. An analysis of the reducing gas is shown in Table 2. The hydrogen (H_2) content of the reducing gas produces a specific output of 3.5–5.0 tons of direct reduced material per hour per square meter of hearth area.

The system is economic only if low-priced fuels such as natural gas, naphtha, or coke-oven gas are available. About 2.3 to 2.5 Gcal of natural gas is required per ton of DRI product. The furnace is divided into an upper reducing zone and a lower cooling zone. Heavy oil and coal require gasifiers to produce the gas for reforming. The reducing gas from the reformer is introduced into the furnace at about 800–870°C through

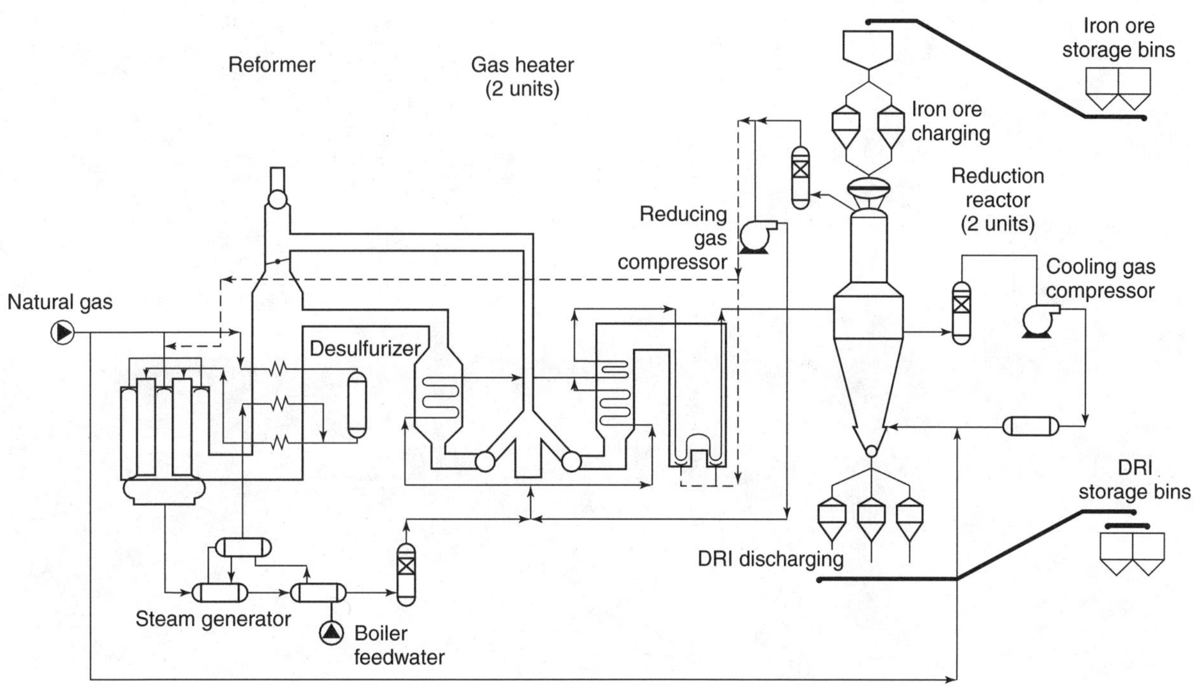

Figure 1. Process Flow Diagram for One Module of Sicartsa's HYL III Plant (Reprinted Courtesy of *Iron & Steelmaker*).

Table 2. Analysis of Reducing Gas

Reformed Gas	
Component	Percent
H_2	73
CO	17
CO_2	7
CH_4	3

numerous openings around the periphery of the furnace. The spent reducing or top gas is passed through a scrubber and cooled to about 40°C. About one-third of the cleaned gas must be purged from the circuit to remove carbon dioxide (CO_2) and nitrogen (N_2). The other two-thirds of the gas is recompressed, mixed with fresh reformed gas, and recirculated back to the bottom of the reactor, where it cools the descending hot material. The pressure inside a Midrex and HYL reactor is 1 bar and 5.5 bar gauge, respectively.

Rotary Kiln Using Coal

As shown in Figure 2, the rotary kiln is another device used to process pellets or lump ore. Flux, ore, or pellets are fed into the furnace countercurrent to the gas flow. The charge must be exposed to temperatures exceeding 900°C for 5–7 h to achieve 92% metallization. The coal can be fed either with the pellets or close to the discharge from the kiln. Coal fed closer to the discharge[2] reduces the coal consumption. Carbon monoxide (CO) is formed from the char/ore bed at the bottom of the hearth from the fixed carbon (C) as well as the volatile fraction of the coal. There is some direct reduction from the contact of the carbon with the ore. Sufficient flux is also fed into the kiln to trap the sulfur (S) from the coal. After cooling, the direct reduction material is separated magnetically from the residual char and ash. The ash and char are separated and the char is fed back with the feed to the kiln.

Costs are lowest for coal-based technologies,[2] but those savings are offset by the difficult process control, higher energy consumption, poor reactor availability, and the presence of char in the product.

Melter Gasifier

As shown in Figure 3, iron ore is partially reduced in a reduction furnace followed by smelting in a melter gasifier. The coal and iron units are fed by separate lock hoppers into the top of the gasifier melter kept at a pressure of 2–5 bar. About 1100 kg of coal per ton of iron is dried and gasified during its fall through the upper part of the furnace. The flow rate in the cylindrical part of the furnace is controlled to maintain a stable fluidized bed. In the lower part of the furnace, about 600 nm^3 of oxygen per metric tonne of hot metal is injected by tuyeres to gasify the coal. Temperatures at the top of the cylindrical section and in the dome produce a reducing gas containing approximately 65–73% CO, 3–4% CO_2, and 20–25% hydrogen. Before the gas is passed into the prereduction section, it is cooled to 850°C and passed through cyclones. It then serves as a reducing gas in the reduction shaft furnace. The flow from the prereduction section is about 1500 m^3/h of gas per ton of DRI containing 48% CO, 34% CO_2, and 16% hydrogen. An additional volume of gas surplus to the needs of the prereduction system is available, depending on the degree of metallization desired.

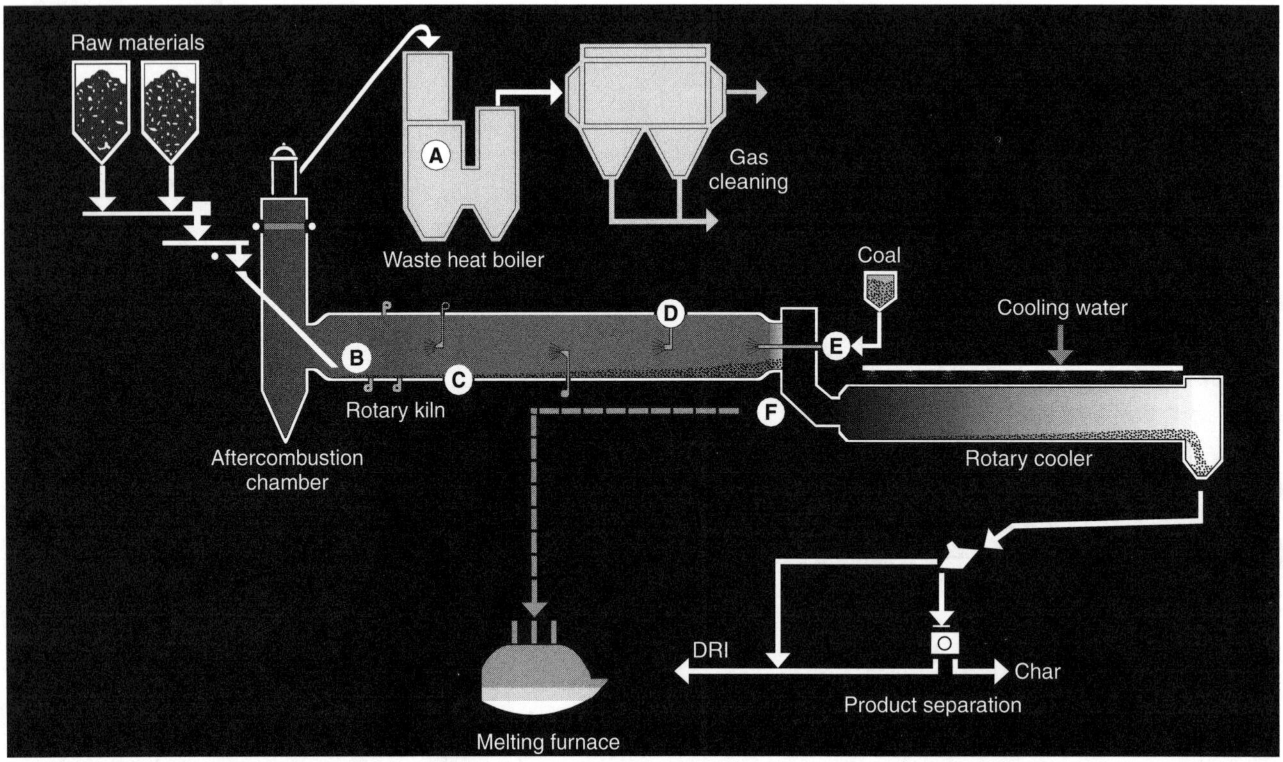

Figure 2. SL/RN Coal-Based Direct Reduction (Reprinted Courtesy Verlag Stahleisen).

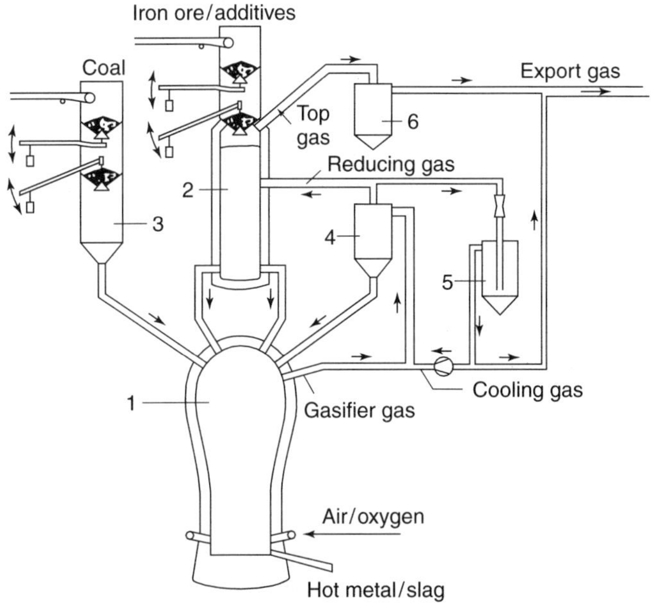

Figure 3. Corex Process: Melter Gasifier (1), Reduction Shaft Furnace (2), Coal Feed System (3), Hot-Dust Cyclone (4), Cooling Gas Scrubber (5), Top Gas Scrubber (6) (Reprinted Courtesy *Iron and Steel Engineer*).

Hot Metal Bath

The fourth variant is the iron bath type. A schematic of a horizontal two-zone countercurrent smelter is shown in Figure 4.[13]

The converter or hearth is first filled with liquid iron, or a heel of liquid from a previous heat is left in. The coal that is fed into the iron bath is gasified by the temperature of the molten iron along with the oxygen that is also injected into the bath. The smelting occurs in the thick, foamy slag layer on top of the hot metal. The oxygen injection provides sufficient agitation to achieve high smelting rates. Either ore or, preferably, hot prereduced wustite-rich ore concentrate is fed into the bath. The reduction and oxidation zones are kept separate by a physical barrier.

Part of the CO evolved is postcombusted above the bath in the hearth and the radiant heat is absorbed by the slag. The off-gases may be directed to a prereduction shaft furnace or used for combustion in boilers and reheat furnaces.

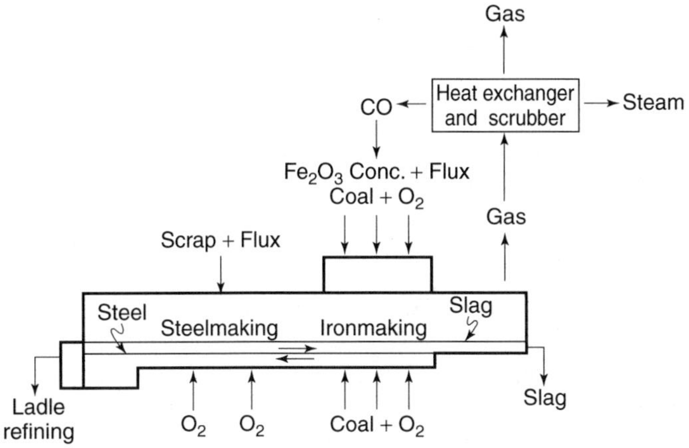

Figure 4. Horizontal Two-Zone Countercurrent Smelter (Reprinted Courtesy of *Iron & Steelmaker*).

AIR EMISSION CHARACTERIZATION

Vertical Shaft Using Reformed Gas as the Reductant

Air emissions from the shaft furnace occur from three major areas. The first is the material-handling system. The emissions from the pellet handling are strongly dependent on the quality of the pellet and the moisture content. Pellets tend to dust at each transfer location. Another possible emission location is the feed to the shaft furnace. If the devices sealing the lock hopper are a problem, then the pressurized gases inside the shaft may escape.

Fuel (usually natural gas) is required to produce the feed to make the carbon monoxide–hydrogen reformed gas in the reformer, as well as the combustion energy to heat the reformer catalyst tubes externally. Some of the purged process gas is used as the combustion fuel. Air emissions[3,4] from the Midrex and HYL processes are illustrated in Tables 3 and 4, respectively.

The oxides of nitrogen (NO_x) emissions[3] from a Midrex plant are much lower than would be expected considering the flame temperature and air-to-fuel ratios used. Ammonia is formed as a trace by-product in the reformer furnace by the reaction of nitrogen and hydrogen over the catalyst. The ammonia reduces the NO_x. The quantity of purged gas[5] may be reduced by installing a scrubber to remove the CO_2 from the recirculated top gas. The amount of hydrogen and CO removed from the circuit to the combustion section is reduced.

If more than 1 ppm of sulfur is present in the fuel to be reformed, it must be removed to avoid poisoning the reformer catalyst. The process gas will remove sulfur from the ores in

Table 3. Typical Emission Levels

	Kilograms Per Ton of Product (Concentration, mg/m^3)		
	Particulate	SO_2	NO_x
Reformer	0.012	0.035	0.172
	(14)	(40)	(196)
Shaft furnace	<0.001	<0.001	<0.002
	(25)	(40)	(196)
Dust collection[a]	0.016	0.000	<0.001

[a] A Midrex facility typically contains three dust-collection systems.

Table 4. Emissions to the Air

	Kilograms Per Tonne of DRI			
	Material Handling	Combustion Systems	Dust Collection	Total
Particulates	0.001	0.006	0.032	0.039
SO_2	—	0.0002	—	0.0002
NO_x	—	0.050	—	0.050
CO	—	0.004	—	0.004
HC	—	0.001	—	0.001
Total	0.001	0.0612	0.032	0.0942

Table 5. Air Pollution Control Measures

	Contaminant Source		
	Process Gas	Raw-Material Feed	Product Discharge
Vertical shaft, reformed gas	Purged gas from reformer can be cleaned using a high-energy scrubber or (ESP). Moisture is too high for fabric filters. Any sulfur removed from the reformer circuit may be recovered or vented to the atmosphere.	Pellet transfer points must be collected and scrubbed.	Product is hot and dusty. Tight hoods are required at product transfer locations.
Rotary kiln	The medium-heat-content kiln exhaust gas can be combusted in an afterburner or boiler. Use of a boiler would allow use of an ESP for particulate control. An afterburner would require a high-energy scrubber for particulate. Adequate control or hydrocarbons could be attained with close control of the CO in the combustion exhaust. Medium-heat-content gases do not produce high NO_x. The SO_x and NO_x control technology would be the same as with standard utility boilers.	Lime, coal, and oxide transfer points require collection and cleaning. Fabric filters would be preferred.	Product is hot and dusty. Same control as above.
Hot metal bath	The particulate emissions from the HMB system would be similar to those from a BOF or combined blown steel-making vessel. PAH emissions may not be destroyed as the temperature at the mouth of suppressed combustion hoods may only reach 840°C. The medium-heat-content gas from the bath would be used for prereduction or combusted in a boiler. Much of the SO_x wold be retained by the oxides in the prereduction section. Further SO_x control after combustion could be attained using standard utility boiler technologies.	Same as for BOF.	The product is molten hot metal or steel. Control techniques applicable to BOF tapping are applicable.
Melter gasifier	A low-to medium-heat-content gas is produced from the gaisfier. The gases would either be used in a prereduction stage or be used for combustion in a boiler. Standard utility boiler cleanup systems could be used for NO_x and SO_x control.	Same as for blast furnace.	The product is hot metal. Control techniques are as discussed under blast furnaces.

the reduction zone and deposit it on the reduced iron in the cooling zone. A limited amount of sulfur from the ores will be emitted to the atmosphere when the purge gas is burned as a combustion fuel.

Rotary Kiln/Hearth

Rotary unit air emissions result from raw materials handling and from the process gas exhaust and the product screening and carbon char return to the kiln. The total dust collected[6] from the raw materials and the product-handling system for a kiln amount to from 200 to 300 kg per tonne of DRI produced. The waste gas from the kilns generates 600 kg of dust per tonne of DRI handled. The volume of gas from a kiln is partially dependent on the volatile content of the coal used and the efficient utilization of the coal. An optimum fill[7] must be achieved, and the bed must be kept in a rolling mode to minimize the segregation of iron ore and reductant so that coal consumption is minimized.

Changing the coal feed to the discharge end[7] reduced the energy consumption from 20 to 15 GJ per ton (coal use was reduced from 690 to 530 kg of coal per ton of DRI). The waste gas contains CO, hydrogen, and hydrocarbons. The excess air–oxygen conditions in the kiln will influence whether the sulfur is present as oxides or hydrogen sulfide (H_2S). Twenty-seven percent of the sulfur in the feed[7] reverts to the gas as sulfur dioxide (SO_2). The remainder is tied up in the product and the waste flux. The gas volume from a kiln[8] varies from 3300 to 6100 nm^3 per ton of DRI. The waste gas from the kiln contains from 20%[2] to 65 V of the energy input (3 GJ per ton of DRI at the 20% level). The retention time of gases[9] in a kiln is from 5 to 9 sec at temperatures varying from 1200°C at the burner end to 500°C at the exit end, favoring destruction of the polycyclic aromatic hydrocarbons (PAHs). Coal fed closest to the product discharge may also minimize PAH emissions. The medium-heat-content gas from the kiln may be passed through an afterburner or combusted in a boiler.

Melter Gasifier

The melter gasifier generates about 1500 m^3 per ton of DRI or gas per ton of hot metal. The calorific value of the gas (termed export gas) is approximately 7600 kJ/nm^3. Therefore, this gas is similar to coke oven gas and would be used as a fuel for reheating furnaces or boilers. Much of the sulfur from the coal would have an affinity for the slag. Part of the sulfur would leave the smelter as H_2S, most of which would be removed by the oxides in the prereduction shaft as its level[10] in the export gas is less than 50 ppm.

Hot Metal Bath

The emissions from the HMB systems are similar to those from basic oxygen furnaces (BOFs). The major difference is the influence of the coal on the volatile and nonvolatile organic emissions from the furnaces.

In a case[11] where the ratio of the slag cavity formed by the top oxygen jet (L_s) to the foaming slag thickness (L_{so}) exceeded 1, the ferrous content in the dust increased to 100 g/nm^3. At ratios of less than 1 and low mixing intensities, the ferrous content in the dust decreased to less than 10 g/nm^3. Control of the slag V ratio[11] and the suspension of coal char or coke in the slag is important for the control of foaming. The size range of particulates from the HMB would be expected to be similar to that from a top-blown or combined blown-oxygen furnace. Uniform coal feeding rates should result in the destruction of volatile matter fed to the furnace.

Processes[12] that operate at extremely high temperatures with gases that contain nitrogen from infiltration air and the nitrogen content of the oxygen used for postcombustion have a high probability of producing high NO_x concentrations. Postcombustion using low-nitrogen-containing oxygen and limiting the ingress of air into the hoods or ducts above the furnace will limit NO_x emissions. Most of the sulfur from the coal and the metallic feed would stay with the slag and metal.

AIR POLLUTION CONTROL MEASURES

See Table 5.

REFERENCES

1. A. Chatterjee, "State of Direct Reduction Ironmaking in India," *Steel Technol. Int.* **49**, 1990–91.
2. W.-D. Hausler, "Improvements in the Codir Direct Reduction Process," *Steel Technol. Int.*, 1990–91.
3. T. W. Hoffman, "The MIDREX Direct Reduction Process: An Environmentally Sound Route to Steel Making," *UNEP Industry Environ.*, 1984.
4. J. M. Pena, "Environmental Aspects of the HYL III Direct Reduction Process," *UNEP Industry Environ.* **9**, 1984.
5. P. Heinrich, K. Knop, and R. Madlinger, "Status Report and Potentials for Development of HYL-III Direct Reduction Technology," *Metallurg. Plant Technol. Int.* (2), 1990.
6. S. Das Gupta, "Environmental Aspects of Electric Arc Furnace and Direct Reduction Plants in India," *UNEP Industry Environ.*, 1984.
7. S. P. Mehrotra and J. K. Brimacombe, "Kinetics of Gasification of Coal Char with CO_2 in a Rotary Reactor," *Ironmaking Steelmaking* **17**(2), 93, 1990.
8. L. Formanek, H. Eichberger, and H. Serbent, "Energy Utilisation in Direct Reduction Using Rotary Kilns," *Metallurg. Plant Technol.* **2**, 9, 1988.
9. B. R. Niihawan, "Environmental Aspects of the Direct Reduction Route to Steel Making," *UNEP Industry Environ.* **25**, 1984.
10. R. Steffen, "The Corex Process: First Operating Results in Hot-Metal Making," *Metallurg. Plant Technol. Int.* **2**, 20, 1990.
11. T. lbaraki, M. Kanemoto, S. Ogata, et al., "Development of Smelting Reduction of Iron Ore — An Approach to Commercial Ironmaking," *Iron Steelmaking* **30**, 1990.
12. J. C. Agarwal, "Strategic Considerations in Direct Steelmaking," *Iron Steelmaking* **18**(3), 30, 1991.
13. P. E. Queneau, "Direct Steelmaking — Quo Modo," *Iron Steelmaking*, 76, 1990.

ROLLING

Michael T. Unger

Rolling steel involves working metal to attain permanent deformation by applying mechanical forces to the surface. The objective may be to produce a desired shape or size (mechanical shaping) or to improve the physical properties of the metal (mechanical treating).

Rolling operations are usually subdivided into hot- and cold-working. Forces required to deform metal in hot rolling are sensitive to the rate of applied loads and to temperature variations. The resultant strength of the metal after deformation is changed very little. During cold rolling, the deformation forces are relatively insensitive to load application and to temperature variations, but the basic strength of the cold-worked metal is increased.

Semifinished forms (slabs, blooms, and billets) are made by the ingot process or the newer continuous-casting process. In the ingot method, the finishing stages of steel-making begin when ingots are lowered into furnaces called soaking pits that reheat them to an even temperature for rolling. The ingot is rolled in a primary mill into semifinished slabs or into rectangular shapes called blooms. In the newer continuous-casting process, steel is poured directly into a tundish, which feeds a curved mold where the steel is solidified directly into semifinished products. The end product can be slabs, blooms, or billets, depending on the final proportions.

Shaping by rolling consists of passing the steel between two rolls that revolve at the same speed, but in opposite directions, as shown in Figure 1. If spaced so that the distance between the rolls is less than the height of the section entering them, the rolls will grip the piece of metal and deliver it reduced in the cross-sectional area and increased in length.

PROCESS DESCRIPTION

During hot rolling, steel heated to 2300°C is placed onto the end of the rolling line. For example, a slab's metallurgical analysis, size, weight, and intended end use are fed into the mill's computer. Sensing devices continuously record the slab's thickness, width, and temperature, and the squeezing force of the rolls as the slab moves through the roughing and finishing stands, as shown in Figure 1. The computer instantaneously resets controls or changes the speed of rolling stands to ensure a properly finished product. Steel rolled on the hot strip mill can be sold in this form or further processed into cold-rolled, galvanized, or aluminized steel.

Air emissions from hot-rolling mills generally consist of water vapor from scale breaking or metallurgical water sprays and are typically uncontrolled. However, there are potential emissions from rolling solution aerosols that are localized close to the rolling stands, as well as small pieces of scale that may become airborne. Emissions from combustion sources for the soaking pits and reheat ovens are frequently controlled.

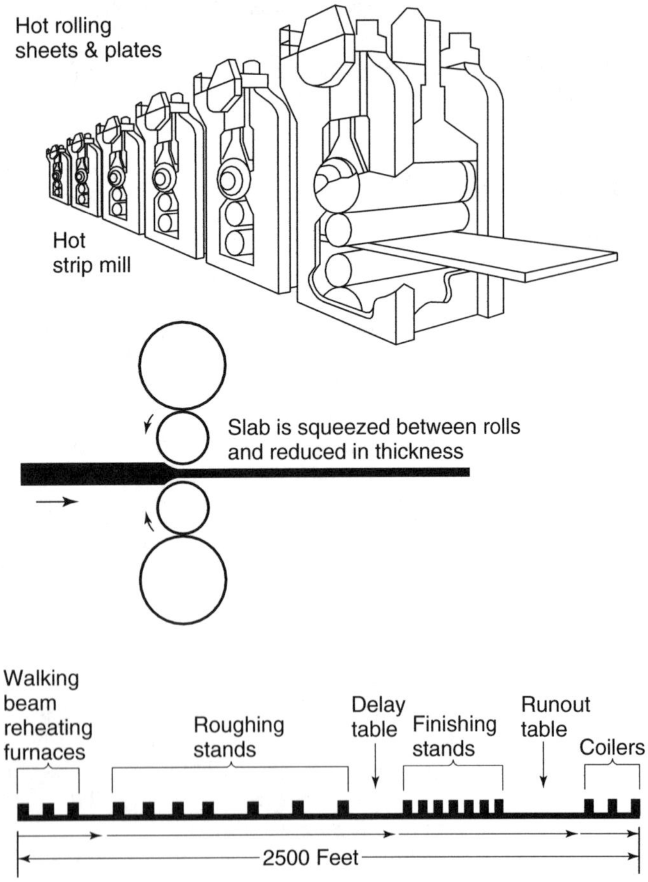

Figure 1. Hot Rolling Schematic.

Hot-band or steel rolled on the hot strip mill can be fed into a multistand cold-rolling mill. Without reheating the steel, the cold mill rolls steel under great pressure to precision thickness and produces specific surfaces and develops controlled mechanical properties. A schematic of a cold-rolling mill started in 1990 is shown in Figure 2. This mill exhibits best available control technology (BACT) and is expected to release minimal quantities of air pollutants because of the high level of pollution control equipment.

AIR EMISSIONS CHARACTERIZATION AND CONTROL MEASURES

During state-of-the-art cold rolling, steel is introduced onto the line and passes through the major processing steps shown in Table 1. Uncontrolled emission estimates, calculated controlled emissions, the types of air pollution control utilized, and the techniques' expected control efficiencies are also summarized in Table 1.

In both hot and cold rolling, one can expect nitrogen oxides (NO_x) emissions resulting from combustion in reheat furnaces, in annealing furnaces, or from boilers. The primary means of controlling NO_x emissions is through combustion modification or selective catalytic reduction. Nitrogen oxides formed from combustion air constituents are referred to as thermal NO_x. The general technique to control thermal NO_x is to suppress the natural gas combustion temperature to below 2600°F. Above this temperature, NO_x formation is exponential, while below this temperature, it is linear at a very limited rate. Methods of controlling NO_x emissions by reduced flame temperatures include:

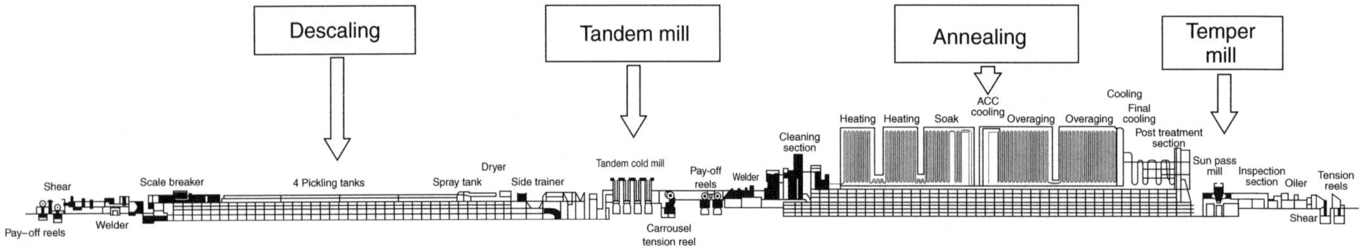

Figure 2. Cold Rolling Schematic.

Table 1. Continuous Cold Mill with BACT Controls

Source	Purpose	Emissions Type	Expected Uncontrolled Emissions (mg/nm^3)	Control Technology	Calculated Controlled Emissions (mg/nm^3)	Percent Control
Pinch roll leveler	Flattens steel	Particulate matter	1000	Fabric filter	10	99
Flash butt welder	Connects coils	Particulate matter	360	Fabric filter	3.6	99
Tension leveler	Removes oxidation	Particulate matter	1000	Fabric filter	10	99
Descaling (pickling)	Cleans surface	HCl vapor	3250	Countercurrent packed-tower scrubber with mist eliminator	6.5	99
Tendem cold rolling	Reduces thickness of strip	Roll coolant spray (water and oil)	100	Baffle plate collision mist eliminator	20	80
Electrolytic cleaning	Removes oil from strip	Alkali mist	100	Horizontal air washer	5	95
Posttreatment picking	Removes light oxide	HCl vapor	840	Gas washing tower	1.7	99
Temper mill	Tempers/restores flatness	Negligible		None		
Electrostatic oiler	Applies protective oil coat	Oil aerosol	13	Mist eliminator	1.3	90

1. A low heat-release furnace, which rapidly absorbs energy from the flame
2. Low-turbulence, laminar burners, which radiate heat rapidly, resulting in low flame temperatures
3. Flue gas recirculation from the boiler outlet into the wind box, which absorbs heat and inhibits combustion, both of which will lower flame temperature
4. Low combustion air temperature, which will lower combustion efficiency, but can reduce NO_x generation
5. Low excess airfiring, which limits the oxygen available and, in general, will decrease nitrogen oxide formation.

Selective catalytic reduction (SCR) is an expensive post-combustion NO_x-control technique used in Japan,[1,2] but not on industrial combustors in the United States. Ammonia is used as a reducing agent in a gas-phase reaction with oxides of nitrogen in the presence of a catalyst to form nitrogen and water.

REFERENCES

1. Prevention of Significant Deterioration Permit Application, United Engineers and Constructors, 1987.
2. *BACT/LAER Clearinghouse—A Compilation of Control Technology Determinations*, EPA-450/3-85-016B, U.S. Environmental Protection Agency, Research Triangle Park, NC, 1985.

BIBLIOGRAPHY

AISE, *The Making, Shaping and Treating of Steel*, 10th ed., Herbich & Held, Pittsburgh, 1985.

W. L. Roberts, *The Cold Rolling of Steel*, Marcel Dekker, New York, 1978.

W. L. Roberts, *Hot Rolling of Steel*, Marcel Dekker, New York, 1983.

SCARFING

S. J. Manganello

Scarfing in the steel mill[1,2] consists of surface removal with an oxy-fuel-gas flame process. The flame from a hot-scarfing machine is used rapidly to melt the steel surface thermochemically between rolling stages while the high-pressure oxygen stream propels the oxidized product from the surface. Subsequent scarfing, if required, is done by hand torches in the conditioning area.

The tip of a hand-scarfing torch is somewhat similar to the oxygen-gas torch used to cut steel. The oxygen not only promotes the chemical reaction that forms the melted iron oxide, but its kinetic energy must force the liquid, oxidized metal from the path of the torch, thereby removing surface defects from the semifinished steel. In addition to oxygen, fuel gas is necessary to the functioning of the hand-scarfing torch. The gas (commonly natural gas) serves as a preheating agent to elevate a spot on the surface to such a temperature that the oxygen and the steel will begin to combine chemically. The starting rod, a small steel rod that extends into the flame and is heated until a small drop of melted rod falls onto the steel surface, is often used to kindle the reaction quickly.

PROCESS DESCRIPTION[1,2]

The mechanical scarfer, installed directly in-line on the rolling mill, is composed of a number of (usually fixed) scarfing units across the steel, designed in such a way that they form a pass on the mill. The hot-scarfing machine is usually placed before the slab/bloom shears or after the roughing or first billet mill. The high temperature of the product being rolled is enough to obtain quick starting action. The machine scarfer can remove defects up to 4.5 mm (0.18 in.)—which is usually sufficient to remove rolled seams, light scabs, and checks. General practice is to desurface each steel grade to the depth that strikes the right economic balance between loss in yield from metal removal and processing costs, on the one hand, and savings in lessened rejection loss, on the other. This depth varies with different materials, but the metal loss is generally 1.5–2.5% of the product. Four-sided scarfing machines are now in common use, as are automated spot-scarfing machines. Water jets are employed to help control the slag.

AIR EMISSIONS AND POLLUTION CONTROL[2–4]

Hand Scarfing

Scarfing generates small particles, mainly iron oxides, that may remain suspended in the atmosphere for a considerable time. The concentration in a closed area can build up over time, and particles will eventually settle on the walls, floor, and equipment. The size of the enclosure, the size of the scarfing operation, and the ventilation affect the concentration. Adequate ventilation is the key, whether it be natural, mechanical, or respirator, and there are safety criteria for each type. General ventilation of the work site is often adequate, but local ventilation may also be required.

Machine Scarfing

Exhaust hoods are commonly used for machine scarfers. They must have sufficient airflow and velocity to keep contaminants at or below permissible limits to control air pollution, which are recommended/stated by the manufacturer.

The scarfing of steel results in the generation of appreciable quantities of smoke, the quantity and density depending on such variables as steel analysis, scarfing oxygen pressure, and efficiency of the slag water jets. It is necessary to collect this smoke and discharge it to the atmosphere outside the building with a suitable exhaust system, consisting of a hood over the roll table, a duct system, a fan, and a stack. Systems are normally designed for maximum gas velocities of 3000 fpm, based on free air movement at average ambient mill temperatures.

Scarfing-machine smoke contains some solids in addition to steam and gases. The solids are mainly oxides of iron with traces of the alloying elements present in the scarfed steel. Much of the solid matter in the smoke is submicrometer in size, with occasional particles up to the maximum that can be airborne at the exhaust velocities. An average four-sided scarfing machine equipped with cross-fire water jets and an efficient smoke-collection system will produce approximately 36 grains (7000 grains equal 1 pound) of smoke solids per pound of metal removed under normal scarfing conditions. The large number of submicrometer particles give the smoke a yellowish-brown color. Smoke gases on a typical installation have been analyzed to contain 0.22% CO_2, 23.0% O_2, 0.12% H_2, and 76.6% N_2. The smoke is a nuisance and can irritate the eyes. Also, the iron in the smoke may cause short circuits if allowed to accumulate on electrical apparatus. Thus consideration should be given to the most suitable location for discharge to the atmosphere. The smoke can be cleaned so that the stack discharge has no visible color by passing it through an electrostatic precipitator. The construction materials used for the smoke-removal system should be corrosion- and abrasion-resistant because of the nature of the smoke, which may vary with each installation in accordance with the composition of the steel being scarfed. Bag-type units have not been used because of the moisture content of the air.

REFERENCES

1. *The Making, Shaping, and Treating of Steel*, 10th ed., 1985, pp. 734–37, edited by William T. Lankford et al., Association of Iron and Steel Engineers, Pittsburgh, PA.
2. Private Correspondence with R. C. Smith, L-TEC, Florence, SC, March 19, 1990.
3. *Welding Handbook, Vol. 1, Welding Technology*, 8th ed., 1987, Chapter 16, ed. Charlotte Weisman, American Welding Society, Miami, FL.
4. *L-TEC Scarfing Machine Bulletin*: SM13-04.

SINTER PLANTS

S. S. Felton and L. M. Stuart

Sintering is one of a number of iron ore agglomeration processes used in the steel industry. It was developed in the early 1900s to utilize the large amount of blast furnace dusts being generated each year and the millions of tons of hematite ore fines that were being accumulated in the Mesabi range. The sintering process fuses iron-bearing fines into a hard clinker suitable for use in blast furnaces. It is capable of converting a wide variety of iron ore fines (iron-bearing blast furnace dusts and sludges, mill scales, and steelmaking dusts and iron-bearing dusts and sludges collected from air and water pollution control processes) into a high-quality blast furnace burden material.

PROCESS DESCRIPTION

The sintering process includes a material storage, blending, and feeding system for the burden materials; a traveling-grate hearth called the sinter strand to produce the sinter; a sinter breaking, screening, and cooling system to size and cool the sinter properly and to recycle undersized sinter to become part of the feed-burden materials; and a variety of environmental control facilities. Figure 1 provides a flow diagram of a modern sinter plant.[1]

Storage, Blending, and Feeding System

In modern sinter plants, the bedding and blending system takes the ores and other materials from storage bins and transfers them in predetermined amounts to a traveling stacker, which deposits them in thin layers down the length of a large bedding pile. As many as 500 layers of materials may be deposited in a single bedding pile, and a pile may contain enough material to operate a sinter plant for 6–12 days. The blended feed material is reclaimed from the pile to large feed storage bins at the sinter plant. Sinter plants without a bedding system meter burden materials from large storage

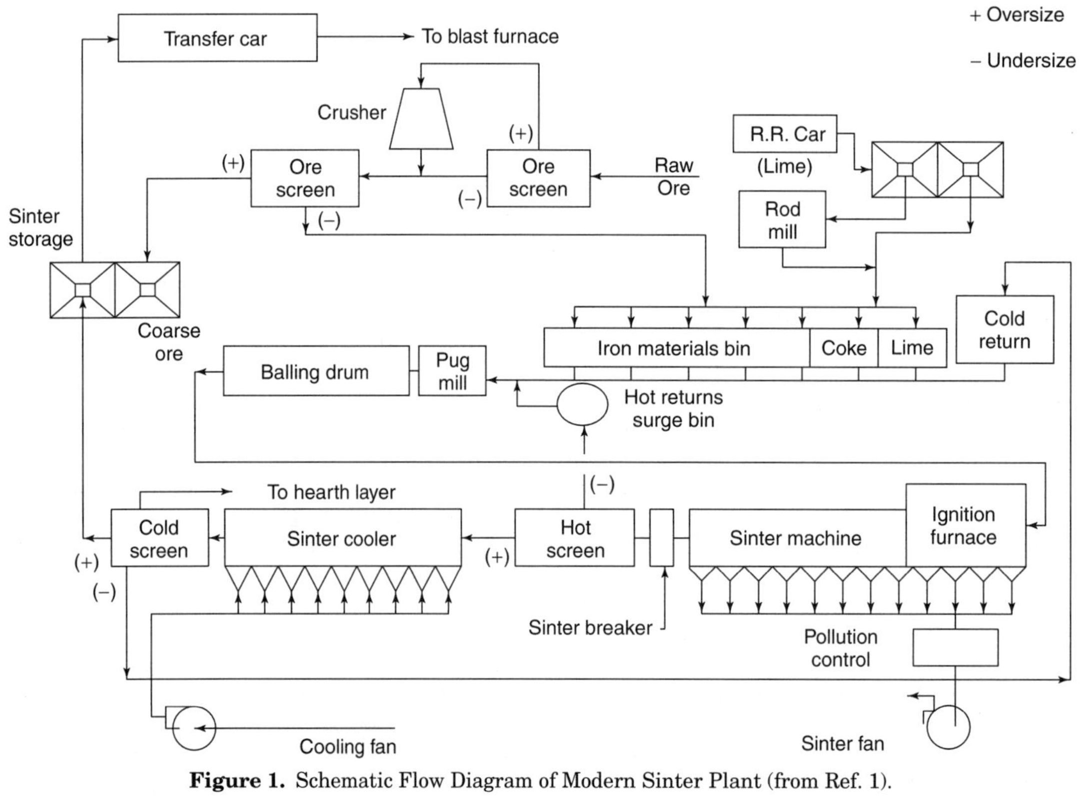

Figure 1. Schematic Flow Diagram of Modern Sinter Plant (from Ref. 1).

bins onto conveyor belts going directly to the main sinter feed conveyor. Careful and homogeneous blending of the raw feed materials produces a sinter product that is consistently uniform in chemical and physical properties.

Feeding Materials to the Sinter Strand

Conversion of the raw material mix into a sinter product takes place on a sinter strand. The feed to the sinter strand consists of the raw materials from the blending pile, fluxes, solid fuel (usually coke breeze), and recycled sinter fines from the sinter screening process. Each of these materials is added in closely controlled amounts to a mixing-reroll drum for further blending and control of moisture content. The final mix is then transferred onto the sintering strand in a uniform depth by a roll-feeder-mix-level control system.

Ignition

The sinter feed mix travels under an ignition hood where the hot combustion gases ignite the solid fuel in the sinter mix to start the sintering process. The fuels used in the ignition hood are usually gaseous fuels, such as natural gas or coke-oven gas. The combustion products from the ignition zone are drawn through the sinter mix into the windbox gas collection main located under the sinter strand.

Sintering

Once the fuel in the sinter mix is ignited, the sintering process has begun. As the sinter strand moves out of the ignition zone, air is drawn downward through the bed into the windboxes under the traveling grate. Air movement is provided by large fans located downstream of the windboxes. When the air is drawn through the sinter bed, the solid fuel in the mix burns at temperatures sufficient to fuse the raw materials into a hard clinker material. The flame front of burning fuel in the mix continues to move downward through the mix as the strand moves forward. Completion, or burn-through, is controlled to occur in the first section of the last windbox at the end of the strand. All the combustion products, excess air, and other gaseous materials exiting the sinter bed during the sinter process are captured in the windboxes under the sinter strand.

Sinter Sizing and Cooling

The sinter leaves the sinter strand in large, hot chunks that must be properly sized and cooled before they can be utilized in the blast furnace. This is accomplished by a series of breakers, screens, and crushers. The screening processes remove the less than 1/4-in. sinter and recycle it into the raw feed mix ahead of the mix-reroll drum.

Types of Sinter

Three general types of sinter are produced by sinter plants: acid sinter, self-fluxed sinter, and superfluxed sinter.[2] The term basicity, which is the ratio of $(CaO + MgO)/(SiO_2 + Al_2O_3)$ found in the sinter, is used to distinguish one type of sinter from another. Acid sinter, normally with a basicity of less than 1.0, is produced primarily from iron-bearing materials. With this type of sinter, all the fluxes needed by the blast furnace must be added in the blast-furnace burden. The use of this type of sinter is declining and most sinter plants are producing either self-fluxing or superfluxed sinters. Self-fluxed sinter is produced by adding sufficient fluxing materials to adjust the basicity to 1.0 to 1.5. This type of sinter provides only the flux needed by the sinter. The fluxes needed for the other blast-furnace burden materials must also be added to the blast furnace. Superfluxed sinter has a basicity of 2.0 or higher and is produced with sufficient additional flux materials so that no additional flux is needed in the blast-furnace burden. All the flux needed by the blast furnace is provided by the sinter.

AIR EMISSIONS CHARACTERIZATION

Sinter-plant windbox emissions contain particulate emissions and combustion gases generated during the sintering process. Tables 1 and 2 show particulate emissions factors and size data for sinter plants based on data collected by the U.S. Environmental Protection Agency (EPA), mostly in the 1970s. It is important to note that many of these data may no longer be applicable since there have been major changes in sintering practices and gas-cleaning system technology. Tables 3 and 4 present emission data collected in the late 1970s at sinter plants equipped with wet scrubbers for treating windbox gases.

Particulate Matter

Chemical and mineralogical examinations of the inorganic fractions of many particulate catches from sinter plant windboxes show that the principal component, after the cyclones, is potassium chloride (KCl).

Hydrocarbon Content

The amount of hydrocarbon particulate generated in the windbox exhaust is a function of the amount of hydrocarbon contained in the sinter burden. Most of this material evolves from the mill scale as a part of the burden. Mill scales are generated during the rolling of hot steel and consist of a mixture of iron and iron oxides. It is a premium burden material since it helps to produce a hard, dense sinter. However, mill scales contain various types of oils and greases. During the sintering process, most of these oils and greases are burned but some of them vaporize into the windbox gases ahead of the high-temperature advancing flame front and then condense at the cooler windbox temperatures. Visible emissions from the windbox collection device containing significant hydrocarbons tend to be bluish.

Gaseous Constituents

Table 5 shows the concentration of oxygen, sulfur dioxide, combustibles (principally carbon monoxide), and nitrogen oxides in the stack gases from a large sinter plant equipped with wet scrubbers. The sulfur content of the stack gases will vary greatly, depending on the sulfur content of the sinter burden materials and the fuels used.

AIR POLLUTION CONTROL MEASURES

Particulate and gaseous emissions are generated at all phases of the sintering process. Each operation has different types

Table 1. Particulate Emission Factors for Sinter Plants

Source	Units	Emission Factors		Emission Factor Rating[a,b]	Size Data
Sintering					
Windbox emissions	kg/Mg (lb/ton) finished sinter				
Uncontrolled					
Leaving grate		5.56	(11.1)	B	Yes
After coarse particulate removal		4.35	(8.7)	A	
Controlled by dry ESP[c]		0.8	(1.6)	B	
Controlled by wet ESP[c]		0.085	(0.17)	B	Yes
Controlled by venturi scrubber		0.235	(0.47)	B	Yes
Controlled by cyclone		0.5	(1.0)	B	Yes
Sinter discharge (breaker and hot screens)	kg/Mg (lb/ton) finished sinter				
Uncontrolled		3.4	(6.8)	B	
Controlled by baghouse		0.05	(0.1)	B	Yes
Controlled by venturi scrubber		0.295	(0.59)	A	
Windbox and discharge	kg/Mg (lb/ton) finished sinter				
Controlled by baghouse		0.15	(0.3)	A	

[a]The following ratings were applied to each emission factor.

$A =$ *excellent*. Developed from A-rated test data taken from many randomly chosen facilities in the industry population. The source category is specific enough to minimize variability within the source category population.

$B =$ *above average*. Developed only from A-rated test data from a reasonable number of facilities. Although no specific bias is evident, it is not clear whether the facilities tested represent a random sample of the industries. As with the A rating, the source category is specific enough to minimize variability within the source category population.

$C =$ *average*. Developed only from A-and B-rated test data from a reasonable number of facilities. Although no specific bias is evident, it is clear whether the facilities tested represent a random sample of the industry. As with the A rating, the source category is specific enough to minimize variability within the source category population.

$D =$ *below average*. The emission factor was developed from A- and B-rated test data, and there may be suspect that these facilities do not represent a random sample of the industry. There also may be evidence of variability within the source category population. Limitations on the use of the emission factor are footnoted in the emission factor table.

$E =$ *poor*. The emission factor was developed from C- and D-rated test data, and there may be reason to suspect that the facilities tested do not represent a random sample of the industry. There also may be evidence of variability within the source category population. Limitations on the use factors are always footnoted.

[b]The data were rated according to the following criteria.

A = Tests performed by a sound methodology and reported in enough detail for adequate validation. These tests are not necessarily EPA reference method tests, although such reference methods are used as a guide.

B = Tests that are performed by a generally sound methodology, but lack enough documentation for adequate validation.

C = Tests that are based on an unproven methodology or lack a significant amount of documentation.

D = Tests that are based on a generally unacceptable method, but may provide an order-of-magnitude value for the source.

[c]Electrostatic precipitator

Source: Ref. 3.

of particulate and gaseous emissions and requires a different type of emission control technology.

Feed Preparation Process

Handling the large quantity and variety of feed materials in storage bins, feeders, screens, and transfer chutes generates dusts and requires that dust emission control hoods and exhaust systems be installed at several points in the feed preparation process. The emissions from these operations consist primarily of particulate matter. These emissions are collected by baghouses or low-energy wet scrubbers.

Sinter Sizing and Cooling

The hot, dirty gases collected during the sizing and cooling of hot sinter also contain primarily particulate matter. These are generally fairly coarse particles and can be removed by cyclones or baghouses. The collected dust is normally recycled directly to the sintering process.

The dusts generated at the cold-screening operations and conveyor transfer points and during sinter crushing are generally collected with a baghouse, although low-energy wet scrubbers have also been used.

Sinter Windbox Gases

The cleaning of the sintering process gases is the most difficult gas cleaning problem in the sinter plant and one of the most difficult and costly gas cleaning problems in the steel industry. Many sinter plants are equipped with cyclone collectors followed by the primary emission control equipment. The primary emission control equipment is located after the induced-draft fans and includes either (1) electrostatic precipitators, (2) wet scrubbers, or (3) baghouses.

Cyclone Separators

The gases leaving the sintering grate contain a significant amount of coarse particles. Most sinter plants are equipped with a series of gravity separators or dropout chambers followed by mechanical collectors, such as cyclones or multiclones. The cyclone collectors are installed ahead of the induced-draft fans and are designed to remove particles with diameters of

Table 2. Size-Specific Emission Factors for Sinter Plants

Source	Emission Factor Rating (A–E)[a]	Particle Size (μm)	Cumulative Mass (% Less Than Stated Size)	Cumulative Mass *Emission Factor* kg/Mg Alloy	(lb/ton Alloy)
Sintering					
Windbox emissions					
Uncontrolled leaving grate	D	0.5	4	0.22	(0.44)
		1.0	4	0.22	(0.44)
		2.5	5	0.28	(0.56)
		5.0	9	0.50	(1.00)
		10	15	0.83	(1.67)
		15	20	1.11	(2.22)
		*	100	5.56	(11.1)
Controlled by wet electrostatic precipitator	C	0.5	18	0.015	(0.03)
		1.0	25	0.021	(0.04)
		2.5	33	0.028	(0.06)
		5.0	48	0.041	(0.08)
		10	59	0.050	(0.10)
		15	69	0.059	(0.12)
		*	100	0.085	(0.17)
Controlled by venturi scrubber	C	0.5	55	0.129	(0.26)
		1.0	75	0.176	(0.35)
		2.5	89	0.209	(0.42)
		5.0	93	0.219	(0.44)
		10	96	0.226	(0.45)
		15	98	0.230	(0.46)
		*	100	0.235	(0.47)
Controlled by cyclone	C	0.5	25	0.13	(0.25)
		1.0	37	0.19	(0.37)
		2.5	52	0.26	(0.52)
		5.0	64	0.32	(0.64)
		10	74	0.37	(0.74)
		15	80	0.40	(0.80)
		*	100	0.5	(1.0)
Controlled by baghouse	D	0.5	3	0.005	(0.009)
		1.0	9	0.014	(0.027)
		2.5	27	0.041	(0.081)
		5.0	47	0.071	(0.141)
		10	69	0.104	(0.207)
		15	79	0.119	(0.237)
		*	100	0.150	(0.30)
Sinter discharge (breaker and hot screens) controlled by baghouse	C	0.5	2	0.001	(0.002)
		1.0	4	0.002	(0.004)
		2.5	11	0.006	(0.011)
		5.0	20	0.010	(0.020)
		10	32	0.016	(0.032)
		15	42	0.021	(0.042)
		*	100	0.05	(0.1)

[a] See Table 1, footnotes *a* and *b*.

Source: Ref. 3.

25 μm or greater. This is done to protect the induced-draft fans from severe wear by these abrasive particles.

Electrostatic Precipitators

In the 1960s, most sinter plants were equipped with dry electrostatic precipitators that operated with removal efficiencies of up to 99% of the windbox emissions produced during the production of "acid" sinters. In recent years, there has been a trend toward the manufacture of "fluxed" sinters—which has been detrimental to the performance of electrostatic precipitators since the dust generated has increased resistivity.

With the increased use of oil-bearing mill scales in the sinter feed materials, there are increased amounts of condensed hydrocarbon particulates in the windbox gases. Dry electrostatic precipitators are not efficient collection devices

Table 3. Performance of Sinter Plant Windbox Gas Collection Systems

	Sinter Plant No.	
	1	2
Cyclones—pounds per net ton sinter		
Inlets	4.6	2.03
Outlets	2.4	1.4
Percent removal	48	33
Wet scrubbers—pounds per net ton sinter		
Inlets	2.7	1.1
Outlets	0.14	0.29
Percent removal	95	74

Source: Ref. 2.

Table 4. Particle Size Distribution of Sinter Windbox Particulates

Particle Size Distribution	Scrubber Inlet	Scrubber Outlet	Percent Removal
Percent less than:			
1 μm	67	90	64
2 μm	70	95	64
3 μm	72	97	65
5 μm	75	98	—
40 μm	85	100	—
			All sizes 74

Source: Ref. 2.

Table 5. Chemical Composition of Sinter Plant Windbox Stack Gases, after Scrubber

Fuels Used (% Btu)	Sinter Plant 1 Coke Oven Gas 10% Coke Breeze 90%	Sinter Plant 2 Coke Oven Gas 19% Coke Breeze 81%
Stack gas constituents		
Oxygen (%)	15.3	16.6
Combustibles (%)	0.8	1.9
Sulfur dioxide, ppm	72	—
NO_x, ppm	88	95
Pounds per ton sinter	0.71	0.66
Pounds per million Btu	0.56	0.56
Temperature (°F)	—	131

Source: Ref. 2.

for hydrocarbons. Also, the presence of hydrocarbons in precipitators can present explosion and fire hazards.

Baghouses

Baghouse installations have been applied to windbox emission controls in several sinter plants where high-basicity sinter is made. Their efficient operation is sensitive to condensation of moisture or hydrocarbon vapors. Condensed moisture and sulfur oxides in the gas stream will react with lime in the dust or in the precoat material on the bags to form calcium sulfites and sulfates on the surface of the bags, thus blinding them.

Scrubbers

High-energy wet scrubbers followed by efficient mist eliminators are very efficient in treating sinter plant windbox gases. The wet scrubber must be designed to remove very fine particulate matter with diameters of 1 μm or less. This usually requires operating the wet scrubber at pressure drops across the scrubber of 30 to 70 in. of water. Since there are large volumes of sinter plant windbox gases, the electric power requirements for these gas cleaning systems are very high. Wet scrubbers are also capable of capturing a significant portion of the condensible hydrocarbons and other fine particulates in the windbox gases.

Much of the water used in the wet scrubbers is recirculated to reduce makeup water requirements and the amount of water that must be blown down from the system. However, the percent recirculation rate will vary from 30% to 80%, depending on the mineral scaling potential of the waters in a given system. Treatment of the scrubber waters requires neutralizing the acidic waters and adding chemical coagulants to enhance the clarification of the scrubber discharge waters prior to recirculation.

REFERENCES

1. W. R. Mura, *An Evaluation of Sinter Plant Air Pollution Control Technology*, Mellon Institute, Pittsburgh, March 1982.
2. L. M. Stuart, Bethlehem Steel Corp., written communication to J. A. Grantz, Armco Steel Co., May 11, 1990.
3. *Compilation of Air Pollutant Emission Factors*, U.S. Environmental Protection Agency, AP-42 October 1986.

BIBLIOGRAPHY

D. F. Ball, A. F. Bradley, and A. Grieve, "Environmental Control in Iron Ore Sintering," Paper 7, presented at Symposium on Minerals and the Environment, Institution of Mining and Metallurgy, London, England, June 4–7, 1974.

B. H. Carpenter et al., *Pollution Effects of Abnormal Operations in Iron and Steel Making, Vol. II, Sintering, Manual of Practice*, EPA-600/2-78-118b, U.S. Environmental Protection Agency, Research Triangle Park, NC, June 1978.

T. A. Cuscino, Jr., *Particulate Emission Factors Applicable to the Iron and Steel Industry*, EPA-450/4-79-028 (NTIS PB81145914), September 1971.

R. G. Genton, "Steel Mill Sinter Plant," Paper No. 72-81, presented at Annual Meeting of the Air Pollution Control Association, Miami Beach, FL, June 18–22, 1972.

Iron and Steel Industry Particulate Emissions: Source Category Report, EPA-600/7-86-036, October 1984, U.S. Environmental Protection Agency.

The Making, Shaping and Treating of Steel, 9th ed. H. B. McGannon, Ed., U.S. Steel Corp., 1971.

D. A. Pengidore, *Sinter Plant Windbox Gas Recirculation System Demonstration—Phase 1. Engineering and Design*, EPA-600/ 2-75-014 U.S. Environmental Protection Agency, Research Triangle Park, NC, August 1975.

B. A. Steiner, "Air Pollution Control in the Iron and Steel Industry," *Int. Metals Rev.* **21**, 171, 1976.

B. A. Steiner, "Ferrous Metallurgical Operations," Chapter 21 in *Air Pollution, Vol. IV*, A. C. Stem, Ed., Academic Press, New York, 1977.

Metal Producing, Vol. 27, No. 5, May 1989.

J. Varga, *Control of Reclamation (Sinter) Plant Emissions Using Electrostatic Precipitators*, EPA-600/2-76-002, U.S. Environmental Protection Agency, Research Triangle Park, NC, January 1976.

PRIMARY LEAD SMELTING

Adapted from AP-42, *Compilation of Air Pollutant Emission Factors*, U.S. EPA, with update by editor

Lead is found naturally as a sulfide ore containing small amounts of copper, iron, zinc, precious metals, and other trace elements. The lead in this ore, typically after being concentrated at or near the mine, is processed into metallurgical lead at four facilities in the United States (2 smelters/refineries in Missouri), one smelter in Montana, and one refinery in Nebraska). Demand for lead from these primary sources is expected to remain relatively stable in the early 1990s, due in large part to storage battery recycling programs being implemented by several states. Significant emissions of sulfur dioxide (SO_2), particulate matter, and especially lead have caused much attention to be focused on identifying, and quantifying emissions from, sources within these facilities.[15]

PROCESS DESCRIPTION[15,16]

The processing of lead concentrate into metallurgical lead involves three major steps: sintering, reduction, and refining. A diagram of a typical facility, with particle and gaseous emission sources indicated, is shown in Figure 1.

SINTERING

The primary purpose of the sinter machine is the reduction of sulfur content of the feed material. This feed material typically consists of the following:

1. Lead concentrates, including pyrite concentrates that are high in sulfur content, and concentrates that are high in impurities such as arsenic, antimony, and bismuth, as well as relatively pure high-lead concentrates
2. Lime rock and silica, incorporated in the feed to maintain a desired sulfur content
3. High-lead-content sludge byproducts from other facilities
4. Undersized sinter recycled from the roast exiting the sinter machine

The undersized sinter return stream mixes with the other feed components, or green feed, as the two streams enter a rotary pelletizing drum. A water spray into the drum enhances the formation of nodules in which the sinter returns form a core rich in lead oxide and the green feed forms a coating rich in lead sulfide. The smaller nodules are separated out and conveyed through an ignition furnace, then covered with the remaining nodules on a moving grate and conveyed through the sinter machine, which is essentially a large oven. Excess air is forced upward through the grate, facilitating combustion,

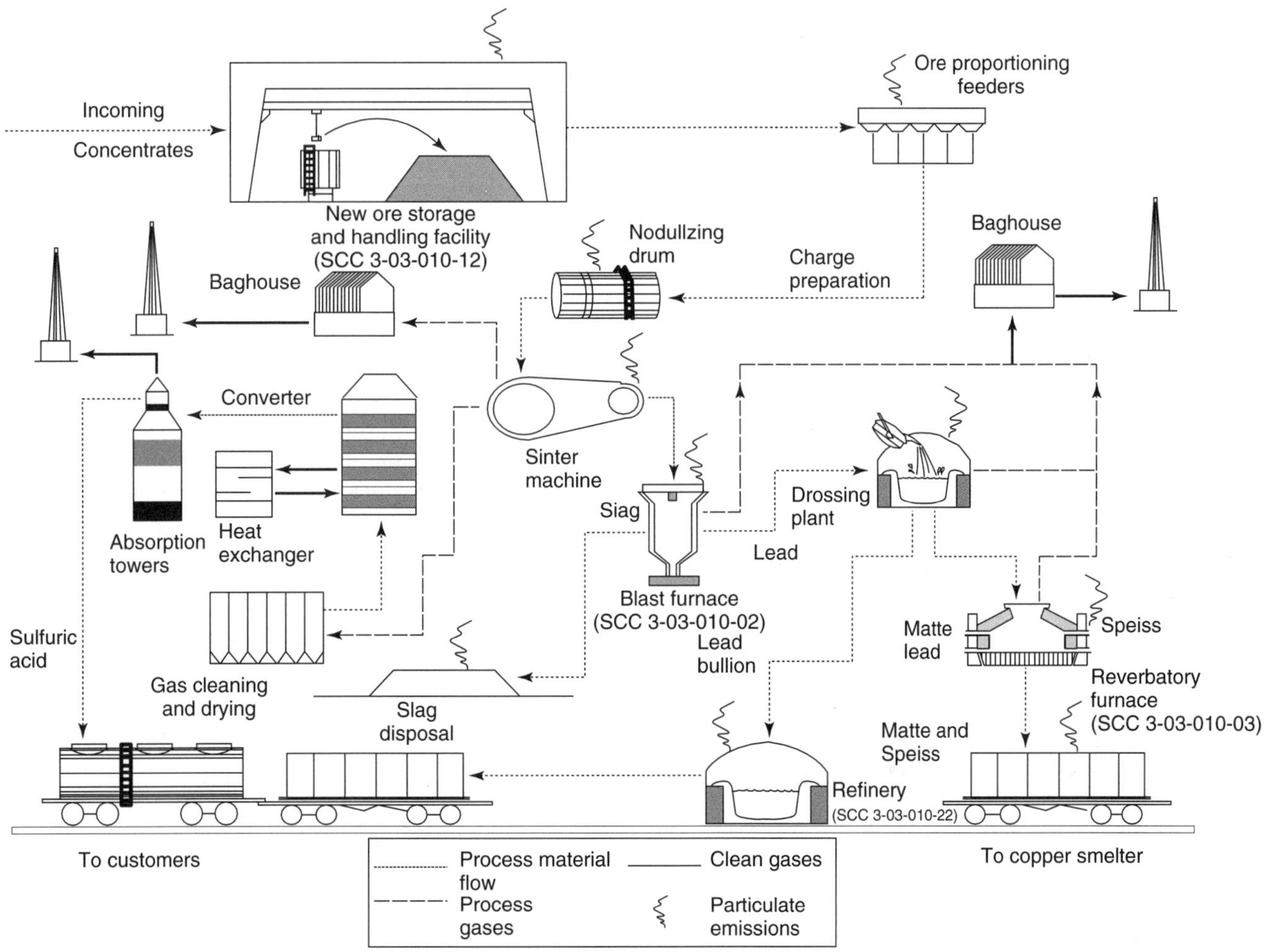

Figure 1. A Typical Primary Lead Smelting and Refining Facility (Source Classification Code in Parentheses).

releasing SO_2, and oxidizing the lead sulfide to lead oxide. The "strong gas" from the front end of the sinter machine, containing 2.5–4% SO_2, is vented to gas cleaning equipment before possibly being piped to a sulfuric plant. Gases from the rear part of the sinter machine are recirculated up through the moving grate and are typically vented to a baghouse. That portion of the product that is undersized, usually due to insufficient desulfurization, is filtered out and recycled through the sinter; the remaining sinter roast is crushed before being transported to the blast furnace.

REDUCTION

The sinter roast is then conveyed to the blast furnace in charge cars along with coke, ores containing high amounts of precious metals, slags and by-product dusts from other smelters, and by-product dusts from baghouses and various other sources within the facility. Iron scrap is often added to the charge to aid heat distribution and to combine with the arsenic in the charge. The blast furnace process rate is controlled by the proportion of coke in the charge and by the air flow through the tuyeres in the floor of the furnace. The charge descends through the furnace shaft into the smelting zone, where it becomes molten, and is tapped into a series of settlers that allow the separation of lead from slag. The slag is allowed to cool before being stored, and the molten lead of roughly 85% purity is transported in pots to the dross building.

REFINING

The drossing area consists of a variety of interconnected kettles, heated from below by natural gas combustion. The lead pots arriving from the blast furnace are poured into receiving kettles and allowed to cool to the point at which copper dross rises to the top of the top and can be skimmed off and transferred to a reverbatory furnace. The remaining lead dross is transferred to a finishing kettle, where such materials as wood chips, coke fines, and sulfur are added and mixed to facilitate further separation, and this sulfur dross is also skimmed off and transferred to the reverbatory furnace. To the drosses in the reverbatory furnace are added tetrahedrite ore, which is high in silver content but low in lead and may have been dried elsewhere within the facility, coke fines, and soda ash. When heated in the same fashion as the kettles, the dross in the reverbatory furnace separates into three layers. Lead bullion settles to the bottom and is tapped back to the receiving kettles; matte (copper sulfide and other metal sulfides), which rises to the top; and speiss (high in arsenic and antimony content). The matte and speiss are both typically forwarded to copper smelters.

The third and final phase in the processing of lead ore to metallurgical lead, the refining of the bullion in cast iron kettles, occurs in five steps: (1) removal of antimony, tin, and arsenic; (2) removal of precious metals by the Parke's Process, in which zinc combines with gold and silver to form an insoluble intermetallic at operating temperatures; (3) vacuum removal of zinc; (4) removal of bismuth by the Betterson Process, in which calcium and magnesium are added to form an insoluble compound with the bismuth that is skimmed from the kettle; and (5) removal of remaining traces of metal impurities through the addition of $NaOH$ and $NaNO_3$. The final refined lead, from 99.990 to 99.999% pure, is typically cast into 45-kg (100-pound) pigs for shipment.

EMISSIONS AND CONTROLS[15–17]

Emission of lead and particulate occur in varying amounts from nearly every process and process component within primary lead smelter/refineries, and SO_2 is also emitted from several sources. The lead and particulate emissions point, volume, and area sources may include:

1. The milling, dividing, and fire assaying of samples of incoming concentrates and high-grade ores
2. Fugitive emissions within the crushing mill area, including the loading and unloading of ores and concentrates from rail cars onto conveyors
3. The ore crushers and associated transfer points, which may be controlled by baghouses
4. Fugitives emissions from the unloading, storage, and transfer of by-product dusts, high-grade ores, residues, coke, lime, silica, and any other materials stored in outdoor piles
5. Strong gases from the front end of the sinter machine, which are typically vented to an electrostatic precipitator (ESP), one or more scrubbers, and a wet ESP for sulfuric acid mist elimination, but during shutdowns of the acid plant may bypass the ESP
6. Weak gases from the back end of the sinter machine, which are high in lead dust content but typically pass through cyclones and a baghouse
7. Fugitive emissions from the sinter building, including leaks in the sinter machine and the sinter cake crusher
8. Gases exiting the top of the blast furnace, which are typically controlled with a baghouse
9. Fugitive emissions from the blast furnace, including leaks from the furnace covers and the bottoms of charge cars, dust from the charge car bottom dump during normal operation, and escaping gases when blow holes develop in the shaft and must be "shot" with explosives
10. Lead fumes from the molten lead and slag leaving the blast furnace area
11. Fugitive leaks from the tapping of the kettles and settlers
12. The hauling and dumping of slag, at both the handling and cooling area and the slag storage pile
13. The combustion of natural gas, as well as the creation of lead-containing fumes at the kettles and reverbatory furnace, all of which are typically vented to a baghouse at the drossing building
14. Fugitive emissions from the various pouring, pumping, skimming, cooling, and tapping operations within the drossing building
15. The transporting, breaking, granulating, and storage of speiss and matte
16. The loading, transferring, and drying of tetrahedrite ore, which is typically controlled with cyclones and a baghouse
17. The periodic cleanout of the blast and reverbatory furnaces
18. Dust caused by wind erosion and plant vehicular traffic, which are normally estimated with factors from Section 13.2 of AP-42, but are addressed herein due to the high lead content of the dust at primary lead smelting and refining facilities

Table 1. Emission Factors for Primary Lead Smelting[a] (Metric Units)

Emission Factor Rating: E

Process	Particulate[b]	PM-10[c]	Lead	SO_2
Ore crushing[d] (SCC 3-03-010-04)	0.023	0.118	0.001	NA
Ore screening[e] (SCC 3-03-010-27)	0.004	0.005	0.001	NA
Tetrahedrite drier[f] (SCC 3-03-010-28)	0.012	0.013	0.0003	NA
Sinter machine (weak gas)[g] (SCC 3-03-010-29)	0.051	0.052	0.009	275[h]
Sinter building fugitives[g] (SCC 3-03-010-25)	0.018	0.058	0.016	NA
Sinter storage[j] (SCC 3-03-010-30)	NA	NA	NA	NA
Blast furnace[k] (SCC 3-03-010-02)	0.21	0.43	0.034	23[h]
Speiss pit[m] (SCC 3-03-010-31)	NA	NA	NA	NA

[a]Most of the processes are controlled by baghouses; otherwise it is noted. SCC, source classification code; ND, not applicable.

[b]Filterable particulate only.

[c]Filterable and condensable particulate; ≤ 10 µm mean diameter.

[d]Entire ore crushing building at one facility, including transfer points; kg/Mg of ore, except lead, which is kg/Mg of lead in ore.

[e]Tests at one facility; kg/Mg ore.

[f]kg/Mg dried; tests at one facility.

[g]kg/Mg sinter produced; tests at one facility. The sinter machine is controlled by ESP and scrubbers.

[h]Uncontrolled emission factor from 1971 tests on two facilities (Ref. 5 and 6).

[j]kg/Mg throughput; includes charge car loading; from tests at one facility.

[k]kg/Mg of bullion, includes dross kettles; from tests at one facility.

[m]kg/Mg granulated; from tests at one facility.

Table 1 presents particulate, PM-10, lead, and SO_2 emission factors for primary lead smelting. Emission factors in lb/ton can be determined by multiplying the values in Table 1 by a factor of 2.

NESHAP FOR PRIMARY LEAD SMELTERS

The U.S. EPA issued a final rule on the National Emission Standard for Hazardous Air Pollutants (NESHAP) for primary lead smelters with an effective date of June 4, 1999, based on the initial proposal issued in April 1998. Compliance for existing sources must be achieved by June 4, 2001. As summarized in the Fact sheet[18] and the Final Rule[19]:

- There are three existing primary lead smelters operating in the United States — an ASARCO, Inc., facility located in Glover Missouri, an ASARCO, Inc., facility located in East Helena, Montana, and a Doe Run Company facility located in Herculaneum, Missouri.
- All three facilities are listed as "major" sources of metallic HAP, including compounds of lead, arsenic, antimony, and cadmium. A "major" source is an industrial ("stationary") source that has the potential to emit 10 tons/year or more of a listed HAP, or 25 tons/year or more of a combination of HAP.
- This final rule establishes a "plantwide" emission limit of 500 grams of lead per megagram of lead produced (1.0 lb/ton) from the aggregation of emissions discharged from nine identified sources that emit hazardous air pollutants. The process sources include sinter machines, blast furnaces, and dross furnaces, whereas the process fugitive emission sources include dross furnace charging location, blast and dross furnace tapping locations, sinter machine charging locations, sinter machine discharge end, sinter crushing and sizing equipment, and sinter machine area.
- This final rule also requires primary smelters to develop and operate according to a Standard Operating Procedure (SOP) Manual for the control of fugitive dust sources and for the operation and maintenance of control devices. Each smelter's SOP Manual will be reviewed and subject to approval by the EPA Administrator or designated authority.

Detailed information on the final rule for this NESHAP can be found on the U.S. EPA Technology Transfer Network on the Unified Air Toxics Web (UATW) bulletin board (*http://www.epa.gov/ttn/uatw/*).

REFERENCES

1. C. Darvin and F. Porter, *Background Information for New Source Performance Standards: Primary Copper, Zinc and Lead Smelters, Volume I*, EPA-450/2-74-002a, U.S. Environmental protection Agency, Research Triangle Park, NC, October 1974.
2. A. E. Vandergrift et al., *Particulate Pollutant System Study, Vol. I: Mass Emissions*, APTD-0743, U.S. Environmental Protection Agency, Research Triangle Park, NC, May 1971.
3. A. Worcester and D. H. Beilstein, "The State of the Art: Lead Recovery," presented at the 10th Annual Meeting of the Metallurgical Society, AIME, New York, March 1971.
4. *Environmental Assessment of the Domestic Primary Copper, Lead and Zinc Industries* (Prepublication), EPA Contract No. 68-03-2537, Pedco Environmental, Cincinnati, OH, October 1978.
5. T. J. Jacobs, *Visit to St. Joe Minerals Corp. Lead Smelter, Herculaneum, MO*, Office of Air Quality Planning and Standards, U.S. Environmental Protection Agency, Research Triangle Park, NC, October 21, 1971.
6. T. J. Jacobs, *Visit to Amax Lead Company, Boss, MO*, Office of Air Quality Planning, and Standards, U.S. Environmental Protection Agency, Research Triangle Park, NC, October 28, 1971.
7. R. B. Paul, American Smelting and Refining Co., Glover, MO, written communication to Regional Administrator, U.S. Environmental Protection Agency, Kansas City, MO, April 3, 1973.

8. *Emission Test No. 72-MM-14*, Office of Air Quality Planning and Standards, U.S. Environmental Protection Agency, Research Triangle Park, NC, May 1972.
9. *Source Sampling Report: Emissions from Lead Smelter at American Smelting and Refining Company, Glover, MO, July 1973*, EMB-73-PLD-1, Office of Air Quality Planning and Standards, U.S. Environmental Protection Agency, Research Triangle Park, NC, August 1974.
10. *Sample Fugitive Lead Emissions from Two Primary Lead Smelters*, EPA-450/3-77-03 1, U.S. Environmental Protection Agency, Research Triangle Park, NC, October 1977.
11. *Silver Valley/Bunker Hill Smelter Environmental Investigation* (Interim Report), Contract No. 68-02-1343, Pedco Environmental, Durham, NC, February 1975.
12. R. E. Iverson, Meeting with U.S. Environmental Protection Agency and AISI on Steel Facility Emission Factors, Office of Air Quality Planning and Standards, U.S. Environmental Protection Agency, Research Triangle Park, NC, June 1976.
13. G. E. Spreight, "Best Practicable Means in the Iron and Steel Industry," *Chem. Eng.* **271**, 132–139, March 1973.
14. *Control Techniques for Lead Air Emissions*, EPA-450/2-77 -012, U.S. Environmental Protection Agency, Research Triangle Park, NC, January 1978.
15. *Mineral Commodity Summaries*, 1992, U.S. Department of the Interior, Bureau of Mines.
16. *Task 2 Summary Report: Revision and Verification of Lead Inventory Source List*, North American Weather Consultants, Salt Lake City, UT, June 1990.
17. *Task 5 Summary Report: ASARCO East Helena Primary Lead Smelter Lead Emission Inventory, Volume 1: Point Source Lead Emission Inventory*, North American Weather Consultants, Salt Lake City, UT, April 1991.
18. *Fact Sheet on Final Air Toxics Rule for Primary Lead Smelters*, U.S. EPA, July 1, 1999.
19. *40 CFR Part 63: NESHAP for Primary Lead Smelters: Final Rule*, June 4, 1999.

ZINC SMELTING

James R. Donnelly

Zinc metal has been known since very early times, when it was mined and smelted for use in making bronze and brass. Zinc is still used in these alloys but is more commonly associated with galvanizing of sheet iron. Because of its resistance to corrosion and light weight, zinc also has a number of other uses, such as: in electrical apparatus, household utensils, automobiles, castings, as a reducing agent, for deozidizing bronze, purifying fats for soaps, and numerous other uses.[1] Commercial forms of zinc include: ingots, lumps, powder, sheets, shot, strips, and wire, with slab zinc the most common starting form for production of the others.

Economic minerals found in zinc ores include sphalerite (zinc blende), smithsonite (zinc spar), franklinite, and others. Zinc sulfide ores are commonly found in conjunction with lead and copper mineralization (Galena and Chalcopyrite).[2] Beneficiation at the mine (crushing, grinding, gravity separation, and flotation) is used to upgrade the ores to zinc-rich concentrates. Zinc concentrates typically contain up to 40–60% zinc, with the balance primarily sulfur (approximately 30%) and iron (approximately 10%). Zinc values will be lower for combined lead–zinc concentrates. Other metals contained in zinc concentrates include antimony, arsenic, cadmium, copper, lead, mercury, and silver.[2] Silver values found in zinc concentrates can represent a significant additional revenue source from processing of the concentrate.

PROCESS DESCRIPTION

Zinc concentrates are processed to metallic slab zinc via roasting and either pyro- or hydrometallurgical purification processes. A generalized process flow schematic of these two process routes is presented in Figure 1.[3] The pyrometallurgical route utilizes typical primary nonferrous smelting techniques, including roasting, sintering, and retorting. The hydrometallurgical route incorporates roasting, leaching, purification, and electrolysis steps.

Roasting

Roasting is the first process step of both pyro- and hydrometallurgical routes. Roasting is the partial oxidation of the zinc sulfide concentrates with air. Its purpose, in the pyrometallurgical route, is to utilize the heats of roasting to dry and heat the charge as well as drive off the majority of sulfur as sulfur dioxide (SO_2), producing an impure form of zinc oxide for further pyroprocessing. In the hydrometallurgical route roasting serves to convert the insoluble sulfide concentrate

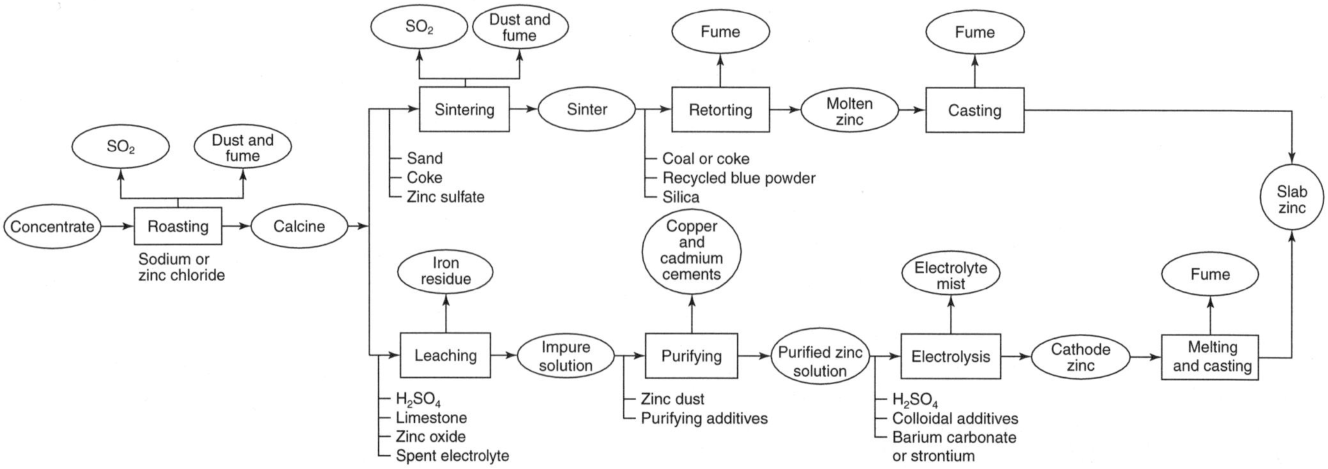

Figure 1. Typical Primary Zinc Smelting Process.

into a more readily leached oxide form. The product produced by roasting is called "Calcine."

Three different types of roasters are used in zinc processing—multiple hearth, suspension, or fluidized bed. Multiple-hearth roasters are the oldest type used in the United States, while fluidized-bed roasters are the most modern.

Primary zinc roasting reactions occurs between 850°C and 1000°C (1300°F and 1800°F), depending on the type of roaster used. Oxygen in the air reacts with sulfur in the concentrate to form sulfur dioxide and to a lesser extent sulfur trioxide as per the following reactions:

$$2ZnS + 3O_2 \rightarrow 2ZnO + 2SO_2 \quad (1)$$

$$2SO_2 + O_2 \rightarrow 2SO_3 \quad (2)$$

Multiple-hearth roasters are the oldest type of the roasters now used in zinc processing and are the least energy efficient. In multiple-hearth roasters, the concentrate is added at the top of a series of nine or more hearths stacked inside a brick-lined cylindrical column. As the feed concentrate drops through the furnace, it is first dried by hot gases passing up through the hearths and then oxidized to produce calcine. The reactions are slow and are sustained by the addition of fuel. These roasters operate at atmospheric pressure and temperatures in the range of 680–700°C. They have the capability of producing very high-purity calcines. Off gas from multiple-hearth roasters typically contain between 4.5 and 6.5% SO_2.

Suspension roasters employ a combustion chamber similar to that of a pulverized coal furnace and achieve more rapid desulfuization than multiple-hearth roasters. The feed is blown into the combustion chamber, where the majority of the calcination takes place. This requires additional grinding, beyond that required for a multiple-hearth furnace, to assure that heat transfer to the material is sufficiently rapid for the desulfurization and oxidation reactions to occur in the furnace chamber. Hearths at the bottom of the roaster capture larger particles, which require additional time within the furnace to complete the desulfurization reaction. Suspension roasters also operate at atmospheric pressure, but at higher temperatures, than multiple-hearth roasters, on the order of 970–990°C. Off-gas from suspension roasters typically contains 10–13% SO_2.

Fluid-bed roasters operate with a pneumatically supported bed of feedstock wherein the desulfurization reactions take place. The heat of reaction is self-sustaining, after ignition, and no additional fuel is required. In a fluidized-bed roaster, finely ground sulfide concentrates are added to the top of the turbulent suspended bed, where they rapidly heat and mix with oxidizing air. Fluidized-bed roasters achieve the lowest sulfide sulfur content calcine of the three designs (about 0.3%). Off-gas from fluidized-bed roaster typically contain on the order of 7–12% SO_2.

Suspension and fluidized-bed roasters offer a number of advantages over multiple-hearth roasters. They are more energy efficient than multiple-hearth roasters, with faster reaction rates that allow for higher specific process rates (throughput per specific roaster volume). The higher off-gas sulfur dioxide (SO_2) content permits more efficient and economical use of acid plants to control SO_2 emissions.

Pyrometallurgical Processing

Pyrometallurgical processing, as the name implies, utilizes heat to accomplish the conversion of calcine to metallic zinc. The typical product from pyroprocessing of calcine is slab metallic zinc. The pyrometallurgical processing route has the advantage over the hydrometallurgical route in that it can process a much wider range of zinc-containing feedstocks, including secondary zinc feedstocks such as electric arc furnace (EAF) dust. Steps in pyroprocessing include roasting, sintering, and retorting.

Sintering

Sintering is the first stage of the pyrometallurgical reduction of calcine to slab zinc. Calcine is mixed with recycled sinter, fluxes, and coke or coal fuel and fed to the sintering machine, where low-boiling point cadmium and lead oxide impurities are volatilzed and subsequently recovered using particulate control devices. Sintering produces an agglomerated permeable mass suitable for feed to retorting furnaces. Typical sinter product contains approximately 48% zinc, 8% iron, 5% aluminum, 4% silicon, and 2% calcium. Smaller amounts of lead, magnesium, and other metals are also present.

Down-draft sintering machines are the most commonly used in the industry. Grate pallets are joined to form a continuous conveyor system. Combustion air is drawn down through the grate pallets where it ignites the fuel present and drives off the sinter impurities. The impurity-laden air is then routed to a particulate control device (typically a bag filter) prior to discharge to the atmosphere.[2]

Retorting

A retort is used to first reduce the zinc oxide to metallic zinc and then volatilize the relatively low-melting-point metallic zinc, 906°C (1663°F), to separate it from other metals found in the sintered feed. Sintered zinc oxide feed and coke are added to the retort, where they are heated, in a reducing atmosphere, to temperatures in the range of 900–1500°C(1650–2600°F). Under these conditions, zinc oxide is simultaneously reduced and volatilized to gaseous zinc:

$$ZnO + CO \rightarrow Zn\ (\text{vapor}) + CO_2 \quad (3)$$

Carbon monoxide is regenerated by the following reaction:

$$CO_2 + C \rightarrow 2CO \quad (4)$$

Zinc vapor and carbon monoxide produced pass from the main furnace to a condenser, where zinc is recovered in a molten zinc bath.

Both electrically and hot combustion gas heated retorts are used, although electrically heated retorts are the most common due to their greater thermal efficiency. Electrically heated retorts typically use electric resistance heating. In a retort furnace, preheated coke and sinter, silica flux, and miscellaneous zinc-bearing materials are fed continuously into the top of the furnace. Feed coke serves as the principal electrical conductor, producing heat, and also provides carbon for the production of the carbon monoxide required for zinc oxide reduction.

The molten zinc metal collected in the condenser is sometimes purified further. The molten zinc finally is cast into small slabs, 27 kg (60 pounds), or large slabs, 640–1000 kg (1400–2400 pounds).

Hydrometallurgical Processing

Hydrometallurgical processing of zinc has the advantage over pyroprocessing in that it generally results in a product with less impurities. This is especially true for concentrates that contain high levels of arsenic, antimony, bismuth, and cadmium. The steps in the hydrometallurgical route include roasting, leaching, purification, electrowinning, melting, and casting.

Leaching

Leaching of calcine (zinc oxide) with sulfuric acid proceeds according to the following equation. Zinc oxide is converted to aqueous zinc sulfate, in an electrolyte solution containing sulfuric acid.

$$ZnO + H_2SO_4 \rightarrow Zn^{2+}(aq) + SO_4^{2-}(aq) + H_2O \quad (5)$$

Typically double-leach methods are used wherein the calcine is first leached in a slightly acidic solution to solublize sulfates and a small fraction of the zinc oxide. This step is followed by more aggressive leaching using recycled barren (zinc-depleted) strong electrolyte. Zinc oxide is dissolved, as are many of the impurities, especially iron, combined as zinc ferrite. The somewhat neutralized product from this step is used as the leach solution for the first stage of the calcine leaching. This two-stage approach also serves as a purification step, in that much of the dissolved iron precipitates out of the solution as goetite, jarosite, or hematite. Hot acid baths and in some cases pressurized leaching (autoclaving) is sometimes employed to maximize the dissolution of the zinc contained in the calcine. The zinc-containing solution from these steps is known as the pregnant solution or pregnant electrolyte.

Purification

The pregnant electrolyte is treated to remove impurities prior to the electrowinning step. The purification steps can be very complex depending on the type and level of impurities present. Purification steps vary by smelter and are considered proprietary by the smelter operators. Typically a major step is the addition of zinc dust to the pregnant solution to force impurities to precipitate followed by filtration to remove the solid precipitates.

Metallic impurities such as arsenic, antimony, bismuth, cobalt, germanium, nickel, and thallium can interfere severely with the electrolyte deposition of zinc. It is necessary to limit their final concentrations to less than 0.05 mg/L. Cadmium is typically recovered as metal, copper as a 60% copper cake and the balance of impurities as a concentrated slime.

Electrowinning

The electrowinning process entails the application of an electrical potential between an inert anode (made of a lead–silver alloy) and an aluminum cathode both immersed in a zinc-bearing solution.[2] In electrowinning the following three major reactions occur within the electrowinning cells. Oxygen gas is released at the anode, metallic zinc is deposited at the cathode, and sulfuric acid is regenerated within the electrolyte. The specific energy consumption for zinc electrowinning is on the order of 3000 kWh per ton.

Electrolytic zinc smelters typically contain several hundred cells. Electrolyte is continuously circulated through the cells and cooling towers, both to lower its temperature and to evaporate water, thus maintaining the electrolyte strength. Periodically, each cell is shut down and the zinc is stripped from the aluminum cathodes.

Melting and Casting

The final stage of electrolytic zinc smelting is the melting and casting of the cathode zinc slabs. Melting is typically carried out in electric furnaces, and the zinc is cast as either small slabs, 27 kg (60 pounds), or large slabs, 640–1100 kg (1400–2400 pounds).

AIR EMISSIONS CHARACTERIZATION

Both pyro- and hydrometallurgical zinc smelting processes generate emissions along the various process steps. Roasting is the major source of both SO_2 and particulate emissions.

Sulfur Dioxide Emissions

Nearly 99% of the potential SO_2 generated from zinc concentrates is released during the roasting process. Concentrations of SO_2 in the exhaust gases vary with the roaster type (from about 4.5 up to about 13%). These levels of SO_2 in the off-gas are sufficient to allow the use of a sulfuric acid plant as the control device. Modern double-contact–double-absorption acid plants can reduce the SO_2 emissions to below 100 ppm.

Additional SO_2 is emitted from the sinter plant, the quantity depending on the sulfur content of the calcine feedstock. SO_2 concentrations of sinter plant exhaust gases ranges from 0.1% to 2.4%. In most cases no SO_2 control is applied to this gas stream. The SO_2 levels are too low for use of an acid plant but can be reduced using either caustic of lime-based wet scrubbers, depending on the off-gas SO_2 concentration and the control required.

Extensive desulfurization of the zinc prior to retorting results in practically no SO_2 emissions from these devices.

Particulate Emissions

As with sulfur dioxide, the majority of particulate emissions are generated during the roaster step. Emissions from the roaster may represent from about 3 up to 60% of the concentrate feed. When expressed in terms of zinc production, emissions are estimated to be 133 kg/Mg (266 pounds per ton) for a multiple-hearth roaster and 1000 kg/Mg (2000 pounds per ton) for a fluidized-bed roaster, expressed in terms of zinc production.[3] For suspension and fluid bed roaster cyclones are typically used as the primary particulate control for the purpose of recycling the product to the roaster. Roasters equipped with an acid plant for SO_2 control, use electrostatic precipitators and/or wet scrubbers to clean the off-gases prior to the sulfuric acid plant. Particulate emissions following the acid plant are very low and are associated primarily with sulfuric acid mist. Particulates, in the form of total SO_3 and H_2SO_4, can be controlled to levels of about 40 mg/Nm^3.

Sinter plant emissions are the next significant source of particulate emissions in pyroprocessing of zinc. Uncontrolled particulate emissions from sinter plants can represent from about 3 to 8% of the feed. Sinter plant particulate emissions are typically controlled using fabric filters.

Table 1. Particulate Emission Factors for Zinc Smelting[a]

Process	Emission Factor (kg/Mg)	Emission Factor Rating	Process	Emission Factor (kg/Mg)	Emission Factor Rating
Roasting			*Sinter plant*		
Multiple hearth			Uncontrolled	62.5	E
Uncontrolled	113	E	W/Cyclone	24.1	D
Suspension			W/Cyclone and ESP	8.25	D
Uncontrolled	1000	E	Vertical retort		
Controlled	4	E	Uncontrolled	7.15	D
Fluidized bed			Electric retort		
Uncontrolled	1083	E	Uncontrolled	10.0	E
			Electrowinning		
			Uncontrolled	3.3	E

[a]From AP-42, *Compilation of Air Pollutant Emission Factors*, U.S. EPA.

Note: Data between controlled and uncontrolled emissions not necessarily compatible. D emission factor rating, data of below average quality, limitations on use may prevail; E emission factor rating, data of poor quality, limitations on use noted.

Particulate emissions from retorts are generally low, usually less than 1% of the feed. These emissions are generally controlled using fabric filters or wet scrubber.

The hydrometallurgical process steps emit much less particulate than does pyrometallurgical reduction. During electrowinning small quantities of sulfuric acid mist are generated. These emissions are exhausted directly to the atmosphere or are controlled using cross-flow ventilation systems and fabric mist eliminators.

Table 1 presents controlled and uncontrolled emission factors for point sources within a zinc smelting plant.

The U.S. Environmental Protection Agency has estimated fugitive emission factors for the zinc smelting industry. These are presented in Table 2. These emission factors are based on similar operations in the steel, lead, and copper industries.

Table 2. Uncontrolled Fugitive Particulate Emission Factors for Zinc Smelting[a]

Process	Emission Factor[b,c] (kg/Mg)
Roasting	Negligible
Sinter plant[d]	
Wind box	0.12–0.55
Discharge and screens	0.28–1.22
Retort building	1.0–2.0
Casting	1.26

[a]From AP-42, *Compilation of Air Pollutant Emission Factors*, U.S. EPA.
[b]Emission Factor Rating E (poor) for all factors presented.
[c]Based on quantity of slab zinc produced except as noted.
[d]Based on quantity of sinter produced.

AIR POLLUTION CONTROL MEASURES

Control measures used in zinc smelting operations are similar to the types of controls used in copper and lead smelting. They include control measures to limit SO_2, acid mist, particulates, and metals emissions. The major emission source is the roasting operation with the highest levels of uncontrolled emissions. The typical control approach is to install a metallurgical acid plant to recover the SO_2 in the roaster off-gas as sulfuric acid. The efficient and reliable operation of an acid plant depends on a clean, relatively dry feed gas to the acid plant drying tower. Figure 2 presents a

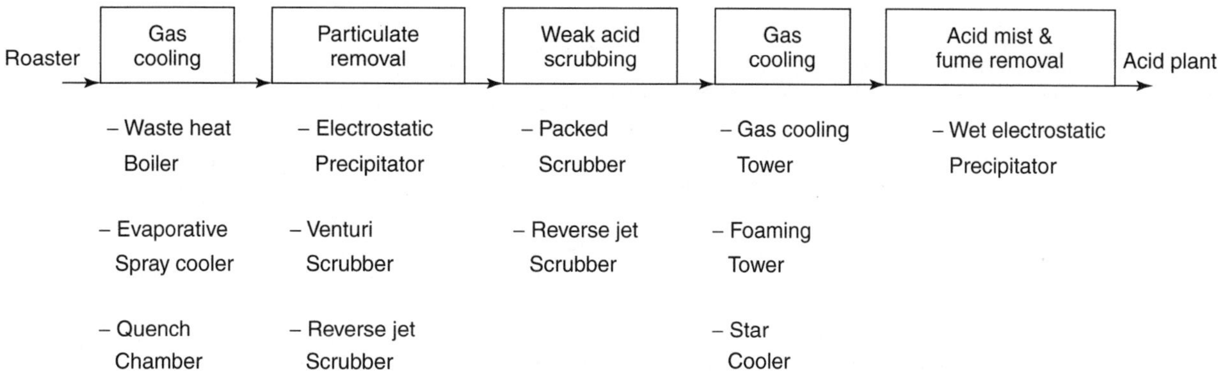

Figure 2. Metallurgical Acid Plant–Gas Cleaning System; Unit Operations and Equipment Utilized.

schematic of a metallurgical acid plant gas cleaning system. This system functions to remove particulate matter, SO_3, and metal fumes to low levels and predries the gas entering the acid plant.

Gas cooling is needed to reduce the roaster off-gas temperatures to levels acceptable for use of a hot electrostatic precipitator (350–400°C) and/or to condense out volatile metals. For large suspension and fluid-bed roasters where an electrostatic precipitator (ESP) is used for particulate removal, a waste heat boiler is often used in order to recover heat energy in the form of steam. For other applications an evaporative gas cooler is used. Where particulate removal will be accomplished using a wet scrubber, a quench chamber is used to saturate and cool the gas to its adiabatic saturation temperature.

Particulate removal is most often accomplished using an ESP. Metallurgical sulfuric acid plant gas cleaning ESPs are typically of the rigid frame or rigid electrode type with three to five electrical fields and specific collecting areas on the order of 100–200 m^2/m^3/sec. Outlet particulate loadings are generally on the order of 25–150 mg/Nm^3.

Wet scrubbers are also used for particulate removal. These are typically high-energy venturi scrubbers (1200–2500 mm pressure drop) or other particulate scrubbers, such as reverse-jet scrubbers. Wet scrubbers are more often used when higher particulate loading is allowable to the acid plant and/or when wet handling/treatment of the collected dust is specified.

Weak acid scrubbers function to remove additional particulate, sulfuric acid mist and condense volatile metals. Packed-bed and reverse-jet scrubbers are typically used for this purpose. These scrubbers typically operate at a pH of 1 or less and therefore do not remove SO_2.

Gas coolers lower the gas temperature further, below the adiabatic saturation temperature, through the use of either gas/liquid heat exchangers or recirculating chilled scrubber fluid. Gas temperatures on the order of 35–40°C are common. These coolers function to condense out additional volatile metals and SO_3 and also reduce the moisture content of the gas. A number of different gas cooler designs are used.

Wet ESPs remove acid mist and condensed metal fumes from the gas stream. These ESPs may be of the wet tube or plate type. In some applications two wet ESPs may be used in series to achieve very high acid mist and particulate removal. The gas entering the acid plant typically is relatively dry, containing low levels of SO_3, metals, and particulate (on the order of 10–50 mg/Nm^3).

In some instances where the concentrate contains high levels of mercury and or very high acid quality is required, a mercury removal step follows the wet ESPs. Both calomel and sulfide-based technologies are used.

Gas coolers (either evaporative or air dilution type) followed by fabric filters are typically used to capture particulate from the sinter machine off gas. High-energy venturi wet scrubbers are also used. Fabric filters are of the pulse cleaned design using off-line cleaning. Low air-to-cloth (filtering) ratios are used due to the fine and sticky nature of the cadmium, lead, and zinc fume collected in these devices.

Off gases from retorting is often discharged directly to the atmosphere or treated as for the sinter off gas. Fugitive emissions are controlled using enclosures, sprays, and dust filters.

REFERENCES

1. *Merck Index*, 10th Edition, Merck & Co., 1983.
2. N. L. Weiss, *SME Mineral Processing Handbook*, Section 15, "Lead and Zinc," Society of Mining Engineers, 1985.
3. AP-42, *Compilation of Air Pollution Emission Factors*, U.S. Environmental Protection Agency.
4. A. J. Bounicore and W. T. Davis, *Air Pollution Engineering Manual*, Van Nostrand Reinhold, New York, 1992.
5. *Pollution Prevention and Abatement Handbook — Toward Cleaner Production*, draft document, World Bank, 1997.
6. *Environmental Assessment of the Domestic Primary Copper, Lead and Zinc Industries*, EPA-600/2-82-066, U.S. Environmental Protection Agency, Cincinnati, OH, October 1978.
7. *Systems Study for Control of Emissions in the Primary Nonferrous Smelting Industry*, Vol. 1, APTD-1280, U.S. Environmental Protection Agency, Research Triangle Park, NC, June 1969.
8. *Encyclopedia of Chemical Technology*, John Wiley & Sons, New York, 1967.
9. *Background Information for New Source Performance Standards: Primary Copper, Lead and Zinc Smelters*, EPA-450/ 2-74-002a, U.S. Environmental Protection Agency, Research Triangle Park, NC, October 1974.
10. J. A. Dutrizac et al., *Zinc and Lead Processing*, The Metallurgical Society of CIM, 1998.

SECONDARY ALUMINUM

Charles A. Licht

PROCESS DESCRIPTION

The development of secondary aluminum operations have grown from the days when the process revolved around the sweating of aluminum in scrap yards where scrap was melted, separated from high-melting-point materials, and poured into ingot or sows. This type of operation still goes on. Scrap aluminum required for recycling is generated in the form of "home" scrap (never considered "secondary" when processed in the facility that generated it), new or manufactured scrap, obsolete automotive scrap components, obsolete appliances and tools, obsolete housing elements such as siding and wiring, cans (UBC), and other varied sources. At present scrap aluminum is used to make "prime" equivalent alloys for extrusions and sheet, where previously most scrap was recycled into the foundry and die casting industries. The alloys needed for extrusion and sheet forms must have low iron, silicon, and copper content, while foundry alloys may contain levels of iron at 1%, silicon ranging from 5% to 15%, zinc ranging from 1% to 3%, and copper ranging from 1% to 3+%.

Foundry alloys are manufactured in plants that utilize the sows and ingot generated in scrap yard sweating furnaces and scrap coming from other sources as noted above. Over the years, techniques have been developed for the processing of the slags and skimmings (drosses) from various melting operations, which include skimmings from primary facilities, foundries and die casters, and those generated in secondary aluminum facilities (see Figures 1 and 2).

For many years, the term secondary aluminum had pejorative overtones, fostered in part by the primary aluminum producers, who had great surpluses of primary capacity, who

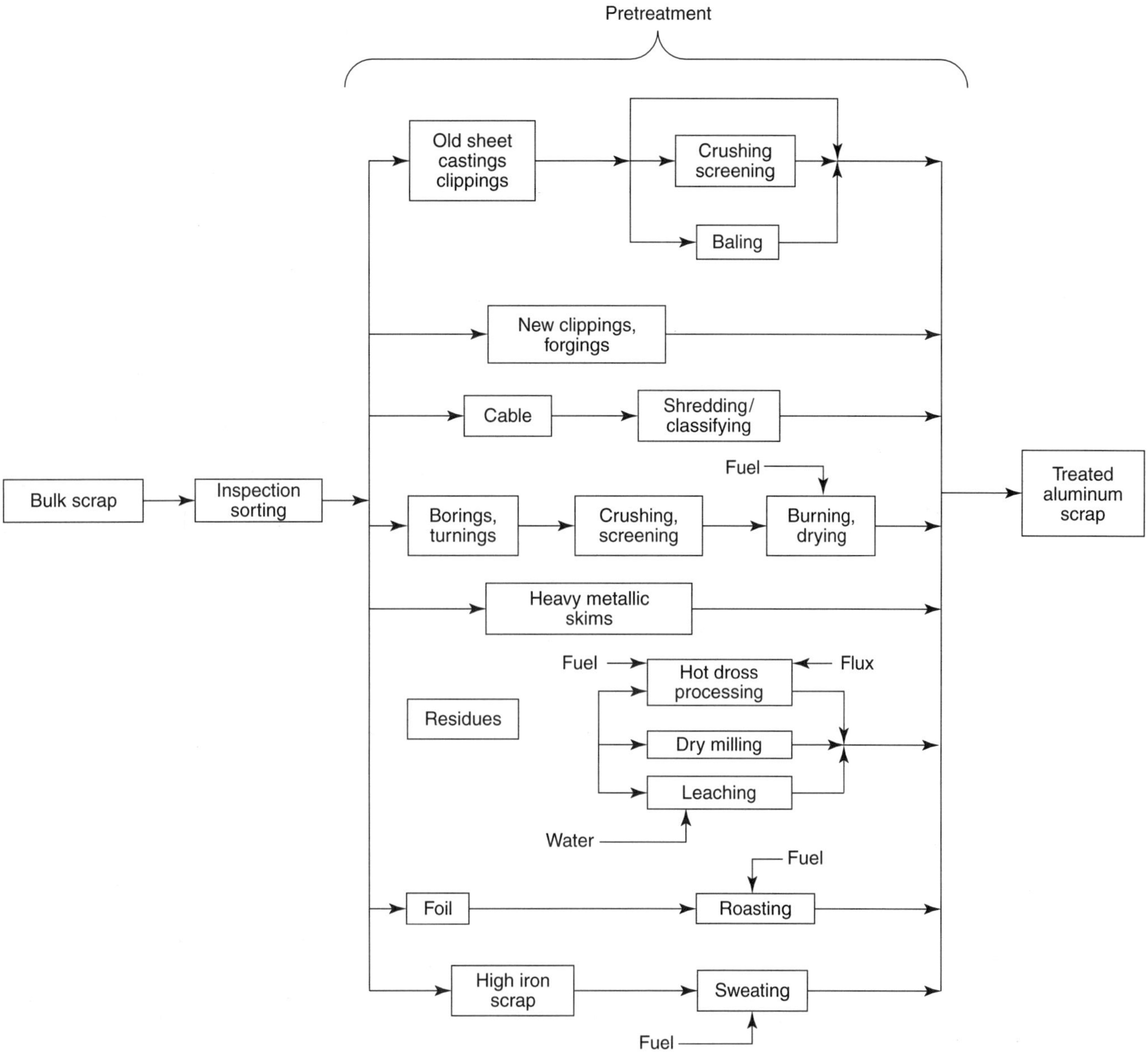

Figure 1. Partial Process Flow Diagram.

"talked down" the recycling of aluminum in order to keep their potlines going. These issues are now historic, since not only is aluminum remelted by the foundry alloy manufacturers, but the foundry alloy producers must now vie with primary companies in the purchase of scrap materials to supplement their primary sources. They must also compete with high-end secondary operations, where scrap is reprocessed into sheet and extrusion form.

The foundry industry consumes a vast percentage of the alloys produced by the secondary aluminum smelters. The secondary aluminum ingot makers, by virtue of both market forces and availability, generally take lower-purity and dirtier materials to recycle into casting alloys. In part, this is due to the broader specifications available in the casting alloy spectrum. The presence of iron, silicon, and other elements makes certain materials impossible to extrude or roll. They do not create the same problem when making castings. Metals, such as silicon and iron, tend to make the materials brittle. Materials that are going to be extruded or rolled cannot have brittleness, but must be ductile and tough.

A number of terms must be defined. These include:

1. *Alloying*. Alloying is a process wherein a molten bath is modified by the addition of alloying elements to bring the alloy into specification. In a typical foundry alloy, the addition of silicon, copper, and certain other minor elements is usually required.
2. *Degassing*. Degassing is a process wherein various gases and/or fluxes are utilized to remove absorbed gases from the molten metal. This is required since the alloyed molten metal will contain much more gas than is possible to be held when the metal solidifies; thus if the gases are not removed from the molten metal as they solidify, holes will develop. This porosity leads to inferior castings and makes it impossible to roll or extrude salable products.

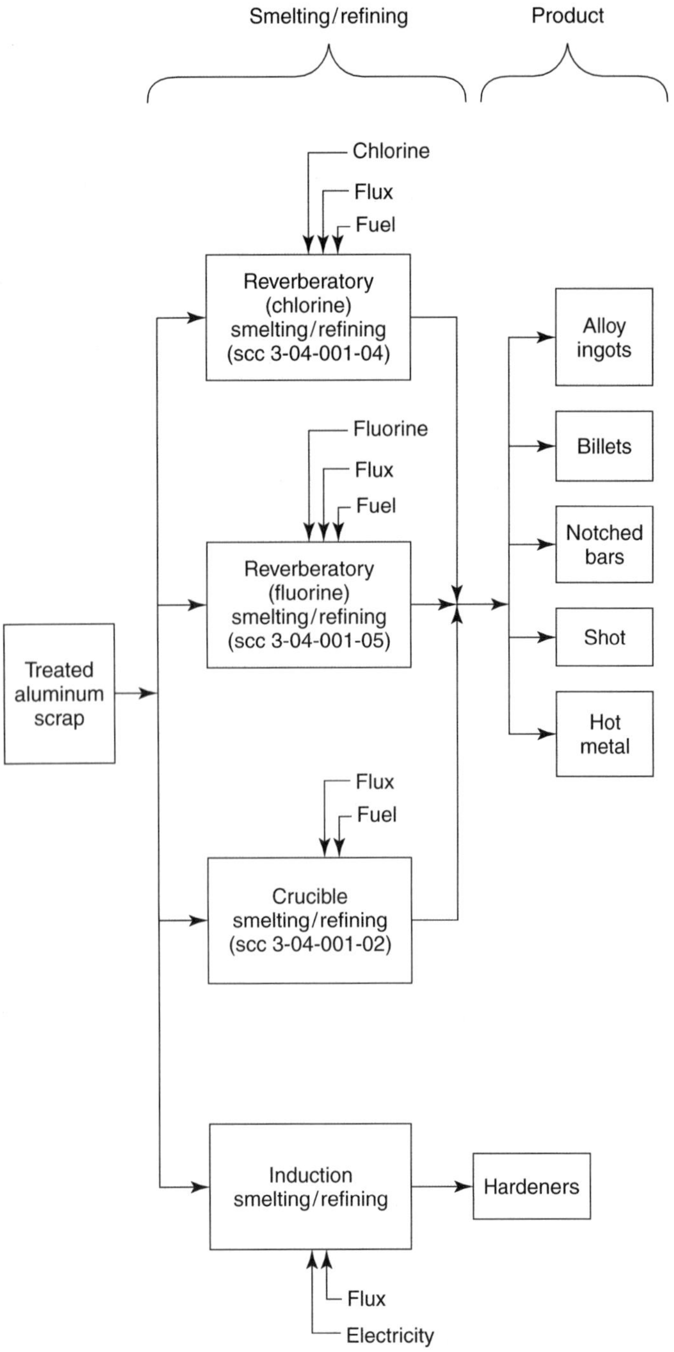

Figure 2. Partial Process Flow Diagram.

3. *Demagging*. Essentially the only technique that is available for refining aluminum, demagging allows the removal of magnesium from the molten bath by reacting the bath with either a fluoride salt or chlorine.
4. *Fluxing*. This term covers a very broad array of processes. In melting aluminum scrap in open charging well furnaces (to be described later), a salt–potash–fluoride flux is used on the open surface of the bath to eliminate the formation of aluminum oxides and to cause the agglomeration of small beads of aluminum that will form in the oxide particles, thus allowing these particles to develop sufficient mass to flow into the bath. Additionally, certain fluxes are used in reverberatory furnaces that do not have charging wells. Basically, these fluxes are used to remove the buildup of oxides from the walls or to eliminate and/or reduce oxidation. Fluxes are also used in the degassing and demagging processes.

RAW MATERIALS PROCESSING

Among the major sources of aluminum materials available to the secondary casting alloy producer are manufactured scrap in the form of chips and turnings. To a lesser extent, clippings or skeletons from stamping operations are available, as well as cutoffs from heavier sections from extrusion operation. The availability of this portion of the scrap market to the secondary foundry ingot makers is minimal because of the high cost of these materials, which are coopted by primary and secondary sheet rollers and extruders.

Scrap is obtained from the automobile shredding process after the nonferrous portions are separated from the ferrous portions. Ultimately, the nonferrous portions themselves are split so that high-quality aluminum scrap can be obtained.

Another source of aluminum scrap is aluminum copper radiators used in many heat exchanger applications, which are a source not only of aluminum, but of inexpensive copper. Another very large source is obsolete scrap. This can be in the form of old sheet and siding, painted and unpainted. Occasionally, old sheet and siding carry insulating materials. Old castings from a variety of consumer products, such as automatic transmission housings and engines from automobiles, lawn chairs, lawn mower housings and motors, outboard motors, aluminum doors and windows, and aluminum screens, are also available.

Often, die casters and foundries will have a secondary smelter process their gates and risers and will return specification alloy ingot to the foundry. Many times, this scrap is sold outright to the smelters.

Used beverage containers (UBCs) represent a very significant volume of metallics. Because the material is essentially one family of alloy, the optimum use of this material is to recycle into can sheet stock. The process for handling cans is similar to other processes in the aluminum cycle in that cans are generally shredded, run through a delacquering system, which is equivalent to a chip dryer, and then processed in either rotary furnaces or specially equipped reverberatory furnaces. Some can scrap is processed as received and melted in rotary furnaces.

Direct melting of can scrap in the rotary furnace is beneficial in that the coatings are generally burned in the furnace. Rotary salt furnaces have adequate air pollution controls so that even partially burned carbonaceous material is collected in the baghouse. The use of delacquered cans allows for open-well or induction furnace operation without the need for substantial air pollution controls.

One area that is now in the proposed MACT rules relates to various chlorine-bearing compounds used as interior coatings and sealants. Hydrochloric gas emission potential from melting can scrap requires control.

CHIP DRYING

The aluminum chips and turnings are a manufactured product generated in the turning, milling, boring, and machining of aluminum. These chips generally contain varying amounts of coolants or lubricants, ranging from 2% to 20% of the inbound weight. Machining operations usually utilize miscible compounds for cooling and lubricating, and thus a substantial proportion of the coolant adhering to the chips is water. However, the oil content can create problems. Additionally, charging water-wet material into a molten bath is to be avoided at any cost.

Because of this potential for problems, many secondary aluminum and some primary aluminum facilities have chip-drying equipment. These systems are generally set up so that a small crusher (hammer mill or ring mill) breaks the particles down to uniform size below $\frac{1}{4}$ to $1\frac{1}{4}$ inches. The materials are then fed into a storage bin and withdrawn from that storage bin by a metering conveyor so that uniform feed can be achieved. The chips are fed through air locks into the drying system, where the chips are heated and the oil and moisture driven off. The chips then leave the system via air locks and magnetic separators to remove "free" iron. They are then available for melting. Obviously, the optimum situation is when these hot chips are taken to the furnaces directly.

The fumes from the process are ducted to a mechanical separator, such as a cyclone, a multiple cyclone, or a "dropout box." The fumes are then ducted into fume incinerators, where the combustible components are removed.

There are several dryer designs that are commercially available, although many "standard" designs are custom engineered. The standard dryer has an open flame within the rotary drum. While this would appear to cause oxidation, in fact, only trivial oxidation is noted. This is so because the drum is sealed so that leakage air is minimized and the oxygen content in the drum is only sufficient to sustain a small flame.

Another technique that is used is to heat the drum externally. This has the advantage of requiring substantially smaller pollution control systems since the combustion gases do not need to be controlled. Only those gases generated by the evaporation and/or partial burning of the "moisture" are then drawn off into mechanical separation and fume-burning equipment. The direct-fired unit has a greater throughput for physical size, thus offsetting the somewhat higher cost of operating the pollution control system.

Another system utilizes a "black" drum wherein no flame occurs. The hot gases are recirculated through mechanical separators and an afterburner. A portion of the "burned-out" gas is exhausted. The remaining gas is recirculated as the dryer heat source.

CAN (UBC) DELACQUERING

UBCs are delivered to processors after having been partially processed at scrap yards and recycling centers. The cans may have been shredded or compressed into briquettes or small or large bales. The compressed bales are broken prior to processing. The shredded UBC is processed as received. The ideal condition for melting of UBC is to have removed the coatings, prior to melting. Many processors do not remove coatings, and thus the potential for forming hydrochloric acid gas, dioxins and furans exists. This requires gas neutralization. UBCs can be processed in a closed circuit rotary drum system. A patented concept is used extensively in can delacquering (Figure 3).

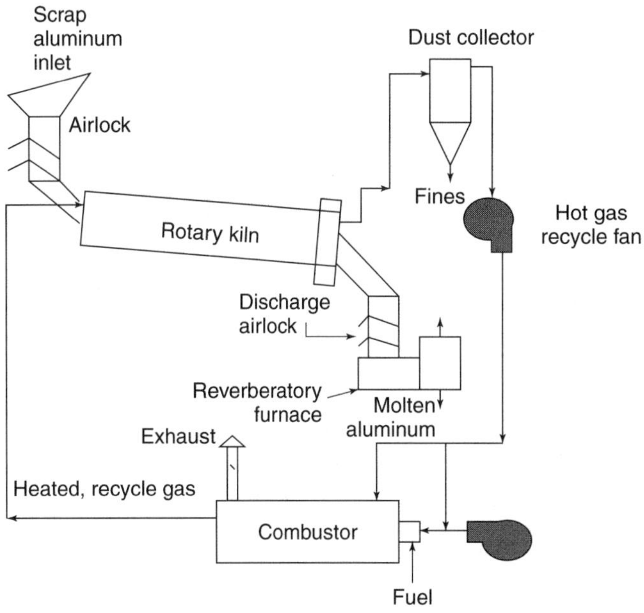

Figure 3. APROS Delacquering System.

SCRAP SHREDDING

Aluminum scrap material, such as painted siding, bus bodies, insulated pieces of all sorts, and a great variety of materials that were processed by "sweating" (to be discussed later), are being processed by large shredding systems. These systems generally utilize hammer mills or ring mills, usually in the range of 60 in. in diameter by 60 in. wide, although larger units are in service. Scrap is fed into the system after having been picked to eliminate oversized materials and remove nonaluminum components such as zinc, magnesium, stainless steel, etc.

Depending on the installation, scrap may simply drop into the shredding device or may be fed with a feeder that controls the feed rate based on motor amperage. Systems range from 400 to over 1000 hp, with throughput ranges from 5000 to over 50,000 pounds an hour, depending on the grade of scrap, the amount of picking that has to be done, and the type of material handling in use. The scrap coming from this process is dramatically cleaned so that the amount of smoke and fume generated in the charging well is substantially reduced. Additionally, free iron and combustibles that can cause oxidation on the charging well are removed.

The hammer mill "windage" is ventilated to a mechanical pollution control system (cyclone or dropout box), which removes large amounts of particulate loading. In some instances, it is necessary to go to secondary pollution controls such as a baghouse, to deal with extremely fine particulate from the mill.

SWEAT FURNACE

These direct-fired furnaces are generally small. They are used to separate low-melting-point from higher-melting-point materials. In the case of aluminum, sweating is done on lawn mowers and other materials that contain steel components. Large hammermill systems allow mechanical, rather than thermal, processing with increased metal recovery.

Automatic transmissions with the steel shafts and gearing are sweated, although typically these are disassembled and then hammermilled. Specific materials often are sweated because of the nature of their configuration. Pistons may be sweated because ferrous components cast into the head of the piston cannot be removed by breaking the piston in a hammermill.

DROSS PROCESSING

Aluminum skimmings and drosses produced in primary operations, foundries, and secondary smelters can be processed in several ways.

A technique that was popular earlier, particularly with the high-salt-content secondary smelter skimmings, was to wash the drosses in large volumes of water. Typically, this was carried on near a large river so that many hundreds of gallons per minute of water could be sluiced through a long drum, generally raising the salt content by 100 to 200 ppm of the water flow going through the system and raising the stream levels substantially less than that. Exiting from the end of the system would be a melange of aluminum oxide and aluminum particles. Some systems screened these materials at 20 mesh, salvaging most of the aluminum particles, which were then dried and melted, while the oxides were allowed to go into settling ponds, where they settled out, with the overflow returning to the stream. These ponds had to be dredged periodically and the oxides taken off to landfill. The use of wet systems has been tried over the past several decades with varying levels of success. Small-scale wet recovery systems were in use in the 1920s and 1930s. Development work on more sophisticated systems was initiated in the 1950s. These systems were not financially viable because of very low land fill costs. In the 1990s several wet systems have been installed where the potential liability of failed landfills became a financial issue to corporate management. The systems wash the salt-bearing materials. The brines are separated, cleaned, and processed to form salt solids for reuse as flux. The oxides are clean enough to be used for exothermic purposes if sufficient metal content is present. They can be utilized in other processes because of the low chloride content. What metal values are left in the drosses after washing (generally plus 20 mesh), are remelted, in either rotary or reverberatory furnaces.

In the middle 1950s, improvements were developed in dry milling dross to the point where facilities were capable of processing a million pounds per day of skimmings and drosses. This process involved primary breaking of large chunks to smaller sizes. The smaller chunks were processed through a series of hammermills and screens, separating out the coarser materials that contained higher metallic components and melting the concentrates. The finer materials that were low in metallics and not readily melted were separated by various air-classification techniques and developed into a product for the ferrous exothermic market. The extremely fine materials were sold to cement companies for their alumina content or discarded to landfill.

Processes were installed to wash the fine materials to develop an exothermic product. This was carried out by at least one facility in the middle 1960s, generating a product that was widely sought after by the ferrous exothermic manufacturers. Since 1975, a number of companies have started up that melt skimmings and drosses in rotary furnaces, adding appropriate salt fluxes and separating the metals from the oxides thermally. The metals are melted and then decanted. The salt cake is poured into molds and, after cooling, is taken to landfill.

FURNACES

Small crucible or pot furnaces are typically not used in the secondary aluminum industry. These small furnaces are primarily used in small sand and permanent mold foundries and as holding units at die casting machines.

OPEN-WELL REVERBERATORY FURNACES

The furnaces (Figure 4) used to melt aluminum scrap in secondary foundry alloy facilities have open-well charging hearths. These furnaces have capacities of 30,000–300,000 pounds. Melting occurs primarily in the open charging well. The closed combustion chamber functions as a heater, although heavy scrap such as sows and ingot may be charged into this chamber. These furnaces can range from 15 to 20 ft wide by 15 to 30 ft long. In addition, a charging well, 5 to 6 ft wide by the length of the furnace, is used.

This type of furnace allows easy degassing, demagging, and fluxing. All these activities occur in the well, which is appropriately hooded to remove fumes that are generated. The process starts with heating the furnace to about 1400°F and charging a substantial amount of heavy section scrap into the combustion chamber. This material, which can be sow, ingot, heavy castings, or plate is used to form a heel. It will represent 10–20% of the furnace capacity. When the heel is molten, additional scrap is charged into the well of the open-well furnaces. Puddling and/or circulation by means of various types of equipment is done so that the hot molten bath inside the furnace combustion zone can be circulated into the charging well to melt the scrap being charged.

The ideal technique for charging scrap into the charging well is by means of a mechanism such as a vibrating feeder or feeder and conveyor so that the scrap is charged at a steady rate. By utilizing mechanical charging techniques, it is possible for the air pollution control hoods to be fitted closely to the furnace allowing minimal potential loss of smoke and fume from the charging well. This type of furnace cannot have "dirty" material charged into the combustion chamber. The emissions from the combustion chamber will primarily be products of combustion (POC) of fuels with only minor traces of particulate from the bath. In the case of direct-charged reverberatory furnaces, substantially more scrap will be charged initially and will be allowed to melt down. Substantial amounts of flux are used in the melting process on the charging well that require added thermal input to heat and melt.

In order to assist in the melting process, a variety of techniques are utilized. Pumps made of extremely high-temperature components are used to circulate the bath. These pumps can also be used with appropriate fitting to do degassing and/or demagging with chlorine. Additionally, mechanical puddling devices are commercially available that can be used both to puddle the scrap in the open charging wells and to "work out" the skimmings and mechanically skim them from the furnace.

The open charging well furnace, because the charging well is controlled by an air pollution control system, can take advantage of considerably "dirtier" scrap — dirtier in the sense

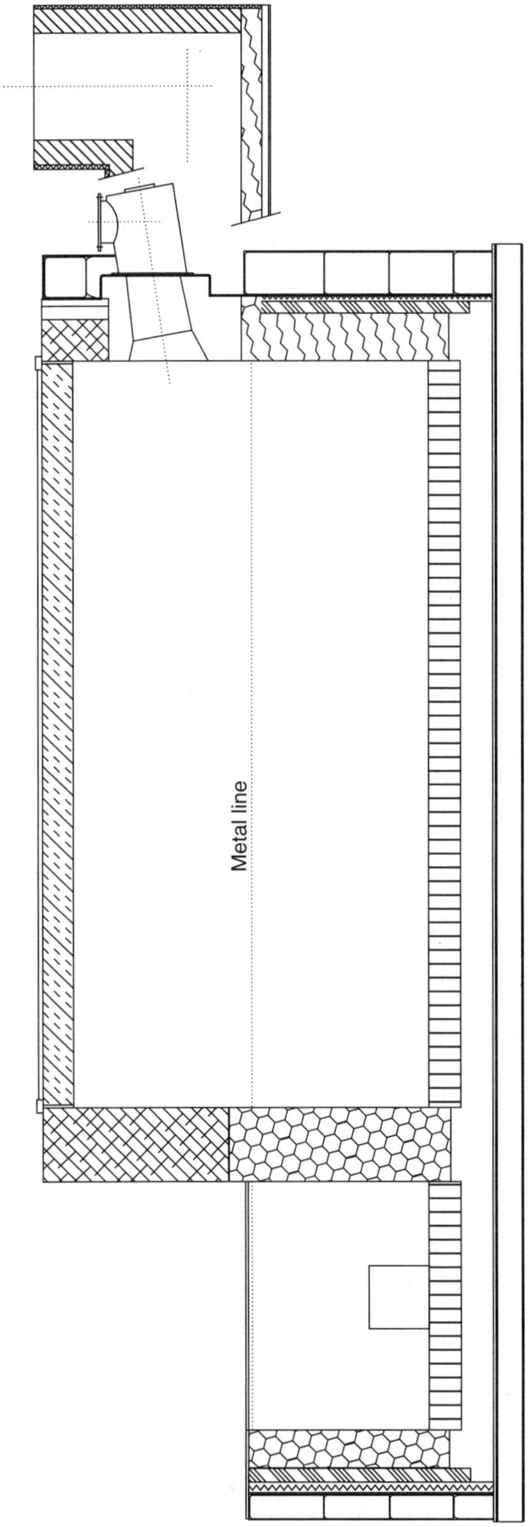

Figure 4. Open-Well Reverberatory Furnace.

of having combustible materials on the surface such as paint, oil, or grease. The "dirt" from this material, if charged into a direct-charge furnace, would only be partially combusted in the combustion zone and can result in hydrocarbon emissions. Substantial amounts of unburned material and fine particulate would be exhausted, thus requiring air pollution control on the high-temperature exhaust stream, which is expensive as compared with the relatively low temperatures that can be utilized in controlling the smoke and fume from a charging well furnace.

As the batch proceeds, silicon and copper are added as required to meet the final specifications. If a pump is not used, then, toward the end of the batch, chlorine or aluminum fluoride are introduced to the bath to reduce the magnesium content. If a pump is used, chlorine can be introduced at a low input level over the last half of the melt cycle. This can be done only in an open charging well furnace, which is one of the reasons secondary casting alloy producers utilize this type of furnace almost exclusively. Magnesium is removed by reaction of the magnesium with chlorine. This process is very efficient with the use of pumps. Chlorine utilization with results at or below stoichiometric reaction ratios have been measured frequently. Until recently the wet scrubber used as part of a chlorination "bell" was used in a few plants.

REVERBERATORY FURNACE

Most primary operations and independent extruders and sheet rollers remelt purchased and home scrap using furnaces where scrap is charged directly into the combustion zone. With the clean scrap that these companies use due to alloy considerations, the scrap can be melted without excessive oxidation. In the proposed MACT Rules this type of furnace may not require air pollution controls. This type of furnace is not amenable to charging auto shredder scrap, turnings and chips, or light-gauge materials.

These furnaces are gas or oil fired. Gas firing is preferred since there are no particulate or SO_x products of combustion.

The process starts by heating the furnace to about 1400°F and charging a substantial amount of heavy section scrap into the combustion chamber. This material, which can be sow, ingot, heavy castings, or plate, is used to form a heel. It will represent 20–30% of the furnace capacity. When the heel is molten, additional scrap, ingot, or sow is charged. Particular car must be taken to assure that *all* charge material is *dry*.

When molten, the furnace may be skimmed and an additional charge placed in the furnace. Depending on the equipment available for doing the charging and the size of the furnace, charges of 10,000 to 60,000 pounds can be made at a time. The direct-charge furnace is more thermally efficient than the charging well furnace because of better heat transfer and in part to the omission of salt fluxes.

While the fluxes require energy, they are very effective in removing the aluminum content of the oxide formations; the amount of aluminum that is skimmed from the furnace is substantially reduced as compared with that when the skimming is done in a direct-charge reverberatory furnace without fluxing. Direct-charge furnaces cannot use melting fluxes since the vaporized slat would generate massive emissions (Figure 5).

After a proportion of the furnace bath is melted, the bath is stirred and sampled. The analysis is compared with the precalculated analysis at that stage. If the analysis is within reasonable limits of the precalculations, then the previously allotted scrap will be added to the furnace. If it is not, an opportunity exists rapidly to recalculate and change the analysis of the material to be charged. By doing this, the chances of having a full furnace come in out of "spec" are reduced.

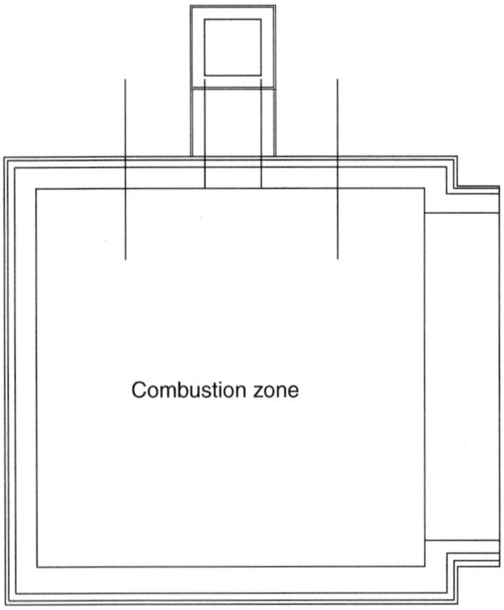

Figure 5. Recerberatory Furnace.

When the alloy is in specification, it can be poured into large 15,000- to 40,000-pound ladles, which can be trucked to foundries and die casters, or it can be tapped into transfer ladles to be taken to another location in the plant for pouring into sow or ingot or tapped via troughs into sow and/or ingot pouring systems adjacent to the furnace.

ROTARY FURNACES

Rotary furnaces are used to melt drosses and can scrap (UBC). They are directly fired with gas or oil. The fumes and POC are vented into baghouses to control fumes. Salt or salt/potash flux is used. The exhausts generally contain salt fumes, oxides, and carbonaceous material. If cans are melted, the exhaust may also contain hydrochloric acid gas. European technology uses rotary furnaces for making alloy. Typical European rotary furnaces are substantially bigger than U.S. equipment.

FLUXES

Generally fluxes fall into four categories, as follows:

1. *Cover fluxes*. These fluxes prevent oxidation of the molten bath and cause the agglomeration of metal droplets, which sink back into the bath. These are used on open-well furnaces only.
2. *Solvent fluxes*. These fluxes generally cause suspended particles and oxides to float to the top of the bath to form a dross. This dross can be moved out into the charging well for fluxing down with a cover flux prior to skimming, or in the case of direct-charge reverberatory furnaces, the oxide flux mixture can be skimmed from the furnace and cooled.
3. *Degassing fluxes*. Generally, these fluxes are used to remove soluble gases, such as hydrogen, from the molten aluminum. If the bath is to be demagged using chlorine, this is generally sufficient to do an adequate job, although other fluxes, including a number of proprietary mixtures, are used for degassing purposes.
4. *Magnesium reducing fluxes*. Basically, there are two materials that are used for the removal of magnesium, chlorine, and aluminum fluoride. The melting point of aluminum fluoride is substantially higher than that of the bath, and thus reactions can be difficult to achieve. When aluminum fluoride is used to remove magnesium from a bath, it is generally mixed with the "smelter's flux" (47.5% calcium chloride, 47.5% sodium chloride, 5% fluoride salt such as cryolite) to reduce the melting point.

Although the literature is replete with descriptions of a great variety of flux materials used for cover, solvent, and degassing, most secondary aluminum alloy producers use "smelter's" flux for their cover and depend on chlorine or chlorine and nitrogen mixtures for their solvent and degassing fluxes.

Uncontrolled utilization of chlorine can generate enormous volumes of air pollutants. Chlorine will react preferentially with magnesium to form magnesium chloride, which, at bath operating temperatures, is a molten salt. Chlorine will also react with aluminum to form aluminum chloride, which is a gas above 368°F (187°C). When cooled, it will react with atmospheric moisture to form aluminum oxide and hydrochloric acid gas, causing an acrid visible plume.

With the use of demagging metal pumps, the flow of chlorine into the system can be sustained at a level low enough that the reaction process takes place without the formation of visible aluminum chloride. The use of a chlorination bell restrains aluminum chloride under the bell and allows it to vent through a scrubbing section so that there is no loss of aluminum chloride to the environment.

A patented (Derham) process for magnesium removal makes use of a pump that circulate metal from the charging well into a secondary chamber, where a thick molten-salt flux bath is maintained with molten aluminum below it. Chlorine is introduced into the chamber well below the bath surface. The magnesium chloride and aluminum chloride are absorbed in the salt bed above the aluminum. The aluminum underflows and returns to the furnace, where it is reheated and recirculated until magnesium specification levels are obtained.

AIR EMISSIONS CHARACTERIZATION

In the secondary alloy producer's operations particularly, but even in sheet rolling and extrusion facilities, scrap being remelted has surface coatings. The obvious paint, dirt, oil, and grease are present, but even when melting new die casting home scrap, parting agents can be on the surface of the casting. In the case of extrusions and sheet rolling, lubricants and coolants remain on the surface. When remelted, the surface coating will cause the formation of visible emissions and VOC emissions from the charging well. According to the tests run in preparation of the MACT rules, chlorine, and chlorine-containing materials such as PVC and neoprene, are present in sufficient quantity to form dioxins and furans, which will require process modification or controls.

Table 1, from AP-42, the U.S. Environmental Protection Agency's *Compilation of Air Pollutant Emission Factors*, presents emission factors for the principal emission sources in secondary aluminum operations.

Table 1. Particulate Emission Factors for Secondary Aluminum Operations[a] (English Units)

All Emission Factors are in lb/ton of Product

Operation	SCCs	Uncontrolled		Baghouse		Electrostatic Precipitator	
Sweating furnace[b]	3-04-001-01	14.5	E	3.3	E		
Smelting							
Crucible furnace[b]	3-04-001-02	1.9	E				
Reverberatory[c]	3-04-001-03	4.3	E	1.3[e]	E	1.3	E
Chlorine demagging[d]	3-04-001-04	1000	E	50	E		

[a]Ref. 2. Emission factors for sweating and smelting furnaces expressed as units per unit weight of metal processed. For chlorine demagging, emission factor is kg/Mg (lb/ton) of chlorine used.
[b]Based upon averages of two source tests.
[c]Uncontrolled, based on averages of ten source tests. Standard deviation of uncontrolled emission factor is 3.5 lb/ton, that of controlled factor is 0.3 lb/ton.
[d]Based on average of ten source tests. Standard deviation of uncontrolled emission factor is 430 lb/ton; of control factor, 36 lb/ton.
[e]This factor may be lower if a coated baghouse is used.

Table 2. Particle Size Distribution and Size-Specific Emission Factors for Uncontrolled Reverberatory Furnaces in Secondary Aluminum Operations[a] (English Units)

Aerodynamic Particle Diameter (μm)	Particle Size Distribution[b]		Size-specific Emission Factor[c], (lb/ton)			
	Chlorine Demagging	Refining	Chlorine Demagging	Rating	Refining	Rating
2.5	19.8	50.0	398	E	2.12	E
6.0	36.9	53.4	738	E	2.3	E
10.0	53.2	60.0	1,064	E	2.6	E

[a]Refs. 3 and 4.
[b]Cumulative weight percent is less than the aerodynamic particle diameter, μm.
[c]Size-specific emission factor equals total particulate emission factor multiplied by particle size distribution, (percent)/100. From Table 2.3-1, total particulate emission factor for chloride demagging is 1000 lb/ton chlorine used, and for refining, 4.3 lb/ton aluminum processed.

Table 2, from AP-42, presents particle size distributions and corresponding emission factors for uncontrolled chlorine demagging and metal refining in secondary aluminum reverberatory furnaces. Test runs for the MACT Program indicate that the controlled emissions from lime coated baghouses are considerably less than the table indicates.

AIR POLLUTION CONTROL MEASURES

The direct-charge reverberatory furnace is generally charged only with clean scrap, sows, T-bar, or pig, resulting in the exhaust flue being clear and generally well within the allowable process weight rate and opacity limitations. MACT rules will *not* require controls *if the charge is clean*. On the other hand, the scrap charged into an open charging well, since lower temperatures exist and the scrap is generally much "dirtier," will generate volumes of smoke and fume. Proper design and application of air pollution control hoods with appropriate closures to furnaces, the use of appropriate doors and the use of charging devices causing small continuous scrap additions minimizes "puffs" of smoke that are difficult to control. Control of large charging wells can be sustained with relatively small amounts of air by using the above techniques. Reverberatory furnace operations on similar furnaces (150,000-pound) capacity are being controlled with air flows that range between 22,000 and 45,000 acfm. Fumes are successfully controlled through adequate design of the hoods and proper control of the charge. Precoating of the dust collector bags utilizing hydrated lime, bentonite, or proprietary compounds has been known technology for many decades. The use of continuous additions of appropriate additives is a technology that has worked effectively for the past 25 years. The additives react with active chemicals in the gas stream forming solids that are collected in the control system. Appropriate reactions between hydrochloric acid gas with an alkali results in the conversion of the gas to an innocuous particulate that can be gathered in the baghouse. Typically, the particulate collected in secondary aluminum casting alloy facilities that employ continuous additive additions to the baghouses produce opacities ranging from 0 to 10%. The dusts collected are generally nonhazardous, passing Toxicity Characteristics Leaching Procedure (TCLP) tests.

BIBLIOGRAPHY

R. S. McClain, G. V. Sullivan, *Beneficiation of Aluminum Plant Residues*, U.S. Department of the Interior, Bureau of Mines, RI #6219, 1963.

G. Bracale, U. Lecis, and M. Leoni, *Degassing & Chemical Refining of Aluminum Melts for D.C. Casting Rolling Slabs*, TMS paper #A70-33, 1970.

M. V. Brant, D. C. Bone, and E. F. Emley, *Fumeless In-Line Degassing & Cleaning of Liquid Aluminum*, TMS paper #A70-51, 1970.

Donald L. Siebert, *Impact of Technology on the Commercial Secondary Aluminum Industry*, U.S. Department of the Interior, Bureau of Mines, IC #8445, 1970.

Fred J. Francis, *Secondary Aluminum Smelter Air Pollution Control Using a Chromatographic Coated Baghouse—A Technically New and Economic Solution*, Air Pollution Control Association, 1972.

Aluminum, Materials Shortages, Permanent Subcommittee on Investigations of the Committee on Government Operations, U.S. Senate, 1974.

Steven M. Gerstein and Mark E. Franza, *Control Technology for Secondary Aluminum Smelters*, Air Pollution Control Association, 1975.

Aluminum Industry Energy Conservation Workshop Papers, The Aluminum Association, Inc., 1976.

J. E. Dore, J. C. Yarwood, and J. A. Ford, *A New Low Emission Process for Degassing & Treating Aluminum Alloy Melts*, TMS paper #A76-48.

H. Spoel, *The Recovery of Metal & Chemical Values from Aluminum Drosses*, TMS paper #A76-54

Environmental Considerations of Selected Energy Conserving Manufacturing Process Options: Vol. VII. Alumina/Aluminum Industry Report, U.S. EPA Office of Research & Development, EPA-600/7-76-034h, 1976.

Robert F. Jones, Thomas H. Ginsburg, and Charles A. Licht, *Secondary Aluminum Plant Particulate Control—A "Nuts & Bolts" Solution*, TMS paper #A76-24.6.

J. C. Cole and W. F. Fleming, *Application of a Baghouse to an Aluminum Remelt Facility*, TMS paper #A76-49.

C. L. Brooks, *Loss & Metal Conservation*, TMS paper #LM78-42, 1978.

J. W. Stamper and H. F. Kurtz, *Aluminum*, Mineral Commodity Profiles #14, Bureau of Mines, U.S. Department of the Interior, 1978.

Anthony Pultz and Charles A. Licht, *"New Source" Pollution Controls at a Secondary Aluminum Plant*, Air Pollution Control Association, 1978.

Warren S. Peterson, Editor, *Guidelines for Handling Molten Aluminum*, The Aluminum Association, Inc., SGE/1500/1280/1-2, 1980.

S. Freti, J. D. Bornand, and K. Buxmann, *Metallurgy of Dross Formation on Aluminum Melts*, Light Metal Age, June 1982.

Charles A. Licht, *Fearless Forecast For Aluminum*, Die Casting Management, July/August 1983.

Compilation of Air Pollutant Emission Factors, Volume 1: Stationary Point and Area Sources, AP-42, U.S. EPA, 1985.

Aluminum Industry Energy Conservation Workshop IX, The Aluminum Association, Inc., April 1986.

C. A. Sorrell, J. G. Groetsch, Jr., and D. M. Soboroff, *Aluminum Fluxing Salts: A Critical Review of the Chemistry and Structures of Alkali Aluminum Halides*, U.S. Department of the Interior, Bureau of Mines Information Circular #9069, 1986.

Edward T. Brookman, *Screening Study on Feasibility of Standards of Performance for Secondary Aluminum Manufacturing*, The Research Corporation of New England, 1987.

Guidelines for Aluminum Scrap Receiving & Inspection Based on Safety & Health Considerations, The Aluminum Association, 1990.

G. J. Kulik and J. C. Daley, *Aluminum Dross Processing in the 90's*, TMS, 1990.

Warren S. Peterson, Editor, *Guidelines for Handling Molten Aluminum*, The Aluminum Association, 1990.

David R. Wilburn and Lorie A. Wagner, *Aluminum Availability & Supply, A Minerals Availability Appraisal*, U.S. Department of the Interior, Bureau of Mines Circular #9371, 1993.

Ray D. Peterson, *Issues in the Melting & Reclamation of Aluminum Scrap*, Journal of Metals, February 1995.

Aluminum Source Book, Alumitech '97, The Aluminum Association, Inc., 1996.

Collection of Papers, VI Aluminum Industry Technology Seminar, IV International Aluminum Recycling Seminar September 1997.

M. Ali Khan, *Control of Emission from a Secondary Aluminum Furnace*, Air Pollution Control Association, June 1997.

Molten Aluminum Processing, International American Foundrymen's Society, Inc., Nov. 1998.

W. C. Shirley, *Secondary Aluminum Industry Emission Control*, report for American Smelting Research Institute.

C. L. Brooks, "Melt Loss and Metal Conservation," LM 78-42, Metallurgical Society of AIME, New York, 1978.

Air Pollution Engineering Manual, AP-40, J. A. Danielson, Ed., U.S. Environmental Protection Agency, Research Triangle Park, NC, 1973, pp. 283–92.

R. F. Jones, T. H. Ginsburg, and C. A. Licht, "Secondary Aluminum Plant Particulate Control," Paper 76-24.6, 69th Annual Meeting of the Air Pollution Control Association, Portland, 1976.

C. A. Licht, "Demagging," *Aluminum Alloys Using an Innovative Technology*, A 77-76, Metallurgical Society of AIME, New York, 1977.

C. A. Licht, "Energy Conservation in a Large Aluminum Melter," *Proceedings of the Third International Aluminum Extrusion Technology Seminary, Vol. 2*, Aluminum Association, 1984.

A. Pultz and C. A. Licht, "New Source Pollution Controls at a Secondary Aluminum Plant," 78-61.6, presented at 71st Annual Meeting of Air Pollution Control Association, Houston, TX, 1978.

E. L. Rooy and J. H. L. VanLinden, "Recycling of Aluminum," *Metals Handbook*, Vol. 2, 10th ed., ASM International, 1990, pp. 1205–13.

D. L. Siebert, *Impact of Technology on the Commercial Secondary Aluminum Industry*, IC 8445, U.S. Bureau of Mines, Washington, D.C., 1970.

SECONDARY BRASS AND BRONZE MELTING PROCESSES

Charles A. Licht

Brass is zinc-alloyed copper. Bronze is tin-alloyed copper. Other copper alloys are identified by the alloying metals, such as aluminum bronze and silicon bronze. Some alloys called bronze are actually brass. Manganese bronze is a high-zinc brass. Because of their high strength, workability, corrosion resistance, color, and other desirable physical characteristics, the copper-base alloys have found wide use in hardware, radiator cores, condensers, jewelry, musical instruments, plumbing fittings, electric equipment, ship propellers, and many other applications.

The remelting of nearly pure copper or bronze causes minor air pollution as a result of the high boiling points of the materials involved. With good melting practice, total emissions to the air should not exceed 0.5% of the process weight. Brasses containing 15–40% zinc are poured at temperatures above the boiling points of zinc; so vaporization and combustion of desirable elements, particularly zinc, occur. Air emissions may vary from less than 0.5% to 6% or more of the total metal charge. A substantial percentage of the zinc content may be lost through fuming, depending on the composition of the alloy, the type of furnace used, and the melting practice.

PROCESS DESCRIPTION

Furnace Types

Brass and bronze shapes for working, such as slabs and billets, are usually produced in large gas- or oil-fired reverberatory furnaces. Most secondary smelters use stationary or rotary reverberatory furnaces for reclaiming and refining scrap metal, casting the alloyed metal into pigs. Brasses and bronzes used to make commercial castings are often melted in induction furnaces in the larger foundries and in crucible, fuel-fired, furnaces in smaller job foundries.

AIR EMISSIONS CHARACTERIZATION

Air contaminants emitted from brass furnaces consist of products of combustion from the fuel and particulate matter in the form of dusts, metallics, and oxide fumes. The particulate matter composing the dust and fume load varies according to the fuel, alloy composition, melting temperature, type of furnace, and other operating factors. Particulate is made up of fly ash, carbon, and mechanically produced dust. Furnace emissions also contain fumes resulting from condensation and oxidation of the more volatile elements, including zinc and lead.

Air pollution resulting from volatilization of metals during melting of nearly pure copper and bronze is not too serious because of the high-boiling-point temperatures of copper, tin, nickel, aluminum, and even lead. Alloys containing zinc ranging up to 7% can be successfully processed with a minimum of fume emission when an inert slag cover is used. This nominal figure is subject to some variation, depending on the composition of the alloy, temperatures, operation procedures, and other factors.

Copper-base alloys containing 20–40% zinc have low boiling points of approximately 2100°F (1150°C) and melting temperatures of approximately 1700°F (925°C) to 1900°F (1040°C). These zinc-rich alloys are poured at approximately 1900°F (1040°C) to 2000°F (1100°C), which is only slightly below their boiling points. Pure zinc melts at 787°F (420°C) and boils at 1663°F (906°C). Even within the pouring range, therefore, fractions of high-zinc alloys usually boil and flash to zinc oxide. The zinc oxide formed is submicrometer in size. Its escape to the atmosphere can be prevented only by collecting the fumes and venting through efficient air pollution control equipment.

The difficulty of controlling metallic fumes from brass furnaces lies in the physical characteristics of these fumes. The particle sizes of zinc oxide fumes vary from 0.03 to 0.3 μm. Lead oxide fumes are in this same range of particle sizes. The collection of these very small particles requires high-efficiency control devices. These metallic fumes also produce very opaque emissions, since particles of 0.2–0.6 μm in diameter produce maximum light scattering.

In brass foundries, as much as 98% of the particulate matter contained in furnace stack gases may be zinc oxide and lead oxide, depending on the composition of the alloy. Constituents of fumes include zinc, lead, tin, copper, cadmium, silicon, and carbon. They are present in varying amounts, depending on the composition of the alloy and foundry practice.

Investigations prove conclusively that the most troublesome fumes consist of particles of zinc and lead compounds submicrometer in size, and that air pollution control equipment capable of collecting particulate matter from 1.0 down to about 0.03 μm is required.

Factors Causing Large Concentrations of Zinc Fumes

According to the Air Pollution Engineering Manual, 1973 Edition, four principal factors cause relatively large concentrations of zinc fumes in brass furnace gases:

1. *Alloy composition.* The rate of loss of zinc is approximately proportional to the zinc percentage in the alloy.
2. *Pouring temperature.* For a given percentage of zinc, an increase of 100°F (50°C) increases the rate of loss of zinc about three times.
3. *Type of furnace.* Direct-fired furnaces produce larger fume concentrations than the crucible type, other conditions being constant. Operation of direct-fired furnaces will require emission control equipment.
4. *Poor foundry practice.* Excessive emissions result from improper combustion, overheating the charge, addition of zinc at maximum furnace temperature, flame impingement on the metal charged, heating the metal charged, heating the metal too fast, and insufficient flux cover. Excessive superheating of the molten metal is to be avoided for metallurgical, economic, and pollution control reasons. From an air pollution emission viewpoint, the early addition of zinc is preferable to gross additions at maximum furnace temperatures.

In any fuel-fired furnace, the internal atmosphere is of prime importance since there exists a constant flow of combustion gases through the melting chamber in contact with the metal. A reducing atmosphere is undesirable from both the metallurgical and air emission viewpoints. With too little oxygen, metal is exposed to a reducing atmosphere of unburned fuel and water vapor, which usually results in gassy metal. Incomplete combustion, especially with oil firing, produces carbon soot. To prevent these difficulties, the atmosphere should be slightly oxidizing. Excess oxygen content should be greater than 0.1%, otherwise castings will be affected by gas porosity. Excess oxygen content must be less than 0.5% to prevent excessive metal oxidation. The need for close control of the internal furnace atmosphere requires careful regulation of the fuel and air input and frequent checking of the combustion gases.

CRUCIBLE FURNACE—PIT AND TILT TYPE

Indirect-fired crucible furnaces are used extensively in foundries requiring small- and medium-sized melts. The liftout type of crucible is frequently employed in small furnaces. According to LA APC data, tests have demonstrated that, with careful practice and use of slag covers, the crucible furnace is capable of low-fume operation on red brasses but not on yellow brass.

The slag cover, consisting mainly of silicates, is not used as a refining flux, but as an inert, cohesive slag of sufficient thickness to keep the molten metal covered. If the quantity of slag is carefully controlled, minimum emissions result. Before any metal is added to the crucible, the flux should be added so that, as melting takes place, a cover is formed to keep the molten metal separated from the atmosphere.

ELECTRIC FURNACE—LOW-FREQUENCY INDUCTION

Induction furnaces have a number of desirable characteristics for melting brasses. The heating is rapid and uniform. Metal temperature can be accurately controlled. Contamination from combustion gases is eliminated. High-frequency induction furnaces are well adapted to copper- and nickel-rich alloys, but are not widely used for zinc-rich alloys. During melting of clean metal, use of a suitable flux cover over the metal prevents excessive fuming, except during back-charging and pouring phases of the heat. The usual flux covers borax, soda ash, and others are destructive to furnace walls. Charcoal is used with satisfactory results.

CUPOLA FURNACE

The cupola furnace is used for reduction of copper-base alloy slag and residues. The residues charged have a recoverable metallic content of 25% to 30%. The balance of the unrecoverable material consists of nonvolatile gangue, mainly silicates. In addition to the residues, coke and flux are charged to the furnace. Periodically, the recovered metal is tapped from the furnace. The slag produced in the cupola is discharged through a slag tap located slightly above the metal tap.

In addition to the usual metallic fumes, the cupola also discharges carbon particles and fly ash. Collection of these emissions is required at the cupola stack, the charge door, and the metal tap spout. With no control equipment, emissions of 60–100% opacity can be expected from the charge door and stack. The opacity of the fumes emitted at the tap varies from 60% to 80%.

The stag discharged from the cupola is rich in zinc oxide. It does not volatilize since its boiling point is 3590°F (1975°C). The discharged slag is usually cooled by water. The emissions from the slag-tapping operation have low opacity.

Hooding and Ventilation Requirements

Regardless of the efficiency of the control device, air pollution control is not complete unless all the fumes generated by the furnace are captured. Since different problems are encountered with the various types of furnaces, each will be discussed separately.

REVERBERATORY FURNACE—STATIONARY TYPE

In an open-hearth reverberatory furnace, the products of combustion and metallic fumes are normally vented directly from the furnace through a cooling device to a baghouse. Auxiliary hoods are required over the charge door, rabble (or slag) door, and taphole. These may be vented to the baghouse serving the furnace and help cool the hot combustion gases by dilution.

The furnace burners should be turned down or off during periods when the furnace is open for charging, rabbling, lancing, removing slag, or adding or pouring metal. Otherwise, the exhaust system may not have sufficient capacity to handle the products of combustion and the additional air required to capture the fumes. Since no two of these operations occur simultaneously, the required air volume for collection may be reduced by the use of properly placed dampers within the exhaust system.

Well-designed hoods, properly located, with an indraft velocity of 200 fpm, adequately capture furnace emissions. If the hood is placed too high for complete capture or is improperly shaped and poorly fitted, higher indraft velocities are required.

The rabble or slag door permits (1) mixing the charge, (2) removing slag from the metal surface, and (3) lancing to refine the metal. Emissions from the furnace may be of 50–90% opacity during these operations, even with the burners partially throttled. Up to 200 fpm indraft velocity is recommended for properly designed hoods.

After the slag has been removed, metal, usually zinc, must be added to bring the brass within specifications. The furnace metal is at a temperature well above the boiling point of zinc and is no longer covered by the tenacious slag cover. Hence, voluminous emissions of zinc oxide result. The addition of slab zinc can produce 100% opaque fumes in great quantity, while a brass addition may generate fumes of 50% opacity. A well-designed hood is required over the charge door or rabble door through which the metal is charged.

Perhaps the most critical operation from the standpoint of air pollution occurs when the furnace is tapped. Nearly continuous emissions of 90–100% opacity may be expected. Much planning is required to design a hood that completely captures the emissions and yet permits sufficient working room and visibility of the molten metal. Again, the burners should be turned off or throttled as much as possible to reduce the quantity of fumes emitted.

The fluxes used in reverberatory furnaces normally present no air pollution problems. Only nonvolatile fluxes such as borax, soda ash, and iron oxide mill scale are used.

ROTARY TYPE

Cylindrical-type reverberatory furnaces present all the collection problems of the open-hearth type with the additional complication of furnace rotation. The cylindrical furnace may be rotated up to 90° for charging, slag removing, and metal tapping. With hoods installed in a fixed position, the source of emissions may be several feet from the hood, and thus no fumes would be collected. Either a hood attached to the furnace and venting to the control device through flexible ductwork or an oversized close-fitting hood covering all possible locations of the emission source is required. A close-fitting hood and high indraft velocities are often necessary.

A cylindrical furnace rotates on its longitudinal axis. A tight breeching is mandatory at the gas discharge end of the furnace. Adequate indraft velocity must be maintained through the breeching connection to prevent the escape of fumes.

The exhaust system for the cylindrical furnace, as well as for all types of reverberatory furnaces, must be designed to handle the products of combustion at the maximum fuel rate. Any lesser capacity results in a positive pressure within the furnace during periods of maximum firing, with resultant emissions from all furnace openings.

TILTING TYPE

The tilting-type furnace differs from the reverberatory furnaces previously discussed in that the exhaust stack is an integral part of the furnace and rotates with the furnace during charging, skimming, and pouring. One type of tilting furnace is charged through the stack, and skimming and pouring are accomplished through a small taphole in the side of the furnace. Another type has a closable charge door and a small port through which the furnace gases escape. These two furnace openings may describe a full 180° of arc during the various phases of a heat.

ROTARY TILTING TYPE

The rotary-tilting-type furnace not only tilts for charging and pouring, but rotates during the melting period to improve heat transfer. Two types are common. One is charged through the burner end and is poured from the exhaust port of the furnace, opposite the burner. The other has a side charge door at the center of the furnace through which charging, slagging, and pouring operations are conducted.

Because of the various movements of this type of furnace, direct connection to the control device is not feasible. The furnace is under positive pressure throughout the heat. Fumes are emitted through all furnace openings.

Hooding a rotary-tilting-type reverberatory furnace for complete capture of fumes is difficult. Complete collection is seldom achieved. These furnaces are undoubtedly the most difficult type of brass furnace to control. To hood them effectively requires a comprehensive design. The major source of emissions occurs at the furnace discharge. Capture of fumes is accomplished by a closely fitted hood.

Hooding is sometimes installed at the burner end of a furnace to capture emissions that may escape from openings during melting or during the time the furnace is tilted to pour. Because both ends of the furnace are open, a venting action is created during the pour, causing fume emissions to be discharged from the elevated end of the furnace. Close hooding is not practical because the operator must observe the conditions within the furnace through the open ends. An overhead canopy hood is usually installed.

Additional heavy emissions may be expected during charging, alloying, and slagging. High overhead canopy hoods are used. These overhead hoods are, however, unsatisfactory unless they cover a large area, and a high indraft velocity is provided along with very large control air volumes.

The need for numerous hoods and large air volumes, with the resultant larger control device, makes the tilting open-flame furnace expensive to control. This type of furnace is being gradually replaced by more easily controlled equipment.

CRUCIBLE FURNACES

Tilting crucible furnaces can be controlled by an exhaust system to control emissions during pouring. An exhaust system may vent several furnaces to a baghouse. The hooding collects all the fumes during pouring without interfering with the furnace operation in any way. The hood can be equipped with a damper that is closed when the furnace is in the normal firing position. A linkage system can open the damper when pouring begins. After the furnace is partially tilted, the damper is fully opened, remaining open for the rest of the pour. It swings shut automatically when the furnace is returned to the firing position. The ductwork leading from the hood pivots when the furnace is tilted. The entire hood is fixed to the furnace with a few bolts, which permits rapid removal for repairs to the furnace lining and crucible replacement.

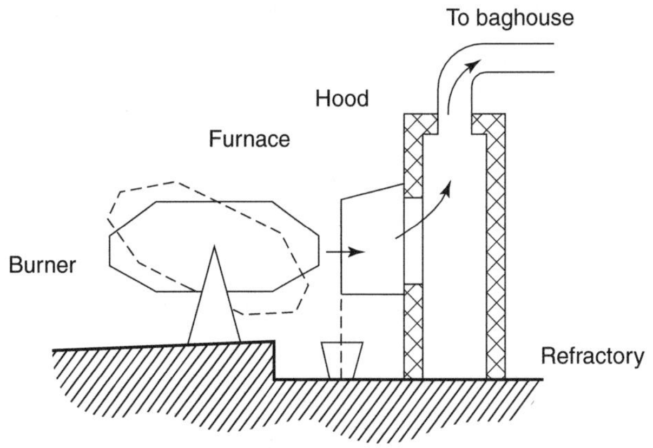

Figure 1. Rotary-Tilting-Type Brass Melting Furnace.

Emissions resulting from pouring molten metal from a ladle into molds can be controlled by one of several devices. One alternative is a fixed pouring station that is hooded so that the emissions from the ladle and molds are captured during the pouring (Figure 1). A second solution, for smaller foundries, is a hood attached to the pouring ladle and vented to the control system through flexible ductwork. One variation places a small baghouse on the crane holding the ladle so that the ladle is vented over the whole pouring floor. Another variation is a stationary baghouse and sufficient flexible ductwork to allow the hood to travel from mold to mold.

A third solution employs a hood attached to the crucible bale, a short section of flexible ductwork, and two ducts — one mounted on the bridge crane and one attached to the stationary crane rails. The ducts are interconnected with a transfer box, which is sealed by a continuous loop of rubberized belt. This allows complete freedom of movement for the ladle within the area served by the overhead crane.

LOW-FREQUENCY INDUCTION FURNACE

The control of the emissions from an induction furnace is much more expensive and difficult if only turnings are charged to the furnace. In addition to the fumes common to brass melting, great clouds of 100% opacity black smoke are generated when the oily shavings contact the molten heel within the furnace. Adequate hooding enclosing the furnace is required, and a large volume of air is necessary to capture the smoke and fumes.

CUPOLA FURNACE

An exhaust system to control a cupola must have sufficient capacity to remove the products of combustion, collect the emissions from the metal tap spout, and provide a minimum indraft velocity of 250 fpm through the charge door. In addition, side curtains may be required around the charge door to shield adverse crosscurrents. A canopy hood is recommended for the metal tap spout. The air requirement for this hood is a function of its size and proximity to the source of emissions.

AIR POLLUTION CONTROL MEASURES

Baghouses

Baghouses with tubular bag filters are used to control the emissions from brass furnaces. This type of collector is available in many useful and effective forms. Wool, cotton, and synthetic filter media effectively separate submicrometer -sized particulate matter from gases because of the filtering action of the "mat" of particles previously collected.

The gases leaving a reverberatory furnace may be 100°F to 200°F hotter than the molten metal and must be cooled before reaching the filter cloth. Direct cooling, by spraying water into the hot combustion gases, is practiced, with maximum water content being controlled to less than 10% by weight. Some plants prefer not to use water because

1. There is increased corrosion of the ductwork and collection equipment;
2. The water vapor increases the exhaust gas volume, necessitating a correspondingly larger baghouse; and
3. The temperature of the gases in the baghouse must be kept above the dew point to prevent condensation of water on the bags. The exhaust gases may be cooled by dilution with cold air, but this increases the size of the control equipment and the operating costs of the exhaust system.

One type of cooling system employed consists of a water-jacketed cooler followed by air-cooled radiation–convection columns, as shown in Figure 2. The radiation–convection coolers reduce the temperature to protect the fabric of the filter medium. Figure 2 depicts an installation showing the cooling columns and baghouse.

Treated polyester or Nomex® has replaced glass cloth as the most favored high-temperature fabric. Although glass bags withstand higher temperatures, the periodic shaking of the bags gradually breaks the glass fibers and results in higher maintenance costs.

Probably the most critical design factor for a tubular baghouse is the filtering velocity. A filtering velocity of 2.5 fpm or less is recommended for collecting the fumes from brass furnaces with relatively small concentrations of fume. Larger concentrations of fume require a lower filtering velocity. A higher filtering velocity requires more frequent shaking to maintain a pressure drop through the baghouse within reasonable limits. Excessive bag wear results from frequent shaking and higher filtering velocities. A pressure drop of 2–5 in. of water column is normal. High-pressure differentials across the bags are to be avoided.

The baghouse should be completely enclosed to protect the bags from inclement weather and water condensation. The exhaust fan should be placed downstream from the baghouse to prevent blade abrasion and imbalance. Broken bags are easily detected when the exhaust system discharges to the atmosphere through one opening.

Electrostatic Precipitators

Electrostatic precipitators are extremely effective collectors for many substances in any size range from 200 mesh (74 μm) to 0.001 μm, wet or dry, ambient or up to 1200°F (650°C). This equipment has not proved satisfactory on lead and zinc fumes. High-voltage precipitators have not been available in the small

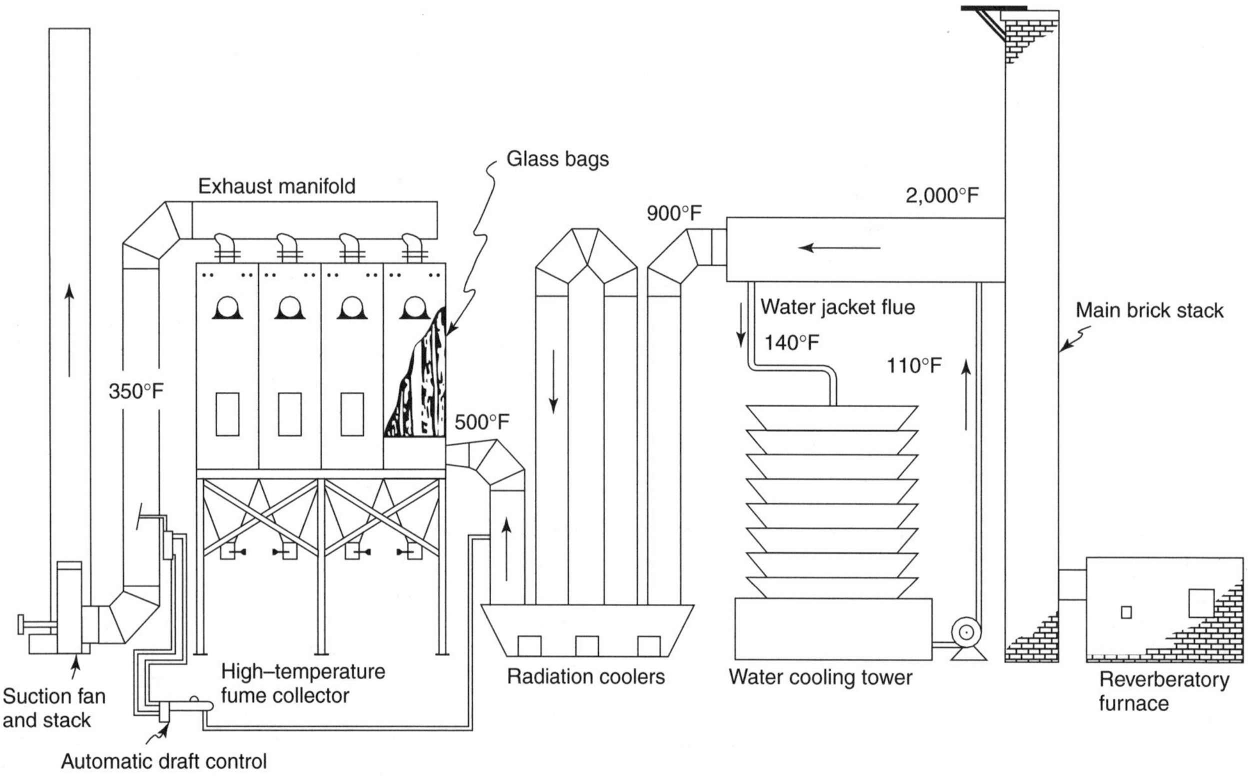

Figure 2. Sketch of Small Baghouse for Zinc Fume (Allen et al., 1952).

units suitable for small nonferrous foundry use. First cost is usually prohibitive.

Scrubbers

Scrubbers or mechanical washers have proved in some applications to be effective from 10 to 1 μm. They are not effective in the submicrometer range unless extremely high pressure drops are used. They have the disadvantage of high power consumption and mechanical wear. They require separation of the metallic fumes and other particulate matter from the circulating water. A scrubber system will require an elaborate water pollution control system, basically transferring an air problem into a much worse water pollution control problem.

REFERENCES

1. *Air Pollution Engineering Manual*, AP-40, J. A. Danielson, Ed.; U.S. Environmental Protection Agency, Research Triangle Park, NC, 1973, pp. 269–83.

BIBLIOGRAPHY

D. V. Neff and R. F. Schmidt, "Recycling of Copper," *Metals Handbook*, Vol. 2, 10th ed., ASM International, 1990, pp. 1213–16.

IRON FOUNDRIES

American Foundrymen's Society Air Quality Committee (10-E)*

Control of the air pollution that results from iron foundries may be considered according to the methods of melting, the handling of sand, the types of molten metals and other materials, and the cleaning of finished castings. The air pollutant characteristics are affected by a number of factors, including the type of melting unit, material-handling and hooding systems, and emissions control systems. The air pollution problem then becomes one of capturing the smoke, dust, and fumes at the furnace and other sources and transporting these contaminants to suitable control devices.

PROCESS DESCRIPTION

Mold and Core Production

Molds are forms used to shape the exterior of castings. The green sand mold, the most common type, uses moist sand mixed with 3–20% clay and 2–5% water, depending on the process. Additives to prevent casting defects include organic material, such as seacoal (a pulverized high-volatility, low-sulfur bituminous coal), wood or corn flour, oat hulls, or similar organic matter.

Cores are molded sand shapes used to form the internal voids in castings. They are made by mixing sand with various binders, shaping it into a core, and curing the core with a variety of processes.

The Melting Process

Electric Furnace (General). In the electric furnace, the basic process operations are (1) furnace charging, in which metal, scrap, alloys, carbon, and flux are added to the furnace; (2) melting, during which the furnace remains closed; (3) back-charging, which involves the addition of more metal and alloys; (4) refining and treating, during which the chemical composition is adjusted to meet product specifications; (5) slag removal; and (6) tapping molten metal in a ladle or directly into molds.

Induction Furnaces. Electric induction furnaces are either horizontal or vertical, cylindrical, refractory-lined vessels. Heating and melting occur when the charge is energized with a low-, medium-, or high-frequency alternating current. Induction furnaces also may be used for holding and superheating.

Electric induction furnaces generally have lower emissions per ton of metal melted than the other furnace types. As a result, in spite of a generally lower unit capacity, induction furnaces have supplanted cupolas in many foundries.

Electric Arc Furnaces. Electric arc melting furnaces are large, welded-steel cylindrical vessels equipped with a removable roof through which three carbon electrodes are inserted. The electrodes are energized by three-phase alternating current, creating arcs that melt the metallic charge material. Additional heat is generated by the electrical resistance of the metal to the current between the arc paths. The most common method of charging an arc furnace is by removing the roof and introducing the charge material directly. Alternatives include charging through a roof chute or side charging door. Once the melting cycle is complete, the metal is tapped by tilting the furnace and pouring the metal into a ladle.

Cupola. The cupola is a vertical, cylindrical shaft furnace and may use pig iron, scrap iron, scrap steel, and coke as the charge components. The mechanism by which melting is accomplished in the cupola is heat release through the combustion of coke—the reaction between oxygen in the air and carbon in the fuel—that is in direct contact with the metallic portion of the charge and the fluxes.

One of the advantageous features of such a furnace is that counterflow preheating of the charge material is an inherent part of the melting process. The upward-flowing hot gases come into close contact with the descending burden, allowing direct and efficient heat exchange to take place. The running or charge coke is also preheated, which aids in the combustion process as it reaches the combustion zone to replenish fuel consumed.

Greater understanding of these features accounts, in part, for the continued popularity of the cupola as a melting unit. However, recent design improvements, such as cokeless, plasma-fired types that alter emission characteristics are now encountered.

* Charles E. Brown, Chairman; Charles H. Borcherding, CIH, CSP, Vice Chairman; William A. Baker, Secretary; J. Peter Aldred, Charles M. Davis, LeRoy E. Euvrard, Jr., JD, Tom Godbey, Jeff Haworth, William B. Huelsen, Walter M. Kiplinger, Fredrick H. Kohloff, Barry Komegay. Gerry Lanham, P.E., Gary E. Mosher, CIH, George W. Nicholas, Gene Odenreider, Jack C. Shih, Craig M. Wehr.

Casting, Cooling, and Finishing

After melting, molten metal is tapped from the furnace and poured into a ladle or directly into molds. If poured into a ladle, the molten iron may be treated with a variety of alloying agents predetermined by the desired metallurgical properties. It then is ladled into molds, where it solidifies and is allowed to cool further before separation of the casting from the mold (shakeout).

In larger, more mechanized foundries, the molds are conveyed automatically through a cooling tunnel before they are placed on a vibrating grid to shake the mold and core sand loose from the casting. In some foundries, molds are placed in an open floor space and molten iron is poured into the molds and allowed to cool. Molding and core sand are separated from the casting(s) either manually or mechanically.

Used sand from casting shakeout is usually returned to the sand preparation area and cleaned, screened, and processed to make new molds. Because of process losses and potential contamination, additional makeup sand may be required.

When castings have cooled, any unwanted appendages, such as sprues, gates, and risers, are removed by an oxygen torch, abrasive saw, friction cutting tool, or hand hammer. The castings then may be subjected to abrasive blast cleaning and/or tumbling to remove any remaining mold sand or scale.

AIR EMISSION CHARACTERIZATION

Mold and Core Production

The major pollutant emitted in mold and core production operations is particulate matter from sand preparation, mold core forming, and curing. In addition, volatile organic compounds (VOCs), carbon monoxide, and particulate matter may be emitted during core and mold curing or drying (see Figure 1).

Melting

The melting process begins with the handling of charge materials going into the melting furnace. Emissions from the raw materials handling are fugitive particulate matter generated from the receiving, unloading, storage, and conveying operations. Scrap preparation and preheating may emit one or more of the following: fumes, organic compounds, carbon monoxide, or coarse particulate matter. Scrap preparation with solvent degreasers may emit VOCs.

Induction and Arc Melting

The highest concentrations of furnace emissions occur during charging, back-charging, alloying, slag removal, and tapping operations. These emissions primarily include particulate (metal oxides) and possibly organics, depending on the scrap quality and pretreatment.

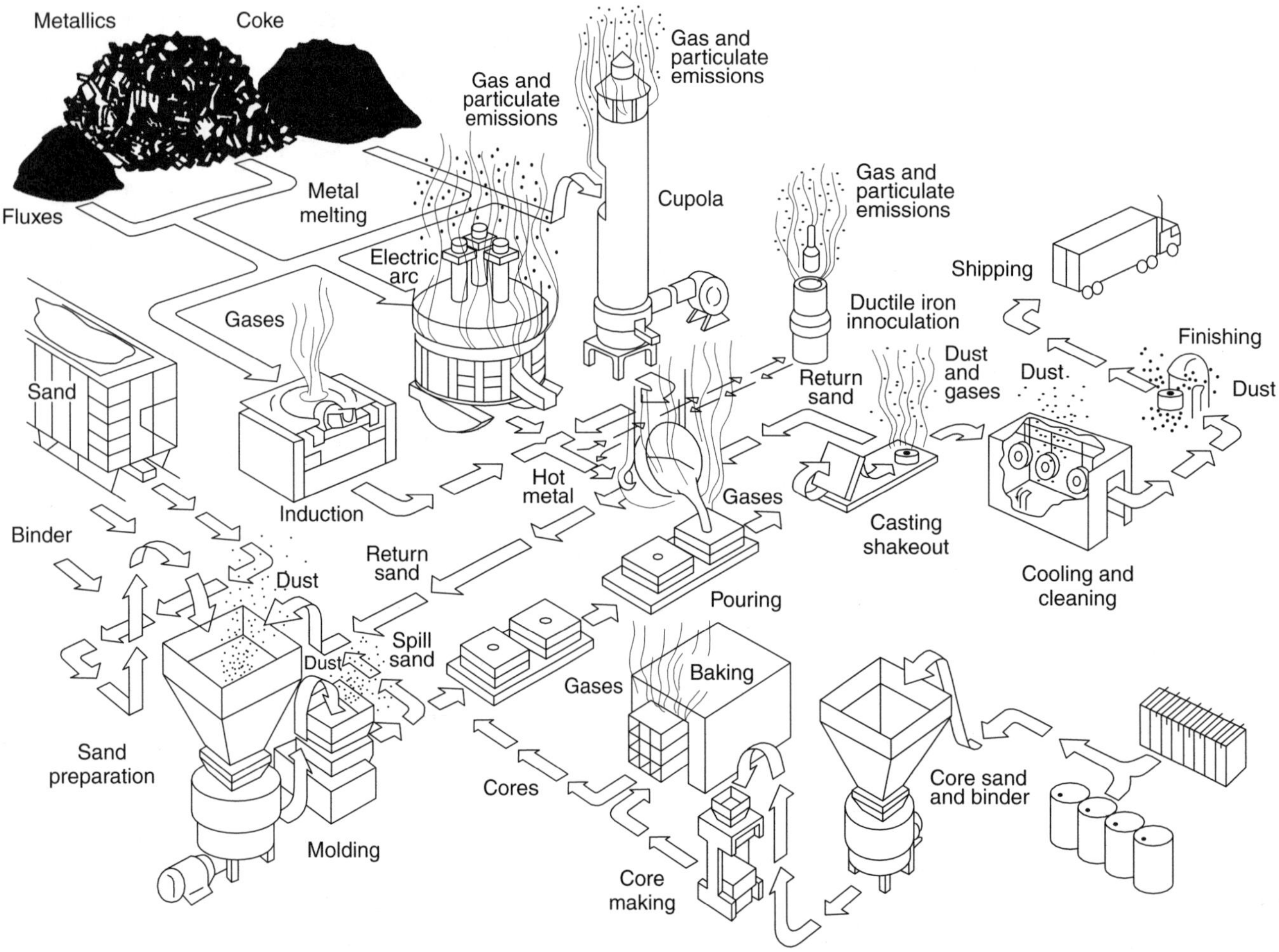

Figure 1. Emission Points in a Typical Iron Foundry.

Typical dust loading from electric arc furnaces can range from 10 to 15 pounds per ton melted. However, electric induction furnaces may emit particulate matter at one-tenth of that value.

Cupola Melting

The quantity and composition of particulate emissions vary among cupolas, and even at intervals in the same cupola. Causes include changes in iron-to-coke ratios, air volumes per ton melted, stack velocity, and the quality of the scrap melted.

Where oily scrap is charged, the raw emissions potentially not only will be greater in quantity, but can be much more visible. The American Foundrymen's Society compiled a survey of cupola emissions and found that an average emission from an uncontrolled cupola was approximately 13 to 17 pounds of particulate per ton melted. Eighty-five percent of the emissions may be greater than 10 g in size.

Each cupola will have varying airflows at different phases in the melt process. This will affect the grains per standard cubic foot in emitted stack gases if all other factors are equal.

Dust composition and amounts vary from cupola to cupola. The source of the raw charge materials also will have a significant impact on dust composition and quantity. The dust could include some or all of the following materials:

- Iron oxide
- Magnesium oxide
- Manganese oxide
- Zinc oxide
- Silicon dioxide
- Calcium oxide
- Lead
- Cadmium

In addition, gases and organic compounds may be emitted as part of the melting process. These include carbon monoxide, sulfur oxides, lead, and organic emissions. Both sulfur and organic emissions may be affected by the amount of oil or grease on the scrap.

The quantity of sulfur oxides also is large enough to be a definite consideration in the corrosion of air pollution control equipment. There are a number of instances of rapid deterioration of dust collectors on cupolas where corrosion protection was not considered.

Where fluorspar is used as an additive, the fluorine driven off can cause a corrosion problem with dust collection equipment. Fluorine also has the potential to dissolve glass bags. The carbonic acid formed when carbon dioxide reacts with water vapor may cause corrosion problems as well.

Pouring, Casting, Cooling, and Finishing

Particulate emissions can be generated during the treatment and inoculation of molten iron before pouring. For example, the addition of magnesium to molten metal to produce ductile iron causes a very violent reaction. This reaction is accompanied by various emissions of magnesium oxides and metallic fumes, depending on the method of treatment. Some methods, such as the tundish method, result in significantly lower emissions than others. Emissions from pouring consist of metal fumes, carbon monoxide, organic compounds, and particulate matter evolved when the mold and core materials are contacted by the molten iron. Emissions continue as the molds cool and during the shakeout operation, although at a much lower rate.

Finishing operations emit particulate matter during the removal of burrs, risers, and gates and during shotblast cleaning. This particulate matter consists primarily of iron, iron oxide, and abrasive media. The painting of castings also can lead to a variety of VOC emissions.

AIR POLLUTION CONTROL MEASURES

There are two primary collection methods for foundry particulate matter—wet and dry. Wet scrubbers include low- and high-energy types. Dry collection includes baghouses, mechanical collectors, and electrostatic precipitators.

In addition, the control of organic compounds may require incineration or afterburners. Air toxics merit special consideration, requiring careful selection of the collection method.

Wet Scrubbers

For particulate collection, the mechanisms used in a wet-type collector are inertial impaction and direct interception. These are used either separately or in combination.

In studying wet collector performance, independent investigators developed the contact power theory, which states that for a well-designed wet-scrubber collection, efficiency is a function of the energy consumed in the air-to-water contact process and is independent of the collector design. On this basis, well-designed collectors operating at or near the same pressure drop can be expected to exhibit comparable performance.

All wet collectors have a fractional efficiency characteristic—that is, their cleaning efficiency varies directly with the size of the particle being collected. In general, collectors operating at a very low pressure loss will remove only medium to coarse particles. High-efficiency collection of fine particles requires increased energy input, which will be reflected in higher collector pressure loss.

Some high-energy scrubbers are designed to use as little as 8 gal/min of scrubbing water per 1000 cfm of saturated gas at the throat. The water is usually recirculated, and a small amount, 0.5–1 gal/min per 1000 cfm, is bled from the system to remove collected dust. This stream is subject to effluent guidelines for metal molding and casting.

Some sand system scrubbers may operate with as little as 8–15 in. w.c. pressure drop. However, when encountering extremely fine particles, such as emissions from melting operations, a high-energy scrubber is required. For example, a typical cupola wet collector operation could require 30–100 in. w.c. pressure drop.

In addition, gas scrubbers may be used to control odors and toxic and sulfur dioxide emissions. In this case, acids, bases, or oxidizing agents may have to be added to the scrubbing liquid. This stream also would be subject to effluent guidelines for metal molding and casting.

Dry Collectors

The most frequently encountered equipment for the removal of solid particulate matter from an air stream or gas stream is the fabric dust collector or baghouse. With a mass median size of 0.5 μm, a collection efficiency of 98–99+% can be expected.

As the filter medium becomes coated in a fabric collector, the collection efficiency rises. However, as material continues to build on the bag surface, higher pressure drops occur, which could result in a significant reduction in airflow. To maintain design flows, the bags must be cleaned periodically by mechanical shaking or with pulsed air.

Filter media are now available for hot corrosive atmospheres, such as furnace emissions. Operating inlet temperatures up to 500°F (260°C) are not uncommon. High humidity can be a problem if no provision is included for the condensation of free moisture. Free moisture and acid dew point are the worst enemies of all fabric collectors. It is important to have the following design information in order to select the proper fabric and the quantity of bags required:

- Gas flow rate
- Temperature and dew point
- Acid dew point
- Particle size and distribution
- Concentration of solids
- Chemical and physical properties of solids

Teflon®-coated woven-glass-fiber bags have been used on a large majority of cupola installations because of their high temperature resistance. If fluorspar is used, Nomex® bags are generally installed. The temperature of the gases entering the baghouse is then reduced to a maximum of 400°F (204°C). Use of these lower-temperature bags creates a potential corrosion hazard because of the acid dew point problem. For reverse-air and mechanical shake collectors, air-to-cloth ratios vary between 1.5 and 2.5 to 1.

Pulse-jet and cartridge collectors also can be used to collect pollutants from sand systems and casting cleaning operations. With either type of unit, care must be taken to select the proper air–cloth ratio (maxima of 15:1 with pulse jet and 1.5:1 with cartridge). In general, these types of collectors will have only marginal results with furnace and inoculation emissions. If considered, they should be employed at a very low air–cloth ratio. In addition, moisture introduced with compressed air may be significant and cause system failure.

Incineration

Afterburners may be used in some processes to control emissions, particularly when oily scrap or hydrocarbons in any form are charged into the furnaces or scrap preheat systems. Afterburning is required for below-the-door takeoff cupola emission systems. If afterburners are not used, carbon monoxide and oil vapors may be emitted through the discharge stack of the air pollution equipment. In order to achieve the required incineration, sufficient retention time (a minimum of 0.6 sec) and ignition temperatures must be maintained.

In general, in the selection of collection devices for all processes, moisture, temperature, and the presence of corrosive materials must be considered. The temptation to operate at higher air–cloth ratios in baghouses must be avoided. Similarly, claims that lower pressure drops in scrubbers create high efficiencies have been proved to be false.

Absorption

Charcoal absorption has been used in conjunction with other control devices for VOC control.

BIBLIOGRAPHY

1. V. H. Baldwin, Jr., *Environmental Assessment of Iron Casting*, EPA-600/2-80-021, Research Triangle Institute, Research Triangle Park, NC, 1980.
2. R. Bethea, *Air Pollution Control Technology*, Van Nostrand Reinhold Environmental Engineering Series, 1978, New York.
3. G. S. Cole, *Odor Pollution Control*, Scientific Research Staff, Ford Motor Co., American Foundrymen's Society Air Quality Committee (10-E), 1976.
4. *Compilation of Air Pollutant Emission Factory, Vol. 1, Stationery Point and Area Sources*, 4th ed., AP-42, Supplement A, U.S. Environmental Protection Agency, Research Triangle Park, NC, October 1986.
5. *Control of External Air Pollution*, American Foundrymen's Society Air Quality Committee (10-E), 1976.
6. *Cupola Handbook*, 5th ed., American Foundrymen's Society, Des Plaines, IL, 1984.
7. G. Engles, *The Nature and Characteristics of Cupola Effluents*, Verein Deutscher Gies sere ifachleute, Dusseldorf, Germany, 1969.
8. *General Principles of Foundry Ventilation*, American Foundrymen's Society Air Quality Committee (10-E), 1972.
9. *Legal Aspects of Air Pollution Control*, American Foundrymen's Society Air Quality Committee (10-E), 1976.
10. *Systems Analysis of Emissions and Emissions Control in the Iron Foundry Industry*, A. T. Kearney and Co., Chicago, IL, 1971.

SECONDARY LEAD SMELTING

Charles A. Licht

INDUSTRY OVERVIEW

The secondary lead smelting industry produces lead and lead alloys. Over 60% of the raw materials for secondary smelters comes from automobile batteries, while a large part of the remainder is derived from lead-covered cable, lead pipe, scrap solder, scrap babbitt, and skimmings and drosses from lead and solder operations.

Reduction of scrap to produce elemental lead requires high-temperature furnacing operations (2300°F), while production of alloys can be achieved at a temperature just above the melting point of the alloy (700°F). Secondary lead may be refined to the purity of primary lead by proper fluxing and oxidizing. The NESHAP–MACT document describing the industry indicated a different grouping for the several areas of secondary lead processing separating the lead alloy makers who remelt the scrap from the lead smelters who refine the lead at high temperature.

The operating systems for the different products are very different. The alloy maker who does little refining can operate a kettle furnace or a reverberatory furnace, depending on lot size and throughput. The process that refines lead will use "blast" furnaces (cupolas using air or oxygen as the oxidizer and coke as the fuel), rotary, or reverberatory furnaces. The typical remelter will use battery plates, lead cable sheathing, lead pipe, etc., but not process the lead chemicals (sulfates and oxides) present in a lead acid battery. These lead chemicals must be processed in high-temperature refining furnaces. Many battery manufacturers operate captive processors that can deal with the entire battery, processing the lead compounds and lead in appropriate furnaces. Some use the cases and separators for fuel.

Electrowinning and electrorefining processes have been investigated. These processes have the benefit of reduced air emissions. At this time no commercial processors are in operation.

PROCESS DESCRIPTION

There are several alternative systems in use depending on the source of the lead. Where lead acid batteries are the source the process is initiated by breaking the batteries. Breaking is usually by sawing or shearing to remove the head but also by using hammer mills to disintegrate the battery. The next step is to separate the lead-bearing components from the plastic and acid. After separation, the lead content is remelted while the lead compounds are processed through a smelting furnace. The final steps are to refine and alloy the lead to meet specifications (see Figure 1).

Processing using shears or saws is followed by tumbling devices to separate components. Cases may still require using hammer mills to remove the posts. When hammer mills are used, it is usually necessary to follow the process with a sink/float separation system to separate the lead from the plastic components. Polypropylene is rewashed and sold for reuse. Hard rubber cases are used as fuel.

The plates, posts, and separators undergo further processing to separate the paste from the plates for desulfurization. Paste desulfurization involves chemical removal of the sulfur content. Removing the sulfur content prior to heating reduces SO_2 emissions.

Due to the impact of environmental regulations and the economies of scale, the small refiners are going out of business. Small remelters are surviving by finding niche markets.

Control of the air pollution resulting from the secondary smelting and reclaiming of lead scrap relates directly to the type of furnaces being used. The reverberatory, blast, and pot furnaces (kettles) are most commonly used. There is reportedly one electric arc furnace in use. The electric furnace, since no fuel is used, can be controlled with substantially smaller air emission control equipment. Smelting Processes are described in Figure 2. Refining Processes are described in Figure 3.

REVERBERATORY FURNACES

Reverberatory furnaces are rectangular refractory-lined, fuel-fired units. Roofs and floors are usually arched. The floor is pitched to the tapping station. Reverberatory furnaces (Figure 4) are usually operated continuously. The furnace charge includes battery grids and desulfurized paste, battery scrap, rerun reverberatory slags, flue dust, drosses, iron silica, and coke. Often the charge is dried in a fuel-fired rotary drum to eliminate the chance of an explosion caused by liquid being

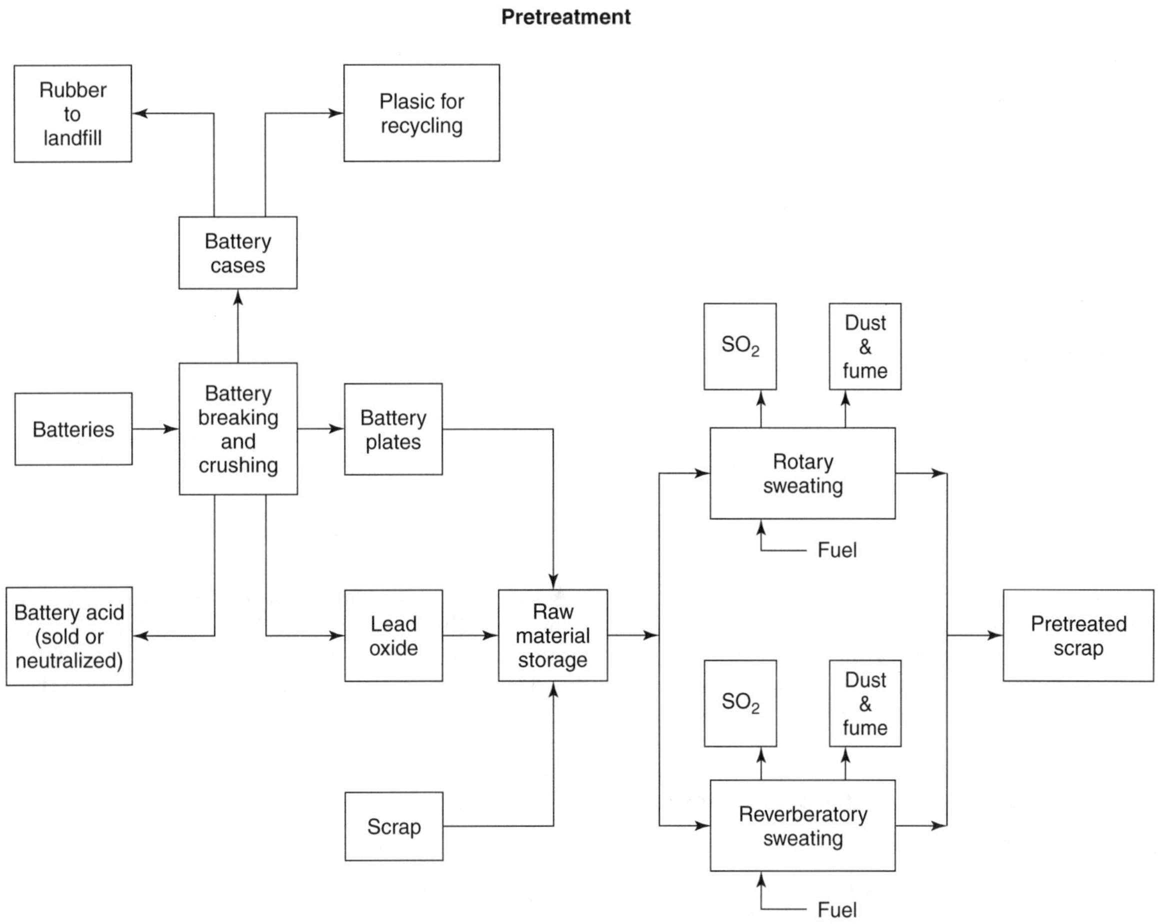

Figure 1. Pretreatment Process Flow Diagram.

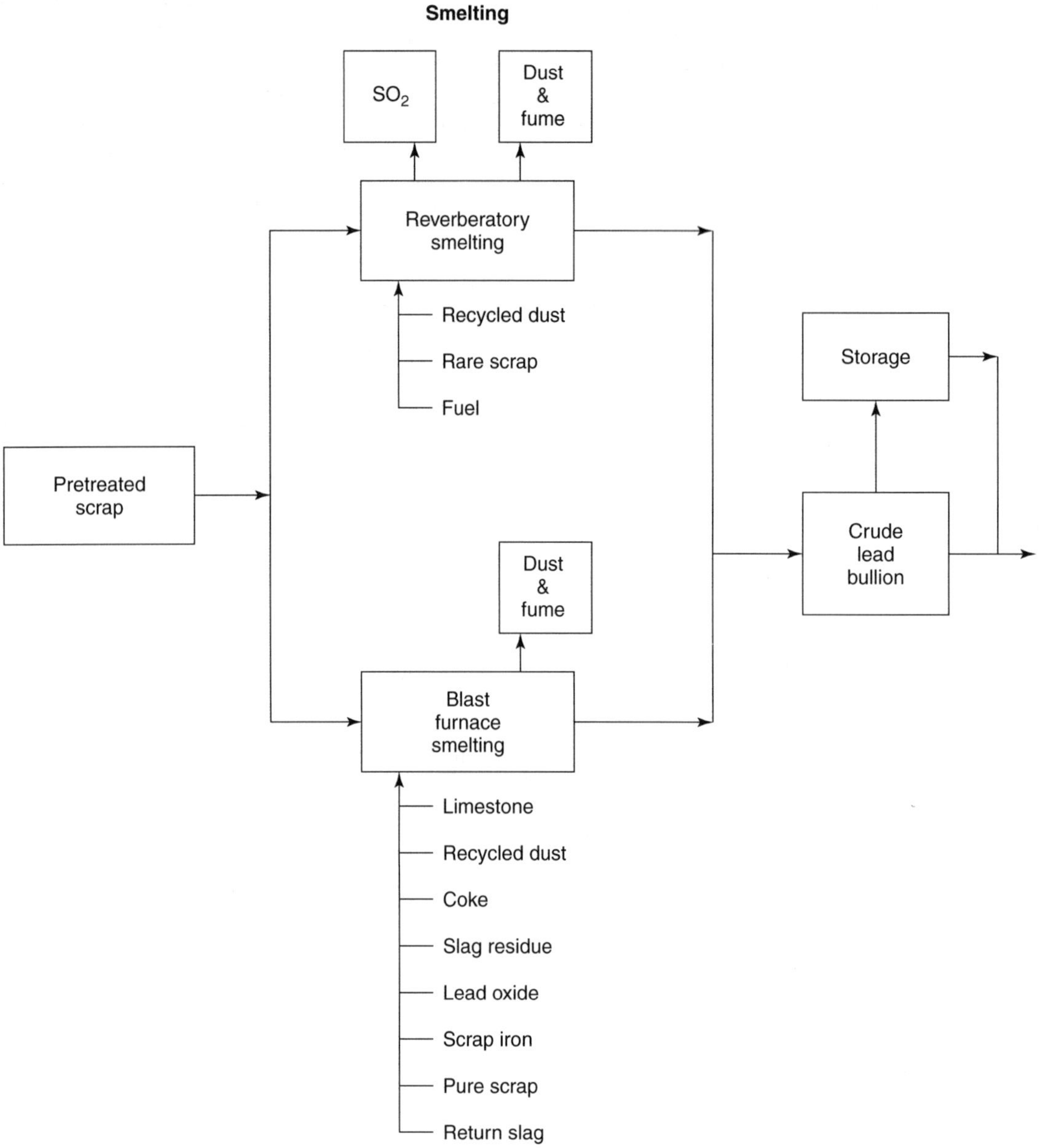

Figure 2. Smelting Process Flow Diagram.

present. Reverberatory furnaces produce soft lead and lead-bearing slag by removing alloying elements and impurities. These furnaces usually operate under negative pressure.

PROCESS DESCRIPTION

Sweating operations can be conducted in a reverberatory furnace or rotary tube. Materials for both sweating and reducing, such as lead scrap, battery plates, oxides, drosses, and lead residues, are charged to a reverberatory furnace. The charges are made up of a mixture of these materials and put into the furnace in such a manner as to keep a very small mound of unmelted material on top of the bath. As the mound becomes molten, more material is charged.

This furnace may be gas or oil fired or a combination of both. Oxygen enrichment may be used to increase melt rates. Sufficient draft is pulled to remove the smoke and fumes but allow the retention of heat over the hearth. Metal is tapped off at intervals as a semisoft lead as the level of the metal rises. Slagging is usually continuous.

AIR EMISSIONS CHARACTERIZATION

A fairly high percentage of sulfur can be present in various forms in the reverberatory furnace charge. The temperature maintained is sufficiently high to "kill" the sulfides and result in the formation of sulfur dioxide and sulfur trioxide in the exit gases. Also present in the smoke and fumes produced are oxides, sulfides, and sulfates of lead, tin, arsenic, copper, and antimony. An overall material balance reported on the product side is approximately 47% recovery of metal, 46% recovery of slag, sometimes called "litharge," and 7% of smoke and fumes.

The unagglomerated particulate emitted from secondary lead-smelting operations have a particle size range of 0.07–0.4 μm, with a mean of about 0.3 μm. The particles are nearly spherical and have a tendency to agglomerate. The

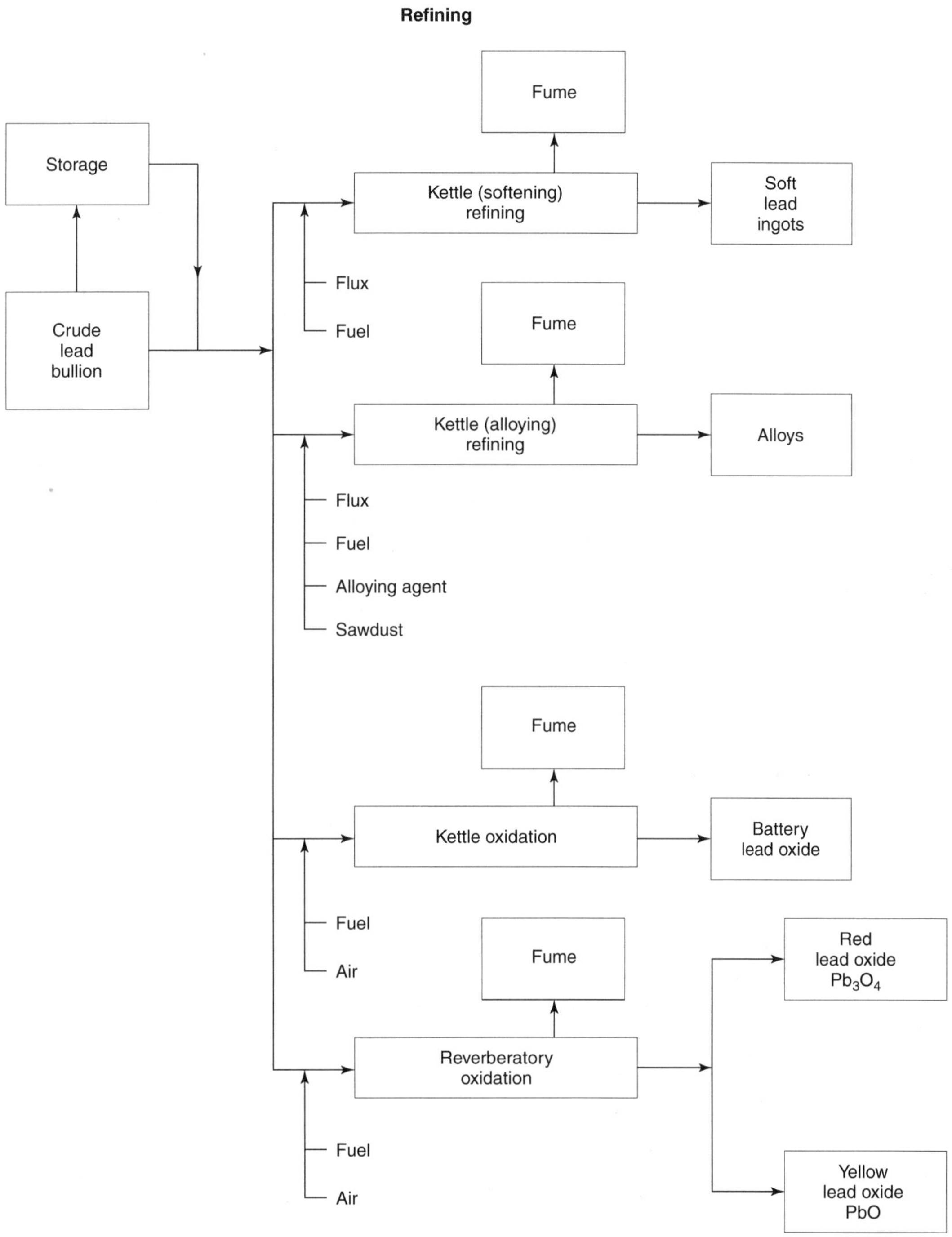

Figure 3. Refining Process Flow Diagram.

concentration of particulate matter in stack gases ranges from 1.4 to 4.5 g/ft^3.

Hooding and Ventilation Requirements

All the smoke and fumes produced by the reverberatory furnace must be collected. These gases are combined with the products of combustion. The entire volume emitted from the furnace must pass through the collector. It is undesirable to draw cold air into these furnaces through the charge doors, inspection ports, or other openings to keep air contaminants from escaping from them. External hoods are used to capture these emissions. The ventilating air for these hoods, as well as for the hoods venting slagging stations, must also pass through the collector. In large furnaces, this represents a considerable volume of gases at high temperatures.

AIR POLLUTION CONTROL MEASURES

The only control systems found to operate satisfactorily use a baghouse followed by a scrubber as a final collector. These systems may include auxiliary items such as gas-cooling devices and settling chambers.

A continuously operating negative-pressure baghouse unit is required. This allows offline cleaning of a compartment.

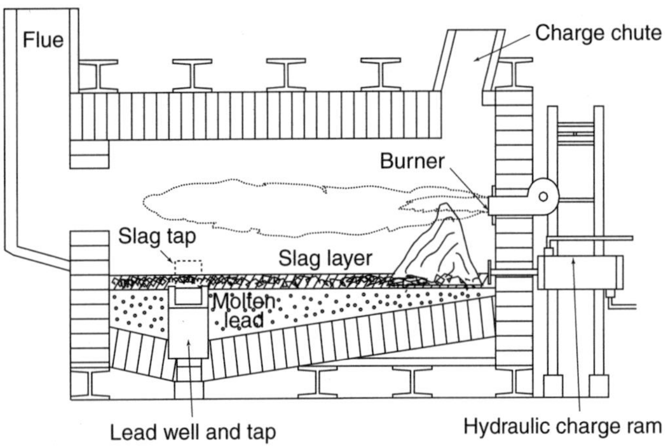

Figure 4. Reverberatory Furnace.

The bags may cleaned by a standard mechanical shaking mechanism.

Provision should be made to prevent sparks and burning materials from contacting the filter cloth. Baghouse inlet temperature must be controlled by use of radiant cooling ducts, water jacketed cooling ducts, or other suitable devices. The type of cloth selected depends on such parameters as the temperature and corrosivity of the entering gases and the permeability and abrasion or stress-resisting characteristics of the cloth. Dacron bags are being used successfully in this service. Air-to-cloth ratios not exceeding 2:1 (preferably much lower) should be applied. Scrubbers are used *after* the baghouse has removed particulate. The purpose of the scrubber is to remove acid gases.

LEAD BLAST FURNACES

Blast furnaces (Figure 5) are cupola furnaces. They are vertical cylindrical units that are top charged with scrap, slag, iron, limestone, sand, and coke. The process involves periodically loading a preweighed charge of proper proportions. Coke is used as the primary fuel, but other fuel (gas or oil) may be used to supplement the coke. Periodic charges are made as the material reduces in volume as it melts. Metal is continuously tapped to holding kettles for later pumping to refining kettles. Slag is tapped periodically. Blast furnaces produce "hard' lead containing alloying elements such as antimony.

PROCESS DESCRIPTION

The materials forming the usual charge for the blast furnace and a typical percentage composition are 4.5% rerun slag, 4.5% scrap cast iron, 3% limestone, 5.5% coke, and 82.5% drosses, oxides, and reverberatory slags. The rerun slag is a silicate slag from previous blast furnace runs. The drosses are miscellaneous drosses consisting of copper drosses, caustic drosses, and dry drosses obtained from refining processes in the pot furnaces. Combustion air is introduced near the bottom of the furnace through tuyeres at about 8–12 oz/in.^{2}. Hard lead is charged into the cupola at the start of the operation to provide molten metal to fill the crucible. Normal charges are added as the material melts down. The limestone and iron form the flux that floats on top of the molten lead and retards its oxidation.

As the level of molten material rises, the slag is tapped at intervals while the molten lead flows from the furnace at a more or less continuous rate. The lead product is "hard" or "antimonial." Approximately 70% of the molten material is tapped off as hard lead and the remaining 30% as slag. About 5% of the slag is retained for rerun later.

AIR EMISSIONS CHARACTERIZATION

Combustion air from the tuyeres passing vertically upward through the charge in a blast furnace conveys oxides, smoke, bits of coke, and particulate present in the charge. A typical material balance based on the charge to a blast furnace in which battery groups are being processed is 70% recovery of lead, 8% slag, 10% matte (sulfur compounds formed with slag), 5% water (moisture contained in charge), and 7% dust (lead oxide and other particulate discharged from stack of furnace with gaseous products of combustion). Particulate matter loading in blast furnace gases is exceedingly heavy, up to 4 g/ft^3.

Blast furnace stack gas temperatures range from 1200°F to 1350°F. In addition to the particulate matter, which consists of smoke, oil vapor, fume, and dust, the blast furnace stack gases contain carbon monoxide. An afterburner is necessary to control the gaseous, liquid, and solid combustible material in the effluent.

Hooding and Ventilation Requirements

The only practical way to capture the contaminants discharged from a lead blast furnace is to seal the furnace and vent all the gases to a control system.

AIR POLLUTION CONTROL MEASURES

The control system for a lead blast furnace requires the elimination of the CO and other combustibles prior to collection requiring an afterburner.

Afterburner

The afterburner must raise the temperature of the combustibles, inspirated air, and cupola gases to at least 1300°F. The secondary combustion zone have a retention time of at least $\frac{1}{4}$ sec (per LA). (The author suggests that time up to 1.5 sec will accomplish better results.) Enough turbulence must be created in the gas stream for thorough mixing of combustibles and air. In large-diameter cupola furnaces, stratification of the gas stream may make this a problem. One device, proved successful in promoting mixing in large-diameter cupolas, is the inverted cone. The combustion air is inspirated through the charging door and, if necessary, also through openings strategically located in the cupola circumference, above the charging opening. The rapid ignition of the combustible effluent by the afterburner frequently results in a pulsating or puffing emission discharge from the charging door. This can be eliminated by installing a burner below the level of the charging door, which ignites and partially burns the combustible effluent.

A cupola afterburner need not be operated through the entire furnace cycle. Even without an afterburner, an active flame can be maintained in the upper portion of the cupola.

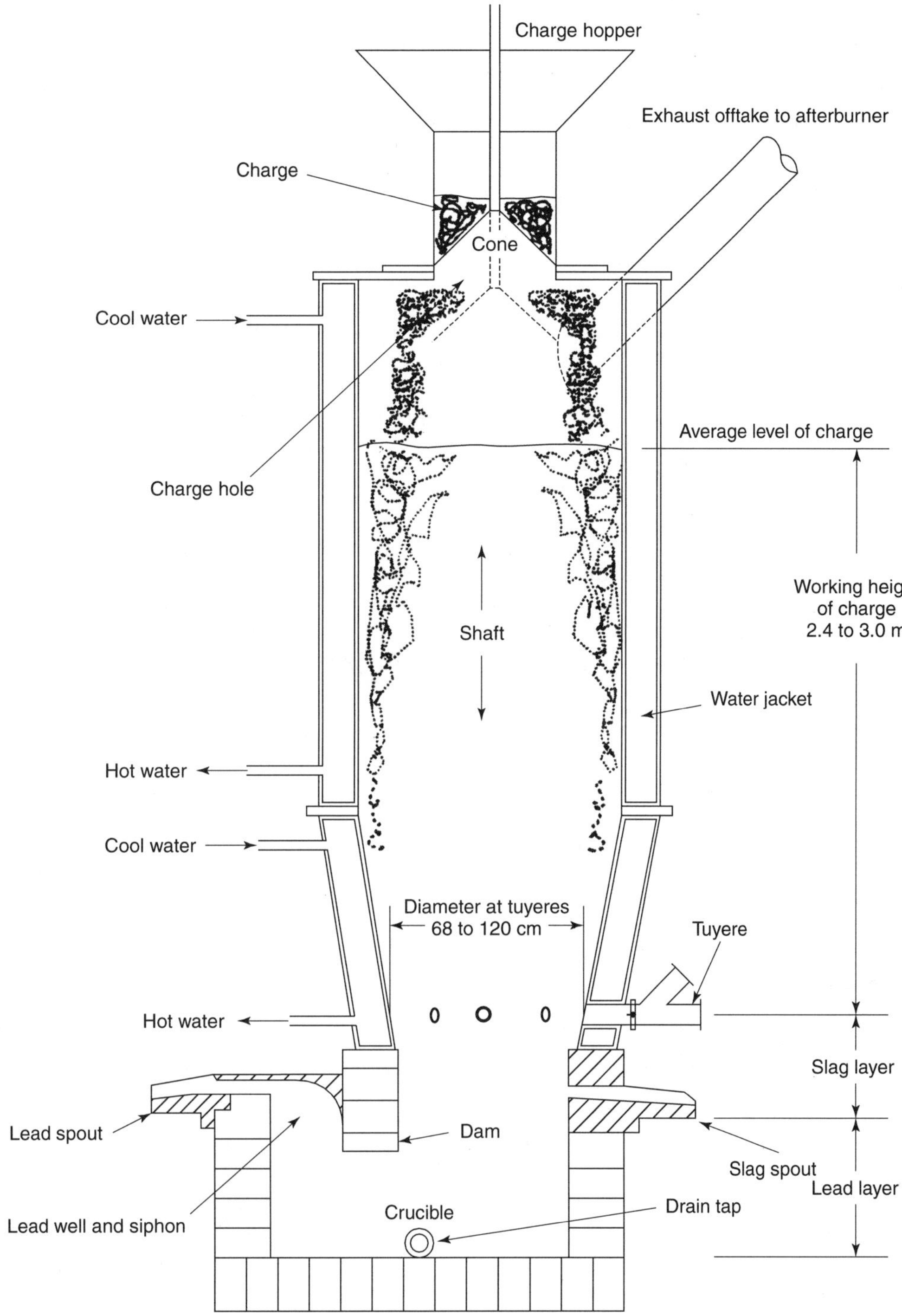

Figure 5. Blast Furnace.

This requires control of the materials charged, and control of combustion air and mixing. The afterburner must be in operation during the furnace light-off procedure.

Baghouse Dust Collectors

The temperature of the gas stream discharged from the top of a cupola may be as high as 2200°F. If a baghouse is used as a control device, these gases must be cooled to prevent bag damage. Cooling can be effected by radiant cooling columns, by evaporative water coolers, or by dilution with ambient air.

For satisfactory baghouse operation, when metallurgical fumes are to be collected, filtering velocity should not exceed 2 fpm, net. In actual practice levels of 1.0–1.5 fpm are routinely found. Provisions for cleaning collected material from the bags require compartmented baghouses so that one section may be isolated and the bags shaken while the remainder of the system is in operation while maintaining a low air-to-cloth ratio. The gas temperature through the baghouse must not fall below the acid dew point to eliminate condensation in the baghouse, which may cause the particles on the bag to agglomerate, deteriorate the cloth, and corrode the baghouse enclosure. A bypass control must also be installed. If the cooling system fails, the bypass must open. This discharges the effluent gas stream to the atmosphere and prevents damage to the bags. Properly designed and maintained baghouses normally can be expected to have efficiencies ranging upward from 95% to as high as 99%.

Scrubbers following the baghouse reduce the acid content to levels to meet requirements. In some operations the use of limestone fluxes and other neutralizing materials are used in the melter to reduce the acid formation in the process. This eliminates the need for a scrubber with its attendant water cleanup circuits and water pollution problems.

ROTARY (REVERBERATORY) FURNACES

Rotary furnaces are used in the United States (Figure 6). The rotary furnace has advantages over the stationery reverberatory furnace in that, since it is operated in batches, the fluxes can be better adjusted. The rotating action promotes better mixing and more uniform product. The charge to a rotary furnace is made up of battery components, drosses, and flue dusts. Fluxes included iron, sand, soda ash, limestone, and coke. These fluxes produce a low-lead slag that does not require further processing before disposal as a nonhazardous waste.

AIR EMISSIONS CHARACTERIZATION

Sulfur is usually present in various forms in the charge to the reverberatory furnace. The temperature is maintained high enough to "kill" the sulfides and result in the formation of sulfur dioxide and sulfur trioxide in the exit gases. Also present in the smoke and fumes produced are oxides, sulfides, and sulfates of lead, tin, arsenic, copper, and antimony. An overall material balance shows on the product side approximately 47% recovery of metal, 46% recovery of slag ("litharge"), and 7% of smoke and fumes.

The unagglomerated particulate emitted from secondary lead-smelting operations has been found to have a particle size range from 0.07 to 0.4 μm with a mean of about 0.3 μm. The particles are nearly spherical and have a tendency to agglomerate. The concentration of particulate matter in stack gases ranges from 1.4 to 4.5 g/ft^3.

Hooding and Ventilation Requirements

All the smoke and fumes produced by the rotary furnace must be collected. These gases are combined with the products of combustion. The entire volume emitted from the furnace must pass through the collector. The flue gases must be ducted to cooling equipment prior to combining with the flow from hoods located to the charging and pouring stations.

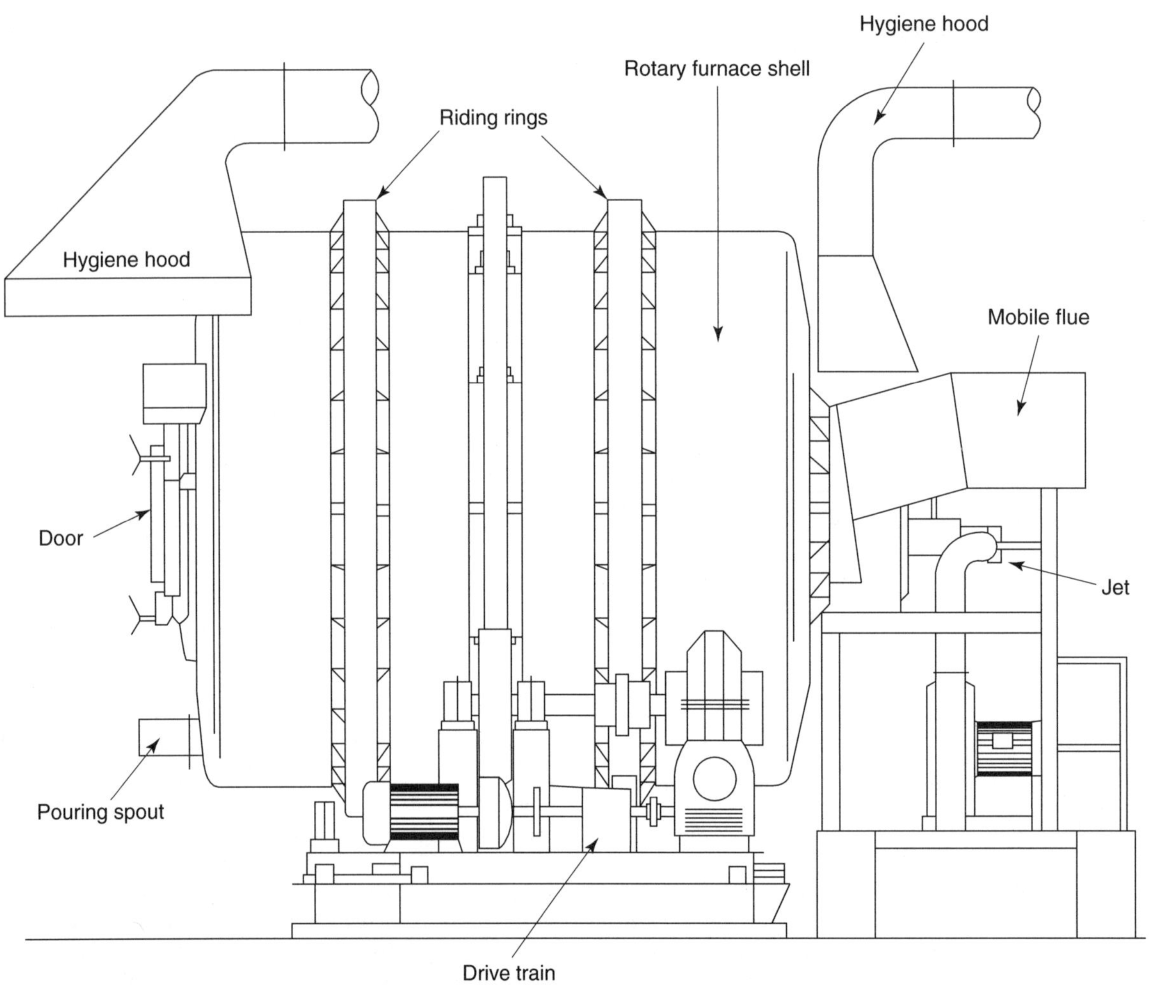

Figure 6. Rotary (Reverberatory) Furnace.

AIR POLLUTION CONTROL MEASURES

The only control systems found to operate satisfactorily use a baghouse, and perhaps a scrubber as final collector, as control devices. These systems may include auxiliary items such as gas-cooling devices and settling chambers.

A continuously operating negative-pressure baghouse unit is required. Allowing offline cleaning of a compartment at a time works well. The bags may cleaned by a standard mechanical shaking mechanism.

Provision should be made to prevent sparks and burning materials from contacting the filter cloth. Baghouse inlet temperature must be controlled by use of radiant cooling ducts, water-jacketed cooling ducts, or other suitable devices. The type of cloth selected depends on such parameters as the temperature and corrosivity of the entering gases and the permeability and abrasion or stress-resisting characteristics of the cloth. Dacron bags are being used successfully in this service. Air-to-cloth ratios not exceeding 2:1 should be applied. Many commercial operations operate with air-to-cloth ratios as low as 1:1.

KETTLES (POT FURNACES)—PROCESS DESCRIPTION

Kettles are used for remelting, alloying, and refining processes (Figure 7). Remelting is usually done in small kettles (1–15 ton capacity). Materials charged are usually alloys in ingot form or scrap with known analysis that do not require any further processing except to be melted for casting operations.

The kettles used in secondary smelters for refining range from 1 to 50 tons capacity or more. These kettles are usually gas fired. Various refining and alloying operations are carried on in these kettles. Alloying usually begins with a metal lower in the percentage of alloying materials than desired. The percent desired is calculated, and that amount is added. Antimony, tin, arsenic, copper, and nickel are the most common alloying elements used.

The refining processes most commonly employed are those for the removal of copper and antimony to produce soft lead and for the removal of arsenic, copper, and nickel to produce hard lead. For copper removal, the temperature of the molten lead is allowed to drop to 620°F and sulfur is added. The mixture is agitated and copper sulfide is skimmed off as dross. This is known as "copper dross" and is charged into the blast furnace.

When aluminum is added to molten lead, it reacts preferentially with copper, antimony, and nickel to form complex compounds that can be skimmed from the surface of the metal. The antimony content can also be reduced to about 0.02% by bubbling air through the molten lead. It can be further reduced by adding a mixture of sodium nitrate and sodium hydroxide and skimming the resulting dross from the surface of the metal.

Another common refining procedure, "dry drossing," consists of introducing sawdust into the agitated mass of molten metal. This forms carbon, which aids in separating the globules of lead suspended in the dross and reduces some of the lead oxide to elemental lead.

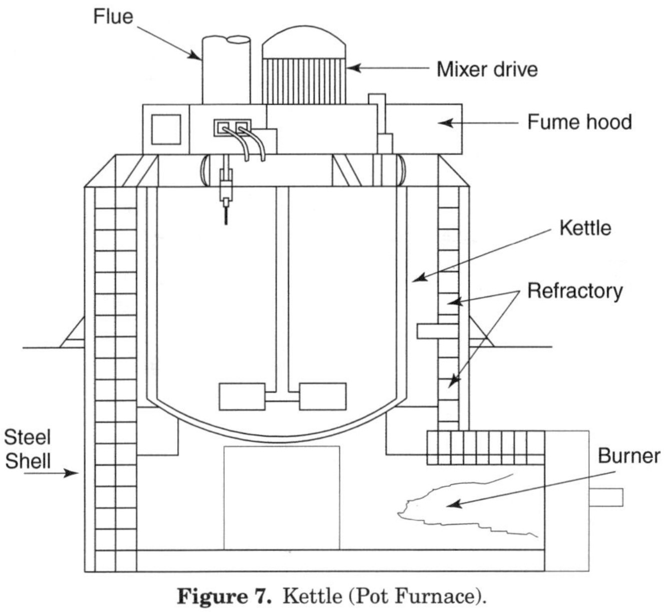

Figure 7. Kettle (Pot Furnace).

AIR EMISSIONS CHARACTERIZATION

Although the quantity of air contaminants discharged from kettles as a result of remelting, alloying, and refining is much less than that from reverberatory or blast furnaces and the flue gases can be vented without control, the capture and control of these contaminants are equally important in order to prevent periodic violations of air pollution regulations and to protect the health of the employees.

Problems of industrial hygiene are inherent in this industry. People working with this equipment frequently inhale and ingest lead oxide fumes, which are cumulative, systemic poisons. Frequent medical examinations are necessary for all employees, and a mandatory compliance with OSHA regulation 40 CFR 1910.1025 is required.

Hooding and Ventilation Requirements

The canopy-type hood is readily adaptable to capturing the fumes from a kettle. Recommended in-draft velocities vary depending on the furnace hood geometry, cross-drafts, and temperatures involved. Hood design is critical to proper and safe operation. Hood inrush velocity can be maintained by minimizing the opening in the hood and closing doors that are not immediately needed for operations. The agitators and other equipment placed within the hood must have auxiliary cooling to minimize damage to the tools used inside the hood.

AIR POLLUTION CONTROL MEASURES

The control systems for kettles, as with the other lead furnaces, require the use of a baghouse as the final collector. Typically they do not need scrubbers. The temperature of the gases from the kettles is much lower than gases from the other type furnaces. Gas-cooling devices, if needed, will be much smaller. Afterburners typically are not required.

The use of multicompartmented shaker-type baghouses is recommended. The compartments should be cleaned with the cleaning compartment off line. The net air-to-cloth ratio should not exceed 2:1, preferably less.

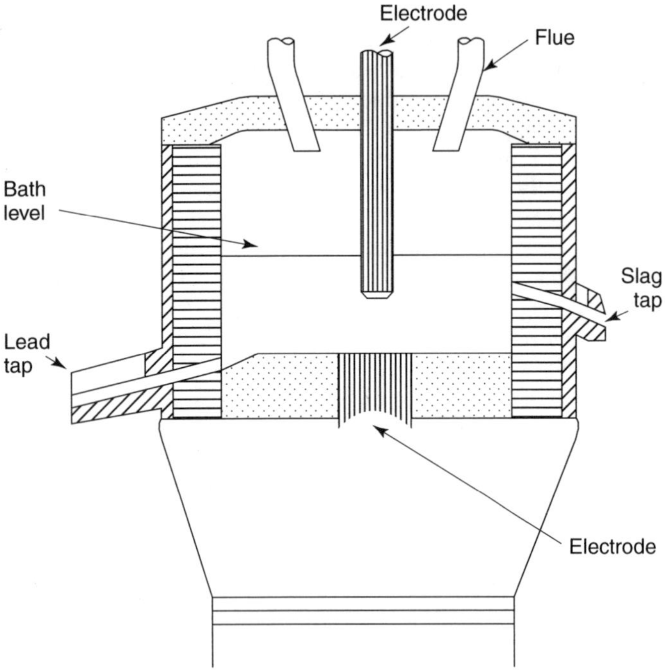

Figure 8. Electric Arc Furnace.

ELECTRIC ARC FURNACE

Electric arc furnaces are seldom used in the United States (Figure 8). They are used to rerun reverberatory slags. They require much less air emission control, since there are no fuels used. The slag product contains very low levels of lead, which allows disposal to land as nonhazardous waste.

The only arc furnace reported operating in the United States makes use of a baghouse to control the emissions from the melting and pouring stations on the furnace. There are no reported hydrocarbon or chlorine/hydrochloric acid gas emissions.

FUGITIVE PROCESS EMISSIONS

The NESHAP–MACT controls for process sources were identified as emanating from smelting furnace charging and tapping locations, flue dust agglomerating furnaces, refining kettles, and dryers. The recommended procedure is to install adequate hooding at these stations and ventilate the hoods to a baghouse. The U.S. EPA also recommends that wetting or use of surfactants, binding agents, or sealers be required on storage piles coupled with ventilated enclosures. The use of materials handling equipment can cause fugitive emissions to be generated within the storage areas. The enclosures require ventilation to dust collectors.

Battery breaking areas were also found to generate fugitive emissions. The use of enclosures, power washing systems, and wetting of storage piles is required. The entire area is to be ventilated to a baghouse.

Furnace and lead refining and casting areas must either be totally enclosed or partially enclosed and have the floors vacuumed or power washed routinely.

Materials storage areas require pavement. The vehicles leaving the areas must be washed. The area requires partial enclosure with wet suppression or total enclosure with ventilation to a baghouse.

Roadways in plants must be paved. They must be cleaned by vacuuming or power washing periodically.

Table 1. Summary of Typical Uncontrolled Organic Emissions from Smelting Furnaces (Mg/yr)

Smelting Furnace Type	Organic HAPs	THC	CO
Blast	100	280	1,800
Rotary	6	34	20
Reverberatory	N/A	8	8
Reverberatory/blast with combined flow	1	2	12
Electric	N/A	N/A	10

N/A, no data available.
MACT Background Document, Table 2-2.
Source:

PROCESS SOURCES OF FUGITIVE EMISSIONS

Process fugitive sources of emissions are smelting furnace charging, lead tapping, slag tapping, refining operations, agglomerating furnaces, and casting machines. Blast furnace charging is a source of organic emissions. Process fugitive emissions may be combined with process emissions to lower

Table 2. Design and Operating Parameters of the Process Baghouses Tested by the EPA

Facility ID	Type of Gas Treated	Type of Baghouse	Air Flow (scfm)	Temp. (°F)	Bag Material	No. of Compartments	Pressure Drop (in. H_2O)	A/C Ratio (ft/min)
Tejas Resources, TX	Process, rotary furnace	Shaker	8,500	250	acrylic	8	0.2–5.0	1.1
East Penn Manufacturing Co., PA	Process, blended reverberatory and blast gas stream	Shaker	26,000	250	Gore-Tex©	6	0–4.9	1.05
Schuylkill Metals, MO	Process, blast furnace, plus process fugitive	Reverse air, sonic	25,000	250	Teflon B© on fiberglass	6	5.0–9.5	1.3

MACT Background Document, Table 3-2.
Source:

Table 3. Summary of EPA Data for Uncontrolled and Controlled THC Emissions

Source Type	Company	Sampling Location	Run[a]	Concentration[b] (ppmv as Propane at 4% CO_2)	Rate (kg/h)	Ref.
Blast	Schuylkill Metals, MO	Furnace outlet	1	3,538	63	12
			2	2,950	54	
			3	3,227	59	
Blast	Schuylkill Metals, MO	Afterburner outlet	1	296	4.4	12
			2[c]	2,300	34.4	
			3	364	5.4	
Reverberatory/blast	East Penn Manufacturing Co., PA	Blast furnace outlet	1	3,133	40.8	13
			2	2,412	27.8	
			3	3,085	36.5	
Reverberatory/blast	East Penn Manufacturing Co., PA	Mixing chamber outlet	1	5	0.5	13
			2	20	1.3	
			3	3	0.3	
Reverberatory	RSR-NY	Furnace outlet (no addon controls)	1	10	0.9	26
			2	9	0.9	
			3	10	0.9	
				11	0.9	
Rotary	Tejas Resources, TX	Furnace outlet (no addon controls)	1	167	3.2	11
			2	89	1.8	
			3	137	2.8	
			4	72	1.4	
			5	110	2.0	
			6	35	0.6	

[a]Individual runs were 3 h at Schuylkill and East Penn; at RSR-NY, run 1 was 5 h and runs 2 to 4 were 1 h; at Tejas, each run was for a complete batch cycle of 15–24 h.
[b]Uncontrolled blast furnace concentrations are not corrected to a constant CO_2 concentration because they are measured at the furnace outlet before dilution. Reverberatory and rotary furnace THC concentrations are corrected to 4% CO_2 to account for dilution. Data collected using EPA reference method 25A.
[c]Afterburner malfunctioned during this run.

Source: MACT Background Document, Table 3–6.

Table 4. Operating Temperatures and Residence Times for Afterburners Controlling Blast Furnaces at Secondary Lead Smelters

Facility ID	Furnace ID	Operating Temperature (°C)	Residence Time (sec)
2	1	670	N/A
		670	N/A
		670	N/A
		670	N/A
3	1	650	N/A
6	1	730	0.7
	2	730	0.7
8	1	650	N/A
12	1	870	1.5
13	1	590	2.5
	2	590	2.5
16	1	700	2.5
27	1	590	N/A
28	1	650	2.0

N/A, Residence time not available.
MACT Background Document, Table 3–6.
Source:

temperatures and reduce the auxiliary cooling equipment requirements.

Hoods are required to control process fugitive emissions at tapping, slagging, and other process stations. The hoods must be designed to operate with minimum openings to reduce the exhaust air needed. The hood opening face velocity must overcome air currents and prevent the escape of contaminated air to the plant. OSHA recommends face velocities of 250–500 ft/min (75–150 m/min).

PRODUCTS

Various grades of lead metal, along with the oxides, are produced by the secondary lead industry. The grade of product desired determines the type of equipment selected for its manufacture. The most common grades of lead produced are soft, semisoft, and hard. By starting with one of these grades and using accepted refining and alloying techniques, any special grade of lead or lead alloy can be made.

Soft lead may be designated as corroding, chemical, acid copper, or common desilverized lead. These four types are high-purity leads. Their chemical requirements are presented in Table 1. These leads are the products of the pot furnace (kettle) after a considerable amount of refining has been done.

Semisoft lead is the product of the reverberatory furnace. It contains from 0.3% to 0.4% antimony and up to 0.05% copper.

Hard lead is made in the blast furnace. A typical composition of hard lead is 5–12% antimony, 0.2–0.6% arsenic, 0.5–1.2% tin, 0.05–0.15% copper, and 0.001–0.01% nickel. In developing the MACT document, the U.S. EPA developed a number of data reporting information which is reported in Tables 1–4.

BIBLIOGRAPHY

1. G. L. Allen, F. H. Viets, and L. C. McCabe, "Control of Metallurgical and Mineral Dusts and Fumes in Los Angeles County,

California," Bureau of Mines Information Circular 7627, U.S. Department of Interior, Washington, D.C., April, 1952.

2. J. E. Williamson, J. F. Neuzell, W. E. Zwincher, *A Study of Five Source Tests on Lead Emissions from Secondary Lead Smelters*, L.A. County Air Pollution Control District, EPA Order No. 2PO-63-02-3326, 11 February 1972.
3. Battelle, Columbus Laboratories, *A Study to Identify Opportunities for Increased Solid Waste Utilization*, National Association of Secondary Material Industries, Inc., June 1972.
4. P. R. Webb, H. E. Carlton, and R. B. Engdahl, Battelle, Columbus Laboratories, *Research Report on Secondary Lead Plant Stack Emission Sampling at N. L. Industries Plant, McCook, Illinois*, Contract #68-02-0230, Task Order #1 to EPA Office of Air Programs, July 19, 1972.
5. Paul E. Webb, H. E. Carlton, and R. B. Engdahl, Battelle, Columbus Laboratories, *Research Report on Secondary Lead Plant Stack Emission Sampling at General Battery Corporation, Reading, PA*, Contract #68-02-0230, Task Order #1 to EPA Office of Air Programs, July 25, 1972.
6. P. R. Webb, H. E. Carlton, and R. B. Engdahl, Battelle, Columbus Laboratories, *Research Report on Secondary Lead Plant Stack Emission Sampling at Revere Smelting & Refining Plant, Newark, NJ*, Contract #68-02-0230, Task Order #1 to EPA Office of Air Programs, August 11, 1972.
7. P. R. Webb, H. E. Carlton, and R. B. Engdahl, *Secondary Lead Plant Stack Emission Sampling at N.L. Industries Plant, McCook, IL*, BATTELLE, Contract No. 68-02-0230 to U.S. EPA Office of Air Programs, 19 August 1972.
8. P. R. Webb, H. E. Carlton, and R. B. Engdahl, *Secondary Lead Plant Stack Emission Sampling at Revere Smelting and Refining Plant, Newark, NJ*, BATTELLE, Contract No. 68-02-0230 to U.S. EPA Office of Air Programs, 19 August 1972.
9. P. R. Webb, H. E. Carlton, and R. B. Engdahl, *Secondary Lead Plant Stack Emission Sampling at General Battery Corporation, Reading, PA*, BATTELLE, Contract No. 68-02-0230 to U.S. EPA Office of Air Programs, 19 August 1972.
10. P. R. Webb, H. E. Carlton, and R. B. Engdahl, Battelle, Columbus Laboratories, *Research Report on Secondary Lead Plant Stack Emission Sampling at Quemetco, Incorporated, Industry, California*, Contract #68-02-0230, Task Order #1 to EPA Office of Air Programs, August 19, 1972.
11. P. R. Webb, H. E. Carlton, and R. B. Engdahl, *Secondary Lead Plant Stack Emission Sampling at Quemetco, Inc., Industry, CA*, BATTELLE, Contract No. 68-02-0230 to U.S. EPA Office of Air Programs, 19 August 1972.
12. U.S. EPA Air Pollution Technical Information Center, *Lead And Air Pollution: A Bibliography with Abstracts*, January 1974.
13. U.S. Department of the Interior, *Lead—A Chapter from Mineral Facts and Problems*, Bureau of Mines Preprint from Bulletin 667, 1975.
14. U.S. Department of the Interior, Bureau of Mines Bulletin 667, *Lead—A Chapter from Mineral Facts and Problems*, 1975.
15. M. Dean High, Engineering-Science, Inc., *Inspection Manual for Enforcement of New Source Performance Standards: Secondary Lead Smelters*, U.S. Department of Commerce, NTIS (PB-269-958) (EPA/340/1-77/001), January 1977.
16. David M. Augenstein, PEDCo-Environmental, Inc., *Control Techniques for Lead Air Emission, Volume II, Chapter 4, Appendix B (Final Report)*, U.S. Department of Commerce, NTIS (PB 80-197551) December 1977; J. W. Watson and K. J. Brooks, *A Review of Standards of Performance for New Stationary Sources—Secondary Lead Smelters*, The Mitre Corporation, MTR-7871, January 1979.
17. J. W. Watson and K. J. Brooks, *A Review of Standards of Performance for New Stationary Sources—Secondary Lead Smelters*, Metrek Division of the MITRE Corporation, McLean, VA., Contract Sponsor, EPA, Contract #68-02-2526, Project No. 1510L, Dept. W56, January 1979.
18. H. J. Taback, KVB, Inc., *Fine Particle Emission from Stationary & Miscellaneous Sources in the South Coast Air, Etc.*, February 1979.
19. J. E. Howes, Jr., Battelle, Columbus Labs, *Evaluation of Stationary Source Particulate Measurement Methods. Volume V. Secondary Lead Smelters*, U.S. Department of Commerce, NTIS (PB-300 337), June 1979.
20. R. T. Coleman, Jr., and R. Vandervort, *Demonstration of Fugitive Emission Controls at a Secondary Lead Smelter*, Lead-Zinc-Tin 80 Symposium, The Metallurgical Society of AIME, 1980.
21. *Lead-Zinc-Tin '80*, Proceedings of a World Symposium on Metallurgy & Environmental Control sponsored by the TMS-AIME Lead, Zinc, & Tin Committee at the February 24–28, 1980, annual meeting, edited by John M. Cigan, St. Joe Minerals Corp., Thomas S. Mackey, Key Metals & Minerals Engineering Corp., & Thomas J. O'Keefe, University of MO—Rolla, MO.
22. Ben B. Ewing, George Provenzano, and Keturah Reinbold, Institute for Environmental Studies, Univ. of Illinois, Urbana–Champaign, *Economic Impact of a Proposed Change in Lead Effluent Standards*, (Document #80-02) IL Institute of Natural Resources, Environmental Management Division, February 1980.
23. Klaus Schwitzgebel and Robert Vandervort, *Emission and Emission Controls at a Secondary Lead Smelter*, Radian Corporation, Austin, TX, January 1981.
24. E. R. Cole, Jr., A. Y. Lee, and D. L. Paulson, *Electrolytic Method for Recovery of Lead From Scrap Batteries*, U.S. Department of the Interior, Bureau of Mines Report of Investigations (RI-8602), 1981.
25. Ernest C. Crosby and Vittorio K. Argento, *Environmental Concentration of Lead in the Area of Secondary Lead Smelters*, University of Texas, for Presentation at 75th Annual Meeting of the Air Pollution Control Association, New Orleans, LA, June 20–25, 1982.
26. Ernest C. Crosby and Vittorio K. Argento, University of Texas, *Environmental Concentration of Lead in the Area of Secondary Lead Smelters & a Battery Manufacturing Plant in Dallas, Texas*, Air Pollution Control Association Annual Meeting, June 20–25, 1982.
27. Industrial Environmental Research Laboratory, *Environmental Controls at a Secondary Lead Smelter*, U.S. EPA Capsule Report EPA 625/2-82-030, July 1983.
28. E. R. Cole, A. Y. Lee, and D. L. Paulson, *Recovery of Lead from Battery Sludge by Electrowinning*, Journal of Metals, August 1983.
29. E. R. Cole, A. Y. Lee, and D. L. Paulson, *Recovery of Lead from Battery Sludge by Electrowinning*, Journal of Metals, August 1983.
30. PEDCo Environmental, Inc., *Inspecting and Operating and Maintenance Guidelines for Secondary Lead Smelter Air Pollution Control*, Contract No. 68-03-2924, U.S. EPA, Industrial Pollution Control Division, January 1984.
31. Josef Blanderer, *The Refining of Lead by Oxygen*, JOMA, December 1984.
32. Y. Lee, E. R. Cole, Jr., and D. L. Paulson, *Electrolytic Method of Recovery of Lead from Scrap Batteries—Scale-Up Study Using 20-Liter Multielectrode Cell*, U.S. Department of the Interior, Bureau of Mines Report of Investigations (RI-8857), 1984.
33. E. R. Cole, A. Y. Lee, and D. L. Paulson, "Update on Recovering Lead from Scrap Batteries," *JOM*, February, 1985.
34. E. R. Cole, Jr., A. Y. Lee, and D. L. Paulson, "Update of Recovering Lead From Scrap Batteries," *Journal of Metals*, February, 1985.

35. *Automatic Dross Removal from Lead Refining Kettles*, W. R. D. Couper, H. Forrest, TMS Technical Paper No. A86-29, Metallurgical Society of AIME.
36. Craig Smith and Michael R. Daley, *Domestic Secondary Lead Industry*, IC9156, U.S. Department of the Interior Bureau of Mines, 1987.
37. Dean M. Peters, "Secondary Lead Gets Mixed Reviews," *Recycling Today*, October, 1987.
38. *Lead-Zinc '90*, Edited by Thomas S. Makey, Key Metals & Materials Engineering Corp., & R. David Prengaman, RSR Corporation, Proceedings of a world symposium on metallurgy & environmental control, sponsored by the TMS Lead, Zinc, & Tin Committee annual meeting, February 18–21, 1990.
39. G. Rath, T. Vlajcic, and O. Metelman, "Lead Smelting in a Submerged Arc Furnace," *JOM*, June, 1990.
40. N. A. Esmen, Pittsburgh University, Department of Industrial Environmental Health Sciences, *Fundamental Investigation of Exhaust Hoods* (Final Report), U.S. Department of Commerce, NTIS (PB91-173393), August, 1990.
41. U.S. EPA Superfund Engineering Issue, "Treatment of Lead Contaminated Soils," April, 1991.
42. U.S. EPA Superfund Engineering Issue, "Treatment of Lead-Contaminated Soils," April, 1991.
43. J. E. Murphy and M. F. Chambers, *Production of Lead Metal by Molten-Salt Electrolysis with Energy-Efficient Electrodes*, U.S. Department of the Interior, Bureau of Mines Report of Investigations (RI-9335), 1991.
44. William D. Woodbury, *Lead*, U.S. Department of the Interior, Bureau of Mines, April, 1992.
45. Tullio Corsini, *Improved CX Process for Spent Lead Acid Batteries Treatment*, EPD Congress 1993, Edited by J. P. Hager, The Minerals, Metals & Materials Society, 1992.
46. Richard A. Lelby, Jr., *Secondary Lead Smelting at East Penn Manufacturing Co., Inc.*, Edited by J. P. Hager, The Minerals, Metals & Materials Society, 1992.
47. Paul B. Queneau, Barry J. Hansen, and D. Erik Spiller, *Recycling Lead and Zinc in the United States*, Hazen Research, Inc., Golden, CO, 1992.
48. L. C. George, Bureau of Mines, Rolla Research Center, Rolla, MO, *Waste Battery Acid: Use or Disposal.* (Final Report), U.S. Department of Commerce, NTIS (PB92-176155), 1992.
49. Tullio Corsini, Ed. J. P. Hager, The Minerals, Metals, & Materials Society, *Improved CX Process for Spent Lead Acid Batteries Treatment*, EPD Congress, 1993.
50. U.S. EPA Office of Air Quality Planning and Standards, *Secondary Lead Smelting Background Information Document for Proposed Standards, Volume 1*, Emission Standards Division, June, 1994.
51. U.S. EPA Risk Reduction Engineering Laboratory, *BESCORP Soil Washing System for Lead Battery Site Treatment*, EPA/540/AR-93/503, January, 1995.
52. U.S. EPA, Super Innovative Technology Evaluation, *BESCORP Soil Washing System for Lead Battery Site Treatment*, Applications Analysis Report, January, 1995.
53. Acurex Environmental Corporation, *Reclamation of Automotive Batteries: Assessment of Health Impacts...*, for California Air Resources Board, Contract No. 93-323, March, 1995.
54. Nahed A. Al-Mahallawy and Robert M. Wilson, *Technical Analysis Lead Smelter Component Cairo Air Improvement Project*, submitted by Datex, Inc., Arlington, VA, June, 1995.
55. *Recycling of Metals & Engineered Materials*, Proceedings of a Symposium Organized by the Recycling Committee of the Extraction & Processing Division and the Light Metals Division of The Minerals, Metals & Materials Society, November 12–15, 1995.
56. Gerald Smith, *Lead In November 1995*, U.S. Dept. of the Interior, Mineral Industry Surveys, 9 February 1996.
57. Gerald Smith, *Lead In December 1995*, U.S. Dept. of the Interior, Mineral Industry Surveys, 19 March 1996.
58. John G. Feller, *OSHA, EPA and Liability Concerns* (April 1996), Consultant Volume 10, #2, Boelter Environmental Consultants, April, 1996.
59. Laurel Staley, *Lead Reclamation from Superfund Waste*, National Risk Management Laboratory, EPA Tech Trends Issue No. 23, June, 1996.
60. Gerald Smith, *Lead in March 1996*, U.S. Dept. of the Interior, Mineral Industry Surveys, 7 June 1996.
61. Laurel Staley, *Lead Reclamation from Superfund Waste*, National Risk Management, June, 1996. Research Laboratory, EPA Tech Trends, Issue #23.
62. Gerald Smith, *Lead in May 1996*, U.S. Dept. of the Interior, Mineral Industry Surveys, 17 July 1996.
63. U.S. Department of Interior, U.S. Geological Survey, Mineral Industry Surveys, *Lead in May 1996*, July 17, 1996.
64. Gerald Smith, *Lead in June 1996*, U.S. Dept. of the Interior, Mineral Industry Surveys, 19 August, 1996.
65. U.S. Department of Interior, U.S. Geological Survey, Mineral Industry Surveys, *Lead in June 1996*, August 19, 1996.
66. Kathleen Fackelmann, "Hypertension's Lead Connection," *Science News* **149**, 15 June 1998.

SECONDARY ZINC

Charles A. Licht

Secondary zinc scrap reclamation operations are carried out using a number of techniques. Among the end products that are generated are zinc powders (for paints and precious metal precipitation), zinc oxide (for use in tire manufacturing, coated-paper manufacturing, paints, etc.), zinc alloy (for manufacturing castings and galvanizing), and various low-grade zinc alloys for off-specification noncritical castings and zinc compounds (used as additives in a variety of products, including fertilizers and animal feeds).

The raw materials are scrap die castings from postconsumer products, including home scrap such as bad castings and gates and risers from foundry operations. Flashing and trim scraps are used, as are edge trim and sheet scrap, reject battery cases, old zinc roof sheet, and galvanizers' dross. Skimmings and drosses are used to reclaim metal and zinc oxide.

Iron or contaminated zinc scrap is processed in sweating furnaces, although occasionally hammermills are used to break the scrap to allow separation. The typical sweating furnace is an unlined rotary furnace, although refractory-lined units are in use. The rotary furnace has the advantage of keeping the material moving, exposing all surfaces of the scrap to heat, and allowing for separation so that the molten zinc is able to run free to the discharge point. Properly operated zinc rotary sweat furnaces are used to separate zinc from aluminum, since there is about a 400°F (200°C) temperature differential between the two melting points.

Molten zinc generated from a rotary sweat furnace is poured into a cast-iron holding pot, where the alloy is checked and modified or refined prior to pouring into a usable end product. Typically, this zinc is poured into blocks or slabs. If the material is alloyed, it is poured in either 50-pound (22-kg) slabs or 20-pound (9-kg) bars.

In the manufacture of zinc powder, zinc scrap, drosses, and carbon-content material are placed in retort furnaces, which are heated externally. Zinc oxide reacts with carbon to form zinc metal. The zinc metal is heated above its boiling point (1668°F, 910°C), causing it to boil over into a condenser. Air is kept out of the system by tight seals. This system can create air pollution problems due to retort seal failure or when a batch is completed, and the seal must be broken to move the condenser in order to empty the retort.

Another end product of retort operations is zinc oxide manufactured utilizing retort technology. In this case, zinc is allowed to boil and exit the retort to burn in air. It is then collected in an air pollution/product control system. The pollution control device collects a salable product.

Another technique employed in the manufacture of zinc oxide is the use of a muffle furnace. In a muffle furnace heating is done indirectly. The furnace is built so that molten zinc can be added to the boiling bath continuously or periodically. The bath is operated at temperatures in excess of the boiling point (1668°F, 910°C) of zinc. This allows the zinc vapors to leave the furnace and enter a chamber, where they are burned in air to form zinc oxide particles. The zinc oxide is air conveyed to a baghouse system whose primary purpose is to collect a salable product, but which also does an excellent job of controlling potential emissions (see Figure 1).

Zinc skimmings and drosses are melted in refractory-lined rotary furnaces, which are filled, heated, refilled, and reheated. The oxide materials are allowed to overflow into coolers while the metal fills the furnace. When the furnace is filled with metal, the metal is tapped into blocks for resale to galvanizers or alloyers. The oxides generated in the dross reclamation operations are generally used as micronutrients in fertilizers and animal feed.

High-purity zinc is made from scrap by utilizing retorts. Zinc metal from a retort goes into a second retort, where it is condensed to form a molten bath. The bath is tapped after all the metal in the first retort has been boiled off.

A number of secondary operations exist that are secondary only in the sense that they involve remelting of primary zinc. The manufacture of zinc casting alloys is frequently carried out in iron pots, although some companies use refractory-lined reverberatory furnaces. In the process of alloy manufacture, special high-grade (99.99%-pure) zinc metal is melted, and minor additions of aluminum, magnesium, and copper are made to meet alloy specifications. The metal is poured off as slabs or 20-pound (9-kg) bars, occasionally in the form of blocks. Other processes include the melting of special high-grade zinc and pouring, without alloying, into special configurations, such as balls used in plating.

The processes utilized in the zinc metal remelt operations generate almost no fumes. On the other hand, the "wet" and "dry" rotary sweating furnaces generate substantial amounts of fumes, as do the scrap remelting furnaces. In these cases, the furnaces must be controlled by adequate air pollution control systems, such as shaker-type baghouses. The products collected in these baghouses, while a lower-grade material than the zinc oxides manufactured for sale, are usually salable.

Fluxing is done in some of the processes, particularly in the reclamation of zinc scrap in iron pots. In this process, zinc chloride is utilized for cleaning the metal. This requires that the zinc pots be ventilated to baghouses because of the periodic high fume generation.

Muffle furnaces have much greater vaporizing capacity than any other process available. They are operated for months-long campaigns. The heat for vaporization, which is supplied by fuel-fired burners, is forced through a silicon carbide arch that separates the zinc vapors from the products of combustion. Molten zinc from a melting pot or sweat furnace is charged through a feed well, which is also the air lock. The zinc vapors are carried to a condenser where the purified zinc is collected

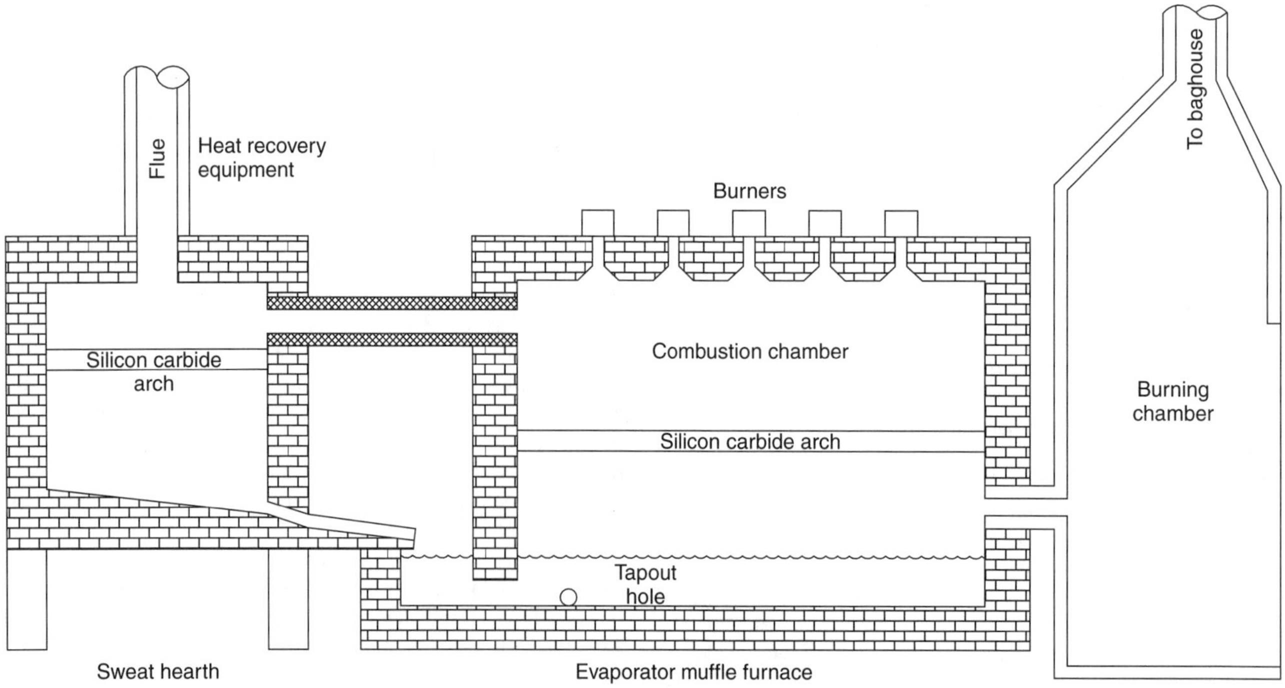

Figure 1. Zinc Muffle Furnace with Integral Sweat Furnace.

or vented and burned in air to generate oxides. The zinc fume is collected in a baghouse.

When scrap is sweated to generate the molten metal necessary for the muffle furnace, the sweating furnace will create fumes. Muffle furnaces are constructed as sweat furnaces. The products of combustion used to boil zinc are utilized to heat the scrap in the sweating furnace and then are vented through a heat exchanger system that allows the use of preheated combustion air in the primary chamber. The only contaminated vent is on the sweating furnace at the charge door. The charge door is enclosed by a hood that controls the fumes to an air pollution control system. This secondary fume will not meet product specifications and is used in lower-grade applications, such as fertilizer and animal feed.

REDUCTION RETORT FURNACES

Belgian Retorts

The Belgian retort furnace is one of several horizontal retort furnaces that had been the most common device for the reduction of zinc. The horizontal retort process is now being replaced by other methods capable of handling larger volumes of metal per retort and by the electrolytic process for the reduction of zinc ore. The reduction process is used to reclaim zinc from the dross formed in zinc-melting operations, from contaminated zinc scrap, from zinc oxide collected by air pollution control systems serving zinc alloy melting operations, and from contaminated zinc oxide from zinc oxide plants.

A typical Belgian retort (Figure 2) is about 8 in. (20 cm) in internal diameter and from 48 to 60 in. (120–150 cm) long. One end is closed. A conical clay condenser from 18 to 24 in. (45–60 cm) long is attached to the open end. The retorts are arranged in banks with rows four to seven high and as many retorts in a row as are needed to obtain the desired production. The retorts are generally gas fired today, although previously many units were coal fired.

The retorts are charged with a mixture of zinc scrap, zinc oxide, and powdered coke. If these materials are powdered, water is added to facilitate charging and to allow the mixture to be packed tightly into the retort. From three to four times more carbon is used than is needed for the reduction reaction.

After the charging, the condensers are replaced and their mouths stuffed with a porous material. A small hole is left through the stuffing to allow moisture and unwanted volatile materials to escape. About 3 h are needed to expel all the undesirable volatile materials from the retort. About 6 h after charging is completed, zinc vapors appear. The charge in the retort is brought up to 1832°F (1000°C) to 2012°F (1100°C) for about 8 h, after which it may rise slowly to a maximum of 2280°F (1250°C). The temperature on the outside of the retorts ranges from 2375°F (1300°C) to 2550°F (1400°C). The condensers are operated at from 780°F (415°C) to 1020°F (550°C), a temperature range above the melting point of zinc, but one at which the vapor pressure is so low that a minimum of zinc vapor is lost.

The reduction reaction of zinc oxide can be summarized by the following reaction.

$$ZnO + C \rightleftharpoons Zn + CO \tag{1}$$

Very little, if any, zinc oxide is actually reduced by the solid carbon in the retort. A series of reactions results in an atmosphere rich in carbon monoxide, which does the actual reducing. The reactions are reversible, but by the use of an excess of carbon, they are forced toward the right. The reactions probably get started by the oxidation of a small portion of the coke by the oxygen in the residual air in the retort. The oxygen is quickly used, but the carbon dioxide formed reacts with the

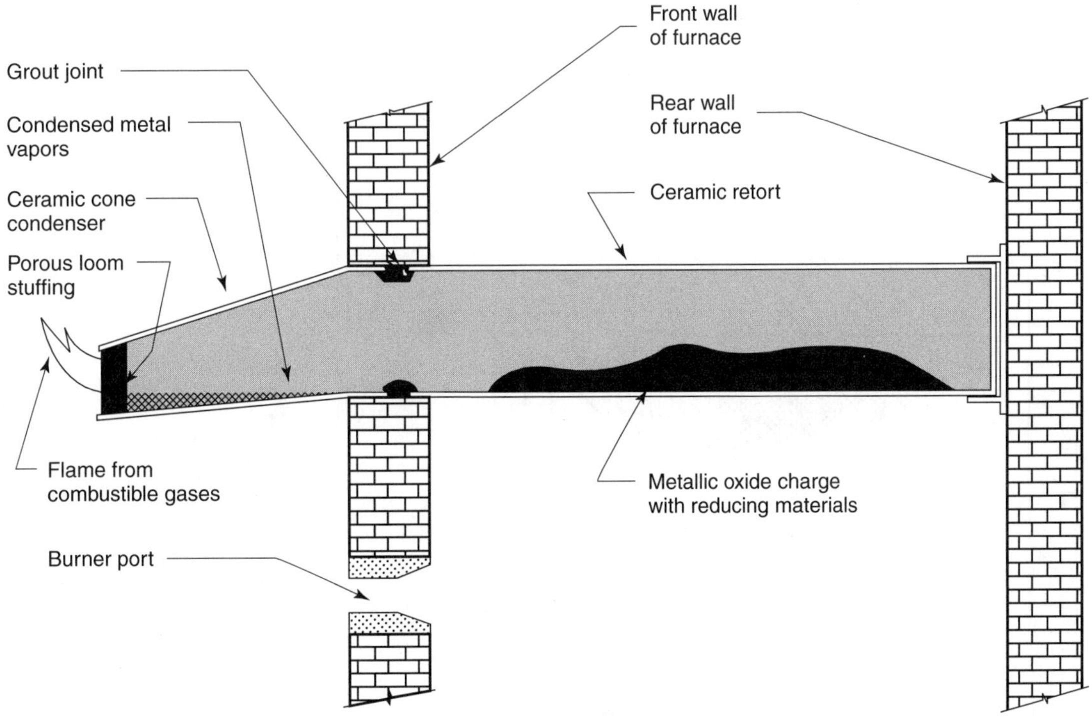

Figure 2. Diagram Showing One Bank of a Belgian Retort Furnace.

carbon to form carbon monoxide, according to the following equation.

$$CO_2 + C \rightleftharpoons 2CO \quad (2)$$

The carbon monoxide in turn reacts with zinc oxide to produce zinc and carbon dioxide:

$$CO + ZnO \rightleftharpoons Zn + CO_2 \quad (3)$$

Carbon monoxide is regenerated by use of Eq. (2), and the reduction of the zinc oxide proceeds.

About 8 h after the first zinc begins to be discharged, the heat needed to maintain production begins to increase and the amount of zinc produced begins to decrease. Although zinc can still be produced, the amount of heat absorbed by the reduction reaction decreases and the temperature of the retort and its contents increases. Care must be taken not to damage the retort or fuse its charge. As a result, a 24-h cycle is usually found to be an economical operation. The zinc values still in the spent charge are recovered by recycling with fresh charges. A single-pass recovery yields 65–70% of the zinc charged, but, by recycling, an overall recovery of 95% may be obtained.

Air Emissions Characterization

The air contaminants emitted vary in composition and concentration during the operating cycle of Belgian retorts. During charging operations, very low concentrations are emitted. The feed is moist and not dusty. As the retorts are heated, steam is emitted. After zinc begins to form, both carbon monoxide and zinc vapors are discharged. These emissions burn to form gaseous carbon dioxide and solid zinc oxide. During the heating cycle zinc is poured from the condensers about three times at 6–7-h intervals. The amount of zinc vapors discharged increases during the tapping operation. Before the spent charge is removed from the retorts, the temperature of the retorts is lowered, but zinc fumes and dust from the spent charge are discharged to the atmosphere.

Hooding and Ventilation Requirements

Air contaminants are discharged from each retort. In one installation reported in Los Angeles, a furnace has 240 retorts arranged in five horizontal rows with 48 retorts per row. The face of the furnace measures 70 ft (21.3 m) long by 8 ft (2.4 m) high; therefore, the air contaminants are discharged from 240 separate openings and over an area of 560 sq ft (15.7 sq m). A hood 2 ft (0.6 m) wide by 70 ft (21.3 m) long positioned immediately above the front of the furnace is used to collect the air contaminants. The hood indraft is 175 fpm (53 m/min).

DISTILLATION RETORT FURNACES

The distillation retort furnace (Figure 3) consists of a pear-shaped graphite retort, which may be 5 ft (1.5 m) long by 2 ft (0.6 m) in diameter at the closed end by 1.5 ft (0.45 m) in diameter at the open end and 3 ft (1 m) in diameter at its widest cross section. Normally, the retort is placed in a brick-lined furnace with only the open end of the retort protruding. It is heated externally with gas- or oil-fired burners. The retorts are charged with molten impure zinc through the open end. A condenser is attached to the opening to receive and condense

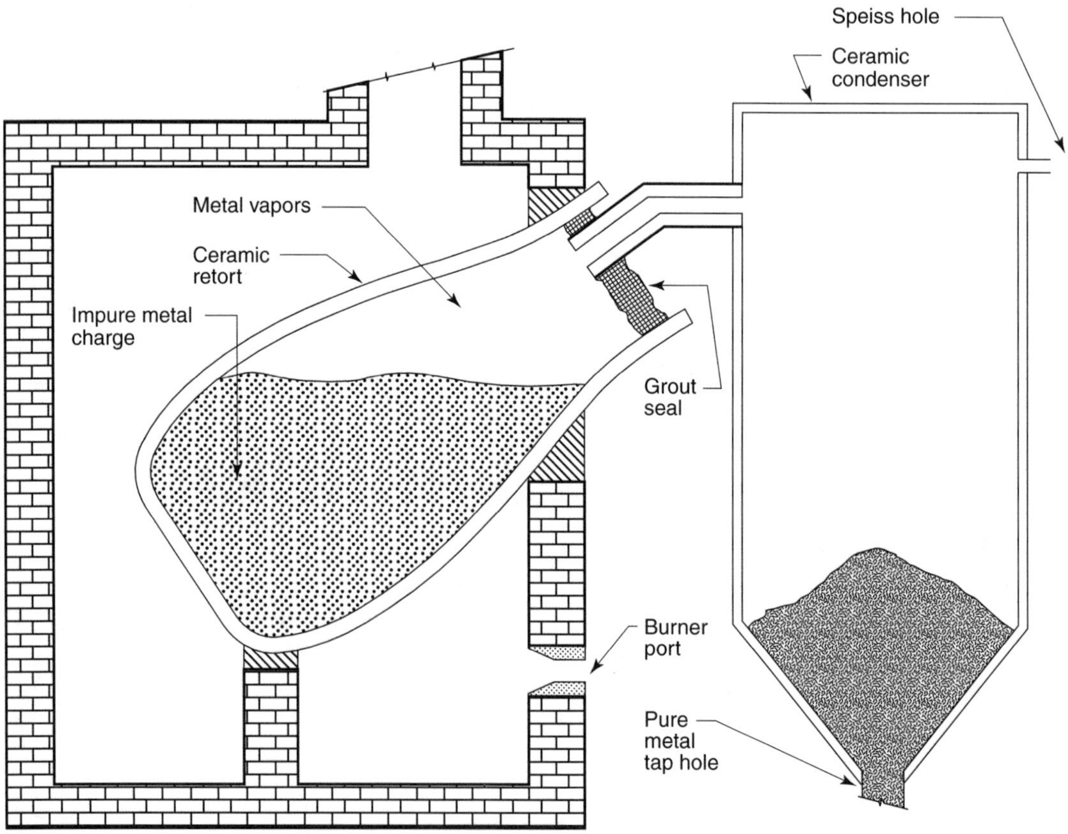

Figure 3. Diagram of a Distillation-Type Retort Furnace.

zinc vapors. After distillation is completed, the condenser is moved away, residue is removed from the retort, and a new batch is started.

The vaporized zinc is either conducted to a condenser or discharged through an orifice into a stream of air. Two types of condensers are used: a brick-lined steel condenser operated at from 780°F to 1010°F (415–545°C) to condense the vapor to liquid zinc, or a larger, unlined steel condenser that cools the vapor to solid zinc. The latter condenser is used to manufacture powered zinc. The condensers must be operated at a slight positive pressure to keep air from entering and oxidizing the zinc. To ensure that there is a positive pressure, a small hole, called a "speiss" hole, is provided through which a small amount of zinc vapor is allowed to escape continuously into the atmosphere. The vapor burns with a bright flame, indicating that there is a pressure in the condenser. If the flame gets too large, the pressure is too high. If it goes out, the pressure is too low. In either case, the proper adjustments are made to obtain the desired condenser pressure.

When it is desired to make zinc oxide, the vapor from a retort is discharged through an orifice into a stream of air where zinc oxide is formed by burning inside a refractory-lined chamber. The combustion gases and air carry the oxide particles to a baghouse collector where the powdered oxide is collected.

Air Emissions Characterization

During the 24-h cycle of the distillation retorts, zinc vapors escape from the retort

1. When the residue from the preceding batch is removed from the retort and a new batch is charged; and
2. When the second charge is added to the retort. As the zinc vapors mix with air, they oxidize to form a dense cloud of fumes. Air contaminants are discharged for about 1 h each time the charging hole is open. When the zinc is actually being distilled, no fumes escape from the retort; however, a small amount of zinc oxide escapes from the "speiss" hole in the condenser. Although the emission rate is low, air contaminants are discharged for about 20 h/d.

Hooding and Ventilation Requirements

To capture the emissions from a distillation retort furnace, simple canopy hoods placed close to and directly over the sources of emissions are sufficient.

The retorts are fuel fired. The products of combustion do not mix with the emissions from the retort or the condenser. The exhausted gases are heated slightly by the combustion of zinc and from radiation and convection losses from the retort, but the amount of heating is so low that no cooling is necessary.

MUFFLE FURNACES

Muffle furnaces (Figure 4) are continuously fed retort furnaces. They have a much greater vaporizing capacity than Belgian retorts or bottle retorts. They are operated continuously for many months at a time. Heat for vaporization is supplied by gas- or oil-fired burners by conduction and radiation through a silicon carbide arch that separates the zinc vapors and the products of combustion. Molten zinc from either a melting pot or a sweat furnace is charged through a feed well that also acts as an air lock. The zinc vapors are conducted to a condenser, where purified liquid zinc is collected, or the condenser is bypassed and the vapors are discharged through an orifice into a stream of air, where zinc oxide is formed.

The furnace, including the feed well and sweating chamber, is heated indirectly with a combination gas- or oil-fired burner. The combustion chamber, located directly over the vaporizing chamber, is heated to about 2400°F (1315°C). On leaving the combustion chamber, the products of combustion are conducted through the sweating chamber to supply the heat needed for melting the zinc alloys from the scrap charged and for heating the zinc prior to discharge into the feed well at 900°F (480°C).

Zinc vapors can be conducted from the vaporizing section into a multiple-chamber condenser. When zinc oxide is the desired product, condensers are not used. The vapors are allowed to escape through an orifice into a burning chamber.

Air Emissions Characterization

Dust and fumes are created by the sweating operation. Scrap is charged into the sweating chamber through a door. After

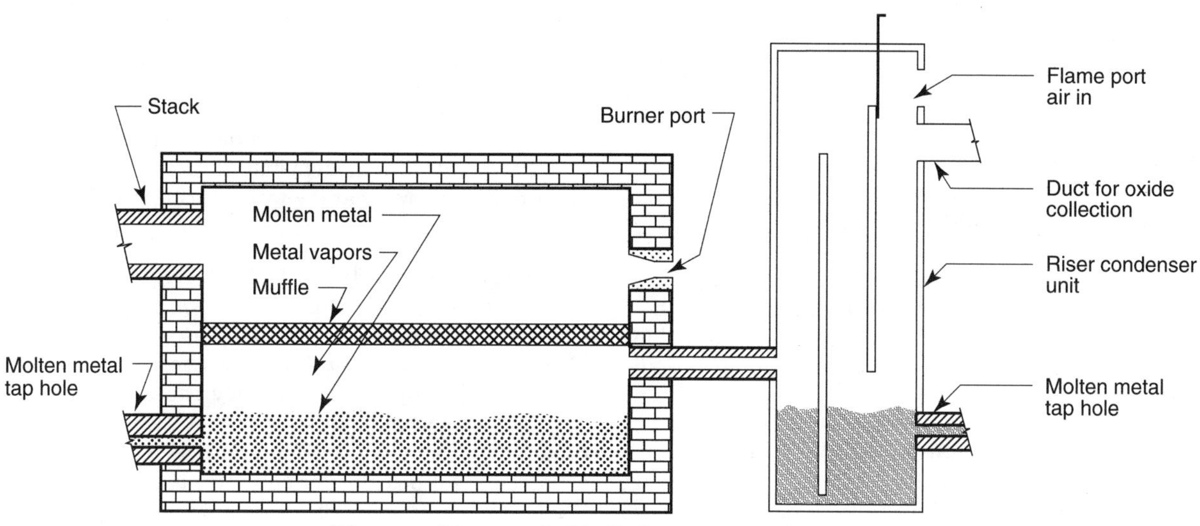

Figure 4. Diagram of a Muffle Furnace and Condenser.

the zinc alloys have been melted, the residue is pushed or pulled out of the chamber into refractory-lined pots. The pots are then taken to a dross stirrer to separate metal from oxide. Dust and fumes are created at the rake-out point and at the dross stirrer, requiring air pollution controls at both locations.

The zinc alloys charged into the vaporizing section contain copper, aluminum, iron, lead, and other impurities. As zinc is distilled from the metals, the concentration of the impurities increases until continued distillation becomes impractical. After periods of operation, the residue, containing 10–50% zinc, must be removed. When tapped, the temperature of the residue is about 1900°F (1055°C), hot enough to release zinc oxide fumes when the residue metal is poured. The molds collecting the residue metal are often arranged so that the metal overflows from one mold to another; however, the metal cools so rapidly that fumes are released only from the pouring spout and the first two or three molds. The fumes, almost entirely zinc oxide, are 100% opaque from the pouring spout and the first mold. At the third mold, the opacity decreases to 10%.

Any discharge of the zinc vapor from the condenser forms zinc oxide of product purity; therefore, the condenser vents into the intake hood of a product-collecting exhaust system. Since some zinc oxide is always produced, even when the condenser is set to produce a maximum of liquid zinc, the product collecting exhaust system is always in operation to prevent air contaminants from escaping from the condenser to the atmosphere.

Hooding and Ventilation Requirements

The dust and fumes created by charging scrap and sweating zinc alloys from the scrap originate inside the sweat chamber. Thermal drafts cause emissions to escape from the upper portion of the sweat chamber doors. Hoods are placed over the doors to collect the emissions. The charging door hood extends to the furnace wall and covers a little more than the width of the door. When possible, hood sides are installed to the plant floor.

A hood enclosing the rake-out door can be controlled with an inlet velocity of 250 fpm (76 m/min), which is sufficient to capture all emissions escaping from both the furnace and the screen.

The ductwork joining the hoods to the control devices is manifolded and dampered so that any or all hoods can be opened or closed. Exhaust systems can provide sufficient ventilation to control the fumes created by multiple furnaces in operation at the same time. Factors for uncontrolled point source and fugitive particulate emissions are listed in Tables 1 and 2, respectively.

Table 1. Uncontrolled Particulate Emission Factors for Secondary Zinc Smelting[a] (Emission Factor Rating: C)

Operation	Emissions kg/Mg	Emissions lb/ton
Reverberatory sweating		
Clean metallic scrap	Negligible	Negligible
General metallic scrap	6.5	13
Residual scrap	16	32
Rotary sweating	5.5–12.5	11–25
Muffle sweating	5.4–16	10.8–32
Kettle sweating		
Clean metallic scrap	Negligible	Negligible
General metallic scrap	5.5	11
Residual scrap	12.5	25
Electric resistance sweating	<5	<10
Crushing/screening	0.5–3.8	1.0–7.5
Sodium carbonate leaching		
Crushing/screening	0.5–3.8	1.0–7.5
Calcining	44.5	89
Kettle (pot) melting	0.05	0.1
Crucible melting	DNA	DNA
Reverberatory melting	DNA	DNA
Electric induction melting	DNA	DNA
Alloying	DNA	DNA
Retort and muffle distillation		
Pouring	0.2–0.4	0.4–0.8
Casting	0.1–0.2	0.2–0.4
Muffle distillation	22.5	45
Graphite rod distillation	Negligible	Negligible
Retort distillation/oxidation[b]	10–20	20–40
Muffle distillation/oxidation[b]	10–20	20–40
Retort reduction	23.5	47
Galvanizing	2.5	5

[a]Expressed as units per unit weight of feed material processed for crushing/screening, skimming/residues processed; for kettle (pot) melting and retort and muffle distillation operations, metal product. Galvanizing factor expressed in units per unit weight of zinc used. DNA: Data not available.
[b]Factor units per unit weight of zinc oxide produced. The product zinc oxide dust is totally carried over in the exhaust gas from the furnace and is recovered with 98–99% efficiency.

Source: Ref. 1.

AIR POLLUTION CONTROL MEASURES

Pollution control can be achieved on reduction retort furnaces, distillation retort furnaces, and muffle furnaces with a baghouse. In some installations on muffle furnaces, a low-efficiency cyclone and a baghouse are used to control the emissions from the sweating chambers and residue pouring operations. Although the cyclone has a low collection efficiency with the particle size involved, it does collect from 5% to 10% of the dust load and provides cooling. Often a cyclone was in existence before the baghouse was installed. Recent systems do not have cyclones, only baghouses. Filtration velocity is usually about 1.5 fpm (0.5 m/min), although some older systems are operating successfully at 3 fpm (1 m/min).

Dust collectors for other zinc-melting and zinc-vaporizing furnaces are very similar to the ones already described. Glass bags have been found adequate when gas temperatures exceed the limits of synthetics, although they are seldom used today. Filtering velocities of 3 fpm (1 m/min) have been employed and found adequate, but modem installations realize better bag life and product recovery at a lower filtering velocity.

AIR POLLUTION CONTROL EQUIPMENT

Generally, zinc processing furnaces are controlled at the point of fume generation by hoods that encompass the source and act as the pickup point to convey the zinc oxide fumes into baghouses. Typical baghouses are multicompartment, shaker-type units operating at temperatures below 275°F (135°C), thus making it possible for polyester bags to be utilized rather than more expensive fabrics. A few applications use reverse-jet baghouses. The volume of zinc oxide collected from muffle

Table 2. Fugitive Particulate Uncontrolled Emission Factors for Secondary Zinc Smelting[a] (Emission Factor Rating: E)

Operation	Particulate kg/Mg	Particulate lb/ton
Reverberatory sweating[b]	0.63	1.30
Rotary sweating[b]	0.45	0.90
Muffle sweating[b]	0.54	1.07
Kettle (pot) sweating[b]	0.28	0.56
Electric resistance sweating[b]	0.25	0.50
Crushing/screening[c]	2.13	4.25
Sodium carbonate leaching	DNA	DNA
Kettle (pot) melting furnace[b]	0.0025	0.005
Crucible melting furnace[d]	0.0025	0.005
Reverberatory melting furnace[b]	0.0025	0.005
Electric induction melting[b]	0.0025	0.005
Alloying retort distillation	DNA	DNA
Retort and muffle distillation	1.18	2.36
Casting[b]	0.0075	0.015
Graphite rod distillation	DNA	DNA
Retort distillation/oxidation	DNA	DNA
Muffle distillation/oxidation	DNA	DNA
Retort reduction	DNA	DNA

[a]Expressed as units per end product, except factors for crushing/screening and electric resistance furnaces, which are expressed as units per unit of scrap processed. DNA: Data not available.

[b]Estimate based on stack emission factor given in Ref. 1, assuming fugitive emissions to be equal to 5% of stack emissions.

[c]Ref. 1. Average of reported emission factors.

[d]Engineering judgment, assuming fugitive emissions from crucible melting furnace to be equal to fugitive emissions from kettle (pot) melting furnace.

Source: Ref. 1.

furnaces will range to more than 3000 pounds (1360 kg) an hour. Retorts operate at a much lower rate. The baghouses are typically set up to shut down compartment by compartment for cleaning on a cyclical basis. In some plants, pulsed-air cleaning dust collectors using either bags or cartridges are used. The product generated in this process is usually withdrawn continuously via a screw conveyor to storage bins, where further processing and/or packaging is accomplished.

REFERENCES

1. *Compilation of Air Pollutant Emission Factors*, Volume 1: Stationary Point and Area Sources, AP-42, U.S. EPA, 1985.

BIBLIOGRAPHY

1. *Air Pollution Engineering Manual*, AP-40; J. A. Danielson, Ed.; U.S. Environmental Protection Agency, Research Triangle Park, NC, 1973, pp. 293–299.
2. M. Bess, "Recycling of Zinc," *Metals Handbook*, Vol. 2, 10th ed., ASM International, 1990, pp. 1223–1224.
3. H. E. Brown, *Zinc Chemicals: Applications, International Lead Zinc Research Organization*, Research Triangle Park, NC, June 1990.
4. B. C. Hafford, W. E. Pepper, and T. B. Lloyd, *Zinc Dust and Zinc Powder: Their Production, Properties, and Applications*, International Lead Zinc Research Organization, Research Triangle Park, NC, 1982.
5. B. Baum and D. S. Carr, *Zinc Oxide: A Weathering Stabilizer for Plastics*, International Lead Zinc Research Organization, and Zinc Institute, New York, rev. ed. June 1984.
6. H. E. Brown, *Zinc Oxide, Properties and Applications*, International Lead Zinc Research Organization, New York, 1976, rev. ed. 1986.
7. *Zinc Oxide and Rubber*, Zinc Institute, Inc., New York, 1984.

STEEL FOUNDRIES

Souad A. Benromdhane and Mackenzie L. Davis

PROCESS DESCRIPTION

Although each foundry accomplishes the same basic tasks (melting, mold and core making, and casting), specific materials, equipment, and processes vary. This section describes the steel foundry processes in their general aspects. It is important to realize that one method of melting steel or a particular molding method used by one manufacturer may or may not be duplicated by another manufacturer.

The major production steps are shown in Figure 1: (1) raw materials handling and preparation; (2) mold and core production; (3) metal melting; (4) back-charging (addition of more metal and possibly alloys); (5) refining by single (oxidizing) slagging or double (oxidizing and reducing) slagging operations; (6) oxygen lancing to adjust the chemistry of the metal and speed up the melt; (7) tapping the molten metal into ladle or directly into molds; and (8) finishing (which includes annealing, grinding, and shot blasting).[1]

Raw Materials Handling, Storage, and Preparation

Handling operations include receiving, unloading, storing, and conveying raw materials for both furnace charging and mold and core preparation. Raw materials are received in railroad cars, trucks, and containers, then transferred by truck, loaders and conveyors to open piles. When needed, the raw materials are transferred from storage to process areas by similar means.

Metallics and fluxes are the major materials required for furnace charging. Metallic raw materials include steel scrap and foundry returns. Fluxes include carbonates (limestone, dolomite), fluoride (fluospar), and carbide compounds (calcium carbide), which are added to the steel to improve the quality of the cast.

The foundry sets specifications for scrap material in accordance with their part-order specifications. The steel designated as plain steel generally contains less than 0.8% C, minimum amounts of P and S, and suitably adjusted amounts of Si and Mn. Other metals may be added in various percentages to make low- and high-alloy steels. In all steel-making processes melting temperatures range between, 1550 and 1650°C.[2]

In addition to the raw materials used to produce the molten metal, a variety of materials are needed to prepare the sand cores and molds that form the steel castings. Lake sand, recycled sand and chemical additives are combined in a sand-handling system that consists of receiving areas, conveyors, storage silos, mixers (sand mullers), core- and mold-making machines, shakeout grates, sand cleaners, and sand screeners.

Used sand from castings shakeout is recycled to the sand preparation area and cleaned to remove any clay or carbonaceous buildup. The sand is then screened and reused to make new molds. Even though sand can be reclaimed to the quality required for molding, current technology may not permit reclaiming it to the quality required for core

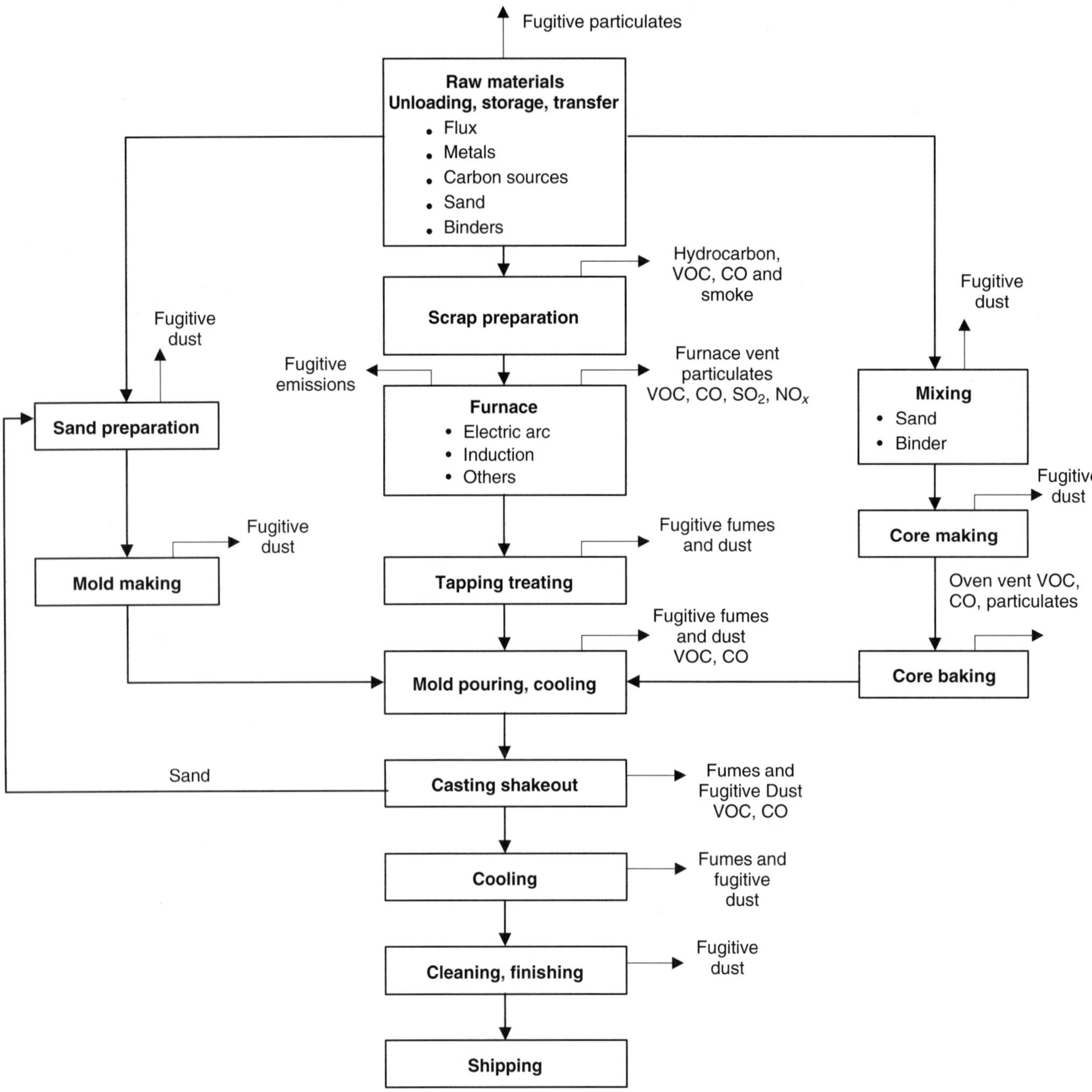

Figure 1. Flow Diagram of a Typical Steel Foundry.

production. Makeup sand is added to replace process losses and contaminated sand that must be discarded.

Steel Melting

The three most common furnaces used for steel melting are open hearth, electric arc (direct and indirect), and electric induction furnaces (EIFs). Among these, the indirect arc and the induction furnaces account for the most tonnage being produced. The lowest tonnage producer is the direct arc furnace. All three types of furnaces may be operated in either acid or basic mode.

The open-hearth furnace is used for large tonnage, and the electric arc furnace is used for small lot production with widely differing analyses. Special compositions of expensive steel are produced in smaller quantities in a high-frequency induction furnace.[3]

The indirect arc furnace (Figure 2) is known for its high-quality melting capabilities. It is highly mechanized, and allows efficient supervision and control. The furnace is rocked at predetermined intervals to allow more exposure of the lining area. This delivers more homogeneous melt and increases the life of the lining. The furnace is equipped with one stationary and one movable electrode. Automatic control of the electrode permits the input of a fixed amount of power for a known alloy.

The direct arc furnace (Figure 3) has three vertical electrodes (graphite or amorphous carbon), and is enclosed in a refractory-lined cylinder that may be tipped forward and backward. The furnace roof is built into a separate steel ring and rests on the side walls of the furnace shell.

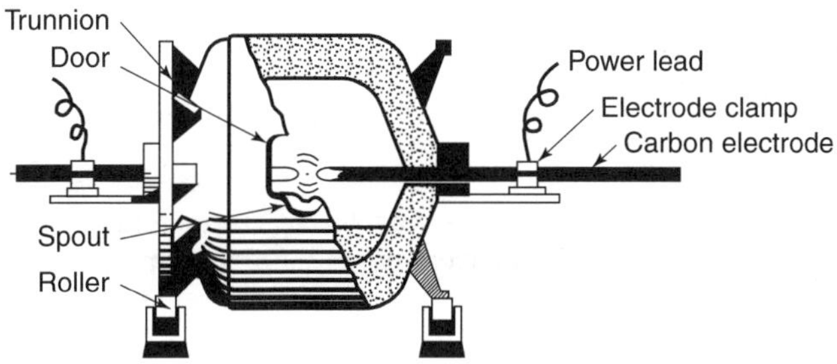

Figure 2. Indirect Arc Furnace "Rocking Arc".

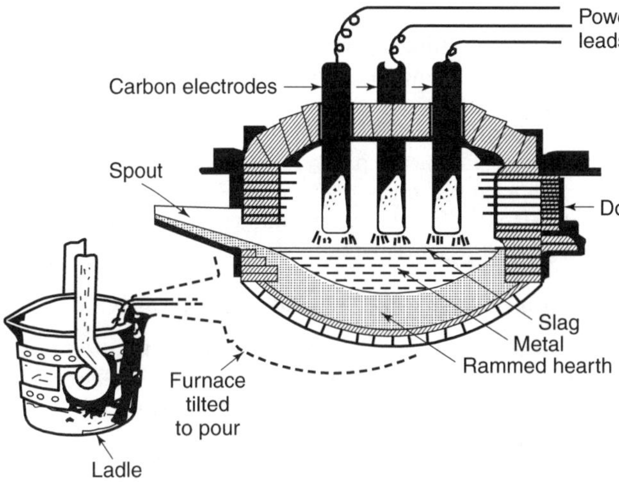

Figure 3. The Direct Arc Furnace (Three Phase).

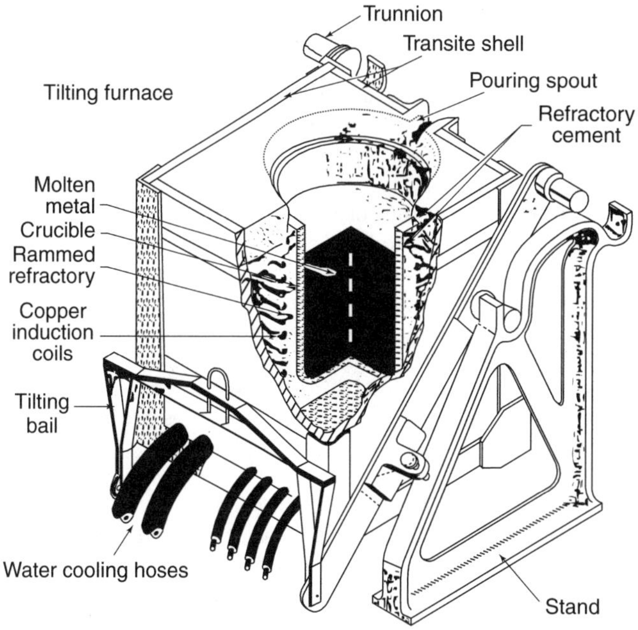

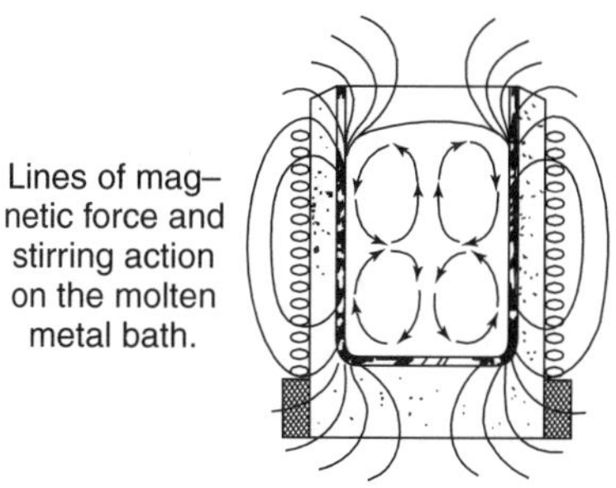

Figure 4. Coreless Induction Furnace.

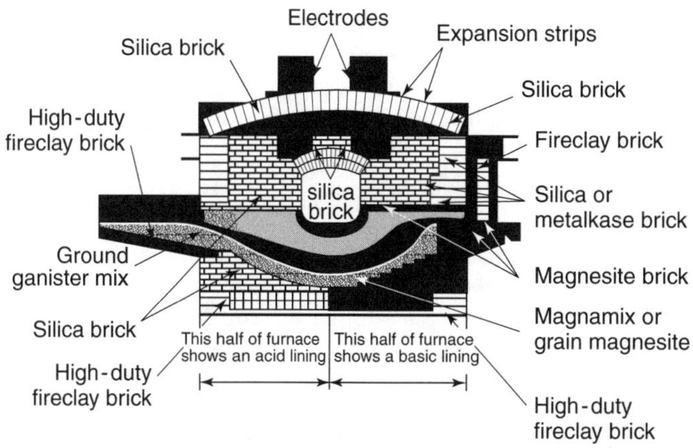

Figure 5. Section of Electric-Arc-Furnace Showing Refractories for Either Acid or Basic Operation (*Modern Refractory Practice*, Harbison–Walker Refractories Co.).

The electrical induction furnaces are cylindrical-shaped refractory-lined vessels surrounded by electrical coils. There are two kinds of EIFs: coreless and channel. A cross section of a coreless EIF is shown in Figure 4. When energized with a high-frequency alternating current, a fluctuating electromagnetic field is produced that heats the metal charge in approximately 2 min. For safety reasons, the scrap metal added to the furnace charge is cleaned and heated before being introduced into the furnace. Oil or moisture on the scrap metal may cause an explosion in the furnace. EIFs are kept closed except when charging or tapping. A permanent opening in the top of the furnace allows oxygen to enter, and is a continuous source of emissions.[3,4]

Foundries that have access to good scrap, and therefore do not need to reduce the phosphorus and sulfur content of the steel to meet the specifications, usually prefer to use furnaces lined with silica refractories (acid lined). The acid practice leads to a siliceous slag covering the steel. Neither phosphorus nor sulfur is eliminated from the steel during steel-making operations; hence, the process is limited to the melting of raw materials low in these two elements. Other foundries use basic-type refractories (magnesite, dolomite, etc.) for the furnace lining. The basic slags, high in lime content, facilitate the removal of phosphorus and sulfur from the steel. Figure 5 shows the differences in refractories for basic or acid practice.[2,5]

For the production of steel from a pure metallic charge, steel scrap with low sulfur and phosphorus content is used with small additions of pure pig iron for the purpose of increasing the carbon content. High-quality carbon steel can be produced if the materials used are sufficiently pure. After fusion is effected, molten steel is held in the furnace to remove oxides and obtain a metal that does not froth. At the correct temperature, steel is poured into molds for production of the steel casting.

Mold and Core Production

Molds are forms used to shape the exterior of the castings. They are prepared using a mixture of wet sand, clay, and organic additives, and are usually dried with hot air. Green-sand mold, the most common type, uses moist sand mixed with 4–6% clay (bentonite) for bonding. The mixture has a 4–5% water content. About 5% sea coal (a pulverized high-volatility bituminous coal) is added to the mixture to prevent casting defects from sand expansion when the hot metal is poured. Potential organic substitutes for sea coal include wood flour, oat hulls, pitch, or similar organic matter.

Cores are molded sand shapes used to make the internal voids in castings. They are made by mixing sand (mulling) with organic binders and resins and molding the sand into a core. Unlike the green sand molding process, cores are chemically bonded by resins.[6] Resins are made to withstand heat so that the sand particles are held tightly together until the iron is shaped. The resin then decomposes and gases are released.

Two core-making processes are used:

Hot-box core: The sand binder is typically 3–5% furan resin with a phosphoric acid activator. It is cured as a solid core in a heated metal pattern at 205–315°C for 0.5–1.5 min. Typically, for furan binders, SO_2 gas is injected as a catalyst to induce rapid cure,[7,8] and

Cold-box core: The sand binder is typically 1–3% of each of two resins, activated by a nitrogen-diluted gas. It hardens after curing for 10–30 sec. when the green core is gassed with polyisocyanate. For phenolic isocyanate binder, triethylamine gas is injected as a catalyst.[7,8]

A computer that ensures the mixing of the proper ratio of binders and resins controls each of these processes. Several binders and resins are used due to variations in humidity, the size and number of cores or molds required production rates, and equipment.

The organic binders undergo thermal decomposition when exposed to the very high temperatures of castings. At these temperatures it is likely that pyrolysis of the chemical binder will produce a complex mixture of free radicals, which will recombine to form a wide range of chemical compounds that have widely differing concentrations.

Modern techniques include molds that do not use sand.

Pouring, Cooling, and Shakeout

Traditional casting involves pouring the molten metal into the molds. Alternative casting methods involve drawing up of the metal into the mold (countergravity).

Several methods are available for countergravity casting into nonmetal permeable molds, using ceramic investment molds and low-temperature bonded sand molds. These are widely used and are known by acronyms such as CLA (countergravity low-pressure casting of air-melted alloys), CLV (countergravity low-pressure casting of vacuum-melted alloys), CV (check valve casting), and CLAS (countergravity low-pressure air-melted sand casting). Other methods include squeeze casting and directional monocrystal solidification.[9]

The cooling molds are mechanically conveyed (partially covered) to an enclosed room to circulate on an open conveyor until the casting solidifies in the mold.

Shakeout is the process of separating the casting from the sand molds and cores. After the castings have solidified, they are separated from the cores and molds by means of a vibrating grid. The hot metal pieces (with small amounts of sand still attached) are then cooled under ambient conditions. The sand, which can be recycled, is returned to the mold-making station by elevator. Waste sand is transferred outside to open storage piles for disposal.

Finishing

To match the contour of the casting, burrs, risers, and gates are broken or ground off in the cleaning and finishing process. The casting can be then shot-blasted to remove remaining mold sand and scale.

STEEL FOUNDRY EMISSIONS

No data on criteria pollutant emissions have been reported. Lead emissions are linked primarily to contamination in scrap used in the EAF. However, no production rates from steel manufacturing were available for lead or SO_2 emissions. U.S. EPA reported emissions of VOCs, NO_x, as well as particulate matter based on furnace capacity.[1]

No specific data on emission rates of hazardous air pollutants (HAPs) for the steel foundry are available. Very limited data on PM_{10} are available (see Table 1).[1]

Emissions from Handling and Preparation of Raw Materials

Emissions generated during handling and preparation of the raw materials include fugitive particulate matter generated from the receiving, unloading, storage, and conveying of raw materials.

Emissions from Melting Furnaces

Steel manufacturing generates one of largest hazardous waste streams nationwide. Electric arc furnace (EAF) dust is a listed hazardous waste by the EPA because of the leachable Pb, Cd, and Cr and is designated as K061 under the Resource Conservation and Recovery Act (RCRA).

Emissions from the furnace are generated in nearly all phases of its operation. These are subdivided in two categories. Primary emissions are defined as those occurring during the melting and refining cycles of the furnace operation. Secondary emissions are defined as those occurring during activities other than melting or refining, specifically during periods of charging, tapping, and slagging. The latter account for about 10% of the total emissions by weight.

The major contaminants evolved during melting are particulate matter and carbon monoxide. Depending on the scrap chemistry and the furnace additives, trace contaminants such as hydrocarbons, sulfur oxides, and fluorides may also be present.

Table 1. PM_{10} Emission Factors for Steel Foundries[1]

Process	PM_{10} in kg/Mg of Metal Processed
Melting electric induction furnace	0.045
Sand grinding/handling	0.27
Mold and core making	3.0
Core ovens	0.45–1.11
Pouring and casting	1.4
Casting cleaning	0.85
Charge handling	0.18
Casting cooling	0.7

The injection of oxygen during refining periods gives rise to the greatest concentration of particulate matter, primarily in the form of metallic oxides. These are due to oxidation and volatilization in the presence of intense heat and turbulence.

Electric furnace dust evolution has been reported in the range of 4.5–22.5 kg/mt (10–50 lb/ton) of steel produced.[9] Dust concentrations in the exhaust gas often exceed 23,000 mg/m^3 (10 gr/scf).[9] The ranges of chemical composition of electric furnace dusts are shown in Table 2. The wide variations are attributed to the type of steel produced.

Emissions from Core and Mold Operations

On a mass basis, the major pollutant emitted in mold and core production operations is particulate matter. Emissions of volatile organic compounds (VOCs) (i.e., acrolein, benzene, toluene, and xylene) begin with their initial exposure to the atmosphere. This occurs from open containers of binder material, the addition of binders to sand, and the mixing of sand and binders. Emissions continue during mold preparation, storage, pouring, and cooling. Organic compounds, carbon monoxide and particulates are emitted from core baking, and organic emissions result from mold drying.[1]

Emissions from Pouring Cooling and Shakeout Operations

Emissions generated from pouring consist of hot metal fumes, carbon monoxide, organic compounds and particulates released from the mold and core materials contacting the molten metal. CO emissions peak during pouring and shakeout (the highest concentration has been measured during pouring). The highest concentrations of hydrocarbons are emitted during shakeout, but also peak during the pouring. Emissions continue as the molds cool.[1]

During pouring, 95% of the particulate matter emissions are less than 5 μm in diameter and during shakeout 50% of PM emissions are less than 5 μm in diameter.[10] Organic compounds contained in the mold and the core materials are expected to sorb to the particulate matter.

Particulate matter emissions from green sand molds peak during shakeout. These emissions typically contain carbonaceous material from burning organic matter in the sand, silica fines, and clay from the molding aggregate, metallic fumes, and other fines present in the molds. During shakeout, a peak of organic emissions is observed, when hot sand near the metal and the metal itself contact cooler sand containing residual organics. Any VOCs in the sand will continue to volatilize during the sand reclaiming process.

Table 2. Composition of Electric Furnace Dusts from Steel-Making[9]

Component	Range of Reported Values (wt %)
Fe_2O_3	19–65
FeO	4–11
Fe	5–36
SiO_2	1–9
Al_2O_3	1–13
CaO	2–22
MgO	2–15
MnO	1–12
Cr_2O_3	0–12
NiO	0–3
PbO	0–4
ZnO	0–44
P	0–1
S	0–1
C	0–4
Alkalis	1–11

AIR POLLUTION CONTROL MEASURES

Raw material handling emissions can be controlled using filters. Scrap preparation using degreasing and heating, on the other hand, requires catalytic incinerators and afterburners that ensure approximately 95% control efficiency for CO and organics removal. This also applies for the core- and mold-making process. The major concern is with the furnace emissions at the melting stage. Emissions from pouring are usually uncontrolled. Cyclones are used to control particulates at the finishing step.[1,9]

Methods of Fume Capture and Control of Primary Emissions from Furnaces

Controls for fugitive furnace emissions include updraft ventilation to bag filters, cyclones, and wet scrubbers.

Various methods of fume capture and all conventional types of gas cleaning equipment have been employed to capture emissions near the furnace doors and taps and route them to pollution control equipment over the years. Primary emissions associated with melting and refining have been the thrust of air pollution control. Three methods of fume capture are available: (1) direct furnace evacuation, (2) roof-mounted hoods, and (3) canopy hoods.[9]

Direct shell evacuation employs an induced draft, which withdraws emissions from the furnace through a special hole in the furnace roof. Consequently, it is often referred to as a fourth-hole system. Generally, a water-cooled duct system, suitably flanged to permit roof swing and furnace tilting, is used to transport the gases to a point at which further cooling is accomplished by evaporative cooling, dilution air cooling, or radiant-convective gas cooling. Air is induced at gaps in the ductwork to assure complete combustion of the carbon monoxide.[9]

Direct shell evacuation has the advantages of: (1) better control of furnace evacuation rates, (2) least system volume, (3) most effective CO control, and (4) best control of particulate emissions during melting and refining.[9]

The most common practice of the roof-mounted hoods is the side draft or lateral exhaust hood. These exhaust arrangements consist of a hood on the furnace roof that draws fumes laterally as they escape the electrode ports of the furnace. They have the advantages of (1) no effects on furnace conditions, (2) the absence of special cooling equipment and furnace pressure controls, and (3) general simplicity of system design.[9]

Canopy hoods consist of large hoods located well above the furnace roof in the melt shop building roof trusses. Fumes are permitted to escape from the furnace uncontrolled and are withdrawn through the canopy hood to a fabric filter system. These have the advantages of: (1) complete freedom of interference with furnace operations, (2) simplicity of system design, (3) the absence of special cooling equipment, and (4) at least partial control of charging and tapping emissions.[9]

REFERENCES

1. AP-42, Background Report, Section 12.13, "Steel Foundries." U.S. Environmental Protection Agency, Pacific Environmental Services, Inc., P.O. Box 12077, Research Triangle Park, NC.
2. McGraw Hill, *Encyclopedia of Engineering*, S. P. Parker, Ed., 1982, pp. 1035–43.
3. Steel Founders Society of America, *Steel Casting Handbook*, Third Edition, Charles W. Briggs, The Electric Printing Company, Cleveland, Ohio, 1960.
4. Alfred Stansfield, *The Electric Furnace for Iron and Steel*, McGraw-Hill Book Company, Inc., New York, 370 Seventh Avenue, p. 453, First Edition, 1923.
5. ASM International, *Metals Handbook*, Ninth Edition, *Casting*, Vol. 15, Ohio, 1988.
6. P. R. Carey, "Today Resin Binder Processes for Coremaking," *Foundry Management & Technology*, 32–39 (1990).
7. P. R. Carey, "Updating Resin Binder Processes—Part VI," *Foundry Management & Technology*, July 1986, 84–88.
8. P. R. Carey, "Updating Resin Binder Processes—Part VII," *Foundry Management & Technology*, September 1986, 90–98.
9. Iron and Steel Society, Inc., *Electric Furnace Steel-Making*, Book Crafters, Inc., Chelsea, MI, 1985.
10. W. D. Scott, and C. E. Bates, *Measurement of Iron Foundry Fugitive Emissions*, Symposium on Fugitive Emissions Measurement and Control. Hartford, CT, U.S. EPA, 1976.

BIBLIOGRAPHY

1. American Foundrymen's Society, "Symposium on Interpretation and Application of Sand Test Data for the Production of Quality Steel Castings," Chicago, 1951.
2. O. S. Fishman, "Technological Advances in Induction Melting of Iron and Steel," Electric Furnace Conference Proceeding, pp. 149–59, 1997.
3. M. Hans, and D. Dart, "Environmental Self-Audits and Enforcement in Steel Minimills, An EPA Initiative," Electric Furnace Conference Proceeding, Vol. 55, pp. 557–60, Chicago, 1997.
4. L. J. Heaslip, and J. M. Svoboda, "Controlling the Behavior of Basic Electric Arc Furnace with Oxygen Injection in the Foundry," Electric Furnace Conference Proceeding, Vol. 55, pp. 51–55, Chicago, 1997.
5. D. R. Meadgen, and H. D. Hunt, "An Update on Project S.T.A.R.: Zero Waste Program at Chaparral Steel," Electric Furnace Conference Proceedings, Vol. 55, pp. 561–66, Chicago, 1997.
6. C. A. Sanders, *Foundry Sand Practice*, Sixth Edition, American Colloid Company, Skokie, IL, 1972.
7. R. J. Schmitt, C. L. Kusik, R. O'Neil, P. M. Cowx, and C. T. Schade, "Minimizing Electric Arc Furnace Dust Generation," Electric Furnace Conference Proceedings, Vol. 55, pp. 567–81, Chicago, 1997.
8. United States Steel Corporation, *The Making, Shaping and Treatment of Steel* Herbick & Held, Pittsburgh, Pennsylvania, 1971.
9. Z. Q. Zhang, "Cleaner Cast Steels and Refining Process," Electric Furnace Conference Proceedings, Vol. 55, pp. 85–91, Chicago, 1997.

15

MINERAL PRODUCTS INDUSTRY

HOT MIX ASPHALT MIXING FACILITIES

Mark J. Miller, George P. Partridge, Jr., and R. Scott Huebner

The top course of roughly 93% of the two million miles of paved roads in the United States is made of asphalt.[1] In 1996, approximately 500 million tons of hot-mix asphalt were produced at the 3,600 (estimated) asphalt plants in the United States.[2] Hot-mix asphalt paving materials are a combination of 2–6 aggregates, often including reclaimed asphalt pavement, coated with hot liquid asphalt cement (with or without additives). The main purpose of a hot-mix asphalt facility is to combine the correct proportions of aggregate, remove the free moisture from the aggregate, heat the aggregate, and coat it with the liquid asphalt cement (bitumen).

A hot-mix asphalt facility can be a permanent plant, a skid-mounted plant (for easy relocation), or a portable plant.

PROCESS DESCRIPTION

There are four types of hot mix asphalt facilities: batch plants, parallel-flow drum plants, counter-flow drum plants, and continuous-mix plants. Very few continuous-mix plants are still in service. No new continuous-mix plants have been built in years; therefore, they are not discussed further.

While there are significant differences between batch plants, parallel-flow drum plants, and counter-flow drum plants, the basic process is the same. Figure 1 is a process flow diagram. The feeding system is loaded with aggregates from stockpiles with various properties, and these aggregates are mixed in accordance with product specifications. The aggregate mix goes through moisture removal and is then heated to approximately 135–170°C (275–340°F).[3] Fine particles (dust waste) from the aggregate are removed through the exhaust system and may be collected. The aggregate is then coated with the hot (usually 300°F or 150°C) asphalt cement.[4] The mixture is temporarily stored in a hot storage silo until dispatched for use.

Batch Mixing

The process of producing hot-mix asphalt in a batch plant starts with loading aggregate from storage piles into cold bins. The quantity of the aggregate is measured as it exits the bins onto the cold conveyor belt that then dumps the aggregate into the top of the dryer. The rotary dryer is set on a slight incline with the burner at the lower end. The burner provides the heat energy to evaporate the moisture from the aggregate and to heat the aggregate to about 20–25°C degrees above the desired temperature of the final product.[3] The hot aggregate is discharged from the dryer at the burner end into an enclosed hot bucket elevator. The elevator takes the hot aggregate to the top of the tower unit. At the top of the tower unit is a set of vibrating screens that are designed to separate the aggregate into a number of sizes. Hot dust that is collected from the exhaust gas of the dryer is added to the aggregate here. The different-sized aggregates are stored in individual hot bins. The hot bins are insulated and/or heated. The hot

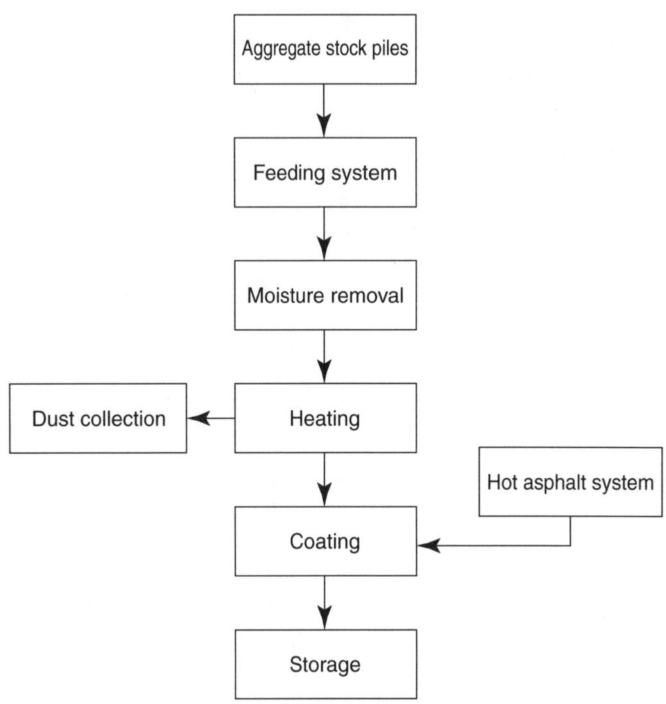

Figure 1. General Process Flow Diagram of Hot Mix Asphalt Plant.

aggregate from these storage bins is dropped into a weigh hopper on a set of scales to accurately proportion the aggregate mix. The weighing off of the hot, dry stone alleviates the need to calculate moisture content and compensate accordingly. Reclaimed asphalt pavement (if used) is added at this point. The hot aggregate is then dumped into the pug mill, and the hot asphalt cement is pumped from a hot storage tank to an asphalt weigh bucket and sprayed into the pug mill mixer after a few seconds of dry mixing. The total mixing time is usually less than 60 sec.[2] The hot-mix asphalt is then dumped into a storage silo or directly into a truck for transport to the job site.

Figure 2 shows the process flow and emissions points for a batch mix facility.

Parallel Flow Drum Mixing

The process of producing hot-mix asphalt in a parallel-flow drum plant also starts with loading aggregate from storage piles into cold bins. The aggregates are proportioned according to the job mix formula specifications as they leave the cold bins, accounting for the theoretical moisture content of each aggregate type. A cold conveyor dumps the aggregates into the top of the dryer. The burner of the dryer is at the top end of the inclined drum. The flow of the exhaust gas is parallel to the flow of the mix materials. The major difference between this type of plant and a batch plant is that the dryer not only dries and heats the aggregate, it also is used to mix the heated aggregates with the hot liquid asphalt cement. The flow of the asphalt cement is electronically controlled based on the virgin aggregate and reclaimed asphalt pavement weigh scales. The reclaimed asphalt pavement and asphalt cement are added midway or more down the drum in a lower temperature zone. Generally the drying and heating is accomplished in the first $\frac{2}{3}$ of the drying drum, and the mixing and coating is accomplished in the last $\frac{1}{3}$. Because the mixing of the aggregates with reclaimed asphalt pavement and asphalt cement occurs in the heated drum, organic emissions may be greater than in other processes. However, this process also traps some of the dust or fines from the exhaust in the asphalt mixture.

Some newer plants use an outer "coater" drum placed around the dryer drum. Aggregates fall into the outer drum and are then mixed with reclaimed asphalt pavement and hot liquid asphalt cement. Some of the gas from the outer drum is mixed in with the combustion air for the burner. This process is

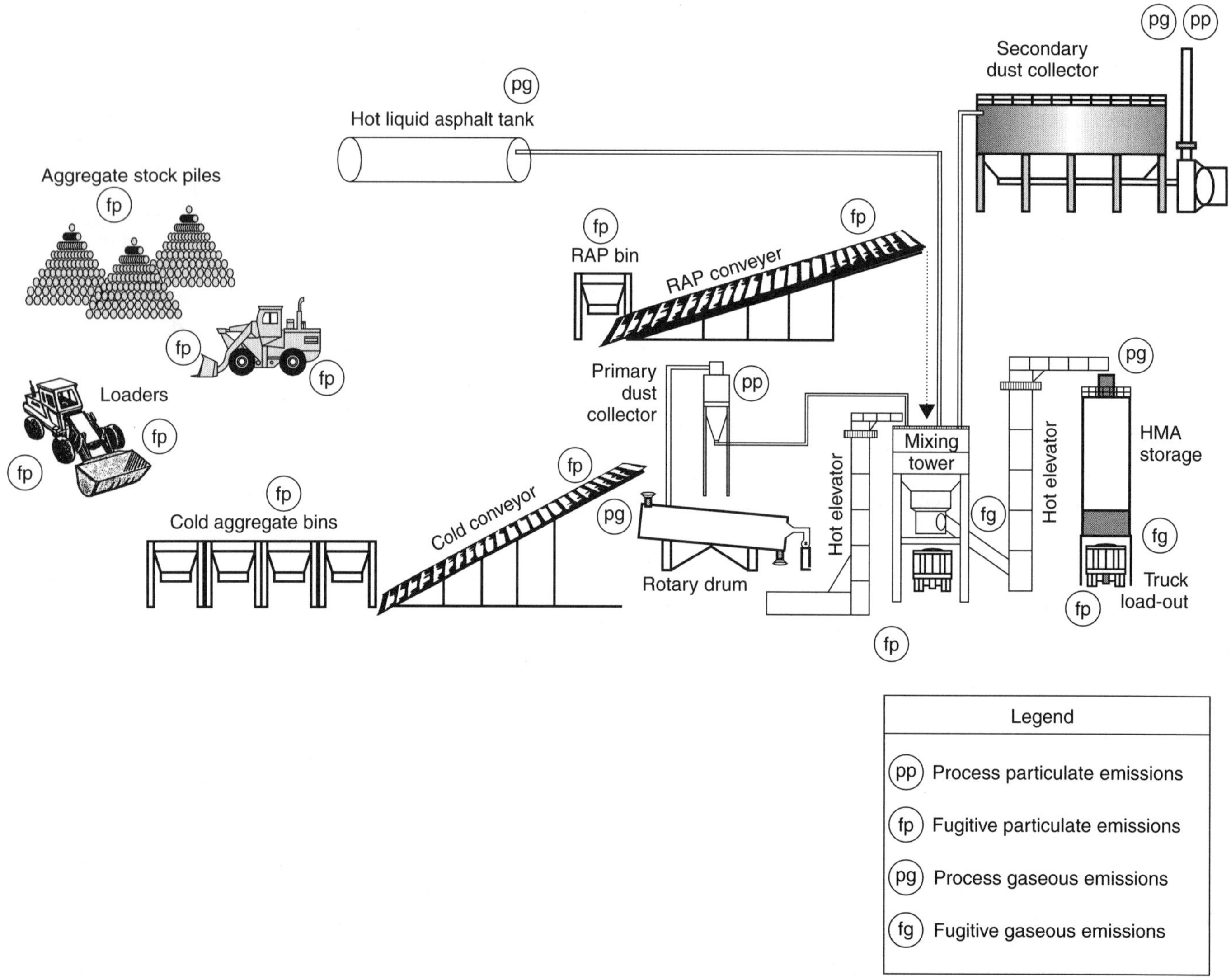

Figure 2. Batch Mix Hot Mix Asphalt Mixing Facility General Process Flow Diagram and Air Pollution Source Diagram.

used to decrease smoke (liquid aerosol) and gaseous emissions from the plant.[5]

The hot-mix asphalt is then dumped into a storage silo. The exhaust gases exit the drum and go to a particulate collection system.

Figure 3 shows the process flow and emissions points for a parallel-flow drum facility.

Counter-Flow Drum Mixing

The process of producing hot-mix asphalt in a counter-flow drum plant also starts with loading aggregate from storage piles into cold bins. The aggregates are proportioned according to the job mix formula specifications as they leave the cold bins, accounting for the theoretical moisture content of each aggregate type. A cold conveyor dumps the aggregates into the top of the dryer. The burner is located near the bottom of the slightly inclined dryer drum; the aggregate moves counter to the direction of the exhaust gas from the burner. The reclaimed asphalt pavement and hot liquid asphalt cement are added behind the burner flame zone, thus removing them from direct contact with the exhaust gases and minimizing organic emissions. Some newer plants use a system of steel rings around the burner flame to allow radiant energy to be transferred to the reclaimed asphalt pavement mixing area.[5]

The hot-mix asphalt is then dumped into a storage silo. The exhaust gases exit the drum and go to a particulate collection system.

Figure 4 shows the process flow and emissions points for a counter-flow drum facility.

Burners

Cost and efficiency in a hot-mix asphalt facility is dependent on the combustion process. Burners are the devices used to produce the heat for the drying and heating processes. A direct-fired burner is the most effective means of providing the aggregate with radiant and convective heat from the flame and the exhaust gases. The job of the burner is to set the appropriate air-to-fuel ratio, to atomize the fuel, and to mix the atomized fuel with the air to obtain complete combustion. The burner may also be used to burn off some of the organics released from the asphalt cement or reclaimed asphalt pavement during the hot mixing.

The burner in hot-mix asphalt facilities is usually a hybrid of a forced-air burner and an induced-draft burner.[6] In a forced-air burner, all the combustion air is forced by a blower through a constricted area to the fuel introduction point. The pressurized fuel is atomized and dispersed through the air, and then ignited by a pilot light. In an induced-draft burner, the combustion air is pulled past the burner by an exhaust fan.

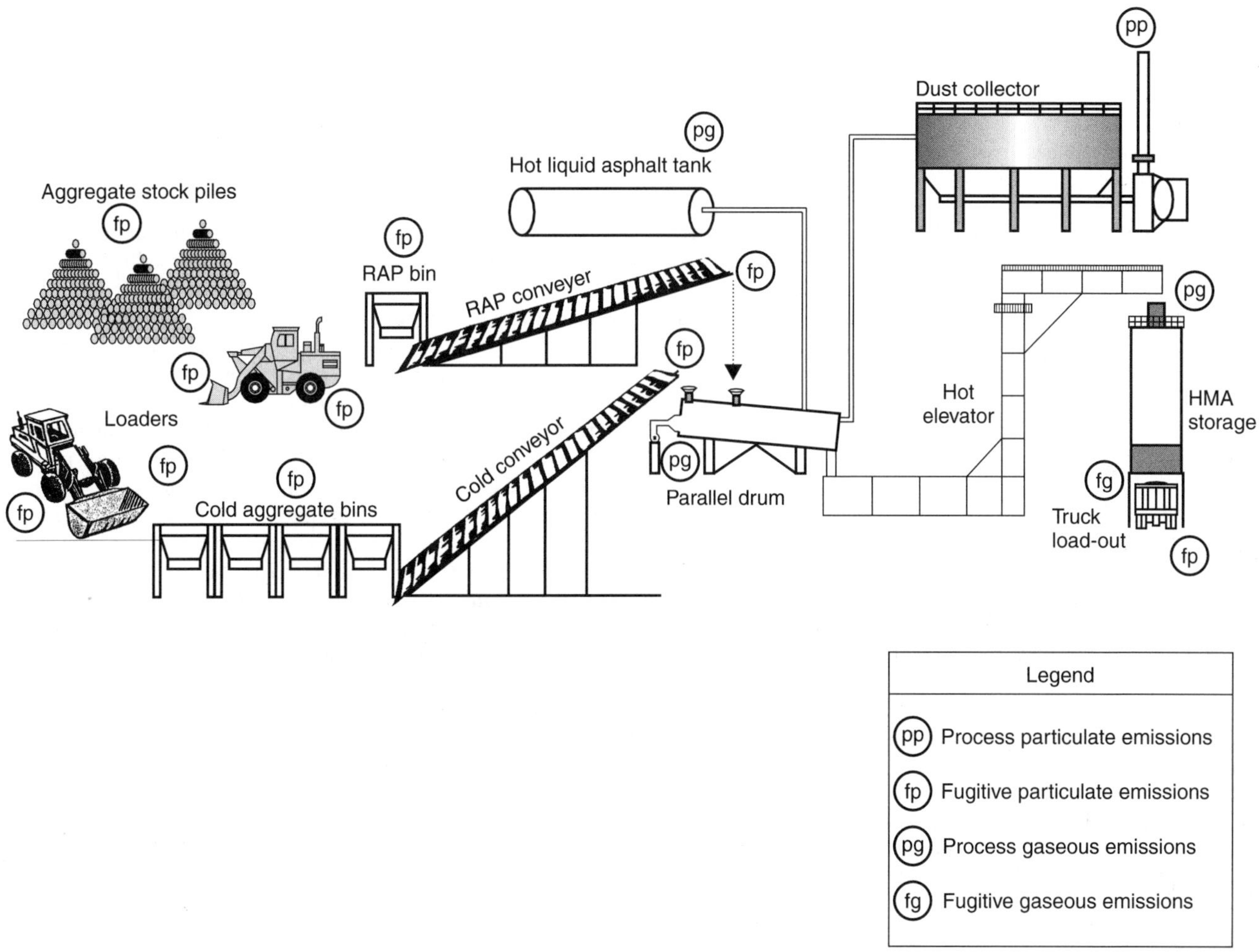

Figure 3. Parallel Flow Drum Mix Hot Mix Asphalt Mixing Facility General Process Flow Diagram and Air Pollution Source Diagram.

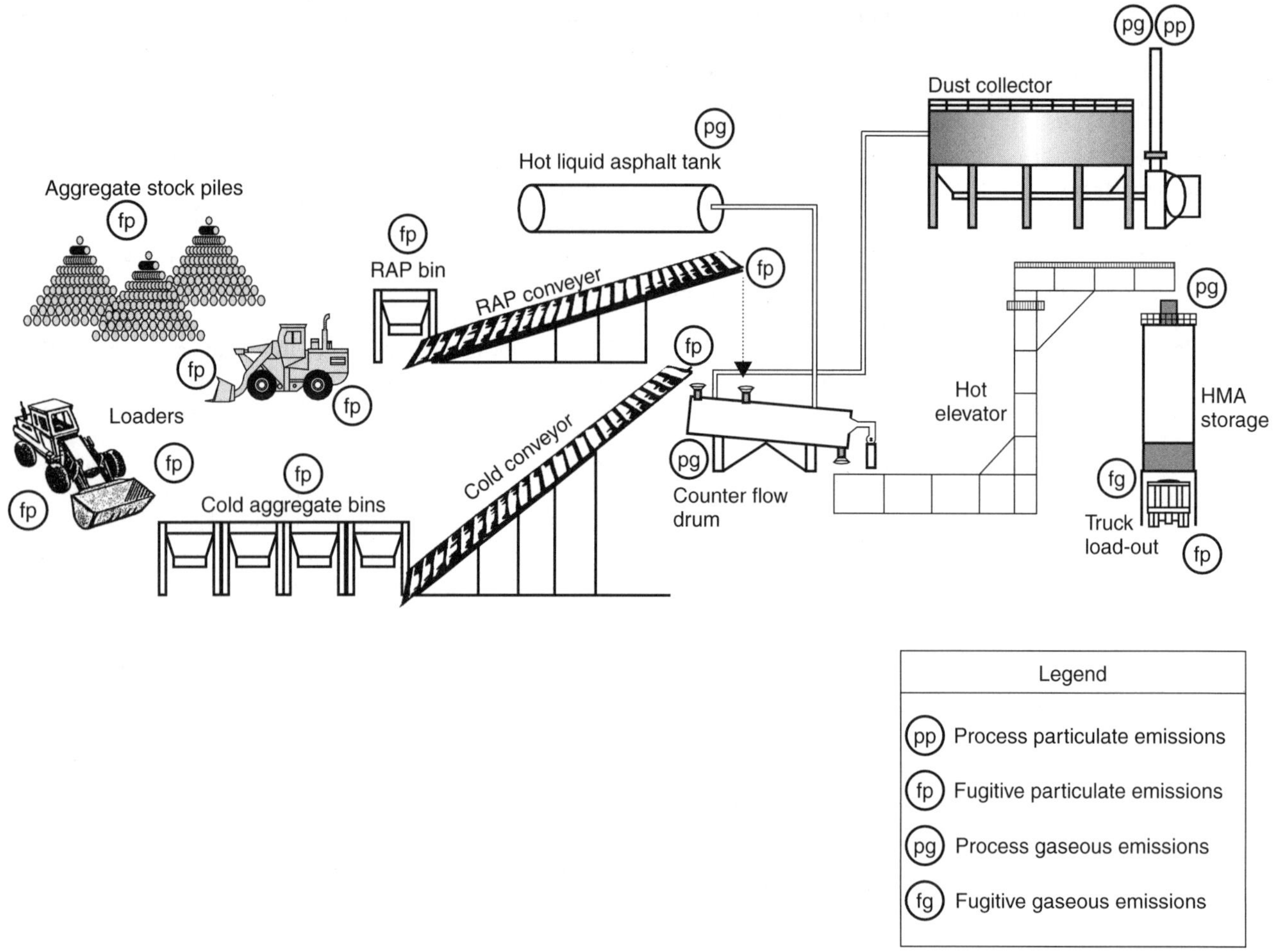

Figure 4. Counter-Flow Drum Mix Hot Mix Asphalt Mixing Facility General Process Flow Diagram and Air Pollution Source Diagram.

In a typical hot-mix asphalt facility, 30% of the combustion air is introduced by a forced-air system, and the remaining 70% of the combustion air, plus excess air, is introduced by the exhaust fan. Typically, 25% excess air is provided.[6] The burners can be designed to burn a variety of fuels—natural gas, fuel oils, diesel, liquefied petroleum gas, and coal are common.

Recycling Waste Materials in Hot-Mix Asphalt

The hot-mix asphalt industry is currently recycling three different waste materials. They are reclaimed asphalt pavement, old tires (crumb rubber), and the dust/fines from the exhaust system.

In recent years, the use of reclaimed asphalt pavement has become commonplace in the hot-mix asphalt industry. Reclaimed asphalt pavement significantly reduces the amount of new (virgin) rock and asphalt cement needed to produce hot-mix asphalt. In the reclamation process, old asphalt pavement is removed from the road base. This material is then transported to the plant, and is crushed and screened to the appropriate size for further processing. In some cases, the process is done in place by an equipment train that removes existing asphalt, crushes it, mixes it with new asphalt and aggregates, and puts it back on the pavement surface in a single continuous process. Depending on the process used and the particular job, asphalt recycling can reduce costs by 25–50%. It is generally agreed that using recycled asphalt does not detract from performance.[7] It is estimated that more than 80% of the asphalt pavement removed in the United States is reused in highway applications.[8]

At very high temperatures, reclaimed asphalt pavement releases organic pollutants, and its constituent asphalt cement may be degraded. This makes it difficult to process reclaimed asphalt pavement through the combustion and heating zones of the dryer drums; therefore, the reclaimed asphalt pavement is introduced into the mix later and relies on conductive heat transfer from the aggregates to reach an acceptable mix temperature.[6] The required superheating of the new (virgin) aggregate to provide the heat needed by the reclaimed asphalt pavement limits the amount of reclaimed asphalt pavement that can be contained in an asphalt mixture. Reclaimed asphalt pavement content of 25% by weight is common.

The use of scrap rubber in asphalt has received a lot of attention. One reason for this is the more than 242 million used tires that are disposed of annually in the United States. Section 1038 of ISTEA (Dec. 18, 1991), "Use of Recycled Paving Material" contained mandatory minimum utilization requirements for crumb rubber modified asphalt until 1995. As of 1995, 48 states had utilized crumb rubber in some asphalt pavement projects.[9]

Crumb rubber is made by shredding tires, magnetically removing steel fragments, and grinding to pebble size. Modified asphalt paving products can be made using crumb rubber by a wet process or a dry process. In the wet process, the rubber is added to the asphalt cement prior to its use as cement at the hot-mix asphalt facility. Typically, rubber may make up 20% of the asphalt cement by weight.[7] The dry process adds the crumb rubber to the aggregate in the hot-mix asphalt facility prior to the addition of asphalt cement (just as reclaimed asphalt pavement is added). The rubber usually represents 2–3% of the aggregate by weight.[10]

It has been common practice to add fines/dust from the hot-mix asphalt production process to the hot-mix asphalt. Batch plants are better at proportioning the amount of fines since they are added in the tower and do not combine in the dryer drum.

Historically, the particulates collected have been returned to the hot-mix asphalt mixing process. The future of this practice may be uncertain, as recent studies are indicating that the use of small aggregates weakens the asphalt and leads to rutting of the road surface under heavy use.[11]

Increasing social and political pressure in the resource recovery area is causing hot-mix asphalt providers to look at use of materials such as roofing shingles, fly ash, contaminated soils, and crushed glass in hot-mix asphalt mixtures.[1]

Heat Recovery

Energy considerations are encouraging a move towards greater waste heat utilization by new and refurbished hot-mix asphalt facilities. The recovered waste heat is used to predry the reclaimed asphalt pavement, and in some cases to predry the aggregates. Total energy savings in excess of 50% are possible, thus reducing the emissions of combustion products.[4]

AIR EMISSIONS CHARACTERIZATION

Hot-mix asphalt facilities, like many other facilities, have two major categories of emissions: point sources (those ducted through a stack, vent, or pipe to the atmosphere) and nonpoint sources (those emitted directly to the atmosphere), also called fugitive sources.

The most significant point source in the hot-mix asphalt mixing process is the rotary drum dryer.[2] The dryer emits steam, particulate matter (PM), carbon dioxide (CO_2), carbon monoxide (CO), nitrogen oxides (NO_x), sulfur oxides (SO_x), volatile organic compounds (VOCs), methane, formaldehyde, and various hazardous air pollutants (HAP). Batch plants have a second significant point source—the hot mixing tower (hot conveyor, hot screens, and pug mill mixer).

The nonpoint, or fugitive, sources at hot-mix asphalt facilities include vehicular traffic, the transport and handling of the hot-mix asphalt and aggregates, as well as the loadout operations to the delivery trucks.

The air pollution from the hot-mix asphalt mixing process is subject to the "U.S. EPA NSPS Subpart I—Standards of Performance for Hot Mix Asphalt Facilities." Since the NSPS regulation that applies to hot-mix asphalt facilities was promulgated in the 1970s, fugitive emissions must be considered when determining whether Title V applicability thresholds are exceeded. For purposes of major source applicability, the pollutants of greatest interest are PM_{10}, CO, and SO_2.[12] Emission factors for other criteria pollutants, such as NO_x and VOCs, are significantly smaller than those for these three pollutants. Hazardous air pollutants (metals, polycyclic aromatic hydrocarbons (PAHs), benzene, ethylbenzene, toluene, xylene, and formaldehyde) are emitted in relatively small quantities relative to criteria pollutants. Generally, hot-mix asphalt facilities may be subject to Title V if they produce 250,000 tons of products in a 12-month period.[12] The average hot-mix asphalt facility produces about 140,000 tons of products per year. Emissions points for each of the three types of facilities are indicated in Figures 2–4.

Process Particulate Emissions

Stone dust, fly ash, soot, and unburned particles of fuel comprise the main source of particulates in hot-mix asphalt facilities. Stone dust is emitted in the dryer drum as the aggregates fracture. Fly ash and soot are emitted from the burner of the dryer drum. Unburned fuel particles may be emitted due to poor combustion management.

Batch plants also have process emissions from the hot elevator, hot screens, and the pug mill mixer. The mixing tower is usually controlled with auxiliary air handling. These emissions are mostly stone dust. Emissions from the pug mill mixer may contain a fine aerosol of condensed organic particles.

Drum plants have lower total uncontrolled particulate emissions than batch plants since some of the particles will adhere to the asphalt cement in the drum plant. Uncontrolled emissions factors for process particulates for both types of plants can be found in Table 1.

Fugitive Particulate Emissions

Stone dust is emitted from the aggregate stockpiles, loader, cold feed aggregate bins, cold conveyors, reclaimed asphalt pavement bins and conveyors, and hot-mix discharge at loadout. Other dust particulates are emitted from the vehicular traffic on plant grounds and roadways. Silt content (particles equal to or less than 75 μm in diameter) is typically 5% for hot-mix asphalt aggregates.[2]

Emissions of fugitive particulates from aggregate stockpiles, loaders, and vehicular traffic can be estimated using emissions factors found in Section 13.2 of AP-42. The EPA is currently researching emissions during the truck loadout operation.[13]

Nonpoint source particulate emissions include PM_{10} and $PM_{2.5}$.

Oxides of Sulfur Emissions

SO_x emissions from hot-mix asphalt facilities are causing increasing environmental concern. One reason for this concern is the low stack height at hot-mix asphalt plants, which allows only minimal dispersion. The combustion of fuel containing sulfur results in the creation of SO_x; therefore, SO_x is a concern in oil- and coal-fired facilities.

Table 1. Uncontrolled Emission Factors for Process Particulates in Hot-Mix Asphalt Plants[a]

	Total PM	PM_{10}	$PM_{2.5}$
Batch mix	32	4.5	0.27
Drum mix	19	4.3	1.05

[a] Units, lb/ton of hot-mix asphalt.

Source: Compilation of Air Pollution Emission Factors (AP-42), U.S. EPA, 1995 January.

Carbon Monoxide

Carbon monoxide emissions in hot-mix asphalt facilities result from incomplete combustion of the fuel in the burner of the dryer drum and in the hot storage tank heater. CO emissions are affected by several variables, including burner design and maintenance, drum design and maintenance, exhaust fan design and maintenance, fuel pressure, and the moisture content of the aggregate. Carbon monoxide emissions are potentially the largest regulated gaseous emissions from hot-mix asphalt facilities.

Carbon Dioxide

Carbon dioxide is formed as part of the combustion process at the burner in the dryer drum and the hot storage tank heater. The amount of CO_2 formed depends on the type and amount of fuel burned. Emissions of CO_2 can be estimated based on fuel use and the fuel combustion emission factors in AP-42, Chapter 1. Emissions of carbon dioxide are not currently regulated.

Nitrogen Oxide

Nitrogen oxides are formed when air (composed of nitrogen and oxygen) is exposed to high heat. NO_x can be formed at the dryer drum and at the heater for the hot storage silo.

Hydrocarbons

Hydrocarbon emissions from hot-mix asphalt facilities include methane, formaldehyde, and more than 50 different volatile organic compounds. These emissions may be process-related (point source) or fugitive (nonpoint source). When asphalt cement is heated, hydrocarbons are released. This applies both to the hot liquid asphalt cement and to the reclaimed asphalt pavement that may be introduced in the hot-mix asphalt mixing process.

Some of the process emissions may be the result of poor combustion at the dryer drum and at the heater for the hot storage silo. VOC emissions are likely to be higher in parallel-flow drum facilities, because liquid asphalt cement and reclaimed asphalt pavement (when used) are in contact with the hot exhaust gas stream. The use of crumb rubber is also likely to increase VOC emissions.

Fugitive hydrocarbon emissions may occur at the hot storage silos, the asphalt cement storage tanks, and the truck loadout area. Fugitive emissions at the truck loadout area are currently a major issue. The U.S. EPA is currently studying these fugitive emissions.

Toxic Emissions

Toxic emissions from hot-mix asphalt facilities include small amounts of hazardous air pollutants and some other metals. Between 1976 and 1993, 19 epidemiological studies examined the potential carcinogenic effects on asphalt workers. Exposure was significantly associated with excess lung cancer, bladder cancer, urethra and pelvic cancer, stomach cancer, and colon cancer.[14] EPA is scheduled to promulgate new rules for hazardous air pollutants at hot-mix asphalt facilities by November 15, 2000.

AP-42 Emission Factors

Emission factors are fundamental tools for air quality management. An emission factor is a representative value that attempts to relate the quantity of a pollutant released to the atmosphere with an activity. The U.S. EPA's *Compilation of Air Pollution Emission Factors* (AP-42) provides emission factors that represent an average of a range of emission rates—about 50% of facilities will have emissions higher than the emission factor, and about 50% of facilities will have emissions lower than the emission factor. Section 11.1 of AP-42 contains the emission factors for hot-mix asphalt plants. Section 13.2 of AP-42 contains the emissions factors for the vehicular traffic, aggregate handling, and storage piles. The EPA is currently working on a revision to the AP-42 emission factors for hot-mix asphalt facilities. The emissions factors listed in Section 11.1 of AP-42 of January 1995 are presented in Tables 2 and 3.

Current AP-42 information can be found on the EPA website at: *http://www.epa.gov/ttnchie1/ap42c11.html*.

AIR POLLUTION CONTROL MEASURES

Particulate Control Equipment

Traditionally, the air pollution control system in most hot-mix asphalt facilities includes one or more types of particulate collectors. The types of particulate collectors used include gravity settling chambers (knockout boxes), cyclone separators, wet collectors (scrubbers), and fabric filters (baghouses).

A gravity settling chamber or a cyclone separator is often used in conjunction with a wet collector. These dry collectors remove the majority of the larger particulates, thus reducing the volume of sludge that is produced in the scrubber.

The wet scrubber also removes some of the hydrocarbon emissions in the exhaust gas of the dryer drum. Wet scrubbers have several disadvantages that have limited their popularity—sludge disposal, water discharge permits, and high energy costs.

Fabric filters, commonly known as baghouses, are the most common type of particulate control equipment currently found at hot-mix asphalt facilities. The volume of air that must be

Table 2. Emission Factors for Batch Mix Hot-Mix Asphalt Plants[a]

Process	CO	CO_2	NO_x	SO_2	TOC	PM	PM_{10}
Natural gas-fired dryer	0.1700	17	0.0130	0.0025	0.0084	16	2.2
Oil-fired dryer	0.0350	19	0.0840	0.1200	0.0230	16	2.2

[a]Factors are in kg/Mg of product. Factors are for uncontrolled emissions. TOC represents the sum of all organics as methane.

Table 3. Emission Factors for Drum Mix Hot Mix Asphalt Plants[a]

Process	CO	CO_2	NO_x	SO_2	TOC	PM	PM_{10}
Natural gas-fired dryer	0.0280	14	0.0150	0.0017	0.0250	19	4.3
Oil-fired dryer	0.0180	19	0.0380	0.0280	0.0350	19	4.3

[a]Factors are in kg/Mg of product. Factors are for uncontrolled emissions. TOC represents the sum of all organics as methane.

filtered determines the size of a baghouse. An air-to-cloth ratio between 5.0:1 and 6.2:1 has been found to be effective in hot-mix asphalt facilities.[6] A typical baghouse for the exhaust is of the pulse jet type, containing between 300 and 1,600 separate bags.

An emissions monitor in the collector's exhaust stack measures the dust level escaping to the atmosphere. Moisture condensing on the probes of conductive-probe-type monitors often causes false alarms. A monitor using a constant light beam and self-compensating photocell amplifier has been shown to be reliable in hot-mix asphalt facilities.[15]

In batch plants, cyclone separators have typically been used to return larger particulates (larger than 200 mesh or 75 μm) from the dryer drum waste gas to the mixing tower. The exhaust from the cyclones is generally routed to a baghouse.

Crusting agents have been effective at limiting emissions of stone dust from aggregate stockpiles in high winds. Watering of the aggregate stockpiles is not used, since part of the hot-mix asphalt process requires the drying of the aggregate. Partial enclosure of the stock is also effective.

Traditionally, watering the road has controlled dust from unpaved driving surfaces. Paving the road is becoming a more common approach to controlling the dust from driving surfaces.

Gaseous Air Pollutant Control Measures

Prior to the Clean Air Act Amendments of 1990, no pressure was exerted on the hot-mix asphalt industry with regard to the control of gaseous air pollutants. As Title V is being implemented, more attention is being given to this subject. The two gaseous pollutants of greatest concern in hot-mix asphalt facilities are carbon monoxide and sulfur dioxide.

The main approach to carbon monoxide control at hot-mix asphalt plants is proper maintenance and adjustment of the burner in the dryer drum. Several industry studies have shown that CO emissions can be reduced by 50–90% with a good maintenance program. These studies have shown that it is the maintenance program, not the type of burner, that has the greatest effect on CO emissions. Newer spin-stabilized burners and older refractory burners are both capable of low emissions.[16]

Since CO emissions increase with fuel use, energy conservation techniques (such as heat recovery) can significantly reduce emissions.

Control of sulfur dioxide emissions has been approached by two different methods in hot-mix asphalt plants. Both dry limestone injection into the dryer drum and the use of low-sulfur fuel have been found to be effective at reducing SO_2 emissions.

Dry limestone injection is a process based on the absorbent reaction between limestone and the sulfur oxides in the flue gas. To make this process efficient, fine limestone dust particles are sprayed pneumatically to the end of the burner flame in the dryer drum. A calcium/sulfur ratio of 92 (mol/mol) has been found to be effective.[17] While SO_2 emissions can be reduced by 99.9% using this process, 90% reduction efficiency is the norm.

Toxic Air Pollutant Control Measures

Public pressure for the control of hazardous air pollutants and other toxic air pollutants has recently become a major issue among grassroots community, environmental, and health organizations. Control of HAPs in batch mix plants is accomplished by recycling the air from the auxiliary exhaust system of the mixing tower into the combustion air for the burner of the dryer drum. New/refurbished parallel-flow plants can have the asphalt cement added in an area that is not directly in the exhaust stream. The exhaust from the asphalt cement can be used as the air for the burner. This essentially burns off the hydrocarbons, including most of the HAP and the VOCs in the exhaust stream.[18]

Concerns about the fugitive emissions from the hot storage and loadout areas have led some to suggest enclosure of these areas for emissions control. Good information regarding these concerns is not currently available. The EPA has made a study of these fugitive emissions a research priority.

FOR FURTHER INFORMATION

The following websites are provided to help you find the most current information.

AP-42 Emissions Factors for HMA	*http://www.epa.gov/ttnchie1/ap42c11.html*
Asphalt Institute	*http://www.asphaltinstitute.org*
Hot Mix Asphalt Database	*http://www.eng.auburn.edu center/ncat/databs.htm*
Crumb Rubber Asphalt	*http://www/rubberizedasphalt. org*
Asphalt Contractor On-line	*http://www.asphalt.com*
Pennsylvania Transportation Institute	*http://www.pti.psu.edu*
National Asphalt Pavement Association	*http://www.hotmix.org/main.htm*
National Quality Initiative	*http://www.nqi.org*
American Association of State Highway and Transportation Officials	*http://www.aashto.org*
U.S. EPA technical information	*http://www.epa.gov/ttn*

REFERENCES

1. Mike Acott, "This Is Not Your Father's Toll Road, Hot Mix Asphalt Changes Are on the Horizon," *American City & County* **105**(1), 42 (1990 January).
2. *Compilation of Air Pollution Emission Factors* (AP-42), U.S. EPA, 1995 January.
3. Kathryn O'C Gunkel, *The Fundamentals of the Operation and Maintenance of the Exhaust Gas System in a Hot Mix Asphalt Facility*, National Asphalt Pavement Association, 1987.
4. B. W. Ang, T. F. Fwa, and T. T. Ng, "Analysis of Process Energy Use of Asphalt Mixing Plants," *Energy* **18**(7), 769 (July 1993).
5. Lynn Landberg, "Asphalt Plants Solving Old Problems; Asphalt Plant Emissions," *Construction Equipment* **86**(3) (September 15, 1992).
6. Roberts et al., *Hot Mix Asphalt Materials, Mixture Design and Construction*, National Asphalt Pavement Association Education Foundation, Lanham, Maryland, 1991.
7. John Prendergast, "Anatomy of Asphalt, What Goes into Making Asphalt Pavements Could Change a Lot by the End of the Decade," *Civil Engineering* **61**(7), 56 (July 1991).

8. Bob Drake, "Limited Study Finds No Fault with Crumb-Rubber Asphalt," *Pit & Quarry* **86**(10), 38 (April 1994).
9. "Statement of the National Asphalt Pavement Association for the Subcommittee on Transportation and Infrastructure, U.S. Senate," *Federal Document Clearing House*, March 30, 1995.
10. Jon A. Epps, *Uses of Recycled Rubber Tires in Highways*, Transportation Research Board, 1994.
11. Seth Shulman, "The Lowdown on Blacktop," *Technology Review* **98**(2), 18–19 (February 1995).
12. Tim Smith, *Technical Support Document for Potential to Emit Guidance Memo*, USEPA/OAQPS, April 1998.
13. William J. Angelo, "New Haze over Asphalt Plants," *Engineering News-Record* **240**(3), 10 (January 19, 1998).
14. Pat A. Cook and James E. Krause, "Asphalt Fumes—A Growing Concern," *Constructioneer* **51**(20), 96 (October 1997).
15. "Emissions Monitor for an Asphalt Manufacturer's Dust Collector Gives Reliable Results," *Powder and Bulk Engineering* **7**(4), 38 (April 1993).
16. Bob Frank, "Regulations More Stringent on Gaseous Pollutants," *Plant Production, The Asphalt Contractor Online* (February 1999).
17. K. Kuuspalo, "Reduction of SO_2 Emissions in Asphalt Plants," *Journal of Aerosol Science* **22**, Suppl. 1, p. S471 (1991).
18. John Speers, "South Jersey Asphalt Plant Refurbished," *Constructioneer* **49**(7), 14 (April 1995).

BIBLIOGRAPHY

S. H. Mellor, "Modular Plants—The Road Ahead for Asphalt Production," *Mine and Quarry* **19**(12), 14 (December 1990).

"Hot Mix Asphalt in Practice in the U.S.," *AASHTO Quarterly Magazine (American Association of State Highway and Transportation Officials)* **69**(2), 12 (April 1990).

Robert F. Baker, "Mix Designs and Air Quality Emissions Tests of Crumb Rubber Modified Asphalt Concrete," *Transportation Research Record*, No. 1515, p. 18 (1995).

"Asphalt Plant Emissions," *Audubon Magazine* **98**(5), 14 (September–October 1996).

"ALANCO Achieves Environmental Compliance for Asphalt Plant," *Business Wire*, p. 4031078 (April 3, 1997).

"Clean Air: Asphalt Industry Cited for Fugitive Emissions," *Greenwire* (November 13, 1997).

Richard Korman, "Paving the Way to Cleaner Air," *Engineering News-Record* (October 14, 1996).

"Being Responsible and Responsive to the Public," *Engineering News-Record* **240**(3), 92 (January 19, 1998).

PORTLAND CEMENT

Walter L. Greer, Ann Dougherty, and Douglas M. Sweeney

INTRODUCTION

Portland cement is a fine, gray powder that consists of a mixture of the hydraulic cement minerals, tricalcium silicate, dicalcium silicate, tricalcium aluminate, and tetracalcium aluminoferrite, to which one or more forms of calcium sulfate have been added. Portland cement accounts for about 93% of the cement production in the United States. Blended cements are about 2% and masonry cement about 5% of domestic cement production. These cementitious materials are also produced in portland cement plants and contain portland cement as an ingredient.

Portland, blended, and masonry cements are produced in several different types or formulations for specific purposes or properties. Chemical and physical specifications for the types of portland and masonry cements are written by several agencies, of which the most widely used are those provided by the American Society for Testing and Materials. The most common types of portland cement are designated by the Roman numerals I through V. Blended cements usually have a letter designation following a Roman numeral to indicate the nature of the blend (i.e., Type I-P for a general purpose portland–pozzolan blend). Types of masonry cement are designated by the letters N, S, and M.

The production of portland cement is a four-step process: (1) acquisition of raw materials, (2) preparation of the raw materials for pyroprocessing, (3) pyroprocessing of the raw materials to form portland cement clinker, and (4) grinding of the clinker to portland cement. In a portland cement plant, the pyroprocessing operation is almost always the limiting factor for productive capacity. Figure 1 is a basic flow diagram of the portland cement process. Figure 2 presents a layout of a cement plant recently built in the United States, and Figure 3 shows another modern cement plant. While the various unit operations and unit processes in portland cement plants accomplish the same end result, no single flow diagram can fully represent all plants. Each plant is unique in layout and appearance owing to variations in climate, location, topography, raw materials, fuels, and preferences of equipment vendors and owners. These plants are capital intensive. In 1997, there were 105 plants in the United States producing approximately 76 million tons of portland cement. Portland cement plants can run 24 hours per day for extended periods, for example, six months or more with only minor downtime for maintenance is not unusual.

Raw materials are selected, crushed, ground, and proportioned so that the resulting mixture has the desired fineness and chemical composition for delivery to the pyroprocessing system. The major chemical constituents of portland cement are calcium, silicon, aluminum, iron, and oxygen. Carbon is a major constituent of the cement raw mix, but that element is eliminated during processing. Minor constituents, generally in a total amount of less than 5% by weight of the mixture, include magnesium, sulfur, sodium, and potassium. And since raw materials for portland cement usually come from the earth's crust, a wide variety of trace elements can be found in the cement, although these generally total less than 1% by weight of the mixture. Some of these naturally occurring trace elements can affect the performance of portland cement and/or appear in particulate emissions and process residues from cement plants. Most often, however, they harmlessly substitute for the four major metals in the crystalline matrix of the portland cement.

The more than 30 raw materials that are known to be used in the manufacture of portland cement can be divided into four categories: lime (calcareous), silica (siliceous), alumina (argillaceous), and iron (ferriferous). Limestone or another form of calcium carbonate ($CaCO_3$) will predominate in the mixture of raw materials. One or more quarries are usually associated with a portland cement plant. The terms slurry, raw meal, raw mix, and kiln feed are synonymous in naming the prepared raw materials or product of the raw mill department. At least 1575 kg (3465 pounds) of dry raw materials are required to produce 1000 kg (2200 pounds) of cement clinker. This ratio of feed to product can increase by several pounds due

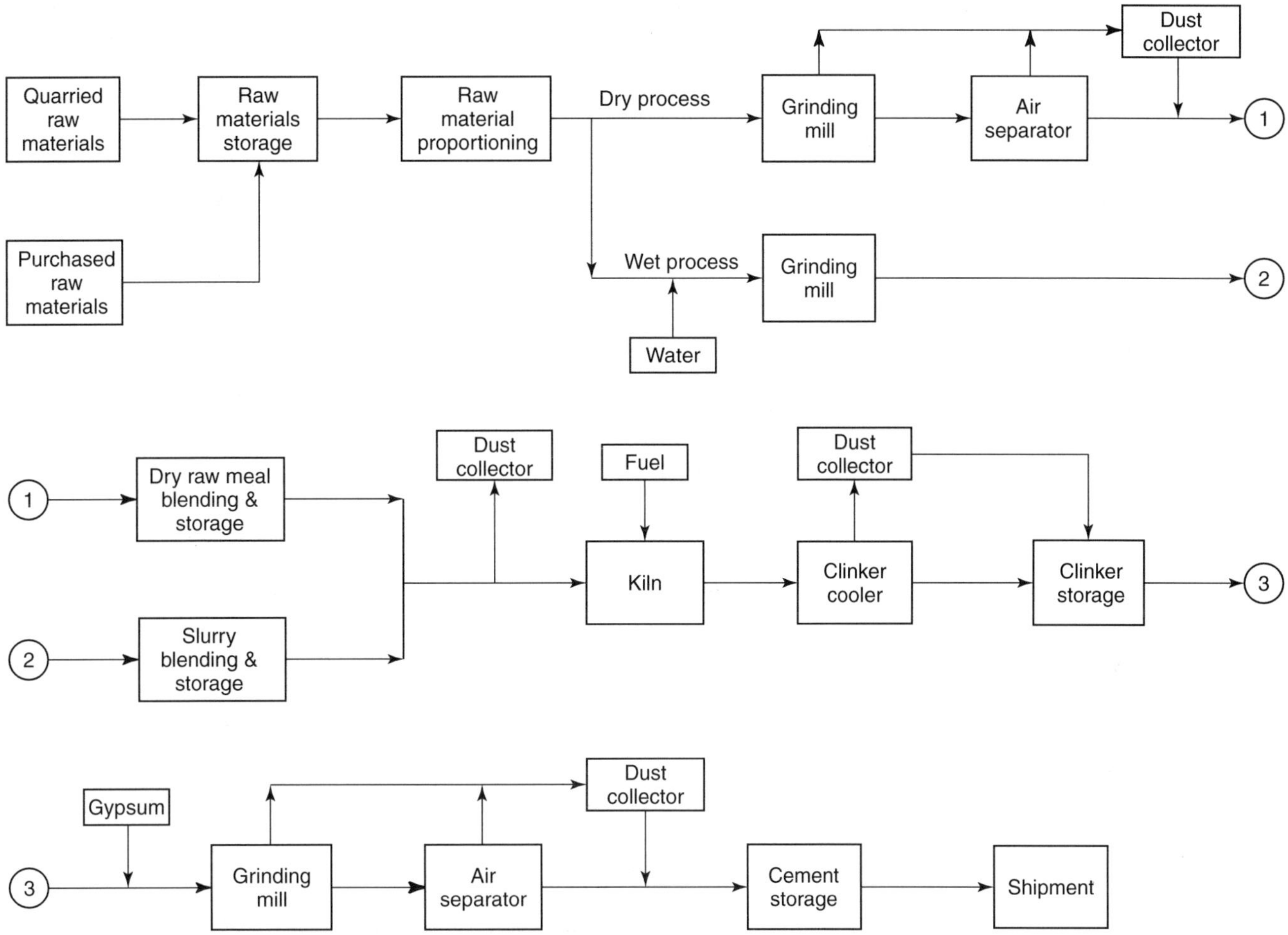

Figure 1. Basic Flow Diagram of the Portland Cement Manufacturing Process.

to raw mix composition and dust removal. Most of the weight lost between raw mix and clinker is carbon dioxide (CO_2) that is calcined from the $CaCO_3$ and emitted to the atmosphere during pyroprocessing. About one ton of CO_2 is emitted per ton of portland cement clinker produced, with about one-half of this amount resulting from the calcination of $CaCO_3$. The balance results from the combustion of fuel. More efficient kilns produce less CO_2 per ton of clinker than less efficient kilns. Fuels containing hydrogen (e.g., natural gas) result in lower CO_2 emissions than coal or petroleum coke.

Standard industry practice is to report the chemical analyses of raw materials, process intermediates, by-products, and portland cement as metal oxides, even though the constituents are rarely present in that form. If desired, the theoretical quantities of minerals in the cement matrix are calculated from the oxide analysis using specified formulas. Actual quantities of minerals may be determined by X-ray diffraction.

There are wet-process and dry-process portland cement plants. In the wet process, the ground raw materials are suspended in sufficient water to form a pumpable slurry. In the dry process, they are dried to a flowable powder. New portland cement plants in the United States have exclusively used the dry process because of its lower thermal energy requirement. The Portland Cement Association estimated in 1997 that the average thermal energy used to produce a ton of cement in the United States was about 4.0 million Btu. Thermal energy consumption ranged from about 2.7 to 7.3 million Btu per ton, depending on the age and design of the plant. Average electric energy consumption is about 0.4 million Btu (117 kWh) per ton of cement.

The prepared raw materials are fed to one of several pyroprocessing systems in the kiln or burning department. Each system accomplishes the same result via the following basic steps: evaporation of free water, evolution of combined water, calcination of the carbonate constituents (decarbonization), and formation of the portland cement minerals (clinkerization). The wet process uses rotary kilns exclusively. The dry process can also employ simple rotary kilns. Thermal efficiency can be improved, however, through the use of one or more cyclone-type preheater vessels that are arranged vertically, in series, ahead of the rotary kiln in the material flow path. It can be further improved by diverting up to 60% of the thermal energy (i.e., fuel) required by the pyroprocessing system to a special calciner vessel located between the preheater vessels and the rotary kiln.

The rotary kiln is the heart of the portland cement process since the several and complex chemical reactions necessary to produce portland cement minerals take place there. The portland cement kiln is a slightly inclined, slowly rotating steel tube that is lined with appropriate refractory materials. The rotation of the kiln causes the solid materials to be slowly transported downhill from the feed end. Fuel is supplied at the lower or discharge end of the kiln. Many fuels can be used

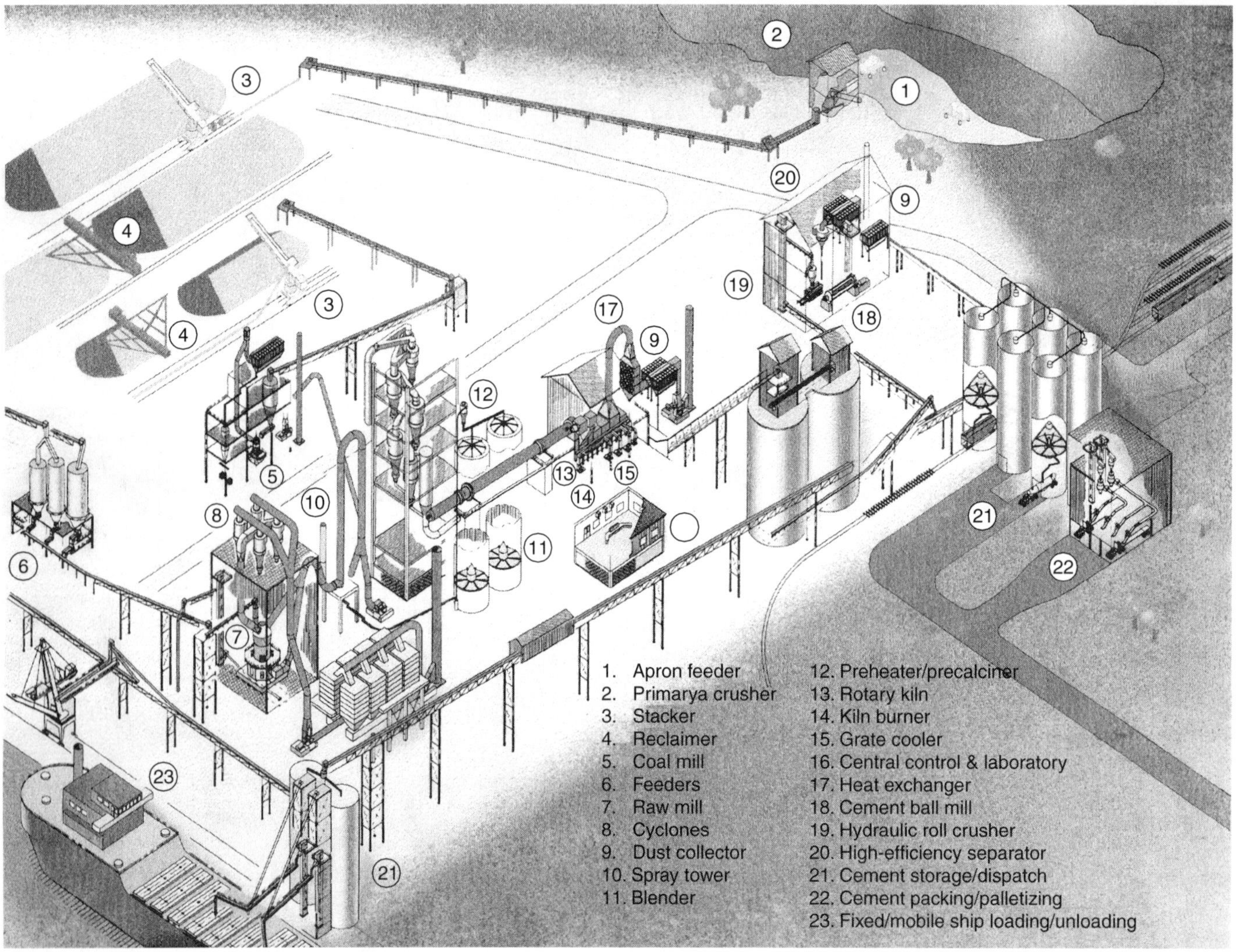

Figure 2. Cement Plant Layout.

Figure 3. Modern Cement Plant.

in the kiln, but coal has predominated in the United States since the mid-1970s. The choice of fuel is based on economics and availability. The hot, gaseous combustion products move countercurrent to the material flow, thereby transferring heat to the solids in the kiln load.

Flame temperatures in excess of 3400°F result in the material temperatures of 2700–2800°F that are required to produce the hydraulic calcium and aluminum silicates. In effect, the cement pyroprocess converts natural sedimentary rocks into synthetic igneous rocks.

The product of the rotary kiln is known as clinker. Heat from just produced clinker is recuperated in a clinker cooling device and returned to the pyroprocess by heating combustion air for the kiln and/or calciner.

The cooled clinker is mixed with a form of calcium sulfate, usually gypsum, and ground in ball or tube mills in the finish mill department to produce portland cement. Masonry cement is similarly produced from portland cement clinker, gypsum, and one or more calcareous materials.

Portland cements are shipped from the packhouse or shipping department in bulk or in paper bags by truck, rail, barge, or ship. Masonry cements are shipped primarily in paper bags.

Except for the quarry and rock crushing operations, and indirect-fired coal mill systems, the New Source Performance Standards (NSPS) that apply to a new or modified portland cement plant constructed after August 17, 1971, are contained in 40 CFR 60, Subpart F, *Standards of Performance for Portland Cement Plants.*

Emission factors for portland cement plants are contained in Section 11.6 of the United States Environmental Protection Agency (EPA) publication, *Compilation of Air Pollutant Emission Factors*, AP-42, Fifth Edition, January 1995. Tables of these emission factors occupy several pages in AP-42 and have not been included here.

An explanation of the operation of the common dust collection devices used in the cement industry, such as multiclones, fabric filters, and electrostatic precipitators (ESPs), is beyond the scope of this chapter but can be found elsewhere in this manual.

ACQUISITION OF RAW MATERIALS

Process Description

The initial step in the manufacture of portland cement is the acquisition of raw materials. The industry is considered an extractive industry since nearly all the required raw materials are obtained from the earth's crust by mining or quarrying. Most cement plants are located near a source of $CaCO_3$, which is most often limestone. Since about one-third of the weight of the limestone is lost as CO_2 during pyroprocessing, process economics dictate that this lost weight be transported as short a distance as possible. Those plants that are not immediately associated with a limestone quarry often have a source of limestone or other form of $CaCO_3$ (e.g., aragonite) that is available by less expensive water transportation. However, there are a few exceptions to these generalizations on plant location.

Calcium is the metallic element of highest concentration in portland cement. The calcareous raw materials can include limestone, chalk, marl, aragonite, and an impure limestone known in the industry as natural cement rock. Limestone, chalk, and cement rock are most often extracted from open-face quarries, but underground mining can be employed. Dredging and underwater mining techniques are used to develop deposits of calcareous raw materials in the ocean or below the water table. Gypsum and/or natural anhydrite (i.e., forms of calcium sulfate) from quarries or mines are calcium-bearing constituents of portland cement that are introduced as part of the final stage of its manufacture, finish grinding. It is rare for a cement plant to have a captive source of gypsum or anhydrite and these materials are usually purchased.

Silicon, aluminum, and iron are the next most prevalent metallic elements in normal portland cement and are listed in descending order of concentration. These metals are found in various siliceous, argillaceous, and ferriferous ores and minerals, such as sand, shale, clay, and iron ore. Although usually extracted in open-face quarries or pits, these raw materials can be dredged or excavated from underwater deposits. They can be obtained from captive sources adjacent to or away from the portland cement plant; however, it is often necessary or economical for the cement manufacturer to purchase them from outside sources.

The wastes and by-products of other industries are successfully employed as portland cement raw materials. Such materials include, but are not limited to, power plant fly ash, steel mill scale, and metal smelting slags.

The cement manufacturing process and the performance of portland cement are sometimes affected by trace elements that are found in virgin raw materials or wastes. Care must be exercised in selecting these raw materials to assure that trace elements will not be present in high enough concentrations to cause problems in the plant or with the product.

AIR EMISSIONS CHARACTERIZATION

Quarries at cement plants are similar to other stone quarries. The necessary operations include rock drilling, blasting, excavation, loading, hauling, crushing, screening, materials handling, stockpiling, and storing. There are many different operating methods, types of equipment, and equipment brands that are used to accomplish these tasks. Particulate matter is the primary air pollutant associated with quarry operations. In some locations (e.g., an underground mine), exhaust emissions from mobile equipment may be of concern. There are usually no atmospheric air pollution problems at underground mines or underwater operations.

The NSPS that apply to quarry and crushing operations at portland cement plants are contained in 40 CFR 60, Subpart 000, *Standards of Performance for Nonmetallic Mineral Processing*. These standards are applicable to those affected facilities that commenced construction, reconstruction, or modification after August 31, 1983.

Raw materials can also be the source of some environmentally undesirable emissions from the kiln stack found later in the process. If the raw materials contain naturally occurring hydrocarbons, such as petroleum or kerogens, these materials can evaporate or pyrolize in the relatively cooler portions of the pyroprocessing system and appear at the stack exit as a visible "blue haze" or as invisible organic constituents. Sulfur and chlorine from the raw materials can participate in reactions with the small amount of ammonia sometimes found in cement kiln emissions to form a "detached plume" of ammonium sulfate or ammonium chloride. Nitrogenous constituents of the raw materials can possibly contribute to emissions of nitrogen oxides (NO_x) emissions that are unrelated to combustion. Sulfides in raw materials contribute to sulfur dioxide (SO_2) emissions under the process conditions found in all preheater or precalciner kilns.

AIR POLLUTION CONTROL MEASURES

Control measures for particulate emissions in quarries include water sprays with and without surfactants, foams, chemical dust suppressants, wind screens, equipment enclosures, paving, mechanical collectors and fabric filters on operating equipment, and material storage buildings, enclosures, bins, and silos with and without exhaust venting to fabric filters. Collected dust is returned to the process.

Typical fabric filters found in the quarry are pulse jet types in newer plants and reverse air or shaker types in the older plants (see Table 1).

Purchased raw materials, including coal or petroleum coke used for fuel, can also generate particulate emissions as a result of vehicle loading and unloading, material handling, stockpiling, and haulage. The particulate emission control measures for purchased materials are the same as those listed for quarries.

RAW MILLING

Process Description

The second step in the manufacture of portland cement is the preparation of the raw materials for pyroprocessing. This operation in the raw mill department combines the blending of appropriate raw materials for proper chemical composition with particle-size reduction through grinding.

Grinding is required to achieve optimum fuel efficiency in the cement kiln and to develop maximum strength and durability potential in portland cement concrete. Typically, the raw material in the kiln feed is ground to about 85% passing a 200-mesh (74-μm) sieve or 90% passing a 170-mesh (88-μm) sieve. Usually less than 1% of the material is retained on a 50-mesh (297-μm) sieve. The actual fineness that is required depends on the reactivity of the raw material components. Excessive grinding of raw materials wastes energy and reduces the productive capacity of the raw mill. Raw milling processes are either wet or dry, depending on the type of pyroprocessing system(s) at the plant.

When raw materials are dried before grinding or when the physical properties of the moist materials permit handling, the raw materials are usually proportioned with weigh feeder systems located in the process flow ahead of a mill feed bin or the raw mill itself. If required and justified by process economics, raw materials can also be proportioned and blended in large (e.g., 1000-ft-long) linear or circular stacker–reclaimer systems that are frequently located in closed buildings.

Cement raw materials are received in the raw mill department with a moisture content varying from 2% to 35%. In the dry process, this moisture is usually reduced to less than 1% before or during grinding. Drying prior to grinding is accomplished in impact dryers, drum dryers, paddle-equipped rapid dryers, air separators, or autogenous mills. Drying can also be carried out during grinding of the raw mix in ball-and-tube mills or roller mills. Thermal energy for drying can be supplied by separate, direct-fired coal, oil, or gas burners that heat the airstream that passes through the drying apparatus or mill. The most efficient and popular source of heat for drying is the hot exit gases from the pyroprocessing system. These gases can come from the kiln, the clinker cooler, the alkali bypass system, or a combination of these sources. Unless the hot gases are supplied solely from the clinker cooler, the gases passing through dryers and raw mills will contain products of combustion, as well as solid particles. The selection of the drying method depends on the physical properties of the raw materials, the type of pyroprocessing system in the plant, the availability and cost of energy, and the preferences of owners, managers, and vendors.

Ball-and-tube mills (i.e., long ball mills) are rotating, horizontal steel tubes that contain steel balls and are used to provide comminution or grinding of the raw materials. Air separators are frequently used in conjunction with these mills in the dry process to separate materials of adequate fineness from the coarse particles that must be returned to the grinding mill for additional work (i.e., closed-circuit grinding). The design and operation of these mills and air separators are discussed in the following description of finish milling. The hot gases required for simultaneous drying and grinding in a ball mill system can enter the feed end of the mill and flow concurrently with the raw materials. Otherwise, the unground raw materials and the hot gases are introduced simultaneously into the air separator. Some operators feel that this latter procedure provides for more efficient drying, easier operation of the mill circuit, and early removal from the mill system of those materials that are already sufficiently ground. The separator does, however, experience additional wear.

Table 1. Fabric Filters in Quarries

	Pulse Jet			Reverse Air/Shaker
Acfm	5,000–25,000	5,000–25,000	5,000–25,000	5,000–25,000
Filter design	Tubular bag	Pleated	PTFE membrane on tubular bag	Tubular bag
Filter type	Polyester	Spun-bonded polyester	Felt with PTFE membrane	Polyester
Temp.(°F)	<275	≤ 265	≤ 275	<275
A/C ratio[a]	$\leq 5:1$	$\leq 3.5:1$	$\leq 5.5:1$	$\leq 2.5:1$
Inlet loading (gr/acf)	5–40	5–40	5–40	5–40
Outlet emission (gr/acf)	<0.02	<0.01	<0.005	0.02

[a] Air-to-cloth ratio (acfm/ft^2).

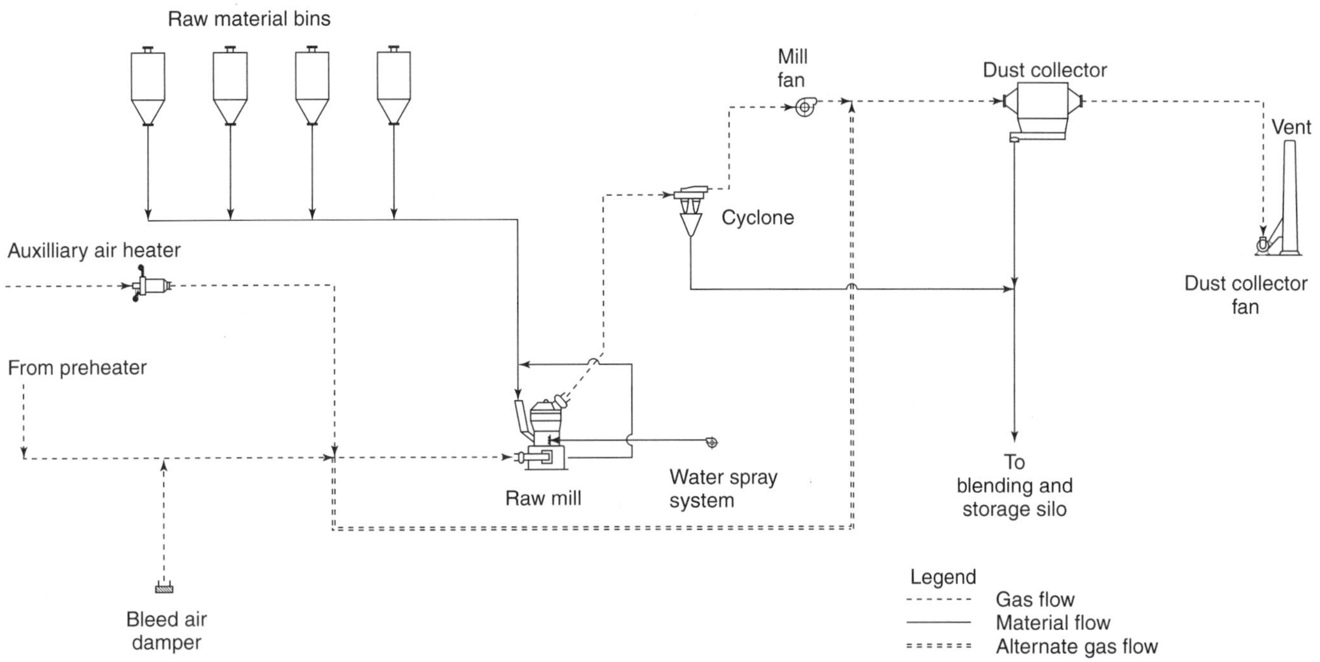

Figure 4. Raw Mill Schematic Diagram.

Vertical roller mills are very popular in new, dry-process portland cement plants because of their relative simplicity and high efficiency. The principle of operation of these mills is similar to that of a mortar and pestle. In this case, the pestle (roll) is stationary and the mortar (table) rotates. Raw materials are dropped on the rotating table to be crushed and ground between the rolls and the table. Hot gases enter the mill through an annular duct at table height. As ground material is forced off the table into the hot gas stream, it is entrained in the gases, dried, and transported upward to internal separators from which coarse material is returned to the mill by gravity. Product is captured in the air pollution control device (e.g., fabric filter). Figure 4 is a process diagram of a typical vertical roller mill raw milling circuit. Figure 5 shows an installed roller mill.

Materials are transported to, within, and away from dry raw milling systems by a variety of mechanisms, including screw conveyors, belt conveyors, drag conveyors, bucket elevators, air slide conveyors, and pneumatic conveying systems.

The dry raw mix is pneumatically blended and stored in specially constructed silos until it is fed to the pyroprocessing system.

In the wet process, water is added to the raw mill during the grinding of the raw materials in ball or tube mills, thereby producing a pumpable slip or slurry of approximately 65% solids. The slurry is agitated, blended, and stored in various kinds and sizes of cylindrical tanks or slurry basins until it is fed to the pyroprocessing system. Until recently, the advantage of the wet process was that the chemical composition of the kiln feed could be controlled more closely since slurries blend more easily than powders. Modern equipment can now blend raw meal powders satisfactorily.

Figure 5. Vertical Roller Mill.

AIR EMISSIONS CHARACTERIZATION

The raw material feeders, stackers, blenders, and reclaimers can produce fugitive dust emissions. Transfer points on belt conveyor systems and bucket elevators that serve to transport raw materials from storage to the raw mill department can also generate fugitive dust emissions.

The dry raw mills and the auxiliary equipment are all designed to run under negative pressure to suppress particulate emissions. Nevertheless, poorly designed or maintained seals and closures throughout the system can result in fugitive dust emissions. If these systems experience positive pressure through a fan failure or other cause, short-term particulate emissions can be expected until the system can be shut down.

During colder weather, the vents from dryers, raw mills, and air separators may exhibit a steam plume that is sometimes confused with particulate emissions. The condensate will dissipate within a few feet of the emission point. Fabric filters in the vent circuits for dryers, raw mills, and air separators must be insulated to prevent internal moisture condensation and the resultant blinding of bags.

There are no particulate emissions from the wet grinding process, except for the materials-handling systems ahead of the mills.

AIR POLLUTION CONTROL MEASURES

Dust collecting devices in the raw mill and raw mix storage areas include mechanical cyclones, fabric filters, and, rarely, electrostatic precipitators (ESPs). When employed, mechanical collectors are used in series with one of the other, more efficient dust collection devices. The collected dust is returned to the mill system or raw mix stream.

Typical fabric filters found in the raw mill area are pulse-jet types in the newer or upgraded plants and reverse-air or shaker types in the older plants. Cartridge-type filters can be found on materials-handling equipment (see Table 2). Vertical mills are most often closely coupled to the pyroprocessing system. The air pollution control measures for these mills will be found in the following section on pyroprocessing.

PYROPROCESSING

Process Description

The third step in the manufacture of portland cement is the pyroprocessing of the raw mix into portland cement clinker. Clinkers are gray-colored, glass-hard, spherical-shaped nodules that generally range from 1/8 to 2 in. in diameter. The clinkers are predominantly composed of the cement minerals, tricalcium silicate, dicalcium silicate, calcium aluminate, and tetracalcium aluminoferrite, which result from chemical reactions between the cement raw materials that are completed at the temperature of incipient fusion. The chemical reactions and physical changes that describe the transformation are very complex. A simplified version of the major sequential events is as follows:

- Evaporation of free water
- Evolution of combined water in the argillaceous components
- Calcination of the $CaCO_3$ to calcium oxide (CaO)
- Reaction of CaO with silica to form dicalcium silicate
- Reaction of CaO with the aluminum and iron-bearing constituents to form the liquid phase (i.e., aluminate and aluminoferrite)
- Formation of the clinker nodules
- Evaporation of volatile constituents (e.g., sodium, potassium, chlorides, and sulfates)
- Reaction of excess CaO with dicalcium silicate to form tricalcium silicate

The pyroprocessing system is generally described as containing three steps or zones: (1) drying or preheating, (2) calcining, and (3) burning or sintering. The pyroprocessing is accomplished in the burning or kiln department. The word *burning* is jargon that is used in the cement industry to describe the intense heat in this zone of the kilns. None of the constituents of the cement raw mix actually combusts during pyroprocessing.

The raw mix is fed to the pyroprocessing system as a slurry in the wet process, as a powder in the dry process. A rotary kiln is the common element in all pyroprocessing systems, and it will always contain the burning zone and all or part of the calcining zone. All the pyroprocessing steps occur in the rotary kiln in wet-process and long, dry-process (i.e., no preheater) systems. The application of chemical engineering principles to cement pyroprocessing has resulted in equipment additions to the rotary kiln system that can accomplish preheating and most of the calcining more quickly and efficiently outside the kiln. Rotary kilns are rotating, cylindrical steel tubes with length-to-diameter ratios in the approximate range of 15:1 to 35:1. The size of the kiln and its relative proportions are determined by the type and capacity of the pyroprocessing system. Wet-process kilns of over 700 ft in length and 23 ft in diameter are in operation. However, many wet and all dry-process kilns in the United States are smaller. Dry-process kilns that are equipped with preheaters are shorter yet. The kiln rotates about the longitudinal axis, which is slightly inclined to the horizontal, at a speed of from 1 to 3.5 rpm.

Table 2. Fabric Filters for Raw Mill Systems

	Pulse Jet			Reverse Air/Shaker
Acfm	10,000–50,000	10,000–50,000	10,000–50,000	10,000–50,000
Filter design	Tubular bag	Pleated	PTFE membrane on tubular bag	Tubular bag
Filter type	Polyester	Spun-bonded polyester	Felt with PTFE membrane	Polyester
Temp.(°F)	<275	≤ 265	≤ 275	<275
A/C ratio[a]	$\leq 5.0:1$	$\leq 2.5:1$	$\leq 5.5:1$	$\leq 2.5:1$
Inlet loading (gr/acf)	5–20	5–20	5–20	5–20
Outlet emission (gr/acf)	<0.02	<0.01	<0.005	0.02

[a] Air-to-cloth ratio (acfm/ft^2).

Refractory material lines the kiln to protect the steel shell from the intense heat and to retain heat within the kiln. The inclination and rotation of the tube result in the transport of solid materials from the upper or feed end to the lower or discharge end. The solids (i.e., load) occupy no more than 15–20% of the internal volume of the rotary kiln inside the refractory. There will be hundreds of tons of material within the kiln at any particular time. Material transit time is measured in hours for long kilns and in fractions of hours for preheater or calciner kilns. Heat energy is supplied at the discharge end of the kiln by the combustion of a variety of fuels. The flow of hot, gaseous combustion products is, therefore, countercurrent to the material flow. Heat is transferred from the flame and hot gases to the solid materials to provide the driving force for the required chemical reactions. The solid material is heated to about 2700°F by flame temperatures in excess of 3400°F.

Wet-process and long, dry-process pyroprocessing systems consist solely of the simple rotary kiln. Heavy chains are suspended in the drying or preheat zone at the feed end of the kiln to improve heat transfer from the hot gases to the solid materials. These chains are attached to the inside of the kiln shell in various patterns. As the kiln rotates, the chains are raised and exposed to the hot gases. Further kiln rotation causes the hot chains to fall into the cooler materials at the bottom of the kiln, thereby transferring the heat to the load.

Dry-process pyroprocessing systems have been improved in thermal efficiency and productive capacity through the addition of one or more cyclone-type preheater vessels in the gas stream after the rotary kiln. The vessels are arranged vertically, in series, and are supported by a structure known as the preheater tower. Hot exhaust gases from the rotary kiln pass countercurrent through the downward-moving raw materials in the preheater vessels. Compared with the simple rotary kiln, the heat transfer rate is significantly increased, the degree of heat utilization is more complete, and the process time is markedly reduced owing to the intimate contact of the solid particles with the hot gases. The required length and material retention time of the rotary kiln is thereby reduced.

The hot gases from the preheater tower are often used as a source of heat for the drying of raw materials in the raw mill. The mechanical collectors, fabric filters, and/or ESPs that follow the raw mill are production machines capturing valuable product as well as pollution control devices.

Additional thermal efficiencies and productivity gains have been achieved by diverting some fuel to a calciner vessel at the base of the preheater tower. At least 40% of the thermal energy is required in the rotary kiln. The amount of fuel that is introduced to the calciner is determined by the availability and source of the oxygen for combustion in the calciner. If available and allowed by environmental regulations, calciner systems can use lower-quality fuels (e.g., less volatile matter) and thereby further improve process economics.

In preheater and calciner kiln systems, it is possible efficiently to remove undesirable volatile constituents through a bypass system located between the feed end of the rotary kiln and the preheater tower. Otherwise, the volatile constituents would condense somewhere in the preheater tower and subsequently recirculate to the kiln. Buildups of these condensed materials can also restrict process flows by blocking gas and material passages. In a bypass system, a portion of the kiln exit gas stream is withdrawn and quickly cooled by air or water to condense the volatile constituents to fine particles. The solid particles are removed from the gas stream by fabric filters or ESPs.

Figure 6 is a flow diagram of a four-stage preheater with a calciner pyroprocessing system that is equipped with an alkali bypass and a reciprocating grate clinker cooler. Figure 7 shows

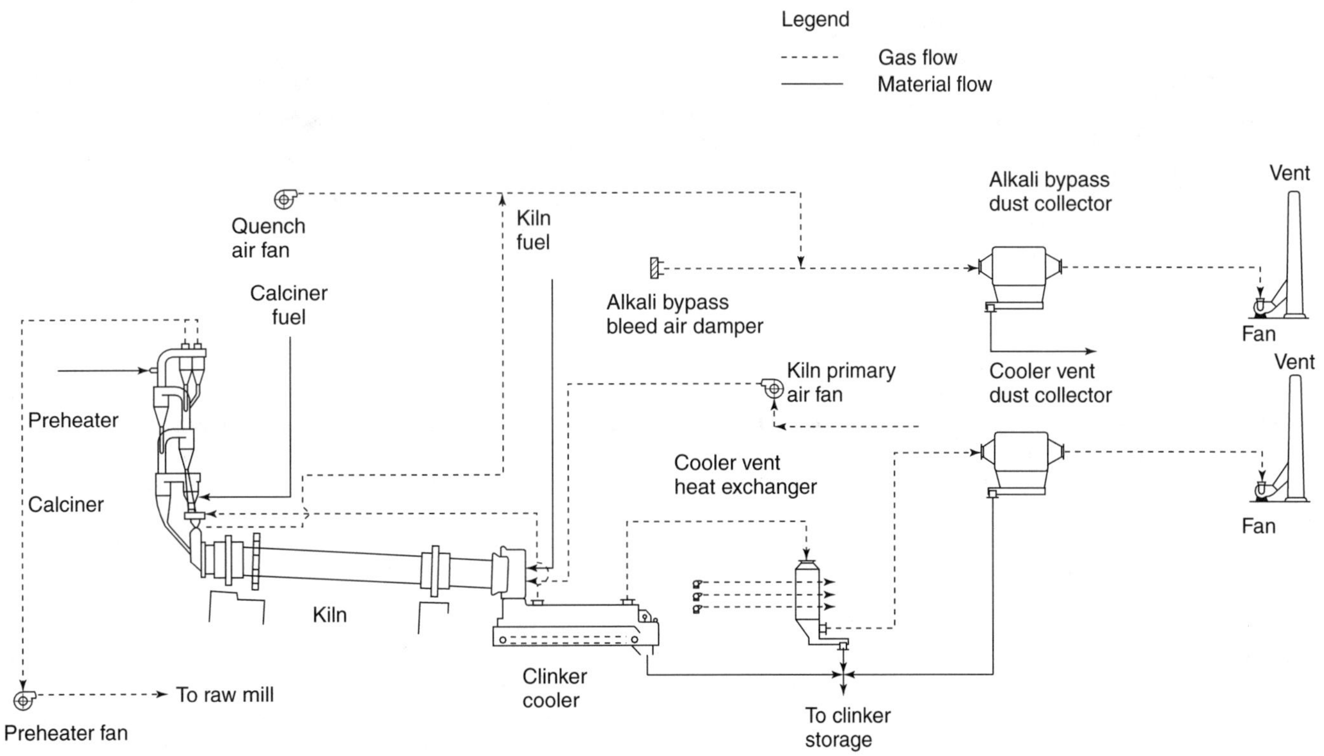

Figure 6. Pyroprocessing System Schematic Diagram.

Figure 7. Preheater and Rotary Kilns.

a four-stage preheater kiln system next to a traditional rotary kiln.

Solid fossil fuels (i.e., coal and petroleum coke) are ground to a fine powder [e.g., 80% passing a 200-mesh (74-μm) sieve] before being blown into the rotary kiln or calciner for combustion. Two classes of coal preparation systems are employed, direct and indirect firing. Direct-firing systems were exclusively found in older plants, but many of these systems have been converted to or replaced by indirect firing. New plants are equipped with indirect firing systems. Indirect firing systems are covered by NSPS contained in 40 CFR 60, Subpart Y, *Standards of Performance for Coal Preparation Plants*. Because direct-fired systems do not vent directly to the atmosphere, these standards do not apply to them.

Direct-fired coal/coke systems obtain hot air for drying and material transport from the clinker cooler or kiln hood. Solid fuel is fed to the coal mill, dried, ground, and transported directly to the burning zone of the rotary kiln without storage. The relatively large quanity of air required for drying results in more primary air in the flame than is desirable. Excess primary air contributes to inefficient combustion and unnecessary NO_x formation. Direct-fired systems are unsuitable for firing coal in calciners or in a riser duct for NO_x control.

Most often, indirect-fired coal/coke systems obtain the hot air for drying and material transport from the pyroprocessing system exhaust. Hot gases from the clinker coller can be used if proper safety precautions are taken. The ground coal is collected in a fabric filter and temporarily stored in feed bins prior to introduction into the burning zone of the rotary kiln, the calciner, or the riser duct. Relatively small quantities of ambient air are required to transport the independently metered quantities of solid fuel from the storage bin to the selected firing point. Indirect firing is a prerequisite for the installation of low-NO_x burners in the rotary kiln. Storing and handling powdered coal has the potential for fires and explosions; therefore, safety systems are installed to minimize the chance of catastrophic events.

Air Emissions Characterization

In simple rotary kiln systems, some finely divided particles of raw mix, calcined kiln feed, clinker dust, and volatile constituents (e.g., potassium sulfate) are entrained in the exiting gas stream. These particles are almost entirely removed from the gas stream before the combustion products are vented to the atmosphere. Affected pyroprocessing systems always meet or exceed the NSPS for particulate emissions from portland cement plants. Even those plants built prior to 1971 that are not subject to NSPS usually meet these standards for particulate emissions.

The powder that is collected from the kiln exhaust gases is known as cement kiln dust (CKD). Most plants return all or a portion of the CKD to the process; others completely remove it from the process. The chemical composition and physical state of the CKD depend on the type of pyroprocessing system, the chemical composition of the raw materials and fuel, and the state of the process at any given time. The composition of the relatively coarse CKD that is caught in the first field of an ESP is similar in chemical and physical characteristics to the raw mix, and in the last field it is very similar to that of the finer particulate emissions from the kiln stack. The same generalization cannot be made about CKD caught in a fabric filter, since there is almost no segregation of particles in a fabric filter. Specifications for portland cement often contain limitations on the quantity of sodium and potassium. Since the volatile oxides and salts of these metals tend to migrate or partition to the CKD, a portion or all of the CKD is sometimes removed from the pyroprocessing system to meet product quality standards. Bypass CKD is rich in sodium and potassium and is not returned to the process. The dust emanating from a preheater tower has the same general chemical and physical composition as the kiln feed and is returned to the process. The CKD that is removed from the system can be used for a variety of beneficial purposes (e.g., waste stabilization) or is managed at the cement plant in a monofill. The handling, storage, and deposition of CKD can result in fugitive dust emissions.

The cleaned bypass gases may be used in the raw mill, vented through a separate stack, or combined with kiln gases in the main kiln stack. The preheater gases may be vented to the atmosphere after particulate removal or used in the raw mill with no treatment other than temperature control. If the plant was built or modified after August 17, 1971, the preheater and bypass exit gases must meet the NSPS opacity limit of 20% and a mass emission limit of 0.15 kg/mg of particulate emissions per metric ton (0.30 lb/ton) of dry kiln feed on a combined basis from all emission points, regardless of the treatment or use of combustion products and tempering air from the pyroprocessing system. The principal gaseous emissions from the pyroprocessing system in a typical descending order by volume are nitrogen, CO_2, water, oxygen, NO_x, SO_2, CO, and hydrocarbons. The volumetric composition range of these constituents is from about 73% to less than 10 ppm. CO_2 and the last four listed gases are the primary constituents of environmental concern.

Emission rates of SO_2 and NO_x display a wide range of values throughout the industry. It is impossible to characterize the industry for gaseous emissions of SO_2 and NO_x with a single number or narrow numerical range. Each individual pyroprocessing system has its own emission characteristics, and the SO_2 and NO_x emissions from proposed or untested pyroprocessing systems are sometimes difficult to predict accurately. Extensive continuous monitoring of several cement plants has shown that SO_2 and NO_x emissions from a single source will normally vary over a rather large range (e.g., an order of magnitude). Short-term tests, such as EPA Methods 6 and 7, can lead to very erroneous conclusions regarding SO_2 and NO_x emissions, since these methods represent nearly instantaneous process conditions.

Sulfur input to a pyroprocessing system is only from feed and fuel. The relative amounts of sulfur in the feed and fuel, the system design, the chemical form of the input sulfur, and the process conditions, such as the presence of an oxidizing or reducing atmosphere in the kiln, are the variables that determine the quantity of SO_2 emissions at any given time. Oxides of nitrogen result primarily from the combustion of fuel, although nitrogenous constituents in the raw mix may make a contribution to NO_x emissions. The two primary sources of NO_x from fuel combustion are known as fuel and thermal NO_x. Nitric oxide (NO) predominates among the oxides of nitrogen that are emitted from cement pyroprocessing systems.

The current NSPS for cement plants recognize the uncertainty about SO_2 and NO_x emission rates from cement kilns and the absence of widely used control technology through the absence of any emission standards for these pollutants. It is often necessary, however, to include air pollution permit limitations on SO_2 and NO_x emissions from cement plants to meet prevention of significant deterioration (PSD) regulatory requirements and/or ambient air quality standards. Attempts to control emissions of SO_2 and NO_x from cement plants are becoming more prevalent and are meeting with success.

Indirect-firing coal systems can divert approximately 10% of the pyroprocess exit gases to the coal mill to dry the solid fossil fuel. These gases contain the products of combustion and calcination. If the coal mill is vented through a stack with a relatively low elevation, the constituents in these gases may dominate the ambient dispersion modeling for the plant. It may be necessary to recombine the coal mill vent gases with the balance of the kiln gases in the main stack to better disperse the coal mill vent gases. If the gases are combined, the more stringent NSPS standard for coal preparation plants will apply to the main stack. Fabric filters on coal mill applications generally have electrically grounded filter bags to eliminate static charges on the filter medium. If electrical sparks occur, a coal-dust explosion or fire could result. Explosion venting is normally included the fabric filter design to reduce the damage to the filter bags and/or the fabric filter enclosure if an explosion were to occur. Explosion venting also provides a pressure relief so that fire does not damage in-line equipment. The fabric filters used in coal mill applications are made from common filter media such as polyester, acrylic, and spun-bonded polyester. PTFE membrane is sometimes added to the filter medium to reduce the amount of coal dust retained on the filter and to lower the pressure drop across the dust collector. Air pollution control equipment on the indirect coal systems is pulse-jet fabric filters (see Table 3).

Table 3. Fabric Filters for Indirect-Fired Coal Systems

	Pulse Jet		
Acfm	10,000–50,000	10,000–50,000	10,000–50,000
Filter design	Tubular bag	Pleated[a]	PTFE membrane on tubular bag
Filter type	Polyester	Spun-bonded polyester	Felt with PTFE membrane
Temp.(°F)	$\leq$275	$\leq$265	$\leq$275
A/C ratio[b]	$\leq$4:1	$\leq$2.0:1	$\leq$4:1
Inlet loading (gr/acf)	10–55	10–55	10–55
Outlet emission (gr/acf)	<0.02	$\leq$0.01	$\leq$0.005

[a] Not recommended when blending petroleum coke.
[b] Air-to-cloth ratio (acfm/ft^2).

AIR POLLUTION CONTROL MEASURES

Air pollution control equipment on the kiln system includes reverse air fabric filters, pulse-jet fabric filters, and ESPs. Acoustic horns are sometimes used in both devices to assist in cleaning (see Table 4). Cement kiln systems have highly alkaline internal environments that can absorb up to 95% of potential SO_2 emissions. Exceptions to the generalization are found in systems that have sulfide sulfur (pyrites) in the kiln feed. Without unique design considerations or changes in raw materials, the overall sulfur absorption rate may be as low as 70%. Historically, the cement kiln system itself has been determined to be best available control technology (BACT) for SO_2 emissions. In fabric filters, there must be an absorbing reagent (e.g., CaO) in the filter cake for SO_2 capture to occur. Without the presence of water, which is undesirable in the operation of a fabric filter, $CaCO_3$ is not an SO_2 absorbing reagent. It has been observed that as much as 50% of the SO_2 can be removed from the pyroprocessing system exhaust gases when this gas stream is used in an in-line raw mill for heat recovery and drying. In this case, moisture and calcium carbonate are simultaneously present for sufficient time to accomplish the chemical reaction with SO_2. Vendors of cement pyroprocessing systems are developing methods to recirculate calcined raw mix or CKD to the process to provide calcium oxide as a reagent for additional internal scrubbing of SO_2. When ambient air quality standards have been threatened, a small number of proposed cement kilns are installing tail pipe wet scrubbers that use calcium carbonate as the reagent for SO_2 removal.

Energy-efficient pyroprocessing systems have the potential to emit less SO_2 than inefficient systems because of the lower sulfur input resulting from reduced fuel consumption. In addition, preheater and calciner kilns have the capacity to absorb relatively large quantities of fuel-derived SO_2. Similarly, raw materials with the lowest content of sulfide sulfur can result in the lowest SO_2 emissions. Selective quarrying or a change in raw materials can lower the input of sulfur to the pyroprocessing system. As noted previously, pyroprocessing systems with in-line raw mills are able to absorb up to half of the potential SO_2 emissions when the mill is in operation.

Several mechanisms for the control of NO_x emissions from cement kilns are in existence. They meet with varying degrees of success. Stable kiln operation, such as is often found in a precalciner system, has been shown to minimize cumulative, long-term NO_x emissions. However, short-term spikes of NO_x emissions during process upsets are unavoidable, since a higher than normal input of thermal energy from the combustion source is required to restore the process to equilibrium. Several equipment vendors sell burner configurations for the rotary kiln that are alleged to reduce NO_x emissions. A form of staged combustion can be used on preheater or precalciner kilns to reduce NO_x emissions. Fuel is burned under reducing conditions in the riser duct from the rotary kiln to the preheater or precalciner to generate CO. This CO chemically reduces the NO_x generated in the kiln to elemental nitrogen. The oxygen-deficient gases thereby generated can be supplied to the calciner to further reduce NO_x generation in that low-temperature combustion source. Pyroprocessing system vendors offer other, similar methods to achieve NO_x reduction through combustion with oxygen-deficient gases or reaction of NO_x with CO. It is possible to inject ammonia or urea into a preheater tower at a point where

Table 4. Kiln System Dust Collectors

	Reverse Air		Pulse Jet	Precipitator
Acfm	50,000–300,000	50,000–300,000	50,000–300,000	50,000–300,000
Filter design	Tubular bag	Tubular bag	Tubular bag	N/A
Filter type	Fiberglass	Fiberglass with PTFE membrane	Fiberglass with PTFE membrane	N/A
Temp.(°F)	350–500	350–500	350–650	350–650
A/C ratio[a]	1.5:1 net	2.0:1	$\leq$3.5:1 net	SCA: 350–500[b]
Inlet loading (gr/acf)	4–18	4–18	4–18	4–18
Outlet emission (gr/acf)	0.02	<0.005	0.005	0.02

[a] Air-to-cloth ratio (acfm/ft^2).
[b] Specific collecting area (ft^2/1000 acfm).

the gas temperature is about 1800°F to achieve a beneficial reaction between ammonia and NO_x. To date, this technology, selective noncatalytic reduction (SNCR), has not been used on a continuous basis in the United States. Ammonia injection does not appear to be possible in pyroprocessing systems with only a rotary kiln, since the point of optimum temperature is not accessible through the rotating kiln shell. Other possibilities for NO_x emissions reduction exist in the recirculation of flue gas as oxygen-deficient primary air in the rotary kiln, alternative or low-nitrogen fuels and/or raw materials, and mid-kiln injection of used tires or other combustible solids as a supplementary fuel.

CLINKER COOLING

Process Description

The clinker produced in a rotary kiln is cooled in a device called a clinker cooler. This process step recoups up to 30% of the heat input to the kiln system, locks in desirable product qualities by freezing mineralogy, and makes it possible to handle the cooled clinker with conventional conveying equipment.

Depicted in Figure 8 is the predominant type of clinker cooler, the reciprocating grate, in which clinker is cooled from about 2000°F to 250°F by ambient air passing through a horizontal bed of clinker. A portion of the resulting hot air is used as combustion air in the kiln. Cooling air not utilized in the rotary kiln for combustion is vented to the atmosphere, used for drying coal or raw materials, or used as a source of heated combustion air in a precalciner. If necessary to further reduce clinker temperature, secondary coolers with direct or indirect air to clinker contact may be employed. If necessary, air from these clinker coolers is vented to atmosphere through fabric filters.

The reciprocating grate cooler consists of a horizontal box of rectangular cross section that houses horizontal rows of fixed and movable grate plates that bisect the cross section. A typical grate plate has multiple passages for air and is about 1 foot square. A row of grate plates is perpendicular to the flow of the clinker and may consist of 6–12 grate plates, depending on the cooler capacity. The clinker cooler is normally oriented so that the clinker continues its flow in line with the longitudinal axis of the kiln as it moves along the top of the grates. Typically, the reciprocating movement of every second row of grates forces the clinker through the cooler. Other methods to move the clinker through the cooler are being introduced by equipment vendors. Ambient air is forced through the grates and the bed of clinker from the chamber below by a series of fans along the length of the cooler.

Only a portion of the air supplied to a reciprocating grate cooler will be used for combustion air in the kiln, combustion air in the calciner, or hot air for the coal or raw mills. The excess air must be cleaned of clinker dust before it is vented to the atmosphere.

Air Emissions Characterization

The collected dust from clinker coolers is fairly coarse, with only about 0–15% of it finer than 10 μm. This abrasive dust

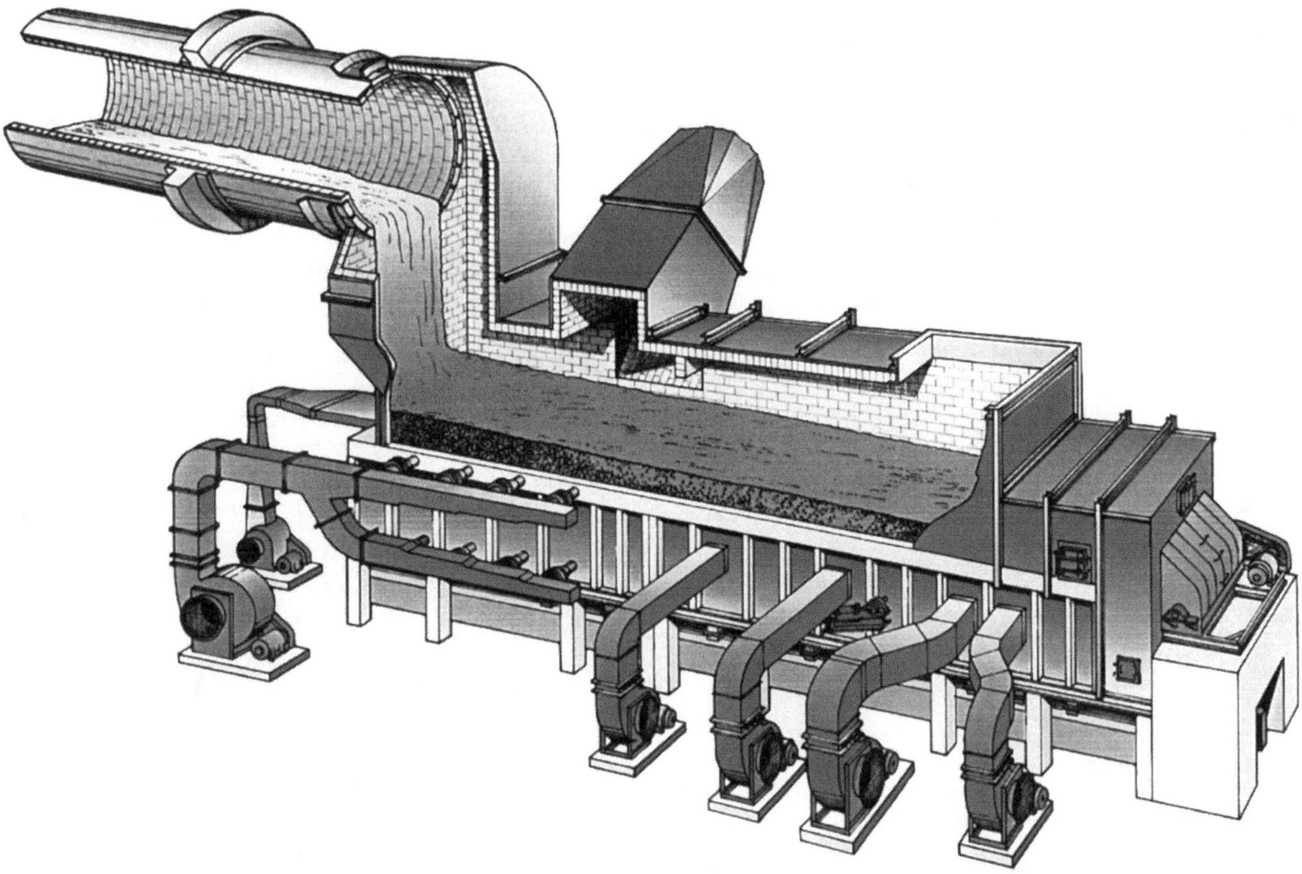

Figure 8. Reciprocating Grate Clinker Cooler.

consists solely of cement minerals and is returned to the process.

The quantity of air used for cooling is about 1–3 pounds per pound of clinker, depending on the efficiency of cooling and the desired temperature of the clinker and vent gas. If some of the gases are used for the drying of coal or for other purposes, then the volume of gas to be cleaned at the cooler vent may be reduced by 10–100%.

The dust content of the cooler exhaust gases is affected by the granular distribution of the clinker, the degree of burning of the clinker, the bulk density of the clinker (i.e., liter weight), and the flow rate of the cooling air. Frequently, a clinker breaker (e.g., hammer mill) is located at the discharge of the cooler and may increase the dust burden.

If applicable, NSPS specify a mass emission limit on the cooler vent stack of 0.05 kg/mg of particulates per metric ton (0.10 lb/ton) of dry kiln feed. An opacity limit of 10% also applies to the cooler stack. If the cooler gases are used for drying in the raw mill, the more stringent mass emission and opacity limits for the cooler stack rather than the kiln stack apply to the raw mill vent.

Air Pollution Control Measures

Upsets in the kiln can rapidly increase the vent gas temperature to 1000°F and the dust load to 13–50 gr/acf. In older plants, there may be bypass arrangements to vent these gases directly to the atmosphere until the upset is over. These particulate emission excursions are not permitted in newer plants. Gas temperatures are controlled to protect the dust collector through the use of tempering bleed air, water sprays, or an air-to-air heat exchanger. All these methods have costs and limitations, and there is no clear universal solution. In a few plants with air-to-air heat exchangers, cooled excess air is recirculated to the cooler, thereby eliminating the need for a clinker cooler vent stack.

The dust collectors used on reciprocating grate clinker coolers are most often fabric filters. ESPs and gravel bed filters are infrequently used. Sometimes clinker cooler dust collectors are preceded in the gas flow stream by a mechanical cyclone or multiclone dust collector. Typical fabric filters on clinker coolers are pulse jets or pulsed plenums in the newer plants and reverse-air types in the older plants (see Table 5).

CLINKER STORAGE

Process Description

To allow for necessary operational flexibility, a cement plant is usually able to store from 5% to 25% of its annual clinker production capacity. The storage requirement largely depends on the shipping cycle. Northern plants usually manufacture and store clinker during the inactive winter months for grinding during the summer shipping season.

The material-handling equipment used to transport clinker from the clinker coolers to storage and then to the finish mill department is similar to that used to transport raw materials (i.e., belt conveyors, screw conveyors, deep-bucket conveyors, and bucket elevators). Drag chains are sometimes used because they are less sensitive to abrasion and high temperatures during upset conditions. Gravity drops and transfer points in the conveying and storage systems are normally enclosed and vented to dust collectors.

Older plants were typically designed to store clinker in partially enclosed buildings and storage halls or in outside piles. Newer and modernized plants store at least some clinker in fully enclosed storage halls or cylindrical, vertical silos. Clinker storage domes are gaining in popularity.

Air Emissions Characterization

Dust in the clinker has a tendency to become airborne during handling. The character of the dust varies by plant and existing process conditions. Dust caught in the clinker cooler exhaust dust collector is often returned to the clinker stream and can result in reentrainment of this material in air during subsequent handling. Clinker dust is normally a small proportion of clinker production and is relatively coarse, but some kilns normally produce dusty clinker. During process upsets when the kiln falls below clinkering temperatures and runs "raw," material that is discharged from the kiln is said to be "unburned" (i.e., not fused into clinker) and is very dusty.

Air Pollution Control Measures

The air pollution control measures and equipment used in clinker handling systems are similar to those described for raw milling.

The free fall of clinker onto storage piles usually creates visible, fugitive particulate emissions. This dust generation can be reduced by discharging the clinker to piles through a simple device known as a rock ladder or by using variable-height, stacker belt conveyor systems. Water sprays are incompatible with clinker dust. However, fugitive dust emissions from the surface of open clinker storage piles are mitigated by precipitation, which causes a crust to form on the piles. Wind breaks and pile covers (e.g., tarpaulins) have also been used to minimize fugitive clinker dust with mixed success. Clinker in open piles is usually reclaimed with mobile equipment, such as front-end loaders. Clinker in storage halls is frequently

Table 5. Fabric Filters on Clinker Coolers

	Pulsed Plenum/Pulse Jet	Reverse Air	Precipitator
Acfm	20,000–100,000	20,000–100,000	20,000–100,000
Filter design	Tubular bag	Tubular bag	N/A
Filter type	Nomex, polyester	Nomex, fiberglass	N/A
Temp.(°F)	≤400, <275	≤400, <500	350–600
A/C ratio[a]	≤5:1 net	2:1 net	SCA: 350–500[b]
Inlet loading (gr/acf)	5–10	5–10	5–10
Outlet emission (gr/acf)	0.02	0.02	0.02

[a] Air-to-cloth ratio (acfm/ft^2).

[b] Specific collecting area (ft^2/1000 acfm).

handled with overhead bucket cranes. Fugitive clinker dust originating from material handling operations around open storage piles is often observed and is very difficult to control.

FINISH MILLING

Process Description

The final step in the manufacture of portland cement is the grinding of clinker to a fine powder. Up to 5% by weight of gypsum and/or natural anhydrite is added to the clinker during grinding to control the setting time of the cement. In the industry, this step is called finish grinding or milling and is accomplished in the finish mill department. Small amounts of various other chemicals may be added to the cement during finish grinding to function as processing additions (e.g., grinding aids) or to impart special properties to the cement or resulting concrete (e.g., air entrainment). Small amounts of water are often sprayed into a cement finish mill to aid in cooling the mill and the cement. Other specification and nonspecification cements with unique properties and constituents can be prepared in the finish mill department. For example, pozzolans or blast furnace slag can be mixed with portland cement clinker and gypsum during the finish grinding process to produce blended cements. In the United States, a common speciality cement derived from portland cement clinker is masonry cement. Typically, a masonry cement is composed of equal portions of portland cement clinker and limestone to which 2–4% by weight of gypsum is added. In addition, chemicals that impart the properties of air entrainment, plasticity, and water repellency to a mortar are added. Each manufacturer of masonry cement has a proprietary formula for its product.

Finish milling is almost exclusively accomplished in ball or tube mills. These mills are rotating, horizontal steel cylinders containing slightly less than half their volume in steel alloy balls, which are called grinding media. These balls can range in size from 4 to $\frac{1}{2}$ in. in diameter. Clinker and gypsum are introduced into the feed end of the mill, and partially ground portland cement exits from the discharge end. A finish mill might be divided into two or more internal compartments in which the grinding media are segregated by size. The larger balls are at the feed end of the mill. The compartments of the mills are formed by slotted division heads that are perpendicular to the mill's longitudinal axis and cover the entire circular cross section of the mill. The slots are small enough to retain the grinding media in the proper compartment, but large enough to allow the partially ground cement to flow toward the discharge end. At the discharge end of the mill, there is a similar slotted barrier called a discharge grate that serves to keep the balls in the mill while allowing partially ground cement to exit. A given particle of cement remains in the mill for 3–7 min. The ends (heads) and sides (shells) of the mills are lined with replaceable alloy steel plates or castings that undergo the wear and abrasion of the grinding process. Mill shell linings are sometimes designed so that the balls are segregated by size during mill operation, thereby eliminating the need for division heads.

Cement is usually ground in a closed circuit with an air separator. This continuously operating device is used to separate particles of cement of acceptable size in the material discharged from the mill from those particles that have not been fully ground. The large particles (i.e., tailings) are returned to the mill and reintroduced to the feed end along with new feed. A figure that is 100 times the ratio of the weight of the returned tailings to the weight of new feed is called the circulating load and is expressed in percent. Circulating loads in the range of 200–500% are typical, but higher and lower circulating loads are found in acceptable mill circuits. Air separators are mechanical devices that use centrifugal force, gravity, and an ascending air current to separate the cement particles. Older air separators have a relatively low separation efficiency. Equipment manufacturers now offer high-efficiency separators that are included in most new finish mill projects and are popular retrofit items because of the increased efficiency, lower operating costs, and improved product performance.

Another device of increasing popularity in the finish mill department is the roll crusher. This device accomplishes the initial size reduction of the clinker and the gypsum outside and prior to the ball mill. The efficiency and/or the productive capacity of a given ball mill is thereby increased at potentially lower grinding temperatures.

For a variety of reasons, cement customers usually demand cement that is at temperatures of 100–150°F when delivered. Cement grinding temperatures can reach 350°F. The high-efficiency separator circuits, with their associated high volumes of mill vent air, provide better cooling of the cement than conventional mill circuits. Water-supplied cement coolers (i.e., heat exchangers) are often installed in the material flow path following the finish mill to reduce the cement temperatures prior to product storage. No air pollution problems are associated with the fully closed cement coolers.

Figure 9 is a process flow sheet for a finish mill circuit that includes a roll crusher and a high-efficiency air separator. Figure 10 shows a two-compartment ball mill in finish mill service as viewed from the feed end of the mill.

Air Emissions Characterization

Particulate emissions from mill vents, air separator vents, and material-handling system vents constitute the air pollution concerns in the finish mill department.

About 30–40% of the particles of ordinary Type I portland cement are finer than 10 μm. For Type III, high-early-strength portland cement, the percentage of particles finer than 10 μm increases to the 45–65% range. Typically, about 90% of portland cement will pass a 325-mesh (44-μm) sieve. The potential air pollution problems associated with the manufacture, handling, and transportation of portland cement have their origin in the large proportion of very fine particles in the product.

Air Pollution Control Measures

Emissions from finish mills are controlled adequately by fabric filters. The fabric filters most often found on new or upgraded plants are the pulse-jet and pulsed-plenum types. Reverse air/shaker fabric filters are typically found in older plants. In almost all cases, pulse-jet or pulsed-plenum fabric filters are installed in conjunction with high-efficiency separators (see Tables 6 and 7).

The cement dust caught in a fabric filter is returned to the process. In colder weather, the water that is used for internal mill cooling can produce a steam plume at the mill baghouse vent. This plume is sometimes confused with excessive particulate emissions, but it will dissipate within

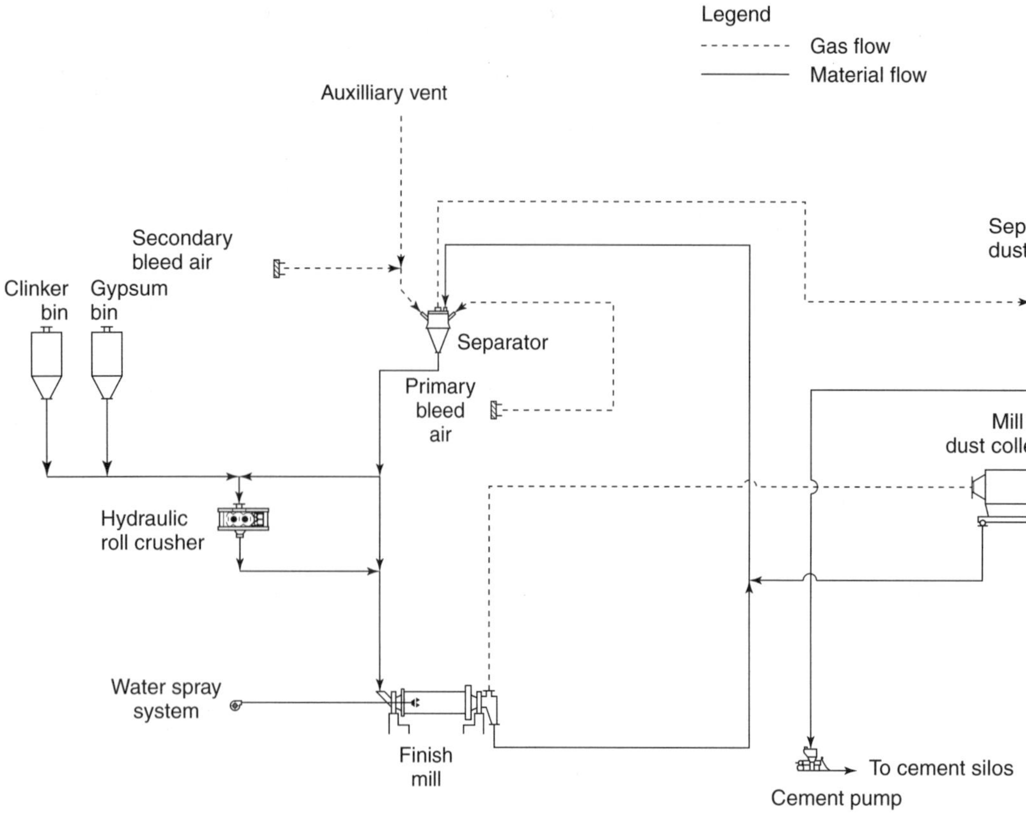

Figure 9. Finish Mill Schematic Diagram.

Figure 10. Finish Mill.

Table 6. Fabric Filters for Finish Mill Systems

	Reverse Air/Shaker		Pulse Jet		Pulse Plenum
Acfm	10,000–30,000	10,000–30,000	10,000–30,000	10,000–30,000	10,000–30,000
Filter design	Tubular bag	Pleated	Tubular bag	PTFE membrane on tubular bag	Tubular bag
Filter type	Polyester	Spun-bonded Polyester	Polyester	Polyester with PTFE membrane	Polyester
Temp.(°F)	≤275	≤265	≤275	≤275	≤275
A/C ratio[a]	≤2.5:1	≤2.5:1	≤4:1	<5:1	4:1
Inlet loading (gr/acf)	5–20	5–100	5–100	5–100	5–100
Outlet emission (gr/acf)	0.02	<0.01	<0.02	0.005	0.02

[a]Air-to-cloth ratio (acfm/ft^2)

Table 7. Fabric Filters for High-Efficiency Separators

	Pulse Jet			Pulse Plenum
acfm	40,000–60,000	40,000–60,000	40,000–60,000	40,000–60,000
Filter design	Pleated	Tubular bag	PTFE membrane on tubular bag	Tubular bag
Filter type	Spun-bonded polyester	Polyester felt	Polyester with PTFE membrane	Polyester
Temp.(°F)	<265	<275	<275	<275
A/C Ratio[a]	<2.0:1	<4.0:1	<5.0:1	≤3.5:1
Inlet loading (gr/acf)	150–300	150–300	150–300	150–300
Outlet emission (gr/acf)	<0.01	<0.02	<0.005	0.02

[a]Air-to-cloth ratio (acfm/ft^2).

a few feet of the vent opening. Fabric filters on finish mill systems that use cooling water must be well insulated to prevent condensation within the baghouse and subsequent blinding of the bags.

PACKING AND LOADING

Process Description

Portland cement is pneumatically conveyed from the finish mill department to large, vertical, cylindrical concrete storage silos in the packhouse or shipping department. Storage domes are coming into use. Mechanical transfer systems, such as bucket elevators, belt conveyors, screw conveyors, and air-slide conveyors, supplement the pneumatic system.

The number and capacity of the storage silos or domes depend on the capacity of the plant, the number (i.e., types) of cements in the product mix, the marketing strategy of the company, and the weather-driven shipping pattern.

Portland cement is withdrawn from the storage silos or domes by a variety of feeding devices and conveyed to loading stations in the plant or directly to transport vehicles using the same kinds of material-transfer systems that were used to put the cement into the silos. Most of the portland cement is shipped from the plant in bulk by rail or truck transport. Those plants located adjacent to water transportation routes usually serve some customers or distribution terminals by barge or ship.

Portland cement is also shipped in multiwall paper bags with a capacity of 94 pounds. These bags are filled on automatic or semiautomatic packing machines. During filling, each bag is vented to a bag filter to allow the escape of displaced air. The filled bags are then manually or mechanically palletized for shipment.

Masonry cement is almost totally shipped in multiwall paper bags. Bag weights range from 70 to 80 pounds, depending on the type of masonry cement in the bag. The packing, palletizing, and dust suppression operations are identical to those used for portland cement.

There are remote distribution terminals associated with some cement plants. Bulk or packaged cement is shipped in advance from the plant to the terminal to provide subsequent timely distribution to customers. Terminals are most often supplied by rail or barge shipments, although trucks and ships are sometimes used. The handling and loading of bulk portland cement at distribution terminals are carried out by the same kinds of pneumatic and mechanical conveying systems as are used at the plant.

Air Emissions Characterization

Particulate emissions from the silo openings, cement-handling equipment, bulk and package loading operations, and the fabric filters constitute the air pollution concerns in the shipping department.

Air Pollution Control Measures

Active and passive fabric filters are used to remove dust from the exhaust airstreams from the silos, domes, transport systems, and packaging operations. The cement dust is returned to the product.

The dust generated during the loading of trucks, railcars, barges, and ships is controlled by venting the transport vessel to a fabric filter. The collected dust is returned to the shipment of cement. Flexible loading spouts with concentric pipes are among the devices that are successfully used for dust-free loading. In a loading spout, the cement flows to the transport vessel by gravity through a central pipe, while air displaced from the transport vessel is drawn through an annular space.

Dust is controlled at distribution terminals through the venting of silos, bins, and transfer points to fabric filters. The captured cement dust is returned to the product.

The typical fabric filters used in the packing and loading departments of newer plants are of the pulse-jet type. Reverse-air or shaker-type fabric filters are found in older plants. Occasionally, a cartridge-type fabric filter will be employed (see Table 8).

SUPPLEMENTAL FUELS AND RAW MATERIALS

The recycling of wastes of other industries in portland cement kilns as fuel and raw material substitutes is a reliable and proven technology. This technology offers a safe, cost-effective and environmentally sound method of beneficial recovery of the energy and chemical values of selected wastes, thereby enabling a portland cement manufacturer to operate at lower cost.

The energy-bearing, ignitable wastes that are currently used in the portland cement industry as fossil fuel substitutes are waste oils, spent organic solvents, paints, organic sludges, and other combustible residues from the chemical, manufacturing, and petroleum industries. Smaller amounts of other waste streams are also being successfully recycled into cement kilns as fuel substitutes. Some waste streams require preparation so that they can be effectively introduced into the kiln. For example, liquids with high or variable chlorine levels are blended with other waste liquids to provide a lower or more consistent chlorine concentration in the waste-derived fuel (WDF). Sludges are liquefied, solidified, or encapsulated to provide better material-handling properties. Solids also may be ground to facilitate blending into liquid fuels. Materials such as sawdust can be handled with the same equipment as coal and are readily consumed in a cement kiln. Other high-energy waste streams, such as used rubber tires, are finding increased acceptance as fossil fuel substitutes.

The greatest economic and societal benefits from utilizing wastes in cement manufacturing are derived from the replacement of fossil fuel. Coal and petroleum coke provide about four-fifths of the thermal energy for manufacturing cement in the United States. In 1997, the portland cement industry obtained about 28 trillion Btu or 9.6% of its thermal energy requirements from waste-derived fuels. Higher substitution rates are technically possible.

Older, less fuel-efficient cement plants often derive the greatest economic benefit from waste fuel substitution because of their higher demand for fuel. These older facilities can successfully compete with more modern facilities as a result of lower operating costs and improved cash flow when fossil fuel is replaced with a cheaper substitute.

Cement kilns have several important characteristics that contribute to the effective destruction of organic waste materials. The gas residence time in the burning zone of the kiln at 3000°F or higher is approximately 3 sec. Temperatures in excess of 2000°F exist for as long as 6 sec. Test burns have repeatedly demonstrated destruction and removal efficiencies (DREs) of 99.99–99.9999% for even the most stable organic compounds.

Constituents of ash resulting from incombustible constituents in any waste (e.g., metals) primarily becomes chemically incorporated into the clinker crystal matrix or is caught with the CKD in the air pollution control device prior to the kiln stack. The more volatile metals (e.g., lead) migrate to the CKD, while refractory metals (e.g., chromium) are mostly found in the clinker. Since cement raw materials come from the earth's crust, lead and chromium, and other naturally occurring trace metals, are found in all cement and CKD. The use of waste-derived fuels and waste raw materials may increase the metal content of cement and CKD by small increments. No matter the source, however, metals emissions from cement kilns can be controlled to meet applicable environmental standards. The chemical and physical specifications for the performance of portland cement also make it necessary for a cement manufacturer to monitor the input of several trace elements that can be found in substitute fuels and raw materials.

An excessive chlorine input can contribute to material buildups within a pyroprocessing system (e.g., kiln rings), deterioration of ESP performance, and excessive corrosion of pyroprocessing system components. A cement manufacturer quickly learns the maximum chlorine feed rate that a

Table 8. Fabric Filters in Packing and Loading Department

	Pulse Jet			Pulse Plenum
Acfm	3,000–10,000	3,000–10,000	3,000–10,000	3,000–10,000
Filter design	Pleated	Tubular bag	PTFE membrane on tubular bag	Tubular bag
Filter type	Spun-bonded polyester	Polyester	Felt with PTFE membrane	Polyester
Temp.(°F)	≤265	≤275	≤275	<275
A/C ratio[a]	≤3.5:1	≤5.0:1	≤5.5:1	≤2.5:1
Inlet loading (gr/acf)	5–40	5–40	5–40	5–40
Outlet emission (gr/acf)	<0.01	<0.02	<0.005	0.02

[a] Air-to-cloth ratio (acfm/ft^2).

pyroprocessing system will tolerate and limits chlorine input from natural fuel and raw materials or waste substitutes to less than that amount. In addition, the feed rate of chlorine to each kiln using hazardous WDF is limited by regulatory constraints. Regulations for burning hazardous WDF in cement kilns were promulgated under RCRA by the EPA on February 21, 1991. The air emissions of pollutants of concern (i.e., CO, hydrocarbons, hydrogen chloride, and heavy metals) during waste burning for energy recovery are regulated under these regulations.

The cement-making process offers many unique opportunities for the utilization of noncombustible solid wastes. Silicon, aluminum, and iron are needed to react chemically with the calcium in the cement raw mix. Materials such as spent cracking catalyst, diatomaceous-earth filter media, foundry sand, and steel mill scale have high concentrations of these elements and are used to replace the conventional siliceous, argillaceous, and ferriferous components of the raw mix.

HAZARDOUS AIR POLLUTANTS

Two National Emission Standards for Hazardous Air Pollutants (NESHAPs) affecting the portland cement manufacturing industry are expected to be promulgated in 1999. The first of these standards applies to all operations in all cement plants and can be found at 40 CFR 63, Subpart LLL. The second standard only applies to the pyroprocessing emissions from those plants burning hazardous WDF and can be found at 40 CFR 63, Subpart EEE.

PROCESS AND QUALITY CONTROL

Modern cement plants are exclusively controlled by digital computers from central control rooms. Process variables in all manufacturing departments are continuously monitored. Usually, process control actions are initiated by the process control computer, but manual intervention is possible during process upsets, equipment malfunctions, or emergency conditions. Older cement plants use analog control systems in either central control rooms or departmental control stations.

Modern cement plants are usually equipped with continuous opacity monitors on the kiln and clinker cooler stacks. At many plants, gaseous emissions of oxygen, CO, NO_x, and SO_2 from the kiln stacks are also continuously monitored. These monitoring devices are reliable. Nevertheless, equipment redundancy may be required if there are to be minimal data gaps in a compliance-monitoring scheme. The location of these devices in the process is often hot and dirty, thereby complicating the monitoring task.

The portland cement process involves rather complex chemistry and requires close process control. Plant laboratories are staffed around the clock. Frequent chemical and physical tests are made on raw materials, raw mix, clinker, and cement. The procedures may range from elementary wet chemistry to more sophisticated testing by X-ray fluorescence. The newest cement plants are equipped with automatic sampling and analytical systems. The operation of the pyroprocessing system receives particularly close attention since product quality is largely determined in the kiln. If proper process conditions and kiln temperatures are not maintained, the complex chemical reactions that take place in the kiln are incomplete and the clinker is unacceptable. Catastrophic and expensive failure of the process can also occur.

REFERENCES

1. U.S. and Canadian Labor-Energy Input Survey, Portland Cement Association, January 1999.
2. 40 Code of Federal Regulations, Part 60, Subpart F, *Standards of Performance for Portland Cement Plants*, U.S. Government Printing Office, Washington, D.C. Last revised by 54 FR 6666, February 14, 1989.
3. 40 Code of Federal Regulations, Part 60, Subpart 000, *Standards of Performance for Nonmetallic Minerals Processing Plants*, U.S. Government Printing Office, Washington, D.C. Last revised by 62 FR 31359, June 9, 1997.
4. 40 Code of Federal Regulations, Part 60, Subpart Y, *Standards of Performance for Coal Preparation Plants*, U.S. Government Printing Office, Washington, D.C. Last revised by 54 FR 6671, February 14, 1989.
5. Compilation of Air Pollutant Factors (AP-42), Section 11.6, *Portland Cement Manufacturing*, Fifth Edition, U.S. Environmental Protection Agency, Research Triangle Park, NC, September 1995.
6. W. Greer, "The Calcining Zone of the Kiln and Other Considerations," Paper No. 11, Portland Cement Association, Kiln Optimization Short Course, Chicago, 1980.
7. W. Greer, "SO2/NO_x Control, Compliance with Environmental Regulations," *IEEE Trans. Industry Applications* **25**(3) (May/June 1989).

BIBLIOGRAPHY

R. H. Bogue, *The Chemistry of Portland Cement*, 2nd ed., Reinhold Publishing Corp., New York, 1955.

W. H. Duda, *Cement Data Book*, 3rd ed., Bauverlag GmbH, Weisbaden and Berlin, 1985.

O. Labalm and B. Kohlhaas, *Cement Engineers Handbook*, 4th Ed., Bauverlag GmbH, Wiesbaden and Berlin, 1983.

K. E. Peray, *Cement Manufacturers Handbook*, Chemical Publishing Co., New York.

K. E. Peray, and J. J. Waddell, *The Rotary Cement Kiln*, Chemical Publishing Co., New York, 1972.

VDZ Congress '85, Process Technology of Cement Manufacturing, Bauverlag GmbH, Wiesbaden and Berlin, 1987.

CERAMIC AND BRICK MANUFACTURING

Dennis A. Brosnan

Ceramic and brick products are considered "traditional" ceramics because of their long history of manufacture using well-known techniques of powder processing and a high-temperature processing step, known as firing, with the latter process to develop a permanent bond between residual particles in the composition. Traditional ceramic products contain clay minerals as a significant part of their composition. The clay functions in the product-forming stage to provide plasticity and cohesion in the powder mass and in the firing step as a precursor to formation of vitreous (glassy) phases, which, on cooling, provide for a permanent bond and for the useful properties of the product.

Ceramic products include wall and floor tile, which are generally made by powder pressing techniques, and whiteware products, which are generally made by slip casting techniques. The term *ceramic product* is used to describe a number of products including tile, toilets, decorative ceramics, dinnerware, porcelain products, and clay-based electrical insulators.

Whiteware products derive their name because the material is generally white in color in the fired and unglazed state facilitating coverage with light- or dark-colored glazes (glassy coatings). Whiteware products may also be produced by plastic-forming techniques such as plastic pressing and jiggering.

Brick and other structural clay products (quarry tile and roofing tile) have almost 100% of their composition as local clays, which have characteristics of plasticity and a reasonable firing range (range in temperatures allowing for desired properties). Approximately 95% of brick made in North America are produced by stiff extrusion followed by drying and firing. The balance of brick (~5%) are made by molding (pressing) of a wet clay mass — a process usually called the "soft mud" process. In Europe, by contrast, the predominant brick-forming processes are soft extrusion and molding. Quarry tile are produced by extrusion, and roofing tile usually employ a tandem process of extrusion followed by plastic pressing prior to drying and firing.

The firing step in producing traditional ceramics involves processing temperatures in the "red heat" range of 950–1250°C. The product is heated to these temperatures on a desired schedule involving specific preheating rates, dwell times, and cooling rates known to produce products meeting customer acceptance or performance-based specifications.

Traditional ceramic products are technically alumino-silicate ceramics since clays contain minerals with alumina (Al_2O_3) and silica (SiO_2) in their composition along with various trace components. These materials vitrify (form a significant amount of glass) at temperatures exceeding about 950°C. Clays that are used include kaolin, ball clay, shale, illite, and montmorillonite. Nonclay minerals such as chlorites may make up a major part of the raw material. Other constituents of traditional ceramic batches may include minerals such as potter's flint (ground silica), feldspar (sodium or potassium alumino-silicate), pyrophyllite (alumino-silicate), talc (magnesium alumino-silicate), and wollastonite (calcium silicate).

Product specifications for traditional ceramic products used in North America are given in ASTM Standards.[1] These product specifications usually provide for minimum values of physical properties that are believed to provide for acceptable performance in a majority of product applications. International product standards are increasingly applied to products made in North America, and efforts are underway to provide for uniform international standards in fields like ceramic tile and whitewares.

There remain significant differences between use of some products, such as brick, in different parts of the world. In Europe, for example, much of the market for brick is in "structural products" used in load-bearing walls covered with stucco (where product color is of small concern). In the United States, customer demands for certain colors in facing brick dictate differences in processing as compared to Europe. This simply means that certain technology developed offshore may not be directly or completely usable for brick production in the United States.

PROCESS DESCRIPTION

The three processes used in forming traditional ceramic products by powder processing methods are (1) slurry processing (slip casting), (2) plastic forming (extrusion and plastic molding, etc.) and (3) dry pressing. Each forming process is preceded by powder preparation steps and followed by drying and firing steps.

Slip Casting

The process of slip casting is used in manufacturing many whiteware products, including most sanitary ware products. Typical batch compositions for slip cast products are shown in Table 1. Whiteware producers usually process prepared raw materials from vendors instead of mining and processing (crushing/grinding) local materials.

All the solid ingredients are typically smaller than 200 mesh (<75 μm) and usually <10 μm. Particle size distributions are carefully monitored along with slip specific gravity and slip rheology to maintain desired casting rates. Additives include deflocculants, flocculants, and binders. Deflocculants may be sodium silicate or sodium polyacrylate. Binders in compositions containing little clay (such as alumina porcelains) may be forms of carboxymethylcellulose.

The slip casting process is described by Wood,[2] Reed,[3] and others.[4] The general processing steps of batching of raw materials to a prescribed mix composition, mixing, forming, drying, and firing are shown in Figure 1. Some raw materials, such as kaolin and ball clay, may be delivered in slurry form, that is, mixed with water by the manufacturer and shipped in tank cars (usually by rail).

Slip casting involves preparation of slurries in a mixer called a "blunger." The slurry is charged to molds that, by suction, draw the slurry to the mold surface, forming a cake that is the product shape being formed. When sufficient wall thickness is obtained, the mold is drained. When the part has dewatered sufficiently to have strength to be handled, it is removed from the mold to be sent for drying and firing. "Pressure casting" involves subjecting the slurry to an external pressure to accelerate the development of the cake against a porous mold.

Plastic Forming

Plastic forming involves the addition of water to a powdered raw material mixture such that the material develops plasticity and cohesion. The definition of plasticity in ceramic powders is "the characteristic property of moist powders which permits

Table 1. Typical Batch Compositions for Whiteware Ceramics

Material	Weight Percentage	
	Whiteware	Al_2O_3 Porcelain
Alumina	—	70–80
Kaolin clay	15–25	—
Ball clay	15–20	<5
Feldspar	10–20	—
Potters flint	15–25	<10
Water	20–25	20–25
Additives	<1	<1

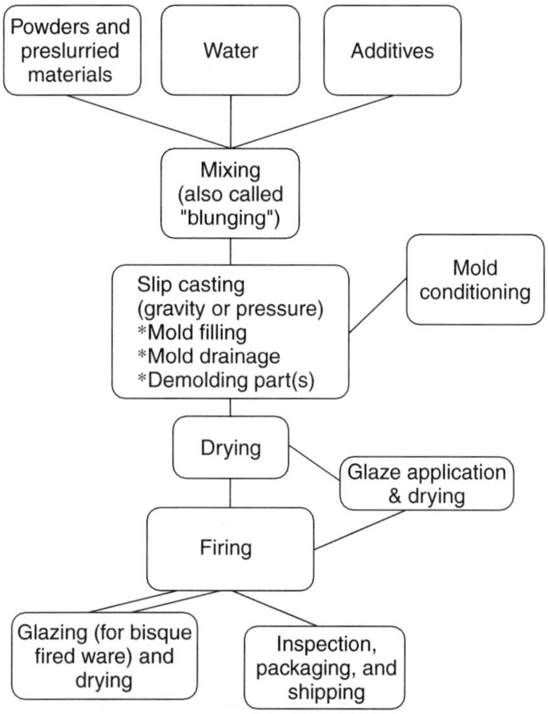

Figure 1. Slip Casting Flow Diagram.

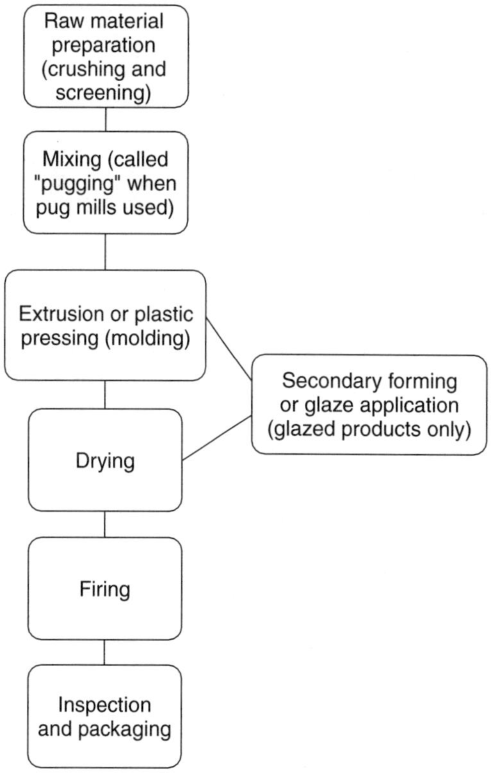

Figure 2. Plastic Forming Flow Diagram.

it to be deformed without cracking and retain its new shape when the deforming stress is removed."[5]

The amount of plasticity developed is dependent on the water content of the mixture and the inherent plasticity of the raw materials. In the case of nonplastic raw materials, binders, such as forms of carboxymethylcellulose or waxes, may be used to facilitate plastic forming.

The flow diagram for the plastic-forming process is shown in Figure 2. In the case of structural clay products like brick, quarry tile, and roofing tile, a single clay or multiple clays may be used to make a particular product. Clays are usually mined and prepared (by crushing and screening) by the manufacturers of the structural clay product.

Clays with high iron oxide content, such as shale, may fire to a desirable red color, and fireclays typically produce light brown or buff colors. Kaolin base products typically fire to white colors. Manufacturers may blend different clays to produce particular fired colors. In addition, coloring oxides including iron oxide may be added to the composition to produce specific colors in the fired product.

The chemical analysis of heavy clay products depends on the local raw material deposit. In general, ceramic producers report their chemical analyses on an oxide weight basis instead of an elemental weight percentage. Thus a brick may exhibit an analysis such as 10–30% Al_2O_3, 0.5–8% Fe_2O_3, 0.1–2% Mn_2O_3, 0.2–3% TiO_2, 0.2–2% Na_2O, 0.2–4% K_2O, and balance SiO_2.

The majority, or all, of the SiO_2 is present in a "combined" form in the minerals present and is not "free" or "available" as a discrete species in either the raw material or fired product. Clay and fired brick may contain a trace of free SiO_2 as quartz ranging up to a few weight percent. Normally a brick producer has difficulty to obtain reasonable product yields in manufacturing if the clay contains a substantial amount of quartz.

The batch ingredients typically contain up to 10% or more of moisture from the as-mined or storage condition. Raw material preparation usually consists of the steps of crushing and screening (scalping) to a particular "top size" specification. In manufacture of brick, the top size or largest particle may be on the order of 8 mesh (2.4 mm) to 6 mesh (3.4 mm), but thinner products, such as quarry tile or roofing tile, may have a top size of 20 mesh (0.8 mm).

The crushed and screened clay is mixed with water in a continuous mixer known as a "pug mill." Then the clay is sent to an extrusion machine that typically has a vacuum deairing chamber. The clay is forced through a die to form the desired shapes. The die may contain internal elements to produce a cored (partially hollow) shape. Both petroleum-based and water-based die lubricants may be used. The clay column produced in an extrusion process in brick manufacturing is shown exiting the extrusion die in Figure 3.

The consistency of the extruded clay is measured using a penetrometer device, which records the force required to make an impression of a standard diameter to a fixed depth in the clay column. In stiff extrusion the penetrometer readings typically exceed 3 kgf/cm^2. In soft extrusion, as in manufacturing roofing tiles, penetrometer readings are less than 2 kgf/cm^2. Further information on brick manufacturing is found in the *Brick and Tile Making Handbook* by Bender and Händle[6] and in a discussion by Robinson.[7]

In producing molded products, brick are made by machines (presses) that "throw" or insert plastic clay slugs into multiple cavities in mold boxes followed by compaction at low pressures (typically <40 MPa). The molds may be coated with sand or water to facilitate release from the mold (producing sand or water "struck" brick, respectively). The bricks are placed on trays and conveyed to the dryer.

Figure 3. Extrusion Operation Showing Exit of Extrusion Die.

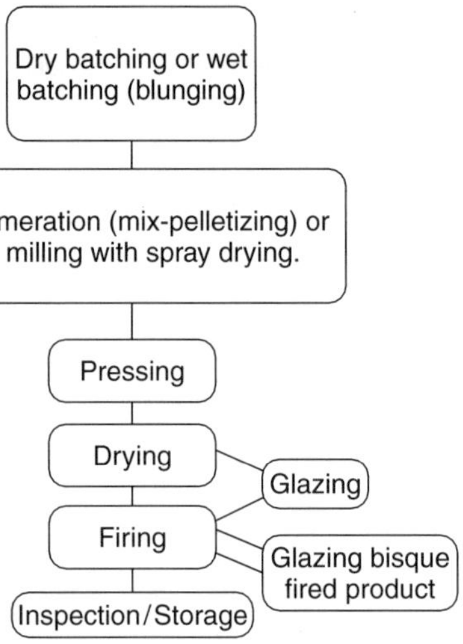

Figure 4. Flow Chart in Dry Pressing of Ceramic Tile.

In many structural clay products, the surface is coated with wet or dry powders to create specific colors or textures. Heavy wet powder coatings are called "engobes." These powders are typically mixtures of clay and colorants. Various sands may be used to create certain colors or textures.

In extrusion, the product may be cut into "slugs" of a convenient length, and they are further cut into brick using reel or "push through" cutters. The product is then loaded on kiln cars for drying and firing. Many new plants use industrial robots to handle the product (stack brick on the kiln cars).

In the case of roofing tile, cut slugs of plastic clay are fed to a press for molding of the final shape. The as-pressed shapes may be loaded onto ceramic supports (called kiln furniture or "cartridges") and further loaded onto cars for passage through the dryer and kiln. Glazes may be applied prior to passage through the kiln or they may be applied to the fired product, called "bisque" fired product, that is subsequently "refired" to vitrify the glaze.

Dry Pressing

Dry pressing is a technique of molding "near dry" powders into a precision shape under pressures on the order of 70–140 MPa. Compositions for tile products include talc–clay mixtures, clay–potters flint mixtures, and clay–wollastonite mixtures where most raw materials are ground or purchased as less than 200 mesh (<75 μm). The powders are usually mixed or otherwise prepared to a "free-flowing" consistency by mix-pelletizing or by spray drying. The flow diagram for dry pressing is shown in Figure 4.

The object of agglomeration is to form spherical "pellets" of diameter ~0.2–0.8 mm. Mix-pelletizing is a preparation operation where water is added to the dry powders in an intensive mixer followed by drying and screening (scalping) of the pellets prior to storage. Spray drying is a process in which batch ingredients are mixed with water (typically in ball mills) prior to drying of the raw material slurry in a spray dryer. Spray dryers typically are heated using natural gas.

In the pressing operation, the agglomerated powder is fed to the press that typically has a multiple cavity mold (Figure 5). In the case of ceramic tile, the mold contains an insert in the mold to allow for isostatic compaction. The pressed tiles are conveyed to drying and firing.

Figure 5. Powder Pressing Operation for Ceramic Tile (Power feed and tile removal mechanisms are not shown).

Dryers and Kilns

Dryers used with traditional ceramic products are typically fed by waste heat drawn from the cooling zone of the kiln. They are technically classed as "continuous counter-current" production dryers because convection is the dominant heat transfer mechanism and because the drying product typically moves in a direction opposite to the movement of air (the draft) in the dryer.

The hot air drawn from the cooling zone of the kiln typically contains no products of combustion. The air movement in the kiln is countercurrent to the direction of movement of the product in the kiln. The pressure in the kiln is maintained so that there is no possibility of air movement from the hot zone of the kiln concurrent with the product movement, and subsequently no products of combustion are drawn into the dryer supply air duct.

The dryer maintains a temperature and relative humidity profile aimed at minimizing the rate of drying during the critical shrinkage period for the product (called Stage I of drying). Excessive drying rates can result in cracking of the product. The entrance end of dryers where product is inserted is typically held at a specified relative humidity and low temperature (typically 50°C or higher), while the exit end condition of the dryer is at high temperature (typically >175°C). The dryer exhaust air exits near the product entrance end through a port called the dryer stack.

Some drying is accomplished within the plant building during the dwell period between extrusion and the entrance to the dryer. Special areas called "holding rooms" are used to accomplish this "predrying," and in some processes, a substantial amount of water is removed in predrying.

Kilns used in the ceramic industry are classified as (1) periodic (batch) and (2) continuous (tunnel or roller hearth). Batch kilns are used in a minority of operations with traditional ceramic manufacturing, and they are usually employed when production volumes are small. Many modern batch kilns are fired (heated) using natural gas burners.

Tunnel kilns are used in most heavy clay operations. A typical gas-fired tunnel kiln is shown in Figure 6. In this view, the ductwork on top of the kiln is visible and burners are located on the lower sides of the kiln.

Tunnel kilns can alternately be fired with burners on the kiln crown (top) or the burners can fire into channels underneath the ware (product). The kiln exhaust stack is typically near the entrance end of the kiln with stack temperatures on the order of 175°C. The exit end of a tunnel kiln producing brick is shown in Figure 7.

Some new tunnel kilns have a product setting height much less than shown in Figure 7. These "low-profile" kilns may be more adaptable to automated loading and unloading of the brick to or from the kiln cars. Low-profile kilns may also allow for faster drying and firing cycles.

Ceramic tiles are fired in either tunnel kilns or in roller hearth kilns. The roller hearth kiln is increasingly preferred for tile production. It uses metal and ceramic rollers arranged across the kiln cross section and protruding from the sides of the kiln. The tiles are placed on the rollers and traverse the length of the kiln being propelled by the motion of the rollers. The kiln draft is countercurrent to the direction of the travel of tile. Some roller hearth kilns have multiple levels of rollers. Other ceramic products are also being fired in roller hearth kilns.

Figure 6. Partial View of a Gas-Fired Tunnel Kiln Producing Brick (the "quick" cooling zone is in center of field; the waste heat duct carrying dryer air is perpendicular to the kiln as shown in the upper right).

Figure 7. Exit End of Gas-Fired Tunnel Kiln Producing Brick.

Brick and tile producers use reduction firing, or flashing, when manufacturing certain products. Reduction firing changes the oxidation state of iron compounds such that magnetite (Fe_3O_4) is formed near the surface of the product.

In addition to firing with natural gas, other fuels used in manufacturing brick include coal and sawdust (wood waste). Oil is no longer used as a primary fuel in firing brick, but either oil or propane may be used as a "backup" fuel during periods when natural gas use is curtailed.

AIR EMISSIONS CHARACTERIZATION

Introduction

In manufacturing of ceramic products, the main pollutant species may be fugitive emissions of particulate matter (in raw material preparation), emissions of hazardous air pollutants (HAPs), and emissions of criteria pollutants (species listed in the national Ambient Air Quality Standards) with the gaseous pollutants originating from the firing process. The main HAPs that can be emitted in firing is hydrogen fluoride (HF), although much lower quantities of hydrogen chloride (HCl) may be emitted. The main criteria pollutant, which may be emitted, is SO_x, but a minority of brick plants emits measurable quantities of SO_x. Trace quantities of volatile organic compounds (VOCs) and metals (in the form of metal oxides) may also be emitted.

It is important to state that clay brick have been manufactured for at least two thousand years. This means that brick manufacturing is one of the oldest manufacturing operations for human beings. As such, brick manufacturing has never been associated with increased incidence of disease over the overall population group. Only in a very limited number of recorded cases has there been environmental damage to vegetation around brick plants.

Ceramic Products Manufacturing—Particulate Matter Emissions

Emission factors published for particulate matter emissions in manufacturing of ceramic products are shown in Table 2. It should be noted that ceramic manufacturers applying glazes to their products are particularly sensitive to ambient dust. This is because glaze defects occur (as pinholes, etc.) if the product has loose particulate on the surface on which glaze is applied or if dust comes into contact with the glazed product in

Table 2. Selected Emission Factors for Filterable Particulate Matter in Ceramic Manufacturing[8] (also Ref. 4)

Source	Emission Factor Filterable PM (lb/ton of fired product)
Crushing and screening with fabric filter (quarry tile processing)	0.12
Crushing and screening (wall tile)	0.12
Spray dryer with fabric filter (gas fired)	0.06
Spray dryer with venturi scrubber (gas fired)	0.19
Glaze spray booth (no controls)	19
Glaze spray booth (wet scrubber)	1.8
Tunnel kiln exhaust (gas fired kiln)[a]	0.49
Roller hearth kiln (gas fired)	0.065
Refiring (glaze on fired bisque ceramic)	0.067

[a]No information on controls available for this source.

the kiln while the glaze is in a fused state. To minimize these problems, the portion of the manufacturing facility handling dried and/or fired ware where glaze application occurs is kept particularly clean and dust generation is minimized.

The data in Table 2 are limited in scope, and application of the emission factors should be made with this limitation in mind. For example, there are more than 100 manufacturers in the United States producing glazed whiteware ceramics, but there are only nine data points in this table. Further study of the references for the data would be required to understand completely if the data are applicable to a particular manufacturing situation.

Ceramic Products Manufacturing— Kiln Stack Emissions

Acid gas emissions originate from fluorine contained in the crystal structure of clays and other hydrous minerals and from sulfur sources within the raw materials. The fluorine in clays is insoluble in water and in most reagents, and it is only released when the clay is heated to at least 500°C. In a like manner, sulfur salts in the form of pyrite, FeS_2, and sulfates, notably calcium sulfate, occur in clays. Only a few clays contain sulfur sources.

Clay minerals usually contain from 80 to about 800 ppm of fluorine (0.008–0.080% by weight). This fluorine is released on decomposition of the clay crystals on heating in a process called "dehydroxylation." With kaolinite minerals, dehydroxylation occurs in the range of 500–600°C, but higher temperatures are observed for dehydroxylation with montmorillonite (600–750°C), pyrophyllite (820°C), illite (600–900°C), and talc (~910°C+).

When fluorine is released from clay materials, it usually undergoes immediate hydrolysis, forming HF gas. In atmospheres deficient in water vapor, it is possible for SiF_4 gas to form, but research has shown that SiF_4 gas will undergo hydrolysis subsequent to emission into the atmosphere. In gas-fired kilns, HF is the only fluorine species usually observed. Fluorine evolution can continue to higher temperatures. Brosnan and Sanders have hypothesized that some fluorine after dehydroxylation is adsorbed on clay surfaces to be released as temperatures are increased in the kiln.[9]

Emissions of SO_x are due to decomposition of pyrite or sulfates in the clay.[10] White body compositions usually do not contain pyrite, as the iron oxide formed on oxidation would cause iron spotting. Pyrite is usually associated with products made from some sources of shale and/or fireclay. The decomposition of pyrite has been shown to go through two steps with the initial oxidation occurring at about 400°C and subsequent oxidation continuing up to about 600°C. As in the case of HF, some sulfur species are held in the ceramic body to be released gradually as temperatures are increased in the kiln.

Ceramic producers like to avoid raw materials containing sulfates because of problems with efflorescence in service (efflorescence is the release of soluble salts on raining/drying cycles causing discoloration). However, certain raw materials essential to the production of whitewares, including ball clays, contain a trace of sulfate.

Emissions of carbon monoxide and carbon dioxide in ceramic materials have their origin in the combustion of fuels and in the oxidation of carbonaceous materials entrained in clays. Incomplete oxidation of carbon within the formed ceramic part ("in the body") leads to a condition known as black coring in the fired product.

Emissions of NO_x are primarily due to the combustion of fuels such as natural gas. It is unusual for clay to contain a substantial amount of nitrogen. Considerable progress has been made in NO_x reduction through new burner designs.

Emission factors for acid gas emissions for ceramic manufacturing are shown in Table 3. As in the case of Table 2,

Table 3. Emission Factors for Acid Gas Emissions in Kiln Stacks in Ceramic Manufacturing (1b/ton of fired product) (Data from Refs. 4 and 8)

Source	HF	Total Fluorides	SO_2	NO_x	CO	CO_2	VOC
Tunnel kiln (gas fired)	0.46	0.56	6.5 See note	0.54	3.3	780	0.43
Roller hearth kiln (gas fired)	0.017	NA	0.078	0.29	NA	190	NA
Glazing kiln (gas fired)	0.019	NA	NA	NA	NA	97	NA

Note: Reference 8 provides an emission factor formula depending on the sulfur content of the raw material. For sulfur content exceeding 0.07%, the emission factor for SO_2 in 1b/ton of fired product is 44 times the sulfur content of the raw material (in weight percent). For sulfur content less than or equal to 0.07%, the emission factor for SO_2 in 1b. per ton of fired product is 9.5 times the sulfur content of the raw material (in weight percent). These factors may only provide a rough estimate of SO_2 emissions.

Table 4. Selected Emission Factors for Filterable Particulate Matter in Brick Manufacturing[11]

Source	Species	Emission Factor Range (1b/ton processed)	Proposed Average Emission Factor (1b/ton)
Grinding area with clay @ 14% moisture content	Filterable PM	NA	0.025
Grinding area with clay @ 14% moisture content	Filterable PM_{10}	NA	0.0023
Grinding area with clay @ 3.9% moisture content	Filterable PM	NA	8.5
Grinding area with clay @ 3.9% moisture content	Filterable PM_{10}	NA	0.53
Extrusion line with fabric filter	Filterable PM_{10}	NA	0.0036
Brick dryer	Filterable PM	NA	0.077
Natural-gas-fired kiln	Filterable PM	NA	0.37
Natural-gas-fired kiln	Filterable PM_{10}	NA	0.28
Sawdust-fired kiln and sawdust dryer	Filterable PM	NA	1.3
Sawdust-fired kiln and sawdust dryer	Filterable PM_{10}	NA	0.25
Coal-fired kiln	Filterable PM	0.67–2.3	1.2
Coal-fired kiln	Filterable PM_{10}	0.42–1.4	0.76

the data must be viewed as limited in scope. In view of the origins of acid gas species, actual emissions are specific to certain raw materials.

Brick Manufacturing—Particulate Matter Emissions

Emission factors for particulate matter emissions in brick manufacturing are shown in Table 4. Data for the grinding areas and for the extrusion line are specific to emissions within the plant buildings. The plant building may serve as an additional "control device," considerably limiting atmospheric fugitive dust emissions. Measurements of the reduction in emissions by the plant building, called the "building efficiency," show reductions exceeding 95%. Therefore, emission factors applied to grinding are not directly usable in calculations such as in emissions inventories. It is clear that emissions within the grinding room are related to the moisture content of the clay raw material. Emissions of fugitive dust in brick manufacturing have been shown to be much lower than in other ceramic fields, such as refractories manufacturing, where dry raw materials are processed.

Information on stack emissions from one dryer and from kilns, using different fuels in firing brick, is shown in Table 4. These data must be viewed only from the perspective of estimates because there are no exact two duplicate kilns in the industry. Kilns vary widely as to length, burner arrangement, product throughput, and firing schedule. These are all factors that may influence particulate matter and acid gas emissions.

Limited information is published on particle size distribution of particulate matter emissions in brick plant kiln stacks (Ref. 11). This information is shown in Table 5. The references indicate that data originate from two sawdust-fired brick plants, one of which was subsequently converted to natural gas firing. The references further indicate that the data originate from two coal-fired brick plants, but one of these plants converted from coal to natural gas in 1980. Because there are more than 15 coal-fired tunnel kilns in brick plants, the data from only one plant are not considered as representative of current operations.

Brick Manufacturing—Kiln Stack Emissions (HAPs and Criteria Pollutants)

Emission factors for hazardous air pollutants in brick kiln stacks (without controls) are shown in Table 6. The data in this table represent testing at a very limited number of sources, limiting the use of these data only to estimations of emissions. In brick manufacturing, fluorine and sulfur emissions are directly related to the content of fluorine and sulfur in raw materials (and to a lesser extent firing variables). Since raw material fluorine contents differ by one order of magnitude (factor of 10), it is reasonable to consider that the emission factors in Table 6 are useful only for estimating purposes.

Experience has shown that the ratio of HF to total fluorides is in the range 0.85–1.00. In this respect, the information in Table 6 is likely specific only to the particular kilns tested.

In a like manner, SO_2 emissions are related to the pyrite content of the raw materials, and most raw materials used

Table 5. Average Particle Size Distribution for Filterable Particulate Matter from a Limited Number of Brick Kilns (Ref. 11)

Source	Aerodynamic Particle Diameter (μm)	Percent of Particulate Matter Less than the Stated Particle Diameter
Sawdust-fired kiln	10	75
	2.5	48
	1	44
Coal-fired kiln	10	63
	2.5	23
	1	9.8

Table 6. Hazardous Air Pollutants and Criteria Pollutants in Brick Kiln Stacks (lb/ton of fired product, Ref. 11)

Source	HF	Total Fluorides	SO_2	CO	CO_2	NO_x	HCl
Gas-fired kiln	0.37	0.59	0.67 see note	1.2	400	0.35	0.17
Coal-fired kiln	0.17	NA	1.2	0.80	300	0.51	NA
Sawdust-fired kiln	0.37	0.59	0.67	1.6	490	0.37	0.17

Note: Reference 11 provides an emission factor for SO_2 for an "abnormally high sulfur" raw material at 5.1 lb per ton of fired product. This factor is based on extremely limited test information.

in brick manufacturing contain no pyrite. Therefore, the SO_2 results in Table 6 can lead to erroneous estimates.

With respect to HCl emissions, the HCl originates from chlorine salts entrained in the raw materials. Some clays may have been in contact with salt water during geological time scales, providing a reason that a few clays contain chlorine-bearing salts. However, this is not true of all clays, and the data on HCl emissions in Table 6 are not representative of all brick manufacturing operations.

The data for CO, CO_2, and NO_x may be fairly representative of brick manufacturing operations due to the negligible influence of the raw materials and to the similarities in fuel compositions and consumption in the United States.

Brick Manufacturing—Kiln Stack Emissions (VOCs)

Emission factors for VOCs in brick plant kiln stack exhaust are shown in Table 7. It is important to note that the data in Table 7 are for tests at only one plant per fuel type. Therefore, it is unreasonable to apply these data generally to other brick plants. It is also interesting that many of the organic "species" exhibit a difference in the gas-fired plant of about one order of magnitude higher than data for that species for the coal or sawdust-fired plants. This is surprising considering the relative cleanliness of natural gas.

Further research is needed to determine if the species listed in Table 7 are actually present or if the sampling technique actually only detected molecular fragments from depolymerization processes of organic species in the clays. More definitive techniques would be required to answer this question.

Table 7. Volatile Organic Compound Emission Factors for Brick Plant Stacks (Ref. 11)

Species	Gas-Fired Kiln	Coal-Fired Kiln	Sawdust-Fired Kiln with Sawdust Dryer
1,1-dichloroethane	NA	5×10^{-6}	NA
1,1,1-trichloroethane	4.7×10^{-6}	<DL	<DL
1,4-dichlorobenzene	4.8×10^{-5}	3.2×10^{-6}	NA
2-butanone	2.2×10^{-4}	2.5×10^{-4}	2.2×10^{-4}
2-hexanone	8.5×10^{-5}	<DL	<DL
2-methylnaphthelene	5.7×10^{-5}	1.7×10^{-6}	NA
2-methylphenol	NA	<DL	<DL
Acetone	1.7×10^{-3}	6.8×10^{-4}	1×10^{-3}
Benzene	2.9×10^{-3}	2.9×10^{-4}	5.6×10^{-4}
Carbon disulfide	4.3×10^{-5}	2.3×10^{-6}	1.8×10^{-5}
Carbon tetrachloride	NA	<DL	<DL
Ethylbenzene	4.4×10^{-5}	2.1×10^{-5}	1×10^{-5}
Methylene chloride	NA	8.0×10^{-7}	6.2×10^{-5}
Naphthelene	6.5×10^{-5}	6.9×10^{-6}	<DL
Phenol	8.6×10^{-5}	3.5×10^{-5}	1×10^{-4}
Tetrachloroethene	2.8×10^{-6}	<DL	<DL
Trichloroethane	NA	<DL	<DL

Brick Manufacturing—Kiln Stack Emissions (Metals)

Emission factors for "metals" in brick kiln stacks are given in Table 8. As in the case of VOC emissions, data for each fuel type for metals emissions are based on a single test. Therefore, the same limitations apply to estimates using these data as have been mentioned in previous cases. Some of the data included in the reference were indicated for "brick kilns" without specifying the fuel. Such data are indicated with an "*a*" superscript in Table 8.

AIR POLLUTION CONTROL MEASURES

Introduction

Air pollution control devices were first applied to a brick plant in Utah in the 1950s when a scrubber was installed to reduce SO_x emissions in a plant in a suburban location. In the 1970s, a dry sorption scrubber was installed on the kiln exhaust at a Canadian plant. Further installations occurred in the 1990s in response to regulatory actions in a few states.

Because of implementation of environmental regulations on brick plants in the 1970s in Germany, much of today's scrubber technology is of German origin. Two types of scrubbers have been used in Germany for brick plants: (1) dry sorption scrubbers employing limestone, $Ca(CO_3)$, as the reagent and (2) lime injection/baghouse scrubbers using either limestone or hydrated lime, $Ca(OH)_2$, as the reagent.

Dry Sorption Scrubbers

The cross section of a dry sorption scrubber is shown in Figure 8. Limestone is held in a reservoir at the top of the scrubber, and it is allowed to proceed by gravitational flow through the scrubber at the required rate. The kiln exhaust

Table 8. Metals Emission Factors for Brick Plant Stacks (Ref. 11)

Metal	Gas-Fired Kiln	Coal-Fired Kiln	Sawdust-Fired Kiln with Sawdust Dryer
As	3.1×10^{-5}	1.3×10^{-4}	2.1×10^{-5}
Be	4.2×10^{-7}	1.6×10^{-5}	3.1×10^{-7}
Cd	1.5×10^{-5}[a]	NA	2.2×10^{-5}
Co	2.1×10^{-6}[a]	NA	NA
Cr	5.1×10^{-5}[a]	NA	4.8×10^{-5}
Hg	7.5×10^{-6}	9.6×10^{-5}	1.1×10^{-5}
Mn	2.9×10^{-4}	NA	4.8×10^{-4}
Ni	7.2×10^{-5}[a]	NA	3.4×10^{-5}
P	NA	9.8×10^{-4}	5.5×10^{-4}
Pb	1.5×10^{-4}[a]	NA	1.2×10^{-4}
Sb	2.7×10^{-5}[a]	NA	2.8×10^{-6}
Se	2.3×10^{-4}[a]	NA	4.7×10^{-5}

[a] Reference indicates "brick kiln" without specifying fuel.

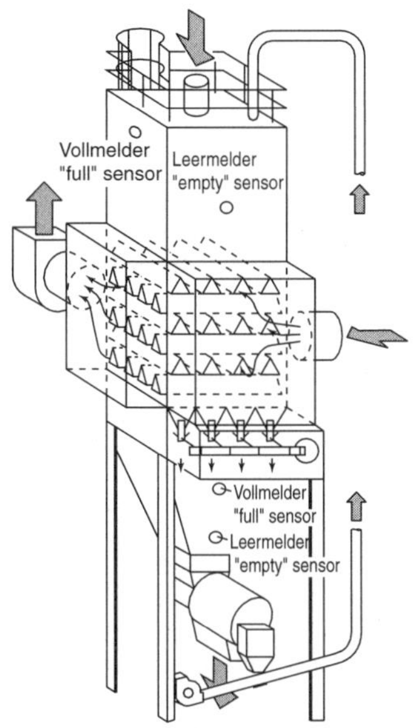

Figure 8. Dry Sorption Scrubber Used in Brick Plants.

enters a plenum on the side of the scrubber (from the right in Figure 8). Gases pass under angled "shields" and through the limestone bed prior to entering an exhaust plenum.

A brick plant installation is shown in Figure 9. The kiln exhaust can be seen on the top of the plant building (to the right). The exhaust plenum is located on the right side of the scrubber tower connected to the induced draft fan supplying the scrubber stack.

Research work has shown that limestone properties may affect scrubber performance. These properties include chemical analysis, density, pore size distribution, and crystal size.[12]

Figure 9. Dry Sorption Scrubber at a Brick Plant.

Lime Injection/Baghouse Scrubbers

In the event that higher levels of SO_x is present in the kiln exhaust along with HF, lime injection scrubbers may be employed at brick plants. These scrubbers work by entraining the reagent in kiln exhaust in an external reactor where neutralization occurs. The reagent is captured in a conventional baghouse. Some scrubbers employ a preconditioning of the reagent, which causes some agglomeration and influences the reagent reactivity, and usage of reagent may be more efficient.

A portion of a lime injection scrubber used on brick plant kiln exhaust is shown in Figure 10. The kiln exhaust enters the venturi reactor (upper left) through a vertical pipe to the left center of the photograph. Effluent gases from the scrubber enter the baghouse to the right of the photograph. A portion of the reagent captured in the baghouse is mixed with fresh reagent in a paddle-type mixer (not shown) and reused.

Recent work has suggested that surface area of fine reagents may be one way of predicting reagent reactivity (Ref. 12). A reactivity index has been proposed for evaluation of the potential neutralizing power of reagents.

Figure 10. Lime Injection/Baghouse Scrubber at a Brick Plant.

MACT and the Ceramic and Brick Industries

Under implementation of the Clean Air Act, the regulatory category "Clay Products Manufacturing" was among the last industry segments to be regulated. No proposed MACT criteria relative to air emissions were published when this chapter was prepared.

SUMMARY AND CONCLUSIONS

The ceramic and brick industries are included in a broad category called "traditional ceramics" by those familiar with the art. These ceramics have vastly different production methods and product specifications, but they have in common clay minerals as a major constituent of their composition.

Ceramic products include toilets, decorative ceramics, dinnerware, tile, and clay-based electrical insulators. This industry segment has not been subjected to the scrutiny of tests sponsored by the U.S. Environmental Protection Agency, making data on emissions limited in scope.

The U.S. Environmental Protection Agency has studied the brick industry only to a limited extent. This has resulted in some data relative to particulate emissions, emissions of hazardous air pollutants, emissions of volatile organic compounds, and metals emissions. The studies have revealed that the brick industry is a relative nonpolluter compared to many other industries.

In the evolution of environmental regulations, it is possible that emissions will become viewed in terms of the "life cycle" of products related to pollution created in their production. When viewed from this perspective, ceramic manufacturers will be considered among the lowest-polluting industries.

Acknowledgment

The author thanks Mr. David McKeown of Richtex Corporation (Columbia, SC), Mr. Terry Schimmel of Boral Bricks, Inc. (Phenix City, AL), and Mr. Mike Vickers of Acme Brick (Denton, TX) for their review of the manuscript of this chapter.

REFERENCES

1. American Society for Testing and Materials, West Conshohocken, PA.
2. Russell K. Wood, "Ceramic Whiteware," in *Engineered Materials Handbook*, Vol. 4 (Ceramics and Glass), ASM International, pp. 931–36 (1991).
3. James S. Reed, *Principles of Ceramic Processing*, 2nd Edition, John Wiley & Sons, pp. 492–534 (1995).
4. "Ceramic Products Manufacturing," *www.epa.gov.80/ttnichie/lap42pdf/c11so7.pdf.*
5. After the definition in *Dictionary of Ceramics*, A. E. Dodd, Littlefield, Adams, and Company (London).
6. "Brick and Tile Making," Bauverlag GmbH, Berlin, Germany (1982).
7. G. C. Robinson, "Structural Clay Products," in *Engineered Materials Handbook*, Vol. 4 (Ceramics and Glass), ASM International, pp. 943–52 (1991).
8. *Op. cit.*, Ref. 4, and "Emission Factor Documentation For AP-42, Section 11.7," Draft Report, U.S. Environmental Protection Agency, June 1994.
9. Denis A. Brosnan, "HF Emissions In Manufacturing Traditional Ceramics," *Ceramic Bulletin* **76**(1), 59–61 (1997).
10. Denis A. Brosnan, "SO_x Emissions in Firing Ceramics," *Ceramic Bulletin* **76**(3), 53–33 (1997).
11. "Emission Factor Documentation for AP-42, Brick and Structural Clay Product Manufacturing," August, 1997, Available through: *http://www.epa.gov:80/ttnchie1/ap42pdf/c11s03.pdf.*
12. "How to Select Solid Reagents for Acid Gas Scrubbers," Denis A. Brosnan, *Ceramic Bulletin* **78**(2), 51–54 (1999).

COAL PROCESSING

Larry L. Simmons and Lisa E. Lambert
(with update by editor and excerpts from AP-42)

This discussion focuses on the emissions from sources at coal-processing plants, which are considered to include all coal-preparation plants and coal-handling facilities.

A coal-preparation plant is any facility (excluding underground mining operations) that prepares coal by one or more of the following processes: breaking, crushing, screening, wet or dry cleaning, or thermal drying. Generally, coal-preparation plants can be classified as:

- Those that clean both coarse and fine coal
- Those that clean only coarse coal
- Those that crush coal to a specific size

Coal-handling facilities include any facility (excluding those associated with mining) that processes coal solely by use of one or more of the following operations: transferring, conveying, loading, unloading, storing.

PROCESS DESCRIPTION

Coal-processing plants can vary according to size, complexity, and purpose. For example, the type of coal-processing plant can range from a simple coal-loading station that handles only a few tons of coal per hour to a complex coal-washing plant processing over 1000 tons of coal per hour. The type of coal processing performed depends on the requirements of the end user. Figure 1 illustrates a typical coal-processing plant.

As shown in Figure 1, this description of particulate matter emissions from coal-processing plants begins at the point where

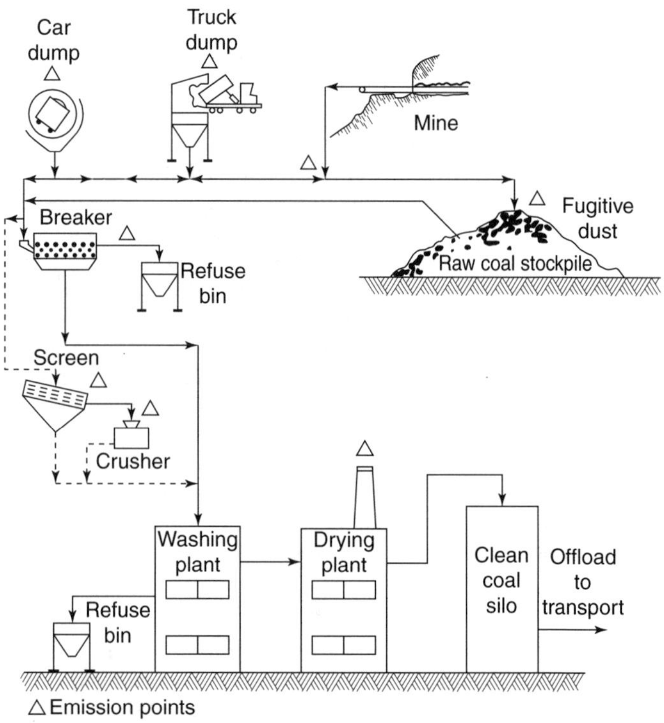

Figure 1. Typical Coal-Processing Plant Schematic (Revised from Ref. 2).

coal is brought to the processing plant via truck, railcar, or conveyor and is unloaded. It concludes at the point where the processed coal is loaded into transportation equipment (truck, railcar, conveyor, barge) for shipment to the end user.

At the processing plant, the coal is unloaded and dumped into a receiving hopper that discharges to a primary crusher, a tipple, or a feeder. A feeder subsequently empties onto a conveyor for transfer to an open storage pile or enclosed silo. Coal that is carried directly to the processing plant by conveyors is unloaded by dropping onto open storage piles or into enclosed silos.

Coal is normally stored this way to allow for optimum scheduling of processing and transportation equipment. The coal from open storage piles is transferred to either crushing, screening, or loading operations by front-end loaders or self-feeding tunnel conveyors. The coal that is stored in silos is transferred to processing equipment with belt conveyors or is loaded into trucks or railroad cars by gravity. Because of the large amount of material involved, belt conveyors that have large carrying capacities are the most common method of transporting material at a coal-processing plant.

Coal at any preparation plant undergoes at least a crushing operation. At many smaller plants, the coal that is processed by crusher is loaded directly into a truck, a railcar, or a barge for shipment to the user. Most of the larger mining operations will also provide secondary crushing, screening, wet or dry cleaning, and drying of the crushed coal. The remainder of this process description will discuss such processing at a typical coal-preparation plant.

The cleaning or beneficiation of coal at preparation plants is performed for a number of reasons. One such reason is to improve the coal quality by removing undesirable impurities. This increases the heating value of the coal and provides a better fuel for the user. In fact, coal cleaning is often necessary to market coal, since mined coal may contain up to 60% of reject material.[3]

Another reason for cleaning coal is that air-pollution control requirements on the user often dictate the partial removal of pyrites with the ash in order to reduce the sulfur content. Also, ash content must often be monitored and reduced to levels stipulated in sales contracts. However, a minimum ash content must be maintained to ensure optimum combustion characteristics.[4]

Last, substantial savings in freight costs for shipping coal may be achieved by the removal of impurities before loading. Moreover, it is generally easier to dispose of the impurities at the mining site than at the combustion site.

At coal-preparation plants, the initial process operations consist of "tramp iron" removal and size reduction (Figure 1).[5] Coal is first exposed to a high-intensity magnet, which is usually suspended over the incoming belt conveyor, and the iron impurities are extracted from the coal. This high-intensity magnet may also be located after the breaker, but it is always located before the screen and crusher.

The coal is next conveyed to a breaker, which consists of a cylindrical shell with perforated holes (2–8 in. diameter) and interior lifting blades. The perforated shell allows the smaller coal to pass through. The breaker rotates on a horizontal axis and breaks the tumbling coal, which is fed into the breaker at one end. The soft coal is broken in the breaker to a size small enough to pass through the shell, while the hard, larger, unbroken material (reject) passes out through the other end of the breaker and into a refuse bin. The coal (usually less than 4 in. in size) that passes through the breaker shell is then transferred to the cleaning plant.[5]

Instead of entering the breaker, some coal is diverted to a scapling screen. From the scalping screen, the oversized material (>4 in.) falls into a crusher and is reduced in size to less than 4 in. This material is then combined with the screenings from the scalping screen and transferred to the cleaning plant. This alternative flow is used more often than the breaker circuit despite the disadvantage of exposing the crusher to large pieces of material. A heavy-duty, single-roll crusher with tramp iron protection is most often used for this process.[6]

As in the case of crushing equipment, coal-preparation plants also use a variety of screening equipment for coal sizing. The types generally used are the grizzly, shaker, and vibrating screens. Grizzly screens, which size coal by gravity only, are usually used on coal preceding a crusher or belt conveyor loading operation.[7] Shaker screens are used infrequently and rarely provide a separation of less than 2 in. Vibrating screens are the most common type of separating device. They are used in both dry and wet processing plants.[8]

After crushing and screening, the raw coal at a preparation plant is usually stored on an open pile or in a silo prior to washing to allow for the smooth, efficient operation of the cleaning plant.[9] At the cleaning plant, the type of mechanical cleaning equipment used depends on the size range of the coal entering the plant and the desired coal size. For example, coal that is larger than 8 in. is usually crushed; however, if lump coal is desired, the large fraction above 8 in. is cleaned by slate pickers.

Coal of a size range of less than $\frac{3}{8}$ in. is often cleaned with pneumatic cleaning devices or air tables. These devices have a perforated bottom plate over which a layer of coal passes. A current of air is passed upward through the bed, which removes the finer particles. The fines are eventually removed from the airstream by cyclones and fabric filters. Generally, air tables are not very efficient with respect to their ability to remove ash from coal. One source reports that their efficiency of ash removal is limited to 2–3%.[10]

Jig-table washing plants use jigs to clean coals greater than $\frac{1}{4}$ in. in size and Diester tables to clean the $\frac{1}{4}$-in. to 28-mesh size range. This equipment is often used with Froth cells and/or thermal dryers. In a jig-table wash plant, the raw coal (<8 mesh) is first separated on a wet screen (usually $\frac{1}{4}$-in. mesh). The larger coal enters the jig, while the remaining coal is transferred to another, separate cleaning circuit. The coal exiting the jig is dewatered on screens and in centrifuges, crushed, and loaded or stored.

The finer coal (<$\frac{1}{4}$ in.) is mixed with water and then poured into the tables, where the refuse is separated from the coal. Water is removed from the refuse by screening, and the refuse is deposited into a bin for storage until it may be hauled to a disposal area. The washing coal is then dewatered by using a stationary gravity screen or "sieve bend," where the fines are removed and sent to a centrifuge for dewatering and extraction of the fines. Finally, the washed coal is loaded or conveyed to a thermal dryer.[11]

A digester table has a flat, riffled surface (about 12 sq. ft in area) that oscillates perpendicularly to the riffles toward the flow of coal. The heavy reject material falls off one end of the table, while the light coal is discharged off the opposite end. The remaining material is distributed in between.

A heavy-media wash plant cleans coal by floating it in a medium with a selected specific gravity, in which a dispersion of finely ground magnetite (Fe_3O_4) is maintained. In this type of plant, the raw coal is first separated at $\frac{1}{4}$ in. on an inclined screen. The oversized fraction is transferred to a flat, wet screen, where the finer particles are sprayed off the $>\frac{1}{4}$-in. coal. The oversized material from this wet screen is then discharged into a heavy-media bath, where the refuse is separated from the coal. The refuse is then dewatered by discharging to a screen. The medium, which is removed from the bath, is divided into two parts, one returning to circulation via the heavy-medium sump and the other being pumped to the magnetite recovery system. The refuse is discharged from the screen for disposal. The coal is next removed from the washer to a rinse screen, where the coal is dewatered.

The thermal drying of coal is occasionally used at coal processing facilities employing fine-coal washing equipment. Generally, coal is dried via a thermal dryer for one or more of the following reasons.[12,31]

- To avoid freezing difficulties and to facilitate handling during shipment, storage, and transfer
- To maintain high pulverizer capacity
- To improve the quality of the coal used for coking
- To decrease transportation costs

Thermal dryers are generally used to dry the $\frac{1}{4}$- by 0-in. fraction. However, sometimes the plus-$\frac{1}{4}$-in. portion is dried for ease of screening.

Thermal dryers are simply contacting devices where the exhaust gases from the combustion process in the dryer are used to heat the wet coal and to evaporate the surface moisture. Seven basic types of thermal dryers are currently used: rotary, screen, cascade, continuous carrier, flash or suspension, multilouver, and fluidized bed.

The most prevalent type of dryer in use today is the fluidized-bed dryer (see Figure 2). This dryer contains a perforated plate in a negative-pressure fluidizing chamber above which coal is suspended by a rising column of the gases. The dried coal exits from the dryer at an overflow weir. A flash or suspension dryer is the second most popular type of dryer. In this dryer, the hot gases generated in the combustion furnace of the dryer transport the wet coal up a riser. The turbulence created in the riser provides an excellent drying environment. This type of dryer is used for extremely fine coal, with the top size not exceeding $\frac{1}{8}$ in.

After washing and drying, mines using unit-train shipment usually store enough clean coal to fill a train. Silos an often used for this purpose. Furthermore, silo storage prevents the accumulation of moisture and exposure to wind. Some mines employ open storage, using conveyors for loading. At other mines, railroad cars or barges are loaded directly as the coal is processed.

AIR EMISSION CHARACTERIZATION

The estimated emission factors for coal-processing-plant fugitive dust sources are summarized in Table 1, and fugitive dust control is discussed in the following paragraphs.

The major emission source in the fine or coarse coal processing phases is the air exhaust from the air separation processes (air tables). For the dry cleaning process, these emissions are generated when the coal is stratified by pulses of air. Particulate matter emissions from this source are normally controlled with cyclones followed by fabric filters. Potential emissions from wet cleaning processes are very low.[43]

The major source of emissions from the final preparation phase is the thermal dryer exhaust. This emission stream contains coal particles entrained in the drying gases and volatile organic compounds (VOC) released from the coal, in addition to the standard products of coal combustion resulting from burning coal to generate the hot gases [including carbon monoxide (CO), carbon dioxide (CO_2), VOC, sulfur dioxide (SO_2), and nitrogen oxides (NO_x)]. Table 2 shows emission factors for particulate matter. Emission factors for SO_2, NO_x, VOC, and CO_2 are presented in Table 3.[43]

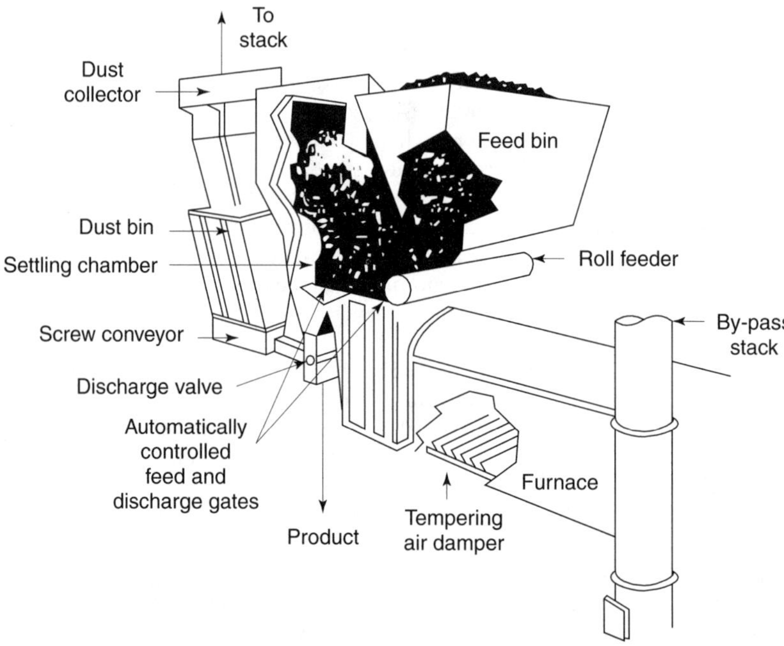

Figure 2. Fluidized-Bed Thermal Coal Dryer (Revised from Ref. 15).

Table 1. Fugitive Dust Emission Factors for Coal-Processing Plants

Unloading	
Truck	0.02 lb/ton unloaded[a]
Railcar	0.40 lb/ton unloaded[b]
Primary crushing	0.02 lb/ton crushed[c]
Secondary crushing/screening	0.16 lb/ton crushed/screened[c]
Transfer and conveying	0.20 lb/ton transferred or conveyed[d]
Cleaning	Negligible[e]
Storage	
Loading onto pile	0.08 lb/ton loaded[f]
Vehicular traffic	0.16 lb/ton stored[f]
Loading out	0.10 lb/ton stored[f]
Wind erosion	0.09 lb/ton stored[f]
Loading	
Truck	0.02 lb/ton loaded[a]
Railcar	0.40 lb/ton loaded[b]
Barge	0.40 lb/ton loaded[b]

[a]Ref. 16.
[b]Ref. 17 and 18.
[c]Ref. 19.
[d]Ref. 20.
[e]Ref. 21 (see Table 2 for dryer emissions).
[f]Ref. 22.

Table 2. PM Emission Factors for Coal Cleaning[a] Emission Factor Rating: D (Except as Noted)

	Filterable PM[b]			Condensible PM[c]	
Process	PM	$PM_{2.5}$	$PM_{1.0}$	Inorganic	Organic
Multilouvered dryer[d] (SCC 3-05-010-03)	3.7	ND	ND	0.057	0.018
Fluidized bed dryer[e] (SCC 3-05-010-01)	26[f]	3.8[g]	1.1[g]	0.034[h]	0.0075[h]
Fluidized bed dryer with venturi scrubber[j] (SCC 3-05-010-01)	0.17	ND	ND	0.043	0.0048
Fluidized bed dryer with venturi scrubber and tray scrubber[k] (SCC 3-05-010-01)	0.025	ND	ND	ND	ND
Air tables with fabric filter[m] (SCC 3-05-010-13)	0.032[n]	ND	ND	0.033[p]	0.0026[q]

[a]Emission factor units are lb/ton of coal feed, unless noted. 1 lb/ton = 2 kg/Mg. SCC, Source Classification Code; ND, no data. Table taken from AP-42, Chapter 11.10, "Coal Processing."
[b]Filterable PM is that PM collected on or prior to the filter of an EPA Method 5 (or equivalent) sampling train.
[c]Condensible PM is that PM collected in the impinger portion of a PM sampling train.
[d]Ref. 35. Alternate SCC is 3-05-310-03, which corresponds to units of lb/thousand tons of coal feed. To determine the emission factor for this alternate SCC, multiply the factor in this table by 1,000.
[e]Alternate SCC is 3-05-310-01, which corresponds to units of lb/thousand tons of coal feed. To determine the emission factor for this alternate SCC, multiply the factor in this table by 1,000.
[f]Refs. 36, 38.
[g]Refs. 36, 38. Emission factor rating: E. Particle size data from Ref. 38 used in conjunction with filterable PM data from Refs. 36 and 38. Actual cut size of $PM_{2.5}$ data was 2.7 μm.
[h]Ref. 36
[j]Refs. 27, 28, 36, 38, 41. See footnote *e* for alternate SCC.
[k]Ref. 42. Tray scrubber using NaOH as the scrubbing liquid. See footnote *e* above for alternate SCC.
[m]Alternate SCC is 3-05-310-13, which corresponds to units of lb/thousand tons of coal feed. To determine the emission factor for this alternate SCC, multiply the factor in this table by 1,000.
[n]Refs. 39, 40.
[p]Ref. 40.
[q]Ref. 39.

A number of inorganic hazardous air pollutants are found in trace quantities in coal. These include arsenic, beryllium, cadmium, chromium, copper, mercury, manganese, nickel, lead, thorium, and uranium. It is likely that many of these are emitted in trace amounts from crushing, grinding, and drying operations.[43]

The new source performance standards (NSPS) for coal preparation plants were promulgated in January 1976 (40 CFR Subpart Y). These standards specify emission limits for PM from coal cleaning thermal dryers and pneumatic cleaning equipment sources, and opacity limits for fugitive emissions from coal processing and conveying equipment, coal storage systems, and coal transfer and loading systems.

AIR POLLUTION CONTROL MEASURES

This section presents the fugitive dust control methods that may be used on a case-by-case basis by the coal-processing industry followed by a discussion of control technologies used in the processing phase.[17,22] Application of fugitive dust control measures would be to meet the National Ambient Air Quality Standards or to reduce the nuisance potential beyond the property line of the coal-processing facility.

The fugitive dust emissions from a truck or railcar may be controlled by one or a combination of the following:

- Total enclosure with ventilation to a fabric filter
- A partial enclosure
- A water spray system
- A wet suppression system using water and chemical wetting agents or foams

For primary and secondary crushing and screening operations, control methods consist of enclosures with ventilation to a fabric filter and wet-suppression systems utilizing a chemical wetting agent or foams.

For fugitive dust emissions from conveying operations, the control methods generally used are partial (top) enclosure, total enclosure, or wet suppression. Also, fugitive dust emissions created by the droppings from the return belt conveyors may be controlled through the use of dribble pans.

Fugitive dust emissions from transfer points may be controlled through the use of total enclosure, enclosure with ventilation to a fabric filter, or wet-suppression systems using chemical wetting agents.

The fugitive dust emissions from storage piles consist of four sources: loadin, wind disturbance, vehicular traffic, and loadout. The control methods available for those sources vary with the type of source. For loadin at storage piles, control methods include (1) enclosures, (2) wind guards, (3) telescopic chutes, (4) wet suppression, and (5) operating practices.

Fugitive dust emissions from wind disturbances at storage piles may be controlled through the use of wet suppression

Table 3. Gaseous Pollutant Emission Factors for Coal Cleaning[a] Emission Factor Rating: D (Except as Noted)

Process	VOC[b]	SO_2	NO_x	CO_2
Multilouvered dryer[c] (SCC 3-05-010-03)	ND	ND	ND	160
Fluidized bed dryer[d] (SCC 3-05-010-01)	ND	1.4[e]	0.16[f]	30[g]
Fluidized bed dryer with venturi scrubber[h] (SCC 3-05-010-01)	0.098[j]	—[k]	0.16[f]	30[g]
Fluidized bed dryer with venturi scrubber and tray scrubber[m] (SCC 3-05-010-01)	ND	0.072[n]	0.16[f]	30[g]

[a]Emission factor units are lb/ton of coal feed, unless noted. 1 lb/ton = 2 kg/Mg. SCC, Source Classification Code; ND, no data. Table taken from AP-42, Section 11.10, "Coal Processing."
[b]VOC as methane, measured with an EPA Method 25A sampling train. Measurement may include compounds designated as nonreactive.
[c]Ref. 35. Emission factor rating: E. Alternate SCC is 3-05-310-03, which corresponds to units of lb/thousand tons of coal feed. To determine the emission factor for this alternate SCC, multiply the factor in this table by 1,000.
[d]Alternate SCC is 3-05-310-01, which corresponds to units of lb/thousand tons of coal feed. To determine the emission factor for this alternate, SCC, multiply the factor in this table by 1,000.
[e]Refs. 36, 37, 26. Emission factor rating: E.
[f]Refs. 36, 37, 42. Includes NO_x measurements before and after control devices that are not expected to provide control of NO_x emissions.
[g]Refs. 28, 36–38, 41. Includes CO_2 measurements before and after control devices that are not expected to provide control of CO_2 emissions.
[h]See footnote *d* above for alternate SCC.
[j]Refs. 28, 37.
[k]Venturi scrubbers may achieve between 0 and 95% control of SO_2 emissions. The use of a neutralizing agent in the scrubber water increases the SO_2 control efficiency.
[m]Venturi scrubber followed by tray scrubber using a NaOH solution as the scrubbing liquid. See footnote *d* above for alternate SCC.
[n]Ref. 42.

and the application of surface crusting agents. Also, operating practices such as orienting the storage piles perpendicularly to prevailing winds to reduce the exposed surface are helpful in reducing wind erosion.

For loading out from coal storage piles, control measures consist of the installing under-pile conveyor systems, bucket-wheel reclaimer systems, and wet suppression.

For coal storage in silos or bins, fugitive dust emissions are normally controlled by covering the conveyors that transport and dump the coal. The fugitive dust in the displaced air from silos and bins may be controlled by the addition of either bin vent filters or exhaust fabric filters.

Control methods for loading operations consist of telescopic chutes and wet suppression. For operations that use wet suppression in subsequent processes, generally only telescopic chutes are required for dust control if the coal has been adequately treated with a wetting agent. A summary of the control measures and their respective efficiencies is presented in Table 4.

The most common technology used to control dryer emissions is venturi scrubbers and mist eliminators downstream from the product recovery cyclones. The control efficiency of these techniques for filterable PM ranges from 98 to 99.9%. Scrubbers also may achieve between 0 and 95% control of SO_2 emissions. The use of a neutralizing agent (such as NaOH) in the scrubber water increases the SO_2 removal efficiency of the scrubber.

Table 4. Summary of Control Alternatives and Their Control Efficiencies

Fugitive Dust Sources	Control Alternatives	CE[a] (%)
Unloading coal		
Railcar	Enclosure, vent to fabric filter	99
	Enclosure	70
	Wet suppression with chemicals	80
	Watering	50
Truck	Enclosure vent to fabric filter	99
	Enclosure	70
	Wet suppression with chemicals	80
	Watering	50
Primary crushing	Enclosure, vent to fabric filter	99
	Wet suppression with chemicals	90
Secondary crushing/screening	Enclosure, vent to fabric filter	99
	Wet suppression with chemicals	90
Transfer and conveying	Enclosure of conveyors and transfer points, vent to fabric filter	99
	Enclosure of conveyors and transfer points	70
	Wet suppression with chemicals	90
Cleaning	Enclosure	100
Storage		
Loading onto piles	Enclosure	80
	Telescopic chutes	75
	Wet supression with chemicals	75
	Wind guards	50
Wind erosion	Enclosure	100
	Wet suppression with chemicals	99
Loading out	Under-pile conveyor	80
	Wet suppression with chemicals	95
	Bucket wheel reclaimer	80
Loading		
Railcar/barge	Wet suppression with chemicals	80
	Telescopic chutes	75
Truck	Wet suppression with chemicals	80
	Telescopic chutes	75

[a]CE, control efficiency.

Source: Adapted from references 17 and 22.

REFERENCES

1. M. Szabo, *Environmental Assessment of Coal Transportation*, PEDCo Environmental, Inc., EPA-600/7-78-081, U.S. Environmental Protection Agency, Industrial Environmental Research Laboratory, Office of Research and Development. Cincinnati, OH, Contract No. 68-02-132 1, May 1978, p. 59.

2. *1977 Division of Mines Report*, State of Ohio, Department of Industrial Relations, pp. 4–6.
3. *Inspection Manual for the Enforcement of New Source Performance Standards: Coal Preparation Plants*, PEDCo Environmental, Inc., EPA-340/1-77-022, U.S. Environmental Protection Agency, Division of Stationary Source Enforcement, Washington, D.C., Contract No. 68-01-3150, November 1977, p. 3–2.
4. Ref. 3, p. 3–3.
5. Ref. 3, p. 4–5.
6. Ref. 3, p. 4–8.
7. Ref. 3, p. 6–7.
8. M. Weisburg, *Field Operations and Enforcement Manual for Air Pollution Control. Vol. III: Inspection Procedures for Specific Industries*, System Development Corp., prepared for U.S. Environmental Protection Agency, Office of Air Programs, Research Triangle Park, NC, Control No. CPA 70122, APTD- 1102, August 1972, pp. 6–8.
9. Ref. 3, p. 4–8.
10. Ref. 3, p. 4–9.
11. Ref. 3, p. 4–12.
12. P. N. Formica, *Controlled and Uncontrolled Emission Rates and Applicable Limitations for Eighty Processes*, prepared for U.S. Environmental Protection Agency, Research Triangle Park, NC, Control No. 68-02-1382, Task Order No. 12, Report No. EPA-450/3-77-016 (PB 266 978), September 1976, p. VIII-20.
13. Ref. 12, p. VIII-21.
14. Ref. 3, p. 4–25.
15. Ref. 3, p. 4723.
16. G. Jutze, K. Axetell, and R. Amick, *Evaluation of Fugitive Dust Emissions from Mining*, PEDCo Environmental, Inc., prepared for U.S. Environmental Protection Agency, Industrial Environmental Research Laboratory, Office of Research and Development, Cincinnati, OH, Contract No. 6802-1321, Task No. 36, EPA-600/9-76-001, June 1976, pp. 38, 40.
17. *Reasonably Available Control Measures for Fugitive Dust*, Environmental Protection Agency, Ohio, p. 2–175.
18. Ref. 1, p. 63.
19. *Technical Guidance for Control of Industrial Process Fugitive Particulate Emissions*, PEDCo Environmental, Inc., prepared for U.S. Environmental Protection Agency, Office of Air and Waste Management, Office of Air Quality Planning and Standards, Research Triangle Park, NC, Contract No. 68-02-1375, Task No. 33, Project No. 3155-GG, EPA-450/3-77-010, March 1977, p. 2–241.
20. Ref.16, p. 46.
21. Ref. 17, p. 2–237.
22. C. Cowherd, G. E. Muleski, and J. S. Kinsey, *Control of Open Dust Sources*, EPA-450/3-88-008, U.S. Environmental Protection Agency, September 1988.
23. "Stack Test Results on Thermal Coal Dryers" (unpublished), Bureau of Air Pollution Control, Pennsylvania Department of Health, Harrisburg, PA.
24. "Amherst's Answer to Air Pollution Laws," *Coal Mining Proc.* **7**(2), 26–29 (1970).
25. E. Northcott, "Dust Abatement at Bird Coal," *Mining Congr.* J. **53**, 26–29 (November 1967).
26. *Report by York Research Corporation on Emissions from the Island Creek Coal Company Coal Processing Plant, Vansant, Virginia*, EMB Report 72-CCL-6, U.S. Environmental Protection Agency, Research Triangle Park, NC, February 1972.
27. *Coal Preparation Plant Emission Tests: Consolidation Coal Company, Bishop, West Virginia*, EMB Report 73-CCL-19, U.S. Environmental Protection Agency, Research Triangle Park, NC, November 1972.
28. *Coal Preparation Plant Emission Tests, Westmoreland Coal Company, Wentz Plant*, EMB Report 72-CCL-22, U.S. Environmental Protection Agency, Research Triangle Park, NC, April 1972.
29. *Background Information for Standards of Performance: Coal Preparation Plants, Vol. 2: Test Data Summary*, EPA-450/274-021b, U.S. Environmental Protection Agency, Research Triangle Park, NC, October 1974.
30. *Background Information for Establishment of National Standards of Performance for New Sources: Coal Cleaning Industry*, Environmental Engineering, Inc., Gainesville, FL, EPA Contract No. CPA-70-142, July 1971.
31. *1986 Keystone Coal Industry Manual*, McGraw-Hill, New York.
32. *Toxic Air Pollutant Emission Factors: A Compilation for Selected Air Toxic Compounds and Sources*, 2nd ed., EPA-450/2-90-011, U.S. Environmental Protection Agency, October 1990.
33. *Second Review of New Source Performance Standards for Coal Preparation Plants*, EPA-450/3-88-001, U.S. Environmental Protection Agency, Research Triangle Park, NC, February 1988.
34. *Estimating Air Toxic Emissions from Coal and Oil Combustion Sources*, EPA-450/2-89-001, U.S. Environmental Protection Agency, Research Triangle Park, NC, April 1989.
35. *Emission Testing Report: Bureau of Mines, Grand Forks, North Dakota*, EMB Report 73-CCL-5, U.S. Environmental Protection Agency, Research Triangle Park, NC, April 1973.
36. *Coal Preparation Plant Emission Tests, Consolidation Coal Company, Bishop, West Virginia*, EMB Report 72-CCL-19A, U.S. Environmental Protection Agency, Research Triangle Park, NC, February 1972.
37. *Emission Test Report, U.S. Steel #50, Pineville, West Virginia*, EMB Report 73-CCL-1, U.S. Environmental Protection Agency, Research Triangle Park, NC, September 1972.
38. *Emission Test Report, Westmoreland Coal Company, Quinwood, West Virginia*, EMB Report 75-CCL-7, U.S. Environmental Protection Agency, Research Triangle Park, NC, May 1976.
39. *Report by York Research Corporation on Emissions from the Florence Mining Company Coal Processing Plant, Seward, Pennsylvania*, EMB Report 72-CCL-4, U.S. Environmental Protection Agency, Research Triangle Park, NC, February 1972.
40. *Coal Preparation Plant Emission Tests: Eastern Associates Coal Company, Keystone, West Virginia*, EMB Report 72-CCL-13, U.S. Environmental Protection Agency, Research Triangle Park, NC, February 1972.
41. *Coal Preparation Plant Emission Tests: Island Creek Coal Company, Vansant, Virginia*, EMB Report 73-CCL-2, U.S. Environmental Protection Agency, Research Triangle Park, NC, September 1972.
42. *Report on Compliance Testing, Performed for Peabody Coal Company, Hawthorne Mine, Carlisle, Indiana*, Clean Air Engineering, Palatine, IL, May 6, 1993.
43. *Compilation of Air Pollution Emission Factors*, AP-42: Chapter 11.10, "Coal Processing," U.S. EPA, 1995.

CONCRETE BATCHING

Michael C. Schum, Michael D. Errickson, and John E. Yocom

Concrete is basically a mixture of two components: aggregate and paste. The paste, composed of cement and water, binds the aggregates into a rocklike mass as the paste hardens. Aggregates are generally divided into two groups: fine and coarse. Fine aggregates consist of natural or manufactured sand with particle sizes ranging up to $\frac{3}{8}$ in. Coarse aggregates are those with particles ranging up to 6 in.

Figure 1. Overview of a Concrete Batch Plant.

The paste is composed of Portland cement, water, and entrapped air. Cement paste typically constitutes about 25–40% of the total volume of concrete, with aggregates making up the remaining volume.

U.S. production of concrete is estimated to be 310 million cubic yards of concrete per year.[1] Concrete is widely used in the construction of roads and buildings. There are approximately 4,000 plants located throughout the United States. Due to the perishable nature of concrete, these plants are distributed throughout areas of heavy population. An overview of a typical concrete batching plant is shown in Figure 1.

PROCESS DESCRIPTION

Raw materials are typically delivered to a plant by rail, truck, or barge. Often, the plant is collocated with a rock crushing plant and/or a sand and gravel plant because of the ready availability of raw materials. Cement is commonly transferred to elevated storage silos pneumatically or by bucket elevator. Sand and coarse aggregate are transferred to elevated bins by a front-end loader, clamshell crane, belt conveyor, or bucket elevator. From these elevated bins, the constituents are fed by gravity or screw conveyor to weigh hoppers, which combine the proper amounts of each material. A typical process description is shown as Figure 2.

There are two types of concrete plants: transit mixed and central mix. Approximately 70% of U.S. plants are transit-mixed. At a transit-mixed concrete plant, sand, aggregate, cement, and water are all fed into the mixer truck. The concrete is then mixed on the way to the delivery site. In a central-mix plant, the sand, aggregate, cement, and water are mixed at the plant and subsequently transferred to a truck to be transported to the delivery site. The delivery truck is most commonly an agitator type; however, it may also be an open-bed dump truck. Shrink-mixed concrete is concrete that is partially mixed at a central-mix plant and then completely mixed in an agitator truck on the way to the delivery site.

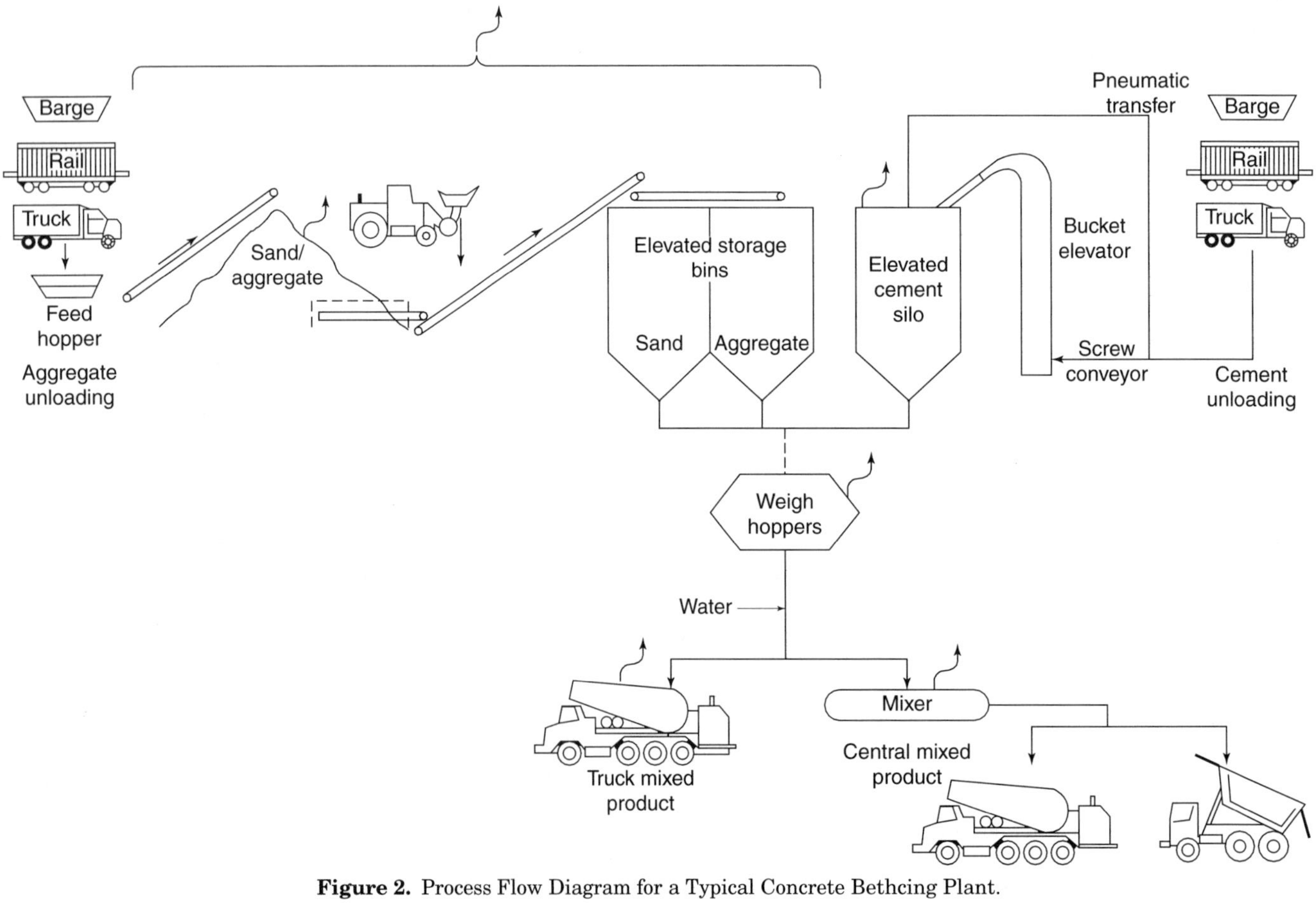

Figure 2. Process Flow Diagram for a Typical Concrete Bethcing Plant.

EMISSION CHARACTERIZATION

The only pollutant of concern from concrete batch plants is particulate matter. Particulate matter emissions consist primarily of cement dust, but also includes fine particles of aggregate and sand. Most emission points are fugitive sources such as material transfer points, loading operations, vehicle traffic, and wind erosion from storage piles. The only potential point source of emissions is from the cement silos.

The current emission factors for concrete batching are summarized in Table 1.[2] Where available, data have been provided for both controlled and uncontrolled operations. The term *controlled* applies to the use of hoods and shrouds in conjunction with dust collectors. With regard to paved roads, the term *controlled* apply to the use of wet suppression techniques. The "uncontrolled" data apply to plants operating without wet suppression techniques or dust collectors. AP-42 emission factor ratings have also been included in Table 1. The scale for emission factor ratings ranges from A (highest confidence level) to E. Particulate matter emission factors provided in Table 1 have a rating of D or E, which provide an order-of-magnitude emissions estimate. The poor emission factor rating is indicative of variability in test data.

The emission factors provided are for total particulate matter and PM_{10} (particles less than or equal to 10 μm). The EPA no longer maintains an air quality standard for total suspended particulate matter (TSP). However, they recently adopted a new air quality standard for $PM_{2.5}$ (particles less than or equal to 2.5 μm). There currently are no data available for $PM_{2.5}$ emissions from concrete batching operations.

Site-specific estimates of particulate matter emissions from unpaved roads can be obtained by utilizing the following equation from AP-42 Section 13.2.2[2]:

$$E = \frac{k(s/12)^a (W/3)^b}{(M/0.2)^c}$$

where E is the size-specific emission factor (lb/vehicle mile traveled); s, the surface material silt content (%); W, the mean vehicle weight (tons); and M, the surface material moisture content (%). The constants k, a, b, and c are defined in the following table:

Variable	$PM_{2.5}$	PM_{10}	PM_{30} (TSP)
k	0.38	2.6	10
a	0.8	0.8	0.8
b	0.4	0.4	0.5
c	0.3	0.3	0.4
Rating	C	B	B

Paved road emissions may also be calculated through the use of the empirical equation:

$$E = k(sL/2)^{0.65}(W/3)^{1.5}$$

where E is the particulate emission factor (having units matching the units of k); sL, the road surface silt loading (g/m^2); W, the average weight (tons) of the vehicles traveling the road; and k, the base emission factor for particle size range and units of interest (see following table)

	Multiplier k		
Size Range	Grams per vehicle kilometer traveled	Grams per vehicle mile traveled	Pounds per vehicle mile traveled
$PM_{2.5}$	1.1	1.8	0.0040
PM_{10}	4.6	7.3	0.016
PM_{15}	5.5	9.0	0.020
PM_{30} (TSP)	24	38	0.082

Further information on calculating paved road emissions can be found in AP-42, Section 13.2.1.

Table 1. Summary of Concrete Batching Emission Factors[2]

Source	Operation	Rating	Total P.M. (#/yd^3)	PM_{10} (#/yd^3)
Aggregate transfer to elevated bin	Uncontrolled	D	0.0069	0.0033
	Controlled		ND	ND
Sand transfer to elevated bin	Uncontrolled	D	0.0021	0.00099
	Controlled		ND	ND
Cement unloading to elevated storage silo (pneumatic)	Uncontrolled	E	0.48	0.39
	Controlled	D	0.00099	0.00034
Cement supplement unloading to elevated storage silo (pneumatic)	Uncontrolled		ND	ND
	Controlled	E	0.0089	0.0049
Weigh hopper loading	Uncontrolled	D	0.0051	0.0024
	Controlled		ND	ND
Mixer loading (central-mix plant)	Uncontrolled	E	0.22	0.078
	Controlled	E	0.011	0.0038
Truck loading (truck-mix plant)	Uncontrolled	D	0.61	0.15
	Controlled	D	0.21	0.051
Vehicle traffic (paved roads)	Uncontrolled	D	9.3	1.8
	Controlled	E	6.9	1.4
Vehicle traffic (unpaved roads)	Uncontrolled	D	5.8	2.6
	Controlled		ND	ND

AIR POLLUTION CONTROL MEASURES

There are three approaches to reducing the quantity of particulate emissions from concrete batching plants: (1) Minimize the amount of dust that is released into the atmosphere by plant process changes; (2) prevent particles from becoming airborne by various suppression techniques such as wetting; (3) capture and remove particles that would otherwise become airborne.[3] Several techniques for minimizing dust emissions include:

- Fine tuning the batch sequence to deliver a smooth, controlled flow of raw material into the mixer.
- Enclose aggregate transfer points and storage piles.
- Enclose the batch plant to protect it from the wind.
- Use of vacuum sweepers to reduce dust accumulation on roadways.

The goal of these techniques is to reduce the amount of fugitive emissions from the plant.

Another approach to reduce particulate emissions is the use of water to prevent dust from becoming airborne. Several areas where dust suppression can be effective include:

- Use of a spray bar to rinse down the charge hopper at the truck mixer load point (transit mix plants).
- Spray down paved roadways.
- Spray aggregate storage piles.
- Use spray bars at aggregate transfer points.

The most effective method to reduce emissions is the use of dust collection systems in conjunction with baghouses. Collection systems include the use of hoods and shrouds at dust generation sources, such as the truck mixer loading point (transit mix plant), the cement weigh hopper, and the central mixer (central mix plant). These dust collectors should be vented to a baghouse for removal. Baghouses or fabric filters should also be installed at the vents on top of the cement storage silos to capture dust generated as the silos are filled or drawn down. Figure 3 shows a central mixer with dust collection hood at the mixer outlet.

Figure 4 is a general view of a central mixer and associated baghouse dust collector.

Figure 3. Central Mixer Dust Collection Hood.

Figure 4. Mixer Dust Collector.

Periodic maintenance is crucial to maintain the effectiveness of dust control devices such as water sprays and baghouses.[4] An inspection schedule specific to the concrete batch facility should be developed to reduce lapses in dust control efforts. At a minimum, the following items should be inspected:

- Check filter bags for holes, blinding, and short-circuiting.
- Confirm that filter bag shakers are operating correctly.
- Observe the spray pattern of water sprays to ensure even moisture distribution.
- Ensure dust collection hoods are properly located near the dust generation source.
- Observe the cement silo loading. Excessive rates of portland cement loading reduce the efficiency of dust control equipment.

REFERENCES

1. Telephone Conversation between Michael Errickson, HRP Associates, Inc., and Colin Lovell of the National Ready Mixed Concrete Association, Silver Spring, MD, February 23, 1999.
2. *Compilation of Air Pollutant Emission Factors*, AP-42, U.S. Environmental Protection Agency, Research Triangle Park, NC, 1995.
3. *Environmental Management Practices*, Chapter 6, National Ready Mixed Concrete Association, Silver Spring, MD.
4. Written Communication from Vic Copelan, North Carolina Department of Environmental Health and Natural Resources, Washington, NC, to Concrete Plant Owners, December 5, 1995.

FIBERGLASS OPERATIONS

Aaron J. Teller and Joseph Y. Hsieh

The production of fiberglass consists of two different forms of product — continuous-filament fiberglass or textile products and fiberglass blown wool or insulation products.

The general-purpose textile fiberglass, which is moisture- and alkali-resistant with good electrical and physical properties, is also called E- (electrical) Glass. The major applications of the general-purpose textile fiberglass are in the production

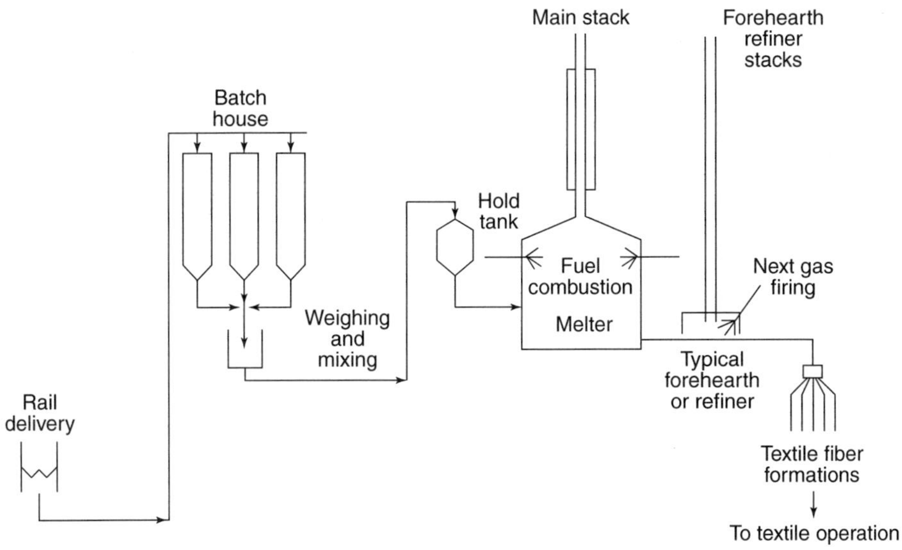

Figure 1. Fiberglass Manufacturing Process.

of fireproof cloth, fiberglass-reinforced plastics (FRP), and composites. It is used in thermal and acoustical insulation because the small cell of air entrained in the wool or blanket prevents the movement of air and sound waves. The manufacturing process consists of (Figure 1): materials blending and transport, melting and refining, and fiberforming and textile operation.

As in the manufacturing of glass, the major air emission problem where acid-gas recovery in addition to particulate is required is related to the melting and refining furnace operation. The fiber-forming and textile operations produce, primarily, particulate emissions and some volatile organic compounds (VOCs). The textile product manufacture presents a more complex emissions control problem than does insulation manufacture because of the presence of boron and fluorine in the most flexible E-Glass product.

PROCESS DESCRIPTION

The composition of textile and insulation fiberglass, although varied as a function of the producer, has the general composition[1] shown in Table 1.

Table 1. Composition of Textile and Insulation Fiberglass

Component	Textile (% wt)	Insulation (% wt)
SiO_2	54–55	59–65
Al_2O_3	13–14	4–8
Na_2O	0.5	15–18
CaO	22–24	4.5–9
MgO	0.3–0.4	0–4
B_2O_3	5.5–6.8	0–2
Fe_2O_3	0.3	0.1–5
F	0.7	
K_2O		0.3
MnO		0–4
BaO		0–5
SO_3		$\leq$0.6

The raw materials are unloaded from freight cars or trucks and transported to specific silos in the batch house. The materials are then withdrawn to automatic weight machines and blended. The mix is transported by air conveying to the holding vessel at the melter and then fed as a batch to the furnace or melter.

The melters are reverberatory with gas firing, although oil may be used. There is a significant difference from the container or float-glass furnace. Thermal recovery is achieved by a recuperator in a continuous rather than cyclical manner. A part of the reasoning is that the fiberglass (especially textile) melt is extremely sensitive to very small pressure variations and that boron emissions can clog the brick-lattice regenerator.

Because of the time–temperature sensitivity of the final product, liquid glass flows through channels with zones of refiners and forehearths, in addition to the melter. These forehearths and refiners are generally fired with natural gas. The flue gas from the melter flows upward from the combustion zone through the stack containing an annular zone heat exchanger. Combustion air flows in the annular zone of the recuperator for preheating. The flue gas at 500–800°C is emitted, generally by passing through an air damper, to the stack. The forehearth and refiner flue gases are emitted through individual stacks. These flue gases are often combined with the melter flue gas.

AIR EMISSIONS CHARACTERIZATION

The air emissions for the fiberglass operation can be characterized according to the three phases of manufacturing processes. Emission factors for the fiberglass operations are given in Tables 2 and 3.

Materials Blending and Transport

The primary air pollution problem in this phase of the operation is that of containment and fugitive dust handling inasmuch as fine particulates are not involved. The raw materials are generally in the size range of 20–500 μm.

Table 2. Emission Factors for Glass Fiber Manufacturing without Controls[a] Emission Factor Rating: B

	Particulates		SO_x		CO		NO_x		VOC[b]		Fluorides	
	lb/ton	kg/Mg	lb/ton	kg/Mg	lb/ton	kg/Mg	lb/ton	kg/Mg	lb/ton	kg/Mg	lb/ton	kg/Mg
Unloading and conveying[c]	3.0	1.5	*d*	*d*	*d*	*d*	*d*	*d*	*d*	*d*	*d*	*d*
Storage bins[c]	0.2	0.1	*d*	*d*	*d*	*d*	*d*	*d*	*d*	*d*	*d*	*d*
Mixing and weighing[c]	0.6	0.3	*d*	*d*	*d*	*d*	*d*	*d*	*d*	*d*	*d*	*d*
Crushing and batch charging[c]	Neg	Neg	*d*	*d*	*d*	*d*	*d*	*d*	*d*	*d*	*d*	*d*
Glass furnace—wool												
Electric	0.5	0.25	0.04	0.02	0.05	0.025	0.27	0.14	*e*	*e*	0.002	0.001
Gas—regenerative	22	11	10	5	0.25	0.13	5	2.5	*e*	*e*	0.12	0.06
Gas—recuperative	25–30	13–15	10	5	0.25	0.13	1.7	0.85	*e*	*e*	0.11	0.06
Gas—unit melter	9	4.5	0.6	0.3	0.25	0.13	0.3	0.15	*e*	*e*	0.12	0.06
Glass furnace—textile												
Recuperative	2	1	3	1.5	0.5	0.25	20	10	*d*	*d*	2	1
Regenerative	16	8	30	15	1	0.5	20	10	*d*	*d*	2	1
Unit melter	6	3	*e*	*e*	0.9	0.45	20	10	*d*	*d*	2	1
Forming—wool												
Flame attenuation	2	1	*d*	*d*	*d*	*d*	*d*	*d*	0.3	0.15	*e*	*e*
Forming—textile	1	0.5	*d*	*d*	*d*	*d*	*d*	*d*	Neg	Neg	*d*	*d*
Oven curing—wool												
Flame attenuation	6	3	*e*	*e*	3.5	1.8	2	1	7	3.5	*e*	*e*
Oven curing and cooling—textile	1.2	0.6	*d*	*d*	1.5	0.75	2.6	1.3	Neg	Neg	*d*	*d*

[a] Expressed as units per unit weight of raw material processed. Neg, negligible.
[b] Includes primarily phenols and aldehydes, and to a lesser degree, methane.
[c] Ref. 10.
[d] Not applicable.
[e] No data are available.

Table 3. Uncontrolled Emission Factors for Rotary Spin Wool Glass Fiber Manufacturing[a]

	Particulate			Organic Compounds[b]		
Products	Front Half	Back Half	Total	Phenolics[c]	Phenol	Formaldehyde
R-19	17.81 (36.21)	4.25 (8.52)	22.36 (44.72)	3.21 (6.92)	0.96 (1.92)	0.75 (1.50)
R-11	19.61 (39.21)	3.19 (6.37)	22.79 (45.59)	6.21 (12.41)	0.92 (1.84)	1.23 (2.46)
Ductboard	27.72 (55.42)	8.55 (17.08)	36.26 (72.50)	10.66 (21.31)	3.84 (7.68)	1.80 (3.61)
Heavy density	4.91 (9.81)	1.16 (2.33)	6.07 (12.14)	0.88 (1.74)	0.53 (1.04)	0.43 (0.85)

[a] Ref. 7. Expressed kg/Mg (lb/ton) of finished product. Gas stream did not pass through any added primary control device (wet electrostatic precipitator, venturi scrubber, etc.)
[b] Included in total particulate catch. These organics are collected as condensible particulate matter and do not necessarily represent the entire organics present in the exhaust gas stream.
[c] Includes phenol.

Melting and Refining

The emissions from the melting–refining furnace during the textile fiberglass operation contain:

1. Fine particulates, including calcium carbonate, sodium fluoride, sodium fluosilicate, silica, calcium fluoride, aluminum silicate, sodium sulfate, boron oxides–boric acid.
2. Gases, including fluorides—hydrogen fluoride, silicon tetrafluoride, sulfur oxides, nitrogen oxides, boric acid, carbon dioxide, water vapor.

The solid particulates emitted in the melter flue gas are classed as fine particulates, with most of the particle size below 10 μm. Another class of particulates is formed downstream of the stack for the untreated emissions. This consists primarily of sublimed boric acid. It can form a very opaque, long-lasting white plume resulting from sublimation after cooling. This "formed particulate" is created in the emissions from the melter, forehearths, and refiners. About 70% of the boron and fluorine emissions are emitted by the melter.

The concentration of the particulates and condensible vapor boric acid in the effluent gases as follows:

Solid particulates	200–1500 mg/nm^3
Boron oxide, solid plus equivalent boric acid	100–1000 mg/nm^3

The gaseous components of the emissions are hydrogen fluoride, silicon tetrafluoride, sulfur oxides, and the boric acid considered in the discussion of particulates. They are present in concentrations ranging as follows:

HF, SiF_4 (equivalent HF)	20–160 ppm
SO_x	40–80 ppm

The emissions from the melting–refining furnace in the insulation fiberglass operation contain only small quantities of boron and no fluoride. The emissions from the fiber forming and textile operations include particulate and VOC emissions. The VOC emissions are lower for textile fiberglass than for insulation fiberglass because of the small quantities of lubricant and VOC used in the textile fiberglass operation.

AIR POLLUTION CONTROL MEASURES

Materials Blending and Transport

In the materials storage and transport area, the railroad discharge is generally partially sleeved and under either positive or negative pressure. The silos are provided with bin vent filters for collecting the fugitive dust. The blending vessels are vented with airflow directed to bag filters. These are generally pulse jets operating at an air-to-cloth ratio in the range of 4–7 fpm. The solid discharge is recycled to the blending zone.

Melting and Refining

A major consideration in the control of emissions from E-Glass furnaces is the suppression by capture of the boron compounds. If this is not achieved, then the prospect of opaque plumes resulting from sublimation of boric acid and the potential for forming a complex of HF and H_3BO_3 such as HBF_4 both exist.

A significant factor in the design of the air pollution control system is the maintenance of the melter pressure, with the permissible variation, as requested by furnace operators, ± 0.01 in. wc. Both wet and dry scrubbing methods have been utilized in controlling the emissions from the melting and refining operation.

Wet Scrubbing

Either the venturi–packed-bed scrubber or a nucleation cross-flow scrubber can be used. The venturi system is a modification of the conventional packed-tower–venturi–cyclone arrangement for recovering the particulate and acid gases. The treatment of blow-down stream required appears to have limited its application.

A low-energy (less than 15 in. wc) nucleation cross-flow scrubber was operated in a Canadian facility.[8] The system used a cooling tower to circumvent the steam plume appearance. The major problems encountered were the dual alkali transfer for (Na → Ca), resulting in excess solubility of calcium ions. This often caused scrubber plugging by deposition of insoluble calcium salts.

Dry Scrubbing

Two types of systems have been installed in the United States for E-Glass production.

1. Spray drier-electrostatic precipitator (ESP) (Figure 2)
2. Quench reactor-dry venturi-baghouse (Figure 3)

The latter has also been installed in Russia and Taiwan.

Spray Dryer–Electrostatic Precipitator

The gases are cooled by a water–lime slurry quench to about 100°C and then flow to the ESP, the system fan, and out of the stack. Although gaseous fluoride was reduced to the 5-ppm range, there was evidence of significant plume opacity, reflecting the presence of boron compounds in the emitted gases. Problems were reported with condensation and adhesion melting of particulate on the ESP electrodes.[2]

Quench Reactor–Dry Venturi–Baghouse[3,4]

This system is operated with either water quench or lime slurry quench as a function of requirements for reduction of SO_2 and HF. Where the only SO_x reduction requirement is related to opacity, and SO_3 only is to be removed, water quench in the quench reactor is utilized.

The quenched gas, at about 100°C, proceeds to the dry venturi, where limestone is introduced into the gas stream as both reagent and cake modifier.[3] The gas is then further cooled by air dilution to the range of 60–70°C. The limestone acts to neutralize the fluorides and aid in the formation of a low-compressibility layer in the baghouse. A significant part of the neutralization reaction occurs in equivalent renewable "fixed-bed" cake on the bags. The cleaning cycle is approximately 6 h.

The reduction of temperature to 60–70°C results in the sublimation of the boric acid. This boric acid is deposited on the surface of the baghouse cake along with the crystalline reagents. Filtration in a temperature region only 20°F greater than the dew point is maintained with a system pressure drop in the range of 8 in. wg. The performance of the system, with greater than 98% on-line reliability, is given in Table 4.

For the insulation fiberglass operation, inasmuch as the emissions from the melter contains no fluoride and only small quantities of boron, the emission control can be achieved by a spray dryer–ESP (Figure 2) or spray dryer–baghouse.

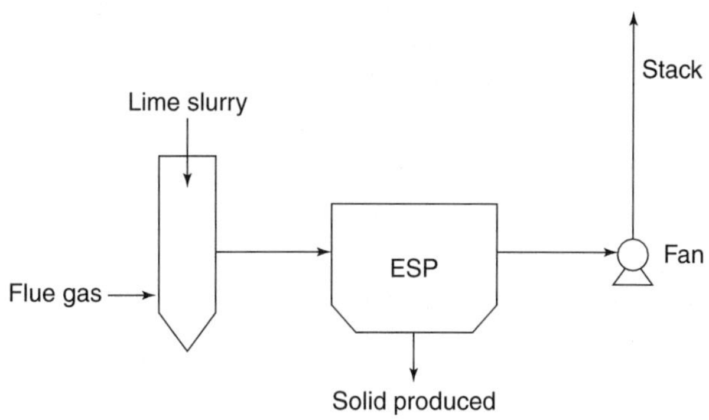

Figure 2. Spray Dryer—Electrostatic Precipitator System.

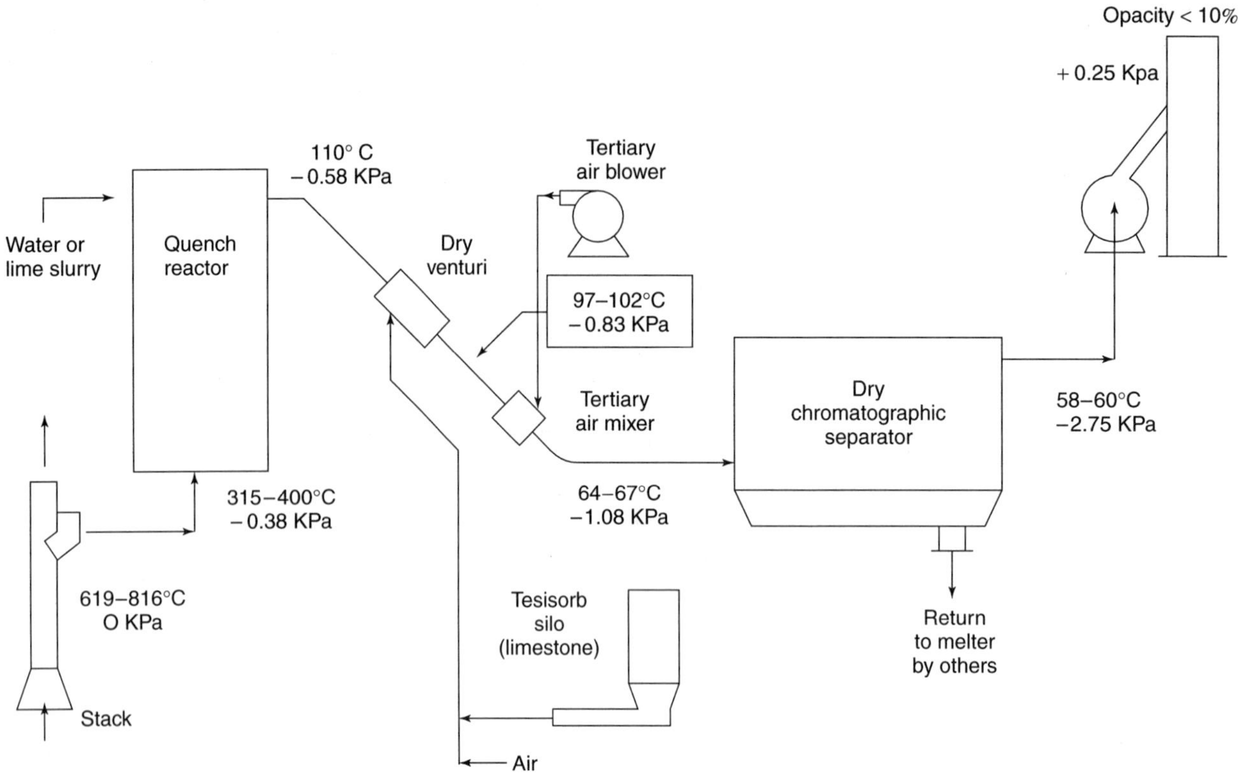

Figure 3. Typical Teller Dry Chromatographic System for the Fiberglass Industry—HF, SO_x, and Particulate.

Table 4. Performance of Semidry System Fiberglass Emission Control

		Outlet Composition					
		Fluoride			Boron		
Dry Gas Flow (sdcfm)	Particulate (mg/nm^3)	Solid (lb/h)	Gas (lb/h)	(ppm)	Solid (lb/h)	Gas (lb/h)	(mg/nm^3)
18,500	1.5	0.015	0.11	2.12	0.03	0.92	13.7
14,600	2.3	0.004	0.009	0.28	0.030	0.475	9.2
15,500	2.1	0.005	0.017	0.45	0.032	0.568	10.3
14,800	4.5	0.003	0.007	0.21	0.019	0.208	4.1
15,100	1.8	0.003	0.002	0.10	0.03	0.52	9.8

Source: Ref. 9.

Fiber Forming and Textile Operation

Fiber formation does not present emission control difficulties. However, the spraying of the mat with lubricants and/or binders to minimize erosion of the fibers can create either VOC or inorganic particulate emissions.

The predominantly inorganic discharges are treated in a cross-flow scrubber with particulate emissions reduced to less than 25 mg/nm^3.[6] The organic emissions are treated by condensation processes for high-molecular-weight emissions and, if necessary, adsorption for low-molecular-weight compounds.

NESHAP FOR WOOL FIBERGLASS MANUFACTURING

Under the Clean Air Act Amendments of 1990, EPA is required to regulate emissions of 188 listed toxic air pollutants. On July 16, 1992, EPA published a list of industrial source categories that emit one or more of these air toxics. For listed categories of "major" sources (those that emit 10 tons/year or more of a listed pollutant or 25 tons/year or more of a combination of pollutants), the Clean Air Act required EPA to develop standards that require the application of stringent air pollution controls, known as maximum achievable control technology (MACT).

EPA's published list of industry groups (known as "source categories") to be regulated includes major sources that manufacture wool fiberglass. The proposed regulation was published in the *Federal Register* on March 31, 1997, and became effective in May 1999. A supplement to the proposed regulation was proposed in the *Federal Register* on February 12, 1999 (64 FR 7149). The EPA was considering all comments on the supplemental proposal at the time of this publication and will amend the final regulation in a future action as appropriate. The following summarizes the rule and the affected facilities[11]:

Table 5. Summary of Final Rule Emission Limits for New and Existing Glass-Melt Furnaces and RS and FA Manufacturing Lines in Wool Fiberglass Plants

Process	Emission Limit	
	Existing	New
Glass Furnace	0.25 kg of PM per Mg of glass pulled	0.25 kg of PM per Mg of glass pulled
RS Manufacturing Line	0.6 kg of formaldehyde per Mg of glass pulled	0.4 kg of formaldehyde per Mg of glass pulled
	Pipe Insulation	Pipe Insulation
FA Manufacturing Line	3.4 kg of formaldehyde per Mg of glass pulled	3.4 kg of formaldehyde per Mg of glass pulled
	Heavy Density	Heavy Density
	None	3.9 kg of formaldehyde per Mg of glass pulled

- There are a total of 21 wool fiberglass manufacturing plants nationwide that are affected by the final rule (there are 27 plants nationwide currently in operation; however, six of them are not major sources or do not have manufacturing lines subject to the standard and therefore are not subject to the regulation).
- The rule applies to new and existing glass manufacturing furnaces; new and existing rotary-spin (RS) manufacturing lines producing building insulation; new and existing flame-attenuation (FA) manufacturing lines producing pipe products, and new FA manufacturing lines producing heavy-density insulation products. Existing FA manufacturing lines are not affected, since they are already in compliance with the regulation and EPA anticipated that no new FA manufacturing lines will be built in the future.[11]
- The hazardous air pollutants (HAPs) emitted by the facilities covered by this rule include compounds of three metals (arsenic, chromium, lead) and three organic HAPs (formaldehyde, phenol, and methanol)
- The rule sets limits for formaldehyde emissions from most RS manufacturing lines and certain FA manufacturing lines. Formaldehyde is a toxic air pollutant and is a surrogate for phenol and methanol, which are also emitted from RS and FA manufacturing lines producing bonded wool fiberglass products.
- The rule also sets emission limits for PM emissions from new and existing glass melting furnaces. PM emissions are surrogate for metal hazardous air pollutant (compounds of arsenic, chromium, and lead) emissions. The limits are based on the use of well-designed and-operated baghouses or electrostatic precipitators (ESP) to collect PM emissions including metal hazardous air pollutants. Since these technologies are considered the best technologies available, they are required for both new and existing facilities.
- The MACT standard requires that each affected source perform an initial compliance test to demonstrate compliance with the emission limits. For air pollution control devices and process equipment used to comply with the emission limits, the initial compliance tests are used to establish levels of control device parameters and process parameters used to monitor compliance. The rule requires that these control device parameters and process parameters be monitored on a regular basis in order to determine that the control device or process equipment is operating properly. The rule also specifies requirements for notifications, reporting, and recordkeeping.

The MACT limits are summarized in Table 5 for new and existing glass-melting furnaces and RS and FA manufacturing lines.

REFERENCES

1. Duffy, "Glass Technology," *Chem. Tech. Rev.*, **184** (1981).
2. Noll, *Glass Tech.* **25**(2), 91–97 (1984).
3. A. J. Teller, U.S. Patent No. 2,018,157, 1972.
4. A. J. Teller, U.S. Patent No. 4,319,890, 1982.
5. A. J. Teller, *Glass Indus.* 18–26 (June 1982).
6. Private communications, Ceilcote.
7. Wool Fiberglass Insulation Manufacturing Industry, *Background Information for Proposed Standards*, EPA-450/3-83022a, U.S. Environmental Protection Agency, Research Triangle Park, NC, December 1983.
8. A. J. Teller, *Glass Indus.*, 15–19 (February 1976).
9. Private communications, PPG Industries.
10. J.R. Schorr et al., *Source Assessment: Pressed and Blown Glass Manufacturing Plants*, EPA-600/2-77-005, U.S. Environmental Protection Agency, Research Triangle Park, NC, January 1977.
11. *Fact Sheet: Final Air Toxics for Wool Fiberglass Manufacturing*, U.S. EPA, May 13, 1999.
12. *National Emission Standard for Hazardous Air Pollutants for Source Categories: Wool Fiberglass Manufacturing: Proposed Rule*, 40 CFR Part 63, March 31, 1997.

GLASS MANUFACTURING

Aaron J. Teller and Joseph Y. Hsieh

The production of glass products, although based on an old art, has emerged as sophisticated and automated batch and continuous processes. The major forms of glass produced are container glass and flat glass, about 75% of the total production. Lesser amounts produced as pressed and blown glasses include lead and borosilicate glass and frit for ceramic coatings.

A typical flow diagram for the production of container glass (Figure 1) establishes three major components of the

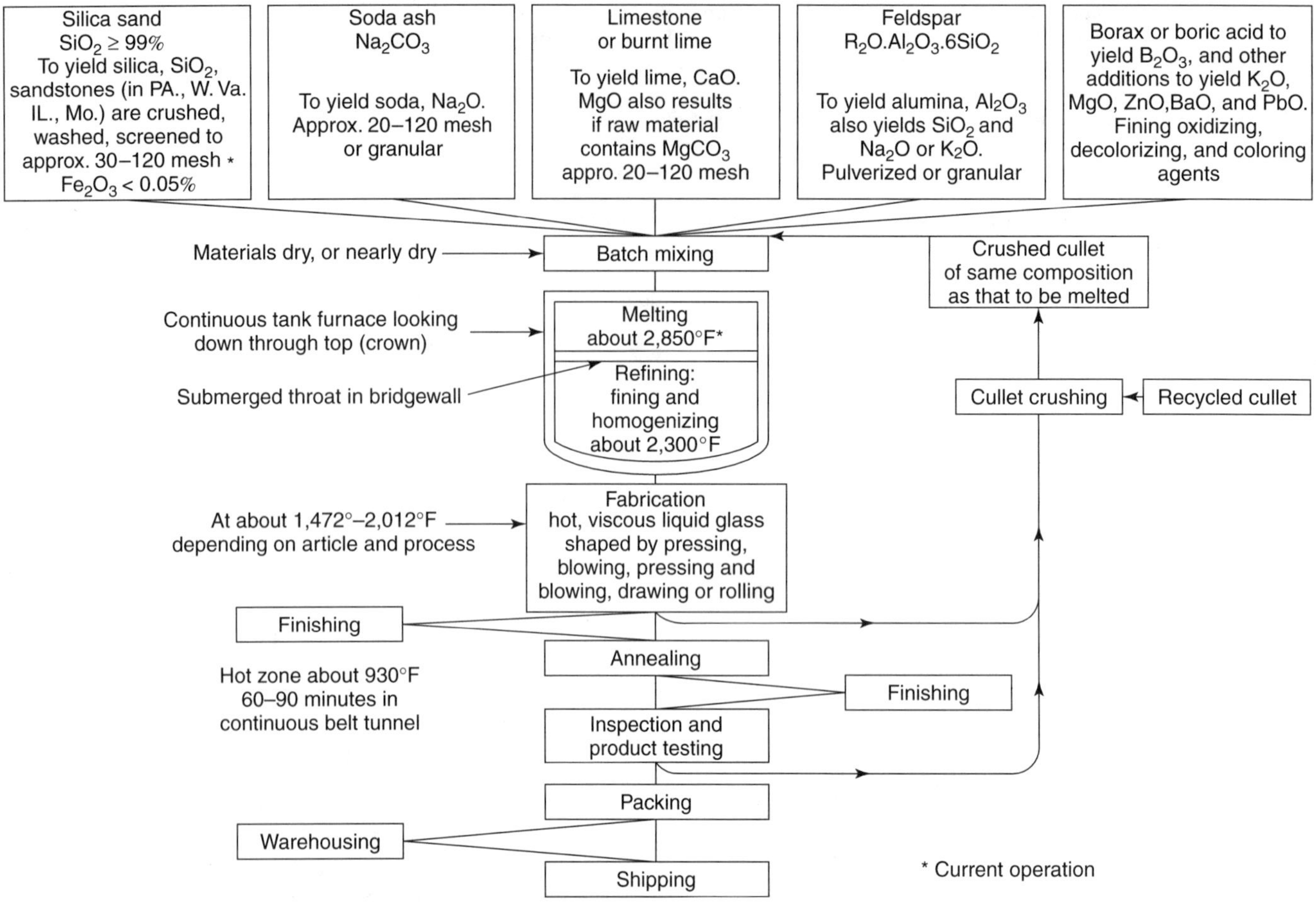

Figure 1. Flow Diagram for Soda-Lime Glass Manufacture (Kird and Othmer, 1947).

manufacturing process: raw-material blending and transport, melting, and forming and finishing. The major difference between container glass and flat glass is in the forming operation. Container glass is formed by blow molding. Flat glass floats the molten glass continuously on a liquid alloy surface with simultaneous cooling to sheet glass. Thus flat glass is sometimes referred to as float glass. The lead and borosilicate glasses and frit may be produced in continuous reverberatory furnaces, but because of small size production, are often conducted in batch, rotary (including tilting), crucible, and hearth furnaces.

The air emissions from glass manufacturing operation are in three zones:

1. Raw-material blending and transport
2. Melting
3. Forming and finishing

The majority of air emissions are in the melting furnace operation.

One of the major concerns in the glass industry was that the addition of an emission control system would adversely affect the pressure stability in the melter and that parallel operation of melters with an emission control system would destabilize the operation. Several years of operation with an emission control system on glass furnaces[8] has dispelled this concern and has resulted in even greater melter stability.

PROCESS DESCRIPTION

The major materials used in the production of glass are cullet (recycled glass), silica sand, limestone, and soda ash.

Additional components used in minor quantities for both container and flat glass, and for specialty glasses, are salt cake; alumina; barium oxide; boron compounds for borosilicate glass; lead compounds for both lead glass and borosilicate glass; cryolite or fluorospar for frit glass; opacifiers (tin oxide, sodium antimonate, etc.)—frit; and color materials (iron oxide, nickel oxide, copper oxide, cobalt oxide, and so on). These materials are stored in batch house silos and are automatically weighed and blended in batches and transported to the holding zone at the furnace, where they are added to the melt. These materials are in the size range of 10–500 μm with a median in the 300-μm range.

Melting for container and flat glass is generally conducted in a continuous reverberatory furnace (Figures 2 and 3) fired by natural gas or fuel oil. Electric boost furnaces have been introduced in some operations to minimize flue gas emissions (Figure 4).

The reverberatory melters, both end-port and side-port designs, utilize regenerative heat recovery. These are commonly constructed of checker brick in two separate arrays. The flue gas and combustion gases are alternated between the two checker arrangements, with the flue gas heating one checker while the combustion air flows through the other checker, recovering a portion of the thermal energy of the flue gas.

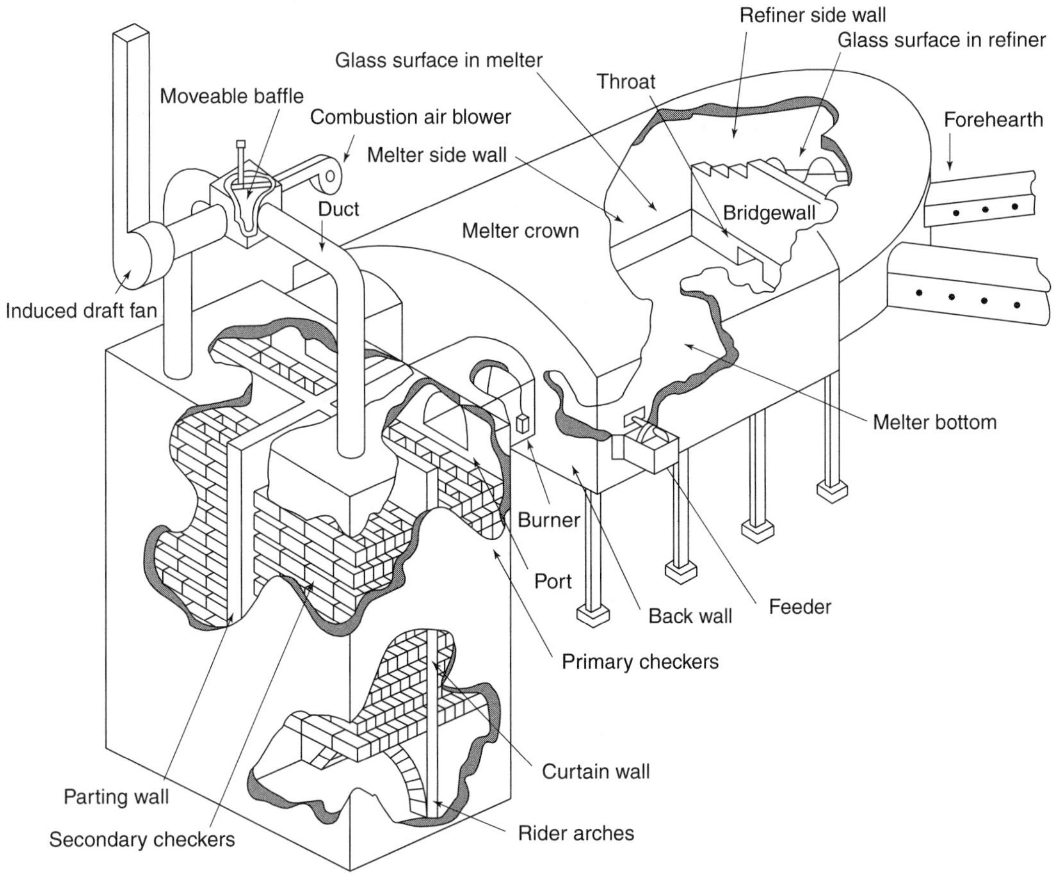

Figure 2. Regenerative and Port Glass-Melting Furnace.

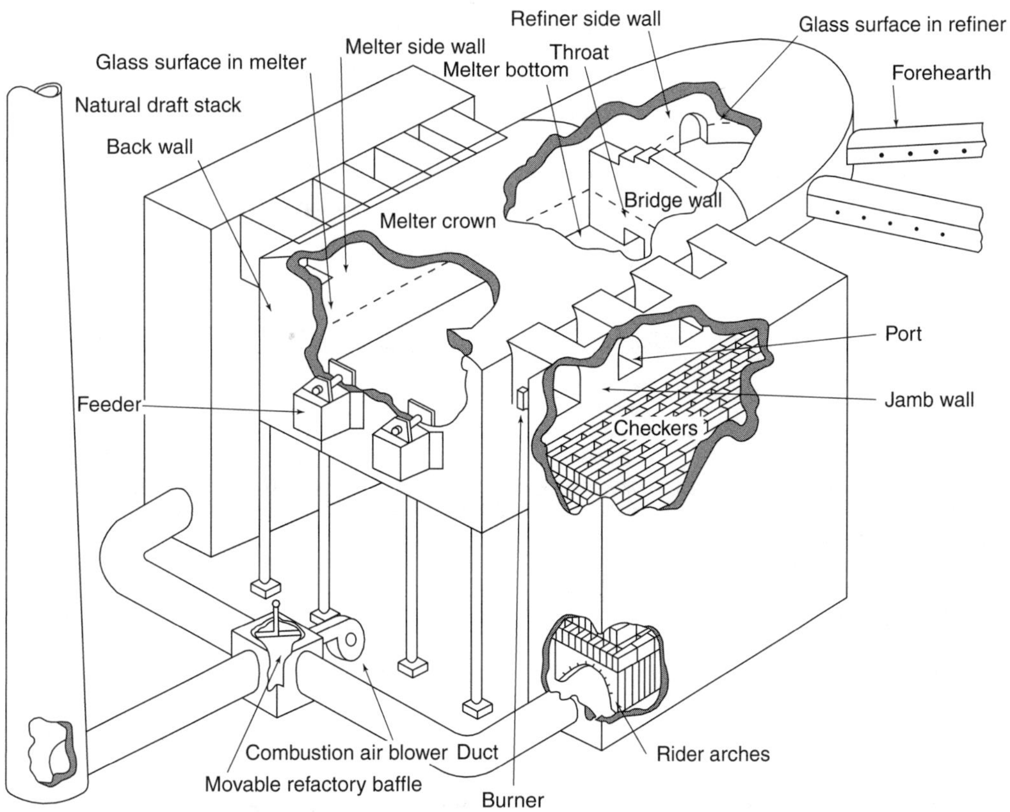

Figure 3. Regenerative Side Port Glass-Melting Furnace.

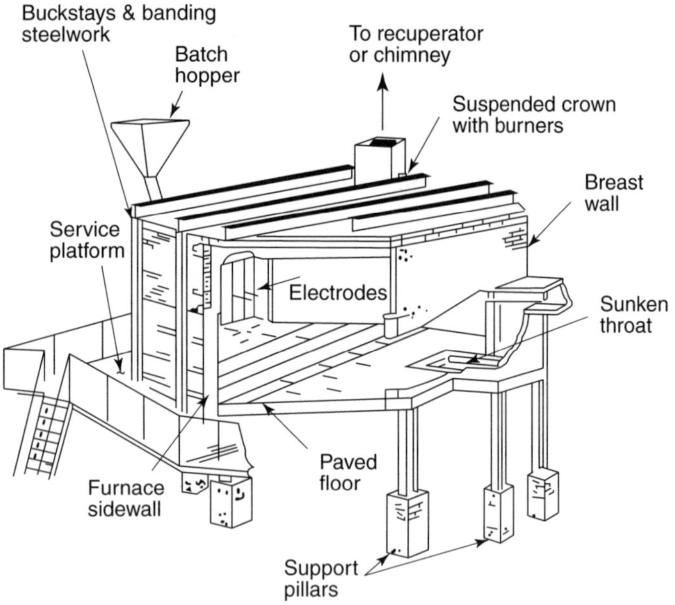

Figure 4. Electric Boost Furnace.

The diversion of the gases causes spikes in the system pressure and temperature at the time of reversal; the durational cycle is 15–25 min.[8] The furnace temperature is maintained in the range of 1200–1575°C. The flue gas, after leaving the regenerative checkers, exhibits a temperature range of 450–500°C.

AIR EMISSIONS CHARACTERIZATION

Raw Material Blending and Transport

Fugitive dust is produced by the preparation transport process. Modern plants contain the fugitive dust in an enclosed system and use air transport of the batch feed. Thus conventional fabric filters are used on all silos and the transport system to confine the particulate emissions.

The emissions consist of particulates in the size ranges of 10, 50, and 300 μm. They are characterized as fugitive dust emissions.

Melting

Two sources contribute to air emissions from melting furnace operations, namely, the combustion of fuel and the evaporation or dissociation of raw materials.

The emissions from glass melting, in the flue gas, consist of particulate matter and sulfur oxides in container and flat glass manufacture. They may contain boron oxides and fluorides in the frit and some specialty glass product manufacture. Almost all the particulate matter emitted from the melting furnace is less than 1 μm in size (Figure 5).

The characteristics of uncontrolled emissions from glass manufacture are presented in Table 1.

Forming and Finishing

The emissions from forming and finishing result from thermal decomposition of lubricants for the gob shears and gob delivery systems.

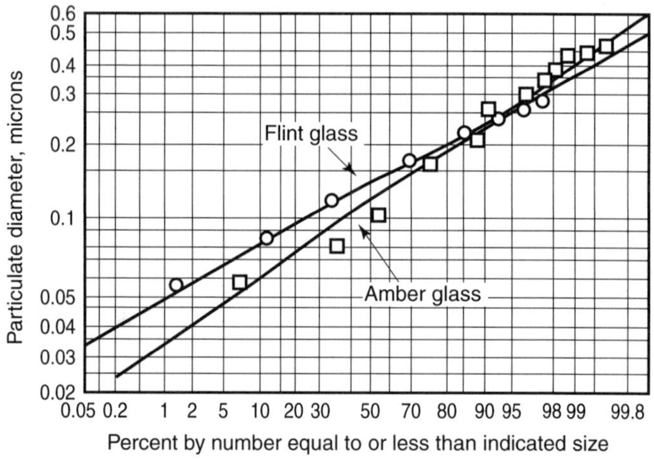

Figure 5. Log-Probability Distribution of Particle Sizes Present in Glass Furnace Effluent.

Table 1. Characteristics of Emissions from Glass Manufacturers

	Container	Float	Opal (Borosilicate)	Frit
Opacity (%)	10–50	20–50	20–100	10–60
Particulate (gr/dscf)	0.02–0.35	0.02–0.35	0.02–0.35	0.02–0.35
Particulate composition (primary composition)	Na_2SO_4, SO_3	Na_2SO_4, SO_3 NaF, Na_2SiF_6	B_2O_3 P_4O_{10}	Na_2SO_4, NaF
Particulate size (Figure 1) (μm)	<5	<5	<5	<5
Gaseous emissions	SO_x, NO_x	SO_x, NO_x	HF, NO_x	SO_x, HF, NO_x
Gas emissions—gas fired (ppm)				
SO_x	50–300	150–400		100–300
HF			400–1000	100–300
NO_x	100–600	200–600	200–600	200–600
SO_3 (% of SO_x)	2–15	2–15		2–15
SO_x when oil-fired (ppm)	400–800	400–800	350–500	400–800

Source: Ref. 7.

AIR POLLUTION CONTROL MEASURES

Raw-Material Blending and Transport

Bin vent filters on the storage silos are standard units supplied by the silo vendors for batch house operation. Fabric filters are used for collecting particulates emitted at the weigh-blending station. These are primarily pulse jet baghouses operating at an air–cloth ratio of 4–7 fpm.

Typical practice provides hooding at the blending and receiving zones. The railroad hopper car discharge is connected by sleeves to the air transport system. The ventilation rate on hooding is of the order of 8–15 cfm/ft^3 of mixer capacity.

Melting

Particulate and Acid Gases. The control of particulate matter and acid gas emissions from glass furnaces has utilized the following systems:

Wet scrubbing — venturi and packed bed
Baghouse
Electrostatic precipitator (ESP)
Electric boost
Dry scrubbing
Semidry scrubbing

These have been specifically related to the removal of particulate and acid gases. No NO_x reduction has been reported for any of these systems.

A comparison of the reported information regarding performance and operating cost is given in Table 2.

Wet Scrubbing

The system consisting of a venturi and packed-bed scrubber with a pressure drop of 20–30 in. wc was earlier built to control the emission from container and opal glass operation. The high energy consumption and the complexity of treatment of liquid discharge required have appeared to limit its application.

Baghouse

Both reverse-air cleaning and shaker-type baghouses with an air-to-cloth ratio of 2–4 fpm have been utilized for the particulate control. The high-temperature baghouse, utilizing fiberglass bags operating in the range of 200°C, functions only as a particulate remover. Because of the fine particulate, a rapid rise in pressure drop is encountered, resulting in short cleaning cycles. A rise in the "clean bag" pressure drop increases over time, coupled with an operating pressure drop of 10–15 in. wg. The recovered product can be recycled to the production process.

Electrostatic Precipitator

The ESP has been the predominant air pollution control system in the glass industry. It functions only as a particulate separator and, in that type of service, has provided reliable performance for container glass operators. The particulate emissions are low, in the range of 20 mg/nm^3.

It has been reported that the collected particulate does not adhere properly to the electrodes in borosilicate glass applications.[4] The material recovered by the ESP can be recycled to the glass-making process.

Dry Scrubbing

Dry scrubbing (Figure 6) is used primarily for to control both particulate and SO_3, while maintaining low pressure drop and long cleaning cycles in the baghouse. The system consists of injecting nepheline syenite, with or without limestone addition, into the dry venturi preceding the baghouse. The system is generally operated in the range of 150°C. These Tesisorbs™ permit the formation of a thick cake in the baghouse with low pressure drop, thus enhancing absorption and fine particulate removal. At an operating pressure drop of 4 in. wg in the

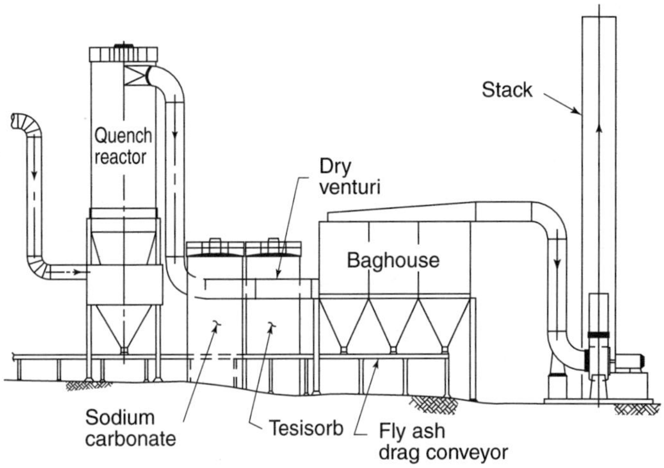

Figure 6. Glass Emission Control.

Table 2. System Comparison-150 TPD-Container Glass

	Wet Scrubbing	Baghouse	ESP	Dry Scrubbing	Semidry Scrubbing	Electric Boost
Outlet						
Opacity (%)	Steam	<10	<10	<5	<5	<20
Particulate (mg/nm^3)	50	20	22	20	14	22–40
SO_x removal (%)	>95	0	0	20	>95	0
Product disposal	Treatment req.	Recycle	Recycle	Recycle	Recycle	Loss
Power consumption						
Installed horsepower	250	90	25	60	70	0
Power cons (kWh/ton)	30	12	3	8	9	0
Electrical use (kWh/ton)	0	0	3	0	0	950
Total power use (kWh/ton)	30	12	6	8	9	950

baghouse, with a 4-h cleaning cycle, and an air-to-cloth ratio of 2 fpm, the SO_3 concentration was reduced from 25–30 to 0.2–2 ppm, in addition to providing particulate emissions in the range of 10–20 mg/nm^3.

Semidry Scrubbing

This system, operating in the glass industry, is installed in two configurations: quench reactor–dry venturi–baghouse (QR-DV-BH) (Figure 6), and spray dryer–ESP (SD-ESP). These systems control particulate, SO_2, and SO_3 emissions and permit the use of fuel oil as the combustion fuel in the melter.

The performance reported for the SD-ESP type of system showed a reduction of SO_x by greater than 90%, while no reduction in conventional ESP performance was indicated.[5] The performance for the QR-DV-BH type of system[8] for the control of three furnaces in parallel, with sodium carbonate solution as the reagent in the quench reactor and limestone as the Tesisorb reagent in the dry venturi, resulted in the following reduction in SO_x:

Stoichiometric Ratio	SO_x Removal (%)
0.39	55.8
0.52	89.1
1.0	>98

Particulate emission was 9.2–20.6 mg/nm^3.

No "brown plume" was noticed during operation, even though the major neutralization was achieved by reaction with Na_2CO_3 solution. The entire baghouse catch is recycled to the batch house inasmuch as the reagents are merely diverted from the melter batch feed. The sulfite–sulfates formed replace the salt cake normally purchased.

Electric Boost

The electric boost process maintains a cool surface in the melt, thus minimizing the emissions of the disassociated products, Na_2O, SO_2, and SO_3 and that of particulate emissions from both disassociation and entrainment (Figure 4). The consumption of energy, however, is quite high (Table 2), thus affecting the economics of glass manufactured.

The reported data on emissions[6] of 1050°C from an electric boost furnace indicated fluoride emissions less than 0.5 mg/nm^3. Contrary to expectations, it was reported that replacement of electric boost by oil firing in an opal furnace reduced metal emissions As, Se, Cd, Sb, and Pb.[3]

Nitrogen Oxides. The reduction of NO_x emissions from glass manufacturing processes is an emerging technology. The systems now being offered, planned, or in the process of installation are as follows:

1. Homogeneous catalytic reaction
2. Selective catalytic reduction (SCR)

The homogeneous catalytic reaction is based on the introduction of ammonia into the flue gas stream in the temperature range of 700–800°C (known as Exxon De NO_x process), or the introduction of ethyl alcohol, benzene, or kerosene into the flue gas stream at temperatures exceeding 700°C.[9]

The ammonia injection generally achieves a 50–75% reduction in NO_x before excessive slip of ammonia occurs in other industrial applications. In glass furnace emission control, the concentration of the "slip" becomes critical because of the potential for white particulate NH_4HSO_3 formation downstream of the stack. The NO_x reduction by the EtOH injection process is claimed to be from 200 to 30–50 ppm.

Selective catalytic reduction for NO_x control requires a clean stream to protect the catalyst. Thus it must be installed downstream of the particulate control unit. It requires a temperature of 300–500°C to achieve a 75–80% reduction in NO_x with minimum ammonia slip for a single-stage SCR. The catalysts used are generally noble or combinations of metal oxides on a ceramic surface.

A report on a molecular sieve catalyst[2] installed in a glass plant indicates that subsequent to acid-gas removal (Figure 7), an 80% reduction in NO_x was achieved with minimum ammonia slip (10–50 ppm) at an operating temperature

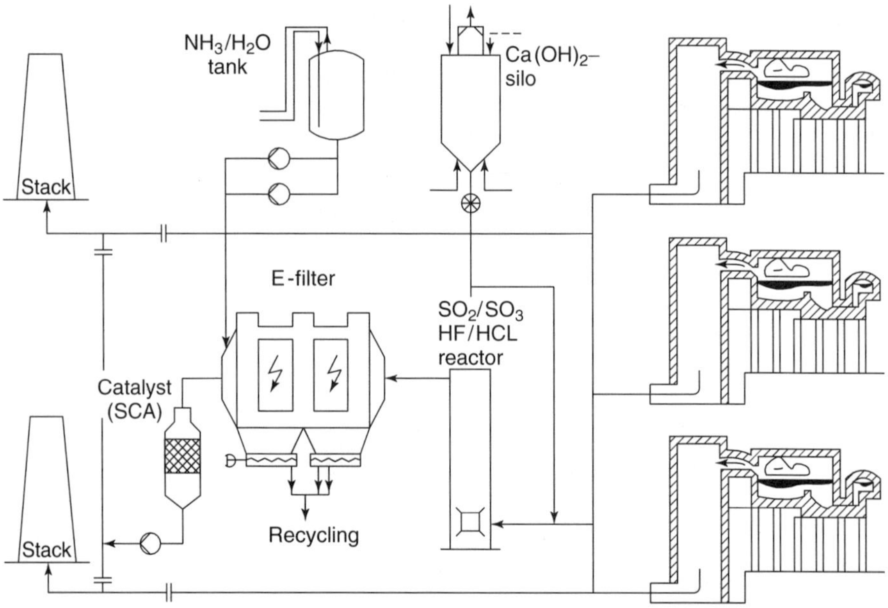

Figure 7. Glass Manufacturing Facility with Emission Control.

of 340–360°C. It should be noted, although not indicated in the flowsheet, that removal of acid gases generally occurs below 200°C and that reheat of the gases must precede SCR treatment.

Forming and Finishing

The development of newer lubricants, including silicone–water emulsions, has decreased the emissions owing to the thermal decomposition of the lubricants. The solution appears to be related to continual improvement of the lubricants. If necessary, a system utilizing a nonplugging condensation–adsorption system can be installed. Such a system has been tested for oily condensibles and volatile organic compound (VOC) recovery in a proprietary dry scrubber system in the cement industry.

REFERENCES

1. Chikamoto et al., *Kyoto-f-u Eisel Koqai Kenleyushoi Nempo* **23**, 124–126 (1978).
2. Grove et al., 49th Conference on Glass Problems, American Ceramic Society, 1988.
3. Mamuru et al., *Ann. Rep. Radiat. Cent. Osaka Prof.* **20**, 29–35 (1979).
4. Noll, *Glass Technol.* **24**(2), 91–97 (1984).
5. Spirina et al., *Steklo Kaam* (6), 6–7 (1979).
6. Tarapore et al., *Glass* **59**(12), 470–473 (1982).
7. Teller, *Glass Indus.* 15–23 (February 1976).
8. Teller, Hsieh, and van Saun, *Proceedings 49th Conference on Glass Problems*, American Ceramic Society, 1988, pp. 196–208.
9. Uetsuki et al., *Osaka Gijutsu Shikensho Kitio* **36**(3), 162–169 (1985)
10. Private communication, R-C Co.
11. Teller, *Proceedings 42nd Conference on Glass Problems*, American Ceramic Society, November 1981, pp. 1–8.

LIME MANUFACTURING

(Adapted from *AP-42, Compilation of Air Pollution Emission Factors)*

PROCESS DESCRIPTION[1–5]

Lime is the high-temperature product of the calcination of limestone. Although limestone deposits are found in every state, only a small portion is pure enough for industrial lime manufacturing. To be classified as limestone, the rock must contain at least 50% calcium carbonate. When the rock contains 30–45% magnesium carbonate, it is referred to as dolomite or dolomitic limestone. Lime can also be produced from aragonite, chalk, coral, marble, and sea shells. The Standard Industry Classification (SIC) code for lime manufacturing is 3274. The six-digit Source Classification Code (SCC) for lime manufacturing is 3-05-016.

Lime is manufactured in various kinds of kilns by one of the following reactions:

$$CaCO_3 + \text{heat} \rightarrow CO_2 + CaO \text{ (high-calcium lime)}$$

$$CaCO_3 \cdot MgCO_3 + \text{heat} \rightarrow 2CO_2 + CaO \cdot MgO \text{ (dolomitic lime)}$$

In some lime plants, the resulting lime is reacted (slaked) with water to form hydrated lime. The basic processes in the production of lime are: (1) quarrying raw limestone; (2) preparing limestone for the kilns by crushing and sizing; (3) calcining limestone; (4) processing the lime further by hydrating; and (5) miscellaneous transfer, storage, and handling operations. A generalized material flow diagram for a lime manufacturing plant is given in Figure 1. Note that some operations shown may not be performed in all plants.

The heart of a lime plant is the kiln. The prevalent type of kiln is the rotary kiln, accounting for about 90% of all lime production in the United States. This kiln is a long, cylindrical, slightly inclined, refractory-lined furnace, through which the limestone and hot combustion gases pass countercurrently. Coal, oil, and natural gas may all be fired in rotary kilns. Product coolers and kiln feed preheaters of various types are commonly used to recover heat from the hot lime product and hot exhaust gases, respectively.

The next most common type of kiln in the United States is the vertical, or shaft, kiln. This kiln can be described as an upright heavy steel cylinder lined with refractory material. The limestone is charged at the top and is calcined as it descends slowly to discharge at the bottom of the kiln. A primary advantage of vertical kilns over rotary kilns is higher average fuel efficiency. The primary disadvantages of vertical kilns are their relatively low production rates and the fact that coal cannot be used without degrading the quality of the lime produced. There have been few recent vertical kiln installations in the United States because of high-product quality requirements.

Other, much less common, kiln types include rotary hearth and fluidized-bed kilns. Both kiln types can achieve high production rates, but neither can operate with coal. The "calcimatic" kiln, or rotary hearth kiln, is a circular kiln with a slowly revolving doughnut-shaped hearth. In fluidized-bed kilns, finely divided limestone is brought into contact with hot combustion air in a turbulent zone, usually above a perforated grate. Because of the amount of lime carryover into the exhaust gases, dust collection equipment must be installed on fluidized-bed kilns for process economy.

Another alternative process that is beginning to emerge in the United States is the parallel-flow regenerative (PR) lime kiln. This process combines two advantages. First, optimum heating conditions for lime calcining are achieved by concurrent flow of the charge material and combustion gases. Second, the multiple-chamber regenerative process uses the charge material as the heat transfer medium to preheat the combustion air. The basic PR system has two shafts, but three-shaft systems are used with small-size grains to address the increased flow resistance associated with smaller feed sizes.

In the two-shaft system, the shafts alternate functions, with one shaft serving as the heating shaft and the other as the flue gas shaft. Limestone is charged alternatively to the two shafts and flows downward by gravity flow. Each shaft includes a heating zone, a combustion/burning zone, and a cooling zone. The two shafts are connected in the middle to allow gas flow between them. In the heating shaft, combustion air flows downward through the heated charge material. After being preheated by the charge material, the combustion air combines with the fuel (natural gas or oil), and the air/fuel mixture is fired downward into the combustion zone. The hot combustion gases pass from the combustion zone in the heating shaft to the combustion zone in the flue gas shaft. The

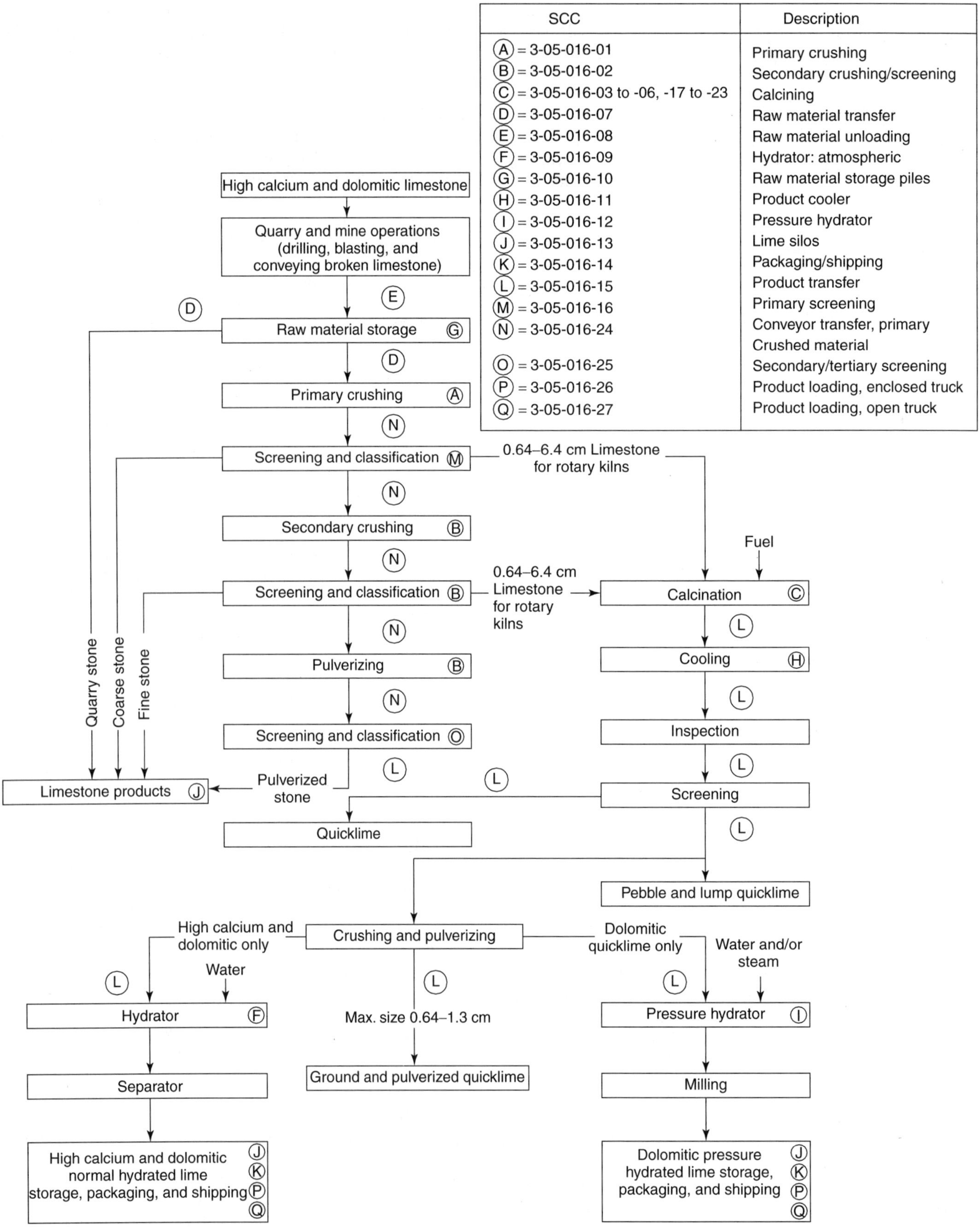

SCC	Description
Ⓐ = 3-05-016-01	Primary crushing
Ⓑ = 3-05-016-02	Secondary crushing/screening
Ⓒ = 3-05-016-03 to -06, -17 to -23	Calcining
Ⓓ = 3-05-016-07	Raw material transfer
Ⓔ = 3-05-016-08	Raw material unloading
Ⓕ = 3-05-016-09	Hydrator: atmospheric
Ⓖ = 3-05-016-10	Raw material storage piles
Ⓗ = 3-05-016-11	Product cooler
Ⓘ = 3-05-016-12	Pressure hydrator
Ⓙ = 3-05-016-13	Lime silos
Ⓚ = 3-05-016-14	Packaging/shipping
Ⓛ = 3-05-016-15	Product transfer
Ⓜ = 3-05-016-16	Primary screening
Ⓝ = 3-05-016-24	Conveyor transfer, primary Crushed material
Ⓞ = 3-05-016-25	Secondary/tertiary screening
Ⓟ = 3-05-016-26	Product loading, enclosed truck
Ⓠ = 3-05-016-27	Product loading, open truck

Figure 1. Process Flow Diagram for Lime Manufacturing (Ref. 4; SCC, Source Classification Code).

heated exhaust gases flow upward through the flue gas shaft combustion zone and into the preheating zone, where they heat the charge material. The function of the two shafts reverses on a 12-min cycle. The bottom of both shafts is a cooling zone. Cooling air flows upward through the shaft countercurrently to the flow of the calcined product. This air mixes with the combustion gases in the crossover area, providing additional combustion air. The product flows by gravity from the bottom of both shafts.

About 15% of all lime produced is converted to hydrated (slaked) lime. There are two kinds of hydrators: atmospheric and pressure. Atmospheric hydrators, the more prevalent type, are used in continuous mode to produce high-calcium and dolomitic hydrates. Pressure hydrators, on the other hand, produce only a completely hydrated dolomitic lime and operate only in batch mode. Generally, water sprays or wet scrubbers perform the hydrating process and prevent product loss. Following hydration, the product may be milled and then conveyed to air separators for further drying and removal of coarse fractions.

The major uses of lime are metallurgical (aluminum, steel, copper, silver, and gold industries), environmental (flue gas desulfurization, water softening, pH control, sewage-sludge destabilization, and hazardous waste treatment), and construction (soil stabilization, asphalt additive, and masonry lime).

AIR EMISSIONS CHARACTERIZATION[1–4,6]

Potential air pollutant emission points in lime manufacturing plants are indicated by SCC in Figure 1. Except for gaseous pollutants emitted from kilns, particulate matter (PM) is the only dominant pollutant. Emissions of filterable PM from rotary lime kilns constructed or modified after May 3, 1977, are regulated to 0.30 kg/Mg (0.60 lb/ton) of stone feed under 40 CFR Part 60, subpart HH.

The largest ducted source of particulate is the kiln. The properties of the limestone feed and the ash content of the coal (in coal-fired kilns) can significantly affect PM emission rates. Of the various kiln types, fluidized beds have the highest levels of uncontrolled PM emissions because of the very small feed rate combined with the high air flow through these kilns. Fluidized-bed kilns are well controlled for maximum product recovery. The rotary kiln is second worst in uncontrolled PM emissions because of the small feed rate and relatively high air velocities and because of dust entrainment caused by the rotating chamber. The calcimatic (rotary hearth) kiln ranks third in dust production primarily because of the larger feed rate and the fact that, during calcination, the limestone remains stationary relative to the hearth. The vertical kiln has the lowest uncontrolled dust emissions due to the large lump feed, the relatively low air velocities, and the slow movement of material through the kiln. In coal-fired kilns, the properties of the limestone feed and the ash content of the coal can significantly affect PM emissions.

Some sort of particulate control is generally applied to most kilns. Rudimentary fallout chambers and cyclone separators are commonly used to control the larger particles. Fabric and gravel bed filters, wet (commonly venturi) scrubbers, and electrostatic precipitators are used for secondary control.

Carbon monoxide (CO), carbon dioxide (CO_2), sulfur dioxide (SO_2), and nitrogen oxides (NO_x) are all produced in kilns. Sulfur dioxide emissions are influenced by several factors, including the sulfur content of the fuel, the sulfur content and mineralogical form (pyrite or gypsum) of the stone feed, the quality of lime being produced, and the type of kiln. Due to variations in these factors, plant-specific SO_2 emission factors are likely to vary significantly from the average emission factors presented here. The dominant source of sulfur emissions is the kiln's fuel, and the vast majority of the fuel sulfur is not emitted because of reactions with calcium oxides in the kiln. Sulfur dioxide emissions may be further reduced if the pollution equipment uses a wet process or if it brings CaO and SO_2 into intimate contact.

Carbon dioxide is emitted from the kiln as a result of the carbonate in the limestone being reduced to CO_2 gas, and the carbon in the fuel oxidizing. If CO_2 emissions from the fuel combustion are estimated using data from Chapter 1 (External Combustion Sources) only noncombustion CO_2 emission factors should be used [915 kg/Mg (1830 lb/ton) lime produced for dolomitic limestone and 785 kg/Mg (1570 lb/ton) lime produced for calcitic limestone]. These estimates are theoretical, based on the production of two moles of CO_2 for each mole of limestone produced. In some facilities a portion of the CO_2 generated is recovered for use in sugar refining.

In sugar refining, a suspension of hydrated lime in water is used to adjust the pH of the product stream and precipitate colloidal impurities.[7] The lime is then removed by reaction with carbon dioxide.[7]

Product coolers are emission sources only when some of their exhaust gases are not recycled through the kiln for use as combustion air. The trend is away from the venting of product cooler exhaust, however, to maximize fuel use efficiencies. Cyclones, baghouses, and wet scrubbers have been used on coolers for particulate control.

Hydrator emissions are low because water sprays or wet scrubbers are usually installed to prevent product loss in the exhaust gases. Emissions from pressure hydrators may be higher than from the more common atmospheric hydrators because the exhaust gases are released intermittently, making control more difficult.

Other particulate sources in lime plants include primary and secondary crushers, mills, screens, mechanical and pneumatic transfer operations, storage piles, and roads. If quarrying is a part of the lime plant operation, particulate emissions may also result from drilling and blasting. Emission factors for some of these operations are presented in *AP-42, Compilation of Air Pollution Emission Factors*, Sections 11.19 and 13.2 on the TTNWeb.

Table 1 (metric units) presents emission factors for PM emissions from lime manufacturing calcining, cooling, and hydrating. Table 2 (metric units) includes emission factors for the mechanical processing (crushing, screening, and grinding) of limestone and for some materials handling operations. Section 11.19, *Construction Aggregate Processing* of AP-42, also includes stone processing emission factors that are based on more recent testing, and, therefore, may be more representative of emissions from stone crushing, grinding, and screening. In addition, Section 13.2, *Fugitive Dust Sources* of AP-42, includes emission factors for materials handling that may be more representative of materials handling emissions than the emission factors in Table 2.

Emission factors for emissions of SO_2, NO_x, CO, and CO_2 from lime manufacturing are presented in Table 3. Particle size distribution for rotary lime kilns is provided in Table 4.

Table 1. Emission Factors for Lime Manufacturing Calcining, Cooling, and Hydrating[a] (Metric Units)

	Filterable[b]				Condensable PM[c]			
Source	PM	Emission Factor Rating	PM_{10}	Emission Factor Rating	Inorganic	Emission Factor Rating	Organic	Emission Factor Rating
Coal-fired rotary kiln (SCC 3-05-016-18)	180[d]	D	22[e]	D	0.67[f]	D	0.51[g]	E
Coal-fired rotary kiln with large diameter cyclone (SCC 3-05-016-18)	60[h]	D	ND		ND		ND	
Coal-fired rotary kiln with fabric filter (SCC 3-05-016-18)	0.14[j]	D	0.77[k]	D	0.19[m]	E	ND	
Coal-fired rotary kiln with ESP (SCC 3-05-016-18)	4.3[h]	D	2.2[n]	D	ND		ND	
Coal-fired rotary kiln with venturi scrubber (SCC 3-05-016-18)	0.72[p]	D	ND		0.14[q]	D	ND	
Gas-fired rotary kiln with ESP (SCC 3-05-016-19)	0.086[r]	E	ND		0.11[r]	E	ND	
Gas-fired rotary kiln with gravel bed filter (SCC 3-05-016-19)	0.51[s]	E	ND		0.24[s]	E	ND	
Coal- and gas fired rotary kiln (SCC 3-05-016-20)	40[t]	E	ND		ND		ND	
Coal- and gas-fired rotary kiln with venturi scrubber (SCC 3-05-016-20)	0.44[t]	D	ND		0.041[t]	D	ND	
Coal- and coke-fired rotary kiln with venturi scrubber (SCC 3-05-016-21)	0.83[u]	D	ND		ND		ND	
Coal-fired rotary preheater kiln with multiclone (SCC 3-05-016-22)	42[v]	E	ND		0.040[v]	E	ND	
Coal-fired rotary preheater kiln with gravel bed filter (SCC 3-05-016-22)	0.59[w]	E	ND		ND		ND	
Coal-fired rotary preheater kiln with multiclone, water spray, and fabric filter (SCC 3-05-016-22)	0.56[x]	E	ND		0.57[x]	E	0.076[x]	E
Gas-fired calcimatic kiln (SCC 3-05-016-05)	48[y]	E	ND		0.14[y]	E	ND	
Gas-fired parallel flow regenerative kiln with fabric filter (SCC 3-05-016-23)	0.051[z]	D	ND		ND		ND	
Atmospheric hydrator with wet scrubber (SCC 3-05-016-09)	0.033[aa]	D	ND		0.0067[aa]	D	ND	
Product cooler (SCC 3-05-016-11)	3.4[y]	E	ND		0.011[y]	E	ND	

[a]Factors represent uncontrolled emissions unless otherwise noted. All emission factors in kg/Mg of lime produced unless noted. ND, no data; SCC, Source Classification Code.
[b]Filterable PM is that PM collected on or prior to the filter of an EPA Method 5 (or equivalent) sampling train.
[c]Condensable PM is that PM collected in the impinger portion of a PM sampling train.
[d]Refs. 11 and 12.
[e]Refs. 4, 11 and 12.
[f]Refs. 11 and 13.
[g]Refs. 11.
[h]Refs. 12.
[j]Refs. 12, 20, 31 and 33.
[k]Refs. 4, 12, 20, 31 and 33.
[m]Refs. 9, 20–23 and 33.
[n]Refs. 4 and 12.
[p]Refs. 8, 28 and 29.
[q]Refs. 8, 15 and 16.
[r]Refs. 14.
[s]Refs. 17 and 32.
[t]Refs. 19.
[u]Refs. 30.
[v]Refs. 12.
[w]Refs. 18.
[x]Refs. 34.
[y]Refs. 25.
[z]Refs. 35.
[aa]Ref. 24; units are kg/Mg of hydrated lime produced.

Table 2. Emission Factors for Lime Manufacturing Raw Material and Product Processing and Handling[a] (Metric Units)

Source	Filterable[b]			
	PM	Emission Factor Rating	PM_{10}	Emission Factor Rating
Primary crusher[c] (SCC 3-05-016-01)	0.0083	E	ND	
Scalping screen and hammermill (secondary crusher)[c] (SCC 3-05-016-02)	0.31	E	ND	
Primary crusher with fabric filter[d] (SCC 3-05-016-01)	0.00021	D	ND	
Primary screen with fabric filter[e] (SCC 3-05-016-16)	0.0030	D	ND	
Crushed material conveyor transfer with fabric filter[f] (SCC 3-05-016-24)	4.4×10^{-5}	D	ND	
Secondary and tertiary screen with fabric filter[g] (SCC 3-05-016-25)	6.5×10^{-5}	D	ND	
Product transfer and conveying (SCC 3-05-016-15)[h]	1.1	E	ND	
Product loading, enclosed truck (SCC 3-05-016-26)[h]	0.31	D	ND	
Product loading, open truck (SCC 3-05-016-27)[h]	0.75	D	ND	

[a]Factors represent uncontrolled emissions unless otherwise noted. Factors are kg/Mg of material processed unless noted. ND, no data; SCC, Source Classification Code.
[b]Filterable PM is that PM collected on or before the filter of an EPA Method 5 (or equivalent) sampling train.
[c]Ref. 8; units of kg/Mg of stone processed.
[d]Ref. 35. Emission factors in units of kg/Mg of material processed. Includes scalping screen, scalping screen discharges, primary crusher, primary crusher discharges, and ore discharge.
[e]Ref. 35. Emission factors in units of kg/Mg of material processed. Includes primary screening, including the screen feed, screen discharge, and surge bin discharge.
[f]Ref. 35. Emission factors in units of kg/Mg of material processed. Based on average of three runs each of emissions from two conveyor transfer points on the conveyor from the primary crusher to the primary stockpile.
[g]Ref. 35. Emission factors in units of kg/Mg of material processed. Based on sum of emissions from two emission points that include conveyor transfer point for the primary stockpile underflow to the secondary screen, secondary screen, tertiary screen, and tertiary screen discharge.
[h]Ref. 12; units of kg/Mg of product loaded.

Table 3. Emission Factors for Lime Manufacturing[a] (Metric Units)

Source	SO_2[b]	Emission Factor Rating	SO_3	Emission Factor Rating	NO_x	Emission Factor Rating	CO	Emission Factor Rating	CO_2[c]	Emission Factor Rating
Coal-fired rotary kiln (SCC 3-05-016-18)	2.7[d]	D	ND		1.6[e]	C	0.74[f]	D	1,600[g]	C
Coal-fired rotary kiln with fabric filter (SCC 3-05-016-18)	0.83[h]	D	ND		ND		ND		ND	
Coal-fired rotary kiln with wet scrubber (SCC 3-05-016-18)	0.15[j]	D	0.11[k]	E	ND		ND		ND	
Gas-fired rotary kiln (SCC 3-05-016-19)	ND		ND		1.7[m]	E	1.1[m]	E	ND	
Coal- and gas-fired rotary kiln with venturi scrubber (SCC 3-05-016-20)	ND		ND		1.4[n]	D	0.41[n]	D	1,600[n]	D
Coal- and coke-fired rotary kiln with venturi scrubber (SCC 3-05-016-21)	ND		ND		ND		ND		1,500[p]	D
Coal-fired rotary preheater kiln with dry PM controls (SCC 3-05-016-22)	1.1[q]	E	ND		ND		ND		ND	
Coal-fired rotary preheater kiln with multiclone, water spray, and fabric filter (SCC 3-05-016-22)	3.2[r]	E	ND		ND		3.2[r]	E	1,200[r]	E
Gas-fired calcimatic kiln (SCC 3-05-016-05)	ND		ND		0.076[s]	D	ND		1,300[s]	E
Gas-fired parallel flow regenerative kiln with fabric filter (SCC 3-05-016-23)	0.0060[t]	D	ND		0.12[t]	D	0.23[t]	D	ND	
Product cooler (SCC 3-05-016-11)	ND	ND			ND		ND		3.9[s]	E

[a]Factors represent uncontrolled emissions unless otherwise noted. Factors are kg/Mg of lime produced unless noted. ND, no data; SCC, Source Classification Code.
[b]Mass balance on sulfur may yield a more representative emission factor for a specific facility.
[c]Mass balance on carbon may yield a more representative emission factor for a specific facility.
[d]Refs. 11 and 20.
[e]Refs. 11, 13, 20, 31, and 33.
[f]Refs. 20 and 27.
[g]Refs. 10, 11, 26–29, 31.
[h]Refs. 20, 31, and 33.
[j]Ref. 27.
[k]Ref. 15.
[m]Ref. 14.
[n]Ref. 19.
[p]Ref. 30.
[q]Refs. 18 and 26.
[r]Ref. 34.
[s]Ref. 25.
[t]Ref. 35.

Table 4. Average Particle Size Distribution for Rotary Lime Kilns[a]

Particle Size (μm)	Cumulative Mass Percent Less than Stated Particle Size			
	Uncontrolled Rotary Kiln	Rotary Kiln with Multiclone	Rotary Kiln with ESP	Rotary Kiln with Fabric Filter
2.5	1.4	6.1	14	27
5.0	2.9	9.8	ND	ND
10.0	12	16	50	55
15.0	31	23	62	73
20.0	ND	31	ND	ND

[a]Ref. 4, Table 4–28; based on A- and C-rated particle size data. Source Classification Codes 3-05-016-04, and -18 to -21. ND, no data.

Because of differences in the sulfur content of the raw material and fuel and in process operations, a mass balance on sulfur may yield a more representative emission factor for a specific facility than the SO_2 emission factors presented in Table 3. In addition, CO_2 emission factors estimated using a mass balance on carbon may be more representative for a specific facility than the CO_2 emission factors presented in Table 3. Additional information on estimating emission factors for CO_2 emissions from lime kilns can be found in the background report for this AP-42 section, which can be found on the CHIEF website as part of the TTNWeb at *http://www.epa.gov/ttn*.

REFERENCES

1. *Screening Study for Emissions Characterization from Lime Manufacture*, EPA Contract No. 68-02-0299, Vulcan-Cincinnati, Inc., Cincinnati, OH, August 1974.
2. *Standards Support and Environmental Impact Statement, Volume I: Proposed Standards of Performance for Lime Manufacturing Plants*, EPA-450/2-77-007a, U.S. Environmental Protection Agency, Research Triangle Park, NC, April 1977.
3. National Lime Association, Lime Manufacturing, *Air Pollution Engineering Manual*, Anthony J. Buonicore, and Wayne T. Davis (eds.), Air and Waste Management Association, Van Nostrand Reinhold, New York, 1992.
4. J. S. Kinsey, *Lime and Cement Industry — Source Category Report, Volume I: Lime Industry*, EPA-600/7-86-031, U.S. Environmental Protection Agency, Cincinnati, OH, September 1986.
5. Written communication from J. Bowers, Chemical Lime Group, Fort Worth, TX, to R. Marinshaw, Midwest Research Institute, Cary, NC, October 28, 1992.
6. Written communication from A. Seeger, Morgan, Lewis & Bockius, to R. Myers, U.S. Environmental Protection Agency, Research Triangle Park, NC, November 3, 1993.
7. *Minerals Yearbook: Lime Annual Report*, U.S. Department of Interior, Washington, D.C., 1993.
8. *Air Pollution Emission Test, J. M. Brenner Company, Lancaster, PA*, EPA Project No. 75-STN-7, U.S. Environmental Protection Agency, Office of Air Quality Planning and Standards, Research Triangle Park, NC, November 1974.
9. D. Crowell et al., *Test Conducted at Marblehead Lime Company, Bellefonte, PA, Report on the Particulate Emissions from a Lime Kiln Baghouse*, Marblehead, Lime Company, Chicago, IL, July 1975.
10. *Stack Sampling Report of Official Air Pollution Emission Tests Conducted on Kiln No. 1 at J. E. Baker Company, Millersville, OH*, Princeton Chemical Research, Inc., Princeton, NJ, March 1975.
11. W. R. Feairheller and T. L. Peltier, *Air Pollution Emission Test, Virginia Lime Company, Ripplemead, VA*, EPA Contract No. 68-02-1404, Task 11 (EPA, Office of Air Quality Planning and Standards), Monsanto Research Corporation, Dayton, OH, May 1975.
12. G. T. Cobb et al., *Characterization of Inhalable Particulate Matter Emissions from a Lime Plant, Vol. I*, EPA-600/X-85-342a, Midwest Research Institute, Kansas City, MO, May 1983.
13. W. R. Feairheller et al., *Source Test of a Lime Plant, Standard Lime Company, Woodville, OH*, EPA Contract No. 68-02-1404, Task 12 (EPA, Office of Air Quality Planning and Standards), Monsanto Research Corporation, Dayton, OH, December 1975.
14. *Air Pollution Emission Test, Dow Chemical, Freeport, TX*, Project Report No. 74-LIM-6, U.S. Environmental Protection Agency, Office of Air Quality Planning and Standards, Research Triangle Park, NC, May 1974.
15. J. B. Schoch, *Exhaust Gas Emission Study, J. E. Baker Company, Millersville, OH*, George D. Clayton and Associates, Southfield, MI, June 1974.
16. *Stack Sampling Report of Official Air Pollution Emission Tests Conducted on Kiln No. 2 Scrubber at J. E. Baker Company, Millersville, OH*, Princeton Chemical Research, Inc., Princeton, NJ, May 1975.
17. R. L. Maurice and P. F. Allard, *Stack Emissions on No. 5 Kiln, Paul Lime Plant, Inc.*, Engineers Testing Laboratories, Inc., Phoenix, AZ, June 1973.
18. R. L. Maurice and P. F. Allard, *Stack Emissions Analysis, U.S. Lime Plant, Nelson, AZ*, Engineers Testing Laboratories, Inc., Phoenix, AZ, May 1975.
19. T. L. Peltier, *Air Pollution Emission Test, Allied Products Company, Montevallo, AL*, EPA Contract No. 68-02-1404, Task 20 (EPA, Office of Air Quality Planning and Standards), Monsanto Research Corporation, Dayton, OH, September 1975.
20. T. L. Peltier, *Air Pollution Emission Test, Martin-Marietta Corporation, Calera, AL* (Draft), EMB Project No. 76-LIM-9, U.S. Environmental Protection Agency, Office of Air Quality Planning and Standards, Research Triangle Park, NC, September 1975.
21. *Report on the Particulate Emissions from a Lime Kiln Baghouse* (Exhibit 1 supplied by the National Lime Association), August 1977.
22. *Report on the Particulate Emissions from a Lime Kiln Baghouse* (Exhibit 2 supplied by the National Lime Association), May 1977.
23. *Report on the Particulate Emissions from a Lime Kiln Baghouse* (Exhibit 3 supplied by the National Lime Association), May 1977.
24. *Air Pollution Emission Test, U.S. Lime Division, Flintkote Company, City of Industry, CA*, Report No. 74-LIM-5, U.S. Environmental Protection Agency, Office of Air Quality Planning and Standards, Research Triangle Park, NC, October 1974.
25. T. L. Peltier and H. D. Toy, *Particulate and Nitrogen Oxide Emission Measurements from Lime Kilns*, EPA Contract No. 68-02-1404, Task No. 17 (EPA, National Air Data Branch, Research Triangle Park, NC), Monsanto Research Corporation, Dayton, OH, October 1975.
26. *Air Pollution Emission Test, Kilns 4, 5, and 6, Martin-Marietta Chemical Corporation, Woodville, OH*, EMB Report No. 76-LIM-12, U.S. Environmental Protection Agency, Office of Air Quality Planning and Standards, Research Triangle Park, NC, August 1976.
27. *Air Pollution Emission Test, Kilns 1 and 2, J. E. Baker Company, Millersville, OH*, EMB Project No. 76-LIM-11, U. S. Environmental Protection Agency, Office of Air Quality Planning and Standards, Research Triangle Park, NC, August 1976.

28. *Particulate Emission Tests Conducted on the Unit #2 Lime Kiln in Alabaster, Alabama, for Allied Products Corporation*, Guardian Systems, Inc., Leeds, AL, October 1990.
29. *Particulate Emission Tests Conducted on #1 Lime Kiln in Alabaster, Alabama, for Allied Products Corporation*, Guardian Systems, Inc., Leeds, AL, October 1991.
30. *Mass Emission Tests Conducted on the #2 Rotary Lime Kiln in Saginaw, Alabama, for SI Lime Company*, Guardian Systems, Inc., Leeds, AL, October 1986.
31. *Flue Gas Characterization Studies Conducted on the #4 Lime Kiln in Saginaw, Alabama, for Dravo Lime Company*, Guardian Systems, Inc., Leeds, AL, July 1991.
32. R. D. Rovang, *Trip Report, Paul Lime Company, Douglas, AZ*, U.S. Environmental Protection Agency, Office of Air Quality Planning and Standards, Research Triangle Park, NC, January 1973.
33. T. E. Eggleston, *Air Pollution Emission Test, Bethlehem Mines Corporation, Annville, PA*, EMB Test No. 74-LIM-1, U.S. Environmental Protection Agency, Office of Air Quality Planning and Standards, Research Triangle Park, NC, August 1974.
34. *Air Pollution Emission Test, Marblehead Lime Company, Gary, Indiana*, Report No. 74-LIM-7, U.S. Environmental Protection Agency, Office of Air Quality Planning and Standards, Research Triangle Park, NC, 1974.
35. *Emissions Survey Conducted at Chemstar Lime Company, Located in Bancroft, Idaho*, American Environmental Testing Company, Inc., Spanish Fork, Utah, February 26, 1993.

SAND AND GRAVEL PROCESSING

John Richards and Brian Palm

U.S. companies produce more than 900,000,000 tons of sand and gravel per year.[1,2] These products are used as a raw material in the manufacturing of materials such as concrete and asphalt for both road and building construction. There are more than 8,000 sand and gravel facilities located throughout all 50 states. Because sand and gravel are high-volume, low-cost commodities, transportation is a significant part of the total delivered cost. Therefore, natural aggregate, including sand and gravel, is commonly used within 30–50 miles of the place of extraction. The industry providing sand and gravel products includes many vertically integrated companies that may also operate stone crushing plants, asphaltic concrete plants, or ready-mixed concrete plants.

Sand and gravel is typically produced from unconsolidated deposits formed by the action of water in the form of rivers, lakes, and glaciers. Deposits are mined, washed to remove impurities such as clay, screened to generate the required size ranges, and stockpiled. In some cases, crushing of larger gravel is necessary. Sand and gravel plants used screening units and other processing equipment similar to that described in the section on "Stone Crushing."

PROCESS DESCRIPTION

Construction Sand and Gravel

The general process for the preparation of sand and gravel is illustrated in Figure 1. Processing sand and gravel starts with the removal of material from naturally occurring deposits of unconsolidated granular material. Some of these deposits occur at or immediately below ground level at the bottoms of streams, lakes, and oceans. More than 300 million tons of sand and gravel are recovered from these underwater deposits. Plants processing these granular, noncohesive materials use a variety of special excavation systems such as floating clamshell dredges, floating bucket ladder dredges, draglines, underwater wheel excavators, and suction pumps.[3,4] In some facilities, submersible pumps with close-coupled drive units are used to optimize delivery of the excavated high-density slurry to the plant. In other plants, the material can be transported by barges or conveyors. The mining area where sand and gravel products are extracted from deposits at or near ground level is referred to as a pit.[3]

Depending on the location and characteristics of the deposit, sand and gravel are removed by power shovels, front-end loaders, or bulldozers equipped with rippers. Many of the pit deposits have inherently high moisture levels. The extracted sand and gravel are transported to the processing plant by means of trucks or conveyors.

After recovery of the mined material from a stockpile, it is processed in a grizzly feeder for removal of oversized material. In some plants, the oversized material is conveyed to a stockpile for use as fill or sub-base material. In other cases, the oversized material is processed in one or more crushers. Each of the crushers is associated with a screen in a loop configuration to allow undersized material to bypass the crusher. The reduction ratios of the crushers are selected to minimize generation of fugitive dust and to provide adequate throughput.

The undersized material from the grizzly feeder is conveyed to one or more sets of vibrating inclined or vibrating horizontal screens for size classification of the material. The large material separated in this process step is stockpiled and sold as gravel. The small material passing through the screens is

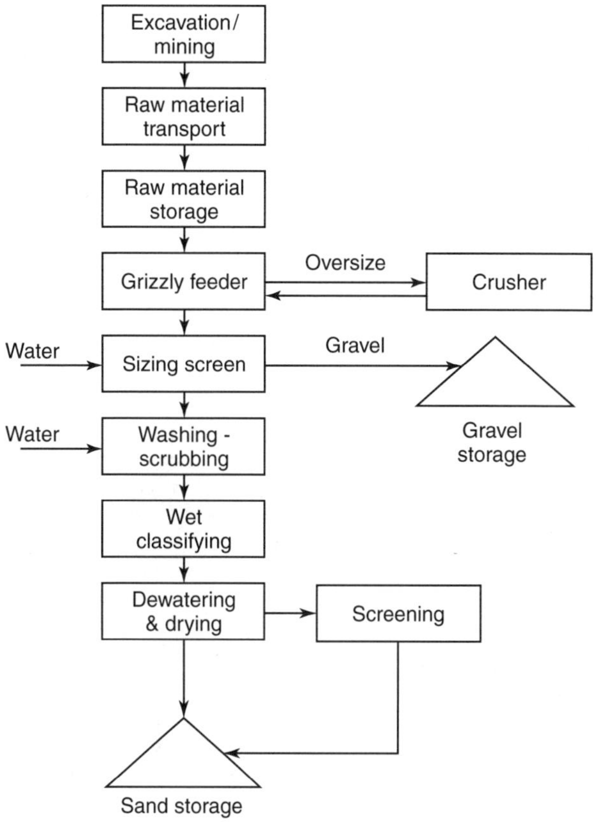

Figure 1. Example Flowchart of a Sand and Gravel Plant.

washed to remove clay particles and organic material present with the sand. The sand is often further separated by size in a set of water classifiers and then dewatered. The sand is stockpiled or stored in bins.

AIR EMISSIONS CHARACTERIZATION

The main pollutant emission from sand and gravel and stone crushing operations is particulate matter. Since 1987, EPA has regulated ambient particulate matter as PM_{10}, particles having an aerodynamic diameter (equivalent diameter of a spherical particle with a density of 1.0 cm^3) equal to or less than 10 μm. In 1997, EPA modified the PM_{10} National Ambient Air Quality Standards (NAAQS) and added new NAAQS for $PM_{2.5}$. This is particulate matter with an aerodynamic diameter equal to or less than 2.5 μm. Total suspended particulate (aerodynamic diameters up to approximately 45 μm) is no longer an EPA regulated air pollutant.

Mineral particulate matter is formed due to physical attrition of the sand and gravel during crushing, screening, conveying, and loading. Physical attrition processes inherently yield particles that have aerodynamic diameters that are predominately larger than 10 μm. Only a small fraction of the mineral particulate is in the PM_{10} size range. Due to the large amount of energy needed to subdivide particles by physical attrition, even less of the mineral particulate emissions from sand and gravel operations are in the $PM_{2.5}$ size range.

Particles formed by physical attrition have chemical compositions identical to the parent material undergoing size reduction. Natural crustal elements are the main constituents of mineral particulate emissions from both crushed stone and sand and gravel processing. These emissions are similar to particulate emitted by agricultural and construction operations and a variety of natural sources.

Potential emission sources for particulate matter in a typical sand and gravel plant include the following:

- Haul roads
- Crushers
- Conveyor transfer points
- Screens
- Driers
- Product loading operations

The extent of particulate emission from each of these sources depends, in part, on the material size distribution and the moisture level. Sand and gravel facilities handling wet material have essentially negligible emissions with the possible exception of sand driers.

The available emissions data for stone crushing plant processing equipment are summarized in Table 1. These data have been compiled from Sections 11.19.1[5] and 11.19.2[6] of the U.S. EPA AP-42 Emission Factor Compilation.

Sand and gravel and crushed stone are used in a variety of similar construction applications. Emission factors for crushed stone processing (AP-42 Section 11.19.2) may be used for sand and gravel processing because of the similarities in geology, extraction methods, and processing methods.

Relatively few sand and gravel plants operate sand driers. Table 1 summarizes the limited data available concerning total particulate emissions. There are no data available concerning PM_{10} or $PM_{2.5}$ emissions.

Emission factor data are provided for two general operating conditions: uncontrolled and controlled. The term *uncontrolled* applies to processes that are operating without (1) wet suppression techniques, or (2) hood and air pollution control systems. Unless otherwise stated, the controlled emission factor data apply to a wet suppression equipment system. The term *controlled* applies to sand plants using conventional wet suppression systems or air pollution control equipment. For plants using wet suppression, the term *controlled* applies to stone moisture levels that are equal to or greater than 1.5% moisture. Additional information concerning the rationale for using the 1.5% value is provided in the section of this manual concerning stone crushing.

The PM_{10} emission factor data for screening operations, crushers, and conveyor transfer points summarized in Table 1 have been compiled by EPA and the National Stone Association in a series of tests conducted over the 1991 to 1997 time period.[7–15] These are included in Section 11.19.2 of EPA's AP-42 data compilation.[6] Due to the lack of data

Table 1. Summary of Processing Equipment Emission Factors[a]

Source	Operating Condition	Total Particulate Matter as "TSP" (lbs/ton)	PM_{10} (lbs/ton)	$PM_{2.5}$ (lbs/ton)
Sand drier	Uncontrolled	2.0 [b]	ND	ND
	Controlled, wet scrubber	0.039 [b]	ND	ND
	Controlled, fabric filter	0.010 [b]	ND	ND
Primary, secondary, or tertiary crushers	Uncontrolled	ND	0.0024 [c]	ND
	Controlled	ND	0.00059 [c]	0.00009 [c]
Fines crushers	Uncontrolled	ND	0.015 [d]	ND
	Controlled	ND	0.0021 [d]	0.00007 [d]
Conveyor transfer points	Uncontrolled	ND	0.0014 [d]	ND
	Controlled	ND	0.000048 [d]	0.000013 [d]

[a]Particulate emissions factors for sand and gravel operations handling wet materials are not included because emissions are considered negligible.

[b]Emission factors are based on the Section 11.19.1 , U.S. EPA Publication AP-42, Fifth Edition, July 1995.

[c]Emissions data for crushers are based on the emission factors presented in Section 11.19.2 for stone crushing plant tertiary crushers.

[d]Emissions data for sand and gravel conveying are based on the emission factors presented in Section 11.19.2 for stone crushing plant conveyors.

specifically applicable to sand and gravel operations, EPA has recommended in AP-42 that emissions from sand and gravel operations be based on the Section 11.19.2 data.

The $PM_{2.5}$ emission factor data[16] have been obtained by the National Stone Association and have been provided to EPA for inclusion in AP-42. All the EPA and NSA sponsored tests have been conducted using state-of-the-art fugitive emission capture techniques and U.S. EPA Reference Method emission testing procedures.[17,18] The EPA and NSA sponsored tests included comprehensive quality assurance procedures and full documentation of process operating conditions. The initial set of PM_{10} tests was conducted using EPA Method 201A. Starting in 1996, tests for PM_{10} and $PM_{2.5}$ were conducted using a draft emission testing procedure that has recently been designated as EPA Method 201B.

There are very limited total particulate matter data for sand and gravel operations. In the case of stone crushing plants, EPA[5] has recommended the use of a multiplying factor of 2.1 to convert from the available PM_{10} data to corresponding total particulate matter data. A similar approach can be taken for sand and gravel operations.

Emissions during the transport of wet material within sand and gravel facilities are negligible. Particulate emission factor data for haul roads at sand and gravel facilities can be estimated using Eq. 1 for PM_{10} and Eq. 2 for $PM_{2.5}$. These equations have been derived based on tests at three stone crushing plants.[19,20]

$$E = (s/3)^{0.8}(M/2)^{-0.9} \quad (1)$$

$$E = 0.25(s/3)^{0.8}(M/2)^{-0.9} \quad (2)$$

Where E is the PM_{10} emissions from haul roads (lbs/VMT); s, the haul road silt content (%); and M, the haul road moisture content (%). These equations indicate that the variables affecting dust emissions from stone crushing plant haul roads are the silt content and the moisture content. Silt is defined as material that has a maximum size of 74 μm and can pass through a 200-mesh screen. Silt and moisture inherent in the material are measured by techniques described in Appendix C of EPA's AP-42 emission factor compilation.[4]

The moisture content of the haul road depends on the frequency of watering, the water application rates, and the evaporation rates. The silt levels typically range from 5 to 10% by weight. The moisture levels are usually in the range of 4 to 8%. These equations do not apply to wintertime conditions, when there is snow or ice on the road surfaces. Equations 1 and 2 also do not apply to facilities using chemical dust suppressants on haul road surfaces.

A comparison of Eqs. 1 and 2 with EPA's general unpaved road equation is provided in the section concerning stone crushing. Emissions from quarry haul roads are slightly lower than predicted by EPA's general equation provided in AP-42.

Sand and gravel facilities using fossil-fuel-fired sand driers will have slight emissions of nitrogen oxides, sulfur oxides, and volatile organic compounds. The emission rate for nitrogen oxides is estimated at 0.031 pounds per ton of sand dried based on data in Section 11.19.1 of EPA's AP-42 compilation.[5] The emission rate of sulfur dioxide can be estimated based on the sulfur content of the fuel. The sulfur dioxide emissions can range from near-negligible levels for natural gas and No. 2 fuel oil to modest emission rates for No. 6 oil. Emission factors for fossil fuels can be found in Section 1 of EPA's AP-42 compilation. The emission rates of volatile organic compounds for drier burners are very small. Limited data concerning the appropriate emission factors for VOC emissions are provided in Section 11.19.1 of AP-42.[5]

AIR POLLUTION CONTROL MEASURES

Sand and gravel operations handle inherently wet materials with negligible fugitive dust emissions. Additional air pollution control measures are not needed for these wet operations.

The air pollution control measures for sand and gravel operations handling drier material from pits at or near ground level include (1) conventional wet suppression, (2) hooding with either wet scrubber or fabric filter control, (3) minimizing drop heights to storage piles, (4) covered conveyors, and (5) good plant housekeeping. The effectiveness of these techniques is similar to applications in the stone crushing industry discussed in another section of this publication.

REFERENCES

1. D. Meyer, and S. Zelnak, Jr., "Structure and Economics." Chapter 2 in *The Aggregates Handbook*, National Stone Association, Washington, D.C., 1992.
2. V. Tepordei, "Construction Sand and Gravel Statistical Compendium," United States Geological Survey Minerals Information, USGS Reston, Virginia, May 5, 1997.
3. R. Archibald, "Extraction Principles." Chapter 7 in *The Aggregates Handbook*, National Stone Association, Washington, D.C., 1992.
4. Ohio Department of Natural Resources, Division of Geological Survey, "Sand and Gravel," GeoFacts No. 19, Columbus, Ohio, 1996.
5. U.S. EPA, "Sand and Gravel Processing," Section 11.19.1 of EPA, *AP-42, Compilation of Emission Factors*, July 1995.
6. U.S. EPA, "Sand and Gravel Processing," Section 11.19.2 of EPA, *AP-42, Compilation of Emission Factors*, July 1995.
7. J. Richards, T. Brozell, and W. Kirk. "PM_{10} Emission Factors for a Stone Crushing Plant Deister Vibrating Screen," Report to the U.S. EPA, Research Triangle Park, NC, Contract No. 68-D1-0055, Task 2.84, February 1992.
8. J. Richards, T. Brozell, and W. Kirk. "PM_{10} Emission Factors for a Stone Crushing Plant Tertiary Crusher," Report to the U.S. EPA, Research Triangle Park, NC, Contract No. 68-D1-0055, Task 2.84, February 1992.
9. W. Kirk, T. Brozell, and J. Richards, "PM_{10} Emission Factors for a Stone Crushing Plant Deister Vibrating Screen and Crusher," Report to the National Stone Association, December 1992.
10. T. Brozell, J. Richards, and W. Kirk. "PM_{10} Emission Factors for a Stone Crushing Plant Tertiary Crusher and Vibrating Screen," Report to the U.S. EPA, Research Triangle Park, NC, Contract No. 68-D0-0122, December 1992.
11. T. Brozell, "PM_{10} Emission Factors for Two Transfer Points at a Granite Stone Crushing Plant," Report to the U.S. EPA, Research Triangle Park, NC, Contract No. 68-D0-0122, January 1994.
12. T. Brozell, "PM_{10} Emission Factors for a Stone Crushing Plant Transfer Point," Report to the U.S. EPA, Research Triangle Park, NC, Contract No. 68-D0-0122, February 1993.
13. T. Brozell, and J. Richards. "PM_{10} Emission Factors for a Limestone Crushing Plant Vibrating Screen and Crusher, Bristol Tennessee," Report to the U.S. EPA, Research Triangle Park, NC, Contract No. 68-D2-0163, July 1993.
14. T. Brozell, and J. Richards. "PM_{10} Emission Factors for a Limestone Crushing Plant Vibrating Screen and Crusher,

Maryville Tennessee," Report to the U.S. EPA, Research Triangle Park, NC, Contract No. 68-D2-0163, July 1993.

15. T. Brozell, T. Holder, and J. Richards, "Measurement of PM_{10} & $PM_{2.5}$ Emission Factors at a Stone Crushing Plant," Final Report to the National Stone Association, December 1996.
16. T. Brozell, "$PM_{2.5}$ and PM_{10} Emission Factor Testing at Two Stone Crushing Plant Quarries," Presented at the Environmental and Safety Meeting, National Stone Association, October 1997.
17. J. Richards, and T. Brozell. "PM_{10} Emission Factors for Sloped, Vibrating Screens and Tertiary Crushers at Five Stone Crushing Plants," Paper presented at the National Stone Association Annual Meeting Concerning Health, Safety, and Environment, Atlanta, Georgia, November, 1993.
18. T. Brozell, "PM_{10} Emission Factors for a Haul Road at a Granite Stone Crushing Plant," Final report to the National Stone Association, December 1994.
19. J. Richards, and T. Brozell, "PM_{10}, $PM_{2.5}$, and PM_1 Emission Factors for Haul Roads at Two Stone Crushing Plants," Final report to the National Stone Association, Washington D.C., November 1995.
20. J. Richards, and T. Brozell. "A Comparison of NSA Quarry Haul Road Emission Factors with EPA's Unpaved Road Equation," Presented at the Environmental and Safety Meeting, National Stone Association, October 1997.

STONE CRUSHING

John Richards, Todd Brozell, and Brian Palm

The Stone Crushing Industry in the United States produces more than 2.4 billion tons of stone per year. During the past 25 years, production has increased at an average annual rate of approximately 3%. There is considerable variability in the size of stone crushing operations. There are small portable plants with capacities of less than 100,000 tons per year and permanent plants with capacities ranging from 300,000 to more than 12,000,000 tons per year.[1] A typical permanent facility has a capacity in the range of 300,000 to 800,000 tons per year.[1]

The stone products are used primarily in road construction, building construction, asphaltic concrete production, and cement production. The distance between the stone crushing plant and the point of stone use must be kept short because the transportation costs of stone are major components of the overall delivered cost. For this reason, the stone crushing industry is one of the most widely dispersed industries in the country with approximately 3,200 quarry operations located in 49 states.

The most common types of stone produced include granite, limestone, dolomite, and sandstone. Other types of stone produced include quartzite, slate, marl, and shell.[2] Of the various types of stone, granite, limestone, and dolomite account for a large majority of U.S. stone production.[2]

PROCESS DESCRIPTION

The production of crushed stone starts with blasting to loosen rock from the quarry wall. Blasting is usually conducted between one and three times per week to provide sufficient inventory of fragmented stone for the processing equipment. Figure 1 shows a typical quarry with fragmented stone in the center right.

Figure 1. Loading of Fragmented Stone in a Quarry.

The fragmented stone is loaded into large-haul road trucks using a front-end loader, as shown in Figure 1, or hydraulic shovels.

These trucks transport the stone to a grizzly feeder and a primary crusher. The distance from the loading point to the primary crusher is usually a fraction of a mile.

The capacities of the haul road trucks range from 30 to 75 tons. The size of the rock ranges from several inches to more than 3 ft. Figure 2 is a general flowchart for a typical stone crushing plant. The material throughputs for this example plant are based on a typical overall production rate of 500 tons per hour, a level equivalent to an annual plant capacity of approximately 1,000,000 tons per year.

The grizzly feeder and the primary crusher are usually located in or near the quarry pit to minimize haul truck transport time. Stone is reduced from a maximum size of approximately 3 ft to a maximum size of approximately 12 in. in the primary crusher. This stone is stored temporarily in a primary surge pile adjacent to the primary crusher.

Stone recovered from the primary surge pile is transported by a conveyor to the primary screen. This vibrating, multideck screen separates the stone into size categories that include (1) oversized material exceeding 3 in., (2) moderately sized material between 1 and 3 in., and (3) undersized material smaller than approximately 1 in. The oversized material passes into one or more secondary crushers to reduce the stone size to a top size less than approximately 3 in. Both compression and impact crushers are used for secondary crushing.

The moderately sized stone from the primary screen and the stone that is processed in the secondary crushers are combined and sent to the secondary screen. This is usually a vibrating inclined multideck screen. The screen mesh sizes on each of the two or three decks can be varied to produce streams of stone having the desired maximum sizes. The smallest stone falls through all the screen decks and often has a top size of approximately $\frac{3}{16}$ in. diameter. The stone that falls through the initial deck but is too large to penetrate the lower decks is directed to several product storage piles or recirculates for additional size reduction. These products have top sizes that are usually 0.5 to 1 in. diameter. The oversized material from the secondary screen is transported by a conveyor to a surge bin feeding one or more tertiary crushers. All the crushed stone from the tertiary crushers is redirected back to the secondary

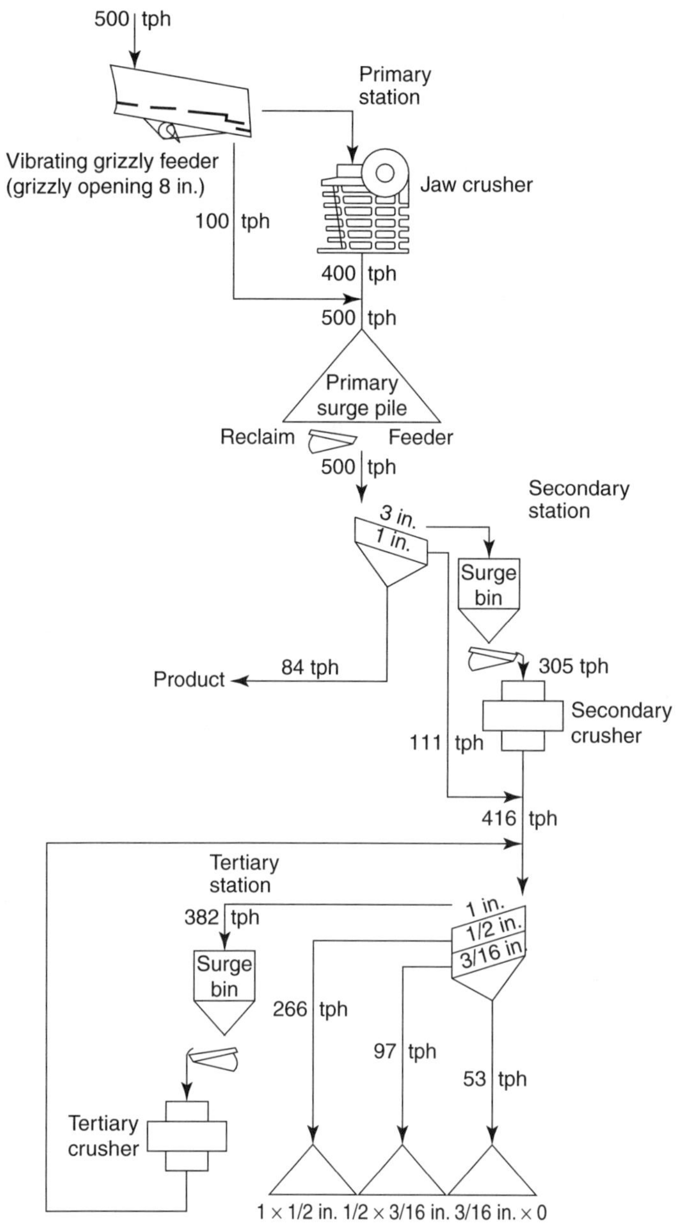

Figure 2. Example Plant Flowchart (Reprinted courtesy of the National Stone Association).

screens. Material from the tertiary crusher is approximately 0.5 in. or less.

In some plants, there is additional treatment of the fines (material less than $\frac{3}{16}$ in. diameter) to produce manufactured sand. In these plants, one or more additional crusher-screening operations are used to further reduce the size of the rock. Manufactured sand is defined as material that will pass through a No. 8 sieve screen.

Throughout the entire process, crushed rock of various size ranges is stockpiled awaiting shipment or further processing. Front-end loaders are often used for loading customer trucks. Conveyors are used for loading barges. At some large facilities, bins are used to facilitate loading customer trucks and rail cars.

Stone crushing plants use several techniques to minimize their effects on surrounding communities. Landscaped berms and trees are often used to improve the appearance and to reduce noise. When a quarry site is no longer needed, the quarry pit is often converted into a lake for recreational use or for community water supplies.

AIR EMISSIONS CHARACTERIZATION

The only pollutant emission of concern from stone crushing operations is particulate matter. Since 1987, EPA has regulated ambient particulate matter as PM_{10}. In 1997, the EPA modified the PM_{10} National Ambient Air Quality Standards (NAAQS) and added new NAAQS for $PM_{2.5}$. Total suspended particulate (material less 45 μm) is no longer an EPA regulated air pollutant.

Mineral particulate matter is formed due to physical attrition of the stone during crushing, screening, conveying, and loading. Physical attrition processes, such as the crusher shown in Figure 3, inherently yield particles that have aerodynamic diameters that are predominately larger than 10 μm. Only a small fraction of the mineral particulate is in the PM_{10} size range. Due to the large energies needed to subdivide particles by physical attrition, even less of the mineral particulate are in the $PM_{2.5}$ size range. Particles formed by physical attrition have chemical compositions identical to the parent material undergoing size reduction. The main constituents of mineral particulate are natural crustal

Figure 3. Crusher.

elements. These emissions are similar to particulate emitted by agricultural and construction operations and a variety of natural sources.

Potential emission points for particulate matter in a typical stone crushing plant include the following. The extent of particulate emission from each of these sources depends, in part, on stone size distribution and the moisture level of the small particulate matter.

- Quarry drilling and blasting
- Fragmented stone loading
- Haul roads
- Primary crushing
- Secondary crushing
- Conveyor transfer points
- Screening
- Tertiary crushing
- Fines crushing
- Product loading

The available emissions data for stone crushing plant processing equipment are summarized in Table 1. Data are provided for two general operating conditions: uncontrolled and controlled. The term "uncontrolled" applies to processes that are operating without (1) wet suppression techniques, or (2) hood and air pollution control systems. The term "controlled" applies to plants using conventional wet suppression systems, chemical dust suppressants, or air pollution control equipment. For plants using wet suppression, the term controlled applies to stone moisture levels that are equal to or greater than 1.5% moisture.

The 1.5% moisture level is an arbitrary value used to separate dry from wet conditions. This value continues to be used for consistency with EPA reports and emission factor compilations prepared in the 1970s and 1980s. Caution is needed in interpreting and using uncontrolled ("dry") emission factors. The potential for dust release from stone surfaces is moderate when the moisture levels are near zero and decreases rapidly as the moisture content increases to approximately the 1.0% to 1.5% range. The addition of even 0.1% moisture can have a significant beneficial impact on particulate emissions when stone moisture levels are in the 0.1 to 0.5% range. Due to the differences in particulate emissions as moisture decreases from 1.5%, it is difficult to assign a single emission factor value for the "dry" conditions.

The PM_{10} emission factor data for screening operations, tertiary crushers, fines crushers, and conveyor transfer points summarized in Table 1 have been compiled by EPA and the National Stone Association in a series of tests conducted from 1991 to 1997[5–12] and are included in EPA's AP-42 data compilation.[4] The $PM_{2.5}$ emission factor data[13] have been obtained by the National Stone Association and have been provided to EPA for later inclusion in AP-42. The EPA and NSA sponsored tests have been conducted using state-of-the-art fugitive emission capture techniques and U.S. EPA Reference Method emission testing procedures.[14–16] The EPA and NSA sponsored tests included comprehensive quality-assurance procedures and full documentation of process operating conditions. The initial set of PM_{10} tests was conducted using EPA Method 201A. Starting in 1996, tests for PM_{10} and $PM_{2.5}$ were conducted using a draft emission testing procedure that has recently been designated as EPA Method 201B.

There are no particulate emissions factor data available for blasting. These emissions are very difficult to measure due

Table 1. Summary of Processing Equipment Emission Factors

Source	Operating Condition	Total Particulate Matter as "TSP" (lbs/ton)	PM_{10} (lbs/ton)	$PM_{2.5}$ (lbs/ton)
Screening	Uncontrolled	ND	0.015[a]	ND
	Controlled	ND	0.00084[a]	0.00005[c]
Fines screening	Uncontrolled	ND	0.071[a]	ND
	Controlled	ND	0.0021[a]	ND
Primary crushing	Uncontrolled	0.0007[a]	ND[b]	ND
	Controlled	ND	ND[b]	ND
Secondary crushing	Uncontrolled	ND	ND[b]	ND
	Controlled	ND	ND[b]	ND
Tertiary crushing	Uncontrolled	ND	0.0024[a]	ND
	Controlled	ND	0.00059[a]	0.00009[c]
Fines crushing	Uncontrolled	ND	0.015[a]	ND
	Controlled	ND	0.0021[a]	0.00007[c]
Conveyor transfer points	Uncontrolled	ND	0.0014[a]	ND
	Controlled	ND	0.000048[a]	0.000013[c]
Wet drilling (unfragmented stone)	Controlled	ND	0.000080[a]	ND
Truck loading (fragmented stone)	Uncontrolled	ND	0.000016[a]	ND
Truck loading (crushed stone)	Controlled	ND	0.00010[a]	ND

[a]Emission factors presented in the U.S. EPA Publication AP-42, Fifth Edition, July 1995.
[b]Emissions data for tertiary crushers used to estimate the maximum possible emissions from primary and secondary crushers.
[c]Emission factor data presented in Ref. 13.

primarily to the low blasting frequency and the very short duration of plume generation. Testing is also difficult due to the inability to contain the dispersing plume and MSHA safety requirements relating to the use of electrically powered instruments close to explosives. Despite the lack of emission factor data, blasting emissions are considered small due to the short duration of the plume.

It is apparent that there are very limited total particulate matter data for stone crushing plant operations. EPA[4] has recommended the use of a multiplying factor of 2.1 to convert from the available PM_{10} data to corresponding total particulate matter data.

Particulate emission factor data for quarry haul roads can be estimated using Eq. 1 for PM_{10} and Eq. 2 for $PM_{2.5}$. These equations have been derived based on tests at three stone crushing plants.[17,18]

$$E = (s/3)^{0.8}(M/2)^{-0.9} \quad (1)$$

$$E = 0.25(s/3)^{0.8}(M/2)^{-0.9} \quad (2)$$

where E is the PM_{10} emissions from haul roads (lbs/VMT); s, the haul road silt content (%); and M, the haul road moisture content (%). These equations indicate that the variables affecting dust emissions from stone crushing plant haul roads are the silt content and the moisture content. Silt is defined as material that has a maximum size of 74 μm and can pass through a 200-mesh sieve. Silt and moisture are measured by techniques described in Appendix C of EPA's AP-42 emission factor compilation.[4]

The moisture content of the haul road depends on the frequency of watering, the water application rates, and the evaporation rates. The silt levels typically range from 5 to 10% by weight. The moisture levels are usually in the range of 4 to 8%. These equations do not apply to wintertime conditions when there is snow or ice on the road surfaces. Equations 1 and 2 also do not apply to facilities using chemical dust suppressants on haul road surfaces.

Equation 1 estimates emissions of PM_{10} that are lower than EPA's unpaved road equation included in Section 11.2 of AP-42.[4] The EPA equation was first published in 1974 and has evolved through seven different forms over a 21-year period. The version shown in Eq. 3 was included in the 1995 edition of AP-42.

$$E = k5.9(s/12)(S/30)(W/3)^{0.7}(w/4)^{0.5}[(365-p)/365] \quad (3)$$

where E is the haul road particulate emission factor (lbs/VMT); k, the particulate size multiplier, PM_{10} (0.36) and TSP (0.80); s, the haul road silt content (%); S, the average vehicle speed (mph); W, the average vehicle weight (tons); w, the average number of wheels; and p, the number of days with at least 0.01 in. of rainfall. EPA's unpaved road equation was developed primarily for rural unpaved roads and iron and steel mine unpaved roads.[4,19] Haul roads at stone crushing plants are considerably different from these unpaved road surfaces. Emissions are lower from quarry haul roads primarily because they are watered and compacted on a routine basis. Haul road trucks in stone quarries move at speeds of 15 to 25 mph, well below the speeds on public unpaved roads and level industrial unpaved roads. The wheels factor included in the general unpaved road equation is not applicable to stone crushing plant quarries.

Figure 4. Road Clearance for a Haul Road Truck and Light Duty Truck.

The road clearances designed into haul road trucks are considerably higher than the clearance of automobiles and light duty trucks, as illustrated in Figure 4. The combined effect of low vehicle speeds and high road clearances reduces the turbulence and air speeds under the vehicles where particulate can be retained. The different tread designs used on heavy-duty off-road trucks also reduces the entrainment of dust and increases compaction of the road.

AIR POLLUTION CONTROL MEASURES

Control measures for reducing or eliminating PM_{10} and $PM_{2.5}$ emissions from quarries and processing plants include the following:

- Wet suppression
- Chemical suppressants
- Covering open operations to prevent dust entrainment by the wind
- Reducing the conveyor drop height
- Using hoods, industrial ventilation systems, and dust collectors (fabric filters or low static pressure drop wet scrubbers) on processes not amenable to wet suppression

Table 2 lists typical operations in a rock quarry and processing plant and shows applicable air pollution control measures. Wet dust suppression is a common and effective method of controlling dust emissions.[24] Emission factor tests conducted by EPA and NSA under controlled (wet suppression) and uncontrolled conditions indicated PM_{10} emission suppression efficiencies of 90 to 95% for tertiary crushers, screening operations, and conveyor transfer points.[14] The effectiveness of wet suppression has been enhanced during the past 10–15 years due to the development of improved atomizing nozzles and the use of wetting agents in some cases. Foaming agents have also been used to enhance PM_{10} emission suppression.[20]

Drilling and Blasting

Quarries must drill and blast to obtain material in sizes that can be transferred and conveyed in an efficient manner to the

Table 2. Particulate Emission Sources and Typical Control Measures for Rock Quarrying and Processing

Operation or source	Control options
Drilling	Liquid injection (water or water plus a wetting agent)
	Capturing and venting emissions to a control device
Blasting	Adoption of good blasting practices
Loading of fragmented stone	Wet suppression
Haul roads	Wet suppression
	Treatment of roads with surface agents
	Paving customer haul roads
Crushing	Wet suppression
	Capturing and venting emissions to a control device
Screening	Wet suppression
	Capturing and venting emissions to a control device
Conveyor transfer points	Wet suppression
	Capturing and venting emissions to a control device
Stockpiling	Stacker conveyors (reduced drop heights)
	Water sprays at conveyor discharge
Storage bins	Capturing and venting to a control device
Windblown dust from roads and plant yard	Wet suppression
	Surface-active agents
	Soil stabilization
	Paving
Loading product into railroad cars, trucks, and ships	Wet suppression

primary crusher. Fugitive dust controls on drilling equipment can efficiently control emissions. These controls include wet suppression and dry collection. The wet suppression technique involves pumping water into the air blow back line leading down to the drill bit. The dry system uses a small hood and cyclone to capture the dust entrained during drilling.

Since the late 1980s, there have been a number of significant improvements in blasting technology. These have been adopted to minimize vibration effects and nuisance dust problems. One of the most significant of these improvements has been the development and implementation of precision detonation techniques. These procedures involve the detonation of the explosive charges located in adjacent drilling holes with precisely timed millisecond delays. This technique results in a number of small detonations instead of one large blast. Precision detonation reduces vibration, noise, and dust generation.

Blast control plugs and stemming materials are inert materials placed on top of the explosive charge to minimize fly rock, air blast, noise, and fugitive dust. Blast control plugs and stemming material oppose the explosive energy from vertical displacement and cause the explosive energy to break through the rock face.[21,22,23]

Loading Fragmented Stone

PM_{10} and $PM_{2.5}$ emissions from loading fragmented stone are inherently low due to the large particle size distribution of the material being handled. Emissions of particulate matter are also minimized by carefully dumping the stone into the bed of the haul road truck. As shown in Figure 1, the drop height is kept very low. In addition to minimizing dust emissions, this practice reduces the maintenance requirements for the haul road trucks.

Quarry Haul Roads

Particulate emissions from quarry haul roads are controlled by the use of wet suppression and/or chemical dust suppressants. The water application rates for quarry haul roads are usually in the range of 0.3 to 0.5 gallons per square yard. The frequency of water application is highly dependent on weather conditions and usually ranges from once every two to once every six hours of quarry operating time. The application frequency is adjusted as necessary to account for seasonal differences in the moisture evaporation rates. Water application is not needed during and immediately after rainfall. Water application is also not needed when there is snow cover on the road surface. Under these wintertime conditions, there is sufficient moisture inherently present on the road surface to minimize dust emissions.

The types of chemical suppressants include both organic binders and hygroscopic materials. The organic binders partially seal the road surface to minimize reentrainment of silt particles. The hygroscopic materials, such as calcium chloride ($CaCl_2$), retain moisture on the road surface and thereby minimize the release of particles. The application frequencies for chemical supressants range from monthly to quarterly depending on the specific characteristics of the additive.

Diesel particulate are emitted from heavy-duty trucks, including the types used in stone quarries. These emissions are most common when the truck is shifting or when it is climbing the sloped surface of the quarry haul road. The rate of diesel emissions is primarily dependent on the design characteristics of the engine; however, normal maintenance of the vehicle engine can minimize these emissions.

Primary Crusher

Figure 5 shows a side view of a quarry haul truck dumping fragmented stone into a grizzly feeder in a quarry of a stone crushing plant. The inlet to the primary crusher is located to the left of the grizzly feeder. Because of the scale of the

Figure 5. Side View of a Grizzly Feeder and a Primary Crusher Showing Water Spraying During Truck Dumping.

operation and the need for accessibility to the crusher opening, enclosure and venting to a control device are impractical. Water sprays, as shown in Figure 5, are used to wet the material and reduce dust emissions.

Due to the large size of the rock, relatively small quantities of water are needed to suppress the dust emissions. Excessive quantities of water in winter are undesirable because they can cause freezing on the belt conveyor that transports the 12 inch topsize material to the primary surge pile.

Secondary and Tertiary Crushing

Secondary and tertiary crushers are used to produce stone products with the appropriate size distributions for use in road and building construction. Each of the crushers is selected and operated to provide the necessary reduction ratio. This is the ratio between the feed size and the crusher outlet material size. Crushers inherently generate less particulate matter when the reduction ratios are in the moderate range. By spreading the crushing requirements over both a secondary and tertiary crusher with a moderate reduction ratio, the overall emissions are lower than if a single crusher with a very high reduction ratio were used.

There are two different categories of crushers used for secondary and tertiary duty. Compression crushers reduce stone sizes by forcing the feed material to pass through a fixed (but adjustable) area. Impact crushers reduce stone sizes by applying a high-speed impact force.

Particulate control of both categories of crushers can usually be achieved by wet suppression systems located throughout the stone crushing plant. Moisture applied from the grizzly feeder and various points along the material path to the crushers reduces the potential for particulate emissions. Water sprays on the feed conveyors at the inlet of the crusher are used as part of the overall plant wet suppression system. All the water spray nozzles along the material flow path are not needed all the time. Excessive moisture does not provide any dust control benefits.

Screening and Conveyor Transfer Points

The types of screening equipment used at stone crushing plants include (1) vibrating inclined, (2) stationary inclined, (3) vibrating grizzly feeder, (4) vibrating horizontal, and (5) rotary. The vibrating inclined screens similar to the unit shown in Figure 6 are the most common. These screens usually have two or more decks to separate the stone feed into three or more streams of different sizes. The size ranges in each of the material streams from the multideck vibrating inclined screen depend on the size of the openings in the wire mesh or rubber media. The ability to produce stones of the necessary size range also depends on the material throughput.

Emissions from screens can be effectively controlled by the plant-wide wet suppression system. One part of this system is a set of nozzles located at the feed side of the screen. Screening operations can also be controlled by enclosures and air pollution control systems. Both fabric filters and low-pressure-drop wet scrubbers can be used as the particulate control system.

Conveyor transfer points are locations where a stream of stone makes an abrupt change in direction or elevation as it is discharged from one conveyor to another. The potential for dust emissions is low because the drop heights are low. Emissions are usually minimized by partial enclosures that block the wind at the point of discharge. Control of particulate emissions

Figure 6. Vibrating Inclined Screen.

can be obtained by the plant-wide wet suppression system or with full enclosures with fabric filters or wet scrubbers.

Stockpiling and Crushed Stone Loading

Dust emissions from stockpiles and stone loading operations are usually small. Dust emissions from stockpiling are not usually amenable to enclosure; therefore, wetting the material is most commonly used. The stockpiles usually develop a crusty surface that minimizes dust emissions from the stored stone products. The emissions that occur as the stone drops from the conveyor to the stockpile are minimized by keeping the drop heights as low as practical.

Emissions of particulate matter during the loading of crushed stone into trucks, barges, or rail cars is minimized by maintaining proper moisture levels in the product stockpiles. The plant-wide wet suppression system provides most of the moisture necessary to control particulate emissions. If necessary, additional moisture can be sprayed on the stockpile to prevent excessive drying. Particulate emissions can also be reduced by minimizing the drop heights from the conveyor or front end loader to the truck, barge, or rail car being used to ship the product.

Summary

The overall impact of stone crushing operations on ambient levels of PM_{10} and $PM_{2.5}$ is small. This is due (1) partially to the fact that relatively small quantities of mineral particulate are formed in these small size ranges because of the relatively low energy levels of the processing equipment and (2) partially to the effectiveness of conventional particulate control techniques. Recent upwind–downwind studies conducted using $PM_{2.5}$ ambient monitoring instruments indicate that there is a near-negligible impact of stone crushing plant operations on ambient $PM_{2.5}$ concentrations.[25,26]

REFERENCES

1. D. Meyer, and S. P. Zelnak, Jr., Chapter 2, "Structure and Economics," in *The Aggregates Handbook*, p. 2–2, National Stone Association, Washington D.C., 1991.
2. R. S. Huhta, Chapter 1, "Introduction to the Aggregate Industry," in *The Aggregates Handbook*, p. 1–4, National Stone Association, Washington D.C., 1991.
3. U.S. EPA, $PM_{2.5}$ Criteria Document, U.S. EPA, July 17, 1997.
4. Compilation of Air Pollutant Emission Factors, AP-42, Fifth Edition, U.S. EPA, Research Triangle Park, N.C., July 1995.
5. J. Richards, T. Brozell, and W. Kirk. "PM_{10} Emission Factors for a Stone Crushing Plant Deister Vibrating Screen," Report to the U.S. EPA, Research Triangle Park, N.C., Contract No. 68-D1-0055, Task 2.84, February 1992.
6. J. Richards, T. Brozell, and W. Kirk. "PM_{10} Emission Factors for a Stone Crushing Plant Tertiary Crusher," Report to the U.S. EPA, Research Triangle Park, NC, Contract No. 68-D1-0055, Task 2.84, February 1992.
7. W. Kirk, T. Brozell, and J. Richards, "PM_{10} Emission Factors for a Stone Crushing Plant Deister Vibrating Screen and Crusher," Report to the National Stone Association, December 1992.
8. T. Brozell, J. Richards, and W. Kirk, "PM_{10} Emission Factors for a Stone Crushing Plant Tertiary Crusher and Vibrating Screen," Report to the U.S. EPA, Research Triangle Park, NC, Contract No. 68-D0-0122, December 1992.
9. T. Brozell, "PM_{10} Emission Factors for Two Transfer Points at a Granite Stone Crushing Plant," Report to the U.S. EPA, Research Triangle Park, NC, Contract No. 68-D0-0122, January 1994.
10. T. Brozell, "PM_{10} Emission Factors for a Stone Crushing Plant Transfer Point," Report to the U.S. EPA, Research Triangle Park, NC, Contract No. 68-D0-0122, February 1993.
11. T. Brozell, and J. Richards, "PM_{10} Emission Factors for a Limestone Crushing Plant Vibrating Screen and Crusher, Bristol, Tennessee," Report to the U.S. EPA, Research Triangle Park, NC, Contract No. 68-D2-0163, July 1993.
12. T. Brozell, and J. Richards, "PM_{10} Emission Factors for a Limestone Crushing Plant Vibrating Screen and Crusher, Maryville, Tennessee," Report to the U.S. EPA, Research Triangle Park, NC, Contract No. 68-D2-0163, July 1993.
13. T. Brozell, T. Holder, and J. Richards, "Measurement of PM_{10} & $PM_{2.5}$ Emission Factors at a Stone Crushing Plant," Final Report to the National Stone Association, December 1996.
14. J. Richards, and T. Brozell, "AP-42 and the Cooperative EPA-NSA Emission Factor Test Program," Paper presented at the NSA Environment, Safety, and Health Learning Forum, Nashville, Tennessee, October 22–24, 1995.
15. T. Brozell, "$PM_{2.5}$ and PM_{10} Emission Factor Testing at Two Stone Crushing Plant Quarries," Presented at the Environmental and Safety Meeting, National Stone Association, October 1997.
16. J. Richards, and T. Brozell, "PM_{10} Emission Factors for Sloped, Vibrating Screens and Tertiary Crushers at Five Stone Crushing Plants," Paper presented at the National Stone Association Annual Meeting Concerning Health, Safety, and Environment, Atlanta, Georgia, November, 1993.
17. T. Brozell, "PM_{10} Emission Factors for a Haul Road at a Granite Stone Crushing Plant," Final report to the National Stone Association, December 1994.
18. J. Richards, and T. Brozell, "PM_{10}, $PM_{2.5}$, and PM_1 Emission Factors for Haul Roads at Two Stone Crushing Plants," Final report to the National Stone Association, Washington D.C., November 1995.
19. J. Richards, and T. Brozell, "A Comparison of NSA Quarry Haul Road Emission Factors with EPA's Unpaved Road Equation," Presented at the Environmental and Safety Meeting, National Stone Association, October 1997.
20. M. Kestner, "Using Suppressants to Control Dust Emissions," Parts I and II, Powder and Bulk Engineering, February 1989 and March 1989.
21. M. D. Wilkins, and P. N. Worsey, "Stemming Technique for Loading Angled Holes Charged with ANFO," Paper presented at the 1997 Meeting of the International Society of Explosive Engineers.
22. C. Deacon, P. Duniam, and M. Jones. "Improved Blast Control through the Use of Programmable Delay Detonators," Paper presented at the 1997 Meeting of the International Society of Explosive Engineers.
23. L. Workman, "Selecting the Right Explosives," *Pit & Quarry*, July 1997, pp. 43–45.
24. Midwest Research Institute, "Inspection Manual for PM_{10} Emissions from Paved/Unpaved Roads and Storage Piles," Final Report to the U.S. EPA, Office of Air Quality Planning and Standards, Stationary Source Compliance Division, EPA Contract No. 68-02-4463, October 1989.
25. J. Hayden, S. Whitt, J. Richards, T. Brozell, and B. Palm. "$PM_{2.5}$ Ambient Air Monitoring at a Stone Crushing Plant," Aggregates Manager, March 1998, pp. 41–46.
26. J. Richards, "$PM_{2.5}$ is Different," *Stone Review*, December–January 1999.

16

PHARMACEUTICAL INDUSTRY

Industry Characterization
Michael J. Barboza, Theodore V. Chleboski, Richard V. Crume, and Jeffrey W. Portzer
Bulk Materials Processing—Bulk Pharmaceutical Production
Michael J. Barboza, Theodore V. Chleboski, Richard V. Crume, and Jeffrey W. Portzer
Pharmaceutical Product Formulation
Michael J. Barboza and Theodore V. Chleboski

INDUSTRY CHARACTERIZATION

MICHAEL J. BARBOZA, THEODORE V. CHLEBOSKI, RICHARD V. CRUME, AND JEFFREY W. PORTZER

INTRODUCTION

The pharmaceutical industry consists of manufacturing, packaging, and sales of chemicals used as medication for humans and animals including both "ethical" (i.e., prescription) and "proprietary" (i.e., nonprescription) drugs. Additionally, the industry manufactures related chemicals for nonpharmaceutical purposes, including preventive medicine and health-enhancing products, medicated and nonmedicated cosmetics, and food additives. Some facilities also manufacture medical apparatus, such as surgical and medical instruments and appliances, dental equipment and supplies, and various products used in diagnostic laboratories.

The predominant characteristic of the industry is diversity. Not only are a wide variety of products manufactured, but a significant number of distinct chemical and manufacturing processes are used in producing these products. Consequently, no two manufacturing facilities are alike, thus making attempts to define the industry in terms of specific products and/or processes extremely complicated. The industry definition is further complicated by the manufacture of nonpharmaceutical products by some pharmaceutical facilities and pharmaceutical products, reagents, and intermediates by certain specialty synthetic chemical manufacturers whose operations would not ordinarily be classified as pharmaceutical. The question of what characteristics best define the pharmaceutical industry is not easily answered, because some pharmaceutical products have multiple end uses or are used in conjunction with nonpharmaceutical products. A more traditional definition of pharmaceutical manufacturing, which is limited to chemicals intended for use as medications (including both ethical and proprietary drugs) and as preventive medicine and health-enhancing agents, has been used. This approach allows classification of pharmaceutical manufacturing processes into a few clearly defined categories for purposes of air emissions control.

Operations

Most pharmaceutical manufacturing operations tend to be relatively clean and closely controlled, because of the importance of product purity and the presence of toxic and biologically active constituents. Controlling air emissions is considered important to avoid inadvertent contamination of products and to prevent occupational and community exposure to potentially harmful substances. Using emission controls also is considered important in recovering expensive solvents. For these reasons, the air emissions from many pharmaceutical-manufacturing facilities are well controlled, and serious air pollution from these facilities is relatively rare.

Nevertheless, problems that occur are usually associated with (1) odorous emissions from solvent extraction and fermentation operations and (2) volatile organic compound (VOC) emissions related to general chemical synthesis operations. Other potential emission sources include product formulation and packaging operations, materials handling and storage, wastewater treatment, spent solvent treatment for recycling, and the regeneration of spent adsorbent carbon. Emissions include both VOCs (including toxic materials) and particulates. Emission controls are available for most pharmaceutical operations, although applying controls is not always cost effective due to the high-volume, low-concentration airstreams that must be handled. Additionally, because manufacturing many pharmaceuticals involves batch operations, emission controls also must be operated in batch mode or operated continuously while handling intermittent, variable feed streams. These methods of operation can reduce the effectiveness of emission control systems, limit the availability of control system options, and increase control system operating costs.

The batch nature of pharmaceutical operations lends itself to decentralized emissions control, where control systems are applied to individual process units rather than to entire production lines. Consequently, regulatory actions tend to focus on controlling specific process units (e.g., reactors, distillation units, crystallizers, centrifuges, dryers, filters, and storage tanks) rather than on limiting emissions from an entire facility. Nevertheless, economical emission reduction for an entire facility is sometimes achieved by using collection manifolds and central emission control devices.

Industry Classification[1]

The Standard Industrial Classification (SIC) code for the Pharmaceutical Industry is 283. The Pharmaceutical Industry (SIC 283) consists of establishments that are primarily involved in fabricating or processing medicinal chemicals and pharmaceutical products. The industry also includes establishments that formulate pharmaceutical products and are involved in the grinding, grading, and milling of botanical products. The industry is further categorized by the following four 4-digit SIC codes:

Medicinals and Botanicals (SIC 2833)
Pharmaceutical Preparations (SIC 2834)
In Vivo and In Vitro Diagnostic Substances (2835)
Biological Products, except diagnostics (SIC 2836).

According to the *U.S. Census of Manufacturers*, in 1992 the Medicinals and Botanicals and the Pharmaceutical

Preparations categories accounted for 64% of establishments and 81% of the value of shipments in the industry. In comparison, the In Vitro and In Vivo Diagnostic Products and Biologicals Products categories are relatively small. Together they accounted for the remaining 36% of the establishments and 19% of the value of the shipments in the industry. In general the processes and subsequent environmental impacts of the In Vitro and In Vivo Diagnostic Products and Biological Products categories are different from those of the Medicinals and Botanicals and Pharmaceutical preparations categories. Therefore, focus of these discussions is on the two larger categories (SIC 2833 and 2834).

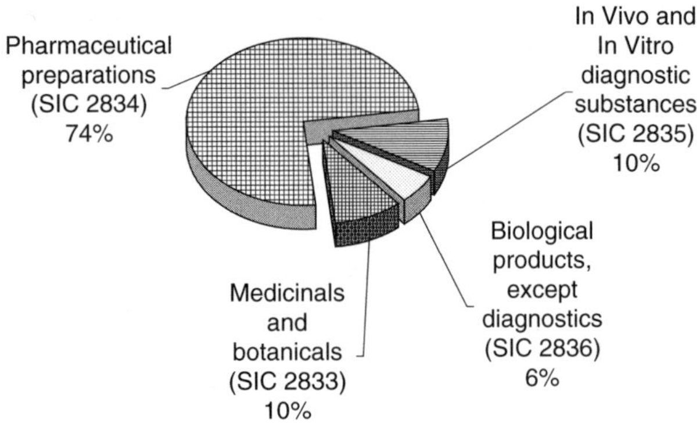

Figure 1. Value of U.S. Pharmaceutical Shipments by Sector.

Industry Size[1]

The *U.S. Census of Manufacturers* identified that in 1992 there were a total of 1,425 Pharmaceutical (SIC 283) establishments employing approximately 194,000 people (excluding Puerto Rico). Table 1 presents summary statistics for the Pharmaceutical industry indicating the number of establishments, number of companies, value of shipments, and employment for the United States and Puerto Rico. The value of total shipments was over $67 billion. Pharmaceutical Preparations (SIC 2834) was the largest sector in terms of number of facilities (48%), employment (63%), and value of shipments (75%). The remaining facilities, employment, and value of shipments were divided evenly among the remaining sectors within the industry. One exception is the In Vivo and In Vitro Diagnostic Products sector (SIC 2835), which claims a higher portion of employment than SIC 2833 and SIC 2836 industry sectors. Figure 1 displays the value of pharmaceutical shipments by industry sector, while the employment by industry sector is presented in Figure 2.

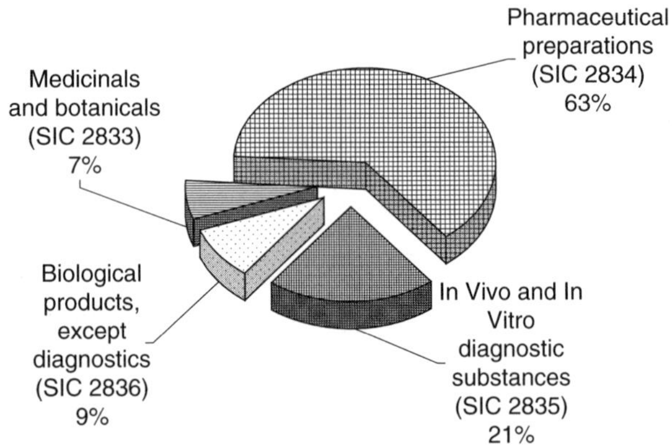

Figure 2. Pharmaceutical Industry Employment by Sector.

A relatively significant number of pharmaceutical establishments are located in Puerto Rico due in part to tax incentives offered to manufacturers to locate new plants in Puerto Rico to encourage creation of jobs. These tax incentives are being phased out over the next 10 years by a 1996 tax law.

The effects of this tax incentive are illustrated by the concentration of pharmaceutical plants in Puerto Rico. According to the 1992 Economic Census of Outlying Areas, which covers statistics for Puerto Rico, there were a total of 88 establishments in Puerto Rico. Of these 88, 74 establishments were in the Pharmaceutical Preparations industry, 8 were in the Medicinals and Botanicals industry, and the remaining 6 establishments were in the In Vitro and In Vivo Diagnostic Products industry, and the Biological Products, except diagnostic substances industry. The total value of shipments of the 88 establishments located in Puerto Rico was about $12 billion. Pharmaceutical Preparations accounted for about 92% of this. The pharmaceutical industry in Puerto Rico employed about 25,000 people in the 88 establishments in 1992.

Table 2 presents a summary of the Medicinals and Botanicals SIC 2833 and Pharmaceutical Preparations SIC 2834 establishments according to number of employees. Note that 20% of the establishments with 250 employees or more are responsible for approximately 88% of the total value of shipments.

Table 1. Summary Statistics for the Pharmaceutical Industry

	50 States				Puerto Rico		
Industry	Number of Establishments	Number of Companies[a]	Value of Shipments (millions of dollars)[b]	Employment (000s)	Number of Establishments	Value of Shipments (millions of dollars)[b]	Employment (000s)
SIC 2833	225	208	6,438	13	8	N/A[c]	N/A[c]
SIC 2834	691	585	50,418	123	74	11,097	22
SIC 2835	234	205	6,838	40	5	477	1
SIC 2836	275	193	3,974	18	1	N/A[c]	N/A[c]
Total	1,425	1,191	67,668	194	88	11,924	25

Source: Ref. 1.
[a]Defined as a business organization consisting of one establishment or more under common ownership or control.
[b]Value of all products and services sold by establishments in the pharmaceuticals industry.
[c]Certain census data are not available for Puerto Rico. Information is withheld to avoid disclosing data for individual facilities.

Table 2. Employment Size Distribution for Medicinals and Botanicals and Pharmaceutical Preparations Establishments[a]

	Medicinals and Botanicals			Pharmaceutical Preparations		
Number of Employees	Number of Facilities	Percent of Facilities	Percent of Value of Shipments	Number of Facilities	Percent of Facilities	Percent of Value of Shipments
<10	104	46	N/A[b]	225	33	0.4
10–49	76	34	N/A[b]	221	30	2
50–249	35	16	N/A[b]	142	21	10
250 or more	10	4	N/A[b]	113	16	88
Total	225	100	100	691	100	100

Source: Ref. 1.
[a]Not including Puerto Rico.
[b]Information has been withheld to avoid disclosing individual establishment data.

Competitive pressures are forcing many companies to restructure and form mergers and strategic alliances. Increasing competition from both domestic and foreign firms, as well as from the generic drug market, has forced mergers between the larger pharmaceutical companies and midsized pharmaceutical companies. Such mergers and alliances allow companies to share their expertise, bring products to market more rapidly, and more effectively market products once they are approved by the Food and Drug Administration (FDA).

Geographic Distribution[1]

The geographic distribution of pharmaceutical facilities in the United States and Puerto Rico is presented in Figure 3. The U.S. pharmaceuticals industry has traditionally been concentrated in New Jersey, California, and New York. These three states account for about one-third of the facilities, employees, and value of shipments. Other states such as Massachusetts, North Carolina, and Maryland have seen recent growth in the pharmaceuticals industry, especially in biotechnology and research and development.

Research and Development[1]

Discovery of new compounds followed by further research and development (R&D) is one of the primary functions of the industry. The pharmaceutical production process starts with an extensive research stage, which can last several years. Following the discovery of a new drug that appears to have efficacy in treating or preventing illness, preclinical tests and clinical trials are conducted. Then a new drug application is submitted to the FDA for approval. Market distribution of

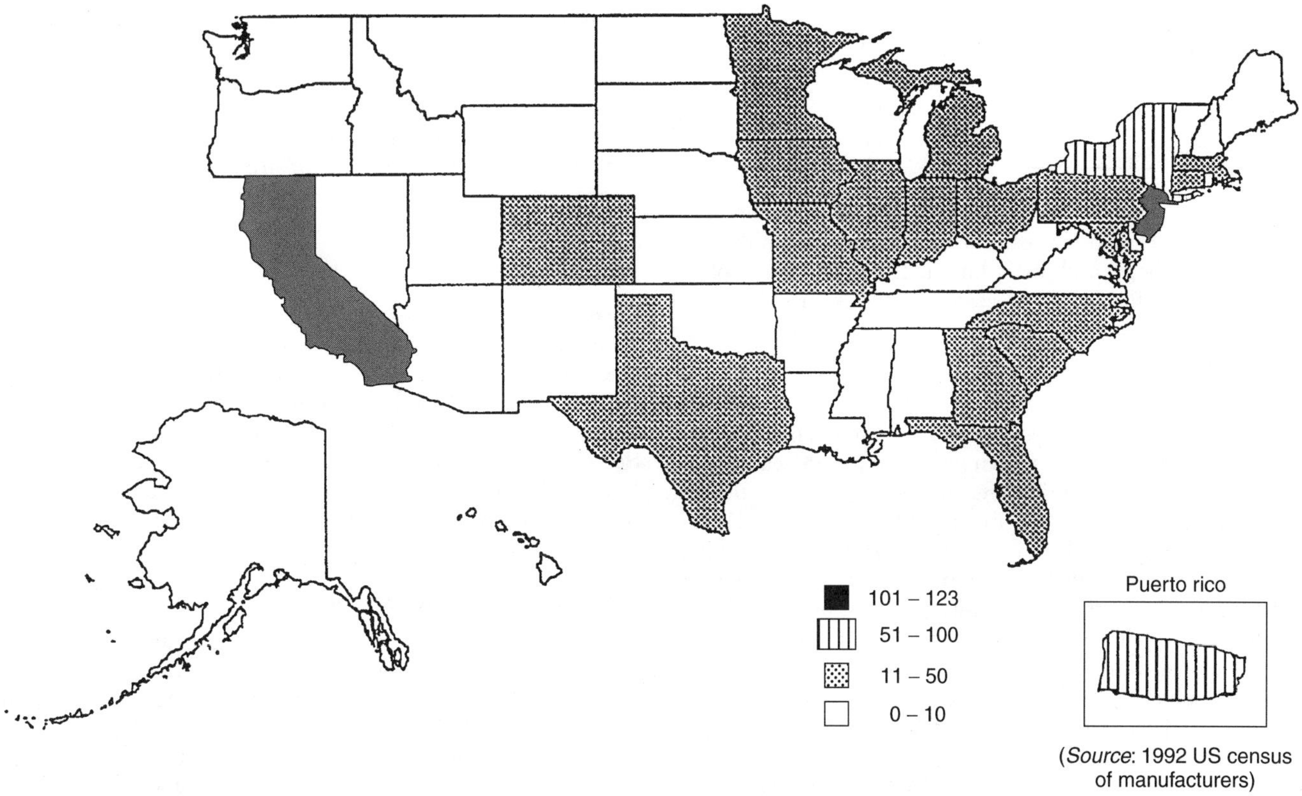

Figure 3. Geographic Distribution of Pharmaceutical Facilities (SIC 2833 and 2834).

a new drug in the United States can begin only after FDA approval. It takes an average of 15 years to bring a new drug to market, from time of discovery to approval. A significant proportion (over 21%) of industry sales are reinvested into research and development.

PROCESS DESCRIPTION

Pharmaceutical production (after R&D activities) can be broken into two major stages:

1. Bulk pharmaceutical production
2. Pharmaceutical product formulation

Bulk pharmaceutical production involves the conversion of organic and natural substances into bulk pharmaceutical chemicals (BPC) (i.e., active ingredients) through fermentation, extraction, and/or chemical synthesis. Pharmaceutical product formulation involves mixing, compounding, and/or formulating operations to convert manufactured BPC substances into final usable forms. These two main production stages are discussed in more detail in subsequent sections.

The manufacturing processes and operations that are most common in the pharmaceutical industry can be divided into the following four main manufacturing process categories: chemical synthesis, fermentation, extraction, and formulation. A breakdown of U.S. pharmaceutical manufacturing facilities by these process categories is presented in the following table[1]:

Pharmaceutical Manufacturing Process Category	Percent of Total U.S. Pharmaceutical Manufacturing Facilities (% of total)
Chemical synthesis	10
Fermentation	1
Extraction	5
Formulation	59
Combination	25

Note that 25% of pharmaceutical facilities utilize a combination of these process categories. The most common combination is chemical synthesis and formulation. These four main pharmaceutical manufacturing process categories are described below.

Chemical Synthesis

Chemical synthesis processes use a variety of batch operations to produce high-purity, pharmaceutical-grade chemicals. Common production equipment include reactors, dryers, distillation units, filters, centrifuges, and crystallizers. Unless a very-high-volume product is manufactured, the equipment is generally campaigned. In this case, the equipment may be used to produce a number of different products over the course of a year and may be dedicated to a specific product for only several weeks at a time.

Organic solvents (including chlorinated compounds) are commonly used in reactors and crystallizers. For example, in synthesis reactions, the solvent serves to: dissolve the reactants, provide dilution to control reaction rate, and control reaction temperature by removing the heat of the reaction through refluxing (returning condensed solvent vapors to the reaction mixture). Additionally, solvents are required for product purification.

Fermentation

Fermentation involves the combination of various species or strains of fungus or bacterium with selected raw materials and the reliance on biological processes to produce desired compounds. The principal pharmaceuticals produced by fermentation are antibiotics, vitamins, and steroids. The fermentation process takes place in the three steps described below.

Inoculum and Seed Preparation. Spores from a master stock are activated with water, nutrients, and heat. As the mass grows, it is transferred to a seed tank with a typical capacity of 350–7500 L (100–2000 gal), where further growth occurs.

Fermentation. Seeds from the foregoing step are combined with raw materials in a fermentation vessel, typically of 18,000–380,000 L (5,000–100,000 gal) in capacity. Air is sparged through the vessel to help sustain growth, and the action of the microorganisms on the raw materials results in the desired product. At the end of the fermentation step, filtration may be used to separate the microorganisms from the product. Fermentation typically requires a period of 12 h to 1 week for completion.

Product Recovery. Product recovery and purification are usually achieved through solvent extraction, direct precipitation, ion exchange, or adsorption.

Extraction

The pharmaceutical extraction process is unique in its low ratio of product weight to raw material weight (often less than 1%). Solvents are used to extract desired compounds from a variety of materials, including roots, leaves, animal glands, parasite fungi, and human blood. In extracting these compounds, solvents also are used to remove and separate fats and oils. After extraction occurs several different types of separation devices (e.g., centrifuges, filters, evaporators, or dryers) can then be used to isolate the desired compounds. Numerous pharmaceuticals are produced by extraction, including insulin, morphine, digitalis derivatives, and allergy medications.

Formulation

Formulation processes convert bulk chemicals into refined products that include: tablets, capsules, liquids, and ointments (i.e., the final dose forms). Typical formulation operations include mixing, blending, granulating, drying, coating, polishing, tablet pressing, capsule filling, sorting, and packaging. Solvents are typically used during mixing and coating operations and to clean the equipment. Most of the dust generated by formulation operations is captured and sometimes returned to the process.

Other Operations and Facilities

A number of operations and facilities, although not considered pharmaceutical processes by definition, are necessary components of many pharmaceutical manufacturing plants, especially some of the larger ones. The need for these operations varies from plant to plant. These other operations and facilities, which can contribute to emissions from pharmaceutical manufacturing plants, include: material storage and transfer, power and steam generation, waste disposal, as well as wastewater treatment.

Storage and Transfer. Volatile compounds are often stored in tank farms, process storage tanks, and storage drums. These storage devices are susceptible to breathing and working losses, the magnitude of which depends on the type of compound stored, the size and design of the tank, ambient temperature and diurnal temperature changes, and tank throughput. Emissions may also be associated with manual material transfer operations within the facility and with the transfer of liquids from tanker cars, trucks, and rail tank cars.

Power and Steam Generation. Pharmaceutical process plants require electricity as well as process heating. Many plants utilize industrial boilers, some with fairly large capacities to generate steam for space heating and process heating needs. In addition, plants may include standby boilers for emergency use as well as emergency generators for emergency electrical power. Some plants may also utilize cogeneration facilities for power and heating needs.

Boilers and process heaters represent an opportunity to burn a waste stream while simultaneously recovering the stream's heating value. Typical applications are combusting high-fuel-value liquid waste streams (uncommon in pharmaceutical operations) or using a pollutant-laden airstream as a source of combustion air. A fairly high-fuel-value vent gas stream is required for boilers or process heaters to function without added fuel. As with flares, boilers and process heaters operating with waste streams more typically are found in petroleum refining or large chemical complexes.

Waste Disposal. Pharmaceutical manufacturing generates wastes that may consist of both hazardous and nonhazardous materials. These wastes can include off-spec or obsolete raw materials or products, spent solvents, reaction residues, used filter media, still bottoms, used chemical reagents, dusts from filtration or air pollution control equipment, raw material packaging wastes, laboratory wastes, spills, as well as wastes generated during packaging of formulated products.

Filter cakes and spent raw materials (plants, roots, animal tissues, etc.) from fermentation and natural product extraction are two of the largest sources of residual wastes in the pharmaceutical industry. Other wastes include reaction residues and filtrates from chemical synthesis processes. These wastes may be stripped of any solvents remaining in them, and then disposed of as either hazardous or nonhazardous wastes.

Facilities may perform shredding of wastes such as outdated or off-spec products. Mechanical shredders may be used to shred dry nonhazardous industrial solid waste material for destruction and volume reduction. Typically, solid wastes are shipped offsite for disposal or incineration. In some instances, facilities may utilize incinerators or waste heat boilers to dispose of solid wastes onsite. This allows recovery of potential heating value from some wastes.

The industry has been implementing practices to reduce waste generation and material losses. Typical practices include process optimization, production scheduling, materials tracking and inventory control, special material handling and storage procedures, preventive maintenance programs, and wastes stream segregation.

Wastewater Treatment Facilities[1]. Pharmaceutical manufacturers use water for process operations, as well as for other nonprocess purposes. However, the use and discharge practices and the characteristics of the wastewater will vary depending on the operations performed. In some cases, water may be formed as part of chemical reactions. Process water includes any water that, during manufacturing or processing, comes into direct contact with or results from the use of any raw material or production of an intermediate, finished product, byproduct, or waste. Process wastewater includes water that was used or formed during the reaction, water used to clean process equipment and floors, and pump seal water. Nonprocess wastewater includes noncontact cooling water (e.g., used in heat exchangers), noncontact ancillary water (e.g., boiler blowdown, bottle washing), sanitary wastewater, and wastewater from other sources (e.g., storm water runoff). Based on a 1990 EPA study comprising 244 pharmaceutical facilities, the average daily wastewater generation by the pharmaceutical manufacturing industry was estimated to be 266 million gallons. Additionally, more than half of the facilities in this study have implemented water conservation measures. Such measures include: careful monitoring of water use, installation of automatic monitoring and alarm systems or in-plant discharges, implementation of alternative production processes, reuse of noncontact water as process makeup water and treatment of contact cooling water to allow reuse. Pharmaceutical manufacturers generate process wastewater containing a variety of conventional parameters [e.g., biological oxygen demand (BOD), total suspended solids (TSS), and pH] and other chemical constituents. The top four chemicals discharged by the pharmaceutical industry (as reported to EPA) are methanol, ethanol, acetone, and isopropanol.

Most process wastewater receives some treatment, either in-plant at the process unit prior to commingling with other facility wastewater or prior to discharge to a permitted outfall. Table 3 provides a trend analysis (from an EPA study) of wastewater treatment technologies used by the

Table 3. Wastewater Treatment Technology Trends

Treatment Technology	Percentage of Facilities Using Technology in 1989/1890
Neutralization	44
Equalization	29
Activated sludge	21
Settleable solids removal	NA[a]
Primary sedimentation	NA
Aerated lagoon	5
Primary clarification	10
Chlorination	3
Polishing ponds	NA
Waste stabilization pond	3
Trickling filter	2
Multimedia filtration	6
Stream stripping	5.7
Evaporation	NA
Secondary clarification	20.9
Granular activated carbon	3.3
Oxidation	2
Dissolved air flotation	NA
pH adjustment	50
Phase separation	12

Source: Ref. 1.
Note: Total percentage is not 100 because facilities may use multiple treatment technologies.
[a]NA, not available.

pharmaceutical industry that indicates since 1986, the use of neutralization, equalization, activated sludge, primary clarification, multimedia filtration, steam stripping, secondary clarification, granular activated carbon, and oxidation have all increased, while the use of aerated lagoons, chlorination, waste stabilization ponds, and trickling filters have decreased slightly. More than half of the surveyed facilities provide pH adjustment or neutralization to adjust the pH prior to discharge. Additionally, because wastewater treatment can be sensitive to spikes of high flow or high constituent concentration, many treatment systems include equalization. Advanced biological treatment is used to treat BOD, chemical oxygen demand (COD), and TSS as well as various organic constituents. Biological systems can be divided into two basic types: aerobic (treatment takes place in the presence of oxygen) and anaerobic (treatment takes places in the absence of oxygen). Very few pharmaceutical facilities use anaerobic treatment. However, more than 30% use aerobic systems

Table 4. 1995 Releases for Pharmaceutical Facilities (SIC 2833 & 2834) in TRI, by Number of Facilities Reporting (releases reported in pounds/year)[a]

Chemical Name	# Reporting Chemical	Fugitive Air (lb/yr)	Point Air (lb/yr)	Total Air[b] (lb/yr)	Water Discharges (lb/yr)	Underground Injection (lb/yr)	Land Disposal (lb/yr)	Total Releases (lb/yr)	Average Releases Per Facility (lb/yr)
Dichloromethane	63	2,386,889	4,611,794	6,998,683	21,635	83,000	5	7,103,323	112,751
Methanol	104	1,396,868	2,100,445	3,497,313	841,250	5,820,000	5	10,159,933	97,692
Ammonia	42	772,824	380,822	1,153,646	1,665,336	0	2,413	3,051,395	72,652
Toluene	54	498,932	593,839	1,092,771	65,336	0	2,413	3,051,395	97,692
n-Butyl alcohol	14	145,024	476,734	621,758	255	6,600	0	628,613	44,901
Acid aerosols (HCl)	62	68,269	532,143	600,412	10	0	5	600,427	9,684
n-hexane	18	201,267	258,124	459,391	2,384	5,300	5	467,080	25,949
Methyl isobutyl ketone	14	273,952	109,175	383,127	15,000	6,500	0	404,627	28,902
Acetonitrile	25	206,608	106,670	313,278	1,405	219,000	5	533,688	21,348
Dichlorodiflouoromethane	8	22,610	195,178	217,788	0	0	0	217,788	27,224
Strong-acid process only, no supplies	2	61,250	140,250	201,500	0	0	0	201,500	100,750
Cyclohexane	9	47,574	147,052	194,626	700	33,000	0	228,326	25,370
Chloroform	14	55,536	88,826	144,362	3,105	0	0	147,467	10,533
1,1,1-Trichlorethane	1	76,500	52,500	129,000	0	0	0	129,000	129,000
Chloromethane	6	28,840	97,844	126,684	44,000	0	0	170,684	28,447
Trichlorofluoromethane	6	59,306	61,801	121,107	0	0	0	121,107	20,185
Xylene	14	10,712	107,105	117,817	0	0	0	117,817	8,416
N,N-Dimethylformamide	20	63,972	10,598	74,570	69,005	1,000,000	1,700	1,145,275	27,264
sec-Butyl alcohol	1	250	71,799	72,049	0	0	0	72,049	72,049
Methyl ethyl ketone	7	20,624	51,120	71,744	50	31,000	0	102,794	14,685
Chlorodifluoromethane	4	31,484	30,009	61,493	0	0	0	61,493	15,373
tert-Butyl alcohol	7	26,713	19,473	46,186	2,400	36,000	0	84,586	12,084
Freon 113	2	3,500	38,119	41,619	0	0	0	41,619	20,810
Dimethylamine	2	23,500	15,250	38,750	250	0	250	39,250	19,625
Triethylamine	17	22,262	15,957	38,219	10,030	5,900	5	54,154	3,186
Certain glycol ethers	7	1,310	27,944	29,254	5	0	0	29,259	4,180
Sulfuric acid	11	22,283	3,091	25,374	0	0	0	25,374	2,307
Formic acid	13	21,550	3,173	24,723	5,160	1,400	5	31,288	2,407
Ethylene glycol	30	21,721	2,638	24,359	20,200	0	500	45,059	1,502
Carbon disulfide	1	2,450	21,000	23,450	0	0	0	23,450	23,450
Methyl *tert*-Butyl ether	11	4,061	18,449	22,510	0	12,000	0	34,510	3,137
Ethylene oxide	3	12,143	9,550	21,693	0	0	0	21,693	7,231
Nitric acid	13	8,029	12,928	20,957	10	0	0	20,967	1,613
2-Methoxyethanol	5	9,130	9,455	18,585	0	0	0	18,585	3,717
Tetrachloroethylene	2	2,239	14,000	16,239	0	0	0	16,239	8,120
Chlorine	19	4,315	9,036	13,351	16,633	0	5	29,989	1,578
Zinc compounds	16	765	11,169	11,934	73,686	100,000	121,500	307,120	19,195
Phosphoric acid	31	5,194	5,160	10,354	20	0	5	10,379	335
Hydrogen fluoride	2	250	8,350	8,600	0	0	0	8,600	4,300
Dichlorotetrafluoroethane (CFC-114)	3	4,978	2,260	7,238	0	0	0	7,238	2,413
Trifluralin	1	6,900	250	7,150	0	0	0	7,150	7,150
Formaldehyde	9	2,662	3,772	6,434	2,000	0	0	8,434	937
Pyridine	7	2,820	3,093	5,193	5	13,000	0	18,918	2,703
Aniline	3	3,896	1,173	5,069	0	0	0	5,069	1,690

Source: Adapted from Ref. 1.
[a]List of compounds represents consolidation of reported 1995 Toxic Release Inventory (TRI) compounds with reported air releases greater than 5000 lb/yr.
[b]Total air emissions represents the sum of TRI fugitive and point air emissions.

Table 5. BPC Process HAP Emissions

Process	Batch		Continuous	
	Uncontrolled (lb/yr)	Controlled (lb/yr)	Uncontrolled (lb/yr)	Controlled (lb/yr)
Chemical synthesis	52,513,671	8,075,164	9,226,728	573,785
Fermentation	6,365,364	1,194,771	2,021,964	413,294
Extraction	832,998	129,940	612,954	202,090
Formulation	7,160,546	2,149,151	673,986	335,428
Total	66,872,579	11,549,026	12,535,632	1,524,597

Source: Ref. 3.

such as activated sludge, aerated lagoons, trickling filter, and rotating biological contactors (RBC).

AIR EMISSIONS CHARACTERIZATION

According to 1995 Toxic Release Inventory (TRI) data, the reporting pharmaceutical facilities released via the air, water, or land without treatment and transferred (shipped off site) a total of 177 million pounds of pollutants, made up of 104 different chemicals. Of these environmental releases, 57% were released to the air as shown in the following table[1]:

Release Category	1995 Release Distribution (% of total)
Air (fugitive and point)	57
Water discharges	17
Underground injection	25
Land disposal	1

The most frequently reported toxic air chemicals on the 1995 TRI list display a prevalence of volatile chemicals, which indicates the air-intensive toxic chemical loading of the pharmaceutical industry. Many of the compounds reported for the 1995 TRI with total air releases greater than 100,000 lbs/yr are highly volatile (i.e., methanol, dichloromethane, toluene, xylene, and acetonitrile). These are primarily solvents used to extract active ingredients and for cleaning equipment. The primary means of release to the environment are from both fugitive and point-source emissions. Other commonly reported chemicals released and transferred are acids (hydrochloric and sulfuric), which can be used for pH control or as catalysts.[1]

Table 4 presents a summary of the 1995 TRI environmental releases by the reporting pharmaceutical facilities with reported air releases greater than 5000 lbs/yr.

EPA has recently compiled HAP emissions profile data for pharmaceutical batch and continuous processes to establish baseline emission data for processes subject to the Pharmaceutical Maximum Achievable Control Technology (MACT) Standard (40 CFR 61, Subpart GGG). Table 5 presents a summary of uncontrolled and controlled emission of HAPs for chemical synthesis, fermentation, extraction, and formulation processes. The average control of HAP emissions from both batch and continuous BPC processes is approximately 85%. The total controlled HAP emissions for batch processes are approximately ten (10) times greater than the total controlled HAP emissions for the continuous processes. HAP emissions are greatest from the chemical synthesis processes for both batch and continuous processes, which is approximately 70 and 40%, respectively, of the total controlled HAP emissions reported for the BPC processes (i.e., chemical synthesis, fermentation, extraction, and formulation).

AIR POLLUTION CONTROL MEASURES

Emission control systems are commonly used in the BPC manufacturing industry to reduce emissions of VOCs (including odors) and particulates during batch and continuous operations. A variety of emission control systems are available for use in the BPC industry, depending on the particular application and on process operating parameters. Because of the batch nature of many BPC manufacturing facilities, it is common for critical emission parameters, such as temperature, pressure, flow rate, material composition and concentration, and particulate loading, to vary throughout a process operating cycle. To control these variable emissions streams effectively, emission controls are designed to handle both peak and nonpeak flow conditions. Specific emission controls available for use at BPC manufacturing facilities include: condensation, absorption, carbon adsorption, thermal destruction, vapor containment, and dust collection. These emission controls are discussed in more detail in the following sections.

REFERENCES

1. *Profile of the Pharmaceutical Manufacturing Industry*, EPA Office of Compliance Sector Notebook Project (EPA/310-R-97-005, September 1997).
2. R. V. Crume, *Air Pollution*, Vol. 7, A. C. Stern, Ed, Academic Press, Orlando, FL, 1986, pp. 478–486.
3. *Pharmaceutical Production NESHAP,* Basis and Purpose Document, USEPA-Emissions Standards Division, January 1997.

BIBLIOGRAPHY

Chemical Process Industries, 4th ed., R. N. Shreve and J. A. Brink, Jr., Eds., McGraw-Hill, New York, 1977, pp. 753–788.

Control of Volatile Organic Emissions from Manufacture of Synthesized Pharmaceutical Products, EPA-450/2-78-029, U.S. Environmental Protection Agency, Research Triangle Park, NC, 1979.

Control of Volatile Organic Compound Emissions from Batch Processes, U.S. EPA Guideline Series, Research Triangle Park, NC. November, 1993 (EPA-453/R-93-017).

B. S. Lane, *Industrial Pollution Control Handbook*, H. F. Lund, Ed., McGraw-Hill, New York, 1971, pp. 17-1–17-34.

U.S. Census of Manufacturers, Industry Series 1992: Drugs, U.S. Department of Commerce, Bureau of the Census, 1995 and 1992 Economic Census of Outlying Areas, Manufacturers: Puerto Rico, U.S. Department of Commerce, Bureau of the Census, 1994.

BIBLIOGRAPHY — INTERNET

Pharmaceutical Online, *http//www.pharmaceutical online.com.*

Pharmaceutical Research and Manufacturers of America (PhRMA), *http//www.phrma.org.*

United States Environmental Agency (U.S. EPA), *http//www.epa.gov.*

BULK MATERIALS PROCESSING — BULK PHARMACEUTICAL PRODUCTION

Michael J. Barboza, Theodore V. Chleboski, Richard V. Crume, and Jeffrey W. Portzer

INTRODUCTION

Bulk pharmaceutical production involves the conversion of organic and natural substances into bulk pharmaceutical chemicals (BPC) (i.e., active ingredients) for use as raw materials for product formulation. This is accomplished through fermentation, extraction, and/or chemical synthesis processes.

Production of BPCs usually involves reaction and subsequent purification to produce pure drug substances, whereas finished drug products usually are the result of the formulation of these drug substances into specific physical forms that may be administered to patients.

Most BPCs are crystalline solid salts, organic bases, or acids with well-defined pharmacological actions. Typical pharmaceutical active ingredients include analgesics, sedatives, antidepressants, anticonvulsants, andrenergics, antihistaminics, cytotoxics, and antibiotics. Raw materials include many organic solvents, intermediates, synthetic chemicals, plant and animal matter, inorganic acids, and bases. Complex metallic or polymetallic compounds are used as reactants and catalysts.[1]

The Federal Food and Drug Administration (FDA) and the International Society of Pharmaceutical Engineers (ISPE) have developed Good Management Practices (GMPs) for manufacture of BPCs. The definition of a drug in the Food, Drug, and Cosmetic Act encompasses BPCs and Section 501(a)(2)(B) of the act requires that all drugs be manufactured, processed, packed, and held in accordance with GMPs. The Code of Federal Regulations (CFR) 210 and 211, apply to finished pharmaceuticals, but are used as a general guide for BPCs.[1]

BPCs may be produced by fermentation, multistep chemical synthesis, enzymatic reactions, recombinant DNA technology, recovery from natural materials, or combination of these processes. A typical synthetic process may involve several different reactions such as condensation, hydrogenation, acylation, or alkylations.[1]

A typical process train for multistep synthesis is composed of a reactor with charge tanks or hoppers, followed by separation and solvent handling vessels. There generally is a processing step for filtration of the solution reaction liquors containing the target compound prior to isolation of the solids or further reaction of the solution or slurry, especially at the final step.[1]

Once isolated, typically on a centrifuge or filtration unit, the solids are washed (with solvent or a solvent/water mixture) and then dried. Dryers typically are forced air/nitrogen, vacuum tray, rotary tumble vacuum, or ribbon. Associated with the general multipurpose trains are various ancilliary systems such as distillation systems, heating and cooling systems, gas-charging systems, and environmental control systems. The isolated product usually is dried, blended, and packaged, and may undergo additional physical operations, such as particle size manipulation (micronizing) before final bulk packaging. These ancillary systems are constructed of chemically resistant materials such as glass, polymers, stainless steel, and other metallic alloys.[1]

BPC manufacturers use many varied chemical intermediates. Many of these intermediates or excipients (binders, disintegrants, fillers, diluents, lubricants, colors, flavoring agents, etc.) are manufactured for the drug industry by fine chemicals (custom) manufacturers.[1]

PROCESS DESCRIPTION

Bulk pharmaceutical substances typically consist of structurally complex organic chemical compounds that are manufactured via a series of intermediate steps and reactions under precise conditions. These substances are manufactured by:

1. Chemical synthesis
2. Fermentation
3. Extraction isolation/recovery from natural sources

Examples of different pharmaceutical products manufactured by each of these processes are presented in Table 1.[2]

Batch Processes[2]

Most pharmaceutical substances are manufactured utilizing batch processes. In a batch process, a particular substance or intermediate is manufactured in a campaign for periods ranging from a few days to several months until sufficient material is manufactured to satisfy the projected sales demand. At the end of the manufacturing campaign, another pharmaceutical intermediate or substance is made. The same equipment with potentially different configurations and the same operating personnel are often used to make a different intermediate or substance, utilizing different raw materials, executing different processes, and generating different waste streams. When the same equipment is used for manufacturing different intermediates and/or different bulk substances, the equipment is thoroughly cleaned and validated prior to its

Table 1. Pharmaceutical Dosage Forms

Chemical Synthesis	Natural Product Extraction	Fermentation
Antibiotics	Antineoplastic agents	Antibiotics
Antihistamines	Enzymes and digestive aids	Antineoplastic agents
Cardiovascular agents	CNS depressants	Therapeutic nutrients
Central nervous system (CNS)	Hematological agents	Vitamins
Stimulants	Insulin	Steroids
CNS depressants	Vaccines	
Hormones		
Vitamins		

Source: Ref. 2.

reuse. Where cleaning of a specific type of equipment is difficult or where a sufficient volume of a certain intermediate or bulk substance is made every year, the equipment may be dedicated to the batch manufacturing of a particular intermediate or bulk substance. Where the equipment is dedicated to the production of successive batches of the same intermediate or bulk substance, the equipment may not be washed and cleaned between batches. Instead, the cleaning schedule will depend on whether there is a potential for carryover of contaminants or degraded materials that could affect the final product.

The specific methods and materials (e.g., water, steam, detergents, and/or organic solvents) used to clean the equipment are based on the ability of the cleaning process to remove residues of raw materials, intermediates, precursors, degradation products, and isomers. Raw materials are checked for their identity and quality before use in the manufacturing processes.

Additionally, in-process testing, as well as quality assurance/quality control (QA/QC) testing in onsite laboratories, is performed during drug product manufacturing. In-process testing may include simple pH measurements or checks on color, while QA/QC testing typically includes more sophisticated analyses such as chromatography. Upon completion of the manufacturing operation, batch production records are checked by competent and responsible personnel for actual yield against theoretical yield of a batch and to ensure that each step has been performed.

Representative Design and Operation

Bulk pharmaceutical production facilities can differ greatly in design and operation; however, pharmaceutical manufacturing, in general, can be illustrated by a description of a representative process such as used for the manufacture of aspirin.

Aspirin production is just one of a variety of pharmaceutical manufacturing processes. However, many pharmaceutical production lines resemble aspirin manufacturing in the use of a series of batch process unit operations that involve reactors, filters, crystallizers, centrifuges, and dryers. Aspirin manufacture differs from other pharmaceutical production processes in that the equipment used tends to be dedicated and the processes may be highly automated. In contrast, other production processes may use campaigned equipment that is used in the production of a number of different products over the course of a year and where a high degree of automation may not be feasible.

A flow diagram for an aspirin manufacturing facility is shown in Figure 1 and can be considered a representative pharmaceutical process.

Several of the raw materials used in aspirin manufacturing are acetic anhydride and salicylic acid. These materials are charged into a reaction vessel and heat is applied for up to 2–6 h. The resulting solution is filtered and then crystallized at about 0°C (32°F). The crystals are separated from the mother

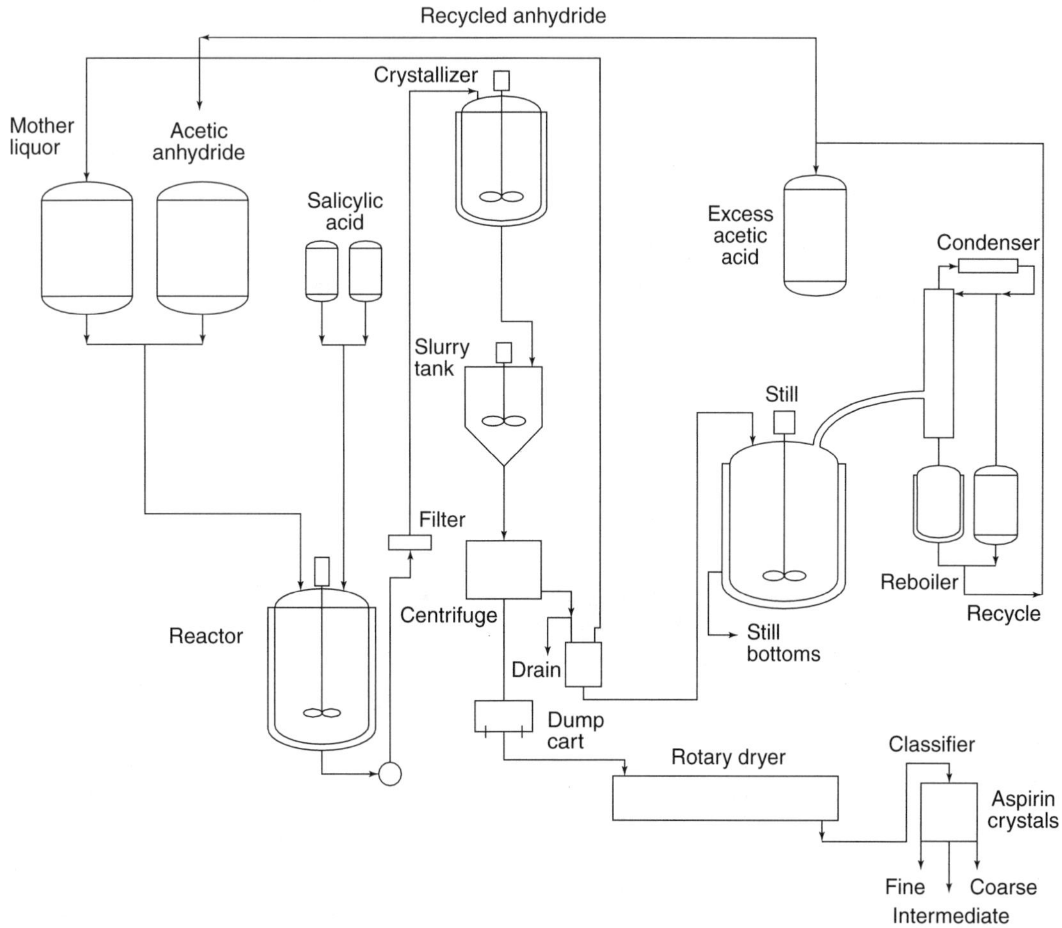

Figure 1. Representative Process Flow Diagram for an Aspirin Manufacturing Facility.

liquor by centrifuge or vacuum filter. Next they are washed and dried for product formulation, where they are mixed with starch and other fillers, and compressed into tablets. A still may be used to remove excess acetic acid and anhydride, and other recovery operations may be used to reclaim solvents used in the process. Solvents such as glacial acetic acid, xylene, methanol, benzene, and toluene may be used during the reaction process or to wash crystals following separation.

Process Operations

Pharmaceutical manufacturing typically consists of several batch processes in the following: reactors, distillation units, extractors, fermentors, centrifuges, filters, crystallizers, dryers, and storage and transfer equipment. These process unit operations are described below.

Reactors. Many pharmaceutical manufacturing processes are initiated with the addition of reactants to glass-lined or stainless steel stirred-tank reactors. These are closed, pressurized vessels typically equipped with agitation capability, a heating–cooling jacket, a vapor vent with reflux condenser, a pressure-relief (safety) valve, a manway, a sight glass, and multiple nozzles to add reactant and/or remove product. Such a vessel may also function as a mixer, preheater, holding tank, crystallizer, evaporator, or distillation kettle. Vessels of this type range in capacity from as little as 10 L (3 gal) for specialty products to 10,000 L (2500 gal) or more for high-volume production. Automatic controls range from simple temperature control to fully automated computer batch process control. Operating pressures range from full vacuum to several hundred thousand kilograms per cubic meter (several hundred pounds per square inch). Figure 2 presents a kettle-type batch reactor. Table 2 provides a summary of typical solvents that may be used during chemical synthesis (reaction) process.

Distillation Units. Batch reactors, small stills, and large columns may all be used for distillation operations. However, pharmaceutical facilities tend to utilize small equipment relative to that employed in petrochemical operations and refineries. Two methods of distillation are in use: (1) boiling the liquid mixture and condensing the vapors, without returning any liquid to the still (essentially evaporation); and (2) the same process, but with part of the condensate (called the reflux) returned to the still and brought into contact with the vapors. The largest distillation equipment in pharmaceutical facilities handles no more than about 3,000 kg/h (6,600 lb/h) of feed material.

Extractors. Extraction is used to separate the components of liquid mixtures. It relies on differences in the solubilities of these components and on the use of a solvent that preferentially combines with one of the components. The resulting mixture consists of an extract (containing the preferentially combined material) and a raffinate (containing the residual phase). Extraction may take place in an agitated reaction vessel (essentially a mixer–settler), in a vertical cylinder (where the

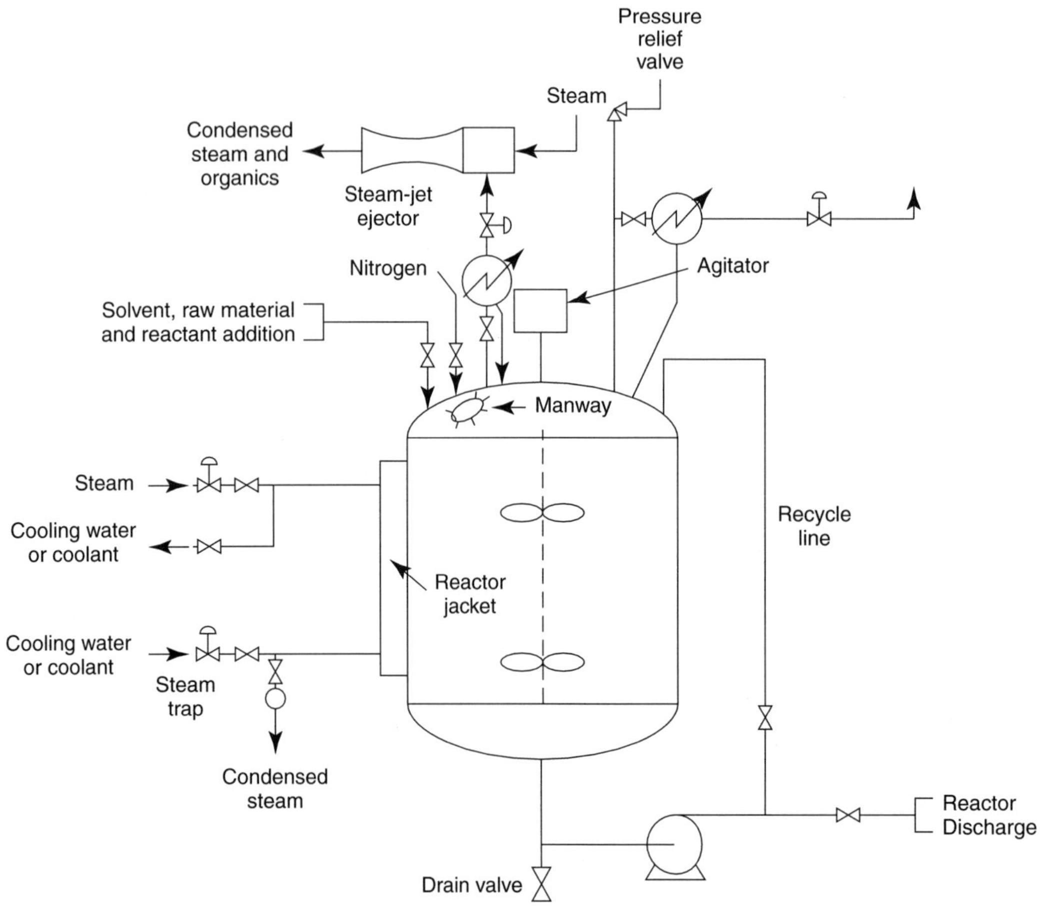

Figure 2. Basic Design of a Kettle-Type Batch Reactor.

Table 2. Solvent Used in Select BPC Manufacturing Processes

Chemical	Chemical Synthesis	Fermentation	Extraction	Hazardous Air Pollutant Under the Clean Air Act
Acetone	X	X	X	
Acetonitrile	X	X	X	X
Ammonia (aqueous)	X	X	X	
n-Amyl acetate	X	X	X	
Amyl alcohol	X	X	X	
Aniline	X		A	X
Benzene	X			X
2-Butanone (MEK)	X			X
n-Butyl acetate	X			X
n-Butyl alcohol	X	X	X	
Chlorobenzene	X			X
Chloroform	X	X	X	X
Chloromethane	X			X
Cyanide	X			
Cyclohexane	X			
o-Dichlorobenzene (1,2-dichlorobenzene)	X			
1,2-Dichlorobenzene	X			
1,2-Dichloroethane			X	
Diethylamine	X		X	
Diethyl ether	X		X	
N,N-Dimethyl acetamide	X			
Diethylamine	X			
N,N-Dimethyl-aniline	X			X
N,N-Dimethylformamide	X	X	X	X
Dimethyl sulfoxide	X		X	
1,4-Dioxane	X		X	X
Ethanol	X	X	X	
Ethyl acetate	X	X	X	
Ethylene glycol	X		X	X
Formaldehyde	X	X	X	X
Formamide	X			
Furfural	X			
n-Heptane	X	X	X	
n-Hexane	X	X	X	
Isobutyraldehyde	X			
Isopropanol	X	X	X	
Isopropyl acetate	X	X	X	
Isopropyl ether	X		X	
Methanol	X	X	X	X
Methylamine	X			
Methyl cellulose	X	X		
Methylene chloride	X	X	X	X
Methyl formate	X			
Methyl isobutyl ketone (MiBK)	X	X		X
2-Methylpyridine	X			
Petroleum naphtha	X	X	X	
Phenol	X	X	X	X
Polyethylene glycol 600	X			
n-Propanol	X		X	
Pyridine	X		X	
Tetrahydrofuran	X		X	
Toluene	X	X	X	X
Trichlorofluoromethane	X			
Triethylamine	X	X		X
Xylenes	X			X

Source: Ref. 2.

solvent flows upward or downward through the liquid mixture), or in a column with internals to mechanically enhance the contact between the two liquid phases. Table 2 provides a summary of typical solvents that may be used during extraction processes.

Fermentation. Fermentation involves the combination of various species or strains of fungus or bacterium with selected raw materials and the reliance on biological processes to produce desired compounds.

Spores from a master stock are activated with water, nutrients, and heat. As the mass grows, it is transferred to a seed tank with a typical capacity of 350–7,500 L (100–2,000 gal), where further growth occurs. Seeds from the foregoing step are combined with raw materials in a fermentation vessel, typically of 18,000–380,000 L (5,000–100,000 gal) in capacity. Air is sparged through the vessel to help sustain growth, and the action of the microorganisms on the raw materials results in the desired product. At the end of the fermentation step, filtration may be used to separate the microorganisms from the product. Fermentation typically requires a period of 12 h to 1 week for completion. Product recovery and purification are usually achieved through solvent extraction, direct precipitation, ion exchange, or adsorption. Figure 3 presents a fermentor vessel using axial impellers. Table 2 provides a summary of typical solvents that may be used during fermentation processes.

Centrifuges. Centrifuging is a process used to separate solid material suspended in a liquid from the liquid. Centrifuges are primarily used to obtain solid filter cake material from the liquid solvents.

Where centrifuging is desired for a solid–liquid separation, the pharmaceutical industry typically uses vertical basket centrifuges. In these devices, the solids are retained on a filter cloth, which is supported by a perforated metal basket. Filtration can usually be improved by establishing a "precoat" of solids on the cloth at the start of the centrifuge cycle or leaving a "heel" after removing the filter cake at the end of the cycle. Centrifuges are blanketed or purged with an inert gas such as nitrogen, especially when flammable solvents are used.

Centrifuges are cylindrical chambers that contain a rotating basket at the base of a vertical shaft. Slurry enters the chamber through an inlet pipe as the basket rotates. Centrifugal force pushes the mixture toward the wall of the basket. The liquid passes through the filter medium and drains through a pipe. The solid particles form a filtercake on the sides of the basket.

A wash liquid is introduced to force the remaining slurry liquid through the cake and filter medium, after the slurry has been fed to the chamber. The basket continues to spin to remove any residual liquid.

The filter cake is unloaded automatically or manually, at the completion of the centrifuging cycle. The automatic method utilizes an adjustable knife that is engaged to scrape the side of the basket and dislodge the filtercake. The cake material falls to the bottom of the chamber, where it is discharged through an opening in the basket. The manual unloading method requires the operators to "scoop out" the filtercake product into a transfer vessel. Figure 4 presents a schematic of a centrifuge.

Filters. A variety of types of filters are in use in the pharmaceutical industry, including pressure, cartridge, plate and frame, atmospheric, and vacuum filters. Similar to centrifuges, filters are used to remove solids from a liquid. These solids can be impurities or contaminants (e.g., catalysts or carbon particles), or they can be process intermediates or final products.

Crystallizers. Products of very high purity can be manufactured via a crystallization process. Crystallization is achieved by creating a supersaturated solution in which the desired compounds form crystals. Supersaturation can result from cooling the solution, especially where solubility is a strong function of temperature. Supersaturation can also result from evaporating a portion of the solvent or by reducing solubility through the addition of a third component.

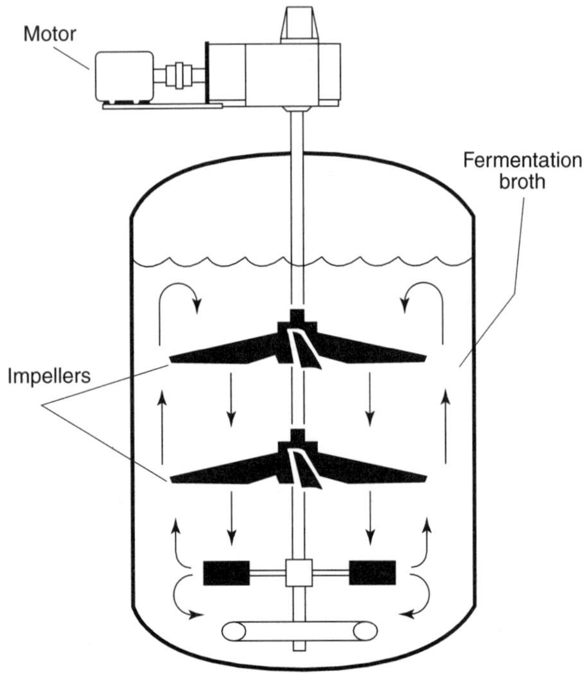

Figure 3. Fermentor Mixer.

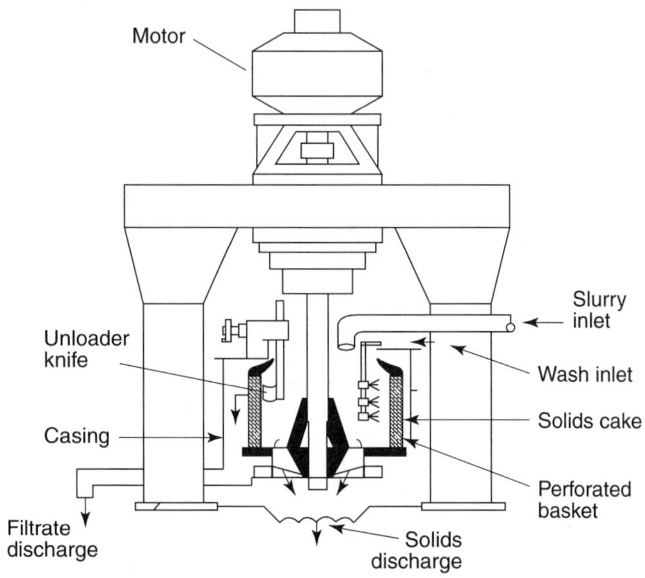

Figure 4. Centrifuge.

Dryers. Drying is often necessary to remove excess solvents remaining in products following centrifuging, filtering, and related operations. There are two types of dryers, defined on the basis of their principle of operation: convection and conduction. In a convective dryer, the heat for drying is provided by a large flow of dry air (or inert gas), which also acts to transport the volatilized solvent out of the dryer vent. In a conductive dryer, the heat for drying is supplied by the walls or trays of the dryer and is thus transferred to the wet product by conduction. The flow rate of gas, plus vaporized solvents from the vents of conductive dryers, is much lower than that from convective dryers.

The mechanical configuration of dryers varies greatly, with batch tray, rotary tray, rotary drum, moving belt, and fluidized-bed designs among the possible types. Some types, such as the rotary tray dryer, can be designed for use as either a convective or a conductive dryer. Vacuum tray dryers are typically conductive dryers.

Storage and Transfer. Volatile organic compounds are stored in tank farms, 55-gal drums, and sometimes in process holding tanks. Storage tanks in tank farms range in size from about 20,000 to 110,000 L (5,000–30,000 gal). Most are horizontal tanks, although vertical tanks are also used. Process holding tanks are smaller and range in size from 2,000 to 20,000 L (500–5,000 gal).

Solvent Recovery. Considerations such as environmental protection, pollution control, and cost of fresh solvent often require solvent(s) to be recovered and reused. Relatively pure wash liquors from solid–liquid separation stages and solvents obtained by condensing vapors from drying may be recycled without any further purification.

When solvents need to be purified, batch or continuous distillation is normally employed. Separate tanks for virgin and recovery solvents is common practice. Recovered solvent reuse from one product to the other is usually avoided. Whether or not recovered solvent from one stage of the process can be reused in other stages of the same product can only be determined from knowledge of the synthesis.[3]

AIR EMISSIONS CHARACTERIZATION

This section describes the characteristics of emissions associated with BPC manufacturing processes. Subsections presented include: emissions sources, types of emissions, emission estimates, and estimation techniques. Guidance is provided on estimating emission rates.

Emission Sources

A summary of typical material inputs and emissions for BPC manufacturing processes is presented in Table 3. Sources of emissions at BPC manufacturing plants vary widely from facility to facility, depending on the products manufactured. Sources of emissions associated with the major BPC manufacturing unit process operations are described below.

Reactors. Emissions from reactors may occur by several mechanisms, including (1) displacement of air during the charging of reactants, (2) inert gas flows during the reaction cycle, (3) opening of the reactor during operation to collect samples or charge reactants, (4) venting of emissions during refluxing, (5) solvent vaporization during cleanup operations, and (6) the relieving of pressure between cycles (where a reactor is operated under pressure). Emissions are maximized when reactions involve highly volatile chemicals and elevated temperatures. In contrast, emissions are minimized with less volatile chemicals, low temperatures, and/or sometimes subatmospheric pressures and when water-based reactions are involved.

Table 3. Summary of Typical Material Inputs and Outputs in BPC Processes

Process	Inputs (examples of some commonly used chemicals provided)	Air Emissions
Chemical Synthesis Reaction	Solvent, catalysts, reactants, e.g., benzene, chloroform, methylene chloride, toluene, methanol, ethylene glycol, methyl isobutyl ketone (MiBK), xylenes, hydrochloric acid, etc.	VOC emissions from reactor vents, manways, material loading and unloading, acid gases (halogen acids, sulfur dioxide, nitrous oxides); fugitive emissions, from pumps, sample collections, valves, tanks
Separation	Separation and extraction solvents, e.g., methanol, toluene, hexanes, etc.	VOC emissions from filtering systems that are not contained; fugitive emissions from valves, tanks and centrifuges
Purification	Purification solvents, e.g., methanol, toluene, hexanes, etc.	Solvent vapors from purification tanks; fugitive emissions
Drying	Finished active drug(s) or intermediates	VOC emissions from manual loading and unloading of dryers
Natural product extraction	Plants, roots, animal tissues, extraction solvents, e.g., ammonia, chloroform, phenol, toluene, etc.	Solvent vapors and VOCs from extraction chemicals
Fermentation	Inoculum, sugars, starches, nutrients, phosphates, fermentation solvents, e.g., ethanol, amyl alcohol, methanol, MiBK, acetone, etc.	Odoriferous gases, extraction solvent vapors, particulates

Source: Ref. 2.

Distillation Units. Emissions occur from the condensers used to recover the evaporated liquids and from the vents as a result of vapor displacement during startup.

Extractors. Emissions from extractors may occur from the (1) displacement of air while charging, emptying, and cleaning the extractor; (2) agitation of liquids during the extraction process; and (3) filling and emptying of associated surge tanks.

Fermentors. Emissions occur during the fermentation process as a result of displacement of large volumes of solvent from one vessel to another, and recovery of product from the concentrated solvent by crystallization, filtration, and drying of the solid product.

Centrifuges. A significant source of emissions is the vaporization of organic compounds from the "wet" solids as they are removed from the centrifuge and transported to the next operation. In addition, emissions can occur as the centrifuge is opened, as the filtrate is discharged into a holding tank, and as inert gases are vented. Some older facilities may have open-top centrifuges, where direct vaporization to the ambient air can occur.

Filters. Emissions typically occur when a filter is opened to remove the collected solids or when the filter is purged with steam or inert gas prior to cleaning.

Crystallizers. Emissions from crystallization operations are usually not significant, unless supersaturation is achieved through solvent evaporation. These emissions may be vented directly to the atmosphere or through a capture device (e.g., a condenser).

Dryers. In the pharmaceutical industry, drying is the unit operation with generally the highest VOC emissions. With a convective dryer, the exhaust gas is a high-flow, relatively dilute stream, which traditionally has often been vented directly to the atmosphere without emission control. With a conductive dryer, the vent gas is much more concentrated; so the flow rate is low. In the case of vacuum drying, VOC emissions occur downstream from the vacuum pump, where the air in-leakage is exhausted from the system.

Storage and Transfer. The vapor space in a tank will in time become saturated with the stored organics. During tank filling, vapors are displaced causing an emission or a "working loss." Some vapors also are displaced as the temperature of the stored VOC rises, such as from solar radiation, or as atmospheric pressure drops; these are "breathing losses." The amount of loss depends on several factors: type of VOC stored, size of tank, type of tank, diurnal temperature changes, and tank throughput. Working and breathing losses can be estimated using the EPA AP-42 Tanks Program.[4]

In-plant transfer of VOC is done mainly by pipeline, but also may be done manually (e.g., loading or unloading 55-gal drums). Raw materials are delivered to the plant by tank truck, rail car, or in 55-gal drums. These storage devices are susceptible to breathing and working losses, the magnitude of which depends on the type of compound stored, the size and design of the tank, ambient temperature and diurnal temperature changes, and tank throughput. Emissions may also be associated with manual material transfer operations within the facility and with the transfer of liquids from tanker cars, trucks, and rail tank cars.

Types of Emissions

Pharmaceutical manufacturing facilities are similar to many chemical-manufacturing operations in that both gaseous and particulate emissions can be significant.

VOCs. Usually of greatest concern at pharmaceutical facilities are emissions of VOCs. Many of these compounds are photochemically reactive and contribute to tropospheric (lower-level) ozone formation (a concern for U.S. facilities located in ozone nonattainment areas), while others may deplete the stratospheric (upper-level) ozone layer. An added concern is that some compounds are air toxics or HAPs (see Table 2), categorized by the EPA as possible or probable human carcinogens, thereby making long-term human exposure undesirable. Finally, some compounds used in the industry may be acute toxicants, where short-term exposure, including emergency releases, can lead to adverse health effects.

Particulates. Emissions of particulates also may be of concern. One reason is that particulate emissions, especially those associated with formulation and packaging operations, may include biologically active ingredients. Additionally, some of these emissions may be in the respirable size range. Within this range, a significant fraction of the particulates may be inhaled directly into the lungs, thereby enhancing the likelihood of being absorbed into the body and damaging lung tissues. Fortunately, because of concern about worker health and product contamination, particulate emissions are often closely controlled, especially where toxic compounds are in use. Consequently, particulate emissions at most pharmaceutical manufacturing facilities are insignificant.

Odors. Additionally, odors can be a problem, especially where fermentation takes place. While odors can create a community nuisance, they generally do not represent a serious air pollution problem.

Other. Another class of gaseous emissions is acid gases; these are principally sulfur dioxide, nitrous oxides, and hydrochloric acid. The oxides of sulfur and nitrogen are typically associated with the large-scale combustion of fossil fuels are utilized for power generation or steam production by large chemical facilities. Because of the generally smaller-scale operations of the pharmaceutical industry, acid gases are usually not an issue. The exception may be halogen acids (e.g., hydrochloric), which are formed as by-products of chemical synthesis or as products of the combustion of halogenated compounds.

Emission Estimation

It is very difficult to characterize specific pharmaceutical operations in terms of representative emission rates. This difficulty is due to the batch nature of the industry, where the following characteristics of process units can vary widely:

- Size
- Annual and daily throughput

- Number of cycles per year and operational duration
- Types of raw materials used and products manufactured, including the production of multiple products over the course of a year

The EPA has ranked pharmaceutical process unit operations according to their VOC emission potentials as indicated in the following (in decreasing order).[5]

Emission Potential Rank	Pharmaceutical Manufacturing Process Unit Operation
1	Dryer
2	Reactor
3	Distillation
4	Storage and transfer
5	Filter
6	Extractor
7	Centrifuge
8	Crystallizer

The first four process unit operations generally account for the majority of emissions at many facilities. However, because of the wide variation in process unit design and operation from facility to facility, the emissions associated with each process unit must be evaluated on a case-by-case basis.

Emission Estimation Techniques

Determining realistic emission factors for pharmaceutical manufacturing can be extremely difficult, because pharmaceutical manufacturing unit processes vary extensively in size, throughput, operation, and product. The most common approaches for determining emission rates involve estimating emissions by using (1) material balances and/or (2) theoretical calculations. Emission testing can be difficult and expensive for batch processes.

Material Balance. Material balance computations can be performed by obtaining records on total solvent purchases for a particular facility and assuming that these purchases represent replacements for solvents used and/or lost through evaporation (and thus emitted to the atmosphere). However, other routes by which solvents can be lost also must be taken into account. These routes include discharges to the sewer, incineration, residuals within solid waste, and incorporation into products.

Emission calculations, using a material balance methodology, typically are based on historical production records, pharmaceutical process step specifications, and quality-control requirements. Quantity of the material input and output (dry pharmaceutical blends and liquid solvents) and material losses to disposal, process cleanup, material residual in equipment, or rejection during quality-control inspection over the time the material is processed are used to develop a material balance.

Theoretical Calculations. It is possible to estimate emission rates from pharmaceutical manufacturing processes using theoretical calculations, based on such factors as vapor pressure, ideal gas behavior, and various simplifying assumptions. The EPA has published a series of VOC emission calculation procedures for charging, evacuation, nitrogen or air sweep, heating, gas evolution, vacuum distillation, and drying operations.[5] When using these and similar calculations to determine hourly, daily, or annual emission rates, correction factors that account for the actual duration of the process must be incorporated into the calculations.

It is important to bear in mind when applying theoretical calculations that the simplifying assumptions inherent in calculations of this type can lead to inaccuracies. These inaccuracies usually overstate emissions and can lead to emission estimates that are from several percent to as much as an order of magnitude higher than the actual value. Thus caution is required in interpreting the results of these calculations.

Many of the BPC manufacturing processes involve several different types of unit operations. Chemical equations may be used to estimate emissions from some of these unit operations. These unit operations are discussed below.

- Filling
- Purging or sweep
- Heating
- Depressurization
- Vacuum
- Gas evolution
- Solids drying
- Storage tank

Filling. When a liquid is charged into a vessel, vapors are discharged through the vent by simple volume displacement. These vapors are from the incoming stream and the vessel contents (if present). The emission for each compound is estimated from its effective vapor-space concentration and the total displacement volume. The chemical composition and temperature of the vessel's liquid contents and inlet stream at the beginning of the operation are used for the vapor-pressure calculations. If the filling also has a gas sweep during the operation, then the emission level is a function of both activities—the volume change and the gas sweep displacement.[6]

Purging or Sweep. In many processing operations, a purge gas (such as nitrogen) is admitted to the process vessel to remove vapors and provide an inert atmosphere. Volatile compounds in the vessel's gas space accompany the purge gas as it exits the equipment vent. Solvent losses are estimated from the flow rate of purge gas and the concentration of each volatile compound.

If the vessel contains process material, then the exit gas from the process vent is normally saturated with the volatile liquid components (nonempty vessel purge). In this case, the component vapor pressure in the vessel's gas space remains constant throughout the purge operation. The total emissions for the vessel are calculated from the amount of purge gas and the constant component vapor pressures.

If the liquid contents have been transferred to another tank and the original vessel is now empty, then the concentration of volatile organic vapors reduces over time through dilution (empty vessel purge). The component vapor pressure is based on the number of volume changes that the vessel undergoes throughout the purge operation.[6]

Heating. If a liquid in a vessel is heated to a higher temperature, vent emissions result from the thermal expansion of the vessel gas space and from an increase in the liquid's vapor pressure. The emission for each volatile component is estimated from this thermal displacement and the average vapor concentration.[6]

Depressurization. Many processes involve operations that are below atmospheric pressure. When a vacuum operation is performed by reducing a vessel's system pressure, a discharge of volatile compounds results. Emissions are estimated by the total amount and composition of gas removed from the vessel's free space.[6]

Vacuum. After a vessel's pressure has been reduced to vacuum, the objective is to maintain the vacuum for the duration of the process operation. Air leaking into the evacuated system through flanges and around gaskets must be continuously removed by a vacuum pump. Vent losses are calculated based on an estimated leak rate and the vessel's vapor-space concentration for each volatile compound.[6]

Gas Evolution. Some chemical reactions generate gases as a chemical by-product. If these gaseous chemicals are not retained by the process solvents, they evolve from the reaction mixture and leave the system through the vessel vent. Solvent vapors are swept out of the vessel as these reaction gases leave the system. Emissions from a reaction are calculated from the amount of gas that has evolved and the concentration of the different compounds in the vessel vapor space.[6]

Solids Drying. Many chemical processes involve some form of solids drying as part of a larger manufacturing process. When the initially wet solids are placed in direct contact with air (such as with air transport drying), the liquid portion of the wet product cake is evaporated and the dried solids are retained. Solvent losses from drying are estimated by simply calculating the solvent loss from the product cake.[6]

Storage Tank. Several fixed-roof storage tank models are available, including cone roof, dome roof, horizontal, underground, and isothermal vessels. Emissions from storage tanks originate from both working losses due to transfer operations when shipments are received and breathing losses due to daily fluctuations in temperature. Breathing losses are calculated using the average gas space volume in the storage tank during the period of time that the tank is in use, molecular weight of liquid, climate conditions, and vapor saturation. Working losses are calculated from the liquid volume throughput, vapor molecular weight, pressure, and frequency of filling operations.[6]

Computer Software. Use of commercial software in some cases may be helpful to assist in developing VOC emission estimates for certain BPC processes or unit operations. Although this approach may be a significant expense for the casual user, the commercial software available today is capable of utilizing chemical equations or material balances to estimate emissions, based on inputting information related to type and quantity of materials used, temperatures, pressures, and flow rates. Available commercial software that may be helpful to estimate emissions via theoretical calculations or material balances include Emission Master[6].

AIR POLLUTION CONTROL MEASURES

Information on air pollution control measures that apply to pharmaceutical processes has been divided into the following three subsections: emission control considerations, process design considerations, and operational practices.

Emission control systems are commonly used to reduce emissions of VOCs (including odors) and particulates. Because some VOCs and particulates may contain toxic components (HAPs), these control systems are also effective in reducing emissions of air toxics. Indeed, air-toxics control practices have been followed in the industry for many years as a result of the biologically active and high-value materials handled by some operations. As discussed under "Air Emissions Characterization," VOC emissions are of primary concern at pharmaceutical facilities and are thus our focus here.

The primary focus of the remainder of this discussion is the control of VOC emissions.

A variety of control systems are available for use in the pharmaceutical industry, depending on the particular application and on process operating parameters. A listing of the types of controls that are in use or available for use at pharmaceutical facilities is presented below.

Control Category	Available Control Systems
Add-on controls	Condensers
	Absorbers
	Adsorption units
	Thermal destruction
	Vapor containment
Process design	VOC minimization
	Solvent substitution
Operational practices	Cleanup operations

Information is presented in the following on the use of these pollution control concepts: add-on controls, process design, and operational practices.

Emission Control Considerations

Because of the batch nature of many pharmaceutical process unit operations, it is common for critical emission parameters to vary throughout a process operating cycle. These parameters include temperature, pressure, flow rate, VOC composition and concentration, and particulate loading.

To control these variable emissions streams effectively, emission controls must be designed to handle both peak and nonpeak flow conditions. Additionally, the controls must be designed to be started up and stopped numerous times over a year and to be effective over short operational cycles. Consequently, it is very important that control equipment be sized correctly and designed to handle a range of operating parameters and conditions.

Add-on controls available for use at pharmaceutical manufacturing facilities include condensation, absorption, carbon adsorption, thermal destruction, and vapor containment.

Condensation. The principle of operation of a condenser in emission control service is that a gas stream containing a

VOC can be cooled to below the saturation temperature ("dew point"), forming a second phase consisting of the VOC liquid. Separating the liquid from the gas (e.g., by gravity in the condenser combined with a demister pad to trap entrainment) removes a fraction of the VOC from the emission stream. The amount of VOC that remains in the gas stream as a vapor is a function of the temperature and the vapor–liquid equilibrium of the VOC species. In general, the lower the temperature, the lower will be the VOC content in the exit gas stream. In theory, any desired level of VOC removal (control efficiency) approaching 100% is possible with a condenser operating at a low enough temperature. There are some very real practical limitations, however.

Low condensation temperatures are required to control high-volatility VOCs or low concentrations of VOCs with any significant degree of volatility. In the latter case, most of the cooling duty of the condenser is to cool the noncondensible gas to the saturation temperature. Because the heat-transfer coefficient is low under these circumstances, the required heat-transfer area is large, increasing the cost of the condenser.

Producing low temperatures requires energy to drive the refrigeration system. Lower temperatures have progressively higher energy consumption (and thus cost) per unit quantity of heat removed. One scheme to raise the required condensation temperature is to compress the vent stream to a higher pressure so that the dew-point temperature will be raised. This approach is particularly useful where high control efficiencies of low-boiling-point materials (e.g., methylene chloride) are required. It is sometimes possible to raise the dew-point temperature sufficiently so that the problem with water vapor to be described can be avoided.

The presence of water vapor in the emission stream along with the VOC is common in actual practice. If room air sweeps or air purges are used, as is often the case in pharmaceutical manufacture, normal humidity will be present. If the process contains water as well as organic solvents, then even isolated vent streams will contain water vapor. Even conditioned "plant air" has traces of water, since the design operation of typical plant air dryers is to lower the water content to a dew point of only about −40°C (−40°F). If a gas stream containing water passes through a condenser that is operating below the freezing point, frost (rime) will form on the tubes, thereby effectively decreasing the capacity of the condenser by blocking efficient heat transfer.

Low-temperature condensation of volatiles from a gas stream containing water can be accomplished with a series of two or more condensers with decreasing temperatures. The first condenser(s) removes the water vapor, while successive, lower-temperature units remove the organic compounds. A concern with this scheme is that the condensate containing the water may also have a substantial organic content (by condensation) and thus may pose a water pollution control problem.

Because a vent condenser can be a modest cost-control device, it is often recommended. Based on the foregoing discussion, there are clearly some situations that are well suited to condensation. For example, when there is little or no water vapor in the stream and a reasonable concentration of a medium- or high-boiling-point VOC (10% or more), the use of a condenser is particularly appropriate. When the VOC is a low-boiling-point material, when there is water vapor present, or when the VOC concentration is low (less than 1%), the applicability of condensers is marginal, and multiple units operated at low temperatures may be required. When the VOC concentration is quite low (at the fractional percent level), condensers are probably not appropriate, as the achievable level of control will be unacceptably low.

Absorption. Absorption is the selective transfer of one or more components of a gas mixture (solute) into a solvent liquid. For any given solvent, solute, and set of operating conditions (pressure and temperature), there is an equilibrium ratio of solute concentration in the gas mixture-to-solute concentration in the solvent. If the solute concentration is higher than the equilibrium value, there is a "driving force" for mass transfer of that solute into the solvent, leading to its removal from the gas stream. If the equilibrium concentration in the gas is low compared with the concentration in the liquid solvent, the solute is said to have high solubility. Conversely, if a high concentration of the solute in the gas phase is in equilibrium with the liquid, the solute is said to have low solubility.

The apparent solubility of a solute species may be enhanced if there is a chemical reaction between it and components of the solvent solution. This reaction has the practical effect of minimizing the concentration of the actual solute species in the liquid by changing the solute species into another, more soluble species.

In emission control service for VOC removal, the solvents are chosen for their high organic solubility. These solvents often include water, high-boiling-point mineral oils, and sometimes-aqueous solutions of sodium carbonate and sodium hydroxide.

Emission control devices that are based on the principle of absorption are commonly called scrubbers and include spray towers, venturi scrubbers, packed columns, and trayed (plate) columns. The basic design concept of all these devices is to provide the most economical means of achieving a high degree of gas–liquid contact in order to promote the desired mass transfer. The degree of control achieved (i.e., the fraction of the solute that is transferred from the gas to the liquid) is a function of residence time, mass transfer area, and physical and thermodynamic properties of the VOC species involved.

Spray towers are inherently simple devices consisting of an open column up which the gas stream flows and a group of atomization nozzles to distribute the solvent as a spray of fine drops. Spray towers are versatile in that particulate matter also can be removed, but they have low mass transfer coefficients and typically are used only with high-solubility gases.

Venturi scrubbers, which are slightly more complex mechanically than spray towers, have a high degree of gas–liquid mixing and high particulate removal efficiency, but have relatively short residence times that limit the mass transfer efficiency. They are typically used only with high-solubility gases.

For controlling organic compound emissions, packed and trayed columns are generally preferred. Packed columns are usually more economical than trayed columns for small-diameter (less than 0.6-m) units and can be readily designed for low-gas-phase pressure drop and for handling corrosive materials and liquids that tend to foam or plug. Trayed columns are preferred for larger-diameter units, where internal cooling is required or where low liquid flow rates would be insufficient to wet the packing.

Scrubbers have been applied particularly to control water-soluble inorganic gases (e.g., sulfur dioxide, hydrogen chloride, hydrogen sulfide, and ammonia) in airstreams using water as the solvent. In the case of acid gases, using an alkaline scrubbing medium increases the effectiveness even more.

In the pharmaceutical industry, scrubbers are also used to remove VOCs. Solvent selection, equipment design parameters, and operating conditions all determine the control efficiency. To absorb organic compounds that have relatively high water solubility (e.g., most alcohols, organic acids, aldehydes, ketones, amines, and glycols), water is the preferred solvent. For organic compounds with low water solubility, another organic liquid (usually one with low vapor pressure) may be chosen. However, in the case of low solubility, water may still be selected, and the equipment will be designed to handle the high flow rate required.

Increasing the depth of the packing or the number of trays increases the control efficiency by increasing the amount of mass transfer area. Decreasing the operating temperature may result in a more favorable gas–liquid equilibrium and achieve higher control efficiency with a given piece of equipment.

Absorption can be used effectively with a wide range of concentrations, ranging from several percent to as low as 200 or 300 ppm. Control efficiencies typically range from 60% to 95% or better. However, because of the hydraulic behavior of packing and trays, a relatively steady gas flow is required to maintain good efficiency. Turndown ratios of approximately 50% are standard, thus limiting the ability of absorption to handle the intermittent flow of vents from batch processes.

Adsorption. The principle of operation for adsorption and the application of adsorption to VOC removal from gas streams are discussed in the following.

Principle of Operation. Gas adsorption is the selective transfer of one or more components (solutes) of a gas mixture onto the surfaces of the pores of a microcrystalline solid sorbent. The selectivity is most pronounced in a monomolecular layer next to the solid surface, although selectivity may persist to a height of three or four molecules. Both natural and synthetic materials are suitable for use as sorbents if they have a microcrystalline structure. The large number of pores in these materials results in extremely high surface areas (e.g., more than a square kilometer per kilogram, or 0.2 square mile per pound, of solid in some instances). Adsorbents in large-scale commercial use include carbons, silicates, aluminas, and aluminosilicates (molecular sieves).

Generally, adsorption is reversible: The effective adsorption capacity (performance) of a sorbent for a particular solute increases with the concentration of the solute in the gas stream and decreases with increasing temperature. This means that the sorbent may be repeatedly "regenerated"; that is to say, the solute can be desorbed by passing a lower-concentration or higher-temperature gas stream over the bed. Reversible adsorption is desired for VOC removal because it allows for reuse of the adsorbent. However, in some situations, the adsorption of a solute from a gas is irreversible. The capacity of a solid sorbent is nearly constant and independent of the gas stream composition, but because the solute is so tightly bonded to the sorbent, it cannot be reused.

Activated carbon is the preferred absorbent material for removing VOCs from gas streams. It has a high affinity for nonpolar compounds, with an even greater capacity for high-molecular-weight materials. The high-molecular-weight compounds can pose some problems, as they tend to be irreversibly adsorbed and effectively decrease the capacity of the carbon in a reversible system. A further problem is that high-molecular-weight compounds can displace lower-molecular-weight materials already adsorbed.

The performance of a particular type of activated carbon for a particular solute species must be determined from equilibrium behavior. Where only one species of solute is present in the gas stream, the equilibrium behavior is conveniently represented by a simple curve that plots solute concentration in the solid phase as a function of solute concentration in the gas phase. These curves are usually only valid at a single temperature and are known as *isotherms*. Isotherms for common organic vapors being adsorbed on activated carbon are available in the open literature and from carbon suppliers. They may represent strictly experimental data or, more commonly, experimental data fitted to established algebraic formulas (e.g., Langmuir or Freundlich isotherms). In some rare cases, they may represent purely theoretical calculations of equilibrium based on the molecular statistics of the solute–surface interactions.

Application to VOC Removal from Gas Streams. In designing a carbon adsorption pollution control device in which the inlet VOC concentration is known and a certain outlet concentration is desired, it is possible to determine from the isotherm and material balance calculations how much carbon will be required. The process design is finalized by applying experience factors in terms of "aging" of the carbon beds (i.e., loss of capacity with time as a result of pore plugging or some other problem) and safety factors. The desired outlet concentration of the solute in the gas phase is used to determine the concentration of the solute on the activated carbon. When the entire mass of carbon reaches the "saturation" level, the solute no longer will be adsorbed from the gas stream to the desired level, and "breakthrough" is said to have occurred. The carbon then must be either regenerated (in the case of reversible adsorption) or disposed of (in the case of irreversible adsorption).

To allow vent streams or other gas streams containing VOCs to come into contact with the activated carbon, the carbon granules are usually arranged in a fixed-bed arrangement in either a vertical or horizontal cylindrical vessel. Small units are manufactured with the carbon already in place (e.g., a canister) with the intent that the entire unit will be replaced when the capacity limit of the carbon is reached. The supplier may regenerate the unit, or it may be disposed of. Larger units are constructed so that the carbon granules are loaded after installation of the vessel. In these units, the carbon may be periodically regenerated in place, or it can be removed and shipped off to a central location for regeneration on a contract basis.

Fixed-bed units are operated "batchwise"; that is, the carbon is gradually saturated with the VOC and must be periodically regenerated, usually in place. This regeneration in place is done by having two or more beds installed and manifolded together so that one bed is in use while the other bed is undergoing regeneration. For very large adsorption operations, "continuous" processes have been designed in which the carbon bed is continuously moved (by a fluidized bed, conveyor belt, or other mechanical means) from an adsorption zone through a regeneration zone and back to the adsorption zone. Continuous adsorption generally does not apply to the smaller-scale operations at pharmaceutical plants.

Activated carbon beds in VOC removal service usually are regenerated using steam. The steam serves as a gas stream with zero concentration of the VOC solute (so that the solute tends to be desorbed) and acts to raise the temperature of the bed so that the equilibrium for desorption will be favored. A much smaller volume of steam is required for regeneration compared with the volume of gas that is treated during the adsorption portion of the cycle. Following desorption, the steam is condensed (further reducing its volume) and sent to a wastewater treatment system to remove the organic content. In some cases, depending on the species and loading on the carbon bed, the VOC concentration in the water is sufficient such that two liquid layers form with the steam condensate. The second layer, with the higher organic compound concentration, can be readily decanted and either recycled or disposed of in a more efficient manner than wastewater.

In the pharmaceutical industry, because of the final product purity requirements, there are limited opportunities for in-plant recycling of recovered solvents. However, where practiced, recovery of a separate organic stream is desirable, since there may be opportunities to sell recovered solvents to another industry or to incinerate them for in plant heat recovery.

Activated carbon adsorption is currently in use in pharmaceutical plants to remove VOCs from vent streams, back up condensers, and control odors. Removal efficiencies of 95% are common, with values as high as 98–99% seen in selected cases. In odor-control applications, the malodorous organic compounds are often fairly large molecules, and thus carbon adsorption may be substantially irreversible. In these cases, the use of disposable carbon devices (such as canisters) may be favored. A related issue is the problem of displacement: If the carbon bed is used to adsorb the larger molecules, it is unlikely that it also will be very effective at simultaneously removing the lighter-molecular-weight compounds. Thus the removal efficiency of the lighter-weight compounds will be poor.

It should be noted that potential fire hazards are associated with using carbon adsorption on air vents or vent streams containing oxygen and flammable hydrocarbons. Whereas the gas streams may be sufficiently dilute in flammables as to be well below the lower flammability limit, the carbon bed concentrates the flammable materials, so that a potentially flammable mixture may be formed inside the bed. Furthermore, the carbon itself, with a high surface area, is combustible. For these reasons, special precautions for early fire detection or bed cooling may be required in some instances or an alternative to carbon adsorption may need to be used.

Thermal Destruction. There are four main types of thermal destruction systems in general use: flares, boilers and process heaters, thermal incinerators, and catalytic (oxidizers) incinerators.

Flares. Flares burn flammable or combustible vapors with an open flame and can be either tower or ground mounted. Nozzle design, steam injection, and air injection are some of the design aspects used to enhance destruction efficiency. Flares are well suited for use with gas streams having a significant fuel value (more than 7.45 MJ/nm^3 or 200 Btu/scf) and are particularly well suited for handling a large, intermittent flow such as might occur from an emergency vent header. They have only limited capability for handling chlorine-containing compounds owing to corrosion of the burner elements and the lack of postcombustion emission control devices (for acid gas removal). The use of flares in the pharmaceutical industry for emission control is very limited; more typical applications are in petroleum refining or large chemical complexes.

Boilers and Process Heaters. Boilers and process heaters represent an opportunity to burn a waste stream while simultaneously recovering the stream's heating value. Typical applications are combusting a high-fuel-value liquid waste stream (uncommon in pharmaceutical operations) or using a pollutant-laden airstream as a source of combustion air. A fairly high-fuel-value vent gas stream is required for the boiler or process heater to function without added fuel. As with flares, boilers and process heaters operating with waste streams are more typically found in petroleum refining or large chemical complexes.

Thermal Incinerators. Thermal incinerators are controlled combustion devices where fuel and air are added to a combustion chamber to maintain a high minimum operating temperature. Because of these high temperatures, thermal incinerators are able to handle dilute vent streams with high destruction efficiency. Gases with heating values below about 1.86 MJ/nm^3 (50 Btu/scf) can be handled with the addition of fuel. Combustion chamber temperatures range from 700°C to 1300°C (1300–2400°F). Packaged, single-unit thermal incinerators are available in a wide range of sizes and are able to handle flow rates ranging from 0.1 nm^3/sec (200 scfm) to about 24 nm^3/sec (50,000 scfm).

The control efficiency achieved by thermal incineration is typically 98% destruction or an exit gas concentration of 20 ppm by volume, whichever is less stringent. Thus an inlet stream with a VOC concentration of 2000 ppm by volume or higher (which, with the typical 1:1 air dilution, becomes 1000 ppm by volume at the inlet to the combustion chamber) will have 98% of the incinerator inlet VOC content removed. Inlet streams with lower VOC concentrations will result in outlet gas concentrations of 20 ppm by volume of unburned organics, but with lower destruction efficiencies. Thermal incinerators can handle chlorine and sulfur containing compounds, but may require a scrubber on the outlet to control acid gases.

Due to the wide range of sizes available and the ability to handle a wide range of compounds in vent gas streams, thermal incineration is well suited for emission control in the pharmaceutical industry. It is, however, a capital- and energy-intensive approach.

Catalytic Incinerators. Catalytic incineration is a variation of thermal incineration. A catalyst is used to promote oxidation of the inlet gas stream at lower temperatures than are required in standard thermal combustion. Catalytic units usually operate over a range of 320–650°C (600–1200°F), and these lower operating temperatures reduce energy requirements. Catalytic incineration units may be smaller than standard thermal incineration units, but the cost of the catalyst may tend to offset any potential savings in the capital investment. Two additional design constraints are that (1) high organic concentrations may cause catalyst failure and (2) the catalyst may be poisoned by sulfur-containing compounds, heavy metals, and halogens.

The control efficiency of catalytic incineration is equivalent to that of thermal incineration, and a wide range of packaged unit sizes is available. As with thermal incineration, catalytic incineration is well suited for emission control in the pharmaceutical industry.

Vapor Containment. Transferring volatile organic liquids from delivery tank trucks and tank cars to storage tanks and from storage tanks to process vessels displaces gas from the headspaces that contain some fraction of VOC vapor. While these intermittent vent streams may be treated by one of the add-on control devices described above, vapor containment may in some cases be the preferred option. In this technology, additional piping is provided so that as the liquid is transferred from the supply tank to the receiving tank, the displaced gas and vapor from the receiving tank are returned in a separate line to the supply tank. With a properly designed system, very little gas/vapor escapes into the atmosphere, and VOC emissions are minimized. This concept can be extended to filling reactors from drums by using a drum pump and vapor return line rather than picking up the drum and pouring.

Other unit operations that can be improved in terms of vapor containment are those designed to operate at atmospheric pressure. Because of the relatively unrestricted mixing of the process vapors with the atmosphere, reactor and condenser vents open to the atmosphere are difficult-to-control sources of emissions. An option is to operate them as closed systems, at a slight pressure or vacuum, so that the flow of noncondensible gas (air or inert gas) is restricted. This approach is likely to decrease emissions, and the restricted flow rate stream that occurs is easier to control.

Bulk quantities of volatile organic liquids typically are stored in atmospheric pressure tanks with vents to relieve excessive pressure and to break any vacuum that may form. However, diurnal temperature fluctuations will result in "breathing losses" where the headspace vapor, approximately saturated with volatile organic vapors, is expelled when the temperature rises. When the temperature drops, fresh air (or an inert blanketing gas) is drawn into the tank, from which it may later be expelled with some degree of saturation. Installing "conservation vents" raises, by a small amount, the pressure or vacuum threshold at which this event occurs, thereby reducing the emission stream volume. Isolated, especially high-pressure, rises or unusually low vacuums will still open the conservation vents, but normal daily fluctuation will not, so that the total volume vented to the atmosphere is substantially reduced.

Process Design Considerations

To use add-on controls, including vapor containment, a certain amount of process design must be undertaken. However, in the case of new products or revamps of existing processes, it may be possible to use a different pollution control concept — designing a process that has an inherently lower potential for air emissions. One strategy is to minimize the use of VOCs, to substitute lower-volatility (and/or less toxic) compounds, and to keep these compounds contained.

For example, in the case of drying emissions in pharmaceutical manufacture, a modified process design could use and transfer intermediates as solutions or slurries, rather than as dry products. Where solvents are required, a lower-vapor-pressure solvent would tend to minimize evaporative losses and increase recovery efficiency when using condensers. With some processes, a reaction or formulation step traditionally conducted in a solvent possibly could be carried out in an aqueous medium. One ultimate goal may be the practice of a "solventless" synthesis, where the reaction is conducted "neat." In this case, a customized reactor design may be required to handle potential heat transfer and rheological problems associated with the resulting more concentrated reaction mixture.

Although there are often physical or chemical reasons why a process design must incorporate a particular volatile compound, there are many other instances where, with ingenuity, VOC emissions can be minimized by eliminating the offending material altogether.

Another strategy in a modified process design is to minimize the transfer of materials from process vessels by having a particular unit perform more than one unit operation. An example is a filter–dryer combination. These units, increasingly being used in the pharmaceutical industry, are batch pressure filters that, instead of discharging a wet filter cake, remain closed and convectively dry the cake with a recirculating gas stream.

For maximum effectiveness with some add-on controls, process redesign may be required. An example is the use of closed-loop drying systems. Acknowledging that a convective dryer is potentially a large source of emissions, a condenser, or a condenser combined with gas compression, is often a good choice. Instead of venting the drying gas, a further refinement of this approach is to recirculate the gas from the discharge of the condenser through the heater and back to the dryer. During the drying cycle there are, other than fugitives, essentially no emissions from the closed loop.

Operational Practices

This last control concept deals with the potential for changing workplace practices in such a way as to minimize emissions. Much of the everyday operation of a pharmaceutical synthesis facility is governed by standard operating procedures (SOPs). With proper engineering and management review, the SOPs can sometimes be revised and modified to reduce emissions. One area where emission reductions are particularly possible is in the use of solvents in cleanup practices. Typical procedures may call for the filling and subsequent emptying and air drying of process vessels. The air-drying step results in significant emissions regardless of the vapor pressure of the solvent involved. As an improved SOP, it may be feasible to eliminate the drying step, leaving residual solvent present before the start of the next cycle. Alternatively, a low-VOC aqueous cleaner might be substituted. A similar emission source involves flushing and blowing transfer lines (piping). Again, it may be possible to change the SOPs so that the lines are left filled with solvent rather than blown dry to the atmosphere.

REFERENCES

1. International Society for Pharmaceutical Engineering/Food Drug Administration, *Baseline Pharmaceutical Engineering Guides for New Facilities, Volume 1: Bulk Pharmaceutical Chemicals*, First Edition, June 1996.
2. *Profile of the Pharmaceutical Manufacturing Industry*, EPA Office of Compliance Sector Notebook Project (EPA/310-R-97-005), September 1997.
3. *Pharmaceutical Engineering*, January/February 1996, Michael P. Bracklebank and Pradeep V. Deo, GMP Issues for Bulk Pharmaceutical Chemical Plants.
4. *Compilation of Air Pollutant Emission Factors, Vol. 1: Stationary Point and Area Sources*, EPA Publication No. AP-42, Fourth Edition, U.S. Environmental Protection Agency, Research Triangle Park, NC, September 1985, Supplement F, April 1993.

5. *Control of Volatile Organic Emissions from Manufacture of Synthesized Pharmaceutical Products*, EPA-450/2-78-029, U.S. Environmental Protection Agency, Research Triangle Park, NC, 1979, App. B, p. 2–5.
6. Mitchell Scientific, Inc., *Emission Master Demonstration Software — Getting Started*, June 1997.

BIBLIOGRAPHY

Control of Volatile Organic Compound Emissions from Batch Processes, U.S. EPA Guideline Series, Research Triangle Park, NC. November, 1993 (EPA-453/R-93-017).

Draft EPA Guideline Document, *Control of Volatile Organic Compound Emissions from Batch Processes*, February 1993.

Development Document for Proposed Effluent Limitations Guidelines and Standards for the Pharmaceutical Manufacturing Point Source Category, U.S. EPA, Washington, D.C., February 1995.

Kirk–Othmer Encyclopedia of Chemical Technology, Fourth Edition, John Wiley and Sons, New York, 1994.

Pharmaceutical Production NESHAP, Basis and Purpose Document, U.S. EPA-Emissions Standards Division, January 1997.

BIBLIOGRAPHY — INTERNET

Good Manufacturing Institute, *http//www.gmpinstitiute.com*.

International Society for Pharmaceutical Engineering, *http//www.ispe.org*.

Pharmaceutical Online, *http//www.pharmaceutical online.com*.

Pharmaceutical Research and Manufacturers of America (PhRMA), *http//www.phrma.org*.

PHARMACEUTICAL PRODUCT FORMULATION

Michael J. Barboza and Theodore V. Chleboski

INTRODUCTION

Pharmaceutical product formulation processes convert manufactured bulk pharmaceutical chemicals (BPC) (i.e., active ingredients) into final, usable dosage forms for consumer use. A variety of types of formulation processes are used to convert the BPC into these final dosage forms. Typical formulation processes include mixing, blending, granulating, drying, coating, polishing, tablet pressing, capsule filling, sorting, and packaging. Many of these final dosage forms are produced with formulation processes in batch operations using various equipment and raw materials, utilizing a campaign regimen. At the end of a production campaign, another product may be formulated and packaged using the same equipment by the same personnel. Additionally, formulation and packaging is performed in accordance with Food and Drug Administration (FDA) and good manufacturing practices (GMP).[1]

Following formulation, the finished product may be packaged at the same site or may be transferred to another site. Packaging includes: placing the final formulated products into containers, labeling, and preparing for shipping. Packaging components of pharmaceuticals formulations is vital to safe and effective use of these products. Besides serving the patient as a convenient unit of use, the composite package (unit container, labeling, and shipping components) provides appropriate identification and necessary information for proper use, including warnings and precautions as well as preservation of the product's chemical and physical integrity.[1]

Dosage Forms[1]

BPCs are manufactured into several common dosage forms, such as tablets, capsules, liquids, creams, and ointments, as well as aerosols, patches, and injectable dosages, using various formulation processes. Table 1 provides a summary of some common pharmaceutical dosage forms and their uses. These common dosage forms are discussed below.

Tablets. Tablets account for the majority of solid medications taken orally in the United States. Tablets can be made to achieve rapid drug release or to produce delayed, repeated, or prolonged therapeutic action. Tablets can be compressed or molded, and may be coated. To prepare a tablet, the active pharmaceutical ingredient is combined with a filler, such as sugar or starch, and a binder, such as corn syrup or starch. The filler is added to ensure that the active ingredient is diluted to the proper concentration. A binder is needed to bind tablet particles together. A lubricant, such as magnesium sterate or polyethylene glycol, may be added to facilitate equipment operation, or to slow disintegration or dissolution of the active ingredient.

Tablets are produced by compression of powder blends or granulations. In direct compression, the ingredients are blended and then compressed into final tablets without modifying the physical nature of the material itself.

Wet Granulation. The most widely used and most general method of tablet preparation is the wet granulation method. In wet granulation, the active ingredient is powdered and mixed with the filler. This mixture is then wetted (often with solvents) and blended with the binder, forming a solution. Coarse granules form which are mixed with lubricants such as magnesium stearate and then compressed into tablets.

Dry Granulation. Slugging or dry granulation is used when tablet ingredients are sensitive to moisture or temperatures associated with drying or when the tablet ingredients have sufficient inherent binding or cohesive properties. Dry granulation includes weighing, mixing, slugging, dry screening, lubrication, and compression. Slugging requires large, heavy presses to compress larger tablets, between 20 and 30 grams in weight. These large tablets are then ground and screened to a desired mesh size then recompressed into final tablets.

Coating may be used to offer protection from moisture, oxygen, or light, to mask unpleasant taste or appearance, and to impart distinctive colors to facilitate patient recognition. Enteric coatings are used to delay the release of the active ingredient in the stomach and prolong therapeutic activity. The latter are used for drugs that are unstable to gastric pH or enzymes, cause nausea and vomiting or irritation to the stomach, or should be present in high concentrations in the intestines. Coating is done in a rotary drum. The coating solution (often solvent based) is poured or sprayed onto the tablets. Aqueous coating solutions can be used instead of solvent-based (usually methylene chloride) solutions. As the drum rotates, the tablets become coated. Once coated, they are dried in the drum and may be sent to another rotary drum for polishing. Polishing works by the friction created when the tablets rotate and rub against each other. Uncoated tablets may also be polished. Once the tablets pass quality-control

Table 1. Pharmaceutical Dosage Forms

Dosage Form	Constituents, Properties	Uses
	Solids	
Powders, bulk	Comminuted or blended, dissolved or mixed with water	External, internal
Effervescent	CO_2-releasing base ingredients	Oral
Insufflation	Insufflator propels medicated powder into body cavity	Body cavities
Lyophilized	Reconstitution by pharmacist of unstable products	Various uses including parenteral and oral
Capsules	Small-dose bulk powder enclosed in gelatin shell, active ingredient plus diluent	Internal
Troches, lozenges	Prepared by piping and cutting or disk candy technology; compounded with glycerogelatin	Slow dissolution in mouth
Compressed tablets	Dissolved or mixed with water; great variety of shapes and formulations	Oral and external
Pellets	For prolonged action	Implantation
Coated tablets	Coating protective, slow release	Oral
	Liquid Solutions	
Syrups	Sweetener, solvent, medicinal agent	Flavoring agent, medicinal
Spirits	Alcohol, water, volatile substances	Flavor or medicinal
Collodions	Pyroxylin in ether, medicinal agent (castor oil, camphor)	External for corns and bunions
Parenteral solutions	Sterile, pyrogen-free, isotonic, pH close to that of blood; oily or aqueous solution	Intravenous, intramuscular, subcutaneous injection
Ophthalmic	Sterile, isotonic, pH close to that of tears; viscosity builder	Eye treatment
Nasal	Aqueous, isotonic, pH close to that of nasal fluids; sprays or drops	Nose treatment
Mouthwash, gargles	Aqueous, antiseptic	Refreshment, short-term bacterial control
Inhalations	Administered with mechanical devices	Medication of trachea or bronchioles
	Liquid Dispersions	
Suspensions	Powder suspended in water, alcohol, glycol, or an oil	Oral dosing, skin application
Emulsions, lotions	Oil-in-water or water-in-oil	Oral, external, or injection
	Semisolid and Plastic Dispersions	
Ointments	Hydrocarbon (oily), adsorptive water-washable, or water-soluble bases; emulsifying agents, glycols, medicating agent	External
Pastes and cerates	Ointments with high dispersed solids and waxes, respectively	External
Suppositories	Theobroma oil, glycinerated gelatin, or polyethylene glycol base plus medicinal agent	Insertion into body cavity

Source: Ref. 1.

requirements, they may be held or sent directly to packaging. Coated tablets are stamped with identifying information (e.g., brand name, code number) in rotary ink presses.

Capsules. After tablets, the most common solid oral dosage form is the capsule. Capsules come in soft- and hard-shelled varieties. Hard capsules or dry-filled capsules are formed by dipping metal pins into a solution of gelatin of a specific temperature. The temperature controls the viscosity of the gelatin and hence the thickness of the capsule walls. When the pins are removed from the solution, a hard coating of gelatin forms on the pins. The coating is dried and trimmed. These capsules are filled by introducing the powdered material into the longer end or body of the capsule and then slipping on the cap. Soft-shelled capsules are formed by placing two continuous gelatin films between rotary die plates. As the plates are brought together, the two gelatin films join and seal, forming the two halves of the capsule. As the two halves join, the ingredients, which can be a liquid, paste, or powder, are injected into the capsules. Commercially filled soft gelatin capsules come in a wide choice of sizes and shapes; they may be round, oval, oblong, tube, or suppository-shaped.

Liquid Dosage. In formulating a liquid product, the ingredients are first weighed and then dissolved in an appropriate liquid. The solutions are mixed in glass-lined or stainless steel vessels, after which they are stored in tanks before final packaging. Preservatives may be added to prevent mold and bacterial growth. If the liquid will be used for injection or ophthalmic use, sterilization is required. In this case, either the container, which has also been previously sterilized/depyrogenated, is filled with liquid that has been rendered sterile by aseptic filtration in a sterile environment, and/or the entire container and its contents are terminally heat sterilized in an autoclave.

Ointments and Creams. Ointments are usually made by blending the bulk active ingredient with a base, such as a petroleum derivative or wax. The mixture is cooled, rolled out, and poured into tubes by machines and packaged. Creams are semisolid emulsions and are either oil-in-water or water-in-oil, rather than being petroleum based. Generally, the ingredients of the two phases are heated separately, then are mixed and stirred vigorously to achieve emulsification. As with all other dosage forms, equipment is washed and cleaned based on batch record requirements. However, because of the greasy nature of ointment and cream production, cleaning often is done with detergents.

Product Quality Control[1]

The pharmaceutical industry uses batch production records and/or processing procedures to describe each manufacturing step in detail. Raw materials are checked prior to use in a process, and the final dosage forms require a myriad of tests to assure therapeutic benefit. For example, the content uniformity, color, homogeneity, dissolution, stability, identity, and potency of the product must be determined and meet stated ranges. Representative samples are collected at the end of the formulation stage and submitted to the chemical and/or microbiological laboratories for final assaying. Representative samples are also collected during packaging operations. Quality-control checks are utilized at various stages in the formulation and packaging process. The quality-control unit of the pharmaceutical manufacturing company has the responsibility and authority to approve or reject raw materials, in-process materials, packaging materials including containers, closures, and labeling materials, as well as the final product. The equipment used to formulate and package the final product is cleaned, maintained, and sanitized at appropriate intervals. Actual maintenance and cleaning schedules and results are documented. The methods, equipment, and materials used (e.g., water wash, steam, detergents, organic solvents) to clean the equipment are specified on a per product basis.

PROCESS DESCRIPTION

Formulation processes used to convert manufactured bulk pharmaceutical chemical substances into final usable forms include: blending, mixing, milling, granulation, drying, tablet forming, capsule filling, and coating processes. These processes are described below.

Blending

Active drug and excipients are preblended together for homogeneity prior to the actual granulation process. The blending process is intended to achieve a homogeneous distribution of pharmaceutical materials for improved compression capabilities. Blending operations are performed using conventional ribbon, twin-shell, or double-cone blenders. For many years, conventional blending operations for noncritical operations were performed by pushing the material with blades or paddles (i.e., ribbon blenders). Other types of conventional blending equipment are: mullers for blending putty, sigma-blade mixer for dough, and planetary for pastes. The blender is charged with the appropriate ingredients from some type of transfer container and tumbled to create the desired homogeneous distribution. Today, depending on the desired consistency, pharmaceutical materials are blended with a V-Blender configuration with a twin cylinder. The V-Blender tumbles and rotates pharmaceutical material in varying forms (i.e., liquid–solids or solids–solids) to blend it uniformly. The twin shell uses an inclined cylinder with intermeshing action when the two such cylinders combine their flows.

Mixing

Pharmaceutical materials are mixed with variable-speed mixers in a large mixing bowl. The conventional mixing equipment includes a large "cake" mixing bowl and a beater that is variable speed. The mixing bowl can be mounted on dollies for easy transport of material for loading and unloading operations. Figure 1 presents a typical mixer that may be used in formulation.

Milling

Pharmaceutical materials are ground and shaped to desired product specifications with various milling equipment. Obtaining a uniform particle size is an important processing step in

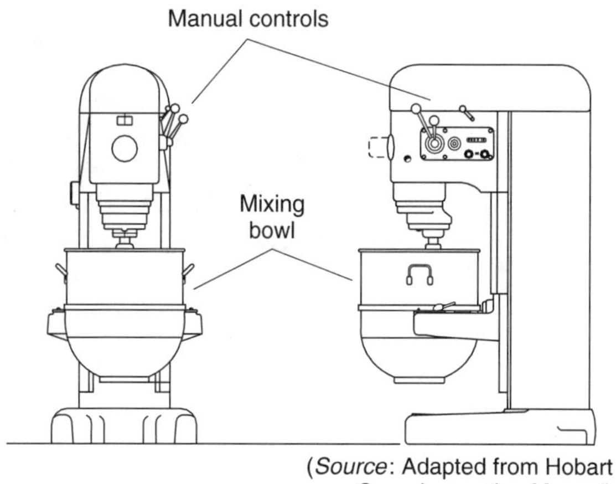

Figure 1. Mixer.

the formulation manufacturing process. Wet material is run through a sieving or milling system to break up oversized agglomerates prior to the drying process. The dried granules are sieved or milled again for final sizing prior to compression. One of the most common sizing devices in the pharmaceutical industry is the cone mill. The cone mill uses a conical basket and rotating rotor. The material to be sized is gravity fed into the basket and forced through the basket holes. The size and layout of the holes determines the product size.

Granulation[2]

Granulation is basically the creation of a final product from the combination of excipients, active drugs, and binders. Wet granulation used in the production of solid dosage forms is a concern when using solvent materials.

Wet Granulation. Wet granulation includes: blending (pre- and postgranulation), classification or milling (wet and dry), granulation/drying, and compression/coating/packaging. Wet granulation is a process where a granulate is formed from a mixture of preblended dry ingredients consisting of active drug, excipients, and other materials. The mixture is then wetted (often with solvents) to create granules with desired characteristics. The formulation is then dried, classified (milled and sized), compressed into tablets, coated, and packaged for distribution. Several methods can be used to perform the wet granulation process which involve the use of a vertical high-shear granulator or a fluidized bed powder/granulator. During the vertical high-shear granulator method, raw materials are formulated and charged into a vertical high-shear granulation bowl. Liquids (usually solvents such as IPA) are added to create the desired granular form. The high-shear granulator incorporates a main blade designed to promote rapid material turnover and movement in the bowl and a horizontal chopper blade designed to impart shear and break up agglomerates. The semiwet granulate is then discharged from the vertical granulator and wet milled to create a uniform particle size. During the next unit step, the wet-milled product is then charged into a fluid bed dryer and dried to remove the moisture and a free-flowing, low-dust granular product. Figure 2 presents a schematic of a granulation system that may be used in formulation processes.

The compression/coating/packaging is the final unit process operation in the granulation process. This process involves the actual dose taken by the user, which is usually in the form of tablets, capsules, and sachets. Tablets are produced by compressing the granules into the desired dosage form. A pinch roller system is used to prevent segregation, granule breakage, or bridging. The pinch roller system is typically used with lay flat polyethylene tubing connected between the granule container and the tablet press inlet. The roller system is allowed to travel freely up and down the tubing, pinching and releasing to control the flow of material.

Drying

Drying is used to dry pharmaceutical materials and product intermediates. Two major types of pharmaceutical materials dried are aqueous based and solvent based. Drying of pharmaceutical material usually is accomplished by circulating air which is heated by steam. Drying processes commonly used in pharmaceutical batch processing are conductive drying and convective drying. In conductive drying, heat is transmitted to the material to be dried by contact with heated surfaces; in convective drying, heat is transmitted by hot gases that are in contact with the material. Conductive drying may occur under vacuum conditions or at atmospheric pressure and in several types of dryers, including tray dryers, tumble

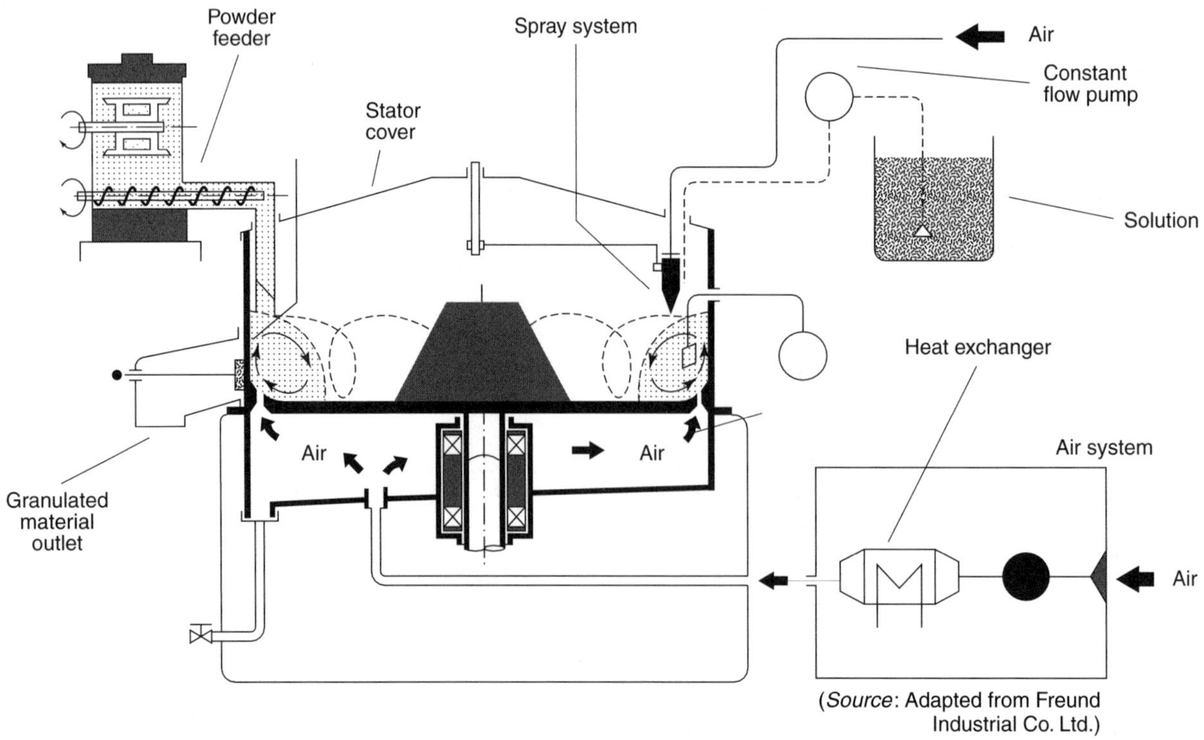

Figure 2. Granulator System.

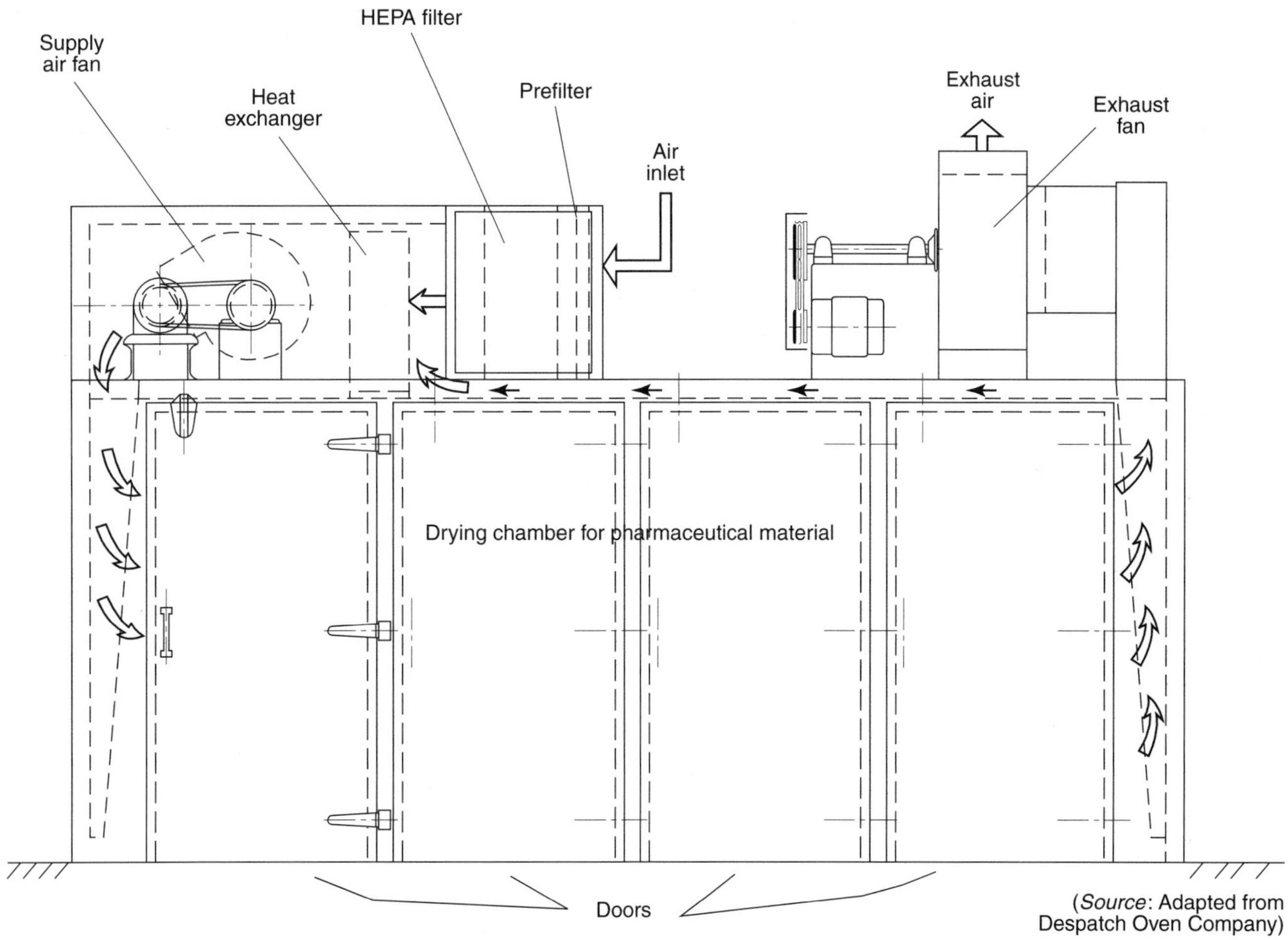

Figure 3. Convective Tray Dryer.

(double-cone) dryers, pan dryers, and rotary dryers. Convective drying occurs at atmospheric conditions. Convective dryers include tray, rotary, fluid bed, and spray. Figures 3 and 4 present examples of convective and vacuum tray dryers, respectively.

Drying is a labor-intensive process because of the necessary manual loading and unloading. A description of a tray drying system operating procedure is as follows:

- The interior and exterior of the tray drying system is visually inspected to ensure proper cleanliness.
- Temperature recording devices, steam supply system, and ducting system are checked to see if properly lined up.
- Equipment and product logs are reviewed.
- Proper airflow rates, drying times, and temperature are set to the prescribed settings.
- Pharmaceutical material intended for drying is placed on the trays or shelves that are stacked on pharmaceutical carts.
- These trays are filled to the desired capacity and placed in carts in the drying oven chamber.
- The drying unit is operated for the prescribed drying time, which will vary depending on the pharmaceutical material being dried.

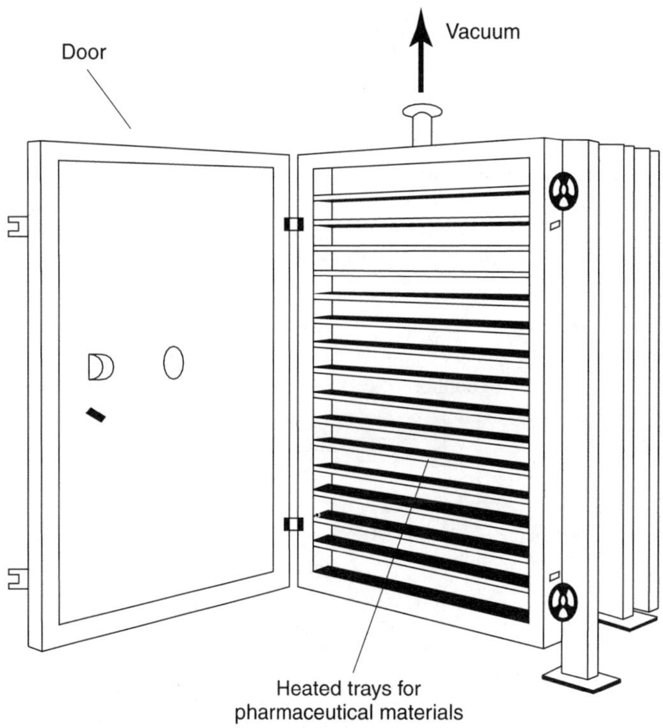

Figure 4. Vacuum Tray Dryer.

- Upon completion of the drying process, the dried pharmaceutical material is removed from the drying unit chamber.
- A review of the time and temperature requirements are made to ensure product integrity.

Other types of drying equipment include rotary and tumble dryers. Rotary dryers consist of a revolving cylinder that is slightly inclined to the horizontal. Feed enters the elevated end and is transported through the dryer by the rotation and slope of the cylinder. The tumble dryer usually is a double-coned shell configuration. These double-coned dryers are operated under vacuum, where heat is transferred to the pharmaceutical material being dried is via conduction from the heated dryer surface.

Tablet Forming[1]

Dry granulated pharmaceutical materials are formed into tablets to desired specification using a tablet press machine and blending active ingredients, filler, and binder. The purpose of the filler is to dilute the active ingredient to the proper concentration, and the binder is used to hold the contents of the tablet together.

Capsule Filling

Capsule filling involves filling gelatin capsules with powdered or granulated pharmaceutical material and inspecting the filled capsules for weight. After capsule filling, a cleaning/polishing step is performed to remove dust from the capsules as well as to polish them, prior to packaging.

Coating

Pharmaceutical materials are coated with a coating material and dried. The coating material may either be water or solvent based. Solvent is used to form coatings on tablet buildup materials in certain solid granulated pharmaceutical material processes. Coating may be applied either in a coating pan, where the coating is sprayed on the tablets, or in a fluidized bed, where the tablets remain suspended while the coating is applied. Figure 5 presents coating equipment that may be used during formulation processes.

Coating pans often have warm air blowing across them as coating is being applied to the pharmaceutical material. The coating sprays are directed at the material with very little overspray. The coating and drying operations in this case occur in the same equipment. Aqueous-based coatings are available for most applications, including some of the functional coating applications such as time-release and enteric-release that have traditionally warranted the use of solvent-based coatings.

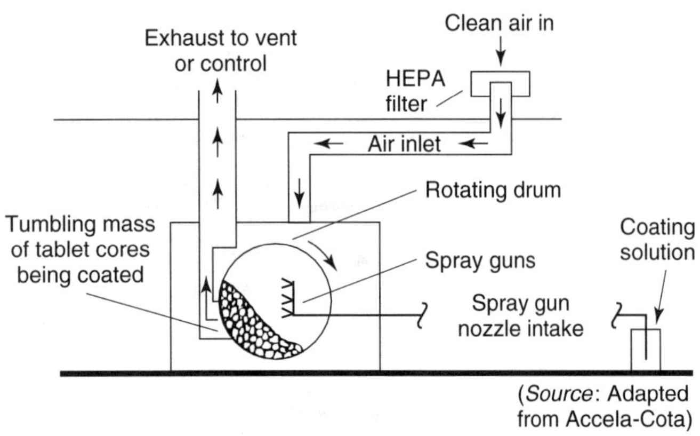

Figure 5. Coating Equipment.

The coating material, such as solvents and tablet makeup materials, are introduced into a fluidized bed reactors. The tablet buildup materials are fluidized to allow the solvent to form a coating. The fluidized bed reactor equipment usually is a specially designed rotor and a large stationary cylinder. The material to be coated is introduced along the wall of the cylinder in a ropelike twisted fashion and suspended in air, which is created by centrifugal force, rotating speed, and gravity. The pharmaceutical material is uniformly coated when the coating material is sprayed onto this fluidized bed of tablets.

AIR EMISSIONS CHARACTERIZATION

Air emissions from pharmaceutical product formulation processes are a function of the emission sources and types of emission substances.

Emissions

There are several types of emissions that may be emitted during product formulation processes. These types of emissions include VOCs, HAPs, and particulates.

VOCs/HAPs. Emissions of VOC/HAPs can occur during product formulation processes when various solvents are used, such as granulating, coating, and especially drying. It is difficult to estimate quantitative emission rates for VOCs/HAPs during formulation processes due to the batch nature of these processes and the various types of raw materials used and products manufactured. Methods to estimate VOCs and HAPs due to solvents during the drying, granulating, and coating processes are based on assuming that 100% volatilization of the VOC/HAP (solvent) occurs during the formulation activity and a knowledge of the quantity of solvent introduced into the process obtained from batch production records. Typical solvents used can include toluene, isopropanol, and methanol.

Particulate. Particulate emissions may occur during blending, milling, mixing, tablet pressing, granulating, and coating processes. Particulate emissions from these processes can be considered relatively insignificant since the pharmaceutical industry places a heavy emphasis on containing particulate through the use good housekeeping practices and use of dust collectors during formulation activities. In some cases, most of the dust generated by formulation operations is captured and sometimes returned to the process.

A material balance of the dry material input (identified in the batch production records), losses of dry pharmaceutical material (due to quality-control material rejection), and material residual in equipment during these processes can be used to account for the small quantity of particulate emissions during these processes.

Emission Sources

Sources of emissions associated with product formulation process operations are described below.

Drying. Drying during pharmaceutical product formulation is the unit operation with generally the highest VOC emissions. With a convective dryer, the exhaust gas is a high-flow, relatively dilute stream, which traditionally has often been vented directly to the atmosphere without emission control. With a conductive dryer, the larger vent gas flow rate results in much more concentrated exhaust streams. In the case of vacuum drying, VOC emissions occur downstream from the vacuum pump, where the air in-leakage is exhausted from the system.

Estimates of the air emissions from drying typically are based on 100% volatilization of the solvent material during the drying operation. A methodology used to estimate emissions of tray drying operations as well as for conductive and convective dryers during drying of granular pharmaceutical materials is discussed below.

In general, the rate of drying of the material depends on many factors associated with the specific drying situation (i.e., moisture or solvent content of material to be dried, heat and mass transfer parameters, drying period, etc.). The drying rate is the highest at the beginning of the drying process; it decreases with time so that a *large percentage* of the total solvent is removed from the pharmaceutical material during the initial portion of the drying cycle.[3]

Studies on the theory of drying of solids usually relate drying rate to moisture content of the solid. (In dryer nomenclature, moisture can also refer to organic solvent content, not just water content.) Three distinct periods of drying can be observed[3]:

- Constant-rate period where surface moisture is evaporated
- Funicular state where capillarity of the liquid in the pores influences the drying rate
- Pendular state where the moisture content is so low that capillary action ceases and the liquid must diffuse through the pores of the solid.

Each of these three periods of drying has successively lower rates of drying; the final drying rate, when the moisture content is zero, is of course also zero.[3]

The drying curve for solids can be related to rate of drying expressed as a function of residence time. Laboratory or pilot-scale experiments are often needed to establish the correct dryer size, operating temperature, gas flow rate, and cycle time. After a dryer is installed and operating correctly, the only data usually readily available are the cycle time, gas flow rates, and moisture content of the solid at the start and finish of the cycle. If very dry solids are produced (i.e., zero moisture content), it is clear that the drying rate at the end of the cycle will be asymptotic to zero. This endpoint condition, and a knowledge of the total solvent removed, can be used to estimate the emission characteristics of an existing, installed dryer. Figure 6 presents an example of an emission profile for a dryer using solvent material.

The solvent content prior to drying is given. At the end of the drying cycle the solvent content is reduced to the percent or fractional percent level, as required by the process. From a mass balance, the total amount of solvent emitted can be calculated (or assumed to be the total solvent prior to drying, i.e., 100% given off).

When the average emission rate must be determined, a reasonable estimate can be obtained by assuming that the drying rate is linear over the length of the drying cycle, starting from the initial, highest value and declining to zero at the end of the cycle. From a material balance of solvent removed, the average drying rate can be calculated knowing the length of the drying cycle.

The average emission rate can be determined by the total quantity of solvent emitted (i.e., 100% volatilized) during the drying operation divided by the drying cycle time. A theoretical maximum can be estimated by using twice the average emission rate, assuming that the emission release rate was linear. However, this theoretical maximum would not account for most drying processes, since most of the solvent is released during the initial portion of the drying cycle (e.g., 1–2 h). Therefore, it is important when designing emission control equipment and for air-permitting purposes that this be considered, since the initial drying emission rate can be substantially greater during the initial part of the drying cycle when compared to the emission rate at the latter stages of the cycle. Emission control equipment should be designed to accommodate this initial peak emission rate.

The calculation of total VOC content emitted during the drying cycle uses a mass balance from initial VOC concentration in the solids to the final concentration. The estimation of the peak dryer emission rate may be used for sizing of emission control equipment. The flow rate of

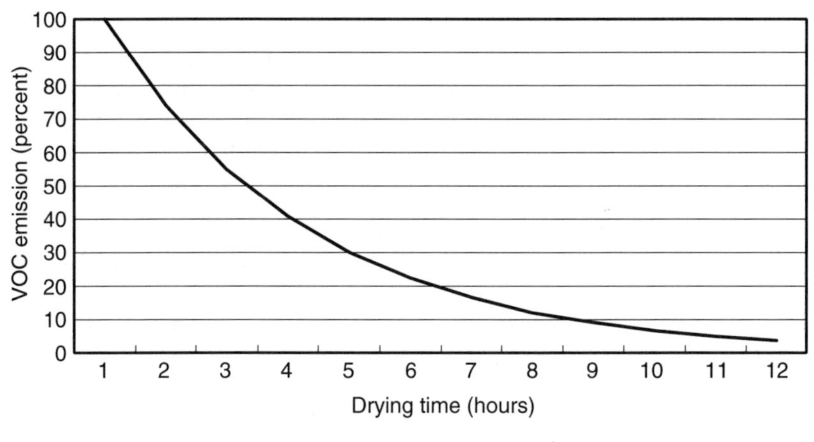

Figure 6. Dryer Emission Profile.

gas through the system is the dryer exhaust gas rate. To provide a conservative emission estimate for the worst-case operating scenario for purposes of permitting and air pollution control equipment sizing, peak hourly solvent emission rate is assumed to occur at the initial drying rate. Laboratory or pilot-scale experiments could be used to verify the actual maximum solvent emission rate from drying.

A conservative approach for tray drying emission calculation for peak emission rate may be calculated based on a maximum loading of the VOC solvent in the pharmaceutical material before drying and the assumption that 100% of the VOC solvent is volatilized (released) within the initial 1–2 h of the drying cycle (e.g., 6–12 h). This could yield 6–7 times the average emission rate. This approach is very conservative and could be used to assess peak emission rates for air-permitting purposes and designing emission control systems.

Particle or dust emissions are usually not anticipated for drying operations. Particle or dust emissions are not probable because of the following:

- Geometry of the pharmaceutical material is in the form of spheres, tablets, and capsules, for which carryover is not likely.
- Air flows are not turbulent enough to agitate fine material to make it airborne.

Granulation. Estimation of total solvent emitted during granulation processes is based on the initial weight of granulated pharmaceutical material, the initial solvent concentration in the pharmaceutical material at the beginning of the granulation process, and a 100% volatilization of the solvent during granulation. However, this is usually very conservative, since it is likely that much of the solvent is contained in the product and will be emitted over time during subsequent drying steps.

Coating. A large source of emissions from formulation activities can be coating operations. The coating material may either be water or solvent based, which is applied either in a coating pan, where the coating is sprayed onto the pharmaceutical materials, or in a fluidized bed, where the material is suspended while the coating is applied.

Emissions of VOC/HAPs can occur from coating pans or fluidized beds if the coating material contains VOC/HAP solvents. Because these operations are typically operated at elevated temperatures (30 to 80°C), there is potential for VOC/HAPs to be emitted from these unit operations. 100% volatilization of the solvent material used in coating is assumed to be emitted during the coating operation for conservatively estimating emissions. Typical solvents used in the coating operations are methanol and methylene chloride.

Aqueous-based coatings are also available for applications, including some of the functional coating applications such as time-release and enteric-release that have traditionally warranted the use of solvent-based coatings.

Blending, Milling, Mixing, Tablet Press. Particulate emissions from blending, milling, mixing, and tablet press processes can be considered relatively insignificant, since the pharmaceutical industry places a heavy emphasis on good housekeeping practices during formulation activities. In some cases, most of the dust generated by formulation operations is captured and sometimes returned to the process.

Historical production records and quality-control requirements of the dry material input as well as losses of dry pharmaceutical material due to quality-control material rejection and material residual in equipment during the process can be used to estimate particulate emissions during these processes.

AIR POLLUTION CONTROL MEASURES

Because of the batch nature of the formulation process unit operations, it is common for critical emission parameters to vary throughout a process operating cycle. These parameters include temperature, pressure, flow rate, VOC composition and concentration, and loading. To control these variable emissions streams effectively, emission controls must be designed to handle both peak and nonpeak flow conditions. Additionally, the controls must be designed to be started up and stopped numerous times over a year and to be effective over short operational cycles. Consequently, it is very important that control equipment be sized correctly and designed to handle a range of operating parameters and conditions.

Volatile Organic Compounds

Condensation. Condensers are often considered the first devices to be used, especially for some drying operations when assessing VOC removal control technology. The principle of operation of a condenser in emission control service is that a gas stream containing a VOC can be cooled to below the saturation temperature ("dew point"), forming a second phase consisting of the VOC liquid. There are clearly some situations that are well suited to condensation. For example, when there is little or no water vapor in the stream and a reasonable concentration of a medium- or high-boiling-point VOC (10% or more), the use of a condenser is particularly appropriate. In the pharmaceutical industry, because of the final product purity requirements, there are limited opportunities for in-plant recycling of the condensed recovered solvents.

Absorption. In the pharmaceutical industry for formulation activities (such as granulating and coating operations), scrubbers are used to remove VOCs. Gas absorption (scrubbing) is a widely used technique for controlling both organic and inorganic air emissions. Gas absorption control technology for VOCs that are highly soluble in water can provide effective VOC control. The effectiveness of scrubbers for VOC control is affected by the availability of a suitable scrubbing liquid and the thermodynamic properties of the compounds to be absorbed (e.g., vapor pressure and solubility).

Scrubbing of VOCs that have high solubility in water, which is a readily available absorbing medium, are options for this type of control technology. Solvent selection, equipment design parameters, and operating conditions all determine the control efficiency. In addition to providing high VOC removal efficiencies, scrubbing offers the further advantage of not producing products of combustion (e.g., NO_x, CO) that are associated with other methods of VOC control. These features, combined with the attractive economics and low maintenance of packed tower absorption, make it an appropriate VOC control technology for certain applications.

To absorb organic compounds that have relatively high water solubility (e.g., most alcohols, organic acids, aldehydes, ketones, amines, and glycols), water is the preferred solvent. For organic compounds with low water solubility, another organic liquid (usually one with low vapor pressure) may be chosen. However, in the case of low solubility, water may still be selected, and the equipment will be designed to handle the higher flow rates of scrubbing water required.

Increasing the depth of the packing or the number of trays increases the control efficiency by increasing the amount of mass transfer area. This will also increase the pressure drop through the device. Decreasing the operating temperature may result in more favorable gas–liquid equilibrium and achieve higher control efficiency with a given piece of equipment.

Absorption can be used effectively with a wide range of concentrations, ranging from several percent to as low as 200 or 300 ppm. Control efficiencies typically range from 60% to 95% or better. However, because of the hydraulic behavior of packing and trays, a relatively steady gas flow is required to maintain good efficiency. Turndown ratios of approximately 50% are standard, thus limiting the ability of absorption to handle the intermittent flow of vents from batch processes.

Adsorption. Activated carbon adsorption is currently in use in pharmaceutical plants to remove VOCs from vent streams, backup condensers, and control odors. Removal efficiencies of 95% are common, with values as high as 98 to 99% seen in selected cases. In odor-control applications, the malodorous organic compounds are often fairly large molecules, and thus carbon adsorption may be substantially irreversible. In these cases, the use of disposable carbon devices (such as canisters) may be favored.

Particulates

Dust Collection. Dust emissions of pharmaceutical material may be generated during formulation processes. Dust collectors or fabric filters may control the dust emissions of pharmaceutical material. Dust collector removal efficiency of 99% may be achieved, depending on particle size, loading, and other operating parameters. In some cases in formulation processes, dust collectors are used as a precontrol device to remove particulate prior to the VOC air streams entering scrubbing or condensing equipment.

Dust collectors typically require plant compressed air and power to operate. During normal operation, air enters the dust collector through the top inlet and passes through the filter elements. Dust is collected on the outside of the elements, and filtered air flows through the center of the filter elements into the clean air plenum, where it exits the dust collector and is exhausted to the atmosphere via an exhaust stack. Filter elements are purged when a solid-state control timer automatically selects a pair of filter elements to be cleaned and activates a solenoid valve that opens an air diaphragm valve that directs high-pressure air pulses into the center of the selected filter elements, blowing the collected dust off the filter elements into the hopper.

REFERENCES

1. *Profile of the Pharmaceutical Manufacturing Industry*, EPA Office of Compliance Sector Notebook Project (EPA/310-R-97-005), September 1997.
2. D. DiProspero, "Use of Intermediate Bulk Container (IBC) Systems in the Wet Granulation Manufacturing Process," *Pharmaceutical Engineering*, May/June 1998.
3. W. McCabe and J. Smith, "Unit Operations of Chemical Engineering," *Chemical Eng.,* Third Edition, 1976.

BIBLIOGRAPHY

Control of Volatile Organic Compound Emissions from Batch Processes, U.S. EPA Guideline Series, Research Triangle Park, NC, November, 1993 (EPA-453/R-93-017).

Draft EPA Guideline Document, *Control of Volatile Organic Compound Emissions from Batch Processes*, February 1993.

U.S. EPA 1993, *Draft EPA Guideline Document for Control of Volatile Organic Compound Emissions from Batch Processes*, February 1993.

International Society for Pharmaceutical Engineering/Food Drug Administration, *Baseline Pharmaceutical Engineering Guides for New Facilities, Vol. 1: Bulk Pharmaceutical Chemicals*, First Edition, June 1996.

Kirk–Othmer Encyclopedia of Chemical Technology, Fourth Edition, John Wiley and Sons, New York, 1994.

Pharmaceutical Production NESHAP, Basis and Purpose Document, U.S. EPA-Emissions Standards Division, January 1997.

BIBLIOGRAPHY — INTERNET

Pharmaceutical Online, *http/ /www.pharmaceutical online.com.*

Pharmaceutical Research and Manufacturers of America (PhRMA), *http/ /www.phrma.gov.*

United States Environmental Protection Agency (USEPA), *http/ /www.epa.gov.*

17

THE PETROLEUM INDUSTRY

J. Eldon Rucker and Robert P. Strieter
(with update by editor, including adaptations from AP-42, *Compilation of Air Pollution Emission Factors*)

OVERVIEW OF OPERATIONS

The petroleum industry is organized into the following four broad segments:

1. Exploration and production
2. Transportation
3. Refining
4. Marketing

These segments encompass the operations that produce crude oil and natural gas (petroleum industry raw materials), transform these raw materials into refined products, transport both the raw materials and refined products from one place to another, and market some of the refined products (such as gasoline) to consumers. Although the industry segments are interrelated, each segment has unique functions involving different operations and equipment. Figure 1 shows schematically the interrelationships between the four industry segments and the flow of petroleum industry products.

Petroleum industry operations begin with exploration to locate new sources of crude oil and natural gas. When potential sources are located, wells are drilled to confirm the presence of oil or gas and to determine whether the reserves are sufficient to support production.

During production, crude oil and/or natural gas is recovered from wells and prepared for transportation from the field. Domestic crude oil is transported from the field to refineries by a complex network of pipelines. Alaskan and imported crude oil are transported to refineries by tanker. Natural gas, which may be produced alone or in combination with crude oil, often must be processed at a gas plant to make it suitable for consumer use.

At the refinery, crude oil is converted into a large variety of products, such as hydrocarbon fuels and feedstocks for the petrochemical industry. Refining operations include physical separation of the components in the crude oil and chemical conversion and treating processes to generate the desired products. The refined petroleum products are transported by pipeline, rail, marine vessel, and tank truck to distribution outlets and sometimes directly to the users. The distribution outlets, referred to as bulk terminals and bulk plants, are part of the marketing segment. From these distribution, or marketing, outlets, tank trucks transport the petroleum refinery products directly to large commercial users or, in the case of gasoline, to service stations.

Because the four petroleum industry segments differ significantly in function, they will be discussed separately.

EXPLORATION AND PRODUCTION

Introduction

The first segment of the petroleum industry, exploration and production (E&P), involves the recovery of natural gas and crude oil from underground reservoirs. This industry segment encompasses the following major activities: (1) exploration and well-site preparation, (2) drilling, (3) crude oil and gas processing, and (4) enhanced recovery. There are an estimated 575,000 crude oil wells, 252,000 natural gas wells, and 724 gas plants in the United States.[1] The leading oil- and gas-producing states are Alaska, Texas, Louisiana, California, Oklahoma, New Mexico, and Kansas.[2,3]

Process Description

Seismic and other geophysical methods are used to locate subterranean formations that signal the potential presence of oil and gas reservoirs. When a likely formation is located, drilling is the only way to confirm that oil and gas are present.

Since exploration and well-site preparation activities do not contribute significantly to air emissions, these activities are not discussed in further detail. Instead, this section will concentrate on drilling and production operations, which are more significant sources of air pollutant emissions.

Drilling Operations

Drilling operations include the activities necessary to bore through the earth's crust to access crude oil and natural gas resources. During drilling operations, specially formulated muds are circulated through the hole to remove cutting from around the drill bit, to provide lubrication for the drill string, to protect the walls of the hole, and to control down-hole pressure. Internal-combustion (IC) engines often driven by diesel and natural gas, provide energy to rotate the drill bit, hoist the drill string, and circulate the mud, and these engines are sources of hydrocarbon air emissions and criteria pollutants [carbon monoxide (CO) particulate matter, nitrogen oxides (NO_x), and sulfur dioxide (SO_2)].

Cuttings are separated from the mud at the well surface as the mud is passed through shale shakers, desanders, desilters, and degassers.[4] The mud flows to a tank for recycling, and the cuttings, which may be contaminated, with hydrocarbons, are pumped to a waste pit for disposal. Because the waste pits may be open to the atmosphere, they are a potential source of hydrocarbon emissions. One other potential air emission source during drilling is blowouts. However, in the 10-year period from 1960 to 1970, blow outs occurred at only 106 of 273,000 wells that were drilled in eight states.[5]

Once a well has been completed and is producing crude oil or natural gas, an arrangement of high-pressure valves termed a "Christmas tree" is installed to control production. As the well ages, an artificial lift device may be needed to help bring product to the well surface. Figure 2 shows examples of a "Christmas tree" and a pumping unit for artificial lift. Pumping units are typically powered by electric motors, but are sometimes powered by IC engines. Gas-fired turbines

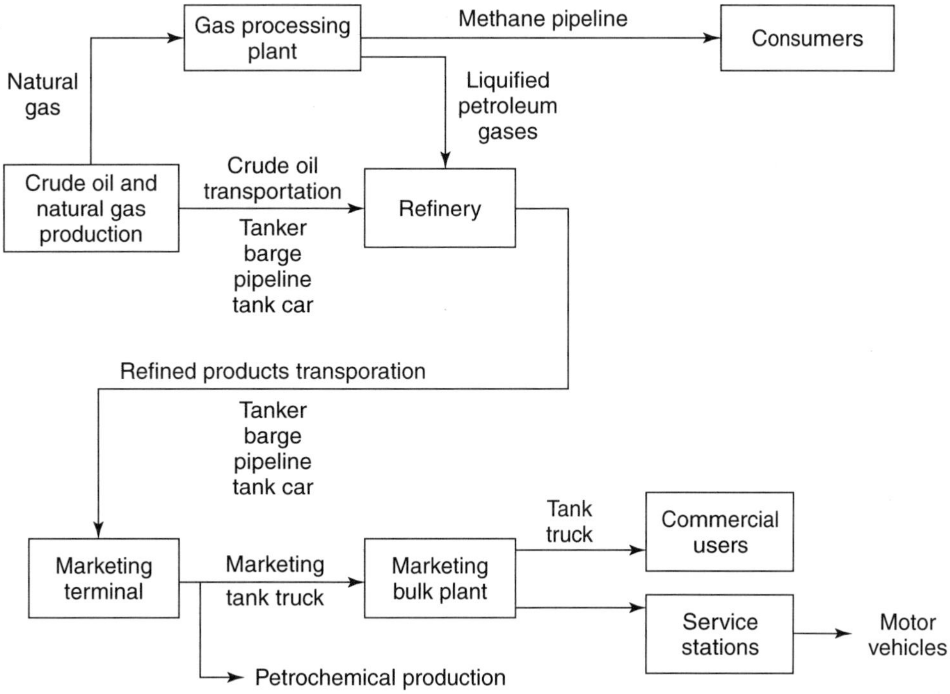

Figure 1. Flowsheet of Petroleum Industry Production, Transportation, Refining, and Marketing.

Figure 2. Typical Piping and Valve Arrangement at Oil and Gas Production Facility.

are another power source typically associated with steam generation or gas processing.

Oil and Gas Production Facilities

The overall objective of the production facility is to prepare the oil and gas for pipeline transport. The oil is usually transferred to a pipeline terminal and eventually to a refinery, while the gas may be sold to a pipeline company for further sale or transported directly to a gas plant for additional treatment. A variety of types of process equipment may be required at a production facility, depending on the quality of the well-head product and whether or not both oil and gas are produced by the well. Figure 3 shows a production facility that is producing and processing both crude oil and natural gas. Not all of the processing steps shown in Figure 3 are needed at all wells. Also, depending on local conditions, the equipment may be used in different combinations or additional equipment not shown may be necessary.

The first processing step employed at many production facilities involves separating the oil, gas, and water produced by the well. The separators provide only one stage of separation, and, in many cases, additional water separation from the oil and gas streams may be required. Water in the oil can form an emulsion. This emulsion is broken using heat in heater treaters or electric energy in devices such as electrostatic coalescers. Cleaned oil flows from the emulsion breakers to crude oil storage tanks, which store the crude oil before it is transferred to the pipeline. The water that is recovered during emulsion breaking is often recycled through skimmers to recover remaining oil, is filtered, and then is stored in water tanks prior to underground injection or other discharge.

Natural gas exiting the first-stage separators is composed primarily of methane, with smaller quantities of ethane, propane, butane, and some natural gasoline. It may also contain such impurities as nitrogen, helium, carbon dioxide

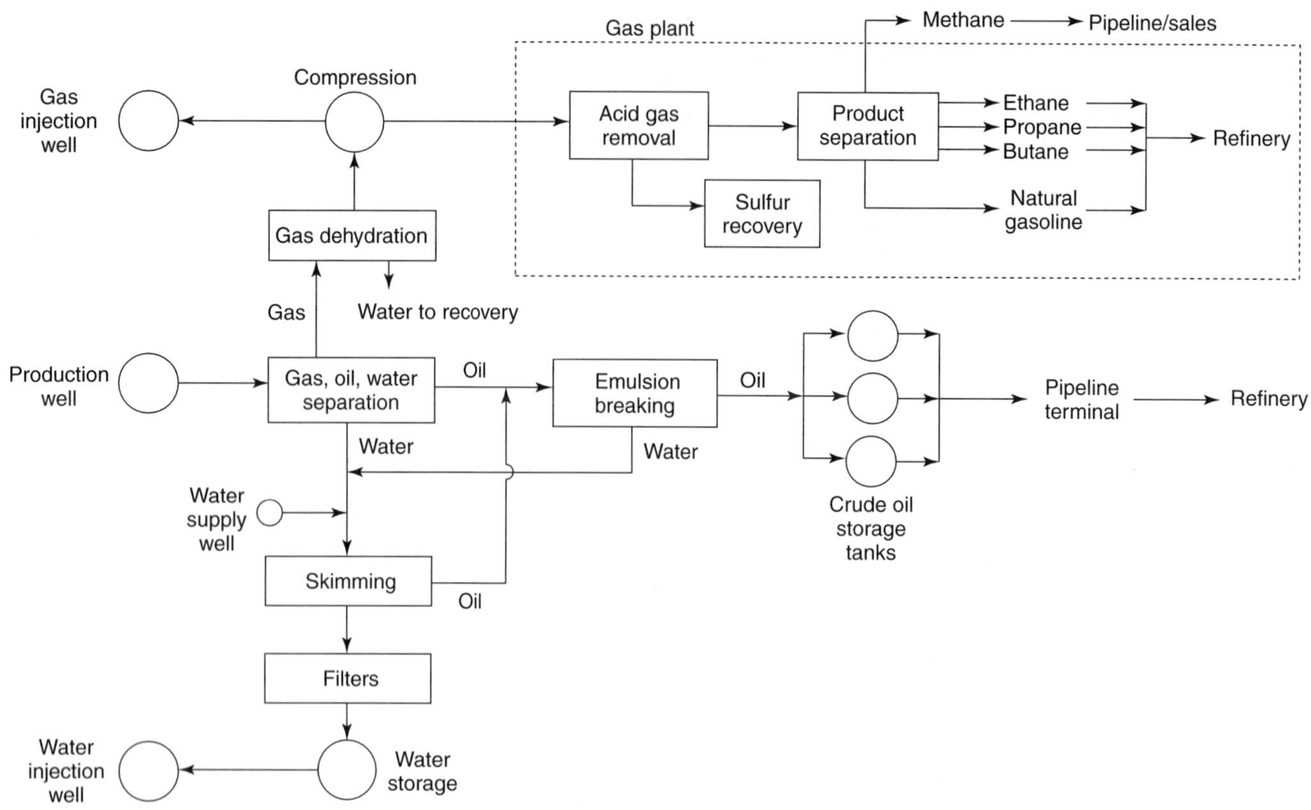

Figure 3. Oil and Gas Production Facility.

(CO_2), hydrogen sulfide (H_2S), mercaptans, and water. Condensible hydrocarbons are recovered from the gas using absorption, refrigeration, or gas expansion processes.[6] Water is removed by passing the gas through equipment such as glycol dehydrators, where the water is absorbed by the glycol. This water is ultimately vented as steam from the glycol reboiler.

The gas-sweetening and product-separation steps shown in Figure 3 are often conducted at a gas plant that is not located at the oil and gas production facility. To sweeten "sour" natural gas, H_2S and CO_2 are usually scrubbed from the gas using an amine solution in an absorption tower.[7] The acid gas stream may be flared, incinerated, or processed further in a sulfur-recovery facility, and the scrubbed gas stream is further processed to separate products. The majority of gas plants use one of the following four processes to separate the hydrocarbon constituents contained in the natural gas stream.[8]

- Absorption
- Refrigerated absorption
- Refrigeration
- Compression

Refrigerated absorption is the process in widest use. In this process, all hydrocarbons except methane are absorbed by an oil operating at low temperature. The methane is transferred to a gas pipeline company, and the heavier hydrocarbon components are separated from the absorber oil and each other in fractionating columns.[9] After separation, these hydrocarbons are transferred to a refinery for additional processing or, in the case of propane and butane, sold directly to users.

Air emissions may occur from a variety of sources at oil and gas production facilities and gas plants. At oil wells, "casing-head gas" may be vented directly to the atmosphere. Hydrocarbons may also be lost with water vapor that is vented from glycol dehydrators and with amines vented from absorbers at gas plants. Crude oil and water are stored in tanks at production facilities, and air emissions occur as a result of working and breathing losses from these tanks. Also, at gas plants, natural gasoline may be stored in tanks at atmospheric pressure, and emissions are expected from these storage tanks. The other hydrocarbon products separated at gas plants are stored in pressure vessels, but fugitive emissions may arise from relief valves on these vessels. Fugitive emissions also may occur from valves, flanges, pumps, and compressors located at both gas plants and production facilities.

In addition to hydrocarbon emissions, criteria pollutants are emitted at gas plants and production facilities where IC engines are used to drive pumps, compressors, and other prime movers. Fired vessels such as heater treaters, which often burn natural gas, are another source of criteria pollutants, as are steam generators. A small number of production facilities and all gas plants operate flares, and these devices are an additional source of criteria air pollutants (e.g., SO_2 emissions during acid gas flaring).

Enhanced Recovery

A number of techniques are used to increase the ultimate production of oil and gas from a reservoir. They include water injection, gas injection with CO, thermal processes, and acid treatment and fracturing. Water injection is the most widely applied of these enhanced recovery techniques. Water recovered with oil and gas from production wells is combined with water pumped from water supply wells and then injected into the oil and gas reservoir. This technique forces oil and gas from the reservoir toward the producing wells. Similarly, gas may be injected into a reservoir to create or strengthen a gas cap, which increases the pressure energy in this type of reservoir.[10]

Thermal processes include the injection of steam or heated water, as well as the in situ burning of a portion of the oil in the reservoir.[11] This technique, which is more common in California than in other states, reduces the viscosity of the oil and improves flow characteristics.

Acid treatment and fracturing are two separate techniques, but both increase production by increasing the porosity or permeability of the underground formations.[12] Acid treatment is effective for carbonate rock formations. Hydrochloric acid is usually used to dissolve the carbonate rock. In fracturing, a sand and fluid suspension is pumped down the well at high pressure to crack open sandstone formations, enabling oil and gas to flow more freely to the producing well.

Pumps are used to inject water and other fluids during water flooding, fracturing, and acid treatment. Air emissions will result if IC engines are used to power these pumps. Also, compressors are used during gas injection, and some of these compressors may be driven by IC engines.

Air Emissions Characterization

Volatile organic compound (VOC) air emissions have been characterized for sources in the E&P industry segment using the 1985 National Acid Precipitation Assessment Program (NAPAP) emissions inventory of the U.S. Environmental Protection Agency (EPA). The principal constituents of VOC emissions include benzene, toluene, xylene(s), 1,3-butadiene, and trimethylbenzene(s). Unfortunately, estimates of emission rates are not well developed in any database, and national emission levels cannot be accurately estimated.

Emission estimates have been developed for specific criteria pollutant emissions associated with the E&P industry segment. Nationwide SO_2 emissions were estimated to be 550,000 tons per year in 1979.[13] These estimates were based on data available for 82 sulfur-recovery facilities performing gas-sweetening operations on "sour" gases (i.e., gas containing H_2S). With respect to NO_x emissions, it was estimated that 111,760 tons of NO_x per year were emitted during drilling operations in 1975.[14] These emissions occurred from IC engines that were powering 1580 mobile drilling rigs during the drilling of 40,248 onshore wells. Total NO_x emissions from oil and gas production during 1975 totaled 618,640 tons per year; major sources of these emissions were IC engines (67%), steam generators (29%), and heaters (4%).[15]

Air Pollution Control Measures

Principal sources of emissions from the E&P segment include well casing vents, fired vessels, IC engines, oil and water storage tanks, glycol venting, and fugitive (equipment leak) sources. Control measures for each of these sources of emissions are discussed in the following paragraphs.

Vent Collection System

Collection systems are used to capture gases discharged from well casings during oil production. The most common disposition of collected gases is to a gas plant or dehydration facility. The collected gases also may be compressed for underground reinjection, used as a fuel for prime movers, or fed to a flare for destruction. Collection systems usually include a knockout pot or other device to remove condensible liquids. The control efficiency of the collection system depends on the capture efficiency and how the collected gas is used or destroyed. For example, flares operate at a destruction efficiency of about 98%.[16]

Air–Fuel Controllers

Criteria and hazardous air pollutant emissions from combustion sources can be reduced by the use of these controls. Proper air-to-fuel adjustment and control can reduce emissions of VOCs and particulate matter (PM) through more efficient combustion. Precombustion chambers vary the air-to-fuel ratio in stages to reduce NO_x emissions. In addition, low-NO_x burners are also used to reduce NO_x formation in combustion. The efficiency of air-to-fuel-ratio controllers is estimated to be about 50%.

Carbon Filter

Carbon adsorption can be used to control organic emissions from vent streams, such as from glycol vents. Organic constituents are adsorbed on the active surface sites of the carbon particles. When the carbon capacity to adsorb organics is exhausted, the carbon must be disposed of or regenerated. Disposable cannisters are sometimes used for low-organic-loading vent streams.

Refrigeration

Low-temperature condensers can be used to recover organic constituents from vent streams. The temperature of the vent stream is reduced to below the dew point of the condensibles. Chilled water, brine, and low-boiling-point gases can be used as coolants. Condenser efficiencies range between 85% and 95%.[17]

Fixed and Floating Roofs

A fixed-roof tank consists of a cylindrical steel shell with a permanently affixed roof that may vary from cone or dome in shape to flat. Fixed roof tanks commonly operate at a slight internal pressure or slight vacuum. Emissions occur from these tanks as a result of working and breathing losses. As a tank is filled, the increasing liquid level forces saturated vapors to vent from the tank; these losses are referred to as working losses. Breathing losses are caused by diurnal temperature changes that influence the vapor pressure of the liquid in the tank.

Fixed-roof tanks may be equipped with conservation vents that relieve pressure to prevent tank damage during filling. However, a vent recovery system can be installed to collect the vent emissions for recovery or destruction. The control efficiency of this type of system will depend on the ultimate recovery or destruction technique, but usually ranges up to 85% or more, depending on the tank size and application.

An open-topped (external) floating roof consists of a cylindrical steel shell equipped with a roof that floats on the liquid surface as it rises and falls. A seal, which is attached to the roof, contacts the tank wall at the annular space between the roof and the wall. As with internal floating roofs, air emissions occur from external floating roofs as a result of liquid that leaks through seals and stands on top of the roof.

However, with a properly designed seal system, the efficiency of an external floating roof ranges from about 95% to 99%.[19] Additional details on the design and emissions from storage tanks can be found in the current edition of *AP-42 Compilation of Air Pollution Emission Factors*, Chapter 7.[18]

Inspection and Maintenance

Fugitive emissions due to equipment leaks from pump and compressor seals, valves, flanges, and other equipment connections can be reduced by an active inspection and maintenance program. Leaking equipment is identified during periodic inspections with a vapor-detection device. The leaking equipment is logged on a maintenance schedule, and mechanical adjustments (e.g., tightening valve stems, packing) are made to repair the leaks. The efficiency of this control procedure is affected by how often the inspections are conducted. In addition, leakless equipment has been developed to reduce fugitive emission losses from such equipment as valves and pump seals.

TRANSPORTATION

Introduction

This segment of the petroleum industry encompasses the transportation operations used to move raw materials and refined products among the various other interrelated segments. It includes the transportation of oil and gas from well-field production facilities to refineries and gas plants and the transportation of refined products by means other than tank trucks (tank truck transport is covered in the marketing section of this discussion).

Process Description

Figure 4 presents a schematic of the transportation network showing the different modes used to transport petroleum products among segments of the industry. Most domestic crude oil is transported from well-field production facilities to refineries through a complex network of pipelines (Figure 5). The crude oil is introduced into these pipelines by pumping stations that are powered by electric motors, reciprocating engines, or gas turbines. There were approximately 3400 pump stations nationwide in 1990. Crude oil is also transported from some production facilities by barges, railroad tank cars, and tank trucks. Unlike domestic crude oil, imported and Alaskan crude oil enter the continental United States by tanker through 52 U.S. marine terminals. Once the crude oil has been processed at refineries, the refined products are transported from the refineries to marketing terminals by barges, railroad tank cars, and tank trucks.

The transportation segment involves many distinct operations that are potential air emission sources. Transportation vessels and vehicles are sources of evaporative hydrocarbon losses from crude and refined products. These losses occur during loading, while in transit, and from marine vessel ballasting operations. Also, pipeline pump stations are sources of fugitive emissions from equipment leaks and storage tanks can be sources of VOC emissions as a result of working and breathing losses.

Air Emission Characterization

Volatile organic compound air emissions occur from a variety of sources during the transportation of petroleum liquids.

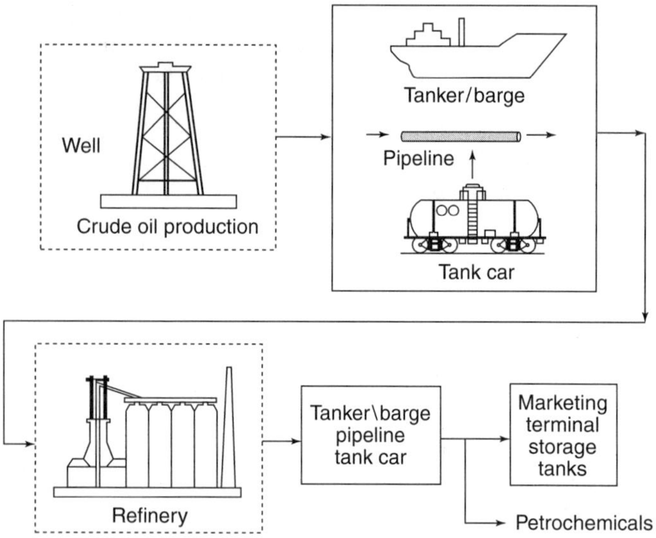

Figure 4. Petroleum Transportation and Distribution Systems.

Figure 5. View of Trans-Alaska Pipeline.

There are emission losses of VOCs and criteria pollutants from transportation vehicles, from storage tanks, and from leaks in equipment used in the transportation segment.

Table 1 presents estimates for the 1985 annual VOC emissions from each of these sources and provides estimates for the number of sources in each category. The emission estimates shown in Table 1 were derived from the EPA NAPAP emissions inventory.

Table 1. Transportation Segment Emission Sources of VOC Emissions

Source	Number of Sources
Loading (marine terminal)	52
Barges (product/crude)	3,800
Carrier (crude)	125
Carrier (product)	150
Equipment leaks at pump stations	
Pumps	8,500
Valves/flanges	76,800
Pump station storage tanks	3,523

Source: Ref. 1.

Emissions from Loading, Transit, and Ballasting Operations

Air emissions from railroad tank cars, tank trucks, and marine vessels are of three types: loading, transit, and marine vessel ballasting losses.

Loading Losses

Loading losses are the primary source of evaporative emissions from rail tank car, tank truck, and marine vessel operations. Loading losses occur as organic vapors in "empty" cargo tanks are displaced to the atmosphere by the liquid being loaded into the tanks. These vapors are a composite of (1) vapors formed in the empty tank by evaporation of residual product from previous loads, (2) vapors transferred to the tank in vapor balance systems as product is being unloaded, and (3) vapors generated in the tank as the new product is being loaded. The quantity of evaporative losses from loading operations is, therefore, a function of the following parameters:

- Physical and chemical characteristics of the previous cargo
- Method of unloading the previous cargo
- Operations to transport the empty carrier to a loading terminal
- Method of loading the new cargo
- Physical and chemical characteristics of the new cargo

The principal methods of cargo carrier loading are illustrated in Figures 6–8. In the splash loading method, the fill pipe dispensing the cargo is lowered only part way into the cargo tank. Significant turbulence and vapor–liquid contact occur during the splash loading operation, resulting in high levels of vapor generation and loss. If the turbulence is great enough, liquid droplets will be entrained in the vented vapors.

A second method of loading is submerged loading. Two types are the submerged fill pipe method and the bottom loading method. In the submerged fill pipe method, the fill pipe extends almost to the bottom of the cargo tank. In the bottom loading method, a permanent fill pipe is attached to the cargo tank bottom. During most of submerged loading by both methods, the fill pipe opening is below the liquid surface level. Liquid turbulence is controlled significantly during submerged loading, resulting in much lower vapor generation than encountered during splash loading.

The recent loading history of a cargo carrier is just as important a factor in loading losses as the method of loading.

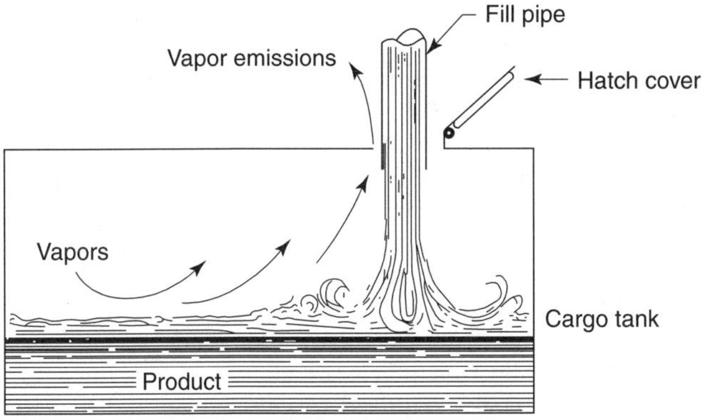

Figure 6. Splash Loading Method.

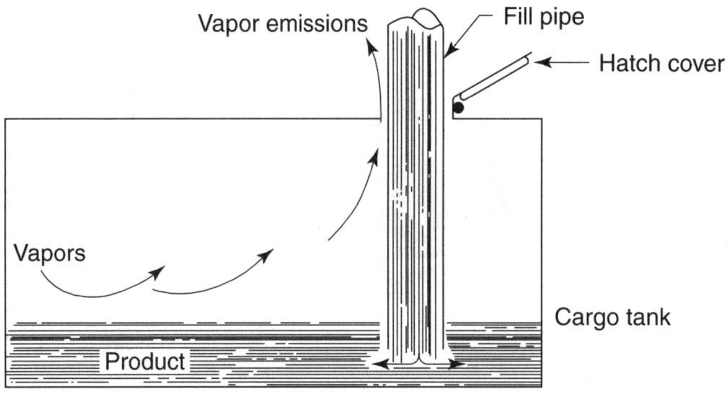

Figure 7. Submerged Fill Pipe.

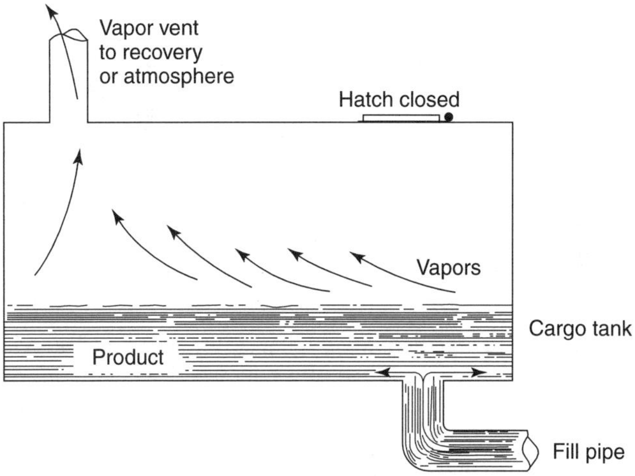

Figure 8. Bottom Loading.

If the carrier has carried a nonvolatile liquid such as fuel oil, or has just been cleaned, it will contain vapor-free air. If it has just carried gasoline and has not been vented, the air in the carrier tank will contain volatile organic vapors, which will be expelled during the loading operation along with newly generated vapors.

Cargo carriers are sometimes designated to transport only one product, and in such cases are practicing "dedicated service." Dedicated gasoline cargo tanks return to a loading terminal containing air fully or partially saturated with vapor from the previous load. Cargo tanks may also be "switch loaded" with various products, so that a nonvolatile product being loaded may expel the vapors remaining from a previous load of a volatile product such as gasoline. These circumstances vary with the type of cargo tank and with the ownership of the carrier, the petroleum liquids being transported, geographic location, and season of the year.

One control measure for vapors displaced during liquid loading is called "vapor balance service," in which the cargo tank retrieves the vapors displaced during product unloading at bulk plants or service stations and transports the vapors back to the loading terminal. Figure 9 shows a tank truck in vapor balance service filling a service station underground tank and taking on displaced gasoline vapors for return to the terminal. A cargo tank returning to a bulk terminal in vapor balance service normally is saturated with organic vapors, and the presence of these vapors at the start of submerged loading of the tanker truck results in greater loading losses than

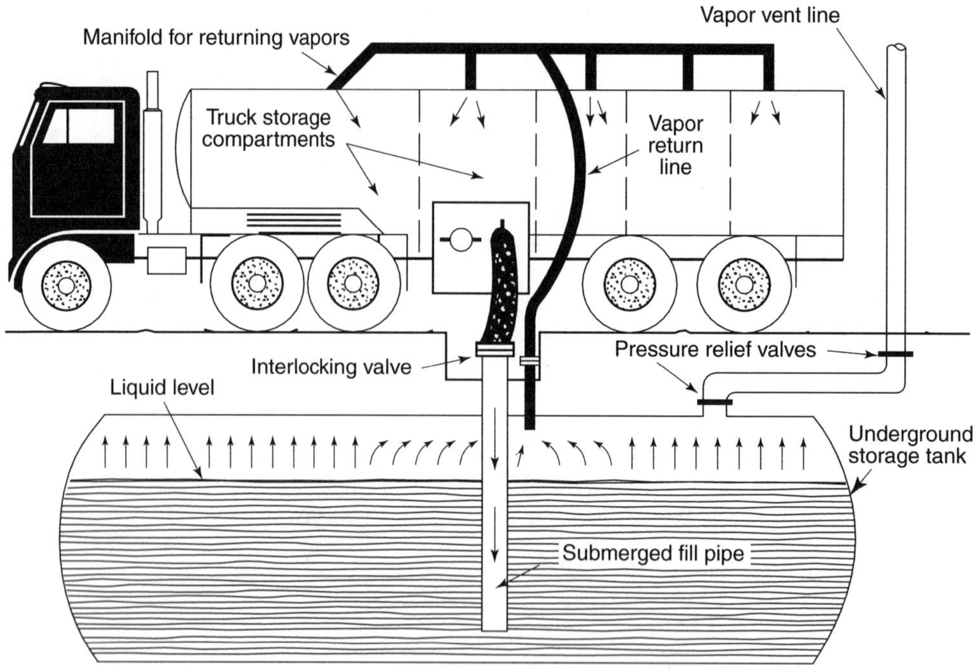

Figure 9. Tank Truck Unloading into a Service Station Underground Storage Tank and Practicing "Vapor Balance" form of Emission Control.

encountered during nonvapor balance, or "normal," service. Vapor balance service is usually not practiced with marine vessels, although some vessels practice emission control by means of vapor transfer within their own cargo tanks during ballasting operations, discussed below.

Emissions from loading petroleum liquid can be estimated (with a probable error of ±30%)[36] using the following expression:

$$L_{\mathrm{L}} = 12.46\frac{SPM}{T} \tag{1}$$

where L_{L} is the loading loss [pounds per 1000 gallons (lb/10^3 gal) of liquid loaded L]; S, a saturation factor (see Table 2); P, the true vapor pressure of liquid loaded [pounds per square inch absolute (psia) (see AP42-Chapter 7.1 for vapor pressures of petroleum products)]; M, the molecular weight of vapors [pounds per pound mole (lb/lb mole)]; and T, the temperature of bulk liquid loaded, °R (°F + 460).

The saturation factor, S, represents the expelled vapor's fractional approach to saturation, and it accounts for the variations observed in emission rates from the different unloading and loading methods. Table 2 lists suggested saturation factors.

Emissions from controlled loading operations can be calculated by multiplying the uncontrolled emission rate calculated in Eq. 1 by an overall reduction efficiency term:

$$\left(1 - \frac{\text{eff}}{100}\right)$$

where eff is the efficiency in percent.

The overall reduction efficiency should account for the capture efficiency of the collection system as well as both the control efficiency and any downtime of the control device. Measures to reduce loading emissions include selection of alternate loading methods and application of vapor recovery equipment.

Table 2. Saturation (*S*) Factors for Calculating Petroleum Liquid Loading Losses

Cargo Carrier	Mode of Operation	*S* Factor
Tank trucks and rail tank cars	Submerged loading of a clean cargo tank	0.50
	Submerged loading: dedicated normal service	0.60
	Submerged loading: dedicated vapor balance service	1.00
	Splash loading of a clean cargo tank	1.45
	Splash loading: dedicated normal service	1.45
	Splash loading: dedicated vapor balance service	1.00
Marine vessels[a]	Submerged loading: ships	0.2
	Submerged loading: barges	0.5

[a]For products other than gasoline and crude oil. For marine loading of gasoline, use factors from Table 3. For marine loading of crude oil, use Eqs. 2 and 3 and Table 4.

The latter captures organic vapors displaced during loading operations and recovers the vapors by the use of refrigeration, absorption, adsorption, and/or compression. The recovered product is piped back to storage. Vapors can also be controlled through combustion in a thermal oxidation unit, with no product recovery. Figure 10 demonstrates the recovery of gasoline vapors from tank trucks during loading operations at bulk terminals. Control efficiencies for the recovery units range from 90% to over 99%, depending on both the nature of the vapors and the type of control equipment used.[37,38] However, only 70–90% of the displaced vapors reach the control device, because of leakage from both the tank truck and collection system.[38] The collection efficiency should be assumed to be 90% for tanker trucks required to pass an annual leak test. Otherwise, 70% should be assumed.

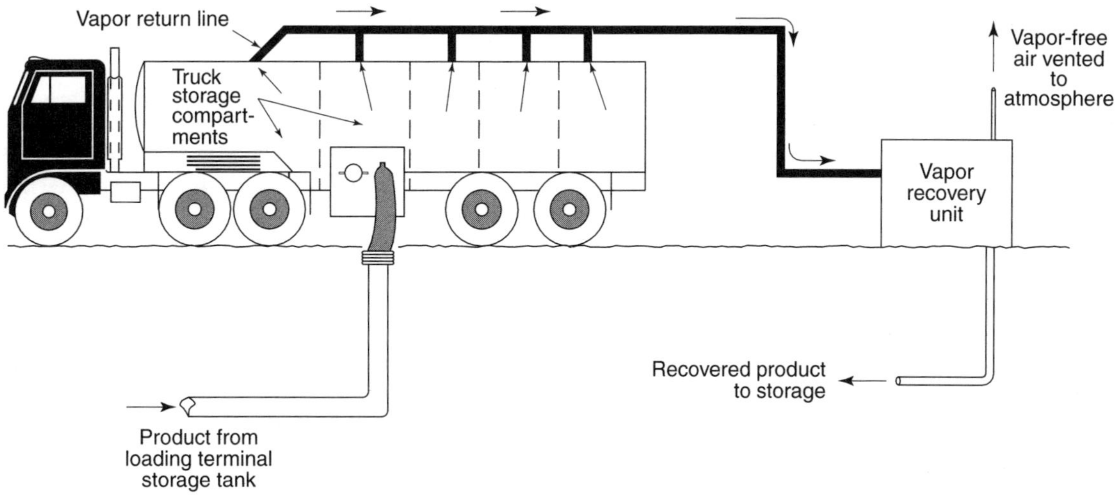

Figure 10. Tank Truck Loading with Vapor Recovery.

Sample Calculation

Loading losses (L_L) from a gasoline tank truck in dedicated vapor balance service and practicing vapor recovery would be calculated as follows, using Eq. 1:

Design basis:

Cargo tank volume is 8000 gal
Gasoline Reid vapor pressure (RVP) is 9 psia
Product temperature is 80°F
Vapor recovery efficiency is 95%
Vapor collection efficiency is 90% (for vessels passing annual leak test)

Loading loss equation:

$$L_L = 12.46 \frac{SPM}{T} \left(1 - \frac{\text{eff}}{100}\right)$$

where S is the saturation factor (see Table 2)—1.00; P, the true vapor pressure of gasoline = 6.6 psia; M, the molecular weight of gasoline vapors = 66; T, the temperature of gasoline = 540°C; and eff, the overall reduction efficiency (95% control × 90% collection) = 85%.

$$L_L = 12.46 \frac{(1.00)(6.6)(66)}{540} \left(1 - \frac{85}{100}\right) = 1.5 \text{ lb}/10^3 \text{ gal}$$

Total loading losses are:

$$(1.5 \text{ lb}/10^3 \text{ gal})(8.0 \times 10^3 \text{ gal}) = 12 \text{ lb}$$

Measurements of gasoline loading losses from ships and barges have led to the development of emission factors for these specific loading operations.[39] These factors are presented in Table 3 and should be used instead of Eq. 1 for gasoline loading operations at marine terminals. Factors are expressed in units

Table 3. (Metric And English Units). Volatile Organic Compound (VOC) Emission Factors for Gasoline Loading Operations at Marine Terminals[a]

		Ships/Ocean Barges[b]		Barges[b]	
Vessel Tank Condition	Previous Cargo	mg/L Transferred	lb/10^3 gal Transferred	mg/L Transferred	lb/10^3 gal Transferred
Uncleaned	Volatile[c]	315	2.6	465	3.9
Ballasted	Volatile	205	1.7	—[d]	—[d]
Cleaned	Volatile	180	1.5	ND	ND
Gas-freed	Volatile	85	0.7	ND	ND
Any condition	Nonvolatile	85	0.7	ND	ND
Gas-freed	Any cargo	ND	ND	245	2.0
Typical overall situation[e]	Any cargo	215	1.8	410	3.4

[a] Refs. 34, 40. Factors are for both VOC emissions (which excludes methane and ethane) and total organic emissions, because methane and ethane have been found to constitute a negligible weight fraction of the evaporative emissions from gasoline. ND, no data.
[b] Ocean barges [tank compartment depth about 12.2 m (40 ft)] exhibit emission levels similar to tank ships. Shallow draft barges [compartment depth 3.0–3.7 m (10–12 ft)] exhibit higher emission levels.
[c] Volatile cargoes are those with a true vapor pressure greater than 10 kilopascals (kPa) (1.5 psia).
[d] Barges are usually not ballasted.
[e] Based on observation that 41% of tested ship compartments were uncleaned, 11% ballasted, 24% cleaned, and 24% gas-freed. For barges, 76% were uncleaned.

of milligrams per liter (mg/L) and pounds per 1000 gallons ($lb/10^3$ gal).

In addition to Eq. 1, which estimates emissions from the loading of petroleum liquids, Eq. 2 has been developed specifically for estimating emissions from the loading of crude oil into ships and ocean barges:

$$C_L = C_A + C_G \tag{2}$$

where C_L is the total loading loss ($lb/10^3$ gal of crude oil loaded); C_A, the arrival emission factor, contributed by vapors in the empty tank compartment before loading ($lb/10^3$ gal loaded; see Note below); and C_G, the generated emission factor, contributed by evaporation during loading ($lb/10^3$ gal loaded). Note: Values of C_A for various cargo tank conditions are listed in Table 4.

This equation was developed empirically from test measurements of several vessel compartments.[39] The quantity C can be calculated using:

$$C_G = 1.84(0.44P - 0.42)\frac{MG}{T} \tag{3}$$

where P is the true vapor pressure of loaded crude oil (psia); M, the molecular weight of vapors (lb/lb mole); G, the vapor growth factor = 1.02 (dimensionless); and T, the temperature of vapors, °R (°F + 460).

Emission factors derived from Eq. 3 and Table 4 represent total organic compounds. Volatile organic compound emission factors (which exclude methane and ethane because they are exempted from the regulatory definition of "VOC") for crude oil vapors have been found to range from approximately 55 to 100 weight percent of these total organic factors. When specific vapor composition information is not available, the VOC emission factor can be estimated by taking 85% of the total organic factor.[35]

Ballasting Losses

Ballasting operations are a major source of evaporative emissions associated with the unloading of petroleum liquids at marine terminals. It is common practice to load several cargo tank compartments with sea water after the cargo has been unloaded. This water, termed "ballast," improves the stability of the empty tanker during the subsequent voyage. Although ballasting practices vary, individual cargo tanks are ballasted typically about 80%, and the total vessel 15–40%, of capacity. Ballasting emissions occur as vapor-laden air in the "empty" cargo tank is displaced to the atmosphere by ballast water being pumped into the tank. Upon arrival at a loading port, the ballast water is pumped from the cargo tanks before the new cargo is loaded. The ballasting of cargo tanks reduces the quantity of vapors returning in the empty tank, thereby reducing the quantity of vapors emitted during subsequent tanker loading. Regulations administered by the U.S. Coast Guard require that, at marine terminals located in ozone nonattainment areas, large tankers with crude oil washing systems contain the organic vapors from ballasting.[41] This is accomplished principally by displacing the vapors during ballasting into a cargo tank being simultaneously unloaded. In other areas, marine vessels emit organic vapors directly to the atmosphere.

Table 4. (English Units). Average Arrival Emission Factors, C_A, for Crude Oil Loading Emission Equation[a]

Ship/Ocean Barge Tank Condition	Previous Cargo	Arrival Emission Factor ($lb/10^3$ gal)
Uncleaned	Volatile[b]	0.86
Ballasted	Volatile	0.46
Cleaned or gas-freed	Volatile	0.33
Any condition	Nonvolatile	0.33

[a]Arrival emission factors (C_A) to be added to generated emission factors (C_G) calculated in Eq. 3 to produce total crude oil loading loss (C_L). Factors are for total organic compounds; VOC emission factors average about 15% lower, because VOC does not include methane or ethane.

[b]Volatile cargoes are those with a true vapor pressure greater than 10 kPa (1.5 psia).

Equation 4 has been developed from test data to calculate the ballasting emissions from crude oil ships and ocean barges:[39]

$$L_B = 0.31 + 0.20P = 0.01PU_A \tag{4}$$

where L_B is the ballasting emission factor ($lb/10^3$ gal of ballast water); P, the true vapor pressure of discharged crude oil (psia); and U_A, the arrival cargo true ullage, before dockside discharge, measured from the deck (ft); (the term *ullage* here refers to the distance between the cargo surface level and the deck level).

Table 5 lists average total organic emission factors for ballasting into uncleaned crude oil cargo compartments. The first category applies to "full" compartments wherein the crude oil true ullage just before cargo discharge is less than 1.5 m (5 ft). The second category applies to lightered, or short-loaded, compartments (part of cargo previously discharged, or original load a partial fill), with an arrival true ullage greater than 1.5 m (5 ft). It should be remembered that these tabulated emission factors are examples only, based on average conditions, to be used when crude oil vapor pressure is unknown. Equation 4 should be used when information about crude oil vapor pressure and cargo compartment condition is available. The following sample calculation illustrates the use of Eq. 4.

Sample Calculation

Ballasting emissions from a crude oil cargo ship would be calculated as follows, using Eq. 4:

Design basis:

Vessel and cargo description	80,000 dead-weight-ton tanker, crude oil capacity 500,000 barrels (bbl); 20% of the cargo capacity is filled with ballast water after cargo discharge. The crude oil has an RVP of 6 psia and is discharged at 75°F.
Compartment conditions	70% of the ballast water is loaded into compartments that had been fully loaded to 2 ft ullage, and 30% is loaded into compartments that had been lightered to 15 ft ullage before arrival at dockside.

Table 5. (Metric and English Units). Total Organic Emission Factors for Crude Oil Ballasting[a]

	Average Emission Factors			
	By Category		Typical Overall[b]	
Compartment Condition Before Cargo Discharge	mg/L Ballast Water	$lb/10^3$ gal Ballast Water	mg/L Ballast Water	$lb/10^3$ gal Ballast Water
Fully loaded[c]	111	0.9	129	1.1
Lightered or previously short loaded[d]	171	1.4		

[a]Assumes crude oil temperature of 16°C (60°F) and RVP of 34 kPa (5 psia). VOC emission factors average about 85% of these total organic factors, because VOCs do not include methane or ethane.
[b]Based on observation that 70% of tested compartments had been fully loaded before ballasting. May not represent average vessel practices.
[c]Assumed typical arrival ullage of 0.6 m (2 ft).
[d]Assumed typical arrival ullage of 6.1 m (20 ft).

Ballasting emission equation:

$$L_B = 0.31 + 0.20P + 0.01PU_A$$

where P is the true vapor pressure of crude oil (4.6 psia) (see AP42, Chapter 7.1); U_A, the true cargo ullage for the full compartments, = 2 ft, and true cargo ullage for the lightered compartments = 15 ft; and

$$\begin{aligned} L_B &= 0.70[0.31 + (0.20)(4.6) + (0.01)(4.6)(2)] \\ &\quad + 0.30[0.31 + (0.20)(4.6) + (0.01)(4.6)(15)] \\ &= 1.5 \text{ lb}/10^3 \text{ gal} \end{aligned}$$

Total ballasting emissions are:

$$(1.5 \text{ lb}/10^3 \text{ gal})(0.20)(500{,}000 \text{ bbl})(42 \text{ gal/bbl}) = 6{,}300 \text{ lb}$$

Since VOC emissions average about 85% of these total organic emissions, emissions of VOCs are about: (0.85)(6,300 lb) = 5,360 lb.

Transit Losses

In addition to loading and ballasting losses, losses occur while the cargo is in transit. Transit losses are similar in many ways to breathing losses associated with petroleum storage. Experimental tests on ships and barges[36] have indicated that transit losses can be calculated using:

$$L_T = 0.1\,PW \tag{5}$$

where L_T is the transit loss from ships and barges (lb/week-10^3 gal transported); P, the true vapor pressure of the transported liquid (psia); and W, the density of the condensed vapors (lb/gal).

Emissions from gasoline truck cargo tanks during transit have been studied by a combination of theoretical and experimental techniques, and typical emission values are presented in Table 6.[42,43] Emissions depend on the extent of venting from the cargo tank during transit, which in turn depends on the vapor tightness of the tank, the pressure relief valve settings, the pressure in the tank at the start of the trip, the vapor pressure of the fuel being transported, and the degree of fuel vapor saturation of the space in the tank. The emissions are not directly proportional to the time spent in transit. If the vapor leakage rate of the tank increases, emissions increase up to a point, and then the rate changes as other determining factors take over. Truck tanks in dedicated vapor balance service usually contain saturated vapors, and this leads to lower emissions during transit because no additional fuel evaporates to raise the pressure in the tank to cause venting. Table 6 lists "typical" values for transit emissions and "extreme" values that could occur in the unlikely event that all determining factors combined to cause maximum emissions.

In the absence of specific inputs for Eqs. 1–5, the typical evaporative emission factors presented in Tables 6 and 7 should be used. It should be noted that, although the crude oil used to calculate the emission values presented in these tables has an RVP of 5, the RVP of crude oils can range from less than 1 up to 10. Similarly, the RVP of gasolines ranges from 7 to 13. In areas where loading and transportation sources are major factors affecting air quality, it is advisable to obtain the necessary parameters and to calculate emission estimates using Eqs. 1–5.[18]

Storage Tank Emissions

Storage tanks at pipeline pump stations are sources of VOC emissions. Storage tank design, emissions, and emission controls are discussed in the E&P section. In general, storage tank emissions can be attributed to breathing and working losses. Breathing losses result from vapor volume and pressure variations caused by diurnal temperature changes. Working losses result from tank-filling operations where hydrocarbon vapors are displaced from the tank because of the rising liquid surface.

Fugitive Emissions

Pump station pumps, seals, valves, and flanges are source of VOC fugitive emissions, which are discussed in the refinery section.

Table 6. (Metric and English Units). Total Uncontrolled Organic Emission Factors for Petroleum Liquid Rail Tank Cars and Tank Trucks

Emission Source	Gasoline[a]	Crude Oil[b]	Jet Naphtha (JP-4)	Jet Kerosene	Distillate Oil No. 2	Residual Oil No. 6
Loading operations[c]						
Submerged loading—dedicated normal service[d]						
mg/L transferred	590	240	180	1.9	1.7	0.01
lb/10^3 gal transferred	5	2	1.5	0.016	0.014	0.0001
Submerged loading—vapor balance service[d]						
mg/L transferred	980	400	300	—[e]	—[e]	—[e]
lb/10^3 gal transferred	8	3	2.5	—[e]	—[e]	—[e]
Splash loading—dedicated normal service						
mg/L transferred	1,430	580	430	5	4	0.03
lb/10^3 gal transferred	12	5	4	0.04	0.03	0.0003
Splash loading—vapor balance service						
mg/L transferred	980	400	300	—[e]	—[e]	—[e]
lb/10^3 gal transferred	8	3	2.5	—[e]	—[e]	—[e]
Transit losses						
Loaded with product						
mg/L transported						
Typical	0–1.0	ND	ND	ND	ND	ND
Extreme	0–9.0	ND	ND	ND	ND	ND
lb/10^3 gal transported						
Typical	0–0.01	ND	ND	ND	ND	ND
Extreme	0–0.08	ND	ND	ND	ND	ND
Return with vapor						
mg/L transported						
Typical	0–13.0	ND	ND	ND	ND	ND
Extreme	0–44.0	ND	ND	ND	ND	ND
lb/10^3 gal transported						
Typical	0–0.11	ND	ND	ND	ND	ND
Extreme	0–0.37	ND	ND	ND	ND	ND

[a]Ref. 30. Gasoline factors represent emissions of VOC as well as total organics, because methane and ethane constitute a negligible weight fraction of the evaporative emissions from gasoline. VOC factors for crude oil can be assumed to be 15% lower than the total organic factors, to account for the methane and ethane content of crude oil evaporative emissions. All other products should be assumed to have VOC factors equal to total organics. The example gasoline has an RVP of 69 kPa (10 psia). ND, no data.
[b]The example crude oil has an RVP of 34 kPa (5 psia).
[c]Loading emission factors are calculated using Eq. 1 for a dispensed product temperature of 16°C (60°F).
[d]Ref. 30.
[e]Not normally used.

Fuel Combustion Emissions

Pumps, tankers, and barges are often powered by IC engines. These engines emit VOC, NO_x, SO_2, CO, and particulate matter as a result of fuel combustion.

Air Pollution Control Measures

Table 8 presents transportation-segment emission source and control options, as further discussed in the following. While specific emission factors are provided in Tables 2–7 for some of these sources, considerable uncertainty still exists for other sources due to the wide range of estimates and certainties for these emission control measures that precludes accurate estimation of control efficiencies in some cases.

Splash loading, submerged fill pipe loading, bottom loading, and vapor balance systems were discussed earlier. Additional techniques that can be utilized for control or capture are presented in the following paragraphs.

Refrigeration

Vapor recovery by refrigeration recovers hydrocarbon vapors by condensing at cryogenic temperatures and atmospheric pressure. Product vapors are first cooled in a dehydrator by direct contact with cooling coils, where most of the vapors are condensed and recovered. From the dehydrator, vapors are further cooled and condensed. The vapor-recovery efficiency of the refrigeration unit depends on the composition and concentration of the hydrocarbon vapors and the operating temperature of the condenser. Control efficiencies range from 85%[5] to 95%.[20]

Absorption

An absorption vapor-recovery unit recovers hydrocarbons by absorption into a lean oil stream. Vapors are compressed and fed to a packed-bed absorber where the hydrocarbons in the vapor are absorbed by the lean oil. The lean oil, which is fed

Table 7. (Metric and English Units). Total Organic Emission Factors for Petroleum Marine Vessel Sources[a]

Emission Source	Gasoline[b]	Crude Oil[c]	Jet Naphtha (JP-4)	Jet Kerosene	Distillate Oil No. 2	Residual Oil No. 6
Loading operations						
Ships/ocean barges						
mg/L transferred	—[d]	73	60	0.63	0.55	0.004
$lb/10^3$ gal transferred	—[d]	0.61	0.50	0.005	0.005	0.00004
Barges						
mg/L transferred	—[d]	120	150	1.60	1.40	0.011
$lb/10^3$ gal transferred	—[d]	1.0	1.2	0.013	0.012	0.00009
Tanker ballasting						
mg/L ballast water	100	—[e]	ND	ND	ND	ND
$lb/10^3$ gal ballast water	0.8	—[e]	ND	ND	ND	ND
Transit						
mg/wk-L transported	320	150	84	0.60	0.54	0.003
lb/wk-10^3 gal transported	2.7	1.3	0.7	0.005	0.005	0.00003

[a]Factors are for a dispensed product of 16°C (60°F). ND, no data.
[b]Factors represent VOC as well as total organic emissions, because methane and ethane constitute a negligible fraction of gasoline evaporative emissions. All products other than crude oil can be assumed to have VOC factors equal to total organic factors. The example gasoline has an RVP of 69 kPa (10 psia).
[c]VOC emission factors for a typical crude oil are 15% lower than the total organic factors shown, in order to account for methane and ethane. The example crude oil has an RVP of 34 kPa (5 psia).
[d]See Table 3 for these factors.
[e]See Table 5 for these factors.

Table 8. Transportation Segment Emission Sources and Emission Controls

Emission Source	Emission Control
Loading (marine terminal)	Submerged loading Refrigeration Compression absorption Thermal oxidation
Barges (product/crude)	Submerged loading Refrigeration
Carrier (crude)	
Ballasting losses	Vapor displacement into cargo tank
Transit losses	Refrigeration
Carrier (product)	
Ballasting losses	Vapor displacement into cargo tank
Transit losses	Refrigeration
Equipment leaks (pumps, flanges, valves)	Quarterly inspection and maintenance Monthly inspection and maintenance
Pump station storage tanks	Fixed roofs with conservation vents Internal floating roofs External floating roofs Refrigeration

Sources: Refs. 18, 20, 28–30.

countercurrently to the vapor stream, is cooled by refrigeration to improve efficiency. The removal efficiencies of absorbers depend on the ratio of the lean oil flow rate to the vapor flow rate. For gasoline loading vapors, the efficiency of vapor removal can be as high as 95%.[20]

Thermal Oxidation

Incineration operates by combusting hydrocarbon vapors to CO_2 and water. Incineration can be used to control vapors from product loading at marine terminals, but these systems require sophisticated instrumentation and can be very expensive to install and operate. The efficiency of incineration for controlling vapors at marine terminals is 99%.[20]

Compression

Compression–refrigeration–condensation units recover hydrocarbon vapors by condensation at low temperature and moderate pressure. Such units can pose a significant safety risk in marine loading of gasoline.

Inspection and Maintenance — Leakless Equipment

Pipeline pump station valves, pumps, and flanges are sources of emission losses. These losses can be controlled by regularly scheduled inspection and maintenance. Equipment is inspected for leaks with a vapor-detection device and leaking equipment is repaired. Also, new valves have been developed that significantly reduce emissions from valve stem packing.

Pump Station Storage Tank Controls

Pump station storage tanks are sources of VOC emissions. Storage tank emission controls are discussed in detail in the E&P section. In general, replacing fixed roofs on product storage tanks with internal floating roofs can reduce vapor losses by 60–99%, depending on the type of roof and seals installed and on the type of organic liquid in the storage tank.

REFINING

Introduction

The petroleum refining industry (SIC Code 29) involves numerous processes that convert crude oil into more than

2500 products, including liquified petroleum gas, gasoline, kerosene, jet fuel, diesel fuel, other fuel oils, lubricating oils, and feedstock for the petrochemical industry.[18] Petroleum refinery activities include the storage of crude at the refinery, petroleum handling and refining operations, and storage of the refined products prior to shipment (Figure 11).

As of January 1990, there were 189 operating refineries in the United States with a total crude capacity of 15.4 million barrels per calendar day.[21] Refineries are generally located in coastal areas, with the majority of the capacity (54%) located in Texas, Louisiana, and California. Refineries also are concentrated in the Chicago, Philadelphia, and Puget Sound areas. A geographical distribution of the U.S. refineries and crude capacities is presented in Table 9.

Process Description

Refining operations consist of separation processes, conversion processes, treating processes, feedstock and product handling, and associated auxiliary operations. The flow scheme at a particular refinery is determined by the composition of the crude oil and the chosen slate of products. An example of a complex refinery flow scheme is presented in Figure 12.

The operations associated with petroleum refining are described more fully in the following sections. Refining operations (separation, conversion, and treating) are discussed first, followed by a brief discussion of auxiliary operations, including wastewater treatment and cooling towers. Handling operations are discussed last and include equipment (tanks, pipes, pumps, and valves) employed throughout the refinery. Potential emission sources from each segment of refining are discussed and are summarized in Table 10. In general, these emission sources are either those resulting from the petroleum products (namely, VOC emissions) or those resulting from combustion sources at the refinery.

Volatile organic compound emissions from refinery operations can be characterized as of two types: process point source emissions and fugitive emissions. Process point source emissions are those emissions directly associated with or generated by a process unit. Process vents are an example of a point source emission. Fugitive emission sources are VOC emission sources not specifically generated by a particular process unit. Such emission sources are found throughout a refinery and may or may not be associated with a process unit. They include valves, flanges, pump and compressor seals, cooling towers, storage tanks, transfer operations, and wastewater treatment systems. Fugitive emissions also result from the evaporation of leaked or spilled hydrocarbon liquid and gases. Combustion sources at refineries result in emissions of SO_2, NO_x, and PM.

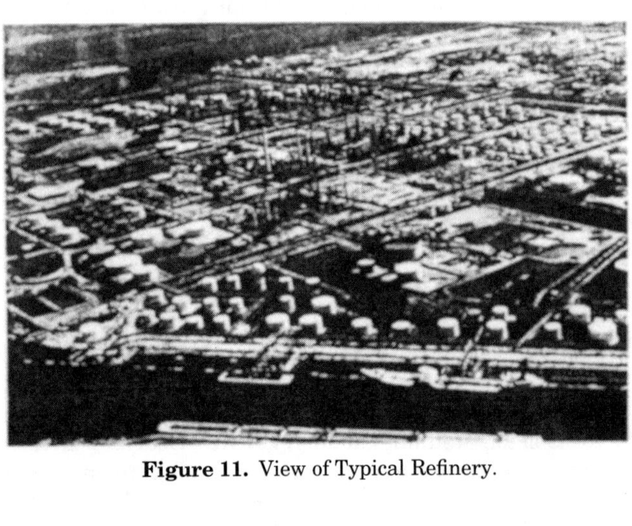

Figure 11. View of Typical Refinery.

Table 9. U.S. Refinery Capacity by State as of January 1990

State	Number of Plants	Crude Capacity (barrels per calendar day)
Texas	30	3,872,750
Louisiana	17	2,272,415
California	30	2,222,248
Illinois	6	920,600
Pennsylvania	9	856,000
New Jersey	6	494,250
Washington	7	492,675
Ohio	4	482,650
Indiana	4	426,900
Oklahoma	6	391,500
Mississippi	5	358,600
Kansas	8	348,125
Minnesota	2	285,600
Alaska	6	231,000
Kentucky	2	218,900
Wyoming	5	164,600
Utah	6	154,500
Delaware	1	140,000
Alabama	3	139,250
Montana	4	138,900
Hawaii	2	129,800
Michigan	4	123,700
Colorado	3	91,200
New Mexico	3	80,300
Tennessee	1	60,000
Arkansas	3	58,570
North Dakota	1	58,000
Virginia	1	53,000
New York	1	42,500
Georgia	2	35,500
Wisconsin	1	32,000
West Virginia	2	29,680
Oregon	1	15,000
Arizona	2	13,710
Nevada	1	4,500
Total	189	15,438,923

Source: Ref. 21.

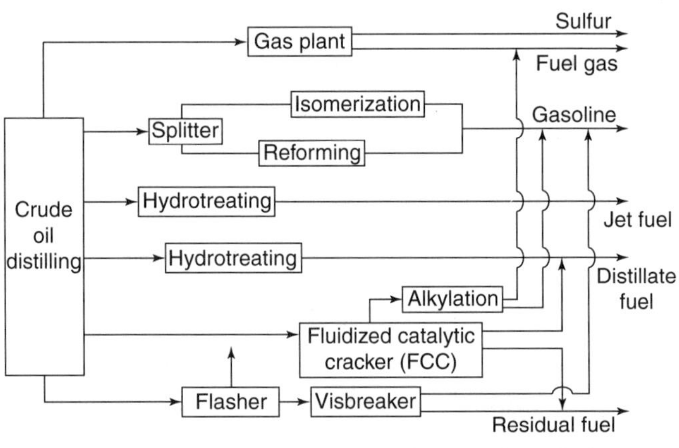

Figure 12. Flow Scheme for a Complex Petroleum Refinery.

Table 10. Emission Sources for Petroleum Refining

Source Category	Potential Emission Sources[a] Process Point	Process Fugitive	Area Fugitive
Crude separation	G,J,L	F,H,M,N	I
Light hydrocarbon processing	O,G	F,H	I
Middle and heavy distillate processing	G,O,P,R	F,H	Q
Residual hydrocarbon processing	B,G,K,O,R	H	I
Auxiliary processes	G	F,H	I

[a]Key to emission sources: B—visbreaker furnace; F—wastewater disposal (process drain, blow-down, cooling water); G—flare, incinerator process heater, boiler; H—storage, transfer, and handling; I—pumps, valves, compressors, fittings, etc.; J—absorber; K—process vent; L—distillation/fractionation; M—hot wells; N—steam ejectors; O—catalyst regeneration; P—evaporation; Q—catalytic cracker; R—stripper.
Source: Ref. 20.

Refining Operations

Separation Processes. Crude oil is a mixture of paraffinic, naphthenic, and aromatic hydrocarbon compounds with small amounts of impurities, including sulfur, nitrogen, and metals. Refinery separation processes separate the crude into common boiling-point fractions. The three petroleum separation processes are atmospheric distillation, vacuum distillation, and light ends recovery.

Refineries operate some process equipment under a vacuum, as in vacuum distillation, where heavy feedstocks (usually the bottom fraction from atmospheric distillation) are distilled under high temperatures. A steam-driven vacuum jet, sometimes coupled with a barometric condenser, is frequently used to produce and maintain a vacuum (see Figure 13). Light volatile compounds, such as H_2S, can be formed during vacuum distillation owing to the high temperatures used in the process. Those compounds that do not condense in the barometric condenser are discharged with the exhaust stream.

Conversion Processes. These processes are used to meet the specifications for high-octane gasoline, jet fuel, and diesel fuel, and to upgrade components such as residual oils and light ends to gasoline. Cracking, coking, and visbreaking processes are used to break large molecules into smaller molecules. Isomerization and reforming processes rearrange the structure of molecules to produce higher-value molecules of a similar molecular size. Polymerization and alkylation processes are used to combine small molecules into larger ones.

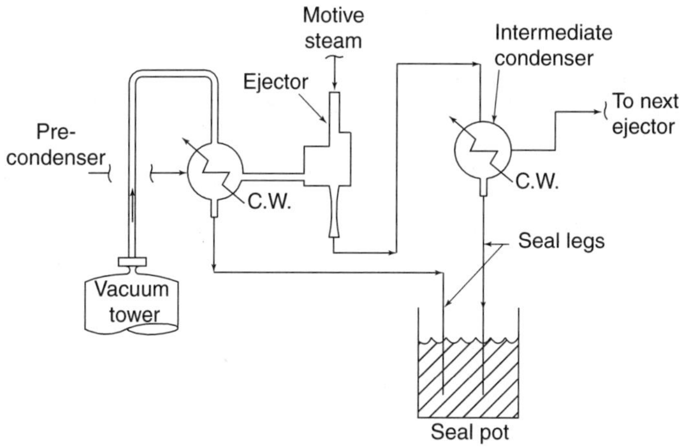

Figure 13. Vacuum System.

Coking is a low-pressure thermal cracking process that produces petroleum coke and gas oil from crude oil residue, decanted oils, and/or tar pitch products. Delayed coking is the most widely used process. In delayed coking, the process stream is initially fed through a fractionator, where the lighter material is flashed off. The bottoms subsequently are heated to 900–1100°F, and the vaporized material is sent to a coke drum. In the drum, some of the process stream condenses into coke, while more volatile components are returned to the fractionator for separation. Once the coke drum is filled with coke, the removal process is initiated. First the drum is cooled, and then the coke is cut from the drum using a high-pressure water jet, after which it is physically removed from the drum. Significant VOC and PM air emissions can occur during coke removal. Substances expected to be present in these emissions include benzene, toluene, xylene, H_2S, metals, and polycyclic aromatic hydrocarbons (PAH).

Catalytic cracking is commonly used to convert a gas oil feed stream into fuel gas, liquified petroleum gas (LPG), high-octane gasoline, and distillate fuel (see Figure 14). The fluid catalytic cracking (FCC) process uses a catalyst in the form of very fine particles that act as a fluid when charged with a vapor. Fresh feed is preheated in a process heater and introduced into the bottom of a vertical transfer line or riser as the catalyst vaporizes the feed, bringing both to the desired reaction temperature, 880–980°F. The high activity of modem catalysts causes most of the cracking reactions to take place in the riser as the catalyst and oil mixture flows upward into the reactor. The hydrocarbon vapors are separated from the catalyst particles by cyclones in the reactor. The reaction products are sent to a fractionator for separation. The spent catalyst falls to the bottom of the reactor and is steam stripped to remove adsorbed hydrocarbons as it exists the reactor bottom. The spent catalyst is then conveyed to a regenerator, where coke deposited on the catalyst during cracking reactions is burned off in a controlled combustion process with preheated air. Regeneration temperature is usually 1100°F to 1250°F. The catalyst is then recycled to be mixed with fresh hydrocarbon feed.

There are two sources of air emissions from catalytic cracking: process heater and catalyst regenerator. Emissions from the catalyst regenerator include hydrocarbons, oxides

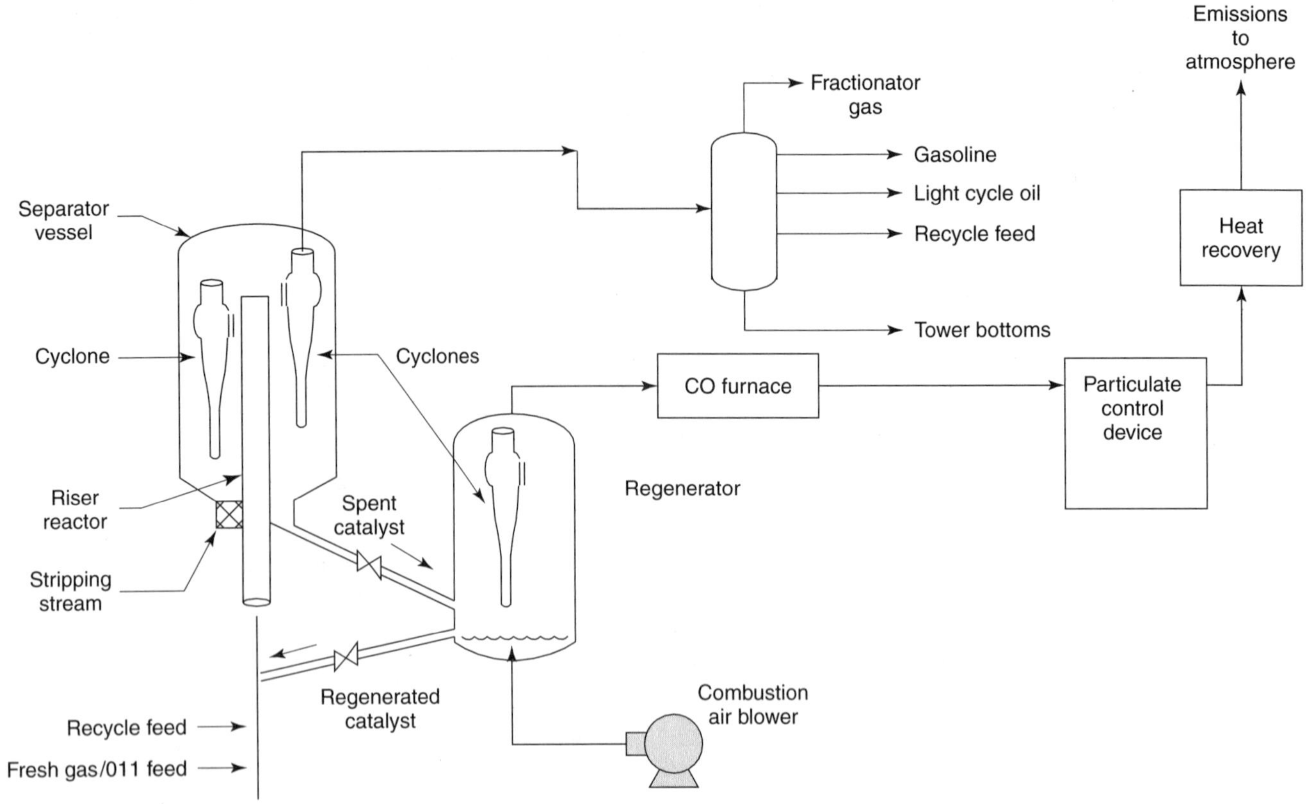

Figure 14. Schematic Diagram of a Fluid Catalytic Cracking Unit.

of sulfur, ammonia, aldehydes, NO_x, cyanides, CO, metals, and PM.

Catalytic reformers utilize catalysts to cause the rearrangement of existing molecules into higher-value molecules. Reformers are classified by method and frequency of catalyst regeneration. In the cyclic method, regeneration frequencies of one to two reactors per day are typical. In semiregeneration, all reactors are regenerated simultaneously at the end of a run, which may range from 3 to 24 months in length. In continuous regeneration, the catalyst is not regenerated in the reactor itself. A portion of the catalyst is continually routed to a separate regeneration system and returned to the top of the reactor. Regeneration is done at 200–500 psig and 900–1050°F. Bimetallic catalysts are the most commonly used, with platinum–rhenium on alumina the most popular combination. Catalysts used in reforming become contaminated with carbon, sulfur, nitrogen, water, and metals and must be regenerated to restore activity. During regeneration, the carbon is oxidized to mixed carbon oxides (CO, CO_2, COS) and hydrocarbons are combusted. Compounds that may be emitted during regeneration include ammonia, carbon tetrachloride, dibenzofurans, dioxins, metals, chlorides, and cyanides. Catalyst fines also may be emitted in small quantities, but this is generally minimized owing to the high cost of platinum catalyst. Continuous reforming has the highest potential for emissions.

Treating Processes. These processes are used to stabilize and upgrade petroleum products by separating them from less desirable products and by removing impurities. Deasphalting is a treating process used primarily for the separation of asphaltic and resinous materials from petroleum products. Treatment of products to remove other undesirable elements, such as sulfur, nitrogen, and oxygen, is achieved by hydrodesulfurization, hydrotreating, chemical sweetening, and acid gas removal. Desalting processes remove salt, minerals, grit, and water from crude oil prior to distillation. Asphalt blowing is used for polymerizing and stabilizing asphalt.

Hydrotreating is a commonly used treatment process that refers to a variety of processes. It is used for viscosity index improvement, desulfurization, denitrification, demetallization, removal of gum-forming compounds, and color improvement. In the hydrotreating process, the oil feed mixed with hydrogen is charged to a fixed-bed catalyst reactor. The catalyst generally has a cobalt or nickel–molybdenum/alumina base. When the catalyst is exhausted, it is regenerated at 700–1200°F using air with steam or flue gas. Common emissions are ammonia, H_2S, nickel, and VOCs. Catalyst handling operations produce negligible particulate emissions of coke fines and catalyst dust.

Sulfur-recovery processes are used to minimize SO_x emissions from refinery fuels and products. Elemental sulfur is produced as a by-product and is often sold commercially. In the widely used Claus process, H_2S is oxidized to form SO_2 and water:

$$H_2S + 1.5O_2 \rightarrow SO_2 + H_2O$$

Additional H_2S reacts with this SO_2 to form elemental sulfur and water:

$$2H_2S + SO_2 \rightarrow 3S + 2H_2O$$

Tail gas from a Claus sulfur-recovery unit contains a variety of pollutants, including SO_2, H_2S, other reduced sulfur compounds (COS and CS_2), CO, and VOCs.

Asphalt blowing refers to the blowing of air through vacuum distillation residuum to polymerize asphalt by oxidation and increase the hardness and melting point. Ferric chloride or phosphorus pentoxide sometimes is added to increase the reaction rate and to create a very-high-melting-point product. The feed is typically preheated to 400–600°F. The oils, containing a large quantity of polycyclic aromatic compounds, are oxidized by blowing heated air through a batch mixture or, in a continuous process, by passing hot air countercurrent to the airflow. The reaction is exothermic and quench steam is sometimes used for temperature control. Air emissions from asphalt blowing are primarily hydrocarbon vapors, including acetaldehyde, benzene, formaldehyde, metals, PAH, toluene, and sulfur compounds.

Distillate sweetening is accomplished by the conversion of mercaptans to alkyl disulfides in the presence of a catalyst. Sulfur is added to the sour distillate with a small amount of caustic and air. The mixture is then passed upward through a fixed-bed catalyst counter to the flow of caustic. Conversion may be followed by disulfide removal in an extraction process. In the conversion and extraction process, the sour distillate is washed with caustic and is then contacted in the extractor with a solution of catalyst and caustic. The extracted distillate is contacted with air to convert mercaptans to disulfides. After oxidation, the distillate is gravity settled, inhibitors are added, and the distillate is sent to storage. The major emissions are VOCs resulting from the contact between the distillate product and air in the "air-blowing" step.

Auxiliary Operations. A wide assortment of auxiliary operations not directly involved in the refining of crude oil is used in functions vital to the operation of the refinery. Examples include boilers, wastewater treatment facilities, cooling towers, and combustion devices.

Wastewater Treatment. Process wastewater containing H_2S, ammonia, and oil is termed *sour water*. It is stripped with steam to remove H_2S, ammonia, and light gases, which are then recovered or incinerated. Because of the low heating value of the process stream to be incinerated, auxiliary fuel is required to maintain the incineration combustion process. The most common emissions from the process are H_2S and ammonia.

All refineries employ some form of wastewater treatment prior to discharge to the environment or reuse in the refinery. The design of wastewater treatment systems is dictated by the diversity of refinery pollutants, including oil, phenols, sulfides, dissolved solids, and toxic chemicals. Although the exact configurations of wastewater treatment processes employed by refineries vary greatly, drain systems, oil–water separators, and air flotation systems are generally included. Drain systems consist of individual process drains, where oily water from various sources is collected, and junction boxes, which receive the oily water from several drains. Oil–water separators generally represent the first step in the treatment of refinery wastewater. The separation and removal of oil from the water are accomplished through density differences that cause the oil to rise to the top and enable it to be skimmed off. Air flotation usually follows the oil–water separator and is used to remove remaining oil (free and emulsified) and solids (colloidal and suspended) by introducing air bubbles into the wastewater by mechanical means. The majority of the air emissions, including VOCs and dissolved gases, evolved by the wastewater treatment are the result of evaporation from the surfaces of wastewater in open drains, separators, and aerated basins. The factors influencing emissions from these systems are wastewater composition, equipment design, and climatic factors.

Cooling Towers. Refineries use large quantities of water for cooling throughout the refining process. The cooling water absorbs heat from process streams in noncontact heat exchangers. Before the water can be reused, it must be cooled. Cooling towers are used to transfer waste heat from the cooling water to the atmosphere by allowing the water to cascade through a series of decks and slat-type grids. Water that enters the tower may contain hydrocarbons from leaking heat exchangers. Atmospheric emissions from the cooling tower can consist of VOCs and gases stripped from the cooling water as the air and water come into contact. They also may contain cooling water treatment chemicals such as chromium and chlorine, although the use of chromium is largely being phased out.

Combustion Sources. There are a number of different types of combustion sources in a typical refinery. In general, these sources can be categorized on the basis of the combustion device and the fuel. Refinery combustion sources include process heaters, boilers, gas turbines, and engines. Refinery fuels are typically residual oil, distillate oil, refinery gas, or natural gas. All the criteria pollutants, as well as VOCs and metals, are emitted from combustion sources. The quantity of these emissions is a function of the fuel type, as well as of the combustor type and size.

Boilers and process heaters are used throughout refineries to produce steam and to raise the temperature of feed materials to meet reaction or distillation requirements, respectively. In addition to the typical refinery fuels, CO-rich regenerator flue gas also may be used as a fuel. Gas compressors often are driven by IC engines and usually are fueled by natural or refinery gas. Gas turbines can fire a variety of fuels and typically are used to produce electricity. They also are used as cogeneration units that produce electricity and steam for process needs.

Handling Operations. The refinery feedstock and product handling operations consist of unloading, storage, blending, and loading activities. The types of equipment used in these operations include tanks, unloading and transfer equipment, valves, flanges, pumps, and compressors. Valves, flanges, pumps, and compressors also are used throughout the refining operations.

All refineries have a feedstock and product storage area, termed a *tank farm*, which provides surge storage capacity to assure smooth, uninterrupted refinery operations. Individual storage tank capacities range from less than 1000 to more than 500,000 barrels.[18] Storage tank, emissions, and emission control technologies are discussed under "Exploration and Production."

Valves and Flanges. Valves and flanges are found throughout a refinery. Under the influence of heat, pressure, vibration, friction, and corrosion, valves and flanges may develop leaks. Valves are subject to product leakage from the valve stem. These leaks may be liquid, vapor, or both, depending on the product carried and on the temperature.

A single pipeline connected through a manifold to a number of feeder lines may be used to carry several different products. To avoid the possibility of product contamination, it is customary to blind off the unused feeder lines. Blinding involves loosening a flanged pipe connection, inserting a flat, solid metal plate between the flanges, and retightening the connection. The installation or removal of blinds results in the spillage of some product. Depending on product volatility, temperature, drainage, and flushing facilities, a certain amount of the spilled product evaporates.

Transfer Operations. Although most refinery feedstocks and products are transported by pipeline, some are transported by truck, railcars, and marine vessels. The emissions from transfer operations and applicable control technologies are discussed under "Marketing" and "Transportation."

Pumps and Compressors. Liquid and gases in the refinery are moved about via pumps and compressors. These pieces of equipment can leak product through their seals, the point of contact between the moving shaft and the stationary casing.

Sampling. The operation of process units is periodically checked throughout the refinery by routine analyses of feedstocks and products. To obtain representative samples for these analyses, sampling lines must be purged, resulting in possible hydrocarbon vapor emissions.

Pressure-Relief Valves. Pressure-relief valves are used to prevent excessive pressures from developing in process vessels and lines. Leaks may develop because of the corrosive nature of the product or the valve may reseal improperly after blowoff. As routine maintenance and observation are often difficult because of a lack of accessibility and the operational difficulties involved with repair, relief valve installations can leak substantially before repair is completed.

The blowdown system allows for the safe disposal of liquid and gaseous hydrocarbon discharged from pressure-relief devices. Refinery equipment subject to planned and unplanned hydrocarbon discharges is manifolded into a collection unit. The blowdown is separated into liquid and vapor, with the liquid being recycled into the refinery and the vapor either recycled or flared. Uncontrolled blowdown emissions primarily consist of hydrocarbons, but also can include any of the other criteria pollutants.

Turnarounds and Tank Cleaning. The periodic maintenance and repair of process equipment and storage tanks are essential to refinery operations. To provide for safe vessel entry, all hydrocarbon liquid and vapor must be removed. Once the liquid has been cooled and pumped to storage facilities, the vessel is depressurized and/or purged with steam, nitrogen, or water to remove any remaining hydrocarbon vapors and odorous material.

Air Emissions Characterization

This section will quantify, where possible, refinery air emissions. The following are 1987 estimates of criteria pollutant emissions based on the 1970 level of control for the entire refining industry.[22] The use of 1970 emission control assumptions undoubtedly results in an overestimation of actual emission rates by perhaps an order of magnitude or more for some pollutants, especially VOCs. More up-to-date emission rate data are not available because even recent estimates rely largely on outdated emission factors to estimate releases.

Pollutant	Thousands of Tons per Year
PM	771
SO_2	1322
NO_x	220
VOC	991
CO	2533

Potential hazardous air pollutant emission sources from refining operations are presented in Table 11. Table 12 summarizes estimated benzene, toluene, xylene, butadiene, and trimethylbenzene emissions from refineries, as reported under SARA Title III community right-to-know requirements.

Air Pollution Control Measures

Control of refinery air emissions can be accomplished by process change, equipment change, procedure change, installation of control equipment, improved housekeeping, increased monitoring, and better equipment maintenance. Specific control devices are addressed in the following for the refining sector. Some combination of these often proves to be the most effective solution. Several of these controls also result in some form of savings.

Vapor Control Systems

Vapor control systems involve the collection of process or fugitive vapor emissions and their recycling or recovery for hydrocarbon content or fuel value or their destruction. In some instances, where economically feasible, the emission stream itself can be recycled to refinery fuel gas production or the fuel value of the emission stream can be recovered via combustion. Hydrocarbons in vapors can be destroyed via flares or incinerators. The specific configuration of vapor control systems depends on the specific configuration and economics of the refinery.

Flares

Flares are commonly used for the disposal of waste gases during process upsets (e.g., startup, shutdown) and emergencies. They are basically safety devices that also are used to destroy organic constituents in waste emission streams. Flares can be used for controlling almost any nonhalogenated VOC emission stream. They are designed and operated to handle large fluctuations in flow rate and VOC content. There are several different types of flares, but the prevalent refinery flare, illustrated in Figure 15, is an elevated steam-assisted flare. Flaring generally is considered a control option when the heating value of the emission stream cannot be recovered because of uncertain or intermittent flow, as in process upsets or emergencies. If the waste gas to be flared does not have sufficient heating value to sustain combustion, auxiliary fuel may be required. According to studies conducted by the EPA, 98% destruction efficiency can be achieved by steam-assisted flares when controlling emission streams with heat contents greater than 300 Btu/scf.

Table 11. Potential Hazardous Air Pollutants (HAPs) for Petroleum Refining

Process	Potential HAPs			
	Organic		Inorganic	
	Vapor	Particulate	Vapor	Particulate
Refining operations[a]				
Crude separation	a,b,d,e,f,g,h,i,j,k,l,m,o,A,B,C,D,E,F,J	o	c,m,t,u,v,x,y,L	p,I,Q,R
Light hydrocarbon processing	g,h,i,n,N,O,P	R	t,v	G,H,Q
Middle and heavy distillate processing	a,d,e,f,g,h,i,j,k,l,F,J,K,O,P,S,T	o,R	m,t,u,v,x,y,L	p,q,G,H,I,Q,U
Residual hydrocarbon processing	a,d,e,f,g,h,i,j,k,l,n,F,J,M,N,P,S,T	o,R	m,s,t,u,v,x,y,L	p,q,G,H,I,Q,U
Auxiliary processes	a,b,d,e,f,g,h,i,j,k,l,n,A,B,C,D,J,K,M,T	o,R	c,m,s,u,y,L	p,q,r,z,I
Combustion processes[b]				
Oil combustion	14	19	13,17,27	1,2,5,6,8,9, 10,11,15,16,18, 20,22,24
Natural gas combustion	14	19		
Gasoline combustion	12,14	12,19	17	15
Diesel combustion	12	12,19		6,18

[a]Pollutant key: a—maleic anhydride, b—benzoic acid, c—chlorides, d—ketones, e—aldehydes, f—heterocyclic compounds (e.g., pyridines) g—benzene, h—toluene, i—xylene, j—phenols, k—organic compounds containing sulfur (sulfonates, sulfones), l—cresol, m—inorganic sulfides, n—mercaptans, o—polynuclear compounds (benzo pyrene, anthracene, etc.), p—vanadium, q—nickel, r—lead, s—sulfuric acid, t—hydrogen sulfide, u—ammonia, v—carbon disulfide, x—carbonyl sulfide, y—cyanides, z—chromates, A—acetic acid, B—formic acid, C—methylethylamine, D—diethylamine, E—thiosulfide, F—methyl mercaptan, G—cobalt, H—molybdenum, I—zinc, J—cresylic acid, K—xylenols, L—thiophenes, M—thiophenol, N—nickel carbonyl, O—tetraethyl lead, P—cobalt carbonyl, Q—catalyst fines, R—coke fines, S—formaldehyde, T—aromatic amines, U—copper.

[b]Pollutant key: 1—arsenic, 2—antimony, 5—barium, 6—beryllium, 8—cadmium, 9—chromium, 10—cobalt, 11—copper, 12—dioxin, 13—fluoride, 14—formaldehyde, 15—lead, 16—manganese, 17—mercury, 18—nickel, 19—polycyclic organic matter (POM), 20—vanadium, 22—radionuclides, 24—zinc, 27—chlorine.

Source: Ref. 20.

Table 12. Estimated Emissions from Petroleum Refinery

Chemical	Point (tons/year)	Fugitive (tons/year)	Total (tons/year)
Benzene	114	29	142
Toluene	437	111	548
Xylene (total)	31	1751	1782
Butadiene	3	1	4
Trimethyl-benzene (1,2,4)	310	141	452

Source: Ref. 32.

Incinerators

There are two types of incinerators—thermal and catalytic. Thermal incinerators are used to control a wide variety of continuous VOC emission streams. Thermal incineration is preferable to flaring when halogenated or sulfur-bearing compounds are present owing to the corrosive properties of these compounds. Destruction efficiencies up to 98–99+% are achievable with thermal incineration. Although they accommodate minor fluctuations in flow, thermal incinerators are not well suited to streams with highly variable flow because of reduced residence time and poor mixing during increased flow conditions. Thermal incineration is typically applied to emission streams that are dilute mixtures of VOCs and air. In such cases, due to safety considerations, the concentration of the VOCs is limited in some instances to 25% of the lower explosive limit (LEL). Thus if the VOC concentration is high, dilution may be required. When emission streams controlled by thermal incineration are dilute (i.e., low heat content), supplementary fuel is required to maintain the desired combustion temperature. Fuel requirements may be reduced by recovering the energy contained in the hot flue gases from the incinerator.

Catalytic incinerators are similar to thermal incinerators in design and operation, but they employ a catalyst to increase the reaction rate. Since the catalyst allows the reaction to take place at lower temperatures, significant fuel savings are possible. Destruction efficiencies of up to 95% are generally achieved with catalytic incineration. Catalytic incineration is not as broadly applicable as thermal incineration because of catalyst sensitivity to pollutants and process conditions.

Refinery Fuel Gas

Existing boilers or process heaters can be used to control emission streams by recovering the fuel value while destroying the VOCs. This is accomplished by diverting the streams to the refinery fuel gas system or directly into the firebox. When used as emission control devices, boilers or process heaters can provide destruction efficiencies of greater than 98% at a small capital cost. Operating costs are reduced as the recovery of the emission stream heat content reduces fuel requirements. Typically, emission streams are controlled in boilers or process heaters and used as supplemental fuel only if they have sufficient heating value (greater than 150 Btu/scf). In some instances, emission streams with high heat content may be the main fuel to the process heater or boiler. Note that emission streams with low heat content can also be burned in boilers or process heaters when the flow rate of the emission stream is small compared with that of the fuel–air mixture. There are some limitations on the application of boilers or process

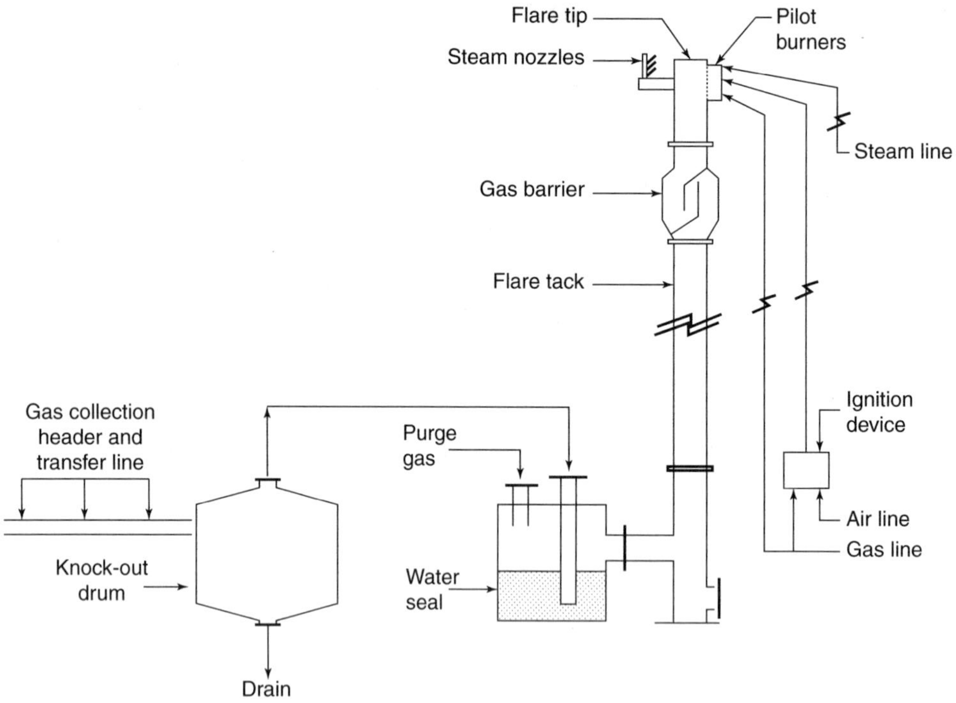

Figure 15. Steam-Assisted Elevated Flare System.

heaters as emission control devices. Because these combustion devices are essential to the operation of the refinery, only those emission streams that will not reduce their performance or reliability can be controlled. Streams not suitable for control include those with varying flow rate and/or heating value, high-volume/low-heating-value streams, and streams with corrosive compounds.

Hydrocarbon Reuse

The ultimate vapor-recovery process recycles the emission stream to the light ends recovery unit (gas processing), where it is recovered as a salable product, usable feedstock, or gasoline blending component. This requires compressing and separating, by distillation, the stream into its respective components.

Covers

Covering a liquid surface or the vapor space above a liquid surface suppresses VOC emissions to the atmosphere. Examples of control by covering are floating roof tanks (external and internal) and wastewater systems. Tank applications are discussed in detail under "Exploration and Production." An equipment-based New Source Performance Standard was promulgated in November 1988 for refinery wastewater systems. The control of wastewater treatment systems requires covering sources where emission generation is greatest, namely, process drains, junction boxes, and oil–water separators. By suppressing emissions through the separator, more hydrocarbon can be recovered as a liquid and recycled back to the refinery. Covering a drain (Figure 16) involves either a physical cover at ground level or a liquid seal in the drain pipe. Emission reductions of 40–50%[23] are achievable by drain seals. Junction boxes (Figure 17) require venting to prevent siphoning and vapor locks.

Vents on junction boxes should be at least 4 in. in diameter and 3 ft in length. To minimize VOC emissions, the vent pipe must also have a water seal. Control efficiencies can be assumed to be equal to those of drain seals.

Any VOC emissions from oil–water separators are controlled by installing a fixed or floating roof. A vapor space under a fixed roof may constitute an explosion or fire hazard. In order to eliminate this problem, the vapor space can be blanketed with either plant gas or an inert gas such as nitrogen. Floating roofs eliminate much of the vapor space, thus greatly reducing the potential for volatilization from the oil layer. Emission reduction estimates based on qualitative information range from 90% to 98%.[23] Data developed by Litchfield indicate that 85% is representative of the emission reduction achievable by a simple fixed or floating roof.[23] The most obvious factor affecting performance is the degree of maintenance.

In addition to covering components in the wastewater system, the system can be completely closed and vented to a combustion device, such as a flare. Emission reductions depend on the efficiency of the control device; for instance, flares would result in 98% reduction.

Inspection and Maintenance

Improved maintenance—including scheduled inspection and monitoring, improved housekeeping, and improved employee training—is a very practical method for reducing hydrocarbon emissions and alleviating odor problems. Moreover, it is often the only control method for some sources, such as valves, relief valves, seals, cooling towers (heat exchanger leaks), and sampling operations. For nearly all sources, especially process drains, wastewater separators, treating units, blind changing, and loading operations, employee awareness of the problem will reduce emissions.

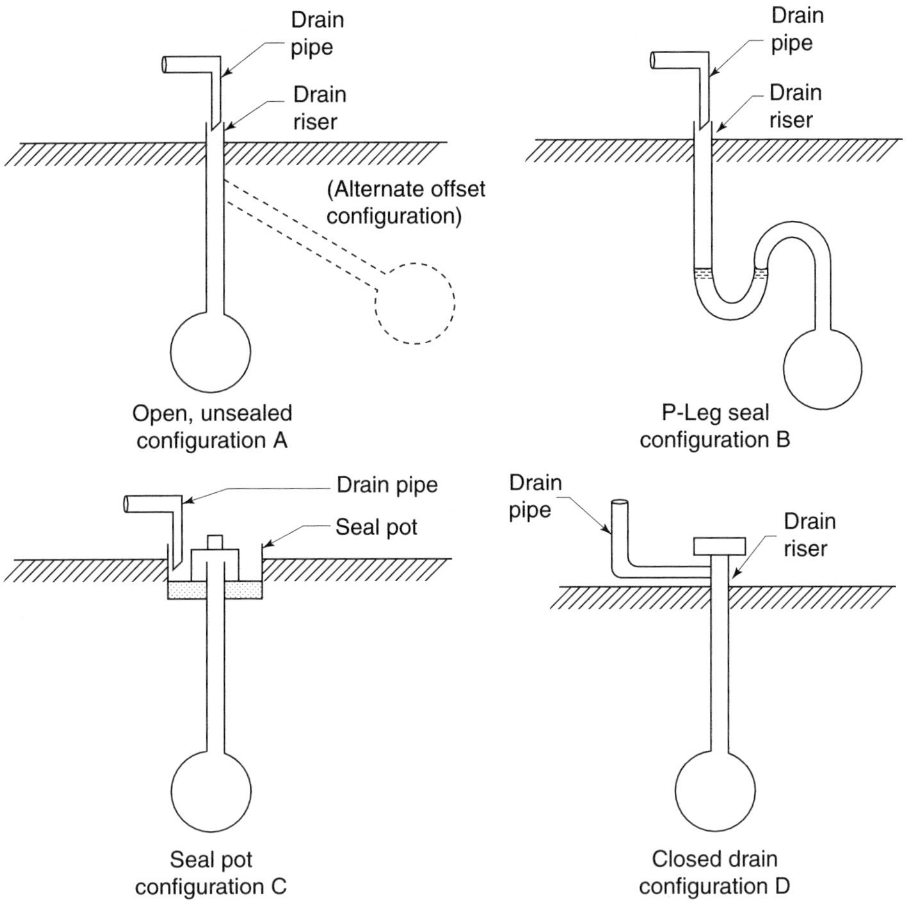

Figure 16. Types of Individual Refinery Drains for Oily Wastewater.

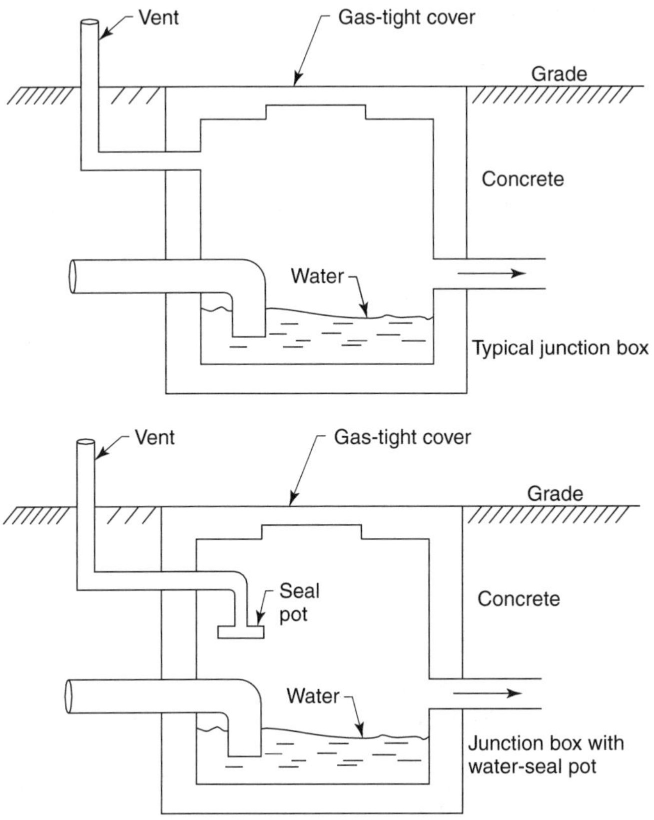

Figure 17. Refinery Drain System Junction Boxes.

In-Line Sampler

In-line sampling allows the collection of a representative process sample without having to flush any hydrocarbon to the process drain. Typically, small piping/tubing is run parallel to the process piping. A three-way valve is used for sample collection, thus eliminating any "dead legs."

Rupture Disks

Rupture disks as a control device are used to protect relief valves from the corrosive process environment. They are typically thin metal disks located on the pressure side of the relief valve. They are designed to burst at the relief valve setting. Owing to their "one-time" use, rupture disks are applicable for relief valves that are expected to be vented in emergencies only. They also can be used in place of relief valves in certain applications.

New Valve Technology

New valve technology has been developed that can significantly reduce pollutant emissions from the valve stem packing. It includes new types of packing material for valve stems and new seal technologies. New valve designs have been developed for some applications to prevent leaks by eliminating packing material leak sources around valve stems or by encasing potential leak sources in a "bellows" structure to trap emissions. The new "leakless" valve designs do have limitations of applicability. Some designs cannot withstand the high temperatures and pressures of many refinery process streams. Other designs are limited to valve sizes of 6–8 in. in diameter and, therefore, are applicable only to smaller line sizes in refineries. For refineries, the most widely used control technique to minimize leakage from valves is an effective schedule of inspection and preventative maintenance.

Mechanical Seals

Mechanical seals are used to control product emissions from centrifugal pump and compressor glands. A simple mechanical seal consists of two rings with wearing surfaces at right angles to the shaft. One ring is stationary, while the other is attached to the shaft and rotates with it. A spring and the action of the fluid pressure keep the two faces in contact. Lubrication of the wearing faces is effected by a thin film of the material being pumped. The wearing faces are precisely finished to ensure perfectly flat surfaces.

A pressure seal can further reduce emissions. A liquid that is less volatile or dangerous than the product being pumped is introduced between a dual set of seals. Since this liquid is maintained at a higher pressure than the product, some of it passes by the seal and into the product. The pressure differential prevents the product from leaking outward and the sealing liquid provides lubrication. As some of the sealing liquid passes the outer seal (hence the need for low volatility), a means should be provided for its disposal.

Quick Change Blind/Manifold Design

Several devices have been developed to reduce spillage, such as Hamer and Greenwood blinds. These "line" blinds do not require a complete break of the flange connection as "slip" blinds do, but rely on a gear mechanism to release the plate. Combinations of these devices in conjunction with gate valves allow changing of the line blind while the line is under pressure from either direction.

In addition, during design stages, the manifold can be designed to minimize the need for blinding, as well as the quantity of liquid that will be spilled. With highly volatile products, water or nitrogen can be used to displace the product, precluding any product spillage during the blind change.

Wetting

Wetting down the coke is used as a PM control in the decoking process.

New Gaskets, Bolts, and Welding

Replacement of leaking gaskets on existing pipe and valve flanges and minimizing the number of flanges with the use of welded pipe are methods of reducing VOC emissions from flanges.

Steam Stripping

Steam stripping is an effective control for removing VOCs from process streams. Two examples within a refinery include FCC catalyst stripping and sour-water treating. Spent FCC catalyst is steam stripped to remove absorbed hydrocarbons at the reactor exit. Process wastewater is stripped with steam to remove H_2S, ammonia, and light gases. The process is generally conducted in the sour-water stripper. The sour water and steam are fed into the column. The stripped water and condensed steam are routed to the wastewater system and the undesirable constituents are incinerated.

Carbon Monoxide Boiler

Depending on the refinery, the catalytic cracker may or may not have a CO boiler. Carbon monoxide boilers are used to recover the energy contained in the catalyst regeneration off-gases. The recovered energy is used as process heat for various refinery processes. The fuel used in the CO boiler consists of the process gas from the catalyst regenerator and an auxiliary fuel source. The process gas may contain up to 5–10% CO.[24] Combusting the gas in the boiler produces heat and reduces emissions of CO and VOCs to negligible levels.

Cyclone

Cyclone separators are used for catalyst dust collection in the upper section of both FCC unit reactors and regenerators. The cyclones are employed as a single unit or in multiple two- or three-stage series units. In general, high-efficiency cyclones have dust collection efficiencies of over 90% for particle sizes of more than 15 μm.[25] The efficiency drops off rapidly for particles less than 10 μm.

Electrostatic Precipitators

Electrostatic precipitator (ESP) particle removal is accomplished by charging the particles, collecting the particles, and transporting the collected particles into a hopper. An ESP is very sensitive to the aerosol density and the electrical resistivity of the material, but is less sensitive to particle size. The electrical resistivity of the particles influences the drift velocity or the attraction between the particles and the collecting plate. A high resistivity will cause a low drift velocity, which will decrease the overall collection efficiency. Used to control PM emissions on FCC units, ESPs achieve efficiencies as high as 80–85%.[25]

Scrubbers

Scrubbers are used in refineries to separate and purify gaseous streams containing high concentrations of VOCs, SO_2, and PM. Examples of scrubbing within a refinery include applications on the asphalt blowing airstream and on the FCC regenerator off-gas. A common system on FCC regenerators is a caustic scrubber, where particulate removal and sulfur oxides absorption take place in a venturi scrubber. Particulate is removed by inertial impaction of the scrubbing liquid with the entrained particles. The sulfur oxides absorption reactions that take place are as follows:

$$2NaOH + SO_2 \rightarrow Na_2SO_3 + H_2O$$

Sulfur dioxide and PM removal efficiencies of 93–98% have been recorded.

Combustion Controls

All the criteria pollutants and several potential hazardous air pollutants (HAPs) are emitted from combustion sources within refineries. Sulfur oxides can be controlled by fuel selection, fuel desulfurization, or flue gas treatment. Carbon monoxide, hydrocarbons, and PM can be minimized by more efficient combustion. Nitrous oxides can be controlled by combustion modification or by flue gas treatment. Controls specific to each refinery combustion source are described below.

Internal-Combustion Engine

The VOC, CO, and NO_x emissions can be controlled by a three-way catalyst (similar to a catalytic converter in a motor vehicle) on a rich-burn engine (fuel-rich combustion). The NO_x emissions can be controlled by selective catalytic reduction (SCR) on a lean-burn engine. Catalyst systems generally are designed for 80% reduction efficiencies. Precombustion chambers, which control the air-to-fuel ratio in stages, are another NO_x-reduction technique for gas-fired engines. Levels of NO_x as low as 1.5 g/bhp h have been reported. Injection timing retard also reduces NO_x production. The magnitude of the reduction may vary considerably among engine types. In general, reductions of 20–34%, with corresponding 1–4% fuel consumption penalty and slight increases of VOCs and CO, are obtained.

Combustion Turbines

Wet injection using either water or steam is the most prevalent NO_x-reduction technique for turbines. Depending on turbine type and size, NO_x emission levels of 25–50 ppm at 15% oxygen are obtainable. Selective catalytic reduction is also applicable to turbines for NO_x reductions of generally 80%. Coupling SCR with wet injection can result in NO_x emission levels as low as 5 ppm.

Heaters and Boilers

In addition to three-way catalysts and SCR, selective noncatalytic reduction processes can be applied to process heater and boiler flue gas for NO_x control. The process using ammonia can achieve 40–60% reduction, and that using urea can achieve 20–80% reduction, depending on temperature. Low-NO_x. burners, which stage the mixing of air and fuel to reduce flame temperature, result in NO_x reductions of 30–40% over conventional burners.

MARKETING

Introduction

The marketing of petroleum products at wholesale and retail facilities involves an extensive distribution network. Of the more than 2500 products produced by the petroleum industry, the major products distributed in bulk include motor gasoline and distillates such as diesel fuel, jet fuel, and heating oil.

Gasoline and other large-volume transportation fuels are distributed primarily by pipeline, ship, and barge to bulk terminals. From bulk terminals, tank trucks generally bring the product to a bulk plant, service station, or commercial customer. Specialty products, such as solvents, thinners, and lubricating oils, as well as some fuels, may be transported by railroad tank cars directly from the refinery. In addition, many products, such as lube oils and solvents, may be put in drums and shipped by truck to warehouses, dealers, or customers.

Process Description

Gasoline and distillate products are made at refineries and sent by pipeline, ship, and barge to bulk storage terminals. The products are stored in large tanks before being dispensed into tank trucks for delivery to customers. Facilities handling more than 20,000 gallons of gasoline per day are considered bulk terminals, but the largest facilities can throughput more than 1 million gallons of petroleum products per day. There were an estimated 637 bulk terminals in the United States in 1987.[26]

Some products loaded at bulk terminals are delivered by truck to bulk plants, which are small storage facilities located close to their customers. Bulk plants have limited storage capacity and are equipped with a small loading rack for dispensing products into local delivery trucks. They deliver fuel to such customers as service stations, commercial accounts, and farmers. There were an estimated 9391 bulk plants[26] nationwide (1987), with a typical daily throughput of 5000 gallons per day.

Tank trucks deliver motor fuels to service stations. The fuels are stored in underground tanks and then pumped directly into motor vehicle fuel tanks using a metering dispenser equipped with a flexible hose and nozzle. There are an estimated 160,000 service stations[27] in the United States in 1987.

The general distribution network in the petroleum marketing sector is outlined in Figure 18. During operations at marketing facilities, emissions of VOCs can occur as a result of product transfer (e.g., loading), during product storage, and

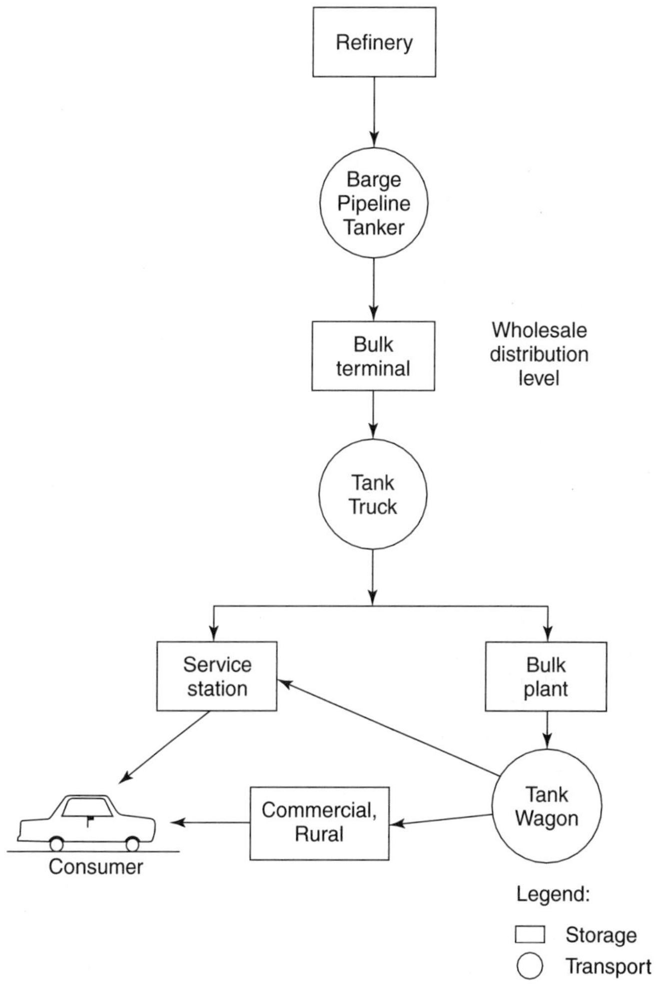

Figure 18. Gasoline Marketing in the United States.

from pump, valve, and flange seals and product spillage. Combustion sources in the marketing segment include vehicles transporting products that emit NO_x, SO_2, CO, and PM.

Air Emission Characterization

In this section, air emissions of VOCs and criteria pollutants are discussed separately. Few data are available to speciate VOC emissions into components of toxic air pollutants that may be emitted.

Volatile Organic Compounds

Emission of VOCs during marketing operations can be broken down into three major categories: storage tanks, truck loading, and service stations.

Storage Tanks

Storage tank design, emissions, and emission controls are discussed in detail in the E&P section. The greatest potential for VOC emissions comes from motor gasoline storage. Diesel and heating fuels have relatively low vapor pressures and are not as significant sources of releases from storage tanks.

Truck Loading Operations

Vapors containing a mixture of VOCs and air are emitted when tank trucks are loaded with motor gasoline or if a truck contains gasoline vapors from a previous load. As product is loaded, it displaces the vapors in the tank truck, and these vapors are vented to the atmosphere or to a vapor processing system.

Gasoline tank trucks are normally divided into compartments, with a loading dome hatchway at the top of each compartment. The method of loading affects the amount of emissions generated. Loading can be by top splash or submerged fill through the dome using a fill tube or by bottom filling. Vapor collection equipment for tank trucks includes piping to capture vapors from the tank being filled back into the tank truck. The vapors are then discharged into the vapor collection systems at bulk terminals when the tank track returns for product loading.

Service Station Operations

There are emissions from three sources at service stations: during the filling of underground tanks, underground tank breathing and emptying losses, and vehicle refueling losses. Splash filling losses and uncontrolled displacement losses during vehicle refueling are the largest emission sources at service station operations.

Tank trucks load gasoline into underground storage tanks at service stations. This is called Stage I refueling. Typically, splash or submerged filling methods are used. Vapor balance is utilized to control emissions by returning displaced vapors to the tank truck rather than allowing them to be emitted to the atmosphere.

A second source of VOC emissions from gasoline service stations is breathing losses from underground tanks. Breathing losses vary by the frequency with which gasoline is withdrawn from the tank and also with barometric pressure. Stage II refueling is the process of loading fuel into vehicles. Vehicle refueling produces evaporative emissions (i.e., venting of vapors) when vapors are displaced from the automobile tank by dispensed liquid gasoline and from spillage.

Air Toxics Emissions

The EPA has identified potential hazardous air pollutants that it is reviewing for possible emissions regulations. The agency compiled emissions profiles of these chemicals from various emissions sources by speciating the VOC emissions from these sources. A synopsis of the distribution of these emissions is presented in Table 13. Currently, little information is available on potential hazardous air pollutants from the marketing sector.

Criteria Pollutant Emissions

Criteria pollutant emissions (NO_x, SO_x, and CO) are from combustion sources related to motor vehicle transport in the marketing sector. Motor vehicles transporting product to bulk plants, bulk terminals, or service stations are sources of emissions from IC engines.

Air Pollution Control Measures

Table 13 presents marketing segment emissions sources and their control options. Several controls are available and in use to control VOC emissions from marketing segment sources.

Storage Tank Controls

Covering a liquid surface or the vapor space above a liquid surface suppresses VOC emissions to the atmosphere. Both internal and external floating roofs provide emission control for storage tanks. Floating roofs rise and fall with the liquid level, preventing formation of a large vapor space. Fixed roof tanks use pressure-vacuum vents to control breathing losses, and some tanks are equipped with vapor-recovery systems to control filling and emptying losses. Tank emission controls were discussed in more detail in the E&P section.

Loading Controls — Submerged Loading

A major source of evaporative emissions is the filling of underground gasoline storage tanks at service stations.

Table 13. Marketing Sector Emission Sources and Emission Controls

Emission Source	Emission Control
Storage tanks (bulk plants/ terminals)	Fixed roof with conservation vent Internal floating roof External floating roof
Loading racks (bulk plants/ terminals)	Submerged loading Refrigeration Adsorption and/or compression Thermal oxidation Vapor balance system
Service station tank truck unloading	Vapor balance system/vapor recovery
Service station refueling	Onboard controls Stage II vapor-recovery controls Vapor balance Hybrid Vacuum assist

Source: Refs. 18, 20, 28–30.

Gasoline is usually delivered to service stations in 30,000-L (8,000-gal) tank trucks or smaller account trucks. Emissions are generated when gasoline vapors in the underground storage tank are displaced to the atmosphere by the gasoline being loaded into the tank. As with other loading losses, the quantity of loss in service station tank filling depends on several variables, including the method and rate of filling, the tank configuration, and the gasoline temperature, vapor pressure, and composition. In splash loading, the gasoline is introduced into a tank at the top and above the liquid level in the tank. Submerged fill methods eliminate splash-induced vaporization and reduce vapors generated during the loading of gasoline into tank trucks and storage tanks. During submerged fill, the gasoline is introduced into the tank being filled, with the transfer line outlet situated below the liquid surface. Submerged filling minimizes droplet entrainment, evaporation, and turbulence, and can significantly reduce vapor loss over splash filling. An average emission rate for submerged filling is 880 mg/L (7.3 lb/1000 gal) of transferred gasoline, and the rate for splash filling is 1380 mg/L (11.5 lb/1000 gal) transferred gasoline (see Table 14).[18,37]

Vapor Balance Systems

During filling operations, vapor balance systems can be used to transfer vapor displaced by liquid in the tank being filled into the tank being emptied. Vapor balance prevents the compression and expansion of vapor spaces that would otherwise occur during filling. The vapor balance system employs a hose that returns gasoline vapors displaced from the underground tank to the tank truck cargo compartments being emptied. The control efficiency of the balance system ranges from 93 to 100%. Organic emissions from underground tank filling operations at a service station employing a vapor balance system and submerged filling are not expected to exceed 40 mg/L (0.3 lb/1000 gal) of transferred gasoline.[18,28]

A second source of vapor emissions from service stations is underground tank breathing. Breathing losses occur daily and are attributable to gasoline evaporation and barometric pressure changes. The frequency with which gasoline is withdrawn from the tank, allowing fresh air to enter to enhance evaporation, also has a major effect on the quantity of these emissions. An average breathing emission rate is 120 mg/L (1.0 lb/1000 gal) of throughput, as shown in Table 14.[18]

Refrigeration — Absorption — Adsorption

Refrigeration, absorption, and adsorption are means of recovering vapors as product. Refrigeration recovers hydrocarbon vapors by condensing vapors at cryogenic temperatures and atmospheric pressure. It can achieve reductions of 85–95%.[28] Absorption vapor-recovery units recover hydrocarbon from gasoline vapors by absorption into a lean oil stream. Carbon adsorption units can be used to recover hydrocarbon vapors by passing the gasoline loading vapors through a carbon bed, where the hydrocarbon constituents are adsorbed onto the carbon.

Service Station Operations

Gasoline transfer operations at service stations are sources of hydrocarbon emissions. The transfer takes place in two stages, referred to as Stage I and Stage II. Stage I refers to loading gasoline into underground storage tanks at the station, while Stage II refers to vehicle refueling.

Vehicle refueling emissions come from vapors displaced from the automobile tank by dispensed gasoline and from spillage. The quantity of displaced vapors depends on gasoline temperature, auto tank temperature, gasoline RVP, and dispensing rate. The following equation can be used to estimate uncontrolled displacement losses from vehicle refueling for a particular set of conditions[45]:

$$E_R = 264.2[(-5.909) - 0.0949(\Delta T) + 0.0884(T_D + 0.485(\mathrm{RVP})] \quad (6)$$

where E_R is the refueling emission (mg/L); ΔT, the difference between temperature of fuel in vehicle tank and temperature of dispensed fuel (°F); T_D, the temperature of dispensed fuel (°F); and RVP, the Reid vapor pressure (psia).
Note that this equation and the spillage loss factor are incorporated into the U.S. EPA mobile source emissions model, referred to as the MOBILE model. The MOBILE model allows for disabling of this calculation if it is desired to include these emissions in the stationary area source

Table 14. (Metric and English Units). Evaporative Emissions from Gasoline Service Station Operations[a]

Emission Source	Emission Rate mg/L Throughput	Emission Rate lb/10^3 gal Throughput
Filling underground tank (Stage I)		
Submerged filling	880	7.3
Splash filling	1,380	11.5
Balanced submerged filling	40	0.3
Underground tank breathing and emptying[b]	120	1.0
Vehicle refueling operations (Stage II)		
Displacement losses (uncontrolled)[c]	1,320	11.0
Displacement losses (controlled)	132	1.1
Spillage	80	0.7

[a]Factors are for VOC as well as total organic emissions, because the methane and ethane content of gasoline evaporative emissions is negligible.
[b]Includes any vapor loss between underground tank and gas pump.
[c]Based on Eq. 6, using average conditions.

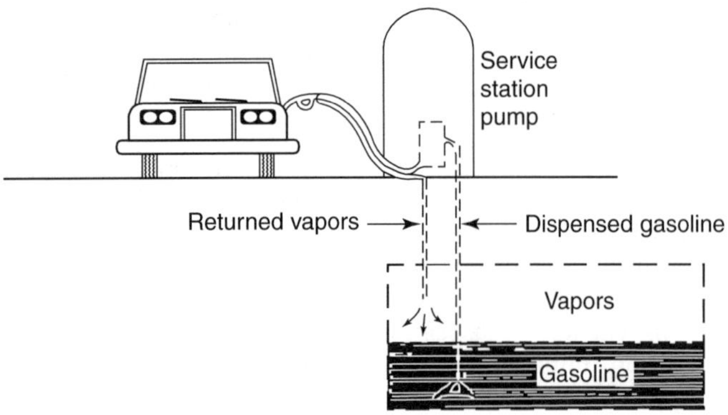

Figure 19. Automobile Refueling Vapor Recovery System.

portion of an inventory rather than in the mobile source portion. It is estimated that the uncontrolled emissions from vapors displaced during vehicle refueling average 1320 mg/L (11.0 lb/1000 gal) of dispensed gasoline.[37,44]

Spillage loss is made up of contributions from prefill and postfill nozzle drip and from spit-back and overflow from the vehicles's fuel tank filler pipe during filling. The amount of spillage loss can depend on several variables, including service station business characteristics, tank configuration, and operator techniques. An average spillage loss is 80 mg/L (0.7 lb/1000 gal) of dispensed gasoline.[18,37,44]

Control methods for vehicle refueling emissions are based on conveying the vapors displaced from the vehicle fuel tank to the underground storage tank vapor space through the use of a special hose and nozzle, as depicted in Figure 19 (termed Stage II vapor control). In "balance" vapor control systems, the vapors are conveyed by natural pressure differentials established during refueling. In "vacuum assist" systems, the conveyance of vapors from the auto fuel tank to the underground storage tank is assisted by a vacuum pump. Tests on a few systems have indicated overall systems control efficiencies in the range of 88–92%.[37,44] When inventorying these emissions as an area source, rule penetration and rule effectiveness should also be taken into account. *Procedures for Emission Inventory Preparation, Volume IV: Mobile Sources*, EPA-450/4-81-026d, provides more detail on this.[18]

To control emissions, Stage I vapor balance is usually used between the tank truck and the underground storage tank at the service station. Vapor balance systems were described above. Stage II controls can be vapor balanced between the service station underground tank and the vehicle gasoline tank. Vacuum-assist systems differ from vapor balance systems in that they use a vacuum pump to provide extra negative pressure at the nozzle-fill-neck interface to increase vapor recovery. They require some venting of excess vapors to the atmosphere. Hybrid systems combine vapor balance and vacuum assist. They enhance vapor recovery at the nozzle-fill neck by vacuum, while keeping vacuum low enough to minimize vapor returned to the vapor-recovery system. The other control technique for vehicle refueling is the use of onboard controls. Here, a vapor seal in the vehicle fill neck forces vapors being displaced from the tank into a carbon cannister on the vehicle, where they are adsorbed.[29–31] The use of on-board control of refueling losses is being required on automobiles in the future and is already being used on some models. This may eventually decrease the importance and or requirement for Stage II controls at service stations.

REFERENCES

1. *An Analysis of Petroleum Industry Costs Associated with Air Toxic Amendments to the Clean Air Act*, Interim Final Report, Delta Management Group, 1989, p. 13.
2. SO_2 Emissions in Natural Gas Production Industry—Background Information for Proposed Standards, EPA-450/3-82-023a, U.S. Environmental Protection Agency, Research Triangle Park, NC, 1983, p. 9-3.
3. *Industrial Process Profiles for Environmental Use: Chapter 2. Oil and Gas Production Industry*, EPA-600/2-77-023b, U.S. Environmental Protection Agency, Cincinnati, OH, 1977, p. 1.
4. *Industrial Process Profiles for Environmental Use*, p. 16.
5. *Industrial Process Profiles for Environmental Use*, p. 14
6. *Industrial Process Profiles for Environmental Use*, p. 34.
7. *SO_2 Emissions in Natural Gas Production Industry*, p. 3-3.
8. *NO_x Emissions from Petroleum Industry Operations*, API Publication 4311, American Petroleum Institute, Washington, D.C., 1979, p. 13.
9. *Industrial Process Profiles for Environmental Use*, p. 43.
10. *Introduction to Oil and Gas Production*, Book I of the Vocational Training Series, American Petroleum Institute, Dallas, 1983, p. 51.
11. *Introduction to Oil and Gas Production*, p. 52.
12. *Industrial Process Profiles for Environmental Use*, pp. 53, 55.
13. *SO_2 Emissions in Natural Gas Production Industry*, p. 3-9.
14. *NO_x Emissions from Petroleum Industry Operations*, pp. 57, 58.
15. *NO_x Emissions from Petroleum Industry Operations*, p. 66.
16. *Evaluation of Control Technologies for Hazardous Air Pollutants*, Final Report, Radian Corp., Research Triangle Park, NC, 1985, p. 3-9.
17. *Evaluation of Control Technologies for Hazardous Air Pollutants*, p. 3-14.
18. *Compilation of Air Pollutant Emission Factors, AP-42, Chapter 5.2, Transportation and Marketing of Petroleum Liquids*, 1995, U.S. EPA, available on the CHIEF bulletin board of the Technology Transfer Network. (http://www.epa.gov/ttn).
19. *Volatile Organic Compound (VOC) Emissions from Volatile Organic Liquid Storage Tanks—Background Information for Proposed Standards*, EPA-450/3-81-003a, U.S. Environmental Protection Agency, Research Triangle Park, NC, 1984, pp. 4-17–4-20.
20. *Background Information on Hydrocarbon Emissions from Marine Terminal Operations*, EPA-450/3-76-038a and 038-b, U.S. Environmental Protection Agency, Research Triangle Park, NC, 1976.
21. L. A. Thrash, *Oil Gas J.* **88**(13), 78 (1990).
22. *National Air Pollutant Emission Estimates 1940–1987*, EPA-450/4-88-022, U.S. Environmental Protection Agency, Office of Air Quality Planning and Standards, Research Triangle Park, NC, March 1989.
23. *VOC Emissions from Petroleum Refinery Wastewater Systems—Background Information for Proposed Standards*, EPA-450/3-85-001a, U.S. Environmental Protection Agency, Research Triangle Park, NC, 1985.
24. *Emission Estimation Techniques for Petroleum Refineries and Bulk Terminals*, Radian Corp., Research Triangle Park, NC, 1989.
25. *Air Pollution Engineering Manual*, 2nd ed., U.S. Environmental Protection Agency, Washington, D.C., 1973.
26. *Census of Wholesale Trade Geographic Area Series U.S., 1987*, WC87-A-52, Bureau of Census, U.S.D.O.C., 1989, US-14.
27. *Gasoline Marketing in the 1980's: Structure, Practice and Public Policy*, Temple, Barker, and Sloane, Inc., 1988.

28. *Evaluation of Air Pollution Regulatory Strategies for Gasoline Marketing Industry*, EPA-450/3-84-012a, U.S. Environmental Protection Agency, Washington, D.C., 1984.
29. *Fugitive Emission Sources of Organic Compounds Additional Information on Emissions, Emission Reductions, and Costs*, EPA-405/3-82-010, U.S. Environmental Protection Agency, Research Triangle Park, NC, 1982.
30. *Draft Regulatory Impact Analysis: Proposed Refueling Emission Regulations for Gasoline-Fueled Motor Vehicles. Vol. I—Analysis of Gasoline Marketing Regulatory Strategies*, EPA-450/3-87-001a, U.S. Environmental Protection Agency, Washington, D.C., 1987.
31. *Petroleum Supply Annual 1988*: Vol. 1, DOE/EIA0340(88)1, U.S. Department of Energy, Washington, D.C., 1988, p. 35.
32. *The Toxics-Release Inventory: A National Perspective, 1987*, U.S. Environmental Protection Agency, Office of Toxic Substances, Economics and Technology Division, Washington, D.C., June 1989.
33. C. E. Burklin and R. L. Honercamp, *Revision of Evaporative Hydrocarbon Emission Factors*, EPA-450/3-76-039, U.S. Environmental Protection Agency, Research Triangle Park, NC, August 1976.
34. G. A. LaFlam et al., *Revision of Tank Truck Loading Hydrocarbon Emission Factors*, Pacific Environmental Services, Inc., Durham, NC, May 1982.
35. G. A. LaFlam, *Revision of Marine Vessel Evaporative Emission Factors*, Pacific Environmental Services, Inc., Durham, NC, November 1984.
36. *Evaporation Loss from Tank Cars, Tank Trucks and Marine Vessels*, Bulletin No. 2514, 1/95 Petroleum Industry 5.2-17, American Petroleum Institute, Washington, D.C., 1959.
37. C. E. Burklin, et al., *A Study of Vapor Control Methods for Gasoline Marketing Operations*, EPA-450/3-75-046A and -046B, U.S. Environmental Protection Agency, Research Triangle Park, NC, May 1975.
38. *Bulk Gasoline Terminals—Background Information for Proposed Standards*, EPA-450/3-80-038a, U.S. Environmental Protection Agency, Research Triangle Park, NC, December 1980.
39. *Atmospheric Hydrocarbon Emissions from Marine Vessel Transfer Operations*, Publication 2514A, American Petroleum Institute, Washington, D.C., 1981.
40. C. E. Burklin, et al., *Background Information on Hydrocarbon Emissions from Marine Terminal Operations*, EPA-450/3-76-038a and -038b, U.S. Environmental Protection Agency, Research Triangle Park, NC, November 1976.
41. *Rules for the Protection of the Marine Environment Relating to Tank Vessels Carrying Oil in Bulk*, 45 FR 43705, June 30, 1980.
42. R. A. Nichols, *Analytical Calculation of Fuel Transit Breathing Loss*, Chevron USA, Inc., San Francisco, CA, March 21, 1977.
43. R. A. Nichols, *Tank Truck Leakage Measurements*, Chevron USA, Inc., San Francisco, CA, June 7, 1977.
44. *Investigation of Passenger Car Refueling Losses: Final Report, 2nd Year Program*, APTD-1453, U.S. Environmental Protection Agency, Research Triangle Park, NC, September 1972.
45. *Refueling Emissions from Uncontrolled Vehicles*, EPA-AA-SDSB-85-6, U.S. Environmental Protection Agency, Ann Arbor, MI, June 1985.

18

WOOD PROCESSING INDUSTRY

Arun V. Someshwar and John E. Pinkerton

The processing of wood to yield such varied products as paper, paperboard, lumber, plywood, and reconstituted building board ranks among the ten largest industrial activities in the United States. To produce paper or paperboard, the wood is first pulped. Pulps can be produced through chemical or mechanical means or by a combination of both. The pulp may then be bleached to various degrees of brightness. Finally, bleached or unbleached pulp is processed into thick sheets of paperboard or paper. The manufacture of building boards, with a few exceptions, involves processing the wood itself into sheets or boards, with the aid of adhesives.

The following material is intended to provide a broad overview of selected segments of the wood processing industry. The production process, atmospheric emissions, and emission control technologies employed are described for (1) several chemical wood pulping processes, (2) mechanical wood pulping processes, (3) recovered paper pulping, and (4) paper and paperboard manufacture. It should be recognized that this manual only provides information on the most common manufacturing processes and control technologies. There are a large number of variations of the basic processes and control systems; thus the descriptions here are necessarily generic in nature. Also, this presentation does not cover solid wood products manufacturing operations.

CHEMICAL WOOD PULPING

Arun V. Someshwar and John E. Pinkerton

Wood can be pulped with chemical or mechanical methods. Chemical wood pulping involves cooking wood chips or sawdust in an aqueous solution of pulping chemicals, resulting in the extraction of cellulose from the wood by dissolving the lignin that binds the cellulose fibers together. The pulping chemicals may be alkaline, acidic, or neutral. Three principal chemical wood pulping processes currently in use are (1) kraft, (2) acid sulfite, and (3) semichemical. Of these three, kraft pulping accounts for nearly 80% of the chemical pulp produced in the United States. Mechanical pulping represents only 10% of the U.S. wood pulping capacity, while semichemical pulping, a combined chemical/mechanical process, represents around 6%. The remaining pulping capacity is either sulfite or soda chemical processes. Of all the pulping processes, kraft mills have the most air pollution concerns because of reduced sulfur compounds, extensive chemical recovery processes, and, at about 75% of the mills, bleaching operations. At the end of 1996, 116 kraft, 11 sulfite, and 14 other chemical pulp mills were in operation in this country.[1]

KRAFT PULPING

The dominant wood pulping process today is the kraft process. The term kraft or sulfate has been in use since 1879, when sodium sulfate replaced sodium carbonate (soda process) as the makeup chemical. The kraft process is expected to continue as the dominant chemical wood pulping process for reasons that include its comparative simplicity, rapidity of the process, its insensitivity to variations in wood condition, its applicability to all wood species, as well as the valuable properties of the pulp produced (Figure 1).[2]

Process Description

The production of pulp by the kraft process can be divided into five areas, namely, (1) wood preparation, (2) pulping, (3) pulp washing, (4) oxygen delignification and/or bleaching of pulp and (5) chemical recovery. Figure 1 provides a schematic representation of the kraft pulping and chemical recovery process.

Wood Preparation. The primary fiber source for the U.S. paper industry is wood. There are a few mills that use nonwood fibers, principally flax and cotton, for specialty paper production, but their output is minuscule compared to wood pulp. Recovered paper is also extensively used as a fiber source, but again the origin of the paper is wood fibers.

Wood pulp can be produced from logs, chips, or sawdust. Logs can be transported by truck or rail to the mill for the production of chips, or the chips can be produced at another location. Sawdust usually comes from off-site, unless the pulp mill is located adjacent to a wood products plant. Mills using tree-length logs cut them into shorter pieces with slasher saws. Bark is almost always removed from the cut logs prior to further use. Most mills use drum debarkers, which are rotating cylinders that tumble the logs to remove the bark. A few mills perform hydraulic debarking, where bark is removed with high-pressure water sprays.

With the exception of the groundwood pulping process where whole logs are fed directly to a grinder, logs are chipped after bark removal. Chips are screened to remove those that are too large or too small. The large chips are reduced to an acceptable size, while too small chips are generally used as boiler fuel. Accepted chips are transferred by conveyer or pneumatically to storage piles, which are generally outdoors.

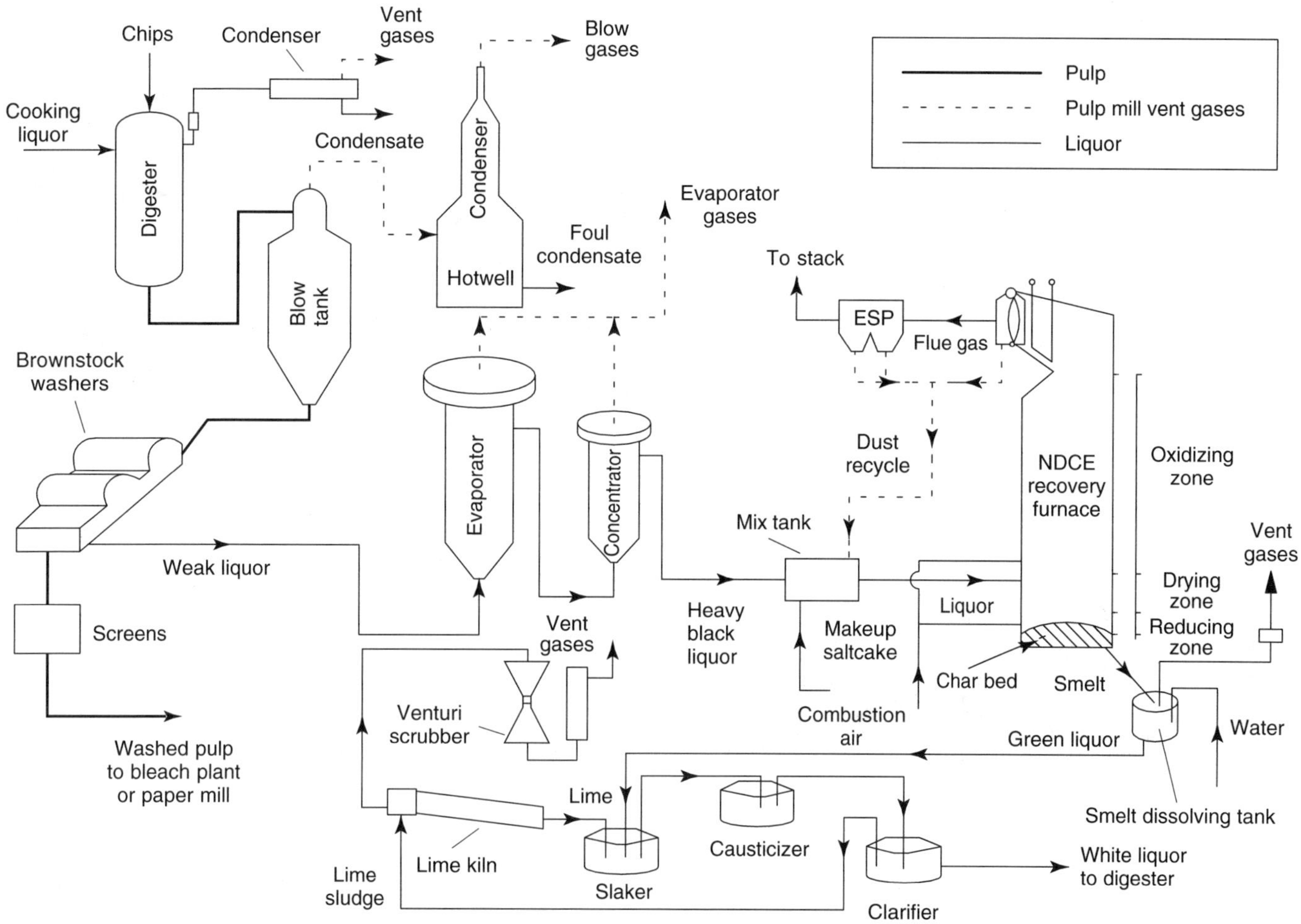

Figure 1. Schematic of the Kraft Pulping and Recovery Process.

Pulping. In the kraft process, wood chips are cooked at an elevated temperature (340–360°F) and pressure (100–135 psig) in an aqueous solution of sodium hydroxide and sodium sulfide (also called white liquor). The sodium sulfide in the cooking liquor serves to buffer and sustain the cooking reaction, while the sodium hydroxide is consumed by reaction with the lignin and carbohydrates in the wood. Once the cooking is complete, the wood is broken down into two phases: a soluble phase containing the lignin and alkali-soluble hemicellulose and an insoluble phase containing the alpha cellulose or pulp. The wood chips are cooked in large vertical vessels called digesters, which may be operated in either the "batch" or the "continuous" mode.

In the case of batch digesters, air trapped with the chips and gases formed during digestion are relieved intermittently during cooking. In softwood pulping, the relief gases are condensed for turpentine recovery before venting. The cooking period ranges from 2 to 6 hours. At the end of the cook, the digester contents are transferred to an atmospheric tank, called the blow tank. Gases leaving the blow tanks pass through a condenser to remove moisture, and the uncondensed gases are incinerated in a combustion device.

In continuous digesters, which are extremely tall and large cooking vessels (see Figure 2), an uninterrupted flow of wood chips and cooking liquor enter from the top. Pulp is withdrawn continuously from the bottom into a blow tank, while the spent liquor is drawn off and transferred to a flash tank. Steam from the flash tank is used to presteam the wood chips, after which it passes through a condenser. Noncondensible gases (NCGs) are vented, and in the case of softwoods, turpentine is decanted from the condensate. Newer digesters utilize a "cold-blow" as compared with the conventional "hot blow." In the cold-blow process, the pulp leaves the digester at a much lower temperature, resulting in a significant reduction in blow gas emissions.

Pulp Washing. Wood pulps made by chemical means generally go through several steps where knots, shives, and unacceptably short fibers are removed; spent cooking chemicals are separated from the pulp; and the consistency of the pulp slurry is adjusted for subsequent processing steps. Removal of knots and other oversize wood material from chemical pulps is generally practiced where the pulp will be bleached. Knot removal is normally accomplished prior to washing using vibrating screens or enclosed pressurized screens. The general objectives of washing are to remove undesirable materials from the pulp that would adversely affect the subsequent bleaching or paper-making process and to recover pulping chemicals. Washing involves displacement

Figure 2. Continuous Digester.

of the "dirty" liquid in the pulp slurry with a "cleaner" liquid. Emissions from washing operations are a function of what compounds are present in the "dirty" and "cleaner" liquids. They also depend upon the extent of contact between these liquids and air that must ultimately be exhausted from the washing system.

There are many different types of pulp washing equipment. Rotary vacuum drum washers are the most common type (Figure 3). They have a rotating drum that is partially immersed in a vat of pulp with a consistency on the order of 1%. An internal vacuum is applied to pick up the pulp from the vat. Above the vat, shower water is sprayed onto the pulp mat to displace the liquid already present. After the shower, the vacuum is stopped and the pulp is removed from the drum just before the vat. The liquid removed from the pulp is piped to a filtrate tank located below the vat. This tank has a vent to release air drawn in with the liquid. The drum portion of the washer is normally underneath a hood, which serves to exhaust steam and vapors from around the drum and vat.

There are many other types of washers besides vacuum drum types. Rotary pressure and compaction baffle washers are somewhat similar to vacuum drum washers, except that the rotating drum is enclosed and put under positive pressure so that the washing liquid is forced through the pulp rather than drawn through the pulp by vacuum. The "dirty" fluid then flows to a filtrate tank, which is vented. The gases in this vent will contain compounds volatilized from the liquids similar to the vacuum drum systems.

Diffusion washers are another type; they often follow continuous digesters at chemical pulp mills. These washers

Figure 3. Brown-Stock Washers (Vacuum Drum Type).

are essentially large tanks where spent cooking liquor is forced from pulp by replacement with a "cleaner" washing liquid. These washers may be pressurized. Vent gas flows from atmospheric diffusion washers are quite small, and thus emissions are very low compared to vacuum drum washer hoods. Pressurized systems do not have a vent on the washer. However, both types of diffusion washer systems have a filtrate tank with a vent, and the vent gas will have volatile organic compounds, TRS (at kraft mills), and SO_2 (at sulfite mills). Again, the sources of these compounds are the spent cooking liquor and washing liquid.

The horizontal belt washer is a relatively new type of washer where pulp is spread onto a moving screen and wash water is applied in a series of showers. The "dirty" liquid moves down through the screen to a tank and then is taken from the tank to be used as "clean" shower water on the preceding stage. Like a series of vacuum drum washers, the wash water is circulated countercurrent to the pulp flow. The belt washer is enclosed under slight positive pressure, but there can be outward air leakage from the pulp entrance and exit points. The only vent in this washer system is typically on the final filtrate tank, although there may also be a purge air vent. Air flows are very small, which tends to minimize the amount of volatile organic compounds released from these washers.

Another new type of washer is the displacement press, which is generally enclosed and vented by an exhaust fan. Filtrate squeezed from the pulp flows to filtrate tanks that are vented. As with other systems, the emissions will be a function of compounds present in the "dirty" and "clean" liquids and the degree of contact between the liquids and air.

At chemical pulp mills, screening and cleaning takes places after washing. Screening is most often accomplished in gravity centrifugal or pressurized devices, which are enclosed and not vented. Cleaning is generally done in a series of enclosed centrifugal cleaners. While these screening and cleaning devices are not vented, tanks which receive accepted and rejected material typically have vents to the atmosphere, and pumps that move the stock have atmospheric exhausts. There are still some open vibrating screen systems in use as well. Any emissions from these systems result from the presence of volatile organic compounds present in the pulp/liquid mixture being processed.

Generally pulp is thickened after screening and cleaning. Further pulp washing may also be done at the same time. Devices for increasing pulp consistency include deckers, vacuum filters, multidisc filters, screw extractors, and presses. These devices provide differing degrees of dewatering. Deckers and vacuum filters are somewhat similar to vacuum drum washers in that a rotating drum in a vat of pulp slurry is used and the removed water flows through a screen to the interior of the drum and then drops by gravity into a filtrate tank below the drum. A hood may be placed over the rotating drum in order to exhaust the steam and vapors to outside the building, and there is a vent on the filtrate tank.

Multidisc filters consist of rotating discs in a vat of pulp that pick up pulp as they turn and remove water by vacuum. Screw extractors and presses remove water by squeezing. These devices can be operated under pressure or at atmospheric pressure. The equipment can be directly or indirectly (through building exhausts) vented to the atmosphere. Besides equipment vents, there can be vents on associated feed tanks, stock storage tanks, filtrate tanks, and vacuum pumps.

Like pulp washing operations, emissions from pulp thickening operations are a function of what compounds are present in the liquids accompanying the pulp and the extent of contact these liquids have with air that is exhausted from the thickening equipment and associated tankage.

Oxygen Delignification and/or Pulp Bleaching. Oxygen delignification is one means to reduce the lignin content of pulp prior to addition of other bleaching chemicals. Although in theory oxygen delignification may be applied to any type of pulp, it is most often used on kraft wood pulps in the United States. Typical major system components are a continuous reactor, blow tank, and one or more washers and associated filtrate tanks. The reactor is operated under elevated temperature and pressure. Steam, caustic (usually in the form of oxidized white liquor), and oxygen are added to the pulp before it enters the reactor. Following the reactor, the pulp goes to a blow tank and then to the post-oxygen washers. A wide variety of washer types are used, including vacuum drum, compaction baffle and rotary pressure washers, and roll presses. Water flow is countercurrent to pulp flow, with the "cleanest" water being added at the final washing stage as shower water. Washer filtrates are cycled back to previous post-oxygen washing stages and finally to the brown stock washer system.

Bleaching of kraft pulps is accomplished in a multistage bleach plant, normally with chlorine-containing compounds. Each stage consists of a bleaching tower followed by a washer. The bleaching tower is used to promote sufficient contact of the bleaching agents with the pulp, and the pulp can flow upward or downward, depending on design. Washers remove the unwanted soluble organic and inorganic materials. Vacuum drum washers are the most common type of washers, although diffusion washers, pressurized washers, and roll presses are also used in the bleach plant (see section on pulp washing for descriptions of this equipment).

Traditionally chlorine was used in the first bleaching stage. Due to concerns about formation of unwanted chlorinated compounds resulting from chlorine use, it is being rapidly replaced with chlorine dioxide. In fact, use of chlorine for bleaching will likely be limited to the production of dissolving pulps by the year 2001. In the second stage, usually referred to as the first extraction stage, sodium hydroxide is added to raise pH and dissolve unwanted organics, which are subsequently removed by washing. Chlorine dioxide is applied in the third stage for further lignin removal and brightening; sodium hypochlorite use in this stage has been largely phased out due to concerns about chloroform formation. An additional extraction and final chlorine dioxide stage often follow the third stage.

Recovery of Cooking Chemicals. An integral part of chemical pulping processes is the recovery of spent cooking chemicals for both economic and environmental reasons. The basic steps are evaporation to reduce the water content of spent liquor, combustion of concentrated spent liquor, and recovery of chemicals from combustion products. The kraft process has an additional step involving regeneration of lime that is used for recovering sodium. The overall chemical recovery process is more complicated for kraft pulping as compared to sulfite and semichemical pulping.

Following cooking, spent cooking liquor, referred to as weak black liquor, is washed from the pulp. Weak black liquor consists of about 85% water and 15% solid material. The solids are a complex mixture of sulfur and sodium-containing organic and inorganic compounds. To recover the sulfur and sodium, the water content of weak black liquor is reduced by evaporation to the point where the solids content is high enough to support combustion, typically in the range of 65 to 80%. The liquor is burned in a recovery furnace, a specialized combustion device designed to produce a molten smelt at the bottom of its combustion chamber. The smelt contains most of the sodium and sulfur present in the liquor burned. The smelt is drained into a dissolving tank and mixed with calcium hydroxide. The resulting mixture is processed in a series of clarifiers to obtain white liquor (a solution of sodium hydroxide and sodium sulfide) and a calcium carbonate precipitate. The precipitate is ultimately sent to a lime kiln, where it is calcined to obtain calcium oxide. This lime is slaked to remake calcium hydroxide. Obviously the kraft chemical recovery process has the potential to be a substantial source of atmospheric emissions.

There are different types of evaporators (falling film, rising film, forced circulation) and numerous ways to route the liquor through the multiple effects. In general, fresh steam enters the first evaporation effect, which is above atmospheric pressure. This is the last liquor evaporation stage, after which the liquor contains about equal amounts of water and solids. The water vapor released from the liquor in one stage is sent to the succeeding stage, where it is used under a lower pressure to evaporate additional water. The final effects of the evaporator are under vacuum. Condensates will be generated in each effect as the vapors transfer heat to the liquor. However, volatile gaseous compounds will also be generated, and these must ultimately be vented from the evaporator system.

Increasing the solids content of liquor from 50 to 65% or greater is customarily viewed as a separate evaporation step, which is accomplished either in a concentrator through indirect contact with steam or in a direct contact evaporator (DCE) with flue gas. There are many concentrator designs, but the vapors from the concentrator are normally routed to the first effect of the evaporator train. These vapors are relatively free from

TRS and volatile organic compounds, since these compounds are released during the earlier evaporation steps.

The direct contact evaporator (DCE) brings hot flue gas exiting the recovery furnace into direct contact with liquor at about 50% solids to evaporate additional water. The evaporated water is carried off in the flue gas. The direct contact takes place either in a cyclone or cascade. An undesirable consequence of this contact is the generation and release of H_2S due to the acidifying action of CO_2 present in the flue gas. Volatile organic compounds present in the liquor may be stripped as well. H_2S formation is minimized by oxidizing sodium sulfide in the liquor to sodium thiosulfate prior to sending the liquor to the DCE. Black liquor oxidation (BLO) is widely used in the United States in order to meet TRS emission requirements for recovery furnaces.

Most BLO systems use air to achieve the oxidation, although a few employ oxygen to avoid foaming problems with liquors from resinous softwood species. In the air-based systems, air is thoroughly mixed with black liquor in an oxidation tower or agitated tank, in either one or two steps. Although oxidation can be performed on weak black liquor prior to any evaporation, it is usually done after the multiple-effect evaporators.

DCEs and associated BLO systems are still used in about one-third of the U.S. kraft recovery furnaces, although the number of these units has been steadily declining for the past 25 years. Only one new unit has been installed in the past 20 years. Compared to DCE/BLO systems, the nondirect contact furnace design has more favorable economics, lower overall energy use, and lower total emissions.

Once the liquor has a minimum solids concentration of 60%, it will support combustion so that it can be burned in the recovery furnace (Figures 4 and 5). The recovery furnace is a unique device designed to burn the organic material in the liquor and capture the inorganic material as a molten smelt in the bottom of the furnace. The lower part of the furnace is run in an oxygen-deficient condition, while excess oxygen is supplied to the upper part. Black liquor is fired into the furnace cavity in the form of droplets with one or more liquor guns. The droplets first undergo drying and then pyrolysis. Pyrolysis results in formation of combustion gases and char. The char drops to the bottom of the furnace, where the carbon burns and the inorganic sodium compounds (sulfide, sulfate, carbonate) drain off as smelt. In the upper oxidizing zone of the furnace, complex reactions occur among sodium particulates and fumes, carbon monoxide, and sulfur-containing gases. The details of the combustion process are complicated and beyond the scope of this text. A good reference source is the text edited by Hough.[3]

The major elements in black liquor are carbon, hydrogen, oxygen, sulfur, and sodium. Potassium, nitrogen, and chlorine are significant trace elements that originate mainly in the wood, although potassium and chlorine can also be introduced through the use of makeup chemicals such as sodium sulfate or process water. In addition, liquor will contain several other metals as trace impurities. The relative amounts of these materials present in the liquor fired affect the emission characteristics to a large degree, but the design characteristics and method of operation of the recovery furnace also have a major impact on its emissions.

Recovery of sodium sulfate from the flue gas for reuse in the liquor cycle is worthwhile provided the cost of capture is less than its economic value. Historically the break point has been considered to be around 90% capture efficiency. Today, capture rates far above 90% are needed to meet environmental requirements. With a single exception, electrostatic precipitators (ESPs) are used in the United States on kraft recovery furnaces. A few furnaces are equipped with scrubbers as well; the scrubbers can achieve some incremental particulate removal. ESP collection efficiency (on a mass basis) ranges from over 90 in the oldest units to over 99.5% for the

Figure 4. Kraft Recovery Furnace.

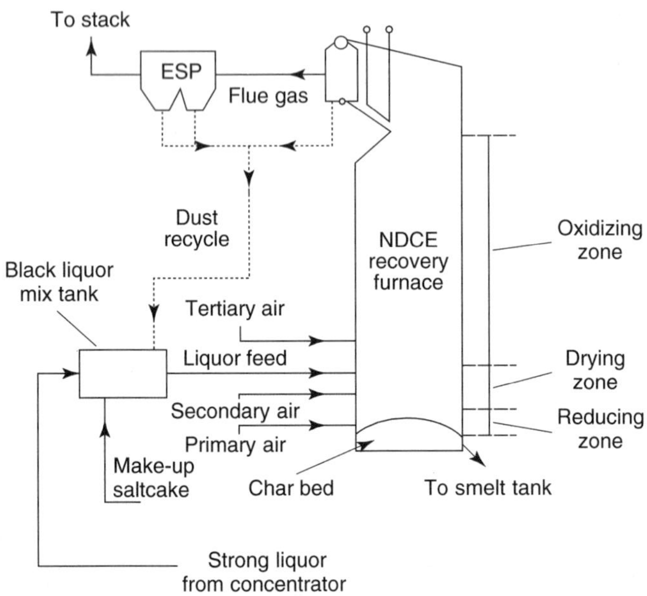

Figure 5. Kraft Recovery Furnace Schematic.

newest units. The captured particulate material is typically sent to a tank, where it is mixed with incoming liquor prior to burning in the recovery furnace.

Molten smelt is drained from the furnace hearth into a tank located below the furnace. Each recovery furnace has one or two of these smelt dissolving tanks. As the smelt drains, steam shatter jets are used to break up the smelt. The smelt particles drop into an agitated solution of weak wash, which is obtained from displacement washing of calcium carbonate precipitate. The mixture of smelt and weak wash is called green liquor, and this is pumped from the dissolving tank to a green liquor clarifier for solids removal.

A great deal of turbulence occurs in the smelt dissolving tank because of the exothermic reactions between sodium and water. Steam, gases, and entrained particles from the tank are vented through a stack. Particulates are removed by a wet scrubbing device, such as a venturi scrubber, often followed by a mist eliminator to capture any entrained water droplets. The scrubber makeup water is typically weak wash, and the scrubber water is recycled back to the dissolving tank.

In the lime kiln, lime mud (mainly calcium carbonate and water) is calcined to form lime (CaO) and CO_2. The source of heat for this reaction can be natural gas, fuel oil, or petroleum coke. In the standard inclined rotating kiln, combustion gases move countercurrent to the lime mud. Calcium oxide from the kiln and green liquor from the smelt dissolving tank are mixed in the slaker to obtain sodium hydroxide and calcium carbonate. The slaker must be vented through a stack to discharge the large amounts of steam generated. There are numerous other pieces of equipment in the causticizing area of a mill. These sources are associated with the processing of green liquor (clarifiers, storage and surge tanks, dregs filters); white liquor (causticizer tanks, clarifiers, pressure filters, storage tanks); and lime mud (mix tanks, dilution tanks, storage tanks, pressure filters, precoat filters, filter vacuum pump exhausts).

Air Emissions Characterization

Historically, odor and visible particulate emissions from kraft pulp mills have received considerable attention. Initial state and federal emission regulations addressed TRS and particulate emissions. The characteristic odor of a kraft pulp mill results from TRS compound emissions, which are the by-products of reactions between the wood lignin and sulfide ions in kraft liquors. Hydrogen sulfide (H_2S) and three organic sulfur compounds—methyl mercaptan (CH_3SH), dimethyl sulfide [$(CH_3)_2S_2$], and dimethyl disulfide [$(CH_3)_2S$]—constitute the major TRS compounds.

Wood Preparation. Wood preparation activities can result in the release of particulate matter. These particulate releases are referred to as "fugitive" emissions because they are not released from a stack or vent. As such, they are extremely difficult to measure. Even estimating the emissions is very challenging due to the numerous factors that influence emission rates. Thus the paucity of information about fugitive emissions from wood preparation activities is not surprising. Most of the particulates released are comprised of wood or bark, although there can also be some soil-based particulates that result from soil becoming dislodged from the bark. An important consideration in evaluating emission rates is particle size, since relatively large particles will tend to settle quickly, while smaller particles will remain airborne longer. Particles that stay airborne long enough to travel beyond the mill property line are of concern because they can create a nuisance dust problem and may contribute to ambient PM_{10} and $PM_{2.5}$ particulate concentrations. Unfortunately there is no available information on particle size distributions for sawing, chipping, screening, unloading, transfer, or chip storage piles.

Another type of fugitive emission associated with wood preparation activities is gaseous. A portion of the extractive component of wood consists of volatile organic compounds that are continuously escaping from the wood in gaseous form. Living trees continuously release these compounds and the releases continue after tree harvesting, although the rates of release decline over time. The major types of compounds released are terpenes (softwoods) and isoprenes (hardwoods). These compounds are classified by EPA as volatile organic compounds (VOCs) because they are expected to participate in photochemical reactions that lead to the formation of ground-level ozone. There have been some limited laboratory and field studies of VOC emissions from chip storage piles,[4,5] suggesting that emission rates from softwood chips are significantly larger (roughly an order of magnitude) than those from hardwoods. The rates were found to vary considerably, being a function of wood species, chip age, pile geometry, and weather conditions. Observed total releases over the study periods varied from less than 0.1 to over 7 lb carbon/ton oven dry chips.[5]

Pulping. Besides the four main TRS compounds mentioned above, the kraft cooking process results in the formation and release of volatile organic compounds from the wood chips. Methanol, acetone, methyl ethyl ketone, and acetaldehyde are the main compounds generated in both hardwood and softwood pulping. A variety of terpenes ($C_{10}H_{16}$) are released when softwoods are pulped, while acetic acid is released from hardwoods. Once formed, all these compounds exit the digester either in gaseous or liquid form. Relief gases formed during cooking are routed through condensers to remove water and condensable compounds. Where softwood species are pulped, turpentine is recovered in the condensers. The gases that remain after the condensers are referred to as noncondensible gases (NCGs). Because the digester NCGs contain significant concentrations of odorous TRS compounds, virtually all kraft mills now collect these gases and route them to an incineration device (lime kiln, thermal oxidizer, boiler, or recovery furnace). The reduced sulfur compounds are oxidized to sulfur dioxide, while any hydrocarbons are combusted to yield carbon dioxide and water.

In continuous digester systems, the steam remaining after the relief gases are sent through flash tanks for condensing is typically used to preheat and condition the chips prior to their entry to the digester. This steam often contains considerable amounts of TRS and other organic gases. These compounds can be released to the atmosphere through chip feeding valves or open chip steaming bins. Many mills collect these gases for later incineration. Batch digesters have blow gases in addition to relief gases. Blow gases are normally condensed in a blow heat recovery system either by direct or indirect condensers. The remaining NCGs are collected for later incineration.

The condensates formed from cooling of digester relief and blow gases contain TRS and organic compounds. These condensates can either be sent to the wastewater treatment system or reused, typically for unbleached pulp washing or in the recausticizing area. Prior to reuse, many mills steam strip the condensates to remove the undesirable compounds and the stripper off-gases (SOGs) are burned in a combustion device. If these compounds are not removed, they can later volatilize when they come into contact with air, such as in a vacuum drum brownstock washer, and result in increased emissions.

Table 1 provides estimates of concentrations for TRS and several VOCs of significance in *uncontrolled* kraft pulp mill and evaporator NCGs and kraft stripper off-gases. As noted earlier, at virtually all kraft mills, the NCGs and SOGs are burned in a combustion device. Based upon the mean TRS concentrations shown in Table 1, the burning of NCGs and SOGs in a combustion device is expected to result in uncontrolled SO_2 emissions from TRS compound oxidation of about 5.70 lb/ADTUBP $[(1.79 + 1.06) \times 2]$. When NCGs are burned in kraft lime kilns or wood residue-fired boilers, the lime or bark ash alkalinity, respectively, serves to capture a significant portion of the SO_2 released from TRS oxidation.[6] Other "standalone" thermal oxidizers are typically equipped with alkaline scrubbers for SO_2 removal.

Pulp Washing. The pulp processing steps of washing, cleaning, screening, and deknotting do not result in the formation of any new pollutants. However, volatile compounds present in the pulp-water mixture transferred from the digester area and compounds present in recirculated water used for washing and pulp dilution can be released from these operations. The points of release, the compounds released, and the amounts released are very mill-specific.

Vibrating screen systems are not enclosed, and any volatile compounds released from the pulp-water-spent liquor mixture being processed can escape into the building air. Volatile compounds released can include reduced sulfur compounds, terpenes, and methanol. Most systems do have hoods above the screens to exhaust the air out of the building for worker exposure reasons. The pressurized screens are not directly vented, but these systems generally have storage tanks for accepts and rejects that are vented. These vents provide an opportunity for volatile organic compounds present in the liquid–pulp mixture to be emitted. There may be other handling equipment in the knot removal system such as knot bucket elevators that represent potential emission sources.

In rotary vacuum drum washers, the hood exhaust and filtrate tank vent gases will contain reduced sulfur and volatile organic compounds, including methanol, acetaldehyde, acetone, and methyl ethyl ketone, with methanol dominating. Terpenes are present where softwoods are pulped. These compounds enter the vacuum drum washer system with the pulp–spent liquor slurry coming from the digesters and any dilution water used to achieve the correct consistency. The compounds can also be present in the shower water when condensates are reused for pulp washing. As a rule, the higher the concentrations of TRS in the shower water, the higher the concentrations of TRS and organic compounds in the hood and filtrate tank vent gases. Typically, there are a wide variety of water use practices in washing systems, ranging from the use of only fresh water to use of only condensates in the showers on the last washing stage. Where there are multiple washers in series, countercurrent water use is practiced where the "cleanest" shower water is applied in the final washing stage. The filtrates from the final stage are then used to as shower water on the preceding stage, and so on. Studies of methanol emissions from kraft mill vacuum drum washer systems have shown a fairly good relationship between emissions and the concentration of methanol in the shower water of the final washing stage.[9] Earlier studies showed TRS emissions were significantly higher from washer systems that used condensates versus fresh water in the showers.[10]

Table 2 gives estimates of emissions from kraft pulp mill brownstock washers and deckers for TRS compounds and several VOCs of significance including total hydrocarbons (THCs). For VOCs, emissions from both vacuum drum type washers and all other type washers (such as diffusion, pressure, belt, and compaction baffle washers) are shown separately.

Oxygen Delignification. In mills that practice oxygen delignification, volatile organic compounds present in the incoming pulp slurry, oxidized white liquor, and washer shower water can be released in blow tank, washer hood, washer filtrate tank, and pulp storage tank vent gases. The amounts of these compounds present in the various process liquids will depend upon the water reuse practices of the mill, particularly if condensates are reused in the post-oxygen washers. As previously explained, releases of volatile organic compounds from washer systems also are a function of the extent to which exhaust air and process liquids are in contact, which in turn depends upon equipment design and ventilation practices. Studies have shown that methanol is the dominant

Table 1. Uncontrolled Noncondensible Gas Emissions from Kraft Pulping

	Pulping & Evap. NCGs				Stripper Off-Gases			
			Range	Mean			Range	Mean
	No.[b]	Ref.	(lb/ADTUBP[a])		No.	Ref.	(lb/ADTUBP)	
TRS as S	3	7	0.67–4.18	1.79	1	7	—	1.06
Methanol	4	7	ND–0.52	9.5E−02	10[c]	8	6.34–19.67	10.46
Acetone	3	7	ND–0.045	1.4E−02	10[c]	8	0.00–2.79	0.68
Acetaldehyde	1	7	ND[1.3E−05][e]	—	10[c]	8	0.02–0.20	0.11
MEK	4	7	ND–8.0E−03	2.5E−03	10[c]	8	0.01–0.15	0.05
Terpenes	1[d]	7	ND–0.24	1.4E−01	1[d]	7	0.028–0.040	0.032

[a]ADTUBP, air-dried ton of unbleached pulp.
[b]No., number of sources tested.
[c]Estimated from stripper feed concentrations assuming 100% stripping efficiency.
[d]α − pinene + β-pinene.
[e]ND[x] - not detected at concentration x.

Table 2. TRS and VOC Emissions from Kraft Pulp Washers and Deckers

		Vacuum Drum Washers			Other Washers			Pulp Deckers		
			Range	Mean		Range	Mean		Range	Mean
	Ref.	No.[b]	(lb/ADTUBP[a])		No.	(lb/ADTUBP)		No.	(lb/ADTUBP)	
TRS as S	7	20	ND[c]–1.62	0.25	7	ND–0.63	0.11	6	ND–0.089	0.025
Methanol	7	30	0.087–6.1	9.4E–01	7	8E–04–0.25	9.5E–02	6	0.022–1.7	3.3E–01
Acetone	7	26	ND–0.27	6.0E–02	7	ND–0.10	1.8E–02	6	ND–0.028	9.7E–03
Acetaldehyde	7	18	1E–03–0.10	2.0E–02	7	ND–0.069	1.1E–02	5	ND–0.014	4.6E–03
Formaldehyde	7	14	ND–0.046	3.6E–03	6	ND–8E–04	1.1E–04	5	ND–0.01	3.3E–03
MEK	7	27	ND–0.067	1.2E–02	7	ND–0.007	2.4E–03	6	ND–0.008	2.8E–03
Terpenes	7	14	ND–6.0	1.5E+00	7	4E–05–0.30	8.7E–02	2	0.015–0.10	5.8E–02
THCs as C[d]	7	10	0.002–2.0	4.8E–01	6	2E–04–5.1	9.8E–01	3	5E–02–0.09	7.3E–02

[a]ADTUBP, air-dried ton of unbleached pulp.
[b]No., number of sources tested.
[c]ND, not detected.
[d]As measured by EPA M25A.

organic compound found in all of these vent gases,[11] with considerably lesser amounts of acetone, acetaldehyde, and methyl ethyl ketone. Terpenes and reduced sulfur compounds were generally at trace amounts or below sampling method detection limits. For methanol, a reasonably good relationship was found between total system methanol emissions and the methanol concentration in the final post-oxygen washer shower water.

Carbon dioxide and smaller amounts of carbon monoxide are formed in the reactor as the delignification proceeds. The majority of these gases will be released from the blow tank vent. Measurements have shown average carbon monoxide emission rates range from 0.01 to 0.45 lb/ODTP from blow tank vents,[12] with no readily apparent relationship to either the amount of oxygen applied or to the reduction in kappa number accomplished. It is not known if organic compounds of regulatory significance (e.g. methanol) are also formed in the reactor.

Table 3 gives estimates of emissions from kraft oxygen delignification system vents for TRS and several VOCs of significance including THCs.

Pulp Bleaching. For mills that bleach pulp, the bleaching sequences and bleaching conditions employed may vary considerably. Chlorination, oxidation, and alkali extraction may each make up a separate bleaching stage. Figure 6 gives a much-simplified diagram of a typical bleaching stage. Gas-phase emissions may occur from each stage's tower, washer hood, and seal tank vent. The various emission points may or may not be combined into one or more common vents. Scrubbers utilizing various scrubbing fluids, such as bleach-plant extraction-stage filtrate, sodium hydroxide solution, sodium bisulfite solution, alkaline wash water from causticizing operations (weak wash), white liquor, and chilled water, have been installed for the control of Cl_2 and ClO_2 air emissions.

Table 3. TRS and VOC Emissions from Oxygen Delignification and Bleach Plant Vents

	O_2 Delignification System Vents				Kraft Bleach Plant Vents			
			Range	Mean			Range	Mean
	No.	Ref.	(lb/ADTUBP[a])		No.	Ref.	(lb/ADTBP)	
TRS as S[d]	3	7	ND–0.0048	1.6E–03	26	7	ND–0.30	1.6E–02
Methanol	10	7	0.32–1.91	9.8E–01	51	7	ND–4.05	1.2E–01
Acetone	10	7	0.0022–0.22	5.4E–02	46	7	ND–0.16	7.2E–03
Acetaldehyde	7	7	0.008–0.16	4.1E–02	33	7	ND–0.031	3.3E–03
Formaldehyde	7	7	ND–0.004	1.7E–03	24	7	ND–0.059	2.9E–03
MEK	10	7	0.0003–0.011	3.6E–03	44	7	ND–0.021	9.9E–04
Chloroform	—	—	—	—	33	7	ND–0.77	1.2E–01
Chlorine[e]	—	—	—	—	12	7	ND–0.66	1.4E–01
Chlorine Dioxide[e]	—	—	—	—	12	7	ND–1.43	1.7E–01
Terpenes	3	7	0.0077–0.044	2.6E–02	19	7	ND–0.24	2.2E–02
THCs as C[f]	8	7	0.033–1.1	4.7E–01	32	7	0.001–0.48	1.1E–01

[a]ADT(U)BP, air-dried ton of (un)bleached pulp.
[b]No., number of sources tested.
[c]ND[x], not detected at concentration x.
[d]TRS excluding H_2S.
[e]Uncontrolled emissions from bleach plants are 0.0–10.4 lb/ADTUBP for Cl_2 and 0.03–25.5 lb/ADTUBP for ClO_2.[41]
[f]As measured by EPA M25A.

In the bleach plant, exhaust gases from the tower vents, washer hoods (if present), and washer filtrate tanks (see Figure 6) will contain volatile organic compounds, unreacted bleaching chemicals, and inadvertently formed compounds. Most of the nonchlorinated volatile organic compounds are carried into the bleach plant with the unbleached pulp slurry, although methanol also can enter with the chlorine dioxide bleaching solution if a methanol-based chlorine dioxide generating process is used. Additional methanol can be formed in the bleaching towers as a result of delignification reactions, although amounts tend to decrease with increasing levels of chlorine dioxide substitution in the initial bleaching stage. Fresh water or paper machine whitewater is usually used for the final stages of bleached pulp washing; so little methanol enters via this route.

Similar to other unit processes, the amount of volatile organic compounds in the various vent gases will be a function of the concentrations present in the process liquid streams, the volatility of the compounds at the temperatures encountered in the process, and the degree of contact between exhaust air and these liquids. For example, measurement studies have shown that methanol emissions tend to be higher from bleach plants with vacuum drum washer systems than from those with pressure or diffusion washers.[13]

Chlorine dioxide can be present in bleach plant vent gases if it is not completely consumed in the bleaching tower. It may also be found in the vent gases of absorbers where the chlorine dioxide gas from the generator is absorbed into chilled water prior to use in bleaching. Since chlorine is an impurity formed in some chlorine dioxide generating processes, it may be also be present in small amounts in the bleach plant and chlorine dioxide generator gases. Of course, chlorine can also be found in bleach plant vent gases if it is directly used in the first bleaching stage. In most bleach plants, scrubbers are employed to remove chlorine dioxide and chlorine from these vent gases. Where used as a bleaching chemical, anecdotal evidence suggests hydrogen peroxide is sometimes found in vent gases.

Chloroform is the primary chlorinated organic compound inadvertently formed when chlorinated bleaching agents are used. The use of hypochlorite results in the most chloroform formation. Hypochlorite was commonly introduced into the first extraction stage or added in a separate hypochlorite bleaching stage. Hypochlorite use at kraft and sulfite mills has been largely eliminated due to chloroform and AOX concerns, although it is still employed for the production of some dissolving pulps. Chlorine use yields lesser amounts of chloroform than hypochlorite. The formation of chloroform in bleach plants using only chlorine dioxide is very low, less than 0.02 lb/ton pulp, and is most likely due to the presence of chlorine in on-site generated chlorine dioxide. Once formed, chloroform will be present as a gas and in the process liquids. The fraction of chloroform leaving the bleach plant in sewered filtrates versus vent gases is highly variable, but more tends to leave in the filtrates when washer vent gases flows are small.[14] Other chlorinated organic compounds may be formed in the bleach plant, but only minimal concentrations have been observed in field studies.[13]

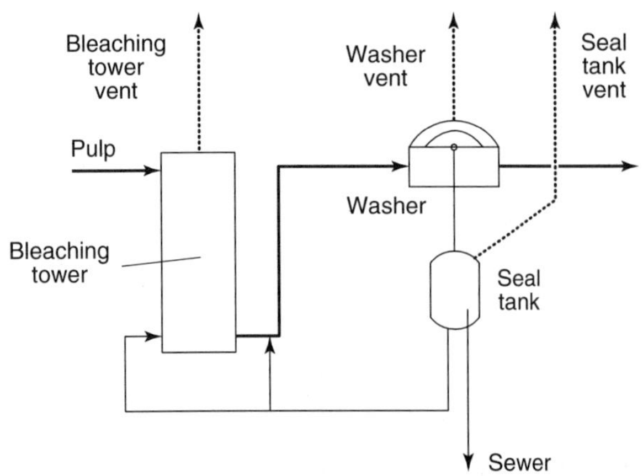

Figure 6. Typical Venting from a Bleaching Stage.

In bleaching of chemical pulps, carbon monoxide is generated mainly in the first bleaching stage where the bulk of delignification takes place. Measurements indicate over 90% of total bleach plant CO emissions come from bleaching towers where chlorine dioxide and chlorine are applied. Temporal variability is quite large, and mill-to-mill total CO emissions have been found to range from 0.2 to 1.7 lb/ODTP.[12] Carbon dioxide is also formed in the bleaching towers, but data are lacking on the levels present in the vent gases.

Table 3 also gives estimates of emissions from kraft pulp mill bleach plant vents for TRS, $CHCl_3$, Cl_2, ClO_2, several VOCs of significance, and THCs.

Recovery of Cooking Chemicals. The first recovery furnace of the current industry-wide design was installed over 70 years ago. Since then there have been many incremental improvements and refinements to this design that have lowered emissions. TRS and particulate emissions received all the attention until recent years, because of their readily observable nature (odor and visibility).

Until recent times, the recovery furnace was the predominant source of TRS emissions at every kraft mill. Although the recovery furnace still can be the largest TRS emission source, emissions have been greatly reduced over the past 25 years.[15] Three factors have contributed to significant emission reductions, namely, (1) avoidance of firing black liquor solids at rates far above the furnace design capacity, which caused excessive H_2S emissions, (2) widespread adoption of black liquor oxidation to minimize H_2S pickup across direct contact evaporators, and (3) the increasing number of NDCE-type furnace installations. In addition, there have been notable design improvements that have led to better combustion conditions in the furnace. Average concentrations of H_2S in recovery furnace flue gases now range from less than 1 to about 10 ppm for NDCEs, and from 5 to 40 ppm for DCEs with black liquor oxidation. In the United States, TRS concentrations are normally reported at standard conditions of 20°C, 0% moisture and 8% O_2 concentration rather than at actual stack conditions. Even with low TRS concentrations, the recovery furnace can still have significant mass emissions because of the high flue gas volumes.

Particulates generated in the recovery furnace are composed mainly of sodium sulfate, with lesser amounts of sodium carbonate and sodium chloride. Similar potassium compounds are also generated, but in much lower amounts. Trace amounts of other metal compounds, for example, magnesium, calcium, and zinc, can be present. The sodium compounds originate from condensation of the gaseous sodium fume released from the smelt bed and from ash generated during combustion of liquor droplets carried upward in the furnace. A significant portion of the particulate material is submicrometer in size, which makes removal with add-on control devices more difficult. The material also has a "sticky" characteristic that promotes adhesion to boiler steam tubes and other surfaces; soot blowing is required to dislodge this material.

In addition to TRS and particulates, recovery furnaces emit SO_2, oxides of nitrogen (NO_x), CO, volatile organic compounds, and products of incomplete combustion. Other potential emissions include sulfur trioxide, gaseous sulfuric acid, hydrogen chloride, and ammonia. Black liquor contains a significant amount of sulfur, nominally 3 to 5% by weight of the dissolved solids. While the vast majority of this sulfur leaves the furnace in the smelt, a small fraction (generally under 1%) can escape in gaseous or particulate form. Average SO_2 concentrations in stack gases can range from nearly 0 to 500 ppm. Factors that influence SO_2 levels are liquor sulfidity, liquor solids content, stack gas oxygen content, furnace load, auxiliary fuel use, and furnace design. None of these factors has exhibited a consistent relationship to SO_2 emissions.[16] Gaseous sulfuric acid may exist in the recovery furnace as a result of a reaction between sulfur trioxide and water vapor; the SO_3 being formed in small amounts from further oxidation of SO_2. Gaseous H_2SO_4 concentrations are small, on the order of 1 ppm.[17]

NO and NO_2 arise mainly from the presence of nitrogen in the black liquor (referred to as fuel NO_x), but limited amounts of NO can be formed from oxidation of N_2 in the combustion air (referred to as thermal NO_x). Fuel NO_x formation is related to black liquor nitrogen content, which can range from 0.05 to 0.25% of the liquor solids content[18] and oxygen concentrations present in and above the combustion zone.[19] Temperature governs the amount of thermal NO_x formed, although it may also influence the amount of fuel NO_x generated. NO amounts are 20 to 30 times higher than NO_2 amounts, although standard practice is to report $NO + NO_2$ emissions as NO_x using the molecular weight of NO_2 when reporting mass emission rates. Stack gas NO_x concentrations are normally in the 50 to 100 ppm range.[20]

In the reducing zone of the furnace, small amounts of ammonia (NH_3) can be formed from the nitrogen in the black liquor droplets. Most of the NH_3 is oxidized as the combustion gas passes upward through the oxidizing zone, but there is limited evidence suggesting NH_3 may still be present in the flue gas as it exits the upper furnace.[21] Additional study is needed on this phenomenon.

Carbon monoxide is a product of incomplete combustion. Complete combustion would result in all the organic carbon in the black liquor being converted to CO_2. CO emissions from recovery furnaces have been found to fluctuate markedly with time, and long-term mean values vary considerably from furnace to furnace. CO levels are affected by swings in liquor firing rates and liquor solids content. Empirical evidence suggests a loose positive correlation between CO and TRS emissions.

Volatile organic compounds are emitted in small amounts from recovery furnaces. The source of these compounds may be incomplete combustion or the liquor itself when it comes into contact with combustion gases. The most obvious contact between liquor and flue gas is in the DCE, where methanol and other volatiles can be transferred from the liquor to the flue gas. Less obvious contact occurs in the bottom of electrostatic precipitators, where black liquor is used to collect the captured ash, and in furnaces that use liquor to transfer ash removed from the upper furnace areas to the salt cake mix tank.

Chlorides are present in black liquor. Sources of chlorides are wood chips, purchased caustic, purchased salt cake, mill water, and spent chlorine dioxide generator acid. When the liquor is combusted, the chlorides will partition to the smelt, to particulate matter, and to gaseous form. Studies indicate roughly 75% of the chlorides will be retained in the smelt,[22] with most of the remainder being carried out of the furnace in particulate form, mainly as NaCl with some KCl. The particulates are captured by the ESP and DCE (if present) and returned to the liquor being fired. Anywhere from 0 to 8% of the chlorides present in the as-fired liquor will exit as gaseous HCl. Concentrations in stack flue gases range from 0.1 to about 50 ppm (at standard conditions). The amount of HCl in the flue gas has been found to correlate closely with the concentration of SO_2. Furnaces with very low SO_2 emissions have very low HCl emissions.[22]

The possibility exists for the formation of other chlorinated compounds in the recovery furnace, although the amounts are expected to be extremely small. Comprehensive emission measurements have shown concentrations of gaseous chlorinated organic compounds (e.g., chloroform) to be below method detection limits in nearly all cases.[7] EPA has focused a great deal of attention on chlorinated dioxins and furans over the past decade, but measurements have shown kraft recovery furnaces to have minimal emissions of these compounds.[23]

It should be noted that kraft recovery furnaces have auxiliary fuel (natural gas, residual oil, or distillate oil) burning capability for the purposes of furnace startup and shutdown. Auxiliary fuel may also be used to stabilize combustion when there are problems with the liquor supply and to maintain steam production if liquor firing is inadequate. Firing of auxiliary fuel can result in additional emissions of NO_x. SO_2 generated from fuel oil combustion is captured within the furnace to a significant extent by the sodium fume when the oil is burned in conjunction with black liquor or when the smelt bed is still present at the bottom of the furnace.[24] This capture does not take place during furnace startup on oil, since there is no smelt bed present. Oil also contains trace amounts of many metals that can contribute to emissions of metals compounds.

Table 4 gives estimates of emissions from kraft recovery furnaces for TRS, THCs, SO_2, NO_x, CO, PM, acid gases H_2SO_4 and HCl, and several VOCs of significance. The data for DCE and NDCE furnaces are provided separately. Table 5 gives estimates of emissions from black liquor oxidation systems for TRS, THCs, and several VOCs of significance. All data correspond to oxidation of strong liquors.

The significant emissions from the dissolving tank vent are TRS compounds and particulate matter. TRS arises principally from the sulfide present in the smelt and weak wash. If TRS-containing condensates are used in the recausticizing area, there is a possibility for higher levels of TRS in the weak wash and greater potential for stripping of these compounds. H_2S is the principal compound present in vent gases, with typical concentrations in the range of 5 to 20 ppm. It is thought to be mainly generated by the smelt shattering.[25] Methyl mercaptan, dimethyl sulfide, and dimethyl disulfide can be present in the vent gases if they are present in the weak wash as a result of condensate reuse. Scrubbers using weak wash can reduce vent gas concentrations of H_2S and methyl mercaptan, but not dimethyl sulfide and dimethyl disulfide. Caustic and oxidized

Table 4. TRS, SO_2, NO_x, CO, PM, Acid Gas and VOC Emissions from Kraft Recovery Furnaces

	DCE Kraft Recovery Furnace				NDCE Kraft Recovery Furnace			
			Range	Mean			Range	Mean
	No.[b]	Ref.	lb/ton BLS[a]		No.	Ref.	lb/ton BLS	
TRS as S[d]	13	7	ND–0.18[c]	0.025	11	7	ND–0.051	0.0066
SO_2	21	20	0.025–10.8	2.06	13	20	0.01–54.1	2.63
NO_x	11	20	1.06–2.44	1.45	16	20	0.63–2.30	1.20
CO	5	26	0.25–26.3	6.63	5	26	0.25–26.3	6.63
PM[e]	10	27	0.41–0.98	0.46	19	28	0.31–1.66	1.04
H_2SO_4	5	7	ND–0.047	9.3E–03	4	7	ND–0.16	3.9E–02
HCl	17	7	ND–0.55	1.6E–01	21	7	ND–1.23	2.6E–01
THCs as C[f]	10	7	0.01–1.51	4.8E–01	16	7	ND–0.98	1.6E–01
Methanol	18	7	ND–1.35	3.0E–01	14	7	ND–0.27	6.9E–02
Acetone	16	7	ND–0.113	1.9E–02	12	7	ND–0.048	5.5E–03
Acetaldehyde	8	7	ND–0.071	2.1E–02	13	7	ND–0.05	5.3E–03
Formaldehyde	7	7	ND–0.0096	2.6E–03	9	7	ND–0.091	1.5E–02
Phenol	7	7	ND–0.08	1.2E–02	5	7	ND[5.8E–02]	—
Terpenes	7	7	ND–0.538	6.0E–02	8	7	ND–0.092	1.8E–02

[a]Ton BLS, ton black liquor solids.
[b]No., number of sources tested.
[c]ND[x], not detected at concentration x.
[d]TRS excluding H_2S; H_2S is the dominant TRS compound emitted; average concentrations of H_2S in recovery furnace flue gases range from <1 to ≈ 10 ppm (or ≈ 0.015 to 0.15 lb S/ton BLS) for NDCEs, and from <5 to ≈ 40 ppm (or ≈ 0.073 to 0.58 lb S/ton BLS) for DCEs with black liquor oxidation.
[e]Particulate emissions from ESPs on recovery furnaces were converted from units of gr/acf and gr/dscf to lb/ton BLS using conversion factors of 400 acfm/TPD and 301 scfm/TPD and 1.6 ton BLS/ADTP.
[f]As measured by EPA M25A.

Table 5. TRS and VOC Emissions from Black Liquor Oxidation Systems

			Range	Mean
	No.	Ref.	(lb/ton BLS[a])	(lb/ton BLS)
TRS as S	9	7	ND–0.40	0.074
THCs as C[d]	6	7	0.012–0.33	1.4E–01
Methanol	16	7	0.035–2.10	4.4E–01
Acetone	13	7	ND–0.097	2.4E–02
Acetaldehyde	9	7	0.002–0.17	4.5E–02
MEK	13	7	ND–0.084	1.2E–02
Terpenes	5	7	ND–0.33	9.3E–02

[a]Ton BLS, ton black liquor solids.
[b]No., number of sources tested.
[c]ND[x], not detected at concentration x.
[d]As measured by EPA M25A.

white liquor are more effective in removing all of the TRS compounds.[25]

As with the recovery furnace, particulates are composed of mainly sodium compounds with much lesser amounts of potassium compounds and some other trace metal compounds. The dominant compound is sodium carbonate, followed by sodium sulfate. Roughly 90% (by weight) of the particles have equivalent aerodynamic diameters under 10 μm, and 50% have diameters under 1 μm.[29]

Volatile organic compounds such as methanol can be released from the weak wash in both the dissolving tank and the wet scrubber particulate control device. In a study of emissions from four smelt dissolving tanks, NCASI found methanol emissions were closely related to the methanol content of the weak wash used in the scrubbers.[30] Ammonia has been found in the smelt dissolving tank exhaust, although amounts appear to be highly variable with measured emission rates ranging from around 0.02 to 3.8 lb/ton BLS.[31]

Table 6 gives estimates of emissions from smelt dissolving tanks for TRS, THCs, SO_2, NO_x, CO, PM, and several VOCs of significance.

For lime kilns, the lime mud is the primary source of H_2S, which can be released when the CO_2-rich flue gas comes into contact with sodium sulfide in the lime mud. Removal of as much as possible of the sodium sulfide from the lime mud prior to its introduction into the kiln is main method for minimizing H_2S release. If a scrubber is used, the scrubber makeup water can be a potential source of TRS compounds if condensates or other TRS-containing liquids are used; organic reduced sulfur compounds can be stripped as the flue gas passes through the scrubber. Minimizing concentrations of reduced sulfur compounds in the scrubber makeup water will reduce the potential for this stripping. Addition of caustic to the scrubber makeup can serve to reduce TRS emissions. Many mills use lime kilns for the burning of concentrated non-condensable gases (NCGs) that contain high concentrations of TRS compounds. Under normal operating conditions, the TRS compounds are oxidized to SO_2 in the kiln. Under upset conditions, it is possible that a portion of the TRS compounds will not be oxidized and thus be emitted.

While passing through the kiln, the combustion gases pick up a good deal of particulate matter that must be removed before the gases exit to the atmosphere. Mechanical devices such as dust chambers or cyclones are generally used to remove larger particles, which are mainly calcium-containing. A wet scrubber or electrostatic precipitator follows for removal of

Table 6. TRS, SO_2, NO_x, CO, PM, and VOC Emissions from Kraft Smelt Dissolving Tank Vents and Lime Kilns

	Smelt Dissolving Tank Vents				Kraft Lime Kilns			
			Range	Mean			Range	Mean
	No.[c]	Ref.	(lb/ton BLS[a])		No.	Ref.	(lb/ton CaO[b])	
TRS as S	22	7	ND–0.084[a]	0.005	22	7	ND–0.102	0.023
SO_2	9	20	ND–0.075	0.016	16	20	0.01–1.63	0.23
NO_x	10	20	ND–0.151	0.033	14	20	0.08–9.0	2.19
CO	—	—	—	—	4	26	0.15–0.44	0.25
PM[e]	10	32	0.04–0.18	0.13	5	32	0.84–4.0	1.82
THCs as C[f]	11	7	ND–0.25	6.3E−02	17	7	ND–1.40	1.3E−01
Methanol	26	7	ND–0.58	7.8E−02	25	7	ND–1.07	7.0E−02
MEK	23	7	ND–0.0043	2.6E−04	22	7	ND–0.17	4.9E−03
Acetone	15	7	ND–0.007	1.6E−03	10	7	ND–0.039	4.5E−03
Acetaldehyde	9	7	ND–0.004	6.8E−04	13	7	ND–0.047	7.1E−03
Ammonia	19	31	0.02–3.80	4.1E−01				
Formaldehyde	2	7	2E−03–0.016	5.2E−03	10	7	ND–0.042	9.9E−03
Terpenes	14	7	ND–0.015	1.5E−03	13	7	ND–0.10	2.1E−02

[a]Ton BLS, ton black liquor solids.
[b]Ton CaO, ton of reburned lime product.
[c]No., number of sources tested.
[d]ND[x] not detected at concentration x.
[e]For smelt tanks and lime kilns with venturi scrubbers.
[f]As measured by EPA M25A.

smaller particulates, which are mainly sodium sulfate and sodium carbonate and have aerodynamic diameters less than 10 μm. Particulate matter from fuel oil and petroleum coke burning will be captured as well.

SO_2 emissions from lime kilns are generally very low because of the highly alkaline atmosphere inside the kiln that is conducive to capture of the SO_2 as calcium and sodium sulfates. SO_2 will be formed when NCG gases are burned in the kiln or when fuel oil or petroleum coke is used as the fuel. Emission test data summarized by NCASI[20] for 20 kilns showed average SO_2 concentrations in stack gases were under 10 ppmdv for 16 of the units, and did not appear to be related to the fuel type or NCG burning in the kiln. On the other hand, 3 out of the 20 kilns had SO_2 concentrations in the 40 to 240 ppmdv range. There was no apparent reason for these higher concentrations.

NO_x emissions result from fossil fuel burning and possibly from combustion of NCGs (if burned in the kiln). Available data indicate NO_x emissions vary by an order of magnitude among kilns, and suggest sizeable contributions may come from the burning of NCGs containing substantial amounts of ammonia.[33] The available data do not show higher NO_x emissions from kilns burning oil versus natural gas, as might be expected based on the higher emissions from boilers burning oil versus natural gas.[20]

Lime kilns emit CO and VOCs. CO and some VOCs (e.g., formaldehyde) are mainly a product of incomplete fuel combustion, and levels are highly variable among kilns. Some volatile organic compounds enter the kiln with the liquid component of lime mud and are released as the mud is heated. Volatile organic compounds present in scrubber makeup water can be stripped by the flue gas exiting the kiln. Methanol has been found to be the dominant organic compound present in lime kiln exhaust gases.[30] Table 6 also gives estimates of emissions from kraft pulp mill lime kilns for TRS, THCs, SO_2, NO_x, CO, PM, and several VOCs of significance.

Steam leaving a slaker vent contains a considerable amount of particulate matter, which is largely calcium and sodium carbonates and sulfates. Scrubbers are generally employed to capture this particulate matter. Ammonia has been found in slaker stacks, with emission rates ranging from less than 0.1 to about 0.8 lb/ton CaO.[31] As with smelt dissolving tanks, the origin of the ammonia is uncertain.

The smelt dissolving tank and slaker vents are the major emission sources in the causticizing area, but there are numerous other pieces of equipment that are vented to the atmosphere. These sources are associated with the processing of green liquor (clarifiers, storage and surge tanks, dregs filters); white liquor (causticizer tanks, clarifiers, pressure filters, storage tanks); and lime mud (mix tanks, dilution tanks, storage tanks, pressure filters, precoat filters, filter vacuum pump exhausts). These sources typically have small gas flow rates and very low concentrations of gaseous organic compounds, with methanol being the primary one.[30] The amounts of organic compounds in the vent gases will be a function of the process liquid concentrations, temperature of the process, and vent gas flow rates. Total VOC emissions from all these sources are on the order of 0.3 lb/ton CaO or less. Causticizer vents have been found to contain small amounts of ammonia,[31] but the amounts are highly variable, with emission rates ranging from less than 0.01 to almost 0.7 lb/ton CaO.

Miscellaneous Kraft Pulping Area Emission Sources. Turpentine and tall oil are two by-products of the kraft pulping process that many mills using softwoods, especially Southern pine species, recover. Turpentine can be obtained by condensing digester relief and blow gases and sending the condensates to a decanter to separate the turpentine from the water. Both the turpentine and water have considerable amounts of reduced sulfur compounds, and the water will contain other pulping by-products, including alcohols and organic acids. Thus the vents on turpentine decanters can contain TRS and volatile organic compounds. Because of odor concerns, decanter vent gases are typically collected for incineration. Following the decanter, most mills route the turpentine to a storage tank. Storage tank vents may contain small amounts of TRS and VOCs.

Crude tall oil can be obtained from processing soap removed from black liquor. Normally, the primary source of soap is skimmings from black liquor at an intermediate point in the multiple-effect evaporators. Lesser amounts of soap can be obtained from foam tanks associated with brownstock washers and from weak black liquor storage tanks. The soap skimmings are sent to a storage tank prior to further processing. Either a batch or continuous acidulation process is used to produce crude tall oil. In the batch process, soap is introduced to a reactor containing a hot sulfuric acid solution. The mixture is agitated for about 1 h, and is then allowed to settle for a few hours to separate the crude tall oil from the spent acid (brine). The crude tall oil and brine are piped to separate storage tanks. Crude tall oil is shipped off-site for distillation and purification. In the continuous process, the reactor and subsequent separators are run continuously.

Because soap skimmings contain some black liquor, H_2S will be formed in the acidulation process from a reaction between the acid and Na_2S. TRS and organic compounds (e.g., methanol) present in the black liquor may also be released during acidulation. The reactor vent is the primary emission point for these compounds, with only minor emissions coming from subsequent settling and storage tanks. H_2S concentrations in batch reactor vents tend to peak about half-way through the acidulation process, and can be a few thousand ppm. Concentrations are much lower in continuous reactor vents.[34] H_2S is the dominant compound present in the reactor vent. Due to odor concerns, scrubbers are frequently installed on reactor vents for H_2S removal.

Over one-third of the U.S. kraft mills have steam strippers. These devices are employed to remove TRS and organic compounds from selected digester and evaporator area condensates. Reasons for stripper installation include reduction of mill odor, water conservation, and reduction of TRS and organic loading to the wastewater treatment plant. Stripped condensates are often reused for pulp washing or in the recausticizing area, although some mills send all or a fraction of the treated condensates to the wastewater treatment plant. Steam stripping transfers volatile compounds from the liquid to the gas phase. Stripper off-gases thus contain very high levels of TRS compounds and methanol, along with lesser amounts of other volatiles, including terpenes (at softwood mills), ethanol, acetone, acetaldehyde, and methyl ethyl ketone. These gases must be incinerated to prevent unacceptable odor impacts.

All kraft mills collect noncondensible gases (NCGs) from the pulping operations for subsequent incineration. Burning of these TRS-containing gases results in the oxidation of the reduced sulfur compounds to SO_2. While burning of NCGs in lime kilns or recovery furnaces does not normally yield any increase in SO_2 emissions because of the large excess of alkaline material present in these devices, burning in boilers or a standalone thermal oxidizer will result in SO_2 emission increases.

Thermal oxidizers are used at over one-third of U.S. kraft mills for the combustion of so-called concentrated NCGs from the digester and evaporator area. Many mills also burn stripper off-gases in these units. These oxidizers are generally operated at temperatures in a range of 1200 to 1600°F to ensure complete oxidation of the TRS compounds to SO_2 and conversion of hydrocarbons to H_2O and CO_2. SO_2 is the primary emission of concern, and most thermal oxidizers not used as a backup incineration device have scrubbers to remove the SO_2. Uncontrolled SO_2 emissions are generally in the range of 2 to 8 lb/ADTP, with the higher values being for units burning stripper off-gases. Average controlled emissions are less than 0.1 lb/ADTP. NO_x emissions occur due to the combustion of auxiliary fuel (normally gas, but occasionally propane or distillate fuel oil) or due to the presence of ammonia in stripper off-gases; median NO_x emission rates are on the order of 0.1 lb/ADTP. CO and VOC emissions are under 0.1 lb/ADTP.

Air Pollution Control Measures

Kraft pulp mills that began construction or modification after September 24, 1976, are subject to NSPS for particulate and TRS compound emissions as shown in Table 7.[35] In 1979, the U.S. Environmental Protection Agency (EPA) issued retrofit emission guidelines for the control of TRS emissions at existing facilities not subject to NSPS, and these are also shown in Table 7.[36] In addition to particulate and TRS, many "preconstruction" air quality permits for kraft pulp mill sources contain limits for PM_{10}, SO_2, NO_x, CO, and VOCs.

Control of TRS Emissions. The collection and treatment of NCGs from kraft pulp mill sources have been practiced for over five decades. Most early systems were designed to control TRS emissions from the digester and multiple-effect evaporator NCGs only. During the past two decades, efforts have been made to include such sources as the brown-stock washer hood vents, condensate stripper off-gases, turpentine decanter vents, and a number of other minor sources.[37] Incineration in existing combustion devices, namely, lime kilns and power boilers, is the treatment most commonly used, although thermal oxidizers dedicated for this purpose are becoming increasingly prevalent. However, the oxidation of TRS results in SO_2 formation and a caustic scrubber is often installed following the thermal oxidizer for SO_2 removal. Other combustion devices, such as kraft recovery furnaces and flares, are also used. Alternative disposal practices employed include venting with the bleach-plant chlorine or chlorine dioxide–stage washer vent gases for chemical oxidation of the sulfur compounds, and, during process upsets, discharge of selected streams to stacks for improved dispersion, thus minimizing the potential for worker exposure to high concentrations of TRS.[37]

The noncondensible gases in the kraft pulping process are of two types — low-volume, high-concentration (LVHC) or "concentrated" NCGs (CNCG) and high-volume, low-concentration (HVLC) or "dilute" NCGs (DNCG). The CNCGs include those from the digester area, condensate strippers, evaporators, and so on, while the DNCGs include those from brown-stock washers, pulp knotters, washer seal tanks, etc. The CNCGs are typically burned in lime kilns, boilers, or thermal oxidizers, while the DNCGs are burned in boilers or recovery furnaces that can accept such large gas volumes.

Table 7. NSPS and Emission Guidelines for Kraft Pulp Mills

Source	Particulate Matter	Total Reduced Sulfur	
		NSPS	Guidelines[a]
Recovery furnace			
New design	0.044 gr/dscf @ 8% O_2	5 ppmdv[b] @ 8% O_2	5 ppmdv @ 8% O_2
Old design	—	—	20 ppmdv @ 8% O_2
Cross recovery	0.044 gr/dscf @ 8% O_2	25 ppmdv @ 8% O_2	25 ppmdv @ 8% O_2
Smelt dissolving tank	0.20 lb/ton bls	0.032 lb/ton bls	—
Lime kiln			
Gas	0.067 gr/dscf @ 10% O_2	8 ppmdv @ 10% O_2	20 ppmdv @ 10% O_2
Liquid	0.13 gr/dscf @ 10% O_2	8 ppmdv @ 10% O_2	20 ppmdv @ 10% O_2
Digester system	—	5 ppmdv @ 10% O_2	No control
BSW system	—	5 ppmdv @ 10% O_2	No control
BLO system	—	No control	No control
MEE system	—	5 ppmdv @ 10% O_2	5 ppmdv @ 10% O_2
Condensate stripper system	—	5 ppmdv @ 10% O_2	5 ppmdv @ 10% O_2

[a] EPA guidelines for TRS emission from existing kraft pulp mills not subject to NSPS[36].
[b] ppmdv, ppm dry on volume basis.

Vent gases from BLO systems are not required to be controlled under NSPS due to the prohibitive cost-effectiveness of control and the declining use of BLO.[38] TRS emissions from the newer diffusion brown-stock washers are extremely small as compared with the older vacuum-drum-type washing systems.[39] The use of fresh water or stripped condensate as a washing medium in brown-stock washers (as opposed to dirty condensate) also leads to reduced TRS emissions. TRS emissions from smelt-dissolving tanks are similarly most effectively controlled by the choice of a water (both for smelt dissolving and particulate control) which contains minimal amounts of reduced sulfur compounds.

TRS emissions from kraft recovery furnaces are most efficiently controlled by maintaining sufficient oxygen, residence time, and turbulence, and by avoiding overloading. Furnaces with DCEs rely additionally on BLO for TRS emission control. By oxidizing the Na_2S in the liquor to $Na_2S_2O_3$ before it enters the DCE, the reactions between the combustion gases and black liquor in a DCE that generate H_2S are inhibited.

Lime kiln TRS control mainly involves achieving a high degree of lime mud washing. Lime mud is the precipitate resulting from the causticizing reaction when Na_2CO_3 in the green liquor is converted to NaOH and lime is converted to $CaCO_3$. Sodium sulfide entrained in the lime mud reacts with CO_2 in the cold end of the kiln, giving rise to H_2S emissions. Proper operation of the lime mud filter to prevent Na_2S from entering the lime kiln and allow for oxidation of residual Na_2S to $Na_2S_2O_3$ is usually sufficient to keep H_2S levels below about 8 ppm. The use of sulfide-free streams such as fresh water or clean condensate for makeup in the scrubber also prevents H_2S formation by contact with kiln gases.

Control of Particulate Emissions. Particulate control on furnaces with DCEs and on NDCE furnaces is achieved predominantly by ESPs. Electrostatic precipitator particulate-removal efficiencies range from about 90% in older installations to well over 99% in newer units. The range and average particulate emission factors for furnaces with ESPs are shown in Table 4. Direct-contact evaporators used to concentrate black liquor also serve to scrub the particulates leaving the furnace, removing from 20% to 50% of the particulate load prior to the ESP. A few scrubbers have been installed following older ESPs to obtain satisfactory levels of particulate removal.

Particulate control of smelt-dissolving tank vent gases is accomplished by installing demister pads, packed towers, or venturi scrubbers. Most lime kilns are controlled by venturi scrubbers, with pressure drops ranging from 17 to 34 in. of water, although ESPs are increasingly being used in new installations. Typical emission factors for smelt-dissolving tank vents and lime kilns with control are also shown in Table 5.

Fugitive emissions from sources or areas of importance in a kraft mill include coal piles, paved and unpaved roads, bulk materials handling (lime, limestone, starch, etc.), and wood handling. Control strategies include wetting; the use of chemical agents, building enclosures, and windscreens; paving or wetting roads; and modifying handling equipment.[40]

Control of SO_2, NO_x, CO, and VOC Emissions. Besides power boilers in which sulfur-containing fuels are fired, SO_2 emissions from a kraft mill occur principally through kraft-recovery-furnace flue gases. Unlike power boilers firing fossil fuel, the combustion of black liquor in a kraft recovery furnace results in SO_2 emissions that are extremely variable. These emissions depend on a variety of factors, which include (1) liquor properties such as sulfidity (or sulfur-to-sodium ratio), heat value, and solids content; (2) combustion air and liquor firing patterns; (3) furnace design features; and (4) other operational parameters.[16] Table 5 gives SO_2 emission factors for DCE and NDCE furnaces developed from averaging long-term, continuous emission monitoring SO_2 data.[16] Strategies to (1) lower liquor sulfidity and (2) optimize liquor and combustion air properties and firing patterns so as to yield maximum and uniform temperatures in the lower furnace have been used to minimize kraft-recovery-furnace SO_2 emissions. Flue gas desulfurization is capital and energy intensive and its efficacy is uncertain, considering the generally low concentrations and rapidly fluctuating levels of SO_2 in the furnace flue gases. Sulfur dioxide may be formed in the lime kiln when fuel oil is combusted. The regenerated quicklime in the kiln acts as an in situ scrubbing agent and the venturi scrubber that normally follows the kiln augments the SO_2 removal process. Limited SO_2 measurements following ESPs on fuel-oil-burning kilns suggest that over 90% of the SO_2 is captured by the time the flue gas exits the ESP.[20]

NO_x emissions from kraft recovery furnaces and lime kilns result from black liquor and fossil fuel combustion, respectively. The emissions from kraft recovery furnaces are mainly attributed to "fuel NO_x," resulting from partial oxidation of the black liquor nitrogen content. Kraft recovery furnaces operate with a reducing zone in the lower part of the furnace (reduction of Na_2SO_4 to Na_2S) and with an oxidizing zone farther up in the region of the liquor spray guns and secondary air. This "staged combustion" is a natural deterrent to excessive NO_x formation, with the result that most existing kraft recovery furnaces emit less than about 100 ppm NO_x. However, the current trend for new units or existing units undergoing upgrade is to burn liquors with increasingly higher solids content (>70% solids). Available emission factors for NO_x from kraft recovery furnaces (shown in Table 4) are based on tests on older units burning liquors with <70% solids. At present, the impact of burning higher solids liquors on NO_x emissions is not clear. Some have observed an increase in NO_x emissions, while others have concluded that the liquor N and not the percent solids in the liquor is the dominant factor that determines furnace NO_x emissions. Coincidentally, higher temperatures in the lower furnace have been documented to yield significant other environmental (such as lower SO_2 and TRS) and furnace operational (such as smelt-bed stability) advantages.[16] Higher temperatures in the lower furnace zone can result from several factors, including firing higher-solids liquor, firing liquor with higher heat content, and operating the furnace at higher than the design load.

The VOC and CO emissions from recovery furnaces and lime kilns result from incomplete combustion of the organic matter in the fuel. Lime kiln CO and VOC emissions are generally small, as seen from Table 5. Recovery-furnace CO and VOC emissions are a function of the level of excess air used and the degree of mixing achieved within the furnace. Control strategies to minimize CO and VOC emissions involve increasing the residence time, oxygen content, temperature, and level of turbulence in the furnace combustion zone. Unfortunately, an increase in excess air, residence time, and temperature has the opposite effect on NO_x formation.

Control of Bleach Plant Emissions. As shown in Table 3, the uncontrolled Cl_2 and ClO_2 emissions from bleach plants exhibit a very broad range. These emissions are expected from bleach towers, washer hood vents, and seal tank vents of the chlorine and chlorine dioxide stages of bleaching, respectively. Emissions of Cl_2 and ClO_2 are also expected from the ClO_2 generator absorbers.

Studies have indicated that maintaining low-bleaching chemical residuals in the pulp leaving the bleaching towers results in minimal Cl_2 and ClO_2 emissions from washer hoods and seal tanks.[41] Application rates for Cl_2 and ClO_2 govern the Cl_2 and ClO_2 emission rates from their respective tower vents. Packed-tower scrubbers using various scrubbing fluids are used to control Cl_2 and ClO_2 emissions from washer hoods and seal tanks. Smaller scrubbers are used to control emissions from tower vents, especially those of an upflow–downflow design.[41] These smaller scrubbers are designed with chemical recovery in mind. Some mills use the larger packed-bed scrubbers to scrub combined vent gases from tower vents, washer hoods, and seal tanks. Average removal efficiencies for Cl_2 and ClO_2, respectively, across packed-bed scrubbers were determined to be: 72% and 65% using E-filtrate; 85% and 50% using caustic; 92% and 90% using sodium bisulfite; 99% and 99% using weak wash; 99% and 99% using white liquor; 90% and 98% using chilled water and caustic.[41]

The formation of by-product chloroform ($CHCl_3$) during pulp bleaching is a much more complex phenomenon. Laboratory and field studies have shown that $CHCl_3$ may be formed in the chlorination (C), extraction (E), and hypochlorite (H) stages of bleaching.[42] The use of hypochlorite in the bleaching sequence seems to have the most impact on $CHCl_3$ formation. Elimination of hypochlorite from the bleaching sequence has been used to reduce $CHCl_3$ generation[42] at a number of bleach plants. A reduction in the chlorine factor, defined as percent chlorine applied/kappa number, where kappa number is a measure of the residual lignin content, also helps in minimizing chloroform by-product formation. Less than 0.02 lb/ton pulp of $CHCl_3$ is formed when only ClO_2 is used in the bleaching sequence (often the case at the present time). The feasibility of gas-phase scrubbing of $CHCl_3$ from bleach plant vents has not yet been demonstrated. A reduction in $CHCl_3$ by-product formation would also lead to decreased emissions from the wastewater treatment area. On an average, slightly over one half of the $CHCl_3$ formed in the bleach plant is expected to be released through the bleach plant vents.[42] Thus the use of ClO_2 in pulp bleaching is perhaps the best method for minimizing emissions of $CHCl_3$.

ACID SULFITE PULPING

The sulfite pulping process has been largely replaced by the kraft process for the production of chemical wood pulps. Sulfite pulp production has been declining for several decades, and there were only 11 U.S. mills still producing sulfite pulp at the end of 1999.

Process Description

Like the kraft process, wood chips are cooked in batch or continuous digesters at elevated temperature (260–320°F) and pressure (90–100 psi). The cooking liquor is an acid solution containing bisulfite ions of calcium, magnesium, sodium, or ammonium. During cooking, SO_2 will be formed and released with the relief and blow gas streams. Organic compounds including methanol, acetone, furfural, and terpenes (only from softwoods) will be formed in the digester. These compounds will partition to both the gas and liquid phases.

Digester relief and blow gases are routed to absorbers to recover the gaseous SO_2 for reuse in making cooking liquor. Some of the organic compounds in the gases may be absorbed as well, depending on the relative concentrations in the gas and liquid phases in the absorber. Additional SO_2 for cooking liquor makeup is obtained by burning sulfur and routing the gas to an SO_2 absorber. Since SO_2 capture will not be 100% in the absorbers, the exiting gas streams are normally scrubbed to remove the residual SO_2.

Sulfite mills also use vacuum drum washers for pulp washing. With respect to emissions, the same considerations apply as for kraft mills, although there are no TRS compounds in the sulfite. However, SO_2 will be present in the vent gases at sulfite mills.

As for kraft pulps, bleaching of sulfite pulps is also accomplished in a multistep process by using chemicals to oxidize and solubilize the lignin compounds. Oxidizing chemicals include oxygen, ozone, chlorine, chlorine dioxide, sodium hypochlorite, and hydrogen peroxide. Bleaching is accomplished in a multistage bleach plant, normally with chlorine-containing compounds. Each stage consists of a bleaching tower followed by a washer.

In sulfite pulping, all the chemical recovery processes involve capture of SO_2 liberated from the digester during cooking of the wood chips in a sulfurous acid solution. As noted earlier, the base cation for sulfite pulping can be sodium, calcium, magnesium, or ammonium. The magnesium-based process is currently the most widely used form of sulfite pulping. A typical Mg-based sulfite pulping and recovery process is shown schematically in Figure 7. Sodium and magnesium are captured for reuse, but it is not practical to recover ammonium. In the United States, the three operating calcium-based sulfite mills do not recover calcium, although recovery processes for calcium do exist.[43] However, the cost of calcium recovery normally exceeds its value.

Liquor Evaporation. After removal from the digesters, the pulp is separated from the spent sulfite pulping liquor (referred to as red liquor). This weak liquor, which has a solids content of about 10%, is then sent to multiple-effect evaporators to increase its solids content. In the case of calcium bases, water evaporation is performed only to the extent necessary to produce lignin-based products such as vanillin. For the other bases, the solids content is increased to around 50% so that the liquor can be burned in a combustion device.

Liquor Combustion. At ammonium and magnesium base mills, the concentrated liquor is most often burned in a specialized recovery furnace, although a fluidized bed combustor or normal boiler can also be used. Sodium base mills have additional combustion options, but no sodium base mills are operating in the United States. Although sulfite liquor at 50% solids is capable of supporting combustion, auxiliary gas or oil is often co-fired with the liquor to ensure stable combustion conditions in a recovery furnace. Fluidized bed units are capable of combusting spent liquors with lower solids content than recovery furnaces.

The flue gas resulting from liquor combustion contains significant amounts of SO_2 that can be captured and used for making new cooking liquor. This is typically done by routing the flue gas through an absorption unit, where the liquid used for SO_2 capture is either an alkaline solution containing ammonia (e.g., ammonium hydroxide) or a slightly acid solution containing magnesium. The resulting bisulfite solution is then sent to a fortification tower to increase the concentration of sulfite and base ions to acceptable strength for pulping. Some mills employ a scrubber following the absorption unit for further reductions in SO_2 concentrations.

Sulfur Burning. To prepare acid cooking liquor, sulfite mills burn elemental sulfur to get a concentrated SO_2 gas. This gas is first routed through a cooling tower or quench to reduce its temperature. At ammonium and magnesium base mills, the

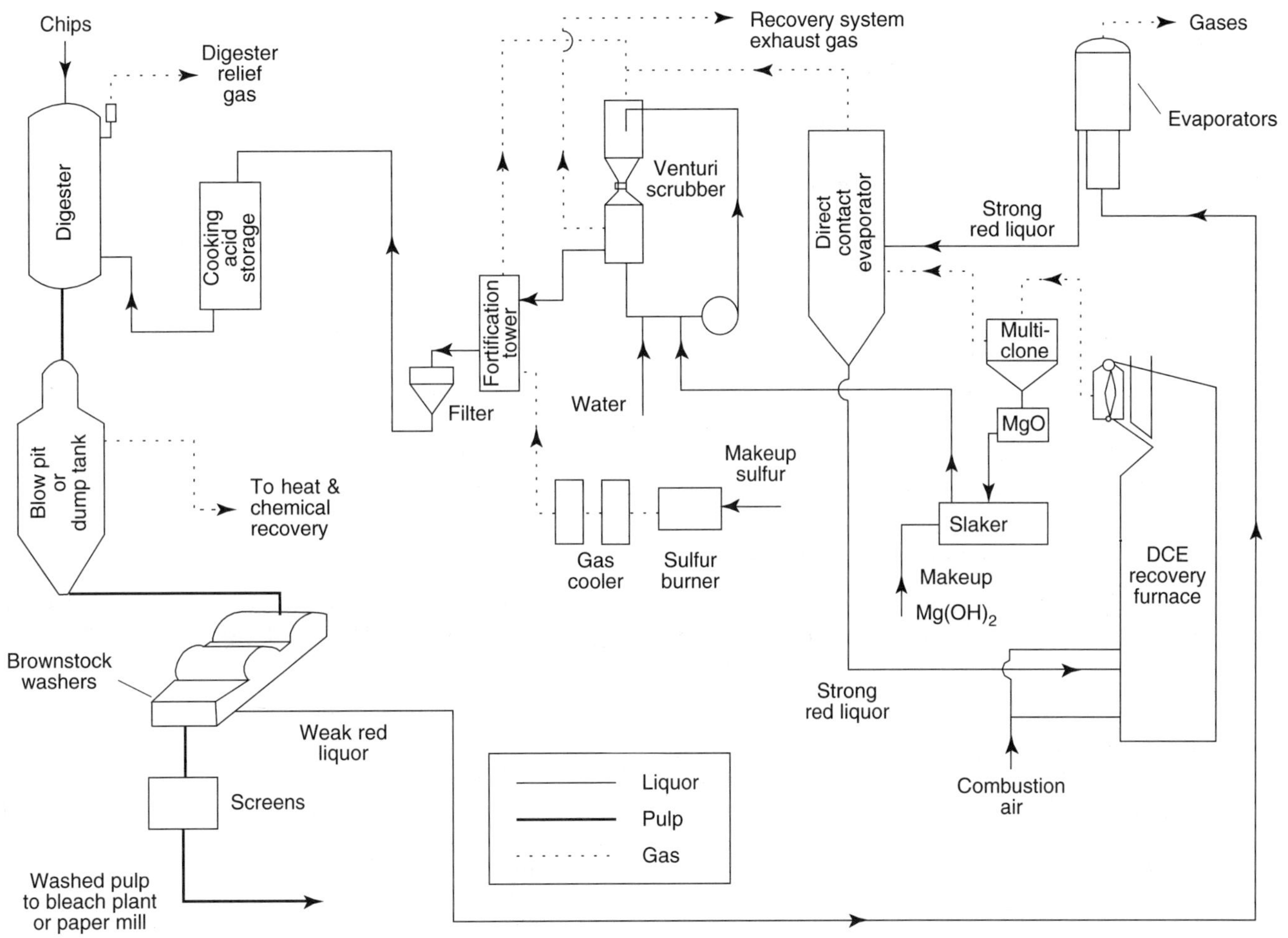

Figure 7. Pulping and Recovery in a Typical Magnesium-Based Sulfite Process.

gas is then sent to the acid fortification tower, where the SO_2 is absorbed into the weak bisulfite solution.

Air Emissions Characterization and Control Measures

Sulfur dioxide is considered the major pollutant of concern from sulfite pulping. The characteristic odor of kraft pulping is generally absent, as TRS compounds are not formed during the lignin–bisulfite reactions. The liquor evaporation process will release SO_2 and volatile organic compounds, notably methanol, from the liquor. The noncondensible portion of these gases will typically be exhausted from the evaporator set at the vacuum system vent. These exhaust gases are normally ducted to a scrubber or absorber for SO_2 capture. The captured SO_2 can be used to reformulate the cooking acid.

The digester and blow tank areas are a major source of SO_2. The relief gases from the digester and NCG gases from the presteaming vessel and flash evaporators are normally all returned to the acid-preparation system for SO_2 recovery. Considerable quantities of SO_2 (ranging from 10 to 70 pounds per ton of pulp[44] could be flashed during a hot blow, with uncontrolled emissions comprising nearly 95% water vapor, 3% SO_2, and 2% CO_2.[45] The quantity of SO_2 actually evolved and emitted in the blow gases is usually much less, depending on the pH of the cooking liquor and the pressure at which the digester contents are discharged. Raising the pH of the digester contents before blow lowers the free SO_2 in solution. Lowering the digester pressure before blow (to as low as 3 psi) and pumping instead of blowing out the contents are added measures for reducing SO_2 emissions. The SO_2 released to the atmosphere, however, depends on the effectiveness of the heat-recovery and SO_2-absorption systems employed for SO_2 recovery. The SO_2 in blow gases can be scrubbed with an alkaline solution of the base and returned to acid preparation, with recoveries as high as 97%.[45] This method is viable with sodium and NH_3 bases. However, magnesium and calcium bases require slurry scrubbers, which are less practical.[45] The SO_2 released during a cold blow from a blow tank is of a lesser magnitude (4–20 pounds SO_2 per ton of pulp),[43] which makes scrubbing impractical.

Many sulfite mills currently utilize displacement or diffusion washers, quite similar to those in kraft mills. Uncontrolled emissions from acid bisulfite washers and screens can be as high as 16 pounds SO_2 per ton of pulp,[45] although typical losses range from 1 to 4 lb/ton.[46] Scrubbing of SO_2 from the washers is generally accomplished by hooding and directing the collected gases to a direct-contact scrubber.

There are no direct SO_2 emissions from the sulfur burning process at these mills. At calcium base mills, the SO_2 gas is routed through packed limestone towers for preparation of the cooking acid. The vent gas from these towers will contain small amounts of SO_2; the vent gases may be scrubbed to remove the SO_2.

Liquor combustion will generate SO_2, particulate matter, NO_x, CO, and products of incomplete combustion. Particulate emissions are a concern for both ammonium and magnesium base liquor burning. The principal issue at ammonium base mills is the formation of very small ammonium salt particles (such as ammonium sulfate) in the furnace and absorber. These particles can result in a visible plume from the recovery furnace stack, sometimes referred to as "blue haze." Removal of the fine particles is accomplished by passing the flue gas through a series of fabric mesh pads called "candles" that are installed after the absorption unit. Particulate emissions following these devices are generally under 0.05 gr/dscf at 8% O_2.[47]

At magnesium base mills, the recovery furnace or fluidized bed reactor is operated such that magnesium oxide is formed and carried out of the unit in the flue gas. The flue gas is then passed through a mechanical collector, typically multiple cyclones, to capture most of the MgO particles. Exhaust gas from the collectors is then passed through a series of three or four packed tower absorbers for SO_2 capture. Some mills use a cooling tower prior to the absorption towers to reduce the temperature of the flue gas to improve SO_2 capture in the absorbers. Some mills use venturi scrubbers in addition to the packed towers, and some mills have a final SO_2 scrubber after the absorption towers to further reduce SO_2 and particulate concentrations. Particulate emission levels after all control devices typically range from 0.05 to 0.1 gr/dscf, corrected to 8% O_2.

If spent ammonium-based liquor is being burned, there will be considerable amounts of NO_x in the flue gas due to the high nitrogen content of the spent liquor. Measurements on operating U.S. recovery furnaces showed average NO_x emission rates of 8.8 lb/ADTP, about three times the reported average of 2.5 lb/ADTP for units burning spent magnesium-base liquor.[48]

There will also be products of incomplete combustion in the flue gas from sulfite liquor burning units. These include carbon monoxide and trace amounts of organic gases such as aldehydes and polycyclic aromatic hydrocarbons. Emissions data on these compounds are relatively scarce. Volatile organic compounds such as methanol can be stripped from the scrubber liquids; emissions will be a function of the concentration present in the incoming scrubber fluid.

Available data on volatile organic compounds emitted during pulping, pulp washing, pulp bleaching, and chemical recovery in the sulfite process are summarized in Tables 8 and 9.[7] Table 8 includes significant VOC emissions measured from a combined blow pit vent, 2 redstock washers, and 1 bleach plant and Table 9 shows the same for 5 sulfite recovery furnaces and 2 so-called "nuisance scubbers."[7]

SEMICHEMICAL PULPING

Semichemical pulping is a process for obtaining high yields of pulps with characteristics suitable for certain end uses. Conventional chemical pulping achieves yields of 50% to 55%, whereas semichemical pulping can achieve from 60% to 80% yields. The yield is defined as the fraction of wood that results in unbleached pulp, both on a dry basis. The lignin is removed only partially during the cook and a second stage involving mechanical disintegration is necessary. Various types of semichemical pulps are produced by the neutral sulfite, kraft, soda, and cold soda pulping processes. The major process difference from conventional chemical pulping lies in the use of lower temperatures, more dilute cooking liquor or shorter cooking time, and mechanical disintegration.[49]

Process Description

In this process, hardwood chips are usually cooked for a few minutes in a continuous digester and then are sent to a refiner for pulp production. Several different types of cooking liquor can be used. Mills that make both kraft and semichemical pulp often use the kraft green liquor, which is a solution of sodium carbonate and sodium sulfide, for semichemical pulp. Other

Table 8. Sulfite Mill VOC Emissions from Pulping, Pulp Washing, and Bleaching

	Combined Blow Pit			Redstock Washers			Bleach Plant	
			Mean		Range	Mean		Mean
	Ref.	No.[b]	(lb/ODTP[a])	No.	(lb/ODTP)		No.	(lb/ODTP)
Methanol	7	1	7.9E+00	2	0.22–0.43	3.3E−01	1	6.2E−01
Acetone	7	1	1.4E−01	2	2E−03–6E−03	4.2E−03	1	1.6E−01
Acetaldehyde	7	1	2.1E−03	2	1E−03–9E−03	4.9E−03	1	1.1E−01
MEK	7	1	4.6E−03	2	ND−0.0009	9.3E−04	1	4.6E−03
Formaldehyde	7	1	ND[c]	2	ND−0.014	7.2E−03	1	1.6E−02
Terpenes	7	1	Sat[d]	2	0.037–0.26	1.5E−01	1	8.5E−02
THCs as C[e]	7	1	2.5E−01	2	0.072–0.16	1.2E−01	1	2.5E−01

[a]ODTP, oven-dried ton of pulp.
[b]No., number of sources tested.
[c]ND, not detected.
[d]Sat, saturated (above quantitation limit).
[e]As measured by EPA M25A.

Table 9. VOC Emissions from Sulfite Nuisance Scrubbers and Recovery Furnaces

	Nuisance Scrubbers				Recovery Furnaces		
			Range	Mean		Range	Mean
	Ref.	No.[b]	(lb/ODTP[a])		No.	lb/ODTP	
Methanol	7	2	0.011–0.91	4.6E−01	4	0.19–6.3	2.2E+00
Acetone	7	2	7E−04–2E−02	9.8E−03	5	ND−0.034[c]	1.9E−02
Acetaldehyde	7	2	9E−04–5E−03	2.9E−03	4	ND−0.20	5.9E−02
MEK	7	2	—	ND	4	ND−0.014	3.6E−03
Formaldehyde	7	2	5E−04–4E−03	ND	4	ND−0.12	3.0E−02
Terpenes	7	2	5.2E−02–sat[d]	>5.2E−02	2	0.17–0.38	2.8E−01
THCs as C[e]	7	2	0.076–2.0	1.04	2	0.33–2.60	1.5E+00

[a]ODTP, oven-dried ton of pulp.
[b]No., number of sources tested.
[c]ND, not detected.
[d]Sat, saturated (above quantitation limit).
[e]As measured by EPA M25A.

cooking processes are the neutral sulfite semichemical (NSSC), which uses a neutral solution of sodium sulfite buffered with sodium carbonate, and the nonsulfur process, which uses a cooking liquor with sodium carbonate and sodium hydroxide.

The chemical recovery process for semichemical pulping liquors depends upon whether or not the semichemical operation is co-located with a kraft pulping operation. As noted earlier, kraft green liquor and neutral sulfite semichemical (NSSC) liquor are used to produce semichemical pulps. Green liquor pulping only is practiced at kraft mills, while NSSC pulping is done at either a kraft mill or as a standalone operation. However, there are only three standalone NSSC mills still operating in the United States. At kraft mills, the green liquor pulping process dominates the NSSC process.

At kraft mills, spent liquor from the semichemical pulping process is mixed with spent kraft black liquor and processed in the same way as black liquor, that is, evaporation and burning in a kraft recovery furnace. The green liquor used for the semichemical cooking is obtained from the green liquor clarifier. At kraft mills with NSSC pulping, the cooking liquor is generally prepared by burning sulfur and passing the concentrated SO_2 gas through a solution of sodium carbonate in an absorption tower.

At standalone NSSC mills, the spent cooking liquor is sent to evaporators to raise its solids content to 40 to 50%. This liquor can be burned in a fluidized bed combustor. The flue gas from the combustor is routed through control devices to capture the sodium particulates and gaseous SO_2. The captured material is used to make new cooking liquor. The cooking liquor is normally fortified with additional SO_2 from a sulfur burner. Emissions of concern include particulates, SO_2, and products of incomplete combustion. Due to the paucity of operating combustors burning spent NSSC liquor, emissions data are lacking.

Liquor recovery operations at standalone semichemical pulp mills that use sodium carbonate and sodium hydroxide for cooking liquor are quite similar to those at stand-alone NSSC mills. Spent liquor is sent to evaporators where its solids content is increased to 30 to 50%; additional evaporation is sometimes accomplished by passing flue gas through the liquor in a venturi scrubber or packed tower. The concentrated liquor is most often burned in a fluidized bed reactor. The flue gas is passed through a mechanical collector to remove the sodium carbonate particulate matter; this material is then dissolved in water or caustic to make new cooking liquor. Sodium carbonate pellets are also withdrawn from the combustor for cooking liquor manufacture. Following the mechanical collector, a scrubber or packed tower is used to remove additional particulate matter, although the collected material is generally not incorporated into the liquor cycle.

The fluidized bed combustor does not emit SO_2 because the pulping liquor does not contain sulfur. There will be some NO_x due to the presence of nitrogen in the liquor, although the amounts will be minimal. Depending on the unit operating conditions, there can be significant amounts of products of incomplete combustion such as carbon monoxide, aldehydes, benzene, styrene, xylenes, toluene, and polycyclic aromatic hydrocarbons. Volatile organic compounds can enter the flue gas if the gas is passed through pulping liquor to achieve additional evaporation or if scrubber makeup water contains strippable compounds such as methanol. Limited data suggest improvement in combustion conditions and careful control of liquor firing rates can significantly reduce VOC emissions, especially those that are products of incomplete combustion.[50]

Spent liquor can also be combusted in recovery furnaces or rotary kilns. These devices generally require co-firing with oil or natural gas to stabilize combustion conditions. In a recovery furnace, the sodium carbonate is removed as a smelt from the bottom of the furnace. The smelt is then dissolved in water or sodium hydroxide solution to make new cooking liquor. In a kiln, the sodium carbonate is removed as an ash and then dissolved. Exhaust gases from these devices may be used for additional liquor evaporation in a direct contact evaporator. Particulate control is most often accomplished with ESPs, wet ESPs, or venturi scrubbers. Emission data from recovery furnaces and kilns are rather limited. NO_x, CO, products of incomplete combustion, and volatile organic compounds present in scrubber liquids or pulping liquor (mainly methanol) being evaporated with flue gas can be expected.

Exhaust gases from dissolving tanks that receive sodium carbonate would contain particulate matter and volatile organic compounds present in the liquid. Some mills use scrubbers to remove most of the particulate matter.

Air Emissions Characterization

Because of short and relatively mild cooking conditions, relief gas volumes from semichemical digesters are considerably less than for kraft continuous digesters. The organic content of the gases is much lower than for kraft digesters as well. Where green liquor is used, the gases will contain some TRS compounds and are generally collected to minimize odor impacts. Digester gases are not collected at other semichemical mills since TRS compounds are not present. However, new EPA regulations will require collection and incineration of these gases to minimize emissions of methanol and other hazardous air pollutants.

Particulate emissions are a potential problem only when recovery furnaces are involved and are similar to those for kraft recovery furnaces. Fluidized-bed reactors (Copeland) are also utilized to burn semichemical spent liquor when a kraft furnace is not available. When recovered in a furnace, the combustion of sulfur-containing semichemical liquor can result in TRS and SO_2 emissions that are somewhat higher than those from kraft black liquors. Lower NSSC liquor heat content resulting from lesser amounts of organic matter extracted from the wood and higher ratios of sulfur to sodium are believed to be responsible. With proper instrumentation and control of the furnace operation, Galeano et al.[51] have shown that NSSC recovery furnaces should be able to operate just as efficiently (with respect to air emissions) as comparable kraft recovery furnaces.

Emissions of methanol and other HAPs have been measured and reported for one kraft green liquor and one sodium carbonate semichemical pulp mill.[50] The results show that, excluding the Copeland reactor, the two highest sources of HAPs (mostly methanol) and VOCs were the pulping area vents and the paper machine vents. The HAP and VOC (EPA M25A) emissions from the pulping system averaged 0.89 lb/oven dried ton of pulp (ODTP) and 0.46 lb C/ODTP, respectively. These emissions from the paper machine averaged 0.45 lb/air dried ton of finished product (ADTFP) and 0.25 lb C/ADTFP, respectively. Emissions from the Copeland reactor were seen to be highly dependent on the manner it was operated.[50]

REFERENCES

1. *Lockwood Post's Directory of the Pulp, Paper and Allied Trades, 1996*, Miller Freeman Publications, San Francisco, CA, 1996.
2. S. A. Rydholm, *Pulping Processes*, 1st ed., Interscience Publishers, New York, 1965.
3. G. Hough (ed.), *Chemical Recovery in the Alkaline Pulping Processes*, TAPPI Press, 1985.
4. H. Axelsson, C. Bostrom, D. Cooper, and U. Svedberg, "Measurements of Turpentine Emissions from Wood Chip Piles Using Fourier Transform Infrared Spectroscopy (FTIR)," *Nordic Pulp and Paper Research Journal* **3**, 155 (1992).
5. *Laboratory and Limited Field Measurements of VOC Emissions from Wood Residuals*, Technical Bulletin No. 723, National Council of the Paper Industry for Air and Stream Improvement, Research Triangle Park, N.C., 1996.
6. *Sulfur Capture in Combination Bark Boilers*, Technical Bulletin No. 640, National Council of the Paper Industry for Air and Stream Improvement, Research Triangle Park, NC, 1992.
7. *Compilation of "Air Toxic" and Total Hydrocarbon Emissions Data for Sources at Chemical Wood Pulp Mills*, Technical Bulletin No. 701, National Council of the Paper Industry for Air and Stream Improvement, Research Triangle Park, NC, 1995.
8. *Acetaldehyde, Acetone, Methanol and Methyl Ethyl Ketone Contents of Kraft Mill Condensates*, Technical Bulletin No. 702, National Council of the Paper Industry for Air and Stream Improvement, Research Triangle Park, NC, 1995.
9. *Volatile Organic Emissions from Pulp and Paper Mill Sources, Part IV—Kraft Brownstock Washing, Screening and Rejects Refining Sources*, Technical Bulletin No. 678, National Council of the Paper Industry for Air and Stream Improvement, Research Triangle Park, NC, 1994.
10. *Atmospheric Emissions from the Pulp and Paper Manufacturing Industry—Report of NCASI-EPA Cooperative Study Project*, Atmospheric Quality Improvement Technical Bulletin No. 69, National Council of the Paper Industry for Air and Stream Improvement, Research Triangle Park, NC, 1974.
11. *Volatile Organic Emissions from Pulp and Paper Mill Sources, Part I—Oxygen Delignification Systems*, Technical Bulletin No. 675, National Council of the Paper Industry for Air and Stream Improvement, Research Triangle Park, NC, 1994.
12. *Carbon Monoxide Emissions from Oxygen Delignification and Chlorine Dioxide Bleaching of Wood Pulp*, Technical Bulletin No. 761, National Council of the Paper Industry for Air and Stream Improvement, Research Triangle Park, NC, 1998.
13. *Volatile Organic Emissions from Pulp and Paper Mill Sources, Part V—Kraft Mill Bleach Plants*, Technical Bulletin No. 679, National Council of the Paper Industry for Air and Stream Improvement, Research Triangle Park, NC, 1994.

14. P. Wiegand, W. Thacker, and R. Miner, "Effluent Quality at Kraft Mills that Use Complete Substitution Bleaching", *TAPPI J.*, **82**(4) 135 (1999).
15. J. Pinkerton, "Trends in U.S. Kraft Mill TRS Emissions", *TAPPI J.*, **82**(4), 1999.
16. *Summary of Long-Term Continuous Emission Monitoring System SO_2 Data and Review of Factors Affecting Sulfur Dioxide Emissions from Kraft Recovery Furnaces*, Technical Bulletin No. 604, National Council of the Paper Industry for Air and Stream Improvement, Research Triangle Park, NC, 1991.
17. *A Study of SO_x Measurement Procedures and their Use at Kraft Recovery Furnaces*, Atmospheric Quality Improvement Technical Bulletin No. 106, National Council of the Paper Industry for Air and Stream Improvement, Research Triangle Park, NC. 1980.
18. K. M. Nichols, L. Thompson, and H. J. Empie, "A Review of NO_x Formation Mechanisms in Recovery Furnaces," *TAPPI J.*, **76**(1), 1993.
19. *An Analysis of Kraft Recovery Furnace NO_x Emissions and Related Parameters*, Technical Bulletin No. 636, National Council of the Paper Industry for Air and Stream Improvement, Research Triangle Park, NC, 1992.
20. *Emission Factors for SO_2, NO_x, and Volatile Organic Compounds for Boilers, Kraft Pulp Mills, and Bleach Plants*, Technical Bulletin No. 646, National Council of the Paper Industry for Air and Stream Improvement, Research Triangle Park, NC, 1995.
21. R. Lovblad, J. Malmstrom, H. Stripple, and D. Cooper, "NO_x Reduction from a Recovery Boiler by Injection of an Enhanced Urea Solution," in *Proc. TAPPI Environmental Conference*, TAPPI, 1991, pp. 1071–1075.
22. *A Study of Kraft Recovery Furnace Hydrochloric Acid Emissions*, Technical Bulletin No. 674, National Council of the Paper Industry for Air and Stream Improvement, Research Triangle Park, NC, 1994.
23. *Locating & Estimating Air Emissions from Sources of Dioxins and Furans*, EPA-454/R-97-003, U.S. Environmental Protection Agency, Research Triangle Park, NC, 1997.
24. *Impact of Burning Oil as Auxiliary Fuel in Kraft Recovery Furnaces upon SO_2 Emissions*, Technical Bulletin No. 592, National Council of the Paper Industry for Air and Stream Improvement, Research Triangle Park, NC, 1990.
25. W. Frederick, "Control of TRS Emissions from Dissolving Tank Vent Stacks," *TAPPI J.*, **79**(6), 144 (1996).
26. *Carbon Monoxide Emissions from Selected Combustion Sources Based on Short-Term Monitoring Records*, Technical Bulletin No. 416, National Council of the Paper Industry for Air and Stream Improvement, Research Triangle Park, NC, 1984.
27. J. S. Henderson, "Final Survey Results for Direct Contact Evaporator Recovery Boiler Electrostatic Precipitators," *TAPPI J.*, **65**(3), 91 (1982).
28. J. S. Henderson, "Final Survey Results for Non-Contact Recovery Boiler Electrostatic Precipitators," *TAPPI J.*, **63**(12), 71 (1980).
29. J. Pinkerton and R. Blosser, "Characterization of Kraft Pulp Mill Particulate Emissions—A Summary of Existing Measurements and Observations," *Atmospheric Environment* **15**(10), (1981).
30. *Volatile Organic Emissions from Pulp and Paper Mill Sources, Part II—Lime Kilns, Smelt Dissolving Tanks, and Miscellaneous Causticizing Area Vents*, Technical Bulletin No. 676, National Council of the Paper Industry for Air and Stream Improvement, Research Triangle Park, NC, 1994.
31. *Ammonia Emissions from Kraft Smelt Dissolving Tanks, Slaker Vents and Causticizer Vents*, Technical Bulletin 789, National Council of the Paper Industry for Air and Stream Improvement, Research Triangle Park, NC, 1999.
32. H. Modetz and M. Murtiff, *Kraft Pulp Industry Particulate Emissions: Source Category Report*, EPA/600/7-87/006, U.S. Environmental Protection Agency, Research Triangle Park, NC, 1987.
33. *Effect of Stripper Off-Gas Burning on NO_x Emissions*, Technical Bulletin 802, National Council of the Paper Industry for Air and Stream Improvement, Research Triangle Park, NC, 2000.
34. J. Drew and M. Propst, "Tall Oil," Pulp Chemicals Association, 1981.
35. *Standards of Performance for New Stationary Sources: Kraft Pulp Mills*, U.S. Environmental Protection Agency (*Federal Register* **51**, 18544), May 20, 1986.
36. *Kraft Pulping—Control of TRS Emissions from Existing Mills*, EPA-450/2-78-003b, U.S. Environmental Protection Agency, Research Triangle Park, NC, 1979.
37. *Collection and Burning of Kraft Non-Condensible Gases Current Practices, Operating Experience and Important Aspects of Design and Operation*, Technical Bulletin No. 469, National Council of the Paper Industry for Air and Stream Improvement, Research Triangle Park, NC, 1985.
38. *Review of New Source Performance Standards for Kraft Pulp Mills*, EPA-450/3-83-017, U.S. Environmental Protection Agency, Research Triangle Park, NC, 1983.
39. *Emission of Reduced Sulfur Compounds from Kraft Process Brownstock Diffusion Washer Vents*, Technical Bulletin No. 406, National Council of the Paper Industry for Air and Stream Improvement, Research Triangle Park, NC, 1983.
40. *Fugitive Dust Emission Factors and Control Methods Important to Forest Products Industry Manufacturing Operations*, Technical Bulletin No. 424, National Council of the Paper Industry for Air and Stream Improvement, Research Triangle Park, NC, 1984.
41. *Bleach Plant Chlorine and Chlorine Dioxide Emissions and Their Control,* Technical Bulletin No. 616, National Council of the Paper Industry for Air and Stream Improvement, Research Triangle Park, NC, 1991.
42. *Results of Field Measurements of Chloroform Formation and Release from Pulp Bleaching*, Technical Bulletin No. 558, National Council of the Paper Industry for Air and Stream Improvement, Research Triangle Park, NC, 1988.
43. *Environmental Pollution Control: Pulp and Paper Industry, Part I, Air*, EPA-625/7-76-001, U.S. Environmental Protection Agency, Washington, D.C., October 1976.
44. *Compilation of Air Pollutant Emission Factors, Volume I: Stationary Point and Area Sources (AP-42)*, U.S. Environmental Protection Agency, Research Triangle Park, NC, 1995, Chapter 10.1.
45. H. Edde, *Environmental Control for Pulp and Paper Mills, Pollution Technology Review No. 108*, Noyes Publication, Park Ridge, NJ, 1984, pp. 148–156.
46. A. L. Caron, "Practices Used by the Sulphite Pulping Industry in the Handling and Treatment of Sulfur Dioxide from Miscellaneous Sources," *Proceedings of the 1976 NCASI Central-Lake States Regional Meeting*, Special Report No. 77-02, National Council of the Paper Industry for Air and Stream Improvement, Research Triangle Park, NC, 1977.
47. *Chemical Recovery Combustion Sources at U.S. Sulfite Pulp Mills*, U.S. Environmental Protection Agency, Research Triangle Park, NC, 1996.
48. *Trends in Emissions of Sulfur Dioxide and Nitrogen Oxides from Pulp and Paper Mills, 1980–1995*, Special Report 97-02, National Council of the Paper Industry for Air and Stream Improvement, Research Triangle Park, NC, 1997.
49. M. Benjamin, I. B. Douglas, G. A. Hansen, et al., "A General Description of Commercial Wood Pulping and Bleaching Processes," *J. Air Pollut. Contr. Assoc.* **19**(3), 155–161 (1969).

50. *Volatile Organic Emissions from Pulp and Paper Mill Sources, Part IX—Semichemical Mills*, Technical Bulletin No. 683, National Council of the Paper Industry for Air and Stream Improvement, Research Triangle Park, NC, 1994.
51. S. F. Galeano, D. C. Kahn, and R. A. Mack, "Air Pollution: Controlled Operation of a NSSC Recovery Furnace," *TAPPI J.*, **54**(5), 741 (1971).

MECHANICAL PULPING

Arun V. Someshwar and John E. Pinkerton

Mechanical pulping, as the name implies, relies mainly on mechanical energy to convert wood to pulp. Mechanical pulping dates back to 1840, and the invention of the pulpwood grinder[1] and the stone-ground wood (SGW) process. In addition to the SGW process, current mechanical pulp manufacturing processes comprise several high-energy refining systems for the production of pulp from chips. These include the refiner mechanical pulping (RMP) process, the thermomechanical pulping (TMP) process, the chemimechanical pulping (CMP) process, and the chemithermomechanical pulping (CTMP) process.

PROCESS DESCRIPTION

In the RMP process, chips are refined directly at atmospheric pressure. Chemicals are sometimes added at the various stages of refining. In the TMP process, the chips are usually steamed under a pressure of 20–40 psi for 2–4 min prior to refining. In some modifications of this process, the refiners are operated under pressure. In the mid-1970s, chemically modified versions of RMP and TMP were introduced. Up until the present, the chemical treatment used has been almost exclusively sulfonation.[1] This may be carried out in several ways, including the treatment of wood chips prior to refining, the treatment of coarse pulp between refining stages, the treatment of completely refined pulp, and the treatment of long fiber. Chips are usually cooked (softened) in a sodium sulfite solution (pH between 4 and 9) for about 30 min at temperatures between 270°F and 320°F. In the CTMP process, the chips are treated with chemicals (sodium sulfite and sodium hydroxide) for softening and refined under pressure. The usual level of addition of sodium sulfite is kept between 1% and 4% Na_2SO_3 on a bone-dry pulp basis.[2]

The CMP and CTMP processes reduce power consumption at the refiners. The RMP and TMP processes are known for their high consumption of electric energy. Pulp yields of 95% or higher are common with the TMP and RMP processes, whereas the CMP and CTMP processes exhibit yields of about 90%.

Bleaching agents are applied to mechanical pulps for the purpose of decolorizing lignin, rather than lignin removal. Either hydrogen peroxide (an oxidant) or sodium hydrosulfite (a reductant), or both, may be used for decolorization. These chemicals can be added at the refiners, prior to bleaching towers, or to pulp stock chests. Studies have not been carried out to determine if application of these bleaching agents leads to formation of any volatile organic compounds such as methanol. Such studies would most likely have to be conducted under laboratory conditions, given the presence of numerous wood-based organic compounds in pulp streams that originate in the grinding and refining operations and the recirculation of these waters throughout the mill.

AIR EMISSIONS CHARACTERIZATION AND CONTROL MEASURES

Large quantities of steam are generated from mechanical and chemimechanical pulping processes. Moisture emission rates are reported to range from 4000 to 7500 lb/ton of pulp for systems not equipped with heat recovery.[3] As wood contains a fair amount of organic material, volatile organic compounds (VOCs) are likely to be released along with the steam during the cooking and refining process.

Heat recovery from steam emissions is extensively practiced. A study[3] by the National Council of the Paper Industry for Air and Steam Improvement on estimating VOC emissions from TMP processes showed the following: (1) Emissions of VOCs from the TMP process operated on Western white wood species ranged from 1.09 to 1.73 and averaged 1.4 pounds of carbon per ton of pulp; from western pine species, it ranged from 0.83 to 3.84 and averaged 1.9 pounds of carbon per ton; and from Southern pine species, it ranged from 2.2 to 7.6 pounds of carbon per ton. (2) Emission rates for VOCs were proportional

Table 1. VOC Emissions from Mechanical Pulping Operations

			Stone Groundwood			Pressurized Groundwood			Thermomechanical	
			Range	Mean		Range	Mean		Range	Mean
	Ref.	No.[b]	(lb/ODTP[a])		No.	(lb/ODTP)		No.	lb/ODTP	
Methanol	4	2	0.022–0.044	3.3E−02	1	—	9.1E−02	3	0.030–0.17	1.2E−01
Limonene	4	2	0.064–0.38	2.2E−01	1	—	4.4E−01	3	0.003–0.074	3.5E−02
Phenol	4	2	ND−0.0053[c]	2.7E−03	1	—	6.7E−02	3	ND−0.012	6.5E−03
Acetaldehyde	4	2	0.017–0.019	1.8E−02	1	—	4.2E−02	3	0.002–0.032	1.9E−02
Formaldehyde	4	2	4E−04–0.003	1.5E−03	1	—	3.6E−03	3	ND−0.017	6.1E−03
Alpha-pinene	4	2	1.4–3.0	2.2E+00	1	—	5.1E+00	3	0.026–1.8	8.0E−01
Beta-pinene	4	2	0.32–3.0	1.7E+00	1	—	3.2E+00	3	0.005–0.70	2.7E−01
THCs as C[d]	4	2	1.3–3.3	2.3E+00	1	—	5.8E+00	3	0.064–1.0	5.6E−01

[a] ODTP, oven-dried ton of pulp.
[b] No., number of sources tested.
[c] ND, not detected.
[d] As measured by EPA M25A.

to moisture emission rates, indicating that those heat-recovery systems that drop the exhaust gas temperature well below the boiling point of water would also be expected to reduce VOC emissions by an unknown amount. No emissions other than VOCs are expected from mechanical pulping processes.

The results of recent testing for hazardous air pollutants and THCs from several mechanical pulping operations at mills are summarized in Table 1.[4] Methanol and terpenes are the major organic compounds emitted. The sources tested included two SGW mills, 1 pressurized groundwood mill, and 3 TMP mills.

REFERENCES

1. D. Atack, "Mechanical Pulping—Some Highlights of the Second 75 years," *Pulp Paper Canada* **90**, 10 (1989).
2. D. M. Mackie and J. S. Taylor, "Review of the Production and Properties of Alphabet Pulps," *Pulp Paper Canada* **89**(2), 58 (1988).
3. *TGNMO Emissions from the Thermomechanical Pulping Process*, Technical Bulletin No. 410, National Council of the Paper Industry for Air and Stream Improvement, Research Triangle Park, NC, 1983.
4. *Volatile Organic Compound Emissions from Non-Chemical Pulp and Paper Mill Sources—Part III—Mechanical Pulping*, Technical Bulletin No. 738, National Council of the Paper Industry for Air and Stream Improvement, Research Triangle Park, NC, 1997.

RECOVERED PAPER PULPING

Arun V. Someshwar and John E. Pinkerton

Since 1980, global paper production from recycled fiber has grown by an average of 6% per year versus about 3% for virgin fiber.[1] In the United States, the collection level of recovered fiber is currently at about 45%, with 50% targeted by the year 2000. Newsprint is the biggest user of recovered paper, but both uncoated and coated magazine grades will likely use some recovered fiber in the future.[1] Over 300 U.S. mills use purchased recovered paper as a fiber source for pulp and paper making. At nearly all mills, "broke" (partly or completely manufactured paper or paperboard from any point in the manufacturing or finishing process) is repulped as well.

PROCESS DESCRIPTION

Repulping of recovered paper and broke most often takes place in an open vessel (commonly called a hydrapulper), where bales of paper are pushed into the vessel filled with water and selected chemicals. Sodium hydroxide is frequently used for pH control. Makeup water for the repulper may come from other areas in the mill, and thus could contain chemicals introduced at the paper machine or from a bleaching/brightening operation (if present). Screening and cleaning in enclosed units takes place following repulping of recovered paper furnish.

If the stock is to be deinked, some type of surfactant (detergent, dispersant, foaming agent) is added in the repulper to promote separation of the ink from the fiber. Most surfactants are chemically modified mineral oils. Following the repulper, deinking is continued using washing, flotation, or a combination of the two. Washing typically takes place in enclosed equipment consisting of centrifugal cleaners, screens, and thickeners. Flotation is conducted in cells that may be closed or open.

Several different bleaching agents can be used on recycled pulps, including chlorine, chlorine dioxide, hypochlorite, ozone, hydrogen peroxide, sodium hydrosulfite, and formamidine sulfinic acid (FAS). The choice is based on the recycled fiber type (mechanical versus chemical), level of contaminants, product brightness requirements, and economics. Use of chlorinated compounds is generally avoided on fibers produced by mechanical means (e.g., old newsprint). Hypochlorite is most often used to decolorize fibers from chemical pulping processes that have also been dyed.

AIR EMISSIONS CHARACTERIZATION AND CONTROL MEASURES

Very few deinking mills use chlorine or chlorine dioxide. Where used, they are added in a mixer just prior to a bleaching tower. The pulp may be washed following the tower. Hypochlorite bleaching can also take place in a tower, although direct addition of hypochlorite solution to repulpers is more common. Like bleaching equipment for kraft or sulfite pulps, use of chlorine or chlorine dioxide may result in these compounds being present in the tower vents, washer hoods, and washer filtrate tank vents. Hypochlorite use will generate chloroform, with the amounts depending on the hypochlorite application rate and type of furnish being bleached. Measurements of chloroform generation rates range from 0.1 to 0.8 lb/ODTP for recycled fiber mills using hypochlorite without chlorine.[2] Only a fraction of these amounts will be released from vents; the remainder will be ultimately discharged through process sewers. However, because of extensive water reuse, the chloroform can be released from many different vents, including bleaching towers, washer hoods, stock chests, filtrate tanks, and building exhausts.

Emissions associated with nonchlorinated bleaching agents have received only limited study. Results from one investigation suggested methanol and acetaldehyde emissions could be a function of the amount of hydrogen peroxide applied to the deinked pulp.[2] At one mill applying 4% H_2O_2 and using a bleaching tower, over 90% of methanol and acetaldehyde emissions from the repulping and deinking area were found in the bleaching tower vent.[2] However, these emissions were only on the order of 0.15 lb/ODTP. Methanol and acetaldehyde emissions also appeared to be greater at mills using hydrogen peroxide versus hypochlorite or sodium hydrosulfite.

Any emissions from repulpers or open deinking units are expected to be the result of chemical additives. Chemicals may be either added directly to the repulper makeup water or be present in water that is recycled from other areas of the mill back to the repulper or deinking units. No direct emission measurements are available for these units due to the difficulty of sampling involved. However, some limited sampling of vents on buildings with repulpers and open deinking equipment suggests emissions of volatile organic compounds from these units are very low.[3,4] The results of recent testing for hazardous air pollutants and THCs from one recycled paperboard and one old corrugated containerboard (OCC) stock preparation system and five deinking operations (with bleaching) are summarized in Table 1.[3,4] Most emissions are seen to be fairly negligible.

Table 1. VOC Emissions from Recycled Paperboard and OCC Stock Preparation and Deinking Operations

			Recycled Paperboard Stock Preparation		OCC Stock Preparation		Deinking w/Bleaching Operations	
			Mean		Mean		Range	Mean
	Ref.	No.	(lb/ODTP[a])	No.[b]	lb/ODTP	No.	lb/ODTP	
Methanol	3,4	1	3.9E−03	1	1.7E−03	5	0.002−0.12	3.5E−02
Limonene	3,4	1	ND[4.3E−04][c]	1	7.6E−05	5	ND−8.2E−04	1.9E−04
Phenol	3,4	1	ND[5.9E−03]	1	3.4E−04	5	ND−7.2E−03	1.8E−03
Acetaldehyde	3,4	1	1.8E−03	1	7.8E−04	5	ND−0.043	1.4E−02
Formaldehyde	3,4	1	1.4E−04	1	1.7E−04	5	ND−4.3E−04	1.2E−04
Chloroform	3,4	1	5.5E−05	1	ND[5.4E−04]	5	9.7E−05−0.14	5.0E−02
Alpha-pinene	2,3	1	ND[5.7E−04]	1	ND[6.3E−04]	5	ND−6.2E−05	1.2E−05
Beta-pinene	3,4	1	ND[4.5E−04]	1	ND[7.4E−04]	5	ND−1.1E−06	2.2E−07
THCs as C[d]	2,3	1	1.2E−02	1	1.0E−02	5	0.032−0.25	1.2E−01

[a] ODTP, oven-dried ton of pulp.
[b] No., number of sources tested.
[c] ND, not detected.
[d] As measured by EPA M25A.

REFERENCES

1. J. O'Brien, "Paper Recycling Grows Worldwide,"*Paper Age*, July 1999, p. 7.
2. *An Investigation of Factors that Influence Chloroform Generation from Recovered Paper Fiber Bleaching with Hypochlorite*, Technical Bulletin No. 674, National Council of the Paper Industry for Air and Stream Improvement, Research Triangle Park, NC, 1993.
3. *Volatile Organic Compound Emissions from Non-Chemical Pulp and Paper Mill Sources—Part II—Recycled Paperboard and Old Corrugated Container Stock Preparation*, Technical Bulletin No. 737, National Council of the Paper Industry for Air and Stream Improvement, Research Triangle Park, NC, 1997.
4. *Volatile Organic Compound Emissions from Non-Chemical Pulp and Paper Mill Sources—Part IV—Deinking Processes*, Technical Bulletin No. 739, National Council of the Paper Industry for Air and Stream Improvement, Research Triangle Park, NC, 1997.

Figure 1. A Typical Paper Machine.

PAPER AND PAPERBOARD MANUFACTURE

Arun V. Someshwar and John E. Pinkerton

Pulping operations using chemical or mechanical processes were discussed in earlier sections of this chapter. Following the washing operation (and bleaching, if performed), the pulp or stock is pumped to high-density storage tanks. From the pulp slurry, thin sheets (paper) or mats (paperboard) are manufactured in the paper mill area on a paper machine (see Figure 1). Individual pulp mills may or may not have paper machines, and mills with standalone operations may purchase baled pulp for paper and paperboard making. To obtain paper or board sheet from a pulp suspension, three operations are performed: stock preparation, sheet formation, and drying. Further operations, finishing and converting, yield a final product. Converting is often carried out in standalone plants located in the consuming districts.

PROCESS DESCRIPTION

Stock preparation involves mixing of various types of pulps and additives and beating of the pulp fibers. Pulps that are derived from virgin wood are termed "virgin" pulp. In recent times, there has been a large increase in the use of "secondary" fiber, derived from recycling paper, for pulp and paper products. Post- or pre-consumer recycled paper is commonly grouped into five categories: deinked, mixed paper, pulp substitutes, newspaper, and corrugated.[1] Depending on the category, one or more of the following processes—asphalt dispersion, cleaning, deinking, pulping, and screening—are used for pulp production.

Figure 2 shows a typical layout of a paper and paperboard mill. Stock from the high-density storage tanks is mixed in blending chests to control consistency and the mixed pulp is sent to a system of "beaters" or "refiners," which further treat the fibers to obtain the desired finished product characteristics, such as paper strength. After stock preparation, the stock is pumped to the head boxes of the paper machine prior to which it is diluted to about 0.5% consistency. From the head boxes, the dilute stock is uniformly distributed across the "wire" of the paper machine. About 98% of the water is removed by gravity and vacuum while the stock is on the wire. The newly formed sheet, still very soft and wet, then passes through the press section of the paper machine, where the sheet is smoothed and additional moisture is removed. The sheet then enters the "dryer" section, where steam-heated drying cylinders are used

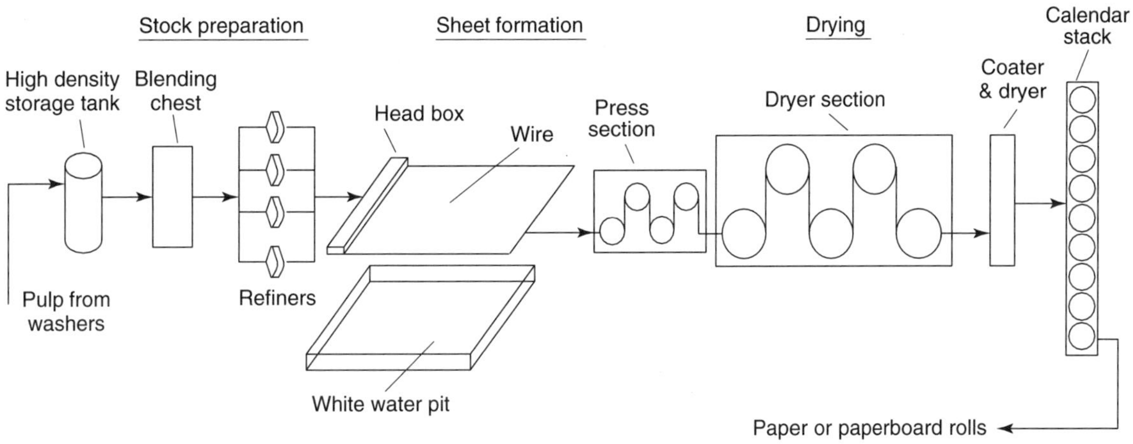

Figure 2. Typical Layout of a Paper Mill.

to evaporate the rest of the moisture in the sheet. After drying, many printing grades of paper and paperboard are surface coated with an aqueous suspension of pigments (such as clay) in adhesives (such as starch) in order to improve surface properties. After coating or sizing, the solvent, usually water, is removed from the coating by evaporative drying, first by air impingement or radiant heating and then over steam-heated drums. Finally, the coated sheet is "ironed" in the calendar stack and then wound in large rolls, ready for shipment.

AIR EMISSIONS CHARACTERIZATION AND CONTROL MEASURES

Emissions from the paper machine consist mainly of water vapor; little or no particulate matter is emitted from the dryers.[2] Paper machines and pulp dryers can, however, be a source of volatile organic compound emissions (VOCs). VOCs present in the water carrying the pulp to the paper machine or dryer can be released as the water is removed from the sheet. Volatile organic compounds are sometimes present in papermaking additives (defoamers, slimicides, retention aids, wet strength agents, wire and felt cleaners, etc.) and these can also be released during the papermaking process. On some paper machines, especially machines used for tissue production, direct-fired dryers burning natural gas, distillate oil or propane are used for additional drying capacity and these dryers are a source of combustion products such as NO_x and CO. Application of coatings to the paper can also be a source of volatile organic compound emissions, although this activity may take place either on the paper machine or in a separate coating operation.

For convenience, paper machines can be viewed as having two distinct parts — the wet end and the dry end. The wet end is where the paper is formed and water is removed mechanically. In the dry end, additional water is removed from the paper web by evaporation. Since paper machines are located inside buildings, the water evaporated and heat must be exhausted from the building. This is done through multiple roof vents, typically fan-assisted, located over the machine. The volumes of air exhausted are large. There are also exhausts associated with the vacuum pumps used to remove water on the wet end of the machine, although the gas volumes are much lower than for the roof vents.

Studies of paper machine emissions have been performed by NCASI.[3,4] Results indicate volatile organic compounds present in the pulp slurry being fed to the machine will be found in the roof vent and vacuum system exhaust gases. The most often detected compound is methanol, a by-product of chemical and mechanical pulping and bleaching processes. However, methanol is also found in recycled fiber slurries, presumably as a result of residual lignin breakdown. Methanol emissions from paper machines at unbleached kraft and semichemical mills ranged from 0.3 to 1.5 lb/ton paper, while paper machines and pulp dryers at bleached kraft mills had emissions under 0.1 lb/ton paper. At mechanical pulp and recycled pulp mills, emissions were found to range from less than 0.01 to 0.3 lb/ton paper. Emissions were proportional to the concentration of methanol present in the paper machine whitewater, which is in turn dependent on the amount of methanol being sent to the machine in the pulp slurry. Other compounds found in considerably lower amounts were acetaldehyde, biphenyl, phenol, and chloroform. However, these compounds were not found consistently at all mills. Because of the extremely large gas flow rates in the roof vents, even very low concentrations of compounds can result in relatively large emission rates. Thus method detection limits become an important issue in the measurement of these emissions.

Measurements of total VOC emissions from paper machines, using an EPA Method 25A analyzer, ranged from 0.1 to 1.2 lb/ton paper.[3] Method 25A analyzers have several limitations and provide no information on the individual compounds present in the gas sample. Thus the contribution of individual organic compounds to total VOC emissions for the various paper machines cannot be determined. Furthermore, the source of these compounds could be either the water carrying the pulp to the machine or additives.

Table 1 provides vent emissions measured by NCASI[3] for certain significant VOCs and THCs corresponding to two unbleached kraft pulp mill (both linerboard), two bleached kraft pulp mill (one pulp dryer and one bleached coated machine), and one semichemical (corrugating medium) pulp mill paper machines. Table 2 provides the same for thirteen "nonchemical pulp mill" paper machines.[4] The "nonchemical pulp mill" paper machines corresponded to four categories, namely (1) 100% secondary fiber furnish (seven), (2) virgin mechanical and chemical pulp furnish (four), (3) nonintegrated

Table 1. Selected VOC and THC Emissions from Kraft and Semichemical Pulp Mill Paper Machines

			Unbleached Linerboard			Bleached Kraft[d]			Semichemical
			Range	Mean		Range	Mean		Mean
	No.	Ref.	(lb/ADTFP[a])		No.[b]	lb/ADTFP			lb/ADTFP
THCs as C[e]	2	3	0.42–0.60	5.1E−01	2	0.038–0.10	6.9E−02	1	2.5E−01
Methanol	2	3	0.92–1.50	1.2E+00	2	0.041–0.10	7.0E−02	1	2.9E−01
Acetaldehyde	2	3	0.0091–0.0098	9.5E−03	2	0.0018–0.003	2.4E−03	1	7.0E−03
Phenol	2	3	0.034–0.053	4.4E−02	2	ND–0.012[c]	6.0E−03	1	2.5E−02
Acetaldehyde	2	3	0.041–0.18	9.2E−02	2	0.017–0.048	7.1E−03	1	7.7E−02
Chloroform	2	3	0.0054–0.0069	6.2E−03	2	0.0013–0.0018	1.6E−03	1	ND
Formaldehyde	1	3	—	2.2E−02	2	ND–0.0095	4.8E−03	1	ND
Terpenes	2	3	0.25–1.5	8.8E−01	2	0.037–0.088	6.3E−02	1	3.6E−02

[a]ADTFP, air dried ton of finished product.
[b]No., number of sources tested.
[c]ND[x], not detected at concentration x.
[d]One bleached coated machine and one pulp dryer.
[e]As measured by EPA M25A.

Table 2. Selected VOC and THC Emissions from Nonchemical Pulp Mill Paper Machines

			100% Secondary Fiber			Virgin Mech. + Chem. Pulp		Non-Integ.	SF + Other[d]
			Range	Median		Range	Mean	Mean	Mean
	Ref.	No.[b]	(lb/ADTFP[a])		No.[b]	(lb/ADTFP)		(lb/ADTFP)	(lb/ADTFP)
THCs as C[e]	4	6	0.23–0.77	3.7E−01	4	0.17–1.2	6.5E−01	2.7E−01	1.0E−01
Methanol	4	7	1.6E−04–0.31	1.0E−01	4	0.019–0.053	3.8E−02	3.3E−02	2.7E−02
Acetaldehyde	4	7	5.7E−04–0.035	1.7E−02	4	8.9E−03–0.044	1.9E−02	3.0E−02	4.1E−03
Biphenyl	4	7	ND–0.098	2.7E−02	4	4.2E−04–0.044	2.4E−02	5.4E−03	2.2E−04
Phenol	4	7	ND–0.021	9.9E−03	4	ND–0.075	5.7E−02	1.1E−02	ND
Chloroform	4	7	ND–0.015	4.7E−03	4	ND–0.047	1.7E−02	4.9E−04	ND
Formaldehyde	4	7	2.8E−05–0.011	4.6E−03	4	7.1E−04–0.013	4.6E−03	2.8E−03	ND

[a]ADTFP, air dried ton of finished product.
[b]No., number of sources tested.
[c]ND[x], not detected at concentration *x*.
[d]Secondary fiber and "other," other includes virgin chemical, groundwood, and thermomechanical pulp.
[e]As measured by EPA M25A.

mill machine (one), and (4) secondary fiber and other (including virgin chemical, groundwood, and TMP pulp) furnish (one).[4]

The burning of natural gas, propane, or distillate fuel oil to generate hot air for impingement drying of paper or to heat infrared dryers will result in NO_x and CO emissions. Distillate oil combustion will also give SO_2 emissions. Data on emissions from these sources are lacking, but the amounts would be small since typical heat input capacities are less than 10×10^6 Btu/h and considerable amounts of excess air are used.

Solvent-based coatings will result in the release of the solvent as the paper web is dried. Commonly used solvents include toluene, xylenes, and acetone. It should be noted EPA does not consider acetone to be a volatile organic compound because of its low photochemical reactivity. In most areas of the country where paper coating operations are located, emissions from solvent-borne coatings must be controlled because of ozone nonattainment concerns. This is one of the reasons there has been a shift away from the use of solvent-borne coatings towards water-borne coatings.

Control techniques for paper machine vents are considered impractical because of the high moisture content and high volume of the vent exhaust gases and the minimal pollutant concentrations. Waste minimization techniques, such as reduced chemical usage and modifications in the mode and location of chemical addition, may be considered on a site-specific basis where reductions are desirable.

REFERENCES

1. L. A. Broeren, "New Technology, Economic Benefits Give Boost to Secondary Fiber Use,"*Pulp Paper*, 69 (November 1989).
2. *Compilation of Air Pollutant Emission Factors, Volume I: Stationary Point and Area Sources* (AP-42), U.S. Environmental Protection Agency, Research Triangle Park, NC, 1995, Chapter 10.2.
3. *Volatile Organic Compound Emissions from Pulp and Paper Mill Sources, Part VII—Pulp Dryers and Paper Machines at Integrated Chemical Pulp Mills*, Technical Bulletin No. 681, National Council of the Paper Industry for Air and Stream Improvement, Research Triangle Park, NC, 1994.
4. *Volatile Organic Compound Emissions from Non-Chemical Pulp and Paper Mill Sources—Part V—Paper Machines*, Technical Bulletin No. 740, National Council of the Paper Industry for Air and Stream Improvement, Research Triangle Park, NC, 1997.

19

SOIL/SOLID WASTE/WATER SOURCES

GROUNDWATER AND SOIL TREATMENT PROCESSES

ROBERT D. FOX, DUANE K. ROOT, AND EDWARD S. ALPERIN

INTRODUCTION

Groundwater and soil contaminated with industrial organic chemicals represent a major environmental remediation challenge in the United States and globally. A wide array of treatment technologies have been used, and more technologies are being developed to address the challenges presented. These technologies provide treatment of organic contamination through physical separation, adsorption, or destruction. Regardless of the technique, a major component of the treatment systems is controlling air emissions related to the treatment process.

Physical separation, or treatment, technologies transfer the organic contaminant from the contaminated matrix to another medium (e.g., water to air, water to solid, soil to air, or soil to water). The physical separation process requires a capture or destruction step for the contaminant, which can result in concentrating the contaminant for recovery, reuse or disposal, or direct destruction. The most common physical separation technologies employed include: soil vapor extraction (SVE), air stripping/sparging, soil flushing/washing, thermal desorption, and carbon adsorption.

Toxic organic contaminants, unlike toxic metals or other inorganics, can be destroyed in treatment. They can be oxidized to carbon dioxide, or they can be reduced/transformed into simpler, more benign structures. These oxidations, reductions, or transformations are accomplished by thermal, chemical, or biological processes.

This discussion focuses on treatment processes for organically contaminated groundwater and soil that have the most significant air emissions issues and air pollution control components. These technologies include air stripping/sparging treatment methods for groundwater, and SVE, aeration and thermal treatment methods for soil.

Incineration by thermal oxidation is the conventional destruction technology for organic wastes in general. However, incineration has been saddled with a reputation as an undesirable source of toxicant emissions and recent trends are toward less costly, less imposing, and less invasive remediation options. Therefore, incineration as treatment for soil is not included in this discussion, but hazardous waste incineration is described elsewhere in the manual.

In situ chemical oxidation methods for treatment of groundwater and soil have increased in utilization and have potential air emission issues, but to date emissions have not been thoroughly characterized and air pollution control measures have not been utilized.

A potentially significant source of organic air emissions also exists at sites where volatile organic compound (VOC) contamination is present and treatment for metals contamination by stabilization/solidification technology is being performed. Soil excavation and reagent/soil mixing operations can provide high-surface-area exposure for VOCs and potential for significant loss to the atmosphere. However, these emissions are difficult to characterize, and air pollution control measures are site specific and often procedural in nature. As such, this topic is also not discussed in further detail.

Most of the treatment technologies described are also discussed in Ref. 1, prepared by the Federal Remediation Technologies Roundtable.[1] Similarly, detailed reports from the U.S. Environmental Protection Agency's Superfund Innovative Technology Evaluation (SITE) program are available on most of the technologies. Information is also available from the U.S. EPA Technology Innovation Office's Innovative Treatment Technology (ITT) Annual Status Report, and Vendor Information System for Innovative Treatment Technology (VISITT) database.

PROCESS DESCRIPTIONS

Groundwater Treatment

Air Stripping. The conventional saturated zone groundwater/soil treatment technology is pump-and-treat, where groundwater is pumped to the surface for treatment and discharged. It relies on the diffusive transport of the contaminant from the soil particles to the groundwater or solubility of a NAPL into the groundwater. This type of transport is generally very slow and results in pumping large volumes (e.g., hundreds of gal/min) of water with very low contaminant concentrations. Typically groundwater contaminated with volatile organic compounds is treated at the surface using air stripping. Treatment is performed by physical separation of the contaminant from a dissolved phase in the groundwater to air in high-surface-area contact with the

Table 1. Typical Air Emission Values

Remedial Option	Typical Operation Rate	Uncontrolled VOC Emissions
Air stripping	3500 L/min	5–50 kg/day[a]
Soil vapor Extraction	0.15–0.85 m^3/min[b]	1–110 kg/day

[a]Assumed 1–10 mg/L pollitant.
[b]Exhaust gas rate per recovery well.
Source: Reference 1.

water using a packed tower. In the packed tower, contaminated water is typically sprayed at the top of the tower and flows down through an inert packing while air passes through the packing in a countercurrent upflow manner. The packing media acts to retard the flow of water (increase liquid holdup) and increase the effective water–air interface surface area. Organic contaminants are transferred to the air phase in a manner predicted by Henry's Law and are either vented from the top of the tower or sent to a vapor treatment system.

As an alternative to conventional groundwater pump-and-treat, several technologies have been developed that offer improved removal performance.

Air Sparging. In situ volatilization of VOCs can be accomplished in saturated soils by sparging air through the groundwater/soils in the contaminated zone. This technology removes VOCs sorbed to the soil particles and dissolved in the groundwater phase, thereby treating both soils and groundwater in the saturated zone. The vaporized VOCs pass to the vadose zone, where they are collected by an SVE system. Air sparging aerates the soil and enhances the aerobic biodegradation of contaminants, especially semivolatile organic compounds (SVOCs) and nonvolatile organics. This type of remediation is called *bioventing*. Thus air sparging is seldom found as a standalone treatment, but is combined with SVE and biodegradation.

In air sparging systems, the injected air flows through the water column past soil particles, transferring VOCs to the air bubbles, which rise into the vadose zone, where they can be captured by an SVE system. The air injection and SVE collection points can be either vertical or horizontal wells.

Horizontal wells can be the only method available to remediate VOC-contamination soil under buildings or tanks. The effectiveness of horizontal wells was tested in a full-scale demonstration on soils contaminated with chlorinated solvents at the DOE's Savannah River Plant.[2]

This in situ air sparging concept uses two parallel horizontal wells: one below the water table and one in the unsaturated (vadose) zone as shown in Figure 1. The deeper well is used as a delivery system for air injection. VOCs are stripped from the groundwater into the vapor phase and are collected and removed from the subsurface by vacuum extraction through the shallower well in the unsaturated zone.

Reported case studies demonstrate that the use of air sparging in conjunction with vapor extraction is effective for cleanup of gasoline and solvent-contaminated sites if the proper site-specific conditions exist, particularly soil permeability and lithological homogeneity.

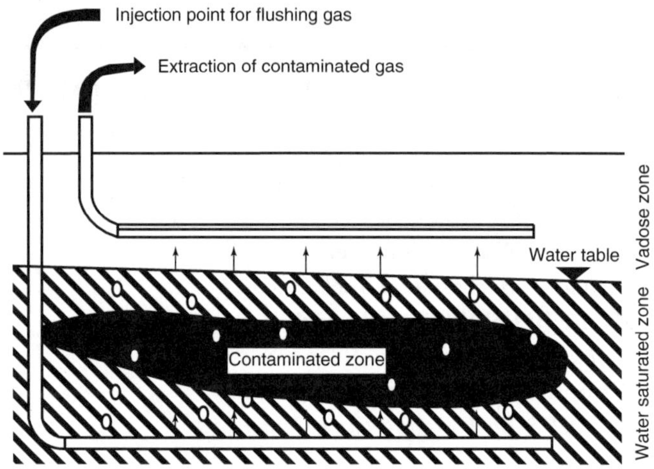

Figure 1. Schematic Diagram of in Situ Air Stripping Using Horizontal Wells, a Flushing Gas, and Vacuum Recovery.

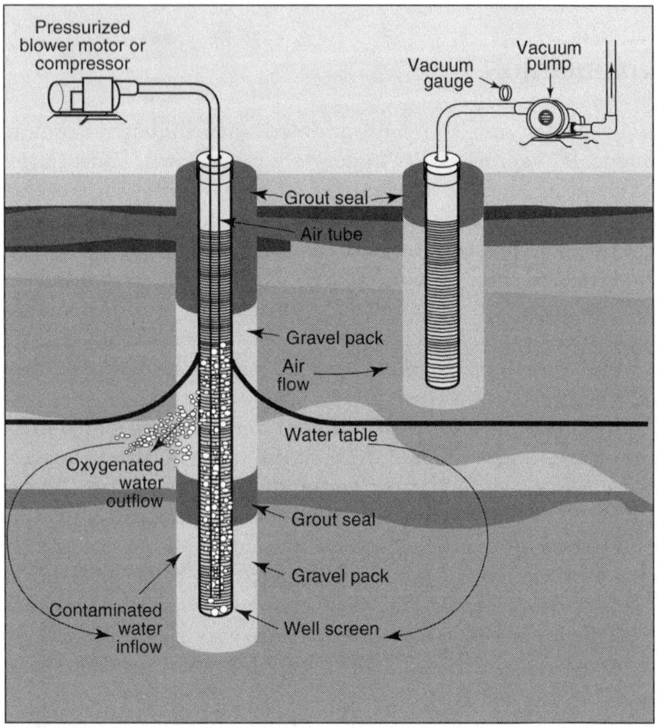

Figure 2. Simplified Schematic of Well-Bore Air Stripping System with Air Vented to Vadose Zone Collected by Soil Vapor Extraction.

Well-Bore Air Stripping. This technology involves the use of air stripping in a specially constructed well to remove VOCs from groundwater that is drawn into the bottom of the well through a lower screened area. The air-stripped groundwater rises in the well and is then allowed to gravity flow back into the formation at another screened point above the contaminated zone. In this process, shown schematically in Figure 2, groundwater is treated in situ and returned to the formation to displace contaminated groundwater. This creates a circulation system in the formation around the well. The stripping air is either collected at the top of the well for surface treatment of the VOCs or allowed to vent into the vadose zone, where biodegradation can destroy the VOCs.

Several variations of this type of well-bore air stripping system have been reported.[3–5] These systems are similar to groundwater pump-and-treat in using water flow through the contaminated saturated soil to transfer the VOC (or contaminant) to the water phase. Their main advantage is that the water is not removed from the well, and can be returned to the formation. The key issues with these systems is where the groundwater goes when it re-enters the groundwater zone and how long it takes to complete the recirculation route. Short-circuiting lessens the sphere of influence. High permeability soils are required to minimize treatment time.

Soil Treatment

Soil Vapor Extraction. Soil vapor extraction (SVE) is an in situ treatment technology, which relies on the volatility of the organic chemical contaminants to enable their separation from the vadose zone soil by vaporization.

SVE uses the application of vacuum to the contaminated unsaturated subsurface area to induce the controlled flow of air through the contamination zone and remove VOCs and some SVOCs. The vacuum is applied principally to vertical wells screened at the contamination zone. It is not uncommon to use monitoring wells for SVE systems. SVE can be used on contamination zones at depths from 5 ft to several hundred feet. Vacuums applied are those necessary to induce adequate airflow and can involve the full range of vacuum pump capabilities, depending on soil permeability. Use of additional wells for air inlet to the contamination zone are sometimes necessary to assure passage of air through the zone of contamination.

SVE can also employ horizontal wells to extract VOCs from the vadose zone. This is especially useful for remediation of soils underlying buildings and active industrial facilities. Subsidence and subsurface obstructions become concerns when using horizontal wells in this type of situation.

Other than the wells, the equipment required to carryout SVE is simple and consists of interconnective piping, a vacuum pump(s), vapor–liquid disengaging tank, appropriate valving, monitors, and controls, and most likely an air emission control system. The contaminated air and VOC vapors withdrawn by the vacuum system may require treatment before discharge of the air to the atmosphere. The cost of SVE is significantly impacted by the need to treat this air stream.

The typical method of operating SVE systems is periodically to monitor the VOC concentration and flow of the air extracted from the subsurface. When the concentration of VOCs drops to a low level and remains low after a period of shutdown, soil sampling is done to verify that the contaminant concentrations in the zone of concern are at acceptable levels.

Enhancements to SVE. A number of technologies are being developed to enhance the performance of SVE for in situ soil remediation. These are air injection, steam sparging, radio frequency heating, hot air heating, and electrical (joule) heating. These enhancements seek to speed the vaporization of VOCs for collection by the SVE system and/or to vaporize SVOCs and other low-volatility contaminants. All but ambient air injection involve heating the soil. A major advantage that is claimed for these thermal-based enhancements is the vaporization of soil moisture. As the water evaporates, soil permeability increases, facilitating improved contaminant removals.

Ambient Air Injection. Ambient air injection enhances SVE by adding air injection wells to the vacuum extraction well network to provide additional air movement through the contamination zone to speed the removal of VOCs. A test of the ability of an air injection–vacuum extraction technology to remediate vadose zone soil was completed under the SITE program and reported in 1995.[6] After 1 year of operation the average mass reduction in VOC and SVOC contamination at the site was 80% from an initial average concentration of 340 mg/kg. The reductions over the aerial extent of the site, as determined from individual boreholes, ranged from 71 to 99%.

Another advantage of the addition of air to subsurface contamination zones is the stimulation of aerobic biological degradation activity by the naturally occurring organisms present in the soil. This technology is called bioventing; it speeds the removal of contamination, but is only effective on biodegradable organics.

Hot Air Injection. The injection of heated air into the vadose zone provides a means of transferring heat energy to a contaminated soil to raise its temperature. The rising temperature of the soil results in changes to several physical characteristics of the contaminant, including increased vapor pressure, decreased viscosity, and decreased interfacial tension.

When the vapor pressure of volatile and semivolatile contaminants adsorbed to the soil increases, mass transfer of the contaminants from the adsorbed phase to the vapor phase also increases. Reduction of the viscosity and adsorptive capacity of semivolatile and nonvolatile contaminants results in a significant mobility increase of the separate phase in the vadose zone and on the capillary fringe. The end result is that significantly more mass is present in the mobile vapor and separate phases and therefore available for removal by standard SVE and separate phase hydrocarbon systems.

Steam Sparging. This technology was originally developed in the petroleum industry as a secondary or tertiary enhanced oil recovery technique, which relies on viscosity reduction and injection pressure to mobilize oil for extraction. As with hot air injection systems, the injection of steam into contaminated soil horizons is used to alter the contaminant vapor pressure, viscosity, and interfacial tension characteristics advantageously. The evaporation rate of both volatile and semivolatile contaminants and the fluid mobility of semivolatile and nonvolatile materials are increased significantly. Additionally, since the heat energy transfer capability of steam is so much greater than hot air due to the latent heat of condensation, steam injection becomes a more efficient means of heating large soil masses. In general, steam injection becomes more cost effective for sites with plume areas of 0.5 acres or greater. The injection of steam coupled with vacuums produced by SVE, results in substantial pressure gradients and resulting driving forces for both vapor and liquid phases within the soil.

At the initiation of the steam injection process, steam exiting from the injection well contacts the soil and immediately condenses. During the condensation process heat energy is transferred to the soil by conduction. A cylindrical zone of heated soil with a condensation front at the outer perimeter expands radially outward from the injection well. As a result of the steam condensation, the soil moisture content will increase until the soil temperature within the heated soil cylinder is raised to steam temperature. At this point the steam no longer condenses and drying of the soil takes place due to the boiling of the pore water along with the evaporating hydrocarbons (steam stripping). Typical operating soil temperatures are 200–220°F.

A technology demonstration project was conducted at the Puget Sound Naval Shipyard in 1991 which demonstrated the effectiveness of steam sparging to mitigate impacts of heavy fuel (No. 6 and diesel) due to leakage from underground storage tanks.[7] Steam was sparged to raise the temperature of the vadose zone and the upper portion of the groundwater to approximately 210°F. In the first 6 months of operation 30,000 gallons of the fuel oil mixture was removed or degraded. In another demonstration Dynamic Underground Stripping (DUS) was used at Lawrence Livermore National Laboratory (LLNL) in 1992–1993 by the developers, LLNL, and the University of California, Berkeley. Steam was injected above the water table to recover 7000 gallons of gasoline (LNAPL) contamination from the site and meet cleanup standards.[8]

Creosote DNAPL is being remediated using steam at the Visalia Pole Yard in California. In continuing operations, a 4.3 acre area (0.5 million cubic yds) has been treated, and nearly 300,000 lbs of NAPL have been either recovered or destroyed. Destruction of a portion of the DNAPL being treated is reported in what is termed hydrous pyrolysis oxidation (HPO) based on levels of carbon dioxide and chloride products that are observed.[9]

Radio-Frequency Heating. Radio-frequency (RF) heating is an in situ treatment process that uses electromagnetic energy to heat soil and enhance SVE. RF heating is done by placing exciter electrode or antenna arrays in the soil either vertically or horizontally. Horizontal insertion is through a horizontal well borehole. Application of energy causes heating to begin near the antenna and proceed outward. Soil can be heated to 250–300°C. Water and contaminants are vaporized. Vapor collection devices are built into the antenna arrays on the surface or are standard SVE systems to collect vapors generated for further treatment and recovery/disposal.

RF heating has been conducted at the Rocky Mountain Arsenal (RMA) and Kirtland AFB sites.[10] At RMA the concentration of organochlorine pesticide contamination in 50 cubic yards of soil was reduced 97–99% by achieving temperatures over 250°C. At Kirtland AFB a maximum of only 139°C was achieved over 42 d of heating on 30 cubic yards to remove gasoline range organics ($<C_{12}$). The heavier diesel-range organics (C_{12}–C_{20}), however, were not successfully removed.

RF heating has also been used to accelerate bioremediation by elevating soil temperatures to the optimal range of 15–35°C. In a pilot test at Fort Wainwright, AK, 22,375 kilowatt-hours of power have been supplied to vadose soils over a 9-month period to provide up to a sixfold increase in biodegradation rates.[11]

Electric Heating. In situ electrical heating of contaminated soil uses common AC electricity to heat soils to enhance SVE removal of organics. Two versions have been under development. The first uses electrodes inserted in the contaminated area of the soil with a technique employing six-phase heating of the soil.[12] This technique is based on the ability to split conventional three-phase electricity into six separate electrical phases. Each phase is connected to one of six separate electrodes placed in a hexagonal array as shown in Figure 3. Because each electrode is at a separate phase, each one conducts to all the others. Voltage gradients in the six-phase array are relatively uniform, thereby facilitating uniform heating throughout the soil bounded by the electrodes. A seventh electrically neutral pipe is inserted in the center of the hexagon and connected to a vacuum system to capture vapors generated during heating.

As part of the DOE Integrated Demonstration Program for VOCs in nonarid soils, a full-scale demonstration was conducted at the Savannah River Site on clayey soils contaminated with PCE and TCE. The electrodes were positioned on a 30-ft-diameter hexagonal array and had electrical contact with the soil between 23 and 44 ft. The clay zone was heated up to ~100°C in 8 d and held there for 17 d. Soil sampling results showed contaminant removal efficiency from the clay zone was 99.7% within the electrode array. Soil moisture was also removed, and this improved soil permeability.[13] The developer, Current Environmental Solutions, is currently performing their first large-scale remediation of DNAPL under a building in Chicago.

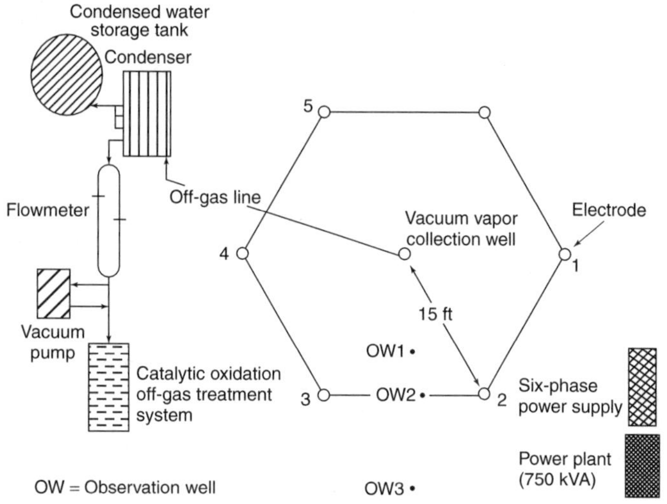

Figure 3. Surface Equipment for in Situ Six-Phase Heating of Soil to Vaporize Organic Contaminants (Electrode Array and Other Wells Drawn to Scale, Other Equipment not Drawn to Scale.

The second method of heating soil is called in situ thermal desorption (ISTD) by the developer. There are two primary forms of ISTD technology; thermal blankets for removal of surfacial contamination down to about 2 ft, and thermal wells, which can be placed to virtually any depth. Overall thermal conduction accounts for over 80% of the heat transfer. A significant feature of ISTD process is the creation of a zone of very high temperature (>1000°F) near the heaters, which oxidizes most of the contaminants before they exit the soil.

This thermal blanket technology was tested on a pilot scale at a site in New York State with surface contamination by PCBs at concentrations over 500 mg/kg.[14] Three 10 ft by 10 ft grids were heated over several days; the PCB concentration in the top 3 in. of the 3–6 in. zone was reduced to below the target level of 2 mg/kg. This type of heating is capable of raising the temperature of surface and near-surface soils to several hundred degrees centigrade, enabling improved vaporization of nonvolatile, semivolatile, as well as volatile contaminants.

Ex Situ Aeration. The full range of organic contaminants in soil, including SVOCs and nonvolatiles, can be physically removed from the soil by heating the soil to temperatures sufficient to vaporize them and collecting them as a concentrate for further treatment or disposal. This vaporization step can be accomplished by aeration, vacuum, steam, or heating, from moderate energy input to heating to several hundred degrees centigrade, and combinations thereof.

Aeration. This physical treatment process involves the placement of VOC-contaminated soil in containers and aerating the soil, similar to in situ SVE. The process is simple and low cost, and uses inexpensive equipment such as rolloff boxes or Baker tanks. The usual practice is to accelerate the process by either heating the air or the soil in the container and/or applying a vacuum to the container. Commercial units are available that utilize one or more of these technologies.[15–17] Not only does the use of heat and vacuum speed the volatilization of VOCs, but it also increases the probability of getting to the desired cleanup levels. It can also aid in the removal of some SVOCs. Systems that heat soil require more energy. Energy requirements increase as the treatment temperature approaches 100°C, where soil moisture evaporation occurs. The accompanying large increase in steam vapors complicates the vapor collection and disposal system and the air emission control systems.

The key issues with volatilization treatment of VOC-contaminated, excavated soils are:

- Control of air emissions during soil excavation and handling. The simplicity of the soil aeration process minimizes volatile losses once the soil is loaded in the container and sealed.
- The ability of the process to achieve the soil cleanup levels that are imposed on excavated soils. As contrasted to in situ volatilization treatment by SVE, where a significant fraction of the VOCs is left on the soil,[18] ex situ treatment will have much more stringent cleanup requirements, such as <2 ppm for PCBs or <50 ppb for TCE.

Thermal Desorption. Thermal desorption separates organic contaminants from soils by heating the soil to a sufficient temperature to vaporize the contaminant. When the heating device is an indirectly heated chamber and the vaporized organics are condensed and/or collected as a concentrate for further treatment or disposal, thermal desorption is a physical treatment technology.

A variety of devices have been developed as the primary heating chamber for thermal desorption: rotary dryers, thermal screws, paddle dryers, anaerobic thermal processors, belt conveyor systems, batch vacuum systems, and mercury retorts.[19] Variations of the rotary dryer, an indirectly heated rotating metal alloy cylinder, are the most prevalent for RCRA/TSCA contaminants. The organic contaminants are volatilized at temperatures in the range of 250–550°C in a low-oxygen purge gas and conveyed as gases to a condensation/scrubbing system that removes the contaminants from the purge gas. The condensed contaminants are collected as concentrates for recovery or further treatment. Further treatment by thermal destruction or chemical destruction can be either on-site or off-site. A schematic of low-temperature thermal desorption with condensate collection is shown in Figure 4.

The commercial literature frequently includes references to low-temperature thermal desorption systems that utilize an afterburner to destroy the volatilized organics, rather than a condensing system to collect them as a condensate. This type of system is in reality an incinerator, and such systems are typically required to meet Resource Conservation and Recovery Act (RCRA) incinerator standards when treating sediments or soils contaminated with hazardous waste constituents.

Direct-fired systems with afterburners are frequently used for the cleanup of non-RCRA contaminated soils such as soil contaminated with gasoline or fuel oil from leaking underground storage tanks. Under these circumstances the units are typically permitted under the respective state air pollution control division regulations and do not have to comply with RCRA permitting requirements.

The performance of a low-temperature thermal desorber in treating sediments or soils depends on the time–temperature conditions achieved in the primary chamber. In general, the higher the temperature, the shorter time that the solids have to be at that temperature to achieve cleanup standards. Lower temperatures require longer residence times in the primary chamber and therefore reduce throughput. For contaminants

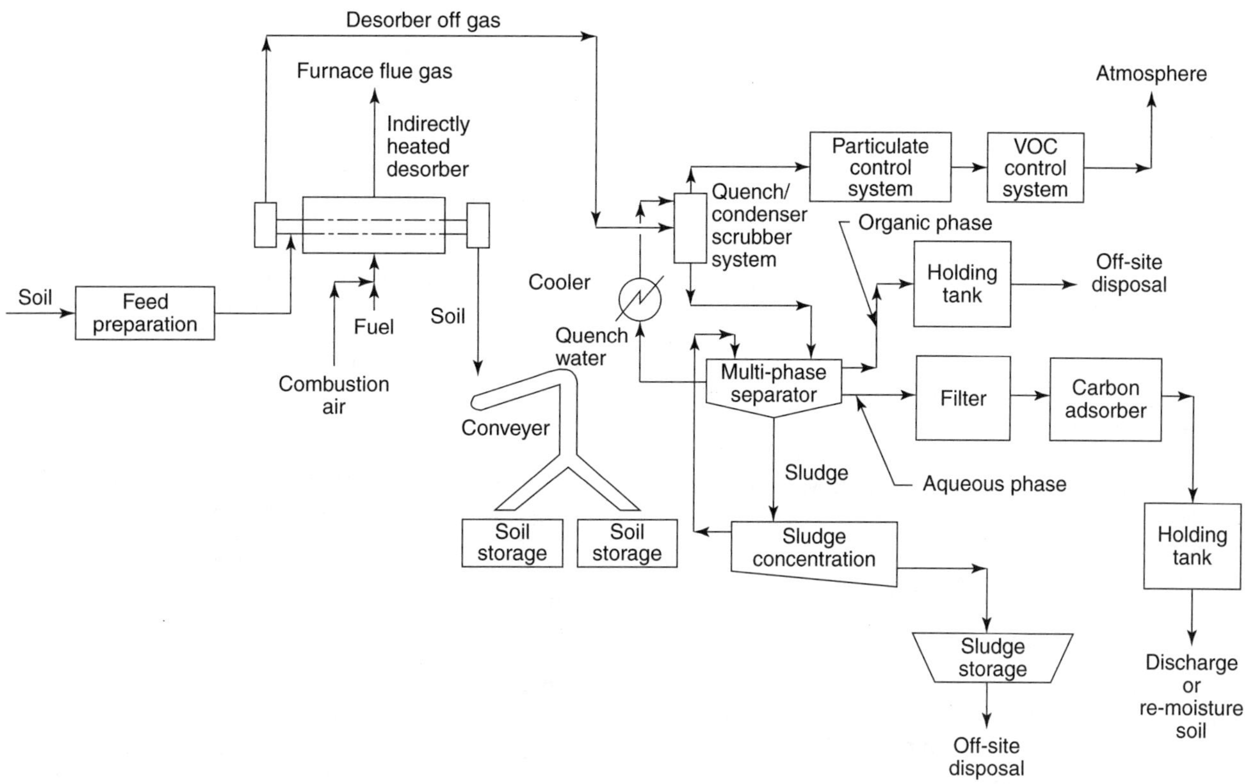

Figure 4. Thermal Desorber with Condenser/Scrubber System Used to Treat Soils and Solids to Remove SVOCs, Pesticides, PCBs, and PCDD/PCDFs.

such as polychlorinated aromatics (e.g., PCBs) and polycyclic aromatic hydrocarbons (PAHs), processing temperatures at or below the normal boiling point of the contaminant will not meet stringent cleanup standards with practical, useful residence times of 1 h or less.

Thermal desorption has been demonstrated at commercial scale for the cleanup of soils, sediments, and sludges contaminated with: VOCs, pesticides, PCBs, PAHs, mercury, and total petroleum hydrocarbons (TPH).[20] The key factors affecting the cost of thermal desorption are throughput requirements, quantity of sediment/soil to be treated, water content of feed, on-stream reliability, and disposal of residuals.

Thermal desorption is a proven and effective remediation technology to achieve stringent cleanup standards in treating organically contaminated sediments and soils. When large quantities of material are involved, the cost of using these thermal technologies in on-site remediation is competitive with other alternatives.

In Situ Vitrification. Vitrification is the process of thermally treating contaminated soil to the point of melting the soil media to form a solidified monolith. The most common method for soil treatment is in situ vitrification by joule heating.

In situ vitrification is accomplished by passing electricity between electrodes implanted in soil so that an electrical path is established through a zone of contamination, usually at depths of less than 20 ft. Sufficient power is utilized in passing electricity through the soil and waste to melt the soil. A highly conductive starter path is used to start the melting process. Continued flow of electricity causes the surrounding media to melt and once molten the flow of electricity results in joule heating between the electrodes. This typically requires media temperatures in the range of 1600–2000°C. The media being treated must be capable of forming a melt with adequate electrical conductivity and most soils meet this criteria; however, additives can be used to aid in the process. During treatment a hood is placed over the treatment zone and process off-gases are collected and passed through a treatment system. Upon cooling the treated zone forms a vitrified monolith. Often there is a volume reduction that occurs, which requires backfill to bring the area to the desired grade.

Vitrification is applicable at sites where semivolatile and nonvolatile contamination is significant. This includes sites contaminated with the following: glycols, PCBs, dioxins/furans, pentachlorophenol (PCP), metals, or radioactive isotopes. Most of the organic and some inorganic contamination is destroyed by thermal decomposition or pyrolysis in oxygen-starved areas of the treatment zone while some are oxidized, especially at the outer areas of the treatment zone where oxygen is available. Most inorganic contamination is encased in the vitrified monolith. Power usage for the process is within the range of 600–900 kWh/ton.

AIR EMISSIONS CHARACTERIZATION

Ambient Physical Treatment Processes

Air emissions from ambient physical treatment processes are primarily characterized by the contaminant compounds being treated, since the treatment is accomplished by separating the contamination from its media. Most air stripping/sparging and SVE processes are included in this category. Data collected by the EPA and summarized in Table 2 show the most common

Table 2. Frequencies of the Most Common Organic Contaminants at NPL Sites with RODs

Compound	No. of Sites Found (Total of 944 Sites)
Trichloroethylene (TCE)	473 (50%)
Benzene	408 (43%)
Toluene	337 (36%)
Tetrachloroethylene (PCE)	226 (36%)
Vinyl chloride	272 (29%)
PCBs	340 (36%)

Source: Ref. 21

Table 3. Typical Air Emission Values

Remedial Option	Typical Operation Rate	Uncontrolled Organic Emissions
Air stripping	570–5700 L/min	8.2–82 kg/d[a]
Soil vapor extraction	1.4–425 m^3/min	6.0–1800 kg/d
Thermal desorption	2700–90,800 kg/h	6.5–220 kg/d[b]

[a]Assumed 10 mg/L pollutant.
[b]Assumed 100 mg/L pollutant.

Source: Ref. 22.

organic contaminants and frequency of occurrence at National Priority List (NPL) sites with Records of Decision (RODs).

Typical uncontrolled air emission values for physical treatment processes used at Superfund sites are shown in Table 3.

Air emissions from these systems also include combustion products from thermal oxidizers that may be used as a secondary treatment. This treatment is used to directly destroy organics separated from the contaminated media. The emissions from thermal processes are described in more detail below. In addition, fugitive emissions exist from handling of secondary waste streams such as recovered groundwater and spent carbon.

Thermal Processes

Thermal processes such as thermally enhanced SVE, thermal desorption, and vitrification provide differing degrees of thermal destruction of organic compounds through oxidation or degradation during treatment. Emissions of primary concern from these processes are the hazardous organic compounds for which treatment is being performed, similar to that for ambient physical treatment processes. Additionally, emissions from thermal processes may include particulate and gaseous oxidation/degradation products such as carbon dioxide, carbon monoxide, water vapor, and acid gases (hydrochloric acid) as well as partially oxidized hydrocarbons (POHCs), degradation products, and hazardous by-products such as dioxins and furans. The degree and amount of these emissions are dependent upon the characteristics of the waste being treated, the type and design of treatment system being used, and the operating conditions of the treatment system.

Air emissions also include those from heating systems that are used to provide the thermal energy to the contaminated media for treatment, combustion products from thermal oxidizers that may be used on the system, and fugitive emissions from handling of secondary waste streams. Secondary waste streams on these systems may include; particulate (fines), condensate/scrubber liquid, scrubber sludge, and spent carbon.

Area Emissions from ex Situ Soil Treatment

Fugitive emissions from area sources during ex situ soil treatment may contribute significantly to the total air emissions from a remediation site. These emissions include those from soil handling operations. Of these the largest source is usually excavation of the contaminated soil. Other potential sources include soil preparation operations such as drying, classifying, etc., and system feed operations (conveyor and feed hopper components).

AIR POLLUTION CONTROL MEASURES

Activated-Carbon Adsorption

Activated-carbon adsorbers are often used to control emissions from air strippers and SVE systems. They are also often used as a polishing step in air pollution control (APC) systems in which thermal oxidizers are not used. Activated-carbon adsorbers are easy to use, but contaminant concentrations in the vent stream need to be monitored to determine when breakthrough occurs, and contaminant loadings on carbon are compound and conditions specific. Carbon adsorption typically achieves greater than 90% efficiency with contaminant concentrations over a few hundred parts per million by volume (ppmv) and the efficiency is typically higher as contaminant concentrations increase. VOC emissions leaving a carbon adsorber are usually less than 100 ppmv. There are a number of factors related to the exhaust gas stream that must be taken into account when designing a carbon adsorption system or predicting adsorption efficiency. Among these are the temperature and water content of the gas stream, which both adversely affect the adsorptive capacity of carbon.

Incineration

Thermal and catalytic oxidizers (incinerators) have also been used to control emissions from air strippers, SVE systems, and many other groundwater and soil treatment systems. These systems destroy organic contaminants by oxidation to carbon dioxide, water vapor, and hydrochloric acid (chlorinated compounds). Typical VOC removal efficiencies with these systems are in the range of 90–98%. A major factor that must be considered with the use of combustion techniques is the generation of hydrochloric and other acid gases from the oxidation of halogenated organics, sulfur-containing compounds, etc. Depending upon feed characteristics, operating conditions, and regulatory requirements, a scrubber downstream from the oxidizer may be needed to neutralize the acid gases.

Thermal oxidation is usually performed at temperatures in the range of 1200–2000°F and requires combustion of a fuel source to maintain oxidation chamber temperature. A residence time usually in the range 0.2–2 sec is also required for complete oxidation of organic contaminants. Thermal oxidizers are effective and efficient but require substantial energy input to destroy dilute gas-phase contaminants.

Catalytic oxidation is performed at much lower temperatures, typically in the range of 650–800°F, and can offer substantial energy savings. As the name implies, a catalyst is used in these oxidizers to promote oxidation at the lower temperatures. The catalyst can become poisoned by halogens, halogenated and sulfur compounds, metals, and particulate, and as a result usually requires periodic cleaning to maintain efficient operation.

Alternate Oxidation Methods

Carbon adsorption and thermal oxidation are by far the most common emission control technologies for air strippers and SVE systems, although some newer innovative oxidation systems are being investigated. The new systems include flameless oxidizers and photocatalytic oxidation (PCO) systems. Both of these systems produce oxidation products of organic contaminants and require consideration of acid gas emissions depending on the feed characteristics.

Flameless oxidizers employ an inert ceramic packing or matrix heated to 1000–1400°F to promote oxidation of organic contaminants. Heat energy can be supplied using either an electrical heater or a gas burner. The ceramic matrix maintains thermal inertia due to its heat capacity and absorbs the exothermic heat released from organic oxidation. The use of the energy content of the waste stream by the flameless oxidation process can result in low operating costs. Fume streams in the range of 10–40 BTU/scf may not require any additional heat input during fume processing. Hydrocarbon and chlorinated compound destruction efficiency is reportedly greater than 99.99%.[23]

PCO systems, which may be usable on more dilute gas streams, operate at near ambient temperatures by adsorbing VOCs onto a UV-active substrate such as titanium dioxide (TiO_2), and the VOCs are oxidized by UV irradiation. In one PCO system offered by Process Technologies, Inc., a patented replaceable reagent panel is used to neutralize acid gases produced by the system.[24]

Air Pollution Control for Thermal Treatment Processes

Some thermally enhanced SVE systems, which are designed for higher-temperature treatment of vadose zone soil have APC systems that are custom designed for the process. The electrically conductive heating element process of Terra-Therm uses a cyclone solids separator followed by a flameless thermal oxidizer, a heat exchanger to reduce the gas temperature, and carbon adsorber beds to treat the vapors drawn from their ISTD process.[25]

The Six-Phase Heating™ process of Current Environmental Solutions is applied to contamination in the saturated zone to remediate DNAPL. SVE is used to collect the steam and contaminant vapors. The SVE vapor stream is passed through a condenser to collect a water phase for further treatment, and the noncondensed gases are sent to a thermal oxidizer.[26]

Many ex situ aeration or thermal desorption systems can operate either by condensing media contaminants for recovery or off-site treatment, or by directing the system off-gases to a thermal or catalytic oxidizer for contaminant destruction. Typical APC systems for thermal desorber off-gases in which contaminants are condensed include provisions for particulate removal, acid gas scrubbing, condensing of organics and water vapor, mist elimination, HEPA filtering, and carbon adsorption. There is a wide variety of thermal desorption systems available from vendors. Their processes are generally proprietary, and the APC systems are tailored to the particular design of the vendor's process. As such, each has its own individual APC system.

Weston's low-temperature thermal treatment (LT3) system includes a baghouse dust collector for particulate emissions control, two condensers (to reduce the volume of water vapor), and a carbon adsorption system. An afterburner or thermal oxidizer can be used with the system and is designed for 1800°F operation and a 1-sec gas residence time. If the waste feed contains sulfides and/or chlorides, the LT3 system is configured to include a caustic wet scrubber following the afterburner.[27]

OHM's X*TRAX™ system uses a scrubber to remove particulate and then condenses the gas to less than 5°C to collect water and organics removed from the soil. The majority of the gas (90–95%) exiting the condenser is reheated and recycled to the process and the remainder is passed through a HEPA filter and carbon adsorber before being vented.[28]

The CleanSoil™ process by Soil Solutions, Inc., uses steam to thermally desorb organics from soil. The off-gas steam and organics are passed through a cooler unit to recover the water and higher boiling organics for recycle and the noncondensed gases are vented through carbon adsorption beds. The carbon adsorption beds are regenerated with steam between each batch.[29]

The APC system for Geosafe Corporation's in situ vitrification process resembles that of similar thermal oxidation/destruction processes. The system consists of a quencher to lower the temperature of the off-gas, a scrubber to remove acid gases and large particulate, a demister and a heater to dewater and dry the stream, a filter to remove fine particulate, and finally, either an activated carbon adsorption system or a thermal oxidation unit.[30]

REFERENCES

1. *Remediation Technologies Screening Matrix and Reference Guide*, EPA/542/B-94/013, Oct. 1994.
2. D. S. Koback, et al., *Innovative Ground Water and Soil Remediation: In Situ Air Stripping Using Horizontal Wells*, Proceedings of the 1991 Outdoor Action Conference, National Water Well Association, Las Vegas, NV.
3. SITE Technology Capsule, *Unterdruck–Verdampfer–Brunnen Technology (UVB) Vacuum Vaporizing Well*, EPA/540/R-95/500a, July 1995.
4. Superfund Innovative Technology Evaluation Program, Technology Profiles, Seventh Edition, EPA/540/R-94/526, p. 180.
5. T. W. Schrauf, et al., "Alternative Method of Groundwater Sparging for Petroleum Hydrocarbon Remediation," Remediation, Winter, 1993/94, p. 93.
6. *Subsurface Volatilization and Ventilation System (SVVS®)*, EPA/540/R-94/529, July 1995.
7. J. F. Dablow and D. Rodgers, "Steam Sparging and Enhanced Bioremediation for Heavy Fuel Oil Remediation," *In Situ and On Site Case Histories: In Situ and On Site Bioremediation*, Vol. V, edited by B. Alleman and A. Leason, Battelle Press, 1997.
8. John C. Fountain, "Technologies for Dense Nonaqueous Phase Liquid Source Zone Remediation," Technology Evaluation Report, Ground-Water Remediation Technologies Analysis Center (GWR-TAC), December 1998, p. 36.
9. Reference 8.
10. Reference 8, p. 39.
11. R. S. Kasevich, M. C. Marley, and M. Johnson, KAI Technologies, Inc., Proceedings from U.S. EPA TIO OSWER Conference on In Situ Thermal Treatment, Atlanta, GA, Dec. 1998.
12. P. A. Gaughtz, et al., "Six Phase Soil Heating Accelerates VOC Extraction From Clay Soil," Proceedings of Spectrum '94, International Nuclear and Hazardous Waste Management, Atlanta, GA, Aug. 14–18, 1994.
13. Reference 8, p. 39.
14. Project Status Fact Sheet, *Field Treatability Study: In-Situ Thermal Desorption*, General Electric Co., Mar. 1994.
15. C. Reece, "VAC*TRAX-Mobile Vacuum/Thermal Treatment System," Proceedings of the Information Exchange Meeting on Waste Retrieval, Treatment and Processing, Houston, TX, Mar. 15–17, 1993, p. 81.
16. McLaren-Hart, technical literature.
17. Superfund Innovative Technology Evaluation Program, Technology Profiles, Seventh Edition, EPA/540/R-94/526, Nov. 1994, p. 87.
18. C. C. Travis, J. M. MacInnis, "Vapor Extraction of Organics from Subsurface Soils — Is It Effective?" *Envir. Sci. & Tech.* **26**(10), 1885, 1992.
19. *Innovative Site Remediation Technology: Design and Application — Thermal Desorption*, William C. Anderson, editor, American Academy of Environmental Engineers, 1997, p. 2–1.
20. Reference 19, Table 5–1.
21. U.S. EPA, Office of Emergency and Remedial Response, ROD Information Directory, December 1995.
22. U.S. EPA, Air/Superfund National Technical Guidance Study — Models for Estimating Air Emission Rates from Superfund Remedial Actions, EPA-451/R-93-001, March 1993.
23. U.S. EPA Information Sources for Innovative Remediation and Site Characterization Technologies, Technology Innovation Office (TIO), OSWER, EPA 542-C-98-003, December 1997.
24. Reference 23.
25. Dr. Harold Vinegar and Dr. George L. Stegemeier, Terra-Therm Environmental Services, Inc., Proceedings from U.S. EPA TIO OSWER Conference on In Situ Thermal Treatment, Atlanta, GA, Dec. 1998.
26. Greg Beyke, Current Environmental Solutions, Inc., Proceedings from the U.S EPA TIO OSWER Conference on In Situ Thermal Treatment, Atlanta, GA, Dec. 1998.
27. R. M. Leuser et al., "Low Temperature Thermal Treatment of Contaminated Soil," Paper No. 89-2.2, 82nd Annual Meeting of the Air and Waste Management Association, Anaheim, CA, June 1989; and U.S. EPA Information Sources TIO, OSWER, EPA 542-C-98-003.
28. Reference 23.
29. Reference 23.
30. James E. Hansen, Geosafe Corporation, Proceedings from the U.S EPA TIO OSWER Conference on In Situ Thermal Treatment, Atlanta, GA, Dec. 1998.

DRINKING WATER TREATMENT

Michael J. Barboza and Jack H. Van Sciver

Drinking water treatment is performed to provide a supply of water that is fit for human consumption and use, that is, free from toxic or otherwise harmful materials, is safe from disease-causing bacteria, viruses and other disease-causing organisms, and free of colors, odors, turbidity, and objectionable tastes. Sources of drinking water include surface water (rivers, streams, and lakes), and groundwater wells. Treatment steps used to provide safe drinking water include prefiltration and filtration for the removal of turbidity and color, disinfection for the control of potentially disease causing organisms (*Giardia* cysts, coliform bacteria, viruses), organic contaminant removal, and inorganic contaminant removal. The selection of water treatment technology alternatives is generally made based on the following considerations:

- Desired effluent characteristics
- Influent water characteristics
- Current treatment system configuration
- Cost/budget
- System operational requirements
- Waste management
- Anticipated future needs of the surrounding area

Regulatory control over drinking water supplies in the United States was significantly strengthened and expanded by the 1986 Amendments to the Safe Drinking Water Act (SDWA). A major focus of the amendments was the establishment of a stringent schedule for the promulgation of drinking water standards that use numerical criteria, such as the Maximum Contaminant Levels (MCLs) for 83 contaminants or treatment technique requirements. Prior to the 1986 amendments, only 22 MCLs had been established. The regulatory requirements stemming from the SDWA include the following:

- *Maximum contaminant levels* (MCLs) are the maximum permissible levels of contaminants delivered to a user of public water supplies.
- *Maximum contaminant level goals* (MCLG) are nonenforceable concentration limits that indicate the concentration of a contaminant that does not cause any known or anticipated adverse human health effects. Formerly termed recommended maximum contaminant levels (RMCLs).
- *Secondary maximum contaminant levels* (SMCLs) are nonenforceable goals for preserving the aesthetic qualities of drinking water.

The SDWA requires that EPA establish treatment technology requirements in lieu of MCLs when it is not economically or technically feasible for water suppliers to determine the level of contaminants. Additionally, the SDWA requires EPA to establish best available technology (BAT) for the purposes of complying with the National Primary Drinking Water Regulations. In accordance with these requirements, EPA established granular activated carbon (GAC) as best available technology for synthetic organic chemical removal. Regulations require that use of other water treatment technologies or control strategies must be at least as effective at the BAT. Additional provisions of the SDWA include monitoring, filtration, disinfection, and requirements concerning the use of lead-containing materials in drinking water systems. Table 1 lists contaminants regulated under the SDWA of 1986.

Table 1. Drinking Water Contaminants Regulated under the SDWA Amendments of 1986

Volatile Organic Compounds	
Trichlorethylene	Benzene
Tetrachloroethylene	Chlorobenzene
Carbon tetrachloride	Dichlorobenzene
1,1,1-Trichloroethane	Trichlorobenzene
Vinyl chloride	1,1-Dichloroethylene
Methylene chloride	*trans*-1,2-Dichloroethylene
	cis-1,2-Dichloroethylene
Synthetic Organic Compounds	
Endrin	1,1,2-Trichloroethane
Lindane	Vydate
Methoxychlor	Simazine
Toxaphene	PAHs
2,4-D	PCBs
2,4,5-TP	Atrazine
Aldicarb	Phthalates
Chlordane	Acrylamide
Dalapon	Dibromochloropropane (DBCP)
Diquat	1,2-Dichloropropane
Endothall	Pentachlorophenol
Glyphosphate	Pichloram
Carbofuran	Dinoseb
Alachlor	Ethylene dibromide (EDB)
Epichlorohydrin	Dibromomethane
Toluene	Xylene
Adipates	Hexachlorocyclopentadiene
2,3,7,8-Chlorinated dibenzo furans	Total trihalomethanes

Source: Ref. 1.

PROCESS DESCRIPTION

Drinking water treatment technologies include:

- Filtration
- Disinfection
- Inorganic compound removal
- Organic compound removal

These drinking water treatment technologies are described below.

Filtration

The objectives of water filtration are to control turbidity, color, and potential microorganism contamination. Key considerations in evaluating water filtration systems include the volume of the resultant sludge and its potential toxicity, the desired frequency of the filter cleaning cycle, the chemical additives consumption rate, and operation simplicity/complexity. These factors must be evaluated on a site-specific basis. Typically, mixing, flocculation, and sedimentation operations are performed prior to actual filtration. Technology alternatives for water filtration include:

- Conventional filtration
- Direct filtration
- Slow sand filtration
- Diatomaceous earth filtration
- Reverse osmosis
- Cartridge filtration

Disinfection

Drinking water disinfection systems are of two types: *primary disinfectants* and *secondary disinfectants*. Primary disinfectants are used for the inactivation of *Giardia* cysts, microorganisms, and viruses. Secondary disinfectants are used for the suppression of biological agents in the drinking water distribution system. Common primary disinfectants include chlorine, chlorine dioxide, ozone, and ultraviolet (UV) radiation. Secondary disinfectants include chlorine, chlorine dioxide, and chloramines. If ozone is selected as a disinfecting agent, care must be taken to ensure that excess ozone from the water treatment contact unit is destroyed. Typical ozone concentrations in contact units exceed 1 g/m^3 (500 ppmv),[1] well in excess of the OSHA Permissible Exposure Limit of 0.0002 g/m^3 (0.1 ppmv). Available technologies for the removal of ozone include thermal incineration (heating the waste ozone to 300–350°C for 3 sec), combination thermal/catalytic incineration, and catalytic destruction. Activated carbon absorption has been used extensively in Europe, particularly at small water treatment plants (i.e., less than 2 MGD).

Inorganic Compound Removal

Numerous treatment techniques are available for the removal of the many potential inorganic contaminants found in drinking water, including radionuclides. Typically, these inorganic treatment processes are effective only for a limited set of potential inorganic contaminants. Inorganic treatment technology alternatives include:

- Conventional treatment with coagulation
- Lime softening
- Cation exchange
- Anion exchange
- Reverse osmosis
- Powdered activated carbon (PAC)
- Granular activated carbon (GAC)
- Activated alumina

Organic Compound Removal

Treatment of organic compounds potentially found in drinking water sources is a function of chemical and physical characteristics of the compounds that must be removed. Classes of organic compounds potentially requiring treatment include aromatic hydrocarbons, halogenated aliphatic and aromatic compounds, polychlorinated biphenyls (PCBs), polynuclear aromatics, and pesticides.

Sometimes it may be preferable to perform oxidative pretreatment of drinking water sources prior to the primary organic compound removal step. Available oxidants for the pretreatment of drinking water include chlorine, ozone, hydrogen peroxide, potassium permanganate, chlorine dioxide, and UV radiation. Organic compounds present in raw drinking water can be oxidized completely to carbon dioxide and water, or partially oxidized into intermediate oxidation products. Drinking water pretreatment with an oxidant may also enhance organic compound removal by forming oxidation compounds with lower aqueous solubilities, causing some precipitation to occur.

Processes employed in the treatment of drinking water for the removal of organic compounds include granular activated carbon, powdered activated carbon with conventional treatment, diffused aeration, and multiple tray aeration. Table 2 presents a number of organic compound treatment technologies for drinking water; along with an indication of acceptability relative to BAT, size suitability, and effectiveness. Table 3 presents typical organic compound removal efficiencies of a number of drinking water treatment technologies for various organic contaminants.

Granular Activated Carbon. Granular activated carbon water treatment systems work on the principle of adsorption. Dissolved contaminants in the water phase are transferred to the porous surface of the activated carbon particles. Activated carbon is an excellent adsorbent for many organic materials due to its very large specific surface area (on the order of 0.8–1.6 km^2/kg) and high degree of porosity. Additionally, the adsorption process is mainly physical in nature and thus easily reversed. The capability of desorbing VOC compounds from GAC and subsequently reusing the GAC is an essential factor in evaluating the usefulness of GAC systems since this facilitates the reuse or recycling of the carbon.

Table 2. Technologies for Organic Contamination Control of Drinking Water

Treatment Technology	Stage of Acceptability	Size Suitability	Comments
Granular activated carbon	BAT	All	Highly effective; potential waste disposal issues
Packed column aeration	BAT	All	Highly effective for volatile compounds; potential air emission issues
Powdered activated carbon	Established	Large	Requires conventional treatment process train for application
Diffused aeration	Established	All	Variable removal effectiveness
Multiple tray aeration	Established	All	Variable removal effectiveness
Oxidation	Experimental	All	By-product concerns
Reverse osmosis	Emerging	Small to medium	Variable removal effectiveness—expensive
Mechanical aeration	Experimental	All	Mostly for wastewater treatment; high energy requirements, easy to operate
Catenary grid	Experimental	All	Performance data are scarce; potential air emissions issues
Higee aeration	Experimental	Small	Compact, high energy requirements; potential air emission issues

Source: Ref. 1

Table 3. Water Treatment Technology Removal Effectiveness Reported for Organic Contaminants (percent)[a]

Contaminant	Coagulation/ Filtration	GAC	PCA	PAC	Diffused Aeration	Oxidation	Reverse Osmosis
Acrylamide	5	NA	0–29	13	NA	NA	0–97
Alachlor	0–49	70–100	70–100	36–100	NA	70–100	70–100
Aldicarb	NA	NA	0–29	NA	NA	NA	94–99
Benzene	0–29	70–100	70–100	NA	NA	70–100	0–29
Carbofuran	54–79	70–100	0–29	45–75	11–20	70–100	70–100
Carbon tetrachloride	0–29	70–100	70–100	0–25	NA	0–29	70–100
Chlordan	NA	70–100	70–100	NA	NA	NA	NA
Chlorobenzene	0–29	70–100	70–100	NA	NA	30–96	70–100
2,4-D	0–29	70–100	70–100	69–100	NA	W[b]	0–65
1,2-Dichloroethane	0–29	70–100	70–100	NA	42–77	0–29	15–70
1,2-Dichloropropane	0–29	70–100	70–100	NA	12–79	0–29	10–100
Dibromochloropropane	0–29	70–100	70–100	NA	NA	0–29	NA
Dichlorobenzene	NA	70–100	70–100	NA	NA	NA	NA
o-Dichlorobenzene	0–29	70–100	70–100	38–95	14–72	30–88	30–69
p-Dichlorobenzene	0–29	70–100	70–100	NA	NA	30–69	0–10
1,1-Dichloroethylene	0–29	70–100	70–100	NA	97	70–100	NA
cis-1,2-Dichloroethylene	0–29	70–100	70–100	NA	32–85	70–100	0–30
trans-1,2- Dichloroethylene	0–29	70–100	70–100	NA	37–96	70–100	0–30
Epichlorohydrin	NA	NA	0–29	NA	NA	0–29	NA
Ethylbenzene	0–29	70–100	70–100	33–99	24–89	70–100	0–30
Ethylene dibromide	0–29	70–100	70–100	NA	NA	0–29	37–100
Heptachlor	64	70–100	70–100	53–97	NA	70–100	NA
Heptachlor epoxide	NA	NA	NA	NA	NA	26	NA
Hydrocarbons[b]	NA	W	NA	NA	NA	NA	NA
Lindane	0–29	70–100	0–29	82–97	NA	0–100	50–75
Methoxychlor	NA	70–100	NA	NA	NA	NA	>90
Monochlorobenzene	NA	NA	NA	14–99	14–85	86–98	50–100
Natural organic material	P	P	NA	P	NA	W	P[c]
PCBs	NA	70–100	70–100	NA	NA	NA	95
Phenol and chlorophenols	NA	W	NA	NA	NA	W	NA
Pentachlorophenol	NA	70–100	0	NA	NA	70–100	NA
Styrene	0–29	NA	NA	NA	NA	70–100	NA
Tetrachloroethylene	NA	70–100	NA	NA	73–95	W	70–90
Trichloroethylene	0–29	70–100	70–100	NA	53–95	30–69	0–100
Trichloroethane	NA	70–100	NA	NA	NA	NA	NA
1,1,1-Trichloroethane	0–29	70–100	70–100	40–65	58–90	0–29	15–100
Toluene	0–29	70–100	70–100	0–67	22–89	70–100	NA
2,4,5-TP	63	70–100	NA	82–89	NA	30–69	NA
Toxaphene	0–29	70–100	70–100	40–99	NA	NA	NA
Vinyl chloride	0–29	70–100	70–100	NA	NA	70–100	NA
Xylenes	0–29	70–100	70–100	60–99	18–89	70–100	10–85

[a]W, well removed; P, poorly removed; NA, not available.
[b]High-molecular-weight hydrocarbons (i.e., gasoline, dyes).

Source: Ref. 1.

The physical and chemical characteristics of the pollutants significantly affect the adsorbent characteristic of GAC systems. Table 4 presents the suitability of a number of organic compounds to treatment by GAC. Generally, GAC systems have greater effectiveness for organic compounds with the following characteristics:

- Branch chained, rather than straight chained
- High molecular weight
- Low aqueous solubility
- Nonpolar
- Present in high concentrations

Raw inlet water properties can also affect GAC adsorbent properties. The most significant characteristic is the presence of competing organic compounds and dissolved solids, which can adversely affect adsorption. Important process design characteristics of GAC water treatment systems include:

- GAC type
- Surface loading rate of the GAC filter
- Empty bed contact time (EBCT)
- Contaminant type and concentration
- Carbon depth and usage

The most significant system design characteristic of GAC water treatment systems with regard to air emissions is whether on-site GAC regeneration will be performed. GAC systems with carbon exhaustion rates in

Table 4. Organic Compounds That Are Readily and Poorly Adsorbed by Granular Activated Carbon

Readily Adsorbed Organics
Aromatic solvents
Chlorinated aromatics
Phenols and chlorophenols
Polynuclear aromatics
Pesticides and herbicides
Chlorinated aliphatics
High-molecular-weight hydrocarbons
Poorly Adsorbed Organics
Alcohols
Low-molecular-weight ketones, acids, and aldehydes
Sugars and starches
Very high-molecular-weight or colloidal organics
Low-molecular-weight aliphatics

Source: Ref. 1.

Table 5. Henry's Law Constants for Selected Compounds

Compound Type	Henry's Law Constant (dimensionless)
Organic Compounds	
Vinyl chloride	265
Trichloroethylene	0.41
Tetrachloroethylene	0.82
cis-1,2-Dichloroethylene	0.32
Pesticides	
Aldicarb	0.00000017
Chlordane	0.004
Dibromochloropropane	0.01
Chlorinated Aromatics	
Polychlorinated biphenols	0.059
Dichlorobenzene	0.081

Source: Ref. 2.

the range of 500–2000 lb carbon/d will generally utilize off-site regeneration, typically provided by a carbon system vendor. Systems exhausting greater than 2000 lb carbon/d may utilize on-site regeneration, which generally utilizes steam to volatilize adsorbed compounds from the carbon.

Packed Column Aeration (Air Stripping). The goal of air stripping is the mass transfer of organic contaminants from the water phase into the contacting air phase. The volatilized organic compounds are either released directly to the atmosphere, or treated through some air pollution control device, and then emitted to the atmosphere. Packed column aeration (PCA) is a mass transfer unit operation in which the water to be treated and air are mixed (typically in a countercurrent configuration). A critical factor in evaluating the effectiveness of air stripping as a VOC removal technique is the Henry's Law coefficient. The Henry's Law coefficient is a measure of a particular contaminants volatility and its ability to partition from the water phase into the air phase. Compounds with high Henry's Law coefficients are effectively removed by air stripping, while those with low coefficients are more difficult to remove from water by air stripping. Table 5 presents Henry's Law Constants for a number of volatile organic compounds. PCA system design considerations include:

- Type of VOC contaminants(s)
- Concentration of VOC contaminants(s)
- Column packing material
- Height of packing material
- Air/water ratio
- Water temperature
- Water flow rate

The major equipment items required for a PCA system are the column, the column packing material, air blower, booster pump (if required), and associated instrumentation. Columns can be fabricated from fiberglass-reinforced plastic (FRP), stainless steel, or aluminum. The column internals include mist eliminators to prevent the entrainment of water droplets, liquid distributors, and support structures to support the packing material. A packed-tower aeration system schematic is presented in Figure 1.

Specific contaminants require different designs to remove them effectively to a given level. Column packing materials are designed to provide a high specific surface area for mass transfer while providing for low pressure drop. The contaminant removal level and the air/water ratio will determine packing height. The air/water ratio in an air stripper is a function of both the water temperature and the required level of contaminant removal. Air/water ratios typically range from 30:1 to 100:1. Column size (diameter) specifications are derived from the required hydraulic (water flow rate)

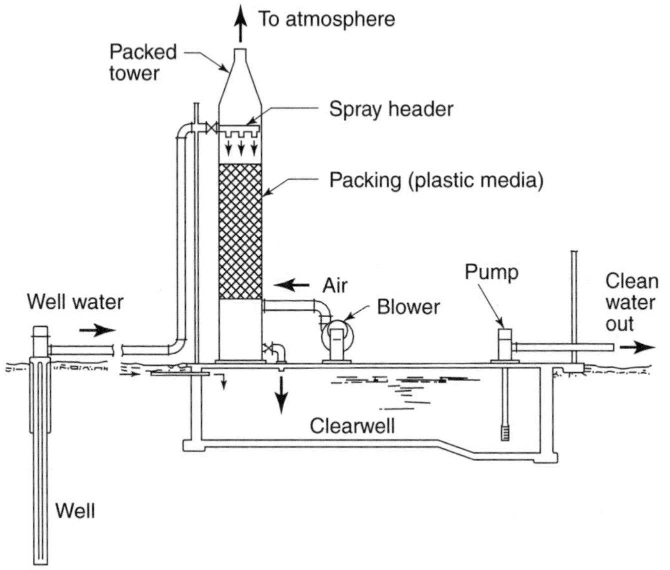

Figure 1. Packed-Tower Aeration System.

requirements. These process design factors all affect the VOC removal efficiency.

Diffused Aeration. In diffused aeration systems, water is treated by bubbling air through the water in a contact chamber. The diffuser is typically located near the bottom of the contact chamber. Air introduced into the contacting chamber produces fine bubbles, which rise through the chamber, producing some mixing turbulence as the bubbles rise. Diffused aeration systems are selected mainly for simplicity and cost considerations, as whether they can be constructed from existing structures; however, diffused aeration is generally less effective that PCA systems. Table 6 presents a listing of organic compounds along with removal efficiencies for diffused aeration water treatment systems.

Tray Aeration. Tray aeration units operate by directing water downward through a series of trays. The trays contain openings made of wire mesh, perforations, or slats. Tray aeration units have less surface area susceptible to clogging or fouling, a particular problem with water containing iron and manganese. Tray aeration is not as effective as PCA systems, and can experience biological growth and corrosion problems. Factors to be considered in the design and selection of tray aeration units are tray type and height, air-to-water ratio, and whether the unit will operate under positive pressure. A typical air-to-water ratio is approximately 30:1. Slat tray aeration units with an air-to-water ratio of 30:1 and a tray height of 12–16 ft have achieved 30–90% reductions of trichloroethylene and 20–85% reductions of tetrachloroethylene.[1]

AIR EMISSIONS CHARACTERIZATION

Air emission concerns associated with drinking water treatment are usually related to the potential transfer of VOC contaminants removed from the water to the air. Drinking water treatment systems with potential to cause air emissions are:

- GAC regeneration
- PCA air strippers

Table 6. Typical VOC Removal Performance of Diffused Aeration Water Treatment

Compound	% Removal
Trichloroethylene	53–95
Tetrachloroethylene	73–95
1,2-Dichloroethane	42–77
1,1-Dichloroethylene	97
1,1,1-Trichloroethane	58–90
1,2-Dichloropropane	12–79
cis-1,2-Dichloroethylene	32–85
trans-1,2-Dichloroethylene	37–96
o-Dichlorobenzene	14–72
Ethylbenzene	24–89
Monochlorobenzene	14–85
Toluene	22–89
Xylenes	18–89

Source: Ref. 1.

GAC Systems

GAC water treatment systems with on-site regeneration have the potential to emit VOCs to the atmosphere during desorption from the carbon. Air emissions from the carbon regeneration process can potentially be significant.

Air Stripper PCA Systems

Air strippers/PCA systems use air to remove VOCs from water and usually emit them directly to the atmosphere. These emissions are usually of VOCs that are readily stripped from water by air.

VOC Emission Rate Estimation. Uncontrolled VOC emissions to the atmosphere from air strippers can be calculated with the following general formula:

$$E_i = (C_i)(L_R)(SE/100)(1.67 \times 10^{-5})$$

where E_i is the emission rate of species i (g/sec); C_i, the concentration of species i in influent water (mg/l); L_R, the influent water flow rate (l/min); and SE, the stripping efficiency (%).[2]

For conservative air emission estimation purposes, it is common to assume that air strippers achieve 100% removal of VOC contaminants in the influent water (i.e., VOC stripping efficiency of 100%). Assumption of 100% VOC removal is also commonly required by regulatory agencies for air permitting purposes. Depending on the magnitude of air emissions from an air stripping tower, vapor-phase VOC emission control may be required. The estimated air stripper VOC emission rate must be evaluated with respect to applicable state, local, and federal air quality regulations, as well as site-specific considerations, such as distance to property lines, and the potential presence of nearby "sensitive" receptors (i.e., schools hospitals, etc.).

AIR POLLUTION CONTROL MEASURES

Methods available for use for the control of vapor-phase VOC compounds from water treatment systems include:

- Thermal oxidation
- Catalytic oxidation
- Ozone oxidation
- Carbon adsorption

Thermal oxidation, catalytic oxidation, and ozone oxidation are not used extensively in water treatment applications. Thermal oxidation is very effective at VOC destruction, but is expensive due to high energy consumption. Catalytic oxidation reduces the energy required, but is not widely used because it is not effective in achieving high destruction efficiencies with low inlet concentrations of chlorinated organic compounds. A catalytic incineration process schematic is presented in Figure 2.

Adsorption of VOC from air strippers can be accomplished with vapor-phase GAC adsorption units. Currently, GAC is the most frequently used treatment technique for the control of vapor-phase VOC from air stripping processes. A schematic of a vapor phase GAC system schematic is presented in Figure 3.

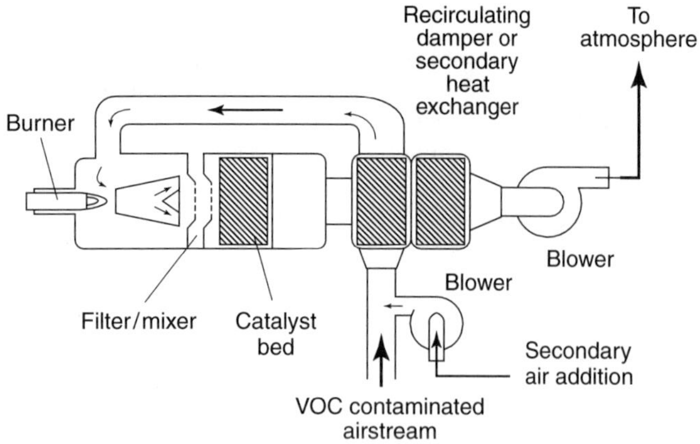

Figure 2. Schematic of Catalytic Incineration Process.

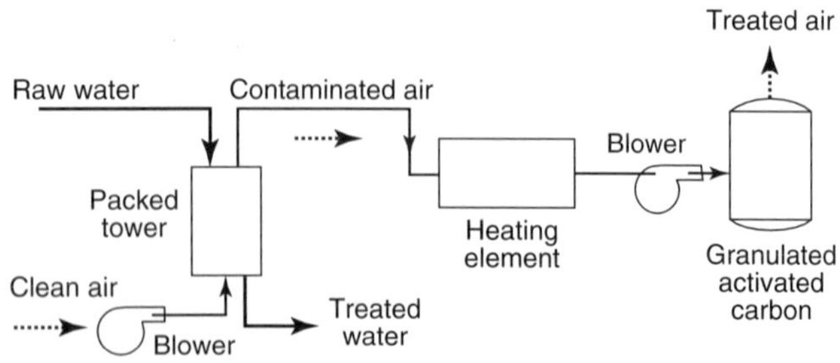

Figure 3. Vapor-Phase GAC System for Treatment of VOC-Laden Airstream.

REFERENCES

1. *Technologies for Upgrading Existing or Designing New Drinking Water Treatment Facilities*, EPA/625/4-89/023, U.S. Environmental Protection Agency, Office of Drinking Water, Cincinnati, OH, March 1990.
2. *Air/Superfund National Technical Guidance Study Series, Estimation of Air Impacts for Air Stripping of Contaminated Water*, EPA/450/1-91/002, U.S. Environmental Protection Agency, Office of Air Quality Planning and Standards, Research Triangle Park, NC, May 1991.

BIBLIOGRAPHY

Louis J. Thibodeaux, *Environmental Chemodynamics*, Wiley Interscience, New York, 1996.

H. Zhu, T. Keener, P. Bishop, T. Orton, and M. Wang, "Emissions of Hazardous Air Pollutants from Aeration Tanks," *Environmental Progress* (Fall 1998).

M. J. Hammer, *Water and Wastewater Technology*, John Wiley and Sons, New York, 1975.

J. N. Murrell and A. D. Jenkins, *Properties of Liquids and Solutions*, John Wiley and Sons, New York, 1994.

R. C. Reid, J. M. Prausnitz, and B. E. Poling, *The Properties of Gases and Liquids*, McGraw-Hill, New York, 1987.

Technical Manual 1203: Air Stripper, New Jersey Department of Environmental Protection, Bureau of New Source Review, Trenton, NJ, March 1994.

Small System Compliance Technology List for the Surface Water Treatment Rule, EPA/815-R/97/002, U.S. Environmental Protection Agency, Office of Drinking Water, Cincinnati, OH, August 1997.

Water Treatment Plant Design, American Water Works Association, American Society of Civil Engineers, Third Edition, McGraw–Hill, New York, 1990.

WEB SITES

Water Environment Federation: *http://www.wef.org*.

American Society of Civil Engineers: *http://www.asce.org*.

U.S. Environmental Protection Agency Office of Groundwater & Drinking Water: *http://www.epa.gov/OGWDW*.

American Water Works Association: *http://www.awwa.org*.

Center for Drinking Water Optimization: *http://www.cdwo.org*.

MUNICIPAL SOLID WASTE LANDFILL GAS EMISSIONS

Michael J. Barboza

Landfilling of municipal solid waste (MSW) has been a generally acceptable means of disposal for many years. MSW landfills (LFs) receive primarily household and/or commercial waste. Current practice is to spread the waste in layers, compacting and covering it with soil. The compacted layers compose the landfill building blocks called *cells*. The buried waste decomposes biologically and chemically to produce solid, liquid, and gaseous products.

MSW landfills are potential sources of emissions of gas mixtures generated from the natural decomposition of organic wastes and vapors from volatile compounds present in the wastes. The concerns associated with LF gas commonly involve odors, combustion/explosion hazards, and possible toxic effects. MSW LFs are different from many other emissions source categories (i.e., manufacturing facilities) because LFs will continue to generate LF gas emissions for many years after closure. Landfill operators must consider subsurface gas migration, gas collection, control and recovery systems, and ambient air quality impacts including odors. Studies of landfill gas emissions have been performed to support landfill permits, landfill closures, permitting and design of gas collection, control, and recovery systems as well as to assess potential impacts and site acceptability for alternate uses. Figure 1 provides an overview that illustrates the gamut of potential activities, emission sources, collection systems, recovery facilities, and control devices at MSW landfills.

NEW SOURCE PERFORMANCE STANDARDS[1]

The U.S. Environmental Protection Agency (EPA) has adopted New Source Performance Standards (NSPS), Subpart WWW and Emission Guidelines, that address uncontrolled emissions of LF gas based on studies that indicate that uncontrolled MSW LF gas emissions can, at certain levels, have adverse effects on both public health and welfare.

The emissions of concern from MSW landfills are nonmethane organic compounds (NMOC) and methane. NMOCs include volatile organic compounds (VOC), hazardous air pollutants (HAPs), and odorous compounds. VOC emissions contribute to ozone formation, which can result in adverse health effects to human health and vegetation. Ozone can penetrate in different regions of the respiratory tract and be absorbed through the respiratory system. The health effects of exposure to HAPs can include cancer, respiratory irritation, and damage to the nervous system. Methane emissions

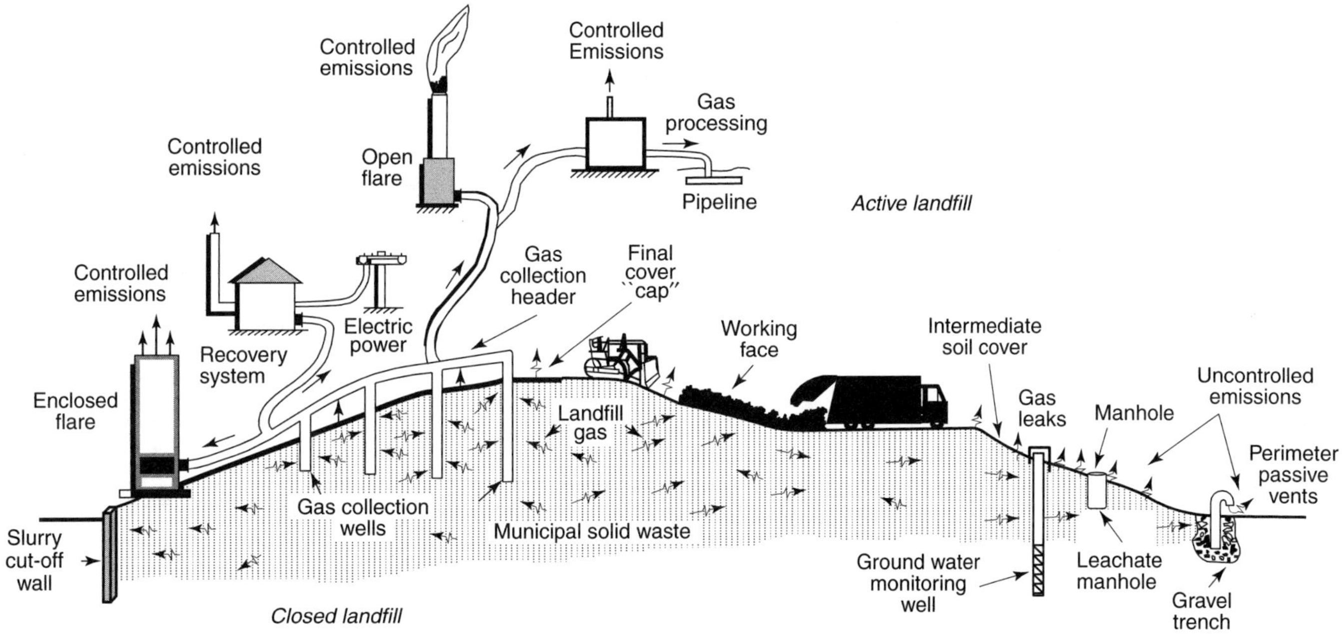

Figure 1. Municipal Solid Waste Landfill Gas Emissions.

contribute to global climate change and can result in fires or explosions if it accumulates in structures on or off landfill sites.

The intended effect of the standards and guidelines is to require certain municipal solid waste landfills to control emissions to the level achievable by the best demonstrated system of continuous emission reduction, considering costs, non-air-quality health, and environmental and energy impacts.

According to New Source Performance Standards (NSPS), Subpart WWW requires use of a control system designed and operated to reduce NMOC by 98% by weight; or when an enclosed combustion device (i.e., not an open flare) is used for control, to either reduce NMOC by 98% by weight; or reduce the outlet NMOC concentration to less than 20 parts per million by volume, dry basis as hexane at 3% oxygen. NMOC emissions are to be reduced through the installation of well-designed and -operated LF gas collection systems and control devices capable of meeting 98% destruction efficiency.

PROCESS DESCRIPTION[2]

There are an estimated 6,000 active MSW landfills in the United States, receiving around 209 million megagrams (Mg) (230 million tons) of waste annually. Most (around 93%) active LFs have a design capacity of 5 million Mg (5.5 tons) or less. Many landfills with capacities of 1–20 million Mg (1.1–22 million tons) are around 44–29% filled, respectively. Average acceptance rates of landfills range from 50 Mg/day (55 tons/day) for smaller LFs (less than 1 Mg design capacity) to around 4,000 Mg/day (4,405 tons/day) for larger LFs (greater than 20 Mg design capacity). The average waste acceptance rate is around 31,780 Mg/year or 122 Mg/d (35,000 tons/year or 135 tons/d). The median acceptance rate, however, is 2,724 Mg/year or 10.4 Mg/d (3,000 tons/year 11.5 tons/d) because of a small number of large landfills.

Cells are usually around 8 ft high. The working face can extend across the width of the cell. Usually 1 day's refuse is covered at the end of each day with at least 6 in. of soil cover. One ft of intermediate and 2 ft of final soil cover are commonly used. Liners made of either soil (i.e., clay) or a combination of soil and synthetic (i.e., high-density polyethylene) are often used beneath the landfill to contain leachate (the liquid produced from waste decomposition) to prevent groundwater contamination and gas migration.

Composition of MSW[2]

Municipal waste composition can be described as containing household wastes (71.97%), commercial wastes (17.19%), construction demolition wastes (5.83%), industrial process wastes (2.73%), sewage sludge (0.51%), and lesser amounts of incinerator ash, small quantity generator hazardous wastes, infectious wastes, asbestos-containing waste materials, and other wastes.

LF Gas Composition

Landfill gas can be generated at both active and inactive LFs. Natural biological processes occurring in LFs transform the waste's constituents, producing leachate and gas (sometimes referred to as biogas). Anaerobic decomposition of buried refuse produces relatively high concentrations of methane and carbon dioxide. The major components of LF gas and various phases of gas generation are shown in Figure 2.[2] Decomposition of waste can reach the anaerobic steady methanogenic phase in around 2–4 years.[3]

Initially, decomposition is aerobic until the oxygen supply is exhausted also characterized by an increase in carbon dioxide (CO_2). The time and duration is difficult to identify for the various phases of waste decomposition and gas generation, which varies with landfill conditions such as waste composition, moisture content, temperature, etc. (weather in that part of the country). This initial phase begins shortly after the placement of the refuse and can be expected to last on the order of between several hours and week. During Phase 2, oxygen reserves are depleted and anaerobic conditions begin. Phase 2 can typically last for several months. Phase 3 is marked by the transformation of complex materials such as cellulose,

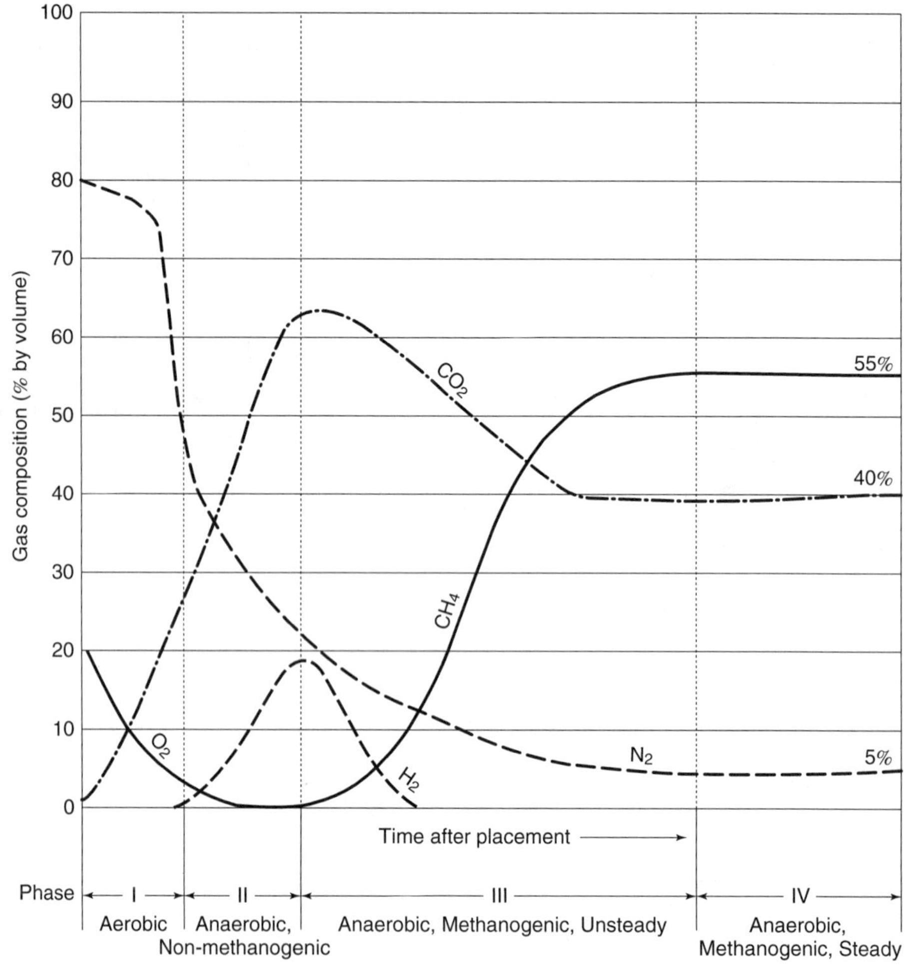

Figure 2. Typical Landfill Gas Generation and Composition.

fats, proteins, and carbohydrates into simple organic materials such as acetic acid. Phase 3 may last for a period of several months to several years. During Phase 4, the acids formed during Phase 3 are consumed by anaerobic methanogenic bacteria and converted into methane and carbon dioxide, the major components of LF gas. During the fifth and final phase, LF gas generation declines. During Phase 5, most of the nutrients required to sustain the methanogenic bacterial population have been depleted, thereby reducing the amount of methane generated. Both Phases 4 and 5, however, may last for decades.

It should be noted that for a given LF, especially an old active one, these various phases of waste decomposition and gas generation will be occurring at the same time, depending on the times and locations of waste placement.

Landfill gas consists of approximately 50% methane and 50% carbon dioxide (by volume). CO_2 is considered a greenhouse gas but is of much less concern at LFs because it has a potency on the order of 20–30 times less than that of methane on a molecule per molecule basis. Other constituents of LF gas can include hydrogen, ammonia, hydrogen sulfide, nitrogen, hydrogen chloride, carbon monoxide, and a variety of volatile organic compounds.

Methane, the major constituent of LF gas, is a colorless odorless gas that is only slightly soluble in water and burns readily in air. It is generally very stable; however, when mixed with air at concentrations of around 5–15% by volume, it is highly explosive. Methane is considered a simple asphyxiant and thus is of lesser concern from a toxicity point of view, except in situations where it may displace oxygen (such as confined spaces). With the exception of methane and carbon dioxide, the other gases are usually present in trace amounts and not of concern with regard to an explosive hazard.

The heating value of landfill gas is derived mostly from its methane content. The higher heating value (HHV) and lower heating value (LHV) of methane are 1,010 and 991 Btu/ft^3 respectively, making the heating value of LF gas around 3,523–4,937 KC/m^3 (396–555 Btu/ft^3), approximately half that of natural gas, depending on the actual methane content (40–55%) of the LF gas. This heating value makes disposal of LF gas practical and efficient by burning. It also can be recovered for use as a fuel for combustion in engines or boilers for generation of electric power or cleaned up to pipeline quality for consumer use.

Typically less than 1% (by volume) of LF gas is NMOCs, including VOCs.[2] A number of different NMOCs that have been detected in LF gas are of regulatory concern, of toxicological or other significance.[4] In addition to HAPs, some air toxics compounds are regulated by many states, such as California[5] and New Jersey.[6] Table 1 lists compounds that have been found in LF gas along with typical concentrations.[7]

Gas Generation

The MSW deposited in landfills decays chemically and biologically to produce solid, liquid, and gaseous products. The

Table 1. Default Concentrations for LFG Constituents

Compound	Molecular Weight	Default Concentration (ppmv)	Emission Factor Rating
1,1,1-Trichloroethane (methyl chloroform)[a]	133.42	0.48	B
1,1,2,2-Tetrachloroethane[a]	167.85	1.11	C
1,1-Dichloroethane (ethylidene dichloride)[a]	98.95	2.35	B
1,1-Dichloroethene (vinylidene chloride)[a]	96.94	0.20	B
1,2-Dichloroethane (ethylene dichloride)[a]	98.96	0.41	B
1,2-Dichloropropane (propylene dichloride)[a]	112.98	0.18	D
2-Propanol (isopropyl alcohol)	60.11	50.1	E
Acetone	58.08	7.01	B
Acrylonitrile[a]	53.06	6.33	D
Bromodichloromethane	163.83	3.13	C
Benzene[a]	78.11	1.91[e]	B
Butane	58.12	5.03	C
Carbon disulfide[a]	76.13	0.58	C
Carbon monoxide[b]	28.01	141	E
Carbon tetrachloride[a]	153.84	0.004	B
Carbonyl sulfide[a]	60.07	0.49	D
Chlorobenzene[a]	112.56	0.25	C
Chlorodifluoromethane	86.47	1.30	C
Chloroethane (ethyl chloride)[a]	64.52	1.25	B
Chloroform[a]	119.39	0.03	B
Chloromethane	50.49	1.21	B
Dichlorobenzene[c]	147	0.21	E
Dichlorodifluoromethane	120.91	15.7	A
Dichlorofluoromethane	102.92	2.62	D
Dichloromethane (methylene chloride)[a]	84.94	14.3	A
Dimethyl sulfide (methyl sulfide)	62.13	7.82	C
Ethane	30.07	889	C
Ethanol	46.08	27.2	E
Ethyl mercaptan (ethanethiol)	62.13	2.28	D
Ethylbenzene[a]	106.16	4.61	B
Ethylene dibromide	187.88	0.001	E
Fluorotrichloromethane	137.38	0.76	B
Hexane[a]	86.18	6.57	B
Hydrogen sulfide	34.08	35.5	B
Mercury (total)[a,d]	200.61	0.000292	E
Methyl ethyl ketone[a]	72.11	7.09	A
Methyl isobutyl ketone[a]	100.16	1.87	B
Methyl mercaptan	48.11	2.49	C
NMOC (as hexane)	86.18	595[f]	B
Pentane	72.15	3.29	C
Perchloroethylene (tetrachloroethylene)[a]	165.83	3.73	B
Propane	44.09	11.1	B
t-1,2-dichloroethene	96.94	2.84	B
Toluene[a]	92.13	39.3[e]	A
Trichloroethylene (trichloroethene)[a]	131.38	2.82	B
Vinyl chloride[a]	62.50	7.34	B
Xylenes[a]	106.16	12.1	B

Note: This is not an all-inclusive list of potential LFG constituents, only those for which test data were available at multiple sites.

[a]Hazardous air pollutants listed in Title III of the 1990 Clean Air Act Amendments.

[b]Carbon monoxide is not a typical constituent of LFG, but does exist in instances involving landfill (underground) combustion. Therefore, this default value should be used with caution. Of 18 sites where CO was measured, only 2 showed detectable levels of CO.

[c]Source tests did not indicate whether this compound was the para- or ortho-isomer. The para isomer is a Title III–listed HAP.

[d]No data were available to speciate total Hg into the elemental and organic forms.

[e]Concentrations at sites with no or unknown co-disposal of hazardous waste. For sites with hazardous waste co-disposal, concentrations are 11.1 ppmv for benzene and 165 ppmv for toluene.

[f]For NSPS/Emission Guideline compliance purposes, the default concentration for NMOC as specified in the final rule must be used. The default NMOC concentration shown is for sites with no or unknown co-disposal of hazardous waste. For sites with hazardous waste co-disposal, the default NMOC concentration is 2420 ppmv. For purposes not associated with NSPS/Emission Guideline compliance, the default VOC content at co-disposal sites = 85% by weight (2060 ppmv as hexane); at sites with no or unknown co-disposal = 39% by weight (235 ppmv as hexane).

Source: Ref. 7.

rate of LF gas generation varies and is affected by a number of factors, such as:

- Type and composition of waste
- Amount of biodegradable materials
- Age of waste
- Moisture content and pH
- Temperature

Under optimal growth conditions, the methanogenic bacteria reproduce and multiply for several years. The bacteria population (and associated LF gas generation) is considered to decline exponentially as the energy value of the refuse is depleted. The methane generation potential of waste in landfills is often considered to be minimal after approximately 20–30 years after placement. Methods to compute gas generation rates typically use exponential decay functions to simulate reduction of waste due to decomposition and depletion of the methane source bacteria population. First-order exponential decay relations tend to overstate gas generation for older waste, since first-order exponential functions approach zero as the age increases and can mathematically simulate gas generation after long periods of time, even though many of the factors required to sustain the physical processes have changed such as depletion of waste substrate and methanogenic bacteria.

Precipitation can affect methane generation. Precipitation can enter and accumulate when a landfill is open before a final cap is in place. Differences can be found (based on boring logs) at various locations within the same landfill, where borings in one location can have very dry conditions with little or no degradation of waste, while another location at the same landfill can have saturated conditions with completely decayed wastes. These differences can be attributed, in part, to the moisture in the waste itself, but is more likely related to whether or not it was raining the day the refuse was placed in the landfill and covered.[8]

The decay process is accelerated with warm temperatures and moisture. Anaerobic decomposition can produce heat with internal LF temperatures that range from 29 to 37°C (84 to 98.6°F). Moisture content of waste can vary with the highest gas production occurring when moisture content reaches around 60–78%.[9]

Waste Acceptance Rate/Waste in Place. The waste acceptance rate or the amount of in-place waste of a landfill can be used to estimate the landfill gas generation rate. Historical disposal records if available can be useful, or a fill history can be estimated for a landfill based on monthly or annual waste receipts, or in-place volume estimates based on topographic surveys and aerial photographs of the landfill. The in-place volume in cubic yards can be converted to tons using an assumed in-place waste density (i.e., 1485 pounds per cubic yard depending on the age of the LF and the amount of cover used, etc.).

LF Gas Model. Projections of LF gas generation rates can be used to estimate LF gas emissions and to support design of gas collection, control, and recovery systems, such as by aiding in sizing collection system components and identifying required flare capacities. A number of different LF gas generation rate (flow) models have been used, some based on stoichiometric chemically limiting formulas and empirical relationships. A review of available landfill gas generation rate models indicates the potential for estimating a fairly broad range of gas generation rates. A number of landfill gas flow models have been used to estimate LF gas generation/production from MSW LFs. These models include: Scholl Canyon, Palos Verde, and Lockman Associates.[2,10]

The maximum expected LF gas generation flow rates for sites with (known or estimated) year-to-year solid waste acceptance rates can be estimated with the following equation based on the NSPS EPA LF Gas Model [Subpart WWW, Section 60.755(a)(1)(ii)], along with appropriate values for the kinetic factors of k and L_0[1]:

$$Q_{\mathrm{M}} = \sum_{i=1}^{n} (1/C_{\mathrm{m}}) k L_0 M_i (e^{-kt_i})$$

where Q_{M} is the maximum expected gas generation flow rate (cubic meters per year); K, the methane generation rate constant (year^{-1}); L_0, the methane generation potential (cubic meters per megagram solid waste); M_i, the mass of solid waste in the ith section (megagrams); t_i, the age of the ith section (years); and C_{m}, the fraction of methane in LF gas (decimal). If not known, use $C_{\mathrm{m}} = 0.50$ (for methane concentration $= 50\%$). The value for mass of solid waste (M_i) used for each year of landfill operation is the waste acceptance for each year. The values for the kinetic factors k and L_0 are those from the most recent version of the Compilation of Air Pollutant Emission Factors (AP-42) (USEPA, 1997). The value for k is the recommended AP-42 default value, $k = 0.04$/year, which is appropriate for areas with normal or above-normal precipitation. A value of 0.02/year is considered more appropriate for landfills with drier waste. The AP-42 recommended value for L_0 is 100 m^3/Mg (3531 ft^3/Mg) refuse, which is considered appropriate for most landfills.[7]

Because of variability and uncertainties in factors affecting gas production, projections can be made for average or typical conditions and worst-case or maximum anticipated gas generation by varying the kinetic factors. Any onsite data available, such as methane generation potential (i.e., from onsite testing) should be used to refine the model to site-specific conditions.

If the year-to-year solid waste acceptance rates are not available but the total quantity of waste in place is known and the years of waste acceptance are known, then the gas generation rate can be estimated with the following equation based on the NSPS EPA LF Gas Model [Subpart WWW Section 60.755(a)(1)(i)], along with appropriate values for the kinetic factors of k and L_0[1]:

$$Q_{\mathrm{M}} = (1/C_{\mathrm{m}}) L_0 R (e^{-kc} - e^{-kt})$$

where Q_{M} is the maximum expected gas generation flow rate (cubic meters per year); k, the methane generation rate constant (year^{-1}); L_0, the methane generation potential (cubic meters per megagram solid waste); R, the average annual acceptance rate (megagrams per year); t_i, the age of the LF at equipment installation plus the time the owner or operator intends to use the gas mover equipment or active life of the LF, whichever is less (if equipment is installed after closure, t is the age of the LF at installation, years); c, the time since closure (years, for active LF $c = 0$ and $e^{-kc} = 1$); and C_{m}, the fraction of methane in LF gas (decimal; if not known use $C_m = 0.50$, for methane concentration = 50%).

The values for the kinetic factors k and L_0 are those from the most recent version of the Compilation of Air Pollutant Emission Factors (AP-42) (USEPA, 1997).

AIR EMISSIONS CHARACTERIZATION

Emissions from MSW LFs are commonly associated with direct releases of landfill gas. Releases of LF gas to the atmosphere can be from a number of different points at a LF. The major gas emission points are from surface releases, venting from various on-site structures, cracks in soil cover surfaces, or venting from gas collection and venting systems.

The emissions of individual gas constituents can be estimated based on the concentrations of constituents in the gas and using gas generation projections and gas collection system efficiencies to estimate gas released. Other methods include estimating emissions from the various source categories with a number of different techniques such as diffusion equations and leak estimation.[4,11,12]

Gas generation often results in pressure differential with higher pressures within the LF. This pressure gradient acts as a driving force for gas movement either laterally or upward. Additionally, normal variations in atmospheric pressure results in cyclical changes in gas pressure within landfills, that tends to lag changes in atmospheric pressure. These changes in differential pressure within LFs can cause atmospheric pumping action, resulting in increases in gas releases.

A concentration of NMOCs in gas from MSW LFs is 595 ppm as hexane, while a value for LFs with co-disposal of MSW and nonresidential waste is 2420 ppm.[7]

Odors are frequently associated with air emissions from MSW LFs.

Landfill Surfaces

Methane is lighter (around 0.544 times) than air and tends to rise within the LF cell, where it may be released from vents, and cracks or openings in the landfill surface and may migrate through the soil cover. There is a tendency for settling to occur at LFs. Settlement at LFs can cause cracks and openings in the surface cover and thus result in gas releases. Ninety percent of the settlement usually occurs during the first 5 years.[2]

Three mechanisms that contribute to gas-phase transport of chemical species to the air from landfills via cover soils are:

- Molecular diffusion
- Biogas generation
- Atmospheric pressure pumping

Molecular diffusion is the dominant mechanism when no biogas is present within the landfill. Biogas can enhance molecular diffusion transport by 3–10 times. Time-varying atmospheric pressure fluxuations can result in pumping action, which enhances the vapor-phase diffusion process. Fluxuations in atmospheric pressure influence cell pressure in a nonlinear fashion, as can the biogas generation rate. Atmospheric pressure fluxuations have been found to enhance molecular diffusion by around 13%. It is suggested that the most critical parameter for controlling vapor emission rates is the cover (cap) porosity.[12]

Structures

MSW landfills often contain many structures, including leachate collection manholes, bedrock wells, groundwater monitoring and withdrawal wells, piezometers, inclinometers, gas headers, and LF gas collection wells that are installed through the landfill cover and, thus, create potential for landfill gas to escape to the atmosphere. There can be large numbers (over a hundred on some sites) of these structures on a given LF site. Gas can accumulate in structures and may be released to the atmosphere from openings or leaks. The accumulation of gas in some structures may pose explosive hazards.

Other Emissions

Other sources of air emissions at landfills include emissions from heavy-duty vehicles on site and sometimes fugitive dust. Additionally there are emissions from trucks delivering waste and soil cover material to the LF. These sources are usually of minor concern. Emissions from vehicles are controlled by equipment design, manufacture, and periodic equipment maintenance. Fugitive dust may be generated during landfilling activities, although concentrations may be easily controlled with dust suppression techniques (i.e. water spraying of roads). Dust is usually of concern as a nuisance.

Controlled Emissions

Combustion is the primary method of controlling LF gas emissions. Uncontrolled emissions of constituents may be reduced with combustion controls but results in emissions of combustion products, such as nitrogen oxides, carbon dioxide, carbon monoxide, and unburned NMOCs. Emissions of sulfur dioxide (SO_2) may result from the burning of sulfur compounds present in the gas, such as hydrogen sulfide (H_2S) and mercaptans.

AIR POLLUTION CONTROL MEASURES

Effective control of MSW landfill gas emissions requires a combination of gas containment, sometimes venting, collection, and effective destruction of organics in collected gas by flaring or recovery systems. The NSPS and Subpart Cc Emission Guidelines require that new and some existing LFs with a design capacity of greater than 2.5 million cubic meters and air emissions in excess of 50 Mg per year of NMOCs to collect LF gas and control emissions of NMOCs with a destruction efficiency of 98%.[1]

Gas Containment

Containment of LF gas is most often used in combination with venting and collection systems. The use of low-permeability soil for cover and slurry walls can impede LF gas movement. Gas containment and control of surface emissions can be accomplished by using a cover soil with low permeability and low porosity a process referred to as *capping*. Capping is commonly used to reduce infiltration of water from precipitation but can also reduce gas emissions from the LF surface cover as well as enhance the effectiveness of LF gas collection systems by preventing infiltration of air into the LF. Slurry walls are commonly used for containment of contaminated groundwater at LFs. These slurry walls can serve as barriers to lateral subsurface gas migration. To remain effective, the clay material in slurry walls must be kept moist and not allowed to dry out. LFs with caps should have either active collection or venting systems or passive vents to allow release of LF gas and prevent excess pressure under the cap due to gas buildup, resulting in potential damage to the cap.

Gas Venting

A number of passive and active venting systems have been used to control lateral subsurface migration of gas beyond LF boundaries. Methods sometimes used in combination with venting systems include use of bentonite slurry walls, concrete grout curtains, and plastic membranes. The effectiveness of these methods is often limited, since they may not be totally impermeable. Methods of passive venting include open trenches and gravel-filled trenches.

Active venting systems can include trenches with collection piping systems and gas extraction wells with mechanical blowers. These systems are more effective, since they provide a positive means of gas extraction under conditions where natural venting alone may not be sufficient, depending on physical and site-specific factors. Also, it allows for active conveyance of gas to a given location.

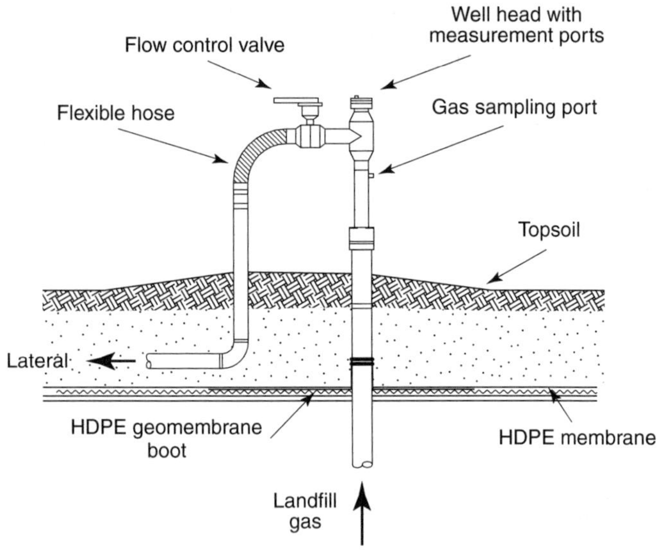

Figure 3. Typical Landfill Gas Extraction Well Head.

Gas Collection

Two basic types of LF gas collection systems are active and passive. Passive collection systems are typically employed for atmospheric venting of LF gas to prevent lateral migration or reduce potential explosion hazards. Passive collection systems emit uncontrolled LF gas. Active systems are generally employed to collect LF gas for control by combustion or for energy recovery. Active systems employ mechanical blowers or compressors to create a pressure gradient and extract the LF gas.

The NSPS, Subpart Cc, Emission Guidelines, and Appendix E prescribe various specifictions and guidance for LF gas collection systems. LF gas collection systems consist of a series of vertical or horizontal recovery wells that withdraw gas from within the LF and convey it via a piping header system to control devices such as flares or recovery systems. These collection systems are used to reduce the amount of uncontrolled LF gas that can escape to the atmosphere. LF gas collection systems must be adjusted to optimize the gas collection process while minimizing the potential for air infiltration into the LF. This is accomplished by adjusting the flow/vacuum of each of the gas extraction wells with the well head control valve. Figure 3 shows a typical well head with control valve, measurement ports, and gas sample port.

A header system (commonly of PVC or HDPE) conveys collected LF gas from the wells or trenches to vents, flares, or recovery systems, with blowers or compressors. The size and the type of blower or compressor depends on total gas flow rate, total system pressure drop, and vacuum required. Centrifugal blowers allow easy throttling throughout their operational range. Blowers can accommodate system pressure drops of up to 50 in. H_2O and flow rates of 100–100,000 cfm. For lower flow rates and higher pressures regenerative (combination of axial and centrifugal) blowers can be used.[2] Gas collection systems are sometimes installed along the perimeter of LFs to prevent lateral subsurface gas migration, while others are placed in the LF refuse for gas venting, control, or recovery.

The efficiency of LF gas collection systems can vary widely from system to system and LF to LF. Gas collection efficiencies can range for active LFs from around 40 to 60% of gas production and for closed LFs with gas collection and disposal using state-of-the-art well and piping systems from 85 to 90%. The South Coast Air Quality Management District (SCAQMD), a district with a large population of LFs with gas controls, estimate greater than 70% gas collection efficiency for both operational and postclosure periods. Waste Management, Inc., a large operator of controlled LF facilities in the nation, estimated on the order of 80% average over periods considering portions of LFs where waste volumes afforded rather complete enclosure of gas collection systems.[10] In summary, typical gas collection efficiency can be 70–80% collected during operation and 90% following closure. Gas collection efficiencies are related to the density of gas collection wells, withdrawal rates, and differential pressures. Excessive vacuum and gas withdrawal can lead to air intrusion, resulting in reduced biological activity (waste decomposition), reduction in methane content (gas combustibility), and creation of potential fire or explosion hazards.

Condensate Handling. Condensate consists of liquids that are the result of moisture and vapors condensing in the collection system piping. The quantity and quality of LF gas condensate depends on the constituents in the decomposing wastes and LF gas generation rate, in addition to other factors such as LF gas temperature, ambient temperature, LF gas collection system design, and amount of LF gas compression induced during conveyance. Condensate is produced as LF gas is extracted from the landfill, as LF gas undergoes natural or artificial cooling, or as a result of physical processes such as LF gas volume adjustments, such as compression or expansion.

Collection system design must consider that piping will convey dual-phase media, specifically, LF gas under a negative pressure and liquid condensate by gravity flow. Wellfield piping systems must be designed with adequate pipe slopes (3%) and perimeter header pipes designed with elevation transitions to promote proper flow.

USTs are needed to collect condensate from topographic low points along perimeter headers. USTs with double-wall construction with provisions for pumps and interstitial monitoring may be required. The capacity of USTs must be designed to accommodate the number of LF gas extraction wells a particular UST will serve.

Condensate management systems are required prior to gas combustion at flare stations and include demisters, knockout drums, and above-ground condensate storage tanks (ASTs). A sump structure may be required to provide tertiary spill containment for condensate ASTs. Collected condensate must be removed from wellfield condensate USTs and condensate ASTs, such as by pumps or by suction tanker. Because of the corrosive properties and biological hazards of collected LF gas condensate, it should be transported to properly licensed treatment and disposal facilities.

Gas Combustion

The high percentage of methane in LF gas and the resultant heat content make it suitable for disposal by combustion in flares, often without auxiliary fuel, and also for use in energy recovery systems as fuel. Table 2 presents emission factors for NO_x, CO, and PM for controlling LF gas with combustion devices (i.e., flares, boilers, gas turbines, and IC engines).[7] Emissions of SO^2 can result from sulfur compounds (such as hydrogen sulfide) in the LF gas. SO^2 emissions can be estimated by assuming that all the sulfur contained in various compounds in the LF gas will be converted to SO^2.

Flares. Flares are devices used to dispose of large quantities of unwanted flammable gases and vapors, such as gas streams that contain potentially odorous or harmful constituents, by combustion. Flares ideally burn waste gas completely and smokelessly. Thus emissions of LF gas and its combustible constituens (including VOCs) can be controlled. Use of combustion techniques requires removal of water from LF gas by condensate traps, knockout drums, or filters.

Two types of flares used at LFs for gas disposal are open (elevated) flares and enclosed (ground) flares. The major components of a flare are gas burner, stack, liquid trap, controls, pilot burner, and ignition system. Some are equipped with automatic pilot ignition systems, temperature sensors, and air and combustion controls.

Table 2. Air Emissions from Control of MSW Landfill Gas

Control Device	Secondary Air Emission[a] (lb/h/DSCFM methane)		
	NO_x	CO	PM[b]
Enclosed flare[c]	0.0024 (C)	0.045 (C)	0.001 (C)
Boiler[d]	0.002 (E)	0.00034 (E)	0.00049 (E)
Gas turbine	0.0052 (D)	0.014 (D)	0.0013 (D)
IC engine	0.015 (D)	0.028 (D)	0.0029 (D)

[a]EPA emission factor rating in parentheses (A through E, with A the best). See Ref. 7 for methods to estimate emissions of CO_2, SO_2, and HCl from total carbon, sulfur, and chloride content of landfill gas.

[b]Based on data for other combustion sources, most of the particulate matter will be less than 2.5 μm in diameter. Hence, this emission rate can be used to provide estimates of PM_{10} or $PM_{2.5}$ emissions.

[c]Where information on equipment was given in the original reference, test data were taken from enclosed flares. Control efficiencies are assumed to be equally representative of open flares.

[d]All source tests were conducted on boilers; however, emission factors should also be representative of steam turbines. Emission factors are representative of boilers equipped with low-NO_x burners and flue gas recirculation. No data were available for uncontrolled NO_x emissions.

Source: Ref. 7.

Elevated flares used for LF gas have a flare tip with no obstruction to flow. The flare tip is the same diameter of the stack. Open flares resemble large bunsen burners with candlelike flames. Combustion and mixing of air and gas take place above the flare with the flame in the atmosphere, as shown in Figure 4.

Enclosed flares are composed of multiple gas burner heads placed at ground level in an enclosure that is usually refractory lined. A diagram of a typical enclosed flare is presented in Figure 5. Some enclosed ground flares are equipped with automatic damper controls that regulate supply air depending on combustion temperature, which is monitored within the flare stack around 3 ft below the outlet. Air pollution control regulatory agencies are tending to require the use of enclosed flares because they are easier to perform emission tests on and have better combustion control than open flares. Requirements can include limits on carbon monoxide emissions (i.e., less than 100 ppm in exhaust gas) and residence time/combustion temperature of at least 0.3 sec and 1000°F.

Flare combustion efficiency is related to flame temperature, residence time of gases in the combustion zone, turbulent mixing of the combustion zone, and amount of oxygen available for combustion. Flares can be quite effective in destruction of hydrocarbons and provide approximately 98% or better control of hydrocarbons depending on design and operating parameters.[2,13]

Emissions from flares are related to LF gas flow rates, combustibility of gas (use of auxiliary fuel), temperature and retention time, and composition of gas feed. The emissions from flares can be estimated from combustion calculations used to estimate the volume and constituents of exhaust gases based on the makeup of the feed gas, flare operating parameters, and combustion efficiency.

The temperature of exhaust gases from flares can range from 1000 to 2000°F. These high temperatures can cause relatively high plume rise and good dispersion, and coupling this with high combustion efficiency can result in relatively low concentration impacts, depending on gas constituents. The plume rise associated with open flares is enhanced due to the height of the open flame. This, however, can be offset by the

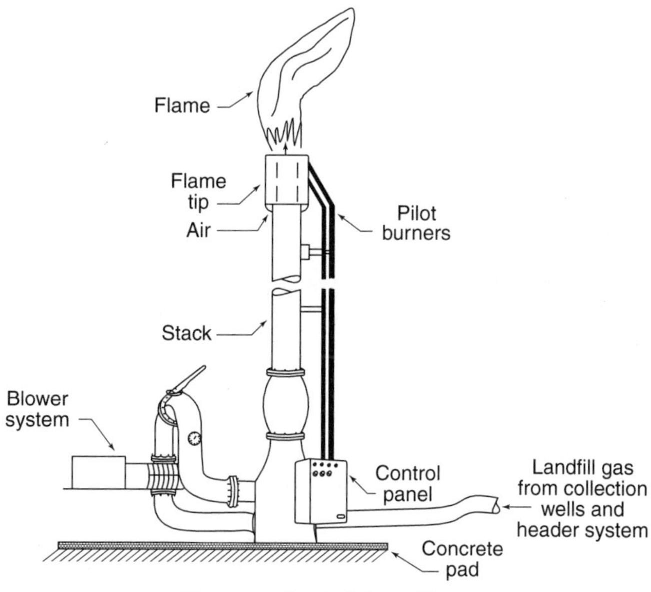

Figure 4. Typical Open Flare.

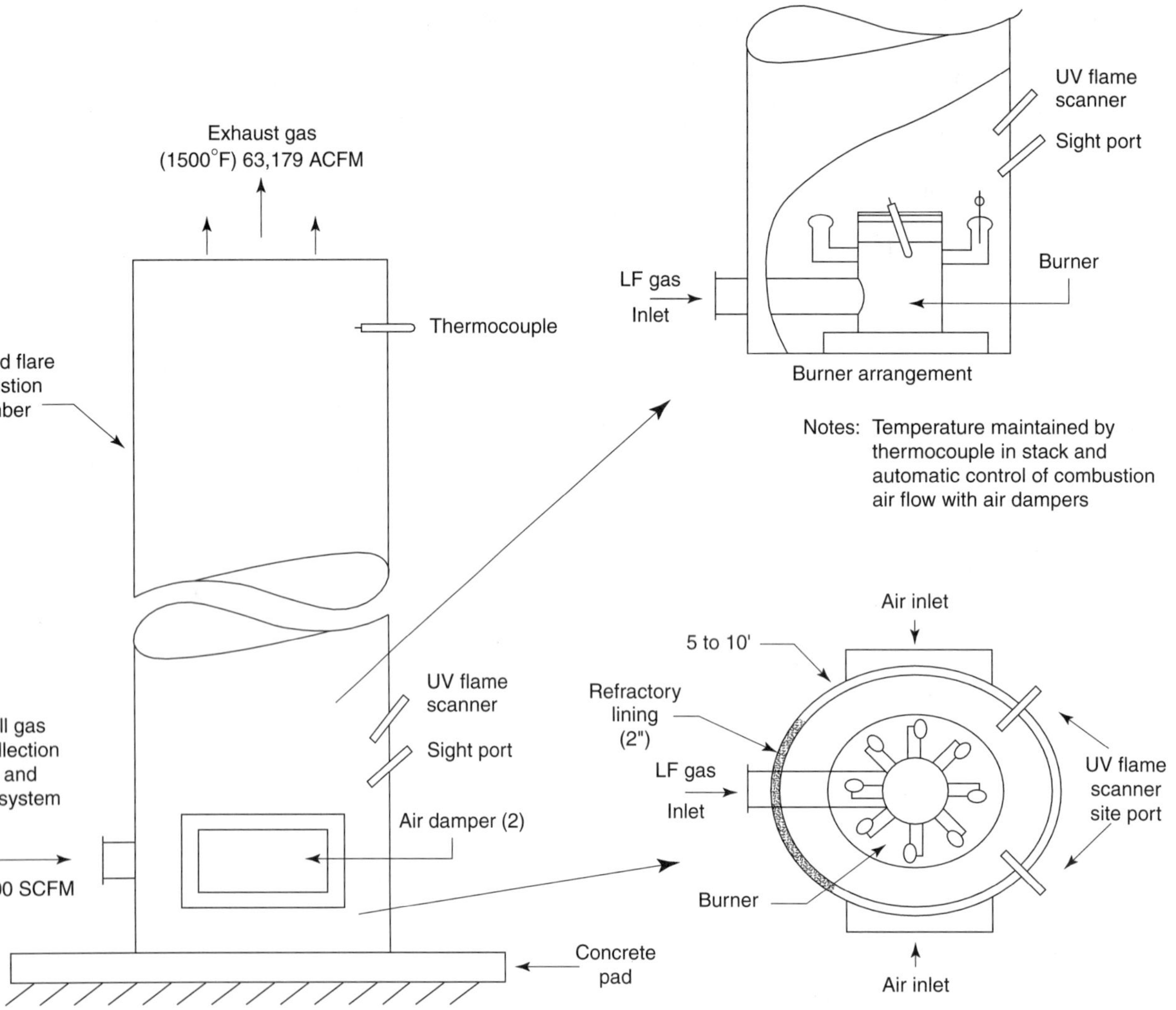

Figure 5. Typical Enclosed Flare. *Source*: Adapted from John Zinc Inc. General Arrangement Drawing Model ZTOF Enclosed Flare.

lack of control of combustion air and the quenching effect of radiant heat loss from the atmosphere and the open flame. Consideration should be given during placement to potential hazards associated with radiant heat from open flares. This is of less concern with enclosed flares since the flames are enclosed in a shell with refractory lining and have better temperature, excess air, and combustion control. It is also easier to perform emission tests on them.

Use of flares requires additional equipment to operate. Figure 6 presents a diagram of a flare station with two flares along with blowers, demisters, knockout drum, condensate AST, and control building. Flare stations similar to that shown in Figure 6 have been installed at the Fresh Kills Landfill in Staten Island, New York. The enclosed flares are John Zinc, Inc., Model ZTOF with LF gas capacity of 5000 scfm for each flare.[14]

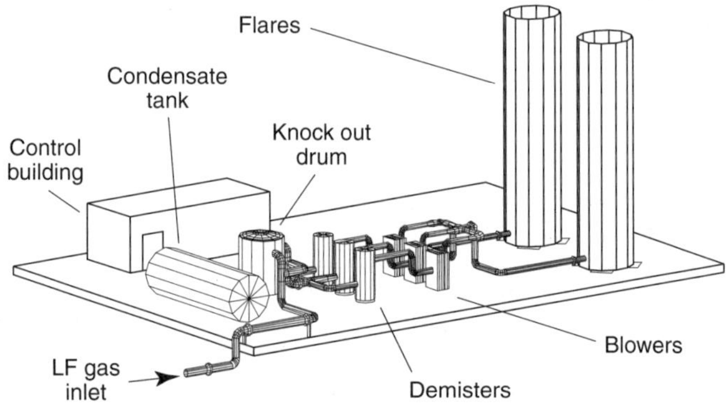

Figure 6. Landfill Gas Flare Station.

Recovery Systems. Gas recovery systems utilize LF gas for beneficial uses instead of just disposing of it as with flares. Recovery systems use combustion techniques that destroy organics and recover energy from the combustion process. The most common energy recovery systems used at MSW LFs include internal combustion engines, boilers with steam turbines, and gas turbines. They are often used to produce electricity. Boilers may be used to produce steam for heating rather than for production of electricity. Factors needed to make LF gas recovery systems economically feasibility are[2]:

- Greater than 2 million tons (1.8 million Mg) of refuse
- Greater than 35 ft depth of waste
- Greater than 35 acres of LF area
- Refuse conducive to gas production

Table 3. Control Efficiencies for LFG Constituents

Control Device	Constituent[a]	Control Efficiency (%)		
		Typical	Range	Rating[b]
Boiler/steam turbine	NMOC	98.0	96–99+	D
	Halogenated species	99.6	87–99+	D
	Nonhalogenated species	99.8	67–99+	D
Flare[c]	NMOC	99.2	90–99+	B
	Halogenated species	98.0	91–99+	C
	Nonhalogenated species	99.7	38–99+	C
Gas turbine	NMOC	94.4	90–99+	E
	Halogenated species	99.7	98–99+	E
	Nonhalogenated species	98.2	97–99+	E
IC engine	NMOC	97.2	94–99+	E
	Halogenated species	93.0	90–99+	E
	Nonhalogenated species	86.1	25–99+	E

[a] Halogenated species are those containing atoms of chlorine, bromine, fluorine, or iodine. For any equipment, the control efficiency for mercury should be assumed to be 0.
[b] EPA rating of emission factor reliability and accuracy (A through E, with A the best).
[c] Where information on equipment was given in the original reference, test data were taken from enclosed flares. Control efficiencies are assumed to be equally representative of open flares.

Source: Ref. 7.

- Continued LF operation
- Short time elapsed after closing for closed LFs

The relative combustion efficiencies (approximate or typical destruction efficiencies) of these recovery systems are presented in Table 3.[7] Emissions from internal combustion engines have the highest by product emissions (e.g., NO_x, CO, etc.) when compared to other recovery techniques (boilers, gas turbines).

Landfill Gas Recovery[15]

A beneficial use of LF gas is to process the gas in a gas recovery plant to produce a pipeline-quality gas. Processing LF gas in a recovery plant serves as a means of disposition of the gas while recovering the inherent energy value of the gas. This requires purification of the LF gas by removing water, certain constituents such as sulfur compounds (i.e., hydrogen sulfide), and other impurities (i.e., various hydrocarbons, NMOCs, VOCs). In addition, producing a pipeline-quality gas requires increasing the heating value by removing carbon dioxide.

A landfill gas recovery facility is usually composed of the three primary unit operations described below.

- Compression and cooling
- Feed gas pretreatment
- Carbon dioxide absorption

Compression and Cooling. After passing LF gas through a separator to physically (cooling process) remove liquids from the gas stream, the feed gas is then conveyed to the plant and is compressed (400–450 psi) in order to condense any remaining liquids from the LF gas. Liquid condensate is sent to storage tanks and later disposed of.

Feed Gas Pretreatment. The feed gas pretreatment system is designed to remove hydrogen sulfide and trace components of hydrocarbons from the LF gas, before entering the carbon dioxide removal process. Hydrogen sulfide removal can be achieved using several types of gas cleaning technologies (a common one is called the iron sponge). Hydrocarbons are initially removed by condensation. Feed gas chillers cool the gas down to approximately 10°F. Additional hydrocarbons and moisture are then separated from the LF gas by absorption into a solvent medium (e.g., glycol-based solutions, Selexol). The solvent is usually regenerated, resulting in off-gases that are usually destroyed by thermal oxidation.

Carbon Dioxide Absorption. The bulk of carbon dioxide contained in the LF gas is removed by solvent absorption towers. Carbon dioxide is absorbed into a solvent at high pressure, resulting in a gas that is high-purity methane. The carbon dioxide is then removed from the solvent by passing the solution through a high vacuum. The carbon dioxide stream can also be a high-purity product or may be treated as off-gas. The resulting purified gas (approximately 99% methane) can be delivered to a pipeline to supplement local gas utilities.

A LF gas recovery plant has operated at the Fresh Kills Landfill located in Staten Island, New York, since 1982 and has recovered and processed more than seven billion cubic feet of gas. The methane produced during this period and sold to Brooklyn Union Gas Company has an energy equivalent of over 2.5 million barrels of oil.

Surface Monitoring for Methane[1]

The NSPS includes a surface methane concentration operational limit, which requires LF operators to periodically monitor surface methane concentrations. The NSPS provisions require the use of an organic vapor analyzer to measure methane concentrations at 30-m intervals over the entire surface of the LF's refuse footprint. The objective is to identify if there are any specific locations or areas of the LF surface where methane concentrations exceed 500 parts per million (ppm). Concentrations that exceed 500 ppm are an indication of inadequate gas collection or breaks in the cover and would require LF operators to make repairs to the LF cover and/or make adjustments to the vacuum of nearby gas wells to increase LF gas collection in the area.

Practices

A number of practices may be used to ensure that LF gas control systems are working properly to control gas emissions and minimize odor complaints. An effective maintenance, repair, and monitoring program includes:

- A crack sealing program to maintain effectiveness of the cover soil.
- Natural soil barriers such as clay in caps and slurry walls must be kept saturated to maintain high efficiency.
- Repair and sealing of gas leaks in structures and maintenance of gas leaks in collection systems.
- Use of good cover material for containment of LF gas.
- Monitoring of migration control systems such as with combustible gas meters and probes placed around the LF perimeter.
- Monitoring for potential explosion hazards.

REFERENCES

1. U.S. EPA, 40 CFR Part 60, *Standards of Performance for New Stationary Sources (Subpart WWW) and Guidelines for Control of Existing Sources (Cc): Municipal Solid Waste Landfills*.
2. *Air Emissions from Municipal Solid Waste Landfills—Background Information for Proposed Standards and Guidelines*, Draft EIS (Preliminary Draft), U.S. EPA Office of Air Quality Planning and Standards, Emission Standards Division, Research Triangle Park, NC 27711, March 12, 1990.
3. D. J. Hagerty, J. L. Pavoni, and J. E. Heers, *Solid Waste Management*, Van Nostrand Reinhold, New York, NY, 1973.
4. M. J. Barboza, "An Integrated Study of Air Toxics Emissions from an MSW Landfill," American Society of Civil Engineers Proceedings of the 1991 National Conference on Environmental Engineering, Reno, NV, July 8–10, 1991.
5. Air Resources Board, State of California, *The Landfill Gas Testing Program: A Report to the Legislature*, prepared by Stationary Source Division Toxic Air Contaminants Identification Branch, 1988.
6. Title 7, New Jersey Administrative Code, Subchapter 17, *Control and Prohibition of Air Pollution by Toxic Substances*.
7. U.S. EPA, *Compilation of Air Pollutant Emission Factors (AP-42). Section 2.4: Municipal Solid Waste Landfills*. U.S. Environmental Protection Agency, Research Triangle Park, NC, November 1998.
8. U.S. EPA, USEPA-600-91-055, *Analysis of Factors Affecting Methane Gas Recovery from Six Landfills*, Prepared by Radian Corporation, dated September 1991.
9. *ADA Landfill Gas Control at Military Installations*, U.S. Army ADA 140-190, Jan. 1984.
10. A. Eschenroeder, D. Burmaster, S. Wolff, and A. Taylor, *Health Risk Assessment of a Proposed Landfill for MSW in Douglas, Massachusetts*, for Douglas Environmental Associates, Inc., Allanova, Lincoln, MA, June 1990.
11. L. W. Baker, and K. P. Mackay, "Screening Models for Estimating Toxic Air Pollution Near a Hazardous Waste LF," *JAPCA* (November 1985).
12. L. J. Thibodeaux, C. Springer, and L. M. Riley, "Models of Mechisms for the Vapor Phase Emission of Hazardous Chemicals from Landfills," *Journal of Hazardous Materials* **7**, 63–74 (1982).
13. *Handbook—Control Technologies for Hazardous Air Pollutants*, EPA/625/6-86/014, U.S. EPA Air and Energy Engineering Research Laboratory, Research Triangle Park, NC, 27711, September 1986.
14. New York City Department of Sanitation, *Fresh Kills Landfill—Landfill Gas Flares Air Permit Application*, prepared by Paulus, Sokolowski, and Sartor, Inc. P.C., April 1998.
15. M. Kammerer, GSF/ECOGAS, personal communication.

BIBLIOGRAPHY

Sanitary Landfill Design and Operation, U.S. Environmental Protection Agency, D. R. Brunner, and D. J. Keller, 1972.

Engineering Science, Austin, Texas, *A report on a Flare Efficiency Study, Volume I, Chemical Manufacturers Association*, Washington, D.C., March 1983.

Pacey, John G., *Controlling Landfill Gas Waste Age*, March 1981, pp. 32–36.

Landfill Methane Recovery, Ed. by M. M. Schunacher, Noyes Data Corporation, Park Ridge, New Jersey, 1983.

The Landfill Gas Testing Program: A Second Report to the California Legislature, State of California Air Resources Board, Stationary Source Division, June 9, 1989, C. Ceni and C. Emerson, Toxics Program Support Section.

U.S. EPA, *Compilation of Air Pollutant Emission Factors (AP-42). Emission Factor Documentation for AP-42 Section 2.4—Municipal Solid Waste Landfills (Revised)*, U.S. Environmental Protection Agency, Research Triangle Park, NC, August 1997.

U.S. EPA, *Emission Inventory Improvement Program, Document Series Volume III: Chapter 15. Landfills*, U.S. Environmental Protection Agency, Research Triangle Park, NC, EPA-454/R-97-04, September 1997.

Environment Canada. *Guidance Document for Landfill Gas Management*. Environment Canada, Waste Treatment Division, Hazardous Waste Branch, EPS 2/UP/5E, March 1996.

WASTEWATER TREATMENT FACILITIES

Michael J. Barboza and Gary Comerford

Wastewater treatment plants (WWTPs) receive and process community sanitary sewerage and industrial wastewaters. Many of these facilities are publicly owned treatment works (POTWs). WWTPs provide various levels and types of wastewater treatment that involve passing wastewater through a series of unit processes that remove solids and biologically or chemically modify or destroy many soluble organic compounds as well as pathogens. This results in a treated wastewater that can be discharged to receiving waters (i.e., rivers, oceans). In some cases treated wastewater may be reused, such as for irrigation. Solids removed from wastewater as sludges are usually given additional treatment (e.g., biological, chemical, and/or physical) prior to disposal or use.

The role of WWTPs has been well understood as one of preventing and controlling pollution of land and waters from discharges of municipal and industrial wastewater. However, WWTPs have emissions to the atmosphere (i.e., air toxic compounds), and odors from wastewater treatment facilities have been a concern for many years, usually associated with sulfur compounds, with focus primarily on hydrogen sulfide or ammonia. Odor control at WWTPs has been practiced for many years; however, control of other emissions from WWTPs is more recent.

Many different industries generate wastewater streams that contain organic compounds. Nearly all these streams undergo collection, contaminant treatment, and/or storage operations before they are finally discharged into either receiving waters or POTWs for further treatment. Industrial wastewater operations can range from pretreatment to full-scale treatment processes. In typical pretreatment facilities, process and/or sanitary wastewater and/or stormwater runoff may be collected, equalized, and/or neutralized and then discharged to POTWs for further treatment.[1]

Typical industrial wastewater operations include the following:

- Mining and milling operations
- Oil and gas industry
- Textile and leather industry
- Chemical and allied products industry
- Primary metals production

PROCESS DESCRIPTION

WWTPs utilize a number of unit processes for treating wastewaters. Although the design of individual plants differ, the major unit processes commonly used can be classified into the following major process categories that are described below:

- Wastewater collection systems
- Wastewater treatment operations
- Digestion of solids (biogas generation, use, and disposal)
- Sludge handling, dewatering, and disposal

Wastewater Collection Systems[1,2]

Wastewater collection systems typically contain lift stations, trenches, junction boxes, and manholes. Wastewater is received into collection systems through open sewer lines from all sources of influent wastewater, which can include sanitary and pretreated or untreated industrial wastewaters and/or storm water runoff.

Wastewater collection system components to be described may be found in industrial and POTW wastewater treatment facilities. The inclusion, arrangement, and sequences of these components are facility-specific.

Wastewater streams normally are introduced into collection systems through individual or area drains, which can be open to the atmosphere or sealed to prevent wastewater contact with the atmosphere. In industry, individual drains may be dedicated to a single source or piece of equipment. Area drains will serve several sources and are located centrally among the sources or pieces of equipment they serve.

Manholes into sewer lines permit service, inspection, and cleaning of lines. They may be located where sewer lines intersect or where there is a significant change in direction, grade, or sewer line diameter.

Trenches or channels may be used to transport industrial wastewater from point of generation to collection units such as junction boxes and lift stations, from one process area of an industrial facility to another, or from one treatment unit to another. POTWs also use trenches or channels to transport wastewater from one treatment unit to another. Trenches and channels are often either open or covered with safety gratings.

Junction boxes typically serve several process sewer lines, which meet at the junction box to combine multiple wastewater streams into one. Junction boxes normally are sized to suit the total flow rate of the entering streams.

Sumps are used typically for collection and equalization of wastewater flow from trenches or sewer lines before treatment or storage. They are usually quiescent and open to the atmosphere.

Lift stations are usually the last collection unit before the treatment system, accepting wastewater from one or several sewer lines. Their main function is to lift the collected wastewater to a treatment and/or storage system, usually by pumping or by use of a hydraulic lift, such as a screw.

Weirs can act as open channel dams, or they can be used to discharge cleaner effluent from settling basins, such as clarifiers. When used as a dam, the face of the weir is normally aligned perpendicular to the bed and walls of the channel. Water from the channel usually flows over the weir and falls to the receiving body of water.[1]

Wastewater Treatment Operations

Wastewater treatment operations may include three stages: primary, secondary, and tertiary treatment. At the beginning of the treatment process, physical operations remove floatable and settleable solids from the influent wastewater. Primary treatment includes oil/water separators, primary clarification, equalization basins, and primary treatment tanks. Secondary treatment systems include chemical and biological processes that remove organic matter from the wastewater. Examples of secondary treatment include aeration basins and secondary clarifiers. Some WWTPs use tertiary treatment, which includes nutrient removal prior to discharge of effluent wastewater.[1]

WWTPs separate wastewater (sewage) into liquid and solids (sludge) and then separately process each of these. A typical treatment facility will use bar screens to remove solids or garbage. From here, the sewage undergoes its initial wastewater/sludge separation phase in primary settling tanks. Additionally, grease is skimmed off the surface of the wastewater. After settling, wastewater begins secondary treatment, flowing into aeration tanks where bacteria act to reduce solids. More sludge is allowed to settle out of the wastewater prior to flowing to final settling tanks where additional skimming takes place. Finally, prior to discharge to local waters, the wastewater may be treated with disinfectant [e.g., sodium hypochlorite (chlorination)] to kill pathogens. A process schematic diagram of a typical POTW is presented in Figure 1.[3]

Headworks. Headworks are raw sewage influent channels where large clumpy items, such as rags, are removed with barscreens. The screens have "fingers" to catch large materials. A mechanical "rake" moves in between the "fingers," taking off the waste and placing the material on a conveyor belt, which empties into storage bins. Bins are loaded onto trucks, which landfill these wastes.

Some headworks are enclosed, but many are open channels. Open channels are potential sources of fugitive emissions. Newer plants may have enclosed headworks with air being vented to odor/emission control systems such as caustic scrubbers.

Grit Chambers. Downstream from the headworks are grit chambers. Grit chambers are used at many WWTPs to remove dense, inorganic solids present in wastewater, such as: sand, gravel, glass, bone chips, egg shells, coffee grounds, or other heavy, solid, generally nonbiodegradable materials that have settling velocities greater than the degradable organic materials also present in wastewater. Grit removal

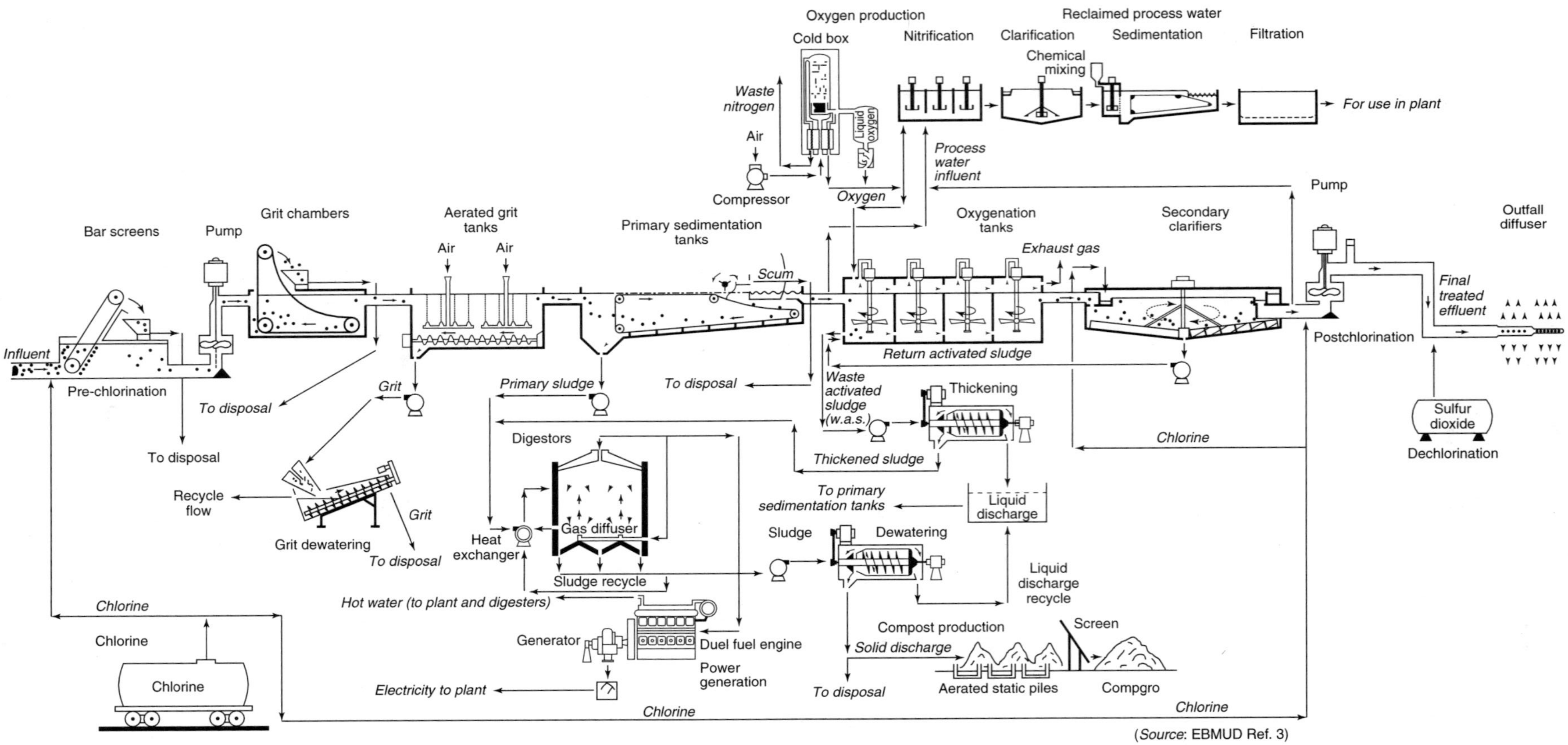

Figure 1. Wastewater Treatment Plant Process Schematic.

is necessary to protect moving mechanical equipment (such as pumps) from abrasion, to keep grit from settling in conveyance channels, and to reduce the frequency of digester cleaning. Grit chambers are usually installed after headworks and before primary sedimentation tanks but can be installed ahead of inlet works pumping stations.[4]

Grit chambers usually include two types: horizontal flow and aerated. Horizontal flow grit chambers depend on gravity to effect inorganic solids removal, while aerated grit chambers work by imparting a helical flow pattern to the wastewater by aerating it along one wall of the chamber. Grit accumulates along the aerated side, and is scooped up by a traveling chain into a grit hopper located on the surface. Collected grit is usually hauled off-site to landfills.[4]

Primary Sedimentation Tanks (Clarifiers). WWTPs include some form of primary sedimentation. Primary sedimentation tanks, sometimes referred to as primary clarifiers, are used to remove easily settleable material from raw sewage. This material includes slower settling organic solid matter as well as fast-settling grit if the WWTP has no grit removal upstream of the primary sedimentation tanks. Primary sedimentation tanks also serve to remove floatable solids comprised mostly of fats and greases commonly known as scum.[4]

Primary sedimentation tanks may be circular or rectangular tanks and include a system of scrapers for collecting the sludge that has settled to the bottom of the tank. The sludge is collected in a sump and pumped to another tank for further treatment. Scum is retained in the primary sedimentation tanks upstream of a surface baffle located near the tank's downstream edge. A surface skimmer collects the scum in a trough; the scum is pumped to another process for further treatment. Clarified wastewater leaves primary sedimentation tanks through a series of weirs and channels.[4]

Activated Sludge (Aeration Tanks). One of the most common types of biological waste water treatment processes is activated sludge.

Diffused Air Activated Sludge. The diffused air activated sludge process is an aerobic, biological process that uses diffused air to provide oxygen for metabolic reactions of microorganisms suspended in the flow. Diffused air activated sludge systems use compressed ambient air to supply oxygen in contrast to pure oxygen or mechanically mixed systems. After the air is diffused into the bottom of an aeration tank, it rises turbulently to the surface. After aeration, the sludge solids are separated from the clarified wastewater in secondary settling.[4]

Mechanically Aerated Activated Sludge. Surface aerators can be grouped into four general categories: radial flow, axial flow, aspirating devices, and horizontal rotors. Each is widely used and has distinct advantages and disadvantages, depending on the application.[5] In all cases mechanically driven turbulence provides oxygen to the system using ambient air.

Trickling Filters. Trickling filters are a form of aerobic, attached-growth biological treatment used to remove organic matter from wastewater. A bed of permeable media may consist of rocks 1–4 in. in diameter and 3–8 ft deep or, in more modern installations, the bed is constructed with plastic media 20–40 ft deep. Forced air may be required to aerate the taller towers. Often trickling filters are the only form of secondary treatment at WWTPs, but they may also precede additional secondary treatment such as activated sludge.[4]

Wastewater is distributed over the trickling filter by rotating distributor arms that sprinkle the water evenly on the surface of the bed. The water cascades through the media, flowing over the biological slime layer that adsorbs and utilizes much of the suspended, colloidal, and dissolved organic matter. Cell growth and respiration occur; respiration converts organics to water and CO_2 while cell growth converts organics into biomass that sloughs off the media and is removed downstream. Water is collected at the base of the trickling filter, with some percentage recirculating prior to traveling to the next process.[4]

Oxidation Ponds. The structure and function of oxidation ponds vary widely from plant to plant; however, oxidation ponds depend on aerobic microorganisms to remove soluble organic pollutants. These aerobic microorganisms can only exist in the presence of dissolved oxygen. Most of the dissolved oxygen in oxidation ponds is supplied by algae, which use the energy from sunlight and the nutrients and carbon dioxide in the wastewater (photosynthesis) to produce more algae and release oxygen into the water. Oxidation ponds generally range from 2 to 5 ft in depth and can cover many acres. They can be characterized by their organic loading rates in pounds of BOD per acre, or per acre foot, per day. Most oxidation ponds follow primary treatment and serve the same secondary treatment function as activated sludge. There are a number of oxidation pond variations.[4]

Secondary Clarifiers. Wastewater with high pollutant concentrations, and in particular high-flow sanitary wastewaters, are typically treated using an activated sludge system where biotreatment is followed by secondary clarification. In this system, settled solids containing biomass are recycled from clarifier sludge to the biotreatment system. This creates a high biomass concentration and therefore allows biodegradation to occur over a shorter residence time.[1]

Tertiary Filters. Tertiary filters are located downstream of secondary clarifiers, and typically use a granular medium to strain fine, unsettled particulates from wastewater. Alum or polymer is often added to the wastewater prior to filtration to form a floc and thus increase particulate removal. Chlorine may also be added upstream of the filter (prechlorination) to help control bacterial growth on the media. Tertiary filters may be enclosed (pressure) filters or open (gravity) filters.[4]

Chlorine Contact Tanks. Disinfection refers to the selective destruction of disease-causing organisms. The most common method of wastewater disinfection is by the addition of chlorine, a strong oxidant, in the form of chlorine gas or sodium hypochlorite. Because of its great oxidation potential, chlorine will oxidize a variety of materials (both organic and inorganic) having a lower oxidation potential. Contact time is normally provided in a chlorine contact tank and typically ranges from 15 min to 2 h. Chlorine contact tanks are typically the last unit process in a plant's flow stream (unless dechlorination is required).[4]

Dechlorination Facilities. Dechlorination is the practice of removing the total combined chlorine residual that exists after chlorination to protect aquatic life from chlorine-induced toxicity. The most practical method of dechlorination is by sulfur dioxide and/or aqueous solutions of sulfite compounds such as sodium bisulfite. Other methods use granular activated carbon or hydrogen peroxide.[4] The objective of wastewater dechlorination systems is to produce a zero-chlorine residual in the effluent before it leaves the plant.

Digestion of Solids (Biogas Generation, Use, and Disposal)

Primary sludge is pumped through a cyclone degritter that separates out particles of sand and gravel. It then passes into thickening tanks. Although there are several ways of treating sludge at this point of the process, the simplest is to let it settle. The settled/thickened sludge is then pumped into closed digestion tanks, which are heated to facilitate the anaerobic bacteria breakdown of organic matter into the by-products of water, carbon dioxide, and methane gas (biogas or digester gas). The digested sludge may be stored prior to disposal.

Anaerobic Digesters. Anaerobic digestion is a biological process that operates in the absence of free oxygen, which degrades organic substances to methane and carbon dioxide, and which stabilizes sludges for further processing by killing pathogens in the sludge and by making the sludges easier to dewater. Sludges generated in primary sedimentation and by secondary treatment are amenable to anaerobic digestion. Combined, these sludges carry with them roughly two-thirds of a WWTPs organic pollutant loading. The anaerobic digestion process can convert more than 50% of these pollutants to methane gas and carbon dioxide. The bioconversion process is very slow, and the biomass doing the work cannot be retained on a fixed media, nor can it be concentrated and recycled as in the secondary treatment reactors. Therefore, the hydraulic retention time must be large to prevent washout of the active biomass, which, in turn, increases reactor volume substantially. Anaerobic digesters may be cylindrical, rectangular, or egg-shaped. The most common design is a circular digester with a diameter ranging from 20 to 125 ft, and a side water depth between 20 and 40 ft. The conversion process is optimized at 35°C, by establishing good mixing inside the unit, and by sealing off the unit from air.[4]

Digester Gas Combustion. Digester gas generated in anaerobic digesters may be used at WWTP sites in combustion processes to produce steam or make electricity or fuel engines that provide power for pumps, fans/blowers (for aeration), and process heating for digesters. In many WWTPs, digester gas may be burned in flares to dispose of it.

Dissolved Air Flotation Thickeners. Dissolved air flotation (DAF) is used in many WWTPs for thickening waste-activated sludges and combinations of primary and waste-activated sludges. DAF systems are used when solid particles tend to float rather than sink or if the sludge resulting from settling processes is insufficiently compacted. The goal of the flotation thickening is to cause solids–liquid separation in an upward direction by attaching small air bubbles to suspended solid particles. The air is forced into solution through a pressurization system. Because the air bubbles have a slightly negative charge, the positively charged solid particles attach and result in a particle that has a specific gravity less than water, tending to rise and float. The floated sludge is then skimmed from the DAF tank surface. Tank configuration may be circular, with center feed, or rectangular, with end feed.[4]

Gravity Sludge Thickeners. Gravity thickeners are similar to primary sedimentation tanks and are used to concentrate sludge prior to additional treatment. The sludge is allowed to settle and compact and is collected at the bottom by scraper blades. Thickened sludge is pumped to digestion or to dewatering processes. Gravity thickening is most effective on sludge from the primary sedimentation tanks, though it is also used for waste-activated sludge thickening. Gravity thickeners may be open or covered tanks.[4]

Sludge Handling, Dewatering, and Disposal

Sludge Centrifuges. Centrifuges are used at many wastewater treatment facilities for a number of solids–liquid separation processes. Typical applications include thickening of waste-activated sludge solids generated in biological unit processes and dewatering of digested solids from primary and secondary treatment processes.[4]

Support equipment for centrifuges include: sludge pumping and storage systems that allow for close control of the feed rate to the machine, chemical conditioning systems for the addition of organic polymers for flocculation, and cake discharge/handling/storage systems. Methods of cake handling may include: direct discharge into a wetwell for subsequent pumping; direct-discharge into an open-throat, progressing cavity pump; discharge onto a belt conveyor system that carries the cake to loading stations or to storage bins; or discharge to horizontal screw conveyors that carry the cake to bins, sumps, pumps, or belt conveyors.[4]

Filter Belt Press. Filter belt presses are a type of sludge dewatering technology that uses gravity and positive pressure to separate liquid wastewater from solid wastes contained in the sludge. The belt is perforated to allow filtration of chemically conditioned sludge. Drainage takes place immediately at the system feed (intake) by means of gravity. As the belt system delivers the sludge farther down the line, a series of rollers apply pressure to the sludge, which produces a dry cake at the system end. Among mechanical processes currently in use, this method of dewatering yields the driest end product.

Sludge Drying Beds. Sewage sludge discharged to drying beds may consist of a combination of primary, secondary, or chemical sludges, depending on the type and level of treatment at any given WWTP. The sludge is typically aerobically or anaerobically digested, and may receive additional mechanical dewatering prior to solar drying. Once the sludge is ready for further dewatering, it is transported to the drying beds generally via a closed system of piping.[4]

Drying beds may be constructed with or without tile underdrains to catch leachate from the sludge while the bottoms of the basins may be asphalt, native soil, or cement. Some drying beds are covered with a greenhouse-type cover to enhance drying capabilities; however, the vast majority are open to the atmosphere. With asphalt bottom beds, the installation of an underdrain system is common practice to enhance dewatering, but as with native soil or sand-bottom beds, this underdrain system is optional. Percolation with a soil bottom takes place for only a short period of time after fresh sludge is applied to the beds with the balance of dewatering accomplished through evaporation.[4]

Sludge Disposal. Sludge may be disposed of by land filling, land application (sometimes as fertilizer), incineration, or composting. Sludge is often transported by truck to landfills. There are a number of processes for stabilizing sludge (such as by mixing with lime) before disposal at landfills and may be used to supplement soil cover material for landfill maintenance. Some WWTPs may dispose of sludge by incineration (such as fluidize bed incinerators, multiple-hearth furnaces, cyclone furnaces, and electric or radiant heat furnaces).[6]

AIR EMISSIONS CHARACTERIZATION

Typical air quality concerns of WWTPs have long been associated with odor emissions generated by wastewater collection and treatment systems. The focus of air quality (odor) concerns were usually placed on the detection and control of ammonia, amines, and sulfur compounds such as mercaptans and hydrogen sulfide (H_2S). More recently, the emissions of criteria pollutants from combustion equipment and the emissions of air toxics, mostly volatile organic compounds (VOCs) from WWTPs have increased in concern.

Air pollutants within wastewater collection and treatment systems may be removed, transformed, or generated, or may pass through the system unchanged. The following primary mechanisms are involved in the fate of air pollutants within WWTPs and may act simultaneously:

- Volatilization
- Degradation
- Adsorption
- Transmission (pass through entire system)
- Generation (as a result of chlorination or as by-products of degradation)

The relative health risks posed by air emissions from various process categories have been rated by Card[7] in terms of source toxicity. Liquid treatment processes are the predominant source of risk at WWTPs. Combustion sources also carry a fair amount pollutants, but owing to better dispersion of stack-emitted gases, they usually pose less risk. Of the liquid treatment processes, secondary treatment usually dominates the risk, followed by the headworks and primary treatment.

Types of Emission Sources

Air emissions characteristics differ for the following various major process categories of wastewater treatment facilities:

- Wastewater collection systems
- Wastewater treatment operations
- Biogas generation, use, and disposal
- Sludge handling, dewatering, and disposal
- Combustion equipment

Wastewater Collection Systems[2]. Studies indicate that emissions of VOC from collection systems may be significant, particularly in sewer lines serving industrial and commercial establishments that discharge large quantities of organic chemicals. These emissions may be of greatest concern where combined sanitary and storm sewers are employed. Gas exchange with the atmosphere takes place at manholes, most of which are vented through pick holes, building vents, and at major pumping stations where wastewater accumulates in vented wet wells.[8] Unfortunately the lack of existing sample data does not allow for meaningful emissions estimates. This is an area where future studies would be of great value to reduce the uncertainties associated with the relative significance of collection systems as VOC emission sources.

Wastewater Treatment Operations. VOCs are currently the focus of regulatory concern regarding emissions from WWTPs, since they may be readily stripped from wastewater during wastewater treatment processes and emitted to the air. Additionally, many of these compounds are of concern due to their potential for carcinogenic or noncarcinogenic health effects.

Headworks. Raw sewage lines are usually combined to form one stream before entering WWTPs at the headworks. The sewage is usually treated with ferric chloride ($FeCl_3$) and liquid chlorine (Cl_2). Ferric chloride storage tank vents can be potential sources of emissions during tank fills and tank breathing. Chlorine is added for odor control (to reduce H_2S emissions). There are usually no emissions from the chlorination process, since bulk liquid Cl_2 is mixed with a stream of plant water, which in turn is injected into the raw influent in a closed system. VOC emissions from combined headworks (bar screens, comminutors, and meters) appear to be of the same magnitude as emissions from aerated grit chambers and sedimentation tank weirs.[8] Enclosed headworks often draw air from the enclosure by negative pressure and convey it to caustic scrubbers for control prior to venting to the atmosphere.

Grit Chambers. Potential emissions of toxic air contaminants from grit chambers is greatest from aerated grit chambers. Air emissions could arise from the aeration process, or there could be fugitive emissions from the collection and transport of grit.[4]

Sedimentation Tanks (Clarifiers). The turbulence created as the flow falls over the weirs and into collecting channels is the primary source of air emissions from primary and secondary sedimentation tanks in addition to evaporative losses from quiescent liquid surfaces. Emission rates are higher for primary sedimentation tanks due to the higher VOC loading. Sometimes sedimentation tanks are covered and include odor scrubbing equipment; many WWTPs, however, have uncovered sedimentation tanks.

Activated Sludge (Aeration Tanks). Due to the turbulence created by aeration systems, air emissions from activated sludge tend to be high relative to other processes at WWTPs. Organisms in the sludge can strip odorous compounds from the air and the air bubbles can strip VOCs from the liquid phase depending upon the concentrations and partial pressures of the specific substances.[4] Mechanical aerators usually produce greater air emissions than diffused air systems. Coarse-bubble-diffused air usually produces greater air emissions than fine-bubble-diffused air systems. Pure oxygen systems produce the lowest air emissions because sufficient amounts of oxygen are transferred to the sludge at much lower aeration rates.

Trickling Filters. The cascading effect of water passing through the trickling filter media is expected to volatilize VOCs present in the wastewater. In addition, VOCs may be biodegraded as well as adsorbed.[4]

Oxidation Ponds. The potential for toxic air emissions from these treatment units will vary depending upon the location of the oxidation pond in the train of the treatment process. Those ponds that receive raw influent should have the highest level of emissions. Oxidation ponds that constitute the secondary treatment units receiving primary effluent should have an intermediate level of emissions. Those ponds receiving secondary effluent will have the lowest level of emissions, since any volatile compounds present are likely to evolve or oxidize during primary and secondary treatment prior to discharge to the ponds.[4]

Tertiary Filters. Except for the brief periods when gravity filters are backwashed, gravity tertiary filters have quiescent surfaces, and little or no VOC release would be expected. Pressure filters are totally enclosed, and no air emissions can occur.[4]

Chlorine Contact Tanks. Depending upon the depth of fall and flow rate per foot of weirs, the turbulence at the weir overflow can cause release of VOCs, if present. Although water surfaces are generally quiescent, most chlorine contact tanks have weirs at the end to control water levels in the tank.[4]

Digestion of Solids (Biogas Generation, Use, and Disposal). Air emissions relating to the digestion of solids are discussed below.

Anaerobic Digesters. Volatile contaminants in the raw sludge may be volatilized in the digester and be carried along with the digester gas to digester gas burning devices on site. Furthermore, there can be production of some VOCs and destruction of others by the biological activity occurring inside the digester.[4]

Digesters are essentially closed containers under a slight pressure. Under normal operation there should be no direct emissions to the atmosphere. Digesters destroy and create compounds through biological processes. The result of these changes in composition of sludges will be reflected in the composition of digester gas and its combustion by products or in the further processing of digested sludge.[4]

Digester Gas Combustion. Digester gas, mostly methane, is produced by anaerobic digestion of solids. Digester gas is often used to fuel internal combustion (IC) engines and boilers to provide power for pumps, fans/blowers (for aeration), and process heating for digesters or is commonly flared in waste gas burners.

Dissolved Air Flotation Thickeners. In dissolved air flotation thickeners, the enhancement of VOC emissions caused by the air stripping effects of flotation air bubbles is small compared to the reduced emissions (by evaporation) caused by the layer of solid flocculants and oil film floating at the surface.[9] However, VOC emissions are still present at these units.

Gravity Sludge Thickeners. Gravity sludge thickener tanks are usually covered and may include odor scrubbing equipment.

Sludge Handling, Dewatering, and Disposal. Air emissions relating to sludge processes are discussed below.

Sludge Centrifuges. Potential emissions of air contaminants could come from feed sludge and central storage wetwells as well as from open cake conveying and storage systems and polymer mixing and storage tanks. The rotating assembly of the centrifuge itself is enclosed by a case with discharge openings for dewatered sludge cake and centrate. Air emissions may occur since centrifuges are usually vented, sometimes through control equipment. The filling of centrifuges can also result in emissions due to displacement of internal volume.

Filter Belt Press. Emissions from filter belt presses are similar to sludge centrifuges except that they usually vent to the room or building that they are enclosed in, while centrifuges are usually vented to the outside and sometimes via controls such as carbon systems.

Sludge Drying Beds. Whether the mixing process results in more or less total emissions of VOCs during a drying cycle is not yet known, even though the rate of emission must certainly be higher if VOCs are present in the sludge.[4] Most concern is related to odor emissions from drying beds, especially during initial pouring of the sludge on the beds.

Sludge Disposal. Emissions from disposed sludge are usually expected to be low and are usually not an issue for WWTPs (except incinerators), since disposal most often occurs at off-site locations.

Emission Points. Wastewater treatment plants have both stack and fugitive emission sources. Stack emissions include combustion sources and liquid processes that are covered and ducted. Examples include head works scrubber vents, sludge handling building vents, and internal combustion engine exhausts. Fugitive emissions sources include large uncovered areas such as settling basins, weirs, and open channels. Many emission sources are controlled; however, some remain uncontrolled. Emissions of VOCs from liquid treatment processes are usually those of greatest concern.

Compounds of Concern

A wide range of suspected and confirmed compounds or classes of compounds are emitted by WWTPs as air pollutants. Table 1 lists some of the compounds of concern and their likely emission points. Because of the recent regulatory focus on air toxics, WWTPs have gained considerable attention as sources of air pollution, especially VOC.

Selection of Target Compounds[10]. Since VOC emissions are the most difficult to quantify yet are of most concern, the

Table 1. Compounds of Concern

Criteria pollutants	Sources
Nitrogen oxides	Boilers, flares, engines, incinerators
Sulfur dioxide	
Carbon monoxide	
Particulates	
Air Toxics	Sources
Volatile organic compounds (e.g., chloroform, methylene chloride, etc.)	Liquid processes
Toxic Metals	Sources
Cadmium	Incinerators
Arsenic	
Odors	Sources
Ammonia	Liquid processes
Hydrogen sulfide	
Mercaptans	
Amines	
Carbon disulfide	
Dimethyl sulfide	
Dimethyl disulfide	

task of selecting target compounds for evaluation demands a rational procedure that considers physical and toxicological principles, limitations of analytical procedures for sampling studies and an emphasis on conservatism to adequately protect human health. The objective is to select compounds that are of the greatest concern from the aspects of quantity (expected wastewater concentrations), volatility (expected transfer from wastewater to the air), and toxicity (based on available health criteria). Reducing the total number of compounds by eliminating compounds of little or no concern results in an effective analytical program and allocates time and financial resources to better identify and quantify the compounds of most concern.

The following steps are suggested to develop a focused list of target compounds for liquid process sources at WWTPs.

1. Identify compounds typically detected in wastewater influent.
2. Add compounds of concern resulting from biodegradation in treatment operations based on literature search.
3. Compile health criteria data for preliminary target compound list.
4. Tabulate volatility data (e.g., vapor pressure, boiling point) for preliminary target compound list.
5. Apply an index value to each compound using the formula below and prioritize based on the computed value. In this formula, the lower the index value for the compound, the higher priority for sampling or evaluation.

$$\text{index} = \text{TR} \times \text{vol/IC}$$

where index is an indication of relative quantity, volatility, and toxicity; TR, the toxicity rating (e.g., allowable ambient limit/unit risk factor); vol, the volatility (e.g., boiling point/vapor pressure/Henry's constant); and IC, the concentration of compound in influent wastewater.

Table 2 provides a list of target compounds developed using the index equation for a wastewater sampling program as detailed in Ref. 10.

Table 2. Typical Target Compounds from WWTPs

1	Vinyldene chloride (1,1, DCE)
2	Ethylene dichloride (1,2, DCA)
3	Vinyl chloride
4	Tetrachloroethene
5	Methylene chloride
6	Tetrachloroethane-1,1,2,2
7	Chloroform
8	Bromomethane
9	Benzene
10	Trichloroethene
11	Styrene
12	Xylene; total M,O, & P
13	Toluene
14	Acetone
15	Butanone MEK (2-butanone)
16	Chlorobenzene
17	Methylphenol-4
18	Naphthalene
19	Hexone (MIBK)
20	Phenol
21	Dichlorobenzene-1,2
22	Ethyl alcohol (ethanol)
23	*Trans*-1,2-Dichloroethene
24	Ethylbenzene
25	Trichloroethane-1,1,1
26	Ethanol, 2 methoxy
27	Carbon disulfide
28	Ethane, 1,1-oxybis
29	Methylphenol-2 (*o*-cresol)

Source: Ref. 10.

Emission Estimation Techniques for Volatile Organic Compounds

Estimation of VOC emissions from WWTP processes can be complex. Techniques to estimate fugitive emissions from liquid surfaces include direct emission measurements (source testing) emission factors (based on test data), calculation using mass balance equations, and use of the more complex mass transfer models. Source testing is difficult and expensive and could be inaccurate if emissions fluctuate in between sampling events. However, all methods are subject to imprecision in data quality. Emission rates can vary from plant to plant, from process to process within a single plant, and day to day due to changes in discharge to the collection system. Emission estimation is further complicated by the number of compounds of concern and the number of different wastewater processes that are involved.[8]

Contaminant emission rates are influenced by many variables. Among them are environmental factors such as wind speed, ambient and wastewater temperatures, liquid surface characteristics (i.e., oil film layer), and the concentrations of individual contaminants in the wastewater influent; plant design characteristics such as exposed surface area and physical structures that influence effective windspeed, liquid depth, degree and type of aeration, and wastewater flow rate; and the physical properties of each contaminant such as vapor pressure, solubility, ability to adsorb onto solids, and susceptibility to biodegradation.

The major emission estimation techniques are described below.

Flux Chambers. Direct measurement of gaseous emission rates from open liquid surfaces can be performed using an emission isolation flux chamber. A flux chamber is a surface enclosure designed to capture gaseous compounds emitted by the surface area beneath them. Flux chambers should be designed to accommodate sweep air across quiescent liquid surfaces in order to simulate a slight wind velocity. This procedure will prevent an artificial buildup of concentrations in the air–liquid boundary layer. The roof of the flux chamber should be made of a transparent material to allow sunlight to penetrate down to the liquid surface. As with sweep air, this procedure is needed to avoid altering the boundary layer environment. Aerated liquid surfaces create additional challenges to collecting representative samples. Because of the high gaseous flow rates associated with aerated basins, the flux chamber must be designed to balance the pressure within the chamber to avoid situations involving positive pressure inside. After the concentration of each target compound is determined, the emission rate per unit area is computed using the following equation[11]:

$$E_{\mathrm{f}} = \frac{CQ}{A}$$

where E_f is the emission flux of compound (μg/m^2 min); C, the measured concentration of compound (μg/m^3); Q, the air flow rate through the flux chamber (m^3/min); and A, the surface area enclosed by the flux chamber (m^2).

Emission Factors. A common approach used for estimating emissions from many types of sources has been the use of emission factors. However, emission factors are most often established from testing data and material balances. Because concern for air emissions from WWTPs is relatively recent, not as much information on emissions is available (i.e., few test data are available and the diversity of plants and their influent characteristics limits using these data for other plants). Therefore, air emissions testing is often needed to collect information to better quantify emissions from WWTPs and develop more accurate emission factors. Some emission factors are available for obtaining gross estimates of average annual emissions but are even more limited when trying to estimate short-term emissions from specific unit processes.

Mass Balance Equations. A mass balance can be performed for any compound that is measured in the WWTP influent and effluent channels. A crude estimate can be calculated simply by assuming that the mass difference between influent and effluent has escaped to the atmosphere. This method is not very accurate because it usually cannot account for internal processes that form sources and sinks for many compounds. These processes include solids adsorption, biodegradations, and chemical transformations and reactions. In general, mass balance equations tend to overestimate emissions; however, they can be useful for providing a sanity check for other methods of estimation.

Mass Transfer Equations. Volatilization of VOCs occurs naturally across open water surfaces and is further intensified by increasing turbulence (e.g., mechanical surface aeration, diffused air systems, and weirs). Competing removal mechanisms for VOCs entering the atmosphere include adsorption and biodegradation within the wastewater. Volatilization is also inhibited by conditions such as oil films at the wastewater surface. The rate of volatilization for VOCs can be calculated using mass transfer equations. These equations estimate VOC emission rates based on the physical properties of each contaminant and can account for the special circumstances that induce or inhibit volatilization. A simple form of these equations[8] is written as:

$$R = KCA$$

where R is the rate of mass transfer between air and wastewater (mg/sec); K the overall mass transfer coefficient (m/sec); C, the VOC concentration in wastewater (mg/m^3); and A, the interfacial contact area between air and wastewater (m^2).

It is only applied to open surfaces, where gas-phase concentrations of VOCs are negligible and competing mechanisms are absent. Many more equations exist that estimate individual liquid and gas-phase mass transfer coefficients, overall mass transfer coefficients, and VOC emission rates. These equations increase in complexity as more and more special circumstances are considered within the WWTP. The accuracy of these equations increases with the use of site-specific information. The more site-specific data available, the more accurate the estimate will be.

Because there are so many individual unit operations and processes at a WWTP, it may be necessary to use several equations to develop an emissions inventory. An extensive list of mass transfer correlations and emissions equations are provided in Refs. 1 and 8. Since these equations can be quite complicated, most have been assimilated into computer models that generate emission estimates based on user-supplied inputs.

Computer-Based Models. Calculating VOC emissions for a WWTP is time consuming, and performing the calculations by hand may also increase the chance of errors. To avoid these problems, a variety of personal computer programs have been developed for use in estimating VOC emissions from WWTPs. The available models vary from program to program and were developed with different types of WWTPs in mind. Therefore, it is important to review each model carefully and choose one that is appropriate for each individual application. The following are commonly used models for estimating emissions of VOCs from WWTPs:

- WATER8
- CHEMDAT8
- BASTE
- TOXCHEM

These models are discussed below.

WATER8. *The Wastewater Treatment Compound Property Processor and Air Emissions Estimator* program (or WATER8) is a U.S. EPA–approved model that consists of analytical expressions for estimating air emissions of individual waste constituents in wastewater treatment, a database for listing many of the common organic compounds, and procedures for obtaining printouts of emission estimates. In the WATER8 program, separate emission estimates are made for each individual compound identified as a constituent of the waste. The emission estimates are based on the properties of the compound and its concentration in the waste. The identification of compounds can be made by selecting them from the database that accompanies the program or by entering new information describing the properties of a compound not contained in the database. Estimates of total emissions from a waste can be obtained by summing the estimates for the individual compounds.[12] The WATER8 program supersedes use of the *Surface Impoundment Modeling System* (SIMS), which is no longer supported as an emission estimation tool by the U.S. EPA.

CHEMDAT8. CHEMDAT8 is a Lotus 1,2,3 spreadsheet that includes several analytical models for estimating volatile organic compound emissions from treatment, storage, and disposal facility (TSDF) processes under user-specified input parameters. The available models for wastewater treatment include aeration and nonaeration impoundment processes. CHEMDAT8 is also a U.S. EPA–approved model.[13]

BASTE. The *Bay Area Sewage Toxic Emissions* (BASTE) model was developed in California to provide emission estimates as a part of the state-wide program to evaluate potential "air toxic hot spots." The BASTE model incorporates chemical concentration data for volatile species commonly found in wastewater, along with the specific engineering design features of the plant. Using the concentration and design data, the BASTE model can predict emissions for existing and future plant process configurations. The BASTE model accounts for the biodegradation, solids sorption, air stripping, and liquid discharge mechanisms at WWTPs for compounds included in the database.[14]

The BASTE model also has a flexible building block approach compared with the fixed process flows of other models. The ability to predict emissions from individual process units is one advantage of this model. The BASTE model is also based on municipal wastewater treatment systems, unlike many other models that are based on industrial treatment.

TOXCHEM. The TOXCHEM model has the ability to simulate the dynamic response of treatment systems, unlike other steady-state models. TOXCHEM also computes the fate of VOCs in aerobic and anaerobic digesters. Like the BASTE model, it also includes a temperature correction for Henry's Law Coefficient. This becomes important when volatilization is the dominant removal mechanism because significant error in the predicted VOC removal can occur over a small temperature range.[8]

Other Models. Several other models exist that perform similar functions as those mentioned above, and each model has its own advantages and disadvantages; however, the ones described above are the most favored.

POLLUTION CONTROL MEASURES[8,15]

There are a number of approaches available for reducing emissions of air toxics from WWTPs. These include source control and industrial pretreatment to reduce discharge of toxics to POTW influent. Many of the technologies potentially available for controlling emissions have been used extensively for other applications but have not been fully demonstrated as effective for controlling emissions of toxics from WWTPs. Some technologies considered for controlling emissions from WWTPs include use of containment structures, adsorption, thermal destruction (flaring), and scrubbing systems.

At this time there are no air pollution control devices specifically designed for VOC control at wastewater treatment facilities. Traditional equipment for industrial VOC control is expensive to use on the wet, relatively dilute air streams that typically occur at POTWs. In addition, industrial VOC control equipment costs significantly more than the traditional odor control equipment currently used at WWTPs. Controlling emissions from existing WWTPs also requires retrofitting considerations such as space, accessibility, safety, reliability, and maintenance.

Containment

Containment of fugitive emissions from open basins has been recently applied for control of odor emissions from such sources as primary clarifiers. Covers contain fugitive emissions, which are usually directed to control devices such as scrubbers. The focus of these systems has been for control of odor emissions. Covers also contain VOC emissions as well; however, control of these compounds depends on the type of control and the efficiency of the specific control device. Cover options include fixed roof covers, flexible floating covers, rigid floating covers, fixed covers in contact with the liquid surface, air-supported structures, and floating hollow spheres.

Adsorption

Gas-phase solid adsorption processes are used for control of a wide range of compounds at relatively low levels. They are usually applied to relatively dilute concentration gas streams. Adsorbent materials include both polar and nonpolar solids. Polar adsorbents have high selectivity for polar compounds, especially water, and are not effective with humid gas streams such as those found at WWTPs.

Activated carbon is effective in removing all but the most volatile of organic compounds from humid gas streams. Since carbon is nonspecific, it tends to adsorb all trace vapors in relative proportion to their concentrations, until its sorptive capacity is approached. Because of competitive adsorption, capacity must be provided for those compounds present that are not intended to be adsorbed.

Activated carbon systems can be either disposable or regenerable systems. Disposable systems are often used where gas flows are low or intermittent; where concentrations are low; or when recovery of the sorbent is not economically feasible.

Flares

Flares are generally a low-cost method for disposal of relatively large amounts of gas containing combustible components. Flares are well suited for processes that handle hydrocarbons, hydrogen, ammonia, and other toxic or dangerous gases that require the immediate release of large volumes of gases. Efficiencies range from 70% up to 100%, with most of them having destruction efficiencies above 90%. Flare efficiency depends on flare head design, gas composition, and flare stability, which depends on the gas exit velocity, and the heating value of the gas.

Flares are used at WWTPs to dispose of excess digester gas. Constituents of gases such as sulfur and halogenated compounds will be oxidized to principally SO_2 and HCl, and then discharged into the atmosphere. Products of incomplete combustion can also be produced due to flame instability or low flare flame temperatures. In general, flares are used for waste gas streams with high heat content, unlike those expected in air emissions from water treatment processes at WWTPs.

Odor Control Systems

Odor has been an issue at wastewater treatment facilities since 1929. Most odor control at wastewater treatment facilities focuses on hydrogen sulfide and a selection of odor-causing, but not necessarily toxic, organic compounds. In some instances, ammonia is also of concern.

Liquid-phase control usually deals exclusively with hydrogen sulfide. The technologies used for liquid-phase control include operational consideration, oxygen control, oxidation, precipitation, pH control, and cleaning. Vapor-phase odor control technology includes dry scrubbing systems, wet scrubbing systems, and biological scrubbing systems. These control technologies can control a fairly broad spectrum of substances. Odor-control technologies are usually most concerned with controlling hydrogen sulfide.

Sludge Incineration Systems

The major source of toxicity from most sludge incineration systems are metals and polynuclear aromatic hydrocarbons (PAHs) that are associated with the particulate emissions.

Particulate removal systems usually consist of an air/water venturi scrubber. Fluidized bed systems usually precede the venturi with a baghouse, to collect the bulk of the ash. Other particulate removal systems are electrostatic precipitators (wet or dry) and impinger trays. VOCs can be easily controlled by maintaining furnace temperature. If VOCs cannot be controlled by furnace temperature, then afterburners may be used.

Biofiltration

Biodegradation of pollutants in waste gases using biofilters and soil filters has been applied to control odors from WWTPs and industrial processes since 1953. More recently, these processes have been used for air emissions removal from chemical and process industries. Other processes mentioned in the literature that employ biological treatment of waste gases include bioscrubbers and trickling filters.

Biodegradation of air pollutants is roughly analogous to the biological treatment processes that are used to degrade organic compounds in water. The pollutant serves as the carbon or energy source for the microbial population. Favorable conditions in temperature, pH, nutrients, oxygen, and water must be provided for growth and maintenance. The majority of pollutants are then metabolized by the microorganisms into harmless oxidation products such as carbon dioxide and water.

The term *biofilter* is used to define processes that use compost, peat, bark, soil, or mixtures of these substances as the filter medium. These media serve as a support system for a microbial population. Filter media is underlaid with a gas distribution system, commonly perforated pipe. Gases flow through the bed, where the pollutants are adsorbed to the filter media.

In bioscrubbers, the microbial population is freely dispersed in the liquid phase. Water-soluble components present in the gases are transferred to the liquid phase in a contact apparatus such as a spray tower. The absorbed substances are then degraded by microbial processes in an activated sludge tank. In a trickling filter (or packed column), the microbial population grows as a biofilm on an inert media in a column. Water is applied at the top of the column, and the waste gases flow countercurrently with the water.

REFERENCES

1. U.S. EPA, *Compilation of Air Pollutant Emission Factors (AP-42). Volume I: Stationary Point and Area Sources—Supplement D, Emission Factors for Municipal Solid Waste Landfills*, U.S. Environmental Protection Agency, Research Triangle Park, NC, September 1997.
2. Chang, et al., *Emissions of Volatile and Potentially Toxic Organic Compounds from Sewage Treatment Plants and Collection Systems*, Report to the California Air Resources Board, Contract A5-127-32, July 1987.
3. EBMUD, *Air Toxics Monitoring Program Report, Phase 1.* East Bay Municipal Utility District Wastewater Department, October 1989.
4. PEEP, *Pooled Emission Estimation Program (PEEP) for POTWs*, Joint Powers Agencies for PEEP, December 1990.
5. WEF, *Design of Wastewater Treatment Plants, Manual of Practice No. 8.* Water Environmental Federation and American Society of Civil Engineers, 1991.
6. C. A. Brunner, "Waste Incinerator Sources—Sewage Sludge," *Air and Waste Management Air Pollution Engineering Manual*, Van Nostrand Reinhold, New York, 1992.
7. T. R. Card and B. Desing, *Quantification of Air Emissions from Wastewater Treatment Plants*, Water Environment Federation, 65th Annual Conference, New Orleans, September 1992.
8. WEF, *Toxic Air Emissions from Wastewater Treatment Facilities*, Special Publication of the Water Environment Federation and American Society of Civil Engineers, 1994.
9. Witherspoon, *VOC and Toxic Emissions from Dissolved Air Flotation Units in the Bay Area*, Air and Waste Management Association, 82nd Annual Conference, Anaheim, California, June 1989.
10. M. J. Barboza, *Air Emissions Testing of Air Toxics at WWTPs*, Proceedings from Water Forum 1992, American Society of Civil Engineers, August 1992.
11. B. Eklund, "Practical Guidance for Flux Chamber Measurements of Fugitive Volatile Organic Emission Rates," *J. Air and Waste Management Assoc.* **42**, December 1992.
12. U.S. EPA, *Users Guide for Wastewater Treatment Compound Property Processor and Air Emissions Estimator (WATER8)*, U.S. Environmental Protection Agency, Office of Air Quality Planning and Standards, Research Triangle Park, NC, EPA-453/C-94-090C, November 1994.
13. U.S. EPA, *CHEMDAT8 User's Guide*, U.S. Environmental Protection Agency, Office of Air Quality Planning and Standards, Research Triangle Park, NC, EPA-453/C-94-090B, November 1994.
14. T. R. Card et al., *Air Permitting of a Wastewater Treatment Plant for Total and Toxic VOC Emissions*, Water Environment Federation, 65th Annual Conference, New Orleans, September 1992.
15. W. Bishop et al., *VOC Vapor Phase Control Technology*, Water Pollution Control Federation Research Foundation Report No. 9-2, WPCF, Alexandria, VA, 1990.

BIBLIOGRAPHY

Gaudes, R. J., G. T. Hunt, and R. C. Porter, Identification, *Control and Dispersion of Air Toxics and Odorous Emissions at a Wastewater Treatment Facility*, Presented at the 63rd Annual Conference of the Water Pollution Control Federation, 1989.

Industrial Wastewater Volatile Organic Compounds Emissions—Background Information for BACT/LAER Determination, Office of Air Quality Planning and Standards, Environmental Protection Agency, Research Triangle Park, NC, EPA-450/3-90-004, January 1990.

Lindstrom, K. P. (Ed.), "Workshop Report and Proceedings: Air Toxic Emissions and POTWs," jointly sponsored by the U.S. Environmental Protection Agency, Alexandria, Virginia, 1989, K. P. Lindstrom, Inc., Sacramento, California, 1990.

Vesilind, P. A., et al., *Sludge Management and Disposal for the Practicing Engineer*, Lewis Publishers, Inc., 1986.

U.S. EPA, *Emission Inventory Improvement Program, Document Series Volume II: Chapter 5.* Preferred and Alternative Methods for estimating Air Emissions from Wastewater Collection and Treatment, U.S. Environmental Protection Agency, Research Triangle Park, NC, EPA-454/R-97-004, March 1997.

20

ACCIDENTAL AND CATASTROPHIC RELEASES

R. Scott Huebner, Kevin S. McLeary, George P. Partridge, Jr., William M. Stayer, and James F. White

INTRODUCTION

The increasing population density in urban areas has increased the number of individuals living in proximity to the chemical process industry (CPI). Residential areas and urban growth have built up around industrial complexes and have therefore increased the concern that a major plant accident could result in significant risk to the surrounding community. The Clean Air Act Amendments of 1990 (CAAA-90) included a new air regulatory concept to address large-scale airborne releases of toxic chemicals from accidental releases that could have catastrophic effects on the surrounding community.[1]

The issues related to process safety and accident prevention have ushered in a whole new field in the air-quality arena. These issues are related to addressing potential gas-phase emissions to the atmosphere in terms of quantifying the release, evaluating the impact, and addressing new air pollution control techniques in the area of release mitigation. Traditionally one had been concerned about routine or planned releases as part of industrial process operations. Now the concern has extended to unplanned releases. Unplanned releases can occur from causes ranging from equipment failure to terrorist activity and warfare.

The objective of this chapter is to address the issues related to a "catastrophic accident," which include prevention, modeling the impact, and mitigation. A catastrophic accident involves a major uncontrolled emission, fire, or explosion, releasing airborne chemicals or toxics into the atmosphere with a resulting effects zone that extends offsite into the surrounding community.

Regulatory Requirements

The 1990 CAAA referred to "extremely hazardous substances" and required facilities to both identify hazards related to accidental chemical releases and to address process safety for accident prevention as well as minimize the consequences of releases when they occur. EPA has not identified "extremely hazardous substances" in relation to the CAAA but has stated that the definition is not limited to the list of substances regulated under the CAAA 112(r) in 40 CFR 68.130. Additional details on 40 CFR Part 68 can be found at *www.epa.gov/docs/fedrgstr*. The Accidental Release Prevention Requirements: Risk Management Programs Under Clean Air Act Section 112 (r)(7) can also be found at that Web site. The EPA Chemical Emergency Preparedness and Prevention Office (CEPPO) has a WEB site listing the CAA 112(r) list of regulated substances at *http://www.epa.gov/ceppo/caalist.htm*. The EPA Office of Solid Waste and Emergency Response has published a consolidated list of chemical substances regulated under Title III of the Clean Air Act.[2] EPA created three different program categories in which to place sources based on the consequences of a worst-case scenario accident. Program 1 applies when public or environmental receptors would not be affected by low-hazard sources. Program 2 addresses medium-hazard sources that are not addressed by Programs 1 and 3. Program 3 is designed to target potentially high-hazard chemical processes. The goals of the tiers in the program are to recognize the different risk levels involved and facilitate emergency response planning.[3]

PROCESS DEFINED

The regulatory rules only focus on stationary sources and address "processes." Process is defined in 40 CFR §68.3 as follows: "any activity involving a regulated substance, including any use, storage, manufacturing, handling, or on-site movement of such substances, or combination of these activities. This is illustrated in Figure 1. For the purposes of this definition, any group of vessels that are interconnected, or separate vessels that are located such that a regulated substance could be involved in a potential release, shall be considered a single process."[4] Figure 2 presents a schematic representation of what is defined as a process or multiple processes.

Plant design seeks to minimize hazards in physical operation. These hazards include:

- Vaporization and diffusion/advective transport of toxic/flammable liquids and gases
- Aerosol formation from toxic/flammable liquids
- Dispersion of combustible/toxic dusts
- Mixing of highly reactive chemicals
- Temperature control of unstable chemicals
- Friction or shock of unstable chemicals
- Pressure control in vessels, and loss of diluents

A major airborne chemical release can occur due to loss of contaminant, illustrated in Figure 3, involving process equipment for causes which include the following[5]:

- Rupture of vessel due to overpressurization
- Rupture of vessel due to brittle fracture
- Flange/gasket failure/seals/plugs
- Weld failures/casting failure
- Overstressing of containment shell
- Vibration
- Corrosion/erosion
- External loading/impact
- Internal explosion

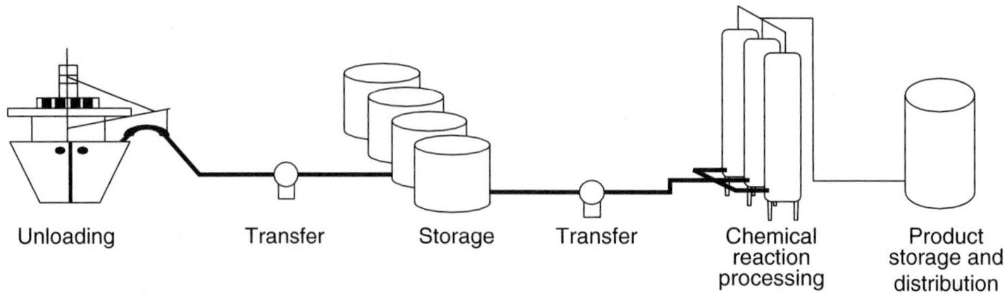

Figure 1. Processes Addressed under the Clean Air Act Amendments of 1990.

Schematic Representation	Description	Interpretation
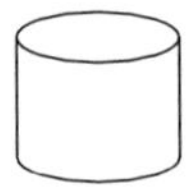	1 vessel 1 regulated substance above TQ	1 process
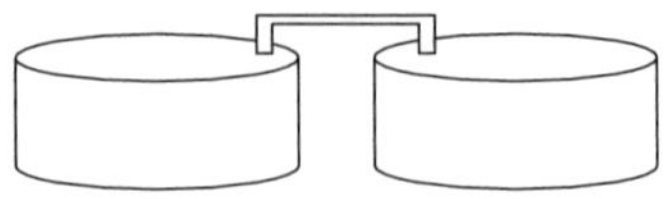	2 or more connected vessels *same* regulated substance above TQ	1 process
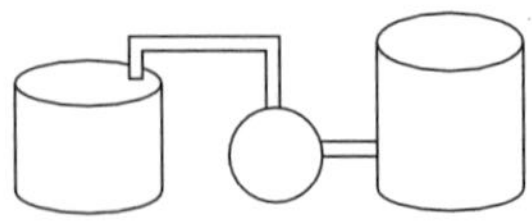	2 or more connected vessels *different* regulated substances each above TQ	1 process
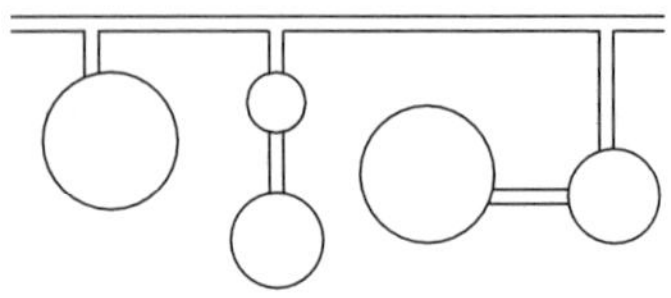	pipeline feeding multiple vessels total above TQ	1 process
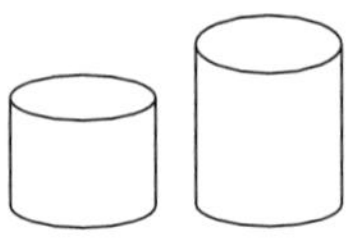	2 or more vessels co-located *same* substance total above TQ	1 process
	2 or more vessels co-located *different* substance each above TQ	1 process
	2 vessels, located so they won't be involved in a single release *same* or *different* substances each above TQ	2 processes
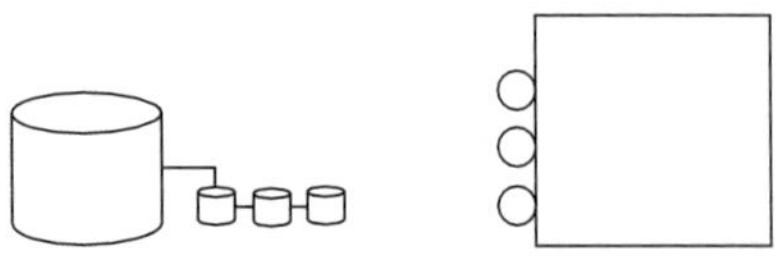	2 locations with regulated substances each above TQ	1 or 2 processes depending on distance
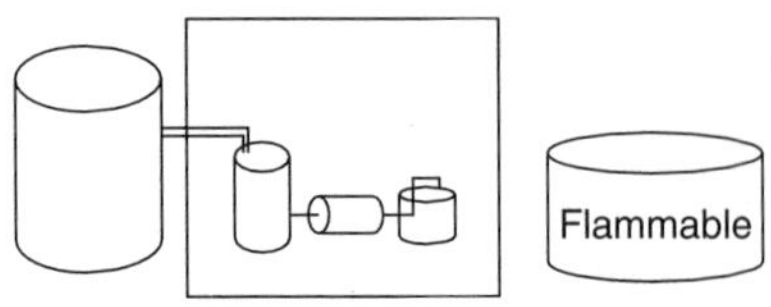	1 series of interconnected vessels *same* or *different* substances above TQs *plus* a co-located storage vessel containing flammables	1 process

Figure 2. Schematic Representation of a Process for a Risk Management Plan (RMP) (*Source*: Ref. 5).

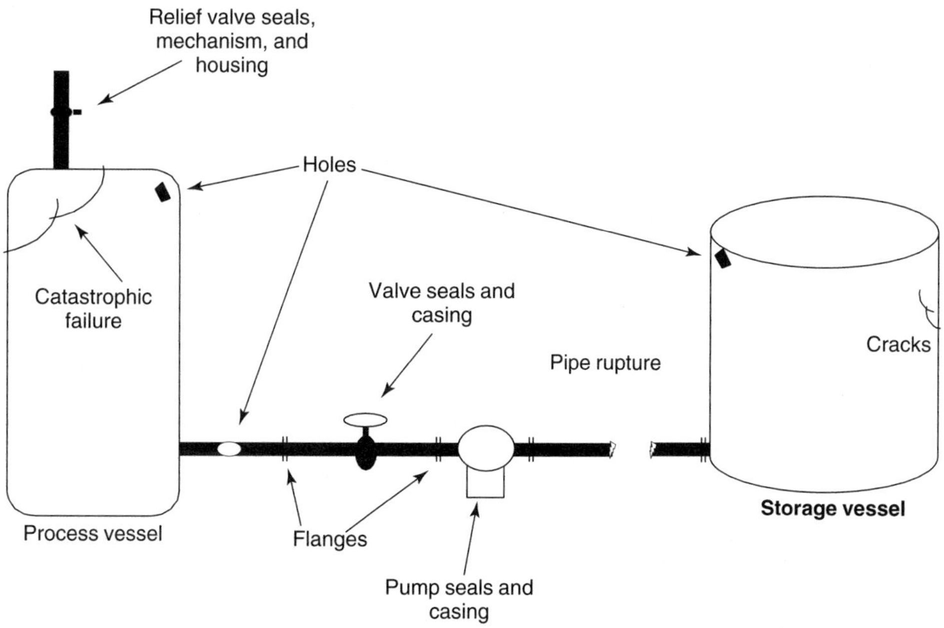

Figure 3. Examples of Release Sources.

Additional contributing events to vapor-phase releases include[6]:

- Overfilling of a vessel or knockout drum
- Excess vapor flow into a vapor destruction system
- Tube rupture in heat exchangers
- Line rupture
- Operator error

Minimizing the Impacts of Potential Releases

The first step in minimizing the potential impact of a catastrophic chemical release prior to any postrelease mitigation action is in the area of engineering design for process safety. Included in this area are basic plant design, individual equipment design, and materials selection.[5] Though relatively infrequent, material failures can be extremely severe, resulting in a catastrophic chemical release. The approaches to minimizing the risk of material failure involves understanding the internal process, the external environment, and failure modes along with proper selection of materials for the intended application. Also important are fabrication techniques, maintenance, inspection, and repair procedures. Mechanical failures are more predictable than material failures and are addressed by controlling the physical operating parameters of the process to minimize their occurrence. Mechanical failures include: brittle fracture, mechanical fatigue, thermal fatigue, creep, thermal runaway exceeding temperature and pressure design limits, stress rupture, catastrophic oxidation, and carbonization. Potential sources of process failures also include the piping and transfer lines. Velocity criteria become important to minimize hazards due to the following conditions: corrosion, erosion, vibration, noise, fluid flow hammer, and static electricity. Process conditions within reaction vessels can lead to runaway reactions and potential ignition if flammable substances are involved. Examples of process conditions that need to be considered in the area of process safety include: catalysis supported by process contamination, reaction with powerful oxidants, thermite reactions, formation of unstable species (such as peroxides, acetylides, nitro compounds), and formation of pyrophoric materials.

Hazard Identification

A hazard can be defined as "a characteristic of the system/plant/process that represents a potential for an accident." It can be thought of as a combination of a hazardous material and an operating environment such that certain unplanned events could lead to an accident.[6] The structured evaluation of chemical process safety is a continuous process that includes: problem identification, solution development, evaluation of solutions against established criteria, and solution implementation. The first step is hazard identification. This step is followed by a preliminary risk assessment, development of a mitigation or control method, risk evaluation, followed by a check to see if the mitigation requirement has been met. If not, then the next step is postrelease mitigation or control.[7]

The American Institute of Chemical Engineers — Center for Chemical Process Safety (AIChE-CCPS) provides documents that present three methods for identifying and displaying the relationships between sequential events contributing to an accident scenario. Included in the method are techniques for estimating both the frequency or probability of initiating events and the sequence of events leading to a specific accident scenario.[6] The three methods are: fault-tree analysis, event-tree analysis, and cause–consequence analysis.

Fault–tree analysis is a deductive-graphic technique for identifying the basic causes of a particular accident event. Examples of causes include equipment failure, human errors, and external events. It can become quite complicated for anything but the simplest chemical plants and can involve thousands of process events. Fortunately the approach can be set up using a variety of computer programs commercially available to draw fault trees from an interactive session. The approach is initiated with a well-defined accident or top event and works backward toward the different potential scenarios that could cause the accident. The logic transfer events incorporated in a fault tree include: AND Gate, OR Gate, INHIBIT Event, BASIC event, INTERMEDIATE Event, UNDEVELOPED Event, EXTERNAL Event, and TRANSFER Symbols. The main disadvantage to using fault trees is that the process becomes very complicated as the details of the process are increased. Fault trees can result in "thousands of gates or logic transfer steps" and require a considerable time to complete. A major advantage to the fault-tree approach is that the process is initialized with the top event. This is opposed to the "event-tree" approach listed below where the events resulting from a single failure might address the events desired by the user.[8] Fault-tree analysis can be utilized to estimate the frequency or probability of an accident event sequence.

Event-tree analysis is an inductive-graphic technique for evaluating or describing the outcomes such as accident event sequences and accident scenarios based on an initiating event. It considers the successes and/or failures of any mitigation technology. It is useful for estimating the probability of each outcome.[6]

Cause–consequence analysis is a combination of fault-tree analysis and event-tree analysis. Risk analysis proceeds with a series of steps starting with hazard identification and proceeding from very quantitative estimates of consequences, probability, and the effectiveness of various mitigation technologies. The Center for Chemical Process Safety of the American Institute of Chemical Engineers has developed flow charts outlining the steps in the risk analysis process. They are summarized in Table 1.[6]

The methods for hazard evaluations are presented in greater detail in a document by the AIChE-CCPS entitled "Guidelines for Chemical Process Quantitative Risk Analysis."

AIR EMISSIONS CHARACTERIZATION

Consequence Analysis

Receptors of interest are defined as "public receptor" and "environmental receptor."[4] A public receptor generally means any place where people live, work, or gather, excluding roadways. By definition this includes areas such as offices, businesses, recreational areas, and residential areas. Environmental receptor refers to a limited number of areas that are officially designated by the state or Federal government. Environmental receptors include natural areas, such as national or state parks, forests, or monuments; officially designated wildlife sanctuaries, preserves, refuges, or areas; and Federal wilderness areas. The distance to an endpoint from a worst-case release is the distance a toxic vapor cloud, fire, or explosion from an accidental release will travel before being dispersed to a degree that serious injuries from short-term exposure will no longer occur.[4]

Table 1. Step in the Risk Analysis Process (from Ref. 6, p. 119)

Step	Objective
1	Identify hazards
2	Estimate consequences
3	Identify opportunities to reduce consequences
4	Identify accident-initiating events
5	Estimate probabilities of initiating events
6	Identify opportunities to reduce probabilities of initiating events
7	Identify accident event sequences
8	Estimate probabilities and consequences of event sequences
9	Identify opportunities to reduce probabilities and/or consequences of event sequences
10	Proceed to more quantitative hazard evaluation analysis if required

According to 40 CFR §68.165 offsite consequence analysis, the owner or operator shall submit in the risk management plan (RMP) the following data: (1) chemical name, (2) physical state (toxics only), (3) basis of results (give model name if used), (4) scenario (explosion, fire, toxic gas release, or liquid spill and vaporization), (5) quantity released in pounds, (6) release rate, (7) release duration, (8) wind speed and atmospheric stability class (toxics only), (9) topography (toxics only), (10) distance to endpoint, (11) public and environmental receptors within the distance, (12) passive mitigation considered, and (13) active mitigation considered (alternative releases only).[4]

The consequences of accidental vapor cloud releases with the greatest potential for adverse acute health hazards are usually due to toxic releases, fires, and explosions. Flammable vapor clouds can result in fires, fireballs, flash fires, or explosions. Ignition of a continuous or instantaneous release can cause a "vapor cloud explosion," which is also referred to as an unconfined vapor cloud explosion (UVCE).[6]

Source and Dispersion Modeling

Release sources include: storage vessels, piping, process vessels, transfer hoses, valves, pumps, joints, and other miscellaneous sources that could result in a release event.[4] Typical scenarios that would be modeled include: liquid spill and subsequent evaporation, two-phase pressurized release, and a gas release.[9] These are illustrated in Figure 4.

When modeling vapor cloud dispersion, there is a sequence of steps to be followed with various pathways and options. A summary of the steps and options to consider are summarized in Table 2.[9]

The input data required for modeling include:

- Source data
- Site characteristics
- Meteorological data and algorithms for calculating input parameters
- Receptor information

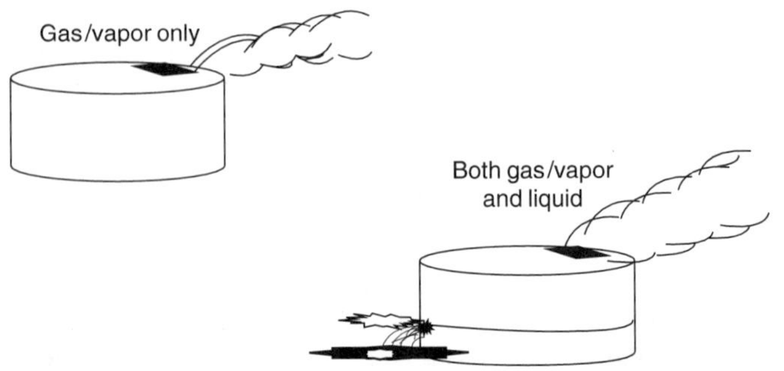

Figure 4. Vapor and Liquid Can Escape from Processes as a Single or Two-Phase Release.

Table 2. Categories of Information to be Considered When Conducting Vapor Cloud Modeling

Scenario description	Liquid spill Pressurized two-phase jet Vapor release
Definition of input parameters	Chemical and physical properties of release Receptor properties Environmental properties
Source emission model	Catastrophic vessel rupture Evaporation from spill Liquid jet Two-phase jet Gas jet Rain out
Vapor cloud dispersion model	Entrainment Thermodynamics Removal Special effects (downwash, averaging time, etc.)
Concentration prediction	Form of model output (tables or graphs) Heights and receptor locations Concentrations as a function of averaging times Ability to superimpose concentration profiles on community maps

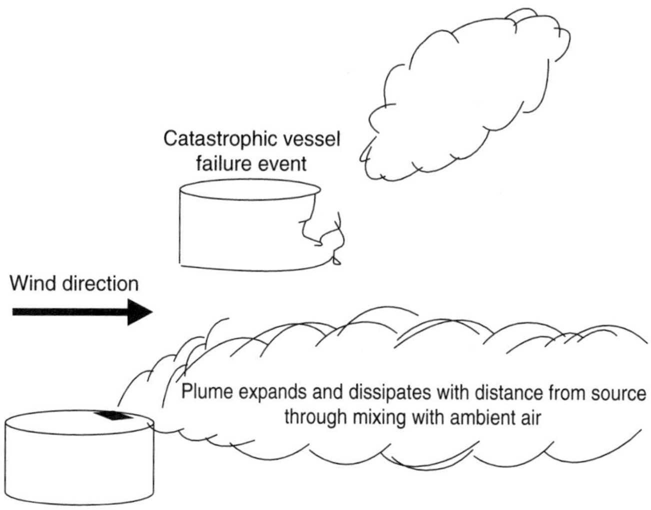

Figure 5. Plumes from Chemical Releases Can Be Viewed as Instantaneous (Puffs) or Continuous, Depending on the Duration of the Release.

Source Data. Source data include the physical and chemical properties of the released material, geometry of the release point, characteristics of spill surface, time variation of release (continuous or instantaneous, as illustrated in Figure 5), and the presence of gas or aerosol in jet releases. Chemical composition along with property data including the following may be important: molecular weight, density, molecular diffusivity, conductivity, boiling point, vapor pressure, heat capacities, heat of vaporization, surface tension, and explosive limits. The release rate is a function of the geometry of the source. Geometry includes the dimensions of the tank, pipe, stack, or orifice that becomes the point of release. Releases of hazardous materials cover a wide range depending on their atmospheric boiling point and the process operating temperature. The thermal conductivity, density, specific heat, moisture content, porosity, and contours of the spill surface will affect pool spreading and heat transfer characteristics that will affect the evaporation rate. The rate of chemical release is typically time varying since flow rates from pressurized containment vessels decrease exponentially with time. The structure of the release point, particularly if it is an orifice, will influence the liquid breakup, turbulence, or shearing effects as the fluid passes through the release opening and experiences atmospheric conditions.

The source release modeling is the most crucial step in the modeling process. It provides the source term or emission rate for the downwind dispersion. Emission calculations for accidental releases may include: multicomponent chemical mixtures, two-phase (gas and liquid) releases, varying emission rates, and dense or cold gases/vapors.[9] A combination of release scenarios may need to be considered resulting from different causes of process failure such as: corrosion, thermal fatigue, overpressurization, or vessel/pipe external impact.

Site Characteristics. Site characteristics include: potential chemical source release locations, buildings, roads, and categories of land use. Another site characteristic that is important is surface roughness, since it influences the intensity of atmospheric turbulence. A further related site concern is meteorological conditions. Real-time meteorological, historical, or hypothetical conditions can be selected. The majority of modeling is conducted with historical or hypothetical data, but it is important to include in the modeling plan "worst-case" meteorological conditions.

Receptor Data. Receptor data include fixed receptors such as monitoring sites, population centers, ignition sources, sensitive crops, and property boundaries.[9] Typical modeling procedures determine the downwind concentrations as a function of distance for various meteorological conditions.

In selecting the dispersion model the primary objectives include:

- Distances to concentrations of interest based on toxicological data
- Concentration contour plots
- Flammable and/or explosive regions involving hazardous materials

A critical determination regarding release characteristics is deciding if a dense gas model should be selected. The estimation of the excess cloud density, along with volume flux and the surrounding ambient conditions, will determine its initial behavior near the point of release. Excess cloud density affects the entrainment rate of ambient air into the top of the cloud and the rate of lateral cloud spreading. The features of the vapor cloud during its initial transport phase can be characterized by the Richardson number.[9]

Model Selection. Once the type of model needed is identified, selection of the appropriate model becomes an important task in itself. There are hundreds of accidental release and dispersion models available.[9] The choices of models can be distinguished by the following categories: planning versus real-time models and public versus proprietary models. Nonproprietary models are available through bulletin boards or at a minimum cost required to copy and distribute the information. Proprietary models can range significantly in cost from less than $1,000 to $15,000–25,000 depending on features. Numerous sources of proprietary models can be located using various search instruments via the Internet. An excellent source for models in the public domain is found at the EPA Support Center for Regulatory Air Models (SCRAM) found on the Technology Transfer Network at *www.epa.gov/ttn/scram*. Several EPA models are available on the Internet, including TSCREEN, RMP*Comp, and DEGADIS.[10] The model ALOHA can be ordered from the National Safety Council. To summarize, it is important to included the parameters that represent modeling of the worst-case scenario as well as a series of sensitivity runs. The sensitivity runs include: variations in humidity, release duration, effect of surface roughness, and averaging time.

The required parameters for modeling worst-case scenarios are presented in Table 3. Alternative modeling scenarios for a process would represent a scenario that is more likely to occur than the worst-case scenario and that reaches an endpoint offsite.

Table 3. Required Parameters for Modeling Worst-Case Scenarios (from Ref. 4, p. 4–6)

Parameter	Description
Endpoints	For toxic substances, use the endpoint specified in part 68, Appendix A For flammable substances, use the endpoint of an over pressure of 1 pound per square inch (psi) for vapor cloud explosions
Wind speed/stability	Use wind speed of 1.5 m/sec and F stability class unless you can demonstrate that local meteorological data applicable to the site show a higher minimum wind speed or less stable atmosphere at all times during the previous 3 years. If you can demonstrate a higher minimum wind speed or less stable atmosphere over 3 years, these minima may be used
Ambient temperature/humidity	For toxic substances, use the highest daily maximum temperature during the past 3 years and average humidity for the site
Height of release	For toxic substances, assume a ground-level release
Topography	Use urban or rural topography, as appropriate
Dense or neutrally buoyant gases	Tables or models used for dispersion of regulated toxic substances must appropriately account for gas density
Temperature of release substance	For liquids (other than gases liquefied by refrigeration), use the highest daily maximum temperature, based on data for the previous 3 years, or at process temperature, whichever is higher Assume gases liquefied by refrigeration at atmospheric pressure are released at their boiling points

AIR POLLUTION CONTROL MEASURES

Prevention Program and Release Control Technologies

Measures to prevent accidents include:

- Improved/upgraded equipment
- Revised maintenance procedures
- Employee and contract employee training programs
- Revised operating procedures
- New process controls
- Mitigation systems to limit the severity of an accidental release
- Emergency response planning
- Process changes
- Strengthening process structures to minimize storm damage
- Chemical inventory reduction
- Process chemical complexation to alter chemical release properties[11]

After taking measures to enhance process safety through design and operational procedures, additional levels of protection can be incorporated to further minimize a catastrophic release to the atmosphere. There are several approaches to address an "out of control" situation where a process has exceeded its safe operating limits or engineering design limitations. Approaches to be considered include not only emergency shutdown, but also safe disposal of emergency releases. An approach may include diverting the release to a control system or to a means of containment followed by treatment or cleanup technologies. Those measures include emergency relief–treatment systems, emergency transfer of materials, double containment, enclosures and walls, dikes, curbs, trenches, and impoundments.[6] The important objective is to seek secondary containment or control for the resulting process failure. Dump systems can serve to divert a reactive mixture to a safer location to advert an overpressurized reactor situation. Quench systems serve similarly to "dump" systems with the addition of a quenching substance to slow or stop the reactions. The reaction can be stopped by changing the chemical reaction, deactivating a catalyst, and by cooling/diluting the process fluid.

Emergency Relief Systems. Emergency relief devices minimize accidental releases of hazardous materials caused by equipment overpressurization and vacuum. In essence one is applying process depressurization or a controlled release to avoid a sudden catastrophic large-scale release. Overpressurization can result from several scenarios, which include: fire, blocked outlet, operation failure, equipment failure (hardware failure such as tube rupture or control system failure), process upset (exothermic runaway reactions), and process causes such as imbalance of fluid flow rates and utility failures.[5] Important factors to consider in the design of emergency relief systems is whether single-phase or two-phase flow exists. Two-phase flow results quite often in a combination of a gas-phase discharge in the presence of a flashing liquid resulting in aerosol/mist formation. Common methods of addressing overpressurization include: safety relief valves and/or rupture disks. The discharge can be directed into containment systems or disposal/treatment systems to prevent or minimize discharge of chemicals to the atmosphere.

Selections of the appropriate control technology includes the following considerations: physical state of effluent, quantitative factors, hazardous properties, and location of disposal system. Physical factors include defining the release phase(s) such as vapor/liquid/particulate or a combination, pressure and temperature, and boiling point. Quantitative factors include flow rate, duration of discharge, and total quantity of material to be treated. Hazardous properties include toxicity and flammability. The location of the disposal system needs to address meteorological conditions, local populations, as well as local regulations and ordances.[5]

Control systems typically involve the following technologies: blowdown, flare, incineration, and vapor control systems. "Blowdown" systems include chemical sewers, closed drain header systems for flammable liquids or special materials, and quench blowdown drums. Blowdown systems are used to handle condensable vapors or contaminated liquid effluents and will minimize volatilization of substances that have significant vapor pressures.

Flares. Flares have frequently been used to control the release of hazardous substances if they are combustible. A flare is suitable for disposing of flammable, toxic, or corrosive gaseous effluents by burning them under controlled conditions. Categories of flares include elevated flares, ground flares, and low-pressure flares. Elevated flares consist of a stack, flare tip, pilot burners, pilot ignition, and supporting facilities. Ground flares can be used when flare luminosity and noise levels are minimal and nontoxic combustion products are produced. Low-pressure flares treat off-gases from units that are discharging VOCs at pressures on slightly above atmospheric. Flaring, when operated properly, is smokeless. Smokeless flares are designed for turbulent combustion conditions created by vortex action or steam injection.

Incineration. Incineration is the burning of the process discharge in a closed system under carefully controlled conditions. Basic design considerations for incineration systems include: state of waste (gas, liquid, slurry, solid), viscosity, mass loading, density, ultimate analysis (C, N, O, H), compositional or phase analysis (percent water, ash, inerts, alkali, heavy metals, sulfur, Cl, F, Fr), and heat of combustion. Incineration technology is continually being developed. Incinerator designs to be considered for disposal of process releases include: fluidized-bed incinerators and liquid and fume incinerators. They may be based on thermal or catalytic systems. The fluidized-bed type of incinerator operates at temperatures in the range of 870–1200°C (1600–2200°F). Gases, liquids, slurries, and solids can be destroyed efficiently, including those waste materials that have a high moisture content, significant inert materials, acidic compounds, or a low heating value. Liquid and fume incinerators are primarily used to destroy VOCs from low-pressure sources. Quite often supplemental fuel is required when the waste has a low heating value or is present at low concentrations. Design criteria for incinerator systems include: organic destruction efficiencies of 99.99% and a destruction efficiency of 99.9999% for wastes containing dioxins, furans, and PCBs, minimum combustion zone temperatures, maximum feed rates for waste and Btu output, continuous emissions monitoring, stack temperature and flow rate, and a negative draft in the incinerator.

Vapor Control Systems — Scrubbers. Vapor control systems are designed for collection of vapor during transfer or storage. Vapor control systems include absorption using lean oil for hydrocarbons and water for acetone or methanol, scrubbers, adsorption utilizing carbon beds, refrigeration, and vapor balancing.[5] A scrubber provides good contact between the vapor/gas and the liquid used to capture the vapor. Scrubber configurations include packed towers, plate columns, spray chambers, wetted-wall columns, and stirred or sparged vessels. The flow pattern is usually countercurrent with the gas entering at the bottom and the liquid at the top. If the gases or vapors are soluble in water, then water is commonly used as the absorbing liquid. Caustic solutions (NaOH, etc.) are used to neutralize acid vapors. Organic liquids with low vapor pressures are typically selected to absorb organic vapors. Packed towers are discouraged from applications where particle levels may be significant, resulting in plugging or fouling of the packing. If the scrubber is treating gases and vapor from emergency release systems including pressure relief devices or pressure control valves, or where multiple valves are used and routed to a scrubber system, then scrubber design should be based on the largest possible release scenario. Standby emergency scrubbers can be designed to operate in the "batch mode." It is critical to design the scrubber to cope with emergency flows. Care should be taken in the design process to include the possibility of two-phase flow reaching the scrubber system and resulting in localized boiling from severe temperature increases. Flooding and breakthrough can occur at peak emergency release flows and must be accounted for in the design process. A batch scrubber sometimes is also referred to as a "passive scrubber," "scrub tank," or "guard tank," since it is intended to treat a relatively large quantity of vapor release during a short duration. This sometimes eliminates the traditional components in an absorption system being present such as recirculation pumps. The absorbing solution would commonly be dumped following an episodic treatment situation. Also a scrubber can be designed for total vapor removal or for selective removal of certain chemical constituents in a multicomponent mixture.[6] Discharge from a passive scrubber can be routed to an "active scrubber" (one designed for continuous operation) or flare if further vapor removal is necessary.

Vapor Control Systems — Adsorption. Adsorption technology is practical in a limited number of emergency situations when the objective is to recover relatively small quantities or low concentrations of leaking materials that would otherwise escape the process. The most widely used adsorptive material is activated carbon, which is suitable for adsorbing numerous organic compounds. Care must be taken when adsorbing oxygenated compounds such as ketones or aldehydes because of the possible fire hazard. Other adsorbents include silica gel, alumina, and molecular sieves. Any "catalytic" properties of the adsorbent must be addressed because of possible ignition of flammable vapors, polymerization of monomers, or decomposition of unstable–reactive materials. Design bed velocities are less than 100 ft/min with temperature limited to 50°C.[6]

Vapor Control Systems—Condenser Technology. Another approach to vapor control is the application of "condenser" technology. Condensers are especially applicable to materials that do not have elevated boiling points and are present in the gas phase at high concentrations. The objective of condenser technology is to cool the vapor below the dew point and recover the condensate. If the boiling point of the chemical being captured is greater than 80°F, water is commonly used as the coolant. For chemicals with low boiling points refrigeration systems are utilized. A large differential can lead to shock cooling and mist generation. Scrubbers or flares are appropriate to apply downstream of condenser technology.[6]

Discharge Point. Whatever the vapor treatment device, there still is a need for a designated discharge point from the treatment system. Traditionally that discharge point has been a stack to disperse the relatively low concentrations of gases, vapors, and particulates that may remain after treatment. In the context of accidental releases, stacks should be looked upon only as a last resort for emergency situations. Stacks provide significant advantages over low-elevation releases such as a ground-level release. These advantages include: plume rise for stack design with significant release velocities, jet entrainment of air, dispersion of contaminants downwind before reaching sensitive ground-level receptors, avoiding wake or downwash effects due to topography or buildings/structures surrounding the release point, and the advantage of higher upper-air wind speeds.[6]

Spill Containment. Spill containment controls the release by containing the chemical(s) when the primary process containment has been breached. Spill containment can serve two purposes: to prevent escape or to minimize the vapor release rate by controlling the air–liquid interface where volatilization occurs. Double containment is included in process design by essentially using double-walled (concentric) piping and double-walled tanks. The entire process or selected chemical storage facilities can be surrounded using enclosures or walls. Secondary containment using enclosures has been applied around tank car and tank truck loading or unloading facilities. Process building enclosures require specific precautions to protect plant workers since the air space is confined around the process. Exhaust ventilation and associated vapor/gas control systems should be utilized as appropriate. Walls can aid in diverting dense-gas releases to a safe area or minimizing the spread of the dense vapor cloud. Dikes, curbs, trenches, and impoundments can contain the spill and reduce the surface area of the spill. Theses approaches are effective for reducing the rate of evaporation or volatilization since the interfacial areas (air/liquid) is reduced and vapor boundary layer formation is encouraged if the walls are above the liquid surface level. Companies have used low-density concrete for wall or dike construction to reduce the rate of heat transfer to the spilled material and thereby reduce its vaporization rate.

Techniques to Suppress Liquid Flashing. Another approach that resembles process containment has been evaluated for application to hydrofluoric acid alkylation facilities in the petroleum industry.[12] The technique incorporates the use of partial enclosures around major process units including reaction vessels and acid inventory tanks. The enclosure resembles a second wall but is open to the atmosphere. If there is an inner-wall failure, the two-phase flashing, chemical release is expanding in a zone where air entrainment is minimized. As a result, the flashing vapor quickly reduces the temperature in the air space and quenches additional vapor flashing, thereby forming a liquid-phase release. The liquid-phase release can then be discharged to secondary containment, where measures are taken to reduce additional volatilization. Those measures include neutralization, dilution, and covers to reduce the interfacial area of the liquid/air interface.

Postrelease Mitigation Technology

Major chemical accidents that have occurred have usually resulted in severe health and environmental impacts due to uncontrolled releases of toxic and flammable vapors in significant quantities. The goal of mitigation is to minimize the release of dangerous amounts of toxic and flammable vapors and the severity of their impact. There are several alternatives to choose from depending on the particular hazard being addressed, the quantity of material involved, location of the facility, and processes involved.[6,13] The hierarchy of approaches has been previously alluded to in the section on process safety. After pursuing accident prevention through process and equipment design and plant operational procedures, the next step is to add an additional level of protection by including postrelease mitigation technologies.

Mitigation can be defined as "reducing the size, duration, and consequences of vapor cloud releases."[6] Mitigation technology can be classified as active or passive. Passive mitigation as defined in §68.3 is "equipment, devices, or technologies that function without human, mechanical, or other energy input."[4] Passive mitigation systems would therefore include containment systems such as dikes or building enclosures. Fire sprinkler systems, water curtains, valves, scrubbers, or flares are considered active mitigation systems since they require energy input to function.

Selection of appropriate mitigation technologies can be simplified by categorizing the operation into chemical unloading, transfer, storage, reaction, and separation processes. The objective of control is to minimize vapor cloud formation. *Vapor* refers to any chemical(s) in the gaseous state, while *cloud* indicates that the vapors represent an unconfined release to the atmosphere rather than being contained in a building or enclosure. Cloud types are characterized by their toxic, flammable, and buoyancy properties. The form of the vapor release can involve: noncondensing gases, condensing gases, two-phase flow systems, and nonflashing liquids. A noncondensing vapor refers to the release of a substance that has a boiling point well below ambient temperature. A condensing vapor is a substance having a boiling point above ambient temperature where some condensate can form an aerosol/mist and potentially rain out. Two-phase flow involves a flashing liquid which is stored or released at process conditions above its boiling point. A nonflashing liquid is released at a temperature below its boiling point and will form a pool on the ground where evaporation or volatilization may occur. Aerosol formation can occur in a pressurized release of a nonflashing liquid.[6] Vapor release events are caused by loss of containment including secondary containment of volatile materials.

Examples of methods for mitigating the effects of toxic vapor clouds include limiting the release amount and/or the rate of release, minimizing liquid pool vaporization, enhancing dispersion of the released vapor, and chemical substitution to one of less concern. Limiting the inventory of the chemical may reduce the consequences if a loss of containment were to occur. Screening methodologies for determining safe inventory levels are based on "worst-case" release and atmospheric dispersion modeling scenarios. Chemical substitution replaces a hazardous chemical with a less hazardous one. Characteristics for identifying a suitable replacement include the fate and transport properties of the chemical(s) being considered. Process modifications can reduce temperatures and pressures at which the chemical(s) are handled to minimize the release rate to the vapor phase if process containment fails. Buffer zones between a chemical process and surrounding populations can be effective for explosion hazards and thermal radiation from fires, and to provide a period of cloud dispersion before reaching sensitive receptors. Buffer zone distance requirements can be significant if the goal is to reduce the impact of a toxic vapor cloud traveling downwind when one considers that every 100 ft is equivalent to less than 15 sec travel time at a typical wind speed of 5 miles per hour.

Toxic volatile or flammable liquids that are not successfully prevented from escaping the process system by passive mitigation techniques must be mitigated using various postrelease mitigation technologies. Various postmitigation technologies add another level of environmental protection in the hierarchy of levels of protection between the process and the surrounding environment. The countermeasures employed will depend on the nature of the release. The scenario can range from a volatile pool, boiling pool, two-phase flashing chemical release, to a total vapor-phase release. The vapor cloud or plume can behave as a buoyant, neutrally buoyant, or a dense gas cloud as it migrates from the process area. Countermeasures may be needed to deal with the pool as well as the resulting vapor cloud.

Mitigation techniques can be categorized as vapor-release countermeasures or liquid release countermeasures.[6] Vapor-release countermeasures include: water sprays, water curtains, steam curtains, and deliberate ignition. Liquid-release countermeasures include: dilution, neutralization, covers, and foams.[6]

Water Sprays. Water sprays potentially can absorb water-soluble gases or vapors as well as dilute the vapor cloud. The technology had been demonstrated on ammonia vapor clouds as well as acid vapor clouds such as hydrofluoric acid. The water spray needs to consist of a fine spray at high flow rates upwind and toward the release source. It is also crucial that a minimal time lag occurs between the moment of release and the activation of the water sprays.

Water Curtains. Water curtains can be used in a similar way as water sprays and are effective for isolating the process releasing the vapor cloud and a point of ignitions.[14] This is particularly important when potentially flammable or explosive vapor clouds are formed. The curtains are configured using nozzles mounted in horizontal piping or manifolds. Air is entrained by the water sprays, resulting in turbulence generated by their mixing. Two concerns that need to be considered in their application are vapor flow over and/or around the water curtain when attempting to mitigate dense vapor clouds, and reduced effectiveness in applications where strong winds prevail. Examples of applications to date include flammable vapor clouds from ethylene releases and toxic vapor clouds from hydrogen chloride and hydrofluoric acid releases.[6] In cold climates where temperatures are below freezing, water curtains may pose operational problems and alternate approaches such as "dry powder curtains" need to be considered.[7] The technology for applying dry powders has been developed by the fire fighting equipment industry.

Steam Curtains. Steam curtains are operated similar to water sprays and water curtains. Since steam elevates the temperature of the vapor cloud, the absorption effects of the spray are minimized. This is offset by the potential increase in buoyancy of the vapor cloud that can be an important attribute of this technology in terms of downwind transport/dispersion characteristics. Significant amounts of steam are required. To reduce concentrations of the vapors below the lower flammable limit, a rate of 0.2 pound of steam is recommended per pound of contaminant.[6] Any potential concerns about static electricity as an ignition source generated by the spray can be prevented by proper design and grounding.

Vapor Cloud Ignition. The use of deliberate ignition of a vapor cloud as a mitigation measure can be applied to both toxic as well as flammable materials. Two benefits of igniting the vapor cloud result: oxidation of the toxic vapors before they can travel off-site to sensitive receptors, and creating a buoyant release to minimize ground-level concentrations.[6] Ignition sources include electric igniters and propane burners. Care must be taken that ignition occurs downwind of the process at a safe distance that does not incur additional process damage. It is recommended that igniters be placed 4 ft above the ground at 3.6 degree intervals around the plant.

Water Dilution. Water dilution is an effective countermeasure for mitigating water-miscible or water-soluble spills. By diluting the spill, the vapor pressure of the mixture is reduced and volatilization of spill components as an emission source is minimized. Initially the addition of water may increase the volatilization rate due to the heat of solution. Larger quantities of water may thus be required and can be estimated using enthalpy diagrams. If the spill contains reactive liquids or liquefied gases, mitigation using dilution is not recommended. For reactive liquids or cases where the sudden temperature rise during the initial addition of water results in volatilization of toxic gases, one may want to consider the use of nonreactive barriers to reduce the airborne emissions.[6] Chemicals where dilution should not be considered are illustrated by chemicals that include chlorine, acetone cyanhydrin, phosphorus, certain hydrides and silanes, and methyl isocyanate.

The technique of neutralization seeks to react the contents of the spill material with a substance to form a less volatile chemical or ester. For the technique to be effective, addition rates much greater than stoichiometric quantities are required. Care must be taken on the addition methodology to not aggravate the vapor-release hazard.

Covers. The use of covers can range from lighter immiscible nonreactive liquids to foams and solid granular materials.[6] The objective of the cover is to minimize the interfacial area of the liquid/air interface to minimize the path of volatilization to the atmosphere. Any material that will cover the surface of the spill and is stable during the cleanup or remediation period is effective. The cover material should not pose an environmental hazard in itself. Examples of cover materials that have been utilized include water, low-density aliphatic oils, and foam materials manufactured by fire protection equipment manufacturers. Covers are particularly effective when the spill spreading has been contained by the use of dikes or other barriers.

Monitoring. Monitoring equipment or fixed-point detectors needs to be incorporated to sense a chemical release situation and activate the appropriate mitigation technology. Detector technology includes: catalytic sensors, infrared sensors, electrochemical sensors, solid-state sensors, and flame ionization detectors.[7]

Risk Management Plan

Risk management strategies include those inherent in the design (materials selection), passive (design of equipment beyond limits of potential process conditions such as temperature and pressure), active (overpressure release–containment–treatment system), and procedural (training and operations).[15] Training should include all personnel working at a facility as well as emergency response training of the off-site population. The facility should inform the local emergency planning committee (LEPC) along with fire and police departments of potential hazards the could occur from the facility and develop with them effective response plans.[7]

EPA's Risk Management Program (RMP) regulation was finalized on June 30, 1993. The rule was required by the Clean Air Act Amendments (CAAA) of 1990 in response to the chemical accidents that occurred in Bhopal, India, and Institute, West Virginia.[1,16,17] The RMP is significant in that it requires information on plant processes and potential hazards to be made public. The deadline for RMP submission was June 21, 1999. Estimates of the number of facilities impacted by the regulation range from 66,000 to 140,000. Any facility under EPA jurisdiction that processes, uses, stores, or handles threshold quantities of listed chemicals is required to submit an RMP.

The elements of the RMP vary depending on the number and program level of the covered processes at the facility. Any facility with one or more covered processes is required to include the following in its RMP: (1) an executive summary (§68.155), (2) the registration for the facility (§68.160), (3) certification statement (§68.85), (4) a worst-case scenario for each Program 1 process; at least one worst-case scenario to cover all Program 2 and 3 processes involving toxic substances; at lease one worst-case scenario to cover all Program 2 and 3 processes involving regulated flammables [§68.165(a)], (5) five-year accident history for each process (§68.168), and (6) a summary of the emergency response program for the facility (§68.180). Any facility with at least one covered process in Program 2 or 3 must also include in the RMP the following: (1) at least one alternative release scenario for each regulated toxic substance in Program 2 or 3 processes and at least one alternative release scenario to cover all regulated flammables in Program 2 or 3 processes [§68.165(b)], (2) summary of the prevention program for each Progam 2 process (§68.170), and (3) summary of the prevention program for each Program 3 process (§68.175).[4]

Software such as RMP*Submit by the EPA is available for use on personal computers for facilities to use in submitting RMPs. Sections of the RMP become public information. There is increasing concern over the impact of public availability of key components of the Section 112(r) RMPs.[18] There is concern that information required under this section of the CAAA by the CPI could be used by terrorists to target U.S. chemical facilities.[19] The requirements to communicate this information to the public has created concern by both small business and large firms. Issues related to concerns range from safety to public reaction. The large chemical firms are typically in the best position to comply, while small businesses can use the requirement as an opportunity to improve the social franchise under which they operate.[20] The communication techniques will depend on location and size of the community and firm.

REFERENCES

1. David Reed, "The Clean Air Act Section 112(r) Risk Management Plans: Compliance Requirements and Alternatives," *Federal Facilities Environmental Journal*, Autumn 1998, pp. 69–78.
2. *Consolidated List of Chemicals Subject to the Emergency Planning and Community Right-To-Know Act (EPCRA) and Section 112(r) of the Clean Air Act as Amended*, EPA 550-B-98-017, *www.epa.gov/ceppo*.
3. William Jorden, "Note: Accidental Release Prevention Requirements: Risk Management Plans under the Clean Air Act Section 112(r)(7)," *The Environmental Lawyer* **3**(2), 515 (1997).
4. *General Guidance for Risk Management Programs (40 CFR Part 68)*, United States Environmental Protection Agency, Office of Solid Waste and Emergency Response, EPA 550-B-98-003 (July 1998).
5. *Guidelines for Engineering Design for Process Safety*, Center for Chemical Process Safety of the American Institute of Chemical Engineers, NY, 1993.
6. Richard W. Prugh and Robert W. Johnson, *Guidelines for Vapor Release Mitigation*, Center for Chemical Process Safety of the American Institute of Chemical Engineers, NY, 1988.
7. *Guidelines for Postrelease Mitigation Technology in the Chemical Process Industry*, Center for Chemical Process Safety of the American Institute of Chemical Engineers, NY, 1997.
8. Daniel A. Crowl and Joseph F. Louvar, *Chemical Process Safety: Fundamentals with Applications*, Prentice Hall International Series in the Physical and Chemical Engineering Sciences, PTR Prentice Hall, New Jersey, 1990.
9. *Guidelines for Use of Vapor Cloud Dispersion Models*, 2nd Edition, Center for Chemical Process Safety of the American Institute of Chemical Engineers, NY, 1996.
10. Robert C. Carroll and David L. Russell, "Gear Up to Meet EPA's Risk Management Program Deadline," *Chemical Engineering Progress*, February 1999.
11. Kenneth R. Comey III, Lee K. Gilmer, George P. Partridge, and David W. Johnson, "Aerosol Reduction from Episodic Releases of Anhydrous Hydrofluoric Acid by Modifying the Acid Catalyst with Liquid Onium Poly(Hydrogen Fluorides)," Proceedings of the *AIChE 1993 Summer National Meeting*; Paper 60C in session entitled "Mitigation of Hazardous Chemical Releases (III); Design Changes."
12. George P. Partridge, Jr., Kenneth R. Comey III, Lee K. Gilmer, and Jame Mudra IV, "Hydrogen Fluoride Alkylation Apparatus and Vapor Recovery Method," U.S. Patent No. 5,277,881, January 11, 1994, Assignee Texaco, Inc., White Plains, NY.
13. Vasilis M. Fthenakis, "Mitigation Options for Accidental Releases of Hazardous Gases," Proceedings of the *International Conference and Workshop on Modeling and Mitigating the Consequences*

of Accidental Releases of Hazardous Materials, New Orleans, Louisiana, September 26–29, 1995 (published by the American Institute of Chemical Engineers), 1995, pp. 881–903.

14. Jean-Philippe Lopez, Joseph Lieto, and Roger Grollier-Baron, "Dilution and Absorption of Unconfined Hazardous Gases by Liquid Spray Curtains," Proceedings of the *International Conference and Workshop on Modeling and Mitigating the Consequences of Accidental Releases of Hazardous Materials*, New Orleans, Louisiana, May 20–24, 1991 (Published by the American Institute of Chemical Engineers), 1991, pp. 453–62.
15. Robert E. Bollinger, David G. Clark, Arthur M. Dowell III, Rodger M. Ewbank, Dennis C. Hendershot, William K. Lutz, Steven I. Meszaros, Donald E. Park, Everett D. Wixom, *Inherently Safer Chemical Processes—A Life Cycle Approach*, Center of Chemical Process Safety of the American Institute of Chemical Engineers, NY, 1996.
16. B. Bowonder and T. Miyake, "Managing Hazardous Facilities: Lessons from the Bophal Accident," *Journal of Hazardous Materials* **19**, 237–69 (1988).
17. "10 Years After Bhopal—Charting a Decade of Change," *Chemical Week*, December 7, 1994, pp. 25–27.
18. David P. Novello and Robert J. Martineau, Jr., "The New Accidental Release Prevention Program under the Clean Air Act," *Natural Resources & Environment* **13**(2), 424 (1998).
19. CSMA Digest, "Clean Air Act Could Increase Threat of Terrorism," *Chemical Times & Trends*, **21**(2) (1998).
20. Patrick J. McNulty, Leon C. Schaller, and Karen R. Chinander, "Communicating under Section 122(r) of the Clean Air Act Amendments," *Risk Analysis* **18**(2), 191–97 (1998).

BIBLIOGRAPHY

EPA Chemical Accident Prevention and Risk Management Planning, *http://www.epa.gov/swercepp/acc-pre.html*. Site contains regulatory information and technical guidance.

Support Center for Regulatory Air Models (SCRAM), *http://www.epa.gov/ttn/scram/*. Site contains EPA models and user guides.

OSHA Process Safety Management Page, *http://www.osha-sic.gov/SLTC/processsafetymanagement/index.html*. The Process Safety Management homepage on the Internet provides numerous documents in downloadable versions related to process safety management. Site contains PSM regulation (29 CFR 1910.100) and related information.

Safety Online, *http://safetyonline.net/hot/rule/*. Site contains CAA 112 (r) rule, data elements, and off-site consequence analysis.

EPA Regulations, *http://www.gov/epahome/rules.html*. Site contains EPA regulations, proposed regulations, and notices.

EPA's Chemical Emergency Preparedness and Prevention Office, *http://www.epa.gov/ceppo/*. Site contains materials related to EPA's risk management program.

EPCRA/Superfund/RCRA/CAAHotline, (800) 424-9346, *http://www.epa.gov/epaoswer/hotline/index.htm*. The EPA's RCRA, Superfund, and EPCRA Hotline provides up-to-date information on EPA programs. The Hotline responds to factual questions developed under the Clean Air Act section 112(r).

Office of Air Quality Planning and Standards, *http://www.epa.gov/oar/oaqps*. The Office of Air Quality Planning and Standards administers EPA's operating permit program.

Center for Chemical Process Safety, *http://www.aiche.org/docs/research/ccps.htm*. This site provides numerous publications and books for purchase in the areas of process safety, catastrophic dispersion modeling, and vapor-release mitigation.

Guidelines for Chemical Process Quantitative Risk Analysis, Center for Chemical Process Safety of the American Institute of Chemical Engineers, NY, 1989.

Risk Management Program Guidance for Wastewater Treatment Plants (40 CFR Part 68), EPA Office of Solid Waste and Emergency Response, EPA 550-B-98-010.

RMP Offsite Consequence Analysis Guidance, EPA 550-B-96-014.

Index